Differentiation

$(cu)' = cu'$ $\quad$ (c constant)

$(u + v)' = u' + v'$

$(uv)' = u'v + v'u$

$\left(\dfrac{u}{v}\right)' = \dfrac{u'v - v'u}{v^2}$

$\dfrac{du}{dx} = \dfrac{du}{dy} \cdot \dfrac{dy}{dx}$ $\quad$ (Chain rule)

$(x^n)' = nx^{n-1}$

$(e^x)' = e^x$

$(a^x)' = a^x \ln a$

$(\sin x)' = \cos x$

$(\cos x)' = -\sin x$

$(\tan x)' = \sec^2 x$

$(\cot x)' = -\csc^2 x$

$(\sinh x)' = \cosh x$

$(\cosh x)' = \sinh x$

$(\ln x)' = \dfrac{1}{x}$

$(\log_a x)' = \dfrac{\log_a e}{x}$

$(\arcsin x)' = \dfrac{1}{\sqrt{1 - x^2}}$

$(\arccos x)' = -\dfrac{1}{\sqrt{1 - x^2}}$

$(\arctan x)' = \dfrac{1}{1 + x^2}$

$(\operatorname{arccot} x)' = -\dfrac{1}{1 + x^2}$

Integration

$\int uv' \, dx = uv - \int u'v \, dx$

$\int x^n \, dx = \dfrac{x^{n+1}}{n+1} + c \quad (n \neq -1)$

$\int \dfrac{1}{x} \, dx = \ln |x| + c$

$\int e^{ax} \, dx = \dfrac{1}{a} e^{ax} + c$

$\int \sin x \, dx = -\cos x + c$

$\int \cos x \, dx = \sin x + c$

$\int \tan x \, dx = -\ln |\cos x| + c$

$\int \cot x \, dx = \ln |\sin x| + c$

$\int \sec x \, dx = \ln |\sec x + \tan x| + c$

$\int \csc x \, dx = \ln |\csc x - \cot x| + c$

$\int \dfrac{dx}{x^2 + a^2} \, dx = \dfrac{1}{a} \arctan \dfrac{x}{a} + c$

$\int \dfrac{dx}{\sqrt{a^2 - x^2}} = \arcsin \dfrac{x}{a} + c$

$\int \dfrac{dx}{\sqrt{x^2 + a^2}} = \sinh^{-1} \dfrac{x}{a} + c$

$\int \dfrac{dx}{\sqrt{x^2 - a^2}} = \cosh^{-1} \dfrac{x}{a} + c$

$\int \sin^2 x \, dx = \tfrac{1}{2}x - \tfrac{1}{4}\sin 2x + c$

$\int \cos^2 x \, dx = \tfrac{1}{2}x + \tfrac{1}{4}\sin 2x + c$

$\int \tan^2 x \, dx = \tan x - x + c$

$\int \cot^2 x \, dx = -\cot x - x + c$

$\int \ln x \, dx = x \ln x - x + c$

$\int e^{ax} \sin bx \, dx = \dfrac{e^{ax}}{a^2 + b^2}(a \sin bx - b \cos bx) + c$

$\int e^{ax} \cos bx \, dx = \dfrac{e^{ax}}{a^2 + b^2}(a \cos bx + b \sin bx) + c$

ADVANCED ENGINEERING MATHEMATICS

SIXTH EDITION

ADVANCED ENGINEERING MATHEMATICS

ERWIN KREYSZIG
Professor of Mathematics
Ohio State University
Columbus, Ohio

JOHN WILEY & SONS
New York Chichester Brisbane Toronto Singapore

Publisher: Don Ford
Mathematics Editor: Robert Pirtle
Editorial Assistant: Jean Gazis
Designer: Madelyn Lesure
Cover Design: Edward A. Burke, Hudson River Studio

Production Supervisor: Susan Ingrao
Production by: Lorraine Burke, Hudson River Studio
Editing Supervisor: Susan Winick
Illustration Manager: John Balbalis
Manufacturing Manager: Bob Ballinger

Compositor: General Graphic Services

Library of Congress Cataloging in Publication Data

Kreyszig, Erwin.
  Advanced engineering mathematics.

  Accompanied by instructor's manual.
  Includes bibliographical references and index.
  1. Mathematical physics. 2. Engineering mathematics.
I. Title.
QA401.K7 1988   510'.2462   87-23038
ISBN 0-471-85824-2

Printed in the United States of America

10 9 8 7 6 5 4 3 2

Preface

Purpose of the Book

This book introduces students of engineering, physics, mathematics and computer science to those areas of mathematics which, from a modern point of view, are most important in connection with practical problems.

The content and character of mathematics needed in applications are changing rapidly. Linear algebra—especially matrices—and numerical methods for computers are of increasing importance. Statistics and graph theory play more prominent roles. Real analysis (ordinary and partial differential equations) and complex analysis remain indispensable. The material in this book is arranged accordingly, in seven independent parts (see also the diagram on the next page):

A Ordinary Differential Equations (Chaps. 1–5)
B Linear Algebra, Vector Calculus (Chaps. 6–9)
C Fourier Analysis and Partial Differential Equations (Chaps. 10, 11)
D Complex Analysis (Chaps. 12–17)
E Numerical Methods (Chaps. 18–20)
F Optimization, Graphs (Chaps. 21, 22)
G Probability and Statistics (Chaps. 23, 24)

This is followed by

References (App. 1)
Answers to Problems (App. 2)
Auxiliary Material (App. 3 and inside of covers)
Tables of Functions (App. 4).

This book has helped to pave the way for the present development and will prepare students for the present situation and the future by a modern approach to the areas listed above and the ideas—some of them computer-related—that are presently causing basic changes: Many methods have become obsolete. New ideas are emphasized, for instance stability, error estimation and structural problems of algorithms, to mention just a few. Trends are driven by supply and demand: supply of powerful new mathematical and computational methods and of enormous computer capacities, demand to solve problems of growing complexity and size, arising from more and more sophisticated systems or production processes, from extreme physical conditions (e.g., those in space travel), from materials with unusual properties (plastics, alloys, superconductors, etc.), or from entirely new tasks in computer vision, robotics and other new fields.

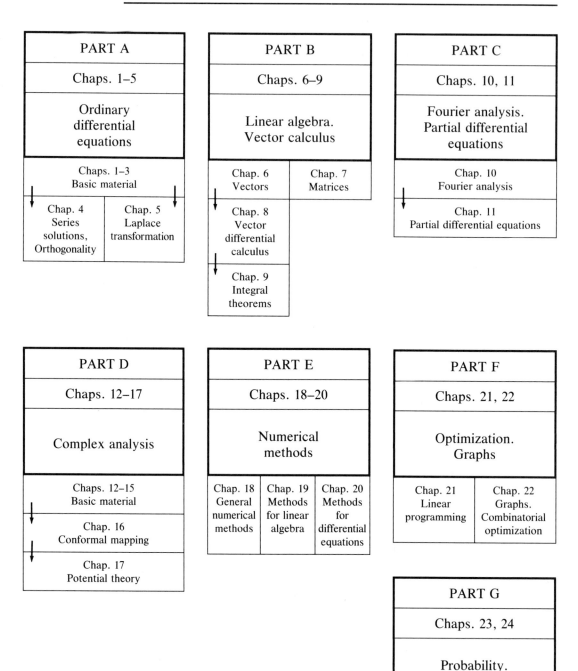

**Parts of the Book
and
Corresponding Chapters**

The general trend seems clear. Details are more difficult to predict. Accordingly, students need solid knowledge of basic principles, methods and results, and a clear perception of what engineering mathematics is all about, in all three phases of solving problems:

- **Modeling:** Translating given physical or other information and data into mathematical form, into a mathematical *model* (a differential equation, a system of equations or some other expression).

- **Solving:** Obtaining the solution by selecting and applying suitable mathematical methods, and in most cases doing numerical work on a computer.

- **Interpreting:** Understanding the meaning and the implications of the mathematical solution for the original problem in terms of physics— or whereever the problem comes from.

It would make no sense to overload students with all kinds of little things that might be of occasional use. Instead, it is important that students become familiar with ways to think mathematically, recognize the need for applying mathematical methods to engineering problems, realize that mathematics is a systematic science built on relatively few basic concepts and involving powerful unifying principles, and get a firm grasp for the interrelation between theory, computing and experiment.

The rapid ongoing developments just sketched have led to many changes and new features in the present edition of this book, causing it to differ very substantially from previous editions.

Changes and New Features Throughout the Book

The book has been simplified by rewriting various sections in a more detailed and leisurely fashion and by placing more emphasis on applications, algorithms and examples.

We first list some major changes and additions pertaining to the book as a whole and then some of the many changes and additions in individual chapters.

- **Problem sets** changed and expanded to contain *over 6000 carefully selected problems,* including more applied problems and more routine problems

- **Chapter review problems** added, to give students practice in choosing a method from the great variety of methods in a whole chapter

- **Worked-out examples** increased to *over 600,* for help in problem solving and better understanding of the text

- **Key formulas** *boxed*

- **Chapter summaries** added, for quick orientation and survey of the most important facts in each chapter

Changes in chapters are listed on the next page.

Changes and New Features in Chapters

- **Ordinary differential equations** (Chaps. 1–4): More systematical treatment of *integrating factors* (Sec. 1.6). *Linear differential equations* (Chap. 2) cast in simpler and more logical form. *Frobenius method* (Sec. 4.4) greatly simplified.

- **Laplace transformation** (Chap. 5). New: shifted data problems, impulsive forces, **Dirac's delta,** list of general formulas, (in addition to the list of transforms)

- **Matrices** (Chap. 7): More *applications* (Markov processes, Leslie matrices, etc.). More on *eigenvalues* and *diagonalization*. Additional modern numerical methods (see below)

- **Vector differential and integral calculus** (Chaps. 8, 9) streamlined by omitting some material of minor interest or making it optional. Grad, div, curl now close together; their forms in curvilinear coordinates (new). Greater emphasis on the types of integrals needed in the integral theorems in Chap. 9.

- **Fourier transformation, Fourier sine and cosine transformations** (Secs. 10.10–10.12, new) with applications to partial differential equations (Sec. 11.14)

- **Complex analysis** (Chaps. 12–17) reorganized to make it more teachable: **1.** *Mappings* by elementary functions added to Chap. 12 (Sec. 12.9). **2.** *Conformal mapping* moved to Chap. 16, to have it close to its applications in Chap. 17 on potential theory, which has been extended by stationary heat problems, etc.
  3. The lengthy introductory chapter on series now reduced to two sections that precede the discussion of power, Taylor and Laurent series.
  4. More on evaluating real integrals by complex integration.

- **Numerical methods** (Chaps. 18–20) modernized throughout, by adding new and more detailed **algorithms** and discussing more worked-out examples, by including **computer-related aspects,** on operations count, pivoting, numerical stability, rounding errors etc.; by giving more extensive treatments of *Newton interpolation, splines, LU-factorization* (Doolittle, Crout, Cholesky), and adding new material, such as *matrix norms, condition numbers, matrix deflation* and *tridiagonalization, QR, spectral shift,* etc.

- **Graph theory:** A new self-contained chapter (Chap. 22) on *graphs* and *digraphs* and their application in **combinatorial optimization** (traveling salesman and other *shortest path problems, shortest spanning trees, network flows, matching,* etc.).

- **Probability and statistics** (Chaps. 23, 24) reorganized by moving sections on sampling to Chap. 24.

- **References** (App. 1) updated and extended, notably those on numerical methods and optimization

- **Auxiliary material added:** Review of partial derivatives (App. 3.2), real series (App. 3.3), first-aid kits of differentiation formulas and integrals, conversion table, Greek alphabet (all on the inside covers).

Suggestions for Courses: A Four-Semester Sequence

The material may be taken in sequence and is suitable for four consecutive semester courses, meeting 3–5 hours a week:

First semester.   Ordinary differential equations (Chaps. 1–5)
Second semester.   Linear algebra and vector analysis (Chaps. 6–9)
Third semester.   Complex analysis (Chaps. 12–17)
Fourth semester.   Numerical methods (Chaps. 18–20)

For the remaining chapters, see below. Possible interchanges are obvious; for instance, numerical methods could precede complex analysis, etc.

Suggestions for Courses: Independent One-Semester Courses

The book is also suitable for various independent one-semester courses meeting 3 hours a week; for example:

Introduction to ordinary differential equations (Chaps. 1, 2)
Laplace transformation (Chap. 5)
Vector algebra and calculus (Chaps. 6, 8)
Matrices and systems of linear equations (Chap. 7)
Fourier series and partial differential equations (Chaps. 10, 11, Secs. 20.4–20.7)
Introduction to complex analysis (Chaps. 12–15)
Numerical analysis (Chaps. 18, 20)
Numerical linear algebra (Chap. 7 for review, Chap. 19)
Optimization (Chaps. 21, 22)
Graphs and combinatorial optimization (Chap. 22)
Probability and statistics (Chaps. 23, 24)

General Features of This Edition

The selection, arrangement and presentation of the material has been made with greatest care, based on past and present teaching, research and consulting experience. Some major features of the book are these:

The book is **self-contained,** except for a few clearly marked places where a proof would be beyond the level of a book of the present type and a reference is given instead. Hiding difficulties or oversimplifying would be of no real help to students.

The presentation is **detailed,** to avoid irritating readers by frequent references to details in other books.

The examples are **simple,** to make the book teachable—why choose complicated examples when simple ones are as instructive or even better?

The notations are **modern and standard,** to help students read articles in journals or other *modern* books and understand other mathematically oriented courses.

The chapters are largely **independent,** providing flexibility in teaching special courses (see above).

The end of a proof is marked by ∎. This sign is also used at the end of some of the definitions and at the end of examples followed by further text.

Acknowledgment

I am indebted to many of my former teachers, colleagues and students who directly or indirectly helped me in preparing this book, in particular the present edition of it. Various parts of the manuscript were distributed to my classes in mimeographed form and returned to me with suggestions for improvement. Discussions with engineers and mathematicians (as well as written comments) were of great help to me; I want to mention particularly Professors S. L. Campbell, J. T. Cargo, P. L. Chambré, V. F. Connolly, A. Cronheim, J. Delany, J. W. Dettman, D. Dicker, D. Ellis, W. Fox, R. G. Helsel, W. N. Huff, J. Keener, E. C. Klipple, V. Komkow, H. Kuhn, G. Lamb, H. B. Mann, I. Marx, K. Millet, J. D. Moore, W. D. Munroe, J. N. Ong, Jr., P. J. Pritchard, H.-W. Pu, W. O. Ray, P. V. Reichelderfer (who helped me very much with the new Chap. 22) J. T. Scheick, H. A. Smith, J. P. Spencer, J. Todd, H. Unz, A. L. Villone, H. J. Weiss, A. Wilansky, C. H. Wilcox, L. Zia, A. D. Ziebur, all from this country, Professors H. S. M. Coxeter and R. Vaillancourt and Mr. H. Kreyszig (whose computer expertise was of great help in Chaps. 18–20) from Canada, and Professors H. Florian, F. Hohenberg, M. Kracht, F. Reutter, C. Schmieden, H. Unger, H. Wielandt, all from Europe. I can offer here only an inadequate acknowledgment of my appreciation.

Furthermore, I wish to thank John Wiley and Sons (see the list on p. iv), Mr. and Mrs. E. A. Burke of Hudson River Studio, and General Graphic Services, in particular Mr. D. Berkheimer and Ms. C. Latshaw, for their effective cooperation and great care in preparing this edition.

Suggestions of many readers were evaluated in preparing the present edition. Any further comment and suggestion for improvement of the book will be gratefully received.

ERWIN KREYSZIG

CONTENTS

PART A

ORDINARY
DIFFERENTIAL EQUATIONS

Differential equations are of fundamental importance in engineering mathematics because many physical laws and relations appear mathematically in the form of such equations. In the present part, which consists of five chapters, we shall consider various physical and geometrical problems that lead to differential equations, and we shall explain the most important standard methods for solving such equations.

We shall pay particular attention to the derivation of differential equations from given physical situations. This transition from the physical problem to a corresponding "mathematical model" is called **modeling.** It is of great practical importance to the engineer and physicist, and will be illustrated by typical examples.

Differential equations are particularly convenient for modern computers. Corresponding **numerical methods** for obtaining approximate solutions of differential equations will be explained in Chap. 20, which is independent of the other chapters of Part E on numerical methods.

Chapter 1

Differential Equations of the First Order

In the present chapter we begin our program of studying ordinary differential equations and their applications by considering the simplest of these equations. These are called differential equations of the *first* order since they involve only the *first* derivative of the unknown function. In this chapter and the following chapters, one of our main goals is to equip the student with methods for solving differential equations, concentrating on those which are of practical importance.

Prerequisite for this chapter: integral calculus.
Sections that may be omitted in a shorter course: 1.8–1.11.
References: Appendix 1, Part A.
Answers to Problems: Appendix 2.

1.1 Basic Concepts and Ideas

Roughly, differential equations are equations that contain derivatives of an unknown function, which we call $y(x)$ and which we want to determine from the equation. Differential equations arise in many engineering and other applications, as mathematical models of various physical and other systems. The simplest of them can be solved by remembering elementary calculus. For example, if a population (of humans, animals, bacteria, etc.) grows at a rate $y' = dy/dx$ (x = time) equal to the population $y(x)$ present, the population model is $y' = y$, a differential equation. If we remember from calculus that $y = e^x$ (or more generally $y = ce^x$) has the property that $y' = y$, we have obtained a solution of our problem.

As another example, if we drop a stone, then its acceleration $y'' = d^2y/dx^2$ is equal to the acceleration of gravity g (a constant). Hence the model of this problem of "free fall" is $y'' = g$, in good approximation, since the air resistance will not matter too much in this case. By integration we get the velocity $y' = dy/dx = gx + v_0$, where v_0 is the initial velocity with which the motion started (e.g., $v_0 = 0$). Integrating once more, we get the distance traveled $y = \frac{1}{2}gx^2 + v_0x + y_0$, where y_0 is the distance from 0 at the beginning (e.g., $y_0 = 0$).

More complicated models, such as those illustrated in Fig. 1 and in many other figures, need for their solution and discussion more refined methods. This calls for a systematic approach, which we shall develop in Chaps. 1–5. In this section we begin by defining and explaining the basic concepts that

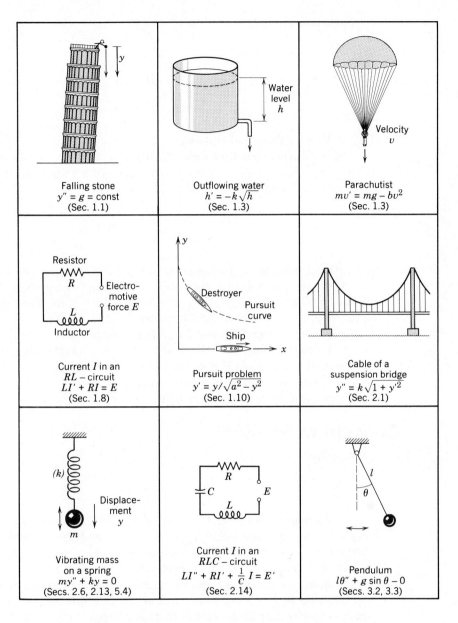

Fig. 1. Some applications of differential equations

are essential in connection with differential equations, and we illustrate these concepts by examples. Then we shall consider two simple practical problems taken from physics and geometry. This will give us a first idea of the nature and purpose of differential equations and their applications.

Ordinary Differential Equation. Its Order

An **ordinary differential equation** is an equation that involves one or several derivatives of an unspecified function y of x; the equation may also involve y itself, given functions of x, and constants.

For example,

(1)
$$y' = \cos x,$$

(2)
$$y'' + 4y = 0,$$

(3)
$$x^2 y''' y' + 2e^x y'' = (x^2 + 2)y^2$$

are ordinary differential equations.

The word **ordinary** distinguishes such an equation from a **partial** *differential equation,* which involves partial derivatives of an unspecified function of two or more independent variables. For example,

$$\frac{\partial^2 u}{\partial x^2} + \frac{\partial^2 u}{\partial y^2} = 0$$

is a partial differential equation. In the present chapter and the next four chapters we shall consider only ordinary differential equations.

We say that a differential equation is of **order** n if the nth derivative of y with respect to x is the highest derivative in the equation.

Thus, a **first-order differential equation** contains only y', and may contain y and given functions of x. Equation (1) is an example. Equation (2) is of the second order, and (3) is of the third order.

In the present chapter we shall consider first order equations. Equations of second and higher order will be discussed in Chaps. 2–5.

Concept of Solution

A function

(4)
$$y = g(x)$$

is called a **solution** of a given first-order differential equation on some interval, say, $a < x < b$ (perhaps infinite) if $g(x)$ is defined and differentiable throughout that interval and is such that the equation becomes an identity when y and y' are replaced by g and g', respectively.

The principal task in differential equations and their applications is to find all solutions of given equations and investigate their properties. We shall discuss various standard methods that have been developed for that purpose.

EXAMPLE 1. Solution
Verify that the function $y = g(x) = x^2$ is a solution of the first-order differential equation

$$xy' = 2y$$

for all x.

To verify this, we differentiate g, finding

$$g' = 2x.$$

We now insert g and g' into the differential equation. Then we see that the equation reduces to an identity $x(2x) = 2x^2$. This completes the verification.   ∎

Sometimes a solution of a differential equation will appear as an implicit function, that is, implicitly given in the form

$$G(x, y) = 0,$$

and is then called an *implicit solution,* in contrast to the *explicit solution* (4).

EXAMPLE 2. Implicit solution
The function y of x implicitly given by

$$x^2 + y^2 - 1 = 0 \qquad\qquad (y > 0)$$

(representing a semicircle) is an implicit solution of the differential equation

$$yy' = -x$$

on the interval $-1 < x < 1$, as the student may verify by differentiation.   ∎

We next observe that a differential equation may have many solutions, as the following example shows.

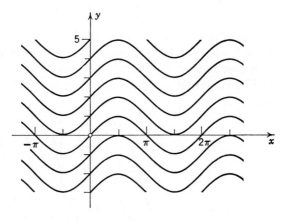

Fig. 2. Solutions of $y' = \cos x$

EXAMPLE 3
Solve the first-order differential equation (1),

$$y' = \cos x.$$

Solution. By calculus, we can immediately integrate,

(5) $y = \sin x + c,$

where c is a constant. c is arbitrary. Choosing $c = 0$, we have the solution $y = \sin x$; for $c = 1.5$ we get $y = \sin x + 1.5$, and so on. And from calculus we know that (5) with arbitrary c represents the totality of *all* solutions of the differential equation (1). See Fig. 2.   ∎

This example is typical. It illustrates that a first-order differential equation may (and, in general, will) have more than one solution, even infinitely many solutions, which can be represented by a single formula involving an arbitrary constant c. It is customary to call such a function which contains an arbitrary[1] constant a **general solution** of the corresponding differential equation of the first order. If we assign a definite value to that constant, then the solution so obtained is called a **particular solution.**

Thus, (5) is a general solution of $y' = \cos x$, and $y = \sin x$, $y = \sin x - 2$ etc. are particular solutions.

EXAMPLE 4. General solution and particular solutions

The student may verify by differentiation that the function

$$(6) \qquad\qquad y = ce^x \qquad\qquad (c \text{ any constant})$$

satisfies the differential equation

$$(7) \qquad\qquad y' = y$$

for all x. Since it involves an arbitrary constant c, it is a general solution of this equation. It represents a family of curves, some of which are shown in Fig. 3. Particular solutions obtained by choosing special c are, for instance,

$$y = e^x, \qquad y = 2e^x, \qquad y = -\tfrac{8}{3}e^x$$

etc. Each of these represents one of those curves.   ∎

In the following sections we shall develop various methods for obtaining general solutions of first-order equations. For a given equation, a general solution obtained by such a method is unique, except for notation, and will then be called *the* general solution of that differential equation.

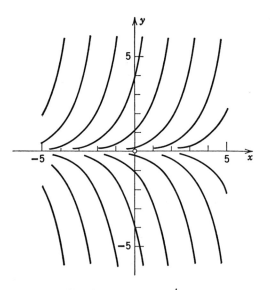

Fig. 3. Solutions of $y' = y$

---

[1]The range of the constant may have to be restricted in some cases to avoid imaginary expressions or other degeneracies.

REMARK. Singular solutions

We mention that in some cases there may be further solutions of a given equation that cannot be obtained by assigning a definite value to the arbitrary constant in the general solution; such a solution is then called a **singular solution** of the equation.

For example, the equation

(8)
$$y'^2 - xy' + y = 0$$

has the general solution

$$y = cx - c^2,$$

as the reader can verify by differentiation and substitution. This represents a family of straight lines, where each line corresponds to a definite value of c. Substitution also shows that a further solution is

$$y = \frac{x^2}{4}.$$

This is a singular solution of (8), since we cannot obtain it by assigning a definite value to c in the general solution. Obviously, each particular solution represents a tangent to the parabola represented by the singular solution (Fig. 4).

Singular solutions will rarely occur in engineering problems.

We mention that in some books, a *general solution* means a formula that includes *all* solutions of an equation, that is, both the particular and the singular ones. We do not adopt this definition for two reasons. First of all, it is frequently quite difficult to prove that a formula includes *all* solutions, hence that definition of a general solution is rather useless in practice. Furthermore, we shall see later that a large and very important class of equations (the so-called *linear* differential equations) do not have singular solutions, and our definition of a general solution can be easily generalized to equations of higher order so that the resulting notion includes all solutions of a differential equation that is linear.

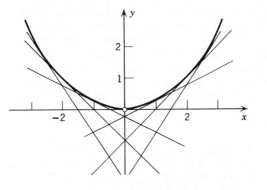

Fig. 4. Singular solution (representing a parabola) and particular solutions of equation (8)

We shall see that the conditions under which a given differential equation has solutions are fairly general. But we should note that there are simple equations that do not have solutions at all, and others that do not have a general solution. For example, the equation $y'^2 = -1$ does not have a solution for real y. (Why?) The equation $|y'| + |y| = 0$ does not have a general solution, because its only solution is $y \equiv 0$.

Modeling. Physical and Geometrical Applications

Differential equations are of great importance in engineering, because many physical laws and relations appear mathematically in the form of differential equations.

Let us consider a simple physical example that will illustrate the typical steps of **modeling,** that is, the steps that lead from the physical situation (*physical system*) to a mathematical formulation (*mathematical model*) and solution, and to the physical interpretation of the result. This may be the easiest way of obtaining a first idea of the nature and purpose of differential equations and their applications.

EXAMPLE 5. Radioactivity, exponential decay

Experiments show that a radioactive substance decomposes at a rate proportional to the amount present. Starting with a given amount of substance, say, 2 grams, at a certain time, say, $t = 0$, what can be said about the amount available at a later time?

Solution. 1st Step. Setting up a mathematical model (a differential equation) of the physical process. We denote by $y(t)$ the amount of substance still present at time t. The rate of change is dy/dt. According to the physical law governing the process of radiation, dy/dt is proportional to y:

(9)
$$\frac{dy}{dt} = ky.$$

Here k is a definite physical constant whose numerical value is known for various radioactive substances. (For example, in the case of radium $_{88}Ra^{226}$ we have $k \approx -1.4 \cdot 10^{-11}$ sec^{-1}.) Clearly, since the amount of substance is positive and decreases with time, dy/dt is negative, and so is k. We see that the physical process under consideration is described mathematically by an ordinary differential equation of the first order. Hence this equation is the mathematical model of that physical process. *Whenever a physical law involves a rate of change of a function, such as velocity, acceleration, etc., it will lead to a differential equation. For this reason differential equations occur frequently in physics and engineering.*

2nd Step. Solving the differential equation. At this early stage of our discussion no systematic method for solving (9) is at our disposal. However, (9) tells us that if there is a solution $y(t)$, its derivative must be proportional to y. From calculus we remember that exponential functions have this property. In fact, the function e^{kt} or more generally

(10)
$$y(t) = ce^{kt}$$

where c is any constant, is a solution of (9) for all t, as can readily be verified by substituting (10) into (9). Since (10) involves an arbitrary constant, it is a general solution of the first order equation (9). [We shall see later (in Sec. 1.2) that (10) includes *all* solutions of (9); that is, (9) does not have singular solutions.]

3rd Step. Determination of a particular solution. It is clear that our physical process has a unique behavior. Hence we can expect that by using further given information we shall be able to select a definite numerical value of c in (10) so that the resulting particular solution will describe the process properly. The amount of substance $y(t)$ still present at time t will depend on the initial amount of substance given. This amount is 2 grams at $t = 0$. Hence we have to specify the value of c so that $y = 2$ when $t = 0$. This condition is called an **initial condition,** since it refers to the initial state of the physical system. By inserting this condition

(11)
$$y(0) = 2$$

in (10) we obtain

Fig. 5. Radioactivity (exponential decay)

$$y(0) = ce^0 = 2 \qquad \text{or} \qquad c = 2.$$

If we use this value of c, then the solution (10) takes the particular form

$$(12) \qquad\qquad\qquad\qquad y(t) = 2e^{kt}.$$

This particular solution of (9) characterizes the amount of substance still present at any time $t \geqq 0$. The physical constant k is negative, and $y(t)$ decreases, as is shown in Fig. 5.

4th Step. Checking. From (12) we have

$$\frac{dy}{dt} = 2ke^{kt} = ky \qquad \text{and} \qquad y(0) = 2e^0 = 2.$$

We see that the function (12) satisfies the equation (9) as well as the initial condition (11).

The student should never forget to carry out this important final step, which shows whether the function is (or is not) the solution of the problem. ∎

Let us show that geometrical problems also lead to differential equations.

EXAMPLE 6. A geometrical application

Find the curve through the point $(1, 1)$ in the xy-plane having at each of its points the slope $-y/x$.

Solution. The function giving the desired curve must be a solution of the differential equation

$$(13) \qquad\qquad\qquad\qquad y' = -\frac{y}{x}.$$

We shall soon learn how to solve such an equation. Meanwhile the student may verify that

$$y = \frac{c}{x}$$

is a solution of (13) for every value of the constant c. Some of the corresponding curves are shown in Fig. 6. Since we are looking for the curve that passes through $(1, 1)$, we must have $y = 1$ when $x = 1$. This yields $c = 1$. Hence the solution of our problem is

$$y = \frac{1}{x}. \qquad\qquad\qquad\qquad\qquad\qquad ∎$$

From a practical point of view, the most basic question is how to solve a given differential equation. In the next sections we discuss the most important methods for solving first-order differential equations, beginning in Sec. 1.2 with the so-called *method of* **separating variables,** which has many applications. Some typical physical applications of this method will be shown in Sec. 1.3 and the corresponding problem set. In Sec. 1.4 we shall see that the method can be extended to further classes of differential equations.

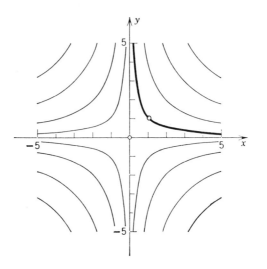

Fig. 6. Solutions of $y' = -y/x$

Problems for Sec. 1.1

In each case state the order of the differential equation and verify that the given function is a solution.

1. $y' + 8y = 0,$ $y = ce^{-8x}$

2. $y'' = e^x,$ $y = e^x + ax + b$

3. $y'' + 16y = 0,$ $y = A \cos 4x + B \sin 4x$

4. $y' - y \cos x = 0,$ $y = c \sin x$

5. $y'' + 4y' + 5y = 0,$ $y = e^{-2x}(A \cos x + B \sin x)$

In each case verify that the given function is a solution of the differential equation. Graph the corresponding curves for some values of the constant c.

6. $y' + y = 0, \quad y = ce^{-x}$ **7.** $y' = -x/y, \quad x^2 + y^2 = c$

8. $y' + y = 5, \quad y = ce^{-x} + 5$ **9.** $y'^2 = 4y, \quad y = 0, \quad y = (x + c)^2$

Solve the following differential equations.

10. $y' = x^3$ **11.** $y' = e^{-x/2}$ **12.** $y' = \cos 3x$ **13.** $y'' = \sin 2x$

Find a first-order differential equation involving both y and y' for which the given function is a solution.

14. $y = e^{-2x}$ **15.** $y = \cos x$ **16.** $y = e^{-x^2}$ **17.** $y = x^4$

In each case verify that the given function is the general solution of the corresponding differential equation and determine c so that the resulting particular solution satisfies the given condition. Graph this solution.

18. $y' + y = 5,$ $y = ce^{-x} + 5.$ $y = 4$ when $x = 0$

19. $y' = 2xy,$ $y = ce^{x^2},$ $y = \frac{1}{2}e$ when $x = 1$

20. $y' = y \cot x,$ $y = c \sin x,$ $y = 2$ when $x = \pi/2$

21. $xy' = 3y,$ $y = cx^3,$ $y = 4$ when $x = -2$

22. $4yy' + x = 0,$ $x^2 + 4y^2 = c,$ $y = 1$ when $x = 0$

Modeling, Applications

23. (Half-life) The *half-life* of a radioactive substance is the time in which half of a given amount will disappear. What is the half-life of $_{88}Ra^{226}$ (in years) in Example 5?

24. (Half-life) Radium $_{88}Ra^{224}$ has a half-life of 3.64 days. Given 1 gram, how much will still be present after 1 day? After 1 year?

25. (Falling body) Experiments show that if a body falls in vacuum due to the action of gravity, then its acceleration is constant (equal to $g = 9.80$ meters/sec^2 = 32.1 ft/sec^2; this is called the *acceleration of gravity*). State this law as a differential equation for $s(t)$, the distance fallen as a function of t. Show that by solving this equation one can obtain the familiar law

$$s(t) = \frac{g}{2} t^2.$$

(In practice, this also applies to the free fall in air if we can neglect the air resistance, for instance, if we drop a stone or an iron ball.)

26. (Falling body, general initial conditions) If in Prob. 25 the body starts at $t = 0$ from initial position $s = s_0$ with initial velocity $v = v_0$, show that then the solution is

$$s(t) = \frac{g}{2} t^2 + v_0 t + s_0.$$

27. (Rocket) A rocket is shot straight up. During the initial stages of flight it has acceleration $7t$ meters/sec^2. The engine cuts out at $t = 10$ sec. How high will the rocket go? (Neglect air resistance.)

28. (Airplane takeoff) An airplane taking off from a landing field has a run of 2 kilometers. If the plane starts with speed 10 meters/sec, moves with constant acceleration and makes the run in 50 sec, with what speed does it take off?

29. (Exponential decay; atmospheric pressure) Observations show that the rate of change of the atmospheric pressure y with altitude x is proportional to the pressure. Assuming that the pressure at 6000 meters (about 18,000 ft) is half its value y_0 at sea level, find the formula for the pressure at any height. *Hint.* Remember from calculus that if $y = e^{kx}$, then $y' = ke^{kx} = ky$.

30. (Exponential growth; a population model. Malthus's law) If relatively small populations (of humans, animals, bacteria, etc.) are left undisturbed, they often grow according to *Malthus's law*,[2] which states that the time rate of growth is proportional to the population present. Formulate this as a differential equation. Show that the solution is $y(t) = y_0 e^{kt}$. For the United States, observed values, in millions, are as follows.

t	0	30	60	90	120	150	180
Year	1800	1830	1860	1890	1920	1950	1980
Population	5.3	13	31	63	105	150	230

Use the first two columns for determining y_0 and k. Then calculate values for 1860, 1890, $\cdots$, 1980 and compare them with the observed values. Comment.

---

[2]THOMAS ROBERT MALTHUS (1766—1834), English social scientist, one of the leaders in classical national economy.

1.2 Separable Differential Equations

Many first-order differential equations can be reduced to the form

$$(1) \qquad\qquad g(y)y' = f(x)$$

by algebraic manipulations. Since $y' = dy/dx$, we find it convenient to write

$$(2) \qquad\qquad \boxed{g(y)\ dy = f(x)\ dx,}$$

but we keep in mind that this is merely another way of writing (1). Such an equation is called an *equation with separable variables,* or a **separable equation,** because in (2) the variables x and y are *separated* so that x appears only on the right and y appears only on the left. By integrating on both sides of (2) we obtain

$$(3) \qquad\qquad \boxed{\int g(y)\ dy = \int f(x)\ dx + c.}$$

If we assume that f and g are continuous functions, the integrals in (3) will exist, and by evaluating these integrals we obtain the general solution of (1).

EXAMPLE 1
Solve the differential equation

$$9yy' + 4x = 0.$$

Solution. By separating variables we have

$$9y\ dy = -4x\ dx.$$

By integrating on both sides we obtain the general solution

$$\frac{9}{2}y^2 = -2x^2 + \tilde{c} \qquad \text{or} \qquad \frac{x^2}{9} + \frac{y^2}{4} = c \qquad\qquad \left(c = \frac{\tilde{c}}{18}\right).$$

The solution represents a family of ellipses.

EXAMPLE 2
Solve the differential equation

$$y' = 1 + y^2.$$

Solution. By separating variables and integrating we obtain

$$\frac{dy}{1 + y^2} = dx, \qquad \text{arc tan } y = x + c, \qquad y = \tan(x + c).$$

It is of great importance to introduce the constant of integration immediately when the integration is carried out.

EXAMPLE 3

Solve the differential equation

$$y' = -2xy.$$

Solution. By separating variables we have

$$\frac{dy}{y} = -2x \, dx \qquad\qquad (y \neq 0).$$

Integration yields

(4) $$\ln |y| = -x^2 + \tilde{c}.$$

In fact, the result on the left may be verified by differentiation as follows. When $y > 0$, then $(\ln y)' = y'/y$. When $y < 0$, then $-y > 0$ and $(\ln (-y))' = -y'/(-y) = y'/y$. Now $y = |y|$ when $y > 0$ and $-y = |y|$ when $y < 0$. Hence we may combine both formulas, finding $(\ln |y|)' = y'/y$.

Furthermore, from (4) we obtain

$$|y| = e^{-x^2 + \tilde{c}}.$$

Noting that $e^{a+b} = e^a e^b$ and setting $e^{\tilde{c}} = c$ when $y > 0$ and $e^{\tilde{c}} = -c$ when $y < 0$, we finally have

$$y = ce^{-x^2}.$$

This solution represents a family of so-called **bell-shaped curves** (Fig. 7). We may also admit $c = 0$, which gives the solution $y \equiv 0$.   ∎

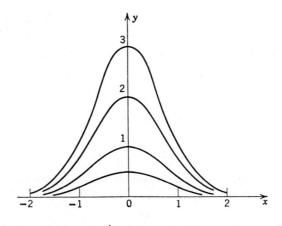

Fig. 7. Solutions of $y' = -2xy$ ("bell-shaped curves") in the upper half-plane ($y > 0$)

Initial Value Problems

In many engineering applications we are not interested in the general solution of a given differential equation but only in the particular solution $y(x)$ satisfying a given **initial condition,** say, the condition that at some point x_0 the solution $y(x)$ has a prescribed value y_0, briefly

(5) $$\boxed{y(x_0) = y_0.}$$

A first-order differential equation together with an initial condition is called an **initial value problem.** To solve such a problem, we must find the particular solution of the equation that satisfies the given initial condition.

The term "initial value problem" is suggested by the fact that the independent variable often is time, so that (5) prescribes the initial situation at some instant and the solution of the problem shows what happens later.

EXAMPLE 4

Solve the initial value problem

$$(x^2 + 1)y' + y^2 + 1 = 0, \qquad y(0) = 1.$$

Solution. 1st Step. Solving the differential equation. Separating variables, we find

$$\frac{dy}{1 + y^2} = -\frac{dx}{1 + x^2}.$$

By integration we obtain

$$\text{arc tan } y = -\text{arc tan } x + c$$

or

$$\text{arc tan } y + \text{arc tan } x = c.$$

Taking the tangent on both sides, we have

(6) $$\tan(\text{arc tan } y + \text{arc tan } x) = \tan c.$$

Now the addition formula for the tangent is

$$\tan(a + b) = \frac{\tan a + \tan b}{1 - \tan a \tan b}.$$

For $a = \text{arc tan } y$ and $b = \text{arc tan } x$ this becomes

$$\tan(\text{arc tan } y + \text{arc tan } x) = \frac{y + x}{1 - xy}.$$

Consequently, (6) may be written

(7) $$\frac{y + x}{1 - xy} = \tan c.$$

2nd Step. Use of the initial condition. We determine c from the initial condition. Setting $x = 0$ and $y = 1$ in (7), we have $1 = \tan c$, so that

$$\frac{y + x}{1 - xy} = 1 \qquad \text{or} \qquad y = \frac{1 - x}{1 + x}.$$

3rd Step. Checking. Check the result.   ∎

Many physical and engineering problems have as their model a differential equation that is separable. We show this in the next section by discussing a careful selection of some typical problems of this nature.

Problems for Sec. 1.2

1. Why is it important to introduce the constant of integration immediately when the integration is performed?

2. Sometimes a student asks whether the derivation of (3) from (2) is correct, since one seems to integrate with respect to different variables on both sides. What would you answer?

Find the general solution of the following equations (where a, b, n are constants).

3. $y' = ky$

4. $y' = \sec y$

5. $y' = y \cot x$

6. $xy' = 5y$

7. $y' = 3x^2y$

8. $xy' = ny$

9. $y' + ay + b = 0 \ (a \neq 0)$

10. $(y - b)y' = a - x$

11. $y' = e^x y^3$

12. $y' = \pi(1 + y^2)$

13. $y' = \sqrt{1 - y^2}$

14. $y' + y^2 = 1$

15. $(x \ln x)y' = y$

16. $y' = x^{n-1}(1 + y^2) \ (n \neq 0)$

17. $y' \sin 2x = y \cos 2x$

18. $y' + \csc y = 0$

Solve the following initial value problems (where L and R are constants).

19. $xy' + y = 0, \quad y(1) = 1$

20. $y' = -x/y, \quad y(2) = \sqrt{5}$

21. $y' = x/y, \quad y(2) = 0$

22. $yy' = \sin^2 x, \quad y(0) = \sqrt{3}$

23. $dr \sin \theta = 2r \cos \theta \, d\theta, \quad r(\pi/4) = -2$

24. $dr/dt = -2tr, \quad r(0) = 4.6$

25. $y' = y \tanh x, \quad y(0) = 1.4$

26. $y^3y' + x^3 = 0, \quad y(2) = 0$

27. $L(dI/dt) + RI = 0, \quad I(0) = I_0$

28. $v(dv/dx) = g = const, \quad v(x_0) = v_0$

29. An initial value problem is usually solved by first determining the general solution of the equation and then the particular solution. Using (3), show that the particular solution of (1) satisfying (5) can also be obtained directly from

(8)
$$\int_{y_0}^{y} g(y^*) \, dy^* = \int_{x_0}^{x} f(x^*) \, dx^*.$$

Use this formula for confirming the result in Example 4.

30. Using (8), find the solution of the differential equation in Example 4 satisfying $y(0) = 0$.

1.3   Modeling: Separable Equations

Modeling means setting up mathematical models of physical or other systems. In this section we consider systems that can be modeled in terms of a separable differential equation.

EXAMPLE 1. (Newton's law of cooling[3])

A copper ball is heated to a temperature of 100°C. Then at time $t = 0$ it is placed in water that is maintained at a temperature of 30°C. At the end of 3 min the temperature of the ball is reduced to 70°C. Find the time at which the temperature of the ball is reduced to 31°C.

---

[3]Sir ISAAC NEWTON (1642—1727), great English physicist and mathematician, became a professor at Cambridge in 1669 and Master of the Mint in 1699. He and the German mathematician and philosopher GOTTFRIED WILHELM LEIBNIZ (1646—1716) invented (independently) the differential and integral calculus. Newton discovered many basic physical laws and created the method of investigating physical problems by means of calculus. His *Philosophiae naturalis principia mathematica* (*Mathematical Principles of Natural Philosophy*, 1687) contain the development of classical mechanics. His work is of greatest importance to both mathematics and physics.

Physical information. Experiments show that the time rate of change of the temperature T of the ball is proportional to the difference between T and the temperature of the surrounding medium (*Newton's law of cooling*).

Experiments also show that heat flows so rapidly in copper that at any time the temperature is practically the same at all points of the ball.

Solution. *1st Step. Modeling.* For our physical system, the mathematical formulation of Newton's law of cooling is

(1)
$$\frac{dT}{dt} = -k(T - 30)$$

where we denoted the constant of proportionality by $-k$ in order that $k > 0$.

2nd Step. General solution. The general solution of (1) is obtained by separating variables; we find

$$T(t) = ce^{-kt} + 30.$$

3rd Step. Use of initial condition. The given initial condition is $T(0) = 100$. The particular solution satisfying this condition is

$$T(t) = 70e^{-kt} + 30.$$

4th Step. Use of further information. The constant k can be determined from the given information $T(3) = 70$. We obtain

$$T(3) = 70e^{-3k} + 30 = 70 \quad \text{or} \quad k = \tfrac{1}{3} \ln \tfrac{7}{4} = 0.1865.$$

Using this value of k we see that the temperature $T(t)$ of the ball is

$$T(t) = 70e^{-0.1865t} + 30.$$

It follows that the value $T = 31°C$ is reached when

$$70e^{-0.1865t} = 1 \quad \text{or} \quad 0.1865t = \ln 70, \quad t = \frac{\ln 70}{0.1865} = 22.78,$$

that is, after approximately 23 minutes.

5th Step. Checking. Check the result.

EXAMPLE 2. Flow of water through orifices (Torricelli's law[4])

A cylindrical tank 1.50 meters high stands on its circular base of diameter 1.00 meter and is initially filled with water. At the bottom of the tank there is a hole of diameter 1.00 cm, which is opened at some instant, so that the water starts draining under the influence of gravity (Fig. 8). Find the height $h(t)$ of the water in the tank at any time t. Find the times at which the tank is one-half full, one-quarter full, and empty.

Solution. *1st Step. Modeling.* We set up the differential equation as a mathematical model of the physical problem. The volume of water which flows out during a short interval of time Δt is

$$\Delta V = Av\Delta t$$

where $A = 0.500^2\pi$ cm^2 is the cross-sectional area of the outlet and v is the velocity of the outflowing water. Torricelli's law states that the velocity with which water issues from an orifice is

$$v = 0.600\sqrt{2gh}$$

where $g = 980$ cm/sec^2 (about 32.17 ft/sec^2) is the acceleration of gravity at the surface of the

---

[4]EVANGELISTA TORRICELLI (1608—1647), Italian physicist and mathematician, pupil and later successor of GALILEO GALILEI (1564—1642) at Florence.

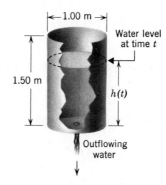

Fig. 8. Tank in Example 2

earth and h is the instantaneous height of the water above the orifice. This formula looks reasonable. Indeed, $\sqrt{2gh}$ is the speed a body acquires if it falls a distance h and the air resistance is so small that it can be neglected. The factor 0.600 is introduced[5] since the cross section of the outflowing stream is somewhat smaller than that of the orifice.

ΔV must equal the change ΔV^* of the volume of water in the tank. Clearly,

$$\Delta V^* = -B\Delta h$$

where B is the cross-sectional area of the tank and Δh is the decrease of the height $h(t)$ of the water. The minus sign appears since the volume of water in the tank decreases. Since $\Delta V = \Delta V^*$, we have

$$A v \Delta t = 0.600 A \sqrt{2gh}\,\Delta t = -B\Delta h.$$

Division by Δt yields

$$\frac{\Delta h}{\Delta t} = -\frac{0.600 A \sqrt{2gh}}{B}.$$

Letting $\Delta t \to 0$, we obtain the differential equation

$$(2) \qquad \frac{dh}{dt} = -\frac{0.600 A \sqrt{2g}}{B}\,\sqrt{h}.$$

Here, $A = 0.500^2 \pi$ cm^2 (see above), the cross-sectional area of the tank is $B = 50.0^2 \pi$ cm^2, and $\sqrt{2g} = \sqrt{2 \cdot 980} = 44.3$ [cm$^{1/2}$/sec]. Substitution of these numbers into (2) and simplification yields

$$\frac{dh}{dt} = -0.00266 h^{1/2}.$$

The initial condition is

$$h(0) = 150 \text{ cm},$$

where $t = 0$ is the instant at which the hole is opened.

2nd Step. General solution, particular solution. We solve the differential equation. Separation of variables yields

$$h^{-1/2}\,dh = -0.00266\,dt.$$

By integration,

$$2h^{1/2} = -0.00266t + \tilde{c}.$$

---

[5]As suggested by J. C. BORDA in 1766.

Dividing and squaring, we have, writing $\bar{c}/2 = c$,

$$h(t) = (c - 0.00133t)^2.$$

By the initial condition, $h(0) = c^2 = 150$. Hence the particular solution corresponding to our problem is

$$h(t) = (12.25 - 0.00133t)^2.$$

3rd Step. To answer the remaining questions, we express t in terms of h:

$$12.25 - 0.00133t = \sqrt{h}, \qquad \text{hence} \qquad t = \frac{12.25 - \sqrt{h}}{0.00133}.$$

This shows that the tank is one-half full at

$$t = \frac{12.25 - \sqrt{75.0}}{0.00133} = 2.70 \cdot 10^3 \text{ sec} = 45.0 \text{ min},$$

one-quarter full at $t = 76.8$ min, and empty at $t = 154$ min.

4th Step. Checking. Check the result.

EXAMPLE 3. Velocity of escape from the earth

Find the minimum initial velocity of a projectile that is fired in radial direction from the earth and is supposed to escape from the earth. Neglect the air resistance and the gravitational pull of other celestial bodies.

Solution. 1st Step. Modeling. According to Newton's law of gravitation, the gravitational force, and therefore also the acceleration a of the projectile, is proportional to $1/r^2$, where r is the distance from the center of the earth to the projectile. Thus

$$a(r) = \frac{dv}{dt} = \frac{k}{r^2}$$

where v is the velocity and t is the time. Since v is decreasing, $a < 0$ and thus $k < 0$. Let R be the radius of the earth. When $r = R$ then $a = -g$, the acceleration of gravity at the surface of the earth. Note that the minus sign occurs because g is positive and the attraction of gravity acts in the negative direction (the direction toward the center of the earth). Thus

$$-g = a(R) = \frac{k}{R^2} \qquad \text{and} \qquad a(r) = -\frac{gR^2}{r^2}.$$

Now $v = dr/dt$, and differentiation yields

$$a = \frac{dv}{dt} = \frac{dv}{dr}\frac{dr}{dt} = \frac{dv}{dr}v.$$

Hence the differential equation for the velocity is

$$\frac{dv}{dr}v = -\frac{gR^2}{r^2}.$$

2nd Step. Solving the differential equation. Separating variables and integrating, we obtain

$$(3) \qquad v\, dv = -gR^2\frac{dr}{r^2} \qquad \text{and} \qquad \frac{v^2}{2} = \frac{gR^2}{r} + c.$$

3rd Step. c in terms of the initial velocity. On the earth's surface, $r = R$ and $v = v_0$, the initial velocity. For these values of r and v, formula (3) becomes

$$\frac{v_0{}^2}{2} = \frac{gR^2}{R} + c, \qquad \text{and} \qquad c = \frac{v_0{}^2}{2} - gR.$$

By inserting this expression for c into (3) we have

(4) $$v^2 = \frac{2gR^2}{r} + v_0{}^2 - 2gR.$$

4th Step. Determination of the velocity of escape. If $v^2 = 0$, then $v = 0$, the projectile stops, and it is clear from the physical situation that the velocity will change from positive to negative, so that the projectile will return to the earth. Consequently, we have to choose v_0 so large that this cannot happen. If we choose

(5) $$v_0 = \sqrt{2gR},$$

then in (4) the expression $v_0{}^2 - 2gR$ is zero and v^2 is never zero for any r. However, if we choose a smaller value for the initial velocity v_0, then $v = 0$ for a certain r. The expression in (5) is called the *velocity of escape* from the earth. Since $R = 6372$ km $= 3960$ mi and $g = 9.80$ meters/sec$^2 = 0.00980$ km/sec$^2 = 32.17$ ft/sec$^2 = 0.00609$ mi/sec^2, we have

$$v_0 = \sqrt{2gR} \approx 11.2 \text{ km/sec} \approx 6.95 \text{ mi/sec}.$$

5th Step. Checking. Check the result.

EXAMPLE 4. Skydiver

Suppose that a skydiver falls from rest toward the earth and the parachute opens at an instant, call it $t = 0$, when the skydiver's speed is $v(0) = v_0 = 10.0$ meters/sec. Find the speed $v(t)$ of the skydiver at any later time t. Does $v(t)$ increase indefinitely?

Physical assumptions and laws. Suppose that the weight of the man plus the equipment is $W = 712$ nt (read "newtons"; about 160 lb), the air resistance U is proportional to v^2, say, $U = bv^2$ nt, where the constant of proportionality b depends mainly on the parachute, and we assume that $b = 30.0$ nt $\cdot$ sec^2/meter$^2 = 30.0$ kg/meter.

Solution. 1st Step. Modeling. We set up the mathematical model (the differential equation) of the problem. **Newton's second law** is

$$\text{Mass} \times \text{Acceleration} = \text{Force}$$

where "force" means the resultant of the forces acting on the skydiver at any instant. These forces are the weight W and the air resistance U.

The weight is $W = mg$, where $g = 9.80$ meters/sec^2 (about 32.2 ft/sec^2) is the acceleration of gravity at the earth's surface. Hence the mass of the man plus the equipment is $m = W/g = 72.7$ kg (about 5.00 slugs). The air resistance U acts upward (against the direction of the motion), so that the resultant is

$$W - U = mg - bv^2.$$

The acceleration a is the time derivative of v, that is, $a = dv/dt$. By Newton's second law,

$$m\frac{dv}{dt} = mg - bv^2.$$

This is the differential equation of our problem. By assumption, $v = v_0 = 10$ meters/sec when $t = 0$. Hence we are looking for the solution $v(t)$ satisfying the *initial condition*

$$v(0) = v_0 = 10.$$

2nd Step. Solving the differential equation. Division by m gives

$$\frac{dv}{dt} = -\frac{b}{m}(v^2 - k^2) \qquad\qquad k^2 = \frac{mg}{b}.$$

Separating variables, we have

(6)
$$\frac{dv}{v^2 - k^2} = -\frac{b}{m} \, dt.$$

On the left,

$$\frac{1}{v^2 - k^2} = \frac{1}{2k} \left(\frac{1}{v - k} - \frac{1}{v + k} \right),$$

as can be readily verified. This is a representation in terms of partial fractions which is convenient for integration:

$$\int \frac{dv}{v^2 - k^2} = \frac{1}{2k} [\ln (v - k) - \ln (v + k)] = \frac{1}{2k} \ln \frac{v - k}{v + k}.$$

From (6) we thus obtain

$$\frac{1}{2k} \ln \frac{v - k}{v + k} = -\frac{b}{m} t + \tilde{c}.$$

Multiplying by $2k$ and taking exponents, we get

(7)
$$\frac{v - k}{v + k} = ce^{-pt} \qquad\qquad p = \frac{2kb}{m},$$

where $c = e^{2k\tilde{c}}$. Solving this for v, we have

(8)
$$v(t) = k \frac{1 + ce^{-pt}}{1 - ce^{-pt}}.$$

We see that $v(t) \to k$ as $t \to \infty$; that is, $v(t)$ does not increase indefinitely but approaches a limit, k. It is interesting to note that this limit is independent of the initial condition $v(0) = v_0$.

3rd Step. c in terms of the initial velocity. We determine c in (8) such that we obtain the particular solution satisfying the initial condition $v(0) = v_0$. From (7) with $t = 0$ we immediately have

$$c = \frac{v_0 - k}{v_0 + k}.$$

With this c, formula (8) represents the solution we are looking for.

4th Step. Use of numerical values. Note that we have not yet made use of the given numerical values. Very often it is a good principle to hunt for general formulas first and to substitute numerical data later. In this way one often gets a better idea of what is going on and, moreover, saves work if one wants to obtain various solutions corresponding to different sets of data.

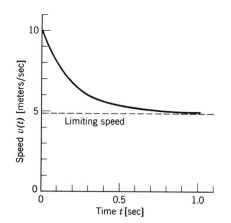

Fig. 9. Speed $v(t)$ of the skydiver in Example 4

In the present case we now compute (see before)

$$k^2 = \frac{mg}{b} = \frac{W}{b} = \frac{712}{30.0} = 23.7 \text{ [meters}^2/\text{sec}^2].$$

Hence $k = 4.87$ meters/sec. This is the limiting speed. Practically speaking, this is the speed of the skydiver after a sufficiently long time.

To $v(0) = v_0 = 10.0$ meters/sec there corresponds

$$c = \frac{v_0 - k}{v_0 + k} = 0.345.$$

Finally,

$$p = \frac{2kb}{m} = \frac{2 \cdot 4.87 \cdot 30.0}{72.7} = 4.02 \text{ [sec}^{-1}].$$

This altogether yields the result (Fig. 9)

(9)
$$v(t) = 4.87 \frac{1 + 0.345e^{-4.02t}}{1 - 0.345e^{-4.02t}}.$$

5th Step. Checking. Check the result.

EXAMPLE 5. Radiocarbon dating

If a fossilized bone contains 25% of the original amount of radioactive carbon $_6C^{14}$, what is its age?

Idea of the method to be used. In the atmosphere, the ratio of radioactive carbon $_6C^{14}$ and ordinary carbon $_6C^{12}$ is constant, and the same holds for *living* organisms. When an organism dies, the absorption of $_6C^{14}$ by breathing and eating terminates. Hence one can estimate the age of a fossil by comparing the carbon ratio in the fossil with that in the atmosphere. This is W. Libby's idea of radiocarbon dating (Nobel Prize for chemistry, 1960). The half-life of $_6C^{14}$ is 5730 years (*CRC Handbook of Chemistry and Physics,* 54th ed., 1973, p. B251).

Solution. As in Example 5 in Sec. 1.1, the mathematical model of the process of radioactive decay is

$$y' = ky, \qquad \text{solution} \qquad y(t) = y_0 e^{kt}.$$

Here, y_0 is the initial amount of $_6C^{14}$. Using the given half-life of $_6C^{14}$, we determine k from

$$\frac{1}{2} = e^{kt} = e^{5730k}, \qquad k = \frac{\ln 1/2}{5730} = -0.000\ 121.$$

The time after which 25% of the original amount of $_6C^{14}$ is still present can now be computed from

$$\frac{1}{4} = e^{kt} = e^{-0.000\ 121t}, \qquad t = \frac{\ln 1/4}{-0.000\ 121} = 11460.$$

The answer is 11460 years. This is twice the half-life of $_6C^{14}$. Is this just by chance? Or would there be a shorter way of getting the answer?

Actually, the experimental determination of the half-life of $_6C^{14}$ involves an error of about 40 years. Also, a comparison with other methods shows that radiocarbon dating tends to give values that are too small, perhaps because the ratio of $_6C^{14}$ to $_6C^{12}$ may have changed over long periods of time. Hence 12 000 or 13 000 years is probably a more realistic answer to our present problem.   ∎

Of course, not every first-order differential equation is separable. However, it will sometimes be possible to convert an equation to separable form by introducing new variables. We discuss this method in the next section.

Problems for Sec. 1.3

1. **(Velocity of escape)** At the earth's surface the velocity of escape is 11.2 km/sec; cf. Example 3. If the projectile is carried by a rocket and is separated from the rocket at a distance of 1000 km from the earth's surface, what would be the minimum velocity at this point sufficient for escape from the earth?

2. **(Velocity of escape)** If in Example 3 the initial velocity is half the velocity of escape, what maximum height above the earth's surface would the projectile reach? First guess, then calculate.

3. **(Boyle-Mariotte's law for ideal gases[6])** Experiments show that for a gas at low pressure p (and constant temperature) the rate of change of the volume $V(p)$ equals $-V/p$. Solve the corresponding differential equation.

4. **(Exponential decay) Lambert's law of absorption[7]** states that the absorption of light in a very thin transparent layer is proportional to the thickness of the layer and to the amount incident on that layer. Formulate this in terms of a differential equation and solve it.

5. Show that a ball thrown vertically upward with initial velocity v_0 takes twice as much time to return as to reach the highest point. Find the velocity upon return. (Air resistance is assumed negligible.)

6. **(Exponential decay, half-life)** Experiments show that most radioactive substances disintegrate at a rate proportional to the amount of substance instantaneously present. The time in which 50% of a given amount disappears is called the *half-life* of the substance. (See also Sec. 1.1.) Uranium $_{92}U^{232}$ has half-life $H = 74$ years. What percentage will disappear within 1 year? Within 10 years? Within 37 years?

7. **(Exponential decay, half-life)** What percentage of a radioactive substance will still be present at time $H/2$, where H is the half-life of the substance? At time $2H$?

8. **(Half-life formula)** Show that the half-life H of a radioactive substance can be determined from two measurements $y_1 = y(t_1)$ and $y_2 = y(t_2)$ of the amounts present at times t_1 and t_2 by the formula

$$H = \frac{(t_2 - t_1) \ln 2}{\ln (y_1/y_2)} .$$

9. **(Exponential growth)** If in a culture of yeast the rate of growth is proportional to the population $p(t)$ present at time t and the population doubles in 1 day, how much can be expected after 1 week at the same rate of growth?

10. **(Evaporation)** Suppose that a mothball loses volume by evaporation at a rate proportional to its instantaneous area. If the diameter of the ball decreases from 2 cm to 1 cm in 3 months, how long will it take until the ball has practically gone, say, until its diameter is 1 mm?

11. **(Evaporation)** Experiments show that a wet porous substance in the open air loses its moisture at a rate proportional to the moisture content. If a sheet hung in the wind loses half its moisture during the first hour, when will it be practically dry, say, when will it have lost 99.9% of its moisture, weather conditions remaining the same?

---

[6]ROBERT BOYLE (1627—1691), English physicist and chemist, one of the founders of the Royal Society; EDMÉ MARIOTTE (about 1620—1684), French physicist and prior of a monastery near Dijon.

[7]JOHANN HEINRICH LAMBERT (1728—1777), German physicist and mathematician, known for his contributions to cartography and astronomy.

12. **(Evaporation)** Two liquids are boiling in a vessel. It is found that the ratio of the quantities of each passing off as vapor at any instant is proportional to the ratio of the quantities x and y still in the liquid state. Show that $y' = ky/x$ and solve this differential equation.

13. **(Sugar inversion)** Experiments show that the rate of inversion of cane sugar in dilute solution is proportional to the concentration $y(t)$ of unaltered sugar. Let the concentration be 1/100 at $t = 0$ and 1/300 at $t = 4$ hours. Find $y(t)$.

14. **(Flywheel)** A flywheel of moment of inertia I is rotating with a relatively small constant angular speed ω_0 (radians/sec). At some instant, call it $t = 0$, the power is shut off and the motion starts slowing down because of friction. Let the friction torque be proportional to $\sqrt{\omega}$, where $\omega(t)$ is the angular speed. Using **Newton's second law** (Moment of inertia × Angular acceleration = Torque), find $\omega(t)$ and the times t_1 at which the wheel is rotating with $\omega_0/2$ and t_2 when it comes to rest.

15. **(Newton's law of cooling)** A thermometer, reading 15°C, is brought into a room whose temperature is 23°C. One minute later the thermometer reading is 19°C. How long does it take until the reading is practically 23°C, say, 22.9°C?

16. **(Torricelli's law)** A conical tank (Fig. 10) of circular cross section whose angle at the apex is 60° has an outlet of cross-sectional area of 0.5 cm². The tank contains water. At time $t = 0$ the outlet is opened and the water flows out. Determine the time when the tank will be empty, assuming that the initial height of water is $h(0) = 1$ meter. *Hint.* You can use (2) (why?), where B is the cross-sectional area of the tank at height $h(t)$, so that B now depends on h.

17. The time to empty the tank in Example 2 is greater than twice the time at which the tank is one-half full. Is this physically understandable?

18. Suppose that the tank in Prob. 16 is hemispherical, of radius R, is initially full of water, and has an outlet of 5 cm² cross-sectional area at the bottom. The outlet is opened at some instant. Find the time it takes to empty the tank (*a*) for any given R, (*b*) for $R = 1$ meter.

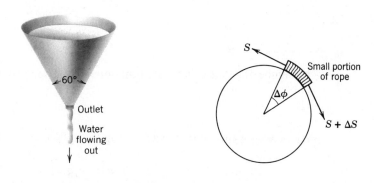

Fig. 10. Conical tank in Prob. 16   **Fig. 11.** Problem 19

19. **(Rope)** If we wind a rope around a rough cylinder which is fixed on the ground, we need only a small force at one end to resist a large force at the other. Let S be the force in the rope. Experiments show that the change ΔS of S in a small portion of the rope is proportional to S and to the small angle $\Delta\phi$ in Fig. 11. Show that the differential equation for S is $dS/d\phi = \mu S$. If $\mu = 0.2$ radian⁻¹, how many times must the rope be snubbed around the cylinder in order that a man holding one end of the rope can resist a force one thousand times greater than he can exert?

20. (Skydiver) How do the equation and solution in Example 4 change if we assume the air resistance to be proportional to v (instead of v^2), say, $U = \tilde{b}v$ where $\tilde{b} = 7.3$ kg/sec? Is this model still reasonable?

21. To what height of free fall does the limiting speed (4.87 meters/sec) in Example 4 correspond?

Geometrical applications. In each case find all curves in the xy-plane having the indicated property.

22. The normal at each point (x, y) intersects the x-axis at $(x + \frac{1}{2}, 0)$.

23. The tangents all pass through the origin.

24. The normals all pass through the origin.

25. The distance from (x, y) to the origin is the same as the distance from (x, y) to the point of intersection of the normal at (x, y) and the x-axis.

1.4 Reduction to Separable Form

Certain first-order differential equations are not separable but can be made separable by a simple change of variables. This holds for equations of the form[8]

$$(1) \qquad \boxed{y' = g\!\left(\frac{y}{x}\right)}$$

where g is any given function of y/x, for example $(y/x)^3$, $\sin(y/x)$, etc. The form of the equation suggests that we set

$$\frac{y}{x} = u,$$

remembering that y and u are functions of x. Then $y = ux$. By differentiation,

$$(2) \qquad y' = u + u'x.$$

By inserting this into (1) and noting that $g(y/x) = g(u)$ we have

$$u + u'x = g(u).$$

Now we may separate the variables u and x, finding

$$\frac{du}{g(u) - u} = \frac{dx}{x}.$$

If we integrate and then replace u by y/x, we obtain the general solution of (1).

---

[8]These equations are sometimes called **homogeneous equations.** We shall not use this terminology but reserve the term ''homogeneous'' for a much more important purpose (cf. Sec. 1.7).

EXAMPLE 1
Solve

$$2xyy' - y^2 + x^2 = 0.$$

Solution. Dividing by x^2, we have

$$2\frac{y}{x}y' - \left(\frac{y}{x}\right)^2 + 1 = 0.$$

If we set $u = y/x$ and use (2), the equation becomes

$$2u(u + u'x) - u^2 + 1 = 0 \quad \text{or} \quad 2xuu' + u^2 + 1 = 0.$$

Separating variables, we find

$$\frac{2u\,du}{1 + u^2} = -\frac{dx}{x}.$$

By integration

$$\ln(1 + u^2) = -\ln|x| + c^* \quad \text{or} \quad 1 + u^2 = \frac{c}{x}.$$

Replacing u by y/x, we finally obtain

$$x^2 + y^2 = cx \quad \text{or} \quad \left(x - \frac{c}{2}\right)^2 + y^2 = \frac{c^2}{4}.$$

EXAMPLE 2. An initial value problem
Solve the initial value problem

$$y' = \frac{y}{x} + \frac{2x^3 \cos x^2}{y}, \qquad y(\sqrt{\pi}) = 0.$$

Solution. We set $u = y/x$. Then $y = xu$, $y' = xu' + u$, and the equation becomes

$$xu' + u = u + \frac{2x^2 \cos x^2}{u}.$$

We simplify algebraically and then integrate:

$$uu' = 2x \cos x^2, \qquad \tfrac{1}{2}u^2 = \sin x^2 + c.$$

Since $u = y/x$, this gives

$$y = ux = x\sqrt{2 \sin x^2 + 2c}.$$

Since $\sin \pi = 0$, the initial condition yields $c = 0$. Hence the answer is

$$y = x\sqrt{2 \sin x^2}.$$

Sometimes the form of a given differential equation suggests other simple substitutions, as the following example illustrates.

EXAMPLE 3
Solve

$$(2x - 4y + 5)y' + x - 2y + 3 = 0.$$

Solution. We set $x - 2y = v$. Then $y' = \tfrac{1}{2}(1 - v')$ and the equation takes the form

$$(2v + 5)v' = 4v + 11.$$

Separating variables and integrating, we find

$$\left(1 - \frac{1}{4v + 11}\right) dv = 2\, dx \quad \text{and} \quad v - \frac{1}{4}\ln|4v + 11| = 2x + c^*.$$

Since $v = x - 2y$, this may be written

$$4x + 8y + \ln|4x - 8y + 11| = c. \qquad \blacksquare$$

Further simple substitutions are illustrated by the equations in Probs. 17–26.

In the next section we consider another standard method for solving a large class of differential equations, the so-called **exact equations.**

Problems for Sec. 1.4

Find the general solution of the following equations.

1. $xy' = x + y$

2. $xy' - 2y = 3x$

3. $xy' = (y - x)^3 + y$

4. $xy' = y^2 + y$

5. $x^2y' = x^2 - xy + y^2$

6. $xy' - y = 3x^4 \cos^2 (y/x)$

7. $xy' = y + x^2 \sec (y/x)$

8. $xyy' = y^2 + (x^2 + 1)^{-1}x^3$

9. $xy' = y + x^5e^x/4y^3$

10. $xy' - y - x^2 \tan (y/x) = 0$

Solve the following initial value problems.

11. $xy' = x + y, \; y(1) = -7.4$

12. $xy' = 2x + 2y, \; y(0.5) = 0$

13. $yy' = x^3 + y^2/x, \; y(2) = 6$

14. $xyy' = 2y^2 + 4x^2, \; y(2) = 4$

15. $y' = \dfrac{y + x}{y - x}, \; y(0) = 2$

16. $y' = \dfrac{y - x}{y + x}, \; y(1) = 1$

Using the indicated transformations, find the general solution of the following equations.

17. $xy' = e^{-xy} - y \quad (xy = v)$

18. $y' = \cot (y + x) - 1 \quad (y + x = v)$

19. $y' = (y - x)^2 \quad (y - x = v)$

20. $e^yy' = k(x + e^y) - 1 \quad (x + e^y = v)$

21. $y' = \dfrac{y - x + 1}{y - x + 5} \quad (y - x = v)$

22. $y' = \dfrac{1 - 2y - 4x}{1 + y + 2x} \quad (y + 2x = v)$

23. Consider $y' = f(ax + by + k)$, where f is continuous. If $b = 0$, the solution is immediate. (Why?) If $b \neq 0$, show that one obtains a separable equation by using $u(x) = ax + by + k$ as a new dependent variable.

Using Prob. 23, find the general solution of the following equations.

24. $y' = (x + y - 4)^2$

25. $y' = (5y + 2)^4$

26. $y' = (6y - y^2 - 8)^{1/2}$

27. Find the curve that passes through the point $(x, y) = (\sqrt{e}, \sqrt{e})$ and has the slope $x/y + y/x$.

28. Find the curve $y(x)$ that passes through $(1, 1/2)$ and is such that at each point (x, y) the intercept of the tangent on the y-axis is equal to $2xy^2$.

29. Show that a straight line through the origin intersects all solution curves of a given differential equation $y' = g(y/x)$ at the same angle.

30. The positions of four battle ships on the ocean are such that the ships form the vertices of a square of length l. At some instant each ship fires a missile that directs its motion steadily toward the missile on its right. Assuming that the four missiles fly horizontally and with the same speed, find the path of each.

1.5   Exact Differential Equations

A first-order differential equation of the form

(1) $$\boxed{M(x, y)\, dx\, +\, N(x, y)\, dy\, =\, 0}$$

is said to be **exact** if its left-hand side is the total or exact differential

(2) $$du\, =\, \frac{\partial u}{\partial x}\, dx\, +\, \frac{\partial u}{\partial y}\, dy$$

of some function $u(x, y)$. Then the differential equation (1) can be written

$$du\, =\, 0.$$

By integration we immediately obtain the general solution of (1) in the form

(3) $$\boxed{u(x, y)\, =\, c.}$$

Comparing (1) and (2), we see that (1) is exact if there is some function $u(x, y)$ such that

(4)   (a) $\dfrac{\partial u}{\partial x} = M,$ (b) $\dfrac{\partial u}{\partial y} = N.$

Suppose that M and N are defined and have continuous first partial derivatives in a region in the xy-plane whose boundary is a closed curve having no self-intersections. Then from (4) (cf. Appendix 3.2 for notation)

$$\frac{\partial M}{\partial y} = \frac{\partial^2 u}{\partial y\, \partial x},$$

$$\frac{\partial N}{\partial x} = \frac{\partial^2 u}{\partial x\, \partial y}.$$

By the assumption of continuity the two second derivatives are equal. Thus

(5) $$\boxed{\frac{\partial M}{\partial y} = \frac{\partial N}{\partial x}.}$$

This condition is not only necessary but also sufficient[9] for $M\, dx\, +\, N\, dy$ to be a total differential.

----

[9]We shall prove this fact at another occasion (in Sec. 9.9); the proof can also be found in some books on elementary calculus; cf. Ref. [18] in Appendix 1.

If (1) is exact, the function $u(x, y)$ can be found by guessing or in the following systematic way. From (4a) we have by integration with respect to x

$$(6) \qquad\qquad u = \int M \, dx + k(y);$$

in this integration, y is to be regarded as a constant, and $k(y)$ plays the role of a "constant" of integration. To determine $k(y)$, we derive $\partial u / \partial y$ from (6), use (4b) to get dk/dy, and integrate.

Formula (6) was obtained from (4a). Instead of (4a) we may equally well use (4b). Then instead of (6) we first have

$$(6^*) \qquad\qquad u = \int N \, dy + l(x).$$

To determine $l(x)$ we derive $\partial u / \partial x$ from (6*), use (4a) to get dl/dx, and integrate.

EXAMPLE 1. An exact equation
Solve

$$xy' + y + 4 = 0.$$

Solution. We write this equation in the form (1), that is,

$$(y + 4) \, dx + x \, dy = 0.$$

We now see that

$$M = y + 4 \qquad \text{and} \qquad N = x.$$

Hence (5) holds, so that the equation is exact. From (6*) we have

$$u = \int N \, dy + l(x) = \int x \, dy + l(x) = xy + l(x).$$

To determine $l(x)$, we differentiate this formula with respect to x and use formula (4a), finding

$$\frac{\partial u}{\partial x} = y + \frac{dl}{dx} = M = y + 4.$$

Thus $dl/dx = 4$ and

$$l = 4x + c^*.$$

Hence we obtain the general solution of our equation in the form

$$u = xy + l(x) = xy + 4x + c^* = const.$$

Division by x gives the explicit form

$$y = \frac{c}{x} - 4.$$

Intelligent students, who know that thinking is important in mathematics, write the given equation in the form

$$y \, dx + x \, dy = -4 \, dx,$$

recognize the left side as the total differential of xy, and integrate, finding $xy = -4x + c$, which is equivalent to our previous result.

EXAMPLE 2. An exact equation
Solve

$$2x \sin 3y \, dx + (3x^2 \cos 3y + 2y) \, dy = 0.$$

Solution. By applying (5) we see that the equation is exact. From (6) we obtain

$$u = \int 2x \sin 3y \, dx + k(y) = x^2 \sin 3y + k(y).$$

From this and the given differential equation,

$$\frac{\partial u}{\partial y} = 3x^2 \cos 3y + \frac{dk}{dy} = 3x^2 \cos 3y + 2y, \qquad \text{hence} \qquad \frac{dk}{dy} = 2y, \qquad k = y^2 + c^*.$$

The general solution is $u = const$ or $x^2 \sin 3y + y^2 = c$.   ∎

Caution! Note well that the present method gives the solution in implicit form, $u(x, y) = c = const$, not in explicit form, $y = f(x)$. For *checking,* we can differentiate $u(x, y) = c$ implicitly and see whether this leads to $dy/dx = -M/N$ or $M \, dx + N \, dy = 0$, the given equation.

EXAMPLE 3. An initial value problem
Solve the initial value problem

$$\sin x \cosh y - y' \cos x \sinh y = 0, \qquad y(0) = 0.$$

Solution. The student may verify that the equation is exact. From (6) we obtain

$$u = \int \sin x \cosh y \, dx + k(y) = -\cos x \cosh y + k(y).$$

From this, $\partial u/\partial y = -\cos x \sinh y + dk/dy$. Hence $dk/dy = 0$, and $k = const$. The general solution is $u = const$, that is, $\cos x \cosh y = c$. The initial condition gives $\cos 0 \cosh 0 = 1 = c$. Hence the answer is $\cos x \cosh y = 1$.

EXAMPLE 4. Warning: Breakdown in the case of nonexactness
Consider the equation

$$y \, dx - x \, dy = 0.$$

We see that $M = y$, $N = -x$, hence $\partial M/\partial y = 1$ but $\partial N/\partial x = -1$. Hence the equation is not exact. Let us show that in such a case, the present method does not work. From (6),

$$u = \int M \, dx + k(y) = xy + k(y).$$

From this,

$$\frac{\partial u}{\partial y} = x + k'(y).$$

This should equal $N = -x$. But this is impossible, since $k(y)$ can depend only on y. Try (6*): it will also fail. Solve the equation by another method we have discussed.   ∎

If an equation is exact, you can destroy exactness by dividing by some (suitable) function. For example, $x \, dx + y \, dy = 0$ is exact, but division by y gives the nonexact equation $(x/y) \, dx + dy = 0$. (Check for nonexactness.) Hence it should be possible to make certain nonexact equations exact by multiplying them by a suitable function. This is the idea of the method in the next section.

Problems for Sec. 1.5

Given $u(x, y)$, find the exact differential equation $du = 0$. What kind of curves are the solution curves $u(x, y) = const$?

1. $u = x^2 + 9y^2$

2. $u = y/x$

3. $u = x^2 - y^2$

4. $u = \cos(y^2 - x)$

5. $u = e^{xy}$

6. $u = (y - 3x + 2)^3$

7. $u = y/x^2$

8. $u = 1/(x^2 + y^2)$

9. $u = e^{x^2/y}$

10. $u = \tan(y^2 - x^3)$

Show that the following differential equations are exact and solve them.

11. $y\,dx + x\,dy = 0$

12. $(x\,dy - y\,dx)/x^2 = 0$

13. $y^2\,dx + 2xy\,dy = 0$

14. $\cos x\,dx + y\,dy = 0$

15. $[(x + 1)e^x - e^y]\,dx - xe^y\,dy = 0$

16. $(\cot y + x^2)\,dx = x\operatorname{cosec}^2 y\,dy$

17. $\sinh x \cos y\,dx = \cosh x \sin y\,dy$

18. $(x\,dx + y\,dy)/(x^2 + y^2)^2 = 0$

19. $e^{-\theta}\,dr - re^{-\theta}\,d\theta = 0$

20. $\tfrac{1}{4}e^{4\theta}\,dr + re^{4\theta}\,d\theta = 0$

Are the following equations exact? Find the general solution.

21. $x\,dy + 3y^2\,dx = 0$

22. $ye^{xy}\,dx + (1 + xe^{xy})\,dy = 0$

23. $2\sin 2x \sinh y\,dx = \cos 2x \cosh y\,dy$

24. $ye^x\,dx + (2y + e^x)\,dy = 0$

25. $2\,dx + x^{-1}\,dy = 0$

26. $(1 + x^2)\,dy + (1 + y^2)\,dx = 0$

27. $2y^{-1}\cos 2x\,dx = y^{-2}\sin 2x\,dy$

28. $2\sin \omega y\,dx + \omega \cos \omega y\,dy = 0$

29. $\sinh x\,dx + y^{-1}\cosh x\,dy = 0$

30. $(y - b)\,dx + (x - a)\,dy = 0$

31. Solve the equation in Example 4.

32. If an equation is separable, show that it is exact. Is the converse true?

33. Under what conditions is $(ax + by)\,dx + (kx + ly)\,dy = 0$ exact? (Here, a, b, k, l are constants.) Solve the exact equation.

34. Under what conditions is $[f(x) + g(y)]\,dx + [h(x) + p(y)]\,dy = 0$ exact?

35. Under what conditions is $f(x, y)\,dx + g(x)h(y)\,dy = 0$ exact?

To see that a differential equation can sometimes be solved by several methods, solve (*a*) by the present method, (*b*) by separation or inspection:

36. $b^2 x\,dx + a^2 y\,dy = 0$

37. $3x^{-4}y\,dx = x^{-3}\,dy$

38. $\sin y \sin x\,dx = \cos y \cos x\,dy$

39. $2x\,dx + x^{-2}(x\,dy - y\,dx) = 0$

Solve the following initial value problems.

40. $4x\,dx + 9y\,dy = 0$, $y(3) = 0$

41. $(y + 3)\,dx + (x - 2)\,dy = 0$, $y(1) = -7$

42. $(2xy\,dx + dy)e^{x^2} = 0$, $y(0) = 2$

43. $3x^2(y - 1)^2\,dx + 2x^3(y - 1)\,dy = 0$, $y(-2) = 2.5$

44. $-3x^{-4}y^2\,dx + 2x^{-3}y\,dy = 0$, $y(4) = 8$

45. $\cos \pi x \cos 2\pi y\,dx = 2\sin \pi x \sin 2\pi y\,dy$, $y(3/2) = 1/2$

46. $[e^x \cos y + 2(x - y)]\,dx = [e^x \sin y + 2(x - y)]\,dy$, $y(0) = \pi$

Find the curve having the given slope y' and passing through the given point (x, y).

47. $y' = -\dfrac{y - 2}{x - 2}$, $(0, 0)$

48. $y' = \dfrac{x + 2}{y + 1}$, $(-3, -1)$

49. $y' = \dfrac{x}{4 - 4y}$, $(0, 2)$

50. $y' = \dfrac{1 - x}{1 + y}$, $(1, 0)$

1.6   Integrating Factors

The differential equation $y^{-1} \, dx + 2x \, dy = 0$ is not exact (verify this!), but if we multiply it by $F(x, y) = y/x$, we get the exact equation $x^{-1} \, dx + 2y \, dy = 0$, which we can readily solve, obtaining $\ln |x| + y^2 = c$. This illustrates that sometimes a given differential equation

$$(1) \qquad\qquad P(x, y) \, dx + Q(x, y) \, dy = 0$$

is not exact but can be made exact by multiplying it by a suitable function $F(x, y)$ ($\neq 0$). This function is then called an **integrating factor** of (1). With some experience, integrating factors can be found by inspection. For this purpose the student should keep in mind the differentials listed in Example 2 below. In some important special cases, integrating factors can be determined in a systematic way, as we shall see below.

EXAMPLE 1. Integrating factor
Solve

$$x \, dy - y \, dx = 0.$$

Solution. The differential equation is not exact. An integrating factor is $F = 1/x^2$, and we obtain

$$F(x)(x \, dy - y \, dx) = \frac{x \, dy - y \, dx}{x^2} = d\left(\frac{y}{x}\right) = 0, \qquad y = cx.$$

EXAMPLE 2. Integrating factors
Find further integrating factors of the equation $x \, dy - y \, dx = 0$ in Example 1.

Solution. Since

$$d\left(\frac{x}{y}\right) = \frac{y \, dx - x \, dy}{y^2}, \qquad d\left(\ln \frac{y}{x}\right) = \frac{x \, dy - y \, dx}{xy}, \qquad d\left(\text{arc tan} \frac{y}{x}\right) = \frac{x \, dy - y \, dx}{x^2 + y^2}$$

the functions $1/y^2$, $1/xy$, and $1/(x^2 + y^2)$ are such factors. The corresponding solutions

$$\frac{x}{y} = c, \qquad \ln \frac{y}{x} = c, \qquad \text{arc tan} \frac{y}{x} = c$$

are essentially the same, because each represents a family of straight lines through the origin.
   This illustrates the following. If we have *one* integrating factor F of (1), we can always get *many* factors since then $FP \, dx + FQ \, dy$ is the differential du of some function u, and for any $H(u)$ another differential is

$$H(FP \, dx + FQ \, dy) = H(u) \, du.$$

This shows that $H(u)F(x, y)$ is another integrating factor of (1). ∎

If $F(x, y)$ is an integrating factor of (1), then

$$FP \, dx + FQ \, dy = 0$$

is an exact equation. Hence the exactness condition $\partial M/\partial y = \partial N/\partial x$ (Sec. 1.5) now becomes

(2)
$$\frac{\partial}{\partial y}(FP) = \frac{\partial}{\partial x}(FQ).$$

This is more complicated than the given equation (1) to be solved and thus practically useless. But let us see whether we get something of interest by looking for an integrating factor F depending only on *one* variable, say x. Then (2) becomes

$$F\frac{\partial P}{\partial y} = \frac{dF}{dx}Q + F\frac{\partial Q}{\partial x}.$$

Dividing by FQ and reordering terms, we have

(3)
$$\frac{1}{F}\frac{dF}{dx} = \frac{1}{Q}\left(\frac{\partial P}{\partial y} - \frac{\partial Q}{\partial x}\right).$$

This proves

Theorem 1 (Integrating factor depending only on one variable)
If (1) *is such that the right-hand side of* (3) *depends only on x, then* (1) *has an integrating factor $F(x)$, which is obtained by solving* (3).

Can you find a similar statement for an integrating factor $F(y)$? (See Prob. 15.)

EXAMPLE 3. Integrating factor F(x)
Solve

(4)
$$(4x + 3y^2)\, dx + 2xy\, dy = 0.$$

Solution. We have $P = 4x + 3y^2$, $\partial P/\partial y = 6y$, $Q = 2xy$, $\partial Q/\partial x = 2y$. Since $\partial P/\partial y \neq \partial Q/\partial x$, this equation is not exact. The right-hand side of (3) is $(6y - 2y)/2xy = 2/x$, a function of x only, so that (4) has an integrating factor $F(x)$. By (3),

$$\frac{1}{F}\frac{dF}{dx} = \frac{2}{x}, \qquad \ln|F| = 2\ln|x|, \qquad F(x) = x^2.$$

Multiplying (4) by x^2 gives the exact equation

$$4x^3\, dx + (3x^2y^2\, dx + 2x^3y\, dy) = 0.$$

By inspection or by the method in Sec. 1.5 we obtain the general solution

$$x^4 + x^3y^2 = c. \qquad\qquad ∎$$

The most important application of the method of integrating factors will be considered in the next section, where we shall solve so-called **linear differential equations,** that is, equations of the form

$$y' + p(x)y = r(x).$$

Problems for Sec. 1.6

Show that the given function is an integrating factor and solve:

1. $2y\,dx + x\,dy = 0,\quad x$ **2.** $2y\,dx + 3x\,dy = 0,\quad xy^2$

3. $x\,dy - y\,dx = 0,\quad 1/x^2$ **4.** $3(y + 1)\,dx = 2x\,dy,\quad (y + 1)/x^4$

5. $\sin y\,dx + \cos y\,dy = 0,\quad e^x$ **6.** $(a + 1)y\,dx + (b + 1)x\,dy = 0,\quad x^a y^b$

7. $y^2\,dx + (1 + xy)\,dy = 0,\quad e^{xy}$

8. $2\sinh x\,dx + \cosh x\,dy = 0,\quad e^y \cosh x$

In each case find an integrating factor F and solve:

9. $2\,dx - e^{y-x}\,dy = 0$ **10.** $3y\,dx + 2x\,dy = 0$

11. $x\cosh y\,dy - \sinh y\,dx = 0$ **12.** $\sin y\,dx + \cos y\,dy = 0$

13. $(y + 1)\,dx - (x + 1)\,dy = 0$ **14.** $2\cos x\cos y\,dx - \sin x\sin y\,dy = 0$

15. (Integrating factor $F(y)$) Show that if $(\partial Q/\partial x - \partial P/\partial y)/P$ depends only on y, then (1) has an integrating factor $F(y)$.

Using Theorem 1 or Prob. 15, solve:

16. $2xy\,dx + 3x^2\,dy = 0$ **17.** $\cos x\,dx + \sin x\,dy = 0$

18. $x\csc y\,dx + \cos y\,dy = 0$ **19.** $2\cosh x\cos y\,dx = \sinh x\sin y\,dy$

20. $3y^2\,dx + 2xy\,dy = 0$ **21.** $(2\cos y + 4x^2)\,dx = x\sin y\,dy$

22. $3x^2 y\,dx + 2x^3\,dy = 0$ **23.** $(3xe^y + 2y)\,dx + (x^2 e^y + x)\,dy = 0$

24. $2\cos y\,dx = \sin y\,dy$ **25.** $2x\tan y\,dx + \sec^2 y\,dy = 0$

In each case find conditions such that F is an integrating factor of (1). *Hint.* Assume $F(P\,dx + Q\,dy) = 0$ to be exact and apply (5) in Sec. 1.5.

26. $F = y^b$ **27.** $F = x^a$ **28.** $F = x^a y^b$ **29.** $F = e^y$

30. Using the result of Prob. 27, solve Probs. 11, 21 and 23.

In each case show that the given function is an integrating factor and find further integrating factors.

31. $2y\,dx + 3x\,dy = 0,\quad F = xy^2$ **32.** $4y\,dx = x\,dy,\quad F = x^3/y^2$

33. $3\,dx - 2x\cot y\,dy = 0,\quad F = x^2\csc^2 y$

34. (Checking) Checking of solutions is always important. In connection with the present method it is particularly essential, since one may have to exclude the function $y(x)$ given by $F(x, y) = 0$. To see this, consider

$$(xy)^{-1}\,dy - x^{-2}\,dx = 0;$$

show that an integrating factor is $F = y$ and leads to $d(y/x) = 0$, hence $y = cx$, where c is arbitrary, but $F = y = 0$ is not a solution of the original equation.

35. It is interesting to note that if $F(x, y)$ is an integrating factor, $1/F(x, y) = 0$ may sometimes yield an additional solution of the corresponding equation. To see this, consider

$$x\,dy - y\,dx = 0;$$

show that an integrating factor is $F = 1/x^2$ and leads to $y = cx$. Show that $1/F = x^2 = 0$ yields the additional solution $x = 0$ of the given equation which is not included in $y = cx$.

1.7   Linear Differential Equations

A first-order differential equation is said to be **linear** if it can be written

(1)
$$\boxed{y' + p(x)y = r(x).}$$

The characteristic feature of this equation is that it is linear in y and y', whereas p and r on the right may be *any* given functions of x.

If the right-hand side $r(x)$ is zero for all x in the interval in which we consider the equation (written $r(x) \equiv 0$), the equation is said to be **homogeneous;** otherwise it is said to be **nonhomogeneous.**

Let us find a formula for the general solution of (1) in some interval I, assuming that p and r are continuous in I. For the homogeneous equation

(2)
$$y' + p(x)y = 0$$

this is very simple. Indeed, by separating variables we have

$$\frac{dy}{y} = -p(x)\,dx \quad \text{and thus} \quad \ln|y| = -\int p(x)\,dx + c^*$$

or

(3)
$$y(x) = ce^{-\int p(x)\,dx} \qquad (c = \pm e^{c^*} \text{ when } y \gtrless 0);$$

here we may also take $c = 0$ and obtain the *trivial solution* $y \equiv 0$.

To solve the nonhomogeneous equation (1), let us write it in the form

$$(py - r)\,dx + dy = 0.$$

This is $P\,dx + Q\,dy = 0$, where $P = py - r$ and $Q = 1$. Hence (3) in Sec. 1.6 becomes simply

$$\frac{1}{F}\frac{dF}{dx} = p(x).$$

Theorem 1 in Sec. 1.6 now implies that our equation has an integrating factor $F(x)$ depending only on x. We get F by separating variables, $dF/F = p\,dx$, and integrating:

$$\ln|F| = \int p(x)\,dx.$$

Hence

$$F(x) = e^{h(x)} \qquad \text{where} \qquad h(x) = \int p(x)\,dx.$$

From this, $h' = p$. Hence (1) multiplied by $F = e^h$ can be written

$$e^h(y' + h'y) = e^h r.$$

But $(e^h y)' = e^h y' + e^h h'y$ by the product and chain rules, so that we have

$$(e^h y)' = e^h r.$$

We can now integrate,

$$e^h y = \int e^h r \, dx + c$$

Dividing both sides by e^h we finally arrive at the desired formula

(4)
$$y(x) = e^{-h} \left[\int e^h r \, dx + c \right], \qquad h = \int p(x) \, dx.$$

This represents the general solution of (1) in the form of an integral. (The choice of the value of the constant of integration in $\int p \, dx$ is immaterial; cf. Prob. 2 below.)

EXAMPLE 1
Solve the linear differential equation

$$y' - y = e^{2x}.$$

Solution. Here

$$p = -1, \qquad r = e^{2x}, \qquad h = \int p \, dx = -x$$

and from (4) we obtain the general solution

$$y(x) = e^x \left[\int e^{-x} e^{2x} \, dx + c \right] = e^x [e^x + c] = ce^x + e^{2x}.$$

Alternatively, we may multiply the given equation by $e^h = e^{-x}$, finding

$$(y' - y)e^{-x} = (ye^{-x})' = e^{2x}e^{-x} = e^x$$

and integrate on both sides, obtaining the same result as before:

$$ye^{-x} = e^x + c, \qquad \text{hence} \qquad y = e^{2x} + ce^x.$$

EXAMPLE 2
Solve

$$xy' + y + 4 = 0.$$

Solution. We write this equation in the form (1), that is,

$$y' + \frac{1}{x} y = -\frac{4}{x}.$$

Hence $p = \dfrac{1}{x}$, $r = -\dfrac{4}{x}$ and, therefore,

$$h = \int p \, dx = \ln|x|, \qquad e^h = x, \qquad e^{-h} = \frac{1}{x}.$$

From this and (4) we obtain the general solution

$$y(x) = \frac{1}{x} \left[\int x \left(-\frac{4}{x} \right) dx + c \right] = \frac{c}{x} - 4,$$

in agreement with Example 1 of Sec. 1.5.

Of course, in simple cases such as Examples 2, 3 and 5 (below) we may solve the equation without using (4). For instance, in Example 2 the student may write the given equation in the form $(xy)' + 4 = 0$ and can now obtain the solution by integration.

EXAMPLE 3

Solve the linear equation

$$xy' + y = \sin x.$$

Solution. We can write the equation in the form

$$(xy)' = \sin x$$

and integrate on both sides, finding

$$xy = -\cos x + c \qquad \text{or} \qquad y = \frac{1}{x}(c - \cos x).$$

Initial Value Problems

EXAMPLE 4. Initial value problem

Solve the initial value problem

$$y' + y \tan x = \sin 2x, \qquad y(0) = 1.$$

Solution. Here $p = \tan x$, $r = \sin 2x = 2 \sin x \cos x$, and

$$\int p \, dx = \int \tan x \, dx = \ln |\sec x|.$$

From this we see that in (4),

$$e^h = \sec x, \qquad e^{-h} = \cos x, \qquad e^h r = (\sec x)(2 \sin x \cos x) = 2 \sin x,$$

and the general solution of our equation is

$$y(x) = \cos x \left[2 \int \sin x \, dx + c \right] = c \cos x - 2 \cos^2 x.$$

According to the initial condition, $y = 1$ when $x = 0$, that is,

$$1 = c - 2 \qquad \text{or} \qquad c = 3.$$

Hence the solution of our initial value problem is

$$y = 3 \cos x - 2 \cos^2 x.$$

EXAMPLE 5. Initial value problem

Solve the initial value problem

$$x^2 y' + 2xy - x + 1 = 0, \qquad y(1) = 0.$$

Solution. The equation may be written as $(x^2 y)' = (x - 1)$. By integration we obtain

$$x^2 y = \frac{1}{2} x^2 - x + c \qquad \text{or} \qquad y(x) = \frac{1}{2} - \frac{1}{x} + \frac{c}{x^2}.$$

From this and the initial condition we have $y(1) = \frac{1}{2} - 1 + c = 0$ or $c = \frac{1}{2}$. Thus

$$y(x) = \frac{1}{2} - \frac{1}{x} + \frac{1}{2x^2}.$$

∎

Very often the solutions of differential equations in engineering applications are not elementary functions. For instance, the integral in (4) may not be elementary. If it is not a tabulated function,[10] we may develop the integrand in a power series and integrate term by term, or we may apply a method of numerical integration (cf. Sec. 18.5) or a numerical method for solving first-order differential equations (cf. Secs. 20.1 and 20.2).

Input and Output

In applications the independent variable x will often be time; the function $r(x)$ on the right-hand side of (1) may represent a force, and the solution $y(x)$ a displacement, a current, or some other variable physical quantity. In engineering mathematics $r(x)$ is frequently called the **input,** and $y(x)$ is called the **output** or *response to the input* (and the initial conditions). For instance, in electrical engineering the differential equation may govern the behavior of an electric circuit and the output $y(x)$ is obtained as the solution of that equation corresponding to the input $r(x)$. We shall discuss this idea in terms of typical examples in the next section and for second-order equations in Secs. 2.6, 2.13, 2.14.

In the next section we show that first-order linear differential equations have important applications to electric circuits.

Problems for Sec. 1.7

1. Show that $e^{-\ln x} = 1/x$ (but not $-x$) and $e^{-\ln (\sec x)} = \cos x$.
2. Show that the choice of the value of the constant of integration in $\int p\, dx$ [cf. (4)] is immaterial (so that we may choose it to be zero).

Find the general solutions of the following differential equations.

3. $y' - y = 2$	**4.** $y' + y = 10$	**5.** $y' - 2y = 1 - 2x$
6. $y' \tan x = y$	**7.** $xy' + y = 2x$	**8.** $y' + xy = 2x$
9. $y' + ky = e^{-kx}$	**10.** $y' + y = 2 \sin x$	**11.** $y' = 2y/x + x^2 e^x$
12. $y' + 3y = e^{2x} + 6$	**13.** $y' - 4y = x - 2x^2$	**14.** $y' + 2y = \cos x$
15. $xy' + 2y = 2e^{x^2}$	**16.** $(x^2 - 1)y' = xy$	**17.** $y' + 2xy = -6x$

Initial value problems. Solve:

18. $y' - y = e^x, \quad y(1) = 0$
19. $y' + y = (x + 1)^2, \quad y(0) = 0$
20. $y' - x^3 y = -4x^3, \quad y(0) = 6$
21. $y' - 2y = 2 \cosh 2x + 4, \quad y(0) = -1.25$
22. $y' - y \cot x = 2x - x^2 \cot x, \quad y(\frac{1}{2}\pi) = \frac{1}{4}\pi^2 + 1$
23. $y' - (1 + 3x^{-1})y = x + 2, \quad y(1) = e - 1$
24. $y' + y \tan x = 2x \cos x, \quad y(0) = -1$
25. $y' = 2(y - 1) \tanh 2x, \quad y(0) = 4$
26. $xy' = (1 + x)y, \quad y(2) = 6e^2$

---

[10]Whether tables of a certain function have been computed can be found in the useful book Ref. [8] listed in Appendix 1.

27. Suppose that p and r in (1) are continuous. Why can you conclude from (4) that an initial value problem consisting of (1) and an initial condition $y(x_0) = y_0$ has a unique solution?

General properties of linear differential equations. The *linear* differential equations (1) and (2) have certain important properties. In the next chapter (Secs. 2.3 and 2.9) we shall see that the same is true for *linear* differential equations of higher order. This fact is quite important, since we can use it to obtain new solutions from given ones. Indeed, prove and illustrate with an example that the **homogeneous equation** (2) has the following properties.

28. $y \equiv 0$ is a solution of (2), called the *trivial solution.*

29. If y_1 is a solution of (2), then $y = cy_1$ (c any constant) is a solution of (2).

30. If y_1 and y_2 are solutions of (2), then their sum $y = y_1 + y_2$ is a solution of (2).

Prove and illustrate with an example that the **nonhomogeneous equation (1)** has the following properties.

31. If y_1 is a solution of (1) and y_2 is a solution of (2), then $y = y_1 + y_2$ is a solution of (1).

32. The difference $y = y_1 - y_2$ of two solutions y_1 and y_2 of (1) is a solution of (2).

33. If y_1 is a solution of (1), then $y = cy_1$ is a solution of $y' + py = cr$.

34. If y_1 is a solution of $y_1' + py_1 = r_1$ and y_2 is a solution of $y_2' + py_2 = r_2$ (with the same p), then $y = y_1 + y_2$ is a solution of $y' + py = r_1 + r_2$.

35. If $p(x)$ and $r(x)$ in (1) are constant, say, $p(x) = p_0$ and $r(x) = r_0$, then (1) can be solved by separating variables and the result will agree with that obtained from (4).

Some Applications

36. **(Motion of a boat)** Two men are riding in a motorboat, the combined weight of the men and the boat being 4900 nt (about 1100 lb). Suppose that the motor exerts a constant force of 200 nt (about 45 lb) and the resistance R of the water is proportional to the speed v, say, $R = kv$ nt, where $k = 10$ nt · sec/meter. Set up the differential equation for $v(t)$, using Newton's second law

$$\text{Mass} \times \text{Acceleration} = \text{Force.}$$

Find $v(t)$ satisfying $v(0) = 0$. Find the maximum speed v_∞ at which the boat will travel (practically after a sufficiently long time). If the boat starts from rest, how long will it take to reach $0.9v_\infty$ and what distance does the boat travel during that time?

37. **Newton's law of cooling** leads to the differential equation (cf. Sec. 1.3)

$$\frac{dT}{dt} = -k(T - T_1)$$

where $T(t)$ is the temperature of a body placed in a medium which is kept at a constant temperature T_1. Solve the equation by the method developed in the present section, assuming the initial temperature of the body to be $T(0) = T_0$.

38. **(Atomic waste disposal)** The Atomic Energy Commission put atomic waste into sealed containers and dumped them in the ocean. It is important that the containers do not break when they hit the bottom of the ocean. Assume that this is the case as long as the speed is less than 12 meters/sec. Show that Newton's second law yields the equation of motion (Fig. 12)

$$m \frac{dv}{dt} = W - B - kv, \qquad v(0) = 0$$

where the drag force $D = -kv$ is assumed to be proportional to the velocity v. (This is experimentally confirmed for velocities that are not too large.) Solve the equation to obtain $v(t)$. Integrate to obtain $y(t)$ such that $y(0) = 0$. Determine the critical time t_{crit} when the container reaches the critical speed $v_{\text{crit}} = 12$ meters/sec, assuming that $W = 2254$ nt (about 500 lb), $B = 2090$ nt (about 460 lb), $k = 0.637$ kg/sec. Show that the container will break if it is dumped at points where the ocean is deeper than 105 meters, approximately.

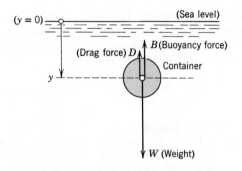

Fig. 12. System and forces in Prob. 38

39. The speed $v(t)$ in Prob. 38 approaches a limiting value as $t \to \infty$. How can this limit be obtained from the differential equation without actually solving it? What would happen if there were no drag force?

40. **(Air resistance)** An extended object falling downward is known to experience a resistive force of the air (called *drag*). We assume the magnitude of this force to be proportional to the speed v. Using Newton's second law, show that

$$m \frac{dv}{dt} = -kv - mg$$

where $g \ (= 9.80$ meters/sec^2) is the acceleration of gravity at the surface of the earth. Solve this equation (a) by the present method, (b) by separating variables, assuming that $v(0) = v_0$. (c) Integrating the result, find $y(t)$, the distance at time t measured from the starting point $y_0 = y(0)$.

Differential equations linear in dx/dy and x. Taking x as the dependent variable, solve:

41. $(e^{-3y} - 3x)y' = 1$ 42. $y' = 1/(2 \sin y - x)$

43. $(e^y + x)y' = 1$ 44. $(1 - 2y + 2y^{-1}x)y' = \frac{1}{2}y^{-1}$

45. $y' = y/(2x + y^3 e^y)$ 46. $y'(\sinh 3y - 2xy) = y^2$

Reduction of nonlinear equations to linear form. The problems on the next page illustrate that *certain nonlinear first-order differential equations may be reduced to linear form by a suitable change of the dependent variable.*

47. (Bernoulli equation) The differential equation

$$y' + p(x)y = g(x)y^a \qquad (a \text{ any real number})$$

is called the **Bernoulli**[11] **equation.** For $a = 0$ and $a = 1$ the equation is linear, and otherwise it is nonlinear. Set $[y(x)]^{1-a} = u(x)$ and show that the equation assumes the linear form

$$u' + (1 - a)p(x)u = (1 - a)g(x).$$

Solve the following Bernoulli equations.

48. $y' + y = y^2$ **49.** $y' + y = xy^{-1}$

50. $y' + x^{-1}y = xy^2$ **51.** $3y' + y = (1 - 2x)y^4$

52. $y' = x^3y^2 + xy$ **53.** $y' + xy = xy^{-1}$

54. (A population model. The logistic law) *Malthus's law* states that the time rate of change of a population $y(t)$ is proportional to $y(t)$ (cf. Prob. 30, Sec. 1.1). This holds for many populations as long as they are not too large. A more refined model is the *logistic law* given by

$$\frac{dy}{dt} = ay - by^2 \qquad (a > 0, b > 0),$$

where the "braking term" $-by^2$ has the effect that the population cannot grow indefinitely. Solve this Bernoulli equation. What is the limit of $y(t)$ as $t \to \infty$?

For the United States, Verhulst predicted in 1845 the values $a = 0.03$ and $b = 1.6 \cdot 10^{-4}$, where t is measured in years and $y(t)$ in millions. Find the particular solution satisfying $y(0) = 5.3$ (corresponding to the year 1800) and compare the values of this solution with some actual values

1800	1830	1860	1890	1920	1950	1980
5.3	13	31	63	105	150	230

Applying suitable substitutions, reduce to linear form and solve:

55. $y' \cos y + x \sin y = 2x$

56. $y' - 1 = e^{-y} \sin x$

57. $2xyy' + (x - 1)y^2 = x^2e^x$

58. $e^xy' + 2e^xy + x = 0$

59. Show that the transformation $x = \phi(t)$ (where ϕ has a continuous derivative) transforms (1) into a linear differential equation involving y and dy/dt.

60. Show that the transformation $y = a(x)u + b(x)$ with differentiable $a \neq 0$ and b transforms (1) into a linear differential equation involving u and du/dx.

---

[11]JAKOB BERNOULLI (1654—1705), Swiss mathematician, professor at Basel, also known for his contributions to elasticity theory and mathematical probability. The method for solving Bernoulli's equation was discovered by Leibniz in 1696. Jakob Bernoulli's students include his nephew NIKLAUS BERNOULLI (1687—1759), who contributed to probability theory and infinite series, and his youngest brother JOHANN BERNOULLI (1667—1748), who had profound influence on the development of calculus, became Jakob's successor at Basel, and had among his students GABRIEL CRAMER (cf. Sec. 7.9) and LEONHARD EULER (cf. Sec. 2.7). His son DANIEL BERNOULLI (1700—1782) is known for his basic work in fluid flow and the kinetic theory of gases.

1.8 Modeling: Electric Circuits

Linear first-order differential equations have various applications in physics and engineering. To illustrate this, let us consider some standard examples in connection with electric circuits. These and similar considerations are important, because they help *all* students to learn how to set up **models,** that is, express physical situations in terms of mathematical relations. This transition from the physical system to a corresponding mathematical model is called **modeling.** This is always the first step in engineering mathematics. Modeling requires experience and training, which can be gained only by considering typical examples from various fields.

The simplest electric circuit is a series circuit in which we have a source of electric energy (**electromotive force**) such as a generator or a battery, and a resistor, which uses energy, for example an electric light bulb (Fig. 13). If we close the switch, a current I will flow through the resistor, and this will cause a **voltage drop,** that is, the electric potential at the two ends of the resistor will be different; this potential difference or voltage drop can be measured by a voltmeter. Experiments show that the following law holds.

The **voltage drop E_R across a resistor** *is proportional to the instantaneous current I,* say,

(1)
$$E_R = RI$$
(Ohm's law)

where the constant of proportionality R is called the **resistance** of the resistor. The current I is measured in *amperes,* the resistance R in *ohms,* and the voltage E_R in *volts.*[12]

The other two important elements in more complicated circuits are *inductors* and *capacitors.* An inductor opposes a change in current, having an inertia effect in electricity similar to that of mass in mechanics; we shall consider this analogy later (Sec. 2.14). Experiments yield the following law.

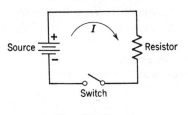

Fig. 13. Circuit

[12]These and the subsequent units are named after ANDRÉ MARIE AMPÈRE (1775—1836), French physicist; GEORG SIMON OHM (1789—1854), German physicist; ALESSANDRO VOLTA (1745—1827), Italian physicist; JOSEPH HENRY (1797—1878), American physicist; MICHAEL FARADAY (1791—1867), English physicist; and CHARLES AUGUSTIN DE COULOMB (1736—1806), French physicist.

The **voltage drop E_L across an inductor** *is proportional to the instantaneous time rate of change of the current I*, say,

$$\text{(2)} \qquad \boxed{E_L = L \frac{dI}{dt}}$$

where the constant of proportionality L is called the **inductance** of the inductor and is measured in *henrys;* the time t is measured in seconds.

A capacitor is an element which stores energy. Experiments yield the following law.

The **voltage drop E_C across a capacitor** *is proportional to the instantaneous electric charge Q on the capacitor*, say,

$$\text{(3*)} \qquad \boxed{E_C = \frac{1}{C} Q}$$

where C is called the **capacitance** and is measured in *farads;* the charge Q is measured in *coulombs.* Since

$$\text{(3')} \qquad I(t) = \frac{dQ}{dt}$$

this may be written

$$\text{(3)} \qquad E_C = \frac{1}{C} \int_{t_0}^{t} I(t^*) \, dt^*.$$

The current $I(t)$ in a circuit may be determined by solving the equation (or equations) resulting from the application of the following physical law.

Kirchhoff's voltage law (KVL)[13]

The algebraic sum of all the instantaneous voltage drops around any closed loop is zero, or the voltage impressed on a closed loop is equal to the sum of the voltage drops in the rest of the loop.

EXAMPLE 1. *RL-circuit*
For the "RL-circuit" in Fig. 14 we obtain from Kirchhoff's voltage law and (1), (2)

$$\text{(4)} \qquad L \frac{dI}{dt} + RI = E(t).$$

Case A (Constant electromotive force). If in (4) we have $E = E_0 = const$, then from (4) in Sec.

---

[13]GUSTAV ROBERT KIRCHHOFF (1824—1887), German physicist. Later we shall also need **Kirchhoff's current law (KCL):**
At any point of a circuit, the sum of the inflowing currents is equal to the sum of the outflowing currents.

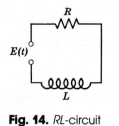

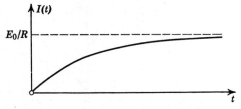

Fig. 14. *RL*-circuit

Fig. 15. Current in an *RL*-circuit due to a constant electromotive force

1.7 we obtain the general solution

$$I(t) = e^{-\alpha t}\left[\frac{E_0}{L}\int e^{\alpha t}\,dt + c\right] \qquad (\alpha = R/L)$$

(5)

$$= \frac{E_0}{R} + ce^{-(R/L)t}.$$

The last term approaches zero as t tends to infinity, so that $I(t)$ tends to the limiting value E_0/R; after a sufficiently long time, I will practically be constant, its value being independent of c, hence independent of an initial condition which we may want to impose. The particular solution corresponding to the initial condition $I(0) = 0$ is (Fig. 15)

(5*)
$$I(t) = \frac{E_0}{R}(1 - e^{-(R/L)t}) = \frac{E_0}{R}(1 - e^{-t/\tau_L})$$

where $\tau_L = L/R$ is called the **inductive time constant** of the circuit (cf. Probs. 5, 6, 10).

Case B (Periodic electromotive force). If $E(t) = E_0 \sin \omega t$, then, by (4) in Sec. 1.7, the general solution of (4) in this current section is

$$I(t) = e^{-\alpha t}\left[\frac{E_0}{L}\int e^{\alpha t}\sin \omega t\,dt + c\right] \qquad (\alpha = R/L).$$

Integration by parts yields

$$I(t) = ce^{-(R/L)t} + \frac{E_0}{R^2 + \omega^2 L^2}(R \sin \omega t - \omega L \cos \omega t).$$

This may be written [cf. (14) in Appendix 3]

(6)
$$I(t) = ce^{-(R/L)t} + \frac{E_0}{\sqrt{R^2 + \omega^2 L^2}}\sin(\omega t - \delta), \qquad \delta = \arctan \frac{\omega L}{R}.$$

The exponential term will approach zero as t approaches infinity. This means that after a sufficiently long time the current $I(t)$ executes practically harmonic oscillations. (Cf. Fig. 16.)

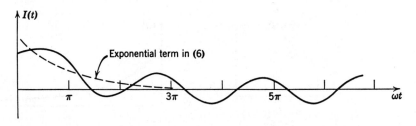

Fig. 16. Current in an *RL*-circuit due to a sinusoidal electromotive force, as given by (6) (with $\delta = \pi/4$)

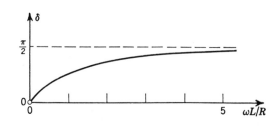

Fig. 17. Phase angle δ in (6) as a function of $\omega L/R$

Figure 17 shows the phase angle δ as a function of $\omega L/R$. If $L = 0$, then $\delta = 0$, and the oscillations of $I(t)$ are in phase with those of $E(t)$. ∎

An electrical (or dynamical) system is said to be in the **steady state** when the variables describing its behavior are periodic functions of time or constant, and it is said to be in the **transient state** (or *unsteady state*) when it is not in the steady state. The corresponding variables are called *steady-state functions* and *transient functions,* respectively.

In Example 1, Case A, the function E_0/R is a steady-state function or **steady-state solution** of (4), and in Case B the steady-state solution is represented by the last term in (6). Before the circuit (practically) reaches the steady state it is in the transient state. It is clear that such an interim or transient period occurs because inductors and capacitors store energy, and the corresponding inductor currents and capacitor voltages cannot be changed instantly. Practically, this transient state will last only a short time.

EXAMPLE 2. *RC-circuit*
By applying Kirchhoff's voltage law and (1), (3) to the *RC*-circuit in Fig. 18 we obtain the equation

$$(7) \qquad\qquad RI + \frac{1}{C} \int I \, dt = E(t).$$

To get rid of the integral we differentiate the equations with respect to t, finding

$$(8) \qquad\qquad R \frac{dI}{dt} + \frac{1}{C} I = \frac{dE}{dt}.$$

According to (4) in Sec. 1.7 this differential equation has the general solution

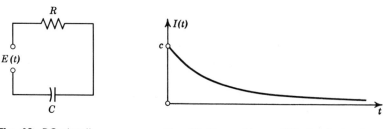

Fig. 18. *RC*-circuit

Fig. 19. Current in an *RC*-circuit due to a constant electromotive force

(9) $$I(t) = e^{-t/(RC)} \left(\frac{1}{R} \int e^{t/(RC)} \frac{dE}{dt} \, dt + c \right).$$

Case A *(Constant electromotive force).* If $E = const$, then $\dfrac{dE}{dt} = 0$, and (9) assumes the simple form

(10) $$I(t) = ce^{-t/(RC)} = ce^{-t/\tau_C} \qquad \text{(Fig. 19)}$$

where $\tau_C = RC$ is called the **capacitive time constant** of the circuit.

Case B *(Sinusoidal electromotive force).* If $E(t) = E_0 \sin \omega t$, then

$$\frac{dE}{dt} = \omega E_0 \cos \omega t.$$

By inserting this in (9) and integrating by parts we find

$$I(t) = ce^{-t/(RC)} + \frac{\omega E_0 C}{1 + (\omega RC)^2} (\cos \omega t + \omega RC \sin \omega t)$$

(11)

$$= ce^{-t/(RC)} + \frac{\omega E_0 C}{\sqrt{1 + (\omega RC)^2}} \sin (\omega t - \delta),$$

where $\tan \delta = -1/(\omega RC)$. The first term decreases steadily as t increases, and the last term represents the steady-state current, which is sinusoidal. The graph of $I(t)$ is similar to that in Fig. 16. ∎

More complicated circuits and the analogy between electrical and mechanical vibrations will be considered in connection with second-order differential equations in Sec. 2.14.

These more complicated circuits will be called ***RLC*-circuits** since they contain resistors, inductors and capacitors. Their mathematical model will be a second-order differential equation obtained from Kirchhoff's laws.

Electrical networks will be considered in Secs. 3.1, 5.8, 7.5 and in the Review Problems to Chap. 7.

In the next section we discuss another basic application of differential equations, namely, the determination of curves that intersect given curves at right angles. This task arises quite frequently in physics and geometry, as our examples will illustrate.

Problems for Sec. 1.8

RL-Circuits

1. Derive (5) from (4). Derive (5*) from (5).

2. Obtain from (5) the solutions satisfying the initial conditions $I(0) = 0.1E_0/R$, E_0/R, $3E_0/R$. Sketch the corresponding curves.

3. In (5) the solution approaches E_0/R as $t \to \infty$. Could this be seen directly from (4) with $E(t) = E_0$, without solving (4)?

4. In Prob. 2, the current is an increasing function when $I(0) < E_0/R$, and a decreasing function when $I(0) > E_0/R$. Could this be seen directly from (4) with $E(t) = E_0$, without solving (4)?

5. Show that the inductive time constant $\tau_L = L/R$ has the (physical) dimensions of time and is the time t at which the current (5*) reaches about 63% of its final value.

6. Show that the current (5*) builds up to half its theoretical maximum value in $\tau_L \ln 2 \approx 0.7 \, \tau_L$ sec.

7. What L should we choose in (4) with $E = E_0 = 100$ volts and $R = 1000$ ohms if we want the current to grow from 0 to 25% of its final value within 10^{-4} sec?

8. In Example 1, Case A, let $R = 12$ ohms, $L = 0.02$ millihenry $(2 \cdot 10^{-5}$ henry) and $I(0) = 0$. Find the time when the current reaches 99.9% of its final value.

9. Will the time in Prob. 8 decrease or increase if we change $I(0) = 0$ to some positive value less than E_0/R?

10. In Example 1, Case A, let $E_0 = 400$ volts, $R = 2000$ ohms and $L = 6$ henrys. Find τ_L. Graph (5*) and the voltage drops E_R and E_L [cf. (1) and (2)].

11. In Example 1, case A, let $R = 100$ ohms, $L = 2.5$ henrys, $E_0 = 110$ volts and $I(0) = 0$. Find the time constant and the time necessary for the current to rise from 0 to 0.6 ampere.

12. If $L = 10$ henrys, what R should we choose in order that (5*) reach 99% of its final value at $t = 1$ sec?

13. Derive (6) from (4) and check it by substitution.

14. Derive the steady-state solution in (6) by substituting $I_p = A \cos \omega t + B \sin \omega t$ into (4) with $E(t) = E_0 \sin \omega t$ and determining A and B by equating the cosine and sine terms in the resulting equation. (Note that in this way one avoids integration bv parts.)

15. How does the phase angle δ in (6) depend on L? Is this physically understandable?

16. For what initial condition does (6) yield the steady-state solution?

17. Solve (4) with electromotive force $E(t) = t$.

18. Solve (4) with $E(t) = e^{-t}$ when (a) $R \neq L$, (b) $R = L$.

19. Find the current in the RL-circuit [cf. (4)] with $R = 1$ ohm, $L = 1$ henry and $E(t) = 1$ if $0 \leq t \leq 3$ sec, $E(t) = 0$ if $t > 3$ sec and $I(0) = \frac{1}{2}$ ampere. *Hint.* Use that $I(t)$ must be continuous at $t = 3$.

20. Periodic electromotive forces other than pure sine or cosine functions occur quite frequently in applications. An example is shown in Fig. 20, where the discontinuities (jumps) are convenient mathematical approximations of very abrupt changes of $E(t)$ from 0 to 10 and conversely. Assume that this $E(t)$ is applied to an RL-circuit with $R = 10$ ohms and $L = 200$ henrys. Assuming $I(0) = 0$, find the current $I(t)$.

21. It is worth noting that periodicity of the input does not entail periodicity of the output since, in general, there will be a transient state. But there are obvious exceptions. To illustrate this, show that $I(t)$ in Prob. 20 satisfies $I(t + 20) = I(t)$ for all $t \geq 0$ if and only if we choose the initial condition

$$I(0) = (e^{1/2} - 1)/(e - 1) \approx 0.378.$$

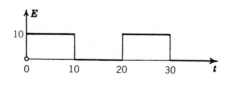

Fig. 20. Problem 20

22. Obtain from (6) the particular solution satisfying the initial condition $I(0) = 0$.

23. Sketch the solutions in Probs. 20 and 21.

RC-Circuits

24. (Discharge of a capacitor) Show that (7) can also be written

(12)
$$R \frac{dQ}{dt} + \frac{1}{C} Q = E(t).$$

Solve this equation with $E(t) = 0$, assuming $Q(0) = Q_0$. Find the time when the capacitor has lost 99.9% of its initial charge.

25. Write the general solution of (12) in the form of an integral and derive from it formula (9) by differentiation and subsequent integration by parts.

26. In Prob. 24, let $R = 10$ ohms, $C = 0.1$ farad and $E(t)$ be exponentially decaying, say, $E(t) = 30e^{-3t}$ volts. Assuming $Q(0) = 0$, find and graph $Q(t)$. Also determine the time when $Q(t)$ reaches a maximum and that maximum charge.

27. Verify that (11) is a solution of (8) with $E = E_0 \sin \omega t$.

28. Obtain from (11) the particular solution satisfying the initial condition $I(0) = 0$.

29. A capacitor ($C = 0.1$ farad) in series with a resistor ($R = 200$ ohms) is charged from a source ($E_0 = 12$ volts); see Fig. 18 with $E(t) = E_0$. Find the voltage $V(t)$ on the capacitor, assuming that at $t = 0$ the capacitor is completely uncharged.

30. Find the current $I(t)$ in the *RC*-circuit shown in Fig. 18, assuming that $E = 100$ volts, $C = 0.25$ farad, R is variable according to $R = (100 - t)$ ohms when $0 \le t \le 100$ sec, $R = 0$ when $t > 100$ sec, and $I(0) = 1$ ampere.

31. Solve (12) when $R = 500$ ohms, $C = 10^{-3}$ farad and $E(t) = 1 - e^{-t}$ volt.

Find the steady-state solution of (12) when $R = 50$ ohms, $C = 0.04$ farad and $E(t)$ equals:

32. $125 \sin t$ **33.** $110 \cos 314t$

34. $100 \cos 2t + 25 \sin 2t + 200 \cos 4t + 25 \sin 4t$

35. $50e^{-t} + 1012 \sin \pi t$

1.9   Families of Curves. Orthogonal Trajectories

In this section we shall learn how to use differential equations for finding curves that intersect given curves at right angles, a task that arises rather often in applications.

If for each fixed real value of c the equation

(1)
$$F(x, y, c) = 0$$

represents a curve in the xy-plane and if for variable c it represents infinitely many curves, then the totality of these curves is called a **one-parameter family of curves,** and c is called the *parameter* of the family.

EXAMPLE 1. Families of curves

The equation

(2) $$F(x, y, c) = x + y + c = 0$$

represents a family of parallel straight lines; each line corresponds to precisely one value of the parameter c. The equation

(3) $$F(x, y, c) = x^2 + y^2 - c^2 = 0$$

represents a family of concentric circles of radius c with center at the origin.   ∎

The general solution of a first-order differential equation involves a parameter c and thus represents a family of curves. This yields a possibility for representing many one-parameter families of curves by first-order differential equations. The practical use of such representations will become obvious from our further considerations.

EXAMPLE 2. Differential equations of families of curves

By differentiating (2) we see that

$$y' + 1 = 0$$

is the differential equation of that family of straight lines. Similarly, the differential equation of the family (3) is obtained by differentiation, $2x + 2yy' = 0$, that is,

$$y' = -x/y.$$ ∎

If the equation obtained by differentiating (1) still contains the parameter c, then we have to eliminate c from this equation by using (1). Let us illustrate this by a simple example.

EXAMPLE 3. Elimination of the parameter of a family

The differential equation of the family of parabolas

(4) $$y = cx^2$$

is obtained by differentiating (4),

(5) $$y' = 2cx,$$

and by eliminating c from (5). From (4) we have $c = y/x^2$, and by substituting this into (5) we find the desired result

(6) $$y' = 2y/x.$$

Note that we may also proceed as follows. We solve (4) for c, finding $c = y/x^2$, and differentiate this equation with respect to x, obtaining

$$0 = \frac{y'}{x^2} - \frac{2y}{x^3}.$$

Hence, as before,

$$y' = \frac{2y}{x}.$$ ∎

Orthogonal Trajectories

In many engineering and other applications, a family of curves is given, and it is required to find another family whose curves intersect each of the given curves at right angles.[14] Then the curves of the two families are said to be *mutually orthogonal,* they form an *orthogonal net,* and the curves of the family to be obtained are called the **orthogonal trajectories** of the given curves (and conversely); cf. Fig. 21.

Let us mention some familiar examples. The meridians on the earth's surface are the orthogonal trajectories of the parallels. On a map the curves of steepest descent are the orthogonal trajectories of the contour lines. In electrostatics the equipotential lines and the lines of electric force are orthogonal trajectories of each other. An illustrative example is shown in Fig. 22. We shall see later that orthogonal trajectories are important in various fields of physics, for example, in hydrodynamics and heat conduction.

Given a family of curves $F(x, y, c) = 0$ that can be represented by a differential equation

$$(7) \qquad\qquad y' = f(x, y)$$

we may find the corresponding orthogonal trajectories as follows. From (7) we see that a curve of the given family that passes through a point (x_0, y_0) has the slope $f(x_0, y_0)$ at this point. The slope of the orthogonal trajectory through (x_0, y_0) at this point should be the negative reciprocal of $f(x_0, y_0)$, that is, $-1/f(x_0, y_0)$, because this is the condition for the tangents of the two curves at (x_0, y_0) to be perpendicular. Consequently, the differential equation of the orthogonal trajectories is

$$(8) \qquad\qquad \boxed{\; y' = -\dfrac{1}{f(x, y)} \;}$$

and the trajectories are obtained by solving this new differential equation.

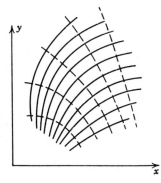

Fig. 21. Curves and their orthogonal trajectories

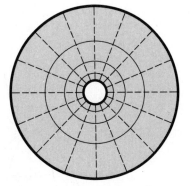

Fig. 22. Equipotential lines and lines of electric force (dashed) between two concentric cylinders

[14]Remember that the **angle of intersection** of two curves is defined to be the angle between the tangents of the curves at the point of intersection.

EXAMPLE 4. Orthogonal trajectories

Find the orthogonal trajectories of the parabolas in Example 3.

Solution. From (6) we see that the differential equation (8) of the orthogonal trajectories is

$$y' = -\frac{1}{2y/x} = -\frac{x}{2y}.$$

By separating variables and integrating we find that the orthogonal trajectories are the ellipses

$$\frac{x^2}{2} + y^2 = c^* \qquad \text{(Fig. 23)}.$$

EXAMPLE 5. Orthogonal trajectories

Find the orthogonal trajectories of the circles

$$(9) \qquad\qquad x^2 + (y - c)^2 = c^2.$$

Solution. We first determine the differential equation of the given family. By differentiating (9) with respect to x we obtain

$$(10) \qquad\qquad 2x + 2(y - c)y' = 0.$$

We must eliminate c. Solving (9) for c, we have

$$(11) \qquad\qquad c = \frac{x^2 + y^2}{2y}.$$

By inserting this into (10) and simplifying we get

$$x + \frac{y^2 - x^2}{2y}y' = 0 \qquad \text{or} \qquad y' = \frac{2xy}{x^2 - y^2}.$$

From this and (8) we see that the differential equation of the orthogonal trajectories is

$$y' = -\frac{x^2 - y^2}{2xy} \qquad \text{or} \qquad 2xyy' - y^2 + x^2 = 0.$$

The orthogonal trajectories obtained by solving this equation (cf. Example 1 in Sec. 1.4) are the circles (Fig. 24)

$$(x - \tilde{c})^2 + y^2 = \tilde{c}^2. \qquad\qquad ■$$

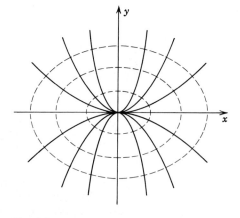

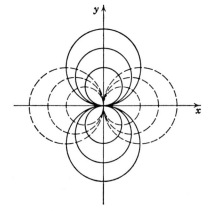

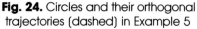

Fig. 23. Parabolas and their orthogonal
trajectories in Example 4

Fig. 24. Circles and their orthogonal
trajectories (dashed) in Example 5

In the next section we motivate and discuss two methods of obtaining approximate solutions without actually solving a given differential equation. The first method, called the *method of* **direction fields,** can relatively easily produce a general picture of all solutions (with limited accuracy) and is of great practical interest. The second method, **Picard's iteration,** is more theoretical; its practical value is limited, since it involves integrations.

Problems for Sec. 1.9

What families of curves are represented by the following equations? Sketch some of the curves.

1. $4y - x + c = 0$
2. $(x - c)^2 + y^2 = 4$
3. $y + 2(x - c)^2 = 0$
4. $(x - c)^2 + y^2 = c^2/2$
5. $y^2 - (x - c)^3 = 0$
6. $y = ce^x$
7. $xy = c$
8. $x^2 - y^2 = c$
9. $c^2 x^2 + y^2 = 1$

Represent the following families of curves in the form (1). Sketch some of the curves.
10. All nonvertical straight lines through the point $(4, -1)$.
11. All ellipses with foci -2 and 2 on the y-axis.
12. All circles of radius 3 with centers on the parabola $y = x^2$
13. All circles through the points -1 and 1 on the x-axis
14. The catenaries obtained by translating the catenary $y = \cosh x$ in the direction of the straight line $y = -x$.

Represent the following families of curves by differential equations.
15. $y = ce^x$
16. $xy = c$
17. $y = cx^3$
18. $y = e^{cx}$
19. $y = \tan(x + c)$
20. $c^2 x^2 + y^2 = c^2$
21. $y = c \sin x$
22. $y = c/x + x$
23. $\cos y = c/\cosh x$

Using differential equations, find the orthogonal trajectories of the following curves. Graph some of the curves and the trajectories.
24. $y = 2x + c$
25. $y = -\frac{1}{2}x^2 + c$
26. $y = \ln |x| + c$
27. $x^2 + 2y^2 = c$
28. $y = cx^3$
29. $y = ce^x$
30. $xy = c$
31. $y^2 - x^2 = c$
32. $y = ce^{x^2}$
33. $x^2 + y^2 = c^2$
34. $y = cx^{3/2}$
35. $y = c\sqrt{x}$

36. Find the orthogonal trajectories of the family $y = cx^n$, where n is any real number.

Applications of orthogonal trajectories

37. **(Electric field)** If an electrical current is flowing in a wire along the z-axis, the resulting *equipotential lines* in the xy-plane are concentric circles about the origin, and the *electric lines of force* are the orthogonal trajectories of these circles. Find the differential equation of these trajectories and solve it.

38. **(Electric field)** Experiments show that the electric lines of force of two opposite charges of the same strength at $(-1, 0)$ and $(1, 0)$ are the circles through $(-1, 0)$ and $(1, 0)$. Show that these circles can be represented by the equation $x^2 + (y - c)^2 = 1 + c^2$. Show that the equipotential lines (orthogonal trajectories) are the circles $(x + c^*)^2 + y^2 = c^{*2} - 1$, which are dashed in Fig. 25 on the next page.

39. **(Fluid flow)** Another area in which orthogonal trajectories play a role is fluid flow. Here the path of a particle of fluid is called a **streamline,** and the orthogonal trajectories of the streamlines are called **equipotential lines** (for reasons to be discussed in some other context, in Sec. 17.4). Suppose that the streamlines are $xy = c$. Show that the walls of the channel in Fig. 26 are streamlines, so that the flow may be regarded as a flow around a corner. Find and graph the equipotential lines.

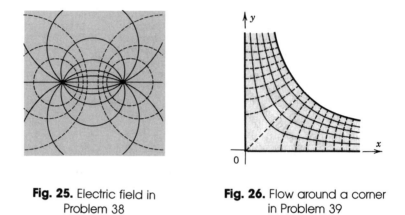

Fig. 25. Electric field in   **Fig. 26.** Flow around a corner
         Problem 38   in Problem 39

40. **(Temperature field)** Curves of constant temperature $T(x, y) = const$ in a temperature field are called **isotherms.** Their orthogonal trajectories are the curves along which heat will flow (in regions that are free of sources or sinks of heat and are filled with a homogeneous medium). If the isotherms are given by $y^2 + 2xy - x^2 = c$, what are the curves of heat flow?

Other forms of the differential equations. Isogonal trajectories

41. Show that (8) may be written in the following form and use this result for determining the orthogonal trajectories of the curves $y = \sqrt{x + c}$.

$$\frac{dx}{dy} = -f(x, y).$$

42. Show that the orthogonal trajectories of a given family $g(x, y) = c$ can be obtained from the differential equation

$$\frac{dy}{dx} = \frac{\partial g/\partial y}{\partial g/\partial x}.$$

Using this equation, find the orthogonal trajectories of the curves $ye^{2x} = c$.

43. **(Cauchy–Riemann equations)** Show that if the equation $u(x, y) = c = const$ represents a family of curves, a representation of its orthogonal trajectories of the form $v(x, y) = c^* = const$ can be obtained from

$$\frac{\partial u}{\partial x} = \frac{\partial v}{\partial y}, \qquad \frac{\partial u}{\partial y} = -\frac{\partial v}{\partial x}.$$

(These so-called *Cauchy–Riemann differential equations* are basic in complex analysis and will be considered in Chap. 12.)

44. **Isogonal trajectories** of a given family of curves are curves that intersect the given curves at a constant angle θ. Show that at each point the slopes m_1 and m_2 of the tangents to the corresponding curves satisfy the relation

$$\frac{m_2 - m_1}{1 + m_1 m_2} = \tan \theta = const.$$

Using this formula, find the curves which intersect the circles $x^2 + y^2 = c$ at an angle of 45°.

45. Using the Cauchy–Riemann equations (Prob. 43), find the orthogonal trajectories of $e^x \cos y = c$.

1.10 Approximate Solutions: Direction Fields, Iteration

In applications, it will often be impossible or not feasible or not necessary to solve differential equations exactly. Indeed, there are various differential equations, even of the first order, for which one cannot obtain formulas for solutions.[15] There are other differential equations for which such formulas can be derived, but they are so complicated that they are practically useless. Finally, since a differential equation is a model of a physical or other system, and in modeling we disregard factors of minor influence in order to keep the model simple, the differential equation will describe the given situation only approximately, and an approximate solution will often be practically as informative as an exact solution.

Approximate solutions of differential equations can be obtained by **numerical methods.** These are discussed in Secs. 20.1 and 20.2. At present we shall consider the *method of direction fields,* which is a geometric procedure, and then the so-called *Picard iteration,* which gives *formulas* for approximate solutions.

Method of Direction Fields

In this method we get a rough picture of all solutions of a given differential equation

(1)
$$y' = f(x, y)$$

without actually solving the equation. The idea is quite natural and simple, as follows.

We assume that the function f is defined in some region of the xy-plane, so that at each point in that region it has one (and only one) value. The

---

[15]Reference [A11] in Appendix 1 includes more than 1500 important differential equations and their solutions, arranged in systematic order and accompanied by numerous references to original literature.

solutions of (1) can be plotted as curves in the xy-plane. We do not know the solutions, but we see from (1) that a solution passing through a point (x_0, y_0) must have the slope $f(x_0, y_0)$ at this point. This suggests the following method.

1st Step (Isoclines). We graph some of the curves in the xy-plane along which $f(x, y)$ is constant. These curves

$$f(x, y) = k = const$$

are called *curves of constant slope* or **isoclines.** Here the value of k differs from isocline to isocline. So these are not yet the solution curves of (1), but just auxiliary curves.

2nd Step (Direction field). Along the isocline $f(x, y) = k$ we draw a number of parallel short line segments **(lineal elements)** with slope k, which is the slope of solution curves of (1) at any point of that isocline. This we do for all isoclines which we graphed before. In this way we obtain a field of lineal elements, called the **direction field** of (1).

3rd Step (Approximate solution curves). With the help of the lineal elements we can now easily graph approximation curves to the (unknown) solution curves of the given equation (1) and thus obtain a qualitatively correct picture of these solution curves.

It suffices to illustrate the method by a simple equation that can be solved exactly, so that we get a feeling for the accuracy of the method.

EXAMPLE 1. Isoclines, direction field
Graph the direction field of the first-order differential equation

(2)
$$y' = xy$$

and an approximation to the solution curve through the point $(1, 2)$. Compare with the exact solution.

Solution. The isoclines are the equilateral hyperbolas $xy = k$ together with the two coordinate axes. We graph some of them. Then we draw lineal elements by sliding a triangle along a fixed ruler. The result is shown in Fig. 27, which also shows an approximation to the solution curve passing through the point $(1, 2)$.

By separating variables, $y = ce^{x^2/2}$. The initial condition is $y(1) = 2$. Hence $2 = ce^{1/2}$, and the exact solution is

$$y = 2e^{(x^2-1)/2}. \qquad\blacksquare$$

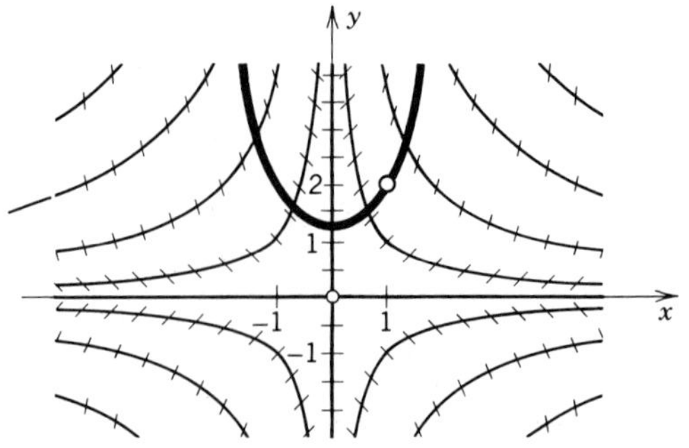

Fig. 27. Direction field of the differential equation (2)

Picard's Iteration Method[16]

This method gives approximate solutions of an initial value problem

$$(3) \qquad \boxed{y' = f(x, y), \quad y(x_0) = y_0}$$

which is assumed to have a unique solution in some interval on the x-axis containing x_0. Picard's method is of great theoretical value in connection with Picard's existence and uniqueness theorem, which we shall discuss in the next section. Its practical value is limited because it involves integrations that may be complicated.

The basic idea of Picard's method is very simple. By integration we see that (3) may be written in the form

$$(4) \qquad \boxed{y(x) = y_0 + \int_{x_0}^{x} f[t, y(t)]\, dt}$$

where t denotes the variable of integration. In fact, when $x = x_0$ the integral is zero and $y = y_0$, so that (4) satisfies the initial condition in (3); furthermore, by differentiating (4) we obtain the differential equation in (3).

To find approximations to the solution $y(x)$ of (4) we proceed as follows. We substitute the crude approximation $y = y_0 = const$ on the right; this yields the presumably better approximation

$$y_1(x) = y_0 + \int_{x_0}^{x} f(t, y_0)\, dt.$$

In the next step we substitute the function $y_1(x)$ in the same way to get

$$y_2(x) = y_0 + \int_{x_0}^{x} f[t, y_1(t)]\, dt,$$

etc. The nth step of this iteration gives an approximating function

$$(5) \qquad \boxed{y_n(x) = y_0 + \int_{x_0}^{x} f[t, y_{n-1}(t)]\, dt.}$$

In this way we obtain a sequence of approximations

$$y_1(x), \qquad y_2(x), \cdots, \qquad y_n(x), \cdots,$$

and we shall see in the next section that the conditions under which this sequence converges to the solution $y(x)$ of (3) are relatively general.

[16]EMILE PICARD (1856—1941), French mathematician, professor in Paris since 1881, also known for his important contributions to complex analysis (see Sec. 14.10 for his famous theorem).

An **iteration method** is a method that yields a sequence of approximations to an (unknown) function, say, y_1, y_2, $\cdots$, where the nth approximation, y_n, is obtained in the nth step by using one (or several) of the previous approximations, and the operation performed in each step is the same. This is a practical advantage, for instance in programming for numerical work.

In the simplest case, y_n is obtained from y_{n-1}; denoting the operation by T, we may write

$$y_n = T(y_{n-1}).$$

Picard's method is of this type, because (5) may be written

$$y_n(x) = T(y_{n-1}(x)) = y_0 + \int_{x_0}^{x} f(t, y_{n-1}(t))\, dt.$$

To illustrate the method, let us apply it to an equation we can readily solve exactly, so that we may compare the approximations with the exact solution. The example to be discussed will also illustrate that the question of the convergence of the method is of practical interest.

EXAMPLE 2. Picard iteration

Find approximate solutions to the initial value problem

$$y' = 1 + y^2, \qquad y(0) = 0.$$

Solution. In this case, $x_0 = 0$, $y_0 = 0$, $f(x, y) = 1 + y^2$, and (5) becomes

$$y_n(x) = \int_0^x [1 + y_{n-1}^2(t)]\, dt = x + \int_0^x y_{n-1}^2(t)\, dt.$$

Starting from $y_0 = 0$, we thus obtain (cf. Fig. 28)

$$y_1(x) = x + \int_0^x 0\, dt = x$$

$$y_2(x) = x + \int_0^x t^2\, dt = x + \tfrac{1}{3}x^3$$

$$y_3(x) = x + \int_0^x \left(t + \frac{t^3}{3}\right)^2 dt = x + \tfrac{1}{3}x^3 + \tfrac{2}{15}x^5 + \tfrac{1}{63}x^7$$

etc. Of course, we can obtain the exact solution of our present problem by separating variables (see Example 2 in Sec. 1.2), finding

(6) $$y(x) = \tan x = x + \tfrac{1}{3}x^3 + \tfrac{2}{15}x^5 + \tfrac{17}{315}x^7 + \cdots \qquad \left(-\frac{\pi}{2} < x < \frac{\pi}{2}\right).$$

The first three terms of $y_3(x)$ and the series in (6) are the same. The series in (6) converges for $|x| < \pi/2$, and all we may expect is that our sequence y_1, y_2, $\cdots$ converges to a function that is the solution of our problem for $|x| < \pi/2$. This illustrates that the study of convergence is of practical importance. ∎

The next section, the last of Chap. 1, concerns the problems of **existence** and **uniqueness** of solutions of first-order differential equations. These problems are of greater relevance to engineering applications than one would at first be inclined to believe. This is so because modeling involves the discarding of minor factors, and in more complicated situations it is often difficult to see whether some physical factor will have a minor or major effect, so that one may not be sure whether a model is faithful and does have a solution, or a unique solution, even though the physical system can be expected to behave reasonably. The matter becomes even more crucial in connection with numerical methods: make sure that the solution exists before you try to compute it.

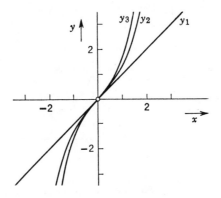

Fig. 28. Approximate solutions in Example 2

Problems for Sec. 1.10

Direction Fields

In each case draw a good direction field. Plot several approximate solution curves. Then solve the equation analytically and compare, to get a feeling for the accuracy of the present method.

1. $y' = -y/x$ 2. $y' = -x/y$ 3. $y' = -xy$

4. $y' = x + y$ 5. $4yy' + x = 0$ 6. $y' = (y - x)/(y + x)$

7. **(Skydiver)** Graph the direction field of the differential equation

$$\frac{dv}{dt} = 10 - 0.4v^2$$

and draw from it the following conclusions. The isoclines are horizontal straight lines. The isocline $v = 5$ is at the same time a solution curve. All solution curves in the upper half-plane ($v > 0$) seem to approach the line $v = 5$ as $t \to \infty$; they are monotone increasing if $0 < v(0) < 5$ and monotone decreasing if $v(0) > 5$. Sketch the solution curves for which $v(0) = 0$ and $v(0) = 7$ and compare with the exact solutions obtained from (8), Sec. 1.3.

8. **(A pursuit problem, tractrix)** In a pursuit problem, an object or *target* moves along a given curve and a second object or *pursuer* follows or *pursues* the target, that is, moves in the direction of the target at all times. Assume the target to be a ship S that moves along a straight line and the pursuer to be a destroyer D that moves so that the distance a from D to S is constant, say, $a = 1$ (nautical mile). Show that (Fig. 29)

$$y' = -y/\sqrt{a^2 - y^2}.$$

Draw a direction field (using $a = 1$) and plot an approximate solution curve corresponding to $y(0) = a = 1$; see Fig. 29. (Such a curve is called a *tractrix*, from Latin *trahere* = to pull.) Separating variables, show that

$$x = -\int y^{-1}\sqrt{a^2 - y^2}\, dy = -\sqrt{a^2 - y^2} + a \ln|y^{-1}(a + \sqrt{a^2 - y^2})| + c.$$

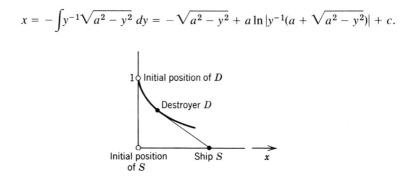

Fig. 29. Tractrix and notations in Problem 8

9. **(Verhulst population model)** Draw the direction field of the differential equation in Prob. 54 of Sec. 1.7, with $a = 0.03$ and $b = 1.6 \cdot 10^{-4}$ and use it to discuss the general behavior of solutions corresponding to initial conditions greater and smaller than 187.5.

Picard Iteration

10. Show that if f in (3) does not depend on y, then the approximations obtained by Picard's method are identical with the exact solution. Why?

11. Apply Picard's method to $y' = y$, $y(0) = 1$, and show that the successive approximations tend to $y = e^x$, the exact solution.

12. Apply Picard's method to $y' = 2xy$, $y(0) = 1$. Graph y_1, y_2, y_3 and the exact solution for $0 \leq x \leq 2$.

13. In Prob. 12, compute the values $y_1(1)$, $y_2(1)$, $y_3(1)$ and compare them with the exact value $y(1) = e = 2.718 \cdots$.

Apply Picard's method to the following initial value problems. Determine also the exact solution. Compare.

14. $y' = xy$, $y(0) = 1$

15. $y' = 2y$, $y(0) = 1$

16. $y' = x + y$, $y(0) = 1$

17. $y' = y^2$, $y(0) = 1$

18. $y' = y^2 + 4$, $y(0) = 0$

19. $y' = xy + 2x - x^3$, $y(0) = 0$

20. $y' - xy = 1$, $y(0) = 1$

1.11 Existence and Uniqueness of Solutions

The initial value problem

$$|y'| + |y| = 0, \qquad y(0) = 1$$

has no solution because $y \equiv 0$ is the only solution of the differential equation. (Why?) The initial value problem

$$y' = x, \qquad y(0) = 1$$

has precisely one solution, namely, $y = \frac{1}{2}x^2 + 1$. The initial value problem

$$xy' = y - 1, \qquad y(0) = 1$$

has infinitely many solutions, namely, $y = 1 + cx$ where c is an arbitrary constant. From these three examples we see that an initial value problem

(1)
$$\boxed{y' = f(x, y), \quad y(x_0) = y_0}$$

may have none, precisely one, or more than one solution. This leads to the following two fundamental questions.

Problem of existence. *Under what conditions does an initial value problem of the form* (1) *have at least one solution?*

Problem of uniqueness. *Under what conditions does that problem have a unique solution, that is, only one solution?*

Theorems that state such conditions are called **existence theorems** and **uniqueness theorems,** respectively.

Of course, our three examples are so simple that we can find the answer to those two questions by inspection, without using any theorems. However, it is clear that in more complicated cases—for example, when the equation cannot be solved by elementary methods—existence and uniqueness theorems will be of great importance. As a matter of fact, a more advanced course in differential equations consists mainly of considerations concerning the existence, uniqueness and general behavior of solutions of various types of differential equations, and many of these considerations have far-reaching practical consequences. For example, results about the existence and uniqueness of periodic solutions of certain differential equations, which were obtained by Poincaré for entirely theoretical reasons about a hundred years ago, are nowadays the base of many practical investigations in nonlinear mechanics.

Uniqueness is of importance, for instance, if we attempt to predict the future behavior of a physical system governed by an initial value problem. Our model may be complicated, so that we have to apply a numerical method for obtaining an approximate solution. But before doing so, we should make sure that the model will yield a unique solution.

We want to mention that in many branches of mathematics existence and uniqueness theorems are of similar importance. Let us illustrate this by a familiar example in linear algebra. The three systems of linear equations

$$\text{(a)} \quad \begin{aligned} x + y &= 1 \\ x + y &= 0 \end{aligned} \qquad \text{(b)} \quad \begin{aligned} x + y &= 1 \\ x - y &= 0 \end{aligned} \qquad \text{(c)} \quad \begin{aligned} x + y &= 1 \\ 2x + 2y &= 2 \end{aligned}$$

have, respectively, none, precisely one, and infinitely many solutions. This is immediately clear, and we need no theorems. However, if we have a system of many equations in many unknowns, there is no immediate answer to the question of whether the system has a solution at all and, if so, how many solutions there are. Then the need for existence and uniqueness theorems becomes obvious.

In connection with differential equations, a student who is exclusively interested in applications and does not like theoretical considerations (an attitude that will prevent him from having good success in the applied area) may be inclined to reason in the following manner. The physical problem corresponding to a certain differential equation has a unique solution, and the same must therefore be true for the differential equation. In simpler cases he may be right. However, he should keep in mind that the differential equation is merely an abstraction of the reality obtained by disregarding certain physical facts that seem to be of minor influence, and in complicated physical situations there may be no way of judging a priori the importance of various factors. In such a case there will then be no a priori guarantee that the differential equation is a faithful model, that is, leads to a faithful picture of the reality. This is part of the art of modeling: to develop a feeling for the relative importance of various factors in a given physical, chemical, biological or other system.

These remarks were intended to motivate our task at hand, which will be the consideration of the existence and uniqueness problem in connection with differential equations of the first order.

In the case of an initial value problem of the form

(1)
$$\boxed{y' = f(x, y), \quad y(x_0) = y_0}$$

there are simple conditions for the existence and uniqueness of the solution. If f is continuous in some region of the xy-plane containing the point (x_0, y_0), then the problem (1) has at least one solution. If, moreover, the partial derivative $\partial f / \partial y$ exists and is continuous in that region, then the problem (1) has precisely one solution. This solution can then be obtained by Picard's iteration method. Let us formulate these statements in a precise way.

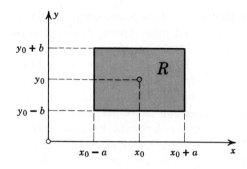

Fig. 30. Rectangle R in the existence and
uniqueness theorems

Existence theorem

If $f(x, y)$ is continuous at all points (x, y) in some rectangle (Fig. 30)

$$R: \qquad |x - x_0| < a, \qquad |y - y_0| < b$$

and bounded[17] in R, say,

(2) $$|f(x, y)| \leqq K \qquad \text{for all } (x, y) \text{ in } R,$$

then the initial value problem (1) *has at least one solution $y(x)$. This solution is defined at least for all x in the interval $|x - x_0| < \alpha$ where α is the smaller of the two numbers a and b/K.*

Uniqueness theorem

If $f(x, y)$ and $\partial f/\partial y$ are continuous for all (x, y) in that rectangle R and bounded, say,

(3) $\qquad$ (a) $\;\; |f| \leqq K,$ $\qquad$ (b) $\;\; \left| \dfrac{\partial f}{\partial y} \right| \leqq M \qquad$ *for all (x, y) in R,*

then the initial value problem (1) *has only one solution $y(x)$. This solution is defined at least for all x in that interval $|x - x_0| < \alpha$. It can be obtained by Picard's method, that is, the sequence $y_0, y_1, \cdots, y_n, \cdots$, where*

$$y_n(x) = y_0 + \int_{x_0}^{x} f[t, y_{n-1}(t)] \, dt, \qquad n = 1, 2, \cdots,$$

converges to that solution $y(x)$.

---

[17] A function $f(x, y)$ is said to be **bounded** when (x, y) varies in a region in the xy-plane, if there is a number K such that $|f(x, y)| \leqq K$ when (x, y) is in that region. For example, $f = x^2 + y^2$ is bounded, with $K = 2$ if $|x| < 1$ and $|y| < 1$. The function $f = \tan(x + y)$ is not bounded for $|x + y| < \pi/2$.

Since the proofs of these theorems require familiarity with uniformly convergent series and other concepts to be considered later in this book, we shall not present these proofs at this time but refer the student to Ref. [A10] in Appendix 1. However, we want to include some remarks and examples that may be helpful for a good understanding of the two theorems.

Since $y' = f(x, y)$, the condition (2) implies that $|y'| \leq K$, that is, the slope of any solution curve $y(x)$ in R is at least $-K$ and at most K. Hence a solution curve which passes through the point (x_0, y_0) must lie in the shaded region in Fig. 31 bounded by the lines l_1 and l_2 whose slopes are $-K$ and K, respectively. Depending on the form of R, two different cases may arise. In the first case, shown in Fig. 31a, we have $b/K \geq a$ and therefore $\alpha = a$ in the existence theorem, which then asserts that the solution exists for all x between $x_0 - a$ and $x_0 + a$. In the second case, shown in Fig. 31b, we have $b/K < a$. Therefore $\alpha = b/K$, and all we can conclude from the theorems is that the solution exists for all x between $x_0 - b/K$ and $x_0 + b/K$. For larger or smaller x's the solution curve may leave the rectangle R, and since nothing is assumed about f outside R, nothing can be concluded about the solution for those larger or smaller x's; that is, for such x's the solution may or may not exist—we don't know.

Let us illustrate our discussion by a simple example.

EXAMPLE 1

Consider the problem

$$y' = 1 + y^2, \qquad y(0) = 0$$

(cf. Example 2 in Sec. 1.10) and take R: $|x| < 5$, $|y| < 3$. Then $a = 5$, $b = 3$ and

$$|f| = |1 + y^2| \leq K = 10, \qquad |\partial f/\partial y| = 2|y| \leq M = 6, \qquad \alpha = b/K = 0.3 < a.$$

In fact, the solution $y = \tan x$ of the problem is discontinuous at $x = \pm \pi/2$, and there is no continuous solution valid in the entire interval $|x| < 5$ from which we started.   ∎

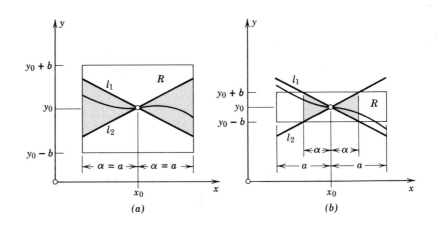

Fig. 31. The condition (2) of the existence theorem.
(a) First case. (b) Second case

The conditions in the two theorems are sufficient conditions rather than necessary ones, and can be lessened. For example, by the mean value theorem of differential calculus we have

$$f(x, y_2) - f(x, y_1) = (y_2 - y_1) \frac{\partial f}{\partial y}\bigg|_{y = \tilde{y}}$$

where (x, y_1) and (x, y_2) are assumed to be in R, and $\tilde{y}$ is a suitable value between y_1 and y_2. From this and (3b) it follows that

(4)
$$|f(x, y_2) - f(x, y_1)| \leqq M|y_2 - y_1|,$$

and it can be shown that (3b) may be replaced by the weaker condition (4) which is known as a **Lipschitz condition.**[18] However, continuity of $f(x, y)$ is not enough to guarantee the *uniqueness* of the solution. This may be illustrated by the following example.

EXAMPLE 2. Nonuniqueness
The initial value problem

$$y' = \sqrt{|y|}, \qquad y(0) = 0$$

has the two solutions

$$y \equiv 0 \qquad \text{and} \qquad y^* = \begin{cases} x^2/4 & \text{when} \quad x \geqq 0 \\ \\ -x^2/4 & \text{when} \quad x < 0 \end{cases}$$

although $f(x, y) = \sqrt{|y|}$ is continuous for all y. The Lipschitz condition (4) is violated in any region which includes the line $y = 0$, because for $y_1 = 0$ and positive y_2 we have

(5)
$$\frac{|f(x, y_2) - f(x, y_1)|}{|y_2 - y_1|} = \frac{\sqrt{y_2}}{y_2} = \frac{1}{\sqrt{y_2}}, \qquad\qquad (\sqrt{y_2} > 0)$$

and this can be made as large as we please by choosing y_2 sufficiently small, whereas (4) requires that the quotient on the left-hand side of (5) should not exceed a fixed constant M.   ∎

This is the end of Chap. 1, on first-order differential equations and their applications. An even greater role in engineering (and elsewhere) is played by second-order differential equations, which we consider in Chap. 2. It is fair to state that not much could be said about *nonlinear* differential equations in general, whereas for *linear* differential equations (that is, equations linear in the unknown function and its derivatives) we shall develop a satisfactory and applicable general theory. We shall also see that this theory readily extends to higher order differential equations that are linear.

Numerical methods for first-order differential equations are presented in Secs. 20.1 and 20.2, which are independent of the other sections on numerical methods and may be studied now.

---

[18]RUDOLF LIPSCHITZ (1832—1903), German mathematician, professor at Bonn, who also contributed to algebra, number theory, potential theory and mechanics.

Problems for Sec. 1.11

1. Show that the initial value problem $xy' = 3y$, $y(0) = 1$ has no solution. Does this contradict our present existence theorem?

2. Determine the largest set S in the xy-plane such that through each point of S there passes one and only one solution curve of $x \, dy = y \, dx$.

3. Find all solutions of the initial value problem $xy' = 2y$, $y(0) = 0$. Graph some of them. Does your answer contradict our present theorems?

4. Consider the equation $xy' = y$ and find all initial conditions $y(x_0) = y_0$ such that the resulting initial value problem has (a) no solution, (b) more than one solution, (c) precisely one solution. Does your answer contradict our present theorems?

5. Find all solutions of the initial value problem $y' = 2\sqrt{y}$, $y(1) = 0$. Which of them do we obtain by Picard's method if we start from $y_0 = 0$?

6. Perform the same task as in Prob. 4, for the equation $(x^2 - x)y' = (2x - 1)y$.

7. Show that if $y' = f(x, y)$ satisfies the assumptions of the present theorems in a rectangle R and y_1 and y_2 are two solutions of the equation whose curves lie in R, then these curves cannot have a point in common (unless they are identical).

8. If the assumptions of the present theorems are satisfied not merely in a rectangle R but even in a vertical strip given by $|x - x_0| < a$, show that then the solution of (1) exists for all x in the interval $|x - x_0| < a$.

9. It is worth noting that the solution of an initial value problem may exist in an interval larger than that in the theorem, the latter being $|x - x_0| < \alpha$ with α defined in the theorem. This is illustrated by Example 1 in the text. Consider that example. Find α in terms of general a and b. What is the maximum possible α that we can achieve by a suitable choice of a and b?

10. Find the largest α for which the present theorems guarantee the existence of the problem $y' = y^2$, $y(1) = 1$. For what x does the solution actually exist?

11. Perform the same task as in Prob. 10, for $y' - y^2 = 4$, $y(0) = 0$.

12. **(Linear differential equation)** Write $y' + p(x)y = r(x)$ in the form (1). If p and r are continuous for all x such that $|x - x_0| \leq a$, show that $f(x, y)$ in this equation satisfies the conditions of our present existence and uniqueness theorems, so that a corresponding initial value problem has a unique solution. [This also follows directly from (4) in Sec. 1.7, so that for the *linear* differential equation, we do not need these theorems.]

13. Show that $f(x, y)$ in Prob. 11 satisfies a Lipschitz condition in R (cf. Fig. 30).

14. **(Linear differential equation)** Show that in Prob. 12, a Lipschitz condition holds. [Hence for the *linear* differential equation, the *continuity* of $f(x, y)$ already guarantees the *uniqueness* of solutions of initial value problems. This is not generally true for nonlinear differential equations.]

15. Find all solutions of the initial value problem $y' = 2\sqrt{y}$, $y(1) = 0$. Which of them do we obtain by Picard's method if we start from $y_0 = 0$?

16. Does $2\sqrt{y}$ in Prob. 15 satisfy a Lipschitz condition?

17. Show that $f(x, y) = |\sin y| + x$ satisfies a Lipschitz condition (4), with $M = 1$, on the whole xy-plane, but $\partial f/\partial y$ does not exist when $y = 0$.

18. Does $f(x, y) = |x| + |y|$ satisfy a Lipschitz condition in the xy-plane? Does $\partial f/\partial y$ exist?

19. Find all solutions of $y' = x|y|$.

20. Does $f(x, y) = x \, |y|$ in Prob. 19 satisfy a Lipschitz condition in a rectangle? Does $\partial f/\partial y$ exist?

Review Problems for Chapter 1

1. What is the difference between an ordinary and a partial differential equation?
2. How many arbitrary constants does a general solution of a first-order differential equation contain? How many conditions are needed to determine these constants?
3. Does every first-order differential equation have a solution? A general solution? (Give reason.)
4. What do we mean by a particular solution of a first-order differential equation? By a singular solution?
5. What gives us the right to talk about *the* general solution of a first-order differential equation?
6. What do we mean by ''orthogonal trajectories''? Why do they lead to differential equations? In what practical problems do they occur?
7. For what reasons do electric circuits lead to differential equations?
8. In what case does a geometrical problem lead to a differential equation?
9. For what reasons may a problem in mechanics lead to a differential equation?
10. What do we mean by exponential growth and exponential decay? What is the corresponding differential equation? In what applications does it occur?

Find the general solution, using one of the methods discussed in this chapter.

11. $xy' - 4y + 12 = 0$
12. $y' \sin y = \frac{1}{2} - x^2$
13. $2 \sin 2y \, dy + \cos 2y \, dx = 0$
14. $\sinh y \tan x \, dy + \cosh y \, dx = 0$
15. $y' = (1 + x)(1 + y^2)$
16. $dr \cos 2\theta = -4r \sin 2\theta \, d\theta$
17. $y' + xy = x/y$
18. $3y \, dx + x \, dy = 0$
19. $x^2y' + 2xy = \sinh 3x$
20. $2xy' = 10x^3y^5 + y$
21. $2dx + \sec x \cos y \, dy = 0$
22. $yy' = 2x \exp(y^2)$
23. $e^y y' - e^y = 2x - x^2$
24. $y' = y/[x + (y + 1)^2]$
25. $y' = (y + x)^2$

Solve the following initial value problems.

26. $y' + 2.5y = 8.75e^x,$ $y(0) = 5.7$
27. $xy' = 3y + x^4(e^x + \cos x) - 2x^2,$ $y(\pi) = \pi^2(\pi e^\pi + 2)$
28. $(x^2 + 1)yy' = 1,$ $y(0) = -3$
29. $4 \sec y \, dx + \sec x \, dy = 0,$ $y(0) = 0$
30. $(2x + e^y) \, dx + xe^y \, dy = 0,$ $y(1) = 0$
31. $y' \sin x = y \cos x,$ $y(\frac{1}{4}\pi) = \sqrt{2}$
32. $(2x + y^4)y' = y,$ $y(24) = 2$
33. $y' + \csc y = 0,$ $y(4.1) = -6.3$
34. $xy' + y = 4(x^{-1} + e^x),$ $y(1) = 2 + 4e \approx 12.873$
35. $xy' + y = x^2y^2,$ $y(0.5) = 0.5$

Find the orthogonal trajectories of the given family of curves. Sketch some of these curves and their orthogonal trajectories.

36. $y = cx$
37. $y = 1 - cx^2$
38. $x^2/a^2 + y^2/b^2 = c$
39. $(x - 1)(y - 1) = c$
40. $2x^2 + 3y^2 = c$

41. If the growth rate of a culture of bacteria is proportional to the number of bacteria present and after 1 day is 1.5 times the original number, within what interval of time will the number of bacteria (a) double, (b) triple?

42. A metal bar whose temperature is 20°C is placed in boiling water. How long does it take to heat the bar to practically 100°C, say, to 99.9°C, if the temperature of the bar after 1 min of heating is 51.5°C? First guess, then calculate.

43. If in a reactor, uranium $_{92}U^{237}$ loses 10% of its weight within 1 day, what is its half-life? How long would it take for 99% of the original amount to disappear?

44. How should one choose R and L in an RL-circuit connected to a battery of 48 volts if one wants the steady-state current to be 10 amp and the time to practically reach this value (say, 9.99 amp) to be 10^{-2} sec after the instant when the circuit is connected to the battery?

45. (**Heart pacemaker**) Figure 32 shows a heart pacemaker consisting of a capacitor of capacitance C, a battery of voltage E_0 and a switch that is periodically moved from A (charging period $t_1 < t < t_2$ of the capacitor) to B (discharging period $t_2 < t < t_3$ during which the capacitor sends an electric stimulus to the heart, which acts as a resistor of resistance R). Find the current during the discharging period, assuming that the charging period is such that the charge on the capacitor is 99% of its maximum possible value.

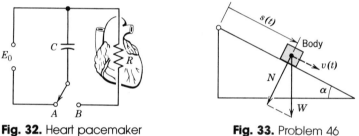

Fig. 32. Heart pacemaker   **Fig. 33.** Problem 46

46. (**Friction**) If a body moves (slides) on a surface, it experiences a force in the direction against the motion. This force is called *friction*. *Coulomb's law of kinetic friction without lubrication*[19] is

$$|F| = \mu|N|$$

where F is the friction, N is the normal force (force that holds the two surfaces together; see Fig. 33), and the constant of proportionality μ is called the *coefficient of kinetic friction*. Using Newton's second law

$$\text{Mass} \times \text{Acceleration} = \text{Force}$$

set up the differential equation in Fig. 33. Find the velocity when the body reaches the bottom of the slide, assuming that the body weighs 45 nt (about 10 lb), $\mu = 0.10$, $\alpha = 30°$, the length of the slide is 10 meters, the air resistance is negligible, and the initial velocity is zero. (For the conversion of newtons, see the front cover.)

47. Assume the body in Prob. 46 to be such that the air resistance is no longer negligible but causes a force of magnitude $0.5v$. Find $v(t)$, the distance $s(t)$ and $s(2.22)$, where $t = 2.22$ sec is the time to cover the whole distance of 10 meters if the air resistance is neglected.

48. If you throw a ball vertically upward with initial velocity 15 meters/sec, what maximum height will it reach? (Neglect air resistance.)

[19]See footnote 8 in Sec. 8.8.

49. (Linear accelerator) Linear accelerators are used in physics for accelerating charged particles. Suppose that an alpha particle enters an accelerator and undergoes a constant acceleration which increases the speed of the particle from 10^3 meters/sec to 10^4 meters/sec in 10^{-3} sec. Find the acceleration a and the distance traveled during this period of 10^{-3} sec.

50. What are the lines of electrical force between two elliptical copper plates $x^2 + 2y^2 = 1$ and $\frac{1}{4}x^2 + \frac{1}{2}y^2 = 1$ that are kept at different electric potentials?

51. If in a thin plate in the xy-plane, heat flows along the curves $xy = const$, what are the isotherms in the plate?

52. (Chemical reaction) The **law of mass action** states that, if the temperature is kept constant, the velocity of a chemical reaction is proportional to the product of the concentrations of the substances which are reacting. A bimolecular reaction $A + B \rightarrow M$ combines a moles per liter of a substance A and b moles of a substance B. If y is the number of moles per liter that have reacted after time t, the rate of reaction is given by

$$\frac{dy}{dt} = k(a - y)(b - y).$$

Solve this equation, assuming that $a \neq b$.

53. Find all curves in the xy-plane such that the slope of the tangent at any point (x, y) is half the slope of the line from the origin to the point.

54. Find all curves such that the y-axis bisects the segment of the normal at (x, y) whose endpoints are (x, y) and the point of intersection of the normal and the x-axis.

55. (Lemniscate) Solve the initial value problem

$$\frac{dr}{d\theta} + \frac{a^2}{r} \sin 2\theta = 0, \qquad r^2(0) = a^2 \qquad (a \neq 0),$$

where r and θ are polar coordinates. The solution curve is called a *lemniscate*.

Summary of Chapter 1
Differential Equations
of the First Order

This chapter concerns **first-order differential equations** and their applications; these are equations

(1) $F(x, y, y') = 0$ or in explicit form $y' = f(x, y)$

involving the derivative y' of an unknown function y, given functions of x, and, perhaps, y itself. We started with the basic concepts (Sec. 1.1), then considered methods of solution (Secs. 1.2–1.9) and, finally, ideas on approximation and existence of solutions (Secs. 1.10, 1.11).

Ordinarily such an equation has a **general solution,** that is, a solution involving an arbitrary constant, which we denoted by c. In most applications, however, one has to find a solution satisfying a given condition, resulting from the physical or other system of which the equation is a mathematical model. This leads to the concept of an **initial value problem**

$$(2) \qquad y' = f(x, y), \qquad y(x_0) = y_0 \qquad (x_0, y_0 \text{ given numbers})$$

in which the **initial condition** $y(x_0) = y_0$ is used to determine a **particular solution,** that is, a solution obtained from the general solution by specifying a value of c. Geometrically, a general solution represents a family of curves, and each particular solution corresponds to a curve of this family.

Perhaps the simplest equations are the **separable equations,** those we can put into the form $g(y) \, dy = f(x) \, dx$, so that we can solve them by integrating on both sides (Secs. 1.2, 1.3). This method can be extended to certain other equations by first applying suitable substitutions (Sec. 1.4).

An **exact equation**

$$M(x, y) \, dx + N(x, y) \, dy = 0$$

is one for which $M \, dx + N \, dy$ is an exact differential $du = (\partial u/\partial x) \, dx + (\partial u/\partial y) \, dy$, so that we get the implicit solution $u(x, y) = c$ (Sec. 1.5). This method extends to nonexact equations that can be made exact by multiplying them by some function F, called an **integrating factor** (Sec. 1.6).

Linear equations (Sec. 1.7)

$$(3) \qquad\qquad y' + p(x)y = r(x)$$

are of that type. For them an integrating factor is $F(x) = \exp(\int p(x) \, dx)$ and leads to the solution formula (4), Sec. 1.7. Certain nonlinear equations can be reduced to linear form by substituting new variables. This holds for the **Bernoulli equation** $y' + p(x)y = g(x)y^a$ (Sec. 1.7, Probs. 47–54).

Applications are included at various places. Sections entirely devoted to applications are 1.3 on separable equations, 1.8 on linear equations applied to **electrical circuits,** and 1.9 on **orthogonal trajectories,** that is, curves that intersect given curves at right angles.

Direction fields (Sec. 1.10) help in sketching families of solutions curves, for instance, in order to gain an impression of their general behavior.

Picard's iteration method (Sec. 1.10) gives approximate solutions of initial value problems by iteration. It is important as the theoretical basis of **Picard's existence and uniqueness theorems** (Sec. 1.11).

Chapter 2

Linear Differential Equations

The ordinary differential equations may be divided into two large classes, the so-called **linear equations** and the **nonlinear equations.** While nonlinear equations (of the second and higher orders) are rather difficult, linear equations are much simpler in many respects, because various properties of their solutions can be characterized in a general way, and standard methods are available for solving many of these equations. In the present chapter we shall consider linear differential equations and their applications. These equations play an important role in engineering mathematics, for example, in connection with mechanical vibrations and electric circuits and networks. We shall consider

homogeneous linear equations of the second order in Secs. 2.1–2.8,

homogeneous linear equations of higher order in Secs. 2.9, 2.10,

nonhomogeneous linear equations in Secs. 2.11–2.16.

We shall devote much space to linear equations of the *second order,* for two reasons. The first reason is that in applications, linear equations of the *second* order are much more important than those of higher order. The second reason is that the necessary theoretical considerations will become easier if we first concentrate upon second-order equations. Once the student understands what is needed to handle second-order equations, he will easily become familiar with the generalizations of the concepts, methods and results to differential equations of the third and higher orders.

Important engineering applications in connection with mechanical and electrical oscillations will be considered in Secs. 2.6 and 2.13–2.15. This will include analogies between mechanical and electrical systems (Sec. 2.14).

Numerical methods for solving second order differential equations are included in Sec. 20.3, which is independent of the other sections in Chap. 20.

(Legendre's, Bessel's, and the hypergeometric equations will be considered in Chap. 4.)

Prerequisite for this chapter: Chap. 1, in particular Sec. 1.7.
Sections that may be omitted in a shorter course: 2.8–2.10, 2.15, 2.16.
References: Appendix 1, Part A.
Answers to problems: Appendix 2.

2.1 Homogeneous Linear Equations of the Second Order

The student has already met with linear differential equations of the first order (Secs. 1.7, 1.8), and we shall now define and consider linear differential equations of the second order.

Linear Differential Equation of the Second Order

A second-order differential equation is said to be **linear** if it can be written

(1)
$$y'' + p(x)y' + q(x)y = r(x).$$

The characteristic feature of this equation is that it is linear in the unknown function y and its derivatives, whereas p and q as well as r on the right may be any given functions of x. If the first term is, say, $f(x)y''$, we have to divide by $f(x)$ to get the "standard form" (1), with y'' as the first term, which is practical.

If $r(x) \equiv 0$ (that is, $r(x) = 0$ for all x considered), then (1) becomes simply

(2)
$$y'' + p(x)y' + q(x)y = 0$$

and is said to be **homogeneous.** If $r(x) \not\equiv 0$, then (1) is said to be **nonhomogeneous.** This is similar to Sec. 1.7. For example,

$$y'' + 4y = e^{-x} \sin x$$

is a nonhomogeneous linear differential equation, whereas

$$(1 - x^2)y'' - 2xy' + 6y = 0$$

is a homogeneous linear differential equation.

The functions p and q in (1) and (2) are called the **coefficients** of the equations.

Any differential equation of the second order that cannot be written in the form (1) is said to be **nonlinear.** For example, nonlinear equations are

$$y''y + y' = 0$$

and

$$y'' = \sqrt{y'^2 + 1}.$$

Linear differential equations of the second order play a basic role in many engineering problems. We shall see that some of these equations are very simple because their solutions are elementary functions. Others are more complicated, their solutions being important higher functions, such as Bessel and hypergeometric functions.

Concept of Solution. Superposition Principle

In all our considerations we assume that x varies in some given finite interval or on the entire x-axis. All our assumptions and statements refer to such an interval, which we need not specify in each case, or to the x-axis.

We call a function

$$y = \phi(x)$$

a **solution** of a (linear or nonlinear) differential equation of the second order *on some interval* (perhaps infinite), if $\phi(x)$ is defined and twice differentiable throughout that interval and is such that the equation becomes an identity when we replace the unspecified function y and its derivatives by ϕ and its corresponding derivatives.

EXAMPLE 1. Solutions of a homogeneous linear differential equation

The functions $y = \cos x$ and $y = \sin x$ are solutions of the homogeneous linear differential equation

$$y'' + y = 0$$

for all x, since for $y = \cos x$ we obtain

$$(\cos x)'' + \cos x = -\cos x + \cos x = 0$$

and similarly for $y = \sin x$. We can even go an important step further. If we multiply the first solution by a constant, for example, by 3, the resulting function $y = 3 \cos x$ is also a solution because

$$(3 \cos x)'' + 3 \cos x = 3[(\cos x)'' + \cos x] = 0.$$

It is clear that instead of 3 we may choose any other constant, for example, -5 or $\frac{2}{9}$. We may even multiply $\cos x$ and $\sin x$ by different constants, say, by 2 and -8, respectively, and add the resulting functions, obtaining

$$y = 2 \cos x - 8 \sin x,$$

and this function is also a solution of our homogeneous equation for all x, because we have

$$(2 \cos x - 8 \sin x)'' + 2 \cos x - 8 \sin x = 2[(\cos x)'' + \cos x] - 8[(\sin x)'' + \sin x] = 0. \blacksquare$$

This example illustrates the very important fact that in the case of the *homogeneous linear* equation (2), we can obtain new solutions from known solutions by multiplication by constants and by addition. Of course, this is a great practical and theoretical advantage, because this property will enable us to generate more complicated solutions from simple solutions. This is often called the **superposition principle**[1] and can be stated as follows.

Fundamental Theorem 1 for the homogeneous equation (2)

*If a solution of the **homogeneous linear** differential equation (2) on some interval I is multiplied by any constant, the resulting function is also a solution of (2) on I. The sum of two solutions of (2) on I is also a solution of (2) on that interval.*

---

[1]Or **linearity principle**.

Proof. We assume that $\phi(x)$ is a solution of (2) on some interval I and show that then $y = c\phi(x)$ is also a solution of (2) on I. If we substitute $y = c\phi(x)$ into (2), the left-hand side of (2) becomes

$$(c\phi)'' + p(c\phi)' + qc\phi = c[\phi'' + p\phi' + q\phi].$$

Since ϕ satisfies (2), the expression in brackets is zero, and the first statement of the theorem is proved. The proof of the last statement is quite simple, too, and is left to the reader.   ∎

An expression of the form

(3) $y = c_1 y_1 + c_2 y_2$ (c_1, c_2 arbitrary constants)

is called a **linear combination** of y_1 and y_2. Combining the two statements in Theorem 1, we conclude that *a linear combination of solutions of* (2) *on I is a solution of* (2) *on I.*

Warning. The student should always remember the highly important Theorem 1, but he should not forget that *the theorem **does not hold** for nonhomogeneous linear equations or nonlinear equations,* as is illustrated by the following two examples.

EXAMPLE 2. A nonhomogeneous linear differential equation
Substitution shows that the functions $y = 1 + \cos x$ and $y = 1 + \sin x$ are solutions of the nonhomogeneous linear differential equation

$$y'' + y = 1,$$

but the following functions are *not* solutions of this differential equation:

$$2(1 + \cos x) \qquad \text{and} \qquad (1 + \cos x) + (1 + \sin x)$$

EXAMPLE 3. A nonlinear differential equation
Substitution shows that the functions $y = x^2$ and $y = 1$ are solutions of the nonlinear differential equation

$$y''y - xy' = 0,$$

but the following functions are *not* solutions of this differential equation:

$$-x^2 \qquad \text{and} \qquad x^2 + 1 \qquad\qquad ∎$$

In the next section we discuss linear differential equations with **constant coefficients** and show that these can readily be solved without integration, the solutions being exponential, cosine and sine functions. We show later (in Secs. 2.6, 2.13, 2.14) that these equations have important engineering applications in connection with mechanical vibrating systems and electric circuits.

Another class of differential equations that can also be solved without integration will be considered in Sec. 2.7.

Problems for Sec. 2.1

Important general properties of homogeneous and nonhomogeneous linear differential equations. Prove the following statements, which refer to any fixed interval I, and illustrate them with examples. Here we assume $r(x) \neq 0$ in (1).

1. $y \equiv 0$ is a solution of (2) (called the *"trivial solution"*) but not of (1).
2. The sum of two solutions of (2) is a solution of (2).
3. A multiple $y = cy_1$ of a solution of (2) is a solution of (2).
4. A "linear combination" $y = c_1 y_1 + c_2 y_2$ of solutions y_1 and y_2 of (2) is a solution of (2).
5. A multiple $y = cy_1$ of a solution of (1) is **not** a solution of (1), unless $c = 1$.
6. The sum of two solutions of (1) is **not** a solution of (1).
7. The difference $y = y_1 - y_2$ of two solutions of (1) is a solution of (2).
8. The sum $y = y_1 + y_2$ of a solution y_1 of (1) and y_2 of (2) is a solution of (1).

Second-order differential equations reducible to the first order. Problems 9–30 illustrate that certain second-order differential equations can be reduced to the first order.

9. If in a second-order equation the dependent variable y does not appear explicitly, the equation is of the form $F(x, y', y'') = 0$. Show that by setting $y' = z$ we obtain a first-order differential equation in z and from its solution the solution of the original equation by integration.

Reduce to the first order and solve:

10. $2xy'' = 3y'$
11. $y'' = y'$
12. $y'' + y' = x + 1$
13. $y'' = y' \tanh x$
14. $y'' = 1 + y'^2$
15. $xy'' + y' = y'^2$

16. **(Reduction to first order)** Another type of equations reducible to first order is $F(y, y', y'') = 0$, in which the independent variable x does not appear explicitly. Using the chain rule, show that $y'' = (dz/dy)z$, where $z = y'$, so that we obtain a first-order equation with y as the *independent* variable.

Reduce to first order and solve:

17. $yy'' = 2y'^2$
18. $yy'' + y'^2 = 0$
19. $y'' + e^y y'^3 = 0$
20. $y'' + 2y'^2 = 0$
21. $y'' + y'^3 \cos y = 0$
22. $y'' + (1 + y^{-1})y'^2 = 0$

23. A particle moves on a straight line so that its acceleration is equal to three times its velocity. At $t = 0$ its displacement from the origin is 1 ft and its velocity is 1.5 ft/sec. Find the time when the displacement is 10 ft.
24. A particle moves on a straight line so that its acceleration equals its velocity. At time $t = 0$ its distance from the origin is 1 meter and its velocity is 2 meters/sec. Find its position and velocity at $t = 5$ sec.
25. A particle moves on a straight line so that the product of its velocity and acceleration is constant, say, 1 meter^2/sec^3. If at $t = 0$ the distance from the origin is 2 meters and the velocity is 2 meters/sec, what are distance and velocity at $t = 6$ sec?

26. **(Hanging cable)** It can be shown that the curve $y(x)$ of an inextensible flexible homogeneous cable hanging between two fixed points is obtained by solving $y'' = k \sqrt{1 + y'^2}$, where the constant k depends on the weight. This curve is called a *catenary* (from Latin *catena* = the chain). Find and graph $y(x)$, assuming $k = 1$ and those fixed points are $(-1, 0)$ and $(1, 0)$ in a vertical xy-plane.

27. Find the curve $y(x)$ through the origin for which $y'' = y'$ and the tangent at the origin is $y = x$.

28. Find the curve $y(x)$ through $(0, 0)$ and $(1, 1)$ for which $2y''y = y'^2$.

29. Find the curve $y(x)$ through the origin for which $y'' = 12\sqrt{y}$ and the tangent at the origin is the x-axis.

30. Given $y'' + ay'^2 + f(y) = 0$, where a is a constant. Show that the substitutions $y' = z$, $u = z^2$ lead to the linear first-order equation

$$\frac{du}{dy} + 2au + 2f(y) = 0.$$

2.2 Homogeneous Equations with Constant Coefficients

We shall now consider linear differential equations of the second order whose coefficients are constant. These equations have important engineering applications, especially in connection with mechanical and electrical vibrations, as we shall see in Secs. 2.6, 2.13 and 2.14.

We start with homogeneous equations, that is, equations of the form

(1)
$$\boxed{y'' + ay' + by = 0}$$

where a and b are constants. *We assume that a and b are real* and the range of x considered is the entire x-axis.

How to solve equation (1)? We remember that the solution of the *first-order* homogeneous linear equation with constant coefficients

$$y' + ky = 0$$

is an exponential function, namely,

$$y = ce^{-kx}.$$

We thus conjecture that

(2)
$$y = e^{\lambda x}$$

might be a solution of the differential equation (1) if λ is properly chosen. Substituting the function (2) and its derivatives

$$y' = \lambda e^{\lambda x} \qquad \text{and} \qquad y'' = \lambda^2 e^{\lambda x}$$

into our equation (1), we obtain

$$(\lambda^2 + a\lambda + b)e^{\lambda x} = 0.$$

Hence (2) is a solution of (1), if λ is a solution of the quadratic equation

(3) $$\boxed{\lambda^2 + a\lambda + b = 0.}$$

This equation is called the **characteristic equation** (or *auxiliary equation*) of (1). Its roots are

(4) $$\lambda_1 = \tfrac{1}{2}(-a + \sqrt{a^2 - 4b}), \qquad \lambda_2 = \tfrac{1}{2}(-a - \sqrt{a^2 - 4b}).$$

Our derivation shows that the functions

(5) $$y_1 = e^{\lambda_1 x} \qquad \text{and} \qquad y_2 = e^{\lambda_2 x}$$

are solutions of (1). The student may check this by substituting (5) into (1).

From elementary algebra we know that, since a and b are real, the characteristic equation may have

> (**Case I**) *two distinct real roots,*
> (**Case II**) *two complex conjugate roots, or*
> (**Case III**) *a real double root.*

These cases will be discussed in detail in Sec. 2.4. For the time being let us illustrate each case by a simple example.

EXAMPLE 1. Distinct real roots
Find solutions of the equation

$$y'' + y' - 2y = 0.$$

Solution. The characteristic equation is

$$\lambda^2 + \lambda - 2 = 0.$$

The roots are 1 and -2. Hence we obtain the two solutions

$$y_1 = e^x \qquad \text{and} \qquad y_2 = e^{-2x}.$$

EXAMPLE 2. Complex conjugate roots
Find solutions of the equation

(6) $$y'' + y = 0.$$

Solution. The characteristic equation is

$$\lambda^2 + 1 = 0.$$

The roots are $i \, (= \sqrt{-1})$ and $-i$. Hence we obtain the two solutions

$$y_1 = e^{ix} \qquad \text{and} \qquad y_2 = e^{-ix}.$$

In Sec. 2.4 we shall define such "complex exponential functions" and see that it is quite simple to obtain real solutions from such "complex solutions." Meanwhile the student may show that $\cos x$ and $\sin x$ are solutions of (6).

EXAMPLE 3. A real double root
Find solutions of the equation

$$y'' - 2y' + y = 0.$$

Solution. The characteristic equation

$$\lambda^2 - 2\lambda + 1 = 0$$

has the real double root 1, and we obtain at first only one solution

$$y_1 = e^x.$$

The case of a real double root will also be discussed in Sec. 2.4. Meanwhile the student may show that another solution is $y_2 = xe^x$ and wonder how we got it, and he may use y_1 and y_2 to find the solution satisfying the "initial conditions" $y(0) = 3$, $y'(0) = 5$.
    [Ans. $y = (3 + 2x)e^x$.]   ∎

Our discussion of constant-coefficient equations is not yet finished. Our goal is to find "all" solutions of such an equation in each of the three cases, so that we can solve any "initial value problem" consisting of the differential equation and two "initial conditions," prescribing at a given x the value of the solution and of its first derivative. (Physically, if x is time, these may be the initial position and the initial speed of a moving body.) This idea leads to the very important concepts of a **general solution** and a **basis of solutions,** which we discuss next (in Sec. 2.3). We introduce these concepts immediately for variable-coefficient equations, for which we shall need them throughout this chapter, and then apply them to constant-coefficient equations in a separate section (Sec. 2.4).

Problems for Sec. 2.2

Find solutions of the following equations. Check the answer by substitution.

1. $y'' - y = 0$	**2.** $4y'' - 9y = 0$	**3.** $y'' - 9y = 0$
4. $y'' - 4y' + 3y = 0$	**5.** $8y'' + 2y' - y = 0$	**6.** $y'' + 3y' + 2y = 0$
7. $y'' - \pi^2 y = 0$	**8.** $y'' + 6y' + 8y = 0$	**9.** $y'' - 0.2y' - 0.08y = 0$
10. $y'' + 2.2y' + 1.17y = 0$	**11.** $y'' + 4y = 0$	**12.** $y'' + 36y = 0$
13. $y'' + 4y' + 5y = 0$	**14.** $y'' + \omega^2 y = 0$	**15.** $y'' - y' = 0$

16. Show that $y = \cosh x$ and $y = \sinh x$ are solutions of $y'' - y = 0$. How can these be obtained from the answer to Prob. 1?

17. Find solutions of $y'' + 4y' = 0$ (a) by the present method, (b) by reduction to the first order.

18. Show that Cases I, II and III occur if and only if $a^2 > 4b$, $a^2 < 4b$ and $a^2 = 4b$, respectively.

19. Show that a and b in (1) can be expressed in terms of λ_1 and λ_2 by the formulas $a = -\lambda_1 - \lambda_2$, $b = \lambda_1 \lambda_2$.

Find a differential equation of the form (1) for which the following functions are solutions. Check your answer by substituting the functions into the equation.

20. e^x, e^{-3x}	**21.** 1, e^{2x}	**22.** e^{2ix}, e^{-2ix}
23. $e^{(-1+2i)x}$, $e^{(-1-2i)x}$	**24.** e^{kx}, e^{lx}	**25.** $e^{-(\alpha+i\omega)x}$, $e^{-(\alpha-i\omega)x}$

2.3 General Solution. Basis. Initial Value Problem

From now on until Sec. 2.8 we shall always be concerned with the **homogeneous linear differential equation of the second order.**

In this section we consider

$$(1) \qquad \boxed{y'' + p(x)y' + q(x)y = 0}$$

and introduce the concept of a "general solution" of such an equation. General solutions will be important throughout this chapter. Their form for equations with constant coefficients (cf. Sec. 2.2) will be discussed in the next section.

Definition (General solution, basis, particular solution)

A **general solution** of (1) on some open[2] interval I is a function of the form

$$(2) \qquad \boxed{y(x) = c_1 y_1(x) + c_2 y_2(x)} \qquad (c_1, c_2 \text{ arbitrary constants}[3])$$

where y_1 and y_2 form a **basis**[4] (or **fundamental system**) of solutions of (1) on I, that is, y_1 and y_2 are solutions of (1) on I which are not proportional on I.

A **particular solution** of (1) on I is obtained if we assign specific values to c_1 and c_2 in (2). ∎

Here we call y_1 and y_2 *proportional* on I if[5]

$$(3) \qquad \text{(a)} \quad y_1 = k y_2 \qquad \text{or} \qquad \text{(b)} \quad y_2 = l y_1$$

holds for all x on I, where k and l are numbers, zero or not.

Note that (2) is a solution of (1), by Fundamental Theorem 1 in Sec. 2.1.

---

[2]A (finite) interval I is said to be **open** if its two endpoints are not regarded as points belonging to I. (The interval I is said to be **closed** if its endpoints are regarded as points of I.) In addition, by definition there are three types of *infinite open intervals*, namely, the real line and intervals of the form $a < x < \infty$ or $-\infty < x < b$.

[3]The range of the constants may have to be restricted in some cases to avoid imaginary expressions or other degeneracies.

[4]Readers familiar with the notion of a vector space (Sec. 6.4) will notice that, by Fundamental Theorem 1 in Sec. 2.1, the solutions of (1) on I form a vector space, and a basis y_1, y_2 is a basis of that vector space in the sense of linear algebra.

[5]Note that if (3a) holds for a $k \neq 0$, we may divide by k, finding $y_2 = y_1/k$, so that in this case (3a) implies (3b) (with $l = 1/k$). However, if $y_1 \equiv 0$, then (3a) holds for $k = 0$, but does not imply (3b); in fact, in this case (3b) will not hold unless $y_2 \equiv 0$, too.

Remark. Linear independence. The definition just stated can also be formulated in terms of "linear independence." We call two functions $y_1(x)$ and $y_2(x)$ **linearly independent** on an interval I where they are defined if

(4) $\qquad k_1 y_1(x) + k_2 y_2(x) = 0 \qquad$ on I implies $\qquad k_1 = 0, k_2 = 0,$

and we call them **linearly dependent** on I if this equation also holds for some constants k_1, k_2 not both zero. Then, if $k_1 \neq 0$ or $k_2 \neq 0$, we can divide and solve, obtaining

$$y_1 = -\frac{k_2}{k_1} y_2 \qquad \text{or} \qquad y_2 = -\frac{k_1}{k_2} y_1.$$

Hence then y_1 and y_2 are proportional, whereas in the case of linear independence, they are not proportional. Hence we have the following result.

y_1 and y_2 form a basis and y in (2) is a general solution of (1) on an interval I if and only if y_1, y_2 are linearly independent solutions of (1) on I.

EXAMPLE 1. Basis. General solution. Particular solution
Find a general solution of the homogeneous equation

$$y'' + 5y' + 6y = 0$$

and the particular solution satisfying the "initial conditions" $y(0) = 1.6$, $y'(0) = 0$.

Solution. 1st Step. The characteristic equation (Sec. 2.2) is

$$\lambda^2 + 5\lambda + 6 = 0.$$

Its roots are $\lambda_1 = -2$ and $\lambda_2 = -3$. This gives the solutions

$$y_1 = e^{\lambda_1 x} = e^{-2x}$$

and

$$y_2 = e^{\lambda_2 x} = e^{-3x}.$$

Since their quotient y_1/y_2 is not a constant (but e^x), they are not proportional. Hence they form a basis. The corresponding general solution is

$$y(x) = c_1 y_1(x) + c_2 y_2(x) = c_1 e^{-2x} + c_2 e^{-3x}.$$

2nd Step. By the first initial condition,

$$y(0) = c_1 + c_2 = 1.6.$$

By differentiation,

$$y'(x) = -2c_1 e^{-2x} - 3c_2 e^{-3x}.$$

Hence by the second initial condition,

$$y'(0) = -2c_1 - 3c_2 = 0.$$

From this, $c_1 = -1.5c_2$. Hence $c_1 + c_2 = -1.5c_2 + c_2 = 1.6$. This gives $c_2 = -3.2$ and $c_1 = -1.5(-3.2) = 4.8$. The particular solution satisfying the two initial conditions is

$$y = 4.8e^{-2x} - 3.2e^{-3x}.$$

EXAMPLE 2. Solutions that are proportional

The functions $y_1 = e^x$ and $y_2 = 3e^x$ are solutions of the equation in Example 3 of Sec. 2.2, but since $y_2 = 3y_1$, they are proportional, so that they do not form a basis.   ∎

Initial Value Problem

In most applications one needs particular solutions rather than a general solution. Now for a *first-order* equation, a general solution contained *one* arbitrary constant, and we needed *one* condition for obtaining a particular solution. Presently we have *two* arbitrary constants and need *two* conditions. In many applications these conditions are of the type

(5)
$$y(x_0) = K_0, \quad y'(x_0) = K_1$$

where $x = x_0$ is a given point and K_0 and K_1 are given numbers. Hence we are looking for a particular solution of (1) whose value at x_0 is K_0 and whose first derivative at x_0 has the value K_1. The conditions in (5) are called **initial conditions.** Equation (1) and conditions (5) together constitute what is known as an **initial value problem.** (These names are suggested by the fact that in applications, x often is time; then (5) describes the initial state of a physical or other system, and the solution shows what happens later.)

So what we have just solved in Example 1 is an initial value problem. Here is another one:

EXAMPLE 3. Initial value problem

Solve the initial value problem

$$y'' + y' - 2y = 0, \qquad y(0) = 4, \qquad y'(0) = 1.$$

Solution. 1st Step. Solutions are $y_1 = e^x$ and $y_2 = e^{-2x}$; cf. Example 1 in the previous section. Since they are not proportional, they form a basis. The corresponding general solution is

$$y(x) = c_1 e^x + c_2 e^{-2x}.$$

2nd Step. $y(0) = c_1 + c_2 = 4$ by the first initial condition. By differentiation,

$$y'(x) = c_1 e^x - 2c_2 e^{-2x}.$$

Hence $y'(0) = c_1 - 2c_2 = 1$ by the second initial condition. Together, $c_1 = 3$, $c_2 = 1$. The solution is

$$y(x) = 3e^x + e^{-2x}.$$ ∎

These problems should also show the reason why we had to exclude proportionality of y_1 and y_2 in the definition of a general solution. Indeed, if (3a) holds, then $y_1 = ky_2$ and we can reduce (2) to the form

$$y = c_1 y_1 + c_2 y_2 = (c_1 k + c_2) y_2 = C y_2$$

involving only *one* arbitrary constant, and we would not be able to impose *two* initial conditions.

A General Solution of (1) Includes All Solutions

To complete our present theory, we raise the remaining questions whether (1) always has a general solution and whether such a solution is really "most general." The answer to both questions is yes:

Theorem 1 (General solution, initial value problem)

Suppose that the homogeneous linear equation (1) *has continuous coefficients* $p(x)$ *and* $q(x)$ *on some open interval I. Then* (1) *has a general solution* $y = c_1 y_1(x) + c_2 y_2(x)$ *on I, and every solution of* (1) *on I involving no arbitrary constants can be obtained by assigning suitable values to* c_1 *and* c_2. *Furthermore, every* initial value problem *on I consisting of the equation* (1) *and two initial conditions* $y(x_0) = K_0$, $y'(x_0) = K_1$ *[with given* x_0 *in I and constants* K_0, K_1] *has a unique solution on I.*

Thus (1) has no "*singular solutions*," that is, solutions *not* obtainable from a general solution.

The proof of these basic facts is not just a matter of a few lines. We therefore postpone it (to Secs. 2.8 and 2.11) and return to practical problems, in order to first gain more practical skill and see applications.

How to Obtain a Basis When One Solution Is Known (Reduction of Order)

One can often obtain a solution y_1 (not identically zero) of a given differential equation (1) by guessing or by some method. We show that then a second independent solution y_2 can be found by solving a *first-order* differential equation. For this reason, this is called **reduction of order.**[6] In this method we substitute $y(x) = u(x)y_1(x)$ and its derivatives

$$y' = u'y_1 + uy_1', \qquad y'' = u''y_1 + 2u'y_1' + uy_1''$$

into (1) and collect terms, obtaining

$$u''y_1 + u'(2y_1' + py_1) + u(y_1'' + py_1' + qy_1) = 0.$$

Since y_1 is a solution of (1), the expression in the last parentheses is zero. We divide the remaining formula by y_1 and write $u' = U$. Then $u'' = U'$ and we have

$$U' + \left(\frac{2y_1'}{y_1} + p\right)U = 0.$$

Separating variables and integrating, we obtain

---

[6]Usually attributed to the French mathematician JEAN-LE-ROND D'ALEMBERT (1717—1783), who is known for his important work in mechanics. Reduction for higher order equations, see in the problem set for Sec. 2.10.

$$\ln U = -2 \ln y_1 - \int p \, dx + \tilde{c}.$$

Taking exponents gives

(6) $$U = \frac{c}{y_1^{\,2}} e^{-\int p \, dx}.$$

Here $U = u'$. Hence the desired second solution is $y_2 = uy_1 = y_1 \int U \, dx$. Since $y_2/y_1 = u = \int U \, dx$ cannot be a constant (why?), we see that y_1 and y_2 form a basis.

EXAMPLE 4. Constant coefficient equation: Case of a double root

From Sec. 2.2 we know that the equation with constant coefficients

(7) $$y'' + ay' + by = 0$$

has solutions $y = e^{\lambda x}$, where λ is a root of the characteristic equation

$$\lambda^2 + a\lambda + b = 0.$$

This equation has a double root $\lambda = -\frac{1}{2}a$ if and only if $a^2 - 4b = 0$. Then $b = \frac{1}{4}a^2$, so that (7) becomes

(7*) $$y'' + ay' + \tfrac{1}{4}a^2 y = 0,$$

and we have one solution

$$y_1 = e^{-ax/2}.$$

We determine a second independent solution. In (6) we need

$$\int p \, dx = \int a \, dx = ax \qquad \text{and} \qquad 1/y_1^{\,2} = e^{ax}.$$

This gives

$$U = ce^{ax}e^{-ax} = c. \qquad \text{Hence} \qquad u = \int U \, dx = cx + k,$$

where c and k are arbitrary. Taking simply $u = x$, we have as a second independent solution $y_2 = uy_1 = xy_1$, that is,

$$y_2 = xe^{-ax/2}.$$

Hence in the case of a double root (and only in this case!), a general solution of (7) [which then has the form (7*)!] is

(8) $$\boxed{y = (c_1 + c_2 x)e^{\lambda x}} \qquad (\lambda = -\tfrac{1}{2}a). \quad \blacksquare$$

We are now well prepared for continuing our discussion of differential equations with constant coefficients.

Problems for Sec. 2.3

Are the following functions linearly dependent or independent on the given interval?

1. e^x, e^{-x}, any interval
2. $\cos x$, $\sin x$, any interval
3. x, $x + 1$ $(0 < x < 1)$
4. $\ln x$, $\ln x^2$ $(x > 1)$
5. $\sin 2x$, $\sin x \cos x$, any interval
6. $e^{\lambda x}$, $xe^{\lambda x}$, any interval
7. 1, e^{4x} $(x < 0)$
8. $x^2 - 4$, $-3x^2 + 12$ $(x > 0)$
9. $|x|x$, x^2 $(0 < x < 1)$
10. $|x|x$, x^2 $(-1 < x < 1)$

Find a general solution of the following differential equations on any interval. Verify that the two functions involved form a basis (are not proportional).

11. $y'' - 25y = 0$ **12.** $y'' - 3y' + 2y = 0$

13. $y'' - y' - 2y = 0$ **14.** $2y'' - y' = 0$

15. $y'' - 8y' + 16y = 0$ **16.** $y'' + 0.4y' + 0.04y = 0$

17. $y'' + 9y = 0$ **18.** $y'' + \omega^2 y = 0$

19. $9y'' - 6y' + y = 0$ **20.** $y'' + 0.12y' - 0.31y = 0$

Solve the following initial value problems. *Hint.* First find a general solution.

21. $y'' - 4y = 0,$ $y(0) = 1,$ $y'(0) = 10$

22. $y'' - 4y' - 5y = 0,$ $y(0) = 0.4,$ $y'(0) = 2.0$

23. $8y'' + 6y' + y = 0,$ $y(-2) = 3e,$ $y'(-2) = -\frac{3}{2}e$

24. $y'' + 1.1y' + 0.1y = 0,$ $y(0) = 0,$ $y'(0) = 1.8$

25. $y'' + 3y' + 2.25y = 0,$ $y(0) = 2,$ $y'(0) = 0$

26. $y'' - 4y' + 3y = 0,$ $y(0) = -1,$ $y'(0) = -5$

27. $y'' - 2y' + \frac{3}{4}y = 0,$ $y(1) = 0.8\sqrt{e} \approx 1.319,$ $y'(0) = 0.4\sqrt{e} \approx 0.659$

28. $10y'' + 2y' + 0.1y = 0,$ $y(10) = -40/e \approx -14.72,$ $y'(10) = 0$

29. $2y'' + y' - y = 0,$ $y(4) = e^2 - e^{-4} \approx 7.371,$ $y'(4) = \frac{1}{2}e^2 + e^{-4} \approx 3.713$

30. $4y'' - 49y = 0,$ $y(0) = 0,$ $y'(0) = 0.84$

Find a differential equation $y'' + ay' + by = 0$ for which the two given functions are solutions and verify that these functions form a basis of solutions on any interval.

31. e^{-3x}, e^{2x} **32.** $1, e^{kx}$

33. $e^{i\omega x}, e^{-i\omega x}$ $(\omega \neq 0)$ **34.** $\cosh 5x, \sinh 5x$

35. $e^{(-3+i)x}, e^{(-3-i)x}$ **36.** $e^{-0.25x}, xe^{-0.25x}$

Reduction of order. Show that the given function y_1 is a solution of the given equation for all positive x and find y_2 such that y_1, y_2 form a basis of solutions for these x.

37. $y'' - \dfrac{1}{x}y' + \dfrac{1}{x^2}y = 0,$ $y_1 = x$

38. $y'' - 4y' + 4y = 0,$ $y_1 = e^{2x}$

39. $y'' - \dfrac{2}{x+1}y' + \dfrac{2}{(x+1)^2}y = 0,$ $y_1 = x + 1$

40. $y'' - \dfrac{x}{x-1}y' + \dfrac{1}{x-1}y = 0,$ $y_1 = e^x$

41. $y'' + \dfrac{2}{x}y' + y = 0,$ $y_1 = \dfrac{\sin x}{x}$

42. $y'' + \dfrac{1}{x}y' + \left(1 - \dfrac{1}{4x^2}\right)y = 0,$ $y_1 = \dfrac{\cos x}{\sqrt{x}}$

43. If y_1, y_2 is a basis for (1) on an interval I, show that $y_3 = y_1 + y_2, y_4 = y_1 - y_2$ is a basis for (1) on I. Give examples.

44. If y_1, y_2 is a basis for (1) on an interval I, show that $y_3 = \alpha y_1 + \beta y_2,$ $y_4 = \gamma y_1 + \delta y_2$ is a basis for (1) on I if and only if $\alpha\delta \neq \beta\gamma$.

45. Show that a general solution of (1) on some interval cannot be reduced to a form containing only *one* arbitrary constant.

2.4 Real Roots, Complex Roots, Double Root of the Characteristic Equation

We shall now see how to obtain a general solution of a homogeneous second-order linear differential equation with real constant coefficients. Such an equation is of the form

(1)
$$y'' + ay' + by = 0,$$

where a and b are constant. From Sec. 2.2 we know that a function

(2)
$$y = e^{\lambda x}$$

is a solution of (1) if λ is a root of the **characteristic equation**

(3)
$$\lambda^2 + a\lambda + b = 0.$$

These roots are

(4) $\lambda_1 = \tfrac{1}{2}(-a + \sqrt{a^2 - 4b})$, $\lambda_2 = \tfrac{1}{2}(-a - \sqrt{a^2 - 4b})$.

Since a and b are real, the characteristic equation may have

> **(Case I)** *two distinct real roots,*
> **(Case II)** *two complex conjugate roots, or*
> **(Case III)** *a real double root.*

This is the point we reached in Sec. 2.2. We can now discuss these cases separately and in each case obtain a general solution.

Case I. Two distinct real roots

This case arises when the *discriminant* $a^2 - 4b$ of (3) is positive, since then the square root in (4) is real (and not zero). Then a basis on any interval is

$$y_1 = e^{\lambda_1 x}, \qquad y_2 = e^{\lambda_2 x}.$$

Indeed, y_2/y_1 is not constant, so that these solutions are not proportional. The corresponding general solution is

(5)
$$y = c_1 e^{\lambda_1 x} + c_2 e^{\lambda_2 x}.$$

EXAMPLE 1. General solution in the case of distinct real roots
Solve

$$y'' + y' - 2y = 0.$$

Solution. The characteristic equation $\lambda^2 + \lambda - 2 = 0$ has the roots 1 and -2, so that we obtain the general solution

$$y = c_1 e^x + c_2 e^{-2x}.$$ ∎

Case II. Complex roots

This case occurs when $a^2 - 4b$ is negative, and (4) shows that then the roots are complex conjugates,

$$\lambda_1 = -\tfrac{1}{2}a + i\omega, \qquad \lambda_2 = -\tfrac{1}{2}a - i\omega,$$

where $\omega = \sqrt{b - \tfrac{1}{4}a^2}$. We claim that in this case, a basis on any interval is

$$y_1 = e^{-ax/2} \cos \omega x, \qquad y_2 = e^{-ax/2} \sin \omega x.$$

Indeed, that these are solutions follows by differentiation and substitution. Also, $y_2/y_1 = \tan \omega x$ is not constant, since $\omega \neq 0$, so that y_1 and y_2 are not proportional. The corresponding general solution is

(6)
$$\boxed{y = e^{-ax/2}(A \cos \omega x + B \sin \omega x).}$$

Complex Exponential Function

Case II (complex roots) has been settled, and nothing is left to prove. We just want to see how one can get the *idea* that y_1 and y_2 might be solutions in this case. We show that this follows from the complex exponential function.

The **complex exponential function** e^z of a complex variable[7] $z = s + it$ is defined by

(7)
$$\boxed{e^z = e^{s+it} = e^s(\cos t + i \sin t).}$$

In Sec. 12.6 we discuss e^z in full. It presently suffices to motivate this definition as follows. For real $z = s$ we have $t = 0$, $\cos 0 = 1$, $\sin 0 = 0$, and e^z becomes the familiar real exponential function e^s of calculus. Furthermore, e^z has properties quite similar to those of the real exponential function: it can be shown (Sec. 12.6) that it is differentiable, its derivative being e^z, and that it satisfies $e^{z_1+z_2} = e^{z_1}e^{z_2}$.

Now for $z = \lambda_1 x = -\tfrac{1}{2}ax + i\omega x$ and $z = \lambda_2 x = -\tfrac{1}{2}ax - i\omega x$ we obtain directly from (7), with $s = -\tfrac{1}{2}ax$ and $t = \omega x$ and $t = -\omega x$, respectively, the functions

---

[7]We write $z = s + it$ instead of $z = x + iy$ since x and y are used as variable and unknown function in our equations.

$$Y_1 = e^{\lambda_1 x} = e^{-ax/2}(\cos \omega x + i \sin \omega x)$$

$$Y_2 = e^{\lambda_2 x} = e^{-ax/2}(\cos \omega x - i \sin \omega x).$$

Note that in the second formula we used that $\sin(-\alpha) = -\sin \alpha$. By addition and subtraction we obtain the above y_1 and y_2:

$$\frac{1}{2}(Y_1 + Y_2) = e^{-ax/2} \cos \omega x = y_1,$$

$$\frac{1}{2i}(Y_1 - Y_2) = e^{-ax/2} \sin \omega x = y_2.$$

EXAMPLE 2. General solution in the case of complex conjugate roots
Find a general solution of the equation

$$y'' - 2y' + 10y = 0.$$

Solution. The characteristic equation is

$$\lambda^2 - 2\lambda + 10 = 0.$$

It has the complex conjugate roots

$$\lambda_1 = 1 + \sqrt{1 - 10} = 1 + 3i, \qquad \lambda_2 = 1 - 3i.$$

This yields the basis

$$y_1 = e^x \cos 3x, \qquad y_2 = e^x \sin 3x.$$

The corresponding general solution is

$$y = e^x(A \cos 3x + B \sin 3x).$$

EXAMPLE 3. An initial value problem
Solve the initial value problem

$$y'' - 2y' + 10y = 0, \qquad y(0) = 4, \qquad y'(0) = 1.$$

Solution. The equation is the same as in the previous example. Hence we can take the general solution just obtained and its derivative

$$y'(x) = e^x(A \cos 3x + B \sin 3x - 3A \sin 3x + 3B \cos 3x).$$

From y, y' and the initial conditions,

$$y(0) = A = 4$$

$$y'(0) = A + 3B = 1.$$

Hence $A = 4$, $B = -1$, and the answer is

$$y = e^x(4 \cos 3x - \sin 3x).$$

EXAMPLE 4
A general solution of the equation

$$y'' + \omega^2 y = 0 \qquad\qquad (\omega \text{ constant, not zero})$$

is

$$y = A \cos \omega x + B \sin \omega x. \qquad\blacksquare$$

Case III. Double root

This is sometimes called the **critical case.** From (4) we see that it arises when the discriminant is zero, that is, $a^2 - 4b = 0$. The root is $\lambda = -\frac{1}{2}a$. Then a basis on any interval is (cf. Example 4, Sec. 2.3)

$$e^{-ax/2}, \qquad xe^{-ax/2}.$$

The corresponding general solution is

(8)
$$\boxed{y = (c_1 + c_2 x)e^{-ax/2}.}$$

Warning. If λ is a *simple* root of (3), then $(c_1 + c_2 x)e^{\lambda x}$ is **not** a solution of (1).

Remark. Our present result can, of course, be derived directly, without using (6) in Sec. 2.3, but using the same idea, as follows. We have $y_1 = e^{-ax/2}$ and substitute $y_2 = uy_1$ and its derivatives

$$y_2{}' = u'y_1 + uy_1{}'$$
$$y_2{}'' = u''y_1 + 2u'y_1{}' + uy_1{}''$$

into (1). Collecting terms, we then obtain

$$u''y_1 + u'(2y_1{}' + ay_1) + u(y_1{}'' + ay_1{}' + by_1) = 0$$

The expression in the last parentheses is zero, since y_1 is a solution of (1). The expression in the first parentheses is zero, too, since

$$2y_1{}' = -ae^{-ax/2} = -ay_1.$$

We are thus left with $u''y_1 = 0$. Hence $u'' = 0$. By two integrations, $u = c_1 x + c_2$. To get a second independent solution $y_2 = uy_1$, we can simply take $u = x$. Then $y_2 = xy_1$, as above.

EXAMPLE 5. General solution in the case of a double root

Solve
$$y'' + 8y' + 16y = 0.$$

Solution. The characteristic equation has the double root $\lambda = -4$. Hence a basis is

$$e^{-4x} \qquad \text{and} \qquad xe^{-4x}$$

and the corresponding general solution is

$$y = (c_1 + c_2 x)e^{-4x}.$$

EXAMPLE 6. An initial value problem in the case of a double root

Solve the initial value problem

$$y'' - 4y' + 4y = 0, \qquad y(0) = 3, \qquad y'(0) = 1.$$

Solution. A general solution of the differential equation is

$$y(x) = (c_1 + c_2 x)e^{2x}.$$

By differentiation we obtain

$$y'(x) = c_2 e^{2x} + 2(c_1 + c_2 x)e^{2x}.$$

From this and the initial conditions it follows that

$$y(0) = c_1 = 3, \qquad y'(0) = c_2 + 2c_1 = 1.$$

Hence $c_1 = 3$, $c_2 = -5$, and the answer is

$$y = (3 - 5x)e^{2x}. \qquad \blacksquare$$

This completes the discussion of all three cases, and we may sum it up:

Case	Roots of (3)	Basis of (1)	General Solution of (1)
I	Distinct real λ_1, λ_2	$e^{\lambda_1 x}, \; e^{\lambda_2 x}$	$y = c_1 e^{\lambda_1 x} + c_2 e^{\lambda_2 x}$
II	Complex conjugate $\lambda_1 = -\frac{1}{2}a + i\omega$, $\lambda_2 = -\frac{1}{2}a - i\omega$	$e^{-ax/2}\cos \omega x$ $e^{-ax/2}\sin \omega x$	$y = e^{-ax/2}(A \cos \omega x + B \sin \omega x)$
III	Real double root $\lambda = -\frac{1}{2}a$	$e^{-ax/2}, \; xe^{-ax/2}$	$y = (c_1 + c_2 x)e^{-ax/2}$

Boundary Value Problem

Applications sometimes also lead to conditions of the type

(9) $$y(P_1) = k_1, \qquad y(P_2) = k_2.$$

These are known as *boundary conditions,* since they refer to the endpoints P_1, P_2 (*boundary points* P_1, P_2) of an interval I on which the equation (1) is considered. Equation (1) and conditions (9) together constitute what is known as a **boundary value problem.** We confine ourselves to the discussion of a typical example.

EXAMPLE 7. Boundary value problem
Solve the boundary value problem

$$y'' + y = 0, \qquad y(0) = 3, \qquad y(\pi) = -3.$$

Solution
1st Step. A basis of solutions is $y_1 = \cos x$, $y_2 = \sin x$. The corresponding general solution is

$$y(x) = c_1 \cos x + c_2 \sin x.$$

2nd Step. The left boundary condition gives $y(0) = c_1 = 3$. The right boundary condition gives $y(\pi) = c_1 \cos \pi + c_2 \cdot 0 = -3$. Now $\cos \pi = -1$ and $c_1 = 3$, so that this equation holds, and we see that it yields no condition for c_2. Hence a solution of the problem is

$$y = 3 \cos x + c_2 \sin x.$$

Here c_2 is still arbitrary. This is a surprise. Of course, the reason is that $\sin x$ is zero at 0 and π. The reader may conclude and prove (Prob. 45) that the solution of a boundary value problem (1), (9) is unique if and only if no solution $y \neq 0$ of (1) satisfies $y(P_1) = y(P_2) = 0$. $\blacksquare$

The next section on **differential operators** is optional and may be omitted without interrupting the flow of ideas. In Sec. 2.6 we discuss vibrating mass–spring systems (vibrations of a mass fixed at the end of an elastic spring) as a basic application of second-order homogeneous linear differential equations with constant coefficients.

Problems for Sec. 2.4

Verify that the following functions are solutions of the given differential equations and obtain from them a real-valued general solution of the form (6).

1. $y = c_1 e^{5ix} + c_2 e^{-5ix}$, $\quad y'' + 25y = 0$

2. $y = c_1 e^{i\pi x} + c_2 e^{-i\pi x}$, $\quad y'' + \pi^2 y = 0$

3. $y = c_1 e^{(1+i)x} + c_2 e^{(1-i)x}$, $\quad y'' - 2y' + 2y = 0$

4. $y = c_1 e^{(-0.1 + 2.5i)x} + c_2 e^{(-0.1 - 2.5i)x}$, $\quad y'' + 0.2y' + 6.26y = 0$

5. $y = c_1 e^{(k + 2\pi i)x} + c_2 e^{(k - 2\pi i)x}$, $\quad y'' - 2ky' + (k^2 + 4\pi^2)y = 0$

General Solution. State whether the given equation corresponds to Case I, Case II or Case III and find a general solution involving real-valued functions.

6. $y'' + 4y = 0$ **7.** $y'' - 4y = 0$

8. $y'' + 4y' + 4y = 0$ **9.** $10y'' + 6y' + 10.9y = 0$

10. $10y'' + 28y' - 6y = 0$ **11.** $y'' + 4y' + 4.25y = 0$

12. $y'' + 4y' - 5y = 0$ **13.** $4y'' + 36y' + 81y = 0$

14. $y'' - 2ky' + (k^2 + 16)y = 0$ **15.** $y'' - 2.56y = 0$

Initial Value Problems. Solve the following initial value problems.

16. $y'' - 4y = 0$, $\quad y(0) = 1$, $\quad y'(0) = 6$

17. $16y'' - 8y' + y = 0$, $\quad y(1) = 0$, $\quad y'(1) = -\sqrt[4]{e}$

18. $y'' + 2y' + 2y = 0$, $\quad y(\tfrac{1}{2}\pi) = 0$, $\quad y'(\tfrac{1}{2}\pi) = -2e^{-\pi/2}$

19. $y'' - 0.2y' + 100.01y = 0$, $\quad y(0) = 1$, $\quad y'(0) = -9.9$

20. $5y'' + 16y' + 12.8y = 0$, $\quad y(0) = 0$, $\quad y'(0) = -2.3$

21. $y'' + 3.7y' = 0$, $\quad y(-2) = 4$, $\quad y'(-2) = 0$

22. $16y'' + y = 0$, $\quad y(0) = -2$, $\quad y'(0) = -1$

23. $y'' + 20y' + 100y = 0$, $\quad y(0.1) = 3.2/e \approx 1.177$, $\quad y'(0.1) = -30/e \approx -11.04$

24. $y'' - 6y' + 9y = 0$, $\quad y(0) = 2$, $\quad y'(0) = 8$

25. $y'' - 2y' + (\pi^2 + 1)y = 0$, $\quad y(\tfrac{1}{4}) = 0$, $\quad y'(\tfrac{1}{4}) = -\pi\sqrt[4]{4e} \approx -5.705$

26. $y'' + 4y' + 4.25y = 0$, $\quad y(0) = 0$, $\quad y'(0) = \tfrac{1}{2}$

27. $4y'' + 15y' - 4y = 0$, $\quad y(0) = 6$, $\quad y'(0) = -7$

28. $y'' + 14y' + 49y = 0$, $\quad y(0) = \sqrt{2}$, $\quad y'(0) = -6\sqrt{2}$

29. $4y'' - 4y' - 3y = 0$, $\quad y(-2) = e$, $\quad y'(-2) = -\tfrac{1}{2}e$

30. $y'' + 4y' + (\omega^2 + 4)y = 0$, $\quad y(\pi/4\omega) = 0$, $\quad y'(\pi/4\omega) = -\omega\sqrt{2}e^{-\pi/2\omega}$

Double Root. Different Bases in Initial Value Problems

31. Apply reduction of order to $y'' - 2y' + y = 0$, using $y_1 = e^x$.

32. Apply reduction of order to $y'' + 10y' + 25y = 0$, using $y_1 = e^{-5x}$.

33. Verify directly that in the case of a double root, $xe^{\lambda x}$ with $\lambda = -a/2$ is a solution of (1).

34. Verify that $y = e^{-3x}$ is a solution of $y'' + 5y' + 6y = 0$, but $y = xe^{-3x}$ is not. Explain.

35. Verify that if (1) does not have a double root, then $xe^{\lambda x}$ is **not** a solution of (1).

36. **Different bases** lead to the same solution. To illustrate this, solve $y'' - 9y = 0$, $y(0) = 4$, $y'(0) = -6$, using (a) e^{3x}, e^{-3x}, (b) $\cosh 3x$, $\sinh 3x$.

Boundary Value Problems. Solve the following boundary value problems.

37. $y'' - 16y = 0$, $y(0) = 3$, $y(\tfrac{1}{4}) = 3e$

38. $y'' - 4y = 0$, $y(-2) = y(2) = \cosh 4$

39. $y'' - 2y' = 0$, $y(0) = 0$, $y(\tfrac{1}{2}) = e - 1$

40. $y'' - y' - 2y = 0$, $y(0) = 3$, $y(1) = 2e^2 + 1/e$

41. $y'' + y' - 2y = 0$, $y(0) = 0$, $y(1) = e - 1/e^2$

42. $3y'' - 8y' - 3y = 0$, $y(-3) = e$, $y(3) = 1/e$

43. $y'' - 2y' + 2y = 0$, $y(0) = -3$, $y(\tfrac{1}{2}\pi) = 0$

44. $y'' + 4y' + 5y = 0$, $y(\tfrac{1}{2}\pi) = 14e^{-\pi} \approx 0.6050$, $y(\tfrac{3}{2}\pi) = -14e^{-3\pi} \approx -0.0011$

45. Show that the solution of a boundary value problem (1), (9) is unique if and only if no solution $y \neq 0$ of (1) satisfies $y(P_1) = y(P_2) = 0$.

2.5 Differential Operators

In this section we give an introduction to differential operators.

By an **operator** we mean a transformation that transforms a function into another function. Operators and corresponding techniques, which are called **operational methods,** play an increasing role in engineering mathematics.

Differentiation suggests an operator as follows. Let D denote differentiation with respect to x, that is, write

$$\boxed{Dy = y'.}$$

D is an operator; it transforms y (assumed differentiable) into its derivative y'. For example,

$$D(x^2) = 2x, \quad D(\sin x) = \cos x.$$

Applying D twice, we obtain the second derivative $D(Dy) = Dy' = y''$. We simply write $D(Dy) = D^2y$, so that

$$Dy = y', \qquad D^2y = y'', \qquad D^3y = y''', \quad \cdots.$$

More generally,

(1) $$L = P(D) = D^2 + aD + b$$

is called a **second-order differential operator.** Here a and b are constant. P

suggests "polynomial." L suggests "linear" (explained below). When L is applied to a function y (assumed twice differentiable), it produces

(2)
$$L[y] = (D^2 + aD + b)y = y'' + ay' + by.$$

L is a *linear* operator. By definition this means that we have

$$L[\alpha y + \beta w] = \alpha L[y] + \beta L[w]$$

for any constants α and β and any functions y and w that are twice differentiable.

The homogeneous linear differential equation $y'' + ay' + by = 0$ may now be simply written

(3)
$$L[y] = P(D)[y] = 0.$$

For example,

(4)
$$L[y] = (D^2 + D - 6)y = y'' + y' - 6y = 0.$$

Since

$$D[e^{\lambda x}] = \lambda e^{\lambda x}, \qquad D^2[e^{\lambda x}] = \lambda^2 e^{\lambda x},$$

we have from (2) and (3)

(5)
$$P(D)[e^{\lambda x}] = (\lambda^2 + a\lambda + b)e^{\lambda x} = P(\lambda)e^{\lambda x} = 0.$$

This confirms our result of the last section that $e^{\lambda x}$ is a solution of (3) if and only if λ is a solution of the characteristic equation $P(\lambda) = 0$. If $P(\lambda)$ has two different roots, we obtain a basis. If $P(\lambda)$ has a double root, we need a second independent solution. To obtain that solution, we differentiate

$$P(D)[e^{\lambda x}] = P(\lambda)e^{\lambda x}$$

[cf. (5)] on both sides with respect to λ and interchange differentiation with respect to λ and x, obtaining

$$P(D)[xe^{\lambda x}] = P'(\lambda)e^{\lambda x} + P(\lambda)xe^{\lambda x}$$

where $P' = dP/d\lambda$. For a double root, $P(\lambda) = P'(\lambda) = 0$, so that we have $P(D)[xe^{\lambda x}] = 0$. Hence $xe^{\lambda x}$ is the desired second solution. This agrees with Sec. 2.4.

$P(\lambda)$ is a polynomial in λ, in the usual sense of algebra. If we replace λ by D, then we obtain the "operator polynomial" $P(D)$. The point of this **"operational calculus"** is that $P(D)$ can be treated just like an algebraic quantity. In particular, we can factor it.

EXAMPLE 1. Factorization, solution of a differential equation

Factor $P(D) = D^2 + D - 6$ and solve $P(D)y = 0$.

Solution. $D^2 + D - 6 = (D + 3)(D - 2)$. Now $(D - 2)y = y' - 2y$ by definition. Hence

$$(D + 3)(D - 2)y = (D + 3)[y' - 2y]$$
$$= y'' - 2y' + 3y' - 6y$$
$$= y'' + y' - 6y.$$

Hence our factorization is "permissible," that is, yields the correct result. Solutions of $(D + 3)y = 0$ and $(D - 2)y = 0$ are $y_1 = e^{-3x}$ and $y_2 = e^{2x}$. This is a basis of $P(D)y = 0$ on any interval. The student should verify that our method in Sec. 2.4 gives the same result. This is not unexpected, since we factored $P(D)$ in the same way as we factor the characteristic polynomial $P(\lambda) = \lambda^2 + \lambda - 6$. ∎

Without going into details at this time, we want to mention that operational methods can also be used for operators $M = D^2 + fD + g$ with *variable* coefficients $f(x)$ and $g(x)$ but are more difficult and require more care. For example, $xD \neq Dx$ because

$$xDy = xy' \qquad \text{but} \qquad Dxy = (xy)' = y + xy'.$$

If operational calculus were limited to the simple situations illustrated in this section, it would perhaps not be worth mentioning. Actually, the power of the linear operator approach comes out in more complicated engineering problems, as we shall see in Chap. 5.

Problems for Sec. 2.5

In each case apply the given operator to each of the given functions.

1. $D^2 + 3D$; $\cosh 3x$, $e^{-x} + e^{2x}$, $10 - e^{-3x}$
2. $D - 4$; $4x^2 - 3x$, $6e^{4x}$, $\cos x - \sin x$
3. $(D - 2)(D + 1)$; e^{2x}, xe^{2x}, e^{-x}, xe^{-x}
4. $(D + 3)^2$; $x^3 + \sin \pi x$, e^{-3x}, xe^{-3x}

Find a general solution of the following equations.

5. $(D^2 - D - 2)y = 0$ 6. $(4D^2 + 12D + 13)y = 0$
7. $(6D^2 - D - 1)y = 0$ 8. $(4D^2 - 12D + 9)y = 0$
9. $(D^2 + 2D - 0.44)y = 0$ 10. $(D^2 + 7D + 14.5)y = 0$
11. $(D^2 - 1.8D + 5.22)y = 0$ 12. $(10D^2 + 12D + 3.6)y = 0$
13. $(\pi^2 D^2 - 4\pi D + 4)y = 0$ 14. $(D^2 + 2kD + k^2 + 3)y = 0$

Solve the following initial value problems.

15. $(D^2 + 2D + 2)y = 0$, $y(0) = 1$, $y'(0) = -1$
16. $(D^2 + 4D + 4 + \pi^2)y = 0$, $y(1) = -1/e^2$ $y'(1) = (2 + 2\pi)/e^2$
17. $(9D^2 + 6D + 1)y = 0$, $y(-3) = 10e \approx 27.18$, $y'(-3) = -\frac{19}{3}e \approx -17.22$
18. $(D^2 - 0.2D + 100.01)y = 0$, $y(0) = 0$, $y'(0) = 40$
19. $(8D^2 + 2D - 1)y = 0$, $y(0) = 5/2$, $y'(0) = -13/8$
20. $(D^2 - 0.1D - 3.8)y = 0$, $y(0) = -3.9$, $y'(0) = 7.8$
21. $(D^2 + D + 0.25)y = 0$, $y(2) = 50/e \approx 18.39$, $y'(2) = 0$

22. Illustrate the linearity of L in (2) by taking $\alpha = 3$, $\beta = -2$, $y = e^{-2x}$ and $w = \sin x$.

23. Prove that the operator L in (2) is linear.

24. If $(D^2 + aD + b)y = 0$ is such that the characteristic equation has distinct roots λ and μ, show that a particular solution is

$$y = \frac{e^{\mu x} - e^{\lambda x}}{\mu - \lambda}.$$

Obtain from this a solution of the form $xe^{\lambda x}$ by letting $\mu \to \lambda$ and applying l'Hôpital's rule to y, considered as a function of μ.

25. Verify by direct calculation that $D^2[e^{-kx}f(x)] = e^{-kx}(D - k)^2 f(x)$. Choosing $e^{-kx}f(x) = c_1 + c_2 x$, show that $f(x)$ is a general solution of (3) in the case of a double root.

2.6 Modeling: Free Oscillations

Linear homogeneous differential equations with constant coefficients have important engineering applications. In this section we shall consider such an application, which is fundamental. It is taken from mechanics, but we shall see later that it has a complete analog in electric circuits. In our discussion we shall gain some more experience in **modeling,** that is, setting up mathematical models of physical systems. We shall also see that the mathematical distinction between the three cases (I, II, III) in Sec. 2.4 is of physical significance because these cases correspond to three different types of motion.

Setting up the Model

We take an ordinary spring that resists compression as well as extension and suspend it vertically from a fixed support (Fig. 34). At the lower end of the spring we attach a body of mass m. We assume m to be so large that we may disregard the mass of the spring. If we pull the body down a certain distance and then release it, it undergoes a motion. We assume that the body moves strictly vertically.

We want to determine the motion of our mechanical system. For this purpose we consider the forces[8] acting on the body during the motion. This will lead to a differential equation, and by solving this equation we shall then obtain the displacement as a function of time.

We may choose the *downward direction* as the positive direction and regard forces which act downward as positive and upward forces as negative.

The most obvious force acting on the body is the *attraction of gravity*

---

[8]**Systems of units** and conversion factors see inside of front cover.

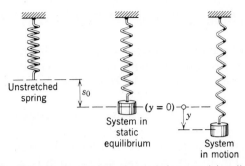

Fig. 34. Mechanical system under consideration

(1) $$F_1 = mg$$

where m is the mass of the body and g ($= 980$ cm/sec^2) is the acceleration of gravity.

We next consider the *spring force* F_2 acting on the body. Experiments show that within reasonable limits, its magnitude is proportional to the change in the length of the spring. Its direction is upward if the spring is extended and downward if the spring is compressed. Thus,

(2) $$F_2 = -ks \qquad \textbf{(Hooke's[9] law)},$$

where s is the vertical displacement of the body (recall that the upper end of the spring is fixed), the constant of proportionality k is called the *spring modulus,* and the minus sign makes F_2 negative (upward) for positive s (extension of the spring) and positive (downward) for negative s (compression of the spring).

If $s = 1$, then $F_2 = -k$. The stiffer the spring, the larger k.

When the body is at rest (motionless), gravitational force and spring force are in equilibrium, their resultant is the zero force,

(3) $$F_1 + F_2 = mg - ks_0 = 0$$

where s_0 is the extension of the spring corresponding to this position, which is called the *static equilibrium position.*

We denote by $y = y(t)$ [t time] the displacement of the body from the static equilibrium position ($y = 0$), with the positive direction downward (Fig. 34). This displacement causes an additional force $-ky$ on the body, by Hooke's law. Hence the resultant of the forces on the body at position $y(t)$ is [cf. (3)]

(4) $$F_1 + F_2 - ky = -ky.$$

[9]ROBERT HOOKE (1635—1703), English physicist, a forerunner of Newton with respect to the law of gravitation.

Undamped System: Equation and Solution

If the damping of the system is so small that it can be disregarded, then (4) is the resultant of *all* the forces acting on the body. The differential equation will now be obtained by the use of **Newton's second law**

$$\text{Mass} \times \text{Acceleration} = \text{Force}$$

where *force* means the resultant of the forces acting on the body at any instant. In our case, the acceleration is $y'' = d^2y/dt^2$ and that resultant is given by (4). Thus

$$my'' = -ky.$$

Hence the motion of our system is governed by the linear differential equation with constant coefficients

(5)
$$\boxed{my'' + ky = 0.}$$

By the method in Sec. 2.4 we obtain the general solution

(6)
$$\boxed{y(t) = A \cos \omega_0 t + B \sin \omega_0 t} \qquad \omega_0 = \sqrt{k/m}.$$

The corresponding motion is called a **harmonic oscillation.** Figure 35 shows typical forms of (6) corresponding to some positive initial displacement $y(0)$ [which determines $A = y(0)$ in (6)] and different initial velocities $y'(0)$ [each of which determines a value of B in (6), since $y'(0) = \omega_0 B$].

By applying the addition formula for the cosine, the student may verify that (6) can be written [cf. also (13) in Appendix 3]

(6*) $$y(t) = C \cos(\omega_0 t - \delta) \qquad \left(C = \sqrt{A^2 + B^2}, \quad \tan \delta = \frac{B}{A} \right).$$

Since the period of the trigonometric functions in (6) is $2\pi/\omega_0$, the body executes $\omega_0/2\pi$ cycles per second. The quantity $\omega_0/2\pi$ is called the **frequency**

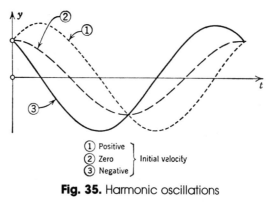

Fig. 35. Harmonic oscillations

of the oscillation and is measured in cycles per second. Another name for cycles/sec is hertz (Hz).[10]

EXAMPLE 1. Undamped system. Harmonic oscillations

If an iron ball of weight $W = 89.00$ nt (about 20 lb) stretches a spring 10.00 cm (about 4 inches), how many cycles per minute will this mass-spring system execute? What will its motion be if we pull down the weight an additional 15.00 cm (about 6 inches)?

Solution. From (2) we obtain the value $k = 89.00/0.1000 = 890.0$ [nt/meter]. The mass is $m = W/g = 89.00/9.8000 = 9.082$ [kg]. This gives the frequency

$$\omega_0/2\pi = \sqrt{890.0/9.082}/2\pi = 9.899/2\pi = 1.576 \text{ Hz}$$

or 94.5 cycles per minute. From (6) and the initial conditions, $y(0) = A = 0.1500$ [meter] and $y'(0) = \omega_0 B = 0$. Hence the motion is

$$y(t) = 0.1500 \cos 9.899t \text{ [meters]} \quad \text{or} \quad 0.492 \cos 9.899t \text{ [ft]}.$$

If you have a chance of experimenting with a mass-spring system, don't miss it. You will be surprised about the good agreement between theory and experiment, usually within a fraction of one percent if you measure carefully.

Damped System: Equation and Solutions

If we connect the mass to a dashpot (Fig. 36), then we have to take the corresponding viscous damping into account. The corresponding damping force has the direction opposite to the instantaneous motion, and we shall assume that it is proportional to the velocity $y' = dy/dt$ of the body. This is generally a good approximation, at least for small velocities. Thus the damping force is of the form

$$F_3 = -cy'.$$

Let us show that the *damping constant c* is positive. If y' is positive, the body moves downward (in the positive y-direction) and $-cy'$ must be an

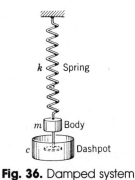

Fig. 36. Damped system

---

[10]HEINRICH HERTZ (1857—1894), German physicist, who discovered electromagnetic waves and made important contributions to electrodynamics.

upward force, that is, by agreement, $-cy' < 0$, which implies $c > 0$. For negative y' the body moves upward and $-cy'$ must represent a downward force, that is, $-cy' > 0$, which implies $c > 0$.

The resultant of the forces acting on the body is now [cf. (4)]

$$F_1 + F_2 + F_3 = -ky - cy'.$$

Hence, by Newton's second law,

$$my'' = -ky - cy',$$

and we see that the motion of the damped mechanical system is governed by the linear differential equation with constant coefficients

(7)
$$\boxed{my'' + cy' + ky = 0.}$$

The corresponding characteristic equation is

$$\lambda^2 + \frac{c}{m}\lambda + \frac{k}{m} = 0.$$

The roots are

$$\lambda_{1,2} = -\frac{c}{2m} \pm \frac{1}{2m}\sqrt{c^2 - 4mk}.$$

Using the short notations

(8)
$$\alpha = \frac{c}{2m} \quad \text{and} \quad \beta = \frac{1}{2m}\sqrt{c^2 - 4mk},$$

we can write

$$\lambda_1 = -\alpha + \beta \quad \text{and} \quad \lambda_2 = -\alpha - \beta.$$

The form of the solution of (7) will depend on the damping, and, as in Secs. 2.2 and 2.4, we now have the following three cases:

Case I.	$c^2 > 4mk$.	*Distinct real roots λ_1, λ_2.*	(*Overdamping*)
Case II.	$c^2 < 4mk$.	*Complex conjugate roots.*	(*Underdamping*)
Case III.	$c^2 = 4mk$.	*A real double root.*	(*Critical damping*)

Let us discuss these three cases separately.

Case I. Overdamping
When the damping constant c is so large that $c^2 > 4mk$, then λ_1 and λ_2 are distinct real roots, the general solution of (7) is

(9)
$$\boxed{y(t) = c_1 e^{-(\alpha - \beta)t} + c_2 e^{-(\alpha + \beta)t}}$$

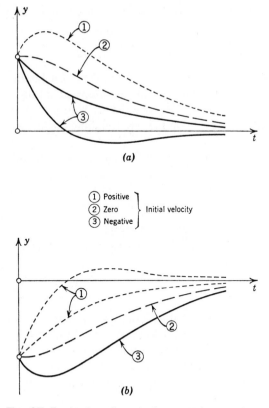

(a)

① Positive
② Zero   } Initial velocity
③ Negative

(b)

Fig. 37. Typical motions in the overdamped case
(a) Positive initial displacement
(b) Negative initial displacement

and we see that in this case the body does not oscillate. For $t > 0$ both exponents in (9) are negative because $\alpha > 0$, $\beta > 0$, and $\beta^2 = \alpha^2 - k/m < \alpha^2$. Hence both terms in (9) approach zero as t approaches infinity. Practically speaking, after a sufficiently long time the mass will be at rest in the static equilibrium position ($y = 0$). This is understandable since the damping takes energy from the system and there is no external force that keeps the motion going. Figure 37 shows (9) for some typical initial conditions.

Case II. Underdamping

This is the most interesting case. If the damping constant c is so small that $c^2 < 4mk$, then β in (8) is pure imaginary, say,

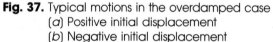

$$(10) \quad \beta = i\omega^* \quad \text{where} \quad \omega^* = \frac{1}{2m} \sqrt{4mk - c^2} = \sqrt{\frac{k}{m} - \frac{c^2}{4m^2}} \quad (> 0).$$

(We write ω^* to reserve ω for Sec. 2.13.) The roots of the characteristic equation are complex conjugate,

$$\lambda_1 = -\alpha + i\omega^*, \qquad \lambda_2 = -\alpha - i\omega^*,$$

[with α given in (8)] and the general solution is

(11) $\boxed{y(t) = e^{-\alpha t}(A \cos \omega^* t + B \sin \omega^* t) = Ce^{-\alpha t} \cos(\omega^* t - \delta)}$

where $C^2 = A^2 + B^2$ and $\tan \delta = B/A$ [as in (6*)].

This solution represents damped oscillations. Since $\cos(\omega^* t - \delta)$ varies between -1 and 1, the curve of the solution lies between the curves $y = Ce^{-\alpha t}$ and $y = -Ce^{-\alpha t}$ in Fig. 38, touching these curves when $\omega^* t - \delta$ is an integer multiple of π.

The frequency is $\omega^*/2\pi$ cycles per second. From (10) we see that the smaller c (> 0) is, the larger is ω^* and the more rapid the oscillations become. As c approaches zero, ω^* approaches the value $\omega_0 = \sqrt{k/m}$ corresponding to the harmonic oscillation (6).

Case III. Critical damping

If $c^2 = 4mk$, then $\beta = 0$, $\lambda_1 = \lambda_2 = -\alpha$, and the general solution is

(12) $\boxed{y(t) = (c_1 + c_2 t)e^{-\alpha t}.}$

Since the exponential function is never zero and $c_1 + c_2 t$ can have at most one positive zero, it follows that the motion can have at most one passage through the equilibrium position ($y = 0$). If the initial conditions are such that c_1 and c_2 have the same sign, there is no such passage at all. This is similar to Case I. Figure 39 shows typical forms of (12).

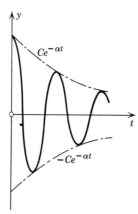

Fig. 38. Damped oscillation in Case II

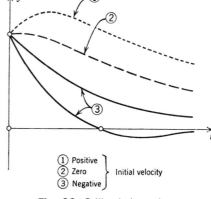

① Positive ⎫
② Zero   ⎬ Initial velocity
③ Negative ⎭

Fig. 39. Critical damping

EXAMPLE 2. The three cases of damped motion

How does the motion in Example 1 change if the system has damping given by

   (I) $c = 200.0$ kg/sec,
   (II) $c = 100.0$ kg/sec,
  (III) $c = 179.8$ kg/sec?

Solution. It is instructive to study and compare these cases with the behavior of the system in Example 1.

(I)   The problem is

$$9.082y'' + 200.0y' + 890.0y = 0, \qquad y(0) = 0.1500 \text{ [meter]}, \qquad y'(0) = 0.$$

The characteristic equation has the roots $\lambda_{1,2} = -\alpha \pm \beta = -11.01 \pm 4.822$, thus $\lambda_1 = -6.190$, $\lambda_2 = -15.83$. From (9) and the initial conditions, $c_1 + c_2 = 0.1500$, $\lambda_1 c_1 + \lambda_2 c_2 = 0$. The solution is

$$y(t) = 0.2463e^{-6.190t} - 0.0963e^{-15.83t}.$$

It approaches 0 as $t \to \infty$. The approach is very rapid. After a few seconds, it is practically 0; that is, the ball is at rest.

(II)   The problem is as before, with $c = 100$ instead of 200. The roots are complex conjugates, $\lambda_{1,2} = -\alpha \pm i\omega^* = -5.506 \pm 8.227i$. From (11) and the initial conditions, $A = 0.1500$, $-\alpha A + \omega^* B = 0$ or $B = 0.1004$. This gives the solution

$$y(t) = e^{-5.506t}(0.1500 \cos 8.227t + 0.1004 \sin 8.227t)$$

These damped oscillations are slower than the harmonic oscillations in Example 1 by about 17%.

(III)   The problem is as before, with $c = 179.8$. Since $c^2 = 4mk$, we get the double root $\lambda = -9.899$. From (12) and the initial conditions, $c_1 = 0.1500$, $c_2 + \lambda c_1 = 0$, $c_2 = 1.485$. The solution is

$$y(t) = (0.150 + 1.485t)e^{-9.899t}.$$

It decreases to zero rapidly and in a monotone fashion.   ∎

This whole section concerned **free motions** of mass–spring systems. These are governed by homogeneous differential equations, as we have seen. **Forced motions** under the influence of a "driving force" lead to nonhomogeneous equations and will be studied in Sec. 2.13, after we have learned how to solve such equations.

We have seen that homogeneous linear constant-coefficient differential equations can be solved by algebra (by solving the characteristic equation; cf. Sec. 2.4). Another class of differential equations also solvable by algebra consists of the so-called **Euler–Cauchy equations,** which we consider next. In engineering, these are of lesser importance.

Problems for Sec. 2.6

Harmonic oscillations (Undamped motion)

1. Show that the harmonic oscillation (6) starting from initial displacement y_0 with initial velocity v_0 is

$$y(t) = y_0 \cos \sqrt{\frac{k}{m}}\, t + v_0 \sqrt{\frac{m}{k}} \sin \sqrt{\frac{k}{m}}\, t.$$

Represent this in the form (6*).

2. To find the modulus k of a spring S, take a body B, determine its weight W, attach B to S (whose upper end is fixed), let B come to rest and measure the stretch s_0 of S. If $W = 69.77$ nt [= 15.68 lb] and $s_0 = 0.441$ meter, what is k and how many cycles per minute will B execute when allowed to vibrate?

3. If a spring is such that a weight of 20 nt (about 4.5 lb) would stretch it 2 cm, what would the frequency of the corresponding harmonic oscillation be? The period?

4. Show that the frequency of a harmonic oscillation of a body on a spring is $(\sqrt{g/s_0})/2\pi$, so that the period is $2\pi\sqrt{s_0/g}$, where s_0 is the elongation in Fig. 34.

5. How does the frequency of the harmonic oscillation change if we (i) double the mass, (ii) take a stiffer spring, (iii) change the initial conditions?

6. Consider (6), assuming that $m = 10$ kg and $k = 10$ nt/meter. Determine A and B such that the initial displacement is 0.01 meter and the initial velocity is (a) -0.01 meter/sec $= -1$ cm/sec, (b) 0, (c) 0.01 meter/sec $= 1$ cm/sec. Represent these functions in the form (6*) and graph their curves.

7. Graph the harmonic oscillation (6) for various values of the spring modulus, say, $k = 1, 4, 9, 16$, assuming that the mass is 1 and the motion starts from $y = 1$ with initial velocity 0. What are the frequencies?

8. At 20.0 nt weight (about 4.5 lb) stretches a spring 9.8 cm, and we pull the weight on the spring down 5 cm from rest and give it an upward velocity of 30.0 cm/sec. Find the resulting motion $y(t)$, assuming no damping.

9. What are the frequencies of vibration of a mass $m = 5$ kg (i) on a spring with modulus $k_1 = 10$ nt/m, (ii) on a spring with $k_2 = 20$ nt/m, (iii) on the two springs in parallel? Cf. Fig. 40.

10. Answer the same question as in Prob. 9, assuming that the mass hangs on the first spring, which in turn hangs on the second spring whose upper end is fixed.

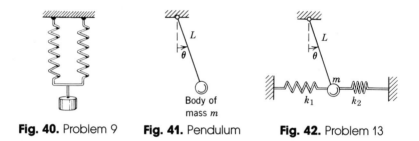

Fig. 40. Problem 9   **Fig. 41.** Pendulum   **Fig. 42.** Problem 13

11. (Pendulum) Determine the frequency of oscillation of the pendulum of length L in Fig. 41. Neglect air resistance and the weight of the rod. Assume that θ is small enough that $\sin\theta \approx \theta$.

12. A clock has a 1 meter pendulum. The clock ticks once for each time the pendulum completes a swing, returning to its original position. How many times a minute does the clock tick?

13. Suppose that the system in Fig. 42 consists of a pendulum as in Prob. 11 and two springs with constants k_1 and k_2 attached to the vibrating body and two vertical walls such that $\theta = 0$ remains the position of static equilibrium and $\theta(t)$ remains small during the motion. Find the period T.

14. (Flat spring) Our present equation $my'' + k_0 y = 0$ governs also the (undamped) vibrations of a body attached to a flat spring (of negligible mass) whose other end is horizontally clamped (Fig. 43); here k_0 is the spring constant in Hooke's law $F = -k_0 s$. What is the motion if the body weighs 4 nt (about 0.9 lb), the system has its equilibrium 1 cm below the horizontal line and we let it start from this position with downward initial velocity 20 cm/sec? When will the body reach its highest position for the first time?

15. (Torsional vibrations) Undamped torsional vibrations (rotations back and forth) of a wheel attached to an elastic thin rod or wire (Fig. 44) are governed by the equation

$$I_0\theta'' + K\theta = 0,$$

Fig. 43. Problem 14   **Fig. 44.** Problem 15
(Flat spring)   (Torsional vibrations)

where θ is the angle measured from the state of equilibrium, I_0 the polar moment of inertia of the wheel about its center and K the torsional stiffness of the rod. Solve the equation for $K/I_0 = 13.69 \text{ sec}^{-2}$, initial angle $15°$ ($= 0.2618$ rad) and initial angular velocity $10° \text{ sec}^{-1}$ ($= 0.1745$ rad $\cdot$ sec^{-1}).

16. A cylindrical buoy 60 cm in diameter stands in water with its axis vertical (Fig. 45). When depressed slightly and released, it is found that the period of vibration is 2 sec. Determine the weight of the buoy. *Hint.* By **Archimedes' principle,** the buoyancy force equals the weight of the water displaced by the body (assumed partly or totally submerged).

17. If 1 liter of water is allowed to vibrate up and down in a U-shaped tube 2 cm in diameter (Fig. 46), what is the frequency? Neglect friction.

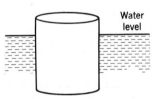

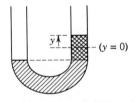

Fig. 45. Buoy in Problem 16   **Fig. 46.** Tube in Problem 17

18. **(Damping)** Determine and graph the motion $y(t)$ of the mechanical system described by (7) for increasing damping, say, $c = 0, 1, 2, 3$. Assume $m = 1, k = 1$, initial displacement 1 and initial velocity 0.

Case I. Overdamped motion

19. Show that for (9) to satisfy initial conditions $y(0) = y_0$ and $v(0) = v_0$ we must have

$$c_1 = [(1 + \alpha/\beta)y_0 + v_0/\beta]/2 \text{ and } c_2 = [(1 - \alpha/\beta)y_0 - v_0/\beta]/2.$$

20. Show that in the overdamped case, the body can pass through $y = 0$ at most once (Fig. 37).

21. In Prob. 20 find conditions for c_1 and c_2 such that the body does not pass through $y = 0$ at all.

22. Show that an overdamped motion with zero initial displacement cannot pass through $y = 0$.

Case II. Underdamped motion (Damped oscillation)

23. Show that the damped oscillation satisfying the initial conditions $y(0) = y_0$, $v(0) = v_0$ is

$$y = e^{-\alpha t}(y_0 \cos \omega^* t + \omega^{*-1}(v_0 + \alpha y_0) \sin \omega^* t).$$

24. Find and graph the three damped oscillations of the form

$$y = e^{-t}(A \cos t + B \sin t) = Ce^{-t} \cos (t - \delta)$$

starting from $y = 1$ with initial velocity $-1, 0, 1$, respectively.

25. Show that the frequency $\omega^*/2\pi$ of the underdamped motion decreases as the damping increases.

26. Show that for small damping,

$$\omega^* \approx \omega_0 \left(1 - \frac{c^2}{8mk}\right).$$

27. For what c (in terms of m and k) is $\omega^*/\omega_0 = 99\%$? 95%? Calculate (a) exactly, (b) by the formula in Prob. 26.

28. Show that the maxima and minima of an underdamped motion occur at equidistant values of t, the distance between two consecutive maxima being $2\pi/\omega^*$.

29. Determine the values of t corresponding to the maxima and minima of the oscillation $y(t) = e^{-t} \sin t$. Check your result by graphing $y(t)$.

30. **(Logarithmic decrement)** Prove that the ratio of two consecutive maximum amplitudes of a damped oscillation (11) is constant, the natural logarithm of this ratio being $\Delta = 2\pi\alpha/\omega^*$. ($\Delta$ is called the *logarithmic decrement* of the oscillation.) Find Δ in the case of $y = e^{-t} \cos t$ and determine the values of t corresponding to the maxima and minima.

31. Consider an underdamped motion of a body of mass $m = 0.5$ kg. If the time between two consecutive maxima is 1 sec and the maximum amplitude decreases to half its initial value after 10 cycles, what is the damping constant of the system?

Case III. Critical Damping

32. Under what conditions does (12) have a maximum or minimum at some instant $t > 0$?

33. Find the critical motion (12) which starts from y_0 with initial velocity v_0.

34. Represent the maximum or minimum amplitude in Prob. 32 in terms of the initial values y_0 and v_0.

35. **(Buckling of a rod)** Equations $y'' + \gamma^2 y = 0$ also have applications in which the independent variable is a space variable (instead of time t). Figure 47 shows a famous example, the buckling of a rod under a constant vertical force F, which is applied at the upper end of the rod. It is shown in mechanics that the curve $y(x)$, which the rod assumes under the load, is a solution of the second-order differential equation

$$EIy'' = -Fy$$

E is the modulus of elasticity of the material of the rod and I is the moment of

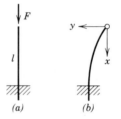

Fig. 47. Rod in Problem 35

inertia of the cross section (assumed circular) about an axis through the center of the circle. We assume E and I to be constant, the lower end of the rod to be clamped and the upper end free. Show that with respect to the coordinates in Fig. 47b this leads to the conditions

$$y(0) = 0, \qquad y'(l) = 0$$

where l is the length of the rod and y is assumed to be small. A solution for any F is $y = 0$ (Fig. 47a), but one knows that if F exceeds a critical value, the equilibrium in Fig. 47a is no longer stable, that is, if slightly displaced, the rod will not return but become curved. Show that the critical force is $F_{\text{crit}} = (\pi/2l)^2 EI$ and the corresponding $y(x)$ in Fig. 47b is a portion of a sine curve.

2.7 Euler–Cauchy Equation

The so-called **Euler–Cauchy equation**[11]

(1)
$$\boxed{x^2 y'' + axy' + by = 0} \qquad (a, b \text{ constant})$$

can also be solved by purely algebraic manipulations. Indeed, substituting

(2)
$$y = x^m$$

and its derivatives into the differential equation (1), we find

$$x^2 m(m - 1)x^{m-2} + axmx^{m-1} + bx^m = 0.$$

Omitting x^m, which is not zero if $x \neq 0$, we obtain the auxiliary equation

(3)
$$\boxed{m^2 + (a - 1)m + b = 0.}$$

[11]LEONHARD EULER (1707—1783) was an enormously creative Swiss mathematician. He studied in Basel under JOHANN BERNOULLI and in 1727 became a professor of physics (and later of mathematics) in St. Petersburg, Russia. In 1741 he went to Berlin as a member of the Berlin Academy. In 1766 he returned to St. Petersburg. He contributed to almost all branches of mathematics and its applications to physical problems, even after he became totally blind in 1771; we mention his fundamental work in differential and difference equations, Fourier and other infinite series, special functions, complex analysis, the calculus of variations, mechanics, and hydrodynamics. He is the exponent of a very rapid expansion of analysis. (So far, the first seventy (!) volumes of his Collected Works have appeared.)

This development was followed by a period characterized by greater rigor, dominated by the great French mathematician AUGUSTIN-LOUIS CAUCHY (1789—1857), the father of modern analysis. Cauchy studied and taught mainly in Paris. He is the creator of complex analysis and exercised a great influence on the theory of infinite series and ordinary and partial differential equations (see the Index to this book for some of these contributions). He is also known for his work in elasticity and optics. Cauchy published nearly 800 mathematical research papers, many of them of basic importance.

If the roots m_1 and m_2 of this equation are different, then the two functions

$$y_1(x) = x^{m_1} \quad \text{and} \quad y_2(x) = x^{m_2}$$

constitute a basis of solutions of the differential equation (1) for all x for which these functions are defined. The corresponding general solution is

(4)
$$\boxed{y = c_1 x^{m_1} + c_2 x^{m_2}}$$
$(c_1, c_2$ arbitrary).

EXAMPLE 1. General solution in the case of different real roots
Solve

$$x^2 y'' - 1.5xy' - 1.5y = 0.$$

Solution. The auxiliary equation is

$$m^2 - 2.5m - 1.5 = 0.$$

The roots are $m_1 = -0.5$ and $m_2 = 3$. Hence a basis of real solutions for all positive x is

$$y_1 = \frac{1}{\sqrt{x}}, \quad y_2 = x^3$$

and the corresponding general solution for all those x is

$$y = \frac{c_1}{\sqrt{x}} + c_2 x^3. \qquad \blacksquare$$

If the roots m_1 and m_2 of (3) are complex, they are conjugate, say $m_1 = \mu + i\nu$, $m_2 = \mu - i\nu$. We claim that in this case, a basis of solutions of (1) for all positive x is

(5) $y_1 = x^\mu \cos (\nu \ln x), \qquad y_2 = x^\mu \sin (\nu \ln x).$

Indeed, that these are solution follows by differentiation and substitution, and linear independence from the fact that these functions are not proportional. The corresponding general solution is

(6) $y = x^\mu [A \cos (\nu \ln x) + B \sin (\nu \ln x)].$

This proves everything and settles the case.

Another question is how we got the *idea* that these might be solutions. To answer it, we state that the formula $x^k = (e^{\ln x})^k = e^{k \ln x}$ extends from real to complex $k = i\nu$ and, together with the Euler formulas (Sec. 2.4) yields

$$x^{i\nu} = e^{i\nu \ln x} = \cos (\nu \ln x) + i \sin (\nu \ln x),$$

$$x^{-i\nu} = e^{-i\nu \ln x} = \cos (\nu \ln x) - i \sin (\nu \ln x).$$

Now multiply by x^μ and add and subtract. This gives $2y_1$ and $2iy_2$, respectively. From this, dividing by 2 and $2i$, we have (5).

EXAMPLE 2. General solution in the case of complex conjugate roots
Solve

$$x^2y'' + 7xy' + 13y = 0.$$

Solution. The auxiliary equation (3) is $m^2 + 6m + 13 = 0$. The roots of this equation are $m_{1,2} = -3 \pm \sqrt{9 - 13} = 3 \pm 2i$. By (6) the answer is

$$y = x^{-3}[A \cos (2 \ln x) + B \sin (2 \ln x)]. \qquad \blacksquare$$

The auxiliary equation (3) has a *double root* $m_1 = m_2$ if and only if

$$b = \frac{1}{4}(1 - a)^2, \qquad \text{and then} \qquad m_1 = m_2 = \frac{1 - a}{2}.$$

In this *critical case* we may obtain a second solution by applying the method of reduction of order. The procedure is similar to that in Sec. 2.4 and yields

$$y_2 = uy_1 = (\ln x)y_1,$$

as the student may verify. Thus, writing m in place of m_1, we see that

(7) $$y_1 = x^m \quad \text{and} \quad y_2 = x^m \ln x \qquad \left(m = \frac{1 - a}{2} \right)$$

are solutions of (1) in the case of a double root m of (3). Since these solutions are linearly independent, they constitute a basis of real solutions of (1) for all positive x, and the corresponding general solution of (1) is

(8) $$y = (c_1 + c_2 \ln x)x^m \qquad (c_1, c_2 \text{ arbitrary}).$$

EXAMPLE 3. General solution in the case of a double root
Solve

$$x^2y'' - 3xy' + 4y = 0.$$

Solution. The auxiliary equation has the double root $m = 2$. Hence a basis of real solutions for all positive x is x^2, $x^2 \ln x$, and the corresponding general solution is

$$y = (c_1 + c_2 \ln x)x^2. \qquad \blacksquare$$

In the next section we return to general second-order homogeneous linear differential equations, completing our discussion of **general solutions, bases** of solutions and **initial value problems** which we started in Sec. 2.4. We first applied these concepts to constant-coefficient equations (Secs. 2.5, 2.6) for practical reasons, but this order also was intended to impart a firm grasp and deeper understanding of the nature and role of these concepts. Such comprehension will be helpful in the next section. There we show that if the coefficients of a homogeneous linear differential equation are *continuous* (not necessarily constant), then this equation has a general solution, which includes all possible solutions of the equation and provides a unique solution of any corresponding initial value problem.

Problems for Sec. 2.7

1. Verify by substitution that y_1 and y_2 in (5) are solutions of (1) for all positive x.
2. Derive y_2 in (7) by the method of reduction of order.
3. Verify directly by substitution that y_2 in (7) is a solution of (1) if (3) has a double root, but $x^{m_1} \ln x$ and $x^{m_2} \ln x$ are **not** solutions of (1) if the roots m_1 and m_2 of (3) are different.

Find a general solution of the following differential equations.

4. $x^2 y'' - 4xy' + 6y = 0$
5. $x^2 y'' - 20y = 0$
6. $x^2 y'' - xy' + 2y = 0$
7. $x^2 y'' - 7xy' + 16y = 0$
8. $x^2 y'' + 6.2xy' + 6.76y = 0$
9. $4x^2 y'' + 12xy' + 3y = 0$
10. $25x^2 y'' + 12.5xy' + y = 0$
11. $x^2 y'' + 3xy' + 5y = 0$
12. $x^2 y'' - 0.4xy' + 0.49y = 0$
13. $10x^2 y'' + 46xy' + 32.4y = 0$
14. $(x^2 D^2 + xD - 1)y = 0$
15. $(x^2 D^2 + 0.7xD - 0.1)y = 0$
16. $(4x^2 D^2 + 8xD - 15)y = 0$
17. $(x^2 D^2 + 7xD + 9)y = 0$
18. $(x^2 D^2 + 1.25)y = 0$
19. $(2x^2 D^2 + 5xD - 9)y = 0$
20. $(x^2 D^2 - 5xD + \frac{82}{9})y = 0$
21. $(x^2 D^2 - 0.2xD + 0.36)y = 0$

Solve the following initial value problems.

22. $x^2 y'' - 2xy' + 2y = 0$, $y(1) = 1.5$, $y'(1) = 1$
23. $x^2 y'' - xy' - 3y = 0$, $y(2) = -1$, $y'(2) = \frac{1}{2}$
24. $x^2 y'' - 3xy' + 4y = 0$, $y(1) = 0$, $y'(1) = 3$
25. $x^2 y'' + xy' + 9y = 0$, $y(1) = 2$, $y'(1) = 0$
26. $4x^2 y'' + 12xy' + 3y = 0$, $y(4) = \frac{3}{4}$, $y'(4) = -\frac{5}{32}$
27. $(x^2 D^2 + xD - 0.01)y = 0$, $y(1) = 1$, $y'(1) = 0.1$
28. $(x^2 D^2 + 3xD + 1)y = 0$, $y(1) = 3$, $y'(1) = -4$
29. $(x^2 D^2 - xD + 2)y = 0$, $y(1) = -1$, $y'(1) = -1$

30. **(Potential between two spheres)** Consider a capacitor consisting of two concentric spheres S_1 and S_2 of radii $r_1 = 4$ cm and $r_2 = 8$ cm, respectively. Suppose that the electrostatic potential on S_1 is $v_1 = 110$ volts and the potential on S_2 is $v_2 = 0$. Find the potential $v(r)$ between the two spheres. *Physical information:* $v(r)$ is a solution of $(rD^2 + 2D)v = 0$, where $D = d/dr$.

31. **(Equations with constant coefficients)** Setting $x = e^t$ $(x > 0)$, transform the Euler–Cauchy equation (1) into the equation $\ddot{y} + (a - 1)\dot{y} + by = 0$, whose coefficients are constant. Here, dots denote derivatives with respect to t.

32. Transform the equation in Prob. 31 back into (1).

33. Show that if we apply the transformation in Prob. 31, then (2) yields an expression of the form (2), Sec. 2.4, and (8) yields an expression of the form (8), Sec. 2.4, except for notation.

Reduce to the form (1) and solve:

34. $(z + 1)^2 y'' + 5(z + 1)y' + 3y = 0$
35. $(2z - 3)^2 y'' + 7(2z - 3)y' + 4y = 0$

2.8 Existence of Solutions, Uniqueness

Content of the Remaining Portion of Chap. 2

In this chapter, up to now we were mainly concerned with *special homogeneous* second-order equations that can be solved by elementary methods, and with their applications. We shall now discuss existence and uniqueness of solutions of *general homogeneous* equations, starting with the case of second order and proceeding to arbitrary order n in the next section. After this, we shall see in Sec. 2.10 that our previous method of solving equations with *constant* coefficients generalizes from second order to nth order. The remaining sections (2.11-2.16) will be devoted to *nonhomogeneous* equations and applications.

Existence and Uniqueness

We first recall that an **initial value problem** *for a second-order differential equation* consists of the equation and two **initial conditions,** one for the solution $y(x)$ and the other for its first derivative $y'(x)$; cf. Sec. 2.3. We now consider initial value problems consisting of a second-order homogeneous linear differential equation

(1a)
$$y'' + p(x)y' + q(x)y = 0$$

and two initial conditions

(1b)
$$y(x_0) = K_0, \qquad y'(x_0) = K_1.$$

Here, x_0 is a given value of x, and K_0, K_1 are given constants. We write (1) to mean (1a), (1b) together. Our important result will be that the continuity of the coefficients $p(x)$ and $q(x)$ is sufficient for the existence and uniqueness of the solution of the problem (1), as follows.

Theorem 1 (Existence and uniqueness theorem)
If $p(x)$ and $q(x)$ are continuous functions on an open interval I and x_0 is in I, then the initial value problem (1) has a unique solution $y(x)$ on the interval I.

The proof of existence uses the same prerequisites as those of the existence theorem in Sec. 1.11 and will not be presented here; it can be found in Ref. [A10] listed in Appendix 1. Uniqueness proofs are usually simpler than existence proofs. But in the present case, even the uniqueness proof is long, and we give it as an optional proof on p. 148 near the end of this chapter.

Linear Independence. Wronskian. Existence of General Solutions

Our goal is to derive very important consequences of Theorem 1 with respect to general solutions

$$(2) \qquad y(x) = c_1 y_1(x) + c_2 y_2(x) \qquad (c_1, c_2 \text{ arbitrary constants})$$

of the homogeneous linear equation (1a). Recall first from Sec. 2.3 that (2) is a **general solution** of (1a) on some open interval I if y_1, y_2 form a **basis** of solutions of (1a) on I, that is, y_1 and y_2 are linearly independent on I. A solution obtained from (2) by assigning specific values to the constants c_1 and c_2 is called a **particular solution** of (1a) on I.

Remember further that we call y_1 and y_2 **linearly independent** on I if

$$k_1 y_1(x) + k_2 y_2(x) = 0 \quad \text{on } I \qquad \text{implies} \qquad k_1 = 0, k_2 = 0;$$

and we call y_1 and y_2 **linearly dependent** on I if this equation also holds for k_1, k_2 not both zero. In this case, and only in this case, y_1 and y_2 are proportional on 1, that is (cf. Sec. 2.3),

$$(3) \qquad \text{(a)} \quad y_1 = k y_2 \qquad \text{or} \qquad \text{(b)} \quad y_2 = l y_1.$$

Our goal is to show that if (1a) has continuous coefficients $p(x)$ and $q(x)$ on I, then (1a) always has a general solution on I and, second, such a general solution includes *all* solutions of (1a) on I; thus the *linear* equation (1a) does not have "*singular solutions*," that is, solutions *not* obtainable by assigning specific values to the constants in a general solution.

This is our program. As a first step, we derive a useful criterion for linear dependence and independence of solutions. This criterion uses the so-called *Wronski determinant*[12] or, briefly, the **Wronskian**, of two solutions y_1 and y_2 of (1a), defined by

$$(4) \qquad \boxed{W(y_1, y_2) = \begin{vmatrix} y_1 & y_2 \\ y_1' & y_2' \end{vmatrix} = y_1 y_2' - y_2 y_1'.}$$

Theorem 2 (Linear dependence and independence of solutions)
Suppose that (1a) *has continuous coefficients* $p(x)$ *and* $q(x)$ *on an open interval* I. *Then two solutions* y_1 *and* y_2 *of* (1a) *on* I *are linearly dependent on* I *if and only if their Wronskian* W *is zero at some* x_0 *in* I. *Furthermore, if* $W = 0$ *for* $x = x_0$, *then* $W \equiv 0$ *on* I; *hence if there is an* x_1 *in* I *at which* $W \neq 0$, *then* y_1, y_2 *are linearly independent on* I.

---

[12]Introduced by I. M. HÖNE (1778—1853), Polish mathematician, who changed his name to Wrónski. Second-order determinants should be familiar from elementary calculus; otherwise, consult Sec. 7.9, which is independent of the other sections in Chap. 7.

Proof. (a) If y_1 and y_2 are linearly dependent on I, then (3a) or (3b) holds on I and gives for (3a)

$$W(y_1, y_2) = W(ky_2, y_2) = \begin{vmatrix} ky_2 & y_2 \\ ky_2{}' & y_2{}' \end{vmatrix} = ky_2 y_2{}' - y_2 ky_2{}' \equiv 0;$$

similarly when (3b) holds.

(b) Conversely, we assume that $W(y_1, y_2) = 0$ for some $x = x_0$ in I and show that then y_1, y_2 are linearly dependent. We consider the system of linear equations

(5)
$$k_1 y_1(x_0) + k_2 y_2(x_0) = 0$$
$$k_1 y_1{}'(x_0) + k_2 y_2{}'(x_0) = 0$$

in the unknowns k_1, k_2. Now this system is homogeneous and its determinant is just the Wronskian $W[y_1(x_0), y_2(x_0)]$, which is zero by assumption. Hence the system has a solution k_1, k_2 where k_1 and k_2 are not both zero (cf. Sec. 7.9). Using these numbers k_1, k_2, we introduce the function

$$y(x) = k_1 y_1(x) + k_2 y_2(x).$$

By Fundamental Theorem 1 in Sec. 2.1 the function $y(x)$ is a solution of (1a) on I. From (5) we see that it satisfies the initial conditions $y(x_0) = 0$, $y'(x_0) = 0$. Now another solution of (1a) satisfying the same initial conditions is $y^* \equiv 0$. Since p and q are continuous, Theorem 1 applies and guarantees uniqueness, that is, $y \equiv y^*$, written out,

$$k_1 y_1 + k_2 y_2 \equiv 0$$

on I. Now since k_1 and k_2 are not both zero, this means linear dependence of y_1, y_2 on I.

(c) We prove the last statement of the theorem. If $W = 0$ at an x_0 in I, we have linear dependence of y_1, y_2 on I by part (b), hence $W \equiv 0$ by part (a) of this proof. Hence $W \neq 0$ at an x_1 in I cannot happen in the case of linear dependence, so that $W \neq 0$ at x_1 implies linear independence.   ∎

EXAMPLE 1. Application of Theorem 2
Show that $y_1 = \cos \omega x$, $y_2 = \sin \omega x$ form a basis of solutions of $y'' + \omega^2 y = 0$, $\omega \neq 0$, on any interval.

Solution. Substitution shows that they are solutions, and linear independence follows from Theorem 2, since

$$W(\cos \omega x, \sin \omega x) = \begin{vmatrix} \cos \omega x & \sin \omega x \\ -\omega \sin \omega x & \omega \cos \omega x \end{vmatrix} = \omega(\cos^2 \omega x + \sin^2 \omega x) = \omega.$$

EXAMPLE 2. Application of Theorem 2

Show that $y = (c_1 + c_2 x)e^x$ is a general solution of $y'' - 2y' + y = 0$ on any interval.

Solution. Substitution shows that $y_1 = e^x$ and $y_2 = xe^x$ are solutions, and Theorem 2 implies linear independence, since

$$W(e^x, xe^x) = \begin{vmatrix} e^x & xe^x \\ e^x & (x + 1)e^x \end{vmatrix} = (x + 1)e^{2x} - xe^{2x} = e^{2x} \neq 0.$$ ∎

Our next step is to show that general solutions always exist:

Theorem 3 (Existence of a general solution)

If the coefficients $p(x)$ and $q(x)$ of (1a) *are continuous on some open interval I, then* (1a) *has a general solution on I.*

Proof. By Theorem 1, equation (1a) has a solution $y_1(x)$ on I satisfying the initial conditions

$$y_1(x_0) = 1, \qquad y_1'(x_0) = 0$$

and a solution $y_2(x)$ on I satisfying the initial conditions

$$y_2(x_0) = 0, \qquad y_2'(x_0) = 1.$$

From this we see that the Wronskian $W(y_1, y_2)$ has at x_0 the value 1. Hence y_1, y_2 are linearly independent on I, by Theorem 2; they form a basis of solutions of (1a) on I, and $y = c_1 y_1 + c_2 y_2$ with arbitrary c_1, c_2 is a general solution of (1a) on I. ∎

A General Solution of (1a) Includes All Solutions

We now reach the final goal of this section by proving that a general solution of (1a) is as general as it can be, namely, it includes *all* solutions of (1a):

Theorem 4 (General solution)

Suppose that (1a) *has continuous coefficients $p(x)$ and $q(x)$ on some open interval I. Then every solution $y = Y(x)$ of* (1a) *on I is of the form*

$$(6) \qquad Y(x) = C_1 y_1(x) + C_2 y_2(x),$$

where y_1, y_2 form a basis of solutions of (1a) *on I and C_1, C_2 are suitable constants.*

Proof. By Theorem 3, our equation has a general solution

$$(7) \qquad y(x) = c_1 y_1(x) + c_2 y_2(x)$$

on I. We have to find suitable values of c_1, c_2 such that $y(x) = Y(x)$ on I. We choose any fixed x_0 in I and show first that we can find c_1, c_2 such that

$$y(x_0) = Y(x_0), \qquad y'(x_0) = Y'(x_0)$$

written out

(8)
$$c_1 y_1(x_0) + c_2 y_2(x_0) = Y(x_0),$$
$$c_1 y_1{}'(x_0) + c_2 y_2{}'(x_0) = Y'(x_0).$$

In fact, this is a system of linear equations in the unknowns c_1, c_2. Its determinant is the Wronskian of y_1 and y_2 at $x = x_0$. Since (7) is a general solution, y_1 and y_2 are linearly independent on I, and from Theorem 2 it follows that their Wronskian is not zero. Hence the system has a unique solution $c_1 = C_1$, $c_2 = C_2$ which can be obtained by Cramer's rule (Sec. 7.9). By using these constants we obtain from (7) the particular solution

$$y^*(x) = C_1 y_1(x) + C_2 y_2(x).$$

Since C_1, C_2 are solutions of (8), from (8) we now see that

$$y^*(x_0) = Y(x_0), \qquad y^*{}'(x_0) = Y'(x_0).$$

From this and the uniqueness theorem (Theorem 1) we conclude that y^* and Y must be equal on I, and the proof is complete.   ∎

In the next section we show that our present discussion and theorems readily extend to nth-order homogeneous linear differential equations with continuous coefficients. This will require neither new ideas nor new concepts.

Problems for Sec. 2.8

The following pairs of functions occurred as bases of solutions in Secs. 2.4 and 2.7. Using Theorem 2, verify the linear independence of these functions on any interval. (In Probs. 5 and 6, assume $x > 0$.)

1. $e^{\lambda_1 x}$, $e^{\lambda_2 x}$, $\lambda_1 \neq \lambda_2$ 2. $e^{\lambda x}$, $x e^{\lambda x}$

3. $e^{-ax/2} \cos \omega x$, $e^{-ax/2} \sin \omega x$ 4. x^{m_1}, x^{m_2}, $m_1 \neq m_2$

5. $x^\mu \cos (\nu \ln x)$, $x^\mu \sin (\nu \ln x)$ 6. x^m, $x^m \ln x$

In each case find a second-order homogeneous linear differential equation for which the given functions are solutions. Show linear independence on any open interval by Theorem 2. (In Probs. 8, 11, 14, assume $x > 0$.)

7. e^x, $x e^x$ 8. x^2, $x^{1/2}$

9. $\cos 4x$, $\sin 4x$ 10. 1, x^2

11. x^2, $x^2 \ln x$ 12. 1, e^{-x}

13. $\cosh 2x$, $\sinh 2x$ 14. $\sqrt{x}$, $1/\sqrt{x}$

15. $\sin x \cos x$, $\cos 2x$

16. Suppose that (1a) has continuous coefficients on I. Show that two solutions of (1a) on I that are zero at the same point in I cannot form a basis of solutions of (1a) on I.

17. Suppose that (1a) has continuous coefficients on I. Show that two solutions of (1a) on I that have maxima or minima at the same point in I cannot form a basis of solutions of (1a) on I.

18. Suppose that y_1, y_2 constitute a basis of solutions of a differential equation satisfying the assumptions of Theorem 2. Show that $z_1 = a_{11}y_1 + a_{12}y_2$, $z_2 = a_{21}y_1 + a_{22}y_2$ is a basis of that equation on the interval I if and only if the determinant of the coefficients a_{jk} is not zero.

19. Illustrate Prob. 18 with $y_1 = e^x$, $y_2 = e^{-x}$, $z_1 = \cosh x$, $z_2 = \sinh x$.

20. (**Euler–Cauchy equation**) Show that $x^2y'' - 4xy' + 6y = 0$ (Sec. 2.7) has $y_1 = x^2$, $y_2 = x^3$ as a basis of solutions for all x. Show that $W(x^2, x^3) = 0$ at $x = 0$. Does this contradict Theorem 2? Can you think of other Euler–Cauchy equations with this property?

2.9 Homogeneous Linear Equations of Arbitrary Order *n*

Before we turn to *nonhomogeneous* linear differential equations (beginning with the second order in Sec. 2.11), let us show how the discussion in Sec. 2.8 extends to homogeneous linear differential equations of any order *n*. With Sec. 2.8 still fresh in our mind (hopefully!), we shall see that this extension is based on the same ideas as the second-order case just completed. Let us first define those equations.

A differential equation of *n*th order is said to be **linear** if it can be written in the form

(1) $$y^{(n)} + p_{n-1}(x)y^{(n-1)} + \cdots + p_1(x)y' + p_0(x)y = r(x)$$

where *r* on the right-hand side and the *coefficients* $p_0, p_1, \cdots, p_{n-1}$ are any given functions of *x*, and $y^{(n)}$ is the *n*th derivative of *y* with respect to *x*, etc. Any differential equation of order *n* which cannot be written in the form (1) is said to be **nonlinear.**

As in Sec. 2.1 for $n = 2$, it is practical to use the "normal form" (1), with $y^{(n)}$ as the first term [instead of a term $f(x)y^{(n)}$].

If $r(x) \equiv 0$, equation (1) becomes

(2) $$\boxed{y^{(n)} + p_{n-1}(x)y^{(n-1)} + \cdots + p_1(x)y' + p_0(x)y = 0}$$

and is said to be **homogeneous.** If $r(x)$ is not identically zero, the equation is said to be **nonhomogeneous.**

Solution, General Solution, Linear Independence

A function $y = \phi(x)$ is called a **solution** of a (linear or nonlinear) differential equation of *n*th order *on an interval I*, if $\phi(x)$ is defined and *n* times differentiable on *I* and is such that the equation becomes an identity when we replace the unspecified function *y* and its derivatives in the equation by ϕ and its corresponding derivatives.

A **general solution** of (2) on some open interval *I* is a function of the form

$$(3) \qquad \boxed{y(x) = c_1 y_1(x) + \cdots + c_n y_n(x)} \qquad (c_1, \cdots, c_n \text{ arbitrary}[13])$$

where $y_1, \cdots, y_n$ form a **basis** (or **fundamental system**) of solutions of (2) on I; that is, they are solutions of (2) on I that are linearly independent on I, as defined below.

A **particular solution** of (2) on I is obtained if we assign specific values to the n constants $c_1, \cdots, c_n$ in (3).

Here, n ($\geqq 2$) functions $y_1(x), \cdots, y_n(x)$ are said to be **linearly independent** *on some interval I* where they are defined if the equation

$$(4) \qquad k_1 y_1(x) + \cdots + k_n y_n(x) = 0 \quad \text{on } I$$

implies that all $k_1, \cdots, k_n$ are zero. These functions are called **linearly dependent** on I if this equation also holds on I for some $k_1, \cdots, k_n$ not all zero.

If and only if $y_1, \cdots, y_n$ are linearly dependent on I, we can express (at least) one of these functions on I as a **"linear combination"** of the other $n - 1$ functions, that is, as a sum of those functions, each multiplied by a constant (zero or not). This motivates the term "linearly dependent." For instance, if (4) holds with $k_1 \neq 0$, we can divide by k_1 and express y_1 as the linear combination

$$y_1 = -\frac{1}{k_1}(k_2 y_2 + \cdots + k_n y_n).$$

Note that when $n = 2$, these concepts reduce to those defined in Sec. 2.3.

EXAMPLE 1. Linear dependence

Show that the functions $y_1 = x$, $y_2 = 3x$, $y_3 = x^2$ are linearly dependent on any interval.

Solution. $y_2 = 3y_1 + 0y_3$.

EXAMPLE 2. Linear independence

Show that $y_1 = x$, $y_2 = x^2$, $y_3 = x^3$ are linearly independent on any interval, for instance, on $-1 \leqq x \leqq 2$.

Solution. Equation (4) is $k_1 x + k_2 x^2 + k_3 x^3 = 0$. Taking $x = -1, 1, 2$, we get

$$-k_1 + k_2 - k_3 = 0, \qquad k_1 + k_2 + k_3 = 0, \qquad 2k_1 + 4k_2 + 8k_3 = 0,$$

respectively, which implies $k_1 = 0$, $k_2 = 0$, $k_3 = 0$, that is, linear independence.   ∎

Initial Value Problem, Existence and Uniqueness

An **initial value problem** for equation (2) now consists of (2) and n **initial conditions**

$$(5) \qquad y(x_0) = K_0, \qquad y'(x_0) = K_1, \qquad \cdots, \qquad y^{(n-1)}(x_0) = K_{n-1},$$

where x_0 is some fixed point in the interval I considered.

In extension of Theorem 1 in Sec. 2.8 we now have:

---

[13]Cf. footnote 3 in Sec. 2.3.

Theorem 1 (Existence and uniqueness theorem)

If $p_0(x), \cdots, p_{n-1}(x)$ in (2) are continuous functions on an open interval I, then the initial value problem consisting of (2) and the initial conditions (5) [with x_0 in I] has a unique solution $y(x)$ on I.

Existence is proved in Ref. [A10] in Appendix 1, and uniqueness can be proved by a slight generalization of the uniqueness proof on p. 148 near the end of this chapter.

Independence Criterion.
Existence of General Solution

The test for linear dependence and independence of solutions (Theorem 2 in Sec. 2.8) can be generalized to nth-order equations as follows.

Theorem 2 (Linear dependence and independence of solutions)

Suppose that the coefficients $p_0(x), \cdots, p_{n-1}(x)$ of (2) are continuous on an open interval I. Then n solutions $y_1, \cdots, y_n$ of (2) on an open interval I are linearly dependent on I if and only if their **Wronskian**

$$(6) \qquad W(y_1, \cdots, y_n) = \begin{vmatrix} y_1 & y_2 & \cdots & y_n \\ y_1' & y_2' & \cdots & y_n' \\ \cdot & \cdot & \cdots & \cdot \\ y_1^{(n-1)} & y_2^{(n-1)} & \cdots & y_n^{(n-1)} \end{vmatrix}$$

is zero for some $x = x_0$ in I. Furthermore, if $W = 0$ for $x = x_0$, then $W \equiv 0$ on I; hence if there is an x_1 in I at which $W \neq 0$, then $y_1, \cdots, y_n$ are linearly independent on I.

The idea of the proof is similar to that of Theorem 2 in Sec. 2.8.

EXAMPLE 3. Test for basis and general solution

Show that $y = c_1 + c_2 e^x + c_3 e^{-x}$ is a general solution of $y''' - y' = 0$ on any open interval I.

Solution. Substitution shows that $y_1 = 1$, $y_2 = e^x$, $y_3 = e^{-x}$ are solutions on any I. Theorem 2 implies that they form a basis on I since their Wronskian is not zero:

$$W(1, e^x, e^{-x}) = \begin{vmatrix} 1 & e^x & e^{-x} \\ 0 & e^x & -e^{-x} \\ 0 & e^x & e^{-x} \end{vmatrix} = 1 \cdot \begin{vmatrix} e^x & -e^{-x} \\ e^x & e^{-x} \end{vmatrix} = e^x e^{-x} + e^{-x} e^x = 2.$$

A systematic discussion of higher order equations with constant coefficients follows in the next section.

The solution of initial value problems is quite similar to that for second-order differential equations, as the following example illustrates.

EXAMPLE 4. Initial value problem
Solve the initial value problem

$$y''' - y' = 0, \qquad y(0) = 3.4, \qquad y'(0) = -5.4, \qquad y''(0) = -0.6.$$

Solution. 1st Step. By Example 3, a general solution is

$$y(x) = c_1 + c_2 e^x + c_3 e^{-x}.$$

2nd Step. We differentiate $y(x)$ and use the initial conditions, finding

$$\text{(a)} \quad y(0) = c_1 + c_2 + c_3 = \quad 3.4,$$

$$\text{(b)} \quad y'(0) = \qquad c_2 - c_3 = -5.4,$$

$$\text{(c)} \quad y''(0) = \qquad c_2 + c_3 = -0.6.$$

(a) minus (c) gives $c_1 = 3.4 + 0.6 = 4$. Then (b) plus (c) gives $2c_2 = -6$, $c_2 = -3$. Finally $c_3 = 2.4$ from (b). The answer is

$$y(x) = 4 - 3e^x + 2.4e^{-x}. \qquad \blacksquare$$

Equation (2) always has general solutions. Indeed, Theorem 3 in Sec. 2.8 extends as follows.

Theorem 3 (Existence of a general solution)

If the coefficients $p_0(x)$, $\cdots$, $p_{n-1}(x)$ of (2) are continuous on some open interval I, then (2) has a general solution on I.

The idea of the proof is similar to that of Theorem 3 in Sec. 2.8.

Finally, a general solution of (2) includes *all* solutions of (2). Indeed, using the idea of the proof of Theorem 4 in Sec. 2.8, one can establish the following extension of that theorem.

Theorem 4 (General solution)

Suppose that (2) has continuous coefficients $p_0(x)$, $\cdots$, $p_{n-1}(x)$ on some open interval I. Then every solution $y = Y(x)$ of (2) on I is of the form

$$\text{(7)} \qquad Y(x) = C_1 y_1(x) + \cdots + C_n y_n(x),$$

where y_1, $\cdots$, y_n forms a basis of solutions of (2) on I and C_1, $\cdots$, C_n are suitable constants.

As for $n = 2$, this theorem means that the *linear* equation (2) has no "*singular solutions*," that is, solutions not obtainable from a general solution, but every solution involving no arbitrary constants is a particular solution as defined above.

In the next section we give a systematic discussion of nth-order homogeneous linear differential equations whose coefficients are constant. This will extend the method considered in Sec. 2.2.

Problems for Sec. 2.9

Show that the given functions form a basis of solutions of the given differential equations on any open interval, verifying linear independence by Theorem 2. (In Probs. 2 and 12, assume $x > 0$.)

1. $1, x, x^2, x^3,$ $y^{IV} = 0$

2. $1, x^2, x^4,$ $x^2 y''' - 3xy'' + 3y' = 0$

3. $e^x, e^{-x}, \cos x, \sin x,$ $y^{IV} - y = 0$

4. $\cosh x, \sinh x, \cos x, \sin x,$ $y^{IV} - y = 0$

5. $e^{-x}, e^x, xe^x,$ $y''' - y'' - y' + y = 0$

6. $e^{-x}, xe^{-x}, x^2 e^{-x},$ $y''' + 3y'' + 3y' + y = 0$

7. $e^x, xe^x, x^2 e^x,$ $y''' - 3y'' + 3y' - y = 0$

8. $e^{-2x}, e^x, xe^x,$ $(D^3 - 3D + 2)y = 0$

9. $e^x, e^{-x}, e^{2x}, e^{-2x},$ $(D^4 - 5D^2 + 4)y = 0$

10. $e^x \cos x, e^x \sin x, e^{-x} \cos x, e^{-x} \sin x,$ $(D^4 + 4)y = 0$

11. $\cos x, \sin x, \cos 2x, \sin 2x,$ $(D^4 + 5D^2 + 4)y = 0$

12. $x^{1/2}, x^{-1/2}, 1,$ $(x^2 D^3 + 3xD^2 + 0.75D)y = 0$

13. $\cos x, \sin x, e^{-x},$ $(D^3 + D^2 + D + 1)y = 0$

14. $x, x^2, e^x,$ $[(x^2 - 2x + 2)D^3 - x^2 D^2 + 2xD - 2]y = 0$

Solve the following initial value problems. *Hint.* For finding a general solution, look at one of Probs. 1–14 if necessary.

15. $y''' = 0,$ $y(2) = 12,$ $y'(2) = 16,$ $y''(2) = 8$

16. $y''' - 3y'' + 3y' - y = 0,$ $y(0) = 2,$ $y'(0) = 2,$ $y''(0) = 10$

17. $x^2 y''' + 3xy'' + 0.75y' = 0,$ $y(1) = 0,$ $y'(1) = 1.5,$ $y''(1) = -1.75$

18. $y''' - y'' - y' + y = 0,$ $y(0) = 2,$ $y'(0) = 1,$ $y''(0) = 0$

19. $y^{IV} + 3y'' - 4y = 0,$ $y(0) = 0,$ $y'(0) = -10,$ $y''(0) = 0,$ $y'''(0) = 40$

20. $y^{IV} - y = 0,$ $y(0) = -1,$ $y'(0) = 7,$ $y''(0) = -1,$ $y'''(0) = 7$

Linear Independence and Dependence

Since these concepts are of *general* importance, far beyond our present study, we add a few more problems on them.

Are the following functions linearly independent or dependent on the positive x-axis?

21. $1, x, x^2$ **22.** $\cos x, \sin x, 1$ **23.** $x, 1/x, 1$

24. $\ln x, \ln x^2, \ln x^3$ **25.** $x, x - 1, x + 1$ **26.** $\cosh^2 x, \sinh^2 x, 1$

27. $\sin x, \sin 2x, \sin 3x$ **28.** $\sinh 2x, e^{2x}, e^{-2x}$ **29.** $(x - 1)^2, (x + 1)^2, x$

30. How would you prove that the functions in Example 2 are linearly independent on the interval $0 < x < 1$?

31. If n functions are linearly dependent on an interval I, prove that they are also linearly dependent on any subinterval of I.

32. Find an interval I on which the three functions $x^3, |x|^3,$ and 1 are linearly independent and a subinterval of I on which these functions are linearly dependent. What fact does this illustrate?

33. If one of the functions of a set of functions on an interval I is a constant multiple of another function of the set on I, prove that the set is linearly dependent on I.

34. If a set of p functions is linearly dependent on an interval I, prove that a set of n ($\geqq p$) functions on I containing that set is linearly dependent on I.

35. If $y \equiv 0$ is a function of a set of functions on an interval I, prove that the set is linearly dependent on I.

2.10 Equations of Order n with Constant Coefficients

The method in Sec. 2.2 may easily be extended to a homogeneous linear equation of any order n with constant coefficients

$$
\text{(1)} \qquad \boxed{y^{(n)} + a_{n-1}y^{(n-1)} + \cdots + a_1 y' + a_0 y = 0.}
$$

Substitution of $y = e^{\lambda x}$ and its derivatives gives the **characteristic equation**

$$
\text{(2)} \qquad \boxed{\lambda^n + a_{n-1}\lambda^{n-1} + \cdots + a_1 \lambda + a_0 = 0}
$$

of (1). If this equation has n **distinct roots** $\lambda_1, \cdots, \lambda_n$, then the n solutions

$$
\text{(3)} \qquad y_1 = e^{\lambda_1 x}, \cdots, y_n = e^{\lambda_n x}
$$

constitute a basis for all x, and the corresponding general solution of (1) is

$$
\text{(4)} \qquad y = c_1 e^{\lambda_1 x} + \cdots + c_n e^{\lambda_n x}.
$$

Indeed, the solutions in (3) are linearly independent on any open interval, since their Wronskian (Sec. 2.9) is nowhere zero; we omit the proof (which is involved), but note that verification for a given equation should be no problem.

For higher n, a practical problem is the actual determination of the roots of (2). Sometimes one can find a root λ_1 by inspection and then divide (2) by $\lambda - \lambda_1$ to get a polynomial of degree $n - 1$. In general, one will have to apply a numerical method, such as Newton's method or fixed-point iteration (Sec. 18.2).

EXAMPLE 1. Distinct real roots
Solve the differential equation

$$
y''' - 2y'' - y' + 2y = 0.
$$

Solution. The roots of the characteristic equation

$$
\lambda^3 - 2\lambda^2 - \lambda + 2 = 0
$$

are -1, 1 and 2, and the corresponding general solution (4) is

$$
y = c_1 e^{-x} + c_2 e^x + c_3 e^{2x}. \qquad \blacksquare
$$

Complex Simple Roots

If complex roots occur, they must occur in conjugate pairs since the coefficients of (1) are real. Thus, if $\lambda = \gamma + i\omega$ is a simple root of (2), so is the conjugate $\bar{\lambda} = \gamma - i\omega$, and two corresponding linearly independent solutions are (as in Sec. 2.4, except for notation)

$$y_1 = e^{\gamma x} \cos \omega x, \qquad y_2 = e^{\gamma x} \sin \omega x.$$

EXAMPLE 2. Complex conjugate simple roots
Solve the initial value problem

$$y''' - 2y'' + 2y' = 0, \qquad y(0) = 0.5, \qquad y'(0) = -1, \qquad y''(0) = 2.$$

Solution. A root of $\lambda^3 - 2\lambda^2 + 2\lambda = 0$ is $\lambda_1 = 0$. A corresponding solution is $y_1 = e^{0x} = 1$. Division by λ gives

$$\lambda^2 - 2\lambda + 2 = 0.$$

The roots are $\lambda_2 = 1 + i$ and $\lambda_3 = 1 - i$. Corresponding solutions are $y_2 = e^x \cos x$ and $y_3 = e^x \sin x$. The corresponding general solution and its derivatives are

$$y = c_1 + e^x[A \cos x + B \sin x],$$

$$y' = e^x[(A + B) \cos x + (B - A) \sin x],$$

$$y'' = e^x[2B \cos x - 2A \sin x],$$

as follows by differentiation. From this and the initial conditions we obtain

$$y(0) = c_1 + A = 0.5, \qquad y'(0) = A + B = -1, \qquad y''(0) = 2B = 2.$$

Hence $B = 1$, $A = -2$, $c_1 = 2.5$. The answer is

$$y = 2.5 + e^x(-2 \cos x + \sin x). \qquad \blacksquare$$

Multiple Roots

If a **double root** occurs, say, $\lambda_1 = \lambda_2$, then $y_1 = y_2$ in (3) and we take y_1 and $y_2 = xy_1$ as two linearly independent solutions corresponding to this root; this is as in Sec. 2.4.

If a **triple root** occurs, say, $\lambda_1 = \lambda_2 = \lambda_3$, then $y_1 = y_2 = y_3$ in (3) and three linearly independent solutions corresponding to this root are

(5) $$y_1, \qquad xy_1, \qquad x^2 y_1;$$

these solutions can be obtained by the method of reduction of order. More generally, *if λ is a **root of order** m, then m corresponding linearly independent solutions are*

(6) $$e^{\lambda x}, \qquad xe^{\lambda x}, \qquad \cdots, \qquad x^{m-1}e^{\lambda x}.$$

These functions are linearly independent on any open interval since $1, x, \cdots, x^{m-1}$ are linearly independent; this follows from Theorem 2 in Sec. 2.9, since these functions are solutions of $y^{(m)} = 0$. [Can you write down their Wronskian and verify that it equals $(m - 1)!(m - 2)! \cdots 3!2!$?]

We show that the functions (6) are solutions of (1) in the present case. The corresponding formulas become a little simpler if we use operator notation (Sec. 2.5), writing (1) in the form

$$L[y] = P(D)[y] = [D^n + a_{n-1}D^{n-1} + \cdots + a_0]y = 0.$$

By assumption,

(7) $$P(D) = Q(D)(D - \lambda)^m$$

where $Q(D)$ is a differential operator of order $n - m$ and corresponds to other roots that the characteristic equation of (1) may have. Now for the functions $x^k e^{\lambda x}$ ($k = 0, \cdots, m - 1$) in (6) we obtain by differentiation

$$(D - \lambda)(x^k e^{\lambda x}) = kx^{k-1}e^{\lambda x} + \lambda x^k e^{\lambda x} - \lambda x^k e^{\lambda x} = kx^{k-1}e^{\lambda x}.$$

Hence by applying $D - \lambda$ to this we have

$$(D - \lambda)^2(x^k e^{\lambda x}) = (D - \lambda)[kx^{k-1}e^{\lambda x}] = k(k - 1)x^{k-2}e^{\lambda x}$$

and so on. Finally

$$(D - \lambda)^k(x^k e^{\lambda x}) = k!e^{\lambda x},$$

so that

$$(D - \lambda)^{k+1}(x^k e^{\lambda x}) = k!(D - \lambda)e^{\lambda x} = 0.$$

Since $k = 0, \cdots, m - 1$ in (6), we have $k + 1 \leqq m$; hence

$$(D - \lambda)^m(x^k e^{\lambda x}) = 0 \qquad k = 0, \cdots, m - 1.$$

Because of (7) this shows that the functions in (6) are solutions of (1) in the case of a root of order m.   ∎

EXAMPLE 3. Real double and triple roots
Solve the differential equation

$$y^{\text{V}} - 3y^{\text{IV}} + 3y''' - y'' = 0.$$

Solution. The characteristic equation

$$\lambda^5 - 3\lambda^4 + 3\lambda^3 - \lambda^2 = 0$$

has the roots $\lambda_1 = \lambda_2 = 0$ and $\lambda_3 = \lambda_4 = \lambda_5 = 1$, and the answer is

(8) $$y = c_1 + c_2 x + (c_3 + c_4 x + c_5 x^2)e^x.$$ ∎

Complex Multiple Roots

In this case, real solutions are obtained as for complex simple roots above. Thus, if $\lambda = \gamma + i\omega$ is a **complex double root**, so is the conjugate $\bar{\lambda} = \gamma - i\omega$. Corresponding linearly independent solutions are

(9) $e^{\gamma x} \cos \omega x$, $e^{\gamma x} \sin \omega x$, $x e^{\gamma x} \cos \omega x$, $x e^{\gamma x} \sin \omega x$.

The first two of these result from $e^{\lambda x}$ and $e^{\bar{\lambda} x}$ as before, and the second two from $xe^{\lambda x}$ and $xe^{\bar{\lambda} x}$ in the same fashion.

For *complex triple roots* (which hardly ever occur in applications), one would obtain two more solutions $x^2 e^{\gamma x} \cos \omega x$, $x^2 e^{\gamma x} \sin \omega x$, and so on.

EXAMPLE 4. Complex double roots
Solve

$$y^{(7)} + 18y^{(5)} + 81y''' = 0.$$

Solution. The characteristic equation

$$\lambda^7 + 18\lambda^5 + 81\lambda^3 = \lambda^3(\lambda^4 + 18\lambda^2 + 81)$$

$$= \lambda^3(\lambda^2 + 9)^2$$

$$= \lambda^3[(\lambda + 3i)(\lambda - 3i)]^2 = 0$$

has a triple root 0 and double roots $-3i$ and $3i$. Hence, by (9) with $\gamma = 0$ and $\omega = 3$, a general solution is

$$y = c_1 + c_2 x + c_3 x^2 + A_1 \cos 3x + B_1 \sin 3x + x(A_2 \cos 3x + B_2 \sin 3x). \qquad \blacksquare$$

Attention! Sections 2.2–2.10 concerned homogeneous equations. *The remaining sections of Chap. 2 concern nonhomogeneous equations.*

In the first of these sections (Sec. 2.11) we introduce the concepts of *general* and *particular solutions* of nonhomogeneous linear differential equations.

Problems for Sec. 2.10

See also Probs. 1–20 for Sec. 2.9.

Find an equation of the form (1) for which the given functions constitute a basis.

1. $1, x, e^{3x}, xe^{3x}$ **2.** $1, x, x^2$

3. $e^{-x}, xe^{-x}, e^x, xe^x$ **4.** $e^{-2x}, e^{-x}, e^x, e^{2x}, 1$

5. $e^{-2x}, xe^{-2x}, x^2e^{-2x}$ **6.** $\cos x, \sin x, x \cos x, x \sin x$

7. $\cosh x, \sinh x, x \cosh x, x \sinh x$ **8.** $\sin x, \sin 2x, \cos x, \cos 2x$

Find a general solution.

9. $y''' - y' = 0$ **10.** $y^{IV} + 3y'' - 4y = 0$

11. $y^{IV} - 5y'' + 4y = 0$ **12.** $y^{IV} - 13y'' + 36y = 0$

13. $y^{IV} - y = 0$ **14.** $y^{IV} + 8y'' + 16y = 0$

15. $y''' + 3y'' + 3y' + y = 0$ **16.** $y^{IV} + 2\pi^2 y'' + \pi^4 y = 0$

Solve the following initial value problems.

17. $y''' + 9y' = 0$, $y(0) = 4$, $y'(0) = 0$, $y''(0) = 9$

18. $y''' + 3y'' + 3y' + y = 0$, $y(0) = 12$, $y'(0) = -12$, $y''(0) = 6$

19. $y^{IV} - y = 0$, $y(0) = 5$, $y'(0) = 2$, $y''(0) = -1$, $y'''(0) = 2$

20. $y^{IV} - 20y'' + 64y = 0$, $y(0) = 2$, $y'(0) = 16$, $y''(0) = -16$, $y'''(0) = 160$

21. $y^{IV} + 3y'' - 4y = 0$, $y(0) = 0$, $y'(0) = -1$, $y''(0) = -5$, $y'''(0) = -1$

22. $y''' - 2y'' - y' + 2y = 0$, $y(0) = 3$, $y'(0) = 0$, $y''(0) = 3$

Reduction of order extends from second-order (cf. Sec. 2.3) to higher order equations. It is particularly simple in the case of constant coefficients, since then we simply have to divide the characteristic equation by $\lambda - \lambda_1$, where $y_1 = e^{\lambda_1 x}$ is a known solution. Reduce and solve the following equations, using the given y_1.

23. $y''' - y'' + 4y' - 4y = 0$, $y_1 = e^x$ **24.** $y''' + 4y'' + 6y' + 4y = 0$, $y_1 = e^{-2x}$

25. $4y''' + 8y'' + 5y' + y = 0$, $y_1 = e^{-x}$ **26.** $y''' - 2y'' - 5y' + 6y = 0$, $y_1 = e^{3x}$

27. $y''' + y'' + 4y' + 30y = 0$, $y_1 = e^{-3x}$

28. $y^{IV} - y''' + y'' - 11y' + 10y = 0$, $y_1 = e^{-x}\cos 2x$

29. Reduction of order is more complicated for an equation with *variable* coefficients $y''' + p_1(x)y'' + p_2(x)y' + p_3(x)y = 0$. If $y_1(x)$ is a solution of it, show that another solution is $y_2(x) = u(x)y_1(x)$ with $u(x) = \int z(x)\,dx$ and z obtained from

$$y_1 z'' + (3y_1' + p_1 y_1)z' + (3y_1'' + 2p_1 y_1' + p_2 y_1)z = 0.$$

30. Solve $x^3 y''' - 3x^2 y'' + (6 - x^2)xy' - (6 - x^2)y = 0$ by reducing the order, using $y_1 = x$.

31. (Euler–Cauchy equation) The *Euler–Cauchy equation of the third order* is

$$x^3 y''' + ax^2 y'' + bxy' + cy = 0.$$

Generalizing the method of Sec. 2.7, show that $y = x^m$ is a solution of the equation if and only if m is a root of the auxiliary equation

$$m^3 + (a - 3)m^2 + (b - a + 2)m + c = 0.$$

Solve:

32. $x^3 y''' + x^2 y'' - 2xy' + 2y = 0$ **33.** $xy''' + 3y'' = 0$

34. $x^3 y''' - 3x^2 y'' + 6xy' - 6y = 0$ **35.** $x^2 y''' + 3xy'' - 3y' = 0$

2.11 Nonhomogeneous Equations

Beginning in this section, we turn from homogeneous *to* nonhomogeneous equations

(1) $$\boxed{y^{(n)} + p_{n-1}(x)y^{(n-1)} + \cdots + p_1(x)y' + p_0(x)y = r(x)}$$

where $r(x) \not\equiv 0$. How to solve such an equation? Before we consider methods, let us first explore what we actually need for proceeding from the corresponding homogeneous equation

(2) $$y^{(n)} + p_{n-1}(x)y^{(n-1)} + \cdots + p_1(x)y' + p_0(x)y = 0$$

to the nonhomogeneous equation (1). The key that relates (1) to (2) and gives us a plan for solving (1) is the following.

Theorem 1 [Relations between solutions of (1) and (2)]

   (a) *The difference of two solutions of* (1) *on some open interval I is a solution of* (2) *on I.*

   (b) *The sum of a solution of* (1) *on I and a solution of* (2) *on I is a solution of* (1) *on I.*

Proof. (a) Denote the left-hand side of (1) by $L[y]$. Let y and $\tilde{y}$ be any solutions of (1) on I. Then $L[y] = r(x)$, $L[\tilde{y}] = r(x)$, and since we have $(y - \tilde{y})' = y' - \tilde{y}'$ etc., we obtain the first assertion,

$$L[y - \tilde{y}] = L[y] - L[\tilde{y}] = r(x) - r(x) \equiv 0.$$

(b) Similarly, for y as before and any solution y^* of (2) on I,

$$L[y + y^*] = L[y] + L[y^*] = r(x) + 0 = r(x). \qquad \blacksquare$$

This situation suggests the following concepts.

Definition (General solution, particular solution)

A **general solution** of the nonhomogeneous equation (1) on some open interval I is a solution of the form

(3) $$y(x) = y_h(x) + y_p(x),$$

where $y_h(x) = c_1 y_1(x) + \cdots + c_n y_n(x)$ is a general solution of the homogeneous equation (2) on I and $y_p(x)$ is any solution of (1) on I containing no arbitrary constants.

A **particular solution** of (1) on I is a solution obtained from (3) by assigning specific values to the arbitrary constants $c_1, \cdots, c_n$ in $y_h(x)$. $\blacksquare$

If the coefficients in (1) and $r(x)$ are continuous functions on I, then (1) has general solutions on I (proof for $n = 2$ in Sec. 2.16; for general n, see e.g. Ref. [A8]); also, an initial value problem consisting of (1) and n initial conditions (5), Sec. 2.9, has a unique solution on I.

Furthermore, justifying the terminology, we now prove that a general solution of (1) includes *all* solutions of (1); hence the situation is the same as for the homogeneous equation:

Theorem 2 (General solution)

Suppose that the coefficients and $r(x)$ in (1) are continuous on some open interval I. Then every solution of (1) on I is obtained by assigning suitable values to the arbitrary constants in a general solution (3) of (1) on I.

Proof. Let $\tilde{y}(x)$ be any solution of (1) on I. Let (3) be any general solution of (1) on I; this solution exists because of our continuity assumption. Theorem 1(a) implies that the difference $Y(x) = \tilde{y}(x) - y_p(x)$ is a solution of the *homogeneous* equation (2). By Theorem 4 in Sec. 2.9, this solution $Y(x)$ is obtained from $y_h(x)$ by assigning suitable values to the arbitrary constants $c_1, \cdots, c_n$. From this and $\tilde{y}(x) = Y(x) + y_p(x)$, the statement follows. $\blacksquare$

Practical conclusion

To solve the nonhomogeneous equation (1) or an initial value problem for (1), we have to solve the homogeneous equation (2) and find any particular solution y_p of (1). Methods for this and applications will be the subject for the remaining sections of Chap. 2.

EXAMPLE 1. A nonhomogeneous equation

Solve the nonhomogeneous equation

$$y'' - 4y' + 3y = 10e^{-2x}.$$

Solution. The characteristic equation $\lambda^2 - 4\lambda + 3 = 0$ has the roots 1 and 3. Hence a general solution of the homogeneous equation is

$$y_h = c_1 e^x + c_2 e^{3x}.$$

We need a solution y_p of the nonhomogeneous equation. Since e^{-2x} has derivatives e^{-2x} times some constants, we try

$$y_p = ke^{-2x}.$$

Then $y_p' = -2ke^{-2x}$, $y_p'' = 4ke^{-2x}$. Substitution thus gives

$$4ke^{-2x} - 4(-2ke^{-2x}) + 3ke^{-2x} = 10e^{-2x}.$$

Hence $4k + 8k + 3k = 10$, and $k = 2/3$. The answer is

$$y = y_h + y_p = c_1 e^x + c_2 e^{3x} + \tfrac{2}{3} e^{-2x}.$$

EXAMPLE 2. An initial value problem

Solve the initial value problem

$$y''' - 2y'' - y' + 2y = 2x^2 - 6x + 4, \qquad y(0) = 5, \qquad y'(0) = -5, \qquad y''(0) = 1.$$

Solution. A general solution of the homogeneous equation is $y_h = c_1 e^{-x} + c_2 e^x + c_3 e^{2x}$; cf. Example 1 in Sec. 2.10. We need a solution y_p. The form of the right-hand side suggests that we try

$$y_p = Kx^2 + Mx + N. \qquad \text{Then} \qquad y_p' = 2Kx + M, \qquad y_p'' = 2K, \qquad y_p''' = 0.$$

Substitution into the equation yields

$$-2 \cdot 2K - (2Kx + M) + 2(Kx^2 + Mx + N) = 2x^2 - 6x + 4.$$

By equating like powers,

$$2Kx^2 = 2x^2, \qquad (-2K + 2M)x = -6x, \qquad -4K - M + 2N = 4.$$

Hence $K = 1$, $M = -2$, $N = 3$. This gives the general solution

$$y = y_h + y_p = c_1 e^{-x} + c_2 e^x + c_3 e^{2x} + x^2 - 2x + 3.$$

For determining the constants from the initial conditions we also need the derivatives

$$y' = -c_1 e^{-x} + c_2 e^x + 2c_3 e^{2x} + 2x - 2,$$

$$y'' = c_1 e^{-x} + c_2 e^x + 4c_3 e^{2x} + 2.$$

Setting $x = 0$ and using the initial conditions, we obtain

$$\text{(a)} \quad y(0) = c_1 + c_2 + c_3 + 3 = 5,$$
$$\text{(b)} \quad y'(0) = -c_1 + c_2 + 2c_3 - 2 = -5,$$
$$\text{(c)} \quad y''(0) = c_1 + c_2 + 4c_3 + 2 = 1.$$

(a) plus (b) gives $2c_2 + 3c_3 = -1$, and (c) minus (a) gives $3c_3 = -3$. Hence $c_3 = -1$, $c_2 = 1$, and $c_1 = 2$ from (a). The answer is

$$y = 2e^{-x} + e^x - e^{2x} + x^2 - 2x + 3.$$

Check that this function satisfies the equation as well as the three initial conditions. ∎

As already announced, in the next section we discuss a method for obtaining a **particular solution** y_p of a nonhomogeneous linear differential equation, as is needed for obtaining a general solution of such an equation.

Problems for Sec. 2.11

In each case verify that $y_p(x)$ is a solution of the given differential equation and find a general solution.

1. $y'' + 3y' + 2y = 2x^2$, $y_p = x^2 - 3x + 3.5$
2. $y'' - 2y' + 5y = 5x^4 - 8x^3 + 2x^2 + 8x - 4$, $y_p = x^4 - 2x^2$
3. $y'' + y = -20 \sin 3x$, $y_p = 2.5 \sin 3x$
4. $y'' + 4y' + 3y = -105e^{2x}$, $y_p = -7e^{2x}$
5. $8y'' - 6y' + y = 6e^x + 3x - 16$, $y_p = 2e^x + 3x + 2$
6. $(D^2 - 4D - 1)y = -\frac{2}{3} \cos x + \frac{1}{2} \sin x$, $y_p = \frac{1}{6} \cos x + \frac{1}{12} \sin x$
7. $(D^2 + 3D - 4)y = 10e^x$, $y_p = 2xe^x$
8. $(D^2 - 4D + 3)y = 2e^{3x}$, $y_p = xe^{3x}$
9. $(D^2 - 6D + 9)y = 2e^{3x}$, $y_p = x^2e^{3x}$
10. $(D^2 + 3D - 4)y = -6.8 \sin x$, $y_p = \sin x + 0.6 \cos x$

Verify that y_p is a solution of the given equation and solve the given initial value problem.

11. $y'' + y = 2x$, $y_p = 2x$, $y(0) = -1$, $y'(0) = 8$
12. $(D^2 - 1)y = 2 \cos x$, $y_p = -\cos x$, $y(0) = 0$, $y'(0) = -0.2$
13. $y'' - 3y' + 2y = 2x^2 - 6x + 2$, $y_p = x^2$, $y(0) = 4$, $y'(0) = 3$
14. $y'' + y = 2e^x$, $y_p = e^x$, $y(0) = 1.75$, $y'(0) = 1$
15. $y'' - y = 2e^x$, $y_p = xe^x$, $y(0) = -1$, $y'(0) = 0$
16. $y'' + 4y = -12 \sin 2x$, $y_p = 3x \cos 2x$, $y(0) = 1.8$, $y'(0) = 5.0$
17. $(D^2 + 1)y = 2 \sin x$, $y_p = -x \cos x$, $y(\frac{1}{2}\pi) = 1$, $y'(\frac{1}{2}\pi) = \frac{1}{2}\pi$
18. $(4x^2D^2 + 1)y = (1 - x^2) \cos 0.5x$,
    $y_p = \cos 0.5x$, $y(1) = 3 + \cos 0.5 \approx 3.88$, $y'(1) = 5.5 - 0.5 \sin 0.5 \approx 5.26$
19. $(x^2D^2 - 3xD + 3)y = 3 \ln x - 4$, $y_p = \ln x$, $y(1) = 0$, $y'(1) = 1$
20. $(x^2D^2 - 2xD + 2)y = (3x^2 - 6x + 6)e^x$, $y_p = 3e^x$, $y(0) = 3$, $y'(0) = 7$

21. Show that if y_1 is a solution of (1) with $r = r_1$ and y_2 is a solution of (1) with $r = r_2$, then $y = y_1 + y_2$ is a solution of (1) with $r = r_1 + r_2$.

22. As an illustration of Prob. 21, find a general solution of the differential equation $y'' + 3y' - 4y = 10e^x - 6.8 \sin x$. *Hint.* Use Probs. 7 and 10.

23. Show that $y_1 = e^x$ is a solution of $y'' + y = 2e^x$ and $y_2 = x \sin x$ is a solution of $y'' + y = 2 \cos x$. Using Prob. 21, find a general solution of the equation $y'' + y = 2e^x + 2 \cos x$.

24. To illustrate that the choice of y_p in (3) is immaterial, show that $y_{p1} = -\cos x$ and $y_{p2} = e^x - \cos x$ are particular solutions of $y'' - y = 2 \cos x$, find the corresponding general solutions (3), and show that one of them can be expressed in terms of the other.

25. Show that two general solutions (3) of (1) that correspond to different y_p can be transformed into each other.

2.12 Nonhomogeneous Equations: Solving by the Method of Undetermined Coefficients

A general solution of a nonhomogeneous linear equation is a sum of the form $y = y_h + y_p$, where y_h is a general solution of the corresponding homogeneous equation and y_p is any particular solution of the nonhomogeneous equation. This has just been shown. Hence our main task is to discuss methods for finding such y_p. There is a general method for this which always works and which we shall consider in Sec. 2.16. There also is a much simpler special method which we discuss now.[14] It is called the **method of undetermined coefficients** and applies to equations with constant coefficients and a right-hand side $r(x)$ that is an exponential function, a polynomial, a cosine or sine, or a sum of such functions. We explain this method for a second-order equation

(1)
$$y'' + ay' + by = r(x)$$

but it will become obvious that it works equally for equations of higher order.

The key idea of the method is to assume for y_p an expression similar to that of $r(x)$, involving unknown coefficients that one must determine by substitution of this y_p into the equation. This should work for functions $r(x)$ whose derivatives are of a similar kind as $r(x)$ itself, which is the case for the functions just mentioned. Note that this is what we just did in the examples in the last section.

Rules for the Method of Undetermined Coefficients

(A) Basic Rule. *If $r(x)$ in (1) is one of the functions in the first column in Table 2.1 (see p. 126), choose the corresponding function y_p in the second column and determine its undetermined coefficients by substituting y_p and its derivatives into (1).*

(B) Modification Rule. *If $r(x)$ is a solution of the homogeneous equation corresponding to (1), then multiply your choice of y_p by x (or by x^2 if this solution corresponds to a double root of the characteristic equation of the homogeneous equation).*

(C) Sum Rule. *If $r(x)$ is a sum of functions listed in several lines of Table 2.1, first column, then choose for y_p the sum of the functions in the corresponding lines of the second column.*

---

[14]The student will see in Sec. 2.16 that our present method actually results from the general method.

Table 2.1
Method of Undetermined Coefficients

Term in $r(x)$	Choice for y_p
$ke^{\gamma x}$	$Ce^{\gamma x}$
$kx^n \ (n = 0, 1, \cdots)$	$K_n x^n + K_{n-1} x^{n-1} + \cdots + K_1 x + K_0$
$k \cos \omega x$	$\left.\vphantom{\begin{array}{c}a\\b\end{array}}\right\}\ K \cos \omega x + M \sin \omega x$
$k \sin \omega x$	

Rule (A) is illustrated by the two examples in the last section. Let us consider a third one.

EXAMPLE 1. Application of Rule (A)

Solve the nonhomogeneous equation

$$(2) \qquad\qquad y'' + 4y = 8x^2.$$

Solution. Table 2.1 suggests the choice

$$y_p = K_2 x^2 + K_1 x + K_0. \qquad \text{Then} \qquad y_p'' = 2K_2.$$

Substitution gives

$$2K_2 + 4(K_2 x^2 + K_1 x + K_0) = 8x^2.$$

Equating the coefficients of x^2, x, and x^0 on both sides, we have $4K_2 = 8$, $4K_1 = 0$, $2K_2 + 4K_0 = 0$. Thus $K_2 = 2$, $K_1 = 0$, $K_0 = -1$. Hence $y_p = 2x^2 - 1$, and a general solution of (2) is

$$y = y_h + y_p = A \cos 2x + B \sin 2x + 2x^2 - 1.$$

Note well that although $r(x) = 8x^2$, a trial $y_p = K_2 x^2$ would fail. Try it. Can you see why it fails?

EXAMPLE 2. Modification Rule (B) in the case of a simple root

Solve

$$(3) \qquad\qquad y'' - 3y' + 2y = e^x.$$

Solution. The characteristic equation $\lambda^2 - 3\lambda + 2 = 0$ has the roots 1 and 2. Hence $y_h = c_1 e^x + c_2 e^{2x}$. Ordinarily, our choice would be $y_p = Ce^x$. But we see that $r(x) = e^x$ is a solution of the homogeneous equation corresponding to a simple root (namely, 1). Hence Rule (B) suggests

$$y_p = Cxe^x. \qquad \text{We need} \qquad y_p' = C(e^x + xe^x), \qquad y_p'' = C(2e^x + xe^x).$$

Substitution gives

$$C(2 + x)e^x - 3C(1 + x)e^x + 2Cxe^x = e^x.$$

The xe^x-terms cancel out, and $-Ce^x = e^x$ remains. Hence $C = -1$. A general solution is

$$y = c_1 e^x + c_2 e^{2x} - xe^x.$$

Check it! Try $y_p = Ce^x$ to convince yourself that it does not work.

EXAMPLE 3. Modification Rule (double root) and Sum Rule

Solve

$$(4) \qquad\qquad y'' - 2y' + y = (D - 1)^2 y = e^x + x.$$

Solution. The characteristic equation has the double root $\lambda = 1$. Hence $y_h = (c_1 + c_2 x)e^x$. We determine a particular solution y_p. By Table 2.1, the term x indicates a particular solution choice

$$K_1 x + K_0.$$

Since 1 is a double root of the characteristic equation $(\lambda - 1)^2 = 0$, by the Modification Rule the term e^x calls for the particular solution

$$Cx^2 e^x \qquad\qquad \text{(instead of } Ce^x\text{)}.$$

Together,

$$y_p = K_1 x + K_0 + Cx^2 e^x.$$

Substituting this into (4), we obtain

$$y_p{}'' - 2y_p{}' + y_p = 2Ce^x + K_1 x - 2K_1 + K_0 = e^x + x.$$

Hence $C = \frac{1}{2}$, $K_1 = 1$, $K_0 = 2$, and a general solution of (4) is

$$y = y_h + y_p = (c_1 + c_2 x)e^x + \tfrac{1}{2}x^2 e^x + x + 2.$$

EXAMPLE 4. Another application of the Sum Rule

Solve

$$(5) \qquad\qquad y'' + 2y' + 5y = 16e^x + \sin 2x.$$

Solution. The characteristic equation $\lambda^2 + 2\lambda + 5 = 0$ has the complex roots $-1 + 2i$ and $-1 - 2i$. Hence $y_h = e^{-x}(A \cos 2x + B \sin 2x)$. By Table 2.1 and differentiation,

$$y_p = Ce^x + K \cos 2x + M \sin 2x,$$
$$y_p{}' = Ce^x - 2K \sin 2x + 2M \cos 2x,$$
$$y_p{}'' = Ce^x - 4K \cos 2x - 4M \sin 2x.$$

We substitute this into (5) and collect terms, finding

$$8Ce^x + (-4K + 4M + 5K) \cos 2x + (-4M - 4K + 5M) \sin 2x = 16e^x + \sin 2x.$$

Hence $8C = 16$, $K + 4M = 0$, $-4K + M = 1$, so that $C = 2$, $K = -4/17$, $M = 1/17$. The answer is

$$y = e^{-x}(A \cos 2x + B \sin 2x) + 2e^x - \tfrac{4}{17} \cos 2x + \tfrac{1}{17} \sin 2x. \qquad\blacksquare$$

Our rules are simple and take care of the cases listed in Table 2.1, which are particularly important to the engineer. For completeness we now prove those rules, using operators (Sec. 2.5) as a convenient tool. The reader may skip the proof or enjoy it as an application of operator methods.

For line 1 of Table 2.1, equation (1) is of the form

$$(6) \qquad\qquad (D - \lambda_1)(D - \lambda_2)y = ke^{\gamma x}.$$

Applying $D - \gamma$ on both sides, we obtain

$$(7) \qquad (D - \gamma)(D - \lambda_1)(D - \lambda_2)y = k(D - \gamma)[e^{\gamma x}] = 0.$$

Hence every solution of (6) satisfies (7). Accordingly, every solution of (6) can be obtained from a general solution of (7). Equation (7) is a homogeneous third-order equation. The form of its general solutions depends on whether the roots are different or equal. General solutions of (7) are

(8)

$$\text{(a)} \quad y = (c_1 e^{\lambda_1 x} + c_2 e^{\lambda_2 x}) + Ce^{\gamma x} \qquad \text{if } \gamma \neq \lambda_1 \neq \lambda_2 \neq \gamma$$

$$y = (c_1 e^{\lambda x} + c_2 x e^{\lambda x}) + Ce^{\gamma x} \qquad \text{if } \lambda_1 = \lambda_2 = \lambda \neq \gamma$$

$$\text{(b)} \quad y = (c_1 e^{\lambda_1 x} + c_2 e^{\gamma x}) + Cxe^{\gamma x} \qquad \text{if } \lambda_1 \neq \lambda_2 = \gamma$$

$$\text{(c)} \quad y = (c_1 e^{\gamma x} + c_2 x e^{\gamma x}) + Cx^2 e^{\gamma x} \qquad \text{if } \lambda_1 = \lambda_2 = \gamma.$$

The functions in parentheses are general solutions of the homogeneous equation corresponding to (6). Hence the last term with suitably determined C is a particular solution y_p of (6). This shows that (8a) proves the Basic Rule, formula (8b) the Modification Rule for a simple root, and (8c) for a double root when $r(x) = ke^{\gamma x}$. For lines 2 and 3 of Table 2.1 the idea of proof is the same; instead of $D - \gamma$ we then have to apply D^{n+1} in line 2, which gives $D^{n+1}[x^n] = 0$, and $D^2 + \omega^2$ in line 3, which gives $(D^2 + \omega^2)[\cos \omega x] = 0$ and $(D^2 + \omega^2)[\sin \omega x] = 0$. ∎

In the next section we discuss a basic engineering application of nonhomogeneous linear differential equations, namely, the motions of a mass–spring system under an **external force** $r(t)$ including **resonance.**

Problems for Sec. 2.12

Find a particular solution of the following differential equations.

1. $y'' + y = x^2 + x$
2. $y'' + 2y' + y = 2x^2$
3. $y'' + y = 2 \sin x$
4. $y'' + y' - 2y = 3e^x$
5. $y'' + y' - 6y = 52 \cos 2x$
6. $y'' - y' - 2y = \sin x$
7. $y'' + 5y' + 6y = 9x^4 - x$
8. $y'' - 2y' + 2y = 2e^x \cos x$
9. $y''' + 2y'' - y' - 2y = 1 - 4x^3$
10. $y^{IV} - 5y'' + 4y = 10 \cos x$

Find a general solution of the following differential equations.

11. $y'' + y = -x - x^2$
12. $y'' + 4y = e^{-x}$
13. $y'' - y = e^x$
14. $y'' - y' - 2y = 4 \sin x$
15. $y'' - 4y' + 3y = e^{3x}$
16. $y'' + y = \sin x$
17. $y'' + y' - 2y = e^x$
18. $y'' - 2y' + 2y = 2e^x \cos x$
19. $y''' - y'' - 4y' + 4y = 6e^{-x}$
20. $y''' + 2y'' - y' - 2y = 1 - 4x^3$

Solve the following initial value problems.

21. $y'' + 25y = 5x$, $\quad y(0) = 5$, $\quad y'(0) = -4.8$
22. $y'' - y = x$, $\quad y(2) = e^2 - 2 \approx 5.389$, $\quad y'(2) = e^2 - 1 \approx 6.389$
23. $y'' - 2y' + y = 2x^2 - 8x + 4$, $\quad y(0) = 0.3$, $\quad y'(0) = 0.3$
24. $y'' - y' - 2y = 10 \sin x$, $\quad y(\frac{1}{2}\pi) = -3$, $\quad y'(\frac{1}{2}\pi) = -1$
25. $y'' - y' - 2y = 3e^{2x}$, $\quad y(1) = 1/e \approx 0.3679$, $\quad y'(1) = -1/e + e^2 \approx 7.021$
26. $(D^2 + 0.5D - 0.5)y = 3 \cos x + \sin x + e^x$, $\quad y(0) = 0$, $\quad y'(0) = 1.5$
27. $(D^2 + 4)y = 8e^{-2x} + 4x^2 + 2$, $\quad y(0) = 2$, $\quad y'(0) = 2$
28. $(D^2 - 4D + 3)y = 2 \sin x - 4 \cos x$, $\quad y(\frac{1}{4}\pi) = 1/\sqrt{2}$, $\quad y'(\frac{1}{4}\pi) = 1/\sqrt{2}$
29. $(D^2 + 2D + 10)y = 10x^2 + 4x + 2$, $\quad y(0) = 1$, $\quad y'(0) = -1$
30. $(4D^2 - 4D + 65) = 64e^{x/2} + 65x - 4$, $\quad y(0) = 1$, $\quad y'(0) = 5.5$

31. Give the details of the proof of the Modification Rule in the case of line 2 in Table 2.1.

32. Perform the same task as in Prob. 31, for line 3 in Table 2.1.

First-Order Equations. The method of undetermined coefficients can also be applied to certain first-order linear differential equations, and may sometimes be simpler than the usual method (Sec. 1.7). Using both methods, solve:

33. $y' + y = x^4$ **34.** $y' - 2y = \sin 4x$ **35.** $y' + y = 2 \cos x$

2.13 Modeling: Forced Oscillations. Resonance

In Sec. 2.6 we considered free oscillations of a body on a spring as shown in Fig. 48. These motions are governed by the homogeneous equation

$$(1) \qquad\qquad my'' + cy' + ky = 0$$

where m is the mass of the body, c is the damping constant, and k is the spring modulus. We shall now extend our consideration, assuming that a variable force $r(t)$ acts on the system. In this way we shall become familiar with further interesting facts that are fundamental in engineering mathematics, in particular with resonance.

Setting up the Model

We remember that the equation (1) was obtained by considering the forces acting on the body and using Newton's second law. From this it is immediately clear that the differential equation corresponding to the present situation is obtained from (1) by adding the force $r(t)$; this yields

$$\boxed{my'' + cy' + ky = r(t).}$$

$r(t)$ is called the **input** or **driving force,** and a corresponding solution is called

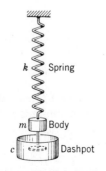

Fig. 48. Mass on a spring

an **output** or a **response** *of the system to the driving force.* (Cf. also Sec. 1.7.) The resulting motion is called a **forced motion,** in contrast to the **free motion** corresponding to (1), which is a motion in the absence of an external force $r(t)$.

Of particular interest are periodic inputs. We shall consider a sinusoidal input, say,

$$r(t) = F_0 \cos \omega t \qquad (F_0 > 0, \; \omega > 0).$$

More complicated periodic inputs will be considered later in Sec. 10.7.
The differential equation now under consideration is

(2)
$$\boxed{my'' + cy' + ky = F_0 \cos \omega t.}$$

Solving the Equation

We want to determine and discuss the general solution of (2), which represents the general output of our vibrating system. Since a general solution of the corresponding homogeneous equation (1) is known from Sec. 2.6, we must now determine a particular solution $y_p(t)$ of (2).

This can be done by the method of undetermined coefficients (Sec. 2.12), starting from

(3)
$$\boxed{y_p(t) = a \cos \omega t + b \sin \omega t.}$$

By differentiating this function we have

$$y_p' = -\omega a \sin \omega t + \omega b \cos \omega t,$$

$$y_p'' = -\omega^2 a \cos \omega t - \omega^2 b \sin \omega t.$$

We substitute these expressions into (2) and collect the cosine and sine terms:

$$[(k - m\omega^2)a + \omega cb] \cos \omega t + [-\omega ca + (k - m\omega^2)b] \sin \omega t = F_0 \cos \omega t.$$

By equating the coefficients of the cosine and sine terms on both sides we have

(4)
$$\begin{aligned}(k - m\omega^2)a + \;\; \omega cb \;\; &= F_0 \\[4pt] -\omega ca \;\; + (k - m\omega^2)b &= 0.\end{aligned}$$

This is a system of two linear algebraic equations in the two unknowns a and b. The solution is obtained in the usual way by elimination or by Cramer's rule (if necessary, see Sec. 7.9). We find

$$a = F_0 \frac{k - m\omega^2}{(k - m\omega^2)^2 + \omega^2 c^2}, \qquad b = F_0 \frac{\omega c}{(k - m\omega^2)^2 + \omega^2 c^2}$$

provided the denominator is not zero. If we set $\sqrt{k/m} = \omega_0 \, (> 0)$ as in Sec. 2.6, this becomes

(5)
$$a = F_0 \frac{m(\omega_0{}^2 - \omega^2)}{m^2(\omega_0{}^2 - \omega^2)^2 + \omega^2 c^2} \, , \; b = F_0 \frac{\omega c}{m^2(\omega_0{}^2 - \omega^2)^2 + \omega^2 c^2} \cdot$$

We thus obtain the general solution

(6)
$$y(t) = y_h(t) + y_p(t),$$

where y_h is a general solution of (1) and y_p is given by (3) with coefficients (5).

Discussion of Types of Solutions

We shall now discuss the behavior of the mechanical system, distinguishing between the two cases $c = 0$ (no damping) and $c > 0$ (damping). These cases will correspond to two different types of output.

Case 1. *Undamped forced oscillations*

If there is no damping, then $c = 0$. We first assume that $\omega^2 \neq \omega_0{}^2$ (where $\omega_0{}^2 = k/m$, as in Sec. 2.6). This is essential. We then obtain from (3) and (5)

(7)
$$y_p(t) = \frac{F_0}{m(\omega_0{}^2 - \omega^2)} \cos \omega t = \frac{F_0}{k[1 - (\omega/\omega_0)^2]} \cos \omega t.$$

From this and (6*) in Sec. 2.6 we have the general solution

(8)
$$y(t) = C \cos (\omega_0 t - \delta) + \frac{F_0}{m(\omega_0{}^2 - \omega^2)} \cos \omega t.$$

*This output represents a superposition of two harmonic oscillations; the frequencies are the "**natural frequency**" $\omega_0/2\pi$ [cycles/sec] of the system (that is, the frequency of the free undamped motion) and the frequency $\omega/2\pi$ of the input.*

From (7) we see that the maximum amplitude of y_p is

(9)
$$a_0 = \frac{F_0}{k} \rho \qquad \text{where} \qquad \rho = \frac{1}{1 - (\omega/\omega_0)^2} \cdot$$

It depends on ω and ω_0. As $\omega \to \omega_0$, the quantities ρ and a_0 tend to infinity. This phenomenon of excitation of large oscillations by matching input and natural frequencies ($\omega = \omega_0$) is known as **resonance,** and is of basic importance in the study of vibrating systems (see below). The quantity ρ is called the *resonance factor* (Fig. 49 on the next page). From (9) we see that ρ/k is the ratio of the amplitudes of the function y_p and the input.

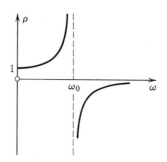

Fig. 49. Resonance factor $\rho(\omega)$

In the case of resonance, equation (2) becomes

(10)
$$y'' + \omega_0^2 y = \frac{F_0}{m} \cos \omega_0 t.$$

From the Modification Rule in the preceding section we conclude that a particular solution of (10) is of the form

$$y_p(t) = t(a \cos \omega_0 t + b \sin \omega_0 t).$$

By substituting this into (10) we find $a = 0$, $b = F_0/2m\omega_0$, and (Fig. 50)

(11)
$$y_p(t) = \frac{F_0}{2m\omega_0} t \sin \omega_0 t.$$

We see that y_p becomes larger and larger. In practice, this means that systems with very little damping may undergo large vibrations that can destroy the system; we shall return to this practical aspect of resonance later in this section.

Another interesting and highly important type of oscillation is obtained when ω is close to ω_0. Take, for example, the particular solution [cf. (8)]

(12)
$$y(t) = \frac{F_0}{m(\omega_0^2 - \omega^2)} (\cos \omega t - \cos \omega_0 t) \qquad (\omega \neq \omega_0)$$

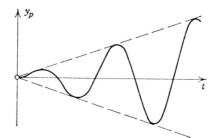

Fig. 50. Particular solution in the case of resonance

corresponding to the initial conditions $y(0) = 0$, $y'(0) = 0$. This may be written [cf. (12) in Appendix 3]

$$y(t) = \frac{2F_0}{m(\omega_0^2 - \omega^2)} \sin \frac{\omega_0 + \omega}{2} t \sin \frac{\omega_0 - \omega}{2} t.$$

Since ω is close to ω_0, the difference $\omega_0 - \omega$ is small, so that the period of the last sine function is large, and we obtain an oscillation of the type shown in Fig. 51.

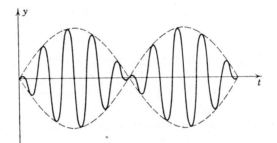

Fig. 51. Forced undamped oscillation when the difference of the input and natural frequencies is small ("beats")

Case 2. *Damped forced oscillations*

If there is damping, then $c > 0$, and we know from Sec. 2.6 that the general solution y_h of (1) is

$$y_h(t) = e^{-\alpha t}(A \cos \omega^* t + B \sin \omega^* t) \qquad \left(\alpha = \frac{c}{2m} > 0\right)$$

and this solution approaches zero as t approaches infinity (practically, after a sufficiently long time); that is, the general solution (6) of (2) now represents the **transient solution** and tends to the **steady-state solution** y_p. Hence, *after a sufficiently long time, the output corresponding to a purely sinusoidal input will practically be a harmonic oscillation whose frequency is that of the input. This is what happens in practice, because no physical system is completely undamped.*

   While in the undamped case the amplitude of y_p approaches infinity as ω approaches ω_0, this will not happen in the damped case: *in this case the amplitude will always be finite,* but may have a maximum for some ω, depending on c. This may be called **practical resonance.** It is of great importance because it shows that some input may excite oscillations with such a large amplitude that the system can be destroyed. Such cases happened in practice, in particular in earlier times when less was known about resonance. Machines, cars, ships, airplanes and bridges are vibrating mechanical systems, and it is sometimes rather difficult to find constructions which are completely free of undesired resonance effects.

Amplitude of y_p

To study the amplitude of y_p as a function of ω, we write (3) in the form

(13)
$$y_p(t) = C^* \cos(\omega t - \eta)$$

where, according to (5),

(14)
$$C^*(\omega) = \sqrt{a^2 + b^2} = \frac{F_0}{\sqrt{m^2(\omega_0{}^2 - \omega^2)^2 + \omega^2 c^2}},$$

$$\tan \eta = \frac{b}{a} = \frac{\omega c}{m(\omega_0{}^2 - \omega^2)}.$$

Let us determine the maximum of $C^*(\omega)$. By setting $dC^*/d\omega = 0$ we find

$$[-2m^2(\omega_0{}^2 - \omega^2) + c^2]\,\omega = 0.$$

The expression in brackets is zero when

(15)
$$c^2 = 2m^2(\omega_0{}^2 - \omega^2).$$

For sufficiently large damping ($c^2 > 2m^2\omega_0{}^2 = 2mk$) equation (15) has no real solution, and C^* decreases in a monotone way as ω increases (Fig. 52). If $c^2 \leqq 2mk$, equation (15) has a real solution $\omega = \omega_{max}$, which increases as c decreases and approaches ω_0 as c approaches zero. The amplitude $C^*(\omega)$ has a maximum at $\omega = \omega_{max}$, and by inserting $\omega = \omega_{max}$ into (14) we find

(16)
$$C^*(\omega_{max}) = \frac{2mF_0}{c\sqrt{4m^2\omega_0{}^2 - c^2}}.$$

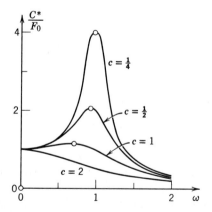

Fig. 52. Amplification C^*/F_0 as a function of ω for $m = 1$, $k = 1$, and various values of the damping constant c

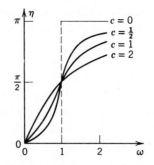

Fig. 53. Phase lag η as a function of ω for $m = 1$, $k = 1$, and various values of the damping constant c

We see that $C^*(\omega_{max})$ is finite when $c > 0$. Since $dC^*(\omega_{max})/dc < 0$ when $c^2 < 2mk$, the value of $C^*(\omega_{max})$ increases as c ($\leq \sqrt{2mk}$) decreases and approaches infinity as c approaches zero, in agreement with our result in Case 1. Figure 52 shows the **amplification** C^*/F_0 (ratio of the amplitudes of output and input) as a function of ω for $m = 1$, $k = 1$, and various values of the damping constant c.

The angle η in (14) is called the **phase angle** or **phase lag** (Fig. 53) because it measures the lag of the output with respect to the input. If $\omega < \omega_0$, then $\eta < \pi/2$; if $\omega = \omega_0$, then $\eta = \pi/2$, and if $\omega > \omega_0$, then $\eta > \pi/2$.

In the next section we show that the mass–spring system just considered has a strictly quantitative analog in electrical engineering, namely, the **RLC-circuit**. This will be an impressive demonstration of the unifying power of mathematics, illustrating that entirely different physical systems may have the same mathematical model and can then be treated and solved by the same methods.

Problems for Sec. 2.13

Find the steady-state oscillations of the vibrating systems governed by the following equations.

1. $y'' + 9y = \sin t$ **2.** $y'' + y = 3 \cos 2t$

3. $y'' + 3y' + 2y = 10 \sin t$ **4.** $y'' + 5y' + 6y = 6.29 \sin \frac{1}{2} t$

5. $(D^2 + 4D + 3)y = 65 \cos 2t$ **6.** $(D^2 + D + 1.5)y = \cos 3t - \sin t$

7. $(3D^2 + D + 2)y = 2 \cos t - 52 \sin 2t$ **8.** $(D^2 + D + 1)y = \cos t + 13 \cos 2t$

Find the transient motions of the vibrating systems governed by the following equations.

9. $y'' + 2y' + 2y = 85 \cos 3t$ **10.** $y'' + 4y' + 3y = 26 \cos 2t$

11. $2y'' + 2y' + 3y = 87 \cos 3t$ **12.** $y'' + 2y' + y = 50 \sin 3t$

13. $(D^2 + 4D + 20)y = \sin t + \frac{4}{19} \cos t$ **14.** $(D^2 + 4D + 5)y = 37.7 \sin 4t$

15. $(D^2 + 1)y = \cos \omega t$, $\omega^2 \neq 1$ **16.** $(D^2 + 6D + 9)y = \cos t$

In each case, the given differential equation is the mathematical model of a vibrating system. Find the motion of the system corresponding to the given initial displacement and initial velocity.

17. $y'' + 25y = 24 \sin t$, $\quad y(0) = 1$, $\quad y'(0) = 1$

18. $(D^2 + 2D + 2)y = \cos t$, $\quad y(0) = 1.2$, $\quad y'(0) = 1.4$

19. $(D^2 + 0.5D + 2)y = 5 \cos t$, $\quad y(0) = 4$, $\quad y'(0) = \frac{3}{2}\sqrt{31} + 2 \approx 10.35$

20. $(D^2 + 4)y = \sin t + \frac{1}{3} \sin 3t + \frac{1}{5} \sin 5t$, $\quad y(0) = 1$, $\quad y'(0) = \frac{3}{35}$

21. $(D^2 + D + 0.25)y = 2 \cos t - \frac{3}{2} \sin t - 2 \cos 2t + 3.75 \sin 2t$,
$\quad y(0) = 0$, $\quad y'(0) = 1.5$

22. (Gun barrel) Solve

$$y'' + y = \begin{cases} 1 - t^2/\pi^2 & \text{if } 0 \leq t \leq \pi \\ 0 & \text{if } t > \pi \end{cases} \qquad y(0) = y'(0) = 0.$$

This may be interpreted as an undamped system on which a force F acts during some interval of time (see Fig. 54), for instance, the force acting on a gun barrel when a shell is fired, the barrel being braked by heavy springs (and then damped by a dashpot, which we disregard for simplicity). *Hint.* At $t = \pi$ both y and y' must be continuous.

Fig. 54. Problem 22

23. In (12) let ω approach ω_0. Show that this leads to a solution of the form (11).

24. For what initial conditions $y(0) = y_0$, $y'(0) = v_0$ does the solution of Prob. 15 represent an oscillation whose frequency equals the frequency of the input?

25. Solve the initial value problem $y'' + y = \cos \omega t$, $\omega^2 \neq 1$, $y(0) = 0$, $y'(0) = 0$. Graph the maximum amplitude as a function of ω. Show that the solution can be written

$$y(t) = \frac{2}{1 - \omega^2} \sin \left[\frac{1}{2}(1 + \omega)t \right] \sin \left[\frac{1}{2}(1 - \omega)t \right].$$

Plot good graphs of $y(t)$ with $\omega = 0.5, 0.9, 1.1, 2$.

2.14 Modeling of Electric Circuits

The last section was devoted to the study of a mechanical system that is of great practical interest. We shall now consider a similarly important *electrical system*, which may be regarded as a basic building block in electrical networks. This consideration will also provide a striking example of the important fact that *entirely different physical systems may correspond to the same mathematical model*—in the present case: to the same differential equation. This will illustrate the role mathematics plays in unifying various phenomena of entirely different physical natures. We shall obtain a *correspondence between mechanical and electrical systems which is not merely qualitative but strictly quantitative* in the sense that to a given mechanical system we can construct an electric circuit whose current will give the exact values of the displacement in the mechanical system when suitable scale factors are introduced. The practical importance of such an ***analogy between mechanical and electrical systems*** is almost obvious. The analogy may be used for constructing an ''electrical model'' of a given mechanical system; in many cases this will be an essential simplification, because electric circuits are easy to assemble and currents and voltages are easy to measure, whereas the construction of a mechanical model may be complicated and expensive, and the measurement of displacements will be time-consuming and inaccurate.

Setting up the Model

We consider the *RLC*-circuit in Fig. 55, in which an Ohm's resistor of resistance R [ohms], an inductor of inductance L [henrys] and a capacitor of capacitance C [farads] are connected in series to a source of electromotive force $E(t)$ [volts], where t is time. The equation for the current $I(t)$ [amperes] in the *RLC*-circuit is obtained by considering the three voltage drops

$$E_L = LI'$$ across the inductor,

$$E_R = RI$$ across the resistor (*Ohm's law*), and

$$E_C = \frac{1}{C} \int I(t)\, dt$$ across the capacitor.

Their sum equals the electromotive force $E(t)$. This is **Kirchhoff's voltage law** (Sec. 1.8), the analog of Newton's second law (Sec. 2.6) for mechanical systems. For a sinusoidal $E(t) = E_0 \sin \omega t$ (E_0 constant), this law gives

(1′) $$LI' + RI + \frac{1}{C} \int dt = E(t) = E_0 \sin \omega t.$$

This process of modeling is the same as that in Sec. 1.8. Indeed, if we add $E_L = LI'$ to the equation (7) in Sec. 1.8 for the *RC*-circuit, we obtain our present equation (1′) for the *RLC*-circuit.

To get rid of the integral in (1′), we differentiate with respect to t, obtaining

(1) $$\boxed{LI'' + RI' + \frac{1}{C} I = E_0 \omega \cos \omega t.}$$

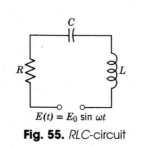

$$E(t) = E_0 \sin \omega t$$

Fig. 55. *RLC*-circuit

This is of the same form as (2), Sec. 2.13. Hence our RLC-circuit is the electrical analog of the mechanical system in Sec. 2.13. The corresponding analogy of electrical and mechanical quantities is shown in Table 2.2.

Table 2.2
**Analogy of Electrical and Mechanical Quantities
in (1), This Section, and (2), Sec. 2.13**

Electrical System	Mechanical System
Inductance L	Mass m
Resistance R	Damping constant c
Reciprocal $1/C$ of capacitance	Spring modulus k
Derivative $E_0\omega \cos \omega t$ $\Big\}$ of electromotive force	Driving force $F_0 \cos \omega t$
Current $I(t)$	Displacement $y(t)$

Comment. Recalling from Sec. 1.8 that $I = Q'$, we have $I' = Q''$ and $Q = \int I\, dt$ in (1′). Hence we obtain from (1′) as the differential equation for the charge Q on the capacitor

$$(1'') \qquad\qquad LQ'' + RQ' + \frac{1}{C} Q = E_0 \sin \omega t.$$

In most practical problems, the current $I(t)$ is more important than $Q(t)$, and for this reason we shall concentrate on (1) rather than on (1″).

Solving Equation (1), Discussion of Solution

To obtain a particular solution of (1) we may proceed as in Sec. 2.13. By substituting

$$(2) \qquad\qquad I_p(t) = a \cos \omega t + b \sin \omega t$$

into (1) we obtain

$$(3) \qquad\qquad a = \frac{-E_0 S}{R^2 + S^2}, \qquad b = \frac{E_0 R}{R^2 + S^2}$$

where S is the so-called **reactance,** given by the expression

$$(4) \qquad\qquad S = \omega L - \frac{1}{\omega C}.$$

In any practical case, $R \neq 0$ so that the denominator in (3) is not zero. The result is that (2), with a and b given by (3), is a particular solution of (1).
   Using (3), we may write I_p in the form

$$(5) \qquad\qquad I_p(t) = I_0 \sin (\omega t - \theta)$$

where [cf. (14) in Appendix 3]

$$I_0 = \sqrt{a^2 + b^2} = \frac{E_0}{\sqrt{R^2 + S^2}}, \qquad \tan \theta = -\frac{a}{b} = \frac{S}{R}.$$

The quantity $\sqrt{R^2 + S^2}$ is called the **impedance.** Our formula shows that the impedance equals the ratio E_0/I_0, which is somewhat analogous to $E/I = R$ (Ohm's law).

A general solution of the homogeneous equation corresponding to (1) is

$$I_h = c_1 e^{\lambda_1 t} + c_2 e^{\lambda_2 t},$$

where λ_1 and λ_2 are the roots of the characteristic equation

$$\lambda^2 + \frac{R}{L}\lambda + \frac{1}{LC} = 0,$$

which we can write in the form $\lambda_1 = -\alpha + \beta$ and $\lambda_2 = -\alpha - \beta$, where

$$\alpha = \frac{R}{2L}, \qquad \beta = \frac{1}{2L}\sqrt{R^2 - \frac{4L}{C}}.$$

As in Sec. 2.13 we conclude that if $R > 0$ (which, of course, is true in any practical case), the general solution $I_h(t)$ of the homogeneous equation approaches zero as t approaches infinity (practically: after a sufficiently long time). Hence, the transient current $I = I_h + I_p$ tends to the steady-state current I_p, and *after some time* **the output will practically be a harmonic oscillation,** *which is given by* (5) *and whose frequency is that of the input.*

EXAMPLE 1. *RLC*-circuit
Find the current $I(t)$ in an *RLC*-circuit with $R = 100$ ohms, $L = 0.1$ henry, $C = 10^{-3}$ farad, which is connected to a source of voltage $E(t) = 155 \sin 377t$ (hence 60 Hz = 60 cycles/sec), assuming zero charge and current when $t = 0$.

Solution. Equation (1) is

$$0.1I'' + 100I' + 1000I = 155 \cdot 377 \cos 377t.$$

We calculate the reactance $S = 37.7 - 1/0.377 = 35.0$ and the steady-state current

$$I_p(t) = a \cos 377t + b \sin 377t$$

where

$$a = \frac{-155 \cdot 35.0}{100^2 + 35^2} = -0.483, \qquad b = \frac{155 \cdot 100}{100^2 + 35^2} = 1.381.$$

Then we solve the characteristic equation

$$0.1\lambda^2 + 100\lambda + 1000 = 0.$$

The roots are $\lambda_1 = -10$ and $\lambda_2 = -990$. Hence the general solution is

$$\text{(6)} \qquad I(t) = c_1 e^{-10t} + c_2 e^{-990t} - 0.483 \cos 377t + 1.381 \sin 377t.$$

We determine c_1 and c_2 from the initial conditions $Q(0) = 0$ and $I(0) = 0$. By the second condition,

$$(7) \qquad\qquad I(0) = c_1 + c_2 - 0.483 = 0.$$

How to use $Q(0) = 0$? Solving (1') algebraically for I', we have

$$(8) \qquad\qquad I'(t) = \frac{1}{L}\left[E(t) - RI(t) - \frac{1}{C}Q(t) \right]$$

since $\int I\,dt = Q$. Here $E(0) = 0$, $I(0) = 0$, and $Q(0) = 0$, so that $I'(0) = 0$. Differentiating (6), we thus obtain

$$(9) \qquad\qquad I'(0) = -10c_1 - 990c_2 + 1.381 \cdot 377 = 0.$$

The solution of (7) and (9) is $c_1 = -0.043$, $c_2 = 0.526$. From (6) we thus have the answer

$$I(t) = -0.043e^{-10t} + 0.526e^{-990t} - 0.483\cos 377t + 1.381\sin 377t.$$

The first two terms will die out rapidly, and after a very short time the current will practically execute harmonic oscillations of frequency 60 Hz = 60 cycles/sec, which is the frequency of the impressed voltage. Note that by (5) we can write the steady-state current in the form

$$I_p(t) = 1.463\sin(377t - 0.34). \qquad\blacksquare$$

The next section concerns a **complex method** for solving equations with cosine or sine on the right and is optional.

Problems for Sec. 2.14

RLC-circuits. Find the steady-state current in the *RLC*-circuit in Fig. 56 where
1. $R = 20$ ohms, $L = 10$ henrys, $C = 0.05$ farad, $E = 50\sin t$ volts
2. $R = 240$ ohms, $L = 40$ henrys, $C = 10^{-3}$ farad, $E = 369\sin 10t$ volts
3. $R = 40$ ohms, $L = 10$ henrys, $C = 0.02$ farad, $E = 800\cos 5t$ volts

Find the transient current in the *RLC*-circuit in Fig. 56 where
4. $R = 200$ ohms, $L = 100$ henrys, $C = 0.005$ farad, $E = 2500\sin t$ volts
5. $R = 20$ ohms, $L = 5$ henrys, $C = 0.01$ farad, $E = 850\sin 4t$ volts
6. $R = 16$ ohms, $L = 8$ henrys, $C = 0.125$ farad, $E = 300\cos 2t$ volts

Using (8), find the current in the *RLC*-circuit in Fig. 56, assuming zero initial current and charge, and
7. $R = 40$ ohms, $L = 10$ henrys, $C = 0.02$ farad, $E = 300$ volts
8. $R = 160$ ohms, $L = 20$ henrys, $C = 0.002$ farad, $E = 481\sin 10t$ volts
9. $R = 6$ ohms, $L = 1$ henry, $C = 0.04$ farad, $E = 24\cos 5t$ volts

10. What are the conditions for an *RLC*-circuit to be overdamped (Case I), underdamped (Case II), and critically damped (Case III)? In particular, what is the critical resistance R_{crit} (the analog of the critical damping constant $c_{crit} = 2\sqrt{mk}$)?
11. Derive (3) in two ways, namely, (a) directly by substituting (2) into (1), (b) from (5) in Sec. 2.13, using Table 2.2 and taking $E_0\omega$ instead of F_0.
12. It was claimed in the text that if $R > 0$, then the transient current approaches $I_p(t)$ as $t \to \infty$. How can this be proved?

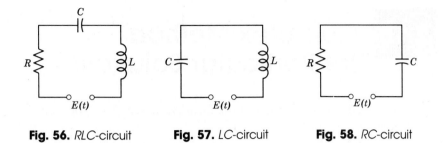

Fig. 56. *RLC*-circuit   **Fig. 57.** *LC*-circuit   **Fig. 58.** *RC*-circuit

13. (**Tuning**) In tuning a radio to a station we turn a knob on the radio which changes C (or perhaps L) in an *RLC*-circuit (Fig. 56) so that the amplitude of the steady-state current becomes maximum. For what C will this be the case?

LC-circuits. Find the current $I(t)$ in the *LC*-circuit in Fig. 57, where
14. $L = 0.4$ henry, $C = 0.1$ farad, $E = 110 \sin \omega t$ volts, $\omega^2 \neq 25$
15. $L = 0.2$ henry, $C = 0.05$ farad, $E = 100$ volts
16. $L = 2.5$ henrys, $C = 10^{-3}$ farad, $E = 10t^2$ volts

Find the current $I(t)$ in the *LC*-circuit in Fig. 57, assuming zero initial current and charge, and
17. $L = 10$ henrys, $C = 0.004$ farad, $E = 250$ volts
18. $L = 1$ henry, $C = 0.25$ farad, $E = 90 \cos t$ volts
19. $L = 2$ henrys, $C = 0.005$ farad, $E = 210 \sin 4t$ volts
20. $L = 10$ henrys, $C = 0.1$ farad, $E = 10t$ volts

21. Show that if $E(t)$ in Fig. 57 has a jump of magnitude J at $t = a$, then $I'(t)$ has a jump of magnitude J/L at $t = a$, whereas $I(t)$ is continuous at $t = a$.

Using the result of Prob. 21, find the current $I(t)$ in the *LC*-circuit in Fig. 57, assuming $L = 1$ henry, $C = 1$ farad, zero initial current and charge, and
22. $E = t$ when $0 < t < a$ and $E = 0$ when $t > a$
23. $E = 1 - e^{-t}$ when $0 < t < \pi$ and $E = 0$ when $t > \pi$
24. $E = 1$ when $0 < t < a$ and $E = 0$ when $t > a$

25. Find the natural frequency (= frequency of free oscillations) of an *LC*-circuit (a) directly, (b) from Sec. 2.6 by means of Table 2.2.

RC-circuits
26. Show that if the initial charge in the capacitor in Fig. 58 is $Q(0)$, the initial current in the *RC*-circuit is $I(0) = E(0)/R - Q(0)/RC$.
27. Show that if $E(t)$ in Fig. 58 has a jump of magnitude J at $t = a$, then the current $I(t)$ in the *RC*-circuit has a jump of magnitude J/R at $t = a$.

Using the result of Prob. 27, find the current $I(t)$ in the *RC*-circuit in Fig. 58, assuming $R = 1$ ohm, $C = 1$ farad, zero initial charge on the capacitor, and
28. $E = t$ if $0 < t < a$ and $E = a$ if $t > a$
29. $E = t$ if $0 < t < a$ and $E = 0$ if $t > a$
30. $E = t + 1$ if $0 < t < a$ and $E = a + 1$ if $t > a$

2.15 Complex Method for Particular Solutions

Given an equation of the form (1) in Sec. 2.14, for example,

(1) $$I'' + I' + 3I = 5 \cos t,$$

we know that we can obtain a particular solution $I_p(t)$ by the method of undetermined coefficients, that is, by substituting $I_p(t) = a \cos t + b \sin t$ into (1) and determining a and b. The result will be

$$I_p(t) = 2 \cos t + \sin t.$$

Engineers often prefer a simple and elegant complex method for obtaining $I_p(t)$. In this method we note that $5 \cos t$ in (1) is the real part of

$$5e^{it} = 5(\cos t + i \sin t)$$

(cf. the Euler formula in Sec. 2.4) and, instead of (1), we consider the differential equation

(2) $$I'' + I' + 3I = 5e^{it} \qquad (i = \sqrt{-1}).$$

We determine a complex particular solution of the form

(3) $$I_p^*(t) = Ke^{it}.$$

By substituting the function and its derivatives

$$I_p^{*\prime} = iKe^{it}, \qquad I_p^{*\prime\prime} = -Ke^{it}$$

into (2) we have

$$(-1 + i + 3)Ke^{it} = 5e^{it}.$$

Solving for K, we obtain

$$K = \frac{5}{2 + i} = \frac{5(2 - i)}{(2 + i)(2 - i)} = \frac{10 - 5i}{5} = 2 - i.$$

Hence

$$I_p^*(t) = (2 - i)e^{it} = (2 - i)(\cos t + i \sin t)$$

is a solution of (2). The real part of I_p^* is

$$I_p(t) = 2 \cos t + \sin t$$

and this function is a solution of the real part of the differential equation (2), that is, of the given differential equation (1). In fact, the function is identical with that obtained above. This illustrates the practical procedure of the complex method.

Our equation (1) is a particular case of the equation [cf. (1), Sec. 2.14]

$$(4) \qquad LI'' + RI' + \frac{1}{C} I = E_0 \omega \cos \omega t,$$

which we shall now consider, assuming that $R \neq 0$. The corresponding complex equation is

$$(5) \qquad LI'' + RI' + \frac{1}{C} I = E_0 \omega e^{i\omega t}.$$

The function on the right suggests a particular solution of the form

$$(6) \qquad I_p{}^*(t) = K e^{i\omega t} \qquad\qquad (i = \sqrt{-1}).$$

By substituting this function and its derivatives

$$I_p{}^{*\prime} = i\omega K e^{i\omega t}, \qquad I_p{}^{*\prime\prime} = -\omega^2 K e^{i\omega t}$$

into (5) we have

$$\left(-\omega^2 L + i\omega R + \frac{1}{C} \right) K e^{i\omega t} = E_0 \omega e^{i\omega t}.$$

Dividing by ω on both sides and solving for K, we obtain

$$(7) \qquad K = \frac{E_0}{-\omega L + iR + \dfrac{1}{\omega C}} = \frac{E_0}{iZ}$$

where Z is the so-called **complex impedance,** given by (use $1/i = -i$)

$$(8) \qquad Z = R + i \left(\omega L - \frac{1}{\omega C} \right).$$

It follows that (6) with K given by (7) is a solution of (5).

We see that the imaginary part of Z is the reactance S defined by (4), Sec. 2.14, and $|Z|$ is the impedance defined in connection with (5), Sec. 2.14. Thus (cf. Fig. 59 on the next page)

$$Z = |Z| e^{i\theta} \qquad \text{where} \qquad \tan \theta = \frac{S}{R}.$$

Consequently, formula (6) with K according to (7) can be written in the form

$$I_p{}^*(t) = \frac{E_0}{iZ} e^{i\omega t} = -i \frac{E_0}{|Z|} e^{i(\omega t - \theta)}.$$

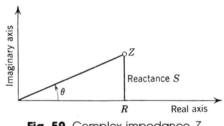

Fig. 59. Complex impedance Z

The real part is

$$(9) \qquad I_p(t) = \frac{E_0}{|Z|} \sin(\omega t - \theta) = \frac{E_0}{\sqrt{R^2 + S^2}} \sin(\omega t - \theta),$$

and this is a solution of the real part of the differential equation (5), that is, $I_p(t)$ is a solution of (4). We see that this function is identical with (5) in the previous section. Since

$$\sin(\omega t - \theta) = \sin \omega t \cos \theta - \cos \omega t \sin \theta$$

and, furthermore,

$$\cos \theta = \frac{\operatorname{Re} Z}{|Z|} = \frac{R}{\sqrt{R^2 + S^2}}, \qquad \sin \theta = \frac{\operatorname{Im} Z}{|Z|} = \frac{S}{\sqrt{R^2 + S^2}},$$

the solution can be written in the form

$$(10) \qquad I_p(t) = \frac{E_0}{R^2 + S^2}[R \sin \omega t - S \cos \omega t].$$

This is identical with (2) in the previous section, where a and b are given by (3) in that section.

In the last section of this chapter we discuss a general method for solving nonhomogeneous linear equations, the method of **variation of parameters.** This method is more powerful than the method of undetermined coefficients (Sec. 2.12), but the latter is generally simpler and should be preferred in the cases to which it applies, which are indicated in Sec. 2.12.

Problems for Sec. 2.15

Using the complex method, find the steady-state current $I_p(t)$ in the circuit governed by (4), where

1. $R = 20, L = 10, C = 0.05, E_0 = 5, \omega = 2$
2. $R = 10, L = 20, C = 0.1, E_0 = 4, \omega = 2$
3. $R = 25, L = 15, C = 0.05, E_0 = 100, \omega = 4$
4. $R = 100, L = 100, C = 10^{-3}, E_0 = 3, \omega = 4$
5. $R = 50, L = 25, C = 0.01, E_0 = 500, \omega = 3$

6. Find the complex impedance Z and the reactance S in Prob. 3.

Using the complex method, find the steady-state output of the following equations.

7. $y'' + y' + 4y = 8 \sin 2t$ **8.** $y'' + 3y' + 16y = 24 \cos 4t$

9. $y'' + y' + 9y = -3 \sin 3t$ **10.** $y'' + 5y' + \frac{1}{8}y = 25 \cos 10t$

2.16 Nonhomogeneous Equations: Solving by the Method of Variation of Parameters

The method of undetermined coefficients in Sec. 2.12 is a simple procedure for obtaining particular solutions of nonhomogeneous linear differential equations, and we have seen in Secs. 2.13–2.15 that it has important applications in connection with vibrations. However, it is limited to certain simpler types of differential equations. The method to be discussed in this section will be completely general but more complicated than that method.

We consider linear differential equations of the form

$$(1) \qquad\qquad y'' + p(x)y' + q(x)y = r(x)$$

assuming that p, q, and r are continuous on an open interval I. We shall obtain a particular solution of (1) by using the method of **variation of parameters**[15] as follows. We know that the corresponding homogeneous differential equation

$$(2) \qquad\qquad y'' + p(x)y' + q(x)y = 0$$

has a general solution $y_h(x)$ on the interval I, which is of the form

$$y_h(x) = c_1 y_1(x) + c_2 y_2(x).$$

The method of variation of parameters consists in replacing c_1 and c_2 by functions $u(x)$ and $v(x)$ to be determined so that the resulting function

$$(3) \qquad\qquad y_p(x) = u(x)y_1(x) + v(x)y_2(x)$$

is a particular solution of (1) on I. By differentiating (3) we obtain

$$y_p{}' = u'y_1 + uy_1{}' + v'y_2 + vy_2{}'.$$

---

[15]Credited to the great mathematician JOSEPH LOUIS LAGRANGE (1736—1813), who was born in Turin, of French extraction, got his first professorship when he was 19 (at the Military Academy of Turin), became director of the mathematical section of the Berlin Academy in 1766, and moved to Paris in 1787. His important major work was in the calculus of variations, celestial mechanics, general mechanics (*Mécanique analytique*, Paris 1788), differential equations, approximation theory, algebra and number theory.

Now (3) contains *two* functions u and v, but the requirement that y_p satisfy (1) imposes only *one* condition on u and v. Hence it seems plausible that we may impose a second arbitrary condition. Indeed, our further calculation will show that we can determine u and v such that y_p satisfies (1) and u and v satisfy as a second condition the relationship

(4) $$u'y_1 + v'y_2 = 0.$$

This reduces the expression for y_p' to the form

(5) $$y_p' = uy_1' + vy_2'.$$

By differentiating this function we have

(6) $$y_p'' = u'y_1' + uy_1'' + v'y_2' + vy_2''.$$

By substituting (3), (5) and (6) into (1) and collecting terms containing u and terms containing v we readily obtain

$$u(y_1'' + py_1' + qy_1) + v(y_2'' + py_2' + qy_2) + u'y_1' + v'y_2' = r.$$

Since y_1 and y_2 are solutions of the homogeneous equation (2), this reduces to

$$u'y_1' + v'y_2' = r.$$

Equation (4) is $$u'y_1 + v'y_2 = 0.$$

This is a system of two linear algebraic equations for the unknown functions u' and v'. The solution is obtained by Cramer's rule (cf. Sec. 7.9). We find

(7) $$u' = -\frac{y_2 r}{W}, \qquad v' = \frac{y_1 r}{W},$$

where

$$W = y_1 y_2' - y_1' y_2$$

is the Wronskian of y_1 and y_2 (cf. Sec. 2.8). Clearly, $W \neq 0$ since y_1, y_2 constitute a basis of solutions. Integration of (7) gives

$$u = -\int \frac{y_2 r}{W}\, dx, \qquad v = \int \frac{y_1 r}{W}\, dx.$$

These integrals exist because $r(x)$ is continuous. Substituting these expressions for u and v into (3), we obtain the desired solution of (1),

(8) $$\boxed{y_p(x) = -y_1 \int \frac{y_2 r}{W}\, dx + y_2 \int \frac{y_1 r}{W}\, dx.}$$

Note that if the constants of integration in (8) are left arbitrary, then (8) represents a general solution of (1).

Caution! Before applying (8), make sure that your equation is written in the standard form (1), with y'' as the first term; divide by $h(x)$ if it starts with $h(x)y''$.

EXAMPLE 1

Solve the differential equation

$$(9) \qquad\qquad y'' + y = \sec x.$$

Solution. The method of undetermined coefficients (Sec. 2.12) cannot be applied. The functions

$$y_1 = \cos x, \qquad y_2 = \sin x$$

constitute a basis of solutions of the homogeneous equation. Their Wronskian is

$$W(y_1, y_2) = \cos x \cos x - (-\sin x) \sin x = 1.$$

From (8) we thus obtain the following particular solution of (9):

$$y_p = -\cos x \int \sin x \sec x \, dx + \sin x \int \cos x \sec x \, dx$$

$$= \cos x \ln |\cos x| + x \sin x.$$

The corresponding general solution of the differential equation (9) is

$$y = y_h + y_p = [c_1 + \ln |\cos x|] \cos x + (c_2 + x) \sin x. \qquad\blacksquare$$

This is the end of Chap. 2. In Chap. 1 we were concerned with first-order differential equations, in Chap. 2 with second and higher order equations, and the next step is the extension to **systems of differential equations** (*simultaneous differential equations*). This will be done in Chap. 3 without the use of matrices and in Sec. 7.16 with the use of matrices.

Problems for Sec. 2.16

Find a general solution of the following equations.

1. $y'' - 4y' + 4y = e^{2x}/x$

2. $y'' + y = \csc x + x$

3. $y'' - 2y' + y = e^x \sin x$

4. $y'' + 9y = \sec 3x$

5. $y'' + 2y' + y = e^{-x} \cos x$

6. $y'' + 6y' + 9y = e^{-3x}/(x^2 + 1)$

7. $(D^2 - 2D + 1)y = x^{3/2}e^x$

8. $(D^2 + 2D + 1)y = e^{-x} \ln x$

9. $(D^2 + 4D + 4)y = e^{-2x}/x^2$

10. $(D^2 - 4D + 5)y = e^{2x}/\sin x$

11. $(D^2 - 4D + 4)y = 6x^{-4}e^{2x}$

12. $(D^2 + 2D + 2)y = e^{-x}/\cos^3 x$

13. $(D^2 - 2D + 1)y = e^x/x^3$

Nonhomogeneous Euler–Cauchy Equations. Find a general solution of the following equations. *Caution!* First divide the equation by x^2 to get the standard form (1).

14. $x^2y'' - 2y = 3x^2$

15. $x^2y'' + xy' - y = 4$

16. $x^2y'' + xy' - 0.25y = 1/x$

17. $x^2y'' - 2xy' + 2y = 1/x^2$

18. $x^2y'' - 4xy' + 6y = x^4 \sin x$

19. $x^2y'' - 2xy' + 2y = x^4$

20. $(x^2D^2 - 2xD + 2)y = x^3 \cos x$

21. $(x^2D^2 - xD)y = 2x^3e^x$

22. $(x^2D^2 + xD - 1)y = x^3e^x$

23. $(x^2D^2 - 4xD + 6)y = 1/x^4$

24. $(x^2D^2 - 2xD + 2)y = 6/x$

25. $(x^2D^2 + xD - 1)y = 16x^3$

Further Proof in Chapter 2

Proof of Theorem 1 (Uniqueness) in Sec. 2.8, page 107[16]

Assuming that the problem consisting of the differential equation

(1a) $$y'' + p(x)y' + q(x)y = 0$$

and the two initial conditions

(1b) $$y(x_0) = K_0, \qquad y'(x_0) = K_1$$

has two solutions $y_1(x)$ and $y_2(x)$ on the interval I in the theorem, we show that their difference

$$y(x) = y_1(x) - y_2(x)$$

is identically zero on I; then $y_1 \equiv y_2$ on I, which implies uniqueness.

Since the equation in (1) is homogeneous and linear, y is a solution of that equation on I, and since y_1 and y_2 satisfy the same initial conditions, y satisfies the conditions

(9) $$y(x_0) = 0, \qquad y'(x_0) = 0.$$

We consider the function

$$z(x) = y(x)^2 + y'(x)^2$$

and its derivative

$$z' = 2yy' + 2y'y''.$$

From the differential equation we have

$$y'' = -py' - qy.$$

By substituting this in the expression for z' we obtain

(10) $$z' = 2yy' - 2py'^2 - 2qyy'.$$

Now, since y and y' are real,

$$(y \pm y')^2 = y^2 \pm 2yy' + y'^2 \geq 0.$$

From this we immediately obtain the two inequalities

(11) (a) $2yy' \leq y^2 + y'^2 = z,$ (b) $-2yy' \leq y^2 + y'^2 = z.$

---

[16]This proof was suggested by my colleague, Prof. A. D. Ziebur. In this proof we use formula numbers that have not yet been used in Sec. 2.8.

From (11b) we have $2yy' \geqq -z$. Together, $|2yy'| \leqq z$. For the last term in (10) we now obtain

$$-2qyy' \leqq |-2qyy'| = |q||2yy'| \leqq |q|z.$$

Using this result as well as $-p \leqq |p|$ and applying (11a) to the term $2yy'$ in (10), we find

$$z' \leqq z + 2|p|y'^2 + |q|z.$$

Since $y'^2 \leqq y^2 + y'^2 = z$, we obtain from this

$$z' \leqq (1 + 2|p| + |q|)z$$

or, denoting the function in parentheses by h,

(12a) $z' \leqq hz$ for all x on I.

Similarly, from (10) and (11) it follows that

(12b)
$$-z' = -2yy' + 2py'^2 + 2qyy'$$
$$\leqq z + 2|p|z + |q|z = hz.$$

The inequalities (12a) and (12b) are equivalent to the inequalities

(13) $z' - hz \leqq 0,$ $z' + hz \geqq 0.$

Integrating factors for the two expressions on the left are

$$F_1 = e^{-\int h(x)\,dx} \qquad \text{and} \qquad F_2 = e^{\int h(x)\,dx}.$$

The integrals in the exponents exist because h is continuous. Since F_1 and F_2 are positive, we thus have from (13)

$$F_1(z' - hz) = (F_1 z)' \leqq 0 \qquad \text{and} \qquad F_2(z' + hz) = (F_2 z)' \geqq 0,$$

which means that $F_1 z$ is nonincreasing and $F_2 z$ is nondecreasing on I. Since $z(x_0) = 0$ by (9), we thus obtain when $x \leqq x_0$

$$F_1 z \geqq (F_1 z)_{x_0} = 0, \qquad F_2 z \leqq (F_2 z)_{x_0} = 0$$

and similarly, when $x \geqq x_0$,

$$F_1 z \leqq 0, \qquad F_2 z \geqq 0.$$

Dividing by F_1 and F_2 and noting that these functions are positive, we altogether have

$$z \leqq 0, \qquad z \geqq 0 \qquad \text{for all } x \text{ on } I.$$

This implies $z = y^2 + y'^2 \equiv 0$ on I. Hence $y \equiv 0$ or $y_1 \equiv y_2$ on I.   ∎

Review Problems for Chapter 2

1. What is the superposition principle? Does it hold for nonlinear equations? For nonhomogeneous linear equations? For homogeneous linear equations?

2. How many arbitrary constants does a general solution of a nonhomogeneous second-order linear equation involve? How many additional conditions does one need to determine them?

3. How would you practically determine whether two solutions of a differential equation are linearly independent? Why is this important and relevant in this chapter?

4. Does it make sense to talk about linear dependence of two functions at a single point? Explain.

5. If we know two solutions of a nonhomogeneous linear equations on the same interval, can we find from them a particular solution of the corresponding homogeneous equation? A general solution of this homogeneous equation?

6. What does an initial value problem for a third-order equation look like?

7. What is the difference between an initial value problem and a boundary value problem? Why did we make no such distinction in the case of a first-order equation?

8. What is a particular solution? Why are particular solutions generally more common as final answers to practical problems than general solutions?

9. Why did we always first determine a general solution, even when we needed just a particular solution?

10. Can a nonhomogeneous linear equation have the trivial solution 0? A homogeneous linear equation?

11. We considered two large classes of differential equations that can essentially be solved by "algebraic" manipulations. What were they?

12. In modeling, one generally prefers linear over nonlinear differential equations whenever one can hope to get a faithful picture of the reality from a linear equation. What is the reason for this?

13. Give some examples in modeling in which a linear differential equation would not be appropriate. (Don't forget first-order equations in this connection.)

14. For second-order constant-coefficient equations, we distinguished three cases. What are they? What is their significance in connection with mass–spring systems? In *RLC*-circuits?

15. What do we mean by "resonance"? Where and under what conditions does it occur?

Find a general solution of the following equations.

16. $y'' + 16y = e^x + x$

17. $(D^2 - 2D + 1)y = e^x/x^3$

18. $y^{IV} + 1.75y'' - 9y = e^{2x} - \sin \pi x$

19. $(D^2 + 4D + 4)y = 9 \cosh x$

20. $y'' + 4y = \sec 2x$

21. $(x^2D^2 - 2xD + 2)y = 10x^3 \cos x$

22. $y'' + 2y' + 2y = 3 \cos 2x$

23. $(4x^2D^2 - 8xD + 5)y = 0$

24. $y^{IV} - 81y = 12x^{-5} - 40.5x^{-1}$

25. $(x^2D^2 - 5xD + 9)y = 0$

26. $y'' + 2y' + 3.56y = \cos 1.56x$

27. $(D^2 + 4D + 3)y = 2 \cos x + \sin x$

28. $y'' - 4y' + 4y = (6x^2 + 4)e^x$

29. $(x^2D^2 - 2xD + 2)y = 48/x^2$

30. $y'' + y' + y = 2 \cos x + 26 \cos 2x$

Solve the following initial value problems.

31. $y'' + 2.89y = 34.68,$ $y(0) = 16,$ $y'(0) = -5.44$

32. $(xD^2 - D)y = x^2 e^x,$ $y(1) = 2,$ $y'(1) = e - 4 \approx -1.282$

33. $y'' + y' + 4y = -10 \sin 2x,$ $y(0) = 5,$ $y'(0) = 0.5$

34. $(D^2 + 2D + 1)y = 4e^{-x} \ln x,$ $y(1) = 0,$ $y'(1) = -e^{-1}$

35. $y^{IV} - 3y'' - 4y = 0,$ $y(0) = -1,$ $y'(0) = 0,$ $y''(0) = 6,$ $y'''(0) = 0$

36. $(xD^2 - D)y = (3 + x)x^2 e^x,$ $y(-1) = 2 + e^{-1} \approx 2.368,$
    $y'(-1) = -2 - e^{-1} \approx -2.368$

37. $y''' + 2y'' + y' = 2x + 4,$ $y(0) = 5,$ $y'(0) = -3,$ $y''(0) = 6$

38. $(D^2 + 2D + 5)y = 16 \sin x,$ $y(0) = -0.6,$ $y'(0) = 0.2$

39. $x^2 y'' + xy' - y = 26,$ $y(0.8) = 0,$ $y'(0.8) = -7.5$

40. $(D^2 - 4D + 4)y = 6 + e^{2x}/x,$ $y(1) = 0,$ $y'(1) = e^2 - 3 \approx 4.389$

41. $y'' + y' - 2y = 2x^2 - 2x - 14,$ $y(0) = 0,$ $y'(0) = 0$

42. $(x^2 D^2 + xD - 1)y = 4x^3 e^x,$ $y(0.5) = 14\sqrt{e} \approx 23.08,$
    $y'(0.5) = -30\sqrt{e} \approx -49.46$

43. $y'' + 2y' + 2y = 2e^{-x} \sec^3 x,$ $y(0) = 0,$ $y'(0) = 0$

44. $(D^3 + 3D^2 + 3D + 1)y = x^2,$ $y(2) = 4 + 6e^{-2} \approx 4.812,$
    $y'(2) = -2 - 3e^{-2} \approx -2.406,$ $y''(2) = 2$

45. $y^{IV} - 5y'' + 4y = 20 \cos x,$ $y(0) = 1,$ $y'(0) = 14,$ $y''(0) = -6,$
    $y'''(0) = 56$

46. Using the Wronskian, find out whether $y_1 = e^x + e^{-2x},$ $y_2 = e^{2x} - e^x,$ $y_3 = \cosh 2x$ are linearly dependent or independent on the x-axis.

47. **(Comparison of methods)** Whenever the method of undetermined coefficients (Sec. 2.12) is applicable, it should be used, since it is simpler than the method in Sec. 2.16. To illustrate this, solve $y'' + 4y' + 3y = 65 \cos 2x$ by both methods.

48. Find the steady-state current in the *RLC*-circuit in Fig. 60, assuming that $L = 1$ henry, $R = 2000$ ohms, $C = 4 \cdot 10^{-3}$ farad and $E(t) = 110 \sin 415t$ (66 cycles/sec).

49. Find a general solution of the homogeneous equation corresponding to the equation in Prob. 48.

50. Find the current in the *RLC*-circuit in Fig. 60 when $R = 20$ ohms, $L = 0.1$ henry, $C = 1.5625 \cdot 10^{-3}$ farad, $E(t) = 160t$ volts if $0 < t < 0.01$, $E(t) = 1.6$ volt if $t > 0.01$ sec, assuming that $I(0) = 0,$ $I'(0) = 0$.

51. Find the steady-state current in the *RLC*-circuit in Fig. 60 when $R = 50$ ohms, $L = 30$ henrys, $C = 0.025$ farad, $E(t) = 200 \sin 4t$ volts.

52. Find the motion of the mass–spring system in Fig. 61 with mass 0.125 kg, damping 0, spring constant 1.125 kg/sec^2 and driving force $\cos t - 4 \sin t$ nt, assuming zero initial displacement and velocity. For what frequency of the driving force would you get resonance?

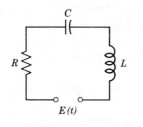

Fig. 60. *RLC*-circuit

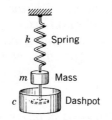

Fig. 61. Mass–spring system

53. Find the steady-state solution of the system in Fig. 61 when $m = 1$, $c = 2$, $k = 6$ and the driving force is $\sin 2t + 2 \cos 2t$.

54. In Fig. 61, let $m = 1$, $c = 4$, $k = 24$ and $r(t) = 10 \cos \omega t$. Determine ω such that you get the steady-state vibration of maximum possible amplitude. Determine this amplitude. Then find the general solution with this ω and check whether the results are in agreement.

55. In Prob. 53, find the solution corresponding to initial displacement 1 and initial velocity 0.

Summary of Chapter 2
Linear Differential Equations

A *first-order* homogeneous linear equation is $y' + p(x)y = 0$ (Sec. 1.7). A **second-order homogeneous linear equation** is an equation that can be written

$$(1) \qquad\qquad y'' + p(x)y' + q(x)y = 0.$$

It has the very important property that a linear combination of solutions is again a solution (**Superposition Principle,** Sec. 2.1). Two linearly independent solutions y_1, y_2 of (1) form a **basis** of solutions and $y = c_1 y_1 + c_2 y_2$ with arbitrary constants c_1, c_2 a **general solution.** From it we obtain a **particular solution** if we specify numerical values of c_1 and c_2, for instance, by imposing two **initial conditions** (Sec. 2.3)

$$(2) \qquad y(x_0) = K_0, \qquad y'(x_0) = K_1 \qquad (x_0,\ K_0,\ K_1 \text{ given numbers}).$$

Together, (1) and (2) constitute an **initial value problem** for (1). If p and q are continuous on some interval I, then (1) has a general solution on I and (1), (2) has a unique solution (which is a particular solution). Cf. Secs. 2.3, 2.8.

Sections 2.1–2.10 concern exclusively *homogeneous* linear equations, Secs. 2.1–2.8 for order $n = 2$ [the above equation (1)] and Secs. 2.9, 2.10 for any n. For arbitrary n, a **general solution** is a linear combination $y = c_1 y_1 + \cdots + c_n y_n$ of n linearly independent solutions $y_1, \cdots, y_n$, called a **basis** of solutions, with arbitrary constants $c_1, \cdots, c_n$ as coefficients. Linear independence may now be tested by means of the **Wronski determinant** (Sec. 2.9). Note that for $n = 2$, linear independence simply means that y_1 and y_2 are not proportional.

The study of **nonhomogeneous linear equations**

$$(3) \qquad\qquad y'' + p(x)y' + q(x)y = r(x) \qquad\qquad r(x) \not\equiv 0$$

begins in Sec. 2.11. A **general solution** of (3) is

$$y = y_h + y_p,$$

y_h a general solution of the corresponding homogeneous equation (1) and y_p a particular solution of (3). Hence the additional practical problem in solving (3) is the determination of such a y_p. For this we give a general method (Sec. 2.16), a simpler special method of undetermined coefficients (Sec. 2.12), valid for constant p, q and special r (powers of x, sine, cosine, etc.), and a complex method for sinusoidal driving forces (Sec. 2.15).

If $p(x)$ and $q(x)$ in (1) or (3) are *variable*, the solutions will in general be higher functions. We study the most important ones of them in Chap. 4. If $p(x)$ and $q(x)$ are constant, we write $p(x) = a$, $q(x) = b$ and obtain from (1) an equation

(4) $$y'' + ay' + by = 0.$$

This equation can be solved by substituting $y = e^{\lambda x}$. Then λ is a root of the **characteristic equation**

$$\lambda^2 + a\lambda + b = 0.$$

Hence there are three cases (Sec. 2.4):

Case	Type of Roots	General Solution
I	Distinct real λ_1, λ_2	$y = c_1 e^{\lambda_1 x} + c_2 e^{\lambda_2 x}$
II	Complex $-\frac{1}{2}a \pm i\omega$	$y = e^{-ax/2}(A \cos \omega x + B \sin \omega x)$
III	Double $-\frac{1}{2}a$	$y = (c_1 + c_2 x)e^{-ax/2}$

This simple algebraic method can be readily extended to homogeneous linear equations of arbitrary order n (Sec. 2.10).

Equation (4) and the nonhomogeneous equation

(5) $$y'' + ay' + by = r(x) \qquad (a, b \text{ constant})$$

have basic applications in mechanics (Secs. 2.6, 2.13) and electrical engineering (Sec. 2.14).

Another large class of equations solvable by an ''algebraic'' method consists of the **Euler–Cauchy equations** (Sec. 2.7)

$$x^2 y'' + axy' + by = 0.$$

If we substitute $y = x^m$, we can determine m from the auxiliary equation

$$m^2 + (a - 1)m + b = 0.$$

Chapter 3

Systems of
Differential Equations,
Phase Plane, Stability

Section 3.1 is devoted to systems of differential equations. This is an elementary introduction, without the use of matrices. (An appetizer for students knowing 2×2 matrices is included in Example 3. The matrix approach to systems will be presented in Sec. 7.16).

In Sec. 3.2 we consider differential equations in the phase plane.

Section 3.3 is devoted to stability in connection with differential equations, using the phase plane. Stability is a concept of increasing importance in modern engineering mathematics, for instance, in connection with feedback and controls. It is suggested by physics, where it means that, roughly speaking, small changes of a physical system at some instant cause only small changes in the behavior of the system at all later times.

Prerequisite for this chapter: Chaps. 1 and 2.
References: Appendix 1, Part A.
Answers to problems: Appendix 2.

3.1 Systems of Differential Equations

Systems of differential equations have important applications. For instance, they arise quite frequently as models of mechanical or electrical systems which are combinations of the simple systems discussed in Secs. 1.8, 2.6, 2.13 and 2.14.

The present section contains an elementary[1] approach to systems of differential equations, and we shall consider the method of **solution by elimination.** In this method, unknown functions and their derivatives are successively eliminated until one arrives at a single higher order differential equation containing only one unknown function and its derivatives. This equation is solved, and then the other unknown functions are found in turn. Let us explain the details in terms of a typical application.

---

[1]That is, without the use of vectors and matrices. The vector approach is included in the chapter on matrices (in Sec. 7.16). Of course, readers familiar with vectors and matrices may immediately study Sec. 7.16.

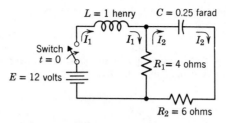

Fig. 62. Electrical network in Example 1

EXAMPLE 1. Model of an electrical network

Find the currents $I_1(t)$ and $I_2(t)$ in the network shown in Fig. 62, assuming that all charges and currents are zero when the switch is closed at $t = 0$.

Solution. 1st Step. Setting up the mathematical model. The mathematical model of this network is obtained from Kirchhoff's voltage law, as in Secs. 1.8 and 2.14 (where we considered single circuits). The left-hand loop yields

$$I_1' + 4(I_1 - I_2) = 12,$$

where I_1 is the current in the left-hand loop and I_2 the current in the right-hand loop, and $4(I_1 - I_2)$ is the voltage drop across the resistor because I_1 and I_2 flow through the resistor in opposite directions. Similarly, for the right-hand loop we obtain

$$6I_2 + 4(I_2 - I_1) + 4 \int I_2 \, dt = 0.$$

Rewriting the first equation and differentiating the second, we see that the currents I_1 and I_2 in the network are governed by the system

(1)
    (a) $I_1' + 4I_1 - 4I_2 = 12$
    (b) $-4I_1' + 10I_2' + 4I_2 = 0.$

This is the mathematical model of the electrical network in Fig. 62. It is a system of two first-order ordinary linear differential equations.

2nd Step. Solving the system of differential equations. Using (1a), we express I_2 and its derivative in terms of I_1 and its derivatives. The result we substitute into (1b), which then only contains I_1 and its derivatives. We solve this equation. Finally we go back to (1a) and use it to express I_2 in terms of the solution just obtained. The details are as follows.

From (1a) we have

(2) $$I_2 = \tfrac{1}{4}I_1' + I_1 - 3.$$

By differentiation,

$$I_2' = \tfrac{1}{4}I_1'' + I_1'.$$

Substituting this and (2) into (1b), we obtain

$$-4I_1' + 10(\tfrac{1}{4}I_1'' + I_1') + 4(\tfrac{1}{4}I_1' + I_1 - 3) = 0.$$

Simplification yields

$$I_1'' + \tfrac{14}{5}I_1' + \tfrac{8}{5}I_1 = \tfrac{24}{5}.$$

A general solution of this equation is

(3a) $$I_1 = c_1 e^{-2t} + c_2 e^{-0.8t} + 3.$$

Its derivative is given by

$$I_1' = -2c_1 e^{-2t} - 0.8 c_2 e^{-0.8t}.$$

We can now calculate I_2 from this and (2), finding

(3b)
$$I_2 = \tfrac{1}{2}c_1 e^{-2t} + \tfrac{4}{5}c_2 e^{-0.8t}.$$

3rd Step. Determination of a particular solution from the initial conditions. The initial conditions are $I_1(0) = 0$ and $I_2(0) = 0$. Hence by (3),

$$I_1(0) = c_1 + c_2 + 3 = 0$$

$$I_2(0) = \tfrac{1}{2}c_1 + \tfrac{4}{5}c_2 = 0.$$

This gives $c_1 = -8$ and $c_2 = 5$. Hence the solution of our physical problem is

$$I_1 = -8e^{-2t} + 5e^{-0.8t} + 3$$

$$I_2 = -4e^{-2t} + 4e^{-0.8t}.$$

We see that I_1 approaches the value of 3 amperes, whereas I_2 approaches 0 as $t \to \infty$. This had to be expected. Why?   ∎

This example is typical. It illustrates the practical importance of systems of differential equations as well as a standard method for solving them, namely, solution by elimination. The key idea of this method is the derivation of a single higher order differential equation that contains only one of the unknown functions (and its derivatives), and the solution of this equation. This method applies to systems of the form

(4)
$$\text{(a)} \qquad x' = a_1 x + b_1 y + f_1(t)$$
$$\text{(b)} \qquad y' = a_2 x + b_2 y + f_2(t)$$

where f_1 and f_2 are given functions and a_1, b_1, a_2, b_2 are constants. x and y are unknown functions. By a **solution** of (4) on some open interval I of the t-axis we mean a pair of functions x, y of t which are defined and differentiable on I and when substituted into (4) reduce these two equations (4) to an identity.[2]

If $b_1 = a_2 = 0$, the system breaks up into two separate first-order linear differential equations which can be solved by the method of Sec. 1.7. If one of these two constants, say b_1, is not zero, we may proceed as follows. From (4a) we first have

(5)
$$b_1 y = x' - a_1 x - f_1.$$

We now differentiate (4a). In the resulting equation we substitute y' as given by (4b) and then $b_1 y$ as given by (5). This yields

$$x'' = a_1 x' + b_1 y' + f_1'$$
$$\quad = a_1 x' + b_1(a_2 x + b_2 y + f_2) + f_1'$$
$$\quad = a_1 x' + b_1 a_2 x + b_2(x' - a_1 x - f_1) + b_1 f_2 + f_1'.$$

---

[2]Note that in our present method we also have to assume the existence of second derivatives of x or y.

Collecting terms, we have

$$x'' - (a_1 + b_2)x' + (a_1 b_2 - b_1 a_2)x = r(t)$$

where

$$r(t) = b_1 f_2(t) - b_2 f_1(t) + f_1'(t).$$

By solving this equation we get x. Then y is obtained from (5).

EXAMPLE 2. Solution of a nonhomogeneous system by elimination
Solve

(6)   (a) $x' = -2x + y$

                                 (b) $y' = -4x + 3y + 10 \cos t$

Solution. From (6a),

(7) $y = x' + 2x.$

Differentiating (6a) and substituting y' from (6b) and then y from (7), we obtain

$$\begin{aligned}
x'' &= -2x' + y' \\
&= -2x' - 4x + 3y + 10 \cos t \\
&= -2x' - 4x + 3(x' + 2x) + 10 \cos t.
\end{aligned}$$

By collecting terms we have

$$x'' - x' - 2x = 10 \cos t.$$

The characteristic equation $\lambda^2 - \lambda - 2 = 0$ has the roots -1 and 2. The method in Sec. 2.12 yields as a particular solution $-3 \cos t - \sin t$, as the student should verify. This gives the general solution

$$x(t) = c_1 e^{-t} + c_2 e^{2t} - 3 \cos t - \sin t.$$

By (7),

$$y(t) = c_1 e^{-t} + 4c_2 e^{2t} - 7 \cos t + \sin t.$$

Check the result by substitution.

EXAMPLE 3. Solution by elimination. Solution by using 2 × 2 matrices
Solve the system of differential equations

(8)   (a) $x' = 4x - 2y$

                                 (b) $y' = x + y$

by elimination. Students having some knowledge of 2 × 2 matrices should also study the solution by matrices and compare the two methods. (The matrix method will be considered systematically in Sec. 7.16.)

Solution by elimination. From (8a) we have

(9) $y = -\tfrac{1}{2}x' + 2x.$

We differentiate (8a). In the resulting equation we first substitute y' as given by (8b) and y as given by (9). This yields

$$\begin{aligned}
x'' &= 4x' - 2y' \\
&= 4x' - 2(x + y) \\
&= 4x' - 2x - 2(-\tfrac{1}{2}x' + 2x).
\end{aligned}$$

Ordering terms and simplifying, we have

$$x'' - 5x' + 6x = 0.$$

A general solution is

(10a) $$x = c_1 e^{3t} + c_2 e^{2t}.$$

By differentiation,

$$x' = 3c_1 e^{3t} + 2c_2 e^{2t}.$$

Hence from (9),

(10b) $$y = \tfrac{1}{2} c_1 e^{3t} + c_2 e^{2t}.$$

Warning. Checking by substitution is particularly important in the present method, because one may introduce constants of integration, and substitution may show that sometimes these constants are not completely arbitrary. To see the point, substitute (10a) into (8b) to get

$$y' - y = c_1 e^{3t} + c_2 e^{2t}.$$

This linear equation can be solved as explained in Sec. 1.7. Its general solution is

$$y = \tfrac{1}{2} c_1 e^{3t} + c_2 e^{2t} + c e^t.$$

This is a solution of (8b) for every c, but substitution in (8a) shows that we must have $c = 0$. Hence checking is important!

Solution by the matrix method. 1st Step. Vector representation of the system. We write $x = y_1$ and $y = y_2$. Then the given system (8) can be written

(11) $$\mathbf{y}' = \mathbf{Ay}, \quad \text{where} \quad \mathbf{A} = \begin{bmatrix} 4 & -2 \\ 1 & 1 \end{bmatrix} \quad \text{and} \quad \mathbf{y} = \begin{bmatrix} y_1 \\ y_2 \end{bmatrix}.$$

2nd Step. Solving the vector equation. If this were a single differential equation (with A being a number), the solution would be an exponential function. So let us try

$$\mathbf{y} = \mathbf{z} e^{\lambda t}.$$

Substitution of $\mathbf{y}$ and $\mathbf{y}' = \lambda \mathbf{z} e^{\lambda t}$ into (11) yields

$$\lambda \mathbf{z} e^{\lambda t} = \mathbf{A} \mathbf{z} e^{\lambda t}.$$

Canceling out $e^{\lambda t}$, we have

(12) $$\mathbf{Az} = \lambda \mathbf{z} \quad \text{or} \quad (\mathbf{A} - \lambda \mathbf{I})\mathbf{z} = \mathbf{0}$$

(**I** the 2×2 unit matrix), written out

(13) $$\begin{bmatrix} 4 - \lambda & -2 \\ 1 & 1 - \lambda \end{bmatrix} \begin{bmatrix} z_1 \\ z_2 \end{bmatrix} = \mathbf{0} \quad \text{or} \quad \begin{aligned} (4 - \lambda)z_1 - 2z_2 &= 0 \\ z_1 + (1 - \lambda)z_2 &= 0 \end{aligned}$$

We determine λ. Equation (12) has the *trivial solution* $\mathbf{z} = \mathbf{0}$ for *every* λ. This is of no interest. (Why?) What we want is a solution $\mathbf{z} \neq \mathbf{0}$. Since the system in (13) is homogeneous, for a solution $\mathbf{z} \neq \mathbf{0}$ to exist, the coefficient determinant of that system must be zero:

(14) $$\begin{vmatrix} 4 - \lambda & -2 \\ 1 & 1 - \lambda \end{vmatrix} = (4 - \lambda)(1 - \lambda) + 2 = \lambda^2 - 5\lambda + 6 = 0.$$

The roots are $\lambda_1 = 3$ and $\lambda_2 = 2$.

We determine a solution $\mathbf{z}$ of (13) with $\lambda = \lambda_1 = 3$. For $\lambda = 3$, the equations in (13) are

$$z_1 - 2z_2 = 0, \qquad z_1 - 2z_2 = 0.$$

A solution is $z_1 = 2$, $z_2 = 1$. Similarly, for $\lambda = \lambda_2 = 2$ that system (13) is

$$2z_1 - 2z_2 = 0, \qquad z_1 - z_2 = 0.$$

A solution is $z_1 = 1$, $z_2 = 1$. Hence solutions of (11) are

$$\begin{bmatrix} 2 \\ 1 \end{bmatrix} e^{3t} \qquad \text{and} \qquad \begin{bmatrix} 1 \\ 1 \end{bmatrix} e^{2t}.$$

Any linear combination of these is also a solution, since (11) is homogeneous; thus

$$\mathbf{y} = a_1 \begin{bmatrix} 2 \\ 1 \end{bmatrix} e^{3t} + a_2 \begin{bmatrix} 1 \\ 1 \end{bmatrix} e^{2t} \qquad \text{or} \qquad \begin{aligned} y_1 &= 2a_1 e^{3t} + a_2 e^{2t} \\ y_2 &= a_1 e^{3t} + a_2 e^{2t} \end{aligned}$$

This agrees with the solution obtained by the elimination method, where $c_1 = 2a_1$ and $c_2 = a_2$.

The present method involves the following concepts. A value of λ for which (12) has a solution $\mathbf{z} \neq \mathbf{0}$ is called an *eigenvalue* of the matrix $\mathbf{A}$, and this $\mathbf{z}$ is called an *eigenvector* of $\mathbf{A}$ corresponding to this eigenvalue λ. Equation (14) is called the *characteristic equation* of $\mathbf{A}$. Hence the eigenvalues of $\mathbf{A}$ are the roots of the characteristic equation of $\mathbf{A}$. More on eigenvalue problems in Secs. 7.12–7.16.   ∎

Elimination Method for Higher Order Equations

The elimination method can be extended to systems of equations of higher order. Let us explain this idea in terms of an important application in mechanics.

EXAMPLE 4. Model of a mechanical system of two masses on two springs

Figure 63 shows a mechanical system consisting of two masses on two springs. Set up a mathematical model and solve it; that is, find the displacements $y_1(t)$ and $y_2(t)$ of the masses from their positions of static equilibrium ($y_1 = 0$ and $y_2 = 0$) when the whole system is at rest, under the assumptions made in Sec. 2.6 (vertical motion, no damping, masses of springs neglected).

Solution. 1st Step. Setting up the mathematical model. As in Sec. 2.6, the differential equations governing our mechanical system are obtained from Newton's second law. The upper mass is connected to *two* springs, and Hooke's law gives an upward spring force $-k_1 y_1 = -3y_1$ and a downward spring force $k_2(y_2 - y_1)$. Note that $y_2 - y_1$ is the difference of the displacements of the two masses from equilibrium, and this gives the net change in length for the lower spring. Similarly, the lower mass is connected to one spring only, and Hooke's law gives an upward

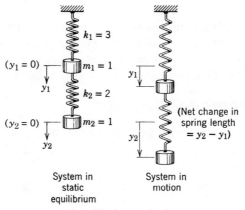

Fig. 63. Example 4

spring force $-k_2(y_2 - y_1) = -2(y_2 - y_1)$. Since the masses are $m_1 = m_2 = 1$, for simplicity, Newton's second law thus gives

$$y_1'' = -3y_1 + 2(y_2 - y_1)$$
$$y_2'' = -2(y_2 - y_1).$$

This can be written

(15) \qquad (a) \quad $y_1'' = -5y_1 + 2y_2$

\qquad\qquad (b) \quad $y_2'' = \ \ 2y_1 - 2y_2.$

This system of two second-order linear equations is the mathematical model of our physical mass–spring system.

2nd Step. Solving the system of differential equations. From (15a) we have

(16) $$y_2 = \tfrac{1}{2}y_1'' + \tfrac{5}{2}y_1.$$

In Examples 1–3 we differentiated once. Here we differentiate (15a) twice. In the resulting equation we substitute y_2'' as given by (15b) and y_2 as given by (16). This yields

$$\begin{aligned}
y_1^{iv} &= -5y_1'' + 2y_2'' \\
&= -5y_1'' + 2(2y_1 - 2y_2) \\
&= -5y_1'' + 4y_1 - 4(\tfrac{1}{2}y_1'' + \tfrac{5}{2}y_1).
\end{aligned}$$

Ordering terms, we have

(17) $$y_1^{iv} + 7y_1'' + 6y_1 = 0.$$

The characteristic equation $\lambda^4 + 7\lambda^2 + 6 = 0$ is quadratic in $k = \lambda^2$. The roots are $k_1 = -1$ and $k_2 = -6$. Hence λ has the four values $i, -i, \sqrt{6}\, i, -\sqrt{6}\, i$ which are complex conjugate in pairs. As in Sec. 2.4 we thus obtain the general solution of (17),

(18a) \qquad $y_1 = a_1 \cos t + b_1 \sin t + a_2 \cos \sqrt{6}t + b_2 \sin \sqrt{6}t.$

Differentiating twice, we have

$$y_1'' = -a_1 \cos t - b_1 \sin t - 6a_2 \cos \sqrt{6}t - 6b_2 \sin \sqrt{6}t.$$

Using this and (18a), we obtain from (16)

(18b) \qquad $y_2 = 2a_1 \cos t + 2b_1 \sin t - \tfrac{1}{2}a_2 \cos \sqrt{6}t - \tfrac{1}{2}b_2 \sin \sqrt{6}t.$

This solution (18) of (15) represents harmonic oscillations of the two masses of our mechanical system. ∎

Further applications are considered in the problem set.

In the next section we consider phase-plane methods for discussing solutions of systems of differential equations. The **phase plane** is the yv-plane (v the velocity), and our discussion includes the famous van der Pol equation.

Problems for Sec. 3.1

Solve the following systems of differential equations (where $D = d/dt$).

1. $x' = y$
    $y' = x$

2. $x' = 2x + 2y$
    $y' = 5x - y$

3. $x' = 2x + 3y$
    $y' = \tfrac{1}{3}x + 2y$

4. $x' = x + y$
    $y' = 3x - y$

5. $x' = x + y$
    $y' = 4x + y$

6. $x' = 2x + 5y$
    $2y' = -x - 3y$

7. $Dx + Dy = \cos t$
    $Dx - Dy = \sin t$

8. $x' = 4x - 2y$
    $y' = 3x - y - 2e^{3t}$

9. $(D + 4)x + 6y = 0$
    $(D - 1)y - x = 0$

10. $Dx = x + y$
    $Dy = x - y + e^t$

11. $D(x + y) = \cosh 4t$
    $D(x - y) = \sinh 4t$

12. $(D - 2)x + 2Dy = 2 - 4e^{2t}$
    $(2D - 3)x + (3D - 1)y = 0$

13. $Dx + y = \cos t - \sin t$
    $Dy + x = \cos t + \sin t$

14. $(D^2 + 1)x = t$
    $D(x + y) = \sin t$

15. $D(x + y) = x + t$
    $D^2 y = Dx$

16. $D^2 x = y + 1$
    $D^2 y = x + t$

Find the solutions of the following systems satisfying the given conditions.

17. $x' = y,\quad y' = x,\quad x(0) = 0,\quad y(0) = 1$

18. $x' = -y,\quad y' = x,\quad x(\tfrac{1}{2}\pi) = -4.5,\quad y(\tfrac{1}{2}\pi) = 0.8$

19. $x' + y' = 4 \cos 2t,\quad x' - y' = 4 \sin 2t,\quad x(0) = 1.2,\quad y(0) = 0$

20. $3x' = 6x + y,\quad y' = 3x + 2y,\quad x(0) = -1,\quad y(0) = 9$

21. $Dx = x + 2y,\quad Dy = -8x + 11y,\quad x(0) = 1,\quad y(0) = 1$

22. $Dx = x + y,\quad Dy = x - y,\quad x(\sqrt{2}) = e^2 \approx 7.39,$
    $y(\sqrt{2}) = (\sqrt{2} - 1)e^2 \approx 3.06$

23. $Dx = 3x + 4y,\quad Dy = 4x - 3y,\quad x(0) = 1,\quad y(0) = 3$

24. $Dx = 5x + 4y - 5t^2 + 6t + 25,\quad Dy = x + 2y - t^2 + 2t + 4,$
    $x(0) = 0,\quad y(0) = 0$

25. $Dx = 2x + 3y + 2e^{2t},\quad Dy = x + 4y + 3e^{2t},\quad x(0) = -2/3,$
    $y(0) = 1/3$

26. Set up a model for the mechanical system in Fig. 64. Solve it under the physical assumptions made in Example 4. In particular, find, graph and compare the solutions corresponding to zero initial velocities and initial displacements

  (i) $y_1(0) = 1,\ y_2(0) = 1$ (ii) $y_1(0) = 1,\ y_2(0) = -1$.

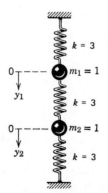

Fig. 64. Problem 26 (System in static equilibrium)

Set up a model (system of differential equations) of the following electrical networks and solve it. Determine also the steady-state currents.

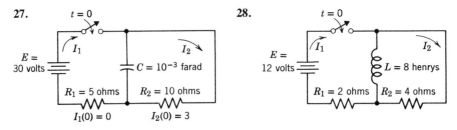

27.   **28.**

$I_1(0) = 0$ $I_2(0) = 3$

29. Solve the model for the network in Fig. 62, page 155, assuming that $E = 100$ volts, the other data being as before. Compare with Example 1 and comment.

30. The solution (18) in Example 4 is an interesting superposition of two harmonic oscillations. Show that two special cases are

$$(i) \quad y_1 = \sin t, \quad y_2 = 2 \sin t \qquad (ii) \quad y_1 = \sin \sqrt{6}t, \quad y_2 = -\tfrac{1}{2} \sin \sqrt{6}t.$$

Show that in (i) at each instant the masses are moving both upward or both downward, so that, at any given time, the springs are both compressed or both extended. What is going on in (ii)? Is it plausible that (ii) has a higher frequency than (i) does?

3.2 Phase Plane

In this section we show that solutions of certain general second-order differential equations that are reducible to first order can be investigated in terms of curves in the plane (the "phase plane," see below). Indeed, engineers make frequent use of this powerful technique. We first state the equations to be considered and reduce them.

A physical system is said to be **autonomous** if its differential equation does not contain the independent variable (time t, say) explicitly. Hence if this differential equation is of second order, it is of the form

$$(1) \qquad\qquad\qquad F(y, y', y'') = 0.$$

Here $y' = dy/dt = v$ is the velocity. By the chain rule,

$$(2) \qquad\qquad\qquad y'' = v' = \frac{dv}{dt} = \frac{dv}{dy}\frac{dy}{dt} = \frac{dv}{dy}v.$$

We thus obtain a first-order differential equation for v as a function of the variable y, which now becomes the *independent* variable. (Cf. also the problem set of Sec. 2.1.) Solutions of this new differential equation represent curves in the yv-plane. The yv-plane is called the **phase plane.** This term is suggested by mechanics, where "phase plane" means the yp-plane, $p = mv$ being the momentum and m the mass of a moving particle.

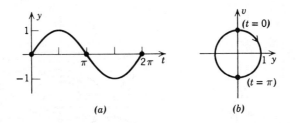

Fig. 65. Graphs of a harmonic oscillation
(a) in the *ty*-plane, (b) in the phase plane

EXAMPLE 1. Model of free undamped harmonic oscillations

The autonomous vibrating system governed by

$$y'' + y = 0, \qquad y(0) = 0, \qquad y'(0) = 1$$

(cf. Sec. 2.6) has the solution (Fig. 65a)

$$y = \sin t.$$

Using (2), we obtain

$$\frac{dv}{dy} v + y = 0, \qquad v(0) = 1.$$

Hence

$$v\, dv + y\, dy = 0, \qquad \text{and} \qquad v^2 + y^2 = 1.$$

This shows that in the phase plane a harmonic oscillation appears as a circle (Fig. 65b). Note that to each t there corresponds a point on the circle with coordinates $y(t)$, $v(t)$. If we wish, we can label the points as shown in Fig. 65b for $t = 0$ and $t = \pi$ and indicate the sense of increasing t by an arrow.   ∎

The new equation obtained by means of (2) is of the first order. Sometimes, one will be able to solve it and then investigate solution curves in the phase plane, as in Example 2 (below). But even if this is not possible, one can still discuss the general behavior of solutions $v(y)$ in a useful fashion. Compared with numerical work, the method has the advantage that one can study many solutions at once, not just a single solution, and in this way one can discover general patterns of behavior.

EXAMPLE 2. Free undamped pendulum

Figure 66 on the next page shows a pendulum consisting of a body of mass m (the bob) and a rod of length L. Study the motion of the pendulum by using the phase plane, assuming that the mass of the rod and air resistance are negligible.

Solution. 1st Step. Setting up the mathematical model. Let θ denote the angular displacement, measured counterclockwise from the equilibrium position. The weight of the bob is mg (g the acceleration of gravity). It causes a restoring force $mg \sin \theta$ tangent to the curve of motion (circular arc) of the bob. By Newton's second law, at each instant this force is balanced by the force of acceleration $mL\theta''$, where $L\theta''$ is the acceleration; hence the resultant of these two forces is zero, and we obtain as the mathematical model

(3) $$mL\theta'' + mg \sin \theta = 0.$$

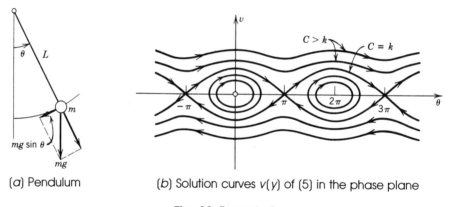

(a) Pendulum   (b) Solution curves $v(y)$ of (5) in the phase plane

Fig. 66. Example 2

2nd Step. Solutions $v(y)$ in the phase plane. Dividing (3) by mL, we have

(4)
$$\theta'' + k \sin \theta = 0 \qquad \left(k = \frac{g}{L} \right).$$

When θ is very small, we could approximate $\sin \theta$ rather accurately by θ and obtain as an *approximate* solution $A \cos \sqrt{k}t + B \sin \sqrt{k}t$, but the *exact* solution for any θ is not an elementary function.

We set $v = \theta'$ (the angular velocity). Then $\theta'' = v' = (dv/d\theta)v$, as in (2) (except for notation), so that from (4) we obtain

$$\frac{dv}{d\theta} v = -k \sin \theta.$$

We separate variables, $v \, dv = -k \sin \theta \, d\theta$, and integrate:

(5)
$$\tfrac{1}{2}v^2 = k \cos \theta + C \qquad \text{(C constant).}$$

3rd Step. Discussion of result, types of motion. The three terms in (5) are proportional to energies. To see this, we multiply (5) by mL^2, finding

$$\tfrac{1}{2}m(Lv)^2 - mL^2 k \cos \theta = mL^2 C.$$

Since v is the angular velocity, Lv is the velocity; hence the first term is the kinetic energy. The second term (including the minus sign) is the potential energy of the pendulum, and $mL^2 C$ is its total energy, which is constant, as expected from the law of conservation of energy, since there is no damping. The type of motion depends on the total energy, hence on C, as follows.

Figure 66b shows a sketch of portions of the curves $v(y)$ for various values of C; these graphs continue periodically with period 2π to the left and to the right. (Why?) We see that some of the curves are ellipse-like and closed, others are wavy, and there are two curves (passing through $\cdots, -\pi, \pi, 3\pi, \cdots$) that separate those two types of curves. From (5) we see that the smallest possible C is $C = -k$; then $v = 0$, and $\cos \theta = 1$, so that the pendulum is at rest. The pendulum will change its direction of motion if there are points at which $v = 0$. Then $k \cos \theta + C = 0$ by (5). If $\theta = \pi$, then $\cos \theta = -1$ and $C = k$. Hence if $-k < C < k$, then the pendulum reverses its direction for a $|\theta| < \pi$, and for these values of C with $|C| < k$ the pendulum oscillates; this corresponds to the closed curves in the figure. However, if $C > k$, then $v = 0$ is impossible and the pendulum makes a whirly motion which appears as a wavy curve in the phase plane. Finally, the value $C = k$ corresponds to the two "separating curves" in Fig. 66b.   ∎

Our next example concerns a phase plane discussion of a more complicated nonlinear equation for which $v(y)$ cannot be obtained exactly. This is the famous van der Pol equation.

EXAMPLE 3. Self-sustained oscillations, van der Pol equation

There are physical systems such that for small oscillations, energy is fed into the system, whereas for large oscillations, energy is taken from the system. In other words, large oscillations will be damped, whereas for small oscillations there is "negative damping" (feeding of energy into the system). For physical reasons we expect such a system to approach a periodic behavior, which will thus appear as a closed curve in the phase plane, called a **limit cycle**. A differential equation describing such vibrations is the famous **van der Pol equation**[3]

(6)
$$y'' - \mu(1 - y^2)y' + y = 0 \qquad (\mu > 0).$$

It occurs in the study of electrical circuits containing vacuum tubes. For $\mu = 0$ this is $y'' + y = 0$ and we obtain harmonic oscillations. Let $\mu > 0$. The damping term has the coefficient $-\mu(1 - y^2)$. This is negative for small oscillations, namely, $y^2 < 1$, so that we have "negative damping," is zero for $y^2 = 1$ (no damping) and is positive if $y^2 > 1$ (positive damping, loss of energy). If μ is small, we expect a limit cycle that is almost a circle because then our equation differs but little from $y'' + y = 0$. If μ is large, the limit cycle will probably look different.

Using $y'' = v \, dv/dy$, we have from (6)

$$\frac{dv}{dy} v - \mu(1 - y^2)v + y = 0.$$

The isoclines are $dv/dy = k = const$, that is,

(7)
$$\frac{dv}{dy} = \mu(1 - y^2) - \frac{y}{v} = k.$$

This yields

$$v = \frac{y}{\mu(1 - y^2) - k}.$$

These curves are a bit complicated. Figure 67 shows some of them for a small $\mu = 0.1$ as well

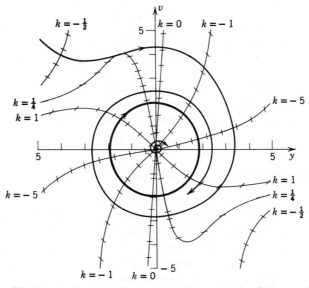

Fig. 67. Lineal element diagram for the van der Pol equation with $\mu = 0.1$ in the phase plane, showing also the limit cycle and two solution curves

[3]BALTHASAR VAN DER POL (1889—1959), Dutch physicist.

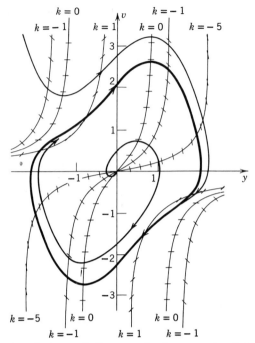

Fig. 68. Lineal element diagram for the van der Pol equation with $\mu = 1$ in the phase plane, showing also the limit cycle and two solution curves approaching it

as the limit cycle (almost a circle) and two solution curves approaching the limit cycle, one from the outside and one from the inside. The latter is a narrow spiral and only the initial part of it is shown in the figure. For larger μ the situation changes and the limit cycle no longer resembles a circle. Figure 68 illustrates this for $\mu = 1$. Note that the approach of solution curves $v(y)$ to the limit cycle is much more rapid than for $\mu = 0.1$. ∎

In the next section we extend our present method to more general systems of differential equations and include a systematic discussion of **"critical points"** of such systems and their stability.

Problems for Sec. 3.2

1. In Example 1, a general solution is $y = A \cos t + B \sin t$. What is the radius of the corresponding circle in the phase plane?

2. **(Harmonic oscillations)** Solve $y'' + \omega_0^2 y = 0$ and indicate the type of corresponding curves in the phase plane.

3. **(Damped oscillations)** Consider the damped oscillation $y(t) = e^{-t} \sin t$. Plot some points of the corresponding curve in the phase plane to obtain the impression that this must be a spiral.

4. **(Free damped motions)** From Sec. 2.6 we know that there are three types of free motions of a damped system. Using the figures of solution curves in Sec. 2.6, give a rough qualitative description and sketch of the corresponding curves in the phase plane. (Assume zero initial velocity, for simplicity.)

5. **(Forced oscillations)** What curves in the phase plane correspond to steady-state forced oscillations of a damped system (Sec. 2.13) under a sinusoidal driving force?

6. **(Falling body)** Graph the solution of $y'' = 32$ [ft/sec^2], $y(0) = 0$, $y'(0) = 0$ in the phase plane.

Solve the following equations and indicate the type of corresponding curves in the phase plane.

7. $4y'' + y = 0$ 8. $y'' + 16y = 0$

9. $y'' - 9y = 0$ 10. $y'' + 4y = 12$

11. $y' = 1 + y^2$ 12. $y'' - k^2y = 0$

Reduce to first order, solve and graph some of the solution curves in the phase plane.

13. $y'' = 4y'$ 14. $y'' + 5y' = 0$

15. $yy'' + y'^2 = 0$ 16. $y'' + y'^2 = 0$

17. $y'' + y = 0$ 18. $y'' = 2yy'$

19. $2yy'' + y'^2 = 0$

20. **(Van der Pol equation)** Show that if $\mu \to 0$, the isoclines in Example 3 approach straight lines through the origin, of slope $-1/k$. Why is this to be expected?

21. Figure 67 shows that some of the isoclines have a maximum or minimum. Show that these points lie on the hyperbola $vy = -1/2\mu$.

22. Consider a generalization of the van der Pol equation, namely,

$$y'' - \mu(1 - y^2 - \gamma^2 y'^2)y' + y = 0 \qquad (\gamma \text{ constant}).$$

Show that this equation describes self-sustained oscillations. Find the equation of the isoclines. Verify that for $\gamma = 1$ the given equation has the periodic solution $y = \sin t$.

23. **(Rayleigh[4] equation)** Show that the so-called *Rayleigh equation*

$$Y'' - \mu(1 - \tfrac{1}{3}Y'^2)Y' + Y = 0 \qquad (\mu > 0)$$

also describes self-sustained oscillations and that by differentiating it and setting $y = Y'$ one obtains the van der Pol equation.

24. **(Duffing equation)** The Duffing equation is

$$y'' + \omega_0^2 y + \beta y^3 = 0$$

where usually $|\beta|$ is small, thus characterizing a small deviation of the restoring force from linearity. $\beta > 0$ and $\beta < 0$ are called the cases of a *hard* and a *soft spring*, respectively. Find the equation of the solution curves in the phase plane. (For $\beta > 0$ all these curves are closed.)

25. **(Pendulum)** To what state (location, speed, direction of motion) of the pendulum do the four points of intersection of a closed curve in Fig. 66 with the axes correspond? The point of intersection of a wavy curve with the v-axis?

---

[4]LORD RAYLEIGH (JOHN WILLIAM STRUTT) (1842—1919), English physicist and mathematician, professor at Cambridge and London, known by his important contributions to various branches of applied mathematics and theoretical physics, in particular the theory of waves, elasticity and hydrodynamics.

3.3 Critical Points. Stability

We consider systems of differential equations of the form

(1)
$$\begin{aligned} x' &= F(x, y) \\ y' &= G(x, y) \end{aligned}$$
$(' = d/dt)$

where the unknown functions x and y depend on a variable t. A system, such as (1), in which the independent variable does not occur explicitly is called an **autonomous system.** In this section we shall see that solutions of an autonomous system can be investigated in the xy-plane. Indeed, this approach yields a characterization of various general properties of solutions, without actually solving the equations. Accordingly, the more complicated the equations (1) are, the more important this approach becomes. Furthermore, we shall be able to study whole families of solutions, not just a single solution. This generally is an advantage over numerical methods, which yield one (approximate) solution at a time (although with much greater accuracy).

The present approach will include, in a rather natural way, stability considerations. Stability concepts are suggested by physics, where **stability** means, roughly speaking, that a small change (small disturbance) of a physical system at some instant changes the behavior of the system only slightly at all future times.

A solution $x(t)$, $y(t)$ of (1) represents a curve C in the xy-plane (or a point as a degenerate case). This curve is called a *solution curve* or **path** (sometimes a *trajectory*) of (1). The sense of increasing t is called the *positive sense* on C and can be marked by an arrowhead. This defines an *orientation* on C. If t is time and C the path of a moving body, the positive sense is the sense in which the body moves along C as time progresses.

From (1) we see that the slope of a path passing through a point P: (x, y) is

(2)
$$\frac{dy}{dx} = \frac{dy/dt}{dx/dt} = \frac{G(x, y)}{F(x, y)}.$$

Note that (2) gives no information about the orientation of a path. Note further that we must have $F(x, y) \neq 0$ at P. If $F(x, y) = 0$ but $G(x, y) \neq 0$ at P, we can take $dx/dy = F(x, y)/G(x, y)$ instead of (2) and conclude from $dx/dy = 0$ that the tangent of C at P is vertical. However, what can we do if both F and G are zero at some point? This problem is the main topic of the present section and will lead to interesting results of practical importance.

Before we go into details, we should make two observations. In the first place, (2) is of the form $dy/dx = f(x, y)$, whose solutions we considered geometrically in the xy-plane in Sec. 1.10. Hence our present study extends that in Sec. 1.10.

Second, if a second-order differential equation

$$y'' = G(y, y')$$
$(' = d/dt)$

is given (as in Chap. 2 and also in Sec. 3.2), we can write it as a system by setting $y' = v$. Then $y'' = v'$, and the system is

$$y' = v$$
$$v' = G(y, v)$$

and a solution $y(t)$, $v(t)$ of this system represents a curve in the yv-plane, the phase plane. This system is a special case of (1), except for the notation (which we can adjust to the notation above if we replace y with x, and v with y). Because of this fact, our present xy-plane is often called the **phase plane** of (1), although y is now any variable, not necessarily x'.

Critical Points, Their Stability

A point P_0: (x_0, y_0) at which both F and G are zero is called a **critical point** of (1).

There are different critical points, and we can classify them in two different ways, once with respect to their stability and on the other hand with respect to the general geometric shape of the paths in their neighborhood. We discuss this in more detail. In doing so we assume that P_0 is a critical point of (1) which is *isolated,* that is, P_0 is the only critical point of (1) in some circular disk about P_0. We begin with the first classification.

P_0 is called a **stable**[5] **critical point** of (1) if, roughly speaking, all paths of (1) that at some instant are sufficiently close to P_0 remain close to P_0 at all future times; precisely: if for every disk D_ϵ of radius $\epsilon > 0$ about P_0 there is a disk D_δ of radius $\delta > 0$ about P_0 such that every path of (1) which has a point P_1 (corresponding to $t = t_1$, say) in D_δ has all its points corresponding to $t \geq t_1$ in D_ϵ. See Fig. 69.

P_0 is called a **stable and attractive critical point**[6] of (1) if P_0 is stable and every path which has a point in D_δ approaches P_0 as $t \to \infty$. See Fig. 70.

P_0 is called **unstable** if P_0 is not stable.

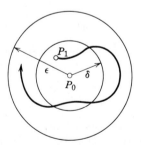

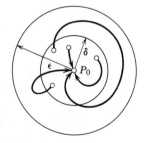

Fig. 69. Stable critical point P_0 of (1) (The path initiating at P_1 stays in the disk of radius ϵ.)

Fig. 70. Stable and attractive critical point P_0 of (1)

---

[5]More precisely: **stable in the sense of Liapunov.** There exist other definitions of stability, but Liapunov's definition—the only we shall consider—is probably the most useful one.

[6]Or an **asymptotically stable critical point.**

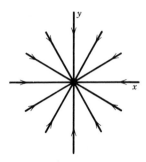

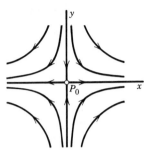

Fig. 71. Stable and attractive proper node

Fig. 72. Saddle point

Critical Points: Node, Saddle, Center, Spiral Point

We turn to the second classification of a critical point P_0 of (1), starting with some simple systems, to see what may happen.

EXAMPLE 1. The four types of critical points illustrated by systems (3)–(6)
Node:

(3) $x' = -x, \quad y' = -y,$ solution $x = Ae^{-t}, \quad y = Be^{-t}$ (Fig. 71).

P_0: (0, 0) is an example of a **node;** since $Ay = Bx$, the paths at P_0 have all possible directions, P_0 is stable and attractive.
Saddle point:

(4) $x' = x, \quad y' = -y,$ solution $x = Ae^t, \quad y = Be^{-t}$ (Fig. 72).

P_0: (0, 0) is an example of a **saddle point;** we have two incoming paths (the positive and negative y-axes, corresponding to $A = 0$), two outgoing paths (in the positive and negative x-directions, when $B = 0$), and infinitely many paths, the hyperbolas $xy = AB$ ($A \neq 0, B \neq 0$), which bypass P_0. A saddle point is unstable.
Center:

(5) $x' = y', \quad y = -2x,$ solution $2x^2 + y^2 = c$ (Fig. 73).

P_0: (0, 0) is an example of a **center;** the paths are closed curves (ellipses) around P_0, and P_0 is stable.
Spiral point:

(6) $x' = -x + y, \quad y' = -x - y,$ solution $r = c_0 e^{\theta}$ (Fig. 74)

P_0: (0, 0) is an example of a **spiral point;** the paths are spirals about P_0: here $r = (x^2 + y^2)^{1/2}$ and $\theta = \text{arc tan } (y/x)$ are polar coordinates. The reader may show that a general solution in terms of x and y is

(7) $x = e^{-t}(A \cos t + B \sin t), \quad y = e^{-t}(B \cos t - A \sin t).$

From this and $\cos^2 t + \sin^2 t = 1$ we obtain by straightforward calculation and simplification

$$r^2 = x^2 + y^2 = (A^2 + B^2)e^{-2t}$$

which implies the solution given in (6) with $c_0 = \sqrt{A^2 + B^2}$ and $t = -\theta$.   ∎

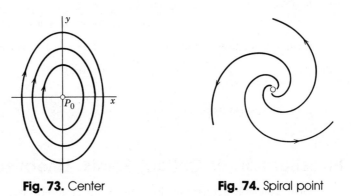

Fig. 73. Center   **Fig. 74.** Spiral point

The definitions of the concepts used in this example (with a further sub-classification of nodes) are as follows. An isolated critical point P_0 of (1) is:

(a) a **proper node** if every path approaches P_0 in a definite direction as $t \to +\infty$ or $t \to -\infty$ and, given any direction, there is a path that approaches P_0 in this direction,

(b) an **improper node** if every path, with the possible exception of one pair of paths, has the same limiting direction at P_0,

(c) a **saddle point** if finitely many paths approach P_0 as $t \to +\infty$ and finitely many paths approach P_0 as $t \to -\infty$,

(d) a **center** if the paths form closed curves containing P_0 in the region enclosed by any path,

(e) a **spiral point** if the paths form spirals about P_0 with P_0 as the asymptotic point.

EXAMPLE 2. Improper nodes

The system

(8) $$x' = -x, \qquad y' = -2y$$

has a stable and attractive improper node at the origin (Fig. 75). The system

(9) $$x' = x, \qquad y' = x + y$$

has an unstable improper node at $(0, 0)$ such that *all* paths have the same limiting direction as $t \to -\infty$ (Fig. 76). Indeed, the reader may show that a general solution is

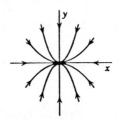

Fig. 75. Stable and attractive improper node

Fig. 76. Unstable improper node

$$x = Ae^t, \qquad y = (At + B)e^t.$$

For $A = 0$ this gives vertical rays ($x = 0$), and for $A \neq 0$ we obtain

$$\frac{y'}{x'} = \frac{(At + B + A)e^t}{Ae^t} \quad \rightarrow \quad -\infty \qquad \text{as } t \rightarrow -\infty,$$

that is, the paths approach the origin in vertical direction as $t \rightarrow -\infty$. ∎

Investigation of Critical Points. Linearization

How can one determine the type of an isolated critical point P_0 of a given system (1)? In most practical cases this can be done by **linearization,** that is, by studying a certain *linear* system, as follows.

We assume P_0 to be (0, 0), without loss of generality. We also assume that F and G in (1) are continuous and have continuous partial derivatives in a neighborhood of P_0. Since P_0 is a critical point, we have $F(0, 0) = 0$ and $G(0, 0) = 0$. Hence F and G have no constant terms. Their linear terms we write explicitly. Then (1) takes the form

(10)
$$x' = a_1 x + b_1 y + F_1(x, y)$$
$$y' = a_2 x + b_2 y + G_1(x, y).$$

One can prove that if $a_1 b_2 - b_1 a_2 \neq 0$, then the type and stability of P_0 is the same as that of the critical point (0, 0) of the *linear system* obtained by **linearization,** that is, by dropping F_1 and G_1 from (10):

(11)
$$\boxed{\begin{aligned} x' &= a_1 x + b_1 y \\ y' &= a_2 x + b_2 y \end{aligned}}$$
$$(' = d/dt).$$

Indeed, the above assumption on the derivatives implies that F_1 and G_1 are small near P_0. **Two exceptions** occur: for equal λ_1, λ_2 (below) or pure imaginary λ_1, λ_2, in addition to the type of points of the *linear* system, the nonlinear system may also have a spiral point. For proofs, see Ref. [A5], pp. 375–388.

We discuss (11). Substituting $x = Ae^{\lambda t}$, $y = Be^{\lambda t}$ into (11) and omitting the factor $e^{\lambda t}$, we obtain the homogeneous algebraic equations

$$(a_1 - \lambda)A + \qquad b_1 B = 0$$
$$a_2 A + (b_2 - \lambda)B = 0$$

This has a nontrivial solution A, B if and only if the determinant of the coefficients is zero, that is,

(12) $$\lambda^2 - (a_1 + b_2)\lambda + a_1 b_2 - b_1 a_2 = 0.$$

Note that by eliminating y from (11) we obtain (cf. Sec. 3.1)

(13) $$x'' - (a_1 + b_2)x' + (a_1 b_2 - b_1 a_2)x = 0.$$

Hence (12) is also the characteristic equation of (13) as used in Chap. 2 (Sec. 2.2, etc.), except for notation.

We now introduce the standard notation

(14) $$p = a_1 + b_2, \qquad q = a_1 b_2 - b_1 a_2, \qquad \Delta = p^2 - 4q.$$

Then, if λ_1, λ_2 denote the roots of (12), we have from (12)

$$\lambda^2 - p\lambda + q = (\lambda - \lambda_1)(\lambda - \lambda_2) = \lambda^2 - (\lambda_1 + \lambda_2)\lambda + \lambda_1 \lambda_2 = 0.$$

Hence p is the sum and q the product of the roots, and Δ is the discriminant. Since p, q determine the roots, which in turn determine the behavior of solutions near P_0, we can characterize P_0 in terms of p, q, Δ, as follows. P_0 is:

(15)

> *stable and attractive* if $p < 0$ and $q > 0$,
>
> *stable* if $p \leqq 0$ and $q > 0$,
>
> *unstable* if $p > 0$ or $q < 0$.

Furthermore, P_0 is a:

(16)

> *node* if $q > 0$ and $\Delta \geqq 0$,
>
> *saddle point* if $q < 0$
>
> *center* if $p = 0$ and $q > 0$,
>
> *spiral point* if $p \neq 0$ and $\Delta < 0$.

This is summarized in the **stability chart** in Fig. 77. In this chart, the region of instability is shaded.

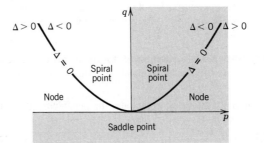

Fig. 77. Stability chart of the system (1) written in the form (10) with p, q, Δ defined in (14). Stable and attractive: The second quadrant without the q-axis. Stability also on the positive q-axis (which corresponds to centers). Unstable: Shaded region

We indicate how to obtain these criteria. If $q = \lambda_1 \lambda_2 > 0$, both roots are positive or both negative, or complex conjugates. If also $p = \lambda_1 + \lambda_2 < 0$, both are negative or have a negative real part. Hence P_0 is stable and attractive. The reasoning for the other two lines in (15) is similar.

If $\Delta < 0$, the roots are complex conjugates, say, $\lambda_1 = \alpha + i\beta$ and $\lambda_2 = \alpha - i\beta$. If also $p = \lambda_1 + \lambda_2 = 2\alpha < 0$, this gives a spiral point that is stable and attractive.

If $p = 0$, then $\lambda_2 = -\lambda_1$ and $q = \lambda_1 \lambda_2 = -\lambda_1^2$. If also $q > 0$, then $\lambda_1^2 = -q < 0$, so that λ_1, and thus λ_2, must be pure imaginary. This gives periodic solutions, their paths being closed curves around $P_0 = (0, 0)$, which is a center. For further details, see Ref. [A5] in Appendix 1.

Further Examples and Applications

EXAMPLE 3. Application of the criteria (15) and (16)

In (3) (above) we have $a_1 = -1$, $b_1 = 0$, $a_2 = 0$, $b_2 = -1$, hence $p = -2$, $q = 1$, $\Delta = 0$, and the criteria indicate a node at $(0, 0)$ that is stable and attractive.

Similarly, in (4) we have $p = 0$, $q = -1$, $\Delta = 4$ and, accordingly, a saddle point, which is always unstable.

The reader may test (5), (6), (8), (9) in a similar way.

EXAMPLE 4. Critical points in the case of free motions of a mass on a spring

What type of critical points does the model of a mass on an elastic spring have?

Solution. In our present notation, equation (7), Sec. 2.6, is

$$x'' + \frac{c}{m} x' + \frac{k}{m} x = 0.$$

As a system, with $y = x'$, we have

(17)
$$x' = y$$
$$y' = -\frac{k}{m} x - \frac{c}{m} y.$$

We see that this system has a single critical point, which is at $(0, 0)$. For our criteria (15), (16) we need $p = -c/m$, $q = k/m$ and $\Delta = (c/m)^2 - 4k/m$. This yields the following results.

No damping. $c = 0$, $p = 0$, $q > 0$, a center.
Underdamping. $c^2 < 4mk$, $p < 0$, $q > 0$, $\Delta < 0$, a stable and attractive spiral point.
Critical damping. $c^2 = 4mk$, $p < 0$, $q > 0$, $\Delta = 0$, a stable and attractive node.
Overdamping. $c^2 > 4mk$, $p < 0$, $q > 0$, $\Delta > 0$, a stable and attractive node.

EXAMPLE 5. Linearization of the damped pendulum equation

Investigate the critical points of the model of a pendulum without damping and with damping proportional to the velocity.

Solution. In our present notation, the pendulum equation (4), Sec. 3.2, with a damping term cx' added is

(18)
$$\boxed{x'' + cx' + k \sin x = 0.}$$

As a system,

(19)
$$x' = y$$
$$y' = -k \sin x - cy$$

We see that the critical points are $(0, 0)$, $(\pm\pi, 0)$, $(\pm 2\pi, 0)$, $\cdots$. We consider $(0, 0)$. By linearization, since $\sin x = x - x^3/6 + - \cdots$,

$$(20) \qquad \begin{aligned} x' &= y \\ y' &= -kx - cy \end{aligned}$$

This is identical with (17), except for the (positive!) factor m (and except for the physical meaning of x); hence for $c = 0$ (no damping) we have a center, for small damping a spiral point, and so on.

We consider the critical point $(\pi, 0)$. Since $\sin(\pi + x) = -\sin x = -x + x^3/6 - + \cdots$, linearization now yields

$$(21) \qquad \begin{aligned} x' &= y \\ y' &= kx - cy \end{aligned}$$

For our criteria (15), (16) we calculate $p = -c$, $q = -k$ and $\Delta = c^2 + 4k$. This yields the following results.

No damping. $c = 0, p = 0, q < 0, \Delta > 0$, a saddle. Cf. Fig. 66 in Sec. 3.2 (where $x = \theta$ and $y = x' = v$).

Damping. $c > 0, p < 0, q < 0, \Delta > 0$, a saddle.

Since $\sin x$ is periodic with period 2π, the critical points $(\pm 2\pi, 0)$, $(\pm 4\pi, 0)$, $\cdots$ are of the same type as $(0, 0)$, and the critical points $(-\pi, 0)$, $(\pm 3\pi, 0)$, $\cdots$ are of the same type as $(\pi, 0)$, so that our task is finished.

Figure 78 shows the paths in the case of damping. What we see agrees with our physical intuition. Indeed, damping means loss of energy; hence instead of the closed paths of periodic solutions in Fig. 66 we now have paths spiraling around one of the critical points $(0, 0)$, $(\pm 2\pi, 0)$, $\cdots$. Even the wavy paths corresponding to whirly motions eventually spiral around one of these points; furthermore, there are no more paths that connect critical points (as there were in the undamped case).   ∎

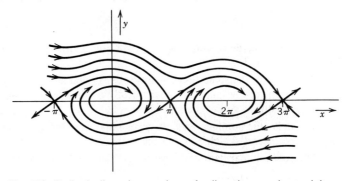

Fig. 78. Paths in the phase plane for the damped pendulum; cf. (18) and (19)

This is the end of Chap. 3. Looking back, we see that the *theory* developed in Chaps. 2 and 3 was quite general, whereas the equations actually solved were rather simple, most of them with constant coefficients. Accordingly, we should direct our further effort to the development of methods for solving differential equations whose coefficients are variable. We shall do this in the next chapter, which includes equations of basic importance in physics and engineering (**Bessel's** and **Legendre's equations**, the **hypergeometric equation**, etc.).

Problems for Sec. 3.3

1. Show that in (3) for every choice of A and B one obtains a ray which is directed toward the origin (so that this critical point is stable).
2. Give a derivation of the solution of (4).
3. Derive the solution of (5).
4. Obtain the solution given in (6) directly from the equation in (6).
5. Derive the solution (7) of (6).
6. Solve (8). What curves are the paths?
7. In (8) let t represent time. Introduce $\tau = -t$ as a new independent variable and solve. What does that transformation mean in terms of mechanics?
8. Derive the solution of (9).

Solve the following systems of differential equations and determine the type and stability behavior of their critical points.

9. $x' = x, \quad y' = 2y$
10. $x' = x, \quad y' = y$
11. $x' = y, \quad y' = -x$
12. $x' = -x, \quad y' = x - y$
13. $x' = 2x + 3y$
    $y' = \frac{1}{3}x + 2y$
14. $x' = x + y$
    $y' = 3x - y$
15. $x' = x + y$
    $y' = 4x + y$
16. $x' = 2x + 5y$
    $y' = -\frac{1}{2}x - \frac{3}{2}y$

17. Derive (13) from (11).
18. What type of critical point do we obtain in Example 4 when $c < 0$? What would this mean mechanically?
19. Determine the location of all critical points of the equation

$$x'' - 9x + x^3 = 0.$$

20. **(Linearization)** Applying linearization, determine the type of the critical points in Prob. 19.
21. Perform the same tasks as in Probs. 19 and 20, for $x'' - 4x + x^3 = 0$. Can you interpret the equation mechanically as the model of an undamped motion, the "restoring force" being $-4x + x^3$?
22. Solve the nonlinear system corresponding to the equation in Prob. 21 in the phase plane and sketch some of the paths.
23. Add a linear damping term to the equation in Prob. 21, to get

$$x'' + x' - 4x + x^3 = 0.$$

Determine the type of each critical point by linearization. First guess (using mechanical arguments and the answer to Prob. 21), then calculate.
24. **(Van der Pol equation)** Determine the type of the critical point $(0, 0)$ of the van der Pol equation (Sec. 3.2) when $\mu > 0$, $\mu = 0$ and $\mu < 0$.
25. What type of critical point would we obtain at $(0, 0)$ for the pendulum equation (cf. Example 5) with "negative damping" $c < 0$?

Review Problems for Chapter 3

1. What is the phase plane? Without looking into the text, state the key idea by which we obtained from a *second-order* equation with unknown function $y(t)$ a *first-order* equation in the phase plane.

2. In this chapter, we graphed a harmonic oscillation as a circle in the phase plane. Can you think of a motion that appears as an ellipse in the phase plane?

3. What do we mean by a path? By a limit cycle? What is a critical point?

4. Sketch the form of solutions near a saddle point. Near a spiral point.

5. When do we call a system of equations *autonomous?* Why was this property essential to some of our considerations?

6. Without looking into the text, write down the formulas that show that a second-order equation $y'' = G(y, y')$ can be written as a system of the type considered in Sec. 3.3.

7. What is a self-sustained oscillation?

8. What factor in the terms of the van der Pol equation is responsible for the fact that this equation governs self-sustained oscillations? Could you replace this factor by some other function that would also give self-sustained oscillations? Give an example of such a function.

9. Is it plausible that one can study critical points of *nonlinear* systems by considering a corresponding *linear* system?

10. Does the linear system in Prob. 9 change if we go from one critical point to another critical point of the same system?

Solve the following systems of differential equations; in Probs. 17–20, take into account the given initial conditions.

11. $x' = 2x + \frac{1}{3}y$
$y' = 3x + 2y$

12. $x' = 2x + y$
$y' = 5x - 2y$

13. $x' = -3x + y$
$y' = -20x + 6y$

14. $x' = x + 2y$
$y' = 2x + y$

15. $x' = x + y$
$y' = -2x - y + t + 1$

16. $x' = 3x - 4y + 10\cos t$
$y' = x - 2y$

17. $x' = x + y,\quad y' = x - y,\quad x(0) = 2,\quad y(0) = -2(\sqrt{2} + 1) \approx -4.828$

18. $x' = 3y,\quad y' = 3x,\quad x(0) = 3,\quad y(0) = -1$

19. $x' = 2x - y + \sin t,\quad y' = x + 3\cos t,\quad x(0) = 1,\quad y(0) = -1$

20. $x'' = 4y + 5e^t,\quad y'' = 4x - 5e^t,\quad x(0) = 2,\quad x'(0) = 7,\quad y(0) = 0,\quad y'(0) = -3$

21. Set up the model of the undamped vibrating system in Fig. 79 for general m_1, m_2, k_1, k_2. Solve it, assuming that $m_1 = 1.5$, $m_2 = 2$, $k_1 = 4.5$, $k_2 = 6$.

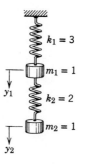

System in
static
equilibrium

Fig. 79. Vibrating system in Problems 21–23

22. In Prob. 21, find the solution if the system starts from initial displacement $y_1(0) = 2$, $y_2(0) = 7$ with initial velocity 0.

23. What motion will the system in Prob. 21 perform if it starts from the static equilibrium position with initial velocity $y_1{}'(0) = 2$, $y_2{}'(0) = 3$?

24. Find the currents I_1 and I_2 in the network in Fig. 80, where $R = 6\frac{2}{3}$ ohms, $L = 1$ henry, $C = 5 \cdot 10^{-3}$ farad, $E = 66\frac{2}{3}$ volts and $I_1(0) = I_2(0) = 0$.

25. Find the currents I_1 and I_2 in the network in Fig. 80, where $R = 2.5$ ohms, $L = 1$ henry, $C = 0.04$ farad, $E(t) = 169 \sin t$ volts and $I_1(0) = I_2(0) = 0$.

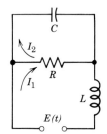

Fig. 80. Network in Problems 24, 25

Determine the type of the critical point (0, 0) of the following systems.

26. $x' = y$
$y' = -\sin x - y$

27. $x' = x - y + x^3$
$y' = x + y - xy$

28. $x' = -4x + 5 \tan y$
$y' = -6x + 4y - y^2$

29. $x' = 3x + y + y^2$
$y' = x + \frac{1}{2}y$

30. $x' = 2x + y^2$
$y' = x^2 - y$

Summary of Chapter 3
Systems of Differential Equations, Phase Plane, Stability

Whereas single electric circuits or single mass–spring systems are governed by single differential equations as their mathematical model (Chap. 2), networks consisting of several circuits, mechanical systems of several masses and springs and other problems of engineering interest lead to **systems of differential equations** (Sec. 3.1) in which we simultaneously deal with several unknown functions, representing the currents in the various circuits, the displacements of those masses, etc. In Sec. 3.1 it is illustrated how linear systems can be solved by reducing them to a single equation, or by simple matrix methods. (The latter will be considered in full generality in Sec. 7.16.)

The **phase plane** (Sec. 3.2) is the plane with displacement y and velocity v of a motion as rectangular coordinates. It is a powerful tool for studying general properties of solutions, particularly for **nonlinear differential equations,** frequently without actually obtaining solutions in the usual form $y = y(t)$, where t is time. Typical equations accessible in this way are the **pendulum equation** (Sec. 3.2)

$$\theta'' + \frac{g}{L} \sin \theta = 0$$

whose exact solutions are nonelementary functions, the pendulum equation with damping (Sec. 3.3) and the famous **van der Pol equation**

$$y'' - \mu(1 - y^2)y' + y = 0$$

governing self-sustained oscillations.

Extension leads to the study of systems (Sec. 3.3)

$$x' = F(x, y), \qquad y' = G(x, y).$$

Points at which both F and G are zero are called **critical points.** The investigation of the general behavior of solutions near such points is very important and leads to a classification of these points and their stability, as is graphically summarized in the **stability chart** in Sec. 3.3. It is quite surprising how much information the phase plane reveals, even in more complicated situations.

Chapter 4

Series Solutions
of Differential Equations.
Orthogonal Functions

In Chap. 2 we saw that a homogeneous linear differential equation whose coefficients are *constant* can be solved by algebraic methods, and the solutions are elementary functions known from calculus. However, if those coefficients are not constant but depend on x, the situation is more complicated and the solutions may be nonelementary functions. Bessel's equation, Legendre's equation and the hypergeometric equation are of this type. Since these and other equations and their solutions play an important role in engineering mathematics, we shall now consider a method for solving such equations. The solutions will appear in the form of power series. For this reason the method is called the **power series method.** We shall also consider some basic properties of the solutions. This will help the student to get acquainted with these "higher transcendental functions" and with some standard procedures used in connection with special functions, in particular for the purpose of establishing properties of the functions and relations between them, some of which are basic in numerical computations.

Actually, some of those differential equations (Bessel's equation, for instance) have solutions of a more general form, namely, power series times a single power of x (whose exponent need not be an integer and may even be complex), sometimes still multiplied by a logarithm. In Sec. 4.4 we show that the power series method can be generalized to such equations. This is known as the **extended power series method** or **Frobenius method.**

The last three sections (Secs. 4.7–4.9) are devoted to the *orthogonality* of Legendre polynomials, Bessel functions, and other sets of solutions of certain boundary value problems, which are known as **Sturm-Liouville problems.**

This chapter can also be studied *immediately after Chap. 2,* since it does not presuppose material from Chap. 3.

Prerequisite for this chapter: Chap. 2.
Sections that may be omitted in a shorter course: 4.2, 4.6, 4.9.
References: Appendix 1, Part A.
Answers to problems: Appendix 2.

4.1 Power Series Method

We shall consider solving differential equations by the so-called *power series method,* which yields solutions in the form of power series. This is a very efficient standard procedure in connection with linear differential equations whose coefficients are variable. It is of great practical importance, since power series may be used for computing numerical values, characterizing various general properties of solutions and obtaining other types of representations of solutions, as we shall see.

Power Series

We first remember that a **power series**[1] (in powers of $x - x_0$) is an infinite series of the form

$$(1) \qquad \sum_{m=0}^{\infty} a_m(x - x_0)^m = a_0 + a_1(x - x_0) + a_2(x - x_0)^2 + \cdots .$$

$a_0, a_1, a_2, \cdots$ are constants, called the **coefficients** of the series. x_0 is a constant, called the **center** of the series, and x is a variable.

If in particular $x_0 = 0$, we obtain a *power series in powers of x*

$$(2) \qquad \sum_{m=0}^{\infty} a_m x^m = a_0 + a_1 x + a_2 x^2 + a_3 x^3 + \cdots .$$

We shall assume in this section that all variables and constants are real.

Familiar examples of power series are the Maclaurin series

$$\frac{1}{1 - x} = \sum_{m=0}^{\infty} x^m = 1 + x + x^2 + \cdots \qquad (|x| < 1, \textit{geometric series}),$$

$$e^x = \sum_{m=0}^{\infty} \frac{x^m}{m!} = 1 + x + \frac{x^2}{2!} + \frac{x^3}{3!} + \cdots ,$$

$$\cos x = \sum_{m=0}^{\infty} \frac{(-1)^m x^{2m}}{(2m)!} = 1 - \frac{x^2}{2!} + \frac{x^4}{4!} - + \cdots ,$$

$$\sin x = \sum_{m=0}^{\infty} \frac{(-1)^m x^{2m+1}}{(2m + 1)!} = x - \frac{x^3}{3!} + \frac{x^5}{5!} - + \cdots .$$

[1]The term "power series" alone usually refers to a series of the form (1), but does not include series of negative powers of x such as $a_0 + a_1 x^{-1} + a_2 x^{-2} + \cdots$ or series involving fractional powers of x. Note that in (1) we write, for convenience, $(x - x_0)^0 = 1$, even when $x = x_0$.

We use m as the summation letter, reserving n as a standard notation for the parameters in the Legendre and Bessel equations for integer values (cf. Secs. 4.3 and 4.5).

Idea of the Power Series Method

The general idea of the power series method for solving differential equations is very simple and natural. We shall describe the practical procedure and illustrate it by simple examples, postponing the mathematical justification of the method to the next section.

A differential equation being given, we first represent all given functions in the equation by power series in powers of x (or in powers of $x - x_0$, if solutions in the form of power series in powers of of $x - x_0$ are wanted). In many practical cases the given functions will be polynomials, and in such a case, nothing need be done in this first step. Then we assume a solution in the form of a power series, say,

$$(3) \qquad y = \sum_{m=0}^{\infty} a_m x^m = a_0 + a_1 x + a_2 x^2 + a_3 x^3 + \cdots$$

and insert this series and the series obtained by termwise differentiation,

$$(4) \qquad \text{(a)} \quad y' = \sum_{m=1}^{\infty} m a_m x^{m-1} = a_1 + 2a_2 x + 3a_3 x^2 + \cdots$$

$$\text{(b)} \quad y'' = \sum_{m=2}^{\infty} m(m-1) a_m x^{m-2} = 2a_2 + 3 \cdot 2 a_3 x + 4 \cdot 3 a_4 x^2 + \cdots$$

etc., into the equation. Then we collect like powers of x and equate the sum of the coefficients of each occurring power of x to zero, starting with the constant terms, the terms containing x, the terms containing x^2, etc. This gives relations from which we can determine the unknown coefficients in (3) successively.

Let us illustrate the procedure for some simple equations that can also be solved by elementary methods.

EXAMPLE 1
Solve

$$y' - y = 0.$$

Solution. In the first step, we insert (3) and (4a) into the equation:

$$(a_1 + 2a_2 x + 3a_3 x^2 + \cdots) - (a_0 + a_1 x + a_2 x^2 + \cdots) = 0.$$

Then we collect like powers of x, finding

$$(a_1 - a_0) + (2a_2 - a_1)x + (3a_3 - a_2)x^2 + \cdots = 0.$$

Equating the coefficient of each power of x to zero, we have

$$a_1 - a_0 = 0, \quad 2a_2 - a_1 = 0, \quad 3a_3 - a_2 = 0, \cdots.$$

Solving these equations, we may express $a_1, a_2, \cdots$ in terms of a_0 which remains arbitrary:

$$a_1 = a_0, \quad a_2 = \frac{a_1}{2} = \frac{a_0}{2!}, \quad a_3 = \frac{a_2}{3} = \frac{a_0}{3!}, \cdots.$$

With these values (3) becomes

$$y = a_0 + a_0 x + \frac{a_0}{2!} x^2 + \frac{a_0}{3!} x^3 + \cdots,$$

and we see that we have obtained the familiar general solution

$$y = a_0 \left(1 + x + \frac{x^2}{2!} + \frac{x^3}{3!} + \cdots \right) = a_0 e^x.$$

EXAMPLE 2
Solve

$$y' = 2xy.$$

Solution. We insert (3) and (4a) into the equation:

$$a_1 + 2a_2 x + 3a_3 x^2 + \cdots = 2x(a_0 + a_1 x + a_2 x^2 + \cdots).$$

We must perform the multiplication by $2x$ on the right and can write the resulting equation in the convenient form

$$a_1 + 2a_2 x + 3a_3 x^2 + 4a_4 x^3 + 5a_5 x^4 + 6a_6 x^5 + \cdots$$
$$= \qquad 2a_0 x + 2a_1 x^2 + 2a_2 x^3 + 2a_3 x^4 + 2a_4 x^5 + \cdots.$$

From this we can immediately conclude that $a_1 = 0$, which implies $a_3 = 0$, $a_5 = 0$, $\cdots$, and for the coefficients with even subscript we obtain

$$a_2 = a_0, \qquad a_4 = \frac{a_2}{2} = \frac{a_0}{2!}, \qquad a_6 = \frac{a_4}{3} = \frac{a_0}{3!}, \cdots.$$

a_0 remains arbitrary. With these values (3) becomes

$$y = a_0 \left(1 + x^2 + \frac{x^4}{2!} + \frac{x^6}{3!} + \frac{x^8}{4!} + \cdots \right) = a_0 e^{x^2}.$$

The student may check this by separating variables.

EXAMPLE 3
Solve

$$y'' + y = 0.$$

Solution. By inserting (3) and (4b) into the equation we obtain

$$(2a_2 + 3 \cdot 2a_3 x + 4 \cdot 3a_4 x^2 + \cdots) + (a_0 + a_1 x + a_2 x^2 + \cdots) = 0.$$

Collecting like powers of x, we find

$$(2a_2 + a_0) + (3 \cdot 2a_3 + a_1)x + (4 \cdot 3a_4 + a_2)x^2 + \cdots = 0.$$

Equating the coefficient of each power of x to zero, we have

$$2a_2 + a_0 = 0 \qquad\qquad\qquad \text{coefficient of } x^0$$
$$3 \cdot 2a_3 + a_1 = 0 \qquad\qquad\qquad \text{coefficient of } x^1$$
$$4 \cdot 3a_4 + a_2 = 0 \qquad\qquad\qquad \text{coefficient of } x^2$$

etc. Solving these equations, we see that a_2, a_4, $\cdots$ may be expressed in terms of a_0; and a_3, a_5, $\cdots$ may be expressed in terms of a_1:

$$a_2 = -\frac{a_0}{2!}, \qquad a_3 = -\frac{a_1}{3!}, \qquad a_4 = -\frac{a_2}{4 \cdot 3} = \frac{a_0}{4!}, \cdots;$$

a_0 and a_1 are arbitrary. With these values (3) becomes

$$y = a_0 + a_1 x - \frac{a_0}{2!} x^2 - \frac{a_1}{3!} x^3 + \frac{a_0}{4!} x^4 + \frac{a_1}{5!} x^5 + \cdots .$$

This can be written

$$y = a_0 \left(1 - \frac{x^2}{2!} + \frac{x^4}{4!} - + \cdots \right) + a_1 \left(x - \frac{x^3}{3!} + \frac{x^5}{5!} - + \cdots \right)$$

and we recognize the familiar general solution

$$y = a_0 \cos x + a_1 \sin x.$$ ∎

In the next section we discuss the basic concepts and facts on power series that we need in connection with the power series method.

Problems for Sec. 4.1

Apply the power series method to the following differential equations.

1. $y' = 3y$ **2.** $y' + 2y = 0$ **3.** $y' = ky$
4. $y' = 2xy$ **5.** $(1 - x)y' = y$ **6.** $y' = xy$
7. $(1 + x)y' = y$ **8.** $(1 - x^2)y' = y$ **9.** $y'' = y$
10. $y'' + 9y = 0$ **11.** $y'' = 4y$ **12.** $y'' = y'$

(More problems of this type are included at the end of the next section.)

Theory of the Power Series Method

We have seen that the power series method yields solutions of differential equations in the form of power series. In the present section we shall give a mathematical justification of the operations occurring in this method, such as differentiation, addition and multiplication of power series. Some of these facts may already be known to the student from elementary calculus, whereas others are usually not considered in elementary classes. [Proofs and further details (not to be used in this chapter) can be found in Secs. 14.3 and 14.4.]

Basic Concepts

A **power series** is an infinite series of the form

$$(1) \qquad \sum_{m=0}^{\infty} a_m(x - x_0)^m = a_0 + a_1(x - x_0) + a_2(x - x_0)^2 + \cdots ,$$

and, as in the previous section, we assume that the variable x, the **center** x_0 and the **coefficients** $a_0, a_1, \cdots$ are real.
  The expression

$$(2) \qquad s_n(x) = a_0 + a_1(x - x_0) + a_2(x - x_0)^2 + \cdots + a_n(x - x_0)^n$$

(n a positive integer) is called the nth **partial sum** of the series (1). Clearly, if we omit the terms of s_n from (1), the remaining expression is

$$(3) \qquad R_n(x) = a_{n+1}(x - x_0)^{n+1} + a_{n+2}(x - x_0)^{n+2} + \cdots.$$

This expression is called the **remainder** *of* (1) *after the term* $a_n(x - x_0)^n$.

For example, in the case of the geometric series

$$1 + x + x^2 + \cdots + x^n + \cdots$$

we have

$$s_1 = 1 + x, \qquad R_1 = x^2 + x^3 + x^4 + \cdots,$$

$$s_2 = 1 + x + x^2, \qquad R_2 = x^3 + x^4 + x^5 + \cdots, \qquad \text{etc.}$$

In this way we have now associated with (1) the sequence of the partial sums $s_1(x), s_2(x), \cdots$. It may happen that for some $x = x_1$ this sequence converges, say,

$$\lim_{n \to \infty} s_n(x_1) = s(x_1).$$

Then we say that the series (1) **converges,** or *is convergent, at* $x = x_1$; the number $s(x_1)$ is called the **value** or *sum* of (1) at x_1, and we write

$$s(x_1) = \sum_{m=0}^{\infty} a_m(x_1 - x_0)^m.$$

If that sequence is divergent at $x = x_1$, then the series (1) is said to **diverge,** or to *be divergent,* at x_1.

We remember that a *sequence* $s_1, s_2, \cdots$ is said to *converge* to a number s, or to *be convergent* with the *limit* s, if for every given positive number ϵ (no matter how small, but not zero) we can find a number N such that

$$(4) \qquad |s_n - s| < \epsilon \qquad \text{for every } n > N.$$

This inequality (4) means that all s_n with $n > N$ lie between $s - \epsilon$ and $s + \epsilon$ (Fig. 81). Of course, N will in general depend on the choice of ϵ.

Now in our case, $s = s_n + R_n$, hence $R_n = s - s_n$. This gives

$$|s_n - s| = |R_n|$$

in (4), and convergence at $x = x_1$ means that we can make $|R_n(x_1)|$ as small as we please, by taking n large enough. In other words, in the case of convergence, $s_n(x_1)$ is an approximation to $s(x_1)$, and the error $|R_n(x_1)|$ of the approximation can be made smaller than any preassigned positive number ϵ by taking n sufficiently large.

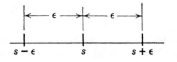

Fig. 81. Inequality (4)

Convergence Interval. Radius of Convergence

1. The series (1) converges at $x = x_0$, because then all its terms except for the first, a_0, are zero. In exceptional cases this may be the only x for which (1) converges. Such a series is of no practical interest.

2. If there are further values of x for which the series converges, these values form an interval, called the **convergence interval.** If this interval is finite, it has the midpoint x_0, so that it is of the form

$$(5) \qquad\qquad |x - x_0| < R$$

and the series (1) converges for all x such that $|x - x_0| < R$ and diverges for all x such that $|x - x_0| > R$. The number R is called the **radius**[2] **of convergence** of (1). It can be obtained from either of the formulas

$$(6) \quad (a) \quad R = 1 \Big/ \lim_{m \to \infty} \sqrt[m]{|a_m|} \qquad (b) \quad R = 1 \Big/ \lim_{m \to \infty} \left| \frac{a_{m+1}}{a_m} \right|$$

provided these limits exist and are not zero. [If they are infinite, then (1) converges only at the center x_0.]

3. The convergence interval may sometimes be infinite, that is, (1) converges for all x. For instance, if the limit in (6a) or (6b) is zero, this case occurs. One then writes $R = \infty$, for convenience. (Proofs of all these facts can be found in Sec. 14.3.)

For each x for which (1) converges, it has a certain value $s(x)$. We say that (1) **represents** the function $s(x)$ in the convergence interval and write

$$s(x) = \sum_{m=0}^{\infty} a_m (x - x_0)^m \qquad\qquad (|x - x_0| < R).$$

Let us illustrate the three possibilities by typical examples.

EXAMPLE 1. The useless Case 1 of convergence only at the center
In the case of the series

$$\sum_{m=0}^{\infty} m! x^m = 1 + x + 2x^2 + 6x^3 + \cdots$$

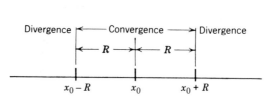

Fig. 82. Convergence interval (5)
of a power series with center x_0

---

[2]Because for a *complex* power series, one has convergence in an open disk of *radius R*. No general statement about convergence or divergence can be made for $x - x_0 = R$ or $x - x_0 = -R$.

we have $a_m = m!$, and in (6b),

$$\frac{a_{m+1}}{a_m} = \frac{(m+1)!}{m!} = m + 1 \to \infty \qquad \text{as } m \to \infty.$$

Thus this series converges only at the center $x = 0$. Such a series is useless.

EXAMPLE 2. Case 2. The geometric series

For the **geometric series** we have

$$\frac{1}{1-x} = \sum_{m=0}^{\infty} x^m = 1 + x + x^2 + \cdots \qquad (|x| < 1).$$

In fact, $a_m = 1$ for all m, and from (6) we obtain $R = 1$, that is, the geometric series converges and represents $1/(1 - x)$ when $|x| < 1$.

EXAMPLE 3. Case 3

In the case of the series

$$e^x = \sum_{m=0}^{\infty} \frac{x^m}{m!} = 1 + x + \frac{x^2}{2!} + \cdots$$

we have $a_m = 1/m!$. Hence in (6b),

$$\frac{a_{m+1}}{a_m} = \frac{1/(m+1)!}{1/m!} = \frac{1}{m+1} \to 0 \qquad \text{as } m \to \infty,$$

so that the series converges for all x.

EXAMPLE 4. A hint for some of the problems

Find the radius of convergence of the series

$$\sum_{m=0}^{\infty} \frac{(-1)^m}{8^m} x^{3m} = 1 - \frac{x^3}{8} + \frac{x^6}{64} - \frac{x^9}{512} + - \cdots.$$

Solution. This is a series in powers of $t = x^3$ with coefficients $a_m = (-1)^m/8^m$, so that in (6b),

$$\left| \frac{a_{m+1}}{a_m} \right| = \frac{8^m}{8^{m+1}} = \frac{1}{8}.$$

Thus $R = 8$. Hence the series converges for $|t| < 8$, that is, $|x| < 2$. ∎

Operations on Power Series

We shall now consider the operations on power series that are used in connection with the power series method. The following three operations are permissible, in the sense explained in each case. We also list a condition concerning the vanishing of all coefficients, which is a basic tool of the power series method.

Termwise differentiation

A power series may be differentiated term by term. More precisely: if

$$y(x) = \sum_{m=0}^{\infty} a_m(x - x_0)^m$$

converges for $|x - x_0| < R$, where $R > 0$, then the series obtained by differentiating term by term also converges for those x and represents the derivative y' of y for those x, that is,

$$y'(x) = \sum_{m=1}^{\infty} m a_m (x - x_0)^{m-1}.$$

(Proof in Sec. 14.4, Theorems 3 and 5.)

Termwise addition

Two power series may be added term by term. More precisely: if the series

$$(7) \qquad \sum_{m=0}^{\infty} a_m (x - x_0)^m \qquad \text{and} \qquad \sum_{m=0}^{\infty} b_m (x - x_0)^m$$

have positive radii of convergence and their sums are $f(x)$ and $g(x)$, then the series

$$\sum_{m=0}^{\infty} (a_m + b_m)(x - x_0)^m$$

converges and represents $f(x) + g(x)$ for each x which lies in the interior of the convergence interval of each of the given series. (Proof in Sec. 14.4.)

Termwise multiplication

Two power series may be multiplied term by term. More precisely: Suppose that the series (7) have positive radii of convergence and let $f(x)$ and $g(x)$ be their sums. Then the series obtained by multiplying each term of the first series by each term of the second series and collecting like powers of $x - x_0$, that is,

$$\sum_{m=0}^{\infty} (a_0 b_m + a_1 b_{m-1} + \cdots + a_m b_0)(x - x_0)^m$$
$$= a_0 b_0 + (a_0 b_1 + a_1 b_0)(x - x_0) + (a_0 b_2 + a_1 b_1 + a_2 b_0)(x - x_0)^2 + \cdots$$

converges and represents $f(x)g(x)$ for each x in the interior of the convergence interval of each of the given series. (See Sec. 14.4.)

Vanishing of all coefficients

If a power series has a positive radius of convergence and a sum that is identically zero throughout its interval of convergence, then each coefficient of the series is zero. (Proof in Sec. 14.4, Theorem 2.)

These properties of power series form the theoretical basis of the power series method. The remaining question is whether a given differential equation has solutions representable by power series at all. We may answer this question by using the following concept.

A function $f(x)$ is said to be **analytic** *at a point* $x = x_0$ if it can be represented by a power series in powers of $x - x_0$ with radius of convergence $R > 0$.

Using this notion, we may formulate a basic criterion of the desired type:

Theorem 1 (Existence of power series solutions)

If the functions p, q and r in the differential equation

$$(8) \qquad\qquad y'' + p(x)y' + q(x)y = r(x)$$

are analytic at $x = x_0$, then every solution $y(x)$ of (8) is analytic at $x = x_0$ and can thus be represented by a power series in powers of $x - x_0$ with radius of convergence[3] $R > 0$.

The proof requires advanced methods of complex analysis and can be found in Ref. [A10] in Appendix 1.

Caution! In applying this theorem it is important to write the linear equation in the form (8), with 1 as the coefficient of y''.

The examples and problems considered so far were intended to familiarize us with the power series method and to gain confidence and technical skill, and they concerned equations we could solve otherwise, to permit comparison of the results. In the next section we consider the first "big" equation of physics, the **Legendre equation,** and its solutions, the *Legendre functions* and **Legendre polynomials,** which are new functions not considered in calculus. Their applications to electrostatics follows in Sec. 11.12.

Problems for Sec. 4.2

Apply the power series method to the following differential equations.

1. $xy' - 3y = 6$

2. $xy' = (x + 1)y$

3. $(1 - x^2)y' = 2xy$

4. $xy' - (x + 2)y = -2x^2 - 2x$

5. $(x + 1)y' - (2x + 3)y = 0$

6. $x(x + 1)y' - (2x + 1)y = 0$

7. $(x - 3)y' - xy = 0$

8. $y'' - y = x$

9. $y'' - 3y' + 2y = 0$

10. $y'' - 4xy' + (4x^2 - 2)y = 0$

11. $(1 - x^2)y'' - 2xy' + 2y = 0$

12. $y'' - xy' + y = 0$

13. Show that $y' = (y/x) + 1$ cannot be solved for y as a power series in x. Solve this equation for y as a power series in powers of $x - 1$. (*Hint.* Introduce $t = x - 1$ as a new independent variable and solve the resulting equation for y as a power series in t.) Compare the result with that obtained by the appropriate elementary method.

Solve for y as a power series in powers of $x - 1$:

14. $y' = ky$

15. $y'' + y = 0$

16. $y'' - y = 0$

Radius of convergence. Find the radius of convergence of the following series.

17. $\displaystyle\sum_{m=0}^{\infty} \frac{x^m}{3^m}$

18. $\displaystyle\sum_{m=0}^{\infty} \frac{(-1)^m x^m}{m!}$

19. $\displaystyle\sum_{m=0}^{\infty} (-1)^m x^{2m}$

20. $\displaystyle\sum_{m=1}^{\infty} \frac{x^m}{m}$

21. $\displaystyle\sum_{m=2}^{\infty} \frac{m(m - 1)}{3^m} x^m$

22. $\displaystyle\sum_{m=0}^{\infty} \frac{1}{2^m} (x - 3)^{2m}$

[3]R is at least equal to the distance between the point $x = x_0$ and that point (or those points) closest to $x = x_0$ at which one of the functions p, q, r, *as functions of a complex variable,* is not analytic. (Note that that point may not lie on the x-axis, but somewhere in the complex plane.)

23. $\displaystyle\sum_{m=0}^{\infty} \frac{(-1)^m}{k^m} x^{2m}$

24. $\displaystyle\sum_{m=1}^{\infty} \frac{(4m)!}{(m!)^4} x^m$

25. $\displaystyle\sum_{m=0}^{\infty} \frac{(-1)^m x^{2m+1}}{(2m+1)!}$

26. $\displaystyle\sum_{m=2}^{\infty} \frac{m(m-1)}{4^m} x^m$

27. $\displaystyle\sum_{m=0}^{\infty} \left(\frac{7}{5}\right)^m x^{2m}$

28. $\displaystyle\sum_{m=0}^{\infty} \frac{x^{2m+1}}{(2m+1)m!}$

29. **(Shift of summation index)** In the power series method it will sometimes be helpful to make simple shifts of summation indices. To get used to this, show that

$$\sum_{m=2}^{\infty} m(m-1)a_m x^{m-2} = \sum_{p=1}^{\infty} (p+1)pa_{p+1}x^{p-1} = \sum_{s=0}^{\infty} (s+2)(s+1)a_{s+2}x^s.$$

What is the relation between m, p and s?

Shift of summation index. Shift the index so that the power under the summation sign is x^m. Check your result by writing the first few terms explicitly. Also determine the radius of convergence.

30. $\displaystyle\sum_{n=1}^{\infty} \frac{(-1)^{n+1}}{3n} x^{n+2}$

31. $\displaystyle\sum_{s=2}^{\infty} \frac{s(s+1)}{s^2+1} x^{s-1}$

32. $\displaystyle\sum_{p=2}^{\infty} p(p-1)x^{p-2}$

33. $\displaystyle\sum_{n=0}^{\infty} \frac{(n+1)^3}{(n+2)!} x^{n+1}$

34. $\displaystyle\sum_{k=3}^{\infty} \frac{(-1)^{k+1}}{6^k} x^{k-3}$

35. $\displaystyle\sum_{s=0}^{\infty} \frac{5^{s+2}}{s+3} x^{s+2}$

4.3 Legendre's Equation. Legendre Polynomials $P_n(x)$

Legendre's differential equation[4]

(1)
$$\boxed{(1-x^2)y'' - 2xy' + n(n+1)y = 0}$$

arises in numerous physical problems, particularly in boundary value problems for spheres, as we shall see in connection with partial differential equations in Sec. 11.12. The parameter n in (1) is a given real number. Any solution of (1) is called a **Legendre function.**

Dividing (1) by $1 - x^2$, we obtain the standard form (8), Sec. 4.2, and we see that the coefficients of the resulting equation are analytic at $x = 0$, so that we may apply the power series method. Substituting

(2)
$$y = \sum_{m=0}^{\infty} a_m x^m$$

and its derivatives into (1) and denoting the constant $n(n+1)$ by k we obtain

[4]ADRIEN MARIE LEGENDRE (1752—1833), French mathematician, who became a professor in Paris in 1775 and made important contributions to special functions, elliptic integrals, number theory and the calculus of variations. His book *Éléments de géométrie* (1794) became very famous and had 12 editions in less than 30 years.

$$(1 - x^2) \sum_{m=2}^{\infty} m(m - 1)a_m x^{m-2} - 2x \sum_{m=1}^{\infty} ma_m x^{m-1} + k \sum_{m=0}^{\infty} a_m x^m = 0.$$

By writing the first expression as two separate series, we have the equation

$$(1^*) \quad \sum_{m=2}^{\infty} m(m - 1)a_m x^{m-2} - \sum_{m=2}^{\infty} m(m - 1)a_m x^m - 2\sum_{m=1}^{\infty} ma_m x^m + k\sum_{m=0}^{\infty} a_m x^m = 0,$$

written out

$$2 \cdot 1a_2 + 3 \cdot 2a_3 x + 4 \cdot 3a_4 x^2 + \cdots + (s + 2)(s + 1)a_{s+2} x^s + \cdots$$
$$- 2 \cdot 1a_2 x^2 - \cdots \qquad - s(s - 1)a_s x^s - \cdots$$
$$- 2 \cdot 1a_1 x - 2 \cdot 2a_2 x^2 - \cdots \qquad - 2sa_s x^s - \cdots$$
$$+ ka_0 + \quad ka_1 x + \quad ka_2 x^2 + \cdots \qquad + ka_s x^s + \cdots = 0.$$

Since this must be an identity in x if (2) is to be a solution of (1), the sum of the coefficients of each power of x must be zero; since $k = n(n + 1)$, this gives

(3a)
$$2a_2 + n(n + 1)a_0 = 0, \qquad \text{coefficients of } x^0$$
$$6a_3 + [-2 + n(n + 1)]a_1 = 0, \qquad \text{coefficients of } x^1$$

and in general, when $s = 2, 3, \cdots$,

(3b) $\quad (s + 2)(s + 1)a_{s+2} + [-s(s - 1) - 2s + n(n + 1)]a_s = 0.$

Now the expression in brackets $[\cdots]$ can be written $(n - s)(n + s + 1)$, as the student may readily verify. We thus obtain from (3)

(4) $\qquad a_{s+2} = -\dfrac{(n - s)(n + s + 1)}{(s + 2)(s + 1)} a_s \qquad (s = 0, 1, \cdots).$

This is called a **recurrence relation** or **recursion formula**. It gives each coefficient in terms of the second one preceding it, except for a_0 and a_1, which are left as arbitrary constants. We find successively

$$a_2 = -\frac{n(n + 1)}{2!} a_0 \qquad\qquad a_3 = -\frac{(n - 1)(n + 2)}{3!} a_1$$

$$a_4 = -\frac{(n - 2)(n + 3)}{4 \cdot 3} a_2 \qquad\qquad a_5 = -\frac{(n - 3)(n + 4)}{5 \cdot 4} a_3$$

$$= \frac{(n - 2)n(n + 1)(n + 3)}{4!} a_0 \qquad\qquad = \frac{(n - 3)(n - 1)(n + 2)(n + 4)}{5!} a_1$$

etc. By inserting these values for the coefficients into (2) we obtain

$$(5) \qquad\qquad y(x) = a_0 y_1(x) + a_1 y_2(x)$$

where

$$(6) \quad y_1(x) = 1 - \frac{n(n+1)}{2!} x^2 + \frac{(n-2)n(n+1)(n+3)}{4!} x^4 - + \cdots$$

and

$$(7) \quad y_2(x) = x - \frac{(n-1)(n+2)}{3!} x^3 + \frac{(n-3)(n-1)(n+2)(n+4)}{5!} x^5 - + \cdots .$$

These series converge for $|x| < 1$. Since (6) contains even powers of x only, while (7) contains odd powers of x only, the ratio y_1/y_2 is not a constant, so that y_1 and y_2 are not proportional and thus linearly independent solutions. Hence (5) is a general solution of (1) on the interval $-1 < x < 1$.

Legendre Polynomials

In many applications the parameter n in Legendre's equation will be a non-negative integer. Then the right-hand side of (4) is zero when $s = n$, and, therefore, $a_{n+2} = 0$, $a_{n+4} = 0$, $a_{n+6} = 0$, $\cdots$. Hence, if n is even, $y_1(x)$ reduces to a polynomial of degree n. If n is odd, the same is true for $y_2(x)$. These polynomials, multiplied by some constants, are called **Legendre polynomials.** Since they are of great practical importance, let us consider them in more detail. For this purpose we solve (4) for a_s, obtaining

$$(8) \qquad\qquad a_s = - \frac{(s+2)(s+1)}{(n-s)(n+s+1)} a_{s+2} \qquad (s \leqq n-2)$$

and may then express all the nonvanishing coefficients in terms of the coefficient a_n of the highest power of x of the polynomial. The coefficient a_n is at first still arbitrary. It is customary to choose $a_n = 1$ when $n = 0$ and

$$(9) \qquad\qquad a_n = \frac{(2n)!}{2^n (n!)^2} = \frac{1 \cdot 3 \cdot 5 \cdots (2n-1)}{n!}, \qquad n = 1, 2, \cdots .$$

The reason is that for this choice of a_n all those polynomials will have the value 1 when $x = 1$; this follows from (14) in Prob. 14. We then obtain from (8) and (9)

$$(9^*) \qquad
\begin{aligned}
a_{n-2} &= - \frac{n(n-1)}{2(2n-1)} a_n = - \frac{n(n-1)(2n)!}{2(2n-1)2^n(n!)^2} \\[2mm]
&= - \frac{n(n-1)2n(2n-1)(2n-2)!}{2(2n-1)2^n n(n-1)! \, n(n-1)(n-2)!} ,
\end{aligned}$$

that is,

$$a_{n-2} = -\frac{(2n-2)!}{2^n(n-1)!\,(n-2)!}.$$

Similarly,

$$a_{n-4} = -\frac{(n-2)(n-3)}{4(2n-3)}\,a_{n-2} = \frac{(2n-4)!}{2^n 2!\,(n-2)!\,(n-4)!}$$

etc., and in general, when $n - 2m \geqq 0$,

$$(10) \qquad a_{n-2m} = (-1)^m \frac{(2n-2m)!}{2^n m!\,(n-m)!\,(n-2m)!}.$$

The resulting solution of Legendre's differential equation (1) is called the **Legendre polynomial** *of degree* n and is denoted by $P_n(x)$. From (10) we obtain

$$(11) \qquad \boxed{\begin{aligned} P_n(x) &= \sum_{m=0}^{M} (-1)^m \frac{(2n-2m)!}{2^n m!\,(n-m)!\,(n-2m)!}\,x^{n-2m} \\ &= \frac{(2n)!}{2^n(n!)^2}\,x^n - \frac{(2n-2)!}{2^n 1!\,(n-1)!\,(n-2)!}\,x^{n-2} + - \cdots, \end{aligned}}$$

where $M = n/2$ or $(n-1)/2$, whichever is an integer. In particular (Fig. 83),

$$P_0(x) = 1, \qquad\qquad P_1(x) = x,$$

$$(11') \quad P_2(x) = \tfrac{1}{2}(3x^2 - 1), \qquad P_3(x) = \tfrac{1}{2}(5x^3 - 3x),$$

$$P_4(x) = \tfrac{1}{8}(35x^4 - 30x^2 + 3), \qquad P_5(x) = \tfrac{1}{8}(63x^5 - 70x^3 + 15x),$$

etc.

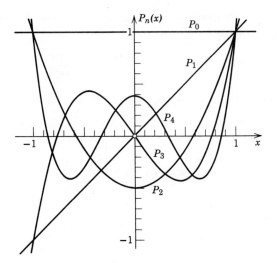

Fig. 83. Legendre polynomials

The so-called *orthogonality* of the Legendre polynomials will be considered in Sec. 4.9.

Whereas the power series method applies to the Legendre equation, as we have just seen, there are other important equations, notably Bessel's equation, which require a more general method of solution, the so-called **Frobenius method** or **extended power series method.** The simple reason is that these equations have no power series solutions but solutions that are a power series times a fractional power of x, sometimes still multiplied by a logarithm. We discuss this more general method in the next section and its application to Bessel's equation in Secs. 4.5 and 4.6.

Problems for Sec. 4.3

1. Using (11′), verify by substitution that $P_0, \cdots, P_5$ satisfy Legendre's equation.
2. Find and graph $P_6(x)$.
3. Derive (11′) from (11).
4. Show that we can get (3) from (1*) more quickly if we write $m - 2 = s$ in the first sum in (1*) and $m = s$ in the other sums, obtaining

$$\sum_{s=0}^{\infty} \{(s + 2)(s + 1)a_{s+2} - [s(s - 1) + 2s - k]a_s\}x^s = 0.$$

5. Show that the equation in Prob. 11, Sec. 4.2, is a special Legendre equation and obtain a general solution from (5), this section.
6. Show that for any n for which the series (6) or (7) do not reduce to a polynomial, these series have radius of convergence 1.
7. Show that if $n = 0$, then $y_2 = x + \frac{1}{3}x^3 + \frac{1}{5}x^5 + \cdots = \frac{1}{2} \ln [(1 + x)/(1 - x)]$.
8. Verify by substitution that y_2 in Prob. 7 is a solution of (1) with $n = 0$.
9. Obtain the result in Prob. 7 by solving (1) with $n = 0$ by an elementary method.
10. Show that for $n = 1$ we have $y_1(x) = 1 - \frac{1}{2}x \ln [(1 + x)/(1 - x)]$.
11. Using (11), show that $P_n(-x) = (-1)^n P_n(x)$ and $P_n'(-x) = (-1)^{n+1}P_n'(x)$.
12. **(Rodrigues' formula**[5]**)** Applying the binomial theorem to $(x^2 - 1)^n$, differentiating n times term by term, and comparing with (11), show that

(12) $$P_n(x) = \frac{1}{2^n n!} \frac{d^n}{dx^n} [(x^2 - 1)^n] \qquad (Rodrigues'\ formula).$$

13. Using (12) and integrating n times by parts, show that

(13) $$\int_{-1}^{1} P_n^2(x)\ dx = \frac{2}{2n + 1} \qquad (n = 0, 1, \cdots).$$

14. **(Generating function)** Show that

(14) $$\frac{1}{\sqrt{1 - 2xu + u^2}} = \sum_{n=0}^{\infty} P_n(x)u^n.$$

[5]OLINDE RODRIGUES (1794—1851), French mathematician and economist.

The function on the left is called a *generating function* of the Legendre polynomials. *Hint.* Start from the binomial expansion of $1/\sqrt{1-v}$, set $v = 2xu - u^2$, multiply the powers of $2xu - u^2$ out, collect all the terms involving u^n, and verify that the sum of these terms is $P_n(x)u^n$.

15. Let A_1 and A_2 be two points in space (Fig. 84, $r_2 > 0$). Using (14), show that

$$\frac{1}{r} = \frac{1}{\sqrt{r_1{}^2 + r_2{}^2 - 2r_1 r_2 \cos \theta}} = \frac{1}{r_2} \sum_{m=0}^{\infty} P_m(\cos \theta) \left(\frac{r_1}{r_2}\right)^m$$

This formula has applications in potential theory.

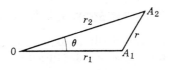

Fig. 84. Problem 15

Using (14), show that

16. $P_n(1) = 1$

17. $P_n(-1) = (-1)^n$

18. $P_{2n+1}(0) = 0$

19. $P_{2n}(0) = 1 \cdot 3 \cdots (2n-1)/2 \cdot 4 \cdots (2n)$.

20. (Bonnet's recursion[6]) Differentiating (14) with respect to u, using (14) in the resulting formula, and comparing coefficients of u^n, obtain the *Bonnet recursion*

$$(15) \qquad (n+1)P_{n+1}(x) = (2n+1)xP_n(x) - nP_{n-1}(x), \qquad n = 1, 2, \cdots.$$

21. (Computation) Formula (15) is useful for computations, the loss of significant figures being small (except at zeros). Using (15), compute $P_2(2.6)$ and $P_3(2.6)$.

22. Using (15) and (11'), find P_6.

23. (Associated Legendre functions) Consider

$$(1-x^2)y'' - 2xy' + \left[n(n+1) - \frac{m^2}{1-x^2} \right] y = 0.$$

Substituting $y(x) = (1-x^2)^{m/2} u(x)$, show that u satisfies

$$(16) \qquad (1-x^2)u'' - 2(m+1)xu' + [n(n+1) - m(m+1)]u = 0.$$

Starting from (1) and differentiating it m times, show that a solution of (16) is

$$u = \frac{d^m P_n}{dx^m}.$$

The corresponding $y(x)$ is denoted by $P_n{}^m(x)$. It is called an *associated Legendre function* and plays a role in quantum physics. Thus

---

[6]OSSIAN BONNET (1819—1892), French mathematician, whose main work was in the differential geometry of surfaces.

$$P_n{}^m(x) = (1 - x^2)^{m/2} \frac{d^m P_n}{dx^m} .$$

24. Find $P_1{}^1(x)$, $P_2{}^1(x)$, $P_2{}^2(x)$, $P_4{}^2(x)$.

25. Show that $P_n{}^m(x) = 0$ when $m > n$.

4.4 Extended Power Series Method. Indicial Equation

Why an Extension of the Method is Needed

The power series method takes care of linear differential equations which, when written in the standard form

$$y'' + p(x)y' + q(x)y = 0$$

[with y'' having coefficient 1] have coefficients p and q that are analytic (cf. Sec. 4.2) for all x for which the equation is considered. However, some equations of practical importance have a **"singular point,"** that is, a point x at which p or q is no longer analytic (or several such points), as opposed to the other points which are called **regular points** of the equation. For instance, the famous Bessel equation

$$y'' + \frac{1}{x} y' + \left(1 - \frac{\nu^2}{x^2}\right) y = 0 \qquad (\nu \text{ a given number})$$

(to be discussed in Secs. 4.5 and 4.6) has $x = 0$ as a singular point. Now if $x = 0$ (or $x = x_0$) is a singular point, the equation may not have a power series solution in powers of x (or $x - x_0$, respectively), so that the power series method would not work. Fortunately, often the behavior of the coefficients of the equation at a singular point is "not too bad," but such that the following theorem applies.

Theorem 1 (Frobenius method)

Any differential equation of the form

(1) $$y'' + \frac{b(x)}{x} y' + \frac{c(x)}{x^2} y = 0,$$

where the functions $b(x)$ and $c(x)$ are analytic at $x = 0$, has at least one solution which can be represented in the form

(2) $$y(x) = x^r \sum_{m=0}^{\infty} a_m x^m = x^r(a_0 + a_1 x + a_2 x^2 + \cdots) \qquad (a_0 \neq 0)$$

where the exponent r may be any (real or complex) number (and r is chosen *so that $a_0 \neq 0$).*[7]

The equation also has a second solution (such that these two solutions are linearly independent) which may be similar to (2) (with a different r and different coefficients) or may contain a logarithmic term (details in Theorem 2, below).

The point is that in (2) we have a power series times a single power of x whose exponent r is not restricted to be a nonnegative integer. (The latter restriction would make the whole expression a power series, by definition; see footnote 1 in Sec. 4.1.)

The proof of the theorem requires advanced methods of complex analysis and can be found in Ref. [A10] listed in Appendix 1.

The subsequent method for solving (1) is based on this theorem and is called the **Frobenius method**[8] or **extended power series method.**

Indicial Equation, Indicating the Form of Solutions

To solve (1), we write it in the somewhat more convenient form

(1′)
$$x^2 y'' + x b(x) y' + c(x) y = 0.$$

We first expand $b(x)$ and $c(x)$ in power series,

$$b(x) = b_0 + b_1 x + b_2 x^2 + \cdots, \qquad c(x) = c_0 + c_1 x + c_2 x^2 + \cdots.$$

Then we differentiate (2) term by term, finding

(2*)
$$y'(x) = \sum_{m=0}^{\infty} (m + r) a_m x^{m+r-1} = x^{r-1}[r a_0 + (r + 1) a_1 x + \cdots],$$

$$y''(x) = \sum_{m=0}^{\infty} (m + r)(m + r - 1) a_m x^{m+r-2}$$

$$= x^{r-2}[r(r - 1) a_0 + (r + 1) r a_1 x + \cdots].$$

By inserting all these series into (1′) we readily obtain

(3)
$$x^r[r(r - 1) a_0 + \cdots] + (b_0 + b_1 x + \cdots) x^r (r a_0 + \cdots)$$
$$+ (c_0 + c_1 x + \cdots) x^r (a_0 + a_1 x + \cdots) = 0.$$

---

[7]In this theorem, we may replace x by $x - x_0$, where x_0 is any number. Note that the condition $a_0 \neq 0$ is no restriction of generality; it simply means that we factor out the highest possible power of x.

The singular point of (1) at $x = 0$ is sometimes called a **regular singular point,** a term, confusing to the student, which we shall not use.

[8]GEORG FROBENIUS (1849—1917), German mathematician, who also made important contributions to the theory of matrices and groups.

We now equate the sum of the coefficients of each power of x to zero, as before. This yields a system of equations involving the unknown coefficients a_m. The smallest power is x^r, and the corresponding equation is

$$[r(r - 1) + b_0 r + c_0]a_0 = 0.$$

Since by assumption $a_0 \neq 0$, the expression in the brackets must be zero. This gives

(4)

$$\boxed{r^2 + (b_0 - 1)r + c_0 = 0.}$$

This important quadratic equation is called the **indicial equation** of the differential equation (1). Its role is as follows.

Our method will yield a basis of solutions. One of the two solutions will always be of the form (2), where r is a root of (4). The form of the other solution will be indicated by the indicial equation; depending on the roots, there are three possible cases as stated in the following theorem.[9]

Theorem 2 (Frobenius method. Form of the second solution)

Suppose that the differential equation (1) satisfies the assumptions in Theorem 1. Let r_1 and r_2 be the roots of the indicial equation (4). Then we have the following three cases.

Case 1. Distinct roots not differing by an integer.[10] *A basis is*

(5)
$$y_1(x) = x^{r_1}(a_0 + a_1 x + a_2 x^2 + \cdots)$$

and

(6)
$$y_2(x) = x^{r_2}(A_0 + A_1 x + A_2 x^2 + \cdots)$$

with coefficients obtained successively from (3) with $r = r_1$ and $r = r_2$, respectively.

Case 2. Double root $r_1 = r_2 = r$. *A basis is*

(7)
$$y_1(x) = x^r(a_0 + a_1 x + a_2 x^2 + \cdots) \qquad r = \tfrac{1}{2}(1 - b_0)$$

(of the same general form as before) and

(8)
$$y_2(x) = y_1(x) \ln x + x^r(A_1 x + A_2 x^2 + \cdots) \qquad\qquad (x > 0).$$

---

[9]A general theory of convergence of the occurring series will not be presented here, but in each individual case convergence may be tested in the usual way.

[10]Note that this case includes complex conjugate roots r_1 and $r_2 = \bar{r}_1$, because we have $r_1 - r_2 = r_1 - \bar{r}_1 = 2i \operatorname{Im} r_1 \neq \pm 1, \pm 2, \cdots$.

Case 3. Roots differing by an integer. *A basis is*

(9) $$y_1(x) = x^{r_1}(a_0 + a_1x + a_2x^2 + \cdots)$$

(of the same general form as before) and

(10) $$y_2(x) = ky_1(x) \ln x + x^{r_2}(A_0 + A_1x + A_2x^2 + \cdots),$$

where the roots are so denoted that $r_1 - r_2 > 0$ *and* k *may turn out to be zero.*

Proofs are given on p. 236 near the end of the chapter. Note that in Case 2 we *must* have a logarithm whereas in Case 3 we may or may not.

Typical Applications

EXAMPLE 1. Euler–Cauchy equation
The Euler–Cauchy equation (Sec. 2.7)

$$x^2y'' + b_0xy' + c_0y = 0 \qquad\qquad (b_0, c_0 \text{ constant})$$

satisfies the assumptions in Theorem 1. It is solved by substituting $y = x^r$ and its derivatives. After dropping the common power x^r there remains

$$r^2 + (b_0 - 1)r + c_0 = 0.$$

We see that this is the indicial equation (4) [and $y = x^r$ is a very special form of (2)!]. If the roots r_1, r_2 are different, a basis of solutions is

$$y_1 = x^{r_1}, \qquad y_2 = x^{r_2},$$

as in Case 1 or in Case 3 without a logarithmic term. For a double root $r_1 = r_2 = r$ a basis of solutions is

$$y_1 = x^r, \qquad y_2 = x^r \ln x.$$

This is Case 2. Interesting is that for this simple equation, Case 3 plays no extra role. Thus the equation foreshadows some of the Frobenius theory but not all of it.

EXAMPLE 2. An equation that leads to Case 1
Solve the differential equation

$$y'' + \frac{1}{2x} y' + \frac{1}{4x} y = 0$$

Solution. This equation satisfies the assumption of Theorems 1 and 2, so that the Frobenius method applies. We write the equation more conveniently as

$$4xy'' + 2y' + y = 0.$$

Substitution of (2) and of its derivatives (2*) gives

(11) $$4 \sum_{m=0}^{\infty} (m + r)(m + r - 1)a_m x^{m+r-1} + 2 \sum_{m=0}^{\infty} (m + r)a_m x^{m+r-1} + \sum_{m=0}^{\infty} a_m x^{m+r} = 0.$$

If you feel more comfortable by writing this out, go ahead:

$$4r(r - 1)a_0 x^{r-1} + 4(r + 1)ra_1 x^r + 4(r + 2)(r + 1)a_2 x^{r+1} + \cdots$$

$$+ 2ra_0 x^{r-1} + 2(r + 1)\, a_1 x^r + \qquad 2(r + 2)\, a_2 x^{r+1} + \cdots$$

$$+ \qquad a_0 x^r + \qquad\qquad a_1 x^{r+1} + \cdots = 0.$$

By equating the sum of the coefficients of x^{r-1} to zero we obtain the indicial equation

$$4r(r - 1) + 2r = 0, \qquad \text{thus} \qquad r^2 - \tfrac{1}{2}r = 0.$$

The roots are $r_1 = \tfrac{1}{2}$ and $r_2 = 0$. This is Case 1.

By equating the sum of the coefficients of x^{r+s} in (11) to zero we obtain (take $m + r - 1 = r + s$, thus $m = s + 1$ in the first two series and $m = s$ in the last series)

$$4(s + r + 1)(s + r)a_{s+1} + 2(s + r + 1)a_{s+1} + a_s = 0.$$

We see that this can be written

$$4(s + r + 1)(s + r + \tfrac{1}{2})a_{s+1} + a_s = 0.$$

We solve this for a_{s+1} in terms of a_s:

(12) $$a_{s+1} = - \frac{a_s}{(2s + 2r + 2)(2s + 2r + 1)} \qquad (s = 0, 1, \cdots).$$

First solution. We determine a first solution $y_1(x)$ corresponding to $r_1 = \tfrac{1}{2}$. For $r = r_1$, formula (12) becomes

$$a_{s+1} = - \frac{a_s}{(2s + 3)(2s + 2)} \qquad (s = 0, 1, \cdots).$$

From this we get successively

$$a_1 = - \frac{a_0}{3 \cdot 2}, \qquad a_2 = - \frac{a_1}{5 \cdot 4}, \qquad a_3 = - \frac{a_2}{7 \cdot 6}, \qquad \text{etc.}$$

In many practical situations an explicit formula for a_m will be rather complicated. Here it is simple: by successive substitution we get

$$a_1 = - \frac{a_0}{3!}, \qquad a_2 = \frac{a_0}{5!}, \qquad a_3 = - \frac{a_0}{7!}, \qquad \cdots$$

and in general, taking $a_0 = 1$,

$$a_m = \frac{(-1)^m}{(2m + 1)!} \qquad\qquad m = 0, 1, \cdots.$$

Hence the first solution is

$$y_1(x) = x^{1/2} \sum_{m=0}^{\infty} \frac{(-1)^m}{(2m + 1)!} x^m = \sqrt{x}\left(1 - \frac{1}{6}x + \frac{1}{120}x^2 - + \cdots\right).$$

Second solution. We now determine a second solution $y_2(x)$ corresponding to $r_2 = 0$, of the form (6). For $r = r_2 = 0$, formula (12) [with A_{s+1} and A_s instead of a_{s+1} and a_s] becomes

$$A_{s+1} = - \frac{A_s}{(2s + 2)(2s + 1)} \qquad (s = 0, 1, \cdots).$$

From this we get successively

$$A_1 = - \frac{A_0}{2 \cdot 1}, \qquad A_2 = - \frac{A_1}{4 \cdot 3}, \qquad A_3 = - \frac{A_2}{6 \cdot 5},$$

and by successive substitution we have

$$A_1 = -\frac{A_0}{2!}, \qquad A_2 = \frac{A_0}{4!}, \qquad A_3 = -\frac{A_0}{6!}, \qquad \cdots$$

and in general, taking $A_0 = 1$,

$$A_m = \frac{(-1)^m}{(2m)!}.$$

Hence the second solution, of the form (6) with $r_2 = 0$, is

$$y_2(x) = \sum_{m=0}^{\infty} \frac{(-1)^m}{(2m)!} x^m = 1 - \tfrac{1}{2}x + \tfrac{1}{24}x^2 - + \cdots.$$

y_1 and y_2 are linearly independent on the positive x-axis since their quotient is not constant. Hence these solutions form a basis for all $x > 0$.

EXAMPLE 3. Illustration of Case 2 (Double root)

Solve the differential equation

$$(13) \qquad\qquad x(x-1)y'' + (3x-1)y' + y = 0.$$

(This is a special hypergeometric differential equation, as we shall see in the problem set.)

Solution. Writing this equation in the standard form (1), we see that it satisfies the assumptions in Theorem 1. By inserting (2) and its derivatives (2*) into (13) we obtain

$$(14) \quad \begin{aligned} &\sum_{m=0}^{\infty} (m+r)(m+r-1)a_m x^{m+r} - \sum_{m=0}^{\infty} (m+r)(m+r-1)a_m x^{m+r-1} \\ &+ 3\sum_{m=0}^{\infty} (m+r)a_m x^{m+r} - \sum_{m=0}^{\infty} (m+r)a_m x^{m+r-1} + \sum_{m=0}^{\infty} a_m x^{m+r} = 0. \end{aligned}$$

The smallest power is x^{r-1}; by equating the sum of its coefficients to zero we have

$$[-r(r-1) - r]a_0 = 0 \qquad \text{or} \qquad r^2 = 0.$$

Hence this indicial equation has the double root $r = 0$.

First solution. We insert this value into (14) and equate the sum of the coefficients of the power x^s to zero, finding

$$s(s-1)a_s - (s+1)sa_{s+1} + 3sa_s - (s+1)a_{s+1} + a_s = 0$$

or $a_{s+1} = a_s$. Hence $a_0 = a_1 = a_2 = \cdots$, and by choosing $a_0 = 1$ we obtain the solution

$$y_1(x) = \sum_{m=0}^{\infty} x^m = \frac{1}{1-x}.$$

Second solution. We now substitute $y_2 = uy_1$ and its derivatives into (13), finding

$$x(x-1)(u''y_1 + 2u'y_1' + uy_1'') + (3x-1)(u'y_1 + uy_1') + uy_1 = 0.$$

Since y_1 is a solution of (13), this reduces to

$$x(x-1)(u''y_1 + 2u'y_1') + (3x-1)u'y_1 = 0.$$

Inserting the expressions for y_1 and y_1', we first have

$$x(x-1)\left(u''\frac{1}{1-x} + 2u'\frac{1}{(1-x)^2}\right) + (3x-1)u'\frac{1}{1-x} = 0.$$

Simplifying this algebraically, we readily obtain

$$xu'' + u' = 0; \qquad \text{hence} \qquad \frac{u''}{u'} = -\frac{1}{x}.$$

By integrating twice,

$$\ln u' = -\ln x = \ln \frac{1}{x}, \qquad u' = \frac{1}{x}, \qquad u = \ln x.$$

Hence a second independent solution of (13) is

$$y_2 = uy_1 = \frac{\ln x}{1 - x}.$$

EXAMPLE 4. Case 3, second solution without logarithmic term.
This case is illustrated by the Euler–Cauchy equation with r_1 and r_2 differing by an integer; cf. Example 1.

EXAMPLE 5. Case 3, second solution with logarithmic term
Solve

$$(x^2 - 1)x^2y'' - (x^2 + 1)xy' + (x^2 + 1)y = 0.$$

Solution. By substituting (2) and its derivatives into the differential equation we first have

$$(x^2 - 1) \sum_{m=0}^{\infty} (m + r)(m + r - 1)a_m x^{m+r} - (x^2 + 1) \sum_{m=0}^{\infty} (m + r)a_m x^{m+r} + (x^2 + 1) \sum_{m=0}^{\infty} a_m x^{m+r} = 0.$$

Performing the indicated multiplications and simplifying, we obtain

$$(15) \qquad \sum_{m=0}^{\infty} (m + r - 1)^2 a_m x^{m+r+2} - \sum_{m=0}^{\infty} (m + r + 1)(m + r - 1)a_m x^{m+r} = 0.$$

By equating the coefficient of x^r to zero we see that the indicial equation is

$$(r + 1)(r - 1) = 0.$$

The roots $r_1 = 1$ and $r_2 = -1$ differ by an integer. Hence we are in Case 3.

First solution. By equating the coefficient of x^{r+1} in (15) to zero we have

$$-(r + 2)ra_1 = 0.$$

For $r = r_1$ (as well as for $r = r_2$) this implies $a_1 = 0$. By equating the sum of the coefficients of x^{s+r+2} in (15) to zero we find

$$(16) \qquad (s + r - 1)^2 a_s = (s + r + 3)(s + r + 1)a_{s+2} \qquad (s = 0, 1, \cdots).$$

Inserting $r = r_1 = 1$ and solving for a_{s+2}, we see that

$$a_{s+2} = \frac{s^2}{(s + 4)(s + 2)} a_s \qquad (s = 0, 1, \cdots).$$

Since $a_1 = 0$, it follows that $a_3 = 0$, $a_5 = 0$, etc. For $s = 0$ we obtain $a_2 = 0$, and from this by taking $s = 2, 4, \cdots$, we have $a_4 = 0$, $a_6 = 0$, etc. Hence to the larger root $r_1 = 1$ there corresponds the solution

$$y_1 = a_0 x.$$

Second solution. If we insert the smaller root $r = r_2 = -1$ into (16), we get

$$(s - 2)^2 a_s = (s + 2)s a_{s+2} \qquad\qquad (s = 0, 1, \cdots).$$

With $s = 0$ this becomes $4a_0 = 0$, which implies $a_0 = 0$. This contradicts our initial assumption $a_0 \neq 0$. In fact, we shall now see that our equation does not have a second independent solution $y_2(x)$ of the form (2) corresponding to the smaller root r_2.

To obtain a second solution, we could start from (10) with $r_2 = -1$ and determine the coefficients by substituting (10) into our present equation. It is simpler to use the method of reduction of order directly; that is, since x is a solution, we may set

$$y_2(x) = xu(x).$$

Substituting this into the present differential equation and simplifying, we find

$$(x^3 - x)u'' + (x^2 - 3)u' = 0.$$

Using partial fractions, we obtain

$$\frac{u''}{u'} = \frac{3 - x^2}{x^3 - x} = -\frac{3}{x} + \frac{1}{x + 1} + \frac{1}{x - 1}.$$

Integrating on both sides, we have

$$\ln u' = -3 \ln x + \ln (x + 1) + \ln (x - 1) = \ln \frac{x^2 - 1}{x^3}.$$

Taking exponentials and integrating again, we find

$$u = \ln x + \frac{1}{2x^2}.$$

This yields as a second independent solution

$$y_2(x) = xu(x) = x \ln x + \frac{1}{2x}. \qquad\qquad \blacksquare$$

The Frobenius method solves the **hypergeometric equation,** whose solutions include many known functions as special cases (see the problem set). In the next section we use this method for solving **Bessel's equation.**

Problems for Sec. 4.4

Find a basis of solutions of the following differential equations. Try to identify the series obtained by the Frobenius method as expansions of known functions.

1. $xy'' + 2y' + 4xy = 0$ 2. $xy'' + 2y' + xy = 0$
3. $xy'' + (1 - 2x)y' + (x - 1)y = 0$ 4. $y'' + xy' + (1 - 2x^{-2})y = 0$
5. $(x - 1)xy'' + (4x - 2)y' + 2y = 0$ 6. $x^2y'' + 6xy' + (6 - 4x^2)y = 0$
7. $x^2y'' - 5xy' + 9y = 0$ 8. $xy'' - y = 0$
9. $x(1 - x)y'' + \frac{1}{2}(x + 1)y' - \frac{1}{2}y = 0$ 10. $y'' + x^{-1}y' - y = 0$
11. $(x + 2)^2 y'' + (x + 2)y' - y = 0$ 12. $xy'' + 3y' + 4x^3 y = 0$
13. $(x - 1)^2 y'' + (x - 1)y' - 4y = 0$ 14. $(x + \frac{1}{2})^2 y'' - (x + \frac{1}{2})y' + y = 0$
15. $x(1 - x)y'' + (\frac{3}{2} - \frac{7}{2}x)y' - \frac{3}{2}y = 0$ 16. $x^2y'' + xy' + (x^2 - \frac{1}{4})y = 0$
17. $(1 + x)x^2 y'' - (1 + 2x)xy' + (1 + 2x)y = 0$
18. $xy'' + (2 - 2x)y' + (x - 2)y = 0$
19. $2x(1 - 2x)y'' + (12x^2 - 4x + 1)y' - 2(4x^2 + 1)y = 0$
20. $x^2(x + 1)^2 y'' - (5x^2 + 8x + 3)xy' + (9x^2 + 11x + 4)y = 0$

Hypergeometric equation, hypergeometric series, hypergeometric functions
21. Gauss's hypergeometric differential equation[11] is

(17)
$$\boxed{x(1-x)y'' + [c - (a + b + 1)x]y' - aby = 0}$$

where a, b, c are constants. Show that the corresponding indicial equation has the roots $r_1 = 0$ and $r_2 = 1 - c$. Show that for $r_1 = 0$ the Frobenius method yields

(18)
$$y_1(x) = 1 + \frac{ab}{1!\,c}x + \frac{a(a+1)b(b+1)}{2!\,c(c+1)}x^2$$
$$+ \frac{a(a+1)(a+2)b(b+1)(b+2)}{3!\,c(c+1)(c+2)}x^3 + \cdots$$

where $c \neq 0, -1, -2, \cdots$. This series is called the **hypergeometric series**. Its sum $y_1(x)$ is denoted by $F(a, b, c; x)$ and is called the **hypergeometric function**.

Using (18), prove:
22. The series (18) converges for $|x| < 1$.
23. $F(1, b, b; x) = 1 + x + x^2 + \cdots$, the geometric series.
24. If a or b is a negative integer, (18) reduces to a polynomial.
25. $\dfrac{dF(a, b, c; x)}{dx} = \dfrac{ab}{c}F(a+1, b+1, c+1; x)$,

$$\frac{d^2F(a, b, c; x)}{dx^2} = \frac{a(a+1)b(b+1)}{c(c+1)}F(a+2, b+2, c+2; x), \qquad \text{etc.}$$

Many elementary functions are special cases of $F(a, b, c; x)$. Prove
26. $\dfrac{1}{1-x} = F(1, 1, 1; x) = F(1, b, b; x) = F(a, 1, a; x)$
27. $(1 + x)^n = F(-n, b, b; -x)$, $(1 - x)^n = 1 - nxF(1 - n, 1, 2; x)$
28. $\arctan x = xF(\tfrac{1}{2}, 1, \tfrac{3}{2}; -x^2)$, $\arcsin x = xF(\tfrac{1}{2}, \tfrac{1}{2}, \tfrac{3}{2}; x^2)$
29. $\ln(1 + x) = xF(1, 1, 2; -x)$, $\ln\dfrac{1+x}{1-x} = 2xF(\tfrac{1}{2}, 1, \tfrac{3}{2}; x^2)$

30. (Second solution) Show that for $r_2 = 1 - c$ in Prob. 21 the Frobenius method yields the following solution (where $c \neq 2, 3, 4, \cdots$):

(19)
$$y_2(x) = x^{1-c}\left(1 + \frac{(a - c + 1)(b - c + 1)}{1!\,(-c + 2)}x\right.$$
$$\left. + \frac{(a - c + 1)(a - c + 2)(b - c + 1)(b - c + 2)}{2!\,(-c + 2)(-c + 3)}x^2 + \cdots\right)$$

---

[11]CARL FRIEDRICH GAUSS (1777–1855), great German mathematician. He already made the first of his great discoveries as a student at Helmstedt and Göttingen. In 1807 he became a professor and director of the Observatory at Göttingen. His work was of basic importance in algebra, number theory, differential equations, differential geometry, non-Euclidean geometry, complex analysis, numerical analysis, astronomy, geodesy, electromagnetism and theoretical mechanics. He also paved the way for a general and systematic use of complex numbers.

31. Show that in Prob. 30,

$$y_2(x) = x^{1-c}F(a - c + 1, b - c + 1, 2 - c; x).$$

32. Show that if $c \neq 0, \pm 1, \pm 2, \cdots$, the functions (18) and (19) constitute a basis of solutions of (17).

33. Consider the differential equation

(20) $$(t^2 + At + B)\ddot{y} + (Ct + D)\dot{y} + Ky = 0$$

where A, B, C, D, K are constants, $\dot{y} = dy/dt$, and $t^2 + At + B$ has distinct zeros t_1 and t_2. Show that by introducing the new independent variable

$$x = \frac{t - t_1}{t_2 - t_1}$$

the equation (20) becomes the hypergeometric equation, where the parameters are related by $Ct_1 + D = -c(t_2 - t_1), C = a + b + 1, K = ab$.

Solve, in terms of hypergeometric functions, the following equations.

34. $2x(1 - x)y'' + (1 - 5x)y' - y = 0$ **35.** $2x(1 - x)y'' - (1 + 5x)y' - y = 0$

36. $5x(1 - x)y'' + (4 - x)y' + y = 0$ **37.** $4x(x - 1)y'' + 2(4x - 1)y' + y = 0$

38. $2x(x - 1)y'' + (6x + 1)y' + 2y = 0$ **39.** $2x(1 - x)y'' + (3 - 7x)y' - 3y = 0$

40. $4x(1 - x)y'' + (1 - 10x)y' - 2y = 0$ **41.** $3x(1 - x)y'' + (1 - 5x)y' + y = 0$

42. $3x(x - 1)y'' + (1 + 9x)y' + 3y = 0$ **43.** $3t(1 + t)\ddot{y} + t\dot{y} - y = 0$

44. $2(t^2 - 5t + 6)\ddot{y} + (2t - 3)\dot{y} - 8y = 0$ **45.** $4(t^2 - 3t + 2)\ddot{y} - 2\dot{y} + y = 0$

4.5 Bessel's Equation. Bessel Functions of the First Kind

One of the most important differential equations in applied mathematics is **Bessel's differential equation**[12]

(1) $$\boxed{x^2y'' + xy' + (x^2 - \nu^2)y = 0}$$

where the parameter ν is a given number. This equation appears in problems on vibrations, electrostatic fields, heat conduction, etc., in most cases when the problem exhibits cylindrical symmetry. The reader may want to take a quick look at the modeling of the vibrations of a circular membrane, a typical problem that leads to the Bessel equation (Sec. 11.10).

We assume that the parameter ν in (1) is real and nonnegative. We see that Bessel's equation is of the type characterized in Theorem 1, Sec. 4.4. Hence it has solutions of the form

---

[12]FRIEDRICH WILHELM BESSEL (1784–1846), German astronomer and mathematician, started out as an apprentice of a trade company, studying astronomy on his own in his spare time, later became an assistant at a small private observatory, and finally director of the new Königsberg observatory. His paper on the Bessel functions (dated 1824) appeared in 1826.

(2) $$y(x) = \sum_{m=0}^{\infty} a_m x^{m+r} \qquad (a_0 \neq 0).$$

Substituting this and its derivatives into Bessel's equation we have

$$\sum_{m=0}^{\infty} (m + r)(m + r - 1)a_m x^{m+r} + \sum_{m=0}^{\infty} (m + r)a_m x^{m+r}$$

$$+ \sum_{m=0}^{\infty} a_m x^{m+r+2} - \nu^2 \sum_{m=0}^{\infty} a_m x^{m+r} = 0.$$

We equate the sum of the coefficients of x^{s+r} to zero. Note that this power x^{s+r} corresponds to $m = s$ in the first, second and fourth sums, and to $m = s - 2$ in the third sum. Hence for $s = 0$ and $s = 1$, the third sum does not contribute since $m \geq 0$. For $s = 2, 3, \cdots$ all four sums contribute, so that we get a general formula for all these s. We find

    (a) $\qquad\qquad r(r - 1)a_0 + ra_0 - \nu^2 a_0 = 0 \qquad\qquad (s = 0)$

(3) (b) $\qquad\quad (r + 1)ra_1 + (r + 1)a_1 - \nu^2 a_1 = 0 \qquad (s = 1)$

    (c) $\quad (s + r)(s + r - 1)a_s + (s + r)a_s + a_{s-2} - \nu^2 a_s = 0 \quad (s = 2, 3, \cdots).$

From (3a) we obtain the **indicial equation**

(4) $$(r + \nu)(r - \nu) = 0.$$

The roots are $r_1 = \nu \; (\geq 0)$ and $r_2 = -\nu$.

Solution Corresponding to the Root $r_1 = \nu$

For $r = r_1 = \nu$, equation (3b) yields $a_1 = 0$. Equation (3c) may be written

$$(s + r + \nu)(s + r - \nu)a_s + a_{s-2} = 0,$$

and for $r = \nu$ this takes the form

(5) $$(s + 2\nu)sa_s + a_{s-2} = 0.$$

Since $a_1 = 0$ and $\nu \geq 0$, it follows that $a_3 = 0$, $a_5 = 0, \cdots$, successively. If we set $s = 2m$ in (5), we obtain for the other coefficients

(6) $$a_{2m} = - \frac{1}{2^2 m(\nu + m)} a_{2m-2}, \qquad m = 1, 2, \cdots$$

and may determine these coefficients $a_2, a_4, \cdots$ successively. a_0 remains arbitrary. It is customary to put

(7) $$a_0 = \frac{1}{2^\nu \Gamma(\nu + 1)}$$

where $\Gamma(\nu + 1)$ is the **gamma function.** Some basic formulas for this important function are included in Appendix 3. For the present purpose it suffices to know that $\Gamma(\alpha)$ is defined by the integral

$$(8) \qquad\qquad \Gamma(\alpha) = \int_0^\infty e^{-t} t^{\alpha-1} \, dt \qquad\qquad (\alpha > 0).$$

By integration by parts we obtain

$$\Gamma(\alpha + 1) = \int_0^\infty e^{-t} t^\alpha \, dt = \left. -e^{-t} t^\alpha \right|_0^\infty + \alpha \int_0^\infty e^{-t} t^{\alpha-1} \, dt.$$

The first expression on the right is zero, and the integral on the right is $\Gamma(\alpha)$. This yields the basic relation

$$(9) \qquad\qquad \Gamma(\alpha + 1) = \alpha \Gamma(\alpha).$$

Since

$$\Gamma(1) = \int_0^\infty e^{-t} \, dt = 1,$$

we conclude from (9) that

$$\Gamma(2) = \Gamma(1) = 1!, \qquad \Gamma(3) = 2\Gamma(2) = 2!, \cdots$$

and in general

$$(10) \qquad\qquad \Gamma(k + 1) = k! \qquad\qquad (k = 0, 1, \cdots).$$

This shows that the gamma function may be regarded as a generalization of the factorial function known from elementary calculus.

Returning to our problem, we obtain from (6), (7) and (9)

$$a_2 = -\frac{a_0}{2^2(\nu + 1)} = -\frac{1}{2^{2+\nu} 1! \, \Gamma(\nu + 2)}$$

$$a_4 = -\frac{a_2}{2^2 2(\nu + 2)} = \frac{1}{2^{4+\nu} 2! \, \Gamma(\nu + 3)}$$

and so on; in general,

$$(11) \qquad\qquad a_{2m} = \frac{(-1)^m}{2^{2m+\nu} m! \, \Gamma(\nu + m + 1)}.$$

By inserting these coefficients in (2) and remembering that $a_1 = 0$, $a_3 = 0$, $\cdots$ we obtain a particular solution of Bessel's equation which is denoted by $J_\nu(x)$:

(12)

$$J_\nu(x) = x^\nu \sum_{m=0}^{\infty} \frac{(-1)^m x^{2m}}{2^{2m+\nu} m! \, \Gamma(\nu + m + 1)} \cdot$$

This solution of (1) is known as the **Bessel function of the first kind** *of order* ν. We see that it has the form of a product, namely, x^ν times a power series. This series converges for all x, as one can verify by the ratio test.

Integer values of ν are frequently denoted by n. Thus, for $n \geqq 0$,

(13)

$$J_n(x) = x^n \sum_{m=0}^{\infty} \frac{(-1)^m x^{2m}}{2^{2m+n} m! \, (n + m)!} \cdot$$

Solution $J_{-\nu}$ of the Bessel Equation

Replacing ν by $-\nu$ in (12), we have

(14) $$J_{-\nu}(x) = x^{-\nu} \sum_{m=0}^{\infty} \frac{(-1)^m x^{2m}}{2^{2m-\nu} m! \, \Gamma(m - \nu + 1)} \cdot$$

Since Bessel's equation involves ν^2, the functions J_ν and $J_{-\nu}$ are solutions of the equation for the same ν. If ν is not an integer, they are linearly independent, because the first term in (12) and the first term in (14) are finite nonzero multiples of x^ν and $x^{-\nu}$, respectively. This yields the following result.

Theorem 1 (General solution of Bessel's equation)
If ν is not an integer, a general solution of Bessel's equation for all $x \neq 0$ is

(15) $$y(x) = c_1 J_\nu(x) + c_2 J_{-\nu}(x).$$

But if ν is an integer, then (15) is not a general solution. This follows from

Theorem 2 (Linear dependence of Bessel functions J_n and J_{-n})
For integer $\nu = n$ the Bessel functions $J_n(x)$ and $J_{-n}(x)$ are linearly dependent, because

(16) $$\boxed{J_{-n}(x) = (-1)^n J_n(x)} \qquad (n = 1, 2, \cdots).$$

Proof. We use (14) and let ν approach a positive integer n. Then the gamma functions in the coefficients of the first n terms become infinite (cf. Fig. 545 in Appendix 3), the coefficients become zero, and the summation starts with $m = n$. Since in this case $\Gamma(m - n + 1) = (m - n)!$ [cf. (10)], we obtain

$$J_{-n}(x) = \sum_{m=n}^{\infty} \frac{(-1)^m x^{2m-n}}{2^{2m-n} m! \, (m - n)!} = \sum_{s=0}^{\infty} \frac{(-1)^{n+s} x^{2s+n}}{2^{2s+n} (n + s)! s!}$$

where $m = n + s$ and $s = m - n$. From (13) we see that the last series represents $(-1)^n J_n(x)$. This completes the proof.   ∎

A general solution of the Bessel equation with integer $\nu = n$ will be obtained in the next section. **Numerical values** of J_0 and J_1 are included in Appendix 4.

EXAMPLE 1. Vibrating cable or chain (Fig. 85)

Applications of Bessel functions occur in connection with the so-called wave equation and will be discussed systematically in Sec. 11.10. For the time being, let us consider another interesting application, namely, vibrations of a flexible hanging cable (or chain) that is fixed at the upper end ($x = 0$ in Fig. 85) and is allowed to make *small* vibrations in a vertical plane. Let L be the length of the cable and ρ its constant density (mass per unit length). Then the weight of the portion of the cable below a point x is $W(x) = \rho g(L - x)$. Assuming small angles α in the displacement, we can regard $W(x)$ to be approximately equal to the tension acting tangentially in the moving cable. The horizontal component of this tension gives the restoring force $F(x)$ $= W \sin \alpha \approx W \tan \alpha = W u_x$, where $u_x = \partial u/\partial x$ and $u(x, t)$ is the displacement of the cable from its vertical equilibrium position, t being time. Hence for a small portion of the cable between x and $x + \Delta x$, the difference in force is (by the mean value theorem of calculus)

$$F(x + \Delta x) - F(x) \approx \Delta x \, (W u_x)_x.$$

By Newton's second law, this equals the mass $\rho \Delta x$ times the acceleration $u_{tt} = \partial^2 u/\partial t^2$ of this portion:

$$\rho \, \Delta x \, u_{tt} = \Delta x \, (W u_x)_x = \Delta x \, \rho g[(L - x)u_x]_x.$$

Expecting periodic motions in time, we try $u(x, t) = y(x) \cos (\omega t + \delta)$ Substituting this and dividing by $\rho \Delta x$, we have

$$-\omega^2 y \cos (\omega t + \delta) = g[(L - x)y']' \cos (\omega t + \delta).$$

Dropping the cosine factor, performing the indicated differentiation and ordering terms, we get

$$(L - x)y'' - y' + \lambda^2 y = 0, \qquad \lambda^2 = \omega^2/g.$$

Setting $L - x = z$, we have $dx = -dz$ and by the chain rule

$$z \frac{d^2 y}{dz^2} + \frac{dy}{dz} + \lambda^2 y = 0.$$

The only really tricky step is to show that this can be cast into the form of Bessel's equation with parameter $\nu = 0$. The substitution that will work is $s = 2\lambda z^{1/2}$. Then by the chain rule

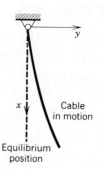

Fig. 85. Vibrating cable or chain in Example 1

$$\frac{dy}{dz} = \frac{dy}{ds}\lambda z^{-1/2}, \qquad \frac{d^2y}{dz^2} = \frac{d^2y}{ds^2}\lambda^2 z^{-1} - \frac{1}{2}\frac{dy}{ds}\lambda z^{-3/2}.$$

Substitution gives

$$\lambda^2\frac{d^2y}{ds^2} + \left(-\frac{1}{2}\lambda z^{-1/2} + \lambda z^{-1/2}\right)\frac{dy}{ds} + \lambda^2 y = 0.$$

Dividing by λ^2 and remembering that $s = 2\lambda z^{1/2}$, we finally obtain

$$\frac{d^2y}{ds^2} + \frac{1}{s}\frac{dy}{ds} + y = 0.$$

This is the Bessel equation with parameter $\nu = 0$. Since $s = 2\lambda z^{1/2} = (2\omega/\sqrt{g})\sqrt{L - x}$, a solution is

$$y(x) = J_0(2\omega\sqrt{(L - x)/g}).$$

Since the upper end ($x = 0$) of the cable is fixed, $y(0) = J_0(2\omega\sqrt{L/g}) = 0$. Hence possible frequencies $\omega/2\pi$ of the vibrating cable are those for which $s = 2\omega\sqrt{L/g}$ is a zero of J_0 (see Fig. 86 in the problem set), or superpositions of such "*normal modes*." Figure 85 shows the *first* of these normal modes, the "fundamental mode," obtained when $x = 0$ corresponds to the *first* positive zero of J_0. That the cable does not look exactly like the first portion of the curve of J_0 in Fig. 86 results from the square root in the solution. Can you imagine what the second normal mode looks like? The third?

Experimental result. A cable of length $L = 228$ cm gave 24.0, 24.5, 24.0, 24.0, 24.0 vibrations per minute, hence 24.1 on the average. From $2\omega\sqrt{L/g} = 2.405$ (first zero of J_0; cf. Table A2 in Appendix 4) we compute the frequency

$$\frac{\omega}{2\pi} = \frac{2.405}{2 \cdot 2\pi\sqrt{L/g}} = \frac{2.405}{4\pi\sqrt{2.28/9.80}} = 0.397 \ [\text{sec}^{-1}],$$

that is, $0.397 \cdot 60 = 23.8$ cycles per minute. The error is 1.3%, less than I expected, since the cable "wiggled" its very end. Try it yourself, it is fun! I was unable to get the higher modes.

History. This problem was considered in 1733 by DANIEL BERNOULLI (1700—1782) and marks the first appearance of Bessel functions. Bessel's classical paper is of 1824. ∎

Recall from Theorem 1 that we have a general solution of Bessel's equation (1) only when ν is not an integer. To complete our discussion, in the next section we introduce **Bessel functions of the second kind** (which are infinite at 0) and use them to get a general solution valid for *all* values of the parameter ν.

Problems for Sec. 4.5

1. Show that the series in (12) converge for all x.

2. Show that $J_n(x)$ is an even function for even n and an odd function for odd n.

Bessel functions J_0 and J_1. J_0 and J_1 are of particular importance in certain engineering problems. Problems 3–7 illustrate some of their properties.

3. Show that (cf. Fig. 86)

$$(17) \qquad J_0(x) = 1 - \frac{x^2}{2^2(1!)^2} + \frac{x^4}{2^4(2!)^2} - \frac{x^6}{2^6(3!)^2} + - \cdots$$

$$(18) \qquad J_1(x) = \frac{x}{2} - \frac{x^3}{2^3 1! \, 2!} + \frac{x^5}{2^5 2! \, 3!} - \frac{x^7}{2^7 3! \, 4!} + - \cdots.$$

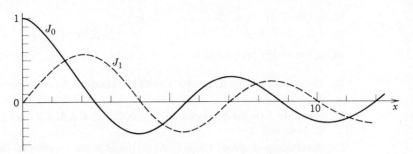

Fig. 86. Bessel functions of the first kind

4. The series (17) and (18) converge very rapidly (why?), so that they are useful in computations. To illustrate this, determine how many terms of (17) one needs to compute $J_0(1)$ with an error less than 1 unit of the fifth decimal place. How many terms would be needed to compute ln 2 from the Maclaurin series of ln $(1 + x)$ with the same degree of accuracy?

5. Show that for small $|x|$ we have $J_0(x) \approx 1 - 0.25x^2$. Using this formula, compute $J_0(x)$ for $x = 0.1, 0.2, \cdots, 1.0$ and determine the relative error by comparing with Table A2 in Appendix 4.

6. **(Behavior for large x)** It can be shown that for large x,

(19)
$$J_{2n}(x) \approx (-1)^n(\pi x)^{-1/2}(\cos x + \sin x)$$
$$J_{2n+1}(x) \approx (-1)^{n+1}(\pi x)^{-1/2}(\cos x - \sin x).$$

Using (19), graph $J_0(x)$ for large x, compute approximate values of the first five positive zeros of $J_0(x)$, and compare them with the more accurate values 2.405, 5.520, 8.654, 11.792, 14.931.

7. Using (19), compute approximate values of the first four positive zeros of $J_1(x)$ and determine the relative error, using the more exact values 3.832, 7.016, 10.173, 13.324.

Relations between Bessel functions. In applying Bessel functions, the engineer must know that these functions satisfy various relations. (Ref. [A14], the standard book on Bessel functions, and Ref. [7] in Appendix 1 contain an incredibly large number of formulas.) Some important relations are included in the following problems. Formulas (22) and (23) are useful for computing tables.

8. Using (12) and (9), show that

(20)
$$[x^\nu J_\nu(x)]' = x^\nu J_{\nu-1}(x).$$

Using (12), show that

(21)
$$[x^{-\nu} J_\nu(x)]' = -x^{-\nu} J_{\nu+1}(x).$$

9. Derive from (20) and (21) the **recurrence relation**

(22)
$$J_{\nu-1}(x) + J_{\nu+1}(x) = \frac{2\nu}{x} J_\nu(x).$$

10. Derive from (20) and (21) the **recurrence relation**

(23)
$$J_{\nu-1}(x) - J_{\nu+1}(x) = 2J_\nu{}'(x).$$

11. Derive the Bessel equation (1) from (20) and (21).

Using (20)–(23), show that

12. $J_0'(x) = -J_1(x)$

13. $J_1'(x) = J_0(x) - x^{-1}J_1(x)$

14. $J_2'(x) = \frac{1}{2}[J_1(x) - J_3(x)]$

15. $J_2'(x) = (1 - 4x^{-2})J_1(x) + 2x^{-1}J_0(x)$

16. Using Table A2 in Appendix 4 and (22), compute $J_2(x)$ for the values $x = 0, 0.1, 0.2, \cdots, 1.0$.

17. Compute $J_3(x)$ for the values $x = 2.0, 2.2, 2.4, 2.6, 2.8$ from (22) and Table A2 in Appendix 4.

18. **(Interlacing of zeros)** Using (20), (21), and Rolle's theorem, show that between two consecutive zeros of $J_0(x)$ there is precisely one zero of $J_1(x)$.

19. Show that between any two consecutive positive zeros of $J_n(x)$ there is precisely one zero of $J_{n+1}(x)$.

Integrals involving Bessel functions can often be evaluated or at least simplified by the use of (20)–(23). Show that

20. $\int x^{\nu}J_{\nu-1}(x)\,dx = x^{\nu}J_{\nu}(x) + c$

21. $\int x^{-\nu}J_{\nu+1}(x)\,dx = -x^{-\nu}J_{\nu}(x) + c$

22. $\int J_{\nu+1}(x)\,dx = \int J_{\nu-1}(x)\,dx - 2J_{\nu}(x)$

Using the formulas in Probs. 20–22 and, if necessary, integration by parts, evaluate

23. $\int J_3(x)\,dx$

24. $\int x^{-3}J_4(x)\,dx$

25. $\int J_5(x)\,dx$

26. $\int x^3 J_0(x)\,dx$

(In other cases, one may be left with the integral of $J_0(x)$, which cannot be evaluated in finite form, but has been tabulated; cf. Ref. [A14] in Appendix 1.)

Elementary functions. For certain values of ν the Bessel function J_ν can be expressed in terms of cosine and sine:

27. Using (12) and the relation $\Gamma(\frac{1}{2}) = \sqrt{\pi}$, show that

(24) $$J_{1/2}(x) = \sqrt{\frac{2}{\pi x}} \sin x, \qquad J_{-1/2}(x) = \sqrt{\frac{2}{\pi x}} \cos x.$$

Conclude from this and (22) that the *Bessel functions $J_\nu(x)$ of order $\nu = \pm\frac{1}{2}$, $\pm\frac{3}{2}, \pm\frac{5}{2}, \cdots$ are elementary functions.*

28. Using (22) and (24), show that

$$J_{3/2}(x) = \sqrt{\frac{2}{\pi x}} \left(\frac{\sin x}{x} - \cos x \right), \qquad J_{-3/2}(x) = -\sqrt{\frac{2}{\pi x}} \left(\frac{\cos x}{x} + \sin x \right).$$

Elimination of first derivative

29. Substitute $y(x) = u(x)v(x)$ into $y'' + p(x)y' + q(x)y = 0$ and show that for obtaining a second-order differential equation for u not containing u', we must take

$$v(x) = \exp\left(-\frac{1}{2} \int p(x)\,dx \right).$$

30. Show that for the Bessel equation the substitution in Prob. 29 is $y = ux^{-1/2}$ and gives

(25) $$x^2 u'' + (x^2 + \tfrac{1}{4} - \nu^2)u = 0.$$

Solve this equation with $\nu = \tfrac{1}{2}$ and compare the result with Prob. 27. Comment. Verify that the functions in Prob. 28 satisfy (25).

Differential equations reducible to Bessel's equation: see the problem set for Sec. 4.6.

4.6 Bessel Functions of the Second Kind

For integer $\nu = n$ the Bessel functions $J_n(x)$ and $J_{-n}(x)$ are linearly dependent (Sec. 4.5), so that they do not form a basis of solutions. This poses the problem of obtaining a second linearly independent solution when ν is an integer n. This is the topic of the present section.

$n = 0$: Bessel Function of the Second Kind $Y_0(x)$

We first consider the case $n = 0$. Then Bessel's equation can be written

(1) $$xy'' + y' + xy = 0,$$

and the indicial equation (4), Sec. 4.5, has a double root $r = 0$. This is Case 2 in Sec. 4.4, and (8), Sec. 4.4, shows that the desired solution must have the form

(2) $$y_2(x) = J_0(x) \ln x + \sum_{m=1}^{\infty} A_m x^m.$$

We substitute y_2 and its derivatives

$$y_2' = J_0' \ln x + \frac{J_0}{x} + \sum_{m=1}^{\infty} m A_m x^{m-1}$$

$$y_2'' = J_0'' \ln x + \frac{2J_0'}{x} - \frac{J_0}{x^2} + \sum_{m=1}^{\infty} m(m-1) A_m x^{m-2}$$

into (1). Then the logarithmic terms disappear because J_0 is a solution of (1), the other two terms containing J_0 cancel, and we find

$$2J_0' + \sum_{m=1}^{\infty} m(m-1) A_m x^{m-1} + \sum_{m=1}^{\infty} m A_m x^{m-1} + \sum_{m=1}^{\infty} A_m x^{m+1} = 0.$$

From (13) in Sec. 4.5 we obtain the power series of J_0' in the form

$$J_0'(x) = \sum_{m=1}^{\infty} \frac{(-1)^m 2m x^{2m-1}}{2^{2m}(m!)^2} = \sum_{m=1}^{\infty} \frac{(-1)^m x^{2m-1}}{2^{2m-1} m! \, (m-1)!}.$$

By inserting this series we have

$$\sum_{m=1}^{\infty} \frac{(-1)^m x^{2m-1}}{2^{2m-2}m!\,(m-1)!} + \sum_{m=1}^{\infty} m^2 A_m x^{m-1} + \sum_{m=1}^{\infty} A_m x^{m+1} = 0.$$

We first show that the A_m with odd subscripts are all zero. The coefficient of the power x^0 is A_1, and so, $A_1 = 0$. By equating the sum of the coefficients of the power x^{2s} to zero we have

$$(2s+1)^2 A_{2s+1} + A_{2s-1} = 0, \qquad s = 1, 2, \cdots.$$

Since $A_1 = 0$, we thus obtain $A_3 = 0$, $A_5 = 0$, $\cdots$, successively.

We now equate the sum of the coefficients of x^{2s+1} to zero. This gives

$$-1 + 4A_2 = 0 \qquad \text{or} \qquad A_2 = \tfrac{1}{4} \qquad\qquad (s = 0)$$

and for the other values $s = 1, 2, \cdots$

$$\frac{(-1)^{s+1}}{2^{2s}(s+1)!s!} + (2s+2)^2 A_{2s+2} + A_{2s} = 0.$$

For $s = 1$ this yields

$$\tfrac{1}{8} + 16A_4 + A_2 = 0 \qquad \text{or} \qquad A_4 = -\tfrac{3}{128}$$

and in general

$$(3) \qquad A_{2m} = \frac{(-1)^{m-1}}{2^{2m}(m!)^2} \left(1 + \frac{1}{2} + \frac{1}{3} + \cdots + \frac{1}{m} \right), \qquad m = 1, 2, \cdots.$$

Using the short notation

$$(4) \qquad\qquad h_m = 1 + \frac{1}{2} + \cdots + \frac{1}{m}$$

and inserting (3) and $A_1 = A_3 = \cdots = 0$ into (2), we obtain the result

$$y_2(x) = J_0(x)\ln x + \sum_{m=1}^{\infty} \frac{(-1)^{m-1} h_m}{2^{2m}(m!)^2} x^{2m}$$

$$(5)$$

$$= J_0(x)\ln x + \tfrac{1}{4}x^2 - \tfrac{3}{128}x^4 + - \cdots.$$

Since J_0 and y_2 are linearly independent functions, they form a basis of (1). Of course, another basis is obtained if we replace y_2 by an independent particular solution of the form $a(y_2 + bJ_0)$ where $a\ (\neq 0)$ and b are constants. It is customary to choose $a = 2/\pi$ and $b = \gamma - \ln 2$, where the number $\gamma = 0.577\,215\,664\,90 \cdots$ is the so-called **Euler constant**, which is defined as the limit of

$$1 + \frac{1}{2} + \cdots + \frac{1}{s} - \ln s$$

as s approaches infinity. The standard particular solution thus obtained is known as the **Bessel function of the second kind** *of order zero* (Fig. 87) or **Neumann's function**[13] *of order zero* and is denoted by $Y_0(x)$. Thus [cf. (4)]

(6)
$$Y_0(x) = \frac{2}{\pi}\left[J_0(x)\left(\ln\frac{x}{2} + \gamma \right) + \sum_{m=1}^{\infty} \frac{(-1)^{m-1}h_m}{2^{2m}(m!)^2}x^{2m} \right],$$

For small $x > 0$ the function $Y_0(x)$ behaves about like $\ln x$ (cf. Fig. 87; why?), and $Y_0(x) \to -\infty$ as $x \to 0$.

Bessel Functions of the Second Kind $Y_n(x)$

For $\nu = n = 1, 2, \cdots$ a second solution can be obtained by manipulations similar to those for $n = 0$, starting from (10), Sec. 4.4. It turns out that in these cases the solution also contains a logarithmic term.

The situation is not yet completely satisfactory, because the second solution is defined differently, depending on whether the order ν is an integer. To provide uniformity of formalism and numerical tabulation, it is desirable to adopt a form of the second solution that is valid for all values of the order. This is the reason for introducing a standard second solution $Y_\nu(x)$ defined for all ν by the formula

(7)

(a)
$$Y_\nu(x) = \frac{1}{\sin \nu\pi}[J_\nu(x)\cos \nu\pi - J_{-\nu}(x)]$$

(b)
$$Y_n(x) = \lim_{\nu \to n} Y_\nu(x).$$

This function is known as the **Bessel function of the second kind** *of order* ν or **Neumann's function**[14] *of order* ν.

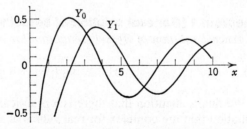

Fig. 87. Bessel functions of the second kind.
(For a small table, see Appendix 4.)

[13]CARL NEUMANN (1832—1925), German mathematician and physicist, became a professor at Leipzig in 1868. His work on potential theory sparked the development in the field of integral equations by VITO VOLTERRA (1860—1940) of Rome, ERIC IVAR FREDHOLM (1866—1927) of Stockholm, whose famous 1901-1903 papers were a sensation to the mathematical world of his time, and DAVID HILBERT (1862—1943) of Göttingen (cf. the footnote in Sec. 6.6).

[14]See the previous footnote. The solutions $Y_\nu(x)$ are sometimes denoted by $N_\nu(x)$; in Ref. [A14] they are called Weber's functions, Euler's constant in (6) is often denoted by C or $\ln \gamma$.

We discuss the linear independence of J_ν and Y_ν.

For noninteger order ν, the function $Y_\nu(x)$ is evidently a solution of Bessel's equation because $J_\nu(x)$ and $J_{-\nu}(x)$ are solutions of that equation. Since for those ν the solutions J_ν and $J_{-\nu}$ are linearly independent and Y_ν involves $J_{-\nu}$, the functions J_ν and Y_ν are linearly independent. Furthermore, it can be shown that the limit in (7b) exists and Y_n is a solution of Bessel's equation for integer order; cf. Ref. [A14] in Appendix 1. We shall see that the series development of $Y_n(x)$ contains a logarithmic term. Hence $J_n(x)$ and $Y_n(x)$ are linearly independent solutions of Bessel's equation. The series development of $Y_n(x)$ can be obtained if we insert the series (12) and (14), Sec. 4.5, for $J_\nu(x)$ and $J_{-\nu}(x)$ into (7a) and then let ν approach n; for details see Ref. [A14]; the result is

$$
\textbf{(8)} \quad
\begin{aligned}
Y_n(x) = {} & \frac{2}{\pi} J_n(x) \left(\ln \frac{x}{2} + \gamma \right) + \frac{x^n}{\pi} \sum_{m=0}^{\infty} \frac{(-1)^{m-1}(h_m + h_{m+n})}{2^{2m+n} m! \, (m+n)!} x^{2m} \\
& - \frac{x^{-n}}{\pi} \sum_{m=0}^{n-1} \frac{(n-m-1)!}{2^{2m-n} m!} x^{2m}
\end{aligned}
$$

where $x > 0$, $n = 0, 1, \cdots$, and

$$
h_0 = 0, \qquad h_s = 1 + \frac{1}{2} + \frac{1}{3} + \cdots + \frac{1}{s} \qquad (s = 1, 2, \cdots),
$$

and when $n = 0$ the last sum in (8) is to be replaced by 0. For $n = 0$ the representation (8) takes the form (6). Furthermore, it can be shown that

$$
Y_{-n}(x) = (-1)^n \, Y_n(x).
$$

We may formulate our main result as follows.

Theorem 1 (General solution of Bessel's equation)

A general solution of Bessel's equation for all values of ν is

$$
\textbf{(9)} \qquad\qquad y(x) = C_1 J_\nu(x) + C_2 Y_\nu(x).
$$

We finally mention that there is a practical need for solutions of Bessel's equation that are complex for real values of x. For this reason the solutions

$$
\textbf{(10)} \qquad
\begin{aligned}
H_\nu^{(1)}(x) &= J_\nu(x) + i Y_\nu(x) \\
H_\nu^{(2)}(x) &= J_\nu(x) - i Y_\nu(x)
\end{aligned}
$$

are frequently used and have been tabulated (cf. Ref. [8] in Appendix 1). These linearly independent functions are called **Bessel functions of the third kind** *of order ν* or *first and second* **Hankel functions**[15] *of order ν.*

---

[15]HERMANN HANKEL (1839—1873), German mathematician.

In the next section we turn to the concept of the **orthogonality** of functions whose importance in applications can hardly be overestimated. It fits our present circle of ideas, since orthogonal functions—Legendre polynomials and Bessel functions among them—arise quite naturally as solutions of certain boundary value problems (*"Sturm–Liouville problems"*), which we discuss in Sec. 4.8.

Problems for Sec. 4.6

Differential equations reducible to Bessel's equation. There are various differential equations which can be reduced to Bessel's equation. To illustrate this, use the indicated substitutions and find a general solution of the given equations in terms of Bessel functions.

1. $x^2 y'' + xy' + (x^2 - 4)y = 0$
2. $x^2 y'' + xy' + (\lambda^2 x^2 - \nu^2)y = 0$ ($\lambda x = z$)
3. $xy'' + y' + \frac{1}{4}y = 0$ ($\sqrt{x} = z$)
4. $4x^2 y'' + 4xy' + (x - n^2)y = 0$ ($\sqrt{x} = z$)
5. $x^2 y'' + xy' + (4x^4 - \frac{1}{4})y = 0$ ($x^2 = z$)
6. $xy'' + (1 + 2n)y' + xy = 0$ ($y = x^{-n}u$)
7. $x^2 y'' - 3xy' + 4(x^4 - 3)y = 0$ ($y = x^2 u,\ x^2 = z$)
8. $x^2 y'' + (1 - 2\nu)xy' + \nu^2(x^{2\nu} + 1 - \nu^2)y = 0$ ($y = x^\nu u,\ x^\nu = z$)
9. $x^2 y'' + \frac{1}{4}(x + \frac{3}{4})y = 0$ ($y = u\sqrt{x},\ \sqrt{x} = z$)
10. $y'' + xy = 0$ ($y = u\sqrt{x},\ \frac{2}{3}x^{3/2} = z$)
11. $y'' + x^2 y = 0$ ($y = u\sqrt{x},\ \frac{1}{2}x^2 = z$)
12. $y'' + k^2 xy = 0$ ($y = u\sqrt{x},\ \frac{2}{3}kx^{3/2} = z$)
13. $y'' + k^2 x^2 y = 0$ ($y = u\sqrt{x},\ \frac{1}{2}kx^2 = z$)
14. $y'' + k^2 x^4 y = 0$ ($y = u\sqrt{x},\ \frac{1}{3}kx^3 = z$)

15. Show that for small $x > 0$ we have $Y_0(x) \approx 2\pi^{-1}(\ln \frac{1}{2}x + \gamma)$. Using this formula, compute an approximate value of the smallest positive zero of $Y_0(x)$ and compare it with the more accurate value 0.9.

16. It can be shown that for large x,

    (11) $Y_n(x) \approx \sqrt{2/(\pi x)} \sin (x - \frac{1}{2}n\pi - \frac{1}{4}\pi)$.

    Using (11) and Table A2 in Appendix 4, graph $Y_0(x)$ and $Y_1(x)$ for $0 < x \leq 15$. Using (11), compute approximate values of the first three positive zeros of $Y_0(x)$ and compare these values with the more accurate values 0.89, 3.96 and 7.09.

17. Show that the Hankel functions (10) constitute a basis of solutions of Bessel's equation for any ν.

Modified Bessel functions

18. The function $I_\nu(x) = i^{-\nu}J_\nu(ix)$, $i = \sqrt{-1}$, is called the *modified Bessel function of the first kind of order ν*. Show that $I_\nu(x)$ is a solution of the differential equation

    (12) $x^2 y'' + xy' - (x^2 + \nu^2)y = 0$

    and has the representation

    (13) $I_\nu(x) = \sum_{m=0}^{\infty} \frac{x^{2m+\nu}}{2^{2m+\nu}m!\,\Gamma(m + \nu + 1)}$.

19. Show that $I_\nu(x)$ is real for all real x (and real ν), $I_\nu(x) \neq 0$ for all real $x \neq 0$, and $I_{-n}(x) = I_n(x)$, where n is any integer.

20. Show that another solution of the differential equation (12) is the so-called *modified Bessel function of the third kind* (sometimes: *of the second kind*)

$$(14) \qquad K_\nu(x) = \frac{\pi}{2 \sin \nu\pi} [I_{-\nu}(x) - I_\nu(x)].$$

4.7   Orthogonal Sets of Functions

Content of the Remainder of This Chapter

Legendre polynomials (Sec. 4.3) and Bessel functions enjoy a property that is called **orthogonality** and is of general importance in engineering and mathematics. In the present section we introduce the corresponding concepts and notations. In Sec. 4.8 we shall consider certain boundary value problems ("Sturm–Liouville problems") whose solutions form orthogonal sets of functions, and in Sec. 4.9 we shall apply the results of Sec. 4.8 to Legendre polynomials and Bessel functions. This is our program for the remaining portion of this chapter.

So let us first define orthogonality.

Orthogonality of Functions

Let $g_m(x)$ and $g_n(x)$ be two real-valued functions that are defined on an interval $a \leq x \leq b$ and are such that the integral of the product $g_m(x)g_n(x)$ over that interval exists. We shall denote this integral by (g_m, g_n). This is a simple standard notation, which is widely used. Thus

$$(1) \qquad (g_m, g_n) = \int_a^b g_m(x)g_n(x) \, dx.$$

g_m and g_n are said to be **orthogonal** *on the interval* $a \leq x \leq b$ if the integral (1) is zero, that is,

$$(2) \qquad \boxed{(g_m, g_n) = \int_a^b g_m(x)g_n(x) \, dx = 0} \qquad (m \neq n).$$

A set of real-valued functions $g_1(x), g_2(x), g_3(x), \cdots$ is called an **orthogonal set** *of functions on an interval* $a \leq x \leq b$ if these functions are defined on that interval and if all the integrals (g_m, g_n) exist and are zero for all pairs of distinct functions in the set.

The nonnegative square root of (g_m, g_m) is called the **norm** of $g_m(x)$ and is generally denoted by $\|g_m\|$; thus

(3)
$$\|g_m\| = \sqrt{(g_m, g_m)} = \sqrt{\int_a^b g_m^2(x)\, dx}.$$

Throughout our discussion we shall make the following

General assumption

All occurring functions are bounded and are such that the occurring integrals exist, and every function has nonzero norm.

Clearly, an orthogonal set $g_1, g_2, \cdots$ on an interval $a \leqq x \leqq b$ whose functions have norm 1 satisfies the relations

(4)
$$(g_m, g_n) = \int_a^b g_m(x)g_n(x)\, dx = \begin{cases} 0 & \text{when } m \neq n \\ 1 & \text{when } m = n \end{cases} \qquad \begin{aligned} m &= 1, 2, \cdots \\ n &= 1, 2, \cdots . \end{aligned}$$

Such a set is called an **orthonormal set** *of functions on the interval* $a \leqq x \leqq b$.

Formula (4) can be written more briefly $(g_m, g_n) = \delta_{mn}$ with the so-called **Kronecker delta**[16] defined by

(5)
$$\delta_{mn} = \begin{cases} 0 & \text{if } m \neq n \\ 1 & \text{if } m = n \end{cases} \qquad (m, n = 1, 2, \cdots).$$

Obviously, from an *orthogonal* set we may obtain an *orthonormal* set by dividing each function by its norm.

EXAMPLE 1

The functions $g_m(x) = \sin mx$, $m = 1, 2, \cdots$ form an orthogonal set on the interval $-\pi \leqq x \leqq \pi$, because for $m \neq n$ we obtain [cf. (11) in Appendix 3]

$$(g_m, g_n) = \int_{-\pi}^{\pi} \sin mx \sin nx\, dx = \frac{1}{2} \int_{-\pi}^{\pi} \cos (m - n)x\, dx - \frac{1}{2} \int_{-\pi}^{\pi} \cos (m + n)x\, dx = 0.$$

The norm $\|g_m\|$ equals $\sqrt{\pi}$, because

$$\|g_m\|^2 = \int_{-\pi}^{\pi} \sin^2 mx\, dx = \pi \qquad (m = 1, 2, \cdots).$$

Hence the corresponding orthonormal set consists of the functions

$$\frac{\sin x}{\sqrt{\pi}}, \qquad \frac{\sin 2x}{\sqrt{\pi}}, \qquad \frac{\sin 3x}{\sqrt{\pi}}, \qquad \cdots .$$

[16]LEOPOLD KRONECKER (1823—1891), German mathematician at Berlin, who made important contributions to algebra, group theory and number theory.

EXAMPLE 2

The functions

$$1, \qquad \cos x, \qquad \sin x, \qquad \cos 2x, \qquad \sin 2x, \qquad \cdots$$

form an orthogonal set on the interval $-\pi \leq x \leq \pi$. This follows from the formula for (g_m, g_n) in Example 1, the analogous relation for the cosine functions, and

$$\int_{-\pi}^{\pi} \cos mx \sin nx \, dx = \frac{1}{2} \int_{-\pi}^{\pi} \sin (m + n)x \, dx - \frac{1}{2} \int_{-\pi}^{\pi} \sin (m - n)x \, dx = 0$$

for all $m, n = 0, 1, \cdots$. The corresponding orthonormal set is

$$\frac{1}{\sqrt{2\pi}}, \qquad \frac{\cos x}{\sqrt{\pi}}, \qquad \frac{\sin x}{\sqrt{\pi}}, \qquad \frac{\cos 2x}{\sqrt{\pi}}, \qquad \frac{\sin 2x}{\sqrt{\pi}}, \qquad \cdots \qquad \blacksquare$$

Series of Orthogonal Functions

Orthogonal sets yield important types of series developments in a relatively simple fashion. In fact, let $g_1(x), g_2(x), \cdots$ be any orthogonal set of functions on an interval $a \leq x \leq b$ and let $f(x)$ be a given function that can be represented in terms of the g_j's by a convergent series

$$(6) \qquad f(x) = \sum_{n=1}^{\infty} a_n g_n(x) = a_1 g_1(x) + a_2 g_2(x) + \cdots .$$

This series is called a **generalized Fourier series** of $f(x)$ and its coefficients $a_1, a_2, \cdots$ are called the **Fourier constants** of $f(x)$ *with respect to that orthogonal set of functions.*

Because of the orthogonality, these constants can be determined in a very simple fashion. In fact, multiplying both sides of (6) by $g_m(x)$ (m fixed), integrating over $a \leq x \leq b$, and assuming that term-by-term integration is permissible,[17] we have

$$(f, g_m) = \int_a^b f g_m \, dx = \int_a^b \left(\sum_{n=1}^{\infty} a_n g_n \right) g_m \, dx = \sum_{n=1}^{\infty} a_n (g_n, g_m).$$

The integral for which $n = m$ equals $(g_m, g_m) = \|g_m\|^2$, whereas all the other integrals on the right are zero, because of the orthogonality. Hence

$$(7') \qquad (f, g_m) = a_m \|g_m\|^2,$$

and the desired formula for the Fourier constants is

$$(7) \qquad a_m = \frac{(f, g_m)}{\|g_m\|^2} = \frac{1}{\|g_m\|^2} \int_a^b f(x) g_m(x) \, dx \qquad m = 1, 2, \cdots .$$

---

[17]This is justified, for instance, in the case of uniform convergence (Theorem 3, Sec. 14.8).

EXAMPLE 3. Fourier series

In the case of the orthogonal set in Example 2, representation (6) may be written

$$(8) \qquad f(x) = a_0 + \sum_{n=1}^{\infty} (a_n \cos nx + b_n \sin nx)$$

and (7) becomes

$$a_0 = \frac{1}{2\pi} \int_{-\pi}^{\pi} f(x) \, dx$$

$$(9) \qquad a_n = \frac{1}{\pi} \int_{-\pi}^{\pi} f(x) \cos nx \, dx$$

$$(n = 1, 2, \cdots)$$

$$b_n = \frac{1}{\pi} \int_{-\pi}^{\pi} f(x) \sin nx \, dx$$

The series (8) is called the **Fourier series** of $f(x)$. Its coefficients are called the **Fourier coefficients** of $f(x)$ and formulas (9) are called the **Euler formulas** for these coefficients. *Since Fourier series are particularly important in engineering mathematics, we shall devote a whole chapter to them (Chap. 10) and present some basic applications in Chap. 11 in connection with partial differential equations.* ∎

Orthogonality with Respect to a Weight Function

Some important sets of functions g_1, g_2, $\cdots$ occurring in applications are not orthogonal but have the property that for some nonnegative function $p(x)$,

$$(10) \qquad \int_a^b p(x)g_m(x)g_n(x) \, dx = 0 \qquad \text{when} \quad m \neq n.$$

Such a set is then said to be *orthogonal with respect to the* **weight function** $p(x)$ *on the interval* $a \leq x \leq b$. The *norm* of g_m is now defined as

$$(11) \qquad \|g_m\| = \sqrt{\int_a^b p(x)g_m^2(x) \, dx}$$

and if the norm of each function g_m is 1, the set is said to be *orthonormal* on that interval with respect to $p(x)$.

If we set $h_m = \sqrt{p} \, g_m$, then (10) becomes

$$\int_a^b h_m(x)h_n(x) \, dx = 0 \qquad (m \neq n),$$

that is, the functions h_m form an orthogonal set in the usual sense. Important examples will be considered in the following sections.

If $g_1(x)$, $g_2(x)$, $\cdots$ is an orthogonal set on an interval $a \leq x \leq b$ with respect to a weight function p, and if a given function $f(x)$ can be represented by a **generalized Fourier series** [cf. (6)]

(12) $$f(x) = a_1 g_1(x) + a_2 g_2(x) + \cdots,$$

then the Fourier constants a_1, a_2, $\cdots$ of $f(x)$ with respect to that set can be determined as before; the only difference is that we now have to multiply that series by pg_m (instead of g_m) before integrating. The other steps are as before and give

(13) $$a_m = \frac{1}{\|g_m\|^2} \int_a^b p(x)f(x)g_m(x)\,dx \qquad (m = 1, 2, \cdots),$$

where the norm is now defined by (11).

Completeness of Orthonormal Sets

In practice, one uses only orthonormal sets which consist of "sufficiently many" functions, so that one can represent large classes of functions—certainly all continuous functions on an interval $a \leq x \leq b$—by a generalized Fourier series (6). These orthonormal sets are called "*complete*" (in the set of functions considered; definition below). For instance the orthonormal set in Example 2 is complete in the set of continuous[18] functions on the interval $-\pi \leq x \leq \pi$, and so are the sets of Legendre polynomials and Bessel functions in Sec. 4.9 (on intervals given there).

In this connection, convergence is *convergence in the norm* (also called *mean-square convergence* or *mean convergence*); that is, a sequence of functions f_n is called **convergent** with the limit f if

(14*) $$\lim_{n \to \infty} \|f_n - f\| = 0,$$

written out by (3) (where we can drop the square root)

(14) $$\lim_{n \to \infty} \int_a^b [f_n(x) - f(x)]^2\,dx = 0.$$

Accordingly, (6) converges and represents f if

(15) $$\lim_{n \to \infty} \int_a^b [s_n(x) - f(x)]^2\,dx = 0$$

where s_n is the nth partial sum of (6),

---

[18] Actually, piecewise continuous and much more general functions, but a full discussion of completeness would need prerequisites not required in this book. See Ref. [13], Secs. 3.4-3.7 listed in Appendix 1. (There "complete sets" are called "total sets", a more modern term.)

$$(16) \qquad s_n(x) = \sum_{m=1}^{n} a_m g_m(x).$$

By definition, an orthonormal set g_1, g_2, $\cdots$ on an interval $a \leqq x \leqq b$ is **complete** *in a set of functions S* defined on $a \leqq x \leqq b$ if we can approximate every f belonging to S arbitrarily closely by a linear combination $a_1 g_1 + a_2 g_2 + \cdots + a_n g_n$; technically: if for every $\epsilon > 0$ we can find constants $a_1, \cdots, a_n$ (with n large enough) such that

$$(17) \qquad \| f - (a_1 g_1 + \cdots + a_n g_n) \| < \epsilon.$$

An interesting and basic consequence of the integral in (15) is obtained as follows. Performing the square and using (16), we first have

$$\int_a^b [s_n(x) - f(x)]^2 \, dx = \int_a^b s_n^2 \, dx - 2 \int_a^b f s_n \, dx + \int_a^b f^2 \, dx$$

$$= \int_a^b \left[\sum_{m=1}^{n} a_m g_m \right]^2 dx - 2 \sum_{m=1}^{n} a_m \int_a^b f g_m \, dx + \int_a^b f^2 \, dx.$$

The integral in the second sum on the right equals a_m, by (7) with $\|g_m\|^2 = 1$. The first integral on the right equals $\Sigma \, a_m^2$ because $\int g_m g_k \, dx = 0$ for $m \neq k$, and $\int g_m^2 \, dx = 1$. Hence the right-hand side reduces to

$$- \sum_{m=1}^{n} a_m^2 + \int_a^b f^2 \, dx.$$

This is nonnegative because the integrand on the left and thus the integral on the left are nonnegative. This proves **Bessel's inequality**

$$(18) \qquad \boxed{\sum_{m=1}^{n} a_m^2 \leqq \|f\|^2 = \int_a^b f(x)^2 \, dx.} \qquad n = 1, 2, \cdots.$$

Here we can let $n \to \infty$, because the left-hand sides form a monotone increasing sequence which is bounded by the right-hand side, so that we have convergence by the familiar Theorem 1 in Appendix 3 (see A3.3). Hence

$$(19) \qquad \sum_{m=1}^{\infty} a_m^2 \leqq \|f\|^2.$$

Furthermore, if g_1, g_2, $\cdots$ is complete in a set of functions S, then (15) holds for every f belonging to S. By (17) this implies equality in (18) with $n \to \infty$; hence in the case of completeness the so-called **Parseval's equality**

(20)
$$\sum_{m=1}^{\infty} a_m^2 = \|f\| = \int_a^b f(x)^2 \, dx$$

holds for every f in S.

As a consequence of (20) we prove that in the case of *completeness* there is no function orthogonal to *every* function of the orthonormal set, with the trivial exception of a function of zero norm:

Theorem 1 (Completeness)

Let g_1, g_2, $\cdots$ be a complete orthonormal set on a $\leqq x \leqq b$ in a set of functions S. Then if a function f belongs to S and is orthogonal to every g_m, it must have norm zero. In particular, if f is continuous, it must be identically zero.

Proof. From the orthogonality assumption we see that the left-hand side of (20) must be zero. This proves the first statement. If f is continuous, then $\|f\| = 0$ implies $f(x) \equiv 0$, as can be seen directly from (3), with f instead of g_m. ∎

EXAMPLE 4. Fourier series

The orthonormal set in Example 2 is complete in the set of continuous functions on $-\pi \leqq x \leqq \pi$. Verify directly that $f(x) \equiv 0$ is the only continuous function orthogonal to all the functions of that set.

Solution. Let f be any continuous function. By the orthogonality (we can omit $\sqrt{2\pi}$ and $\sqrt{\pi}$),

$$\int_{-\pi}^{\pi} 1 \cdot f(x) \, dx = 0, \qquad \int_{-\pi}^{\pi} f(x) \cos nx \, dx = 0, \qquad \int_{-\pi}^{\pi} f(x) \sin nx \, dx = 0.$$

Hence $a_n = 0$ and $b_n = 0$ in (9) for all n, so that (6) reduces to $f(x) \equiv 0$. ∎

In the next section we show that sets of orthogonal functions are obtained as solutions of boundary value problems for second-order homogeneous linear equations satisfying certain conditions. These are called **Sturm–Liouville problems** and include many problems of practical importance in physics.

Problems for Sec. 4.7

To get used to the notation, evaluate (g_1, g_2), $\|g_1\|$, $\|g_2\|$ and $(g_1 + g_2, g_1 - g_2)$, where in (1) and (3), $a = 0$, $b = 1$ and

1. $g_1(x) = x$, $g_2(x) = x^2$
2. $g_1(x) = e^x$, $g_2(x) = e^{-x}$
3. $g_1(x) = \sin 2\pi x$, $g_2(x) = \cos 2\pi x$
4. $g_1(x) = 1$, $g_2(x) = 1 - x$

In each case show that the given set is orthogonal on the given interval I and determine the corresponding orthonormal set.

5. $1, \cos x, \cos 2x, \cos 3x, \cdots,$ $I: 0 \leqq x \leqq 2\pi$
6. $\sin x, \sin 2x, \sin 3x, \cdots,$ $I: 0 \leqq x \leqq \pi$
7. $\sin \pi x, \sin 2\pi x, \sin 3\pi x, \cdots,$ $I: -1 \leqq x \leqq 1$
8. $1, \cos 2x, \cos 4x, \cos 6x, \cdots,$ $I: 0 \leqq x \leqq \pi$

9. $1, \cos \pi x, \cos 2\pi x, \cos 3\pi x, \cdots,$ $I: -1 \leqq x \leqq 1$

10. $1, \cos \dfrac{n\pi}{L} x$ $(n = 1, 2, \cdots),$ $I: 0 \leqq x \leqq 2L$

11. $\sin \dfrac{n\pi}{L} x$ $(n = 1, 2, \cdots),$ $I: -L \leqq x \leqq L$

12. $1, \cos \dfrac{\pi}{L} x, \sin \dfrac{\pi}{L} x, \cos \dfrac{2\pi}{L} x, \sin \dfrac{2\pi}{L} x, \cdots,$ $I: 0 \leqq x \leqq 2L$

13. $P_0(x), P_1(x), P_2(x),$ $I: -1 \leqq x \leqq 1$ (cf. Sec. 4.3)

14. Determine constants $a_0, b_0, \cdots, c_2$ so that the functions $g_0 = a_0, g_1 = b_0 + b_1 x,$ $g_2 = c_0 + c_1 x + c_2 x^2$ form an orthonormal set on the interval $-1 \leqq x \leqq 1$. Compare the result with that of Prob. 13.

15. Show that if the functions $g_1(x), g_2(x), \cdots$ form an orthogonal set on an interval $a \leqq x \leqq b$, then the functions $g_1(ct + k), g_2(ct + k), \cdots, c > 0$, form an orthogonal set on the interval $(a - k)/c \leqq t \leqq (b - k)/c$.

16. Using the result of Prob. 15, derive the orthogonality property of the set in Prob. 10 from that in Prob. 5.

17. Using the result of Prob. 15, derive the orthogonality property of the set in Prob. 7 from that in Prob. 6.

Orthogonality with respect to a weight function. In each case, verify that the given functions are orthogonal on the given interval with respect to the given weight function $p(x)$ and have the indicated norm.

18. $T_0(x) = 1, T_1(x) = x, T_2(x) = 2x^2 - 1, -1 \leqq x \leqq 1, p(x) = (1 - x^2)^{-1/2},$ $\|T_0\| = \sqrt{\pi}, \|T_1\| = \|T_2\| = \sqrt{\pi/2}$. *Hint.* Set $x = \cos \theta$.

19. $L_0(x) = 1, L_1(x) = 1 - x, L_2(x) = 1 - 2x + \frac{1}{2}x^2, 0 \leqq x < \infty, p(x) = e^{-x},$ norm 1

20. $U_0(x) = 1, U_1(x) = 2x, U_2(x) = 4x^2 - 1, -1 \leqq x \leqq 1, p(x) = (1 - x^2)^{1/2},$ norm $\sqrt{\pi/2}$.

4.8 Sturm–Liouville Problem

In engineering mathematics, various important orthogonal sets of functions arise as solutions of linear second-order differential equations of the form

(1)
$$\boxed{[r(x)y']' + [q(x) + \lambda p(x)]y = 0}$$

on some given interval $a \leqq x \leqq b$, satisfying conditions of the form

(2)

(a) $\boxed{k_1 y(a) + k_2 y'(a) = 0}$

(b) $\boxed{l_1 y(b) + l_2 y'(b) = 0.}$

Here λ is a parameter. k_1 and k_2 are given real constants not both zero, and so are l_1 and l_2.

In connection with (1) and (2) one uses the following terms. Equation (1) is called a **Sturm–Liouville equation.**[19] We shall see that Legendre's, Bessel's and other equations can be written in the form (1). Conditions (2) are called **boundary conditions,** since they refer to the endpoints (boundary points) $x = a$ and $x = b$ of that interval. A differential equation together with boundary conditions constitutes what is known as a **boundary value problem.** Our boundary value problem (1), (2) is called a **Sturm–Liouville problem.**

These problems occur in various engineering applications, for instance, in connection with vibrations of strings and membranes (Secs. 11.3, 11.8) and in heat conduction (Sec. 11.5).

Eigenvalues, Eigenfunctions

We see directly from (1) and (2) that for every number λ, the problem has the trivial solution $y \equiv 0$, that is, $y(x) = 0$ for all x in that interval. A number λ for which the problem has a solution $y \not\equiv 0$ [if such a number exists] is called an **eigenvalue** of the problem, and this solution y is called an **eigenfunction** of the problem *corresponding to this eigenvalue* λ.

EXAMPLE 1. Vibrating elastic string

Find the eigenvalues and eigenfunctions of the Sturm–Liouville problem

(3)   (a) $y'' + \lambda y = 0$ (b) $y(0) = 0, y(\pi) = 0$.

This problem arises, for instance, if an elastic string (a violin string, for example) is stretched a little and then fixed at its ends $x = 0$ and $x = \pi$ and allowed to vibrate. Then $y(x)$ is the "space function" of the deflection $u(x, t)$ of the string, assumed in the form $u(x, t) = y(x)w(t)$, where t is time. (This model will be discussed in Secs. 11.2–11.4.)

Solution. For negative $\lambda = -\nu^2$ a general solution of the equation is

$$y(x) = c_1 e^{\nu x} + c_2 e^{-\nu x}.$$

From (3b) we obtain $c_1 = c_2 = 0$ and $y \equiv 0$, which is not an eigenfunction. For $\lambda = 0$ the situation is similar. For positive $\lambda = \nu^2$ a general solution is

$$y(x) = A \cos \nu x + B \sin \nu x.$$

From the first boundary condition we obtain $y(0) = A = 0$. The second boundary condition then yields

$$y(\pi) = B \sin \nu\pi = 0 \qquad \text{or} \qquad \nu = 0, \pm 1, \pm 2, \cdots.$$

For $\nu = 0$ we have $y \equiv 0$. For $\lambda = \nu^2 = 1, 4, 9, 16, \cdots$, taking $B = 1$, we obtain

$$y(x) = \sin \nu x \qquad\qquad \nu = 1, 2, \cdots.$$

Hence the eigenvalues of the problem are $\lambda = \nu^2$ where $\nu = 1, 2, \cdots$, and corresponding eigenfunctions are $y(x) = \sin \nu x$ where $\nu = 1, 2, \cdots$.   ∎

---

[19]JACQUES CHARLES FRANÇOIS STURM (1803—1855), was born and studied in Switzerland and then moved to Paris, where he later became the successor of Poisson in the chair of mechanics at the Sorbonne.

JOSEPH LIOUVILLE (1809—1882), French mathematician and professor in Paris, contributed to various fields in mathematics and is particularly known by his important work in complex analysis (Liouville's theorem, cf. Sec. 13.6), special functions, differential geometry, and number theory.

Orthogonality of Eigenfunctions

It can be shown that under rather general conditions on the functions p, q and r in (1) the Sturm–Liouville problem (1), (2) has infinitely many eigenvalues; the corresponding rather complicated theory can be found in Refs. [A2] and [A10] listed in Appendix 1. (For instance, sufficient conditions are those in Theorem 1, below, together with $p(x) > 0$ and $r(x) > 0$ on $a < x < b$.)

Furthermore, we have the following important theorem.

Theorem 1 (Orthogonality of eigenfunctions)

Suppose that the functions p, q, r and r' in the Sturm–Liouville equation (1) are real-valued and continuous on the interval $a \leqq x \leqq b$. Let $y_m(x)$ and $y_n(x)$ be eigenfunctions of the Sturm–Liouville problem (1), (2) which correspond to different eigenvalues λ_m and λ_n, respectively. Then y_m, y_n are orthogonal on that interval with respect to the weight function p.

If $r(a) = 0$, then (2a) can be dropped from the problem. If $r(b) = 0$, then (2b) can be dropped. If $r(a) = r(b)$, then (2) can be replaced by

$$(4) \qquad y(a) = y(b), \qquad y'(a) = y'(b).$$

Proof. By assumption, y_m satisfies

$$(ry_m')' + (q + \lambda_m p)y_m = 0,$$

and y_n satisfies

$$(ry_n')' + (q + \lambda_n p)y_n = 0.$$

Multiplying the first equation by y_n, the second by $-y_m$ and adding, we get

$$(\lambda_m - \lambda_n)py_my_n = y_m(ry_n')' - y_n(ry_m')'$$

$$= [(ry_n')y_m - (ry_m')y_n]'$$

where the last equality can be readily verified by performing the indicated differentiation of the last expression in brackets. This expression is continuous on $a \leqq x \leqq b$ since r and r' are continuous by hypothesis and y_m, y_n are solutions of (1). Integrating over x from a to b, we thus obtain

$$(5) \qquad (\lambda_m - \lambda_n) \int_a^b py_my_n \, dx = \left[r(y_n'y_m - y_m'y_n) \right]_a^b.$$

The expression on the right equals

$$(6) \qquad \begin{aligned} &r(b)[y_n'(b)y_m(b) - y_m'(b)y_n(b)] \\ &-r(a)[y_n'(a)y_m(a) - y_m'(a)y_n(a)]. \end{aligned}$$

Case 1. If $r(a) = 0$ and $r(b) = 0$, then the expression in (6) is zero. Hence the expression on the left-hand side of (5) must be zero, and since λ_m and λ_n are distinct, we obtain the desired orthogonality

$$(7) \qquad\qquad \int_a^b p(x)y_m(x)y_n(x)\,dx = 0 \qquad\qquad (m \ne n)$$

without the use of the boundary conditions (2).

Case 2. Let $r(b) = 0$, but $r(a) \ne 0$. Then the first line in (6) is zero. We consider the remaining expression in (6). From (2a) we have

$$k_1 y_n(a) + k_2 y_n{}'(a) = 0,$$

$$k_1 y_m(a) + k_2 y_m{}'(a) = 0.$$

Let $k_2 \ne 0$. Then by multiplying the first equation by $y_m(a)$ and the last by $-y_n(a)$ and adding, we have

$$k_2[y_n{}'(a)y_m(a) - y_m{}'(a)y_n(a)] = 0.$$

Since $k_2 \ne 0$, the expression in brackets must be zero. This expression is identical with that in the last line of (6). Hence (6) is zero, and from (5) we obtain (7). If $k_2 = 0$, then by assumption $k_1 \ne 0$, and the argument of proof is similar.

Case 3. If $r(a) = 0$, but $r(b) \ne 0$, the proof is similar to that in Case 2, but instead of (2a) we now have to use (2b).

Case 4. If $r(a) \ne 0$ and $r(b) \ne 0$, we have to use both boundary conditions (2) and proceed as in Cases 2 and 3.

Case 5. Let $r(a) = r(b)$. Then (6) takes the form

$$r(b)[y_n{}'(b)y_m(b) - y_m{}'(b)y_n(b) - y_n{}'(a)y_m(a) + y_m{}'(a)y_n(a)].$$

We may use (2) as before and conclude that the expression in brackets is zero. However, we immediately see that this would also follow from (4), so that we may replace (2) by (4). Hence (5) yields (7), as before. This completes the proof of Theorem 1.   ∎

EXAMPLE 2. Vibrating elastic string

The differential equation in Example 1 is of the form (1) where $r = 1$, $q = 0$, and $p = 1$. From Theorem 1 it follows that the eigenfunctions are orthogonal on the interval $0 \le x \le \pi$.

EXAMPLE 3. Fourier series

The reader may show that the functions

$$1, \quad \cos x, \quad \sin x, \quad \cos 2x, \quad \sin 2x, \cdots$$

in the Fourier series in Example 3, Sec. 4.7, are the eigenfunctions of the Sturm–Liouville problem

$$y'' + \lambda y = 0, \qquad y(\pi) = y(-\pi), \qquad y'(\pi) = y'(-\pi).$$

Hence from Theorem 1 it follows that these functions form an orthogonal set on the interval $-\pi \le x \le \pi$. Note that the boundary conditions of our present problem are of the form (4).

Further examples will be included in the next section.   ∎

A generalized Fourier series (cf. Sec. 4.7) in which the orthogonal set is a set of eigenfunctions is called an **eigenfunction expansion.**

Reality of Eigenvalues

The eigenvalues of Sturm–Liouville problems have the following interesting property, which the physicist would probably expect, because eigenvalues may be related to frequencies, energies and other physical quantities that must be real.

Theorem 2 (Real eigenvalues)

If the Sturm–Liouville problem (1), (2) *satisfies the condition in Theorem* 1 *and p is positive in the whole interval* $a \leqq x \leqq b$ *(or negative everywhere in that interval), then all the eigenvalues of the problem are real.*

Proof. Let $\lambda = \alpha + i\beta$ be an eigenvalue of the problem and let

$$y(x) = u(x) + iv(x)$$

be a corresponding eigenfunction; here α, β, u, and v are real. Substituting this into (1) we have

$$(ru' + irv')' + (q + \alpha p + i\beta p)(u + iv) = 0.$$

This complex equation is equivalent to the following pair of equations for the real and imaginary parts:

$$(ru')' + (q + \alpha p)u - \beta p v = 0,$$
$$(rv')' + (q + \alpha p)v + \beta p u = 0.$$

Multiplying the first equation by v, the second by $-u$ and adding, we get

$$-\beta(u^2 + v^2)p = u(rv')' - v(ru')'$$
$$= [(rv')u - (ru')v]'.$$

The expression in brackets is continuous on $a \leqq x \leqq b$, for reasons similar to those in the previous proof. Integrating over x from a to b, we thus obtain

$$-\beta \int_a^b (u^2 + v^2)p \, dx = \left[r(uv' - u'v) \right]_a^b.$$

Because of the boundary conditions the right-hand side is zero; this is as in the previous proof. Since y is an eigenfunction, $u^2 + v^2 \not\equiv 0$. Since y and p are continuous and $p > 0$ (or $p < 0$) on the interval $a \leqq x \leqq b$, the integral on the left is not zero. Hence, $\beta = 0$, which means that $\lambda = \alpha$ is real. This completes the proof. ∎

Theorem 2 is illustrated by Examples 2 and 3, and further examples are included in the next section and the corresponding problem set.

Fourier series (cf. Example 3) are by far the most important eigenfunction expansions. We shall devote to them a whole chapter (Chap. 10) and consider numerous applications of these series in Chap. 11. Other Sturm–Liouville problems include problems for the *Legendre* and *Bessel equations,* which we discuss in the next section. The Sturm–Liouville theory then guarantees the orthogonality of the solution sets (Legendre polynomials, Bessel functions). This will be vital to our solution of the problem of the vibrating circular membrane (drumhead) in Sec. 11.10.

Problems for Sec. 4.8

1. Carry out the proof of Theorem 1 for the special Sturm–Liouville problem in Example 1.
2. Carry out the details of the proof of Theorem 1 in Cases 3 and 4.
3. **(Normalization of eigenfunctions)** Show that if $y = y_0$ is an eigenfunction of (1), (2) corresponding to some eigenvalue $\lambda = \lambda_0$, then $y = \alpha y_0$ ($\alpha \neq 0$, arbitrary) is an eigenfunction of (1), (2) corresponding to λ_0. (Note that this property can be used to "normalize" eigenfunctions, that is, to obtain eigenfunctions of norm 1.)

Find the eigenvalues and eigenfunctions of the following Sturm–Liouville problems.

4. $y'' + \lambda y = 0,$ $\quad y(0) = 0,$ $\quad y(\tfrac{1}{2}\pi) = 0$
5. $y'' + \lambda y = 0,$ $\quad y(0) = 0,$ $\quad y'(1) = 0$
6. $y'' + \lambda y = 0,$ $\quad y(0) = 0,$ $\quad y(L) = 0$
7. $y'' + \lambda y = 0,$ $\quad y(0) = 0,$ $\quad y'(L) = 0$
8. $y'' + \lambda y = 0,$ $\quad y'(0) = 0,$ $\quad y(\pi) = 0$
9. $y'' + \lambda y = 0,$ $\quad y'(0) = 0,$ $\quad y'(\pi) = 0$
10. $y'' + \lambda y = 0,$ $\quad y'(0) = 0,$ $\quad y(L) = 0$
11. $y'' + \lambda y = 0,$ $\quad y'(0) = 0,$ $\quad y'(L) = 0$
12. $y'' + \lambda y = 0,$ $\quad y(0) = y(2\pi),$ $\quad y'(0) = y'(2\pi)$
13. $y'' + \lambda y = 0,$ $\quad y(0) = y(\pi),$ $\quad y'(0) = y'(\pi)$
14. $(xy')' + \lambda x^{-1}y = 0,$ $\quad y(1) = 0,$ $\quad y(e) = 0.$ *Hint.* Set $x = e^t.$
15. $(xy')' + \lambda x^{-1}y = 0,$ $\quad y(1) = 0,$ $\quad y'(e) = 0$
16. $(e^{2x}y')' + e^{2x}(\lambda + 1)y = 0,$ $\quad y(0) = 0,$ $\quad y(\pi) = 0$
17. $(x^{-1}y')' + (\lambda + 1)x^{-3}y = 0,$ $\quad y(1) = 0,$ $\quad y(e) = 0$

18. Show that the eigenvalues of the Sturm–Liouville problem $y'' + \lambda y = 0,$ $y(0) = 0, y(1) + y'(1) = 0$ are obtained as solutions of the equation $\tan k = -k,$ where $k = \sqrt{\lambda}.$ Show graphically that this equation has infinitely many solutions $k = k_n$ and the eigenfunctions are $y_n = \sin k_n x$ ($k_n \neq 0$). Show that the positive k_n are of the form $k_n = \tfrac{1}{2}(2n + 1)\pi + \delta_n,$ where δ_n is positive and small and $\delta_n \to 0$ as $n \to \infty.$ Compute k_0 and k_1 (by Newton's method, Sec. 18.2).

Find a Sturm–Liouville problem whose eigenfunctions are

19. $1, \cos \dfrac{n\pi}{L}x, \sin \dfrac{n\pi}{L}x, (n = 1, 2, \cdots)$ 20. $1, \cos x, \cos 2x, \cos 3x, \cdots$

4.9   Orthogonality of Bessel Functions and Legendre Polynomials

EXAMPLE 1. Legendre polynomials

Legendre's equation (1) in Sec. 4.3 can be written

$$[(1 - x^2)y']' + \lambda y = 0, \qquad \lambda = n(n + 1).$$

Hence it is a Sturm–Liouville equation (1), Sec. 4.8, with $r = 1 - x^2$, $q = 0$, and $p = 1$. Since $r = 0$ when $x = \pm 1$, no boundary conditions are needed to form a Sturm–Liouville problem on the interval $-1 \leq x \leq 1$. We know that, for $n = 0, 1, \cdots$, the Legendre polynomials $P_n(x)$ are solutions of the problem. Hence these are eigenfunctions, and from Theorem 1 in Sec. 4.8 it follows that they are orthogonal on that interval, that is,

(1)
$$\int_{-1}^{1} P_m(x)P_n(x)\, dx = 0 \qquad\qquad (m \neq n).$$

The norm is (cf. Prob. 13 in Sec. 4.3)

(2)
$$\boxed{\; \|P_m\| = \sqrt{\int_{-1}^{1} P_m{}^2(x)\, dx} = \sqrt{\frac{2}{2m + 1}} \;} \qquad m = 0, 1, \cdots.$$

EXAMPLE 2. Bessel functions

The orthogonality of the Bessel functions J_n is quite important in engineering applications, for example in connection with vibrations of circular membranes (to be considered in Sec. 11.10). $J_n(s)$ satisfies the Bessel equation (Sec. 4.5)

$$s^2 \ddot{J}_n + s \dot{J}_n + (s^2 - n^2)J_n = 0$$

where the dots denote derivatives with respect to s. We assume that n is a nonnegative integer. Setting $s = \lambda x$ where λ is a constant, not zero, we have $dx/ds = 1/\lambda$, and, by the chain rule,

$$\dot{J}_n = J_n{}'/\lambda, \qquad \ddot{J}_n = J_n{}''/\lambda^2,$$

where primes denote derivatives with respect to x. By substitution,

$$x^2 J_n{}''(\lambda x) + x J_n{}'(\lambda x) + (\lambda^2 x^2 - n^2)J_n(\lambda x) = 0.$$

Dividing by x, we may write this equation in the form

(3)
$$[x J_n{}'(\lambda x)]' + \left(-\frac{n^2}{x} + \lambda^2 x \right)J_n(\lambda x) = 0.$$

We see that for each fixed n this is a Sturm–Liouville equation (1), Sec. 4.8, with the parameter written as λ^2 instead of λ, and

$$p(x) = x, \qquad q(x) = -n^2/x, \qquad r(x) = x.$$

Since $r(x) = 0$ at $x = 0$, it follows from Theorem 1 in the last section that those solutions of (3) on a given interval $0 \leq x \leq R$ which satisfy the boundary condition

(4)
$$J_n(\lambda R) = 0$$

form an orthogonal set on that interval with respect to the weight function $p(x) = x$. (Note that for $n \neq 0$ the function q is discontinuous at $x = 0$, but this does not affect the proof of that theorem.) It can be shown (Ref. [A14]) that $J_n(s)$ has infinitely many real zeros; let $\alpha_{1n} < \alpha_{2n} < \alpha_{3n} \cdots$ denote the positive zeros of $J_n(s)$. Then (4) holds for

(5) $\lambda R = \alpha_{mn}$ or $\lambda = \lambda_{mn} = \dfrac{\alpha_{mn}}{R}$ $(m = 1, 2, \cdots)$,

and we obtain the following result.

Theorem 1 (Orthogonality of Bessel functions)

For each fixed $n = 0, 1, \cdots$ the Bessel functions $J_n(\lambda_{1n}x)$, $J_n(\lambda_{2n}x)$, $J_n(\lambda_{3n}x)$, $\cdots$, with λ_{mn} as in (5), form an orthogonal set on the interval $0 \leqq x \leqq R$ with respect to the weight function $p(x) = x$, that is,

(6)
$$\int_0^R xJ_n(\lambda_{mn}x)J_n(\lambda_{kn}x)\,dx = 0 \qquad\qquad k \neq m.$$

Hence we have obtained infinitely many orthogonal sets, each corresponding to one of the fixed values of n.

Since $p(x) = x$, we see from (11)–(13) in Sec. 4.7 that if a given function $f(x)$ has an eigenfunction expansion in terms of one of those orthogonal sets, the expansion is

(7) $f(x) = a_1 J_n(\lambda_{1n}x) + a_2 J_n(\lambda_{2n}x) + \cdots$.

This is called a **Fourier–Bessel series.** We shall prove that

(8) $\|J_n(\lambda_{mn}x)\|^2 = \displaystyle\int_0^R xJ_n{}^2(\lambda_{mn}x)\,dx = \dfrac{R^2}{2}J_{n+1}^2(\lambda_{mn}R)$,

where $\lambda_{mn} = \alpha_{mn}/R$. By (13) in Sec. 4.7 this implies that the Fourier constants a_m of f in (7) are

(9) $a_m = \dfrac{2}{R^2 J_{n+1}^2(\alpha_{mn})}\displaystyle\int_0^R xf(x)J_n(\lambda_{mn}x)\,dx$, $m = 1, 2, \cdots$.

We prove (8). Multiplying (3) by $2xJ_n{}'(\lambda x)$, we obtain an equation that can be written

$$\{[xJ_n{}'(\lambda x)]^2\}' + (\lambda^2 x^2 - n^2)\{J_n{}^2(\lambda x)\}' = 0.$$

Integrating over x from 0 to R we find

(10) $[xJ_n{}'(\lambda x)]^2\ \Big|_0^R = -\displaystyle\int_0^R (\lambda^2 x^2 - n^2)\{J_n{}^2(\lambda x)\}'\,dx$.

From (21) in Prob. 8, Sec. 4.5, writing s and n instead of x and v, and denoting the derivative with respect to s by a dot, we have

$$-ns^{-n-1}J_n(s) + s^{-n}\dot{J}_n(s) = -s^{-n}J_{n+1}(s).$$

Multiplying by s^{n+1} and setting $s = \lambda x$, we obtain

$$\lambda xJ_n{}'(\lambda x)\frac{1}{\lambda} = nJ_n(\lambda x) - \lambda xJ_{n+1}(\lambda x),$$

where the prime denotes the derivative with respect to x. Hence the left-hand side of (10) equals

$$\left[[nJ_n(\lambda x) - \lambda xJ_{n+1}(\lambda x)]^2\right]_{x=0}^R .$$

If $\lambda = \lambda_{mn}$ then $J_n(\lambda R) = 0$, and since $J_n(0) = 0$ $(n = 1, 2, \cdots)$, that left-hand side becomes

(11) $\lambda_{mn}{}^2 R^2 J_{n+1}^2(\lambda_{mn}R)$.

If we integrate by parts on the right-hand side of (10), we see that this right-hand side becomes

$$-\left[(\lambda^2 x^2 - n^2)J_n^2(\lambda x)\right]_0^R + 2\lambda^2 \int_0^R xJ_n^2(\lambda x)\, dx.$$

When $\lambda = \lambda_{mn}$, the first expression is zero at $x = R$. It is also zero at $x = 0$ because $\lambda^2 x^2 - n^2 = 0$ when $n = x = 0$, and $J_n(\lambda x) = 0$ when $x = 0$ and $n = 1, 2, \cdots$. From this and (11) we readily obtain (8). This completes the proof.   ∎

If p or r in (1), Sec. 4.8, vanish at a or b (or at both a and b), then (1), (2), Sec. 4.8, is called a **singular Sturm–Liouville problem.** Our present examples are of this type.

This is the end of Chap. 4 on the power series method, the Frobenius method and orthogonality. In the next and final chapter of Part A on ordinary differential equations, we consider an **operational method** for solving such equations, the **Laplace transformation,** which from the practical point of applications in physics and engineering is the most important operational method. (**Fourier transformations** will be considered in the chapter on Fourier series, in Secs. 10.10 and 10.11.)

Problems for Sec. 4.9

Legendre polynomials. Represent the following polynomials in terms of Legendre polynomials.

1. $1, x, x^2, x^4$

2. $3x^2 + x$

3. $5x^3 - 3x^2 - x - 1$

4. $70x^4 - 45x^2 - 3$

In each case, obtain the first few terms of the expansion of $f(x)$ in terms of Legendre polynomials and graph the first three partial sums. *Hint.* Use (6), (7) in Sec. 4.7.

5. $f(x) = \begin{cases} 0 & \text{if } -1 < x < 0 \\ x & \text{if } \;\; 0 < x < 1 \end{cases}$

6. $f(x) = \begin{cases} 0 & \text{if } -1 < x < 0 \\ 1 & \text{if } \;\; 0 < x < 1 \end{cases}$

7. $f(x) = |x|$ if $-1 < x < 1$

8. $f(x) = e^x$ if $-1 < x < 1$

9. Show that the functions $P_n(\cos \theta)$, $n = 0, 1, \cdots$, form an orthogonal set on the interval $0 \leqq \theta \leqq \pi$ with respect to the weight function $\sin \theta$.

Hermite polynomials[20]. The functions

$$He_0 = 1, \qquad He_n(x) = (-1)^n e^{x^2/2} \frac{d^n}{dx^n}(e^{-x^2/2}), \qquad n = 1, 2, \cdots$$

are called *Hermite polynomials.*

Remark. As is true for many special functions, the literature contains more than one notation, and one sometimes defines as Hermite polynomials the functions

$$H_0^* = 1, \qquad H_n^*(x) = (-1)^n e^{x^2} \frac{d^n e^{-x^2}}{dx^n}.$$

This differs from our definition, which is preferably used in applications.

[20]CHARLES HERMITE (1822—1901), French mathematician, is known by his work in algebra and number theory. The great HENRI POINCARÉ (1854—1912) was one of his students.

10. Show that

$$He_1(x) = x, \quad He_2(x) = x^2 - 1, \quad He_3(x) = x^3 - 3x, \quad He_4(x) = x^4 - 6x^2 + 3.$$

11. Show that the Hermite polynomials are related to the coefficients of the Maclaurin series

$$e^{tx-t^2/2} = \sum_{n=0}^{\infty} a_n(x)t^n$$

by the formula $He_n(x) = n!a_n(x)$. *Hint.* Note that $tx - t^2/2 = x^2/2 - (x - t)^2/2$. (The exponential function is called the **generating function** of the He_n.)

12. Show that the Hermite polynomials satisfy the relation

$$He_{n+1}(x) = xHe_n(x) - He_n'(x).$$

13. Differentiating the generating function in Prob. 11 with respect to x, show that

$$He_n'(x) = nHe_{n-1}(x).$$

Using this and the formula in Prob. 12 (with n replaced by $n - 1$), prove that $He_n(x)$ satisfies the differential equation

$$y'' - xy' + ny = 0.$$

14. Using the equation in Prob. 13, show that $w = e^{-x^2/4}He_n(x)$ is a solution of **Weber's equation**[21]

$$w'' + (n + \tfrac{1}{2} - \tfrac{1}{4}x^2)w = 0 \qquad (n = 0, 1, \cdots).$$

15. Show that the Hermite polynomials are orthogonal on the x-axis $-\infty < x < \infty$ with respect to the weight function $p(x) = e^{-x^2/2}$.

Chebyshev polynomials.[22] The functions

$$T_n(x) = \cos(n \text{ arc } \cos x), \qquad U_n(x) = \frac{\sin[(n + 1) \text{ arc } \cos x]}{\sqrt{1 - x^2}} \qquad n = 0, 1, \cdots$$

are called *Chebyshev polynomials of the first and second kind,* respectively.

16. Show that

$$T_0 = 1, \qquad T_1(x) = x, \qquad T_2(x) = 2x^2 - 1, \qquad T_3(x) = 4x^3 - 3x,$$
$$U_0 = 1, \qquad U_1(x) = 2x, \qquad U_2(x) = 4x^2 - 1, \qquad U_3(x) = 8x^3 - 4x.$$

17. Show that the Chebyshev polynomials $T_n(x)$ are orthogonal on the interval $-1 \leq x \leq 1$ with respect to the weight function $p(x) = 1/\sqrt{1 - x^2}$. *Hint.* To evaluate the integral, set arc $\cos x = \theta$.

18. Show that $T_n(x)$ is a solution of the differential equation

$$(1 - x^2)T_n'' - xT_n' + n^2T_n = 0.$$

---

[21]HEINRICH WEBER (1842—1913), German mathematician.

[22]PAFNUTI CHEBYSHEV (1821—1894), Russian mathematician, is known by his work in approximation theory and the theory of numbers. Another transliteration of the name is TCHE-BICHEF.

Laguerre polynomials.[23] The functions

$$L_0 = 1, \qquad L_n(x) = \frac{e^x}{n!} \frac{d^n(x^n e^{-x})}{dx^n}, \qquad n = 1, 2, \cdots$$

are called *Laguerre polynomials.*

19. Show that

$$L_1(x) = 1 - x, \quad L_2(x) = 1 - 2x + x^2/2, \quad L_3(x) = 1 - 3x + 3x^2/2 - x^3/6.$$

20. Verify by direct integration that L_0, $L_1(x)$, $L_2(x)$ are orthogonal on the positive axis $0 \le x < \infty$ with respect to the weight function $p(x) = e^{-x}$.

21. Prove that the set of all Laguerre polynomials is orthogonal on $0 \le x < \infty$ with respect to the weight function $p(x) = e^{-x}$.

22. Show that

$$L_n(x) = \sum_{m=0}^{n} \frac{(-1)^m}{m!} \binom{n}{m} x^m = 1 - nx + \frac{n(n-1)}{4} x^2 - + \cdots + \frac{(-1)^n}{n!} x^n.$$

23. $L_n(x)$ satisfies Laguerre's differential equation $xy'' + (1 - x)y' + ny = 0$. Verify this fact for $n = 0, 1, 2, 3$.

Bessel functions

24. Graph $J_0(\lambda_{10}x)$, $J_0(\lambda_{20}x)$, $J_0(\lambda_{30}x)$, and $J_0(\lambda_{40}x)$ for $R = 1$ in the interval $0 \le x \le 1$. (Use Table A2 in Appendix 4.)

Develop the following functions $f(x)$ $(0 < x < R)$ in a **Fourier–Bessel series**

$$f(x) = a_1 J_0(\lambda_{10}x) + a_2 J_0(\lambda_{20}x) + a_3 J_0(\lambda_{30}x) + \cdots$$

and graph the first few partial sums.

25. $f(x) = 1$. *Hint.* Use (20), Sec. 4.5.   **26.** $f(x) = \begin{cases} 1 & \text{if} \quad 0 < x < R/2 \\ 0 & \text{if} \quad R/2 < x < R \end{cases}$

27. $f(x) = \begin{cases} k & \text{if} \quad 0 < x < a \\ 0 & \text{if} \quad a < x < R \end{cases}$ **28.** $f(x) = \begin{cases} 0 & \text{if} \quad 0 < x < R/2 \\ k & \text{if} \quad R/2 < x < R \end{cases}$

29. $f(x) = 1 - x^2$ $(R = 1)$. *Hint.* Use (20), Sec. 4.5, and integration by parts.[24]

30. $f(x) = R^2 - x^2$ **31.** $f(x) = x^2$

32. $f(x) = x^4$

33. Show that $f(x) = x^n$ $(0 < x < 1, n = 0, 1, \cdots)$ can be represented by the Fourier–Bessel series

$$x^n = \frac{2 J_n(\alpha_{1n}x)}{\alpha_{1n} J_{n+1}(\alpha_{1n})} + \frac{2 J_n(\alpha_{2n}x)}{\alpha_{2n} J_{n+1}(\alpha_{2n})} + \cdots.$$

34. Find a representation of $f(x) = x^n$ $(0 < x < R, n = 0, 1, \cdots)$ similar to that in Prob. 33.

35. Represent $f(x) = x^3$ $(0 < x < 2)$ by a Fourier–Bessel series involving J_3.

---

[23]EDMOND LAGUERRE (1834—1886), French mathematician, who did research work in geometry and the theory of infinite series.

[24]This problem will be needed in Example 1 of Sec. 11.10.

Further Proof in Chapter 4

Proof of Theorem 2 in Sec. 4.4, page 198

The formula numbers in this proof are the same as in the text of Sec. 4.4. An additional formula not appearing in Sec. 4.4 will be called (A) (see below).

The differential equation in Theorem 2 is

(1) $$y'' + \frac{b(x)}{x} y' + \frac{c(x)}{x^2} y = 0,$$

where $b(x)$ and $c(x)$ are analytic functions. We can write it

(1') $$x^2 y'' + x b(x) y' + c(x) y = 0.$$

The indicial equation of (1) is

(4) $$r^2 + (b_0 - 1)r + c_0 = 0.$$

The roots r_1, r_2 of this quadratic equation determine the general form of a basis of solutions of (1), and there are three possible cases as follows.

Case 1 (Roots not differing by an integer). A first solution of (1) of the form

(5) $$y_1(x) = x^{r_1}(a_0 + a_1 x + a_2 x^2 + \cdots)$$

can be determined as in the power series method. For a proof that in this case, equation (1) has a second independent solution of the form

(6) $$y_2(x) = x^{r_2}(A_0 + A_1 x + A_2 x^2 + \cdots),$$

see Ref. [A10] listed in Appendix 1.

Case 2 (Double root). The indicial equation (4) has a double root r if and only if $(b_0 - 1)^2 - 4c_0 = 0$, and then $r = \frac{1}{2}(1 - b_0)$. A first solution

(7) $$y_1(x) = x^r(a_0 + a_1 x + a_2 x^2 + \cdots) \qquad r = \tfrac{1}{2}(1 - b_0)$$

can be determined as in Case 1. We show that a second independent solution is of the form

(8) $$y_2(x) = y_1(x) \ln x + x^r(A_1 x + A_2 x^2 + \cdots) \qquad (x > 0).$$

We use the method of reduction of order (as in Sec. 2.3), that is, we determine $u(x)$ such that

$$y_2(x) = u(x) y_1(x)$$

is a solution of (1). By inserting this and the derivatives

$$y_2' = u'y_1 + uy_1', \qquad y_2'' = u''y_1 + 2u'y_1' + uy_1''$$

into the differential equation (1') we obtain

$$x^2(u''y_1 + 2u'y_1' + uy_1'') + xb(u'y_1 + uy_1') + cuy_1 = 0.$$

Since y_1 is a solution of (1'), the sum of the terms involving u is zero, and this equation reduces to

$$x^2 y_1 u'' + 2x^2 y_1' u' + xby_1 u' = 0.$$

By dividing by $x^2 y_1$ and inserting the power series for b we obtain

$$u'' + \left(2\frac{y_1'}{y_1} + \frac{b_0}{x} + \cdots\right) u' = 0.$$

Here and in the following the dots designate terms which are constant or involve positive powers of x. Now from (7) it follows that

$$\frac{y_1'}{y_1} = \frac{x^{r-1}[ra_0 + (r+1)a_1 x + \cdots]}{x^r[a_0 + a_1 x + \cdots]}$$

$$= \frac{1}{x}\left(\frac{ra_0 + (r+1)a_1 x + \cdots}{a_0 + a_1 x + \cdots}\right) = \frac{r}{x} + \cdots.$$

Hence the previous equation can be written

(A) $$u'' + \left(\frac{2r + b_0}{x} + \cdots\right) u' = 0.$$

Since $r = (1 - b_0)/2$ the term $(2r + b_0)/x$ equals $1/x$, and by dividing by u' we thus have

$$\frac{u''}{u'} = -\frac{1}{x} + \cdots.$$

By integration we obtain

$$\ln u' = -\ln x + \cdots, \qquad \text{hence} \qquad u' = \frac{1}{x} e^{(\cdots)}.$$

Expanding the exponential function in powers of x and integrating once more, we see that u is of the form

$$u = \ln x + k_1 x + k_2 x^2 + \cdots.$$

Inserting this into $y_2 = uy_1$, we obtain for y_2 a representation of the form (8).

Case 3 (Roots differing by an integer). We write $r_1 = r$ and $r_2 = r - p$ where p is a *positive* integer. A first solution

$$(9) \qquad y_1(x) = x^{r_1}(a_0 + a_1 x + a_2 x^2 + \cdots)$$

can be determined as in Cases 1 and 2. We show that a second independent solution is of the form

$$(10) \qquad y_2(x) = k y_1(x) \ln x + x^{r_2}(A_0 + A_1 x + A_2 x^2 + \cdots)$$

where we may have $k \neq 0$ or $k = 0$. As in Case 2 we set $y_2 = u y_1$. The first steps are literally as in Case 2 and give equation (A),

$$u'' + \left(\frac{2r + b_0}{x} + \cdots \right) u' = 0.$$

Now by elementary algebra, the coefficient $b_0 - 1$ of r in (4) equals minus the sum of the roots,

$$b_0 - 1 = -(r_1 + r_2) = -(r + r - p) = -2r + p.$$

Hence $2r + b_0 = p + 1$, and division by u' gives

$$\frac{u''}{u'} = -\left(\frac{p + 1}{x} + \cdots \right).$$

The further steps are as in Case 2. Integrating, we find

$$\ln u' = -(p + 1) \ln x + \cdots, \qquad \text{thus} \qquad u' = x^{-(p+1)} e^{(\cdots)}$$

where dots stand for some series of nonnegative integer powers of x. By expanding the exponential function as before we obtain a series of the form

$$u' = \frac{1}{x^{p+1}} + \frac{k_1}{x^p} + \cdots + \frac{k_{p-1}}{x^2} + \frac{k_p}{x} + k_{p+1} + k_{p+2} x + \cdots.$$

We integrate once more. Writing the resulting logarithmic term first, we get

$$u = k_p \ln x + \left(-\frac{1}{p x^p} - \cdots - \frac{k_{p-1}}{x} + k_{p+1} x + \cdots \right).$$

Hence, by (9) we get for $y_2 = u y_1$ the formula

$$y_2 = k_p y_1 \ln x + x^{r_1 - p} \left(-\frac{1}{p} - \cdots - k_{p-1} x^{p-1} + \cdots \right)(a_0 + a_1 x + \cdots).$$

But this is of the form (10) with $k = k_p$ since $r_1 - p = r_2$ and the product of the two series involves nonnegative integer powers of x only.   ∎

Review Problems for Chapter 4

1. Can a power series in x contain negative powers of x? Fractional powers? A constant term?

2. What do we mean by saying that a function is analytic at some point? Why did this concept arise in this chapter?

3. Why did we generalize the power series method? Give an example of an equation to which the Frobenius method applies whereas the usual power series method does not.

4. Can a power series solution ever reduce to a polynomial?

5. Can the series in the Frobenius method ever reduce to a power series?

6. In this chapter, we always took $x_0 = 0$ and considered power series in x. Explain why this is no essential restriction of generality.

7. What do we mean by the indicial equation? For what purpose did we use it? How did it arise?

8. In the Frobenius method we distinguished three cases. What were they? Which of them does the Euler–Cauchy equation reveal and which not?

9. To which case of the Frobenius method do complex conjugate roots of the indicial equation correspond?

10. For which of the two equations, the Legendre equation or the Bessel equation, did we need the Frobenius method? (Give reason.)

11. Is J_ν a solution of the Bessel equation for every ν? Why did we also introduce the Bessel functions Y_ν? What is the main difference in the behavior of J_ν and Y_ν ($\nu > 0$) near $x = 0$?

12. Let $g_1(x)$, $g_2(x)$, $\cdots$ be orthogonal on the interval $1 \leqq x \leqq 3$ with respect to some weight function $p(x)$. What functions are then orthogonal on $1 \leqq x \leqq 3$ in the usual sense?

13. If we want to represent a given function by a linear combination or an infinite series of other functions, what advantage do we have in the determination of the coefficients if these functions are orthogonal? What do we mean by saying that an orthonormal set of functions is complete? Why is this concept of practical importance?

14. Without looking into the text, write down what you can remember about Sturm–Liouville problems (the definition, properties of solutions).

15. Can a Sturm–Liouville problem involve boundary conditions of the form $y(2) + y'(5) = 0$ or of the form $3y(2) - \frac{1}{2}y'(2) = 1$? (Give reasons.)

Using one of the two methods discussed in this chapter, solve the following equations. Try to identify the series obtained as expansions of known functions.

16. $y'' + y = 2x^2 + x$

17. $y'' + 4y = 0$

18. $(x - 1)y'' - xy' + y = 0$

19. $(x^2 + 2x)y'' + (x^2 - 2)y' - (2x + 2)y = 0$

20. $x^2y'' - 2xy' + (x^2 + 2)y = 0$

21. $xy'' - y' - 4x^3y = 0$

22. $(x - 1)^2y'' + (x - 1)y' - y = 0$

23. $x^2y'' - 2xy' + (2 - x^2)y = 0$

24. $xy'' + 2y' + xy = 0$

25. $(2x^3 + x^2)y'' + (6x^2 + 4x)y' + 2y = 0$

Using the indicated substitutions, solve, in terms of Bessel functions:

26. $x^2y'' + xy' + 4(x^4 - \nu^2)y = 0$ $(x^2 = z)$

27. $xy'' - y' + xy = 0$ $(y = xu)$

28. $x^2y'' + (x^2 + \frac{1}{4})y = 0$ $(y = u\sqrt{x})$

29. $xy'' + 3y' + xy = 0$ $(y = u/x)$

30. $y'' - xy = 0$ $(y = u\sqrt{x}, \frac{2}{3}ix^{3/2} = z; i = \sqrt{-1};$ **Airy's equation**)

Show orthogonality on the given interval and determine the corresponding ortho-normal set of functions.

31. $\sin x, \sin 2x, \sin 3x, \cdots, \quad -\pi \leq x \leq \pi$

32. $1, \cos \omega x, \cos 2\omega x, \cos 3\omega x, \cdots, \quad 0 \leq x \leq 2\pi/\omega$

33. $1, \cos 4nx, \sin 4nx, \quad n = 1, 2, \cdots, \quad 0 \leq x \leq \frac{1}{2}\pi$

34. $\sin kn\pi x, \quad n = 1, 2, \cdots, \quad -1/k \leq x \leq 1/k$

35. $P_0(\frac{1}{2}x), P_1(\frac{1}{2}x), P_2(\frac{1}{2}x), \quad -2 \leq x \leq 2;$ cf. Sec. 4.3

Find the eigenvalues and eigenfunctions of the following problems.

36. $(xy')' + \lambda y/x = 0$ $(\lambda > 0),$ $y(1) = 0, \quad y(e^\pi) = 0.$ *Hint.* Set $x = e^t$.

37. $y'' + \lambda y = 0,$ $\quad y(0) = 0, \quad y(\pi) = 0$

38. $y'' + \lambda y = 0,$ $\quad y(-\frac{1}{2}\pi) = 0, \quad y(\frac{1}{2}\pi) = 0$

39. $(xy')' + \lambda y/x = 0$ $(\lambda > 0),$ $y(e^{1/2}) = 0, \quad y(e^{3/2}) = 0$

40. $y'' + \lambda y = 0,$ $\quad y'(0) = 0, \quad y'(\frac{1}{2}\pi) = 0$

Summary of Chapter 4
Series Solutions of Differential Equations. Orthogonal Functions

The **power series method** is a general method for solving linear differential equations

(1) $$y'' + p(x)y' + q(x)y = 0$$

with variable $p(x)$ and $q(x)$; it also applies to nonhomogeneous and higher order equations. It gives solutions in the form of power series; this motivates the name. In this method, one substitutes a power series (with any center x_0, e.g., $x_0 = 0$)

(2) $$y(x) = a_0 + a_1(x - x_0) + a_2(x - x_0)^2 + \cdots$$

into (1), similarly $y'(x) = a_1 + 2a_2(x - x_0) + \cdots$ and $y''(x)$. In this way, one determines the undetermined coefficients a_m in (2), as explained in Sec. 4.1 and other sections. This gives solutions y represented by power series. If $p(x)$ and $q(x)$ are **analytic** at $x = x_0$ (Sec. 4.2), then (1) has solutions of this form.

If $p(x)$ and $q(x)$ are not analytic at x_0, we call them **singular** at x_0. If such singularities are "not too bad," namely, such that (1) can be written

$$(3) \qquad y'' + \frac{a(x)}{x - x_0} y' + \frac{b(x)}{(x - x_0)^2} y = 0$$

with $a(x)$ and $b(x)$ analytic at x_0, then (3) has at least one solution of the form

$$(4) \qquad y(x) = x^r [a_0 + a_1(x - x_0) + a_2(x - x_0)^2 + \cdots],$$

where r can be a fraction or even a complex number and is determined by substituting (4) into (3); this also gives the a_m's. A second independent solution may be of a similar form (with different r and a_m's) or may involve a logarithmic term. This method is called the **Frobenius method** or the **extended power series method** (Sec. 4.4).

If an equation (1) or (3) and its series solutions occur frequently in applications, one gives them a name and introduces special symbols. Of this kind, and particularly useful to the engineer and physicist, are **Legendre's equation** and the Legendre polynomials $P_0(x)$, $P_1(x)$, $P_2(x)$, $\cdots$ (Sec. 4.3), the **hypergeometric equation** and the hypergeometric functions $F(a, b, c; x)$ (Sec. 4.4), and **Bessel's equation** and the Bessel functions J_ν and Y_ν (Secs. 4.5, 4.6). Indeed, second-order linear differential equations are one of the two main sources of such "higher functions". (The other source originates from nonelementary integrals, such as those listed in Appendix 3.)

Modeling involving differential equations in most cases leads to initial value (Chap. 2) or boundary value problems. Many of the latter can be written in the form of **Sturm–Liouville problems** (Sec. 4.8). These are **eigenvalue problems** involving a parameter λ, which in applications may be related to frequencies, energies or other physical quantities. Solutions of these problems, called *eigenfunctions*, have many general properties in common, notably the highly important orthogonality (Sec. 4.8). Section 4.9 shows this for Legendre polynomials and Bessel functions. This leads to series developments called **eigenfunction expansions**, Fourier series (Chap. 10) being a prototype of these, as we shall see in more detail in Chap. 11 on partial differential equations.

Chapter 5

Laplace Transformation

The Laplace transformation is a method for solving differential equations and corresponding initial and boundary value problems. The process of solution consists of three main steps:

1st step.　　The given "hard" problem is transformed into a "simple" equation (*subsidiary equation*).

2nd step.　　The subsidiary equation is solved by purely algebraic manipulations.

3rd step.　　The solution of the subsidiary equation is transformed back to obtain the solution of the given problem.

In this way the Laplace transformation reduces the problem of solving a differential equation to an algebraic problem. The third step is made easier by tables, whose role is similar to that of integral tables in integration. (These tables are also useful in the first step.) Such a table is included at the end of the chapter.

This switching from operations of calculus to *algebraic* operations on transforms is called **operational calculus,** a very important area of applied mathematics, and the Laplace transformation is practically the most important method for this purpose. (For another such method, the Fourier transformation, see Sec. 10.11.) Indeed, the Laplace transformation is widely used in engineering mathematics, where it has numerous applications. It is particularly useful in problems where the (mechanical or electrical) driving force has discontinuities, for instance, acts for a short time only, or is periodic but is not merely a sine or cosine function. Another advantage is that it solves problems directly. Indeed, initial value problems are solved without first determining a general solution. Similarly, nonhomogeneous equations are solved without first solving the corresponding homogeneous equation.

In this chapter we consider the Laplace transformation from a practical point of view and illustrate its use by important engineering problems, many of them related to *ordinary* differential equations.

Partial differential equations can also be treated by the Laplace transformation. An introduction to this field is given in Sec. 11.13.

Section 5.9 contains a list of general formulas and Sec. 5.10 a list of transforms F(s) and corresponding functions f(t).

Prerequisite for this chapter: Chap. 2.
Sections that may be omitted in a very short course: 5.5–5.7.
References: Appendix 1, Part A.
Answers to problems: Appendix 2.

5.1 Laplace Transform. Inverse Transform. Linearity

Let $f(t)$ be a given function that is defined for all $t \geqq 0$. We multiply $f(t)$ by e^{-st} and integrate with respect to t from zero to infinity. Then, if the resulting integral exists, it is a function of s, say, $F(s)$:

$$F(s) = \int_0^\infty e^{-st} f(t) \, dt.$$

The function $F(s)$ of the variable s is called the **Laplace transform**[1] of the original function $f(t)$, and will be denoted by $\mathcal{L}(f)$. Thus

(1)
$$F(s) = \mathcal{L}(f) = \int_0^\infty e^{-st} f(t) \, dt.$$

The operation just described, which yields $F(s)$ from a given $f(t)$, is called the **Laplace transformation.**

Furthermore, the original function $f(t)$ in (1) is called the *inverse transform* or **inverse** of $F(s)$ and will be denoted by $\mathcal{L}^{-1}(F)$; that is, we shall write

$$f(t) = \mathcal{L}^{-1}(F).$$

Notation

Original functions are denoted by lowercase letters and their transforms by the same letters in capitals, so that $F(s)$ denotes the transform of $f(t)$, and $Y(s)$ denotes the transform of $y(t)$, and so on.

EXAMPLE 1

Let $f(t) = 1$ when $t \geqq 0$. Find $F(s)$.

Solution. From (1) we obtain by integration

$$\mathcal{L}(f) = \mathcal{L}(1) = \int_0^\infty e^{-st} \, dt = -\frac{1}{s} e^{-st} \bigg|_0^\infty;$$

hence, when $s > 0$,

$$\mathcal{L}(1) = \frac{1}{s}.$$

---

[1]PIERRE SIMON MARQUIS DE LAPLACE (1749—1827), great French mathematician, was professor in Paris. He developed the foundation of potential theory and made important contributions to celestial mechanics, astronomy in general, special functions, and probability theory. Napoleon Bonaparte was his student for a year. For Laplace's interesting political involvements, see Ref. [2], p. 260, listed in Appendix 1.

Our notation in the first line on the right is convenient, but we should say a word about it. The interval of integration in (1) is infinite. Such an integral is called an *improper integral* and, by definition, is evaluated according to the rule

$$\int_0^\infty e^{-st} f(t) \, dt = \lim_{T \to \infty} \int_0^T e^{-st} f(t) \, dt.$$

Hence our convenient notation means

$$\int_0^\infty e^{-st} \, dt = \lim_{T \to \infty} \left[-\frac{1}{s} e^{-st} \right]_0^T = \lim_{T \to \infty} \left[-\frac{1}{s} e^{-sT} + \frac{1}{s} e^0 \right] = \frac{1}{s} \qquad (s > 0).$$

We shall use that notation throughout this chapter.

EXAMPLE 2

Let $f(t) = e^{at}$ when $t \geq 0$, where a is a constant. Find $\mathscr{L}(f)$.

Solution. Again by (1),

$$\mathscr{L}(e^{at}) = \int_0^\infty e^{-st} e^{at} \, dt = \frac{1}{a - s} e^{-(s-a)t} \Big|_0^\infty ;$$

hence, when $s - a > 0$,

$$\mathscr{L}(e^{at}) = \frac{1}{s - a} .$$ ∎

Must we go on in this fashion and obtain the transform of one function after another directly from the definition? The answer is no. And the reason is that the Laplace transformation has many general properties that are helpful for that purpose. A very important property is that the Laplace transformation is a linear operation, just as differentiation and integration. By this we mean the following.

Theorem 1 (Linearity of the Laplace transformation)

The Laplace transformation is a linear operation; that is, for any functions $f(t)$ and $g(t)$ whose Laplace transform exists and any constants a and b,

$$\boxed{\mathscr{L}\{af(t) + bg(t)\} = a\mathscr{L}\{f(t)\} + b\mathscr{L}\{g(t)\}.}$$

Proof. By the definition,

$$\mathscr{L}\{af(t) + bg(t)\} = \int_0^\infty e^{-st}[af(t) + bg(t)] \, dt$$

$$= a \int_0^\infty e^{-st} f(t) \, dt + b \int_0^\infty e^{-st} g(t) \, dt$$

$$= a\mathscr{L}\{f(t)\} + b\mathscr{L}\{g(t)\}.$$ ∎

EXAMPLE 3

Let $f(t) = \cosh at = (e^{at} + e^{-at})/2$. Find $\mathscr{L}(f)$.

Solution. From Theorem 1 and Example 2 we obtain

$$\mathscr{L}(\cosh at) = \frac{1}{2} \mathscr{L}(e^{at}) + \frac{1}{2} \mathscr{L}(e^{-at}) = \frac{1}{2} \left(\frac{1}{s - a} + \frac{1}{s + a} \right);$$

that is, when $s > a \ (\geqq 0)$,

$$\mathscr{L}(\cosh at) = \frac{s}{s^2 - a^2}.$$

EXAMPLE 4. A simple partial fraction reduction

Let $F(s) = \dfrac{1}{(s-a)(s-b)}$, $a \neq b$. Find $\mathscr{L}^{-1}(F)$.

Solution. The inverse of a linear transformation is linear (cf. Prob. 45). By partial fraction reduction we thus obtain from Example 2

$$\mathscr{L}^{-1}(F) = \mathscr{L}^{-1}\left\{\frac{1}{a-b}\left(\frac{1}{s-a} - \frac{1}{s-b}\right)\right\}$$

$$= \frac{1}{a-b}\left[\mathscr{L}^{-1}\left(\frac{1}{s-a}\right) - \mathscr{L}^{-1}\left(\frac{1}{s-b}\right)\right] = \frac{1}{a-b}(e^{at} - e^{bt}).$$

This proves formula 11 in the table in Sec. 5.10, p. 299.

EXAMPLE 5. Derivation of formula 12 in the table in Sec. 5.10

Let $F(s) = \dfrac{s}{(s-a)(s-b)}$, $a \neq b$. Find $\mathscr{L}^{-1}(F)$.

Solution. Using the idea of Example 4, we obtain

$$\mathscr{L}^{-1}\left\{\frac{s}{(s-a)(s-b)}\right\} = \mathscr{L}^{-1}\left\{\frac{1}{a-b}\left(\frac{a}{s-a} - \frac{b}{s-b}\right)\right\} = \frac{1}{a-b}(ae^{at} - be^{bt}). \quad \blacksquare$$

A short list of some important elementary functions and their Laplace transforms is given in Table 5.1, and a more extensive list in Sec. 5.10. Cf. also the references in Appendix 1, Part A.

Once we know the transforms in Table 5.1, nearly all the transforms we shall need can be obtained through the use of some simple general theorems which we consider in the subsequent sections.

Formulas 1, 2 and 3 in Table 5.1 are special cases of formula 4. Formula 4 follows from formula 5 and $\Gamma(n+1) = n!$, where n is a nonnegative integer

Table 5.1
Some Functions $f(t)$ and Their Laplace Transforms $\mathscr{L}(f)$
(Cf. formulas 1–7 and 13–16 in the table in Sec. 5.10)

	$f(t)$	$\mathscr{L}(f)$		$f(t)$	$\mathscr{L}(f)$
1	1	$1/s$	6	e^{at}	$\dfrac{1}{s-a}$
2	t	$1/s^2$	7	$\cos \omega t$	$\dfrac{s}{s^2 + \omega^2}$
3	t^2	$2!/s^3$	8	$\sin \omega t$	$\dfrac{\omega}{s^2 + \omega^2}$
4	t^n ($n = 1, 2, \cdots$)	$\dfrac{n!}{s^{n+1}}$	9	$\cosh at$	$\dfrac{s}{s^2 - a^2}$
5	t^a (a positive)	$\dfrac{\Gamma(a+1)}{s^{a+1}}$	10	$\sinh at$	$\dfrac{a}{s^2 - a^2}$

[cf. (10) in Sec. 4.5 or (26) in Appendix 3]. Formula 5 can be proved by starting from the definition

$$\mathcal{L}(t^a) = \int_0^\infty e^{-st} t^a \, dt,$$

setting $st = x$, and using (8) in Sec. 4.5 [(24) in Appendix 3]; then

$$\mathcal{L}(t^a) = \int_0^\infty e^{-x} \left(\frac{x}{s}\right)^a \frac{dx}{s} = \frac{1}{s^{a+1}} \int_0^\infty e^{-x} x^a \, dx = \frac{\Gamma(a + 1)}{s^{a+1}} \qquad (s > 0).$$

Formula 6 was proved in Example 2. To prove the formulas 7 and 8, we set $a = i\omega$ in formula 6. Then

$$\mathcal{L}(e^{i\omega t}) = \frac{1}{s - i\omega} = \frac{s + i\omega}{(s - i\omega)(s + i\omega)} = \frac{s + i\omega}{s^2 + \omega^2} = \frac{s}{s^2 + \omega^2} + i\frac{\omega}{s^2 + \omega^2}.$$

On the other hand, by Theorem 1,

$$\mathcal{L}(e^{i\omega t}) = \mathcal{L}(\cos \omega t + i \sin \omega t) = \mathcal{L}(\cos \omega t) + i\mathcal{L}(\sin \omega t).$$

Equating the real and imaginary parts of these two equations, we obtain the formulas 7 and 8. Avoiding working in complex, the reader may prove formulas 7 and 8 by using the defining integrals (1) and integration by parts. Another derivation of these formulas in real will be given in the next section.

Formula 9 was proved in Example 3, and formula 10 can be proved in a similar manner.   ∎

Existence of Laplace Transforms

In conclusion of this introductory section we should say something about the existence of the Laplace transform. Roughly and intuitively speaking, the situation is as follows. For a fixed s the integral in (1) will exist if the whole integrand $e^{-st} f(t)$ goes to zero fast enough as $t \to \infty$, say, at least like an exponential function with a negative exponent. This motivates the inequality (2) in the subsequent existence theorem. $f(t)$ need not be continuous. This is of practical importance since discontinuous inputs (driving forces) are just those for which the Laplace transformation becomes particularly useful. It suffices to require that $f(t)$ be piecewise continuous on every finite interval in the range $t \geq 0$.

By definition, a function $f(t)$ is **piecewise continuous** on a finite interval $a \leq t \leq b$ if $f(t)$ is defined on that interval and is such that the interval can be subdivided into finitely many intervals, in each of which $f(t)$ is continuous and has finite limits as t approaches either endpoint of the interval of subdivision from the interior.

It follows from this definition that finite jumps are the only discontinuities that a piecewise continuous function may have; these are known as *ordinary discontinuities*. Figure 88 shows an example. Clearly, the class of piecewise continuous functions includes every continuous function.

Fig. 88. Example of a piecewise continuous function $f(t)$.
(The dots mark the function values at the jumps.)

Theorem 2 (Existence theorem for Laplace transforms)

*Let $f(t)$ be a function that is piecewise continuous on every finite interval
in the range $t \geq 0$ and satisfies*

$$(2) \qquad\qquad |f(t)| \leq Me^{\gamma t} \qquad\qquad \text{for all } t \geq 0$$

*and for some constants γ and M. Then the Laplace transform of $f(t)$ exists
for all $s > \gamma$.*

Proof. Since $f(t)$ is piecewise continuous, $e^{-st}f(t)$ is integrable over any
finite interval on the t-axis. From (2), assuming that $s > \gamma$, we obtain

$$|\mathcal{L}(f)| = \left| \int_0^\infty e^{-st}f(t)\, dt \right| \leq \int_0^\infty |f(t)|e^{-st}\, dt \leq \int_0^\infty Me^{\gamma t}e^{-st}\, dt = \frac{M}{s - \gamma}$$

where the condition $s > \gamma$ was needed for the existence of the last integral.
This completes the proof. ∎

The conditions in Theorem 2 are sufficient for most applications, and it
is easy to find out whether a given function satisfies an inequality of the
form (2). For example,

$$(3) \qquad \cosh t < e^t, \qquad t^n < n!\, e^t \quad (n = 0, 1, \cdots) \qquad \text{for all } t > 0,$$

and any function that is bounded in absolute value for all $t \geq 0$, such as the
sine and cosine functions of a real variable, satisfies that condition. An
example of a function that does not satisfy a relation of the form (2) is the
exponential function e^{t^2}, because, no matter how large we choose M and γ
in (2),

$$e^{t^2} > Me^{\gamma t} \qquad\qquad \text{for all } t > t_0$$

where t_0 is a sufficiently large number, depending on M and γ.

It should be noted that the conditions in Theorem 2 are sufficient rather
than necessary. For example, the function $1/\sqrt{t}$ is infinite at $t = 0$, but its
transform exists; in fact, from the definition and $\Gamma(\tfrac{1}{2}) = \sqrt{\pi}$ [cf. (30) in
Appendix 3] we obtain

$$\mathcal{L}(t^{-1/2}) = \int_0^\infty e^{-st}t^{-1/2}\, dt = \frac{1}{\sqrt{s}} \int_0^\infty e^{-x}x^{-1/2}\, dx = \frac{1}{\sqrt{s}} \Gamma(\tfrac{1}{2}) = \sqrt{\frac{\pi}{s}}.$$

If the Laplace transform of a given function exists, it is uniquely determined. Conversely, it can be shown that if two functions (both defined on the positive real axis) have the same transform, these functions cannot differ over an interval of positive length, although they may differ at various isolated points (cf. Ref. [A15] in Appendix 1). Since this is of no importance in applications, we may say that the inverse of a given transform is essentially unique. In particular, if two *continuous* functions have the same transform, they are completely identical. Of course, this *is* of practical importance. Why? (Remember the introduction to the chapter.)

Concluding comment. This section contained the basic concepts, some transforms we shall use quite frequently and a discussion on the existence of Laplace transforms. In the next section we discuss and apply the most crucial property of the Laplace transformation, namely, that, roughly speaking, differentiation of functions corresponds to the multiplication of transforms by s, and integration of functions corresponds to the division of transforms by s. Hence *the Laplace transformation replaces operations of calculus by operations of algebra on transforms.* This, in a nutshell, is Laplace's basic idea, for which we should admire him.

Problems for Sec. 5.1

Find the Laplace transforms of the following functions, where a, b, c, ω, θ are constants.

1. $t + 4$ **2.** $at + b$ **3.** $a + bt + ct^2$ **4.** $4t^3 + t^2$

5.   **6.**   **7.**   **8.**

9. $(t^2 + \frac{1}{2})^2$ **10.** ce^{-at+b} **11.** $\sin \pi t$ **12.** $3 \cos 0.6t$

13. $\cos (\omega t + \theta)$ **14.** $\sin (\omega t + \theta)$ **15.** $\cos^2 t$ **16.** $\sin^2 t$

17. $\cos^2 \omega t$ **18.** $\sin^2 \omega t$ **19.** $\sinh^2 2t$ **20.** $\cosh^2 3t$

Given $F(s)$, find the inverse transform $f(t) = \mathcal{L}^{-1}(F)$.

21. $\dfrac{1}{s^2 + 9}$ **22.** $\dfrac{2s + 1}{s^2 + 4}$ **23.** $\dfrac{s - 4}{s^2 - 4}$ **24.** $\dfrac{a_1}{s} + \dfrac{a_2}{s^2} + \dfrac{a_3}{s^3}$

25. $\dfrac{0.8}{s + 1.3}$ **26.** $\dfrac{-2.1}{s - \frac{1}{2}\pi}$ **27.** $\dfrac{2}{s} + \dfrac{1}{s + 2}$ **28.** $\dfrac{1}{(s+1)(s+2)}$

29. $\dfrac{1}{s^5} + \dfrac{1}{s^2}$ **30.** $\dfrac{4(s + 1)}{s^2 - 16}$ **31.** $\dfrac{1}{s(s + 1)}$ **32.** $\dfrac{0.15}{s - 7} - \dfrac{0.4}{s^3}$

33. $\dfrac{1}{s^2 + 3s}$ **34.** $\dfrac{n\pi L}{L^2 s^2 + n^2 \pi^2}$ **35.** $\dfrac{1}{(s - a)(s - b)}$ $(a \neq b)$

36. Obtain the answer to Prob. 8 from Probs. 5 and 7.

37. Solve Prob. 16 by using Prob. 15 and $\cos^2 t + \sin^2 t = 1$.

38. Using Prob. 5, find $\mathcal{L}(f)$, where $f(t) = 0$ if $t \leq 4$, $f(t) = 1$ if $t > 4$.

39. Obtain formula 1 in Table 5.1 from formula 6.

40. Obtain formulas 9 and 10 in Table 5.1 from formula 6.
41. Obtain formula 6 in Table 5.1 from formulas 9 and 10.
42. Derive formulas 7 and 8 in Table 5.1 by integration by parts.
43. Prove formula 4 in Table 5.1 by induction.
44. Prove (3).
45. **(Linearity of the inverse Laplace transform)** Show that $\mathcal{L}^{-1}$ is linear. *Hint.* Use that $\mathcal{L}$ is linear.

5.2 Laplace Transforms of Derivatives and Integrals

Probably the most important property of the Laplace transformation is its linearity (Theorem 1 in the previous section). Next in the order of importance comes the fact that, roughly speaking, differentiation of a function $f(t)$ corresponds simply to multiplication of the transform $F(s)$ by s. This permits replacing operations of calculus by simple algebraic operations on transforms.

Furthermore, since integration is the inverse operation of differentiation, we expect it to correspond to division of transforms by s. This is in fact so. Accordingly, our program for this section is as follows. Theorem 1 concerns the differentiation of $f(t)$, Theorem 2 the extension to higher derivatives, and Theorem 3 the integration of $f(t)$. We include examples as well as a first application to a differential equation.

Theorem 1 [Laplace transform of the derivative of $f(t)$]
Suppose that $f(t)$ is continuous for all $t \geq 0$, satisfies (2), p. 247, for some γ and M, and has a derivative $f'(t)$ that is piecewise continuous on every finite interval in the range $t \geq 0$. Then the Laplace transform of the derivative $f'(t)$ exists when $s > \gamma$, and ∎

(1)
$$\boxed{\mathcal{L}(f') = s\mathcal{L}(f) - f(0)}$$
$(s > \gamma)$.

Proof. We first consider the case when $f'(t)$ is continuous for all $t \geq 0$. Then, by definition and by integrating by parts,

$$\mathcal{L}(f') = \int_0^\infty e^{-st}f'(t)\,dt = \left[e^{-st}f(t)\right]\Big|_0^\infty + s\int_0^\infty e^{-st}f(t)\,dt.$$

Since f satisfies (2), Sec. 5.1, the integrated portion on the right is zero at the upper limit when $s > \gamma$, and at the lower limit it is $-f(0)$. The last integral is $\mathcal{L}(f)$, the existence for $s > \gamma$ being a consequence of Theorem 2 in Sec. 5.1. This proves that the expression on the right exists when $s > \gamma$, and is equal to $-f(0) + s\mathcal{L}(f)$. Consequently, $\mathcal{L}(f')$ exists when $s > \gamma$, and (1) holds.

If $f'(t)$ is merely piecewise continuous, the proof is quite similar; in this case, the range of integration in the original integral must be broken up into parts such that f' is continuous in each such part. ∎

Remark. This theorem may be extended to piecewise continuous functions $f(t)$, but in place of (1) we then obtain the formula (1*) in Prob. 41 at the end of the current section.

By applying (1) to the second-order derivative $f''(t)$ we obtain

$$\mathcal{L}(f'') = s\mathcal{L}(f') - f'(0)$$

$$= s[s\mathcal{L}(f) - f(0)] - f'(0);$$

that is,

(2)
$$\boxed{\mathcal{L}(f'') = s^2\mathcal{L}(f) - sf(0) - f'(0).}$$

Similarly,

(3)
$$\mathcal{L}(f''') = s^3\mathcal{L}(f) - s^2 f(0) - sf'(0) - f''(0),$$

etc. By induction we thus obtain the following extension of Theorem 1.

Theorem 2 (Laplace transform of the derivative of any order *n*)

Let $f(t)$ and its derivatives $f'(t)$, $f''(t)$, $\cdots$, $f^{(n-1)}(t)$ be continuous functions for all $t \geq 0$, satisfying (2), p. 247, for some γ and M, and let the derivative $f^{(n)}(t)$ be piecewise continuous on every finite interval in the range $t \geq 0$. Then the Laplace transform of $f^{(n)}(t)$ exists when $s > \gamma$, and is given by the formula

(4)
$$\boxed{\mathcal{L}(f^{(n)}) = s^n \mathcal{L}(f) - s^{n-1} f(0) - s^{n-2} f'(0) - \cdots - f^{(n-1)}(0).}$$

EXAMPLE 1

Let $f(t) = t^2$. Find $\mathcal{L}(f)$.

Solution. Since $f(0) = 0$, $f'(0) = 0$, $f''(t) = 2$, and $\mathcal{L}(2) = 2\mathcal{L}(1) = 2/s$, we obtain from (2)

$$\mathcal{L}(f'') = \mathcal{L}(2) = \frac{2}{s} = s^2\mathcal{L}(f), \quad \text{hence} \quad \mathcal{L}(t^2) = \frac{2}{s^3},$$

in agreement with Table 5.1. The example is typical: it illustrates that in general there are several ways of obtaining the transforms of given functions.

EXAMPLE 2

Derive the Laplace transforms of $\cos \omega t$ and $\sin \omega t$.

Solution. Let $f(t) = \cos \omega t$. Then $f''(t) = -\omega^2 \cos \omega t = -\omega^2 f(t)$. Also $f(0) = 1$, $f'(0) = 0$. From this and (2),

$$-\omega^2\mathcal{L}(f) = \mathcal{L}(f'') = s^2\mathcal{L}(f) - s, \quad \text{hence} \quad \mathcal{L}(f) = \mathcal{L}(\cos \omega t) = \frac{s}{s^2 + \omega^2}.$$

Similarly for $g(t) = \sin \omega t$. Then $g(0) = 0$, $g'(0) = \omega$, and

$$-\omega^2\mathcal{L}(g) = \mathcal{L}(g'') = s^2\mathcal{L}(g) - \omega, \quad \text{hence} \quad \mathcal{L}(g) = \mathcal{L}(\sin \omega t) = \frac{\omega}{s^2 + \omega^2}.$$

EXAMPLE 3

Let $f(t) = \sin^2 t$. Find $\mathcal{L}(f)$.

Solution. We have $f(0) = 0$, $f'(t) = 2 \sin t \cos t = \sin 2t$, and (1) gives

$$\mathcal{L}(\sin 2t) = \frac{2}{s^2 + 4} = s\mathcal{L}(f) \quad \text{or} \quad \mathcal{L}(\sin^2 t) = \frac{2}{s(s^2 + 4)} \,.$$

EXAMPLE 4

Let $f(t) = t \sin \omega t$. Find $\mathcal{L}(f)$.

Solution. We have $f(0) = 0$ and

$$f'(t) = \sin \omega t + \omega t \cos \omega t, \qquad f'(0) = 0,$$

$$f''(t) = 2\omega \cos \omega t - \omega^2 t \sin \omega t$$

$$= 2\omega \cos \omega t - \omega^2 f(t),$$

so that by (2),

$$\mathcal{L}(f'') = 2\omega\mathcal{L}(\cos \omega t) - \omega^2 \mathcal{L}(f) = s^2\mathcal{L}(f).$$

Using the formula for the Laplace transform of $\cos \omega t$, we thus obtain

$$(s^2 + \omega^2)\mathcal{L}(f) = 2\omega\mathcal{L}(\cos \omega t) = \frac{2\omega s}{s^2 + \omega^2} \,.$$

Hence the result is

$$\mathcal{L}(t \sin \omega t) = \frac{2\omega s}{(s^2 + \omega^2)^2} \,.$$

EXAMPLE 5. A differential equation. Initial value problem

Solve the initial value problem

$$y'' + 4y' + 3y = 0, \qquad y(0) = 3, \quad y'(0) = 1.$$

Solution.   1st Step. Setting up the subsidiary equation. Let $Y(s) = \mathcal{L}(y)$ be the Laplace transform of the (unknown) solution $y(t)$. Then by Theorems 1 and 2 and the initial conditions,

$$\mathcal{L}(y') = sY - y(0) \qquad\qquad = sY - 3,$$

$$\mathcal{L}(y'') = s^2 Y - sy(0) - y'(0) = s^2 Y - 3s - 1.$$

We substitute this into the Laplace transform of the given differential equation, finding

$$s^2 Y + 4sY + 3Y = 3s + 1 + 4 \cdot 3.$$

The equation for the transform $Y(s)$ of the unknown function $y(t)$ is called the **subsidiary equation** of the given differential equation.

2nd Step. Solution of the subsidiary equation. Our present subsidiary equation can be written

$$(s + 3)(s + 1)Y = 3s + 13.$$

Solving algebraically for Y and using partial fractions, we obtain

$$Y = \frac{3s + 13}{(s + 3)(s + 1)} = \frac{-2}{s + 3} + \frac{5}{s + 1} \,.$$

3rd Step. Solution of the given problem. From Table 5.1 we see that

$$\mathcal{L}^{-1}\left\{\frac{1}{s + 3}\right\} = e^{-3t}, \qquad \mathcal{L}^{-1}\left\{\frac{1}{s + 1}\right\} = e^{-t}.$$

Using the linearity (Theorem 1 in Sec. 5.1), we see that the solution of our problem is

$$y(t) = -2e^{-3t} + 5e^{-t}.$$

Summary of the approach. In our approach we assume that the unknown solution $y(t)$ has a transform $Y(s)$ and Theorems 1 and 2 are applicable. Once the solution has been found, these assumptions should be justified. Practically speaking, we find it simpler and more natural to check by substitution whether $y(t)$ satisfies the given equation and initial conditions. This is the case.

Let us summarize our approach:

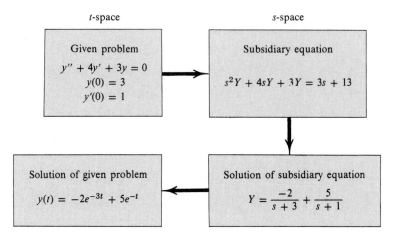

Did we gain anything, compared with the method in Secs. 2.2-2.4? The initial conditions were automatically accounted for. So we gained something but not much, since the example is too simple. However, it shows the typical steps of the new method. Applications illustrating substantial gains will be considered after we know further properties and techniques which make the Laplace transformation powerful and flexible. We start this program in the next section with so-called shifting processes and the unit step function.   ∎

Shifted data problems is a short name for initial value problems in which the initial conditions refer to some later instant instead of $t = 0$. We explain the idea of solving such a problem by the Laplace transformation in terms of a simple example.

EXAMPLE 6. Shifted data problem
Solve the initial value problem

$$y'' + y = 2t, \qquad y(\tfrac{1}{4}\pi) = \tfrac{1}{2}\pi, \qquad y'(\tfrac{1}{4}\pi) = 2 - \sqrt{2}$$

Solution. We see by inspection that $y = A \cos t + B \sin t + 2t$ is a general solution, and we know how we would have to go from here to take care of the initial conditions. What we want to learn is how the Laplace transformation can proceed although $y(0)$ and $y'(0)$ are unknown.

1st Step. Setting up the subsidiary equation. From (2) and Table 5.1 (Sec. 5.1) we obtain

$$s^2 Y - s y(0) - y'(0) + Y = 2/s^2.$$

2nd Step. Solution of the subsidiary equation. Solving algebraically and using partial fractions, we get

$$Y = \frac{2}{(s^2 + 1)s^2} + y(0)\frac{s}{s^2 + 1} + y'(0)\frac{1}{s^2 + 1}$$

$$= 2\left(\frac{1}{s^2} - \frac{1}{s^2 + 1}\right) + y(0)\frac{s}{s^2 + 1} + y'(0)\frac{1}{s^2 + 1}.$$

3rd Step. Solution of the given problem. From Table 5.1 we obtain $y = \mathcal{L}^{-1}(Y)$ in the form

$$y(t) = 2t + y(0) \cos t + [y'(0) - 2]\sin t$$

$$= 2t + A \cos t + B \sin t,$$

(where $A = y(0)$ and $B = y'(0) - 2$, but this is of no further interest). The first initial condition gives

$$y(\tfrac{1}{4}\pi) = \tfrac{1}{2}\pi + A/\sqrt{2} + B/\sqrt{2} = \tfrac{1}{2}\pi,$$

hence $B = -A$. By differentiation,

$$y'(t) = 2 - A \sin t + B \cos t.$$

By the second initial condition,

$$y'(\tfrac{1}{4}\pi) = 2 - A/\sqrt{2} + B/\sqrt{2} = 2 - \sqrt{2}.$$

This gives $A = 1$, $B = -1$ and the answer

$$y(t) = \cos t - \sin t + 2t.$$ ∎

Laplace Transform of the Integral of a Function

We conclude the present section with integration of $f(t)$, the inverse operation of differentiation; we expect it to correspond to division of the transform by s, since division is the inverse operation of multiplication.

Theorem 3 [Integration of $f(t)$]

If $f(t)$ is piecewise continuous and satisfies an inequality of the form (2), *Sec. 5.1, then*

(5)
$$\boxed{\mathcal{L}\left\{\int_0^t f(\tau)\, d\tau\right\} = \frac{1}{s}\,\mathcal{L}\{f(t)\}} \qquad (s > 0,\quad s > \gamma).$$

Proof. Suppose that $f(t)$ is piecewise continuous and satisfies (2), Sec. 5.1, for some γ and M. Clearly, if (2) holds for some negative γ, it also holds for positive γ, and we may assume that γ is positive. Then the integral

$$g(t) = \int_0^t f(\tau)\, d\tau$$

is continuous, and by using (2) in Sec. 5.1 we obtain

$$|g(t)| \le \int_0^t |f(\tau)|\, d\tau \le M \int_0^t e^{\gamma\tau}\, d\tau = \frac{M}{\gamma}(e^{\gamma\tau} - 1) \qquad (\gamma > 0).$$

Also $g'(t) = f(t)$, except for points at which $f(t)$ is discontinuous. Hence $g'(t)$ is piecewise continuous on each finite interval, and, by Theorem 1,

$$\mathcal{L}\{f(t)\} = \mathcal{L}\{g'(t)\} = s\mathcal{L}\{g(t)\} - g(0) \qquad (s > \gamma).$$

Here, clearly, $g(0) = 0$, so that $\mathcal{L}(f) = s\mathcal{L}(g)$. This implies (5) and completes the proof. ∎

Equation (5) has a useful companion, which we obtain by writing $\mathscr{L}\{f(t)\} = F(s)$, interchanging the two sides and taking the inverse transform on both sides. Then

(6)
$$\boxed{\mathscr{L}^{-1}\left\{\frac{1}{s}F(s)\right\} = \int_0^t f(\tau)\,d\tau.}$$

EXAMPLE 7

Let $\mathscr{L}(f) = \dfrac{1}{s(s^2 + \omega^2)}$. Find $f(t)$.

Solution. From Table 5.1 in Sec. 5.1 we have

$$\mathscr{L}^{-1}\left(\frac{1}{s^2 + \omega^2}\right) = \frac{1}{\omega}\sin \omega t.$$

From this and Theorem 3 we obtain the answer

$$\mathscr{L}^{-1}\left\{\frac{1}{s}\left(\frac{1}{s^2 + \omega^2}\right)\right\} = \frac{1}{\omega}\int_0^t \sin \omega\tau\,d\tau = \frac{1}{\omega^2}(1 - \cos \omega t).$$

This proves formula 19 in the table in Sec. 5.10.

EXAMPLE 8

Let $\mathscr{L}(f) = \dfrac{1}{s^2(s^2 + \omega^2)}$. Find $f(t)$.

Solution. Applying Theorem 3 to the answer in Example 7, we obtain the desired formula

$$\mathscr{L}^{-1}\left\{\frac{1}{s^2}\left(\frac{1}{s^2 + \omega^2}\right)\right\} = \frac{1}{\omega^2}\int_0^t (1 - \cos \omega\tau)\,d\tau = \frac{1}{\omega^2}\left(t - \frac{\sin \omega t}{\omega}\right).$$

This proves formula 20 in the table in Sec. 5.10. ∎

So far we discussed the linearity (Sec. 5.1) and the basic operational properties (Theorems 1, 2, 3) of the Laplace transformation. But there are further general properties that make the transformation even more flexible and powerful. As the next of these, we discuss in Sec. 5.3 the processes of **"shifting"** (replacement of s by $s - a$, or replacement of t by $t - a$).

Problems for Sec. 5.2

Using (1) or (2), find the transform $\mathscr{L}(f)$ of the given function $f(t)$.

1. $\cosh^2 t$ **2.** $\sinh^2 t$ **3.** te^t **4.** te^{-5t}

5. $\sin^2 \omega t$ **6.** te^{at} **7.** $t \cos t$ **8.** $\cos^2 \pi t$

Further transforms. Using Theorems 1 and 2, derive the following transforms that occur in applications (in connection with resonance, etc.).

9. $\mathscr{L}(t \cos \omega t) = \dfrac{s^2 - \omega^2}{(s^2 + \omega^2)^2}$ **10.** $\mathscr{L}(t \sin \omega t) = \dfrac{2\omega s}{(s^2 + \omega^2)^2}$

11. $\mathscr{L}(t \cosh at) = \dfrac{s^2 + a^2}{(s^2 - a^2)^2}$ **12.** $\mathscr{L}(t \sinh at) = \dfrac{2as}{(s^2 - a^2)^2}$

Using the formulas in Probs. 9 and 10, show that

13. $\mathscr{L}^{-1}\left(\dfrac{1}{(s^2 + \omega^2)^2}\right) = \dfrac{1}{2\omega^3}(\sin \omega t - \omega t \cos \omega t)$

14. $\mathcal{L}^{-1}\left(\dfrac{s^2}{(s^2 + \omega^2)^2}\right) = \dfrac{1}{2\omega}(\sin \omega t + \omega t \cos \omega t)$

Derivation by different methods is possible for various formulas, and is typical of the Laplace transformation. Probs. 15–20 illustrate this point.

15. Using (1), derive $\mathcal{L}(\sin \omega t)$ from $\mathcal{L}(\cos \omega t)$.

16. Find $\mathcal{L}(\cos^2 t)$ (a) by the use of the result of Example 3, (b) by the method used in Example 3, (c) by expressing $\cos^2 t$ in terms of $\cos 2t$.

17. Check the result in Example 3, using $\sin^2 t = \frac{1}{2}(1 - \cos 2t)$.

18. Check the answer to Prob. 1 by writing $\cosh t$ in terms of exponential functions.

19. Using (2), derive formula 7 in Table 5.1 (Sec. 5.1).

20. Using (2), derive formula 8 in Table 5.1 (Sec. 5.1).

Application of Theorem 3. Find $f(t)$ if $\mathcal{L}(f)$ equals

21. $\dfrac{1}{s^2 + s}$ **22.** $\dfrac{1}{s(s - 2)}$ **23.** $\dfrac{10}{s(s^2 + 9)}$ **24.** $\dfrac{4}{s(s^2 - 1)}$

25. $\dfrac{1}{s^2(s + 1)}$ **26.** $\dfrac{1}{s^4 - 2s^3}$ **27.** $\dfrac{1}{s^2}\left(\dfrac{s - 1}{s + 1}\right)$ **28.** $\dfrac{1}{s^2}\left(\dfrac{s - 2}{s^2 + 4}\right)$

29. $\dfrac{54}{s^3(s - 3)}$ **30.** $\dfrac{2s - \pi}{s^3(s - \pi)}$ **31.** $\dfrac{1}{s^2}\left(\dfrac{s + 1}{s^2 + 1}\right)$ **32.** $\dfrac{1}{s^5 - s^3}$

Initial value problems. Using the Laplace transformation, solve:

33. $y'' + 9y = 0$, $y(0) = 0$, $y'(0) = 2$

34. $4y'' + \pi^2 y = 0$, $y(0) = 2$, $y'(0) = 0$

35. $y'' + \omega^2 y = 0$, $y(0) = A$, $y'(0) = B$, $(\omega \neq 0)$

36. $y'' + 25y = t$, $y(0) = 1$, $y'(0) = 0.04$

37. $y'' - 2y' - 3y = 0$, $y(0) = 1$, $y'(0) = 7$

38. $y'' - 4y' + 3y = 2t - \frac{8}{3}$, $y(0) = 0$, $y'(0) = -\frac{16}{3}$

39. $4y'' + y = 0$, $y(0) = 1$, $y'(0) = -2$

40. $y'' + 2y' - 8y = 0$, $y(0) = 1$, $y'(0) = 8$

41. (Extension of Theorem 1) The following extension of Theorem 1 is of practical interest in applications. Show that if $f(t)$ is continuous, except for an ordinary discontinuity (finite jump) at $t = a$ (> 0), the other conditions remaining the same as in Theorem 1, then (cf. the figure)

(1*) $\mathcal{L}(f') = s\mathcal{L}(f) - f(0) - [f(a + 0) - f(a - 0)]e^{-as}$.

Graph the following functions. Using (1*), find their Laplace transform.

42. $f(t) = t$ if $0 \le t < 1$, $f(t) = 1$ if $1 < t < 2$, $f(t) = 0$ otherwise

43. $f(t) = k$ if $1 < t < 2$, $f(t) = 0$ otherwise

44. $f(t) = 0$ if $0 \le t < a$, $f(t) = k$ if $t > a$

45. $f(t) = 3$ if $0 \le t < 5$, $f(t) = 1$ if $t > 5$

5.3 Shifting on the *s*-Axis, Shifting on the *t*-Axis, Unit Step Function

What is the state we have reached, and what is our next goal? We know that the Laplace transformation is linear (Theorem 1, Sec. 5.1), that differentiation of $f(t)$ roughly corresponds to the multiplication of $\mathscr{L}(f)$ by s (Theorems 1 and 2, Sec. 5.2), and that this property is essential in solving differential equations. A first illustration of the technique is given in Example 5, Sec. 5.2. But there the solution may easily be found by the usual methods.

To provide applications such that the Laplace transformation can show its real power, we first have to derive some further general properties of the transformation. Two very important properties concern the shifting on the *s*-axis and the shifting on the *t*-axis, as expressed in the two shifting theorems (Theorems 1 and 2 of this section).

Shifting on the *s*-Axis: Replacing *s* by *s* − *a* in *F(s)*

Theorem 1 (First shifting theorem; shifting on the *s*-axis)

If $f(t)$ has the transform $F(s)$ where $s > \gamma$, then $e^{at}f(t)$ has the transform $F(s - a)$ where $s - a > \gamma$; thus, if $\mathscr{L}\{f(t)\} = F(s)$, then

(1)
$$\mathscr{L}\{e^{at}f(t)\} = F(s - a).$$

*Hence if we know the transform $F(s)$ of $f(t)$, we get the transform of $e^{at}f(t)$ by "**shifting on the s-axis**" [i.e., by replacing s with $s - a$, to get $F(s - a)$]. See Fig. 89.*

Remark. By taking the inverse transform on both sides and interchanging the left and right sides we obtain from (1)

(1*)
$$\mathscr{L}^{-1}\{F(s - a)\} = e^{at}f(t)$$

Proof of Theorem 1. By definition, $F(s) = \displaystyle\int_0^\infty e^{-st}f(t)\,dt$ and, therefore,

$$F(s - a) = \int_0^\infty e^{-(s-a)t}f(t)\,dt = \int_0^\infty e^{-st}[e^{at}f(t)]\,dt = \mathscr{L}\{e^{at}f(t)\}. \quad \blacksquare$$

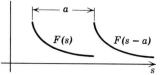

Fig. 89. First shifting theorem, shifting on the *s*-axis

EXAMPLE 1

By applying Theorem 1 to the formulas 4, 7, and 8 in Table 5.1 we obtain the following results.

$f(t)$	$\mathcal{L}(f)$
$e^{at}t^n$	$\dfrac{n!}{(s - a)^{n+1}}$
$e^{at} \cos \omega t$	$\dfrac{s - a}{(s - a)^2 + \omega^2}$
$e^{at} \sin \omega t$	$\dfrac{\omega}{(s - a)^2 + \omega^2}$

This proves formulas 8, 9, 17, 18 in the table in Sec. 5.10.

EXAMPLE 2. Damped free vibrations

A small body of mass $m = 2$ is attached at the lower end of an elastic spring whose upper end is fixed, the spring modulus being $k = 10$. Let $y(t)$ be the displacement of the body from the position of static equilibrium. Determine the free vibrations of the body, starting from the initial position $y(0) = 2$ with the initial velocity $y'(0) = -4$, assuming that there is damping proportional to the velocity, the damping constant being $c = 4$.

Solution. The motion is described by the solution $y(t)$ of the initial value problem

$$y'' + 2y' + 5y = 0, \qquad y(0) = 2, \qquad y'(0) = -4,$$

cf. (7) Sec. 2.6. Using (1) and (2) in Sec. 5.2, we obtain the subsidiary equation

$$s^2 Y - 2s + 4 + 2(sY - 2) + 5Y = 0.$$

The solution is

$$Y(s) = \frac{2s}{(s + 1)^2 + 2^2} = 2 \frac{s + 1}{(s + 1)^2 + 2^2} - \frac{2}{(s + 1)^2 + 2^2}.$$

Now

$$\mathcal{L}^{-1} \left(\frac{s}{s^2 + 2^2} \right) = \cos 2t, \qquad \mathcal{L}^{-1} \left(\frac{2}{s^2 + 2^2} \right) = \sin 2t.$$

From this and Theorem 1 we obtain the expected type of solution

$$y(t) = \mathcal{L}^{-1}(Y) = e^{-t}(2 \cos 2t - \sin 2t). \qquad\blacksquare$$

Shifting on the t-Axis: Replacing t by t − a in f(t)

The first shifting theorem (Theorem 1) concerns shifting on the s-axis: the replacement of s in $F(s)$ by $s - a$ corresponds to the multiplication of the original function $f(t)$ by e^{at}. We shall now state the second shifting theorem (Theorem 2), which concerns shifting on the t-axis: the replacement of t in $f(t)$ by $t - a$ corresponds roughly to the multiplication of the transform $F(s)$ by e^{-as}; the precise formulation of this is as follows.

Theorem 2 (Second shifting theorem; shifting on the t-axis)

If $f(t)$ has the transform $F(s)$, then the function

$$(2) \qquad \tilde{f}(t) = \begin{cases} 0 & \text{if } t < a \\ f(t - a) & \text{if } t > a \end{cases}$$

with arbitrary $a \geqq 0$ has the transform $e^{-as}F(s)$. Hence if we know that *transform F(s) of f(t), we get the transform of the function (2), whose variable has been shifted ("**shifting on the t-axis**"), by multiplying F(s) by e^{-as}.* (Proof and applications follow below.)

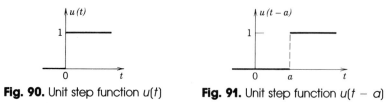

Fig. 90. Unit step function $u(t)$ **Fig. 91.** Unit step function $u(t - a)$

Unit Step Function $u(t - a)$

By definition, $u(t - a)$ is 0 for $t < a$, has a jump of size 1 at $t = a$ (where we can leave it undefined) and is 1 for $t > a$:

(3)
$$u(t - a) = \begin{cases} 0 & \text{if } t < a \\ 1 & \text{if } t > a \end{cases} \qquad (a \geqq 0).$$

Figure 90 shows the special case $u(t)$, which has the jump at zero, and Fig. 91 the general case $u(t - a)$ for an arbitrary positive a. The unit step function is also called the **Heaviside function.**[2]

The unit step function $u(t - a)$ is a basic building block of various functions, as we shall see, and it greatly increases the usefulness of Laplace transform methods. At present we can use it to write $\tilde{f}(t)$ in (2) in the form $f(t - a)u(t - a)$. Figure 92 shows an example. And the second shifting theorem (Theorem 2) asserts that if $\mathcal{L}\{f(t)\} = F(s)$, then

(4)
$$\mathcal{L}\{f(t - a)u(t - a)\} = e^{-as}F(s).$$

Furthermore, taking the inverse transform on both sides of (4) and interchanging sides, we obtain

(4*)
$$\mathcal{L}^{-1}\{e^{-as}F(s)\} = f(t - a)u(t - a).$$

Proof of Theorem 2. From the definition we have

$$e^{-as}F(s) = e^{-as} \int_0^\infty e^{-s\tau}f(\tau) \, d\tau = \int_0^\infty e^{-s(\tau+a)}f(\tau) \, d\tau.$$

Substituting $\tau + a = t$ in the integral, we obtain

$$e^{-as}F(s) = \int_a^\infty e^{-st}f(t - a) \, dt.$$

[2]OLIVER HEAVISIDE (1850—1925), English electrical engineer, who is known by his important work in operational calculus ("Heaviside calculus") and its engineering applications.

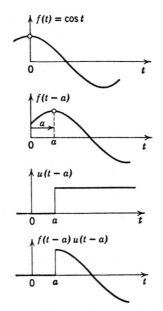

Fig. 92. $f(t - a)u(t - a)$, where $f(t) = \cos t$

We can write this as an integral from 0 to ∞ if we make sure that the integrand is zero for all t from 0 to a. We may easily accomplish this by multiplying the present integrand by the step function $u(t - a)$, thereby obtaining (4) and completing the proof:

$$e^{-as}F(s) = \int_0^\infty e^{-st}f(t - a)u(t - a)\, dt = \mathcal{L}\{f(t - a)u(t - a)\}. \quad \blacksquare$$

It is fair to say that we are already approaching the stage where we can attack problems for which the Laplace transformation is preferable to the usual classical method, as the examples in the next section will illustrate. In this connection we need the transform of the unit step function $u(t - a)$,

(5)
$$\mathcal{L}\{u(t - a)\} = \frac{e^{-as}}{s} \qquad (s > 0).$$

This formula follows directly from the definition because

$$\mathcal{L}\{u(t - a)\} = \int_0^\infty e^{-st}u(t - a)\, dt$$

$$= \int_0^a e^{-st}0\, dt + \int_a^\infty e^{-st}1\, dt = -\frac{1}{s}e^{-st}\Big|_a^\infty .$$

Let us consider two simple examples. Further applications follow in the problem set and in the next sections.

EXAMPLE 3

Find the inverse transform of e^{-3s}/s^3.

Solution. Since $\mathscr{L}^{-1}(1/s^3) = t^2/2$ (cf. Table 5.1 in Sec. 5.1), Theorem 2 gives (Fig. 93)

$$\mathscr{L}^{-1}(e^{-3s}/s^3) = \tfrac{1}{2}(t - 3)^2 u(t - 3).$$

EXAMPLE 4

Find the transform of the function (Fig. 94)

$$f(t) = \begin{cases} 1 & \text{if } 0 < t < \pi \\ 0 & \text{if } \pi < t < 2\pi \\ \sin t & \text{if } \quad t > 2\pi. \end{cases}$$

Solution. We can represent $f(t)$ in terms of unit step functions:

$$f(t) = u(t) - u(t - \pi) + u(t - 2\pi) \sin t.$$

Here $\sin t = \sin(t - 2\pi)$ because of the periodicity. From (5) and (4) and Table 5.1 (Sec. 5.1),

$$\mathscr{L}(f) = \frac{1}{s} - \frac{e^{-\pi s}}{s} + \frac{e^{-2\pi s}}{s^2 + 1}. \qquad \blacksquare$$

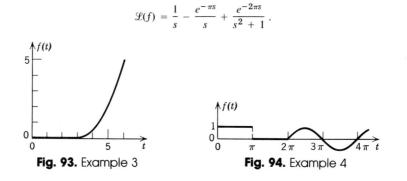

Fig. 93. Example 3   **Fig. 94.** Example 4

In the next section we increase the usefulness of the Laplace transformation still further by introducing the so-called **Dirac delta function** and considering some typical examples.

Problems for Sec. 5.3

Applications of the First Shifting Theorem

Find the Laplace transforms of the following functions.

1. $0.4te^{2.5t}$ **2.** $\tfrac{1}{2}t^2 e^{-3t}$ **3.** $e^t \cos t$ **4.** $e^{-t/2} \sin \tfrac{1}{4}t$

5. $e^{-2t} \sin n\pi t$ **6.** $e^{-t} \sin(\omega t + \theta)$ **7.** $e^{-t} \cosh t$ **8.** $e^t(a + bt)$

9. $e^{-3t}(\cos 2t - \tfrac{3}{2} \sin 2t)$ **10.** $e^{-\alpha t}(A \cos \beta t + B \sin \beta t)$

Find $f(t)$ if $\mathscr{L}(f)$ equals

11. $\dfrac{3}{(s + 1)^2}$ **12.** $\dfrac{8}{(s - \tfrac{1}{2})^3}$ **13.** $\dfrac{0.3}{s^2 + 2s + 2}$ **14.** $\dfrac{s - 2}{s^2 - 4s + \tfrac{17}{4}}$

15. $\dfrac{6}{s^2 - 4s - 5}$ **16.** $\dfrac{1.4s}{s^2 - 2s + 2}$ **17.** $\dfrac{10}{(s + \pi)^4}$ **18.** $\dfrac{2}{s^2 + s + \tfrac{1}{2}}$

19. $\dfrac{s + 1 - 2\omega}{s^2 + 2s + \omega^2 + 1}$ **20.** $\dfrac{2s + 2\alpha}{s^2 + 2\alpha s + \alpha^2 - 16}$

Representing the hyperbolic functions in terms of exponential functions and applying the first shifting theorem, show that

21. $\mathcal{L}(\cosh at \cos at) = \dfrac{s^3}{s^4 + 4a^4}$ **22.** $\mathcal{L}(\cosh at \sin at) = \dfrac{a(s^2 + 2a^2)}{s^4 + 4a^4}$

23. $\mathcal{L}(\sinh at \cos at) = \dfrac{a(s^2 - 2a^2)}{s^4 + 4a^4}$ **24.** $\mathcal{L}(\sinh at \sin at) = \dfrac{2a^2 s}{s^4 + 4a^4}$

Initial value problems. Using the Laplace transformation, solve the following problems. (Initial value problems involving step functions follow in the problem set for the next section.)

25. $y'' - 4y' + 4y = 0$, $y(0) = 0$, $y'(0) = 2$
26. $y'' + 2y' + 10y = 0$, $y(0) = 0$, $y'(0) = 3$
27. $y'' + 4y' + 4y = 0$, $y(0) = 2$, $y'(0) = -3$
28. $9y'' - 6y' + y = 0$, $y(0) = 3$, $y'(0) = 1$
29. $y'' + y' + 1.25y = 0$, $y(0) = 1$, $y'(0) = -0.5$
30. $4y'' - 4y' + 37y = 0$, $y(0) = 3$, $y'(0) = 1.5$

Unit step function. Represent the following functions in terms of unit step functions and find their Laplace transforms.

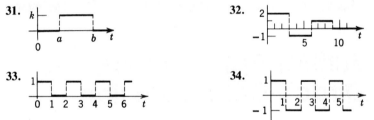

31.   **32.**

33.   **34.**

Applications of the Second Shifting Theorem

Graph the following functions and find their Laplace transforms.

35. $(t - 1)u(t - 1)$ **36.** $tu(t - 1)$ **37.** $(t - 1)^2 u(t - 1)$ **38.** $t^2 u(t - 1)$
39. $u(t - \pi) \cos t$ **40.** $u(t - \tfrac{1}{2}\pi) \sin t$ **41.** $e^{-2t} u(t - 1)$ **42.** $e^{kt} u(t - a)$

In each case graph the given function, which is assumed to be zero outside the given interval, and find its Laplace transform.

43. t $(0 < t < 1)$ **44.** t $(1 < t < 4)$
45. t^2 $(0 < t < 1)$ **46.** t^2 $(0 < t < 3)$
47. $K \sin t$ $(2\pi < t < 4\pi)$ **48.** $\sin \omega t$ $(0 < t < \pi/\omega)$
49. $K \cos \omega t$ $(0 < t < \pi/\omega)$ **50.** $1 - e^{-t}$ $(0 < t < \pi)$

Find and graph $f(t)$ if $\mathcal{L}(f)$ equals

51. $(e^{-3s} - e^{-s})/s$ **52.** e^{-s}/s^2
53. e^{-2s}/s **54.** e^{-s}/s^3
55. $e^{-2s}/(s - 2)$ **56.** $e^{-\pi s}/(s^2 + 2s + 2)$
57. $e^{-s}/(s^2 + \pi^2)$ **58.** $s(1 + e^{-\pi s})/(s^2 + 1)$
59. $e^{-s}/(s - 3)$ **60.** $(1 - e^{-\pi s})/(s^2 + 4)$

Models of electric circuits

61. A capacitor of capacitance C is charged so that its potential is V_0. At $t = 0$ the switch in Fig. 95 is closed and the capacitor starts to discharge through the resistor of resistance R. Using the Laplace transformation, find the charge $q(t)$ on the capacitor.

62. Find the current $i(t)$ in the circuit in Fig. 96, assuming that no current flows when $t \leqq 0$ and the switch is closed at $t = 0$.

Find the current $i(t)$ in the *LC*-circuit shown in Fig. 97, assuming $L = 1$ henry, $C = 1$ farad, zero initial current and charge on the capacitor and $v(t)$ as follows.

63. $v = 1$ when $0 < t < a$ and 0 otherwise

64. $v = t$ when $0 < t < 1$ and $v = 1$ when $t > 1$

65. $v = 1 - e^{-t}$ when $0 < t < \pi$ and 0 otherwise

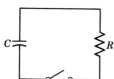

Fig. 95. Problem 61

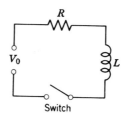

Fig. 96. Problem 62

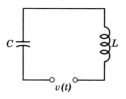

Fig. 97. Problems 63—65

5.4 Further Applications. Dirac's Delta Function

EXAMPLE 1. Response of an *RC*-circuit to a single square wave

Find the current $i(t)$ in the circuit in Fig. 98 if a single square wave with voltage V_0 is applied. The circuit is assumed to be quiescent before the square wave is applied.

Solution. The equation of the circuit is (cf. Sec. 1.8)

$$Ri(t) + \frac{q(t)}{C} = Ri(t) + \frac{1}{C}\int_0^t i(\tau)\, d\tau = v(t)$$

where $v(t)$ can be represented in terms of two unit step functions:

$$v(t) = V_0[u(t - a) - u(t - b)].$$

Using Theorem 3 in Sec. 5.2 and formula (5) in Sec. 5.3, we obtain the subsidiary equation

$$RI(s) + \frac{I(s)}{sC} = \frac{V_0}{s}[e^{-as} - e^{-bs}].$$

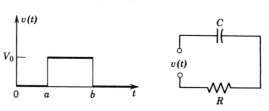

Fig. 98. Example 1

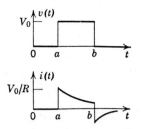

Fig. 99. Voltage and current in Example 1

Solving this equation algebraically for $I(s)$, we get

$$I(s) = F(s)(e^{-as} - e^{-bs}) \qquad \text{where} \qquad F(s) = \frac{V_0/R}{s + 1/RC}.$$

From Table 5.1 in Sec. 5.1 we have

$$\mathcal{L}^{-1}(F) = \frac{V_0}{R} e^{-t/(RC)}.$$

Hence Theorem 2 in Sec. 5.3 yields the solution (Fig. 99)

$$i(t) = \mathcal{L}^{-1}(I) = \mathcal{L}^{-1}\{e^{-as}F(s)\} - \mathcal{L}^{-1}\{e^{-bs}F(s)\}$$

$$= \frac{V_0}{R}\,[e^{-(t-a)/(RC)}u(t-a) - e^{-(t-b)/(RC)}u(t-b)];$$

that is, $i = 0$ if $t < a$, and

$$i(t) = \begin{cases} K_1 e^{-t/(RC)} & \text{if } a < t < b \\ (K_1 - K_2)e^{-t/(RC)} & \text{if } t > b \end{cases}$$

where $K_1 = V_0 e^{a/(RC)}/R$ and $K_2 = V_0 e^{b/(RC)}/R$.

EXAMPLE 2. Response of an undamped system to a single square wave

Solve the initial value problem

$$y'' + 2y = r(t), \qquad y(0) = 0, \qquad y'(0) = 0,$$

where $r(t) = 1$ if $0 < t < 1$ and 0 otherwise (Fig. 100).

Solution. From Theorem 2 in Sec. 5.2, the initial conditions and (3) and (5) in Sec. 5.3 we obtain the subsidiary equation

$$s^2 Y + 2Y = \frac{1}{s} - \frac{e^{-s}}{s}.$$

The solution is

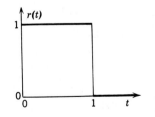

Fig. 100. Input in Examples 2 and 3

$$Y(s) = \frac{1}{s(s^2 + 2)} - \frac{e^{-s}}{s(s^2 + 2)} .$$

On the right we use partial fractions:

$$\frac{1}{s(s^2 + 2)} = \frac{1}{2}\left(\frac{1}{s} - \frac{s}{s^2 + 2}\right).$$

Hence by Table 5.1 (Sec. 5.1) and Theorem 2 in Sec. 5.3,

$$y(t) = \tfrac{1}{2}[1 - \cos \sqrt{2}t] - \tfrac{1}{2}[1 - \cos \sqrt{2}(t - 1)]u(t - 1),$$

that is,

$$y(t) = \begin{cases} \tfrac{1}{2} - \tfrac{1}{2}\cos \sqrt{2}t & \text{if } 0 \leq t < 1 \\ \tfrac{1}{2}\cos \sqrt{2}(t - 1) - \tfrac{1}{2}\cos \sqrt{2}t & \text{if } \quad t > 1. \end{cases}$$

We see that $y(t)$ represents a composite of harmonic oscillations.

EXAMPLE 3. Response of a damped vibrating system to a single square wave
Determine the response of the damped vibrating system corresponding to

$$y'' + 3y' + 2y = r(t), \qquad y(0) = 0, \qquad y'(0) = 0$$

with $r(t)$ as in Example 2 (Fig. 100).

Solution. By the same method as before we obtain the subsidiary equation

$$s^2 Y + 3sY + 2Y = \frac{1}{s}(1 - e^{-s}).$$

Solving for Y, we have

$$Y(s) = F(s)(1 - e^{-s}) \qquad \text{where} \qquad F(s) = \frac{1}{s(s + 1)(s + 2)} .$$

In terms of partial fractions,

$$F(s) = \frac{1/2}{s} - \frac{1}{s + 1} + \frac{1/2}{s + 2} .$$

Hence by Table 5.1 (Sec. 5.1),

$$f(t) = \mathcal{L}^{-1}(F) = \tfrac{1}{2} - e^{-t} + \tfrac{1}{2}e^{-2t}.$$

Therefore, by Theorem 2 in Sec. 5.3 we have

$$\mathcal{L}^{-1}\{e^{-s}F(s)\} = f(t - 1)u(t - 1) = \begin{cases} 0 & (0 \leq t < 1) \\ \tfrac{1}{2} - e^{-(t-1)} + \tfrac{1}{2}e^{-2(t-1)} & (t > 1). \end{cases}$$

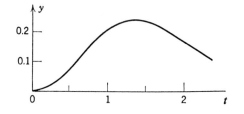

Fig. 101. Output in Example 3

This yields the solution (cf. Fig. 101)

$$y(t) = \mathcal{L}^{-1}(Y) = f(t) - f(t-1)u(t-1) = \begin{cases} \frac{1}{2} - e^{-t} + \frac{1}{2}e^{-2t} & (0 \leq t < 1) \\ K_1 e^{-t} - K_2 e^{-2t} & (t > 1) \end{cases}$$

where $K_1 = e - 1$ and $K_2 = (e^2 - 1)/2$. ∎

Short Impulses. Dirac's Delta Function

Phenomena of an impulsive nature, such as the action of very large forces (or voltages) over very short intervals of time, are of great practical interest, since they arise in various applications. This situation occurs, for instance, when a tennis ball is hit, a system is given a blow by a hammer, an airplane makes a "hard" landing, a ship is hit by a high single wave, and so on. Our present goal is to show how to solve problems involving short impulses by the Laplace transformation.

In mechanics, the **impulse** of a force $f(t)$ over a time interval $a \leq t \leq a + k$ is defined to be the integral of $f(t)$ from a to $a + k$. The analog for an electric circuit is the integral of the electromotive force applied to the circuit, integrated from a to $a + k$. Of particular practical interest is the case of a very short k (and its limit $k \to 0$), that is, the impulse of a force acting only for an instant. To handle the case, we consider the function (Fig. 102)

$$(1) \qquad f_k(t) = \begin{cases} 1/k & \text{if } a \leq t \leq a + k, \\ 0 & \text{otherwise.} \end{cases}$$

Its impulse I_k is 1, since the integral evidently gives the area of the rectangle in Fig. 102:

$$(2) \qquad I_k = \int_0^\infty f_k(t)\, dt = \int_a^{a+k} \frac{1}{k}\, dt = 1.$$

We can represent $f_k(t)$ in terms of two unit step functions (Sec. 5.3), namely,

$$f_k(t) = \frac{1}{k}\, [u(t-a) - u(t - (a+k))].$$

From (5) in Sec. 5.3 we obtain the Laplace transform

$$(3) \qquad \mathcal{L}\{f_k(t)\} = \frac{1}{ks}\, [e^{-as} - e^{-(a+k)s}] = e^{-as}\, \frac{1 - e^{-ks}}{ks}.$$

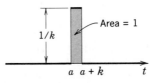

Fig. 102. The function $f_k(t)$ in (1)

The limit of $f_k(t)$ as $k \to 0$ is denoted by

$$\delta(t - a)$$

and is called the **Dirac delta function**[3] (sometimes the **unit impulse function**). The quotient in (3) has the limit 1 as $k \to 0$, as follows by l'Hôpital's rule. Thus

(4)
$$\boxed{\mathscr{L}\{\delta(t - a)\} = e^{-as}.}$$

We note that $\delta(t - a)$ is not a function in the ordinary sense as used in calculus, but a so-called *"generalized function,"*[4] because (1) and (2) with $k \to 0$ imply

$$\delta(t - a) = \begin{cases} \infty & \text{if } t = a \\ 0 & \text{otherwise} \end{cases}$$

and

$$\int_0^\infty \delta(t - a)\, dt = 1,$$

but an ordinary function which is everywhere 0 except at a single point must have the integral 0. Nevertheless, in impulse problems it is convenient to operate on $\delta(t - a)$ as though it were an ordinary function.

EXAMPLE 4. Response of a damped vibrating system to a unit impulse

Determine the response of the damped mass–spring system (cf. Sec. 2.13) governed by

$$y'' + 3y' + 2y = \delta(t - a), \qquad y(0) = 0, \qquad y'(0) = 0.$$

Thus the system is initially at rest and at time $t = a$ is suddenly given a sharp hammerblow.

Solution. By (4) we obtain the subsidiary equation

$$s^2 Y + 3sY + 2Y = e^{-as}.$$

Solving for Y, we have

$$Y(s) = F(s)e^{-as}, \qquad \text{where} \qquad F(s) = \frac{1}{(s + 1)(s + 2)} = \frac{1}{s + 1} - \frac{1}{s + 2}.$$

Taking the inverse transform, we obtain

$$f(t) = \mathscr{L}^{-1}(F) = e^{-t} - e^{-2t}.$$

Hence by the second shifting theorem (Sec. 5.3) we have

---

[3]PAUL DIRAC (1902—1984), English physicist, was awarded the Nobel Prize [jointly with ERWIN SCHRÖDINGER (1887—1961)] in 1933 for his work in quantum mechanics.

[4]Or *"distribution."* A systematic theory of generalized functions was created in 1936 by the Soviet mathematician, SERGEI L'VOVICH SOBOLEV (born 1908) and in 1945, under wider aspects, by the French mathematician, LAURENT SCHWARTZ (born 1915).

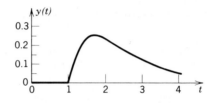

Fig. 103. Output in Example 4 with $a = 1$

$$y(t) = \mathcal{L}^{-1}\{e^{-as}F(s)\} = f(t - a)u(t - a) = \begin{cases} 0 & \text{if } 0 \leqq t < a, \\ e^{-(t-a)} - e^{-2(t-a)} & \text{if } t > a. \end{cases}$$

Figure 103 shows this solution for $a = 1$. The reader may compare this with the output in Example 3 (Fig. 101) and comment. ∎

In Sec. 5.2 we have seen the effects on transforms corresponding to the differentiation or integration of *functions*. In the next section we discuss another useful general property: the effects on functions corresponding to the differentiation or integration of *transforms*.

Problems for Sec. 5.4

RC-circuit. Find the current in the *RC*-circuit in Fig. 104 with $R = 100$ ohms, $C = 0.1$ farad and electromotive force $v(t)$ [volts] as follows. Assume that the circuit is quiescent before $v(t)$ is applied. *Hint.* Use partial fractions if necessary.

1. $v(t) = 100$ if $1 < t < 2$ and 0 otherwise
2. $v(t) = 10000$ if $1 < t < 1.01$ and 0 otherwise
3. $v(t) = e^{-t}$ if $t > 2$ and 0 otherwise
4. $v(t) = 50(t - 3)$ if $t > 3$ and 0 otherwise
5. $v(t) = 10(t - 1)$ if $t > 1$ and 0 otherwise
6. $v(t) = 200t$ if $0 < t < 1$ and 0 if $t > 1$

Undamped vibrating system. Using the Laplace transformation, solve the initial value problem $y'' + y = r(t)$, $y(0) = 0$, $y'(0) = 0$, where $r(t)$ equals

7. $r(t) = k$ if $0 < t < a$ and 0 otherwise $(a > 0)$
8. $r(t) = t - 1$ if $t > 1$ and 0 if $t < 1$
9. $r(t) = \sin 2t$ if $0 < t < \pi$ and 0 otherwise
10. $r(t) = e^{-(t-2\pi)}$ if $t > 2\pi$ and 0 if $t < 2\pi$

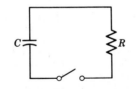

Fig. 104. *RC*-circuit in Probs. 1–6

Effect of the delta function on vibrating systems and in other initial value problems.
Solve:

11. $y'' + 9y = \delta(t - 1)$, $\quad y(0) = 0$, $\quad y'(0) = 0$

12. $y'' + 4y = \delta(t - \pi)$, $\quad y(0) = 2$, $\quad y'(0) = 0$

13. $y'' + y = \delta(t - \pi) - \delta(t - 2\pi)$, $\quad y(0) = 0$, $\quad y'(0) = 1$

14. $y'' - y = \sin t + \delta(t - \frac{1}{2}\pi)$, $\quad y(0) = 3.0$, $\quad y'(0) = -3.5$

15. $y'' + 2y' + 2y = \delta(t - 2\pi)$, $\quad y(0) = 1$, $\quad y'(0) = -1$

16. $y'' + 4y' + 5y = \delta(t - 1)$, $\quad y(0) = 0$, $\quad y'(0) = 3$

17. $y'' + 2y' - 3y = -8e^{-t} - \delta(t - \frac{1}{2})$, $\quad y(0) = 3$, $\quad y'(0) = -5$

18. $y'' + 5y' + 6y = u(t - 1) + \delta(t - 2)$, $\quad y(0) = 0$, $\quad y'(0) = 1$

19. $y'' + 2y' + 5y = 25t - \delta(t - \pi)$, $\quad y(0) = -2$, $\quad y'(0) = 5$

20. $y'' + 4y' + 4y = 1 + t + \delta(t - 1)$, $\quad y(0) = 0$, $\quad y'(0) = 2.25$

5.5 Differentiation and Integration of Transforms

The Laplace transformation has surprisingly many general properties, which we can use for obtaining transforms or inverse transforms. Indeed, methods for that purpose based on those properties are direct integration (Sec. 5.1), use of the linearity (Sec. 5.1), shifting (Sec. 5.3), and differentiation or integration of original functions $f(t)$ (Sec. 5.2). But this is not all: in this section we consider differentiation and integration of transforms $F(s)$ and find out the corresponding operations for original functions $f(t)$. And in the next section we shall consider multiplication of transforms and find out the corresponding operation for original functions.

Differentiation of Transforms

It can be shown that if $f(t)$ satisfies the conditions of the existence theorem in Sec. 5.1, then the derivative of the corresponding transform

$$F(s) = \mathcal{L}(f) = \int_0^\infty e^{-st} f(t)\, dt$$

with respect to s can be obtained by differentiating under the integral sign with respect to s (proof in Ref. [5] listed in Appendix 1); thus

$$F'(s) = -\int_0^\infty e^{-st}[t f(t)]\, dt.$$

Consequently, if $\mathcal{L}(f) = F(s)$, then

(1)
$$\boxed{\mathcal{L}\{t f(t)\} = -F'(s);}$$

differentiation of the transform of a function corresponds to the multiplication of the function by $-t$.

This property of the Laplace transformation enables us to obtain new transforms from given ones.

EXAMPLE 1

We shall derive the following three formulas (formulas 21-23 in the table in Sec. 5.10):

	$\mathcal{L}(f)$	$f(t)$
(2)	$\dfrac{1}{(s^2 + \beta^2)^2}$	$\dfrac{1}{2\beta^3}(\sin \beta t - \beta t \cos \beta t)$
(3)	$\dfrac{s}{(s^2 + \beta^2)^2}$	$\dfrac{t}{2\beta}\sin \beta t$
(4)	$\dfrac{s^2}{(s^2 + \beta^2)^2}$	$\dfrac{1}{2\beta}(\sin \beta t + \beta t \cos \beta t)$

Solution. From (1) and formula 8 (with $\omega = \beta$) in Table 5.1, Sec. 5.1, we obtain

$$\mathcal{L}(t \sin \beta t) = \frac{2\beta s}{(s^2 + \beta^2)^2}.$$

By dividing by 2β we obtain (3). From (1) and formula 7 (with $\omega = \beta$) in Table 5.1, Sec. 5.1, we find

$$(5) \qquad \mathcal{L}(t \cos \beta t) = -\frac{(s^2 + \beta^2) - 2s^2}{(s^2 + \beta^2)^2} = \frac{s^2 - \beta^2}{(s^2 + \beta^2)^2}.$$

From this and formula 8 (with $\omega = \beta$) in Table 5.1, Sec. 5.1, we have

$$\mathcal{L}\left(t \cos \beta t \pm \frac{1}{\beta}\sin \beta t\right) = \frac{s^2 - \beta^2}{(s^2 + \beta^2)^2} \pm \frac{1}{s^2 + \beta^2}.$$

On the right we now take the common denominator. Then for the plus sign the numerator is $s^2 - \beta^2 + s^2 + \beta^2 = 2s^2$, so that (4) follows. For the minus sign the numerator takes the form $s^2 - \beta^2 - s^2 - \beta^2 = -2\beta^2$, and we obtain (2).   ∎

Integration of Transforms

Similarly, if $f(t)$ satisfies the conditions of the existence theorem in Sec. 5.1 and the limit of $f(t)/t$, as t approaches 0 from the right, exists, then

$$(6) \qquad \boxed{\mathcal{L}\left\{\frac{f(t)}{t}\right\} = \int_s^\infty F(\tilde{s})\, d\tilde{s}} \qquad (s > \gamma);$$

in this manner, *integration of the transform of a function $f(t)$ corresponds to the division of $f(t)$ by t.*

In fact, from the definition it follows that

$$\int_s^\infty F(\tilde{s})\, d\tilde{s} = \int_s^\infty \left[\int_0^\infty e^{-\tilde{s}t}f(t)\, dt\right] d\tilde{s},$$

and it can be shown (cf. Ref. [5] in Appendix 1) that under the above assumptions we may reverse the order of integration, that is

$$\int_s^\infty F(\tilde{s}) \, d\tilde{s} = \int_0^\infty \left[\int_s^\infty e^{-\tilde{s}t} f(t) \, d\tilde{s}\right] dt = \int_0^\infty f(t) \left[\int_s^\infty e^{-\tilde{s}t} \, d\tilde{s}\right] dt.$$

The integral over $\tilde{s}$ on the right equals e^{-st}/t when $s > \gamma$, and, therefore,

$$\int_s^\infty F(\tilde{s}) \, d\tilde{s} = \int_0^\infty e^{-st} \frac{f(t)}{t} \, dt = \mathcal{L}\left\{\frac{f(t)}{t}\right\} \qquad (s > \gamma).$$

EXAMPLE 2

Find the inverse transform of the function $\ln\left(1 + \dfrac{\omega^2}{s^2}\right)$.

Solution. By differentiation,

$$-\frac{d}{ds} \ln\left(1 + \frac{\omega^2}{s^2}\right) = \frac{2\omega^2}{s(s^2 + \omega^2)} = \frac{2}{s} - 2\frac{s}{s^2 + \omega^2},$$

where the last equality can be readily verified by direct calculation. This is our present $F(s)$. From Table 5.1 in Sec. 5.1 we obtain

$$f(t) = \mathcal{L}^{-1}(F) = \mathcal{L}^{-1}\left\{\frac{2}{s} - 2\frac{s}{s^2 + \omega^2}\right\} = 2 - 2\cos \omega t.$$

This function satisfies the conditions under which (6) holds. Therefore,

$$\mathcal{L}^{-1}\left\{\ln\left(1 + \frac{\omega^2}{s^2}\right)\right\} = \mathcal{L}^{-1}\left\{\int_s^\infty F(\tilde{s}) \, d\tilde{s}\right\} = \frac{f(t)}{t}.$$

Our result is

$$\mathcal{L}^{-1}\left\{\ln\left(1 + \frac{\omega^2}{s^2}\right)\right\} = \frac{2}{t}(1 - \cos \omega t).$$

This proves formula 42 in the table in Sec. 5.10.

EXAMPLE 3

Reasoning as in Example 2, we obtain (cf. formula 43 in the table in Sec. 5.10)

$$\mathcal{L}^{-1}\left\{\ln\left(1 - \frac{a^2}{s^2}\right)\right\} = \frac{2}{t}(1 - \cosh at). \qquad \blacksquare$$

There is still another "big" theorem, the *convolution theorem*, which tells us what operation on functions corresponds to the multiplication of their transforms. This operation, called **convolution,** is some sort of multiplication and integration, which we explain in the next section, along with applications to the solution of differential equations and **integral equations.**

Problems for Sec. 5.5

Using (1), find the Laplace transform of

1. te^t	**2.** $t^2 e^{2t}$	**3.** $t^2 e^{-t}$	**4.** $t \sin 3t$
5. $t \sinh 2t$	**6.** $t^2 \sinh 2t$	**7.** $t^2 \cos t$	**8.** $te^{-t} \cos t$
9. $te^{-t} \cosh 2t$	**10.** $te^{-2t} \sin t$	**11.** $t^2 \cos \omega t$	**12.** $te^{-2t} \sin \omega t$

Using (6) or (1), find $f(t)$, if $\mathcal{L}(f)$ equals

13. $\dfrac{2s}{(s^2 + 1)^2}$ **14.** $\dfrac{s}{(s^2 + 4)^2}$ **15.** $\dfrac{6s}{(s^2 - 9)^2}$ **16.** $\dfrac{2s + 6}{(s^2 + 6s + 10)^2}$

17. $\ln \dfrac{s + 5}{s - 5}$ **18.** $\ln \dfrac{s + a}{s + b}$ **19.** $\ln \dfrac{s}{s - 1}$ **20.** $\ln \dfrac{s^2 + 1}{(s - 1)^2}$

21. arc cot (s/ω) **22.** arc cot $(s + 1)$

23. Solve Probs. 1–3 by using the first shifting theorem.

24. Find $\mathcal{L}(t^n e^{at})$ by repeated application of (1), choosing $f(t) = e^{at}$.

25. Carry out the derivation in Example 3.

5.6 Convolution. Integral Equations

Another important general property of the Laplace transformation has to do with products of transforms. It often happens that we are given two transforms $F(s)$ and $G(s)$ whose inverses $f(t)$ and $g(t)$ we know, and we would like to calculate the inverse of the product $H(s) = F(s)G(s)$ from those known inverses $f(t)$ and $g(t)$. This inverse $h(t)$ is written $(f * g)(t)$, which is a standard notation, and is called the **convolution** of f and g. How to find h from f and g? This is stated in the following theorem. Since the situation and task just described rather frequently arise in applications, this theorem is of considerable practical importance.

Theorem 1 (Convolution theorem)
*Let $f(t)$ and $g(t)$ satisfy the hypothesis of the existence theorem (Sec. 5.1) Then the product of their transforms $F(s) = \mathcal{L}(f)$ and $G(s) = \mathcal{L}(g)$ is the transform $H(s) = \mathcal{L}(h)$ of the **convolution** $h(t)$ of $f(t)$ and $g(t)$, written $\mathcal{L}(f * g)(t)$ and defined by*

(1)
$$h(t) = (f * g)(t) = \int_0^t f(\tau)g(t - \tau)\, d\tau.$$

Proof. By the definition of $G(s)$ and the second shifting theorem, for each fixed τ ($\tau \geq 0$) we have

$$e^{-s\tau}G(s) = \int_0^\infty e^{-st}g(t - \tau)u(t - \tau)\, dt = \int_\tau^\infty e^{-st}g(t - \tau)\, dt$$

where $s > \gamma$. From this and the definition of $F(s)$ we obtain

$$F(s)G(s) = \int_0^\infty e^{-s\tau}f(\tau)G(s)\, d\tau = \int_0^\infty f(\tau) \int_\tau^\infty e^{-st}g(t - \tau)\, dt\, d\tau$$

where $s > \gamma$. Here we integrate over t from τ to ∞ and then over τ from 0 to ∞; this corresponds to the shaded wedge-shaped region extending to infinity in the $t\tau$-plane shown in Fig. 105. Our assumptions on f and g are such that the order of integration can be reversed. (A proof requiring the knowledge of uniform convergence is included in Ref. [A4] listed in Appendix 1.) We then integrate first over τ from 0 to t (cf. Fig. 105) and then over t from 0 to ∞; thus

$$F(s)G(s) = \int_0^\infty e^{-st} \int_0^t f(\tau)g(t - \tau)\, d\tau\, dt$$

$$= \int_0^\infty e^{-st} h(t)\, dt = \mathscr{L}(h)$$

where h is given by (1). This completes the proof.　■

Using the definition, the reader may show that the convolution $f * g$ has the properties

$$f * g = g * f \qquad \text{(commutative law)}$$

$$f * (g_1 + g_2) = f * g_1 + f * g_2 \qquad \text{(distributive law)}$$

$$(f * g) * v = f * (g * v) \qquad \text{(associative law)}$$

$$f * 0 = 0 * f = 0,$$

just as the multiplication of numbers. But $1 * g \neq g$ in general. For example if $g(t) = t$, then

$$(1 * g)(t) = \int_0^t 1 \cdot (t - \tau)\, d\tau = \frac{t^2}{2}.$$

Another unusual property is that $(f * f)(t) \geqq 0$ may not hold. An illustration of this fact will be suggested in Prob. 28, below, and further examples could be given without difficulty.

We shall now illustrate that convolution is useful for obtaining inverse transforms and solving differential equations.

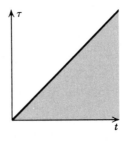

Fig. 105. Region of integration in the $t\tau$-plane in the proof of Theorem 1

EXAMPLE 1

Let $H(s) = 1/[s^2(s - a)]$. Find $h(t)$.

Solution. From Table 5.1 (Sec. 5.1) we know that

$$\mathcal{L}^{-1}\left\{\frac{1}{s^2}\right\} = t, \qquad \mathcal{L}^{-1}\left\{\frac{1}{s - a}\right\} = e^{at}.$$

We use Theorem 1 and integrate by parts, finding

$$h(t) = t * e^{at} = \int_0^t \tau e^{a(t-\tau)}\, d\tau = e^{at} \int_0^t \tau e^{-a\tau}\, d\tau$$

$$= \frac{1}{a^2}(e^{at} - at - 1).$$

EXAMPLE 2. Model of a forced vibrating system, resonance

Solve the initial value problem

$$my'' + ky = K_0 \sin pt, \qquad y(0) = 0, \quad y'(0) = 0.$$

Solution. From Sec. 2.13 we know that this equation is the mathematical model of forced oscillations of a body of mass m attached at the lower end of an elastic spring whose upper end is fixed (Fig. 106). k is the spring modulus, and $K_0 \sin pt$ is the driving force or input. Setting $\omega_0 = \sqrt{k/m}$, we may write the equation in the form

$$y'' + \omega_0{}^2 y = K \sin pt \qquad\qquad \left(K = \frac{K_0}{m}\right).$$

The subsidiary equation is

(2) $$s^2 Y + \omega_0{}^2 Y = K\,\frac{p}{s^2 + p^2}.$$

The solution is

(3) $$Y(s) = \frac{Kp}{(s^2 + \omega_0{}^2)(s^2 + p^2)}.$$

Now

$$\mathcal{L}^{-1}\left\{\frac{1}{s^2 + \omega_0{}^2}\right\} = \frac{1}{\omega_0}\sin \omega_0 t$$

$$\mathcal{L}^{-1}\left\{\frac{1}{s^2 + p^2}\right\} = \frac{1}{p}\sin pt.$$

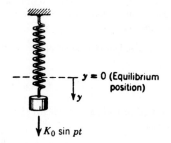

$y = 0$ (Equilibrium position)

y

$K_0 \sin pt$

Fig. 106. Vibrating system. Forced oscillations

Hence by the convolution theorem,

(4)
$$y(t) = \frac{Kp}{\omega_0 p} \sin \omega_0 t * \sin pt = \frac{K}{\omega_0} \int_0^t \sin \omega_0 \tau \sin (pt - p\tau) \, d\tau.$$

The integrand equals [cf. (11) in Appendix 3]

(5)
$$\tfrac{1}{2}[-\cos (pt + (\omega_0 - p)\tau) + \cos (pt - (\omega_0 + p)\tau)].$$

No resonance. If $p^2 \neq \omega_0^2$, using (5), we obtain by integration with respect to τ

$$y(t) = \frac{K}{2\omega_0} \left[-\frac{\sin (pt + (\omega_0 - p)\tau)}{\omega_0 - p} + \frac{\sin (pt - (\omega_0 + p)\tau)}{-(\omega_0 + p)} \right]_0^t$$

$$= \frac{K}{2\omega_0} \left[\frac{\sin pt - \sin \omega_0 t}{\omega_0 - p} + \frac{\sin pt + \sin \omega_0 t}{\omega_0 + p} \right].$$

Forming the common denominator and simplifying, we finally obtain the result

$$y(t) = \frac{K}{p^2 - \omega_0^2} \left[\frac{p}{\omega_0} \sin \omega_0 t - \sin pt \right].$$

This represents a superposition of two harmonic oscillations whose frequencies are the natural frequency of the freely vibrating system and the frequency of the driving force.

Resonance. If $p = \omega_0$, then (5) is simply

$$\tfrac{1}{2}[-\cos \omega_0 t + \cos (\omega_0 t - 2\omega_0 \tau)].$$

Integration with respect to τ yields

$$y(t) = \frac{K}{2\omega_0} \left[-\tau \cos \omega_0 t - \frac{1}{2\omega_0} \sin (\omega_0 t - 2\omega_0 \tau) \right]_0^t$$

$$= \frac{K}{2\omega_0^2} [\sin \omega_0 t - \omega_0 t \cos \omega_0 t].$$

This shows that we now have resonance as studied in Sec. 2.13.

EXAMPLE 3. Response of an undamped system to a single square wave

We reconsider the model in Sec. 5.4, Example 2, that is,

$$y'' + 2y = r(t), \qquad y(0) = 0, \qquad y'(0) = 0$$

where $r(t) = 1$ if $0 < t < 1$ and 0 otherwise. We shall solve this problem by the convolution technique. A main point will be to see how the convolution approach works for *inputs that act for some time only*.

Solution. The subsidiary equation $(s^2 + 2)Y = \mathcal{L}(r)$ has the solution

$$Y = \frac{1}{s^2 + 2} \mathcal{L}(r) = \mathcal{L}\left(\frac{1}{\sqrt{2}} \sin \sqrt{2} t \right) \mathcal{L}(r);$$

here we used formula 8 of Table 5.1 (Sec. 5.1). Hence, by the convolution theorem, we obtain the solution

$$y(t) = \left(\frac{1}{\sqrt{2}} \sin \sqrt{2} t \right) * r$$

Now we must be careful and remember that $r(t) = 1$ if $0 < t < 1$ but 0 if $t > 1$. Hence for

$t < 1$ the convolution integral is

$$r * \left(\frac{1}{\sqrt{2}} \sin \sqrt{2}t \right) = \frac{1}{\sqrt{2}} \int_0^t \sin (\sqrt{2}t - \sqrt{2}\tau) \, d\tau = \frac{1}{2} (1 - \cos \sqrt{2}t)$$

but for $t > 1$ we have to integrate from 0 to 1 only, since $r(t) = 0$ when $t > 1$; the result will be the same as in Sec. 5.4, Example 2; indeed,

$$r * \left(\frac{1}{\sqrt{2}} \sin \sqrt{2}t \right) = \frac{1}{\sqrt{2}} \int_0^1 \sin (\sqrt{2}t - \sqrt{2}\tau) \, d\tau$$

$$= \tfrac{1}{2} \cos (\sqrt{2}t - \sqrt{2}\tau) \Big|_{\tau=0}^{1}$$

$$= \tfrac{1}{2}[\cos (\sqrt{2}t - \sqrt{2}) - \cos \sqrt{2}t]. \qquad \blacksquare$$

Integral Equations

An **integral equation** is an equation in which the unknown function, call it $y(t)$, occurs in the integrand of an integral (and may also occur outside the integral). For instance, (4) in Sec. 1.10, appearing in Picard's method, is a special integral equation. Since integral equations are of practical importance, but are often difficult to solve, it is worth noting that some of them [namely, those whose integral can be written as a convolution] can be solved by the Laplace transformation. However, we should admit that this is a relatively small class among all integral equations of practical interest, and it will certainly be enough to explain the method in terms of a typical example and a handful of problems.

EXAMPLE 4. Integral equation
Solve the integral equation

$$y(t) = t + \int_0^t y(\tau) \sin (t - \tau) \, d\tau.$$

Solution. 1st Step. Equation in terms of convolution. We see that the given equation can be written

$$y = t + y * \sin t.$$

2nd Step. Application of the convolution theorem. We write $Y = \mathscr{L}(y)$. By the convolution theorem,

$$Y(s) = \frac{1}{s^2} + Y(s) \frac{1}{s^2 + 1} \; .$$

Solving for $Y(s)$, we obtain

$$Y(s) = \frac{s^2 + 1}{s^4} = \frac{1}{s^2} + \frac{1}{s^4} \; .$$

3rd Step. Taking the inverse transform. This gives the solution

$$y(t) = t + \tfrac{1}{6}t^3.$$

The reader may check this by substitution and evaluating the integral by repeated integration by parts (which will need patience). $\blacksquare$

In algebraically solving the subsidiary equation of a differential equation, the transform $Y(s)$ of the unknown function $y(t)$ usually appears as a quotient $Y(s) = F(s)/G(s)$, as we have seen in a number of cases in the previous sections, and the unknown $y(t)$ is then usually obtained directly or with the help of a simple partial fraction reduction, without much difficulty. For the sake of completeness and for reference if somebody gets stuck, we give in the next section a systematic discussion of partial fractions and their transforms, together with an application to **systems of differential equations.**

Problems for Sec. 5.6

Find the following convolutions. [*Hint.* In Probs. 7–9, 13 use (11) in Appendix 3.]

1. $1 * 1$ **2.** $1 * \sin t$ **3.** $1 * \cos t$

4. $e^t * e^{-t}$ **5.** $e^{kt} * e^{kt}$ **6.** $e^{at} * e^{bt}$ $(a \neq b)$

7. $\sin \omega t * \sin \omega t$ **8.** $\sin \omega t * \cos \omega t$ **9.** $\cos \omega t * \cos \omega t$

10. $t * e^{at}$ **11.** $1 * e^t$ **12.** $t^2 * e^{at}$

13. $\sin t * \sin 2t$ **14.** $u(t - 1) * t$ **15.** $u(t - \pi) * \cos t$

Application of the Convolution Theorem. Find $h(t)$ by the convolution theorem if $H(s) = \mathcal{L}\{h(t)\}$ equals

16. $\dfrac{1}{(s - a)^2}$ **17.** $\dfrac{1}{s(s - 2)}$ **18.** $\dfrac{1}{(s - a)(s - b)}$ $(a \neq b)$

19. $\dfrac{12}{(s - 0.4)^3}$ **20.** $\dfrac{1}{s(s^2 + 1)}$ **21.** $\dfrac{1}{(s^2 + 1)^2}$

22. $\dfrac{1}{s(s^2 + \omega^2)}$ **23.** $\dfrac{1}{s^2(s^2 + \omega^2)}$ **24.** $\dfrac{1}{s^2(s - a)}$

25. $\dfrac{s}{(s^2 + \omega^2)^2}$ **26.** $\dfrac{s^2}{(s^2 + 1)^2}$ **27.** $\dfrac{s^2 - \omega^2}{(s^2 + \omega^2)^2}$

General properties of convolution

28. Give an example showing that $(f * f)(t) \geqq 0$ may not hold. (Take $f(t) = \cos t$.)

29. Prove the commutative law $f * g = g * f$.

30. Prove the associative law $(f * g) * v = f * (g * v)$.

31. Prove the distributive law $f * (g_1 + g_2) = f * g_1 + f * g_2$.

32. How can we obtain Theorem 3 in Sec. 5.2 from the convolution theorem?

33. **(Dirac's delta function)** Using the convolution theorem and treating Dirac's delta function (Sec. 5.4) as though it were an ordinary function, show that

$$(\delta * f)(t) = f(t).$$

34. Derive the formula in Prob. 33 by using f_k with $a = 0$ (Sec. 5.4) and applying the mean value theorem for integrals.

35. Using the convolution theorem, prove by induction that

$$\mathcal{L}^{-1}\left\{\frac{1}{(s - a)^n}\right\} = \frac{1}{(n - 1)!} t^{n-1} e^{at}.$$

Application of convolution to initial value problems

36. (Forced vibrations) Consider the model of an undamped mass–spring system

$$y'' + \omega^2 y = r(t), \qquad y(0) = K_1, \quad y'(0) = K_2, \qquad (\omega \neq 0).$$

Leaving the *driving force r(t) unspecified*, show that the solution can be written in convolution form

$$y = \frac{1}{\omega} \sin \omega t * r(t) + K_1 \cos \omega t + \frac{K_2}{\omega} \sin \omega t.$$

37. Show that the solution of the initial value problem

$$y'' + 3y' + 2y = r(t),$$

$$y(0) = 0, \quad y'(0) = 1$$

with arbitrary continuous driving force $r(t)$ can be written

$$y = r * (e^{-t} - e^{-2t}) + e^{-t} - e^{-2t}.$$

Applying the convolution theorem, solve the following initial value problems.

38. $y'' + y = 4.7t, \qquad y(0) = 0, \quad y'(0) = 0$

39. $y'' + y = 1.5 \sin 2t, \qquad y(0) = 0, \quad y'(0) = 0$

40. $y'' + 4y = -6 \cos t, \qquad y(0) = 1, \quad y'(0) = 0$

41. $y'' + 25y = 5.2e^{-t}, \qquad y(0) = 1.2, \quad y'(0) = -10.2$

42. $y'' + 4y = u(t - 1), \qquad y(0) = 0, \quad y'(0) = 0$

43. $y'' + 3y' + 2y = 1 - u(t - 1), \qquad y(0) = 0, \quad y'(0) = 1$

44. The gun barrel problem (Prob. 22 in Sec. 2.13).

Integral equations

Using the Laplace transformation, solve

45. $y(t) = 1 + \displaystyle\int_0^t y(\tau) \, d\tau$

46. $y(t) = 1 - \displaystyle\int_0^t (t - \tau)y(\tau) \, d\tau$

47. $y(t) = \sin 2t + \displaystyle\int_0^t y(\tau) \sin 2(t - \tau) \, d\tau$

48. $y(t) = te^t - 2e^t \displaystyle\int_0^t e^{-\tau} y(\tau) \, d\tau$

49. $y(t) = 1 - \sinh t + \displaystyle\int_0^t (1 + \tau)y(t - \tau) \, d\tau$

50. $y(t) = t + e^t - \displaystyle\int_0^t y(\tau) \cosh (t - \tau) \, d\tau$

5.7 Partial Fractions. Systems of Differential Equations

Most of the applications considered so far in this chapter led to a subsidiary equation whose solution $Y(s)$ was of the form

$$Y(s) = \frac{F(s)}{G(s)}$$

where $F(s)$ and $G(s)$ are polynomials in s. In such a case, $y(t) = \mathcal{L}^{-1}(Y)$, the solution of the problem, can be determined by first expressing $Y(s)$ in terms of partial fractions. Actually, this is what we did in an ad hoc fashion and without difficulty. In fact, one may proceed in that way, using one's "favorite" method of determining the constant coefficients of partial fractions. In the present section we shall discuss partial fractions and their inverse transforms in a systematic, case-by-case fashion, so that the reader can look for help at a single spot—this section—should it become necessary. Throughout this section we make the following

Assumption

$F(s)$ and $G(s)$ have real coefficients and no common factors. The degree of $F(s)$ is lower than that of $G(s)$.

The form of partial fractions depends on the type of factor, and we shall see that the corresponding inverse transforms are already known to us. We shall consider the following cases:

- **(Case 1)** Unrepeated factor $s - a$.
- **(Case 2)** Repeated factor $(s - a)^m$.
- **(Case 3)** Complex factors $(s - a)(s - \bar{a})$.
- **(Case 4)** Repeated complex factors $[(s - a)(s - \bar{a})]^2$.

We leave aside higher powers $[(s - a)(s - \bar{a})]^m$, which are of minor practical interest. What we must know in each case is the general form of the partial fraction or fractions and the corresponding inverse transforms. We also list formulas for the coefficients of the fractions one may use if the specific situation offers no simpler possibility. Examples follow.

Case 1. Unrepeated factor $s - a$
To this factor corresponds in $Y = F/G$ a fraction

(1a)
$$\frac{A}{s - a}.$$

From Table 5.1 in Sec. 5.1 we see that its inverse transform is

(1b)
$$Ae^{at}.$$

We prove below that the constant A is given by

(1c) $\qquad A = \lim_{s \to a} \dfrac{(s - a)F(s)}{G(s)} \qquad$ or by $\qquad A = \dfrac{F(a)}{G'(a)}.$

Case 2. Repeated factor $(s - a)^m$

To this factor corresponds in $Y = F/G$ a sum of m fractions

(2a) $\qquad \dfrac{A_m}{(s - a)^m} + \dfrac{A_{m-1}}{(s - a)^{m-1}} + \cdots + \dfrac{A_1}{s - a}.$

From Table 5.1 and the first shifting theorem (Sec. 5.3) it follows that its inverse transform is

(2b) $\qquad e^{at} \left(A_m \dfrac{t^{m-1}}{(m - 1)!} + A_{m-1} \dfrac{t^{m-2}}{(m - 2)!} + \cdots + A_2 \dfrac{t}{1!} + A_1 \right).$

We prove below that the constant A_m is given by

(2c) $\qquad\qquad\qquad A_m = \lim_{s \to a} \dfrac{(s - a)^m F(s)}{G(s)}$

and the other constants are given by

(2d) $\qquad A_k = \dfrac{1}{(m - k)!} \lim_{s \to a} \dfrac{d^{m-k}}{ds^{m-k}} \left[\dfrac{(s - a)^m F(s)}{G(s)} \right], \qquad k = 1, \cdots, m - 1.$

Case 3. Unrepeated complex factors $(s - a)(s - \bar{a})$

Here $a = \alpha + i\beta$, say, and $\bar{a} = \alpha - i\beta$ is the complex conjugate of a. We could work in complex, but it is preferable to stay in real. Then to

$$(s - a)(s - \bar{a}) = s^2 - (a + \bar{a})s + a\bar{a}$$
$$= s^2 - 2\alpha s + \alpha^2 + \beta^2$$
$$= (s - \alpha)^2 + \beta^2$$

corresponds the partial fraction

(3a) $\qquad \dfrac{As + B}{(s - \alpha)^2 + \beta^2}, \qquad$ which we can write $\qquad \dfrac{A(s - \alpha) + \alpha A + B}{(s - \alpha)^2 + \beta^2}.$

From Table 5.1 and the first shifting theorem we obtain the inverse transform

(3b) $\qquad\qquad\qquad e^{\alpha t} \left(A \cos \beta t + \dfrac{\alpha A + B}{\beta} \sin \beta t \right).$

We prove below that A is the imaginary part and $(\alpha A + B)/\beta$ the real part of

(3c)
$$Q_a = \frac{1}{\beta} \lim_{s \to a} \frac{[(s - \alpha)^2 + \beta^2]F(s)}{G(s)}.$$

Case 4. Repeated complex factors $[(s - a)(s - \bar{a})]^2$

To these factors correspond in $Y = F/G$ the partial fractions

(4a)
$$\frac{As + B}{[(s - \alpha)^2 + \beta^2]^2} + \frac{Cs + D}{(s - \alpha)^2 + \beta^2}.$$

The second fraction is as in (3a). For the first fraction we can use (3) and (2) in Sec. 5.5 and the first shifting theorem. Since we can write $As + B = A(s - \alpha) + (\alpha A + B)$, this gives as the inverse transform of both fractions

(4b)
$$e^{\alpha t} \left[\frac{A}{2\beta} t \sin \beta t + \frac{\alpha A + B}{2\beta^3} (\sin \beta t - \beta t \cos \beta t) \right]$$
$$+ e^{\alpha t} \left[C \cos \beta t + \frac{\alpha C + D}{\beta} \sin \beta t \right].$$

We prove below that the constants in (4b) are given by the formulas

(4c)
$$A = \mathrm{Im}\, R_a / \beta, \qquad\qquad \alpha A + B = \mathrm{Re}\, R_a,$$
$$C = (A - \mathrm{Re}\, S_a)/2\beta^2, \qquad \alpha C + D = \mathrm{Im}\, S_a / 2\beta$$

where Re denotes the real part and Im the imaginary part,

$$R_a = \lim_{s \to a} R(s), \qquad S_a = \lim_{s \to a} R'(s)$$

and

$$R(s) = \frac{[(s - \alpha)^2 + \beta^2]^2 F(s)}{G(s)}.$$

Typical Examples

EXAMPLE 1. Unrepeated factors

Find the inverse transform of

$$Y(s) = \frac{F(s)}{G(s)} = \frac{s + 1}{s^3 + s^2 - 6s}.$$

Solution. Since $G(s) = s(s - 2)(s + 3)$ has three distinct linear factors, Y has the representation

$$Y(s) = \frac{A_1}{s} + \frac{A_2}{s - 2} + \frac{A_3}{s + 3}.$$

We determine A_1, A_2, A_3. Multiplication by $G(s)$ gives

$$s + 1 = (s - 2)(s + 3)A_1 + s(s + 3)A_2 + s(s - 2)A_3.$$

Taking $s = 0$, $s = 2$, $s = -3$, we obtain

$$1 = -2 \cdot 3A_1, \qquad 3 = 2 \cdot 5A_2, \qquad -2 = -3(-5)A_3.$$

Hence $A_1 = -1/6$, $A_2 = 3/10$, $A_3 = -2/15$. The answer is

$$\mathcal{L}^{-1}(Y) = -\tfrac{1}{6} + \tfrac{3}{10}e^{2t} - \tfrac{2}{15}e^{-3t}.$$

EXAMPLE 2. Repeated factor. An initial value problem

Solve the initial value problem

$$y'' - 3y' + 2y = 4t \qquad\qquad y(0) = 1, \qquad y'(0) = -1.$$

Solution. From Table 5.1, Sec. 5.1, and (1) and (2) in Sec. 5.2 we see that the subsidiary equation is

$$s^2 Y - s + 1 - 3(sY - 1) + 2Y = \frac{4}{s^2}.$$

Collecting the terms in Y on the left and the others on the right, we have

$$(s^2 - 3s + 2)Y = \frac{4}{s^2} + s - 4 = \frac{4 + s^3 - 4s^2}{s^2}.$$

Since $s^2 - 3s + 2 = 0$ has the roots 2 and 1, this gives the partial fraction representation

$$Y(s) = \frac{F(s)}{G(s)} = \frac{s^3 - 4s^2 + 4}{s^2(s - 2)(s - 1)} = \frac{A_2}{s^2} + \frac{A_1}{s} + \frac{B}{s - 2} + \frac{C}{s - 1}.$$

We first determine the constants A_1 and A_2 in the partial fractions corresponding to the root $a = 0$. From (2c) with $m = 2$,

$$A_2 = \lim_{s \to 0} \frac{s^2(s^3 - 4s^2 + 4)}{s^2(s^2 - 3s + 2)} = \left. \frac{s^3 - 4s^2 + 4}{s^2 - 3s + 2} \right|_{s=0} = 2.$$

From (2d) with $m = 2$ and $k = 1$,

$$A_1 = \lim_{s \to 0} \frac{d}{ds}\left(\frac{s^3 - 4s^2 + 4}{s^2 - 3s + 2} \right) = 3.$$

We determine B and C. From the first formula in (1c),

$$B = \left. \frac{F(s)}{s^2(s - 1)} \right|_{s=2} = -1, \qquad C = \left. \frac{F(s)}{s^2(s - 2)} \right|_{s=1} = -1.$$

From this we obtain the answer

$$y(t) = \mathcal{L}^{-1}(Y) = \mathcal{L}^{-1}\left\{ \frac{2}{s^2} + \frac{3}{s} - \frac{1}{s - 2} - \frac{1}{s - 1} \right\} = 2t + 3 - e^{2t} - e^t.$$

EXAMPLE 3. Forced oscillations, resonance. Complex factors

Solve the initial value problem in Example 2 of Sec. 5.6 by the method of partial fractions.

No resonance. If $p^2 \neq \omega_0^2$, the solution $Y(s)$ of the subsidiary equation (2), Sec. 5.6, and its representation in terms of partial fractions are

$$Y(s) = \frac{Kp}{(s^2 + \omega_0^2)(s^2 + p^2)} = \frac{As + B}{s^2 + \omega_0^2} + \frac{Ms + N}{s^2 + p^2}.$$

We first determine A and B in the fraction corresponding to $a_1 = \alpha_1 + i\beta_1 = i\omega_0$. In (3c) we thus have $\alpha = \alpha_1 = 0$, $\beta = \beta_1 = \omega_0$ and $F(s) = Kp$, so that

$$Q_{a_1} = \frac{1}{\omega_0} \lim_{s \to a_1} \frac{(s^2 + \omega_0^2)Kp}{(s^2 + \omega_0^2)(s^2 + p^2)} = \left. \frac{Kp}{s^2 + p^2} \right|_{s=a_1} = \frac{Kp}{p^2 - \omega_0^2}.$$

Since this Q_{a_1} is real, $A = 0$ in (3b), and Q_{a_1} is the other coefficient, $(\alpha A + B)/\beta$ in (3b), with $\beta = \omega_0$. Hence (3b) gives the inverse

$$\frac{Kp}{\omega_0(p^2 - \omega_0^2)} \sin \omega_0 t.$$

We determine M and N in the last fraction, which corresponds to $a_2 = \alpha_2 + i\beta_2 = ip$. In (3c) we now have

$$Q_{a_2} = \frac{Kp}{s^2 + \omega_0^2}\bigg|_{s=a_2} = \frac{Kp}{\omega_0^2 - p^2},$$

and (3b) now gives the transform

$$\frac{Kp}{p(\omega_0^2 - p^2)} \sin pt.$$

By adding these two transforms we obtain the desired solution

$$y(t) = \mathcal{L}^{-1}(Y) = \frac{K}{p^2 - \omega_0^2}\left(\frac{p}{\omega_0}\sin \omega_0 t - \sin pt\right).$$

Resonance. If $p = \omega_0$, the solution of the subsidiary equation is

$$Y(s) = \frac{K\omega_0}{(s^2 + \omega_0^2)^2}.$$

The denominator has the double roots $a = \alpha + i\beta = i\omega_0$ and $\bar{a} = -i\omega_0$, that is, $\alpha = 0$ and $\beta = \omega_0$. This denominator cancels out in $R(s)$ in (4c) and we simply have

$$R(s) = K\omega_0, \qquad R_a = K\omega_0, \qquad S_a = 0.$$

This gives $\alpha A + B = K\omega_0$ in (4c), whereas the other expressions in (4c) are zero. Hence (4b) yields the solution

$$y(t) = \mathcal{L}^{-1}(Y) = \frac{K}{2\omega_0^2}(\sin \omega_0 t - \omega_0 t \cos \omega_0 t).$$

Our present results agree with those in Example 2 of Sec. 5.6.

EXAMPLE 4. Unrepeated complex and real factors
Find the inverse transform of

$$Y(s) = \frac{F(s)}{G(s)} = \frac{1}{s^4 - k^4}.$$

Solution. By inspection,

(5) $$Y(s) = \frac{1}{(s^2 + k^2)(s^2 - k^2)} = \frac{1}{2k^2}\left(\frac{1}{s^2 - k^2} - \frac{1}{s^2 + k^2}\right).$$

From this and Table 5.1,

(6) $$\mathcal{L}^{-1}\left\{\frac{1}{s^4 - k^4}\right\} = \frac{1}{2k^3}(\sinh kt - \sin kt).$$

This proves formula 27 in the table in Sec. 5.10.

Our example illustrates an important point. If one does not see shortcuts, such as the one in (5), one can still get **help from a systematic partial fraction reduction:**

$$Y(s) = \frac{1}{(s^2 + k^2)(s - k)(s + k)} = \frac{A_1 s + A_2}{s^2 + k^2} + \frac{B}{s - k} + \frac{C}{s + k}.$$

In (3c) we have $a = ik$. Hence $\alpha = 0$ and $\beta = k$. Thus,

$$Q_a = \frac{1}{k} \cdot \frac{1}{(s-k)(s+k)} \bigg|_{s=ik} = \frac{1}{k(s^2 - k^2)} \bigg|_{s=ik} = -\frac{1}{2k^3}.$$

This is real. Hence $(\alpha A + B)/\beta = Q_a$ and $A = 0$. From (3b) we now obtain the inverse transform

$$-\frac{1}{2k^3} \sin kt.$$

Also, by (1c),

$$B = \frac{1}{(s^2 + k^2)(s+k)} \bigg|_{s=k} = \frac{1}{4k^3}, \qquad C = \frac{1}{(s^2 + k^2)(s-k)} \bigg|_{s=-k} = -\frac{1}{4k^3}.$$

Together, this gives as the answer an expression that agrees with (6):

$$-\frac{1}{2k^3} \sin kt + \frac{1}{4k^3} (e^{kt} - e^{-kt}). \qquad \blacksquare$$

Proofs of Formulas (1c), (2c), (2d), (3c), (4c)

In these proofs, $W(s)$ always denotes the sum of the other partial fractions of $Y(s) = F(s)/G(s)$.

Proof of (1c) in Case 1
In this case,

$$Y(s) = \frac{F(s)}{G(s)} = \frac{A}{s-a} + W(s)$$

Multiplication by $s - a$ gives

$$\frac{(s-a)F(s)}{G(s)} = A + (s-a)W(s).$$

The first formula in (1c) now follows by letting $s \to a$ and noting that $(s-a)W(s) \to 0$, since $W(s)$ has no factor that could cancel $s - a$. The second formula in (1c) is obtained by writing

$$\frac{(s-a)F(s)}{G(s)} = \frac{F(s)}{G(s)/(s-a)}$$

and letting $s \to a$. Then $F(s) \to F(a)$ in the numerator, while the denominator appears as an indeterminate of the form 0/0; evaluating it in the usual way, we get the denominator of the second formula in (1c):

$$\lim_{s \to a} \frac{G(s)}{s-a} = \lim_{s \to a} \frac{G'(s)}{(s-a)'} = G'(a). \qquad \blacksquare$$

Proof of (2c) and (2d) in Case 2
In this case,

$$Y(s) = \frac{F(s)}{G(s)} = \frac{A_m}{(s-a)^m} + \frac{A_{m-1}}{(s-a)^{m-1}} + \cdots + \frac{A_1}{s-a} + W(s).$$

Multiplication by $(s - a)^m$ gives

$$\frac{(s - a)^m F(s)}{G(s)} = A_m + (s - a)A_{m-1} + (s - a)^2 A_{m-2} + \cdots + (s - a)^m W(s).$$

Letting $s \to a$, we obtain (2c). By differentiation,

$$\frac{d}{ds}\left[\frac{(s - a)^m F(s)}{G(s)}\right] = A_{m-1} + \text{further terms containing factors } s - a.$$

Letting $s \to a$, we obtain A_{m-1} as given by (2d) with $k = m - 1$. Differentiating again, we obtain A_{m-2}, and so on. This proves (2d).   ∎

Proof of (3c) in Case 3
For simplicity we write in this proof $p(s) = (s - a)(s - \bar{a}) = (s - \alpha)^2 + \beta^2$. Then by (3a)

$$Y(s) = \frac{F(s)}{G(s)} = \frac{As + B}{p(s)} + W(s) \qquad (A, B \text{ real}).$$

We multiply this by $p(s)$,

$$\frac{p(s)F(s)}{G(s)} = As + B + p(s)W(s).$$

We now let $s \to a$. Since $p(a) = 0$ and $W(s)$ has no factor that could cancel $s - a$, and since $a = \alpha + i\beta$, we get

$$Q_a = \lim_{s \to a} \frac{p(s)F(s)}{G(s)} = Aa + B = (\alpha A + B) + i\beta A.$$

Dividing by β and separating the real and the imaginary parts on both sides, we obtain the statement involving (3c).   ∎

Proof of (4c) in Case 4
Writing $p(s)$ as before, from (4a) we have

$$Y(s) = \frac{F(s)}{G(s)} = \frac{As + B}{p^2(s)} + \frac{Cs + D}{p(s)} + W(s).$$

Multiplying this by $p^2(s)$, we get

$$R(s) = \frac{p^2(s)F(s)}{G(s)} = As + B + p(s)(Cs + D) + p^2(s)W(s).$$

We let $s \to a$. Then we get, since $p(a) = 0$,

$$R_a = \lim_{s \to a} R(s) = Aa + B = (\alpha A + B) + i\beta A.$$

Separating the real and the imaginary parts, we obtain the first two formulas in (4c). By differentiation,

$$R'(s) = A + p'(s)(Cs + D) + \text{terms containing } p(s).$$

Here $p(s) = (s - \alpha)^2 + \beta^2$ gives $p'(s) = 2(s - \alpha)$. We now let $s \to a$. Since $p'(a) = 2(a - \alpha) = 2i\beta$, we then obtain

$$S_a = \lim_{s \to a} R'(s) = A + 2i\beta[C(\alpha + i\beta) + D] = (A - 2\beta^2 C) + 2i\beta(\alpha C + D).$$

Separation of the real and the imaginary parts gives the other two formulas in (4c).   ∎

Systems of Differential Equations

The Laplace transformation may also be used for solving systems of differential equations. We explain the method in terms of a typical example.

EXAMPLE 5. Model of two masses on springs

The mechanical system in Fig. 107 consists of two masses on three springs and is governed by the differential equations

(7)
$$y_1'' = -ky_1 + k(y_2 - y_1)$$
$$y_2'' = -k(y_2 - y_1) - ky_2,$$

where k is the spring modulus of each of the three springs, y_1 and y_2 are the displacements of the masses from their position of static equilibrium; the masses of the springs and the damping are neglected. The derivation of (7) is similar to that of the differential equation in Sec. 2.6. Cf. also Sec. 3.1.

We shall determine the solution corresponding to the initial conditions $y_1(0) = 1$, $y_2(0) = 1$, $y_1'(0) = \sqrt{3k}$, $y_2'(0) = -\sqrt{3k}$. Let $Y_1 = \mathcal{L}(y_1)$ and $Y_2 = \mathcal{L}(y_2)$. Then from (2) in Sec. 5.2 and the initial conditions we obtain the subsidiary equations

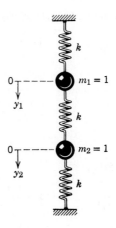

Fig. 107. Example 5

$$s^2 Y_1 - s - \sqrt{3k} = -kY_1 + k(Y_2 - Y_1)$$

$$s^2 Y_2 - s + \sqrt{3k} = -k(Y_2 - Y_1) - kY_2.$$

This system of linear algebraic equations in the unknowns Y_1 and Y_2 may be written

$$(s^2 + 2k)Y_1 - kY_2 = s + \sqrt{3k}$$

$$-kY_1 + (s^2 + 2k)Y_2 = s - \sqrt{3k}.$$

Cramer's rule (Sec. 7.9) or elimination yields the solution

$$Y_1 = \frac{(s + \sqrt{3k})(s^2 + 2k) + k(s - \sqrt{3k})}{(s^2 + 2k)^2 - k^2}$$

$$Y_2 = \frac{(s^2 + 2k)(s - \sqrt{3k}) + k(s + \sqrt{3k})}{(s^2 + 2k)^2 - k^2}.$$

The representations in terms of partial fractions are

$$Y_1 = \frac{s}{s^2 + k} + \frac{\sqrt{3k}}{s^2 + 3k}, \qquad Y_2 = \frac{s}{s^2 + k} - \frac{\sqrt{3k}}{s^2 + 3k}.$$

Hence the solution of our initial value problem is

$$y_1(t) = \mathcal{L}^{-1}(Y_1) = \cos \sqrt{k}t + \sin \sqrt{3k}t$$

$$y_2(t) = \mathcal{L}^{-1}(Y_2) = \cos \sqrt{k}t - \sin \sqrt{3k}t. \qquad \blacksquare$$

Our concluding section of text (Sec. 5.8) contains further applications. This is followed in Sec. 5.9 by a list of important **general properties** of the Laplace transformation and in Sec. 5.10 by a **table of Laplace transforms.**

Problems for Sec. 5.7

Using partial fractions, find $f(t)$ if $\mathcal{L}(f)$ equals

1. $\dfrac{1}{(s - a)(s - b)}$ **2.** $\dfrac{6s - 4}{s^2 - s}$ **3.** $\dfrac{s^2 + 9s - 9}{s^3 - 9s}$

4. $\dfrac{4s + 4}{s^2 + 16}$ **5.** $\dfrac{s}{(s + 1)^2}$ **6.** $\dfrac{-s + 4.5}{s^2 + 2.25}$

7. $\dfrac{s + 2}{(s + 1)^2}$ **8.** $\dfrac{s}{s^2 + 2s + 2}$ **9.** $\dfrac{s^2 + 2s}{(s^2 + 2s + 2)^2}$

10. $\dfrac{3s^2 - 6s + 7}{(s^2 - 2s + 5)^2}$ **11.** $\dfrac{s^2 + s - 2}{(s + 1)^3}$ **12.** $\dfrac{s^3 + 6s^2 + 14s}{(s + 2)^4}$

13. $\dfrac{2s^2 - 3s}{(s - 2)(s - 1)^2}$ **14.** $\dfrac{s^4 + 3(s + 1)^3}{s^4(s + 1)^3}$ **15.** $\dfrac{s^2 - 6s + 7}{(s^2 - 4s + 5)^2}$

16. $\dfrac{s^3 + 3s^2 - s - 3}{(s^2 + 2s + 5)^2}$ **17.** $\dfrac{s^3 - 7s^2 + 14s - 9}{(s - 1)^2(s - 2)^3}$ **18.** $\dfrac{s^3 - 3s^2 + 6s - 4}{(s^2 - 2s + 2)^2}$

19. Solve Prob. 1 by convolution.

20. Check the result in Example 1 by (i) working backward, (ii) using Theorem 3, Sec. 5.2, (iii) convolution methods.

Some inverses in terms of hyperbolic functions. Show that

21. $\mathcal{L}^{-1}\left\{\dfrac{1}{s^4 + 4a^4}\right\} = \dfrac{1}{4a^3}(\cosh at \sin at - \sinh at \cos at)$

22. $\mathcal{L}^{-1}\left\{\dfrac{s}{s^4 + 4a^4}\right\} = \dfrac{1}{2a^2}\sinh at \sin at$

23. $\mathcal{L}^{-1}\left\{\dfrac{s^2}{s^4 + 4a^4}\right\} = \dfrac{1}{2a}(\cosh at \sin at + \sinh at \cos at)$

24. $\mathcal{L}^{-1}\left\{\dfrac{s^3}{s^4 + 4a^4}\right\} = \cosh at \cos at$

Initial value problems. Using the Laplace transformation, solve:

25. $y'' + y = 2 \cos t, \quad y(0) = 2, \ y'(0) = 0$

26. $y'' + y = -2 \sin t, \quad y(0) = 0, \ y'(0) = 1$

27. $y'' - 4y' + 3y = 0, \quad y(0) = 3, \ y'(0) = 7$

28. $y'' + 3y' + 2y = u(t - 2), \quad y(0) = 0, \ y'(0) = 0$

29. $y'' - 4y = 8t^2 - 4, \quad y(0) = 5, \ y'(0) = 10$

30. $y'' + 2y' + y = 2 \cos t, \quad y(0) = 3, \ y'(0) = 0$

31. $y'' - 3y' + 2y = 6e^{-t}, \quad y(0) = 3, \ y'(0) = 3$

32. $y'' + 2y' + 5y = t - 0.04\,\delta(t - \pi), \quad y(0) = -0.08, \ y'(0) = 0.2$

33. $y'' - 2y' + 5y = 8 \sin t - 4 \cos t, \quad y(0) = 1, \ y'(0) = 3$

34. $y'' + 2y' - 3y = 10 \sinh 2t, \quad y(0) = 0, \ y'(0) = 4$

Systems of differential equations. Solve the following initial value problem by means of the Laplace transformation.

35. $y_1' = -y_2, \ y_2' = y_1, \quad y_1(0) = 1, \ y_2(0) = 0$

36. $y_1' + y_2 = 2 \cos t, \ y_1 + y_2' = 0, \quad y_1(0) = 0, \ y_2(0) = 1$

37. $y_1'' = y_1 + 3y_2, \ y_2'' = 4y_1 - 4e^t,$
$\qquad y_1(0) = 2, \ y_1'(0) = 3, \ y_2(0) = 1, \ y_2'(0) = 2$

38. $y_1'' + y_2 = -5 \cos 2t, \ y_2'' + y_1 = 5 \cos 2t,$
$\qquad y_1(0) = 1, \ y_1'(0) = 1, \ y_2(0) = -1, \ y_2'(0) = 1$

39. $y_1' + y_2' = 2 \sinh t, \ y_2' + y_3' = e^t, \ y_3' + y_1' = 2e^t + e^{-t},$
$\qquad y_1(0) = 1, \ y_2(0) = 1, \ y_3(0) = 0$

40. $2y_1' - y_2' - y_3' = 0, \ y_1' + y_2' = 4t + 2, \ y_2' + y_3 = t^2 + 2,$
$\qquad y_1(0) = y_2(0) = y_3(0) = 0$

5.8 Periodic Functions. Further Applications

Periodic functions appear in many practical problems, and in most cases they are more complicated than single cosine or sine functions. This justifies the topic of the present section, which is a systematic approach to the transformation of periodic functions. The text and the problem set also include further applications.

Let $f(t)$ be a function that is defined for all positive t and has the period p (> 0), that is,

$$f(t + p) = f(t) \qquad \text{for all } t > 0.$$

If $f(t)$ is piecewise continuous on an interval of length p, then its Laplace transform exists, and we can write the integral from zero to infinity as the series of integrals over successive periods:

$$\mathcal{L}(f) = \int_0^\infty e^{-st} f(t)\, dt = \int_0^p e^{-st} f\, dt + \int_p^{2p} e^{-st} f\, dt + \int_{2p}^{3p} e^{-st} f\, dt + \cdots.$$

If we substitute $t = \tau + p$ in the second integral, $t = \tau + 2p$ in the third integral, $\cdots$, $t = \tau + (n - 1)p$ in the nth integral, $\cdots$, then the new limits in every integral are 0 and p. Since

$$f(\tau + p) = f(\tau), \qquad f(\tau + 2p) = f(\tau),$$

etc., we thus obtain

$$\mathcal{L}(f) = \int_0^p e^{-s\tau} f(\tau)\, d\tau + \int_0^p e^{-s(\tau + p)} f(\tau)\, d\tau + \int_0^p e^{-s(\tau + 2p)} f(\tau)\, d\tau + \cdots.$$

The factors that do not depend on τ can be taken out from under the integral signs; this gives

$$\mathcal{L}(f) = [1 + e^{-sp} + e^{-2sp} + \cdots] \int_0^p e^{-s\tau} f(\tau)\, d\tau.$$

The series in brackets $[\cdots]$ is a geometric series whose sum is $1/(1 - e^{-ps})$. This establishes the following result.

Theorem 1 (Transform of periodic functions)

The Laplace transform of a piecewise continuous periodic function $f(t)$ with period p is

(1)
$$\boxed{\mathcal{L}(f) = \frac{1}{1 - e^{-ps}} \int_0^p e^{-st} f(t)\, dt} \qquad (s > 0).$$

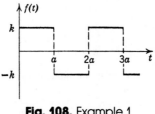 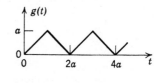

Fig. 108. Example 1   **Fig. 109.** Example 2

EXAMPLE 1. Periodic square wave

Find the transform of the square wave shown in Fig. 108.

Solution. We use (1). Since $p = 2a$, we obtain by direct integration and simplification

$$\mathscr{L}(f) = \frac{1}{1 - e^{-2as}} \left(\int_0^a ke^{-st}\, dt + \int_a^{2a} (-k)e^{-st}\, dt \right)$$

$$= \frac{k}{s} \frac{1 - 2e^{-as} + e^{-2as}}{(1 + e^{-as})(1 - e^{-as})} = \frac{k}{s} \left(\frac{1 - e^{-as}}{1 + e^{-as}} \right)$$

$$= \frac{k}{s} \frac{e^{-as/2}(e^{as/2} - e^{-as/2})}{e^{-as/2}(e^{as/2} + e^{-as/2})} .$$

Hence the result is

$$\mathscr{L}(f) = \frac{k}{s} \tanh \frac{as}{2} .$$

We can also obtain a less elegant but often more useful form of the result if we write

$$\mathscr{L}(f) = \frac{k}{s} \left(\frac{1 - e^{-as}}{1 + e^{-as}} \right) = \frac{k}{s} \left(1 - \frac{2e^{-as}}{1 + e^{-as}} \right) = \frac{k}{s} \left(1 - \frac{2}{e^{as} + 1} \right).$$

EXAMPLE 2. Periodic triangular wave

Find the transform of the periodic function shown in Fig. 109.

Solution. We see that $g(t)$ is the integral of the function $f(t)$ with $k = 1$ in Example 1. Hence, by Theorem 3 in Sec. 5.2,

$$\mathscr{L}(g) = \frac{1}{s} \mathscr{L}(f) = \frac{1}{s^2} \tanh \frac{as}{2} .$$

EXAMPLE 3. Half-wave rectifier

Find the transform of the following function $f(t)$ with period $p = 2\pi/\omega$:

$$f(t) = \begin{cases} \sin \omega t & \text{if} \quad 0 < t < \pi/\omega, \\ 0 & \text{if} \quad \pi/\omega < t < 2\pi/\omega. \end{cases}$$

Note that this function is the half-wave rectification of $\sin \omega t$ (Fig. 110).

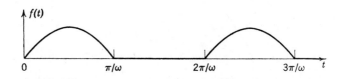

Fig. 110. Half-wave rectification of $\sin \omega t$

Solution. From (1) we obtain

$$\mathscr{L}(f) = \frac{1}{1 - e^{-2\pi s/\omega}} \int_0^{\pi/\omega} e^{-st} \sin \omega t \, dt.$$

Using $1 - e^{-2\pi s/\omega} = (1 + e^{-\pi s/\omega})(1 - e^{-\pi s/\omega})$ and integrating by parts or noting that the integral is the imaginary part of the integral

$$\int_0^{\pi/\omega} e^{(-s+i\omega)t} \, dt = \frac{1}{-s + i\omega} e^{(-s+i\omega)t} \Big|_0^{\pi/\omega} = \frac{-s - i\omega}{s^2 + \omega^2} (-e^{-s\pi/\omega} - 1)$$

we obtain the result

$$\mathscr{L}(f) = \frac{\omega(1 + e^{-\pi s/\omega})}{(s^2 + \omega^2)(1 - e^{-2\pi s/\omega})} = \frac{\omega}{(s^2 + \omega^2)(1 - e^{-\pi s/\omega})}.$$

EXAMPLE 4. Sawtooth wave

Find the Laplace transform of the function (Fig. 111).

$$f(t) = \frac{k}{p} t \quad \text{when } 0 < t < p, \quad f(t + p) = f(t).$$

Solution. Integration by parts yields

$$\int_0^p e^{-st} t \, dt = -\frac{t}{s} e^{-st} \Big|_0^p + \frac{1}{s} \int_0^p e^{-st} \, dt$$

$$= -\frac{p}{s} e^{-sp} - \frac{1}{s^2} (e^{-sp} - 1),$$

and from (1) we thus obtain the result

$$\mathscr{L}(f) = \frac{k}{ps^2} - \frac{ke^{-ps}}{s(1 - e^{-ps})} \qquad (s > 0).$$

EXAMPLE 5. Staircase function

Find the Laplace transform of the staircase function (Fig. 112)

$$g(t) = kn \qquad\qquad [np < t < (n + 1)p, \quad n = 0, 1, 2, \cdots].$$

Solution. Since $g(t)$ is the difference of the functions $h(t) = kt/p$ (whose transform is k/ps^2) and $f(t)$ in Example 4, we obtain

$$\mathscr{L}(g) = \mathscr{L}(h) - \mathscr{L}(f) = \frac{ke^{-ps}}{s(1 - e^{-ps})} \qquad (s > 0).$$

The next example illustrates the application of the Laplace transformation to the model of an oscillating system with a rather general periodic function as the driving force.

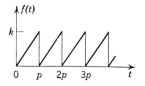

Fig. 111. Sawtooth wave

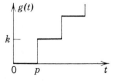

Fig. 112. Staircase function

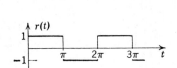

Fig. 113. Input $r(t)$ in Example 6

EXAMPLE 6. Forced oscillations

Solve the differential equation

(2) $$y'' + 2y' + 10y = r(t)$$

where (Fig. 113)

$$r(t) = \begin{cases} 1 & (0 < t < \pi) \\ -1 & (\pi < t < 2\pi) \end{cases} \qquad r(t + 2\pi) = r(t).$$

Solution. The subsidiary equation of (2) is

$$s^2 Y - s y(0) - y'(0) + 2[sY - y(0)] + 10Y = R(s)$$

where $R(s)$ denotes the transform of $r(t)$. By solving for Y we obtain

(3) $$Y(s) = \frac{(s + 2)y(0) + y'(0)}{s^2 + 2s + 10} + \frac{R(s)}{s^2 + 2s + 10}.$$

The roots of the denominator are

$$a = -1 + 3i \qquad \text{and} \qquad \bar{a} = -1 - 3i.$$

The inverse transform of the first term in (3) is

(4)
$$\mathscr{L}^{-1}\left\{\frac{(s + 2)y(0) + y'(0)}{s^2 + 2s + 10}\right\} = \mathscr{L}^{-1}\left\{\frac{(s + 1)y(0)}{(s + 1)^2 + 9} + \frac{y(0) + y'(0)}{(s + 1)^2 + 9}\right\}$$

$$= e^{-t}\left(y(0)\cos 3t + \frac{y(0) + y'(0)}{3}\sin 3t\right).$$

This is the solution of the homogeneous equation corresponding to (2). We consider the last term in (3) and determine $R(s)$. We can represent $r(t)$ in terms of step functions (Sec. 5.3):

$$r(t) = u(t) - 2u(t - \pi) + 2u(t - 2\pi) - 2u(t - 3\pi) + - \cdots.$$

From (5) in Sec. 5.3 we obtain the transform

$$R(s) = \mathscr{L}(r) = \frac{1}{s}[1 - 2e^{-\pi s} + 2e^{-2\pi s} - 2e^{-3\pi s} + - \cdots].$$

Hence

$$\frac{R(s)}{s^2 + 2s + 10} = \frac{1}{s[(s + 1)^2 + 9]}\{1 - 2e^{-\pi s} + - \cdots\}.$$

Now we know from the first shifting theorem (Sec. 5.3) that

$$\mathscr{L}^{-1}\left\{\frac{1}{(s + 1)^2 + 9}\right\} = \frac{1}{3}e^{-t}\sin 3t.$$

Therefore, according to Theorem 3 in Sec. 5.2,

(5) $\qquad \mathscr{L}^{-1}\left\{\dfrac{1}{s}\left(\dfrac{1}{(s+1)^2+9}\right)\right\} = \dfrac{1}{3}\displaystyle\int_0^t e^{-\tau}\sin 3\tau\, d\tau = \dfrac{1}{10}[1 - h(t)]$

where

(6) $\qquad\qquad h(t) = e^{-t}(\cos 3t + \tfrac{1}{3}\sin 3t).$

Denoting the function on the right-hand side of (5) by $f(t)$, that is,

$$f(t) = \frac{1}{10} - \frac{h(t)}{10},$$

and applying the second shifting theorem (Sec. 5.3), we thus obtain

(7) $\quad \mathscr{L}^{-1}\left\{\dfrac{R(s)}{s^2 + 2s + 10}\right\} = f(t) - 2f(t - \pi)u(t - \pi) + 2f(t - 2\pi)u(t - 2\pi) - + \cdots.$

Now in the terms on the right,

$$h(t - \pi) = e^{-(t-\pi)}[\cos 3(t - \pi) + \tfrac{1}{3}\sin 3(t - \pi)] = -h(t)e^{\pi}$$

and similarly,

$$h(t - 2\pi) = h(t)e^{2\pi}, \qquad h(t - 3\pi) = -h(t)e^{3\pi},$$

etc. It follows that in (7),

$$f(t - \pi) = \frac{1}{10} - \frac{h(t - \pi)}{10} = \frac{1}{10} + \frac{h(t)}{10}e^{\pi}$$

$$f(t - 2\pi) = \frac{1}{10} - \frac{h(t - 2\pi)}{10} = \frac{1}{10} - \frac{h(t)}{10}e^{2\pi}$$

etc., and the right-hand side of (7) equals

$$\frac{1}{10} - \frac{h(t)}{10} \qquad\qquad\qquad\qquad \text{if } 0 < t < \pi,$$

$$-\frac{1}{10} - \frac{h(t)}{10} - \frac{h(t)}{5}e^{\pi} \qquad\qquad \text{if } \pi < t < 2\pi,$$

$$\frac{1}{10} - \frac{h(t)}{10} - \frac{h(t)}{5}(e^{\pi} + e^{2\pi}) \qquad \text{if } 2\pi < t < 3\pi,$$

$$\frac{(-1)^n}{10} + \frac{h(t)}{10} - \frac{h(t)}{5}(1 + e^{\pi} + \cdots + e^{n\pi}) \quad \text{if } n\pi < t < (n + 1)\pi.$$

Summing the finite geometric progression in the parentheses, we see that the expression in the last line may be written

$$\frac{(-1)^n}{10} + \frac{h(t)}{10} - \frac{h(t)}{5}\left(\frac{e^{(n+1)\pi} - 1}{e^{\pi} - 1}\right).$$

If we use (6), this becomes

(8)
$$\left[\frac{1}{10} + \frac{1}{5(e^{\pi} - 1)}\right]e^{-t}\left(\cos 3t + \frac{1}{3}\sin 3t\right)$$

$$+ \frac{(-1)^n}{10} - \frac{1}{5(e^{\pi} - 1)}e^{-[t-(n+1)\pi]}\left(\cos 3t + \frac{1}{3}\sin 3t\right),$$

and the solution of (2) on the interval $n\pi < t < (n + 1)\pi$ is the sum of (8) and the right-hand side of (4).

Clearly the function in (4) and the function in the first line of (8) approach zero as t approaches infinity. The function in the last line of (8) is a periodic function of t with period 2π and thus represents the *steady-state solution* of our problem. To see this more distinctly we denote that function by $y_2(t)$ and set $t - n\pi = \tau$. Then $t = \tau + n\pi$, and τ ranges from 0 to π as t ranges from $n\pi$ to $(n + 1)\pi$. Also

$$\cos 3t = \cos (3\tau + 3n\pi) = \cos 3\tau \cos 3n\pi = (-1)^n \cos 3\tau,$$

$$\sin 3t = \sin (3\tau + 3n\pi) = \sin 3\tau \cos 3n\pi = (-1)^n \sin 3\tau,$$

and the function y_2 becomes

(9)
$$y_2 = (-1)^n \left[\frac{1}{10} - \frac{e^{-\tau + \pi}}{5(e^\pi - 1)} \left(\cos 3\tau + \frac{1}{3} \sin 3\tau \right) \right]$$

$$\approx (-1)^n [0.1 - 0.22 e^{-\tau} \cos (3\tau - 0.32)]$$

where $0 < \tau < \pi$ and $t = \tau + n\pi$, $n = 0, 1, \cdots$. Figure 114 shows this function; the graph for $0 < t < \pi$ is obtained from (9) with $n = 0$ and $t = \tau$, and for $\pi < t < 2\pi$ it is obtained from (9) with $n = 1$ and $t = \tau + \pi$.

We have the surprising result that the output y_2 tends to oscillate more rapidly than the input $r(t)$. The reason for this unexpected behavior of our system will become obvious from our later consideration of Fourier series in Sec. 10.7. ∎

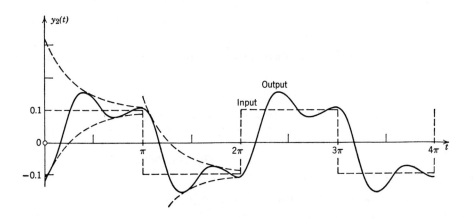

Fig. 114. Steady-state output (9) in Example 6

This is the end of Chap. 5 (except for the tables in Secs. 5.9 and 5.10) as well as of Part A on ordinary differential equations. **Partial differential equations** will be taken up, along with Fourier series, in Chaps. 10 and 11, and **numerical methods** for both ordinary and partial differential equations in Chap. 20.

Problems for Sec. 5.8

Laplace transforms of periodic functions. Graph the following functions, which are assumed to have the period 2π, and find their transforms.

1. $f(t) = t \quad (0 < t < 2\pi)$ 2. $f(t) = 2\pi - t \quad (0 < t < 2\pi)$
3. $f(t) = e^t \quad (0 < t < 2\pi)$ 4. $f(t) = K \sin(t/2) \quad (0 < t < 2\pi)$
5. $f(t) = t^2 \quad (0 < t < 2\pi)$ 6. $f(t) = 4\pi^2 - t^2 \quad (0 < t < 2\pi)$
7. $f(t) = -1$ if $0 < t < \pi$, $f(t) = 1$ if $\pi < t < 2\pi$
8. $f(t) = t$ if $0 < t < \pi$, $f(t) = \pi - t$ if $\pi < t < 2\pi$
9. $f(t) = 0$ if $0 < t < \pi$, $f(t) = t - \pi$ if $\pi < t < 2\pi$
10. $f(t) = t$ if $0 < t < \pi$, $f(t) = 0$ if $\pi < t < 2\pi$

11. Apply Theorem 1 to the function $f(t) = 1$, which is periodic with any period p.
12. How can the answer to Prob. 8 be obtained from the answers to Probs. 9 and 10?
13. Solve Prob. 9 by applying the second shifting theorem (Sec. 5.3) to Prob. 10.

Half-wave and full-wave rectifiers
14. Find the Laplace transform of the half-wave rectification of $-\sin \omega t$ (Fig. 115).

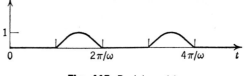

Fig. 115. Problem 14

15. Find the Laplace transform of the full-wave rectification of $\sin \omega t$ (Fig. 116).

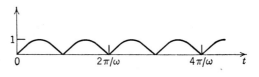

Fig. 116. Problem 15

16. Find the Laplace transform of $|\cos \omega t|$.
17. Solve Prob. 14 by applying Theorem 2, Sec. 5.3, to the result of Example 3.
18. Check the answer to Prob. 15 by using the results of Example 3 and Prob. 14.

Models of mechanical systems
19. **(Automatic pressure control)** Figure 117 shows a system for automatic control of pressure. $y(t)$ is the displacement, where $y = 0$ corresponds to the equilibrium position due to a given constant pressure. We make the following assumptions. The damping of the system is proportional to the velocity y'. For $t < 0$, the system is at rest. At $t = 0$ the pressure is suddenly increased in the form of a unit step function. Show that the corresponding differential equation is

$$my'' + cy' + ky = Pu(t)$$

(m = effective mass of the moving parts, c = damping constant, k = spring

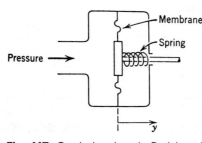

Fig. 117. Control system in Problem 19

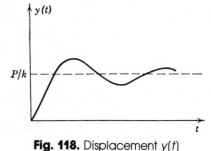
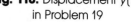

Fig. 118. Displacement $y(t)$
in Problem 19

modulus, P = force due to the increase of pressure at $t = 0$). Using the Laplace transformation, show that if $c^2 < 4mk$, then (Fig. 118)

$$y(t) = \frac{P}{k} [1 - e^{-\alpha t} \sqrt{1 + (\alpha/\omega^*)^2} \cos (\omega^* t + \theta)],$$

where $\alpha = c/2m$, $\omega^* = \sqrt{(k/m) - \alpha^2}$ (> 0), $\tan \theta = -\alpha/\omega^*$.

20. Solve Prob. 19 when $c^2 > 4mk$.

21. Solve Prob. 19 when $c^2 = 4mk$.

22. (**Flywheels**) Two flywheels (moments of inertia M_1 and M_2) are connected by an elastic shaft (moment of inertia negligible), and are rotating with constant angular velocity ω. At $t = 0$ a constant retarding couple P is applied to the first wheel. Find the subsequent angular velocity $v(t)$ of the other wheel.

23. Same conditions as in Prob. 22, but the retarding couple is applied during the interval $0 < t < 1$ only. Find $v(t)$.

Models of electric circuits

24. Find the ramp-wave response of the *RC*-circuit in Fig. 119 (the current $i(t)$ in the circuit), assuming that the circuit is quiescent at $t = 0$.

25. Solve Prob. 24 without the use of the Laplace transformation.

26. In Prob. 24, explain the reason for the jump of $i(t)$ at $t = 1$.

27. Find the current $i(t)$ in the *RC*-circuit in Fig. 119 when $v(t) = \sin \omega t$ ($0 < t < \pi/\omega$), $v(t) = 0$ ($t > \pi/\omega$), and $i(0) = 0$.

28. A capacitor ($C = 1$ farad) is charged to the potential $V_0 = 100$ volts and discharged starting at $t = 0$ by closing the switch in Fig. 120 on the next page. Find the current in the circuit and the charge on the capacitor.

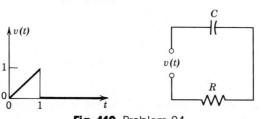

Fig. 119. Problem 24

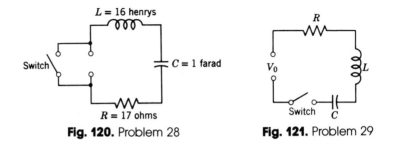

Fig. 120. Problem 28   **Fig. 121.** Problem 29

29. Using the Laplace transformation, show that the current $i(t)$ in the *RLC*-circuit in Fig. 121 (constant electromotive force V_0, zero initial current and charge) is

$$i(t) = \begin{cases} (K/\omega^*)e^{-\alpha t}\sin\omega^* t & \text{if } \omega^{*2} > 0 \\ Kte^{-\alpha t} & \text{if } \omega^{*2} = 0 \\ (K/\beta)e^{-\alpha t}\sinh\beta t & \text{if } \omega^{*2} = -\beta^2 < 0; \end{cases}$$

here $K = V_0/L$, $\alpha = R/2L$, $\omega^{*2} = (1/LC) - \alpha^2$.

30. Find the current in the circuit in Prob. 29, assuming that the electromotive force applied at $t = 0$ is $V_0\sin pt$, and the current and charge at $t = 0$ are zero.

31. Find the current in the *RLC*-circuit in Fig. 121 if a battery of electromotive force V_0 is connected to the circuit at $t = 0$ and short-circuited at $t = a$. Assume that the initial current and charge are zero, and ω^{*2}, as defined in Prob. 29, is positive.

32. Find the steady-state current in the circuit in Fig. 122.

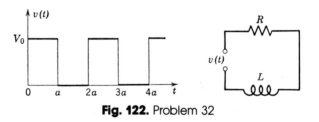

Fig. 122. Problem 32

33. Find the current $i(t)$ in the circuit in Fig. 123, assuming that $i(0) = 0$.

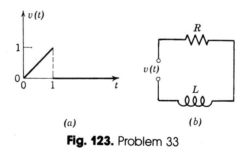

(a)   *(b)*

Fig. 123. Problem 33

34. Find the steady-state current in the circuit in Fig. 123*b*, if $v(t) = t$ when $0 < t < 1$ and $v(t + 1) = v(t)$ as shown in Fig. 124.

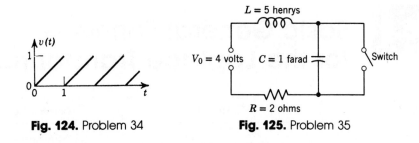

Fig. 124. Problem 34   **Fig. 125.** Problem 35

35. Steady current is flowing in the circuit in Fig. 125 with the switch closed. At $t = 0$ the switch is opened. Find the current $i(t)$.

36. The circuits in Fig. 126 are coupled by mutual inductance M, and at $t = 0$ the currents $i_1(t)$ and $i_2(t)$ are zero. Applying Kirchhoff's voltage law (Sec. 1.8), show that the differential equations for the currents are

$$L_1 i_1' + R_1 i_1 = M i_2' + V_0 u(t)$$

$$L_2 i_2' + R_2 i_2 = M i_1'$$

where $u(t)$ is the unit step function. Setting $\mathcal{L}(i_1) = I_1(s)$ and $\mathcal{L}(i_2) = I_2(s)$, show that the subsidiary equations are

$$L_1 s I_1 + R_1 I_1 = M s I_2 + \frac{V_0}{s}$$

$$L_2 s I_2 + R_2 I_2 = M s I_1.$$

Assuming that $A \equiv L_1 L_2 - M^2 > 0$, show that the expression for I_2, obtained by solving these algebraic equations, may be written

$$I_2 = \frac{K}{(s + \alpha)^2 + \omega^{*2}}$$

where $K = A^{-1} V_0 M$, $\alpha = (2A)^{-1}(R_1 L_2 + R_2 L_1)$, $\omega^{*2} = A^{-1} R_1 R_2 - \alpha^2$. Show that $i_2(t) = \mathcal{L}^{-1}(I_2)$ is of the same form as $i(t)$ in Prob. 29, where the constants K, α, and ω^* are now those defined in the present problem.

37. Show that if $A = 0$ in Prob. 36, then $i_2(t) = K_0 e^{-at}$ where $K_0 = BV_0 M$, $a = BR_1 R_2$, $B = (R_1 L_2 + R_2 L_1)^{-1}$.

38. Suppose that in the first circuit in Fig. 126 the switch is closed and a steady current V_0/R_1 is flowing in it. At $t = 0$ the switch is opened. Find the secondary current $i_2(t)$. *Hint.* Note that $i_1(0) = V_0/R_1$ and $i_1 = 0$ when $t > 0$.

39. Find $i_1(t)$ in Prob. 38, assuming that $\omega^{*2} > 0$.

40. Show that the solution y_2 in Example 6 is continuous at 0 and π.

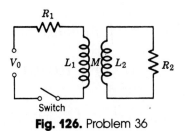

Fig. 126. Problem 36

Basic General Formulas for the Laplace Transformation

Formula	Name, Comments	Sec.
$F(s) = \mathcal{L}\{f(t)\} = \displaystyle\int_0^\infty e^{-st} f(t)\, dt$ $f(t) = \mathcal{L}^{-1}\{F(s)\}$	Definition of Transform Inverse Transform	5.1
$\mathcal{L}\{af(t) + bg(t)\} = a\mathcal{L}\{f(t)\} + b\mathcal{L}\{g(t)\}$	Linearity	5.1
$\mathcal{L}(f') = s\mathcal{L}(f) - f(0)$ $\mathcal{L}(f'') = s^2\mathcal{L}(f) - sf(0) - f'(0)$ $\mathcal{L}(f^{(n)}) = s^n\mathcal{L}(f) - s^{n-1}f(0) - \cdots$ $\cdots - f^{(n-1)}(0)$ $\mathcal{L}\left\{\displaystyle\int_0^t f(\tau)\, d\tau\right\} = \dfrac{1}{s}\mathcal{L}(f)$	Differentiation of Function Integration of Function	5.2
$\mathcal{L}\{e^{at}f(t)\} = F(s - a)$ $\mathcal{L}^{-1}\{F(s - a)\} = e^{at}f(t)$	s-Shifting (1st Shifting Theorem)	5.3
$\mathcal{L}\{f(t - a)u(t - a)\} = e^{-as}F(s)$ $\mathcal{L}^{-1}\{e^{-as}F(s)\} = f(t - a)u(t - a)$	t-Shifting (2nd Shifting Theorem)	5.3
$\mathcal{L}\{tf(t)\} = -F'(s)$ $\mathcal{L}\left\{\dfrac{f(t)}{t}\right\} = \displaystyle\int_s^\infty F(\tilde{s})\, d\tilde{s}$	Differentiation of Transform Integration of Transform	5.5
$(f * g)(t) = \displaystyle\int_0^t f(\tau)g(t - \tau)\, d\tau$ $= \displaystyle\int_0^t f(t - \tau)g(\tau)\, d\tau$ $\mathcal{L}(f * g) = \mathcal{L}(f)\mathcal{L}(g)$	Convolution	5.6
$\mathcal{L}(f) = \dfrac{1}{1 - e^{-ps}}\displaystyle\int_0^p e^{-st}f(t)\, dt$	f Periodic with Period p	5.8

5.10 Table of Laplace Transforms

For more extensive tables, see Refs. [A7] and [A13] in Appendix 1.

	$F(s) = \mathcal{L}\{f(t)\}$	$f(t)$	Cf. Sec.
1	$1/s$	1	
2	$1/s^2$	t	
3	$1/s^n, \quad (n = 1, 2, \cdots)$	$t^{n-1}/(n-1)!$	5.1
4	$1/\sqrt{s}$	$1/\sqrt{\pi t}$	
5	$1/s^{3/2}$	$2\sqrt{t/\pi}$	
6	$1/s^a \quad (a > 0)$	$t^{a-1}/\Gamma(a)$	
7	$\dfrac{1}{s-a}$	e^{at}	5.1
8	$\dfrac{1}{(s-a)^2}$	te^{at}	
9	$\dfrac{1}{(s-a)^n} \quad (n = 1, 2, \cdots)$	$\dfrac{1}{(n-1)!} t^{n-1}e^{at}$	5.3
10	$\dfrac{1}{(s-a)^k} \quad (k > 0)$	$\dfrac{1}{\Gamma(k)} t^{k-1}e^{at}$	
11	$\dfrac{1}{(s-a)(s-b)} \quad (a \neq b)$	$\dfrac{1}{(a-b)}(e^{at} - e^{bt})$	5.1
12	$\dfrac{s}{(s-a)(s-b)} \quad (a \neq b)$	$\dfrac{1}{(a-b)}(ae^{at} - be^{bt})$	
13	$\dfrac{1}{s^2 + \omega^2}$	$\dfrac{1}{\omega}\sin \omega t$	
14	$\dfrac{s}{s^2 + \omega^2}$	$\cos \omega t$	5.1
15	$\dfrac{1}{s^2 - a^2}$	$\dfrac{1}{a}\sinh at$	
16	$\dfrac{s}{s^2 - a^2}$	$\cosh at$	
17	$\dfrac{1}{(s-a)^2 + \omega^2}$	$\dfrac{1}{\omega}e^{at}\sin \omega t$	5.3
18	$\dfrac{s-a}{(s-a)^2 + \omega^2}$	$e^{at}\cos \omega t$	
19	$\dfrac{1}{s(s^2 + \omega^2)}$	$\dfrac{1}{\omega^2}(1 - \cos \omega t)$	5.2
20	$\dfrac{1}{s^2(s^2 + \omega^2)}$	$\dfrac{1}{\omega^3}(\omega t - \sin \omega t)$	
21	$\dfrac{1}{(s^2 + \omega^2)^2}$	$\dfrac{1}{2\omega^3}(\sin \omega t - \omega t \cos \omega t)$	
22	$\dfrac{s}{(s^2 + \omega^2)^2}$	$\dfrac{t}{2\omega}\sin \omega t$	5.5
23	$\dfrac{s^2}{(s^2 + \omega^2)^2}$	$\dfrac{1}{2\omega}(\sin \omega t + \omega t \cos \omega t)$	
24	$\dfrac{s}{(s^2 + a^2)(s^2 + b^2)} \quad (a^2 \neq b^2)$	$\dfrac{1}{b^2 - a^2}(\cos at - \cos bt)$	

Table of Laplace Transforms (*continued*)

	$F(s) = \mathscr{L}\{f(t)\}$	$f(t)$	Cf. Sec.
25	$\dfrac{1}{s^4 + 4k^4}$	$\dfrac{1}{4k^3}(\sin kt \cosh kt - \cos kt \sinh kt)$	
26	$\dfrac{s}{s^4 + 4k^4}$	$\dfrac{1}{2k^2}\sin kt \sinh kt$	5.7
27	$\dfrac{1}{s^4 - k^4}$	$\dfrac{1}{2k^3}(\sinh kt - \sin kt)$	
28	$\dfrac{s}{s^4 - k^4}$	$\dfrac{1}{2k^2}(\cosh kt - \cos kt)$	
29	$\sqrt{s-a} - \sqrt{s-b}$	$\dfrac{1}{2\sqrt{\pi t^3}}(e^{bt} - e^{at})$	
30	$\dfrac{1}{\sqrt{s+a}\,\sqrt{s+b}}$	$e^{-(a+b)t/2}I_0\left(\dfrac{a-b}{2}t\right)$	4.6
31	$\dfrac{1}{\sqrt{s^2 + a^2}}$	$J_0(at)$	4.5
32	$\dfrac{s}{(s-a)^{3/2}}$	$\dfrac{1}{\sqrt{\pi t}}e^{at}(1 + 2at)$	
33	$\dfrac{1}{(s^2 - a^2)^k}\quad (k > 0)$	$\dfrac{\sqrt{\pi}}{\Gamma(k)}\left(\dfrac{t}{2a}\right)^{k-1/2}I_{k-1/2}(at)$	4.6
34	$e^{-as/s}$	$u(t - a)$	5.3
35	e^{-as}	$\delta(t - a)$	5.4
36	$\dfrac{1}{s}e^{-k/s}$	$J_0(2\sqrt{kt})$	4.5
37	$\dfrac{1}{\sqrt{s}}e^{-k/s}$	$\dfrac{1}{\sqrt{\pi t}}\cos 2\sqrt{kt}$	
38	$\dfrac{1}{s^{3/2}}e^{k/s}$	$\dfrac{1}{\sqrt{\pi k}}\sinh 2\sqrt{kt}$	
39	$e^{-k\sqrt{s}}\quad (k > 0)$	$\dfrac{k}{2\sqrt{\pi t^3}}e^{-k^2/4t}$	
40	$\dfrac{1}{s}\ln s$	$-\ln t - \gamma \quad (\gamma \approx 0.5772)$	4.6
41	$\ln\dfrac{s-a}{s-b}$	$\dfrac{1}{t}(e^{bt} - e^{at})$	
42	$\ln\dfrac{s^2 + \omega^2}{s^2}$	$\dfrac{2}{t}(1 - \cos \omega t)$	5.5
43	$\ln\dfrac{s^2 - a^2}{s^2}$	$\dfrac{2}{t}(1 - \cosh at)$	
44	$\arctan\dfrac{\omega}{s}$	$\dfrac{1}{t}\sin \omega t$	
45	$\dfrac{1}{s}\operatorname{arc cot} s$	$\mathrm{Si}(t)$	Appendix 3

Review Problems for Chap. 5

1. What do we mean by saying that the Laplace transformation is a *linear* operation? Why is this practically important?

2. Does every continuous function have a Laplace transform? (Give reason or counterexample.)

3. Can a discontinuous function have a Laplace transform? (Give reason for your answer.)

4. Why is it of practical importance that two continuous functions that are different also have different Laplace transforms?

5. Is $\mathcal{L}\{f(t)g(t)\} = \mathcal{L}\{f(t)\}\mathcal{L}\{g(t)\}$? Or what?

6. Is $\displaystyle\int_0^t f(\tau)g(t - \tau)\,d\tau = \int_0^t f(t - \tau)g(\tau)\,d\tau$? (Give reason for your answer.)

7. If you know $f(t) = \mathcal{L}^{-1}\{F(s)\}$, how would you find $\mathcal{L}^{-1}\{F(s)/s^2\}$?

8. Obtain the Laplace transform of t^n from Theorem 2 in Sec. 5.2.

9. What is the difference in the shifting by the first shifting theorem and by the second shifting theorem?

10. Show that $e^{-as}\mathcal{L}\{f(t + a)\} = \mathcal{L}\{f(t)u(t - a)\}$. (Use a shifting theorem.)

Find the Laplace transform of the given function.

11. $e^{-t} \cos 4t$ 12. $e^{-2t} \sin \pi t$ 13. $tu(t - 3)$

14. $t * e^{-t}$ 15. $\cosh^2 t$ 16. $t \sin 4\pi t$

17. $t^{-1} \sin t$ 18. $t^3 u(t - 1)$ 19. $\sin t + t \cos t$

20. $K \cos \omega t$ if $0 < t < 2\pi/\omega$ and 0 otherwise

Find the inverse Laplace transform of the given function.

21. $\dfrac{1}{s^2 + s - 12}$ 22. $\dfrac{s + 2}{s^2 + 2}$ 23. $\dfrac{1}{s^3 - 4s}$

24. $\dfrac{s^2 + 2}{s^3 + 4s}$ 25. $\dfrac{8}{s^4 - 2s^3}$ 26. $\dfrac{e^{-\pi s}}{s^2 + 1}$

27. $\dfrac{s^2 - 6s + 4}{s^3 - 3s^2 + 2s}$ 28. $\dfrac{3s^2 - 2s - 1}{(s - 3)(s^2 + 1)}$ 29. $\dfrac{e^{-4s}}{s(s^2 + 3)}$

30. $\dfrac{s^4 - 7s^3 + 13s^2 + 4s - 12}{s^2(s - 3)(s^2 - 3s + 2)}$

Using the Laplace transformation, find $y(t)$ satisfying the given equation and conditions.

31. $y'' + 9y = 1.8u(t - 3)$, $y(0) = 0$, $y'(0) = 0$

32. $y'' + y = \delta(t - 2)$, $y(0) = 2.5$, $y'(0) = 0$

33. $y'' + 3y' + 2y = e^{-3t}$, $y(0) = 0$, $y'(0) = 0$

34. $y'' + 4y' + 4y = 4 \cos t + 3 \sin t$, $y(0) = 1$, $y'(0) = 0$

35. $y'' + y' - 2y = 7 + t - t^2$, $y(0) = 0$, $y'(0) = 0$

36. $y'' + 4y = \delta(t - \pi) - \delta(t - 2\pi)$, $y(0) = 0$, $y'(0) = 2$

37. $y'' - 2y' + 2y = 2e^t \cos t$, $y(0) = 1.5$, $y'(0) = 1.5$

38. $y'' - 3y' + 2y = 4t + e^{3t}$, $y(0) = 1$, $y'(0) = -1$

39. $y'' - 4y' + 4y = (3t^2 + 2)e^t$, $y(0) = 20$, $y'(0) = 34$

40. $y'' + 4y' + 3y = r(t)$, $r(t) = t - 1$ if $1 < t < 2$ and 0 otherwise,
$y(0) = 2$, $y'(0) = -2$

41. $y(t) = \sin t + \int_0^t y(\tau) \sin(t - \tau)\, d\tau$

42. $y(t) = t^2 + \frac{1}{60}t^6 - \int_0^t y(\tau)(t - \tau)^3\, d\tau$

43. $y(t) = \cosh 3t - 3e^{3t} \int_0^t y(\tau)e^{-3\tau}\, d\tau$

Models of Circuits, Networks, Mass–Spring Systems. Solve the following problems by the Laplace transformation.

44. Find the current $i(t)$ in the *RC*-circuit in Fig. 127, where $R = 10$ ohms, $C = 0.1$ farad, $e(t) = 10t$ volts if $0 < t < 4$, $e(t) = 40$ volts if $t > 4$, and the initial charge on the capacitor is 0.

45. Find the current $i(t)$ in the *RLC*-circuit in Fig. 128, where $R = 160$ ohms, $L = 20$ henrys, $C = 0.002$ farad, $e(t) = 37 \sin 10t$ volts, assuming zero initial current and charge.

46. Find the charge $q(t)$ and the current $i(t)$ in the *LC*-circuit in Fig. 129, assuming $L = 1$ henry, $C = 1$ farad, $e(t) = 1 - e^{-t}$ if $0 < t < \pi$, $e(t) = 0$ if $t > \pi$, and zero initial current and charge.

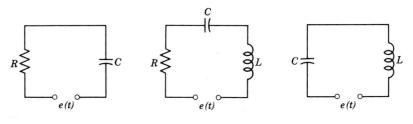

Fig. 127. *RC*-circuit   **Fig. 128.** *RLC*-circuit   **Fig. 129.** *LC*-circuit

47. Show that the model of the system in Fig. 130 (no friction, no damping) is

$$m_1 y_1'' = -k_1 y_1 + k_2(y_2 - y_1)$$
$$m_2 y_2'' = -k_2(y_2 - y_1) - k_3 y_2.$$

48. In Prob. 47, let $m_1 = m_2 = 10$ kg, $k_1 = k_3 = 20$ kg/sec², $k_2 = 40$ kg/sec². Find the solution satisfying the initial conditions $y_1(0) = y_2(0) = 0$, $y_1'(0) = 1$ meter/sec, $y_2'(0) = -1$ meter/sec.

49. Solve Prob. 48, assuming that $y_1(0) = y_2(0) = 1$ meter, $y_1'(0) = y_2'(0) = 0$, all the other data being as before. Compare the solution (frequency, type of motion) with that in Prob. 48.

50. Solve Prob. 48 by classical methods (cf. Sec. 3.1).

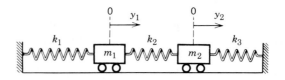

Fig. 130. System in Problems 47–50

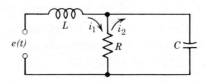

Fig. 131. Network in Problems 51–54

51. Show that, by Kirchhoff's voltage law (Sec. 1.8), the currents in the network in Fig. 131 are obtained from the system

$$Li_1' + R(i_1 - i_2) = e(t)$$

$$R(i_2' - i_1') + \frac{1}{C}i_2 = 0$$

52. Solve the system in Prob. 51, where $R = 10$ ohms, $L = 20$ henrys, $C = 0.05$ farad, $e = 20$ volts, $i_1(0) = 0$, $i_2(0) = 2$ amperes.

53. Show that in Prob. 52, we have $i_1(t) \to 2$, $i_2(t) \to 0$ as $t \to \infty$. Can you conclude this directly from the network?

54. Solve the system in Prob. 51, where $R = 0.8$ ohm, $L = 1$ henry, $C = 0.25$ farad, $e(t) = \frac{4}{5}t + \frac{21}{25}$ volt, $i(0) = 1$ ampere, $i_2(0) = -3.8$ ampere.

55. Set up the model of the network in Fig. 132 and find the solution, assuming that all charges and currents are 0 when the switch is closed at $t = 0$. Find the limits of $i_1(t)$ and $i_2(t)$ as $t \to \infty$, (i) from the solution, (ii) directly from the given network.

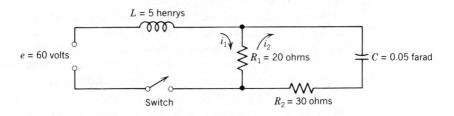

Fig. 132. Network in Problem 55

Summary of Chapter 5
Laplace Transformation

The main purpose of the Laplace transformation is the solution of differential equations and systems of such equations, as well as corresponding initial value problems. The **Laplace transform** $F(s) = \mathcal{L}(f)$ of a function $f(t)$ is defined by (Sec. 5.1)

(1)
$$F(s) = \mathcal{L}(f) = \int_0^\infty e^{-st} f(t)\, dt.$$

This definition is motivated by the property that the differentiation of f with respect to t corresponds to the multiplication of the transform F by s; precisely (Sec. 5.2):

(2)
$$\mathscr{L}(f') = s\mathscr{L}(f) - f(0),$$
$$\mathscr{L}(f'') = s^2\mathscr{L}(f) - sf(0) - f'(0),$$

etc. Hence by taking the transform of a given differential equation

(3)
$$y'' + ay' + by = r(t)$$

and writing $\mathscr{L}(y) = Y(s)$, we obtain the **subsidiary equation**

(4)
$$(s^2 + as + b)Y = \mathscr{L}(r) + sf(0) + f'(0) + af(0).$$

Here, in obtaining the transform $\mathscr{L}(r)$ we can get help from the small table in Sec. 5.1 or the big table in Sec. 5.10. In the second step we solve the subsidiary equation *algebraically* for $Y(s)$. In the third step we determine the inverse transform $y(t) = \mathscr{L}^{-1}(Y)$, that is, the solution of the problem. This is generally the hardest step, and in it we may again use one of those two tables. $Y(s)$ will often be a rational function, so that we can obtain the inverse $\mathscr{L}^{-1}(Y)$ by partial fraction reduction (Sec. 5.7) if we see no simpler way.

Given initial values can be substituted directly into (4) on the right, so that the determination of a general solution is avoided. Two more facts account for the practical importance of the Laplace transformation. First, it has some basic properties and resulting techniques that simplify the determination of transforms and inverses. The most important of these properties are listed in Sec. 5.9, together with references to the corresponding sections. More on the use of unit step functions and Dirac's delta can be found in Secs. 5.3 and 5.4, and on convolution in Sec. 5.6. Second, due to these properties, the present method is particularly suitable for handling right-hand sides $r(t)$ given by different expressions over different intervals of time, for instance, when $r(t)$ is a square wave or an impulse or of the form $r(t) = \cos t$ if $0 \leq t \leq 4\pi$ and 0 elsewhere.

For the application of the Laplace transformation to *partial* differential equations, see Sec. 11.13. Other important transformations (Fourier transformation, etc.) will be considered in Secs. 10.10 and 10.11.

PART B
LINEAR ALGEBRA, VECTOR CALCULUS

Two main factors have affected the development of engineering mathematics during the past two decades. These factors are the widespread application of computers to engineering problems and the increasing use of linear algebra and linear analysis, for instance, in handling large problems in systems analysis.

The first two chapters of the present part are devoted to **linear algebra,** consisting of the theory and application of vectors and vector spaces (Chap. 6) and of matrices (Chap. 7). This also includes the application of matrices in solving systems of differential equations (Sec. 7.16).

Numerical methods in linear algebra are presented in Chap. 19, which is independent of the other chapters in Part E on numerical methods. Thus it can be studied immediately after Chap. 7 if desired.

The last two chapters of the present part are devoted to **linear analysis,** usually called **vector calculus.** Chapter 8 concerns vector *differential* calculus (vector fields, curves, velocity, directional derivative, gradient, divergence, curl), and Chap. 9 covers vector *integral* calculus (line, surface and triple integrals and their transformation by the integral theorems of Green, Gauss and Stokes).

Chapter 6

Vectors

In the present chapter we shall consider basic concepts and methods of vector algebra and its applications to physical and geometrical problems. The usefulness of vectors in engineering mathematics results from the fact that many physical quantities—for example, forces and velocities—may be represented by vectors, and in several respects the rules of vector calculation are as simple as the rules governing the system of real numbers.

It is true that any problem that can be solved by the use of vectors can also be treated by nonvectorial methods, but vector analysis is a shorthand that simplifies many calculations considerably. Furthermore, it is a way of visualizing physical and geometrical quantities and relations between them. For all these reasons, vector methods are used extensively throughout modern applied mathematics.

> *Prerequisite for this chapter:* In the last two sections, determinants of second and third order will be needed.
> *Section that may be omitted in a shorter course:* Secs. 6.4, 6.6.
> *References:* Appendix 1, Part B.
> *Answers to problems:* Appendix 2.

6.1 Scalars and Vectors

Length, area, temperature, pressure and voltage are examples of quantities each of which can be specified by a single number (number of units on a suitable scale of measurement; e.g., 5.37 meters, 23.4 °C, 110 volts, etc.). We call them **scalars.**[1] This name is introduced to distinguish them from other physical or geometrical quantities that have a direction as well as a magnitude. We call such quantities **vectors** (precise definition below). Important examples of vectors are forces and velocities. A *force* has a magnitude (small or large, measured according to some scale) and a direction in which it acts. Figure 133 shows the force of attraction for the earth's motion around the sun. A *velocity* of a moving body has a magnitude (fast or slow, usually called *speed*) and points in the instantaneous direction of motion. Figure 133 shows the instantaneous velocity of the earth. Forces or velocities may be graphically represented by *arrows,* or **directed line segments.**

---

[1]Clearly, if a number characterizes a scalar that has a physical or geometrical meaning, this number should be independent of the choice of coordinates; since this point will not matter in the present chapter, we discuss it later (in Sec. 8.1).

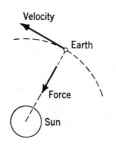

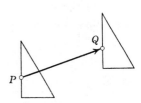

Fig. 133. Force and velocity   **Fig. 134.** Translation

As a third example, if we translate a triangle (displace it without rotating it), each point of it moves from its original position (P in Fig. 134) to its new position (Q in Fig. 134), and we can indicate this motion by an arrow $\overrightarrow{PQ}$. If we do this for each point of the triangle, we get directed line segments that have the same length and the same direction (that is, they are parallel and are directed in the same sense). We may say that each of these directed line segments "carries" a point of the triangle from its original position to its new position. The word "vector" comes from Latin and means "carrier."

This situation suggests the following

Definition of a vector

A directed line segment is called a **vector.** Its length is called the **length** of the vector and its direction is called the **direction** of the vector. Two vectors are equal if and only if they have the same length and the same direction.

The length of a vector is also called the **Euclidean norm** or **magnitude** of the vector.

The initial point of that directed segment is called the *initial point* of the vector, and the terminal point of the segment the *terminal point* of the vector.   ∎

Vectors will be denoted by lowercase boldfaced letters,[2] such as **a, b, v,** and the length of a vector **a** by $|\mathbf{a}|$.

Of course, if we want, we may combine the two parts of this definition and define a vector to be the collection of all directed line segments having a given length and a given direction.

From the definition we see that a vector may be arbitrarily translated (that is, displaced without rotation) or, what amounts to the same thing, its initial point may be chosen in an arbitrary fashion. Clearly, if we choose a certain point as the initial point of a given vector, the terminal point of the vector is uniquely determined.

If two vectors **a** and **b** are equal, we write

$$\mathbf{a} = \mathbf{b},$$

and if they are different, we may write

$$\mathbf{a} \neq \mathbf{b}.$$

---

[2]This is customary in printed work; in handwritten work one may characterize vectors by arrows, for example $\vec{a}$ (in place of **a**), $\vec{b}$, etc.

Equal vectors,
a = b

Vectors having
the same length
but different
direction

Vectors having
the same direction
but different
length

Vectors having
different length
and different
direction

Fig. 135. Vectors

Any vector may be represented graphically as an arrow of suitable length and direction, as is illustrated in Fig. 135. Then the tail is the initial point and the tip the terminal point of the vector.

A vector of length one is called a **unit vector.**

For the sake of completeness, we mention that in physics and geometry there are situations where we want to impose restrictions on the position of the initial point of a vector. For example, as is known from mechanics, a force acting on a rigid body may be equally well applied to any point of the body on its line of action. This suggests the concept of a **sliding vector,** that is, a vector whose initial point can be any point on a straight line that is parallel to the vector. A force acting on an elastic body is a vector whose initial point cannot be changed at all. In fact, if we choose another point of application of the force, the effect of the force will in general be different. This suggests the notion of a **bound vector,** that is, a vector having a certain fixed initial point (*point of application*). Since these concepts occur in the literature, the student should know about them, but they will not be of particular importance in our further work.

Our definition of a vector is "geometric." This has the advantage that it is immediately clear that this definition is independent of the choice of coordinates. In the next section we discuss vectors "algebraically," by defining their components with respect to a given coordinate system.

6.2 Components of a Vector

A point in three-dimensional space is a geometric object, but if we introduce a coordinate system, we may describe it by (or even regard it as) an ordered triple of numbers (called its coordinates). Similarly, in the previous section we defined a vector in a geometrical fashion (using the notion of a directed line segment), but if we use a coordinate system, we may describe vectors in algebraic terms.

In fact, let us introduce a coordinate system in space whose axes are three mutually perpendicular straight lines. On all three axes we choose the same scale. Then the three *unit points* on the axes, whose coordinates are $(1, 0, 0)$, $(0, 1, 0)$, and $(0, 0, 1)$, have the same distance from the *origin,* the point of intersection of the axes. The rectangular coordinate system thus obtained is called a **Cartesian coordinate system** in space (cf. Fig. 136).

We now consider a vector **a** obtained by directing a line segment PQ such that P is the initial point and Q is the terminal point (Fig. 137). Let (x_1, y_1, z_1) and (x_2, y_2, z_2) be the coordinates of P and Q, respectively. Then the numbers

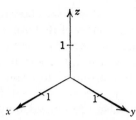

Fig. 136. Cartesian coordinate system

(1)
$$a_1 = x_2 - x_1, \qquad a_2 = y_2 - y_1, \qquad a_3 = z_2 - z_1$$

are called the **components** of the vector **a** with respect to that coordinate system, and we write simply

$$\mathbf{a} = [a_1, a_2, a_3].$$

Similarly, for a vector **v** with components v_1, v_2, v_3 we may write $\mathbf{v} = [v_1, v_2, v_3]$, and so on.

By definition, the length $|\mathbf{a}|$ of the vector $\mathbf{a} = [a_1, a_2, a_3]$ is the distance $\overline{PQ}$, and from (1) and the theorem of Pythagoras it follows that

(2)
$$|\mathbf{a}| = \sqrt{a_1{}^2 + a_2{}^2 + a_3{}^2}.$$

EXAMPLE 1. Components and length of a vector

The vector **a** with initial point P: $(3, 1, 4)$ and terminal point Q: $(1, -2, 4)$ has the components

$$a_1 = 1 - 3 = -2, \qquad a_2 = -2 - 1 = -3, \qquad a_3 = 4 - 4 = 0.$$

Hence $\mathbf{a} = [-2, -3, 0]$, so that (2) gives the length

$$|\mathbf{a}| = \sqrt{(-2)^2 + (-3)^2 + 0^2} = \sqrt{13}.$$

If we choose $(-1, 5, 8)$ as the initial point of **a**, the corresponding terminal point is $(-3, 2, 8)$. If we choose the origin $(0, 0, 0)$ as the initial point of **a**, the corresponding terminal point is $(-2, -3, 0)$; its coordinates equal the components of **a**.   ∎

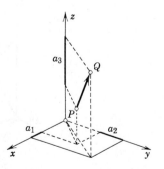

Fig. 137. Components of a vector

From (1) we see that if we choose the origin as the initial point of a given vector, the coordinates of its terminal point are equal to the components of the vector, and the vector is then called the **position vector** of the terminal point (with respect to our coordinate system) and is usually denoted by **r**. Cf. Fig. 138. For instance, in Example 1, the point A: $(-2, -3, 0)$ has the position vector $\mathbf{r} = \mathbf{a} = [-2, -3, 0]$.

From (1) we also see immediately that *the components a_1, a_2, a_3 of the vector $\mathbf{a}$ are independent of the choice of the initial point of $\mathbf{a}$*, because if we translate **a**, then corresponding coordinates of P and Q are altered by the same amount. *Hence, a fixed Cartesian coordinate system being given, each vector is uniquely determined by the ordered triple of its components with respect to that coordinate system.*

We now introduce the *null vector or* **zero vector 0,** defined as the vector with components 0, 0, 0. Then any ordered triple of real numbers, including the triple 0, 0, 0, can be chosen as the components of a vector.

The coordinate system being fixed, the correspondence between the ordered triples of real numbers and the vectors in space is one-to-one, that is, to each such triple there corresponds a vector in space that has those three numbers as its components with respect to that coordinate system, and conversely.

It follows that two vectors $\mathbf{a} = [a_1, a_2, a_3]$ *and* $\mathbf{b} = [b_1, b_2, b_3]$ *are equal* if and only if corresponding components of these vectors are equal.

Consequently, *a vector equation*

$$\mathbf{a} = \mathbf{b}$$

is equivalent to the three equations for the components of **a** *and* **b** *with respect to any fixed Cartesian coordinate system, that is,*

$$a_1 = b_1, \qquad a_2 = b_2, \qquad a_3 = b_3.$$

There is yet another consequence of that one-to-one correspondence between all ordered triples of real numbers and all vectors as defined in Sec. 6.1 in a geometric fashion. That correspondence implies that a vector could equally well be *defined* as an ordered triple of real numbers (called the *components* of the vector). Starting from this definition, we would then

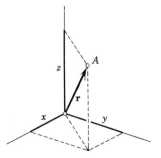

Fig. 138. Position vector **r** of a point A: (x, y, z)

obtain the geometrical interpretation of a vector that we used as a definition in Sec. 6.1.

Clearly, if a vector is a physical or geometric object, its length and direction should be independent of the particular choice of a coordinate system, and this requirement implies a certain transformation behavior of the components under the transition to another coordinate system. We shall discuss this problem later (in the proofs after Sec. 8.11).

Later (in Sec. 6.4) we shall also generalize the concept of a vector to include ordered n-tuples of numbers, where n is any fixed positive integer.

In the next section we define an **addition** of vectors and a **multiplication by scalars** (that is, real numbers), which are easy to perform since they have properties similar to the addition and multiplication of real numbers.

Problems for Sec. 6.2

Find the components of the vector **v** with given initial point $P: (x_1, y_1, z_1)$ and terminal point $Q: (x_2, y_2, z_2)$. Find $|\mathbf{v}|$. Graph **v**.

1. $P: (1, 0, 2)$, $Q: (3, 2, 1)$
2. $P: (0, 0, 0)$, $Q: (\frac{1}{2}, -4, 0)$
3. $P: (12, -2, 3)$, $Q: (-2, 3, 12)$
4. $P: (a, b, c)$, $Q: (0, 0, 0)$
5. $P: (0, 2, 0)$, $Q: (3, 0, -7)$
6. $P: (\frac{1}{4}, \frac{1}{2}, -3)$, $Q: (0, \frac{3}{2}, 5)$
7. $P: (4, -1, 2)$, $Q: (0, 0, 0)$
8. $P: (0.1, 0.3, 0.2)$, $Q: (1.3, 0, 2.8)$
9. $P: (4, -1, \frac{1}{2})$, $Q: (-4, 1, -\frac{1}{2})$
10. $P: (-4, -1, -6)$, $Q: (4, 8, 12)$

Find the terminal point of the given vector $[v_1, v_2, v_3]$ corresponding to the given initial point P. Find the length of the vector.

11. $[2, 3, 0]$, $P: (4, 9, 0)$
12. $[8, 0, -\frac{1}{2}]$, $P: (\frac{1}{2}, 0, \frac{1}{2})$
13. $[0.2, 0.4, 2.0]$, $P: (1.4, 2.3, 0.1)$
14. $[-3, 0, -8]$, $P: (2, 1, -1)$
15. $[12, -15, -4]$, $P: (0, 0, 3)$
16. $[0, 0, 0]$, $P: (3, -8, 4)$
17. $[4, 4, 4]$, $P: (4, 4, 4)$
18. $[0, a, b]$, $P: (1, 0, 2)$
19. $[5, 1, \frac{1}{2}]$, $P: (-2, 1, \frac{1}{2})$
20. $[\frac{3}{4}, 2, -\frac{1}{2}]$, $P: (0, 0, 0)$

Find the initial point of the given vector corresponding to the given terminal point Q. Find the length of the vector.

21. $[\frac{1}{2}, \frac{1}{3}, \frac{1}{4}]$, $Q: (-1, 0, \frac{1}{4})$
22. $[0, 1, 2]$, $Q: (4, 5, 7)$
23. $[13, 1, 0]$, $Q: (13, 1, 0)$
24. $[2, 2, 2]$, $Q: (0, -2, 4)$
25. $[4, 0, -2]$, $Q: (0, 0, 0)$
26. $[\frac{1}{2}, -1, \frac{1}{2}]$, $Q: (0, -1, -\frac{1}{2})$
27. $[3, 6, 7]$, $Q: (-3, -6, -7)$
28. $[a, b, c]$, $Q: (-a, -b, -c)$
29. $[3, -5, 1]$, $Q: (2, 4, -6)$
30. $[0, 4, \frac{1}{4}]$, $Q: (4, \frac{1}{4}, 0)$

From the length and direction of a vector **v** in the plane we can determine the components of **v**. For instance, if $|\mathbf{v}| = 3$ and the angle with the positive x-axis is $\alpha = 30°$, then $v_1 = 3 \cos 30° = 2.60$ and $v_2 = 3 \sin 30° = 1.50$. Sketch **v** and find its components, if

31. $|\mathbf{v}| = 1.6$, $\alpha = 60°$
32. $|\mathbf{v}| = 4$, $\alpha = \frac{3}{4}\pi$
33. $|\mathbf{v}| = \sqrt{6}$, $\alpha = -\frac{1}{4}\pi$
34. $|\mathbf{v}| = 10$, $\alpha = 10°$
35. $|\mathbf{v}| = 8.3$, $\alpha = 165°$

6.3 | Addition of Vectors, Multiplication by Scalars

Since we want to do calculations with vectors, we shall now introduce two algebraic operations, called addition of vectors and multiplication of vectors by scalars. Vector addition is motivated by the familiar parallelogram law for getting the resultant of forces (cf. Fig. 139). This suggests the following

Definition of vector addition
Given two vectors **a** and **b**, put the initial point of **b** at the terminal point of **a**; then the **sum** of **a** and **b** is defined as the vector **c** drawn from the initial point of **a** to the terminal point of **b** (Fig. 140), and we write

$$\mathbf{c} = \mathbf{a} + \mathbf{b}. \qquad\blacksquare$$

From this definition we conclude the following. If with respect to any fixed Cartesian coordinate system, $\mathbf{a} = [a_1, a_2, a_3]$ and $\mathbf{b} = [b_1, b_2, b_3]$, then

(1)
$$\mathbf{a} + \mathbf{b} = [a_1 + b_1, \quad a_2 + b_2, \quad a_3 + b_3];$$

that is, *the components of a sum of vectors are obtained by adding corresponding components of these vectors.* Figure 141 shows this for the plane, and in space the situation is similar.

From the definition or from (1) we see that vector addition has the following properties. For any vectors (cf. Figs. 142, 143)

(2)

(a)	$\mathbf{a} + \mathbf{b} = \mathbf{b} + \mathbf{a}$	(*commutativity*)
(b)	$(\mathbf{u} + \mathbf{v}) + \mathbf{w} = \mathbf{u} + (\mathbf{v} + \mathbf{w})$	(*associativity*)
(c)	$\mathbf{a} + \mathbf{0} = \mathbf{0} + \mathbf{a} = \mathbf{a}$	
(d)	$\mathbf{a} + (-\mathbf{a}) = \mathbf{0}$	

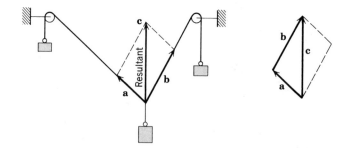

Fig. 139. Resultant of two forces (parallelogram law)

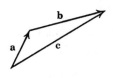
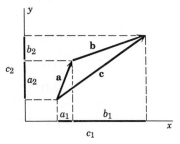

Fig. 140. Vector addition   **Fig. 141.** Vector addition in terms of components

where $-\mathbf{a}$ denotes the vector having the length $|\mathbf{a}|$ and the direction opposite to that of $\mathbf{a}$.

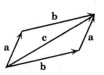
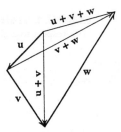

Fig. 142. Commutativity   **Fig. 143.** Associativity
of vector addition   of vector addition

In (2b) we may simply write $\mathbf{u} + \mathbf{v} + \mathbf{w}$, and similarly for sums of more than three vectors. Instead of $\mathbf{a} + \mathbf{a}$ we also write $2\mathbf{a}$, and so on. This (and the notation $-\mathbf{a}$ used before) suggests that we define the second algebraic operation for vectors, the multiplication of a vector by an arbitrary real number (called a **scalar**), as follows.

Multiplication of vectors by scalars (numbers)

Let $\mathbf{a}$ be any vector and c any real number. Then the vector $c\mathbf{a}$ is defined as follows.

> The length of $c\mathbf{a}$ is $|c||\mathbf{a}|$.
> If $\mathbf{a} \neq \mathbf{0}$ and $c > 0$, then $c\mathbf{a}$ has the direction of $\mathbf{a}$.
> If $\mathbf{a} \neq \mathbf{0}$ and $c < 0$, then $c\mathbf{a}$ has the direction opposite to $\mathbf{a}$.
> If $\mathbf{a} = \mathbf{0}$ or $c = 0$ (or both), then $c\mathbf{a} = \mathbf{0}$.   ∎

From this definition it follows that in terms of components, if $\mathbf{a} = [a_1, a_2, a_3]$, then

(3)
$$c\mathbf{a} = [ca_1, ca_2, ca_3].$$

Simple examples are shown in Fig. 144 on the next page. Furthermore, from the definition or from (3) we see that multiplication by scalars has the following basic properties.

(4)

(a)	$c(\mathbf{a} + \mathbf{b})$	$= c\mathbf{a} + c\mathbf{b}$	
(b)	$(c + k)\mathbf{a}$	$= c\mathbf{a} + k\mathbf{a}$	
(c)	$c(k\mathbf{a})$	$= (ck)\mathbf{a}$	(written $ck\mathbf{a}$)
(d)	$1\mathbf{a}$	$= \mathbf{a}.$	

The reader may prove that (2) and (4) imply that for any **a**,

(5)

(a) $0\mathbf{a} = \mathbf{0}$

(b) $(-1)\mathbf{a} = -\mathbf{a}.$

Instead of $\mathbf{b} + (-\mathbf{a})$ we simply write $\mathbf{b} - \mathbf{a}$ (Fig. 145).

EXAMPLE 1. Vector addition. Multiplication by scalars
With respect to a given coordinate system, let

$$\mathbf{a} = [4, 0, 1] \qquad \text{and} \qquad \mathbf{b} = [2, -5, \tfrac{1}{3}].$$

Then $-\mathbf{a} = [-4, 0, -1]$, $7\mathbf{a} = [28, 0, 7]$, $\mathbf{a} + \mathbf{b} = [6, -5, \tfrac{4}{3}]$ and

$$2(\mathbf{a} - \mathbf{b}) = 2[2, 5, \tfrac{2}{3}] = [4, 10, \tfrac{4}{3}] = 2\mathbf{a} - 2\mathbf{b}. \qquad ▮$$

Fig. 144. Multiplication of vectors by scalars (numbers)

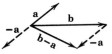

Fig. 145. Difference of vectors

The Unit Vectors i, j, k

With a given Cartesian coordinate system, we may associate three unit vectors **i, j, k** that have the positive directions of the coordinate axes, as shown in Fig. 146. Then we have the possibility of representing a given vector $\mathbf{a} = [a_1, a_2, a_3]$ as the sum of three vectors parallel to the coordinate axes (cf. Fig. 146),

(6)

$$\boxed{\mathbf{a} = [a_1, a_2, a_3] = a_1\mathbf{i} + a_2\mathbf{j} + a_3\mathbf{k}.}$$

For instance, $\mathbf{a} = [\tfrac{1}{2}, -3, 0] = \tfrac{1}{2}\mathbf{i} - 3\mathbf{j}$, $\mathbf{b} = [2, 0, 1] = 2\mathbf{i} + \mathbf{k}$, etc. This is convenient in certain applications, as we shall see.

Figure 146 shows **i, j, k** when the origin is chosen as their common initial point. Since the coordinate axes are perpendicular, so are these vectors. Instead of *perpendicular,* the term **orthogonal** is also used quite frequently, and we say that **i, j, k** form a *triple of orthogonal unit vectors*. This triple is also known as the *fundamental orthogonal triad* associated with that

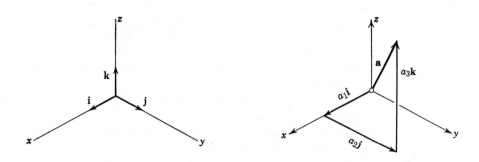

Fig. 146. The unit vectors **i, j, k** and the representation (6)

coordinate system. In terms of components,

(7) $\mathbf{i} = [1, 0, 0]$, $\mathbf{j} = [0, 1, 0]$, $\mathbf{k} = [0, 0, 1]$.

The next section can be *skipped* in a first reading, without destroying the continuity later on. In it we explain and illustrate the concepts of a **vector space,** which is defined by axioms suggested by basic properties of vector addition and multiplication by scalars discussed in the present section.

Problems for Sec. 6.3

Let $\mathbf{a} = [1, 2, -3]$, $\mathbf{b} = [2, 1, 4]$, $\mathbf{c} = [0, -5, 0]$. Find

1. $\mathbf{a} + \mathbf{b}, \mathbf{b} + \mathbf{a}$ **2.** $3\mathbf{a} - 3\mathbf{b}, 3(\mathbf{a} - \mathbf{b})$

3. $(\mathbf{a} + \mathbf{b}) + \mathbf{c}, \mathbf{a} + (\mathbf{b} + \mathbf{c})$ **4.** $\mathbf{a} + 2\mathbf{b} - 3\mathbf{c}$

5. $-4\mathbf{a} + 8\mathbf{c}, -4(\mathbf{a} - 2\mathbf{c})$ **6.** $|\mathbf{a} + \mathbf{b}|, |\mathbf{a}| + |\mathbf{b}|$

7. $|\mathbf{a} - \mathbf{c}|, |\mathbf{a}| - |\mathbf{c}|$ **8.** $(1/|\mathbf{a}|)\mathbf{a}, (1/|\mathbf{b}|)\mathbf{b}, (1/|\mathbf{c}|)\mathbf{c}$

9. $|\mathbf{a} - \mathbf{b} + 2\mathbf{c}|, |-\mathbf{a} + \mathbf{b} - 2\mathbf{c}|$ **10.** $|\mathbf{a} + \mathbf{b}| + |\mathbf{c}|, |\mathbf{a}| + |\mathbf{b} + \mathbf{c}|$

11. $|\mathbf{a}|(\mathbf{b} - \mathbf{c})$ **12.** $|\mathbf{a}||\mathbf{b}|, |\mathbf{a}|\mathbf{b}$

13. What laws do Probs. 1, 2, 3 illustrate?

Forces. In each case, find the resultant (its components) and its magnitude.
14. $\mathbf{p} = [1, 3, -1]$, $\mathbf{q} = [5, 0, -2]$, $\mathbf{u} = [0, -1, 3]$
15. $\mathbf{p} = [3, 4, 2]$, $\mathbf{q} = [-2, -5, -3]$, $\mathbf{u} = [0, 0, 1]$
16. $\mathbf{p} = [3, 2, 1]$, $\mathbf{q} = [4, -1, -1]$, $\mathbf{u} = [-7, -1, 0]$
17. $\mathbf{p} = [-1, 0, 2]$, $\mathbf{q} = [0, 4, 13]$, $\mathbf{u} = [1, 2, 3]$, $\mathbf{v} = [3, -8, 0]$
18. $\mathbf{p} = [-1, -1, 3]$, $\mathbf{q} = [4, 0, -6]$, $\mathbf{u} = [-3, 1, 3]$

19. Determine a force $\mathbf{p}$ such that $\mathbf{p}, \mathbf{q} = [1, 1, -2]$ and $\mathbf{u} = [-1, 2, 0]$ are in equilibrium.
20. Determine all forces $\mathbf{p}$ such that the resultant of $\mathbf{p}, \mathbf{q} = [2, -4, 3]$ and $\mathbf{u} = [-4, 2, -16]$ is parallel to the xy-plane.
21. Determine two forces $\mathbf{p}, \mathbf{q}$ in the direction of the coordinate axes in the plane such that $\mathbf{p}, \mathbf{q}, \mathbf{u} = [3, -4]$, $\mathbf{v} = [-1, 2]$ are in equilibrium.

22. Determine three forces **p**, **q**, **u** in the direction of the coordinate axes such that **p**, **q**, **u**, **v** = [3, −2, 1], **w** = [0, 0, −2] are in equilibrium. Are **p**, **q**, **u** uniquely determined?

23. If $|\mathbf{p}| = 2$ and $|\mathbf{q}| = 3$, what can be said about the length and direction of the resultant?

24. Using vectors, show that the midpoints of the sides of any quadrilateral form the vertices of a parallelogram.

25. Show that the sum of the vectors drawn from the center of a regular polygon to its vertices is the zero vector.

6.4 Vector Spaces

In the previous sections we introduced vectors of the form $\mathbf{a} = [a_1, a_2, a_3]$, $\mathbf{b} = [b_1, b_2, b_3]$ etc. geometrically as directed line segments and algebraically as ordered triples of numbers. We say that all these vectors form the **vector space** R^3, with the two algebraic operations of vector addition and multiplication by scalars defined by

$$
(1) \qquad
\begin{aligned}
\mathbf{a} + \mathbf{b} &= [a_1 + b_1, \quad a_2 + b_2, \quad a_3 + b_3], \\
c\mathbf{a} &= [ca_1, \quad ca_2, \quad ca_3].
\end{aligned}
$$

Physicists and mathematicians realized about a hundred years ago that there is no need to stop with triples but that it is practical to consider ordered quadruples, quintuples, etc. as vectors. This idea led to the following concept.

Vector space R^n

By definition, R^n consists of all ordered n-tuples of real numbers, called **vectors** and written

$$
\mathbf{a} = [a_1, a_2, \cdots, a_n], \qquad \mathbf{b} = [b_1, b_2, \cdots, b_n],
$$

with the two algebraic operations called *vector addition* and *multiplication of vectors by scalars* and defined by

$$
(2) \qquad
\begin{aligned}
\mathbf{a} + \mathbf{b} &= [a_1 + b_1, \quad a + b_2, \quad \cdots, \quad a_n + b_n], \\
c\mathbf{a} &= [ca_1, \quad ca_2, \quad \cdots, \quad ca_n].
\end{aligned}
$$

The numbers $a_1, \cdots, a_n$ are called the **components** of the vector **a**.

We see that (2) is suggested by (1), to which it reduces when $n = 3$.

EXAMPLE 1. Vector space R^4

The vector space R^4 consists of all ordered quadruples of real numbers, written $\mathbf{a} = [a_1, a_2, a_3, a_4]$, $\mathbf{b} = [b_1, b_2, b_3, b_4]$ and so on. By adding two such vectors we get as their sum another vector in R^4 that is uniquely determined; similarly for the multiplication of a vector in R^4 by a scalar (a real number). For instance,

$$[\tfrac{1}{2}, 3, 0, -6] + [0, -5, 4, \tfrac{1}{3}] = [\tfrac{1}{2}, -2, 4, -\tfrac{17}{3}],$$

$$\tfrac{8}{5}[-1, \tfrac{1}{4}, 10, 2] = [-\tfrac{8}{5}, \tfrac{2}{5}, 16, \tfrac{16}{5}].$$

Of course we cannot "see" and sketch vectors in R^4, R^5, $\cdots$ as we can in R^2 or R^3, but we can imagine various situations in which they are useful. Thus, in $\mathbf{a} = [a_1, a_2, a_3, a_4]$, the first three components may be the coordinates of a moving particle in space and a_4 may be the corresponding time. This idea of "space-time" is used in Einstein's theory of relativity. Or we may associate with a motion a vector $\mathbf{a} = [a_1, \cdots, a_6]$ in R^6, the first three components being the coordinates of the instantaneous position and the other three components the velocity components. Or we may use a vector in R^{23} whose 23 components give the daily production of 23 chemicals in some factory. Similarly for problems on storing, selling or shipping, or on the composition of alloys. Can you think of further examples? What vector space would be appropriate in a three-body problem (three planets, for instance) if you want to describe positions and velocities (disregarding rotations)?

General Vector Spaces

So far we have generalized the vector space concept from R^3 to R^n for any positive integer n. But it pays to go still further and, instead of n-tuples of numbers, to take a set of any elements and define for them an addition and a multiplication by scalars (real numbers) that have properties analogous to those of these operations in R^3 (and R^n). This leads to the following concept, which is important since it applies to many kinds of sets (sets of vectors, matrices, functions, operators, etc.) that play a role in differential equations, functional analysis (abstract modern analysis; cf. Ref. [13] listed in Appendix 1), numerical analysis and other fields of practical interest to the engineer. The reader should not be scared by the many conditions in this definition, since all of them are suggested by the vectors in R^3, and each one expresses just a simple and familiar property that one needs in working with vectors in R^3, so that one does not want to lose it in the present more general situation.

Definition of a real vector space

A nonempty set V of elements $\mathbf{a}$, $\mathbf{b}$, $\cdots$ is called a *real vector space* (or *real linear space*), and these elements are called **vectors**,[3] if in V there are defined two algebraic operations (called *vector addition* and *multiplication by scalars*) as follows.

I. Vector addition associates with every pair of vectors $\mathbf{a}$ and $\mathbf{b}$ of V a unique vector of V, called the *sum* of $\mathbf{a}$ and $\mathbf{b}$ and denoted by $\mathbf{a} + \mathbf{b}$, such that the following axioms are satisfied.

I.1 *Commutativity.* For any two vectors $\mathbf{a}$ and $\mathbf{b}$ of V,

$$\mathbf{a} + \mathbf{b} = \mathbf{b} + \mathbf{a}.$$

I.2 *Associativity.* For any three vectors $\mathbf{u}$, $\mathbf{v}$, $\mathbf{w}$ of V,

$$(\mathbf{u} + \mathbf{v}) + \mathbf{w} = \mathbf{u} + (\mathbf{v} + \mathbf{w}) \quad \text{(written } \mathbf{u} + \mathbf{v} + \mathbf{w}\text{).}$$

---

[3]Regardless of what they actually are, but this convention causes no confusion because in any specific case the nature of those elements is clear from the context.

I.3 There is a unique vector in V, called the *zero vector* and denoted by **0**, such that for every **a** in V,

$$\mathbf{a} + \mathbf{0} = \mathbf{a}.$$

I.4 For every **a** in V there is a unique vector in V that is denoted by $-\mathbf{a}$ and is such that

$$\mathbf{a} + (-\mathbf{a}) = \mathbf{0}.$$

II. Multiplication by scalars. The real numbers are called *scalars*. Multiplication by scalars associates with every **a** in V and every scalar c a unique vector of V, called the *product* of c and **a** and denoted by $c\mathbf{a}$ (or $\mathbf{a}c$) such that the following axioms are satisfied.

II.1 *Distributivity*. For every scalar c and vectors **a** and **b** in V,

$$c(\mathbf{a} + \mathbf{b}) = c\mathbf{a} + c\mathbf{b}.$$

II.2 *Distributivity*. For all scalars c and k and every **a** in V,

$$(c + k)\mathbf{a} = c\mathbf{a} + k\mathbf{a}.$$

II.3 *Associativity*. For all scalars c and k and every **a** in V,

$$c(k\mathbf{a}) = (ck)\mathbf{a} \qquad\qquad \text{(written } ck\mathbf{a}).$$

II.4 For every **a** in V,

$$1\mathbf{a} = \mathbf{a}. \qquad\qquad \blacksquare$$

A *complex vector space* is obtained if, instead of real numbers, we take complex numbers as scalars.

EXAMPLE 2. Vectors in R^3 and in R^n

The space R^3 is a vector space in the sense of the present definition. Indeed, the axioms I.1 to II.4 follow by direct calculation.

The set of all vectors in R^3 whose last component is 0 forms a vector space, as follows by direct verification of the axioms.

The set of all vectors in R^3 whose components are all nonnegative does not form a vector space, since this property is lost under the multiplication by a negative scalar. The set of all vectors in R^3 whose components have the sum 1 is not a vector space. Why?

The space R^n is a vector space since it satisfies I.1 to II.4, as follows by direct calculation.

EXAMPLE 3. Polynomials

The set of all constant, linear, quadratic and cubic polynomials together forms a vector space under the usual addition and multiplication by a real number, since these two operations give polynomials of degree not exceeding 3, and the axioms in our definition follow by direct calculation.

EXAMPLE 4. Second-order homogeneous linear differential equations

The solutions of such an equation on a fixed interval $a < x < b$ form a vector space under the usual addition and multiplication by a number since these two operations give again such a solution, by Fundamental Theorem 1 in Sec. 2.1, and I.1 to II.4 follow by direct calculation. Do the solutions of a nonhomogeneous linear differential equation form a vector space?

Matrices will furnish further important vector spaces, as we shall see in the next chapter. In conclusion we should note that we have obtained the **abstract concept** of a vector space by the principle of defining it in terms of some of the most important properties of our **concrete model,** the set of all vectors in three-dimensional space. This is the principle by which many abstract mathematical concepts are derived from concrete models. Of course, in each case experience is needed in selecting properties to be stated in the form of axioms.

Linear Independence and Dependence

We shall now discuss some important concepts related to vector spaces.

Let $\mathbf{a}_{(1)}, \cdots, \mathbf{a}_{(m)}$ be any vectors[4] in a vector space V. Then an expression of the form

$$c_1 \mathbf{a}_{(1)} + \cdots + c_m \mathbf{a}_{(m)} \qquad (c_1, \cdots, c_m \text{ any scalars})$$

is called a **linear combination** of these vectors. Clearly, this is a vector in V, by the axioms of a vector space. The set S of all these linear combinations is called the **span** of $\mathbf{a}_{(1)}, \cdots, \mathbf{a}_{(m)}$ and is denoted by span $(\mathbf{a}_{(1)}, \cdots, \mathbf{a}_{(m)})$, and we say that S is **spanned** or *generated* by those m vectors.

It is not difficult to see that the span of a set of vectors is again a vector space. (Proof?)

Using the notion of linear combination, we can now introduce the basic concepts of linear independence and linear dependence of any given vectors $\mathbf{a}_{(1)}, \cdots, \mathbf{a}_{(m)}$ in a vector space V. For this purpose we consider the vector equation

$$(3) \qquad \boxed{c_1 \mathbf{a}_{(1)} + c_2 \mathbf{a}_{(2)} + \cdots + c_m \mathbf{a}_{(m)} = \mathbf{0}.}$$

This equation holds if we choose all scalars $c_1, \cdots, c_m$ equal to zero, since then it becomes $\mathbf{0} = \mathbf{0}$. If this is *the only* n-tuple of scalars for which (3) holds, we call the set of vectors $\mathbf{a}_{(1)}, \cdots, \mathbf{a}_{(m)}$ **linearly independent.**

Otherwise, that is, if (3) also holds with scalars $c_1, \cdots, c_m$ not all zero (so that at least one of them is not zero), we call this set of vectors **linearly dependent.** There is a good reason for this name. Namely, in the case of linear dependence, at least one of the vectors can be expressed as a linear combination of the others. Why? Well, if (3) holds with, say, $c_1 \neq 0$, we can solve (3) for $\mathbf{a}_{(1)}$:

$$(4) \qquad \mathbf{a}_{(1)} = l_2 \mathbf{a}_{(2)} + \cdots + l_m \mathbf{a}_{(m)} \qquad \text{where } l_j = -c_j / c_1.$$

Similarly if (3) holds with another $c_j \neq 0$. In contrast, in the case of linear independence, all c_j in (3) must be zero, by definition, and we cannot solve for any of the vectors in terms of the others. To avoid misunderstandings: in (4), some or even all l_j may be zero.

Also note that for a single vector $\mathbf{a}$, equation (3) is $c\mathbf{a} = \mathbf{0}$ and shows that a single vector $\mathbf{a}$ is linearly independent if and only if $\mathbf{a} \neq \mathbf{0}$.

---

[4]If you wish, write simply $\mathbf{a}_1, \cdots, \mathbf{a}_m$, but keep in mind that these are *vectors,* not vector *components.*

EXAMPLE 5. Linearly independent and linearly dependent sets in R^3
The vectors

$$\mathbf{i} = [1, 0, 0], \qquad \mathbf{j} = [0, 1, 0], \qquad \mathbf{k} = [0, 0, 1]$$

(cf. Sec. 6.3) form a linearly independent set in R^3 since

$$c_1\mathbf{i} + c_2\mathbf{j} + c_3\mathbf{k} = \mathbf{0}$$

is equivalent to the three equations $c_1 \cdot 1 = 0$, $c_2 \cdot 1 = 0$, $c_3 \cdot 1 = 0$ for the three components on both sides.

The vectors $\mathbf{a} = [1, 2, 1]$, $\mathbf{b} = [0, 0, 3]$, $\mathbf{d} = [2, 4, 0]$ form a linearly dependent set in R^3. Indeed, $6\mathbf{a} - 2\mathbf{b} - 3\mathbf{d} = \mathbf{0}$. Hence, for instance, $\mathbf{a} = \frac{1}{3}\mathbf{b} + \frac{1}{2}\mathbf{d}$. Geometrically, if we let their initial points coincide, then $\mathbf{a}$ lies in the plane passing though this common initial point and determined by the other two vectors. This is typical: the vectors of a linearly dependent set of three vectors in R^3 with common initial point P lie in the same plane through P or even in the same line. Figure 148 shows the idea.

Similarly, two vectors $\mathbf{a}$ and $\mathbf{b}$ in R^3 (or in R^2) form a linearly dependent set if and only if they are proportional, because (3) is $c_1\mathbf{a} + c_2\mathbf{b} = \mathbf{0}$ and implies $\mathbf{a} = k\mathbf{b}$ or $\mathbf{b} = l\mathbf{a}$; geometrically, if we let their initial points coincide, these vectors lie in the same line (cf. Fig. 147).

Any set of vectors containing the zero vector is linearly dependent. This holds for any vector space.   ∎

If a vector space V is such that it contains a linearly independent set B of n vectors, whereas any set of $n + 1$ or more vectors in V is linearly dependent, then V is said to have **dimension** n (or to be *n-dimensional*), and B is called a **basis** of V. Then every vector $\mathbf{v}$ of V can be represented as a linear combination of the n vectors of the basis in a unique fashion. Any such vector space is said to be **finite dimensional.**

EXAMPLE 6. Dimension, basis
The vector space R^3 has dimension 3. A basis is $\mathbf{i}, \mathbf{j}, \mathbf{k}$ in Example 5. More generally, space R^n has dimension n. A basis is $\mathbf{e}_{(1)}, \cdots, \mathbf{e}_{(n)}$, where $\mathbf{e}_{(j)}$ has the jth component equal to 1 and all others zero. This is often called the **standard basis** in R^n.

The space of solutions in Example 4 has dimension 2, and any basis of solutions as defined in Sec. 2.3 forms a basis in the sense of vector space.

The space in Example 3 has dimension 4, and a basis is $1, x, x^2, x^3$.   ∎

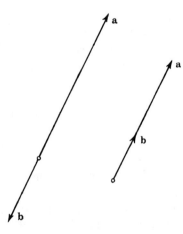

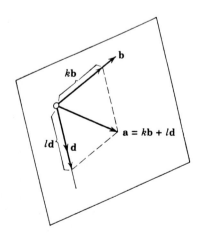

Fig. 147. Sets of two linearly
dependent vectors **a, b**

Fig. 148. Set of three linearly
dependent vectors **a, b, d**

If a vector space V contains linearly independent sets of n vectors for every n, no matter how large, we call V **infinite dimensional**. For example, the vector space of all polynomials of any degree is infinite dimensional since for any n, the polynomials $1, x, x^2, \cdots, x^n$ form a linearly independent set. The vector space of all continuous functions on a given interval $\alpha \leqq x \leqq \beta$ is also infinite dimensional. The study of infinite-dimensional vector spaces would exceed the level of this book; it is done in functional analysis; cf., e.g., Ref. [13] in Appendix 1.

In the next section we return to the vectors in three-dimensional space. For them we introduce a product, called the **inner product** or **dot product**, which is a scalar. This multiplication of vectors by vectors is suggested by the work done by a force and other applications.

Problems for Sec. 6.4

Let $\mathbf{u} = [4, 2, 0, -1]$, $\mathbf{v} = [1, 0, 1, 0]$, $\mathbf{w} = [3, -5, 6, 0]$. Find

1. $2\mathbf{u} - 6\mathbf{v}$, $-2(3\mathbf{v} - \mathbf{u})$
2. $\mathbf{u} + \mathbf{v} - \mathbf{w}$, $\mathbf{w} - \mathbf{u} - \mathbf{v}$
3. $\mathbf{v} + 2\mathbf{w}$, $2\mathbf{w} + \mathbf{v}$
4. $16\mathbf{u} - 32\mathbf{w}$
5. $\mathbf{u} - 2\mathbf{v} + 3\mathbf{w}$
6. $\frac{1}{4}\mathbf{u} + \frac{3}{2}\mathbf{w}$

Show that each of the following sets forms a vector space (the sets of functions taken with the usual addition and multiplication by scalars). Determine its dimension. Find a basis.

7. The set of all ordered triples of real numbers of the form $[v_1, v_2, 0]$.
8. The set of all multiples of $[2, 1, 4]$ in R^3.
9. The set of all vectors in R^2 satisfying $v_1 + 3v_2 = 0$.
10. The set of all vectors in R^5 whose first two components are 0.
11. The set of all vectors in R^3 satisfying $v_1 + v_2 + v_3 = 0$.
12. The set of all vectors in R^4 such that $v_1 + v_2 = 0$, $v_1 + 2v_2 + v_3 - v_4 = 0$.
13. The set of all vectors in R^3 of the form $a[1, 0, 0] + b[0, 1, 1]$ (a, b any scalars).
14. The set of all real numbers.
15. Span $\{[1, 2, 3, 0], [0, 0, 0, 1]\}$.
16. Span $\{[1, 0, 0], [0, 1, 0], [4, -1, 0]\}$.
17. The set of all functions of the form $f(x) = a \cos 2x + b \sin 2x$ (a, b any scalars).
18. The set of all functions of the form $f(x) = (c_1 x + c_2)e^{-x}$ (c_1, c_2 any scalars).
19. The set of all functions of the form $f(x) = ae^x + be^{2x} + ce^{-x}$ (a, b, c any scalars).
20. The set of all polynomials in x, of degree not exceeding 2.

21. **(Subspace)** A nonempty subset W of a vector space V is called a *subspace* of V if W is itself a vector space with respect to the algebraic operations defined in V. Give examples of one- and two-dimensional subspaces of the space of all vectors in R^3.

22. Show that a subset W of a vector space V is a subspace of V if and only if for any $\mathbf{a}$ and $\mathbf{b}$ in W and any scalars c and q the vector $c\mathbf{a} + q\mathbf{b}$ is in W.

In each case state whether the given set (with the usual addition and multiplication by scalars) is a vector space or not. Give reasons for your answer.

23. The set of all vectors $[v_1, v_2]$ such that $v_1 + v_2 = 1$.

24. The set of all vectors in R^5 such that $v_1 + v_3 + v_5 = 0$.

25. The set of all ordered quadruples of positive real numbers.

26. The set of all unit vectors in R^3.

27. The set of all polynomials in x, of degree 2.

28. The set of all functions $f(x)$ defined for $0 \leq x \leq 2$ and satisfying $f(1) = 0$.

29. The set of all forces of the form $\mathbf{p} = \mathbf{i} + a\mathbf{j}$ (a any scalar).

30. The set of all ordered triples $[v_1, v_2, v_3]$ such that $v_1 \leq v_2$.

Geometric applications. Using vectors, prove:

31. The diagonals of a parallelogram bisect each other.

32. The line that joins one vertex of a parallelogram to the midpoint of an opposite side divides the diagonal in the ratio $1:2$.

33. The medians of a triangle meet at a point P that divides each median in the ratio $1:2$.

34. The four diagonals of a parallelepiped meet and bisect each other.

35. The line that passes through the midpoints of adjacent sides of a parallelogram divides one of the diagonals in the ratio $1:3$.

Linear independence, basis

36. Show that a subset of a linearly independent set is itself linearly independent.

37. If a subset S_0 of a set S is linearly dependent, show that S is itself linearly dependent.

38. Find three different bases in R^2.

39. **(Standard basis)** Show that $\{\mathbf{e}_{(1)}, \cdots, \mathbf{e}_{(n)}\}$ is a basis in R^n, the so-called *standard basis;* here, $\mathbf{e}_{(j)}$ has its jth component 1 and all others 0. What is the standard basis in R^3?

40. **(Uniqueness)** Show that the representation $\mathbf{v} = c_1\mathbf{a}_{(1)} + \cdots + c_n\mathbf{a}_{(n)}$ of any given vector $\mathbf{v}$ in an n-dimensional vector space V in terms of a basis $\mathbf{a}_{(1)}, \cdots, \mathbf{a}_{(n)}$ in V is unique.

6.5 Inner Product (Dot Product)

We now return to the vectors in three-dimensional space, as considered in Secs. 6.1–6.3. So far we have defined the addition of these vectors and their multiplication by scalars (real numbers). We shall now introduce a multiplication of vectors by vectors which has many practical applications.

The **inner product, dot product** or **scalar product** of two vectors $\mathbf{a}$ and $\mathbf{b}$ in three-dimensional space is written $\mathbf{a} \cdot \mathbf{b}$ and is defined as

(1)
$$\begin{aligned} \mathbf{a} \cdot \mathbf{b} &= |\mathbf{a}||\mathbf{b}| \cos \gamma && \text{if } \mathbf{a} \neq \mathbf{0}, \mathbf{b} \neq \mathbf{0}, \\ \mathbf{a} \cdot \mathbf{b} &= 0 && \text{if } \mathbf{a} = \mathbf{0} \text{ or } \mathbf{b} = \mathbf{0}. \end{aligned}$$

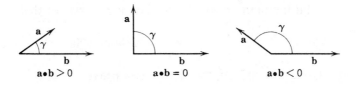

Fig. 149. Angle between vectors and value of inner product

This is read "**a** dot **b**"; here γ $(0 \leqq \gamma \leqq \pi)$ is the angle between **a** and **b** (computed when the vectors have their initial point coinciding, as shown in Fig. 149).

The value of the inner product is a scalar (a real number), and this motivates the name "scalar product." Since the cosine in (1) may be positive, zero or negative, the same is true for the inner product (cf. Fig. 149). Since for γ between 0 and π the cosine is 0 if and only if $\gamma = \pi/2$, we have the following important result.

Theorem 1 (Orthogonality)

Two nonzero vectors are orthogonal (perpendicular) if and only if their inner product (their dot product) is zero.

Setting **b** = **a** in (1), we have $\mathbf{a} \cdot \mathbf{a} = |\mathbf{a}|^2$, and this shows that the length (Euclidean norm) of a vector can be written in terms of the inner product,

(2)
$$|\mathbf{a}| = \sqrt{\mathbf{a} \cdot \mathbf{a}}.$$

From this and (1) we obtain for the angle γ between two nonzero vectors

(3)
$$\cos \gamma = \frac{\mathbf{a} \cdot \mathbf{b}}{|\mathbf{a}||\mathbf{b}|} = \frac{\mathbf{a} \cdot \mathbf{b}}{\sqrt{\mathbf{a} \cdot \mathbf{a}} \sqrt{\mathbf{b} \cdot \mathbf{b}}}.$$

From the definition we also see that the inner product has the following properties. For any vectors **a**, **b**, **c** and scalars q_1, q_2,

(4)

(a) $[q_1\mathbf{a} + q_2\mathbf{b}] \cdot \mathbf{c} = q_1\mathbf{a} \cdot \mathbf{c} + q_2\mathbf{b} \cdot \mathbf{c}$ (*Linearity*)

(b) $\mathbf{a} \cdot \mathbf{b} = \mathbf{b} \cdot \mathbf{a}$ (*Symmetry*)

(c) $\mathbf{a} \cdot \mathbf{a} \geqq 0$
$\mathbf{a} \cdot \mathbf{a} = 0$ if and only if $\mathbf{a} = \mathbf{0}$ $\Bigg\}$ (*Positive-definiteness*).

Hence *dot multiplication is commutative* [cf. (4b)] *and is distributive with respect to vector addition;* in fact, from (4a) with $q_1 = 1$ and $q_2 = 1$ we have

(4a*) $(\mathbf{a} + \mathbf{b}) \cdot \mathbf{c} = \mathbf{a} \cdot \mathbf{c} + \mathbf{b} \cdot \mathbf{c}$ (*Distributivity*).

Furthermore, from (1) and $|\cos \gamma| \leq 1$ we see that

(5) $$|\mathbf{a} \cdot \mathbf{b}| \leq |\mathbf{a}||\mathbf{b}| \qquad (Schwarz^5 \ inequality).$$

Using this and (2), the reader may prove

(6) $$|\mathbf{a} + \mathbf{b}| \leq |\mathbf{a}| + |\mathbf{b}| \qquad \textbf{(Triangle inequality).}$$

A simple direct calculation with inner products shows that

(7) $\quad |\mathbf{a} + \mathbf{b}|^2 + |\mathbf{a} - \mathbf{b}|^2 = 2(|\mathbf{a}|^2 + |\mathbf{b}|^2) \qquad (Parallelogram \ equality).$

If $\mathbf{a}$ and $\mathbf{b}$ are represented in terms of components, say, $\mathbf{a} = [a_1, a_2, a_3]$ and $\mathbf{b} = [b_1, b_2, b_3]$, their inner product is given by the simple basic formula

(8) $$\boxed{\mathbf{a} \cdot \mathbf{b} = a_1 b_1 + a_2 b_2 + a_3 b_3.}$$

We prove (8). Using (6), Sec. 6.3, we can write

$$\mathbf{a} = a_1 \mathbf{i} + a_2 \mathbf{j} + a_3 \mathbf{k} \qquad \text{and} \qquad \mathbf{b} = b_1 \mathbf{i} + b_2 \mathbf{j} + b_3 \mathbf{k}.$$

Since $\mathbf{i}$, $\mathbf{j}$ and $\mathbf{k}$ are unit vectors, we have from (2)

$$\mathbf{i} \cdot \mathbf{i} = 1, \qquad \mathbf{j} \cdot \mathbf{j} = 1, \quad \mathbf{k} \cdot \mathbf{k} = 1,$$

and since they are orthogonal, it follows from Theorem 1 that

$$\mathbf{i} \cdot \mathbf{j} = 0, \qquad \mathbf{j} \cdot \mathbf{k} = 0, \qquad \mathbf{k} \cdot \mathbf{i} = 0.$$

Hence if we substitute those representations of $\mathbf{a}$ and $\mathbf{b}$ into $\mathbf{a} \cdot \mathbf{b}$ and use (4a*) and (4b), we first have a sum of nine inner products,

$$\mathbf{a} \cdot \mathbf{b} = a_1 b_1 \mathbf{i} \cdot \mathbf{i} + a_1 b_2 \mathbf{i} \cdot \mathbf{j} + \cdots + a_3 b_3 \mathbf{k} \cdot \mathbf{k}.$$

Since six of these products are zero, we obtain (8). ∎

EXAMPLE 1. Inner product. Angle between vectors

Find the inner product and the lengths of $\mathbf{a} = [1, 2, 0]$ and $\mathbf{b} = [3, -2, 1]$ as well as the angle between these vectors.

Solution. $\mathbf{a} \cdot \mathbf{b} = 1 \cdot 3 + 2 \cdot (-2) + 0 \cdot 1 = -1$, $|\mathbf{a}| = \sqrt{\mathbf{a} \cdot \mathbf{a}} = \sqrt{5}$, $|\mathbf{b}| = \sqrt{\mathbf{b} \cdot \mathbf{b}} = \sqrt{14}$, and (3) gives the angle

$$\gamma = \text{arc cos} \frac{\mathbf{a} \cdot \mathbf{b}}{|\mathbf{a}| \, |\mathbf{b}|} = \text{arc cos} \, (-0.11952) = 1.69061 = 96.865°.$$

Can you sketch these vectors and convince yourself that they make an angle greater than 90°, so that the inner product comes out negative?

---

[5]HERMANN AMANDUS SCHWARZ (1843—1921), German mathematician, successor of Weierstrass at Berlin, known by his work in complex analysis (conformal mapping), differential geometry and the calculus of variations (minimal surfaces).

EXAMPLE 2. Orthonormal basis

By definition, an *orthonormal basis* in three-dimensional space is a basis {**a**, **b**, **c**} consisting of orthogonal unit vectors. It has the great advantage that the determination of the coefficients in representations

$$v = l_1 a + l_2 b + l_3 c \qquad\qquad (v \text{ a given vector})$$

is very simple. We claim that $l_1 = \mathbf{a} \cdot \mathbf{v}$, $l_2 = \mathbf{b} \cdot \mathbf{v}$, $l_3 = \mathbf{c} \cdot \mathbf{v}$. Indeed, this follows simply by taking the inner products of the representation with **a**, **b**, **c**, respectively, and using the orthonormality, $\mathbf{a} \cdot \mathbf{v} = l_1 \mathbf{a} \cdot \mathbf{a} + l_2 \mathbf{a} \cdot \mathbf{b} + l_3 \mathbf{a} \cdot \mathbf{c} = l_1$, etc.

If we introduce a Cartesian coordinate system with the positive directions on the axes in the directions of the basis vectors, the associated triad is $\mathbf{i} = \mathbf{a}$, $\mathbf{j} = \mathbf{b}$, $\mathbf{k} = \mathbf{c}$. ∎

Before we consider some applications of inner products, let us introduce one more concept. Suppose that **a** and **b** ($\neq \mathbf{0}$) are any given vectors and let γ denote the angle between them. Then the real number

$$p = |a| \cos \gamma$$

is called the **component** *of* **a** *in the direction of* **b** or the **projection** *of* **a** *in the direction of* **b**. If $\mathbf{a} = \mathbf{0}$, then γ is undefined, and we set $p = 0$.

It follows that $|p|$ is the length of the orthogonal projection of **a** on a straight line l in the direction of **b**. p may be positive, zero or negative (Fig. 150). If $p > 0$, then $p\mathbf{b}$ points *in* the direction of **b** and if $p < 0$ in the *opposite* direction.

From this definition we see that in particular the components of a vector **a** in the directions of the unit vectors **i**, **j**, **k** of the fundamental triad associated with a Cartesian coordinate system are the components a_1, a_2, a_3 of **a** as defined in Sec. 6.2. This shows that our present use of the term "component" is merely a slight generalization of the previous one.

From (3) we obtain

$$(9) \qquad\qquad p = |a| \cos \gamma = \frac{a \cdot b}{|b|} \qquad\qquad (b \neq 0)$$

and if in particular **b** is a unit vector, then we simply have

$$(10) \qquad\qquad p = a \cdot b.$$

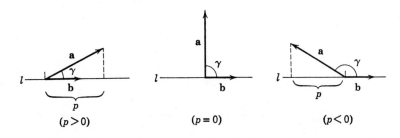

Fig. 150. Component of a vector **a** in the direction of a vector **b**

Applications of the Inner Product (Dot Product)

The following examples may illustrate the usefulness of inner products. Various other applications will be considered later. (More about straight lines will follow in Sec. 8.3, in connection with curves and their applications.)

EXAMPLE 3. Resultant of forces

What is the angle between the resultant $\mathbf{r}$ of the forces $\mathbf{a} = [3, 2, 0]$ and $\mathbf{b} = [-1, 4, 0]$ and the x-axis?

Solution. $\mathbf{r} = \mathbf{a} + \mathbf{b} = [2, 6, 0]$, and

$$\gamma = \text{arc cos } \frac{\mathbf{r} \cdot \mathbf{i}}{|\mathbf{r}|\,|\mathbf{i}|} = \text{arc cos } \frac{2}{\sqrt{40}} = \text{arc cos } 0.31623 = 1.24905 = 71.565°$$

Can you check the result by sketching the forces and $\mathbf{r}$?

EXAMPLE 4. Work done by a force

Consider a particle on which a constant force $\mathbf{a}$ acts. Let the particle be given a displacement $\mathbf{d}$. Then the work W done by $\mathbf{a}$ in the displacement is defined as the product $|\mathbf{d}|$ and the component of $\mathbf{a}$ in the direction of $\mathbf{d}$, that is,

(11)
$$W = |\mathbf{a}|\,|\mathbf{d}| \cos \alpha = \mathbf{a} \cdot \mathbf{d}$$

where α is the angle between $\mathbf{d}$ and $\mathbf{a}$. If $\alpha < 90°$, as in Fig. 151, then $W > 0$. If $\mathbf{a}$ and $\mathbf{d}$ are orthogonal, then the work is zero (why?). If $\alpha > 90°$, then $W < 0$, which means that in the displacement one has to do work against the force (against its component in the direction of $\mathbf{d}$).

EXAMPLE 5. Component of a force in a given direction

What force in the rope in Fig. 152 will hold a car of 5000 lb in equilibrium if the ramp makes an angle of 25° with the horizontal?

Solution. Introducing coordinates as shown, the weight is $\mathbf{w} = [0, -5000]$ and a vector in the direction of the rope is $\mathbf{b} = [-1, \tan 25°] = [-1, 0.46631]$. Hence the force needed is about 2100 lb because by (9),

$$p = \frac{\mathbf{w} \cdot \mathbf{b}}{|\mathbf{b}|} = \frac{-2331.5}{\sqrt{1.2174}} = -2113.1.$$

EXAMPLE 6. Orthogonal straight lines in a plane

Find a representation of the straight line L_1 through the point P: (1, 3) in the xy-plane and perpendicular to the line L_2 represented by $x - 2y + 2 = 0$; cf. Fig. 153.

Solution. Any straight line L_1 in the xy-plane can be represented in the form $a_1 x + a_2 y = c$. Because of (8) we can write this in the form

$$\mathbf{a} \cdot \mathbf{r} = c,$$

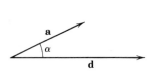

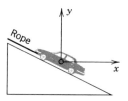

Fig. 151. Work done by a force   **Fig. 152.** Example 5

where $\mathbf{a} = [a_1, a_2]$, and $\mathbf{r} = [x, y]$ is the position vector of a running point on the line L_1. If $c = 0$, then L_1 passes through the origin. If $c \neq 0$, then

$$\mathbf{a} \cdot \mathbf{r} = 0$$

represents a line $L_1{}^*$ through the origin and parallel to L_1. Certainly $\mathbf{a} \neq \mathbf{0}$ and, by Theorem 1, the vector $\mathbf{a}$ is perpendicular to $\mathbf{r}$ and, therefore, perpendicular to the line $L_1{}^*$. It is called a **normal vector** to $L_1{}^*$. Since L_1 and $L_1{}^*$ are parallel, $\mathbf{a}$ is also a normal vector to L_1. Hence two lines L_1 and L_2: $b_1 x + b_2 y = d$ are perpendicular or *orthogonal* if and only if their normal vectors $\mathbf{a}$ and $\mathbf{b} = [b_1, b_2]$ are orthogonal, that is, $\mathbf{a} \cdot \mathbf{b} = 0$. Note that this implies that the slopes of the lines are negative reciprocals.

In our case, $\mathbf{b} = [1, -2]$, and a vector perpendicular to $\mathbf{b}$ is $\mathbf{a} = [2, 1]$. Hence the representation of L_1 must be of the form $2x + y = c$, and by substituting the coordinates of P we obtain $c = 5$. This yields the solution $y = -2x + 5$. (cf. Fig. 153).

EXAMPLE 7. Normal vector to a plane

Find a unit vector perpendicular to the plane $4x + 2y + 4z = -7$.

Solution. Any plane in space can be represented in the form

$$(12) \qquad\qquad a_1 x + a_2 y + a_3 z = c.$$

The position vector of a point in this plane is of the form $\mathbf{r} = [x, y, z]$. Introducing the vector $\mathbf{a} = [a_1, a_2, a_3]$ and using (8), we may write (12) as follows.

$$(13) \qquad\qquad \mathbf{a} \cdot \mathbf{r} = c.$$

Certainly $\mathbf{a} \neq \mathbf{0}$, and the unit vector in the direction of $\mathbf{a}$ is (cf. Fig. 154)

$$\mathbf{n} = \frac{1}{|\mathbf{a}|}\,\mathbf{a}.$$

Dividing by $|\mathbf{a}|$, we obtain from (13)

$$(14) \qquad\qquad \mathbf{n} \cdot \mathbf{r} = p \qquad \text{where} \qquad p = \frac{c}{|\mathbf{a}|}\,.$$

From (10) we see that p is the projection of $\mathbf{r}$ in the direction of $\mathbf{n}$, and this projection has the same constant value $c/|\mathbf{a}|$ for the position vector $\mathbf{r}$ of any point in the plane. Clearly this holds if and only if $\mathbf{n}$ is perpendicular to the plane. $\mathbf{n}$ is called a **unit normal vector** to the plane (the other being $-\mathbf{n}$). Furthermore, from this and the definition of projection it follows that $|p|$ is the distance of the plane from the origin. Representation (14) is called **Hesse's**[6] **normal form** of a plane. In our case, $\mathbf{a} = [4, 2, 4]$, $c = -7$, $|\mathbf{a}| = 6$, $\mathbf{n} = \frac{1}{6}\mathbf{a} = [\frac{2}{3}, \frac{1}{3}, \frac{2}{3}]$, and the plane has the distance $7/6$ from the origin. ∎

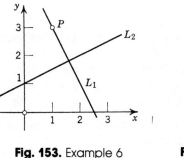

Fig. 153. Example 6

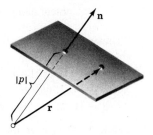
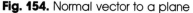

Fig. 154. Normal vector to a plane

[6]LUDWIG OTTO HESSE (1811—1874), German mathematician, who contributed to the theory of curves and surfaces.

By comparing the last two examples we discover a substantial **advantage of vector notation,** namely, that formulas for analogous objects in different dimensions (lines and planes in our case) take the same vector form.

The next section may be *skipped* in a first reading without destroying continuity. It explains the concept of an **inner product space,** which is defined by axioms suggested by basic properties of the dot product in the present section.

Problems for Sec. 6.5

Inner products. Let $\mathbf{a} = [1, 3, 2]$, $\mathbf{b} = [0, -1, 4]$, $\mathbf{c} = [3, -4, -1]$. Find

1. $\mathbf{a} \cdot \mathbf{b}$, $\mathbf{b} \cdot \mathbf{a}$
2. $|\mathbf{a}|$, $|\mathbf{b}|$, $|\mathbf{c}|$
3. $\frac{3}{2}|\mathbf{a}|$, $|\frac{3}{2}\mathbf{a}|$
4. $(\mathbf{a} + \mathbf{b}) \cdot \mathbf{c}$, $\mathbf{a} \cdot \mathbf{c} + \mathbf{b} \cdot \mathbf{c}$
5. $2\mathbf{a} \cdot 3\mathbf{b}$, $6\mathbf{a} \cdot \mathbf{b}$
6. $|4\mathbf{a} + 8\mathbf{b}|$, $4|\mathbf{a} + 2\mathbf{b}|$
7. $\mathbf{a} \cdot (\mathbf{b} - \mathbf{c})$, $\mathbf{a} \cdot (\mathbf{c} - \mathbf{b})$
8. $|\mathbf{a} + \mathbf{c}|$, $|\mathbf{a}| + |\mathbf{c}|$
9. $|\mathbf{a} - 2\mathbf{b}|$, $|2\mathbf{b} - \mathbf{a}|$
10. $|\mathbf{a} + \mathbf{b} - \mathbf{c}|$
11. $-\mathbf{a} \cdot (2\mathbf{c} - \mathbf{b})$
12. $|12\mathbf{a} + 24\mathbf{b} - 12\mathbf{c}|$

13. What laws do Probs. 1 and 4 illustrate?
14. If $\mathbf{a}$ is given and $\mathbf{a} \cdot \mathbf{b} = \mathbf{a} \cdot \mathbf{c}$, can we conclude that $\mathbf{b} = \mathbf{c}$?
15. Find $\mathbf{a} \cdot \mathbf{b}$ where $\mathbf{a} = 2\mathbf{i}$ and $\mathbf{b} = \mathbf{i} + \mathbf{j}$, $\mathbf{b} = \mathbf{j}$, $\mathbf{b} = -\mathbf{i} + \mathbf{j}$, and sketch a figure similar to Fig. 150.

Work. Find the work done by a force $\mathbf{p}$ acting on a body if the body is displaced from a point A to a point B along the straight segment AB, where

16. A: $(0, 0, 0)$, B: $(0, 3, 0)$, $\mathbf{p} = [1, 1, 1]$
17. A: $(2, 2, 2)$, B: $(4, 4, 4)$, $\mathbf{p} = [0, -4, 2]$
18. A: $(4, 1, 3)$, B: $(5, 2, 5)$, $\mathbf{p} = [1, 1, -1]$
19. A: $(0, \frac{1}{2}, 3)$, B: $(\frac{1}{4}, \frac{1}{2}, 4)$, $\mathbf{p} = [8, -12, 4]$
20. A: $(0, -1, 2)$, B: $(0, 0, 0)$, $\mathbf{p} = [0, 0, 1]$

21. Why is the work in Prob. 18 equal to zero?
22. Show that the work done by the resultant of two constant forces $\mathbf{p}$ and $\mathbf{q}$ in the displacement of a body from a point A to a point B along the straight segment $\overrightarrow{AB}$ is the same as the sum of the works done by each of the forces in this displacement.
23. Show that the work done by a constant force $\mathbf{p}$ in the displacement of a body along a straight segment $\overrightarrow{AB}$ and then along $\overrightarrow{BC}$ is the same as in the direct displacement along $\overrightarrow{AC}$.

Component in the direction of a vector. In each case find the component of $\mathbf{a}$ in the direction of $\mathbf{b}$.

24. $\mathbf{a} = [1, 1, -1]$, $\mathbf{b} = [3, 0, 7]$
25. $\mathbf{a} = [1, 1, 2]$, $\mathbf{b} = [0, 0, 6]$
26. $\mathbf{a} = [4, 1, 0]$, $\mathbf{b} = [-1, 4, 3]$
27. $\mathbf{a} = [0, 4, 3]$, $\mathbf{b} = [0, -4, -3]$
28. $\mathbf{a} = [1, 1, 1]$, $\mathbf{b} = [-8, 0, -6]$
29. $\mathbf{a} = [4, 3, 9]$, $\mathbf{b} = [2, \frac{1}{3}, -1]$

30. In what case is the component of $\mathbf{a}$ in the direction of $\mathbf{b}$ equal to twice the component of $\mathbf{b}$ in the direction of $\mathbf{a}$?

Orthogonality of vectors

31. For what values of a_1 are $\mathbf{a} = [a_1, 2, 0]$ and $\mathbf{b} = [3, 4, -1]$ orthogonal?
32. Find all unit vectors $\mathbf{a} = [a_1, a_2]$ orthogonal to $[2, -5]$.

33. Show that $\mathbf{a} = [1, -1, 2]$, $\mathbf{b} = [0, 4, 2]$, $\mathbf{c} = [-10, -2, 4]$ are orthogonal.

34. Find an orthonormal basis $\{\mathbf{a}, \mathbf{b}, \mathbf{c}\}$ in three-dimensional space, where $\mathbf{b} = q_1[3, 4, 0]$, $\mathbf{c} = q_2[4, -3, 0]$, and q_1, q_2 are suitable scalars.

35. Find all vectors $\mathbf{v}$ orthogonal to $\mathbf{a} = [1, 2, 0]$. Do they form a vector space?

36. Find a force $\mathbf{p} = [p_1, p_2, 0]$ such that the resultant of $\mathbf{p}$ and $\mathbf{q} = [2, 6, 4]$ is perpendicular to the xy-plane.

37. Show that the straight lines $4x + 3y = 2$ and $3x - 4y = -1$ are orthogonal.

38. Find a unit normal vector to the plane $5x + y - 4z = 3$.

39. For what c are the planes $x + y + z = 1$ and $2x + cy + 7z = 0$ orthogonal?

40. Using vectors, show that if the diagonals of a rectangle are orthogonal, the rectangle must be a square.

41. Under what conditions will the diagonals of a parallelogram be orthogonal?

42. Are the four diagonals of a cube orthogonal?

43. Show that a parallelogram with diagonals of the same length must be a rectangle.

Angle between vectors. Let $\mathbf{a} = [3, 1, 4]$, $\mathbf{b} = [1, -1, 0]$, $\mathbf{c} = [1, 0, 5]$. Find the cosine of the angle between the following vectors.

44. $\mathbf{a}, \mathbf{c}$ 45. $\mathbf{a}, -\mathbf{c}$ 46. $\mathbf{a}, \mathbf{b} + \mathbf{c}$ 47. $\mathbf{a} + \mathbf{b}, \mathbf{a} - \mathbf{b}$

48. Find the angle between the straight lines $x + y = 1$ and $2x - 3y = 0$.

49. Find the angle between the straight lines $4x - y = 2$ and $x + 4y = 3$.

50. Find the angle between the planes $x + y + z = 1$ and $x - y = 0$.

51. Find the angle between the planes $3x - 4y + 2z = 0$ and $x + 2y - z = 0$.

52. Find the angles of the triangle with vertices A: $(1, 1, 0)$, B: $(3, 1, 0)$, C: $(1, 3, 0)$.

53. Find the angles of the triangle with vertices A: $(0, 0, 0)$, B: $(1, 2, 3)$, C: $(4, -1, 3)$.

54. Find the angle between the diagonals of the rectangle with vertices $(2, 1)$, $(1, 2)$, $(5, 4)$, $(4, 5)$.

55. Find the angle between the diagonals of the parallelogram with vertices $(0, 0)$, $(4, 0)$, $(1, 2)$, $(5, 2)$.

56. Show that if $\mathbf{a}, \mathbf{b}, \mathbf{c}$ are orthogonal unit vectors, then $\mathbf{a} + \mathbf{b} + \mathbf{c}$ makes the same angle with each of the vectors.

57. Deduce the law of cosines by using vectors $\mathbf{a}, \mathbf{b}$ and $\mathbf{a} - \mathbf{b}$.

58. Let $\mathbf{a} = \cos \alpha \, \mathbf{i} + \sin \alpha \, \mathbf{j}$ and $\mathbf{b} = \cos \beta \, \mathbf{i} + \sin \beta \, \mathbf{j}$, where $0 \leq \alpha \leq \beta \leq 2\pi$. Show that $\mathbf{a}$ and $\mathbf{b}$ are unit vectors. Use (8) to obtain the trigonometric identity for $\cos (\beta - \alpha)$.

59. Prove the triangle inequality (6).

60. Prove the parallelogram equality (7).

6.6 Inner Product Spaces

In Sec. 6.4 we obtained the definition of a vector space by taking some of the basic properties of vector addition and multiplication by scalars in three-dimensional space. It is quite interesting that the dot product (Sec. 6.5) may be used in a similar fashion and motivates the concept of a *real inner product space*. By definition, this is a real vector space on which there is defined an inner product satisfying (4), Sec. 6.5:

Definition of a real inner product space

A real vector space V is called a *real inner product space* (or *real pre-Hilbert*[7] *space*) if it has the following property. With every pair of vectors **a** and **b** in V there is associated a real number, which is denoted by (**a**, **b**) and is called the **inner product** of **a** and **b**, such that the following axioms are satisfied.

I. For all scalars q_1 and q_2 and all vectors **a**, **b**, **c** in V,

$$(q_1\mathbf{a} + q_2\mathbf{b}, \mathbf{c}) = q_1(\mathbf{a}, \mathbf{c}) + q_2(\mathbf{b}, \mathbf{c}) \qquad (Linearity).$$

II. For all vectors **a** and **b** in V,

$$(\mathbf{a}, \mathbf{b}) = (\mathbf{b}, \mathbf{a}) \qquad (Symmetry).$$

III. For every **a** in V,

$$(\mathbf{a}, \mathbf{a}) \geqq 0, \quad \text{and}$$
$$(\mathbf{a}, \mathbf{a}) = 0 \quad \text{if and only if} \quad \mathbf{a} = \mathbf{0} \qquad (Positive\text{-}definiteness).$$

Definition of orthogonality

Vectors **a** and **b** in an inner product space V are called *orthogonal* if

$$(\mathbf{a}, \mathbf{b}) = 0. \qquad \blacksquare$$

Note that this definition is suggested and motivated by Theorem 1 in Sec. 6.5.

Using the inner product, with every element **a** in V we may associate a number which is denoted by $\|\mathbf{a}\|$, is defined by

$$\|\mathbf{a}\| = \sqrt{(\mathbf{a}, \mathbf{a})} \qquad (\geqq 0),$$

and is called the **norm** of **a**. From (2), Sec. 6.5, we see that this generalizes the concept of a length. In fact, the dot product is an inner product in the sense of our new definition,

$$(\mathbf{a}, \mathbf{b}) = \mathbf{a} \cdot \mathbf{b}.$$

---

[7]DAVID HILBERT (1862—1943), great German mathematician, taught at Königsberg and (from 1895) at Göttingen, where he developed the famous Göttingen mathematical school; his students and collaborators included H. Bolza, R. Courant, H. B. Curry, A. Haar, E. R. Hedrick, O. D. Kellogg, J. v. Neumann, E. Schmidt, H. Steinhaus, H. Weyl, and many others. Hilbert's work was of basic importance to higher algebra and number theory, integral equations, calculus of variations, functional analysis, and mathematical logic. His "Foundations of Geometry" helped the axiomatic method to gain general recognition and acceptance. Hilbert's famous twenty-three problems (presented in a talk in 1900 at the International Congress of Mathematicians in Paris) had considerable influence on the development of modern mathematics.

If V in this definition is finite dimensional (Sec. 6.4), it is actually a so-called *Hilbert space;* proof in Ref. [13], p. 73, listed in Appendix 1.

In our new notation, formula (2), Sec. 6.5, can be written

$$\|\mathbf{a}\| = |\mathbf{a}| = \sqrt{(\mathbf{a}, \mathbf{a})} = \sqrt{\mathbf{a} \cdot \mathbf{a}}.$$

From the axioms of an inner product and the definition of a norm, one can derive [cf. (5)–(7) in Sec. 6.5]

$$|(\mathbf{a}, \mathbf{b})| \leq \|\mathbf{a}\| \, \|\mathbf{b}\| \qquad (\textit{Schwarz inequality}),$$

from this

$$\|\mathbf{a} + \mathbf{b}\| \leq \|\mathbf{a}\| + \|\mathbf{b}\| \qquad (\textit{Triangle inequality})$$

and by a simple direct calculation

$$\|\mathbf{a} + \mathbf{b}\|^2 + \|\mathbf{a} - \mathbf{b}\|^2 = 2(\|\mathbf{a}\|^2 + \|\mathbf{b}\|^2) \qquad (\textit{Parallelogram equality}).$$

We shall not go into details, because these belong to more advanced courses (on functional analysis, in particular Hilbert spaces, approximation theory etc.; cf. [13] in Appendix 1). However, we want to illustrate the generality of the concept of an inner product space by two important examples.

EXAMPLE 1. *n*-dimensional Euclidean space

The vector space R^n in Sec. 6.5 becomes an inner product space if we define the inner product of two vectors $\mathbf{a} = [a_1, \cdots, a_n]$ and $\mathbf{b} = [b_1, \cdots, b_n]$ by

(1) $$(\mathbf{a}, \mathbf{b}) = a_1 b_1 + a_2 b_2 + \cdots + a_n b_n.$$

This is suggested by (8) in Sec. 6.5, to which it reduces when $n = 3$. It satisfies axioms I-III, in the definition of a real inner product space, above, as one can verify by direct calculation. The space obtained is called *n-dimensional Euclidean space* and is often denoted by E^n or again simply by R^n. From (1) we obtain the norm

$$\|\mathbf{a}\| = \sqrt{a_1^2 + a_2^2 + \cdots + a_n^2}.$$

For $n = 3$ this is the usual length (2) in Sec. 6.5.

EXAMPLE 2. An inner product for functions

The set of all continuous functions $f(x)$, $g(x)$, $\cdots$ on a given interval $\alpha \leq x \leq \beta$ forms a real vector space under the usual addition of functions and multiplication by scalars (real numbers). On this space we can define an inner product by the integral

(2) $$(f, g) = \int_\alpha^\beta f(x)g(x)\, dx$$

Axioms I-III can be verified by direct calculation. Orthogonality $(f, g) = 0$ in this inner product space is identical with orthogonality as defined in Sec. 4.7.   ∎

In the next section we return to the vectors in three-dimensional space. We define for them another product, the **vector product** or **cross product,** which is a vector. This product is suggested by applications such as the moment of a force or a rotation in space, as we shall explain.

6.7 Vector Product (Cross Product)

Dot multiplication of two vectors gives as the product a scalar (cf. Sec. 6.5). Now various applications suggest another kind of multiplication of vectors such that the product of two vectors is again a vector. This so-called **vector product** or **cross product** of two vectors **a** and **b** is written

$$\boxed{\mathbf{a} \times \mathbf{b}}$$ (read "a cross b")

and is a vector **v**, which is defined as follows.

Definition of vector product
If **a** and **b** have the same or opposite direction or one of these vectors is the zero vector, then $\mathbf{v} = \mathbf{a} \times \mathbf{b} = \mathbf{0}$.

In any other case, $\mathbf{v} = \mathbf{a} \times \mathbf{b}$ is the vector whose length is equal to the area of the parallelogram with **a** and **b** as adjacent sides and whose direction is perpendicular to both **a** and **b** and is such that **a**, **b**, **v**, in this order, form a right-handed triple or right-handed triad, as shown in Fig. 155.   ∎

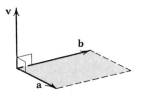

Fig. 155. Vector product

The term **right-handed** comes from the fact that the vectors **a**, **b**, **v**, in this order, assume the same sort of orientation as the thumb, index finger and middle finger of the right hand when these are held as shown in Fig. 156. We may also say that if **a** is rotated into the direction of **b** through the angle $\alpha\,(< \pi)$, then **v** advances in the same direction as a right-handed screw would if turned in the same way (cf. Fig. 157).

The parallelogram with **a** and **b** as adjacent sides has the area $|\mathbf{a}||\mathbf{b}| \sin \gamma$, where γ is the angle between **a** and **b** (cf. Fig. 155). We thus obtain

(1) $$\boxed{|\mathbf{v}| = |\mathbf{a}||\mathbf{b}| \sin \gamma.}$$

Let $\mathbf{a} \times \mathbf{b} = \mathbf{v}$ and let $\mathbf{b} \times \mathbf{a} = \mathbf{w}$. Then, by definition, $|\mathbf{v}| = |\mathbf{w}|$, and in order that **b**, **a**, **w** form a right-handed triple we must have $\mathbf{w} = -\mathbf{v}$. This implies

(2) $$\boxed{\mathbf{b} \times \mathbf{a} = -(\mathbf{a} \times \mathbf{b})}$$ (Fig. 158),

that is, cross multiplication of vectors has the following unusual property.

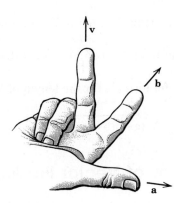

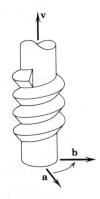

Fig. 156. Right-handed triple of vectors
a, b, v

Fig. 157. Right-handed screw

Cross multiplication of vectors is not commutative but anticommutative. Hence the order of the factors in a vector product is of great importance and must be carefully observed.

From the definition it follows that for any constant k,

$$(3) \qquad (k\mathbf{a}) \times \mathbf{b} = k(\mathbf{a} \times \mathbf{b}) = \mathbf{a} \times (k\mathbf{b}).$$

Furthermore, cross multiplication is distributive with respect to vector addition, that is,

$$(4) \qquad \begin{aligned} \mathbf{a} \times (\mathbf{b} + \mathbf{c}) &= (\mathbf{a} \times \mathbf{b}) + (\mathbf{a} \times \mathbf{c}), \\ (\mathbf{a} + \mathbf{b}) \times \mathbf{c} &= (\mathbf{a} \times \mathbf{c}) + (\mathbf{b} \times \mathbf{c}). \end{aligned}$$

The proof will be given in the next section.

Cross multiplication is not associative. That is, in general,

$$\mathbf{a} \times (\mathbf{b} \times \mathbf{c}) \neq (\mathbf{a} \times \mathbf{b}) \times \mathbf{c}.$$

Indeed, for instance, $\mathbf{i} \times (\mathbf{i} \times \mathbf{j}) = \mathbf{i} \times \mathbf{k} = -\mathbf{j}$, whereas $(\mathbf{i} \times \mathbf{i}) \times \mathbf{j} = \mathbf{0} \times \mathbf{j} = \mathbf{0}$.

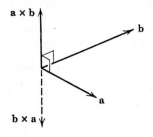

Fig. 158. Anticommutativity of cross multiplication

From (1) in this section and (1) in Sec. 6.5 it follows that

$$|\mathbf{v}|^2 = |\mathbf{a}|^2|\mathbf{b}|^2 \sin^2 \gamma = |\mathbf{a}|^2|\mathbf{b}|^2(1 - \cos^2 \gamma) = (\mathbf{a}\cdot\mathbf{a})(\mathbf{b}\cdot\mathbf{b}) - (\mathbf{a}\cdot\mathbf{b})^2.$$

Taking roots, we obtain a useful formula for the length of a vector product:

(5) $$|\mathbf{a} \times \mathbf{b}| = \sqrt{(\mathbf{a}\cdot\mathbf{a})(\mathbf{b}\cdot\mathbf{b}) - (\mathbf{a}\cdot\mathbf{b})^2}.$$

Two Typical Applications of Vector Products

EXAMPLE 1. Moment of a force
In mechanics the moment m of a force $\mathbf{p}$ about a point Q is defined as the product $m = |\mathbf{p}|d$, where d is the (perpendicular) distance between Q and the line of action L of $\mathbf{p}$ (Fig. 159). If $\mathbf{r}$ is the vector from Q to any point A on L, then $d = |\mathbf{r}| \sin \gamma$ (Fig. 159) and

$$m = |\mathbf{r}| |\mathbf{p}| \sin \gamma.$$

Since γ is the angle between $\mathbf{r}$ and $\mathbf{p}$,

$$m = |\mathbf{r} \times \mathbf{p}|,$$

as follows from (1). The vector

(6)

is called the **moment vector** or **vector moment** of $\mathbf{p}$ about Q. Its magnitude is m, and its direction is that of the axis of the rotation about Q which $\mathbf{p}$ has the tendency to produce.

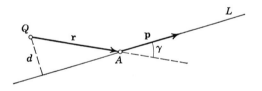

Fig. 159. Moment of a force

EXAMPLE 2. Velocity of a rotating body
A rotation of a rigid body B in space can be simply and uniquely described by a vector $\mathbf{w}$ as follows. The direction of $\mathbf{w}$ is that of the axis of rotation and such that the rotation appears clockwise, if one looks from the initial point of $\mathbf{w}$ to its terminal point. The length of $\mathbf{w}$ is equal to the **angular speed** ω (> 0) of the rotation, that is, the linear (or tangential) speed of a point of B divided by its distance from the axis of rotation.

Let P be any point of B and d its distance from the axis. Then P has the speed ωd. Let $\mathbf{r}$ be the position vector of P referred to a coordinate system with origin 0 on the axis of rotation. Then $d = |\mathbf{r}| \sin \gamma$ where γ is the angle between $\mathbf{w}$ and $\mathbf{r}$. Therefore,

$$\omega d = |\mathbf{w}| |\mathbf{r}| \sin \gamma = |\mathbf{w} \times \mathbf{r}|.$$

From this and the definition of vector product we see that the velocity vector $\mathbf{v}$ of P can be represented in the form (Fig. 160)

(7) $$\boxed{\mathbf{v} = \mathbf{w} \times \mathbf{r}.}$$

This simple formula is useful for determining $\mathbf{v}$ at any point of B.   ∎

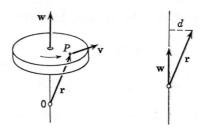

Fig. 160. Rotation of a rigid body

The definition of a vector product shows that this product is independent of the particular choice of coordinates. Of course, in practically working with vector products we often use a coordinate system and must then know how to express such a product in terms of components. The derivation of the corresponding formulas will be our next task.

6.8 Components of Vector Products

In specific applications, vector formulas such as (6) or (7) in the preceding section are not enough, but we must know the components of vector products $\mathbf{v} = \mathbf{a} \times \mathbf{b}$ in terms of the components of their factors. In this connection it is important to note that there are two types of coordinate systems, depending on the orientation of the axes, namely, right-handed and left-handed. The definitions are as follows.

A Cartesian coordinate system is called **right-handed** if the corresponding unit vectors $\mathbf{i}, \mathbf{j}, \mathbf{k}$ in the positive directions of the axes form a right-handed triple (Fig. 161a); it is called **left-handed** if these vectors form a *left-handed triple,* that is, assume the same sort of orientation as the thumb, index finger and middle finger of the left hand (Fig. 161b).

In applications we preferably use right-handed systems.

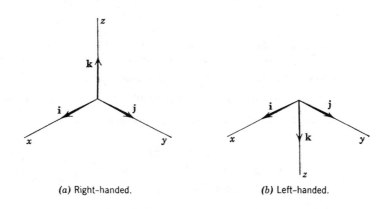

(a) Right-handed.   (b) Left-handed.

Fig. 161. The two types of Cartesian coordinate systems

Theorem 1 (Vector product in terms of components)

*With respect to a **right-handed** Cartesian coordinate system, let* $\mathbf{a} = [a_1, a_2, a_3]$ *and* $\mathbf{b} = [b_1, b_2, b_3]$. *Then* $\mathbf{v} = \mathbf{a} \times \mathbf{b}$ *has the components*

$$v_1 = a_2 b_3 - a_3 b_2, \qquad v_2 = a_3 b_1 - a_1 b_3, \qquad v_3 = a_1 b_2 - a_2 b_1$$

or, written in terms of second-order determinants,[8]

$$(1) \qquad v_1 = \begin{vmatrix} a_2 & a_3 \\ b_2 & b_3 \end{vmatrix}, \qquad v_2 = \begin{vmatrix} a_3 & a_1 \\ b_3 & b_1 \end{vmatrix}, \qquad v_3 = \begin{vmatrix} a_1 & a_2 \\ b_1 & b_2 \end{vmatrix}.$$

Proof on p. 344 near the end of the chapter.

Remark 1

For memorizing, it is useful to note that (1) can be interpreted as the expansion of the determinant

$$(2) \qquad \mathbf{a} \times \mathbf{b} = \begin{vmatrix} \mathbf{i} & \mathbf{j} & \mathbf{k} \\ a_1 & a_2 & a_3 \\ b_1 & b_2 & b_3 \end{vmatrix}$$

by the first row, but we should keep in mind that this is not an ordinary determinant because the entries of the first row are vectors.

Remark 2

With respect to a *left-handed* Cartesian coordinate system, the determinants in (1) are preceded by a minus sign.

EXAMPLE 1

With respect to a right-handed Cartesian coordinate system, let $\mathbf{a} = [4, 0, -1]$ and $\mathbf{b} = [-2, 1, 3]$. Then

$$\mathbf{a} \times \mathbf{b} = \begin{vmatrix} \mathbf{i} & \mathbf{j} & \mathbf{k} \\ 4 & 0 & -1 \\ -2 & 1 & 3 \end{vmatrix} = \mathbf{i} - 10\mathbf{j} + 4\mathbf{k} = [1, -10, 4].$$

EXAMPLE 2. Moment of a force

Find the moment of the force $\mathbf{p}$ in Fig. 162 about the center of the wheel.

Solution. Introducing coordinates as shown in Fig. 162, we have

$$\mathbf{p} = [1000 \cos 30°, \quad 1000 \sin 30°, \quad 0] = [866, \ 500, \ 0], \qquad \mathbf{r} = [0, \ -1.5, \ 0],$$

so that (6) in Sec. 6.7 and (2) in the present section give

$$\mathbf{m} = \mathbf{r} \times \mathbf{p} = \begin{vmatrix} \mathbf{i} & \mathbf{j} & \mathbf{k} \\ 0 & -1.5 & 0 \\ 866 & 500 & 0 \end{vmatrix} = 0\mathbf{i} - 0\mathbf{j} + \begin{vmatrix} 0 & -1.5 \\ 866 & 500 \end{vmatrix} \mathbf{k} = [0, \ 0, \ 1299].$$

---

[8]Second- and third-order determinants are usually considered in elementary calculus. Readers unfamiliar with them may consult Sec. 7.9.

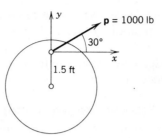

Fig. 162. Moment of a force **p**

This moment vector is normal (perpendicular) to the plane of the wheel; hence it has the direction of the axis of rotation about the center of the wheel which the force has the tendency to produce.

EXAMPLE 3. Plane through three given points
Find the plane through A: $(1, 2, 3)$, B: $(-1, 4, -5)$, C: $(0, 1, -6)$.

Solution. A normal vector to the plane is given by the vector product of $\mathbf{b} = \overrightarrow{AB} = [-2, 2, -8]$ and $\mathbf{c} = \overrightarrow{AC} = [-1, -1, -9]$,

$$\mathbf{b} \times \mathbf{c} = \begin{vmatrix} \mathbf{i} & \mathbf{j} & \mathbf{k} \\ -2 & 2 & -8 \\ -1 & -1 & -9 \end{vmatrix} = -26\mathbf{i} - 10\mathbf{j} + 4\mathbf{k}.$$

This gives the representation

$$-26x - 10y + 4z = k$$

where $k = -26 \cdot 1 - 10 \cdot 2 + 4 \cdot 3 = -34$ because the plane should pass through A. Division by -2 then gives the answer

$$13x + 5y - 2z = 17.$$

EXAMPLE 4. Area of a parallelogram
Find the area of the parallelogram with vertices A, B, C as in Example 3 and D: $(-2, 3, -14)$.

Solution. From (1) in Sec. 6.7 and Example 3 we obtain the area

$$|\mathbf{b} \times \mathbf{c}| = \sqrt{26^2 + 10^2 + 4^2} = \sqrt{792} \approx 28.14. \qquad \blacksquare$$

We can now also prove the **distributive law** (4) in the previous section. From (1) it follows that the first component of the vector $\mathbf{a} \times (\mathbf{b} + \mathbf{c})$ is

$$\begin{vmatrix} a_2 & a_3 \\ b_2 + c_2 & b_3 + c_3 \end{vmatrix} = a_2(b_3 + c_3) - a_3(b_2 + c_2)$$

$$= (a_2 b_3 - a_3 b_2) + (a_2 c_3 - a_3 c_2)$$

$$= \begin{vmatrix} a_2 & a_3 \\ b_2 & b_3 \end{vmatrix} + \begin{vmatrix} a_2 & a_3 \\ c_2 & c_3 \end{vmatrix}.$$

The expression on the right is the first component of $\mathbf{a} \times \mathbf{b} + \mathbf{a} \times \mathbf{c}$. For the other two components of the vector $\mathbf{a} \times (\mathbf{b} + \mathbf{c})$ the consideration is similar. This proves the first of the relations (4) in the previous section, and the second relation in (4) can be proved by the same argument. ∎

The reader may prove the following criterion for linear dependence and independence of two vectors.

Theorem 2 (Linear dependence)
Two vectors form a linearly dependent set if and only if their vector product is the zero vector.

In the next section we discuss the **scalar triple product a•(b × c)** and other products of three and more factors and show that the latter can be expressed in terms of dot products, cross products and scalar triple products.

Problems for Sec. 6.8

With respect to a right-handed Cartesian coordinate system, let $\mathbf{a} = [1, 2, -3]$, $\mathbf{b} = [1, 2, 0]$ and $\mathbf{c} = [-1, 1, 0]$. Find

1. $\mathbf{a} \times \mathbf{b}, \mathbf{b} \times \mathbf{a}, \mathbf{a}\cdot\mathbf{b}, \mathbf{b}\cdot\mathbf{a}$
2. $\mathbf{a} \times \mathbf{c}, |\mathbf{a} \times \mathbf{c}|, \mathbf{a}\cdot\mathbf{c}$
3. $\mathbf{b} \times \mathbf{c}, |\mathbf{b} \times \mathbf{c}|, |\mathbf{c} \times \mathbf{b}|$
4. $\mathbf{a} \times (\mathbf{b} + \mathbf{c}), \mathbf{a} \times \mathbf{b} + \mathbf{a} \times \mathbf{c}$
5. $(\mathbf{c} - \mathbf{a}), \times 2\mathbf{b}, (\mathbf{a} - \mathbf{c}) \times 2\mathbf{b}$
6. $(\mathbf{a} + 2\mathbf{b}) \times \mathbf{c}, (\frac{1}{2}\mathbf{a} + \mathbf{b}) \times 2\mathbf{c}$
7. $(\mathbf{a} + \mathbf{b}) \times \mathbf{c}, \mathbf{a} \times \mathbf{c} - \mathbf{c} \times \mathbf{b}$
8. $\mathbf{a} \times (2\mathbf{b} + 3\mathbf{c}), 2\mathbf{a} \times \mathbf{b} + 3\mathbf{a} \times \mathbf{c}$
9. $\mathbf{a} \times (\mathbf{b} - \mathbf{c}), \mathbf{b} \times \mathbf{a} + \mathbf{a} \times \mathbf{c}$
10. $\mathbf{c} \times \mathbf{c}, \mathbf{a} \times \mathbf{c} + \mathbf{c} \times \mathbf{a}$
11. $(\mathbf{a}\cdot\mathbf{b})\mathbf{c}, (\mathbf{a} \times \mathbf{b})\cdot\mathbf{c}$
12. $(\mathbf{a} \times \mathbf{b}) \times \mathbf{c}, \mathbf{a} \times (\mathbf{b} \times \mathbf{c})$

13. What properties of cross multiplication do Probs. 1, 4 and 6 illustrate?
14. Show that $\mathbf{i} \times \mathbf{j} = \mathbf{k}, \mathbf{k} \times \mathbf{i} = \mathbf{j}, \mathbf{j} \times \mathbf{k} = \mathbf{i}$, assuming right-handed coordinates. Find $(\mathbf{i} \times \mathbf{j}) \times \mathbf{k}, \mathbf{i} \times (\mathbf{j} \times \mathbf{k})$.

Moment of a force. A force $\mathbf{p}$ acts on a line through a point A. Find the moment vector $\mathbf{m}$ of $\mathbf{p}$ about a point Q, where the force, the point A and the point Q are
15. $[3, -1, 2], (0, -1, 4), (3, 0, 2)$
16. $[2, 1, 0], (1, 1, 1), (2, 2, 2)$
17. $[1, 0, 0], (1, 1, 0), (-5, 1, 0)$
18. $[1, 1, 0], (0, 0, 0), (0, 1, 0)$
19. $[3, 0, -6], (1, 2, 1), (4, 0, -1)$
20. $[1, -2, 3], (4, 3, 1), (6, -1, 7)$

Force in a magnetic field. If a particle of electric charge q moves with velocity $\mathbf{v}$ in a magnetic field (for instance, in a cyclotron), it experiences a force $\mathbf{p}$ perpendicular to both its direction of motion and the field, $\mathbf{p} = q\mathbf{v} \times \mathbf{b}$ (**b** the magnetic induction). Let $q = 1$. Find $\mathbf{p}$ if with respect to a right-handed Cartesian coordinate system,
21. $\mathbf{v} = [3, 4, 0], \mathbf{b} = [0, 0, 6]$
22. $\mathbf{v} = [0, 3, 1], \mathbf{b} = [2, 2, 0]$
23. $\mathbf{v} = [0, 4, -4], \mathbf{b} = [3, 0, -5]$
24. $\mathbf{v} = [3, -1, 2], \mathbf{b} = [1, 1, 1]$

Parallel and orthogonal forces. Are the following forces parallel? Orthogonal?
25. $[1, 1, 2], [1, -1, 0]$
26. $[1, -2, 3], [-3, 6, -8]$
27. $[\frac{3}{8}, -2, \frac{21}{5}], [-\frac{3}{14}, \frac{8}{7}, -\frac{12}{5}]$
28. $[\frac{2}{5}, 4, -\frac{1}{2}], [2, \frac{3}{8}, \frac{23}{5}]$
29. $[3, -1, \frac{1}{2}], [\frac{1}{2}, -1, 3]$
30. $[5, -\frac{1}{4}, \frac{1}{2}], [-\frac{4}{3}, \frac{1}{15}, -\frac{2}{15}]$

Planes. Find a representation of the plane through the three given points.
31. $(3, 1, 1), (1, 3, 1), (1, 1, 3)$
32. $(1, 2, 4), (-1, -2, -4), (1, 0, 0)$

33. (1, 1, 1), (5, 4, 3), (10, 8, 6)   **34.** (3, 5, 7), (11, 13, 17), (23, 29, 31)

35. (2, 1, $\frac{1}{2}$), (3, 2, $\frac{3}{2}$), (4, 5, -1)   **36.** (0, 4, 9), (16, 0, 1), (3, 3, 0)

Find the plane through the point (1, 1, 1) and perpendicular to the two given planes.

37. $x + y + z = 1$, $x + 2z = 2$ **38.** $3x - 4y = 1, y + 2z = 5$

39. $x = 4$, $2x - 2y + 3z = 1$ **40.** $\frac{1}{2}x + 4y - \frac{1}{4}z = 0$, $4x - 2y + 2z = 3$

Area of parallelograms and triangles. Find the area of the parallelogram of which the given vectors are adjacent sides.

41. [1, 0, 2], [1, 1, 2]   **42.** [-3, -1, 0], [4, 9, 0]

43. [4, -4, 4], [3, 6, 0]   **44.** [2, 2, 6], [-2, -2, -12]

45. [0, 1, $\frac{1}{2}$], [-4, 3, $\frac{1}{2}$]   **46.** [2, -1, 5], [5, -1, 2]

Find the area of the parallelogram that has the following vertices in the xy-plane.

47. (0, 0), (2, 2), (-1, 1), (1, 3)   **48.** (1, 2), (0, 0), (2, 6), (1, 4)

49. (-4, 2), (-6, 5), (-3, 6), (-5, 9)   **50.** (8, -3), (10, 1), (9, 0), (11, 4)

Find the area of the triangle with vertices

51. (1, 0, 0), (0, 1, 0), (0, 0, 1)   **52.** (1, 1, 0), (0, 5, 6), (11, 0, 6)

53. (0, 0, 0), (0, 1, 0), (1, 1, 0)   **54.** (3, 2, 0), (1, -1, 0), (2, 3, 0)

55. (4, -2, 6), (6, -1, 7), (5, 0, 5)   **56.** (0, 4, 0), (1, 1, 1), (-2, 1, -3)

57. Find two unit vectors perpendicular to both [2, -2, 0] and [0, 4, 8].

58. Find two unit vectors perpendicular to both [1, -2, 3] and [-2, 4, 0].

59. Find a vector **v** parallel to the intersection of the planes $2x + 3y + 4z = 0$ and $x - y + z = 2$.

60. Prove Theorem 2.

Scalar Triple Product.
Other Repeated Products

Repeated products of vectors having three or more factors occur frequently in applications. The most important of these products is the **scalar triple product** or *mixed triple product* **a**•(**b** × **c**) of three vectors, which we shall denote by

$$(\mathbf{a} \quad \mathbf{b} \quad \mathbf{c}).$$

With respect to any right-handed Cartesian coordinate system, let

$$\mathbf{a} = [a_1, a_2, a_3], \qquad \mathbf{b} = [b_1, b_2, b_3], \qquad \mathbf{c} = [c_1, c_2, c_3].$$

Writing $\mathbf{b} \times \mathbf{c} = \mathbf{v} = [v_1, v_2, v_3]$, from (8) in Sec. 6.5 and (1) in Sec. 6.8 (with **b** and **c** instead of **a** and **b**) we obtain

$$\mathbf{a} \cdot (\mathbf{b} \times \mathbf{c}) = a_1 v_1 + a_2 v_2 + a_3 v_3 = a_1 \begin{vmatrix} b_2 & b_3 \\ c_2 & c_3 \end{vmatrix} + a_2 \begin{vmatrix} b_3 & b_1 \\ c_3 & c_1 \end{vmatrix} + a_3 \begin{vmatrix} b_1 & b_2 \\ c_1 & c_2 \end{vmatrix}.$$

The right-hand side is a development of a third-order determinant, so that we have the result

$$(a \quad b \quad c) = a \cdot (b \times c) = \begin{vmatrix} a_1 & a_2 & a_3 \\ b_1 & b_2 & b_3 \\ c_1 & c_2 & c_3 \end{vmatrix}.$$

(1)

Since interchanging of two rows reverses the sign of the determinant, we have

$$(2) \qquad (a \quad b \quad c) = -(b \quad a \quad c), \qquad \text{etc.}$$

Interchanging twice, we obtain

$$(3) \qquad (a \quad b \quad c) = (b \quad c \quad a) = (c \quad a \quad b).$$

Now, by definition

$$(a \quad b \quad c) = a \cdot (b \times c), \qquad (c \quad a \quad b) = c \cdot (a \times b),$$

and since dot multiplication is commutative, the last expression is equal to $(a \times b) \cdot c$. Therefore,

$$(4) \qquad a \cdot (b \times c) = (a \times b) \cdot c.$$

Furthermore, for any constant k,

$$(5) \qquad (ka \quad b \quad c) = k(a \quad b \quad c).$$

Geometrical interpretation. *The absolute value of the scalar triple product* (a b c) *has a simple geometrical interpretation. It is equal to the volume of the parallelepiped P with* **a, b, c** *as adjacent edges.* (A *parallelepiped* is a solid with three pairs of parallel sides; cf. Fig. 163. If the sides are perpendicular, it is a box; and if, moreover, the edges all have the same length, it is a cube.)

Indeed, from (1) in Sec. 6.5 we obtain

$$(a \quad b \quad c) = a \cdot (b \times c) = |a||b \times c| \cos \beta$$

where β is the angle between **a** and the product vector **b** × **c**. Now $|b \times c|$ is the area of the base of P, and the altitude h of P is equal to the absolute value of $|a| \cos \beta$ (Fig. 163). This proves our statement. ∎

From this geometrical consideration it follows that the value of the scalar triple product is a real number that is independent of the choice of right-handed Cartesian coordinates in space. But we should keep in mind that for *left-handed* Cartesian coordinate systems we have a minus sign in front of the determinants in (1), Sec. 6.8, and this leads to a minus sign in front of

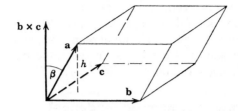

Fig. 163. Geometrical interpretation
of a scalar triple product

the determinant in (1). We may also say that the value of the determinant is **invariant** under transformations of right-handed into right-handed, or left-handed into left-handed, Cartesian coordinate systems, but is multiplied by -1 under a transition from a right-handed to a left-handed system (or conversely).

EXAMPLE 1. Tetrahedron
A tetrahedron is determined by three edge vectors **a**, **b**, **c** as indicateed in Fig. 164.
    Find the volume of the tetrahedron with $\mathbf{a} = [2, 0, 3]$, $\mathbf{b} = [0, 6, 2]$, $\mathbf{c} = [3, 3, 0]$ as edge vectors.

Solution. The volume V of the parallelepiped with these vectors as edge vectors is the absolute value of the scalar triple product

$$(\mathbf{a}\ \ \mathbf{b}\ \ \mathbf{c}) = \begin{vmatrix} 2 & 0 & 3 \\ 0 & 6 & 2 \\ 3 & 3 & 0 \end{vmatrix} = 2\begin{vmatrix} 6 & 2 \\ 3 & 0 \end{vmatrix} + 3\begin{vmatrix} 0 & 6 \\ 3 & 3 \end{vmatrix} = -12 - 54 = -66,$$

that is, $V = 66$. The minus sign indicates that **a**, **b**, **c**, in this order, form a left-handed triple if the Cartesian coordinate system is right-handed (and a right-handed triple if this system is left-handed). The volume of the tetrahedron is $\frac{1}{6}$ of that of the parallelepiped, hence 11.
    Can you sketch the tetrahedron, choosing the origin as the common initial point of the three vectors? What are the coordinates of the four vertices?

EXAMPLE 2. Parallelepiped with given vertices
Find the volume of the parallelepiped determined by the points $(1, 1, 1)$, $(2, 4, 1)$, $(5, 1, 1)$, $(4, 9, 3)$ as vertices.

Solution. Edge vectors are $[2 - 1,\ \ 4 - 1,\ \ 1 - 1] = [1, 3, 0]$, $[4, 0, 0]$, $[3, 8, 2]$. Hence the volume is the absolute value 24 of the scalar triple product

$$\begin{vmatrix} 1 & 3 & 0 \\ 4 & 0 & 0 \\ 3 & 8 & 2 \end{vmatrix} = -4\begin{vmatrix} 3 & 0 \\ 8 & 2 \end{vmatrix} = -24. \qquad \blacksquare$$

Fig. 164. Tetrahedron

From the geometrical interpretation of a scalar triple product we also obtain a useful criterion for linear dependence and independence of three vectors. If we let their initial points coincide, they form a linearly dependent set if and only if they lie in the same plane (or same line); cf. Sec. 6.4, particularly Example 5. This implies

Theorem 1 (Linear dependence)
Three vectors form a linearly dependent set if and only if their scalar triple product is zero.

Other Repeated Products

Other repeated products occurring in applications may be expressed in terms of dot, vector and scalar triple products. An important corresponding formula is

(6) $$\mathbf{b} \times (\mathbf{c} \times \mathbf{d}) = (\mathbf{b} \cdot \mathbf{d})\mathbf{c} - (\mathbf{b} \cdot \mathbf{c})\mathbf{d}$$

(proof below). It implies the **identity of Lagrange**

(7) $$(\mathbf{a} \times \mathbf{b}) \cdot (\mathbf{c} \times \mathbf{d}) = (\mathbf{a} \cdot \mathbf{c})(\mathbf{b} \cdot \mathbf{d}) - (\mathbf{a} \cdot \mathbf{d})(\mathbf{b} \cdot \mathbf{c})$$

(whose proof is left to the reader; cf. Prob. 43), as well as (cf. Prob. 45)

(8) $$(\mathbf{a} \times \mathbf{b}) \times (\mathbf{c} \times \mathbf{d}) = (\mathbf{a} \quad \mathbf{b} \quad \mathbf{d})\mathbf{c} - (\mathbf{a} \quad \mathbf{b} \quad \mathbf{c})\mathbf{d}.$$

Before we prove (6), we note that (6) also implies

$$(\mathbf{b} \times \mathbf{c}) \times \mathbf{d} = -\mathbf{d} \times (\mathbf{b} \times \mathbf{c}) = (\mathbf{d} \cdot \mathbf{b})\mathbf{c} - (\mathbf{d} \cdot \mathbf{c})\mathbf{b}.$$

This shows that in general $\mathbf{b} \times (\mathbf{c} \times \mathbf{d})$ and $(\mathbf{b} \times \mathbf{c}) \times \mathbf{d}$ are different vectors, that is, *cross multiplication is not associative* and the parentheses in (6) are important and cannot be omitted. For instance, with respect to a right-handed Cartesian coordinate system,

$$(\mathbf{i} \times \mathbf{j}) \times \mathbf{j} = \mathbf{k} \times \mathbf{j} = -\mathbf{i} \quad \text{but} \quad \mathbf{i} \times (\mathbf{j} \times \mathbf{j}) = \mathbf{0}.$$

***Proof of* (6).** We choose a right-handed Cartesian coordinate system such that the x-axis has the direction of $\mathbf{d}$ and the xy-plane contains $\mathbf{c}$. Then the vectors in (6) are of the form

$$\mathbf{b} = [b_1, b_2, b_3], \qquad \mathbf{c} = [c_1, c_2, 0], \qquad \mathbf{d} = [d_1, 0, 0].$$

Hence by (2) in Sec. 6.8,

$$\mathbf{c} \times \mathbf{d} = \begin{vmatrix} \mathbf{i} & \mathbf{j} & \mathbf{k} \\ c_1 & c_2 & 0 \\ d_1 & 0 & 0 \end{vmatrix} = -c_2 d_1 \mathbf{k}, \qquad \mathbf{b} \times (\mathbf{c} \times \mathbf{d}) = \begin{vmatrix} \mathbf{i} & \mathbf{j} & \mathbf{k} \\ b_1 & b_2 & b_3 \\ 0 & 0 & -c_2 d_1 \end{vmatrix}.$$

The determinant on the right equals $[-b_2c_2d_1, \quad b_1c_2d_1, \quad 0]$. But also

$$(\mathbf{b} \cdot \mathbf{d})\mathbf{c} - (\mathbf{b} \cdot \mathbf{c})\mathbf{d} = b_1d_1[c_1, c_2, 0] - (b_1c_1 + b_2c_2)[d_1, 0, 0]$$

$$= [-b_2c_2d_1, \quad b_1d_1c_2, \quad 0].$$

This proves (6) for our special coordinate system. Now the length and direction of a vector and a vector product, and the value of an inner product are independent of the choice of the coordinates. Furthermore, the representation of $\mathbf{b} \times (\mathbf{c} \times \mathbf{d})$ in terms of $\mathbf{i}, \mathbf{j}, \mathbf{k}$ will be the same for right-handed and left-handed systems, because of the double cross multiplication. Hence, (6) holds in any Cartesian coordinate system, and the proof is complete. ∎

   This is the end of Chap. 6 devoted to vectors and their algebraic operations. The next chapter concerns **matrices** and vectors and their algebraic operations. This extends the algebraic aspects of the present chapter, although the presence of matrices will introduce many other ideas in connection with systems of linear algebraic equations, eigenvalue problems, quadratic forms and so on. **Differential calculus** for vectors will be discussed in Chap. 8 and **integral calculus** for vectors in Chap. 9.

Problems for Sec. 6.9

Find the value of the scalar triple product of the given vectors (taken in the given order), whose components are referred to right-handed Cartesian coordinates.

1. $\mathbf{i}, \mathbf{j}, \mathbf{k}$
2. $\mathbf{i}, \mathbf{k}, \mathbf{j}$
3. $\mathbf{j}, \mathbf{k}, \mathbf{i}$
4. $3\mathbf{k}, -4\mathbf{j}, 8\mathbf{i}$
5. $[1, 3, 2], [4, 0, -1], [5, 2, -3]$
6. $[3, -1, 1], [0, 1, 1], [-1, 2, -3]$
7. $[4, 2, 1], [1, -1, -1], [3, 2, 4]$
8. $[\frac{1}{2}, 1, 2], [\frac{1}{4}, \frac{1}{8}, \frac{1}{5}], [3, \frac{9}{2}, \frac{44}{5}]$
9. $[1, 1, -1], [1, -1, 1], [-1, 1, 1]$
10. $[k, l, m], [m, k, l], [l, m, k]$

Linear dependence and independence. Are the following vectors linearly dependent or independent?

11. $\mathbf{i} + \mathbf{j}, \mathbf{i} + 2\mathbf{j}, \mathbf{i} + 3\mathbf{j}$
12. $\mathbf{i} - \mathbf{j}, \mathbf{j} - \mathbf{k}, \mathbf{k} - \mathbf{i}$
13. $\mathbf{i} + \mathbf{j}, \mathbf{j} + \mathbf{k}, \mathbf{k} + \mathbf{i}$
14. $[3, \frac{1}{8}, -4], [-8, -\frac{1}{3}, \frac{32}{3}]$
15. $[2, 3], [4, 1], [\frac{1}{2}, \frac{1}{4}]$
16. $\mathbf{j}, 0, \mathbf{k}$
17. $[4, -2, 7], \mathbf{i}, \mathbf{j}, \mathbf{k}$
18. $[1, 3, 6], [0, 0, \frac{1}{4}], [3, \frac{1}{3}, -\frac{1}{3}]$
19. $[1, 2, 3], [1, 2, 0], [1, 0, 0]$
20. $[1, \frac{1}{2}, \frac{1}{4}], [\frac{1}{4}, 1, \frac{1}{2}], [\frac{1}{2}, \frac{1}{4}, 1]$

21. Do the points $(1, 0, 2), (2, 2, 1), (3, 0, 1), (-1, 1, 0)$ lie in a plane?
22. Do the points $(9, 2, 1), (5, 1, 2), (5, -5, 0), (2, 1, 3)$ lie in a plane?
23. Find all λ such that $[1, 3, 2], [2, \lambda, 0], [4, 4, 1]$ are linearly dependent.
24. Determine λ and μ such that the points $(4, 4, -3), (5, \lambda, \mu)$ and $(2, 0, 5)$ lie on a straight line.

Volume of parallelepipeds and tetrahedra. Find the volume of the parallelepiped that has the given vectors as adjacent edges.

25. $4\mathbf{i}, 3\mathbf{j}, -8\mathbf{k}$
26. $\mathbf{i}, \mathbf{i} + \mathbf{j}, \mathbf{i} + \mathbf{j} + \mathbf{k}$
27. $[0, 2, 1], [1, -1, 0], [0, -1, 4]$
28. $[\frac{1}{2}, 1, 0], [-2, 4, 5], [3, 2, \frac{1}{2}]$
29. $[4, 3, 2], [1, 0, -1], [0, -2, -3]$
30. $[7, 6, 4], [1, 9, 3], [8, 15, 6]$

Find the volume of the tetrahedron that has the following vertices.

31. $(0, 1, 1), (1, 0, 0), (2, 2, 3), (-1, 0, 4)$ **32.** $(2, 1, 8), (3, 2, 9), (2, 1, 4), (3, 3, 10)$

33. $(0, 0, 0), (1, 0, 0), (0, 1, 0), (0, 0, 1)$ **34.** $(-1, 0, 1), (4, 4, 5), (0, 1, 0), (2, 2, 0)$

35. $(1, 0, 0), (0, 1, 1), (2, 0, 0), (0, 3, 5)$

36. $(-2, 4, -3), (1, 0, 4), (9, 9, 7), (3, 8, 4)$

Other repeated products. With respect to a right-handed Cartesian coordinate system, let $\mathbf{a} = [1, 2, -1]$, $\mathbf{b} = [3, 0, -4]$, $\mathbf{c} = [-1, 1, 0]$ and $\mathbf{d} = [2, -1, 3]$. Find

37. $(\mathbf{a} \times \mathbf{b}) \times \mathbf{c}, (\mathbf{a} \times \mathbf{b}) \cdot \mathbf{c}$ **38.** $(\mathbf{b} \times \mathbf{c}) \times \mathbf{d}, \mathbf{d} \times (\mathbf{b} \times \mathbf{c})$

39. $(\mathbf{a} \times \mathbf{c}) \times \mathbf{d}, (\mathbf{a} \times \mathbf{d}) \times \mathbf{c}$ **40.** $(\mathbf{b} \times \mathbf{b}) \times \mathbf{c}, \mathbf{b} \times (\mathbf{b} \times \mathbf{c})$

41. $(\mathbf{a} \times \mathbf{b}) \times (\mathbf{c} \times \mathbf{d})$ **42.** $(\mathbf{a} \times \mathbf{c}) \cdot (\mathbf{b} \times \mathbf{d}), (\mathbf{c} \times \mathbf{a}) \cdot (\mathbf{d} \times \mathbf{b})$

43. Prove (7). *Hint.* Take the dot product of **a** and (6).

44. A system of three homogeneous linear equations in three unknowns has a non-trivial solution if and only if the determinant of the coefficients of the system is zero. Using this familiar theorem, prove Theorem 1.

45. Derive (8) from (6).

Further Proof in Chapter 6

Proof of Theorem 1 in Sec. 6.8, page 336

We prove that in right-handed Cartesian coordinates, the vector product

$$\mathbf{v} = \mathbf{a} \times \mathbf{b} = [a_1, \quad a_2, \quad a_3] \times [b_1, \quad b_2, \quad b_3]$$

has the components

$$(1) \quad v_1 = a_2 b_3 - a_3 b_2, \qquad v_2 = a_3 b_1 - a_1 b_3, \qquad v_3 = a_1 b_2 - a_2 b_1$$

[which can easily be memorized by formula (2) in the text].

We need only consider the case $\mathbf{v} \neq \mathbf{0}$. Since $\mathbf{v}$ is perpendicular to both **a** and **b**, Theorem 1 in Sec. 6.5 gives $\mathbf{a} \cdot \mathbf{v} = 0$ and $\mathbf{b} \cdot \mathbf{v} = 0$, in components [cf. (8), Sec. 6.5]

$$(3) \qquad \begin{aligned} a_1 v_1 + a_2 v_2 + a_3 v_3 &= 0, \\ b_1 v_1 + b_2 v_2 + b_3 v_3 &= 0. \end{aligned}$$

Multiplying the first equation by b_3, the last by a_3, and subtracting, we obtain

$$(a_3 b_1 - a_1 b_3) v_1 = (a_2 b_3 - a_3 b_2) v_2.$$

Multiplying the first equation by b_1, the last by a_1, and subtracting, we obtain

$$(a_1 b_2 - a_2 b_1) v_2 = (a_3 b_1 - a_1 b_3) v_3.$$

We can easily verify that these two equations are satisfied by

(4) $v_1 = c(a_2 b_3 - a_3 b_2),$ $v_2 = c(a_3 b_1 - a_1 b_3),$ $v_3 = c(a_1 b_2 - a_2 b_1),$

where c is a constant. The reader may verify by inserting that (4) also satisfies (3). Now each of the equations in (3) represents a plane through the origin in $v_1 v_2 v_3$-space. The vectors **a** and **b** are normal vectors of these planes (cf. Example 7 in Sec. 6.5). Since $\mathbf{v} \neq \mathbf{0}$, these vectors are not parallel and the two planes do not coincide. Hence their intersection is a straight line L through the origin. Since (4) is a solution of (3) and, for varying c, represents a straight line, we conclude that (4) represents L, and every solution of (3) must be of the form (4). In particular, the components of **v** must be of this form, where c is to be determined. From (4) we obtain

$$|\mathbf{v}|^2 = v_1{}^2 + v_2{}^2 + v_3{}^2 = c^2[(a_2 b_3 - a_3 b_2)^2 + (a_3 b_1 - a_1 b_3)^2 + (a_1 b_2 - a_2 b_1)^2].$$

This can be written

$$|\mathbf{v}|^2 = c^2[(a_1{}^2 + a_2{}^2 + a_3{}^2)(b_1{}^2 + b_2{}^2 + b_3{}^2) - (a_1 b_1 + a_2 b_2 + a_3 b_3)^2],$$

as can be verified by performing the indicated multiplications in both formulas and comparing. Using (8) in Sec. 6.5, we thus have

$$|\mathbf{v}|^2 = c^2[(\mathbf{a} \cdot \mathbf{a})(\mathbf{b} \cdot \mathbf{b}) - (\mathbf{a} \cdot \mathbf{b})^2].$$

By comparing this with (5) in Sec. 6.7 we conclude that $c = \pm 1$.

We show that $c = +1$. This can be done as follows.

If we change the lengths and directions of **a** and **b** continuously and so that at the end $\mathbf{a} = \mathbf{i}$ and $\mathbf{b} = \mathbf{j}$ (Fig. 161a in Sec. 6.8), then **v** will change its length and direction continuously, and at the end, $\mathbf{v} = \mathbf{i} \times \mathbf{j} = \mathbf{k}$. Obviously we may effect the change so that both **a** and **b** remain different from the zero vector and are not parallel at any instant. Then **v** is never equal to the zero vector, and since the change is continuous and c can only assume the values $+1$ or -1, it follows that at the end c must have the same value as before. Now at the end $\mathbf{a} = \mathbf{i}$, $\mathbf{b} = \mathbf{j}$, $\mathbf{v} = \mathbf{k}$ and, therefore, $a_1 = 1$, $b_2 = 1$, $v_3 = 1$, and the other components in (4) are zero. Hence from (4) we see that $v_3 = c = +1$. This proves Theorem 1.

For a left-handed coordinate system, $\mathbf{i} \times \mathbf{j} = -\mathbf{k}$ (cf. Fig. 161b in Sec. 6.8), resulting in $c = -1$. This proves Remark 2 in Sec. 6.8.   ∎

Review Problems for Chapter 6

1. Give examples of quantities that are vectors.
2. Why do the components of a vector remain unchanged if we change the initial point of the vector?
3. Is the addition of a vector with two components and a vector with three components defined?
4. Is the addition of a vector and a scalar defined?
5. Show that if $\mathbf{v} \neq \mathbf{0}$, then $(1/|\mathbf{v}|)\mathbf{v}$ is a unit vector.
6. What are typical applications that motivate the inner product and the vector product?

7. In what case will the work done by a constant force in a displacement be zero?

8. What are right-handed and left-handed coordinates? Where in this chapter was this distinction essential?

9. The vector product has two unusual properties. Can you remember them?

10. Vector formulas in two and three dimensions often look the same. Explain this in terms of representations of a circle and a sphere. Can you think of other examples?

11. What do we mean by orthogonality? Why is this concept important, and in what connection?

12. What is wrong with $\mathbf{a} \times \mathbf{b} \times \mathbf{c}$? With $\mathbf{a} \cdot \mathbf{b} \cdot \mathbf{c}$? With $(\mathbf{a} \cdot \mathbf{b}) \times \mathbf{c}$?

13. Give examples of subspaces of R^4.

14. How many different bases does R^3 have?

15. What is an orthonormal basis? What is the advantage of using it?

Let $\mathbf{a} = [3, 0, 2]$, $\mathbf{b} = [2, 2, -1]$, $\mathbf{c} = [1, -2, 3]$, $\mathbf{d} = [2, -1, 5]$. Find the following expressions. (In connection with vector products assume the coordinate system to be right-handed.)

16. $\mathbf{a} + \mathbf{b} + \mathbf{c} + \mathbf{d}$

17. $\mathbf{b} + \mathbf{c} - \mathbf{a}$

18. $\mathbf{a} - \mathbf{b} + \mathbf{c} - \mathbf{d}$

19. $4\mathbf{a} + 8\mathbf{b}, 4(2\mathbf{b} + \mathbf{a})$

20. $5\mathbf{a} + 2\mathbf{b}, |5\mathbf{a} + 2\mathbf{b}|$

21. $|\mathbf{b} - \mathbf{c}|, |\mathbf{b}| - |\mathbf{c}|$

22. $(1/|\mathbf{a}|)\mathbf{a}, (1/|\mathbf{b}|)\mathbf{b}$

23. $(1/|\mathbf{b} - \mathbf{a}|)(\mathbf{b} - \mathbf{a})$

24. $\mathbf{a} \cdot \mathbf{b}, \mathbf{b} \cdot \mathbf{a}$

25. $\mathbf{a} \cdot (\mathbf{b} - \mathbf{c}), \mathbf{a} \cdot (\mathbf{c} - \mathbf{b})$

26. $\mathbf{a} \times \mathbf{b}, \mathbf{b} \times \mathbf{a}$

27. $\mathbf{a} \times (\mathbf{b} - \mathbf{c}), (\mathbf{c} - \mathbf{b}) \times \mathbf{a}$

28. $6\mathbf{c} \cdot \mathbf{d}, 3\mathbf{d} \cdot 2\mathbf{c}$

29. $(\mathbf{a} - \mathbf{b}) \times (\mathbf{c} - \mathbf{d}), \mathbf{a} - (\mathbf{b} \times \mathbf{c}) - \mathbf{d}$

30. $\mathbf{a} \times \mathbf{d}, |\mathbf{a} \times \mathbf{d}|$

31. $(\mathbf{a} \quad \mathbf{b} \quad \mathbf{c}), (\mathbf{b} \quad \mathbf{c} \quad \mathbf{a})$

32. $(\mathbf{b} \quad -\mathbf{b} \quad \mathbf{d}), (\mathbf{b} \quad \mathbf{b} + \mathbf{d} \quad \mathbf{d})$

33. $(\mathbf{b} \quad \mathbf{c} \quad \mathbf{d}), (\mathbf{c} \quad \mathbf{b} \quad 2\mathbf{d})$

34. $\mathbf{a} \times (\mathbf{b} \times \mathbf{c}), (\mathbf{a} \times \mathbf{b}) \times \mathbf{c}$

35. $(\mathbf{a} + 2\mathbf{b}) \times (\mathbf{c} \times \mathbf{d})$

36. Find the resultant of the forces $\mathbf{p} = [3, 2, 1]$, $\mathbf{q} = [4, -4, 2]$ and $\mathbf{u} = [2, -2, 3]$.

37. Find a force $\mathbf{p}$ such that the resultant of $\mathbf{p}$, $\mathbf{q} = [1, -2, 2]$ and $\mathbf{u} = [2, -3, 4]$ is the zero vector.

38. Find the angle between the forces $[2, 2]$ and $[3, 1]$. Make a figure.

39. Find the angle between the forces $[2, 1, 3]$ and $[4, 4, 0]$.

40. Find the angles α, β, γ between the force $\mathbf{p} = [3, 6, 8]$ and the coordinate axes. Verify that $\cos^2 \alpha + \cos^2 \beta + \cos^2 \gamma = 1$. Prove this in general.

Find the work done by a force $\mathbf{p}$ acting on a body if the body is displaced from a point P to a point Q along the straight segment PQ, where P, Q and $\mathbf{p}$ are

41. $(0, 0), (1, 1), [4, 6]$

42. $(3, 5), (5, 3), [7, -1]$

43. $(2, 6), (4, 14), [-8, 2]$

44. $(12, 0, 2), (0, 3, 8), [1, -1, 3]$

45. $(1, 1, 1), (5, 5, 5), [2, 4, 0]$

46. $(-3, 8, 4), (1, 2, 4), [3, 2, 14]$

Find the moment vector $\mathbf{m}$ of a force $\mathbf{p}$ about a point Q if $\mathbf{p}$ acts on a line through a point A, where $\mathbf{p}$, A, Q are

47. $[0, 5, 0], (2, 3, 0), (0, 0, 0)$

48. $[3, 4, 0], (7, 8, 0), (1, 1, 0)$

49. $[0, 8, -1], (0, 2, 2), (0, 1, 1)$

50. $[1, 1, 0], (0, 4, 2), (3, 0, 2)$

51. $[2, 9, -3], (4, 6, 0), (3, 1, 5)$

52. $[8, 6, 4], (2, 2, 3), (10, 8, 7)$

Find the volume of the parallelepiped with edge vectors

53. [4, 0, 2], [1, −3, 1], [5, 2, 0]   **54.** [1, 2, −4], [0, 5, 7], [3, 3, −3]

Find the volume of the tetrahedron with vertices

55. (2, 1, 1), (3, 3, 4), (0, 1, 5), (1, 2, 2)   **56.** (0, 1, 2), (5, 5, 6), (1, 2, 1), (3, 3, 1)

Find the area of the triangle with vertices

57. (1, 1, 1), (2, 3, 4), (0, −1, 2)   **58.** (1, 2), (4, −1), (−3, 1)

59. (1, 0, 1), (0, 1, 1), (0, 0, 2)   **60.** $(-1, \frac{1}{2}, \frac{3}{2})$, $(0, 2, \frac{1}{2})$, $(4, 1, \frac{3}{2})$

Find a representation of the plane through the points

61. (1, 1, 1), (5, 0, −5), (3, 2, 0)   **62.** (−1, 0, 2), (8, 6, 4), (1, 3, 5)

Find the component of a force **p** in the direction of a vector **b**, where

63. **p** = [1, 2], **b** = [4, 0]   **64.** **p** = [−5, 3], **b** = [8, 6]

65. **p** = [4, 0, −2], **b** = [1, −3, 5]   **66.** **p** = $[6, 2, -\frac{1}{2}]$, **b** = [3, 1, 40]

67. Find two forces **p**, **q** in the directions of [1, 1] and [−1, 1] such that **p**, **q**, **u** = [3, −5] and **v** = [4, 8] are in equilibrium.

68. In what case is the component of **a** in the direction of **b** equal to the component of **b** in the direction of **a**?

69. Let the pairs **a**, **b**, and **c**, **d** each determine a plane. Show that these planes are parallel if and only if (**a** × **b**) × (**c** × **d**) = **0**.

70. What does it mean geometrically if in Prob. 69, (**a** × **b**)•(**c** × **d**) = 0?

Summary of Chapter 6
Vectors

Vector algebra, as developed in this chapter, is a powerful shorthand for handling quantities such as forces and velocities, which have a magnitude and a direction. We call them **vectors.** Geometrically, they are arrows (directed line segments); algebraically, they can be handled as ordered triples of real numbers (their *"components"*); notations are

(1) $\mathbf{a} = [a_1, a_2, a_3] = a_1\mathbf{i} + a_2\mathbf{j} + a_3\mathbf{k}$ (Secs. 6.2, 6.3).

For distinction, we call **scalar** any number or quantity such as mass or area, which has a magnitude but no direction (Sec. 6.1).

Vector addition (suggested by taking resultants of forces) and **multiplication of vectors by scalars** (Sec. 6.3) enable us to calculate with vectors almost as easily as with numbers. With these two algebraic operations, the set of all vectors is called the **vector space** R^3. (Section 6.4 includes the idea of generalizing R^3 to R^n and beyond, but further details would belong to more advanced courses.)

Applications suggest two kinds of multiplication of vectors by vectors. The **inner product** or **dot product**

$$\text{(2)} \qquad \mathbf{a} \cdot \mathbf{b} = |\mathbf{a}||\mathbf{b}| \cos \gamma = a_1 b_1 + a_2 b_2 + a_3 b_3 \qquad \text{(Sec. 6.5)}$$

is a scalar. It is suggested, for instance, by the concept of work done by a force. By (2) we have for the length of **a**

$$\text{(3)} \qquad |\mathbf{a}| = \sqrt{\mathbf{a} \cdot \mathbf{a}} = \sqrt{a_1^2 + a_2^2 + a_3^2}$$

as well as a formula for the angle γ between two vectors. If $\mathbf{a} \cdot \mathbf{b} = 0$, we call the vectors **orthogonal**. (Again, this can be generalized; cf. Sec. 6.6.)

The **vector product** or **cross product** $\mathbf{v} = \mathbf{a} \times \mathbf{b}$ is a vector of length

$$\text{(4)} \qquad |\mathbf{a} \times \mathbf{b}| = |\mathbf{a}||\mathbf{b}| \sin \gamma \qquad \text{(Sec. 6.7)}$$

and perpendicular to both **a** and **b** such that **a, b, v** form a *right-handed* triple. In terms of components with respect to right-handed coordinates,

$$\text{(5)} \qquad \mathbf{a} \times \mathbf{b} = \begin{vmatrix} \mathbf{i} & \mathbf{j} & \mathbf{k} \\ a_1 & a_2 & a_3 \\ b_1 & b_2 & b_3 \end{vmatrix} \qquad \text{(Sec. 6.8)}.$$

The vector product is suggested, for instance, by moments of forces or by rotations. Caution! This multiplication is *anti*commutative, $\mathbf{a} \times \mathbf{b} = -\mathbf{b} \times \mathbf{a}$, and is *not* associative (Sec. 6.7).

Repeated products can be expressed in terms of scalar products, vector products and **scalar triple products** (Sec. 6.9)

$$\text{(6)} \quad (\mathbf{a} \ \mathbf{b} \ \mathbf{c}) = \mathbf{a} \cdot (\mathbf{b} \times \mathbf{c}) = (\mathbf{a} \times \mathbf{b}) \cdot \mathbf{c} = \begin{vmatrix} a_1 & a_2 & a_3 \\ b_1 & b_2 & b_3 \\ c_1 & c_2 & c_3 \end{vmatrix}.$$

Chapter 7

Matrices
and Determinants

A matrix is a rectangular array of numbers. Such arrays occur in various branches of applied mathematics. In many cases they form the coefficients of linear transformations (examples in Sec. 7.1) or systems of linear equations (simultaneous linear equations) arising, for instance, from electrical networks, frameworks in mechanics, curve fitting in statistics and transportation problems. Matrices are useful because they enable us to consider an array of many numbers as a single object, denote it by a single symbol, and perform calculations with these symbols in a very compact form. The "mathematical shorthand" thus obtained is very elegant and powerful and is suitable for various practical problems. It entered engineering mathematics more than 60 years ago and is of increasing importance in various branches.

This chapter has four big parts:

> Calculation with matrices, Secs. 7.1—7.4
> Systems of linear equations, Secs. 7.5—7.11
> Eigenvalue problems, Secs. 7.12—7.15
> Systems of differential equations, Sec. 7.16.

After the introduction of matrices and the operations of matrix algebra (Secs. 7.1—7.4), we apply matrices to systems of linear equations, without using determinants (Secs. 7.5—7.8) and later with the use of determinants (Secs. 7.9—7.11). Here we discuss the Gauss elimination (Sec. 7.5), the concepts of rank and inverse (Secs. 7.6, 7.8) and Cramer's rule (Sec. 7.11).

Eigenvalue problems (Secs. 7.12—7.15) include a study of complex matrices (Hermitian, skew-Hermitian, unitary, in Secs. 7.13, 7.14) and the diagonalization of quadratic forms (Sec. 7.15). Section 7.16 is devoted to systems of differential equations. Applications of matrices to practical problems are shown throughout the chapter.

Numerical methods are presented in Chap. 19, which is independent of Chap. 18, so that these methods can be studied immediately after the corresponding material in the present chapter.

Prerequisite for this chapter: Secs. 6.1—6.5.
Sections that may be omitted in a shorter course: 7.13—7.16.
References: Appendix 1, Part B.
Answers to problems: Appendix 2.

7.1 Basic Concepts

A **matrix** is a rectangular array of numbers enclosed in brackets (or parentheses). For instance,

$$(1) \qquad \begin{bmatrix} 2 & 0.4 & 8 \\ 5 & -3 & 0 \end{bmatrix}, \quad \begin{bmatrix} 0 \\ 4 \end{bmatrix}, \quad \begin{bmatrix} a_1 & a_2 & a_3 \end{bmatrix}, \quad \begin{bmatrix} a & b \\ c & d \end{bmatrix}$$

are matrices. The first has two "rows" (horizontal lines) and three "columns" (vertical lines). The next two look like vectors. The last is "square." Just as we calculate with numbers or with vectors, we can calculate with matrices, as we shall see in the next two sections. This makes matrices a powerful tool for systems of linear equations, linear transformations and numerous other applications. Let us first illustrate how matrices appear in linear equations and how we can find good notations that extend to more general problems.

EXAMPLE 1. Matrix occurring in linear equations or linear transformations
Consider

$$(2) \qquad \begin{aligned} 20x_1 - 3x_2 &= y_1 \\ 0.5x_1 + 7x_2 &= y_2. \end{aligned}$$

If y_1 and y_2 are given numbers (for instance, $y_1 = 43$, $y_2 = -6$), this is a system of two linear equations in two unknowns x_1, x_2. The coefficients of this system are $20, -3, 0.5, 7$. To recognize their position, we may arrange them in the way they occur in the equations, that is, in the form of a matrix,

$$(3) \qquad \begin{bmatrix} 20 & -3 \\ 0.5 & 7 \end{bmatrix}.$$

If x_1, x_2 and y_1, y_2 are variable quantities (for instance, coordinates corresponding to two different coordinate systems in the plane), then our pair of formulas (2) represents a linear transformation. This transformation is completely determined by its four coefficients, that is, by the matrix (3). We may denote the matrix by a single letter, for instance **A,** and write it in the form

$$\mathbf{A} = \begin{bmatrix} a_{11} & a_{12} \\ a_{21} & a_{22} \end{bmatrix},$$

where $a_{11} = 20$, $a_{12} = -3$, $a_{21} = 0.5$, $a_{22} = 7$. This notation is better than the notation a, b, c, d in (1). It has the advantage that we can see immediately the position of a coefficient a_{jk} because the first subscript is the number of the row (horizontal line) and the second is the number of the column (vertical line) of a_{jk}. Thus a_{11} stands in the first row and first column, a_{12} in the first row and second column, a_{21} in the second row and first column and a_{22} in the second row and second column. ∎

We shall now define the general concept of a matrix and introduce related concepts and corresponding notations.

Algebraic operations for matrices which will enable us to do calculations with matrices will be defined and explained in the next section.

General Notations and Names

A **matrix** is a rectangular array of (real or complex) numbers of the form

$$
(4) \qquad \mathbf{A} = \begin{bmatrix} a_{11} & a_{12} & \cdots & a_{1n} \\ a_{21} & a_{22} & \cdots & a_{2n} \\ \cdot & \cdot & \cdots & \cdot \\ a_{m1} & a_{m2} & \cdots & a_{mn} \end{bmatrix}.
$$

These mn numbers are called the **entries** in the matrix $\mathbf{A}$ or the *elements* of $\mathbf{A}$. The horizontal lines are called **rows** or *row vectors,* and the vertical lines are called **columns** or *column vectors* of $\mathbf{A}$. A matrix with m rows and n columns, as in (4), is briefly called an $\boldsymbol{m \times n}$ **matrix** (read "m by n matrix"). Thus, the matrices in (1) are $2 \times 3, 2 \times 1, 1 \times 3$ and 2×2.

Matrices will be denoted by capital boldface letters $\mathbf{A}$, $\mathbf{B}$ etc. We shall also denote $\mathbf{A}$ by $[a_{jk}]$, that is, by the general entry in $\mathbf{A}$ enclosed in brackets. Similarly, $\mathbf{B} = [b_{jk}]$ and so on.

*In the **double-subscript notation** for the entries, the first subscript always denotes the **row** and the second the **column** in which the given entry stands.* Thus a_{23} is the entry in the second row and third column.

If all the entries in a matrix $\mathbf{A}$ are *real* numbers, we call $\mathbf{A}$ a **real matrix,** otherwise (that is, if not all entries are real) a *complex matrix.*

If a matrix has only one row, we call it a **row matrix** or *row vector,* and denote it by a lowercase boldface letter; thus

$$
\mathbf{a} = \begin{bmatrix} a_1 & a_2 & \cdots & a_n \end{bmatrix}.
$$

Similarly, a **column matrix** or *column vector* is a matrix with only one column, and will also be denoted by a lowercase boldface letter; thus,

$$
\mathbf{b} = \begin{bmatrix} b_1 \\ b_2 \\ \vdots \\ b_m \end{bmatrix}.
$$

A **square matrix** is a matrix that has as many rows as columns. If it has n rows and n columns, it is an $n \times n$ matrix, and we call n its **order.** The diagonal containing the entries $a_{11}, a_{22}, \cdots, a_{nn}$ in a square matrix $\mathbf{A} = [a_{jk}]$ is called the **main diagonal** or *principal diagonal* of this matrix. Thus the last matrix in (1) is square of order 2, and a and d are on its main diagonal.

Square matrices are of particular importance, as we shall see.

Any matrix obtained by omitting some rows and columns from a given $m \times n$ matrix $\mathbf{A}$ is called a **submatrix** of $\mathbf{A}$.

As a matter of convention and convenience the notion "submatrix" includes **A** itself (as the matrix obtained from **A** by omitting no rows or columns).

EXAMPLE 2. Submatrices of a matrix
The matrix

$$\begin{bmatrix} a_{11} & a_{12} & a_{13} \\ a_{21} & a_{22} & a_{23} \end{bmatrix}$$

contains three 2×2 submatrices, namely,

$$\begin{bmatrix} a_{11} & a_{12} \\ a_{21} & a_{22} \end{bmatrix}, \quad \begin{bmatrix} a_{11} & a_{13} \\ a_{21} & a_{23} \end{bmatrix}, \quad \begin{bmatrix} a_{12} & a_{13} \\ a_{22} & a_{23} \end{bmatrix},$$

two 1×3 submatrices (the two row vectors), three 2×1 submatrices (the column vectors), six 1×2 submatrices, namely,

$$\begin{bmatrix} a_{11} & a_{12} \end{bmatrix}, \quad \begin{bmatrix} a_{11} & a_{13} \end{bmatrix}, \quad \begin{bmatrix} a_{12} & a_{13} \end{bmatrix},$$
$$\begin{bmatrix} a_{21} & a_{22} \end{bmatrix}, \quad \begin{bmatrix} a_{21} & a_{23} \end{bmatrix}, \quad \begin{bmatrix} a_{22} & a_{23} \end{bmatrix},$$

and six 1×1 submatrices, $\begin{bmatrix} a_{11} \end{bmatrix}, \begin{bmatrix} a_{12} \end{bmatrix}, \cdots, \begin{bmatrix} a_{23} \end{bmatrix}$. ∎

In the next section we define two algebraic operations for matrices, the addition of matrices and the multiplication of a matrix by a scalar (a number). Problems follow at the end of the next section.

7.2 Addition of Matrices, Multiplication by Scalars

We begin by defining equality of matrices.

Definition. Equality of matrices
Two matrices $\mathbf{A} = [a_{jk}]$ and $\mathbf{B} = [b_{jk}]$ are **equal** if and only if **A** and **B** have the same number of rows and the same number of columns and corresponding elements are equal, that is,

$$a_{jk} = b_{jk} \qquad \text{for all occurring } j \text{ and } k.$$

Then we write

$$\mathbf{A} = \mathbf{B}.$$

EXAMPLE 1. Equality of matrices
The definition implies that

$$\mathbf{A} = \begin{bmatrix} a_{11} & a_{12} \\ a_{21} & a_{22} \end{bmatrix} = \mathbf{B} = \begin{bmatrix} 4 & 0 \\ 3 & -1 \end{bmatrix} \quad \text{if and only if} \quad \begin{matrix} a_{11} = 4, & a_{12} = 0, \\ a_{21} = 3, & a_{22} = -1. \end{matrix}$$

A cannot be equal to, say, a 2×3 matrix. A column matrix cannot be equal to a row matrix. ∎

We shall now define two algebraic operations for matrices. These are called *addition of matrices* and *multiplication of matrices by scalars* (numbers). These definitions are motivated and suggested by various applications (to be considered later in this chapter) as well as by the corresponding definitions for vectors in Sec. 6.3, which they will include as special cases.

Definition. Addition of matrices

Addition is defined only for matrices with the *same* number of rows and the *same* number of columns, and is then defined as follows. The **sum** of two $m \times n$ matrices $A = [a_{jk}]$ and $B = [b_{jk}]$ is written

$$A + B$$

and is the $m \times n$ matrix obtained by adding corresponding entries in A and B; thus, the entries in $A + B$ are

$$(1) \qquad\qquad a_{jk} + b_{jk} \qquad\qquad \begin{matrix} j = 1, \cdots, m, \\ k = 1, \cdots, n. \end{matrix}$$

Matrices with different numbers of rows or different numbers of columns cannot be added.

EXAMPLE 2. Addition of matrices

If

$$A = \begin{bmatrix} -4 & 6 & 3 \\ 0 & 1 & 2 \end{bmatrix} \text{ and } B = \begin{bmatrix} 5 & -1 & 0 \\ 3 & 1 & 0 \end{bmatrix}, \text{ then } A + B = \begin{bmatrix} 1 & 5 & 3 \\ 3 & 2 & 2 \end{bmatrix}.$$

A in this example and B in Example 1 cannot be added.   ∎

We define the $m \times n$ **zero matrix 0** to be the $m \times n$ matrix with all entries zeros. Then we see from the definition that matrix addition enjoys properties quite similar to those of the addition of real numbers,

$$(2) \quad \begin{aligned} &\text{(a)} && A + B = B + A \\ &\text{(b)} && (U + V) + W = U + (V + W) && \text{(written } U + V + W) \\ &\text{(c)} && A + 0 = A \\ &\text{(d)} && A + (-A) = 0 \end{aligned}$$

where $-A = [-a_{jk}]$ is the $m \times n$ matrix obtained by multiplying every entry in A by -1 and is called the **negative** of A.

Instead of $A + (-B)$ we simply write $A - B$ and call this matrix the **difference** of A and B. Obviously, the entries in $A - B$ are obtained by subtracting corresponding entries in A and B.

Definition. Multiplication of matrices by scalars (numbers)

The **product** of an $m \times n$ matrix $\mathbf{A} = [a_{jk}]$ by a scalar c is written $c\mathbf{A}$ (or $\mathbf{A}c$) and is the $m \times n$ matrix obtained by multiplying each entry in $\mathbf{A}$ by c:

$$
(3) \qquad c\mathbf{A} = \mathbf{A}c = \begin{bmatrix} ca_{11} & ca_{12} & \cdots & ca_{1n} \\ ca_{21} & ca_{22} & \cdots & ca_{2n} \\ \cdot & \cdot & \cdots & \cdot \\ ca_{m1} & ca_{m2} & \cdots & ca_{mn} \end{bmatrix}.
$$

EXAMPLE 3. Multiplication of matrices by scalars (numbers)

If $\mathbf{A} = \begin{bmatrix} 2.7 & -1.8 \\ 0.9 & 3.6 \end{bmatrix}$, then $\mathbf{A} + \mathbf{A} = 2\mathbf{A} = \begin{bmatrix} 5.4 & -3.6 \\ 1.8 & 7.2 \end{bmatrix}$ and $\dfrac{10}{9}\mathbf{A} = \begin{bmatrix} 3 & -2 \\ 1 & 4 \end{bmatrix}$. ∎

From the definitions we see that for any $m \times n$ matrices (with fixed m and n) and any numbers,

$$
(4) \qquad
\begin{aligned}
&\text{(a)} \quad c(\mathbf{A} + \mathbf{B}) = c\mathbf{A} + c\mathbf{B} \\
&\text{(b)} \quad (c + k)\mathbf{A} = c\mathbf{A} + k\mathbf{A} \\
&\text{(c)} \quad c(k\mathbf{A}) = (ck)\mathbf{A} \qquad \text{(written } ck\mathbf{A}) \\
&\text{(d)} \quad 1\mathbf{A} = \mathbf{A}.
\end{aligned}
$$

Note that $(-1)\mathbf{A} = -\mathbf{A}$, the negative of $\mathbf{A}$.

Formulas (2) and (4) express the properties which are characteristic of a vector space (cf. Sec. 6.4). This gives

Theorem 1 (Vector spaces of matrices)

All real (or complex) $m \times n$ matrices (with m and n fixed) form a real (or complex) vector space of dimension mn. A basis consists of the mn matrices $\mathbf{E}_{rs}$ ($r = 1, \cdots, m$; $s = 1, \cdots, n$), where $\mathbf{E}_{rs}$ is the $m \times n$ matrix with $e_{rs} = 1$ (that is, the entry in the rth row and sth column equals 1) and all the other entries zero.

EXAMPLE 4. Vector space of all real 2×2 matrices

This vector space has dimension 4. A basis is

$$
\mathbf{E}_{11} = \begin{bmatrix} 1 & 0 \\ 0 & 0 \end{bmatrix}, \quad \mathbf{E}_{12} = \begin{bmatrix} 0 & 1 \\ 0 & 0 \end{bmatrix}, \quad \mathbf{E}_{21} = \begin{bmatrix} 0 & 0 \\ 1 & 0 \end{bmatrix}, \quad \mathbf{E}_{22} = \begin{bmatrix} 0 & 0 \\ 0 & 1 \end{bmatrix}.
$$

For **B** in Example 1 we obtain $\mathbf{B} = 4\mathbf{E}_{11} + 3\mathbf{E}_{21} - \mathbf{E}_{22}$. ∎

In this section we have defined two algebraic operations for matrices, addition and multiplication by a scalar. In the next section we define a third algebraic operation, the **multiplication of two matrices.** These three operations will then enable us to do calculations with matrices that have useful practical applications to systems of linear equations, linear transformations and so on.

Problems for Secs. 7.1 and 7.2

$$\text{Let}\quad A = \begin{bmatrix} 2 & 1 \\ 1 & 3 \end{bmatrix}, \quad B = \begin{bmatrix} -2 & 0 \\ 3 & 4 \end{bmatrix}, \quad C = \begin{bmatrix} 3 & 2 & 0 \\ 1 & 0 & 4 \end{bmatrix}, \quad D = \begin{bmatrix} 2 & 6 & 0 \\ 3 & 5 & 0 \end{bmatrix}.$$

Find the following expressions or give reasons why they are undefined.

1. $A + B, B + A$
2. $A - B, B - A$
3. $4A - 2B, 2(2A - B)$
4. $A + D, D + A$
5. $6C + 6D, 6(C + D)$
6. $5D - 4C, 4C - 5D$
7. $A + B + C + D$
8. $-C - D, -(C + D)$
9. $\frac{1}{2}A - \frac{3}{4}B, \frac{1}{2}(A - \frac{3}{2}B)$
10. $0B + D, 0A + B$

11. Write **B** in terms of the basis in Example 4.

12. What is c_{12} in **C**? c_{21}? c_{23}? List all submatrices of C.

13. Verify that $9A = 2A + 7A = 11A - 2A$ and $3(4A) = 12A$.

$$\text{Let}\quad M = \begin{bmatrix} 2 & 0 & 0 \\ -1 & 1 & 0 \\ 3 & -4 & 0 \end{bmatrix}, \quad P = \begin{bmatrix} 2 & 1 & 4 \\ 8 & 0 & 0 \\ 3 & -1 & 1 \end{bmatrix}, \quad Q = \begin{bmatrix} 0 \\ 3 \\ 7 \end{bmatrix}, \quad R = \begin{bmatrix} 2 \\ 9 \\ 0 \end{bmatrix}.$$

Find the following expressions or give reasons why they are undefined.

14. $M + P, P + M$
15. $M - P$
16. $3Q - 5R, 6Q - 10R$
17. $-3M + 3P$
18. $P + Q - R$
19. $2M - 4P, 2(M - 2P)$

20. Prove (2) and (4) in Sec. 7.2. Give illustrative examples.

Vector spaces whose elements are matrices

21. Prove that all 3×3 matrices form a vector space. What is its dimension?

22. Do the 2×2 and 3×3 matrices together form a vector space? (Give reason.)

23. **(Positive matrices)** By definition, a *positive matrix* is a real $n \times n$ matrix whose entries are all positive. Do these matrices with fixed n form a vector space? (These matrices have applications in mechanics, economics, etc.)

24. Do the 3×2 matrices with entries in the first column all zero form a vector space?

25. Show that the scalar multiples of any given matrix form a vector space. What is its dimension?

Use of matrices in modeling networks. Matrices have various engineering applications, as we shall see. For instance, they may be used to characterize connections (in electrical networks, in nets of roads connecting cities, in production processes, etc.), as follows.

26. **(Nodal incidence matrix)** Figure 165 shows an electrical network having 6 *branches* (connections) and 4 *nodes* (points where two or more branches come together). One node is the *reference node* (node whose voltage is 0 since the node is

grounded). We number the other nodes and number and direct the branches. This we do arbitrarily. The network can now be described by a matrix $\mathbf{A} = [a_{jk}]$, where

$$a_{jk} = \begin{cases} +1 \text{ if branch } k \text{ leaves node } \textcircled{j} \\ -1 \text{ if branch } k \text{ enters node } \textcircled{j} \\ 0 \text{ if branch } k \text{ does not touch node } \textcircled{j}. \end{cases}$$

$\mathbf{A}$ is called the *nodal incidence matrix* of the network. Show that for the network in Fig. 165, $\mathbf{A}$ has the given form.

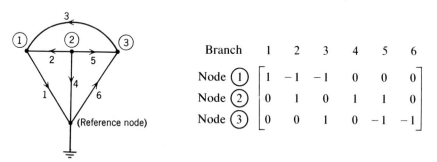

Branch	1	2	3	4	5	6
Node ①	1	−1	−1	0	0	0
Node ②	0	1	0	1	1	0
Node ③	0	0	1	0	−1	−1

Fig. 165. Network and nodal incidence matrix in Problem 26

27. Find the nodal incidence matrix of the electrical network in the figure.

28. Find the nodal incidence matrix in the electrical network in the figure.

29. Methods of electrical network analysis have applications in other fields, too. Determine the analog of the nodal incidence matrix for the net of one-way streets (directions as indicated by the arrows) shown in the figure.

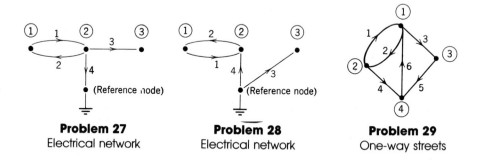

Problem 27
Electrical network

Problem 28
Electrical network

Problem 29
One-way streets

Draw a sketch of the network having the following nodal incidence matrix.

30.
$$\begin{bmatrix} 0 & -1 & 1 & 0 & -1 \\ -1 & 1 & 0 & 0 & 0 \\ 0 & 0 & -1 & 0 & 0 \\ 0 & 0 & 0 & -1 & 0 \end{bmatrix}$$

31.
$$\begin{bmatrix} 1 & 0 & 0 & -1 \\ -1 & 1 & 0 & 0 \\ 0 & -1 & 1 & 0 \end{bmatrix}$$

32.
$$\begin{bmatrix} 0 & 1 \\ 1 & 0 \\ -1 & -1 \end{bmatrix}$$

33. (Mesh incidence matrix) A network can also be characterized by the *mesh incidence matrix* $\mathbf{M} = (m_{jk})$, where

$$m_{jk} = \begin{cases} +1 \text{ if branch } k \text{ is in mesh } \boxed{j} \text{ and has the same orientation} \\ -1 \text{ if branch } k \text{ is in mesh } \boxed{j} \text{ and has the opposite orientation} \\ 0 \text{ if branch } k \text{ is not in mesh } \boxed{j} \end{cases}$$

where a *mesh* is a loop with no branch in its interior (or in its exterior). Here, the meshes are numbered and directed (oriented) in an arbitrary fashion. Show that for the network in Fig. 166, the matrix **M** has the given form.

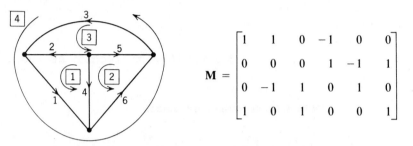

$$\mathbf{M} = \begin{bmatrix} 1 & 1 & 0 & -1 & 0 & 0 \\ 0 & 0 & 0 & 1 & -1 & 1 \\ 0 & -1 & 1 & 0 & 1 & 0 \\ 1 & 0 & 1 & 0 & 0 & 1 \end{bmatrix}$$

Fig. 166. Network and matrix **M** in Problem 33

7.3 Matrix Multiplication

We shall now define the multiplication of matrices by matrices. Although the definition will include the inner product (dot product) of vectors [cf. (8) in Sec. 6.5] as a special case, it may at first look somewhat artificial, but will afterward find its motivation by the use of matrices in connection with linear transformations.

Multiplication of a Matrix by a Matrix

Let $\mathbf{A} = [a_{jk}]$ be an $m \times n$ matrix and $\mathbf{B} = [b_{jk}]$ an $r \times p$ matrix. Then the product $\mathbf{AB}$ (in this order) is defined only when $r = n$, that is,

Number of rows of B = Number of columns of A.

$\mathbf{AB}$ is then the $m \times p$ matrix $\mathbf{C} = [c_{jk}]$ whose entry c_{jk} is the inner product (dot product)

$$c_{jk} = (j\text{th row vector of } \mathbf{A}) \cdot (k\text{th column vector of } \mathbf{B})$$

[see Fig. 167], that is, for $j = 1, \cdots, m$ and $k = 1, \cdots, p$ we have

(1)
$$c_{jk} = \sum_{l=1}^{n} a_{jl} b_{lk} = a_{j1}b_{1k} + a_{j2}b_{2k} + \cdots + a_{jn}b_{nk}.$$

Matrix multiplication is, therefore, conveniently referred to as the *multiplications of rows into columns*.

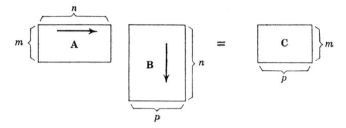

Fig. 167. Matrix multiplication. **AB** = **C**

EXAMPLE 1. Multiplication of matrices

$$\begin{bmatrix} 4 & 2 \\ 1 & 8 \end{bmatrix} \begin{bmatrix} 3 \\ 5 \end{bmatrix} = \begin{bmatrix} 12 + 10 \\ 3 + 40 \end{bmatrix} = \begin{bmatrix} 22 \\ 43 \end{bmatrix} \quad \text{whereas} \quad \begin{bmatrix} 3 \\ 5 \end{bmatrix} \begin{bmatrix} 4 & 2 \\ 1 & 8 \end{bmatrix} \quad \text{is undefined.}$$

EXAMPLE 2. Multiplication of matrices is not commutative

$$\begin{bmatrix} 2 & 3 & 1 \\ 0 & 4 & 5 \end{bmatrix} \begin{bmatrix} 8 & 0 \\ 6 & 1 \\ 7 & 9 \end{bmatrix} = \begin{bmatrix} 41 & 12 \\ 59 & 49 \end{bmatrix}$$

because

$$c_{11} = 2 \cdot 8 + 3 \cdot 6 + 1 \cdot 7 = 41, \qquad c_{12} = 2 \cdot 0 + 3 \cdot 1 + 1 \cdot 9 = 12,$$

$$c_{21} = 0 \cdot 8 + 4 \cdot 6 + 5 \cdot 7 = 59, \qquad c_{22} = 0 \cdot 0 + 4 \cdot 1 + 5 \cdot 9 = 49.$$

Similarly,

$$\begin{bmatrix} 8 & 0 \\ 6 & 1 \\ 7 & 9 \end{bmatrix} \begin{bmatrix} 2 & 3 & 1 \\ 0 & 4 & 5 \end{bmatrix} = \begin{bmatrix} 16 & 24 & 8 \\ 12 & 22 & 11 \\ 14 & 57 & 52 \end{bmatrix}.$$

Hence, even if both **AB** and **BA** are defined, as in this example, then **AB** $\neq$ **BA** in general; that is, matrix multiplication is *not commutative*.

EXAMPLE 3. Products of row and column matrices

$$\begin{bmatrix} 3 & 6 & 1 \end{bmatrix} \begin{bmatrix} 1 \\ 2 \\ 4 \end{bmatrix} = \begin{bmatrix} 19 \end{bmatrix}, \qquad \begin{bmatrix} 1 \\ 2 \\ 4 \end{bmatrix} \begin{bmatrix} 3 & 6 & 1 \end{bmatrix} = \begin{bmatrix} 3 & 6 & 1 \\ 6 & 12 & 2 \\ 12 & 24 & 4 \end{bmatrix}$$

EXAMPLE 4. AB = 0 does not necessarily imply A = 0 or B = 0 or BA = 0

$$\begin{bmatrix} 1 & 1 \\ 2 & 2 \end{bmatrix} \begin{bmatrix} -1 & 1 \\ 1 & -1 \end{bmatrix} = \begin{bmatrix} 0 & 0 \\ 0 & 0 \end{bmatrix}, \qquad \begin{bmatrix} -1 & 1 \\ 1 & -1 \end{bmatrix} \begin{bmatrix} 1 & 1 \\ 2 & 2 \end{bmatrix} = \begin{bmatrix} 1 & 1 \\ -1 & -1 \end{bmatrix}$$

Properties of Matrix Multiplication

Warning. Examples 2 and 4 illustrate that matrix multiplication has some properties that are quite unusual because they have no counterparts in the usual multiplication of numbers. We state these properties in (b) and (c) in the following theorem.

Theorem 1 (Properties of matrix multiplication)

(a) *Matrix multiplication is associative and is distributive with respect to addition of matrices, that is,*

$$(k\mathbf{A})\mathbf{B} = k(\mathbf{AB}) = \mathbf{A}(k\mathbf{B}) \qquad written\ k\mathbf{AB}\ or\ \mathbf{A}k\mathbf{B}$$

$$\mathbf{A}(\mathbf{BC}) = (\mathbf{AB})\mathbf{C} \qquad written\ \mathbf{ABC}$$

(2)

$$(\mathbf{A} + \mathbf{B})\mathbf{C} = \mathbf{AC} + \mathbf{BC}$$

$$\mathbf{C}(\mathbf{A} + \mathbf{B}) = \mathbf{CA} + \mathbf{CB}$$

provided **A, B** *and* **C** *are such that the expressions on the left are defined.* (*k is any number.*)

(b) *Matrix multiplication is not commutative; that is, if* **A** *and* **B** *are matrices such that both* **AB** *and* **BA** *are defined, then*

$$\mathbf{AB} \neq \mathbf{BA} \qquad in\ general.$$

(c) *The cancellation law is not true, in general, that is,*

$$\mathbf{AB} = \mathbf{0} \qquad \textit{does not necessarily imply} \qquad \mathbf{A} = \mathbf{0}\ or\ \mathbf{B} = \mathbf{0}.$$

Proof. The proof of (a) is left as an exercise. (b) can be seen from Example 2 and (c) from Example 4.   ∎

Properties **(b)** *and* **(c)** *are quite unusual* and should be carefully observed. Property (c) will be considered in more detail in Sec. 7.8.

Property (b) implies that *the order of the factors in a matrix product must be carefully observed.* To be precise we say that in the product **AB**, the matrix **B** is **premultiplied** by the matrix **A**, or, alternatively, that **A** is **postmultiplied** by **B**.

Special Matrices

Triangular matrices

A square matrix whose entries above the main diagonal are all zero is called a *lower triangular matrix.* Similarly, an *upper triangular matrix* is a square matrix whose entries below the main diagonal are all zero. For instance,

$$\mathbf{T}_1 = \begin{bmatrix} 1 & 0 & 0 \\ -2 & 3 & 0 \\ 5 & 0 & 2 \end{bmatrix} \quad and \quad \mathbf{T}_2 = \begin{bmatrix} 1 & 6 & -1 \\ 0 & 2 & 3 \\ 0 & 0 & 4 \end{bmatrix}$$

are lower and upper triangular, respectively.

Diagonal matrices

A square matrix $\mathbf{A} = [a_{jk}]$ whose entries above *and* below the main diagonal are all zero, that is, $a_{jk} = 0$ for all $j \neq k$, is called a **diagonal matrix.** For example,

$$\begin{bmatrix} 2 & 0 & 0 \\ 0 & 0 & 0 \\ 0 & 0 & -4 \end{bmatrix} \quad \text{and} \quad \begin{bmatrix} 2 & 0 & 0 \\ 0 & 2 & 0 \\ 0 & 0 & 2 \end{bmatrix}$$

are diagonal matrices.

A diagonal matrix whose entries on the main diagonal are all equal is called a **scalar matrix.** Thus a scalar matrix is of the form

$$\mathbf{S} = \begin{bmatrix} c & 0 & \cdots & 0 \\ 0 & c & \cdots & \cdot \\ \cdot & \cdot & \cdots & \cdot \\ 0 & 0 & \cdots & c \end{bmatrix}$$

where c is any number. The name comes from the fact that an $n \times n$ scalar matrix $\mathbf{S}$ commutes with any $n \times n$ matrix $\mathbf{A,}$ and the multiplication by $\mathbf{S}$ has the same effect as the multiplication by a scalar,

(3) $$\mathbf{AS} = \mathbf{SA} = c\mathbf{A}.$$

In particular, a scalar matrix whose entries on the main diagonal are all 1 is called a **unit matrix** and is denoted by $\mathbf{I}_n$ or simply by $\mathbf{I}.$ For $\mathbf{I},$ formula (3) becomes

(4) $$\mathbf{AI} = \mathbf{IA} = \mathbf{A}.$$

For example, the 3×3 unit matrix is

$$\mathbf{I} = \begin{bmatrix} 1 & 0 & 0 \\ 0 & 1 & 0 \\ 0 & 0 & 1 \end{bmatrix}.$$

Multiplication in Terms of Row and Column Vectors

Matrix multiplication is a multiplication of rows by columns, as we know, and we can restate (1) in terms of inner products. To do this, we write $\mathbf{A}$ in terms of its rows vectors, that is,

$$A = \begin{bmatrix} \mathbf{a}_1 \\ \mathbf{a}_2 \\ \vdots \\ \mathbf{a}_m \end{bmatrix}, \quad \text{where} \quad \begin{aligned} \mathbf{a}_1 &= \begin{bmatrix} a_{11} & a_{12} & \cdots & a_{1n} \end{bmatrix} \\ \mathbf{a}_2 &= \begin{bmatrix} a_{21} & a_{22} & \cdots & a_{2n} \end{bmatrix} \\ &\quad\vdots \\ \mathbf{a}_m &= \begin{bmatrix} a_{m1} & a_{m2} & \cdots & a_{mn} \end{bmatrix} \end{aligned}$$

and we write $\mathbf{B}$ in terms of its column vectors, that is

$$\mathbf{B} = \begin{bmatrix} \mathbf{b}_1 & \mathbf{b}_2 & \cdots & \mathbf{b}_p \end{bmatrix},$$

where

$$\mathbf{b}_1 = \begin{bmatrix} b_{11} \\ b_{21} \\ \vdots \\ b_{n1} \end{bmatrix}, \quad \mathbf{b}_2 = \begin{bmatrix} b_{12} \\ b_{22} \\ \vdots \\ b_{n2} \end{bmatrix}, \quad \cdots, \quad \mathbf{b}_p = \begin{bmatrix} b_{1p} \\ b_{2p} \\ \vdots \\ b_{np} \end{bmatrix}.$$

Then we see from (1) that

$$(5) \qquad \mathbf{C} = \mathbf{AB} = \begin{bmatrix} \mathbf{a}_1 \cdot \mathbf{b}_1 & \mathbf{a}_1 \cdot \mathbf{b}_2 & \cdots & \mathbf{a}_1 \cdot \mathbf{b}_p \\ \mathbf{a}_2 \cdot \mathbf{b}_1 & \mathbf{a}_2 \cdot \mathbf{b}_2 & \cdots & \mathbf{a}_2 \cdot \mathbf{b}_p \\ \cdot & \cdot & \cdots & \cdot \\ \mathbf{a}_m \cdot \mathbf{b}_1 & \mathbf{a}_m \cdot \mathbf{b}_2 & \cdots & \mathbf{a}_m \cdot \mathbf{b}_p \end{bmatrix}.$$

This idea sometimes helps in applications to see more clearly what is going on. Related to this, we observe that $\mathbf{Ab}_1$ is an $m \times 1$ matrix, a column matrix,

$$\mathbf{Ab}_1 = \begin{bmatrix} \mathbf{a}_1 \\ \vdots \\ \mathbf{a}_m \end{bmatrix} \mathbf{b}_1 = \begin{bmatrix} \mathbf{a}_1 \cdot \mathbf{b}_1 \\ \vdots \\ \mathbf{a}_m \cdot \mathbf{b}_1 \end{bmatrix}.$$

We see that this is the first column of $\mathbf{AB}$ in (5). Taking $\mathbf{b}_2$, we get the second column, and so on. Hence we can write the product $\mathbf{AB}$ in terms of its column vectors by a formula that is often useful (and will help us later):

$$(6) \qquad\qquad \mathbf{AB} = \begin{bmatrix} \mathbf{Ab}_1 & \mathbf{Ab}_2 & \cdots & \mathbf{Ab}_p \end{bmatrix}.$$

EXAMPLE 5. Product in terms of row and column vectors
Writing a 2×2 matrix $\mathbf{A}$ in terms of row vectors, say,

$$\mathbf{A} = \begin{bmatrix} a_{11} & a_{12} \\ a_{21} & a_{22} \end{bmatrix} = \begin{bmatrix} \mathbf{a}_1 \\ \mathbf{a}_2 \end{bmatrix} \quad \text{where} \quad \begin{aligned} \mathbf{a}_1 &= \begin{bmatrix} a_{11} & a_{12} \end{bmatrix} \\ \mathbf{a}_2 &= \begin{bmatrix} a_{21} & a_{22} \end{bmatrix} \end{aligned}$$

and a 2×2 matrix $\mathbf{B}$ in terms of column vectors, say,

$$\mathbf{B} = \begin{bmatrix} b_{11} & b_{12} \\ b_{21} & b_{22} \end{bmatrix} = \begin{bmatrix} \mathbf{b}_1 & \mathbf{b}_2 \end{bmatrix} \quad \text{where} \quad \mathbf{b}_1 = \begin{bmatrix} b_{11} \\ b_{21} \end{bmatrix}, \quad \mathbf{b}_2 = \begin{bmatrix} b_{12} \\ b_{22} \end{bmatrix},$$

we see that (5) takes the form

$$\mathbf{AB} = \begin{bmatrix} \mathbf{a}_1 \\ \mathbf{a}_2 \end{bmatrix} \begin{bmatrix} \mathbf{b}_1 & \mathbf{b}_2 \end{bmatrix} = \begin{bmatrix} \mathbf{a}_1 \cdot \mathbf{b}_1 & \mathbf{a}_1 \cdot \mathbf{b}_2 \\ \mathbf{a}_2 \cdot \mathbf{b}_1 & \mathbf{a}_2 \cdot \mathbf{b}_2 \end{bmatrix} = \begin{bmatrix} a_{11}b_{11} + a_{12}b_{21} & a_{11}b_{12} + a_{12}b_{22} \\ a_{21}b_{11} + a_{22}b_{21} & a_{21}b_{12} + a_{22}b_{22} \end{bmatrix}.$$

Also, $\mathbf{AB} = \begin{bmatrix} \mathbf{Ab}_1 & \mathbf{Ab}_2 \end{bmatrix}$ by (6).   ∎

Motivation of Matrix Multiplication

We consider three coordinate systems in the plane, which we denote as the w_1w_2-system, the x_1x_2-system and the y_1y_2-system, and we assume that these systems are related by transformations

$$(7) \qquad \begin{aligned} y_1 &= a_{11}x_1 + a_{12}x_2 \\ y_2 &= a_{21}x_1 + a_{22}x_2 \end{aligned}$$

and

$$(8) \qquad \begin{aligned} x_1 &= b_{11}w_1 + b_{12}w_2 \\ x_2 &= b_{21}w_1 + b_{22}w_2 \end{aligned}$$

which are called *linear transformations*. By substituting (8) into (7) we see that the y_1y_2-coordinates can be obtained directly from the w_1w_2-coordinates by a single linear transformation of the form

$$(9) \qquad \begin{aligned} y_1 &= c_{11}w_1 + c_{12}w_2 \\ y_2 &= c_{21}w_1 + c_{22}w_2. \end{aligned}$$

Now this substitution gives

$$\begin{aligned} y_1 &= a_{11}(b_{11}w_1 + b_{12}w_2) + a_{12}(b_{21}w_1 + b_{22}w_2) \\ y_2 &= a_{21}(b_{11}w_1 + b_{12}w_2) + a_{22}(b_{21}w_1 + b_{22}w_2). \end{aligned}$$

Comparing this with (9), we see that we must have

$$\begin{aligned} c_{11} &= a_{11}b_{11} + a_{12}b_{21} & c_{12} &= a_{11}b_{12} + a_{12}b_{22} \\ c_{21} &= a_{21}b_{11} + a_{22}b_{21} & c_{22} &= a_{21}b_{12} + a_{22}b_{22} \end{aligned}$$

or briefly

$$(10) \qquad c_{jk} = a_{j1}b_{1k} + a_{j2}b_{2k} = \sum_{i=1}^{2} a_{ji}b_{ik} \qquad j, k = 1, 2.$$

This is (1) with $m = n = p = 2$.

What does our calculation show? Essentially two things. First, matrix multiplication is defined in such a way that linear transformations can be written in compact form, using matrices; in our case, (7) becomes

$$(7^*) \quad \mathbf{y} = \mathbf{Ax} \quad \text{where} \quad \mathbf{y} = \begin{bmatrix} y_1 \\ y_2 \end{bmatrix}, \quad \mathbf{A} = \begin{bmatrix} a_{11} & a_{12} \\ a_{21} & a_{22} \end{bmatrix}, \quad \mathbf{x} = \begin{bmatrix} x_1 \\ x_2 \end{bmatrix}$$

and (8) becomes

$$(8^*) \quad \mathbf{x} = \mathbf{Bw} \quad \text{where} \quad \mathbf{x} = \begin{bmatrix} x_1 \\ x_2 \end{bmatrix}, \quad \mathbf{B} = \begin{bmatrix} b_{11} & b_{12} \\ b_{21} & b_{22} \end{bmatrix}, \quad \mathbf{w} = \begin{bmatrix} w_1 \\ w_2 \end{bmatrix}.$$

Second, if we substitute linear transformations into each other, we can obtain the coefficient matrix $\mathbf{C}$ of the composite transformation (the transformation obtained by the substitution) simply by multiplying the coefficient matrices $\mathbf{A}$ and $\mathbf{B}$ of the given transformations, in the right order suggested by the substitution; from (7^*), (8^*) and (9) we get

$$\mathbf{y} = \mathbf{Ax} = \mathbf{A(Bw)} = \mathbf{ABw} = \mathbf{Cw,} \quad \text{where} \quad \mathbf{C} = \mathbf{AB.}$$

For higher dimensions the idea and the result are exactly the same; only the number of variables changes. We then have m variables $y_1, \cdots, y_m$ and n variables $x_1, \cdots, x_n$ and p variables $w_1, \cdots, w_p$. The matrix $\mathbf{A}$ is $m \times n$, the matrix $\mathbf{B}$ is $n \times p$ and $\mathbf{C}$ is $m \times p$, as in Fig. 167. And the requirement that $\mathbf{C}$ be the product $\mathbf{AB}$ leads to formula (1) in its general form. This completely motivates the definition of matrix multiplication.

Linear Transformations

We want to say more about matrices as related to linear transformations.

Let X and Y be any vector spaces (cf. Sec. 6.4). To each vector $\mathbf{x}$ in X we assign a unique vector $\mathbf{y}$ in Y. Then we say that a **mapping** (or **transformation** or **operator**) of X into Y is given. Such a mapping is denoted by a capital letter, say F. The vector $\mathbf{y}$ in Y assigned to a vector $\mathbf{x}$ in X is called the **image** of $\mathbf{x}$ and is denoted by $F(\mathbf{x})$ [or $F\mathbf{x}$, without parentheses].

F is called a **linear mapping** or **linear transformation** if for all vectors $\mathbf{v}$ and $\mathbf{x}$ in X and scalars c,

$$(11) \qquad\qquad F(\mathbf{v} + \mathbf{x}) = F(\mathbf{v}) + F(\mathbf{x})$$
$$F(c\mathbf{x}) = cF(\mathbf{x}).$$

For instance, (7) and (8) are linear transformations in the sense of the present definition.

Linear transformation of space R^n into space R^m

From now on, we let $X = R^n$, the vector space of all ordered n-tuples of real numbers, and $Y = R^m$, the vector space of all ordered m-tuples of real numbers. Then any real $m \times n$ matrix $\mathbf{A} = [a_{jk}]$ yields a transformation of R^n into R^m given by

$$\mathbf{y} = \mathbf{Ax}.$$

Since $\mathbf{A}(\mathbf{u} + \mathbf{x}) = \mathbf{Au} + \mathbf{Ax}$ and $\mathbf{A}(c\mathbf{x}) = c\mathbf{Ax}$, this transformation is linear.

We show that, conversely, every linear transformation F of R^n into R^m can be given in terms of an $m \times n$ matrix $\mathbf{A}$, after a basis in R^n and a basis in R^m have been introduced. Let $\mathbf{e}_{(1)}, \cdots, \mathbf{e}_{(n)}$ be any basis in R^n. Then every $\mathbf{x}$ in R^n has a unique representation

$$\mathbf{x} = x_1 \mathbf{e}_{(1)} + \cdots + x_n \mathbf{e}_{(n)}.$$

Since F is linear, this implies for the image $F(\mathbf{x})$:

$$F(\mathbf{x}) = F(x_1 \mathbf{e}_{(1)} + \cdots + x_n \mathbf{e}_{(n)}) = x_1 F(\mathbf{e}_{(1)}) + \cdots + x_n F(\mathbf{e}_{(n)}).$$

Hence F is uniquely determined by the images of the vectors of a basis in R^n. We now choose in R^n the **"standard basis"**

$$\mathbf{e}_{(1)} = \begin{bmatrix} 1 \\ 0 \\ \vdots \\ 0 \end{bmatrix}, \qquad \mathbf{e}_{(2)} = \begin{bmatrix} 0 \\ 1 \\ \vdots \\ 0 \end{bmatrix}, \qquad \cdots, \qquad \mathbf{e}_{(n)} = \begin{bmatrix} 0 \\ 0 \\ \vdots \\ 1 \end{bmatrix},$$

where $\mathbf{e}_{(j)}$ has its jth component equal to 1 and all others 0. We can now determine an $m \times n$ matrix $\mathbf{A} = [a_{jk}]$ such that for every $\mathbf{x}$ in R^n and image $\mathbf{y} = F(\mathbf{x})$ in R^m,

$$\mathbf{y} = F(\mathbf{x}) = \mathbf{Ax}.$$

Indeed, from the image $\mathbf{y}^{(1)} = F(\mathbf{e}_{(1)})$ of $\mathbf{e}_{(1)}$ we get the condition

$$\mathbf{y}^{(1)} = \begin{bmatrix} y_1^{(1)} \\ y_2^{(1)} \\ \vdots \\ y_m^{(1)} \end{bmatrix} = \begin{bmatrix} a_{11} & \cdots & a_{1n} \\ a_{21} & \cdots & a_{2n} \\ \vdots & \cdots & \vdots \\ a_{m1} & \cdots & a_{mn} \end{bmatrix} \begin{bmatrix} 1 \\ 0 \\ \vdots \\ 0 \end{bmatrix}$$

from which we can determine the first column of $\mathbf{A}$, namely, $a_{11} = y_1^{(1)}$, $a_{21} = y_2^{(1)}, \cdots, a_{m1} = y_m^{(1)}$. Similarly, from the image of $\mathbf{e}_{(2)}$ we get the second column of $\mathbf{A}$, and so on. This completes the proof.   ■

We say that $\mathbf{A}$ **represents** F, or *is a representation of F*, with respect to the bases in R^n and R^m. Quite generally, the purpose of a *"representation"* is the replacement of one object of study by another object whose properties are more readily apparent.

EXAMPLE 6. Linear transformations
Interpreted as transformations of Cartesian coordinates in the plane, the matrices

$$\begin{bmatrix} 0 & 1 \\ 1 & 0 \end{bmatrix}, \quad \begin{bmatrix} 1 & 0 \\ 0 & -1 \end{bmatrix}, \quad \begin{bmatrix} -1 & 0 \\ 0 & -1 \end{bmatrix}, \quad \begin{bmatrix} a & 0 \\ 0 & 1 \end{bmatrix}$$

represent a reflection in the line $x_2 = x_1$, a reflection in the x_1-axis, a reflection in the origin and a stretch (when $a > 1$, or a contraction when $0 < a < 1$) in the x_1-direction, respectively.

EXAMPLE 7. Linear transformations
Our discussion preceding Example 6 is simpler than it may look at first sight. To see this, find the matrix **A** representing the linear transformation that maps (x_1, x_2) onto $(2x_1 - 5x_2, 3x_1 + 4x_2)$.

Solution. Since

$$\begin{bmatrix} 1 \\ 0 \end{bmatrix} \text{ and } \begin{bmatrix} 0 \\ 1 \end{bmatrix} \text{ are mapped onto } \begin{bmatrix} 2 \\ 3 \end{bmatrix} \text{ and } \begin{bmatrix} -5 \\ 4 \end{bmatrix},$$

respectively, we obtain, according to our discussion,

$$\mathbf{A} = \begin{bmatrix} 2 & -5 \\ 3 & 4 \end{bmatrix}.$$

We check this, finding

$$\begin{bmatrix} y_1 \\ y_2 \end{bmatrix} = \begin{bmatrix} 2 & -5 \\ 3 & 4 \end{bmatrix} \begin{bmatrix} x_1 \\ x_2 \end{bmatrix} = \begin{bmatrix} 2x_1 - 5x_2 \\ 3x_1 + 4x_2 \end{bmatrix}.$$

The reader may obtain **A** also at once by writing the given transformation in the form

$$y_1 = 2x_1 - 5x_2$$
$$y_2 = 3x_1 + 4x_2. \qquad \blacksquare$$

In the next section we define an operation called **transposition** of matrices and, related to it, *real symmetric* and *real skew-symmetric matrices,* which are of particular importance in applications, as we shall see later.

Problems for Sec. 7.3

$$\text{Let} \quad \mathbf{a} = \begin{bmatrix} 1 \\ 0 \\ 2 \end{bmatrix}, \quad \mathbf{B} = \begin{bmatrix} 2 & 0 \\ 0 & 3 \\ 1 & -1 \end{bmatrix}, \quad \mathbf{C} = \begin{bmatrix} 1 & -1 & 0 \\ 0 & 2 & 3 \\ 4 & 0 & -1 \end{bmatrix}, \quad \mathbf{d} = \begin{bmatrix} 4 & 3 & 0 \end{bmatrix}.$$

Find those of the following expressions which are defined.

1. ad, da
2. $\mathbf{a}^2$, $\mathbf{B}^2$, $\mathbf{C}^2$
3. BC, CB
4. Bd, dB
5. Cd, dC
6. aC, Ca
7. dCB
8. adC
9. adCa
10. $\mathbf{C}^2$, $\mathbf{C}^4$, $\mathbf{C}^4 - 3\mathbf{C}^2 + 2\mathbf{I}$

11. **(Idempotent matrix)** A matrix A is said to be *idempotent* if $\mathbf{A}^2 = \mathbf{A}$. Give examples of idempotent matrices, different from the zero or unit matrix.
12. Find real 2×2 matrices (as many as you can) whose square is **I**, the unit matrix.
13. Find a 2×2 matrix $\mathbf{A} \neq \mathbf{0}$ such that $\mathbf{A}^2 = \mathbf{0}$.

14. Find two 2×2 matrices **A**, **B** such that $(\mathbf{A} + \mathbf{B})^2 \neq \mathbf{A}^2 + 2\mathbf{AB} + \mathbf{B}^2$.

15. Prove (2) and (4).

16. Show that the product of two upper triangular $n \times n$ matrices is upper triangular.

Use of matrices in linear transformations

17. Represent each of the transformations

$$
\begin{aligned}
y_1 &= 2x_1 - x_2 \\
y_2 &= -x_1 + 2x_2
\end{aligned}
\quad \text{and} \quad
\begin{aligned}
x_1 &= w_1 + w_2 \\
x_2 &= w_1 - w_2
\end{aligned}
$$

by the use of matrices and find the composite transformation that expresses y_1, y_2 in terms of w_1, w_2.

18. Give a geometrical interpretation (as in Example 6) of the linear transformations with matrices

$$
\begin{bmatrix} 1 & 0 \\ 0 & 0 \end{bmatrix}, \begin{bmatrix} 0 & 0 \\ 0 & 1 \end{bmatrix}, \begin{bmatrix} -1 & 0 \\ 0 & 1 \end{bmatrix}, \begin{bmatrix} 1 & 0 \\ 0 & 3 \end{bmatrix}, \begin{bmatrix} a & 0 \\ 0 & a \end{bmatrix} \quad (a > 1).
$$

19. **(Rotation)** Show that the linear transformation $\mathbf{y} = \mathbf{Ax}$ with matrix

$$
\mathbf{A} = \begin{bmatrix} \cos\theta & -\sin\theta \\ \sin\theta & \cos\theta \end{bmatrix} \quad \text{and} \quad \mathbf{x} = \begin{bmatrix} x_1 \\ x_2 \end{bmatrix}, \quad \mathbf{y} = \begin{bmatrix} y_1 \\ y_2 \end{bmatrix}
$$

is a counterclockwise rotation of the Cartesian $x_1 x_2$-coordinate system in the plane about the origin.

20. Show that in Prob. 19,

$$
\mathbf{A}^n = \begin{bmatrix} \cos n\theta & -\sin n\theta \\ \sin n\theta & \cos n\theta \end{bmatrix}.
$$

What does this result mean geometrically?

21. Find **A** such that $\mathbf{y} = \mathbf{Ax}$ represents a clockwise rotation in the plane through an angle of $45°$.

22. Give a geometrical interpretation (as in Example 6) of the linear transformation $\mathbf{x} = \mathbf{Bw}$ with matrix

$$
\mathbf{B} = \begin{bmatrix} 3 & 0 \\ 0 & 1 \end{bmatrix}.
$$

Find the composite transformation $\mathbf{y} = \mathbf{ABw}$ with **A** given in Prob. 19. Show that $\mathbf{AB} \neq \mathbf{BA}$. What does this mean geometrically?

23. **(Computer graphics)** To visualize a three-dimensional object with plane faces (e.g., a cube), we may store the position vectors of the vertices with respect to a suitable $x_1 x_2 x_3$-coordinate system (and a list of the connecting edges) and then obtain a two-dimensional image on a video screen by projecting the object onto a coordinate plane, for instance, onto the $x_1 x_2$-plane by setting $x_3 = 0$. To change the appearance of the image, we can impose a linear transformation on the position vectors stored. Show that a diagonal matrix **D** with main diagonal entries $3, 1, \frac{1}{2}$ gives from an $\mathbf{x} = [x_j]$ the new position vector $\mathbf{y} = \mathbf{Dx}$, where $y_1 = 3x_1$ (stretch in the x_1-direction by a factor 3), $y_2 = x_2$ (unchanged), $y_3 = \frac{1}{2}x_2$ (contraction in the x_3-direction). What effect would a scalar matrix have?

24. **(Rotations in space in computer graphics)** What effect would the following matrices have in the situation described in Prob. 23?

$$\begin{bmatrix} 1 & 0 & 0 \\ 0 & \cos\theta & -\sin\theta \\ 0 & \sin\theta & \cos\theta \end{bmatrix}, \begin{bmatrix} \cos\varphi & 0 & -\sin\varphi \\ 0 & 1 & 0 \\ \sin\varphi & 0 & \cos\varphi \end{bmatrix}, \begin{bmatrix} \cos\psi & -\sin\psi & 0 \\ \sin\psi & \cos\psi & 0 \\ 0 & 0 & 1 \end{bmatrix}$$

25. (Assignment problem) Contractors C_1, C_2, C_3 bid for jobs J_1, J_2, J_3 as the cost matrix (in 100 000 dollar units) shows. What assignment minimizes the total cost (a) under no condition? (b) Under the condition that each contractor be assigned to only one job?

$$\begin{array}{c} \\ C_1 \\ C_2 \\ C_3 \end{array} \begin{array}{ccc} J_1 & J_2 & J_3 \\ \begin{bmatrix} 12 & 2 & 5 \\ 9 & 3 & 6 \\ 8 & 4 & 4 \end{bmatrix} \end{array} \qquad \mathbf{A} = [a_{jk}] = \begin{bmatrix} 20 & 14 & 10 \\ 17 & 13 & 7 \\ 18 & 15 & 10 \end{bmatrix}$$

Cost matrix for Problem 25   Matrix A for Problem 26

26. If worker W_j can do job J_k in a_{jk} hours, as shown by the matrix **A**, and each worker should do one job only, which assignment would minimize the total time?

27. If **A** gives the prices (in cents/lb) of two items at three stores and **x** gives the quantities (lb) somebody wants to buy, what can he see from $\mathbf{y} = \mathbf{Ax}$?

$$\begin{array}{c} \\ \mathbf{A} = \end{array} \begin{array}{cc} \text{Potatoes} & \text{Onions} \\ \begin{bmatrix} 16 & 13 \\ 14 & 17 \\ 10 & 16 \end{bmatrix} & \begin{array}{l} \text{Store I} \\ \text{Store II} \\ \text{Store III} \end{array} \end{array} \qquad \mathbf{x} = \begin{bmatrix} 40 \\ 10 \end{bmatrix} \begin{array}{l} \text{Potatoes} \\ \text{Onions} \end{array}$$

28. If the cost matrix **A** gives the cost/item (in $) of producing chairs and tables, and the production vector **v** gives the daily production, what information does $\mathbf{w} = \mathbf{Av}$ yield?

$$\mathbf{A} = \begin{array}{cc} \text{Chair} & \text{Table} \\ \begin{bmatrix} 16 & 25 \\ 34 & 72 \end{bmatrix} \begin{array}{l} \text{Material} \\ \text{Labor} \end{array} \end{array} \qquad \mathbf{v} = \begin{bmatrix} 1600 \\ 400 \end{bmatrix} \begin{array}{l} \text{Chairs} \\ \text{Tables} \end{array}$$

29. The 2×2 cost matrix **C** shows the production costs per item of two products I and II produced by some company. The 2×4 production matrix **P** shows production figures per quarter. Find $\mathbf{M} = \mathbf{CP}$ and interpret its entries.

	Product I	Product II		Spring	Summer	Fall	Winter
Raw materials	4.0	4.5	I	1000	2000	2000	3000
Labor and other	20.0	25.0	II	3000	1000	2000	1000

30. (Associative algebra) A *real* (or complex) *associative algebra* is defined to be a set A of elements **A, B, C,** $\cdots$ such that: (*a*) A is a real (or complex) vector space (cf. Sec. 6.4). (*b*) A multiplication is defined on the elements of A such that with any ordered pair of elements **A, B** in A, there is associated a unique element of A, written **AB** and called the *product* of **A** and **B,** and this multiplication satisfies (2). Show that the set of all real ($n \times n$) matrices (n fixed) is an associative algebra.

7.4 Transpose of a Matrix

In this short section we want to define another simple operation on a matrix and introduce certain matrices of special types that are of practical importance.

Transposition of a matrix is defined as follows. The **transpose** $\mathbf{A}^\mathsf{T}$ of an $m \times n$ matrix $\mathbf{A} = [a_{jk}]$ is the $n \times m$ matrix that has the columns of $\mathbf{A}$ as rows (that is, the kth *column* of $\mathbf{A}$ becomes the kth *row* of $\mathbf{A}^\mathsf{T}$) and thus the rows of $\mathbf{A}$ as columns (the jth *row* of $\mathbf{A}$ is the jth *column* of $\mathbf{A}^\mathsf{T}$). In formulas,

(1)
$$\mathbf{A}^\mathsf{T} = [a_{jk}] = \begin{bmatrix} a_{11} & a_{21} & \cdots & a_{m1} \\ a_{12} & a_{22} & \cdots & a_{m2} \\ \cdot & \cdot & \cdots & \cdot \\ a_{1n} & a_{2n} & \cdots & a_{mn} \end{bmatrix}.$$

The reader may prove that

(2)
$$(\mathbf{A} + \mathbf{B})^\mathsf{T} = \mathbf{A}^\mathsf{T} + \mathbf{B}^\mathsf{T}.$$

EXAMPLE 1. Transposition of a matrix
If
$$\mathbf{A} = \begin{bmatrix} 5 & -8 & 1 \\ 4 & 0 & 0 \end{bmatrix}, \quad \text{then} \quad \mathbf{A}^\mathsf{T} = \begin{bmatrix} 5 & 4 \\ -8 & 0 \\ 1 & 0 \end{bmatrix}.$$

EXAMPLE 2. Transposition of a row matrix
The transpose of a row matrix is a column matrix and conversely. For instance, if
$$\mathbf{b} = \begin{bmatrix} 7 & 5 & -2 \end{bmatrix}, \quad \text{then} \quad \mathbf{b}^\mathsf{T} = \begin{bmatrix} 7 \\ 5 \\ -2 \end{bmatrix}. \qquad \blacksquare$$

A *real* square matrix $\mathbf{A} = [a_{jk}]$ is said to be **symmetric** if it is equal to its transpose,

(3) $\mathbf{A}^\mathsf{T} = \mathbf{A},$ that is, $a_{kj} = a_{jk}$ $(j, k = 1, \cdots, n).$

A *real* square matrix $\mathbf{A} = (a_{jk})$ is said to be **skew-symmetric** if

(4) $\mathbf{A}^\mathsf{T} = -\mathbf{A},$ that is, $a_{kj} = -a_{jk}$ $(j, k = 1, \cdots, n).$

(Similar notions for complex matrices will be defined later, in Sec. 7.13.) Note that for $k = j$ in (4) we have $a_{jj} = -a_{jj}$, which implies that the entries on the main diagonal of a skew-symmetric matrix are all zero.

Note well and keep in mind that a symmetric or skew-symmetric matrix is always *real*, by definition.

Any real square matrix **A** may be written as the sum of a symmetric matrix **R** and a skew-symmetric matrix **S**, where

$$(5) \qquad \mathbf{R} = \tfrac{1}{2}(\mathbf{A} + \mathbf{A}^\mathsf{T}) \quad \text{and} \quad \mathbf{S} = \tfrac{1}{2}(\mathbf{A} - \mathbf{A}^\mathsf{T}).$$

EXAMPLE 3. Symmetric and skew-symmetric matrices

The matrices

$$\mathbf{A} = \begin{bmatrix} -3 & 1 & 5 \\ 1 & 0 & -2 \\ 5 & -2 & 4 \end{bmatrix} \quad \text{and} \quad \mathbf{B} = \begin{bmatrix} 0 & -4 & 1 \\ 4 & 0 & -5 \\ -1 & 5 & 0 \end{bmatrix}$$

are symmetric and skew-symmetric, respectively. The matrix

$$\mathbf{A} = \begin{bmatrix} 2 & 3 \\ 5 & -1 \end{bmatrix}$$

is neither symmetric nor skew-symmetric. **A** may be written in the form **A** = **R** + **S**, where

$$\mathbf{R} = \tfrac{1}{2}(\mathbf{A} + \mathbf{A}^\mathsf{T}) = \begin{bmatrix} 2 & 4 \\ 4 & -1 \end{bmatrix} \quad \text{and} \quad \mathbf{S} = \tfrac{1}{2}(\mathbf{A} - \mathbf{A}^\mathsf{T}) = \begin{bmatrix} 0 & -1 \\ 1 & 0 \end{bmatrix}$$

are symmetric and skew-symmetric, respectively. ∎

Transposition of a product. *The transpose of a product equals the product of the transposed factors, taken in reverse order,*

$$(6) \qquad \boxed{(\mathbf{AB})^\mathsf{T} = \mathbf{B}^\mathsf{T}\mathbf{A}^\mathsf{T}.}$$

The proof follows from the definition of matrix multiplication and is left to the reader.

In working with matrices, one generally prefers *column* matrices (column vectors) over *row* matrices (row vectors), but transposition quickly provides the switch from one to the other when needed. Let us consider an important application in which matrices play a basic role and that switch is feasible.

EXAMPLE 4. Stochastic matrix. Markov process

Suppose that the 1988 state of land use in a city of 50 square miles of (nonvacant) area is:

I	(Residentially used)	30%
II	(Commercially used)	20%
III	(Industrially used)	50%.

Find the states in 1993 and 1998, assuming that the transition probabilities for 5-year intervals are given by the following matrix **A** = $[a_{jk}]$.

	To I	To II	To III
From I	0.8	0.1	0.1
From II	0.1	0.7	0.2
From III	0	0.1	0.9

Remark. A square matrix with nonnegative entries and row sums all equal to 1 is called a **stochastic matrix.** Thus **A** is a stochastic matrix. A stochastic process for which the probability of entering a certain state depends only on the *last* state occupied (and on the matrix governing the process) is called a **Markov process.** [1] Thus our example concerns a Markov process.

Solution. From the matrix **A** and the 1988 state we can compute the 1993 state

$$\text{I} \quad \text{(Residential)} \quad 0.8 \cdot 30 + 0.1 \cdot 20 + \ \ 0 \cdot 50 = 26 \ [\%]$$

$$\text{II} \quad \text{(Commercial)} \quad 0.1 \cdot 30 + 0.7 \cdot 20 + 0.1 \cdot 50 = 22 \ [\%]$$

$$\text{III} \quad \text{(Industrial)} \quad 0.1 \cdot 30 + 0.2 \cdot 20 + 0.9 \cdot 50 = 52 \ [\%].$$

The sum is 100%, as it should be. We write this in matrix form. Let the column vector **x** denote the 1988 state; thus, $\mathbf{x}^\mathsf{T} = [30 \ \ 20 \ \ 50]$. Let **y** denote the 1993 state. Then

$$\mathbf{y}^\mathsf{T} = \mathbf{x}^\mathsf{T}\mathbf{A} = \begin{bmatrix} 30 & 20 & 50 \end{bmatrix} \begin{bmatrix} 0.8 & 0.1 & 0.1 \\ 0.1 & 0.7 & 0.2 \\ 0 & 0.1 & 0.9 \end{bmatrix} = \begin{bmatrix} 26 & 22 & 52 \end{bmatrix}.$$

Similarly, for the vector **z** of the 1998 state we get, as the reader may verify,

$$\mathbf{z}^\mathsf{T} = \mathbf{y}^\mathsf{T}\mathbf{A} = (\mathbf{x}^\mathsf{T}\mathbf{A})\mathbf{A} = \mathbf{x}^\mathsf{T}\mathbf{A}^2 = \begin{bmatrix} 23 & 23.2 & 53.8 \end{bmatrix}.$$

Answer. In 1993, the residential area will be 26% (13 square miles), the commercial 22% (11 square miles) and the industrial 52% (26 square miles). For 1998, the corresponding figures are 23%, 23.2%, 53.8%. ∎

This is the end of the first portion of Chap. 7 (Secs. 7.1—7.4), which was devoted to the introduction of the basic algebraic operations for matrices. The next portion (Secs. 7.5—7.8) concerns **systems of linear algebraic equations** (without the use of determinants) and includes discussions of the *rank* of a matrix (Sec. 7.6) and the *inverse* of a matrix (Sec. 7.8). *Determinants* follow in Secs. 7.9 and 7.10.

Problems for Sec. 7.4

$$\text{Let} \qquad \mathbf{a} = \begin{bmatrix} 1 \\ 3 \\ 4 \end{bmatrix}, \qquad \mathbf{B} = \begin{bmatrix} 2 & 0 \\ 0 & 4 \\ 3 & 5 \end{bmatrix}, \qquad \mathbf{C} = \begin{bmatrix} 0 & 1 & 2 \\ 2 & 0 & 3 \\ 5 & -1 & 4 \end{bmatrix}.$$

Find those of the following expressions that are defined.

1. $\mathbf{a}^2$, $\mathbf{a}^\mathsf{T}\mathbf{a}$, $\mathbf{a}\mathbf{a}^\mathsf{T}$ 2. $\mathbf{B}^2$, $\mathbf{B}^\mathsf{T}\mathbf{B}$, $\mathbf{B}\mathbf{B}^\mathsf{T}$ 3. $\mathbf{C} + \mathbf{C}^\mathsf{T}$ 4. $\mathbf{C}^2$, $\mathbf{C}^\mathsf{T}\mathbf{C}$, $\mathbf{C}\mathbf{C}^\mathsf{T}$

5. $\mathbf{C}\mathbf{a}$, $\mathbf{C}^\mathsf{T}\mathbf{a}$, $\mathbf{C}\mathbf{a}^\mathsf{T}$ 6. $\mathbf{a}^\mathsf{T}\mathbf{a}\mathbf{C}$, $\mathbf{C}\mathbf{a}\mathbf{a}^\mathsf{T}$ 7. $\mathbf{a}^\mathsf{T}\mathbf{B}$, $\mathbf{B}^\mathsf{T}\mathbf{a}$ 8. $\mathbf{C}\mathbf{B}\mathbf{B}^\mathsf{T}$, $\mathbf{B}\mathbf{B}^\mathsf{T}\mathbf{C}^\mathsf{T}$

9. Verify $(\mathbf{C}\mathbf{a})^\mathsf{T} = \mathbf{a}^\mathsf{T}\mathbf{C}^\mathsf{T}$ for the above **a** and **C**.

10. Represent the above **C** as the sum of a symmetric and a skew-symmetric matrix.

---

[1] ANDREI ANDREJEVITCH MARKOV (1856-1922), Russian mathematician, known for his work in probability theory.

11. Prove that

$$(A + B)^T = A^T + B^T, \qquad (aA)^T = aA^T, \qquad (A^T)^T = A$$

and give an example for each formula.

12. Prove (6).

13. Show that the symmetric 4×4 matrices form a real vector space. What is its dimension?

14. Show that all symmetric $n \times n$ matrices (n fixed) form a vector space, determine the dimension of that space and find a basis.

15. Show that all skew-symmetric $n \times n$ matrices (n fixed) form a vector space. What is its dimension?

16. Show that AA^T is symmetric.

17. Show that the sum and difference of symmetric $n \times n$ matrices are symmetric.

18. Show that the product of symmetric matrices A, B is symmetric if and only if A and B commute,

$$AB = BA.$$

19. Find all real square matrices that are both symmetric and skew-symmetric.

20. How can you arrange the distance between four cities in the form of a 4×4 matrix A? When will A be symmetric?

21. Could you think of arranging the differences in altitudes of five locations in the mountains in the form of a 5×5 skew-symmetric matrix?

22. If an airline serves 46 airports, how can it display the average daily number of persons on the various routes in the form of a 46×46 matrix A? Will A be symmetric? Skew-symmetric? What would an off-diagonal 0 indicate?

23. Verify z in Example 4 by using

$$\text{(a) } z^T = y^T A, \qquad \text{(b) } z^T = x^T A^2.$$

24. For the Markov process with transition matrix $A = [a_{jk}]$, whose entries are

$$a_{11} = a_{12} = 0.5, \quad a_{21} = 0.2, \quad a_{22} = 0.8,$$

and initial state $[0.7 \quad 0.7]^T$, compute the next 4 states. Do they seem to approach a limit vector?

25. Using (6), show that $z = A^T y$ in Example 4 and check the computation of z by this formula.

26. In a production process, let N mean "no trouble" and T "trouble." Let the transition probabilities from one day to the next be 0.8 for $N \to N$, hence 0.2 for $N \to T$, and 0.5 for $T \to N$, hence 0.5 for $T \to T$. If today there is no trouble, what is the probability of trouble 2 days after today? 3 days after today?

27. Show that if A is a stochastic matrix, so is A^2.

28. Show that any power of a stochastic matrix is a stochastic matrix. (Use induction.)

29. Suppose that in some town there are 5000 married and 1000 single women and 20% of the single women get married each year and 10% of the married women get divorced, and the total population of women remains constant. Set up a matrix model and compute the number of married and single women after 1, 2 and 3 years.

30. Under what condition for a stochastic matrix A will the new state $y^T = x^T A$ be equal to the initial state?

7.5 Systems of Linear Equations. Gauss Elimination

A *system of m linear equations* (or *set of m simultaneous linear equations*) *in n unknowns* $x_1, \cdots, x_n$ is a set of equations of the form

(1)

$$
\begin{array}{l}
a_{11}x_1 + \cdots + a_{1n}x_n = b_1 \\
a_{21}x_1 + \cdots + a_{2n}x_n = b_2 \\
\cdots \cdots \cdots \cdots \cdots \cdots \cdots \\
a_{m1}x_1 + \cdots + a_{mn}x_n = b_m
\end{array}
$$

The a_{jk} are given numbers, which are called the **coefficients** of the system. The b_i are also given numbers. If the b_i are all zero, then (1) is called a **homogeneous system.** If at least one b_i is not zero, then (1) is called a **non-homogenous system.**

A **solution** of (1) is a set of numbers $x_1, \cdots, x_n$ which satisfy all the m equations. A **solution vector** of (1) is a vector **x** whose components constitute a solution of (1). If the system (1) is homogeneous, it has at least the **trivial solution** $x_1 = 0, \cdots, x_n = 0$.

From the definition of matrix multiplication we see that the m equations of (1) may be written as a single vector equation

(2)
$$ \mathbf{Ax} = \mathbf{b} $$

where the **coefficient matrix** $\mathbf{A} = [a_{jk}]$ is the $m \times n$ matrix

$$
\mathbf{A} = \begin{bmatrix}
a_{11} & a_{12} & \cdots & a_{1n} \\
a_{21} & a_{22} & \cdots & a_{2n} \\
\cdot & \cdot & \cdots & \cdot \\
a_{m1} & a_{m2} & \cdots & a_{mn}
\end{bmatrix},
\quad \text{and} \quad
\mathbf{x} = \begin{bmatrix} x_1 \\ \cdot \\ \cdot \\ \cdot \\ x_n \end{bmatrix}
\quad \text{and} \quad
\mathbf{b} = \begin{bmatrix} b_1 \\ \cdot \\ \cdot \\ \cdot \\ b_m \end{bmatrix}
$$

are column vectors. We assume that the coefficients a_{jk} are not all zero, so that **A** is not a zero matrix. Note that **x** has n components, whereas **b** has m components. The matrix

$$
\tilde{\mathbf{A}} = \begin{bmatrix}
a_{11} & \cdots & a_{1n} & b_1 \\
\cdot & \cdots & \cdot & \cdot \\
\cdot & \cdots & \cdot & \cdot \\
a_{m1} & \cdots & a_{mn} & b_m
\end{bmatrix}
$$

is called the **augmented matrix** of the system (1). We see that $\tilde{\mathbf{A}}$ is obtained by augmenting $\mathbf{A}$ by the column $\mathbf{b}$. The matrix $\tilde{\mathbf{A}}$ determines the system (1) completely, because it contains all the given numbers appearing in (1).

EXAMPLE 1. Geometric interpretation. Existence of solutions
If $m = n = 2$, we have two equations in two unknowns x_1, x_2

$$a_{11}x_1 + a_{12}x_2 = b_1$$
$$a_{21}x_1 + a_{22}x_2 = b_2.$$

If we interpret x_1, x_2 as coordinates in the x_1x_2-plane, then each of the two equations represents a straight line, and (x_1, x_2) is a solution if and only if the point P with coordinates x_1, x_2 lies on both lines. Hence there are three possible cases:
   (*a*) No solution if the lines are parallel.
   (*b*) Precisely one solution if they intersect.
   (*c*) Infinitely many solutions if they coincide.
For instance,

$x + y = 1$	$x + y = 1$	$x + y = 1$
$x + y = 0$	$x - y = 0$	$2x + 2y = 2$
Case (*a*)	Case (*b*)	Case (*c*)

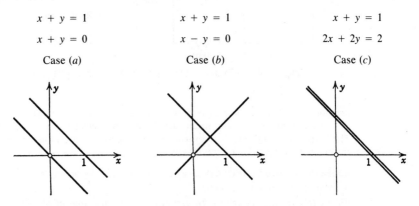

If the system is homogeneous, Case (*a*) cannot happen, because then those two straight lines pass through the origin, whose coordinates 0, 0 constitute the trivial solution. The reader may consider three equations in three unknowns as representations of three planes in space and discuss the various possible cases in a similar fashion.   ∎

Our simple example illustrates that a system (1) may not always have a solution, and relevant problems are as follows. Does a given system (1) have a solution? Under what conditions does it have precisely one solution? If it has more than one solution, how can the set of all solutions be characterized? How can the solutions be obtained?

Gauss Elimination

The Gauss elimination is a standard method for solving systems of linear equations. This is a systematic process of elimination, a method of great importance that works in practice and is reasonable with respect to computing time and storage demand (two aspects we shall consider in Sec. 19.1 on numerical methods). We first explain the method by some typical examples. Since a system of linear equations is completely determined by its augmented matrix, the process of elimination can be performed by merely considering the matrices. To see this correspondence, we shall write systems of equations and augmented matrices side by side.

Solving systems of linear equations is a basic practical task in linear algebra because such systems appear frequently as models of various problems, for instance, in frameworks, electrical networks, traffic flow, production and consumption, assignment of jobs to workers, population growth, statistics, numerical methods for differential equations (Chap. 20), systems of differential equations (Sec. 7.16) and many others, and this accounts for the importance of the Gauss elimination and variants of it. Let us begin with an illustrative example.

EXAMPLE 2. Electrical network. Gauss elimination

Find the currents in the network in Fig. 168.

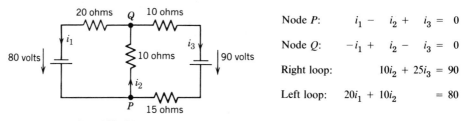

Node P: $i_1 - i_2 + i_3 = 0$

Node Q: $-i_1 + i_2 - i_3 = 0$

Right loop: $10i_2 + 25i_3 = 90$

Left loop: $20i_1 + 10i_2 = 80$

Fig. 168. Network in Example 2 and equations for the currents

Solution. We label the currents as shown, choosing directions arbitrarily; if a current will come out negative, this will simply mean that the current flows against the direction of our arrow. The current entering each battery will be the same as the current leaving it. The equations for the currents result from Kirchhoff's laws:

Kirchhoff's current law (KCL). At any point of a circuit, the sum of the inflowing currents equals the sum of the outflowing currents.

Kirchhoff's voltage law (KVL). In any closed loop, the sum of all voltage drops equals the impressed electromotive force.

This gives the four equations in Fig. 168. We write $x_1 = i_1$, $x_2 = i_2$, $x_3 = i_3$, to get these equations in the form (1). Then we have:

Equations   Augmented Matrix $\tilde{\mathbf{A}}$

$$
\begin{aligned}
\text{Pivot} \longrightarrow \boxed{x_1} - x_2 + x_3 &= 0 \\
\boxed{-x_1} + x_2 - x_3 &= 0 \\
\text{Eliminate} \longrightarrow \qquad 10x_2 + 25x_3 &= 90 \\
\boxed{20x_1} + 10x_2 &= 80
\end{aligned}
\qquad
\begin{bmatrix}
1 & -1 & 1 & 0 \\
-1 & 1 & -1 & 0 \\
0 & 10 & 25 & 90 \\
20 & 10 & 0 & 80
\end{bmatrix}
$$

We solve this system. This system is so simple that we could solve it almost by inspection. This is not the point. The point is to perform a systematic elimination—the Gauss elimination—which will work in general, also for large systems. It is a reduction to *"triangular form"* from which we shall then readily obtain the values of the unknowns by *"back substitution."*

First Step. Elimination of x_1
Call the first equation the **pivot equation** and its x_1-term the **pivot** in this step, and use this equation to eliminate x_1 (get rid of x_1) in the other equations. For this, do these operations:

Subtract[2] -1 times the pivot equation from the second equation.

Subtract 20 times the pivot equation from the fourth equation.

_____

[2]To express operations in terms of "subtraction" rather than "addition" is somewhat better in discussions of numerical algorithms (Sec. 19.1).

This corresponds to row operations on the augmented matrix, which we indicate behind the *new* matrix in (3). The result is:

(3)

$$
\begin{aligned}
x_1 - x_2 + x_3 &= 0 \\
0 &= 0 \\
10x_2 + 25x_3 &= 90 \\
30x_2 - 20x_3 &= 80
\end{aligned}
\qquad
\begin{bmatrix}
1 & -1 & 1 & 0 \\
0 & 0 & 0 & 0 \\
0 & 10 & 25 & 90 \\
0 & 30 & -20 & 80
\end{bmatrix}
\begin{matrix}
\\
\text{Row 2 + Row 1} \\
\\
\text{Row 4} - 20\,\text{Row 1}
\end{matrix}
$$

Second Step. Elimination of x_2

The first equation, which has just served as the pivot equation, remains untouched. We want to take the (new!) second equation as the next pivot equation. Since it contains no x_2-term (needed as the next pivot)—in fact, it is $0 = 0$—we have first to change the order of equations (and corresponding rows of the new matrix) to get a nonzero pivot. We put the second equation ($0 = 0$) at the end and move the third and fourth equations one place up; this is called **partial pivoting**.[3] We get

$$
\begin{aligned}
x_1 - x_2 + x_3 &= 0 \\
\text{Pivot} \longrightarrow \boxed{10x_2}\!+ 25x_3 &= 90 \\
\text{Eliminate} \longrightarrow \boxed{30x_2} - 20x_3 &= 80 \\
0 &= 0
\end{aligned}
\qquad
\begin{bmatrix}
1 & -1 & 1 & 0 \\
0 & 10 & 25 & 90 \\
0 & 30 & -20 & 80 \\
0 & 0 & 0 & 0
\end{bmatrix}
$$

To eliminate x_2, do:

Subtract 3 times the pivot equation from the third equation.

The result is

(4)

$$
\begin{aligned}
x_1 - x_2 + x_3 &= 0 \\
10x_2 + 25x_3 &= 90 \\
- 95x_3 &= -190 \\
0 &= 0
\end{aligned}
\qquad
\begin{bmatrix}
1 & -1 & 1 & 0 \\
0 & 10 & 25 & 90 \\
0 & 0 & -95 & -190 \\
0 & 0 & 0 & 0
\end{bmatrix}
\begin{matrix}
\\
\\
\text{Row 3} - 3\,\text{Row 2} \\
\\
\end{matrix}
$$

Back Substitution. Determination of x_3, x_2, x_1

Working backward from the last to the first equation of this "triangular" system (4), we can now readily find x_3, then x_2 and then x_1:

$$
\begin{aligned}
-95x_3 &= -190, & x_3 &= i_3 = 2 \text{ [amperes]}, \\
10x_2 + 25x_3 &= 90, & x_2 &= \tfrac{1}{10}(90 - 25x_3) = i_2 = 4 \text{ [amperes]}, \\
x_1 - x_2 + x_3 &= 0, & x_1 &= x_2 - x_3 = i_1 = 2 \text{ [amperes]}.
\end{aligned}
$$

This is the answer to our problem. The solution is unique.   ∎

A system (1) is called **overdetermined** if it has more equations than unknowns, as in Example 2, **determined** if $m = n$, as in Example 1, and **underdetermined** if (1) has fewer equations than unknowns. An underdetermined system always has solutions, whereas in the other two cases, solutions may or may not exist. (Details follow in Sec. 7.7.) We want to illustrate next that the Gauss elimination applies to any system, no matter whether it has many solutions, a unique solution or no solutions.

---

[3] As opposed to **total pivoting**, in which also the order of the unknowns is changed. Cf. Sec. 19.1.

EXAMPLE 3. Gauss elimination for an underdetermined system

Solve the system of three linear equations in four unknowns

(5)
$$\begin{aligned}
\boxed{3.0x_1} + 2.0x_2 + 2.0x_3 - 5.0x_4 &= 8.0 \\
\boxed{0.6x_1} + 1.5x_2 + 1.5x_3 - 5.4x_4 &= 2.7 \\
\boxed{1.2x_1} - 0.3x_2 - 0.3x_3 + 2.4x_4 &= 2.1
\end{aligned}
\qquad
\begin{bmatrix}
3.0 & 2.0 & 2.0 & -5.0 & 8.0 \\
0.6 & 1.5 & 1.5 & -5.4 & 2.7 \\
1.2 & -0.3 & -0.3 & 2.4 & 2.1
\end{bmatrix}$$

Solution. As in the previous example, we circle pivots and box terms to be eliminated.

First Step. Elimination of x_1 from the second and third equations by subtracting

$$0.6/3.0 = 0.2 \text{ times the first equation from the second equation,}$$

$$1.2/3.0 = 0.4 \text{ times the first equation from the third equation.}$$

This gives a new system of equations

(6)
$$\begin{aligned}
3.0x_1 + 2.0x_2 + 2.0x_3 - 5.0x_4 &= 8.0 \\
\boxed{1.1x_2} + 1.1x_3 - 4.4x_4 &= 1.1 \\
\boxed{-1.1x_2} - 1.1x_3 + 4.4x_4 &= -1.1
\end{aligned}
\qquad
\begin{bmatrix}
3.0 & 2.0 & 2.0 & -5.0 & 8.0 \\
0 & 1.1 & 1.1 & -4.4 & 1.1 \\
0 & -1.1 & -1.1 & 4.4 & -1.1
\end{bmatrix}$$

and we circle the pivot to be used in the next step.

Second Step. Elimination of x_2 from the third equation of (6) by subtracting

$$-1.1/1.1 = -1 \text{ times the second equation from the third equation.}$$

This gives

(7)
$$\begin{aligned}
3.0x_1 + 2.0x_2 + 2.0x_3 - 5.0x_4 &= 8.0 \\
1.1x_2 + 1.1x_3 - 4.4x_4 &= 1.1 \\
0 &= 0
\end{aligned}
\qquad
\begin{bmatrix}
3.0 & 2.0 & 2.0 & -5.0 & 8.0 \\
0 & 1.1 & 1.1 & -4.4 & 1.1 \\
0 & 0 & 0 & 0 & 0
\end{bmatrix}$$

Back Substitution. From the second equation, $x_2 = 1 - x_3 + 4x_4$. From this and the first equation, $x_1 = 2 - x_4$. Since x_3 and x_4 remain arbitrary, we have infinitely many solutions; if we choose a value of x_3 and a value of x_4, then the corresponding values of x_1 and x_2 are uniquely determined.

EXAMPLE 4. Gauss elimination if a unique solution exists

Solve the system

$$\begin{aligned}
-x_1 + x_2 + 2x_3 &= 2 \\
3x_1 - x_2 + x_3 &= 6 \\
-x_1 + 3x_2 + 4x_3 &= 4
\end{aligned}
\qquad
\begin{bmatrix}
-1 & 1 & 2 & 2 \\
3 & -1 & 1 & 6 \\
-1 & 3 & 4 & 4
\end{bmatrix}$$

First Step. Elimination of x_1 from the second and third equations gives

$$\begin{aligned}
-x_1 + x_2 + 2x_3 &= 2 \\
2x_2 + 7x_3 &= 12 \\
2x_2 + 2x_3 &= 2
\end{aligned}
\qquad
\begin{bmatrix}
-1 & 1 & 2 & 2 \\
0 & 2 & 7 & 12 \\
0 & 2 & 2 & 2
\end{bmatrix}
\begin{array}{l}
\\ \text{Row 2 + 3 Row 1} \\ \text{Row 3 - Row 1}
\end{array}$$

Second Step. Elimination of x_2 from the third equation gives

$$\begin{aligned}
-x_1 + x_2 + 2x_3 &= 2 \\
2x_2 + 7x_3 &= 12 \\
-5x_3 &= -10
\end{aligned}
\qquad
\begin{bmatrix}
-1 & 1 & 2 & 2 \\
0 & 2 & 7 & 12 \\
0 & 0 & -5 & -10
\end{bmatrix}
\begin{array}{l}
\\ \\ \text{Row 3 - Row 2}
\end{array}$$

Back Substitution. Beginning with the last equation, we obtain successively $x_3 = 2$, $x_2 = -1$, $x_1 = 1$. We see that the system has a unique solution.

EXAMPLE 5. Gauss elimination if no solution exists

What will happen if we apply the Gauss elimination to a system of equations that has no solution? The answer is that in this case the method will show this fact by producing a contradiction. For instance, consider

$$
\begin{aligned}
3x_1 + 2x_2 + x_3 &= 3 \\
2x_1 + x_2 + x_3 &= 0 \\
6x_1 + 2x_2 + 4x_3 &= 6
\end{aligned}
\qquad
\begin{bmatrix}
3 & 2 & 1 & 3 \\
2 & 1 & 1 & 0 \\
6 & 2 & 4 & 6
\end{bmatrix}
$$

First Step. Elimination of x_1 from the second and third equations by subtracting

2/3 times the first equation from the second equation,

$6/3 = 2$ times the first equation from the third equation.

This gives

$$
\begin{aligned}
3x_1 + 2x_2 + x_3 &= 3 \\
-\tfrac{1}{3}x_2 + \tfrac{1}{3}x_3 &= -2 \\
-2x_2 + 2x_3 &= 0
\end{aligned}
\qquad
\begin{bmatrix}
3 & 2 & 1 & 3 \\
0 & -\tfrac{1}{3} & \tfrac{1}{3} & -2 \\
0 & -2 & 2 & 0
\end{bmatrix}
$$

Second Step. Elimination of x_2 from the third equation gives

$$
\begin{aligned}
3x_1 + 2x_2 + x_3 &= 3 \\
-\tfrac{1}{3}x_2 + \tfrac{1}{3}x_3 &= -2 \\
0 &= 12
\end{aligned}
\qquad
\begin{bmatrix}
3 & 2 & 1 & 3 \\
0 & -\tfrac{1}{3} & \tfrac{1}{3} & -2 \\
0 & 0 & 0 & 12
\end{bmatrix}
$$

This shows that the system has no solution. ∎

The form of the system and of the matrix in the last step of the Gauss elimination is called the **echelon form.** Thus in Example 5 the echelon forms of the coefficient matrix and the augmented matrix are

$$
\begin{bmatrix}
3 & 2 & 1 \\
0 & -\tfrac{1}{3} & \tfrac{1}{3} \\
0 & 0 & 0
\end{bmatrix}
\quad \text{and} \quad
\begin{bmatrix}
3 & 2 & 1 & 3 \\
0 & -\tfrac{1}{3} & \tfrac{1}{3} & -2 \\
0 & 0 & 0 & 12
\end{bmatrix}.
$$

At the end of the Gauss elimination (before the back substitution) the reduced system will have the form

$$
\begin{aligned}
a_{11}x_1 + a_{12}x_2 + \cdots\cdots\cdots + a_{1n}x_n &= b_1 \\
c_{22}x_2 + \cdots\cdots\cdots + c_{2n}x_n &= b_2{}^* \\
&\;\;\vdots \\
k_{rr}x_r + \cdots + k_{rn}x_n &= \tilde{b}_r \\
0 &= \tilde{b}_{r+1} \\
&\;\;\vdots \\
0 &= \tilde{b}_m
\end{aligned}
$$

(8)

where $r \leq m$ (and $a_{11} \neq 0$, $c_{22} \neq 0$, $\cdots$, $k_{rr} \neq 0$). From this we see that with respect to solutions of this system (8), there are three possible cases:

(a) No solution if $r < m$ and one of the numbers $\tilde{b}_{r+1}$, $\cdots$, $\tilde{b}_m$ is not zero. This is illustrated by Example 5, where $r = 2 < m = 3$ and $\tilde{b}_{r+1} = \tilde{b}_3 = 12$.

(b) Precisely one solution if $r = n$ and $\tilde{b}_{r+1}$, $\cdots$, $\tilde{b}_m$, if present, are zero. This solution is obtained by solving the nth equation of (8) for x_n, then the $(n-1)$th equation for x_{n-1} and so on up the line. See Example 2, where $r = n = 3$ and $m = 4$.

(c) Infinitely many solutions if $r < n$ and $\tilde{b}_{r+1}$, $\cdots$, $\tilde{b}_m$, if present, are zero. Then any of these solutions is obtained by choosing values at pleasure for the unknowns x_{r+1}, $\cdots$, x_n, solving the rth equation for x_r, then the $(r-1)$th equation for x_{r-1} and so on up the line. Example 3 illustrates this case.

Elementary Row Operations

To justify the Gauss elimination as a method of solving systems of linear equations, we first introduce two related concepts.

Elementary operations for equations
  Interchange of two equations
  Multiplication of an equation by a nonzero constant
  Addition of a constant multiple of one equation to another equation

To these correspond the following

Elementary row operations for matrices
  Interchange of two rows
  Multiplication of a row by a nonzero constant
  Addition of a constant multiple of one row to another row

The Gauss elimination consists of the use of the third of these operations (for getting zeros) and of the first (in pivoting).

Now call a system of linear equations S_1 **row-equivalent** to a system of linear equations S_2 if S_1 can be obtained from S_2 by (finitely many!) elementary row operations. Clearly, the system produced by the Gauss elimination at the end is row-equivalent to the original system to be solved. Hence the desired justification of the Gauss elimination as a solution method now follows from the subsequent theorem, which implies that the Gauss elimination gives all solutions of the *original* system.

Theorem 1 (Row-equivalent systems)
Row-equivalent systems of linear equations have the same sets of solutions.

Proof. The interchange of two equations does not alter the solution set. Neither does the multiplication of an equation by a (nonzero!) constant c, because multiplication by $1/c$ produces the original equation. Similarly for the addition of an equation E_1 to an equation E_2, since by adding $-E_1$ (the equation obtained from E_1 by multiplying E_1 by -1) to the equation resulting from the addition we get back the original equation.   ∎

Variants of the Gauss elimination based on factorizing **A** in terms of triangular matrices, $\mathbf{A} = \mathbf{LU}$, such as **Doolittle's, Crout's** and **Cholesky's methods,** are popular in modern numerical work and will be discussed in Sec. 19.2 (which is independent of other sections on numerical methods), along with numerical aspects of the Gauss elimination.

Since a system of linear algebraic equations may have no solutions or precisely one solution or infinitely many solutions, the problems of **existence and uniqueness** of solutions are of considerable practical importance. We discuss them in Sec. 7.7. To prepare for this discussion, in the next section (Sec. 7.6) we introduce the concept of the **rank** of a matrix.

Problems for Sec. 7.5

Solve the following systems of linear equations by the Gauss elimination.

1. $2x + y = 4$
$5x - 2y = 1$

2. $3x + y = 11$
$x - y = 5$

3. $x + y = 0$
$3x - 4y = 1$

4. $4y + 3z = 13$
$x - 2y + z = 3$
$3x + 5y = 11$

5. $-x + y + 2z = 0$
$3x + 4y + z = 0$
$2x + 5y + 3z = 0$

6. $4x - y + z = 0$
$x + 2y - z = 0$
$3x + y + 5z = 0$

7. $x + y + z = -1$
$4y + 6z = 6$
$y + z = 1$

8. $3y - 4z = 1$
$9x - 4y + z = 4$
$x + y + z = 15$

9. $2x + 2y - 4z = -1$
$3y + 2z = 9$
$-4x + y + 3z = 4.5$

10. $-3x + 3y + 7z = 4$
$x - y + 2z = 3$
$4x - 4y + z = 5$

11. $7x - 4y - 2z = -6$
$16x + 2y + z = 3$

12. $3x - 12y = 6$
$-7x + 28y = -14$
$5x - 20y = 10$

13. $2y - z = 1$
$4x - 10y + 3z = 5$
$3x - 3y = 6$

14. $6x - 6y + 2z = 28$
$-x + 9y - 3z = -10$
$\frac{1}{4}x - y + \frac{1}{3}z = \frac{5}{3}$

15. $3x + 4y + 6z = 1$
$-2x + 8y - 4z = 2$
$4x - 8y + 8z = -2$

16. $5x + 3y = 22$
$-4x + 7y = 20$
$9x - 2y = 15$

17. $4y + 7z = -13$
$5x - 3y + 4z = -23$
$-x + 2y - 8z = 29$

18. $6x + 18y - 4z = 20$
$-x - 3y + 8z = 4$
$5x + 15y - 9z = 11$

19. $w + x + y = 3$
$-3w - 17x + y + 2z = 1$
$4w - 17x + 8y - 5z = 1$
$-5x - 2y + z = 1$

20. $w - x + 3y - 3z = 3$
$-5w + 2x - 5y + 4z = -5$
$-3w - 4x + 7y - 2z = 7$
$2w + 3x + y - 11z = 1$

Models of electrical networks
Using Kirchhoff's laws, find the currents in the following networks.

21.

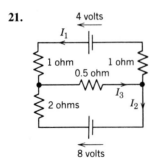

22.

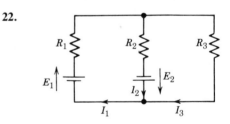

23.

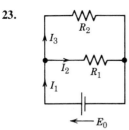

24.

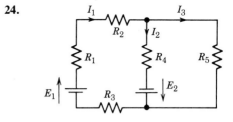

25. (Wheatstone bridge) Show that if $R_x/R_3 = R_1/R_2$ in the figure, then $I = 0$. (R_0 is the resistance of the instrument by which I is measured.)

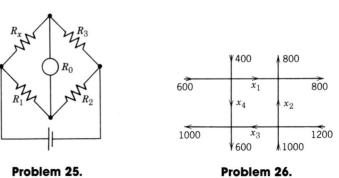

Problem 25.
Wheatstone bridge

Problem 26.
Net of one-way streets

26. (Traffic flow) Methods of electrical circuit analysis have applications to other fields. For instance, applying the analog of Kirchhoff's current law, find the traffic flow (cars per hour) in the net of one-way streets (in the directions indicated by the arrows) shown in the figure. Is the solution unique?

27. (Supply and demand) Determine the equilibrium solution ($D = S$) of the one-commodity market with linear model

$$D = 13 - 2P$$

$$S = 3P - 7$$

where D, S, P mean demand, supply and price of the commodity.

28. Determine the equilibrium solution ($D_1 = S_1$, $D_2 = S_2$) of the two-commodity market with linear model

$$D_1 = 20 - 2P_1 - P_2$$
$$S_1 = 4P_1 - P_2 + 2$$
$$D_2 = 5P_1 - 2P_2 + 8$$
$$S_2 = 3P_2 - 2$$

where D, S, P mean demand, supply, price, and the subscripts 1 and 2 refer to the first and second commodities, respectively.

29. **(Equivalence relation)** By definition, an *equivalence relation* on a set is a relation satisfying three conditions:

    (1) Each element A of the set is equivalent to itself.
    (2) If A is equivalent to B, then B is equivalent to A.
    (3) If A is equivalent to B and B is equivalent to C, then A is equivalent to C.

    For instance, equality is an equivalence relation on the set of real numbers. Show that row equivalence satisfies these three conditions.

30. Show that equality is an equivalence relation on the set of all matrices.

7.6 Rank of a Matrix

In the last section we have seen how to solve a system of linear equations, but we also know that there may be no solutions or a single solution or more than just one solution. So we ask whether we can make general statements about these *problems of existence and uniqueness*. The answer is yes; the main tool will be the concept of the rank of a matrix to be introduced now, and those statements will be given in the next section.

The maximum number of linearly independent row vectors of a matrix $\mathbf{A} = [a_{jk}]$ is called the **rank** of $\mathbf{A}$ and is denoted by

$$\text{rank } \mathbf{A}.$$

The rank of a zero matrix is 0, by definition. (Linear independence of vectors is defined in Sec. 6.4.)

EXAMPLE 1. Rank
The matrix

(1)
$$\mathbf{A} = \begin{bmatrix} 3 & 0 & 2 & 2 \\ -6 & 42 & 24 & 54 \\ 21 & -21 & 0 & -15 \end{bmatrix}$$

has rank $\mathbf{A} = 2$ since the last row is a linear combination of the two others (six times the first row minus $\frac{1}{2}$ times the second), which are linearly independent.   ∎

Note further that rank $\mathbf{A} = 0$ if and only if $\mathbf{A} = \mathbf{0}$. This follows directly from the definition.

In our proposed discussion of the existence and uniqueness of solutions of systems of linear equations we shall need the following very important

Theorem 1 (Rank in terms of column vectors)

The rank of a matrix $\mathbf{A}$ *equals the maximum number of linearly independent* **column** *vectors of* $\mathbf{A}$.

Hence $\mathbf{A}$ *and* $\mathbf{A}^\mathsf{T}$ *have the same rank.*

Proof. Let $\mathbf{A} = [a_{jk}]$ and let rank $\mathbf{A} = r$. Then, by definition, $\mathbf{A}$ has a linearly independent set of r row vectors, call them $\mathbf{v}_{(1)}, \cdots, \mathbf{v}_{(r)}$, and all row vectors $\mathbf{a}_{(1)}, \cdots, \mathbf{a}_{(m)}$ of $\mathbf{A}$ are linear combinations of those independent ones, say,

$$\mathbf{a}_{(1)} = c_{11}\mathbf{v}_{(1)} + c_{12}\mathbf{v}_{(2)} + \cdots + c_{1r}\mathbf{v}_{(r)}$$

$$\mathbf{a}_{(2)} = c_{21}\mathbf{v}_{(1)} + c_{22}\mathbf{v}_{(2)} + \cdots + c_{2r}\mathbf{v}_{(r)}$$

$$\vdots \qquad \vdots \qquad \vdots \qquad \vdots$$

$$\mathbf{a}_{(m)} = c_{m1}\mathbf{v}_{(1)} + c_{m2}\mathbf{v}_{(2)} + \cdots + c_{mr}\mathbf{v}_{(r)}$$

These are vector equations. Each of them is equivalent to n equations for corresponding components. Denoting the components of $\mathbf{v}_{(1)}$ by $v_{11}, \cdots, v_{1n}$, the components of $\mathbf{v}_{(2)}$ by $v_{21}, \cdots, v_{2n}$, etc., and similarly for the vectors on the left, we thus have

$$a_{1k} = c_{11}v_{1k} + c_{12}v_{2k} + \cdots + c_{1r}v_{rk}$$

$$a_{2k} = c_{21}v_{1k} + c_{22}v_{2k} + \cdots + c_{2r}v_{rk}$$

$$\vdots \qquad \vdots \qquad \vdots \qquad \vdots$$

$$a_{mk} = c_{m1}v_{1k} + c_{m2}v_{2k} + \cdots + c_{mr}v_{rk}$$

where $k = 1, \cdots, n$. This can be written

$$
\begin{bmatrix} a_{1k} \\ a_{2k} \\ \vdots \\ a_{mk} \end{bmatrix}
= v_{1k} \begin{bmatrix} c_{11} \\ c_{21} \\ \vdots \\ c_{m1} \end{bmatrix}
+ v_{2k} \begin{bmatrix} c_{12} \\ c_{22} \\ \vdots \\ c_{m2} \end{bmatrix}
+ \cdots + v_{rk} \begin{bmatrix} c_{1r} \\ c_{2r} \\ \vdots \\ c_{mr} \end{bmatrix}
$$

where $k = 1, \cdots, n$. The vector on the left is the kth column vector of $\mathbf{A}$. Hence the equation shows that each column vector of $\mathbf{A}$ is a linear combination of the r vectors on the right. Hence the maximum number of linearly independent column vectors of $\mathbf{A}$ cannot exceed r, which is the maximum number of linearly independent row vectors of $\mathbf{A}$, by the definition of rank.

Now the same conclusion applies to the transpose $\mathbf{A}^\mathsf{T}$ of $\mathbf{A}$. Since the row vectors of $\mathbf{A}^\mathsf{T}$ are the column vectors of $\mathbf{A}$, and the column vectors of $\mathbf{A}^\mathsf{T}$ are the row vectors of $\mathbf{A}$, that conclusion means that the maximum number of linearly independent row vectors of $\mathbf{A}$ (which is r) cannot exceed the maximum number of linearly independent column vectors of $\mathbf{A}$. Hence that number must equal r, and the proof is complete.   ∎

EXAMPLE 2. Illustration of Theorem 1

What does Theorem 1 mean with respect to our matrix $\mathbf{A}$ in (1)? Since we have rank $\mathbf{A} = 2$, the column vectors should contain two linearly independent ones, and the other two should be linear combinations of them. Indeed, the first two column vectors are linearly independent, and

$$\begin{bmatrix} 2 \\ 24 \\ 0 \end{bmatrix} = \frac{2}{3} \begin{bmatrix} 3 \\ -6 \\ 21 \end{bmatrix} + \frac{2}{3} \begin{bmatrix} 0 \\ 42 \\ -21 \end{bmatrix} \quad \text{and} \quad \begin{bmatrix} 2 \\ 54 \\ -15 \end{bmatrix} = \frac{2}{3} \begin{bmatrix} 3 \\ -6 \\ 21 \end{bmatrix} + \frac{29}{21} \begin{bmatrix} 0 \\ 42 \\ -21 \end{bmatrix}.$$

This is easy to verify but not so easy to see. Imagining that $\mathbf{A}^\mathsf{T}$ were given, we realize that the determination of a rank by a direct application of the definition is not the proper way, unless a matrix is sufficiently simple. This suggests asking whether we can "simplify" (transform) a matrix without altering its rank.   ∎

Invariance of Rank Under Row Operations

We claim that *elementary row operations* (Sec. 7.5) *do not alter the rank of a matrix* $\mathbf{A}$.

For the first operation (interchange of two row vectors) this is clear. The second operation (multiplication of a row vector by a nonzero constant) does not alter the rank either, since it does not alter the maximum number of linearly independent row vectors. Finally, the third operation is the addition of c times a row vector $\mathbf{a}_{(j)}$, say, to another row vector, say, $\mathbf{a}_{(i)}$. This produces a matrix that differs from $\mathbf{A}$ only in the ith row vector, which is of the form $\mathbf{a}_{(i)} + c\mathbf{a}_{(j)}$, a linear combination of the row vectors $\mathbf{a}_{(i)}$ and $\mathbf{a}_{(j)}$, so that the number of linearly independent row vectors remains the same. Hence the new matrix has the same rank as $\mathbf{A}$. Remembering from the previous section that *row-equivalent matrices* are those that can be obtained from each other by finitely many elementary row operations, our result is

Theorem 2 (Row-equivalent matrices)

Row-equivalent matrices have the same rank.

This theorem tells us what we can do to determine the rank of a matrix $\mathbf{A}$, namely, we can reduce $\mathbf{A}$ to echelon form (Sec. 7.5), using the technique of the Gauss elimination, because this leaves the rank unchanged, by Theorem 2, and from the echelon form we can recognize the rank directly.

EXAMPLE 3. Determination of rank

For the matrix in Example 1 we obtain successively

$$\mathbf{A} = \begin{bmatrix} 3 & 0 & 2 & 2 \\ -6 & 42 & 24 & 54 \\ 21 & -21 & 0 & -15 \end{bmatrix} \quad \text{(given)}$$

$$\begin{bmatrix} 3 & 0 & 2 & 2 \\ 0 & 42 & 28 & 58 \\ 0 & -21 & -14 & -29 \end{bmatrix} \quad \begin{matrix} \\ \text{Row 2 + 2 Row 1} \\ \text{Row 3 - 7 Row 1} \end{matrix}$$

$$\begin{bmatrix} 3 & 0 & 2 & 2 \\ 0 & 42 & 28 & 58 \\ 0 & 0 & 0 & 0 \end{bmatrix} \quad \begin{matrix} \\ \\ \text{Row 3 + } \tfrac{1}{2} \text{ Row 2} \end{matrix}$$

The last matrix is in echelon form. From the row vectors and Theorem 2 we see immediately that rank $\mathbf{A} \leqq 2$, and rank $\mathbf{A} = 2$ by Theorem 1, since the first two column vectors are certainly linearly independent. ∎

This method of determining rank has practical applications in connection with the determination of linear dependence and independence of vectors. The key to this is the following theorem, which results immediately from the definition of rank.

Theorem 3 (Linear dependence and independence)

p vectors $\mathbf{x}_{(1)}, \cdots, \mathbf{x}_{(p)}$ in vector space R^n (cf. Sec. 6.4) are linearly independent if the matrix with row vectors $\mathbf{x}_{(1)}, \cdots, \mathbf{x}_{(p)}$ has rank p; they are linearly dependent if that rank is less than p.

Since each of those p vectors has n components, that matrix, call it $\mathbf{A}$, has p rows and n columns; and if $n < p$, then by Theorem 1 we must have rank $\mathbf{A} \leqq n < p$, so that Theorem 3 yields the following result, which one should keep in mind.

Theorem 4

p vectors with $n < p$ components are always linearly dependent.

For instance, three or more vectors in the plane are linearly dependent. Similarly, four or more vectors in space are linearly dependent.

The rank of a matrix, as defined in this section, plays a basic role in characterizing the totality of solutions of a system of linear algebraic equations. We show this in the next section, where we discuss the problems of **existence and uniqueness of solutions** of such systems.

Problems for Sec. 7.6

Determine the rank of the following matrices. (Use inspection or the method in Example 3.)

1. $\begin{bmatrix} 2 & 2 \\ 2 & 2 \end{bmatrix}$ **2.** $\begin{bmatrix} 3 & 7 & 5 \\ 0 & 14 & 10 \end{bmatrix}$ **3.** $\begin{bmatrix} \tfrac{1}{2} & 1 & -4 \\ 0 & 0 & 0 \end{bmatrix}$

4. $\begin{bmatrix} 1 & 5 \\ 5 & 1 \end{bmatrix}$ **5.** $\begin{bmatrix} 0 & 4 & 5 \\ 1 & 4 & 5 \end{bmatrix}$ **6.** $\begin{bmatrix} 3 & 2 & 1 \\ 1 & 2 & 3 \end{bmatrix}$

7. $\begin{bmatrix} 0 & 0 \\ 1 & 0 \end{bmatrix}$ **8.** $\begin{bmatrix} a & b & c \\ b & a & c \end{bmatrix}, a \neq \pm b$ **9.** $\begin{bmatrix} 4 & 0 & -2 & 1 \\ -8 & 0 & 4 & -2 \end{bmatrix}$

10. $\begin{bmatrix} 7 & 7 \\ \frac{1}{2} & -\frac{1}{2} \\ 0 & 0 \end{bmatrix}$ **11.** $\begin{bmatrix} 3 & -3 & 0 \\ 1 & 4 & 5 \\ 4 & 4 & 8 \end{bmatrix}$ **12.** $\begin{bmatrix} 6 & 1 & 8 & 3 \\ 2 & 3 & 0 & 2 \\ 4 & -1 & -8 & -3 \end{bmatrix}$

13. $\begin{bmatrix} 4 & 1 \\ 1 & \frac{1}{4} \\ 2 & \frac{1}{2} \end{bmatrix}$ **14.** $\begin{bmatrix} 0 & 1 & 0 \\ 1 & 0 & 0 \\ 0 & 0 & 1 \end{bmatrix}$ **15.** $\begin{bmatrix} 1 & 0 & 0 & a \\ 0 & 1 & 0 & b \\ 0 & 0 & 1 & c \end{bmatrix}$

Linear independence and dependence
Are the given vectors linearly independent or dependent?

16. [3 2 7], [2 4 1], [1 -2 6]   **17.** [1 5 3], [2 4 6], [3 9 11]

18. [2 0 5], [3 0 2], [7 0 -2]   **19.** [0 1 0], [1 0 0], [1 1 0]

20. [9 9 9], [0 -1 2], [$\frac{1}{2}$ $\frac{1}{4}$ 0]   **21.** [1 2 3], [4 5 6], [7 8 9]

22. [1 0 0], [0 2 0], [0 0 3]   **23.** [1 1 1], [1 -1 1], [1 1 -1]

24. Show by an example that rank $\mathbf{A}$ = rank $\mathbf{B}$ does *not* imply rank $\mathbf{A}^2$ = rank $\mathbf{B}^2$.

25. (Rank of the transpose) Show that rank $\mathbf{A}$ = rank $\mathbf{A}^\mathsf{T}$.

26. Prove that if the row vectors of an $n \times n$ matrix $\mathbf{A}$ are linearly independent, then so are the column vectors of $\mathbf{A}$ (and vice versa).

27. Prove that if $\mathbf{A}$ is not square, then either the row vectors or the column vectors of $\mathbf{A}$ are linearly dependent.

28. (Row space and column space) By definition, the *row space* of an $m \times n$ matrix $\mathbf{A}$ is the vector space of all linear combinations of the row vectors. Similarly, the *column space* of $\mathbf{A}$ is the vector space of all linear combinations of the *column vectors* of $\mathbf{A}$. Find a basis of each of these two spaces in the case of the matrix in Prob. 10.

29. Prove the important fact that the dimensions of the row and column spaces of any matrix $\mathbf{A}$ are equal, namely, equal to rank $\mathbf{A}$.

30. Prove that row-equivalent matrices have the same row space.

7.7 Systems of Linear Equations: General Properties of Solutions

Using the concept of rank defined in the previous section, we now state the

Fundamental Theorem for systems of linear equations

(a) *A system of m linear equations*

(1)
$$
\begin{aligned}
a_{11}x_1 + a_{12}x_2 + \cdots + a_{1n}x_n &= b_1 \\
a_{21}x_1 + a_{22}x_2 + \cdots + a_{2n}x_n &= b_2 \\
\cdots\cdots\cdots\cdots\cdots\cdots\cdots\cdots \\
a_{m1}x_1 + a_{m2}x_2 + \cdots + a_{mn}x_n &= b_m
\end{aligned}
$$

in n unknowns $x_1, \cdots, x_n$ has solutions if and only if the coefficient matrix A and the augmented matrix $\widetilde{A}$, that is (cf. also Sec. 7.5)

$$
A = \begin{bmatrix} a_{11} & \cdots & a_{1n} \\ \cdot & \cdots & \cdot \\ \cdot & \cdots & \cdot \\ a_{m1} & \cdots & a_{mn} \end{bmatrix}
\quad and \quad
\widetilde{A} = \begin{bmatrix} a_{11} & \cdots & a_{1n} & b_1 \\ \cdot & \cdots & \cdot & \cdot \\ \cdot & \cdots & \cdot & \cdot \\ a_{m1} & \cdots & a_{mn} & b_m \end{bmatrix},
$$

have the same rank.

(b) *If this rank r equals n, the system (1) has precisely one solution.*

(c) *If $r < n$, the system (1) has infinitely many solutions, all of which are obtained by determining r suitable unknowns (whose submatrix of coefficients must have rank r) in terms of the remaining $n - r$ unknowns, to which arbitrary values can be assigned.*

(d) *If solutions exist, they can all be obtained by the Gauss elimination* (cf. Sec. 7.5). (This elimination may be started without first looking at the ranks of A and $\widetilde{A}$, since it will automatically reveal whether or not solutions exist; see, for instance, Example 5 in Sec. 7.5.)

Proof. (a) We can write the system (1) in the form

(1)
$$ Ax = b $$

or in terms of the column vectors $c_{(1)}, \cdots, c_{(n)}$ of A:

(2)
$$ c_{(1)}x_1 + c_{(2)}x_2 + \cdots + c_{(n)}x_n = b. $$

Since $\widetilde{A}$ is obtained by attaching to A the additional column b, Theorem 1 in Sec. 7.6 implies that rank $\widetilde{A}$ equals rank A or rank A + 1. Now if (1) has a solution $\mathbf{x}$, then (2) shows that $\mathbf{b}$ must be a linear combination of those column vectors. Hence rank $\tilde{\mathbf{A}}$ cannot exceed rank $\mathbf{A}$, so that rank $\tilde{\mathbf{A}}$ = rank $\mathbf{A}$.

Conversely, if rank $\tilde{\mathbf{A}}$ = rank $\mathbf{A}$, then $\mathbf{b}$ must be a linear combination of the column vectors of $\mathbf{A}$, say,

$$\mathbf{b} = \alpha_1 \mathbf{c}_{(1)} + \cdots + \alpha_n \mathbf{c}_{(n)}$$

since otherwise rank $\tilde{\mathbf{A}}$ = rank $\mathbf{A}$ + 1. But this means that (1) has a solution, namely, $x_1 = \alpha_1, \cdots, x_n = \alpha_n$.

(b) If rank $\mathbf{A} = r = n$, then the set $C = \{\mathbf{c}_{(1)}, \cdots, \mathbf{c}_{(n)}\}$ is linearly independent, by Theorem 1 in the previous section. It follows that then the representation (2) of $\mathbf{b}$ is unique because

$$\mathbf{c}_{(1)} x_1 + \cdots + \mathbf{c}_{(n)} x_n = \mathbf{c}_{(1)} \tilde{x}_1 + \cdots + \mathbf{c}_{(n)} \tilde{x}_n$$

would imply

$$(x_1 - \tilde{x}_1) \mathbf{c}_{(1)} + \cdots + (x_n - \tilde{x}_n) \mathbf{c}_{(n)} = \mathbf{0}$$

and $x_1 - \tilde{x}_1 = 0, \cdots, x_n - \tilde{x}_n = 0$ by the linear independence. Hence the scalars $x_1, \cdots, x_n$ in (2) are uniquely determined, that is, the solution of (1) is unique.

(c) If rank $\mathbf{A}$ = rank $\tilde{\mathbf{A}} = r < n$, by Theorem 1, Sec. 7.6, there is a linearly independent set K of r column vectors of $\mathbf{A}$ such that the other $n - r$ column vectors of $\mathbf{A}$ are linear combinations of those vectors. We renumber the columns and unknowns, denoting the renumbered quantities by $\hat{}$, so that $\{\hat{\mathbf{c}}_{(1)}, \cdots, \hat{\mathbf{c}}_{(r)}\}$ is that linearly independent set K. Then (2) becomes

$$\hat{\mathbf{c}}_{(1)} \hat{x}_1 + \cdots + \hat{\mathbf{c}}_{(n)} \hat{x}_n = \mathbf{b},$$

$\hat{\mathbf{c}}_{(r+1)}, \cdots, \hat{\mathbf{c}}_{(n)}$ are linear combinations of the vectors of K, and so are the vectors $\hat{x}_{r+1} \hat{\mathbf{c}}_{(r+1)}, \cdots, \hat{x}_n \hat{\mathbf{c}}_{(n)}$. Expressing these vectors in terms of the vectors of K and collecting terms, we can thus write the system in the form

(3) $$\hat{\mathbf{c}}_{(1)} y_1 + \cdots + \hat{\mathbf{c}}_{(r)} y_r = \mathbf{b}$$

with $y_j = \hat{x}_j + \beta_j$, where β_j results from the terms $\hat{x}_{r+1} \hat{\mathbf{c}}_{(r+1)}, \cdots, \hat{x}_n \hat{\mathbf{c}}_{(n)}$; here, $j = 1, \cdots, r$. Since the system has a solution, there are $y_1, \cdots, y_r$ satisfying (3). These scalars are unique since K is linearly independent. Choosing $\hat{x}_{r+1}, \cdots, \hat{x}_n$ fixes the β_j and corresponding $\hat{x}_j = y_j - \beta_j$, where $j = 1, \cdots, r$.

(d) This was proved in Sec. 7.6 and is restated here as a reminder.   ∎

The theorem is illustrated by the examples in Sec. 7.5: in Example 3 we have rank $\mathbf{A}$ = rank $\tilde{\mathbf{A}}$ = 2 < n = 4 and can choose x_3 and x_4 arbitrarily; in Example 4 there is a unique solution since rank $\mathbf{A}$ = rank $\tilde{\mathbf{A}}$ = n = 3, and in Example 5 there is no solution, since rank $\mathbf{A}$ = 2 < rank $\tilde{\mathbf{A}}$ = 3.

The Homogeneous System

The system in (1) is said to be **homogeneous** if all the b_j's on the right-hand side are zero. Otherwise it is said to be **nonhomogeneous.** (Cf. also Sec. 7.5.) From the Fundamental Theorem we readily obtain the following results.

Theorem 2 (Homogeneous system)
A homogeneous system of linear equations

(4)
$$
\begin{aligned}
a_{11}x_1 + a_{12}x_2 + \cdots + a_{1n}x_n &= 0 \\
a_{21}x_1 + a_{22}x_2 + \cdots + a_{2n}x_n &= 0 \\
&\cdots\cdots\cdots\cdots\cdots \\
a_{m1}x_1 + a_{m2}x_2 + \cdots + a_{mn}x_n &= 0
\end{aligned}
$$

always has the **trivial solution** $x_1 = 0, \cdots , x_n = 0$. *Nontrivial solutions exist if and only if rank* $\mathbf{A} < n$. *If rank* $\mathbf{A} = r < n$, *these solutions form a vector space of dimension* $n - r$ *(cf. Sec. 6.4). In particular, if* $\mathbf{x}_{(1)}$ *and* $\mathbf{x}_{(2)}$ *are solution vectors of* (4), *then* $\mathbf{x} = c_1\mathbf{x}_{(1)} + c_2\mathbf{x}_{(2)}$, *where* c_1 *and* c_2 *are any constants, is a solution vector of* (4). (*This* ***does not hold*** *for nonhomogeneous systems.*)

Proof. The first proposition is obvious and is in agreement with the fact that for a homogeneous system the matrix of the coefficients and the augmented matrix have the same rank. The solution vectors form a vector space because if $\mathbf{x}_{(1)}$ and $\mathbf{x}_{(2)}$ are any of them, then $\mathbf{A}\mathbf{x}_{(1)} = \mathbf{0}$, $\mathbf{A}\mathbf{x}_{(2)} = \mathbf{0}$, and this implies $\mathbf{A}(\mathbf{x}_{(1)} + \mathbf{x}_{(2)}) = \mathbf{A}\mathbf{x}_{(1)} + \mathbf{A}\mathbf{x}_{(2)} = \mathbf{0}$ as well as $\mathbf{A}(c\mathbf{x}_{(1)}) = c\mathbf{A}\mathbf{x}_{(1)} = \mathbf{0}$, where c is arbitrary. If rank $\mathbf{A} = r < n$, the Fundamental Theorem implies that we can choose $n - r$ suitable unknowns, call them $x_{r+1}, \cdots , x_n$, in an arbitrary fashion, and every solution is obtained in this way. It follows that a basis of solutions is $\mathbf{y}_{(1)}, \cdots , \mathbf{y}_{(n-r)}$, where the solution vector $\mathbf{y}_{(j)}, j = 1, \cdots , n - r$, is obtained by choosing $x_{r+j} = 1$ and the other $x_{r+1}, \cdots , x_n$ zero; the corresponding $x_1, \cdots , x_r$ are then determined. This proves that the vector space of all solutions has dimension $n - r$ and completes the proof. ∎

We mention that the vector space of all solutions of (4) is called the **null space** of the coefficient matrix $\mathbf{A}$, because if we muliply any $\mathbf{x}$ in this null space by $\mathbf{A}$ we get $\mathbf{0}$. The dimension of the null space is called the **nullity** of $\mathbf{A}$. In terms of these concepts, Theorem 2 states that

$$\text{rank } \mathbf{A} + \text{nullity } \mathbf{A} = n$$

where n is the number of unknowns (number of columns of $\mathbf{A}$). If we have rank $\mathbf{A} = n$, then nullity $\mathbf{A} = 0$, so that the system has only the trivial solution. If rank $\mathbf{A} = r < n$, then nullity $\mathbf{A} = n - r > 0$, so that we have nontrivial solutions which, together with $\mathbf{0}$, form a vector space of dimension $n - r > 0$.

Note that rank $\mathbf{A} \leqq m$ in (4), by the definition, so that rank $\mathbf{A} < n$ when $m < n$. By Theorem 2 this proves the following theorem, which is of considerable practical importance.

Theorem 3 (System with fewer equations than unknowns)

A homogeneous system of linear equations with fewer equations than unknowns always has nontrivial solutions.

The Nonhomogeneous System

If a nonhomogeneous system of linear equations has solutions, their totality can be characterized as follows.

Theorem 4 (Nonhomogeneous system)

If a nonhomogeneous system of linear equations of the form (1) *has solutions, then all these solutions are of the form*

$$\mathbf{x} = \mathbf{x}_0 + \mathbf{x}_h$$

where $\mathbf{x}_0$ *is any fixed solution of* (1) *and* $\mathbf{x}_h$ *runs through all the solutions of the corresponding homogeneous system* (4).

Proof. Let $\mathbf{x}$ be any given solution of (1) and $\mathbf{x}_0$ an arbitrarily chosen solution of (1). Then $\mathbf{A}\mathbf{x} = \mathbf{b}$, $\mathbf{A}\mathbf{x}_0 = \mathbf{b}$ and, therefore,

$$\mathbf{A}(\mathbf{x} - \mathbf{x}_0) = \mathbf{A}\mathbf{x} - \mathbf{A}\mathbf{x}_0 = \mathbf{0}.$$

This shows that the difference $\mathbf{x} - \mathbf{x}_0$ of any solution $\mathbf{x}$ of (1) and any fixed solution $\mathbf{x}_0$ of (1) is a solution of (4), say, $\mathbf{x}_h$. Hence all solutions of (1) are obtained by letting $\mathbf{x}_h$ run through all the solutions of the homogeneous system (4), and the proof is complete.   ∎

Related to the problem of solving systems of linear algebraic equations is the problem of determining the *inverse* $\mathbf{A}^{-1}$ of a *square* matrix $\mathbf{A}$, a concept we introduce in the next section, along with the *Gauss–Jordan method* for computing the inverse.

7.8 Inverse of a Matrix

In this section we consider exclusively **square** *matrices.*

The **inverse** of an $n \times n$ matrix $\mathbf{A} = [a_{jk}]$ is denoted by $\mathbf{A}^{-1}$ and is an $n \times n$ matrix such that

(1)
$$\mathbf{A}\mathbf{A}^{-1} = \mathbf{A}^{-1}\mathbf{A} = \mathbf{I},$$

where $\mathbf{I}$ is the $n \times n$ unit matrix.

If **A** has an inverse, then **A** is called a **nonsingular matrix.** If **A** has no inverse, then **A** is called a **singular matrix.**

If **A** *has an inverse, the inverse is unique.*

Indeed, if both **B** and **C** are inverses of **A**, then **AB** = **I** and **CA** = **I**, so that we obtain the uniqueness from

$$\mathbf{B} = \mathbf{IB} = (\mathbf{CA})\mathbf{B} = \mathbf{C}(\mathbf{AB}) = \mathbf{CI} = \mathbf{C}.$$

Let us motivate the concept of an inverse and find conditions for the existence. For this purpose we start from **A** and consider a corresponding **linear transformation** (Sec. 7.3)

$$(2) \qquad\qquad\qquad \mathbf{y} = \mathbf{Ax},$$

where the vectors **x** and **y** are column vectors with n components $x_1, \cdots, x_n$ and $y_1, \cdots, y_n$, respectively. If the inverse matrix $\mathbf{A}^{-1}$ exists, we may multiply (2) from the left ("premultiply": cf. Sec. 7.3) by $\mathbf{A}^{-1}$ and have $\mathbf{A}^{-1}\mathbf{y} = \mathbf{A}^{-1}\mathbf{Ax} = \mathbf{Ix} = \mathbf{x}$. That is, we have obtained the **inverse transformation** of (2),

$$(3) \qquad\qquad\qquad \mathbf{x} = \mathbf{A}^{-1}\mathbf{y},$$

which expresses **x** in terms of **y**, and involves the inverse $\mathbf{A}^{-1}$.

For a given **y**, equation (2) may be regarded as a system of n linear equations in n unknowns $x_1, \cdots, x_n$, and we know from the Fundamental Theorem in the previous section that it has a unique solution if and only if **A** has rank n, the greatest possible rank for an $n \times n$ matrix. This proves

Theorem 1 (Existence of the inverse)

For an $n \times n$ matrix **A**, *the inverse* $\mathbf{A}^{-1}$ *exists if and only if* rank **A** = n. *Hence* **A** *is nonsingular if* rank **A** = n, *and is singular if* rank **A** < n.

Determination of the Inverse

We want to show that for practically determining the inverse $\mathbf{A}^{-1}$ of a nonsingular $n \times n$ matrix **A** we can use the Gauss elimination (Sec. 7.5), actually, a variant of it, called the **Gauss–Jordan elimination.**[4] Our idea is as follows. Using **A**, we form the n systems $\mathbf{Ax}_{(1)} = \mathbf{e}_{(1)}, \cdots, \mathbf{Ax}_{(n)} = \mathbf{e}_{(n)}$, where $\mathbf{e}_{(j)}$ has the jth component 1 and the other components 0. Introducing the $n \times n$ matrices $\mathbf{X} = [\mathbf{x}_{(1)} \ \cdots \ \mathbf{x}_{(n)}]$ and $\mathbf{I} = [\mathbf{e}_{(1)} \ \cdots \ \mathbf{e}_{(n)}]$, we combine the n systems into a single matrix equation $\mathbf{AX} = \mathbf{I}$ and the n augmented matrices $[\mathbf{A} \ \mathbf{e}_{(1)}], \cdots, [\mathbf{A} \ \mathbf{e}_{(n)}]$ into a single augmented matrix $\tilde{\mathbf{A}} = [\mathbf{A} \ \mathbf{I}]$. Now $\mathbf{AX} = \mathbf{I}$ implies $\mathbf{X} = \mathbf{A}^{-1}\mathbf{I} = \mathbf{A}^{-1}$, and to solve $\mathbf{AX} = \mathbf{I}$ for **X** we can

---

[4]See W. Jordan, *Handbuch der Vermessungskunde* (7th ed., Stuttgart, 1920), vol. I, Sec. 36. We do **not recommend** it as a method for solving systems of linear equations, since the number of operations in addition to those of the Gauss elimination is larger than that for back substitution, which the Gauss–Jordan elimination avoids. See also Sec. 19.1.

apply the Gauss elimination to $\tilde{\mathbf{A}} = [\mathbf{A} \quad \mathbf{I}]$ to get $[\mathbf{U} \quad \mathbf{H}]$, where $\mathbf{U}$ is upper triangular, since the Gauss elimination triangularizes systems. The Gauss–Jordan elimination now operates on $[\mathbf{U} \quad \mathbf{H}]$ and, by eliminating the entries in $\mathbf{U}$ above the main diagonal, reduces it to $[\mathbf{I} \quad \mathbf{K}]$, the augmented matrix of $\mathbf{IX} = \mathbf{A}^{-1}$. Hence we must have $\mathbf{K} = \mathbf{A}^{-1}$ and can thus read off $\mathbf{A}^{-1}$ at the end.

(A formula for the entries in $\mathbf{A}^{-1}$ in terms of those in $\mathbf{A}$ follows in Sec. 7.11, in connection with determinants.)

EXAMPLE 1. Inverse of a matrix. Gauss-Jordan elimination

Find the inverse $\mathbf{A}^{-1}$ of

$$\mathbf{A} = \begin{bmatrix} -1 & 1 & 2 \\ 3 & -1 & 1 \\ -1 & 3 & 4 \end{bmatrix}.$$

Solution. We apply the Gauss elimination (Sec. 7.5) to

$$[\mathbf{A} \quad \mathbf{I}] = \left[\begin{array}{ccc|ccc} -1 & 1 & 2 & 1 & 0 & 0 \\ 3 & -1 & 1 & 0 & 1 & 0 \\ -1 & 3 & 4 & 0 & 0 & 1 \end{array}\right]$$

$$\left[\begin{array}{ccc|ccc} -1 & 1 & 2 & 1 & 0 & 0 \\ 0 & 2 & 7 & 3 & 1 & 0 \\ 0 & 2 & 2 & -1 & 0 & 1 \end{array}\right] \begin{array}{l} \\ \text{Row } 2 + 3 \text{ Row } 1 \\ \text{Row } 3 - \text{ Row } 1 \end{array}$$

$$\left[\begin{array}{ccc|ccc} -1 & 1 & 2 & 1 & 0 & 0 \\ 0 & 2 & 7 & 3 & 1 & 0 \\ 0 & 0 & -5 & -4 & -1 & 1 \end{array}\right] \begin{array}{l} \\ \\ \text{Row } 3 - \text{ Row } 2 \end{array}$$

This is $[\mathbf{U} \quad \mathbf{H}]$ as produced by the Gauss elimination, and $\mathbf{U}$ agrees with Example 4 in Sec. 7.5. Now follow the additional Gauss–Jordan steps, reducing $\mathbf{U}$ to $\mathbf{I}$, that is, to diagonal form with entries 1 on the main diagonal.

$$\left[\begin{array}{ccc|ccc} 1 & -1 & -2 & -1 & 0 & 0 \\ 0 & 1 & 3.5 & 1.5 & 0.5 & 0 \\ 0 & 0 & 1 & 0.8 & 0.2 & -0.2 \end{array}\right] \begin{array}{l} -\text{Row } 1 \\ 0.5 \text{ Row } 2 \\ -0.2 \text{ Row } 3 \end{array}$$

$$\left[\begin{array}{ccc|ccc} 1 & -1 & 0 & 0.6 & 0.4 & -0.4 \\ 0 & 1 & 0 & -1.3 & -0.2 & 0.7 \\ 0 & 0 & 1 & 0.8 & 0.2 & -0.2 \end{array}\right] \begin{array}{l} \text{Row } 1 + 2 \text{ Row } 3 \\ \text{Row } 2 - 3.5 \text{ Row } 3 \\ \\ \end{array}$$

$$\left[\begin{array}{ccc|ccc} 1 & 0 & 0 & -0.7 & 0.2 & 0.3 \\ 0 & 1 & 0 & -1.3 & -0.2 & 0.7 \\ 0 & 0 & 1 & 0.8 & 0.2 & -0.2 \end{array}\right] \begin{array}{l} \text{Row } 1 + \text{ Row } 2 \\ \\ \\ \end{array}$$

The last three columns constitute $\mathbf{A}^{-1}$. Check:

$$
\begin{bmatrix} -1 & 1 & 2 \\ 3 & -1 & 1 \\ -1 & 3 & 4 \end{bmatrix}
\begin{bmatrix} -0.7 & 0.2 & 0.3 \\ -1.3 & -0.2 & 0.7 \\ 0.8 & 0.2 & -0.2 \end{bmatrix}
=
\begin{bmatrix} 1 & 0 & 0 \\ 0 & 1 & 0 \\ 0 & 0 & 1 \end{bmatrix}.
$$

Hence $\mathbf{AA}^{-1} = \mathbf{I}$. Similarly, $\mathbf{A}^{-1}\mathbf{A} = \mathbf{I}$. ∎

Some Useful Formulas for Inverses

For a nonsingular 2×2 matrix we obtain

(4) $\qquad \mathbf{A} = \begin{bmatrix} a_{11} & a_{12} \\ a_{21} & a_{22} \end{bmatrix}, \qquad \mathbf{A}^{-1} = \dfrac{1}{\det \mathbf{A}} \begin{bmatrix} a_{22} & -a_{12} \\ -a_{21} & a_{11} \end{bmatrix},$

where $\det \mathbf{A} = a_{11}a_{22} - a_{12}a_{21}$ and will be discussed in the next section. Indeed, one can readily verify that (1) holds.

Similarly, for a nonsingular diagonal matrix we simply have

(5) $\qquad \mathbf{A} = \begin{bmatrix} a_{11} & \cdots & 0 \\ \cdot & \cdots & \cdot \\ \cdot & \cdots & \cdot \\ 0 & \cdots & a_{nn} \end{bmatrix}, \qquad \mathbf{A}^{-1} = \begin{bmatrix} 1/a_{11} & \cdots & 0 \\ \cdot & \cdots & \cdot \\ \cdot & \cdots & \cdot \\ 0 & \cdots & 1/a_{nn} \end{bmatrix};$

the elements of $\mathbf{A}^{-1}$ on the main diagonal are the reciprocals of those of $\mathbf{A}$.

EXAMPLE 2. Inverse of a 2 × 2 matrix

$$
\mathbf{A} = \begin{bmatrix} 3 & 1 \\ 2 & 4 \end{bmatrix}, \qquad \mathbf{A}^{-1} = \frac{1}{10}\begin{bmatrix} 4 & -1 \\ -2 & 3 \end{bmatrix} = \begin{bmatrix} 0.4 & -0.1 \\ -0.2 & 0.3 \end{bmatrix}.
$$

EXAMPLE 3. Inverse of a diagonal matrix

$$
\mathbf{A} = \begin{bmatrix} -0.5 & 0 & 0 \\ 0 & 4 & 0 \\ 0 & 0 & 1 \end{bmatrix}, \qquad \mathbf{A}^{-1} = \begin{bmatrix} -2 & 0 & 0 \\ 0 & 0.25 & 0 \\ 0 & 0 & 1 \end{bmatrix} \qquad ∎
$$

The inverse of the inverse is the given matrix $\mathbf{A}$**:**

(6) $\qquad\qquad\qquad\qquad (\mathbf{A}^{-1})^{-1} = \mathbf{A}.$

The simple proof is left to the reader (Prob. 2).

Inverse of a product. The inverse of a product $\mathbf{AC}$ can be calculated by inverting each factor separately and multiplying the results *in reverse order:*

(7) $\qquad\qquad\qquad\qquad \boxed{(\mathbf{AC})^{-1} = \mathbf{C}^{-1}\mathbf{A}^{-1}.}$

To prove (7), we start from (1), with $\mathbf{A}$ replaced by $\mathbf{AC}$, that is,

$$\mathbf{AC(AC)^{-1} = I.}$$

Multiplying this by $\mathbf{A^{-1}}$ from the left and using $\mathbf{A^{-1}A = I}$, we obtain

$$\mathbf{C(AC)^{-1} = A^{-1}.}$$

If we multiply this by $\mathbf{C^{-1}}$ from the left, the result follows.

Of course, (7) may be generalized to products of more than two matrices; by induction we obtain

$$(8) \qquad\qquad \mathbf{(AC \cdots PQ)^{-1} = Q^{-1}P^{-1} \cdots C^{-1}A^{-1}.}$$

Vanishing of Matrix Products. Cancellation Law

We can now obtain more information about the strange fact that for matrix multiplication, the cancellation law is not true, in general; that is, $\mathbf{AB = 0}$ does not necessarily imply that $\mathbf{A = 0}$ or $\mathbf{B = 0}$ (as for numbers), and it does also not imply that $\mathbf{BA = 0}$. These facts were stated in Sec. 7.3 and illustrated with

$$\begin{bmatrix} 1 & 1 \\ 2 & 2 \end{bmatrix} \begin{bmatrix} -1 & 1 \\ 1 & -1 \end{bmatrix} = \begin{bmatrix} 0 & 0 \\ 0 & 0 \end{bmatrix},$$

$$\begin{bmatrix} -1 & 1 \\ 1 & -1 \end{bmatrix} \begin{bmatrix} 1 & 1 \\ 2 & 2 \end{bmatrix} = \begin{bmatrix} 1 & 1 \\ -1 & -1 \end{bmatrix}.$$

Each of these two matrices has rank less than $n = 2$. This is typical, and the situation changes when $n \times n$ matrices have rank n:

Theorem 2 (Cancellation law)
Let $\mathbf{A}, \mathbf{B}, \mathbf{C}$ be $n \times n$ matrices. Then:

(a) *If rank $\mathbf{A} = n$ and $\mathbf{AB = AC}$, then $\mathbf{B = C}$.*

(b) *If rank $\mathbf{A} = n$, then $\mathbf{AB = 0}$ implies $\mathbf{B = 0}$. Hence if $\mathbf{AB = 0}$, but $\mathbf{A \neq 0}$ as well as $\mathbf{B \neq 0}$, then rank $\mathbf{A} < n$ and rank $\mathbf{B} < n$.*

(c) *If $\mathbf{A}$ is singular, so are $\mathbf{AB}$ and $\mathbf{BA}$.*

Proof. (a) Premultiply $\mathbf{AB = AC}$ on both sides by $\mathbf{A^{-1}}$, which exists by Theorem 1.

(b) Take $\mathbf{C = 0}$ in (a).

(c_1) Rank $\mathbf{A} < n$ by Theorem 1. Hence $\mathbf{Ax = 0}$ has nontrivial solutions, by Theorem 2 in Sec. 7.7. Multiplication gives $\mathbf{BAx = 0}$. Hence those solutions also satisfy $\mathbf{BAx = 0}$. Hence rank $\mathbf{(BA)} < n$ by Theorem 2 in Sec. 7.7, and $\mathbf{BA}$ is singular by Theorem 1.

(c₂) $\mathbf{A}^\mathsf{T}$ is singular by Theorem 1 in Sec. 7.6. Hence $\mathbf{B}^\mathsf{T}\mathbf{A}^\mathsf{T}$ is singular by part (c₁). But $\mathbf{B}^\mathsf{T}\mathbf{A}^\mathsf{T} = (\mathbf{AB})^\mathsf{T}$; cf. Sec. 7.4. Hence $\mathbf{AB}$ is singular by Theorem 1 in Sec. 7.6. ∎

The first portion of this chapter (Secs. 7.1—7.4) was concerned with the basic algebraic matrix operations. This is the end of the second portion (Secs. 7.5—7.8), in which we have discussed and solved systems of linear algebraic equations without the use of determinants. The third portion of the chapter (Secs. 7.9—7.11) is on **determinants.** The next section (Sec. 7.9) is mainly for reference. Section 7.10 contains the general definition and properties of determinants, and Sec. 7.11 the application of determinants to systems of linear equations.

Problems for Sec. 7.8

1. Verify (4) and (5) by showing that $\mathbf{AA}^{-1} = \mathbf{A}^{-1}\mathbf{A} = \mathbf{I}$.

2. Prove (6).

Find the inverse and check the result.

3. $\begin{bmatrix} 2 & 1 \\ 5 & 3 \end{bmatrix}$

4. $\begin{bmatrix} 1/\sqrt{2} & 1/\sqrt{2} \\ -1/\sqrt{2} & 1/\sqrt{2} \end{bmatrix}$

5. $\begin{bmatrix} 5 & 0 \\ 0 & 0.4 \end{bmatrix}$

6. $\begin{bmatrix} \cos\theta & -\sin\theta \\ \sin\theta & \cos\theta \end{bmatrix}$

7. $\begin{bmatrix} 0 & a \\ b & 0 \end{bmatrix}$

8. $\begin{bmatrix} 0.5 & -0.4 \\ 0 & 0.2 \end{bmatrix}$

9. $\begin{bmatrix} 4 & 0 & 0 \\ 0 & \frac{1}{2} & 0 \\ 0 & 0 & -5 \end{bmatrix}$

10. $\begin{bmatrix} 5 & -1 & -6 \\ 0 & 2 & -8 \\ 0 & 0 & 10 \end{bmatrix}$

11. $\begin{bmatrix} 1 & 0 & 0 \\ \frac{1}{2} & 1 & 0 \\ 1 & 5 & 2 \end{bmatrix}$

12. $\begin{bmatrix} 0 & -2 & 1 \\ \frac{1}{2} & \frac{1}{2} & -\frac{1}{2} \\ -1 & 2 & 0 \end{bmatrix}$

13. $\begin{bmatrix} 2 & 0 & -1 \\ 5 & 1 & 0 \\ 0 & 1 & 3 \end{bmatrix}$

14. $\begin{bmatrix} 5 & -1 & 5 \\ 0 & 2 & 0 \\ -5 & 3 & -15 \end{bmatrix}$

15. $\begin{bmatrix} 0 & 1 & 0 \\ 1 & 0 & 0 \\ 0 & 0 & 1 \end{bmatrix}$

16. $\begin{bmatrix} 1 & 0 & 0 \\ 0 & 0 & 1 \\ 0 & 1 & 0 \end{bmatrix}$

17. $\begin{bmatrix} 0 & 0 & c \\ 0 & b & 0 \\ a & 0 & 0 \end{bmatrix}$

Find the inverse of the given linear transformation.

18. $x^* = x + 2y$
$y^* = 3x + 4y$

19. $x^* = 3x - y$
$y^* = -5x + 2y$

20. $x^* = x$
$y^* = -\frac{1}{2}x + y$
$z^* = \frac{3}{4}x - \frac{5}{2}y + \frac{1}{2}z$

21. $x^* = x + 2y + 5z$
$y^* = -y + 2z$
$z^* = 2x + 4y + 11z$

22. $x^* = 19x + 2y - 9z$
$\ y^* = -4x - y + 2z$
$\ z^* = -2x + z$

23. $x^* = 2x + 4y + z$
$\ y^* = x + 2y + z$
$\ z^* = 3x + 4y + 2z$

24. $x^* = 6x + 4y + 3z$
$\ y^* = 4x + 3y + 4z$
$\ z^* = 3x + 2y + 2z$

25. $x^* = 3x - y + z$
$\ y^* = -15x + 6y - 5z$
$\ z^* = 5x - 2y + 2z$

26. Show that $(\mathbf{A}^{-1})^\mathsf{T} = (\mathbf{A}^\mathsf{T})^{-1}$.

27. Show that $(\mathbf{A}^2)^{-1} = (\mathbf{A}^{-1})^2$.

28. Find the inverse of the square of the matrix in Prob. 14.

29. Show that the inverse of a nonsingular symmetric matrix is symmetric.

30. Show that the inverse of a nonsingular skew-symmetric matrix is skew-symmetric.

7.9 Determinants of Second and Third Order

This section is independent of the other sections in this chapter. It is for reference in connection with other chapters, and it also serves to motivate the discussions in the next section, where general determinants will be considered.

Determinants of second order can be introduced and used in connection with systems of two linear equations

$$
\begin{aligned}
a_{11}x_1 + a_{12}x_2 &= b_1 \\
a_{21}x_1 + a_{22}x_2 &= b_2
\end{aligned}
\tag{1}
$$

in two unknowns x_1, x_2, if we proceed as follows. To solve this system, we multiply the first equation by a_{22}, the second by $-a_{12}$, and add, finding

$$(a_{11}a_{22} - a_{21}a_{12})x_1 = b_1 a_{22} - b_2 a_{12}.$$

Then we multiply the first equation of (1) by $-a_{21}$, the second by a_{11}, and add again, finding

$$(a_{11}a_{22} - a_{21}a_{12})x_2 = a_{11}b_2 - a_{21}b_1.$$

If $a_{11}a_{22} - a_{21}a_{12}$ is not zero, we may divide and obtain the desired result

$$
x_1 = \frac{b_1 a_{22} - b_2 a_{12}}{a_{11}a_{22} - a_{21}a_{12}}, \qquad
x_2 = \frac{a_{11}b_2 - a_{21}b_1}{a_{11}a_{22} - a_{21}a_{12}}.
\tag{2}
$$

The expression in the denominators is written in the form

$$\begin{vmatrix} a_{11} & a_{12} \\ a_{21} & a_{22} \end{vmatrix}$$

and is called a **determinant of second order;** thus

(3)
$$\begin{vmatrix} a_{11} & a_{12} \\ a_{21} & a_{22} \end{vmatrix} = a_{11}a_{22} - a_{21}a_{12}.$$

The four numbers a_{11}, a_{12}, a_{21}, a_{22} are called the **entries** in the determinant or the *elements* of the determinant. The entries in a horizontal line form a **row** and the entries in a vertical line a form a **column** of the determinant.

We may now write the solution (2) of the system (1) in the form

(4)
$$x_1 = \frac{D_1}{D}, \qquad x_2 = \frac{D_2}{D} \qquad\qquad (D \neq 0)$$

where

$$D = \begin{vmatrix} a_{11} & a_{12} \\ a_{21} & a_{22} \end{vmatrix}, \qquad D_1 = \begin{vmatrix} b_1 & a_{12} \\ b_2 & a_{22} \end{vmatrix}, \qquad D_2 = \begin{vmatrix} a_{11} & b_1 \\ a_{21} & b_2 \end{vmatrix}.$$

This formula is called **Cramer's rule.**[5] Note that D_1 is obtained by replacing the first column of D by the column with entries b_1, b_2, and D_2 is obtained by replacing the second column of D by that column.

If both b_1 and b_2 are zero, the system is said to be **homogeneous.** *In this case it has at least the "trivial solution" $x_1 = 0$, $x_2 = 0$. It has further solutions if and only if $D = 0$.*

If at least b_1 or b_2 is not zero, the system is said to be **nonhomogeneous.** *Then if $D \neq 0$, it has precisely one solution, which is obtained from (4).*

Determinants of Third Order

These occur in connection with systems of three linear equations

(5)
$$a_{11}x_1 + a_{12}x_2 + a_{13}x_3 = b_1$$
$$a_{21}x_1 + a_{22}x_2 + a_{23}x_3 = b_2$$
$$a_{31}x_1 + a_{32}x_2 + a_{33}x_3 = b_3$$

in three unknowns x_1, x_2, x_3. To obtain an equation involving only x_1, we multiply the equations by

[5]GABRIEL CRAMER (1704—1752), Swiss mathematician, also known by his book on the theory of curves, which appeared 1750 in Geneva.

$$a_{22}a_{33} - a_{32}a_{23}, \qquad -(a_{12}a_{33} - a_{32}a_{13}), \qquad a_{12}a_{23} - a_{22}a_{13},$$

respectively. We see that these expressions may be written as second-order determinants:

$$M_{11} = \begin{vmatrix} a_{22} & a_{23} \\ a_{32} & a_{33} \end{vmatrix}, \qquad -M_{21} = -\begin{vmatrix} a_{12} & a_{13} \\ a_{32} & a_{33} \end{vmatrix}, \qquad M_{31} = \begin{vmatrix} a_{12} & a_{13} \\ a_{22} & a_{23} \end{vmatrix}.$$

Adding the resulting equations, we obtain

$$(6) \qquad (a_{11}M_{11} - a_{21}M_{21} + a_{31}M_{31})x_1 = b_1M_{11} - b_2M_{21} + b_3M_{31}.$$

Two further equations containing only x_2 and x_3, respectively, may be obtained in a similar manner.

To simplify our notation, we now define a **determinant of third order** by

$$(7) \quad D = \begin{vmatrix} a_{11} & a_{12} & a_{13} \\ a_{21} & a_{22} & a_{23} \\ a_{31} & a_{32} & a_{33} \end{vmatrix} = a_{11}\begin{vmatrix} a_{22} & a_{23} \\ a_{32} & a_{33} \end{vmatrix} - a_{21}\begin{vmatrix} a_{12} & a_{13} \\ a_{32} & a_{33} \end{vmatrix} + a_{31}\begin{vmatrix} a_{12} & a_{13} \\ a_{22} & a_{23} \end{vmatrix}.$$

We see that

$$D = a_{11}M_{11} - a_{21}M_{21} + a_{31}M_{31},$$

the coefficient of x_1 in (6), and if we write out the second-order determinants in (7), we obtain

$$(8) \qquad D = a_{11}a_{22}a_{33} - a_{11}a_{32}a_{23} + a_{21}a_{32}a_{13} - a_{21}a_{12}a_{33} + a_{31}a_{12}a_{23} - a_{31}a_{22}a_{13}.$$

Obviously, the determinant on the right-hand side of (7) which is multiplied by a_{i1}, $i = 1, 2$ or 3, is obtained from D by omitting the first column and the ith row of D.

We see that (6) may now be written

$$Dx_1 = D_1 \qquad \text{where} \qquad D_1 = \begin{vmatrix} b_1 & a_{12} & a_{13} \\ b_2 & a_{22} & a_{23} \\ b_3 & a_{32} & a_{33} \end{vmatrix}.$$

Similarly, the aforementioned equation containing only x_2 may be written

$$Dx_2 = D_2 \qquad \text{where} \qquad D_2 = \begin{vmatrix} a_{11} & b_1 & a_{13} \\ a_{21} & b_2 & a_{23} \\ a_{31} & b_3 & a_{33} \end{vmatrix}$$

and the equation containing only x_3 may be written

$$Dx_3 = D_3 \quad \text{where} \quad D_3 = \begin{vmatrix} a_{11} & a_{12} & b_1 \\ a_{21} & a_{22} & b_2 \\ a_{31} & a_{32} & b_3 \end{vmatrix}.$$

Note that the entries in D are arranged in the same order as they occur as coefficients in the equations of (5), and D_j, $j = 1, 2$ or 3, is obtained from D by replacing the jth column by the column with entries b_1, b_2, b_3, the expressions on the right-hand sides of the equations of (5).

It follows that if $D \neq 0$, then system (5) has the unique solution

(9) $$x_1 = \frac{D_1}{D}, \qquad x_2 = \frac{D_2}{D}, \qquad x_3 = \frac{D_3}{D} \qquad \text{(Cramer's rule)}.$$

If (5) is homogeneous, that is, $b_1 = b_2 = b_3 = 0$, it has at least the trivial solution $x_1 = x_2 = x_3 = 0$, and nontrivial solutions exist if and only if $D = 0$.

If (5) is nonhomogeneous and $D \neq 0$, it has precisely one solution, which is obtained from (9).

EXAMPLE 1. Cramer's rule
Solve by Cramer's rule:

$$\begin{aligned} 2x_1 - x_2 + 2x_3 &= 2 \\ x_1 + 10x_2 - 3x_3 &= 5 \\ -x_1 + x_2 + x_3 &= -3. \end{aligned}$$

Solution. The determinant of the system is

$$D = \begin{vmatrix} 2 & -1 & 2 \\ 1 & 10 & -3 \\ -1 & 1 & 1 \end{vmatrix} = 46.$$

The determinants in the numerators in (9) are

$$D_1 = \begin{vmatrix} 2 & -1 & 2 \\ 5 & 10 & -3 \\ -3 & 1 & 1 \end{vmatrix} = 92, \quad D_2 = \begin{vmatrix} 2 & 2 & 2 \\ 1 & 5 & -3 \\ -1 & -3 & 1 \end{vmatrix} = 0, \quad D_3 = \begin{vmatrix} 2 & -1 & 2 \\ 1 & 10 & 5 \\ -1 & 1 & -3 \end{vmatrix} = -46.$$

Therefore, $x_1 = 2$, $x_2 = 0$, and $x_3 = -1$. ∎

Remark
Cramer's rule is impractical for the numerical solution of larger systems (to which it can be extended; see Sec. 7.11), but plays a role in the theory, and determinants appear in connection with differential equations, eigenvalue problems and other applications. In **numerical work,** one uses the Gauss elimination or similar methods (Secs. 7.5, 19.1).

General Properties of Third-Order Determinants

We shall now list the most important properties of our determinants. The proofs follow from (7) by direct calculation. (In the next section we shall see that determinants of any order n have quite similar properties.)

(A) *The value of a determinant is not altered if its rows are written as columns in the same order.* Example:

$$(10) \qquad \begin{vmatrix} 1 & 3 & 0 \\ 2 & 6 & 4 \\ -1 & 0 & 2 \end{vmatrix} = \begin{vmatrix} 1 & 2 & -1 \\ 3 & 6 & 0 \\ 0 & 4 & 2 \end{vmatrix} = -12.$$

(B) *If any two rows (or two columns) of a determinant are interchanged, the value of the determinant is multiplied by* -1. Example:

$$(11) \qquad \begin{vmatrix} 2 & 6 & 4 \\ 1 & 3 & 0 \\ -1 & 0 & 2 \end{vmatrix} = - \begin{vmatrix} 1 & 3 & 0 \\ 2 & 6 & 4 \\ -1 & 0 & 2 \end{vmatrix} = 12.$$

The second-order determinant obtained from D [cf. (7)] by deleting one row and one column is called the **minor** of the entry that belongs to the deleted row and column. Example: The minors of a_{21} and a_{22} in D are

$$\begin{vmatrix} a_{12} & a_{13} \\ a_{32} & a_{33} \end{vmatrix} \qquad \text{and} \qquad \begin{vmatrix} a_{11} & a_{13} \\ a_{31} & a_{33} \end{vmatrix},$$

respectively. The **cofactor** of the entry in the ith row and the kth column of D is defined as $(-1)^{i+k}$ times the minor of that entry. Example: the cofactors of a_{21} and a_{22} are

$$- \begin{vmatrix} a_{12} & a_{13} \\ a_{32} & a_{33} \end{vmatrix} \qquad \text{and} \qquad \begin{vmatrix} a_{11} & a_{13} \\ a_{31} & a_{33} \end{vmatrix},$$

respectively. The signs $(-1)^{i+k}$ form a checkerboard pattern:

$$\begin{matrix} + & - & + \\ - & + & - \\ + & - & + \end{matrix}$$

Furthermore, we see that we may now write (7) in the form

$$D = a_{11}C_{11} + a_{21}C_{21} + a_{31}C_{31},$$

where C_{i1} is the cofactor of a_{i1} in D. From this and the properties (**A**) and (**B**) we obtain the following property.

(**C**) *The determinant D may be developed by any row or column; that is, it may be written as the sum of the three entries in any row (or column) each multiplied by its cofactor.* For example, the development of D by its second row is

$$D = -a_{21}\begin{vmatrix} a_{12} & a_{13} \\ a_{32} & a_{33} \end{vmatrix} + a_{22}\begin{vmatrix} a_{11} & a_{13} \\ a_{31} & a_{33} \end{vmatrix} - a_{23}\begin{vmatrix} a_{11} & a_{12} \\ a_{31} & a_{32} \end{vmatrix}.$$

From (**C**) we obtain

(**D**) *A factor of the entries in any row (or column) can be placed before the determinant.* Example:

$$\begin{vmatrix} 4 & 6 & 1 \\ 3 & -9 & 2 \\ -1 & 12 & 5 \end{vmatrix} = \begin{vmatrix} 4 & 2\cdot 3 & 1 \\ 3 & -3\cdot 3 & 2 \\ -1 & 4\cdot 3 & 5 \end{vmatrix} = 3\begin{vmatrix} 4 & 2 & 1 \\ 3 & -3 & 2 \\ -1 & 4 & 5 \end{vmatrix}.$$

From the properties (**B**) and (**D**) we may draw the following conclusion.

(**E**) *If corresponding entries in two rows (or columns) of a determinant are proportional, the value of the determinant is zero.*

The following property is basic for simplifying determinants to be evaluated.

(**F**) *The value of a determinant remains unaltered if the elements of one row (or column) are altered by adding to them any constant multiple of the corresponding elements in any other row (or column).* Example:

$$\begin{vmatrix} -6 & 21 & -30 \\ 1 & -3 & 5 \\ 2 & 7 & -4 \end{vmatrix} = \begin{vmatrix} -6+1\cdot 7 & 21-3\cdot 7 & -30+5\cdot 7 \\ 1 & -3 & 5 \\ 2 & 7 & -4 \end{vmatrix}$$

$$= \begin{vmatrix} 1 & 0 & 5 \\ 1 & -3 & 5 \\ 2 & 7 & -4 \end{vmatrix} = \begin{vmatrix} 1 & 0 & 5 \\ 0 & -3 & 0 \\ 2 & 7 & -4 \end{vmatrix} = -3\begin{vmatrix} 1 & 5 \\ 2 & -4 \end{vmatrix} = 42.$$

(**G**) *If each entry in a row (or column) of a determinant is expressed as a binomial, the determinant can be written as the sum of two determinants.* Example:

$$(12) \qquad \begin{vmatrix} 4x+3 & 2 & 1 \\ x & 3 & 4 \\ 2x-1 & 1 & -1 \end{vmatrix} = \begin{vmatrix} 4x & 2 & 1 \\ x & 3 & 4 \\ 2x & 1 & -1 \end{vmatrix} + \begin{vmatrix} 3 & 2 & 1 \\ 0 & 3 & 4 \\ -1 & 1 & -1 \end{vmatrix}$$

By applying the product rule of differentiation we obtain the following property.

(H) *If the entries in a determinant are differentiable functions of a variable, the derivative of the determinant may be written as a sum of three determinants,*

$$\frac{d}{dx}\begin{vmatrix} f & g & h \\ p & q & r \\ u & v & w \end{vmatrix} = \begin{vmatrix} f' & g' & h' \\ p & q & r \\ u & v & w \end{vmatrix} + \begin{vmatrix} f & g & h \\ p' & q' & r' \\ u & v & w \end{vmatrix} + \begin{vmatrix} f & g & h \\ p & q & r \\ u' & v' & w' \end{vmatrix},$$

where primes denote derivatives with respect to x.

In the next section we define determinants of arbitrary order n and discuss their general properties.

Problems for Sec. 7.9

Evaluate

1. $\begin{vmatrix} 3 & -1 \\ 2 & 4 \end{vmatrix}$

2. $\begin{vmatrix} 0 & 3 \\ 5 & 7 \end{vmatrix}$

3. $\begin{vmatrix} \cos\theta & \sin\theta \\ -\sin\theta & \cos\theta \end{vmatrix}$

4. $\begin{vmatrix} 16 & 89 & 76 \\ 0 & 10 & 37 \\ 0 & 0 & 25 \end{vmatrix}$

5. $\begin{vmatrix} 13 & 24 & 12 \\ 0 & 6 & 0 \\ 12 & 81 & 5 \end{vmatrix}$

6. $\begin{vmatrix} 17 & 2 & 4 \\ 1 & 3 & 8 \\ 16 & -1 & -4 \end{vmatrix}$

7. $\begin{vmatrix} 6 & 13 & -2 \\ 5 & 37 & -5 \\ 1 & 13 & -2 \end{vmatrix}$

8. $\begin{vmatrix} 0.5 & 0.1 & 0.8 \\ 1.5 & 0.3 & 0.6 \\ 1.0 & 0.4 & 0.2 \end{vmatrix}$

9. $\begin{vmatrix} 1.6 & 0.4 & 1.2 \\ 2.2 & -0.3 & 2.5 \\ 0.4 & 0.2 & 0.2 \end{vmatrix}$

10. $\begin{vmatrix} 1 & 2 & 3 \\ 4 & 5 & 6 \\ 7 & 8 & 9 \end{vmatrix}$

11. $\begin{vmatrix} 3 & 5 & -2 \\ 0 & 6 & 1 \\ 2 & 0 & -4 \end{vmatrix}$

12. $\begin{vmatrix} 13 & 1 & 4 \\ 6 & 2 & 4 \\ 4 & 1 & 4 \end{vmatrix}$

13. $\begin{vmatrix} a & b & c \\ c & a & b \\ b & c & a \end{vmatrix}$

14. $\begin{vmatrix} 1 & a & a^2 \\ 1 & b & b^2 \\ 1 & c & c^2 \end{vmatrix}$

15. $\begin{vmatrix} b^2 + c^2 & ab & ca \\ ab & c^2 + a^2 & bc \\ ca & bc & a^2 + b^2 \end{vmatrix}$

16. (Straight line) Show that the straight line through two points P_1: (x_1, y_1) and P_2: (x_2, y_2) in the xy-plane is given by formula (a) (below), and derive from (a) the familiar formula (b).

(a) $\begin{vmatrix} x & y & 1 \\ x_1 & y_1 & 1 \\ x_2 & y_2 & 1 \end{vmatrix} = 0$ (b) $\dfrac{x - x_1}{x_1 - x_2} = \dfrac{y - y_1}{y_1 - y_2}$

7.10 Determinants of Arbitrary Order

A determinant of arbitrary order n is a scalar associated with a square matrix of order n, as explained below. It can be defined in several equivalent ways. One possible way is suggested by a process of solving n linear equations in n unknowns. For $n = 2$ and $n = 3$ this was shown in the previous section, and the generalization to arbitrary n will be discussed in Sec. 7.11.

Determinants have various applications in engineering mathematics, although their importance has decreased since they are generally impractical in numerical work. Indeed, Cramer's rule (Secs. 7.9 and 7.11) is certainly not the *practical* answer to the problem of solving systems of linear equations numerically, and there are numerically better methods (cf. Secs. 7.5 and 19.1—19.3). However, determinants will still retain their place in eigenvalue problems (Sec. 7.12), differential equations, vector algebra (vector and scalar triple products) and so on.

We shall now define determinants of order n in a way that is practical in applications. Then we shall see that the useful properties of determinants of orders $n = 2$ and 3 listed in Sec. 7.9 carry over to determinants of general order n.

Definition. Determinant of order *n*

A *determinant of order n* is a scalar associated with an n-rowed square matrix $\mathbf{A} = [a_{jk}]$, which is written

(1)
$$D = \det \mathbf{A} = \begin{vmatrix} a_{11} & a_{12} & \cdots & a_{1n} \\ a_{21} & a_{22} & \cdots & a_{2n} \\ \cdot & \cdot & \cdots & \cdot \\ \cdot & \cdot & \cdots & \cdot \\ a_{n1} & a_{n2} & \cdots & a_{nn} \end{vmatrix}$$

and is defined for $n = 1$ by

(2)
$$D = a_{11}$$

and for $n \geq 2$ by

(3a) $D = a_{j1}C_{j1} + a_{j2}C_{j2} + \cdots + a_{jn}C_{jn}$ $(j = 1, 2, \cdots, \text{or } n)$

or

(3b) $D = a_{1k}C_{1k} + a_{2k}C_{2k} + \cdots + a_{nk}C_{nk}$ $(k = 1, 2, \cdots, \text{or } n)$.

where

$$C_{jk} = (-1)^{j+k}M_{jk}$$

and M_{jk} is a determinant of order $n - 1$, namely, the determinant of the submatrix of **A** obtained from **A** by deleting the row and column of the entry a_{jk} (the jth row and the kth column).   ∎

In this way, D is defined in terms of n determinants of order $n - 1$, each of which is, in turn, defined in terms of $n - 1$ determinants of order $n - 2$, and so on; we finally arrive at second-order determinants, in which those submatrices consist of single entries whose determinant is defined by (2).

From the definition it follows that *we may* **develop** *D by any row or column,* that is, choose in (3) the entries in any row or column, similarly when developing the C_{jk}'s cofactors in (3), and so on.

This definition is unambiguous, that is, yields the same value for D no matter which columns or rows we choose. A proof is given on p. 448 near the end of the chapter.

Terms used in connection with determinants are taken from matrices: in D we have n^2 **entries** or *elements* a_{jk}, also n **rows** and n **columns,** a **main diagonal** or *principal diagonal* on which $a_{11}, a_{22}, \cdots, a_{nn}$ stand. Two names are new: M_{jk} is called the **minor** *of* a_{jk} *in* D, and C_{jk} the **cofactor** *of* a_{jk} *in* D.

For example, in the third-order determinant

$$\begin{vmatrix} a_{11} & a_{12} & a_{13} \\ a_{21} & a_{22} & a_{23} \\ a_{31} & a_{32} & a_{33} \end{vmatrix}$$

we have the cofactors and minors

$$C_{11} = M_{11} = \begin{vmatrix} a_{22} & a_{23} \\ a_{32} & a_{33} \end{vmatrix}, \qquad C_{32} = -M_{32} = -\begin{vmatrix} a_{11} & a_{13} \\ a_{21} & a_{23} \end{vmatrix}, \qquad \text{etc.}$$

For later use we note that (3) may also be written in terms of minors

(4a) $D = \sum_{k=1}^{n} (-1)^{j+k}a_{jk}M_{jk}$ $(j = 1, 2, \cdots, \text{or } n)$

(4b) $D = \sum_{j=1}^{n} (-1)^{j+k}a_{jk}M_{jk}$ $(k = 1, 2, \cdots, \text{or } n)$.

EXAMPLE 1. Determinant of second order
For

$$D = \det A = \begin{vmatrix} a_{11} & a_{12} \\ a_{21} & a_{22} \end{vmatrix}$$

formula (3) gives four possibilities of developing it, namely, by

$$\text{the first row:} \quad D = a_{11}a_{22} + a_{12}(-a_{21}),$$

$$\text{the second row:} \quad D = a_{21}(-a_{12}) + a_{22}a_{11},$$

$$\text{the first column:} \quad D = a_{11}a_{22} + a_{21}(-a_{12}),$$

$$\text{the second column:} \quad D = a_{12}(-a_{21}) + a_{22}a_{11}.$$

They all give the same value $D = a_{11}a_{22} - a_{12}a_{21}$, which also agrees with that in Sec. 7.9.

EXAMPLE 2. A third-order determinant
Let

$$D = \begin{vmatrix} 1 & 3 & 0 \\ 2 & 6 & 4 \\ -1 & 0 & 2 \end{vmatrix}.$$

The development by the first row is

$$D = 1 \begin{vmatrix} 6 & 4 \\ 0 & 2 \end{vmatrix} - 3 \begin{vmatrix} 2 & 4 \\ -1 & 2 \end{vmatrix} = 1(12 - 0) - 3(4 + 4) = -12.$$

The development by the third column is

$$D = 0 \begin{vmatrix} 2 & 6 \\ -1 & 0 \end{vmatrix} - 4 \begin{vmatrix} 1 & 3 \\ -1 & 0 \end{vmatrix} + 2 \begin{vmatrix} 1 & 3 \\ 2 & 6 \end{vmatrix} = 0 - 12 + 0 = -12,$$

etc.

EXAMPLE 3. Determinant of a triangular matrix
The determinant of any triangular matrix equals the product of all the entries on the main diagonal. To see this, develop by rows if the matrix is lower triangular, and by columns if it is upper triangular. For instance,

$$\begin{vmatrix} -3 & 0 & 0 \\ 6 & 4 & 0 \\ -1 & 2 & 5 \end{vmatrix} = -3 \begin{vmatrix} 4 & 0 \\ 2 & 5 \end{vmatrix} = -3 \cdot 4 \cdot 5 = -60. \qquad \blacksquare$$

General Properties of Determinants

From our definition we may now readily obtain the most important properties of determinants, as follows.

Since the same value is obtained whether we expand a determinant by any row or any column, we have

Theorem 1 (Transposition)
The value of a determinant is not altered if its rows are written as columns, in the same order.   Example: Sec. 7.9, formula (10).

Theorem 2 (Multiplication by a constant)
If all the entries in one row (or one column) of a determinant are multiplied by the same factor k, the value of the new determinant is k times the value of the given determinant.

Proof. Develop the determinant by that row (or column) whose entries are multiplied by k.   ∎

Caution! $\det k\mathbf{A} = k^n \det \mathbf{A}$ (not $k \det \mathbf{A}$). Explain why.

EXAMPLE 4

$$\begin{vmatrix} 1 & 3 & 0 \\ 2 & 6 & 4 \\ -1 & 0 & 2 \end{vmatrix} = 2 \begin{vmatrix} 1 & 3 & 0 \\ 1 & 3 & 2 \\ -1 & 0 & 2 \end{vmatrix} = 6 \begin{vmatrix} 1 & 1 & 0 \\ 1 & 1 & 2 \\ -1 & 0 & 2 \end{vmatrix} = 12 \begin{vmatrix} 1 & 1 & 0 \\ 1 & 1 & 1 \\ -1 & 0 & 1 \end{vmatrix} = -12$$ ∎

From Theorem 2, with $k = 0$, or directly by expanding, we obtain

Theorem 3
If all the entries in a row (or a column) of a determinant are zero, the value of the determinant is zero.

Theorem 4
If each entry in a row (or a column) of a determinant is expressed as a binomial, the determinant can be written as the sum of two determinants, for example,

$$\begin{vmatrix} a_1 + d_1 & b_1 & c_1 \\ a_2 + d_2 & b_2 & c_2 \\ a_3 + d_3 & b_3 & c_3 \end{vmatrix} = \begin{vmatrix} a_1 & b_1 & c_1 \\ a_2 & b_2 & c_2 \\ a_3 & b_3 & c_3 \end{vmatrix} + \begin{vmatrix} d_1 & b_1 & c_1 \\ d_2 & b_2 & c_2 \\ d_3 & b_3 & c_3 \end{vmatrix}.$$

Proof. Expand the determinant by the row (or column) whose entries are binomials.   ∎

Another example is given in (12), Sec. 7.9. The generalization of Theorem 4 to the case of entries that are sums of more than two terms is obvious.

Theorem 5 (Interchange of rows or columns)
If any two rows (or two columns) of a determinant are interchanged, the value of the determinant is multiplied by -1.

Proof. The proof is by induction. We see that the theorem holds for determinants of order $n = 2$. Assuming that it holds for determinants of order $n - 1$, we will show that it holds for determinants of order n.

Let D be of order n and E obtained from D by interchanging two rows. Expand D and E by a row that is not one of those interchanged, call it the jth row. Then, by (4a),

(5) $D = \sum_{k=1}^{n} (-1)^{j+k} a_{jk} M_{jk},$ $E = \sum_{k=1}^{n} (-1)^{j+k} a_{jk} N_{jk}$

where N_{jk} is obtained from the minor M_{jk} of a_{jk} in D by interchanging two rows. Since these minors are of order $n - 1$, the induction hypothesis applies and gives $N_{jk} = -M_{jk}$. Hence $E = -D$ by (5). This proves the statement for *rows*. To get it for *columns*, apply Theorem 1.   ∎

An example is given in formula (11) of the preceding section.

Theorem 6 (Proportional rows or columns)
If corresponding entries in two rows (or two columns) of a determinant are proportional, the value of the determinant is zero.

Proof. Let the entries in the ith and jth rows of D be proportional, say, $a_{ik} = ca_{jk}, k = 1, \cdots, n$. If $c = 0$, then $D = 0$. Let $c \neq 0$. By Theorem 2,

$$D = cB$$

where the ith and jth rows of B are identical. Interchange these rows. Then B goes over into $-B$, by Theorem 5. On the other hand, since the rows are identical, the new determinant is still B. Thus $B = -B$, $B = 0$, and $D = 0$.   ∎

EXAMPLE 5

$$\begin{vmatrix} 3 & 6 & -4 \\ 1 & -1 & 3 \\ -6 & -12 & 8 \end{vmatrix} = 0$$ ∎

Before evaluating a determinant, it is advisable to simplify it. This may be done by Theorem 2 and

Theorem 7 (Addition of a row or column)
The value of a determinant is left unchanged if the entries in a row (or column) are altered by adding to them any constant multiple of the corresponding entries in any other row (or column, respectively).

Proof. Apply Theorem 4 to the determinant that results from the given addition. This yields a sum of two determinants; one is the original determinant and the other contains two proportional rows. According to Theorem 6, the second determinant is zero, and the proof is complete.   ∎

The trick in simplifying determinants is to generate many zeros in a row (or column) and then to develop the given determinant by that row or column in order to express it by a small number of determinants of order one less (or even a single such determinant), which are then handled in the same fashion, and so on. If we do this on a pocket calculator, we can find out by inspection what rows or columns are best in a specific case. By doing some of the problems at the end of the section, the reader may gain skill in this art (which old-fashioned texts emphasize to this day). We proceed to a general method that can be programmed. The method is based on Theorem 7 and modeled after the Gauss elimination. We explain it in terms of a typical example.

EXAMPLE 6. Evaluation of a determinant by reduction to "triangular form"

$$D = \begin{vmatrix} 2 & 0 & -4 & 6 \\ 4 & 5 & 1 & 0 \\ 0 & 2 & 6 & -1 \\ -3 & 8 & 9 & 1 \end{vmatrix}$$

$$= \begin{vmatrix} 2 & 0 & -4 & 6 \\ 0 & 5 & 9 & -12 \\ 0 & 2 & 6 & -1 \\ 0 & 8 & 3 & 10 \end{vmatrix} \begin{matrix} \\ \text{Row 2} - 2 \text{ Row 1} \\ \\ \text{Row 4} + 1.5 \text{ Row 1} \end{matrix}$$

$$= \begin{vmatrix} 2 & 0 & -4 & 6 \\ 0 & 5 & 9 & -12 \\ 0 & 0 & 2.4 & 3.8 \\ 0 & 0 & -11.4 & 29.2 \end{vmatrix} \begin{matrix} \\ \\ \text{Row 3} - 0.4 \text{ Row 2} \\ \text{Row 4} - 1.6 \text{ Row 2} \end{matrix}$$

$$= \begin{vmatrix} 2 & 0 & -4 & 6 \\ 0 & 5 & 9 & -12 \\ 0 & 0 & 2.4 & 3.8 \\ 0 & 0 & 0 & 47.25 \end{vmatrix} \begin{matrix} \\ \\ \\ \text{Row 4} + 4.75 \text{ Row 3} \end{matrix}$$

$$= 2 \times 5 \times 2.4 \times 47.25 = 1134.$$

In work with pencil and paper, one writes down lower order determinants when they appear, instead of carrying along zeros,

$$D = \begin{vmatrix} 2 & 0 & -4 & 6 \\ 4 & 5 & 1 & 0 \\ 0 & 2 & 6 & -1 \\ -3 & 8 & 9 & 1 \end{vmatrix} = \cdots = 2 \begin{vmatrix} 5 & 9 & -12 \\ 2 & 6 & -1 \\ 8 & 3 & 10 \end{vmatrix} = \cdots = 10 \begin{vmatrix} 2.4 & 3.8 \\ -11.4 & 29.2 \end{vmatrix} = 1134. \quad ∎$$

For determinants of products of matrices, there is a very useful formula that has various applications. A proof of this formula will be given in the next section.

Theorem 8 (Determinant of a product of matrices)

For any $n \times n$ matrices $\mathbf{A}$ and $\mathbf{B}$,

$$(6) \qquad\qquad \det (\mathbf{AB}) = \det (\mathbf{BA}) = \det \mathbf{A} \det \mathbf{B}.$$

EXAMPLE 7

$$\begin{vmatrix} 2 & 4 & 3 \\ 6 & 10 & 14 \\ 4 & 7 & 9 \end{vmatrix} \begin{vmatrix} 4 & 0 & 5 \\ -2 & 1 & -1 \\ 3 & 0 & 4 \end{vmatrix} = \begin{vmatrix} 9 & 4 & 18 \\ 46 & 10 & 76 \\ 29 & 7 & 49 \end{vmatrix}$$ ∎

Historically, determinants were introduced for solving systems of linear algebraic equations. In the next section we derive **Cramer's rule,** which expresses the unknowns in such a system as quotients of determinants. For *computing* solutions, one uses the Gauss elimination or variants of it (Secs. 7.5, 19.1, 19.2), not Cramer's rule, which would be very impractical even for moderately large systems.

Problems for Sec. 7.10

Evaluate

1. $\begin{vmatrix} 1 & 3 & 1 & -2 \\ 2 & 0 & -2 & -4 \\ -1 & 1 & 0 & 1 \\ 2 & 5 & 3 & 6 \end{vmatrix}$
2. $\begin{vmatrix} 8 & 2 & 0 & 7 \\ 4 & 4 & 1 & 4 \\ 7 & 4 & 3 & 7 \\ 1 & -2 & 0 & 2 \end{vmatrix}$
3. $\begin{vmatrix} -1 & 0 & 0 & 0 \\ 14 & 17 & 0 & 0 \\ 13 & 12 & 2 & 0 \\ 28 & -7 & 5 & \frac{1}{2} \end{vmatrix}$

4. $\begin{vmatrix} -4 & 1 & 4 & -2 \\ -5 & -8 & 4 & -2 \\ 8 & 16 & -6 & 15 \\ -2 & 8 & -3 & 7 \end{vmatrix}$
5. $\begin{vmatrix} 2 & 7 & 2 & 1 \\ -6 & 4 & 5 & 6 \\ -1 & 7 & 2 & 4 \\ -7 & 4 & 5 & 7 \end{vmatrix}$
6. $\begin{vmatrix} 2 & 5 & 3 & -2 \\ 2 & 8 & 3 & 2 \\ 0 & 3 & 8 & 5 \\ -4 & -7 & 2 & 11 \end{vmatrix}$

7. $\begin{vmatrix} 3 & 2 & 0 & 0 \\ 6 & 8 & 0 & 0 \\ 0 & 0 & 4 & 7 \\ 0 & 0 & 2 & 5 \end{vmatrix}$
8. $\begin{vmatrix} 9 & -4 & 0 & 0 \\ 2 & 8 & 0 & 0 \\ 0 & 0 & 7 & 1 \\ 0 & 0 & 6 & -2 \end{vmatrix}$
9. $\begin{vmatrix} 3 & 2 & 0 & 1 \\ -6 & 0 & 5 & 5 \\ 3 & -2 & -3 & -6 \\ -9 & 2 & 14 & 16 \end{vmatrix}$

10. $\begin{vmatrix} 4 & 0 & 2 & 5 \\ 0 & 3 & 7 & 4 \\ -8 & 0 & -2 & -8 \\ 0 & 6 & 18 & 11 \end{vmatrix}$
11. $\begin{vmatrix} 2 & 1 & 0 & 0 \\ 5 & 3 & 2 & 0 \\ 0 & 7 & 5 & 4 \\ 0 & 0 & 6 & 4 \end{vmatrix}$
12. $\begin{vmatrix} 4 & 3 & 0 & 0 \\ -8 & 1 & 2 & 0 \\ 0 & -7 & 3 & -6 \\ 0 & 0 & 5 & -5 \end{vmatrix}$

13. Show that det $(k\mathbf{A}) = k^n$ det $\mathbf{A}$ (*not* k det $\mathbf{A}$), where $\mathbf{A}$ is any $n \times n$ matrix.

14. Write the product of the determinants in Probs. 7 and 11, Sec. 7.9, as a determinant.

15. Verify that the answer to Prob. 7 equals the product of the determinants of the 2×2 submatrices containing no zero entries. Explain why.

7.11   Rank in Terms of Determinants. Cramer's Rule

The rank of a matrix is a very important concept, as we can see from Sec. 7.7. From Sec. 7.6 we know that the rank of a matrix **A** is the maximum number of linearly independent row or column vectors of **A**. It is remarkable, or perhaps even surprising, that we can use determinants for characterizing the rank of a matrix. This characterization is often used for *defining* the rank. We formulate it as follows, assuming rank **A** > 0 (since rank **A** = 0 if and only if **A** = **0**; cf. Sec. 7.6).

Theorem 1 (Rank in terms of determinants)

*An $m \times n$ matrix $\mathbf{A} = [a_{jk}]$ has rank $r \geqq 1$ if and only if **A** has an $r \times r$ submatrix with nonzero determinant whereas the determinant of every square submatrix with $r + 1$ or more rows that **A** has (or does not have!) is zero.*

*In particular, if **A** is a square matrix, **A** is nonsingular, so that the inverse $\mathbf{A}^{-1}$ of **A** exists, if and only if*

$$\det \mathbf{A} \neq 0.$$

Proof. The key lies in the fact that elementary row operations (Sec. 7.5), which do not alter the rank (by Theorem 2 in Sec. 7.6), also do not alter the property of a determinant of being zero or not zero, since the determinant is multiplied by

(*i*) -1 if we interchange two rows (Theorem 5, Sec. 7.10),
(*ii*) $c \neq 0$ if we multiply a row by $c \neq 0$ (Theorem 2, Sec. 7.10),
(*iii*)   1 if we add a multiple of a row to another row (Theorem 7, Sec. 7.10).

Let $\hat{\mathbf{A}}$ denote the echelon form of **A** (cf. Sec. 7.5). $\hat{\mathbf{A}}$ has r nonzero row vectors (which are the first r row vectors) if and only if rank **A** = r. Let $\hat{\mathbf{R}}$ be the $r \times r$ submatrix of $\hat{\mathbf{A}}$ consisting of the r^2 entries that are simultaneously in the first r rows and the first r columns of $\hat{\mathbf{A}}$. Since $\hat{\mathbf{R}}$ is triangular and has all diagonal entries different from zero, $\det \hat{\mathbf{R}} \neq 0$. Since $\hat{\mathbf{R}}$ is obtained from the corresponding $r \times r$ submatrix **R** of **A** by elementary operations, we have $\det \mathbf{R} \neq 0$. Similarly, $\det \mathbf{S} = 0$ for a square submatrix **S** of $r + 1$ or more rows possibly contained in **A**, since the corresponding submatrix $\hat{\mathbf{S}}$ of $\hat{\mathbf{A}}$ must contain a row of zeros, so that $\det \hat{\mathbf{S}} = 0$ by Theorem 3 in Sec. 7.10. This proves the assertion of the theorem for an $m \times n$ matrix.

If **A** is square, say, an $n \times n$ matrix, the statement just proved implies that rank **A** = n if and only if **A** has an $n \times n$ submatrix with a nonzero determinant; but this means that we must have $\det \mathbf{A} \neq 0$.   ∎

Using this theorem, we shall now derive Cramer's rule, which represents solutions of systems of linear equations as quotients of determinants. Cramer's rule is, in general, only of ***theoretical interest.*** For ***numerical work,*** other methods are preferable (cf. Secs. 7.5, 19.1–19.3), because the fastest known way to calculate $\det \mathbf{A}$ for an arbitrary $n \times n$ matrix **A** is to apply the Gauss elimination to **A** (ignoring the right-hand side).

Cramer's Theorem (Solution of linear equations by determinants)

(a) *If the determinant* $D = \det \mathbf{A}$ *of a system of n linear equations*

(1)

$$a_{11}x_1 + a_{12}x_2 + \cdots + a_{1n}x_n = b_1$$

$$a_{21}x_1 + a_{22}x_2 + \cdots + a_{2n}x_n = b_2$$

$$\cdots\cdots\cdots\cdots\cdots\cdots\cdots\cdots\cdots$$

$$a_{n1}x_1 + a_{n2}x_2 + \cdots + a_{nn}x_n = b_n$$

in the same number of unknowns $x_1, \cdots, x_n$ *is not zero, the system has precisely one solution. This solution is given by the formulas*

(2) $$x_1 = \frac{D_1}{D}, \quad x_2 = \frac{D_2}{D}, \cdots, \quad x_n = \frac{D_n}{D}$$ **(Cramer's rule)**

where D_k *is the determinant obtained from D by replacing in D the kth column by the column with the entries* $b_1, \cdots, b_n$.

(b) *Hence if* (1) *is* **homogeneous** *and* $D \neq 0$, *it has only the trivial solution* $x_1 = 0, x_2 = 0, \cdots, x_n = 0$. *If* $D = 0$, *the homogeneous system also has nontrivial solutions.*

Proof. From Theorem 1 and the Fundamental Theorem in Sec. 7.7 it follows that (1) has a unique solution, since

$$D = \det \mathbf{A} = \begin{vmatrix} a_{11} & \cdots & a_{1n} \\ \cdot & \cdots & \cdot \\ \cdot & \cdots & \cdot \\ a_{n1} & \cdots & a_{nn} \end{vmatrix} \neq 0$$

implies rank $\mathbf{A} = n$. We prove (2). Developing D by the kth column, we obtain

(3) $$D = a_{1k}C_{1k} + a_{2k}C_{2k} + \cdots + a_{nk}C_{nk},$$

where C_{ik} is the cofactor of the entry a_{ik} in D. If we replace the entries in the kth column of D by any other numbers, we obtain a new determinant, say, $\tilde{D}$. Clearly, its development by the kth column will be of the form (3), with $a_{1k}, \cdots, a_{nk}$ replaced by those new numbers and the cofactors C_{ik} as before. In particular, if we choose as new numbers the entries $a_{1l}, \cdots, a_{nl}$ in the lth column of D (where $l \neq k$), then the development of the resulting determinant $\tilde{D}$ becomes

(4) $$a_{1t}C_{1k} + a_{2l}C_{2k} + \cdots + a_{nl}C_{nk} = 0$$ $(l \neq k)$,

because $\tilde{D}$ has two identical columns and is zero (cf. Theorem 6 in Sec. 7.10). If we multiply the first equation in (1) by C_{1k}, the second by $C_{2k}, \cdots$, the last by C_{nk} and add the resulting equations, we first obtain

$$C_{1k}(a_{11}x_1 + \cdots + a_{1n}x_n) + \cdots + C_{nk}(a_{n1}x_1 + \cdots + a_{nn}x_n)$$
$$= b_1 C_{1k} + \cdots + b_n C_{nk}.$$

The expression on the left may be written

$$x_1(a_{11}C_{1k} + \cdots + a_{n1}C_{nk}) + \cdots + x_n(a_{1n}C_{1k} + \cdots + a_{nn}C_{nk}).$$

From (3) we conclude that the factor of x_k in this representation is equal to D, and from (4) it follows that the factor of x_l ($l \neq k$) is zero. Thus

$$x_k D = b_1 C_{1k} + b_2 C_{2k} + \cdots + b_n C_{nk}.$$

Since $D \neq 0$, we may divide, finding

(5) $$x_k = \frac{1}{D}(b_1 C_{1k} + b_2 C_{2k} + \cdots + b_n C_{nk}) = \frac{D_k}{D},$$

where $k = 1, \cdots, n$. This is identical with (2).

Furthermore, if $D \neq 0$ as before and (1) is homogeneous, then we have $D_1 = 0, \cdots, D_n = 0$, so that (2) yields the trivial solution. If $D = 0$ and (1) is homogeneous, then rank $\mathbf{A} < n$ by Theorem 1, so that nontrivial solutions exist by (c) in the Fundamental Theorem (Sec. 7.7). ∎

An example is included in Sec. 7.9.

As an important consequence of Cramer's theorem, we may now express the entries in the inverse of a matrix as follows.

Theorem 3 (Inverse of a matrix)

The inverse of a nonsingular $n \times n$ matrix $\mathbf{A} = [a_{jk}]$ is given by

(6) $$\mathbf{A}^{-1} = \frac{1}{\det \mathbf{A}} \begin{bmatrix} A_{11} & A_{21} & \cdots & A_{n1} \\ A_{12} & A_{22} & \cdots & A_{n2} \\ . & . & \cdots & . \\ A_{1n} & A_{2n} & \cdots & A_{nn} \end{bmatrix},$$

where A_{jk} is the cofactor of a_{jk} in $\det \mathbf{A}$. (Note well that in $\mathbf{A}^{-1}$, the cofactor A_{jk} occupies the same place as a_{kj} (not a_{jk}) does in $\mathbf{A}$.)

Proof. We denote the right-hand side of (6) by $\mathbf{B}$ and show that $\mathbf{BA} = \mathbf{I}$. We write

(7) $$\mathbf{BA} = \mathbf{G} = [g_{kl}].$$

Here, by the definition of matrix multiplication,

(8) $$g_{kl} = \sum_{s=1}^{n} \frac{A_{sk}}{\det \mathbf{A}} a_{sl} = \frac{1}{\det \mathbf{A}} \sum_{s=1}^{n} A_{sk} a_{sl}.$$

For $l = k$ the last sum is the development of $D = \det \mathbf{A}$ by the kth column. Hence

$$g_{kk} = \frac{1}{\det \mathbf{A}} \sum_{s=1}^{n} A_{sk} a_{sk} = \frac{1}{\det \mathbf{A}} \det \mathbf{A} = 1.$$

For $l \neq k$ this is a similar development of the determinant $\tilde{D}$ obtained from D by replacing the kth column of D with the lth column of D, so that $\tilde{D}$ has two identical columns and is zero:

$$g_{kl} = \frac{1}{\det \mathbf{A}} \sum_{s=1}^{n} A_{sk} a_{sl} = \frac{1}{\det \mathbf{A}} \tilde{D} = 0 \qquad (k \neq l).$$

Hence $\mathbf{BA} = \mathbf{G} = \mathbf{I}$ in (7). Similarly, $\mathbf{AB} = \mathbf{I}$. Hence $\mathbf{B} = \mathbf{A}^{-1}$. ∎

Using Theorem 1, we may now also prove the theorem on determinants of matrix products (Theorem 8 in Sec. 7.10), which we first restate.

Theorem 4 (Determinant of a product of matrices)
For any $n \times n$ matrices $\mathbf{A}$ and $\mathbf{B}$,

$$(9) \qquad \det(\mathbf{AB}) = \det(\mathbf{BA}) = \det \mathbf{A} \det \mathbf{B}.$$

Proof. If $\mathbf{A}$ is singular, so is $\mathbf{AB}$ by Theorem 2c in Sec. 7.8. Hence we have $\det \mathbf{A} = 0$, $\det(\mathbf{AB}) = 0$ by Theorem 1, and (9) is $0 = 0$, which holds.

Let $\mathbf{A}$ be nonsingular. Then we can reduce $\mathbf{A}$ to a diagonal matrix $\hat{\mathbf{A}} = [\hat{a}_{jk}]$ by Gauss–Jordan steps (Sec. 7.8). Under these operations, $\det \mathbf{A}$ retains its value, by Theorem 7 in Sec. 7.10, except perhaps for a sign reversal if we have to interchange two rows to get a nonzero pivot (cf. Theorem 5 in Sec. 7.10). But the same operations reduce $\mathbf{AB}$ to $\hat{\mathbf{A}}\mathbf{B}$ with the same effect on $\det(\mathbf{AB})$. Hence it remains to prove (9) for $\hat{\mathbf{A}}\mathbf{B}$, written out

$$\hat{\mathbf{A}}\mathbf{B} = \begin{bmatrix} \hat{a}_{11} & 0 & \cdots & 0 \\ 0 & \hat{a}_{22} & \cdots & 0 \\ & & \ddots & \\ 0 & 0 & \cdots & \hat{a}_{nn} \end{bmatrix} \begin{bmatrix} b_{11} & b_{12} & \cdots & b_{1n} \\ b_{21} & b_{22} & \cdots & b_{2n} \\ & & \vdots & \\ b_{n1} & b_{n2} & \cdots & b_{nn} \end{bmatrix}$$

$$= \begin{bmatrix} \hat{a}_{11}b_{11} & \hat{a}_{11}b_{12} & \cdots & \hat{a}_{11}b_{1n} \\ \hat{a}_{22}b_{21} & \hat{a}_{22}b_{22} & \cdots & \hat{a}_{22}b_{2n} \\ & & \vdots & \\ \hat{a}_{nn}b_{n1} & \hat{a}_{nn}b_{n2} & \cdots & \hat{a}_{nn}b_{nn} \end{bmatrix}.$$

We now take the determinant $\det(\hat{\mathbf{A}}\mathbf{B})$. On the right we can take out a factor $\hat{a}_{11}$ from the first row, $\hat{a}_{22}$ from the second, $\cdots$, $\hat{a}_{nn}$ from the nth. But this product $\widehat{a}_{11}\widehat{a}_{22} \cdots \widehat{a}_{nn}$ equals det $\widehat{\mathbf{A}}$, since $\widehat{\mathbf{A}}$ is diagonal. The remaining determinant is det $\mathbf{B}$, and (9) is proved.   ∎

This finishes our discussion of systems of linear algebraic equations, which we started in Sec. 7.5. The next and last portion of this chapter (Secs. 7.12 to 7.16) concerns **eigenvalue problems,** whose importance in physics and engineering can hardly be overestimated. In the next section (Sec. 7.12) we introduce the basic concepts. In Secs. 7.13 and 7.14 we consider important *complex* matrices (Hermitian, skew-Hermitian and unitary matrices) and their eigenvalues. In Sec. 7.15 we discuss general properties of eigenvectors and the diagonalization of square matrices. Finally, in Sec. 7.16 we show that eigenvalues and eigenvectors are basic in solving systems of differential equations, a task we already considered in Chap. 3 in a more elementary fashion, without the use of matrices.

Problems for Sec. 7.11

Using Theorem 1, find the rank of the following matrices.

1. $\begin{bmatrix} 3 & -5 \\ 7 & 0 \end{bmatrix}$

2. $\begin{bmatrix} 7 & 3 & -1 \\ 1 & 0 & 2 \end{bmatrix}$

3. $\begin{bmatrix} 2 & -4 & 0 & 5 \\ 0 & 1 & 0 & 6 \end{bmatrix}$

4. $\begin{bmatrix} 0.4 & -1.2 \\ 0 & 0 \\ -4.8 & 14.4 \end{bmatrix}$

5. $\begin{bmatrix} 3 & 6 & 9 \\ 0 & 0 & 5 \\ 0 & 0 & 7 \end{bmatrix}$

6. $\begin{bmatrix} 0 & 7 & 8 & 4 \\ 0 & 6 & 5 & 3 \\ 0 & 0 & 1 & 2 \end{bmatrix}$

7. $\begin{bmatrix} 0 & -8 \\ -6 & 1 \\ 2 & 2 \end{bmatrix}$

8. $\begin{bmatrix} 1 & 0 & 0 \\ 0 & 0 & 1 \\ 0 & 1 & 0 \end{bmatrix}$

9. $\begin{bmatrix} 5 & 0 & 2 & 0 \\ 5 & 8 & 2 & 6 \\ 5 & 0 & 2 & 6 \end{bmatrix}$

Using Theorem 3, find the inverse. Check the answer.

10. $\begin{bmatrix} -1 & 5 \\ 2 & 3 \end{bmatrix}$

11. $\begin{bmatrix} -2 & 1 \\ 1.5 & -0.5 \end{bmatrix}$

12. $\begin{bmatrix} \cos\theta & \sin\theta \\ -\sin\theta & \cos\theta \end{bmatrix}$

13. $\begin{bmatrix} 2 & 0 & -1 \\ 5 & 1 & 0 \\ 0 & 1 & 3 \end{bmatrix}$

14. $\begin{bmatrix} 1 & 0 & 0 \\ 2 & 1 & 0 \\ 5 & 4 & 1 \end{bmatrix}$

15. $\begin{bmatrix} 5 & 2 & 4 \\ 0 & 5 & 1 \\ 0 & 0 & 5 \end{bmatrix}$

16. $\begin{bmatrix} a & 0 & 0 \\ 0 & b & 0 \\ 0 & 0 & c \end{bmatrix}$

17. $\begin{bmatrix} 0 & 0 & a \\ 0 & b & 0 \\ c & 0 & 0 \end{bmatrix}$

18. $\begin{bmatrix} 0 & 1 & 0 \\ 1 & 0 & 0 \\ 0 & 0 & 1 \end{bmatrix}$

19. $\begin{bmatrix} -2 & -2 & 7 \\ 4 & 3 & -12 \\ -1 & 0 & 2 \end{bmatrix}$ **20.** $\begin{bmatrix} 19 & 2 & -9 \\ -4 & -1 & 2 \\ -2 & 0 & 1 \end{bmatrix}$ **21.** $\begin{bmatrix} 0 & -2 & 1 \\ \frac{1}{2} & \frac{1}{2} & -\frac{1}{2} \\ -1 & 2 & 0 \end{bmatrix}$

Solve by Cramer's rule and by the Gauss elimination (Sec. 7.5).

22. $-x - 2y = -5$
  $5x + 4y = 1$

23. $x - 3y = -10$
  $10x - 5y = 0$

24. $7x + 3y = 13$
  $3x + 7y = 17$

25. $2x + y + 2z = -1$
  $x \qquad + z = -1$
  $-x + 3y - 2z = 7$

26. $x - y + 2z = 2$
  $3x + y - z = 3$
  $2x + 2y - 3z = 1$

27. $x + y + z = 0$
  $2x + 5y + 3z = 1$
  $-x + 2y + z = 2$

28. Obtain (4) and (5) in Sec. 7.8 from the present Theorem 3.

29. Using $\mathbf{A}^{-1}$ as given by (6), show that $\mathbf{AA}^{-1} = \mathbf{I}$.

30. Verify Theorem 4 for the matrices in Probs. 1 and 10.

31. Verify Theorem 4 for the matrices in Probs. 19 and 21.

32. (a) Show that the product of two $n \times n$ matrices is singular if and only if at least one of the two matrices is singular. (b) Show that the sum of two nonsingular $n \times n$ matrices may be singular, and the sum of two singular matrices may be nonsingular.

33. (**Similar matrices**) An $n \times n$ matrix $\mathbf{B}$ is called *similar* to an $n \times n$ matrix $\mathbf{A}$ if there is a (nonsingular!) matrix $\mathbf{T}$ such that $\mathbf{B} = \mathbf{T}^{-1}\mathbf{AT}$. Show that the determinants of similar matrices have the same value. (Similarity is important, as we shall see in Sec. 7.15.)

Geometrical applications. Using Cramer's theorem, part (b), show:

34. The **plane through three points** (x_1, y_1, z_1), (x_2, y_2, z_2), (x_3, y_3, z_3) in space is given by (a) and (b). [Explain why. Also, derive (b) from (a).]

(a) $\begin{vmatrix} x & y & z & 1 \\ x_1 & y_1 & z_1 & 1 \\ x_2 & y_2 & z_2 & 1 \\ x_3 & y_3 & z_3 & 1 \end{vmatrix} = 0,$ (b) $\begin{vmatrix} x - x_1 & y - y_1 & z - z_1 \\ x_2 - x_1 & y_2 - y_1 & z_2 - z_1 \\ x_3 - x_1 & y_3 - y_1 & z_3 - z_1 \end{vmatrix} = 0.$

35. The **circle through three points** (x_1, y_1), (x_2, y_2), (x_3, y_3) in the plane is given by

$$\begin{vmatrix} x^2 + y^2 & x & y & 1 \\ x_1^2 + y_1^2 & x_1 & y_1 & 1 \\ x_2^2 + y_2^2 & x_2 & y_2 & 1 \\ x_3^2 + y_3^2 & x_3 & y_3 & 1 \end{vmatrix} = 0.$$

36. Find the plane through $(1, 0, 0)$, $(0, 1, 0)$, $(0, 0, 1)$.

37. Find the plane through $(4, 1, 3)$, $(0, -1, 2)$, $(5, 7, -3)$.

38. Find the circle through $(-3, 2)$, $(2, 7)$, $(6, -1)$.

7.12 Eigenvalues, Eigenvectors

From the standpoint of engineering applications, eigenvalue problems are among the most important problems in connection with matrices, and the number of research papers on corresponding numerical methods for computers is enormous. The basic concepts are as follows.

Let $A = [a_{jk}]$ be a given $n \times n$ matrix and consider the vector equation

$$(1) \qquad\qquad \boxed{Ax = \lambda x}$$

where λ is a number.

It is clear that the zero vector $x = 0$ is a solution of (1) for any value of λ. A value of λ for which (1) has a solution $x \neq 0$ is called an **eigenvalue**[6] or **characteristic value** (or *latent root*) of the matrix A. The corresponding solutions $x \neq 0$ of (1) are called **eigenvectors** or **characteristic vectors** of A corresponding to that eigenvalue λ. The set of the eigenvalues is called the **spectrum** of A. The largest of the absolute values of the eigenvalues of A is called the **spectral radius** of A.

The set of all eigenvectors corresponding to an eigenvalue of A, together with 0, forms a vector space, called the **eigenspace** of A corresponding to this eigenvalue.

The problem of determining the eigenvalues and eigenvectors of a matrix is called an *eigenvalue problem*.[7] Problems of this type occur in connection with physical and technical applications. Therefore, the student should know the fundamental ideas and concepts that are important in this field of mathematics. During the past two decades, various new methods for the approximate determination of eigenvalues have been developed and other methods that have been known for still a longer time have been put into a form suitable for computers. Cf. Refs. [B2], [B9] in Appendix 1.

Applications

EXAMPLE 1. Vibrating system of two masses on two springs (Fig. 63 in Sec. 3.1)

Mass–spring systems involving several masses and springs can be treated as eigenvalue problems. For instance, in Example 4 of Sec. 3.1 we obtained as the mathematical model of the mechanical system in Fig. 63 (Sec. 3.1) the system of differential equations

$$(2) \qquad \begin{aligned} y_1'' &= -5y_1 + 2y_2 \\ y_2'' &= 2y_1 - 2y_2. \end{aligned}$$

We show how this leads to an eigenvalue problem of the form (1). First of all, we can write these two equations as a single vector equation

$$y'' = \begin{bmatrix} y_1'' \\ y_2'' \end{bmatrix} = Ay = \begin{bmatrix} -5 & 2 \\ 2 & -2 \end{bmatrix} \begin{bmatrix} y_1 \\ y_2 \end{bmatrix}.$$

Now comes the idea. By analogy with a single constant-coefficient equation we substitute

[6]German: *Eigenwert;* "*eigen*" means "proper"; "*Wert*" means "value."

[7]More precisely: an *algebraic* eigenvalue problem, because there are other eigenvalue problems involving a differential equation (see Secs. 4.8 and 11.3) or an integral equation.

(3) $$\mathbf{y} = \mathbf{x}e^{\omega t}.$$

This yields

$$\omega^2 \mathbf{x} e^{\omega t} = \mathbf{A} \mathbf{x} e^{\omega t}$$

and we recognize that this is an eigenvalue problem

(4) $$\mathbf{A}\mathbf{x} = \lambda\mathbf{x}$$ where $\lambda = \omega^2$.

Hence, for (3) to be a solution of (2), not identically zero, $\omega^2 = \lambda$ must be an eigenvalue of $\mathbf{A}$ and the vector $\mathbf{x}$ in (3) must be a corresponding eigenvector.

How to get from (4) the actual solutions of (2) is shown later, in a separate section on differential equations (Sec. 7.16), after we have gained some experience with eigenvalue problems in general.   ∎

We shall now consider another engineering application that does not involve differential equations but leads directly to an eigenvalue problem, and we shall explain the typical steps of solution, namely, the determination first of the eigenvalues and then of corresponding eigenvectors.

EXAMPLE 2. Stretching of an elastic membrane

An elastic membrane in the $x_1 x_2$-plane with boundary circle $x_1^2 + x_2^2 = 1$ (Fig. 169) is stretched so that a point P: (x_1, x_2) goes over into the point Q: (y_1, y_2) given by

(5) $$\mathbf{y} = \begin{bmatrix} y_1 \\ y_2 \end{bmatrix} = \mathbf{A}\mathbf{x} = \begin{bmatrix} 5 & 3 \\ 3 & 5 \end{bmatrix} \begin{bmatrix} x_1 \\ x_2 \end{bmatrix},$$ in components, $$\begin{aligned} y_1 &= 5x_1 + 3x_2 \\ y_2 &= 3x_1 + 5x_2 \end{aligned}.$$

Find the "*principal directions*" that is, directions of the position vector $\mathbf{x}$ of P for which the direction of the position vector $\mathbf{y}$ of Q is the same or exactly opposite. What shape does the boundary circle take under this deformation?

Solution. We are looking for vectors $\mathbf{x}$ such that $\mathbf{y} = \lambda\mathbf{x}$. Since $\mathbf{y} = \mathbf{A}\mathbf{x}$, this gives $\mathbf{A}\mathbf{x} = \lambda\mathbf{x}$, an equation of the form (1), an eigenvalue problem. In components, $\mathbf{A}\mathbf{x} = \lambda\mathbf{x}$ is

(6) $$\begin{aligned} 5x_1 + 3x_2 &= \lambda x_1 \\ 3x_1 + 5x_2 &= \lambda x_2 \end{aligned}$$ or $$\begin{aligned} (5 - \lambda)x_1 + 3x_2 &= 0 \\ 3x_1 + (5 - \lambda)x_2 &= 0 \end{aligned}$$

This is a homogeneous system of linear equations. It has a nontrivial solution if and only if its coefficient determinant is zero (cf. Cramer's theorem in Sec. 7.11). Hence we must have

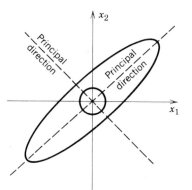

Fig. 169. Undeformed and deformed membrane in Example 2

(7) $\qquad \begin{vmatrix} 5 - \lambda & 3 \\ 3 & 5 - \lambda \end{vmatrix} = (5 - \lambda)^2 - 9 = 0.$ Solutions $\lambda_1 = 8$, $\lambda_2 = 2$.

These are the **eigenvalues** of our problem. We also need eigenvectors. We get them by inserting λ_1 into (6) and then λ_2. For $\lambda = \lambda_1 = 8$, our system (6) becomes

$$-3x_1 + 3x_2 = 0,$$
$$3x_1 - 3x_2 = 0.$$

Solution $x_2 = x_1$, x_1 arbitrary.

For $\lambda_2 = 2$, our system (6) is

$$3x_1 + 3x_2 = 0,$$
$$3x_1 + 3x_2 = 0.$$

Solution $x_2 = -x_1$, x_1 arbitrary.

We thus obtain as **eigenvectors** of **A** the vectors

$$\begin{bmatrix} x_1 \\ x_1 \end{bmatrix}$$ corresponding to λ_1; $$\begin{bmatrix} x_1 \\ -x_1 \end{bmatrix}$$ corresponding to λ_2.

x_1 is arbitrary and we can choose, for instance, $x_1 = 1$. Or we can choose $x_1 = 1/\sqrt{2}$ to get unit vectors. These vectors make 45° and 135° angles with the positive x_1-direction. They give the principal directions, the answer to our problem. Whereas the multiplication of a vector **x** by **A** will generally produce a $\mathbf{y} = \mathbf{Ax}$ of any other direction, for these eigenvectors the multiplication by **A** has the same effect as the multiplication by the scalars λ_1 and λ_2, respectively. These eigenvalues show that in the principal directions the membrane is stretched by factors 8 and 2, respectively; cf. Fig. 169.

The additional question (shape of the deformed membrane) should help us to get a clearer picture of what is going on. We answer it now. The inverse of **A** is [cf. (4), Sec. 7.8]

$$\mathbf{A}^{-1} = \frac{1}{\det \mathbf{A}} \begin{bmatrix} 5 & -3 \\ -3 & 5 \end{bmatrix} = \begin{bmatrix} 5/16 & -3/16 \\ -3/16 & 5/16 \end{bmatrix}.$$

Hence the inverse transformation of (5) is

$$x_1 = \tfrac{5}{16}y_1 - \tfrac{3}{16}y_2,$$
$$x_2 = -\tfrac{3}{16}y_1 + \tfrac{5}{16}y_2.$$

Now the boundary circle is $x_1^2 + x_2^2 = 1$. From the inverse transformation we thus obtain

$$x_1^2 + x_2^2 = \frac{1}{16^2} [(5y_1 - 3y_2)^2 + (-3y_1 + 5y_2)^2] = 1.$$

Writing out the squares and multiplying by 128, we get

$$17y_1^2 - 30y_1y_2 + 17y_2^2 = 128.$$

This is an ellipse with principal semiaxes 8 and 2 (the eigenvalues of **A**) in the directions of the eigenvectors (cf. Fig. 169). Hence we can transform it to principal axes by a 45° rotation

$$y_1 = z_1/\sqrt{2} - z_2/\sqrt{2},$$
$$y_2 = z_1/\sqrt{2} + z_2/\sqrt{2}.$$

After simplification this gives $2z_1^2 + 32z_2^2 = 128$, that is,

$$\frac{z_1^2}{8^2} + \frac{z_2^2}{2^2} = 1.$$ ∎

Determination of Eigenvalues

We shall now discuss (1) with general n, proceeding as in Example 2, where $n = 2$. We first note that if x is any vector, then $\{x, Ax\}$ will in general be a linearly independent set. However, if x is an eigenvector of A, then $\{x, Ax\}$ is a linearly dependent set; corresponding components of x and Ax are then proportional, the factor of proportionality being the eigenvalue λ of A to which the eigenvector corresponds.

Next let us show that *any $n \times n$ matrix has at least 1 and at most n distinct (real or complex) eigenvalues.*

For this purpose we write (1) out:

$$a_{11}x_1 + \cdots + a_{1n}x_n = \lambda x_1$$

$$a_{21}x_1 + \cdots + a_{2n}x_n = \lambda x_2$$

$$\cdots\cdots\cdots\cdots\cdots\cdots\cdots\cdots\cdots$$

$$a_{n1}x_1 + \cdots + a_{nn}x_n = \lambda x_n.$$

Transferring the terms on the right-hand side to the left-hand side we have

$$(8) \qquad \begin{aligned} (a_{11} - \lambda)x_1 + \quad a_{12}x_2 \quad + \cdots + \quad a_{1n}x_n \quad &= 0 \\ a_{21}x_1 \quad + \quad (a_{22} - \lambda)x_2 \quad + \cdots + \quad a_{2n}x_n \quad &= 0 \\ \cdots\cdots\cdots\cdots\cdots\cdots\cdots\cdots\cdots\cdots\cdots\cdots \\ a_{n1}x_1 \quad + \quad a_{n2}x_2 \quad + \cdots + \quad (a_{nn} - \lambda)x_n \quad &= 0. \end{aligned}$$

In matrix notation,

$$\boxed{(A - \lambda I)x = 0.}$$

By Cramer's theorem in Sec. 7.11, this homogeneous system of linear equations has a nontrivial solution if and only if the corresponding determinant of the coefficients is zero:

$$(9) \quad D(\lambda) = \det(A - \lambda I) = \begin{vmatrix} a_{11} - \lambda & a_{12} & \cdots & a_{1n} \\ a_{21} & a_{22} - \lambda & \cdots & a_{2n} \\ \cdot & \cdot & \cdots & \cdot \\ a_{n1} & a_{n2} & \cdots & a_{nn} - \lambda \end{vmatrix} = 0.$$

$D(\lambda)$ is called the **characteristic determinant,** and (9) is called the **characteristic equation** corresponding to the matrix A. By developing $D(\lambda)$ we obtain a polynomial of nth degree in λ. This is called the **characteristic polynomial** corresponding to A. [For instance, in Example 2, the characteristic determinant and equation are shown in (7).]

We have thus obtained the following important result.

Theorem 1 (Eigenvalues)

The eigenvalues of a square matrix $\mathbf{A}$ are the roots of the corresponding characteristic equation (9).

Determination of Eigenvectors

The eigen*values* must be determined first. Once these are known, corresponding eigen*vectors* can be determined from the system (8), where λ is the eigenvalue for which an eigenvector is wanted. This is what we did in Example 2 and shall do again in the examples below.

If $\mathbf{x}$ is an eigenvector of $\mathbf{A}$, then $k\mathbf{x}$ with any nonzero constant k is also an eigenvector of $\mathbf{A}$ corresponding to the same eigenvalue. This follows since the system (8) is homogeneous. (See Theorem 2 in Sec. 7.7 if necessary.)

EXAMPLE 3. Multiple eigenvalues
Find the eigenvalues and eigenvectors of the matrix

$$\mathbf{A} = \begin{bmatrix} -2 & 2 & -3 \\ 2 & 1 & -6 \\ -1 & -2 & 0 \end{bmatrix}.$$

Solution. For determining the eigenvalues as the roots of the characteristic equation, one may have to use Newton's method (Sec. 18.2) or another numerical method explained in Secs. 19.7 or 19.8. Sometimes it may also help to observe that the product and sum of the eigenvalues are the constant term and $(-1)^{n-1}$ times the coefficient of the second highest term, respectively, of the characteristic polynomial (why?). Once an eigenvalue λ_1 has been found, one may divide by $\lambda - \lambda_1$. For our matrix, the characteristic determinant gives the characteristic polynomial

$$-\lambda^3 - \lambda^2 + 21\lambda + 45 = 0.$$

The roots (eigenvalues of $\mathbf{A}$) are $\lambda_1 = 5$, $\lambda_2 = \lambda_3 = -3$. To find eigenvectors, we apply the Gauss elimination (Sec. 7.5) to the system $(\mathbf{A} - \lambda\mathbf{I})\mathbf{x} = \mathbf{0}$, first with $\lambda = 5$ and then with $\lambda = -3$. We find that the vector

$$\mathbf{x}_1 = \begin{bmatrix} 1 \\ 2 \\ -1 \end{bmatrix}$$

is an eigenvector of $\mathbf{A}$ corresponding to the eigenvalue 5, and the vectors

$$\mathbf{x}_2 = \begin{bmatrix} -2 \\ 1 \\ 0 \end{bmatrix} \quad \text{and} \quad \mathbf{x}_3 = \begin{bmatrix} 3 \\ 0 \\ 1 \end{bmatrix}$$

are two linearly independent eigenvectors of $\mathbf{A}$ corresponding to the eigenvalue -3. This agrees with the fact that, for $\lambda = -3$, the matrix $\mathbf{A} - \lambda\mathbf{I}$ has rank 1 and so, by Theorem 2 in Sec. 7.7, a basis of solutions of the corresponding system (8), viz.,

$$x_1 + 2x_2 - 3x_3 = 0$$
$$2x_1 + 4x_2 - 6x_3 = 0$$
$$-x_1 - 2x_2 + 3x_3 = 0$$

consists of two linearly independent vectors. ∎

If an eigenvalue λ of a matrix $\mathbf{A}$ is a root of order M_λ of the characteristic polynomial of $\mathbf{A}$, then M_λ is called the **algebraic multiplicity** of λ, as opposed to the **geometric multiplicity** m_λ of λ, which is defined to be the number of linearly independent eigenvectors corresponding to λ, thus, the dimension of the corresponding eigenspace. Since the characteristic polynomial has degree n, the sum of all algebraic multiplicities equals n. In Example 3, for $\lambda = -3$ we have $m_\lambda = M_\lambda = 2$. In general, $m_\lambda \leq M_\lambda$, as can be shown. We convince ourselves that $m_\lambda < M_\lambda$ is possible:

EXAMPLE 4. Algebraic and geometric multiplicity

The characteristic equation of the matrix

$$\mathbf{A} = \begin{bmatrix} 0 & 1 \\ 0 & 0 \end{bmatrix} \quad \text{is} \quad \det (\mathbf{A} - \lambda \mathbf{I}) = \begin{vmatrix} -\lambda & 1 \\ 0 & -\lambda \end{vmatrix} = \lambda^2 = 0.$$

Hence $\lambda = 0$ is an eigenvalue of algebraic multiplicity 2. But its geometric multiplicity is only 1, since eigenvectors result from $-0x_1 + x_2 = 0$, hence $x_2 = 0$, in the form $[x_1 \quad 0]^\mathsf{T}$.

EXAMPLE 5. Real matrices with complex eigenvalues

Since real polynomials may have complex roots (which then occur in conjugate pairs), a real matrix may have complex eigenvalues and eigenvectors. For instance, the characteristic equation of the skew-symmetric matrix

$$\mathbf{A} = \begin{bmatrix} 0 & 1 \\ -1 & 0 \end{bmatrix} \quad \text{is} \quad \det (\mathbf{A} - \lambda \mathbf{I}) = \begin{vmatrix} -\lambda & 1 \\ -1 & -\lambda \end{vmatrix} = \lambda^2 + 1 = 0$$

and gives the eigenvalues $\lambda_1 = i \ (= \sqrt{-1})$, $\lambda_2 = -i$. Eigenvectors are obtained from $-ix_1 + x_2 = 0$ and $ix_1 + x_2 = 0$, respectively, and we can choose $x_1 = 1$ to get

$$\begin{bmatrix} 1 \\ i \end{bmatrix} \quad \text{and} \quad \begin{bmatrix} 1 \\ -i \end{bmatrix}.$$

The reader may show that, more generally, these vectors are eigenvectors of the matrix

$$\mathbf{A} = \begin{bmatrix} a & b \\ -b & a \end{bmatrix} \qquad (a, b \text{ real})$$

and that $\mathbf{A}$ has the eigenvalues $a + ib$ and $a - ib$.

Further Applications

EXAMPLE 6. Eigenvalue problems arising from Markov processes

As another application, let us show that Markov processes also lead to eigenvalue problems. To see this, let us determine the limit state of the land-use succession in Example 4, Sec. 7.4.

Solution. We recall that Example 4 in Sec. 7.4 concerns a *Markov process* and that such a transition process is governed by a **stochastic matrix** $\mathbf{A} = [a_{jk}]$, that is, a square matrix with nonnegative entries a_{jk} (giving transition probabilities) and all row sums equal to 1. Furthermore, state $\mathbf{y}$ (a column vector) is obtained from state $\mathbf{x}$ according to $\mathbf{y}^\mathsf{T} = \mathbf{x}^\mathsf{T}\mathbf{A}$, equivalently, $\mathbf{y} = \mathbf{A}^\mathsf{T}\mathbf{x}$. A limit is reached if states remain unchanged, thus if $\mathbf{x}^\mathsf{T} = \mathbf{x}^\mathsf{T}\mathbf{A}$ or

(10) $$\mathbf{A}^\mathsf{T}\mathbf{x} = \mathbf{x}.$$

This means that $\mathbf{A}^\mathsf{T}$ should have the eigenvalue 1. But $\mathbf{A}^\mathsf{T}$ has the same eigenvalues as $\mathbf{A}$ (by Theorem 1 in Sec. 7.10); and $\mathbf{A}$ has the eigenvalue 1, with eigenvector $\mathbf{v}^\mathsf{T} = [1 \cdots 1]$, because the row sums of $\mathbf{A}$ equal 1. In our example,

$$\mathbf{Av} = \begin{bmatrix} 0.8 & 0.1 & 0.1 \\ 0.1 & 0.7 & 0.2 \\ 0 & 0.1 & 0.9 \end{bmatrix} \begin{bmatrix} 1 \\ 1 \\ 1 \end{bmatrix} = \begin{bmatrix} 1 \\ 1 \\ 1 \end{bmatrix}$$

as claimed. Hence (10) has a nontrivial solution $\mathbf{x} \neq \mathbf{0}$, which is an eigenvector of $\mathbf{A}^\mathsf{T}$ corresponding to $\lambda = 1$. Now (10) is $(\mathbf{A}^\mathsf{T} - \mathbf{I})\mathbf{x} = \mathbf{0}$, written out,

$$-0.2x_1 + 0.1x_2 \qquad\qquad = 0$$
$$0.1x_1 - 0.3x_2 + 0.1x_3 = 0$$
$$0.1x_1 + 0.2x_2 - 0.1x_3 = 0.$$

A solution is $x^\mathsf{T} = [12.5 \quad 25 \quad 62.5]$. *Answer.* Assuming that the probabilities remain the same as time progresses, we see that the states tend to 12.5% residentially, 25% commercially and 62.5% industrially used area.

EXAMPLE 7. Eigenvalue problems arising from population models. Leslie model

The Leslie model describes age-specified population growth, as follows. Let the oldest age attained by the females in some animal population be 6 years. Divide the population into three age classes of 2 years each. Let the *"Leslie matrix"* be

$$\mathbf{L} = [l_{jk}] = \begin{bmatrix} 0 & 2.3 & 0.4 \\ 0.6 & 0 & 0 \\ 0 & 0.3 & 0 \end{bmatrix}$$

where l_{1k} is the average number of daughters born to a single female during the time she is in age class k, and $l_{j,j-1}$ $(j = 2, 3)$ is the fraction of females in age class $j - 1$ that will survive and pass into class j. (a) What is the number of females in each class after 2, 4, 6 years if each class initially consists of 500 females? (b) For what initial distribution will the number of females in each class change by the same proportion? What is this rate of change?

Solution. (a) Initially, $\mathbf{x}_{(0)}^\mathsf{T} = [500 \quad 500 \quad 500]$. After 2 years,

$$\mathbf{x}_{(2)} = \mathbf{Lx}_{(0)} = \begin{bmatrix} 0 & 2.3 & 0.4 \\ 0.6 & 0 & 0 \\ 0 & 0.3 & 0 \end{bmatrix} \begin{bmatrix} 500 \\ 500 \\ 500 \end{bmatrix} = \begin{bmatrix} 1350 \\ 300 \\ 150 \end{bmatrix}.$$

Similarly, after 4 years we have $\mathbf{x}_{(4)}^\mathsf{T} = (\mathbf{Lx}_{(2)})^\mathsf{T} = [750 \quad 810 \quad 90]$ and after 6 years we have $\mathbf{x}_{(6)}^\mathsf{T} = (\mathbf{Lx}_{(4)})^\mathsf{T} = [1899 \quad 450 \quad 243]$.

(b) Proportional change means that we are looking for a distribution vector $\mathbf{x}$ such that $\mathbf{Lx} = \lambda\mathbf{x}$, where λ is the rate of change (growth if $\lambda > 1$, decrease if $\lambda < 1$). The characteristic equation is

$$\det(\mathbf{L} - \lambda\mathbf{I}) = -\lambda^3 - 0.6(-2.3\lambda - 0.3 \cdot 0.4) = -\lambda^3 + 1.38\lambda + 0.072 = 0.$$

A positive root is found to be (for instance, by Newton's method, Sec. 18.2) $\lambda = 1.2$. A corresponding eigenvector can then be determined from $0.6x_1 - 1.2x_2 = 0, 0.3x_2 - 1.2x_3 = 0$. Thus, $x^\mathsf{T} = [1 \quad 0.5 \quad 0.125]$. To get an initial population of 1500, as before, we multiply x^T by 923. *Answer.* 923 females in class 1, 462 in class 2, 115 in class 3. The growth rate is 1.2. ∎

The present section was devoted to the introduction and explanation of the basic concepts and facts in connection with matrix eigenvalue problems. In the next section we consider, for the first time in this chapter, **complex matrices,** called *Hermitian, skew-Hermitian* and *unitary matrices.* These are the complex analogs of real symmetric, skew-symmetric (cf. Sec. 7.4) and orthogonal matrices. They are important in physics, particularly in quantum mechanics.

Problems for Sec. 7.12

Find the eigenvalues and eigenvectors of the following matrices.

1. $\begin{bmatrix} 8 & -4 \\ 2 & 2 \end{bmatrix}$ 2. $\begin{bmatrix} 3 & 4 \\ 4 & -3 \end{bmatrix}$ 3. $\begin{bmatrix} 1 & 0 \\ 0 & 2 \end{bmatrix}$ 4. $\begin{bmatrix} 1 & 0 \\ 0 & 1 \end{bmatrix}$

5. $\begin{bmatrix} 0 & 0 \\ 0 & -1 \end{bmatrix}$ 6. $\begin{bmatrix} 0 & 0 \\ 0 & 0 \end{bmatrix}$ 7. $\begin{bmatrix} 1.4 & 0.5 \\ -1.0 & -0.1 \end{bmatrix}$

8. $\begin{bmatrix} -1 & 0 \\ -2 & 1 \end{bmatrix}$ 9. $\begin{bmatrix} 1.4 & -1.0 \\ 0.5 & -0.1 \end{bmatrix}$ 10. $\begin{bmatrix} 1 & 2 \\ 0 & 3 \end{bmatrix}$

11. $\begin{bmatrix} 5 & 10 \\ 4 & -1 \end{bmatrix}$ 12. $\begin{bmatrix} 6 & 2 \\ -9 & 0 \end{bmatrix}$ 13. $\begin{bmatrix} 2 & 0 & 0 \\ 0 & 4 & 0 \\ 0 & 0 & 3 \end{bmatrix}$

14. $\begin{bmatrix} a & 0 & 0 \\ 0 & b & 0 \\ 0 & 0 & c \end{bmatrix}$ 15. $\begin{bmatrix} 3 & 1 & 4 \\ 0 & 2 & 6 \\ 0 & 0 & 5 \end{bmatrix}$ 16. $\begin{bmatrix} 2 & -2 & 3 \\ -2 & -1 & 6 \\ 1 & 2 & 0 \end{bmatrix}$

Find the eigenvalues of the following matrices.

17. $\begin{bmatrix} 13 & 0 & -15 \\ -3 & 4 & 9 \\ 5 & 0 & -7 \end{bmatrix}$ 18. $\begin{bmatrix} 26 & 2 & 4 \\ -2 & 21 & 2 \\ 2 & 4 & 28 \end{bmatrix}$ 19. $\begin{bmatrix} 1.5 & -1.0 & -2.0 \\ -0.4 & 1.2 & 0.4 \\ -0.3 & -0.6 & -0.2 \end{bmatrix}$

Some general properties of the spectrum. Let $\lambda_1, \cdots, \lambda_n$ be the eigenvalues of a given matrix $\mathbf{A} = [a_{jk}]$. In each case prove the proposition and illustrate it with an example.

20. If $\mathbf{A}$ is real, the eigenvalues are real or complex conjugates in pairs.
21. If $\mathbf{A}$ is real and n is odd, at least one λ_j is real.
22. **(Trace)** The so-called *trace* of $\mathbf{A}$, given by trace $\mathbf{A} = a_{11} + a_{22} + \cdots + a_{nn}$, is equal to $\lambda_1 + \cdots + \lambda_n$.
23. The constant term of $D(\lambda)$ equals det $\mathbf{A}$.
24. **(Inverse)** The inverse $\mathbf{A}^{-1}$ exists if and only if $\lambda_j \neq 0$ $(j = 1, \cdots, n)$.
25. The inverse $\mathbf{A}^{-1}$ has the eigenvalues $1/\lambda_1, \cdots, 1/\lambda_n$.
26. **(Triangular matrix)** If $\mathbf{A}$ is triangular, the entries on the main diagonal are the eigenvalues of $\mathbf{A}$.
27. **("Spectral shift")** The matrix $\mathbf{A} - k\mathbf{I}$ has the eigenvalues $\lambda_1 - k, \cdots, \lambda_n - k$.
28. The matrix $k\mathbf{A}$ has the eigenvalues $k\lambda_1, \cdots, k\lambda_n$.
29. The matrix $\mathbf{A}^m$ (m a nonnegative integer) has the eigenvalues $\lambda_1^m, \cdots, \lambda_n^m$.

30. (Spectral mapping theorem) The matrix

$$k_m\mathbf{A}^m + k_{m-1}\mathbf{A}^{m-1} + \cdots + k_1\mathbf{A} + k_0\mathbf{I}$$

which is called a *polynomial matrix,* has the eigenvalues

$$k_m\lambda_j^m + k_{m-1}\lambda_j^{m-1} + \cdots + k_1\lambda_j + k_0 \quad (j = 1, \cdots, n).$$

(This proposition is called the *spectral mapping theorem for polynomial matrices.*) The eigenvectors of that matrix are the same as those of $\mathbf{A}$.

Find limit states of the Markov processes governed by the following stochastic matrices.

31. $\begin{bmatrix} 0.4 & 0.6 \\ 0.5 & 0.5 \end{bmatrix}$
     32. $\begin{bmatrix} 0.50 & 0.25 & 0.25 \\ 0.25 & 0.50 & 0.25 \\ 0.25 & 0.25 & 0.50 \end{bmatrix}$
     33. $\begin{bmatrix} 0.6 & 0.4 & 0 \\ 0.1 & 0.1 & 0.8 \\ 0 & 1.0 & 0 \end{bmatrix}$

Find the principal directions and corresponding factors of extension or contraction of the elastic deformation $\mathbf{y} = \mathbf{A}\mathbf{x}$, where $\mathbf{A}$ equals

34. $\begin{bmatrix} 3/2 & 1/\sqrt{2} \\ 1/\sqrt{2} & 1 \end{bmatrix}$
     35. $\begin{bmatrix} 3 & \sqrt{2} \\ \sqrt{2} & 2 \end{bmatrix}$
     36. $\begin{bmatrix} 1 & 0.2 \\ 0.2 & 1 \end{bmatrix}$

Find the growth rate in the Leslie model with Leslie matrix

37. $\begin{bmatrix} 0 & 8 & 0 \\ 0.5 & 0 & 0 \\ 0 & 0.2 & 0 \end{bmatrix}$
     38. $\begin{bmatrix} 0 & 5.2 & 2.125 \\ 0.4 & 0 & 0 \\ 0 & 0.3 & 0 \end{bmatrix}$
     39. $\begin{bmatrix} 0 & 4.5 & 2.5 \\ 0.2 & 0 & 0 \\ 0 & 0.2 & 0 \end{bmatrix}$

40. Show that a Leslie matrix $\mathbf{L}$ with positive $l_{12}, l_{13}, l_{21}, l_{32}$ has a positive eigenvalue. *Hint.* Use Probs. 20—23, (This is a special case of the famous *Perron–Frobenius theorem* in Sec. 19.7, which is difficult to prove in its general form.)

41. (Leontief[8] input–output model) Suppose that three industries are interrelated so that their outputs are used as inputs by themselves, according to the 3×3 **consumption matrix**

$$\mathbf{A} = [a_{jk}] = \begin{bmatrix} 0.2 & 0.5 & 0 \\ 0.6 & 0 & 0.4 \\ 0.2 & 0.5 & 0.6 \end{bmatrix}$$

where a_{jk} is the fraction of the output of industry k consumed (purchased) by industry j. Let p_j be the price charged by industry j for its total output. A problem is to find prices so that for each industry, total expenditures equal total income. Show that this leads to $\mathbf{A}\mathbf{p} = \mathbf{p}$, where $\mathbf{p} = [p_1 \quad p_2 \quad p_3]^T$, and find a solution $\mathbf{p}$ with nonnegative p_1, p_2, p_3.

42. Show that a consumption matrix as considered in Prob. 41 must have column sums 1 and always has the eigenvalue 1.

---

[8]WASSILY LEONTIEF (born 1906), American economist. For his work he was awarded the Nobel Prize in 1973.

43. **(Open Leontief input–output model)** If not the whole output is consumed by the industries themselves (as in Prob. 41), then instead of $\mathbf{Ax} = \mathbf{x}$ we have $\mathbf{x} - \mathbf{Ax} = \mathbf{y}$, where $\mathbf{x} = [x_1 \quad x_2 \quad x_3]^\mathsf{T}$ is produced, $\mathbf{Ax}$ is consumed by the industries and, thus, $\mathbf{y}$ is the net production available for other consumers. Find for what production $\mathbf{x}$ a given $\mathbf{y} = [0.1 \quad 0.3 \quad 0.1]^\mathsf{T}$ can be achieved if the consumption matrix is

$$\mathbf{A} = \begin{bmatrix} 0.1 & 0.4 & 0.2 \\ 0.5 & 0 & 0.1 \\ 0.1 & 0.4 & 0.4 \end{bmatrix}.$$

7.13 Hermitian, Skew-Hermitian and Unitary Matrices

We shall now introduce three classes of *complex* square matrices having important applications, for instance, in quantum physics, and consider their eigenvalues.

In this connection we use the standard notation

$$\overline{\mathbf{A}} = [\overline{a}_{jk}]$$

for the matrix obtained from $\mathbf{A} = [a_{jk}]$ by replacing each entry by its complex conjugate, and we also use the notation

$$\overline{\mathbf{A}}^\mathsf{T} = [\overline{a}_{kj}]$$

for the conjugate transpose. For example, if

$$\mathbf{A} = \begin{bmatrix} 3 + 4i & -5i \\ -7 & 6 - 2i \end{bmatrix} \quad \text{then} \quad \overline{\mathbf{A}}^\mathsf{T} = \begin{bmatrix} 3 - 4i & -7 \\ 5i & 6 + 2i \end{bmatrix}$$

Definitions of Hermitian,[9] skew-Hermitian and unitary matrices
A square matrix $\mathbf{A} = [a_{jk}]$ is called

Hermitian if	$\overline{\mathbf{A}}^\mathsf{T} = \mathbf{A}$,	that is,	$\overline{a}_{kj} = a_{jk}$
skew-Hermitian if	$\overline{\mathbf{A}}^\mathsf{T} = -\mathbf{A}$,	that is,	$\overline{a}_{kj} = -a_{jk}$
unitary if	$\overline{\mathbf{A}}^\mathsf{T} = \mathbf{A}^{-1}$.		

From these definitions we see the following. If $\mathbf{A}$ is Hermitian, the entries on the main diagonal must satisfy $\overline{a}_{jj} = a_{jj}$, that is, they are real. Similarly, if $\mathbf{A}$ is skew-Hermitian, then $\overline{a}_{jj} = -a_{jj}$ or, if we set $a_{jj} = \alpha + i\beta$, then $\alpha - i\beta = -(\alpha + i\beta)$, so that $\alpha = 0$ and a_{jj} is pure imaginary or 0.

[9]Cf. footnote 19 in Sec. 4.9.

EXAMPLE 1. Hermitian, skew-Hermitian and unitary matrices

The matrices

$$\begin{bmatrix} 4 & 1 - 3i \\ 1 + 3i & 7 \end{bmatrix}, \quad \begin{bmatrix} 3i & 2 + i \\ -2 + i & -i \end{bmatrix}, \quad \begin{bmatrix} i/2 & \sqrt{3}/2 \\ \sqrt{3}/2 & i/2 \end{bmatrix}$$

are Hermitian, skew-Hermitian and unitary, respectively, as the reader may verify.   ∎

If a Hermitian matrix is real, then $\overline{\mathbf{A}}^\mathsf{T} = \mathbf{A}^\mathsf{T} = \mathbf{A}$. Hence a real Hermitian matrix is a symmetric matrix (cf. Sec. 7.4). Similarly, a real skew-Hermitian matrix is a skew-symmetric matrix. Hence *Hermitian and skew-Hermitian matrices generalize symmetric and skew-symmetric matrices,* respectively.

Unitary and Orthogonal Matrices

By the above definition, a unitary matrix satisfies

$$(1) \qquad\qquad \overline{\mathbf{A}}^\mathsf{T} = \mathbf{A}^{-1}.$$

A *real* unitary matrix is called an **orthogonal matrix;** then (1) becomes

$$(2) \qquad\qquad \mathbf{A}^\mathsf{T} = \mathbf{A}^{-1},$$

that is, *an orthogonal matrix is a real matrix for which the inverse equals its transpose.* Orthogonal matrices are important, as we shall see below.

A set of vectors $\mathbf{x}_1, \cdots, \mathbf{x}_n$ for which

$$(3) \qquad\qquad \overline{\mathbf{x}}_j{}^\mathsf{T}\mathbf{x}_k = \delta_{jk} = \begin{cases} 0 & \text{if } j \neq k \\ 1 & \text{if } j = k \end{cases} \qquad (j, k = 1, \cdots, n)$$

is called a **unitary system.**

Clearly, if these vectors are real, then $\overline{\mathbf{x}}_j{}^\mathsf{T}\mathbf{x}_k = \mathbf{x}_j{}^\mathsf{T}\mathbf{x}_k$ is the inner product (dot product) of $\mathbf{x}_j$ and $\mathbf{x}_k$ in the elementary sense, and condition (3) means that $\mathbf{x}_1, \cdots, \mathbf{x}_n$ are orthogonal (mutually perpendicular) and are normalized (are unit vectors). For this reason, the real system is then called an **orthonormal system.**

Hence a unitary system of *real* vectors is an orthonormal system, a set of mutually perpendicular unit vectors.

Theorem 1 (Unitary matrix)

A square matrix $\mathbf{A}$ is unitary if and only if its column vectors (and also its row vectors) form a unitary system.

Proof. (a) Let $\mathbf{A}$ be a unitary matrix with column vectors $\mathbf{a}_1, \cdots, \mathbf{a}_n$. Then by (1),

$$(4) \quad \mathbf{A}^{-1}\mathbf{A} = \overline{\mathbf{A}}^\mathsf{T}\mathbf{A} = \begin{bmatrix} \overline{\mathbf{a}}_1{}^\mathsf{T} \\ \vdots \\ \overline{\mathbf{a}}_n{}^\mathsf{T} \end{bmatrix} \begin{bmatrix} \mathbf{a}_1 \cdots \mathbf{a}_n \end{bmatrix} = \begin{bmatrix} \overline{\mathbf{a}}_1{}^\mathsf{T}\mathbf{a}_1 & \overline{\mathbf{a}}_1{}^\mathsf{T}\mathbf{a}_2 & \cdots & \overline{\mathbf{a}}_1{}^\mathsf{T}\mathbf{a}_n \\ \overline{\mathbf{a}}_2{}^\mathsf{T}\mathbf{a}_1 & \overline{\mathbf{a}}_2{}^\mathsf{T}\mathbf{a}_2 & \cdots & \overline{\mathbf{a}}_2{}^\mathsf{T}\mathbf{a}_n \\ \vdots & \vdots & \cdots & \vdots \\ \overline{\mathbf{a}}_n{}^\mathsf{T}\mathbf{a}_1 & \overline{\mathbf{a}}_n{}^\mathsf{T}\mathbf{a}_2 & \cdots & \overline{\mathbf{a}}_n{}^\mathsf{T}\mathbf{a}_n \end{bmatrix} = \mathbf{I}$$

so that

(5)
$$\overline{\mathbf{a}}_j^{\mathsf{T}} \mathbf{a}_k = \begin{cases} 0 & \text{if } j \neq k, \\ 1 & \text{if } j = k. \end{cases}$$

This shows that the column vectors of $\mathbf{A}$ form a unitary system. From (1) it follows that the inverse $\mathbf{A}^{-1}$ of a unitary matrix $\mathbf{A}$ is unitary (cf. Prob. 29) and, furthermore, the column vectors of $\mathbf{A}^{-1}$ are the complex conjugates of the row vectors of $\mathbf{A}$. Hence the row vectors of $\mathbf{A}$ also form a unitary system.

(b) Conversely, if the column vectors of $\mathbf{A}$ satisfy (5), then the off-diagonal entries in the big matrix are zero and the diagonal entries are 1. Hence $\overline{\mathbf{A}}^{\mathsf{T}} \mathbf{A} = \mathbf{I}$, as (4) shows. Similarly, $\mathbf{A}\overline{\mathbf{A}}^{\mathsf{T}} = \mathbf{I}$. This implies $\overline{\mathbf{A}}^{\mathsf{T}} = \mathbf{A}^{-1}$, since we also have $\mathbf{A}^{-1}\mathbf{A} = \mathbf{A}\mathbf{A}^{-1} = \mathbf{I}$ and the inverse is unique. Hence $\mathbf{A}$ is unitary. Similarly when the row vectors form a unitary system, by what has been said at the end of part (a).   ∎

Theorem 2 (Determinant of a unitary matrix)
The determinant of a unitary matrix $\mathbf{A}$ has absolute value 1.

Proof. $1 = \det \mathbf{I} = \det(\mathbf{A}\mathbf{A}^{-1}) = \det(\mathbf{A}\overline{\mathbf{A}}^{\mathsf{T}})$. By Theorem 4 in Sec. 7.11 this equals $\det \mathbf{A} \det \overline{\mathbf{A}}^{\mathsf{T}}$. Now $\det \overline{\mathbf{A}}^{\mathsf{T}} = \det \overline{\mathbf{A}}$ by Theorem 1, Sec. 7.10. Also $\det \overline{\mathbf{A}} = \overline{(\det \mathbf{A})}$ since $\overline{a} + \overline{b} = \overline{(a + b)}$ and $\overline{a}\overline{b} = \overline{(ab)}$ and similarly for products and sums of more than two complex numbers. Our result is

$$1 = (\det \mathbf{A})\overline{(\det \mathbf{A})} = |\det \mathbf{A}|^2, \qquad \text{hence} \qquad |\det \mathbf{A}| = 1. \qquad ∎$$

Since a *real* unitary matrix is an orthogonal matrix [cf. (2)], Theorems 1 and 2 immediately imply the important

Theorem 3 (Orthogonal matrix)
A real square matrix is an orthogonal matrix if and only if its column vectors, call them $\mathbf{a}_1, \cdots, \mathbf{a}_n$, form an orthonormal system, thus

(6)
$$\mathbf{a}_j^{\mathsf{T}} \mathbf{a}_k = \delta_{jk} = \begin{cases} 0 & \text{if } j \neq k \\ 1 & \text{if } j = k \end{cases} \qquad (j, k = 1, \cdots, n)$$

(or its row vectors form such a system).
The determinant of an orthogonal matrix equals $+1$ or -1.

Orthogonal Transformations

A linear transformation $\mathbf{y} = \mathbf{A}\mathbf{x}$ (cf. Sec. 7.3) whose coefficient matrix $\mathbf{A}$ is orthogonal is called an **orthogonal transformation**. Orthogonal matrices are of great practical importance because these transformations have the following property.

Theorem 4 (Orthogonal transformation)

An orthogonal transformation leaves the inner product

$$(\mathbf{x}, \mathbf{y}) = \mathbf{x}^T\mathbf{y} \quad \text{and thus the norm} \quad \|\mathbf{x}\| = \sqrt{\mathbf{x}^T\mathbf{x}}$$

invariant. In particular, in three-dimensional Euclidean space, an orthogonal transformation preserves the length of any vector and the angle between any two vectors; hence it is a rotation (possibly combined with a reflection).

Proof. Let $\mathbf{A}$ be orthogonal and $\mathbf{u} = \mathbf{A}\mathbf{x}$ and $\mathbf{v} = \mathbf{A}\mathbf{y}$. Using (2) in this section and (6) in Sec. 7.4, we obtain

$$(\mathbf{u}, \mathbf{v}) = \mathbf{u}^T\mathbf{v} = (\mathbf{A}\mathbf{x})^T\mathbf{A}\mathbf{y} = \mathbf{x}^T\mathbf{A}^T\mathbf{A}\mathbf{y} = \mathbf{x}^T\mathbf{A}^{-1}\mathbf{A}\mathbf{y} = \mathbf{x}^T\mathbf{I}\mathbf{y} = \mathbf{x}^T\mathbf{y} = (\mathbf{x}, \mathbf{y}). \quad \blacksquare$$

We mention without proof that the coefficient matrix of *every* rotation of a Cartesian coordinate system is an orthogonal matrix. This matrix has determinant $+1$. For a rotation combined with a reflection it has determinant -1.

In the next section we continue our work on Hermitian, skew-Hermitian and unitary matrices by investigating the eigenvalues of these matrices, whose behavior (shown in Fig. 170 in Sec. 7.14) is largely responsible for the importance of these matrices in physics. We also discuss *Hermitian, skew-Hermitian* and *quadratic forms,* which play a role in physics and geometry, for instance, in connection with conic sections and quadratic surfaces.

Problems for Sec. 7.13

Orthogonal matrices
Are the following matrices symmetric? Skew-symmetric? Orthogonal?

1. $\begin{bmatrix} 0 & 1 \\ -1 & 0 \end{bmatrix}$ 2. $\begin{bmatrix} 1/2 & \sqrt{3}/2 \\ \sqrt{3}/2 & -1/2 \end{bmatrix}$ 3. $\begin{bmatrix} \cos\theta & -\sin\theta \\ \sin\theta & \cos\theta \end{bmatrix}$

4. $\begin{bmatrix} 0 & 0 & 1 \\ 0 & 1 & 0 \\ -1 & 0 & 0 \end{bmatrix}$ 5. $\begin{bmatrix} \cos\theta & -\sin\theta & 0 \\ \sin\theta & \cos\theta & 0 \\ 0 & 0 & 1 \end{bmatrix}$ 6. $\begin{bmatrix} \frac{2}{3} & \frac{1}{3} & \frac{2}{3} \\ -\frac{2}{3} & \frac{2}{3} & \frac{1}{3} \\ \frac{1}{3} & \frac{2}{3} & -\frac{2}{3} \end{bmatrix}$

7. Calculate the determinants of the matrices in Probs. 1, 3, 5. Which theorem do the results illustrate?

8. Find all 2×2 matrices that are both orthogonal and skew-symmetric.

9. Do there exist orthogonal skew-symmetric 3×3 matrices? Orthogonal symmetric 3×3 matrices?

10. Interpret the transformation $\mathbf{y} = \mathbf{A}\mathbf{x}$ geometrically, where $\mathbf{A}$ is the matrix in Prob. 3 and the components of $\mathbf{x}$ and $\mathbf{y}$ are Cartesian coordinates.

11. Same task as in Prob. 10, where $\mathbf{A}$ is the matrix in Prob. 5.

12. Find $\mathbf{A}$ such that $\mathbf{y} = \mathbf{Ax}$ is a counterclockwise rotation through $30°$ in the plane.

13. Let $\mathbf{v}^\mathsf{T} = [4 \quad 2]$, $\mathbf{x}^\mathsf{T} = [-2 \quad 1]$, $\mathbf{w} = \mathbf{Av}$, $\mathbf{y} = \mathbf{Ax}$ with $\mathbf{A}$ given in Prob. 3. Find $|\mathbf{v}|$, $|\mathbf{x}|$, $|\mathbf{w}|$, $|\mathbf{y}|$ and the angles between $\mathbf{v}$ and $\mathbf{x}$ and between $\mathbf{w}$ and $\mathbf{y}$. Which theorem do the results illustrate?

14. Show that the inverse of an orthogonal matrix is orthogonal.

15. Show that the product of two orthogonal $n \times n$ matrices is orthogonal.

16. Is the sum of two orthogonal matrices orthogonal?

Hermitian, skew-Hermitian and unitary matrices

17. Show that any square matrix may be written as the sum of a Hermitian matrix and a skew-Hermitian matrix.

18. If $\mathbf{A}$ and $\mathbf{B}$ are $n \times n$ Hermitian matrices and a and b are any *real* numbers, show that $\mathbf{C} = a\mathbf{A} + b\mathbf{B}$ is a Hermitian matrix.

19. Let $\mathbf{A}$ and $\mathbf{B}$ be skew-Hermitian matrices. Under what conditions is $\mathbf{C} = a\mathbf{A} + b\mathbf{B}$ a skew-Hermitian matrix? (a and b are numbers.)

20. Show that the entries on the main diagonal of a skew-Hermitian matrix are pure imaginary or zero.

21. Show that the product of two Hermitian $n \times n$ matrices $\mathbf{A}$ and $\mathbf{B}$ is Hermitian if and only if $\mathbf{AB} = \mathbf{BA}$.

Show that the following matrices are unitary. Find the inverse of each, and verify that $\mathbf{AA}^{-1} = \mathbf{A}^{-1}\mathbf{A} = \mathbf{I}$.

22. $\begin{bmatrix} 1 & 0 \\ 0 & i \end{bmatrix}$
   23. $\begin{bmatrix} 0 & i \\ i & 0 \end{bmatrix}$
   24. $\begin{bmatrix} 1/\sqrt{2} & i/\sqrt{2} \\ -i/\sqrt{2} & -1/\sqrt{2} \end{bmatrix}$

25. $\begin{bmatrix} 0.8i & -0.6i \\ 0.6i & 0.8i \end{bmatrix}$
   26. $\begin{bmatrix} 1/\sqrt{3} & i\sqrt{2/3} \\ -i\sqrt{2/3} & -1/\sqrt{3} \end{bmatrix}$
   27. $\begin{bmatrix} 1/\sqrt{2} & \frac{1}{2} - \frac{1}{2}i \\ \frac{1}{2} + \frac{1}{2}i & -1/\sqrt{2} \end{bmatrix}$

28. Verify that the determinants of the matrices in Prob. 23, 25, 27 have absolute value 1.

29. Show that the inverse of a unitary matrix is unitary.

30. Show that the product of two unitary $n \times n$ matrices is unitary. Verify this for the matrices in Probs. 24 and 26.

31. **(Pauli spin matrices)** Find the eigenvalues and eigenvectors of the so-called *Pauli spin matrices*

$$\mathbf{S}_x = \begin{bmatrix} 0 & 1 \\ 1 & 0 \end{bmatrix}, \quad \mathbf{S}_y = \begin{bmatrix} 0 & -i \\ i & 0 \end{bmatrix}, \quad \mathbf{S}_z = \begin{bmatrix} 1 & 0 \\ 0 & -1 \end{bmatrix}.$$

32. Show that $\mathbf{S}_x\mathbf{S}_y = i\mathbf{S}_z$, $\mathbf{S}_y\mathbf{S}_x = -i\mathbf{S}_z$, $\mathbf{S}_x^2 = \mathbf{S}_y^2 = \mathbf{S}_z^2 = \mathbf{I}$ (cf. Prob. 31).

7.14 Eigenvalues of Hermitian, Skew-Hermitian and Unitary Matrices

It is quite remarkable and in part accounts for the importance of the matrices introduced in Sec. 7.13 that the spectra of these matrices can be characterized in a general way (cf. Fig. 170).

Theorem 1 (Eigenvalues)

(a) *The eigenvalues of a Hermitian matrix are real.*

(b) *The eigenvalues of a skew-Hermitian matrix are pure imaginary or zero.*

(c) *The eigenvalues of a unitary matrix have the absolute value 1.*

Proof. Let λ be an eigenvalue of $\mathbf{A}$ and $\mathbf{x}$ a corresponding eigenvector. Then

$$(1) \qquad\qquad \mathbf{Ax} = \lambda\mathbf{x}.$$

(a) Let $\mathbf{A}$ be Hermitian. Multiplying (1) by $\bar{\mathbf{x}}^\mathsf{T}$ from the left, we obtain

$$\bar{\mathbf{x}}^\mathsf{T}\mathbf{Ax} = \bar{\mathbf{x}}^\mathsf{T}\lambda\mathbf{x} = \lambda\bar{\mathbf{x}}^\mathsf{T}\mathbf{x}.$$

Now $\bar{\mathbf{x}}^\mathsf{T}\mathbf{x} = \bar{x}_1 x_1 + \cdots + \bar{x}_n x_n = |x_1|^2 + \cdots + |x_n|^2$ is real, and is not 0 since $\mathbf{x} \neq \mathbf{0}$. Hence we may divide to get

$$(2) \qquad\qquad \lambda = \frac{\bar{\mathbf{x}}^\mathsf{T}\mathbf{Ax}}{\bar{\mathbf{x}}^\mathsf{T}\mathbf{x}}.$$

We see that λ is real if the numerator is real. Now this numerator is the

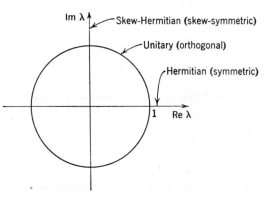

Fig. 170. Location of the eigenvalues of Hermitian, skew-Hermitian and unitary matrices in the complex λ-plane

inner product of the vectors $\bar{x}$ and $\mathbf{Ax}$, hence a number. We prove that this number is real by showing that it is equal to its complex conjugate. In this step we can take transposes, since transposition has no effect on a number. Using (6) in Sec. 7.4 and then $\overline{\mathbf{A}}^{\mathsf{T}} = \mathbf{A}$, we thus obtain

$$(3) \qquad \overline{(\bar{\mathbf{x}}^{\mathsf{T}}\mathbf{Ax})} = \overline{(\bar{\mathbf{x}}^{\mathsf{T}}\mathbf{Ax})}^{\mathsf{T}} = (\mathbf{x}^{\mathsf{T}}\overline{\mathbf{A}}\bar{\mathbf{x}})^{\mathsf{T}} = \bar{\mathbf{x}}^{\mathsf{T}}\overline{\mathbf{A}}^{\mathsf{T}}\mathbf{x} = \bar{\mathbf{x}}^{\mathsf{T}}\mathbf{Ax}.$$

From this and (2) it follows that λ is real.

(b) Let $\mathbf{A}$ be skew-Hermitian. Then $\overline{\mathbf{A}}^{\mathsf{T}} = -\mathbf{A}$, and we get a minus sign in (3),

$$(4) \qquad \overline{(\bar{\mathbf{x}}^{\mathsf{T}}\mathbf{Ax})} = -\bar{\mathbf{x}}^{\mathsf{T}}\mathbf{Ax}.$$

So this is a complex number $c = a + ib$ whose conjugate $\bar{c} = a - ib$ equals $-c = -a - ib$. Hence $a = 0$ and c is pure imaginary or 0. Division by the real $\bar{x}^{\mathsf{T}}x$ in (2) does not change this and proves (b).

(c) Let $\mathbf{A}$ be unitary. We take the conjugate transpose of (1) and use $\overline{\mathbf{A}}^{\mathsf{T}} = \mathbf{A}^{-1}$. Then

$$(\overline{\mathbf{A}}\bar{\mathbf{x}})^{\mathsf{T}} = \bar{\mathbf{x}}^{\mathsf{T}}\overline{\mathbf{A}}^{\mathsf{T}} = \bar{\mathbf{x}}^{\mathsf{T}}\mathbf{A}^{-1} = \bar{\lambda}\bar{\mathbf{x}}^{\mathsf{T}}.$$

We now multiply $\bar{\mathbf{x}}^{\mathsf{T}}\mathbf{A}^{-1} = \bar{\lambda}\bar{\mathbf{x}}^{\mathsf{T}}$ on the left-hand side by $\mathbf{Ax}$ from the right and on the right-hand side by $\lambda\mathbf{x}$ [= $\mathbf{Ax}$; cf. (1)] from the right. This gives

$$\bar{\mathbf{x}}^{\mathsf{T}}\mathbf{A}^{-1}\mathbf{Ax} = \bar{\mathbf{x}}^{\mathsf{T}}\mathbf{x} = \bar{\lambda}\bar{\mathbf{x}}^{\mathsf{T}}\lambda\mathbf{x} = \lambda\bar{\lambda}\bar{\mathbf{x}}^{\mathsf{T}}\mathbf{x}.$$

Since $\bar{\mathbf{x}}^{\mathsf{T}}\mathbf{x} \neq 0$, we may divide on both sides and get $1 = \lambda\bar{\lambda} = |\lambda|^2$, hence $|\lambda| = 1$, as asserted. ∎

EXAMPLE 1. Illustration of Theorem 1

For the matrices in Example 1, Sec. 7.13, we find by direct calculation:

Matrix	Characteristic Equation	Eigenvalues
Hermitian	$\lambda^2 - 11\lambda + 18 = 0$	9, 2
Skew-Hermitian	$\lambda^2 - 2i\lambda + 8 = 0$	$4i$, $-2i$
Unitary	$\lambda^2 - i\lambda - 1 = 0$	$\frac{1}{2}\sqrt{3} + \frac{1}{2}i$, $-\frac{1}{2}\sqrt{3} + \frac{1}{2}i$ ∎

An important special case of Theorem 1 is

Theorem 2 (Symmetric, skew-symmetric and orthogonal matrices)

(a) *The eigenvalues of a symmetric matrix are real.*

(b) *The eigenvalues of a skew-symmetric matrix are pure imaginary or zero.*

(c) *The eigenvalues of an orthogonal matrix have the absolute value* 1 *and are real or complex conjugates in pairs.*

EXAMPLE 2. Illustration of Theorem 2

The matrix in Example 2, Sec. 7.12, is symmetric and has the real eigenvalues 8 and 2. The first matrix in Example 5, Sec. 7.12, is skew-symmetric and has the pure imaginary eigenvalues i and $-i$. The second matrix in that example is *not* skew-symmetric; unless $a = 0$. The matrix

$$\begin{bmatrix} 1/\sqrt{2} & -1/\sqrt{2} \\ 1/\sqrt{2} & 1/\sqrt{2} \end{bmatrix}$$

is orthogonal and has the eigenvalues $\lambda_1 = (1 + i)/\sqrt{2}$ and $\lambda_2 = (1 - i)/\sqrt{2}$, which are conjugate and have absolute value 1. (Verify this!)   ∎

Hermitian and Skew-Hermitian Forms

The numerator $\bar{\mathbf{x}}^\mathsf{T}\mathbf{A}\mathbf{x}$ in (2) is called a **form** in the components $x_1, \cdots, x_n$ of $\mathbf{x}$. For instance, when $n = 2$,

$$\bar{\mathbf{x}}^\mathsf{T}\mathbf{A}\mathbf{x} = \begin{bmatrix} \bar{x}_1 & \bar{x}_2 \end{bmatrix} \begin{bmatrix} a_{11} & a_{12} \\ a_{21} & a_{22} \end{bmatrix} \begin{bmatrix} x_1 \\ x_2 \end{bmatrix} = \begin{bmatrix} \bar{x}_1 & \bar{x}_2 \end{bmatrix} \begin{bmatrix} a_{11}x_1 + a_{12}x_2 \\ a_{21}x_1 + a_{22}x_2 \end{bmatrix}$$

$$= \begin{cases} a_{11}\bar{x}_1 x_1 + a_{12}\bar{x}_1 x_2 \\ + a_{21}\bar{x}_2 x_1 + a_{22}\bar{x}_2 x_2. \end{cases}$$

Similarly for general n,

$$\bar{\mathbf{x}}^\mathsf{T}\mathbf{A}\mathbf{x} = \sum_{j=1}^{n} \sum_{k=1}^{n} a_{jk}\bar{x}_j x_k = \begin{aligned} & a_{11}\bar{x}_1 x_1 + \cdots + a_{1n}\bar{x}_1 x_n \\ & + a_{21}\bar{x}_2 x_1 + \cdots + a_{2n}\bar{x}_2 x_n \\ & + \cdots\cdots\cdots\cdots\cdots\cdots\cdots \\ & + a_{n1}\bar{x}_n x_1 + \cdots + a_{nn}\bar{x}_n x_n. \end{aligned}$$

(5)

This form is called a **Hermitian** or **skew-Hermitian form** if its coefficient matrix $\mathbf{A}$ is Hermitian or skew-Hermitian, respectively.

Theorem 3 (Hermitian and skew-Hermitian forms)
For every choice of the vector $\mathbf{x}$ the value of a Hermitian form is real, and the value of a skew-Hermitian form is pure imaginary or 0.

Proof. In proving (3) and (4), we made no use of the fact that $\mathbf{x}$ was an eigenvector, and the proofs remain valid for any vectors (and Hermitian or skew-Hermitian matrices). From this, our present theorem follows.   ∎

EXAMPLE 3. Hermitian form
If

$$\mathbf{A} = \begin{bmatrix} 3 & 2 - i \\ 2 + i & 4 \end{bmatrix} \quad \text{and} \quad \mathbf{x} = \begin{bmatrix} 1 + i \\ 2i \end{bmatrix},$$

then

$$\bar{\mathbf{x}}^\mathsf{T}\mathbf{A}\mathbf{x} = \begin{bmatrix} 1 - i & -2i \end{bmatrix} \begin{bmatrix} 3 & 2 - i \\ 2 + i & 4 \end{bmatrix} \begin{bmatrix} 1 + i \\ 2i \end{bmatrix}$$

$$= \begin{bmatrix} 1 - i & -2i \end{bmatrix} \begin{bmatrix} 3(1 + i) + (2 - i)2i \\ (2 + i)(1 + i) + 4 \cdot 2i \end{bmatrix} = 34.$$

Quadratic Forms

If the matrix $\mathbf{A}$ in (5) and the vector $\mathbf{x}$ are real, then (5) becomes

$$
\begin{aligned}
\mathbf{x}^T\mathbf{A}\mathbf{x} = \sum_{j=1}^{n}\sum_{k=1}^{n} a_{jk}x_jx_k = \quad & a_{11}x_1^2 + a_{12}x_1x_2 + \cdots + a_{1n}x_1x_n \\[4pt]
& + a_{21}x_2x_1 + a_{22}x_2^2 + \cdots + a_{2n}x_2x_n \\[4pt]
& + \cdots\cdots\cdots\cdots\cdots\cdots\cdots \\[4pt]
& + a_{n1}x_nx_1 + a_{n2}x_nx_2 + \cdots + a_{nn}x_n^2.
\end{aligned}
$$

(6)

and is called a **quadratic form.** Without restriction we may assume the coefficient matrix to be *symmetric,* because we can take off-diagonal terms together in pairs and then write the result as a sum of two equal terms, as the next example (Example 4) illustrates. Quadratic forms occur in physics and geometry, for instance, in connection with conic sections (ellipses $x_1^2/a^2 + x_2^2/b^2 = 1$, etc.) and quadratic surfaces.

EXAMPLE 4. Quadratic form. Symmetric coefficient matrix C
Let

$$
\mathbf{x}^T\mathbf{A}\mathbf{x} = \begin{bmatrix} x_1 & x_2 \end{bmatrix} \begin{bmatrix} 3 & 4 \\ 6 & 2 \end{bmatrix} \begin{bmatrix} x_1 \\ x_2 \end{bmatrix} = 3x_1^2 + 4x_1x_2 + 6x_2x_1 + 2x_2^2 = 3x_1^2 + 10x_1x_2 + 2x_2^2.
$$

Here $4 + 6 = 5 + 5$. From the corresponding *symmetric* matrix $\mathbf{C} = [c_{jk}]$, where $c_{jk} = \frac{1}{2}(a_{jk} + a_{kj})$, thus $c_{11} = 3$, $c_{12} = c_{21} = 5$, $c_{22} = 2$, we get the same result

$$
\mathbf{x}^T\mathbf{C}\mathbf{x} = \begin{bmatrix} x_1 & x_2 \end{bmatrix} \begin{bmatrix} 3 & 5 \\ 5 & 2 \end{bmatrix} \begin{bmatrix} x_1 \\ x_2 \end{bmatrix} = 3x_1^2 + 5x_1x_2 + 5x_2x_1 + 2x_2^2 = 3x_1^2 + 10x_1x_2 + 2x_2^2. \quad \blacksquare
$$

In our discussion of matrix eigenvalue problems so far we have emphasized properties of eigen*values.* In the next section we turn to eigen*vectors* and discuss their general properties, including their use as basis vectors and the **diagonalization** of square matrices.

Problems for Sec. 7.14

     1. Do the computations in Example 1.

     2. Do the computations in Example 2.

Find the eigenvalues of the following matrices. In each case, state which theorems are illustrated (e.g., Theorems 2(b), 2(c) in Prob. 3, etc.).

3. $\begin{bmatrix} 0 & -1 \\ 1 & 0 \end{bmatrix}$ **4.** $\begin{bmatrix} 4 & i \\ -i & 2 \end{bmatrix}$ **5.** $\begin{bmatrix} 0 & a \\ -\bar{a} & 0 \end{bmatrix}$

6. $\begin{bmatrix} 0.6 & 0.8 \\ 0.8 & -0.6 \end{bmatrix}$ **7.** $\begin{bmatrix} \cos\theta & -\sin\theta \\ \sin\theta & \cos\theta \end{bmatrix}$ **8.** $\begin{bmatrix} 2i & 4 \\ -4 & 0 \end{bmatrix}$

9. $\begin{bmatrix} 1/\sqrt{2} & i/\sqrt{2} \\ -i/\sqrt{2} & -1/\sqrt{2} \end{bmatrix}$ **10.** $\begin{bmatrix} 5i & i \\ i & i \end{bmatrix}$ **11.** $\begin{bmatrix} -1 & 5+i \\ 5-i & 3 \end{bmatrix}$

12. $\begin{bmatrix} 0 & 1 & 0 \\ 1 & 0 & 0 \\ 0 & 0 & 1 \end{bmatrix}$ **13.** $\begin{bmatrix} 1/\sqrt{2} & 0 & -1/\sqrt{2} \\ 0 & -1 & 0 \\ 1/\sqrt{2} & 0 & 1/\sqrt{2} \end{bmatrix}$ **14.** $\begin{bmatrix} \frac{2}{3} & \frac{1}{3} & \frac{2}{3} \\ -\frac{2}{3} & \frac{2}{3} & \frac{1}{3} \\ \frac{1}{3} & \frac{2}{3} & -\frac{2}{3} \end{bmatrix}$

Hermitian forms. Find $H = \bar{\mathbf{x}}^T\mathbf{A}\mathbf{x}$, where

15. $\mathbf{A} = \begin{bmatrix} 1 & 0 \\ 0 & 1 \end{bmatrix}$, $\mathbf{x} = \begin{bmatrix} 1+i \\ 1-i \end{bmatrix}$ **16.** $\mathbf{A} = \begin{bmatrix} -1 & 5+i \\ 5-i & 2 \end{bmatrix}$, $\mathbf{x} = \begin{bmatrix} x_1 \\ x_2 \end{bmatrix}$

17. $\mathbf{A} = \begin{bmatrix} 0 & i \\ -i & 0 \end{bmatrix}$, $\mathbf{x} = \begin{bmatrix} 1 \\ i \end{bmatrix}$ **18.** $\mathbf{A} = \begin{bmatrix} a & b \\ \bar{b} & c \end{bmatrix}$, $\mathbf{x} = \begin{bmatrix} 3i \\ 4 \end{bmatrix}$, a, c real

19. $\mathbf{A} = \begin{bmatrix} 1 & -i & 2i \\ i & 1 & 0 \\ -2i & 0 & 1 \end{bmatrix}$, $\mathbf{x} = \begin{bmatrix} 0 \\ 1 \\ i \end{bmatrix}$ **20.** $\mathbf{A} = \begin{bmatrix} 1 & i & 0 \\ -i & 0 & 3 \\ 0 & 3 & 2 \end{bmatrix}$, $\mathbf{x} = \begin{bmatrix} x_1 \\ x_2 \\ x_3 \end{bmatrix}$

Skew-Hermitian forms. Find $S = \bar{\mathbf{x}}^T\mathbf{A}\mathbf{x}$, where

21. $\mathbf{A} = \begin{bmatrix} i & 0 \\ 0 & -i \end{bmatrix}$, $\mathbf{x} = \begin{bmatrix} 1 \\ i \end{bmatrix}$ **22.** $\mathbf{A} = \begin{bmatrix} 5i & 2+i \\ -2+i & i \end{bmatrix}$, $\mathbf{x} = \begin{bmatrix} 2i \\ 3 \end{bmatrix}$

23. $\mathbf{A} = \begin{bmatrix} 2i & 4 \\ -4 & 0 \end{bmatrix}$, $\mathbf{x} = \begin{bmatrix} x_1 \\ x_2 \end{bmatrix}$ **24.** $\mathbf{A} = \begin{bmatrix} i\alpha & \mu + i\nu \\ -\mu + i\nu & i\beta \end{bmatrix}$, $\mathbf{x} = \begin{bmatrix} -i \\ 4 \end{bmatrix}$

Quadratic forms. Find a real symmetric matrix $\mathbf{C}$ such that $Q = \mathbf{x}^T\mathbf{C}\mathbf{x}$, where Q equals

25. $6x_1^2 - 4x_1x_2 + 2x_2^2$ **26.** $(x_1 - x_2)^2$

27. $5x_1^2 - 2x_1x_2 + x_2^2$ **28.** $x_1^2 + 2x_1x_2 + 3x_2^2 + 6x_2x_3 + 2x_3^2$

29. $(x_1 + x_2)^2 - x_3^2$ **30.** $(x_1 - x_2 + 2x_3 - 2x_4)^2$

31. $(x_1 - x_2 + x_3)^2$ **32.** $(x_1 + x_2)^2 + (x_3 + x_4)^2$

33. **(Definiteness)** A real quadratic form $Q = \mathbf{x}^T\mathbf{C}\mathbf{x}$ and its symmetric matrix $\mathbf{C} = (c_{jk})$ are said to be **positive definite** if $Q > 0$ for all $(x_1, \cdots, x_n) \neq (0, \cdots, 0)$. A necessary and sufficient condition for positive definiteness is that all the determinants

$$C_1 = c_{11}, \quad C_2 = \begin{vmatrix} c_{11} & c_{12} \\ c_{21} & c_{22} \end{vmatrix}, \quad C_3 = \begin{vmatrix} c_{11} & c_{12} & c_{13} \\ c_{21} & c_{22} & c_{23} \\ c_{31} & c_{32} & c_{33} \end{vmatrix}, \cdots, \quad C_n = \det \mathbf{C}$$

are positive (cf. Ref. [B3], vol. I, p. 306). Show that the form in Prob. 25 is positive definite.

34. Test the forms in Probs. 26 and 28 for positive definiteness.

35. Consider the quadratic form $Q = \mathbf{x}^T\mathbf{C}\mathbf{x}$ ($\mathbf{C}$ symmetric) in three variables x_1, x_2, x_3. Find $\partial Q/\partial x_1$, $\partial Q/\partial x_2$, $\partial Q/\partial x_3$ and show that these are the components of the vector $2\mathbf{C}\mathbf{x}$.

7.15 Properties of Eigenvectors. Diagonalization

Eigenvectors of $n \times n$ matrices $\mathbf{A}$ have some basic properties of practical interest. They may (or may not!) form a basis in R^n, and if they do, we can use them to *"diagonalize"* $\mathbf{A}$, that is, to transform $\mathbf{A}$ into a diagonal matrix whose entries on the main diagonal are the eigenvalues of $\mathbf{A}$. These are the key issues in this section. We begin with a concept that is of central interest in eigenvalue problems.

Definition of similarity of matrices

An $n \times n$ matrix $\hat{\mathbf{A}}$ is called **similar** to an $n \times n$ matrix $\mathbf{A}$ if

$$(1) \qquad \boxed{\hat{\mathbf{A}} = \mathbf{T}^{-1}\mathbf{A}\mathbf{T}}$$

for some (nonsingular!) $n \times n$ matrix $\mathbf{T}$. This transformation, which gives $\hat{\mathbf{A}}$ from $\mathbf{A}$, is called a **similarity transformation.**

Similarity transformations are important since they preserve eigenvalues:

Theorem 1 (Eigenvalues of similar matrices)

If $\hat{\mathbf{A}}$ is similar to $\mathbf{A}$, then $\hat{\mathbf{A}}$ has the same eigenvalues as $\mathbf{A}$.

Furthermore, if $\mathbf{x}$ is an eigenvector of $\mathbf{A}$, then $\mathbf{y} = \mathbf{T}^{-1}\mathbf{x}$ is an eigenvector of $\hat{\mathbf{A}}$ corresponding to the same eigenvalue.

Proof. Let λ be an eigenvalue of $\mathbf{A}$ and $\mathbf{x}$ a corresponding eigenvector. Then $\mathbf{A}\mathbf{x} = \lambda\mathbf{x}$ ($\mathbf{x} \neq \mathbf{0}$). We multiply this by $\mathbf{T}^{-1}$ from the left to get

$$\mathbf{T}^{-1}\mathbf{A}\mathbf{x} = \lambda\mathbf{T}^{-1}\mathbf{x}.$$

Now comes a trick worth remembering. We can insert $\mathbf{I}$ where we want and write it as the product of a matrix times its inverse. Here we write $\mathbf{I} = \mathbf{T}\mathbf{T}^{-1}$ and get

$$\mathbf{T}^{-1}\mathbf{A}\mathbf{x} = \mathbf{T}^{-1}\mathbf{A}\mathbf{I}\mathbf{x} = \mathbf{T}^{-1}\mathbf{A}\mathbf{T}\mathbf{T}^{-1}\mathbf{x} = \hat{\mathbf{A}}(\mathbf{T}^{-1}\mathbf{x}) = \lambda\mathbf{T}^{-1}\mathbf{x}.$$

This shows that λ is an eigenvalue of $\hat{\mathbf{A}}$ and $\mathbf{T}^{-1}\mathbf{x}$ is a corresponding eigenvector of $\hat{\mathbf{A}}$. Note that $\mathbf{T}^{-1}\mathbf{x} \neq \mathbf{0}$, since $\mathbf{T}^{-1}\mathbf{x} = \mathbf{0}$ would give the contradiction $\mathbf{x} = \mathbf{I}\mathbf{x} = \mathbf{T}\mathbf{T}^{-1}\mathbf{x} = \mathbf{T}\mathbf{0} = \mathbf{0}$. ∎

The next theorem is of interest in itself and of help in connection with bases of eigenvectors.

Theorem 2 (Linear independence of eigenvectors)

*Let $\lambda_1, \lambda_2, \cdots, \lambda_k$ be **distinct** eigenvalues of an $n \times n$ matrix. Then corresponding eigenvectors $x_1, x_2, \cdots, x_k$ form a linearly independent set.*

Proof. Suppose that the conclusion is false. Let r be the largest integer such that $\{x_1, \cdots, x_r\}$ is a linearly independent set. Then $r < k$ and $\{x_1, \cdots, x_{r+1}\}$ is linearly dependent. Thus there are scalars $c_1, \cdots, c_{r+1}$, not all zero, such that

$$(2) \qquad c_1 x_1 + \cdots + c_{r+1} x_{r+1} = 0$$

(cf. Sec. 6.4). Multiplying both sides by A and using $Ax_j = \lambda_j x_j$, we obtain

$$(3) \qquad c_1 \lambda_1 x_1 + \cdots + c_{r+1} \lambda_{r+1} x_{r+1} = 0.$$

To get rid of the last term, we subtract λ_{r+1} times (2) from this, obtaining

$$c_1(\lambda_1 - \lambda_{r+1})x_1 + \cdots + c_r(\lambda_r - \lambda_{r+1})x_r = 0.$$

Here $c_1(\lambda_1 - \lambda_{r+1}) = 0, \cdots, c_r(\lambda_r - \lambda_{r+1}) = 0$ since $\{x_1, \cdots, x_r\}$ is linearly independent. Hence $c_1 = \cdots = c_r = 0$, since all the eigenvalues are distinct. But with this, (2) reduces to $c_{r+1}x_{r+1} = 0$, hence $c_{r+1} = 0$, since $x_{r+1} \neq 0$ (an eigenvector!). This contradicts the fact that not all scalars in (2) are zero. Hence the conclusion of the theorem must hold.   ∎

This theorem immediately implies

Theorem 3 (Basis of eigenvectors)

*If an $n \times n$ matrix A has n **distinct** eigenvalues, then A has a basis of eigenvectors in R^n.*

EXAMPLE 1. Basis of eigenvectors

The matrix

$$A = \begin{bmatrix} 5 & 3 \\ 3 & 5 \end{bmatrix} \qquad \text{has a basis of eigenvectors} \qquad \begin{bmatrix} 1 \\ 1 \end{bmatrix}, \begin{bmatrix} 1 \\ -1 \end{bmatrix}$$

corresponding to the eigenvalues $\lambda_1 = 8$, $\lambda_2 = 2$. (Cf. Example 2 in Sec. 7.12.)

EXAMPLE 2. Basis when not all eigenvalues are distinct. Nonexistence of basis

Even if not all n eigenvalues are different, a matrix A may still provide a basis of eigenvectors in R^n. This is illustrated by Example 3 in Sec. 7.12, where $n = 3$. On the other hand, A may not have enough linearly independent eigenvectors to make up a basis. For instance, the matrix in Example 4, Sec. 7.12,

$$A = \begin{bmatrix} 0 & 1 \\ 0 & 0 \end{bmatrix} \qquad \text{has only one eigenvector} \qquad \begin{bmatrix} k \\ 0 \end{bmatrix},$$

where k is arbitrary, not zero. Hence A does not provide a basis of eigenvectors in R^2.   ∎

Bases of eigenvectors have great advantages in connection with **linear transformations** (Sec. 7.3) as given by $n \times n$ matrices. Indeed, if the eigenvectors $x_1, \cdots, x_n$ of a matrix A corresponding to (not necessarily distinct) eigenvalues $\lambda_1, \cdots, \lambda_n$, respectively, form a basis in R^n, then any vector x in R^n has a unique representation

$$x = c_1 x_1 + c_2 x_2 + \cdots + c_n x_n.$$

Now comes the point. If we apply A, we get

$$
\begin{aligned}
Ax &= A(c_1 x_1 + \cdots + c_n x_n) \\
&= c_1 A x_1 + \cdots + c_n A x_n \\
&= c_1 \lambda_1 x_1 + \cdots + c_n \lambda_n x_n.
\end{aligned}
$$

(4)

This shows the advantage: we have decomposed the complicated action of A on arbitrary vectors x into a sum of simple actions (multiplication by scalars) on the eigenvectors of A.

Bases of eigenvectors exist under conditions much more general than those given in Theorem 3, and for the matrices considered in Secs. 7.12—7.14 we can even choose a unitary system of eigenvectors, hence an orthonormal system in the case of a symmetric, skew-symmetric or orthogonal matrix:

Theorem 4 (Basis of eigenvectors)

A Hermitian, skew-Hermitian or unitary matrix has a basis in R^n that is a unitary system (cf. Sec. 7.13). (Proof in Ref. [B3], vol. 1, pp. 270—272.)

EXAMPLE 3. Orthonormal basis of eigenvectors
The matrix in Example 1 is symmetric, and an orthonormal basis of eigenvectors is $[1/\sqrt{2} \quad 1/\sqrt{2}]^T, [1/\sqrt{2} \quad -1/\sqrt{2}]^T$. ∎

Diagonalization

Bases of eigenvectors also play a central role in the diagonalization of an $n \times n$ matrix A, as the following theorem explains.

Theorem 5 (Diagonalization of a matrix)

If an $n \times n$ matrix A has a basis of eigenvectors, then

(5)
$$\boxed{D = X^{-1} A X}$$

is diagonal, with the eigenvalues of A as the entries on the main diagonal. Here X is the matrix with these eigenvectors as column vectors. Also,

(5*)
$$D^m = X^{-1} A^m X.$$

Proof. Let $x_1, \cdots, x_n$ form a basis of eigenvectors of A in R^n corresponding to eigenvalues $\lambda_1, \cdots, \lambda_n$, respectively, of A. Then $X = [x_1 \cdots x_n]$ has rank n, by Theorem 1 in Sec. 7.6. Hence X^{-1} exists by Theorem 1 in Sec. 7.8. Now (6) in Sec. 7.3 and $A x_j = \lambda_j x_j$ give

$$AX = A[x_1 \cdots x_n] = [A x_1 \cdots A x_n] = [\lambda_1 x_1 \cdots \lambda_n x_n].$$

Direct calculation shows that

$$[\lambda_1 x_1 \;\; \cdots \;\; \lambda_n x_n] = XD \qquad \text{where} \qquad D = \begin{bmatrix} \lambda_1 & & 0 \\ & \ddots & \\ 0 & & \lambda_n \end{bmatrix}.$$

(For instance, if $n = 2$ and $\lambda_1 = 3$, $\lambda_2 = 7$, then

$$\begin{bmatrix} 3x_{11} & 7x_{12} \\ 3x_{21} & 7x_{22} \end{bmatrix} = \begin{bmatrix} x_{11} & x_{12} \\ x_{21} & x_{22} \end{bmatrix} \begin{bmatrix} 3 & 0 \\ 0 & 7 \end{bmatrix}$$

and similarly in the general case.) Together, $AX = XD$. Multiplication by X^{-1} from the left gives the desired result (5), the equation namely, $X^{-1}AX = X^{-1}XD = D$. Also, (5*) follows by noting that

$$D^2 = DD = X^{-1}AXX^{-1}AX = X^{-1}AAX = X^{-1}A^2X, \qquad \text{etc.} \quad \blacksquare$$

EXAMPLE 4. Diagonalization
Calculation as in the examples in Sec. 7.12 etc. shows that the matrix

$$A = \begin{bmatrix} 5 & 4 \\ 1 & 2 \end{bmatrix} \quad \text{has eigenvectors} \quad \begin{bmatrix} 4 \\ 1 \end{bmatrix} \quad \text{and} \quad \begin{bmatrix} 1 \\ -1 \end{bmatrix}. \quad \text{Hence} \quad X = \begin{bmatrix} 4 & 1 \\ 1 & -1 \end{bmatrix}$$

and, using (4) in Sec. 7.8, we obtain

$$X^{-1}AX = \frac{1}{-5}\begin{bmatrix} -1 & -1 \\ -1 & 4 \end{bmatrix}\begin{bmatrix} 5 & 4 \\ 1 & 2 \end{bmatrix}\begin{bmatrix} 4 & 1 \\ 1 & -1 \end{bmatrix} = \begin{bmatrix} 0.2 & 0.2 \\ 0.2 & -0.8 \end{bmatrix}\begin{bmatrix} 24 & 1 \\ 6 & -1 \end{bmatrix} = \begin{bmatrix} 6 & 0 \\ 0 & 1 \end{bmatrix}.$$

The student may show that an interchange of the columns of X results in an interchange of the eigenvalues 6 and 1 in the diagonal matrix.

EXAMPLE 5. Diagonalization
Diagonalize

$$A = \begin{bmatrix} 7.3 & 0.2 & -3.7 \\ -11.5 & 1.0 & 5.5 \\ 17.7 & 1.8 & -9.3 \end{bmatrix}.$$

Solution. The characteristic determinant gives the characteristic equation $-\lambda^3 - \lambda^2 + 12\lambda = 0$. The roots (eigenvalues of A) are $\lambda_1 = 3$, $\lambda_2 = -4$, $\lambda_3 = 0$. By the Gauss elimination applied to $(A - \lambda I)x = 0$ with $\lambda = \lambda_1, \lambda_2, \lambda_3$ we find eigenvectors and then X^{-1} by the Gauss–Jordan elimination (Sec. 7.8, Example 1). The results are

$$\begin{bmatrix} -1 \\ 3 \\ -1 \end{bmatrix}, \quad \begin{bmatrix} 1 \\ -1 \\ 3 \end{bmatrix}, \quad \begin{bmatrix} 2 \\ 1 \\ 4 \end{bmatrix}, \quad X = \begin{bmatrix} -1 & 1 & 2 \\ 3 & -1 & 1 \\ -1 & 3 & 4 \end{bmatrix}, \quad X^{-1} = \begin{bmatrix} -0.7 & 0.2 & 0.3 \\ -1.3 & -0.2 & 0.7 \\ 0.8 & 0.2 & -0.2 \end{bmatrix}$$

Calculating AX and multiplying by X^{-1} from the left, we thus obtain

$$D = X^{-1}AX = \begin{bmatrix} -0.7 & 0.2 & 0.3 \\ -1.3 & -0.2 & 0.7 \\ 0.8 & 0.2 & -0.2 \end{bmatrix}\begin{bmatrix} -3 & -4 & 0 \\ 9 & 4 & 0 \\ -3 & -12 & 0 \end{bmatrix} = \begin{bmatrix} 3 & 0 & 0 \\ 0 & -4 & 0 \\ 0 & 0 & 0 \end{bmatrix}. \quad \blacksquare$$

Transformation of Forms to Principal Axes

This is an important practical task related to the diagonalization of matrices. We explain the idea for quadratic forms (cf. Sec. 7.14)

(6)
$$Q = \mathbf{x}^\mathsf{T}\mathbf{A}\mathbf{x}.$$

Without restriction we can assume that $\mathbf{A}$ is real *symmetric* (cf. Sec. 7.14). Then $\mathbf{A}$ has an orthonormal basis of n eigenvectors, by Theorem 4. Hence the matrix $\mathbf{X}$ with these vectors as column vectors is orthogonal, so that $\mathbf{X}^{-1} = \mathbf{X}^\mathsf{T}$. From (5) we thus have $\mathbf{A} = \mathbf{X}\mathbf{D}\mathbf{X}^{-1} = \mathbf{X}\mathbf{D}\mathbf{X}^\mathsf{T}$. Substitution into (6) gives

$$Q = \mathbf{x}^\mathsf{T}\mathbf{X}\mathbf{D}\mathbf{X}^\mathsf{T}\mathbf{x}.$$

If we set $\mathbf{X}^\mathsf{T}\mathbf{x} = \mathbf{y}$, then, since $\mathbf{X}^\mathsf{T} = \mathbf{X}^{-1}$, we get

(7)
$$\mathbf{x} = \mathbf{X}\mathbf{y}$$

and Q becomes simply

(8)
$$Q = \mathbf{y}^\mathsf{T}\mathbf{D}\mathbf{y} = \lambda_1 y_1{}^2 + \lambda_2 y_2{}^2 + \cdots + \lambda_n y_n{}^2.$$

This proves

Theorem 6 (Principal axes theorem)
The substitution (7) transforms a quadratic form

$$Q = \mathbf{x}^\mathsf{T}\mathbf{A}\mathbf{x} = \sum_{j=1}^{n}\sum_{k=1}^{n} a_{jk}x_j x_k$$

to the principal axes form (8), where $\lambda_1, \cdots, \lambda_n$ are the (not necessarily distinct) eigenvalues of the (symmetric!) matrix $\mathbf{A}$, and $\mathbf{X}$ is an orthogonal matrix with corresponding eigenvectors $\mathbf{x}_1, \cdots, \mathbf{x}_n$, respectively, as column vectors.

EXAMPLE 6. Transformation to principal axes. Conic sections
Find out what type of conic section the following quadratic form represents and transform it to principal axes:

$$Q = 17x_1{}^2 - 30x_1 x_2 + 17x_2{}^2 = 128.$$

Solution. We have $Q = \mathbf{x}^\mathsf{T}\mathbf{A}\mathbf{x}$, where

$$\mathbf{A} = \begin{bmatrix} 17 & -15 \\ -15 & 17 \end{bmatrix}, \qquad \mathbf{x} = \begin{bmatrix} x_1 \\ x_2 \end{bmatrix}.$$

This gives the characteristic equation $(17 - \lambda)^2 - 15^2 = 0$. It has the roots $\lambda_1 = 2$, $\lambda_2 = 32$. Hence (8) becomes

$$Q = 2y_1{}^2 + 32y_2{}^2.$$

We see that $Q = 128$ represents the ellipse $2y_1{}^2 + 32y_2{}^2 = 128$, that is,

$$\frac{y_1^2}{8^2} + \frac{y_2^2}{2^2} = 1.$$

If we want to know the direction of the principal axes in the x_1x_2-coordinates, we have to determine normalized eigenvectors from $(A - \lambda I)x = 0$ with $\lambda = \lambda_1 = 2$ and $\lambda = \lambda_2 = 32$ and then use (7). We get

$$\begin{bmatrix} 1/\sqrt{2} \\ 1/\sqrt{2} \end{bmatrix} \text{ and } \begin{bmatrix} -1/\sqrt{2} \\ 1/\sqrt{2} \end{bmatrix},$$

hence

$$x = Xy = \begin{bmatrix} 1/\sqrt{2} & -1/\sqrt{2} \\ 1/\sqrt{2} & 1/\sqrt{2} \end{bmatrix} \begin{bmatrix} y_1 \\ y_2 \end{bmatrix}, \qquad \begin{array}{l} x_1 = y_1/\sqrt{2} - y_2/\sqrt{2} \\ x_2 = y_1/\sqrt{2} + y_2/\sqrt{2}. \end{array}$$

This is a 45° rotation. Our results agree with those in Sec. 7.12, Example 2, except for the notations. See also Fig. 169 in that example.   ∎

In the next section, the last of Chap. 7 on matrices, we show that eigenvalue problems arise quite naturally in connection with systems of linear **differential equations** and are of great help in solving and discussing corresponding initial value problems, as they result from mass–spring systems, electrical networks and other applications.

Problems for Sec. 7.15

Similarity transformations
Find $\hat{A} = T^{-1}AT$. Find the eigenvalues of $\hat{A}$ and A and verify that they are the same. Find corresponding eigenvectors y of $\hat{A}$, compute $x = Ty$ and verify that they are eigenvectors of A.

1. $A = \begin{bmatrix} 2 & 0 \\ 0 & -1 \end{bmatrix}$, $T = \begin{bmatrix} 0.8 & 0.6 \\ -0.6 & 0.8 \end{bmatrix}$

2. $A = \begin{bmatrix} 3 & 4 \\ 4 & -3 \end{bmatrix}$, $T = \begin{bmatrix} -2 & 1 \\ \frac{3}{2} & -\frac{1}{2} \end{bmatrix}$

3. $A = \begin{bmatrix} 8 & -4 \\ 2 & 2 \end{bmatrix}$, $T = \begin{bmatrix} 5 & 2 \\ 2 & 1 \end{bmatrix}$

4. $A = \begin{bmatrix} 1 & 0 \\ 2 & -1 \end{bmatrix}$, $T = \begin{bmatrix} 10 & 3 \\ 0 & 1 \end{bmatrix}$

5. $A = \begin{bmatrix} 1.4 & 0.5 \\ -1.0 & -0.1 \end{bmatrix}$, $T = \begin{bmatrix} 1 & 1 \\ -1 & -2 \end{bmatrix}$

6. $A = \begin{bmatrix} 5 & 10 \\ 4 & -1 \end{bmatrix}$, $T = \begin{bmatrix} 5 & 1 \\ 2 & -1 \end{bmatrix}$

7. $A = \begin{bmatrix} 2 & 8 \\ 0 & -2 \end{bmatrix}$, $T = \begin{bmatrix} 0.1 & 0.2 \\ -0.3 & 0.4 \end{bmatrix}$

8. $A = \begin{bmatrix} 2 & 1 \\ -2 & -1 \end{bmatrix}$, $T = \begin{bmatrix} 7 & -5 \\ 10 & -7 \end{bmatrix}$

Diagonalization
Find a basis of eigenvectors and diagonalize or indicate that no such basis exists.

9. $\begin{bmatrix} 0 & 4 \\ 9 & 0 \end{bmatrix}$

10. $\begin{bmatrix} 2 & 2 \\ 0 & 0 \end{bmatrix}$

11. $\begin{bmatrix} 1 & 1 \\ 0 & -1 \end{bmatrix}$

12. $\begin{bmatrix} 0 & 1 \\ 0 & 0 \end{bmatrix}$

13. $\begin{bmatrix} 1 & 1 \\ 1 & 1 \end{bmatrix}$ **14.** $\begin{bmatrix} \sqrt{2} & \sqrt{2} \\ \sqrt{2} & -\sqrt{2} \end{bmatrix}$ **15.** $\begin{bmatrix} 0 & 0 & 1 \\ 0 & 2 & 0 \\ 3 & 0 & 0 \end{bmatrix}$ **16.** $\begin{bmatrix} -2 & 0 & 0 \\ -2 & 0 & 0 \\ 0 & 0 & 2 \end{bmatrix}$

Quadratic forms

Find a symmetric matrix **C** such that $Q = \mathbf{x}^T \mathbf{C} \mathbf{x}$, where Q equals

17. $(x_1 - x_2)^2$

18. $2x_1^2 + 4x_1 x_2 + 6x_2^2$

19. $-9x_1 x_2 + 3x_2^2$

20. $4x_1^2 - \frac{1}{2}x_2^2$

21. $(x_1 + x_2 + x_3)^2$

22. $x_1^2 + x_1 x_2 - 2x_2^2 - 6x_2 x_3 - 4x_3^2$

23. $(6x_1 + 2x_2)^2 + 8x_1 x_3$

24. $2x_1 x_2 - 6x_1 x_3 + 12x_2 x_3$

Find out what type of conic section (or pair of straight lines) is represented by the given quadratic form. Transform it to principal axes. Express $\mathbf{x}^T = [x_1 \ \ x_2]$ in terms of the new coordinate vector $\mathbf{y}^T = [y_1 \ \ y_2]$, as in Example 6.

25. $7x_1^2 + 48x_1 x_2 - 7x_2^2 = 25$

26. $3x_1^2 + 4\sqrt{3}x_1 x_2 + 7x_2^2 = 9$

27. $801x_1^2 - 600x_1 x_2 + 1396x_2^2 = 169$

28. $4x_1 x_2 + 3x_2^2 = 1$

29. $x_1^2 + 6x_1 x_2 + 9x_2^2 = 10$

30. $10x_1^2 - 9x_1 x_2 + \frac{25}{4}x_2^2 = 13$

31. $x_1^2 + 24x_1 x_2 - 6x_2^2 = 5$

32. $-3x_1^2 + 8x_1 x_2 + 3x_2^2 = 0$

33. Show that the form in Prob. 18 is positive definite. (See Prob. 33 in Sec. 7.14.)

34. Test the forms in Probs. 22 and 24 for positive definiteness.

Trace of a square matrix

The sum of the entries on the main diagonal of an $n \times n$ matrix $\mathbf{A} = [a_{jk}]$ is called the **trace** of **A**; thus trace $\mathbf{A} = a_{11} + a_{22} + \cdots + a_{nn}$.

35. Show that trace $\mathbf{AB} = \sum_{i=1}^{n} \sum_{l=1}^{n} a_{il} b_{li} =$ trace **BA**, where $\mathbf{A} = [a_{jk}]$ and $\mathbf{B} = [b_{jk}]$ are $n \times n$ matrices.

36. Using Prob. 35, show that similar matrices have equal traces.

37. (**Sum of the eigenvalues**) Show that trace **A** equals the sum of the eigenvalues of **A**, each counted as often as its algebraic multiplicity indicates.

38. Using Prob. 37, show that trace $\hat{\mathbf{A}} =$ trace **A** when $\hat{\mathbf{A}}$ is similar to **A**.

39. Prove Theorem 1 by showing that the characteristic determinants of $\hat{\mathbf{A}}$ and **A** are equal.

40. (**Inheritance**) For pea plants, let gene A mean "red-flowered" and gene a "white-flowered." Offsprings inherit with probability $\frac{1}{2}$ one of the two genes of each parent. Let x_1, x_2, x_3 be the fractions of AA, Aa, aa parents, respectively, and $y_1^{(1)}$, $y_2^{(1)}$, $y_3^{(1)}$ the fractions of AA, Aa, aa offspring in an experiment. Assume that all plants are fertilized with AA plants. Set up a model $\mathbf{y}^{(1)} = \mathbf{A}\mathbf{x}$ for this experiment, where $\mathbf{x} = [x_1 \ \ x_2 \ \ x_3]^T$ and $\mathbf{y}^{(1)} = [y_1^{(1)} \ \ y_2^{(1)} \ \ y_3^{(1)}]^T$. To see what will happen in the long run, find a formula for the fractions in the mth generation, $\mathbf{y}^{(m)} = \mathbf{A}^m \mathbf{x}$. *Hint.* Use (5*).

7.16 Systems of Differential Equations

Systems of differential equations arise in various engineering problems. A first impression of this fact was given in Secs. 3.1 and 7.12. Simpler systems can be treated without the use of matrices (cf. Sec. 3.1). However, matrices, and in particular eigenvalue and diagonalization methods, are in general much more efficient in solving systems of *linear* differential equations, small or large, and for discussing general properties and developing a systematic theory.

In this section we shall see how our techniques and results on eigenvalues and diagonalization have very useful applications to those systems and are generally far superior to other methods.

EXAMPLE 1. Model of a vibrating system. Four methods of solution

Determine the motion of the mechanical system in Fig. 171 satisfying the initial conditions

$$y_1(0) = 1, \qquad y_2(0) = 2, \qquad y_1'(0) = -2\sqrt{6}, \qquad y_2'(0) = \sqrt{6}.$$

Solution. In Sec. 3.1, Example 4, we derived the corresponding mathematical model

(1)
$$\text{(a)} \quad y_1'' = -5y_1 + 2y_2$$
$$\text{(b)} \quad y_2'' = 2y_1 - 2y_2.$$

1st Method. Reduction to a fourth-order differential equation

In Sec. 3.1, Example 4, we have shown that y_1 satisfies

(2)
$$y_1^{\text{iv}} + 7y_1'' + 6y_1 = 0,$$

from which we obtained y_1 and then y_2 from (1a),

(3)
$$y_1 = a_1 \cos t + b_1 \sin t + a_2 \cos \sqrt{6}t + b_2 \sin \sqrt{6}t$$
$$y_2 = 2a_1 \cos t + 2b_1 \sin t - \tfrac{1}{2}a_2 \cos \sqrt{6}t - \tfrac{1}{2}b_2 \sin \sqrt{6}t.$$

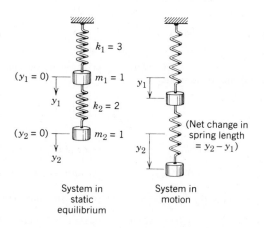

Fig. 171. Mechanical system in Example 1

From this and the initial conditions,

$$y_1(0) = a_1 + a_2 = 1, \quad y_2(0) = 2a_1 - \tfrac{1}{2}a_2 = 2, \quad \text{hence} \quad a_1 = 1, \quad a_2 = 0;$$

$$y_1{}'(0) = b_1 + b_2\sqrt{6} = -2/6, \quad y_2{}'(0) = 2b_1 - \tfrac{1}{2}b_2\sqrt{6} = \sqrt{6}, \quad \text{hence} \quad b_1 = 0, \quad b_2 = -2.$$

This gives the solution

(4)
$$y_1(t) = \quad \cos t - 2 \sin \sqrt{6}t$$
$$y_2(t) = 2 \cos t + \quad \sin \sqrt{6}t.$$

2nd Method. "Triangularization" using operators
We write (1) in the form

$$-2y_2 \quad + (D^2 + 5)y_1 = 0$$
$$(D^2 + 2)y_2 - \quad 2y_1 \quad = 0.$$

As in the Gauss elimination, we "multiply" the first equation by $\tfrac{1}{2}(D^2 + 2)$ [that is, we apply this operator] and add it to the second equation to get the "triangular system"

$$-2y_2 + \quad\quad\quad\quad (D^2 + 5)y_1 = 0$$
$$[\tfrac{1}{2}(D^2 + 2)(D^2 + 5) - 2]y_1 = 0.$$

The second equation, multiplied by 2, gives

$$(D^4 + 7D^2 + 6D)y_1 = 0.$$

This is identical with (2). The remaining steps are as in the first method.

3rd Method. Substitution of $\mathbf{y} = \mathbf{x}e^{\omega t}$
The remaining methods use matrices. We can write (1) as a vector equation

(5)
$$\mathbf{y}'' = \mathbf{A}\mathbf{y}, \quad\quad \text{where} \quad\quad \mathbf{y} = \begin{bmatrix} y_1 \\ y_2 \end{bmatrix} \quad\quad \text{and} \quad\quad \mathbf{A} = \begin{bmatrix} -5 & 2 \\ 2 & -2 \end{bmatrix}.$$

Motivated by the case of a single equation, we substitute $\mathbf{y} = \mathbf{x}e^{\omega t}$. Canceling out $e^{\omega t}$, we obtain

$$\mathbf{A}\mathbf{x} = \lambda\mathbf{x} \quad\quad\quad\quad (\lambda = \omega^2).$$

The characteristic equation is

(6)
$$\det(\mathbf{A} - \lambda\mathbf{I}) = \begin{vmatrix} -5 - \lambda & 2 \\ 2 & -2 - \lambda \end{vmatrix} = \lambda^2 + 7\lambda + 6 = 0.$$

The roots are $\lambda_1 = -1$ and $\lambda_2 = -6$. These are the eigenvalues of $\mathbf{A}$. Corresponding eigenvectors are found to be

(7)
$$\mathbf{x}_1 = \begin{bmatrix} 1 \\ 2 \end{bmatrix}, \quad\quad \mathbf{x}_2 = \begin{bmatrix} 2 \\ -1 \end{bmatrix}.$$

To $\lambda_1 = -1$ and $\lambda_2 = -6$ there correspond $\omega_1 = \pm i$ and $\omega_2 = \pm i\sqrt{6}$. Hence a solution is

(8)
$$\mathbf{y}(t) = a_1\mathbf{x}_1 \cos t + b_1\mathbf{x}_1 \sin t + a_2\mathbf{x}_2 \cos \sqrt{6}t + b_2\mathbf{x}_2 \sin \sqrt{6}t,$$

where a_1, b_1, a_2, b_2 are arbitrary constants. In terms of components this can be written

$$y_1(t) = a_1 \cos t + b_1 \sin t + 2a_2 \cos \sqrt{6}t + 2b_2 \sin \sqrt{6}t$$
$$y_2(t) = 2a_1 \cos t + 2b_1 \sin t - a_2 \cos \sqrt{6}t - b_2 \sin \sqrt{6}t,$$

as can be seen from the form of $\mathbf{x}_1$ and $\mathbf{x}_2$. This agrees with (3), except for notation (our present $2a_2$ and $2b_2$ correspond to a_2 and b_2 there). From (8) and the first two initial conditions $y_1(0) = 1$, $y_2(0) = 2$, written in vector form, we have

$$\mathbf{y}(0) = a_1\mathbf{x}_1 + a_2\mathbf{x}_2 = a_1\begin{bmatrix} 1 \\ 2 \end{bmatrix} + a_2\begin{bmatrix} 2 \\ -1 \end{bmatrix} = \begin{bmatrix} 1 \\ 2 \end{bmatrix}.$$

Hence $a_1 = 1$ and $a_2 = 0$. Substitution of this into (8) and differentiation gives

$$\mathbf{y}'(t) = -\mathbf{x}_1 \sin t + b_1\mathbf{x}_1 \cos t + b_2\sqrt{6}\mathbf{x}_2 \cos \sqrt{6}t.$$

From this and the other two initial conditions $y_1'(0) = -2\sqrt{6}$, $y_2'(0) = \sqrt{6}$ we obtain

$$\mathbf{y}'(0) = b_1\mathbf{x}_1 + b_2\sqrt{6}\,\mathbf{x}_2 = b_1\begin{bmatrix} 1 \\ 2 \end{bmatrix} + b_2\sqrt{6}\begin{bmatrix} 2 \\ -1 \end{bmatrix} = \begin{bmatrix} -2\sqrt{6} \\ \sqrt{6} \end{bmatrix}.$$

Hence $b_1 = 0$ and $b_2 = -1$. The solution of our problem is

$$\mathbf{y}(t) = \mathbf{x}_1 \cos t - \mathbf{x}_2 \sin \sqrt{6}t.$$

In terms of components, as in (4),

$$y_1(t) = \cos t - 2 \sin \sqrt{6}t$$

$$y_2(t) = 2 \cos t + \sin \sqrt{6}t.$$

4th Method. Diagonalization
This method is more systematic, avoiding the ad hoc substitution of the previous method. It makes essential use of Theorem 5 in Sec. 7.15. We diagonalize the system (5) by using the eigenvectors (7) and setting

(9)
$$\mathbf{y} = \mathbf{X}\mathbf{z} = \begin{bmatrix} 1 & 2 \\ 2 & -1 \end{bmatrix}\begin{bmatrix} z_1 \\ z_2 \end{bmatrix} = \begin{bmatrix} z_1 + 2z_2 \\ 2z_1 - z_2 \end{bmatrix}.$$

Then (5) becomes

$$\mathbf{y}'' = \mathbf{X}\mathbf{z}'' = \mathbf{A}\mathbf{y} = \mathbf{A}\mathbf{X}\mathbf{z}.$$

Multiplying by $\mathbf{X}^{-1}$ from the left, noting $\mathbf{X}^{-1}\mathbf{X} = \mathbf{I}$ and using Theorem 5 in Sec. 7.15, we get

$$\mathbf{z}'' = \begin{bmatrix} z_1'' \\ z_2'' \end{bmatrix} = \mathbf{X}^{-1}\mathbf{A}\mathbf{X}\mathbf{z} = \mathbf{D}\mathbf{z} = \begin{bmatrix} -1 & 0 \\ 0 & -6 \end{bmatrix}\begin{bmatrix} z_1 \\ z_2 \end{bmatrix} = \begin{bmatrix} -z_1 \\ -6z_2 \end{bmatrix}.$$

Hence

$$z_1'' = -z_1 \qquad \text{Solution} \qquad z_1 = c_{11} \cos t + c_{12} \sin t$$

$$z_2'' = -6z_2. \qquad \text{Solution} \qquad z_2 = c_{21} \cos \sqrt{6}t + c_{22} \sin \sqrt{6}t.$$

From this and (9) we calculate

$$y_1 = z_1 + 2z_2 = c_{11} \cos t + c_{12} \sin t + 2c_{21} \cos \sqrt{6}t + 2c_{22} \sin \sqrt{6}t$$

$$y_2 = 2z_1 - z_2 = 2c_{11} \cos t + 2c_{12} \sin t - c_{21} \cos \sqrt{6}t - c_{22} \sin \sqrt{6}t.$$

This again agrees with (3), except for the notations of the arbitrary constants. The remaining steps are as before.

We note that in the case of our homogeneous system, it was not necessary to compute the inverse $\mathbf{X}^{-1}$. For nonhomogeneous systems we would have to, as our subsequent discussion shows.   ∎

Single equations as well as systems can usually be converted to systems of first order. For instance, we can do this for our second-order system (1) in Example 1 by setting $y_1{}' = y_3$, $y_2{}' = y_4$. Then $y_1{}'' = y_3{}'$, $y_2{}'' = y_4{}'$ and we have the system

$$y_1{}' = y_3, \qquad y_2{}' = y_4, \qquad y_3{}' = -5y_1 + 2y_2, \qquad y_4{}' = 2y_1 - 2y_2.$$

This explains it why the general theory is usually formulated for first-order systems

$$y_j{}' = \sum_{k=1}^{n} a_{jk}y_k + h_j(t) \qquad\qquad j = 1, \cdots, n.$$

We want to say a few words about a general theory for such systems. Ordinarily the a_{jk}'s depend on t, but, for simplicity, we assume that they are constant. (In Example 1 we have $h_j = 0$.) In matrix form, the system is

(10) $$\mathbf{y}' = \mathbf{Ay} + \mathbf{h}.$$

We assume that $\mathbf{A}$ has a basis of eigenvectors $\mathbf{x}_1, \cdots, \mathbf{x}_n$. Then the matrix $\mathbf{X} = [\mathbf{x}_1 \cdots \mathbf{x}_n]$ is nonsingular. To diagonalize, as in Sec. 7.15 we can now set

(11) $$\mathbf{y} = \mathbf{Xz}.$$

Then $\mathbf{y}' = \mathbf{Xz}'$ and (10) becomes

$$\mathbf{Xz}' = \mathbf{AXz} + \mathbf{h}.$$

Multiplying by $\mathbf{X}^{-1}$ from the left and using Theorem 5, Sec. 7.15, we get, since $\mathbf{X}^{-1}\mathbf{AX} = \mathbf{D}$,

(12) $$\mathbf{z}' = \mathbf{Dz} + \mathbf{X}^{-1}\mathbf{h}.$$

In terms of components, taking $\mathbf{Dz}$ to the left, we can write

(12*) $$z_j{}' - \lambda_j z_j = r_j(t) \qquad\qquad (j = 1, \cdots, n),$$

where, because of the definition of matrix multiplication, $r_j(t)$ is a linear combination, with constant coefficients, of $h_1(t), \cdots, h_n(t)$. By Sec. 1.7, the solution of (12*) is

(13) $$z_j(t) = e^{\lambda_j t} \left(\int e^{-\lambda_j t} r_j(t)\, dt + c_j \right),$$

where c_j is an arbitrary constant. As in Sec. 1.7 we can assign to c_j a unique value by prescribing an initial condition

(14) $$\mathbf{y}(t_0) = \mathbf{y}_0.$$

Indeed, by (11), this determines

$$\mathbf{z}(t_0) = \mathbf{z}_0 = \mathbf{X}^{-1}\mathbf{y}_0$$

or, in terms of components, $z_j(t_0) = z_{0j}$, where z_{0j} is the jth component of the constant vector $\mathbf{z}_0$.

Remembering how we obtained the general solution in the case of a single first-order linear equation in Sec. 1.7, we see that we can formulate our result as follows.

Existence and uniqueness theorem for systems

Let $\mathbf{h}(t)$ in (10) be continuous in an interval $\alpha < t < \beta$, and let t_0 be any given point in that interval. Assume that $\mathbf{A}$ has a linearly independent set of n eigenvectors. Then the initial value problem (10), (14) has a unique solution $\mathbf{y}(t)$ on that interval.

Real symmetric and, more generally, Hermitian matrices satisfy that condition; see Theorem 4 in Sec. 7.15. Since in many applications the occurring matrices are real symmetric or Hermitian, this theorem is of considerable practical importance.

Let us give an example that illustrates the formulas preceding this theorem.

EXAMPLE 2. Diagonalization in the case of a nonhomogeneous system

Solve the initial value problem consisting of the equations

$$y_1' = 5y_1 + 8y_2 + 1$$
$$y_2' = -6y_1 - 9y_2 + t$$

and the initial conditions $y_1(0) = 4$, $y_2(0) = -3$.

Solution. In matrix notation we have $\mathbf{y}' = \mathbf{A}\mathbf{y} + \mathbf{h}$, where

$$\mathbf{A} = \begin{bmatrix} 5 & 8 \\ -6 & -9 \end{bmatrix}, \qquad \mathbf{h} = \begin{bmatrix} 1 \\ t \end{bmatrix}.$$

The characteristic equation yields the eigenvalues $\lambda_1 = -1$ and $\lambda_2 = -3$. We now determine corresponding eigenvectors, finding

$$\mathbf{x}_1 = \begin{bmatrix} 1 \\ -3/4 \end{bmatrix}, \qquad \mathbf{x}_2 = \begin{bmatrix} 1 \\ -1 \end{bmatrix}.$$

Hence

$$\mathbf{X} = [\mathbf{x}_1 \ \mathbf{x}_2] = \begin{bmatrix} 1 & 1 \\ -3/4 & -1 \end{bmatrix}, \qquad \mathbf{X}^{-1} = \begin{bmatrix} 4 & 4 \\ -3 & -4 \end{bmatrix}$$

and (12) takes the form

$$\mathbf{z}' = \mathbf{D}\mathbf{z} + \mathbf{X}^{-1}\mathbf{h},$$

where

$$\mathbf{D} = \begin{bmatrix} -1 & 0 \\ 0 & -3 \end{bmatrix}, \qquad \mathbf{X}^{-1}\mathbf{h} = \begin{bmatrix} 4 + 4t \\ -3 - 4t \end{bmatrix}.$$

We take $\mathbf{Dz}$ to the left, write the equation in terms of components and use (13):

$$z_1' + z_1 = 4 + 4t, \quad \text{solution} \quad z_1 = c_1 e^{-t} + 4t$$

$$z_2' + 3z_2 = -3 - 4t, \quad \text{solution} \quad z_2 = c_2 e^{-3t} - \tfrac{5}{9} - \tfrac{4}{3}t.$$

From this we obtain

$$\mathbf{y} = \mathbf{Xz} = \begin{bmatrix} 1 & 1 \\ -3/4 & -1 \end{bmatrix}\begin{bmatrix} z_1 \\ z_2 \end{bmatrix} = c_1 \begin{bmatrix} 1 \\ -3/4 \end{bmatrix} e^{-t} + c_2 \begin{bmatrix} 1 \\ -1 \end{bmatrix} e^{-3t} + \frac{1}{9}\begin{bmatrix} -5 + 24t \\ 5 - 15t \end{bmatrix}.$$

By the initial conditions,

$$\mathbf{y}(0) = c_1 \begin{bmatrix} 1 \\ -3/4 \end{bmatrix} + c_2 \begin{bmatrix} 1 \\ -1 \end{bmatrix} + \begin{bmatrix} -5/9 \\ 5/9 \end{bmatrix} = \begin{bmatrix} 4 \\ -3 \end{bmatrix}.$$

This yields $c_1 = 4$, $c_2 = 5/9$. Hence the solution of our problem is

$$y_1 = 4e^{-t} + \tfrac{5}{9}e^{-3t} - \tfrac{5}{9} + \tfrac{8}{3}t$$

$$y_2 = -3e^{-t} - \tfrac{5}{9}e^{-3t} + \tfrac{5}{9} - \tfrac{5}{3}t. \qquad \blacksquare$$

In applications, it is often important to know whether the solutions tend to zero as $t \to \infty$. For instance, such a solution may describe the effect of an undesirable disturbance acting on a mechanical or electrical system and we want to know whether it will disappear. Our discussion shows that for $\mathbf{h} = \mathbf{0}$ in (10), hence $\mathbf{X}^{-1}\mathbf{h} = \mathbf{0}$ in (12) we have $z_j = c_j e^{\lambda_j t}$ in (13). This proves

Theorem 2 (Stability)
All solutions of the system (10) *with* $\mathbf{h} = \mathbf{0}$ *approach* $\mathbf{0}$ *as* $t \to \infty$ *if and only if all the eigenvalues of* $\mathbf{A}$ *have negative real parts.*

This is the end of Chap. 7 on matrices. *In the next chapter we return to the vectors in three-dimensional space,* without the involvement of matrices. Algebraic calculations for these vectors were discussed in Chap. 6. A differential calculus for them follows now, in Chap. 8, and an integral calculus for them in Chap. 9, which is the last chapter of Part B on linear algebra and vector calculus.

Problems for Sec. 7.16

Solve the following systems of differential equations.

1. $y_1' = y_1 + y_2$
$\quad\ y_2' = 3y_1 - y_2$

2. $y_1' = -y_2$
$\quad\ y_2' = -y_1$

3. $y_1' = 14y_1 - 10y_2$
$\quad\ y_2' = 5y_1 - y_2$

4. $y_1' = y_1 + 2y_2$
$\quad\ y_2' = -2y_1 + 6y_2$

5. $y_1' = -8y_1 - 6y_2$
$\quad\ y_2' = 11y_1 + 9y_2$

6. $y_1' = -y_1 - 3y_2$
$\quad\ y_2' = y_1 - 5y_2$

7. $y_1' = 3y_1 + y_2$
$\quad\ y_2' = 6y_1 - 2y_2$
$\quad\ y_3' = 2y_3$

8. $y_1' = 3y_1 + y_2 + 4y_3$
$\quad\ y_2' = 2y_2 + 6y_3$
$\quad\ y_3' = 5y_3$

Solve the following initial value problems.

9. $y_1' = 4y_1 - 2y_2$
$\quad y_2' = y_1 + y_2$
$\qquad y_1(0) = 3, \quad y_2(0) = 2$

10. $y_1' = 3y_1 + 4y_2$
$\quad\ y_2' = 4y_1 - 3y_2$
$\qquad y_1(0) = 1, \quad y_2(0) = 3$

11. $y_1' = \quad y_1 + 2y_2$
$\quad\ y_2' = -8y_1 + 11y_2$
$\qquad y_1(0) = 1, \quad y_2(0) = 1$

12. $y_1' = y_2$
$\quad\ y_2' = y_1 + 3y_3 \qquad y_1(0) = 2, \quad y_2(0) = 0, \quad y_3(0) = 2$
$\quad\ y_3' = y_2$

13. $y_1' = 3y_1$
$\quad\ y_2' = 5y_1 + 4y_2 \qquad\qquad y_1(0) = 2, \quad y_2(0) = -10, \quad y_3(0) = -27$
$\quad\ y_3' = 3y_1 + 6y_2 + y_3$

Models of vibrating systems

14. In Example 1, find and graph the solutions satisfying the initial conditions $y_1(0) = 1, y_2(0) = 2, y_1'(0) = 0, y_2'(0) = 0$.

15. Show that for general m_1, m_2, k_1, k_2, the mechanical system in Fig. 171 is governed by $\mathbf{y}'' = \mathbf{A}\mathbf{y}$, where

$$\mathbf{A} = \begin{bmatrix} -\dfrac{k_1 + k_2}{m_1} & \dfrac{k_2}{m_1} \\[2ex] \dfrac{k_2}{m_2} & -\dfrac{k_2}{m_2} \end{bmatrix}.$$

16. Show that, for any positive k_1, k_2, m_1, m_2, the eigenvalues of the matrix in Prob. 15 are negative real. What does this imply with respect to motions of the system? Had this to be expected for physical reasons?

17. Find the solution in Prob. 15 when $m_1 = m_2 = 1, k_1 = 6, k_2 = 4, y_1(0) = 0,$ $y_2(0) = 0, y_1'(0) = \sqrt{2}, y_2'(0) = 2\sqrt{2}$.

18. Find the solution in Prob. 15 when $m_1 = m_2 = 1, k_1 = 9, k_2 = 6, y_1(0) = -2,$ $y_2(0) = 1, y_1'(0) = 0, y_2'(0) = 0$.

19. Show that the vertical vibrations of the mechanical system in Fig. 172 on the next page (masses of springs and damping neglected) are governed by the differential equations

$$y_1'' = -ky_1 + k(y_2 - y_1) \qquad \text{thus} \quad \mathbf{y}'' = \mathbf{A}\mathbf{y} \quad \text{where} \quad \mathbf{A} = \begin{bmatrix} -2k & k \\ k & -2k \end{bmatrix}.$$
$$y_2'' = -k(y_2 - y_1) - ky_2$$

20. To solve the vector equation in Prob. 19, substitute $\mathbf{y} = \mathbf{x}e^{\omega t}$. Show that this leads to the eigenvalue problem $\mathbf{A}\mathbf{x} = \lambda\mathbf{x}$ where $\lambda = \omega^2$. Find the eigenvalues and eigenvectors.

21. In Prob. 20, find the solution of the vector equation that satisfies the initial conditions $y_1(0) = 1, y_2(0) = 1, y_1'(0) = \sqrt{3k}, y_2'(0) = -\sqrt{3k}$.

22. In Prob. 19, the eigenvalues are negative real. Show that this still holds if we replace the springs by springs with any (positive) spring constants k_1, k_2, k_3.

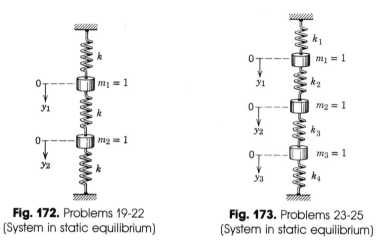

Fig. 172. Problems 19-22
(System in static equilibrium)

Fig. 173. Problems 23-25
(System in static equilibrium)

23. Find the mathematical model (system of differential equations in vectorial form) of the vibrating masses in Fig. 173 (no damping, mass of springs neglected).

24. Verify that the characteristic equation $\lambda^3 + a\lambda^2 + b\lambda + c = 0$ of the matrix in Prob. 23 has positive coefficients and conclude that the eigenvalues of the matrix must be negative real. What does this mean with respect to the motion of the mechanical system?

25. Let $\mathbf{A}$ be the matrix in Prob. 23, with $k_1 = k_2 = k_3 = k_4 = k$. Show that the corresponding matrix $\mathbf{B} = (1/k)\mathbf{A} + 2\mathbf{I}$ has the eigenvalues 0, $+\sqrt{2}$ and $-\sqrt{2}$. Conclude from this that $\mathbf{A}$ has the eigenvalues $-2k$, $(-2 + \sqrt{2})k$ and $(-2 - \sqrt{2})k$.

26. Derive the eigenvalues and eigenvectors in Example 2.

27. Solve Example 2 by reducing the system to a single equation.

28. Give a detailed proof of Theorem 2.

Further Proof in Chapter 7

Proof that the definition of a determinant in Sec. 7.10 is unambiguous
We show that the definition of a determinant

$$
(1) \qquad D = \det \mathbf{A} =
\begin{vmatrix}
a_{11} & a_{12} & \cdots & a_{1n} \\
a_{21} & a_{22} & \cdots & a_{2n} \\
\cdot & \cdot & \cdots & \cdot \\
\cdot & \cdot & \cdots & \cdot \\
a_{n1} & a_{n2} & \cdots & a_{nn}
\end{vmatrix}
$$

as given in Sec. 7.10 is unambiguous, that is, yields the same value of D no matter which rows or columns we choose. (Here we shall use formula numbers not yet used in Sec. 7.10.)

We shall prove first that *the same value is obtained no matter which row is chosen.*

The proof is by induction. The statement is true for a second-order determinant (see Example 1). Assuming that it is true for a determinant of order $n - 1$, we prove that it is true for a determinant D of order n.

For this purpose we expand D, given in the definition, in terms of each of two arbitrary rows, say the ith and the jth, and compare the results. Without loss of generality let us assume $i < j$.

First expansion. We expand D by the ith row. A typical term in this expansion is

$$(7) \qquad a_{ik}C_{ik} = a_{ik} \cdot (-1)^{i+k}M_{ik}.$$

The minor M_{ik} of a_{ik} in D is an $(n - 1)$th order determinant. By the induction hypothesis we may expand it by any row. We expand it by the row corresponding to the jth row of D. This row contains the entries a_{jl} $(l \neq k)$. It is the $(j - 1)$th row of M_{ik}, because M_{ik} does not contain entries of the ith row of D, and $i < j$. We have to distinguish between two cases as follows.

Case I. If $l < k$, then the entry a_{jl} belongs to the lth column of M_{ik} (cf. Fig. 174). Hence the term involving a_{jl} in this expansion is

$$(8) \qquad a_{jl} \cdot (\text{cofactor of } a_{jl} \text{ in } M_{ik}) = a_{jl} \cdot (-1)^{(j-1)+l}M_{ikjl}$$

where M_{ikjl} is the minor of a_{jl} in M_{ik}. Since this minor is obtained from M_{ik} by deleting the row and column of a_{jl}, it is obtained from D by deleting the ith and jth rows and the kth and lth columns of D. We insert the expansions of the M_{ik} into that of D. Then it follows from (7) and (8) that the terms of the resulting representation of D are of the form

$$(9a) \qquad a_{ik}a_{jl} \cdot (-1)^{b}M_{ikjl} \qquad (l < k)$$

where

$$b = i + k + j + l - 1.$$

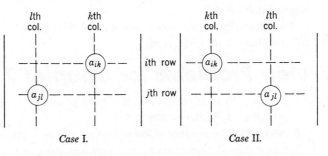

Case I.   Case II.

Fig. 174. Cases I and II of the two expansions of D

Case II. If $l > k$, the only difference is that then a_{jl} belongs to the $(l - 1)$th column of M_{ik}, because M_{ik} does not contain entries of the kth column of D, and $k < l$. This causes an additional minus sign in (8), and, instead of (9a), we therefore obtain

(9b) $$-a_{ik}a_{jl} \cdot (-1)^b M_{ikjl} \qquad\qquad (l > k)$$

where b is the same as before.

Second expansion. We now expand D at first by the jth row. A typical term in this expansion is

(10) $$a_{jl}C_{jl} = a_{jl} \cdot (-1)^{j+l} M_{jl}.$$

By the induction hypothesis we may expand the minor M_{jl} of a_{jl} in D by its ith row, which corresponds to the ith row of D, since $j > i$.

Case I. If $k > l$, the entry a_{ik} in that row belongs to the $(k - 1)$th column of M_{jl}, because M_{jl} does not contain entries of the lth column of D, and $l < k$ (cf. Fig. 169). Hence the term involving a_{ik} in this expansion is

(11) $$a_{ik} \cdot (\text{cofactor of } a_{ik} \text{ in } M_{jl}) = a_{ik} \cdot (-1)^{i+(k-1)} M_{ikjl},$$

where the minor M_{ikjl} of a_{ik} in M_{jl} is obtained by deleting the ith and jth rows and the kth and lth columns of D [and is, therefore, identical with M_{ikjl} in (8), so that our notation is consistent]. We insert the expansions of the M_{jl} into that of D. It follows from (10) and (11) that this yields a representation whose terms are identical with those given by (9a) when $l < k$.

Case II. If $k < l$, then a_{ik} belongs to the kth column of M_{jl}, we obtain an additional minus sign, and the result agrees with that characterized by (9b).

We have shown that the two expansions of D consist of the same terms, and this proves our statement concerning rows.

The proof of the statement concerning *columns* is quite similar; if we expand D in terms of two arbitrary columns, say, the kth and the lth, we find that the general term involving $a_{jl}a_{ik}$ is exactly the same as before. This proves that not only all column expansions of D yield the same value, but also that their common value is equal to the common value of the row expansions of D.

This completes the proof and shows that *our definition of an nth-order determinant is unambiguous.*   ∎

Review Problems for Chapter 7

1. If **A** is a 10 × 20 matrix and **B** a 20 × 10 matrix, is **A** + **B** defined? **A** + **B**$^\mathsf{T}$? **AB**? **BA**? **A**$^\mathsf{T}$**B**? (Give reason.)
2. Show that the product of lower triangular $n \times n$ matrices is lower triangular.
3. Show that all linear combinations of two given 3 × 4 matrices form a vector space.

4. Give an example of a system of linear equations (as small and simple as possible) that has no solutions.

5. How can the rank of a matrix be given in terms of row vectors? Column vectors? Determinants?

6. Why is the rank of a matrix important in connection with solving systems of linear equations?

7. If $\mathbf{A}$ is a 4×10 matrix, can the column vectors of $\mathbf{A}$ be linearly independent? The row vectors? (Give reason.)

8. What do you know about the existence and number of solutions of a nonhomogeneous system of linear equations? A homogeneous system?

9. What is the Gauss elimination good for? What is its basic idea? Why is this elimination generally better than Cramer's rule? What does pivoting mean?

10. What is the inverse of a matrix? When does it exist? How would you practically compute it?

11. What is a matrix eigenvalue problem and in what connections did it arise in this chapter?

12. Do there exist square matrices with no eigenvalues? Can a real matrix have complex eigenvalues? (Give reasons and examples.)

13. Can a matrix with a complex entry have only real eigenvalues?

14. Once you know an eigenvector of a matrix, how can you immediately find the corresponding eigenvalue?

15. Show that the transpose $\mathbf{A}^\mathsf{T}$ of a square matrix $\mathbf{A}$ has the same eigenvalues as $\mathbf{A}$. Do $\mathbf{A}$ and $\mathbf{A}^\mathsf{T}$ have the same eigenvectors?

$$\text{Let } \mathbf{a} = \begin{bmatrix} 1 \\ 0 \\ -2 \end{bmatrix}, \quad \mathbf{b} = \begin{bmatrix} 5 \\ -4 \\ 9 \end{bmatrix}, \quad \mathbf{C} = \begin{bmatrix} 2 & 1 & 0 \\ 1 & 0 & -2 \\ 0 & 3 & 2 \end{bmatrix}, \quad \mathbf{D} = \begin{bmatrix} 6 & -2 & 2 \\ 8 & 3 & 2 \\ 1 & 5 & -9 \end{bmatrix}. \quad \text{Find}$$

16. $\mathbf{a} + 3\mathbf{b}$

17. $10\mathbf{C} - 5\mathbf{D}$

18. $\mathbf{Ca}, \mathbf{a}^\mathsf{T}\mathbf{C}$

19. $\mathbf{Cb}$

20. $(\mathbf{C} - \mathbf{D})(\mathbf{a} - \mathbf{b})$

21. $\mathbf{a} + \mathbf{Db}$

22. $\mathbf{C} + \mathbf{C}^\mathsf{T}$

23. $\mathbf{CC}^\mathsf{T}$

24. $\mathbf{CD}, \mathbf{D}^\mathsf{T}\mathbf{C}^\mathsf{T}$

25. $\mathbf{C}^{-1}$

26. rank $\mathbf{D}$

27. rank $(\mathbf{b}^\mathsf{T}\mathbf{D})$

28. $\mathbf{a}^\mathsf{T}\mathbf{b}, \mathbf{ab}^\mathsf{T}$

29. $\mathbf{aa}^\mathsf{T}, \mathbf{a}^\mathsf{T}\mathbf{a}$

30. $\mathbf{DCab}^\mathsf{T}$

31. $\mathbf{C}^4$

32. Represent $\mathbf{C}$ as the sum of a symmetric and a skew-symmetric matrix.

33. Solve $\mathbf{Cx} = \mathbf{b}$.

34. Solve $\mathbf{Dx} = [36 \quad 39 \quad -46]^\mathsf{T}$.

Solve the following systems of equations or indicate that no solution exists.

35.
$$25x + 40y = -5$$
$$4x - 6y = 24$$
$$-8x - 12y = 0$$

36.
$$y + 6z = 0$$
$$12x + y - z = 0$$
$$-x + 4y - 5z = 0$$

37.
$$4x + 7y - 2z = 0$$
$$-8x + y + 4z = 0$$
$$6x - 9y - 3z = 0$$

38.
$$5x + 3y + 4z = 3$$
$$9x - 7y + z = -7$$
$$4y + 2z = 4$$

39.
$$4x + 2y - 6z = -6$$
$$5x + 3y - 8z = -9$$

40.
$$3x - 2y = -2$$
$$5x + 4y = 26$$
$$x - 3y = 8$$

41.
$$4y + 3z = -5$$
$$4x - 5z = 7$$
$$3x - 5y = 19$$

42.
$$6x + 4y = 4$$
$$8x - 6z = 7$$
$$-8y - 2z = -1$$

43.
$$2x + 5y - 12z = -16$$
$$5x - 3y + z = -9$$

Find the inverse or indicate that it does not exist. Check your result.

44. $\begin{bmatrix} 2 & -1 \\ 4 & 3 \end{bmatrix}$

45. $\begin{bmatrix} 0.6 & -0.4 \\ 0.4 & -0.1 \end{bmatrix}$

46. $\begin{bmatrix} -0.3 & 0.2 \\ 6 & -4 \end{bmatrix}$

47. $\begin{bmatrix} 0.8 & 0.6 \\ -0.6 & 0.8 \end{bmatrix}$

48. $\begin{bmatrix} 0 & 0 & 4 \\ 0 & 8 & 0 \\ 1 & 0 & 0 \end{bmatrix}$

49. $\begin{bmatrix} 1 & 1 & 1 \\ 1 & 1 & 1 \\ 1 & 1 & 1 \end{bmatrix}$

50. $\begin{bmatrix} 1 & 0 & 0 \\ 0 & 0 & 1 \\ 0 & 1 & 0 \end{bmatrix}$

51. $\begin{bmatrix} 1 & 3 & 5 \\ 0 & 1 & 2 \\ 0 & 0 & 1 \end{bmatrix}$

52. $\begin{bmatrix} 2 & 0 & 0 \\ 2 & 4 & 0 \\ 2 & 4 & 1 \end{bmatrix}$

53. $\begin{bmatrix} -0.7 & 0.2 & 0.3 \\ -1.3 & -0.2 & 0.7 \\ 0.8 & 0.2 & -0.2 \end{bmatrix}$

54. $\begin{bmatrix} \frac{1}{2} & -\frac{3}{2} & \frac{1}{2} \\ -\frac{1}{4} & 1 & -\frac{3}{4} \\ 0 & -1 & 3 \end{bmatrix}$

55. $\begin{bmatrix} 2 & 0 & 4 \\ -9 & 3 & 1 \\ -8 & 1 & 7 \end{bmatrix}$

Find the eigenvalues and eigenvectors.

56. $\begin{bmatrix} 6 & -2 \\ 9 & -5 \end{bmatrix}$

57. $\begin{bmatrix} 4 & 6 \\ -2 & -4 \end{bmatrix}$

58. $\begin{bmatrix} 1 & \frac{1}{2} \\ \frac{1}{2} & 1 \end{bmatrix}$

59. $\begin{bmatrix} 0 & 4 \\ 2 & 0 \end{bmatrix}$

60. $\begin{bmatrix} 3 & 2 \\ -5 & -3 \end{bmatrix}$

61. $\begin{bmatrix} 0 & i \\ -i & 0 \end{bmatrix}$

62. $\begin{bmatrix} 0 & i \\ i & 0 \end{bmatrix}$

63. $\begin{bmatrix} 1 & i \\ i & 1 \end{bmatrix}$

64. $\begin{bmatrix} 1 & 2 & 3 \\ 0 & 1 & 3 \\ 0 & 0 & 0 \end{bmatrix}$

65. $\begin{bmatrix} 0 & 2 & 0 \\ 3 & -2 & 3 \\ 0 & 3 & 0 \end{bmatrix}$

66. $\begin{bmatrix} 0 & 1 & 0 \\ 1 & 0 & 0 \\ 0 & 0 & 1 \end{bmatrix}$

67. $\begin{bmatrix} -16i & 0 & 12i \\ 0 & i & 0 \\ -24i & 0 & 18i \end{bmatrix}$

Solve

68. $y_1' = -2y_1 + 2y_2$
$ y_2' = -4y_1 - 8y_2$

69. $y_1' = 7y_2$
$ y_2' = 7y_1$

70. $y_1' = -5y_1 + 3.5y_2$
$ y_2' = -4.5y_1 + 3y_2$

71. $y_1' = -y_1 + 2y_2$
$ y_2' = -2y_1 + 4y_2$

72. $y_1' = 5y_1 + 11y_2$
$ y_2' = -y_1 - 7y_2$

73. $y_1' = 10y_1 - 5y_2$
$ y_2' = 36y_1 - 17y_2$

74. $y_1' = 2y_1 + y_2 + 3y_3$ **75.** $y_1' = 4y_2$ **76.** $y_1' = -9y_1 + 4y_2 - 4y_3$

$y_2' = y_2 + 3y_3$ $y_2' = 6y_1 - 4y_2 + 6y_3$ $y_2' = -15y_1 + 7y_2 - 5y_3$

$y_3' = 3y_3$ $y_3' = 6y_2$ $y_3' = 15y_1 - 6y_2 + 8y_3$

Find the currents in the following networks.

77.

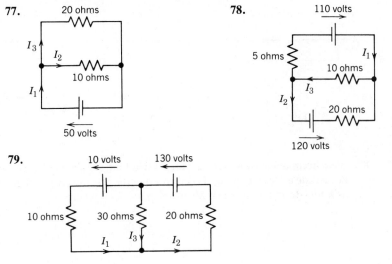

78.

79.

80. (Four-terminal network) Consider a *four-terminal network* (Fig. 175) in which an input signal is applied to one pair of terminals and the output signal is taken from the other pair. Assume that the network is linear, that is, the currents i_1 and i_2 are linear functions of u_1 and u_2, say,

$$
(1) \qquad \mathbf{i} = \begin{bmatrix} i_1 \\ i_2 \end{bmatrix} = \mathbf{Au} = \begin{bmatrix} a_{11} & a_{12} \\ a_{21} & a_{22} \end{bmatrix} \begin{bmatrix} u_1 \\ u_2 \end{bmatrix} = \begin{bmatrix} a_{11}u_1 + a_{12}u_2 \\ a_{21}u_1 + a_{22}u_2 \end{bmatrix}.
$$

Show that the input potential and current may be expressed as linear functions of the output potential and current, say,

$$
(2) \qquad \begin{aligned} u_1 &= t_{11}u_2 + t_{12}i_2, \\ i_1 &= t_{21}u_2 + t_{22}i_2, \end{aligned} \qquad \text{or} \qquad \mathbf{v}_1 = \mathbf{Tv}_2 \qquad \text{where} \qquad \mathbf{v}_1 = \begin{bmatrix} u_1 \\ i_1 \end{bmatrix}, \mathbf{v}_2 = \begin{bmatrix} u_2 \\ i_2 \end{bmatrix},
$$

and the **"transmission matrix"** $\mathbf{T}$ is

$$
\mathbf{T} = \begin{bmatrix} t_{11} & t_{12} \\ t_{21} & t_{22} \end{bmatrix} = \frac{1}{a_{21}} \begin{bmatrix} -a_{22} & 1 \\ -\det \mathbf{A} & a_{11} \end{bmatrix}.
$$

Hint. To verify the last expression, start from (1); express u_1 in terms of u_2 and i_2; then express i_1 in terms of u_2 and i_2; finally, compare the resulting representations with (2).

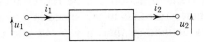

Fig. 175. Four-terminal network

81. Express **A** in terms of the elements of **T**.

In each case, show that the given transmission matrix **T** corresponds to the indicated four-terminal network.

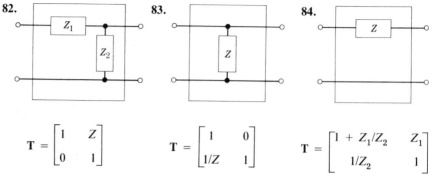

82.
$$\mathbf{T} = \begin{bmatrix} 1 & Z \\ 0 & 1 \end{bmatrix}$$

83.
$$\mathbf{T} = \begin{bmatrix} 1 & 0 \\ 1/Z & 1 \end{bmatrix}$$

84.
$$\mathbf{T} = \begin{bmatrix} 1 + Z_1/Z_2 & Z_1 \\ 1/Z_2 & 1 \end{bmatrix}$$

85. (Four-terminal networks in cascade) If two four-terminal networks are connected in cascade as shown in Fig. 176, prove that the resulting network may be regarded as a four-terminal network for which $\mathbf{v}_1 = \mathbf{T}\mathbf{v}_2$, where

$$\mathbf{v}_1 = \begin{bmatrix} u_1 \\ i_1 \end{bmatrix}, \qquad \mathbf{v}_2 = \begin{bmatrix} u_2 \\ i_2 \end{bmatrix}, \qquad \mathbf{T} = \mathbf{T}_1\mathbf{T}_2.$$

Using this result, obtain the matrix in Prob. 84 from those in Prob. 82 (with $Z = Z_1$) and Prob. 83 (with $Z = Z_2$).

Fig. 176. Four-terminal networks in cascade

Summary of Chapter 7
Matrices and Determinants

An $m \times n$ matrix $\mathbf{A} = [a_{jk}]$ is a rectangular array of numbers ("entries" or "elements") arranged in m rows and n columns. Matrix algebra extends vector algebra: addition and multiplication by scalars (Sec. 7.2) are defined so that the $m \times n$ matrices (m, n fixed!) form a vector space (of dimension mn). **Multiplication** of matrices ("row times column"; Sec. 7.3) is motivated by the composition of **linear transformations**. A matrix **A** defines a linear transformation $\mathbf{y} = \mathbf{A}\mathbf{x}$, and every linear transformation can be represented by matrices that depend on the choice of bases in the space in which **x** and **y** vary. Multiplication is now defined so that, if $\mathbf{y} = \mathbf{A}\mathbf{x}$ and $\mathbf{x} = \mathbf{B}\mathbf{w}$, then $\mathbf{y} = \mathbf{C}\mathbf{w}$, where $\mathbf{C} = \mathbf{A}\mathbf{B}$. Multiplication is associative, but is *not commutative*: if **AB** is defined, **BA** may not be defined, but even if **BA** is defined, **AB** $\neq$ **BA** in general. Also **AB** = **0** may not imply **A** = **0** or **B** = **0** or **BA** = **0** (Secs. 7.3, 7.8):

$$\begin{bmatrix} 1 & 1 \\ 2 & 2 \end{bmatrix} \begin{bmatrix} -1 & 1 \\ 1 & -1 \end{bmatrix} = \begin{bmatrix} 0 & 0 \\ 0 & 0 \end{bmatrix}, \quad \begin{bmatrix} -1 & 1 \\ 1 & -1 \end{bmatrix} \begin{bmatrix} 1 & 1 \\ 2 & 2 \end{bmatrix} = \begin{bmatrix} 1 & 1 \\ -1 & -1 \end{bmatrix}$$

$$\begin{bmatrix} 1 & 2 \end{bmatrix} \begin{bmatrix} 3 \\ 4 \end{bmatrix} = \begin{bmatrix} 11 \end{bmatrix}, \quad \begin{bmatrix} 3 \\ 4 \end{bmatrix} \begin{bmatrix} 1 & 2 \end{bmatrix} = \begin{bmatrix} 3 & 6 \\ 4 & 8 \end{bmatrix}.$$

A main application of matrices concerns **systems of linear equations** (Sec. 7.5)

(1) $$\mathbf{Ax} = \mathbf{b}$$

(m equations in n unknowns $x_1, \cdots, x_n$; **b** given). The most important method of solution is the **Gauss elimination** (Sec. 7.5), which reduces the system to "triangular" form by *elementary row operations,* which leave the set of solutions unchanged. The *Gauss–Jordan elimination* is practical for finding the **inverse** $\mathbf{A}^{-1}$ of a square matrix **A** (Sec. 7.8) but not for solving systems, since it requires more operations than the Gauss method with back substitution. (Numerical aspects and variants, such as *Doolittle's method* are discussed in Secs. 19.1 and 19.2.) *Cramer's rule* (Sec. 7.11) represents the unknowns in (1) as quotients of determinants; for numerical work, it is impractical. **Determinants** (Secs. 7.9, 7.10) have decreased in importance, but will retain their place in eigenvalue problems, elementary geometry, etc.

The **rank** r of a matrix **A** is the maximum number of linearly independent rows or columns of **A**, equivalently, the number of rows of the largest square submatrix of **A** with nonzero determinant (Secs. 7.6, 7.11). The system (1) has solutions if and only if rank **A** = rank [**A** **b**], [**A** **b**] the *augmented matrix* (Fundamental Theorem, Sec. 7.7). The *homogeneous system*

(2) $$\mathbf{Ax} = \mathbf{0}$$

has solutions $\mathbf{x} \neq \mathbf{0}$ ("nontrivial solutions") if and only if rank **A** $< n$, in the case $m = n$ equivalently if and only if det **A** = 0 (Secs. 7.7, 7.11).

Hence the system

(3) $$\mathbf{Ax} = \lambda\mathbf{x} \qquad \text{or} \qquad (\mathbf{A} - \lambda\mathbf{I})\mathbf{x} = \mathbf{0}$$

has solutions $\mathbf{x} \neq \mathbf{0}$ if and only if λ is a root of the *characteristic equation* (Sec. 7.12)

(4) $$\det (\mathbf{A} - \lambda\mathbf{I}) = 0.$$

Such a (real or complex) number λ is called an **eigenvalue** of **A** and that solution $\mathbf{x} \neq \mathbf{0}$ an **eigenvector** of **A** corresponding to this λ. Equation (3) is called an (algebraic) **eigenvalue problem** (Sec. 7.12). Eigenvalue problems are of great importance in physics and engineering, and they also have applications in economics and statistics. They are basic in solving systems of differential equations (Sec. 7.16).

The *transpose* $\mathbf{A}^\mathsf{T}$ of a matrix $\mathbf{A} = [a_{jk}]$ is $\mathbf{A}^\mathsf{T} = [a_{kj}]$, rows become columns and conversely (Sec. 7.4). The *complex conjugate* of **A** is $\overline{\mathbf{A}} = [\overline{a}_{jk}]$. Six classes of square matrices of practical importance arise from this: a real matrix **A** is called *real* **symmetric** if $\mathbf{A}^\mathsf{T} = \mathbf{A}$, *real* **skew-symmetric** if $\mathbf{A}^\mathsf{T} = -\mathbf{A}$, **orthogonal** if $\mathbf{A}^\mathsf{T} = \mathbf{A}^{-1}$ (Secs. 7.4, 7.13); a complex matrix is called **Hermitian** if $\overline{\mathbf{A}}^\mathsf{T} = \mathbf{A}$, **skew-Hermitian** if $\overline{\mathbf{A}}^\mathsf{T} = -\mathbf{A}$, **unitary** if $\overline{\mathbf{A}}^\mathsf{T} = \mathbf{A}^{-1}$. (Sec. 7.13). The eigenvalues of Hermitian (and real-symmetric) matrices are real; those of skew-Hermitian (and real skew-symmetric) are pure imaginary or 0; those of unitary (and orthogonal) matrices have absolute value 1 (Sec. 7.14).

The diagonalization of matrices and the transformation of quadratic forms to principal axes are discussed in Sec. 7.15.

Chapter 8

Vector Differential Calculus

Vector calculus concerns differentiation (this chapter) and integration (Chap. 9) of vector functions. Vector functions represent vector fields (Sec. 8.1), which have various physical and geometrical applications. The basic concepts of differential calculus can be extended to vector functions in a simple and natural fashion (Sec. 8.2). Vector functions are useful for representing and investigating curves (Secs. 8.3—8.6) and their application in mechanics as paths of moving bodies (Sec. 8.5). Physically and geometrically important concepts in connection with scalar and vector fields are the gradient (Sec. 8.8), divergence (Sec. 8.9) and curl (Sec. 8.10). (Corresponding integral theorems will be considered in Chap. 9.) In the last section (Sec. 8.11) we show how the gradient, divergence, curl and Laplacian can be transformed into general curvilinear coordinates.

Prerequisite for this chapter: Chap. 6.
Sections that may be omitted in a shorter course: 8.4—8.6, 8.11.
References: Appendix 1, Part B.
Answers to problems: Appendix 2.

8.1 Scalar Fields and Vector Fields

A **scalar function** is a function that is defined at each point of a certain set of points in space and whose values are real numbers depending only on the points in space but not on the particular choice of the coordinate system. In most applications the domain of definition D of a scalar function f will be a curve, a surface or a three-dimensional region in space. The function f associates with each point in D a scalar, a real number, and we say that a **scalar field** is given in D.

If we introduce coordinates x, y, z, then f may be represented in terms of the coordinates, and we write $f(x, y, z)$, but we should keep in mind that the value of f at any point P is independent of the particular choice of coordinates. To indicate this fact it is also customary to write $f(P)$ instead of $f(x, y, z)$. The function f may also depend on parameters such as time.

EXAMPLE 1. Scalar function

The distance $f(P)$ of any point P from a fixed point P_0 in space is a scalar function whose domain of definition D is the whole space. $f(P)$ defines a scalar field in space. If we introduce a Cartesian coordinate system and P_0 has the coordinates x_0, y_0, z_0, then f is given by the

457

well-known formula

$$f(P) = f(x, y, z) = \sqrt{(x - x_0)^2 + (y - y_0)^2 + (z - z_0)^2}$$

where x, y, z are the coordinates of P. If we replace the given Cartesian coordinate system by another such system, then the values of the coordinates of P and P_0 will in general change, but $f(P)$ will have the same value as before. Hence $f(P)$ is a scalar function. The direction cosines of the line through P and P_0 are not scalars because their values will depend on the choice of the coordinate system.

EXAMPLE 2. Scalar fields
The temperature T within a body B is a scalar function. It defines a scalar field, namely, the temperature field in B. The function T may depend on time or other parameters. Other examples of scalar fields are the pressure within a region through which a compressible fluid is flowing, and the density of the air of the earth's atmosphere.   ∎

Vector Function, Vector Field

A **vector function** $\mathbf{v}(P)$ is a function that is defined on some point set D in space (for instance, on the set of the points of a curve, a surface or a three-dimensional region) and associates with each point P in D a vector $\mathbf{v}(P)$, and we say that a **vector field** is given in D. Some typical illustrative examples are shown in Figs. 177—180.

If we introduce Cartesian coordinates x, y, z, then we may write our vector function in terms of component functions,

$$\mathbf{v}(x, y, z) = [v_1(x, y, z), \quad v_2(x, y, z), \quad v_3(x, y, z)]$$

or, using $\mathbf{i}$, $\mathbf{j}$, $\mathbf{k}$ as in Sec. 6.3,

$$\mathbf{v}(x, y, z) = v_1(x, y, z)\mathbf{i} + v_2(x, y, z)\mathbf{j} + v_3(x, y, z)\mathbf{k},$$

but we should keep in mind that $\mathbf{v}$ depends only on the points of its domain of definition, and at any such point defines the same vector for every choice of the coordinate system. Our notations are the same as in Chap. 6; the only difference is that the components v_1, v_2, v_3 of $\mathbf{v}$ now become functions of x, y, z, since $\mathbf{v}$ is a function of x, y, z.

EXAMPLE 3. Vector field (Velocity field)
At any instant the velocity vectors $\mathbf{v}(P)$ of a rotating body B constitute a vector field, the so-called **velocity field** of the rotation. If we introduce a Cartesian coordinate system having the origin on the axis of rotation, then (cf. Example 2 in Sec. 6.7)

$$(1) \qquad \mathbf{v}(x, y, z) = \mathbf{w} \times [x, \quad y, \quad z] = \mathbf{w} \times (x\mathbf{i} + y\mathbf{j} + z\mathbf{k}),$$

Fig. 177. Field of tangent vectors   **Fig. 178.** Field of normal vectors
of a curve   of a surface

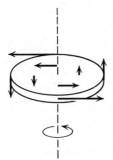

Fig. 179. Velocity field
of a rotating body

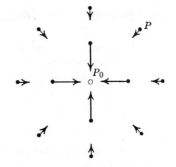

Fig. 180. Gravitational field

where x, y, z are the coordinates of any point P of B at the instant under consideration. If the coordinates are such that the z-axis is the axis of rotation and $\mathbf{w}$ points in the positive z-direction, then $\mathbf{w} = \omega\mathbf{k}$ and

$$\mathbf{v} = \begin{vmatrix} \mathbf{i} & \mathbf{j} & \mathbf{k} \\ 0 & 0 & \omega \\ x & y & z \end{vmatrix} = \omega(-y\mathbf{i} + x\mathbf{j}) = \omega\,[-y, \quad x, \quad 0].$$

An example of a rotating body and the corresponding velocity field are shown in Fig. 179.

EXAMPLE 4. Vector field (Field of force)

Let a particle A of mass M be fixed at a point P_0 and let a particle B of mass m to be free to take up various positions P in space. Then A attracts B. According to **Newton's law of gravitation** the corresponding gravitational force $\mathbf{p}$ is directed from P to P_0, and its magnitude is proportional to $1/r^2$ where r is the distance between P and P_0, say,

$$(2) \qquad\qquad |\mathbf{p}| = \frac{c}{r^2}, \qquad\qquad c = GMm,$$

where $G\ (= 6.67 \cdot 10^{-8}\ \mathrm{cm}^3/\mathrm{gm} \cdot \sec^2)$ is the gravitational constant. Hence $\mathbf{p}$ defines a vector field in space. If we introduce Cartesian coordinates such that P_0 has the coordinates x_0, y_0, z_0 and P has the coordinates x, y, z, then by the Pythagorean theorem,

$$r = \sqrt{(x - x_0)^2 + (y - y_0)^2 + (z - z_0)^2} \qquad (\geqq 0).$$

Assuming that $r > 0$ and introducing the vector

$$\mathbf{r} = [x - x_0, \quad y - y_0, \quad z - z_0] = (x - x_0)\mathbf{i} + (y - y_0)\mathbf{j} + (z - z_0)\mathbf{k},$$

we have $|\mathbf{r}| = r$, and $(-1/r)\mathbf{r}$ is a unit vector in the direction of $\mathbf{p}$; the minus sign indicates that $\mathbf{p}$ is directed from P to P_0 (Fig. 180). From this and (2) we obtain

$$(3) \qquad \mathbf{p} = |\mathbf{p}|\left(-\frac{1}{r}\,\mathbf{r}\right) = -\frac{c}{r^3}\,\mathbf{r} = -c\,\frac{x - x_0}{r^3}\,\mathbf{i} - c\,\frac{y - y_0}{r^3}\,\mathbf{j} - c\,\frac{z - z_0}{r^3}\,\mathbf{k}.$$

This vector function describes the gravitational force acting on B. ∎

Having introduced the concept of a vector function, we define in the next section the concepts of differential calculus for such functions, in particular the **derivative.** This will be a straightforward extension of familiar definitions of calculus.

Problems for Sec. 8.1

Level curves. Determine the *level curves* $T(x, y) = const$ (curves of constant temperature or *isotherms*) of the temperature fields given by the following functions. Sketch some of these isotherms.

1. $T = xy$

2. $T = y$

3. $T = \ln (x^2 + y^2)$

4. $T = x + 3y$

5. $T = 5x - 2y$

6. $T = x^2 - y^2$

7. $T = x^2 + 2x - y^2$

8. $T = x^2 - y^2 + 4y$

9. $T = \text{arc tan } (y/x)$

Consider the scalar field (pressure field) determined by $f(x, y) = 9x^2 + y^2$. Find:

10. The pressure at the points $(1, 1)$, $(-4, \frac{1}{2})$, and $(-8, 2)$.

11. The level curves (curves of constant pressure or *isobars*). Sketch some of them.

12. The pressure at the points of the hyperbola $x^2 - y^2 = 1$.

13. The region in which $1 \leq f(x, y) \leq 9$.

Level surfaces. Find the *level surfaces* $f(x, y, z) = const$ of the scalar fields in space given by the following functions.

14. $f = x + y + z$

15. $f = x^2 + y^2 + z^2$

16. $f = x^2 + 4y^2 + 9z^2$

17. $f = x^2 + y^2$

18. $f = x^2 + y^2 - z$

19. $f = (x^2 + y^2 + z^2)^{-1}$

20. $f = \sin xy$

21. $f = z - \sqrt{x^2 + y^2}$

22. $f = y^2 - z$

Consider the pressure field in space given by $f(x, y, z) = x^2 + 4y^2 + z^2$. Find:

23. The level surfaces. (Sketch some of them.)

24. The level curves in the plane $z = 1$. (Sketch some of them.)

25. A formula for the pressure in the plane $z = x + y$.

26. The region in which $f(x, y, z) > 4$.

Vector fields. Sketch figures (similar to Fig. 180) of the vector fields in the xy-plane given by the following vector functions.

27. $\mathbf{v} = 3\mathbf{j}$

28. $\mathbf{v} = \mathbf{i} - \mathbf{j}$

29. $\mathbf{v} = x\mathbf{i} + \mathbf{j}$

30. $\mathbf{v} = 2x\mathbf{i} + 2y\mathbf{j}$

31. $\mathbf{v} = y\mathbf{i} - x\mathbf{j}$

32. $\mathbf{v} = x^2\mathbf{i} + y^2\mathbf{j}$

33. $\mathbf{v} = -(x^2 + y^2)\mathbf{j}$

34. $\mathbf{v} = -y\mathbf{i} + x\mathbf{j}$

Find and sketch the curves on which $\mathbf{v}$ has constant length and the curves on which $\mathbf{v}$ has constant direction, where

35. $\mathbf{v} = x\mathbf{i} + y\mathbf{j}$

36. $\mathbf{v} = x\mathbf{i} + 2y\mathbf{j}$

37. $\mathbf{v} = x^2\mathbf{i} - y^2\mathbf{j}$

38. $\mathbf{v} = (x + y)(\mathbf{i} + \mathbf{j})$

39. $\mathbf{v} = \mathbf{i} + (x^2 + y^2)\mathbf{j}$

40. $\mathbf{v} = (x - y)\mathbf{i} + (x + y)\mathbf{j}$

Vector Calculus

The basic concepts of calculus, such as convergence, continuity and differentiability, can be introduced to vector analysis in a simple and natural way as follows.

Convergence. An infinite sequence of vectors $\mathbf{a}_{(n)}$, $n = 1, 2, \cdots$, is said to converge if there is a vector $\mathbf{a}$ such that

(1)
$$\lim_{n \to \infty} |\mathbf{a}_{(n)} - \mathbf{a}| = 0.$$

a is called the **limit vector** of that sequence, and we write

$$(2) \qquad \lim_{n \to \infty} \mathbf{a}_{(n)} = \mathbf{a}.$$

Clearly, if a Cartesian coordinate system has been introduced, then that sequence of vectors converges to **a** if and only if the three sequences of the components of the vectors converge to the corresponding components of **a**. The simple proof is left to the reader.

Similarly, a vector function $\mathbf{v}(t)$ of a real variable t is said to have the **limit** *l* as t approaches t_0, if $\mathbf{v}(t)$ is defined in some *neighborhood*[1] of t_0 (possibly except at t_0) and

$$(3) \qquad \lim_{t \to t_0} |\mathbf{v}(t) - l| = 0.$$

Then we write

$$(4) \qquad \lim_{t \to t_0} \mathbf{v}(t) = l.$$

Continuity. A vector function $\mathbf{v}(t)$ is said to be **continuous** at $t = t_0$ if it is defined in some neighborhood of t_0 and

$$(5) \qquad \lim_{t \to t_0} \mathbf{v}(t) = \mathbf{v}(t_0).$$

If we introduce a Cartesian coordinate system, we may write

$$\mathbf{v}(t) = [v_1(t), \quad v_2(t), \quad v_3(t)] = v_1(t)\mathbf{i} + v_2(t)\mathbf{j} + v_3(t)\mathbf{k}.$$

Then $\mathbf{v}(t)$ is continuous at t_0 if and only if its three components are continuous at t_0.

Derivative of a Vector Function

A vector function $\mathbf{v}(t)$ is said to be **differentiable** at a point t if the limit

$$(6) \qquad \boxed{\mathbf{v}'(t) = \lim_{\Delta t \to 0} \frac{\mathbf{v}(t + \Delta t) - \mathbf{v}(t)}{\Delta t}}$$

exists. The vector $\mathbf{v}'(t)$ is called the **derivative** of $\mathbf{v}(t)$. Cf. Fig. 181. (The

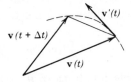

Fig. 181. Derivative of a vector function

---

[1]That is, in some interval on the t-axis containing t_0 as an interior point.

curve in this figure is the locus of the heads of the arrows representing **v** for values of the independent variable in some interval containing t and $t + \Delta t$.)

Suppose that a Cartesian coordinate system has been introduced. Then **v**(t) is differentiable at a point t if and only if its three components are differentiable at that t, that is, if and only if the derivatives

$$v_m{}'(t) = \lim_{\Delta t \to 0} \frac{v_m(t + \Delta t) - v_m(t)}{\Delta t} \qquad (m = 1, 2, 3)$$

exist. Then

(7) $$\mathbf{v}'(t) = [v_1{}'(t), \quad v_2{}'(t), \quad v_3{}'(t)];$$

that is, *to differentiate a vector function one differentiates each component separately.*

The familiar rules of differentiation yield corresponding rules for differentiating vector functions, for example,

$$(c\mathbf{v})' = c\mathbf{v}' \quad (c \text{ constant}), \qquad (\mathbf{u} + \mathbf{v})' = \mathbf{u}' + \mathbf{v}'$$

and in particular

(8) $$(\mathbf{u} \cdot \mathbf{v})' = \mathbf{u}' \cdot \mathbf{v} + \mathbf{u} \cdot \mathbf{v}'$$

(9) $$(\mathbf{u} \times \mathbf{v})' = \mathbf{u}' \times \mathbf{v} + \mathbf{u} \times \mathbf{v}'$$

(10) $$(\mathbf{u} \quad \mathbf{v} \quad \mathbf{w})' = (\mathbf{u}' \quad \mathbf{v} \quad \mathbf{w}) + (\mathbf{u} \quad \mathbf{v}' \quad \mathbf{w}) + (\mathbf{u} \quad \mathbf{v} \quad \mathbf{w}').$$

The simple proofs are left to the reader. In (9), the order of the vectors must be carefully observed because cross multiplication is not commutative.

EXAMPLE 1. Derivative of a vector function of constant length
Let **v**(t) be a vector function whose length is constant, say, $|\mathbf{v}(t)| = c$. Then $|\mathbf{v}|^2 = \mathbf{v} \cdot \mathbf{v} = c^2$, and $(\mathbf{v} \cdot \mathbf{v})' = 2\mathbf{v} \cdot \mathbf{v}' = 0$, by differentiation [cf. (8)]. This yields the following result. *The derivative of a vector function* **v**(t) *of constant length is either the zero vector or is perpendicular to* **v**(t). ∎

Important applications of derivatives will be discussed in the following sections.

Partial Derivatives of a Vector Function

From our considerations, the way of introducing partial differentiation to vector analysis is obvious. Indeed, let the components of a vector function

$$\mathbf{v} = [v_1, \quad v_2, \quad v_3] = v_1\mathbf{i} + v_2\mathbf{j} + v_3\mathbf{k}$$

be differentiable functions of n variables $t_1, \cdots, t_n$. Then the **partial derivative** of **v** with respect to t_l is denoted by $\partial \mathbf{v}/\partial t_l$ and is defined as the vector function

$$\frac{\partial \mathbf{v}}{\partial t_l} = \frac{\partial v_1}{\partial t_l}\mathbf{i} + \frac{\partial v_2}{\partial t_l}\mathbf{j} + \frac{\partial v_3}{\partial t_l}\mathbf{k}.$$

Similarly,

$$\frac{\partial^2 \mathbf{v}}{\partial t_l \partial t_m} = \frac{\partial^2 v_1}{\partial t_l \partial t_m}\, \mathbf{i} + \frac{\partial^2 v_2}{\partial t_l \partial t_m}\, \mathbf{j} + \frac{\partial^2 v_3}{\partial t_l \partial t_m}\, \mathbf{k},$$

and so on.

EXAMPLE 2. Partial derivatives

Let $\mathbf{r}(t_1, t_2) = a \cos t_1\, \mathbf{i} + a \sin t_1\, \mathbf{j} + t_2 \mathbf{k}$. Then

$$\frac{\partial \mathbf{r}}{\partial t_1} = -a \sin t_1\, \mathbf{i} + a \cos t_1\, \mathbf{j}, \qquad \frac{\partial \mathbf{r}}{\partial t_2} = \mathbf{k}.$$

Note that if $\mathbf{r}(t_1, t_2)$ is interpreted as a position vector, it represents a cylinder of revolution of radius a, having the z-axis as axis of rotation. (Representations of surfaces will be considered in Sec. 9.4.)   ∎

The present section contained the basic concepts of differential calculus for vector functions. In the next section we apply these concepts to a discussion of **curves** in space (and in the plane as a special case). The student should study the next section with particular care, since curves are of interest in themselves as well as in mechanics, where they occur as paths of moving bodies (as we show in Sec. 8.5) and in vector integral calculus (Chap. 9), where they take the place of intervals of integration as used in ordinary calculus.

Problems for Sec. 8.2

Derivatives. Find the first and second derivatives of the following vector functions and the lengths of these derivatives.

1. $\mathbf{a} + t\mathbf{b}$
2. $\mathbf{a} + t^2\mathbf{b}$
3. $(1 + t)\mathbf{i} + (3 + 4t)\mathbf{j}$
4. $(5 - t)\mathbf{i} + (-5 + 2t)\mathbf{j}$
5. $t^2\mathbf{i} + 3t^2\mathbf{j}$
6. $2t\mathbf{i} + t^2\mathbf{j}$
7. $\cos t\, \mathbf{i} + \sin t\, \mathbf{j}$
8. $c \cos t\, \mathbf{i} + c \sin t\, \mathbf{j}$
9. $\cos t\, \mathbf{i} + \sin t\, \mathbf{j} + t\mathbf{k}$
10. $4 \cos t\, \mathbf{i} + \sin t\, \mathbf{j}$
11. $t\mathbf{i} + t^2\mathbf{j} + t^3\mathbf{k}$
12. $e^t\mathbf{i} + e^{-t}\mathbf{j}$

Derivatives of products. Let $\mathbf{u} = t\mathbf{i} + 2t^2\mathbf{k}$, $\mathbf{v} = t^3\mathbf{j} + t\mathbf{k}$, $\mathbf{w} = \mathbf{i} + t\mathbf{j} + t^2\mathbf{k}$. Find

13. $(\mathbf{u} \cdot \mathbf{v})'$, $(\mathbf{v} \cdot \mathbf{u})'$
14. $(\mathbf{u} \times \mathbf{v})'$, $(\mathbf{v} \times \mathbf{u})'$
15. $(\mathbf{u}\ \mathbf{v}\ \mathbf{w})'$, $(\mathbf{w}\ \mathbf{u}\ \mathbf{v})'$
16. $[(\mathbf{u} - 2\mathbf{v}) \cdot \mathbf{w}]'$
17. $[(\mathbf{u} + \mathbf{v}) \cdot \mathbf{w}]'$
18. $(4\mathbf{w} \times \mathbf{u})'$, $-4(\mathbf{u} \times \mathbf{w})'$
19. $(2\mathbf{u} \cdot 3\mathbf{w})'$, $6(\mathbf{w} \cdot \mathbf{u})'$
20. $[\mathbf{u} \times (\mathbf{v} \times \mathbf{w})]'$
21. $[(\mathbf{u} \times \mathbf{v}) \times \mathbf{w}]'$

Partial derivatives. Find the first partial derivatives with respect to x, y, z:

22. $yz\mathbf{i} + xz\mathbf{j}$
23. $x\mathbf{i} + 2y\mathbf{j}$
24. $\frac{1}{2}x^2\mathbf{j} - xyz^2\mathbf{k}$
25. $z\mathbf{i} + y \sin x\, \mathbf{j} + e^y\mathbf{k}$
26. $y^2\mathbf{i} + z^2\mathbf{j} + x^2\mathbf{k}$
27. $(x + y)\mathbf{j} + (x - y)\mathbf{k}$
28. $e^x\mathbf{i} + e^{yz}\mathbf{j} + e^{xy}\mathbf{k}$
29. $\cos yz\, \mathbf{i}$
30. $e^x \sin z\, \mathbf{i} - 2z^2\mathbf{k}$

31. Using (8) in Sec. 6.5, prove (8).
32. Prove (9).
33. Derive (10) from (8) and (9).
34. Prove (10) directly.
35. Find formulas similar to (8) and (9) for $(\mathbf{u} \cdot \mathbf{v})''$ and $(\mathbf{u} \times \mathbf{v})''$.
36. Show that if $\mathbf{v}(t)$ is a unit vector and $\mathbf{v}'(t) \neq \mathbf{0}$, then $\mathbf{v}$ and $\mathbf{v}'$ are orthogonal.
37. Show that the equation $\mathbf{v}'(t) = \mathbf{c}$ has the solution $\mathbf{v}(t) = \mathbf{c}t + \mathbf{b}$ where $\mathbf{c}$ and $\mathbf{b}$ are constant vectors.
38. Show that $\mathbf{v}(t) = \mathbf{b}e^{\lambda t} + \mathbf{c}e^{-\lambda t}$ satisfies the equation $\mathbf{v}'' - \lambda^2\mathbf{v} = \mathbf{0}$.
39. Show that $\left(\dfrac{\mathbf{v}}{|\mathbf{v}|}\right)' = \dfrac{\mathbf{v}'(\mathbf{v} \cdot \mathbf{v}) - \mathbf{v}(\mathbf{v} \cdot \mathbf{v}')}{(\mathbf{v} \cdot \mathbf{v})^{3/2}}$.
40. Differentiate $(t\mathbf{i} + t^2\mathbf{j})/|t\mathbf{i} + t^2\mathbf{j}|$.

8.3 Curves

As an important application of vector calculus, let us now consider some basic facts about curves in space. The student will know that curves occur in various problems in calculus as well as in physics, for example, as paths of moving particles. We mention that the study of curves and surfaces in space by means of calculus is an important branch of mathematics which is called **differential geometry.**

A Cartesian coordinate system being given, we may represent a curve C by a vector function (Fig. 182)

(1) $$\boxed{\mathbf{r}(t) = [x(t),\quad y(t),\quad z(t)] = x(t)\mathbf{i} + y(t)\mathbf{j} + z(t)\mathbf{k};}$$

to each value t_0 of the real variable t there corresponds a point of C having the position vector $\mathbf{r}(t_0)$, that is, the coordinates $x(t_0)$, $y(t_0)$, $z(t_0)$.

A representation of the form (1) is called a **parametric representation** of the curve C, and t is called the *parameter* of this representation. This type of representation is useful in many applications, for example, in mechanics, where the variable t may be time.

Other types of representations of curves in space are

(2) $$y = f(x),\qquad z = g(x)$$

and

(3) $$F(x, y, z) = 0,\qquad G(x, y, z) = 0.$$

In (2), the function $y = f(x)$ is the projection of the curve in the xy-plane, and $z = g(x)$ is the projection of the curve in the xz-plane. In (3), each equation represents a surface, and the curve is the intersection of these two surfaces.

A **plane curve** is a curve that lies in a plane in space. A curve that is not a plane curve is called a **twisted curve.**

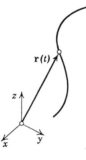

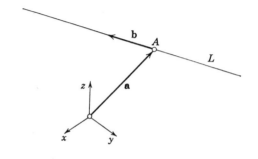

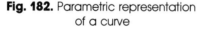

Fig. 182. Parametric representation of a curve

Fig. 183. Parametric representation of a straight line

EXAMPLE 1. Straight line

Any straight line L can be represented in the form

(4) $\mathbf{r}(t) = \mathbf{a} + t\mathbf{b} = (a_1 + tb_1)\mathbf{i} + (a_2 + tb_2)\mathbf{j} + (a_3 + tb_3)\mathbf{k}$

or, in the other notation,

$$\mathbf{r}(t) = [a_1 + tb_1, \quad a_2 + tb_2, \quad a_3 + tb_3],$$

where $\mathbf{a}$ and $\mathbf{b}$ are constant vectors. L passes through the point A with position vector $\mathbf{r} = \mathbf{a}$ and has the direction of $\mathbf{b}$ (Fig. 183). If $\mathbf{b}$ is a unit vector, its components are the **direction cosines** of L, and in this case, $|t|$ measures the distance of the points of L from A. For instance, the straight line in the xy-plane through A: (3, 2) having slope 1 is given by

$$\mathbf{r}(t) = [3, \quad 2, \quad 0] + t[1, \quad 1, \quad 0] = [3 + t, \quad 2 + t, \quad 0].$$

EXAMPLE 2. Ellipse, circle

The vector function

(5) $\mathbf{r}(t) = a \cos t\, \mathbf{i} + b \sin t\, \mathbf{j}$

represents an ellipse in the xy-plane with center at the origin and principal axes in the direction of the x and y axes. In fact, since $\cos^2 t + \sin^2 t = 1$, we obtain from (5)

$$\frac{x^2}{a^2} + \frac{y^2}{b^2} = 1, \qquad z = 0.$$

If $b = a$, then (5) represents a *circle* of radius a. For instance,

$$\mathbf{r}(t) = 4 \cos t\, \mathbf{i} + 2 \sin t\, \mathbf{j} \qquad \text{represents the ellipse} \qquad \tfrac{1}{16}x^2 + \tfrac{1}{4}y^2 = 1$$

with semiaxes 4 and 2.

EXAMPLE 3. Circular helix

The twisted curve C represented by the vector function

(6) $\mathbf{r}(t) = a \cos t\, \mathbf{i} + a \sin t\, \mathbf{j} + ct\mathbf{k}$ $(c \neq 0)$

is called a *circular helix*. It lies on the cylinder $x^2 + y^2 = a^2$. If $c > 0$, the helix is shaped like a right-handed screw (Fig. 184). If $c < 0$, it looks like a left-handed screw (Fig. 185). For instance, the helix

$$\mathbf{r}(t) = \cos t\, \mathbf{i} + \sin t\, \mathbf{j} + t\mathbf{k}$$

is right-handed, lies on the cylinder $x^2 + y^2 = 1$ and has pitch 2π; that is, the z-coordinates of points of the helix vertically above each other differ by integer multiples of 2π.   ∎

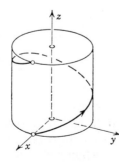

Fig. 184. Right-handed circular helix

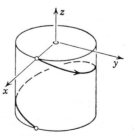

Fig. 185. Left-handed circular helix

The portion between any two points of a curve is often called an **arc** *of a curve*. For the sake of simplicity we shall use the single term "curve" to denote an entire curve as well as an arc of a curve.

A curve may have points at which it intersects or touches itself. Such a point is called a *multiple point* of a curve. Examples are shown in Fig. 186. A curve without multiple points is called a **simple curve.**

EXAMPLE 4. Simple and nonsimple curves

Ellipses and helices are simple curves. The curve represented by

$$\mathbf{r}(t) = (t^2 - 1)\mathbf{i} + (t^3 - t)\mathbf{j}$$

is not simple, since it has a double point at the origin; this point corresponds to the two values $t = 1$ and $t = -1$. Can you sketch this curve?   ∎

Fig. 186. Curves having double and triple points

We finally mention that a given curve C may be represented by various vector functions. For example, if C is given by (1) and we set $t = h(t^*)$, then we obtain a new vector function $\tilde{\mathbf{r}}(t^*)$ representing C, provided $h(t^*)$ takes on all the values of t occurring in (1). For instance, if C is the path of a moving body B and t is time, a change of parameter $t = h(t^*)$ means that the motion of B changes in time, the path C remaining the same.

EXAMPLE 5. Change of parameter

The parabola $y = x^2$ in the xy-plane may be represented by the vector function

$$\mathbf{r}(t) = t\mathbf{i} + t^2\mathbf{j} \qquad\qquad (-\infty < t < \infty).$$

If we set $t = -2t^*$, we obtain another representation of the parabola:

$$\tilde{\mathbf{r}}(t^*) = \mathbf{r}(-2t^*) = -2t^*\mathbf{i} + 4t^{*2}\mathbf{j}.$$

If we set $t = t^{*2}$, we obtain

$$\tilde{\mathbf{r}}(t^*) = t^{*2}\mathbf{i} + t^{*4}\mathbf{j},$$

but this function represents only the portion of the parabola in the first quadrant, because $t^{*2} \geqq 0$ for all t^*.   ∎

In the next section we continue our discussion on curves by exploring the geometric meaning of the **derivative** of a vector function in connection with the tangent of a curve, and by discussing the *length* of a curve. This includes the introduction of the **arc length function** s of a curve, which turns out to be a particularly useful parameter in parametric representations of curves.

Problems for Sec. 8.3

Find a parametric representation of the straight line through a point A in the direction of a vector $\mathbf{b}$, where

1. A: $(0, 0, 0)$, $\mathbf{b} = \mathbf{i} + \mathbf{j}$

2. A: $(4, \frac{1}{2}, 0)$, $\mathbf{b} = [0, 0, 1]$

3. A: $(0, -5, 3)$, $\mathbf{b} = [2, 1, -1]$

4. A: $(2, 0, 6)$, $\mathbf{b} = 2\mathbf{i} + 4\mathbf{j} - \frac{1}{4}\mathbf{k}$

5. A: $(-3, 8, 0)$, $\mathbf{b} = \frac{1}{2}\mathbf{i} - 7\mathbf{k}$

6. A: $(1, 2, 5)$, $\mathbf{b} = [1, -2, 9]$

Find a parametric representation of the straight line through the points A and B, where

7. A: $(0, 0, 0)$, B: $(1, 1, 1)$

8. A: $(3, 2, 4)$, B: $(3, 1, 6)$

9. A: $(0, 1, 3)$, B: $(2, 1, 3)$

10. A: $(4, 3, 2)$, B: $(4, -2, 2)$

11. A: $(4, 2, 0)$, B: $(7, -3, 0)$

12. A: $(1, 5, 3)$, B: $(0, 2, -1)$

Find a parametric representation of the straight line represented by

13. $y = x$, $z = x$

14. $x + y + z = 8$, $2x - y + 3z = 4$

15. $y = x + 1$, $z = 2x - 3$

16. $3x + 6y - z = 2$, $x + y = 0$

What curves are represented by the following parametric representations?

17. $\mathbf{r}(t) = t\mathbf{i} + (2/t)\mathbf{j}$

18. $\mathbf{r}(t) = 2\cos t\, \mathbf{i} - 2\sin t\, \mathbf{j} + \mathbf{k}$

19. $\mathbf{r}(t) = \cos t\, \mathbf{j} + 4\sin t\, \mathbf{k}$

20. $\mathbf{r}(t) = t\mathbf{i} + 2t^2\mathbf{j} - \frac{1}{2}\mathbf{k}$

21. $\mathbf{r}(t) = \cosh t\, \mathbf{i} + \sinh t\, \mathbf{j}$

22. $\mathbf{r}(t) = \cos t\, \mathbf{i} + \sin t\, \mathbf{j} - \pi t\mathbf{k}$

23. $\mathbf{r}(t) = e^t\mathbf{i} + e^{-t}\mathbf{j}$

24. $\mathbf{r}(t) = (a + c\cos t)\mathbf{i} + (b + c\sin t)\mathbf{j}$

Represent the following curves in parametric form.

25. $x^2 + y^2 = 4$, $z = 5$

26. $\frac{1}{4}x^2 + \frac{1}{9}y^2 = 1$, $z = 0$

27. $x^2 - 16y^2 = 1$, $z = 1$

28. $(x - 2)^2 + (y + 1)^2 = 25$, $z = 0$

29. $y = 1 - x^2$, $z = -2$

30. $4(x + 2)^2 + (y - 4)^2 = 4$, $z = 0$

31. $x^2 + y^2 = 36$, $z = x$

32. $y^2 - z^2 = 9$, $x = 4$

33. What curve is represented by $\mathbf{r}(t) = t\mathbf{i} + t^4\mathbf{j}$? Find another representation of it by setting $t = t^{*3}$.

34. If in $\mathbf{r}(t) = (1 - 4t)\mathbf{i} + (3 + 2t)\mathbf{j}$ you set $t = e^{t^*}$, does the resulting representation represent the entire line?

35. Set $t = -t^*$ in $\mathbf{r}(t) = \cos t\, \mathbf{i} + \sin t\, \mathbf{j}$ and show that the sense of increasing t^* is the clockwise sense on the circle.

8.4 Tangent, Arc Length of a Curve

The **tangent** to a curve C at a point P of C is defined as the limiting position of the straight line L through P and another point Q of C as Q approaches P along the curve (Fig. 187 on the next page).

Suppose that C is represented by a continuously differentiable vector function $\mathbf{r}(t)$, where t is any parameter. Let P and Q correspond to t and $t + \Delta t$, respectively. Then L has the direction of the vector

$$\frac{1}{\Delta t}\,[\mathbf{r}(t + \Delta t) - \mathbf{r}(t)].$$

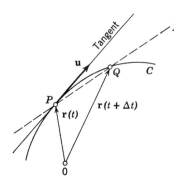

Fig. 187. Tangent to a curve

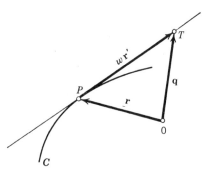

Fig. 188. Representation of the tangent to a curve

Hence, if the derivative of **r**, that is, the vector

(1)
$$\mathbf{r}'(t) = \lim_{\Delta t \to 0} \frac{1}{\Delta t} [\mathbf{r}(t + \Delta t) - \mathbf{r}(t)]$$

is not the zero vector, it has the direction of the tangent to C at P. It points in the direction of increasing values of t, and its sense, therefore, depends on the orientation of the curve. $\mathbf{r}'$ is called a **tangent vector** of C at P. The corresponding unit vector

(2)
$$\mathbf{u} = \frac{1}{|\mathbf{r}'|} \mathbf{r}'$$

is called the **unit tangent vector** to C at P. These vectors $\mathbf{r}'$ and $\mathbf{u}$ point in the direction of increasing values of t.

Now the position vector of a point T on the tangent is the sum of the position vector $\mathbf{r}$ of P and a vector in the direction of the tangent. Hence a parametric representation of the tangent is (Fig. 188)

(3)
$$\mathbf{q}(w) = \mathbf{r} + w\mathbf{r}',$$

where both $\mathbf{r}$ and $\mathbf{r}'$ depend on P and the parameter w is a real variable.

EXAMPLE 1. Tangent to an ellipse

Find the tangent to the ellipse $\frac{1}{4}x^2 + y^2 = 1$ at P: $(\sqrt{2}, 1/\sqrt{2})$.

Solution. $\mathbf{r}(t) = 2 \cos t\, \mathbf{i} + \sin t\, \mathbf{j}$, hence $\mathbf{r}'(t) = -2 \sin t\, \mathbf{i} + \cos t\, \mathbf{j}$, and P corresponds to $t = \pi/4$, since $2 \cos (\pi/4) = \sqrt{2}$ and $\sin (\pi/4) = 1/\sqrt{2}$. Thus $\mathbf{r}'(\pi/4) = [-\sqrt{2}, 1/\sqrt{2}]$. *Answer:*

$$\mathbf{q}(w) = [\sqrt{2}, \ 1/\sqrt{2}] + w[-\sqrt{2}, \ 1/\sqrt{2}] = \sqrt{2}(1 - w)\mathbf{i} + (1/\sqrt{2})(1 + w)\mathbf{j}.$$

To check the result, sketch the ellipse and the tangent. ∎

Tangent vectors will now be helpful in our next step, which is the introduction of two related concepts, the length l of a curve and the arc length s of a curve.

Fig. 189. Length of a curve

Length of a Curve

To define the length of a curve C, we may proceed as follows. We inscribe in C a broken line of n chords joining the two endpoints of C as shown in Fig. 189. This we do for every positive integer n in an arbitrary way but so that the maximum chord-length approaches zero as n approaches infinity. The lengths of these lines of chords can be obtained from the theorem of Pythagoras. If the sequence of these lengths l_1, l_2, $\cdots$ is convergent, with limit l, then C is said to be **rectifiable,** and l is called the **length** of C.

If C can be expressed by a continuously differentiable[2] vector function

$$\mathbf{r} = \mathbf{r}(t) \qquad\qquad (a \leqq t \leqq b),$$

then it can be shown that C is rectifiable, and its length l is given by the integral

(4)
$$l = \int_a^b \sqrt{\mathbf{r}' \cdot \mathbf{r}'}\ dt \qquad\qquad \left(\mathbf{r}' = \frac{d\mathbf{r}}{dt}\right),$$

whose value is independent of the choice of the parametric representation. The proof is quite similar to that for plane curves usually considered in elementary integral calculus (cf. Ref. [18]) and can be found in Ref. [B6] in Appendix 1. The practical evaluation of (4) will be difficult, in general; for some simple cases see the problem set.

Arc Length s of a Curve

If we replace the fixed upper limit b in (4) with a variable upper limit t, the integral becomes a function of t, say, $s(t)$; denoting the variable of integration by $\tilde{t}$, we have

(5)
$$s(t) = \int_a^t \sqrt{\mathbf{r}' \cdot \mathbf{r}'}\ d\tilde{t} \qquad\qquad \left(\mathbf{r}' = \frac{d\mathbf{r}}{d\tilde{t}}\right).$$

This function $s(t)$ is called the *arc length function* or, simply, the **arc length** of C.

From our consideration it follows that, geometrically, for a fixed value $t = t_0 \geqq a$, the arc length $s(t_0)$ is the length of the portion of C between the points corresponding to $t = a$ and $t = t_0$. For $t = t_0 < a$ we have $s(t_0) < 0$ and that length is $-s(t_0)$.

---

[2]*"Continuously differentiable"* means that the derivative exists and is continuous; *"twice continuously differentiable"* means that the first and second derivatives exist and are continuous, and so on.

The constant a in (5) may be replaced by another constant; that is, the point of the curve corresponding to $s = 0$ may be chosen in an arbitrary manner. The sense corresponding to increasing values of s is called the **positive sense** on C; in this fashion any representation $\mathbf{r}(s)$ or $\mathbf{r}(t)$ of C defines a certain **orientation** of C. Obviously, there are two ways of *orienting C*, and it is not difficult to see that the transition from one orientation to the opposite orientation can be effected by a transformation of the parameter whose derivative is negative.

From (5) we obtain by differentiating and squaring

$$(6) \qquad \left(\frac{ds}{dt}\right)^2 = \frac{d\mathbf{r}}{dt} \cdot \frac{d\mathbf{r}}{dt} = \left(\frac{dx}{dt}\right)^2 + \left(\frac{dy}{dt}\right)^2 + \left(\frac{dz}{dt}\right)^2.$$

It is customary to write

$$d\mathbf{r} = [dx, \, dy, \, dz] = dx \, \mathbf{i} + dy \, \mathbf{j} + dz \, \mathbf{k}$$

and

$$(7) \qquad ds^2 = d\mathbf{r} \cdot d\mathbf{r} = dx^2 + dy^2 + dz^2.$$

ds is called the **linear element** of C.

Note that (6) may also be written

$$(8) \qquad \boxed{\left(\frac{ds}{dt}\right)^2 = |\mathbf{r}'(t)|^2.}$$

The arc length s may serve as a parameter in parametric representations of curves. This simplifies various formulas, as we shall see.

As a first and important case, we show that the use of s simplifies the formula (2) for the **unit tangent vector:**

$$(9) \qquad \boxed{\mathbf{u}(s) = \mathbf{r}'(s).}$$

This follows from (8) with $t = s$ since $ds/ds = 1$.

EXAMPLE 2. Circular helix. Circle. Arc length as parameter

For the helix in Example 3, Sec. 8.3,

$$\mathbf{r}(t) = a \cos t \, \mathbf{i} + a \sin t \, \mathbf{j} + ct\mathbf{k}$$

we get

$$\mathbf{r}'(t) = -a \sin t \, \mathbf{i} + a \cos t \, \mathbf{j} + c\mathbf{k}.$$

From this, $\mathbf{r}' \cdot \mathbf{r}' = a^2 + c^2$, so that (5) gives

$$s = \int_0^t \sqrt{a^2 + c^2} \, d\tilde{t} = t\sqrt{a^2 + c^2}.$$

Hence $t = s/\sqrt{a^2 + c^2}$, and a formula for the helix with the arc length s as parameter is

$$\mathbf{r}^*(s) = \mathbf{r}\left(\frac{s}{\sqrt{a^2 + c^2}}\right) = a \cos \frac{s}{\sqrt{a^2 + c^2}} \mathbf{i} + a \sin \frac{s}{\sqrt{a^2 + c^2}} \mathbf{j} + \frac{cs}{\sqrt{a^2 + c^2}} \mathbf{k}.$$

Setting $c = 0$, we have $t = s/a$ and obtain for a circle of radius a the representation

$$\mathbf{r}\left(\frac{s}{a}\right) = a \cos \frac{s}{a} \mathbf{i} + a \sin \frac{s}{a} \mathbf{j}.$$

The circle is oriented in the counterclockwise sense, which corresponds to increasing values of s. Setting $s = -s^*$ and using $\cos(-\alpha) = \cos \alpha$ and $\sin(-\alpha) = -\sin \alpha$, we obtain

$$\mathbf{r}\left(-\frac{s^*}{a}\right) = a \cos \frac{s^*}{a} \mathbf{i} - a \sin \frac{s^*}{a} \mathbf{j};$$

we have $ds/ds^* = -1 < 0$, and the circle is now oriented clockwise.   ∎

In the next section we shall discuss the role of curves in **mechanics** and in Sec. 8.6 some further geometric properties of curves.

Problems for Sec. 8.4

Tangent vectors and tangents. Find (a) a tangent vector $\mathbf{r}'(t)$ and the corresponding unit tangent vector $\mathbf{u}(t)$ of the given curve $C: \mathbf{r}(t)$, (b) $\mathbf{r}'$ and $\mathbf{u}$ at P (sketch it), (c) the tangent at P.

1. $\mathbf{r}(t) = t\mathbf{i} + t^2\mathbf{j}$, $P: (1, 1, 0)$
2. $\mathbf{r}(t) = t\mathbf{i} + t^3\mathbf{j}$, $P: (2, 8, 0)$
3. $\mathbf{r}(t) = (4 + 3t)\mathbf{i} + (2 - t)\mathbf{k}$, $P: (13, 0, 0)$
4. $\mathbf{r}(t) = (1 + \cos t)\mathbf{i} + \sin t\,\mathbf{j}$, $P: (0, 0, 0)$
5. $\mathbf{r}(t) = \cosh t\,\mathbf{i} + \sinh t\,\mathbf{j}$, $P: (\frac{5}{3}, \frac{4}{3}, 0)$
6. $\mathbf{r}(t) = t\mathbf{i} + t^2\mathbf{j} + t^3\mathbf{k}$, $P: (1, 1, 1)$
7. $\mathbf{r}(t) = \cos t\,\mathbf{i} + \sin t\,\mathbf{j} + t\mathbf{k}$, $P: (1, 0, 0)$
8. $\mathbf{r}(t) = 2 \cos t\,\mathbf{i} + \sin t\,\mathbf{j}$, $P: (-\sqrt{2}, 1/\sqrt{2}, 0)$
9. $\mathbf{r}(t) = 3 \cos t\,\mathbf{i} - 3 \sin t\,\mathbf{j}$, $P: (0, 3, 0)$
10. $\mathbf{r}(t) = 3 \cos t\,\mathbf{i} + 3 \sin t\,\mathbf{j} + 4t\mathbf{k}$, $P: (0, 3, 2\pi)$

Length of a curve. Sketch the following curves and find their lengths.

11. **Catenary** $\mathbf{r}(t) = t\mathbf{i} + \cosh t\,\mathbf{j}$ from $t = 0$ to $t = 1$
12. **Circular helix** $\mathbf{r}(t) = a \cos t\,\mathbf{i} + a \sin t\,\mathbf{j} + ct\mathbf{k}$ from $(a, 0, 0)$ to $(a, 0, 2\pi c)$
13. **Semicubical parabola** $\mathbf{r}(t) = t\mathbf{i} + t^{3/2}\mathbf{j}$ from $(0, 0, 0)$ to $(4, 8, 0)$
14. **Four-cusped hypocycloid** $\mathbf{r}(t) = a \cos^3 t\,\mathbf{i} + a \sin^3 t\,\mathbf{j}$, total length
15. **Involute of circle** $\mathbf{r}(t) = (\cos t + t \sin t)\mathbf{i} + (\sin t - t \cos t)\mathbf{j}$ from $t = 0$ to $t = \pi$.
16. $\mathbf{r}(t) = e^t \cos t\,\mathbf{i} + e^t \sin t\,\mathbf{j}$ from $t = 0$ to $t = \pi/2$

17. If a plane curve is represented in the form $y = f(x)$, $z = 0$, using (4) show that its length between $x = a$ and $x = b$ is

$$l = \int_a^b \sqrt{1 + y'^2}\, dx.$$

18. Using the formula in Prob. 17, find the length of a circle of radius a.

19. Show that if a plane curve is represented in polar coordinates $\rho = \sqrt{x^2 + y^2}$ and $\theta = \arctan(y/x)$, then $ds^2 = \rho^2\, d\theta^2 + d\rho^2$ and

$$l = \int_\alpha^\beta \sqrt{\rho^2 + \rho'^2}\, d\theta \qquad\qquad (\rho' = d\rho/d\theta).$$

Using the formula in Prob. 19, find the lengths of the following curves.

20. Circle of radius a, total length   **21.** $\rho = e^{\theta}, 0 \leq \theta \leq \pi$

22. $\rho = \theta^2, 0 \leq \theta \leq \pi/2$ **23.** $\rho = 1 + \cos \theta, 0 \leq \theta \leq \pi/2$

24. Cardioid $\rho = a(1 - \cos \theta)$, total length. [Use (10) in Appendix A3.1.] Sketch this curve.

25. If a curve is represented by a parametric representation, show that a transformation of the parameter whose derivative is negative reverses the orientation.

8.5 Velocity and Acceleration

In the previous section we discussed geometric properties of a curve C in terms of the derivative $\mathbf{r}'(t)$ of the vector function $\mathbf{r}(t)$ representing C. We shall now see that $\mathbf{r}'(t)$ and $\mathbf{r}''(t)$ are also basic in mechanics when C is the path of a moving body, where t is time.

From Sec. 8.4 we know that the vector

(1)
$$\mathbf{v} = \mathbf{r}' = \frac{d\mathbf{r}}{dt}$$

is tangent to C and, therefore, points in the instantaneous direction of motion of P. From (6) in Sec. 8.4 we see that this vector has the length

$$|\mathbf{v}| = \sqrt{\mathbf{r}' \cdot \mathbf{r}'} = \frac{ds}{dt},$$

where s is the arc length, which measures the distance of P from a fixed point ($s = 0$) on C along the curve. Hence ds/dt is the **speed** of P. The vector $\mathbf{v}$ is, therefore, called the **velocity vector**[3] of the motion.

The derivative of the velocity vector is called the **acceleration vector** and will be denoted by $\mathbf{a}$; thus

(2)
$$\mathbf{a}(t) = \mathbf{v}'(t) = \mathbf{r}''(t).$$

EXAMPLE 1. Centripetal acceleration
The vector function

$$\mathbf{r}(t) = R \cos \omega t \, \mathbf{i} + R \sin \omega t \, \mathbf{j} \qquad (\omega > 0)$$

represents a circle C of radius R with center at the origin of the xy-plane and describes a motion of a particle P in the counterclockwise sense. The velocity vector

$$\mathbf{v} = \mathbf{r}' = -R\omega \sin \omega t \, \mathbf{i} + R\omega \cos \omega t \, \mathbf{j}$$

(cf. Fig. 190) is tangent to C, and its magnitude, the speed

---

[3]When no confusion is likely to arise, the word *velocity* is often used to denote the speed, the length of $\mathbf{v}$.

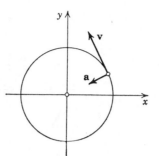

Fig. 190. Centripetal acceleration

$$|\mathbf{v}| = \sqrt{\mathbf{r}' \cdot \mathbf{r}'} = R\omega$$

is constant. The **angular speed** (speed divided by the distance R from the center) is equal to ω. The acceleration vector is

(3) $$\mathbf{a} = \mathbf{v}' = -R\omega^2 \cos \omega t \, \mathbf{i} - R\omega^2 \sin \omega t \, \mathbf{j} = -\omega^2 \mathbf{r}.$$

We see that there is an acceleration of constant magnitude $|\mathbf{a}| = \omega^2 R$ toward the origin. This is called the **centripetal acceleration.** It results from the fact that the velocity vector is changing direction at a constant rate. The **centripetal force** is $m\mathbf{a}$, where m is the mass of P. The opposite vector $-m\mathbf{a}$ is called the **centrifugal force,** and the two forces are in equilibrium at each instant of the motion.   ∎

Tangential Acceleration and Normal Acceleration

It is clear that $\mathbf{a}$ is the time rate of change of $\mathbf{v}$. In Example 1 we have $|\mathbf{v}| = const$, but $|\mathbf{a}| \neq 0$. This illustrates that the magnitude of $\mathbf{a}$ is not in general the rate of change of $|\mathbf{v}|$. The reason is that, in general, $\mathbf{a}$ is not tangent to the path C. In fact, by applying the chain rule to (1) we have

$$\mathbf{v} = \frac{d\mathbf{r}}{dt} = \frac{d\mathbf{r}}{ds} \frac{ds}{dt} = \mathbf{u}(s) \frac{ds}{dt},$$

where $\mathbf{u}(s) = d\mathbf{r}/ds$ is the unit tangent vector of C (Sec. 8.4), and, by differentiating this again,

(4) $$\mathbf{a} = \frac{d\mathbf{v}}{dt} = \frac{d}{dt}\left(\mathbf{u}(s) \frac{ds}{dt}\right) = \frac{d\mathbf{u}}{ds}\left(\frac{ds}{dt}\right)^2 + \mathbf{u}(s) \frac{d^2 s}{dt^2}.$$

Since $\mathbf{u}(s)$ is tangent to C and $d\mathbf{u}/ds$ is perpendicular to $\mathbf{u}(s)$, formula (4) is a decomposition of the acceleration vector into its normal component $(d\mathbf{u}/ds)(ds/dt)^2$, called the **normal acceleration,** and its tangential component $\mathbf{u}(s)d^2 s/dt^2$, called the **tangential acceleration.** (Examples below.) From this we see that if and only if the normal acceleration is zero, $|\mathbf{a}|$ equals the time rate of change of $|\mathbf{v}| = ds/dt$ (except for the sign), because then we have $|\mathbf{a}| = |\mathbf{u}(s)| \, |d^2 s/dt^2| = |d^2 s/dt^2|$.

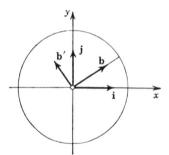

Fig. 191. Motion in Example 2

EXAMPLE 2. Coriolis acceleration[4]

A particle P moves on a disk toward the edge, the position vector being

$$(5) \qquad\qquad \mathbf{r}(t) = t\mathbf{b}$$

where $\mathbf{b}$ is a unit vector, rotating together with the disk with constant angular speed ω in the counterclockwise sense (Fig. 191). Find the acceleration $\mathbf{a}$ of P.

Solution. Because of the rotation, $\mathbf{b}$ is of the form

$$(6) \qquad\qquad \mathbf{b}(t) = \cos \omega t\, \mathbf{i} + \sin \omega t\, \mathbf{j}.$$

Differentiating (5) with respect to t, we obtain the velocity

$$(7) \qquad\qquad \mathbf{v} = \mathbf{r}' = \mathbf{b} + t\mathbf{b}'.$$

Obviously $\mathbf{b}$ is the velocity of P relative to the disk, and $t\mathbf{b}'$ is the additional velocity due to the rotation. Differentiating once more, we obtain the acceleration

$$(8) \qquad\qquad \mathbf{a} = \mathbf{v}' = 2\mathbf{b}' + t\mathbf{b}''.$$

In the last term of (8) we have $\mathbf{b}'' = -\omega^2\mathbf{b}$, as follows by differentiating (6). Hence this acceleration $t\mathbf{b}''$ is directed toward the center of the disk, and from Example 1 we see that this is the centripetal acceleration due to the rotation. In fact, the distance of P from the center is equal to t which, therefore, plays the role of R in Example 1.

The most interesting and probably unexpected term in (8) is $2\mathbf{b}'$, the so-called **Coriolis acceleration,** which results from the interaction of the rotation of the disk and the motion of P on the disk. It has the direction of $\mathbf{b}'$; that is, it is tangential to the edge of the disk and, referred to the fixed xy-coordinate system, it points in the direction of the rotation. If P is a person of mass m_0 walking on the disk according to (5), he will feel a force $-2m_0\mathbf{b}'$ in the opposite direction, that is, against the sense of the rotation.

EXAMPLE 3. Superposition of two rotations, Coriolis acceleration

Find the acceleration of a projectile P moving along a "meridian" M of a rotating sphere (for instance, the surface of our earth) with constant speed relative to the sphere, which also rotates with constant angular speed.

Solution. The motion of P on M can be described in the form

$$(9) \qquad\qquad \mathbf{r}(t) = R \cos \gamma t\, \mathbf{b} + R \sin \gamma t\, \mathbf{k}$$

where R is the radius of the sphere, $\gamma\ (> 0)$ the angular speed of P on M, $\mathbf{b}$ a horizontal unit

---

[4]GUSTAVE GASPARD CORIOLIS (1792—1843), French engineer, who did research in mechanics.

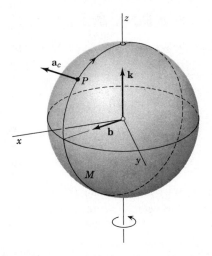

Fig. 192. Superposition of two rotations

vector in the plane of M (Fig. 192), and $\mathbf{k}$ the unit vector in the positive z-direction. Since $\mathbf{b}$ rotates together with the sphere, it is of the form

(10) $$\mathbf{b} = \cos \omega t \, \mathbf{i} + \sin \omega t \, \mathbf{j}$$

where ω (> 0) is the angular speed of the sphere and $\mathbf{i}$ and $\mathbf{j}$ are the unit vectors in the positive x- and y-directions, which are fixed in space. By differentiating (9) we obtain the velocity

(11) $$\mathbf{v} = \mathbf{r}' = R \cos \gamma t \, \mathbf{b}' - \gamma R \sin \gamma t \, \mathbf{b} + \gamma R \cos \gamma t \, \mathbf{k}.$$

By differentiating this again with respect to t we obtain the acceleration

(12) $$\mathbf{a} = \mathbf{v}' = R \cos \gamma t \, \mathbf{b}'' - 2\gamma R \sin \gamma t \, \mathbf{b}' - \gamma^2 R \cos \gamma t \, \mathbf{b} - \gamma^2 R \sin \gamma t \, \mathbf{k},$$

where, by (10),

$$\mathbf{b}' = -\omega \sin \omega t \, \mathbf{i} + \omega \cos \omega t \, \mathbf{j},$$

$$\mathbf{b}'' = -\omega^2 \cos \omega t \, \mathbf{i} - \omega^2 \sin \omega t \, \mathbf{j} = -\omega^2 \mathbf{b}.$$

From (9) we see that the sum of the last two terms in (12) is equal to $-\gamma^2 \mathbf{r}$, and (12) becomes

(13) $$\mathbf{a} = -\omega^2 R \cos \gamma t \, \mathbf{b} - 2\gamma R \sin \gamma t \, \mathbf{b}' - \gamma^2 \mathbf{r}.$$

The first term on the right is the centripetal acceleration caused by the rotation of the sphere, and the last term is the centripetal acceleration resulting from the rotation of P on M. The second term is the **Coriolis acceleration**

(14) $$\mathbf{a}_c = -2\gamma R \sin \gamma t \, \mathbf{b}'.$$

On the "Northern Hemisphere," $\sin \gamma t > 0$ [cf. (9)] and because of the minus sign, $\mathbf{a}_c$ is directed opposite to $\mathbf{b}'$, that is, tangential to the surface of the sphere, perpendicular to M, and opposite to the rotation of the sphere. Its magnitude $2\gamma R \, |\sin \gamma t| \, \omega$ is maximum at the "North Pole" and zero at the equator. If the projectile P has mass m_0, it experiences a force $-m_0 \mathbf{a}_c$, opposite to $m_0 \mathbf{a}_c$; this is quite similar to the force in Example 2. This force tends to let the projectile deviate from the path M to the right. On the "Southern Hemisphere," $\sin \gamma t < 0$ and that force acts in the opposite direction, tending to let the projectile deviate from M to the left. This effect can be observed in connection with missiles, rockets and shells. The flow of air toward an area of low pressure also shows those deviations.   ∎

The next section is optional and may be skipped without breaking continuity.

Problems for Sec. 8.5

Let $\mathbf{r}(t)$ be the position vector of a moving particle, where t ($\geqq 0$) is time. Describe the geometric shape of the path and find the velocity vector, the speed and the acceleration vector.

1. $\mathbf{r} = t\,\mathbf{i}$ **2.** $\mathbf{r} = t^2\mathbf{k}$

3. $\mathbf{r} = (2t - t^2)\mathbf{i}$ **4.** $\mathbf{r} = t^3\mathbf{i}$

5. $\mathbf{r} = 2t\mathbf{i} - 2t\mathbf{j} + t\mathbf{k}$ **6.** $\mathbf{r} = \cos 2t\,\mathbf{i} + \sin 2t\,\mathbf{k}$

7. $\mathbf{r} = e^t\mathbf{i} + e^{-t}\mathbf{j}$ **8.** $\mathbf{r} = \cos t^2\,\mathbf{i} + \sin t^2\,\mathbf{j}$

9. $\mathbf{r} = 3\cos t\,\mathbf{i} + 3\sin t\,\mathbf{j} + 2t\mathbf{k}$ **10.** $\mathbf{r}(t) = 3\cos 2t\,\mathbf{i} + 2\sin 3t\,\mathbf{j}$

11. Obtain (3) by differentiating (7), Sec. 6.7.

12. Find (a) the velocity vector $\mathbf{v}(t)$ and the acceleration vector $\mathbf{a}(t)$ of the motion $\mathbf{r}(t) = \sin t\,\mathbf{j}$, (b) the points at which $\mathbf{v}(t) = \mathbf{0}$, (c) the points at which $|\mathbf{a}(t)|$ is maximum.

13. (Cycloid) Sketch $\mathbf{r}(t) = (R\sin \omega t + \omega Rt)\mathbf{i} + (R\cos \omega t + R)\mathbf{j}$, taking $R = 1$ and $\omega = 1$. This so-called *cycloid* is the path of a point on the rim of a wheel of radius R that rolls without slipping along the x-axis,

14. Find the tangential acceleration and the normal acceleration of the motion given by $\mathbf{r}(t) = t\mathbf{i} - t^2\mathbf{j}$

15. Let a motion be given by $\mathbf{r}(t) = \cos t\,\mathbf{i} + 2\sin t\,\mathbf{j}$. Find the tangential acceleration.

16. Find the centripetal acceleration of the moon toward the earth, assuming that the orbit of the moon is a circle of radius 239,000 miles $= 3.85 \cdot 10^8$ meters and the time for one complete revolution is 27.3 days $= 2.36 \cdot 10^6$ sec.

17. Find the acceleration of the earth toward the sun from the fact that the earth revolves about the sun in a nearly circular orbit with an almost constant speed of 30 km/sec.

18. (Satellite) Find the speed of an artificial earth satellite traveling at an altitude of 80 miles above the earth's surface where $g = 31$ ft/sec^2. (The radius of the earth is 3960 miles.) *Hint.* Cf. Example 1.

19. Find the motion for which the acceleration vector is constant.

20. Find the Coriolis acceleration in Example 2 with (5) replaced by $\mathbf{r} = t^2\mathbf{b}$.

8.6   Curvature and Torsion of a Curve (Optional)

This section is optional. Together with Secs. 8.3 and 8.4, it gives the foundations of the theory of curves in space.

The **curvature** $\kappa(s)$ of a curve C, represented by $\mathbf{r}(s)$ with arc length s as parameter, is defined by

(1)
$$\kappa(s) = |\mathbf{u}'(s)| = |\mathbf{r}''(s)|$$
$(' = d/ds)$.

Here $\mathbf{u}(s) = \mathbf{r}'(s)$ is the unit tangent vector of C (Sec. 8.4), and we have to assume that $\mathbf{r}(s)$ is twice differentiable, so that $\mathbf{r}''(s)$ exists.

Since κ is the length of the rate of change of the unit tangent vector of C with s, we see that κ measures the deviation of C from the tangent at each point of C. Examples are given below and in the problem set. For general parameters t, the formula for the curvature is complicated (cf. Prob. 2).

If $\kappa(s) \neq 0$ (hence > 0), the unit vector $\mathbf{p}(s)$ in the direction of $\mathbf{u}'(s)$ is

(2)
$$\mathbf{p} = \frac{1}{\kappa}\mathbf{u}'$$
$(\kappa > 0)$

and is called the **unit principal normal vector** of C. From Example 1 in Sec. 8.2 we see that $\mathbf{p}$ is perpendicular to $\mathbf{u}$. The vector

(3)
$$\mathbf{b} = \mathbf{u} \times \mathbf{p}$$
$(\kappa > 0)$

is called the **unit binormal vector** of C. From the definition of a vector product it follows that $\mathbf{u}, \mathbf{p}, \mathbf{b}$ constitute a right-handed triple of orthogonal unit vectors (Secs. 6.3 and 6.7). This triple is called the **trihedron** of C at the point under consideration (Fig. 193). The three straight lines through that point in the directions of $\mathbf{u}, \mathbf{p}, \mathbf{b}$ are called the **tangent,** the **principal normal** and the **binormal** of C. Figure 193 also shows the names of the three planes spanned by each pair of those vectors.

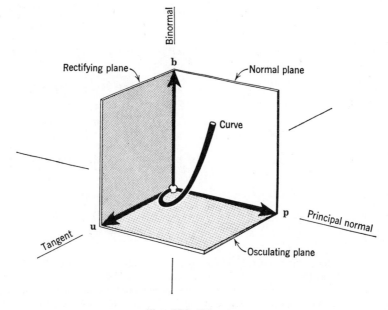

Fig. 193. Trihedron

Torsion of a Curve

To introduce the torsion $\tau(s)$ of a curve C represented by $\mathbf{r}(s)$, we consider the derivative $\mathbf{b}' = d\mathbf{b}/ds$ of the unit binormal vector $\mathbf{b}$ of C. Here we assume that $\mathbf{r}(s)$ is three times differentiable, so that $\mathbf{b}'$ exists. If $\mathbf{b}'(s) \neq \mathbf{0}$, it is perpendicular to $\mathbf{b}$ (cf. Example 1 in Sec. 8.2). We show that $\mathbf{b}'$

is also perpendicular to $\mathbf{u}$. In fact, by differentiating $\mathbf{b} \cdot \mathbf{u} = 0$ we have $\mathbf{b}' \cdot \mathbf{u} + \mathbf{b} \cdot \mathbf{u}' = 0$; hence $\mathbf{b}' \cdot \mathbf{u} = 0$ because $\mathbf{b} \cdot \mathbf{u}' = 0$. Consequently, $\mathbf{b}'$ is of the form $\mathbf{b}' = \alpha \mathbf{p}$ where α is a scalar. It is customary to set $\alpha = -\tau$. Then

$$(4) \qquad\qquad\qquad \mathbf{b}' = -\tau \mathbf{p} \qquad\qquad\qquad (\kappa > 0).$$

The scalar function τ is called the **torsion** of C. Taking the dot product by $\mathbf{p}$ on both sides of (4), we obtain

$$(5) \qquad\qquad\qquad \boxed{\tau(s) = -\mathbf{p}(s) \cdot \mathbf{b}'(s).}$$

The concepts just introduced are basic in the theory and application of curves. Let us illustrate them by a typical example. Further applications are given in the problem set.

EXAMPLE 1. Circular helix. Circle

In the case of the circular helix (6) in Sec. 8.3 we obtain the arc length $s = t\sqrt{a^2 + c^2}$; cf. Example 2 in Sec. 8.4. Hence we may represent the helix in the form

$$\mathbf{r}(s) = a \cos \frac{s}{K} \mathbf{i} + a \sin \frac{s}{K} \mathbf{j} + c \frac{s}{K} \mathbf{k} \qquad \text{where} \qquad K = \sqrt{a^2 + c^2}.$$

It follows that

$$\mathbf{u}(s) = \mathbf{r}'(s) = -\frac{a}{K} \sin \frac{s}{K} \mathbf{i} + \frac{a}{K} \cos \frac{s}{K} \mathbf{j} + \frac{c}{K} \mathbf{k}$$

$$\mathbf{r}''(s) = -\frac{a}{K^2} \cos \frac{s}{K} \mathbf{i} - \frac{a}{K^2} \sin \frac{s}{K} \mathbf{j}$$

$$\kappa = |\mathbf{r}''| = \sqrt{\mathbf{r}'' \cdot \mathbf{r}''} = \frac{a}{K^2} = \frac{a}{a^2 + c^2}$$

$$\mathbf{p}(s) = \frac{1}{\kappa(s)} \mathbf{r}''(s) = -\cos \frac{s}{K} \mathbf{i} - \sin \frac{s}{K} \mathbf{j}$$

$$\mathbf{b}(s) = \mathbf{u}(s) \times \mathbf{p}(s) = \frac{c}{K} \sin \frac{s}{K} \mathbf{i} - \frac{c}{K} \cos \frac{s}{K} \mathbf{j} + \frac{a}{K} \mathbf{k}$$

$$\mathbf{b}'(s) = \frac{c}{K^2} \cos \frac{s}{K} \mathbf{i} + \frac{c}{K^2} \sin \frac{s}{K} \mathbf{j}$$

$$\tau(s) = -\mathbf{p}(s) \cdot \mathbf{b}'(s) = \frac{c}{K^2} = \frac{c}{a^2 + c^2}.$$

Hence the circular helix has constant curvature and torsion. If $c > 0$ (right-handed helix, cf. Fig. 184 in Sec. 8.3), then $\tau > 0$, and if $c < 0$ (left-handed helix, cf. Fig. 185), then $\tau < 0$.

If $c = 0$, we get a circle of radius a, and our formulas then yield $\kappa = 1/a$ (thus the curvature is the reciprocal of the radius) and $\tau = 0$; also $\mathbf{b}(s) = \mathbf{k}$ is constant, namely, perpendicular to the plane of the circle (the xy-plane).   ∎

Since $\mathbf{u}$, $\mathbf{p}$ and $\mathbf{b}$ are linearly independent vectors, we may represent any vector in space as a linear combination of these vectors. Hence if the derivatives $\mathbf{u}'$, $\mathbf{p}'$ and $\mathbf{b}'$ exist, they may be represented in this fashion. The corresponding formulas are the so-called **Frenet formulas**[5]

---

[5]JEAN-FRÉDÉRIC FRENET (1816—1900), French mathematician.

$$\text{(a)} \quad \mathbf{u}' = \qquad\qquad \kappa\mathbf{p}$$

(6) $$\qquad \text{(b)} \quad \mathbf{p}' = -\kappa\mathbf{u} \qquad\qquad + \tau\mathbf{b}$$

$$\text{(c)} \quad \mathbf{b}' = \qquad\qquad -\tau\mathbf{p}$$

Formula (6a) follows from (2), and (6c) is identical with (4). The derivation of (6b) is not very difficult either and is left to the reader (Prob. 20).

It can be shown that continuous functions $\kappa = \kappa(s)$ (> 0) and $\tau = \tau(s)$ determine a curve (with arc length s) uniquely, except for its position in space. This follows by integrating the Frenet formulas and accounts for the fundamental role of curvature and torsion in the differential geometry of space curves. $\kappa = \kappa(s)$ and $\tau = \tau(s)$ are called **natural equations** of C.

So far in this chapter, we have discussed vector functions of a single variable and their applications. In the remaining sections we consider vector functions of several variables. As a first step, in the next section we prove and discuss the **chain rule** and the mean value theorem of differential calculus for such functions.

Problems for Sec. 8.6

1. Show that the curvature of a circle of radius a equals $1/a$.

2. Using (1), show that if a curve is represented by $\mathbf{r}(t)$, where t is any parameter, then its curvature is

$$\textbf{(1')} \qquad \kappa(t) = \frac{\sqrt{(\mathbf{r}'\cdot\mathbf{r}')(\mathbf{r}''\cdot\mathbf{r}'') - (\mathbf{r}'\cdot\mathbf{r}'')^2}}{(\mathbf{r}'\cdot\mathbf{r}')^{3/2}}.$$

3. Using (1'), show that for a curve $y = y(x)$ in the xy-plane,

$$\textbf{(1'')} \qquad \kappa(x) = |y''|/(1 + y'^2)^{3/2} \qquad\qquad (y' = dy/dx, \text{ etc.}).$$

Indicate what kind of curve is represented and find its curvature, using (1') or (1'').

4. $\mathbf{r} = a\cos t\,\mathbf{i} + b\sin t\,\mathbf{j}$ **5.** $y = x^2$

6. $xy = 1$ **7.** $\mathbf{r} = a\cos t\,\mathbf{i} + a\sin t\,\mathbf{j} + ct\mathbf{k}$

8. $\mathbf{r} = \cosh t\,\mathbf{i} + \sinh t\,\mathbf{j}$ **9.** $\mathbf{r} = t\mathbf{i} + t^{3/2}\mathbf{j}$ $(t \geq 0)$

Torsion of a curve

10. Using (3) and (5), show that the torsion $\tau(s)$ of a curve C: $\mathbf{r}(s)$ is

$$\textbf{(5')} \qquad\qquad \tau(s) = (\mathbf{u} \quad \mathbf{p} \quad \mathbf{p}') \qquad\qquad (\kappa > 0).$$

11. Using (2), show that (5') may be written ($' = d/ds$, etc.)

$$\textbf{(5'')} \qquad\qquad \tau(s) = (\mathbf{r}' \quad \mathbf{r}'' \quad \mathbf{r}''')/\kappa^2 \qquad\qquad (\kappa > 0).$$

12. Show that if a curve C is represented by $\mathbf{r}(t)$, where t is any parameter, then (5'') becomes ($' = d/dt$, etc.)

$$\textbf{(5''')} \qquad\qquad \tau(t) = \frac{(\mathbf{r}' \quad \mathbf{r}'' \quad \mathbf{r}''')}{(\mathbf{r}'\cdot\mathbf{r}')(\mathbf{r}''\cdot\mathbf{r}'') - (\mathbf{r}'\cdot\mathbf{r}'')^2} \qquad (\kappa > 0).$$

13. Show that the torsion of a plane curve (with $\kappa > 0$) is identically zero.

14. Find the torsion of the helix $\mathbf{r}(t) = a \cos t\, \mathbf{i} + a \sin t\, \mathbf{j} + ct\mathbf{k}$. Use (5'''). Compare the result with Example 1.

15. Find the torsion of the curve C: $\mathbf{r}(t) = t\mathbf{i} + t^2\mathbf{j} + t^3\mathbf{k}$ (which looks similar to the curve in Fig. 193).

16. What type of curves are the projections of the curve in Prob. 15 in the coordinate planes?

17. Sketch C: $\mathbf{r}(t) = \cos t\, \mathbf{i} + \sin t\, \mathbf{j} + e^t\mathbf{k}$ and find its torsion.

18. What kind of curve is $\mathbf{r}(t) = \cos t\, \mathbf{i} + \sin t\, \mathbf{j} + \sin t\, \mathbf{k}$? Find κ and τ.

19. Can you visualize and sketch C: $\mathbf{r}(t) = \cos t\, \mathbf{i} + \sin t\, \mathbf{j} + \sin 2t\, \mathbf{k}$? Find τ.

20. Prove the Frenet formula (6b). *Hint.* Start by differentiating $\mathbf{p} = \mathbf{b} \times \mathbf{u}$.

8.7 Functions of Several Variables: Chain Rule, Mean Value Theorem

We shall now consider some facts about functions of several variables that will be needed in the following sections. Some of this may already be known to the student from elementary calculus, but we shall keep our presentation self-contained. For simplicity we shall formulate everything for functions of two and three variables; the generalization to functions of more than three variables will be obvious. Familiarity of the student with the concept of a partial derivative will be assumed (cf. Appendix 3).

Continuity

A function $f(x, y)$ is said to be **continuous** *at a point* (x_0, y_0) if f is defined in a *neighborhood*[6] of that point and if for any positive number ϵ (no matter how small, but not zero) we can find a positive number δ such that

$$|f(x, y) - f(x_0, y_0)| < \epsilon$$

for all (x, y) in that neighborhood that satisfy the inequality

$$(x - x_0)^2 + (y - y_0)^2 < \delta^2.$$

Geometrically speaking, continuity of $f(x, y)$ at (x_0, y_0) means that for each interval of length 2ϵ with midpoint $f(x_0, y_0)$ we can find a circular disk with nonzero radius δ and center (x_0, y_0) in that neighborhood such that for every point (x, y) in the disk the corresponding function value $f(x, y)$ lies in that interval (Fig. 194).

Continuity will be essential in the next theorem and proof.

---

[6]That is, in some circular disk $(x - x_0)^2 + (y - y_0)^2 < r^2$, $(r > 0)$, in the xy-plane.

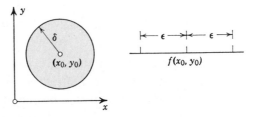

Fig. 194. Continuity of a function of two variables

Chain Rules for Functions of Several Variables

From elementary calculus we know that if w is a differentiable function of x, and x is a differentiable function of t, then

$$\frac{dw}{dt} = \frac{dw}{dx} \frac{dx}{dt} \, .$$

This so-called chain rule of differentiation can be generalized as follows.

Theorem 1 (Chain rule)

Let $w = f(x, y)$ be continuous and have continuous first partial derivatives at every point of a domain[7] D in the xy-plane. Let $x = x(t)$ and $y = y(t)$ be differentiable functions of a variable in some interval T such that, for each t in T, the point $[x(t), y(t)]$ lies in D. Then $w = f[x(t), y(t)]$ is a differentiable function for all t in T, and

(1)
$$\boxed{\frac{dw}{dt} = \frac{\partial w}{\partial x} \frac{dx}{dt} + \frac{\partial w}{\partial y} \frac{dy}{dt} \, .}$$

Proof. We choose a t in T and Δt so small that $t + \Delta t$ is also in T, and set

(2) $\Delta x = x(t + \Delta t) - x(t), \qquad \Delta y = y(t + \Delta t) - y(t),$

and furthermore

$$\Delta w = f(x + \Delta x, y + \Delta y) - f(x, y).$$

By adding and subtracting a term we see that this equation may be written

$$\Delta w = [f(x + \Delta x, y + \Delta y) - f(x, y + \Delta y)] + [f(x, y + \Delta y) - f(x, y)].$$

If we apply the mean value theorem for a function of a single variable (cf. Ref. [18]) to each of the two expressions in the brackets, we obtain

---

[7]A **domain** D is an open connected point set, where "connected" means that any two points of D can be joined by a broken line of finitely many linear segments all of whose points belong to D, and "open" means that every point of D has a neighborhood all of whose points belong to D. For example, the interior of a rectangle or a circle is a domain.

$$\Delta w = \Delta x \left.\frac{\partial f}{\partial x}\right|_{x_1, y + \Delta y} + \Delta y \left.\frac{\partial f}{\partial y}\right|_{x, y_1}$$

where x_1 lies between x and $x + \Delta x$, and y_1 lies between y and $y + \Delta y$. Dividing this equation by Δt on both sides, letting Δt approach 0, and noting that $\partial f/\partial x$ and $\partial f/\partial y$ are assumed to be continuous, we obtain (1).   ∎

Under analogous conditions this theorem remains valid for a function $w = f(x, y, z)$ with x, y, z depending on t; instead of (1) we then have

(3)
$$\boxed{\frac{dw}{dt} = \frac{\partial w}{\partial x}\frac{dx}{dt} + \frac{\partial w}{\partial y}\frac{dy}{dt} + \frac{\partial w}{\partial z}\frac{dz}{dt}.}$$

The idea of the proof is the same. This extends further to the case when x, y, z depend on *two* variables u, v, as follows.

Theorem 2 (Chain rule)

Let $w = f(x, y, z)$ be continuous and have continuous first partial derivatives in a domain D in xyz-space. Let $x = x(u, v)$, $y = y(u, v)$, $z = z(u, v)$ be functions which are continuous and have first partial derivatives in a domain B in the uv-plane, where B is such that for every point (u, v) in B, the corresponding point $[x(u, v), y(u, v), z(u, v)]$ lies in D. Then the function

$$w = f(x(u, v), y(u, v), z(u, v))$$

is defined in B, has first partial derivatives with respect to u and v in B, and

(4)
$$\boxed{\begin{aligned}\frac{\partial w}{\partial u} &= \frac{\partial w}{\partial x}\frac{\partial x}{\partial u} + \frac{\partial w}{\partial y}\frac{\partial y}{\partial u} + \frac{\partial w}{\partial z}\frac{\partial z}{\partial u}, \\[2ex] \frac{\partial w}{\partial v} &= \frac{\partial w}{\partial x}\frac{\partial x}{\partial v} + \frac{\partial w}{\partial y}\frac{\partial y}{\partial v} + \frac{\partial w}{\partial z}\frac{\partial z}{\partial v}.\end{aligned}}$$

The proof follows immediately from the (extended) Theorem 1 [formula (3)] by keeping one of the two variables u and v constant.

Mean Value Theorem

From calculus we know that if a function $f(x)$ is differentiable, then

$$f(x_0 + h) - f(x_0) = h\frac{df}{dx},$$

the derivative being evaluated at a suitable point between x_0 and $x_0 + h$ (cf. Ref. [18] in Appendix 1). This so-called mean value theorem of differential calculus can be extended to functions of two variables as follows.

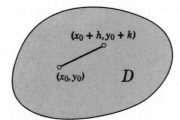

Fig. 195. Mean value theorem

Theorem 3 (Mean value theorem)

Let $f(x, y)$ be continuous and have continuous first partial derivatives in a domain D. Furthermore, let (x_0, y_0) and $(x_0 + h, y_0 + k)$ be points in D such that the line segment joining these points lies in D (Fig. 195). Then

$$(5) \qquad f(x_0 + h, y_0 + k) - f(x_0, y_0) = h\frac{\partial f}{\partial x} + k\frac{\partial f}{\partial y},$$

the partial derivatives being evaluated at a suitable point of that segment.

Proof. Let

$$x = x(t) = x_0 + th, \qquad y = y(t) = y_0 + tk \qquad (0 \le t \le 1)$$

and

$$F(t) = f(x(t), y(t)) = f(x_0 + th, y_0 + tk).$$

Then

$$f(x_0 + h, y_0 + k) = F(1), \qquad f(x_0, y_0) = F(0).$$

By the mean value theorem for a function of a single variable there is a value t_1 between 0 and 1 such that

$$(6) \qquad f(x_0 + h, y_0 + k) - f(x_0, y_0) = F(1) - F(0) = F'(t_1).$$

We now apply Theorem 1 to $F(t)$. Since $dx/dt = h$ and $dy/dt = k$, we obtain

$$(7) \qquad F'(t_1) = \frac{\partial f}{\partial x} h + \frac{\partial f}{\partial y} k,$$

the derivatives on the right being evaluated at the point $(x_0 + t_1 h, y_0 + t_1 k)$, which lies on the line segment with endpoints (x_0, y_0) and $(x_0 + h, y_0 + k)$. By inserting (7) into (6) we obtain (5), and the proof is complete. ∎

For a function $f(x, y, z)$ of three variables, satisfying conditions analogous to those in Theorem 3, the consideration is quite similar and leads to

$$(8) \quad f(x_0 + h, y_0 + k, z_0 + l) - f(x_0, y_0, z_0) = h\frac{\partial f}{\partial x} + k\frac{\partial f}{\partial y} + l\frac{\partial f}{\partial z},$$

the partial derivatives being evaluated at a suitable point of the segment with endpoints (x_0, y_0, z_0) and $(x_0 + h, y_0 + k, z_0 + l)$.

The theorems in the present section are the basis of our further discussion of scalar and vector functions of several variables. This discussion in Secs. 8.8—8.11 (the end of the chapter) centers around the concepts of **gradient** (Sec. 8.8), **divergence** (Sec. 8.9) and **curl** (Sec. 8.10), which are of great importance in vector calculus and its physical applications. Indeed, it is fair to say that without these three concepts, the whole vector calculus would not be of much value to physicists and engineers. **Integral theorems** involving these concepts follow in Chap. 9.

Problems for Sec. 8.7

Find dw/dt by (1) and check the result by substitution and differentiation, where

1. $w = x + 2y$, $x = t^4$, $y = \ln t$ **2.** $w = \sqrt{x^2 + y^2}$, $x = e^{4t}$, $y = e^{-4t}$

3. $w = x/y$, $x = g(t)$, $y = h(t)$ **4.** $w = y^x$, $x = \cos t$, $y = \sin t$

5. Carry out the details of the proof of (3).

Find dw/dt by (3) and check the result by substitution and differentiation, where

6. $w = x^2 + y^2 + z^2$, $x = e^{2t} \cos 2t$, $y = e^{2t} \sin 2t$, $z = e^{4t}$

7. $w = (x^2 + y^2 + z^2)^{-1/2}$, $x = \cos t$, $y = \sin t$, $z = t$

8. $w = x \sin yz$, $x = t^4 + 1$, $y = t^2 + 1$, $z = t^2 - 1$

9. $w = xy - yz + zx$, $x = \cosh t$, $y = e^t$, $z = \sinh t$

10. Prove Theorem 2.

Find $\partial w/\partial u$ and $\partial w/\partial v$, where

11. $w = x^2 + y^2$, $x = u + v$, $y = u - v$

12. $w = x^4 - 4x^2y^2 + y^4$, $x = uv$, $y = u/v$

13. $w = \ln (x^2 + y^2)$, $x = e^u \cos v$, $y = e^u \sin v$

14. $w = xy$, $x = \cosh u^2 \cosh v^2$, $y = \sinh u^2 \sinh v^2$

15. $w = x^2 - y^2$, $x = u^2 - v^2$, $y = 2uv$

16. (Partial derivatives on a surface) Let $w = f(x, y, z)$, and let $z = g(x, y)$ represent a surface S in space. Then on S, the function becomes $\widetilde{w}(x, y) = f[x, y, g(x, y)]$. Show that its partial derivatives are obtained from

$$\frac{\partial \widetilde{w}}{\partial x} = \frac{\partial f}{\partial x} + \frac{\partial f}{\partial z}\frac{\partial g}{\partial x}, \qquad \frac{\partial \widetilde{w}}{\partial y} = \frac{\partial f}{\partial y} + \frac{\partial f}{\partial z}\frac{\partial g}{\partial y} \qquad [z = g(x, y)].$$

Apply this to $f = x^3 + y^3 + z^2$, $g = x^2 + y^2$ and check by substitution and direct differentiation.

17. Let $w = f(x, y)$ and $x = r \cos \theta$, $y = r \sin \theta$. Show that

$$\frac{\partial w}{\partial r} = \frac{\partial w}{\partial x} \cos \theta + \frac{\partial w}{\partial y} \sin \theta, \qquad \frac{\partial w}{\partial \theta} = -\frac{\partial w}{\partial x} r \sin \theta + \frac{\partial w}{\partial y} r \sin \theta$$

18. Let f and x, y be as in Prob. 17. Show that

$$\left(\frac{\partial w}{\partial r}\right)^2 + \frac{1}{r^2}\left(\frac{\partial w}{\partial \theta}\right)^2 = \left(\frac{\partial w}{\partial x}\right)^2 + \left(\frac{\partial w}{\partial y}\right)^2.$$

19. **(Laplacian)** $w_{xx} + w_{yy}$ is called the *Laplacian* of w. Show that for w, x, y as in Prob. 17,

$$w_{xx} + w_{yy} = w_{rr} + \frac{1}{r} w_r + \frac{1}{r^2} w_{\theta\theta}.$$

20. **(Wave equation)** $w_{tt} - c^2 w_{xx} = 0$ is called the *one-dimensional wave equation*. Let $w = f(v, z)$ and $v = x + ct$, $z = x - ct$. Show that

$$w_{tt} - c^2 w_{xx} = -4c^2 w_{vz}.$$

8.8 Directional Derivative. Gradient of a Scalar Field

We consider a scalar field in space given by a function $f(P) = f(x, y, z)$ (cf. Sec. 8.1). We know that the first partial derivatives of f are the rates of change of f in the directions of the coordinate axes. It seems unnatural to restrict attention to these three directions, and we may ask for the rate of change of f in any direction. This simple idea leads to the notion of a directional derivative.

Directional Derivative

To define this derivative we choose a point P in space and a direction at P, given by a vector **b.** Let C be the ray from P in the direction of **b,** and let Q be a point on C, whose distance from P is s (Fig. 196). Then if the limit

(1) $$D_{\mathbf{b}}f = \frac{df}{ds} = \lim_{s \to 0} \frac{f(Q) - f(P)}{s} \qquad (s = \text{distance between } P \text{ and } Q)$$

exists, it is called the **directional derivative** *of f at P in the direction of* **b.** Obviously, this is the rate of change of f at P in the direction of **b.** Both notations $D_{\mathbf{b}}f$ and df/ds are common, but $D_{\mathbf{b}}f$ has the advantage of indicating the direction.

In this way there are now infinitely many directional derivatives of f at P, each corresponding to a certain direction. But, a Cartesian coordinate system being given, we may express any such derivative in terms of the first

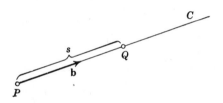

Fig. 196. Directional derivative

partial derivatives of f at P as follows. If P has the position vector $\mathbf{p}_0$ and we now assume that $\mathbf{b}$ is a unit vector, we can represent C in the form

$$(2) \qquad \mathbf{r}(s) = x(s)\mathbf{i} + y(s)\mathbf{j} + z(s)\mathbf{k} = \mathbf{p}_0 + s\mathbf{b} \qquad (s \geqq 0, \quad |\mathbf{b}| = 1).$$

Then $D_{\mathbf{b}}f = df/ds$ is the derivative of the function $f[x(s), y(s), z(s)]$ with respect to the arc length s of C. Hence, assuming that f has continuous first partial derivatives and applying the chain rule (Theorem 1 in the previous section), we obtain

$$(3) \qquad D_{\mathbf{b}}f = \frac{df}{ds} = \frac{\partial f}{\partial x} x' + \frac{\partial f}{\partial y} y' + \frac{\partial f}{\partial z} z'$$

where primes denote derivatives with respect to s (which are evaluated at $s = 0$).

Gradient

From (2) we have

$$\mathbf{r}' = x'\mathbf{i} + y'\mathbf{j} + z'\mathbf{k} = \mathbf{b}.$$

Together with (3) this suggests that we introduce the vector

$$(4) \qquad \boxed{\operatorname{grad} f = \frac{\partial f}{\partial x}\mathbf{i} + \frac{\partial f}{\partial y}\mathbf{j} + \frac{\partial f}{\partial z}\mathbf{k},}$$

which is called the **gradient** of the scalar function f; then we can write (3) in the form of an inner product (dot product):

$$(5) \qquad \boxed{D_{\mathbf{b}}f = \frac{df}{ds} = \mathbf{b} \cdot \operatorname{grad} f} \qquad (|\mathbf{b}| = 1).$$

Attention! If the direction is given by a vector $\mathbf{a}$ of any length ($\neq 0$), then

$$(5') \qquad D_{\mathbf{a}}f = \frac{df}{ds} = \frac{1}{|\mathbf{a}|} \mathbf{a} \cdot \operatorname{grad} f.$$

By introducing the *differential operator*

$$\boxed{\nabla = \frac{\partial}{\partial x}\mathbf{i} + \frac{\partial}{\partial y}\mathbf{j} + \frac{\partial}{\partial z}\mathbf{k}}$$

(read **nabla** or "del") we may write

$$\boxed{\operatorname{grad} f = \nabla f = \frac{\partial f}{\partial x}\mathbf{i} + \frac{\partial f}{\partial y}\mathbf{j} + \frac{\partial f}{\partial z}\mathbf{k}.}$$

The notation ∇f for the gradient is frequently used in engineering.

If in particular $\mathbf{b}$ has the direction of the positive x-axis, then $\mathbf{b} = \mathbf{i}$, and

$$D_{\mathbf{i}}f = \mathbf{i} \cdot \text{grad } f = \mathbf{i} \cdot \frac{\partial f}{\partial x} \mathbf{i} = \frac{\partial f}{\partial x} \mathbf{i} \cdot \mathbf{i} = \frac{\partial f}{\partial x}.$$

Similarly, the directional derivative in the positive y-direction is $\partial f / \partial y$, etc.

EXAMPLE 1. Directional derivative

Find the directional derivative of $f(x, y, z) = 2x^2 + 3y^2 + z^2$ at the point P: (2, 1, 3) in the direction of the vector $\mathbf{a} = \mathbf{i} - 2\mathbf{k}$.

Solution. We obtain

$$\text{grad } f = 4x\mathbf{i} + 6y\mathbf{j} + 2z\mathbf{k}, \quad \text{and at } P, \quad \text{grad } f = 8\mathbf{i} + 6\mathbf{j} + 6\mathbf{k}.$$

From this and (5'),

$$D_{\mathbf{a}}f = \frac{1}{\sqrt{5}}(\mathbf{i} - 2\mathbf{k}) \cdot (8\mathbf{i} + 6\mathbf{j} + 6\mathbf{k}) = \frac{1 \cdot 8 - 2 \cdot 6}{\sqrt{5}} = -\frac{4}{\sqrt{5}} \approx -1.789.$$

The minus sign indicates that f decreases at P in the direction of $\mathbf{a}$.   ∎

Gradient Characterizes Maximum Increase

We first want to show that *the length and direction of* grad f *are independent of the particular choice of Cartesian coordinates.*

This is, of course, not obvious, because (4) involves partial derivatives, which depend on the choice of the coordinates, so that we do not know yet whether the corresponding expression

$$\frac{\partial f}{\partial x^*}\mathbf{i}^* + \frac{\partial f}{\partial y^*}\mathbf{j}^* + \frac{\partial f}{\partial z^*}\mathbf{k}^*$$

with respect to other Cartesian coordinates x^*, y^*, z^* (and corresponding unit vectors $\mathbf{i}^*$, $\mathbf{j}^*$, $\mathbf{k}^*$) will have the same length and direction as (4).

To prove our proposition, we may reason as follows. By the definition of a scalar function, the value of f at a point P depends on P but is independent of the coordinates, and s, the arc length of that ray C, is also independent of the choice of coordinates. Hence $D_{\mathbf{b}}f$ is independent of the particular choice of coordinates. From (5) we obtain

$$D_{\mathbf{b}}f = |\mathbf{b}| \, |\text{grad } f| \cos \gamma = |\text{grad } f| \cos \gamma,$$

where γ is the angle between $\mathbf{b}$ and grad f. We see that $D_{\mathbf{b}}f$ is maximum when $\cos \gamma = 1$, $\gamma = 0$, and then $D_{\mathbf{b}}f = |\text{grad } f|$. This shows that the length and direction of grad f are independent of the coordinates, and we have:

Theorem 1 (Gradient)

Let $f(P) = f(x, y, z)$ be a scalar function having continuous first partial derivatives. Then grad f *exists and its length and direction are independent of the particular choice of Cartesian coordinates in space. If at a point P the gradient of f is not the zero vector, it has the direction of maximum increase of f at P.*

Gradient as Normal Vector to Surfaces

Another important geometrical characterization of the gradient can be obtained as follows. Consider a differentiable scalar function $f(x, y, z)$ in space. Suppose that for each constant c the equation

$$(6) \qquad f(x, y, z) = c = const$$

represents a surface S in space. Then, by letting c assume all values, we obtain a family of surfaces, which are called the **level surfaces** of the function f. Since, by the definition of a function, our function f has a unique value at each point in space, it follows that through each point in space there passes one, and only one, level surface of f. We remember that a curve C in space may be represented in the form (cf. Sec. 8.3)

$$(7) \qquad \mathbf{r}(t) = [x(t), y(t), z(t)] = x(t)\mathbf{i} + y(t)\mathbf{j} + z(t)\mathbf{k}.$$

If we now require that C lie on S, then the functions $x(t)$, $y(t)$, $z(t)$ in (7) must be such that

$$f[x(t), y(t), z(t)] = c;$$

cf. (6). By differentiating this with respect to t and using the chain rule (Sec. 8.7) we obtain

$$(8) \qquad \frac{\partial f}{\partial x} x' + \frac{\partial f}{\partial y} y' + \frac{\partial f}{\partial z} z' = (\text{grad } f) \cdot \mathbf{r}' = 0,$$

where the vector

$$\mathbf{r}' = \frac{d\mathbf{r}}{dt} = [x', y', z'] = x'\mathbf{i} + y'\mathbf{j} + z'\mathbf{k}$$

is tangent to C (cf. Sec. 8.4). If we consider curves on S passing through a point P of S in various directions, their tangents at P will, in general, lie in the same plane that touches S at P. This plane is called the **tangent plane** of S at P. The straight line through P and perpendicular to the tangent plane is called the **normal** of S at P (Fig. 197). From (8) and Theorem 1, Sec. 6.5, we thus obtain the following result.

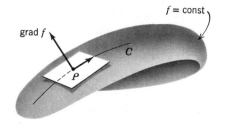

Fig. 197. Level surface and gradient

Theorem 2 (Gradient and surface normal)

Let f be a scalar function that is defined and differentiable in a domain D in space, and let P be any point in D on a level surface S of f. Then if the gradient of f at P is not the zero vector, it is perpendicular to S at P; that is, it has the direction of the normal to S at P.

EXAMPLE 2. Normal to a plane curve

The level curves $f = const$ of $f(x, y) = \ln (x^2 + y^2)$ are concentric circles about the origin. The gradient

$$\text{grad } f = \frac{\partial f}{\partial x}\mathbf{i} + \frac{\partial f}{\partial y}\mathbf{j} = \frac{2x}{x^2 + y^2}\mathbf{i} + \frac{2y}{x^2 + y^2}\mathbf{j}$$

has the direction of the normals to the circles, and its direction corresponds to that of the maximum increase of f. For example, at the point P: (2, 1), we have (cf. Fig. 198a)

$$\text{grad } f = 0.8\mathbf{i} + 0.4\mathbf{j}.$$

EXAMPLE 3. Normal to a surface

Find a unit normal vector $\mathbf{n}$ of the cone of revolution $z^2 = 4(x^2 + y^2)$ at the point P: (1, 0, 2).

Solution. The cone is the level surface $f = 0$ of $f(x, y, z) = 4(x^2 + y^2) - z^2$. Thus

$$\text{grad } f = 8x\mathbf{i} + 8y\mathbf{j} - 2z\mathbf{k} \qquad \text{and at } P, \qquad \text{grad } f = 8\mathbf{i} - 4\mathbf{k}.$$

Hence, by Theorem 2, a unit normal vector of the cone at P is

$$\mathbf{n} = \frac{1}{|\text{grad } f|} \text{grad } f = \frac{2}{\sqrt{5}}\mathbf{i} - \frac{1}{\sqrt{5}}\mathbf{k}$$

(cf. Fig. 198b), and the other one is $-\mathbf{n}$. ∎

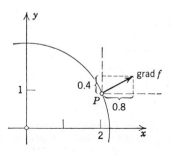

Fig. 198a. Normal to a circle in Example 2

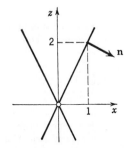

Fig. 198b. Intersection of the xz-plane and the cone in Example 3

Gradient of a Potential

Some of the vector fields occurring in physics are given by vector functions that can be obtained as the gradients of suitable scalar functions. Such a scalar function is then called a *potential function* or **potential** of the corresponding vector field. The use of potentials simplifies the investigation of those vector fields considerably. To obtain a first impression of this approach to vector fields, let us consider an important example.

EXAMPLE 4. Gravitational field. Laplace's equation

In Example 4, Sec. 8.1, we have seen that, according to Newton's law of gravitation, the force of attraction between two particles is

(9)
$$\mathbf{p} = -\frac{c}{r^3}\mathbf{r} = -c\left(\frac{x - x_0}{r^3}\mathbf{i} + \frac{y - y_0}{r^3}\mathbf{j} + \frac{z - z_0}{r^3}\mathbf{k}\right),$$

where

$$r = \sqrt{(x - x_0)^2 + (y - y_0)^2 + (z - z_0)^2}$$

is the distance between the two particles and c is a constant. Observing that

(10a)
$$\frac{\partial}{\partial x}\left(\frac{1}{r}\right) = -\frac{2(x - x_0)}{2[(x - x_0)^2 + (y - y_0)^2 + (z - z_0)^2]^{3/2}} = -\frac{x - x_0}{r^3}$$

and similarly

(10b)
$$\frac{\partial}{\partial y}\left(\frac{1}{r}\right) = -\frac{y - y_0}{r^3}, \qquad \frac{\partial}{\partial z}\left(\frac{1}{r}\right) = -\frac{z - z_0}{r^3},$$

we see that $\mathbf{p}$ is the gradient of the scalar function

$$f(x, y, z) = \frac{c}{r} \qquad\qquad (r > 0);$$

that is, f is a potential of that gravitational field.

By differentiating (10) we find

$$\frac{\partial^2}{\partial x^2}\left(\frac{1}{r}\right) = -\frac{1}{r^3} + \frac{3(x - x_0)^2}{r^5}, \qquad \frac{\partial^2}{\partial y^2}\left(\frac{1}{r}\right) = -\frac{1}{r^3} + \frac{3(y - y_0)^2}{r^5},$$

$$\frac{\partial^2}{\partial z^2}\left(\frac{1}{r}\right) = -\frac{1}{r^3} + \frac{3(z - z_0)^2}{r^5}.$$

Since the sum of the three expressions on the right is zero, we see that the potential $f = c/r$ satisfies the equation

(11)
$$\boxed{\frac{\partial^2 f}{\partial x^2} + \frac{\partial^2 f}{\partial y^2} + \frac{\partial^2 f}{\partial z^2} = 0.}$$

This important partial differential equation is called **Laplace's equation**; it will be considered in detail in Chaps. 11 and 17. The expression on the left is called the **Laplacian** of f and is denoted by $\nabla^2 f$ or Δf. The differential operator

$$\nabla^2 = \Delta = \frac{\partial^2}{\partial x^2} + \frac{\partial^2}{\partial y^2} + \frac{\partial^2}{\partial z^2}$$

(read "nabla squared" or "delta") is called the **Laplace operator**. Using this operator, we may write (11) in the form

$$\boxed{\nabla^2 f = 0.}$$

It can be shown that the field of force produced by any distribution of masses is given by a vector function that is the gradient of a scalar function f, and f satisfies (11) in any region of space that is free of matter.

There are other laws in physics that are of the same form as Newton's law of gravitation. For example, in electrostatics the force of attraction (or repulsion) between two particles of opposite (or like) charges Q_1 and Q_2 is

$$\mathbf{p} = \frac{k}{r^3}\mathbf{r} \qquad \text{(Coulomb's law[8])}$$

where $k = Q_1 Q_2/4\pi\epsilon$, and ϵ is the dielectric constant. Hence $\mathbf{p}$ is the gradient of the potential $f = -k/r$, and f satisfies (11) when $r > 0$.   ∎

If the vector function defining a vector field is the gradient of a scalar function, the field is said to be **conservative,** because, as we shall see in Sec. 9.9, in such a field the work done in displacing a particle from a point P_1 to a point P_2 in the field depends only on P_1 and P_2 but not on the path along which the particle is displaced from P_1 to P_2. We shall also see that not every field is conservative.

We have seen that the gradient is a vector function associated with a given scalar function. The **divergence,** which we discuss next, will be a scalar function associated with a given vector function. It is suggested by physics, for instance, in connection with flow problems.

Problems for Sec. 8.8

Gradient. (a) Find the gradient ∇f. (b) Graph some level curves $f = const$ and indicate ∇f by arrows at some points of these curves.

1. $f = xy$
2. $f = 2x - 3y$
3. $f = x^2 + y^2$
4. $f = \sqrt{x^2 + y^2}$
5. $f = y - 4x^2$
6. $f = \arctan(y/x)$
7. $f = x^2 + 4y^2$
8. $f = y/x^2$
9. $f = \ln(x^2 + y^2)$

Find the gradient ∇f, where f equals

10. $\sin x \cosh y$
11. $e^x \cos y$
12. $\sinh(x^2 + y^2)$
13. xyz
14. $yz + zx + xy$
15. $\frac{1}{4}x^2 - 4y^2 + \frac{1}{2}z^2$
16. $(x + y + z)^2$
17. e^{xyz}
18. $(x^2 + y^2 + z^2)^{-1/2}$

Find a scalar function f whose gradient is

19. $\mathbf{i} + \mathbf{j} + \mathbf{k}$
20. $\mathbf{0}$
21. $2x\mathbf{i} - 6y\mathbf{j} + \mathbf{k}$
22. $(x + y)(\mathbf{i} + \mathbf{j})$
23. $yz\mathbf{i} + xz\mathbf{j} + xy\mathbf{k}$
24. $(x\mathbf{i} + y\mathbf{j})/(x^2 + y^2)$
25. $y\mathbf{i} + (x + z)\mathbf{j} + y\mathbf{k}$
26. $y^{-1}\mathbf{i} - xy^{-2}\mathbf{j}$
27. $e^{xy}(y\mathbf{i} + x\mathbf{j})$

Formulas for the gradient. Granted sufficient differentiability, show that

28. $\nabla(fg) = f\nabla g + g\nabla f$
29. $\nabla(f^n) = nf^{n-1}\nabla f$
30. $\nabla(f/g) = (1/g^2)(g\nabla f - f\nabla g)$
31. $\nabla^2(fg) = g\nabla^2 f + 2\nabla f \cdot \nabla g + f\nabla^2 g$

Directional derivative. Find the directional derivative of f at P in the direction of $\mathbf{a}$, where

32. $f = x - y$, P: (4, 5), $\mathbf{a} = 2\mathbf{i} + \mathbf{j}$
33. $f = x^2 + y^2$, P: (1, 1), $\mathbf{a} = 3\mathbf{i} - 4\mathbf{j}$
34. $f = \ln(x^2 + y^2)$, P: (4, 0), $\mathbf{a} = \mathbf{i} - \mathbf{j}$
35. $f = 1/\sqrt{x^2 + y^2 + z^2}$, P: (3, 0, 4), $\mathbf{a} = \mathbf{i} + \mathbf{j} + \mathbf{k}$
36. $f = x^2 + y^2 + z^2$, P: (1, 0, 1), $\mathbf{a} = -\mathbf{i} - \mathbf{j} - \mathbf{k}$

[8]CHARLES AUGUSTIN DE COULOMB (1736—1806), French physicist and engineer. Coulomb's law was derived by him from his own very precise measurements.

37. $f = xyz,$ $P: (3, 3, 0),$ $\mathbf{a} = 3\mathbf{j} - 2\mathbf{k}$
38. $f = xy,$ $P: (1, 1, -\frac{1}{2}),$ $\mathbf{a} = \mathbf{i} - \mathbf{j} + 7\mathbf{k}$
39. $f = x^2 + 4y^2 + \frac{1}{4}z^2,$ $P: (3, 1, 1),$ $\mathbf{a} = 2\mathbf{i} - \frac{5}{2}\mathbf{j} + 16\mathbf{k}$
40. $f = xy + yz + 25,$ $P: (-2, 4, 7),$ $\mathbf{a} = -\mathbf{i} + \mathbf{j} + 5\mathbf{k}$

Some applications. Find a unit normal vector to the given curve in the xy-plane or surface in space at the given point P.

41. $y = \frac{4}{3}x - \frac{2}{3},$ $P: (2, 2)$ 42. $y = 1 - x^2,$ $P: (1, 0)$
43. $x^2 + y^2 = 25,$ $P: (3, 4)$ 44. $4x^2 + 9y^2 = 36,$ $P: (0, 2)$
45. $y = x^{3/2},$ $P: (1, 1)$ 46. $ax + by + cz + d = 0,$ any P
47. $z = \sqrt{x^2 + y^2},$ $P: (6, 8, 10)$ 48. $z = x^2 + y^2,$ $P: (1, 2, 5)$
49. $x^2 + y^2 + z^2 = 32,$ $P: (4, 4, 0)$ 50. $x^2 + y^2 + 2z^2 = 26,$ $P: (2, 2, 3)$

51. **(Heat flow)** In a temperature field, heat flows in the direction of maximum decrease of temperature T. Find this direction at $P: (2, 1)$ when $T = x^3 - 3xy^2$.
52. Same question as in Prob. 51, when $T = 4x^2 + y^2 - 5z^2$ and $P: (\frac{1}{4}, -2, \frac{1}{2})$.
53. If on a mountain, the elevation above sea level is $z(x, y) = 1500 - 3x^2 - 5y^2$ [meters], what is the direction of steepest ascent at $P: (-0.2, 0.1)$?
54. **(Electric field)** If the potential between two concentric cylinders is $V(x, y) = 110 + 30 \ln (x^2 + y^2)$ [volts], what is the direction of the electric force (the gradient) at $P: (2, 5)$?
55. What is the direction of the equipotential line in Prob. 54 at P?

8.9 Divergence of a Vector Field

Let $\mathbf{v}(x, y, z)$ be a differentiable vector function, where x, y, z are Cartesian coordinates, and let v_1, v_2, v_3 be the components of $\mathbf{v}$. Then the function

$$(1) \qquad \boxed{\text{div } \mathbf{v} = \frac{\partial v_1}{\partial x} + \frac{\partial v_2}{\partial y} + \frac{\partial v_3}{\partial z}}$$

is called the **divergence** *of* $\mathbf{v}$ or the *divergence of the vector field defined by* $\mathbf{v}$. Another common notation for the divergence of $\mathbf{v}$ is $\nabla \cdot \mathbf{v}$,

$$\text{div } \mathbf{v} = \nabla \cdot \mathbf{v} = \left(\frac{\partial}{\partial x}\mathbf{i} + \frac{\partial}{\partial y}\mathbf{j} + \frac{\partial}{\partial z}\mathbf{k} \right) \cdot (v_1\mathbf{i} + v_2\mathbf{j} + v_3\mathbf{k})$$

$$= \frac{\partial v_1}{\partial x} + \frac{\partial v_2}{\partial y} + \frac{\partial v_3}{\partial z},$$

with the understanding that the "product" $(\partial/\partial x)v_1$ in the dot product means the partial derivative $\partial v_1/\partial x$, etc. This is a convenient notation, but nothing more. Note that $\nabla \cdot \mathbf{v}$ means the scalar div $\mathbf{v}$, whereas ∇f means the vector grad f defined in Sec. 8.8.

For example, if

$$\mathbf{v} = 3xz\mathbf{i} + 2xy\mathbf{j} - yz^2\mathbf{k}, \qquad \text{then} \qquad \text{div } \mathbf{v} = 3z + 2x - 2yz.$$

We shall see below that the divergence has an important physical meaning. Clearly the values of a function that characterizes a physical or geometrical property must be independent of the particular choice of coordinates; that is, those values must be invariant with respect to coordinate transformations.

Theorem 1 (Invariance of the divergence)

The values of div **v** *depend only on the points in space* (*and, of course, on* **v**) *but not on the particular choice of the coordinates in* (1), *so that with respect to other Cartesian coordinates* x^*, y^*, z^* *and corresponding components* $v_1{}^*$, $v_2{}^*$, $v_3{}^*$ *of* **v** *the function* div **v** *is given by*

$$(2) \qquad\qquad \text{div } \mathbf{v} = \frac{\partial v_1{}^*}{\partial x^*} + \frac{\partial v_2{}^*}{\partial y^*} + \frac{\partial v_3{}^*}{\partial z^*} .$$

This is proved on p. 506 near the end of the chapter. Presently let us turn to the more immediate task of getting a feeling for the significance of the divergence.

If $f(x, y, z)$ is a twice differentiable scalar function, then

$$\text{grad } f = \frac{\partial f}{\partial x} \mathbf{i} + \frac{\partial f}{\partial y} \mathbf{j} + \frac{\partial f}{\partial z} \mathbf{k}$$

and by (1),

$$\text{div (grad } f) = \frac{\partial^2 f}{\partial x^2} + \frac{\partial^2 f}{\partial y^2} + \frac{\partial^2 f}{\partial z^2} .$$

The expression on the right is the Laplacian of f (cf. Sec. 8.8). Thus

$$(3) \qquad\qquad \boxed{\text{div (grad } f) = \nabla^2 f.}$$

EXAMPLE 1. Gravitational force
The gravitational force **p** in Example 4, Sec. 8.8, is the gradient of the scalar function $f(x, y, z) = c/r$, which satisfies Laplace's equation $\nabla^2 f = 0$. According to (3), this means that div **p** $= 0$ ($r > 0$). ∎

The following example, taken from hydrodynamics, illustrates the physical significance of the divergence of a vector field. More about the physical interpretation of the divergence follows later (in Sec. 9.7).

EXAMPLE 2. Motion of a compressible fluid
We consider the motion of a fluid in a region R having no **sources** or **sinks** in R, that is, no points at which fluid is produced or disappears. The concept of **fluid state** is meant to cover also gases and vapors. Fluids in the restricted sense, or liquids, have a small compressibility, which can be neglected in many problems. Gases and vapors have a large compressibility; that is, their density ρ ($=$ mass per unit volume) depends on the coordinates x, y, z in space (and may depend on time t). We assume that our fluid is compressible.

We consider the flow through a small rectangular parallelepiped W of dimensions[9] Δx, Δy, Δz

---

[9]It is a standard usage to indicate small quantities by Δ; this has, of course, nothing to do with the Laplacian.

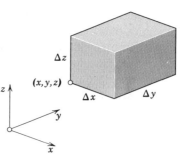

Fig. 199. Physical interpretation of the divergence

with edges parallel to the coordinate axes (Fig. 199). W has the volume $\Delta V = \Delta x \, \Delta y \, \Delta z$. Let

$$\mathbf{v} = [v_1, v_2, v_3] = v_1\mathbf{i} + v_2\mathbf{j} + v_3\mathbf{k}$$

be the velocity vector of the motion. We set

(4) $$\mathbf{u} = \rho\mathbf{v} = [u_1, u_2, u_3] = u_1\mathbf{i} + u_2\mathbf{j} + u_3\mathbf{k}$$

and assume that $\mathbf{u}$ and $\mathbf{v}$ are continuously differentiable vector functions of x, y, z and t. Let us calculate the change in the mass included in W by considering the **flux** across the boundary, that is, the total loss of mass leaving W per unit time. Consider the flow through the left-hand face of W, whose area is $\Delta x \, \Delta z$. The components v_1 and v_3 of $\mathbf{v}$ are parallel to that face and contribute nothing to this flow. Hence the mass of fluid entering through that face during a short time interval Δt is given approximately by

$$(\rho v_2)_y \, \Delta x \, \Delta z \, \Delta t = (u_2)_y \, \Delta x \, \Delta z \, \Delta t,$$

where the subscript y indicates that this expression refers to the left-hand face. The mass of fluid leaving the parallelepiped W through the opposite face during the same time interval is approximately $(u_2)_{y+\Delta y} \, \Delta x \, \Delta z \, \Delta t$, where the subscript $y + \Delta y$ indicates that this expression refers to the right-hand face. The difference

$$\Delta u_2 \, \Delta x \, \Delta z \, \Delta t = \frac{\Delta u_2}{\Delta y} \Delta V \, \Delta t \qquad [\Delta u_2 = (u_2)_{y+\Delta y} - (u_2)_y]$$

is the approximate loss of mass. Two similar expressions are obtained by considering the other two pairs of parallel faces of W. If we add these three expressions, we find that the total loss of mass in W during the time interval Δt is approximately

$$\left(\frac{\Delta u_1}{\Delta x} + \frac{\Delta u_2}{\Delta y} + \frac{\Delta u_3}{\Delta z} \right) \Delta V \, \Delta t,$$

where

$$\Delta u_1 = (u_1)_{x+\Delta x} - (u_1)_x \quad \text{and} \quad \Delta u_3 = (u_3)_{z+\Delta z} - (u_3)_z.$$

This loss of mass in W is caused by the time rate of change of the density and is thus equal to

$$-\frac{\partial \rho}{\partial t} \Delta V \, \Delta t.$$

If we equate both expressions, divide the resulting equation by $\Delta V \, \Delta t$, and let Δx, Δy, Δz and Δt approach zero, then we obtain

$$\operatorname{div} \mathbf{u} = \operatorname{div} (\rho\mathbf{v}) = -\frac{\partial \rho}{\partial t}$$

or

(5) $$\frac{\partial \rho}{\partial t} + \operatorname{div} (\rho\mathbf{v}) = 0.$$

This important relation is called the *condition for the conservation of mass* or the **continuity equation** *of a compressible fluid flow.*

If the flow is **steady,** that is, independent of time, then $\partial \rho / \partial t = 0$ and the continuity equation is

(6) $$\text{div } (\rho \mathbf{v}) = 0.$$

If the density ρ is constant, so that the fluid is incompressible, then equation (6) becomes

(7) $$\text{div } \mathbf{v} = 0.$$

This relation is known as the **condition of incompressibility.** It expresses the fact that the balance of outflow and inflow for a given volume element is zero at any time. Clearly, the assumption that the flow has no sources or sinks in R is essential to our argument.   ∎

Gradient, divergence and curl are three very important concepts in vector calculus. Having introduced the *gradient* in the previous section and the *divergence* in the present section, we continue in the next section with the **curl.** The curl will be a *vector* function associated with a given *vector* function.

Problems for Sec. 8.9

Find the divergence of the following vector functions.

1. $xi + yj$

2. $xi + yj + zk$

3. $(xi + yj)/(x^2 + y^2)$

4. $(x^2 + y^2 + z^2)(i + j + k)$

5. $yzi + xzj + xyk$

6. $v_1(y, z)i + v_2(x, z)j + v_3(x, y)k$

7. $e^x(\cos y \, i + \sin y \, j)$

8. $x^2 i + y^2 j - 2(x + y)zk$

9. $xy(i + j) + xyzk$

10. $(xi + yj + zk)/(x^2 + y^2 + z^2)^{3/2}$

Find $\nabla^2 f$ by (3) and check the result by direct differentiation.

11. $f = \sin x \cosh y$

12. $f = \arctan (y/x)$

13. $f = x^2 + y^2 + z^2$

14. $f = x^3 y z^4$

15. $f = 1 - x^2 - 4y^2 + 2z^2$

16. $f = x^6 + y^6 + z^6 + 30x^2 y^2 z^2$

17. $f = xz/y$

18. $f = e^{xyz}$

Find grad (div **v**), where **v** equals

19. $x^2 i + y^2 j + z^2 k$

20. $e^x i + e^y j + e^z k$

21. $(x^2 - y^2)i + 3k$

22. $xy \sin z \, (i + j + k)$

23. $x^2 yi + yj - 2xyzk$

24. $xyz(i - j + k)$

Find the directional derivative of div **u** at the point P: (4, 4, 2) in the direction of the corresponding outer normal of the sphere $x^2 + y^2 + z^2 = 36$ where

25. $\mathbf{u} = x^4 i + y^4 j + z^4 k$

26. $\mathbf{u} = xzi + yxj + zyk$

Formulas for the divergence. Show that

27. div $(kv) = k$ div **v** (k constant)

28. div $(fv) = f$ div $\mathbf{v} + \mathbf{v} \cdot \nabla f$

29. div $(f \nabla g) = f \nabla^2 g + \nabla f \cdot \nabla g$

30. div $(f \nabla g) -$ div $(g \nabla f) = f \nabla^2 g - g \nabla^2 f$

31. Verify the formula in Prob. 28 when $f = e^{xyz}$ and $\mathbf{v} = axi + byj + czk.$

32. Obtain the answer to Prob. 10 from the formula in Prob. 28.

33. **(Flow)** Consider a steady flow whose velocity vector is $v = y\mathbf{i}$. Show that it has the following properties. The flow is incompressible. The particles that at time $t = 0$ are in the cube bounded by the planes $x = 0$, $x = 1$, $y = 0$, $y = 1$, $z = 0$, $z = 1$ occupy at $t = 1$ the volume 1.

34. **(Flow)** Consider a steady flow having the velocity $v = x\mathbf{i}$. Show that the individual particles have the position vectors $\mathbf{r}(t) = c_1 e^t \mathbf{i} + c_2 \mathbf{j} + c_3 \mathbf{k}$ where c_1, c_2, c_3 are constants, the flow is compressible and e is the volume occupied at $t = 1$ by the particles that at $t = 0$ fill the cube in Prob. 33.

35. **(Rotational flow)** The velocity vector $\mathbf{v}(x, y, z)$ of an incompressible fluid rotating in a cylindrical vessel is of the form $\mathbf{v} = \mathbf{w} \times \mathbf{r}$, where $\mathbf{w}$ is the (constant) rotation vector; cf. Example 2 in Sec. 6.7. Show that div $\mathbf{v} = 0$. Is this plausible from our present Example 2?

8.10 Curl of a Vector Field

Let x, y, z be right-handed Cartesian coordinates in space, and let

$$\mathbf{v}(x, y, z) = v_1\mathbf{i} + v_2\mathbf{j} + v_3\mathbf{k}$$

be a differentiable vector function. Then the function

(1)

$$\operatorname{curl} \mathbf{v} = \nabla \times \mathbf{v} = \begin{vmatrix} \mathbf{i} & \mathbf{j} & \mathbf{k} \\ \dfrac{\partial}{\partial x} & \dfrac{\partial}{\partial y} & \dfrac{\partial}{\partial z} \\ v_1 & v_2 & v_3 \end{vmatrix}$$

$$= \left(\frac{\partial v_3}{\partial y} - \frac{\partial v_2}{\partial z} \right)\mathbf{i} + \left(\frac{\partial v_1}{\partial z} - \frac{\partial v_3}{\partial x} \right)\mathbf{j} + \left(\frac{\partial v_2}{\partial x} - \frac{\partial v_1}{\partial y} \right)\mathbf{k}$$

is called the **curl** *of the vector function* $\mathbf{v}$ or the *curl of the vector field defined by* $\mathbf{v}$. For a left-handed Cartesian coordinate system, the determinant in (1) is preceded by a minus sign, in agreement with Remark 2 in Sec. 6.8.

Instead of curl $\mathbf{v}$ the notation rot $\mathbf{v}$ is also used in the literature.

EXAMPLE 1. Curl of a vector function

With respect to right-handed Cartesian coordinates, let $\mathbf{v} = yz\mathbf{i} + 3zx\mathbf{j} + z\mathbf{k}$. Then (1) gives

$$\operatorname{curl} \mathbf{v} = \begin{vmatrix} \mathbf{i} & \mathbf{j} & \mathbf{k} \\ \partial/\partial x & \partial/\partial y & \partial/\partial z \\ yz & 3zx & z \end{vmatrix} = -3x\mathbf{i} + y\mathbf{j} + (3z - z)\mathbf{k} = -3x\mathbf{i} + y\mathbf{j} + 2z\mathbf{k}. \qquad ∎$$

The curl plays an important role in many applications. Let us illustrate this with a typical basic example. (We shall say more about the role and nature of the curl in Sec. 9.8.)

EXAMPLE 2. Rotation of a rigid body

We have seen in Example 2, Sec. 6.7, that a rotation of a rigid body B about a fixed axis in space can be described by a vector $\mathbf{w}$ of magnitude ω in the direction of the axis of rotation, where ω (> 0) is the angular speed of the rotation, and $\mathbf{w}$ is directed so that the rotation appears clockwise if we look in the direction of $\mathbf{w}$. According to (7), Sec. 6.7, the velocity field of the rotation can be represented in the form

$$\mathbf{v} = \mathbf{w} \times \mathbf{r}$$

where $\mathbf{r}$ is the position vector of a moving point with respect to a Cartesian coordinate system having the origin on the axis of rotation. Let us choose right-handed Cartesian coordinates such that $\mathbf{w} = \omega\mathbf{k}$; that is, the axis of rotation is the z-axis. Then (cf. Example 3 in Sec. 8.1)

$$\mathbf{v} = \mathbf{w} \times \mathbf{r} = -\omega y\mathbf{i} + \omega x\mathbf{j}$$

and, therefore,

$$\text{curl } \mathbf{v} = \begin{vmatrix} \mathbf{i} & \mathbf{j} & \mathbf{k} \\ \dfrac{\partial}{\partial x} & \dfrac{\partial}{\partial y} & \dfrac{\partial}{\partial z} \\ -\omega y & \omega x & 0 \end{vmatrix} = 2\omega\mathbf{k},$$

that is,

$$(2) \qquad\qquad\qquad\qquad \text{curl } \mathbf{v} = 2\mathbf{w}.$$

Hence, in the case of a rotation of a rigid body, the curl of the velocity field has the direction of the axis of rotation, and its magnitude equals twice the angular speed ω of the rotation.

Note that our result does not depend on the particular choice of the Cartesian coordinate system in space.   ∎

For any twice continuously differentiable scalar function f,

$$(3) \qquad\qquad \boxed{\text{curl } (\text{grad } f) = \mathbf{0},}$$

as can easily be verified by direct calculation. *Hence if a vector function is the gradient of a scalar function, its curl is the zero vector.* Since the curl characterizes the rotation in a field, we also say more briefly that *gradient fields describing a motion are irrotational.* (If such a field occurs in some other connection, not as a velocity field, it is usually called *conservative;* cf. at the end of Sec. 8.8.)

EXAMPLE 3

The gravitational field in Example 4, Sec. 8.8, has curl $\mathbf{p} = \mathbf{0}$. The field in Example 2 of the present section is not irrotational. A similar velocity field is obtained by stirring coffee in a cup.   ∎

The curl is defined in (1) in terms of coordinates, but if it is supposed to have a physical or geometrical significance, it should not depend on the choice of these coordinates. This is true, as follows.

Theorem 1 (Invariance of the curl)

The length and direction of curl $\mathbf{v}$ are independent of the particular choice of Cartesian coordinate systems in space.

This theorem will be proved on p. 507.

Problems for Sec. 8.10

Find curl **v** where, with respect to right-handed Cartesian coordinates, **v** equals

1. $y\mathbf{i} - x\mathbf{j}$

2. $xy\mathbf{i} + (x^2 - y^2)\mathbf{j}$

3. $e^{xy}\mathbf{i} + e^{-xy}\mathbf{j}$

4. $\sin y\,\mathbf{i} + \cos x\,\mathbf{j}$

5. $y\mathbf{i} + z\mathbf{j} + x\mathbf{k}$

6. $xyz(x\mathbf{i} + y\mathbf{j} + z\mathbf{k})$

7. $v_1(x)\mathbf{i} + v_2(y)\mathbf{j} + v_3(z)\mathbf{k}$

8. $xz^4\mathbf{i} - zx^4\mathbf{k}$

9. $y \sin z\,\mathbf{j} + z \cos y\,\mathbf{k}$

10. $(x\mathbf{i} + y\mathbf{j} + z\mathbf{k})/(x^2 + y^2 + z^2)^{3/2}$

Fluid motion. In each case the velocity vector **v** of a steady fluid motion is given. Find curl **v**. Is the motion incompressible? Find the paths of the particles.

11. $\mathbf{v} = y^3\mathbf{i}$

12. $\mathbf{v} = x\mathbf{i} + y\mathbf{j}$

13. $\mathbf{v} = y\mathbf{i} - x\mathbf{j}$

14. $\mathbf{v} = -y^2\mathbf{i} + 2\mathbf{j}$

Formulas for curl, div, etc. Show that, granted sufficient differentiability,

15. curl $(\mathbf{u} + \mathbf{v}) = $ curl $\mathbf{u} + $ curl $\mathbf{v}$

16. div (curl $\mathbf{v}$) = 0

17. curl $(f\mathbf{v}) = $ grad $f \times \mathbf{v} + f$ curl $\mathbf{v}$

18. curl (grad f) = **0**

19. div $(\mathbf{u} \times \mathbf{v}) = \mathbf{v} \cdot$ curl $\mathbf{u} - \mathbf{u} \cdot$ curl $\mathbf{v}$

20. div $(g\nabla f \times f\nabla g) = 0$

With respect to right-handed Cartesian coordinates, let $\mathbf{u} = y\mathbf{i} + z\mathbf{j} + x\mathbf{k}$ and $\mathbf{v} = xy\mathbf{i} + yz\mathbf{j} + zx\mathbf{k}$. Find

21. $\mathbf{u} \times$ curl $\mathbf{v}$, $\mathbf{v} \times$ curl $\mathbf{u}$

22. curl $(\mathbf{u} \times \mathbf{v})$, curl $(\mathbf{v} \times \mathbf{u})$

23. grad $(\mathbf{u} \cdot \mathbf{v})$, div [grad $(\mathbf{u} \cdot \mathbf{v})$]

24. div $(\mathbf{u} \times \mathbf{v})$, div $(\mathbf{v} \times \mathbf{u})$

8.11 Grad, Div, Curl in Curvilinear Coordinates (Optional)

In engineering problems we often use coordinates other than Cartesian coordinates. Then we may have to transform the gradient, divergence or curl into these new coordinates. We show how this can be done and give explicit formulas for the cases of cylindrical and spherical coordinates, the two most important coordinates systems in space (other than Cartesian).

For Cartesian coordinates x, y, z we also write

$$x = x_1, \qquad y = x_2, \qquad z = x_3$$

to get simpler formulas later on. **Cylindrical coordinates** r, θ, z are given by

(1)
$$x_1 = r \cos \theta, \qquad x_2 = r \sin \theta, \qquad x_3 = z$$
(Fig. 200c)

and are often used in problems exhibiting cylindrical symmetry.

Spherical coordinates r, θ, ϕ are given by (Fig. 200d)

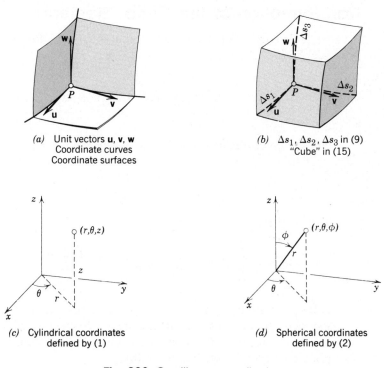

(a) Unit vectors **u**, **v**, **w**
Coordinate curves
Coordinate surfaces

(b) $\Delta s_1, \Delta s_2, \Delta s_3$ in (9)
"Cube" in (15)

(c) Cylindrical coordinates
defined by (1)

(d) Spherical coordinates
defined by (2)

Fig. 200. Curvilinear coordinates

(2) $\qquad x_1 = r \cos \theta \sin \phi, \qquad x_2 = r \sin \theta \sin \phi, \qquad x_3 = r \cos \phi$

and may be good in the case of spherical symmetry. Typical applications will be given in Chap. 11.

Equations (1) and (2) and corresponding equations for other coordinates in space are of the form

(3) $\quad x_1 = x_1(q_1, q_2, q_3), \qquad x_2 = x_2(q_1, q_2, q_3), \qquad x_3 = x_3(q_1, q_2, q_3).$

We assume that every point P is determined by a unique tripel of coordinates q_1, q_2, q_3, so that we can solve (3) in the form

(4) $\quad q_1 = q_1(x_1, x_2, x_3), \qquad q_2 = q_2(x_1, x_2, x_3), \qquad q_3 = q_3(x_1, x_2, x_3).$

Through P there pass three surfaces $q_1 = const$, $q_2 = const$, $q_3 = const$, called **coordinate surfaces,** which are curved in general (Fig. 200a). Their three pairwise intersections are called **coordinate curves** through P. This motivates the name "**curvilinear coordinates**" of the q-coordinates. The coordinate systems used in practice suggest that we assume the q-coordinates to be **orthogonal,** that is, the coordinate curves through P are mutually perpendicular (their tangents at P make right angles, cf. Fig. 200a). For instance, cylindrical and spherical coordinates have this property (except when $r = 0$). We further assume the functions in (3) and (4) to be differentiable.

Transformation of the Linear Element

As a first task, let us transform the square of the linear element [(7) in Sec. 8.4]

$$(5) \qquad ds^2 = \sum_{i=1}^{3} dx_i^2 = dx_1^2 + dx_2^2 + dx_3^2$$

into the q-coordinates. By the chain rule (Sec. 8.7),

$$dx_i = \sum_{j=1}^{3} \frac{\partial x_i}{\partial q_j} dq_j.$$

This implies

$$(6) \qquad ds^2 = \sum_{i=1}^{3} dx_i^2 = \sum_{i=1}^{3} dx_i\, dx_i = \sum_{i=1}^{3} \left[\sum_{j=1}^{3} \frac{\partial x_i}{\partial q_j} dq_j \sum_{k=1}^{3} \frac{\partial x_i}{\partial q_k} dq_k \right]$$

(where we used a different summation letter, k, in the last sum because the two summations are independent). Since the x-coordinates and the q-coordinates are orthogonal, we have in (5) only squares, and similarly in (6). Consequently, we can write (6)

$$(7) \qquad \boxed{ds^2 = h_1^2\, dq_1^2 + h_2^2\, dq_2^2 + h_3^2\, dq_3^2}$$

and determine h_1^2, h_2^2, h_3^2 by collecting the coefficients of dq_1^2, dq_2^2, dq_3^2 in (6). Here, dq_1^2 is obtained in (6) by taking $j = 1$ and $k = 1$; hence h_1^2 has three terms because i goes from 1 to 3. Similarly for h_2^2, h_3^2:

$$(8) \qquad \boxed{h_1^2 = \sum_{i=1}^{3} \left(\frac{\partial x_i}{\partial q_1}\right)^2, \qquad h_2^2 = \sum_{i=1}^{3} \left(\frac{\partial x_i}{\partial q_2}\right)^2, \qquad h_3^2 = \sum_{i=1}^{3} \left(\frac{\partial x_i}{\partial q_3}\right)^2.}$$

EXAMPLE 1. Linear element in cylindrical coordinates

In our present q-notation, cylindrical coordinates $q_1 = r$, $q_2 = \theta$, $q_3 = z$ are defined by (1) in the form

$$x_1 = q_1 \cos q_2, \qquad x_2 = q_1 \sin q_2, \qquad x_3 = q_3.$$

By differentiation we have from (8)

$$h_1^2 = \cos^2 q_2 + \sin^2 q_2 = 1$$

$$h_2^2 = (-q_1 \sin q_2)^2 + (q_1 \cos q_2)^2 = q_1^2$$

$$h_3^2 = 1.$$

Hence, by (7) the square of the linear element in cylindrical coordinates is

$$ds^2 = dq_1^2 + q_1^2\, dq_2^2 + dq_3^2$$

$$= dr^2 + r^2\, d\theta^2 + dz^2.$$

For $z = const$ we have $dz = 0$ and obtain the familiar $ds^2 = dr^2 + r^2 d\theta^2$ in **polar coordinates** in the plane.

EXAMPLE 2. Linear element in spherical coordinates

Spherical coordinates $q_1 = r$, $q_2 = \theta$, $q_3 = \phi$ are defined by [cf. (2)]

$$x_1 = q_1 \cos q_2 \sin q_3, \qquad x_2 = q_1 \sin q_2 \sin q_3, \qquad x_3 = q_1 \cos q_3.$$

Hence (8) gives

$$h_1{}^2 = (\cos q_2 \sin q_3)^2 + (\sin q_2 \sin q_3)^2 + \cos^2 q_3 = 1$$

$$h_2{}^2 = (-q_1 \sin q_2 \sin q_3)^2 + (q_1 \cos q_2 \sin q_3)^2 + 0 = q_1{}^2 \sin^2 q_3$$

$$h_3{}^2 = (q_1 \cos q_2 \cos q_3)^2 + (q_1 \sin q_2 \cos q_3)^2 + (-q_1 \sin q_3)^2 = q_1{}^2.$$

This and (7) give the square of the linear element in spherical coordinates

$$ds^2 = dq_1{}^2 + q_1{}^2 \sin^2 q_3 \, dq_2{}^2 + q_1{}^2 \, dq_3{}^2$$

$$= dr^2 + r^2 \sin^2 \phi \, d\theta^2 + r^2 \, d\phi^2. \qquad\blacksquare$$

Transformation of the Gradient

For the x-coordinates we have (cf. Sec. 8.4, where dx is denoted dr)

$$d\mathbf{x} = dx_1 \, \mathbf{i} + dx_2 \, \mathbf{j} + dx_3 \, \mathbf{k}$$

where $\mathbf{i}$, $\mathbf{j}$, $\mathbf{k}$ are unit vectors in the directions of the coordinate axes, and

$$ds^2 = d\mathbf{x} \cdot d\mathbf{x} = dx_1{}^2 + dx_2{}^2 + dx_3{}^2.$$

Similarly, for the q-coordinates we can write

$$d\mathbf{q} = h_1 \, dq_1 \, \mathbf{u} + h_2 \, dq_2 \, \mathbf{v} + h_3 \, dq_3 \, \mathbf{w}$$

where $\mathbf{u}$, $\mathbf{v}$, $\mathbf{w}$ are orthogonal unit vectors in the directions of the coordinate curves at P (Fig. 200a on p. 499) and we obtain, in agreement with (7)

$$ds^2 = d\mathbf{q} \cdot d\mathbf{q} = h_1{}^2 \, dq_1{}^2 + h_2{}^2 \, dq_2{}^2 + h_3{}^2 \, dq_3{}^2.$$

If we now let q_1 increase by Δq_1, then we move in the $\mathbf{u}$-direction by $\Delta s_1 = h_1 \Delta q_1$ (cf. Fig. 200b). In the limit this gives $\partial q_1 / \partial s_1 = 1/h_1$. Similarly for the other two directions. Together,

$$(9) \qquad \frac{\partial q_1}{\partial s_1} = \frac{1}{h_1}, \qquad \frac{\partial q_2}{\partial s_2} = \frac{1}{h_2}, \qquad \frac{\partial q_3}{\partial s_3} = \frac{1}{h_3}.$$

From this we now get the components of the gradient ∇f of a differentiable scalar function $f(q_1, q_2, q_3)$ as the projections in the $\mathbf{u}$-, $\mathbf{v}$-, $\mathbf{w}$-directions; and on the other hand these are the directional derivatives $\partial f / \partial s_i$ in these three directions. Thus by (9),

$$\frac{\partial f}{\partial s_1} = \frac{\partial f}{\partial q_1} \frac{\partial q_1}{\partial s_1} = \frac{\partial f}{\partial q_1} \frac{1}{h_1}$$

and similarly for the other two components. This gives the following expression for the **gradient** ∇f in orthogonal curvilinear coordinates q_1, q_2, q_3 [with h_1, h_2, h_3 given by (8) and **u, v, w** shown in Fig. 200a]

(10)
$$\boxed{\text{grad } f = \nabla f = \frac{1}{h_1} \frac{\partial f}{\partial q_1} \mathbf{u} + \frac{1}{h_2} \frac{\partial f}{\partial q_2} \mathbf{v} + \frac{1}{h_3} \frac{\partial f}{\partial q_3} \mathbf{w}.}$$

EXAMPLE 3. Gradient

In the cylindrical coordinates r, θ, z given by (1) we have (cf. Example 1)

(11) $h_1 = h_r = 1, \qquad h_2 = h_\theta = q_1 = r, \qquad h_3 = h_z = 1$

and (10) with $q_1 = r, q_2 = \theta, q_3 = z$ gives the gradient

(12)
$$\boxed{\text{grad } f = \nabla f = \frac{\partial f}{\partial r} \mathbf{u} + \frac{1}{r} \frac{\partial f}{\partial \theta} \mathbf{v} + \frac{\partial f}{\partial z} \mathbf{w}.}$$

In the spherical coordinates r, θ, ϕ given by (2) we have (cf. Example 2)

(13) $h_1 = h_r = 1, \qquad h_2 = h_\theta = q_1 \sin q_3 = r \sin \phi, \qquad h_3 = h_\phi = q_1 = r$

and (10) with $q_1 = r, q_2 = \theta, q_3 = \phi$ gives the gradient

(14)
$$\boxed{\text{grad } f = \nabla f = \frac{\partial f}{\partial r} \mathbf{u} + \frac{1}{r \sin \phi} \frac{\partial f}{\partial \theta} \mathbf{v} + \frac{1}{r} \frac{\partial f}{\partial \phi} \mathbf{w}.}$$ ∎

Transformation of the Divergence and the Laplacian

For the **divergence** we obtain in orthogonal curvilinear coordinates q_1, q_2, q_3 [with h_1, h_2, h_3 given by (8)]

(15)
$$\boxed{\text{div } \mathbf{F} = \nabla \cdot \mathbf{F} = \frac{1}{h_1 h_2 h_3} \left[\frac{\partial}{\partial q_1} (h_2 h_3 F_1) + \frac{\partial}{\partial q_2} (h_3 h_1 F_2) + \frac{\partial}{\partial q_3} (h_1 h_2 F_3) \right].}$$

This follows as in Example 2 of Sec. 8.9 by considering the flux through the faces of the "cube" in Fig. 200b, the divergence being the limit of the flux divided by the volume $h_1 \Delta q_1 h_2 \Delta q_2 h_3 \Delta q_3$; here $h_2 \Delta q_2 h_3 \Delta q_3$ is the area of the faces $q_1 = const$ and $q_1 + h_1 \Delta q_1 = const$, so that $\partial(h_2 h_3 F_1)/\partial q_1$ results from the contribution of these faces to the flux. The other two terms in (15) correspond to the other two pairs of faces of the "cube."

EXAMPLE 4. Divergence

In the cylindrical coordinates given by (1) we obtain from (11) and (15) with $q_1 = r, q_2 = \theta$, $q_3 = z$ the divergence

$$\nabla \cdot \mathbf{F} = \frac{1}{r} \left[\frac{\partial}{\partial r} (rF_1) + \frac{\partial F_2}{\partial \theta} + \frac{\partial}{\partial z} (rF_3) \right]$$

that is,

(16)
$$\operatorname{div} \mathbf{F} = \nabla \cdot \mathbf{F} = \frac{1}{r} \frac{\partial}{\partial r} (rF_1) + \frac{1}{r} \frac{\partial F_2}{\partial \theta} + \frac{\partial F_3}{\partial z} \, .$$

In the spherical coordinates given by (2) we obtain from (13) and (15) with $q_1 = r$, $q_2 = \theta$, $q_3 = \phi$ the divergence

$$\nabla \cdot \mathbf{F} = \frac{1}{r^2 \sin \phi} \left[\frac{\partial}{\partial r} (r^2 \sin \phi \, F_1) + \frac{\partial}{\partial \theta} (rF_2) + \frac{\partial}{\partial \phi} (r \sin \phi \, F_3) \right]$$

that is,

(17)
$$\operatorname{div} \mathbf{F} = \nabla \cdot \mathbf{F} = \frac{1}{r^2} \frac{\partial}{\partial r} (r^2 F_1) + \frac{1}{r \sin \phi} \frac{\partial F_2}{\partial \theta} + \frac{1}{r \sin \phi} \frac{\partial}{\partial \phi} (\sin \phi \, F_3).$$

Since $\nabla^2 f = \nabla \cdot \nabla f = \operatorname{div} (\operatorname{grad} f)$ (cf. Sec. 8.9), we also obtain from (15) with $\mathbf{F} = \nabla f$ and ∇f given by (10) the **Laplacian** in orthogonal curvilinear coordinates q_1, q_2, q_3

(18)
$$\nabla^2 f = \frac{1}{h_1 h_2 h_3} \left[\frac{\partial}{\partial q_1} \left(\frac{h_2 h_3}{h_1} \frac{\partial f}{\partial q_1} \right) + \frac{\partial}{\partial q_2} \left(\frac{h_3 h_1}{h_2} \frac{\partial f}{\partial q_2} \right) + \frac{\partial}{\partial q_3} \left(\frac{h_1 h_2}{h_3} \frac{\partial f}{\partial q_3} \right) \right]$$

with h_1, h_2, h_3 given by (8).

EXAMPLE 5. Laplacian

In the cylindrical coordinates $q_1 = r$, $q_2 = \theta$, $q_3 = z$ given by (1) we obtain from (11) and (18) the Laplacian

$$\nabla^2 f = \frac{1}{r} \left[\frac{\partial}{\partial r} \left(r \frac{\partial f}{\partial r} \right) + \frac{\partial}{\partial \theta} \left(\frac{1}{r} \frac{\partial f}{\partial \theta} \right) + \frac{\partial}{\partial z} \left(r \frac{\partial f}{\partial z} \right) \right]$$

hence

$$\nabla^2 f = \frac{1}{r} \frac{\partial}{\partial r} \left(r \frac{\partial f}{\partial r} \right) + \frac{1}{r^2} \frac{\partial^2 f}{\partial \theta^2} + \frac{\partial^2 f}{\partial z^2}$$

or by performing the indicated product differentiation,

(19)
$$\nabla^2 f = \frac{\partial^2 f}{\partial r^2} + \frac{1}{r} \frac{\partial f}{\partial r} + \frac{1}{r^2} \frac{\partial^2 f}{\partial \theta^2} + \frac{\partial^2 f}{\partial z^2} \, .$$

In the spherical coordinates $q_1 = r$, $q_2 = \theta$, $q_3 = \phi$ given by (2) we obtain from (13) and (18) the Laplacian

$$\nabla^2 f = \frac{1}{r^2 \sin \phi} \left[\frac{\partial}{\partial r} \left(r^2 \sin \phi \frac{\partial f}{\partial r} \right) + \frac{\partial}{\partial \theta} \left(\frac{1}{\sin \phi} \frac{\partial f}{\partial \theta} \right) + \frac{\partial}{\partial \phi} \left(\sin \phi \frac{\partial f}{\partial \phi} \right) \right]$$

hence

$$\nabla^2 f = \frac{1}{r^2} \frac{\partial}{\partial r} \left(r^2 \frac{\partial f}{\partial r} \right) + \frac{1}{r^2 \sin^2 \phi} \frac{\partial^2 f}{\partial \theta^2} + \frac{1}{r^2 \sin \phi} \frac{\partial}{\partial \phi} \left(\sin \phi \frac{\partial f}{\partial \phi} \right)$$

or by performing the indicated product differentiations,

(20)
$$\nabla^2 f = \frac{\partial^2 f}{\partial r^2} + \frac{2}{r} \frac{\partial f}{\partial r} + \frac{1}{r^2 \sin^2 \phi} \frac{\partial^2 f}{\partial \theta^2} + \frac{1}{r^2} \frac{\partial^2 f}{\partial \phi^2} + \frac{\cot \phi}{r^2} \frac{\partial f}{\partial \phi} \, .$$

Transformation of the Curl

For curl $\mathbf{F} = \nabla \times \mathbf{F}$ the general formula in *right-handed* orthogonal curvilinear coordinates q_1, q_2, q_3 is the following. (Proof in Ref. [10], p. 311, listed in Appendix 1.)

(21)

$$\operatorname{curl} \mathbf{F} = \nabla \times \mathbf{F} = \frac{1}{h_1 h_2 h_3} \begin{vmatrix} h_1 \mathbf{u} & h_2 \mathbf{v} & h_3 \mathbf{w} \\ \dfrac{\partial}{\partial q_1} & \dfrac{\partial}{\partial q_2} & \dfrac{\partial}{\partial q_3} \\ h_1 F_1 & h_2 F_2 & h_3 F_3 \end{vmatrix}$$

From this and (11) or (13) the curl can now readily be obtained in cylindrical or spherical coordinates.

Because of the practical importance of cylindrical and spherical coordinates we have used them for illustration in our examples, but we want to emphasize that our formulas are *general* and can be used in any orthogonal coordinate system.

We have reached the end of Chap. 8 on vector *differential* calculus and shall start with **vector integral calculus** in the next section (Sec. 9.1).

Further Proofs in Chapter 8

We prove the invariance of the divergence and the curl as stated in Secs. 8.9 and 8.10. For this we need the following

Theorem 1 (Transformation law for vector components)

For any vector $\mathbf{v}$, the components v_1, v_2, v_3 and v_1^*, v_2^*, v_3^* in any two systems of Cartesian coordinates x_1, x_2, x_3 and x_1^*, x_2^*, x_3^*, respectively, are related by

(1)

$$\begin{aligned} v_1^* &= c_{11}v_1 + c_{12}v_2 + c_{13}v_3 \\ v_2^* &= c_{21}v_1 + c_{22}v_2 + c_{23}v_3 \\ v_3^* &= c_{31}v_1 + c_{32}v_2 + c_{33}v_3. \end{aligned}$$

and conversely

(2)

$$\begin{aligned} v_1 &= c_{11}v_1^* + c_{21}v_2^* + c_{31}v_3^* \\ v_2 &= c_{12}v_1^* + c_{22}v_2^* + c_{32}v_3^* \\ v_3 &= c_{13}v_1^* + c_{23}v_2^* + c_{33}v_3^* \end{aligned}$$

with coefficients

$$
\begin{array}{lll}
c_{11} = \mathbf{i}^* \cdot \mathbf{i} & c_{12} = \mathbf{i}^* \cdot \mathbf{j} & c_{13} = \mathbf{i}^* \cdot \mathbf{k} \\
c_{21} = \mathbf{j}^* \cdot \mathbf{i} & c_{22} = \mathbf{j}^* \cdot \mathbf{j} & c_{23} = \mathbf{j}^* \cdot \mathbf{k} \\
c_{31} = \mathbf{k}^* \cdot \mathbf{i} & c_{32} = \mathbf{k}^* \cdot \mathbf{j} & c_{33} = \mathbf{k}^* \cdot \mathbf{k}
\end{array}
$$

(3)

satisfying

(4)
$$
\sum_{j=1}^{3} c_{kj} c_{mj} = \delta_{km} \qquad (k, m = 1, 2, 3),
$$

where the **Kronecker delta**[10] *is given by*

$$
\delta_{km} = \begin{cases} 0 & (k \neq m) \\ 1 & (k = m). \end{cases}
$$

and $\mathbf{i}, \mathbf{j}, \mathbf{k}$ *and* $\mathbf{i}^*, \mathbf{j}^*, \mathbf{k}^*$ *denote the unit vectors in the positive* x_1-, x_2-, x_3- *and* x_1^*-, x_2^*-, x_3^*-*directions, respectively.*

Proof. The representations of $\mathbf{v}$ in the two systems are

(5) (a) $\mathbf{v} = v_1 \mathbf{i} + v_2 \mathbf{j} + v_3 \mathbf{k}$ (b) $\mathbf{v} = v_1^* \mathbf{i}^* + v_2^* \mathbf{j}^* + v_3^* \mathbf{k}^*$

Since $\mathbf{i}^* \cdot \mathbf{i}^* = 1$, $\mathbf{i}^* \cdot \mathbf{j}^* = 0$, $\mathbf{i}^* \cdot \mathbf{k}^* = 0$, we get from (5b) simply $\mathbf{i}^* \cdot \mathbf{v} = v_1^*$ and from this and (5a)

$$
v_1^* = \mathbf{i}^* \cdot \mathbf{v} = \mathbf{i}^* \cdot v_1 \mathbf{i} + \mathbf{i}^* \cdot v_2 \mathbf{j} + \mathbf{i}^* \cdot v_3 \mathbf{k} = v_1 \mathbf{i}^* \cdot \mathbf{i} + v_2 \mathbf{i}^* \cdot \mathbf{j} + v_3 \mathbf{i}^* \cdot \mathbf{k}.
$$

Because of (3), this is the first formula in (1), and the other two formulas are obtained similarly, by considering $\mathbf{j}^* \cdot \mathbf{v}$ and then $\mathbf{k}^* \cdot \mathbf{v}$. Formula (2) follows by the same idea, taking $\mathbf{i} \cdot \mathbf{v} = v_1$ from (5a) and then from (5b) and (3)

$$
v_1 = \mathbf{i} \cdot \mathbf{v} = v_1^* \mathbf{i} \cdot \mathbf{i}^* + v_2^* \mathbf{i} \cdot \mathbf{j}^* + v_3^* \mathbf{i} \cdot \mathbf{k}^* = c_{11} v_1^* + c_{21} v_2^* + c_{31} v_3^*,
$$

and similarly for the other two components.

We prove (4). We can write (1) and (2) briefly as

(6)   (a) $v_j = \displaystyle\sum_{m=1}^{3} c_{mj} v_m^*$,   (b) $v_k^* = \displaystyle\sum_{j=1}^{3} c_{kj} v_j$.

Substituting v_j into v_k^*, we get

---

[10]LEOPOLD KRONECKER (1823—1891), German mathematician at Berlin, who made important contributions to algebra, group theory and number theory.

We shall keep our discussion completely independent of Chap. 7, but readers familiar with matrices should recognize that we are dealing with **orthogonal transformations and matrices,** and that our present theorem follows from Theorem 4 in Sec. 7.13.

$$v_k^* = \sum_{j=1}^{3} c_{kj} \sum_{m=1}^{3} c_{mj} v_m^* = \sum_{m=1}^{3} v_m^* \left(\sum_{j=1}^{3} c_{kj} c_{mj} \right),$$

where $k = 1, 2, 3$. Taking $k = 1$, we have

$$v_1^* = v_1^* \left(\sum_{j=1}^{3} c_{1j} c_{1j} \right) + v_2^* \left(\sum_{j=1}^{3} c_{1j} c_{2j} \right) + v_3^* \left(\sum_{j=1}^{3} c_{1j} c_{3j} \right).$$

For this to hold for *every* vector **v**, the first sum must be 1 and the other two sums 0. This proves (4) with $k = 1$ for $m = 1, 2, 3$. Taking $k = 2$ and then $k = 3$, we obtain (4) with $k = 2$ and 3, for $m = 1, 2, 3$. ∎

The most general transformation of a Cartesian coordinate system into another such system may be decomposed into a transformation of the type just considered and a translation. Under a translation, corresponding coordinates differ merely by a constant. We thus obtain

Theorem 2 (Transformation law for Cartesian coordinates)

The transformation of any Cartesian $x_1 x_2 x_3$-coordinate system into any other Cartesian $x_1^* x_2^* x_3^*$-coordinate system is of the form

$$
\begin{aligned}
x_1^* &= c_{11} x_1 + c_{12} x_2 + c_{13} x_3 + b_1 \\
(7) \qquad x_2^* &= c_{21} x_1 + c_{22} x_2 + c_{23} x_3 + b_2 \\
x_3^* &= c_{31} x_1 + c_{32} x_2 + c_{33} x_3 + b_3
\end{aligned}
$$

with coefficients (3) and constant b_1, b_2, b_3, and conversely,

$$
\begin{aligned}
x_1 &= c_{11} x_1^* + c_{21} x_2^* + c_{31} x_3^* + \tilde{b}_1 \\
(8) \qquad x_2 &= c_{12} x_1^* + c_{22} x_2^* + c_{32} x_3^* + \tilde{b}_2 \\
x_3 &= c_{13} x_1^* + c_{23} x_2^* + c_{22} x_3^* + \tilde{b}_3.
\end{aligned}
$$

Proof of Theorem 1 in Sec. 8.9 (*Invariance of the divergence*)

We write x_1, x_2, x_3 instead of x, y, z. Then by definition,

$$(9) \qquad \text{div } \mathbf{v} = \sum_{j=1}^{3} \frac{\partial v_j}{\partial x_j} = \frac{\partial v_1}{\partial x_1} + \frac{\partial v_2}{\partial x_2} + \frac{\partial v_3}{\partial x_3}.$$

We also write x_1^*, x_2^*, x_3^* instead of x^*, y^*, z^*. We have to show that in these coordinates,

$$(10) \qquad \text{div } \mathbf{v} = \sum_{k=1}^{3} \frac{\partial v_k^*}{\partial x_k^*} = \frac{\partial v_1^*}{\partial x_1^*} + \frac{\partial v_2^*}{\partial x_2^*} + \frac{\partial v_3^*}{\partial x_3^*}.$$

From the chain rule for functions of several variables (Sec. 8.7) we obtain

(11)
$$\frac{\partial v_j}{\partial x_j} = \sum_{m=1}^{3} \frac{\partial v_j}{\partial x_m^*} \frac{\partial x_m^*}{\partial x_j}.$$

In this formula, $\partial x_m^*/\partial x_j = c_{mj}$ by (7); and by (2),

$$\frac{\partial v_j}{\partial x_m^*} = \sum_{k=1}^{3} c_{kj} \frac{\partial v_k^*}{\partial x_m^*}.$$

Substituting all this into (11), summing over j from 1 to 3, and noting (4), we get

$$\text{div } \mathbf{v} = \sum_{j=1}^{3} \frac{\partial v_j}{\partial x_j} = \sum_{j=1}^{3} \sum_{k=1}^{3} \sum_{m=1}^{3} c_{kj} \frac{\partial v_k^*}{\partial x_m^*} c_{mj}$$

$$= \sum_{k=1}^{3} \sum_{m=1}^{3} \delta_{km} \frac{\partial v_k^*}{\partial x_m^*} = \sum_{k=1}^{3} \frac{\partial v_k^*}{\partial x_k^*}.$$

This proves (10). ∎

Proof of Theorem 1 in Sec. 8.10 (*Invariance of the curl*)

We write again x_1, x_2, x_3 instead of x, y, z, and similarly x_1^*, x_2^*, x_3^* for other Cartesian coordinates, assuming that both systems are right-handed. Let a_1, a_2, a_3 denote the components of curl $\mathbf{v}$ in the $x_1 x_2 x_3$-coordinates, as given by (1), Sec. 8.10, with $x = x_1$, $y = x_2$, $z = x_3$. Similarly, let a_1^*, a_2^*, a_3^* denote the components of curl $\mathbf{v}$ in the $x_1^* x_2^* x_3^*$-coordinate system. We prove that the length and direction of curl $\mathbf{v}$ are independent of the particular choice of Cartesian coordinates, as asserted in the theorem. We do this by showing that the components of curl $\mathbf{v}$ satisfy the transformation law (2) which is characteristic of vector components. We consider a_1. We use (6a), then the chain rule for functions of several variables (Sec. 8.7) and then (7). This gives

$$a_1 = \frac{\partial v_3}{\partial x_2} - \frac{\partial v_2}{\partial x_3} = \sum_{m=1}^{3} \left(c_{m3} \frac{\partial v_m^*}{\partial x_2} - c_{m2} \frac{\partial v_m^*}{\partial x_3} \right)$$

$$= \sum_{m=1}^{3} \sum_{j=1}^{3} \left(c_{m3} \frac{\partial v_m^*}{\partial x_j^*} \frac{\partial x_j^*}{\partial x_2} - c_{m2} \frac{\partial v_m^*}{\partial x_j^*} \frac{\partial x_j^*}{\partial x_3} \right)$$

$$= \sum_{m=1}^{3} \sum_{j=1}^{3} (c_{m3} c_{j2} - c_{m2} c_{j3}) \frac{\partial v_m^*}{\partial x_j^*}$$

$$= (c_{33} c_{22} - c_{32} c_{23}) \left(\frac{\partial v_3^*}{\partial x_2^*} - \frac{\partial v_2^*}{\partial x_3^*} \right) + \cdots$$

$$= (c_{33} c_{22} - c_{32} c_{23}) a_1^* + (c_{13} c_{32} - c_{12} c_{33}) a_2^* + (c_{23} c_{12} - c_{22} c_{13}) a_3^*.$$

Note what we did. The double sum had $3 \times 3 = 9$ terms, 3 of which were zero (when $m = j$), and the remaining 6 terms we combined in pairs as we needed them in getting $a_1{}^*, a_2{}^*, a_3{}^*$.

We now use (3), Lagrange's identity (Sec. 6.9) and $\mathbf{k}^* \times \mathbf{j}^* = -\mathbf{i}^*$ and $\mathbf{k} \times \mathbf{j} = -\mathbf{i}$. Then

$$c_{33}c_{22} - c_{32}c_{23} = (\mathbf{k}^* \cdot \mathbf{k})(\mathbf{j}^* \cdot \mathbf{j}) - (\mathbf{k}^* \cdot \mathbf{j})(\mathbf{j}^* \cdot \mathbf{k})$$

$$= (\mathbf{k}^* \times \mathbf{j}^*) \cdot (\mathbf{k} \times \mathbf{j}) = \mathbf{i}^* \cdot \mathbf{i} = c_{11}, \qquad \text{etc.}$$

Hence $a_1 = c_{11}a_1{}^* + c_{21}a_2{}^* + c_{31}a_3{}^*$. This is of the form of the first formula in (2), and the other two formulas of the form (2) are obtained similarly. This proves the theorem for right-handed systems. If the $x_1 x_2 x_3$-coordinates are left-handed, then $\mathbf{k} \times \mathbf{j} = +\mathbf{i}$, but then there is a minus sign in front of the determinant in (1), Sec. 8.10. ∎

Review Problems for Chapter 8

1. Give some examples of vector fields in physics and geometry.
2. Without looking into the text, state the definition of the derivative of a vector function $\mathbf{v}(t)$ and of the first partial derivatives of a vector function $\mathbf{v}(x, y, z)$. What is $\mathbf{v}'(t)$ geometrically?
3. If $\mathbf{r}(t)$ represents a curve, what is $\mathbf{r}'(t)$ in mechanics when t is time?
4. Can a body have constant speed and still have a varying velocity vector?
5. What properties of a curve do curvature and torsion characterize?
6. What do we mean by the directional derivative?
7. What are the directional derivatives of a function $f(x, y, z)$ in the x-, y- and z-directions? In the negative x-, y- and z-directions?
8. Without looking into the text, write down the definitions of grad, div and curl.
9. What properties of a scalar function do the direction and length of the gradient characterize?
10. If f is a scalar function and $\mathbf{v}$ a vector function, both sufficiently often differentiable, which of the following expressions make sense? grad f, f grad f, $\mathbf{v} \cdot$ grad f, div f, div $(f\mathbf{v})$, curl $(f\mathbf{v})$ curl f, f curl $\mathbf{v}$

Let $\quad f = x^2 + 9y^2 + 4z^2, \qquad g = xy^3z^2, \qquad \mathbf{v} = xz\mathbf{i} + (y - z)^2\mathbf{j} + 2xyz\mathbf{k},$
$w = 2y\mathbf{i} + 4z\mathbf{j} + x^2z^2\mathbf{k}$. Find

11. grad f at $(3, -1, 0)$	12. grad g at $(\frac{1}{2}, 2, -4)$
13. $\nabla^2 f$ 14. $\nabla^2 g$	15. $\nabla f \cdot \nabla g$ 16. $w \cdot \nabla g$
17. $\partial^2 g/\partial x \partial y,\ \partial^2 g/\partial y \partial z,\ \partial^2 g/\partial z \partial x$	18. div $\mathbf{v}$ at $(1, 0, 1)$
19. $\nabla f \cdot \mathbf{v}$ 20. $\nabla f \times \nabla g$	21. div w 22. curl $(\text{grad } f)$
23. div $\mathbf{v}$ at $(-2, 2, 7)$	24. curl $\mathbf{v}$ 25. div $(\text{curl } \mathbf{v})$
26. curl w 27. curl $(\mathbf{v} \times \mathbf{k})$	28. $\nabla f \cdot$ curl w 29. $D_v f$ at $(1, 1, 1)$
30. $D_w f$ at $(0, 1, 2)$	31. $D_w g$ at $(3, 0, -2)$
32. $D_v g$ at $(1, 3, 2)$ 33. div $(\mathbf{v} + w)$	34. grad $(\text{div } \mathbf{v})$
35. grad $(\text{div } w)$ 36. grad $(w \cdot w)$	37. div $[g(\mathbf{i} + \mathbf{j} + \mathbf{k})]$
38. curl $(f\mathbf{i})$ 39. curl $(g\mathbf{j})$	40. div (gw)

Indicate the kind of curve represented by $\mathbf{r}(t)$, where t is time, and find the velocity vector, speed and acceleration vector at P, where

41. $\mathbf{r} = t\mathbf{i} + 3t^2\mathbf{j}, \quad P: (2, 12)$

42. $\mathbf{r} = (1 + 2\cos t)\mathbf{i} + 2\sin t\,\mathbf{j}, \quad P: (1, 2)$

43. $\mathbf{r} = \cos 2t\,\mathbf{i} + (3 + \sin 2t)\mathbf{j}, \quad P: (-1, 3)$

44. $\mathbf{r} = (1 + \cos t)\mathbf{i} + 9\sin t\,\mathbf{j}, \quad P: (1, -9)$

45. $\mathbf{r} = t\mathbf{i} + 3t\mathbf{j} + t\mathbf{k}, \quad P: (4, 12, 4)$

46. $\mathbf{r} = -t^2\mathbf{i} + 4t^2\mathbf{j} + t^2\mathbf{k}, \quad P: (-4, 16, 4)$

47. $\mathbf{r} = t\mathbf{i} + t^{-1}\mathbf{j}, \quad P: (2, \frac{1}{2})$

48. $\mathbf{r} = \cosh t\,\mathbf{i} + \sinh t\,\mathbf{j}, \quad P: (\sqrt{2}, 1)$

49. $\mathbf{r} = \cos 2t\,\mathbf{i} + \sin 2t\,\mathbf{j} + t\mathbf{k}, \quad P: (1, 0, \pi)$

50. $\mathbf{r} = \cos t\,\mathbf{i} - \sin t\,\mathbf{j} - 2t\mathbf{k}, \quad P: (0, -1, -\pi)$

Find the curvature and torsion of the following curves.

51. $\mathbf{r} = t\mathbf{i} + t^2\mathbf{j}$ **52.** $\mathbf{r} = t\mathbf{i} + t^3\mathbf{j}$

53. $\mathbf{r} = t\mathbf{i} + t^{-1}\mathbf{j}$ **54.** $\mathbf{r} = t\mathbf{i} + t^{3/2}\mathbf{j} + t\mathbf{k}$

55. $\mathbf{r} = t\mathbf{i} + t^2\mathbf{j} + t^3\mathbf{k}$ **56.** $\mathbf{r} = \cosh t\,\mathbf{i} + \sinh t\,\mathbf{j} + e^t\mathbf{k}$

57. $\mathbf{r} = t\mathbf{i} + t^{-1}\mathbf{j} + t^2\mathbf{k}$ **58.** $\mathbf{r} = \cos t\,\mathbf{i} + \sin t\,\mathbf{j} + \cos t\,\mathbf{k}$

59. $\mathbf{r} = \cos t\,\mathbf{i} + \sin t\,\mathbf{j} + e^{-t}\mathbf{k}$ **60.** $\mathbf{r} = t\mathbf{i} + t^4\mathbf{j} + t^6\mathbf{k}$

Summary of Chapter 8
Vector Differential Calculus

Chapter 8 extends differential calculus to vectors. In Chap. 6 we were concerned with vectors in three-dimensional space (or in the plane) whose components are constant. For these we used the two notations

$$(1) \qquad \mathbf{a} = [a_1, a_2, a_3] \qquad \text{or} \qquad \mathbf{a} = a_1\mathbf{i} + a_2\mathbf{j} + a_3\mathbf{k},$$

($\mathbf{i}, \mathbf{j}, \mathbf{k}$ unit vectors in the positive directions of the axes of the Cartesian xyz-coordinate system), whichever was more convenient, and we introduced the *algebraic* operations for vectors, namely, vector addition and multiplication by scalars (which make the set of these vectors into a vector space), and two kinds of multiplication of vectors by vectors, which give the inner product (dot product) $\mathbf{a} \cdot \mathbf{b}$, which is a scalar, and the vector product $\mathbf{a} \times \mathbf{b}$, which is a vector.

In the present chapter we made the essential step from vector *algebra* to vector *analysis* (vector calculus), beginning with vector *differential* calculus (and continuing with vector *integral* calculus in Chap. 9). This greatly increases the usefulness of vector methods and makes them truly interesting to the engineer and physicist.

This extension begins with the introduction of **vector functions** $\mathbf{v}$ (Sec. 8.1), which may depend on the coordinates x, y, z in space, so that their components v_1, v_2, v_3 depend on x, y, z; in analogy to (1) we then write

$$(2a) \qquad \mathbf{v}(x, y, z) = [v_1(x, y, z), \quad v_2(x, y, z), \quad v_3(x, y, z)]$$

or

$$(2b) \qquad \mathbf{v}(x, y, z) = v_1(x, y, z)\mathbf{i} + v_2(x, y, z)\mathbf{j} + v_3(x, y, z)\mathbf{k}.$$

Such a function $\mathbf{v}$ defines a **vector field** in a region in space; with every point P_0: (x_0, y_0, z_0) in the region, $\mathbf{v}$ associates a vector $\mathbf{v}(P_0) = \mathbf{v}(x_0, y_0, z_0)$. Similarly, $\mathbf{v}$ may be a function of a single variable t, where t varies in some interval $a \leqq t \leqq b$ (or on the entire real axis). Then we write

$$(3) \qquad \mathbf{v}(t) = [v_1(t), v_2(t), v_3(t)] \quad \text{or} \quad \mathbf{v}(t) = v_1(t)\mathbf{i} + v_2(t)\mathbf{j} + v_3(t)\mathbf{k}.$$

The central concept in this chapter is the **derivative** $\mathbf{v}'(t)$ of a vector function $\mathbf{v}(t)$ (Sec. 8.2),

$$(4) \qquad\qquad \mathbf{v}' = \frac{d\mathbf{v}}{dt} = \lim_{\Delta t \to 0} \frac{\mathbf{v}(t + \Delta t) - \mathbf{v}(t)}{\Delta t},$$

a formula that looks quite similar to the familiar formula in calculus. It implies that in terms of components,

$$\mathbf{v}' = [v_1', v_2', v_3'] = v_1'\mathbf{i} + v_2'\mathbf{j} + v_3'\mathbf{k}.$$

Differentiation rules are as in calculus. They imply (Sec. 8.2)

$$(\mathbf{u} \cdot \mathbf{v})' = \mathbf{u}' \cdot \mathbf{v} + \mathbf{u} \cdot \mathbf{v}', \qquad (\mathbf{u} \times \mathbf{v})' = \mathbf{u}' \times \mathbf{v} + \mathbf{u} \times \mathbf{v}'$$

Similarly, partial derivatives of vector functions of several variables are obtained by taking partial derivatives of components; for instance, for $\mathbf{v}(x, y, z)$ we have (Sec. 8.2)

$$\frac{\partial \mathbf{v}}{\partial x} = \left[\frac{\partial v_1}{\partial x}, \ \frac{\partial v_2}{\partial x}, \ \frac{\partial v_3}{\partial x}\right] \qquad\qquad \text{etc.}$$

A **curve** C (Sec. 8.3) can be represented by a vector function $\mathbf{r}(t)$, which associates with each $t = t_0$ in some interval $a < t < b$ the point of C with position vector $\mathbf{r}(t_0)$. The derivative $\mathbf{r}'(t)$ is a **tangent vector** of C (Sec. 8.4). If we choose the **arc length** s (Sec. 8.4) as a parameter, then we get the *unit tangent vector* $\mathbf{r}'(s)$. The length of the derivative of this vector is the **curvature** $\kappa(s) = |\mathbf{r}''(s)|$; cf. Sec. 8.6.

In mechanics, $\mathbf{r}(t)$ may represent the path of a moving body, where t is time. Then $\mathbf{v}(t) = \mathbf{r}'(t)$ is the **velocity vector,** its length is the *speed,* and its derivative $\mathbf{a}(t) = \mathbf{v}'(t) = \mathbf{r}''(t)$ is the **acceleration vector** (Sec. 8.5).

Section 8.7 on the chain rule and a few other facts about functions of several variables prepares us for Secs. 8.8—8.10, in which we discuss the **gradient** of a scalar function f (Sec. 8.8),

$$(5) \qquad \operatorname{grad} f = \nabla f = \left[\frac{\partial f}{\partial x}, \ \frac{\partial f}{\partial y}, \ \frac{\partial f}{\partial z} \right],$$

the **divergence** of a vector function $\mathbf{v}$ (Sec. 8.9),

$$(6) \qquad \operatorname{div} \mathbf{v} = \nabla \bullet \mathbf{v} = \frac{\partial v_1}{\partial x} + \frac{\partial v_2}{\partial y} + \frac{\partial v_3}{\partial z}$$

and the **curl** of $\mathbf{v}$ (Sec. 8.10)

$$(7) \qquad \operatorname{curl} \mathbf{v} = \nabla \times \mathbf{v} = \begin{vmatrix} \mathbf{i} & \mathbf{j} & \mathbf{k} \\ \partial/\partial x & \partial/\partial y & \partial/\partial z \\ v_1 & v_2 & v_3 \end{vmatrix}$$

(with a minus sign in front of the determinant for left-handed coordinates). Some basic formulas for grad, div, curl are (Secs. 8.8—8.10)

$$(8) \qquad \nabla(fg) = f\nabla g + g\nabla f, \qquad \nabla(f/g) = (1/g^2)(g\nabla f - f\nabla g)$$

$$(9) \qquad \operatorname{div} (f\mathbf{v}) = f \operatorname{div} \mathbf{v} + \mathbf{v} \bullet \nabla f, \qquad \operatorname{div} (f\nabla g) = f\nabla^2 g + \nabla f \bullet \nabla g$$

$$(10) \qquad \nabla^2 f = \operatorname{div} (\nabla f), \qquad \nabla^2(fg) = g\nabla^2 f + 2\nabla f \bullet \nabla g + f\nabla^2 g$$

$$(11) \qquad \begin{aligned} \operatorname{curl} (f\mathbf{v}) &= \nabla f \times \mathbf{v} + f \operatorname{curl} \mathbf{v} \\ \operatorname{div} (\mathbf{u} \times \mathbf{v}) &= \mathbf{v} \bullet \operatorname{curl} \mathbf{u} - \mathbf{u} \bullet \operatorname{curl} \mathbf{v} \end{aligned}$$

$$(12) \qquad \operatorname{curl} (\nabla f) = \mathbf{0}, \qquad \operatorname{div} (\operatorname{curl} \mathbf{v}) = 0.$$

The form of grad, div, curl and the Laplacian in general rectangular coordinates is shown in Sec. 8.11.

The directional derivative of f in the direction of a unit vector $\mathbf{b}$ is (Sec. 8.8)

$$(13) \qquad D_{\mathbf{b}}f = \frac{df}{ds} = \mathbf{b} \bullet \nabla f.$$

Chapter 9

Line and Surface Integrals. Integral Theorems

In this chapter we shall define line integrals and surface integrals and consider some important applications of such integrals, which occur frequently in connection with physical and engineering problems. We shall see that a line integral is a natural generalization of a definite integral, and a surface integral is a generalization of a double integral.

Line integrals can be transformed into double integrals (Sec. 9.3) or into surface integrals (Sec. 9.8), and conversely. Triple integrals can be transformed into surface integrals (Sec. 9.6), and conversely. These transformations are of great practical importance. The corresponding formulas of Gauss, Green and Stokes serve as powerful tools in many applications as well as in theoretical problems. We shall see that they also lead to a better understanding of the physical meaning of the divergence and the curl of a vector function.

> *Prerequisites for this chapter:* elementary integral calculus and Chap. 8.
> *Sections that may be omitted in a shorter course:* 9.2, 9.4, 9.7.
> *References:* Appendix 1, Part B.
> *Answers to problems:* Appendix 2.

9.1 Line Integrals

The concept of a line integral is a simple and natural generalization of the concept of a definite integral

$$(1) \qquad \int_a^b f(x)\, dx.$$

In (1) we integrate along the x-axis from a to b and the integrand f is a function defined at each point between a and b. In the case of a line integral we shall integrate along a curve C in space (or in the plane) and the integrand will be a function defined at each point of C. (Hence *curve integral* would be a better name, but *line integral* is the standard term.) The curve C is called the **path of integration.**

We call C a **smooth curve** if C has a representation (cf. Sec. 8.3)

(2) $\mathbf{r}(t) = [x(t), \quad y(t), \quad z(t)] = x(t)\mathbf{i} + y(t)\mathbf{j} + z(t)\mathbf{k}$ $(a \leqq t \leqq b)$

such that $\mathbf{r}(t)$ has a continuous derivative $\mathbf{r}'(t) = d\mathbf{r}/dt$ that is nowhere zero. Geometrically: the curve C has a unique tangent at each of its points whose direction varies continuously as we traverse C. We call A: $\mathbf{r}(a)$ the initial point and B: $\mathbf{r}(b)$ the terminal point of C, and we say that C is now *oriented*, calling the direction from A to B the *positive direction* along C and indicating it by an arrow (as in Fig. 201*a*). The points A and B may coincide (as in Fig. 201*b*); then C is called a *closed curve*.

General assumption

In this book, every path of integration of a line integral is assumed to be **piecewise smooth;** *that is, it consists of finitely many smooth curves.*

Definition and Evaluation of Line Integrals

A **line integral** of a vector function $\mathbf{F}(\mathbf{r})$ over a curve C is defined by

(3)
$$\int_C \mathbf{F}(\mathbf{r}) \cdot d\mathbf{r} = \int_a^b \mathbf{F}(\mathbf{r}(t)) \cdot \frac{d\mathbf{r}}{dt}\, dt$$
[cf. (2)].

In terms of components, with $d\mathbf{r} = [dx, \quad dy, \quad dz]$ as in Sec. 8.4 and $' = d/dt$, formula (3) becomes

(3′)
$$\int_C \mathbf{F}(\mathbf{r}) \cdot d\mathbf{r} = \int_C (F_1\, dx + F_2\, dy + F_3\, dz)$$
$$= \int_a^b (F_1 x' + F_2 y' + F_3 z')\, dt.$$

If the path of integration C in (3) is a *closed* curve, then instead of

$$\int_C \qquad \text{we also write} \qquad \oint_C.$$

$\mathbf{F} \cdot d\mathbf{r}/dt$ with $t = s$ (the arc length of C) is the tangential component of $\mathbf{F}$, and this integral arises naturally in mechanics, where it gives the work done by a force $\mathbf{F}$ in a displacement along C (details and examples below). We

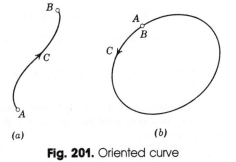

(a)   (b)

Fig. 201. Oriented curve

may thus call the line integral (3) the **work integral.** Other forms of the line integral will be discussed below. We see that the integral in (3) on the right is a definite integral taken over the interval $a \leqq t \leqq b$ on the t-axis in the *positive* direction (the direction of increasing t). This definite integral exists for continuous $\mathbf{F}$ and piecewise smooth C, because this makes $\mathbf{F} \cdot \mathbf{r}'$ piecewise continuous.

EXAMPLE 1. Evaluation of a line integral in the plane

Find the value of the line integral (3) when $\mathbf{F}(\mathbf{r}) = -y\mathbf{i} + xy\mathbf{j}$ and C is the circular arc in Fig. 202 from A to B.

Solution. We may represent C by

$$\mathbf{r}(t) = \cos t \, \mathbf{i} + \sin t \, \mathbf{j} \qquad\qquad (0 \leqq t \leqq \pi/2).$$

Thus $x(t) = \cos t$, $y(t) = \sin t$, so that

$$\mathbf{F}(\mathbf{r}(t)) = -y(t)\mathbf{i} + x(t)y(t)\mathbf{j} = -\sin t \, \mathbf{i} + \cos t \sin t \, \mathbf{j}.$$

Also

$$\mathbf{r}'(t) = -\sin t \, \mathbf{i} + \cos t \, \mathbf{j}$$

so that

$$\int_C \mathbf{F}(\mathbf{r}) \cdot d\mathbf{r} = \int_0^{\pi/2} (-\sin t \, \mathbf{i} + \cos t \sin t \, \mathbf{j}) \cdot (-\sin t \, \mathbf{i} + \cos t \, \mathbf{j}) \, dt$$

$$= \int_0^{\pi/2} (\sin^2 t + \cos^2 t \sin t) \, dt = \frac{\pi}{4} + \frac{1}{3} \approx 1.119$$

[use (10) in Appendix 3.1; set $\cos t = u$ in the second term]. ∎

Two important questions now arise. Does this value depend on the particular choice of a representation of the circular arc C? The answer is no; see Theorem 1 below. Does this value change if we integrate from the same A to the same B as before but along another path? The answer is yes, in general; see Example 2.

EXAMPLE 2. Dependence of a line integral on path (same endpoints)

Evaluate the line integral (3) with $\mathbf{F}(\mathbf{r})$ as in Example 1 from A to B as in Example 1, but along a different path C, namely, along the straight segment from A to B (Fig. 203).

Solution. We may represent C by

$$\mathbf{r}(t) = (1 - t)\mathbf{i} + t\mathbf{j} \qquad\qquad (0 \leqq t \leqq 1)$$

so that A: $\mathbf{r}(0) = \mathbf{i}$ and B: $\mathbf{r}(1) = \mathbf{j}$, as we want it. Then

$$\mathbf{F}(\mathbf{r}(t)) = -y(t)\mathbf{i} + x(t)y(t)\mathbf{j} = -t\mathbf{i} + (1 - t)t\mathbf{j}.$$

Also $\mathbf{r}'(t) = -\mathbf{i} + \mathbf{j}$ by differentiation, and (3) becomes

$$\int_C \mathbf{F}(\mathbf{r}) \cdot d\mathbf{r} = \int_0^1 [-t\mathbf{i} + (1 - t)t\mathbf{j}] \cdot (-\mathbf{i} + \mathbf{j}) \, dt = \int_0^1 [t + (1 - t)t] \, dt = \frac{2}{3}.$$

In Example 1 we obtained a different value. This shows that **the value of a line integral (3) will in general depend** *not only on $\mathbf{F}$ and the endpoints A, B of the path but also on the path along which we integrate from A to B.*

Can we find conditions that guarantee independence? We shall do so in Sec. 9.9, where we shall also explain the importance of this question in physics.

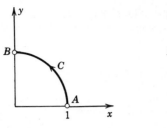
Fig. 202. Example 1

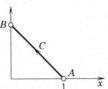
Fig. 203. Example 2

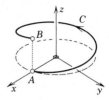
Fig. 204. Example 3

EXAMPLE 3. Line integral in space

To see that the method of computing line integrals in space is the same as that in the plane just considered, find the value of (3) when $\mathbf{F}(\mathbf{r}) = z\mathbf{i} + x\mathbf{j} + y\mathbf{k}$ and C is the helix (Fig. 204)

$$(4) \qquad\qquad \mathbf{r}(t) = \cos t\,\mathbf{i} + \sin t\,\mathbf{j} + 3t\mathbf{k} \qquad\qquad (0 \leqq t \leqq 2\pi).$$

Solution. From (4) we have $x(t) = \cos t$, $y(t) = \sin t$, $z(t) = 3t$. Thus

$$\mathbf{F}(\mathbf{r}(t)) \cdot \mathbf{r}'(t) = (3t\mathbf{i} + \cos t\,\mathbf{j} + \sin t\,\mathbf{k}) \cdot (-\sin t\,\mathbf{i} + \cos t\,\mathbf{j} + 3\mathbf{k}).$$

The dot product is $3t(-\sin t) + \cos^2 t + 3 \sin t$. Hence (3) gives

$$\int_C \mathbf{F}(\mathbf{r}) \cdot d\mathbf{r} = \int_0^{2\pi} (-3t \sin t + \cos^2 t + 3 \sin t)\, dt = 6\pi + \pi + 0 = 7\pi \approx 21.99. \qquad \blacksquare$$

Motivation of the Line Integral (3): Work by a Force

The work W done by a *constant* force $\mathbf{F}$ in the displacement along a *straight* segment $\mathbf{d}$ is $W = \mathbf{F} \cdot \mathbf{d}$; cf. Example 4 in Sec. 6.5. This suggests that we define the work W done by a *variable* force $\mathbf{F}$ in the displacement along a curve C: $\mathbf{r}(t)$ as the limit of sums of works done in displacements along small chords of C, and that we show that this amounts to defining W by the line integral (3).

For this we choose points $t_0 (= a) < t_1 < \cdots < t_n (= b)$. Then the work ΔW_m done by $\mathbf{F}(\mathbf{r}(t_m))$ in the straight displacement from $\mathbf{r}(t_m)$ to $\mathbf{r}(t_{m+1})$ is

$$\Delta W_m = \mathbf{F}(\mathbf{r}(t_m)) \cdot [\mathbf{r}(t_{m+1}) - \mathbf{r}(t_m)] \approx \mathbf{F}(\mathbf{r}(t_m)) \cdot \mathbf{r}'(t_m)\Delta t_m \qquad (\Delta t_m = t_{m+1} - t_m).$$

The sum of these n works is $W_n = \Delta W_0 + \cdots + \Delta W_{n-1}$. If we choose points and consider W_n for every n arbitrarily but so that the greatest Δt_m approaches zero as $n \to \infty$, then the limit of W_n as $n \to \infty$ exists and is the line integral (3) since $\mathbf{F}$ is continuous and C is piecewise smooth, which makes $\mathbf{r}'(t)$ continuous, except at finitely many points where C may have corners or cusps.

EXAMPLE 4. Work done by a variable force

If $\mathbf{F}$ in Example 1 is a force, the work done by $\mathbf{F}$ in the displacement along the quarter-circle is 1.119, measured in suitable units, say, newton-meters (nt·m, also called joules, abbreviation J; see also front cover). Similarly in Examples 2 and 3.

EXAMPLE 5. Work done equals the gain in kinetic energy

Let $\mathbf{F}$ be a force, so that (3) is work. Let t be time, so that $d\mathbf{r}/dt = \mathbf{v}$, velocity. Then we can write (3) as

$$(5) \qquad W = \int_C \mathbf{F} \cdot d\mathbf{r} = \int_a^b \mathbf{F}(\mathbf{r}(t)) \cdot \mathbf{v}(t) \, dt.$$

Now by Newton's second law (force = mass × acceleration),

$$\mathbf{F} = m\mathbf{r}''(t) = m\mathbf{v}'(t),$$

where m is the mass of the body displaced. Substitution into (5) gives [cf. (8), Sec. 8.2]

$$W = \int_a^b m\mathbf{v}' \cdot \mathbf{v} \, dt = \int_a^b m \left(\frac{\mathbf{v} \cdot \mathbf{v}}{2} \right)' dt = \frac{m}{2} |\mathbf{v}|^2 \Big|_{t=a}^{t=b}.$$

On the right, $m|\mathbf{v}|^2/2$ is the kinetic energy. Hence *the work done equals the gain in kinetic energy*. This is a basic law in mechanics.   ∎

Other Forms of Line Integrals

$$(6) \qquad \int_C F_1 \, dx, \qquad \int_C F_2 \, dy, \qquad \int_C F_3 \, dz$$

are special cases of (3) when $\mathbf{F} = F_1\mathbf{i}$ or $F_2\mathbf{j}$ or $F_3\mathbf{k}$, respectively. Another form is

$$(7) \qquad \int_C f(\mathbf{r}) \, dt = \int_a^b f(\mathbf{r}(t)) \, dt$$

with C as in (2). But this definition can also be regarded as a special case of (3), with $\mathbf{F} = F_1\mathbf{i}$ and $F_1 = f/(dx/dt)$, so that $f = F_1 x'$, as in (3'). The evaluation of (7) is similar to that in the previous examples.

EXAMPLE 6. A line integral of the form (7)

Find the value of (7) when $f = (x^2 + y^2 + z^2)^2$ and C is the helix (4) in Example 3.

Solution. From $\mathbf{r}(t) = \cos t \, \mathbf{i} + \sin t \, \mathbf{j} + 3t\mathbf{k}$ we see that on C,

$$(x^2 + y^2 + z^2)^2 = [\cos^2 t + \sin^2 t + (3t)^2]^2 = (1 + 9t^2)^2.$$

This gives

$$\int_C f(\mathbf{r}) \, dt = \int_0^{2\pi} (1 + 9t^2)^2 \, dt = 2\pi + 6(2\pi)^3 + \frac{81}{5}(2\pi)^5 \approx 160\ 135.$$ ∎

General Properties of the Line Integral (3)

From familiar properties of integrals in calculus we obtain corresponding formulas for line integrals (3):

$$(8a) \qquad \int_C k\mathbf{F} \cdot d\mathbf{r} = k \int_C \mathbf{F} \cdot d\mathbf{r} \qquad\qquad (k \text{ constant})$$

(8b)
$$\int_C (\mathbf{F} + \mathbf{G}) \cdot d\mathbf{r} = \int_C \mathbf{F} \cdot d\mathbf{r} + \int_C \mathbf{G} \cdot d\mathbf{r}$$

(8c)
$$\int_C \mathbf{F} \cdot d\mathbf{r} = \int_{C_1} \mathbf{F} \cdot d\mathbf{r} = \int_{C_2} \mathbf{F} \cdot d\mathbf{r} \qquad \text{(Fig. 205)}$$

where in (8c) the path C is subdivided into two arcs C_1 and C_2 that have the same orientation as C (Fig. 205). In (8b) the orientation of C is the same in all three integrals. If the sense of integration along C is reversed, the value of the integral is multiplied by -1.

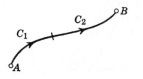

Fig. 205. Formula (8c)

If a line integral (3) is supposed to represent physical quantities, such as work, the choice of one or another representation of a given curve C should not be essential, as long as the positive directions are the same in both cases. This is what we now show.

Theorem 1 (Direction-preserving transformations of parameter)
Any representations of C that give the same positive direction on C also yield the same value of the line integral (3).

Proof. We represent C in (3) using another parameter t^* given by a function $t = \phi(t^*)$ which has a positive derivative and is such that $a^* \leq t^* \leq b^*$ corresponds to $a \leq t \leq b$. Then, writing $\mathbf{r}(\phi(t^*)) = \mathbf{r}^*(t^*)$ and using the chain rule, we have $dt^* = (dt^*/dt)\, dt$ and thus

$$\int_C \mathbf{F}(\mathbf{r}^*) \cdot d\mathbf{r}^* = \int_{a^*}^{b^*} \left[\mathbf{F}(\mathbf{r}^*(t^*)) \cdot \frac{d\mathbf{r}^*}{dt^*} \right] dt^*$$

$$= \int_{a^*}^{b^*} \mathbf{F}(\mathbf{r}(\phi(t^*))) \cdot \frac{d\mathbf{r}}{dt} \frac{dt}{dt^*} \, dt^*$$

$$= \int_a^b \mathbf{F}(\mathbf{r}(t)) \cdot \frac{d\mathbf{r}}{dt} \, dt = \int_C \mathbf{F}(\mathbf{r}) \cdot d\mathbf{r}. \qquad \blacksquare$$

Looking back at the present section, the student will realize that curves and their parametric representations are central to a good understanding of line integrals, and a review of Secs. 8.3 and 8.4 may perhaps be good. In the next section we discuss **double integrals** over regions in the plane, as another concept that will be needed several times in this chapter.

Problems for Sec. 9.1

Compute $\int_C \mathbf{F}(\mathbf{r}) \cdot d\mathbf{r}$, where

1. $\mathbf{F} = 3y\mathbf{i} - x\mathbf{j}$, C: the straight line segment from $(0, 0)$ to $(2, \frac{1}{2})$
2. $\mathbf{F} = x^2\mathbf{i} + y^2\mathbf{j}$, C: $y = 1 - x^2$ from $x = -1$ to $x = 1$
3. $\mathbf{F} = y^2\mathbf{i} - x^4\mathbf{j}$, C: $\mathbf{r} = t\mathbf{i} + t^{-1}\mathbf{j}$, $1 \leq t \leq 3$
4. $\mathbf{F} = \exp(y^{2/3})\mathbf{i} - \exp(x^{3/2})\mathbf{j}$, C: $\mathbf{r} = t\mathbf{i} + t^{3/2}\mathbf{j}$ from $(0, 0)$ to $(0, 1)$
5. $\mathbf{F} = y^2\mathbf{i}$, C: from $(2, 0)$ to $(0, 1)$ counterclockwise along $x^2 + 4y^2 = 4$
6. $\mathbf{F} = -y\mathbf{i} + x\mathbf{j}$, C: $\mathbf{r} = \cosh t\, \mathbf{i} + \sinh t\, \mathbf{j}$, $0 \leq t \leq 10$
7. $\mathbf{F} = x^4\mathbf{i} + y^6\mathbf{j}$, C: from $(2, 0)$ to $(-2, 0)$ counterclockwise along $x^2 + y^2 = 4$
8. $\mathbf{F} = y^4\mathbf{i} + 2y^3\mathbf{j}$, C: $y = e^{-x}$, $0 \leq x \leq 1/4$
9. $\mathbf{F} = 2xy^3\mathbf{i} + 3x^2y^2\mathbf{j}$, C: counterclockwise around the triangle with vertices $(1, 1)$, $(3, 1)$, $(3, 3)$
10. $\mathbf{F} = x(5y + 2)\mathbf{i} + 6y\mathbf{j}$, C: clockwise around the rectangle with vertices $(0, 0)$, $(2, 0)$, $(2, 3)$, $(0, 3)$

Work. Find the work done by the force $\mathbf{F} = x\mathbf{i} - z\mathbf{j} + 2y\mathbf{k}$ in the displacement

11. Along the y-axis from 0 to 1
12. Along the parabola $y = 2x^2$, $z = 2$ from $(0, 0, 2)$ to $(1, 2, 2)$
13. Along the curve $z = y^4$, $x = 1$ from $(1, 0, 0)$ to $(1, 1, 1)$
14. Counterclockwise around the circle $x^2 + y^2 = 4$, $z = 0$
15. Along the closed path consisting of the three straight segments from $(0, 0, 0)$ to $(1, 1, 0)$ to $(1, 1, 1)$ and back to $(0, 0, 0)$
16. Along the straight line $y = x$, $z = x$ from $(1, 1, 1)$ to $(3, 3, 3)$
17. Along the curve $y = x^3$, $z = 2$ from $(1, 1, 2)$ to $(2, 8, 2)$
18. Along the curve $y = x$, $z = x^2$ from $(0, 0, 0)$ to $(1, 1, 1)$
19. Along the helix in Example 3 from $(1, 0, 0)$ to $(1, 0, 12\pi)$
20. Along the helix in Example 3 from $(1, 0, 0)$ to $(-1, 0, 3\pi)$

Orienting C so that the sense of integration becomes the positive sense on C, evaluate $\int_C (x^2 + y^2)\, ds$:

21. Over the path $y = 2x$ from $(0, 0)$ to $(1, 2)$
22. Over the path $y = -x$ from $(1, -1)$ to $(2, -2)$
23. Counterclockwise along the circle $x^2 + y^2 = 4$ from $(2, 0)$ to $(0, 2)$
24. Over the y-axis from $(0, 0)$ to $(0, 1)$, then parallel to the x-axis from $(0, 1)$ to $(1, 1)$
25. Over the x-axis from $(0, 0)$ to $(1, 0)$, then parallel to the y-axis from $(1, 0)$ to $(1, 1)$
26. Counterclockwise around the circle $x^2 + y^2 = 1$ from $(1, 0)$ to $(1, 0)$

27. **(Invariance)** Verify Theorem 1 for the integral in Example 1 and the transformation $t = t^{*3}$.
28. Verify Theorem 1 for $\mathbf{F} = y^2\mathbf{i} + x^2\mathbf{k}$, C: $\mathbf{r}(t) = t\mathbf{i} + t^2\mathbf{j} + t^3\mathbf{k}$, $1 \leq t \leq 2$, and $t = t^{*2}$.

Estimation of line integrals

29. Let $\mathbf{F}$ be a vector function defined at all points of a curve C, and suppose that $|\mathbf{F}|$ is bounded, say $|\mathbf{F}| \leq M$ on C, where M is some positive number. Show that

(9)
$$\left| \int_C \mathbf{F} \cdot d\mathbf{r} \right| \leq ML$$

where L is the length of C.

30. Using (9), find an upper bound for the absolute value of the work W done by the force $\mathbf{F} = x\mathbf{i} + y^2\mathbf{j}$ in the displacement along the straight line from $(0, 0, 0)$ to $(1, 1, 0)$. Find W by integration and compare the results.

9.2 Double Integrals

In our further discussion we shall need the concept of a double integral. Although the reader will be familiar with double integrals from calculus, we now present a brief review.

In the case of a definite integral

$$\int_a^b f(x) \, dx$$

the integrand is a function $f(x)$ that exists for all x in an interval $a \leq x \leq b$ of the x-axis. In the case of a double integral the integrand is a function $f(x, y)$ that is given for all (x, y) in a closed bounded[1] region R of the xy-plane.

The definition of the double integral is quite similar to that of the definite integral. We subdivide the region R by drawing parallels to the x and y axes (Fig. 206). We number the rectangles that are within R from 1 to n. In each such rectangle we choose a point, say, (x_k, y_k) in the kth rectangle, and then we form the sum

$$J_n = \sum_{k=1}^n f(x_k, y_k) \, \Delta A_k$$

where ΔA_k is the area of the kth rectangle. This we do for larger and larger

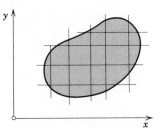

Fig. 206. Subdivision of R

---

[1]"Closed" means that the boundary is part of the region, and "bounded" means that the region can be enclosed in a circle of sufficiently large radius.

positive integers n in a completely independent manner but so that the length of the maximum diagonal of the rectangles approaches zero as n approaches infinity. In this fashion we obtain a sequence of real numbers $J_{n_1}, J_{n_2}, \cdots$. Assuming that $f(x, y)$ is continuous in R and R is bounded by finitely many smooth curves (cf. Sec. 9.1), one can show[2] that this sequence converges and its limit is independent of the choice of subdivisions and corresponding points (x_k, y_k). This limit is called the **double integral** *of* $f(x, y)$ *over the region* R, and is denoted by

$$\iint\limits_{R} f(x, y) \, dx \, dy \quad \text{or} \quad \iint\limits_{R} f(x, y) \, dA.$$

From the definition it follows that double integrals enjoy properties that are quite similar to those of definite integrals. Let f and g be functions of x and y, defined and continuous in a region R. Then

$$\iint\limits_{R} kf \, dx \, dy = k \iint\limits_{R} f \, dx \, dy \qquad (k \text{ constant})$$

$$\textbf{(1)} \qquad \iint\limits_{R} (f + g) \, dx \, dy = \iint\limits_{R} f \, dx \, dy + \iint\limits_{R} g \, dx \, dy$$

$$\iint\limits_{R} f \, dx \, dy = \iint\limits_{R_1} f \, dx \, dy + \iint\limits_{R_2} f \, dx \, dy \qquad (\text{cf. Fig. 207}).$$

Furthermore, there exists at least one point (x_0, y_0) in R such that we have

$$\textbf{(2)} \qquad \iint\limits_{R} f(x, y) \, dx \, dy = f(x_0, y_0)A,$$

where A is the area of R; this is called the **mean value theorem** *for double integrals*.

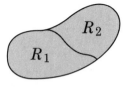

Fig. 207. Formula (1)

Evaluation of Double Integrals

Double integrals over a region R may be evaluated by two successive integrations as follows. Suppose that R can be described by inequalities of

[2]Cf. Ref. [5] in Appendix 1.

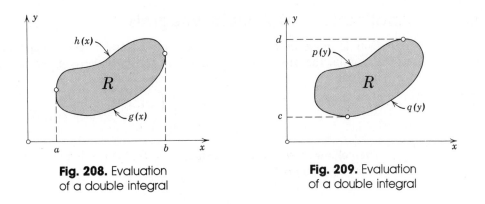

Fig. 208. Evaluation of a double integral

Fig. 209. Evaluation of a double integral

the form

$$a \leqq x \leqq b, \qquad g(x) \leqq y \leqq h(x)$$

(Fig. 208), so that $y = g(x)$ and $y = h(x)$ represent the boundary of R. Then

(3)
$$\iint\limits_{R} f(x, y)\, dx\, dy = \int_{a}^{b} \left[\int_{g(x)}^{h(x)} f(x, y)\, dy \right] dx.$$

We first integrate the inner integral

$$\int_{g(x)}^{h(x)} f(x, y)\, dy.$$

In this integration we keep x fixed, that is, we regard x as a constant. The result of this integration will be a function of x, say, $F(x)$. By integrating $F(x)$ over x from a to b we then obtain the value of the double integral in (3).

Similarly, if R can be described by inequalities of the form

$$c \leqq y \leqq d, \qquad p(y) \leqq x \leqq q(y)$$

(Fig. 209), then we obtain

(4)
$$\iint\limits_{R} f(x, y)\, dx\, dy = \int_{c}^{d} \left[\int_{p(y)}^{q(y)} f(x, y)\, dx \right] dy;$$

we now integrate first over x (treating y as a constant) and then the resulting function of y from c to d.

If R cannot be represented by those inequalities, but can be subdivided into finitely many portions that have that property, we may integrate $f(x, y)$ over each portion separately and add the results; this will give us the value of the double integral of $f(x, y)$ over that region R.

Applications of Double Integrals

Double integrals have various geometrical and physical applications. For example, the **area** A of a region R in the xy-plane is given by the double integral

$$A = \iint_R dx \, dy.$$

The **volume** V beneath the surface $z = f(x, y)$ (> 0) and above a region R in the xy-plane is (Fig. 210)

$$V = \iint_R f(x, y) \, dx \, dy,$$

because the term $f(x_k, y_k) \Delta A_k$ in J_n at the beginning of this section represents the volume of a rectangular parallelepiped with base ΔA_k and altitude $f(x_k, y_k)$.

Let $f(x, y)$ be the density ($=$ mass per unit area) of a distribution of mass in the xy-plane. Then the total **mass** M in R is

$$M = \iint_R f(x, y) \, dx \, dy,$$

the **center of gravity** of the mass in R has the coordinates $\bar{x}, \bar{y}$, where

$$\bar{x} = \frac{1}{M} \iint_R x f(x, y) \, dx \, dy \qquad \text{and} \qquad \bar{y} = \frac{1}{M} \iint_R y f(x, y) \, dx \, dy,$$

the **moments of inertia** I_x and I_y of the mass in R about the x- and y-axes, respectively, are

$$I_x = \iint_R y^2 f(x, y) \, dx \, dy, \qquad I_y = \iint_R x^2 f(x, y) \, dx \, dy,$$

and the **polar moment of inertia** I_0 about the origin of the mass in R is

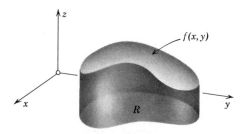

Fig. 210. Double integral as volume

$$I_0 = I_x + I_y = \iint_R (x^2 + y^2) f(x, y) \, dx \, dy.$$

EXAMPLE 1. Applications of the double integral
Let $f(x, y) = 1$ be the density of mass in the region $R: 0 \leqq y \leqq \sqrt{1 - x^2}, 0 \leqq x \leqq 1$ (Fig. 211). Find the center of gravity and the moments of inertia I_x, I_y and I_0.

Solution. The total mass in R is

$$M = \iint_R dx \, dy = \int_0^1 \left[\int_0^{\sqrt{1-x^2}} dy \right] dx = \int_0^1 \sqrt{1 - x^2} \, dx = \int_0^{\pi/2} \cos^2 \theta \, d\theta = \frac{\pi}{4}$$

($x = \sin \theta$), which is the area of R. The coordinates of the center of gravity are

$$\bar{x} = \frac{4}{\pi} \iint_R x \, dx \, dy = \frac{4}{\pi} \int_0^1 \left[\int_0^{\sqrt{1-x^2}} x \, dy \right] dx = \frac{4}{\pi} \int_0^1 x \sqrt{1 - x^2} \, dx = -\frac{4}{\pi} \int_1^0 z^2 \, dz = \frac{4}{3\pi}$$

($\sqrt{1 - x^2} = z$), and $\bar{y} = \bar{x}$, for reasons of symmetry. Furthermore,

$$I_x = \iint_R y^2 \, dx \, dy = \int_0^1 \left[\int_0^{\sqrt{1-x^2}} y^2 \, dy \right] dx = \frac{1}{3} \int_0^1 (\sqrt{1 - x^2})^3 \, dx$$

$$= \frac{1}{3} \int_0^{\pi/2} \cos^4 \theta \, d\theta = \frac{\pi}{16}, \quad I_y = \frac{\pi}{16}, \quad I_0 = I_x + I_y = \frac{\pi}{8} \approx 0.3927.$$

A simpler way of calculating I_x will be shown in Example 2.   ∎

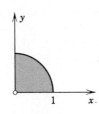

Fig. 211. Example 1

Change of Variables in Double Integrals

It will often be necessary to change the variables of integration in double integrals. We know from calculus that in the case of a definite integral

$$\int_a^b f(x) \, dx$$

a new variable of integration u can be introduced by setting

$$x = x(u)$$

where the function $x(u)$ is continuous and has a continuous derivative in some interval $\alpha \leqq u \leqq \beta$ such that $x(\alpha) = a$, $x(\beta) = b$ [or $x(\alpha) = b$,

$x(\beta) = a$] and $x(u)$ varies between a and b when u varies between α and β. Then

(5)
$$\int_a^b f(x) \, dx = \int_\alpha^\beta f(x(u)) \frac{dx}{du} \, du.$$

For example, let $f(x) = \sqrt{1 - x^2}$, $a = 0$, $b = 1$, and set $x = \sin u$. Then

$$f[x(u)] = \sqrt{1 - \sin^2 u} = \cos u, \qquad \frac{dx}{du} = \cos u, \qquad \alpha = 0, \qquad \beta = \frac{\pi}{2},$$

and

$$\int_0^1 \sqrt{1 - x^2} \, dx = \int_0^{\pi/2} \cos^2 u \, du = \frac{\pi}{4}.$$

In the case of a double integral

$$\iint\limits_R f(x, y) \, dx \, dy$$

new variables of integration u, v can be introduced by setting

$$x = x(u, v), \qquad y = y(u, v);$$

here the functions $x(u, v)$ and $y(u, v)$ are continuous and have continuous first partial derivatives in some region R^* in the uv-plane so that each point (u_0, v_0) in R^* corresponds to a point $[x(u_0, v_0), y(u_0, v_0)]$ in R, where the points corresponding to different points in R are different; and furthermore the **Jacobian**[3]

$$J = \frac{\partial(x, y)}{\partial(u, v)} = \begin{vmatrix} \dfrac{\partial x}{\partial u} & \dfrac{\partial x}{\partial v} \\[2mm] \dfrac{\partial y}{\partial u} & \dfrac{\partial y}{\partial v} \end{vmatrix}$$

is either positive throughout R^* or negative throughout R^*. Then

(6)
$$\iint\limits_R f(x, y) \, dx \, dy = \iint\limits_{R^*} f[x(u, v), y(u, v)] \left| \frac{\partial(x, y)}{\partial(u, v)} \right| du \, dv;$$

that is, the integrand is expressed in terms of u and v, and $dx \, dy$ is replaced by $du \, dv$ times the absolute value of the Jacobian J. For the proof see Ref. [5] in Appendix 1.

---

[3]Named after the German mathematician CARL GUSTAV JACOB JACOBI (1804—1851), professor at Königsberg and Berlin, who made important contributions to elliptic functions, partial differential equations, mechanics, astronomy and the calculus of variations.

For example, **polar coordinates** r and θ can be introduced by setting

$$x = r \cos \theta, \qquad y = r \sin \theta.$$

Then

$$J = \frac{\partial(x, y)}{\partial(r, \theta)} = \begin{vmatrix} \cos \theta & -r \sin \theta \\ \sin \theta & r \cos \theta \end{vmatrix} = r,$$

and

$$(7) \qquad \iint\limits_{R} f(x, y)\, dx\, dy = \iint\limits_{R^*} f(r \cos \theta, r \sin \theta) r\, dr\, d\theta,$$

where R^* is the region in the $r\theta$-plane corresponding to R in the xy-plane.

EXAMPLE 2. Double integral in polar coordinates
Using (7), we obtain for I_x in Example 1

$$I_x = \iint\limits_{R} y^2\, dx\, dy = \int_0^{\pi/2} \int_0^1 r^2 \sin^2 \theta\, r\, dr\, d\theta = \int_0^{\pi/2} \sin^2 \theta\, d\theta \int_0^1 r^3\, dr = \frac{\pi}{4} \cdot \frac{1}{4} = \frac{\pi}{16}.$$

EXAMPLE 3. Change of variables in a double integral
Evaluate the double integral

$$\iint\limits_{R} (x^2 + y^2)\, dx\, dy$$

where R is the square in Fig. 212.

Solution. The shape of R suggests the transformation $x + y = u$, $x - y = v$. Then $x = \frac{1}{2}(u + v)$, $y = \frac{1}{2}(u - v)$, the Jacobian is

$$J = \frac{\partial(x, y)}{\partial(u, v)} = \begin{vmatrix} \frac{1}{2} & \frac{1}{2} \\ \frac{1}{2} & -\frac{1}{2} \end{vmatrix} = -\frac{1}{2},$$

R corresponds to the square $0 \leq u \leq 2$, $0 \leq v \leq 2$, and, therefore,

$$\iint\limits_{R} (x^2 + y^2)\, dx\, dy = \int_0^2 \int_0^2 \frac{1}{2}(u^2 + v^2)\frac{1}{2}\, du\, dv = \frac{8}{3}. \qquad \blacksquare$$

In the next section we discuss **Green's theorem in the plane,** which transforms double integrals into line integrals, and conversely. This is the key theorem in deriving transformation theorems for surface and volume integrals in Secs. 9.6–9.8.

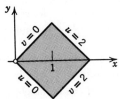

Fig. 212. Region in Example 3

Problems for Sec. 9.2

Evaluate and describe the region of integration.

1. $\displaystyle\int_0^1 \int_0^1 (x^2 + y^2)\, dy\, dx$

2. $\displaystyle\int_0^1 \int_0^{\sqrt{1-x^2}} (x^2 + y^2)\, dy\, dx$

3. $\displaystyle\int_{-\pi}^{\pi} \int_{-1}^1 xy\, dx\, dy$

4. $\displaystyle\int_0^3 \int_{-y}^y (x^2 + y^2)\, dx\, dy$

5. $\displaystyle\int_0^2 \int_{\sinh x^2}^{\cosh x^2} x\, dy\, dx$

6. $\displaystyle\int_0^{\pi/2} \int_0^{y} \frac{\sin y}{y}\, dx\, dy$

7. $\displaystyle\int_0^1 \int_0^{e^x} (x + y^2)\, dy\, dx$

8. $\displaystyle\int_1^2 \int_{-x}^{x} e^y \cosh x\, dy\, dx$

9. $\displaystyle\int_0^1 \int_y^{y^2+1} x^2 y\, dx\, dy$

10. $\displaystyle\int_0^{\pi/2} \int_0^{\cos y} x^2 \sin y\, dx\, dy$

Volume. Find the volume of the following regions in space.

11. The region beneath the plane $z = 6x - y + 10$ and above the rectangle with vertices $(0, 0)$, $(2, 0)$, $(2, 6)$, $(0, 6)$ in the xy-plane

12. The region beneath the plane $x + y + z = 1$ and above the triangle with vertices $(1, 0)$, $(\frac{1}{2}, 0)$, $(\frac{1}{2}, \frac{1}{2})$ in the xy-plane

13. The region beneath $z = x^2 + y^2$ and above the square with vertices $(0, 0)$, $(1, 0)$, $(1, 1)$, $(0, 1)$ in the xy-plane

14. The region beneath $z = x^2 + y^2$ and above the triangle with vertices $(0, 0)$, $(1, 0)$, $(1, 1)$ in the xy-plane

15. The tetrahedron cut from the first octant by the plane $x + y + z = 1$

16. The tetrahedron cut from the first octant by the plane $6x + 3y + 2z = 6$

17. The region beneath $z = y^2$ and above the triangle with vertices $(0, 0)$, $(1, 0)$, $(0, 1)$

18. The first octant region bounded by the coordinate planes and the surfaces $y = 1 - x^2$, $z = 1 - x^2$

19. The first octant section cut from the region inside the cylinder $x^2 + z^2 = 1$ by the planes $y = 0$, $z = 0$, $x = y$

20. The region bounded by the surfaces $y = x^2$, $x = y^2$ and the planes $z = 0$, $z = 3$

Use of polar coordinates. Using polar coordinates, evaluate $\displaystyle\iint_R f(x, y)\, dx\, dy$, where

21. $f = 4x + y$, R: $x^2 + y^2 \leq 9$, $x \geq 0$

22. $f = (x^5 + 2)y$, R: $x^2 + y^2 \leq 1$, $y \geq 0$

23. $f = (x^2 + y^2)^2$, R: $x^2 + y^2 \leq 4$, $y \geq 0$

24. $f = 3 \cos (x^2 + y^2)$, R: $x^2 + y^2 \leq \pi/2$, $x \geq 0$

25. $f = e^{-x^2-y^2}/\pi$, R: the annulus bounded by $x^2 + y^2 = 1$ and $x^2 + y^2 = 4$

Center of gravity. Find the coordinates $\bar{x}$, $\bar{y}$ of the center of gravity of a mass of density $f(x, y)$ in a region R, where

26. $f(x, y) = 1$, R the rectangle $0 \leq x \leq 4, 0 \leq y \leq 10$

27. $f(x, y) = 3xy$, R: the rectangle $0 \leq x \leq 4, 0 \leq y \leq 2$

28. $f(x, y) = 1,$ R: the region $x^2 + y^2 \leq a^2$ in the first quadrant

29. $f(x, y) = x^2 + y^2,$ R: the region $x^2 + y^2 \leq 1$ in the first quadrant

30. $f(x, y) = 1,$ R the triangle with vertices $(0, 0),$ $(b, 0),$ (b, h)

Moments of inertia. Find the moments of inertia I_x, I_y, I_0 of a mass of density $f(x, y) = 1$ in a region R shown in the following figures (which the engineer is likely to need in applications).

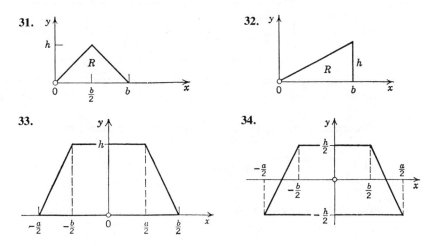

31. **32.**

33. **34.**

Jacobian. Find the Jacobian and give a geometrical reason for your result.

35. Translation $x = u + a, y = v + b$

36. Reflection $x = -u, y = v$

37. Rotation $x = u \cos \phi - v \sin \phi, y = u \sin \phi + v \cos \phi$

38. Expansion in one direction $x = au, y = v$

39. Expansion $x = au, y = bv$

40. Interchange of variables $x = v, y = u$

9.3 Transformation of Double Integrals into Line Integrals (Green's Theorem in the Plane)

Double integrals over a plane region may be transformed into line integrals over the boundary of the region and conversely. This transformation is of practical interest because it may help to make the evaluation of an integral easier. It also helps in the theory whenever one wants to switch from one type of integral to the other. The transformation can be done by the so-called Green's[4] theorem in the plane.

[4]GEORGE GREEN (1793—1841), English mathematician, who was self-educated, started out as a baker and at his death was fellow of Caius College, Cambridge. His work concerned potential theory in connection with electricity and magnetism, vibrations, waves and elasticity theory. It remained almost unknown, even in England, until after his death.

Green's theorem in the plane
(Transformation between double integrals and line integrals)

Let R be a closed bounded region in the xy-plane whose boundary C consists of finitely many smooth curves. Let $F_1(x, y)$ and $F_2(x, y)$ be functions that are continuous and have continuous partial derivatives $\partial F_1/\partial y$ and $\partial F_2/\partial x$ everywhere in some domain containing R. Then

(1)
$$\iint_R \left(\frac{\partial F_2}{\partial x} - \frac{\partial F_1}{\partial y} \right) dx\, dy = \oint_C (F_1\, dx + F_2\, dy),$$

the integration being taken along the entire boundary C of R such that R is on the left as one advances in the direction of integration (cf. Fig. 213).

Comment. Formula (1) can be written in vectorial form

(1′)
$$\iint_R (\text{curl } \mathbf{F}) \cdot \mathbf{k}\, dx\, dy = \oint_C \mathbf{F} \cdot d\mathbf{r} \qquad (\mathbf{F} = F_1 \mathbf{i} + F_2 \mathbf{j}).$$

This follows from (1), Sec. 8.10, which shows that the third component of curl **F** is $\partial F_2/\partial x - \partial F_1/\partial y$.

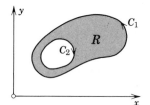

Fig. 213. Region R whose boundary C consists of two parts; C_1 is traversed counterclockwise, while C_2 is traversed clockwise

EXAMPLE 1. Verification of Green's theorem in the plane
Green's theorem in the plane will be quite important in our further work. Before proving it, let us get used to it by verifying it for $F_1 = y^2 - 7y$, $F_2 = 2xy + 2x$ and C the circle $x^2 + y^2 = 1$.

Solution. In (1) on the left we get

$$\iint_R \left(\frac{\partial F_2}{\partial x} - \frac{\partial F_1}{\partial y} \right) dx\, dy = \iint_R [(2y + 2) - (2y - 7)]\, dx\, dy = 9 \iint_R dx\, dy = 9\pi$$

since the circular disk R has area π. On the right in (1) we represent C (oriented counterclockwise!) by

$$\mathbf{r}(t) = [\cos t, \sin t]. \qquad \text{Then} \qquad \mathbf{r}'(t) = [-\sin t, \cos t]$$

and furthermore on C,

$$F_1 = \sin^2 t - 7 \sin t, \qquad F_2 = 2 \cos t \sin t + 2 \cos t.$$

Hence the integral in (1) on the right becomes

$$\oint_C (F_1 x' + F_2 y')\, dt = \int_0^{2\pi} [(\sin^2 t - 7 \sin t)(-\sin t) + 2(\cos t \sin t + \cos t)(\cos t)]\, dt$$

$$= 0 + 7\pi + 0 + 2\pi = 9\pi.$$

This verifies the theorem.   ∎

Proof of Green's theorem. We first prove Green's theorem for a *special region R* that can be represented in both the forms (Fig. 214)

$$a \leqq x \leqq b, \qquad u(x) \leqq y \leqq v(x),$$

and

$$c \leqq y \leqq d, \qquad p(y) \leqq x \leqq q(y)$$

Using (3) in Sec. 9.2, we obtain

(2)
$$\iint_R \frac{\partial F_1}{\partial y}\, dx\, dy = \int_a^b \left[\int_{u(x)}^{v(x)} \frac{\partial F_1}{\partial y}\, dy \right] dx.$$

We integrate the inner integral:

$$\int_{u(x)}^{v(x)} \frac{\partial F_1}{\partial y}\, dy = F_1(x, y) \Big|_{y=u(x)}^{y=v(x)} = F_1[x, v(x)] - F_1[x, u(x)].$$

By inserting this into (2) we find

$$\iint_R \frac{\partial F_1}{\partial y}\, dx\, dy = \int_a^b F_1[x, v(x)]\, dx - \int_a^b F_1[x, u(x)]\, dx$$

$$= -\int_a^b F_1[x, u(x)]\, dx - \int_b^a F_1[x, v(x)]\, dx.$$

Since $y = u(x)$ represents the oriented curve C^* (Fig. 214a) and $y = v(x)$ represents C^{**}, the integrals on the right may be written as line integrals over C^* and C^{**}, and, therefore,

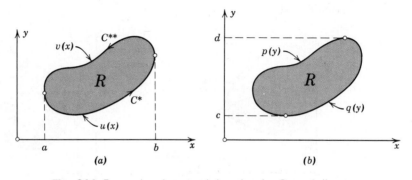

Fig. 214. Example of a special region for Green's theorem

(3)
$$\iint_R \frac{\partial F_1}{\partial y} \, dx \, dy = -\int_{C^*} F_1(x, y) \, dx - \int_{C^{**}} F_1(x, y) \, dx$$
$$= -\oint_C F_1(x, y) \, dx.$$

If portions of C are segments parallel to the y-axis (such as $\tilde{C}$ and $\tilde{\tilde{C}}$ in Fig. 215), then the result is the same as before, because the integrals over these portions are zero and may be added to the integrals over C^* and C^{**} to obtain the integral over the whole boundary C in (3). Similarly, using (4), Sec. 9.2, we obtain (cf. Fig. 214b)

$$\iint_R \frac{\partial F_2}{\partial x} \, dx \, dy = \int_c^d \left[\int_{p(y)}^{q(y)} \frac{\partial F_2}{\partial x} \, dx \right] dy$$
$$= \int_c^d F_2[q(y), y] \, dy + \int_d^c F_2[p(y), y] \, dy$$
$$= \oint_C F_2(x, y) \, dy.$$

From this and (3), the formula (1) follows, and the theorem is proved for special regions.

We now prove the theorem for a region R which itself is not a special region but can be subdivided into finitely many special regions (Fig. 216). In this case we apply the theorem to each subregion and then add the results; the left-hand members add up to the integral over R while the right-hand members add up to the line integral over C plus integrals over the curves introduced for subdividing R. Each of the latter integrals occurs twice, taken once in each direction. Hence these two integrals cancel each other, and we are left with the line integral over C.

The proof thus far covers all regions that are of interest in engineering problems. To prove the theorem for the most general region R satisfying the conditions in the theorem, we must approximate R by a region of the type just considered and then use a limiting process. For details see Ref. [5] in Appendix 1. ∎

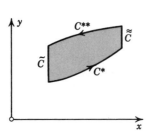

Fig. 215. Proof of Green's theorem

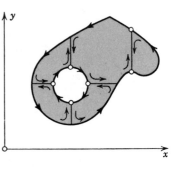

Fig. 216. Proof of Green's theorem

We continue with some further applications of Green's theorem. Later in this chapter we shall use Green's theorem as the basic tool in deriving two very important theorems (by Gauss and Stokes) for transforming integrals.

EXAMPLE 2. Area of a plane region as a line integral over the boundary
In (1), let $F_1 = 0$ and $F_2 = x$. Then

$$\iint_R dx\, dy = \oint_C x\, dy.$$

The integral on the left is the area A of R. Similarly, let $F_1 = -y$ and $F_2 = 0$; then from (1),

$$A = \iint_R dx\, dy = -\oint_C y\, dx.$$

By adding both formulas we obtain

(4)
$$A = \frac{1}{2}\oint_C (x\, dy - y\, dx),$$

the integration being taken as indicated in Green's theorem. This interesting formula expresses the area of R in terms of a line integral over the boundary. It has various applications; for example, the theory of certain **planimeters** (instruments for measuring area) is based on it.

EXAMPLE 3. Area of a plane region in polar coordinates
Let r and θ be polar coordinates defined by $x = r\cos\theta$, $y = r\sin\theta$. Then

$$dx = \cos\theta\, dr - r\sin\theta\, d\theta, \qquad dy = \sin\theta\, dr + r\cos\theta\, d\theta,$$

and (4) becomes a formula which is well known from calculus, namely,

(5)
$$A = \frac{1}{2}\oint_C r^2\, d\theta.$$

As an application of (5), we consider the **cardioid** $r = a(1 - \cos\theta)$, where $0 \leq \theta \leq 2\pi$ (Fig. 217). We find

$$A = \frac{a^2}{2}\int_0^{2\pi} (1 - \cos\theta)^2\, d\theta = \frac{3\pi}{2} a^2.$$

EXAMPLE 4. Transformation of a double integral of the Laplacian of a function into a line integral of its normal derivative
Let $w(x, y)$ be a function that is continuous and has continuous first and second partial derivatives in a domain of the xy-plane containing a region R of the type indicated in Green's theorem. We set $F_1 = -\partial w/\partial y$ and $F_2 = \partial w/\partial x$. Then $\partial F_1/\partial y$ and $\partial F_2/\partial x$ are continuous in R, and

(6)
$$\frac{\partial F_2}{\partial x} - \frac{\partial F_1}{\partial y} = \frac{\partial^2 w}{\partial x^2} + \frac{\partial^2 w}{\partial y^2} = \nabla^2 w,$$

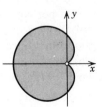

Fig. 217. Cardioid

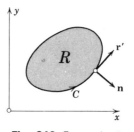

Fig. 218. Example 4

the Laplacian of w (cf. Sec. 8.8). Furthermore, using those expressions for F_1 and F_2, we obtain

$$(7) \quad \oint_C (F_1\, dx + F_2\, dy) = \oint_C \left(F_1 \frac{dx}{ds} + F_2 \frac{dy}{ds}\right) ds = \oint_C \left(-\frac{\partial w}{\partial y}\frac{dx}{ds} + \frac{\partial w}{\partial x}\frac{dy}{ds}\right) ds,$$

where s is the arc length of C, and C is oriented as shown in Fig. 218. The integrand of the last integral may be written as the dot product of the vectors

$$\operatorname{grad} w = \frac{\partial w}{\partial x}\mathbf{i} + \frac{\partial w}{\partial y}\mathbf{j} \qquad \text{and} \qquad \mathbf{n} = \frac{dy}{ds}\mathbf{i} - \frac{dx}{ds}\mathbf{j};$$

that is,

$$(8) \quad -\frac{\partial w}{\partial y}\frac{dx}{ds} + \frac{\partial w}{\partial x}\frac{dy}{ds} = (\operatorname{grad} w)\cdot\mathbf{n}.$$

The vector $\mathbf{n}$ is a unit normal vector to C, because the vector

$$\mathbf{r}'(s) = \frac{d\mathbf{r}}{ds} = \frac{dx}{ds}\mathbf{i} + \frac{dy}{ds}\mathbf{j}$$

(cf. Sec. 8.4) is the unit tangent vector to C, and $\mathbf{r}'\cdot\mathbf{n} = 0$. Furthermore, it is not difficult to see that $\mathbf{n}$ is directed to the *exterior* of C. From this and (5) in Sec. 8.8 it follows that the expression on the right-hand side of (8) is the derivative of w in the direction of the outward normal to C. Denoting this directional derivative by $\partial w/\partial n$ and taking (6), (7) and (8) into account, we obtain from Green's theorem the useful integral formula

$$(9) \quad \boxed{\iint_R \nabla^2 w\, dx\, dy = \oint_C \frac{\partial w}{\partial n}\, ds.}$$

For instance, $w = x^2 - y^2$ satisfies Laplace's equation $\nabla^2 w = 0$. Hence its normal derivative integrated over a closed curve must give 0. Can you verify this directly by integration, say, for the square $0 \le x \le 1, 0 \le y \le 1$? ∎

   Green's theorem in the plane may facilitate the evaluation of integrals and can be used in both directions, depending on the type of integral that is simpler in a concrete case. This is illustrated further in the problem set below. Moreover, and perhaps more fundamentally, Green's theorem will be the essential tool in the proof of Gauss's divergence theorem in Sec. 9.6 (and of Stokes's theorem in Sec. 9.8). Gauss's theorem transforms surface integrals into volume integrals. Hence to get started on Gauss's theorem, we first discuss **surfaces** in the next section and then **surface integrals** in Sec. 9.5.

Problems for Sec. 9.3

Using Green's theorem, evaluate the line integral $\oint_C \mathbf{F(r)} \cdot d\mathbf{r}$ counterclockwise around C, where

1. $\mathbf{F} = 3x^2\mathbf{i} - 4xy\mathbf{j}$, C the boundary of the rectangle $0 \le x \le 4$, $0 \le y \le 1$
2. $\mathbf{F} = y\mathbf{i} - x\mathbf{j}$, C the ellipse $x^2 + 9y^2 = 9$
3. $\mathbf{F} = -y^4\mathbf{i} + xy^2\mathbf{j}$, C the boundary of the triangle with vertices $(0, 0)$, $(1, 0)$, $(1, 1)$
4. $\mathbf{F} = y^2\mathbf{i} + x^2\mathbf{j}$, C the boundary of the square $-1 \le x \le 1$, $-1 \le y \le 1$
5. $\mathbf{F} = y\mathbf{i} - x\mathbf{j}$, C the circle $x^2 + y^2 = 1/4$
6. $\mathbf{F} = x \sin y\ \mathbf{i} + y \cosh x\ \mathbf{j}$, C the boundary of the rectangle $0 \le x \le 2$, $0 \le y \le \pi/2$
7. $\mathbf{F} = (x - y)\mathbf{i} - x^2\mathbf{j}$, C the boundary of the square $0 \le x \le 2$, $0 \le y \le 2$
8. $\mathbf{F} = (3x^2 + y)\mathbf{i} + 4y^2\mathbf{j}$, C the boundary of the triangle with vertices $(0, 0)$, $(1, 0)$, $(0, 2)$
9. $\mathbf{F} = (x^3 - 3y)\mathbf{i} + (x + \sin y)\mathbf{j}$, C as in Prob. 8
10. $\mathbf{F} = (e^x - 3y)\mathbf{i} + (e^y + 6x)\mathbf{j}$, C the ellipse $x^2 + 4y^2 = 4$
11. $\mathbf{F} = xy^2\mathbf{i} - x^2y\mathbf{j}$, C the boundary of the region $x \ge 0$, $0 \le y \le 1 - x^2$
12. $\mathbf{F} = y^3\mathbf{i} + x^3\mathbf{j}$, C the circle $x^2 + y^2 = 1$
13. $\mathbf{F} = (\cos x \sin y - xy)\mathbf{i} + \sin x \cos y\ \mathbf{j}$, C as in Prob. 12
14. $\mathbf{F} = -\cosh y\ \mathbf{i} + \sin x\ \mathbf{j}$, C the boundary of the rectangle $0 \le x \le \pi$, $0 \le y \le 1$
15. $\mathbf{F} = x^{-1}e^y\mathbf{i} + (e^y \ln x + 2x)\mathbf{j}$, C the boundary of the region $1 + x^4 \le y \le 2$

Area. Using one of the formulas in Example 2, find the area of the following plane regions.

16. The region given by $0 \le y \le 1 - x^2$
17. The interior of the **ellipse** $x^2/a^2 + y^2/b^2 = 1$
18. The region in the first quadrant bounded by $y = x$ and $y = x^3$
19. The region under one arch of the **cycloid** $\mathbf{r} = a(t - \sin t)\mathbf{i} + a(1 - \cos t)\mathbf{j}$, $0 \le t \le 2\pi$.
20. The region in the first quadrant bounded by $y = x$, $y = 1/x$, and $y = x/4$

Integral of the normal derivative. Using (9), evaluate $\oint_C \dfrac{\partial w}{\partial n}\ ds$ (counterclockwise), where

21. $w = x^2 + 3y^2$, $C: x^2 + y^2 = 4$
22. $w = e^x + e^y$, C the boundary of the rectangle $0 \le x \le 2$, $0 \le y \le 1$
23. $w = x^3 + 3x^2 - 3xy^2$, C the circle $(x - 4)^2 + (y + 2)^2 = 10$
24. $w = \ln (x^2 + y^2)$, C the boundary of the region $0 \le y \le e^x$, $1 \le x \le 3$
25. $w = 3x^2y - y^3 + y^2$, C the ellipse $25x^2 + y^2 = 25$
26. $w = \cosh x$, C the boundary of the triangle with vertices $(0, 0)$, $(4, 2)$, $(0, 2)$
27. $w = e^x \sin y$, C as in Prob. 26
28. $w = x^2 + y^3$, C the boundary of the region $0 \le y \le \sin x$, $0 \le x \le \pi$
29. $w = (x + y)^2$, C as in Prob. 26

30. (Laplace's equation) If $w(x, y)$ satisfies Laplace's equation $\nabla^2 w = 0$ in a region R, show that (10) with $\partial w/\partial n$ defined as in Example 4 holds. *Hint.* Model your work somewhat after that of Example 4.

$$(10) \qquad \iint\limits_{R} \left[\left(\frac{\partial w}{\partial x}\right)^2 + \left(\frac{\partial w}{\partial y}\right)^2 \right] dx\, dy = \oint_C w\, \frac{\partial w}{\partial n}\, ds$$

31. Show that $w = 2e^x \sin y$ satisfies Laplace's equation and, using (10), integrate $w(\partial w/\partial n)$ counterclockwise around the boundary C of the square $0 \leqq x \leqq 1$, $0 \leqq y \leqq 1$.

Other forms of Green's theorem

32. Show that (1) may be written in the form (11), below, where $\mathbf{n}$ is the outward unit normal vector to the curve C (Fig. 218) and s is the arc length of C. *Hint.* Introduce $\mathbf{F} = F_2\mathbf{i} - F_1\mathbf{j}$.

$$(11) \qquad \iint\limits_{R} \operatorname{div} \mathbf{F}\, dx\, dy = \oint_C \mathbf{F} \cdot \mathbf{n}\, ds$$

33. Verify (11) when $\mathbf{F} = 7x\mathbf{i} - 3y\mathbf{j}$ and C is the circle $x^2 + y^2 = 4$.

34. Show that (1) may be written in the form (12), below, where $\mathbf{k}$ is a unit vector perpendicular to the xy-plane, $\mathbf{r}'$ is the unit tangent vector to C and s is the arc length of C.

$$(12) \qquad \iint\limits_{R} (\operatorname{curl} \mathbf{F}) \cdot \mathbf{k}\, dx\, dy = \oint_C \mathbf{F} \cdot \mathbf{r}'\, ds$$

35. Verify (12) when $\mathbf{F} = y\mathbf{i} + 4x\mathbf{j}$ and C is the boundary of the triangle with vertices at $(0, 0)$, $(2, 0)$, $(2, 1)$.

9.4　Surfaces for Surface Integrals

The idea that will lead to the concept of a surface integral is quite similar to that which led to a line integral in Sec. 9.1. In a line integral we integrate over a curve C and use the representation of C for writing the line integral as a definite integral taken over an interval $a \leqq t \leqq b$ on the t-axis, where t is the parameter in the parametric representation $\mathbf{r}(t)$ of C; see (3), Sec. 9.1.

In a surface integral we shall integrate over a surface S and shall use the representation of S to write the surface integral as a double integral (Sec. 9.2) over some region R in a plane. For this reason we first consider representations of surfaces suitable for this purpose.

Representations of Surfaces

We say briefly "surface," also for a *portion* of a surface, just as we said "curve" for an *arc* of a curve, for simplicity.

Representations of a surface S in xyz-space are

(1) $z = f(x, y)$ or $g(x, y, z) = 0$.

For example, $z = +\sqrt{a^2 - x^2 - y^2}$ or $x^2 + y^2 + z^2 - a^2 = 0 \ (z \geqq 0)$ represents a hemisphere of radius a and center 0.

Now for *curves* C in line integrals, it was more practical and gave greater flexibility to use a *parametric* representation $\mathbf{r} = \mathbf{r}(t)$, where $a \leqq t \leqq b$. This is a mapping of the interval $a \leqq t \leqq b$, located on the t-axis, onto the curve C in xyz-space. It maps every t in that interval onto the point of C with position vector $\mathbf{r}(t)$. See Fig. 219.

Similarly, for surfaces S in surface integrals, it will be more practical to use a *parametric* representation. Surfaces are *two*-dimensional. Hence we need *two* parameters, which we call u and v. Thus a **parametric representation** of a surface S in space is of the form

(2) $$\boxed{\mathbf{r}(u, v) = x(u, v)\mathbf{i} + y(u, v)\mathbf{j} + z(u, v)\mathbf{k}}$$ (u, v) in R

where R is some region in the uv-plane. This mapping (2) maps every point (u, v) in R onto the point of S with position vector $\mathbf{r}(u, v)$. See Fig. 219.

EXAMPLE 1. Parametric representation of a cylinder

The circular cylinder $x^2 + y^2 = a^2$, $-1 \leqq z \leqq 1$, has radius a, height 2 and the z-axis as axis. A parametric representation is (Fig. 220 on the next page)

$$\mathbf{r}(u, v) = a \cos u \, \mathbf{i} + a \sin u \, \mathbf{j} + v\mathbf{k}$$

where the parameters u, v vary in the rectangle R in the uv-plane given by $0 \leqq u \leqq 2\pi$, $-1 \leqq v \leqq 1$. The components of $\mathbf{r}(u, v)$ are

$$x = a \cos u, \qquad y = a \sin u, \qquad z = v.$$

The curves $v = const$ are parallel circles. The curves $u = const$ are vertical straight lines. The point P in Fig. 220 corresponds to $u = \pi/3 = 60°$, $v = 0.7$.

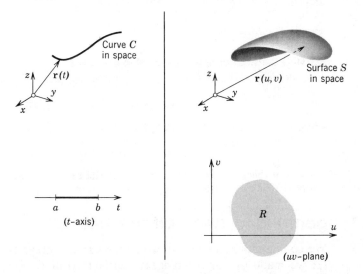

Fig. 219. Parametric representations of a curve and a surface

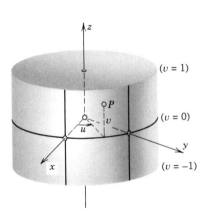

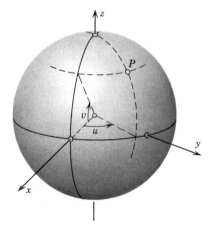

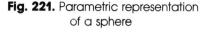

Fig. 220. Parametric representation of a cylinder   **Fig. 221.** Parametric representation of a sphere

EXAMPLE 2. Parametric representation of a sphere

A sphere $x^2 + y^2 + z^2 = a^2$ can be represented in the form

$$(3) \qquad \boxed{\mathbf{r}(u, v) = a \cos v \cos u \, \mathbf{i} + a \cos v \sin u \, \mathbf{j} + a \sin v \, \mathbf{k}}$$

where the parameters u, v vary in the rectangle R in the uv-plane given by the inequalities $0 \leq u \leq 2\pi$, $-\pi/2 \leq v \leq \pi/2$. The components of $\mathbf{r}$ are

$$x = a \cos v \cos u, \qquad y = a \cos v \sin u, \qquad z = a \sin v.$$

The curves $u = const$ and $v = const$ are the "meridians" and "parallels" on S (cf. Fig. 221). *This representation is used in **geography** for measuring the latitude and longitude of points on the globe.*

Another parametric representation of the sphere also used in mathematics is

$$(3^*) \qquad \boxed{\mathbf{r}(u, v) = a \cos u \sin v \, \mathbf{i} + a \sin u \sin v \, \mathbf{j} + a \cos v \, \mathbf{k}}$$

where $0 \leq u \leq 2\pi$, $0 \leq v \leq \pi$.

EXAMPLE 3. Parametric representation of a cone

A circular cone $z = \sqrt{x^2 + y^2}$, $0 \leq z \leq H$ can be represented by

$$\mathbf{r}(u, v) = u \cos v \, \mathbf{i} + u \sin v \, \mathbf{j} + u\mathbf{k}$$

where u, v vary in the rectangle R: $0 \leq u \leq H$, $0 \leq v \leq 2\pi$. The components of $\mathbf{r}(u, v)$ are

$$x = u \cos v, \qquad y = u \sin v, \qquad z = u$$

and we can check that $x^2 + y^2 = z^2$, as it should be. What are the curves $u = const$ and $v = const$? Make a sketch.

Tangent Plane and Surface Normal

We go one more step before we define surface integrals. In the line integral (1) we made use of the tangent vector $\mathbf{r}'(t)$ of C. Similarly, in a surface integral we shall make use of the surface normal, which we get by considering first the tangent plane.

The **tangent plane** $T(P)$ of a surface S at a point P of S is the plane that contains the tangents of all curves on S passing through P, at this point P.

How to find $T(P)$ for a given S represented by $\mathbf{r}(u, v)$? The key to the answer lies in the fact that a curve on S can be represented by a pair of continuous functions

$$u = u(t), \qquad v = v(t)$$

because this gives a curve in the uv-plane and corresponding to it a curve on S, which is then represented in space by

$$\tilde{\mathbf{r}}(t) = \mathbf{r}(u(t), v(t)).$$

For example, the helix

$$\tilde{\mathbf{r}}(t) = a \cos t \, \mathbf{i} + a \sin t \, \mathbf{j} + ct\mathbf{k}$$

lies on the cylinder S in Example 1,

$$\mathbf{r}(u, v) = a \cos u \, \mathbf{i} + a \sin u \, \mathbf{j} + v\mathbf{k}$$

and is given on S by $u = t$, $v = ct$; just substitute this for verification.

Now by the chain rule, the tangent vector of the curve $\tilde{\mathbf{r}}(t) = \mathbf{r}(u(t), v(t))$ is

$$\tilde{\mathbf{r}}'(t) = \frac{d\tilde{\mathbf{r}}}{dt} = \frac{\partial \mathbf{r}}{\partial u} u' + \frac{\partial \mathbf{r}}{\partial v} v'.$$

Fixing a point P on S and considering all curves through P on S, that is, all pairs of functions $u(t)$, $v(t)$, we get all tangent vectors $\tilde{\mathbf{r}}'(t)$ of these curves at P. These will span a plane—which is $T(P)$—provided $\mathbf{r}_u = \partial \mathbf{r}/\partial u$ and $\mathbf{r}_v = \partial \mathbf{r}/\partial v$ exist and are linearly independent, thus (cf. Sec. 6.8, Theorem 2),

(4)
$$\boxed{\mathbf{N} = \mathbf{r}_u \times \mathbf{r}_v \neq \mathbf{0}.}$$

By the definition of a vector product, this vector $\mathbf{N}$ is perpendicular to $T(P)$. It is called a **normal vector** of S at P. The corresponding **unit normal vector** of S at P is (Fig. 222)

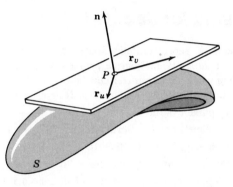

Fig. 222. Tangent plane and normal vector

(5)
$$\mathbf{n} = \frac{1}{|\mathbf{N}|}\mathbf{N} = \frac{1}{|\mathbf{r}_u \times \mathbf{r}_v|}\mathbf{r}_u \times \mathbf{r}_v.$$

Also, if S is represented by $g(x, y, z) = 0$, then, by Theorem 2 in Sec. 8.8,

(5*)
$$\mathbf{n} = \frac{1}{|\text{grad } g|}\text{grad } g$$

We can summarize:

Theorem 1 (Tangent plane and surface normal)
If a surface S is given by (2) with continuous $\mathbf{r}_u = \partial\mathbf{r}/\partial u$ and $\mathbf{r}_v = \partial\mathbf{r}/\partial v$ satisfying (4) at every point of S, then S has everywhere a unique tangent plane $T(P)$, passing through P on S and spanned by $\mathbf{r}_u$ and $\mathbf{r}_v$, and a unique normal whose direction depends continuously on the points of S. Cf. Fig. 222.

Such a surface S is called a **smooth surface.** We call S **piecewise smooth** if it consists of finitely many smooth portions. For instance, a sphere is smooth, and the boundary surface of a cube is piecewise smooth.

EXAMPLE 4. Unit normal vector of a sphere
The sphere $x^2 + y^2 + z^2 - a^2 = 0$ has the unit normal vector
$$\mathbf{n}(x, y, z) = \frac{x}{a}\mathbf{i} + \frac{y}{a}\mathbf{j} + \frac{z}{a}\mathbf{k}.$$

EXAMPLE 5. Unit normal vector of a cone
At the apex of the cone $f(x, y, z) = -z + \sqrt{x^2 + y^2} = 0$ in Example 3 the unit normal vector $\mathbf{n}$ becomes undetermined, since
$$\mathbf{n} = \frac{1}{\sqrt{2}}\left(\frac{x}{\sqrt{x^2 + y^2}}\mathbf{i} + \frac{y}{\sqrt{x^2 + y^2}}\mathbf{j} - \mathbf{k}\right). \qquad \blacksquare$$

What we have learned about surfaces in the present section will be applied to discuss **surface integrals** in the next section. The transformation of surface integrals over closed surfaces into volume integrals (Gauss's theorem) follows in Sec. 9.6.

Problems for Sec. 9.4

Parametric representations and surface normal. To prepare for our discussion of surface integrals, familiarize yourself with parametric representations of surfaces by stating what the **parameter curves** (curves $u = const$ and $v = const$) on the surface are and finding the normal vector $\mathbf{N} = \mathbf{r}_u \times \mathbf{r}_v$ of the surface.

1. xy-plane $\mathbf{r} = u\mathbf{i} + v\mathbf{j}$
2. xy-plane in polar coordinates $\mathbf{r} = u\cos v\,\mathbf{i} + u\sin v\,\mathbf{j}$
3. xy-plane $\mathbf{r} = (u + v)\mathbf{i} + (u - v)\mathbf{j}$
4. Plane $\mathbf{r} = u\mathbf{i} + v\mathbf{j} - c^{-1}(d + au + bv)\mathbf{k}$
5. Elliptic cylinder $\mathbf{r} = 4\cos u\,\mathbf{i} + 2\sin u\,\mathbf{j} + v\mathbf{k}$

6. Hyperbolic cylinder $\mathbf{r} = \cosh u\,\mathbf{i} + \sinh u\,\mathbf{j} + v\mathbf{k}$

7. Cone $\mathbf{r} = u\cos v\,\mathbf{i} + u\sin v\,\mathbf{j} + cu\mathbf{k}$

8. Paraboloid $\mathbf{r} = u\cos v\,\mathbf{i} + u\sin v\,\mathbf{j} + u^2\mathbf{k}$

9. Helicoid $\mathbf{r} = u\cos v\,\mathbf{i} + u\sin v\,\mathbf{j} + v\mathbf{k}$

10. Ellipsoid $\mathbf{r} = a\cos v\cos u\,\mathbf{i} + b\cos v\sin u\,\mathbf{j} + c\sin v\,\mathbf{k}$

11. Ellipsoid of revolution $\mathbf{r} = \cos v\cos u\,\mathbf{i} + \cos v\sin u\,\mathbf{j} + 2\sin v\,\mathbf{k}$

12. Elliptic paraboloid $\mathbf{r} = au\cos v\,\mathbf{i} + bu\sin v\,\mathbf{j} + u^2\mathbf{k}$

13. Hyperbolic paraboloid $\mathbf{r} = au\cosh v\,\mathbf{i} + bu\sinh v\,\mathbf{j} + u^2\mathbf{k}$

14. Hyperboloid $\mathbf{r} = a\sinh u\cos v\,\mathbf{i} + b\sinh u\sin v\,\mathbf{j} + c\cosh u\,\mathbf{k}$

15. Catenoid $\mathbf{r} = v\cos u\,\mathbf{i} + v\sin u\,\mathbf{j} + \cosh^{-1} v\,\mathbf{k}$

Derivation of parametric representations of surfaces. Find a parametric representation of the following surfaces. (The answer gives *one* such representation and there are many others.) Find a normal vector.

16. yz-plane

17. Plane $y = z$

18. Plane $x + y + z = 1$

19. Parabolic cylinder $y = x^2$

20. Cylinder $y^2 + z^2 = 16$

21. Elliptic cylinder $9x^2 + 4z^2 = 36$

22. $z = xy$

23. Paraboloid $z = 4(x^2 + y^2)$

24. Elliptic cone $z = \sqrt{x^2 + 4y^2}$

25. $z = (x^2 + y^2)^2$

26. Show that a representation $z = f(x, y)$ can be written ($f_u = \partial f/\partial u$, etc.)

(6)
$$\mathbf{r}(u, v) = u\mathbf{i} + v\mathbf{j} + f(u, v)\mathbf{k}, \quad \text{and} \quad \mathbf{N} = -f_u\mathbf{i} - f_v\mathbf{j} + \mathbf{k}.$$

27. Write $x = h(y, z)$ in parametric form and find a normal vector.

28. Find the points in Probs. 2, 24 and 25 at which (4) does not hold and indicate whether this is because of the shape of the surface or the choice of the representation.

29. Find a parametric representation of the paraboloid in Prob. 23 such that $\mathbf{N}(0, 0) \neq \mathbf{0}$. Find $\mathbf{N}$.

30. **(Orthogonal parameters on a surface)** Show that the parameter curves $u = const$ and $v = const$ on a surface $\mathbf{r} = \mathbf{r}(u, v)$ intersect at right angles if and only if $\mathbf{r}_u \cdot \mathbf{r}_v = 0$.

Surface normal. Using (5*), find a unit normal vector of the surfaces given by

31. $4x - 5y + 3z = \frac{1}{2}$

32. $z - x^2 - y^2 = 0$

33. $y^2 + z^2 = 25$

34. $x^3 + y^2 + z = a$

35. $x^2 + y^2 + z^2 = a^2$

36. $x^2 + 9y^2 + 4z^2 = 36$

37. $z = xy$

38. $x^2 + y^2 - z^2 = 1$

39. $x^2 - y - z = 0$

40. $x^2 + y^2 = 2\ln z$

Tangent plane. The tangent plane $T(P)$ as such will be of lesser importance in our further work, but one should know how to represent it. Show that:

41. If S: $\mathbf{r}(u, v)$, then $T(P)$: $\mathbf{r}^*(p, q) = \mathbf{r}(P) + p\mathbf{r}_u(P) + q\mathbf{r}_v(P)$.

42. If S: $\mathbf{r}(u, v)$, then $T(P)$: $(\mathbf{r}^* - \mathbf{r} \quad \mathbf{r}_u \quad \mathbf{r}_v) = 0$ (cf. Sec. 6.9).

43. If S: $g(x, y, z) = 0$, then $T(P)$: $(\mathbf{r}^* - \mathbf{r}) \cdot \text{grad } g = 0$.

44. If S: $z = f(x, y)$, then $T(P)$: $z^* - z = (x^* - x)f_x(P) + (y^* - y)f_y(P)$.

Find a representation of the tangent plane of the following surfaces at the point P_0: (x_0, y_0, z_0).

45. $z = xy$, P_0: $(1, 1, 1)$ **46.** $x^2 + y^2 + z^2 = 100$, P_0: $(8, 6, 0)$

47. $z = x^2$, P_0: $(2, 1, 4)$ **48.** $x^2 + y^2 = 8$, P_0: $(2, 2, 3)$

49. $y = x^{3/2}$, P_0: $(4, 8, 6)$ **50.** $4x^2 + y^2 + z^2 = 17$, P_0: $(1, 3, 2)$

9.5 Surface Integrals

Given a piecewise smooth surface S by [cf. (2), Sec. 9.4]

$$(1) \qquad \mathbf{r}(u, v) = x(u, v)\mathbf{i} + y(u, v)\mathbf{j} + z(u, v)\mathbf{k} \qquad (u, v) \text{ in } R$$

with unit normal vector $\mathbf{n} = (1/|\mathbf{N}|)\mathbf{N}$, where [cf. (4), Sec. 9.4]

$$(2) \qquad \mathbf{N} = \mathbf{r}_u \times \mathbf{r}_v,$$

we can now define a **surface integral** of a vector function $\mathbf{F}$ over S by

$$(3) \qquad \boxed{\iint_S \mathbf{F} \cdot \mathbf{n} \, dA = \iint_R \mathbf{F}(\mathbf{r}(u, v)) \cdot \mathbf{N}(u, v) \, du \, dv.}$$

$\mathbf{F} \cdot \mathbf{n}$ is the normal component of $\mathbf{F}$, and this integral arises naturally in flow problems where it gives the *flux* across S (= mass of fluid crossing S per unit time; cf. Sec. 8.9) when $\mathbf{F} = \rho\mathbf{v}$, where ρ is the density of the fluid and $\mathbf{v}$ the velocity vector of the flow (example below). We may thus call the surface integral (3) the **flux integral.** Other forms of surface integrals will be discussed below. We see that the integral in (3) on the right is a double integral (Sec. 9.2) over the region R in the uv-plane corresponding to S and exists for continuous $\mathbf{F}$ and piecewise smooth S, because this makes $\mathbf{F} \cdot \mathbf{N}$ piecewise continuous. Note also that the integrand is a scalar, not a vector, since we take dot products.

In (3) we have $\mathbf{N} \, du \, dv = \mathbf{n}|\mathbf{N}| \, du \, dv$ by (5) in Sec. 9.4. Now, by the definition of vector product, $|\mathbf{N}| = |\mathbf{r}_u \times \mathbf{r}_v|$ is the area of the parallelogram with sides $\mathbf{r}_u$ and $\mathbf{r}_v$. Hence $|\mathbf{N}| \, du \, dv$ is the element of area dA of S, so that $\mathbf{n} \, dA = \mathbf{N} \, du \, dv$, which motivates (3).

Setting $\mathbf{F} = F_1\mathbf{i} + F_2\mathbf{j} + F_3\mathbf{k}$, $\mathbf{n} = \cos\alpha \, \mathbf{i} + \cos\beta \, \mathbf{j} + \cos\gamma \, \mathbf{k}$ and $\mathbf{N} = N_1\mathbf{i} + N_2\mathbf{j} + N_3\mathbf{k}$, we see that (3) can be written

$$(4) \qquad \begin{aligned} \iint_S \mathbf{F} \cdot \mathbf{n} \, dA &= \iint_S (F_1 \cos\alpha + F_2 \cos\beta + F_3 \cos\gamma) \, dA \\ &= \iint_R (F_1 N_1 + F_2 N_2 + F_3 N_3) \, du \, dv. \end{aligned}$$

Here, α, β and γ are the angles between $\mathbf{n}$ and the positive directions of the coordinate axes, since we obtain the dot product $\mathbf{n} \cdot \mathbf{i} = \cos \alpha$ and, on the other hand, from (3) in Sec. 6.5 the same result, $\cos \alpha = \mathbf{n} \cdot \mathbf{i}/|\mathbf{n}||\mathbf{i}| = \mathbf{n} \cdot \mathbf{i}$. Similarly for β and γ.

EXAMPLE 1. Flux through a surface

Compute the flux of water through the parabolic cylinder $S: y = x^2, 0 \leq x \leq 2, 0 \leq z \leq 3$ (Fig. 223) when the velocity vector is $\mathbf{F} = y\mathbf{i} + 2\mathbf{j} + xz\mathbf{k}$, speed being measured in meter/sec.

Solution. Writing $x = u$ and $z = v$, we have $y = x^2 = u^2$. Hence a representation of S is

$$S: \quad \mathbf{r} = u\mathbf{i} + u^2\mathbf{j} + v\mathbf{k} \qquad\qquad (0 \leq u \leq 2, 0 \leq v \leq 3).$$

From this,

$$\mathbf{r}_u = \mathbf{i} + 2u\mathbf{j}$$

$$\mathbf{r}_v = \mathbf{k}$$

$$\mathbf{N} = \mathbf{r}_u \times \mathbf{r}_v = 2u\mathbf{i} - \mathbf{j}.$$

On S,

$$\mathbf{F} = u^2\mathbf{i} + 2\mathbf{j} + uv\mathbf{k}.$$

Hence

$$\mathbf{F} \cdot \mathbf{N} = u^2 \cdot 2u + 2(-1) = 2u^3 - 2.$$

By integration we thus get from (3)

$$\iint_S \mathbf{F} \cdot \mathbf{n} \, dA = \int_0^3 \int_0^2 (2u^3 - 2) \, du \, dv = 3 \int_0^2 (2u^3 - 2) \, du = 12 \; [\text{meter}^3/\text{sec}].$$

Since water has the density $\rho = 1 \text{ gram/cm}^3 = 1 \text{ kg/liter} = 1000 \text{ kg/meter}^3$, the answer is 12 000 kg/sec.

EXAMPLE 2. Surface integral

Evaluate (3) when $\mathbf{F} = x^2\mathbf{i} + 3y^2\mathbf{k}$ and S is the portion of the plane $x + y + z = 1$ in the first octant (Fig. 224).

Solution. Writing $x = u$ and $y = v$, we have $z = 1 - x - y = 1 - u - v$. Hence we can represent the plane $x + y + z = 1$ in the form

$$\mathbf{r}(u, v) = u\mathbf{i} + v\mathbf{j} + (1 - u - v)\mathbf{k}.$$

We obtain the first-octant portion S of this plane by restricting $x = u$ and $y = v$ to the projection R of S in the xy-plane. R is the triangle bounded by the two coordinate axes and the straight line $x + y = 1$; thus $0 \leq x \leq 1 - y, 0 \leq y \leq 1$.

Differentiating $\mathbf{r}(u, v)$, we calculate

$$\mathbf{N} = \mathbf{r}_u \times \mathbf{r}_v = (\mathbf{i} - \mathbf{k}) \times (\mathbf{j} - \mathbf{k}) = \mathbf{i} + \mathbf{j} + \mathbf{k}.$$

Hence $\mathbf{F} \cdot \mathbf{N} = u^2 + 3v^2$, so that (3) gives

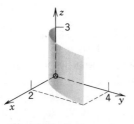

Fig. 223. Surface S
in Example 1

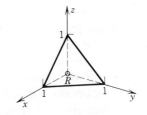

Fig. 224. Portion of a plane
in Example 2

$$\iint_S \mathbf{F} \cdot \mathbf{n} \, dA = \iint_R (u^2 + 3v^2) \, du \, dv = \int_0^1 \int_0^{1-v} (u^2 + 3v^2) \, du \, dv$$

$$= \int_0^1 [\tfrac{1}{3}(1 - v)^3 + 3v^2(1 - v)] \, dv = \tfrac{1}{3}. \quad \blacksquare$$

From (3) or (4) we see that the value of the integral depends on the choice of the unit normal vector **n.** (Instead of **n** we could choose $-\mathbf{n}$.) We express this by saying that such an integral is an *integral over an* **oriented surface** S, that is, over a surface S on which we have chosen one of the two possible unit normal vectors in a continuous fashion. (For a piecewise smooth surface, this needs some further discussion, which we give below.) If we change the orientation of S, which means that we replace **n** by $-\mathbf{n}$, then each component of **n** in (4) is multiplied by -1, so that we have

Theorem 1 (Change of orientation)
The replacement of **n** *by* $-\mathbf{n}$ *(hence of* **N** *by* $-\mathbf{N}$*) corresponds to the multiplication of the integral in* (3) *or* (4) *by* -1.

How to effect such a change of **N** in practice if S is given in the form (1)? The simplest way is to interchange u and v, because then $\mathbf{r}_u$ becomes $\mathbf{r}_v$ and conversely, so that $\mathbf{N} = \mathbf{r}_u \times \mathbf{r}_v$ becomes $\mathbf{r}_v \times \mathbf{r}_u = -\mathbf{r}_u \times \mathbf{r}_v = -\mathbf{N}$, as wanted. Let us illustrate this.

EXAMPLE 3. Change of orientation
In Example 1 we had $\mathbf{F} = y\mathbf{i} + 2\mathbf{j} + xz\mathbf{k}$ and S: $y = x^2$, where $0 \leqq x \leqq 2$, $0 \leqq z \leqq 3$. By setting $x = u$ and $z = v$ we got the representation

$$\mathbf{r} = u\mathbf{i} + u^2\mathbf{j} + v\mathbf{k} \qquad (0 \leqq u \leqq 2, 0 \leqq v \leqq 3)$$

and $+12$ as the value of the integral.
Let us verify that if we interchange u and v, that is, set $x = v$, $z = u$ and

$$\mathbf{r} = v\mathbf{i} + v^2\mathbf{j} + u\mathbf{k} \qquad (0 \leqq u \leqq 3, 0 \leqq v \leqq 2),$$

the integral will be -12. Indeed, by straightforward calculation we get

$$\mathbf{N} = \mathbf{r}_u \times \mathbf{r}_v = \mathbf{k} \times (\mathbf{i} + 2v\mathbf{j}) = \mathbf{j} - 2v\mathbf{i},$$

$$\mathbf{F} = v^2\mathbf{i} + 2\mathbf{j} + uv\mathbf{k} \qquad \text{(on } S\text{)}$$

and from this,

$$\iint_R \mathbf{F} \cdot \mathbf{N} \, du \, dv = \int_0^2 \int_0^3 (-2v^3 + 2) \, du \, dv = \int_0^2 (-6v^3 + 6) \, dv = -12. \quad \blacksquare$$

More about orientation. We consider first a *smooth* surface. If S is smooth and P any of its points, we may choose a unit normal vector **n** of S at P. The direction of **n** is then called the *positive normal direction* of S at P. Obviously there are two possibilities in choosing **n.**

A smooth surface S is said to be **orientable** if the positive normal direction, when given at an arbitrary point P_0 of S, can be continued in a unique and continuous way to the entire surface.

Essential in practice is the fact that *a sufficiently small portion of a smooth surface is always orientable.* From a theoretical point of view it is interesting

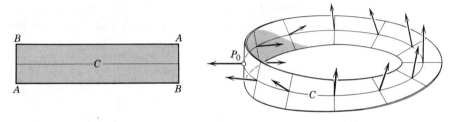

Fig. 225. Möbius strip

that this may not hold in the large. There are nonorientable surfaces. A well-known example of such a surface is the **Möbius strip**[5] shown in Fig. 225. When a normal vector, which is given at P_0, is displaced continuously along the curve C in Fig. 225, the resulting normal vector upon returning to P_0 is opposite to the original vector at P_0. A model of a Möbius strip can be made by taking a long rectangular piece of paper, making a half-twist and sticking the shorter sides together so that the two points A and the two points B in Fig. 225 coincide.

If the boundary of an orientable smooth surface S is a simple closed curve C, then we may associate with each of the two possible orientations of S an orientation of C, as shown in Fig. 226a. Using this simple idea, we may now readily extend the concept of orientation to piecewise smooth surfaces as follows.

A *piecewise smooth* surface S is said to be **orientable** if we can orient each smooth piece of S in such a manner that along each curve C^* which is a common boundary of two pieces S_1 and S_2 the positive direction of C^* relative to S_1 is opposite to the positive direction of C^* relative to S_2.

Figure 226b illustrates the situation for a surface consisting of two smooth pieces.

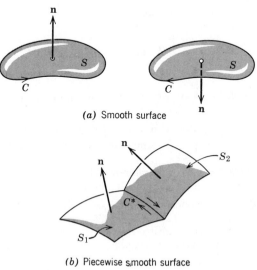

(a) Smooth surface

(b) Piecewise smooth surface

Fig. 226. Orientation of a surface

[5]AUGUST FERDINAND MÖBIUS (1790—1868), German mathematician, student of Gauss, professor of astronomy at Leipzig, known for his important work in the theory of surfaces, projective geometry and mechanics. He also contributed to number theory.

Another notation for the integrals in (4). After this discussion of orientation we can now explain another way of writing (4). It is customary to write in (4) also

(5)

(a) $$\iint_S F_1 \cos \alpha \, dA = \iint_S F_1 \, dy \, dz$$

(b) $$\iint_S F_2 \cos \beta \, dA = \iint_S F_2 \, dz \, dx$$

(c) $$\iint_S F_3 \cos \gamma \, dA = \iint_S F_3 \, dx \, dy$$

and together

(6) $$\iint_S \mathbf{F} \cdot \mathbf{n} \, dA = \iint_S (F_1 \, dy \, dz + F_2 \, dz \, dx + F_3 \, dx \, dy).$$

This is an analog of (3′) in Sec. 9.1 for line integrals. We can use these formulas for evaluating surface integrals by converting them to double integrals over plane regions, but must carefully take into account the orientation of S (the choice of $\mathbf{n}$). We explain this for (5c). If the surface S is given by $z = h(x, y)$ with (x, y) varying in a region R in the xy-plane, and if S is oriented so that $\cos \gamma > 0$, then

(5c′) $$\iint_S F_3 \cos \gamma \, dA = + \iint_R F_3[x, y, h(x, y)] \, dx \, dy,$$

but if $\cos \gamma < 0$, then

(5c″) $$\iint_S F_3 \cos \gamma \, dA = - \iint_R F_3[x, y, h(x, y)] \, dx \, dy.$$

This follows by noting that the element of area $dx \, dy$ in the xy-plane is the projection $|\cos \gamma| \, dA$ of the element of area dA of S, and we have $\cos \gamma = +|\cos \gamma|$ in (5c′), where $\cos \gamma > 0$, but $\cos \gamma = -|\cos \gamma|$ in (5c″), where $\cos \gamma < 0$. Similarly for (5b), (5c) and in (6).

EXAMPLE 4. An application of (6)

Verify the result in Example 1 by (6).

Solution. In Example 1 we have $F_1 = y$, $F_2 = 2$, $F_3 = xz$ and S: $y = x^2$ with $0 \leq x \leq 2$ (thus $0 \leq y \leq 4$) and $0 \leq z \leq 3$. The surface normal is parallel to the xy-plane, thus $\cos \gamma = 0$, and $\mathbf{n}$ points in the positive x-direction, thus $\cos \alpha > 0$, and somewhat down in the negative y-direction, thus $\cos \beta < 0$, causing a minus sign in (5b). (Indeed, $\mathbf{N} = 2u\mathbf{i} - \mathbf{j} = 2x\mathbf{i} - \mathbf{j}$.) We thus obtain from (6)

$$\iint_S \mathbf{F} \cdot \mathbf{n} \, dA = \iint_R (y \, dy \, dz - 2 \, dz \, dx)$$

$$= \int_0^3 \int_0^4 y \, dy \, dz - \int_0^2 \int_0^3 2 \, dz \, dx$$

$$= 3 \cdot \frac{y^2}{2} \Big|_0^4 - 2 \cdot 2 \cdot 3 = 24 - 12 = 12.$$

This confirms the result in Example 1.   ∎

Integrals over Nonoriented Surfaces

Another type of surface integral is

(7)
$$\iint_S G(\mathbf{r}) \, dA = \iint_R G([\mathbf{r}(u, v)]|\mathbf{N}(u, v)| \, du \, dv.$$

Here $dA = |\mathbf{N}| \, du \, dv = |\mathbf{r}_u \times \mathbf{r}_v| \, du \, dv$ is the element of area of the surface S represented by (1) and we disregard the orientation. In particular, taking $G = 1$, we obtain the area A of S given by

(8)
$$A(S) = \iint_S dA = \iint_R |\mathbf{r}_u \times \mathbf{r}_v| \, du \, dv.$$

EXAMPLE 5. Area of a sphere

A sphere of radius a can be represented by (3), Sec. 9.4, that is,

$$\mathbf{r}(u, v) = a \cos v \cos u \, \mathbf{i} + a \cos v \sin u \, \mathbf{j} + a \sin v \, \mathbf{k},$$

where $0 \leqq u \leqq 2\pi$, $-\pi/2 \leqq v \leqq \pi/2$. By direct calculation we obtain (verify!)

$$\mathbf{r}_u \times \mathbf{r}_v = a^2 \cos^2 v \cos u \, \mathbf{i} + a^2 \cos^2 v \sin u \, \mathbf{j} + a^2 \cos v \sin v \, \mathbf{k}.$$

Hence

$$|\mathbf{r}_u \times \mathbf{r}_v| = a^2[\cos^4 v \cos^2 u + \cos^4 v \sin^2 u + \cos^2 v \sin^2 v]^{1/2} = a^2 |\cos v|.$$

With this, (8) gives the familiar formula

$$A(S) = a^2 \int_{-\pi/2}^{\pi/2} \int_0^{2\pi} |\cos v| \, du \, dv = 2\pi a^2 \int_{-\pi/2}^{\pi/2} \cos v \, dv = 4\pi a^2.$$

EXAMPLE 6. Representation and area of a torus surface (doughnut)

A *torus surface* S is obtained by rotating a circle C about a straight line L in space so that C does not intersect or touch L but its plane always passes through L. If L is the z-axis and C has radius b and its center has distance a from L, as in Fig. 227 on the next page, then S can be represented by

$$\mathbf{r}(u, v) = (a + b \cos v) \cos u \, \mathbf{i} + (a + b \cos v) \sin u \, \mathbf{j} + b \sin v \, \mathbf{k}.$$

Thus

$$\mathbf{r}_u = -(a + b \cos v)\sin u \, \mathbf{i} + (a + b \cos v)\cos u \, \mathbf{j}$$

$$\mathbf{r}_v = -b \sin v \cos u \, \mathbf{i} - b \sin v \sin u \, \mathbf{j} + b \cos v \, \mathbf{k}$$

$$\mathbf{r}_u \times \mathbf{r}_v = b(a + b \cos v)[\cos u \cos v \, \mathbf{i} + \sin u \cos v \, \mathbf{j} + \sin v \, \mathbf{k}].$$

Hence $|\mathbf{r}_u \times \mathbf{r}_v| = b(a + b \cos v)$, and (8) gives the total area of the torus

(9)
$$A(S) = \int_0^{2\pi} \int_0^{2\pi} b(a + b \cos v) \, du \, dv = 4\pi^2 ab.$$

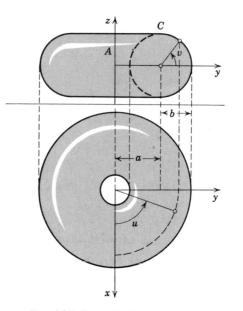

Fig. 227. Torus in Example 6

EXAMPLE 7. Moment of inertia

Find the moment of inertia I of a homogeneous spherical lamina S: $x^2 + y^2 + z^2 = a^2$ of mass M about the z-axis.

Solution. If a mass is distributed over a surface S and $\mu(x, y, z)$ is the density of the mass (= mass per unit area), then the moment of inertia I of the mass with respect to a given axis L is defined by the surface integral

$$(10) \qquad\qquad I = \iint_S \mu D^2 \, dA$$

where $D(x, y, z)$ is the distance of the point (x, y, z) from L. Since, in the present example, μ is constant and S has the area $A = 4\pi a^2$, we have

$$\mu = \frac{M}{A} = \frac{M}{4\pi a^2}.$$

Using for S the representation (3) in Sec. 9.4,

$$\mathbf{r}(u, v) = a \cos v \cos u \, \mathbf{i} + a \cos v \sin u \, \mathbf{j} + a \sin v \, \mathbf{k}$$

we get for the square of the distance of a point (x, y, z) from the z-axis the expression $D^2 = x^2 + y^2 = a^2 \cos^2 v$. Also, by straightforward calculation,

$$dA = |\mathbf{N}| \, du \, dv = |\mathbf{r}_u \times \mathbf{r}_v| \, du \, dv = a^2 \cos v \, du \, dv$$

(verify this!). Hence we obtain the result

$$I = \iint_S \mu D^2 \, dA = \frac{M}{4\pi a^2} \int_{-\pi/2}^{\pi/2} \int_0^{2\pi} a^4 \cos^3 v \, du \, dv = \frac{Ma^2}{2} \int_{-\pi/2}^{\pi/2} \cos^3 v \, dv = \frac{2Ma^2}{3}. \quad \blacksquare$$

When S is given by $z = f(x, y)$, then setting $u = x$, $v = y$, $\mathbf{r} = [u, v, f]$ gives

$$|\mathbf{N}| = |\mathbf{r}_u \times \mathbf{r}_v| = |[1, 0, f_u] \times [0, 1, f_v]|$$

$$= |[-f_u, -f_v, 1]| = \sqrt{1 + f_u^2 + f_v^2}$$

and, since $f_u = f_x$, $f_v = f_y$, formula (7) becomes

$$\textbf{(11)} \quad \boxed{\iint_S G(\mathbf{r}) \, dA = \iint_{R^*} G[x, y, f(x, y)] \sqrt{1 + \left(\frac{\partial f}{\partial x}\right)^2 + \left(\frac{\partial f}{\partial y}\right)^2} \, dx \, dy}$$

where R^* is the projection of S into the xy-plane (Fig. 228) and the normal vector $\mathbf{N}$ on S points *up*. If it points *down*, the integral on the right is preceded by a minus sign.

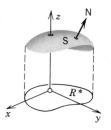

Fig. 228. Formula (11)

From (11) with $G = 1$ we obtain for the **area** $A(S)$ of S: $z = f(x, y)$ the formula

$$\textbf{(12)} \qquad A(S) = \iint_{\overline{S}} \sqrt{1 + \left(\frac{\partial f}{\partial x}\right)^2 + \left(\frac{\partial f}{\partial y}\right)^2} \, dx \, dy$$

where $\overline{S}$ is the projection of S into the xy-plane, as before.

In the next section we first discuss volume integrals. Then we formulate and prove the first "big" integral theorem, which transforms surface integrals into volume integrals. It is called **Gauss's divergence theorem,** since it involves the divergence of a vector function, a concept we discussed in Sec. 8.9 (which the student may perhaps wish to review).

Problems for Sec. 9.5

Flux integrals of the form (3). Evaluate $\iint_S \mathbf{F} \cdot \mathbf{n} \, dA$, where

1. $\mathbf{F} = x\mathbf{i} + y\mathbf{j}$, $S: z = 2x + 3y$, $0 \le x \le 2$, $-1 \le y \le 1$
2. $\mathbf{F} = y\mathbf{i} + 4x\mathbf{j}$, $S: z = -x + 2y$, $1 \le x \le 4$, $2 \le y \le 3$
3. $\mathbf{F} = x^2\mathbf{j} - xz\mathbf{k}$, $S: \mathbf{r} = u\mathbf{i} + u^2\mathbf{j} + v\mathbf{k}$, $0 \le u \le 1$, $-2 \le v \le 2$
4. $\mathbf{F} = e^y\mathbf{i} + ze^x\mathbf{k}$, $S: \mathbf{r} = u\mathbf{i} + 2u\mathbf{j} + v\mathbf{k}$, $-1 \le u \le 1$, $0 \le v \le 5$
5. $\mathbf{F} = \sin x \, \mathbf{i} + z\mathbf{j} + y\mathbf{k}$, $S: y^2 + z^2 = 4$, $-1/2 \le x \le 1/2$, $y \ge 0$, $z \ge 0$

6. $\mathbf{F} = -e^z\mathbf{i} + e^y\mathbf{j} - e^x\mathbf{k}$, $S: x^2 + z^2 = 9$, $x \geq 0$, $0 \leq y \leq 2$, $z \geq 0$

7. $\mathbf{F} = \mathbf{i} + x^2\mathbf{j} + xyz\mathbf{k}$, $S: z = xy$, $0 \leq x \leq y$, $0 \leq y \leq 1$

8. $\mathbf{F} = y\mathbf{i} + z \sin y\,\mathbf{j} + xy\mathbf{k}$, $S: z = 1 - x^2$, $1 \leq y \leq 3$, $z \geq 0$

9. $\mathbf{F} = y^3\mathbf{i} + x^3\mathbf{j} + 3z^2\mathbf{k}$, $S: z = x^2 + y^2$, $x^2 + y^2 \leq 4$

10. $\mathbf{F} = xy\mathbf{i} + z^2\mathbf{j} + z^4\mathbf{k}$,
    $S: \mathbf{r} = u \cos v\,\mathbf{i} + u \sin v\,\mathbf{j} + u\mathbf{k}$, $0 \leq u \leq 2, 0 \leq v \leq \pi$

11. $\mathbf{F} = \cosh x\,\mathbf{i} + \sinh y\,\mathbf{k}$, $S: z = x + y^2$, $0 \leq y \leq x$, $0 \leq x \leq 1$

12. $\mathbf{F} = x\mathbf{i} - y\mathbf{j} + x^2y^2\mathbf{k}$, $S: z = e^{xy}$, $-1 \leq x \leq 1$, $-1 \leq y \leq 1$

13. $\mathbf{F} = e^{\sin x}\mathbf{i} - 2y^2\mathbf{j} + e^x\mathbf{k}$, $S: \mathbf{r} = u\mathbf{i} + v\mathbf{j} + v^2\mathbf{k}$, $1 \leq u \leq 3$, $1 \leq v \leq 2$

14. $\mathbf{F} = y^3\mathbf{i} + x^3\mathbf{j} + z^3\mathbf{k}$, $S: x^2 + 4y^2 = 4$, $x \geq 0$, $y \geq 0$, $0 \leq z \leq h$

15. $\mathbf{F} = y\mathbf{i} - x\mathbf{j} + z^2\mathbf{k}$,
    $S: \mathbf{r} = u \cos v\,\mathbf{i} + u \sin v\,\mathbf{j} + v\mathbf{k}$, $0 \leq u \leq 1$, $0 \leq v \leq \pi/2$

16. $\mathbf{F} = xyz\mathbf{i}$,
    $S: \mathbf{r} = 2 \cos v \cos u\,\mathbf{i} + 2 \cos v \sin u\,\mathbf{j} + 2 \sin v\,\mathbf{k}$, $0 \leq u \leq \frac{1}{2}\pi$, $0 \leq v \leq \frac{1}{2}\pi$

17. $\mathbf{F} = 2xy\mathbf{i} + x^2\mathbf{j}$, $S: \mathbf{r} = \cosh u\,\mathbf{i} + \sinh u\,\mathbf{j} + v\mathbf{k}$, $0 \leq u \leq 2$, $-3 \leq v \leq 3$

18. $\mathbf{F} = xz\mathbf{i} + yz\mathbf{j} + z^2\mathbf{k}$, S as in Prob. 16

19. $\mathbf{F} = x^2yz\mathbf{i} + xy^2z\mathbf{j} + z^3\mathbf{k}$, S as in Prob. 16

20. $\mathbf{F} = \mathbf{i} + \mathbf{j} + \mathbf{k}$, $S: x^2 + y^2 + 4z^2 = 4$, $z \geq 0$

Surface integrals of the form (7). Using (7) or (11), evaluate $\displaystyle\iint_S G(\mathbf{r})\,dA$, where

21. $G = x + y + z$, $S: z = x + y$, $0 \leq y \leq x$, $0 \leq x \leq 1$

22. $G = \cos x + \sin y$, S: the portion of $x + y + z = 1$ in the first octant

23. $G = e^{x+y}$, S as in Prob. 22

24. $G = 8x$, $S: \mathbf{r} = u\mathbf{i} + v\mathbf{j} + u^2\mathbf{k}$, $0 \leq u \leq 2$, $-1 \leq v \leq 2$

25. $G = x + 1$, $S: \mathbf{r} = \cos u\,\mathbf{i} + \sin u\,\mathbf{j} + v\mathbf{k}$, $0 \leq u \leq 2\pi$, $0 \leq v \leq 3$

26. $G = 5xy$, $S: \mathbf{r} = u\mathbf{i} + v\mathbf{j} + uv\mathbf{k}$, $0 \leq u \leq 1$, $0 \leq v \leq 1$

27. $G = x + y + z$, $S: x^2 + y^2 = 1$, $0 \leq z \leq 2$

28. $G = \arctan (y/x)$, $S: z = x^2 + y^2$, $1 \leq z \leq 4$, $x \geq 0$, $y \geq 0$

29. $G = xy$, $S: x^2 + y^2 = 4$, $-1 \leq z \leq 1$

30. $G = 3x^3 \sin y$, $S: \mathbf{r} = u\mathbf{i} + v\mathbf{j} + u^3\mathbf{k}$, $0 \leq u \leq 1$, $0 \leq v \leq \pi$

31. $G = (1 - x^2)y$, $S: \mathbf{r} = u\mathbf{i} + v\mathbf{j} + (1 - v^2)\mathbf{k}$, $-1 \leq u \leq 1$, $0 \leq v \leq 1$

32. $G = xz^2 + 12xy - xy^4$, $S: \mathbf{r} = u\mathbf{i} + v\mathbf{j} + v^2\mathbf{k}$, $0 \leq u \leq 1$, $0 \leq v \leq 1$

33. $G = x^2 + y^2$, $S: z = \sqrt{x^2 + y^2}$, $x^2 + y^2 \leq 4$

34. $G = (x^2 + y^2)^2$, $S: z = (x^2 + y^2)^2$, $x^2 + y^2 \leq 1$

35. $G = x^2 + y^2 + z^2$, S as in Prob. 16

36. **(Potential)** The electrostatic potential at $(0, 0, -a)$ of a charge of constant density σ on the hemisphere $S: x^2 + y^2 + z^2 = a^2, z \geq 0$ is

$$U = \iint_S \frac{\sigma}{\sqrt{x^2 + y^2 + (z + a)^2}}\,dA.$$

Using (3*) in Sec. 9.4, show that $U = 2\pi\sigma a(2 - \sqrt{2})$.

Center of gravity. Moments of inertia

37. **(Center of gravity)** Justify the following formulas for the mass M and the center

of gravity $(\bar{x}, \bar{y}, \bar{z})$ of a lamina S of density (mass per unit area) $\sigma(x, y, z)$ in space:

$$M = \iint_S \sigma \, dA, \quad \bar{x} = \frac{1}{M} \iint_S x\sigma \, dA, \quad \bar{y} = \frac{1}{M} \iint_S y\sigma \, dA, \quad \bar{z} = \frac{1}{M} \iint_S z\sigma \, dA.$$

38. **(Moments of inertia)** Justify the following formulas for the moments of inertia of the lamina in Prob. 37 about the x-, y- and z-axes, respectively:

$$I_x = \iint_S (y^2 + z^2)\sigma \, dA, \quad I_y = \iint_S (x^2 + z^2)\sigma \, dA, \quad I_z = \iint_S (x^2 + y^2)\sigma \, dA.$$

39. Find a formula for the moment of inertia of the lamina in Prob. 37 about the line $y = x, z = 0$.

Find the moment of inertia of a lamina S of density 1 about an axis A, where

40. $S: x^2 + y^2 = 1, \quad 0 \leq z \leq h, \quad A$: the z-axis

41. S as in Prob. 40, $\quad A$: the line $z = h/2$ in the xz-plane

42. $S: x^2 + y^2 = z^2, \quad 0 \leq z \leq h, \quad A$: the z-axis

43. S: the torus in Example 6, $\quad A$: the z-axis

44. **(Steiner's theorem[6])** If I_A is the moment of inertia of a mass distribution of total mass M with respect to an axis A through the center of gravity, show that its moment of inertia I_B with respect to an axis B, which is parallel to A and has the distance k from it, is

$$I_B = I_A + k^2 M.$$

45. Using Steiner's theorem, find the moment of inertia of S in Prob. 41 about the x-axis.

46. Construct a paper model of a Möbius strip. What happens if you cut it along the curve C in Fig. 225?

First fundamental form of a surface. Given a surface $S: \mathbf{r}(u, v)$, the corresponding quadratic differential form

(13) $$ds^2 = E \, du^2 + 2F \, du \, dv + G \, dv^2$$

with coefficients[7]

(14) $$E = \mathbf{r}_u \cdot \mathbf{r}_u, \qquad F = \mathbf{r}_u \cdot \mathbf{r}_v, \qquad G = \mathbf{r}_v \cdot \mathbf{r}_v$$

is called the **first fundamental form** of S. It is basic in the theory of surfaces, since with its help, one can determine lengths (Prob. 48, below), angles (Prob. 49) and areas (Prob. 50) on S, as the following problems illustrate.

47. From (7), Sec. 8.4, we have $ds^2 = d\mathbf{r} \cdot d\mathbf{r}$. Show that for a curve $C: u = u(t)$, $v = v(t)$, $a \leq t \leq b$, on S this yields (13) and (14).

48. Show that C in Prob. 47 has length

[6]JACOB STEINER (1796—1863), Swiss geometer, born in a small village, learned to write only at age 14, became a pupil of Pestalozzi at 18, later studied at Heidelberg and Berlin and, finally, because of his outstanding research, was appointed professor of geometry at Berlin University.

[7]E, F, G are standard notations; of course, they have nothing to do with the functions F and G occurring at some places in this chapter.

$$(15) \qquad l = \int_a^b \sqrt{\mathbf{r}'(t) \cdot \mathbf{r}'(t)} \; dt = \int_a^b \sqrt{Eu'^2 + 2Fu'v' + Gv'^2} \; dt.$$

49. Show that the angle γ between two intersecting curves C_1: $u = g(t)$, $v = h(t)$ and C_2: $u = p(t)$, $v = q(t)$ on S: $\mathbf{r}(u, v)$ is obtained from

$$(16) \qquad\qquad\qquad\qquad \cos \gamma = \frac{\mathbf{a} \cdot \mathbf{b}}{|\mathbf{a}| \, |\mathbf{b}|}$$

where $\mathbf{a} = \mathbf{r}_u g' + \mathbf{r}_v h'$ and $\mathbf{b} = \mathbf{r}_u p' + \mathbf{r}_v q'$ are tangent vectors of C_1 and C_2.

50. Show that

$$(17) \qquad\qquad\qquad |\mathbf{N}|^2 = |\mathbf{r}_u \times \mathbf{r}_v|^2 = EG - F^2,$$

so that the formula (8) for the area $A(S)$ of S becomes

$$(18) \qquad A(S) = \iint_S dA = \iint_R |\mathbf{N}| \; du \; dv = \iint_R \sqrt{EG - F^2} \; du \; dv.$$

51. (Polar coordinates) Show that for polar coordinates $u \,(= r)$ and $v \,(= \theta)$ defined by $x = u \cos v$, $y = u \sin v$ we have $E = 1$, $F = 0$, $G = u^2$, so that

$$ds^2 = du^2 + u^2 \, dv^2 = dr^2 + r^2 \, d\theta^2$$

and calculate from this and (18) the area of a disk of radius a.

52. Obtain (12) from (18).

53. Calculate the area of the torus from (18).

54. Obtain (18) by Lagrange's identity (7), Sec. 6.9.

55. (Theorem of Pappus[8]**)** Show that the area A of the torus Example 6 can be obtained by the *theorem of Pappus*, which states that the area of a surface of revolution equals the product of the length of a meridian C and the length of the path of the center of gravity of C when C is rotated through the angle 2π.

9.6 Triple Integrals. Divergence Theorem of Gauss

The triple integral is a generalization of the double integral introduced in Sec. 9.2. For defining this integral we consider a function $f(x, y, z)$ defined in a bounded closed[9] region T in space. We subdivide this three-dimensional region T by planes parallel to the three coordinate planes. Then we number the parallelepipeds inside T from 1 to n. In each such parallelepiped we choose an arbitrary point, say, (x_k, y_k, z_k) in the kth parallelepiped, and form the sum

---

[8]PAPPUS OF ALEXANDRIA (about 300), Greek mathematician. The theorem is also called *Guldin's theorem,* after the Austrian mathematician HABAKUK GULDIN (1577—1643), professor at Graz and Vienna.

[9]Explained in footnote 1, Sec. 9.2 (with "sphere" instead of "circle").

$$J_n = \sum_{k=1}^{n} f(x_k, y_k, z_k) \, \Delta V_k$$

where ΔV_k is the volume of the kth parallelepiped. This we do for larger and larger positive integers n in an arbitrary manner, but so that the lengths of the edges of the largest parallelepiped of subdivision approach zero as n approaches infinity. In this way we obtain a sequence of real numbers $J_{n_1}, J_{n_2}, \cdots$. We assume that $f(x, y, z)$ is continuous in a domain containing T and T is bounded by finitely many *smooth surfaces* (cf. Sec. 9.4). Then it can be shown (cf. Ref. [5] in Appendix 1) that the sequence converges to a limit which is independent of the choice of subdivisions and corresponding points (x_k, y_k, z_k). This limit is called the **triple integral** *of $f(x, y, z)$ over the region T,* and is denoted by

$$\iiint_T f(x, y, z) \, dx \, dy \, dz \qquad \text{or} \qquad \iiint_T f(x, y, z) \, dV.$$

Triple integrals can be evaluated by three successive integrations. This is similar to the evaluation of double integrals by two successive integrations, as discussed in Sec. 9.2. An example is shown below (in Example 1).

Divergence Theorem of Gauss

We shall now see that triple integrals can be transformed into surface integrals over the boundary surface of a region in space and conversely. This is of practical interest, since in many cases, one of the two kinds of integral is simpler than the other; and it also helps in establishing fundamental equations in fluid flow, heat conduction, etc., as we shall see. The transformation is done by the so-called *divergence theorem* (below), which involves the divergence of a vector function **F**,

(1) $$\text{div } \mathbf{F} = \frac{\partial F_1}{\partial x} + \frac{\partial F_2}{\partial y} + \frac{\partial F_3}{\partial z} \qquad \text{(Sec. 8.9).}$$

Divergence theorem of Gauss
(Transformation between volume integrals and surface integrals)
Let T be a closed[10] bounded region in space whose boundary is a piecewise smooth[11] orientable surface S. Let $\mathbf{F}(x, y, z)$ be a vector function that is continuous and has continuous first partial derivatives in some domain containing T. Then

(2) $$\boxed{\iiint_T \text{div } \mathbf{F} \, dV = \iint_S \mathbf{F} \cdot \mathbf{n} \, dA}$$

*where $\mathbf{n}$ is the **outer** unit normal vector of S.* (Proof on p. 553.)

[10]"Closed" means that the boundary surface S is part of the region.
[11]Cf. Sec. 9.4.

Comment. If we write $\mathbf{F}$ and the outer unit normal vector $\mathbf{n}$ in components,

$$\mathbf{F} = F_1\mathbf{i} + F_2\mathbf{j} + F_3\mathbf{k} \qquad \text{and} \qquad \mathbf{n} = \cos\alpha\,\mathbf{i} + \cos\beta\,\mathbf{j} + \cos\gamma\,\mathbf{k},$$

so that α, β, and γ are the angles between $\mathbf{n}$ and the positive x-, y-, and z-axes, respectively, formula (2) takes the form

$$
\begin{aligned}
\iiint_T \left(\frac{\partial F_1}{\partial x} + \frac{\partial F_2}{\partial y} + \frac{\partial F_3}{\partial z} \right) dx\, dy\, dz \\
= \iint_S (F_1 \cos\alpha + F_2 \cos\beta + F_3 \cos\gamma)\, dA.
\end{aligned}
$$

(3*)

Because of (6) in the last section this may be written

$$
\begin{aligned}
\iiint_T \left(\frac{\partial F_1}{\partial x} + \frac{\partial F_2}{\partial y} + \frac{\partial F_3}{\partial z} \right) dx\, dy\, dz \\
= \iint_S (F_1\, dy\, dz + F_2\, dz\, dx + F_3\, dx\, dy).
\end{aligned}
$$

(3)

EXAMPLE 1. Evaluation of a surface integral by the divergence theorem

Before we prove the divergence theorem, let us show a typical application. By transforming to a triple integral, evaluate

$$I = \iint_S (x^3\, dy\, dz + x^2 y\, dz\, dx + x^2 z\, dx\, dy)$$

where S is the closed surface consisting of the cylinder $x^2 + y^2 = a^2$ $(0 \le z \le b)$ and the circular disks $z = 0$ and $z = b$ $(x^2 + y^2 \le a^2)$.

Solution. In (3) we now have

$$F_1 = x^3, \quad F_2 = x^2 y, \quad F_3 = x^2 z, \qquad \text{hence} \qquad \text{div } \mathbf{F} = 3x^2 + x^2 + x^2 = 5x^2.$$

Hence, if we make use of the symmetry of the region T bounded by S, we get 4 times the integral of div $\mathbf{F}$ over the portion of T in the first octant $x \ge 0$, $y \ge 0$, $z \ge 0$; thus by (3),

$$\iiint_T (3x^2 + x^2 + x^2)\, dx\, dy\, dz = 4 \cdot 5 \int_0^b \int_0^a \int_0^{\sqrt{a^2 - y^2}} x^2\, dx\, dy\, dz.$$

The integral over x equals $\frac{1}{3}(a^2 - y^2)^{3/2}$. Setting $y = a \cos t$, we have

$$dy = -a \sin t\, dt, \qquad (a^2 - y^2)^{3/2} = a^3 \sin^3 t,$$

and the integral over y becomes

$$\frac{1}{3} \int_0^a (a^2 - y^2)^{3/2}\, dy = -\frac{1}{3} a^4 \int_{\pi/2}^0 \sin^4 t\, dt = \frac{\pi}{16} a^4.$$

The integral over z contributes the factor b, and therefore the answer is

$$I = 4 \cdot 5 \frac{\pi}{16} ba^4 = \frac{5\pi}{4} a^4 b.$$

Let us check the integration by using polar coordinates in the xy-plane (thus, cylindrical coordinates in space), setting $x = r \cos \theta$, $y = r \sin \theta$. Then $dx\, dy = r\, dr\, d\theta$ and

$$\iiint_T 5x^2\, dx\, dy\, dz = \int_0^b \int_0^a \int_0^{2\pi} 5r^2 \cos^2 \theta\, r\, dr\, d\theta\, dz = 5b\pi \frac{r^4}{4}\bigg|_0^a = \frac{5\pi}{4} ba^4. \qquad \blacksquare$$

Proof of the divergence theorem. Clearly, (3*) is true if the following three relations hold simultaneously:

(4)
$$\iiint_T \frac{\partial F_1}{\partial x}\, dx\, dy\, dz = \iint_S F_1 \cos \alpha\, dA,$$

(5)
$$\iiint_T \frac{\partial F_2}{\partial y}\, dx\, dy\, dz = \iint_S F_2 \cos \beta\, dA,$$

(6)
$$\iiint_T \frac{\partial F_3}{\partial z}\, dx\, dy\, dz = \iint_S F_3 \cos \gamma\, dA.$$

We first prove (6) for a *special region* T which is bounded by a piecewise smooth orientable surface S and has the property that any straight line parallel to any one of the coordinate axes and intersecting T has at most *one* segment (or a single point) in common with T. This implies that T can be represented in the form

(7)
$$g(x, y) \leqq z \leqq h(x, y)$$

where (x, y) varies in the orthogonal projection R of T in the xy-plane. Clearly, $z = g(x, y)$ represents the "bottom" S_2 of S (Fig. 229), whereas $z = h(x, y)$ represents the "top" S_1 of S, and there may be a remaining vertical portion S_3 of S. (The portion S_3 may degenerate into a curve, as for a sphere.)

To prove (6), we use (7). Since $\mathbf{F}$ is continuously differentiable in some domain containing T, we have

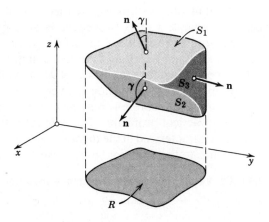

Fig. 229. Example of a special region

(8)
$$\iiint\limits_{T} \frac{\partial F_3}{\partial z} \, dx \, dy \, dz = \iint\limits_{R} \left[\int_{g(x, \, y)}^{h(x, \, y)} \frac{\partial F_3}{\partial z} \, dz \right] dx \, dy.$$

We integrate the inner integral:

$$\int_{g}^{h} \frac{\partial F_3}{\partial z} \, dz = F_3[x, \, y, \, h(x, \, y)] - F_3[x, \, y, \, g(x, \, y)].$$

Hence the left-hand side of (8) equals

(9)
$$\iint\limits_{R} F_3[x, \, y, \, h(x, \, y)] \, dx \, dy - \iint\limits_{R} F_3[x, \, y, \, g(x, \, y)] \, dx \, dy.$$

But the same result is also obtained by evaluating the right-hand side of (6), that is [see also (3)],

$$\iint\limits_{S} F_3 \cos \gamma \, dA = \iint\limits_{S} F_3 \, dx \, dy$$

$$= + \iint\limits_{R} F_3[x, \, y, \, h(x, \, y)] - \iint\limits_{R} F_3[x, \, y, \, g(x, \, y)],$$

where the first integral gets a plus sign because $\cos \gamma > 0$ on S_1 in Fig. 229 [as in (5c′), Sec. 9.5], and the second integral gets a minus sign because $\cos \gamma < 0$ on S_2 [as in (5c″), Sec. 9.5, with g instead of h]. This proves (6).

The relations (4) and (5) now follow by merely relabeling the variables and using the fact that, by assumption, T has representations similar to (7), namely,

$$\widetilde{g}(y, \, z) \leqq x \leqq \widetilde{h}(y, \, z) \qquad \text{and} \qquad \widetilde{\widetilde{g}}(z, \, x) \leqq y \leqq \widetilde{\widetilde{h}}(z, \, x).$$

This establishes the divergence theorem for special regions.

For any region T that can be subdivided into *finitely many* special regions by means of auxiliary surfaces, the theorem follows by adding the result for each part separately; this procedure is analogous to that in the proof of Green's theorem in Sec. 9.3. The surface integrals over the auxiliary surfaces cancel in pairs, and the sum of the remaining surface integrals is the surface integral over the whole boundary S of T; the volume integrals over the parts of T add up to the volume integral over T.

The divergence theorem is now proved for any bounded region that is of interest in practical problems. The extension to the most general region T of the type characterized in the theorem would require a certain limit process; this is similar to the situation in the case of Green's theorem in Sec. 9.3. ∎

EXAMPLE 2. Verification of the divergence theorem

Evaluate $\iint\limits_{S} (7x\mathbf{i} - z\mathbf{k}) \cdot \mathbf{n} \, dA$ over $S: x^2 + y^2 + z^2 = 4$ (a) by (2), (b) directly.

Solution. (a) div $(7x\mathbf{i} - z\mathbf{k}) = 6$. *Answer:* $6 \cdot (4/3)\pi \cdot 2^3 = 64\pi$.
 (b) We can represent S by (3), Sec. 9.4, with $a = 2$, that is,

$$S: \mathbf{r} = 2 \cos v \cos u \, \mathbf{i} + 2 \cos v \sin u \, \mathbf{j} + 2 \sin v \, \mathbf{k}.$$

Then

$$\mathbf{r}_u = -2 \cos v \sin u \, \mathbf{i} + 2 \cos v \cos u \, \mathbf{j}$$

$$\mathbf{r}_v = -2 \sin v \cos u \, \mathbf{i} - 2 \sin v \sin u \, \mathbf{j} + 2 \cos v \, \mathbf{k}$$

$$\mathbf{N} = \mathbf{r}_u \times \mathbf{r}_v = 4 \cos^2 v \cos u \, \mathbf{i} + 4 \cos^2 v \sin u \, \mathbf{j} + 4 \cos v \sin v \, \mathbf{k}.$$

From the representation of S we see that on S,

$$7x\mathbf{i} - z\mathbf{k} = 14 \cos v \cos u \, \mathbf{i} - 2 \sin v \, \mathbf{k}.$$

From this and the formula for $\mathbf{N}$ we thus obtain

$$(7x\mathbf{i} - z\mathbf{k}) \cdot \mathbf{N} = (14 \cos v \cos u)(4 \cos^2 v \cos u) - (2 \sin v)(4 \cos v \sin v)$$

$$= 56 \cos^2 u \cos^3 v - 8 \sin^2 v \cos v$$

$$= 56(\tfrac{1}{2} + \tfrac{1}{2} \cos 2u)(\tfrac{1}{4} \cos 3v + \tfrac{3}{4} \cos v) - 8 \sin^2 v \cos v.$$

We integrate this over u from 0 to 2π and over v from $-\pi/2$ to $\pi/2$, since this corresponds to the entire sphere. This gives

$$56(\pi + 0) \left[\frac{1}{4} \frac{\sin 3v}{3} \Big|_{-\pi/2}^{\pi/2} + \frac{3}{4} \sin v \Big|_{-\pi/2}^{\pi/2} \right] - 8 \cdot 2\pi \frac{\sin^3 v}{3} \Big|_{-\pi/2}^{\pi/2}$$

$$= 56\pi \left[\frac{1}{12}(-1 - 1) + \frac{3}{4}(1 - (-1)) \right] - \frac{16\pi}{3}(1 - (-1)) = 64\pi. \qquad \blacksquare$$

Further applications of the divergence theorem follow in the problem set and in the next section. The examples in the next section shed further light on the nature of the divergence and its applications in physics. There we also derive the **heat equation** governing heat flow (which we discuss further in several sections of Chap. 11), the two famous **Green's formulas** and some general basic properties of solutions of the **Laplace equation** (which we also discuss further in Chap. 11).

Problems for Sec. 9.6

Application of triple integrals. Find the total **mass** of a mass distribution of density σ in a region T in space, where

1. $\sigma = xyz$, T the cube $0 \leq x \leq 1$, $0 \leq y \leq 1$, $0 \leq z \leq 1$
2. $\sigma = x + y + z$, T the box $1 \leq x \leq 2$, $2 \leq y \leq 5$, $0 \leq z \leq 1$
3. $\sigma = x^2 + y^2$, T the cylinder $x^2 + y^2 \leq 4$, $0 \leq z \leq 3$
4. $\sigma = x$, T the tetrahedron with vertices $(0, 0, 0)$. $(1, 0, 0)$, $(0, 1, 0)$, $(0, 0, 1)$
5. $\sigma = x + y + 4z$, T as in Prob. 4
6. $\sigma = xy$, T as in Prob. 4
7. $\sigma = \sin x \cos y$, T the box $0 \leq x \leq \pi$, $0 \leq y \leq \pi/2$, $0 \leq z \leq 2$
8. $\sigma = x^2(y^2 + z^2)$, T the cylinder $-1 \leq x \leq 1$, $y^2 + z^2 \leq 9$
9. $\sigma = (x + y)^2$, T the cube $|x| \leq 1$, $|y| \leq 1$, $|z| \leq 1$
10. $\sigma = z$, T the region in the first octant bounded by $y = 1 - x^2$ and $z = x$

Moment of inertia. Find the moment of inertia $I_x = \displaystyle\iiint_T (y^2 + z^2) \, dx \, dy \, dz$ of a mass of density 1 in T about the x-axis, where T is

11. The box $0 \leq x \leq a$, $-b/2 \leq y \leq b/2$, $-c/2 \leq z \leq c/2$

12. The cube $0 \leq x \leq 1$, $0 \leq y \leq 1$, $0 \leq z \leq 1$

13. The cylinder $y^2 + z^2 \leq a^2$, $0 \leq x \leq h$

14. The cone $y^2 + z^2 \leq x^2$, $0 \leq x \leq h$

15. The ball $x^2 + y^2 + z^2 \leq a^2$

Application of the divergence theorem. Evaluate the surface integral $\iint\limits_{S} \mathbf{F} \cdot \mathbf{n} \, dA$ by the divergence theorem, where

16. $\mathbf{F} = x\mathbf{i} + y\mathbf{j} + z\mathbf{k}$, S the surface of the cube $0 \leq x \leq a$, $0 \leq y \leq a$, $0 \leq z \leq a$

17. $\mathbf{F} = e^x\mathbf{i} + \cosh y \, \mathbf{j} + \sinh z \, \mathbf{k}$, S as in Prob. 16

18. $\mathbf{F} = y^2\mathbf{i} + x^2\mathbf{j} + z^2\mathbf{k}$, S the surface of the cylinder $x^2 + y^2 \leq 4$, $0 \leq z \leq 5$

19. $\mathbf{F} = xy^2\mathbf{i} + y^3\mathbf{j} + 4x^2z\mathbf{k}$, S as in Prob. 18

20. $\mathbf{F} = (x^2 - z)\mathbf{k}$, S the surface of the tetrahedron in Prob. 4

21. $\mathbf{F} = x^6\mathbf{i} + y \cos^3 x \, \mathbf{j} + 2z\mathbf{k}$,
     S the surface of the cylinder $-\pi \leq x \leq \pi$, $y^2 + 4z^2 \leq 4$

22. $\mathbf{F} = e^{y+z}\mathbf{k}$, S the surface of $0 \leq x \leq 1$, $0 \leq y \leq 4$, $0 \leq z \leq 2$

23. $\mathbf{F} = (x^2 + y^2)\mathbf{i} + z^2\mathbf{k}$, S as in Prob. 22

24. $\mathbf{F} = x^2\mathbf{i} + y^2\mathbf{j} + z^2\mathbf{k}$, S the surface of the cube in Prob. 1

25. $\mathbf{F} = \sin^2 x \, \mathbf{i} + z^2\mathbf{j} - z \sin 2x \, \mathbf{k}$, S the surface of $x^2 + y^2 \leq z$, $z \leq 1/2$

26. $\mathbf{F} = (x + z)\mathbf{i} + (y + z)\mathbf{j} + (x + y)\mathbf{k}$, $S: x^2 + y^2 + z^2 = 4$

27. $\mathbf{F} = x^3\mathbf{i} + y^3\mathbf{j} + z^3\mathbf{k}$, S as in Prob. 26

28. $\mathbf{F} = 10y\mathbf{j} + z^3\mathbf{k}$, S the surface of $0 \leq x \leq 6$, $0 \leq y \leq 1$, $0 \leq z \leq y$

29. $\mathbf{F} = x\mathbf{i} + x^2y\mathbf{j} - x^2z\mathbf{k}$, S the surface of the tetrahedron in Prob. 4

30. $\mathbf{F} = x^2\mathbf{i} - (2x - 1)y\mathbf{j} + 4z\mathbf{k}$, S the surface of $x^2 + y^2 \leq z^2$, $0 \leq z \leq 2$

31. $\mathbf{F} = e^x\mathbf{i} - ye^x\mathbf{j} + 3z\mathbf{k}$, S the surface of $x^2 + y^2 \leq c^2$, $0 \leq z \leq h$

32. $\mathbf{F} = x^3\mathbf{i} + y^3\mathbf{j} + 3z(2 - x^2 - y^2)\mathbf{k}$, $S: 9x^2 + y^2 + 9z^2 = 9$

33. $\mathbf{F} = yz\mathbf{i} + zx\mathbf{j} + xy\mathbf{k}$, $S: x^2 + y^2 + z^2 = 25$

34. $\mathbf{F} = xyz\mathbf{i} + x^2z\mathbf{j}$, S the surface of $0 \leq x \leq 3$, $0 \leq y \leq 2$, $0 \leq z \leq 1$

35. $\mathbf{F} = \sin x \, \mathbf{i} + (4 - \cos x)y\mathbf{j}$, S as in Prob. 34

9.7 Further Applications of the Divergence Theorem

The divergence theorem has various applications and important consequences, some of which may be illustrated by the subsequent examples. In these examples, the regions and functions are assumed to satisfy the conditions under which the divergence theorem holds, and in each case, $\mathbf{n}$ is the *outer* unit normal vector of the boundary surface of the region, as before.

EXAMPLE 1. Representation of the divergence independent of coordinates

Dividing both sides of (2) in Sec. 9.6 by the volume $V(T)$ of the region T, we obtain

$$(1) \qquad \frac{1}{V(T)} \iiint\limits_{T} \operatorname{div} \mathbf{F} \, dV = \frac{1}{V(T)} \iint\limits_{S(T)} \mathbf{F} \cdot \mathbf{n} \, dA$$

where $S(T)$ is the boundary surface of T. The basic properties of the triple integral are essentially the same as those of the double integral considered in Sec. 9.2. In particular the **mean value theorem** *for triple integrals* asserts that for any continuous function $f(x, y, z)$ in the region T under consideration there is a point $Q: (x_0, y_0, z_0)$ in T such that

$$\iiint\limits_{T} f(x, y, z) \, dV = f(x_0, y_0, z_0) V(T).$$

Setting $f = \text{div } \mathbf{F}$ and using (1), we have

(2)
$$\frac{1}{V(T)} \iiint\limits_{T} \text{div } \mathbf{F} \, dV = \text{div } \mathbf{F}(x_0, y_0, z_0).$$

Let $P: (x_1, y_1, z_1)$ be any fixed point in T, and let T shrink down onto P, so that the maximum distance $d(T)$ of the points of T from P approaches zero. Then Q must approach P, and from (1) and (2) it follows that the divergence of $\mathbf{F}$ at P is

(3)
$$\boxed{\text{div } \mathbf{F}(x_1, y_1, z_1) = \lim_{d(T) \to 0} \frac{1}{V(T)} \iint\limits_{S(T)} \mathbf{F} \cdot \mathbf{n} \, dA.}$$

This formula is sometimes used as a *definition* of the divergence. While the definition of the divergence in Sec. 8.9 involves coordinates, formula (3) is independent of coordinates. Hence from (3) it follows immediately that *the divergence is independent of the particular choice of Cartesian coordinates.*

EXAMPLE 2. Physical interpretation of the divergence

From the divergence theorem we may obtain an intuitive interpretation of the divergence of a vector. For this purpose we consider the flow of an incompressible fluid (cf. at the end of Sec. 8.9) of constant density $\rho = 1$ which is **stationary** or **steady,** that is, does not vary with time. Such a flow is determined by the field of its velocity vector $\mathbf{v}(P)$ at any point P.

Let S be the boundary surface of a region T in space, and let $\mathbf{n}$ be the outer unit normal vector of S. The mass of fluid that flows through a small portion ΔS of S of area ΔA per unit time from the interior of S to the exterior is equal to $\mathbf{v} \cdot \mathbf{n} \, \Delta A$, where[12] $\mathbf{v} \cdot \mathbf{n}$ is the normal component of $\mathbf{v}$ in the direction of $\mathbf{n}$, taken at a suitable point of ΔS. Consequently, the total mass of fluid that flows across S from T to the outside per unit of time is given by the surface integral

$$\iint\limits_{S} \mathbf{v} \cdot \mathbf{n} \, dA.$$

Hence this integral represents the total flow out of T, and the integral

(4)
$$\frac{1}{V} \iint\limits_{S} \mathbf{v} \cdot \mathbf{n} \, dA$$

where V is the volume of T, represents the average flow out of T. Since the flow is steady and the fluid is incompressible, the amount of fluid flowing outward must be continuously supplied. Hence, if the value of the integral (4) is different from zero, there must be **sources** (*positive sources and negative sources, called* **sinks**) in T, that is, points where fluid is produced or disappears.

If we let T shrink down to a fixed point P in T, we obtain from (4) the *source intensity* at P given by the right-hand side of (3) with $\mathbf{F} \cdot \mathbf{n}$ replaced by $\mathbf{v} \cdot \mathbf{n}$. From this and (3) it follows that *the divergence of the velocity vector* $\mathbf{v}$ *of a steady incompressible flow is the source intensity of the flow at the corresponding point.* There are no sources in T if and only if $\text{div } \mathbf{v} \equiv 0$; in this case,

---

[12]Note that $\mathbf{v} \cdot \mathbf{n}$ may be negative at a certain point, which means that fluid *enters* the interior of S at such a point.

$$\iint_S \mathbf{v} \cdot \mathbf{n} \, dA = 0$$

for any closed surface S in T.

EXAMPLE 3. Modeling of heat flow. Heat equation

We know that in a body heat will flow in the direction of decreasing temperature. Physical experiments show that the rate of flow is proportional to the gradient of the temperature. This means that the velocity $\mathbf{v}$ of the heat flow in a body is of the form

(5) $$\mathbf{v} = -K \, \text{grad} \, U$$

where $U(x, y, z, t)$ is temperature, t is time and K is called the *thermal conductivity* of the body; in ordinary physical circumstances K is a constant. Using this information, set up the mathematical model of heat flow, the so-called **heat equation.**

Solution. Let T be a region in the body and let S be its boundary surface. Then the amount of heat leaving T per unit of time is

$$\iint_S \mathbf{v} \cdot \mathbf{n} \, dA,$$

where $\mathbf{v} \cdot \mathbf{n}$ is the component of $\mathbf{v}$ in the direction of the outer unit normal vector $\mathbf{n}$ of S. This expression is obtained in a fashion similar to that in the preceding example. From (5) and the divergence theorem we obtain [cf. (3), Sec. 8.9]

(6) $$\iint_S \mathbf{v} \cdot \mathbf{n} \, dA = -K \iiint_T \text{div} \, (\text{grad} \, U) \, dx \, dy \, dz = -K \iiint_T \nabla^2 U \, dx \, dy \, dz$$

where $\nabla^2 U = U_{xx} + U_{yy} + U_{zz}$ is the Laplacian of U.

On the other hand, the total amount of heat H in T is

$$H = \iiint_T \sigma \rho U \, dx \, dy \, dz$$

where the constant σ is the specific heat of the material of the body and ρ is the density (= mass per unit volume) of the material. Hence the time rate of decrease of H is

$$-\frac{\partial H}{\partial t} = -\iiint_T \sigma \rho \frac{\partial U}{\partial t} \, dx \, dy \, dz$$

and this must be equal to the above amount of heat leaving T; from (6) we thus have

$$-\iiint_T \sigma \rho \frac{\partial U}{\partial t} \, dx \, dy \, dz = -K \iiint_T \nabla^2 U \, dx \, dy \, dz$$

or

$$\iiint_T \left(\sigma \rho \frac{\partial U}{\partial t} - K \nabla^2 U \right) dx \, dy \, dz = 0.$$

Since this holds for any region T in the body, the integrand (if continuous) must be zero everywhere; that is,

(7) $$\boxed{\frac{\partial U}{\partial t} = c^2 \nabla^2 U}$$ $$c^2 = \frac{K}{\sigma \rho}.$$

This partial differential equation is called the **heat equation;** it is fundamental for heat ...
Methods for solving problems in heat conduction will be considered in Chap. 11.   ...ion.
    If the heat flow does not depend on time t ("**steady-state flow**"), then $\partial U/\partial t = 0$ and the hea...
equation (7) reduces to Laplace's equation $\nabla^2 U = 0$; see below.   ∎

Potential Theory. Harmonic Functions

The theory of solutions of Laplace's equation

(8)
$$\nabla^2 f = \frac{\partial^2 f}{\partial x^2} + \frac{\partial^2 f}{\partial y^2} + \frac{\partial^2 f}{\partial z^2} = 0$$

is called **potential theory.** A solution of (8) that has *continuous* second-order partial derivatives is called a **harmonic function.**[13] We shall consider details of potential theory in Chaps. 11 and 17. At present let us show that the divergence theorem is very useful in potential theory.

EXAMPLE 4. A basic property of solutions of Laplace's equation
Consider the formula in the divergence theorem:

(8*)
$$\iiint_T \text{div } \mathbf{F} \, dV = \iint_S \mathbf{F} \cdot \mathbf{n} \, dA.$$

Assume that $\mathbf{F}$ is the gradient of a scalar function, say, $\mathbf{F} = \text{grad } f$. Then [cf. (3) in Sec. 8.9]

$$\text{div } \mathbf{F} = \text{div } (\text{grad } f) = \nabla^2 f.$$

Furthermore,

$$\mathbf{F} \cdot \mathbf{n} = \mathbf{n} \cdot \text{grad } f,$$

and from (5) in Sec. 8.8 we see that the right side is the directional derivative of f in the outer normal direction of S. If we denote this derivative by $\partial f/\partial n$, formula (8*) becomes

(9)
$$\iiint_T \nabla^2 f \, dV = \iint_S \frac{\partial f}{\partial n} \, dA.$$

Obviously this is the three-dimensional analog of the formula (9) in Sec. 9.3.
    Taking into account the assumptions under which the divergence theorem holds, we immediately obtain from (9) the following result.

Theorem 1 (A basic property of harmonic functions)
Let $f(x, y, z)$ be a harmonic function in some domain D. Then the integral of the normal derivative of the function f over any piecewise smooth[14] *closed orientable surface S in D is zero.*

EXAMPLE 5. Green's theorems
Let f and g be scalar functions such that $\mathbf{F} = f \text{ grad } g$ satisfies the assumptions of the divergence theorem in some region T. Then

$$\text{div } \mathbf{F} = \text{div } (f \text{ grad } g) = f\nabla^2 g + \text{grad } f \cdot \text{grad } g$$

---

[13]This continuity requirement should not be deleted from the definition of a harmonic function, as some older books do.
    [14]Cf. Sec. 9.4.

(cf. Prob. 29 at the end of Sec. 8.9). Furthermore,

$$\mathbf{F \cdot n} = \mathbf{n} \cdot (f \text{ grad } g) = f(\mathbf{n} \cdot \text{grad } g).$$

The expression $\mathbf{n} \cdot \text{grad } g$ is the directional derivative of g in the direction of the outer normal vector $\mathbf{n}$ of the surface S in the divergence theorem. If we denote this derivative by $\partial g/\partial n$, the formula in the divergence theorem becomes

(10)
$$\iiint_T (f\nabla^2 g + \text{grad } f \cdot \text{grad } g) \, dV = \iint_S f \frac{\partial g}{\partial n} \, dA.$$

This formula is called **Green's first formula** or (together with the assumptions) the *first form of Green's theorem.*

Interchanging f and g we obtain a similar formula. Subtracting this formula from (10) we find

(11)
$$\iiint_T (f\nabla^2 g - g\nabla^2 f) \, dV = \iint_S \left(f \frac{\partial g}{\partial n} - g \frac{\partial f}{\partial n} \right) dA.$$

This formula is called **Green's second formula** or (together with the assumptions) the *second form of Green's theorem.*

EXAMPLE 6. Uniqueness of solutions of Laplace's equation

Suppose that f satisfies the assumptions in Theorem 1 and is zero everywhere on a piecewise smooth closed orientable surface S in D. Then, setting $g = f$ in (10) and denoting the interior of S by T, we obtain

$$\iiint_T \text{grad } f \cdot \text{grad } f \, dV = \iiint_T |\text{grad } f|^2 \, dV = 0.$$

Since by assumption $|\text{grad } f|$ is continuous in T and on S and is nonnegative, it must be zero everywhere in T. Hence $f_x = f_y = f_z = 0$, and f is constant in T and, because of continuity, equal to its value 0 on S. This proves

Theorem 2 (Harmonic functions)

If a function $f(x, y, z)$ is harmonic in some domain D and is zero at every point of a piecewise smooth closed orientable surface S in D, then f is identically zero in the region T bounded by S.

This theorem has an important consequence. Let f_1 and f_2 be functions that satisfy the assumptions of Theorem 1 and take on the same values on S. Then their difference $f_1 - f_2$ satisfies those assumptions and has the value 0 everywhere on S. Consequently, from Theorem 2 it follows that $f_1 - f_2 = 0$ throughout T, and we have the following result.

Theorem 3 (Uniqueness theorem for Laplace's equation)

Let T be a region that satisfies the assumptions of the divergence theorem, and let $f(x, y, z)$ be a harmonic function in a domain D that contains T and its boundary surface S. Then f is uniquely determined in T by its values on S.

The problem of determining a solution u of a partial differential equation in a region T such

that u assumes given values on the boundary surface S of T is called the **Dirichlet problem.** [15]
We may thus reformulate Theorem 3 as follows.

Theorem 3* (Uniqueness theorem for the Dirichlet problem)

*Under the assumptions in Theorem 3 the solution of the Dirichlet problem
for the Laplace equation is unique.*

These theorems demonstrate the importance of Gauss's theorem in potential theory.   ■

Looking back at this section and the previous one and at the corresponding
problem sets, the student should realize the great usefulness of the divergence theorem, which is really a key to various areas related to vector
functions and scalar functions and their integrals. Equally useful is the second
"big" theorem in this chapter, **Stokes's theorem,** which we discuss in the
next section. Stokes's theorem transforms surface integrals into line integrals
over the boundary curve of the surface and conversely. It involves the curl
of a vector function as defined in Sec. 8.10 (which the student may want to
review).

Problems for Sec. 9.7

1. Verify Theorem 1 for $f = x^2 + y^2 - 2z^2$ and S the surface of the cube
   $0 \leq x \leq 1, 0 \leq y \leq 1, 0 \leq z \leq 1$.
2. Verify Theorem 1 for $f = x^2 - y^2$ and the surface of the cylinder $x^2 + y^2 \leq 1$,
   $0 \leq z \leq 1$.

Let f and g be harmonic functions in some domain D that contains a region T and
its boundary surface S such that T satisfies the assumptions in the divergence theorem. Show that then:

3. $\displaystyle\iint_S g \frac{\partial g}{\partial n} \, dA = \iiint_T |\text{grad } g|^2 \, dV$

4. If $\partial g/\partial n = 0$ on S, then g is constant in T.
5. If $\partial f/\partial n = \partial g/\partial n$ on S, then $f = g + c$ in T, where c is a constant.
6. $\displaystyle\iint_S \left(f \frac{\partial g}{\partial n} - g \frac{\partial f}{\partial n} \right) dA = 0$.

7. **(Laplacian)** Show that the Laplacian can be represented independently of all
   coordinate systems in the form

$$\nabla^2 f = \lim_{d(T) \to 0} \frac{1}{V(T)} \iint_{S(T)} \frac{\partial f}{\partial n} \, dA$$

where $d(T)$ is the maximum distance of the points of a region T bounded by $S(T)$
from the point at which the Laplacian is evaluated and $V(T)$ is the volume of T.
Hint. Put $\mathbf{F} = \text{grad } f$ in (3) and use (5), Sec. 8.8, with $\mathbf{b} = \mathbf{n}$, the outer unit
normal vector to S.

[15]PETER GUSTAV LEJEUNE DIRICHLET (1805—1859), German mathematician, studied
in Paris under Cauchy and others and succeeded Gauss at Göttingen in 1855. He became known
by his important research on Fourier series (he knew Fourier personally) and in number theory.

8. **(Volume)** Using the divergence theorem, show that the volume V of a region T bounded by a surface S is

$$V = \iint\limits_{S} x \, dy \, dz = \iint\limits_{S} y \, dz \, dx = \iint\limits_{S} z \, dx \, dy$$

$$= \frac{1}{3} \iint\limits_{S} (x \, dy \, dz + y \, dz \, dx + z \, dx \, dy).$$

9. **(Volume)** Using $\mathbf{F} = x\mathbf{i} + y\mathbf{j} + z\mathbf{k}$ in (2), Sec. 9.6, show that a region T with boundary surface S has the volume

$$V = \frac{1}{3} \iint\limits_{S} r \cos \phi \, dA$$

where r is the distance of a variable point P: (x, y, z) on S from the origin O and ϕ is the angle between the directed line OP and the outer normal to S at P.

10. Find the volume of a ball of radius a by means of the formula in Prob. 9.

9.8 Stokes's Theorem

In Sec. 9.3 we saw that double integrals over a region in the plane can be transformed into line integrals over the boundary curve of the region and conversely; this was done by Green's theorem in the plane. We shall now see that, more generally, surface integrals over a surface S with boundary curve C can be transformed into line integrals over C and conversely. This will be done by *Stokes's theorem*, which involves the curl,

$$(1) \qquad \operatorname{curl} \mathbf{F} = \begin{vmatrix} \mathbf{i} & \mathbf{j} & \mathbf{k} \\ \partial/\partial x & \partial/\partial y & \partial/\partial z \\ F_1 & F_2 & F_3 \end{vmatrix} \qquad \text{(cf. Sec. 8.10).}$$

Stokes's theorem
(Transformation between surface integrals and line integrals)
Let S be a piecewise smooth oriented surface in space and let the boundary of S be a piecewise smooth simple closed curve C. Let $\mathbf{F}(x, y, z)$ be a continuous vector function that has continuous first partial derivatives in a domain in space containing S. Then

$$(2) \qquad \boxed{\iint\limits_{S} (\operatorname{curl} \mathbf{F}) \cdot \mathbf{n} \, dA = \oint\limits_{C} \mathbf{F} \cdot \mathbf{r}'(s) \, ds}$$

where $\mathbf{n}$ is a unit normal vector of S and, depending on $\mathbf{n}$, the integration around C is taken in the sense shown in Fig. 230; also, $\mathbf{r}' = d\mathbf{r}/ds$ is the unit tangent vector and s the arc length of C. (Proof on p. 564.)

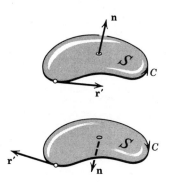

Fig. 230. Stokes's theorem

Comment. This theorem is named after G. G. Stokes.[16] For "piecewise smooth", see Secs. 9.4 and 9.5.

From (3) in Sec. 9.5, with curl **F** instead of **F**, we get (2) in terms of components [cf. (1)]

$$(3) \quad \iint_R \left[\left(\frac{\partial F_3}{\partial y} - \frac{\partial F_2}{\partial z} \right) N_1 + \left(\frac{\partial F_1}{\partial z} - \frac{\partial F_3}{\partial x} \right) N_2 + \left(\frac{\partial F_2}{\partial x} - \frac{\partial F_1}{\partial y} \right) N_3 \right] du \, dv$$

$$= \oint_{\overline{C}} (F_1 \, dx + F_2 \, dy + F_3 \, dz)$$

where R is the region with boundary curve $\overline{C}$ in the uv-plane corresponding to S, which is represented by $\mathbf{r}(u, v)$, and $\mathbf{N} = N_1 \mathbf{i} + N_2 \mathbf{j} + N_3 \mathbf{k} = \mathbf{r}_u \times \mathbf{r}_v$.

EXAMPLE 1. Verification of Stokes's theorem

Before proving Stokes's theorem, let us get used to it by verifying it for $\mathbf{F} = y\mathbf{i} + z\mathbf{j} + x\mathbf{k}$ and S the paraboloid $z = f(x, y) = 1 - (x^2 + y^2)$, $z \geqq 0$ in Fig. 231.

Solution. C: $\mathbf{r}(s) = \cos s \, \mathbf{i} + \sin s \, \mathbf{j}$, $\mathbf{r}'(s) = -\sin s \, \mathbf{i} + \cos s \, \mathbf{j}$, hence

$$\oint_C \mathbf{F} \cdot d\mathbf{r} = \int_0^{2\pi} ([(\sin s)(-\sin s) + 0 + 0] \, ds = -\pi.$$

On the other hand we now consider the integral over S in (2) on the left, using (3), Sec. 9.5, with curl **F** instead of **F** and $u = x$, $v = y$. We need (verify this)

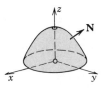

Fig. 231. Surface S in Example 1

[16]Sir GEORGE GABRIEL STOKES (1819—1903), Irish mathematician and physicist, who became a professor in Cambridge in 1849. He is also known for his important contribution to the theory of infinite series and to viscous flow (Navier-Stokes equations), geodesics and optics.

$$\text{curl } \mathbf{F} = -\mathbf{i} - \mathbf{j} - \mathbf{k}, \qquad \mathbf{N} = -f_x\mathbf{i} - f_y\mathbf{j} + \mathbf{k} = 2x\mathbf{i} + 2y\mathbf{j} + \mathbf{k}$$

and calculate from this $(\text{curl } \mathbf{F}) \cdot \mathbf{N} = -2x - 2y - 1$. Now S corresponds to $x^2 + y^2 \leq 1$. Using polar coordinates r, θ, given by $x = r \cos \theta$, $y = r \sin \theta$, we thus obtain from (3), Sec. 9.5,

$$\iint_S (\text{curl } \mathbf{F}) \cdot \mathbf{n} \, dA = \int_0^{2\pi} \int_0^1 (-2r \cos \theta - 2r \sin \theta - 1) r \, dr \, d\theta$$

$$= \int_0^{2\pi} [\tfrac{2}{3}(-\cos \theta - \sin \theta) - \tfrac{1}{2}] \, d\theta = -\pi.$$

Note that $\mathbf{N}$ is an upper normal vector of S (why?) and C is oriented counterclockwise, which is as is required in the theorem. ∎

Proof of Stokes's theorem. Obviously, (2) holds if the integrals of each component on both sides of (3) are equal, that is,

$$(4) \qquad \iint_R \left(\frac{\partial F_1}{\partial z} N_2 - \frac{\partial F_1}{\partial y} N_3 \right) du \, dv = \oint_{\overline{C}} F_1 \, dx$$

$$(5) \qquad \iint_R \left(-\frac{\partial F_2}{\partial z} N_1 + \frac{\partial F_2}{\partial x} N_3 \right) du \, dv = \oint_{\overline{C}} F_2 \, dy$$

$$(6) \qquad \iint_R \left(\frac{\partial F_3}{\partial y} N_1 - \frac{\partial F_3}{\partial x} N_2 \right) du \, dv = \oint_{\overline{C}} F_3 \, dz.$$

We prove this first for a surface S that can be represented simultaneously in the forms

$$(7) \quad \text{(a)} \quad z = f(x, y), \qquad \text{(b)} \quad y = g(x, z), \qquad \text{(c)} \quad x = h(y, z).$$

We prove (4), using (7a). Then, setting $u = x$, $v = y$, we have

$$\mathbf{r}(u, v) = \mathbf{r}(x, y) = x\mathbf{i} + y\mathbf{j} + f\mathbf{k}$$

and in (3), Sec. 9.5, by direct calculation

$$\mathbf{N} = \mathbf{r}_u \times \mathbf{r}_v = \mathbf{r}_x \times \mathbf{r}_y = -f_x\mathbf{i} - f_y\mathbf{j} + \mathbf{k}.$$

Note that $\mathbf{N}$ is an *upper* normal vector of S, since it has a *positive* z-component. Also, $R = S^*$, the projection of S into the xy-plane (Fig. 232). Hence the left-hand side of (4) is

$$(8) \qquad \iint_{S^*} \left[\frac{\partial F_1}{\partial z} (-f_y) - \frac{\partial F_1}{\partial y} \right] dx \, dy.$$

We now consider the right-hand side of (4). We transform this line integral over $\overline{C}$ into a double integral over S^* by applying Green's theorem [formula

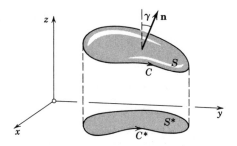

Fig. 232. Proof of Stokes's theorem

(1) in Sec. 9.3 with $F_2 = 0$]. This gives

$$\oint_{\overline{C}} F_1 \, dx = \int\int_{S^*} -\frac{\partial F_1}{\partial y} \, dx \, dy.$$

Here, $F_1 = F_1[x, y, f(x, y)]$. Hence by the chain rule (cf. also Prob. 16 in Sec. 8.7),

$$-\frac{\partial F_1[x, y, f(x, y)]}{\partial y} = -\frac{\partial F_1(x, y, z)}{\partial y} - \frac{\partial F_1(x, y, z)}{\partial z}\frac{\partial f}{\partial y} \qquad [z = f(x, y)]$$

We see that the right-hand side of this equals the integrand in (8). This proves (4). Relations (5) and (6) follow in the same way if we use (7b) and (7c), respectively. By addition we obtain (3). This proves Stokes's theorem for a surface S that can be represented simultaneously in the forms (7a), (7b) and (7c).

As in the proof of the divergence theorem, our result may be immediately extended to a surface S that can be decomposed into finitely many pieces, each of which is of the type just considered. This covers most of the cases of practical interest. The proof in the case of a most general surface S satisfying the assumptions of the theorem would require a limit process; this is similar to the situation in the case of Green's theorem in Sec. 9.3. ∎

EXAMPLE 2. Green's theorem in the plane as a special case of Stokes's theorem
Let $\mathbf{F} = F_1\mathbf{i} + F_2\mathbf{j}$ be a vector function that is continuously differentiable in a domain in the xy-plane containing a simply connected bounded closed region S whose boundary C is a piecewise smooth simple closed curve. Then, according to (1),

$$(\text{curl } \mathbf{F})\cdot\mathbf{n} = (\text{curl } \mathbf{F})\cdot\mathbf{k} = \frac{\partial F_2}{\partial x} - \frac{\partial F_1}{\partial y}.$$

Hence the formula in Stokes's theorem now takes the form

$$\int\int_S \left(\frac{\partial F_2}{\partial x} - \frac{\partial F_1}{\partial y}\right) dA = \oint_C (F_1 \, dx + F_2 \, dy).$$

This shows that Green's theorem in the plane (Sec. 9.3) is a special case of Stokes's theorem.

EXAMPLE 3. Evaluation of a line integral by Stokes's theorem
Evaluate $\int_C \mathbf{F}\cdot\mathbf{r}' \, ds$, where C is the circle $x^2 + y^2 = 4$, $z = -3$, oriented counterclockwise as seen by a person standing at the origin, and, with respect to right-handed Cartesian coordinates,

$$\mathbf{F} = y\mathbf{i} + xz^3\mathbf{j} - zy^3\mathbf{k}.$$

Solution. As a surface S bounded by C we can take the plane circular disk $x^2 + y^2 \leq 4$ in the plane $z = -3$. Then $\mathbf{n}$ in Stokes's theorem points in the positive z-direction; thus $\mathbf{n} = \mathbf{k}$. Hence (curl $\mathbf{F}$)$\cdot\mathbf{n}$ is simply the component of curl $\mathbf{F}$ in the positive z-direction. Since $\mathbf{F}$ with $z = -3$ has the components $F_1 = y$, $F_2 = -27x$, $F_3 = 3y^3$, we thus obtain

$$(\text{curl } \mathbf{F})\cdot\mathbf{n} = \frac{\partial F_2}{\partial x} - \frac{\partial F_1}{\partial y} = -27 - 1 = -28.$$

Hence the integral over S in Stokes's theorem equals -28 times the area 4π of the disk S. This yields the answer $-28 \cdot 4\pi = -112\pi \approx -352$.

The reader may confirm the result by direct calculation, which involves somewhat more work.

EXAMPLE 4. Physical interpretation of the curl

Let S_r be a circular disk of radius r and center P bounded by the circle C_r (Fig. 233), and let $\mathbf{F}(Q) \equiv \mathbf{F}(x, y, z)$ be a continuously differentiable vector function in a domain containing S_r. Then by Stokes's theorem and the mean value theorem for surface integrals,

$$\oint_{C_r} \mathbf{F}\cdot\mathbf{r}' \, ds = \iint_{S_r} (\text{curl } \mathbf{F})\cdot\mathbf{n} \, dA = (\text{curl } \mathbf{F})\cdot\mathbf{n}(P^*)A_r$$

where A_r is the area of S_r and P^* is a suitable point of S_r. This may be written in the form

$$(\text{curl } \mathbf{F})\cdot\mathbf{n}(P^*) = \frac{1}{A_r} \oint_{C_r} \mathbf{F}\cdot\mathbf{r}' \, ds.$$

In the case of a fluid motion with velocity vector $\mathbf{F} = \mathbf{v}$, the integral

$$\oint_{C_r} \mathbf{v}\cdot\mathbf{r}' \, ds$$

is called the **circulation** of the flow around C_r; it measures the extent to which the corresponding fluid motion is a rotation around the circle C_r. If we now let r approach zero, we find

$$(9) \qquad\qquad (\text{curl } \mathbf{v})\cdot\mathbf{n}(P) = \lim_{r\to 0} \frac{1}{A_r} \oint_C \mathbf{v}\cdot\mathbf{r}' \, ds;$$

that is, the component of the curl in the positive normal direction can be regarded as the **specific circulation** (circulation per unit area) of the flow in the surface at the corresponding point.

EXAMPLE 5. Work done in the displacement around a closed curve

Find the work done by the force $\mathbf{F} = 2xy^3 \sin z \, \mathbf{i} + 3x^2y^2 \sin z \, \mathbf{j} + x^2y^3 \cos z \, \mathbf{k}$ in the displacement around the curve of intersection of the paraboloid $z = x^2 + y^2$ and the cylinder $(x - 1)^2 + y^2 = 1$.

Solution. This work is given by the line integral in Stokes's theorem. Now $\mathbf{F} = \text{grad } f$, where $f = x^2y^3 \sin z$ and curl grad $f = \mathbf{0}$ (cf. (3) in Sec. 8.10), so that (curl $\mathbf{F}$)$\cdot\mathbf{n} = 0$ and the work is 0 by Stokes's theorem.

This suggests that Stokes's theorem may be of great help in connection with the independence of path of line integrals. We confirm this in the next section.   ∎

Fig. 233. Example 4

We have seen and emphasized that the value of a line integral generally depends not only on the function to be integrated and the two endpoints P and Q of the path of integration C but also on the particular choice of a path from P to Q. This raises the following basic question. Under what condition will that value depend only on P and Q but not on the choice of C ("**independence of path**")? We discuss and answer this question in the next section. As a tool we shall use Stokes's theorem.

Problems for Sec. 9.8

Evaluate $\displaystyle\iint_S (\text{curl } \mathbf{F}) \cdot \mathbf{n} \, dA$ by direct integration, where

1. $\mathbf{F} = z^2\mathbf{i} + 4x\mathbf{j}$, S the square $0 \le x \le 1$, $0 \le y \le 1, z = 1$
2. $\mathbf{F} = \cos y \, \mathbf{i} + \cosh z \, \mathbf{j} + x\mathbf{k}$, S as in Prob. 1
3. $\mathbf{F} = -y^3\mathbf{i} + x^3\mathbf{j}$, S the circular disk $x^2 + y^2 \le 1$, $z = 0$
4. $\mathbf{F} = e^z\mathbf{i} + e^x\mathbf{k}$, S the rectangle $0 \le x \le 1$, $0 \le y \le 1$, $z = y$
5. $\mathbf{F} = z^2\mathbf{i} + \sin y \, \mathbf{j} - x^2\mathbf{k}$,
   S the triangle with vertices $(0, 0, 0)$, $(2, 0, 0)$, $(0, 0, 2)$
6. $\mathbf{F} = x \cos z \, \mathbf{k}$, $S: x^2 + y^2 = 4$, $y \ge 0$, $0 \le z \le \pi/2$
7. $\mathbf{F} = e^{2z}\mathbf{i} + e^z \sin y \, \mathbf{j} + e^z \cos y \, \mathbf{k}$, $S: 0 \le x \le 2$, $0 \le y \le 1$, $z = y^2$
8. $\mathbf{F} = z^2\mathbf{i} + x^2\mathbf{j} + y^2\mathbf{k}$, $S: z^2 = x^2 + y^2$, $y \ge 0$, $0 \le z \le 2$

9. Verify Stokes's theorem for $\mathbf{F}$ and S in Prob. 1.
10. Verify Stokes's theorem for $\mathbf{F}$ and S in Prob. 3.

Evaluate $\displaystyle\oint_C \mathbf{F} \cdot \mathbf{r}'(s) \, ds$ (clockwise as seen by a person standing at the origin) by Stokes's theorem, where, with respect to right-handed Cartesian coordinates,

11. $\mathbf{F} = \sinh x \, \mathbf{j}$, C the boundary of the triangle with vertices $(1, 0, 0)$, $(0, 1, 0)$, $(0, 0, 1)$
12. $\mathbf{F} = z\mathbf{i} + x\mathbf{j} + y\mathbf{k}$, C as in Prob. 11
13. $\mathbf{F} = y^2\mathbf{i} + x^2\mathbf{k}$, C as in Prob. 11
14. $\mathbf{F} = (x + y)\mathbf{i} + (2x - z)\mathbf{j} + y\mathbf{k}$, C the boundary of the triangle with vertices $(2, 0, 0)$, $(0, 3, 0)$, $(0, 0, 6)$
15. $\mathbf{F} = y^2\mathbf{i} - x^2\mathbf{j} + z^2\mathbf{k}$, C as in Prob. 14
16. $\mathbf{F} = 4z\mathbf{i} - x\mathbf{j} + (2y - ze^{4z})\mathbf{k}$, C as in Prob. 11
17. $\mathbf{F} = -3y\mathbf{i} + 3x\mathbf{j} + z\mathbf{k}$, C the circle $x^2 + y^2 = 1, z = 1$
18. $\mathbf{F} = y^2\mathbf{i} + x^2\mathbf{j} - (x + z)\mathbf{k}$,
   C the boundary of the triangle with vertices $(0, 0, 1)$, $(1, 0, 1)$, $(1, 1, 1)$
19. $\mathbf{F} = 4z\mathbf{i} - 2x\mathbf{j} + 2x\mathbf{k}$, C the intersection of $x^2 + y^2 = 1$ and $z = y + 1$
20. $\mathbf{F} = x^2\mathbf{i} + y^2\mathbf{j} + z^2\mathbf{k}$, C the intersection of $x^2 + y^2 + z^2 = a^2$ and $z = y^2$
21. $\mathbf{F} = yz\mathbf{i} + xz\mathbf{j} + xy\mathbf{k}$, C the intersection of $x^2 + y^2 = 1$ and $z = y^2$
22. $\mathbf{F} = 2y\mathbf{i} + z\mathbf{j} + 3y\mathbf{k}$, C the intersection of $x^2 + y^2 + z^2 = 6z$ and $z = x + 3$
23. $\mathbf{F} = \sin z \, \mathbf{i} - \cos x \, \mathbf{j} + \sin y \, \mathbf{k}$,
   C the boundary of the rectangle $0 \le x \le \pi$, $0 \le y \le 1$, $z = 3$

24. Evaluate $\int_C \mathbf{F} \cdot \mathbf{r}' \, ds$, $\mathbf{F} = (x^2 + y^2)^{-1}(-y\mathbf{i} + x\mathbf{j})$, $C: x^2 + y^2 = 1$, $z = 0$, oriented clockwise. Note that Stokes's theorem cannot be applied. Why?
25. Show that if $\mathbf{F}$ and S satisfy the assumptions of Stokes's theorem and $\mathbf{F} = \text{grad } f$, then $\int_C \mathbf{F} \cdot \mathbf{r}' \, ds = 0$, where C is the boundary curve of S.

Line Integrals Independent of Path

In Sec. 9.1 we saw that the value of a line integral

$$(1) \qquad \int_C (F_1 \, dx + F_2 \, dy + F_3 \, dz)$$

in general depends not only on the endpoints P and Q of the path C but also on C. That is, if we integrate from P to Q along different paths, we shall in general obtain different values of the integral. We shall now see under what conditions that value depends only on P and Q but does not depend on the path C from P to Q. This problem is of great importance. For instance, in mechanics, independence of path may mean that we have to do the same amount of work regardless of the path to the mountain top, be it short and steep or long and gentle, or that we gain back the work done in extending an elastic spring when we release it. Not all forces are of this type—think of swimming in a big whirlpool. Having available Stokes's theorem, we can now discuss the problem in full generality (instead of doing a fraction of it and then waiting for Stokes). We begin with some basic concepts.

Independence of Path. Exact Differential Forms

Let F_1, F_2, F_3 be functions defined and continuous in a domain D in space. Then we call a line integral of the form (1) **independent of path** in D, if for every pair of endpoints P and Q in D the value of the integral is the same for all paths C in D starting from P and ending at Q. This value will then, in general, depend on the choice of the points P and Q but not on the choice of the path joining them.

In this connection we recall that, by definition, a function is a *single-valued* relation, that is, to each point in its domain of definition it assigns a *single* value in its range. This will be essential in our present discussion.

An expression of the form

$$(2) \qquad F_1 \, dx + F_2 \, dy + F_3 \, dz$$

where F_1, F_2, F_3 are functions defined in some domain D, is called a *first-order* **differential form** in three variables. This form is said to be **exact** or *an exact differential* in D if it is the differential

$$(3) \qquad df = \frac{\partial f}{\partial x} \, dx + \frac{\partial f}{\partial y} \, dy + \frac{\partial f}{\partial z} \, dz$$

of a differentiable function $f(x, y, z)$ everywhere in D; that is, if we have

$$(4) \qquad F_1 \, dx + F_2 \, dy + F_3 \, dz = df.$$

By comparing (3) and (4) we see that *the form* (2) *is exact in D if and only if there is a differentiable function $f(x, y, z)$ such that the three relationships*

(5) $$F_1 = \frac{\partial f}{\partial x}, \qquad F_2 = \frac{\partial f}{\partial y}, \qquad F_3 = \frac{\partial f}{\partial z}$$

hold everywhere in D.

In vector language this means that *the form* (2) *is exact in D if and only if the vector function*

$$\mathbf{F} = F_1\mathbf{i} + F_2\mathbf{j} + F_3\mathbf{k}$$

is the gradient of a function f(x, y, z) in D:

(5') $$\mathbf{F} = \text{grad } f = \frac{\partial f}{\partial x}\mathbf{i} + \frac{\partial f}{\partial y}\mathbf{j} + \frac{\partial f}{\partial z}\mathbf{k}.$$

We shall now prove that exactness is a necessary and sufficient condition for independence of path, as follows.

Theorem 1 (Exactness and independence of path)

Let $F_1(x, y, z)$, $F_2(x, y, z)$, $F_3(x, y, z)$ be continuous in a domain D in space. Then the line integral

(6) $$\int_C (F_1 \, dx + F_2 \, dy + F_3 \, dz)$$

is independent of path in D if and only if the differential form under the integral sign is exact in D.

Comment. Note that (6) can be written

(6') $$\int_C \mathbf{F} \cdot d\mathbf{r}$$

where $\mathbf{F} = F_1\mathbf{i} + F_2\mathbf{j} + F_3\mathbf{k}$ and $d\mathbf{r} = dx \, \mathbf{i} + dy \, \mathbf{j} + dz \, \mathbf{k}$.

Proof of the theorem. (a) Suppose that the integral under consideration is independent of path in D. We choose any fixed point P: (x_0, y_0, z_0) and a point Q: (x, y, z) in D. Then we define a function $f(x, y, z)$ by the formula

(7) $$f(x, y, z) = f_0 + \int_P^Q (F_1 \, dx^* + F_2 \, dy^* + F_3 \, dz^*)$$

where f_0 is a constant and we integrate along an arbitrary path in D from P to Q. Since the integral is independent of path and P is fixed, it does indeed depend only on the coordinates x, y, z of the endpoint Q and defines a function $f(x, y, z)$ in D. It remains to show that from (7) the relations (5) follow, which means that $F_1 \, dx + F_2 \, dy + F_3 \, dz$ is exact in D. We prove the first of these three relations. Since the integral is independent of path, we may integrate from P to a point Q_1: (x_1, y, z) and then parallel to the

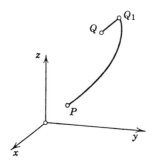

Fig. 234. Proof of Theorem 1

x-axis along the line segment from Q_1 to Q (Fig. 234); here Q_1 is chosen so that the whole segment $Q_1 Q$ lies in D. Then

$$f(x, y, z) = f_0 + \int_P^{Q_1} (F_1 \, dx^* + F_2 \, dy^* + F_3 \, dz^*)$$

$$+ \int_{Q_1}^{Q} (F_1 \, dx^* + F_2 \, dy^* + F_3 \, dz^*).$$

We take the partial derivative with respect to x. Since P and Q_1 do not depend on x, that derivative of the first integral is zero. Since on the segment $Q_1 Q$, both y and z are constant, the last integral may be written as the definite integral

$$\int_{x_1}^{x} F_1(x^*, y, z) \, dx^*.$$

Hence the partial derivative of that integral with respect to x is $F_1(x, y, z)$, and the first of the relations (5) is proved. The other two relations can be proved by the same argument.

(b) Conversely, suppose that $F_1 \, dx + F_2 \, dy + F_3 \, dz$ is exact in D. Then (5) holds in D for some function f. Let C be any path from P to Q in D, and let

$$\mathbf{r}(t) = x(t)\mathbf{i} + y(t)\mathbf{j} + z(t)\mathbf{k} \qquad (t_0 \leq t \leq t_1)$$

be a parametric representation of C such that P corresponds to $t = t_0$ and Q corresponds to $t = t_1$. Then

$$\int_P^Q (F_1 \, dx + F_2 \, dy + F_3 \, dz) = \int_P^Q \left(\frac{\partial f}{\partial x} \, dx + \frac{\partial f}{\partial y} \, dy + \frac{\partial f}{\partial z} \, dz \right)$$

$$= \int_{t_0}^{t_1} \frac{df}{dt} \, dt = f[x(t), y(t), z(t)] \Big|_{t=t_0}^{t=t_1}$$

$$= f(Q) - f(P).$$

This shows that the value of the integral is simply the difference of the values of f at the two endpoints of C and is, therefore, independent of the path C. Theorem 1 is proved.   ∎

The last formula in the proof,

(8)
$$\int_P^Q (F_1\, dx + F_2\, dy + F_3\, dz) = f(Q) - f(P)$$ $[\mathbf{F} = \text{grad } f]$

is the analog of the usual formula

$$\int_a^b g(x)\, dx = G(x) \Big|_a^b = G(b) - G(a)$$ $[\text{where } G'(x) = g(x)]$

for evaluating definite integrals, which is known from elementary calculus. Formula (8) should be applied whenever a line integral is independent of path. The practical use of this formula will be explained near the end of this section, in Example 3.

We remember that the work W done by a force $\mathbf{F} = [F_1, F_2, F_3]$ in the displacement of a particle along a path C in space is given by the line integral (3) in Sec. 9.1,

$$W = \int_C \mathbf{F} \cdot d\mathbf{r} = \int_C (F_1\, dx + F_2\, dy + F_3\, dz),$$

the integration being taken in the sense of the displacement. From Theorem 1 we conclude that W depends only on the endpoints P and Q of the path C if and only if the differential form under the integral sign is exact; and this form is exact if and only if $\mathbf{F}$ is the gradient of a scalar function f. In this case we may find W by using (8). We shall say more about this near the end of the section. For the time being, let us illustrate this application of (8) by a typical example.

EXAMPLE 1. Work done by a variable force
Find the work W done by the force $\mathbf{F} = yz\mathbf{i} + xz\mathbf{j} + xy\mathbf{k}$ in the displacement of a particle along the straight segment C from P: (1, 1, 1) to Q: (3, 3, 2).

Solution. The work is given by the integral

$$W = \int_C (yz\, dx + xz\, dy + xy\, dz).$$

This integral is of the form (1) where

$$F_1 = yz, \qquad F_2 = xz, \qquad F_3 = xy$$

and (5) holds for $f = xyz$. Hence W is independent of path and we may use (8), finding

$$W = f(Q) - f(P) = f(3, 3, 2) - f(1, 1, 1) = 18 - 1 = 17.$$ ∎

Another important characterization of independence of path is given by the following theorem.

Theorem 2 (Independence of path)

Let F_1, F_2, F_3 *be continuous in a domain D of space. Then the line integral*

$$\int_C (F_1 \, dx + F_2 \, dy + F_3 \, dz)$$

is independent of path in D if and only if it is zero around every simple closed path in D.

Proof. (a) Let the integral of

$$(9) \qquad\qquad F_1 \, dx + F_2 \, dy + F_3 \, dz$$

be independent of path in D. Consider any simple closed path C in D. Subdivide C into two arcs C_1 and C_2 as in Fig. 235. Let K denote the value of the integral of (9) from P to Q along C_1. Then, because of independence of path, the integral of (9) from P to Q along C_2 has the same value K, hence the value $-K$ when we reverse the direction of integration and integrate from Q to P along C_2. Together we get the value $K + (-K) = 0$ when we integrate around the closed path C from P to Q along C_1 and then back to P along C_2.

(b) Conversely, let the integral of (9) over any simple closed path in D be 0. Let P and Q be any two points in D and C_1 and C_2 any two paths in D that join P and Q and do not cross, so that together they form a simple closed path C in D. Let K be the value of the integral of (9) from P to Q along C_1. Then the integral of (9) from Q to P along C_2 must have the value $-K$, to ensure that we have $K + (-K) = 0$ as the value of the integral around the closed path C. Reversing the sense of integration along C_2, that is, integrating from P to Q along C_2, we thus obtain the value $+K$. Since P, Q, C_1 and C_2 were arbitrary in D, this proves independence of path, by definition. ∎

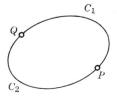

Fig. 235. Proof of Theorem 2

Criterion for Exactness and Independence of Path

To make effective practical use of the simple Theorem 1, we need a criterion by which we can decide whether the differential form under the integral sign is exact. For formulating such a criterion we shall need the following concept.

A domain D is called **simply connected** if every closed curve in D can be continuously shrunk to any point in D without leaving D.

For example, the interior of a sphere or a cube, the interior of a sphere with finitely many points removed, and the domain between two concentric spheres are simply connected, while the interior of a torus (cf. Sec. 9.5) and the interior of a cube with one space diagonal removed are not simply connected.

Theorem 3 (Criterion for exactness and independence of path)

Let $F_1(x, y, z)$, $F_2(x, y, z)$ and $F_3(x, y, z)$ be continuous functions having continuous first partial derivatives in a domain D in space. If the line integral

(10) $$\int_C (F_1 \, dx + F_2 \, dy + F_3 \, dz)$$

is independent of path in D (and, thus $F_1 \, dx + F_2 \, dy + F_3 \, dz$ is exact in D), then

(11) $$\frac{\partial F_3}{\partial y} = \frac{\partial F_2}{\partial z}, \qquad \frac{\partial F_1}{\partial z} = \frac{\partial F_3}{\partial x}, \qquad \frac{\partial F_2}{\partial x} = \frac{\partial F_1}{\partial y}$$

or, in vector language, with $\mathbf{F} = F_1\mathbf{i} + F_2\mathbf{j} + F_3\mathbf{k}$,

(11′) $$\boxed{\operatorname{curl} \mathbf{F} = \mathbf{0}}$$

everywhere in D. Conversely, if D is simply connected and (11) holds everywhere in D, then the integral (10) is independent of path in D (and, consequently, the differential form $F_1 \, dx + F_2 \, dy + F_3 \, dz$ is exact in D).

Proof. (a) Suppose that (10) is independent of path in D. Then Theorem 1 implies that the form $F_1 \, dx + F_2 \, dy + F_3 \, dz$ is exact in D and there is a function f such that

$$\mathbf{F} = F_1\mathbf{i} + F_2\mathbf{j} + F_3\mathbf{k} = \operatorname{grad} f \qquad \text{[cf. (5′)]}$$

in D. From this and (3) in Sec. 8.10 we obtain curl $\mathbf{F}$ = curl (grad f) = **0**.

   (b) Conversely, suppose that D is simply connected and (11′) holds everywhere in D. Let C be any simple closed path in D. Since D is simply connected, we can find a surface S in D bounded by C, Stokes's theorem (Sec. 9.8) is applicable, and we obtain

$$\oint_C (F_1 \, dx + F_2 \, dy + F_3 \, dz) = \oint_C \mathbf{F} \cdot \mathbf{r}' \, ds = \iint_S (\operatorname{curl} \mathbf{F}) \cdot \mathbf{n} \, dA$$

for proper direction on C and normal vector $\mathbf{n}$ on S. Now, regardless of the choice of C, the integral over S is zero because, by (11′), its integrand is identically zero. From this and Theorem 2 it follows that the integral (10) is independent of path in D. This completes the proof.   ∎

We note that in the case of a line integral in the *xy*-plane,

$$\int_C (F_1 \, dx + F_2 \, dy),$$

condition (11) reduces to the single relation

(11*) $$\frac{\partial F_2}{\partial x} = \frac{\partial F_1}{\partial y}.$$

The assumption in Theorem 3 that D be simply connected is essential and cannot be omitted. This can be seen from the following example.

EXAMPLE 2. On the assumption of simple connectedness in Theorem 3
Let

$$F_1 = -\frac{y}{x^2 + y^2}, \qquad F_2 = \frac{x}{x^2 + y^2}, \qquad F_3 = 0.$$

Differentiation shows that (11*) is satisfied in any domain of the xy-plane not containing the origin, for example, in the domain D: $\frac{1}{2} < \sqrt{x^2 + y^2} < \frac{3}{2}$ shown in Fig. 236. Clearly, D is not simply connected. If the integral

$$I = \int_C (F_1 \, dx + F_2 \, dy) = \int_C \frac{-y \, dx + x \, dy}{x^2 + y^2}$$

were independent of path in D, then $I = 0$ on any closed curve in D, for example, on the circle $x^2 + y^2 = 1$. But setting $x = r \cos \theta$, $y = r \sin \theta$, and noting that the circle is represented by $r = 1$, we easily obtain

$$I = \int_0^{2\pi} d\theta = 2\pi,$$

the integration being taken once around the circle in the counterclockwise sense. In fact, since D is not simply connected, we cannot apply Theorem 3 and conclude that I is independent of path in D. ∎

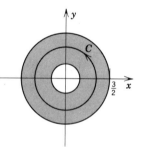

Fig. 236. Example 2

If a line integral is independent of path, it can be evaluated by (8):

EXAMPLE 3. Evaluation of a line integral in the case of independence of path
Evaluate the following integral on any path C from P: $(0, 0, 1)$ to Q: $(1, \pi/4, 2)$.

$$I = \int_C [2xyz^2 \, dx + (x^2z^2 + z \cos yz) \, dy + (2x^2yz + y \cos yz) \, dz]$$

Solution. From Theorem 3 and a little calculation it follows that the integral is independent of path in space. Since $F_1 = 2xyz^2$, from the first relation in (5) we obtain

(12) $$f = \int F_1 \, dx = x^2yz^2 + a(y, z).$$

From this and the second relation in (5) it follows that

$$\frac{\partial f}{\partial y} = x^2z^2 + \frac{\partial a}{\partial y} = F_2 = x^2z^2 + \cos yz.$$

Hence we must have

$$\frac{\partial a}{\partial y} = z \cos yz, \qquad a = \sin yz + c(z).$$

From this, the third relation in (5), and (12) we obtain

$$\frac{\partial f}{\partial z} = 2x^2 yz + y \cos yz + \frac{dc}{dz} = F_3 = 2x^2 yz + y \cos yz.$$

Hence $dc/dz = 0$, $c = const$ and, therefore,

$$f(x, y, z) = x^2 yz^2 + a = x^2 yz^2 + \sin yz + c.$$

The reader should take another look at the integral and try to find this f by inspection. Formula (8) now yields the result

$$I = [x^2 yz^2 + \sin yz + c] \,\bigg|_P^Q = \pi + \sin \frac{\pi}{2} = \pi + 1.$$

Potential Functions and Independence of Path

If a vector function $\mathbf{F}(x, y, z)$ is the gradient of some scalar function f,

(13) $$\mathbf{F} = \operatorname{grad} f,$$

then f is called a *potential function* or, briefly, **potential** (cf. Sec. 8.8), and we speak of **velocity potentials** (if $\mathbf{F} = \mathbf{v}$ is a velocity vector), **electrostatic potentials** (if $\mathbf{F}$ is an electrostatic force) and so on.

If $\mathbf{F} = \operatorname{grad} f$, then $\mathbf{F}$ and the vector field defined by $\mathbf{F}$ are called **conservative** because in this case, mechanical energy is conserved: we do no work in the displacement around a *closed* curve from a point P and back to P:

Theorem 4 (Conservation of mechanical energy)

If $\mathbf{F}$ is conservative, that is, $\mathbf{F} = \operatorname{grad} f$, in a domain D in space, then

$$\int_C \mathbf{F} \cdot \mathbf{n} \, ds$$

is independent of path in D. Hence this integral is 0 around closed paths in D. Physically, this means that in a conservative field, the work in a displacement from a point P and back to P is 0, so that in this process the mechanical energy is conserved.

Proof. This follows from Theorem 3 and curl $\mathbf{F} =$ curl (grad f) = $\mathbf{0}$; cf. (3), Sec. 8.10. ∎

Physically, the kinetic energy of a body B can be interpreted as the ability of B to do work by virtue of its motion, and if B moves in a conservative field of force, after the completion of a round trip the body B will return to its initial position with the same kinetic energy it had originally. For instance, the gravitational force is conservative; if we throw a ball vertically up, it will (if we assume air resistance to be negligible) return to our hand with the same kinetic energy it had when it left our hand.

Friction, air resistance or water resistance always act against the direction of motion, tending to diminish the total mechanical energy of a system (usually converting it into heat or mechanical energy of the surrounding medium, or both), and if in the motion of a body B, these forces are so large that they can no longer be neglected, then the resultant $\mathbf{F}$ of the forces acting on B is no longer conservative. Quite generally, a physical system is called **conservative** if all the forces acting in it are conservative; otherwise it is called **nonconservative** or **dissipative.**

Example 3 illustrates how we can find a potential of a conservative vector $\mathbf{F}$ by integration. Let us add another example:

EXAMPLE 4. (Determination of a potential)

Is the field given by $\mathbf{F} = 2xyz\mathbf{i} + (e^y + x^2z)\mathbf{j} + x^2y\mathbf{k}$ conservative? If it is, find a potential.

Solution. curl $\mathbf{F} = (x^2 - x^2)\mathbf{i} + (2xy - 2xy)\mathbf{j} + (2xz - 2xz)\mathbf{k} = \mathbf{0}$ by direct calculation. Verify this. Hence the answer is yes.

We write $\mathbf{F} = [F_1, F_2, F_3]$ as usual and, as in Example 3, find f such that we obtain $\mathbf{F} = \text{grad } f = [f_x, f_y, f_z]$ by integration; here $f_x = \partial f/\partial x$, etc. We integrate

$$F_1 = 2xyz = f_x \qquad \text{to get} \qquad f = x^2yz + g(y, z),$$

and determine g by equating the other two components, first

$$F_2 = e^y + x^2z = f_y = x^2z + g_y, \qquad \text{hence} \qquad g_y = e^y, \quad g = e^y + h(z).$$

We now have $f = x^2yz + e^y + h(z)$ and determine h from

$$F_3 = x^2y = f_z = x^2y + 0 + h'(z), \qquad \text{hence} \qquad h'(z) = 0, \quad h = c = const.$$

This shows that f is uniquely determined, except for an additive constant c, which we can take 0. Thus

$$f = x^2yz + e^y + c.$$

Check that grad $f = \mathbf{F}$. ∎

This is the end of Chap. 9 as well as of Part B on linear algebra and vector calculus. Part C consists of two chapters and centers around **partial differential equations** (Chap. 11), for which **Fourier series** and integrals (Chap. 10) are basic tools of solution.

Problems for Sec. 9.9

Are the following differential forms exact? In the case of exactness, find f such that the form equals df.

1. $2xy\,dx + x^2\,dy + 3z^2\,dz$
2. $4x^3y\,dx + x^4\,dy$
3. $x^3\,dx + y\,dy + \cos z\,dz$
4. $\sin z\,dy - y\cos z\,dx$
5. $ye^{xy}\,dx + x^2e^{xy}\,dy + dz$
6. $2xy\,dx + (x^2 - 2yz)\,dy - y^2\,dz$
7. $z\,dx - z\,dy + (x - y)\,dz$
8. $e^{xyz}(yz\,dx + xz\,dy + xy\,dz)$
9. $e^z(y\,dx + x\,dy - xy\,dz)$
10. $xz^2\,dx - y\,dy + x^2z\,dz$
11. $x\,dx + (y - z)\,dy + (z - y)\,dz$
12. $\sin(x - y)\,dy - \cos(x - y)\,dz$
13. $(e^y + ze^x)\,dx + xe^y\,dy + e^x\,dz$
14. $\cos z\,dx - x\sin z\,dz$
15. $x^{-2}(y + z)\,dx - x^{-1}(dy + dz)$
16. $(y/z)\,dx + (x/z)\,dy - (xy/z^2)\,dz$
17. $ye^z\,dy + ze^y\,dz$
18. $y\cos xy\,dx + x\cos xy\,dy - dz$
19. $\sin^2 y\,dx + x\sin 2y\,dy$
20. $e^z\,dx + e^y\,dy + e^x\,dz$

Show that the form under the integral sign is exact and evaluate:

21. $\int_{(0,0,0)}^{(1,1,1)} (2\,dx + 2y\,dy + dz)$

22. $\int_{(1,0,1)}^{(4,3,2)} (z\,dx + x\,dz)$

23. $\int_{(2,3,3)}^{(0,5,5)} (e^y\,dy - e^z\,dz)$

24. $\int_{(0,-1,-1)}^{(\pi,3,2)} (\cos x\,dx + z\,dy + y\,dz)$

25. $\int_{(2,1,1)}^{(4,1,0)} (2xy^3\,dx + 3x^2y^2\,dy)$

26. $\int_{(1,3,2)}^{(4,0,1)} [z\,dx - z\,dy + (x - y)\,dz]$

27. $\int_{(1,0,2)}^{(8,2,0)} (\cosh^2 z\,dy + y\sinh 2z\,dz)$

28. $\int_{(2,1,0)}^{(0,1,2)} (ze^{xz}\,dx - dy + xe^{xz}\,dz)$

29. $\int_{(0,0,0)}^{(1,4,2)} (3y^2e^z\,dy + y^3e^z\,dz)$

30. $\int_{(2,1,\pi/2)}^{(2,1,\pi)} (\sin^2 z\,dx + x\sin 2z\,dz)$

Review Problems for Chapter 9

1. What is a smooth curve? A piecewise smooth curve? Are all curves in applications smooth?

2. Why is Theorem 1 in Sec. 9.1 essential in physical applications?

3. Why is independence of path significant in physical applications?

4. How did we use Green's theorem in the plane in the proof of the divergence theorem?

5. Which are the two most important theorems in this chapter? What are they good for? State them from memory as best as you can.

6. What is a smooth surface? A piecewise smooth surface?

7. Are all surfaces in applications smooth, or can you think of some that are merely piecewise smooth?

8. What do we mean by an oriented curve? An oriented surface?

9. How many possibilities of orienting a surface at one of its points do we have? Explain; give an example.

10. Does a square satisfy the conditions in Green's theorem in the plane? Can you think of a curve C that does not satisfy those conditions?

11. State the definition of the divergence of a vector function and explain where it played a role in this chapter.

12. State the definition of the curl of a vector function and explain its significance in this chapter.

13. Our line and surface integrals were made up in terms of a *vector* function **F**, but their integrands are actually *scalar* functions. How did we accomplish this? Why did we proceed in this fashion (which at first sight looks somewhat like a detour)?

14. State Laplace's equation and name some areas in physics in which this equation is basic.

15. What is a harmonic function? What facts about harmonic functions did we discover in this chapter?

Evaluate $\int_C \mathbf{F}(\mathbf{r}) \cdot d\mathbf{r}$ with the force **F** and the curve C as follows.
(*Hint.* In some of these problems, you may save work by using exactness or applying Stokes's theorem.)

16. $\mathbf{F} = x^2\mathbf{i} - 2y^2\mathbf{j}$, C the straight line segment from $(0, 0)$ to $(2, 2)$

17. $\mathbf{F} = y^2\mathbf{i} - x\mathbf{j}$, C as in Prob. 16

Evaluation of $\int_C \mathbf{F(r)} \cdot d\mathbf{r}$ (*continued*)

18. $\mathbf{F} = 3y\mathbf{i} + 6x\mathbf{j}$, C the boundary of $0 \leq x \leq 1, 0 \leq y \leq 1$, counterclockwise

19. $\mathbf{F} = xy^2\mathbf{i} + x^2y\mathbf{j}$, C as in Prob. 18

20. $\mathbf{F} = z\mathbf{i} + 2y\mathbf{j} + x\mathbf{k}$, $C: x^2 + y^2 = 9$, $z = 2$, counterclockwise as seen by a person standing at the origin

21. $\mathbf{F} = z\mathbf{i} + x\mathbf{j} + x\mathbf{k}$, C as in Prob. 20

22. $\mathbf{F} = xy^3\mathbf{i} - e^{xy}z\mathbf{k}$, C as in Prob. 20

23. $\mathbf{F} = xy\mathbf{i} + z^2\mathbf{j}$, $C: y = x^2$, $z = x$ from $(1, 1, 1)$ to $(2, 4, 2)$

24. $\mathbf{F} = ye^x\mathbf{i} + e^x\mathbf{j}$, C as in Prob. 23

25. $\mathbf{F} = 3x^2z\mathbf{i} + x^3\mathbf{k}$, $C: y = x^2, z = x^3$ from $(0, 0, 0)$ to $(2, 4, 8)$

26. $\mathbf{F} = z\mathbf{i} + y^2\mathbf{j} + x^3\mathbf{k}$, C as in Prob. 25

27. $\mathbf{F} = \sin \pi x \, \mathbf{i} + z\mathbf{j}$, C the boundary of the triangle with vertices $(0, 0, 0)$, $(1, 0, 0)$, $(1, 1, 0)$, counterclockwise

28. $\mathbf{F} = \sinh x \, \mathbf{i} + (x + \cosh y)\mathbf{j}$, C as in Prob. 27

29. $\mathbf{F} = 3x^2y^2\mathbf{i} + 2x^3y\mathbf{j}$, C as in Prob. 27

30. $\mathbf{F} = y\mathbf{i} - x\mathbf{j}$, C as in Prob. 18

31. $\mathbf{F} = e^{xyz}(yz\mathbf{i} + xz\mathbf{j} + xy\mathbf{k})$, C as in Prob. 18

32. $\mathbf{F} = y\mathbf{i} + (x + z)\mathbf{j} + x^5\mathbf{k}$, $C: x^2 + y^2 = 1, z = x$ from $(1, 0, 1)$ to $(0, 1, 0)$, shortest way

33. $\mathbf{F} = ye^{xy}\mathbf{i} + xe^{xy}\mathbf{j} + \sinh z \, \mathbf{k}$, C as in Prob. 32

34. $\mathbf{F} = 2xy^3\mathbf{i} + 3x^2y^2\mathbf{j}$, $C: x^2 + y^2 = 1, z = 0$, counterclockwise

35. $\mathbf{F} = 2xy\mathbf{i} + (e^x + x^2)\mathbf{j}$, C as in Prob. 27

36. $\mathbf{F} = z\mathbf{i} + 2x\mathbf{j} + 3y\mathbf{k}$, C as in Prob. 18

37. $\mathbf{F} = x^3\mathbf{i} + e^{3y}\mathbf{j} - e^{-3z}\mathbf{k}$, $C: y^2 + z^2 = 25, x = -1$, counterclockwise as seen by a person standing at the origin

38. $\mathbf{F} = 4(z^2\mathbf{i} + x^2\mathbf{j} + y^2\mathbf{k})$, C the boundary of $0 \leq x \leq 1, 0 \leq y \leq 1, z = 1$, clockwise as seen by a person standing at the origin

39. $\mathbf{F} = z\mathbf{i} + x\mathbf{j} + y\mathbf{k}$, C the ellipse $25x^2 + y^2 = 25, z = 0$, counterclockwise

40. $\mathbf{F} = \cosh yz \, \mathbf{i} + xz \sinh yz \, \mathbf{j} + xy \sinh yz \, \mathbf{k}$, C any path from $(0, 1, 3)$ to $(9, 8, 4)$

Evaluate the following double integrals.

41. $\int_0^1 \int_0^{2\pi} (1 + x^2 + y^2) \, dx \, dy$

42. $\int_0^2 \int_0^x e^{x+y} \, dy \, dx$

43. $\int_0^1 \int_{x^2}^x (1 - xy) \, dy \, dx$

44. $\int_0^{\pi/2} \int_{-1}^1 x^2y^2 \, dx \, dy$

45. $\int_{-1}^2 \int_{x^2}^{x+2} (x + y) \, dy \, dx$

Find the coordinates $\bar{x}$, $\bar{y}$ of the center of gravity of a mass of density $f(x, y)$ in a region R, where

46. $f = 1$, $R: 0 \leq y \leq 1 - x^2$

47. $f = 1$, $R: 0 \leq y \leq \sin x, 0 \leq x \leq \pi$

48. $f = 1$, $R: x^2 + y^2 \leq a^2, y \geq 0$

49. $f = xy$, $R: 0 \leq y \leq x, 0 \leq x \leq 1$

50. $f = e^{x+y}$, $R: 0 \leq y = 1 - x, 0 \leq x \leq 1$

Represent the following surfaces in parametric form $\mathbf{r}(u, v)$ and find a normal vector.

51. $3x + 2y + z = 1$ **52.** $y^2 + 9z^2 = 9$

53. $y = \sqrt{x^2 + z^2}$ **54.** $y^2 + z^2 = 9$

55. $z = x^2$ **56.** $x = 2\sqrt{y^2 + z^2}$

57. $x^2 - 4y^2 = 4$ **58.** $x = 2(y^2 + z^2)$

59. $z = \sqrt{4x^2 + y^2}$ **60.** $x^2 + 4y^2 + z^2 = 4$

Evaluate $\displaystyle\iint\limits_{S} \mathbf{F} \cdot \mathbf{n} \, dA$, where

61. $\mathbf{F} = x^2\mathbf{i} + 2y\mathbf{j},$ $S: z = 1 - x - y,$ $x \geqq 0,$ $y \geqq 0,$ $z \geqq 0$

62. $\mathbf{F} = y\mathbf{i} - x\mathbf{j},$ S the first-octant section of the plane in Prob. 51

63. $\mathbf{F} = y^2\mathbf{i} + x^2\mathbf{j} + z^2\mathbf{k},$ $S: x^2 + y^2 = 16,$ $x \geqq 0,$ $y \geqq 0,$ $0 \leqq z \leqq 5$

64. $\mathbf{F} = x\mathbf{i} + xy\mathbf{j} + e^{yz}\mathbf{k},$ $S: x^2 + y^2 = 1,$ $1 \leqq z \leqq 3$

65. $\mathbf{F} = xyz\mathbf{i} + 3z^2\mathbf{j} + \mathbf{k},$ $S: z = x^2,$ $0 \leqq x \leqq 2,$ $3 \leqq y \leqq 5$

66. $\mathbf{F} = \mathbf{i} + \mathbf{j} + z^2\mathbf{k},$ $S: z = \sqrt{x^2 + y^2},$ $0 \leqq z \leqq 4$

67. $\mathbf{F} = \sin^2 x \, \mathbf{i} - y \sin 2x \, \mathbf{j} + 5z\mathbf{k},$ S the surface of $|x| \leqq 1,$ $|y| \leqq 1,$ $|z| \leqq 1$

68. $\mathbf{F} = 3x^2\mathbf{i} - (6xy + 4yz)\mathbf{j} + 2z(z - 1)\mathbf{k},$ $S: x^2 + y^2 + z^2 = 9$

69. $\mathbf{F} = z\mathbf{i} - z^4\mathbf{k},$ $S: x^2 + 4y^2 + z^2 = 4,$ $x \geqq 0,$ $y \geqq 0,$ $z \geqq 0$

70. $\mathbf{F} = 3z^2x\mathbf{i} - (x^2 + 1)y\mathbf{j} + (x^2 - z^2)z\mathbf{k},$ S the ellipsoid in Prob. 60

Summary of Chapter 9
Line and Surface Integrals.
Integral Theorems

Chapter 8 was concerned with the extension of *differential* calculus to vectors, that is, to vector functions $\mathbf{v}(x, y, z)$ or $\mathbf{v}(t)$, and Chap. 9 extended *integral* calculus to vector functions. This involves *line integrals* (Sec. 9.1), *double integrals* (Sec. 9.2), *surface integrals* (Sec. 9.5) and *triple integrals* (Sec. 9.6) and the three "big" theorems for transforming these integral into one another, the theorems by Green (Sec. 9.3), Gauss (Sec. 9.6) and Stokes (Sec. 9.8).

The analog of the definite integral of calculus is the **line integral**

$$(1) \quad \int_{C} \mathbf{F}(\mathbf{r}) \cdot d\mathbf{r} = \int_{C} (F_1 \, dx + F_2 \, dy + F_3 \, dz) = \int_{a}^{b} \mathbf{F}(\mathbf{r}(t)) \cdot \frac{d\mathbf{r}}{dt} \, dt$$

where $C: \mathbf{r}(t) = x(t)\mathbf{i} + y(t)\mathbf{j} + z(t)\mathbf{k}$ $(a \leqq t \leqq b)$ is a curve in space (or in the plane). Cf. Sec. 9.1.

Double integrals over a region R in the plane (Sec. 9.2) can be transformed into line integrals over the boundary C of R (and conversely) by **Green's theorem in the plane** (Sec. 9.3)

$$(2) \qquad \iint_R \left(\frac{\partial F_2}{\partial x} - \frac{\partial F_1}{\partial y} \right) dx\, dy = \oint_C (F_1\, dx + F_2\, dy).$$

Other forms of this are given in Sec. 9.3. This theorem is also needed as the essential tool in proving the divergence theorem and Stokes's theorem (see below), and is a special case of the latter.

Triple integrals (Sec. 9.6) taken over a region T in space can be transformed into **surface integrals** (Sec. 9.5) over the boundary surface S of T (and conversely) by the **divergence theorem of Gauss** (Sec. 9.6)

$$(3) \qquad \iiint_T \operatorname{div} \mathbf{F}\, dV = \iint_S \mathbf{F} \cdot \mathbf{n}\, dA$$

($\mathbf{n}$ the outer unit normal vector to S). Among other things, this theorem implies **Green's formulas** (Sec. 9.7)

$$(4) \qquad \iiint_T (f\nabla^2 g + \nabla f \cdot \nabla g)\, dV = \iint_S f\, \frac{\partial g}{\partial n}\, dA$$

($\partial g/\partial n$ the directional derivative of g in the direction of $\mathbf{n}$) and

$$(5) \qquad \iiint_T (f\nabla^2 g - g\nabla^2 f)\, dV = \iint_S \left(f\, \frac{\partial g}{\partial n} - g\, \frac{\partial f}{\partial n} \right) dA.$$

Surface integrals over a surface S with boundary curve C can be transformed into line integrals over C (and conversely) by **Stokes's theorem** (Sec. 9.8)

$$(6) \qquad \iint_S (\operatorname{curl} \mathbf{F}) \cdot \mathbf{n}\, dA = \oint_C \mathbf{F} \cdot \mathbf{r}'(s)\, ds.$$

Independence of path of a line integral in a region D means that the integral of a given function over any path C with endpoints P and Q has the same value for all paths from P to Q that lie in D; here P and Q are fixed. An integral

$$(7) \qquad \int_C (F_1\, dx + F_2\, dy + F_3\, dz)$$

is independent of path in D if and only if the differential form $F_1\, dx + F_2\, dy + F_3\, dz$ is exact in D (Sec. 9.9). Also, if curl $\mathbf{F} = \mathbf{0}$, where $\mathbf{F} = [F_1, F_2, F_3]$, has continuous first partial derivatives in a *simply connected* domain D, then the integral (7) is independent of path in D (Sec. 9.9). The proof of this uses Stokes's theorem.

PART C

FOURIER ANALYSIS AND PARTIAL DIFFERENTIAL EQUATIONS

Chapter 10   Fourier Series, Fourier Integrals,
             Fourier Transforms
Chapter 11   Partial Differential Equations

Periodic phenomena occur quite frequently in physics and its engineering applications, and it is an important practical problem to represent the corresponding periodic functions in terms of simple periodic functions such as sine and cosine. This leads to **Fourier series,** whose terms are sine and cosine functions. Their introduction by Fourier (after work by Euler and Daniel Bernoulli) was one of the most important events in the development of applied mathematics. Chapter 10 is primarily concerned with Fourier series. The corresponding ideas and techniques can also be extended to nonperiodic phenomena. This leads to **Fourier integrals** and **Fourier transforms** (Secs. 10.9—10.11), and a common name for the whole area is **Fourier analysis.**

Chapter 11 is concerned with the most important **partial differential equations** of physics and engineering. In this area, Fourier analysis has its most important applications, as a basic tool in solving boundary and initial value problems in mechanics, heat flow, electrostatics and other fields.

Chapter 10

Fourier Series, Fourier Integrals, Fourier Transforms

Fourier series[1] (Sec. 10.2) are series of cosine and sine terms and arise in the important practical task of representing general periodic functions. They constitute a very important tool in solving problems that involve ordinary and partial differential equations.

In the present chapter we discuss basic concepts, facts and techniques in connection with Fourier series. Illustrative examples and some important engineering applications will be included. Further applications will be shown in the next chapter on partial differential equations and initial and boundary value problems.

The *theory* of Fourier series is rather complicated, but the *application* of these series is simple. Fourier series are, in a certain sense, more universal than Taylor series, because many *discontinuous* periodic functions of practical interest can be developed in Fourier series, but, of course, do not have Taylor series representations.

The last three sections of this chapter concern **Fourier integrals** and **Fourier transformations,** which extend the ideas and techniques of Fourier series to nonperiodic functions defined for all *x*. Corresponding applications to partial differential equations will be considered in the next chapter (Sec. 11.14).

Prerequisite for this chapter: elementary integral calculus.
Sections that may be omitted in a shorter course: 10.6—10.11.
References: Appendix 1, Part C.
Answers to problems: Appendix 2.

---

[1]JEAN-BAPTISTE JOSEPH FOURIER (1768—1830), French physicist and mathematician, lived and taught in Paris, accompanied Napoleon to Egypt and was later made prefect of Grenoble. He utilized Fourier series in his main work *Théorie analytique de la chaleur* (*Analytic Theory of Heat,* Paris 1822) in which he developed the theory of heat conduction (heat equation, cf. Sec. 11.5). These new series became a most important tool in mathematical physics and also had considerable influence on the further development of mathematics itself; see Ref. [13] in Appendix 1.

10.1 Periodic Functions. Trigonometric Series

A function $f(x)$ is said to be **periodic** if it is defined for all real x and if there is some positive number p such that

(1) $$f(x + p) = f(x)$$ for all x.

This number p is then called a **period**[2] of $f(x)$. The graph of such a function is obtained by periodic repetition of its graph in any interval of length p (Fig. 237). Periodic phenomena and functions occur in many applications.

Familiar examples of periodic functions are the sine and cosine functions, and we note that the function $f = c = const$ is also a periodic function in the sense of the definition, because it satisfies (1) for every positive p. Examples of functions that are *not* periodic are x, x^2, x^3, e^x and $\ln x$, to mention just a few.

From (1) we have $f(x + 2p) = f[(x + p) + p] = f(x + p) = f(x)$, etc., and for any integer n,

(2) $$f(x + np) = f(x)$$ for all x.

Hence $2p, 3p, 4p, \cdots$ are also periods of $f(x)$. Furthermore, if $f(x)$ and $g(x)$ have period p, then the function

$$h(x) = af(x) + bg(x)$$ $(a, b$ constant$)$

also has the period p.

Our problem in the first few sections of this chapter will be the representation of various functions of period $p = 2\pi$ in terms of the simple functions

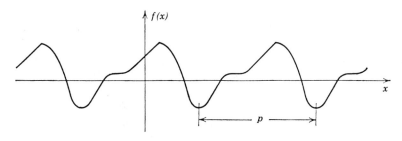

Fig. 237. Periodic function

[2]If a periodic function $f(x)$ has a smallest period p (> 0), this is often called the **primitive period** of $f(x)$. For example, the primitive periods of $\sin x$ and $\sin 2x$ are 2π and π, respectively. Examples of periodic functions without primitive period are $f = const$ and $f(x) = 0$ (x rational), $f(x) = 1$ otherwise.

(3) $1, \quad \cos x, \quad \sin x, \quad \cos 2x, \quad \sin 2x, \cdots, \quad \cos nx, \quad \sin nx, \cdots,$

which have the period 2π (Fig. 238). The series that will arise in this connection will be of the form

(4) $a_0 + a_1 \cos x + b_1 \sin x + a_2 \cos 2x + b_2 \sin 2x + \cdots,$

where $a_0, a_1, a_2, \cdots, b_1, b_2, \cdots$ are real constants. Such a series is called a **trigonometric series,** and the a_n and b_n are called the **coefficients** of the series. Using the summation sign,[3] we may write this series

(4)
$$a_0 + \sum_{n=1}^{\infty} (a_n \cos nx + b_n \sin nx).$$

The set of functions (3) from which we have made up the series (4) is often called the **trigonometric system,** to have a short name for it.

We see that each term of the series (4) has the period 2π. Hence *if the series (4) converges, its sum will be a function of period 2π.*

Periodic functions that occur in engineering problems are often rather complicated, and it is desirable to represent these functions in terms of simple periodic functions. We shall see that almost any periodic function $f(x)$ of period 2π that appears in applications—for example, in connection with vibrations—can be represented by a trigonometric series, and in the next section we shall derive formulas for the coefficients in (4) in terms of $f(x)$ such that (4) converges and has the sum $f(x)$. In Sec. 10.3 we shall extend our results to functions of arbitrary period; this extension will turn out to be quite simple.

In the next section we define **Fourier series,** the most important concept in the whole chapter. These are trigonometric series with coefficients obtained from a given function by integration [by the **Euler formulas** (6) in the next section].

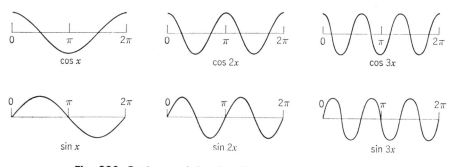

Fig. 238. Cosine and sine functions having the period 2π

---

[3] And inserting parentheses, which yields from a convergent series a convergent series having the same sum.

Problems for Sec. 10.1

Find the smallest positive period p of the following functions.

1. $\cos x$, $\sin x$, $\cos 2x$, $\sin 2x$, $\cos \pi x$, $\sin \pi x$, $\cos 2\pi x$, $\sin 2\pi x$

2. $\cos nx$, $\sin nx$, $\cos \dfrac{2\pi x}{k}$, $\sin \dfrac{2\pi x}{k}$, $\cos \dfrac{2\pi n x}{k}$, $\sin \dfrac{2\pi n x}{k}$

3. If $f(x)$ and $g(x)$ have period p, show that $h = af + bg$ (a, b constant) has the period p. Thus all functions of period p form a vector space.

4. If p is a period of $f(x)$, show that np, $n = 2, 3, \cdots$ is a period of $f(x)$.

5. Show that the function $f(x) = const$ is a periodic function of period p for every positive p.

6. If $f(x)$ is a periodic function of x of period p, show that $f(ax)$, $a \neq 0$, is a periodic function of x of period p/a, and $f(x/b)$, $b \neq 0$, is a periodic function of x of period bp. Verify these results for $f(x) = \cos x$, $a = b = 2$.

Graph the following functions $f(x)$, which are assumed to be periodic of period 2π and, for $-\pi < x < \pi$, are given by the formulas

7. $f(x) = x$ **8.** $f(x) = |x|$

9. $f(x) = |x|^3$ **10.** $f(x) = 2x - x^2$

11. $f(x) = \sin x/2$ **12.** $f(x) = |\cos 2x|$

13. $f(x) = 1 - e^{-|x|}$

14. $f(x) = \begin{cases} x & \text{if} \quad -\pi < x < 0 \\ 0 & \text{if} \quad 0 < x < \pi \end{cases}$ **15.** $f(x) = \begin{cases} 0 & \text{if} \quad -\pi < x < 0 \\ \sin x & \text{if} \quad 0 < x < \pi \end{cases}$

16. $f(x) = \begin{cases} \pi + x & \text{if} \quad -\pi < x < 0 \\ \pi - x & \text{if} \quad 0 < x < \pi \end{cases}$ **17.** $f(x) = \begin{cases} 1 & \text{if} \quad -\pi < x < 0 \\ \cos^2 2x & \text{if} \quad 0 < x < \pi \end{cases}$

Plot accurate graphs of the following functions.

18. $\sin x$, $\sin x + \frac{1}{3}\sin 3x$, $\sin x + \frac{1}{3}\sin 3x + \frac{1}{5}\sin 5x$,

$$f(x) = \begin{cases} -\pi/4 & \text{when} \quad -\pi < x < 0 \\ \pi/4 & \text{when} \quad 0 < x < \pi \end{cases} \quad \text{and} \quad f(x + 2\pi) = f(x)$$

19. $\sin 2\pi x$, $\sin 2\pi x + \frac{1}{3}\sin 6\pi x$, $\sin 2\pi x + \frac{1}{3}\sin 6\pi x + \frac{1}{5}\sin 10\pi x$

20. $\sin x$, $\sin x - \frac{1}{2}\sin 2x$, $\sin x - \frac{1}{2}\sin 2x + \frac{1}{3}\sin 3x$,

$f(x) = x/2$ when $-\pi < x < \pi$ and $f(x + 2\pi) = f(x)$

21. $-\cos x$, $-\cos x + \frac{1}{4}\cos 2x$, $-\cos x + \frac{1}{4}\cos 2x - \frac{1}{9}\cos 3x$,

$f(x) = x^2/4 - \pi^2/12$ when $-\pi < x < \pi$ and $f(x + 2\pi) = f(x)$

Evaluate the following integrals where $n = 0, 1, 2, \cdots$. (These are typical examples of integrals that will be needed in our further work.)

22. $\displaystyle\int_0^{\pi/2} \cos nx \, dx$ **23.** $\displaystyle\int_{-\pi/2}^{\pi/2} x \cos nx \, dx$ **24.** $\displaystyle\int_{-\pi/2}^{\pi/2} x \sin nx \, dx$

25. $\displaystyle\int_0^{\pi} \sin nx \, dx$ **26.** $\displaystyle\int_0^{\pi} x \sin nx \, dx$ **27.** $\displaystyle\int_0^{\pi} e^x \cos nx \, dx$

28. $\displaystyle\int_{-\pi}^{\pi} x \sin nx \, dx$ **29.** $\displaystyle\int_{-\pi}^{0} e^x \sin nx \, dx$ **30.** $\displaystyle\int_0^{\pi} x^2 \cos nx \, dx$

10.2 Fourier Series

Fourier series arise from the task of representing a given periodic function $f(x)$ by a trigonometric series. The Fourier series of $f(x)$ is a trigonometric series whose coefficients are determined from $f(x)$ by certain formulas ["Euler formulas" (6), below] which we shall derive first. Later in this section we take a look at the theory of Fourier series.

Euler Formulas for the Fourier Coefficients

Let us assume that $f(x)$ is a periodic function of period 2π that can be **represented** by a trigonometric series,

$$(1) \qquad f(x) = a_0 + \sum_{n=1}^{\infty} (a_n \cos nx + b_n \sin nx);$$

that is, we assume that this series converges and has $f(x)$ as its sum. Given such a function $f(x)$, we want to determine the coefficients a_n and b_n of the corresponding series (1).

We first determine a_0. Integrating on both sides of (1) from $-\pi$ to π, we have

$$\int_{-\pi}^{\pi} f(x)\, dx = \int_{-\pi}^{\pi} \left[a_0 + \sum_{n=1}^{\infty} (a_n \cos nx + b_n \sin nx) \right] dx.$$

If term-by-term integration of the series is allowed,[4] then we obtain

$$\int_{-\pi}^{\pi} f(x)\, dx = a_0 \int_{-\pi}^{\pi} dx + \sum_{n=1}^{\infty} \left(a_n \int_{-\pi}^{\pi} \cos nx\, dx + b_n \int_{-\pi}^{\pi} \sin nx\, dx \right).$$

The first term on the right equals $2\pi a_0$. All the other integrals on the right are zero, as can be readily seen by performing the integrations. Hence our first result is

$$(2) \qquad a_0 = \frac{1}{2\pi} \int_{-\pi}^{\pi} f(x)\, dx.$$

We now determine $a_1, a_2, \cdots$ by a similar procedure. We multiply (1) by $\cos mx$, where m is any fixed positive integer, and integrate from $-\pi$ to π:

$$(3) \quad \int_{-\pi}^{\pi} f(x) \cos mx\, dx = \int_{-\pi}^{\pi} \left[a_0 + \sum_{n=1}^{\infty} (a_n \cos nx + b_n \sin nx) \right] \cos mx\, dx.$$

---

[4]This is justified, for instance, in the case of uniform convergence (cf. Theorem 3 in Sec. 14.8).

Integrating term by term, we see that the right-hand side becomes

$$a_0 \int_{-\pi}^{\pi} \cos mx \, dx + \sum_{n=1}^{\infty} \left[a_n \int_{-\pi}^{\pi} \cos nx \cos mx \, dx + b_n \int_{-\pi}^{\pi} \sin nx \cos mx \, dx \right].$$

The first integral is zero. By applying (11) in Appendix 3.1 we obtain

$$\int_{-\pi}^{\pi} \cos nx \cos mx \, dx = \frac{1}{2} \int_{-\pi}^{\pi} \cos (n + m)x \, dx + \frac{1}{2} \int_{-\pi}^{\pi} \cos (n - m)x \, dx,$$

$$\int_{-\pi}^{\pi} \sin nx \cos mx \, dx = \frac{1}{2} \int_{-\pi}^{\pi} \sin (n + m)x \, dx + \frac{1}{2} \int_{-\pi}^{\pi} \sin (n - m)x \, dx.$$

Integration shows that the four terms on the right are zero, except for the last term in the first line, which equals π when $n = m$. Since in (3) this term is multiplied by a_m, the right-hand side in (3) is equal to $a_m \pi$, and our second result is

(4) $$a_m = \frac{1}{\pi} \int_{-\pi}^{\pi} f(x) \cos mx \, dx, \qquad m = 1, 2, \cdots.$$

We finally determine $b_1, b_2, \cdots$ in (1). If we multiply (1) by $\sin mx$, where m is any fixed positive integer, and then integrate from $-\pi$ to π, we have

(5) $$\int_{-\pi}^{\pi} f(x) \sin mx \, dx = \int_{-\pi}^{\pi} \left[a_0 + \sum_{n=1}^{\infty} (a_n \cos nx + b_n \sin nx) \right] \sin mx \, dx.$$

Integrating term by term, we see that the right-hand side becomes

$$a_0 \int_{-\pi}^{\pi} \sin mx \, dx + \sum_{n=1}^{\infty} \left[a_n \int_{-\pi}^{\pi} \cos nx \sin mx \, dx + b_n \int_{-\pi}^{\pi} \sin nx \sin mx \, dx \right].$$

The first integral is zero. The next integral is of the type considered before, and we know that it is zero for all $n = 1, 2, \cdots$. For the last integral we obtain

$$\int_{-\pi}^{\pi} \sin nx \sin mx \, dx = \frac{1}{2} \int_{-\pi}^{\pi} \cos (n - m)x \, dx - \frac{1}{2} \int_{-\pi}^{\pi} \cos (n + m)x \, dx.$$

The last term is zero. The first term on the right is zero when $n \neq m$ and is π when $n = m$. Since in (5) this term is multiplied by b_m, the right-hand side in (5) is equal to $b_m \pi$, and our last result is

$$b_m = \frac{1}{\pi} \int_{-\pi}^{\pi} f(x) \sin mx \, dx, \qquad m = 1, 2, \cdots,$$

Writing n in place of m, we altogether have the so-called **Euler formulas**[5]

(6)

$$
\textbf{(a)} \qquad a_0 = \frac{1}{2\pi} \int_{-\pi}^{\pi} f(x) \, dx
$$

$$
\textbf{(b)} \qquad a_n = \frac{1}{\pi} \int_{-\pi}^{\pi} f(x) \cos nx \, dx \qquad n = 1, 2, \cdots,
$$

$$
\textbf{(c)} \qquad b_n = \frac{1}{\pi} \int_{-\pi}^{\pi} f(x) \sin nx \, dx \qquad n = 1, 2, \cdots.
$$

These numbers given by (6) are called the **Fourier coefficients** of $f(x)$. The trigonometric series

(7)

$$
a_0 + \sum_{n=1}^{\infty} (a_n \cos nx + b_n \sin nx)
$$

with coefficients given by (6) is called the **Fourier series** of $f(x)$ (regardless of convergence—we discuss this below).

EXAMPLE 1. Square wave

Find the Fourier coefficients of the periodic function $f(x)$ in Fig. 239a. The analytic representation is

$$
f(x) = \begin{cases} -k & \text{if} & -\pi < x < 0 \\ k & \text{if} & 0 < x < \pi \end{cases} \qquad \text{and} \qquad f(x + 2\pi) = f(x).
$$

Functions of this type may occur as external forces acting on mechanical systems, electromotive forces in electric circuits, etc.

Solution. From (6a) we obtain $a_0 = 0$. This can also be seen without integration, since the area under the curve of $f(x)$ between $-\pi$ and π is zero. From (6b),

$$
a_n = \frac{1}{\pi} \int_{-\pi}^{\pi} f(x) \cos nx \, dx = \frac{1}{\pi} \left[\int_{-\pi}^{0} (-k) \cos nx \, dx + \int_{0}^{\pi} k \cos nx \, dx \right]
$$

$$
= \frac{1}{\pi} \left[-k \frac{\sin nx}{n} \bigg|_{-\pi}^{0} + k \frac{\sin nx}{n} \bigg|_{0}^{\pi} \right] = 0
$$

because $\sin nx = 0$ at $-\pi$, 0 and π for all $n = 1, 2, \cdots$. Similarly, from (6c) we obtain

$$
b_n = \frac{1}{\pi} \int_{-\pi}^{\pi} f(x) \sin nx \, dx = \frac{1}{\pi} \left[\int_{-\pi}^{0} (-k) \sin nx \, dx + \int_{0}^{\pi} k \sin nx \, dx \right]
$$

$$
= \frac{1}{\pi} \left[k \frac{\cos nx}{n} \bigg|_{-\pi}^{0} - k \frac{\cos nx}{n} \bigg|_{0}^{\pi} \right].
$$

Since $\cos(-\alpha) = \cos \alpha$ and $\cos 0 = 1$, this yields

$$
b_n = \frac{k}{n\pi} [\cos 0 - \cos(-n\pi) - \cos n\pi + \cos 0] = \frac{2k}{n\pi} (1 - \cos n\pi).
$$

---

[5]Students familiar with Sec. 4.7 may have noticed that in this derivation of (6) we merely used the *orthogonality* of the trigonometric system (3), Sec. 10.1, on an interval of length 2π.

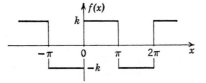

(a) The given function $f(x)$ (Periodic square wave)

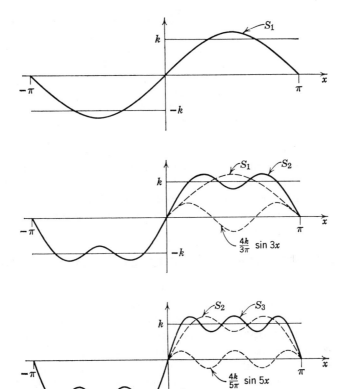

(b) The first three partial sums of the corresponding Fourier series

Fig. 239. Example 1

Now, $\cos \pi = -1$, $\cos 2\pi = 1$, $\cos 3\pi = -1$ etc., in general,

$$\cos n\pi = \begin{cases} -1 & \text{for odd } n, \\ 1 & \text{for even } n, \end{cases} \quad \text{and thus} \quad 1 - \cos n\pi = \begin{cases} 2 & \text{for odd } n, \\ 0 & \text{for even } n. \end{cases}$$

Hence the Fourier coefficients b_n of our function are

$$b_1 = \frac{4k}{\pi}, \qquad b_2 = 0, \qquad b_3 = \frac{4k}{3\pi}, \qquad b_4 = 0, \qquad b_5 = \frac{4k}{5\pi}, \cdots,$$

and since the a_n are zero, the Fourier series of $f(x)$ is

(8)
$$\frac{4k}{\pi} \left(\sin x + \frac{1}{3} \sin 3x + \frac{1}{5} \sin 5x + \cdots \right).$$

The partial sums are

$$S_1 = \frac{4k}{\pi} \sin x, \qquad S_2 = \frac{4k}{\pi}\left(\sin x + \frac{1}{3}\sin 3x\right), \qquad \text{etc.},$$

and their graphs in Fig. 239 seem to indicate that the series is convergent and has the sum $f(x)$, the given function. We notice that at $x = 0$ and $x = \pi$, the points of discontinuity of $f(x)$, all partial sums have the value zero, the arithmetic mean of the values $-k$ and k of our function. Furthermore, assuming that $f(x)$ is the sum of the series and setting $x = \pi/2$, we have

$$f\left(\frac{\pi}{2}\right) = k = \frac{4k}{\pi}\left(1 - \frac{1}{3} + \frac{1}{5} - + \cdots\right)$$

or

$$1 - \frac{1}{3} + \frac{1}{5} - \frac{1}{7} + - \cdots = \frac{\pi}{4}.$$

This is a famous result by Leibniz (obtained about 1673 from geometrical considerations). It illustrates that the values of various series with constant terms can be obtained by evaluating Fourier series at specific points. ∎

In this whole chapter we consider Fourier series from a practical point of view. We shall see that the application of these series is rather simple. In contrast to this, the theory of these series is complicated, and we shall not go into any details of it. Accordingly, we confine ourselves to a theorem on the convergence and the sum of Fourier series, which we present next.

Convergence and Sum of Fourier Series

Suppose that $f(x)$ is any given periodic function of period 2π for which the integrals in (6) exist; for instance, $f(x)$ is continuous or merely piecewise continuous (continuous except for finitely many finite jumps in the interval of integration). Then we can compute the Fourier coefficients (6) of $f(x)$ and use them to form the Fourier series (7) of $f(x)$. It would be nice if the series thus obtained converged and had the sum $f(x)$. Most functions appearing in applications are such that this is true (except at jumps of $f(x)$, which we discuss below). In this case, in which the Fourier series of $f(x)$ does represent $f(x)$, we write

$$f(x) = a_0 + \sum_{n=1}^{\infty} (a_n \cos nx + b_n \sin nx)$$

with an equality sign. If the Fourier series of $f(x)$ does not have the sum $f(x)$ or does not converge, one still writes

$$f(x) \sim a_0 + \sum_{n=1}^{\infty} (a_n \cos nx + b_n \sin nx)$$

with a tilde $\sim$, which indicates that the trigonometric series on the right has the Fourier coefficients of $f(x)$ as its coefficients, so it is the Fourier series of $f(x)$.

The class of functions that can be represented by Fourier series is surprisingly large and general. Corresponding sufficient conditions covering almost any conceivable engineering application are as follows.

Theorem 1 (Representation by a Fourier series)

If a periodic function $f(x)$ with period 2π is piecewise continuous[6] in the interval $-\pi \leq x \leq \pi$ and has a left-hand derivative and right-hand derivative[7] at each point of that interval, then the Fourier series (7) of $f(x)$ [with coefficients (6)] is convergent. Its sum is $f(x)$, except at a point x_0 at which $f(x)$ is discontinuous and the sum of the series is the average of the left- and right-hand limits[7] of $f(x)$ at x_0.

Proof of convergence in Theorem 1 for a continuous function $f(x)$ having continuous first and second derivatives. Integrating (6b) by

parts, we get

$$a_n = \frac{1}{\pi} \int_{-\pi}^{\pi} f(x) \cos nx \, dx = \frac{f(x) \sin nx}{n\pi} \bigg|_{-\pi}^{\pi} - \frac{1}{n\pi} \int_{-\pi}^{\pi} f'(x) \sin nx \, dx.$$

The first term on the right is zero. Another integration by parts gives

$$a_n = \frac{f'(x) \cos nx}{n^2 \pi} \bigg|_{-\pi}^{\pi} - \frac{1}{n^2 \pi} \int_{-\pi}^{\pi} f''(x) \cos nx \, dx.$$

The first term on the right is zero because of the periodicity and continuity of $f'(x)$. Since f'' is continuous in the interval of integration, we have

$$|f''(x)| < M$$

for an appropriate constant M. Furthermore, $|\cos nx| \leq 1$. It follows that

$$|a_n| = \frac{1}{n^2 \pi} \left| \int_{-\pi}^{\pi} f''(x) \cos nx \, dx \right| < \frac{1}{n^2 \pi} \int_{-\pi}^{\pi} M \, dx = \frac{2M}{n^2}.$$

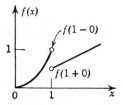

Fig. 240. Left- and right-hand limits

$$f(1 - 0) = 1,$$
$$f(1 + 0) = \tfrac{1}{2}$$

of the function

$$f(x) = \begin{cases} x^2 & \text{if } x < 1 \\ x/2 & \text{if } x > 1 \end{cases}$$

[6]Definition in Sec. 5.1.

[7]The **left-hand limit** of $f(x)$ at x_0 is defined as the limit of $f(x)$ as x approaches x_0 from the left and is frequently denoted by $f(x_0 - 0)$. Thus

$$f(x_0 - 0) = \lim_{h \to 0} f(x_0 - h) \text{ as } h \to 0 \text{ through positive values.}$$

The **right-hand limit** is denoted by $f(x_0 + 0)$ and

$$f(x_0 + 0) = \lim_{h \to 0} f(x_0 + h) \text{ as } h \to 0 \text{ through positive values.}$$

The **left-** and **right-hand derivatives** of $f(x)$ at x_0 are defined as the limits of

$$\frac{f(x_0 - h) - f(x_0 - 0)}{-h} \quad \text{and} \quad \frac{f(x_0 + h) - f(x_0 + 0)}{h},$$

respectively, as $h \to 0$ through positive values. Of course if $f(x)$ is continuous at x_0 the last term in both numerators is simply $f(x_0)$.

Similarly, $|b_n| < 2\,M/n^2$ for all n. Hence the absolute value of each term of the Fourier series of $f(x)$ is at most equal to the corresponding term of the series

$$|a_0| + 2M\left(1 + 1 + \frac{1}{2^2} + \frac{1}{2^2} + \frac{1}{3^2} + \frac{1}{3^2} + \cdots\right)$$

which is convergent. Hence that Fourier series converges and the proof is complete. (Readers already familiar with uniform convergence will see that, by the Weierstrass test in Sec. 14.8, under our present assumptions the Fourier series converges uniformly, and our derivation of (6) by integrating term by term is then justified by Theorem 3 of Sec. 14.8.)

The proofs of convergence in the case of a piecewise continuous function $f(x)$ and of the last statement of Theorem 1 can be found in more advanced texts, for example, in Ref. [C11] listed in Appendix 1. ∎

EXAMPLE 2. Convergence at a jump as indicated in Theorem 1

The square wave in Example 1 has a jump at $x = 0$. Its left-hand limit there is $-k$ and its right-hand limit is k (cf. Fig. 239), so that the average of these limits is 0. The Fourier series (8) of the square wave does indeed converge to this value when $x = 0$ because then all its terms are 0. Similarly for the other jumps. This is in agreement with Theorem 1. ∎

Summary. A Fourier series of a given function $f(x)$ of period 2π is a series of the form (7) with coefficients given by the Euler formulas (6). Theorem 1 gives conditions that are sufficient for this series to converge and at each x to have the value $f(x)$, except at discontinuities of $f(x)$, where the series equals the arithmetic mean of the left-hand and right-hand limits of $f(x)$ at that point.

Most periodic functions in applications will have periods p other than just 2π, but in the next section we show that the transition from period $p = 2\pi$ to general $p = 2L$ (a notation convenient for later applications) is quite simple and direct.

Problems for Sec. 10.2

Find the Fourier series of the function $f(x)$, which is assumed to have the period 2π, and plot accurate graphs of the first three partial sums[8] where $f(x)$ equals

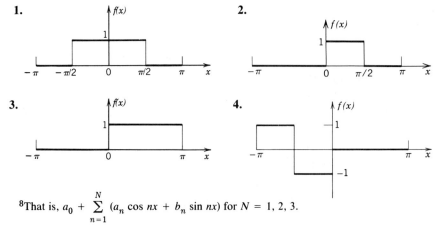

[8]That is, $a_0 + \sum\limits_{n=1}^{N} (a_n \cos nx + b_n \sin nx)$ for $N = 1, 2, 3$.

5. $f(x) = \begin{cases} -k & \text{if} & -\pi/2 < x < \pi/2 \\ k & \text{if} & \pi/2 < x < 3\pi/2 \end{cases}$

6. $f(x) = \begin{cases} -k & \text{if} & 0 < x < \pi/2 \\ 0 & \text{if} & \pi/2 < x < 2\pi \end{cases}$

7. $f(x) = x \quad (-\pi < x < \pi)$

8. $f(x) = \pi - 2|x| \quad (-\pi < x < \pi)$

9. $f(x) = x \quad (0 < x < 2\pi)$

10. $f(x) = x + |x| \quad (-\pi < x < \pi)$

11. $f(x) = x^2 \quad (-\pi < x < \pi)$

12. $f(x) = x^2 \quad (0 < x < 2\pi)$

13. $f(x) = \begin{cases} x & \text{if} & -\pi/2 < x < \pi/2 \\ 0 & \text{if} & \pi/2 < x < 3\pi/2 \end{cases}$

14. $f(x) = \begin{cases} x & \text{if } -\pi/2 < x < \pi/2 \\ \pi - x & \text{if} \quad \pi/2 < x < 3\pi/2 \end{cases}$

15. $f(x) = \begin{cases} -x^2 & \text{if} & -\pi < x < 0 \\ x^2 & \text{if} & 0 < x < \pi \end{cases}$

16. $f(x) = \begin{cases} x^2 & \text{if } -\pi/2 < x < \pi/2 \\ \pi^2/4 & \text{if} \quad \pi/2 < x < 3\pi/2 \end{cases}$

17. Show that if $f(x)$ has the Fourier coefficients a_n, b_n and $g(x)$ has the Fourier coefficients a_n^*, b_n^*, then $kf(x) + lg(x)$ has the Fourier coefficients $ka_n + la_n^*$, $kb_n + lb_n^*$.

18. Using Prob. 17, obtain the Fourier series of $f(x) = |x| \, (-\pi < x < \pi)$ from Probs. 7 and 10.

19. Obtain the Fourier series in Prob. 3 from that in Prob. 1.

20. Verify the last statement in Theorem 1 concerning the discontinuities for the function in Prob. 1.

10.3 Functions of Any Period $p = 2L$

Periodic functions in applications rarely have period 2π but some other period[9] $p = 2L$. If such a function $f(x)$ has a **Fourier series,** we claim that it is of the form

(1)
$$f(x) = a_0 + \sum_{n=1}^{\infty} \left(a_n \cos \frac{n\pi}{L}x + b_n \sin \frac{n\pi}{L}x \right)$$

with the **Fourier coefficients** of $f(x)$ given by the **Euler formulas**

(2)

$$\textbf{(a)} \quad a_0 = \frac{1}{2L} \int_{-L}^{L} f(x) \, dx$$

$$\textbf{(b)} \quad a_n = \frac{1}{L} \int_{-L}^{L} f(x) \cos \frac{n\pi x}{L} \, dx \quad n = 1, 2, \cdots,$$

$$\textbf{(c)} \quad b_n = \frac{1}{L} \int_{-L}^{L} f(x) \sin \frac{n\pi x}{L} \, dx \quad n = 1, 2, \cdots.$$

[9]This notation is practical since in applications, L will be the length of a vibrating string (Sec. 11.2), of a rod in heat conduction (Sec. 11.5), etc.

Proof. The idea is to derive this from Sec. 10.2 by a change of scale. We set $v = \pi x/L$, hence $x = Lv/\pi$. Then $x = \pm L$ corresponds to $v = \pm\pi$. Thus f, regarded as a function of v which we call $g(v)$,

$$f(x) = g(v),$$

has period 2π. Accordingly, by (7) and (6), Sec. 10.2, with v instead of x, this 2π-periodic function $g(v)$ has the Fourier series

$$(3) \qquad g(v) = a_0 + \sum_{n=1}^{\infty} (a_n \cos nv + b_n \sin nv)$$

with coefficients

$$a_0 = \frac{1}{2\pi} \int_{-\pi}^{\pi} g(v)\, dv$$

$$(4) \qquad a_n = \frac{1}{\pi} \int_{-\pi}^{\pi} g(v) \cos nv\, dv$$

$$b_n = \frac{1}{\pi} \int_{-\pi}^{\pi} g(v) \sin nv\, dv.$$

Since $v = \pi x/L$ and $g(v) = f(x)$, formula (3) gives (1). In (4) we introduce $x = Lv/\pi$ as variable of integration. Then the limits of integration $v = \pm\pi$ become $x = \pm L$. Also, $v = \pi x/L$ implies $dv = \pi\, dx/L$. Thus $dv/2\pi = dx/2L$ in a_0. Similarly, $dv/\pi = dx/L$ in a_n and b_n. Hence (4) gives (2). ∎

The interval of integration in (2) may be replaced by any interval of length $p = 2L$ for example, by the interval $0 \leqq x \leqq 2L$.

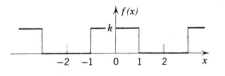

Fig. 241. Example 1

EXAMPLE 1. Periodic square wave
Find the Fourier series of the function (cf. Fig. 241)

$$f(x) = \begin{cases} 0 & \text{if} \quad -2 < x < -1 \\ k & \text{if} \quad -1 < x < 1 \\ 0 & \text{if} \quad 1 < x < 2 \end{cases} \qquad p = 2L = 4, \quad L = 2.$$

Solution. From (2a) and (2b) we obtain

$$a_0 = \frac{1}{4} \int_{-2}^{2} f(x)\, dx = \frac{1}{4} \int_{-1}^{1} k\, dx = \frac{k}{2},$$

$$a_n = \frac{1}{2} \int_{-2}^{2} f(x) \cos \frac{n\pi x}{2} \, dx = \frac{1}{2} \int_{-1}^{1} k \cos \frac{n\pi x}{2} \, dx = \frac{2k}{n\pi} \sin \frac{n\pi}{2}.$$

Thus $a_n = 0$ if n is even and

$$a_n = 2k/n\pi \quad \text{if} \quad n = 1, 5, 9, \cdots, \qquad a_n = -2k/n\pi \quad \text{if} \quad n = 3, 7, 11, \cdots.$$

From (2c) we find that $b_n = 0$ for $n = 1, 2, \cdots$. Hence the result is

$$f(x) = \frac{k}{2} + \frac{2k}{\pi} \left(\cos \frac{\pi}{2}x - \frac{1}{3} \cos \frac{3\pi}{2}x + \frac{1}{5} \cos \frac{5\pi}{2}x - + \cdots \right).$$

EXAMPLE 2. Half-wave rectifier

A sinusoidal voltage $E \sin \omega t$, where t is time, is passed through a half-wave rectifier that clips the negative portion of the wave (Fig. 242). Find the Fourier series of the resulting periodic function

$$u(t) = \begin{cases} 0 & \text{if} \quad -L < t < 0, \\ E \sin \omega t & \text{if} \quad 0 < t < L \end{cases} \qquad p = 2L = \frac{2\pi}{\omega}, \ L = \frac{\pi}{\omega}.$$

Solution. Since $u = 0$ when $-L < t < 0$, we obtain from (2a), with t instead of x,

$$a_0 = \frac{\omega}{2\pi} \int_{0}^{\pi/\omega} E \sin \omega t \, dt = \frac{E}{\pi}$$

and from (2b), by using formula (11) in Appendix 3.1 with $x = \omega t$ and $y = n\omega t$,

$$a_n = \frac{\omega}{\pi} \int_{0}^{\pi/\omega} E \sin \omega t \cos n\omega t \, dt = \frac{\omega E}{2\pi} \int_{0}^{\pi/\omega} [\sin (1 + n)\omega t + \sin (1 - n)\omega t] \, dt.$$

If $n = 1$, the integral on the right is zero, and if $n = 2, 3, \cdots$, we readily obtain

$$a_n = \frac{\omega E}{2\pi} \left[-\frac{\cos (1 + n)\omega t}{(1 + n)\omega} - \frac{\cos (1 - n)\omega t}{(1 - n)\omega} \right]_{0}^{\pi/\omega}$$

$$= \frac{E}{2\pi} \left(\frac{-\cos (1 + n)\pi + 1}{1 + n} + \frac{-\cos (1 - n)\pi + 1}{1 - n} \right).$$

When n is odd, this is equal to zero, and for even n we have

$$a_n = \frac{E}{2\pi} \left(\frac{2}{1 + n} + \frac{2}{1 - n} \right) = -\frac{2E}{(n - 1)(n + 1)\pi} \qquad (n = 2, 4, \cdots).$$

In a similar fashion we find from (2c) that $b_1 = E/2$ and $b_n = 0$ for $n = 2, 3, \cdots$. Consequently,

$$u(t) = \frac{E}{\pi} + \frac{E}{2} \sin \omega t - \frac{2E}{\pi} \left(\frac{1}{1 \cdot 3} \cos 2\omega t + \frac{1}{3 \cdot 5} \cos 4\omega t + \cdots \right). \qquad \blacksquare$$

The function in Example 1 is even, and its Fourier series has only cosine terms, no sine terms. This is typical, and in the next section we show how to save work (and avoid errors) in the case of even or odd periodic functions.

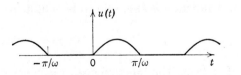

Fig. 242. Half-wave rectifier

Problems for Sec. 10.3

Find the Fourier series of the periodic function $f(x)$, of period $p = 2L$, and plot accurate graphs of $f(x)$ and the first three partial sums.

1. $f(x) = 1$ $(-1 < x < 1)$, $f(x) = 0$ $(1 < x < 3)$, $p = 2L = 4$
2. $f(x) = 1$ $(-1 < x < 0)$, $f(x) = -1$ $(0 < x < 1)$, $p = 2L = 2$
3. $f(x) = 0$ $(-2 < x < 0)$, $f(x) = 1$ $(0 < x < 2)$, $p = 2L = 4$
4. $f(x) = |x|$ $(-2 < x < 2)$, $p = 2L = 4$
5. $f(x) = x$ $(-4 < x < 4)$, $p = 2L = 8$
6. $f(x) = x$ $(0 < x < 1)$, $f(x) = 1 - x$ $(1 < x < 2)$, $p = 2L = 2$
7. $f(x) = 1/2$ $(-1 < x < 0)$, $f(x) = -x$ $(0 < x < 1)$, $p = 2L = 2$
8. $f(x) = (1/2) + x$ $(-1/2 < x < 0)$, $f(x) = (1/2) - x$ $(0 < x < 1/2)$, $p = 2L = 1$
9. $f(x) = x$ $(-\pi/8 < x < \pi/8)$, $f(x) = (\pi/4) - x$ $(\pi/8 < x < 3\pi/8)$, $p = 2L = \pi/2$
10. $f(x) = x^2$ $(0 < x < 2)$, $p = 2L = 2$
11. $f(x) = x^2$ $(-1 < x < 1)$, $p = 2L = 2$
12. $f(x) = x^3$ $(-1 < x < 1)$, $p = 2L = 2$
13. $f(x) = \sin \pi x$ $(0 < x < 1)$, $p = 2L = 1$
14. $f(x) = 1$ $(-2 < x < 0)$, $f(x) = e^{-x}$ $(0 < x < 2)$, $p = 2L = 4$

15. Find the Fourier series of the periodic function obtained by passing the voltage $v(t) = 2 \cos 100\pi t$ through a half-wave rectifier that clips the negative portion of the wave.
16. Obtain the Fourier series in Prob. 3 directly from that in Example 1.
17. Obtain the Fourier series in Prob. 2 directly from that in Example 1, Sec. 10.2.
18. Obtain the Fourier series in Prob. 11 directly from that in Prob. 11, Sec. 10.2.
19. Show that each term in (3) has the period $p = 2L$.
20. Show that in (2) the interval of integration may be replaced by any interval of length $p = 2L$.

10.4 Even and Odd Functions

Unnecessary work (and corresponding sources of errors) in determining Fourier coefficients of a function can be avoided if the function is odd or even.

We first remember that a function $y = g(x)$ is said to be **even** if

$$g(-x) = g(x) \qquad \text{for all } x.$$

The graph of such a function is symmetric with respect to the y-axis (Fig. 243). A function $h(x)$ is said to be **odd** if

$$h(-x) = -h(x) \qquad \text{for all } x.$$

(Cf. Fig. 244). The function $\cos nx$ is even, while the function $\sin nx$ is odd. *If $g(x)$ is an even function, then*

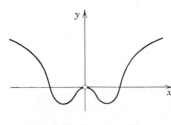

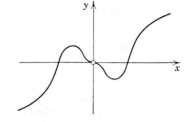

Fig. 243. Even function

Fig. 244. Odd function

$$(1) \qquad \int_{-L}^{L} g(x)\, dx = 2 \int_{0}^{L} g(x)\, dx \qquad\qquad (g \text{ even}).$$

If $h(x)$ is an odd function, then

$$(2) \qquad \int_{-L}^{L} h(x)\, dx = 0 \qquad\qquad (h \text{ odd}).$$

Formulas (1) and (2) are obvious from the graphs of g and h, and we leave the formal proofs to the student.

The product $q = gh$ of an even function g and an odd function h is odd, because

$$q(-x) = g(-x)h(-x) = g(x)[-h(x)] = -q(x).$$

Hence if $f(x)$ is even, then the integrand $f \sin (n\pi x/L)$ in (2c), Sec. 10.3, is odd, and $b_n = 0$. Similarly, if $f(x)$ is odd, then $f \cos (n\pi x/L)$ in (2b), Sec. 10.3, is odd, and $a_n = 0$. From this and (1) we obtain the following theorem.

Theorem 1. (Fourier series of even and odd functions)
*The Fourier series of an even function $f(x)$ of period $2L$ is a "**Fourier cosine series**"*

$$(3) \qquad \boxed{\, f(x) = a_0 + \sum_{n=1}^{\infty} a_n \cos \frac{n\pi}{L}x \,} \qquad\qquad (f \text{ even})$$

with coefficients

$$(4) \qquad a_0 = \frac{1}{L} \int_{0}^{L} f(x)\, dx, \qquad a_n = \frac{2}{L} \int_{0}^{L} f(x) \cos \frac{n\pi x}{L}\, dx, \qquad n = 1, 2, \cdots.$$

*The Fourier series of an odd function $f(x)$ of period $2L$ is a "**Fourier sine series**"*

$$(5) \qquad \boxed{\, f(x) = \sum_{n=1}^{\infty} b_n \sin \frac{n\pi}{L}x \,} \qquad\qquad (f \text{ odd})$$

with coefficients

(6)
$$b_n = \frac{2}{L} \int_0^L f(x) \sin \frac{n\pi x}{L} \, dx.$$

In particular, this theorem implies that the Fourier series of an *even* function $f(x)$ of period $2L = 2\pi$ is a Fourier cosine series

(3*) $f(x) = a_0 + a_1 \cos x + a_2 \cos 2x + a_3 \cos 3x + \cdots$

with coefficients

(4*) $a_0 = \frac{1}{\pi} \int_0^\pi f(x) \, dx, \qquad a_n = \frac{2}{\pi} \int_0^\pi f(x) \cos nx \, dx, \qquad n = 1, 2, \cdots.$

Similarly, the Fourier series of an *odd* function $f(x)$ of period 2π is a Fourier sine series

(5*) $f(x) = b_1 \sin x + b_2 \sin 2x + b_3 \sin 3x + \cdots$

with coefficients

(6*)
$$b_n = \frac{2}{\pi} \int_0^\pi f(x) \sin nx \, dx, \qquad\qquad n = 1, 2, \cdots.$$

For instance, $f(x)$ in Example 1, Sec. 10.2, is odd and, therefore, is represented by a Fourier sine series.

Further simplifications result from

Theorem 2 (Sum of functions)

The Fourier coefficients of a sum $f_1 + f_2$ are the sums of the corresponding Fourier coefficients of f_1 and f_2.

The Fourier coefficients of cf are c times the corresponding Fourier coefficients of f.

EXAMPLE 1. Rectangular pulse

The function $f*(x)$ in Fig. 245 is the sum of the function $f(x)$ in Example 1 of Sec. 10.2 and the constant k. Hence, from that example and Theorem 2 we conclude that

$$f*(x) = k + \frac{4k}{\pi} \left(\sin x + \frac{1}{3} \sin 3x + \frac{1}{5} \sin 5x + \cdots \right).$$

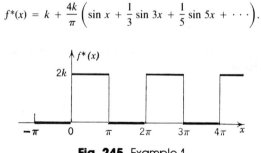

Fig. 245. Example 1

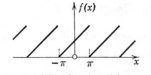

(a) The function $f(x)$.

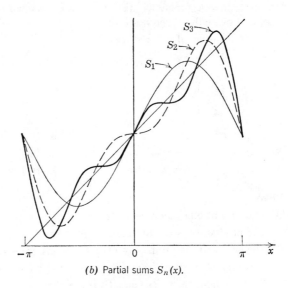

(b) Partial sums $S_n(x)$.

Fig. 246. Example 2

EXAMPLE 2. Saw-toothed wave

Find the Fourier series of the function (Fig. 246a)

$$f(x) = x + \pi \quad \text{if} \quad -\pi < x < \pi \quad \text{and} \quad f(x + 2\pi) = f(x).$$

Solution. We may write

$$f = f_1 + f_2 \quad \text{where} \quad f_1 = x \quad \text{and} \quad f_2 = \pi.$$

The Fourier coefficients of f_2 are zero, except for the first one (the constant term), which is π. Hence, by Theorem 2, the Fourier coefficients a_n, b_n are those of f_1, except for a_0, which is π. Since f_1 is odd, $a_n = 0$ for $n = 1, 2, \cdots$, and

$$b_n = \frac{2}{\pi} \int_0^\pi f_1(x) \sin nx \, dx = \frac{2}{\pi} \int_0^\pi x \sin nx \, dx.$$

Integrating by parts we obtain

$$b_n = \frac{2}{\pi} \left[\frac{-x \cos nx}{n} \bigg|_0^\pi + \frac{1}{n} \int_0^\pi \cos nx \, dx \right] = -\frac{2}{n} \cos n\pi.$$

Hence $b_1 = 2$, $b_2 = -2/2$, $b_3 = 2/3$, $b_4 = -2/4, \cdots$, and the Fourier series of $f(x)$ is

$$f(x) = \pi + 2\left(\sin x - \frac{1}{2} \sin 2x + \frac{1}{3} \sin 3x - + \cdots \right). \qquad \blacksquare$$

As an important application, in the next section we show that a function $f(x)$ defined on an interval $0 \leqq x \leqq L$ can be represented by a Fourier cosine series as well as by a Fourier sine series of period $p = 2L$; these are called **half-range expansions** of f and will be of basic interest in Chap. 11.

Problems for Sec. 10.4

Are the following functions odd, even or neither odd nor even?

1. $|x^3|$, $x \cos nx$, $x^2 \cos nx$, $\cosh x$, $\sinh x$, $\sin x + \cos x$, $x|x|$

2. $x + x^2$, $|x|$, e^x, e^{x^2}, $\sin^2 x$, $x \sin x$, $\ln x$, $x \cos x$, $e^{-|x|}$

Are the following functions $f(x)$, which are assumed to be periodic, of period 2π, even, odd or neither even nor odd?

3. $f(x) = \begin{cases} -x^2 & \text{if} \quad -\pi < x < 0 \\ x^2 & \text{if} \quad 0 < x < \pi \end{cases}$

4. $f(x) = \begin{cases} 0 & \text{if} \quad -\pi < x < 0 \\ x & \text{if} \quad 0 < x < \pi \end{cases}$

5. $f(x) = \begin{cases} 0 & \text{if} \quad -\pi < x < -\pi/2 \\ x & \text{if} \quad -\pi/2 < x < \pi \end{cases}$

6. $f(x) = \begin{cases} x & \text{if} \quad -\pi/2 < x < \pi/2 \\ 0 & \text{if} \quad \pi/2 < x < 3\pi/2 \end{cases}$

7. $f(x) = e^{|x|}$ $(-\pi < x < \pi)$

8. $f(x) = |\sin x|$ $(-\pi < x < \pi)$

9. $f(x) = x|x|$ $(-\pi < x < \pi)$

10. $f(x) = e^{-|x|}$ $(-\pi < x < \pi)$

11. $f(x) = x^2$ $(0 < x < 2\pi)$

12. $f(x) = x$ $(0 < x < 2\pi)$

Prove:

13. The sum of odd functions is odd.

14. The product of two odd functions is even.

15. The sum and the product of even functions are even.

16. If $f(x)$ is even, then $|f(x)|$, $f^2(x)$, and $f^3(x)$ are even.

17. If $f(x)$ is odd, then $|f(x)|$ and $f^2(x)$ are even.

18. If $g(x)$ is any function, defined for all x, then $p(x) = [g(x) + g(-x)]/2$ is even and $q(x) = [g(x) - g(-x)]/2$ is odd, and $g(x) = p(x) + q(x)$.

Represent the following functions as the sum of an even and an odd function.

19. e^x **20.** e^{-2x} **21.** $x/(1 - x)$ **22.** $(1 + x)/(1 - x)$

23. Find all functions that are both even and odd.

24. Prove Theorem 2.

25. Show that the familiar identity $\sin^3 x = \frac{3}{4} \sin x - \frac{1}{4} \sin 3x$ can be interpreted as a Fourier series expansion, and the same holds for $\cos^3 x = \frac{3}{4} \cos x + \frac{1}{4} \cos 3x$.

Find the Fourier series of the following functions, which are assumed to have the period 2π. *Hint.* Use that some of these functions are even or odd.

26. $f(x) = x$ $(-\pi < x < \pi)$ **27.** $f(x) = |x|$ $(-\pi < x < \pi)$

28. $f(x) = x^3$ $(-\pi < x < \pi)$ **29.** $f(x) = x^2/4$ $(-\pi < x < \pi)$

30. $f(x) = |\sin x|$ $(-\pi < x < \pi)$ **31.** $f(x) = |x|^3$ $(-\pi < x < \pi)$

32. $f(x) = x(\pi^2 - x^2)$ $(-\pi < x < \pi)$

33. $f(x) = \begin{cases} 1 & \text{if} \quad -\pi/2 < x < \pi/2 \\ 0 & \text{if} \quad \pi/2 < x < 3\pi/2 \end{cases}$

34. $f(x) = \begin{cases} x - \pi & \text{if} \quad 0 < x < \pi \\ -x & \text{if} \quad \pi < x < 2\pi \end{cases}$

35. $f(x) = \begin{cases} x^2 & \text{if} \quad -\pi/2 < x < \pi/2 \\ \pi^2/4 & \text{if} \quad \pi/2 < x < 3\pi/2 \end{cases}$

36. $f(x) = \begin{cases} -x^2 & \text{if} \quad -\pi < x < 0 \\ x^2 & \text{if} \quad 0 < x < \pi \end{cases}$

Show that

37. $1 - \frac{1}{3} + \frac{1}{5} - \frac{1}{7} + - \cdots = \dfrac{\pi}{4}$ (Use Prob. 33.)

Show that

38. $1 + \frac{1}{4} + \frac{1}{9} + \frac{1}{16} + \frac{1}{25} + \cdots = \dfrac{\pi^2}{6}$ (Euler's famous result.   Use Prob. 29.)

39. $1 - \frac{1}{4} + \frac{1}{9} - \frac{1}{16} + - \cdots = \dfrac{\pi^2}{12}$ (Use Prob. 29.)

40. $1 + \dfrac{1}{3^2} + \dfrac{1}{5^2} + \dfrac{1}{7^2} + \cdots = \dfrac{\pi^2}{8}$ (Use Probs. 38, 39.)

10.5 Half-Range Expansions

In various physical and engineering problems there is a practical need to use Fourier series in connection with functions $f(x)$ that are given merely on some finite interval. Typical applications will arise in the next chapter (Secs. 11.3 and 11.5) in connection with partial differential equations. Then $f(x)$ will be defined on some interval $0 \leq x \leq L$, and on this interval we want to represent $f(x)$ by a Fourier series. We could choose L as the interval of periodicity and go ahead. We can do better by choosing the period $2L$ because then, since $f(x)$ is given only on half of this range (this interval), we can get for $f(x)$ a Fourier cosine series, representing the *even* extension of $f(x)$ to the full range $-L \leq x \leq L$ (Fig. 247b), or we can get for $f(x)$ a Fourier sine series, representing the *odd* extension of $f(x)$ to the full range (Fig. 247c). These two series are called the two **half-range expansions** of the function $f(x)$, which is given only on "half the range" (half the periodicity interval of these series). The form of these series is given in Sec. 10.4. The

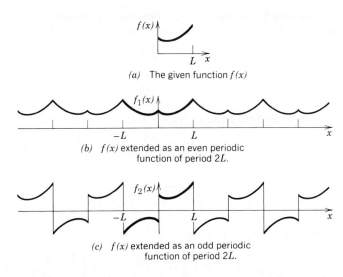

(a)   The given function $f(x)$

(b) $f(x)$ extended as an even periodic function of period $2L$.

(c) $f(x)$ extended as an odd periodic function of period $2L$.

Fig. 247. (a) Function $f(x)$ given on an interval $0 \leq x \leq L$,
(b) its even extension to the full "range" (interval) $-L \leq x \leq L$ (heavy curve)
and the periodic extension of period $2L$ to the x-axis,
(c) its odd extension to $-L \leq x \leq L$ (heavy curve) and the periodic extension
of period $2L$ to the x-axis

cosine half-range expansion is [cf. (3), (4), Sec. 10.4]

(1) $$f(x) = a_0 + \sum_{n=1}^{\infty} a_n \cos \frac{n\pi}{L} x$$

where

(2) $$a_0 = \frac{1}{L} \int_0^L f(x)\, dx, \qquad a_n = \frac{2}{L} \int_0^L f(x) \cos \frac{n\pi x}{L}\, dx, \qquad n = 1, 2, \cdots.$$

The sine half-range expansion is [cf. (5), (6), Sec. 10.4]

(3) $$f(x) = \sum_{n=1}^{\infty} b_n \sin \frac{n\pi}{L} x$$

where

(4) $$b_n = \frac{2}{L} \int_0^L f(x) \sin \frac{n\pi x}{L}\, dx, \qquad\qquad n = 1, 2, \cdots.$$

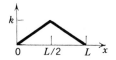

Fig. 248. The given function in Example 1

EXAMPLE 1. "Triangle" and its half-range expansions
Find the two half-range expansions of the function (Fig. 248)

$$f(x) = \begin{cases} \dfrac{2k}{L} x & \text{if } 0 < x < \dfrac{L}{2} \\[2ex] \dfrac{2k}{L}(L - x) & \text{if } \dfrac{L}{2} < x < L \end{cases}$$

Solution. (a) **Even periodic extension.** From (4), Sec. 10.4, we obtain

$$a_0 = \frac{1}{L} \left[\frac{2k}{L} \int_0^{L/2} x\, dx + \frac{2k}{L} \int_{L/2}^L (L - x)\, dx \right] = \frac{k}{2},$$

$$a_n = \frac{2}{L} \left[\frac{2k}{L} \int_0^{L/2} x \cos \frac{n\pi}{L} x\, dx + \frac{2k}{L} \int_{L/2}^L (L - x) \cos \frac{n\pi}{L} x\, dx \right].$$

Now by integration by parts,

$$\int_0^{L/2} x \cos \frac{n\pi}{L} x\, dx = \frac{Lx}{n\pi} \sin \frac{n\pi}{L} x \Big|_0^{L/2} - \frac{L}{n\pi} \int_0^{L/2} \sin \frac{n\pi}{L} x\, dx$$

$$= \frac{L^2}{2n\pi} \sin \frac{n\pi}{2} + \frac{L^2}{n^2\pi^2} \left(\cos \frac{n\pi}{2} - 1 \right).$$

Similarly,

$$\int_{L/2}^{L} (L - x) \cos \frac{n\pi}{L}x \, dx = -\frac{L^2}{2n\pi} \sin \frac{n\pi}{2} - \frac{L^2}{n^2\pi^2} \left(\cos n\pi - \cos \frac{n\pi}{2} \right).$$

By inserting these two results we obtain

$$a_n = \frac{4k}{n^2\pi^2} \left(2 \cos \frac{n\pi}{2} - \cos n\pi - 1 \right).$$

Thus,

$$a_2 = -16k/2^2\pi^2, \quad a_6 = -16k/6^2\pi^2, \quad a_{10} = -16k/10^2\pi^2, \cdots,$$

and $a_n = 0$ if $n \neq 2, 6, 10, 14, \cdots$. Hence the first half-range expansion of $f(x)$ is

$$f(x) = \frac{k}{2} - \frac{16k}{\pi^2} \left(\frac{1}{2^2} \cos \frac{2\pi}{L}x + \frac{1}{6^2} \cos \frac{6\pi}{L}x + \cdots \right).$$

This series represents the even periodic extension of the given function $f(x)$, of period $2L$, shown in Fig. 249a.

(b) Odd periodic extension. Similarly, from (6), Sec. 10.4, we obtain

(5)
$$b_n = \frac{8k}{n^2\pi^2} \sin \frac{n\pi}{2}.$$

Hence the other half-range expansion of $f(x)$ is

$$f(x) = \frac{8k}{\pi^2} \left(\frac{1}{1^2} \sin \frac{\pi}{L}x - \frac{1}{3^2} \sin \frac{3\pi}{L}x + \frac{1}{5^2} \sin \frac{5\pi}{L}x - + \cdots \right).$$

This series represents the odd periodic extension of $f(x)$, of period $2L$, shown in Fig. 249b. ∎

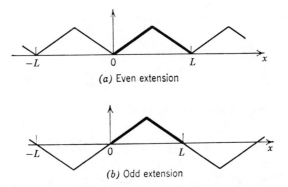

(a) Even extension

(b) Odd extension

Fig. 249. Periodic extensions of $f(x)$ in Example 1

Our main work so far in Secs. 10.2—10.5 has been concerned with finding Fourier coefficients by integration. In the next section we show that in most practical cases, integration can be avoided by applying what we call the **method of jumps,** which is very powerful and of great practical interest.

Problems for Sec. 10.5

Represent the following functions $f(x)$ by a Fourier cosine series and graph the corresponding periodic extension of $f(x)$.

1. $f(x) = k \quad (0 < x < L)$

2. $f(x) = x \quad (0 < x < L)$

3. $f(x) = \begin{cases} 0 & \text{if} \quad 0 < x < L/2 \\ 1 & \text{if} \quad L/2 < x < L \end{cases}$

4. $f(x) = \begin{cases} 1 & \text{if} \quad 0 < x < L/2 \\ 0 & \text{if} \quad L/2 < x < L \end{cases}$

5. $f(x) = x^2 \quad (0 < x < L)$

6. $f(x) = 1 - \dfrac{x}{L} \quad (0 < x < L)$

7. $f(x) = x^3 \quad (0 < x < L)$

8. $f(x) = e^x \quad (0 < x < L)$

9. $f(x) = \sin \dfrac{\pi x}{L} \quad (0 < x < L)$

10. $f(x) = \sin \dfrac{\pi x}{2L} \quad (0 < x < L)$

Represent the following functions $f(x)$ by a Fourier sine series and graph the corresponding periodic extension of $f(x)$.

11. $f(x) = k \quad (0 < x < L)$

12. $f(x) = kx \quad (0 < x < L)$

13. $f(x) = L - x \quad (0 < x < L)$

14. $f(x) = 1 - (2/L)x \quad (0 < x < L)$

15. $f(x) = \begin{cases} 1/2 & \text{if} \quad 0 < x < L/2 \\ 3/2 & \text{if} \quad L/2 < x < L \end{cases}$

16. $f(x) = \begin{cases} 1 & \text{if} \quad 0 < x < L/2 \\ 0 & \text{if} \quad L/2 < x < L \end{cases}$

17. $f(x) = \begin{cases} x & \text{if} \quad 0 < x < L/2 \\ L - x & \text{if} \quad L/2 < x < L \end{cases}$

18. $f(x) = \begin{cases} x & \text{if} \quad 0 < x < L/2 \\ L/2 & \text{if} \quad L/2 < x < L \end{cases}$

19. $f(x) = x^2 \quad (0 < x < L)$

20. $f(x) = x^3 \quad (0 < x < L)$

21. **(Complex form of the Fourier series, complex Fourier coefficients)** Using the formula $e^{i\theta} = \cos \theta + i \sin \theta$ (cf. Sec. 2.4), show that

$$\cos nx = \frac{1}{2}(e^{inx} + e^{-inx}), \qquad \sin nx = \frac{1}{2i}(e^{inx} - e^{-inx})$$

and the Fourier series

$$f(x) = a_0 + \sum_{n=1}^{\infty} (a_n \cos nx + b_n \sin nx)$$

may be written in the form

(6) $$f(x) = c_0 + \sum_{n=1}^{\infty} (c_n e^{inx} + k_n e^{-inx})$$

where $c_0 = a_0$, $c_n = (a_n - ib_n)/2$, $k_n = (a_n + ib_n)/2$, $n = 1, 2, \cdots$. Using (6), Sec. 10.2, show that

$$c_n = \frac{1}{2\pi} \int_{-\pi}^{\pi} f(x)e^{-inx} \, dx, \qquad k_n = \frac{1}{2\pi} \int_{-\pi}^{\pi} f(x)e^{inx} \, dx, \qquad n = 1, 2, \cdots.$$

Introducing the notation $k_n = c_{-n}$, show that (6) may be written

(7) $$\boxed{f(x) = \sum_{n=-\infty}^{\infty} c_n e^{inx}, \qquad c_n = \frac{1}{2\pi} \int_{-\pi}^{\pi} f(x)e^{-inx} \, dx} \qquad n = 0, \pm 1, \pm 2, \cdots.$$

This is the so-called **complex form of the Fourier series,** and the c_n are called the **complex Fourier coefficients** of $f(x)$.

22. Show that the complex Fourier coefficients of an odd function are pure imaginary and those of an even function are real.

Find the complex form of the Fourier series of the following periodic functions of period 2π. Obtain from it the corresponding real Fourier series and compare it with previous results.

23. $f(x) = x \quad (-\pi < x < \pi) \quad$ (cf. Prob. 12).

24. $f(x) = -1 \quad$ if $-\pi < x < 0, \quad f(x) = 1 \quad$ if $0 < x < \pi.$ \quad (Cf. Example 1, Sec. 10.2.)

25. Find the complex form of the Fourier series of $f(x) = e^x \quad (-\pi < x < \pi),$ $f(x + 2\pi) = f(x).$

26. Obtain from the answer to Prob. 25 the corresponding real Fourier series.

10.6 Calculating Fourier Coefficients Without Integration (Method of Jumps)

In several of our previous examples, relatively complicated and lengthy integrations led to relatively simple expressions for the Fourier coefficients a_n and b_n. This raises the question of whether there might be a simpler way of obtaining Fourier coefficients. There is, and we want to show that the Fourier coefficients of a periodic function that is represented by polynomials can be obtained in terms of the jumps of the function and its derivatives. Of course, this is a big advantage, and the corresponding formulas are of great practical importance, because by applying them we can avoid integrations (except for a_0, which must be determined as before).

By a **jump** j of a function $g(x)$ at a point x_0 we mean the difference between the right-hand and left-hand limits (Sec. 10.2) of $g(x)$ at x_0; that is,

(1) $$j = g(x_0 + 0) - g(x_0 - 0).$$

It follows that an upward jump is positive and a downward jump is negative; see Fig. 250.

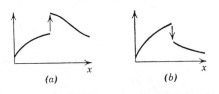

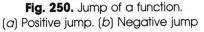

Fig. 250. Jump of a function.
(a) Positive jump. (b) Negative jump

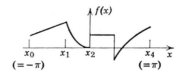

Fig. 251. Example of a representation
of the form (2) (with $m = 4$)

Let $f(x)$ be a function that has period 2π and is represented by polynomials $P_1, \cdots, P_m$ in the interval $-\pi < x < \pi$, say (Fig. 251),

(2)
$$f(x) = \begin{cases} P_1(x) & \text{if } x_0 < x < x_1, \quad (x_0 = -\pi) \\ P_2(x) & \text{if } x_1 < x < x_2, \\ \quad \vdots \\ P_m(x) & \text{if } x_{m-1} < x < x_m \ (= \pi). \end{cases}$$

Then f may have jumps at $x_0, x_1, \cdots, x_m$, and the same is true for the derivatives $f', f'', \cdots$. We choose the following notation.

(3)
$$\begin{aligned} j_s &= \text{jump of } f \text{ at } x_s, \\ j_s' &= \text{jump of } f' \text{ at } x_s, \qquad\qquad (s = 1, 2, \cdots, m) \\ j_s'' &= \text{jump of } f'' \text{ at } x_s, \text{ etc.} \end{aligned}$$

Of course, if f is continuous at x_s, then $j_s = 0$, and for the derivatives the situation is similar, so that some of the numbers $j_s, j_s', \cdots$ in (3) may be zero.

EXAMPLE 1. Jumps of a function and its derivatives
Let
$$f(x) = \begin{cases} 0 & \text{when } -\pi < x < 0, \\ x^2 & \text{when } 0 < x < \pi. \end{cases}$$

We plot graphs of f and its derivatives (Fig. 252),

$$f' = \begin{cases} 0 \\ 2x \end{cases} \qquad f'' = \begin{cases} 0 \\ 2 \end{cases} \qquad f''' = 0.$$

We see that the jumps are

	Jump at $x_1 = 0$	Jump at $x_2 = \pi$
f	$j_1 = 0$	$j_2 = \pi^2$
f'	$j_1' = 0$	$j_2' = -2\pi$
f''	$j_1'' = 2$	$j_2'' = -2$

Note that the jumps at $x = -\pi$ are not listed, because they are taken into account at $x = \pi$, the other end of the interval of periodicity.   ∎

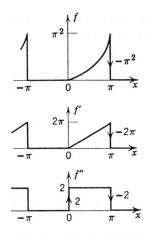

Fig. 252. $f(x)$ and derivatives in Example 1

To get the desired formula for the Fourier coefficients a_1, a_2, $\cdots$ of the function f, given by (2), we start from the Euler formula (6b) in Sec. 10.2:

$$(4) \qquad\qquad \pi a_n = \int_{-\pi}^{\pi} f \cos nx \, dx.$$

Since f is represented by (2), we write the integral as the sum of m integrals:

$$(5) \qquad \pi a_n = \int_{x_0}^{x_1} + \int_{x_1}^{x_2} + \cdots + \int_{x_{m-1}}^{x_m} = \sum_{s=1}^{m} \int_{x_{s-1}}^{x_s} f \cos nx \, dx$$

where $x_0 = -\pi$ and $x_m = \pi$. Integration by parts yields

$$(6) \qquad \int_{x_{s-1}}^{x_s} f \cos nx \, dx = \frac{f}{n} \sin nx \Bigg|_{x_{s-1}}^{x_s} - \frac{1}{n} \int_{x_{s-1}}^{x_s} f' \sin nx \, dx.$$

Now comes an important point: the evaluation of the first expression on the right. $f(x)$ may be discontinuous at x_s (Fig. 253), and we have to take the left-hand limit $f(x_s - 0)$ of f at x_s. Similarly, at x_{s-1} we have to take the right-hand limit $f(x_{s-1} + 0)$. Hence the first expression on the right-hand side of (6) equals

$$\frac{1}{n} [f(x_s - 0) \sin nx_s - f(x_{s-1} + 0) \sin nx_{s-1}].$$

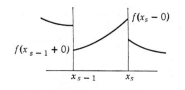

Fig. 253. The formula (6)

Consequently, by inserting (6) into (5) and using the short notations $S_0 = \sin nx_0$, $S_1 = \sin nx_1$, etc., we obtain

$$\pi a_n = \frac{1}{n}[f(x_1 - 0)S_1 - f(x_0 + 0)S_0 + f(x_2 - 0)S_2 - f(x_1 + 0)S_1$$

(7)
$$+ \cdots + f(x_m - 0)S_m - f(x_{m-1} + 0)S_{m-1}]$$

$$- \frac{1}{n}\sum_{s=1}^{m}\int_{x_{s-1}}^{x_s} f' \sin nx \, dx.$$

If we collect terms with the same S, the expression in the brackets becomes

$$-f(x_0 + 0)S_0 + [f(x_1 - 0) - f(x_1 + 0)]S_1$$

(8)
$$+ [f(x_2 - 0) - f(x_2 + 0)]S_2 + \cdots + f(x_m - 0)S_m.$$

The expressions in the brackets in (8) are the jumps of f, multiplied by -1. Furthermore, because of periodicity, $S_0 = S_m$ and $f(x_0) = f(x_m)$, so that we may combine the first and the last term in (8). Thus (8) is equal to

$$-j_1 S_1 - j_2 S_2 - \cdots - j_m S_m,$$

and from (7) we therefore have the intermediate result

(9)
$$\pi a_n = -\frac{1}{n}\sum_{s=1}^{m} j_s \sin nx_s - \frac{1}{n}\sum_{s=1}^{m}\int_{x_{s-1}}^{x_s} f' \sin nx \, dx.$$

Applying the same procedure to the integrals on the right-hand side we find

(10)
$$\sum_{s=1}^{m}\int_{x_{s-1}}^{x_s} f' \sin nx \, dx = \frac{1}{n}\sum_{s=1}^{m} j_s' \cos nx_s + \frac{1}{n}\sum_{s=1}^{m}\int_{x_{s-1}}^{x_s} f'' \cos nx \, dx.$$

Continuing in this fashion, we obtain integrals involving higher and higher derivatives of f. Since f is represented by polynomials and the $(r + 1)$th derivative of a polynomial of degree r is identically zero, we shall reach the point where no integrals are left, and this will happen after finitely many steps. By inserting (10) and the analogous formulas obtained in the further steps into (9) we obtain the desired formula

(11a)
$$\boxed{a_n = \frac{1}{n\pi}\left[-\sum_{s=1}^{m} j_s \sin nx_s - \frac{1}{n}\sum_{s=1}^{m} j_s' \cos nx_s \right. \\ \left. + \frac{1}{n^2}\sum_{s=1}^{m} j_s'' \sin nx_s + \frac{1}{n^3}\sum_{s=1}^{m} j_s''' \cos nx_s - - + + \cdots \right],}$$

where $n = 1, 2, \cdots$ (and a_0 must be obtained by integration as before). In precisely the same fashion we obtain from (6c) in Sec. 10.2 the formula

(11b)

$$
b_n = \frac{1}{n\pi} \left[\sum_{s=1}^{m} j_s \cos nx_s - \frac{1}{n} \sum_{s=1}^{m} j_s' \sin nx_s \right.
$$

$$
\left. - \frac{1}{n^2} \sum_{s=1}^{m} j_s'' \cos nx_s + \frac{1}{n^3} \sum_{s=1}^{m} j_s''' \sin nx_s + - - + + \cdots \right].
$$

To avoid errors, it is practical to graph $f(x)$ and its derivatives and list the jumps in a table as shown in Example 1.

EXAMPLE 2. Periodic square wave

Find the Fourier coefficients of the function (Fig. 254)

$$
f(x) = \begin{cases} -k & \text{if} & -\pi < x < 0, \\ k & \text{if} & 0 < x < \pi. \end{cases}
$$

Solution. We see that $f' \equiv 0$ and the jumps of f are

	Jump at $x_1 = 0$	Jump at $x_2 = \pi$
f	$j_1 = 2k$	$j_2 = -2k$

f is odd. Hence $a_n = 0$, and from (11b),

$$
b_n = \frac{1}{n\pi} [j_1 \cos nx_1 + j_2 \cos nx_2] = \frac{1}{n\pi} [2k \cos 0 - 2k \cos n\pi]
$$

$$
= \frac{2k}{n\pi} (1 - \cos n\pi) = \begin{cases} 4k/n\pi & \text{for odd } n \\ 0 & \text{for even } n \end{cases} \qquad \text{(cf. Example 1, Sec. 10.2).}
$$

EXAMPLE 3

Find the Fourier series of the function in Example 1 of the current section.

Solution. By integration,

$$
a_0 = \frac{1}{2\pi} \int_0^{\pi} x^2 \, dx = \frac{\pi^2}{6}.
$$

From (11a),

$$
a_n = \frac{1}{n\pi} \left[\pi^2 \sin n\pi + \frac{2\pi}{n} \cos n\pi + \frac{1}{n^2} (2 \sin 0 - 2 \sin n\pi) \right] = \frac{2}{n^2} \cos n\pi.
$$

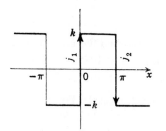

Fig. 254. Example 2

Hence $a_1 = -2/1^2$, $a_2 = 2/2^2$, $a_3 = -2/3^2$, $\cdots$. From (11b) it follows that

$$b_n = \frac{1}{n\pi}\left[-\pi^2 \cos n\pi + \frac{2\pi}{n}\sin n\pi - \frac{1}{n^2}(2\cos 0 - 2\cos n\pi)\right]$$

$$= -\frac{\pi}{n}\cos n\pi + \frac{2}{n^3\pi}(\cos n\pi - 1).$$

Hence,

$$b_1 = \pi - \frac{4}{\pi}, \qquad b_2 = -\frac{\pi}{2}, \qquad b_3 = \frac{\pi}{3} - \frac{4}{3^3\pi}, \qquad b_4 = -\frac{\pi}{4}, \cdots,$$

and the Fourier series is

$$f(x) = \frac{\pi^2}{6} - 2\cos x + \left(\pi - \frac{4}{\pi}\right)\sin x + \frac{1}{2}\cos 2x - \frac{\pi}{2}\sin 2x + \cdots. \qquad ∎$$

Fourier series are very powerful in handling and solving models involving **differential equations.** In the next section we give a first illustration of this fact in terms of a typical vibrational problem. Chapter 11 on partial differential equations will include numerous further such applications of Fourier series.

Problems for Sec. 10.6

1. Obtain the Fourier series in Example 1 by the usual method and compare the amount of work with that in Example 3.
2. Find the Fourier series of $f(x) = x^4$ ($-\pi < x < \pi$), $f(x + 2\pi) = f(x)$ by the use of (6), Sec. 10.2, and (11), this section, and compare the amount of work.
3. Derive (11b) from (6c), Sec. 10.2.
4. Show that in the case of a function $f(x)$ having period $p = 2L$ the formulas corresponding to (11) are

(12a)
$$a_n = \frac{1}{n\pi}\left[-\sum_{s=1}^{m} j_s \sin K_n x_s - \frac{1}{K_n}\sum_{s=1}^{m} j_s' \cos K_n x_s \right. \qquad \left(K_n = \frac{n\pi}{L}\right)$$
$$\left. + \frac{1}{K_n^2}\sum_{s=1}^{m} j_s'' \sin K_n x_s + \frac{1}{K_n^3}\sum_{s=1}^{m} j_s''' \cos K_n x_s - - + + \cdots\right],$$

(12b)
$$b_n = \frac{1}{n\pi}\left[\sum_{s=1}^{m} j_s \cos K_n x_s - \frac{1}{K_n}\sum_{s=1}^{m} j_s' \sin K_n x_s \right.$$
$$\left. - \frac{1}{K_n^2}\sum_{s=1}^{m} j_s'' \cos K_n x_s + \frac{1}{K_n^3}\sum_{s=1}^{m} j_s''' \sin K_n x_s + - - + + \cdots\right].$$

Using (11) or (12), find the Fourier series of the functions f in:

5. Probs. 1—4, Sec. 10.2   6. Probs. 11—14, Sec. 10.2
7. Probs. 4—6, Sec. 10.3   8. Probs. 10—12, Sec. 10.3

Using (11) or (12), find the Fourier sine series of

9. f in Probs. 12 and 17 in Sec. 10.5   10. f in Probs. 14 and 18 in Sec. 10.5
11. $f(x) = x(\pi^2 - x^2)$ $(0 < x < \pi)$ 12. $f(x) = x^3 - x$ $(0 < x < \pi)$

Using (12), find the Fourier cosine series of

13. f in Probs. 2 and 5 in Sec. 10.5   **14.** $f(x) = x^3$ $(0 < x < L)$

15. Can (11) be applied to find the Fourier coefficients of the function $f(x) = e^x$ $(0 < x < 2\pi)$, $f(x + 2\pi) = f(x)$?

10.7 Forced Oscillations

Fourier series have important applications in connection with differential equations. Let us consider an important practical problem involving an ordinary differential equation. (Applications to partial differential equations will be considered in the next chapter.)

From Sec. 2.13 we know that forced oscillations of a body of mass m on a spring (cf. Fig. 255) are governed by the equation

$$(1) \qquad\qquad my'' + cy' + ky = r(t)$$

where k is the spring modulus, c is the damping constant and t is time.

If the external force $r(t)$ is a sine or cosine function and the damping constant c is not zero, the steady-state solution represents a harmonic oscillation having the frequency of the external force.

We shall now see that if $r(t)$ is not a pure sine or cosine function but any other periodic function, then the steady-state solution will represent a superposition of harmonic oscillations having the frequency of $r(t)$ and multiples of this frequency. If the frequency of one of these oscillations is close to the resonant frequency of the vibrating system (cf. Sec. 2.13), then that oscillation may be the dominant part of the response of the system to the external force. Of course, this is quite surprising to an observer not familiar with the corresponding mathematical theory, which is highly important in the study of vibrating systems and resonance. Let us illustrate the situation by a typical example.

EXAMPLE 1. Forced oscillations under a nonsinusoidal periodic driving force
In (1), let $m = 1$ (gm), $c = 0.02$ (gm/sec), and $k = 25$ (gm/sec^2), so that (1) becomes

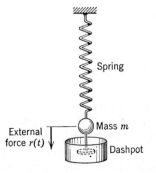

Fig. 255. Vibrating system under consideration

(2) $$y'' + 0.02y' + 25y = r(t)$$

where $r(t)$ is measured in gm $\cdot$ cm/sec^2. Let (Fig. 256)

$$r(t) = \begin{cases} t + \dfrac{\pi}{2} & \text{when} \quad -\pi < t < 0, \\[2mm] -t + \dfrac{\pi}{2} & \text{when} \quad 0 < t < \pi, \end{cases} \qquad r(t + 2\pi) = r(t).$$

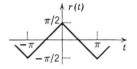

Fig. 256. Force in Example 1

Find the steady-state solution $y(t)$.

Solution. We represent $r(t)$ by a Fourier series, finding

(3) $$r(t) = \frac{4}{\pi}\left(\cos t + \frac{1}{3^2}\cos 3t + \frac{1}{5^2}\cos 5t + \cdots\right).$$

Then we consider the differential equation

(4) $$y'' + 0.02y' + 25y = \frac{4}{n^2\pi}\cos nt \qquad (n = 1, 3, \cdots)$$

whose right-hand side is a single term of the series (3). From Sec. 2.13 we know that the steady-state solution $y_n(t)$ of (4) is of the form

(5) $$y_n = A_n \cos nt + B_n \sin nt.$$

By substituting this into (4) we find that

(6) $$A_n = \frac{4(25 - n^2)}{n^2\pi D}, \qquad B_n = \frac{0.08}{n\pi D}, \qquad \text{where} \qquad D = (25 - n^2)^2 + (0.02n)^2.$$

Since the differential equation (2) is linear, we may expect the steady-state solution to be

(7) $$y = y_1 + y_3 + y_5 + \cdots$$

where y_n is given by (5) and (6). In fact, this follows readily by substituting (7) into (2) and using the Fourier series of $r(t)$, provided that termwise differentiation of (7) is permissible. (Readers already familiar with the notion of uniform convergence [Sec. 14.8] may prove that (7) may be differentiated term by term.)

From (6) we find that the amplitude of (5) is

$$C_n = \sqrt{A_n{}^2 + B_n{}^2} = \frac{4}{n^2\pi\sqrt{D}}.$$

Numerical values are

$$C_1 = 0.0530$$
$$C_3 = 0.0088$$
$$C_5 = 0.5100$$
$$C_7 = 0.0011$$
$$C_9 = 0.0003$$

For $n = 5$ the quantity D is very small, the denominator of C_5 is small, and C_5 is so large that y_5 is the dominating term in (7). This implies that the steady-state motion is almost a harmonic oscillation whose frequency equals five times that of the exciting force (Fig. 257). ∎

The application of Fourier series to more general vibrating systems, heat conduction and other problems follows in Chap. 11. Another area of practical

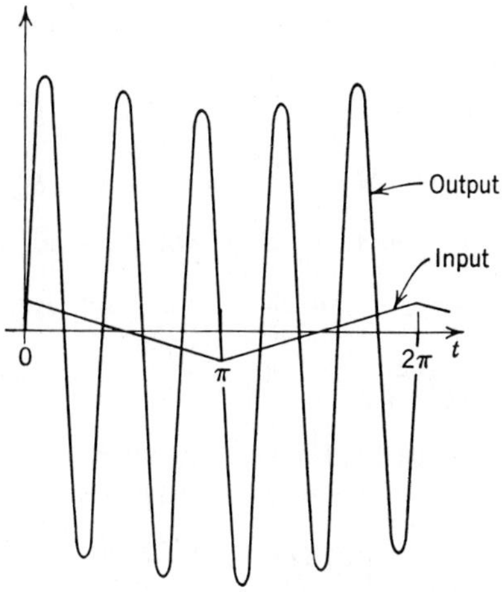

Fig. 257. Input and steady-state output
in Example 1

interest in which Fourier series play a role is the approximation of functions
by simpler functions, known as **approximation theory.** This will be explained
in the next section.

Problems for Sec. 10.7

1. What would happen to the amplitudes C_n in Example 1 (and, accordingly, to the
   form of the vibrations) if we changed the spring constant to the value 9? If we
   took a stiffer spring with $k = 49$? If we increased the damping?

Find a general solution of the differential equation $y'' + \omega^2 y = r(t)$, where

2. $r(t) = \cos \alpha t + \cos \beta t$ $(\omega^2 \neq \alpha^2, \beta^2)$
3. $r(t) = \sin t$, $\omega = 0.5, 0.7, 0.9, 1.1, 1.5, 2.0, 10.0$
4. $r(t) = \sin t + \frac{1}{9} \sin 3t + \frac{1}{25} \sin 5t$, $\omega = 0.5, 0.9, 1.1, 2, 2.9, 3.1, 4, 4.9, 5.1,$
   $6, 8$
5. $r(t) = \sum\limits_{n=1}^{N} b_n \sin nt$, $|\omega| \neq 1, 2, \cdots, N$
6. $r(t) = \sum\limits_{n=1}^{N} a_n \cos nt$, $|\omega| \neq 1, 2, \cdots, N$
7. $r(t) = \dfrac{\pi}{4} |\sin t|$ when $-\pi < t < \pi$ and $r(t + 2\pi) = r(t)$, $|\omega| \neq 0, 2, 4, \cdots$
8. $r(t) = \begin{cases} t + \pi & \text{if} \quad -\pi < t < 0 \\ -t + \pi & \text{if} \quad\;\; 0 < t < \pi \end{cases}$ and $(r(t + 2\pi) = r(t), |\omega| \neq 0, 1, 3, \cdots$
9. $r(t) = \begin{cases} t & \text{if } -\pi/2 < t < \pi/2 \\ \pi - t & \text{if } \quad \pi/2 < t < 3\pi/2 \end{cases}$ and $r(t + 2\pi) = r(t), |\omega| \neq 1, 3, 5, \cdots$

10. Some of the $A(\omega)$ in Prob. 3 are positive and some negative. Is this physically
    understandable? Some are large in absolute value and some small. Is the reason
    the same as in Example 1 of the text?

Find the steady-state oscillation corresponding to $y'' + cy' + y = r(t)$, where $c > 0$ and

11. $r(t) = K \sin t$
12. $r(t) = \sin 3t$

13. $r(t) = a_n \cos nt$
14. $r(t) = \sum_{n=1}^{N} b_n \sin nt$

15. $r(t) = \begin{cases} \pi t/4 & \text{if } -\pi/2 < t < \pi/2 \\ \pi(\pi - t)/4 & \text{if } \pi/2 < t < 3\pi/2 \end{cases}$ and $r(t + 2\pi) = r(t)$

16. $r(t) = \dfrac{t}{12}(\pi^2 - t^2)$ if $-\pi < t < \pi$ and $r(t + 2\pi) = r(t)$

17. (**RLC-circuit**) Find the steady-state current $I(t)$ in the RLC-circuit in Fig. 258, where $R = 100$ ohms, $L = 10$ henrys, $C = 10^{-2}$ farad,

$$E(t) = \begin{cases} 100(\pi t + t^2) & \text{if } -\pi < t < 0 \\ 100(\pi t - t^2) & \text{if } \quad 0 < t < \pi \end{cases} \quad \text{and} \quad E(t + 2\pi) = E(t).$$

Proceed as follows. Develop $E(t)$ in a Fourier series. $I(t)$ will appear in the form of a trigonometric series. Find the general formulas for the coefficients of this series. Compute numerical values of the first few coefficients. Graph the sum of the first few terms of that series.

18. Same task as in Prob. 17 with R, L, C as before, and $E(t) = 200t(\pi^2 - t^2)$ volts if $-\pi < t < \pi$ and $E(t + 2\pi) = E(t)$.

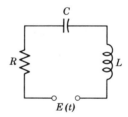

Fig. 258. RLC-circuit in Problems 17 and 18

10.8 Approximation by Trigonometric Polynomials. Square Error

Let $f(x)$ be a given function of period 2π that can be represented by a Fourier series. Then the Nth partial sum of this series is an approximation to $f(x)$:

$$(1) \qquad f(x) \approx a_0 + \sum_{n=1}^{N} (a_n \cos nx + b_n \sin nx).$$

It is natural to ask whether (1) is the "best" approximation to f by a **trigonometric polynomial,** that is, a function of the form

$$(2) \qquad F(x) = \alpha_0 + \sum_{n=1}^{N} (\alpha_n \cos nx + \beta_n \sin nx),$$

with the same fixed N, that is, an approximation for which the "error" is minimum.

Of course, we must first define what we mean by the error E of such an approximation. We want to choose a definition that measures the goodness of agreement between f and F *on the whole interval* $-\pi \leqq x \leqq \pi$. Obviously, the maximum of $|f - F|$ is not suitable for that purpose: in Fig. 259, the function F is a good approximation to f, but $|f - F|$ is large near x_0. We choose

(3)
$$E = \int_{-\pi}^{\pi} (f - F)^2 \, dx.$$

This is called the **total square error** of F relative to the function f on the interval $-\pi \leqq x \leqq \pi$. Clearly, $E \geqq 0$.

N being fixed, we want to determine the coefficients in (2) such that E is minimum. Since $(f - F)^2 = f^2 - 2fF + F^2$, we have

(4)
$$E = \int_{-\pi}^{\pi} f^2 \, dx - 2 \int_{-\pi}^{\pi} fF \, dx + \int_{-\pi}^{\pi} F^2 \, dx.$$

By inserting (2) into the last integral and evaluating the occurring integrals as in Sec. 10.2 we readily obtain

$$\int_{-\pi}^{\pi} F^2 \, dx = \pi(2\alpha_0^2 + \alpha_1^2 + \cdots + \alpha_N^2 + \beta_1^2 + \cdots + \beta_N^2).$$

By inserting (2) into the second integral in (4) we see that the occurring integrals are those in the Euler formulas (6), Sec. 10.2, and we thus obtain

$$\int_{-\pi}^{\pi} fF \, dx = \pi(2\alpha_0 a_0 + \alpha_1 a_1 + \cdots + \alpha_N a_N + \beta_1 b_1 + \cdots + \beta_N b_N).$$

With these expressions (4) becomes

(5)
$$E = \int_{-\pi}^{\pi} f^2 \, dx - 2\pi \left[2\alpha_0 a_0 + \sum_{n=1}^{N} (\alpha_n a_n + \beta_n b_n) \right]$$
$$+ \pi \left[2\alpha_0^2 + \sum_{n=1}^{N} (\alpha_n^2 + \beta_n^2) \right].$$

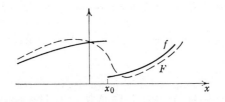

Fig. 259. Error of approximation

If we take $\alpha_n = a_n$ and $\beta_n = b_n$ in (2), then from (5) we see that the square error corresponding to this particular choice of the coefficients of F is given by

(6)
$$E^* = \int_{-\pi}^{\pi} f^2 \, dx - \pi \left[2a_0^2 + \sum_{n=1}^{N} (a_n^2 + b_n^2) \right].$$

By subtracting (6) from (5) we obtain

$$E - E^* = \pi \left\{ 2(\alpha_0 - a_0)^2 + \sum_{n=1}^{N} [(\alpha_n - a_n)^2 + (\beta_n - b_n)^2] \right\}.$$

Since the sum of squares of real numbers on the right cannot be negative,

$$E - E^* \geq 0, \qquad \text{thus} \qquad E \geq E^*,$$

and $E = E^*$ if and only if $\alpha_0 = a_0, \cdots, \beta_N = b_N$. This proves

Theorem 1 (Minimum square error)
The total square error of F [cf. (2), N fixed] relative to f on the interval $-\pi \leq x \leq \pi$ is minimum if and only if the coefficients of F in (2) are the corresponding Fourier coefficients of f. This minimum value is given by (6).

From (6) we see that E^* cannot increase as N increases, but may decrease. Hence, *with increasing N, the partial sums of the Fourier series of f yield better and better approximations to $f(x)$*, considered from the viewpoint of the square error.

Since $E^* \geq 0$ and (6) holds for every N, we obtain from (6) the important **Bessel inequality**[10]

(7)
$$2a_0^2 + \sum_{n=1}^{\infty} (a_n^2 + b_n^2) \leq \frac{1}{\pi} \int_{-\pi}^{\pi} f(x)^2 \, dx$$

for the Fourier coefficients of any function f for which the integral on the right exists.

EXAMPLE 1. Square error for the saw-toothed wave
Compute the total square error of F with $N = 3$ relative to

$$f(x) = x + \pi \quad (-\pi < x < \pi) \qquad \text{(Fig. 246a, Sec. 10.4)}$$

on the interval $-\pi \leq x \leq \pi$.

Solution. $F(x) = \pi + 2 \sin x - \sin 2x + \frac{2}{3} \sin 3x$ by Example 2, Sec. 10.4. From this and (6),

$$E^* = \int_{-\pi}^{\pi} (x + \pi)^2 \, dx - \pi[2\pi^2 + 2^2 + 1^2 + (\tfrac{2}{3})^2],$$

[10]It can be shown that for such a function f even the equality sign in (7) holds. Proof in Ref. [C11] listed in Appendix 1. Formula (7) with the equality sign is called **Parseval's equality**.

hence

$$E^* = \tfrac{8}{3}\pi^3 - \pi(2\pi^2 + \tfrac{49}{9}) \approx 3.567.$$

$F = S_3$ is shown in Fig. 246b, and although $|f(x) - F(x)|$ is large at $x = \pm\pi$ (how large?), where f is discontinuous, F approximates f quite well on the whole interval. How could you get a better approximation with a smaller square error? Will the maximum of $|f(x) - F(x)|$ become smaller for such a better approximation?   ∎

    This brings to an end our discussion of Fourier series, which has emphasized the practical aspects of these series, as needed in applications. In the last four sections of this chapter we show how ideas and techniques in Fourier series can be extended to nonperiodic functions defined on the entire x-axis or at least on the positive x-axis. This leads to the **"Fourier integral"** (Sec. 10.9) and on to **"Fourier transformations"** (Secs. 10.10—10.12).

Problems for Sec. 10.8

1. Let $f(x) = -1$ when $-\pi < x < 0$, $f(x) = 1$ when $0 < x < \pi$, and $f(x + 2\pi) = f(x)$. Find the function $F(x)$ of the form (2) for which the total square error (3) is minimum.

2. Compute the minimum square error in Prob. 1 for $N = 1, 3, 5, 7$. What is the smallest N such that $E^* \leqq 0.2$?

3. Show that the minimum square error (6) is a monotone decreasing function of N.

In each case, find the function $F(x)$ of the form (2) for which the total square error E on the interval $-\pi \leqq x \leqq \pi$ is minimum and compute this minimum value for $N = 1, 2, \cdots, 5$, where, for $-\pi < x < \pi$,

4. $f(x) = |x|$ 5. $f(x) = x$

6. $f(x) = x(\pi^2 - x^2)/12$ 7. $f(x) = x^2$

8. Compare the rapidity of decrease of the square error for the discontinuous function in Prob. 5 and the continuous function in Prob. 7 and comment.

9. Verify Bessel's inequality for the function in Prob. 1. (Use Prob. 40, Sec. 10.4.)

10. Using Bessel's inequality with the equality sign, show that

$$\int_{-\pi}^{\pi} \cos^4 x \, dx = \frac{3\pi}{4}, \qquad \int_{-\pi}^{\pi} \cos^6 x \, dx = \frac{5\pi}{8}.$$

10.9   Fourier Integral

Fourier series are powerful tools in treating various problems involving *periodic* functions. A first illustration of this fact was given in Sec. 10.7, and further important problems in connection with partial differential equations will be considered in Chap. 11. Since, of course, many practical problems do not involve periodic functions, it is desirable to generalize the method of Fourier series to include *nonperiodic functions*. This is our goal, and we proceed as follows.

At first let us consider two simple examples of periodic functions $f_L(x)$ of period $2L$ and see what happens if we let $L \to \infty$. Then we shall consider the Fourier series of an arbitrary function f_L of period $2L$ and again let $L \to \infty$. This will motivate and make plausible the main result of this section, which is an integral representation stated in Theorem 1 (below). Let us get started by first presenting those two introductory examples.

EXAMPLE 1. Square wave

Consider the function $f_L(x)$ (square wave) of period $2L > 2$ given by (Fig. 260, upper part)

$$f_L(x) = \begin{cases} 0 & \text{if} & -L < x < -1 \\ 1 & \text{if} & -1 < x < 1 \\ 0 & \text{if} & 1 < x < L. \end{cases}$$

For $L \to \infty$ we obtain a function $f(x)$ (Fig. 260, lower part) which is no longer periodic, namely,

$$f(x) = \lim_{L \to \infty} f_L(x) = \begin{cases} 1 & \text{if} & -1 < x < 1 \\ 0 & \text{otherwise.} \end{cases}$$

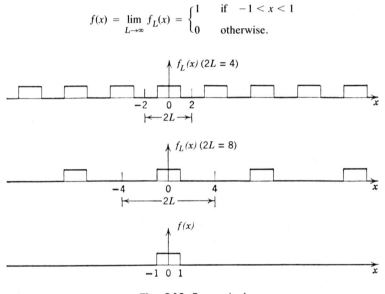

Fig. 260. Example 1

EXAMPLE 2. Exponential function

Figure 261 shows the periodic function $f_L(x)$ of period $2L$ given by

$$f_L(x) = e^{-|x|} \quad \text{if} \ -L < x < L \quad \text{and} \quad f_L(x + 2L) = f_L(x).$$

For $L \to \infty$ we again obtain a function $f(x)$ that is no longer periodic (Fig. 261, lower part),

$$f(x) = \lim_{L \to \infty} f_L(x) = e^{-|x|}. \qquad \blacksquare$$

We now consider any periodic function $f_L(x)$ of period $2L$ which can be represented by a Fourier series. We use the short notation

$$w_n = \frac{n\pi}{L}.$$

Then we can write this Fourier series representation simply as

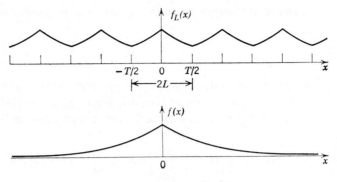

Fig. 261. Example 2

$$f_L(x) = a_0 + \sum_{n=1}^{\infty} (a_n \cos w_n x + b_n \sin w_n x).$$

We want to see what happens if we let $L \to \infty$. As mentioned above, we do this for reasons of motivation as well as for making it plausible that for a nonperiodic function, one should expect an integral (instead of a series) involving $\cos wx$ and $\sin wx$ with w taking *all* values (instead of being restricted to integer multiples $w = w_n = n\pi/L$ of a fixed π/L).

We insert a_n and b_n according to the Euler formulas (2) in Sec. 10.3, denoting the variable of integration by v. Then the Fourier series of $f_L(x)$ becomes

$$f_L(x) = \frac{1}{2L} \int_{-L}^{L} f_L(v)\, dv + \frac{1}{L} \sum_{n=1}^{\infty} \left[\cos w_n x \int_{-L}^{L} f_L(v) \cos w_n v\, dv \right.$$

$$\left. + \sin w_n x \int_{-L}^{L} f_L(v) \sin w_n v\, dv \right].$$

We now introduce

$$\Delta w = w_{n+1} - w_n = \frac{(n+1)\pi}{L} - \frac{n\pi}{L} = \frac{\pi}{L}.$$

Then $1/L = \Delta w/\pi$, and we may write that Fourier series in the form

$$f_L(x) = \frac{1}{2L} \int_{-L}^{L} f_L(v)\, dv + \frac{1}{\pi} \sum_{n=1}^{\infty} \left[(\cos w_n x)\, \Delta w \int_{-L}^{L} f_L(v) \cos w_n v\, dv \right.$$

(1)

$$\left. + (\sin w_n x)\, \Delta w \int_{-L}^{L} f_L(v) \sin w_n v\, dv \right].$$

This representation is valid for any fixed L, arbitrarily large, but finite.

We now let $L \to \infty$ and assume that the resulting nonperiodic function

$$f(x) = \lim_{L \to \infty} f_L(x)$$

is **absolutely integrable** on the x-axis, that is, the following integral exists:

$$(2) \qquad \int_{-\infty}^{\infty} |f(x)| \, dx.$$

Then $1/L \to 0$, and the value of the first term on the right side of (1) approaches zero. Also $\Delta w = \pi/L \to 0$ and it seems *plausible* that the infinite series in (1) becomes an integral from 0 to ∞, which represents $f(x)$, namely,

$$(3) \qquad f(x) = \frac{1}{\pi} \int_0^{\infty} \left[\cos wx \int_{-\infty}^{\infty} f(v) \cos wv \, dv \right.$$
$$\left. + \sin wx \int_{-\infty}^{\infty} f(v) \sin wv \, dv \right] dw.$$

If we introduce the notations

$$(4) \qquad \boxed{A(w) = \frac{1}{\pi} \int_{-\infty}^{\infty} f(v) \cos wv \, dv, \qquad B(w) = \frac{1}{\pi} \int_{-\infty}^{\infty} f(v) \sin wv \, dv.}$$

we can write this in the form

$$(5) \qquad \boxed{f(x) = \int_0^{\infty} [A(w) \cos wx + B(w) \sin wx] \, dw.}$$

This is a representation of $f(x)$ by a so-called **Fourier integral.**

It is clear that our naive approach merely *suggests* the representation (5), but by no means establishes it; in fact, the limit of the series in (1) as Δw approaches zero is not the definition of the integral (3). Sufficient conditions for the validity of (5) are as follows.

Theorem 1 (Fourier integral)

If $f(x)$ is piecewise continuous (cf. Sec. 5.1) in every finite interval and has a right-hand derivative and a left-hand derivative at every point (cf. Sec. 10.2) and if the integral (2) exists, then $f(x)$ can be represented by a Fourier integral. At a point where $f(x)$ is discontinuous the value of the Fourier integral equals the average of the left- and right-hand limits of $f(x)$ at that point (cf. Sec. 10.2). (Proof in Ref. [C2]; cf. Appendix 1.)

The main use of the Fourier integral is in solving differential equations, as we shall see in Sec. 11.14. However, we can also use the Fourier integral in integration and in discussing functions defined by integrals, as the next examples illustrate.

EXAMPLE 3. Single pulse, sine integral

Find the Fourier integral representation of the function (Fig. 262)

$$f(x) = \begin{cases} 1 & \text{when } |x| < 1, \\ 0 & \text{when } |x| > 1. \end{cases}$$

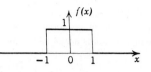

Fig. 262. Example 3

Solution. From (4) we obtain

$$A(w) = \frac{1}{\pi} \int_{-\infty}^{\infty} f(v) \cos wv \, dv = \frac{1}{\pi} \int_{-1}^{1} \cos wv \, dv = \frac{\sin wv}{\pi w} \Big|_{-1}^{1} = \frac{2 \sin w}{\pi w},$$

$$B(w) = \frac{1}{\pi} \int_{-1}^{1} \sin wv \, dv = 0,$$

and (5) gives the answer

(6) $$f(x) = \frac{2}{\pi} \int_0^{\infty} \frac{\cos wx \sin w}{w} \, dw.$$

The average of the left- and right-hand limits of $f(x)$ at $x = 1$ is equal to $(1 + 0)/2$, that is, $1/2$. Furthermore, from (6) and Theorem 1 we obtain

$$\int_0^{\infty} \frac{\cos wx \sin w}{w} \, dw = \begin{cases} \pi/2 & \text{if} \quad 0 \leq x < 1, \\ \pi/4 & \text{if} \quad x = 1, \\ 0 & \text{if} \quad x > 1. \end{cases}$$

We mention that this integral is called **Dirichlet's discontinuous factor**.[11] Let us consider the case $x = 0$, which is of particular interest. When $x = 0$, then

(7) $$\int_0^{\infty} \frac{\sin w}{w} \, dw = \frac{\pi}{2}.$$

We see that this integral is the limit of the so-called **sine integral**

(8) $$\text{Si}(z) = \int_0^z \frac{\sin w}{w} \, dw$$

as $z \to \infty$ (z real). The graph of Si(z) is shown in Fig. 263.

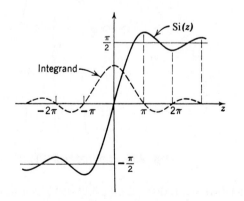

Fig. 263. Sine integral

[11]See footnote 15 in Sec. 9.7.

In the case of a Fourier series the graphs of the partial sums are approximation curves of the curve of the periodic function represented by the series. Similarly, in the case of the Fourier integral (5), approximations are obtained by replacing ∞ by numbers a. Hence the integral

(9)
$$\int_0^a \frac{\cos wx \sin w}{w}\, dw$$

approximates the integral in (6) and therefore $f(x)$; cf. Fig. 264.

Figure 264 shows oscillations near the points of discontinuity of $f(x)$. We might expect that these oscillations disappear as a approaches infinity, but this is not true; with increasing a, they are shifted closer to the points $x = \pm 1$. This unexpected behavior, which also occurs in connection with Fourier series, is known as the **Gibbs phenomenon.**[12] It can be explained by representing (9) in terms of the sine integral as follows. Using (11) in Appendix 3.1, we first have

$$\frac{2}{\pi}\int_0^a \frac{\cos wx \sin w}{w}\, dw = \frac{1}{\pi}\int_0^a \frac{\sin (w + wx)}{w}\, dw + \frac{1}{\pi}\int_0^a \frac{\sin (w - wx)}{w}\, dw.$$

In the first integral on the right we set $w + wx = t$. Then $dw/w = dt/t$, and $0 \le w \le a$ corresponds to $0 \le t \le (x + 1)a$. In the last integral we set $w - wx = -t$. Then $dw/w = dt/t$, and the interval $0 \le w \le a$ corresponds to $0 \le t \le (x - 1)a$. Since $\sin (-t) = -\sin t$, we thus obtain

$$\frac{2}{\pi}\int_0^a \frac{\cos wx \sin w}{w}\, dw = \frac{1}{\pi}\int_0^{(x+1)a} \frac{\sin t}{t}\, dt - \frac{1}{\pi}\int_0^{(x-1)a} \frac{\sin t}{t}\, dt.$$

From this and (8) we see that our integral equals

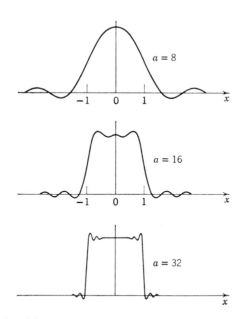

Fig. 264. The integral (9) for $a = 8, 16$ and 32

[12]JOSIAH WILLARD GIBBS (1839—1903), American mathematician, professor of mathematical physics at Yale from 1871, one of the founders of vector calculus, mathematical thermodynamics and statistical mechanics (*Elementary Principles in Statistical Mechanics,* New York, 1902). His work was of great importance to the development of vector analysis and mathematical physics.

$$\frac{1}{\pi} \, \text{Si}(a[x + 1]) - \frac{1}{\pi} \, \text{Si}(a[x - 1]),$$

and the oscillations in Fig. 264 result from those in Fig. 263. The increase of a amounts to a transformation of the scale on the axis and causes the shift of the oscillations.   ∎

Fourier Cosine and Sine Integrals

It is of practical interest to note that if a function is even or odd and can be represented by a Fourier integral, then this representation will be simpler than in the case of an arbitrary function. This follows immediately from our previous formulas, as we shall now see.

If $f(x)$ is an even function, then in (4) we have $B(w) = 0$ and

$$(10) \qquad \boxed{A(w) = \frac{2}{\pi} \int_0^\infty f(v) \cos wv \, dv,}$$

and the Fourier integral (5) reduces to the so-called **Fourier cosine integral**

$$(11) \qquad \boxed{f(x) = \int_0^\infty A(w) \cos wx \, dw} \qquad (f \text{ even}).$$

Similarly, if $f(x)$ is odd, then in (4) we have $A(w) = 0$ and

$$(12) \qquad \boxed{B(w) = \frac{2}{\pi} \int_0^\infty f(v) \sin wv \, dv,}$$

and the Fourier integral (5) reduces to the so-called **Fourier sine integral**

$$(13) \qquad \boxed{f(x) = \int_0^\infty B(w) \sin wx \, dw} \qquad (f \text{ odd}).$$

These simplifications are quite similar to those in the case of a Fourier series discussed in Sec. 10.4.

Evaluation of Integrals

Fourier integral representations may also be used for evaluating integrals. We illustrate this method with a typical example.

EXAMPLE 4. Laplace integrals

Find the Fourier cosine and sine integrals of

$$f(x) = e^{-kx} \qquad\qquad (x > 0, k > 0).$$

Solution. **(a)** From (10) we have

$$A(w) = \frac{2}{\pi} \int_0^\infty e^{-kv} \cos wv \, dv.$$

Now, by integration by parts,

$$\int e^{-kv} \cos wv \, dv = -\frac{k}{k^2 + w^2} e^{-kv} \left(-\frac{w}{k} \sin wv + \cos wv \right).$$

When $v = 0$, the expression on the right equals $-k/(k^2 + w^2)$; when v approaches infinity, it approaches zero because of the exponential factor. Thus

$$(14) \qquad\qquad\qquad\qquad A(w) = \frac{2k/\pi}{k^2 + w^2}.$$

By substituting this into (11) we thus obtain the Fourier cosine integral representation

$$f(x) = e^{-kx} = \frac{2k}{\pi} \int_0^\infty \frac{\cos wx}{k^2 + w^2} \, dw \qquad\qquad (x > 0, \, k > 0).$$

From this representation we see that

$$(15) \qquad\qquad \int_0^\infty \frac{\cos wx}{k^2 + w^2} \, dw = \frac{\pi}{2k} e^{-kx} \qquad\qquad (x > 0, \, k > 0).$$

(b) Similarly, from (12) we have

$$B(w) = \frac{2}{\pi} \int_0^\infty e^{-kv} \sin wv \, dv.$$

By integration by parts,

$$\int e^{-kv} \sin wv \, dv = -\frac{w}{k^2 + w^2} e^{-kv} \left(\frac{k}{w} \sin wv + \cos wv \right).$$

This equals $-w/(k^2 + w^2)$ when $v = 0$, and approaches 0 as $v \to \infty$. Thus

$$(16) \qquad\qquad\qquad\qquad B(w) = \frac{2w/\pi}{k^2 + w^2}.$$

From (13) we thus obtain the Fourier sine integral representation

$$f(x) = e^{-kx} = \frac{2}{\pi} \int_0^\infty \frac{w \sin wx}{k^2 + w^2} \, dw.$$

From this we see that

$$(17) \qquad\qquad \int_0^\infty \frac{w \sin wx}{k^2 + w^2} \, dw = \frac{\pi}{2} e^{-kx} \qquad\qquad (x > 0, \, k > 0).$$

The integrals (15) and (17) are the so-called **Laplace integrals.** ∎

In the next section we shall see that formulas (10)—(13) can be used for defining two integral transformations known as the **Fourier cosine** and **Fourier sine transformations.** This will be quite simple. Somewhat harder work follows in Sec. 10.11, where we introduce and discuss the (complex) **Fourier transformation.** Tables of functions and their transforms are included in Sec. 10.12.

Problems for Sec. 10.9

Using the Fourier integral representation, show that

1. $\displaystyle\int_0^\infty \frac{\cos xw + w \sin xw}{1 + w^2}\, dw = \begin{cases} 0 & \text{if } x < 0 \\ \pi/2 & \text{if } x = 0 \\ \pi e^{-x} & \text{if } x > 0 \end{cases}$ [Use (5).]

2. $\displaystyle\int_0^\infty \frac{1 - \cos \pi w}{w} \sin xw\, dw = \begin{cases} \dfrac{\pi}{2} & \text{if } 0 < x < \pi \\ 0 & \text{if } x > \pi \end{cases}$ [Use (13).]

3. $\displaystyle\int_0^\infty \frac{\cos xw}{1 + w^2}\, dw = \frac{\pi}{2} e^{-x} \quad (x > 0)$ [Use (11).]

4. $\displaystyle\int_0^\infty \frac{\sin w \cos xw}{w}\, dw = \begin{cases} \pi/2 & \text{if } 0 \leqq x < 1 \\ \pi/4 & \text{if } x = 1 \\ 0 & \text{if } x > 1 \end{cases}$ [Use (11).]

5. $\displaystyle\int_0^\infty \frac{\cos (\pi w/2) \cos xw}{1 - w^2}\, dw = \begin{cases} \dfrac{\pi}{2} \cos x & \text{if } |x| < \dfrac{\pi}{2} \\ 0 & \text{if } |x| > \dfrac{\pi}{2} \end{cases}$ [Use (11).]

6. $\displaystyle\int_0^\infty \frac{w^3 \sin xw}{w^4 + 4}\, dw = \frac{\pi}{2} e^{-x} \cos x \quad \text{if } x > 0$ [Use (13).]

7. $\displaystyle\int_0^\infty \frac{\sin \pi w \sin xw}{1 - w^2}\, dw = \begin{cases} \dfrac{\pi}{2} \sin x & \text{if } 0 \leqq x \leqq \pi \\ 0 & \text{if } x > \pi \end{cases}$ [Use (13).]

Represent the following functions $f(x)$ in the form (11).

8. $f(x) = \begin{cases} 1 & \text{if } 0 < x < a \\ 0 & \text{if } x > a \end{cases}$ **9.** $f(x) = \begin{cases} x & \text{if } 0 < x < a \\ 0 & \text{if } x > a \end{cases}$

10. $f(x) = \dfrac{1}{1 + x^2}$ [cf. (15)]   **11.** $f(x) = e^{-x} + e^{-2x} \quad (x > 0)$

12. $f(x) = \begin{cases} x & \text{if } 0 < x < 1 \\ 2 - x & \text{if } 1 < x < 2 \\ 0 & \text{if } x > 2 \end{cases}$ **13.** $f(x) = \begin{cases} x^2 & \text{if } 0 < x < a \\ 0 & \text{if } x > a \end{cases}$

14. Show that $f(x) = 1$ $(0 < x < \infty)$ cannot be represented by a Fourier integral.

If $f(x)$ has the representation (11), show that

15. $f(ax) = \dfrac{1}{a} \displaystyle\int_0^\infty A\left(\dfrac{w}{a}\right) \cos xw\, dw \quad (a > 0)$

16. $x^2 f(x) = \displaystyle\int_0^\infty A^*(w) \cos xw\, dw, \qquad A^* = -\dfrac{d^2 A}{dw^2}$

17. Verify the formula in Prob. 15 for $f(x) = e^{-x}$.

18. Solve Prob. 13 by applying the formula in Prob. 16 to the result of Prob. 8.

19. Show that $xf(x) = \int_0^\infty B^*(w) \sin xw \, dw$, where $B^* = -\dfrac{dA}{dw}$ and A is given by formula (10).

20. Verify the formula in Prob. 19 for $f(x) = 1$ if $0 < x < a$ and $f(x) = 0$ if $x > a$.

10.10 Fourier Cosine Transformation, Fourier Sine Transformation

An **integral transformation** is a transformation that produces from given functions new functions which depend on a different variable and appear in the form of an integral to be evaluated. These transformations are of interest mainly as tools in solving ordinary differential equations, partial differential equations and integral equations, and they often also help in handling and applying special functions. The **Laplace transformation** (Chap. 5) is of this kind and is by far the most important integral transformation in engineering. From the viewpoint of applications, the next in order of importance are perhaps the **Fourier transformations,** although these are somewhat more difficult to handle than the Laplace transformation. We shall see that they can be obtained from the Fourier integral representations in Sec. 10.9. In this section we consider two of them, called the *Fourier cosine* and *sine transformations,* which are real, and in the next section a third one, which is complex.

Fourier Cosine Transformation

For an *even* function $f(x)$, the Fourier integral is the Fourier cosine integral [cf. (10), (11), Sec. 10.9]

(1) (a) $f(x) = \int_0^\infty A(w) \cos wx \, dw$, where (b) $A(w) = \dfrac{2}{\pi} \int_0^\infty f(v) \cos wv \, dv$.

We now set $A(w) = \sqrt{2/\pi} \, \hat{f}_c(w)$, where c suggests "cosine." Then from (1b), writing $v = x$, we have

(2)
$$\hat{f}_c(w) = \sqrt{\frac{2}{\pi}} \int_0^\infty f(x) \cos wx \, dx$$

and from (1a),

(3)
$$f(x) \sqrt{\frac{2}{\pi}} \int_0^\infty \hat{f}_c(w) \cos wx \, dw.$$

Attention! In (2) we integrate with respect to x and in (3) with respect to w. Formula (2) gives from $f(x)$ a new function $\hat{f}_c(w)$, called the **Fourier cosine transform** of $f(x)$. Formula (3) gives us back $f(x)$ from $\hat{f}_c(w)$, and we therefore call $f(x)$ the **inverse Fourier cosine transform** of $\hat{f}(w)$.

The process of obtaining the transform $\hat{f}_c$ from a given f is called the **Fourier cosine transformation.** Similarly for the inverse process.

Fourier Sine Transformation

Similarly, for an *odd* function $f(x)$, the Fourier integral is the Fourier sine integral [cf. (12), (13), Sec. 10.9]

(4)   (a) $f(x) = \displaystyle\int_0^\infty B(w) \sin wx \, dw$,   where   (b) $B(w) = \dfrac{2}{\pi} \displaystyle\int_0^\infty f(v) \sin wv \, dv$.

We now set $B(w) = \sqrt{2/\pi} \, \hat{f}_s(w)$, where s suggests "sine." Then from (4b), writing $v = x$, we have

(5)
$$\hat{f}_s(w) = \sqrt{\frac{2}{\pi}} \int_0^\infty f(x) \sin wx \, dx,$$

called the **Fourier sine transform** of $f(x)$, and from (4a)

(6)
$$f(x) = \sqrt{\frac{2}{\pi}} \int_0^\infty \hat{f}_s(w) \sin wx \, dw,$$

called the **inverse Fourier sine transform** of $\hat{f}_s(w)$. The process of obtaining $\hat{f}_s(w)$ from $f(x)$ is called the **Fourier sine transformation.** Similarly for the inverse process.

Other notations are

$$\mathcal{F}_c(f) = \hat{f}_c, \qquad\qquad \mathcal{F}_s(f) = \hat{f}_s$$

and $\mathcal{F}_c^{-1}$ and $\mathcal{F}_s^{-1}$ for the inverses of $\mathcal{F}_c$ and $\mathcal{F}_s$, respectively.

EXAMPLE 1. Fourier cosine and Fourier sine transforms

Find the Fourier cosine and sine transforms of the function

$$f(x) = \begin{cases} k & \text{if } 0 < x < a \\ 0 & \text{if } x > a. \end{cases}$$

Solution. From the definitions (2) and (5) we obtain by integration

$$\hat{f}_c(w) = \sqrt{\frac{2}{\pi}} k \int_0^a \cos wx \, dx = \sqrt{\frac{2}{\pi}} k \left[\frac{\sin aw}{w} \right],$$

$$\hat{f}_s(w) = \sqrt{\frac{2}{\pi}} k \int_0^a \sin wx \, dx = \sqrt{\frac{2}{\pi}} k \left[\frac{1 - \cos aw}{w} \right].$$

This agrees with formulas 1 in the first two tables in Sec. 10.12 (where $k = 1$).

Note that for $f(x) = k = const$ $(0 < x < \infty)$, these transforms do not exist. (Why?)

EXAMPLE 2. Fourier cosine transform of the exponential function

Find $\mathcal{F}_c(e^{-x})$.

Solution. By integration by parts and recursion,

$$\mathcal{F}_c(e^{-x}) = \sqrt{\frac{2}{\pi}} \int_0^\infty e^{-x} \cos wx \, dx = \sqrt{\frac{2}{\pi}} \frac{e^{-x}}{1 + w^2} (-\cos wx + w \sin wx) \Big|_0^\infty = \frac{\sqrt{2/\pi}}{1 + w^2}.$$

This agrees with formula 3 in Table I, Sec. 10.12, with $a = 1$.   ∎

What have we done in order to introduce the two integral transformations under consideration? Actually not much: We have changed the notations A and B to get a "symmetric" distribution of the constant $2/\pi$ in the original formulas (10)–(13), Sec. 10.9. This redistribution is a standard convenience, but is not essential. One could do without it.

What have we gained? We show next that these transformations have operational properties that permit them to convert differentiations into algebraic operations (just as the Laplace transformation does). This is the key to their application in solving differential equations.

Linearity. Transforms of Derivatives

If $f(x)$ is absolutely integrable (cf. Sec. 10.9) on the positive x-axis and piecewise continuous (cf. Sec. 5.1) on every finite interval, then the Fourier cosine and sine transformations of f exist.

Furthermore, for a function $af(x) + bg(x)$ we have from (2)

$$\mathcal{F}_c(af + bg) = \sqrt{\frac{2}{\pi}} \int_0^\infty [af(x) + bg(x)] \cos wx \, dx$$

$$= a \sqrt{\frac{2}{\pi}} \int_0^\infty f(x) \cos wx \, dx + b \sqrt{\frac{2}{\pi}} \int_0^\infty g(x) \cos wx \, dx.$$

The right-hand side is $a\mathcal{F}_c(f) + b\mathcal{F}_c(g)$. Similarly for $\mathcal{F}_s$, by (5). This shows that the Fourier cosine and sine transformations are **linear operations,**

(7)
$$\text{(a)} \quad \mathcal{F}_c(af + bg) = a\mathcal{F}_c(f) + b\mathcal{F}_c(g),$$
$$\text{(b)} \quad \mathcal{F}_s(af + bg) = a\mathcal{F}_s(f) + b\mathcal{F}_s(g).$$

Theorem 1 (Cosine and sine transforms of derivatives)

Let $f(x)$ be continuous and absolutely integrable (cf. Sec. 10.9) on the x-axis, let $f'(x)$ be piecewise continuous on each finite interval, and let $f(x) \to 0$ as $x \to \infty$. Then

(8)
$$\text{(a)} \quad \mathcal{F}_c\{f'(x)\} = w\mathcal{F}_s\{f(x)\} - \sqrt{\frac{2}{\pi}} f(0),$$
$$\text{(b)} \quad \mathcal{F}_s\{f'(x)\} = -w\mathcal{F}_c\{f(x)\}.$$

Proof. This follows from the definitions by integration by parts, namely,

$$\mathscr{F}_c\{f'(x)\} = \sqrt{\frac{2}{\pi}} \int_0^\infty f'(x) \cos wx \, dx$$

$$= \sqrt{\frac{2}{\pi}} \left[f(x) \cos wx \, \bigg|_0^\infty + w \int_0^\infty f(x) \sin wx \, dx \right]$$

$$= -\sqrt{\frac{2}{\pi}} f(0) + w\mathscr{F}_s\{f(x)\};$$

similarly,

$$\mathscr{F}_s\{f'(x)\} = \sqrt{\frac{2}{\pi}} \int_0^\infty f'(x) \sin wx \, dx$$

$$= \sqrt{\frac{2}{\pi}} \left[f(x) \sin wx \, \bigg|_0^\infty - w \int_0^\infty f(x) \cos wx \, dx \right]$$

$$= 0 - w\mathscr{F}_c\{f(x)\}. \qquad \blacksquare$$

Formula (8a) with f' instead of f gives

$$\mathscr{F}_c\{f''(x)\} = w\mathscr{F}_s\{f'(x)\} - \sqrt{\frac{2}{\pi}} f'(0);$$

hence by (8b),

(9a)
$$\boxed{\mathscr{F}_c\{f''(x)\} = -w^2\mathscr{F}_c\{f(x)\} - \sqrt{\frac{2}{\pi}} f'(0).}$$

Similarly,

(9b)
$$\boxed{\mathscr{F}_s\{f''(x)\} = -w^2\mathscr{F}_s\{f(x)\} + \sqrt{\frac{2}{\pi}} wf(0).}$$

An application of (9) to differential equations will be given in Sec. 11.14. For the time being, we show how (9) can be used to derive transforms.

EXAMPLE 3. An application of the operational formula (9)

Find the Fourier cosine transform of $f(x) = e^{-ax}$, where $a > 0$.

Solution. By differentiation, $(e^{-ax})'' = a^2 e^{-ax}$; thus $a^2 f(x) = f''(x)$. From this and (9a),

$$a^2\mathscr{F}_c(f) = \mathscr{F}_c(f'') = -w^2\mathscr{F}_c(f) - \sqrt{\frac{2}{\pi}} f'(0) = -w^2\mathscr{F}_c(f) + a\sqrt{\frac{2}{\pi}}.$$

Hence $(a^2 + w^2)\mathscr{F}_c(f) = a\sqrt{2/\pi}$. The answer is (cf. Table I, Sec. 10.12)

$$\mathscr{F}_c(e^{-ax}) = \sqrt{\frac{2}{\pi}} \left(\frac{a}{a^2 + w^2} \right) \qquad (a > 0). \qquad \blacksquare$$

Tables of Fourier cosine and sine transforms are included in Sec. 10.12. For more extensive tables, see Ref. [C4] in Appendix 1.

The two transformations considered refer to the positive x-axis. In the next section we shall consider the **Fourier transformation,** which applies to functions defined on the *entire* axis.

Problems for Sec. 10.10

1. Do the Fourier cosine and sine transforms of $f(x) = e^x$ exist?
2. Show that $f(x) = 1$ has no Fourier cosine or sine transform.
3. Derive formula 3 in Table I of Sec. 10.12 by integration.
4. Find $\mathscr{F}_s(e^{-ax})$, $a > 0$, by integration.
5. Obtain $\mathscr{F}_c(1/(1 + x^2))$ from Prob. 3 in Sec. 10.9.
6. Obtain the inverse Fourier cosine transform of e^{-w}.
7. Obtain $\mathscr{F}_s(w^{-1} - w^{-1} \cos \pi w)$. *Hint.* Use Prob. 2, Sec. 10.9.
8. Does the Fourier cosine transform of $(\cos x)/x$ exist? Of $(\sin x)/x$?
9. Find $\mathscr{F}_c\{(\cos \pi x/2)/(1 - x^2)\}$. *Hint.* Use Prob. 5, Sec. 10.9.
10. Obtain formula 10 in Table I of Sec. 10.12 with $a = 1$ from Example 3 in Sec. 10.9.
11. Why did we require $a > 0$ in Example 3?
12. Obtain the answer to Prob. 4 from (9b).
13. Find $\mathscr{F}_s(e^{-x})$ from (8a) and formula 3 of Table I, Sec. 10.12.
14. Using (8b), obtain $\mathscr{F}_s(xe^{-x^2/2})$ from a suitable formula in Table I, Sec. 10.12.
15. Using $\Gamma(\tfrac{1}{2}) = \sqrt{\pi}$, obtain formula 2 in Table II, Sec. 10.12, from formula 4 in that table.

10.11 Fourier Transformation

The previous section concerned two transformations obtained from the Fourier cosine and sine integrals in Sec. 10.9. We now consider a third transformation, which is called the *Fourier transformation* and is obtained from the Fourier integral in complex form. (For a motivation of this transformation, see at the beginning of Sec. 10.10.) We therefore consider first the complex form of the Fourier integral.

Complex Form of the Fourier Integral

The (real) Fourier integral is [cf. (4), (5), Sec. 10.9]

$$f(x) = \int_0^\infty [A(w) \cos wx + B(w) \sin wx] \, dw$$

where

$$A(w) = \frac{1}{\pi} \int_{-\infty}^\infty f(v) \cos wv \, dv, \qquad B(w) = \frac{1}{\pi} \int_{-\infty}^\infty f(v) \sin wv \, dv.$$

Substituting A and B into the integral for f, we have

$$f(x) = \frac{1}{\pi} \int_0^\infty \int_{-\infty}^\infty f(v)[\cos wv \cos wx + \sin wv \sin wx]\, dv\, dw.$$

If we apply the addition formula for the cosine [(6) in Appendix 3.1], we get

(1*)
$$f(x) = \frac{1}{\pi} \int_0^\infty \left[\int_{-\infty}^\infty f(v) \cos (wx - wv)\, dv \right] dw.$$

The integral in brackets is an *even* function of w, call it $F(w)$, because $\cos (wx - wv)$ is an even function of w, the function f does not depend on w, and we integrate with respect to v (not w). Hence the integral of $F(w)$ from $w = 0$ to ∞ is $1/2$ times the integral of $F(w)$ from $-\infty$ to ∞. Thus

(1)
$$f(x) = \frac{1}{2\pi} \int_{-\infty}^\infty \left[\int_{-\infty}^\infty f(v) \cos (wx - wv)\, dv \right] dw.$$

We claim that the corresponding integral of the form (1) with sin instead of cos is zero:

(2)
$$\frac{1}{2\pi} \int_{-\infty}^\infty \left[\int_{-\infty}^\infty f(v) \sin (wx - wv)\, dv \right] dw = 0.$$

This is true since $\sin (wx - wv)$ is an odd function of w, which makes the integral in brackets an odd function of w, call it $G(w)$, so that the integral of $G(w)$ from $-\infty$ to ∞ is zero, as claimed. We now use the **Euler formula**

(3)
$$e^{it} = \cos t + i \sin t$$

for the complex exponential function (review of this function in Sec. 2.4). We set $t = wx - wv$ and add (1) and i times (2), obtaining

(4)
$$f(x) = \frac{1}{2\pi} \int_{-\infty}^\infty \int_{-\infty}^\infty f(v)e^{iw(x-v)}\, dv\, dw \qquad (i = \sqrt{-1}).$$

This is called the **complex Fourier integral.**

It is now a short step from this to the Fourier transform, our present goal.

Fourier Transformation

Writing the exponential function in (4) as a product of exponential functions, we have

(5)
$$f(x) = \frac{1}{\sqrt{2\pi}} \int_{-\infty}^\infty \left[\frac{1}{\sqrt{2\pi}} \int_{-\infty}^\infty f(v)e^{-iwv}\, dv \right] e^{iwx}\, dw.$$

The expression in brackets is a function of w, is denoted by $\hat{f}(w)$ and is called the **Fourier transform** of f; writing $v = x$, we have

(6)
$$\hat{f}(w) = \frac{1}{\sqrt{2\pi}} \int_{-\infty}^{\infty} f(x)e^{-iwx}\,dx.$$

With this, (5) becomes

(7)
$$f(x) = \frac{1}{\sqrt{2\pi}} \int_{-\infty}^{\infty} \hat{f}(w)e^{iwx}\,dw$$

and is called the **inverse Fourier transform** of $\hat{f}(w)$.

Another notation is $\mathcal{F}(f) = \hat{f}(w)$ and $\mathcal{F}^{-1}$ for the inverse.

The process of obtaining the Fourier transform $\mathcal{F}(f) = \hat{f}$ from a given f is called **Fourier transformation.** Similarly for the inverse.

Existence. The following two conditions are sufficient for the existence of the Fourier transform (6) of a function $f(x)$ defined on the x-axis, as we mention without proof.

1. $f(x)$ is piecewise continuous on every finite interval (cf. Sec. 5.1).

2. $f(x)$ is **absolutely integrable** on the x-axis; that is, the following (finite!) limits exist:

$$\lim_{a \to -\infty} \int_a^0 |f(x)|\,dx + \lim_{b \to \infty} \int_0^b |f(x)|\,dx \qquad \left(\text{written } \int_{-\infty}^{\infty} |f(x)|\,dx \right).$$

EXAMPLE 1. Fourier transform
Find the Fourier transform of $f(x) = k$ if $0 < x < a$ and $f(x) = 0$ otherwise.

Solution. From (6) by integration,

$$\hat{f}(w) = \frac{1}{\sqrt{2\pi}} \int_0^a ke^{-iwx}\,dx = \frac{k}{\sqrt{2\pi}} \left(\frac{e^{-iwa} - 1}{-iw} \right) = \frac{k(1 - e^{-iaw})}{iw\sqrt{2\pi}}.$$

This shows that the Fourier transform will in general be a complex-valued function.

EXAMPLE 2. Fourier transform
Find the Fourier transform of e^{-ax^2}, where $a > 0$.

Solution. We use the definition, complete the square in the exponent and pull out the exponential factor that contains no x:

$$\mathcal{F}(e^{-ax^2}) = \frac{1}{\sqrt{2\pi}} \int_{-\infty}^{\infty} \exp\left[-ax^2 - iwx\right] dx$$

$$= \frac{1}{\sqrt{2\pi}} \int_{-\infty}^{\infty} \exp\left[-\left(\sqrt{a}x + \frac{iw}{2\sqrt{a}} \right)^2 + \left(\frac{iw}{2\sqrt{a}} \right)^2 \right] dx.$$

$$= \frac{1}{\sqrt{2\pi}} \exp\left(-\frac{w^2}{4a} \right) \int_{-\infty}^{\infty} \exp\left[-\left(\sqrt{a}x + \frac{iw}{2\sqrt{a}} \right)^2 \right] dx.$$

We denote the integral by I and show that it equals $\sqrt{\pi/a}$. For this we use $\sqrt{a}x + iw/2\sqrt{a} = v$ as a new variable of integration. Then $dx = dv/\sqrt{a}$, so that

$$I = \frac{1}{\sqrt{a}} \int_{-\infty}^{\infty} e^{-v^2}\, dv.$$

We now get the result by the following trick. We square the integral, convert it to a double integral and introduce polar coordinates $r = \sqrt{u^2 + v^2}$ and θ. Since $du\, dv = r\, dr\, d\theta$, we get

$$I^2 = \frac{1}{a} \int_{-\infty}^{\infty} e^{-u^2}\, du \int_{-\infty}^{\infty} e^{-v^2}\, dv = \frac{1}{a} \int_{-\infty}^{\infty} \int_{-\infty}^{\infty} e^{-(u^2+v^2)}\, du\, dv$$

$$= \frac{1}{a} \int_0^{2\pi} \int_0^{\infty} e^{-r^2} r\, dr\, d\theta = \frac{2\pi}{a} \left(-\frac{1}{2} e^{-r^2} \right) \Big|_0^{\infty} = \frac{\pi}{a}.$$

Hence $I = \sqrt{\pi/a}$. From this and the first formula in this solution,

$$\mathscr{F}(e^{-ax^2}) = \frac{1}{\sqrt{2\pi}} \exp\left(-\frac{w^2}{4a} \right) \sqrt{\frac{\pi}{a}} = \frac{1}{\sqrt{2a}} e^{-w^2/4a}.$$

This agrees with formula 9 in Table III, Sec. 10.12.   ∎

Linearity. Fourier Transform of Derivatives

New transforms can be obtained from given ones by

Theorem 1 (Linearity of the Fourier transformation)

The Fourier transformation is a linear operation; that is, for any functions $f(x)$ and $g(x)$ whose Fourier transforms exist and any constants a and b,

(8)
$$\boxed{\mathscr{F}(af + bg) = a\mathscr{F}(f) + b\mathscr{F}(g).}$$

Proof. This is true since integration is a linear operation, so that (6) gives

$$\mathscr{F}\{af(x) + bg(x)\} = \frac{1}{\sqrt{2\pi}} \int_{-\infty}^{\infty} [af(x) + bg(x)]e^{-iwx}\, dx$$

$$= a \frac{1}{\sqrt{2\pi}} \int_{-\infty}^{\infty} f(x)e^{-iwx}\, dx + b \frac{1}{\sqrt{2\pi}} \int_{-\infty}^{\infty} g(x)e^{-iwx}\, dx.$$

$$= a\mathscr{F}\{f(x)\} + b\mathscr{F}\{g(x)\}.\qquad\blacksquare$$

Linearity is also vital to the applicability of the Fourier transformation to the solution of differential equations, which is based on the key property that differentiation of functions corresponds to multiplication of transforms by iw:

Theorem 2 [Fourier transform of the derivative of $f(x)$]

Let $f(x)$ be continuous on the x-axis and $f(x) \to 0$ as $|x| \to \infty$. Furthermore, let $f'(x)$ be absolutely integrable on the x-axis. Then

(9)
$$\boxed{\mathscr{F}\{f'(x)\} = iw\mathscr{F}\{f(x)\}.}$$

Proof. Integrating by parts and using $f(x) \to 0$ as $|x| \to \infty$, we obtain

$$\mathscr{F}\{f'(x)\} = \frac{1}{\sqrt{2\pi}} \int_{-\infty}^{\infty} f'(x)e^{-iwx}\, dx$$

$$= \frac{1}{\sqrt{2\pi}} \left[f(x)e^{-iwx} \Big|_{-\infty}^{\infty} - (-iw) \int_{-\infty}^{\infty} f(x)e^{-iwx}\, dx \right]$$

$$= 0 + iw\, \mathscr{F}\{f(x)\}. \qquad \blacksquare$$

Two successive applications of (9) give

$$\mathscr{F}(f'') = iw\mathscr{F}(f') = (iw)^2 \mathscr{F}(f).$$

Since $(iw)^2 = -w^2$, we have for the transform of the second derivative of f

(10)
$$\boxed{\mathscr{F}\{f''(x)\} = -w^2 \mathscr{F}\{f(x)\}.}$$

Similarly for higher derivatives.

An application of (10) to differential equations will be given in Sec. 11.14. For the time being we show how (9) can be used to derive transforms.

EXAMPLE 3. An application of the operational formula (9)
Find the Fourier transform of xe^{-x^2} from Table III, Sec. 10.12.

Solution. We use (9). By formula 9 in Table III,

$$\mathscr{F}(xe^{-x^2}) = \mathscr{F}\left\{ -\frac{1}{2}(e^{-x^2})' \right\} = -\frac{1}{2}\mathscr{F}\left\{ (e^{-x^2})' \right\}$$

$$= -\frac{1}{2} iw\mathscr{F}(e^{-x^2}) = -\frac{1}{2} iw \frac{1}{\sqrt{2}} e^{-w^2/4} = -\frac{iw}{2\sqrt{2}} e^{-w^2/4}. \qquad \blacksquare$$

Convolution

The **convolution** $f * g$ of functions f and g is defined by

(11) $\quad h(x) = (f * g)(x) = \int_{-\infty}^{\infty} f(p)g(x - p)\, dp = \int_{-\infty}^{\infty} f(x - p)g(p)\, dp.$

The purpose is the same as in the case of the Laplace transformation (Sec. 5.6): the convolution of functions corresponds to the multiplication of their Fourier transforms (except for a factor $\sqrt{2\pi}$):

Theorem 3 (Convolution theorem)
Let $f(x)$ and $g(x)$ be piecewise continuous, bounded and absolutely integrable on the x-axis. Then

(12)
$$\boxed{\mathscr{F}(f * g) = \sqrt{2\pi}\,\mathscr{F}(f)\mathscr{F}(g).}$$

Proof. By the definition and an interchange of the order of integration we have

$$\mathscr{F}(f * g) = \frac{1}{\sqrt{2\pi}} \int_{-\infty}^{\infty} \int_{-\infty}^{\infty} f(p)g(x - p)e^{-iwx} \, dp \, dx$$

$$= \frac{1}{\sqrt{2\pi}} \int_{-\infty}^{\infty} \int_{-\infty}^{\infty} f(p)g(x - p)e^{-iwx} \, dx \, dp.$$

Instead of x we now take $x - p = q$ as a new variable of integration. Then $x = p + q$ and

$$\mathscr{F}(f * g) = \frac{1}{\sqrt{2\pi}} \int_{-\infty}^{\infty} \int_{-\infty}^{\infty} f(p)g(q)e^{-iw(p+q)} \, dq \, dp$$

$$= \frac{1}{\sqrt{2\pi}} \int_{-\infty}^{\infty} f(p)e^{-iwp} \, dp \int_{-\infty}^{\infty} g(q)e^{-iwq} \, dq$$

$$= \sqrt{2\pi}\,\mathscr{F}(f)\mathscr{F}(g). \qquad \blacksquare$$

By taking the inverse Fourier transform on both sides of (12), writing $\hat{f} = \mathscr{F}(f)$ and $\hat{g} = \mathscr{F}(g)$ as before, and noting that $\sqrt{2\pi}$ and $1/\sqrt{2\pi}$ cancel each other, we obtain

(13) $$(f * g)(x) = \int_{-\infty}^{\infty} \hat{f}(w)\hat{g}(w)e^{iwx} \, dw,$$

a formula that will help us in solving partial differential equations (Sec. 11.14).

A **table** of Fourier transforms is included in the next section. For more extensive tables, see Ref. [C4] in Appendix 1.

This is the end of Chap. 10 on Fourier series, Fourier integrals and Fourier transforms. The introduction of Fourier series (and Fourier integrals) was one of the greatest advances ever made in mathematical physics and its engineering applications. In the next chapter, which concerns **partial differential equations,** we shall see why this is so: Fourier series (and Fourier integrals) are probably the most important tool in solving boundary value and initial value problems for partial differential equations. These problems arise in connection with vibrations, heat flow, fluid flow, electrostatic and other potential problems, and so on.

Problems for Sec. 10.11

Find the Fourier transforms of the following functions $f(x)$ (without using Table III, Sec. 10.12).

1. $f(x) = \begin{cases} e^{-x} & \text{if } x > 0 \\ 0 & \text{if } x < 0 \end{cases}$

2. $f(x) = \begin{cases} e^{x} & \text{if } x < 0 \\ 0 & \text{if } x > 0 \end{cases}$

3. $f(x) = \begin{cases} e^{2ix} & \text{if } -1 < x < 1 \\ 0 & \text{otherwise} \end{cases}$

4. $f(x) = \begin{cases} 1 & \text{if } 1 < x < 3 \\ 0 & \text{otherwise} \end{cases}$

5. $f(x) = \begin{cases} e^x & \text{if } -1 < x < 1 \\ 0 & \text{otherwise} \end{cases}$ **6.** $f(x) = \begin{cases} x & \text{if } 0 < x < a \\ 0 & \text{otherwise} \end{cases}$

7. Obtain formula 1 in Table III, Sec. 10.12, from formula 2 in that table.

In Table III, Sec. 10.12, verify:

8. Formula 5   **9.** Formula 2   **10.** Formula 6

11. (Shifting) Show that if $f(x)$ has a Fourier transform, so does $f(x - a)$, and

$$\mathcal{F}\{f(x - a)\} = e^{-iwa}\mathcal{F}\{f(x)\}.$$

12. Using Prob. 11, obtain formula 1 in Table III, Sec. 10.12, from formula 2 (with $c = 3b$).

13. (Shifting on the w-axis) Show that if $\hat{f}(w)$ is the Fourier transform of $f(x)$, then $\hat{f}(w - a)$ is the Fourier transform of $e^{iax}f(x)$.

14. Using Prob. 13, obtain formula 7 in Table III, Sec. 10.12, from formula 1 in that table.

15. Using Prob. 13, obtain formula 8 in Table III, Sec. 10.12, from formula 2 in that table.

10.12 Tables of Fourier Cosine Transforms, Fourier Sine Transforms and Fourier Transforms

For more extensive tables, see Ref. [C4] in Appendix 1.

Table I. Fourier Cosine Transforms

See (2) in Sec. 10.10.

	$f(x)$	$\hat{f}_c(w) = \mathscr{F}_c(f)$	
1	$\begin{cases} 1 & \text{if } 0 < x < a \\ 0 & \text{otherwise} \end{cases}$	$\sqrt{\dfrac{2}{\pi}} \dfrac{\sin aw}{w}$	
2	x^{a-1} $\quad(0 < a < 1)$	$\sqrt{\dfrac{2}{\pi}} \dfrac{\Gamma(a)}{w^a} \cos \dfrac{aw}{2}$	(Cf. Appendix 3.1)
3	e^{-ax} $\quad(a > 0)$	$\sqrt{\dfrac{2}{\pi}} \left(\dfrac{a}{a^2 + w^2} \right)$	
4	$e^{-x^2/2}$	$e^{-w^2/2}$	
5	e^{-ax^2} $\quad(a > 0)$	$\dfrac{1}{\sqrt{2a}} e^{-w^2/4a}$	
6	$x^n e^{-ax}$	$\sqrt{\dfrac{2}{\pi}} \dfrac{n!}{(a^2 + w^2)^{n+1}} \operatorname{Re} (a + iw)^{n+1}$	Re = Real part
7	$\begin{cases} \cos x & \text{if } 0 < x < a \\ 0 & \text{otherwise} \end{cases}$	$\dfrac{1}{\sqrt{2\pi}} \left[\dfrac{\sin a(1 - w)}{1 - w} + \dfrac{\sin a(1 + w)}{1 + w} \right]$	
8	$\cos ax^2$ $\quad(a > 0)$	$\dfrac{1}{\sqrt{2a}} \cos \left(\dfrac{w^2}{4a} - \dfrac{\pi}{4} \right)$	
9	$\sin ax^2$ $\quad(a > 0)$	$\dfrac{1}{\sqrt{2a}} \cos \left(\dfrac{w^2}{4a} + \dfrac{\pi}{4} \right)$	
10	$\dfrac{\sin ax}{x}$ $\quad(a > 0)$	$\sqrt{\dfrac{\pi}{2}} u(a - w)$	(Cf. Sec. 5.3)
11	$\dfrac{e^{-x} \sin x}{x}$	$\dfrac{1}{\sqrt{2\pi}} \arctan \dfrac{2}{w^2}$	
12	$J_0(ax)$ $\quad(a > 0)$	$\sqrt{\dfrac{2}{\pi}} \dfrac{u(a - w)}{\sqrt{a^2 - w^2}}$	(Cf. Secs. 4.5, 5.3)

Table II. Fourier Sine Transforms

See (5) in Sec. 10.10.

	$f(x)$	$\hat{f}_s(w) = \mathscr{F}_s(f)$
1	$\begin{cases} 1 & \text{if } 0 < x < a \\ 0 & \text{otherwise} \end{cases}$	$\sqrt{\dfrac{2}{\pi}} \left[\dfrac{1 - \cos aw}{w} \right]$
2	$1/\sqrt{x}$	$1/\sqrt{w}$
3	$1/x^{3/2}$	$2\sqrt{w}$
4	$x^{a-1} \quad (0 < a < 1)$	$\sqrt{\dfrac{2}{\pi}} \dfrac{\Gamma(a)}{w^a} \sin \dfrac{a\pi}{2}$ (Cf. Appendix 3.1)
5	e^{-x}	$\sqrt{\dfrac{2}{\pi}} \left(\dfrac{w}{1 + w^2} \right)$
6	$e^{-ax}/x \quad (a > 0)$	$\sqrt{\dfrac{2}{\pi}} \arctan \dfrac{w}{a}$
7	$x^n e^{-ax} \quad (a > 0)$	$\sqrt{\dfrac{2}{\pi}} \dfrac{n!}{(a^2 + w^2)^{n+1}} \text{Im}\,(a + iw)^{n+1}$ $\text{Im} = \text{Imaginary part}$
8	$xe^{-x^2/2}$	$we^{-w^2/2}$
9	$xe^{-ax^2} \quad (a > 0)$	$\dfrac{w}{(2a)^{3/2}} e^{-w^2/4a}$
10	$\begin{cases} \sin x & \text{if } 0 < x < a \\ 0 & \text{otherwise} \end{cases}$	$\dfrac{1}{\sqrt{2\pi}} \left[\dfrac{\sin a(1 - w)}{1 - w} - \dfrac{\sin a(1 + w)}{1 + w} \right]$
11	$\dfrac{\cos ax}{x} \quad (a > 0)$	$\sqrt{\dfrac{\pi}{2}} u(w - a)$ (Cf. Sec. 5.3)
12	$\arctan \dfrac{2a}{x} \quad (a > 0)$	$\sqrt{2\pi} \dfrac{\sinh aw}{w} e^{-aw}$

Table III. Fourier Transforms

See (6) in Sec. 10.11.

	$f(x)$	$\hat{f}(w) = \mathscr{F}(f)$
1	$\begin{cases} 1 & \text{if } -b < x < b \\ 0 & \text{otherwise} \end{cases}$	$\sqrt{\dfrac{2}{\pi}} \dfrac{\sin bw}{w}$
2	$\begin{cases} 1 & \text{if } b < x < c \\ 0 & \text{otherwise} \end{cases}$	$\dfrac{e^{-ibw} - e^{-icw}}{iw\sqrt{2\pi}}$
3	$\dfrac{1}{x^2 + a^2} \qquad (a > 0)$	$\sqrt{\dfrac{\pi}{2}} \dfrac{e^{-a|w|}}{a}$
4	$\begin{cases} x & \text{if } 0 < x < b \\ 2x - a & \text{if } b < x < 2b \\ 0 & \text{otherwise} \end{cases}$	$\dfrac{-1 + 2e^{ibw} - e^{-2ibw}}{\sqrt{2\pi}\, w^2}$
5	$\begin{cases} e^{-ax} & \text{if } x > 0 \\ 0 & \text{otherwise} \end{cases} \quad (a > 0)$	$\dfrac{1}{\sqrt{2\pi}(a + iw)}$
6	$\begin{cases} e^{ax} & \text{if } b < x < c \\ 0 & \text{otherwise} \end{cases}$	$\dfrac{e^{(a-iw)c} - e^{(a-iw)b}}{\sqrt{2\pi}(a - iw)}$
7	$\begin{cases} e^{iax} & \text{if } -b < x < b \\ 0 & \text{otherwise} \end{cases}$	$\sqrt{\dfrac{2}{\pi}} \dfrac{\sin b(w - a)}{w - a}$
8	$\begin{cases} e^{iax} & \text{if } b < x < c \\ 0 & \text{otherwise} \end{cases}$	$\dfrac{i}{\sqrt{2\pi}} \dfrac{e^{ib(a-w)} - e^{ic(a-w)}}{a - w}$
9	$e^{-ax^2} \qquad (a > 0)$	$\dfrac{1}{\sqrt{2a}} e^{-w^2/4a}$
10	$\dfrac{\sin ax}{x} \qquad (a > 0)$	$\sqrt{\dfrac{\pi}{2}} \text{ if } |w| < a; \quad 0 \text{ if } |w| > a$

Review Problems for Chapter 10

1. The transition from Fourier series for functions of period 2π to those for functions of arbitrary period was surprisingly simple. Can you write down the corresponding calculations from memory?

2. Can you obtain the Fourier coefficients of a sum of functions from the Fourier coefficients of the functions? (Give reason.)

3. Can a function with discontinuities have a Fourier series? A Taylor series?

4. What simplifications of the Fourier series of a function occur if the function is even? Odd?

5. If f is odd, what can one say about $f^2, f^3, \cdots$?

6. If f is even, what can one say about $f^2, f^3, \cdots$?

7. If $f(x)$ is odd, is $e^{f(x)}$ odd? (Give reason.)

8. If the Fourier series of a function $f(x)$ has both cosine and sine terms, what is the sum of the cosine terms expressed in terms of $f(x)$? The sum of the sine terms?

9. What is the basic difference in the vibrations of a mass–spring system governed by $my'' + cy' + ky = r(t)$ with an arbitrary periodic driving force as opposed to a pure sinusoidal driving force?

10. Using the "jump formulas" (Sec. 10.6), can you discover relations between the rapidity of convergence of a Fourier series and the continuity or discontinuity of a function and its derivatives?

11. Can the "jump method" be applied to $f(x) = \cosh x \, (-\pi < x < \pi)$, $f(x + 2\pi) = f(x)$?

12. What is piecewise continuity? In what connection did it occur in this chapter?

13. What is a trigonometric polynomial? Why did we consider it?

14. What is the mean square error? Why did we consider it?

15. What is the (real) Fourier integral and what can it be used for?

16. What is the Fourier transformation? How did we obtain it from the Fourier integral?

17. Does every continuous function have a Fourier cosine or sine transform?

18. Can a discontinuous function have a Fourier cosine or sine transform?

19. What is the complex form of a Fourier series?

20. Can you think of a way of suggesting the complex form of the Fourier integral by the complex form of the Fourier series?

Find the Fourier series of the following functions, which are assumed to have period 2π.

21. $f(x) = \begin{cases} 0 & \text{if } 0 < x < \pi \\ 1 & \text{if } \pi < x < 2\pi \end{cases}$

22. $f(x) = \begin{cases} 1 & \text{if } -\pi/2 < x < \pi/2 \\ -1 & \text{if } \pi/2 < x < 3\pi/2 \end{cases}$

23. $f(x) = \begin{cases} -1 & \text{if } -\pi < x < -\pi/2 \\ 0 & \text{if } -\pi/2 < x < \pi/2 \\ 1 & \text{if } \pi/2 < x < \pi \end{cases}$

24. $f(x) = \begin{cases} 0 & \text{if } -\pi < x < 0 \\ -1 & \text{if } 0 < x < \pi/2 \\ 1 & \text{if } \pi/2 < x < \pi \end{cases}$

25. $f(x) = |x| \quad (-\pi < x < \pi)$

26. $f(x) = \frac{1}{2}(\pi - x) \quad (0 < x < 2\pi)$

27. $f(x) = -x/2 \quad (-\pi < x < \pi)$

28. $f(x) = x^3 \quad (-\pi < x < \pi)$

29. $f(x) = 1 + x^2 \quad (-\pi < x < \pi)$

30. $f(x) = x^4 \quad (-\pi < x < \pi)$

31. $f(x) = \begin{cases} x^2/2 & \text{if } -\pi/2 < x < \pi/2 \\ \pi^2/8 & \text{if } \pi/2 < x < 3\pi/2 \end{cases}$

32. $f(x) = \begin{cases} 0 & \text{if } -\pi < x < 0 \\ \pi x & \text{if } 0 < x < \pi \end{cases}$

33. $f(x) = \begin{cases} \pi x & \text{if } -\pi/2 < x < \pi/2 \\ 0 & \text{if } \pi/2 < x < 3\pi/2 \end{cases}$

34. $f(x) = \begin{cases} (\pi^2 + \pi x)/2 & \text{if } -\pi < x < 0 \\ -\pi x/2 & \text{if } 0 < x < \pi \end{cases}$

35. $f(x) = \begin{cases} x/a & \text{if } -a < x < a \ (< \pi) \\ \dfrac{\pi - x}{\pi - a} & \text{if } a < x < 2\pi - a \end{cases}$

36. $f(x) = \begin{cases} 0 & \text{if } -\pi < x < 0 \\ x^2 & \text{if } 0 < x < \pi \end{cases}$

37. $f(x) = \begin{cases} 0 & \text{if } -\pi < x < -\pi/2 \\ k & \text{if } -\pi/2 < x < \pi/2 \\ 2k & \text{if } \pi/2 < x < \pi \end{cases}$

38. $f(x) = \begin{cases} -\pi/2 & \text{if } -\pi < x < -\pi/2 \\ x & \text{if } -\pi/2 < x < \pi/2 \\ \pi/2 & \text{if } \pi/2 < x < \pi \end{cases}$

39. $f(x) = \begin{cases} \pi x + x^2 & \text{if } -\pi < x < 0 \\ \pi x - x^2 & \text{if } 0 < x < \pi \end{cases}$

40. $f(x) = \begin{cases} k & \text{if } 0 < x < \pi/2 \\ 0 & \text{if } \pi/2 < x < 3\pi/2 \end{cases}$

Find the Fourier series of the following functions that are assumed to have period $p = 2L$.

41. $f(x) = -1 \ (-1 < x < 0), \quad f(x) = 1 \ (0 < x < 1), \quad p = 2L = 2$

42. $f(x) = 1 \ (-1 < x < 1), \quad f(x) = 0 \ (1 < x < 3), \quad p = 2L = 4$

43. $f(x) = x \ (-1 < x < 1), \quad p = 2L = 2$

44. $f(x) = 0 \ (-1 < x < 0), \quad f(x) = x \ (0 < x < 1), \quad p = 2L = 2$

45. $f(x) = -x \ (-1 < x < 0), \quad f(x) = 0 \ (0 < x < 1), \quad p = 2L = 2$

46. $f(x) = 1 - x^2 \ (-1 < x < 1), \quad p = 2L = 2$

47. $f(x) = -1 \ (-1 < x < 0), \quad f(x) = 2x \ (0 < x < 1), \quad p = 2L = 2$

48. $f(x) = 1 - x^2, \ (0 < x < 2), \quad p = 2L = 2$

49. $f(x) = 1 + x^2 \ (-1 < x < 1), \quad p = 2L = 2$

50. $f(x) = \pi x/2 \ (-\pi/4 < x < \pi/4), \quad f(x) = \pi^2/4 - \pi x/2 \ (\pi/4 < x < 3\pi/4),$
$p = 2L = \pi$

Write the following functions as the sum of an even and an odd function.

51. e^{3x} **52.** $1/(1 - x)^2$ **53.** $\sin \pi x$ **54.** $x/(x + 1)$

55. $\sinh x^2$ **56.** $f(x)$ in Prob. 47

Find the sum of

57. $1 - \frac{1}{3} + \frac{1}{5} - \frac{1}{7} + - \cdots$ (Use Prob. 21.)

58. $1 + \frac{1}{9} + \frac{1}{25} + \frac{1}{47} + \cdots$ (Use Prob. 25.)

59. $1 - 3^{-3} + 5^{-3} - 7^{-3} + - \cdots$ (Use Prob. 39.)

60. Verify that in Prob. 27 the Bessel inequality with the equality sign holds. (Use Prob. 38, Sec. 10.4.)

61. Assuming that in Prob. 21 the Bessel inequality with the equality sign holds, confirm the formula in Prob. 40, Sec. 10.4.

62. Using the Bessel inequality with the equality sign and Prob. 25, show that

$$1 + \frac{1}{3^4} + \frac{1}{5^4} + \frac{1}{7^4} + \cdots = \frac{\pi^4}{96}.$$

63. Assuming that in Prob. 39 the Bessel inequality with the equality sign holds, show that

$$1 + \frac{1}{3^6} + \frac{1}{5^6} + \frac{1}{7^6} + \cdots = \frac{\pi^6}{960}.$$

Find a general solution of $y'' + \omega^2 y = r(t)$, where

64. $r(t) = t^2/4 \quad (-\pi < t < \pi), \quad r(t + 2\pi) = r(t), \qquad |\omega| \neq 0, 1, 2, \cdots$

65. $r(t) = t(\pi^2 - t^2)/12 \quad (-\pi < t < \pi), \quad r(t + 2\pi) = r(t), \qquad |\omega| \neq 1, 2, \cdots$

Summary of Chapter 10
Fourier Series, Fourier Integrals, Fourier Transforms

A **trigonometric series** (Sec. 10.1) is a series of the form

(1) $$a_0 + \sum_{n=1}^{\infty} (a_n \cos nx + b_n \sin nx).$$

The **Fourier series** of a given periodic function $f(x)$ of period 2π is a trigonometric series (1) whose coefficients are the **Fourier coefficients** of $f(x)$ given by the **Euler formulas** (Sec. 10.2)

(2) $$a_0 = \frac{1}{2\pi} \int_{-\pi}^{\pi} f(x)\,dx, \qquad a_n = \frac{1}{\pi} \int_{-\pi}^{\pi} f(x) \cos nx\,dx,$$

$$b_n = \frac{1}{\pi} \int_{-\pi}^{\pi} f(x) \sin nx\,dx.$$

For a function $f(x)$ of any period $p = 2L$ the Fourier series is (Sec. 10.3)

(3) $$f(x) = a_0 + \sum_{n=1}^{\infty} \left(a_n \cos \frac{n\pi}{L}x + b_n \sin \frac{n\pi}{L}x \right)$$

with the Fourier coefficients of $f(x)$ given by the Euler formulas

(4) $$a_0 = \frac{1}{2L} \int_{-L}^{L} f(x)\,dx, \qquad a_n = \frac{1}{L} \int_{-L}^{L} f(x) \cos \frac{n\pi x}{L}\,dx$$

$$b_n = \frac{1}{L} \int_{-L}^{L} f(x) \sin \frac{n\pi x}{L}\,dx$$

These series are fundamental in connection with periodic phenomena, particularly in models involving differential equations (Sec. 10.7, Chap. 11). If $f(t)$ is even [$f(-t) = f(t)$] or odd [$f(-t) = -f(t)$], they reduce to **Fourier cosine** or **Fourier sine series,** respectively (Sec. 10.4). A function $f(t)$ given on an interval $0 \leq t \leq L$ can be developed in a Fourier cosine or sine series of period $2L$; these are called **half-range expansions** of $f(t)$; cf. Sec. 10.5. The set of cosine and sine functions in (4) is called the **trigonometric system.** Its most basic property is its **orthogonality** on an interval of length $2L$, that is, for all integers m and $n \neq m$,

$$\int_{-L}^{L} \cos \frac{m\pi x}{L} \cos \frac{n\pi x}{L} \, dx = 0, \qquad \int_{-L}^{L} \sin \frac{m\pi x}{L} \sin \frac{n\pi x}{L} \, dx = 0$$

and for all integers m and n,

$$\int_{-L}^{L} \cos \frac{m\pi x}{L} \sin \frac{n\pi x}{L} \, dx = 0.$$

This makes Fourier series the model of **orthogonal developments** (Secs. 4.7—4.9).

Partial sums of Fourier series minimize the **square error** (Sec. 10.8). Integration in calculating Fourier coefficients is avoided by the surprising **method of jumps** (Sec. 10.6).

Ideas and techniques of Fourier series extend to nonperiodic functions $f(x)$ defined on the entire real line; this leads to the **Fourier integral** (Sec. 10.9)

$$(5) \qquad f(x) = \int_{0}^{\infty} [A(w) \cos wx + B(w) \sin wx] \, dw,$$

where

$$(6) \quad A(w) = \frac{1}{\pi} \int_{-\infty}^{\infty} f(v) \cos wv \, dv, \qquad B(w) = \frac{1}{\pi} \int_{-\infty}^{\infty} f(v) \sin wv \, dv$$

or, in complex form (Sec. 10.11),

$$(7) \qquad f(x) = \frac{1}{\sqrt{2\pi}} \int_{-\infty}^{\infty} \hat{f}(w) e^{iwx} \, dw \qquad (i = \sqrt{-1}),$$

where

$$(8) \qquad \hat{f}(w) = \frac{1}{\sqrt{2\pi}} \int_{-\infty}^{\infty} f(x) e^{-iwx} \, dx.$$

Formula (8) transforms $f(x)$ into its **Fourier transform** $\hat{f}(w)$. Applications of the Fourier cosine, Fourier sine (Sec. 10.10) and Fourier transformations to partial differential equations will be shown in Sec. 11.14.

Chapter 11

Partial Differential Equations

Partial differential equations arise in connection with various physical and geometrical problems when the functions involved depend on two or more independent variables. It is fair to say that only the simplest physical systems can be modeled by *ordinary* differential equations. Fluid and solid mechanics, heat transfer, electromagnetic theory and other areas of physics are full of problems that must be modeled by *partial* differential equations. Indeed, the range of application of the latter is enormous, compared to that of ordinary differential equations. The independent variables involved may be time and one or several coordinates in space. The present chapter will be devoted to some of the most important partial differential equations occurring in engineering applications. We shall derive these equations as models of physical systems and consider methods for solving **initial** and **boundary value problems,** that is, methods for obtaining solutions of those equations corresponding to the given physical situations.

In Sec. 11.1 we shall define the notion of a solution of a partial differential equation. Sections 11.2—11.4 will be devoted to the one-dimensional wave equation, governing the motion of a vibrating string. The heat equation will be considered in Secs. 11.5 and 11.6, the two-dimensional wave equation (vibrating membranes) in Secs. 11.7—11.10, and Laplace's equation in Secs. 11.11 and 11.12.

In Secs. 11.13 and 11.14 we shall see that partial differential equations can also be solved by the Laplace transformation (cf. Chap. 5) or the Fourier transformation (cf. Secs. 10.10—10.12).

Numerical methods for partial differential equations are presented in Secs. 20.4—20.7.

Prerequisites for this chapter: ordinary linear differential equations (Chap. 2) and Fourier series (Chap. 10).
Sections that may be omitted in a shorter course: 11.6, 11.9, 11.10.
References: Appendix 1, Part C.
Answers to problems: Appendix 2.

11.1 Basic Concepts

An equation involving one or more partial derivatives of an (unknown) function of two or more independent variables is called a **partial differential equation.** The order of the highest derivative is called the **order** of the equation.

Just as in the case of an ordinary differential equation, we say that a partial differential equation is **linear** if it is of the first degree in the dependent variable (the unknown function) and its partial derivatives. If each term of such an equation contains either the dependent variable or one of its derivatives, the equation is said to be **homogeneous;** otherwise it is said to be **nonhomogeneous.**

EXAMPLE 1. Important linear partial differential equations of the second order

(1)
$$\frac{\partial^2 u}{\partial t^2} = c^2 \frac{\partial^2 u}{\partial x^2}$$
One-dimensional wave equation

(2)
$$\frac{\partial u}{\partial t} = c^2 \frac{\partial^2 u}{\partial x^2}$$
One-dimensional heat equation

(3)
$$\frac{\partial^2 u}{\partial x^2} + \frac{\partial^2 u}{\partial y^2} = 0$$
Two-dimensional Laplace equation

(4)
$$\frac{\partial^2 u}{\partial x^2} + \frac{\partial^2 u}{\partial y^2} = f(x, y)$$
Two-dimensional Poisson equation

(5)
$$\frac{\partial^2 u}{\partial x^2} + \frac{\partial^2 u}{\partial y^2} + \frac{\partial^2 u}{\partial z^2} = 0$$
Three-dimensional Laplace equation

Here c is a constant, t is time and x, y, z are Cartesian coordinates. Equation (4) (with $f \not\equiv 0$) is nonhomogeneous, while the other equations are homogeneous. ∎

A **solution** *of a partial differential equation in some region R of the space of the independent variables* is a function that has all the partial derivatives appearing in the equation in some domain containing R and satisfies the equation everywhere in R. (Often one merely requires that that function is continuous on the boundary of R, has those derivatives in the interior of R and satisfies the equation in the interior of R.)

In general, the totality of solutions of a partial differential equation is very large. For example, the functions

(6) $u = x^2 - y^2,$ $u = e^x \cos y,$ $u = \ln (x^2 + y^2),$

which are entirely different from each other, are solutions of (3), as the student may verify. We shall see later that the unique solution of a partial differential equation corresponding to a given physical problem will be obtained by the use of additional information arising from the physical situation. For example, in some cases the values of the required solution of the problem on the boundary of some domain will be given (**"boundary conditions"**); in other cases when time t is one of the variables, the values of the solution at $t = 0$ will be prescribed (**"initial conditions"**).

We know that if an *ordinary* differential equation is linear and homogeneous, then from known solutions we can obtain further solutions by superposition. For a homogeneous linear *partial* differential equation the situation is quite similar. In fact, the following theorem holds.

Fundamental Theorem 1 (Superposition principle)

If u_1 and u_2 are any solutions of a linear homogeneous partial differential equation in some region, then

$$u = c_1 u_1 + c_2 u_2,$$

where c_1 and c_2 are any constants, is also a solution of that equation in that region.

The proof of this important theorem is simple and quite similar to that of Theorem 1 in Sec. 2.1 and is left to the student.

In the next section we shall begin our discussion of the first of the very important equations listed in Example 1, namely, the **one-dimensional wave equation,** where "one-dimensional" indicates that the equation involves one space variable, x. This equation governs the motion of an elastic string, such as a violin string.

Problems for Sec. 11.1

1. Prove Fundamental Theorem 1 for second-order differential equations in two and three independent variables.
2. Verify that the functions (6) are solutions of (3).

Verify that the following functions are solutions of the wave equation (1) for a suitable value of the constant c in (1).

3. $u = x^2 + t^2$ 4. $u = x^2 + 9t^2$ 5. $u = \sin t \sin x$
6. $u = \cos t \sin x$ 7. $u = \cos ct \sin x$ 8. $u = \sin \omega ct \sin \omega x$

Verify that the following functions are solutions of the heat equation (2) for a suitable value of c.

9. $u = e^{-t} \cos x$ 10. $u = e^{-4t} \sin 2x$ 11. $u = e^{-9t} \cos x$
12. $u = e^{-9t} \sin 3x$ 13. $u = e^{-16t} \cos 2x$ 14. $u = e^{-\omega^2 c^2 t} \sin \omega x$

Verify that the following functions are solutions of Laplace's equation.

15. $u = x^2 - y^2$ 16. $u = 3x^2 y - y^3$ 17. $u = e^x \cos y$
18. $u = e^x \sin y$ 19. $u = \arctan (y/x)$ 20. $u = \sin x \cosh y$

21. Show that $u(x, t) = v(x + ct) + w(x - ct)$ is a solution of the wave equation (1); here, v and w are any twice differentiable functions.
22. Verify that $u(x, y) = a \ln (x^2 + y^2) + b$ satisfies Laplace's equation (3) and determine a and b so that u satisfies the boundary conditions $u = 0$ on the circle $x^2 + y^2 = 1$ and $u = 3$ on the circle $x^2 + y^2 = 4$.
23. Show that $u = 1/\sqrt{x^2 + y^2 + z^2}$ is a solution of Laplace's equation (5).
24. Find the electrostatic potential [solution of (5)] between the two concentric spheres $S_1 : x^2 + y^2 + z^2 = 1$ and $S_2 : x^2 + y^2 + z^2 = 4$ when S_1 is at 110 volts and S_2 is grounded [0 volt]. *Hint.* Remember Prob. 23.

25. Show that $u = y - y/(x^2 + y^2)$ satisfies the Laplace equation (3) and is 0 on the x-axis as well as on the circle $x^2 + y^2 = 1$.

Relations to ordinary differential equations
If a partial differential equation involves derivatives with respect to one of the independent variables only, we may solve it like an ordinary differential equation, treating the other independent variables as parameters. Solve the following equations where u is a function of two variables x and y.

26. $u_y = 0$ **27.** $u_x = 0$ **28.** $u_{xx} + u = 0$

29. $u_{xx} = 0$ **30.** $u_{yy} - u = 0$ **31.** $u_{xx} + u_x - 2u = 0$

Setting $u_x = p$, solve

32. $u_{xy} = u_x$ **33.** $u_{xy} = 0$ **34.** $u_{xy} + u_x = 0$

Solve the following systems of partial differential equations.

35. $u_x = 0$, $u_y = 0$ **36.** $u_{xx} = 0$, $u_{yy} = 0$ **37.** $u_{xx} = 0$, $u_{xy} = 0$

38. $u_{xx} = 0$, $u_{xy} = 0$, $u_{yy} = 0$

39. Show that if the level curves $z = const$ of a surface $z = z(x, y)$ are straight lines parallel to the x-axis, then z is a solution of the differential equation $z_x = 0$. Give examples.

40. Show that the solutions $z = z(x, y)$ of $yz_x - xz_y = 0$ represent surfaces of revolution. Give examples. *Hint.* Set $x = r \cos \theta$, $y = r \sin \theta$ and show that the equation becomes $z_\theta = 0$.

11.2 Modeling: Vibrating String. One-Dimensional Wave Equation

As a first important partial differential equation, let us derive the equation governing small transverse vibrations of an elastic string, which is stretched to length L and then fixed at the endpoints. Suppose that the string is distorted and then at a certain instant, say, $t = 0$, is released and allowed to vibrate. The problem is to determine the vibrations of the string, that is, to find its deflection $u(x, t)$ at any point x and at any time $t > 0$; cf. Fig. 265.

When deriving a differential equation corresponding to a given physical problem, we usually have to make simplifying assumptions to ensure that

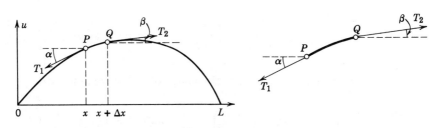

Fig. 265. Vibrating string

the resulting equation does not become too complicated. We know this important fact from our considerations of ordinary differential equations, and for partial differential equations the situation is similar.

In our present case we make the following assumptions.

1. *The mass of the string per unit length is constant ("homogeneous string"). The string is perfectly elastic and does not offer any resistance to bending.*
2. *The tension caused by stretching the string before fixing it at the endpoints is so large that the action of the gravitational force on the string can be neglected.*
3. *The string performs a small transverse motion in a vertical plane; that is, every particle of the string moves strictly vertically and so that the deflection and the slope at every point of the string remain small in absolute value.*

These assumptions are such that we may expect that the solution $u(x, t)$ of the differential equation to be obtained will reasonably well describe small vibrations of the physical "nonidealized" string of small homogeneous mass under large tension.

To obtain the differential equation we consider the forces acting on a small portion of the string (Fig. 265). Since the string does not offer resistance to bending, the tension is tangential to the curve of the string at each point. Let T_1 and T_2 be the tension at the endpoints P and Q of that portion. Since there is no motion in horizontal direction, the horizontal components of the tension must be constant. Using the notation shown in Fig. 265 we thus obtain

$$(1) \qquad T_1 \cos \alpha = T_2 \cos \beta = T = \text{const.}$$

In vertical direction we have two forces, namely, the vertical components $-T_1 \sin \alpha$ and $T_2 \sin \beta$ of T_1 and T_2; here the minus sign appears because that component at P is directed downward. By Newton's second law the resultant of these two forces is equal to the mass $\rho \, \Delta x$ of the portion times the acceleration $\partial^2 u / \partial t^2$, evaluated at some point between x and $x + \Delta x$; here ρ is the mass of the undeflected string per unit length, and Δx is the length of the portion of the undeflected string. Hence

$$T_2 \sin \beta - T_1 \sin \alpha = \rho \, \Delta x \, \frac{\partial^2 u}{\partial t^2} .$$

By using (1) we obtain

$$(2) \qquad \frac{T_2 \sin \beta}{T_2 \cos \beta} - \frac{T_1 \sin \alpha}{T_1 \cos \alpha} = \tan \beta - \tan \alpha = \frac{\rho \, \Delta x}{T} \frac{\partial^2 u}{\partial t^2} .$$

Now $\tan \alpha$ and $\tan \beta$ are the slopes of the curve of the string at x and $x + \Delta x$:

$$\tan \alpha = \left(\frac{\partial u}{\partial x} \right)_x \qquad \text{and} \qquad \tan \beta = \left(\frac{\partial u}{\partial x} \right)_{x + \Delta x} .$$

Here we have to write *partial* derivatives because u also depends on t. Dividing (2) by Δx, we thus have

$$\frac{1}{\Delta x}\left[\left(\frac{\partial u}{\partial x}\right)_{x+\Delta x} - \left(\frac{\partial u}{\partial x}\right)_{x}\right] = \frac{\rho}{T}\frac{\partial^2 u}{\partial t^2}.$$

If we let Δx approach zero, we obtain the linear partial differential equation

(3) $$\boxed{\frac{\partial^2 u}{\partial t^2} = c^2 \frac{\partial^2 u}{\partial x^2}} \qquad\qquad c^2 = \frac{T}{\rho}.$$

This is the so-called **one-dimensional wave equation,** which governs our problem. We see that it is homogeneous and of the second order. The notation c^2 (instead of c) for the physical constant T/ρ has been chosen to indicate that this constant is positive. Solutions of the equation will be obtained in the following section.

11.3 Method of Separating Variables (Product Method)

In the preceding section we showed that the vibrations of an elastic string, such as a violin string, are governed by the one-dimensional wave equation

(1) $$\boxed{\frac{\partial^2 u}{\partial t^2} = c^2 \frac{\partial^2 u}{\partial x^2},}$$

where $u(x, t)$ is the deflection of the string. To find out how the string moves, we solve this equation; more precisely, we determine a solution u of (1) that also satisfies the conditions imposed by the physical system. Since the string is fixed at the ends $x = 0$ and $x = L$, we have the two **boundary conditions**

(2) $$\boxed{u(0, t) = 0, \qquad u(L, t) = 0 \qquad \text{for all } t.}$$

The form of the motion of the string will depend on the initial deflection (deflection at $t = 0$) and on the initial velocity (velocity at $t = 0$). Denoting the initial deflection by $f(x)$ and the initial velocity by $g(x)$, we thus obtain the two **initial conditions**

(3) $$\boxed{u(x, 0) = f(x)}$$

and

(4) $$\boxed{\left.\frac{\partial u}{\partial t}\right|_{t=0} = g(x).}$$

Our problem is now to find a solution of (1) satisfying the conditions (2)—(4). We shall proceed step by step, as follows.

> *First Step.* By applying the so-called **method of separating variables** or *product method,* we shall obtain two *ordinary* differential equations.
> *Second Step.* We shall determine solutions of those two equations that satisfy the boundary conditions.
> *Third Step.* Those solutions will be composed so that the result will be a solution of the wave equation (1) satisfying also the given initial conditions.

The details are as follows.

First Step. Two Ordinary Differential Equations

The product method yields solutions of the wave equation (1) of the form

(5)
$$u(x, t) = F(x)G(t)$$

which are a product of two functions, each depending only on one of the variables x and t. We shall see later that this method has various applications in engineering mathematics. By differentiating (5) we obtain

$$\frac{\partial^2 u}{\partial t^2} = F\ddot{G} \quad \text{and} \quad \frac{\partial^2 u}{\partial x^2} = F''G,$$

where dots denote derivatives with respect to t and primes derivatives with respect to x. By inserting this into our differential equation (1) we have

$$F\ddot{G} = c^2 F''G.$$

Dividing by $c^2 FG$, we find

$$\frac{\ddot{G}}{c^2 G} = \frac{F''}{F}.$$

The expression on the left involves functions depending only on t, while the expression on the right involves functions depending only on x. Hence both expressions must be equal to a constant, say, k, because if the expression on the left is not constant, then changing t will presumably change the value of this expression but certainly not that on the right, since the latter does not depend on t. Similarly, if the expression on the right is not constant, changing x will presumably change the value of this expression but certainly not that on the left. Thus

$$\frac{\ddot{G}}{c^2 G} = \frac{F''}{F} = k.$$

This yields immediately two ordinary linear differential equations, namely,

(6)
$$F'' - kF = 0$$

and

(7)
$$\ddot{G} - c^2 kG = 0.$$

Here, k is still arbitrary.

Second Step. Satisfying the Boundary Conditions

We shall now determine solutions F and G of (6) and (7) so that $u = FG$ satisfies the boundary conditions (2), that is,

$$u(0, t) = F(0)G(t) = 0, \qquad u(L, t) = F(L)G(t) = 0 \qquad \text{for all } t.$$

Clearly, if $G \equiv 0$, then $u \equiv 0$, which is of no interest. Thus $G \not\equiv 0$ and then

(8) $$\text{(a) } F(0) = 0 \qquad \text{(b) } F(L) = 0.$$

For $k = 0$ the general solution of (6) is $F = ax + b$, and from (8) we obtain $a = b = 0$. Hence $F \equiv 0$, which is of no interest because then $u \equiv 0$. For positive $k = \mu^2$ the general solution of (6) is

$$F = Ae^{\mu x} + Be^{-\mu x},$$

and from (8) we obtain $F \equiv 0$, as before. Hence we are left with the possibility of choosing k negative, say, $k = -p^2$. Then equation (6) takes the form

$$F'' + p^2 F = 0.$$

Its general solution is

$$F(x) = A \cos px + B \sin px.$$

From this and (8) we have

$$F(0) = A = 0 \qquad \text{and then} \qquad F(L) = B \sin pL = 0.$$

We must take $B \neq 0$ since otherwise $F \equiv 0$. Hence $\sin pL = 0$. Consequently,

(9) $$pL = n\pi \qquad \text{or} \qquad p = \frac{n\pi}{L} \qquad \text{(} n \text{ integer)}.$$

Setting $B = 1$, we thus obtain infinitely many solutions $F(x) = F_n(x)$, where

(10) $$F_n(x) = \sin \frac{n\pi}{L}x \qquad n = 1, 2, \cdots.$$

These solutions satisfy (8). [For negative integer n we obtain essentially the same solutions, except for a minus sign, because $\sin(-\alpha) = -\sin \alpha$.]

 k is now restricted to the values $k = -p^2 = -(n\pi/L)^2$, resulting from (9). For these values of k, equation (7) takes the form

$$\ddot{G} + \lambda_n{}^2 G = 0 \qquad \text{where} \qquad \lambda_n = \frac{cn\pi}{L}.$$

A general solution is

$$G_n(t) = B_n \cos \lambda_n t + B_n{}^* \sin \lambda_n t.$$

Hence the functions $u_n(x, t) = F_n(x)G_n(t)$, written out

(11) $\qquad \boxed{\; u_n(x, t) = (B_n \cos \lambda_n t + B_n{}^* \sin \lambda_n t) \sin \frac{n\pi}{L} x \;}$ $\qquad (n = 1, 2, \cdots),$

are solutions of (1), satisfying the boundary conditions (2). These functions are called the **eigenfunctions,** or *characteristic functions,* and the values $\lambda_n = cn\pi/L$ are called the **eigenvalues,** or *characteristic values,* of the vibrating string. The set $\{\lambda_1, \lambda_2, \cdots\}$ is called the **spectrum.**

We see that each u_n represents a harmonic motion having the frequency $\lambda_n/2\pi = cn/2L$ cycles per unit time. This motion is called the nth **normal mode** of the string. The first normal mode is known as the *fundamental mode* $(n = 1)$, and the others are known as *overtones;* musically they give the octave, octave plus fifth, etc. Since in (11)

$$\sin \frac{n\pi x}{L} = 0 \qquad \text{at} \qquad x = \frac{L}{n}, \frac{2L}{n}, \cdots, \frac{n-1}{n} L,$$

the nth normal mode has $n - 1$ so-called **nodes,** that is, points of the string that do not move (Fig. 266).

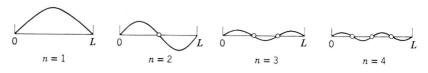

Fig. 266. Normal modes of the vibrating string

Figure 267 shows the second normal mode for various values of t. At any instant the string has the form of a sine wave. When the left part of the string is moving down the other half is moving up, and conversely. For the other modes the situation is similar.

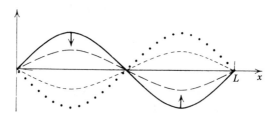

Fig. 267. Second normal mode for various values of t

Third Step. Solution of the Entire Problem

Clearly, a single solution $u_n(x, t)$ will, in general, not satisfy the initial conditions (3) and (4). Now, since the equation (1) is linear and homogeneous, it follows from Fundamental Theorem 1 in Sec. 11.1 that the sum of finitely many solutions u_n is a solution of (1). To obtain a solution that satisfies (3) and (4), we consider the infinite series

$$
(12) \quad u(x, t) = \sum_{n=1}^{\infty} u_n(x, t) = \sum_{n=1}^{\infty} (B_n \cos \lambda_n t + B_n{}^* \sin \lambda_n t) \sin \frac{n\pi}{L} x
$$

where $\lambda_n = cn\pi/L$, as before. From this and the initial condition (3) we obtain

$$
(13) \qquad u(x, 0) = \sum_{n=1}^{\infty} B_n \sin \frac{n\pi}{L} x = f(x).
$$

Hence, for (12) to satisfy (3), the coefficients B_n must be chosen so that $u(x, 0)$ becomes a half-range expansion of $f(x)$, namely, the Fourier sine series of $f(x)$; that is [cf. (4) in Sec. 10.5],

$$
(14) \qquad B_n = \frac{2}{L} \int_0^L f(x) \sin \frac{n\pi x}{L} \, dx, \qquad\qquad n = 1, 2, \cdots.
$$

Similarly, by differentiating (12) with respect to t and using the second initial condition, (4), we find

$$
\left. \frac{\partial u}{\partial t} \right|_{t=0} = \left[\sum_{n=1}^{\infty} (-B_n \lambda_n \sin \lambda_n t + B_n{}^* \lambda_n \cos \lambda_n t) \sin \frac{n\pi x}{L} \right]_{t=0}
$$

$$
= \sum_{n=1}^{\infty} B_n{}^* \lambda_n \sin \frac{n\pi x}{L} = g(x).
$$

Hence, for (12) to satisfy (4), the coefficients $B_n{}^*$ must be chosen so that, for $t = 0$, the derivative $\partial u/\partial t$ becomes the Fourier sine series of $g(x)$; thus, by (4) in Sec. 10.5,

$$
B_n{}^* \lambda_n = \frac{2}{L} \int_0^L g(x) \sin \frac{n\pi x}{L} \, dx
$$

or, since $\lambda_n = cn\pi/L$,

$$
(15) \qquad B_n{}^* = \frac{2}{cn\pi} \int_0^L g(x) \sin \frac{n\pi x}{L} \, dx, \qquad\qquad n = 1, 2, \cdots.
$$

It follows that $u(x, t)$, given by (12) with coefficients (14) and (15), is a solution of (1) that satisfies the conditions (2)—(4), provided that the series

(12) converges and also that the series obtained by differentiating (12) twice (termwise) with respect to x and t, converge and have the sums $\partial^2 u/\partial x^2$ and $\partial^2 u/\partial t^2$, respectively, which are continuous.

Hence the solution (12) is at first a purely formal expression, and we shall now establish it. For the sake of simplicity we consider only the case when the initial velocity $g(x)$ is identically zero. Then the $B_n{}^*$ are zero, and (12) reduces to

$$(16) \qquad u(x, t) = \sum_{n=1}^{\infty} B_n \cos \lambda_n t \sin \frac{n \pi x}{L}, \qquad \lambda_n = \frac{c n \pi}{L}.$$

It is possible to *sum* this series, that is, to write the result in a closed or finite form. For this purpose we use the formula [cf. (11) in Appendix 3.1]

$$\cos \frac{c n \pi}{L} t \sin \frac{n \pi}{L} x = \frac{1}{2} \left[\sin \left\{ \frac{n \pi}{L} (x - ct) \right\} + \sin \left\{ \frac{n \pi}{L} (x + ct) \right\} \right].$$

Consequently, we may write (16) in the form

$$u(x, t) = \frac{1}{2} \sum_{n=1}^{\infty} B_n \sin \left\{ \frac{n \pi}{L} (x - ct) \right\} + \frac{1}{2} \sum_{n=1}^{\infty} B_n \sin \left\{ \frac{n \pi}{L} (x + ct) \right\}.$$

These two series are those obtained by substituting $x - ct$ and $x + ct$, respectively, for the variable x in the Fourier sine series (13) for $f(x)$. Therefore,

$$(17) \qquad \boxed{u(x, t) = \tfrac{1}{2}[f^*(x - ct) + f^*(x + ct)]}$$

where f^* is the odd periodic extension of f with the period $2L$ (Fig. 268). Since the initial deflection $f(x)$ is continuous on the interval $0 \leqq x \leqq L$ and zero at the endpoints, it follows from (17) that $u(x, t)$ is a continuous function of both variables x and t for all values of the variables. By differentiating (17) we see that $u(x, t)$ is a solution of (1), provided $f(x)$ is twice differentiable on the interval $0 < x < L$, and has one-sided second derivatives at $x = 0$ and $x = L$, which are zero. Under these conditions $u(x, t)$ is established as a solution of (1), satisfying (2)—(4).

If $f'(x)$ and $f''(x)$ are merely piecewise continuous (cf. Sec. 5.1), or if those one-sided derivatives are not zero, then for each t there will be finitely many values of x at which the second derivatives of u appearing in (1) do not exist. Except at these points the wave equation will still be satisfied, and we may then regard $u(x, t)$ as a solution of our problem in a broader sense. For example, the case of a triangular initial deflection (Example 1, below) leads to a solution of this type.

Fig. 268. Odd periodic extension of $f(x)$

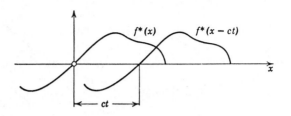

Fig. 269. Interpretation of (17)

Let us mention a very interesting physical interpretation of (17). The graph of $f^*(x - ct)$ is obtained from the graph of $f^*(x)$ by shifting the latter ct units to the right (Fig. 269). This means that $f^*(x - ct)$ $(c > 0)$ represents a wave that is traveling to the right as t increases. Similarly, $f^*(x + ct)$ represents a wave that is traveling to the left, and $u(x, t)$ is the superposition of these two waves.

EXAMPLE 1. Vibrating string if the initial deflection is triangular

Find the solution of the wave equation (1) corresponding to the triangular initial deflection

$$f(x) = \begin{cases} \dfrac{2k}{L}\,x & \text{if } 0 < x < \dfrac{L}{2} \\[2mm] \dfrac{2k}{L}\,(L - x) & \text{if } \dfrac{L}{2} < x < L \end{cases}$$

and initial velocity zero. (Fig. 270 on the next page shows $f(x) = u(x, 0)$ at the top.)

Solution. Since $g(x) \equiv 0$, we have $B_n^* = 0$ in (12), and from Example 1 in Sec. 10.5 we see that the B_n are given by (5), Sec. 10.5. Thus (12) takes the form

$$u(x, t) = \frac{8k}{\pi^2}\left[\frac{1}{1^2}\sin\frac{\pi}{L}x\cos\frac{\pi c}{L}t - \frac{1}{3^2}\sin\frac{3\pi}{L}x\cos\frac{3\pi c}{L}t + - \cdots\right].$$

For plotting the graph of the solution we may use $u(x, 0) = f(x)$ and the above interpretation of the two functions in the representation (17). This leads to the graph shown in Fig. 270. ∎

It is very interesting that the solution (17) can also be obtained very quickly by a suitable transformation of the wave equation, following an ingenious idea by d'Alembert, which we discuss in the next section.

Problems for Sec. 11.3

1. How does doubling the tension affect the pitch of the fundamental tone of a string?

Find the deflection $u(x, t)$ of the vibrating string (length $L = \pi$, ends fixed, and $c^2 = T/\rho = 1$) corresponding to zero initial velocity and initial deflection:

2. $0.01 \sin x$ **3.** $k \sin 2x$ **4.** $k(\sin x + \sin 3x)$

5. $0.01x(\pi - x)$ **6.** $0.01x(\pi^2 - x^2)$ **7.** $k[(\tfrac{1}{2}\pi)^4 - (x - \tfrac{1}{2}\pi)^4]$

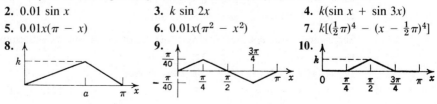

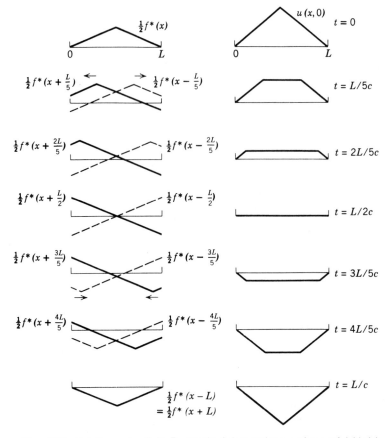

Fig. 270. Solution $u(x, t)$ in Example 1 for various values of t (right part of the figure) obtained as the superposition of a wave traveling to the right (dashed) and a wave traveling to the left (left part of the figure)

11. What is the ratio of the amplitudes of the fundamental mode and the second overtone in Prob. 5? The ratio $a_1{}^2/(a_1{}^2 + a_2{}^2 + \cdots)$? *Hint.* Use (7) in Sec. 10.8, with the equality sign.

12. In what manner does the frequency of the fundamental mode of the vibrating string depend on the length of the string, the tension and the mass of the string per unit length?

Find the deflection $u(x, t)$ of the vibrating string (length $L = \pi$, ends fixed, $c^2 = 1$) if the initial deflection $f(x)$ and the initial velocity $g(x)$ are:

13. $f = 0$, $g(x) = 0.1 \sin 2x$

14. $f = 0$, $g(x) = 0.01x$ if $0 \leq x \leq \frac{1}{2}\pi$, $g(x) = 0.01(\pi - x)$ if $\frac{1}{2}\pi < x \leq \pi$.

15. $f(x) = 0.1 \sin x$, $g(x) = -0.2 \sin x$

Separating variables, find solutions $u(x, y)$ of the following equations.

16. $u_x = yu_y$ 17. $u_x + u_y = 0$ 18. $u_x = u_y$
19. $u_x + u_y = 2(x + y)u$ 20. $ayu_x = bxu_y$ 21. $xu_x = yu_y$
22. $x^2u_{xy} + 3y^2u = 0$ 23. $u_{xy} = u$ 24. $u_{xx} + u_{yy} = 0$

25. Show that a problem (1)—(4) with more complicated boundary conditions, say, $u(0, t) = 0$, $u(L, t) = h(t)$, can be reduced to a problem for a new function v satisfying conditions $v(0, t) = v(L, t) = 0$, $v(x, 0) = f_1(x)$, $v_t(x, 0) = g_1(x)$ but a nonhomogeneous wave equation. *Hint.* Set $u = v + w$ and determine w suitably.

11.4 D'Alembert's Solution of the Wave Equation

It is interesting to note that the solution (17), Sec. 11.3, of the wave equation

$$(1) \qquad \frac{\partial^2 u}{\partial t^2} = c^2 \frac{\partial^2 u}{\partial x^2} \qquad\qquad c^2 = \frac{T}{\rho}$$

can be immediately obtained by transforming (1) in a suitable way, namely, by introducing the new independent variables[1]

$$(2) \qquad v = x + ct, \qquad z = x - ct.$$

Then u becomes a function of v and z, and the derivatives in (1) can be expressed in terms of derivatives with respect to v and z by the use of the chain rule in Sec. 8.7. Denoting partial derivatives by subscripts, we see from (2) that $v_x = 1$ and $z_x = 1$. For simplicity let us denote $u(x, t)$, as a function of v and z, by the same letter u. Then

$$u_x = u_v v_x + u_z z_x = u_v + u_z.$$

Applying the chain rule to the right side and using $v_x = 1$ and $z_x = 1$ we find

$$u_{xx} = (u_v + u_z)_x = (u_v + u_z)_v v_x + (u_v + u_z)_z z_x = u_{vv} + 2u_{vz} + u_{zz}.$$

We transform the other derivative in (1) by the same procedure, finding

$$u_{tt} = c^2(u_{vv} - 2u_{vz} + u_{zz}).$$

By inserting these two results in (1) we obtain (cf. footnote 1 in Appendix 3.1)

$$(3) \qquad \boxed{\; u_{vz} \equiv \frac{\partial^2 u}{\partial z\,\partial v} = 0. \;}$$

Obviously, the point of the present approach is that the resulting equation

---

[1]We mention that the general theory of partial differential equations provides a systematic way for finding this transformation which will simplify the equation. Cf. Ref. [C14] in Appendix 1.

(3) can be readily solved by two successive integrations. In fact, integrating with respect to z, we find

$$\frac{\partial u}{\partial v} = h(v)$$

where $h(v)$ is an arbitrary function of v. Integration with respect to v gives

$$u = \int h(v) \, dv + \psi(z)$$

where $\psi(z)$ is an arbitrary function of z. Since the integral is a function of v, say, $\phi(v)$, the solution u is of the form $u = \phi(v) + \psi(z)$. Because of (2),

$$(4) \qquad\qquad u(x, t) = \phi(x + ct) + \psi(x - ct).$$

This is known as **d'Alembert's solution**[2] of the wave equation (1).

The functions ϕ and ψ can be determined from the initial conditions. Let us illustrate this in the case of zero initial velocity and given initial deflection $u(x, 0) = f(x)$. By differentiating (4) we have

$$(5) \qquad\qquad \frac{\partial u}{\partial t} = c\phi'(x + ct) - c\psi'(x - ct)$$

where primes denote derivatives with respect to the *entire* arguments $x + ct$ and $x - ct$, respectively. From (4), (5) and the initial conditions we obtain

$$u(x, 0) = \phi(x) + \psi(x) = f(x)$$

$$u_t(x, 0) = c\phi'(x) - c\psi'(x) = 0.$$

From the last equation, $\psi' = \phi'$. Hence $\psi = \phi + k$, and from this and the first equation, $2\phi + k = f$ or $\phi = (f - k)/2$. With these functions ϕ and ψ the solution (4) becomes

$$(6) \qquad\qquad u(x, t) = \tfrac{1}{2}[f(x + ct) + f(x - ct)],$$

in agreement with (17) in Sec. 11.3. The student may show that because of the boundary conditions (2) in that section the function f must be odd and must have period $2L$.

If the initial velocity $g(x)$ is not identically zero, instead of (6) we similarly obtain

$$(7) \qquad \boxed{u(x, t) = \tfrac{1}{2}[f(x + ct) + f(x - ct)] + \frac{1}{2c} \int_{x-ct}^{x+ct} g(s) \, ds.}$$

---

[2]JEAN LE ROND D'ALEMBERT (1717—1783), French mathematician, who is also known for his important work in mechanics.

Our result shows that the two initial conditions and the boundary conditions determine the solution uniquely.

The solution of the wave equation by the Laplace transformation and the Fourier transformation will be shown in Secs. 11.13 and 11.14.

In the next section we turn to the **heat equation,** which is another partial differential equation of basic importance.

Problems for Sec. 11.4

Using (6), sketch a figure (of the type of Fig. 270 in Sec. 11.3) of the deflection $u(x, t)$ of a vibrating string (length $L = 1$, ends fixed) starting with initial velocity zero and the following initial deflection $f(x)$, where k is small, say, $k = 0.01$.

1. $f(x) = k \sin \pi x$ **2.** $f(x) = kx(1 - x)$ **3.** $f(x) = k(x - x^3)$

4. $f(x) = k(x^2 - x^4)$ **5.** $f(x) = k(1 - \cos 2\pi x)$ **6.** $f(x) = k \sin^2 \pi x$

Using the indicated transformations, solve the following equations.

7. $u_{xy} = u_{xx}$ $(v = y, z = x + y)$

8. $y u_{xy} = x u_{xx} + u_x$ $(v = y, z = xy)$

9. $u_{xx} - 2u_{xy} + u_{yy} = 0$ $(v = y, z = x + y)$

10. $u_{xx} = u_{yy}$ $(v = y + x, z = y - x)$

11. $u_{yy} + u_{xy} - 2u_{xx} = 0$ $(v = x + y, z = 2y - x)$

12. $u_{xx} - 4u_{xy} + 3u_{yy} = 0$ $(v = x + y, z = 3x + y)$

Types and normal forms of linear partial differential equations. An equation of the form

$$(8) \qquad\qquad A u_{xx} + 2B u_{xy} + C u_{yy} = F(x, y, u, u_x, u_y)$$

is said to be **elliptic** if $AC - B^2 > 0$, **parabolic** if $AC - B^2 = 0$, and **hyperbolic** if $AC - B^2 < 0$. (Here A, B, C may be functions of x and y, and the type of (8) may be different in different parts of the xy-plane.)

13. Show that

> **Laplace's equation** $u_{xx} + u_{yy} = 0$ is elliptic,
>
> the **heat equation** $u_t = c^2 u_{xx}$ is parabolic,
>
> the **wave equation** $u_{tt} = c^2 u_{xx}$ is hyperbolic,
>
> the **Tricomi equation** $y u_{xx} + u_{yy} = 0$ is of mixed type (elliptic in the upper half-plane and hyperbolic in the lower half-plane).

14. If the equation (8) is *hyperbolic,* it can be transformed to the *normal form* $u_{vz} = F^*(v, z, u, u_v, u_z)$ by setting $v = \Phi(x, y)$, $z = \Psi(x, y)$, where $\Phi = const$ and $\Psi = const$ are the solutions $y = y(x)$ of the equation $Ay'^2 - 2By' + C = 0$ (cf. Ref. [C12]). Show that in the case of the wave equation (1),

$$\Phi = x + ct, \qquad \Psi = x - ct.$$

15. If (8) is *parabolic,* the substitution $v = x$, $z = \Psi(x, y)$, with Ψ defined as in Prob. 14, reduces it to the *normal form* $u_{vv} = F^*(v, z, u, u_v, u_z)$. Verify this result for the equation $u_{xx} + 2u_{xy} + u_{yy} = 0$.

16. (Airy equation) Show that by separating variables we can obtain from the **Tricomi equation** the Airy equation $G'' - yG = 0$. (For solutions, see p. 446 of Ref. [1] listed in Appendix 1. See also Review Prob. 30 for Chap. 4.)

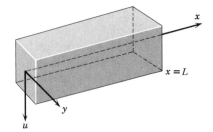

Fig. 271. Undeformed beam in Problem 17

Vibrations of a beam. It can be shown that the small free vertical vibrations of a uniform beam (Fig. 271) are governed by the fourth-order equation

$$(9) \qquad \frac{\partial^2 u}{\partial t^2} + c^2 \frac{\partial^4 u}{\partial x^4} = 0 \qquad \text{(Ref. [C14]),}$$

where $c^2 = EI/\rho A$ (E = Young's modulus of elasticity, I = moment of inertia of the cross section with respect to the y-axis in the figure, ρ = density, A = cross-sectional area).

17. Substituting $u = F(x)G(y)$ into (9) and separating variables, show that

$$F^{(4)}/F = -\ddot{G}/c^2 G = \beta^4 = const,$$

$$F(x) = A \cos \beta x + B \sin \beta x + C \cosh \beta x + D \sinh \beta x,$$

$$G(t) = a \cos c\beta^2 t + b \sin c\beta^2 t.$$

18. Find solutions $u_n = F_n(x)G_n(t)$ of (9) corresponding to zero initial velocity and satisfying the boundary conditions (cf. Fig. 272)
  $u(0, t) = 0$, $u(L, t) = 0$ (ends simply supported for all times t),
  $u_{xx}(0, t) = 0$, $u_{xx}(L, t) = 0$ (zero moments, hence zero curvature, at the ends).

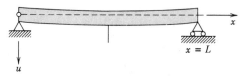

Fig. 272. Beam in Problem 18

19. Find the solution of (9) that satisfies the conditions in Prob. 18 and the initial condition $u(x, 0) = f(x) = x(L - x)$.

20. Compare the results of Prob. 19 and Prob. 5, Sec. 11.3. What is the basic difference between the frequencies of the normal modes of the vibrating string and the vibrating beam?

21. What are the boundary conditions if the beam is clamped at both ends? (Cf. Fig. 273.)

Fig. 273. Beam in Problem 21

Fig. 274. Beam in Problem 24

22. Show that $F(x)$ in Prob. 17 satisfies the conditions in Prob. 21 if βL is a root of the equation

(10) $$\cosh \beta L \cos \beta L = 1.$$

23. Determine approximate solutions of (10).

24. If the beam is clamped at the left and free at the other end (Fig. 274), the boundary conditions are

$$u(0, t) = 0, \qquad u_x(0, t) = 0, \qquad u_{xx}(L, t) = 0, \qquad u_{xxx}(L, t) = 0.$$

Show that $F(x)$ in Prob. 17 satisfies these conditions if βL is a root of the equation

(11) $$\cosh \beta L \cos \beta L = -1.$$

25. Find approximate solutions of (11).

11.5 Heat Flow

The heat flow in a body of homogeneous material is governed by the heat equation (cf. Sec. 9.7)

$$\frac{\partial u}{\partial t} = c^2 \nabla^2 u \qquad\qquad c^2 = \frac{K}{\sigma \rho}$$

where $u(x, y, z, t)$ is the temperature in the body, K is the thermal conductivity, σ is the specific heat and ρ is the density of the material of the body. $\nabla^2 u$ is the Laplacian of u, and with respect to Cartesian coordinates x, y, z,

$$\nabla^2 u = \frac{\partial^2 u}{\partial x^2} + \frac{\partial^2 u}{\partial y^2} + \frac{\partial^2 u}{\partial z^2}.$$

As an important application, let us consider the temperature in a long thin bar or wire of constant cross section and homogeneous material, which is oriented along the x-axis (Fig. 275) and is perfectly insulated laterally, so that heat flows in the x-direction only. Then u depends only on x and time

Fig. 275. Bar under consideration

t, and the heat equation becomes the so-called **one-dimensional heat equation**

$$(1) \qquad \boxed{\frac{\partial u}{\partial t} = c^2 \frac{\partial^2 u}{\partial x^2}.}$$

We shall solve (1) for some important types of boundary and initial conditions. The procedure of solving (1) will be similar to that in the case of the wave equation. The behavior of the solutions will be entirely different from that of the solutions of the wave equation, because (1) involves u_t whereas the wave equation involves u_{tt}. (Hence the classification in Prob. 13 of the last section is not merely a formal matter but has far-reaching consequences with respect to the general behavior of the solutions.)

Let us start with the case when the ends $x = 0$ and $x = L$ of the bar are kept at temperature zero. Then the **boundary conditions** are

$$(2) \qquad u(0, t) = 0, \qquad u(L, t) = 0 \qquad \text{for all } t.$$

Note that this has the same form as (2) in Sec. 11.3. Let $f(x)$ be the initial temperature in the bar. Then the **initial condition** is

$$(3) \qquad u(x, 0) = f(x) \qquad\qquad [f(x) \text{ given}].$$

We shall determine a solution $u(x, t)$ of (1) satisfying (2) and (3).

First step. Applying the method of separation of variables we first determine solutions of (1) that satisfy the boundary conditions (2). We start from

$$(4) \qquad u(x, t) = F(x)G(t).$$

Substituting this expression into (1), we obtain the equation $F\dot{G} = c^2 F''G$, where dots denote derivatives with respect to t and primes denote derivatives with respect to x. To separate variables, we divide this equation by c^2FG:

$$(5) \qquad \frac{\dot{G}}{c^2G} = \frac{F''}{F}.$$

The expression on the left depends only on t, while the right-hand side depends only on x. As in Sec. 11.3, we conclude that both expressions must be equal to a constant, say, k. The student may show that for $k \geqq 0$ the only solution $u = FG$ that satisfies (2) is $u \equiv 0$. For negative $k = -p^2$ we have from (5)

$$\frac{\dot{G}}{c^2G} = \frac{F''}{F} = -p^2.$$

We see that this yields the two linear ordinary differential equations

$$(6) \qquad \boxed{F'' + p^2F = 0}$$

and

(7)
$$\boxed{\dot{G} + c^2 p^2 G = 0.}$$

Second step. We consider (6). The general solution is

(8) $F(x) = A \cos px + B \sin px.$

From the boundary conditions (2) it follows that

$$u(0, t) = F(0)G(t) = 0 \quad \text{and} \quad u(L, t) = F(L)G(t) = 0.$$

Since $G \equiv 0$ implies $u \equiv 0$, we require that $F(0) = 0$ and $F(L) = 0$. By (8), $F(0) = A$. Thus $A = 0$, and, therefore,

$$F(L) = B \sin pL.$$

We must have $B \neq 0$, since otherwise $F \equiv 0$. Hence the condition $F(L) = 0$ leads to

$$\sin pL = 0, \quad \text{hence} \quad p = \frac{n\pi}{L}, \quad n = 1, 2, \cdots.$$

Setting $B = 1$, we thus obtain the following solutions of (6) satisfying (2):

$$F_n(x) = \sin \frac{n\pi x}{L}, \qquad n = 1, 2, \cdots.$$

(As in Sec. 11.3, we need not consider negative integral values of n.)

We now consider the differential equation (7). For the values $p = n\pi/L$ just obtained it takes the form

$$\dot{G} + \lambda_n^2 G = 0 \quad \text{where} \quad \lambda_n = \frac{cn\pi}{L}.$$

The general solution is

$$G_n(t) = B_n e^{-\lambda_n^2 t}, \qquad n = 1, 2, \cdots,$$

where B_n is a constant. Hence the functions

(9) $$\boxed{u_n(x, t) = F_n(x)G_n(t) = B_n \sin \frac{n\pi x}{L} e^{-\lambda_n^2 t}} \qquad n = 1, 2, \cdots.$$

are solutions of the heat equation (1), satisfying (2).

As in the previous section, in order to obtain the solution of the entire problem, we shall now utilize an infinite series.

Third step. To obtain a solution also satisfying (3), we consider the series

(10) $$u(x, t) = \sum_{n=1}^{\infty} u_n(x, t) = \sum_{n=1}^{\infty} B_n \sin \frac{n\pi x}{L} e^{-\lambda_n^2 t} \qquad \left(\lambda_n = \frac{cn\pi}{L} \right).$$

From this and (3),

$$u(x, 0) = \sum_{n=1}^{\infty} B_n \sin \frac{n\pi x}{L} = f(x).$$

Hence, for (10) to satisfy (3), the coefficients B_n must be chosen such that $u(x, 0)$ becomes a half-range expansion of $f(x)$, namely, the Fourier sine series of $f(x)$; that is [cf. (4) in Sec. 10.5],

(11) $$B_n = \frac{2}{L} \int_0^L f(x) \sin \frac{n\pi x}{L} dx \qquad n = 1, 2, \cdots .$$

The solution of our problem can be established, assuming that $f(x)$ is piecewise continuous on the interval $0 \leq x \leq L$ (cf. Sec. 5.1), and has one-sided derivatives[3] at all interior points of that interval; that is, under these assumptions the series (10) with coefficients (11) is the solution of our physical problem. The proof, which requires the knowledge of uniform convergence of series, will be given at a later occasion (Probs. 21, 22 at the end of Sec. 14.8).

Because of the exponential factor all the terms in (10) approach zero as t approaches infinity. The rate of decay increases with n.

EXAMPLE 1. Sinusoidal initial temperature

Find the temperature $u(x, t)$ in a laterally insulated copper bar 80 cm long if the initial temperature is $100 \sin (\pi x/80)$ °C and the ends are kept at 0 °C. How long will it take for the maximum temperature in the bar to drop to 50 °C? First guess, then calculate. Physical data for copper: density 8.92 gm/cm^3, specific heat 0.092 cal/gm °C, thermal conductivity 0.95 cal/cm sec °C.

Solution. The initial condition gives

$$u(x, 0) = \sum_{n=1}^{\infty} B_n \sin \frac{n\pi x}{80} = f(x) = 100 \sin \frac{\pi x}{80} .$$

Hence, by inspection or from (10) we get $B_1 = 100$, $B_2 = B_3 = \cdots = 0$. In (10) we need $\lambda_1^2 = c^2 \pi^2 / L^2$, where $c^2 = K/\sigma\rho = 0.95/0.092 \cdot 8.92 = 1.158$ [cm^2/sec]. Hence we obtain $\lambda_1^2 = 1.158 \cdot 9.870/6400 = 0.001785$ [sec^{-1}]. The solution (10) is

$$u(x, t) = 100 \sin \frac{\pi x}{80} e^{-0.001785t}.$$

Also, $100 e^{-0.001785t} = 50$ when $t = (\ln 0.5)/(-0.001785) = 388$ [seconds] ≈ 6.5 [minutes].

EXAMPLE 2. Speed of decay

Solve the problem in Example 1 when the initial temperature is $100 \sin (3\pi x/80)$ °C and the other data are as before.

---

[3]Cf. footnote 7 in Sec. 10.2.

Solution. In (10), instead of $n = 1$ we now have $n = 3$, and $\lambda_3{}^2 = 3^2\lambda_1{}^2 = 9 \cdot 0.001785 = 0.01607$, so that the solution now is

$$u(x, t) = 100 \sin \frac{3\pi x}{80} e^{-0.01607t}.$$

Hence the maximum temperature drops to 50 °C in $t = (\ln 0.5)/(-0.01607) \approx 43$ [seconds], which is much faster (9 times as fast as in Example 1).

Had we chosen a bigger n, the decay would have been still faster, and in a sum or series of such terms, each term has its own rate of decay, and terms with large n are practically 0 after a very short time. Our next example is of this type, and the curve in Fig. 276 corresponding to $t = 0.5$ looks like a sine curve; that is, it is practically the graph of the first term of the solution.

EXAMPLE 3. "Triangular" initial temperature in a bar

Find the temperature in a laterally insulated bar of length L whose ends are kept at temperature 0, assuming that the initial temperature is

$$f(x) = \begin{cases} x & \text{if} & 0 < x < L/2, \\ L - x & \text{if} & L/2 < x < L. \end{cases}$$

(The upper part of Fig. 276 shows this function for the special $L = \pi$.)

Solution. From (11) we get

$$(12) \qquad B_n = \frac{2}{L} \left(\int_0^{L/2} x \sin \frac{n\pi x}{L} \, dx + \int_{L/2}^L (L - x) \sin \frac{n\pi x}{L} \, dx \right).$$

Integration yields $B_n = 0$ if n is even and

$$B_n = \frac{4L}{n^2 \pi^2} \qquad\qquad (n = 1, 5, 9, \cdots),$$

$$B_n = -\frac{4L}{n^2 \pi^2} \qquad\qquad (n = 3, 7, 11, \cdots).$$

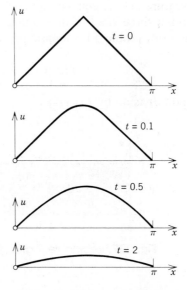

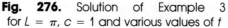

Fig. 276. Solution of Example 3 for $L = \pi$, $c = 1$ and various values of t

Hence the solution is

$$u(x, t) = \frac{4L}{\pi^2} \left[\sin \frac{\pi x}{L} e^{-(c\pi/L)^2 t} - \frac{1}{9} \sin \frac{3\pi x}{L} e^{-(3c\pi/L)^2 t} + \cdots \right].$$

The reader may compare Fig. 276 and Fig. 270 in Sec. 11.3. ■

Steady-State Two-Dimensional Heat Flow

The two-dimensional heat equation is (see at the beginning of this section)

$$\frac{\partial u}{\partial t} = c^2 \nabla^2 u = c^2 \left(\frac{\partial^2 u}{\partial x^2} + \frac{\partial^2 u}{\partial y^2} \right).$$

If the heat flow is **steady** or **stationary** (i.e., time independent), then $\partial u / \partial t = 0$, and the heat equation reduces to **Laplace's equation**[4]

$$(13) \qquad \boxed{\nabla^2 u = \frac{\partial^2 u}{\partial x^2} + \frac{\partial^2 u}{\partial y^2} = 0.}$$

A heat problem then consists of this equation to be considered in some region R of the xy-plane and a given boundary condition on the boundary curve of R. This is called a **boundary value problem.** One calls it

the **Dirichlet problem** if u is prescribed on C,
the **Neumann problem** if the normal derivative $u_n = \partial u / \partial n$ is prescribed on C,
the **mixed problem** if u is prescribed on a portion of C and u_n on the rest of C.

We consider a Dirichlet problem in a rectangle R (Fig. 277), assuming that the temperature $u(x, y)$ equals a given function $f(x)$ on the upper side and 0 on the other three sides of the rectangle R.

We solve this problem by separating variables. Substitution of

$$u(x, y) = F(x)G(y)$$

into (13) and division by FG gives

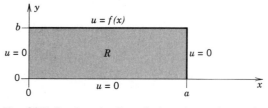

Fig. 277. Rectangle R and given boundary values

---

[4]This very important equation appeared in Sec. 9.7 and will be discussed further in Secs. 11.9, 11.11, 11.12, 12.5 and in Chap. 17.

$$\frac{1}{F} \cdot \frac{d^2F}{dx^2} = -\frac{1}{G} \cdot \frac{d^2G}{dy^2} = -k.$$

From this and the boundary conditions on the left- and right-hand sides of R,

$$\frac{d^2F}{dx^2} + kF = 0, \qquad F(0) = 0, \quad F(a) = 0.$$

This gives $k = (n\pi/a)^2$ and corresponding nonzero solutions

(14) $F(x) = F_n(x) = \sin\frac{n\pi}{a}x,$ $n = 1, 2, \cdots .$

The equation for G then becomes

$$\frac{d^2G}{dy^2} - \left(\frac{n\pi}{a}\right)^2 G = 0.$$

Solutions are

$$G(y) = G_n(y) = A_n e^{n\pi y/a} + B_n e^{-n\pi y/a}.$$

Now the boundary condition $u = 0$ on the lower side of R implies that $G_n(0) = 0$; that is, $G_n(0) = A_n + B_n = 0$ or $B_n = -A_n$. This gives

$$G_n(y) = A_n(e^{n\pi y/a} - e^{-n\pi y/a}) = 2A_n \sinh\frac{n\pi y}{a}.$$

From this and (14), writing $2A_n = A_n^*$, we obtain the **"eigenfunctions"** of our problem

(15) $u_n(x, y) = F_n(x)G_n(y) = A_n^* \sin\frac{n\pi x}{a} \sinh\frac{n\pi y}{a}.$

These satisfy the boundary condition $u = 0$ on the left, right and lower sides of the rectangle R. To obtain the solution of our problem also satisfying the boundary condition

(16) $u(x, b) = f(x)$

on the upper side of R, we consider the infinite series

$$u(x, y) = \sum_{n=1}^{\infty} u_n(x, y).$$

From this, (16) and (15) with $y = b$ we get

$$u(x, b) = f(x) = \sum_{n=1}^{\infty} A_n^* \sin\frac{n\pi x}{a} \sinh\frac{n\pi b}{a}.$$

We can write this in the form

$$u(x, b) = \sum_{n=1}^{\infty} \left(A_n^* \sinh \frac{n\pi b}{a} \right) \sin \frac{n\pi x}{a}.$$

This shows that the expressions in the parentheses must be the Fourier coefficients b_n of $f(x)$; that is, by (4) in Sec. 10.5,

$$b_n = A_n^* \sinh \frac{n\pi b}{a} = \frac{2}{a} \int_0^a f(x) \sin \frac{n\pi x}{a} \, dx.$$

From this and (15) we see that the solution of our problem is

(17)
$$u(x, y) = \sum_{n=1}^{\infty} A_n^* \sin \frac{n\pi x}{a} \sinh \frac{n\pi y}{a}$$

where

(18)
$$A_n^* = \frac{2}{a \sinh (n\pi b/a)} \int_0^a f(x) \sin \frac{n\pi x}{a} \, dx.$$

This solution, obtained formally without regard to the convergence and the sums of the series for u, u_{xx} and u_{yy}, can be established when f and f' are continuous and f'' is piecewise continuous on the interval $0 \leq x \leq a$. The proof is somewhat involved and relies on uniform convergence; it can be found in Ref. [C2] listed in Appendix 1.

Electrostatic Potential

The Laplace equation (13) also governs the electrostatic potential of electrical charges in any region that is free of these charges. Thus our steady-state heat problem can also be interpreted as an electrostatic potential problem, so that (17), (18) is the potential in the rectangle R when the upper side of R is at potential $f(x)$ and the other three sides are grounded.

Actually, in the steady-state case, the two-dimensional wave equation (to be considered in Secs. 11.7, 11.8) also reduces to (13), and (17), (18) then is the displacement of a rectangular elastic membrane (rubber sheet, drumhead) which is fixed along its boundary, with three sides lying in the *xy*-plane and the fourth side given the displacement $f(x)$. This "rubber model" of an electrostatic potential field can be used (and is used) to find the paths of electrons: one can simply let little marbles run down the slope and photograph their paths. Similarly, this would also give the curves along which heat flows in the heat model.

This is an impressive demonstration of the ***unifying power*** of mathematics: physical systems of entirely different natures may have the same mathematical model and can then be treated by the same mathematical methods— and such analogies can be practically used in physical modeling, as indicated.

In the next section we show that our discussion of the heat equation in the first part of the present section can be extended to infinite bars. These are good models of very long bars or wires (for instance, a wire 300 ft long). In these models, the **Fourier integral** (Sec. 10.9) will take the place of the Fourier series.

Problems for Sec. 11.5

1. Compare Figs. 270 and 276 and describe the different behavior of the solutions of the two equations.

2. How does the rate of decay of (9) for fixed n depend on the specific heat, the density and the thermal conductivity of the material?

3. Graph u_1, u_2, u_3 [cf. (9), with $B_n = 1$, $c = 1$, $L = \pi$] as functions of x for the values $t = 0, 1, 2, 3$. Compare the behavior of these functions.

Find the temperature $u(x, t)$ in a bar of silver (length 10 cm, constant cross section of area 1 cm^2, density 10.6 gm/cm^3, thermal conductivity 1.04 cal/cm sec °C, specific heat 0.056 cal/gm °C) that is perfectly insulated laterally, whose ends are kept at temperature 0 °C and whose initial temperature (in °C) is $f(x)$, where

4. $f(x) = \sin 0.2\pi x$ 5. $f(x) = \sin 0.1\pi x$

6. $f(x) = \begin{cases} x & \text{if } 0 < x < 5 \\ 0 & \text{if } 5 < x < 10 \end{cases}$ 7. $f(x) = \begin{cases} x & \text{if } 0 < x < 5 \\ 10 - x & \text{if } 5 < x < 10 \end{cases}$

8. $f(x) = x(100 - x^2)$ 9. $f(x) = x(10 - x)$

10. $f(x) = x$ if $0 < x < 2.5$, $f(x) = 2.5$ if $2.5 < x < 7.5$, $f(x) = 10 - x$ if $7.5 < x < 10$

11. Suppose that a rod satisfies the assumptions in the text and that its ends are kept at different constant temperatures $u(0, t) = U_1$ and $u(L, t) = U_2$. Find the temperature $u_1(x)$ in the rod after a long time (theoretically: as $t \to \infty$).

12. In Prob. 11, let the initial temperature be $u(x, 0) = f(x)$. Show that the temperature for any time $t > 0$ is $u(x, t) = u_1(x) + u_{II}(x, t)$ with u_1 as before and

$$u_{II} = \sum_{n=1}^{\infty} B_n \sin \frac{n\pi x}{L} e^{-(cn\pi/L)^2 t},$$

where

$$B_n = \frac{2}{L} \int_0^L [f(x) - u_1(x)] \sin \frac{n\pi x}{L} dx$$

$$= \frac{2}{L} \int_0^L f(x) \sin \frac{n\pi x}{L} dx + \frac{2}{n\pi} [(-1)^n U_2 - U_1].$$

13. **(Insulated ends, adiabatic boundary conditions)** Find the temperature $u(x, t)$ in a bar of length L that is perfectly insulated, also at the ends at $x = 0$ and $x = L$, assuming that $u(x, 0) = f(x)$. *Physical information:* the flux of heat through the faces at the ends is proportional to the values of $\partial u/\partial x$ there. Show that this situation corresponds to the conditions

$$u_x(0, t) = 0, \qquad u_x(L, t) = 0, \qquad u(x, 0) = f(x).$$

Show that the method of separating variables yields the solution

$$u(x, t) = A_0 + \sum_{n=1}^{\infty} A_n \cos \frac{n\pi x}{L} e^{-(cn\pi/L)^2 t}$$

where, by (2) in Sec. 10.5,

$$A_0 = \frac{1}{L} \int_0^L f(x)\, dx, \qquad A_n = \frac{2}{L} \int_0^L f(x) \cos \frac{n\pi x}{L}\, dx, \quad n = 1, 2, \cdots .$$

14. In Prob. 13, $u \to A_0$ as $t \to \infty$. Does this agree with your physical intuition?

Find the temperature in the bar in Prob. 13 if $L = \pi$, $c = 1$, and:

15. $f(x) = k = const$ **16.** $f(x) = x$

17. $f(x) = \cos 2x$ **18.** $f(x) = 1 - x/\pi$

19. $f(x) = \cos^2 x$ **20.** $f(x) = \sin x$

21. $f(x) = x$ if $0 < x < \pi/2$, $f(x) = 0$ if $\pi/2 < x < \pi$

22. Consider the bar in Probs. 4—10. Assume that the ends are kept at 100 °C for a long time. Then at some instant, say, at $t = 0$, the temperature at $x = L$ is suddenly changed to 0 °C and kept at this value, while the temperature at $x = 0$ is kept at 100 °C. What are the temperatures in the middle of the bar at $t = 1, 2, 3, 10, 50$ sec?

23. **(Radiation at end of rod)** Consider a laterally insulated rod of length π and such that $c = 1$ in (1), whose left end is kept at 0 °C and whose right end radiates freely into air of constant temperature 0 °C. *Physical information.* The "radiation boundary condition" is

$$-u_x(\pi, t) = k[u(\pi, t) - u_0],$$

where $u_0 = 0$ is the temperature of the surrounding air and k is a constant, say $k = 1$ for simplicity. Show that a solution satisfying these boundary conditions is $u(x, t) = \sin px\, e^{-p^2 t}$, where p is a solution of $\tan p\pi = -p$. Show graphically that this equation has infinitely many positive solutions $p_1, p_2, p_3, \cdots$, where $p_n > n - \frac{1}{2}$ and $\lim_{n\to\infty} (p_n - n + \frac{1}{2}) = 0$.

24. **(Nonhomogeneous heat equation)** Consider the problem consisting of

$$u_t - c^2 u_{xx} = Ne^{-\alpha x}$$

and conditions (2), (3). Here the term on the right may represent loss of heat due to radioactive decay in the bar. Show that this problem may be reduced to a problem for the homogeneous equation as follows. Set $u(x, t) = v(x, t) + w(x)$ and determine $w(x)$ so that v satisfies the homogeneous equation and the conditions $v(0, t) = v(L, t) = 0$, $v(x, 0) = f(x) - w(x)$.

25. **(Radiation)** If the rod in the text is free to radiate into the surrounding medium that is kept at constant temperature zero, the equation becomes

$$v_t = c^2 v_{xx} - \beta v.$$

Show that this equation can be reduced to the form (1) by setting $v(x, t) = u(x, t)w(t)$.

Two-Dimensional Problems

26. **(Laplace's equation)** Find the potential in the square $0 \le x \le 3$, $0 \le y \le 3$ if the upper side is kept at the potential $\sin \frac{1}{3}\pi x$ and the other sides are kept at zero.

27. Find the potential in the rectangle $0 \le x \le 20$, $0 \le y \le 40$ whose upper side is kept at potential 110 V and whose other sides are grounded.

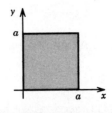

Fig. 278. Square plate

28. **(Heat flow in a plate)** The faces of a thin square plate (Fig. 278, where $a = 12$) are perfectly insulated. The upper side is kept at temperature 100 °C and the other sides are kept at 0 °C. Find the steady-state temperature $u(x, y)$ in the plate.

29. Find solutions u of the two-dimensional heat equation

$$u_t = c^2(u_{xx} + u_{yy})$$

in the thin plate in Fig. 278 with $a = \pi$ such that $u = 0$ on the vertical sides, assuming that the faces and the horizontal sides of the plate are perfectly insulated.

30. Find formulas similar to (17), (18), for the temperature distribution in the rectangle R considered in the text when the lower side of R is kept at temperature $f(x)$ and the three other sides are kept at 0.

11.6 Heat Flow in an Infinite Bar

We shall now consider solutions of the heat equation

$$(1) \qquad\qquad \frac{\partial u}{\partial t} = c^2 \frac{\partial^2 u}{\partial x^2}$$

in the case of a bar that extends to infinity on both sides (and is laterally insulated, as before). In this case we do not have boundary conditions but only the initial condition

$$(2) \qquad\qquad u(x, 0) = f(x) \qquad\qquad (-\infty < x < \infty)$$

where $f(x)$ is the given initial temperature of the bar.

To solve our present problem we start as in the last section; that is, we substitute $u(x, t) = F(x)G(t)$ into (1). This yields the two ordinary differential equations

$$(3) \qquad\qquad F'' + p^2 F = 0 \qquad\qquad \text{[cf. (6), Sec. 11.5]}$$

and

$$(4) \qquad\qquad \dot{G} + c^2 p^2 G = 0 \qquad\qquad \text{[cf. (7), Sec. 11.5]}.$$

Solutions are

$$F(x) = A \cos px + B \sin px \qquad \text{and} \qquad G(t) = e^{-c^2 p^2 t}$$

respectively; here A and B are any constants. Hence a solution of (1) is

$$(5) \qquad u(x, t; p) = FG = (A \cos px + B \sin px)e^{-c^2 p^2 t}.$$

[As in the previous section, we had to choose the constant of separation k negative, $k = -p^2$, because positive values of k lead to an increasing exponential function in (5), which has no physical meaning.]

Any series of functions (5), found in the usual manner by taking p as multiples of a fixed number, would lead to a function that is periodic in x when $t = 0$. However, since $f(x)$ in (2) is not assumed to be periodic, it is natural to use **Fourier integrals** in the present case instead of Fourier series.

Since A and B in (5) are arbitrary, we may consider these quantities as functions of p and write $A = A(p)$ and $B = B(p)$. Since the heat equation is linear and homogeneous, the function

$$(6) \quad u(x, t) = \int_0^\infty u(x, t; p) \, dp = \int_0^\infty [A(p) \cos px + B(p) \sin px] \, e^{-c^2 p^2 t} \, dp$$

is then a solution of (1), provided this integral exists and can be differentiated twice with respect to x and once with respect to t.

From the representation (6) and the initial condition (2) it follows that

$$(7) \qquad u(x, 0) = \int_0^\infty [A(p) \cos px + B(p) \sin px] \, dp = f(x).$$

Using (4) and (5) in Sec. 10.9, we thus obtain

$$(8) \quad A(p) = \frac{1}{\pi} \int_{-\infty}^\infty f(v) \cos pv \, dv, \qquad B(p) = \frac{1}{\pi} \int_{-\infty}^\infty f(v) \sin pv \, dv.$$

According to (1*) in Sec. 10.11, this Fourier integral may be written

$$u(x, 0) = \frac{1}{\pi} \int_0^\infty \left[\int_{-\infty}^\infty f(v) \cos (px - pv) \, dv \right] dp,$$

and (6), this section, thus becomes

$$u(x, t) = \frac{1}{\pi} \int_0^\infty \left[\int_{-\infty}^\infty f(v) \cos (px - pv) \, e^{-c^2 p^2 t} \, dv \right] dp.$$

Assuming that we may invert the order of integration, we obtain

$$(9) \qquad u(x, t) = \frac{1}{\pi} \int_{-\infty}^\infty f(v) \left[\int_0^\infty e^{-c^2 p^2 t} \cos (px - pv) \, dp \right] dv.$$

We evaluate the inner integral by the use of the formula

(10)
$$\int_0^\infty e^{-s^2} \cos 2bs \; ds \; = \; \frac{\sqrt{\pi}}{2} \, e^{-b^2}$$

which will be derived in Sec. 15.4 (Prob. 13). For this purpose we introduce in (10) a new variable of integration p by setting $s = cp\sqrt{t}$ and choosing

$$b \; = \; \frac{x - v}{2c\sqrt{t}} .$$

Then $2bs = (x - v)p$ and $ds = c\sqrt{t}\, dp$, so that (10) becomes

$$\int_0^\infty e^{-c^2 p^2 t} \cos\,(px - pv)\, dp \; = \; \frac{\sqrt{\pi}}{2c\sqrt{t}} \, e^{-(x-v)^2/(4c^2 t)}.$$

By inserting this result into (9) we obtain the representation

(11)
$$u(x, t) \; = \; \frac{1}{2c\sqrt{\pi t}} \int_{-\infty}^\infty f(v) \exp\left\{ - \frac{(x - v)^2}{4c^2 t} \right\} dv.$$

We finally introduce the variable of integration $z = (v - x)/(2c\sqrt{t})$. Then

(12)
$$u(x, t) \; = \; \frac{1}{\sqrt{\pi}} \int_{-\infty}^\infty f(x + 2cz\sqrt{t})\, e^{-z^2}\, dz.$$

If $f(x)$ is bounded for all values of x and integrable in every finite interval, it can be shown (cf. Ref. [C18]) that the function (11) or (12) satisfies (1) and (2). Hence this function is the required solution in the present case.

EXAMPLE 1. Temperature in an infinite bar

Find the temperature in the infinite bar if the initial temperature is (Fig. 279)

$$f(x) \; = \; \begin{cases} U_0 = const & \text{if} \quad |x| < 1. \\ 0 & \text{if} \quad |x| > 1. \end{cases}$$

Solution. From (11) we have

$$u(x, t) \; = \; \frac{U_0}{2c\sqrt{\pi t}} \int_{-1}^1 \exp\left\{ - \frac{(x - v)^2}{4c^2 t} \right\} dv.$$

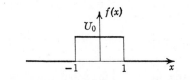

Fig. 279. Initial temperature in Example 1

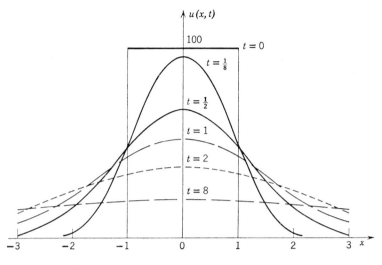

Fig. 280. Solution $u(x, t)$ to Example 1 for $U_0 = 100\,°C$, $c^2 = 1\ cm^2/sec$, and several values of t

If we introduce the above variable of integration z, then the integration over v from -1 to 1 corresponds to the integration over z from $(-1 - x)/2c\sqrt{t}$ to $(1 - x)/2c\sqrt{t}$, and

$$(13) \qquad u(x, t) = \frac{U_0}{\sqrt{\pi}} \int_{-(1+x)/2c\sqrt{t}}^{(1-x)/2c\sqrt{t}} e^{-z^2}\, dz \qquad (t > 0).$$

We mention that this integral is not an elementary function, but can easily be expressed in terms of the error function, whose values have been tabulated. (Table A5 in Appendix 4 contains a few values; larger tables are listed in Ref. [1] in Appendix 1. Cf. also Probs. 11—20, below.) Figure 280 shows $u(x, t)$ for $U_0 = 100\ °C$, $c^2 = 1\ cm^2/sec$, and several values of t. ∎

The equations considered so far (except for the Laplace equation in the second half of Sec. 11.5) have been **"one-dimensional"**; that is, they involve a single space variable x (and time t). In the next section we begin the discussion of a **"two-dimensional"** equation (the two-dimensional wave equation), which involves *two* space variables x and y and governs the vibrations of elastic membranes.

Problems for Sec. 11.6

1. Graph the temperatures in Example 1 (with $U_0 = 100\ °C$ and $c^2 = 1\ cm^2/sec$) at the points $x = 0.5$, 1, and 1.5 as functions of t. Do the results agree with your physical intuition?

2. If $f(x) = 1$ when $x > 0$ and $f(x) = 0$ when $x < 0$, show that (12) becomes

$$u(x, t) = \frac{1}{\sqrt{\pi}} \int_{-x/2c\sqrt{t}}^{\infty} e^{-z^2}\, dz \qquad (t > 0).$$

3. Show that for $x = 0$ the solution in Prob. 2 is independent of t. Can this result be expected for physical reasons?

4. If the bar is semi-infinite, extending from 0 to ∞, the end at $x = 0$ is held at temperature 0 and the initial temperature is $f(x)$, show that the temperature in the bar is

$$(14) \quad u(x, t) = \frac{1}{\sqrt{\pi}} \left[\int_{-x/\tau}^{\infty} f(x + \tau w)e^{-w^2} \, dw - \int_{x/\tau}^{\infty} f(-x + \tau w)e^{-w^2} \, dw \right],$$

where $\tau = 2c\sqrt{t}$.

5. Obtain (14) from (11) by assuming that $f(v)$ in (11) is odd.

6. If $f(x) = 1$ in Prob. 4, show that

$$u(x, t) = \frac{2}{\sqrt{\pi}} \int_{0}^{x/\tau} e^{-w^2} \, dw \qquad (t > 0).$$

7. What form does (14) take if $f(x) = 1$ when $a < x < b$ (where $a > 0$) and $f(x) = 0$ otherwise?

8. Show that the result of Prob. 6 can be obtained from (11) or (12) by using $f(x) = 1$, when $x > 0$, and $f(x) = -1$ when $x < 0$. What is the reason?

9. Show that in Prob. 6 the times required for any two points to reach the same temperature are proportional to the squares of their distances from the boundary at $x = 0$.

10. (Normal distribution) Introducing $w = z\sqrt{2}$ as a new variable of integration, show that (12) becomes

$$u(x, t) = \frac{1}{\sqrt{2\pi}} \int_{-\infty}^{\infty} f(x + cw\sqrt{2t})e^{-w^2/2} \, dw.$$

Students familiar with probability theory will notice that this involves the density $e^{-w^2/2}/\sqrt{2\pi}$ of the normal distribution (cf. Sec. 23.7).

Error function. The error function is defined by the integral

$$\text{erf } x = \frac{2}{\sqrt{\pi}} \int_{0}^{x} e^{-w^2} \, dw.$$

It is important in engineering mathematics. To become familiar with this function, the student may solve the following problems. (A few formulas are included in Appendix 3.1; see (35)—(37). Cf. also Example 1 in Sec. 18.6.)

11. Graph the integrand of erf x (the so-called **bell-shaped curve**).

12. Show that erf x is odd.

13. Show that (13) may be written

$$u(x, t) = \frac{U_0}{2} \left[\text{erf } \frac{1 - x}{2c\sqrt{t}} + \text{erf } \frac{1 + x}{2c\sqrt{t}} \right] \qquad (t > 0).$$

14. Show that in Prob. 6 we have $u(x, t) = \text{erf } (x/2c\sqrt{t})$.

15. It can be shown that erf$(\infty) = 1$. Using this, show that in Prob. 2,

$$u(x, t) = \tfrac{1}{2} + \tfrac{1}{2} \text{erf } (x/2c\sqrt{t}).$$

16. Show that

$$\int_{a}^{b} e^{-w^2} \, dw = \frac{\sqrt{\pi}}{2} (\text{erf } b - \text{erf } a), \qquad \int_{-b}^{b} e^{-w^2} \, dw = \sqrt{\pi} \, \text{erf } b.$$

17. Show that erf $x = \sqrt{\dfrac{2}{\pi}} \displaystyle\int_0^{x\sqrt{2}} e^{-\tau^2/2}\, d\tau$.

18. Obtain the Maclaurin series of erf x by term-by-term integration of that of the integrand.

19. Show that $y = e^{x^2}$ erf x satisfies the differential equation $y' = 2xy + 2/\sqrt{\pi}$.

20. Compute erf x for $x = 0(0.1)0.5$ from Prob. 18 and compare with the 3D-values 0.000, 0.112, 0.223, 0.329, 0.428, 0.520.

11.7 Modeling: Vibrating Membrane. Two-Dimensional Wave Equation

As another important problem from the field of vibrations, let us consider the motion of a stretched membrane, such as a drumhead. The reader will notice that our present considerations are similar to those in the case of the vibrating string in Sec. 11.2.

We make the following assumptions:

1. *The mass of the membrane per unit area is constant ("homogeneous membrane"). The membrane is perfectly flexible and is so thin that it does not offer any resistance to bending.*

2. *The membrane is stretched and then fixed along its entire boundary in the xy-plane. The tension per unit length T caused by stretching the membrane is the same at all points and in all directions and does not change during the motion.*

3. *The deflection u(x, y, t) of the membrane during the motion is small compared to the size of the membrane, and all angles of inclination are small.*

Although these assumptions cannot be realized in practice, small transverse vibrations of a thin physical membrane will satisfy these assumptions relatively accurately.

To derive the differential equation that governs the motion of the membrane, we consider the forces acting on a small portion of the membrane in Fig. 281. Since the deflections of the membrane and the angles of inclination are small, the sides of the portion are approximately equal to Δx and Δy. The tension T is the force per unit length. Hence the forces acting on the sides of the portion are approximately $T \Delta x$ and $T \Delta y$. Since the membrane is perfectly flexible, these forces are tangent to the membrane.

We first consider the horizontal components of the forces. These components are obtained by multiplying the forces by the cosines of the angles of inclination. Since these angles are small, their cosines are close to 1. Hence the horizontal components of the forces at opposite sides are approximately equal. Therefore, the motion of the particles of the membrane in horizontal direction will be negligibly small. From this we conclude that we may regard the motion of the membrane as transversal; that is, each particle moves vertically.

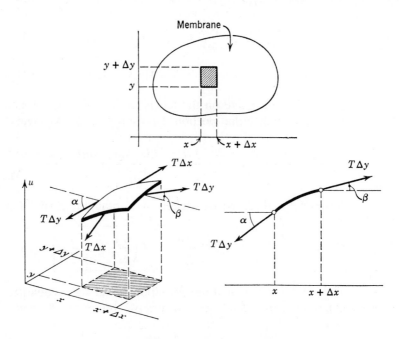

Fig. 281. Vibrating membrane

The vertical components of the forces along the sides parallel to the yu-plane are[5] (Fig. 281)

$$T \, \Delta y \sin \beta \quad \text{and} \quad - T \, \Delta y \sin \alpha;$$

here the minus sign appears because the force on the left side is directed downward. Since the angles are small, we may replace their sines by their tangents. Hence the resultant of those two vertical components is

(1)
$$T \, \Delta y(\sin \beta - \sin \alpha) \approx T \, \Delta y(\tan \beta - \tan \alpha)$$
$$= T \, \Delta y[u_x(x + \Delta x, y_1) - u_x(x, y_2)]$$

where subscripts x denote partial derivatives and y_1 and y_2 are values between y and $y + \Delta y$. Similarly, the resultant of the vertical components of the forces acting on the other two sides of the portion is

(2)
$$T \, \Delta x[u_y(x_1, y + \Delta y) - u_y(x_2, y)]$$

where x_1 and x_2 are values between x and $x + \Delta x$.

By Newton's second law (cf. Sec. 2.6), the sum of the forces given by (1) and (2) is equal to the mass $\rho \, \Delta A$ of that small portion times the acceleration $\partial^2 u/\partial t^2$; here ρ is the mass of the undeflected membrane per unit area and $\Delta A = \Delta x \, \Delta y$ is the area of that portion when it is undeflected. Thus

---

[5]Note that the angle of inclination varies along the sides, and α and β represent values of that angle at a suitable point of the sides under consideration.

$$\rho \, \Delta x \, \Delta y \, \frac{\partial^2 u}{\partial t^2} = T \, \Delta y [u_x(x + \Delta x, y_1) - u_x(x, y_2)]$$

$$+ \, T \, \Delta x [u_y(x_1, y + \Delta y) - u_y(x_2, y)]$$

where the derivative on the left is evaluated at some suitable point $(\tilde{x}, \tilde{y})$ corresponding to that portion. Division by $\rho \, \Delta x \, \Delta y$ yields

$$\frac{\partial^2 u}{\partial t^2} = \frac{T}{\rho} \left[\frac{u_x(x + \Delta x, y_1) - u_x(x, y_2)}{\Delta x} + \frac{u_y(x_1, y + \Delta y) - u_y(x_2, y)}{\Delta y} \right].$$

If we let Δx and Δy approach zero, we obtain the partial differential equation

$$(3) \qquad \boxed{\frac{\partial^2 u}{\partial t^2} = c^2 \left(\frac{\partial^2 u}{\partial x^2} + \frac{\partial^2 u}{\partial y^2} \right)} \qquad\qquad c^2 = \frac{T}{\rho} \, .$$

This equation is called the **two-dimensional wave equation.** The expression in parentheses is the Laplacian $\nabla^2 u$ of u (cf. Sec. 8.8), and (3) can be written

$$(3') \qquad \boxed{\frac{\partial^2 u}{\partial t^2} = c^2 \nabla^2 u.}$$

Solutions will be obtained and discussed in the next section.

11.8 Rectangular Membrane

To solve the problem of a vibrating membrane, we have to determine a solution $u(x, y, t)$ of the two-dimensional wave equation

$$(1) \qquad \frac{\partial^2 u}{\partial t^2} = c^2 \left(\frac{\partial^2 u}{\partial x^2} + \frac{\partial^2 u}{\partial y^2} \right)$$

that satisfies the boundary condition

$$(2) \qquad u = 0 \qquad\qquad \text{on the boundary of the membrane for all } t \geq 0$$

and the two initial conditions

$$(3) \qquad u(x, y, 0) = f(x, y) \qquad\qquad \text{[given initial displacement } f(x, y)]$$

and

$$(4) \qquad \left. \frac{\partial u}{\partial t} \right|_{t=0} = g(x, y) \qquad\qquad \text{[given initial velocity } g(x, y)].$$

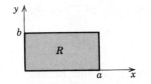

Fig. 282. Rectangular membrane

These conditions are quite similar to those in the case of the vibrating string.

As a first important case, let us consider the rectangular membrane R shown in Fig. 282.

First Step. Three Ordinary Differential Equations

By applying the method of separation of variables we first determine solutions of (1) that satisfy the boundary condition (2). We start from

(5) $$u(x, y, t) = F(x, y)G(t).$$

By substituting this into the wave equation (1) we have

$$F\ddot{G} = c^2(F_{xx}G + F_{yy}G)$$

where subscripts denote partial derivatives and dots denote derivatives with respect to t. To separate the variables, we divide both sides by c^2FG:

$$\frac{\ddot{G}}{c^2G} = \frac{1}{F}(F_{xx} + F_{yy}).$$

Since the functions on the left depend only on t while the functions on the right do not depend on t, the expressions on both sides must be equal to a constant. A little investigation shows that only negative values of this constant will lead to solutions that satisfy (2) without being identically zero; this is similar to the procedure in Sec. 11.3. Denoting this negative constant by $-\nu^2$, we have

$$\frac{\ddot{G}}{c^2G} = \frac{1}{F}(F_{xx} + F_{yy}) = -\nu^2.$$

From this we obtain two equations, the ordinary differential equation

(6) $$\boxed{\ddot{G} + \lambda^2 G = 0}$$ where $\lambda = c\nu$,

and the partial differential equation

(7) $$F_{xx} + F_{yy} + \nu^2 F = 0.$$

Equation (7) is called the two-dimensional **Helmholtz equation.** We consider (7) and apply the method of separating variables once more, that is, we determine solutions of (7) of the form

(8) $$F(x, y) = H(x)Q(y)$$

which are zero on the boundary of the membrane. We substitute (8) into (7):

$$\frac{d^2H}{dx^2} Q = -\left(H \frac{d^2Q}{dy^2} + \nu^2 HQ\right).$$

To separate the variables, we divide both sides by HQ, finding

$$\frac{1}{H} \frac{d^2H}{dx^2} = -\frac{1}{Q}\left(\frac{d^2Q}{dy^2} + \nu^2 Q\right).$$

The functions on the left depend only on x, whereas the functions on the right depend only on y. Hence the expressions on both sides must be equal to a constant. This constant must be negative, say, $-k^2$, because only negative values will lead to solutions that satisfy (2) without being identically zero. Thus

$$\frac{1}{H} \frac{d^2H}{dx^2} = -\frac{1}{Q}\left(\frac{d^2Q}{dy^2} + \nu^2 Q\right) = -k^2.$$

This yields two ordinary linear differential equations for H and Q:

(9) $$\boxed{\frac{d^2H}{dx^2} + k^2 H = 0}$$

and

(10) $$\boxed{\frac{d^2Q}{dy^2} + p^2 Q = 0} \qquad \text{where } p^2 = \nu^2 - k^2.$$

Second Step. Satisfying the Boundary Conditions

The general solutions of (9) and (10) are

$$H(x) = A \cos kx + B \sin kx \qquad \text{and} \qquad Q(y) = C \cos py + D \sin py$$

where A, B, C and D are constants. From (5) and (2) it follows that the function $F = HQ$ must be zero on the boundary, which corresponds to $x = 0$, $x = a$, $y = 0$, and $y = b$; cf. Fig. 282. This yields the conditions

$$H(0) = 0, \qquad H(a) = 0, \qquad Q(0) = 0, \qquad Q(b) = 0.$$

Therefore, $H(0) = A = 0$, and then

$$H(a) = B \sin ka = 0.$$

We must take $B \neq 0$, since otherwise $H \equiv 0$ and $F \equiv 0$. Hence $\sin ka = 0$ or $ka = m\pi$, that is,

$$k = \frac{m\pi}{a} \qquad (m \text{ integer}).$$

In precisely the same fashion we conclude that $C = 0$ and p must be restricted to the values $p = n\pi/b$ where n is an integer. We thus obtain the solutions

$$H_m(x) = \sin \frac{m\pi x}{a} \qquad \text{and} \qquad Q_n(y) = \sin \frac{n\pi y}{b} \qquad \begin{array}{l} m = 1, 2, \cdots, \\ n = 1, 2, \cdots. \end{array}$$

(As in the case of the vibrating string, it is not necessary to consider m, $n = -1, -2, \cdots$ since the corresponding solutions are essentially the same as for positive m and n, except for a factor -1.) It follows that the functions

(11) $\qquad F_{mn}(x, y) = H_m(x)Q_n(y) = \sin \frac{m\pi x}{a} \sin \frac{n\pi y}{b} \qquad \begin{array}{l} m = 1, 2, \cdots, \\ n = 1, 2, \cdots, \end{array}$

are solutions of the equation (7) that are zero on the boundary of the rectangular membrane.

Since $p^2 = \nu^2 - k^2$ in (10) and $\lambda = c\nu$ in (6), we have

$$\lambda = c\sqrt{k^2 + p^2}.$$

Hence to $k = m\pi/a$ and $p = n\pi/b$ there corresponds the value

(12) $\qquad \boxed{\lambda = \lambda_{mn} = c\pi \sqrt{\frac{m^2}{a^2} + \frac{n^2}{b^2}}} \qquad \begin{array}{l} m = 1, 2, \cdots, \\ n = 1, 2, \cdots, \end{array}$

in the differential equation (6). The corresponding general solution of (6) is

$$G_{mn}(t) = B_{mn} \cos \lambda_{mn} t + B_{mn}^* \sin \lambda_{mn} t.$$

It follows that the functions $u_{mn}(x, y, t) = F_{mn}(x, y)G_{mn}(t)$, written out

(13) $\qquad \boxed{u_{mn}(x, y, t) = (B_{mn} \cos \lambda_{mn} t + B_{mn}^* \sin \lambda_{mn} t) \sin \frac{m\pi x}{a} \sin \frac{n\pi y}{b}}$

with λ_{mn} according to (12) are solutions of the wave equation (1) that are zero on the boundary of the rectangular membrane in Fig. 282. These functions are called the **eigenfunctions** or *characteristic functions,* and the numbers λ_{mn} are called the **eigenvalues** or *characteristic values* of the vibrating membrane. The frequency of u_{mn} is $\lambda_{mn}/2\pi$.

It is interesting to note that, depending on a and b, several functions F_{mn} may correspond to the same eigenvalue. Physically this means that there may exist vibrations having the same frequency but entirely different **nodal lines** (curves of points on the membrane that do not move). Let us illustrate this by the following example.

EXAMPLE 1. Eigenvalues and eigenfunctions of the square membrane

Consider the square membrane for which $a = b = 1$. From (12) we obtain the eigenvalues

(14) $$\lambda_{mn} = c\pi \sqrt{m^2 + n^2}.$$

Hence

$$\lambda_{mn} = \lambda_{nm}$$

but for $m \neq n$ the corresponding functions

$$F_{mn} = \sin m\pi x \sin n\pi y \quad \text{and} \quad F_{nm} = \sin n\pi x \sin m\pi y$$

are certainly different. For example, to $\lambda_{12} = \lambda_{21} = c\pi\sqrt{5}$ there correspond the two functions

$$F_{12} = \sin \pi x \sin 2\pi y \quad \text{and} \quad F_{21} = \sin 2\pi x \sin \pi y.$$

Hence the corresponding solutions

and
$$u_{12} = (B_{12} \cos c\pi\sqrt{5}t + B_{12}^* \sin c\pi\sqrt{5}t)F_{12}$$

$$u_{21} = (B_{21} \cos c\pi\sqrt{5}t + B_{21}^* \sin c\pi\sqrt{5}t)F_{21}$$

have the nodal lines $y = \frac{1}{2}$ and $x = \frac{1}{2}$, respectively (cf. Fig. 283). Taking $B_{12} = 1$ and $B_{12}^* = B_{21}^* = 0$, we obtain

(15) $$u_{12} + u_{21} = \cos c\pi\sqrt{5}t\ (F_{12} + B_{21}F_{21})$$

which represents another vibration corresponding to the eigenvalue $c\pi\sqrt{5}$. The nodal line of this function is the solution of the equation

$$F_{12} + B_{21}F_{21} = \sin \pi x \sin 2\pi y + B_{21} \sin 2\pi x \sin \pi y = 0$$

or, since $\sin 2\alpha = 2 \sin \alpha \cos \alpha$,

(16) $$\sin \pi x \sin \pi y\ (\cos \pi y + B_{21} \cos \pi x) = 0.$$

This solution depends on the value of B_{21} (Fig. 284).

From (14) we see that even more than two functions may correspond to the same numerical value of λ_{mn}. For example, the four functions F_{18}, F_{81}, F_{47} and F_{74} correspond to the value $\lambda_{18} = \lambda_{81} = \lambda_{47} = \lambda_{74} = c\pi\sqrt{65}$, because

$$1^1 + 8^2 = 4^2 + 7^2 = 65.$$

This happens because 65 can be expressed as the sum of two squares of natural numbers in several ways. According to a theorem by Gauss, this is the case for every sum of two squares among whose prime factors there are at least two different ones of the form $4n + 1$ where n is a positive integer. In our case, $65 = 5 \cdot 13 = (4 + 1)(12 + 1)$. ∎

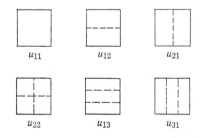

u_{11} u_{12} u_{21}

u_{22} u_{13} u_{31}

Fig. 283. Nodal lines of the solutions $u_{11}, u_{12}, u_{21}, u_{22}, u_{13}, u_{31}$ in the case of the square membrane

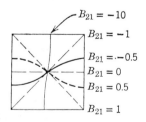

$B_{21} = -10$

$B_{21} = -1$

$B_{21} = -0.5$

$B_{21} = 0$

$B_{21} = 0.5$

$B_{21} = 1$

Fig. 284. Nodal lines of the solution (15) for some values of B_{21}

Third Step. Solution of the Entire Problem

To obtain the solution that also satisfies the initial conditions (3) and (4), we proceed similarly as in Sec. 11.3. We consider the double series[6]

(17)
$$u(x,y,t) = \sum_{m=1}^{\infty} \sum_{n=1}^{\infty} u_{mn}(x,y,t)$$

$$= \sum_{m=1}^{\infty} \sum_{n=1}^{\infty} (B_{mn}\cos\lambda_{mn}t + B^*_{mn}\sin\lambda_{mn}t)\sin\frac{m\pi x}{a}\sin\frac{n\pi y}{b}.$$

From this and (3) we obtain

(18)
$$u(x, y, 0) = \sum_{m=1}^{\infty} \sum_{n=1}^{\infty} B_{mn}\sin\frac{m\pi x}{a}\sin\frac{n\pi y}{b} = f(x, y).$$

This series is called a **double Fourier series.** Suppose that $f(x, y)$ can be developed in such a series.[7] Then the Fourier coefficients B_{mn} of $f(x, y)$ in (18) may be determined as follows. Setting

(19)
$$K_m(y) = \sum_{n=1}^{\infty} B_{mn}\sin\frac{n\pi y}{b}$$

we may write (18) in the form

$$f(x, y) = \sum_{m=1}^{\infty} K_m(y)\sin\frac{m\pi x}{a}.$$

For fixed y this is the Fourier sine series of $f(x, y)$, considered as a function of x, and from (4) in Sec. 10.5 we see that the coefficients of this expansion are

(20)
$$K_m(y) = \frac{2}{a}\int_0^a f(x, y)\sin\frac{m\pi x}{a}\,dx.$$

Furthermore, (19) is the Fourier sine series of $K_m(y)$, and from (4) in Sec. 10.5 it follows that the coefficients are

$$B_{mn} = \frac{2}{b}\int_0^b K_m(y)\sin\frac{n\pi y}{b}\,dy.$$

From this and (20) we obtain the **generalized Euler formula**

---

[6]We shall not consider the problems of convergence and uniqueness.
[7]Sufficient conditions: f, $\partial f/\partial x$, $\partial f/\partial y$, $\partial^2 f/\partial x\,\partial y$ continuous in the rectangle R under consideration.

$$(21) \quad \boxed{B_{mn} = \frac{4}{ab} \int_0^b \int_0^a f(x, y) \sin \frac{m\pi x}{a} \sin \frac{n\pi y}{b} \, dx \, dy} \quad \begin{array}{l} m = 1, 2, \cdots, \\ n = 1, 2, \cdots, \end{array}$$

for the Fourier coefficients of $f(x, y)$ in the double Fourier series (18).

The B_{mn} in (17) are now determined in terms of $f(x, y)$. To determine the B_{mn}^*, we differentiate (17) termwise with respect to t; using (4), we obtain

$$\frac{\partial u}{\partial t}\bigg|_{t=0} = \sum_{m=1}^{\infty} \sum_{n=1}^{\infty} B_{mn}^* \lambda_{mn} \sin \frac{m\pi x}{a} \sin \frac{n\pi y}{b} = g(x, y).$$

Suppose that $g(x, y)$ can be developed in this double Fourier series. Then, proceeding as before, we find

$$(22) \quad B_{mn}^* = \frac{4}{ab\lambda_{mn}} \int_0^b \int_0^a g(x, y) \sin \frac{m\pi x}{a} \sin \frac{n\pi y}{b} \, dx \, dy \quad \begin{array}{l} m = 1, 2, \cdots, \\ n = 1, 2, \cdots, \end{array}$$

The result is that, for (17) to satisfy the initial conditions, the coefficients B_{mn} and B_{mn}^* must be chosen according to (21) and (22).

EXAMPLE 2. Vibrations of a rectangular membrane

Find the vibrations of a rectangular membrane of sides $a = 4$ ft and $b = 2$ ft if the tension is 12.5 lb/ft, the density is 2.5 slugs/ft^2 (as for light rubber), the initial velocity is 0 and the initial displacement is

$$(23) \qquad\qquad f(x, y) = 0.1(4x - x^2)(2y - y^2) \text{ ft.}$$

Solution. $c^2 = T/\rho = 12.5/2.5 = 5$ [ft^2/sec^2]. Also, $B_{mn}^* = 0$ from (22). From (21) and (23),

$$B_{mn} = \frac{4}{4 \cdot 2} \int_0^2 \int_0^4 0.1(4x - x^2)(2y - y^2) \sin \frac{m\pi x}{4} \sin \frac{n\pi y}{2} \, dx \, dy$$

$$= \frac{1}{20} \int_0^4 (4x - x^2) \sin \frac{m\pi x}{4} \, dx \int_0^2 (2y - y^2) \sin \frac{n\pi y}{2} \, dy.$$

Two integrations by parts give for the first integral on the right

$$\frac{128}{m^3\pi^3} [1 - (-1)^m] = \frac{256}{m^3\pi^3} \quad (m \text{ odd})$$

and for the second integral

$$\frac{16}{n^3\pi^3} [1 - (-1)^n] = \frac{32}{n^3\pi^3} \quad (n \text{ odd}).$$

For even m or n we get 0. Together with the factor $1/20$ we thus have $B_{mn} = 0$ if m or n is even and

$$B_{mn} = \frac{256 \cdot 32}{20m^3n^3\pi^6} \approx \frac{0.426\ 050}{m^3n^3} \quad (m \text{ and } n \text{ both odd}).$$

From this and (17) we obtain the answer

$$u(x, y, t) = 0.426\,050 \sum_{m,n\,\text{odd}} \sum \frac{1}{m^3 n^3} \cos\left(\frac{5\pi}{4}\sqrt{m^2 + 4n^2}\right) t \sin\frac{m\pi x}{4} \sin\frac{n\pi y}{2}$$

(24)
$$= 0.426\,050 \left(\cos\frac{5\pi\sqrt{5}}{4} t \sin\frac{\pi x}{4} \sin\frac{\pi y}{2} + \frac{1}{27}\cos\frac{5\pi\sqrt{37}}{4} t \sin\frac{\pi x}{4} \sin\frac{3\pi y}{2}\right.$$

$$\left. + \frac{1}{27}\cos\frac{5\pi\sqrt{13}}{4} t \sin\frac{3\pi x}{4} \sin\frac{\pi y}{2} + \frac{1}{729}\cos\frac{5\pi\sqrt{45}}{4} t \sin\frac{3\pi x}{4} \sin\frac{3\pi y}{2} + \cdots \right).$$

Note how rapidly the coefficients of this series decrease.   ∎

To investigate vibrating membranes of other shapes, the strategy is to introduce a coordinate system in which the boundary of the membrane is given by a simple formula; this is necessary because of the boundary condition ($u = 0$ along the boundary). The practically most important case is the **circular membrane (drumhead).** It calls for polar coordinates and requires us to express the Laplacian in the wave equation in terms of polar coordinates. This we do, once and for all, in the next section.

Problems for Sec. 11.8

1. How does the frequency of a solution (13) change if the tension of the membrane is increased?

2. Determine and graph the nodal lines of the solutions (13) with $m = 1, 2, 3, 4$ and $n = 1, 2, 3, 4$ in the case $a = b = 1$.

3. Same task as in Prob. 2, when $a = 2$ and $b = 1$.

4. Find further eigenvalues of the square membrane with side 1 such that four different eigenfunctions correspond to each such eigenvalue.

5. Find eigenvalues of the rectangular membrane of sides $a = 2$, $b = 1$ such that two or more different eigenfunctions correspond to each such eigenvalue.

6. Show that, among all rectangular membranes of the same area $A = ab$ and the same c, the square membrane is that for which u_{11} [cf. (13)] has the lowest frequency.

7. Find a similar result as in Prob. 6 for the frequency of a solution (13) with arbitrary fixed m and n.

8. Can you visualize $f(x, y)$ in Example 2? Can you see that the first terms of the series (24) already approximates $f(x, y)$ quite well? Can you find a reason why this should be the case?

9. Using integration by parts, verify the calculation of B_{mn} in Example 2.

10. B_{mn} in Example 2 is a product of two integrals. Find out to what functions these integrals correspond and verify their values by the jump method (Sec. 10.6).

11. In what cases will B_{mn} in (21) be a product of two integrals (as in Example 2)?

Double Fourier series. Represent $f(x, y)$ by a double Fourier series of the form (18), where $0 < x < \pi$, $0 < y < \pi$.

12. $f(x, y) = 1$

13. $f(x, y) = x$

14. $f(x, y) = y$

15. $f(x, y) = xy$

16. $f(x, y) = \begin{cases} 1 & \text{if} \quad 0 < x < \pi/2 \\ 0 & \text{if} \quad \pi/2 < x < \pi \end{cases}$

17. $f(x, y) = \begin{cases} 1 & \text{if } \pi/2 < x, y < \pi \\ 0 & \text{otherwise} \end{cases}$

Represent the following functions $f(x, y)$ $(0 < x < a, 0 < y < b)$ by a double Fourier series of the form (18).

18. $f = 1$

19. $f = xy$

20. $f = x + y$

21. $f = xy(a - x)(b - y)$

22. $f = (x + 1)(y + 1)$

23. $f = xy(a^2 - x^2)(b^2 - y^2)$

Find the deflection $u(x, y, t)$ of the square membrane with $a = b = 1$ and $c = 1$, if the initial velocity is zero and the initial deflection is $f(x, y)$, where

24. $f(x, y) = k \sin \pi x \sin \pi y$

25. $f(x, y) = k \sin \pi x \sin 2\pi y$

26. $f(x, y) = k \sin 3\pi x \sin 4\pi y$

27. $f(x, y) = 0.01xy(1 - x)(1 - y)$

28. $f(x, y) = k \sin^2 \pi x \sin^2 \pi y$

29. $f(x, y) = kx(1 - x^2)y(1 - y^2)$

30. **(Forced vibrations of a membrane)** Show that the forced vibrations of a membrane are governed by the equation

$$u_{tt} = c^2 \nabla^2 u + P/\rho$$

where $P(x, y, t)$ is the external force per unit area acting normal to the xy-plane.

11.9 Laplacian in Polar Coordinates

In connection with boundary value problems for partial differential equations, it is a general principle to use coordinates with respect to which the boundary of the region under consideration is given by simple formulas. In the next section we shall consider circular membranes. Then polar coordinates r and θ, defined by

$$x = r \cos \theta, \qquad y = r \sin \theta$$

will be appropriate, because the boundary of the membrane can then be represented by the simple equation $r = const.$

When using r and θ we have to transform the Laplacian

$$\nabla^2 u = \frac{\partial^2 u}{\partial x^2} + \frac{\partial^2 u}{\partial y^2}$$

in the wave equation into these new coordinates.

Transformations of differential expressions from one coordinate system into another are frequently required in applications. Therefore, the student should follow our present consideration with great attention.

As in Sec. 11.4, we use the chain rule. For the sake of simplicity we denote partial derivatives by subscripts and $u(x, y, t)$, as a function of r, θ, t, by the same letter u. By applying the chain rule (4), Sec. 8.7, we obtain

$$u_x = u_r r_x + u_\theta \theta_x.$$

Differentiating this again with respect to x we first have

(1)
$$u_{xx} = (u_r r_x)_x + (u_\theta \theta_x)_x$$
$$= (u_r)_x r_x + u_r r_{xx} + (u_\theta)_x \theta_x + u_\theta \theta_{xx}.$$

Now, by applying the chain rule again, we find

$$(u_r)_x = u_{rr} r_x + u_{r\theta} \theta_x \quad \text{and} \quad (u_\theta)_x = u_{\theta r} r_x + u_{\theta\theta} \theta_x.$$

To determine the partial derivatives r_x and θ_x, we have to differentiate

$$r = \sqrt{x^2 + y^2} \quad \text{and} \quad \theta = \text{arc tan } \frac{y}{x},$$

finding

$$r_x = \frac{x}{\sqrt{x^2 + y^2}} = \frac{x}{r}, \qquad \theta_x = \frac{1}{1 + (y/x)^2}\left(-\frac{y}{x^2}\right) = -\frac{y}{r^2}.$$

Differentiating these two formulas again, we obtain

$$r_{xx} = \frac{r - x r_x}{r^2} = \frac{1}{r} - \frac{x^2}{r^3} = \frac{y^2}{r^3}, \qquad \theta_{xx} = -y\left(-\frac{2}{r^3}\right)r_x = \frac{2xy}{r^4}.$$

We substitute all these expressions into (1). Assuming continuity of the first and second partial derivatives, we have $u_{r\theta} = u_{\theta r}$, and by simplifying,

(2)
$$u_{xx} = \frac{x^2}{r^2} u_{rr} - 2\frac{xy}{r^3} u_{r\theta} + \frac{y^2}{r^4} u_{\theta\theta} + \frac{y^2}{r^3} u_r + 2\frac{xy}{r^4} u_\theta.$$

In a similar fashion it follows that

(3)
$$u_{yy} = \frac{y^2}{r^2} u_{rr} + 2\frac{xy}{r^3} u_{r\theta} + \frac{x^2}{r^4} u_{\theta\theta} + \frac{x^2}{r^3} u_r - 2\frac{xy}{r^4} u_\theta.$$

By adding (2) and (3) we see that the Laplacian of u in polar coordinates is

(4)
$$\boxed{\nabla^2 u = \frac{\partial^2 u}{\partial r^2} + \frac{1}{r}\frac{\partial u}{\partial r} + \frac{1}{r^2}\frac{\partial^2 u}{\partial \theta^2}.}$$

In the next section we shall apply this formula to the investigation of the vibrations of a **drumhead** (circular membrane).

Problems for Sec. 11.9

1. Perform the details of the calculations that lead to (2) and (3).
2. Show that (4) may be written

$$\nabla^2 u = \frac{1}{r}\frac{\partial}{\partial r}\left(r\frac{\partial u}{\partial r}\right) + \frac{1}{r^2}\frac{\partial^2 u}{\partial \theta^2}.$$

3. If u is independent of θ, then (4) reduces to $\nabla^2 u = u_{rr} + u_r/r$. Derive this result directly from the Laplacian in Cartesian coordinates by assuming that u is independent of θ.

4. Transform (4) back into Cartesian coordinates.

5. Show that the only solution of $\nabla^2 u = 0$ depending only on $r = \sqrt{x^2 + y^2}$ is $u = a \ln r + b$.

6. If x and y are Cartesian coordinates, show that $x^* = x \cos \alpha - y \sin \alpha$ and $y^* = x \sin \alpha + y \cos \alpha$ are Cartesian coordinates, and verify by calculation that $\nabla^2 u = u_{x^* x^*} + u_{y^* y^*}$.

7. Express $\nabla^2 u$ in terms of the coordinates $x^* = ax + b$, $y^* = cy + d$, where x, y are Cartesian coordinates.

8. Show that $u_n = r^n \cos n\theta$, $u_n = r^n \sin n\theta$, $n = 0, 1, \cdots$, are solutions of $\nabla^2 u = 0$ with $\nabla^2 u$ given by (4).

9. Find the solution of the Laplace equation in the disk of radius 1 with boundary values $f(\theta) = 10 \cos^2 \theta$.

10. Assuming that termwise differentiation is permissible, show that a solution of the Laplace equation in the disk $R < 1$ satisfying the boundary condition $u(R, \theta) = f(\theta)$ (f given) is

$$u(r, \theta) = a_0 + \sum_{n=1}^{\infty} \left[a_n \left(\frac{r}{R}\right)^n \cos n\theta + b_n \left(\frac{r}{R}\right)^n \sin n\theta \right]$$

where a_n, b_n are the Fourier coefficients of f (cf. Sec. 10.3).

Electrostatic potential. Steady-state heat problems. The electrostatic potential u satisfies Laplace's equation $\nabla^2 u = 0$ in any region free of charges. Also, the heat equation $u_t = c^2 \nabla^2 u$ (cf. Sec. 11.5) reduces to Laplace's equation when the temperature u is independent of time t ("**steady-state case**"). Find the electrostatic potential (equivalently: the steady-state temperature distribution) in the disk $r < 1$ corresponding to the following boundary values.

11. $u(\theta) = \begin{cases} -1 & \text{if } -\pi < \theta < 0 \\ 1 & \text{if } \;\; 0 < \theta < \pi \end{cases}$

12. $u(\theta) = \begin{cases} 110 & \text{if } -\pi/2 < \theta < \pi/2 \\ 0 & \text{if } \;\; \pi/2 < \theta < 3\pi/2 \end{cases}$

13. $u(\theta) = \begin{cases} \theta & \text{if } -\pi/2 < \theta < \pi/2 \\ 0 & \text{if } \;\; \pi/2 < \theta < 3\pi/2 \end{cases}$

14. $u(\theta) = \begin{cases} \theta & \text{if } -\pi/2 < \theta < \pi/2 \\ \pi - \theta & \text{if } \;\; \pi/2 < \theta < 3\pi/2 \end{cases}$

15. $u(\theta) = |\theta| \quad (-\pi < \theta < \pi)$

16. $u(\theta) = \theta^2 \quad (-\pi < \theta < \pi)$

17. Find the electrostatic potential in the semidisk $r < 1$, $0 < \theta < \pi$, which is equal to $110\theta(\pi - \theta)$ on the semicircle $r = 1$ and 0 on the segment $-1 < x < 1$.

18. Find the steady-state temperature u in a semicircular thin plate $r < a$, $0 < \theta < \pi$, if the semicircle $r = a$ is kept at constant temperature u_0 and the bounding segment $-a < x < a$ is kept at $u = 0$. (Use separation of variables.)

19. Find a formula for the potential u on the x-axis in Prob. 15. Use the first four terms of this series for computing u at $x = -0.75, -0.5, -0.25, 0, 0.25, 0.5, 0.75$ (two decimals).

20. Find a formula for the potential u on the y-axis in Prob. 15.

21. Let r, θ, ϕ be *spherical coordinates*, defined by

$$x = r \cos \theta \sin \phi, \qquad y = r \sin \theta \sin \phi, \qquad z = r \cos \phi.$$

If $u(x, y, z)$ is a function of $r = \sqrt{x^2 + y^2 + z^2}$ only, show that

$$\nabla^2 u = u_{rr} + \frac{2}{r} u_r.$$

22. **(Laplacian in spherical coordinates)** Show that the Laplacian in the spherical coordinates r, θ, ϕ defined in Prob. 21 is

$$\nabla^2 u = u_{rr} + \frac{2}{r} u_r + \frac{1}{r^2} u_{\phi\phi} + \frac{\cot \phi}{r^2} u_\phi + \frac{1}{r^2 \sin^2 \phi} u_{\theta\theta}.$$

Show that this can also be written

$$\nabla^2 u = \frac{1}{r^2} \left(\frac{\partial}{\partial r} (r^2 u_r) + \frac{1}{\sin \phi} \frac{\partial}{\partial \phi} (\sin \phi \, u_\phi) + \frac{1}{\sin^2 \phi} u_{\theta\theta} \right).$$

23. If the surface of the ball $r^2 = x^2 + y^2 + z^2 \leqq R^2$ is kept at temperature zero and the initial temperature in the ball is $f(r)$, show that the temperature $u(r, t)$ in the ball is the solution of $u_t = c^2 \left(u_{rr} + \frac{2}{r} u_r \right)$, satisfying the conditions $u(R, t) = 0$, $u(r, 0) = f(r)$.

24. Show that by setting $v = ru$ the formulas in Prob. 23 take the form $v_t = c^2 v_{rr}$, $v(R, t) = 0$, $v(r, 0) = rf(r)$. Include the condition $v(0, t) = 0$ (which holds because u must be bounded at $r = 0$), and solve the resulting problem by separating variables.

25. **(Laplacian in cylindrical coordinates)** Show that the Laplacian in cylindrical coordinates r, θ, z defined by $x = r \cos \theta$, $y = r \sin \theta$, $z = z$ is

$$\nabla^2 u = u_{rr} + \frac{1}{r} u_r + \frac{1}{r^2} u_{\theta\theta} + u_{zz}.$$

11.10 Circular Membrane. Bessel's Equation

We shall now consider vibrations of the circular membrane of radius R (Fig. 285). Using polar coordinates defined by $x = r \cos \theta$, $y = r \sin \theta$, we see from (4) in Sec. 11.9 that the wave equation (3′) in Sec. 11.7 takes the form

$$\frac{\partial^2 u}{\partial t^2} = c^2 \left(\frac{\partial^2 u}{\partial r^2} + \frac{1}{r} \frac{\partial u}{\partial r} + \frac{1}{r^2} \frac{\partial^2 u}{\partial \theta^2} \right).$$

We shall consider solutions $u(r, t)$ of this equation that are radially sym-

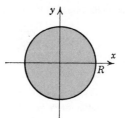

Fig. 285. Circular membrane

metric,[8] that is, do not depend on θ. Then the wave equation reduces to

$$
(1) \qquad \frac{\partial^2 u}{\partial t^2} = c^2 \left(\frac{\partial^2 u}{\partial r^2} + \frac{1}{r} \frac{\partial u}{\partial r} \right).
$$

Since the membrane is fixed along the boundary $r = R$, we have the boundary condition

$$
(2) \qquad u(R, t) = 0 \qquad\qquad \text{for all } t \geq 0.
$$

Solutions not depending on θ will occur if the initial conditions do not depend on θ, that is, if they are of the form

$$
(3) \qquad u(r, 0) = f(r) \qquad\qquad \text{[initial deflection } f(r)\text{]}
$$

and

$$
(4) \qquad \left. \frac{\partial u}{\partial t} \right|_{t=0} = g(r) \qquad\qquad \text{[initial velocity } g(r)\text{]}.
$$

First Step. Two Ordinary Differential Equations

Using the method of separating variables, we first determine solutions of (1) that satisfy the boundary condition (2). We start from

$$
(5) \qquad u(r, t) = W(r)G(t).
$$

By differentiating and inserting (5) into (1) and dividing the resulting equation by $c^2 WG$ we obtain

$$
\frac{\ddot{G}}{c^2 G} = \frac{1}{W} \left(W'' + \frac{1}{r} W' \right)
$$

where dots denote derivatives with respect to t and primes denote derivatives with respect to r. The expressions on both sides must be equal to a constant, and this constant must be negative, say, $-k^2$, in order to obtain solutions that satisfy the boundary condition without being identically zero. Thus,

$$
\frac{\ddot{G}}{c^2 G} = \frac{1}{W} \left(W'' + \frac{1}{r} W' \right) = -k^2.
$$

This yields the two ordinary linear differential equations

$$
(6) \qquad \boxed{\ddot{G} + \lambda^2 G = 0} \qquad\qquad \text{where } \lambda = ck
$$

---

[8]For solutions depending on θ, see the problem set.

and

(7)

$$W'' + \frac{1}{r} W' + k^2 W = 0.$$

Second Step. Satisfying the Boundary Condition

We first consider (7). Introducing the new independent variable $s = kr$, we have $1/r = k/s$, so that, by the chain rule, the derivatives become

$$W' = \frac{dW}{dr} = \frac{dW}{ds} \frac{ds}{dr} = \frac{dW}{ds} k \quad \text{and} \quad W'' = \frac{d^2W}{ds^2} k^2.$$

By substituting this into (7) and omitting the common factor k^2 we obtain

$$\frac{d^2W}{ds^2} + \frac{1}{s} \frac{dW}{ds} + W = 0.$$

This is **Bessel's equation** (1), Sec. 4.5, with $\nu = 0$. A general solution is (cf. Sec. 4.6)

$$W = C_1 J_0(s) + C_2 Y_0(s)$$

where J_0 and Y_0 are the Bessel functions of the first and second kind of order zero. Since the deflection of the membrane is always finite while Y_0 becomes infinite as s approaches zero, we cannot use Y_0 and must choose $C_2 = 0$. Clearly, $C_1 \neq 0$, since otherwise $W \equiv 0$. We may set $C_1 = 1$; then

(8)
$$W(r) = J_0(s) = J_0(kr).$$

On the boundary $r = R$ we must have $u(R, t) = W(R)G(t) = 0$. Since $G \equiv 0$ would imply $u \equiv 0$, we require that

$$W(R) = J_0(kR) = 0.$$

The Bessel function J_0 has (infinitely many) real zeros. Let us denote the positive zeros of $J_0(s)$ by $s = \alpha_1, \alpha_2, \cdots$ (cf. Fig. 286 on the next page). We mention that numerical values (exact to 4 decimal places) are

$$\alpha_1 = 2.4048, \quad \alpha_2 = 5.5201, \quad \alpha_3 = 8.6537, \quad \alpha_4 = 11.7915, \quad \alpha_5 = 14.9309.$$

We see that these zeros are irregularly spaced, and the same is true for the further zeros $\alpha_6, \alpha_7, \cdots$. (More extended tables are included in Ref. [1] in Appendix 1.) Equation (8) now implies

(9)
$$kR = \alpha_m \quad \text{or} \quad k = k_m = \frac{\alpha_m}{R}, \qquad m = 1, 2, \cdots.$$

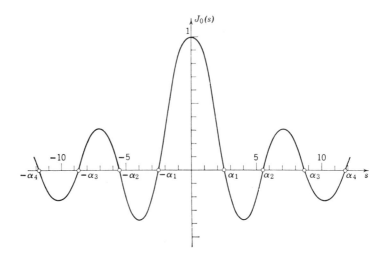

Fig. 286. Bessel function $J_0(s)$

Hence the functions

$$(10) \qquad W_m(r) = J_0(k_m r) = J_0\left(\frac{\alpha_m}{R}\, r\right) \qquad\qquad m = 1, 2, \cdots$$

are solutions of (7) that vanish at $r = R$.

The corresponding general solutions of (6) with $\lambda = \lambda_m = ck_m$ are

$$G_m(t) = a_m \cos \lambda_m t + b_m \sin \lambda_m t.$$

Hence the functions

$$(11) \qquad \boxed{u_m(r, t) = W_m(r)G_m(t) = (a_m \cos \lambda_m t + b_m \sin \lambda_m t)J_0(k_m r)}$$

where $m = 1, 2, \cdots$, are solutions of the wave equation (1), satisfying the boundary condition (2). These are the **eigenfunctions** of our problem, and the corresponding **eigenvalues** are λ_m.

The vibration of the membrane corresponding to u_m is called the mth **normal mode**; it has the frequency $\lambda_m/2\pi$ cycles per unit time. Since the zeros of J_0 are not regularly spaced on the axis (in contrast to the zeros of the sine functions appearing in the case of the vibrating string), the sound of a drum is entirely different from that of a violin. The forms of the normal modes can easily be obtained from Fig. 286 and are shown in Fig. 287. For $m = 1$, all the points of the membrane move up (or down) at the same time. For $m = 2$, the situation is as follows. The function

$$W_2(r) = J_0\left(\frac{\alpha_2}{R}\, r\right)$$

is zero for $\alpha_2 r/R = \alpha_1$ or $r = \alpha_1 R/\alpha_2$. The circle $r = \alpha_1 R/\alpha_2$ is, therefore, a nodal line, and when at some instant the central part of the membrane

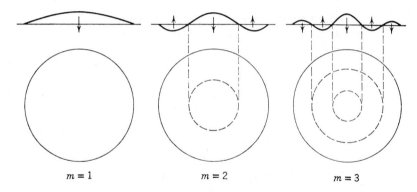

$m = 1$ $m = 2$ $m = 3$

Fig. 287. Normal modes of the circular membrane in the case of vibrations independent of the angle

moves up, the outer part $(r > \alpha_1 R/\alpha_2)$ moves down, and conversely. The solution $u_m(r, t)$ has $m - 1$ nodal lines, which are concentric circles (Fig. 287).

Third Step. Solution of the Entire Problem

To obtain a solution that also satisfies the initial conditions (3) and (4), we may proceed as in the case of the string, that is, we consider the series[9]

$$(12) \quad u(r,t) = \sum_{m=1}^{\infty} W_m(r)G_m(t) = \sum_{m=1}^{\infty} (a_m \cos \lambda_m t + b_m \sin \lambda_m t) J_0\left(\frac{\alpha_m}{R} r\right).$$

Setting $t = 0$ and using (3), we obtain

$$(13) \qquad u(r, 0) = \sum_{m=1}^{\infty} a_m J_0\left(\frac{\alpha_m}{R} r\right) = f(r).$$

Thus for (12) to satisfy (3), the a_m must be the coefficients of the Fourier–Bessel series that represents $f(r)$ in terms of $J_0(\alpha_m r/R)$; that is [cf. (9) in Sec. 4.9],

$$(14) \qquad \boxed{a_m = \frac{2}{R^2 J_1^2(\alpha_m)} \int_0^R rf(r)J_0\left(\frac{\alpha_m}{R} r\right) dr} \qquad m = 1, 2, \cdots.$$

Differentiability of $f(r)$ in the interval $0 \leq r \leq R$ is sufficient for the existence of the development (13), cf. Ref. [A14]. The coefficients b_m in (12) can be determined from (4) in a similar fashion. To obtain numerical values of a_m and b_m, we may apply one of the usual methods of approximate integration, using tables of J_0 and J_1. Sometimes it may not be necessary to apply numerical integration, as the following example illustrates.

---

[9]We shall not consider the problem of convergence and uniqueness.

EXAMPLE 1. Vibrations of a circular membrane
Find the vibrations of a circular drumhead of radius 1 ft and density 2 slugs/ft^2 if the tension is 8 lb/ft, the initial velocity is 0 and the initial displacement is

$$f(r) = 1 - r^2 \text{ [ft]}.$$

Solution. $c^2 = T/\rho = 8/2 = 4$ [ft^2/sec^2]. Also $b_m = 0$, since the initial velocity is 0. From (14) and Prob. 29 in Sec. 4.9, since $R = 1$,

$$a_m = \frac{2}{J_1{}^2(\alpha_m)} \int_0^1 r(1 - r^2)J_0(\alpha_m r) \, dr = \frac{4J_2(\alpha_m)}{\alpha_m{}^2 J_1{}^2(\alpha_m)}$$

Using the tables in Ref. [1], pp. 390—394, 409, we compute

m	α_m	$J_1(\alpha_m)$	$J_2(\alpha_m)$	a_m
1	2.4048	0.51915	0.43170	1.108
2	5.5201	−0.34026	−0.12316	−0.140
3	8.6537	0.27145	0.06262	0.045
4	11.7915	−0.23246	−0.03940	−0.021
5	14.9309	0.20655	0.02762	0.012

Thus

$$f(r) = 1.108J_0(2.4048r) - 0.140J_0(5.5201r) + 0.045J_0(8.6537r) - \cdots.$$

We see that the coefficients decrease relatively slowly. The sum of the explicitly given coefficients in the table is 1.004. The sum of *all* the coefficients should be 1. (Why?)
Since $\lambda_m = ck_m = c\alpha_m/R = 2\alpha_m$, from (12) we thus obtain the solution (with r measured in feet and t in seconds)

$$u(r, t) = 1.108J_0(2.4048r) \cos 4.8096t - 0.140J_0(5.5201r) \cos 11.0402t$$

$$+ 0.045J_0(8.6537r) \cos 17.3074t - \cdots. \quad \blacksquare$$

In the next two sections we extend our discussion of the **Laplace equation,** which we have met before (Secs. 9.7, 11.5) and which is probably the most important partial differential equation in physics and its engineering applications.

Problems for Sec. 11.10

1. What is the reason for using polar coordinates in the present consideration?
2. If the tension of the membrane is increased, how does the frequency of each normal mode (11) change?
3. A smaller drum should have a higher fundamental mode than a large drum, tension and density being the same. How can this be concluded from our formulas?
4. Determine numerical values of the radii of the nodal lines of u_2 and u_3 [cf. (11)] when $R = 1$.
5. Sketch a figure similar to Fig. 287, for u_4, u_5, u_6.
6. Same question as in Prob. 4, for u_4, u_5, u_6.
7. Show that for (12) to satisfy (4),

(15)
$$b_m = \frac{2}{c\alpha_m R J_1{}^2(\alpha_m)} \int_0^R rg(r)J_0(\alpha_m r/R) \, dr, \qquad m = 1, 2, \cdots.$$

8. Why should $a_1 + a_2 + \cdots = 1$ in Example 1?

9. Verify the computation of the a_m in Example 1.

Find the deflection $u(r, t)$ of the circular membrane of radius $R = 1$, if $c = 1$, the initial velocity is zero and the initial deflection is $f(r)$, where

10. $f = 0.1 J_0(\alpha_2 r)$ **11.** $f = k(1 - r^2)$ **12.** $f = k(1 - r^4)$

     Hint. Remember Probs. 25—32, Sec. 4.9.

13. Is it possible that, for fixed c and R, two or more functions u_m [cf. (11)] that have different nodal lines correspond to the same eigenvalue?

Vibrations of a Circular Membrane Depending on Radius r and Angle θ

14. Show that substitution of $u = F(r, \theta)G(t)$ into the wave equation

$$(16) \qquad u_{tt} = c^2 \left(u_{rr} + \frac{1}{r} u_r + \frac{1}{r^2} u_{\theta\theta} \right)$$

leads to

$$(17) \qquad \ddot{G} + \lambda^2 G = 0, \qquad \text{where } \lambda = ck,$$

$$(18) \qquad F_{rr} + \frac{1}{r} F_r + \frac{1}{r^2} F_{\theta\theta} + k^2 F = 0.$$

15. Show that substitution of $F = W(r)Q(\theta)$ in (18) yields

$$(19) \qquad Q'' + n^2 Q = 0,$$

$$(20) \qquad r^2 W'' + rW' + (k^2 r^2 - n^2) W = 0.$$

16. Show that $Q(\theta)$ must be periodic with period 2π, and, therefore, $n = 0, 1, \cdots$ in (19) and (20). Show that this yields the solutions $Q_n = \cos n\theta$, $Q_n{}^* = \sin n\theta$, $W_n = J_n(kr)$, $n = 0, 1, \cdots$.

17. Show that the boundary condition

$$(21) \qquad u(R, \theta, t) = 0$$

leads to the values $k = k_{mn} = \alpha_{mn}/R$, where $s = \alpha_{mn}$ is the mth positive zero of $J_n(s)$.

18. Show that solutions of (16) that satisfy (21) are

$$(22) \qquad \begin{aligned} u_{mn} &= (A_{mn} \cos ck_{mn}t + B_{mn} \sin ck_{mn}t)J_n(k_{mn}r) \cos n\theta, \\ u_{mn}^* &= (A_{mn}^* \cos ck_{mn}t + B_{mn}^* \sin ck_{mn}t)J_n(k_{mn}r) \sin n\theta. \end{aligned}$$

19. Show that $u_{m0}^* \equiv 0$ and u_{m0} is identical with (11) in the current section.

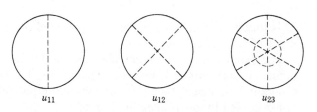

u_{11} u_{12} u_{23}

Fig. 288. Nodal lines of some of the solutions (22)

20. Show that u_{mn} has $m + n - 1$ nodal lines.

Graph the nodal lines of the following solutions.

21. u_{3n}, $n = 1, 2, 3$ **22.** u_{4n}, $n = 1, 2, 3$ **23.** u_{mn}^*, $m, n = 1, 2, 3$

24. Show that the initial condition $u_t(r, \theta, 0) = 0$ leads to $B_{mn} = 0$, $B_{mn}^* = 0$ in (22).

25. Show that u_{11} represents the fundamental mode of a semicircular membrane and find the corresponding frequency when $c^2 = 1$ and $R = 1$.

11.11 Laplace's Equation. Potential

One of the most important partial differential equations in physics is Laplace's equation

(1)
$$\nabla^2 u = 0.$$

Here $\nabla^2 u$ is the Laplacian of u. In Cartesian coordinates x, y, z in space,

(2)
$$\nabla^2 u = \frac{\partial^2 u}{\partial x^2} + \frac{\partial^2 u}{\partial y^2} + \frac{\partial^2 u}{\partial z^2}.$$

The theory of the solutions of Laplace's equation is called **potential theory.** Solutions of (1) that have *continuous* second-order partial derivatives are called **harmonic functions.**

The two-dimensional case, when u depends on two variables only, can most conveniently be treated by methods of complex analysis and will be considered in Sec. 12.5 and Chap. 17.

To illustrate the importance of Laplace's equation in engineering mathematics, let us mention some basic applications.

Laplace's equation occurs in connection with gravitational forces. In Example 4, Sec. 8.8, we have seen that if a particle A of mass M is fixed at a point (X, Y, Z) and another particle B of mass m is at a point (x, y, z), then A attracts B, the gravitational force being the gradient of the scalar function

$$u(x, y, z) = \frac{c}{r}, \qquad c = GMm = const,$$

$$r = \sqrt{(x - X)^2 + (y - Y)^2 + (z - Z)^2} \qquad (> 0).$$

This function of x, y, z is called the **potential** of the gravitational field, and it satisfies Laplace's equation.

The extension to the potential and force due to a continuous distribution of mass is quite direct. If a mass of density $\rho(X, Y, Z)$ is distributed throughout a region T in space, then the corresponding potential u at a point (x, y, z) not occupied by mass is defined to be

(3)
$$u(x, y, z) = k \iiint\limits_T \frac{\rho}{r} \, dX \, dY \, dZ \qquad (k > 0),$$

with r as given by the previous formula. Since $1/r$ $(r > 0)$ is a solution of (1), that is, $\nabla^2(1/r) = 0$, and ρ does not depend on x, y, z, we obtain

$$\nabla^2 u = k \iiint\limits_T \rho \nabla^2 \left(\frac{1}{r}\right) dX \, dY \, dZ = 0;$$

that is, the gravitational potential defined by (3) satisfies Laplace's equation at any point that is not occupied by matter.

In *electrostatics* the electrical force of attraction or repulsion between charged particles is governed by *Coulomb's law* (cf. Sec. 8.8), which is of the same mathematical form as Newton's law of gravitation. From this it follows that the field created by a distribution of electrical charges can be described mathematically by a potential function that satisfies Laplace's equation at any point not occupied by charges.

In Chap. 17 we shall see that Laplace's equation also appears in the theory of incompressible fluid flow.

Furthermore, the basic equation in heat conduction is the heat equation

$$u_t = c^2 \nabla^2 u,$$

(cf. Secs. 9.7 and 11.5) and if the temperature u is independent of time t ("*steady state*"), this equation reduces to Laplace's equation; cf. Sec. 11.5.

In most applications leading to Laplace's equation, it is required to solve a **boundary value problem,** that is, to determine the solution of (1) satisfying given boundary conditions on the boundary surface S of the region T in which the equation is considered. This is called:

(I)   The **first boundary value problem** or **Dirichlet problem** if u is prescribed on S.

(II)   The **second boundary value problem** or **Neumann problem** if the normal derivative $u_n = \partial u/\partial n$ is prescribed on S.

(III)   The **third** or **mixed boundary value problem** if u is prescribed on a portion of S and u_n on the remaining portion of S.

It is then necessary to introduce coordinates in space such that S is given by simple formulas. This requires the transformation of the Laplacian (2) into other coordinate systems. Of course, such a transformation is quite similar to that in the case of the Laplacian of a function of two variables (cf. Sec. 11.9).

From (4) in Sec. 11.9 it follows immediately that the Laplacian of a function u in **cylindrical coordinates**[10] (cf. Fig. 289 on the next page)

---

[10]Observe that θ is not completely determined by the ratio y/x but we must also take the signs of x and y into account.

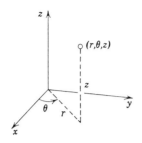

Fig. 289. Cylindrical coordinates

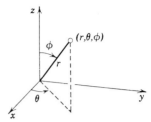

Fig. 290. Spherical coordinates

(4) $r = \sqrt{x^2 + y^2}, \qquad \theta = \arctan \dfrac{y}{x}, \qquad z = z$

is

(5) $\nabla^2 u = \dfrac{\partial^2 u}{\partial r^2} + \dfrac{1}{r}\dfrac{\partial u}{\partial r} + \dfrac{1}{r^2}\dfrac{\partial^2 u}{\partial \theta^2} + \dfrac{\partial^2 u}{\partial z^2}.$

Other important coordinates are **spherical coordinates** r, θ, ϕ, which are related to Cartesian coordinates as follows (cf. Fig. 290):

(6) $x = r \cos \theta \sin \phi, \qquad y = r \sin \theta \sin \phi, \qquad z = r \cos \phi.$

The Laplacian of a function u in spherical coordinates is

(7) $\nabla^2 u = \dfrac{\partial^2 u}{\partial r^2} + \dfrac{2}{r}\dfrac{\partial u}{\partial r} + \dfrac{1}{r^2}\dfrac{\partial^2 u}{\partial \phi^2} + \dfrac{\cot \phi}{r^2}\dfrac{\partial u}{\partial \phi} + \dfrac{1}{r^2 \sin^2 \phi}\dfrac{\partial^2 u}{\partial \theta^2}.$

This may also be written

(7′) $\nabla^2 u = \dfrac{1}{r^2}\left[\dfrac{\partial}{\partial r}\left(r^2 \dfrac{\partial u}{\partial r} \right) + \dfrac{1}{\sin \phi}\dfrac{\partial}{\partial \phi}\left(\sin \phi \dfrac{\partial u}{\partial \phi} \right) + \dfrac{1}{\sin^2 \phi}\dfrac{\partial^2 u}{\partial \theta^2} \right].$

This formula can be derived in a manner similar to that in Sec. 11.9; the details are left as an exercise for the reader.

In the next section we show that the separation of variables in spherical coordinates leads to **Legendre's equation,** which we discussed in Sec. 4.3.

Problems for Sec. 11.11

1. Find the Laplacian in rectangular coordinates $x^* = ax$, $y^* = by$, $z^* = cz$, where x, y, z are Cartesian coordinates.
2. Starting from (2), verify by calculation that with respect to the Cartesian coordinates defined by (7), (8), page 506,

$$\nabla^2 u = u_{x^*x^*} + u_{y^*y^*} + u_{z^*z^*}.$$

Show that the following functions $u = f(x, y)$ satisfy the Laplace equation and plot some of the equipotential lines $u = const.$

3. $x^2 - y^2$ **4.** xy **5.** $y/(x^2 + y^2)$

6. $x^3 - 3xy^2$ **7.** $x/(x^2 + y^2)$ **8.** $(x^2 - y^2)/(x^2 + y^2)^2$

9. Verify that $u = c/r$ satisfies Laplace's equation in spherical coordinates.

10. Show that the only solution of $\nabla^2 u = 0$ depending only on the variable $r = \sqrt{x^2 + y^2 + z^2}$ is $u = c/r + k$; here c and k are constants.

11. Determine c and k in Prob. 10 such that u represents the electrostatic potential between two concentric spheres of radii $r_1 = 2$ cm and $r_2 = 10$ cm kept at the potentials $U_1 = 110$ volts and $U_2 = 0$ volt, respectively.

12. Show that the only solution of the two-dimensional Laplace equation depending only on $r = \sqrt{x^2 + y^2}$ is $u = c \ln r + k$.

13. Find the electrostatic potential between two coaxial cylinders of radii $r_1 = 2$ cm and $r_2 = 10$ cm kept at the potentials $U_1 = 110$ volts and $U_2 = 0$ volt, respectively. Graph and compare the solutions of Probs. 11 and 13.

14. Express the spherical coordinates defined by (6) in terms of Cartesian coordinates.

15. Verify (5) by transforming $\nabla^2 u$ back into Cartesian coordinates.

Find the steady-state (time-independent) temperature distribution:

16. Between two parallel plates $x = x_0$ and $x = x_1$ kept at the temperatures u_0 and u_1, respectively.

17. Between two coaxial circular cylinders of radii r_0 and r_1 kept at the temperatures u_0 and u_1, respectively.

18. Between two concentric spheres of radii r_0 and r_1 kept at the temperatures u_0 and u_1, respectively.

19. (**Helmholtz equation**[11]) Show that the substitution of $u = U(x, y, z)e^{-i\omega t}$ ($i = \sqrt{-1}$) into the three-dimensional wave equation $u_{tt} = c^2 \nabla^2 u$ yields the so-called *Helmholtz equation*

$$\nabla^2 U + k^2 U = 0 \qquad\qquad k = \omega/c.$$

20. Let r, θ, ϕ be spherical coordinates. If $u(r, \theta, \phi)$ satisfies $\nabla^2 u = 0$, show that $v(r, \theta, \phi) = r^{-1}u(r^{-1}, \theta, \phi)$ satisfies $\nabla^2 v = 0$.

11.12 Laplace's Equation in Spherical Coordinates. Legendre's Equation

Let us consider a typical boundary value problem that involves Laplace's equation in spherical coordinates. Suppose that a sphere S of radius R is kept at a fixed distribution of electric potential

$$(1) \qquad\qquad u(R, \theta, \phi) = f(\phi)$$

[11]HERMANN VON HELMHOLTZ (1821—1894), German physicist, known by his important work in thermodynamics, hydrodynamics and acoustics.

where r, θ, ϕ are the spherical coordinates defined in the last section, with the origin at the center of S, and $f(\phi)$ is a given function. We wish to find the potential u at all points in space, which is assumed to be free of further charges. Since the potential on S is independent of θ, so is the potential in space. Thus $\partial^2 u/\partial\theta^2 = 0$, and from (7') in Sec. 11.11 we see that Laplace's equation reduces to

$$(2) \qquad \frac{\partial}{\partial r}\left(r^2 \frac{\partial u}{\partial r}\right) + \frac{1}{\sin\phi}\frac{\partial}{\partial\phi}\left(\sin\phi\,\frac{\partial u}{\partial\phi}\right) = 0.$$

Furthermore, at infinity the potential will be zero; that is, we must have

$$(3) \qquad \lim_{r\to\infty} u(r,\phi) = 0.$$

We shall solve the boundary value problem consisting of the equation (2) and the boundary condition (1) and the condition (3) at infinity, using the method of separation of variables. Substituting a solution of the form

$$u(r,\phi) = G(r)H(\phi)$$

into (2), and dividing the resulting equation by the function GH, we obtain

$$\frac{1}{G}\frac{d}{dr}\left(r^2 \frac{dG}{dr}\right) = -\frac{1}{H\sin\phi}\frac{d}{d\phi}\left(\sin\phi\,\frac{dH}{d\phi}\right).$$

By the usual argument, the two sides of this equation must be equal to a constant, say, k, so that

$$(4) \qquad \frac{1}{\sin\phi}\frac{d}{d\phi}\left(\sin\phi\,\frac{dH}{d\phi}\right) + kH = 0$$

and

$$\frac{1}{G}\frac{d}{dr}\left(r^2 \frac{dG}{dr}\right) = k.$$

The last equation may be written

$$r^2 G'' + 2rG' - kG = 0.$$

This is the Euler–Cauchy equation. From Sec. 2.7 we know that it has solutions of the form $G = r^\alpha$. These solutions will have a particularly simple form if we change our notation and write $n(n + 1)$ for k. Then

$$(5) \qquad r^2 G'' + 2rG' - n(n + 1)G = 0$$

where n is still arbitrary. By substituting $G = r^\alpha$ into (5) we have

$$[\alpha(\alpha - 1) + 2\alpha - n(n + 1)]r^\alpha = 0.$$

The zeros of the expression in brackets are $\alpha = n$ and $\alpha = -n - 1$. Hence we obtain the solutions

(6) $$G_n(r) = r^n \quad \text{and} \quad G_n^*(r) = \frac{1}{r^{n+1}}.$$

Introducing $k = n(n + 1)$ into (4) and setting

$$\cos \phi = w,$$

we have $\sin^2 \phi = 1 - w^2$ and

$$\frac{d}{d\phi} = \frac{d}{dw} \frac{dw}{d\phi} = -\sin \phi \frac{d}{dw}.$$

Consequently, (4) takes the form

(7) $$\frac{d}{dw} \left[(1 - w^2) \frac{dH}{dw} \right] + n(n + 1)H = 0$$

or

(7′) $$\boxed{(1 - w^2) \frac{d^2 H}{dw^2} - 2w \frac{dH}{dw} + n(n + 1)H = 0.}$$

This is **Legendre's equation** (cf. Sec. 4.3).

Solution of Legendre's Equation

For integer[12] $n = 0, 1, \cdots$, the Legendre polynomials

$$H = P_n(w) = P_n(\cos \phi), \qquad n = 0, 1, \cdots,$$

are solutions of Legendre's equation (7). We thus obtain the following two sequences of solutions $u = GH$ of Laplace's equation (2):

(8*) $$u_n(r, \phi) = A_n r^n P_n(\cos \phi), \qquad u_n^*(r, \phi) = \frac{B_n}{r^{n+1}} P_n(\cos \phi)$$

where $n = 0, 1, \cdots$, and A_n and B_n are constants.

To find a solution of (2), valid at points *inside* the sphere and satisfying (1), we consider the series[13]

---

[12]So far, n was arbitrary since k was arbitrary. It can be shown that the restriction of n to integers is necessary to make the solution of (7) continuous, together with its derivative of the first order, in the interval $-1 \leqq w \leqq 1$ or $0 \leqq \phi \leqq \pi$.

[13]Convergence will not be considered. It can be shown that if $f(\phi)$ and $f'(\phi)$ are piecewise continuous in the interval $0 \leqq \phi \leqq \pi$, the series (8) with coefficients (10) can be differentiated termwise twice with respect to r and with respect to ϕ and the resulting series converge and represent $\partial^2 u/\partial r^2$ and $\partial^2 u/\partial \phi^2$, respectively. Hence the series (8) with coefficients (10) is then the solution of our problem inside the sphere.

(8)
$$u(r, \phi) = \sum_{n=0}^{\infty} A_n r^n P_n(\cos \phi).$$

For (8) to satisfy (1) we must have

(9)
$$u(R, \phi) = \sum_{n=0}^{\infty} A_n R^n P_n(\cos \phi) = f(\phi);$$

that is, (9) must be the generalized Fourier series of $f(\phi)$ in terms of Legendre polynomials. From (6) and (7) in Sec. 4.7 and (2) in Sec. 4.9 it follows that

$$A_n R^n = \frac{2n + 1}{2} \int_{-1}^{1} \widetilde{f}(w) P_n(w) \, dw$$

where $\widetilde{f}(w)$ denotes $f(\phi)$ as a function of $w = \cos \phi$. Since $dw = -\sin \phi \, d\phi$, and the limits of integration -1 and 1 correspond to $\phi = \pi$ and $\phi = 0$, respectively, we also have

(10)
$$A_n = \frac{2n + 1}{2R^n} \int_{0}^{\pi} f(\phi) P_n(\cos \phi) \sin \phi \, d\phi, \qquad n = 0, 1, \cdots.$$

Thus the series (8) with coefficients (10) is the solution of our problem for points inside the sphere.

To find the solution *exterior* to the sphere, we cannot use the functions $u_n(r, \phi)$ because these functions do not satisfy (3), but we may use the functions $u_n^*(r, \phi)$, which satisfy (3), and proceed as before. This leads to the solution

(11)
$$u(r, \phi) = \sum_{n=0}^{\infty} \frac{B_n}{r^{n+1}} P_n(\cos \phi) \qquad (r \geq R)$$

with coefficients

(12)
$$B_n = \frac{2n + 1}{2} R^{n+1} \int_{0}^{\pi} f(\phi) P_n(\cos \phi) \sin \phi \, d\phi.$$

EXAMPLE 1. Spherical capacitor

Find the potential inside and outside a spherical capacitor consisting of two metallic hemispheres of radius 1 ft separated by a small slit for reasons of insulation, if the upper hemisphere is kept at 110 volts and the lower is grounded (Fig. 291).

Solution. The given boundary condition is (recall Fig. 290 in Sec. 11.11)

$$f(\phi) = \begin{cases} 110 & \text{if} \quad 0 \leq \phi < \pi/2 \\ 0 & \text{if} \quad \pi/2 < \phi \leq \pi. \end{cases}$$

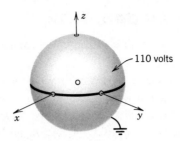

Fig. 291. Spherical capacitor in Example 1

Since $R = 1$, we thus obtain from (10)

$$A_n = \frac{2n + 1}{2} \cdot 110 \int_0^{\pi/2} P_n(\cos \phi) \sin \phi \, d\phi.$$

We set $w = \cos \phi$. Then $P_n(\cos \phi) \sin \phi \, d\phi = -P_n(w) \, dw$ and we integrate from 1 to 0. We get rid of the minus by integrating from 0 to 1. Then from (11) in Sec. 4.3,

$$A_n = 55(2n + 1) \sum_{m=0}^{M} (-1)^m \frac{(2n - 2m)!}{2^n m!(n - m)!(n - 2m)!} \int_0^1 w^{n-2m} \, dw$$

where $M = n/2$ for even n and $M = (n - 1)/2$ for odd n. The integral equals $1/(n - 2m + 1)$. Thus

(13) $$A_n = \frac{55(2n + 1)}{2^n} \sum_{m=0}^{M} (-1)^m \frac{(2n - 2m)!}{m!(n - m)!(n - 2m + 1)!}.$$

Taking $n = 0$, we get $A_0 = 55$ (since $0! = 1$). For $n = 1, 2, 3, \cdots$ we get

$$A_1 = \frac{165}{2} \cdot \frac{2!}{0!1!2!} = \frac{165}{2},$$

$$A_2 = \frac{275}{4} \left(\frac{4!}{0!2!3!} - \frac{2!}{1!1!1!} \right) = 0,$$

$$A_3 = \frac{385}{8} \left(\frac{6!}{0!3!4!} - \frac{4!}{1!2!2!} \right) = -\frac{385}{8},$$

and so on. Hence the potential (8) inside the sphere is (since $P_0 = 1$)

$$u(r, \phi) = 55 + \frac{165}{2} r P_1(\cos \phi) - \frac{385}{8} r^3 P_3(\cos \phi) + \cdots$$

with $P_1, P_3, \cdots$ given by (11'), Sec. 4.3. Since $R = 1$, we see from (10) and (12) in the present section that $B_n = A_n$, and (11) thus gives the potential outside the sphere

$$u(r, \phi) = \frac{55}{r} + \frac{165}{2r^2} P_1(\cos \phi) - \frac{385}{8r^4} P_3(\cos \phi) + \cdots.$$

Partial sums of these series can now be used for computing approximate values of the potential. Also, it is interesting to see that far away from the sphere the potential is approximately that of a point charge, namely, $55/r$ (cf. Example 4 in Sec. 8.8).   ∎

In the last two sections of this chapter we show that problems involving partial differential equations can also be solved by **operational methods** (the Laplace and Fourier transformations).

Problems for Sec. 11.12

1. Verify by substitution that $u_n(r, \phi)$ and $u_n^*(r, \phi)$, $n = 0, 1, 2$, in (8*) are solutions of (2).

2. Find the surfaces on which the functions u_1, u_2, u_3 are zero.

3. Graph the functions $P_n(\cos \phi)$ for $n = 0, 1, 2$, [cf. (11'), Sec. 4.3].

4. Graph the functions $P_3(\cos \phi)$ and $P_4(\cos \phi)$.

5. Derive the values of A_0, A_1, A_2, A_3 in Example 1 from (13).

6. Verify the values in Prob. 5 by integrating the corresponding formulas (11') in Sec. 4.3.

Let r, θ, ϕ be the spherical coordinates used in the text. Find the potential in the interior of the sphere $R = 1$, assuming that there are no charges in the interior and the potential on the surface is $f(\phi)$, where

7. $f(\phi) = 1$ **8.** $f(\phi) = \cos \phi$ **9.** $f(\phi) = \cos 2\phi$

10. $f(\phi) = \cos^2 \phi$ **11.** $f(\phi) = \cos^3 \phi$ **12.** $f(\phi) = \cos 3\phi$

13. $f(\phi) = 10 \cos^3 \phi - 3 \cos^2 \phi - 5 \cos \phi - 1$

14. Show that in Prob. 7, the potential exterior to the sphere is the same as that of a point charge at the origin.

15. Graph the intersections of the equipotential surfaces in Prob. 8 with the xz-plane.

16. Find the potential exterior to the sphere in Probs. 7—13.

17. Verify by substitution that the solution of Prob. 10 satisfies (2).

18. **(Transmission line equations)** Consider a long cable or telephone wire (Fig. 292) that is imperfectly insulated so that leaks occur along the entire length of the cable. The source S of the current $i(x, t)$ in the cable is at $x = 0$, the receiving end T at $x = l$. The current flows from S to T, through the load, and returns to the ground. Let the constants R, L, C and G denote the resistance, inductance, capacitance to ground and conductance to ground, respectively, of the cable per unit length. Show that

$$-\frac{\partial u}{\partial x} = Ri + L\frac{\partial i}{\partial t} \qquad \textbf{(First transmission line equation)}$$

where $u(x, t)$ is the potential in the cable. *Hint.* Apply Kirchhoff's voltage law to a small portion of the cable between x and $x + \Delta x$ (difference of the potentials at x and $x + \Delta x$ = resistive drop + inductive drop).

19. Show that for the cable in Prob. 18,

$$-\frac{\partial i}{\partial x} = Gu + C\frac{\partial u}{\partial t} \qquad \textbf{(Second transmission line equation).}$$

Hint. Use Kirchhoff's current law (difference of the currents at x and

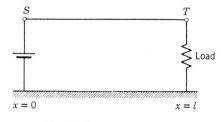

Fig. 292. Transmission line

$x + \Delta x$ = loss due to leakage to ground + capacitive loss).

20. Show that elimination of i from the transmission line equations leads to

$$u_{xx} = LCu_{tt} + (RC + GL)u_t + RGu.$$

21. (**Telegraph equations**) For a submarine cable, G is negligible and the frequencies are low. Show that this leads to the so-called *submarine cable equations* or **telegraph equations**

$$u_{xx} = RCu_t, \qquad i_{xx} = RCi_t.$$

22. Find the potential in a submarine cable with ends ($x = 0$, $x = l$) grounded and initial voltage distribution $U_0 = const.$

23. What problem in heat conduction is the analog of Prob. 22?

24. (**High-frequency line equations**) Show that in the case of alternating currents of high frequencies the equation in Prob. 22 and the analogous equation for the current i can be approximated by the so-called **high-frequency line equations**

$$u_{xx} = LCu_{tt}, \qquad i_{xx} = LCi_{tt}.$$

25. Solve the first high frequency line equation, assuming that the initial potential is $U_0 \sin(\pi x/l)$, $u_t(x, 0) = 0$ and $u = 0$ at the ends $x = 0$ and $x = l$ for all t.

11.13 Laplace Transformation Applied to Partial Differential Equations

Readers familiar with the Laplace transformation (Chap. 5) may wonder whether that method can also be used for solving *partial* differential equations. The answer is yes, in particular if one of the independent variables ranges over the positive axis. The basic idea may be explained in terms of two typical examples (below). The steps of solution are similar to those in Chap. 5; for an equation in two independent variables they are as follows.

1. Take the Laplace transform with respect to one of the two variables, usually t. This gives an *ordinary differential equation* for the transform of the unknown function. This is so since the derivatives of this function with respect to the other variable slip into the transformed equation. The latter also incorporates the given boundary and initial conditions.

2. Solving that ordinary differential equation, obtain the transform of the unknown function.

3. Applying the inverse transformation to that transform, obtain the solution of the given problem.

If the coefficients of the given equation do not depend on t, the transformation will simplify the problem.

EXAMPLE 1. A first-order equation

Solve the problem

$$\frac{\partial w}{\partial x} + x\,\frac{\partial w}{\partial t} = 0, \qquad w(x, 0) = 0, \quad w(0, t) = t \qquad (t \geq 0). \tag{1}$$

We write w since we need u to denote the unit step function (Sec. 5.3).

Solution. We take the Laplace transform of (1) **with respect to t.** By (1) in Sec. 5.2,

$$\mathscr{L}\left\{\frac{\partial w}{\partial x}\right\} + x[s\mathscr{L}\{w\} - w(x, 0)] = 0. \tag{2}$$

Here $w(x, 0) = 0$. In the first term we assume that we may interchange integration and differentiation:

$$\mathscr{L}\left\{\frac{\partial w}{\partial x}\right\} = \int_0^\infty e^{-st}\,\frac{\partial w}{\partial x}\,dt = \frac{\partial}{\partial x}\int_0^\infty e^{-st}w(x, t)\,dt = \frac{\partial}{\partial x}\,\mathscr{L}\{w(x, t)\}. \tag{3}$$

Writing $W(x, s) = \mathscr{L}\{w(x, t)\}$, we thus obtain from (2)

$$\frac{\partial W}{\partial x} + xsW = 0.$$

This may be regarded as an *ordinary* differential equation with x as the independent variable, since derivatives with respect to s do not occur in the equation. The general solution is (Sec. 1.7)

$$W(x, s) = c(s)e^{-sx^2/2}.$$

Since $\mathscr{L}\{t\} = 1/s^2$, the condition $w(0, t) = t$ yields $W(0, s) = 1/s^2$, that is,

$$W(0, s) = c(s) = 1/s^2.$$

Hence

$$W(x, s) = \frac{1}{s^2}\,e^{-sx^2/2}.$$

Now $\mathscr{L}^{-1}\{1/s^2\} = t$, so that the second shifting theorem (Sec. 5.3) with $a = x^2/2$ gives

$$w(x, t) = \left(t - \frac{x^2}{2}\right)u(t - \tfrac{1}{2}x^2) = \begin{cases} 0 & \text{if } t < x^2/2 \\ t - \tfrac{1}{2}x^2 & \text{if } t > x^2/2 \end{cases} \tag{4}$$

Since we proceeded formally, we have to verify that (4) satisfies (1). We leave this to the reader.

EXAMPLE 2. Semi-infinite string

Find the displacement $w(x, t)$ of an elastic string subject to the following conditions.

(*i*) The string is initially at rest on the x-axis from $x = 0$ to ∞ ("*semi-infinite string*").

(*ii*) For time $t > 0$ the left end of the string is moved in a given fashion (Fig. 293)

$$w(0, t) = f(t) = \begin{cases} \sin t & \text{if } 0 \leq t \leq 2\pi \\ 0 & \text{otherwise} \end{cases}$$

(*iii*) Furthermore

$$\lim_{x \to \infty} w(x, t) = 0 \quad \text{for } t \geq 0.$$

Of course there is no infinite string, but our model describes a long string or rope (of negligible weight) with its right end fixed far out on the x-axis. (We again write w since we need u to denote the unit step function.)

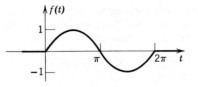

Fig. 293. Motion of the left end of the string
in Example 2 as a function of time t

Solution. We have to solve the wave equation (Sec. 11.2)

$$(5) \qquad\qquad \frac{\partial^2 w}{\partial t^2} = c^2 \frac{\partial^2 w}{\partial x^2} \qquad\qquad c^2 = \frac{T}{\rho}$$

for positive x and t, subject to the "boundary conditions"

$$(6) \qquad\qquad w(0,\, t) = f(t), \qquad \lim_{x \to \infty} w(x,\, t) = 0 \qquad\qquad (t \geqq 0)$$

with f as given above, and the initial conditions

$$(7) \qquad\qquad w(x,\, 0) = 0$$

$$(8) \qquad\qquad \frac{\partial w}{\partial t}\bigg|_{t=0} = 0.$$

We take the Laplace transform **with respect to t.** By (2) in Sec. 5.2,

$$\mathscr{L}\left\{\frac{\partial^2 w}{\partial t^2}\right\} = s^2 \mathscr{L}\{w\} - s w(x,\, 0) - \frac{\partial w}{\partial t}\bigg|_{t=0} = c^2 \mathscr{L}\left\{\frac{\partial^2 w}{\partial x^2}\right\}.$$

Two terms drop out, by (7) and (8). On the right we assume that we may interchange integration and differentiation:

$$\mathscr{L}\left\{\frac{\partial^2 w}{\partial x^2}\right\} = \int_0^\infty e^{-st} \frac{\partial^2 w}{\partial x^2}\, dt = \frac{\partial^2}{\partial x^2} \int_0^\infty e^{-st} w(x,\, t)\, dt = \frac{\partial^2}{\partial x^2}\, \mathscr{L}\{w(x,\, t)\}.$$

Writing $W(x,\, s) = \mathscr{L}\{w(x,\, t)\}$, we thus obtain

$$s^2 W = c^2 \frac{\partial^2 W}{\partial x^2}$$

or

$$\frac{\partial^2 W}{\partial x^2} - \frac{s^2}{c^2}\, W = 0.$$

Since this equation contains only a derivative with respect to x, it may be regarded as an ordinary differential equation for $W(x,\, s)$ considered as a function of x. A general solution is

$$(9) \qquad\qquad W(x,\, s) = A(s) e^{sx/c} + B(s) e^{-sx/c}.$$

From (6) we obtain, writing $F(s) = \mathscr{L}\{f(t)\}$,

$$W(0,\, s) = \mathscr{L}\{w(0,\, t)\} = \mathscr{L}\{f(t)\} = F(s)$$

and, assuming that the order of integrating with respect to t and taking the limit as $x \to \infty$ can be interchanged,

$$\lim_{x \to \infty} W(x,\, s) = \lim_{x \to \infty} \int_0^\infty e^{-st} w(x,\, t)\, dt = \int_0^\infty e^{-st} \lim_{x \to \infty} w(x,\, t)\, dt = 0.$$

This implies $A(s) = 0$ in (9) because $c > 0$, so that for every fixed positive s the function $e^{sx/c}$ increases as x increases. Note that we may assume $s > 0$ since a Laplace transform generally exists for *all* s greater than some fixed γ (Sec. 5.2). Hence we have

$$W(0, s) = B(s) = F(s),$$

so that (9) becomes

$$W(x, s) = F(s)e^{-sx/c}.$$

From the second shifting theorem (Sec. 5.3) with $a = x/c$ we obtain the inverse transform (Fig. 294)

(10) $$w(x, t) = f\left(t - \frac{x}{c}\right) u\left(t - \frac{x}{c}\right),$$

that is,

$$w(x, t) = \sin\left(t - \frac{x}{c}\right) \quad \text{if} \quad \frac{x}{c} < t < \frac{x}{c} + 2\pi \quad \text{or} \quad ct > x > (t - 2\pi)c$$

and zero otherwise. This is a single sine wave traveling to the right with speed c. Note that a point x remains at rest until $t = x/c$, the time needed to reach that x if one starts at $t = 0$ (start of the motion of the left end) and travels with speed c. The result agrees with our physical intuition. Since we proceeded formally, we must verify that (10) satisfies the given conditions. We leave this step to the reader. ∎

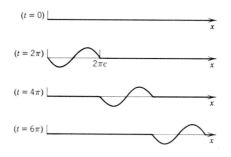

Fig. 294. Traveling wave in Example 2

In the next section we show that besides the Laplace transformation other operational methods, namely, **Fourier transformations** (cf. Secs. 10.10—10.12) can be used for solving partial differential equations.

Problems for Sec. 11.13

1. Sketch a figure similar to Fig. 294 if $c = 1$ and f is "triangular" as in Example 1, Sec. 11.3, with $k = L/2 = 1$.
2. How does the speed of the wave in Example 2 depend on the tension and the mass of the string?
3. Verify the solution in Example 2. What traveling wave do we obtain in Example 2 if we impose a (nonterminating) sinusoidal motion of the left end starting at $t = 0$?

Solve by the Laplace transformation:

4. $\dfrac{\partial u}{\partial x} + 2x\dfrac{\partial u}{\partial t} = 2x$, $u(x, 0) = 1$, $u(0, t) = 1$

5. $x \dfrac{\partial u}{\partial x} + \dfrac{\partial u}{\partial t} = xt$, $u(x, 0) = 0$ if $x \geq 0$, $u(0, t) = 0$ if $t \geq 0$.

6. Solve Prob. 5 by another method.

Find the temperature $w(x, t)$ in a semi-infinite laterally insulated bar extending from $x = 0$ along the x-axis to ∞, assuming that the initial temperature is 0, $w(x, t) \to 0$ as $x \to \infty$ for every fixed $t \geq 0$, and $w(0, t) = f(t)$. Proceed as follows.

7. Set up the model and show that the Laplace transformation leads to

$$sW(x, s) = c^2 \frac{\partial^2 W}{\partial x^2} \qquad\qquad W = \mathcal{L}\{w\}$$

and

$$W(x, s) = F(s)e^{-\sqrt{s}\,x/c} \qquad\qquad F = \mathcal{L}\{f\}.$$

8. Applying the convolution theorem in Prob. 7, show that

$$w(x, t) = \frac{x}{2c\sqrt{\pi}} \int_0^t f(t - \tau)\tau^{-3/2}e^{-x^2/4c^2\tau}\,d\tau.$$

9. Let $w(0, t) = f(t) = u_0(t)$ (Sec. 5.3). Denote the corresponding w, W and F by w_0, W_0 and F_0. Show that then in Prob. 8,

$$w_0(x, t) = \frac{x}{2c\sqrt{\pi}} \int_0^t \tau^{-3/2}e^{-x^2/4c^2\tau}\,d\tau = 1 - \operatorname{erf}\left(\frac{x}{2c\sqrt{t}}\right)$$

with the error function erf as defined in the problem set for Sec. 11.6.

10. (Duhamel's formula[14]) Show that in Prob. 9,

$$W_0(x, s) = \frac{1}{s}\,e^{-\sqrt{s}\,x/c}$$

and the convolution theorem gives *Duhamel's formula*

$$w(x, t) = \int_0^t f(t - \tau)\frac{\partial w_0}{\partial \tau}\,d\tau.$$

11.14 Fourier Transformations Applied to Partial Differential Equations

We shall now discuss a further method for solving partial differential equations, namely, by the application of one of the Fourier transformations. This refers to problems such that initial or boundary data are given on the positive half-axis—then the Fourier cosine or sine transformation (Sec. 10.10) may be appropriate—or on the entire axis—in which case we may use the Fourier

[14]JEAN MARIE CONSTANT DUHAMEL (1797-1872), French mathematician.

transformation in Sec. 10.11. We discuss these techniques in terms of typical applications.

EXAMPLE 1. A heat problem on the x-axis

Find the temperature $u(x, t)$ in a laterally insulated homogeneous rod of constant cross section extending from $x = -\infty$ to $x = \infty$, for time $t > 0$, assuming that the given initial temperature is

(1) $$u(x, 0) = f(x) \qquad (-\infty < x < \infty),$$

and for all $t \geq 0$ the solution and its x derivative satisfy

(2) $$u(x, t) \to 0, \qquad u_x(x, t) \to 0 \qquad \text{as } |x| \to \infty.$$

As a particular case, find $u(x, t)$ when

(3) $\qquad f(x) = U_0 = const \quad$ if $|x| < 1 \qquad$ and $\qquad f(x) = 0 \quad$ if $|x| > 1.$

Solution. We have to solve the **heat equation**

(4) $$u_t = c^2 u_{xx}$$

subject to conditions (1), (2). Our strategy is to take the Fourier transform with respect to x and then solve the resulting ordinary differential equation in t. The details are as follows.

Let $\hat{u} = \mathcal{F}(u)$ denote the Fourier transform of u, *regarded as a function of x*. From (10) in Sec. 10.11 we see that (4) gives

$$\mathcal{F}(u_t) = c^2 \mathcal{F}(u_{xx}) = c^2(-w^2)\mathcal{F}(u) = -c^2 w^2 \hat{u}.$$

On the left, assuming that we may interchange the order of differentiation and integration,

$$\mathcal{F}(u_t) = \frac{1}{\sqrt{2\pi}} \int_{-\infty}^{\infty} u_t e^{-iwx} \, dx = \frac{1}{\sqrt{2\pi}} \frac{\partial}{\partial t} \int_{-\infty}^{\infty} u e^{-iwx} \, dx = \frac{\partial \hat{u}}{\partial t}.$$

Thus

$$\frac{\partial \hat{u}}{\partial t} = -c^2 w^2 \hat{u}.$$

Since this equation involves only a derivative with respect to t but none with respect to w, this is a first-order **ordinary** differential equation with t as the independent variable and w as a parameter. By separating variables (Sec. 1.2) we get the general solution

$$\hat{u}(w, t) = C(w)e^{-c^2 w^2 t},$$

with the arbitrary "constant" $C(w)$ depending on the parameter w. The initial condition (1) gives $\hat{u}(w, 0) = C(w) = \hat{f}(w) = \mathcal{F}(f)$. Our intermediate result is

$$\hat{u}(w, t) = \hat{f}(w)e^{-c^2 w^2 t}.$$

The inversion formula (7), Sec. 10.11, now gives the solution

(5) $$u(x, t) = \frac{1}{\sqrt{2\pi}} \int_{-\infty}^{\infty} \hat{f}(w)e^{-c^2 w^2 t}e^{iwx} \, dw.$$

In this we may insert the Fourier transform

$$\hat{f}(w) = \frac{1}{\sqrt{2\pi}} \int_{-\infty}^{\infty} f(v)e^{-iwv} \, dv.$$

Assuming that we may invert the order of integration, we then obtain

$$u(x, t) = \frac{1}{2\pi} \int_{-\infty}^{\infty} f(v) \left[\int_{-\infty}^{\infty} e^{-c^2 w^2 t}e^{i(wx - wv)} \, dw \right] dv.$$

By the Euler formula (3), Sec. 10.11, the integrand of the inner integral equals

$$e^{-c^2w^2t} \cos (wx - wv) + ie^{-c^2w^2t} \sin (wx - wv).$$

This shows that its imaginary part is an odd function of w, so that the integral[15] of this part is 0, and the real part is even, so that its integral is twice the integral from 0 to ∞:

$$u(x, t) = \frac{1}{\pi} \int_{-\infty}^{\infty} f(v) \left[\int_0^{\infty} e^{-c^2w^2t} \cos (wx - wv) \, dw \right] dv.$$

This agrees with (9), Sec. 11.6 and leads to the further formulas (11) and (13) in Sec. 11.6.

EXAMPLE 2. Problem in Example 1 solved by the method of convolution

Solve the heat problem in Example 1 by the method of convolution.

Solution. The beginning is as before and leads to (5), that is,

$$(5) \qquad u(x, t) = \frac{1}{\sqrt{2\pi}} \int_{-\infty}^{\infty} \hat{f}(w) e^{-c^2w^2t} e^{iwx} \, dw.$$

Now comes the crucial idea. We recognize that this is of the form (13) in Sec. 10.11, that is,

$$(6) \qquad u(x, t) = (f * g)(x) = \int_{-\infty}^{\infty} \hat{f}(w) \hat{g}(w) e^{iwx} \, dw$$

where

$$(7) \qquad \hat{g}(w) = \frac{1}{\sqrt{2\pi}} e^{-c^2w^2t}.$$

Since, by the definition of convolution [(11), Sec. 10.11],

$$(8) \qquad (f * g)(x) = \int_{-\infty}^{\infty} f(p) g(x - p) \, dp,$$

as our next and last step we must determine the inverse Fourier transform g of $\hat{g}$. For this we can use formula 9 in Table III of Sec. 10.12 (which was derived in Example 2 of Sec. 10.11),

$$\mathscr{F}(e^{-ax^2}) = \frac{1}{\sqrt{2a}} e^{-w^2/4a}$$

with a suitable a. With $c^2t = 1/4a$ or $a = 1/4c^2t$, using (7) we obtain

$$\mathscr{F}(e^{-x^2/4c^2t}) = \sqrt{2c^2t} \, e^{-c^2w^2t} = \sqrt{2c^2t} \, \sqrt{2\pi} \, \hat{g}(w).$$

Hence $\hat{g}$ has the inverse

$$\frac{1}{\sqrt{2c^2t}\sqrt{2\pi}} e^{-x^2/4c^2t}.$$

Replacing x with $x - p$ and substituting this into (8) we finally have

$$(9) \qquad u(x, t) = (f * g)(x) = \frac{1}{2c\sqrt{\pi t}} \int_{-\infty}^{\infty} f(p) \exp \left\{ - \frac{(x - p)^2}{4c^2t} \right\} dp.$$

This solution formula of our problem agrees with (11) in Sec. 11.6. We wrote $(f * g)(x)$, without indicating the parameter t with respect to which we did not integrate.

---

[15]Actually, the principal part of the integral; cf. Sec. 15.3.

EXAMPLE 3. Fourier sine transform applied to the heat equation

Find a solution formula in Example 1, assuming that the rod extends from 0 to ∞, the initial temperature is

$$u(x, 0) = f(x) \qquad (0 \leqq x < \infty)$$

and at the left end we have the boundary condition

$$u(0, t) = 0 \qquad (t \geqq 0).$$

Solution. Instead of the Fourier transformation we may now apply the Fourier sine transformation (Sec. 10.10), since x varies from 0 to ∞. Proceeding as in Example 1, we obtain from the heat equation and (9b), Sec. 10.10, since $f(0) = u(0, 0) = 0$,

$$\mathcal{F}_s(u_t) = \frac{\partial \hat{u}_s}{\partial t} = c^2 \mathcal{F}_s(u_{xx}) = -c^2 w^2 \mathcal{F}_s(u) = -c^2 w^2 \hat{u}_s(w, t).$$

The solution of this first-order ordinary differential equation is

$$\hat{u}_s(w, t) = C(w) e^{-c^2 w^2 t}.$$

From the initial condition $u(x, 0) = f(x)$ we have $\hat{u}_s(w, 0) = \hat{f}_s(w) = C(w)$. Hence

$$\hat{u}_s(w, t) = \hat{f}_s(w) e^{-c^2 w^2 t}.$$

Taking the inverse sine transform and substituting

$$f_s(w) = \sqrt{\frac{2}{\pi}} \int_0^\infty f(p) \sin wp \, dp,$$

we obtain the desired solution formula

$$(10) \qquad u(x, t) = \frac{2}{\pi} \int_0^\infty \int_0^\infty f(p) \sin wp \, e^{-c^2 w^2 t} \sin wx \, dp \, dx.$$

EXAMPLE 4. Wave equation on an infinite interval. D'Alembert's solution

Solve the wave equation

$$u_{tt} = c^2 u_{xx} \qquad (-\infty < x < \infty, \quad t > 0)$$

subject to the conditions

$$u(x, 0) = f(x) \qquad \text{(given initial deflection)}$$

$$u_t(x, 0) = 0 \qquad \text{(initial speed zero)}$$

$$u \to 0, \qquad u_x \to 0 \qquad \text{as } |x| \to \infty \text{ for all } t,$$

where f and g are assumed to have a Fourier transform.

Solution. We transform with respect to x, writing $\hat{u} = \mathcal{F}(u)$. Using (10), Sec. 10.11, we get from the wave equation

$$\mathcal{F}(u_{tt}) = \hat{u}_{tt} = c^2 \mathcal{F}(u_{xx}) = -c^2 w^2 \hat{u}.$$

Thus

$$\hat{u}_{tt} + c^2 w^2 \hat{u} = 0.$$

Since no w-derivatives occur, this is a second-order **ordinary** differential equation whose coefficient $c^2 w^2$ is "constant" (that is, independent of t). A general solution is

$$u(w, t) = A(w) \cos cwt + B(w) \sin cwt.$$

For $t = 0$, since $\mathcal{F}\{u(x, 0)\} = \hat{u}(w, 0)$, from the initial conditions we obtain

$$\hat{u}(w, 0) = A(w) = \hat{f}(w)$$

$$\hat{u}_t(w, 0) = cwB(w) = 0.$$

Thus,

$$\hat{u}(w, t) = \hat{f}(w) \cos cwt.$$

We want u. This needs an idea. The idea is to express the cosine in terms of exponential functions and then apply the shifting formula

(11)
$$\boxed{\mathcal{F}\{f(x - a)\} = e^{-iwa}\mathcal{F}\{f(x)\},}$$

which follows directly from the definition of the Fourier transformation by setting $x - a = p$, $x = p + a$, $dx = dp$, so that

$$\mathcal{F}\{f(x - a)\} = \frac{1}{\sqrt{2\pi}} \int_{-\infty}^{\infty} f(x - a)e^{-iwx}\, dx = \frac{1}{\sqrt{2\pi}} \int_{-\infty}^{\infty} f(p)e^{-iw(p+a)}\, dp$$

$$= \frac{1}{\sqrt{2\pi}} e^{-iwa} \int_{-\infty}^{\infty} f(p)e^{-iwp}\, dp.$$

Thus

$$\hat{f}(w) \cos cwt = \tfrac{1}{2} \hat{f}(w)[e^{icwt} + e^{-icwt}]$$

has the inverse Fourier transformation

(12)
$$u(x, t) = \tfrac{1}{2}[f(x - ct) + f(x + ct)].$$

This is **d'Alembert's solution** (6), Sec. 11.4. ∎

This is the end of Chap. 11, in which we concentrated on the most important partial differential equations in physics and engineering. This is also the end of Part C on Fourier analysis and partial differential equations. In Part D (Chaps. 12—17), we again turn to an area of a different nature, **complex analysis,** which is also highly important to the engineer, as our examples and problems will show.

Review Problems for Chapter 11

1. What is the superposition principle? To what kind of equations does it apply?
2. What physical law gave the wave equation, and why does this equation involve the *second* partial derivatives with respect to t? What t-derivative does the heat equation involve?
3. Verify that $u = x^2 + 4t^2$ and $u = x^3 + 3xt^2$ are solutions of the one-dimensional wave equation with a suitable c.
4. What do we mean by d'Alembert's solution of the wave equation?
5. What are the eigenfunctions of a vibrating string? Of a vibrating membrane?
6. In the separation of the one-dimensional wave equation we got only trigonometric functions, whereas for the heat equation we also got an exponential function. What was the reason?
7. How does the exponential function in Prob. 6 change if c^2 in the equation is increased? Is your answer understandable from the physical point of view?
8. How many initial conditions can be given in the case of the wave equation? In the case of the heat equation?
9. Verify that $u = x^4 - 6x^2y^2 + y^4$ and $u = \sin x \sinh y$ are solutions of Laplace's equation.
10. Verify that $u = (x^2 - y^2)/(x^2 + y^2)^2$ and $u = 2x/(x^2 + y^2)^2$ are solutions of Laplace's equation.

11. Verify that $u = x^2 + y^2 - 2z^2$ is a solution of Laplace's equation.

12. Verify that $u = x^2 + y^2 + z^2$ satisfies Poisson's equation.

13. What do we mean by an elliptic equation? What is the most important elliptic equation?

14. What is a hyperbolic equation? A parabolic equation? Give examples.

15. Certain simpler partial differential equations can be solved by methods for ordinary differential equations. Explain. Give examples.

16. What is the error function? In what connection did it appear in this chapter?

17. Why did Bessel functions and Legendre polynomials appear in this chapter?

18. Why did Fourier series play a basic role in this chapter, although the given functions that were physically of interest were not periodic?

19. For what reasons did the Fourier integral occur in this chapter?

20. In Chap. 5, the subsidiary equation was an *algebraic* equation. Why do we obtain an ordinary *differential* equation in solving a partial differential equation by a transform method (an operational method)?

21. Solve $u_{xx} + 9u = 0$.

22. Solve $u_{xy} + u_x + x + y + 1 = 0$.

23. Solve $u_{yy} + 3u_y - 4u = 8$.

24. Find all solutions $u(x, y) = F(x)G(y)$ of Laplace's equation in two variables.

25. Find solutions of $yu_x = xu_y$ by separating variables.

Find the motion of the vibrating string of length π and $c^2 = T/\rho = 1$ starting with initial velocity 0 and deflection

26. $f(x) = \sin x - \frac{1}{2}\sin 2x$ **27.** $f(x) = -0.1 \sin 3x$

28. $f(x) = \pi/2 - |x - \pi/2|$ **29.** $f(x) = \sin^3 x$

30. $f(x) = x$ if $0 < x < \pi/3$, $\quad f(x) = (\pi - x)/2$ if $\pi/3 < x < \pi$

Forced oscillations of an elastic string. The corresponding equation is

$$(1) \qquad\qquad u_{tt} = c^2 u_{xx} + \frac{P}{\rho}.$$

Consider and solve this equation as follows.

31. Show that the forced vibrations of an elastic string are governed by (1), where $P(x, t)$ is the external force per unit length acting perpendicular to the string.

32. Assume the external force to be sinusoidal, say, $P = A\rho \sin \omega t$. Show that

$$P/\rho = A \sin \omega t = \sum_{n=1}^{\infty} k_n(t) \sin \frac{n\pi x}{L}$$

where $k_n(t) = (2A/n\pi)(1 - \cos n\pi) \sin \omega t$; consequently $k_n = 0$ (n even), and $k_n = (4A/n\pi) \sin \omega t$ (n odd). Furthermore, show that substitution of

$$u(x, t) = \sum_{n=1}^{\infty} G_n(t) \sin \frac{n\pi x}{L} \quad \text{into } u_{tt} = c^2 u_{xx} \text{ gives } \quad \ddot{G}_n + \lambda_n^2 G_n = 0$$

where $\lambda = cn\pi/L$.

33. Show that by substituting the expressions for u and P/ρ in Prob. 32 into (1) we obtain

$$\ddot{G}_n + \lambda_n^2 G_n = \frac{2A}{n\pi} (1 - \cos n\pi) \sin \omega t, \qquad \lambda_n = \frac{cn\pi}{L} .$$

Show that if $\lambda_n^2 \neq \omega^2$, the solution is

$$G_n(t) = B_n \cos \lambda_n t + B_n^* \sin \lambda_n t + \frac{2A(1 - \cos n\pi)}{n\pi(\lambda_n^2 - \omega^2)} \sin \omega t.$$

34. Determine B_n and B_n^* in Prob. 33 so that u satisfies the initial conditions $u(x, 0) = f(x)$, $u_t(x, 0) = 0$.

35. Show that in the case of **resonance** ($\lambda_n = \omega$),

$$G_n(t) = B_n \cos \omega t + B_n^* \sin \omega t - \frac{A}{n\pi\omega} (1 - \cos n\pi)t \cos \omega t.$$

Using the indicated transformations, solve the following equations.

36. $u_{xx} - 2u_{xy} + u_{yy} = 0$ $\quad (v = x, \quad z = x + y)$

37. $u_{xx} + 2u_{xy} + u_{yy} = 0$ $\quad (v = x, \quad z = x - y)$

38. $2u_{xx} + 5u_{xy} + 2u_{yy} = 0$ $\quad (v = 2x - y, \quad z = 2y - x)$

39. $xu_{xy} = yu_{yy} + u_y$ $\quad (v = x, \quad z = xy)$

40. $u_{xx} + u_{xy} = 2u_{yy}$ $\quad (v = x + y, \quad z = 2x - y)$

Find the temperature distribution in a laterally insulated thin copper bar ($c^2 = K/\sigma\rho = 1.158$ cm^2/sec), 50 cm long and of constant cross section whose endpoints at $x = 0$ and $x = 50$ are kept at 0 °C and whose initial temperature is

41. $f(x) = \sin (\pi x/50)$ $\qquad$ **42.** $f(x) = 100 \sin (\pi x/25)$

43. $f(x) = \sin^3 (\pi x/10)$ $\qquad$ **44.** $f(x) = x(50 - x)$

45. $f(x) = x$ if $0 < x < 25$, $\quad f(x) = 50 - x$ if $25 < x < 50$

Recall from Sec. 11.5 that for adiabatic boundary conditions the temperature $u(x, t)$ in a laterally insulated rod of length L is

$$u(x, t) = A_0 + \sum_{n=1}^{\infty} A_n \cos \frac{n\pi x}{L} \exp\left[- \left(\frac{cn\pi}{L}\right)^2 t \right].$$

Assuming $L = \pi$ and $c = 1$, find from this the solution satisfying the initial condition

46. $f(x) = 30x^2$ $\qquad$ **47.** $f(x) = 95 \cos 2x$

48. $f(x) = \begin{cases} 10 & \text{if} \quad 0 < x < \pi/2 \\ 0 & \text{if} \quad \pi/2 < x < \pi \end{cases}$ $\qquad$ **49.** $f(x) = \begin{cases} 4x & \text{if} \quad 0 < x < \pi/2 \\ 4(\pi - x) & \text{if} \quad \pi/2 < x < \pi \end{cases}$

50. Using Table A2 in Appendix 4 and linear interpolation and the explicitly given coefficients in Example 1 of Sec. 11.10, show that $f(0.5) \approx 0.7498$. What is the exact value?

Show that the following membranes of area 1 with $c^2 = 1$ have the frequencies of the fundamental mode as given (4-decimal values). Compare.

51. Circle: $\alpha_1/2\sqrt{\pi} = 0.6784$ $\qquad$ **52.** Square: $1/\sqrt{2} = 0.7071$

53. Quadrant of circle: $\alpha_{12}/4\sqrt{\pi} = 0.7244$ ($\alpha_{12} = 5.13562 = $ first positive zero of J_2)

54. Semicircle: $3.832/\sqrt{8\pi} = 0.7644$ $\qquad$ **55.** Rectangle (sides 1:2): $\sqrt{5/8} = 0.7906$

Summary of Chapter 11
Partial Differential Equations

Whereas *ordinary* differential equations (Chaps. 1—5) appear as models of simpler engineering problems involving a single independent variable, problems in which one has two or more independent variables (space variables, or time t and one or several space variables) lead to *partial* differential equations. Hence the importance of these equations to the engineer and physicist can hardly be overestimated.

In this chapter we were mainly concerned with the most important partial differential equations of physics and engineering, namely:

(1) $u_{tt} = c^2 u_{xx}$ **One-dimensional wave equation**
(Secs. 11.2—11.4)

(2) $u_{tt} = c^2(u_{xx} + u_{yy})$ **Two-dimensional wave equation**
(Secs. 11.7—11.10)

(3) $u_t = c^2 u_{xx}$ **One-dimensional heat equation**
(Secs. 11.5, 11.6)

(4) $\nabla^2 u = u_{xx} + u_{yy} = 0$ **Two-dimensional Laplace equation**
(Secs. 11.5, 11.9)

(5) $\nabla^2 u = u_{xx} + u_{yy} + u_{zz} = 0$ **Three-dimensional Laplace equation**
(Secs. 11.11, 11.12).

(1) and (2) are hyperbolic, (3) is parabolic, (4) and (5) are elliptic. (Cf. Problems for Sec. 11.4.)

In practice, one is interested in obtaining the solution of such an equation in a given region satisfying given additional conditions, such as **initial conditions** (conditions at time $t = 0$) or **boundary conditions** (prescribed values of the solution u or some of its derivatives on the boundary surface S, or boundary curve C, of the region) or both. For (1) and (2) one prescribes two initial conditions (initial displacement and initial velocity). For (3) one prescribes the initial temperature distribution. For (4) or (5) one prescribes a boundary condition and calls the resulting problem

> **Dirichlet problem** if u is prescribed on S,
> **Neumann problem** if $u_n = \partial u/\partial n$ is prescribed on S,
> **Mixed problem** if u is prescribed on one part of S and u_n on the other.

Cf. Sec. 11.5.

A general method for solving such problems is the method of **separating variables** or **product method,** in which one assumes solutions in the form of products of functions each depending on one variable only.

Thus equation (1) is solved by setting (Sec. 11.3)

$$u(x, t) = F(x)G(t);$$

similarly for (3) in Sec. 11.5. Substitution into the given equation yields *ordinary* differential equations for F and G, and from these one gets infinitely many solutions $F = F_n$ and $G = G_n$ such that the corresponding functions

$$u_n(x, t) = F_n(x)G_n(t)$$

are solutions of the partial differential equations satisfying the given boundary conditions. These are the **eigenfunctions** of the problem, and the corresponding **eigenvalues** determine the frequency of the vibration (or the rapidity of the decrease of temperature in the case of the heat equation, etc.). To satisfy also the initial condition (or conditions), one must consider infinite series of the u_n, whose coefficients turn out to be the Fourier coefficients of the functions f and g representing the given initial conditions (Secs. 11.3, 11.5). Hence **Fourier series** (and *Fourier integrals*) are of basic importance here (Secs. 11.3, 11.5, 11.6, 11.8).

   Steady-state or **stationary problems** are those in which the solution does not depend on time t. For these, the heat equation $u_t = c^2 \nabla^2 u$ becomes the Laplace equation (Sec. 11.5).

   Before solving an initial or boundary value problem, one often transforms the equation into coordinates in which the boundary of the region considered is given by simple formulas. Thus in polar coordinates given by $x = r \cos \theta$, $y = r \sin \theta$, the **Laplacian** becomes (Sec. 11.9)

$$(6) \qquad \nabla^2 u = u_{rr} + \frac{1}{r} u_r + \frac{1}{r^2} u_{\theta\theta},$$

and for spherical coordinates see Sec. 11.11. If one now applies the separation of variables, one gets **Bessel's equation** from (2) and (6) (vibrating circular membrane, Sec. 11.10) and **Legendre's equation** from (5) transformed to spherical coordinates (Sec. 11.12).

   Operational methods (Laplace transformation, Fourier transformations) are helpful in solving partial differential equations in infinite regions (Secs. 11.12, 11.13), in particular if the coefficients of the equation are constant or depend on one variable only.

PART D
COMPLEX ANALYSIS

Many engineering problems may be treated and solved by methods involving complex numbers and complex functions. Roughly speaking, these problems can be subdivided into two large classes. The first class consists of "elementary problems" for which the knowledge of complex numbers gained in college algebra and calculus is sufficient. For example, many applications in connection with models of electric circuits and mechanical vibrating systems are of this type.

The second class of problems requires a detailed knowledge of the theory of complex analytic functions—**"complex function theory"** or **"complex analysis,"** for short—and of the powerful and elegant methods used in this branch of mathematics. Interesting problems in the theory of heat, in fluid dynamics and in electrostatics belong to this category.

The next six chapters (12—17) will be devoted to complex analysis and its applications. We shall see that the importance of complex analytic functions in engineering mathematics has the following three main roots.

1. *The real and imaginary parts of an analytic function are solutions of Laplace's equation in two independent variables. Consequently, two-dimensional potential problems can be treated by methods developed in connection with analytic functions.*
2. *Many complicated real and complex integrals that occur in applications can be evaluated by methods of complex integration.*
3. *The majority of nonelementary functions appearing in engineering mathematics are analytic functions, and the consideration of these functions for complex values of the independent variable leads to a much deeper and more detailed knowledge of their properties.*

Chapter 12

Complex Numbers.
Complex Analytic Functions

Complex numbers and the complex plane are discussed in Secs. 12.1—12.3. Complex analysis is concerned with complex analytic functions, as defined in Sec. 12.4. In Sec. 12.5 we explain a check for analyticity based on the so-called Cauchy–Riemann equations. The latter are related to Laplace's equation. In the remaining sections of Chap. 12 we study the most important elementary complex functions (exponential function, trigonometric functions, etc.), which generalize familiar real functions known from calculus. In Sec. 12.9 we discuss mappings given by these functions. (In Chaps. 16 and 17 we extend this discussion and show applications to potential problems.)

Prerequisites for this chapter: elementary calculus.
References: Appendix 1, Part D.
Answers to problems: Appendix 2

12.1 Complex Numbers

It was observed early in history that there are equations which are not satisfied by any real number. Examples are

$$x^2 = -3 \qquad \text{or} \qquad x^2 - 10x + 40 = 0.$$

This led to the invention of complex numbers.[1]

Definition

A **complex number** z is an ordered pair (x, y) of real numbers x, y and we write

$$z = (x, y).$$

We call x the **real part** of z and y the **imaginary part** of z and write

---

[1]First to use complex numbers for this purpose was the Italian mathematician GIROLAMO CARDANO (1501—1576), who found the formula for solving cubic equations. The term "complex number" was introduced by the great German mathematician CARL FRIEDRICH GAUSS (cf. the footnote in Sec. 4.4), who also paved the way for a general use of complex numbers.

$$\text{Re } z = x, \qquad \text{Im } z = y.$$

Example: Re $(4, -3) = 4$ and Im $(4, -3) = -3$. Furthermore, we define two complex numbers $z_1 = (x_1, y_1)$ and $z_2 = (x_2, y_2)$ to be **equal** if and only if their real parts are equal and their imaginary parts are equal:

$$z_1 = z_2 \qquad \text{if and only if} \qquad x_1 = x_2 \text{ and } y_1 = y_2.$$

Addition of complex numbers $z_1 = (x_1, y_1)$ and $z_2 = (x_2, y_2)$ is defined by

$$(1) \qquad z_1 + z_2 = (x_1, y_1) + (x_2, y_2) = (x_1 + x_2, \quad y_1 + y_2).$$

Multiplication is defined by

$$(2) \qquad z_1 z_2 = (x_1, y_1)(x_2, y_2) = (x_1 x_2 - y_1 y_2, \quad x_1 y_2 + x_2 y_1).$$

We shall say more about these arithmetical operations and discuss examples below, but we first want to introduce a much more convenient form of writing complex numbers and then learn how to "see" these numbers by plotting them as points in the plane.

Representation in the Form $z = x + iy$

A complex number whose imaginary part is zero is of the form $(x, 0)$. For such numbers we simply have from (1) and (2)

$$(x_1, 0) + (x_2, 0) = (x_1 + x_2, 0)$$

and

$$(x_1, 0)(x_2, 0) = (x_1 x_2, 0),$$

as for real numbers. This suggests that we identify $(x, 0)$ with the real number x. Hence the complex number system is an *extension* of the real number system.

The complex number $(0, 1)$ is denoted by i,

$$i = (0, 1),$$

and is called the **imaginary unit.** We show that it has the property

$$(3) \qquad \boxed{i^2 = -1.}$$

Indeed, from (2) we have $i^2 = (0, 1)(0, 1) = (-1, 0) = -1$.
Furthermore, for every real y we obtain from (2)

$$iy = (0, 1)(y, 0) = (0, y).$$

Combining this with the above $x = (x, 0)$ and using (1), that is,

$$(x, y) = (x, 0) + (0, y),$$

we see that we can write every complex number $z = (x, y)$ in the form

$$\boxed{z = x + iy}$$

or $z = x + yi$. This is done in practice almost exclusively.[2]

EXAMPLE 1. Complex numbers, their real and imaginary parts

$$z = (4, -3) = 4 - 3i, \quad \text{Re } (4 - 3i) = 4, \quad \text{Im } (4 - 3i) = -3$$

$$z = (-\tfrac{1}{2}, 0) = -\tfrac{1}{2}, \quad \text{Re } (-\tfrac{1}{2}) = -\tfrac{1}{2}, \quad \text{Im } (-\tfrac{1}{2}) = 0$$

$$z = (0, \pi) \;\; = \pi i, \quad \text{Re } (\pi i) = 0, \quad \text{Im } (\pi i) = \pi$$ ∎

Complex Plane

This is a geometrical representation of complex numbers as points in the plane. It is of great importance in applications. The idea is quite simple and natural. We choose two perpendicular coordinate axes, the horizontal x-axis, called the **real axis,** and the vertical y-axis, called the **imaginary axis.** On both axes we choose the same unit of length (Fig. 295). This is called a **Cartesian coordinate system.**[3] We now plot $z = (x, y) = x + iy$ as the point P with coordinates x, y. The xy-plane in which the complex numbers are represented in this way is called the **complex plane** or *Argand diagram.*[4] Figure 296 shows an example.

Instead of saying "the point represented by z in the complex plane" we say briefly and simply ***"the point z in the complex plane."*** This will cause no misunderstandings.

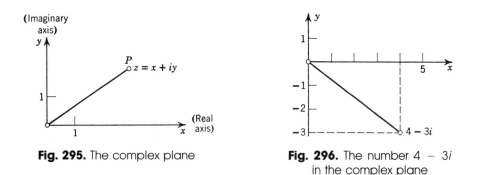

Fig. 295. The complex plane

Fig. 296. The number $4 - 3i$ in the complex plane

---

[2]Electrical engineers often write j for i, to reserve the letter i for the electrical current.

[3]Named after the French philosopher and mathematician RENATUS CARTESIUS (latinized for RENÉ DESCARTES (1596—1650), who invented analytic geometry. His basic work *Géométrie* appeared in 1637, as an appendix to his *Discours de la méthode.*

[4]JEAN ROBERT ARGAND (1768—1822), French mathematician, was born in Geneva and later became a librarian in Paris. His paper on the complex plane appeared in 1806, nine years after a similar memoir by the Norwegian mathematician CASPAR WESSEL (1745—1818), who was surveyor of the Danish Academy of Sciences.

Arithmetic Operations

We can now make use of the notation $z = x + iy$ and of the complex plane.

Addition. The sum of $z_1 = x_1 + iy_1$ and $z_2 = x_2 + iy_2$ can now be written

$$(4) \qquad\qquad z_1 + z_2 = (x_1 + x_2) + i(y_1 + y_2).$$

Example: $(5 + i) + (1 + 3i) = 6 + 4i$. We see that addition of complex numbers is in accordance with the *"parallelogram law"* by which forces are added in mechanics (Fig. 297).

Subtraction is defined to be the inverse operation of addition. That is, the difference $z = z_1 - z_2$ is the complex number z for which $z_1 = z + z_2$. Obviously (cf. Fig. 298)

$$(5) \qquad\qquad z_1 - z_2 = (x_1 - x_2) + i(y_1 - y_2).$$

Example: $(5 + i) - (1 + 3i) = 4 - 2i$.

Multiplication. The product $z_1 z_2$ in (2) can now be written

$$(6) \quad z_1 z_2 = (x_1 + iy_1)(x_2 + iy_2) = (x_1 x_2 - y_1 y_2) + i(x_1 y_2 + x_2 y_1).$$

This is easy to remember since it is obtained formally by the rules of arithmetic for real numbers and using (3), that is, $i^2 = -1$.

    Example: $(5 + i)(1 + 3i) = 5 + 15i + i + 3i^2 = 2 + 16i$.

Division is defined to be the inverse operation of multiplication. That is, the quotient $z = z_1/z_2$ is the complex number $z = x + iy$ for which

$$(7) \qquad\qquad z_1 = z z_2 = (x + iy)(x_2 + iy_2) \qquad\qquad (z_2 \neq 0).$$

We show that for $z_2 \neq 0$ the quotient $z = x + iy = z_1/z_2$ is given by

$$(8^*) \qquad\qquad x = \frac{x_1 x_2 + y_1 y_2}{x_2^2 + y_2^2}, \qquad y = \frac{x_2 y_1 - x_1 y_2}{x_2^2 + y_2^2} \qquad (z_2 \neq 0).$$

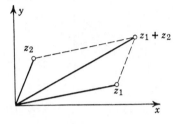

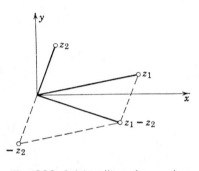

Fig. 297. Addition of complex
numbers

Fig. 298. Subtraction of complex
numbers

The *practical rule* for getting (8*) is the multiplication of both the numerator and the denominator of the quotient z_1/z_2 by $x_2 - iy_2$ and simplification:

$$(8) \qquad z = \frac{x_1 + iy_1}{x_2 + iy_2} = \frac{(x_1 + iy_1)(x_2 - iy_2)}{(x_2 + iy_2)(x_2 - iy_2)} = \frac{x_1 x_2 + y_1 y_2}{x_2^2 + y_2^2} + i\frac{x_2 y_1 - x_1 y_2}{x_2^2 + y_2^2}.$$

Example: if $z_1 = 9 - 8i$ and $z_2 = 5 + 2i$, then

$$\frac{9 - 8i}{5 + 2i} = \frac{(9 - 8i)(5 - 2i)}{(5 + 2i)(5 - 2i)} = \frac{45 - 18i - 40i - 16}{25 + 4} = 1 - 2i.$$

The reader may check this result by showing that

$$zz_2 = (1 - 2i)(5 + 2i) = 9 - 8i = z_1.$$

A proof of (8*) runs as follows. From (6) we see that (7) can be written

$$x_1 + iy_1 = (x_2 x - y_2 y) + i(y_2 x + x_2 y).$$

By the definition of equality the real parts and the imaginary parts on both sides must be equal:

$$x_1 = x_2 x - y_2 y$$

$$y_1 = y_2 x + x_2 y.$$

This is a system of two linear equations in the unknowns x and y. Assuming that x_2 and y_2 are not both zero (briefly written $z_2 \neq 0$), we obtain the unique solution (8*). ∎

Properties of the Arithmetic Operations

From the familiar laws for *real* numbers we obtain for any complex numbers z_1, z_2, z_3, z the following laws (where $-z = -x - iy$):

$$
\left.
\begin{aligned}
z_1 + z_2 &= z_2 + z_1 \\
z_1 z_2 &= z_2 z_1
\end{aligned}
\right\} \quad \textit{(Commutative laws)}
$$

$$
\left.
\begin{aligned}
(z_1 + z_2) + z_3 &= z_1 + (z_2 + z_3) \\
(z_1 z_2)z_3 &= z_1(z_2 z_3)
\end{aligned}
\right\} \quad \textit{(Associative laws)}
$$

(9)

$$z_1(z_2 + z_3) = z_1 z_2 + z_1 z_3 \qquad \textit{(Distributive law)}$$

$$0 + z = z + 0 = z$$

$$z + (-z) = (-z) + z = 0$$

$$z \cdot 1 = z$$

Complex Conjugate Numbers

Let $z = x + iy$ be any complex number. Then $x - iy$ is called the **conjugate** of z and is denoted by $\bar{z}$. Thus,

$$z = x + iy, \qquad \bar{z} = x - iy.$$

Example: the conjugate of $z = 5 + 2i$ is $\bar{z} = 5 - 2i$ (Fig. 299).

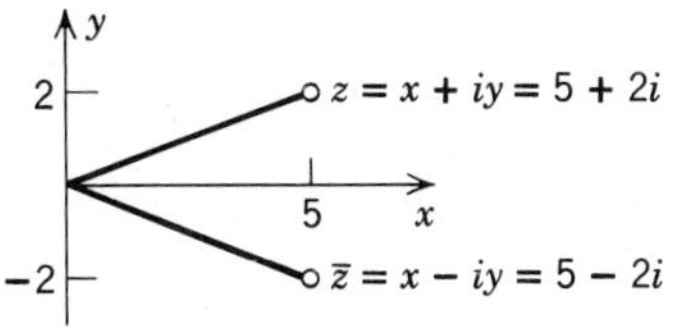

Fig. 299. Complex conjugate numbers

Conjugates are useful since $z\bar{z} = x^2 + y^2$ is real, a property we used in the above division. Moreover, addition and subtraction yields $z + \bar{z} = 2x$, $z - \bar{z} = 2iy$, so that we can express the real part and the imaginary part of z by the important formulas

(10) $\qquad \text{Re } z = x = \dfrac{1}{2}(z + \bar{z}), \qquad \text{Im } z = y = \dfrac{1}{2i}(z - \bar{z}).$

Example: if $z = 6 - 5i$, then we have $\bar{z} = 6 + 5i$ and from (10) we obtain $x = \frac{1}{2}(6 - 5i + 6 + 5i) = 6$ and $y = -5$.

z is real if and only if $y = 0$, hence $\bar{z} = z$ by (10).

z is said to be **pure imaginary** if and only if $x = 0$, hence $\bar{z} = -z$. Then z corresponds to a point on the imaginary axis.

Working with conjugates is easy, since we have

(11)
$$\overline{(z_1 + z_2)} = \bar{z}_1 + \bar{z}_2, \qquad \overline{(z_1 - z_2)} = \bar{z}_1 - \bar{z}_2,$$

$$\overline{(z_1 z_2)} = \bar{z}_1 \bar{z}_2, \qquad \overline{\left(\frac{z_1}{z_2}\right)} = \frac{\bar{z}_1}{\bar{z}_2}.$$

In this section we were mainly concerned with complex numbers, their arithmetic operations and their representation as points in the complex plane, a key idea for great progress in early complex analysis, conceptually and technically. In the complex plane we used rectangular xy-coordinates. In the next section we discuss the use of **polar coordinates** in the complex plane and situations in which polar coordinates are advantageous.

Problems for Sec. 12.1

1. (Powers of the imaginary unit) Show that

(12) $\qquad i^2 = -1, \quad i^3 = -i, \quad i^4 = 1, \quad i^5 = i, \cdots$

$$\frac{1}{i} = -i, \quad \frac{1}{i^2} = -1, \quad \frac{1}{i^3} = i, \cdots.$$

Let $z_1 = 3 + 4i$ and $z_2 = 5 - 2i$. Find (in the form $x + iy$)

2. $(z_1 - z_2)^2$ **3.** z_1/z_2 **4.** $1/z_1{}^2$ **5.** $z_2/2z_1$

Find, in the form $x + iy$,

6. $(7 - 3i) - (-2 + 4i)$ **7.** $(2 + 4i)^2$

8. $(3 + 5i)(3 - 5i)$ **9.** $(5 + 2i)i$

10. $(1 - i)^4$ **11.** $\dfrac{11 + 2i}{4 + 3i}$ **12.** $\dfrac{52.5 - 12.5i}{3 - i}$ **13.** $\dfrac{101}{10 - i}$

Find:

14. $\operatorname{Re} \dfrac{1}{2 + i}$ **15.** $\operatorname{Im} \dfrac{2 + i}{3 + 4i}$ **16.** $\operatorname{Re} \dfrac{(1 + i)^2}{3 + 2i}$ **17.** $\operatorname{Im} \dfrac{2 - i}{4 - 3i}$

18. $\operatorname{Re} z^3$, $(\operatorname{Re} z)^3$ **19.** $\operatorname{Im} z^4$, $(\operatorname{Im} z^2)^2$ **20.** $\operatorname{Im} (1/z)$ **21.** $\operatorname{Re} (z/\bar{z})$

Prove:

22. The distributive law in (9).

23. z is pure imaginary if and only if $\bar{z} = -z$.

24. $\overline{iz} = -i\bar{z}$, $\operatorname{Re} (iz) = -\operatorname{Im} z$, $\operatorname{Im} (iz) = \operatorname{Re} z$.

25. If a product of two complex numbers is zero, at least one factor must be zero.

12.2 Polar Form of Complex Numbers. Powers and Roots

It is often practical to express complex numbers $z = x + iy$ in terms of polar coordinates r, θ. These are defined by

(1)
$$x = r \cos \theta, \qquad y = r \sin \theta.$$

By substituting this we obtain the **polar form** of z,

(2)
$$z = r \cos \theta + ir \sin \theta = r(\cos \theta + i \sin \theta).$$

r is called the **absolute value** or **modulus** of z and is denoted by $|z|$. Hence

(3)
$$|z| = r = \sqrt{x^2 + y^2} = \sqrt{z\bar{z}}.$$

Geometrically, $|z|$ is the distance of the point z from the origin (Fig. 300). Similarly, $|z_1 - z_2|$ is the distance between z_1 and z_2 (Fig. 301).

θ is called the **argument** of z and is denoted by arg z. Thus (Fig. 300)

(4)
$$\theta = \arg z = \arctan \frac{y}{x} \qquad (z \neq 0).$$

Geometrically, θ is the directed angle from the positive x-axis to OP in Fig. 300. Here, as in calculus, *all angles are measured in **radians** and positive in the counterclockwise sense.*

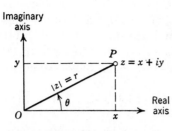

Fig. 300. Complex plane, polar form of a complex number

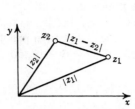

Fig. 301. Distance between two points in the complex plane

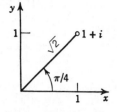

Fig. 302. Example 1

For $z = 0$ this angle θ is undefined. (Why?) For given $z \neq 0$ it is determined only up to integer multiples of 2π. The value of θ that lies in the interval $-\pi < \theta \leq \pi$ is called the **principal value** of the argument of z $(\neq 0)$ and is denoted by Arg z. Thus $\theta = $ Arg z satisfies by definition

$$-\pi < \text{Arg } z \leq \pi.$$

EXAMPLE 1. Polar form of complex numbers. Principal value

Let $z = 1 + i$ (cf. Fig. 302). Then

$$z = \sqrt{2}\left(\cos\frac{\pi}{4} + i\sin\frac{\pi}{4}\right), \qquad |z| = \sqrt{2}, \qquad \arg z = \frac{\pi}{4} \pm 2n\pi \qquad (n = 0, 1, \cdots).$$

The principal value of the argument is Arg $z = \pi/4$. Other values are $-7\pi/4$, $9\pi/4$, etc.

EXAMPLE 2. Polar form of complex numbers. Principal value

Let $z = 3 + 3\sqrt{3}\, i$. Then $z = 6\left(\cos\frac{\pi}{3} + i\sin\frac{\pi}{3}\right)$, the absolute value of z is $|z| = 6$, and the principal value of arg z is Arg $z = \pi/3$.   ∎

Caution! In using (4), we must pay attention to the quadrant in which z lies, since $\tan\theta$ has period π, so that the arguments of z and $-z$ have the same tangent. *Example:* for $\theta_1 = \arg(1 + i)$ and $\theta_2 = \arg(-1 - i)$ we have $\tan\theta_1 = \tan\theta_2 = 1$.

Triangle inequality

For any complex numbers we have the important **triangle inequality**

(5)
$$\boxed{|z_1 + z_2| \leq |z_1| + |z_2|}$$
(Fig. 303)

which we shall use quite frequently. This inequality follows by noting that

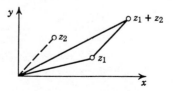

Fig. 303. Triangle inequality

the three points 0, z_1 and $z_1 + z_2$ are the vertices of a triangle[5] (Fig. 303) with sides $|z_1|$, $|z_2|$ and $|z_1 + z_2|$, and one side cannot exceed the sum of the other two sides. A formal proof is left to the reader (Prob. 45).

Example: if $z_1 = 1 + i$ and $z_2 = -2 + 3i$, then (sketch a figure!)

$$|z_1 + z_2| = |-1 + 4i| = \sqrt{17} = 4.123 < \sqrt{2} + \sqrt{13} = 5.020.$$

By induction the triangle inequality can be extended to arbitrary sums:

(6) $$|z_1 + z_2 + \cdots + z_n| \leq |z_1| + |z_2| + \cdots + |z_n|;$$

that is, *the absolute value of a sum cannot exceed the sum of the absolute values of the terms.*

Multiplication and Division in Polar Form

This will give us a better understanding of multiplication and division. Let

$$z_1 = r_1(\cos \theta_1 + i \sin \theta_1) \quad \text{and} \quad z_2 = r_2(\cos \theta_2 + i \sin \theta_2).$$

Then, by (6), Sec. 12.1, the product is at first

$$z_1 z_2 = r_1 r_2[(\cos \theta_1 \cos \theta_2 - \sin \theta_1 \sin \theta_2) + i(\sin \theta_1 \cos \theta_2 + \cos \theta_1 \sin \theta_2)].$$

The addition rules for the sine and cosine [(6) in Appendix 3.1] now yield

(7) $$\boxed{z_1 z_2 = r_1 r_2[\cos (\theta_1 + \theta_2) + i \sin (\theta_1 + \theta_2)].}$$

Taking absolute values and arguments on both sides, we thus obtain the important rules

(8) $$|z_1 z_2| = |z_1| \, |z_2|$$

and

(9) $$\arg (z_1 z_2) = \arg z_1 + \arg z_2 \qquad \text{(up to multiples of } 2\pi\text{).}$$

We now turn to *division*. The quotient $z = z_1/z_2$ is the number z satisfying $z z_2 = z_1$. Hence $|z z_2| = |z| \, |z_2| = |z_1|$, $\arg (z z_2) = \arg z + \arg z_2 = \arg z_1$. This yields

(10) $$\left| \frac{z_1}{z_2} \right| = \frac{|z_1|}{|z_2|} \qquad\qquad (z_2 \neq 0)$$

and

(11) $$\arg \frac{z_1}{z_2} = \arg z_1 - \arg z_2 \qquad \text{(up to multiples of } 2\pi\text{).}$$

---

[5]Which degenerates if z_1 and z_2 lie on the same straight line through the origin.

By combining these two formulas (10) and (11) we also have

(12)
$$\frac{z_1}{z_2} = \frac{r_1}{r_2} [\cos(\theta_1 - \theta_2) + i\sin(\theta_1 - \theta_2)].$$

EXAMPLE 3. Illustration of formulas (8)-(11)

Let $z_1 = -2 + 2i$ and $z_2 = 3i$. Then $z_1 z_2 = -6 - 6i$, $z_1/z_2 = 2/3 + (2/3)i$. Hence

$$|z_1 z_2| = 6\sqrt{2} = 3\sqrt{8} = |z_1| |z_2|, \qquad |z_1/z_2| = 2\sqrt{2}/3 = |z_1|/|z_2|,$$

and for the arguments we obtain $\operatorname{Arg} z_1 = 3\pi/4$, $\operatorname{Arg} z_2 = \pi/2$,

$$\operatorname{Arg} z_1 z_2 = -\frac{3\pi}{4} = \operatorname{Arg} z_1 + \operatorname{Arg} z_2 - 2\pi,$$

$$\operatorname{Arg}(z_1/z_2) = \frac{\pi}{4} = \operatorname{Arg} z_1 - \operatorname{Arg} z_2.$$

Integer powers of z

From (7) and (12) we have

$$z^2 = r^2(\cos 2\theta + i\sin 2\theta),$$

$$z^{-2} = r^{-2}[\cos(-2\theta) + i\sin(-2\theta)]$$

and, more generally, for any integer n,

(13)
$$\boxed{z^n = r^n(\cos n\theta + i\sin n\theta).}$$

EXAMPLE 4. Formula of De Moivre

For $|z| = r = 1$, formula (13) yields the so-called **formula of De Moivre**[6]

(13*)
$$(\cos\theta + i\sin\theta)^n = \cos n\theta + i\sin n\theta.$$

This formula is useful for expressing $\cos n\theta$ and $\sin n\theta$ in terms of $\cos\theta$ and $\sin\theta$. For instance when $n = 2$ and we take the real and imaginary parts on both sides of (13*), we get the familiar formulas
$$\cos 2\theta = \cos^2\theta - \sin^2\theta, \qquad \sin 2\theta = 2\cos\theta\sin\theta.$$

This illustrates the general fact that *complex* methods often simplify the derivation of *real* formulas. ∎

Roots

If $z = w^n$ $(n = 1, 2, \cdots)$, then to each value of w there corresponds one value of z. We shall immediately see that to a given $z \neq 0$ there correspond precisely n distinct values of w. Each of these values is called an **nth root** of z, and we write

(14)
$$w = \sqrt[n]{z}.$$

Hence this symbol is *multivalued,* namely, *n-valued,* in contrast to the usual

---

[6]ABRAHAM DE MOIVRE (1667—1754), French mathematician, who introduced imaginary quantities in trigonometry and contributed to probability theory (cf. Sec. 23.7).

conventions made in *real* calculus. The n values of $\sqrt[n]{z}$ can easily be determined as follows. In terms of polar forms for z and

$$w = R(\cos \phi + i \sin \phi),$$

the equation $w^n = z$ becomes

$$w^n = R^n(\cos n\phi + i \sin n\phi) = z = r(\cos \theta + i \sin \theta).$$

By equating the absolute values on both sides we have

$$R^n = r, \quad \text{thus} \quad R = \sqrt[n]{r}$$

where the root is real positive and thus uniquely determined. By equating the arguments we obtain

$$n\phi = \theta + 2k\pi, \quad \text{thus} \quad \phi = \frac{\theta}{n} + \frac{2k\pi}{n}$$

where k is an integer. For $k = 0, 1, \cdots, n - 1$ we get n *distinct* values of w. Further integers of k would give values already obtained. For instance, $k = n$ gives $2k\pi/n = 2\pi$, hence the w corresponding to $k = 0$, etc. Consequently, $\sqrt[n]{z}$, for $z \neq 0$, has the n distinct values

$$(15) \quad \boxed{\sqrt[n]{z} = \sqrt[n]{r}\left(\cos \frac{\theta + 2k\pi}{n} + i \sin \frac{\theta + 2k\pi}{n}\right)} \quad k = 0, 1, \cdots, n - 1.$$

These n values lie on a circle of radius $\sqrt[n]{r}$ with center at the origin and constitute the vertices of a regular polygon of n sides.

The value of $\sqrt[n]{z}$ obtained by taking the principal value of arg z and $k = 0$ in (15) is called the **principal value** *of* $w = \sqrt[n]{z}$.

EXAMPLE 5. Square root
From (15) it follows that $w = \sqrt{z}$ has the two values

$$(16a) \qquad\qquad w_1 = \sqrt{r}\left(\cos \frac{\theta}{2} + i \sin \frac{\theta}{2}\right)$$

and

$$(16b) \qquad w_2 = \sqrt{r}\left[\cos\left(\frac{\theta}{2} + \pi\right) + i \sin\left(\frac{\theta}{2} + \pi\right)\right] = -w_1$$

which lie symmetric with respect to the origin. For instance, the square root of $4i$ has the values

$$\sqrt{4i} = \pm 2\left(\cos \frac{\pi}{4} + i \sin \frac{\pi}{4}\right) = \pm(\sqrt{2} + i\sqrt{2}).$$

From (16) we can obtain the much more practical formula

$$(17) \qquad\qquad \sqrt{z} = \pm[\sqrt{\tfrac{1}{2}(|z| + x)} + (\text{sign } y)i\sqrt{\tfrac{1}{2}(|z| - x)}]$$

where sign $y = 1$ if $y \geq 0$, sign $y = -1$ if $y < 0$, and all square roots of positive numbers are taken with the positive sign. This follows from (16) if we use the trigonometric identities

$$\cos \tfrac{1}{2}\theta = \sqrt{\tfrac{1}{2}(1 + \cos\theta)}, \qquad \sin \tfrac{1}{2}\theta = \sqrt{\tfrac{1}{2}(1 - \cos\theta)},$$

multiply them by $\sqrt{r}$,

$$\sqrt{r}\cos \tfrac{1}{2}\theta = \sqrt{\tfrac{1}{2}(r + r\cos\theta)}, \qquad \sqrt{r}\sin \tfrac{1}{2}\theta = \sqrt{\tfrac{1}{2}(r - r\cos\theta)},$$

use $r \cos \theta = x$, and finally choose the sign of $\operatorname{Im}\sqrt{z}$ so that sign $[(\operatorname{Re}\sqrt{z})(\operatorname{Im}\sqrt{z})] = \operatorname{sign} y$ (why?).

EXAMPLE 6. Complex quadratic equation

Solve $z^2 - (5 + i)z + 8 + i = 0$

Solution.

$$z = \tfrac{1}{2}(5 + i) \pm \sqrt{\tfrac{1}{4}(5 + i)^2 - 8 - i} = \tfrac{1}{2}(5 + i) \pm \sqrt{-2 + \tfrac{3}{2}i}$$

$$= \tfrac{1}{2}(5 + i) \pm [\sqrt{\tfrac{1}{2}(\tfrac{5}{2} + (-2))} + i\sqrt{\tfrac{1}{2}(\tfrac{5}{2} - (-2))}]$$

$$= \tfrac{1}{2}(5 + i) \pm [\tfrac{1}{2} + \tfrac{3}{2}i] = \begin{cases} 3 + 2i \\ 2 - i \end{cases}$$

EXAMPLE 7. Cube root of a positive real number

If z is *positive real*, then $w = \sqrt[3]{z}$ has the real value $\sqrt[3]{r}$ and the complex conjugate values

$$\sqrt[3]{r}\left(\cos\frac{2\pi}{3} + i\sin\frac{2\pi}{3}\right) = \sqrt[3]{r}\left(-\frac{1}{2} + \frac{\sqrt{3}}{2}i\right)$$

and

$$\sqrt[3]{r}\left(\cos\frac{4\pi}{3} + i\sin\frac{4\pi}{3}\right) = \sqrt[3]{r}\left(-\frac{1}{2} - \frac{\sqrt{3}}{2}i\right).$$

For instance, $\sqrt[3]{1} = 1$, $-\tfrac{1}{2} \pm \tfrac{1}{2}\sqrt{3}\,i$ (Fig. 304). These are the roots of the equation $w^3 = 1$.

EXAMPLE 8. *n*th root of unity

Solve the equation $z^n = 1$.

Solution. From (15) we obtain

$$(18) \qquad \sqrt[n]{1} = \cos\frac{2k\pi}{n} + i\sin\frac{2k\pi}{n}, \qquad k = 0, 1, \cdots, n - 1.$$

If ω denotes the value corresponding to $k = 1$, then the n values of $\sqrt[n]{1}$ can be written as $1, \omega, \omega^2, \cdots, \omega^{n-1}$. These values are the vertices of a regular polygon of n sides inscribed in the unit circle, with one vertex at the point 1. Each of these n values is called an **nth root of unity**. For instance, $\sqrt[4]{1}$ has the values 1, i, -1, and $-i$ (Fig. 305). Figure 306 shows $\sqrt[5]{1}$.

If w_1 is any nth root of an *arbitrary* complex number z, then the n values of $\sqrt[n]{z}$ are

$$w_1, \qquad w_1\omega, \qquad w_1\omega^2, \qquad \cdots, \qquad w_1\omega^{n-1}$$

since multiplying w_1 by ω^k corresponds to increasing the argument of w_1 by $2k\pi/n$. ∎

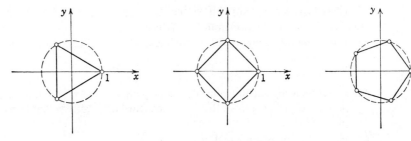

Fig. 304. $\sqrt[3]{1}$ **Fig. 305.** $\sqrt[4]{1}$ **Fig. 306.** $\sqrt[5]{1}$

The student should do the problems related to the polar representation with particular care, since we shall need this representation quite often in our work. In the next section we discuss some curves and regions in the complex plane which we shall also need in the chapters on complex analysis.

Problems for Sec. 12.2

1. **(Multiplication by i)** Show that multiplication of a complex number by i corresponds to a counterclockwise rotation of the corresponding vector through the angle $\pi/2$.

Find

2. $|1 - i|^2$ **3.** $\left|-\frac{3}{2}i\right|$ **4.** $|\cos\theta + i\sin\theta|$ **5.** $|6 - 8i|$

6. $\left|\dfrac{1 + 4i}{4 + i}\right|$ **7.** $\left|\dfrac{z}{\bar{z}}\right|$ **8.** $\left|\dfrac{(3 + 4i)^4}{(3 - 4i)^3}\right|$ **9.** $\left|\dfrac{1}{7 - \pi i}\right|$

Determine the principal value of the arguments of

10. $-2 + 2i$ **11.** -4 **12.** $-3i/2$ **13.** $1 - i\sqrt{3}$

Represent in polar form:

14. $1 + i$ **15.** $-4 + 4i$ **16.** $4i$ **17.** -7

18. $\dfrac{2 + 2i}{1 - i}$ **19.** $\dfrac{i\sqrt{2}}{3 + 3i}$ **20.** $\dfrac{6 + 8i}{4 - 3i}$ **21.** $\dfrac{3 + i\sqrt{2}}{-\sqrt{2} - 2i/3}$

Represent in the form $x + iy$:

22. $4\left(\cos\dfrac{\pi}{2} + i\sin\dfrac{\pi}{2}\right)$ **23.** $\sqrt{8}\left(\cos\dfrac{\pi}{4} + i\sin\dfrac{\pi}{4}\right)$

24. $\pi(\cos\pi + i\sin\pi)$ **25.** $\sqrt{50}\left(\cos\dfrac{3\pi}{4} + i\sin\dfrac{3\pi}{4}\right)$

Find all values of the following roots and plot them in the complex plane.

26. $\sqrt{i}$ **27.** $\sqrt{-i}$ **28.** $\sqrt{-4}$ **29.** $\sqrt{1 - i\sqrt{3}}$

30. $\sqrt[4]{-1}$ **31.** $\sqrt{3 + 4i}$ **32.** $\sqrt{-5 + 12i}$ **33.** $\sqrt{-8 - 6i}$

34. $\sqrt[3]{1 + i}$ **35.** $\sqrt[5]{-1}$ **36.** $\sqrt[6]{-1}$ **37.** $\sqrt[8]{1}$

Find and plot all solutions of the following equations.

38. $z^2 + z + 1 = i$ **39.** $z^2 - 3z + 3 = i$

40. $z^2 - (5 + i)z + 8 + i = 0$ **41.** $z^4 - 3(1 + 2i)z^2 = 8 - 6i$

42. **(Parallelogram equality)** Show that $|z_1 + z_2|^2 + |z_1 - z_2|^2 = 2(|z_1|^2 + |z_2|^2)$. This is called the *parallelogram equality*. Why?

43. Prove the following useful inequalities, which we shall need from time to time.

    (19) $|\operatorname{Re} z| \leq |z|$, $|\operatorname{Im} z| \leq |z|$

44. Let P be a regular polygon of n sides with vertices on the unit circle. Find the product of the lengths of the $n - 1$ straight-line segments that join a fixed vertex of P with the $n - 1$ other vertices.

45. Prove the triangle inequality.

12.3 Curves and Regions in the Complex Plane

In this section we consider some important curves and regions and some related concepts we shall frequently need. This will also help us to become more familiar with the complex plane.

The distance between two points z and a is $|z - a|$. Hence a **circle** C of radius ρ and center at a (Fig. 307) can be represented by

$$(1) \qquad\qquad |z - a| = \rho.$$

In particular, the so-called **unit circle,** that is, the circle of radius 1 and center at the origin $a = 0$ (Fig. 308), is given by

$$|z| = 1.$$

Furthermore, the inequality

$$(2) \qquad\qquad |z - a| < \rho$$

holds for every point z inside C; that is, (2) represents the interior of C. Such a region is called a **circular disk,** or, more precisely, an *open* circular disk, in contrast to the *closed* circular disk

$$|z - a| \leqq \rho,$$

which consists of the interior of C and C itself. The open circular disk (2) is also called a **neighborhood** of the point a. Obviously, a has infinitely many such neighborhoods, each of which corresponds to a certain value of $\rho \; (> 0)$; and a belongs to each of these neighborhoods, that is, a is a point of each of them.[7]

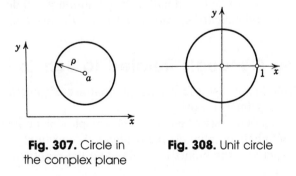

Fig. 307. Circle in the complex plane   **Fig. 308.** Unit circle

[7]More generally, any set that contains an open disk (2) is also called a *neighborhood of a.* For distinction, a disk (2) is often called an *open circular neighborhood of a.*

Similarly, the inequality

$$|z - a| > \rho$$

represents the exterior of the circle C. Furthermore, the region between two concentric circles of radii ρ_1 and ρ_2 ($> \rho_1$) can be represented in the form

(3) $\rho_1 < |z - a| < \rho_2,$

where a is the center of the circles. Such a region is called an *open circular ring* or *open* **annulus** (Fig. 309).

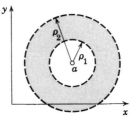

Fig. 309. Annulus
in the complex plane

EXAMPLE 1. Circular disk

Determine the region in the complex plane given by $|z - 3 + i| \leq 4$.

Solution. The inequality is valid precisely for all z whose distance from $a = 3 - i$ does not exceed 4. Hence this is a closed circular disk of radius 4 with center at $3 - i$.

EXAMPLE 2. Unit circle and unit disk

Determine each of the regions

(a) $|z| < 1$ (b) $|z| \leq 1$ (c) $|z| > 1$.

Solution. (a) The interior of the unit circle. This is called the *open unit disk*.
(b) The unit circle and its interior. This is called the *closed unit disk*.
(c) The exterior of the unit circle.   ∎

By the (open) *upper* **half-plane** we mean the set of all points $z = x + iy$ such that $y > 0$. Similarly, the condition $y < 0$ defines the *lower half-plane*, $x > 0$ the *right half-plane* and $x < 0$ the *left half-plane*.

Some Concepts Related to Sets in the Complex Plane

We finally list a few concepts that are of general interest and will be used in our further work.

The term **set of points** in the complex plane means any sort of collection of finitely or infinitely many points. For example, the solutions of a quadratic equation, the points on a line, and the points in the interior of a circle are sets.

A set S is called **open** if every point of S has a neighborhood consisting entirely of points that belong to S. For example, the points in the interior of a circle or a square form an open set, and so do the points of the "right half-plane" Re $z = x > 0$.

An open set S is said to be **connected** if any two of its points can be joined by a broken line of finitely many straight line segments all of whose points belong to S. An open connected set is called a **domain.** Thus an open disk (2) and an open annulus (3) are domains. An open square with a diagonal removed is not a domain since this set is not connected. (Why?)

The **complement** of a set S in the complex plane is defined to be the set of all points of the complex plane that do *not* belong to S. A set S is called **closed** if its complement is open. For example, the points on and inside the unit circle form a closed set ("closed unit disk": cf. Example 2) since its complement $|z| > 1$ is open.

A **boundary point** of a set S is a point every neighborhood of which contains both points that belong to S and points that do not belong to S. For example, the boundary points of an annulus are the points on the two bounding circles. Clearly, if a set S is open, then no boundary point belongs to S; if S is closed, then every boundary point belongs to S.

A **region** is a set consisting of a domain plus, perhaps, some or all of its boundary points. (The reader is warned that some authors use the term "region" for what we call a domain [following the modern standard terminology], and others make no distinction between the two terms.)

So far we have been concerned with complex numbers and the complex plane (just as at the beginning of calculus, one talks about real numbers and the real line). In the next section we start doing complex calculus: we introduce complex functions and **derivatives.** This will generalize familiar concepts of calculus.

Problems for Sec. 12.3

Determine and sketch the sets represented by

1. $|z - 2i| = 2$ **2.** $1 \leqq |z + 1 - i| \leqq 3$ **3.** Re $(z^2) \leqq 1$

4. $|\arg z| < \pi/4$ **5.** $-\pi < \operatorname{Im} z \leqq \pi$ **6.** $|1/z| < 1$

7. $\left|\dfrac{z + 1}{z - 1}\right| = 1$ **8.** $\left|\dfrac{z + 3i}{z - i}\right| = 1$ **9.** $\operatorname{Im} \dfrac{2z + 1}{4z - 4} \leqq 1$

10. $z\bar{z} + (1 + 2i)z + (1 - 2i)\bar{z} + 1 = 0$

12.4 Limit. Derivative. Analytic Function

The functions with which complex analysis is concerned are complex functions that are differentiable. Hence we should first say what we mean by a complex function and then define the concepts of limit and derivative in complex. This discussion will be quite similar to that in calculus.

Complex Function

Recall from calculus that a *real* function f defined on a set S of real numbers (usually an interval) is a rule that assigns to every x in S a real number $f(x)$, called the *value* of f at x.

Now in complex, S is a set of *complex* numbers. And a **function** f defined on S is a rule that assigns to every z in S a complex number w, called the *value* of f at z. We write

$$w = f(z).$$

Here z varies in S and is called a **complex variable.** The set S is called the **domain** *of definition*[8] *of* f.

Example: $w = f(z) = z^2 + 3z$ is a complex function defined for all z; that is, its domain S is the whole complex plane.

The set of all values of a function f is called the *range of* f.

w is complex, and we write $w = u + iv$, where u and v are the real and imaginary parts, respectively. Now w depends on $z = x + iy$. Hence u becomes a real function of x and y, and so does v. We may thus write

$$\boxed{w = f(z) = u(x, y) + iv(x, y).}$$

This shows that a *complex* function $f(z)$ is equivalent to a pair of *real* functions $u(x, y)$ and $v(x, y)$, each depending on the two real variables x and y.

EXAMPLE 1. Function of a complex variable

Let $w = f(z) = z^2 + 3z$. Find u and v and calculate the values of f at $z = 1 + 3i$ and $z = 2 - i$.

Solution. $u = \operatorname{Re} f(z) = x^2 - y^2 + 3x$ and $v = 2xy + 3y$. Also,

$$f(1 + 3i) = (1 + 3i)^2 + 3(1 + 3i) = 1 - 9 + 6i + 3 + 9i = -5 + 15i.$$

This shows that $u(1, 3) = -5$ and $v(1, 3) = 15$. Similarly,

$$f(2 - i) = (2 - i)^2 + 3(2 - i) = 4 - 1 - 4i + 6 - 3i = 9 - 7i.$$

EXAMPLE 2. Function of a complex variable

Let $w = f(z) = 2iz + 6\bar{z}$. Find u and v and the value of f at $z = \frac{1}{2} + 4i$.

Solution. $f(z) = 2i(x + iy) + 6(x - iy)$ gives $u(x, y) = 6x - 2y$ and $v(x, y) = 2x - 6y$. Also

$$f(\tfrac{1}{2} + 4i) = 2i(\tfrac{1}{2} + 4i) + 6(\overline{\tfrac{1}{2} - 4i}) = i - 8 + 3 - 24i = -5 - 23i.$$

Limit, Continuity

A function $f(z)$ is said to have the **limit** l as z approaches a point z_0, written

$$(1) \qquad\qquad \lim_{z \to z_0} f(z) = l,$$

---

[8]This is a standard term. In most cases, a domain of definition will be an open and connected set (a *domain* as defined in Sec. 12.3); exceptions rarely occur in applications.

In the literature on complex analysis, one sometimes uses relations such that to a value of z there may correspond more than one value of w, and it is customary to call such a relation a function (a "multivalued function"). We shall not adopt this convention but assume that all occurring functions are *single-valued* relations, that is, functions in the usual sense: to each z in S corresponds only *one* value $w = f(z)$.

Strictly speaking, $f(z)$ denotes the value of f at z, but it is a convenient abuse of language to talk about *the function* $f(z)$ (instead of *the function* f), thereby exhibiting the notation for the independent variable.

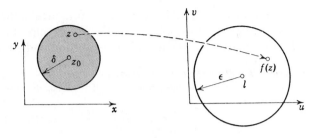

Fig. 310. Limit

if f is defined in a neighborhood of z_0 (except perhaps at z_0 itself) and if the values of f are "close" to l for all z "close" to z_0; that is, in precise terms, for every positive real ϵ we can find a positive real δ such that for all $z \neq z_0$ in the disk $|z - z_0| < \delta$ (Fig. 310) we have

$$(2) \qquad\qquad |f(z) - l| < \epsilon;$$

that is, for every $z \neq z_0$ in that δ-disk the value of f lies in the disk (2).

Formally, this definition is similar to that in calculus, but there is a big difference. Whereas in the real case, x can approach an x_0 only along the real line, here, by definition, z may approach z_0 *from any direction* in the complex plane. This will be quite essential in what follows.

If a limit exists, it is unique. (Cf. Prob. 30.)

A function $f(z)$ is said to be **continuous** at $z = z_0$ if $f(z_0)$ is defined and

$$(3) \qquad\qquad \lim_{z \to z_0} f(z) = f(z_0).$$

Note that by the definition of a limit this implies that $f(z)$ is defined in some neighborhood of z_0.

$f(z)$ is said to be *continuous in a domain* if it is continuous at each point of this domain.

Derivative

The **derivative** of a complex function f at a point z_0 is written $f'(z_0)$ and is defined by

$$(4) \qquad\qquad \boxed{f'(z_0) = \lim_{\Delta z \to 0} \frac{f(z_0 + \Delta z) - f(z_0)}{\Delta z}}$$

provided this limit exists. Then f is said to be **differentiable** at z_0. If we write $\Delta z = z - z_0$, we also have, since $z = z_0 + \Delta z$,

$$(4') \qquad\qquad f'(z_0) = \lim_{z \to z_0} \frac{f(z) - f(z_0)}{z - z_0}.$$

Remember that the definition of a limit implies that $f(z)$ is defined (at least) in a neighborhood of z_0. Also, by that definition, z may approach z_0 from any direction. Hence differentiability at z_0 means that, along whatever path z approaches z_0, the quotient in (4') always approaches a certain value and all these values are equal. This is important and should be kept in mind.

EXAMPLE 3. Differentiability. Derivative

The function $f(z) = z^2$ is differentiable for all z and has the derivative $f'(z) = 2z$ because

$$f'(z) = \lim_{\Delta z \to 0} \frac{(z + \Delta z)^2 - z^2}{\Delta z} = \lim_{\Delta z \to 0} (2z + \Delta z) = 2z.$$ ∎

*The **differentiation rules** are the same as in real calculus,* since their proofs are literally the same. Thus,

$$(cf)' = cf', \quad (f + g)' = f' + g', \quad (fg)' = f'g + fg', \quad \left(\frac{f}{g}\right)' = \frac{f'g - fg'}{g^2},$$

as well as the chain rule and the power rule $(z^n)' = nz^{n-1}$ (n integer) hold.

Also, if $f(z)$ is differentiable at z_0, it is continuous at z_0. (Cf. Prob. 34.)

EXAMPLE 4. $\bar{z}$ not differentiable

It is important to note that there are many simple functions that do not have a derivative at any point. For instance, $f(z) = \bar{z} = x - iy$ is such a function. Indeed, if we write $\Delta z = \Delta x + i\,\Delta y$, we have

$$(5) \qquad \frac{f(z + \Delta z) - f(z)}{\Delta z} = \frac{\overline{(z + \Delta z)} - \bar{z}}{\Delta z} = \frac{\overline{\Delta z}}{\Delta z} = \frac{\Delta x - i\,\Delta y}{\Delta x + i\,\Delta y}.$$

If $\Delta y = 0$, this is $+1$. If $\Delta x = 0$, this is -1. Hence (5) approaches $+1$ along path I in Fig. 311 but -1 along path II. Hence, by definition, the limit of (5) as $\Delta z \to 0$ does not exist at any z.

This example may be surprising, but it merely illustrates that differentiability of a complex function is a rather severe requirement.

The idea of proof (approach from different directions) is basic and will be used again in the next section. ∎

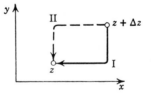

Fig. 311. Paths in (5)

Analytic Functions

These are the functions that are differentiable in some domain, so that we can do "calculus in complex." They are the main concern of complex analysis. Their introduction is our main goal in this section:

Definition (Analyticity)

A function $f(z)$ is said to be *analytic in a domain D* if $f(z)$ is defined and differentiable at all points of D. The function $f(z)$ is said to be *analytic at a point $z = z_0$ in D* if $f(z)$ is analytic in a neighborhood (cf. Sec. 12.3) of z_0.

Also, by an **analytic function** we mean a function that is analytic in *some* domain.

Hence analyticity of $f(z)$ at z_0 means that $f(z)$ has a derivative at every point in some neighborhood of z_0 (including z_0 itself since, by definition, z_0 is a point of all its neighborhoods). This concept is *motivated* by the fact that it is of no practical interest when a function is differentiable merely at a single point z_0 but not throughout some neighborhood of z_0. Problem 28 gives an example.

An older term for *analytic in D* is *regular in D*, and a more modern term is *holomorphic in D*.

EXAMPLE 5. Polynomials, rational functions

The integer powers $1, z, z^2, \cdots$ and, more generally, **polynomials**, that is, functions of the form

$$f(z) = c_0 + c_1 z + c_2 z^2 + \cdots + c_n z^n$$

where $c_0, \cdots, c_n$ are complex constants, are analytic in the entire complex plane.

The quotient of two polynomials $g(z)$ and $h(z)$,

$$f(z) = \frac{g(z)}{h(z)},$$

is called a **rational function**. This f is analytic except at the points where $h(z) = 0$; here we assume that common factors of g and h have been canceled. **Partial fractions**

$$\frac{c}{(z - z_0)^m} \qquad (c \neq 0)$$

(c and z_0 complex, m a positive integer) are special rational functions; they are analytic except at z_0. It is proved in algebra that every rational function can be written as a sum of a polynomial (which may be 0) and finitely many partial fractions. ∎

The concepts discussed in this section extend familiar concepts of calculus. Most important is the concept of an analytic function. Indeed, complex analysis is concerned exclusively with analytic functions, and although many simple functions are not analytic, the large variety of remaining functions will yield a branch of mathematics that is most beautiful from the theoretical point of view and most useful for practical purposes.

Before we consider special analytic functions (exponential functions, cosine, sine, etc.), let us give equations by means of which we can readily decide whether a function is analytic or not. These are the famous **Cauchy–Riemann equations,** which we discuss in the next section.

Problems for Sec. 12.4

Find $f(2 + i)$, $f(3i)$, $f(-4 + i)$ where $f(z)$ equals

1. $3z^2 + z$ **2.** $1/z^2$ **3.** $(z + 1)/(z - 1)$

Find the real and imaginary parts of the following functions.

4. $f(z) = 2z^2 - 3z$ **5.** $f(z) = 1/(1 - z)$ **6.** $f(z) = z/(1 + z)$

Suppose that z varies in a region R in the z-plane. Find the (precise) region in the w-plane in which the corresponding values of $w = f(z)$ lie, and show the two regions graphically.

7. $f(z) = 1/z$, Re $z \geq 0$ **8.** $f(z) = z^2$, $|z| < 8$ **9.** $f(z) = z^4$, $|\arg z| \leq \frac{1}{6}\pi$

In each case, find whether $f(z)$ is continuous at the origin, assuming that $f(0) = 0$ and, for $z \neq 0$, the function $f(z)$ equals

10. $\operatorname{Im} z/|z|$
11. $\operatorname{Re} z/(1 + z)$
12. $\operatorname{Re} z^2/|z|$
13. $(\operatorname{Im} z)^2/|z|^2$
14. $\operatorname{Im} z/|z|^2$
15. $(z \operatorname{Im} z)/|z|$

Differentiate

16. $1/(1 - z)$
17. i/z^5
18. $(z + i)/(z - i)$
19. $(z^2 + i)^3$
20. $z + 1/z$
21. $(iz + 2)/(3z - 6i)$

Find the value of the derivative of

22. $z^3 - 2z$ at $-i$
23. $1/z^4$ at $1 + i$
24. $(z^2 - i)^2$ at $i/2$
25. $(1 + i)/z^4$ at 2
26. i/z^4 at $3i$
27. $(z - i)/2z$ at $2i$

28. Show that $f(z) = |z|^2$ is differentiable only at $z = 0$.; hence it is nowhere analytic. *Hint.* Use the relation $|z + \Delta z|^2 = (z + \Delta z)(\bar{z} + \overline{\Delta z})$.

29. Show that $f(z) = \operatorname{Re} z = x$ is not differentiable at any z.

30. If $\lim_{z \to z_0} f(z)$ exists, show that this limit is unique.

31. Prove that (1) is equivalent to the pair of relations

$$\lim_{z \to z_0} \operatorname{Re} f(z) = \operatorname{Re} l, \qquad \lim_{z \to z_0} \operatorname{Im} f(z) = \operatorname{Im} l.$$

32. If functions f and g have limit l and p, respectively, as z approaches z_0, show that then $f + g$ has the limit $l + p$ and fg has the limit lp.

33. If $z_1, z_2, \cdots$ are complex numbers for which $\lim_{n \to \infty} z_n = a$, and if $f(z)$ is continuous at $z = a$, show that

$$\lim_{n \to \infty} f(z_n) = f(a).$$

34. If $f(z)$ is differentiable at z_0, show that $f(z)$ is continuous at z_0.

35. Prove the product rule $[f(z)g(z)]' = f'(z)g(z) + f(z)g'(z)$.

12.5 Cauchy–Riemann Equations

We shall now derive a very important criterion (a test) for the analyticity of a complex function

$$w = f(z) = u(x, y) + iv(x, y).$$

Roughly, f is analytic in a domain D if and only if the first partial derivatives of u and v satisfy the two equations

(1) $$\boxed{u_x = v_y, \qquad u_y = -v_x}$$

everywhere in D; here $u_x = \partial u/\partial x$ and $u_y = \partial u/\partial y$ (and similarly for v) are the usual notations for partial derivatives. The precise formulation of this statement is given in Theorems 1 and 2 below. The equations (1) are called the **Cauchy–Riemann equations.** They are the most important equations in the whole chapter.[9]

Example: $f(z) = z^2 = x^2 - y^2 + 2ixy$ is analytic for all z, and $u = x^2 - y^2$ and $v = 2xy$ satisfy (1), namely, $u_x = 2x = v_y$ and $u_y = -2y = -v_x$. More examples will follow.

Theorem 1 (Cauchy–Riemann equations)

Let $f(z) = u(x, y) + iv(x, y)$ be defined and continuous in some neighborhood of a point $z = x + iy$ and differentiable at z itself. Then at that point, the first-order partial derivatives of u and v exist and satisfy the Cauchy–Riemann equations (1).

Hence if $f(z)$ is analytic in a domain D, those partial derivatives exist and satisfy (1) at all points of D.

Proof. By assumption, the derivative $f'(z)$ at z exists. It is given by

$$(2) \qquad f'(z) = \lim_{\Delta z \to 0} \frac{f(z + \Delta z) - f(z)}{\Delta z}.$$

The idea of the proof is very simple. By the definition of a limit in complex (cf. Sec. 12.4) we can let Δz approach zero along any path in a neighborhood of z. Thus we may choose the two paths I and II in Fig. 312 and equate the results. By comparing the real parts we shall obtain the first Cauchy–Riemann equation and by comparing the imaginary parts the other equation in (1). The technical details are as follows.

We write $\Delta z = \Delta x + i\Delta y$. In terms of u and v, the derivative in (2) becomes

$$(3) \quad f'(z) = \lim_{\Delta z \to 0} \frac{[u(x + \Delta x, y + \Delta y) + iv(x + \Delta x, y + \Delta y)] - [u(x, y) + iv(x, y)]}{\Delta x + i\Delta y}.$$

We first choose path I in Fig. 312. Thus we let $\Delta y \to 0$ first and then $\Delta x \to 0$.

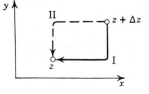

Fig. 312. Paths in (2)

[9]For A.—L. CAUCHY, see the footnote in Sec. 2.7. BERNHARD RIEMANN (1826–1866), German mathematician, studied and received his Ph.D. (in 1851) under Gauss (cf. Sec. 4.4) at Göttingen, where he also taught from 1854 until he died, only 39 years old. He developed what may be called the "geometrical approach" to complex analysis, based on the Cauchy–Riemann equations and conformal mapping, in contrast to the German mathematician KARL WEIERSTRASS (1815–1897; cf. also Sec. 14.8), who based complex analysis on power series. Riemann introduced the concept of the integral as it is used in basic calculus courses, in connection with his work on Fourier series. He created new methods in ordinary and partial differential equations and made basic contributions to number theory and mathematical physics. He also developed the so-called Riemannian geometry, which is the mathematical base of Einstein's theory of relativity. His important work gave the impetus to many ideas in modern mathematics (in particular, in topology and functional analysis). Cf. N. Bourbaki, *Elements of Mathematics, General Topology*, Part 1, pp. 161–166. Paris: Hermann, 1966.

After Δy becomes zero, $\Delta z = \Delta x$. Then (3) becomes, if we first write the two u-terms and then the two v-terms,

$$f'(z) = \lim_{\Delta x \to 0} \frac{u(x + \Delta x, y) - u(x, y)}{\Delta x} + i \lim_{\Delta x \to 0} \frac{v(x + \Delta x, y) - v(x, y)}{\Delta x}.$$

Since $f'(z)$ exists, the two real limits on the right exist. By definition, they are the partial derivatives of u and v with respect to x. Hence the derivative $f'(z)$ of $f(z)$ can be written

(4)
$$\boxed{f'(z) = u_x + iv_x.}$$

Similarly, if we choose path II in Fig. 312, we let $\Delta x \to 0$ first and then $\Delta y \to 0$. After Δx becomes zero, $\Delta z = i\Delta y$, so that from (3) we now obtain

$$f'(z) = \lim_{\Delta y \to 0} \frac{u(x, y + \Delta y) - u(x, y)}{i\,\Delta y} + i \lim_{\Delta y \to 0} \frac{v(x, y + \Delta y) - v(x, y)}{i\,\Delta y}.$$

Since $f'(z)$ exists, the limits on the right exist and yield partial derivatives with respect to y; noting that $1/i = -i$, we obtain

(5)
$$\boxed{f'(z) = -iu_y + v_y.}$$

The existence of the derivative $f'(z)$ thus implies the existence of the four partial derivatives in (4) and (5). By equating the real parts u_x and v_y in (4) and (5) we obtain the first Cauchy–Riemann equation (1). Equating the imaginary parts yields the other. This proves the first statement of the theorem and implies the second because of the definition of analyticity. ∎

Formulas (4) and (5) are also quite practical for calculating derivatives $f'(z)$, as we shall see.

EXAMPLE 1. Cauchy–Riemann equations

$f(z) = z^2$ is analytic for all z. It follows that the Cauchy–Riemann equations must be satisfied (as we have verified above).

For $f(z) = \bar{z} = x - iy$ we have $u = x$, $v = -y$ and see that the second Cauchy–Riemann equation is satisfied, $u_y = -v_x = 0$, but the first is not: $u_x = 1 \neq v_y = -1$. We conclude that $f(z) = \bar{z}$ is not analytic, confirming Example 4 of Sec. 12.4. Note the savings in calculation! ∎

The Cauchy–Riemann equations are fundamental because they are not only necessary but also sufficient for a function to be analytic. More precisely, the following theorem holds.

Theorem 2 (Cauchy–Riemann equations)

If two real-valued continuous functions $u(x, y)$ and $v(x, y)$ of two real variables x and y have continuous first partial derivatives that satisfy the Cauchy–Riemann equations in some domain D, then the complex function $f(z) = u(x, y) + iv(x, y)$ is analytic in D.

The proof of this theorem is more involved than the previous proof; we leave it optional and include it on p. 762 near the end of the chapter.

Theorems 1 and 2 are of great practical importance, since by using the Cauchy–Riemann equations we can now easily find out whether or not a given complex function is analytic.

EXAMPLE 2. Cauchy–Riemann equations

Is $f(z) = z^3$ analytic?

Solution. We find $u = x^3 - 3xy^2$ and $v = 3x^2y - y^3$. Next we calculate

$$u_x = 3x^2 - 3y^2, \qquad v_y = 3x^2 - 3y^2$$

$$u_y = -6xy, \qquad v_x = 6xy.$$

We see that the Cauchy–Riemann equations are satisfied for every z. Hence $f(z) = z^3$ is analytic for every z, by Theorem 2.

EXAMPLE 3. Determination of an analytic function with given real part

We illustrate another class of practical problems that can be solved by the Cauchy–Riemann equations.

Find the most general analytic function $f(z)$ whose real part is $u = x^2 - y^2 - x$.

Solution. We have $u_x = 2x - 1 = v_y$ by the first Cauchy–Riemann equation. This we integrate with respect to y:
$$v = 2xy - y + k(x).$$

As an important point, since we integrated a *partial* derivative with respect to y, the "constant" of integration k may depend on the other variable, x. (To understand this, calculate v_y from this v.) From v and the second Cauchy–Riemann equation,

$$u_y = -v_x = -2y + \frac{dk}{dx}.$$

On the other hand, from the given $u = x^2 - y^2 - x$ we have $u_y = -2y$. By comparison, $dk/dx = 0$, hence $k = const$, which must be real. (Why?) The result is

$$f(z) = u + iv = x^2 - y^2 - x + i(2xy - y + k).$$

This we can express in terms of z, namely, $f(z) = z^2 - z + ik$.

EXAMPLE 4. An analytic function of constant absolute value is constant

The Cauchy–Riemann equations also help to establish certain general properties of analytic functions.

For example, show that if $f(z)$ is analytic in a domain D and $|f(z)| = k = const$ in D, then $f(z) = const$ in D.

Solution. By assumption, $u^2 + v^2 = k^2$. By differentiation,

$$uu_x + vv_x = 0, \qquad uu_y + vv_y = 0.$$

From this and the Cauchy–Riemann equations,

(6) \qquad\qquad (a) $uu_x - vu_y = 0,$ \qquad (b) $uu_y + vu_x = 0.$

To get rid of u_y, multiply (6a) by u and (6b) by v and add. Similarly, to eliminate u_x, multiply (6a) by $-v$ and (6b) by u and add. This yields

$$(u^2 + v^2)u_x = 0, \qquad (u^2 + v^2)u_y = 0.$$

If $k^2 = u^2 + v^2 = 0$, then $u = v = 0$, hence $f = 0$. If $k \neq 0$, then $u_x = u_y = 0$, hence, by the Cauchy–Riemann equations, also $v_x = v_y = 0$. Together, $u = const$ and $v = const$, hence $f = const$. ∎

We mention that if we use the polar form $z = r(\cos \theta + i \sin \theta)$ and set $f(z) = u(r, \theta) + iv(r, \theta)$, then the Cauchy–Riemann equations are

(7)
$$u_r = \frac{1}{r} v_\theta \quad \text{and} \quad v_r = -\frac{1}{r} u_\theta \qquad (r > 0).$$

The derivative can then be calculated from

(8a)
$$f'(z) = (u_r + iv_r)(\cos \theta - i \sin \theta)$$

or from

(8b)
$$f'(z) = (v_\theta - iu_\theta)(\cos \theta - i \sin \theta)/r.$$

EXAMPLE 5. Cauchy–Riemann equations in polar form

Let $f(z) = z^3 = r^3(\cos 3\theta + i \sin 3\theta)$. Then $u = r^3 \cos 3\theta$, $v = r^3 \sin 3\theta$. By differentiation,

$$u_r = 3r^2 \cos 3\theta, \qquad v_\theta = 3r^3 \cos 3\theta,$$

$$v_r = 3r^2 \sin 3\theta, \qquad u_\theta = -3r^3 \sin 3\theta.$$

We see that (7) holds for all $z \neq 0$. This confirms that z^3 is analytic for all $z \neq 0$ (and we know that it is also analytic at $z = 0$). From (8a) we obtain the derivative as expected:

$$f'(z) = 3r^2(\cos 3\theta + i \sin 3\theta)(\cos \theta - i \sin \theta) = 3z^2.$$

Laplace's Equation. Harmonic Functions

One of the main reasons for the great practical importance of complex analysis in engineering mathematics results from the fact that the real part of an analytic function $f = u + iv$ satisfies the so-called **Laplace's equation**

(9)
$$\nabla^2 u = u_{xx} + u_{yy} = 0$$

(∇^2 read "nabla squared"), and the same holds for the imaginary part,

(10)
$$\nabla^2 v = v_{xx} + v_{yy} = 0.$$

Laplace's equation is one of the most important equations in physics, occurring in gravitation, electrostatics, fluid flow, etc. (cf. Chaps. 11, 17). Let us discover why this equation arises in complex analysis.

Theorem 3 (Laplace's equation)

If $f(z) = u(x, y) + iv(x, y)$ is analytic in a domain D, then u and v satisfy Laplace's equation (9) and (10) in D and have continuous second partial derivatives in D.

Proof. Differentiating $u_x = v_y$ with respect to x and $u_y = -v_x$ with respect to y, we obtain

(11)
$$u_{xx} = v_{yx}, \qquad u_{yy} = -v_{xy}.$$

Now the derivative of an analytic function is itself analytic, as we shall prove later (in Sec. 13.6). This implies that u and v have continuous partial derivatives of all orders; in particular, the mixed second derivatives are equal: $v_{yx} = v_{xy}$. By adding (11) we thus obtain (9). Similarly, (10) is obtained by differentiating $u_x = v_y$ with respect to y and $u_y = -v_x$ with respect to x and subtraction, using $u_{xy} = u_{yx}$.   ∎

Solutions of Laplace's equation having *continuous* second-order partial derivatives are called **harmonic functions** and their theory is called **potential theory** (cf. also Sec. 11.11). Hence the real and imaginary parts of an analytic function are harmonic functions.

If two harmonic functions u and v satisfy the Cauchy–Riemann equations in a domain D, they are the real and imaginary parts of an analytic function f in D. Then v is said to be a **conjugate harmonic function** of u in D. (Of course this use of the word "conjugate" has nothing to do with that employed in defining $\bar{z}$, the conjugate of a complex number z.)

A conjugate of a given harmonic function can be obtained from the Cauchy–Riemann equations, as may be illustrated by the following example.

EXAMPLE 6. Conjugate harmonic function

Verify that $u = x^2 - y^2 - y$ is harmonic in the whole complex plane and find a conjugate harmonic function v of u.

Solution. $\nabla^2 u = 0$ by direct calculation. Now $u_x = 2x$ and $u_y = -2y - 1$. Hence a conjugate v of u must satisfy

$$v_y = u_x = 2x, \qquad v_x = -u_y = 2y + 1.$$

Integrating the first equation with respect to y and differentiating the result with respect to x, we obtain

$$v = 2xy + h(x), \qquad v_x = 2y + \frac{dh}{dx}.$$

A comparison with the second equation shows that $dh/dx = 1$. This gives $h(x) = x + c$. Hence $v = 2xy + x + c$ (c any real constant) is the most general conjugate harmonic of the given u. The corresponding analytic function is

$$f(z) = u + iv = x^2 - y^2 - y + i(2xy + x + c) = z^2 + iz + ic.$$

Can you see that the present task was quite similar to that in Example 3? Explain.   ∎

The Cauchy–Riemann equations are the most important equations in this chapter. Their relation to Laplace's equation opens wide ranges of engineering and physical applications, as we show in Chap. 17. In the remainder of this chapter we discuss **elementary functions,** one after the other, beginning with e^z in the next section. Without knowing these functions and their properties we would not be able to do any useful practical work. This is just as in calculus.

Problems for Sec. 12.5

Are the following functions analytic? [Use (1) or (7).]

1. $f(z) = z^4$

2. $f(z) = i|z|^4$

3. $f(z) = e^x(\cos y + i \sin y)$

4. $f(z) = i/z$

5. $f(z) = z\bar{z}$

6. $f(z) = \operatorname{Re} z / \operatorname{Im} z$

7. $f(z) = (1 + i)z^2$

8. $f(z) = 1/z^3$

9. $f(z) = (1 + i)(x + y)^2$

10. $f(z) = \arg z$

11. $f(z) = z - \bar{z}$

12. $f(z) = \ln |z| + i \operatorname{Arg} z$

Are the following functions harmonic? If so, find a corresponding analytic function
$f(z) = u(x, y) + iv(x, y)$.

13. $u = y^2 - x^2$
        14. $u = xy$
        15. $v = y/(x^2 + y^2)$

16. $u = x^2 + y^2$
        17. $v = (x^2 - y^2)^2$
        18. $u = e^x \cos y$

19. $v = x^3 - 3xy^2$
        20. $u = e^x \sin y$
        21. $u = \sin x \cosh y$

Determine conditions on the constants a and b such that the following functions are
harmonic and determine a conjugate harmonic function.

22. $u = x^2 + ay^2$
        23. $u = e^{ax} \cos 2y$
        24. $u = ax^3 + by^3$

25. $u = \cos ax \cosh y$
        26. $u = ax + by$
        27. $u = ax^3 + bxy$

28. Let v be a conjugate harmonic of u in some domain D. Show that then
$h = u^2 - v^2$ is harmonic in D.

29. Show that if u is harmonic and v a conjugate harmonic of u, then u is a conjugate
harmonic of $-v$.

30. Show that if $f(z)$ is analytic and Re $f(z)$ is constant, then $f(z)$ is constant.

Familiarize yourself with (4) and (5), which one needs from time to time; calculate
the derivative by (4) or (5) and verify that the result is as expected:

31. $f(z) = i(z + 2)$
        32. $f(z) = -iz^3$
        33. $f(z) = z^2$

34. Show that, in addition to (4) and (5),

(12)
$$f'(z) = u_x - iu_y, \qquad f'(z) = v_y + iv_x.$$

35. **(Identically vanishing derivative)** Using (4), show that an analytic function whose
derivative is identically zero is a constant.

36. Derive the Cauchy–Riemann equations in polar form (7) from (1).

12.6 Exponential Function

The remaining sections of this chapter will be devoted to the most important
elementary *complex* functions, the exponential function, logarithm, trigonometric functions, etc. We shall see that these complex functions can easily
be defined in such a way that, for real values of the independent variable,
the functions become identical with the familiar real functions. Some of the
complex functions have interesting properties, which do not show when the
independent variable is restricted to real values. The student should follow
the consideration with great care, because these elementary functions will
be frequently needed in applications.

We begin with the complex **exponential function**

$$e^z, \qquad \text{also written} \qquad \exp z,$$

one of the most important analytic functions. The definition of e^z in terms
of the real functions e^x, $\cos y$ and $\sin y$ is

(1)
$$\boxed{e^z = e^x(\cos y + i \sin y).}$$

This definition[10] is motivated by requirements that make e^z a natural extension of the real exponential function e^x, namely,

(a) e^z should reduce to the latter when $z = x$ is real;

(b) e^z should be an **entire function,** that is, analytic for all z, and, resembling calculus, its derivative should be

$$(2) \qquad\qquad (e^z)' = e^z.$$

From (1) we see that (a) holds, since $\cos 0 = 1$ and $\sin 0 = 0$. That e^z is entire is easily verified by the Cauchy–Riemann equations. Formula (2) then follows from (4) in Sec. 12.5:

$$(e^z)' = (e^x \cos y)_x + i(e^x \sin y)_x = e^x \cos y + ie^x \sin y = e^z.$$

e^z has further interesting properties. Let us first show that, as in real, we have the functional relation

$$(3) \qquad\qquad e^{z_1 + z_2} = e^{z_1} e^{z_2}$$

for any $z_1 = x_1 + iy_1$ and $z_2 = x_2 + iy_2$. Indeed, by (1),

$$e^{z_1} e^{z_2} = e^{x_1}(\cos y_1 + i \sin y_1)e^{x_2}(\cos y_2 + i \sin y_2).$$

Since $e^{x_1} e^{x_2} = e^{x_1 + x_2}$ for these *real* functions, by an application of the addition formulas for the cosine and sine functions (similar to that in Sec. 12.2) we find that this equals

$$e^{z_1} e^{z_2} = e^{x_1 + x_2}[\cos (y_1 + y_2) + i \sin (y_1 + y_2)] = e^{z_1 + z_2}.$$

as asserted. An interesting special case is $z_1 = x$, $z_2 = iy$:

$$(4) \qquad\qquad e^z = e^x e^{iy}.$$

Furthermore, for $z = iy$ we have from (1) the so-called **Euler formula**

$$(5) \qquad\qquad e^{iy} = \cos y + i \sin y.$$

Hence the **polar form** of a complex number, $z = r(\cos \theta + i \sin \theta)$, may now be written

$$(6) \qquad\qquad \boxed{z = re^{i\theta}.}$$

From (5) we also see that

---

[10]This definition provides for a relatively simple discussion. We could also define e^z by the familiar series $e^x = \sum_{n=0}^{\infty} x^n/n!$ with x replaced by z but then we would first have to discuss complex series at this early stage. We show the connection in Sec. 14.6.

$$(7) \qquad |e^{iy}| = |\cos y + i \sin y| = \sqrt{\cos^2 y + \sin^2 y} = 1.$$

That is, for pure imaginary exponents the exponential function has absolute value one, a result the student should remember. From (7) and (1),

$$(8) \quad |e^z| = e^x. \qquad \text{Hence} \qquad \arg e^z = y + 2n\pi \quad (n = 0, 1, 2, \cdots),$$

since $|e^z| = e^x$ shows that (1) is actually e^z in polar form.

EXAMPLE 1. Illustration of some properties of the exponential function

Computation of values from (1) provides no problem. For instance, verify that

$$e^{1.4-0.6i} = e^{1.4}(\cos 0.6 - i \sin 0.6) = 4.055(0.825 - 0.565i) = 3.347 - 2.290i,$$

$$|e^{1.4-0.6i}| = e^{1.4} = 4.055, \qquad \text{Arg } e^{1.4-0.6i} = -0.6.$$

Since $\cos 2\pi = 1$ and $\sin 2\pi = 0$, we have from (5)

$$(9) \qquad\qquad\qquad e^{2\pi i} = 1.$$

Furthermore, use (1), (5) or (6) to verify these important special values:

$$(10) \qquad e^{\pi i/2} = i, \qquad e^{\pi i} = -1, \qquad e^{-\pi i/2} = -i, \qquad e^{-\pi i} = -1.$$

To illustrate (3), take the product of

$$e^{2+i} = e^2(\cos 1 + i \sin 1) \qquad \text{and} \qquad e^{4-i} = e^4(\cos 1 - i \sin 1)$$

and verify that it equals

$$e^2 e^4(\cos^2 1 + \sin^2 1) = e^6 = e^{(2+i)+(4-i)}.$$

Finally, conclude from $|e^z| = e^x \neq 0$ in (8) that

$$(11) \qquad\qquad\qquad e^z \neq 0 \qquad \text{for all } z.$$

So here we have an entire function that never vanishes, in contrast to (nonconstant) polynomials, which are also entire (cf. Example 5 in Sec. 12.4) but always have a zero, as is proved in algebra. [Can you obtain (11) from (3)?] ∎

Periodicity of e^z with period $2\pi i$,

$$(12) \qquad\qquad\qquad e^{z+2\pi i} = e^z \qquad\qquad \text{for all } z$$

is a basic property that follows from (1) and the periodicity of $\cos y$ and $\sin y$. [It also follows from (3) and (9).] Hence all the values that $w = e^z$ can assume are already assumed in the horizontal strip of width 2π

$$(13) \qquad\qquad\qquad\qquad -\pi < y \leq \pi \qquad\qquad\qquad \text{(Fig. 313).}$$

This infinite strip is called a **fundamental region** of e^z.

EXAMPLE 2. Solution of an equation

Find all solutions of $e^z = 3 + 4i$.

Solution. $|e^z| = e^x = 5$, $x = \ln 5 = 1.609$ is the real part of all solutions. Furthermore, since $e^x = 5$,

$$e^x \cos y = 3, \qquad e^x \sin y = 4, \qquad \cos y = 0.6, \qquad \sin y = 0.8, \qquad y = 0.927.$$

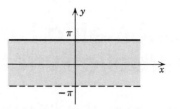

Fig. 313. Fundamental region of the exponential function e^z in the z-plane

Ans. $z = 1.609 + 0.927i \pm 2n\pi i$ $(n = 0, 1, 2, \cdots)$. These are infinitely many solutions (due to the periodicity of e^z). They lie on the vertical line $x = 1.609$ at a distance 2π from their neighbors. ∎

To summarize: many properties of $e^z = \exp z$ parallel those of e^x; an exception is the periodicity of e^z with $2\pi i$, which suggested the concept of a fundamental region and causes the periodicity of $\cos z$ and $\sin z$ with the *real* period 2π, as we shall see in the next section. Keep in mind that e^z is an *entire function*. (Do you still remember what that means?)

Problems for Sec. 12.6

1. Using the Cauchy–Riemann equations, show that e^z is analytic for all z.

Compute e^z (in the form $u + iv$) and $|e^z|$ when z equals

2. $3 + \pi i$ **3.** $-i$ **4.** $-\frac{1}{2} + \frac{3}{2}i$ **5.** $-2 - 3\pi i$

6. $9\pi i/2$ **7.** $e + 5\pi i$ **8.** $\pi - i/2$ **9.** $-1 - 7\pi i/4$

Find the real and imaginary parts of

10. $e^{-z/2}$ **11.** $e^{2\pi z}$ **12.** $e^{z^2}, (e^z)^2$ **13.** $e^{(1+i)z}$

Write the following expressions in the polar form (6).

14. $\sqrt{i}, \sqrt{-i}$ **15.** $-1, -1 + i$ **16.** $\sqrt[n]{z}$ **17.** $6 - 8i$

Find all solutions and plot some of them in the complex plane.

18. $e^z = -3 + 4i$ **19.** $e^z = i$ **20.** $e^{2z} = -2$ **21.** $e^{z^2} = 1$

Find all values of z such that

22. $|e^z| \leq 1$ **23.** Im $e^{2z} = 0$ **24.** $e^{i\bar{z}} = \overline{e^{iz}}$ **25.** $e^{\bar{z}} = \overline{e^z}$

Differentiate

26. $z^4 e^{z^3}$ **27.** $e^{\pi i z}$ **28.** $e^{(a+ib)z^2}$ **29.** $e^{1/z}$ $(z \neq 0)$

30. Show that $u = e^{xy} \cos (x^2/2 - y^2/2)$ is harmonic and find a conjugate.

31. It is interesting that $f(z) = e^z$ is *uniquely* determined by the two properties $f(x + i0) = e^x$ and $f'(z) = f(z)$, where f is assumed to be entire. Prove this. *Hint.* Let g be another entire function with these two properties and show that $(g/f)' = 0$.

32. Prove the statement in Prob. 31, using only the Cauchy–Riemann equations.

33. Let $z = re^{i\theta}$. Write $\bar{z}, -z, -\bar{z}$ in this polar form.

34. Show that $e^{\pi i/4} + e^{-\pi i/4} = 2 \cos \frac{1}{4}\pi = \sqrt{2}$, $e^{\pi i/4} - e^{-\pi i/4} = i\sqrt{2}$.

35. Show $e^{iz} + e^{-iz} = 2 \cos x \cosh y - 2i \sin x \sinh y$.

12.7 Trigonometric Functions, Hyperbolic Functions

Just as e^z extends e^x to complex, we want the *complex* trigonometric functions to extend the familiar *real* trigonometric functions. The idea of making the connection is the use of the Euler formulas (Sec. 12.6)

$$e^{ix} = \cos x + i \sin x, \qquad e^{-ix} = \cos x - i \sin x.$$

By addition and subtraction we obtain

$$\cos x = \frac{1}{2}(e^{ix} + e^{-ix}), \qquad \sin x = \frac{1}{2i}(e^{ix} - e^{-ix}) \qquad (x \text{ real}).$$

This suggests the following definitions for complex values $z = x + iy$:

(1)
$$\cos z = \frac{1}{2}(e^{iz} + e^{-iz}), \qquad \sin z = \frac{1}{2i}(e^{iz} - e^{-iz}).$$

Furthermore, in agreement with the definitions from real calculus we define

(2)
$$\tan z = \frac{\sin z}{\cos z}, \qquad \cot z = \frac{\cos z}{\sin z}$$

and

(3)
$$\sec z = \frac{1}{\cos z}, \qquad \csc z = \frac{1}{\sin z}.$$

Since e^z is entire, $\cos z$ and $\sin z$ are entire functions. $\tan z$ and $\sec z$ are not entire; they are analytic except at the points where $\cos z$ is zero; and $\cot z$ and $\csc z$ are analytic except where $\sin z$ is zero. Formulas for the derivatives follow readily from $(e^z)' = e^z$ and (1)—(3); as in calculus,

(4) $\quad (\cos z)' = -\sin z, \qquad (\sin z)' = \cos z, \qquad (\tan z)' = \sec^2 z,$

etc. Equation (1) also shows that **Euler's formula** *is valid in complex:*

(5)
$$e^{iz} = \cos z + i \sin z \qquad \text{for all } z.$$

Real and imaginary parts of $\cos z$ and $\sin z$ are needed in computing values, and they also help in displaying properties of our functions. We illustrate this by a typical example.

EXAMPLE 1. Real and imaginary parts. Absolute value. Periodicity
Show that

(6)

(a) $\quad \cos z = \cos x \cosh y - i \sin x \sinh y$

(b) $\quad \sin z = \sin x \cosh y + i \cos x \sinh y$

and

(7)

$$\textbf{(a)} \quad |\cos z|^2 = \cos^2 x + \sinh^2 y$$

$$\textbf{(b)} \quad |\sin z|^2 = \sin^2 x + \sinh^2 y$$

and give some applications of these formulas.

Solution. From (1),

$$\cos z = \tfrac{1}{2}(e^{i(x+iy)} + e^{-i(x+iy)})$$

$$= \tfrac{1}{2}e^{-y}(\cos x + i \sin x) + \tfrac{1}{2}e^{y}(\cos x - i \sin x)$$

$$= \tfrac{1}{2}(e^{y} + e^{-y})\cos x - \tfrac{1}{2}i(e^{y} - e^{-y})\sin x.$$

This yields (6a) since, as is known from calculus,

(8) $$\cosh y = \tfrac{1}{2}(e^{y} + e^{-y}), \qquad \sinh y = \tfrac{1}{2}(e^{y} - e^{-y});$$

(6b) is obtained similarly. From (6a) and $\cosh^2 y = 1 + \sinh^2 y$ we obtain

$$|\cos z|^2 = \cos^2 x \,(1 + \sinh^2 y) + \sin^2 x \sinh^2 y.$$

Since $\sin^2 x + \cos^2 x = 1$, this gives (7a), and (7b) is obtained similarly.

For instance, $\cos (2 + 3i) = \cos 2 \cosh 3 - i \sin 2 \sinh 3 = -4.190 - 9.109i$.

From (6) we see that $\cos z$ and $\sin z$ are *periodic with period* 2π, just as in real. Periodicity of $\tan z$ and $\cot z$ with period π now follows.

Formula (7) points to an essential difference between the real and the complex cosine and sine: whereas $|\cos x| \leq 1$ and $|\sin x| \leq 1$, the complex cosine and sine functions are no longer bounded but approach infinity in absolute value as $y \to \infty$, since then $\sinh y \to \infty$.

EXAMPLE 2. Solution of equations. Zeros

Solve (a) $\cos z = 5$ (which has no real solution!), (b) $\cos z = 0$, (c) $\sin z = 0$.

Solution. (a) $e^{2iz} - 10e^{iz} + 1 = 0$ from (1) by multiplication by e^{iz}. This is a quadratic equation in e^{iz}, with solutions (3D-values)

$$e^{iz} = e^{-y+ix} = 5 \pm \sqrt{25 - 1} = 9.899 \text{ and } 0.101.$$

Thus $e^{-y} = 9.899$ or 0.101, $e^{ix} = 1$, $y = \pm 2.292$, $x = 2n\pi$.

Ans. $z = \pm 2n\pi \pm 2.292i$ $(n = 0, 1, 2, \cdots)$. Can you obtain this by using (6a)?

(b) $\cos x = 0$, $\sinh y = 0$ by (7a), $y = 0$. *Ans.* $z = \pm \tfrac{1}{2}(2n + 1)\pi$ $(n = 0, 1, 2, \cdots)$.

(c) $\sin x = 0$, $\sinh y = 0$ by (7b). *Ans.* $z = 2n\pi$ $(n = 0, 1, 2, \cdots)$. Hence the only zeros of $\cos z$ and $\sin z$ are those of the real cosine and sine functions.   ∎

From the definitions it follows immediately that *all the familiar formulas for the real trigonometric functions continue to hold for complex values.* We mention in particular the addition rules

(9)

$$\cos (z_1 \pm z_2) = \cos z_1 \cos z_2 \mp \sin z_1 \sin z_2$$

$$\sin (z_1 \pm z_2) = \sin z_1 \cos z_2 \pm \sin z_2 \cos z_1$$

and the formula

(10) $$\cos^2 z + \sin^2 z = 1.$$

Some further useful formulas are included in the problem set.

Hyperbolic Functions

The complex **hyperbolic cosine** and **sine** are defined by the formulas

(11)
$$\boxed{\cosh z = \tfrac{1}{2}(e^z + e^{-z}), \qquad \sinh z = \tfrac{1}{2}(e^z - e^{-z}).}$$

This is suggested by the familiar definitions for a real variable [cf. (8)]. These functions are entire, with derivatives

(12)
$$(\cosh z)' = \sinh z, \qquad (\sinh z)' = \cosh z,$$

as in calculus. The other hyperbolic functions are defined by

(13)
$$\tanh z = \frac{\sinh z}{\cosh z}, \qquad \coth z = \frac{\cosh z}{\sinh z},$$
$$\operatorname{sech} z = \frac{1}{\cosh z}, \qquad \operatorname{csch} z = \frac{1}{\sinh z}.$$

Complex trigonometric and hyperbolic functions are related. If in (11), we replace z by iz and use (1), we obtain

(14)
$$\cosh iz = \cos z, \qquad \sinh iz = i \sin z.$$

From this, since cosh is even and sinh is odd, conversely

(15)
$$\cos iz = \cosh z, \qquad \sin iz = i \sinh z.$$

Apart from their practical importance, these formulas are remarkable in principle. Whereas in real calculus, the trigonometric and hyperbolic functions are of a different character, in complex these functions are intimately related. Moreover, the Euler formula relates them to the exponential function. This situation illustrates that by working in complex, rather than in real, one can often gain a deeper understanding of **special functions.** This is one of three main reasons for the practical importance of complex analysis, mentioned at the beginning of the chapter.

In the next section we discuss the **complex logarithm,** which differs substantially from the real logarithm (which is simpler), and the student should work the next section with particular care.

Problems for Sec. 12.7

1. Prove that $\cos z$, $\sin z$, $\cosh z$, $\sinh z$ are entire functions.
2. Obtain the differentiation formulas (4) and (12) from (1) and (11).

Differentiate

3. $\sin (1/z)$ **4.** $\cosh^2 \pi z$ **5.** $\sinh z^2$

Compute (in the form $u + iv$)

6. $\cos (3 + 2i)$ **7.** $\sin (3 + 2i)$ **8.** $\sin 3\pi i$
9. $\cos (\pi + i\pi)$ **10.** $\cos i$ **11.** $\sin (\sqrt{2} - 4i)$

Find all solutions of the following equations.

12. $\sinh z = 0$ **13.** $\sin z = \cosh 2$ **14.** $\cos z = 3i$

Find

15. $\operatorname{Re} \tan z$ **16.** $\operatorname{Im} \tan z$ **17.** $\operatorname{Re} \sec z$

18. Show that $\cosh z = \cosh x \cos y + i \sinh x \sin y$.

19. Show that $\sinh z = \sinh x \cos y + i \cosh x \sin y$.

Find (in the form $u + iv$)

20. $\sinh (4 - 3i)$ **21.** $\cosh (1 + \frac{1}{2}\pi i)$ **22.** $\cosh (-\frac{1}{2}\pi + \frac{3}{2}\pi i)$

23. Verify by differentiation that the real and imaginary parts of $\cos z$ and $\cosh z$ are harmonic functions.

24. Find all values of z for which (*a*) $\cos z$, (*b*) $\sin z$ has real values.

25. Prove that $\cos z$ is even, $\cos (-z) = \cos z$, and $\sin z$ is odd, $\sin (-z) = -\sin z$.

26. Show that $\cos \bar{z} = \overline{\cos z}$ and $\sin \bar{z} = \overline{\sin z}$ for all z.

27. Show that $\cos z = \sin (z + \frac{1}{2}\pi)$ and $\sin (\pi - z) = \sin z$.

28. Show that $\sinh z$ and $\cosh z$ are periodic with period $2\pi i$.

29. Prove the addition rules (9).

30. Show that (6) can also be obtained from the addition rules (9).

31. From (9) and (15) derive the addition rules

$$\cosh (z_1 + z_2) = \cosh z_1 \cosh z_2 + \sinh z_1 \sinh z_2$$

$$\sinh (z_1 + z_2) = \sinh z_1 \cosh z_2 + \cosh z_1 \sinh z_2.$$

32. Prove

$$\cos^2 z + \sin^2 z = 1, \qquad \cos^2 z - \sin^2 z = \cos 2z$$

$$\cosh^2 z - \sinh^2 z = 1, \qquad \cosh^2 z + \sinh^2 z = \cosh 2z.$$

33. Solve (a) in Example 2 by (6a).

34. Show that $|\sinh y| \leq |\cos z| \leq \cosh y$ and $|\sinh y| \leq |\sin z| \leq \cosh y$. Conclude that the complex cosine and sine are not bounded in the whole complex plane.

35. Show that $|\cosh z| \leq \cosh x$.

12.8 Logarithm. General Power

The **natural logarithm** of $z = x + iy$ is denoted by $\ln z$ (sometimes also by $\log z$) and is defined as the inverse of the exponential function; that is, $w = \ln z$ is defined for $z \neq 0$ by the relation

$$e^w = z.$$

(Note that $z = 0$ is impossible, since $e^w \neq 0$ for all w; cf. Sec. 12.6.) If we set $w = u + iv$ and $z = re^{i\theta}$, this becomes

$$e^w = e^{u+iv} = re^{i\theta}.$$

Now from Sec. 12.6 we know that e^{u+iv} has the absolute value e^u and the argument v. These must be equal to the absolute value and argument on the right:

$$e^u = r, \qquad v = \theta.$$

$e^u = r$ gives $u = \ln r$, where $\ln r$ is the familiar *real* natural logarithm of the positive number $r = |z|$. Hence $w = u + iv = \ln z$ is given by

(1) $$\boxed{\ln z = \ln r + i\theta} \qquad (r = |z|, \ \theta = \arg z).$$

Now comes an important point (without analog in real calculus): since the argument of z is determined only up to integer multiples of 2π, *the complex natural logarithm* $\ln z$ $(z \neq 0)$ *is infinitely many-valued.*

The value of $\ln z$ corresponding to the principal value Arg z is denoted by Ln z and is called the **principal value** of $\ln z$. Thus

(2) $$\boxed{\text{Ln } z = \ln |z| + i \text{ Arg } z.}$$

The uniqueness of Arg z for given z $(\neq 0)$ implies that Ln z is single-valued, that is, a function in the usual sense. Since the other values of arg z differ by integer multiples of 2π, the other values of $\ln z$ are given by

(3) $$\boxed{\ln z = \text{Ln } z \pm 2n\pi i} \qquad (n = 1, 2, \cdots).$$

They all have the same real part, and their imaginary parts differ by integer multiples of 2π.

If z is positive real, then Arg $z = 0$, and Ln z becomes identical with the real natural logarithm known from calculus. If z is negative real (so that the natural logarithm of calculus is not defined!), then Arg $z = \pi$ and

$$\text{Ln } z = \ln |z| + \pi i.$$

EXAMPLE 1. Natural logarithm. Principal value

$$\ln 1 = 0, \pm 2\pi i, \pm 4\pi i, \cdots \qquad \text{Ln } 1 = 0$$
$$\ln 4 = 1.386\,294 \pm 2n\pi i \qquad \text{Ln } 4 = 1.386\,294$$
$$\ln(-1) = \pm \pi i, \pm 3\pi i, \pm 5\pi i, \cdots \qquad \text{Ln}(-1) = \pi i$$
$$\ln(-4) = 1.386\,294 \pm (2n+1)\pi i \qquad \text{Ln}(-4) = 1.386\,294 + \pi i$$
$$\ln i = \pi i/2, -3\pi i/2, 5\pi i/2, \cdots \qquad \text{Ln } i = \pi i/2$$
$$\ln 4i = 1.386\,294 + \pi i/2 \pm 2n\pi i \qquad \text{Ln } 4i = 1.386\,294 + \pi i/2$$
$$\ln(-4i) = 1.386\,294 - \pi i/2 \pm 2n\pi i \qquad \text{Ln}(-4i) = 1.386\,294 - \pi i/2$$
$$\ln(3-4i) = \ln 5 + i\arg(3-4i) \qquad \text{Ln}(3-4i) = 1.609\,439 - 0.927\,295i$$
$$= 1.609\,438 - 0.927\,295i \pm 2n\pi i$$

The familiar relations for the natural logarithm continue to hold for complex values, that is,

(4)   (a) $\ln (z_1 z_2) = \ln z_1 + \ln z_2$,   (b) $\ln (z_1/z_2) = \ln z_1 - \ln z_2$

but these relations are to be understood in the sense that each value of one side is also contained among the values of the other side.

EXAMPLE 2. Illustration of the functional relations (4) in complex
Let

$$z_1 = z_2 = e^{\pi i} = -1.$$

If we take

$$\ln z_1 = \ln z_2 = \pi i,$$

then (4a) holds provided we write $\ln (z_1 z_2) = \ln 1 = 2\pi i$; it is not true for the principal value, $\mathrm{Ln} (z_1 z_2) = \mathrm{Ln} 1 = 0$. ∎

From (1) and $e^{\ln r} = r$ for positive real r we obtain

(5a) $e^{\ln z} = z$

as expected, but, since arg $(e^z) = y \pm 2n\pi$ is multivalued,

(5b) $\ln (e^z) = z \pm 2n\pi i.$

For every fixed nonnegative integer n, formula (3) defines a function. We prove that each of these functions, in particular the principal value $\mathrm{Ln}\, z$, is analytic except at $z = 0$ and except on the negative real axis (where the imaginary part of such a function is not even continuous but has a jump of magnitude 2π). We do this by proving that

(6) $(\ln z)' = \dfrac{1}{z}$ (z not negative real or zero).

In (1) we have $\ln z = u + iv$, where

$$u = \ln r = \ln |z| = \tfrac{1}{2} \ln (x^2 + y^2), \qquad v = \arg z = \arctan \frac{y}{x} + c,$$

where c is a constant (a multiple of $n\pi$). We calculate the partial derivatives of u and v and find that they satisfy the Cauchy–Riemann equations (Sec. 12.5):

$$u_x = \frac{x}{x^2 + y^2} = v_y = \frac{1}{1 + (y/x)^2} \cdot \frac{1}{x},$$

$$u_y = \frac{y}{x^2 + y^2} = -v_x = -\frac{1}{1 + (y/x)^2}\left(-\frac{y}{x^2}\right).$$

Formula (4) in Sec. 12.5 now yields the desired result:

$$(\ln z)' = u_x + iv_x = \frac{x}{x^2 + y^2} + i\,\frac{1}{1 + (y/x)^2}\left(-\frac{y}{x^2}\right) = \frac{x - iy}{x^2 + y^2} = \frac{1}{z}. ∎$$

General Powers

General powers of a complex number $z = x + iy$ are defined by the formula

$$(7) \qquad\qquad z^c = e^{c \ln z} \qquad\qquad (c \text{ complex}, z \neq 0).$$

Since $\ln z$ is infinitely many-valued, z^c will, in general, be multivalued. The particular value

$$z^c = e^{c \operatorname{Ln} z}$$

is called the *principal value of z^c.*

If $c = n = 1, 2, \cdots$, then z^n is single-valued and identical with the usual nth power of z. If $c = -1, -2, \cdots$, the situation is similar.

If $c = 1/n$ where $n = 2, 3, \cdots$, then

$$z^c = \sqrt[n]{z} = e^{(1/n) \ln z} \qquad\qquad (z \neq 0),$$

the exponent is determined up to multiples of $2\pi i/n$ and we obtain the n distinct values of the nth root, in agreement with the result in Sec. 12.2. If $c = p/q$, the quotient of two positive integers, the situation is similar, and z^c has only finitely many distinct values. However, if c is real irrational or genuinely complex, then z^c is infinitely many-valued.

EXAMPLE 3. General power

$$i^i = e^{i \ln i} = \exp (i \ln i) = \exp \left[i \left(\frac{\pi}{2} i \pm 2n\pi i \right) \right] = e^{-(\pi/2) \mp 2n\pi}.$$

All these values are real, and the principal value ($n = 0$) is $e^{-\pi/2}$. Similarly, by direct calculation and multiplying out in the exponent,

$$(1 + i)^{2-i} = \exp [(2 - i) \ln (1 + i)] = \exp [(2 - i) \{\ln \sqrt{2} + \tfrac{1}{4} \pi i \pm 2n\pi i\}]$$

$$= 2e^{\pi/4 \pm 2n\pi}[\sin (\tfrac{1}{2} \ln 2) + i \cos (\tfrac{1}{2} \ln 2)]. \qquad\qquad \blacksquare$$

It is a *convention* that for real positive $z = x$ the expression z^c means $e^{c \ln x}$ where $\ln x$ is the elementary real natural logarithm (that is, the principal value $\operatorname{Ln} z$ ($z = x > 0$) in the sense of our definition). Also, if $z = e$, the base of the natural logarithm, $z^c = e^c$ is *conventionally* regarded as the unique value obtained from (1) in Sec. 12.6.

From (7) we see that for any complex number a,

$$(8) \qquad\qquad \boxed{a^z = e^{z \ln a}.}$$

We have now introduced and discussed the most important special complex functions. In the next section, which is *optional,* we consider mapping properties of some of these function, with the intention to give the student additional insights into the nature of these functions.[11]

---

[11]A detailed discussion of mappings given by analytic functions follows in Chap. 16.

Problems for Sec. 12.8

1. Verify the calculations in Example 1. Plot some of the values of ln $(3 - 4i)$ in the complex plane.
2. Prove (5a) and (5b).
3. Verify (4) for $z_1 = -i$ and $z_2 = -1$.
4. Prove the analyticity of Ln z by using the Cauchy–Riemann equations in polar form (cf. Sec. 12.5). Obtain the derivative of Ln z from (8b) in Sec. 12.5.

Find the principal value Ln z when z equals

5. -7 6. $\sqrt{2} + i\sqrt{2}$ 7. $(1 - i)^2$ 8. $3 + i\sqrt{27}$

Find all values of the given expressions and plot some of them in the complex plane.

9. ln 2   10. ln (-5) 11. ln $(-i/3)$ 12. ln $(3 + 2i)$
13. ln (e^{2i}) 14. ln $(-\frac{1}{4} + 5i)$ 15. ln $(\pi - i)$ 16. ln $(e + \pi i)$

Solve the following equations for z.

17. ln $z = -\pi i/2$ 18. ln $z = -2 - \frac{3}{2}i$ 19. ln $z = 3 - i$ 20. ln $z = e - \pi i$

Find the principal value of

21. $i^{1/2}$ 22. $(2i)^{2i}$ 23. $(1 + 3i)^i$ 24. $(1 + i)^{1-i}$
25. 3^{4-i} 26. $(3 + 4i)^{1/3}$ 27. $(-5)^{2-4i}$ 28. $(5 - 2i)^{3+\pi i}$

By definition, the **inverse sine** $w = \sin^{-1} z$ is the relation such that $\sin w = z$. The **inverse cosine** $w = \cos^{-1} z$ is the relation such that $\cos w = z$. The **inverse tangent, inverse cotangent, inverse hyperbolic sine,** etc., are defined and denoted in a similar fashion. Using $\sin w = (e^{iw} - e^{-iw})/2i$ and similar representations of $\cos w$ etc., show that

29. $\sin^{-1} z = -i \ln (iz + \sqrt{1 - z^2})$ 30. $\cos^{-1} z = -i \ln (z + \sqrt{z^2 - 1})$
31. $\cosh^{-1} z = \ln (z + \sqrt{z^2 - 1})$ 32. $\sinh^{-1} z = \ln (z + \sqrt{z^2 + 1})$
33. $\tan^{-1} z = \dfrac{i}{2} \ln \dfrac{i + z}{i - z}$ 34. $\tanh^{-1} z = \dfrac{1}{2} \ln \dfrac{1 + z}{1 - z}$

35. Show that $w = \sin^{-1} z$ is infinitely many-valued, and if w_1 is one of these values, the others are of the form $w_1 \pm 2n\pi$ and $\pi - w_1 \pm 2n\pi$, $n = 0, 1, \cdots$. (The *principal value* of $w = u + iv = \sin^{-1} z$ is defined to be the value for which $-\pi/2 \le u \le \pi/2$ when $v \ge 0$ and $-\pi/2 < u < \pi/2$ when $v < 0$.)

12.9 Mapping by Special Functions

This section is *optional*. In it we discuss the elementary functions as mappings. This will supplement the student's understanding of these functions. A discussion of the mapping properties of analytic functions in general follows in Chap. 16, along with applications in Chap. 17.

From calculus we know that a *real* function $y = f(x)$ can be exhibited graphically by plotting it in the xy-plane. This gives a curve when f is continuous. For a *complex* function

$$w = u + iv = f(z) \qquad (z = x + iy)$$

we need two planes, the z-plane in which we plot values of z, and the w-plane in which we plot the corresponding function values $w = f(z)$. In this way, a given function f assigns to each point z in its domain of definition D the corresponding point $w = f(z)$ in the w-plane. We say that f defines a **mapping** of D into the w-plane. For any point z_0 in D we call the point $w_0 = f(z_0)$ the **image** of z_0 with respect to f. More generally, for the points of a curve C in D, the image points form what is called the *image* of C; similarly for other point sets in D.

The next idea is this. By plotting single points and their images we would not gain much insight into the mapping properties of a given function. For this purpose it is much better if we graph and study regions and their images, or families of simple curves (for instance, parallel straight lines or concentric circles) and their images. We shall explain this method for the elementary functions in terms of selected examples of practical importance.

Linear Functions. Positive Integer Powers of z

EXAMPLE 1. Mapping $w = az + b$

Describe, in geometrical terms, the mapping properties of the linear function $w = az + b$ with complex a and b.

Solution. If $a = 1$, this is a **translation** $w = z + b$. It moves (translates) every point in the direction of b through the same distance $|b|$. Hence every figure and its image are congruent. Figure 314 shows an example. $b = 0$ gives the **identity mapping**, which leaves every point fixed.

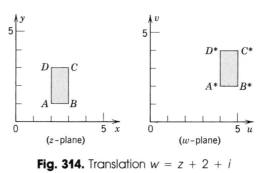

Fig. 314. Translation $w = z + 2 + i$

If $b = 0$, we get $w = az$. For $|a| = 1$ we have $a = e^{i\alpha}$ and $w = e^{i\alpha}z$, which is a rotation through the angle α. Figure 315 shows an example. For real a with $0 < a < 1$ we get a uniform contraction and for real $a > 1$ a uniform expansion. (Explain! What about negative real a?)

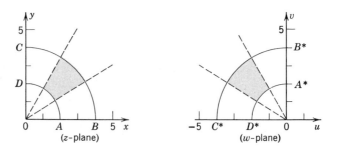

Fig. 315. Rotation $w = iz$. (Angle of rotation $\pi/2$, counterclockwise)

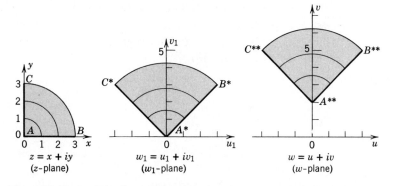

Fig. 316. Linear transformation $w = u + iv = (1 + i)z + 2i$, consisting of the rotation and expansion $w_1 = u_1 + iv_1 = (1 + i)z$ followed by the translation $w = w_1 + 2i$

Combining our results, we conclude that the general linear function $w = az + b$ is a rotation and an expansion (or contraction) $w_1 = az$ followed by a translation $w = w_1 + b$. For instance, Fig. 316 shows a counterclockwise rotation through the angle $\pi/4$ and an expansion by the factor $|1 + i| = \sqrt{2}$ followed by a translation vertically upward.

EXAMPLE 2. Mapping $w = z^2$

Find the images of the circles $|z| = r_0 = const$ and of the rays $\theta = \theta_0 = const$ under the mapping $w = z^2$. Also find the image of the region $1 \le |z| \le 3/2$, $\pi/6 < \theta < \pi/3$.

Solution. In terms of polar coordinates

$$w = Re^{i\phi} = z^2 = r^2 e^{2i\theta} \qquad (z = re^{i\theta}).$$

This shows that any circle $|z| = r_0$ maps onto the circle $|w| = r_0^2$, and any ray $\theta = \theta_0$ onto the ray $\phi = 2\theta_0$. Hence that region is mapped onto the region $1 \le |w| \le 9/4$, $\pi/3 < \phi < 2\pi/3$ in the w-plane (cf. Fig. 317).

The circles and rays considered intersect at right angles, and the mapping $w = z^2$ preserves these angles: the corresponding images also intersect at right angles. This is typical of analytic functions f: they preserve angles between any intersecting curves whatsoever, as we shall see in Chap. 16. An exception occurs at points at which $f'(z) = 0$. For $w = z^2$ we have $w' = 2z = 0$ at $z = 0$; indeed, we see that images of rays do not make the same angle as the rays themselves, but make *twice* this angle. ∎

Example 2 is a model case for all positive integer powers z^n. In terms of polar coordinates,

$$w = Re^{i\phi} = z^n = r^n e^{in\theta}.$$

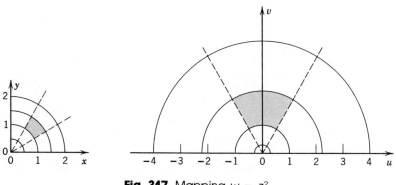

Fig. 317. Mapping $w = z^2$

This shows that $w = z^n$ maps circles $|z| = r_0$ onto circles $|w| = r_0{}^n$, and rays $\theta = \theta_0$ onto rays $\phi = n\theta_0$. For instance, $w = z^3$ triples angles at 0. Hence $w = z^n$ maps the angular region $0 \leqq \theta \leqq \pi/n$ onto the upper half of the w-plane (cf. Fig. 318).

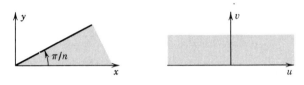

Fig. 318. Mapping defined by $w = z^n$

Exponential Function and Logarithm

As in the previous examples, we shall exhibit various properties of the mappings by considering curves and regions for which we can obtain the corresponding images without computations but simply by using properties of the functions under consideration. This is the general idea underlying our discussion in the whole section.

EXAMPLE 3. Exponential function $w = e^z$
For $w = e^z$, find the image of:

- (a) the straight lines $x = x_0 = const$ and $y = y_0 = const$,
- (b) the rectangle $0 \leqq x \leqq 1, \frac{1}{2} < y < 1$,
- (c) the fundamental region $-\pi < y \leqq \pi$,
- (d) the horizontal strip $0 \leqq y \leqq \pi$.

Solution. (a) Recall from Sec. 12.6 that $|w| = e^x$ and arg $w = y$. Hence $x = x_0$ is mapped onto the circle $|w| = e^{x_0}$, and $y = y_0$ onto the ray arg $w = y_0$.

(b) From (a) we conclude that any rectangle with sides parallel to the coordinate axes is mapped onto a region bounded by portions of rays and circles. Cf. Fig. 319.

(c) This fundamental region is mapped onto the full w-plane (cut along the negative real axis) without the origin 0. More generally, every horizontal strip bounded by two lines $y = c$ and $y = c + 2\pi$ is mapped onto the full w-plane without the origin. This reflects the fact that $w = e^z$ is periodic with the period $2\pi i$.

(d) The strip $0 \leqq y \leqq \pi$ is mapped onto the upper half of the w-plane. Its right half ($x > 0$) is mapped onto the exterior of the unit circle $|w| = 1$ and its left half ($x < 0$) onto the interior without 0 (cf. Fig. 320). ∎

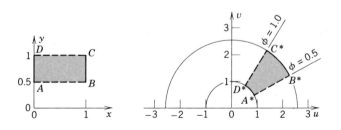

Fig. 319. Mapping by $w = e^z$ [Example 3, part (b)]

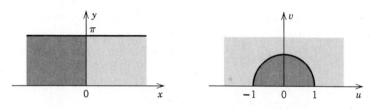

Fig. 320. Mapping by $w = e^z$ [Example 3, part (d)]

We now turn to the mapping given by the **natural logarithm.** In discussing it we use the following basic idea and principle. *The properties of the inverse of a mapping $w = f(z)$ can be obtained from those of the mapping itself by interchanging the roles of the z- and w-planes.* Since the natural logarithm $w = u + iv = \ln z$ is the inverse relation of the exponential function, we conclude from Example 3, part (c), that $w = \text{Ln } z$, $z \ne 0$ (the principal value of $\ln z$) maps the z-plane (cut along the negative real axis and with $z = 0$ omitted) onto the horizontal strip $-\pi < v \leqq \pi$ of the w-plane. Since the mapping

$$w = \text{Ln } z + 2\pi i$$

differs from $w = \text{Ln } z$ by the translation $2\pi i$ (vertically upward), this mapping maps the z-plane (cut as before and 0 omitted) onto the strip $\pi < v \leqq 3\pi$. Similarly for each of the infinitely many mappings

$$w = \ln z = \text{Ln } z \pm 2n\pi i \qquad (n = 0, 1, 2, \cdots).$$

The corresponding horizontal strips of width 2π (images of the z-plane under these mappings) together cover the whole w-plane without overlapping. We invite the reader to interpret the other parts of Example 3 in terms of the mapping $w = \text{Ln } z$, by interchanging the roles of z and w in the example.

Mapping properties of further functions (trigonometric, hyperbolic, fractional linear functions, etc.) will be discussed in Chap. 16.

This is the end of Chap. 12 on analytic functions, roughly, the first stage of complex differential calculus, with more to come. For instance higher derivatives. Why did we not consider them? Well, their existence will appear in the next chapter by arguments in *integral* calculus. This seems strange, but it simply indicates structural differences between *real* calculus and *complex* analysis. Accordingly, in the next chapter we turn to **integration,** which will pleasantly surprise us by its flavor and elegance not present in *real* integral calculus.

Problems for Sec. 12.9

Consider the mapping $w = (1 + i)z - 2$. Find and plot the images of the given curves or regions.

   1. $x = 0, 1, 2, 3$ **2.** $y = -1, 0, 1, 2$ **3.** $|z - 2| \leqq 2$

Plot the images of the following regions under the mapping $w = z^2$.

4. $|z| > 2$ **5.** $|z| \leq 4$ **6.** $|\arg z| < \pi/4$

7. $\frac{1}{2} < |z| < \sqrt{2}$ **8.** $\pi/2 \leq \arg z \leq \pi$ **9.** $-\pi/4 < \arg z < \pi/2$

10. **(Level curves)** Consider $w = u + iv = z^2$. Show that the *level curves* of u, that is, the curves $u = const$ in the z-plane, are hyperbolas with the asymptotes $y = \pm x$. Sketch some of them.

11. Find the level curves $v = const$ in Prob. 10 and sketch some of them.

12. Show that $w = u + iv = z^2$ maps $x = c$ onto the parabola $v^2 = 4c^2(c^2 - u)$ and graph it for $c = 2, 1, \frac{1}{2}, 0$.

13. Find the image of $y = k$ under the mapping $w = u + iv = z^2$.

Find and sketch the images of the following curves and regions under the mapping $w = 1/z$, where $z \neq 0$.

14. $|z| = 1$ **15.** $|\arg z| < \pi/4$ **16.** $|z| \geq 2$

17. Find the image of $x = 1$ under $w = 1/z$. *Hint.* Consider $u^2 + v^2$.

Find and sketch the images of the following regions under the mapping $w = e^z$.

18. $|x| < 1, |y| < \pi/2$ **19.** $x < 1, |y| \leq \pi$ **20.** $0 \leq y \leq \pi/2$

Further Proof in Chapter 12

Proof of Theorem 2 in Sec. 12.5 (Cauchy–Riemann equations)
We prove that the Cauchy–Riemann equations

$$(1) \qquad\qquad u_x = v_y, \qquad\qquad u_y = -v_x$$

are sufficient for a complex function

$$f(z) = u(x, y) + iv(x, y)$$

to be analytic; precisely, if the real part u and the imaginary part v of $f(z)$ satisfy (1) in a domain D in the complex plane and if the partial derivatives in (1) are continuous in D, then $f(z)$ is analytic in D.

In this proof we write $\Delta z = \Delta x + i\Delta y$ and $\Delta f = f(z + \Delta z) - f(z)$. The idea of proof is as follows.

(a) We express Δf in terms of first partial derivatives of u and v, by applying the mean value theorem of Sec. 8.7.

(b) We get rid of partial derivatives with respect to y, by applying the Cauchy–Riemann equations.

(c) We let Δz approach zero and show that then $\Delta f/\Delta z$ as obtained approaches a limit, which is equal to $u_x + iv_x$, the right-hand side of (4) in Sec. 12.5, regardless of the way of approach to zero.

The details are as follows.

(a) Let P: (x, y) be any fixed point in D. Since D is a domain, it contains a neighborhood of P. We can choose a point Q: $(x + \Delta x, y + \Delta y)$ in this neighborhood such that the straight-line segment PQ is in D. Because of our continuity assumptions we may apply the mean value theorem in Sec. 8.7.

This yields

$$u(x + \Delta x, y + \Delta y) - u(x, y) = (\Delta x)u_x(M_1) + (\Delta y)u_y(M_1)$$

$$v(x + \Delta x, y + \Delta y) - v(x, y) = (\Delta x)v_x(M_2) + (\Delta y)v_y(M_2),$$

where M_1 and M_2 ($\neq M_1$ in general!) are suitable points on that segment. The first line is Re Δf and the second is Im Δf, so that

$$\Delta f = (\Delta x)u_x(M_1) + (\Delta y)u_y(M_1) + i[(\Delta x)v_x(M_2) + (\Delta y)v_y(M_2)].$$

(b) Now $u_y = -v_x$ and $v_y = u_x$ by the Cauchy–Riemann equations, so that

$$\Delta f = (\Delta x)u_x(M_1) - (\Delta y)v_x(M_1) + i[(\Delta x)v_x(M_2) + (\Delta y)u_x(M_2)].$$

Also $\Delta z = \Delta x + i\Delta y$, so that we can write $\Delta x = \Delta z - i\Delta y$ in the first term and $\Delta y = (\Delta z - \Delta x)/i = -i(\Delta z - \Delta x)$ in the second term. This gives

$$\Delta f = (\Delta z - i\Delta y)u_x(M_1) + i(\Delta z - \Delta x)v_x(M_1) + i[(\Delta x)v_x(M_2) + (\Delta y)u_x(M_2)].$$

By performing the multiplications and reordering we obtain

$$\Delta f = (\Delta z)u_x(M_1) - i\Delta y\{u_x(M_1) - u_x(M_2)\}$$

$$+ i[(\Delta z)v_x(M_1) - \Delta x\{v_x(M_1) - v_x(M_2)\}].$$

Division by Δz now yields

$$\text{(A)} \quad \frac{\Delta f}{\Delta z} = u_x(M_1) + iv_x(M_1) - \frac{i\Delta y}{\Delta z}\{u_x(M_1) - u_x(M_2)\} - \frac{i\Delta x}{\Delta z}\{v_x(M_1) - v_x(M_2)\}.$$

(c) We finally let Δz approach zero and note that $|\Delta y/\Delta z| \leq 1$ and $|\Delta x/\Delta z| \leq 1$ in (A). Then $Q: (x + \Delta x, y + \Delta y)$ approaches $P: (x, y)$, so that M_1 and M_2 must approach P. Also, since the partial derivatives in (A) are assumed to be continuous, they approach their value at P. In particular, the differences in the braces in (A) approach zero. Hence the limit of the right-hand side of (A) exists and is independent of the path along which $\Delta z \to 0$. We see that this limit equals the right-hand side of (4) in Sec. 12.5. This means that $f(z)$ is analytic at every point z in D, and the proof is complete. ∎

Review Problems for Chapter 12

1. Show that any complex number is equal to the conjugate of its conjugate.
2. The definition of the multiplication of complex numbers looks strange at first sight. What is the motivation for it?
3. What does multiplication of a complex number by i mean geometrically?
4. Prove the associative law of the multiplication of complex numbers.

5. Is a complex number $|z|e^{i\theta}$ uniquely determined if $|z|$ and arc tan (y/x) are given?

6. What do we mean by the principal value of the argument of a complex number?

7. What is the triangle inequality? What does it mean geometrically?

8. Is Im $(1/z) = 1/\text{Im } z$? Is $1/|z| = |1/z|$? Is $|z^2| = |z|^2$?

9. What is the unit circle? An annulus? A half-plane? A domain?

10. Why are the differentiation formulas for complex functions the same as for real functions in calculus?

11. What is the essential difference with respect to the limit in the definition of the derivative in this chapter as compared to that in calculus?

12. What is an analytic function? Can a function be differentiable at a point z_0 without being analytic at z_0?

13. State the Cauchy–Riemann equations from memory. (They are so important that you should memorize them.)

14. What is the idea that led to the Cauchy–Riemann equations?

15. Given a real differentiable function $g(x)$, could you find a function $h(x, y)$ such that $g(x) + ih(x, y)$ is analytic?

16. Show that $\cosh \bar{z} = \overline{\cosh z}$, $\sinh \bar{z} = \overline{\sinh z}$ for all z.

17. Show that $\tanh z$ is periodic with πi.

18. What is the essential difference between the complex logarithm and the logarithm in calculus?

19. What is the principal value of $\ln z$?

20. What do we mean by an entire function? Which of the functions in this chapter are entire and which are not? Is Ln z entire?

Find, in the form $x + iy$,

21. $(3 - 4i)/i$ **22.** i^{26} **23.** $(2 + 3i)^3$ **24.** $e^{\pi i/2}$, $e^{-\pi i/2}$

25. $\dfrac{23 + 2i}{4 - 5i}$ **26.** $\dfrac{5i}{2 + i}$ **27.** $\dfrac{(1 + i)^{10}}{(1 - i)^6}$ **28.** $\dfrac{11 + 2i}{4 + 3i}$

Find

29. Re $\dfrac{(2 - 3i)^2}{2 + 3i}$ **30.** Im $\dfrac{3 + 4i}{7 - i}$ **31.** Re $\dfrac{8 + i}{3 + 2i}$ **32.** Im $\dfrac{24 + 7i}{3 + 4i}$

33. Re $\dfrac{44 + 53i}{2 + 3i}$ **34.** Im $\dfrac{32 + 2i}{17 - 15i}$ **35.** Re $\dfrac{-32 - 48i}{(1 + i)^6}$ **36.** Im $\dfrac{7 - i}{1 - i}$

Represent in polar form

37. $-3 + 3i$ **38.** $17 + 4i$ **39.** $(1 - i)^{12}$ **40.** $20/(6 + 2i)$

Sketch the following curves or regions in the complex plane.

41. $|z - 5i| = 5$ **42.** $1 \leq |z - i| \leq 4$ **43.** $|\arg z| < \pi/4$ **44.** $1 \leq \text{Re } z^2 \leq 4$

Find the value of the derivative of

45. e^{z^2} at $3\pi i$ **46.** $1/z$ at $2 + 1.5i$

47. $z + 1/z$ at $2 + 2i$ **48.** z^5 at $1 - i$

49. Ln z at $4 - 5i$ **50.** $\sin (1/z)$ at i/π

51. $\sinh 3z$ at $1 + \pi i/3$ **52.** $\cos z$ at $3 + 2i$

Find an analytic function $f(z) = u(x, y) + iv(x, y)$ such that

53. $v = 2y(x + 1)$ **54.** $u = 2x/(x^2 + y^2)$

55. $v = e^{x^2 - y^2} \sin 2xy$ **56.** $u = x^4 - 6x^2y^2 + y^4$

Are the following functions harmonic? If so, find a conjugate harmonic.

57. $e^{\pi x} \cos \pi y$ **58.** $x^2 + y^2$ **59.** $x/(x^2 + y^2)$ **60.** $\sin x \cosh y$

Summary of Chapter 12
Complex Numbers.
Complex Analytic Functions

For arithmetic operations with **complex numbers**

(1) $$z = x + iy = re^{i\theta} = r(\cos \theta + i \sin \theta),$$

$r = |z| = \sqrt{x^2 + y^2}$, $\theta = $ arc tan (y/x), and for their representation in the complex plane, see Secs. 12.1 and 12.2.

A complex function $f(z) = u(x, y) + iv(x, y)$ is **analytic** in a domain D if it has a **derivative** (Sec. 12.4)

(2) $$f'(z) = \lim_{\Delta z \to 0} \frac{f(z + \Delta z) - f(z)}{\Delta z}$$

everywhere in D. Also, $f(z)$ is *analytic at a point* $z = z_0$ if it has a derivative in a neighborhood of z_0 (not merely at z_0 itself).

If $f(z)$ is analytic in D, then $u(x, y)$ and $v(x, y)$ satisfy the (very important!) **Cauchy–Riemann equations** (Sec. 12.5)

(3) $$\frac{\partial u}{\partial x} = \frac{\partial v}{\partial y}, \qquad \frac{\partial u}{\partial y} = - \frac{\partial v}{\partial x}$$

everywhere in D. Then u and v also satisfy **Laplace's equation**

(4) $$u_{xx} + u_{yy} = 0, \qquad v_{xx} + v_{yy} = 0$$

everywhere in D. If $u(x, y)$ and $v(x, y)$ are continuous and have *continuous* partial derivatives in D that satisfy (3) in D, then $f(z) = u(x, y) + iv(x, y)$ is analytic in D. Cf. Sec. 12.5.

The complex **exponential function** (Sec. 12.6)

(5) $$e^z = \exp z = e^x (\cos y + i \sin y)$$

is periodic with $2\pi i$, reduces to e^x when $z = x$ $(y = 0)$ and has the derivative e^z. The **trigonometric functions** are (Sec. 12.7)

(6) $$\cos z = \frac{1}{2} (e^{iz} + e^{-iz}) = \cos x \cosh y - i \sin x \sinh y$$

$$\sin z = \frac{1}{2i} (e^{iz} - e^{-iz}) = \sin x \cosh y + i \cos x \sinh y$$

$\tan z = (\sin z)/\cos z$, $\cot z = 1/\tan z$, etc.

The **hyperbolic functions** are (Sec. 12.7)

$$\cosh z = \frac{1}{2}(e^z + e^{-z}) = \cos iz,$$

(7)   (Sec. 12.7)

$$\sinh z = \frac{1}{2}(e^z - e^{-z}) = -i \sin iz$$

etc. An **entire function** is a function that is analytic everywhere in the complex plane. The functions in (5)—(7) are entire.

The **natural logarithm** is (Sec. 12.8)

(8)
$$\ln z = \ln |z| + i \arg z \qquad (\arg z = \theta, z \ne 0)$$

$$= \ln |z| + i \operatorname{Arg} z \pm 2n\pi i \qquad (n = 0, 1, \cdots),$$

where Arg z is the principal value of arg z, that is, $-\pi < \operatorname{Arg} z \leqq \pi$. We see that ln z is infinitely many-valued. Taking $n = 0$ gives the *principal value* Ln z of ln z; thus

(8*) $$\operatorname{Ln} z = \ln |z| + i \operatorname{Arg} z.$$

General powers are defined by (Sec. 12.8)

(9) $$z^c = e^{c \ln z} \qquad (c \text{ complex}, z \ne 0).$$

A complex function $w = f(z)$ defines a mapping of its domain of definition in the z-plane into the w-plane. The mappings given by the functions (5), (8) and some others are discussed in Sec. 12.9.

Chapter 13

Complex Integration

Integration in the complex plane is important for two reasons:

1. In applications there occur real integrals that can be evaluated by complex integration, whereas the usual methods of real integral calculus are not successful.

2. Some basic properties of analytic functions can be established by integration, but would be difficult to prove by other methods. The existence of higher derivatives of analytic functions is a striking property of this type.[1]

In this chapter we define and explain complex integrals. The most fundamental result in the whole chapter is Cauchy's integral theorem (Sec. 13.3). It implies the important Cauchy integral formula (Sec. 13.5). In Sec. 13.6 we prove that if a function is analytic, it has derivatives of all orders. Hence in this respect, complex analytic functions behave much more simply than real-valued functions of real variables.

(Integration by means of residues and applications to real integrals will be considered in Chap. 15.)

Prerequisite for this chapter: Chap. 12.
References: Appendix 1, Part D.
Answers to problems: Appendix 2.

13.1 Line Integral in the Complex Plane

As in real calculus, we distinguish between definite integrals and indefinite integrals or antiderivatives. An **indefinite integral** is a function whose derivative equals a given analytic function in a region. By inverting known differentiation formulas we may find many types of indefinite integrals.

We shall now define *definite integrals,* or **line integrals,** of complex functions $f(z)$, where $z = x + iy$. This needs a short preparation on curves in the complex plane, as follows.

Path of integration. In *real* calculus, a definite integral is taken over an interval (a segment) of the real line. In the case of a *complex* definite integral

---

[1]Proved without integration or equivalent methods only relatively recently, in 1961 [by P. Porcelli and E. H. Connell (*Bulletin of the American Mathematical Society,* vol. 67, pp. 177—181), who make use of a topological theorem by G. T. Whyburn.

we integrate along a curve C in the complex plane,[2] which will be called the *path of integration.*

Now a curve C in the complex plane can be represented in the form

(1)
$$\boxed{z(t) = x(t) + iy(t)}$$
$(a \leqq t \leqq b)$

where t is a real parameter. For example,

$$z(t) = t + 3it$$
$(0 \leqq t \leqq 2)$

represents a portion of the line $y = 3x$ (sketch it!),

$$z(t) = 4\cos t + 4i \sin t$$
$(-\pi \leqq t \leqq \pi)$

represents the circle $|z| = 4$, etc. (More examples below.)

C is called a **smooth curve** if C has a derivative

$$\dot{z}(t) = \frac{dz}{dt} = \dot{x}(t) + i\dot{y}(t)$$

at each of its points which is continuous and nowhere zero. Geometrically this means that C has a continuously turning tangent. This follows directly from the definition (cf. Fig. 321)

$$\dot{z}(t) = \lim_{\Delta t \to 0} \frac{z(t + \Delta t) - z(t)}{\Delta t}$$

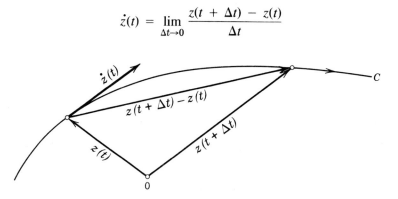

Fig. 321. Tangent vector $\dot{z}(t)$ of a curve C in the complex plane given by $z(t)$. The arrow on the curve indicates the positive sense (sense of increasing t).

Definition of the complex line integral. This will be similar to the method used in calculus. Let C be a smooth curve in the z-plane represented in the form (1). Let $f(z)$ be a continuous function defined (at least) at each point of C. We subdivide ("partition") the interval $a \leqq t \leqq b$ in (1) by points

---

[2]Actually, often only along a portion (an *arc*) of a curve, but we shall say **"curve"** in either case, for simplicity. Readers who have studied Sec. 9.1 will note that our present discussion is similar.

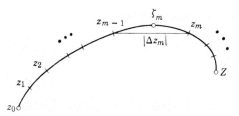

Fig. 322. Complex line integral

$$t_0 \, (= a), \, t_1, \, \cdots, \, t_{n-1}, \, t_n \, (= b)$$

where $t_0 < t_1 < \cdots < t_n$. To this subdivision there corresponds a subdivision of C by points (cf. Fig. 322)

$$z_0, \, z_1, \, \cdots, \, z_{n-1}, \, z_n \, (= Z),$$

where $z_j = z(t_j)$. On each portion of subdivision of C we choose an arbitrary point, say, a point ζ_1 between z_0 and z_1 (that is, $\zeta_1 = z(t)$ where t satisfies $t_0 \leqq t \leqq t_1$), a point ζ_2 between z_1 and z_2, etc. Then we form the sum

$$(2) \qquad\qquad S_n = \sum_{m=1}^{n} f(\zeta_m) \, \Delta z_m$$

where

$$\Delta z_m = z_m - z_{m-1}.$$

This we do for each $n = 2, 3, \cdots$ in a completely independent manner, but in such a way that the greatest $|\Delta z_m|$ approaches zero as n approaches infinity. This gives a sequence of complex numbers $S_2, S_3, \cdots$. The limit of this sequence is called the **line integral** (or simply the *integral*) of $f(z)$ along the oriented curve C and is denoted by

$$(3) \qquad\qquad \boxed{\int_C f(z) \, dz.}$$

The curve C is called the **path of integration.**

C is called a **closed path** if $Z = z_0$, that is, if its terminal point coincides with its initial point. (*Examples:* a circle, a curve shaped like an 8, etc.) Then one also writes

$$\oint_C \qquad\text{instead of}\qquad \int_C.$$

Examples follow in the next section.

General Assumption

All paths of integration for complex line integrals will be assumed to be **piecewise smooth,** *that is, to consist of finitely many smooth curves joined end to end.*

Existence of the Complex Line Integral[3]

From our assumptions that $f(z)$ is continuous and C is piecewise smooth, the existence of the line integral (3) follows. In fact, as in the previous chapter let us write $f(z) = u(x, y) + iv(x, y)$. We also set

$$\zeta_m = \zeta_m + i\eta_m \quad \text{and} \quad \Delta z_m = \Delta x_m + i\Delta y_m.$$

Then (2) may be written

$$(4) \qquad S_n = \sum (u + iv)(\Delta x_m + i\Delta y_m)$$

where $u = u(\zeta_m, \eta_m)$, $v = v(\zeta_m, \eta_m)$ and we sum over m from 1 to n. We may now split up S_n into four sums:

$$S_n = \sum u\,\Delta x_m - \sum v\,\Delta y_m + i\left[\sum u\,\Delta y_m + \sum v\,\Delta x_m\right].$$

These sums are real. Since f is continuous, u and v are continuous. Hence, if we let n approach infinity in the aforementioned way, then the greatest Δx_m and Δy_m will approach zero and each sum on the right becomes a real line integral:

$$(5) \qquad \lim_{n\to\infty} S_n = \int_C f(z)\,dz = \int_C u\,dx - \int_C v\,dy + i\left[\int_C u\,dy + \int_C v\,dx\right].$$

This shows that under our assumptions (f continuous on C and C piecewise smooth) the line integral (3) exists and its value is independent of the choice of subdivisions and intermediate points ζ_m. ∎

Three Basic Properties of Complex Line Integrals

We list three properties of complex line integrals that are quite similar to those of real definite integrals (and real line integrals) and follow immediately from the definition.

1. Integration is a linear operation, that is, a sum of two (or more) functions can be integrated term by term, and constant factors can be taken out from under the integral sign:

$$(6) \qquad \int_C [k_1 f_1(z) + k_2 f_2(z)]\,dz = k_1 \int_C f_1(z)\,dz + k_2 \int_C f_2(z)\,dz.$$

---

[3]If one does not want to refer to real line integrals, one can take (1) in the next section as the *definition* of the complex line integral.

Fig. 323. Subdivision of path [formula (7)]

2. Decomposing C into two portions C_1 and C_2 (Fig. 323), we get

$$(7) \qquad \int_C f(z)\, dz = \int_{C_1} f(z)\, dz + \int_{C_2} f(z)\, dz.$$

3. Reversing the sense of integration, we get the negative of the original value:

$$(8) \qquad \int_{z_0}^{Z} f(z)\, dz = -\int_{Z}^{z_0} f(z)\, dz;$$

here the path C with endpoints z_0 and Z is the same; on the left we integrate from z_0 to Z, and on the right from Z to z_0.

Applications follow in the next section and problems at the end of it.

13.2 Two Integration Methods. Examples

Complex integration is rich in methods for evaluating integrals. We discuss the first two of them, and others will follow later in this chapter and in Chap. 15.

First Method: Use of a Representation of the Path

This method applies to any continuous complex function.

Theorem 1 (Integration by the use of the path)

Let C be a piecewise smooth path, represented by $z = z(t)$, where $a \leq t \leq b$. Let $f(z)$ be a continuous function on C. Then

$$(1) \qquad \boxed{\int_C f(z)\, dz = \int_a^b f[z(t)]\dot{z}(t)\, dt} \qquad \left(\dot{z} = \frac{dz}{dt}\right).$$

Proof. The left-hand side of (1) is given by (5), Sec. 13.1, in terms of real line integrals, and we show that the right-hand side of (1) also equals (5). We have $z = x + iy$, hence $\dot{z} = \dot{x} + i\dot{y}$. We simply write u for $u[x(t), y(t)]$ and v for $v[x(t), y(t)]$. We also have $dx = \dot{x}\, dt$ and $dy = \dot{y}\, dt$. Consequently, in (1),

$$\int_a^b f[z(t)]\dot{z}(t)\, dt = \int_a^b (u + iv)(\dot{x} + i\dot{y})\, dt$$

$$= \int_C [u\, dx - v\, dy + i(u\, dy + v\, dx)],$$

which is the right-hand side of (5), as claimed.   ∎

Steps in applying Theorem 1

(A) Represent the path C in the form $z(t)$ ($a \leqq t \leqq b$).
(B) Calculate the derivative $\dot{z}(t) = dz/dt$.
(C) Substitute $z(t)$ for every z in $f(z)$ (hence $x(t)$ for x and $y(t)$ for y).
(D) Integrate $f[z(t)]\dot{z}(t)$ over t from a to b.

EXAMPLE 1. A basic result: Integral of 1/z around the unit circle

Show that

(2) $$\oint_C \frac{dz}{z} = 2\pi i$$ (C the unit circle, counterclockwise).

This important result will be frequently needed.

Solution. We may represent the unit circle C in the form

$$z(t) = \cos t + i \sin t \qquad (0 \leqq t \leqq 2\pi),$$

so that the counterclockwise integration corresponds to an increase of t from 0 to 2π. By differentiation,

$$\dot{z}(t) = -\sin t + i \cos t.$$

Also $f[z(t)] = 1/z(t)$. Formula (1) now yields the desired result

$$\oint_C \frac{dz}{z} = \int_0^{2\pi} \frac{1}{\cos t + i \sin t}(-\sin t + i \cos t)\, dt = i\int_0^{2\pi} dt = 2\pi i.$$

The Euler formula (Sec. 12.6) helps us to save work by representing the unit circle simply in the form

$$z(t) = e^{it}. \qquad \text{Then} \qquad 1/z(t) = e^{-it}, \qquad dz = ie^{it}\, dt.$$

As before, we now get more quickly

$$\oint_C \frac{dz}{z} = \int_0^{2\pi} e^{-it}ie^{it}\, dt = i\int_0^{2\pi} dt = 2\pi i.$$

EXAMPLE 2. Integral of integer powers

Let $f(z) = (z - z_0)^m$ where m is an integer and z_0 is a constant. Integrate in the counterclockwise sense around the circle C of radius ρ with center at z_0 (cf. Fig. 324).

Solution. We may represent C in the form

$$z(t) = z_0 + \rho(\cos t + i \sin t) = z_0 + \rho e^{it} \qquad (0 \leqq t \leqq 2\pi).$$

Then we have

$$(z - z_0)^m = \rho^m e^{imt}, \qquad dz = i\rho e^{it}\, dt,$$

and we obtain

$$\oint_C (z - z_0)^m \, dz = \int_0^{2\pi} \rho^m e^{imt} i\rho e^{it} \, dt = i\rho^{m+1} \int_0^{2\pi} e^{i(m+1)t} \, dt.$$

By the Euler formula (5) in Sec. 12.6 the right-hand side equals

$$i\rho^{m+1} \left[\int_0^{2\pi} \cos (m + 1)t \, dt + i \int_0^{2\pi} \sin (m + 1)t \, dt \right].$$

When $m = -1$, we have $\rho^{m+1} = 1$, $\cos 0 = 1$, $\sin 0 = 0$ and thus obtain $2\pi i$. For integer $m \neq 1$ each of the two integrals is zero because we integrate over an interval of length 2π, equal to a period of sine and cosine. Hence the result is

(3)
$$\oint_C (z - z_0)^m \, dz = \begin{cases} 2\pi i & (m = -1), \\ 0 & (m \neq -1 \text{ and integer}). \end{cases}$$
∎

Let us now illustrate the following important fact. If we integrate a function $f(z)$ from a point z_0 to a point z_1 along different paths, we generally get different values of the integral. In other words, a complex line integral generally depends not only on the endpoints of the path but also on the geometric shape of the path.

EXAMPLE 3. Integral of a nonanalytic function

Integrate $f(z) = \text{Re } z = x$ from 0 to $1 + i$ (a) along C^* in Fig. 325, (b) along C consisting of C_1 and C_2.

Solution. (a) C^* can be represented by $z(t) = t + it$ $(0 \leq t \leq 1)$. Hence

$$\dot{z}(t) = 1 + i \qquad \text{and} \qquad f[z(t)] = x(t) = t \qquad \text{(on } C^*\text{)}.$$

We now calculate

$$\int_{C^*} \text{Re } z \, dz = \int_0^1 t(1 + i) \, dt = \tfrac{1}{2}(1 + i).$$

(b) C_1 can be represented by $z(t) = t$ $(0 \leq t \leq 1)$. Hence

$$\dot{z}(t) = 1 \qquad \text{and} \qquad f[z(t)] = x(t) = t \qquad \text{(on } C_1\text{)}.$$

C_2 can be represented by $z(t) = 1 + it$ $(0 \leq t \leq 1)$. Hence

$$\dot{z}(t) = i \qquad \text{and} \qquad f[z(t)] = x(t) = 1 \qquad \text{(on } C_2\text{)}.$$

Using (7) in Sec. 13.1, we calculate

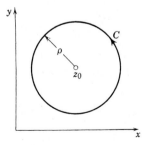

Fig. 324. Path in Example 2

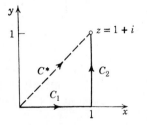

Fig. 325. Paths in Example 3

$$\int_C \text{Re } z \, dz = \int_{C_1} \text{Re } z \, dz + \int_{C_2} \text{Re } z \, dz = \int_0^1 t \, dt + \int_0^1 1 \cdot i \, dt = \tfrac{1}{2} + i.$$

Note that this result differs from the result in (a).   ∎

Second Method: Indefinite Integration

In real calculus, if for given $f(x)$ we know an $F(x)$ such that $F'(x) = f(x)$, then we can apply the formula

$$\int_a^b f(x) \, dx = F(b) - F(a) \qquad\qquad [F'(x) = f(x)].$$

This method extends to complex functions. We shall see that it is simpler than the previous method, but, of course, we have to find an $F(z)$ whose derivative $F'(z)$ equals the given function $f(z)$ that we want to integrate. Clearly, differentiation formulas will often help us in finding such an $F(z)$, so that this method becomes of great practical importance.

Theorem 2 (Indefinite integration of analytic functions)

Let $f(z)$ be continuous in a domain D. Let $F(z)$ be analytic in D and such that[4] $F'(z) = f(z)$. Then for any piecewise smooth path C in D, joining two points z_0 and z_1 in D,

(4)
$$\boxed{\int_{z_0}^{z_1} f(z) \, dz = F(z_1) - F(z_0)} \qquad\qquad [F'(z) = f(z)].$$

(Note that we can write z_0 and z_1 instead of C, since we get the same value for all those C from z_0 to z_1.) Hence for a closed path C in D that encloses only points of D,

(5)
$$\oint_C f(z) \, dz = 0.$$

Proof. Let C be any smooth path from z_0 to z_1, represented by $z = z(t)$ ($a \leqq t \leqq b$). Let $\dot{z} = dz/dt$, as before. Since $f(z) = F'(z)$, we obtain from (1)

$$\int_C f(z) \, dz = \int_C F'(z) \, dz = \int_a^b \frac{dF}{dz} \cdot \frac{dz}{dt} \, dt = \int_a^b \frac{d}{dt} F[z(t)] \, dt$$

$$= F[z(b)] - F[z(a)] = F(z_1) - F(z_0).$$

If C is merely piecewise smooth, consider the integral over each smooth portion of C separately.   ∎

---

[4]To avoid a false impression: this implies that $f(z)$ itself is **analytic** in D (as we shall see in Sec. 13.6) and restricts the method to *analytic* functions.

EXAMPLE 4
$$\int_0^{1+i} z^2 \, dz = \frac{1}{3} z^3 \bigg|_0^{1+i} = \frac{1}{3}(1+i)^3 = -\frac{2}{3} + \frac{2}{3} i$$

EXAMPLE 5
$$\int_{-\pi i}^{\pi i} \cos z \, dz = \sin z \bigg|_{-\pi i}^{\pi i} = 2 \sin \pi i = 2i \sinh \pi = 23.097i$$

EXAMPLE 6
$$\int_{8+\pi i}^{8-3\pi i} e^{z/2} \, dz = 2e^{z/2} \bigg|_{8+\pi i}^{8-3\pi i} = 2(e^{4-3\pi i/2} - e^{4+\pi i/2}) = 0$$

since e^z is periodic with period $2\pi i$.

EXAMPLE 7
Choosing $F(z) = kz$ and $F(z) = z^2/2$, we obtain from Theorem 2 for any path from z_0 to z_1

$$\int_{z_0}^{z_1} k \, dz = k(z_1 - z_0) \qquad \text{and} \qquad \int_{z_0}^{z_1} z \, dz = \tfrac{1}{2}(z_1^2 - z_0^2). \qquad \blacksquare$$

Bound for the Absolute Value of Integrals

There will be a frequent need for estimating the absolute value of complex line integrals. The basic formula is

(6)
$$\left| \int_C f(z) \, dz \right| \leq ML \qquad \text{(\textbf{ML-inequality});}$$

here L is the length of C and M a constant such that $|f(z)| \leq M$ everywhere on C.

Proof. We consider S_n as given by (2), Sec. 13.1. By the generalized triangle inequality (6) in Sec. 12.2 we obtain

$$|S_n| = \left| \sum_{m=1}^{n} f(\zeta_m) \, \Delta z_m \right| \leq \sum_{m=1}^{n} |f(\zeta_m)| \, |\Delta z_m| \leq M \sum_{m=1}^{n} |\Delta z_m|.$$

Now $|\Delta z_m|$ is the length of the chord whose endpoints are z_{m-1} and z_m, cf. Fig. 322 in Sec. 13.1. Hence the sum on the right represents the length L^* of the broken line of chords whose endpoints are $z_0, z_1, \cdots, z_n \; (= Z)$. If n approaches infinity in such a way that the greatest $|\Delta z_m|$ approaches zero, then L^* approaches the length L of the curve C, by the definition of the length of a curve. From this the inequality (6) follows. $\blacksquare$

We cannot see from (6) how close to the bound ML the actual absolute value of the integral is, but this will be no handicap in applying (6). For the time being we explain the practical use of (6) by a simple example.

EXAMPLE 8. Estimation of an integral
Find an upper bound for the absolute value of the integral

$$\int_C z^2 \, dz, \qquad C \text{ the straight-line segment from } 0 \text{ to } 1 + i.$$

Solution. $L = \sqrt{2}$ and $|f(z)| = |z^2| \leq 2$ on C gives by (6)

$$\left| \int_C z^2 \, dz \right| \leq 2\sqrt{2} = 2.8284.$$

The absolute value of the integral is $\left| -\dfrac{2}{3} + \dfrac{2}{3} i \right| = \dfrac{2}{3}\sqrt{2} = 0.9428$ (cf. Example 4).   ∎

In the next section we discuss the most important theorem of the whole chapter, **Cauchy's integral theorem,** which is basic in itself and has far-reaching consequences which we shall explore in Secs. 13.4—13.6, above all the existence of all higher derivatives of an analytic function, which are themselves analytic functions.

Problems for Secs. 13.1 and 13.2

Find a representation $z = z(t)$ of the line segment with endpoints
1. $z = 0$ and $z = 1 + 2i$
2. $z = 0$ and $z = 4 - 8i$
3. $z = 1 + i$ and $z = 3 - 4i$
4. $z = 3i$ and $z = 4 - i$
5. $z = -2 + i$ and $z = -2 + 4i$
6. $z = -4i$ and $z = -7 + 38i$

Represent the following curves in the form $z = z(t)$.
7. $|z - 1 + 2i| = 3$
8. $y = x^2$ from $(0, 0)$ to $(3, 9)$
9. $y = 1/x$ from $(1, 1)$ to $(4, 1/4)$
10. $y = -1 + 3/x$ from $(1, 2)$ to $(3, 0)$
11. $x^2 + 9y^2 = 9$
12. $4(x - 1)^2 + 9(y + 2)^2 = 36$

Find out what curves are represented by the following functions $z(t)$. Calculate $\dot{z}(t)$. Sketch the curves and $\dot{z}(0)$ [as an arrow tangent to the curve at $z(0)$].
13. $3i + 3e^{it}, \quad 0 \leq t \leq \pi$
14. $4 - i + 5e^{-it}, \quad 0 \leq t < 2\pi$
15. $t + 3t^2 i, \quad -1 \leq t \leq 2$
16. $4 + 2e^{it}, \quad -\pi \leq t \leq 0$
17. $\cos t + 2i \sin t, \quad -\pi < t < \pi$
18. $i + t - t^3 i, \quad 0 \leq t \leq 3$

Evaluate $\displaystyle\int_C f(z) \, dz$, where
19. $f(z) = z + 1/z, \quad C$ the unit circle (clockwise)
20. $f(z) = \text{Re } z, \quad C$ the parabola $y = x^2$ from 0 to $1 + i$
21. $f(z) = 3/z, \quad C$ the circular arc from 2 clockwise to $-2i$
22. $f(z) = |z|^2, \quad C$ the unit circle (counterclockwise)
23. $f(z) = (z - i)^{-1}, \quad C$ the circle $|z - i| = 2$ (clockwise)
24. $f(z) = \cos z$ from $1 - i$ to $1 + i$ along any path
25. $f(z) = \text{Re } z, \quad C$ the circle $|z| = r$ (counterclockwise)

Evaluate $\displaystyle\int_C f(z) \, dz$ by the method in Theorem 1 and check the result by the method in Theorem 2:
26. $f(z) = e^z, \quad C$ the line segment from 0 to $1 + \pi i/2$
27. $f(z) = az + b, \quad C$ the line segment from $-1 - i$ to $1 + i$
28. $f(z) = 3z^2, \quad C$ the line segment from $3i$ to $4 - i$
29. $f(z) = z^3, \quad C$ the semicircle $|z| = 2$ from $-2i$ to $2i$ in the right half-plane
30. $f(z) = z^2, \quad C$ the boundary of the triangle with vertices 0, 1, i (clockwise)

31. Verify (6) in Sec. 13.1 for $k_1 f_1 + k_2 f_2 = 3z - z^2$, where C is the upper half of the unit circle from 1 to -1.

32. Prove (6) in Sec. 13.1.

33. Verify (7) in Sec. 13.1 for $f(z) = 1/z$, where C is the unit circle, C_1 its upper half and C_2 its lower half.

34. Verify (8) in Sec. 13.1 for $f(z) = z^2$, where C is the line segment from $-1 - i$ to $1 + i$.

35. Evaluate $\displaystyle\int_C$ Im $(z^2)\, dz$ from 0 to $2 + 4i$ along (a) the line segment, (b) the x-axis

to 2 and then vertically to $2 + 4i$, (c) the parabola $y = x^2$.

36. Evaluate $\displaystyle\int_C (1/\sqrt{z})\, dz$ from 1 to -1 (a) along the upper semicircle $|z| = 1$,

(b) along the lower semicircle $|z| = 1$, where $\sqrt{z}$ is the principal value of the square root.

37. Evaluate $\displaystyle\int_C |z|\, dz$ from A: $z = -i$ to B: $z = i$ along (a) the line segment AB,

(b) the unit circle in the left half-plane, (c) the unit circle in the right half-plane.

Evaluate $\displaystyle\int_C f(z)\, dz$, where

38. $f(z) = e^{2z}$, C the vertical segment from πi to $2\pi i$

39. $f(z) = \sinh z$, C the segment from 0 to $2i$

40. $f(z) = \bar{z}$, C the parabola $y = x^2$ from 0 to $1 + i$

41. $f(z) = z \cos z^2$, C any path from 0 to πi

42. $f(z) = (z - 1)^{-1} + 2(z - 1)^{-2}$, C the circle $|z - 1| = 4$ (clockwise)

43. $f(z) = \sin^2 z$, C the semicircle $|z| = \pi$ from $-\pi i$ to πi in the right half-plane

44. $f(z) = \cosh 3z$, C any path from $\pi i/6$ to 0

45. $f(z) = \frac{1}{4}z^{-4} + \frac{1}{2}z^{-2} - z^{-1} + 3z^4$, C the circle $|z| = 3$ (clockwise)

Using the ML-inequality (6), find upper bounds for the following integrals, where C is the line segment from 0 to $3 + 4i$.

46. $\displaystyle\int_C$ Im $z\, dz$ **47.** $\displaystyle\int_C e^z\, dz$ **48.** $\displaystyle\int_C$ Ln $(z + 1)\, dz$ **49.** $\displaystyle\int_C (z + 1)^{-1}\, dz$

50. Find a better bound in Prob. 47 by decomposing C into two arcs.

13.3 Cauchy's Integral Theorem

Cauchy's integral theorem is very important in complex analysis and has various theoretical and practical consequences. To state this theorem, we shall need the following concepts.

A closed path C (cf. Sec. 13.1) is called a **simple closed path** if C does not intersect or touch itself; cf. Fig. 326 on the next page. For example, a circle is simple, an eight-shaped curve is not.

A domain D (cf. Sec. 12.3) in the complex plane is called a **simply connected domain** if every simple closed path in D encloses only points of D. A domain that is not simply connected is called *multiply connected*.

For instance, the interior of a circle ("circular disk"), ellipse or square is

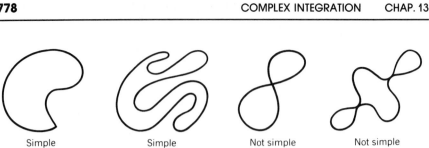

| Simple | Simple | Not simple | Not simple |

Fig. 326. Closed paths

simply connected. More generally, the interior of a simple closed curve is simply connected. A circular ring or annulus (cf. Sec. 12.3) is multiply connected (more precisely: doubly connected[5]). Figure 327 shows further examples.

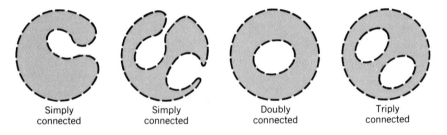

| Simply connected | Simply connected | Doubly connected | Triply connected |

Fig. 327. Simply and multiply connected domains

Recalling that, by definition, a function is a *single-valued* relation (cf. Sec. 12.4), we can now state Cauchy's integral theorem as follows. This theorem is sometimes also called the **Cauchy–Goursat theorem.**

Cauchy's integral theorem

If $f(z)$ is analytic in a simply connected domain D, then for every simple closed path C in D,

(1)
$$\oint_C f(z)\, dz = 0.$$
Cf. Fig. 328.

Proofs. If we make the additional assumption that there is an analytic function $F(z)$ in D such that $F'(z) = f(z)$ in D, then Cauchy's theorem becomes a special case of Theorem 2 in the last section. If we do not assume that there is such an $F(z)$, but if we assume—as Cauchy did—that the derivative $f'(z)$ of $f(z)$ is *continuous* in D (the existence of $f'(z)$ in D being a consequence of analyticity), then Cauchy's theorem follows from a basic theorem

---

[5]A **bounded domain** (i.e., one that lies entirely in some circle about the origin) is said to be *p-fold connected* if its boundary consist of p closed connected sets without common points. For the annulus, $p = 2$, because the boundary consists of two circles having no points in common.

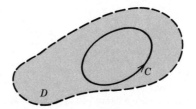

Fig. 328. Cauchy's integral theorem

on real line integrals (proof below). Goursat[6] finally proved Cauchy's theorem without the assumption that $f'(z)$ is continuous (optional proof on p. 796 near the end of the chapter). Before we go into details, let us consider some examples in order to really understand what is going on.

We mention that a simple closed path is sometimes called a *contour* and an integral over such a path a **contour integral.**

EXAMPLE 1

$$\oint_C e^z \, dz = 0, \qquad \oint_C \cos z \, dz = 0, \qquad \oint_C z^n \, dz = 0 \qquad (n = 0, 1, \cdots)$$

for any closed path, since these functions are entire (analytic for all z).

EXAMPLE 2

$$\oint_C \sec z \, dz = 0, \qquad \oint_C \frac{dz}{z^2 + 4} = 0$$

where C is the unit circle. $\sec z = 1/\cos z$ is not analytic at $z = \pm \pi/2, \pm 3\pi/2, \cdots$, but all these points lie outside C; none lies on C or inside C. Similarly for the second integral, whose integrand is not analytic at $z = \pm 2i$ outside C.

EXAMPLE 3

$$\oint_C \bar{z} \, dz = 2\pi i$$

(C the unit circle, counterclockwise) does not contradict Cauchy's theorem, since $f(z) = \bar{z}$ is not analytic, so that the theorem does not apply. (Verify this result!)

EXAMPLE 4

$$\oint_C \frac{dz}{z^2} = 0,$$

where C is the unit circle (cf. Sec. 13.2). This result does *not* follow from Cauchy's theorem, because $f(z) = 1/z^2$ is not analytic at $z = 0$. Hence *the condition that f be analytic in D is sufficient rather than necessary for* (1) *to be true.*

---

[6]EDOUARD GOURSAT (1858—1936), French mathematician. Cauchy published the theorem in 1825. The removal of that condition by Goursat (see *Transactions Amer. Math. Soc.*, vol. 1, 1900) is quite important, for instance, in connection with the fact that derivatives of analytic functions are also analytic, as we shall prove soon. Goursat also made basic contributions to partial differential equations.

EXAMPLE 5

$$\oint_C \frac{dz}{z} = 2\pi i,$$

the integration being taken around the unit circle in the counterclockwise sense (cf. Sec. 13.2). C lies in the annulus $\frac{1}{2} < |z| < \frac{3}{2}$ where $1/z$ is analytic, but this domain is not simply connected, so that Cauchy's theorem cannot be applied. Hence *the condition that the domain D be simply connected is quite essential.*

In other words, by Cauchy's theorem, if $f(z)$ is analytic on a simple closed path C and everywhere in C, with no exception, not even a single point, then (1) holds. The point that causes trouble here is $z = 0$ where $1/z$ is not analytic.

EXAMPLE 6

$$\oint_C \frac{7z - 6}{z^2 - 2z} \, dz = \oint_C \frac{3}{z} \, dz + \oint_C \frac{4}{z - 2} \, dz = 3 \cdot 2\pi i + 0 = 6\pi i$$

(C the unit circle, counterclockwise) by partial fraction reduction, Example 1 in Sec. 13.2, and Cauchy's theorem.

Cauchy's proof under the condition that $f'(z)$ is continuous. From (5) in Sec. 13.1 we have

$$\oint_C f(z) \, dz = \oint_C (u \, dx - v \, dy) + i \oint_C (u \, dy + v \, dx).$$

Since $f(z)$ is analytic in D, its derivative $f'(z)$ exists in D. Since $f'(z)$ is assumed to be continuous, (4) and (5) in Sec. 12.5 imply that u and v have *continuous* partial derivatives in D. Hence Green's theorem (Sec. 9.3) (with u and $-v$ instead of F_1 and F_2) is applicable and gives

$$\oint_C (u \, dx - v \, dy) = \iint_R \left(-\frac{\partial v}{\partial x} - \frac{\partial u}{\partial y} \right) dx \, dy$$

where R is the region bounded by C. The second Cauchy–Riemann equation (Sec. 12.5) shows that the integrand on the right is identically zero. Hence the integral on the left is zero. In the same fashion it follows by the use of the first Cauchy–Riemann equation that the last integral in the above formula is zero. This completes Cauchy's proof. (For Goursat's proof, see p. 796.) ∎

Independence of Path, Deformation of Path

We shall now discuss an important consequence of Cauchy's integral theorem that has great practical interest, proceeding as follows. If we subdivide the path C in Cauchy's theorem into two arcs $C_1{}^*$ and C_2 (Fig. 329), then (1) takes the form

(2′)
$$\int_{C_1{}^*} f \, dz + \int_{C_2} f \, dz = 0.$$

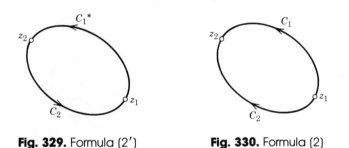

Fig. 329. Formula $(2')$ **Fig. 330.** Formula (2)

If we now reverse the sense of integration along $C_1{}^*$, then the integral over $C_1{}^*$ is multiplied by -1. Denoting $C_1{}^*$ with its new orientation by C_1, we thus obtain from $(6')$ (cf. Fig. 330)

$$(2) \qquad \boxed{\int_{C_2} f(z)\, dz = \int_{C_1} f(z)\, dz.}$$

Hence, if f is analytic in D, and C_1 and C_2 are any paths in D joining two points in D and having no further points in common, then (2) holds.

   If those paths C_1 and C_2 have finitely many points in common (Fig. 331), then (2) continues to hold. This follows by applying the previous result to the portions of C_1 and C_2 between each pair of consecutive points of intersection.

   It is even true that (2) *holds for any paths that join any points z_1 and z_2 and lie entirely in the simply connected domain D in which $f(z)$ is analytic.* To express this we say that *the integral of $f(z)$ is* **independent of path in D.** (Of course, the value of the integral depends on the choice of z_1 and z_2.)

   The *proof* would require an additional consideration of the case in which C_1 and C_2 have infinitely many points of intersection, and is not presented here.

   We may imagine that the path C_2 in (2) was obtained from C_1 by a continuous deformation (Fig. 332). It follows that in a given integral we may impose a continuous deformation on the path of integration (keeping the endpoints fixed); as long as we do not pass through a point where $f(z)$ is not analytic, the value of the line integral will not change under such a deformation. This is often called the **principle of deformation of path.**

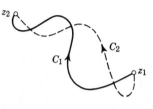

Fig. 331. Paths having finitely many intersections

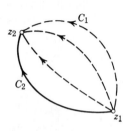

Fig. 332. Continuous deformation of path

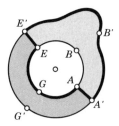

Fig. 333. Unit circle
and path C

EXAMPLE 7. Deformation of path

$$\oint_C \frac{dz}{z} = 2\pi i$$

(counterclockwise integration) now follows from Example 1 in Sec. 13.2 for any simple closed path C whose interior contains 0. Figure 333 gives the idea: first deform ABE continuously into the path $AA'B'E'E$. The heavy curve in Fig. 333 shows the resulting deformed path. Then deform $E'EGAA'$ into $E'G'A'$.

There is a more general and systematic approach to problems of this kind, as we shall now see.   ∎

Cauchy's Theorem for Multiply Connected Domains

A multiply connected domain D^* can be cut so that the resulting domain (that is, D^* without the points of the cut or cuts) becomes simply connected. For a doubly connected domain D^* we need one cut $\tilde{C}$ (Fig. 334). If $f(z)$ is analytic in D^* and at each point of C_1 and C_2 then, since C_1, C_2 and C bound a simply connected domain, it follows from Cauchy's theorem that the integral of f taken over C_1, $\tilde{C}$, C_2 in the sense indicated by the arrows in Fig. 334 has the value zero. Since we integrate along $\tilde{C}$ in both directions, the corresponding integrals cancel out, and we obtain

(3*) $\displaystyle \int_{C_1} f(z)\, dz + \int_{C_2} f(z)\, dz = 0$

where one of the curves is traversed in the counterclockwise sense and the other in the opposite sense. Reversing the sense of integration on one of the curves, we may write this

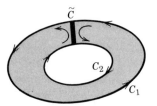

Fig. 334. Doubly connected domain

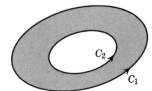
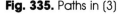

Fig. 335. Paths in (3)

(3)
$$\int_{C_1} f(z)\, dz = \int_{C_2} f(z)\, dz$$

where C_1 and C_2 are now traversed in the same sense (Fig. 335). We remember that (3) holds under the assumption that $f(z)$ is analytic in the domain bounded by C_1 and C_2 and at each point of C_1 and C_2.

Can you see how the result in Example 7 now follows immediately from our present consideration?

For more complicated domains we may need more than one cut, but the basic idea remains the same as before. For instance, for the triply connected domain in Fig. 336,

$$\int_{C_1} f(z)\, dz + \int_{C_2} f(z)\, dz + \int_{C_3} f(z)\, dz = 0$$

where C_2 and C_3 are traversed in the same sense and C_1 is traversed in the opposite sense.

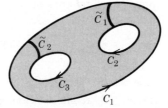

Fig. 336. Triply connected domain

EXAMPLE 8. A basic result: Integral of integer powers
From (3) and Sec. 13.2, Example 2, it now follows that

$$\oint_C (z - z_0)^m\, dz = \begin{cases} 2\pi i & (m = -1) \\ 0 & (m \neq -1 \text{ and integer}) \end{cases}$$

for counterclockwise integration around any simple closed path containing z_0 in its interior. ∎

In the next section, using Cauchy's integral theorem, we prove the existence of indefinite integrals of analytic functions. This is more of theoretical than of practical interest, as the reader can see from the first few lines in the section, which we therefore leave optional. We shall not need it later on.

Problems for Sec. 13.3

1. Verify Cauchy's integral theorem for the integral of z^2 taken over the boundary of the triangle with vertices 0, 2 and $2i$ (counterclockwise).

2. For what contours C will it follow from Cauchy's theorem that

   (a) $\displaystyle\oint_C \frac{dz}{z} = 0,$ (b) $\displaystyle\oint_C \frac{e^{-z}}{z^5 - z}\, dz = 0?$

3. Verify the result in Example 3.

4. If the integral of a function $f(z)$ over the unit circle equals 2 and over the circle $|z| = 2$ equals 1, why can $f(z)$ not be analytic everywhere in the annulus bounded by these circles?

5. Show that the integral of $1/z^3$ around the unit circle is zero. (a) Does this follow from Cauchy's theorem? (b) Can we conclude from this that the integral of $1/z^3$ around the boundary of the square with vertices ± 1, $\pm i$ is zero?

6. Can we conclude from Example 2 and the principle of deformation of path that the integral of $1/(z^2 + 4)$ (a) over $|z - 2| = 2$, (b) $|z - 2| = 3$, is zero? (Give reason.)

In each case find the value of the integral of the given function around the unit circle in the counterclockwise sense and indicate whether Cauchy's theorem may be applied.

7. $f(z) = e^{-z^2}$ 8. $f(z) = \text{Re } z$ 9. $f(z) = \text{Im } z$

10. $f(z) = 1/(\pi z - 1)$ 11. $f(z) = \tan z$ 12. $f(z) = |z|^2$

13. $f(z) = 1/\bar{z}$ 14. $f(z) = 1/(1 - 2z)$ 15. $f(z) = \bar{z}^2$

16. Obtain the answer to Prob. 9 from that to Prob. 8 and Cauchy's theorem applied to $f(z) = z$.

17. Using the principle of deformation of path and

$$\frac{3z + 5}{z^2 + z} = \frac{5}{z} - \frac{2}{z + 1}, \qquad \text{show that} \qquad \int_C \frac{3z + 5}{z^2 + z}\, dz = 6\pi i$$

where C is the path in Fig. 337.

18. Integrate $f(z) = \bar{z}/|z|$ around the circle (a) $|z| = 2$, (b) $|z| = 4$, in the counterclockwise sense. Can the result in (b) be obtained from that in (a) by the principle of deformation of path?

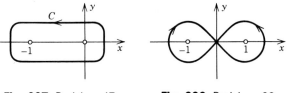

Fig. 337. Problem 17   **Fig. 338.** Problem 20

Evaluate the following integrals. (*Hint.* If necessary, represent the integrand in terms of partial fractions.)

19. $\oint_C \dfrac{z}{z^2 + 1}\, dz$, C: (a) $|z| = 2$, (b) $|z + i| = 1$ (counterclockwise)

20. $\oint_C \dfrac{dz}{z^2 - 1}$, C the contour in Fig. 338

21. $\oint_C \dfrac{dz}{z^2 + 1}$, C: (a) $|z + i| = 1$, (b) $|z - i| = 1$ (counterclockwise)

22. $\oint_C \dfrac{2z^3 + z^2 + 4}{z^4 + 4z^2}\, dz$, C: $|z - 2| = 4$ (clockwise)

23. $\oint_C \dfrac{e^z}{z}\, dz$, C consists of $|z| = 2$ (counterclockwise) and $|z| = 1$ (clockwise)

24. $\oint_C \dfrac{\cos z}{z^2} \, dz,$ $C: |z - 2i| = 1$ (counterclockwise)

25. $\oint_C \dfrac{3z + 1}{z^3 - z} \, dz,$ $C:$ (a) $|z| = 1/2$, (b) $|z| = 2$ (counterclockwise)

26. $\oint_C \mathrm{Re}\,(z^2) \, dz,$ C the boundary of the triangle with vertices at 0, 2 and $2 + i$

27. $\oint_C \mathrm{Ln}\,(1 - z) \, dz,$ C the boundary of the square with vertices $\pm \frac{1}{2} \pm \frac{1}{2}i$

28. $\oint_C \dfrac{dz}{z^3 + 1},$ C the boundary of the quadrangle with vertices $\pm\frac{1}{2}$, $\pm\frac{1}{2}i$

29. $\oint_C \mathrm{Re}\, z \, dz,$ C the contour in Fig. 339

30. $\displaystyle\int_C \dfrac{dz}{z^2(z^2 + 4)},$ C consists of $|z| = 3$ (counterclockwise) and $|z| = 1$ (clockwise)

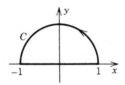

Fig. 339. Problem 29

13.4 Existence of Indefinite Integral

This section includes an application of Cauchy's integral theorem and is *optional*. It relates to Theorem 2 in Sec. 13.2 on the evaluation of line integrals by indefinite integration and substitution of the limits of integration:

(1) $$\int_{z_0}^{z_1} f(z) \, dz = F(z_1) - F(z_0) \qquad [F'(z) = f(z)],$$

where $F(z)$ is an indefinite integral of $f(z)$, that is, $F'(z) = f(z)$, as indicated. In most applications, such an $F(z)$ can be found from differentiation formulas. In Sec. 13.2 we *assumed* the existence of $F(z)$ for given analytic $f(z)$. It is interesting that this assumption is superfluous because we can *prove* the existence of $F(z)$ by Cauchy's theorem:

Theorem 1 (Existence of an indefinite integral)
If $f(z)$ is analytic in a simply connected domain D (cf. Sec. 13.3), then there exists an indefinite integral $F(z)$ of $f(z)$ in D, which is analytic in D and satisfies $F'(z) = f(z)$. Hence this $F(z)$ can be used in evaluating line integrals by (1).

Proof. The conditions of Cauchy's integral theorem are satisfied. Hence the line integral of $f(z)$ from any z_0 in D to any z in D is independent of path in D. We keep z_0 fixed. Then this integral becomes a function of z, which we denote by $F(z)$:

$$(2) \qquad\qquad F(z) = \int_{z_0}^{z} f(z^*)\, dz^*.$$

We show that this $F(z)$ is analytic in D and that $F'(z) = f(z)$. The idea of doing this is as follows. We form the difference quotient

$$\frac{F(z + \Delta z) - F(z)}{\Delta z} = \frac{1}{\Delta z}\left[\int_{z_0}^{z+\Delta z} f(z^*)\, dz^* - \int_{z_0}^{z+\Delta z} f(z^*)\, dz^* \right]$$

$$= \frac{1}{\Delta z} \int_{z}^{z+\Delta z} f(z^*)\, dz^*,$$

subtract $f(z)$ from it and show that expression obtained approaches zero as $\Delta z \to 0$; this is done by using the continuity of $f(z)$, which is a consequence of the analyticity of $f(z)$. We now give the details.

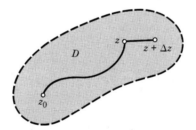

Fig. 340. Path of integration

We keep z fixed. Then we choose $z + \Delta z$ so that the line segment with endpoints z and $z + \Delta z$ is in D. This is possible since D is a domain, hence D contains a neighborhood of z. Cf. Fig. 340. This segment we use as the path of integration in the previous formula. We now subtract $f(z)$. This is a constant, since z is kept fixed. Hence

$$\int_{z}^{z+\Delta z} f(z)\, dz^* = f(z)\int_{z}^{z+\Delta z} dz^* = f(z)\,\Delta z. \quad \text{Thus} \quad f(z) = \frac{1}{\Delta z}\int_{z}^{z+\Delta z} f(z)\, dz^*.$$

This trick permits us to write a single integral:

$$\frac{F(z + \Delta z) - F(z)}{\Delta z} - f(z) = \frac{1}{\Delta z}\int_{z}^{z+\Delta z} [f(z^*) - f(z)]\, dz^*.$$

$f(z)$ is analytic, hence continuous. An $\epsilon > 0$ being given, we can thus find a $\delta > 0$ such that

$$|f(z^*) - f(z)| < \epsilon \qquad \text{when} \qquad |z^* - z| < \delta.$$

Consequently, letting $|\Delta z| < \delta$, we see that the *ML*-inequality (Sec. 13.2) yields

$$\left| \frac{F(z + \Delta z) - F(z)}{\Delta z} - f(z) \right| = \frac{1}{|\Delta z|} \left| \int_z^{z + \Delta z} [f(z^*) - f(z)] \, dz^* \right| \leq \frac{1}{|\Delta z|} \epsilon |\Delta z| = \epsilon;$$

that is, by the definition of a limit and of the derivative,

$$F'(z) = \lim_{\Delta z \to 0} \frac{F(z + \Delta z) - F(z)}{\Delta z} = f(z).$$

Since z is any point in D, this proves that $F(z)$ is analytic in D and is an indefinite integral or antiderivative of $f(z)$ in D, written

$$F(z) = \int f(z) \, dz.$$

Also, if $G'(z) = f(z)$, then $F'(z) - G'(z) \equiv 0$ in D; hence $F(z) - G(z)$ is constant in D (cf. Prob. 35 at the end of Sec. 12.5). That is, two indefinite integrals of $f(z)$ differ only by a constant. The latter drops out in (1). Hence in (1) we can use any indefinite integral of $f(z)$. This proves the theorem. ∎

See Sec. 13.2 for examples and problems on indefinite integration.

The theorem in this section followed from Cauchy's integral theorem. A much more fundamental consequence is **Cauchy's integral formula** for evaluating integrals over closed curves, which we discuss in the next section.

Problems for Sec. 13.4

Evaluate the following integrals.

1. $\displaystyle\int_i^{2+i} z \, dz$ 2. $\displaystyle\int_1^{3i} z^2 \, dz$ 3. $\displaystyle\int_i^{2i} (z^2 - 1)^3 \, dz$ 4. $\displaystyle\int_{2i}^{3i} (z + 4)^2 \, dz$

5. $\displaystyle\int_1^{1+\pi i} e^z \, dz$ 6. $\displaystyle\int_0^i z e^{z^2} \, dz$ 7. $\displaystyle\int_{-2\pi i}^{2\pi i} e^{-z/2} \, dz$ 8. $\displaystyle\int_0^{\pi i} z \cos z \, dz$

9. $\displaystyle\int_{-\pi i}^{\pi i} \sin^2 z \, dz$ 10. $\displaystyle\int_1^{1+i} z^{-2} \, dz$ 11. $\displaystyle\int_{-i}^{i} z \cosh z^2 \, dz$ 12. $\displaystyle\int_{-\pi i}^{\pi i} z \cosh z \, dz$

13. Show by the present method that $\displaystyle\int_C \frac{dz}{z} = \pi i/2$, where $C: z(t) = e^{it}, 0 \leq t \leq \pi/2$.

14. Can we integrate $1/z$ around the unit circle, using the present method?

15. Can we integrate $1/z$ around the circle $|z - 2i| = 1$, using the present method?

13.5 Cauchy's Integral Formula

The most important consequence of Cauchy's integral theorem is Cauchy's integral formula. This formula is useful for evaluating integrals (examples below). More importantly, it plays a key role in proving the surprising fact that analytic functions have derivatives of all orders (Sec. 13.6), in establishing Taylor series representations (Sec. 14.5) and so on. Cauchy's integral formula and its conditions of validity may be stated as follows.

Theorem 1 (Cauchy's integral formula)

Let $f(z)$ be analytic in a simply connected domain D. Then for any point z_0 in D and any simple closed path C in D which encloses z_0 (Fig. 341),

(1)
$$\oint_C \frac{f(z)}{z - z_0} \, dz = 2\pi i f(z_0) \qquad \text{(Cauchy's integral formula),}$$

the integration being taken in the counterclockwise sense.

Proof. By addition and subtraction, $f(z) = f(z_0) + [f(z) - f(z_0)]$. We insert this into (1) on the left and can take the constant factor $f(z_0)$ out from under the integral sign. Then

(2)
$$\oint_C \frac{f(z)}{z - z_0} \, dz = f(z_0) \oint_C \frac{dz}{z - z_0} + \oint_C \frac{f(z) - f(z_0)}{z - z_0} \, dz.$$

The first term on the right equals $f(z_0) \cdot 2\pi i$ (cf. Example 8 in Sec. 13.3, with $m = -1$). This proves the theorem, provided the second integral on the right is zero. This is what we are now going to show. Its integrand is analytic, except at z_0. Hence by the principle of deformation of path (Sec. 13.3) we can replace C by a small circle K of radius ρ and center z_0 (Fig. 342), without altering the value of the integral. Since $f(z)$ is analytic, it is continuous. Hence, an $\epsilon > 0$ being given, we can find a $\delta > 0$ such that

$$|f(z) - f(z_0)| < \epsilon \qquad \text{for all } z \text{ in the disk } |z - z_0| < \delta.$$

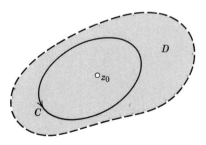

Fig. 341. Cauchy's integral formula

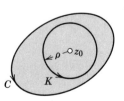

Fig. 342. Proof of Cauchy's integral formula

Choosing the radius ρ of K smaller than δ, we thus have the inequality

$$\left| \frac{f(z) - f(z_0)}{z - z_0} \right| < \frac{\epsilon}{\rho}$$

at each point of K. The length of K is $2\pi\rho$. Hence, by the *ML*-inequality in Sec. 13.2,

$$\left| \oint_K \frac{f(z) - f(z_0)}{z - z_0} \, dz \right| < \frac{\epsilon}{\rho} 2\pi\rho = 2\pi\epsilon.$$

Since ϵ (> 0) can be chosen arbitrarily small, it follows that the last integral on the right-hand side of (2) has the value zero, and the theorem is proved.   ∎

EXAMPLE 1. Cauchy's integral formula

$$\oint_C \frac{e^z}{z - 2} \, dz = 2\pi i e^z \bigg|_{z=2} = 2\pi i e^2$$

for any contour enclosing $z_0 = 2$ (since e^z is entire), and zero for any contour for which $z_0 = 2$ lies outside (by Cauchy's integral theorem).

EXAMPLE 2. Cauchy's integral formula

$$\oint_C \frac{z^3 - 6}{2z - i} \, dz = \oint_C \frac{\frac{1}{2}z^3 - 3}{z - \frac{1}{2}i} \, dz = 2\pi i [\tfrac{1}{2}z^3 - 3] \bigg|_{z=i/2} = \frac{\pi}{8} - 6\pi i \qquad (z_0 = \tfrac{1}{2}i \text{ inside } C).$$

EXAMPLE 3. Integration around different contours
Integrate

$$g(z) = \frac{z^2 + 1}{z^2 - 1}$$

in the counterclockwise sense around a circle of radius 1 with center at the point

(a) $z = 1$ (b) $z = \frac{1}{2}$ (c) $z = -1 + \frac{1}{2}i$,   (d) $z = i$.

Solution. To see what is going on, locate the points where $g(z)$ is not analytic and sketch them along with the contours (Fig. 343). These points are -1 and 1. We see that (b) will give the same result as (a), by the principle of deformation of path. And (d) gives zero, by Cauchy's integral theorem. We consider (a) and afterward (c).
   (a) Here $z_0 = 1$, so that $z - z_0 = z - 1$ in (1). Hence we must write

$$g(z) = \frac{z^2 + 1}{z^2 - 1} = \frac{z^2 + 1}{z + 1} \frac{1}{z - 1}; \qquad \text{thus} \qquad f(z) = \frac{z^2 + 1}{z + 1},$$

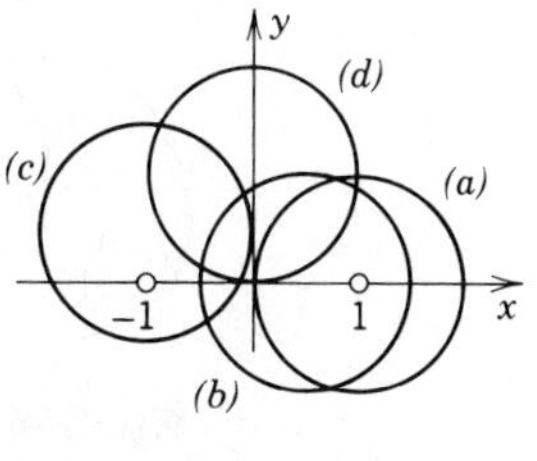

Fig. 343. Example 3

and (1) gives

$$\oint_C \frac{z^2 + 1}{z^2 - 1}\, dz = 2\pi i f(1) = 2\pi i \left[\frac{z^2 + 1}{z + 1}\right]_{z=1} = 2\pi i.$$

(b) $g(z)$ is the same as before, but $f(z)$ changes, because we must now take $z_0 = -1$ (instead of 1). This gives a factor $z - z_0 = z + 1$ in (1). Hence we must now write

$$g(z) = \frac{z^2 + 1}{z - 1}\frac{1}{z + 1}\,; \qquad\text{thus}\qquad f(z) = \frac{z^2 + 1}{z - 1}\,.$$

Compare this for a minute with the previous expression and then go on:

$$\oint_C \frac{z^2 + 1}{z^2 - 1}\, dz = 2\pi i f(-1) = 2\pi i \left[\frac{z^2 + 1}{z - 1}\right]_{z=-1} = -2\pi i.$$

EXAMPLE 4. Use of partial fractions

Integrate $g(z) = (z^2 - 1)^{-1} \tan z$ around the circle $C\colon |z| = 3/2$ (counterclockwise).

Solution. $\tan z$ is not analytic at $\pm\pi/2$, $\pm 3\pi/2$, $\cdots$, but all these points lie outside the contour. $(z^2 - 1)^{-1} = 1/(z - 1)(z + 1)$ is not analytic at 1 and -1. To get integrals of the form (1), with only a *single* point inside C at which the integrand is not analytic, we use partial fractions:

$$\frac{1}{z^2 - 1} = \frac{1}{2}\left(\frac{1}{z - 1} - \frac{1}{z + 1}\right).$$

From this and (1) we obtain

$$\oint_C \frac{\tan z}{z^2 - 1}\, dz = \frac{1}{2}\left[\oint_C \frac{\tan z}{z - 1}\, dz - \oint_C \frac{\tan z}{z + 1}\, dz\right]$$

$$= \frac{2\pi i}{2}\,[\tan 1 - \tan(-1)] = 2\pi i \tan 1 \approx 9.785i. \qquad\blacksquare$$

Multiply connected domains may be handled as in Sec. 13.3. For instance, if $f(z)$ is analytic on C_1 and C_2 and in the ring-shaped domain bounded by C_1 and C_2 (Fig. 344) and z_0 is any point in that domain, then

$$(3) \qquad f(z_0) = \frac{1}{2\pi i}\oint_{C_1} \frac{f(z)}{z - z_0}\, dz + \frac{1}{2\pi i}\oint_{C_2} \frac{f(z)}{z - z_0}\, dz,$$

where the outer integral is taken in the counterclockwise sense and the inner integral (the second) in the clockwise sense, as indicated in Fig. 344. This is the analog of formula (3*) in Sec. 13.3.

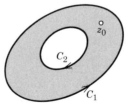

Fig. 344. Formula (3)

Looking back, we point to a chain of basic results. The beginning was Cauchy's integral theorem in Sec. 13.3. From it followed Cauchy's integral formula (1) in this section. From it follows the existence of all **higher derivatives** of an analytic function, in the next section. This is probably the most exciting link of our chain. From it follows in Sec. 14.5 the Taylor series for analytic functions.

Problems for Sec. 13.5

Integrate $z^2/(z^2 + 1)$ in the counterclockwise sense around the circle:

1. $|z + i| = 1$ **2.** $|z - i| = 1/2$ **3.** $|z| = 2$ **4.** $|z| = 1/2$

Integrate $z^2/(z^4 - 1)$ in the counterclockwise sense around the circle:

5. $|z - 1| = 1$ **6.** $|z + i| = 1$ **7.** $|z - i| = 1/2$ **8.** $|z| = 2$

Integrate the following functions in the counterclockwise sense around the unit circle.

9. $1/(4z + i)$ **10.** e^z/z **11.** $e^{z^2}/(2z - i)$ **12.** $(\cos z)/z$
13. $(\sin z)/z$ **14.** $(e^z - 1)/z$ **15.** $(\cosh 3z)/z$ **16.** $(z - 2)^{-1} \sin z$

Integrate the given function around the given contour C (counterclockwise).

17. $1/(z^2 + 4)$, C the ellipse $x^2 + 4(y - 2)^2 = 4$
18. $(2 - \sin z)/(z^2 - z)$, C the boundary of the rectangle with vertices $\frac{1}{2} \pm i$, $-\frac{1}{2} \pm i$
19. $(\tan z)/(z - i)$, C the boundary of the triangle with vertices -1, 1, $2i$
20. $(e^z + i)/(ze^z - 2iz)$, C the boundary of the square with vertices ± 1, $\pm i$
21. $e^{z^3}/[(z - 1 - i)z^2]$, C consists of $|z| = 3$ (counterclockwise) and $|z| = 1$ (clockwise)
22. $[\operatorname{Ln}(z + 1)]/(z^2 + 1)$, C consists of boundary of the the triangle with vertices $1 - \frac{1}{2}i$, $-1 - \frac{1}{2}i$, $2i$ (counterclockwise) and $|z| = \frac{1}{4}$ (clockwise)
23. $[\operatorname{Ln}(z - 1)]/(z - 5)$, C the circle $|z - 4| = 2$

24. Solve Example 4 by applying Cauchy's integral theorem to a suitable triply connected domain, without partial fraction reduction, and then using (1).

25. Show that $\oint_C (z - z_1)^{-1}(z - z_2)^{-1}\, dz = 0$ for a simple closed curve C enclosing

z_1 and z_2, which are arbitrary.

13.6 Derivatives of Analytic Functions

From the assumption that a *real* function of a real variable is once differentiable, nothing follows about the existence of derivatives of higher order. We shall now see that from the assumption that a *complex* function has a first derivative in a domain D, there follows the existence of derivatives of all orders in D. This means that in this respect complex analytic functions behave much more simply than real functions that are once differentiable.

Theorem 1 (Derivatives of an analytic function)

If $f(z)$ is analytic in a domain D, then it has derivatives of all orders in D, which are then also analytic functions in D. The values of these derivatives at a point z_0 in D are given by the formulas

$$(1') \qquad f'(z_0) = \frac{1}{2\pi i} \oint_C \frac{f(z)}{(z - z_0)^2} \, dz,$$

$$(1'') \qquad f''(z_0) = \frac{2!}{2\pi i} \oint_C \frac{f(z)}{(z - z_0)^3} \, dz,$$

and in general

$$(1) \qquad \boxed{f^{(n)}(z_0) = \frac{n!}{2\pi i} \oint_C \frac{f(z)}{(z - z_0)^{n+1}} \, dz} \qquad (n = 1, 2, \cdots);$$

here C is any simple closed path in D that encloses z_0 and whose full interior belongs to D; and we integrate counterclockwise around C (Fig. 345).

Comment. For memorizing (1), it is useful to observe that these formulas are obtained formally by differentiating the Cauchy formula (1), Sec. 13.5, under the integral sign *with respect to z_0.*

Proof of Theorem 1. We prove $(1')$. We start from the definition

$$f'(z_0) = \lim_{\Delta z \to 0} \frac{f(z_0 + \Delta z) - f(z_0)}{\Delta z}.$$

On the right we represent $f(z_0 + \Delta z)$ and $f(z_0)$ by Cauchy's integral formula (1), Sec. 13.5; the two integrals we can combine into a single integral by taking the common denominator and simplifying the numerator (where $z - z_0$ drops out and only $f(z)\Delta z$ remains):

$$\frac{f(z_0 + \Delta z) - f(z_0)}{\Delta z} = \frac{1}{2\pi i \Delta z} \left[\oint_C \frac{f(z)}{z - (z_0 + \Delta z)} \, dz - \oint_C \frac{f(z)}{z - z_0} \, dz \right]$$

$$= \frac{1}{2\pi i} \oint_C \frac{f(z)}{(z - z_0 - \Delta z)(z - z_0)} \, dz.$$

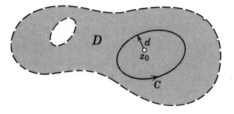

Fig. 345. Theorem 1 and its proof

Clearly, we can now establish (1′) by showing that, as $\Delta z \to 0$, the integral on the right approaches the integral in (1′). To do this, we consider the difference between these two integrals. We can write this difference as a single integral by taking the common denominator and simplifying. This gives

$$\oint_C \frac{f(z)}{(z - z_0 - \Delta z)(z - z_0)}\, dz - \oint_C \frac{f(z)}{(z - z_0)^2}\, dz$$

$$= \oint_C \frac{f(z)\Delta z}{(z - z_0 - \Delta z)(z - z_0)^2}\, dz.$$

We show by the *ML*-inequality (Sec. 13.2) that this difference approaches zero as $\Delta z \to 0$.

Being analytic, the function $f(z)$ is continuous on C, hence bounded in absolute value, say, $|f(z)| \leqq K$. Let d be the smallest distance from z_0 to the points of C (cf. Fig. 345). Then for all $|z|$ on C,

$$|z - z_0|^2 \geqq d^2, \qquad \text{hence} \qquad \frac{1}{|z - z_0|^2} \leqq \frac{1}{d^2}.$$

Furthermore, if $|\Delta z| \leqq d/2$, then for all z on C we also have

$$|z - z_0 - \Delta z| \geqq \frac{d}{2}, \qquad \text{hence} \qquad \frac{1}{|z - z_0 - \Delta z|} \leqq \frac{2}{d}.$$

Let L be the length of C. Then by the *ML*-inequality, if $|\Delta z| \leqq d/2$,

$$\left| \oint_C \frac{f(z)\Delta z}{(z - z_0 - \Delta z)(z - z_0)^2}\, dz \right| \leqq K|\Delta z|\, \frac{2}{d} \cdot \frac{1}{d^2}.$$

This approaches zero as $\Delta z \to 0$. Formula (1′) is proved.

Note that we used Cauchy's integral formula (1), Sec. 13.5, but if all we had known about $f(z_0)$ is the fact that it can be represented by (1), Sec. 13.5, our argument would have established the existence of the derivative $f'(z_0)$ of $f(z)$. This is essential to the continuation and completion of this proof, because it implies that (1″) can be proved by a similar argument, with f replaced by f', and that the general formula (1) then follows by induction. ∎

EXAMPLE 1. Evaluation of line integrals
From (1′), for any contour enclosing the point πi (counterclockwise)

$$\oint_C \frac{\cos z}{(z - \pi i)^2}\, dz = 2\pi i(\cos z)'\bigg|_{z=\pi i} = -2\pi i \sin \pi i = 2\pi \sinh \pi.$$

EXAMPLE 2
From (1″), for any contour enclosing the point $-i$ (counterclockwise)

$$\oint_C \frac{z^4 - 3z^2 + 6}{(z + i)^3}\, dz = \pi i(z^4 - 3z^2 + 6)''\bigg|_{z=-i} = \pi i[12z^2 - 6]_{z=-i} = -18\pi i.$$

EXAMPLE 3

By (1'), for any contour for which 1 lies inside and $\pm 2i$ lie outside (counterclockwise),

$$\oint_C \frac{e^z}{(z-1)^2(z^2+4)} dz = 2\pi i \left(\frac{e^z}{z^2+4}\right)' \bigg|_{z=1} = 2\pi i \frac{e^z(z^2+4) - e^z 2z}{(z^2+4)^2} \bigg|_{z=1} = \frac{6e\pi}{25} i \approx 2.050i. \blacksquare$$

Theorem 1 is also important in deriving general results on analytic functions. Let us show this by proving the converse of Cauchy's integral theorem:

Morera's[7] theorem

If $f(z)$ is continuous in a simply connected domain D and if

$$(2) \qquad \oint_C f(z)\, dz = 0$$

for every closed path in D, then $f(z)$ is analytic in D.

Proof. In Sec. 13.4 it was shown that if $f(z)$ is analytic in D, then

$$F(z) = \int_{z_0}^z f(z^*)\, dz^*$$

is analytic in D and $F'(z) = f(z)$. In the proof we used only the continuity of $f(z)$ and the property that its integral around every closed path in D is zero; from these assumptions we concluded that $F(z)$ is analytic. By Theorem 1, the derivative of $F(z)$ is analytic, that is, $f(z)$ is analytic in D, and Morera's theorem is proved. $\blacksquare$

Theorem 1 also yields a basic inequality that has many applications. To get it, all we have to do is to choose for C in (1) a circle of radius r and center z_0 and apply the *ML*-inequality (Sec. 13.2); with $|f(z)| \le M$ on C we obtain from (1)

$$|f^{(n)}(z_0)| = \frac{n!}{2\pi} \left| \oint_C \frac{f(z)}{(z-z_0)^{n+1}}\, dz \right| \le \frac{n!}{2\pi} M \frac{1}{r^{n+1}} 2\pi r.$$

This yields **Cauchy's inequality**

$$(3) \qquad \boxed{|f^{(n)}(z_0)| \le \frac{n!M}{r^n}.}$$

To gain a first impression of the importance of this inequality, let us prove a famous theorem on entire functions (functions that are analytic for all z; cf. Sec. 12.6).

---

[7]GIACINTO MORERA (1856—1909), Italian mathematician, who worked in Genoa and Turin.

Liouville's theorem

If an entire function $f(z)$ is bounded in absolute value for all z, then $f(z)$ must be a constant.

Proof. By assumption, $|f(z)|$ is bounded, say, $|f(z)| < K$ for all z. Using (3), we see that $|f'(z_0)| < K/r$. Since this is true for every r, we can take r as large as we please and conclude that $f'(z_0) = 0$. Since z_0 is arbitrary, $f'(z) = 0$ for all z, and $f(z)$ is a constant (cf. Prob. 35 at the end of Sec. 12.5). This completes the proof. ∎

This is the end of Chap. 13 on complex integration, which gave us a first impression of methods that have no counterpart in real integral calculus. We have seen that these methods result directly or indirectly from Cauchy's integral theorem (Sec. 13.3). More on integration follows in Chap. 15.

In the next chapter we consider **power series,** which play a great role in complex analysis, and we shall see that the Taylor series of calculus have a complex counterpart, so that e^z, cos z, sin z, etc. have Maclaurin series that are quite similar to those in calculus.

Problems for Sec. 13.6

Integrate the following functions in the counterclockwise sense around the unit circle. (In Probs. 8, 12, 13 n is a positive integer.)

1. $z^2/(2z - 1)^2$ 2. $z^2/(2z - 1)^4$ 3. $z^3/(2z + i)^3$ 4. $z/(4z + i)^2$
5. e^z/z^2 6. $e^{3z}/(z - \pi)^3$ 7. e^{z^2}/z^3 8. e^z/z^n
9. $ze^z/(4z + \pi i)^2$ 10. $z^{-2} \cos z$ 11. $z^{-2} \sin z$ 12. $z^{-2n} \cos z$
13. $z^{-2n-1} \cos z$ 14. $z^{-3} \cosh z$ 15. $z^{-2}e^{-z} \sin z$ 16. $z^{-4} \sinh 2z$

Integrate $f(z)$ around the contour C in the counterclockwise sense (or as indicated).

17. $f(z) = z^{-2} \tan z$, C any contour enclosing 0
18. $f(z) = (z - \frac{1}{2}\pi)^{-2} \cot z$, C the rectangle with vertices $1 \pm i$, $2 \pm i$
19. $f(z) = (z - 2)^{-2} \text{Ln } z$, $C: |z - 3| = 2$
20. $(z + i)^{-1} e^z \tan z$, $C: |z - i| = 1$
21. $f(z) = \dfrac{e^{2z}}{(z - 1)^2(z^2 + 4)}$, C the ellipse $x^2 + 4y^2 = 4$
22. $f(z) = \dfrac{\text{Ln }(z + 3) + \cos z}{(z + 1)^2}$, C the square with vertices ± 2, $\pm 2i$
23. $f(z) = \dfrac{2z^3 - 3}{z(z - 1 - i)^2}$, C consists of $|z| = 2$ (counterclockwise) and $|z| = 1$ (clockwise)
24. $f(z) = \dfrac{(1 + z) \sin z}{(2z - 1)^2}$, $C: |z - i| = 2$
25. $f(z) = \dfrac{e^{z^2}}{z(z - 2i)^2}$, C consists of the square with vertices $\pm 3 \pm 3i$ (counterclockwise) and $|z| = 1$ (clockwise)
26. $f(z) = \dfrac{z^3 + \sin z}{(z - i)^3}$, C the triangle with vertices ± 2, $2i$

27. If $f(z)$ is not a constant and is analytic for all (finite) z, and R and M are any positive real numbers (no matter how large), show that there exist values of z for which $|z| > R$ and $|f(z)| > M$. *Hint.* Use Liouville's theorem.

28. If $f(z)$ is a polynomial of degree $n > 0$ and M an arbitrary positive real number (no matter how large), show that there exists a positive real number R such that $|f(z)| > M$ for all $|z| > R$.

29. Show that $f(z) = e^z$ has the property characterized in Prob. 27, but does not have that characterized in Prob. 28.

30. Prove the **Fundamental theorem of algebra:** If $f(z)$ is a polynomial in z, not a constant, then $f(z) = 0$ for at least one value of z. *Hint.* Assume $f(z) \neq 0$ for all z and apply the result of Prob. 27 to $g = 1/f$.

Further Proof in Chapter 13

Goursat's proof of Cauchy's integral theorem in Sec. 13.3 without the condition that f'(z) is continuous

We start with the case when C is the boundary of a triangle. We orient C counterclockwise. By joining the midpoints of the sides we subdivide the triangle into four congruent triangles (Fig. 346). Let $C_{\text{I}}, C_{\text{II}}, C_{\text{III}}, C_{\text{IV}}$ denote their boundaries. We claim that (cf. Fig. 346)

$$(1) \qquad \oint_C f\,dz = \oint_{C_{\text{I}}} f\,dz + \oint_{C_{\text{II}}} f\,dz + \oint_{C_{\text{III}}} f\,dz + \oint_{C_{\text{IV}}} f\,dz.$$

Indeed, on the right we integrate along each of the three segments of subdivision in both possible directions (Fig. 346), so that the corresponding integrals cancel out in pairs, and the sum of the integrals on the right equals the integral on the left. We now pick an integral on the right which is biggest in absolute value and call its path C_1. Then, by the triangle inequality (Sec. 12.2),

$$\left| \oint_C f\,dz \right| \leqq \left| \oint_{C_{\text{I}}} f\,dz \right| + \left| \oint_{C_{\text{II}}} f\,dz \right| + \left| \oint_{C_{\text{III}}} f\,dz \right| + \left| \oint_{C_{\text{IV}}} f\,dz \right|$$

$$\leqq 4 \left| \oint_{C_1} f\,dz \right|.$$

We now subdivide the triangle bounded by C_1 as before and select a triangle of subdivision with boundary C_2 for which

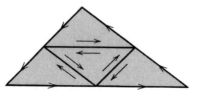

Fig. 346. Proof of Cauchy's integral theorem

$$\left| \oint_{C_1} f\, dz \right| \leq 4 \left| \oint_{C_2} f\, dz \right|. \qquad \text{Then} \qquad \left| \oint_{C} f\, dz \right| \leq 4^2 \left| \oint_{C_2} f\, dz \right|.$$

Continuing in this fashion, we obtain a sequence of triangles T_1, T_2, $\cdots$ with boundaries C_1, C_2, $\cdots$ which are similar and such that T_n lies in T_m when $n > m$, and

$$(2) \qquad\qquad \left| \oint_{C} f\, dz \right| \leq 4^n \left| \oint_{C_n} f\, dz \right|, \qquad\qquad n = 1, 2, \cdots$$

Let z_0 be the point which belongs to all these triangles. Since f is differentiable at $z = z_0$, the derivative $f'(z_0)$ exists. Let $h(z)$ denote the difference between the difference quotient and the derivative:

$$(3) \qquad\qquad h(z) = \frac{f(z) - f(z_0)}{z - z_0} - f'(z_0).$$

Since $f'(z_0)$ is the limit of this difference quotient, we see that $h(z)$ can be made as small in absolute value as we please: for a given positive number ϵ we can find a positive number δ such that

$$(4) \qquad\qquad |h(z)| < \epsilon \qquad \text{when} \qquad |z - z_0| < \delta.$$

We shall need this in a minute. Solving (3) algebraically for $f(z)$, we have

$$f(z) = f(z_0) + (z - z_0) f'(z_0) + h(z)(z - z_0).$$

Integrating this over the boundary C_n of the triangle T_n gives

$$\oint_{C_n} f(z)\, dz = \oint_{C_n} f(z_0)\, dz + \oint_{C_n} (z - z_0) f'(z_0)\, dz + \oint_{C_n} h(z)(z - z_0)\, dz.$$

Since $f(z_0)$ and $f'(z_0)$ are constants and C_n is a closed path, the first two integrals on the right are zero (cf. Example 7 in Sec. 13.2). Thus we have

$$\oint_{C_n} f(z)\, dz = \oint_{C_n} h(z)(z - z_0)\, dz.$$

We may now take n so large that the triangle T_n lies in the disk $|z - z_0| < \delta$. Let L_n be the length of C_n. Then $|z - z_0| < L_n$ for all z on C_n and z_0 in T_n. From this and (4) we have $|h(z)(z - z_0)| < \epsilon L_n$. The ML-inequality in Sec. 13.2 now gives

$$(5) \qquad \left| \oint_{C_n} f(z)\, dz \right| = \left| \oint_{C_n} h(z)(z - z_0)\, dz \right| < \epsilon L_n \cdot L_n = \epsilon L_n^{\,2}.$$

Now let L be the length of C. Then the path C_1 has the length $L_1 = L/2$, the path C_2 has the length $L_2 = L_1/2 = L/4$, etc., and C_n has the length

$$L_n = \frac{L}{2^n} . \qquad \text{Hence} \qquad L_n{}^2 = \frac{L^2}{4^n} .$$

From (2) and (5) we thus obtain

$$\left| \oint_C f \, dz \right| \leq 4^n \left| \oint_{C_n} f \, dz \right| < 4^n \epsilon L_n{}^2 = 4^n \epsilon \frac{L^2}{4^n} = \epsilon L^2.$$

By choosing ϵ (> 0) sufficiently small we can make the expression on the right as small as we please, while the expression on the left is the definite value of an integral. Consequently, this value must be zero, and the proof is complete.

The proof for *the case in which C is the boundary of a polygon* follows from the previous proof by subdividing the polygon into triangles (Fig. 347). The integral corresponding to each such triangle is zero. The sum of these integrals is equal to the integral over C, because we integrate along each segment of subdivision in both directions, the corresponding integrals cancel out in pairs, and we are left with the integral over C.

The case of a general simple closed path C can be reduced to the preceding one by inscribing in C a closed polygon P of chords, which approximates C "sufficiently accurately," and it can be shown that there is a polygon P such that the integral over P differs from that over C by less than any preassigned positive real number ϵ, no matter how small. The details of this proof are somewhat involved and can be found in Ref. [D6] listed in Appendix 1. ∎

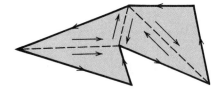

Fig. 347. Proof of Cauchy's integral theorem for a polygon

Review Problems for Chapter 13

1. If a curve C is represented by $z = z(t)$, what is $\dot{z} = dz/dt$ geometrically?
2. What are our general assumptions about paths of integration?
3. What value do you get by integrating $1/z$ counterclockwise around the unit circle? (This is basic; you should memorize it.)
4. Which integration methods in this chapter apply only to analytic functions and which to general continuous complex functions?
5. Which theorem in this chapter do you regard as the most important one? Why? State it from memory.

6. State the inequality that gives upper bounds for the absolute value of an integral. Does it apply only to analytic functions or more generally?

7. What is the principle of deformation of path? Give a typical illustrative example of it.

8. An analytic function $f(z)$ in a domain D is differentiable in D. Does it have higher derivatives in D? Answer the same question for a real function $f(x)$ that is differentiable on some interval of the real line.

9. Is $\operatorname{Re} \int_C f(z)\, dz = \int_C \operatorname{Re} f(z)\, dz$?

10. Using Liouville's theorem, show that e^z and $\cos z$ are not bounded in the complex plane.

Integrate:

11. $z \sinh z^2$ from πi to 0 along any path

12. $6z^2$ from 0 to $1 + i$ along any path

13. $\cos z$ from $-\pi$ to π along the upper semicircle $|z| = \pi$

14. $z^2/(z - \pi)$ counterclockwise around $|z| = 4$

15. $1/\operatorname{Ln} (z + 3i)$ counterclockwise around the unit circle

16. $|z|$ along the straight line segment from 0 to $3 + 3i$

17. $\operatorname{Re} z$ counterclockwise around the boundary of the triangle with vertices 0, 1, $1 + i$

18. $\operatorname{Im} z$ around the contour in Prob. 17

19. $8z - 6$ along the straight line segment from i to $1 + i$

20. $\cos 3z$ from $\pi/3$ along the x-axis to 0 and then along the y-axis to πi

21. $(e^{\pi z} \sin z)/z^2$ counterclockwise around $|z| = 0.1$

22. $(\sin 2\pi z)/(z - i)^3$ counterclockwise around $|z| = 2$

23. $z^3 + 1/z^5$ along the upper arc of the unit circle from 1 to -1

24. $1/z + 1/(z - 2)$ clockwise around the ellipse $(x - 1)^2 + 4y^2 = 4$

25. $\sin z$ along the straight line segment from 0 to i

26. $(\tan \pi z)/(z - 1)^2$ counterclockwise around the ellipse $x^2 + 9y^2 = 9$

27. e^{2z} from $14 + 3\pi i$ to $14 - \pi i$ along any path

28. $\sin \pi z$ from $-i$ to i along the imaginary axis

29. $(z \cosh z^2)/(z - 2i)^3$ clockwise around $|z + 2i| = 1$

30. $z \exp z^2$ from 1 to i counterclockwise along $|z| = 1$

31. $|z| + z$ counterclockwise around the unit circle

32. $\cosh z$ along the straight line segment from 0 to i

33. e^z/z^4 counterclockwise around $|z| = 1/4$

34. $(z + 2i)^{-1} - (z + 2i)^{-2}$ clockwise around the unit circle

35. The function in Prob. 34 clockwise around $|z + 2i| = 1$

36. $\cos iz$ along the real axis from $-\pi$ to π

37. $e^z \cos (e^z)$ from $-\pi i$ to $2\pi i$ along any path

38. $(\operatorname{Ln} z)/(z - 2i)^2$ counterclockwise around $|z - 2i| = \pi/2$

39. $z \cosh z$ from $-i$ to i along the left half of the unit circle

40. $z^{-2n} \sin z$ counterclockwise around the unit circle

Summary of Chapter 13
Complex Integration

The **complex line integral** of a function $f(z)$ taken over a path C is denoted by (Sec. 13.1)

$$\int_C f(z)\, dz \qquad \text{or, if } C \text{ is closed, also by} \qquad \oint_C f(z)\, dz.$$

Such an integral can be evaluated by using the equation $z = z(t)$ of C, where $a \leq t \leq b$ (Sec. 13.2):

$$(1) \qquad \int_C f(z)\, dz = \int_a^b f(z(t))\dot{z}(t)\, dt \qquad \left(\dot{z} = \frac{dz}{dt}\right).$$

As another method, if $f(z)$ is analytic (Sec. 12.4) in a simply connected domain D (cf. Sec. 13.3), then there exists an $F(z)$ in D such that $F'(z) = f(z)$ and for every path C in D from a point z_0 to a point z_1 we have (Secs. 13.2, 13.4)

$$(2) \qquad \int_C f(z)\, dz = F(z_1) - F(z_0) \qquad [F'(z) = f(z)].$$

Cauchy's integral theorem states that if $f(z)$ is analytic in a simply connected domain D, then for every closed path C in D (Sec. 13.3),

$$(3) \qquad \oint_C f(z)\, dz = 0.$$

If $f(z)$ is as in Cauchy's integral theorem, then for any z_0 in D and closed path C in D that contains z_0 in its interior we have **Cauchy's integral formula**

$$(4) \qquad f(z_0) = \frac{1}{2\pi i} \oint_C \frac{f(z)}{z - z_0}\, dz.$$

Furthermore, then $f(z)$ has derivatives of all orders in D that are themselves analytic functions in D and (Sec. 13.6)

$$(5) \qquad f^{(n)}(z_0) = \frac{n!}{2\pi i} \oint_C \frac{f(z)}{(z - z_0)^{n+1}}\, dz \qquad (n = 1, 2, \cdots).$$

Chapter 14

Power Series, Taylor Series, Laurent Series

Sections 14.1 and 14.2 contain the basic concepts and convergence tests for complex series, which are similar to those for real series usually discussed in calculus. Readers reasonably familiar with the latter may use Secs. 14.1 and 14.2 for reference and start with Sec. 14.3, in which we give a thorough discussion of **power series.**

Beginning in Sec. 14.4, we explain why power series play a fundamental role in complex analysis and its applications. The reason is that power series represent analytic functions (Sec. 14.4, Theorem 5) and, conversely, *every* analytic function can be represented by power series (called **Taylor series,** analogs of those in calculus; cf. Sec. 14.5).

Sections 14.6 and 14.7 concern special Taylor series and practical methods for obtaining them. In Sec. 14.8 we discuss uniform convergence of power and other series.

Laurent series (Sec. 14.9) are series of positive *and negative* integer powers of z (or $z - z_0$). These also help in classifying singularities (Sec. 14.10) and lead to an elegant general integration method (to be explained in the next chapter).

Prerequisites for this chapter: Chaps. 12, 13
Sections that may be omitted in a shorter course: Secs. 14.7, 14.8, 14.10 (and use Sec. 14.2 only for reference).
References: Appendix 1, Part D
Answers to Problems: Appendix 2.

14.1 Sequences and Series

In this section we shall define the basic concepts relating to *complex* sequences and series and their convergence and divergence. These concepts are quite similar to those for *real* sequences and series in calculus. *If you feel at home with them and want to take for granted that the ratio test also holds in complex, skip this section and the next, and go to Sec. 14.3.*

Sequences. If to each positive integer n there is assigned a number z_n, then these numbers

$$z_1, z_2, \cdots, z_n, \cdots$$

are called an *infinite sequence* or, briefly, a **sequence** and each of the numbers is called a **term** of the sequence. This sequence is often written $\{z_1, z_2, \cdots\}$ or, more briefly, $\{z_n\}$.

A sequence whose terms are real is called a *real sequence*.

At times it is convenient to number the terms of a sequence starting with 0, with 2, or with some other integer.

A sequence $z_1, z_2, \cdots$ is said to **converge** or to **be convergent** if there is a number c, called the **limit** of the sequence, with the following property. For every $\epsilon > 0$ we can find an integer N such that

(1) $$|z_n - c| < \epsilon \qquad \text{for all } n > N.$$

Then we write

$$\lim_{n \to \infty} z_n = c \qquad \text{or, simply,} \qquad z_n \to c,$$

and say that the sequence *converges to c* or *has the limit c*.

A sequence that is not convergent is said to *be* **divergent** or to **diverge.**

Geometrically, (1) means that all the terms with $n > N$ lie in the open disk of radius ϵ and center c and at most finitely many terms do not lie in this disk (Fig. 348).

For a *real* sequence, (1) means that all the terms with $n > N$ lie in the interval of length 2ϵ and midpoint c (Fig. 349) and at most finitely many terms lie outside this interval on the real line.

EXAMPLE 1. Convergent and divergent sequences

The sequence whose terms are $z_n = 1 + \dfrac{2}{n}$ is $3, 2, \frac{5}{3}, \frac{6}{4}, \frac{7}{5}, \cdots$. It is convergent and has the limit $c = 1$. In fact, in (1)

$$z_n - c = 1 + \frac{2}{n} - 1 = \frac{2}{n} \quad \text{and} \quad \frac{2}{n} < \epsilon \quad \text{when} \quad \frac{n}{2} > \frac{1}{\epsilon} \quad \text{or} \quad n > \frac{2}{\epsilon}.$$

For example, choosing $\epsilon = 0.01$ we have $2/n < 0.01$ when $n > 200$.

The sequence $1, 2, 3, \cdots$ is divergent, and so is the sequence $\frac{1}{4}, \frac{3}{4}, \frac{1}{5}, \frac{4}{5}, \frac{1}{6}, \frac{5}{6}, \cdots$.

EXAMPLE 2. Convergent sequence

The sequence $\{z_n\}$ with $z_n = x_n + iy_n = 2 - \dfrac{1}{n} + i\left(1 + \dfrac{2}{n}\right)$ is

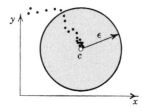

Fig. 348. Convergent complex sequence

Fig. 349. Convergent real sequence

$$1 + 3i, \quad \tfrac{3}{2} + 2i, \quad \tfrac{5}{3} + \tfrac{5}{3}i, \quad \tfrac{7}{4} + \tfrac{3}{2}i, \cdots ,$$

cf. Fig. 350. This sequence is convergent, the limit being $c = 2 + i$. Indeed, in (1) we have

$$|z_n - c| = \left| \frac{2n-1}{n} + i\frac{n+2}{n} - (2+i) \right| = \left| -\frac{1}{n} + \frac{2i}{n} \right| = \frac{\sqrt{5}}{n} < \epsilon \quad \text{when} \quad n > \frac{\sqrt{5}}{\epsilon}.$$

Furthermore, we see that the sequence $\{x_n\}$ of the real parts converges to $2 = \operatorname{Re} c$, and $\{y_n\}$ converges to $1 = \operatorname{Im} c$. This is typical. It illustrates the following theorem by which the convergence of a *complex* sequence can be referred back to that of the two *real* sequences of the real and imaginary parts. ∎

Theorem 1 (Sequences of the real and imaginary parts)
A sequence $z_1, z_2, \cdots , z_n, \cdots$ of complex numbers $z_n = x_n + iy_n$ (where $n = 1, 2, \cdots$) converges to $c = a + ib$ if and only if the sequence of the real parts $x_1, x_2, \cdots$ converges to a and the sequence of the imaginary parts $y_1, y_2, \cdots$ converges to b.

Proof. If $|z_n - c| < \epsilon$, then $z_n = x_n + iy_n$ is within the circle of radius ϵ about $c = a + ib$ so that necessarily (Fig. 351a)

$$|x_n - a| < \epsilon, \qquad |y_n - b| < \epsilon.$$

Thus convergence $z_n \to c$ implies convergence $x_n \to a$ and $y_n \to b$.

Conversely, if $x_n \to a$ and $y_n \to b$ as $n \to \infty$, then for a given $\epsilon > 0$ we can choose N so large that, for every $n > N$,

$$|x_n - a| < \frac{\epsilon}{2}, \qquad |y_n - b| < \frac{\epsilon}{2}.$$

These two inequalities imply that $z_n = x_n + iy_n$ lies in a square with center c and side ϵ. Hence, z_n must lie within a circle of radius ϵ with center c (Fig. 351b). ∎

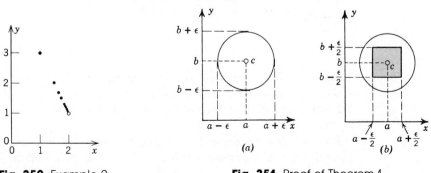

Fig. 350. Example 2

(a)

(b)

Fig. 351. Proof of Theorem 1

Series

Given a sequence $z_1, z_2, \cdots, z_m, \cdots$, we may form the sequence of sums

$$s_1 = z_1$$

$$s_2 = z_1 + z_2$$

(2)

$$\cdots \cdots \cdots \cdots$$

$$s_n = z_1 + z_2 + \cdots + z_n$$

$$\cdots \cdots$$

We call this the sequence of **partial sums** of the *infinite series,* or **series**

(3)
$$\sum_{m=1}^{\infty} z_m = z_1 + z_2 + \cdots.$$

We call the z_m the **terms** of the series. (As a *letter of summation,* we generally use n; and m only if we simultaneously need n in s_n, as in the present case.)

If the sequence of the partial sums converges, say,

$$\lim_{n \to \infty} s_n = s,$$

we say that the series (3) **converges** or *is convergent* and call s its **sum** or *value.* And then we write

$$s = \sum_{m=1}^{\infty} z_m = z_1 + z_2 + \cdots.$$

If that sequence $\{s_n\}$ is not convergent, we call the series (3) **divergent.**

If we omit the terms of s_n from (3), there remains

(4)
$$R_n = z_{n+1} + z_{n+2} + z_{n+3} + \cdots.$$

This is called the **remainder** *of the series* (3) *after the term* z_n. Clearly, if (3) converges and has the sum s, then

$$s = s_n + R_n, \qquad \text{thus} \qquad R_n = s - s_n.$$

Now $s_n \to s$ by the definition of convergence; hence $R_n \to 0$. In applications, when s is unknown and we calculate an approximation s_n of s, then $|R_n|$ is the error, and $R_n \to 0$ means that we can make $|R_n|$ as small as we please, by choosing n large enough.

By applying Theorem 1 to the partial sums of a series we obtain the convergence of a complex series to the convergence of the corresponding two series of the real parts and of the imaginary parts.

Theorem 2 (Real and imaginary parts)

A series (3) with $z_m = x_m + iy_m$ converges with sum $s = u + iv$ if and only if $x_1 + x_2 + \cdots$ converges with the sum u and $y_1 + y_2 + \cdots$ converges with the sum v.

We shall discuss a practically much more important relation between real and complex series in the next section (in connection with "absolute convergence").

The next section will be concerned with **convergence tests** for complex series, which are practically the same as in calculus.

Problems for Sec. 14.1

Find and plot the first few terms of the sequence $z_1, z_2, \cdots, z_n, \cdots$ where z_n equals

1. i^n/n^3　　　　　　**2.** $in/(n + 1)$　　　　　**3.** $i^n n^2/(n + i)$

4. $(1 + i)^n/n^2$　　　　**5.** $e^{in\pi/4}$　　　　　　**6.** $(-1)^n + 2n\pi i$

Are the following sequences $z_1, z_2, \cdots, z_n, \cdots$ convergent? In the case of convergence find the limit.

7. $z_n = i^n/n$　　　　　**8.** $z_n = i^n$　　　　　　**9.** $z_n = n^2/(n + i)$

10. $z_n = in/(n + 1)$　　**11.** $z_n = e^{in\pi/4}$　　　　**12.** $z_n = \frac{1}{2}\pi + e^{in\pi/4}/n\pi$

13. $z_n = (1 + 2i)^n/n!$　**14.** $z_n = \cos(\pi i/n)$　　**15.** $z_n = n\pi/(1 + 2in)$

Problems on series (and some more problems on sequences) follow at the end of the next section.

14.2 Convergence Tests for Series

Before we use a series, we must make sure that it converges.

Divergence can often be shown very simply as follows.

Theorem 1 (Divergence)

If a series $z_1 + z_2 + \cdots$ converges, then

$$\lim_{m \to \infty} z_m = 0.$$

Hence if the series does not satisfy this condition, it diverges.

Proof. If $z_1 + z_2 + \cdots$ converges with the sum s, then, since $z_m = s_m - s_{m-1}$,

$$\lim_{m \to \infty} z_m = \lim_{m \to \infty} (s_m - s_{m-1}) = \lim_{m \to \infty} s_m - \lim_{m \to \infty} s_{m-1} = s - s = 0. \quad \blacksquare$$

Caution! $z_m \to 0$ is *necessary* for convergence but *not sufficient*, as we see from the harmonic series $1 + \frac{1}{2} + \frac{1}{3} + \frac{1}{4} + \cdots$, which satisfies this condition but diverges, as is shown in calculus (cf., e.g., p. 623 in Ref. [18] listed in Appendix 1).

The practical difficulty in proving convergence is that in most cases the sum of a series is unknown. Cauchy overcame this by showing that a series converges if and only if its partial sums eventually get close to each other:

Theorem 2 (Cauchy's convergence principle for series)
A series $z_1 + z_2 + \cdots$ is convergent if and only if for every given $\epsilon > 0$ (no matter how small) we can find an N (which depends on ϵ, in general) such that

(1) $|z_{n+1} + z_{n+2} + \cdots + z_{n+p}| < \epsilon$ *for every $n > N$ and $p = 1, 2, \cdots$.*

The proof of this theorem is somewhat involved; we leave it optional and give it on p. 853 near the end of the chapter.

A series $z_1 + z_2 + \cdots$ is called **absolutely convergent** if

$$\sum_{m=1}^{\infty} |z_m| = |z_1| + |z_2| + \cdots$$

is convergent. The latter is a series of nonnegative real terms, so that absolute convergence provides a relation between complex and real series.

If $z_1 + z_2 + \cdots$ converges but $|z_1| + |z_2| + \cdots$ diverges, then $z_1 + z_2 + \cdots$ is called, more precisely, **conditionally convergent.**

EXAMPLE 1. A conditionally convergent series
The series $1 - \frac{1}{2} + \frac{1}{3} - \frac{1}{4} + - \cdots$ converges, but only conditionally since the harmonic series diverges, as mentioned above (after Theorem 1).   ∎

If a series is absolutely convergent, it is convergent.

This follows readily from Cauchy's principle (Problem 16). This principle also yields the following general convergence test.

Theorem 3 (Comparison test)
If a series $z_1 + z_2 + \cdots$ is given and we can find a converging series $b_1 + b_2 + \cdots$ with nonnegative real terms such that

$$|z_n| \leq b_n \qquad \text{for } n = 1, 2, \cdots,$$

then the given series converges, even absolutely.

Proof. By Cauchy's principle, since $b_1 + b_2 + \cdots$ converges, for any given $\epsilon > 0$ we can find an N such that

$$b_{n+1} + \cdots + b_{n+p} < \epsilon \qquad \text{for every } n > N \text{ and } p = 1, 2, \cdots.$$

From this and $|z_1| \leq b_1, |z_2| \leq b_2, \cdots$ we conclude that for those n and p,

$$|z_{n+1}| + \cdots + |z_{n+p}| \leq b_{n+1} + \cdots + b_{n+p} < \epsilon.$$

Hence, again by Cauchy's principle, $|z_1| + |z_2| + \cdots$ converges, so that $z_1 + z_2 + \cdots$ is absolutely convergent.   ∎

A good comparison series in Theorem 3 is the **geometric series**

$$\sum_{m=0}^{\infty} q^m = 1 + q + q^2 + \cdots.$$

We first recall its convergence behavior:

Theorem 4 (Geometric series)

The geometric series $1 + q + q^2 + \cdots$ converges with the sum $1/(1 - q)$ when $|q| < 1$ and diverges when $|q| \geq 1$.

Proof. If $|q| \geq 1$, then $|q^m| \geq 1$ and Theorem 1 implies divergence.
    Now let $|q| < 1$. The nth partial sum is

$$s_n = 1 + q + \cdots + q^n.$$

From this,

$$q s_n = \quad q + \cdots + q^n + q^{n+1}.$$

On subtraction, most terms on the right cancel in pairs, and we are left with

$$s_n - q s_n = (1 - q)s_n = 1 - q^{n+1}.$$

Now $1 - q \neq 0$ since $q \neq 1$, and we may solve for s_n, finding

$$(2) \qquad s_n = \frac{1 - q^{n+1}}{1 - q} = \frac{1}{1 - q} - \frac{q^{n+1}}{1 - q}.$$

Since $|q| < 1$, the last term approaches zero as $n \to \infty$. Hence the series is convergent and has the sum $1/(1 - q)$. This completes the proof.   ∎

Ratio Test

Using the geometric series as the comparison series $b_1 + b_2 + \cdots$ in the comparison theorem, we can now obtain the ratio test, which will be the most important test in our further work.

Theorem 5 (Ratio test)

If a series $z_1 + z_2 + \cdots$ with $z_n \neq 0$ ($n = 1, 2, \cdots$) has the property that for every n greater than some N,

$$(3) \qquad \left| \frac{z_{n+1}}{z_n} \right| \leq q < 1 \qquad\qquad (n > N)$$

*(note well that q is **fixed** and strictly less than 1), this series converges absolutely. If for every $n > N$,*

$$(4) \qquad \left| \frac{z_{n+1}}{z_n} \right| \geq 1 \qquad\qquad (n > N),$$

the series diverges.

Proof. If (4) holds, then $|z_{n+1}| \geq |z_n|$ for those n, so that divergence of $z_1 + z_2 + \cdots$ follows from Theorem 1.

If (3) holds, then $|z_{n+1}| \leq |z_n|q$ for $n > N$, in particular,

$$|z_{N+2}| \leq |z_{N+1}|q, \qquad |z_{N+3}| \leq |z_{N+2}|q \leq |z_{N+1}|q^2, \qquad \text{etc.}$$

and in general, $|z_{N+p}| \leq |z_{N+1}|q^{p-1}$. Hence

$$|z_{N+1}| + |z_{N+2}| + |z_{N+3}| + \cdots \leq |z_{N+1}|(1 + q + q^2 + \cdots).$$

Absolute convergence of $z_1 + z_2 + \cdots$ now follows from the comparison test and the geometric series, since $q < 1$ (Theorems 3 and 4). ∎

Caution! The inequality (3) implies $|z_{n+1}/z_n| < 1$, but this does **not** imply convergence, as we see from the harmonic series, which satisfies $z_{n+1}/z_n = n/(n + 1) < 1$ for all n but diverges.

Often in practice the sequence of the ratios in (3) and (4) will converge. Then the ratio test takes a more convenient form:

Theorem 6 (Ratio test)

Let a series $z_1 + z_2 + \cdots$ with $z_n \neq 0$ $(n = 1, 2, \cdots)$ have the property that

$$(5) \qquad \qquad \lim_{n \to \infty} \left| \frac{z_{n+1}}{z_n} \right| = L.$$

Then we have the following.
   (a) *If $L < 1$, the series converges absolutely.*
   (b) *If $L > 1$, it diverges.*
   (c) *If $L = 1$, the test fails; that is, no conclusion is possible.*

Proof. (a) Write $k_n = |z_{n+1}/z_n|$ in (5). Let $L = 1 - b < 1$. Then, by the definition of a limit, the k_n must eventually get close to $1 - b$, say, $k_n \leq q = 1 - \frac{1}{2}b < 1$ for all n greater than some N. Convergence of $z_1 + z_2 + \cdots$ now follows from Theorem 5.

(b) Similarly, for $L = 1 + c > 1$ we have $k_n \geq 1 + \frac{1}{2}c > 1$ for all $n > N^*$ (sufficiently large), which implies divergence of $z_1 + z_2 + \cdots$ by Theorem 5.

(c) The harmonic series has $z_{n+1}/z_n = n/(n + 1)$, hence $L = 1$ and diverges. The series

$$1 + \frac{1}{4} + \frac{1}{9} + \frac{1}{16} + \frac{1}{25} + \cdots \qquad \text{has} \qquad \frac{z_{n+1}}{z_n} = \frac{n^2}{(n + 1)^2},$$

hence also $L = 1$, but converges. Convergence follows from (Fig. 352)

$$s_n = 1 + \frac{1}{4} + \cdots + \frac{1}{n^2} \leq 1 + \int_1^n \frac{dx}{x^2} = 2 - \frac{1}{n},$$

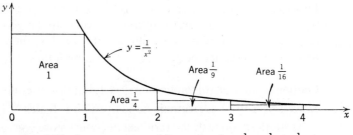

Fig. 352. Convergence of the series $1 + \frac{1}{4} + \frac{1}{9} + \frac{1}{16} + \cdots$

so that $s_1, s_2, \cdots$ is a bounded sequence and is monotone increasing (since the terms of the series are all positive); both properties together are sufficient for the convergence of the real sequence $s_1, s_2, \cdots$. ∎

EXAMPLE 2. Ratio test

Is the following series convergent or divergent? (First guess, then calculate.)

$$\sum_{n=0}^{\infty} \frac{(100 + 75i)^n}{n!} = 1 + (100 + 75i) + \frac{1}{2!}(100 + 75i)^2 + \cdots$$

Solution. By Theorem 6, the series is convergent, since

$$\left|\frac{z_{n+1}}{z_n}\right| = \frac{|100 + 75i|^{n+1}/(n + 1)!}{|100 + 75i|^n/n!} = \frac{|100 + 75i|}{n + 1} = \frac{125}{n + 1} \quad \rightarrow \quad L = 0.$$

EXAMPLE 3. Theorem 5 more general than Theorem 6

Is the following series convergent or divergent?

$$\sum_{n=0}^{\infty} \left(\frac{i}{2^{3n}} + \frac{1}{2^{3n+1}}\right) = i + \frac{1}{2} + \frac{i}{8} + \frac{1}{16} + \frac{i}{64} + \frac{1}{128} + \cdots$$

Solution. The ratios of the absolute values of successive terms are $\frac{1}{2}, \frac{1}{4}, \frac{1}{2}, \frac{1}{4}, \cdots$. Hence convergence follows from Theorem 5. Since the sequence of these ratios has no limit, Theorem 6 is not applicable. ∎

Root Test

The two practically most important tests are the ratio test and the root test. The ratio test is generally simpler; the root test is somewhat more general.

Theorem 7 (Root test)

If a series $z_1 + z_2 + \cdots$ is such that for every n greater than some N,

(6)
$$\sqrt[n]{|z_n|} \leq q < 1 \qquad\qquad (n > N)$$

*(note again that q is **fixed** and strictly less than 1), this series converges absolutely. If for infinitely many n,*

(7)
$$\sqrt[n]{|z_n|} \geq 1,$$

the series diverges.

Proof. If (6) holds, then $|z_n| \leq q^n < 1$ for all $n > N$. Hence the series $|z_1| + |z_2| + \cdots$ converges by comparison with the geometric series, so that $z_1 + z_2 + \cdots$ converges absolutely. If (7) holds, then $|z_n| \geq 1$ for infinitely many n. Divergence of $z_1 + z_2 + \cdots$ now follows from Theorem 1.   ∎

Caution! Equation (6) implies $\sqrt[n]{|z_n|} < 1$, but this does not imply convergence, as we see from the harmonic series, which satisfies $\sqrt[n]{1/n} < 1$ but diverges.

As for the ratio test, the root test becomes more convenient if the sequence in (6) and (7) converges:

Theorem 8 (Root test)

Let the series $z_1 + z_2 + \cdots$ have the property that

$$(8) \qquad \lim_{n \to \infty} \sqrt[n]{|z_n|} = L.$$

Then we have the following.

(a) *If $L < 1$, the series converges absolutely.*
(b) *If $L > 1$, it diverges.*
(c) *If $L = 1$, the test fails; that is, no conclusion is possible.*

Proof. The proof parallels that of Theorem 6.

(a) Let $L = 1 - a^* < 1$. Then by the definition of a limit we have $\sqrt[n]{|z_n|} < q = 1 - \frac{1}{2}a^* < 1$ for all n greater than some (sufficiently large) N^*. Hence $|z_n| < q^n < 1$ for all $n > N^*$. Absolute convergence of the series $z_1 + z_2 + \cdots$ now follows by the comparison with the geometric series.

(b) If $L > 1$, also $\sqrt[n]{|z_n|} > 1$ for all sufficiently large n. Hence $|z_n| > 1$ for those n. Theorem 1 now implies that $z_1 + z_2 + \cdots$ diverges.

(c) Both the *divergent* harmonic series and the *convergent* series $1 + \frac{1}{4} + \frac{1}{9} + \frac{1}{16} + \frac{1}{25} + \cdots$ give $L = 1$. This can be seen from $(\ln n)/n \to 0$ and

$$\sqrt[n]{\frac{1}{n}} = \frac{1}{n^{1/n}} = \frac{1}{e^{(1/n) \ln n}} \;\to\; \frac{1}{e^0}, \qquad \sqrt[n]{\frac{1}{n^2}} = \frac{1}{n^{2/n}} = \frac{1}{e^{(2/n) \ln n}} \;\to\; \frac{1}{e^0}. \quad ∎$$

EXAMPLE 4. Root test

Is the following series convergent or divergent?

$$\sum_{n=0}^{\infty} \frac{(-1)^n}{2^{2n} + 3} (4 - i)^n = \frac{1}{4} - \frac{1}{7}(4 - i) + \frac{1}{19}(4 - i)^2 - + \cdots$$

Solution. By Theorem 8, the series diverges, since

$$\sqrt[n]{\frac{|(4 - i)^n|}{2^{2n} + 3}} = \frac{|4 - i|}{\sqrt[n]{4^n + 3}} = \frac{\sqrt{17}}{\sqrt[n]{4^n + 3}} \;\to\; L = \frac{\sqrt{17}}{4} > 1. \quad ∎$$

This is the end of our discussion of basic concepts and facts on complex series and convergence tests. In the next section we begin our actual work, which will concern **power series.** These are the most important series in complex analysis. The reason for their importance will appear in Secs. 14.4 and 14.5: the sum of such a series (with a positive radius of convergence) is an analytic function and, conversely, every analytic function can be represented by power series. The latter are called **Taylor series** and are quite similar to the familiar Taylor series in calculus.

Problems for Sec. 14.2

Sequences

Are the following sequences $z_1, z_2, \cdots, z_n, \cdots$ bounded? Convergent? Find their limit points.

1. $z_n = (-1)^n + i/n$ **2.** $z_n = e^{in\pi/2}$ **3.** $z_n = i^n \cos n\pi$

4. $z_n = (i^n \cosh n\pi)/n$ **5.** $z_n = (3 + 4i)^n/n!$ **6.** $z_n = (-3i)^n/n^2$

7. $z_n = (1 + i)^{2n}/n$ **8.** $z_n = 1/(1 + i)^n$ **9.** $z_n = (-1)^n + 2i$

10. $z_1 = 1, z_2 = i, z_n = z_{n-2}z_{n-1}$ $(n = 3, 4, \cdots)$

11. Show that a complex sequence $z_1, z_2, \cdots$ is bounded if and only if the two corresponding sequences of the real parts and the imaginary parts are bounded.

12. (**Uniqueness of limit**) Show that if a sequence converges, its limit is unique.

13. If $z_1, z_2, \cdots$ converges with the limit l and $z_1{}^*, z_2{}^*, \cdots$ converges with the limit l^*, show that $z_1 + z_1{}^*, z_2 + z_2{}^*, \cdots$ converges with the limit $l + l^*$.

14. Show that under the assumptions of Prob. 13 the sequence $z_1 z_1{}^*, z_2 z_2{}^*, \cdots$ converges with the limit ll^*.

Series

15. Confirm the result in Example 3 by the ratio test.

16. (**Absolute convergence**) Show that if a series converges absolutely, it is convergent.

Are the following series convergent or divergent?

17. $\displaystyle\sum_{n=0}^{\infty} \frac{(1 + 2i)^n}{n!}$ **18.** $\displaystyle\sum_{n=0}^{\infty} \frac{(3 + i)^{2n}}{(2n)!}$ **19.** $\displaystyle\sum_{n=0}^{\infty} n \left(\frac{i}{2}\right)^n$ **20.** $\displaystyle\sum_{n=0}^{\infty} \left(\frac{9i}{10}\right)^n n^5$

21. $\displaystyle\sum_{n=1}^{\infty} \frac{(2i)^n n!}{n^n}$ **22.** $\displaystyle\sum_{n=1}^{\infty} \frac{n + 1}{2^n n}$ **23.** $\displaystyle\sum_{n=1}^{\infty} \frac{n^{2n} + i^{2n}}{n!}$ **24.** $\displaystyle\sum_{n=0}^{\infty} \frac{i^n}{n^2 + i}$

25. $\displaystyle\sum_{n=1}^{\infty} \frac{(n!)^2}{(2n)!}$ **26.** $\displaystyle\sum_{n=1}^{\infty} \frac{1}{n(n + 1)}$ **27.** $\displaystyle\sum_{n=1}^{\infty} \frac{1}{\sqrt{n}}$ **28.** $\displaystyle\sum_{n=2}^{\infty} \frac{1}{\ln n}$

29. Determine how many terms are needed to compute the sum s of the geometric series $1 + q + q^2 + \cdots$ with an error less than 0.01, when $q = 0.25$, $q = 0.5$, $q = 0.9$.

30. Suppose that $|z_{n+1}/z_n| \le q < 1$, so that the series $z_1 + z_2 + \cdots$ converges by the ratio test (Theorem 5). Show that the remainder $R_n = z_{n+1} + z_{n+2} + \cdots$ satisfies $|R_n| \le |z_{n+1}|/(1 - q)$. (*Hint.* Use the fact that the ratio test is a comparison of the series $z_1 + z_2 + \cdots$ with the geometric series.) Using the result, find how many terms suffice for computing the sum s of the series in Prob. 22 with an error not exceeding 0.05 and compute s to this accuracy.

14.3 Power Series

Power series are the most important series in complex analysis, as was mentioned at the beginning of the chapter and as we shall now see in detail. A **power series** *in powers of* $z - z_0$ is a series of the form

(1)
$$\sum_{n=0}^{\infty} a_n(z - z_0)^n = a_0 + a_1(z - z_0) + a_2(z - z_0)^2 + \cdots$$

where z is a variable, $a_0, a_1, \cdots$ are constants, called the **coefficients** of the series, and z_0 is a constant, called the **center** of the series.

If $z_0 = 0$, we obtain as a particular case a *power series in powers of* z:

(2)
$$\sum_{n=0}^{\infty} a_n z^n = a_0 + a_1 z + a_2 z^2 + \cdots.$$

Convergence Behavior of Power Series

We have made the definitions in the previous section for series of *constant terms*. If the terms of a series are *variable*, say, functions of a variable z (e.g., powers of z, as in a power series), they assume definite values when we fix z, and then all those definitions apply. Clearly, for a series of functions of z, the partial sums, the remainders and the sum will be functions of z. Usually such a series will converge for some z, for instance, throughout some region, and diverge for the other z. In general, such a region may be complicated. For a *power series* it is simple. Indeed, we shall see (in Theorem 1, below) that (1) converges throughout a disk with center z_0—this is the general case—or in the whole complex plane; or it converges merely at the center z_0 [in which case (1) is practically useless]. Let us first illustrate these three possibilities with three typical examples.

EXAMPLE 1. Convergence in a disk. Geometric series
The *geometric series*

$$\sum_{n=0}^{\infty} z^n = 1 + z + z^2 + \cdots$$

converges absolutely when $|z| < 1$ and diverges when $|z| \geq 1$ (cf. Theorem 4 in Sec. 14.2).

EXAMPLE 2. Convergence for every z
The power series

$$\sum_{n=0}^{\infty} \frac{z^n}{n!} = 1 + z + \frac{z^2}{2!} + \frac{z^3}{3!} + \cdots$$

is absolutely convergent for every z. In fact, by the ratio test, for any fixed z,

$$\left| \frac{z^{n+1}/(n+1)!}{z^n/n!} \right| = \frac{|z|}{n+1} \to 0 \quad \text{as} \quad n \to \infty.$$

EXAMPLE 3. Convergence only at the center

The power series

$$\sum_{n=0}^{\infty} n! z^n = 1 + z + 2z^2 + 6z^3 + \cdots$$

converges only at $z = 0$, but diverges for every $z \neq 0$. In fact, from the ratio test we have

$$\left| \frac{(n+1)! z^{n+1}}{n! z^n} \right| = (n+1)|z| \to \infty \quad \text{as} \quad n \to \infty \qquad (z \text{ fixed and} \neq 0). \quad \blacksquare$$

Every power series (1) converges at the center z_0, since $z - z_0 = 0$ when $z = z_0$ and (1) reduces to the single term a_0. The center $z = z_0$ may sometimes be the only point of convergence (as in Example 3). If not—that is, if (1) converges for some $z_1 \neq z_0$—we shall see that then (1) converges for every z that is closer to z_0 than z_1:

Theorem 1 (Convergence of a power series)

If the power series (1) *converges at a point* $z = z_1 \neq z_0$, *it converges absolutely for every* z *closer to* z_0 *than* z_1, *that is*, $|z - z_0| < |z_1 - z_0|$. *Cf. Fig. 353.*

If (1) *diverges at a* $z = z_2$, *it diverges for every* z *farther away from* z_0 *than* z_2.

Proof. Since the series (1) converges for z_1. Theorem 1 in Sec. 14.2 gives

$$a_n(z_1 - z_0)^n \to 0 \quad \text{as} \quad n \to \infty.$$

This implies that for $z = z_1$ the terms of the series (1) are bounded, say,

$$|a_n(z_1 - z_0)^n| < M \qquad \text{for every } n = 0, 1, \cdots.$$

Multiplying and dividing by $(z_1 - z_0)^n$, we obtain from this

$$(3) \qquad |a_n(z - z_0)^n| = \left| a_n(z_1 - z_0)^n \left(\frac{z - z_0}{z_1 - z_0} \right)^n \right| < M \left| \frac{z - z_0}{z_1 - z_0} \right|^n.$$

Now our assumption $|z - z_0| < |z_1 - z_0|$ implies that

$$\left| \frac{z - z_0}{z_1 - z_0} \right| < 1. \qquad \text{Hence the series} \qquad M \sum_{n=0}^{\infty} \left| \frac{z - z_0}{z_1 - z_0} \right|^n$$

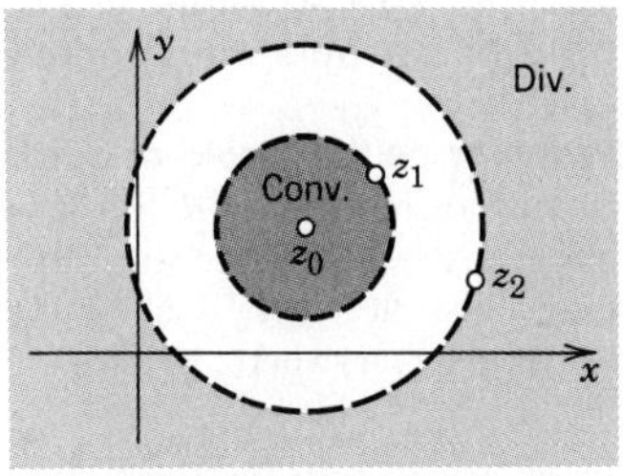

Fig. 353. Theorem 1

is a converging geometric series (cf. Theorem 4 in Sec. 14.2). The convergence of (1) when $|z - z_0| < |z_1 - z_0|$ now follows from (3) and the comparison test in Sec. 14.2.

If the second statement of our theorem were false, we would have convergence at a z_3 such that $|z_3 - z_0| > |z_2 - z_0|$, implying convergence at z_2 by the statement just proved, and contradicting the assumed divergence at z_2. ∎

Radius of Convergence of a Power Series

Examples 2 and 3 illustrate that a power series may converge for all z or only for $z = z_0$. Let us exclude these two cases for a moment. Then, if a power series (1) is given, we may consider all the points z in the complex plane for which the series converges. Let R be the smallest real number such that the distance of each of those points from the center z_0 is at most equal to R. (For instance, $R = 1$ in Example 1.) Then from Theorem 1 it follows that the series converges for all z within the circle of radius R about z_0, that is, for all z for which

(4) $$|z - z_0| < R,$$

and, by the definition of R, the series diverges for all z for which

$$|z - z_0| > R.$$

The circle

$$|z - z_0| = R$$

is called the **circle of convergence,** and its radius R is called the **radius of convergence** of (1). Cf. Fig. 354.

To include those two excluded cases in the notation, we write

$R = \infty$ when the series (1) converges for all z (as in Example 2)

$R = 0$ when (1) converges only at the center $z = z_0$ (as in Example 3).

These are convenient notations, but nothing else.

For a *real* power series (1) in powers of $x - x_0$ with real coefficients and center, formula (4) gives the **convergence interval** $|x - x_0| < R$ of length $2R$ on the real line.

To avoid misunderstandings: no general statements can be made about the convergence of a power series (1) *on the circle of convergence* itself. The series (1) may converge at some or all or none of these points. Details will not be essential to us; hence a simple example may just give us the idea.

EXAMPLE 4. Behavior on the circle of convergence
On the circle of convergence (radius $R = 1$ in all three series),

$\Sigma z^n/n^2$ converges everywhere since $\Sigma 1/n^2$ converges,

$\Sigma z^n/n$ converges at -1 (by Leibniz's test) but diverges at 1,

Σz^n diverges everywhere. ∎

The radius of convergence R of the power series (1) may be determined from the coefficients of the series as follows.

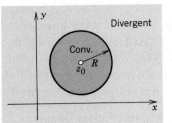

Fig. 354. Circle of convergence

Theorem 2 (Radius of convergence R)

Suppose that the sequence $|a_{n+1}/a_n|$, $n = 1, 2, \cdots$, converges with limit L^*. If $L^* = 0$, then $R = \infty$; that is, the power series (1) *converges for all z. If $L^* \neq 0$ (hence $L^* > 0$), then*

(5)
$$R = \frac{1}{L^*}$$
(Cauchy–Hadamard formula[1]**).**

If $|a_{n+1}/a_n| \to \infty$, then $R = 0$ (convergence only at the center z_0).

Proof. The series (1) has the terms $z_n = a_n(z - z_0)^n$. Hence in the ratio test (Theorem 6 in Sec. 14.2),

$$L = \lim_{n \to \infty} \left| \frac{z_{n+1}}{z_n} \right| = \lim_{n \to \infty} \left| \frac{a_{n+1}(z - z_0)^{n+1}}{a_n(z - z_0)^n} \right| = \lim_{n \to \infty} \left| \frac{a_{n+1}}{a_n} \right| |z - z_0|,$$

that is,

$$L = L^*|z - z_0|.$$

If $L^* = 0$, then $L = 0$ for every z, and the ratio test gives convergence for all z, as asserted. Let $L^* > 0$. If $|z - z_0| < 1/L^*$, then $L = L^*|z - z_0| < 1$, and (1) converges by the ratio test. If $|z - z_0| > 1/L^*$, then $L > 1$, and (1) diverges by the ratio test. By definition, this shows that $1/L^*$ is the radius of convergence R of (1) and proves (5). If $|a_{n+1}/a_n| \to \infty$, then $|z_{n+1}/z_n| \geqq 1$ for every $z \neq z_0$ and all sufficiently large n, so that divergence of (1) for all $z \neq z_0$ now follows from Theorem 5 in Sec. 14.2. ∎

Comment. If $L^* = \lim |a_{n+1}/a_n| \neq 0$, then (5) gives

(6)
$$R = \lim_{n \to \infty} \left| \frac{a_n}{a_{n+1}} \right|.$$

---

[1]Named after the French mathematicians, A. L. CAUCHY (cf. the footnote in Sec. 2.7) and JACQUES HADAMARD (1865—1963). Hadamard made basic contributions to the theory of power series and devoted his lifework to partial differential equations.

This is plausible, because if $a_n \to 0$ rapidly, then $|a_n/a_{n+1}|$ will be large on the average (why?) and the series converges in a large disk. If $a_n \to 0$ slowly, then $|a_n/a_{n+1}|$ will be relatively small and the series converges in a relatively small disk.

Formulas (5) and (6) will not help if L^* does not exist, but extensions of Theorem 2 are still possible, as we discuss in Example 6 below.

EXAMPLE 5. Radius of convergence

Determine the radius of convergence R of the power series

$$\sum_{n=0}^{\infty} \frac{(2n)!}{(n!)^2} (z - 3i)^n.$$

Solution. From (5),

$$L^* = \lim_{n\to\infty} \frac{(2n + 2)!/[(n + 1)!]^2}{(2n)!/(n!)^2} = \lim_{n\to\infty} \frac{(2n + 2)(2n + 1)}{(n + 1)^2} = 4.$$

Hence $R = 1/L^* = 1/4$. The series converges in the open disk $|z - 3i| < 1/4$, of radius 1/4 and center $3i$.

EXAMPLE 6. Extension of Theorem 2

Find the radius of convergence R of the power series

$$\sum_{n=0}^{\infty} \left[1 + (-1)^n + \frac{1}{2^n} \right] z^n = 3 + 2^{-1}z + (2 + 2^{-2})z^2 + 2^{-3}z^3 + (2 + 2^{-4})z^4 + \cdots.$$

Solution. Since the sequence of the ratios $1/6$, $2(2 + 2^{-2})$, $1/2^3(2 + 2^{-2})$, $\cdots$ does not converge, Theorem 2 is of no help. It can be shown that

$$(5^*) \qquad\qquad R = 1/\tilde{L}, \qquad \tilde{L} = \lim_{n\to\infty} \sqrt[n]{|a_n|}.$$

This does still not help here, since $\{\sqrt[n]{|a_n|}\}$ does not converge because inspection shows that it has two limit points $1/2$ and 1. It can further be shown that

$$(5^{**}) \qquad R = 1/\tilde{l}, \qquad \tilde{l} \text{ the greatest limit point of the sequence } \{\sqrt[n]{|a_n|}\}.$$

Here $\tilde{l} = 1$, so that $R = 1$. *Answer.* The series converges for $|z| < 1$.   ∎

This section was concerned with power series and their convergence behavior, which is quite simple: as a rule, power series converge in an open circular disk with center z_0; some of them converge even for every z. (Some of them converge only at z_0, but these are practically useless.) In the next section we show why power series are of fundamental importance in complex analysis: except for the useless ones, they represent **analytic functions;** that is, their sum is an analytic function.

Problems for Sec. 14.3

Find the center and the radius of convergence of the following power series.

1. $\displaystyle\sum_{n=0}^{\infty} (z + 2i)^n$

2. $\displaystyle\sum_{n=1}^{\infty} \frac{z^n}{n}$

3. $\displaystyle\sum_{n=0}^{\infty} n \left(\frac{z}{3}\right)^n$

4. $\displaystyle\sum_{n=0}^{\infty} \frac{(-1)^n}{n!}(z + 1 - i)^n$

5. $\displaystyle\sum_{n=0}^{\infty} \frac{3n + 4}{2^n}(z - 2 - i)^n$

6. $\displaystyle\sum_{n=0}^{\infty} \frac{2^n}{3^n + 1}(z - \tfrac{1}{4})^n$

7. $\displaystyle\sum_{n=2}^{\infty} \left(\frac{3n^2 + n}{n^2 - 1}\right)(z + i)^n$ **8.** $\displaystyle\sum_{n=0}^{\infty} \frac{z^{n+2}}{(n + 1)(n + 2)}$ **9.** $\displaystyle\sum_{n=0}^{\infty} 5^n(z + 5i)^n$

10. $\displaystyle\sum_{n=0}^{\infty} \frac{(-1)^n}{2^{2n}(n!)^2} z^{2n}$ **11.** $\displaystyle\sum_{n=0}^{\infty} \frac{(2n)!}{(n!)^2} z^n$ **12.** $\displaystyle\sum_{n=0}^{\infty} \frac{(3n)!}{2^n(n!)^3}(z + \pi i)^n$

13. $\displaystyle\sum_{n=0}^{\infty} \frac{z^{2n+1}}{(2n + 1)!}$ **14.** $\displaystyle\sum_{n=0}^{\infty} 3^{2n}(z + 1)^{2n}$ **15.** $\displaystyle\sum_{n=1}^{\infty} n^n(z - 3 + \pi i)^n$

16. $\displaystyle\sum_{n=1}^{\infty} \frac{n^n}{n!} (z - 1)^n$. *Hint.* Use $\displaystyle\lim_{n\to\infty} \left(1 + \frac{1}{n}\right)^n = e$.   **17.** $\displaystyle\sum_{n=1}^{\infty} \frac{(2n)!}{n^n} z^n$

18. $\displaystyle\frac{1}{2^2} z^2 + 2^3 z^3 + \frac{1}{2^4} z^4 + 2^5 z^5 + \frac{1}{2^6} z^6 + \cdots$

19. $4^2 z^2 + 2^3 z^3 + 4^4 z^4 + 2^5 z^5 + 4^6 z^6 + 2^7 z^7 + \cdots$

Find the radii of convergence of the given series and of its derived series by some test, thereby verifying that they are equal, as asserted in Theorem 3.

20. $\displaystyle\sum_{n=0}^{\infty} \frac{(-1)^n}{n!} z^{2n}$ **21.** $\displaystyle\sum_{n=0}^{\infty} \frac{1}{(1 - i)^n} (z - i)^n$ **22.** $\displaystyle\sum_{n=0}^{\infty} \frac{1}{n + 1} \left(\frac{z}{5}\right)^{n+1}$

23. $\displaystyle\sum_{n=0}^{\infty} \frac{2^{10n}}{n!} (z - i)^n$ **24.** $\displaystyle\sum_{n=0}^{\infty} n^{100} z^n$ **25.** $\displaystyle\sum_{n=0}^{\infty} \frac{i^n n^2}{3^n} z^n$

26. Show that the sum $s(z)$ of $\displaystyle\sum_{n=0}^{\infty} z^{2n}/n!$ satisfies $s'(z) = 2zs(z)$.

27. Using the geometric series, find the radius of convergence of $\displaystyle\sum_{n=0}^{\infty} \binom{n + m}{n}$. (Apply repeated differentiation.)

28. Does there exist a power series in powers of z that converges at $z = 5 + 4i$ and diverges at $z = 6 - 2i$? (Give reason.)

29. Show that if a power series $\Sigma a_n z^n$ has radius of convergence R (assumed finite), then $\Sigma a_n z^{2n}$ has the radius of convergence $\sqrt{R}$.

30. Show that in Prob. 29, the series $\Sigma a_n^2 z^n$ has the radius of convergence R^2.

14.4 Functions Given by Power Series

To simplify the formulas in this section, we take $z_0 = 0$ and write

(1) $$\sum_{n=0}^{\infty} a_n z^n.$$

This is no restriction, since in a series in powers of $z^* - z_0$ with any center z_0 we can always set $z^* - z_0 = z$ to reduce it to the form (1).

If an arbitrary power series (1) has a nonzero radius of convergence, its sum is a function of z, say, $f(z)$. Then we write

(2) $$f(z) = \sum_{n=0}^{\infty} a_n z^n = a_0 + a_1 z + a_2 z^2 + \cdots \qquad (|z| < R).$$

We say that $f(z)$ *is* **represented** *by the power series* or that *it is* **developed** *in the power series*. For instance, the geometric series represents the function $f(z) = 1/(1 - z)$ in the interior of the unit circle $|z| = 1$. (Cf. Example 1 in Sec. 14.3.)

Our present goal is to show the **uniqueness** of such a representation; that is, *a function $f(z)$ cannot be represented by two different power series with the same center*. If $f(z)$ can at all be developed in a power series with center z_0, the development is unique. This important fact is frequently used in complex and real analysis. This result (Theorem 2, below) will follow from

Theorem 1 (Continuity of the sum of a power series)
The function $f(z)$ in (2) with $R > 0$ is continuous at $z = 0$.

Proof. We must show that

$$\lim_{z \to 0} f(z) = f(0) = a_0,$$

that is, by definition, for a given $\epsilon > 0$ there is a $\delta > 0$ such that $|z| < \delta$ implies $|f(z) - a_0| < \epsilon$. Now (2) converges absolutely for $|z| \le r < R$, by Theorem 1 in Sec. 14.3. Hence the series

$$\sum_{n=1}^{\infty} |a_n| r^{n-1} = \frac{1}{r} \sum_{n=1}^{\infty} |a_n| r^n \qquad (r > 0)$$

converges. Let S be its sum. Then, for $0 < |z| \le r$,

$$|f(z) - a_0| = \left| \sum_{n=1}^{\infty} a_n z^n \right| \le |z| \sum_{n=1}^{\infty} |a_n| |z|^{n-1}$$

$$\le |z| \sum_{n=1}^{\infty} |a_n| r^{n-1} = |z| S.$$

This is less than ϵ for $|z| < \delta$, where $\delta > 0$ is less than both r and ϵ/S. ∎

From this theorem we can now readily obtain the desired uniqueness theorem (assuming $z_0 = 0$ without loss of generality):

Theorem 2 (Identity theorem for power series)
Suppose that the power series

$$\sum_{n=0}^{\infty} a_n z^n \qquad and \qquad \sum_{n=0}^{\infty} b_n z^n$$

both converge for $|z| < R$, where R is positive, and have the same sum for all these z. Then these series are identical, that is,

$$a_n = b_n \qquad\qquad for \ all \ n = 0, 1, \cdots.$$

Proof. We proceed by induction. By assumption,

$$a_0 + a_1 z + a_2 z^2 + \cdots = b_0 + b_1 z + b_2 z^2 + \cdots \qquad (|z| < R).$$

The sums of these two power series are continuous at $z = 0$, by Theorem 1. Hence $a_0 = b_0$: the assertion is true when $n = 0$. Now assume that $a_n = b_n$ for $n = 0, 1, \cdots, m$. Then we may omit on both sides the terms that are equal and divide the result by z^{m+1} ($\neq 0$); this gives

$$a_{m+1} + a_{m+2} z + a_{m+3} z^2 + \cdots = b_{m+1} + b_{m+2} z + b_{m+3} z^2 + \cdots .$$

By Theorem 1, each of these power series represents a function that is continuous at $z = 0$. Hence $a_{m+1} = b_{m+1}$. This completes the proof. ∎

Power Series Represent Analytic Functions

This will be shown as the main goal in this section, after a short preparation. In the next section we shall see that, conversely, *every* analytic function can be represented by power series (called *Taylor series*). This is quite surprising and accounts for the great importance of power series in complex analysis.

Termwise addition or subtraction of two power series with radii of convergence R_1 and R_2 yields a power series with radius of convergence at least equal to the smaller of R_1 and R_2. *Proof.* Add (or subtract) the partial sums s_n and $s_n{}^*$ term by term and use $\lim (s_n \pm s_n{}^*) = \lim s_n \pm \lim s_n{}^*$.

Termwise multiplication of two power series

$$f(z) = \sum_{k=0}^{\infty} a_k z^k = a_0 + a_1 z + \cdots$$

and

$$g(z) = \sum_{m=0}^{\infty} b_m z^m = b_0 + b_1 z + \cdots ,$$

means the multiplication of each term of the first series by each term of the second series and the collection of like powers of z. This gives a power series, which is called the **Cauchy product** of the two series and is given by

$$a_0 b_0 + (a_0 b_1 + a_1 b_0) z + (a_0 b_2 + a_1 b_1 + a_2 b_0) z^2 + \cdots$$

$$= \sum_{n=0}^{\infty} (a_0 b_n + a_1 b_{n-1} + \cdots + a_n b_0) z^n.$$

We mention without proof that this power series converges absolutely for each z within the circle of convergence of each of the two given series and has the sum $s(z) = f(z)g(z)$. For a proof, see p. 113 in vol. 1 of Ref. [D5] listed in Appendix 1 (use the absolute convergence stated in our Theorem 1 of Sec. 14.3).

Termwise differentiation and integration of power series is permissible, as we show next. We call **derived series** *of the power series* (1) the power series obtained from (1) by termwise differentiation, that is,

$$(3) \qquad \sum_{n=1}^{\infty} n a_n z^{n-1} = a_1 + 2a_2 z + 3a_3 z^2 + \cdots .$$

Theorem 3 (Termwise differentiation of a power series)

The derived series of a power series has the same radius of convergence as the original series.

Proof. This follows from (5) in Sec. 14.3 because

$$\lim_{n \to \infty} \frac{(n+1)a_{n+1}}{na_n} = \lim_{n \to \infty} \frac{n+1}{n} \lim_{n \to \infty} \frac{a_{n+1}}{a_n} = \lim_{n \to \infty} \frac{a_{n+1}}{a_n}$$

or, if the last limit does not exist, from (5**) in Sec. 14.3 by noting that $\sqrt[n]{n} \to 1$ as $n \to \infty$. ∎

EXAMPLE 1. An application of Theorem 3

Find the radius of convergence R of the following series by applying Theorem 3.

$$\sum_{n=2}^{\infty} \binom{n}{2} z^n = z^2 + 3z^3 + 6z^4 + 10z^5 + \cdots .$$

Solution. Differentiate the geometric series twice term by term and multiply the result by $z^2/2$. This yields the given series. Hence $R = 1$ by Theorem 3. ∎

Theorem 4 (Termwise integration of power series)

The power series

$$\sum_{n=0}^{\infty} \frac{a_n}{n+1} z^{n+1} = a_0 z + \frac{a_1}{2} z^2 + \frac{a_2}{3} z^3 + \cdots$$

obtained by integrating the series $a_0 + a_1 z + a_2 z^2 + \cdots$ *term by term has the same radius of convergence as the original series.*

The proof is similar to that of Theorem 3.

With Theorem 3 as a tool, we are now ready to establish our main result in this section:

Theorem 5 (Analytic functions. Their derivatives)

A power series with a nonzero radius of convergence R represents an analytic function at every point interior to its circle of convergence. The derivatives of this function are obtained by differentiating the original series term by term. All the series thus obtained have the same radius of convergence as the original series. Hence, by the first statement, each of them represents an analytic function.

Proof. (a) We consider any power series (1) with positive radius of convergence R. Let $f(z)$ be its sum and $f_1(z)$ the sum of its derived series; thus

(4) $$f(z) = \sum_{n=0}^{\infty} a_n z^n \qquad \text{and} \qquad f_1(z) = \sum_{n=1}^{\infty} n a_n z^{n-1}.$$

We show that $f(z)$ is analytic and has the derivative $f_1(z)$ in the interior of the circle of convergence. We do this by proving that for any fixed z with $|z| < R$ and $\Delta z \to 0$ the difference quotient $[f(z + \Delta z) - f(z)]/\Delta z$ approaches $f_1(z)$. By termwise addition we first have from (4)

(5) $$\frac{f(z + \Delta z) - f(z)}{\Delta z} - f_1(z) = \sum_{n=2}^{\infty} a_n \left[\frac{(z + \Delta z)^n - z^n}{\Delta z} - n z^{n-1} \right].$$

Note that the summation starts with 2, since the constant term drops out in taking the difference $f(z + \Delta z) - f(z)$, and so does the linear term when we subtract $f_1(z)$ from the difference quotient.

(b) We claim that the series in (5) can be written

(6) $$\sum_{n=2}^{\infty} a_n \, \Delta z [(z + \Delta z)^{n-2} + 2z(z + \Delta z)^{n-3} + \cdots + (n - 1)z^{n-2}].$$

We give the somewhat technical proof of this on p. 855 near the end of the chapter.

(c) We consider (6). The brackets contain $n - 1$ terms, and the largest coefficient is $n - 1$. Since $(n - 1)^2 < n(n - 1)$, we see that for $|z| \le R_0$ and $|z + \Delta z| \le R_0$, $R_0 < R$, the absolute value of this series cannot exceed

$$|\Delta z| \sum_{n=2}^{\infty} |a_n| n(n - 1) R_0^{n-2}.$$

This series with a_n instead of $|a_n|$ is the second derived series of (2) at $z = R_0$ and converges absolutely by Theorem 3 and Sec. 14.3, Theorem 1. Hence our present series converges. Let $K(R_0)$ be its sum. Then we can write our present result

$$\left| \frac{f(z + \Delta z) - f(z)}{\Delta z} - f_1(z) \right| \le |\Delta z| \, K(R_0).$$

Letting $\Delta z \to 0$ and noting that $R_0 (< R)$ is arbitrary, we conclude that $f(z)$ is analytic at any point interior to the circle of convergence and its derivative is represented by the derived series. From this the statements about the higher derivatives follow by induction.   ∎

The results in this section show that power series are about as nice as we could hope for: we can differentiate and integrate them term by term (Theorems 3 and 4). Theorem 5 accounts for the great importance of power series in complex analysis: the sum of such a series is an analytic function and has derivatives of all orders, which thus are analytic functions. But this is only part of the story. In the next section we show that, conversely, *every* given analytic function $f(z)$ can be represented by power series. These are called **Taylor series** of $f(z)$ and are as in calculus, with the real x replaced by the complex z.

Problems for Sec. 14.4

1. Verify Theorems 3 and 4 for the geometric series and for the series in Example 2, Sec. 14.3.

Applying Theorems 3 or 4 to the geometric series, find the radius of convergence of the following series.

2. $\displaystyle\sum_{n=1}^{\infty} \frac{n}{5^n} (z - i)^n$ **3.** $\displaystyle\sum_{n=1}^{\infty} \frac{4^n}{n} (z + 2i)^n$ **4.** $\displaystyle\sum_{n=1}^{\infty} \frac{(z + 2)^n}{n}$

5. $\displaystyle\sum_{n=0}^{\infty} \left[\binom{n + k}{n} \right]^{-1} z^{n+k}$ **6.** $\displaystyle\sum_{n=k}^{\infty} \binom{n}{k} \left(\frac{z}{2}\right)^n$ **7.** $\displaystyle\sum_{n=0}^{\infty} \left[\binom{n + k}{n} \right]^{-1} \left(\frac{z}{3}\right)^n$

Find the radius of convergence of the following series in two ways, (a) directly by the Cauchy–Hadamard formula (Sec. 14.3), (b) by Theorem 3 or Theorem 4 and a series with simpler coefficients.

8. $\displaystyle\sum_{n=3}^{\infty} \frac{n(n - 1)(n - 2)}{4^n} z^n$ **9.** $\displaystyle\sum_{n=0}^{\infty} \frac{(z - 1)^{n+2}}{(n + 1)(n + 2)}$ **10.** $\displaystyle\sum_{n=2}^{\infty} \frac{n(n - 1)}{2^n} z^{2n}$

11. $\displaystyle\sum_{n=1}^{\infty} \frac{2n}{n!} z^{2n-1}$ **12.** $\displaystyle\sum_{n=1}^{\infty} (-1)^n n z^{2n-1}$ **13.** $\displaystyle\sum_{n=0}^{\infty} \frac{(2n)!}{(n + 1)(n!)^2} (z - i)^{n+1}$

14. Show that $(1 - z)^{-2} = \sum_{n=0}^{\infty} (n + 1) z^n$ (a) by using the Cauchy product, (b) by differenting a suitable series.

15. If $f(z)$ in (1) is even, show that $a_n = 0$ for odd n. (Use Theorem 2.)

16. If $f(z)$ in (1) is odd, show that $a_n = 0$ for even n. Give examples.

17. Applying Theorem 2 to $(1 + z)^p (1 + z)^q = (1 + z)^{p+q}$ (p and q positive integers), show that

$$\sum_{n=0}^{r} \binom{p}{n}\binom{q}{r - n} = \binom{p + q}{r}.$$

18. In the proof of Theorem 3, we claimed that $\sqrt[n]{n} \to 1$ as $n \to \infty$. Prove this. *Hint.* Set $\sqrt[n]{n} = 1 + c_n$, where $c_n > 0$, and show that $c_n \to 0$ as $n \to \infty$.

19. Write out the proof on termwise addition and subtraction of power series indicated in the text.

20. **(Fibonacci[2] numbers)** The *Fibonacci numbers* are recursively defined by $a_0 = a_1 = 1$, $a_n = a_{n-2} + a_{n-1}$ when $n \geq 2$. Show that if a power series $a_0 + a_1 z + \cdots$ represents $f(z) = 1/(1 - z - z^2)$, it must have these numbers as coefficients and conversely. *Hint.* Start from $f(z)(1 - z - z^2) = 1$ and use Theorem 2.

---

[2]LEONARDO OF PISA, called FIBONACCI (= son of Bonaccio), about 1180—1250, Italian mathematician, credited with the first renaissance of mathematics on Christian soil.

14.5 Taylor Series

In the previous section (in Theorem 5) we proved that the sum of every power series (with $R > 0$) is an analytic function. We shall now see that, conversely, every analytic function $f(z)$ can be represented by power series, called **Taylor series** of $f(z)$. These will be of the same form as in calculus, with x replaced by the complex z.

Let us consider a function $f(z)$ that is analytic in a neighborhood of a point $z = z_0$, and let us derive for it Taylor's formula and then the Taylor series with center z_0. The key tool for this is Cauchy's integral formula (1) in Sec. 13.5; writing z and z^* instead of z_0 and z, we have

$$(1) \qquad\qquad f(z) = \frac{1}{2\pi i} \oint_C \frac{f(z^*)}{z^* - z}\, dz^*.$$

z lies inside C, for which we take a circle of radius r with center z_0 in that neighborhood. z^* is the complex variable of integration over C (cf. Fig. 355). Our next idea is now to develop $1/(z^* - z)$ in (1) in powers of $z - z_0$ in a form similar to the sum formula for a finite geometric sum. The latter is

$$(2^*) \qquad 1 + q + \cdots + q^n = \frac{1 - q^{n+1}}{1 - q} = \frac{1}{1 - q} - \frac{q^{n+1}}{1 - q} \qquad (q \ne 1),$$

and we use it in the form

$$(2) \qquad\qquad \frac{1}{1 - q} = 1 + q + \cdots + q^n + \frac{q^{n+1}}{1 - q}.$$

Accordingly, by a standard algebraic manipulation (which is worthwhile remembering) we produce

$$(2^{**}) \qquad \frac{1}{z^* - z} = \frac{1}{z^* - z_0 - (z - z_0)} = \frac{1}{(z^* - z_0)\left(1 - \dfrac{z - z_0}{z^* - z_0}\right)}.$$

For later use we note that since z^* is on C while z is inside C, we have

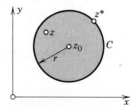

Fig. 355. Cauchy formula (1)

(3)
$$\left| \frac{z - z_0}{z^* - z_0} \right| < 1.$$

Applying (2) with $q = (z - z_0)/(z^* - z_0)$ on the right, we obtain

$$\frac{1}{z^* - z} = \frac{1}{z^* - z_0} \left[1 + \frac{z - z_0}{z^* - z_0} + \left(\frac{z - z_0}{z^* - z_0} \right)^2 + \cdots + \left(\frac{z - z_0}{z^* - z_0} \right)^n \right.$$
$$\left. + \frac{[(z - z_0)/(z^* - z_0)]^{n+1}}{(z^* - z)/(z^* - z_0)} \right].$$

We insert this into (1). Since z and z_0 are constant, we may take the powers of $z - z_0$ out from under the integral sign, and (1) takes the form

$$f(z) = \frac{1}{2\pi i} \oint_C \frac{f(z^*)}{z^* - z_0} dz^* + \frac{z - z_0}{2\pi i} \oint_C \frac{f(z^*)}{(z^* - z_0)^2} dz^* + \cdots$$

(4)
$$\cdots + \frac{(z - z_0)^n}{2\pi i} \oint_C \frac{f(z^*)}{(z^* - z_0)^{n+1}} dz^* + R_n(z)$$

where the last term is given by the formula

(5)
$$R_n(z) = \frac{(z - z_0)^{n+1}}{2\pi i} \oint_C \frac{f(z^*)}{(z^* - z_0)^{n+1}(z^* - z)} dz^*.$$

Using the integral formulas (1) in Sec. 13.6 for the derivatives, we obtain

$$f(z) = f(z_0) + \frac{z - z_0}{1!} f'(z_0) + \frac{(z - z_0)^2}{2!} f''(z_0) + \cdots$$

(6)
$$\cdots + \frac{(z - z_0)^n}{n!} f^{(n)}(z_0) + R_n(z).$$

This representation is called **Taylor's**[3] **formula.** $R_n(z)$ is called the *remainder*. Since the analytic function $f(z)$ has derivatives of all orders, we may take n in (6) as large as we please. If we let n approach infinity, we obtain from (6) the power series

(7)
$$\boxed{f(z) = \sum_{m=0}^{\infty} \frac{f^{(m)}(z_0)}{m!} (z - z_0)^m.}$$

This series is called the **Taylor series** of $f(z)$ *with center* z_0. The particular

---

[3]BROOK TAYLOR (1685—1731), English mathematician, who introduced this formula for functions of a real variable.

case of (7) with center $z_0 = 0$ is called the **Maclaurin**[4] **series** of $f(z)$.

Clearly, the series (7) will converge and represent $f(z)$ if and only if

$$\text{(8)} \qquad \lim_{n \to \infty} R_n(z) = 0.$$

To prove (8), we consider (5). Since z^* is on C while z is inside C (cf. Fig. 355), we have $|z^* - z| > 0$. From this and the analyticity of $f(z)$ inside and on C, it follows that the absolute value of $f(z^*)/(z^* - z)$ is bounded, say,

$$\left| \frac{f(z^*)}{z^* - z} \right| < \tilde{M}$$

for all z^* on C. Also, $|z^* - z_0| = r$, the radius of C, and C has the length $2\pi r$. Hence by applying the *ML*-inequality (Sec. 13.2) to (5) we obtain

$$|R_n| = \frac{|z - z_0|^{n+1}}{2\pi} \left| \int_C \frac{f(z^*)}{(z^* - z_0)^{n+1}(z^* - z)} \, dz^* \right|$$

$$< \frac{|z - z_0|^{n+1}}{2\pi} \tilde{M} \frac{1}{r^{n+1}} 2\pi r = \tilde{M} r \left| \frac{z - z_0}{r} \right|^{n+1}.$$

On the right, $|z - z_0| < r$, since z lies *inside* the circle C. Thus $|z - z_0|/r < 1$, so that the right-hand side approaches zero as $n \to \infty$. This proves (8) for all z inside C. By Theorem 2 in the preceding section, the representation of $f(z)$ in the form (7) is unique in the sense that (7) is the only power series with center at z_0 that represents the given function $f(z)$, we may sum up our result as follows.

Theorem 1 (Taylor's theorem)

Let $f(z)$ be analytic in a domain D, and let $z = z_0$ be any point in D. Then there exists precisely one power series with center z_0 that represents $f(z)$. This series is of the form

$$\text{(9)} \qquad f(z) = \sum_{n=0}^{\infty} a_n(z - z_0)^n \qquad \text{where} \qquad a_n = \frac{1}{n!} f^{(n)}(z_0).$$

This representation is valid in the largest open disk with center z_0 contained in D. The remainders $R_n(z)$ of (9) can be represented in the form (5). The coefficients satisfy the inequality

$$\text{(10)} \qquad |a_n| \leqq \frac{M}{r^n}$$

where M is the maximum of $|f(z)|$ on the circle $|z - z_0| = r$.

---

[4]COLIN MACLAURIN (1698—1746), Scots mathematician, professor at Edinburgh.

Relation (10) follows from Cauchy's inequality (3) in Sec. 13.6.

Examples will be discussed in the next section.

Practically speaking, (8) means that for all z for which (9) converges, the nth partial sum of (9) will approximate $f(z)$ to any assigned degree of accuracy; we just have to choose n large enough.

A point at which a function $f(z)$ ceases to be analytic is called a **singular point** of $f(z)$; we also say that $f(z)$ has a **singularity** at such a point. More precisely: a point $z = c$ is called a *singular point* of $f(z)$, if $f(z)$ is not differentiable at c but if every neighborhood of c contains points at which $f(z)$ is differentiable.

Using this concept, we may say that there is at least one singular point of $f(z)$ on the circle of convergence[5] of the development (9).

One surprising property of complex analytic functions is that they have derivatives of all orders, and now we have discovered the other surprising property that they can always be represented by power series of the form (9). This is not true in general for *real* functions; there are real functions that have derivatives of all orders but cannot be represented by a power series. (Example: $f(x) = \exp(-1/x^2)$ when $x \neq 0$ and $f(0) = 0$; this function cannot be represented by a Maclaurin series since all its derivatives at 0 are zero.)

The relation between our present consideration and that on power series in the preceding section can easily be established as follows.

Theorem 2

Every power series with a nonzero radius of convergence is the Taylor series of the function represented by that power series (more briefly: *is the Taylor series of its sum*).

Proof. Consider any power series with positive radius of convergence R and call its sum $f(z)$; thus,

$$f(z) = a_0 + a_1(z - z_0) + a_2(z - z_0)^2 + \cdots .$$

From Theorem 5 in the previous section it follows that

$$f'(z) = a_1 + 2a_2(z - z_0) + \cdots$$

and more generally

$$f^{(n)}(z) = n!a_n + (n + 1)n \cdots 3 \cdot 2a_{n+1}(z - z_0) + \cdots ;$$

all these series converge in the disk $|z - z_0| < R$ and represent analytic functions. Hence these functions are continuous at $z = z_0$, by Theorem 1

---

[5]The radius of convergence of (9) will in general be equal to the distance from z_0 to the nearest singular point of $f(z)$, but it may be larger; for example, Ln z is singular along the negative real axis, and the distance from $z_0 = -1 + i$ to that axis is 1, but the Taylor series of Ln z with center $z_0 = -1 + i$ has radius of convergence $\sqrt{2}$.

in Sec. 14.4. If we set $z = z_0$, we thus obtain

$$f(z_0) = a_0, \qquad f'(z_0) = a_1, \qquad \cdots, \qquad f^{(n)}(z_0) = n!a_n, \qquad \cdots.$$

Since these formulas are identical with those in Taylor's theorem, the proof is complete. ∎

In the next section we discuss the Maclaurin serin series of complex functions, such as e^z, cos z, etc. In Sec. 14.7 we show practical methods for obtaining Maclaurin series of given functions.

14.6 Taylor Series of Elementary Functions

We shall see that the Maclaurin series of elementary functions look the same as in calculus, with the real x replaced by the complex z.

EXAMPLE 1. Geometric series

Let $f(z) = 1/(1 - z)$. Then we have $f^{(n)}(z) = n!/(1 - z)^{n+1}$, $f^{(n)}(0) = n!$. Hence the Maclaurin expansion of $1/(1 - z)$ is the geometric series

$$(1) \qquad \frac{1}{1 - z} = \sum_{n=0}^{\infty} z^n = 1 + z + z^2 + \cdots \qquad (|z| < 1).$$

$f(z)$ is singular at $z = 1$; this point lies on the circle of convergence.

EXAMPLE 2. Exponential function

We know that the exponential function e^z (Sec. 12.6) is analytic for all z, and $(e^z)' = e^z$. Hence from (9) in the previous section, with $z_0 = 0$, we obtain the Maclaurin series

$$(2) \qquad \boxed{e^z = \sum_{n=0}^{\infty} \frac{z^n}{n!} = 1 + z + \frac{z^2}{2!} + \cdots.}$$

This series is also obtained if we replace x in the Maclaurin series of e^x by z.

Furthermore, by setting $z = iy$ in (2) and separating the series into the real and imaginary parts (cf. Theorem 2, Sec. 14.1) we obtain

$$e^{iy} = \sum_{n=0}^{\infty} \frac{(iy)^n}{n!} = \sum_{k=0}^{\infty} (-1)^k \frac{y^{2k}}{(2k)!} + i \sum_{k=0}^{\infty} (-1)^k \frac{y^{2k+1}}{(2k + 1)!}.$$

Since the series on the right are the familiar Maclaurin series of the real functions cos y and sin y, this shows that we have rediscovered the **Euler formula**

$$(3) \qquad e^{iy} = \cos y + i \sin y.$$

Indeed, one may use (2) for *defining* e^z and derive from (2) the basic properties of e^z. For instance, the differentiation formula $(e^z)' = e^z$ follows readily from (2) by termwise differentiation.

EXAMPLE 3. Trigonometric and hyperbolic functions

By substituting (2) into (1) of Sec. 12.7 we obtain

(4)

$$\cos z = \sum_{n=0}^{\infty} (-1)^n \frac{z^{2n}}{(2n)!} = 1 - \frac{z^2}{2!} + \frac{z^4}{4!} - + \cdots$$

$$\sin z = \sum_{n=0}^{\infty} (-1)^n \frac{z^{2n+1}}{(2n+1)!} = z - \frac{z^3}{3!} + \frac{z^5}{5!} - + \cdots.$$

When $z = x$ these are the familiar Maclaurin series of the real functions $\cos x$ and $\sin x$. Similarly, by substituting (2) into (11), Sec. 12.7, we obtain

(5)

$$\cosh z = \sum_{n=0}^{\infty} \frac{z^{2n}}{(2n)!} = 1 + \frac{z^2}{2!} + \frac{z^4}{4!} + \cdots$$

$$\sinh z = \sum_{n=0}^{\infty} \frac{z^{2n+1}}{(2n+1)!} = z + \frac{z^3}{3!} + \frac{z^5}{5!} \cdots.$$

EXAMPLE 4. Logarithm

From (9) in the previous section it follows that

(6)
$$\text{Ln } (1 + z) = z - \frac{z^2}{2} + \frac{z^3}{3} - + \cdots \qquad (|z| < 1).$$

Replacing z by $-z$ and multiplying both sides by -1, we get

(7)
$$-\text{Ln } (1 - z) = \text{Ln } \frac{1}{1 - z} = z + \frac{z^2}{2} + \frac{z^3}{3} + \cdots \qquad (|z| < 1).$$

By adding both series we obtain

(8)
$$\text{Ln } \frac{1 + z}{1 - z} = 2 \left(z + \frac{z^3}{3} + \frac{z^5}{5} + \cdots \right) \qquad (|z| < 1). \quad \blacksquare$$

In the next section we explain some practical methods of obtaining Taylor series which avoid the cumbersome calculations of the derivatives occurring in the coefficient formulas of Taylor's theorem (Sec. 14.5).

Problems for Sec. 14.6

1. Using (2), prove $(e^z)' = e^z$.
2. Derive (4) and (5) from (2). Obtain (6) from Taylor's theorem.
3. Using (4), show that $\cos z$ is even and $\sin z$ is odd.
4. Using (5), show that $\cosh z \neq 0$ for all real $z = x$.
5. Using (4), show that $\sin z \neq 0$ for all pure imaginary $z = iy \neq 0$.
6. Using (4), show that $(\sin z)' = \cos z$ and $(\cos z)' = -\sin z$.
7. $f(z) = (\sin z)/z$ is undefined at $z = 0$. Define $f(0)$ so that $f(z)$ becomes entire. (Give reason.)
8. Using (4), show that $\cos z + i \sin z$ yields the Maclaurin series for e^{iz} and thus a proof of the Euler formula for complex z.

9. Find the derivative of Ln $(1 + z)$ by differentiating (6).

10. Derive the relations (14) and (15), Sec. 12.7, between the trigonometric and hyperbolic sine and cosine from the present (4) and (5).

Find the Taylor series of the following functions with the given point as center and determine the radius of convergence.

11. e^z, πi 12. e^z, 1   13. e^{-2z}, 0

14. $1/z^2$, 1   15. $1/z$, -1 16. $1/(1 - z)$, i

17. $\cosh(z + i)$, $-i$ 18. $\cos 2z$, 0   19. $\cos z$, $-\pi/4$

20. $1/(z + 2)$, $1 + i$ 21. $(z + i)^3$, $2i$ 22. z^5, -1

23. $\cos z$, $-\pi/2$ 24. $\sin^2 z$, 0   25. $\cos^2 z$, 0

26. $\sin z^2$, 0   27. $\operatorname{Ln} z$, 1   28. $\sinh(z - 4i)$, $4i$

Find the first three terms of the Maclaurin series of the following functions.

29. $\tan z$ 30. $e^z \sin z$ 31. $z \cot z$

32. Obtain the answer to Prob. 24 from Prob. 25 and a relation between $\cos^2 z$ and $\sin^2 z$.

33. Obtain the Maclaurin series of $\cosh^2 z$ from Prob. 25.

34. Obtain the Maclaurin series of $\sinh^2 z$ from Prob. 33 and a relation between $\sinh^2 z$ and $\cosh^2 z$.

Find the Maclaurin series by integrating that of the integrand term by term. (erf z is called the **error function,** Si(z) the **sine integral,** and S(z) and C(z) the **Fresnel integrals.** [6] Cf. also Sec. 18.6 and Appendix 3.)

35. $\displaystyle\int_0^z \frac{e^t - 1}{t}\, dt$ 36. $\displaystyle\int_0^z \frac{1 - \cos t}{t^2}\, dt$ 37. $\operatorname{erf} z = \dfrac{2}{\sqrt{\pi}} \displaystyle\int_0^z e^{-t^2}\, dt$

38. $\operatorname{Si}(z) = \displaystyle\int_0^z \frac{\sin t}{t}\, dt$ 39. $\operatorname{S}(z) = \displaystyle\int_0^z \sin t^2\, dt$ 40. $\operatorname{C}(z) = \displaystyle\int_0^z \cos t^2\, dt$

14.7 Practical Methods for Obtaining Power Series

In most practical cases the determination of the coefficients of a Taylor series by means of the formula in Taylor's theorem will be complicated or time-consuming. There are a number of simpler practical procedures for that purpose, which may be illustrated by the following examples. The uniqueness of the representations thus obtained follows from Theorem 2 in Sec. 14.4.

EXAMPLE 1. Substitution
Find the Maclaurin series of $f(z) = 1/(1 + z^2)$.

Solution. By substituting $-z^2$ for z in (1), Sec. 14.6, we obtain

$$(1) \quad \frac{1}{1 + z^2} = \frac{1}{1 - (-z^2)} = \sum_{n=0}^{\infty} (-z^2)^n = \sum_{n=0}^{\infty} (-1)^n z^{2n}$$

$$= 1 - z^2 + z^4 - z^6 + \cdots \qquad (|z| < 1).$$

[6] AUGUSTIN FRESNEL (1788—1827), French physicist, known by his work in optics.

EXAMPLE 2. Integration
Find the Maclaurin series of $f(z) = \tan^{-1} z$.

Solution. We have $f'(z) = 1/(1 + z^2)$. Integrating (1) term by term and using $f(0) = 0$ we get

$$\tan^{-1} z = \sum_{n=0}^{\infty} \frac{(-1)^n}{2n+1} z^{2n+1} = z - \frac{z^3}{3} + \frac{z^5}{5} - + \cdots \qquad (|z| < 1);$$

this series represents the principal value of $w = u + iv = \tan^{-1} z$, defined as that value for which $|u| < \pi/2$.

EXAMPLE 3. Development by using the geometric series
Develop $1/(c - bz)$ in powers of $z - a$ where $c - ab \neq 0$ and $b \neq 0$.

Solution. To get powers of $z - a$ later on, we first use simple algebra:

$$\frac{1}{c - bz} = \frac{1}{c - ab - b(z - a)} = \frac{1}{(c - ab)\left[1 - \dfrac{b(z - a)}{c - ab}\right]}.$$

To the last expression we apply (1) in Sec. 14.6 with z replaced by $b(z - a)/(c - ab)$, finding

$$\frac{1}{c - bz} = \frac{1}{c - ab} \sum_{n=0}^{\infty} \left[\frac{b(z - a)}{c - ab}\right]^n = \sum_{n=0}^{\infty} \frac{b^n}{(c - ab)^{n+1}} (z - a)^n$$

$$= \frac{1}{c - ab} + \frac{b}{(c - ab)^2} (z - a) + \frac{b^2}{(c - ab)^3} (z - a)^2 + \cdots.$$

This series converges for

$$\left|\frac{b(z - a)}{c - ab}\right| < 1 \qquad \text{that is,} \qquad |z - a| < \left|\frac{c - ab}{b}\right| = \left|\frac{c}{b} - a\right|.$$

EXAMPLE 4. Binomial series, reduction by partial fractions
Find the Taylor series of the following function with center $z_0 = 1$.

$$f(z) = \frac{2z^2 + 9z + 5}{z^3 + z^2 - 8z - 12}$$

Solution. Given a rational function, we may first represent it as a sum of partial fractions and then apply the **binomial series**

(2)
$$\frac{1}{(1 + z)^m} = (1 + z)^{-m} = \sum_{n=0}^{\infty} \binom{-m}{n} z^n$$

$$= 1 - mz + \frac{m(m+1)}{2!} z^2 - \frac{m(m+1)(m+2)}{3!} z^3 + \cdots.$$

Since the function on the left is singular at $z = -1$, the series converges in the disk $|z| < 1$. In our case we obtain

$$f(z) = \frac{1}{(z + 2)^2} + \frac{2}{z - 3} = \frac{1}{[3 + (z - 1)]^2} - \frac{2}{2 - (z - 1)}.$$

This may be written in the form

$$f(z) = \frac{1}{9}\left(\frac{1}{[1 + \frac{1}{3}(z - 1)]^2}\right) - \frac{1}{1 - \frac{1}{2}(z - 1)}.$$

By using the binomial series we obtain

$$f(z) = \frac{1}{9} \sum_{n=0}^{\infty} \binom{-2}{n}\left(\frac{z - 1}{3}\right)^n - \sum_{n=0}^{\infty} \left(\frac{z - 1}{2}\right)^n.$$

We may add the two series on the right term by term. Since the binomial coefficient in the first series equals $(-2)(-3) \cdots (-[n + 1])/n! = (-1)^n(n + 1)$, we find

$$f(z) = \sum_{n=0}^{\infty} \left[\frac{(-1)^n(n + 1)}{3^{n+2}} - \frac{1}{2^n} \right] (z - 1)^n = -\frac{8}{9} - \frac{31}{54} (z - 1) - \frac{23}{108} (z - 1)^2 - \cdots .$$

Since $z = 3$ is the singular point of $f(z)$ that is nearest to the center $z = 1$, the series converges in the disk $|z - 1| < 2$.

EXAMPLE 5. Use of differential equations

Find the Maclaurin series of $f(z) = \tan z$.

Solution. We have $f'(z) = \sec^2 z$ and, therefore, since $f(0) = 0$,

$$f'(z) = 1 + f^2(z), \qquad f'(0) = 1.$$

Observing that $f(0) = 0$, we obtain by successive differentiation

$$f'' = 2ff', \qquad\qquad\qquad f''(0) = 0,$$
$$f''' = 2f'^2 + 2ff'', \qquad\qquad f'''(0) = 2, \qquad f'''(0)/3! = 1/3,$$
$$f^{(4)} = 6f'f'' + 2ff''', \qquad\quad f^{(4)}(0) = 0,$$
$$f^{(5)} = 6f''^2 + 8f'f''' + 2ff^{(4)}, \quad f^{(5)}(0) = 16, \qquad f^{(5)}(0)/5! = 2/15, \qquad\qquad \text{etc.}$$

Hence the result is

(3) $$\tan z = z + \frac{1}{3} z^3 + \frac{2}{15} z^5 + \frac{17}{315} z^7 + \cdots \qquad\qquad \left(|z| < \frac{\pi}{2} \right).$$

EXAMPLE 6. Undetermined coefficients

Find the Maclaurin series of $\tan z$ by using those of $\cos z$ and $\sin z$ (Sec. 14.6).

Solution. Since $\tan z$ is odd, the desired expansion will be of the form

$$\tan z = a_1 z + a_3 z^3 + a_5 z^5 + \cdots .$$

Using $\sin z = \tan z \cos z$ and inserting those developments, we obtain

$$z - \frac{z^3}{3!} + \frac{z^5}{5!} - + \cdots = (a_1 z + a_3 z^3 + a_5 z^5 + \cdots) \left(1 - \frac{z^2}{2!} + \frac{z^4}{4!} - + \cdots \right).$$

Since $\tan z$ is analytic except at $z = \pm \pi/2, \pm 3\pi/2, \cdots$, its Maclaurin series converges in the disk $|z| < \pi/2$, and for these z we may form the Cauchy product of the two series on the right (cf. Sec. 14.4), that is, multiply the series term by term and arrange the resulting series in powers of z. By Theorem 2 in Sec. 14.4 the coefficient of each power of z is the same on both sides. This yields

$$1 = a_1, \qquad -\frac{1}{3!} = -\frac{a_1}{2!} + a_3, \qquad \frac{1}{5!} = \frac{a_1}{4!} - \frac{a_3}{2!} + a_5, \qquad \text{etc.}$$

Hence $a_1 = 1$, $a_3 = \frac{1}{3}$, $a_5 = \frac{2}{15}$, etc., as before.   ∎

In Sec. 14.3 we showed that a power series with center z_0 and radius of convergence R (> 0) converges absolutely in every closed disk of the form $|z - z_0| \leqq r < R$. In the next section we give a thorough discussion of **uniform convergence** and show that the convergence of a power series with center z_0 in that closed disk is not only absolute but also uniform. Thereafter we shall leave power series and (in Sec. 14.9) turn to more general series, called **Laurent series,** which involve positive *and negative* integer powers of $z - z_0$.

Problems for Sec. 14.7

Find the Maclaurin series of the following functions and determine the radius of convergence.

1. $\dfrac{z}{1 + z^3}$

2. $\dfrac{1}{4 - \pi z^3}$

3. $\dfrac{1}{(1 + z^2)^2}$

4. $\dfrac{1 - \cos z^2}{z^4}$

5. $\dfrac{e^{z^4} - 1}{z^3}$

6. $\dfrac{4z^2 + 30z + 68}{(z + 4)^2(z - 2)}$

7. $\dfrac{1}{(1 - z)^3}$

8. $e^{z^2} \displaystyle\int_0^z e^{-t^2}\, dt$

9. $\dfrac{1}{(z + 3 - 4i)^2}$

Find the first few terms of the Maclaurin series of the following functions.

10. $\dfrac{\cos 2z}{1 - 4z^2}$

11. $e^{z^2} \sin z^2$

12. $e^{z^2}/\cos z$

13. $e^{1/(1 - 2z)}$

14. $\cos\left(\dfrac{z}{3 - z}\right)$

15. $e^{(e^z)}$

16. Find a Maclaurin series for which the corresponding function has more than one singularity on the circle of convergence.

17. Had the radius of convergence in Example 3 to be expected from the form of the given function?

Find the Taylor series of the given function with the given point as center. (In Probs. 25 and 26 find only the first four terms.)

18. $\dfrac{1}{2z - i}$, $\ -1$

19. $\dfrac{1}{4 - 3z}$, $\ 1 + i$

20. $\dfrac{1}{z}$, $\ 1$

21. $\dfrac{1}{z}$, $\ 2 + 3i$

22. $\dfrac{1}{(z + i)^2}$, $\ -2i$

23. $\dfrac{1 + z - \sin (z + 1)}{(z + 1)^3}$, $\ -1$

24. $(z + i)^3$, $\ 1 - i$

25. $\tan z$, $\ \dfrac{\pi}{4}$

26. $\dfrac{4 - 6z}{2z^2 - 3z + 1}$, $\ -1$

27. (Euler numbers) The Maclaurin series

$$(4) \qquad\qquad \sec z = E_0 - \frac{E_2}{2!} z^2 + \frac{E_4}{4!} z^4 - + \cdots$$

defines the *Euler numbers* E_{2n}. Show that $E_0 = 1, E_2 = -1, E_4 = 5, E_6 = -61$.

28. (Bernoulli numbers) The Maclaurin series

$$(5) \qquad\qquad \frac{z}{e^z - 1} = 1 + B_1 z + \frac{B_2}{2!} z^2 + \frac{B_3}{3!} z^3 + \cdots$$

defines the *Bernoulli numbers* B_n. Using the method of undetermined coefficients, show that[7]

$$(6) \quad B_1 = -\frac{1}{2}, \quad B_2 = \frac{1}{6}, \quad B_3 = 0, \quad B_4 = -\frac{1}{30}, \quad B_5 = 0, \quad B_6 = \frac{1}{42}, \cdots.$$

---

[7]For tables, see J.W.L. Glaisher, *Quarterly Journal of Mathematics* **29**, 1898, 1—168.

29. Using (1), (2), Sec. 12.7, and (5), show that

$$(7) \quad \tan z = \frac{2i}{e^{2iz} - 1} - \frac{4i}{e^{4iz} - 1} - i = \sum_{n=1}^{\infty} (-1)^{n-1} \frac{2^{2n}(2^{2n} - 1)}{(2n)!} B_{2n} z^{2n-1}.$$

30. Developing $1/\sqrt{1 - z^2}$ and integrating, show that

$$\sin^{-1} z = z + \left(\frac{1}{2}\right)\frac{z^3}{3} + \left(\frac{1 \cdot 3}{2 \cdot 4}\right)\frac{z^5}{5} + \left(\frac{1 \cdot 3 \cdot 5}{2 \cdot 4 \cdot 6}\right)\frac{z^7}{7} + \cdots \qquad (|z| < 1).$$

Show that this series represents the principal value of $\sin^{-1} z$ (defined in Prob. 35, Sec. 12.8).

14.8 Uniform Convergence

We know that power series are *absolutely convergent* (Sec. 14.3, Theorem 1), and before we leave these series let us show that they are *uniformly convergent*. Uniform convergence is of great general interest. To define this concept, we consider a series whose terms are functions $f_0(z)$, $f_1(z)$, $\cdots$:

$$(1) \qquad \sum_{m=0}^{\infty} f_m(z) = f_0(z) + f_1(z) + f_2(z) + \cdots.$$

(For the special $f_m(z) = a_m(z - z_0)^m$ this is a power series.) We assume that this series converges for all z in some region G. We call its sum $s(z)$ and its nth partial sum $s_n(z)$; thus

$$s_n(z) = f_0(z) + f_1(z) + \cdots + f_n(z).$$

Convergence in G means the following. If we pick a $z = z_1$ in G, then, by the definition of convergence at z_1, for given $\epsilon > 0$ we can find an $N_1(\epsilon)$ such that

$$|s(z_1) - s_n(z_1)| < \epsilon \qquad \text{for all } n > N_1(\epsilon).$$

If we pick a z_2 in G, keeping ϵ as before, we can find an $N_2(\epsilon)$ such that

$$|s(z_2) - s_n(z_2)| < \epsilon \qquad \text{for all } n > N_2(\epsilon),$$

and so on. Hence, given an $\epsilon > 0$, to each z in G there corresponds a number $N_z(\epsilon)$. This number tells us how many terms we need (what s_n we need) at a z to make $|s(z) - s_n(z)|$ smaller than ϵ. It measures the speed of convergence. Small $N_z(\epsilon)$ means rapid convergence, large $N_z(\epsilon)$ means slow convergence. Now, if we can find an $N(\epsilon)$ larger than *all* these $N_z(\epsilon)$, we say that the convergence of the series (1) in G is *uniform:*

Definition (Uniform convergence)

A series (1) with sum $s(z)$ is called **uniformly convergent** in a region G if for every $\epsilon > 0$ we can find an $N = N(\epsilon)$, *not depending on z,* such that

$$|s(z) - s_n(z)| < \epsilon \qquad \text{for all } n > N(\epsilon) \text{ *and all z in G.*}$$

Uniformity of convergence is thus a property that always refers to an *infinite set* in the z-plane.

EXAMPLE 1. Geometric series

Show that the geometric series $1 + z + z^2 + \cdots$ is (a) uniformly convergent in any closed disk $|z| \leq r < 1$, (b) not uniformly convergent in its whole disk of convergence $|z| < 1$.

Solution. (a) For z in that closed disk we have $|1 - z| \geq 1 - r$ (sketch it). This implies that $1/|1 - z| \leq 1/(1 - r)$. Hence

$$|s(z) - s_n(z)| = \left| \sum_{m=n+1}^{\infty} z^m \right| = \left| \frac{z^{n+1}}{1 - z} \right| \leq \frac{r^{n+1}}{1 - r}.$$

Since $r < 1$, we can make the right side as small as we want by choosing n large enough, and since the right side does not depend on z (in the closed disk considered), this means uniform convergence.

(b) For given real K (no matter how large) and n we can always find a z in the disk $|z| < 1$ such that

$$\left| \frac{z^{n+1}}{1 - z} \right| = \frac{|z|^{n+1}}{|1 - z|} > K,$$

simply by taking z close enough to 1. Hence no single $N(\epsilon)$ will suffice to make $|s(z) - s_n(z)|$ smaller than a given $\epsilon > 0$ *throughout the whole disk.* By definition, this shows that the convergence of the geometric series in $|z| < 1$ is not uniform.　∎

This example seems to suggest that for a power series, the uniformity of convergence may at most be disturbed near the circle of convergence. This impression is true:

Theorem 1 (Uniform convergence of power series)

A power series

$$\tag{2} \sum_{m=0}^{\infty} a_m (z - z_0)^m$$

with a nonzero radius of convergence R is uniformly convergent in every circular disk $|z - z_0| \leq r$ of radius $r < R$.

Proof. For $|z - z_0| \leq r$ and any positive integers n and p we have

$$\tag{3} \begin{aligned} &|a_{n+1}(z - z_0)^{n+1} + \cdots + a_{n+p}(z - z_0)^{n+p}| \\ &\qquad \leq |a_{n+1}| r^{n+1} + \cdots + |a_{n+p}| r^{n+p}. \end{aligned}$$

Now (2) converges absolutely if $|z - z_0| = r < R$ (cf. Sec. 14.3, Theorem 1). Hence it follows from the Cauchy convergence principle (Sec. 14.2) that, an $\epsilon > 0$ being given, we can find an $N(\epsilon)$ such that

$$|a_{n+1}| r^{n+1} + \cdots + |a_{n+p}| r^{n+p} < \epsilon \qquad \text{for } n > N(\epsilon) \quad \text{and} \quad p = 1, 2, \cdots.$$

From this and (3) we obtain

$$|a_{n+1}(z - z_0)^{n+1} + \cdots + a_{n+p}(z - z_0)^{n+p}| < \epsilon$$

for all z in the disk $|z - z_0| \leqq r$, every $n > N(\epsilon)$, and every $p = 1, 2, \cdots$. Since $N(\epsilon)$ is independent of z, this shows uniform convergence, and the theorem is proved.   ∎

This theorem meets with our immediate need and concern, which is power series. The remainder of this section should provide a deeper understanding of the concept of uniform convergence as such.

Properties of Uniformly Convergent Series (*Optional*)

Uniform convergence derives its main importance from two facts:
1. If a series of *continuous* terms is uniformly convergent, its sum is also continuous (Theorem 2, below).
2. Under the same assumptions, termwise integration is permissible (Theorem 3).

This raises two questions:
1. How can a converging series of continuous terms manage to have a discontinuous sum? (Example 2)
2. How can something go wrong in termwise integration? (Example 3)

Another natural question seems to be:
3. What is the relation between absolute convergence and uniform convergence? The surprising answer: none. (Example 4)

These are the ideas we shall discuss.

If we add *finitely many* continuous functions, we get a continuous function as their sum. Example 2 will show that this is no longer true for an infinite series, even if it converges absolutely. However, if it converges *uniformly*, this cannot happen:

Theorem 2 (Continuity of the sum)
Let the series

$$\sum_{m=0}^{\infty} f_m(z) = f_0(z) + f_1(z) + \cdots$$

be uniformly convergent in a region G. Let F(z) be its sum. Then if each term $f_m(z)$ is continuous at a point z_1 in G, the function F(z) is continuous at z_1.

Proof. Let $s_n(z)$ be the nth partial sum of the series and $R_n(z)$ the corresponding remainder:

$$s_n = f_0 + f_1 + \cdots + f_n, \qquad R_n = f_{n+1} + f_{n+2} + \cdots.$$

Since the series converges uniformly, for a given $\epsilon > 0$ we can find an $n = N(\epsilon)$ such that

$$|R_N(z)| < \frac{\epsilon}{3} \qquad\qquad \text{for all } z \text{ in } G.$$

Since $s_N(z)$ is a sum of finitely many functions that are continuous at z_1, this sum is continuous at z_1. Therefore we can find a $\delta > 0$ such that

$$|s_N(z) - s_N(z_1)| < \frac{\epsilon}{3} \qquad \text{for all } z \text{ in } G \text{ for which } |z - z_1| < \delta.$$

Using $F = s_N + R_N$ and the triangle inequality (Sec. 12.2), for these z we thus obtain

$$|F(z) - F(z_1)| = |s_N(z) + R_N(z) - [s_N(z_1) + R_N(z_1)]|$$

$$\leqq |s_N(z) - s_N(z_1)| + |R_N(z)| + |R_N(z_1)| < \frac{\epsilon}{3} + \frac{\epsilon}{3} + \frac{\epsilon}{3} = \epsilon.$$

This implies that $F(z)$ is continuous at z_1, and the theorem is proved. ∎

EXAMPLE 2. Series of continuous terms with a discontinuous sum

Consider the series

$$x^2 + \frac{x^2}{1 + x^2} + \frac{x^2}{(1 + x^2)^2} + \frac{x^2}{(1 + x^2)^3} + \cdots \qquad (x \text{ real}).$$

Using the formula (2*), Sec. 14.5, for a finite geometric sum with $q = 1/(1 + x^2)$, hence $1/(1 - q) = (1 + x^2)/x^2$, we get for the nth partial sum (Fig. 356)

$$s_n(z) = 1 + x^2 - \frac{1}{(1 + x^2)^n}.$$

We see that when $x \neq 0$, the sum is $s(x) = \lim s_n(x) = 1 + x^2$, but for $x = 0$ we have $s_n(0) = 1 - 1 = 0$ for all n, hence $s(0) = 0$. So we have the surprising fact that the sum is discontinuous (at $x = 0$), although all the terms are continuous and the series converges even absolutely (its terms are nonnegative, thus equal to their absolute value!).

Theorem 2 now tells us that the convergence cannot be uniform in an interval containing $x = 0$. We can verify this also directly. Indeed, for $x \neq 0$ the remainder has the absolute value

$$|R_n(x)| = |s(x) - s_n(x)| = \frac{1}{(1 + x^2)^n}$$

and we see that for a given ϵ (< 1) we cannot find an N depending only on ϵ such that $|R_n| < \epsilon$ for all $n > N(\epsilon)$ and all x, say, in the interval $0 \leqq x \leqq 1$. ∎

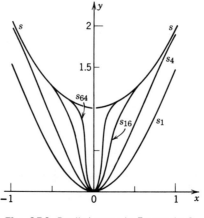

Fig. 356. Partial sums in Example 2

Termwise Integration

This is our second topic in connection with uniform convergence, and we begin with an example to become aware of the danger.

EXAMPLE 3. A series for which termwise integration is not permissible

Let

$$u_m(x) = mxe^{-mx^2}$$

and consider the series

$$\sum_{m=1}^{\infty} f_m(x) \quad \text{where} \quad f_m(x) = u_m(x) - u_{m-1}(x)$$

in the interval $0 \leq x \leq 1$. The nth partial sum is

$$s_n = u_1 - u_0 + u_2 - u_1 + \cdots + u_n - u_{n-1} = u_n - u_0 = u_n.$$

Hence the series has the sum

$$F(x) = \lim_{n \to \infty} s_n(x) = \lim_{n \to \infty} u_n(x) = 0 \qquad (0 \leq x \leq 1).$$

From this we obtain

$$\int_0^1 F(x)\,dx = 0.$$

On the other hand, by integrating term by term,

$$\sum_{m=1}^{\infty} \int_0^1 f_m(x)\,dx = \lim_{n \to \infty} \sum_{m=1}^{n} \int_0^1 f_m(x)\,dx = \lim_{n \to \infty} \int_0^1 s_n(x)\,dx.$$

Now $s_n = u_n$ and the expression on the right becomes

$$\lim_{n \to \infty} \int_0^1 u_n(x)\,dx = \lim_{n \to \infty} \int_0^1 nxe^{-nx^2}\,dx = \lim_{n \to \infty} \frac{1}{2}(1 - e^{-n}) = \frac{1}{2},$$

but not 0. This shows that the series under consideration cannot be integrated term by term from $x = 0$ to $x = 1$. ∎

The series in Example 3 is not uniformly convergent in the interval of integration, and we shall now prove that in the case of a uniformly convergent series of continuous functions we may integrate term by term.

Theorem 3 (Termwise integration)

Let

$$F(z) = \sum_{m=0}^{\infty} f_m(z) = f_0(z) + f_1(z) + \cdots$$

be a uniformly convergent series of continuous functions within a region G. Let C be any path in G. Then the series

$$(4) \qquad \sum_{m=0}^{\infty} \int_C f_m(z)\,dz = \int_C f_0(z)\,dz + \int_C f_1(z)\,dz + \cdots$$

is convergent and has the sum $\int_C F(z)\,dz$.

Proof. From Theorem 2 it follows that $F(z)$ is continuous. Let $s_n(z)$ be the nth partial sum of the given series and $R_n(z)$ the corresponding remainder. Then $F = s_n + R_n$ and

$$\int_C F(z)\ dz = \int_C s_n(z)\ dz + \int_C R_n(z)\ dz.$$

Let L be the length of C. Since the given series converges uniformly, for every given $\epsilon > 0$ we can find a number N such that

$$|R_n(z)| < \frac{\epsilon}{L} \qquad \text{for all } n > N \text{ and all } z \text{ in } G.$$

By applying the *ML*-inequality (Sec. 13.2) we thus obtain

$$\left| \int_C R_n(z)\ dz \right| < \frac{\epsilon}{L} L = \epsilon \qquad \text{for all } n > N.$$

Since $R_n = F - s_n$, this means that

$$\left| \int_C F(z)\ dz - \int_C s_n(z)\ dz \right| < \epsilon \qquad \text{for all } n > N.$$

Hence, the series (4) converges and has the sum indicated in the theorem. This completes the proof. ∎

Theorems 2 and 3 characterize the two most important properties of uniformly convergent series.

Of course, since differentiation and integration are inverse processes, we readily conclude from Theorem 3 that a convergent series may be differentiated term by term, provided the terms of the given series have continuous derivatives and the resulting series is uniformly convergent; more precisely, the following theorem holds.

Theorem 4 (Termwise differentiation)

Let the series $f_0(z) + f_1(z) + f_2(z) + \cdots$ be convergent in a region G and let $F(z)$ be its sum. Suppose that the series $f_0'(z) + f_1'(z) + f_2'(z) + \cdots$ converges uniformly in G and its terms are continuous in G. Then

$$F'(z) = f_0'(z) + f_1'(z) + f_2'(z) + \cdots \qquad \text{for all } z \text{ in } G.$$

The simple proof is left to the student (Prob. 16).

The next example shows that there are series that converge absolutely but not uniformly, and others that converge uniformly but not absolutely. This shows that there is generally no relation between absolute and uniform convergence.

EXAMPLE 4. No relation between absolute and uniform convergence

The series in Example 2 converges absolutely but not uniformly, as we have shown. On the other hand, the series

$$\sum_{m=1}^{\infty} \frac{(-1)^{m-1}}{x^2 + m} = \frac{1}{x^2 + 1} - \frac{1}{x^2 + 2} + \frac{1}{x^2 + 3} - + \cdots \qquad (x \text{ real})$$

converges uniformly on the whole real line but not absolutely.

Proof. By the familiar Leibniz test of calculus (cf. Appendix A3.3) the remainder R_n does not exceed its first term in absolute value, since we have a series of alternating terms whose absolute values form a monotone decreasing sequence with limit zero. Hence, given $\epsilon > 0$, for all x we have

$$|R_n(x)| \leq \frac{1}{x^2 + n + 1} < \frac{1}{n} < \epsilon \qquad \text{when } n > N(\epsilon) = \frac{1}{\epsilon}.$$

This proves uniform convergence, since $N(\epsilon)$ does not depend on x.

The convergence is not absolute because

$$\left| \frac{(-1)^{m-1}}{x^2 + m} \right| = \frac{1}{x^2 + m} > \frac{k}{m}$$

(k a suitable constant) and $k\Sigma 1/m$ diverges.   ∎

Usually uniform convergence is established by a comparison test, which is called the Weierstrass M-test.[8]

Theorem 5 (Weierstrass *M*-test for uniform convergence)

Consider a series of the form (1) *in a region G of the z-plane. Suppose that one can find a convergent series of constant terms,*

$$(5) \qquad\qquad M_0 + M_1 + M_2 + \cdots,$$

such that $|f_m(z)| \leq M_m$ *for all z in G and every* $m = 0, 1, \cdots$. *Then* (1) *is uniformly convergent in G.*

The simple proof is left to the student (Prob. 17).

EXAMPLE 5. Weierstrass *M*-test

Does the following series converge uniformly in the disk $|z| \leq 1$?

$$\sum_{m=0}^{\infty} \frac{z^m + 1}{m^2 + \cosh m|z|}$$

Solution. Uniform convergence follows by the Weierstrass M-test and the convergence of $\Sigma 1/m^2$ (cf. Sec. 14.2, in the proof of Theorem 6) because

$$\left| \frac{z^m + 1}{m^2 + \cosh m|z|} \right| \leq \frac{|z|^m + 1}{m^2} \leq \frac{2}{m^2}.$$ ∎

---

[8]KARL WEIERSTRASS (1815—1897), German mathematician, whose lifework was the development of complex analysis based on the concept of power series (cf. the footnote in Sec. 12.5). He also made basic contributions to the calculus, the calculus of variations, approximation theory and differential geometry. He obtained the concept of uniform convergence in 1841 (published 1894); the first publications on the concept were by G. G. STOKES (cf. Sec. 9.8) in 1847 and PHILIPP LUDWIG VON SEIDEL (1821—1896) in 1848.

By definition, power series involve positive integer powers $z - z_0$, $(z - z_0)^2, \cdots$ (and a constant term). In the next section we discuss **Laurent series,** which involve positive *and* negative integer powers of $z - z_0$. These series will be a useful generalization of Taylor series. They will also help to classify "singular points" (points at which an analytic function ceases to be analytic, to be discussed in Sec. 14.10), and they will provide the basis for another powerful integration method to be developed in Chap. 15.

Problems for Sec. 14.8

Prove that the following series converge uniformly in the given regions.

1. $\displaystyle\sum_{n=0}^{\infty} z^n, \quad |z| \leq 0.99$

2. $\displaystyle\sum_{n=0}^{\infty} \frac{z^n}{n!}, \quad |z| \leq 10^{30}$

3. $\displaystyle\sum_{n=1}^{\infty} \frac{(n!)^2}{(2n)!} z^n, \quad |z| \leq 3$

4. $\displaystyle\sum_{n=1}^{\infty} \frac{z^n}{n^2}, \quad |z| \leq 1$

5. $\displaystyle\sum_{n=1}^{\infty} \frac{1}{n^4} \left(\frac{\pi}{2}\right)^n z^{2n}, \quad |z| \leq 0.79$

6. $\displaystyle\sum_{n=1}^{\infty} \frac{\tanh^n |z|}{n(n+1)}, \quad \text{all } z$

7. $\displaystyle\sum_{n=0}^{\infty} \frac{n(n+1)}{3^n} (z-i)^n, \quad |z-i| \leq 2$

8. $\displaystyle\sum_{n=0}^{\infty} \frac{z^n}{|z|^{2n} + 2}, \quad 2 \leq |z| \leq 3$

9. $\displaystyle\sum_{n=1}^{\infty} \frac{\cos^n |z|}{n^2}, \quad \text{all } z$

10. $\displaystyle\sum_{n=1}^{\infty} \frac{nz^n}{n^3 + |z|}, \quad |z| \leq 1$

Where do the following power series converge uniformly?

11. $\displaystyle\sum_{n=0}^{\infty} \frac{(-1)^n}{2^n} (z-i)^{2n}$

12. $\displaystyle\sum_{n=0}^{\infty} \left(\frac{3n+1}{4n-2}\right)^n z^n$

13. $\displaystyle\sum_{n=0}^{\infty} \frac{\pi^n}{n!} (z+3-i)^n$

14. $\displaystyle\sum_{n=0}^{\infty} \binom{n}{3} (3z+i)^n$

15. If the series (1) converges uniformly in a region G, show that it converges uniformly in any portion of G. Is the converse true?

16. Derive Theorem 4 from Theorem 3.

17. Give a proof of the Weierstrass M-test (Theorem 5).

18. Determine the smallest integer n such that $|R_n| < 0.01$ in Example 1, when $z = x = 0.5, 0.6, 0.7, 0.8, 0.9$. What does the result mean from the viewpoint of computing $1/(1 - x)$ with an absolute error less than 0.01 by means of the geometric series?

19. Show that $x^2 \sum_{m=1}^{\infty} (1 + x^2)^{-m} = 1$ if $x \neq 0$ and 0 if $x = 0$. Verify by computation that the partial sums s_1, s_2, s_3 look as shown in Fig. 357 and compute and graph s_4, s_5, s_6.

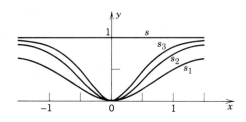

Fig. 357. Sum s and partial sums in Problem 19

20. Find the precise region of convergence of the series in Example 2 with x replaced by a complex variable z.

Heat equation. Show that (10), Sec. 11.5, with coefficients (11) is a solution of the heat equation for $t > 0$, assuming that $f(x)$ is continuous on the interval $0 \leqq x \leqq L$ and has one-sided derivatives at all interior points of that interval. Proceed as follows.

21. Show that $|B_n|$ is bounded, say, $|B_n| < K$ for all n. Conclude that

$$|u_n| < Ke^{-\lambda_n^2 t_0} \qquad \text{when} \qquad t \geqq t_0 > 0$$

and, by the Weierstrass test, the series (10) converges uniformly with respect to x and t when $t \geqq t_0$, $0 \leqq x \leqq L$. Using Theorem 2, show that $u(x, t)$ is continuous when $t \geqq t_0$ and thus satisfies the boundary conditions (2) when $t \geqq t_0$.

22. Show that $|\partial u_n/\partial t| < \lambda_n^2 Ke^{-\lambda_n^2 t_0}$ when $t \geqq t_0$ and the series of the expressions on the right converges, by the ratio test. Conclude from this, the Weierstrass test and Theorem 4 that the series (10) can be differentiated term by term with respect to t and the resulting series has the sum $\partial u/\partial t$. Show that (10) can be differentiated twice with respect to x and the resulting series has the sum $\partial^2 u/\partial x^2$. Conclude from this and the result of Prob. 21 that (10) is a solution of the heat equation for all $t \geqq t_0$. (The proof that (10) satisfies the given initial condition can be found in Ref. [C18] listed in Appendix 1.)

14.9 Laurent Series

In various applications it is necessary to expand a function $f(z)$ around points where $f(z)$ is singular. Taylor's theorem cannot be applied in such cases. A new type of series, known as *Laurent series*[9] is necessary. This will be a representation that is valid in an annulus bounded by two concentric circles C_1 and C_2 such that $f(z)$ is analytic in the annulus and at each point of C_1 and C_2 (Fig. 358). As in the case of the Taylor series, $f(z)$ may be singular at some points outside C_1 and, as the essentially new feature, it may also be singular at some points inside C_2.

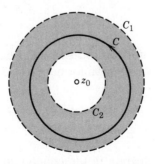

Fig. 358. Laurent's theorem

[9]PIERRE ALPHONSE LAURENT (1813—1854), French engineer and mathematician, who published this theorem in 1843.

Laurent's theorem

*If $f(z)$ is analytic on two concentric circles C_1 and C_2 with center z_0 and in the annulus between them, then $f(z)$ can be represented by the **Laurent series***

(1)
$$f(z) = \sum_{n=0}^{\infty} a_n(z - z_0)^n + \sum_{n=1}^{\infty} \frac{b_n}{(z - z_0)^n}$$

$$= a_0 + a_1(z - z_0) + a_2(z - z_0)^2 + \cdots$$

$$\cdots + \frac{b_1}{z - z_0} + \frac{b_2}{(z - z_0)^2} + \cdots .$$

The coefficients of this Laurent series are given by the integrals[10]

(2) $$a_n = \frac{1}{2\pi i} \oint_C \frac{f(z^*)}{(z^* - z_0)^{n+1}} \, dz^*, \qquad b_n = \frac{1}{2\pi i} \oint_C (z^* - z_0)^{n-1} f(z^*) \, dz^*,$$

each integral being taken in the counterclockwise sense around any simple closed path C that lies in the annulus and encircles the inner circle (Fig. 358).

This series converges and represents $f(z)$ in the open annulus obtained from the given annulus by continuously increasing the circle C_1 and decreasing C_2 until each of the two circles reaches a point where $f(z)$ is singular.

Comment. Obviously, instead of (1), (2) we may write (denoting b_n by a_{-n})

(1′)
$$f(z) = \sum_{n=-\infty}^{\infty} a_n(z - z_0)^n$$

where

(2′)
$$a_n = \frac{1}{2\pi i} \oint_C \frac{f(z^*)}{(z^* - z_0)^{n+1}} \, dz^*.$$

Proof of Laurent's theorem. Let z be any point in the given annulus. Then from Cauchy's integral formula [cf. (3) in Sec. 13.5] it follows that

(3) $$f(z) = \frac{1}{2\pi i} \oint_{C_1} \frac{f(z^*)}{z^* - z} \, dz^* - \frac{1}{2\pi i} \oint_{C_2} \frac{f(z^*)}{z^* - z} \, dz^*,$$

where we integrate in the counterclockwise sense. We transform each of these two integrals as in Sec. 14.5. The first integral is *precisely* as in Sec. 14.5, so we get precisely the same result

---

[10]We denote the variable of integration by z^* because z is used in $f(z)$.

(4)
$$\frac{1}{2\pi i} \oint_{C_1} \frac{f(z^*)}{z^* - z} \, dz^* = \sum_{n=0}^{\infty} a_n (z - z_0)^n$$

with coefficients (counterclockwise integration)

(5)
$$a_n = \frac{1}{2\pi i} \oint_{C_1} \frac{f(z^*)}{(z^* - z_0)^{n+1}} \, dz^*.$$

Here we can replace C_1 by C (cf. Fig. 358), by the principle of deformation of path, since z_0, the point where the integrand in (5) is not analytic, is not a point of the annulus. This proves the formula for the a_n in (2).

We turn to the second integral in (3) and shall get the formula for the b_n from it. Since z lies in the annulus, it lies *outside* the path C_2. Hence the situation differs from that for the first integral. The essential point is that instead of (3) in Sec. 14.5 we now have

(6)
$$\left| \frac{z^* - z_0}{z - z_0} \right| < 1$$

and we must develop $1/(z^* - z)$ in the integrand in powers of $(z^* - z_0)/(z - z_0)$ [instead of the reciprocal of this] to get a *convergent* series. We find

$$\frac{1}{z^* - z} = \frac{1}{z^* - z_0 - (z - z_0)} = \frac{-1}{(z - z_0)\left(1 - \dfrac{z^* - z_0}{z - z_0}\right)}.$$

Compare this for a moment with (2**) in Sec. 14.5, to really understand the difference. Then go on and apply formula (2), Sec. 14.5, for a finite geometric sum, obtaining

$$\frac{1}{z^* - z} = -\frac{1}{z - z_0}\left\{ 1 + \frac{z^* - z_0}{z - z_0} + \left(\frac{z^* - z_0}{z - z_0}\right)^2 + \cdots + \left(\frac{z^* - z_0}{z - z_0}\right)^n \right\}$$

$$- \frac{1}{z - z^*}\left(\frac{z^* - z_0}{z - z_0}\right)^{n+1}.$$

Multiplication by $-f(z^*)/2\pi i$ and integration over C_2 on both sides now yields

$$-\frac{1}{2\pi i} \oint_{C_2} \frac{f(z^*)}{z^* - z} \, dz^* = \frac{1}{2\pi i}\left\{ \frac{1}{z - z_0} \oint_{C_2} f(z^*) \, dz^* + \frac{1}{(z - z_0)^2} \oint_{C_2} (z^* - z_0) f(z^*) \, dz^* + \cdots \right.$$

$$\left. + \frac{1}{(z - z_0)^{n+1}} \oint_{C_2} (z^* - z_0)^n f(z^*) \, dz^* \right\} + R_n^*(z)$$

with the last term on the right given by

(7)
$$R_n{}^*(z) = \frac{1}{2\pi i(z - z_0)^{n+1}} \oint_{C_2} \frac{(z^* - z_0)^{n+1}}{z - z^*} f(z^*)\, dz^*.$$

As before, we can integrate over C instead of C_2 in the integrals on the right. We see that on the right, the power $1/(z - z_0)^n$ is multiplied by b_n as given in (2). This establishes Laurent's theorem, provided

(8)
$$\lim_{n \to \infty} R_n{}^*(z) = 0.$$

We prove (8). The expression $f(z^*)/(z - z^*)$ in (7) is bounded in absolute value, say,

$$\left| \frac{f(z^*)}{z - z^*} \right| < \tilde{M} \qquad\qquad \text{for all } z^* \text{ on } C_2$$

because $f(z^*)$ is analytic in the annulus and on C_2, and z^* lies on C_2 and z outside, so that $z - z^* \neq 0$. From this and the ML-inequality (Sec. 13.2) applied to (7) we get (L = length of C_2)

$$|R_n{}^*(z)| < \frac{1}{2\pi |z - z_0|^{n+1}} |z^* - z_0|^{n+1} \tilde{M} L = \frac{\tilde{M} L}{2\pi} \left| \frac{z^* - z_0}{z - z_0} \right|^{n+1}.$$

From (6) we see that the expression on the right approaches zero as n approaches infinity. This proves (8). The representation (1) with coefficients (2) is now established in the given annulus.

Finally let us prove convergence of (1) in the open annulus characterized at the end of the theorem.

We denote the sums of the two series in (1) by $g(z)$ and $h(z)$, and the radii of C_1 and C_2 by r_1 and r_2, respectively. Then $f = g + h$. The first series is a power series. Since it converges in the annulus, it must converge in the entire disk bounded by C_1, and g is analytic in that disk.

If we set $Z = 1/(z - z_0)$, the last series becomes a power series in Z. The annulus $r_2 < |z - z_0| < r_1$ then corresponds to the annulus $1/r_1 < |Z| < 1/r_2$, the new series converges in this annulus and, therefore, in the entire disk $|Z| < 1/r_2$. Since this disk corresponds to $|z - z_0| > r_2$, the exterior of C_2, the given series converges for all z outside C_2, and h is analytic for all these z.

Since $f = g + h$, it follows that g must be singular at all those points outside C_1 where f is singular, and h must be singular at all those points inside C_2 where f is singular. Consequently, the first series converges for all z inside the circle about z_0 whose radius is equal to the distance of that singularity of f outside C_1 which is closest to z_0. Similarly, the second series converges for all z outside the circle about z_0 whose radius is equal to the maximum distance of the singularities of f inside C_2. The domain common to both of those domains of convergence is the open annulus characterized near the end of the theorem, and the proof of Laurent's theorem is complete. ∎

The Laurent series of a given analytic function $f(z)$ in its annulus of convergence is unique (cf. Prob. 11, this section). *However, $f(z)$ may have different Laurent series in two annuli with the same center;* see the examples below. The uniqueness is essential. As for a Taylor series, to obtain the coefficients of Laurent series, we do not generally use the integral formulas (2) but various other methods, some of which we shall illustrate by our examples. If a Laurent series has been found by any such process, the uniqueness guarantees that it must be *the* Laurent series of the given function in the given annulus.

EXAMPLE 1. Use of Maclaurin series

Find the Laurent series of $z^{-5} \sin z$ with center 0.

Solution. By (4), Sec. 14.6, we obtain

$$z^{-5} \sin z = \sum_{n=0}^{\infty} \frac{(-1)^n}{(2n+1)!} z^{2n-4} = \frac{1}{z^4} - \frac{1}{6z^2} + \frac{1}{120} - \frac{1}{5040} z^2 + - \cdots \qquad (|z| > 0).$$

EXAMPLE 2. Substitution

Find the Laurent series of $z^2 e^{1/z}$ with center 0.

Solution. From (2) in Sec. 14.6 with z replaced by $1/z$ we obtain

$$z^2 e^{1/z} = z^2 \left(1 + \frac{1}{1!z} + \frac{1}{2!z^2} + \cdots \right) = z^2 + z + \frac{1}{2} + \frac{1}{3!z} + \frac{1}{4!z^2} + \cdots \qquad (|z| > 0).$$

EXAMPLE 3

Develop $1/(1 - z)$ (a) in nonnegative powers of z, (b) in negative powers of z.

Solution.

(a)
$$\frac{1}{1-z} = \sum_{n=0}^{\infty} z^n \qquad \text{valid when } |z| < 1$$

(b)
$$\frac{1}{1-z} = \frac{-1}{z(1-z^{-1})} = -\sum_{n=0}^{\infty} \frac{1}{z^{n+1}} = -\frac{1}{z} - \frac{1}{z^2} - \cdots \qquad \text{valid when } |z| > 1$$

EXAMPLE 4. Laurent expansions in different concentric annuli

Find all Laurent series of $1/(z^3 - z^4)$ with center 0.

Solution. Multiplying by $1/z^3$, we get from Example 3

(I)
$$\frac{1}{z^3 - z^4} = \sum_{n=0}^{\infty} z^{n-3} = \frac{1}{z^3} + \frac{1}{z^2} + \frac{1}{z} + 1 + z + \cdots \qquad (0 < |z| < 1),$$

(II)
$$\frac{1}{z^3 - z^4} = -\sum_{n=0}^{\infty} \frac{1}{z^{n+4}} = -\frac{1}{z^4} - \frac{1}{z^5} - \cdots \qquad (|z| > 1).$$

EXAMPLE 5. Use of partial fractions

Find all Taylor and Laurent series of $f(z) = \dfrac{-2z + 3}{z^2 - 3z + 2}$ with center 0.

Solution. In terms of partial fractions,

$$f(z) = -\frac{1}{z-1} - \frac{1}{z-2}.$$

(a) and (b) in Example 3 take care of the first fraction. For the second fraction,

(c)
$$-\frac{1}{z-2} = \frac{1}{2\left(1 - \frac{1}{2}z\right)} = \sum_{n=0}^{\infty} \frac{1}{2^{n+1}} z^n \qquad (|z| < 2)$$

(d)
$$-\frac{1}{z-2} = -\frac{1}{z\left(1 - \frac{2}{z}\right)} = -\sum_{n=0}^{\infty} \frac{2^n}{z^{n+1}} \qquad (|z| > 2).$$

(I) From (a) and (c), valid for $|z| < 1$ (cf. Fig. 359)

$$f(z) = \sum_{n=0}^{\infty} \left(1 + \frac{1}{2^{n+1}}\right) z^n = \frac{3}{2} + \frac{5}{4}z + \frac{9}{8}z^2 + \cdots.$$

(II) From (c) and (b), valid for $1 < |z| < 2$,

$$f(z) = \sum_{n=0}^{\infty} \frac{1}{2^{n+1}} z^n - \sum_{n=0}^{\infty} \frac{1}{z^{n+1}} = \frac{1}{2} + \frac{1}{4}z + \frac{1}{8}z^2 + \cdots - \frac{1}{z} - \frac{1}{z^2} - \cdots.$$

(III) From (d) and (b), valid for $|z| > 2$,

$$f(z) = -\sum_{n=0}^{\infty} (2^n + 1) \frac{1}{z^{n+1}} = -\frac{2}{z} - \frac{3}{z^2} - \frac{5}{z^3} - \frac{9}{z^4} - \cdots.$$

EXAMPLE 6

Find the Laurent series of $f(z) = 1/(1 - z^2)$ that converges in the annulus $1/4 < |z - 1| < 1/2$ and determine the precise region of convergence.

Solution. The annulus has center 1, so that we must develop

$$f(z) = \frac{-1}{(z - 1)(z + 1)}$$

in powers of $z - 1$. We calculate

$$\frac{1}{z + 1} = \frac{1}{2 + (z - 1)} = \frac{1}{2} \frac{1}{\left[1 - \left(-\frac{z - 1}{2}\right)\right]}$$

$$= \frac{1}{2} \sum_{n=0}^{\infty} \left(-\frac{z - 1}{2}\right)^n = \sum_{n=0}^{\infty} \frac{(-1)^n}{2^{n+1}} (z - 1)^n;$$

this series converges in the disk $|(z - 1)/2| < 1$, that is, $|z - 1| < 2$. Multiplication by $-1/(z - 1)$ now gives the desired series

$$f(z) = \sum_{n=0}^{\infty} \frac{(-1)^{n+1}}{2^{n+1}} (z - 1)^{n-1} = \frac{-1/2}{z - 1} + \frac{1}{4} - \frac{1}{8}(z - 1) + \frac{1}{16}(z - 1)^2 - + \cdots.$$

The precise region of convergence is $0 < |z - 1| < 2$; cf. Fig. 360. We confirm this by noting that $1/(z + 1)$ in $f(z)$ is singular at -1, at distance 2 from the center of the series. ∎

If $f(z)$ in Laurent's theorem is analytic inside C_2, the coefficients b_n in (2) are zero by Cauchy's integral theorem, so that the Laurent series reduces to a Taylor series. Examples 3(a) and 5(I) illustrate this.

Of particular practical importance is the case when $z = z_0$ is the only singular point of $f(z)$ inside the circle C_2. We discuss this in the next section, in connection with the classification of **"singularities"** of analytic functions. In the next section we shall also discuss and characterize *zeros* of analytic functions by means of Taylor series.

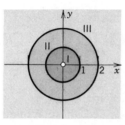

Fig. 359. Regions of convergence
in Example 5

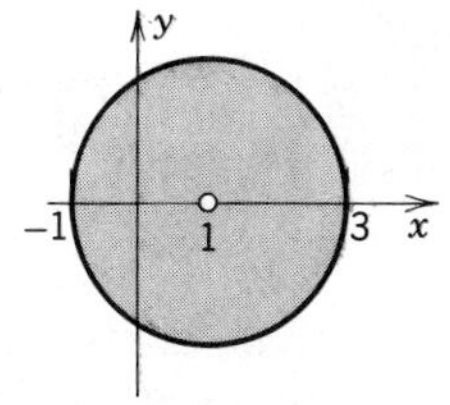

Fig. 360. Region of convergence
in Example 6

Problems for Sec. 14.9

Expand each of the following functions in a Laurent series that converges for
$0 < |z| < R$ and determine the precise region of convergence.

1. $\dfrac{e^{2z}}{z^3}$

2. $\dfrac{1}{z - z^2}$

3. $z \cos \dfrac{1}{z}$

4. $\dfrac{\sinh \pi z}{z^3}$

5. $\dfrac{1}{z^2(z - 3)}$

6. $\dfrac{1}{z^8 + z^4}$

7. $\dfrac{1}{z^2(1 + z)^2}$

8. $\dfrac{3 - z}{z^2 - z^4}$

9. $\dfrac{8 - 2z}{4z - z^3}$

10. Does $\tan (1/z)$ have a Laurent series convergent in a region $0 < |z| < R$?

11. Prove that the Laurent expansion of a given analytic function in a given annulus is unique.

Expand each of the following functions in a Laurent series that converges for
$0 < |z - z_0| < R$ and determine the precise region of convergence.

12. $\dfrac{e^z}{(z - 1)^2}$, $z_0 = 1$

13. $z^3 \cosh \dfrac{1}{z}$, $z_0 = 0$

14. $\dfrac{\cos z}{(z - \pi)^2}$, $z_0 = \pi$

15. $\dfrac{1}{z^2 + 1}$, $z_0 = i$

16. $\dfrac{z^2 - 4}{(z - 1)^2}$, $z_0 = 1$

17. $\dfrac{z + 1 + i}{(z + i)^2}$, $z_0 = -i$

18. $\dfrac{\sin z}{(z - \frac{1}{4}\pi)^3}$, $z_0 = \dfrac{\pi}{4}$

19. $\dfrac{1}{z^2 + 1}$, $z_0 = -i$

20. $\dfrac{1}{1 - z^4}$, $z_0 = -1$

Find all Taylor and Laurent series with center $z = z_0$ and determine the precise region of convergence.

21. $\dfrac{1}{1 - z^2}$, $z_0 = 0$

22. $\dfrac{1}{1 - z^2}$, $z_0 = 1$

23. $\dfrac{1}{z}$, $z_0 = 1$

24. $\dfrac{1}{z^2}$, $z_0 = i$

25. $\dfrac{4}{1 - z^4}$, $z_0 = 0$

26. $\dfrac{z^3 - 2iz^2}{(z - i)^2}$, $z_0 = i$

27. $\dfrac{4z - 1}{z^4 - 1}$, $z_0 = 0$

28. $\dfrac{\sinh z}{(z - 1)^2}$, $z_0 = 1$

29. $\dfrac{\sin z}{z + \frac{1}{2}\pi}$, $z_0 = -\dfrac{1}{2}\pi$

30. Using partial fractions, develop $f(z) = \dfrac{4z^2 + 2z - 4}{z^3 - 4z}$ in a Laurent series that converges in the annulus $2 < |z - 2| < 3$ and determine the precise region of convergence.

14.10 Singularities and Zeros. Infinity

Roughly, a *singularity* of an analytic function $f(z)$ is a z at which $f(z)$ ceases to be analytic, and a *zero* is a z at which $f(z) = 0$. Precise definitions follow below. Singularities may be discussed and classified by means of Laurent series. This is what we do first. Then we show that zeros may be discussed by means of Taylor series.

We say that a function[11] $f(z)$ **is singular** or **has a singularity** at a point $z = z_0$ if $f(z)$ is not analytic (perhaps not even defined) at $z = z_0$, but every neighborhood of $z = z_0$ contains points at which $f(z)$ is analytic.

We call $z = z_0$ an **isolated singularity** of $f(z)$ if $z = z_0$ has a neighborhood without further singularities of $f(z)$. *Example:* tan z has isolated singularities at $\pm \pi/2$, $\pm 3\pi/2$, etc.; tan $(1/z)$ has a nonisolated singularity at 0. (Explain!) Isolated singularities of $f(z)$ at $z = z_0$ can be classified by the Laurent series

$$(1) \qquad f(z) = \sum_{n=0}^{\infty} a_n (z - z_0)^n + \sum_{n=1}^{\infty} \frac{b_n}{(z - z_0)^n} \qquad \text{(Sec. 14.9)}$$

valid in the **immediate neighborhood** of the singular point $z = z_0$, except at z_0 a itself, that is, in a region of the form

$$0 < |z - z_0| < R.$$

The sum of the first series is analytic at $z = z_0$, as we know from the previous section. The second series, containing the negative powers, is called the **principal part** of (1). If it has only finitely many terms, it is of the form

$$(2) \qquad \frac{b_1}{z - z_0} + \cdots + \frac{b_m}{(z - z_0)^m} \qquad (b_m \neq 0).$$

Then the singularity of $f(z)$ at $z = z_0$ is called a **pole,** and m is called its **order.** Poles of the first order are also known as *simple poles.*

If the principal part of (1) has infinitely many terms, we say that $f(z)$ has at $z = z_0$ an **isolated essential singularity.**

We leave aside nonisolated singularities (see the example above).

EXAMPLE 1. Poles. Essential singularities

The function

$$f(z) = \frac{1}{z(z - 2)^5} + \frac{3}{(z - 2)^2}$$

has a simple pole at $z = 0$ and a pole of fifth order at $z = 2$. Examples of functions having an isolated essential singularity at $z = 0$ are

$$e^{1/z} = \sum_{n=0}^{\infty} \frac{1}{n! \, z^n} = 1 + \frac{1}{z} + \frac{1}{2! \, z^2} + \cdots$$

and

$$\sin \frac{1}{z} = \sum_{n=0}^{\infty} \frac{(-1)^n}{(2n + 1)! \, z^{2n+1}} = \frac{1}{z} - \frac{1}{3! \, z^3} + \frac{1}{5! \, z^5} - + \cdots$$

---

[11]We recall that, by definition, a function is a *single-valued* relation. (Cf. Sec. 12.4.)

Section 14.9 provides further examples. For instance, Example 1 shows that $z^{-5} \sin z$ has a fourth-order pole at 0. Example 4 shows that $1/(z^3 - z^4)$ has a third-order pole at 0 and a Laurent series with infinitely many negative powers. This is no contradiction, since this series is valid for $|z| > 1$; it merely tells us that it is quite important to consider the Laurent series valid in the *immediate neighborhood* of a singularity.   ∎

The classification of singularities into poles and essential singularities is not merely a formal matter, because the behavior of an analytic function in a neighborhood of an essential singularity is entirely different from that in the neighborhood of a pole.

EXAMPLE 2. Behavior near a pole

The function $f(z) = 1/z^2$ has a pole at $z = 0$, and $|f(z)| \to \infty$ as $z \to 0$ in any manner.   ∎

This example illustrates

Theorem 1 (Poles)

If $f(z)$ is analytic and has a pole at $z = z_0$, then $|f(z)| \to \infty$ as $z \to z_0$ in any manner. (Cf. Prob. 26.)

EXAMPLE 3. Behavior near an essential singularity

The function $f(z) = e^{1/z}$ has an essential singularity at $z = 0$. It has no limit for approach along the imaginary axis; it becomes infinite if $z \to 0$ through positive real values, but it approaches zero if $z \to 0$ through negative real values. It takes on any given value $c = c_0 e^{i\alpha} \neq 0$ in an arbitrarily small neighborhood of $z = 0$. In fact, setting $z = re^{i\theta}$, we must solve the equation

$$e^{1/z} = e^{(\cos\theta - i \sin\theta)/r} = c_0 e^{i\alpha}$$

for r and θ. Equating the absolute values and the arguments, we have $e^{(\cos\theta)/r} = c_0$, that is,

$$\cos\theta = r \ln c_0, \quad \text{and} \quad \sin\theta = -\alpha r.$$

From these two equations and $\cos^2\theta + \sin^2\theta = 1$ we obtain the formulas

$$r^2 = \frac{1}{(\ln c_0)^2 + \alpha^2} \quad \text{and} \quad \tan\theta = -\frac{\alpha}{\ln c_0}.$$

Hence r can be made arbitrarily small by adding multiples of 2π to α, leaving c unaltered.   ∎

This example illustrates the famous

Theorem 2 (Picard's[12] theorem)

If $f(z)$ is analytic and has an isolated essential singularity at a point z_0, it takes on every value, with at most one exceptional value, in an arbitrarily small neighborhood of z_0.

In Example 3, the exceptional value is $z = 0$. The proof of Picard's theorem is rather complicated; it can be found in Ref. [D12].

Removable singularities. We say that a function $f(z)$ has a *removable singularity* at $z = z_0$ if $f(z)$ is not analytic at $z = z_0$, but can be made analytic there by assigning a suitable value $f(z_0)$. Such singularities are of no interest since they can be removed as just indicated. *Example:* $f(z) = (\sin z)/z$ becomes analytic at $z = 0$ if we define $f(0) = 1$.

---

[12]See the footnote in Sec. 1.10.

Zeros of Analytic Functions

We say that a function $f(z)$ that is analytic in some domain D has a **zero** at a point $z = z_0$ in D if $f(z_0) = 0$. We also say that this zero is of **order** n if not only f but also the derivatives f', f'', $\cdots$, $f^{(n-1)}$ are all zero at $z = z_0$ but $f^{(n)}(z_0) \neq 0$.

A zero of first order is also called a *simple zero;* for it, $f(z_0) = 0$ but $f'(z_0) \neq 0$. For a second-order zero, $f(z_0) = 0$, $f'(z_0) = 0$ but $f''(z_0) \neq 0$, and so on.

EXAMPLE 4. Zeros

The function $1 + z^2$ has simple zeros at $\pm i$. The function $(1 - z^4)^2$ has second-order zeros at ± 1 and $\pm i$. The function $(z - a)^3$ has a third-order zero at $z = a$. The function e^z has no zeros (cf. Sec. 12.6). The function $\sin z$ has simple zeros at 0, $\pm \pi$, $\pm 2\pi$, $\cdots$, and $\sin^2 z$ has second-order zeros at these points. The function $1 - \cos z$ has second-order zeros at 0, $\pm 2\pi$, $\pm 4\pi$, $\cdots$, and $(1 - \cos z)^2$ has fourth-order zeros at these points. ∎

Taylor series at a zero. At an nth-order zero $z = z_0$ of $f(z)$, the derivatives $f'(z_0)$, $\cdots$, $f^{(n-1)}(z_0)$ are zero, by definition. Hence the first few coefficients a_0, $\cdots$, a_{n-1} of the Taylor series (9), Sec. 14.5, are zero, too, whereas $a_n \neq 0$, so that this series takes the form

$$
\begin{aligned}
f(z) &= a_n(z - z_0)^n + a_{n+1}(z - z_0)^{n+1} + \cdots \\
&= (z - z_0)^n[a_n + a_{n+1}(z - z_0) + a_{n+2}(z - z_0)^2 + \cdots]
\end{aligned}
\qquad (a_n \neq 0).
$$

(3)

This is characteristic of such a zero, because if $f(z)$ has such a Taylor series, it has an nth-order zero at $z = z_0$, as follows by differentiation.

Whereas nonisolated singularities may occur, for zeros we have

Theorem 3 (Zeros)

The zeros of an analytic function $f(z)$ ($\not\equiv 0$) are isolated; that is, each of them has a neighborhood that contains no further zeros of $f(z)$.

Proof. In (3), the factor $(z - z_0)^n$ is zero only at $z = z_0$. The power series in the brackets $[\cdot \cdot]$ represents an analytic function (cf. Theorem 5 in Sec. 14.4), call it $g(z)$. Now $g(z_0) = a_n \neq 0$, since an analytic function is continuous, and because of this continuity, also $g(z) \neq 0$ in some neighborhood of $z = z_0$. Hence the same holds for $f(z)$. ∎

This theorem is illustrated by the functions in Example 4.

Poles are often caused by zeros in the denominator. (*Example:* $\tan z$ has poles where $\cos z$ is zero.) This is a major reason for the importance of zeros. The key to the connection is

Theorem 4 (Poles and zeros)

Let $f(z)$ be analytic at $z = z_0$ and have a zero of nth order at $z = z_0$. Then $1/f(z)$ has a pole of nth order at $z = z_0$.

The same holds for $h(z)/f(z)$ if $h(z)$ is analytic at $z = z_0$ and $h(z_0) \neq 0$.

The proof follows from (3) (cf. Prob. 11).

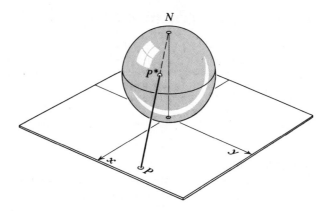

Fig. 361. Riemann number sphere

Riemann Number Sphere. Infinity

The study of the behavior of analytic functions $f(z)$ for large $|z|$ is a natural task of practical importance. When $|z|$ is large, the complex plane becomes somewhat inconvenient. We may then prefer a representation of the complex numbers on a sphere, which was suggested by Riemann and is obtained as follows.

Let S be a sphere of diameter 1 which touches the complex z-plane at the origin (Fig. 361). Let N be the "North Pole" of S (the point diametrically opposite to the point of contact between the sphere and the plane). Let P be any point in the complex plane. Then the straight line segment with endpoints P and N intersects S at a point P^*. We let P and P^* correspond to each other. In this way we obtain a mapping of the complex plane into the sphere S, and P^* is the image point of P with respect to this mapping. The complex numbers, first represented in the plane, are now represented by points on S. To each z there corresponds a point on S.

Conversely, each point on S represents a complex number z, except for the point N, which does not correspond to any point in the complex plane. This suggests that we introduce an additional point, called the **point at infinity** and denoted by the symbol ∞ (*infinity*). The complex plane together with the point ∞ is called the **extended complex plane.** The complex plane without that point ∞ is often called the *finite complex plane*, for distinction, or simply the *complex plane*, as before.

Of course, we now let the point $z = \infty$ correspond to N. Then our mapping becomes a one-to-one mapping of the extended complex plane onto S. The sphere S is called the **Riemann number sphere.** The particular mapping we have used is called a **stereographic projection.**

Obviously, the unit circle is mapped onto the "equator" of S. The interior of the unit circle corresponds to the "Southern Hemisphere" and the exterior to the "Northern Hemisphere." Numbers z whose absolute values are large lie close to the North Pole N. The x and y axes (and, more generally, all the straight lines through the origin) are mapped onto "meridians," while circles with center at the origin are mapped onto "parallels." It can be shown that any circle or straight line in the z-plane is mapped onto a circle on S.

Analytic or Singular at Infinity

If we want to investigate a function $f(z)$ for large $|z|$, we may now set $z = 1/w$ and investigate $f(z) = f(1/w) \equiv g(w)$ in a neighborhood of $w = 0$. We define $f(z)$ to be **analytic or singular at infinity** if $g(w)$ is analytic or singular, respectively, at $w = 0$. We also define

$$(4) \qquad\qquad g(0) = \lim_{w \to 0} g(w)$$

if this limit exists.

Furthermore, we say that $f(z)$ has an *nth-order zero at infinity* if $f(1/w)$ has such a zero at $w = 0$. Similarly for poles and essential singularities.

EXAMPLE 5. Functions analytic or singular at infinity

The function $f(z) = 1/z^2$ is analytic at ∞ since $g(w) = f(1/w) = w^2$ is analytic at $w = 0$, and $f(z)$ has a second-order zero at ∞. The function $f(z) = z^3$ is singular at ∞ and has there a pole of third order since $g(w) = f(1/w) = 1/w^3$ has such a pole at $w = 0$. The function e^z has an essential singularity at ∞ since $e^{1/w}$ has such a singularity at $w = 0$. Similarly, $\cos z$ and $\sin z$ have an essential singularity at ∞.

Recall that an **entire function** is one that is analytic everywhere in the (finite) complex plane. Liouville's theorem (Sec. 13.6) tells us that the only *bounded* entire functions are the constants, hence any nonconstant entire function must be unbounded. Hence it has a singularity at ∞, a pole if it is a polynomial or an essential singularity if it is not. The functions just considered are typical in this respect. ∎

A **meromorphic function** is an analytic function whose only singularities in the finite plane are poles.

EXAMPLE 6. Meromorphic functions

Rational functions with nonconstant denominator, $\tan z$, $\cot z$, $\sec z$ and $\csc z$ are meromorphic functions. ∎

This is the end of Chap. 14 on power series, particularly Taylor series, and Laurent series. We have seen that Taylor and Laurent series of analytic functions have relations to complex integration inasmuch as the coefficients of these series are given by integral formulas, although one almost never uses these formulas in practice, since there are various other methods for developing given functions $f(z)$ in Taylor or Laurent series. Does this make those coefficient formulas useless? On the contrary: what about determining those coefficients in some other way and then using those formulas for evaluating complex integrals? This is the idea we shall put to work in the next chapter. Actually, things will be simpler as they presently look: a single one of those formulas will be sufficient, namely, that for the coefficient of the power $1/(z - z_0)$ in the Laurent series of $f(z)$ that converges in the immediate neighborhood of z_0 (except at z_0 itself). We call this coefficient the **"residue"** of $f(z)$ at z_0, and this gives the name "residue integration method."

Together with Chap. 13 this shows that complex integration is rich in methods that have no counterpart in real integral calculus.

Even more surprising is the fact that residue integration can also be used for evaluating *real* integrals (which are quite complicated).

Problems for Sec. 14.10

Zeros. Determine the location and order of the zeros of the following functions.

1. $z^4 - 1$

2. $(z - 1)^4$

3. $(z + i)^3$

4. $(\sin z - 1)^2$

5. $\cosh^2 z$

6. $(z^2 + 1)(e^z - 1)$

7. $\dfrac{e^z - e^{2z}}{z^2}$

8. $\dfrac{(9 + z^2)^2}{(1 - z^2)^3}$

9. $\dfrac{\sin^3 \pi z}{z^3}$

10. Let $f(z)$ have a zero of order n at $z = z_0$. Show that then $f^2(z)$ has a zero of order $2n$, and the derivative $f'(z)$ has a zero of order $n - 1$ at $z = z_0$, (provided $n > 1$).

11. Prove Theorem 4.

12. Why do we need the condition $h(z_0) \neq 0$ in Theorem 4?

Singularities. Determine the location and type of singularities of the following functions.

13. $z + z^{-1}$

14. $\tan \pi z$

15. $\cot z$

16. $1/[(z^2 + 1)(z + 1)^2]$

17. $e^z + e^{1/z}$

18. $e^{z^2}/(z - i)^4$

19. $\dfrac{z^3 + 3z}{z^2 + 1}$

20. $\dfrac{z^4 - 1}{z(z - 2)^2}$

21. $\cosh^2 \dfrac{1}{z - \pi}$

22. $(\cosh 2z - \sinh z)/z$

23. $(\cos z - \sin z)^{-1}$

24. $e^{1/(z-1)}/(e^z - 1)$

25. Verify Theorem 1 for $f(z) = z^{-3} + z^{-1}$.

26. Prove Theorem 1.

Analytic and singular at infinity. Are the following functions analytic at infinity?

27. e^z, $e^{1/z}$, $ze^{1/z}$

28. $1/(z - 2i)^5$

29. $z^{-3} + z^{-1} + 4i$

30. $(z^3 + 1)/(z^2 - 1)$

31. $(z^3 + i)/(iz^4 + 2)$

32. $\sin (1/z)$, $\csc z$

33. ze^{1/z^2}, $z^{-1} \sin z$

34. $z^{-3} \tan (1/z)$

35. $\cot (1/z)$

Riemann number sphere. Describe and sketch the images of the following regions on the Riemann number sphere.

36. $|z| \leq 1$

37. $\text{Re } z \geq 0$

38. $|z| \geq 5$

39. $\frac{1}{2} \leq |z| \leq 2$

40. $|z| > 1$, $\text{Im } z > 0$

41. $-\pi \leq \text{Im } z \leq \pi$

42. $|\arg z| \leq \pi/4$

43. $\text{Im } z \geq \text{Re } z$

44. $|\arg z| < \pi$

45. Describe the relative positions of z, $-z$, $\bar{z}$, $-\bar{z}$ in the complex plane and on the Riemann sphere.

Further Proofs in Chapter 14

Proof of Theorem 2 in Sec. 14.2 (Cauchy's convergence principle)

(a) In this proof we need two concepts and a theorem, which we list first.

1. A **bounded sequence** $s_1, s_2, \cdots$ is a sequence whose terms all lie in a disk of (sufficiently large, finite) radius K with center at the origin; thus $|s_n| < K$ for all n.

2. A **limit point** a of a sequence s_1, s_2, $\cdots$ is a point such that, given an $\epsilon > 0$, there are infinitely many terms satisfying $|s_n - a| < \epsilon$. (Note that this does *not* imply convergence, since there may still be infinitely many terms that do not lie within that circle of radius ϵ about a.

Example: $\frac{1}{4}, \frac{3}{4}, \frac{1}{8}, \frac{7}{8}, \frac{1}{16}, \frac{15}{16}, \cdots$ has the limit points 0 and 1 and diverges.

3. A bounded sequence in the complex plane has at least one limit point. (Bolzano–Weierstrass theorem; proof below. Recall that "sequence" always means *infinite* sequence.)

(b) We now turn to the actual proof that $z_1 + z_2 + \cdots$ converges if and only if for every $\epsilon > 0$ we can find an N such that

(1) $|z_{n+1} + \cdots + z_{n+p}| < \epsilon$ for every $n > N$ and $p = 1, 2, \cdots$.

Here, by the definition of partial sums,

$$s_{n+p} - s_n = z_{n+1} + \cdots + z_{n+p}.$$

Writing $n + p = r$, we see from this that (1) is equivalent to

(1*) $|s_r - s_n| < \epsilon$ for all $r > N$ and $n > N$.

Suppose that s_1, s_2, $\cdots$ converges. Denote its limit by s. Then for a given $\epsilon > 0$ we can find an N such that

$$|s_n - s| < \frac{\epsilon}{2} \qquad \text{for every } n > N.$$

Hence, when $r > N$ and $n > N$, then by the triangle inequality (Sec. 12.2),

$$|s_r - s_n| = |(s_r - s) - (s_n - s)| \le |s_r - s| + |s_n - s| < \frac{\epsilon}{2} + \frac{\epsilon}{2} = \epsilon,$$

that is, (1*) holds.

(c) Conversely, assume that s_1, s_2, $\cdots$ satisfies (1*). We first prove that then the sequence must be bounded. Indeed, choose a fixed ϵ and a fixed $n = n_0 > N$ in (1*). Then (1*) implies that all s_r with $r > N$ lie in the disk of radius ϵ and center s_{n_0} and only *finitely many* terms $s_1, \cdots, s_N$ may not lie in this disk. Clearly, we can now find a circle so large that this disk and these finitely many terms all lie within this new circle. Hence the sequence is bounded. By the Bolzano–Weierstrass theorem, it has at least one limit point, call it s.

We now show that the sequence is convergent with the limit s. Let $\epsilon > 0$ be given. Then there is an N^* such that $|s_r - s_n| < \epsilon/2$ for all $r > N^*$ and $n > N^*$, by (1*). Also, by the definition of a limit point, $|s_n - s| < \epsilon/2$ for *infinitely many* n, so that we can find and fix an $n > N^*$ such that $|s_n - s| < \epsilon/2$. Together, for *every* $r > N^*$,

$$|s_r - s| = |(s_r - s_n) + (s_n - s)| \le |s_r - s_n| + |s_n - s| < \frac{\epsilon}{2} + \frac{\epsilon}{2} = \epsilon;$$

that is, the sequence s_1, s_2, $\cdots$ is convergent with the limit s.   ∎

Bolzano–Weierstrass theorem[13]

A bounded infinite sequence $z_1, z_2, z_3, \cdots$ in the complex plane has at least one limit point.

Proof. It is obvious that we need both conditions: a finite sequence cannot have a limit point, and the sequence $1, 2, 3, \cdots$, which is infinite but not bounded, has no limit point. To prove the theorem, consider a bounded infinite sequence $z_1, z_2, \cdots$ and let K be such that $|z_n| < K$ for all n. If only finitely many values of the z_n are different, then, since the sequence is infinite, some number z must occur infinitely many times in the sequence, and, by definition, this number is a limit point of the sequence.

We may now turn to the case when the sequence contains infinitely many different terms. We draw a large square Q_0 which contains all z_n. We subdivide Q_0 into four congruent squares, which we number 1, 2, 3, 4. Clearly, at least one of these squares (each taken with its complete boundary) must contain infinitely many terms of the sequence. The square of this type with the lowest number (1, 2, 3 or 4) will be denoted by Q_1. This is the first step. In the next step we subdivide Q_1 into four congruent squares and select a square Q_2 according to the same rule, and so on. This yields an infinite sequence of squares $Q_0, Q_1, Q_2, \cdots, Q_n, \cdots$ with the property that the side of Q_n approaches zero as n approaches infinity, and Q_m contains all Q_n with $n > m$. It is not difficult to see that the number which belongs to all these squares,[14] call it $z = a$, is a limit point of the sequence. In fact, given an $\epsilon > 0$, we can choose an N so large that the side of the square Q_N is less than ϵ and, since Q_N contains infinitely many z_n we have $|z_n - a| < \epsilon$ for infinitely many n. This completes the proof.   ∎

Part (b) of the proof of Theorem 5 in Sec. 14.4

We show that

$$\frac{(z + \Delta z)^n - z^n}{\Delta z} - nz^{n-1}$$

$$= \Delta z[(z + \Delta z)^{n-2} + 2z(z + \Delta z)^{n-3} + \cdots + (n - 1)z^{n-2}]$$

since then (5) implies (6) in Sec. 14.4. That is, writing $z + \Delta z = b$ and $z = a$, thus $\Delta z = b - a$, we have to prove that for any integer $n \geq 2$,

(7a)
$$\frac{b^n - a^n}{b - a} - na^{n-1} = (b - a)A_n,$$

where A_n is the expression in the brackets in (6),

---

[13]BERNARD BOLZANO (1781—1848), Austrian mathematician and professor of religious studies, is a pioneer in the study of point sets, the foundations of analysis and mathematical logic.

For Weierstrass, see footnote 8 in Sec. 14.8.

[14]The fact that such a unique number $z = a$ exists seems to be obvious, but it actually follows from an axiom of the real number system, the so-called *Cantor–Dedekind axiom*. Cf. footnote 2 in Appendix 3.3.

(7b) $A_n = b^{n-2} + 2ab^{n-3} + 3a^2b^{n-4} + \cdots + (n-1)a^{n-2},$

thus, $A_2 = 1$, $A_3 = b + 2a$, etc. We use induction. When $n = 2$, then (7) holds, since then

$$\frac{b^2 - a^2}{b - a} - 2a = \frac{b^2 - a^2 - 2a(b - a)}{b - a} = \frac{(b - a)^2}{b - a} = b - a = (b - a)A_2.$$

Assuming that (7) holds for $n = k$, we show that it holds for $n = k + 1$. By adding and subtracting a term in the numerator and then dividing we get

$$\frac{b^{k+1} - a^{k+1}}{b - a} = \frac{b^{k+1} - ba^k + ba^k - a^{k+1}}{b - a} = b\frac{b^k - a^k}{b - a} + a^k.$$

By the induction hypothesis, the right-hand side equals

$$b[(b - a)A_k + ka^{k-1}] + a^k.$$

Direct calculation shows that this is equal to

$$(b - a)\{bA_k + ka^{k-1}\} + aka^{k-1} + a^k$$

From (7b) with $n = k$ we see that the expression in the braces equals

$$b^{k-1} + 2ab^{k-2} + \cdots + (k - 1)ba^{k-2} + ka^{k-1} = A_{k+1}.$$

Hence our result is

$$\frac{b^{k+1} - a^{k+1}}{b - a} = (b - a)A_{k+1} + (k + 1)a^k.$$

Taking the last term to the left, we obtain (7) with $n = k + 1$. This proves (7) for any integer $n \geq 2$ and completes the proof.   ∎

Review Problems for Chapter 14

1. What is a power series? Why are these series very important in complex analysis?
2. State from memory the ratio test and the root test.
3. State the definition of the radius of convergence of a power series. Can one make general statements about the convergence of such a series on the circle of convergence?
4. Do there exist power series that converge only at the center?
5. What is the Cauchy product of two power series, and what can we say about its convergence?
6. What do we mean by absolute convergence? By uniform convergence?
7. For what z does a power series converge absolutely and uniformly?
8. What is the relation between absolute and uniform convergence? Between absolute and conditional convergence?
9. Is termwise differentiation of a power series permissible? Termwise integration?
10. What is the radius of convergence of a Taylor series of an entire function?

11. Do we obtain an analytic function if we replace x by z in the Maclaurin series of a function $f(x)$?

12. Write down the Maclaurin series of e^z, $\cos z$, $\sin z$, $\cosh z$, $\sinh z$ from memory.

13. Does $\operatorname{Ln} z$ have a Maclaurin series. (Give reason.)

14. Describe some of the methods for obtaining Taylor series in practice. Give examples.

15. What is a Laurent series? What do you know about its convergence?

16. Can a function have more than one Taylor series with a given center? More than one Laurent series with a given center? Explain.

17. Can a Laurent series reduce to a Taylor series and if so, what does this tell us about the corresponding function?

18. What is a singularity of an analytic function? Give examples.

19. Describe briefly how singularities can be classified by using Laurent series.

20. Do we need the integral formulas for the coefficients of a Laurent series, or can we obtain these coefficients by other means? (Keep this in mind; it will be essential in the next chapter.)

Find the radius of convergence of the following power series.

21. $\displaystyle\sum_{n=0}^{\infty} (2z - i)^n$ **22.** $\displaystyle\sum_{n=1}^{\infty} \frac{z^{2n}}{n^2}$ **23.** $\displaystyle\sum_{n=0}^{\infty} \left(\frac{z}{3} + 4\right)^n$ **24.** $\displaystyle\sum_{n=0}^{\infty} \frac{i^n n^2}{4^n} z^n$

25. $\displaystyle\sum_{n=1}^{\infty} \frac{\pi^{2n}}{n^2} z^n$ **26.** $\displaystyle\sum_{n=2}^{\infty} \frac{(z + i)^2}{(2n - 3)!}$ **27.** $\displaystyle\sum_{n=0}^{\infty} \frac{n^{10}}{n!} (z + 2i)^n$ **28.** $\displaystyle\sum_{n=0}^{\infty} \frac{n!}{n^n} z^n$

29. $\displaystyle\sum_{n=0}^{\infty} 3^n (z - 3i)^{2n}$ **30.** $\displaystyle\sum_{n=0}^{\infty} \binom{n + m}{n} z^n$

Find the Taylor series of the following functions with the given point as center and determine the radius of convergence.

31. e^{-z}, 0 **32.** e^z, i **33.** $\operatorname{Ln} z$, 2 **34.** $\cos \pi z^2$, 0

35. $\dfrac{1}{1 + z^4}$, 0 **36.** $\dfrac{1}{1 - z}$, -1 **37.** $\dfrac{1}{1 - z}$, $2i$ **38.** $\dfrac{z^2}{(1 - z^3)^2}$, 0

39. $\displaystyle\int_0^z e^{-t^2}\, dt$, 0 **40.** $\dfrac{4 - 3z}{(1 - z)^2}$, 0

Expand the following functions in a Laurent series that converges for $0 < |z - z_0| < R$ and determine the precise region of convergence as well as the type of singularity of the function at z_0.

41. $\dfrac{1}{z^4 - z^5}$, $z_0 = 0$ **42.** $\dfrac{e^z}{(z - i)^2}$, $z_0 = i$

43. $z^{-8} \cosh \dfrac{1}{z^2}$, $z_0 = 0$ **44.** $\dfrac{1 - z^2}{z^4}$, $z_0 = 0$

45. $\dfrac{\cos z}{(z - \frac{1}{2}\pi)^3}$, $z_0 = \dfrac{\pi}{2}$ **46.** $\dfrac{1}{z^3} \sin 2z^2$, $z_0 = 0$

47. $(z - 1)^{-5} \operatorname{Ln} z$, $z_0 = 1$ **48.** $(z + i)^2 \sin \dfrac{1}{2z + 2i}$, $z_0 = -i$

49. $\dfrac{1}{z^2} \displaystyle\int_0^z \dfrac{e^t - 1}{t}\, dt$, $z_0 = 0$ **50.** $\dfrac{\sinh z + \sin z}{z^2}$, $z_0 = 0$

Summary of Chapter 14
Power Series, Taylor Series,
Laurent Series

Sequences, series and convergence tests are discussed in Secs. 14.1 and 14.2. A **power series** is of the form (Sec. 14.3)

$$(1) \qquad \sum_{n=0}^{\infty} a_n(z - z_0)^n = a_0 + a_1(z - z_0) + a_2(z - z_0)^2 + \cdots,$$

z_0 is its *center*. It converges for $|z - z_0| < R$ and diverges for $|z - z_0| > R$. Some power series converge for all z (then we write $R = \infty$). R is the *radius of convergence*. Also, $R = \lim |a_n/a_{n+1}|$ if this limit exists. The series (1) converges absolutely (Sec. 14.3) and **uniformly** (Sec. 14.8) in every closed disk $|z - z_0| \leq r < R$. If $R > 0$, it represents an analytic function $f(z)$ for $|z - z_0| < R$. The derivatives $f'(z)$, $f''(z)$, $\cdots$ are obtained by termwise differentiation of (1), and these series have the same radius of convergence R as (1). Cf. Sec. 14.4.

Conversely, *every* analytic function $f(z)$ can be represented by power series. These **Taylor series** of $f(z)$ are of the form (Sec. 14.5)

$$(2) \qquad f(z) = \sum_{n=0}^{\infty} \frac{1}{n!} f^{(n)}(z_0)(z - z_0)^n \qquad (|z - z_0| < R),$$

as in calculus. They converge for all z if $f(z)$ is **entire** (= analytic for all z) or in the open disk with center z_0 and radius equal to the distance from z_0 to the nearest **singularity** of $f(z)$ (= point at which $f(z)$ ceases to be analytic, cf. Sec. 14.10). The functions e^z, $\cos z$, $\sin z$, etc. have Maclaurin and Taylor series similar to those in calculus (Sec. 14.6).

A **Laurent series** is of the form (Sec. 14.9)

$$(3) \qquad f(z) = \sum_{n=0}^{\infty} a_n(z - z_0)^n + \sum_{n=1}^{\infty} \frac{b_n}{(z - z_0)^n}$$

or, more briefly written [but this means the same as (3)!]

$$(3^*) \quad f(z) = \sum_{n=-\infty}^{\infty} a_n(z - z_0)^n, \quad a_n = \frac{1}{2\pi i} \oint_C \frac{f(z^*)}{(z^* - z_0)^{n+1}} \, dz^*.$$

This series converges in an annulus (ring) A with center z_0. In A the function $f(z)$ is analytic. At points not in A it may have singularities.

The first series in (3) is a power series. The second series is called the **principal part** of the Laurent series. In a given annulus, a Laurent series of $f(z)$ is unique, but $f(z)$ may have different Laurent series in different annuli with the same center.

If $f(z)$ has an isolated singularity (cf. Sec. 14.10) at $z = z_0$, the Laurent series of $f(z)$ that converges for $0 < |z - z_0| < R$ (R suitable) can be used for classifying this singularity, which is called a **pole** if the principal part of this Laurent series is a finite sum, otherwise an (isolated) **essential singularity.**

A pole is said to be of **order** n if $1/(z - z_0)^n$ is the highest negative power of the principal part in (3). A first-order pole is also called a *simple pole*.

Similarly, the Taylor series (2) can be used for classifying **zeros** of $f(z)$. The function $f(z)$ given by (2) has a zero of **nth order** at z_0 if f and its derivatives f', f'', $\cdots$, $f^{(n-1)}$ are all zero at z_0 whereas $f^{(n)}(z_0) \neq 0$. A first-order zero is also called a *simple zero*. Cf. Sec. 14.10.

Section 14.10 also includes a discussion of the **extended complex plane,** obtained from the complex plane by attaching an improper point ∞ (*"infinity"*).

Chapter 15

Residue Integration Method

Since there are various methods for determining the coefficients of a Laurent series (1), Sec. 14.9, without using the integral formulas (2), Sec. 14.9, we may use the formula for b_1 for evaluating complex integrals in a very elegant and simple fashion. b_1 will be called the **residue** of $f(z)$ at $z = z_0$. This powerful method may also be applied for evaluating certain real integrals, as we shall see in Secs. 15.3 and 15.4.

Prerequisites for this chapter: Chaps. 12–14.
References: Appendix 1, Part D.
Answers to problems: Appendix 2.

15.1 Residues

Let us first explain what a residue is and how it can be used for evaluating integrals

$$\oint_C f(z)\, dz.$$

These will be contour integrals taken around a simple closed path C.

If $f(z)$ is analytic everywhere on C and inside C, such an integral is zero by Cauchy's integral theorem (Sec. 13.3), and we are done.

If $f(z)$ has a singularity at a point $z = z_0$ inside C, but is otherwise analytic on C and inside C, then $f(z)$ has a Laurent series

$$f(z) = \sum_{n=0}^{\infty} a_n(z - z_0)^n + \frac{b_1}{z - z_0} + \frac{b_2}{(z - z_0)^2} + \cdots$$

that converges for all points near $z = z_0$ (except at $z = z_0$ itself), in some domain of the form $0 < |z - z_0| < R$. Now comes the key idea. The coefficient b_1 of the first negative power $1/(z - z_0)$ of this Laurent series is given by the integral formula (2), Sec. 14.9, with $n = 1$, that is,

$$b_1 = \frac{1}{2\pi i} \oint_C f(z)\, dz,$$

but since we can obtain Laurent series by various methods, without using the integral formulas for the coefficients (see the examples in Sec. 14.9), we can find b_1 by one of these methods and then use the formula for b_1 for evaluating the integral:

(1)
$$\oint_C f(z)\ dz\ =\ 2\pi i b_1.$$

Here we integrate in the counterclockwise sense around the simple closed path C that contains $z = z_0$ in its interior.

The coefficient b_1 is called the **residue** of $f(z)$ at $z = z_0$ and we shall denote it by

(2)
$$b_1\ =\ \operatorname*{Res}_{z=z_0} f(z).$$

EXAMPLE 1. Evaluation of an integral by means of a residue

Integrate the function $f(z) = z^{-4} \sin z$ around the unit circle C in the counterclockwise sense.

Solution. From (4) in Sec. 14.6 we obtain the Laurent series

$$f(z) = \frac{\sin z}{z^4} = \frac{1}{z^3} - \frac{1}{3!z} + \frac{z}{5!} - \frac{z^3}{7!} + - \cdots$$

which converges for $|z| > 0$ (that is, for all $z \neq 0$). This series shows that $f(z)$ has a pole of third order at $z = 0$ and that the residue of $f(z)$ at $z = 0$ is $b_1 = -1/3!$. From (1) we thus obtain the answer

$$\oint_C \frac{\sin z}{z^4}\ dz = 2\pi i b_1 = -\frac{\pi i}{3}.$$

EXAMPLE 2. Be careful to use the *right* Laurent series!

Integrate $f(z) = 1/(z^3 - z^4)$ around the circle C: $|z| = 1/2$ in the clockwise sense.

Solution. $z^3 - z^4 = z^3(1 - z)$ shows that $f(z)$ is singular at $z = 0$ and $z = 1$. Now $z = 1$ lies outside C. Hence it is of no interest here. So we need the residue of $f(z)$ at 0. We find it from the Laurent series that converges for $0 < |z| < 1$. This is series (I) in Example 4, Sec. 14.9:

$$\frac{1}{z^3 - z^4} = \frac{1}{z^3} + \frac{1}{z^2} + \frac{1}{z} + 1 + z + \cdots \qquad (0 < |z| < 1).$$

We see from it that this residue is 1. Clockwise integration thus yields

$$\oint_C \frac{dz}{z^3 - z^4} = -2\pi i \operatorname*{Res}_{z=0} f(z) = -2\pi i.$$

Caution! Had we used the wrong series (II) in Example 4, Sec. 14.9,

$$\frac{1}{z^3 - z^4} = -\frac{1}{z^4} - \frac{1}{z^5} - \frac{1}{z^6} - \cdots \qquad (|z| < 1),$$

we would have obtained the wrong answer 0. Explain! ∎

Two Formulas for Residues at Simple Poles

Before we continue integration, we ask the following. To get a residue, a single coefficient of a Laurent series, must we derive the whole series or is there a more economical way? For poles, there is. We shall derive, once and for all, some formulas for residues at poles, so that in this case we no longer need the whole series.

Let $f(z)$ have a simple pole at $z = z_0$. Then the corresponding Laurent series is [cf. (2) with $m = 1$ in Sec. 14.10]

$$f(z) = \frac{b_1}{z - z_0} + a_0 + a_1(z - z_0) + a_2(z - z_0)^2 + \cdots \qquad (0 < |z - z_0| < R).$$

Here $b_1 \neq 0$. (Why?) Multiplying both sides by $z - z_0$ we have

$$(z - z_0)f(z) = b_1 + (z - z_0)[a_0 + a_1(z - z_0) + \cdots].$$

We now let $z \to z_0$. Then the right-hand side approaches b_1. This gives

(3)
$$\boxed{\operatorname*{Res}_{z=z_0} f(z) = b_1 = \lim_{z \to z_0} (z - z_0)f(z).}$$

EXAMPLE 3. Residue at a simple pole

$$\operatorname*{Res}_{z=i} \frac{9z + i}{z(z^2 + 1)} = \lim_{z \to i} (z - i) \frac{9z + i}{z(z + i)(z - i)} = \left[\frac{9z + i}{z(z + i)}\right]_{z=i} = \frac{10i}{-2} = -5i \qquad ∎$$

Another, sometimes simpler formula for the residue at a simple pole is obtained by starting from

$$f(z) = \frac{p(z)}{q(z)}$$

with analytic $p(z)$ and $q(z)$, where we assume that $p(z_0) \neq 0$ and $q(z)$ has a simple zero at $z = z_0$ (so that $f(z)$ has a simple pole at $z = z_0$ as wanted, by Theorem 4 in Sec. 14.10). By the definition of a simple zero, $q(z)$ has a Taylor series of the form

$$q(z) = (z - z_0)q'(z_0) + \frac{(z - z_0)^2}{2!} q''(z_0) + \cdots .$$

This we substitute into $f = p/q$ and then f into (3), finding

$$\operatorname*{Res}_{z=z_0} f(z) = \lim_{z \to z_0} (z - z_0)\frac{p(z)}{q(z)} = \lim_{z \to z_0} \frac{(z - z_0)p(z)}{(z - z_0)[q'(z_0) + (z - z_0)q''(z_0)/2 + \cdots]} .$$

We now see that on the right, a factor $z - z_0$ is canceled and the resulting denominator has the limit $q'(z_0)$. Hence our second formula for the residue at a simple pole is

(4)
$$\boxed{\operatorname*{Res}_{z=z_0} f(z) = \operatorname*{Res}_{z=z_0} \frac{p(z)}{q(z)} = \frac{p(z_0)}{q'(z_0)} .}$$

EXAMPLE 4. Residue at a simple pole calculated by formula (4)

$$\operatorname*{Res}_{z=i} \frac{9z + i}{z(z^2 + 1)} = \left[\frac{9z + i}{3z^2 + 1}\right]_{z=i} = \frac{10i}{-2} = -5i \qquad \text{(cf. Example 3).}$$

EXAMPLE 5. Another application of formula (4)
Find all poles and the corresponding residues of the function

$$f(z) = \frac{\cosh \pi z}{z^4 - 1}.$$

Solution. $p(z) = \cosh \pi z$ is entire, and $q(z) = z^4 - 1$ has simple zeros at $1, i, -1, -i$. Hence $f(z)$ has simple poles at these points (and no further poles). Since $q'(z) = 4z^3$, we see from (4) that the residues equal the values of $(\cosh \pi z)/4z^3$ at those points, that is,

$$\frac{\cosh \pi}{4} \approx 2.8980, \qquad \frac{\cosh \pi i}{4i^3} = \frac{\cos \pi}{-4i} = -\frac{i}{4}, \qquad -\frac{\cosh \pi}{4}, \qquad \frac{\cosh (-\pi i)}{4(-i)^3} = \frac{i}{4}. \quad \blacksquare$$

Formula for the Residue at a Pole of Any Order

Let $f(z)$ be an analytic function that has a pole of any order $m > 1$ at a point $z = z_0$. Then, by the definition of such a pole (Sec. 14.10), the Laurent series of $f(z)$ converging near $z = z_0$ (except at $z = z_0$ itself) is

$$f(z) = \frac{b_m}{(z - z_0)^m} + \frac{b_{m-1}}{(z - z_0)^{m-1}} + \cdots + \frac{b_2}{(z - z_0)^2} + \frac{b_1}{z - z_0}$$

$$+ a_0 + a_1(z - z_0) + \cdots,$$

where $b_m \neq 0$. Multiplying both sides by $(z - z_0)^m$, we have

$$(z - z_0)^m f(z) = b_m + b_{m-1}(z - z_0) + \cdots + b_2(z - z_0)^{m-2} + b_1(z - z_0)^{m-1}$$

$$+ a_0(z - z_0)^m + a_1(z - z_0)^{m+1} + \cdots.$$

We see that the residue b_1 of $f(z)$ at $z = z_0$ is now the coefficient of the power $(z - z_0)^{m-1}$ in the Taylor series of the function

$$g(z) = (z - z_0)^m f(z)$$

on the left, with center $z = z_0$. Thus, by Taylor's theorem (Sec. 14.5),

$$b_1 = \frac{1}{(m - 1)!} g^{(m-1)}(z_0).$$

Hence if $f(z)$ has a pole of mth order at $z = z_0$, the residue is given by

$$(5) \qquad \boxed{\operatorname*{Res}_{z=z_0} f(z) = \frac{1}{(m - 1)!} \lim_{z \to z_0} \left\{ \frac{d^{m-1}}{dz^{m-1}} [(z - z_0)^m f(z)] \right\}.}$$

In particular, for a second-order pole ($m = 2$),

(5*)

$$\text{Res}_{z=z_0} f(z) = \lim_{z \to z_0} \left\{ [(z - z_0)^2 f(z)]' \right\}.$$

EXAMPLE 6. Residue at a pole of higher order

The function

$$f(z) = \frac{50z}{(z + 4)(z - 1)^2}$$

has a pole of second order at $z = 1$, and from (5*) we obtain the corresponding residue

$$\text{Res}_{z=1} f(z) = \lim_{z \to 1} \frac{d}{dz} [(z - 1)^2 f(z)] = \lim_{z \to 1} \frac{d}{dz} \left(\frac{50z}{z + 4} \right) = 8.$$

EXAMPLE 7. Residues from partial fractions

If $f(z)$ is rational, we can also determine its residues from partial fractions. In Example 6,

$$f(z) = \frac{50z}{(z + 4)(z - 1)^2} = \frac{-8}{z + 4} + \frac{8}{z - 1} + \frac{10}{(z - 1)^2}.$$

This shows that the residue at $z = 1$ is 8 (as before), and at $z = -4$ (simple pole) it is -8.

Why is this so? Consider $z = 1$. There the Laurent series has the last two fractions as its principal part and the first fraction as the sum of its other part. This first fraction is analytic at $z = 1$, so that it has a Taylor series with center $z = 1$, as it should be. Similarly, at $z = -4$ the first fraction is the principal part of the Laurent series.

EXAMPLE 8. Integration around a second-order pole

Counterclockwise integration around any simple closed path C such that $z = 1$ is inside C and $z = -4$ outside C yields (cf. Example 6 or 7)

$$\oint_C \frac{z}{(z + 4)(z - 1)^2} \, dz = 2\pi i \, \text{Res}_{z=1} \frac{z}{(z + 4)(z - 1)^2} = 2\pi i \frac{8}{50} \approx 1.0053i. \qquad \blacksquare$$

So far we can evaluate integrals of analytic functions $f(z)$ over closed curves C when $f(z)$ has only *one* singular point inside C. In the next section we show that the residue integration method can be readily extended to the case of several singular points of $f(z)$ inside C.

Problems for Sec. 15.1

Find the residues at the singular points of the following functions.

1. $\dfrac{6}{1 - z}$

2. $\dfrac{\cos 2z}{z^4}$

3. $\dfrac{z + 3}{z + 1}$

4. $\dfrac{e^z}{(z + \pi i)^6}$

5. $\dfrac{1}{(z^2 - 1)^2}$

6. $\dfrac{z^2 + 11z + 1}{(z + 1)^2(z - 2)}$

7. $1/(1 - e^z)$

8. $\sec z$

9. $\tan z$

Find the residues at the singular points that lie inside the circle $|z| = 1.5$.

10. $\dfrac{3z^2}{1 - z^4}$

11. $\dfrac{z - \frac{1}{4}}{z^2 + 3z + 2}$

12. $\dfrac{4 - 3z}{z^3 - 3z^2 + 2z}$

13. $\dfrac{6 - 4z}{z^3 + 3z^2}$

14. $\dfrac{1}{(z^4 - 1)^2}$

15. $\dfrac{-z^2 - 22z + 8}{z^3 - 5z^2 + 4z}$

16. Solve Example 5 by (3) and compare the amount of work with that in Example 5.

17. Show that (3) is a particular case of (5).

18. Another derivation of (5). Obtain (5) without using Taylor's theorem, by $m - 1$ differentiations of the formula for $(z - z_0)^m f(z)$.

Evaluate the following integrals where C is the unit circle (counterclockwise).

19. $\oint_C \cot z \, dz$

20. $\oint_C \csc z \, dz$

21. $\oint_C \tan z \, dz$

22. $\oint_C \dfrac{z^2 - 4}{(z - 2)^4} \, dz$

23. $\oint_C \dfrac{5z}{2z + i} \, dz$

24. $\oint_C \dfrac{4z^3 + 8}{4z + \pi} \, dz$

25. $\oint_C \dfrac{\sin \pi z}{z^4} \, dz$

26. $\oint_C z^n e^{1/z} \, dz$

27. $\oint_C \dfrac{dz}{1 - e^z}$

28. $\oint_C \dfrac{6 \cos z}{2i - 3z} \, dz$

29. $\oint_C \dfrac{z^2 + 1}{z^2 - 2z} \, dz$

30. $\oint_C \dfrac{\tanh^2 (z + \frac{1}{2})}{e^z \sin z} \, dz$

15.2 Residue Theorem

So far we are in a position to evaluate contour integrals whose integrands have only a single isolated singularity inside the contour of integration. We shall now see that our simple method may be extended to the case when the integrand has several isolated singularities inside the contour. This extension is surprisingly simple, as follows.

Residue theorem

Let $f(z)$ be a function that is analytic inside a simple closed path C and on C, except for finitely many singular points $z_1, z_2, \cdots, z_k$ inside C. Then

(1)
$$\oint_C f(z) \, dz = 2\pi i \sum_{j=1}^{k} \operatorname*{Res}_{z = z_j} f(z),$$

the integral being taken in the counterclockwise sense around the path C.

Proof. We enclose each of the singular points z_j in a circle C_j with radius small enough that those k circles and C are all separated (Fig. 362). Then

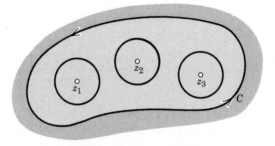

Fig. 362. Residue theorem

$f(z)$ is analytic in the multiply connected domain D bounded by C and $C_1, \cdots, C_k$ and on the entire boundary of D. From Cauchy's integral theorem we have

$$(2) \quad \oint_C f(z)\, dz + \oint_{C_1} f(z)\, dz + \oint_{C_2} f(z)\, dz + \cdots + \oint_{C_k} f(z)\, dz = 0,$$

the integral along C being taken in the counterclockwise sense and the other integrals in the clockwise sense (cf. Sec. 13.3). We now reverse the sense of integration along $C_1, \cdots, C_k$. Then the signs of the values of these integrals change, and we obtain from (2)

$$(3) \quad \oint_C f(z)\, dz = \oint_{C_1} f(z)\, dz + \oint_{C_2} f(z)\, dz + \cdots + \oint_{C_k} f(z)\, dz.$$

All these integrals are now taken in the counterclockwise sense. By (1) in the previous section,

$$\oint_{C_j} f(z)\, dz = 2\pi i \operatorname*{Res}_{z=z_j} f(z),$$

so that (3) yields (1), and the theorem is proved.   ∎

   This important theorem has various applications in connection with complex and real integrals. We shall first consider some complex integrals.

EXAMPLE 1. Integration by the residue theorem

Evaluate the following integral counterclockwise around any simple closed path such that (a) 0 and 1 are inside C, (b) 0 is inside, 1 outside, (c) 1 is inside, 0 outside, (d) 0 and 1 are outside.

$$\oint_C \frac{4 - 3z}{z^2 - z}\, dz$$

Solution. The integrand has simple poles at 0 and 1, with residues [by (3), Sec. 15.1]

$$\operatorname*{Res}_{z=0} \frac{4 - 3z}{z(z - 1)} = \left[\frac{4 - 3z}{z - 1} \right]_{z=0} = -4, \qquad \operatorname*{Res}_{z=1} \frac{4 - 3z}{z(z - 1)} = \left[\frac{4 - 3z}{z} \right]_{z=1} = 1.$$

[Confirm this by (4), Sec. 15.1.] *Ans.* (a) $2\pi i(-4 + 1) = -6\pi i$, (b) $-8\pi i$, (c) $2\pi i$, (d) 0.

EXAMPLE 2. Poles and essential singularities

Evaluate the following integral, where C is the ellipse $9x^2 + y^2 = 9$ (counterclockwise).

$$\oint_C \left(\frac{z e^{\pi z}}{z^4 - 16} + z e^{\pi/z} \right) dz$$

Solution. Since $z^4 - 16 = 0$ at $\pm 2i$ and ± 2, the first term of the integrand has simple poles at $\pm 2i$ inside C, with residues [by (4), Sec. 15.1; note that $e^{2\pi i} = 1$]

$$\operatorname*{Res}_{z=2i} \frac{z e^{\pi z}}{z^4 - 16} = \left[\frac{z e^{\pi z}}{4z^3} \right]_{z=2i} = -\frac{1}{16}, \qquad \operatorname*{Res}_{z=-2i} \frac{z e^{\pi z}}{z^4 - 16} = \left[\frac{z e^{\pi z}}{4z^3} \right]_{z=-2i} = -\frac{1}{16}$$

and simple poles at ± 2, which lie outside C, so that they are of no interest here. The second term of the integrand has an essential singularity at 0, with residue $\pi^2/2$ as obtained from

$$ze^{\pi/z} = z\left(1 + \frac{\pi}{z} + \frac{\pi^2}{2!z^2} + \frac{\pi^3}{3!z^3} + \cdots\right) = z + \pi + \frac{\pi^2}{2}\cdot\frac{1}{z} + \cdots .$$

Ans. $2\pi i(-1/16 - 1/16 + \pi^2/2) = \pi(\pi^2 - 1/4)i = 30.221i$ by the residue theorem.

EXAMPLE 3. Confirmation of an earlier result

Integrate $1/(z - z_0)^m$ (m a positive integer) in the counterclockwise sense around any simple closed path C enclosing the point $z = z_0$.

Solution. $1/(z - z_0)^m$ is its own Laurent series with center $z = z_0$ consisting of this one-term principal part, and

$$\operatorname*{Res}_{z=z_0} \frac{1}{z - z_0} = 1, \qquad \operatorname*{Res}_{z=z_0} \frac{1}{(z - z_0)^m} = 0 \qquad (m = 2, 3, \cdots).$$

In agreement with Example 2, Sec. 13.2, we thus obtain

$$\oint_C \frac{dz}{(z - z_0)^m} = \begin{cases} 2\pi i & \text{if } m = 1 \\ 0 & \text{if } m = 2, 3, \cdots. \end{cases} \qquad \blacksquare$$

It should be very surprising to hear that our present *complex* integration method can be used for evaluating **real integrals** (incidentally, some of them difficult to evaluate by other methods). In the next section we discuss two methods for accomplishing this goal.

Problems for Sec. 15.2

Integrate $\dfrac{3 + 5z}{9z - z^3}$ around the following paths C in the counterclockwise sense.

1. $|z - 1| = 3$ **2.** $|z - 2 + i| = 8$ **3.** $x^2 + 4y^2 = 4$

Integrate $\dfrac{2z + 2}{z(z - 1)(z - 2)}$ in the clockwise sense around the path

4. $|z - i| = 2$ **5.** $|z - 4| = \frac{5}{2}$ **6.** $|z - \frac{3}{2}| = 1$

Evaluate the following integrals, where C is any simple closed path such that all the singularities lie inside C (counterclockwise).

7. $\oint_C \dfrac{z}{4z^2 - 1}\, dz$ **8.** $\oint_C \dfrac{3z}{3z - 1}\, dz$ **9.** $\oint_C \dfrac{3}{z^2 - 2z}$

10. $\oint_C \dfrac{z + 1}{4z^3 - z}\, dz$ **11.** $\oint_C \dfrac{5z}{z^2 + 4}\, dz$ **12.** $\oint_C \dfrac{z + e^z}{z^3 - z}\, dz$

13. $\oint_C \dfrac{\sinh z}{2z - i}\, dz$ **14.** $\oint_C \dfrac{z^2 \sin z}{4z^2 - 1}\, dz$ **15.** $\oint_C \dfrac{z \cosh \pi z}{z^4 + 13z^2 + 36}\, dz$

Evaluate the following integrals, where C is the unit circle (counterclockwise).

16. $\oint_C \dfrac{z + 1}{z^4 - 2z^3}\, dz$ **17.** $\oint_C \tan \pi z\, dz$

18. $\oint_C \dfrac{(z + 4)^3}{z^4 + 5z^3 + 6z^2}\, dz$ **19.** $\oint_C \dfrac{\tan \pi z}{z^3}\, dz$

20. $\oint_C \cot z\, dz$ **21.** $\oint_C \dfrac{6z^2 - 4z + 1}{(z - 2)(1 + 4z^2)}\, dz$

22. $\oint_C \coth z \, dz$

23. $\oint_C \dfrac{\cot z}{z} \, dz$

24. $\oint_C \dfrac{z^5 - 3z^3 + 1}{(2z + 1)(z^2 + 4)} \, dz$

25. $\oint_C \dfrac{z}{1 + 9z^2} \, dz$

26. $\oint_C \dfrac{e^z}{\cos \pi z} \, dz$

27. $\oint_C \dfrac{e^{(z-i)\pi/2}}{\sin \pi z} \, dz$

28. $\oint_C \dfrac{e^{z^2}}{\cos \pi z} \, dz$

29. $\oint_C \dfrac{\cosh z}{z^2 - 3iz} \, dz$

30. $\oint_C \dfrac{30z^2 - 23z + 5}{(2z - 1)^2(3z - 1)} \, dz$

15.3 Evaluation of Real Integrals

We want to show that the residue theorem also yields a very elegant and simple method for evaluating certain classes of complicated *real* integrals.

Integrals of Rational Functions of cos θ and sin θ

We first consider integrals of the type

$$(1) \qquad I = \int_0^{2\pi} F(\cos \theta, \sin \theta) \, d\theta$$

where $F(\cos \theta, \sin \theta)$ is a real rational function of $\cos \theta$ and $\sin \theta$ [for example, $(\sin^2 \theta)/(5 - 4 \cos \theta)$] and is finite on the interval of integration. Setting $e^{i\theta} = z$, we obtain

$$(2) \qquad \boxed{\begin{aligned} \cos \theta &= \frac{1}{2}(e^{i\theta} + e^{-i\theta}) = \frac{1}{2}\left(z + \frac{1}{z}\right) \\ \sin \theta &= \frac{1}{2i}(e^{i\theta} - e^{-i\theta}) = \frac{1}{2i}\left(z - \frac{1}{z}\right) \end{aligned}}$$

and we see that the integrand becomes a rational function of z, say, $f(z)$. As θ ranges from 0 to 2π, the variable z ranges once around the unit circle $|z| = 1$ in the counterclockwise sense. Since $dz/d\theta = ie^{i\theta}$, we have $d\theta = dz/iz$, and the given integral takes the form

$$(3) \qquad I = \oint_C f(z) \frac{dz}{iz},$$

the integration being taken counterclockwise around the unit circle.

EXAMPLE 1. An integral of the type (1)
Show by the present method that

$$\int_0^{2\pi} \frac{d\theta}{\sqrt{2} - \cos \theta} = 2\pi.$$

Solution. We use $\cos \theta = (z + 1/z)/2$ and $d\theta = dz/iz$. Then the integral becomes

$$\oint_C \frac{dz/iz}{\sqrt{2} - \frac{1}{2}\left(z + \frac{1}{z}\right)} = \oint_C \frac{dz}{-\frac{i}{2}(z^2 - 2\sqrt{2}z + 1)} = -\frac{2}{i} \oint_C \frac{dz}{(z - \sqrt{2} - 1)(z - \sqrt{2} + 1)}.$$

We see that the integrand has two simple poles, one at $z_1 = \sqrt{2} + 1$, which lies outside the unit circle $C: |z| = 1$ and is thus of no interest, and the other at $z_2 = \sqrt{2} - 1$ inside C, where the residue is (by (3) in Sec. 15.1)

$$\operatorname*{Res}_{z=z_2} \frac{1}{(z - \sqrt{2} - 1)(z - \sqrt{2} + 1)} = \left[\frac{1}{z - \sqrt{2} - 1}\right]_{z=\sqrt{2}-1} = -\frac{1}{2}.$$

Together with the factor $-2/i$ in front of the integral this yields the desired result $2\pi i(-2/i)(-1/2) = 2\pi$. ∎

Improper Integrals of Rational Functions

We now consider real integrals of the type

$$(4) \qquad \int_{-\infty}^{\infty} f(x)\, dx.$$

Such an integral, for which the interval of integration is not finite, is called an **improper integral,** and it has the meaning

$$(5') \qquad \int_{-\infty}^{\infty} f(x)\, dx = \lim_{a\to-\infty} \int_{a}^{0} f(x)\, dx + \lim_{b\to\infty} \int_{0}^{b} f(x)\, dx.$$

If both limits exist, we may couple the two independent passages to $-\infty$ and ∞, and write[1]

$$(5) \qquad \int_{-\infty}^{\infty} f(x)\, dx = \lim_{R\to\infty} \int_{-R}^{R} f(x)\, dx.$$

We assume that the function $f(x)$ in (4) is a real rational function whose denominator is different from zero for all real x and is of degree at least two units higher than the degree of the numerator. Then the limits in $(5')$ exist, and we may start from (5). We consider the corresponding contour integral

$$(5^*) \qquad \oint_C f(z)\, dz$$

around a path C in Fig. 363 on p. 870. Since $f(x)$ is rational, $f(z)$ has finitely many poles in the upper half-plane, and if we choose R large enough, then

---

[1]The expression on the right-hand side of (4) is called the **Cauchy principal value** of the integral; it may exist even if the limits in $(5')$ do not exist. For instance,

$$\lim_{R\to\infty} \int_{-R}^{R} x\, dx = \lim_{R\to\infty} \left(\frac{R^2}{2} - \frac{R^2}{2}\right) = 0, \quad \text{but} \quad \lim_{b\to\infty} \int_{0}^{b} x\, dx = \infty.$$

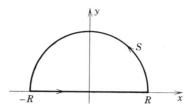

Fig. 363. Path C of the contour integral in (5^*)

C encloses all these poles. By the residue theorem we then obtain

$$\oint_C f(z)\, dz = \int_S f(z)\, dz + \int_{-R}^{R} f(x)\, dx = 2\pi i \sum \operatorname{Res} f(z)$$

where the sum consists of all the residues of $f(z)$ at the points in the upper half-plane at which $f(z)$ has a pole. From this we have

(6) $$\int_{-R}^{R} f(x)\, dx = 2\pi i \sum \operatorname{Res} f(z) - \int_S f(z)\, dz.$$

We prove that, if $R \to \infty$, the value of the integral over the semicircle S approaches zero. If we set $z = Re^{i\theta}$, then S is represented by $R = const$, and as z ranges along S, the variable θ ranges from 0 to π. Since, by assumption, the degree of the denominator of $f(z)$ is at least two units higher than the degree of the numerator, we have

$$|f(z)| < \frac{k}{|z|^2} \qquad\qquad (|z| = R > R_0)$$

for sufficiently large constants k and R_0. By the *ML*-inequality in Sec. 13.2,

$$\left| \int_S f(z)\, dz \right| < \frac{k}{R^2}\, \pi R = \frac{k\pi}{R} \qquad\qquad (R > R_0).$$

Hence, as R approaches infinity, the value of the integral over S approaches zero, and (5) and (6) yield the result

(7) $$\boxed{\int_{-\infty}^{\infty} f(x)\, dx = 2\pi i \sum \operatorname{Res} f(z),}$$

the sum being extended over all the residues of $f(z)$ corresponding to the poles of $f(z)$ in the upper half-plane.

EXAMPLE 2. An improper integral from 0 to ∞
Using (7), show that

$$\int_0^{\infty} \frac{dx}{1 + x^4} = \frac{\pi}{2\sqrt{2}}.$$

Solution. Indeed, $f(z) = 1/(1 + z^4)$ has four simple poles at the points

$$z_1 = e^{\pi i/4}, \qquad z_2 = e^{3\pi i/4}, \qquad z_3 = e^{-3\pi i/4}, \qquad z_4 = e^{-\pi i/4}.$$

The first two of these poles lie in the upper half-plane (Fig. 364). From (4) in Sec. 15.1 we find

$$\operatorname*{Res}_{z=z_1} f(z) = \left[\frac{1}{(1 + z^4)'} \right]_{z=z_1} = \left[\frac{1}{4z^3} \right]_{z=z_1} = \frac{1}{4} e^{-3\pi i/4} = -\frac{1}{4} e^{\pi i/4},$$

$$\operatorname*{Res}_{z=z_2} f(z) = \left[\frac{1}{(1 + z^4)'} \right]_{z=z_2} = \left[\frac{1}{4z^3} \right]_{z=z_2} = \frac{1}{4} e^{-9\pi i/4} = \frac{1}{4} e^{-\pi i/4}.$$

By (1) in Sec. 12.7 and (7) in the current section,

$$\int_{-\infty}^{\infty} \frac{dx}{1 + x^4} = \frac{2\pi i}{4} (-e^{\pi i/4} + e^{-\pi i/4}) = \pi \sin \frac{\pi}{4} = \frac{\pi}{\sqrt{2}} .$$

Since $1/(1 + x^4)$ is an even function, we thus obtain, as asserted,

$$\int_{0}^{\infty} \frac{dx}{1 + x^4} = \frac{1}{2} \int_{-\infty}^{\infty} \frac{dx}{1 + x^4} = \frac{\pi}{2\sqrt{2}} .$$

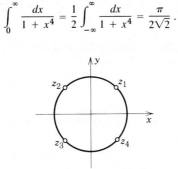

Fig. 364. Example 2

EXAMPLE 3. Another improper integral

Using (7), show that

$$\int_{-\infty}^{\infty} \frac{x^2 - 1}{x^4 + 5x^2 + 4} \, dx = \frac{\pi}{6} .$$

Solution. The degree of the denominator is two units higher than that of the numerator, so that our method again applies. Now

$$f(z) = \frac{p(z)}{q(z)} = \frac{z^2 - 1}{z^4 + 5z^2 + 4} = \frac{z^2 - 1}{(z^2 + 4)(z^2 + 1)}$$

has simple poles at $2i$ and i in the upper half-plane (and at $-2i$ and $-i$ in the lower half-plane, which are of no interest here). We calculate the residues from (4), Sec. 15.1, noting that $q'(z) = 4z^3 + 10z$,

$$\operatorname*{Res}_{z=2i} f(z) = \left[\frac{z^2 - 1}{4z^3 + 10z} \right]_{z=2i} = \frac{5}{12i} , \qquad \operatorname*{Res}_{z=i} f(z) = \left[\frac{z^2 - 1}{4z^3 + 10z} \right]_{z=i} = \frac{-2}{6i} .$$

Ans. $2\pi i(5/12i - 1/3i) = \pi/6$, as asserted.   ∎

Looking back, we realize that the key ideas of our present methods were these. In the first method we mapped the interval of integration on the real axis onto a closed curve in the complex plane (the unit circle). In the second method we attached to an interval on the real axis a semicircle such that we got a closed curve in the complex plane, which we then "blew up." This second method can be applied to further types of integrals, as we show in the next section, the last in this chapter.

Problems for Sec. 15.3

Evaluate the following integrals involving cosine and sine.

1. $\displaystyle\int_0^{2\pi} \frac{d\theta}{2 + \cos\theta}$

2. $\displaystyle\int_0^{2\pi} \frac{d\theta}{25 - 24\cos\theta}$

3. $\displaystyle\int_0^{\pi} \frac{d\theta}{k + \cos\theta}$ $(k > 1)$

4. $\displaystyle\int_0^{2\pi} \frac{d\theta}{5/4 - \sin\theta}$

5. $\displaystyle\int_0^{2\pi} \frac{1 + \sin\theta}{3 + \cos\theta}\,d\theta$

6. $\displaystyle\int_0^{2\pi} \frac{\cos\theta}{17 - 8\cos\theta}\,d\theta$

7. $\displaystyle\int_0^{2\pi} \frac{1 + 4\cos\theta}{17 - 8\cos\theta}\,d\theta$

8. $\displaystyle\int_0^{2\pi} \frac{2\sin^2\theta}{5 - 4\cos\theta}\,d\theta$

9. $\displaystyle\int_0^{2\pi} \frac{\cos\theta}{4 + \frac{4}{3}\sin\theta}\,d\theta$

10. $\displaystyle\int_0^{2\pi} \frac{\cos^2 3\theta}{5 - 4\cos 2\theta}\,d\theta$

11. $\displaystyle\int_0^{2\pi} \frac{\cos^2\theta}{26 - 10\cos 2\theta}\,d\theta$

12. $\displaystyle\int_0^{2\pi} \frac{\sin^2\theta - 2\cos\theta}{2 + \cos\theta}\,d\theta$

13. Using the present method and the binomial theorem applied to $(z + z^{-1})^{2n}$, show that

$$\int_0^{2\pi} \cos^{2n}\theta\,d\theta = \frac{(2n)!\pi}{2^{2n-1}(n!)^2}, \qquad n = 0, 1, 2, \cdots.$$

14. Show directly, without using the result of Prob. 13, that

$$\int_0^{2\pi} \sin^{2n}\theta\,d\theta = \frac{(2n)!}{2^{2n-1}(n!)^2}, \qquad n = 0, 1, 2, \cdots.$$

15. How can you obtain the result in Prob. 14 from that in Prob. 13?

16. Show that

$$\int_0^{2\pi} (1 - \cos\theta)^n \cos n\theta\,d\theta = \frac{(-1)^n\pi}{2^{n-1}}, \quad n = 0, 1, 2, \cdots.$$

Evaluate the following improper integrals.

17. $\displaystyle\int_{-\infty}^{\infty} \frac{dx}{1 + x^2}$

18. $\displaystyle\int_{-\infty}^{\infty} \frac{dx}{x^4 + 16}$

19. $\displaystyle\int_{-\infty}^{\infty} \frac{x}{(x^2 - 2x + 2)^2}\,dx$

20. $\displaystyle\int_{-\infty}^{\infty} \frac{dx}{(1 + x^2)^2}$

21. $\displaystyle\int_{-\infty}^{\infty} \frac{dx}{1 + x^6}$

22. $\displaystyle\int_{-\infty}^{\infty} \frac{x^2}{(x^2 + 1)(x^2 + 4)}\,dx$

23. $\displaystyle\int_{-\infty}^{\infty} \frac{dx}{(1 + x^2)^3}$

24. $\displaystyle\int_{-\infty}^{\infty} \frac{x}{(4 + x^2)^2}\,dx$

25. $\displaystyle\int_{-\infty}^{\infty} \frac{dx}{(x^2 + 1)(x^2 + 9)}$

26. $\displaystyle\int_0^{\infty} \frac{1 + x^2}{1 + x^4}\,dx$

27. $\displaystyle\int_{-\infty}^{\infty} \frac{x^3}{1 + x^8}\,dx$

28. $\displaystyle\int_{-\infty}^{\infty} \frac{dx}{(x^2 + 1)(x^2 + 4)^2}$

29. Solve Prob. 17 by elementary methods.

30. Obtain the answers to Probs. 24 and 27 without calculation.

15.4 Further Types of Real Integrals

There are further classes of real integrals that can be evaluated by applying the residue theorem to suitable complex integrals. In applications such integrals may arise in connection with integral transformations or representations of special functions. In the present section we shall consider two such classes of integrals. One of them is important in problems involving the Fourier integral representation (Sec. 10.9). The other class consists of real integrals whose integrand is infinite at some point in the interval of integration.

Fourier Integrals

Real integrals of the form

(1) $$\int_{-\infty}^{\infty} f(x) \cos sx \, dx \quad \text{and} \quad \int_{-\infty}^{\infty} f(x) \sin sx \, dx \qquad (s \text{ real})$$

occur in connection with the Fourier integral (cf. Sec. 10.9).

If $f(x)$ is a rational function satisfying the assumptions on the degree stated in connection with (4), Sec. 15.3, then the integrals (1) may be evaluated in a way similar to that used for the integral in (4) of the previous section. In fact, we may then consider the corresponding integral

$$\oint_C f(z) \, e^{isz} \, dz \qquad (s \text{ real and positive})$$

over the contour C in Fig. 363 (Sec. 15.3). Instead of (7), Sec. 15.3, we get

(2) $$\int_{-\infty}^{\infty} f(x) \, e^{isx} \, dx = 2\pi i \sum \text{Res} \, [f(z) \, e^{isz}] \qquad (s > 0)$$

where the sum consists of the residues of $f(z)e^{isz}$ at its poles in the upper half-plane. Equating the real and the imaginary parts on both sides of (2), we have

(3) $$\int_{-\infty}^{\infty} f(x) \cos sx \, dx = -2\pi \sum \text{Im Res} \, [f(z)e^{isz}],$$
$$\int_{-\infty}^{\infty} f(x) \sin sx \, dx = 2\pi \sum \text{Re Res} \, [f(z)e^{isz}].$$
$$(s > 0)$$

We remember that (7), Sec. 15.3, was established by proving that the value of the integral over the semicircle S in Fig. 363 approaches zero as $R \to \infty$. To establish (2) we should now prove the same fact for our present contour integral. This can be done as follows. Since S lies in the upper half-plane $y \geq 0$ and $s > 0$, we see that

$$|e^{isz}| = |e^{isx}| |e^{-sy}| = e^{-sy} \leq 1 \qquad (s > 0, \quad y \geq 0).$$

From this we obtain the inequality

$$|f(z) \, e^{isz}| = |f(z)| \, |e^{isz}| \leq |f(z)| \qquad (s > 0, \quad y \geq 0)$$

which reduces our present problem to that in the previous section. Continuing as before, we see that the value of the integral under consideration approaches zero as R approaches infinity. This establishes (2), which implies (3).

EXAMPLE 1. An application of (3)

Show that

$$\int_{-\infty}^{\infty} \frac{\cos sx}{k^2 + x^2} \, dx = \frac{\pi}{k} \, e^{-ks}, \qquad \int_{-\infty}^{\infty} \frac{\sin sx}{k^2 + x^2} \, dx = 0 \qquad (s > 0, \quad k > 0).$$

Solution. In fact, $e^{isz}/(k^2 + z^2)$ has only one pole in the upper half-plane, namely, a simple pole at $z = ik$, and from (4) in Sec. 15.1 we obtain

$$\operatorname*{Res}_{z=ik} \frac{e^{isz}}{k^2 + z^2} = \left[\frac{e^{isz}}{2z} \right]_{z=ik} = \frac{e^{-ks}}{2ik}.$$

Therefore,

$$\int_{-\infty}^{\infty} \frac{e^{isx}}{k^2 + x^2} \, dx = 2\pi i \frac{e^{-ks}}{2ik} = \frac{\pi}{k} \, e^{-ks}.$$

Since $e^{isx} = \cos sx + i \sin sx$, this yields the above results [see also (15) in Sec. 10.9.]   ∎

Other Types of Real Improper Integrals

Another kind of improper integral is a definite integral

$$(4) \qquad \int_A^B f(x) \, dx$$

whose integrand becomes infinite at a point a in the interval of integration,

$$\lim_{x \to a} |f(x)| = \infty.$$

Then the integral (4) means

$$(5) \qquad \int_A^B f(x) \, dx = \lim_{\epsilon \to 0} \int_A^{a-\epsilon} f(x) \, dx + \lim_{\eta \to 0} \int_{a+\eta}^B f(x) \, dx$$

where both ϵ and η approach zero independently and through positive values. It may happen that neither of these limits exists if $\epsilon, \eta \to 0$ independently, but

$$(6) \qquad \lim_{\epsilon \to 0} \left[\int_A^{a-\epsilon} f(x) \, dx + \int_{a+\epsilon}^B f(x) \, dx \right]$$

exists. This is called the **Cauchy principal value** of the integral. It is written

$$\text{pr. v.} \int_A^B f(x)\, dx.$$

For example,

$$\text{pr. v.} \int_{-1}^{1} \frac{dx}{x^3} = \lim_{\epsilon \to 0} \left[\int_{-1}^{-\epsilon} \frac{dx}{x^3} + \int_{\epsilon}^{1} \frac{dx}{x^3} \right] = 0;$$

the principal value exists although the integral itself has no meaning. The whole situation is quite similar to that discussed in the second part of the previous section.

To evaluate improper integrals whose integrands have poles on the real axis, we use a path that avoids these singularities by following small semicircles with centers at the singular points; the procedure may be illustrated by the following example.

EXAMPLE 2. Integrand having a pole on the real axis. Sine integral
Show that

$$\int_0^\infty \frac{\sin x}{x}\, dx = \frac{\pi}{2}.$$

(This is the limit of the sine integral Si(x) as $x \to \infty$; cf. Sec. 10.9.)

Solution. **(a)** We do not consider (sin z)/z because this function does not behave suitably at infinity. We consider e^{iz}/z, which has a simple pole at $z = 0$, and integrate around the contour in Fig. 365. Since e^{iz}/z is analytic inside and on C, Cauchy's integral theorem gives

(7)
$$\oint_C \frac{e^{iz}}{z}\, dz = 0.$$

(b) We prove that the value of the integral over the large semicircle C_1 approaches zero as R approaches infinity. Setting $z = Re^{i\theta}$, we have $dz = iRe^{i\theta}\, d\theta$, $dz/z = i\, d\theta$ and therefore

$$\left| \int_{C_1} \frac{e^{iz}}{z}\, dz \right| = \left| \int_0^\pi e^{iz} i\, d\theta \right| \leq \int_0^\pi |e^{iz}|\, d\theta \qquad (z = Re^{i\theta}).$$

In the integrand on the right,

$$|e^{iz}| = |e^{iR(\cos\theta + i\sin\theta)}| = |e^{iR\cos\theta}| |e^{-R\sin\theta}| = e^{-R\sin\theta}.$$

We insert this, use sin $(\pi - \theta) = \sin\theta$ to get an integral from 0 to $\pi/2$, and then sin $\theta \geq 2\theta/\pi$ (when $0 \leq \theta \leq \pi/2$; see Fig. 366) to get an integral that we can evaluate:

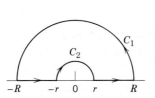

Fig. 365. Contour in Example 2

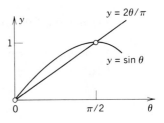

Fig. 366. Inequality in Example 2

$$\int_0^\pi |e^{iz}|\, d\theta = \int_0^\pi e^{-R\sin\theta}\, d\theta = 2\int_0^{\pi/2} e^{-R\sin\theta}\, d\theta$$

$$< 2\int_0^{\pi/2} e^{-2R\theta/\pi}\, d\theta = \frac{\pi}{R}(1 - e^{-R}) \to 0 \qquad \text{as} \qquad R \to \infty.$$

Hence the value of the integral over C_1 approaches 0 as $R \to \infty$.

(c) For the integral over the small semicircle C_2 in Fig. 365 we have

$$\int_{C_2} \frac{e^{iz}}{z}\, dz = \int_{C_2} \frac{dz}{z} + \int_{C_2} \frac{e^{iz} - 1}{z}\, dz.$$

The first integral on the right equals $-\pi i$. The integrand of the second integral is analytic and thus bounded, say, less than some constant M in absolute value for all z on C_2 and between C_2 and the x-axis. Hence by the ML-inequality (Sec. 13.2), the absolute value of this integral cannot exceed $M\pi r$. This approaches 0 as $r \to 0$. Because of part (b), from (7) we thus obtain

$$\int_C \frac{e^{iz}}{z}\, dz = \text{pr. v.} \int_{-\infty}^\infty \frac{e^{ix}}{x}\, dx + \lim_{r \to 0} \int_{C_2} \frac{e^{iz}}{z}\, dz = \text{pr. v.} \int_{-\infty}^\infty \frac{e^{ix}}{x}\, dx - \pi i = 0.$$

Hence this principal value equals πi; its real part is 0 and its imaginary part is

$$(8) \qquad\qquad\qquad \text{pr. v.} \int_{-\infty}^\infty \frac{\sin x}{x}\, dx = \pi.$$

(d) Now the integrand in (8) is not singular at $x = 0$. Furthermore, since for positive x the function $1/x$ decreases, the areas under the curve of the integrand between two consecutive positive zeros decrease in a monotone fashion, that is, the absolute values of the integrals

$$I_n = \int_{n\pi}^{n\pi + \pi} \frac{\sin x}{x}\, dx \qquad\qquad n = 0, 1, \cdots$$

form a monotone decreasing sequence $|I_1|, |I_2|, \cdots$, and $I_n \to 0$ as $n \to \infty$. Since these integrals have alternating sign (why?), it follows from the Leibniz test (in Appendix 3) that the infinite series $I_0 + I_1 + I_2 + \cdots$ converges. Clearly, the sum of the series is the integral

$$\int_0^\infty \frac{\sin x}{x}\, dx = \lim_{b \to \infty} \int_0^b \frac{\sin x}{x}\, dx$$

which therefore exists. Similarly, the integral from 0 to $-\infty$ exists. Hence we need not take the principal value in (8), and

$$\int_{-\infty}^\infty \frac{\sin x}{x}\, dx = \pi.$$

Since the integrand is an even function, the desired result follows.   ∎

In part (c) of Example 2 we avoided the simple pole by integrating along a small semicircle C_2, and then we let C_2 shrink to a point. This process suggests the following.

Theorem 1. Simple poles on the real axis

If $f(z)$ has a simple pole at $z = a$ on the real axis, then (Fig. 367)

$$\lim_{r \to 0} \int_{C_2} f(z)\, dz = \pi i \operatorname*{Res}_{z=a} f(z).$$

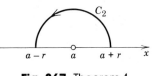

Fig. 367. Theorem 1

Proof. By the definition of a simple pole (Sec. 14.10) the integrand $f(z)$ has at $z = a$ the Laurent series

$$f(z) = \frac{b_1}{z - a} + g(z), \qquad b_1 = \operatorname*{Res}_{z=a} f(z)$$

where $g(z)$ is analytic on the semicircle of integration

$$C_2: \quad z = a + re^{i\theta}, \qquad 0 \leqq \theta = \pi,$$

and for all z between C_2 and the x-axis. By integration,

$$\int_{C_2} f(z)\, dz = \int_0^\pi \frac{b_1}{re^{i\theta}} ire^{i\theta}\, d\theta + \int_{C_2} g(z)\, dz.$$

The first integral on the right equals $b_1 \pi i$. The second cannot exceed $M\pi r$ in absolute value, by the ML-inequality (Sec. 13.2), and $M\pi r \to 0$ as $r \to 0$. ∎

We may combine this theorem with (7) of Sec. 15.3 or (3) in this section. Thus [cf. (7), Sec. 15.3],

(9) $$\text{pr. v.} \int_{-\infty}^{\infty} f(x)\, dx = 2\pi i \sum \operatorname{Res} f(z) + \pi i \sum \operatorname{Res} f(z)$$

(summation over all poles in the upper half-plane in the first sum, and on the x-axis in the second), valid for rational $f(x) = p(x)/q(x)$ with degree $q \geqq$ degree $p + 2$, having simple poles on the x-axis.

This is the end of Chap. 15, which added another powerful general integration method to the methods discussed in the chapter on integration (Chap. 13). Remember that our present residue method is based on Laurent series, which we therefore had to discuss first (in Chap. 14).

In the next chapter we present a systematic discussion of mapping by analytic functions (**"conformal mapping"**). Conformal mapping will then be applied to potential theory in Chap. 17, our last chapter on complex analysis.

Problems for Sec. 15.4

1. Derive (3) from (2).

Evaluate the following real integrals.

2. $\int_{-\infty}^{\infty} \dfrac{\cos x}{(x^2 + 16)^2}\, dx$

3. $\int_{-\infty}^{\infty} \dfrac{\cos 2x}{(x^2 + 4)^2}\, dx$

4. $\displaystyle\int_{-\infty}^{\infty} \frac{\cos 4x}{(x^2 + 1)(x^2 + 4)}\, dx$

5. $\displaystyle\int_{0}^{\infty} \frac{\cos sx}{x^2 + 1}\, dx$

6. $\displaystyle\int_{-\infty}^{\infty} \frac{\sin 3x}{1 + x^4}\, dx$

7. $\displaystyle\int_{-\infty}^{\infty} \frac{\cos x}{x^4 + 13x^2 + 36}\, dx$

8. $\displaystyle\int_{-\infty}^{\infty} \frac{\cos x}{x^4 + 1}\, dx$

9. $\displaystyle\int_{-\infty}^{\infty} \frac{\sin 2x}{x^2 + x + 1}\, dx$

10. $\displaystyle\int_{0}^{\infty} \frac{\cos 2x}{4x^4 + 13x^2 + 9}\, dx$

11. Evaluate $\displaystyle\int_{-\infty}^{\infty} (x^2 + 1)^{-1} \cos ax\, dx$ and obtain from the result the first result in Example 1 by a suitable transformation.

12. Obtain the answer to Prob. 3 from that of Prob. 2 by a suitable transformation of the variable of integration.

13. Integrating e^{-z^2} around the boundary of the rectangle with vertices $-a$, a, $a + ib$, $-a + ib$, letting $a \to \infty$, and using

$$\int_{0}^{\infty} e^{-x^2}\, dx = \frac{\sqrt{\pi}}{2}, \qquad \text{show that} \qquad \int_{0}^{\infty} e^{-x^2} \cos 2bx\, dx = \frac{\sqrt{\pi}}{2} e^{-b^2}.$$

Poles on the real axis. Find the Cauchy principal value of the following integrals.

14. $\displaystyle\int_{-\infty}^{\infty} \frac{dx}{(x + 1)(x^2 + 2)}$ (Fig. 368)

15. $\displaystyle\int_{-\infty}^{\infty} \frac{dx}{x^2 - ix}$

16. $\displaystyle\int_{-\infty}^{\infty} \frac{dx}{x(x - 2 - i)}$

17. $\displaystyle\int_{-\infty}^{\infty} \frac{x}{8 - x^3}\, dx$

18. $\displaystyle\int_{-\infty}^{\infty} \frac{dx}{x^4 - 1}$

19. $\displaystyle\int_{-\infty}^{\infty} \frac{dx}{(x^2 + 1)(x - 1)}$

20. $\displaystyle\int_{-\infty}^{\infty} \frac{\cos \frac{1}{2}\pi x}{x^2 - 1}\, dx$

21. $\displaystyle\int_{-\infty}^{\infty} \frac{\sin \frac{1}{4}\pi x}{2x - x^2}\, dx$

22. $\displaystyle\int_{-\infty}^{\infty} \frac{\sin \pi x}{x - x^5}\, dx$

23. $\displaystyle\int_{-\infty}^{\infty} \frac{x \sin \pi x}{(x^2 + 1)(x^4 - 1)}\, dx$ (*Hint.* Note that the pole at $z = i$ is of second order.)

24. Show that the result in Example 2 remains the same if we replace the upper semicircle C_2 by the corresponding lower semicircle.

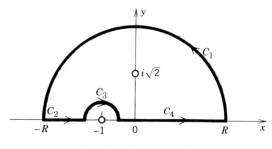

Fig. 368. Problem 14

Review Problems for Chapter 15

1. Without looking into the text, explain briefly the idea of residue integration as applied to complex functions.

2. Why is it important in residue integration to use that Laurent series which converges in a neighborhood of the singular point z_0 (except at z_0 itself)?

3. In which cases does $f(z)$ have only one Laurent series with given center z_0? (Also give an example.)

4. Can we use residue integration for evaluating the integral of tan $(1/z)$ around a closed contour containing $z = 0$ in its interior?

5. If we want to use residue integration, is it *necessary* to first derive some Laurent series? Is it sometimes *advisable* to do so?

6. Can we apply residue integration in the case of a function that is not analytic but merely continuous?

7. If $f(z)$ has a pole of first order at z_0, can its residue at z_0 be zero? Answer the same question for a second-order pole.

8. Without looking into the text, describe briefly the idea of applying residue integration to real integrals whose integrands are rational functions of cosine and sine.

9. Perform the same task as in Prob. 8, for integrals from $-\infty$ to ∞.

10. What is the Cauchy principal value of an integral, and why did it occur in the present chapter?

Integrate the given function over the given path C, using residue integration or one of the methods discussed in Chap. 13, and indicate whether residue integration can be used.

11. e^{4z}/z^2, C the circle $|z| = 2$, counterclockwise

12. $(\cosh \pi z)/z^3$, C the unit circle, counterclockwise

13. $1/(1 + z^2)$, C the ellipse $25x^2 + y^2 = 25$, clockwise

14. $z/|z|^2$, C the unit circle, counterclockwise

15. $(3z^2 - 1)/(z^3 - z)$, C the circle $|z - 1/2| = 1$, counterclockwise

16. $\cot 8z$, C the circle $|z| = 0.3$, counterclockwise

17. $z \sinh 3z^2$, C the quarter-circle $|z| = \sqrt{\pi}$ from $\sqrt{\pi}$ to $i\sqrt{\pi}$

18. $(3z + 4)/z^2$, C the ellipse $4x^2 + 9y^2 = 36$, clockwise

19. $(z + 1) \cos z$, C: from $z = 3$ to $z = 1$ counterclockwise along $|z - 2| = 1$

20. $(z - \pi)^{-2} \sin z$, C the circle $|z - 4| = 1$, counterclockwise

21. $(z - \pi/4)^{-3} \cos 8z$, C the unit circle, counterclockwise

22. $(\text{Re } z + \text{Im } z)/z$, C the unit circle, counterclockwise

23. $(4z^3 + 7z)/\cos z$, C the circle $|z + 1| = 1$, counterclockwise

24. $z^3 \exp z^4$, C any path from $1 + i$ to 1

25. $e^{4/z}$, C the unit circle, counterclockwise

Evaluate by the methods in this chapter:

26. $\displaystyle\int_0^{2\pi} \frac{d\theta}{13 - 5 \sin \theta}$

27. $\displaystyle\int_0^{2\pi} \frac{d\theta}{1 + \frac{1}{2} \cos \theta}$

28. $\displaystyle\int_0^{2\pi} \frac{d\theta}{4 + \sin \theta}$

29. $\displaystyle\int_0^{2\pi} \frac{d\theta}{2\pi + \cos \theta}$

30. $\displaystyle\int_0^{2\pi} \frac{d\theta}{5 - 3 \cos \theta}$

31. $\displaystyle\int_0^{2\pi} \frac{d\theta}{37 - 12 \cos \theta}$

32. $\int_0^{2\pi} \dfrac{\sin^2 3\theta}{1 - \frac{4}{5}\sin 2\theta}\, d\theta$ **33.** $\int_0^{2\pi} \dfrac{1 + 4\sin\theta}{34 - 16\sin\theta}\, d\theta$ **34.** $\int_0^{\infty} \dfrac{dx}{1 + 4x^2}$

35. $\int_{-\infty}^{\infty} \dfrac{dx}{1 + 4x^4}$ **36.** $\int_{-\infty}^{\infty} \dfrac{x}{(1 + x^2)^2}\, dx$ **37.** $\int_{-\infty}^{\infty} \dfrac{x^2}{x^4 + 5x^2 + 4}\, dx$

38. $\int_0^{\infty} \dfrac{dx}{1 + 8x^6}$ **39.** $\int_{-\infty}^{\infty} \dfrac{dx}{1 + k^2 x^2}$, $k > 0$ **40.** $\int_0^{\infty} \dfrac{1 + 2x^2}{1 + 4x^4}\, dx$

Summary of Chapter 15
Residue Integration Method

The **residue** of an analytic function $f(z)$ at a point $z = z_0$ is the coefficient b_1 of the power $1/(z - z_0)$ in that Laurent series

$$f(z) = a_0 + a_1(z - z_0) + \cdots + \frac{b_1}{z - z_0} + \frac{b_2}{(z - z_0)^2} + \cdots$$

of $f(z)$ which converges near z_0 (except at z_0 itself). This residue is given by the integral (Sec. 15.1)

$$(1) \qquad\qquad b_1 = \frac{1}{2\pi i} \oint_C f(z)\, dz$$

but can be obtained in various other ways, so that one can use (1) for evaluating integrals over closed curves. More generally, the **residue theorem** (Sec. 15.2) states that if $f(z)$ is analytic in a domain D except at finitely many points z_j and C is a simple closed path in D such that no z_j lies on C and the full interior of C belongs to D, then

$$(2) \qquad\qquad \oint_C f(z)\, dz = \frac{1}{2\pi i} \sum_j \operatorname*{Res}_{z=z_j} f(z)$$

(summation only over those z_j that lie *inside* C).

This integration method is elegant and powerful. Formulas for the residue at **poles** are (m = order of the pole)

$$(3) \quad \operatorname*{Res}_{z=z_0} f(z) = \frac{1}{(m-1)!} \lim_{z \to z_0} \left(\frac{d^{m-1}}{dz^{m-1}} [(z - z_0)^m f(z)] \right), \quad m = 1, 2 \cdots$$

(cf. Sec. 15.1). Hence for a simple pole ($m = 1$),

$$(3^*) \qquad\qquad \operatorname*{Res}_{z=z_0} f(z) = \lim_{z \to z_0} (z - z_0) f(z).$$

Another formula for the case of a simple pole of $f(z) = p(z)/q(z)$ is

(3**) $$\operatorname*{Res}_{z=z_0} f(z) = \frac{p(z_0)}{q'(z_0)}.$$

Residue integration involves *closed* curves, but the real interval of integration $0 \leqq \theta \leqq 2\pi$ is transformed into the unit circle by setting $z = e^{i\theta}$, so that by residue integration we can integrate **real integrals** of the form (Sec. 15.3)

$$\int_0^{2\pi} F(\cos\,\theta,\,\sin\,\theta)\,d\theta,$$

where F is a rational function of $\cos\,\theta$ and $\sin\,\theta$, such as, for instance, $(\sin^2\,\theta)/(5 - 4\cos\,\theta)$, etc.

Another method of integrating *real* integrals by residues is the use of a closed contour consisting of an interval $-R \leqq x \leqq R$ of the real axis and a semicircle $|z| = R$. From the residue theorem, if we let $R \to \infty$, we obtain for rational $f(x) = p(x)/q(x)$ (with $q(x) \neq 0$ and degree $q \geqq$ degree $p + 2$)

$$\int_{-\infty}^{\infty} f(x)\,dx = 2\pi i \sum \operatorname{Res} f(z) \qquad \text{(Sec. 15.3)}$$

$$\int_{-\infty}^{\infty} f(x)\,\cos\,sx\,dx = -2\pi \sum \operatorname{Im} \operatorname{Res}\,[f(z)e^{isz}]$$

$$\text{(Sec. 15.4)}$$

$$\int_{-\infty}^{\infty} f(x)\,\sin\,sx\,dx = 2\pi \sum \operatorname{Re} \operatorname{Res}\,[f(z)e^{isz}]$$

(sum of all residues at poles in the upper half-plane). In Sec. 15.4 we also · nd this method to real integrals whose integrands become infinite . some point in the interval of integration.

Chapter 16

Conformal Mapping

If a complex function $w = f(z)$ is defined in a domain D of the z-plane, then to each point in D there corresponds a point in the w-plane. In this way we have a *mapping* of D onto the range of values of $f(z)$ in the w-plane. This "geometric approach" to complex analysis helps us to "visualize" the nature of a complex function by considering the manner in which the function maps certain curves and regions. We have seen this for special functions in Sec. 12.9, which the reader may want to review before going on.

This chapter concerns a systematic approach to mappings by general *analytic* functions $w = f(z)$. We show (in Sec. 16.1) that such a mapping is **conformal**; that is, it preserves angles in magnitude and sense, except at *"critical points"* (points at which the derivative $f'(z)$ is zero).

Conformal mapping is important in engineering mathematics, since it is a standard method for solving boundary value problems in two-dimensional potential theory by transforming a given complicated region into a simpler one (details in Chap. 17). For this task of mapping, linear fractional transformations (Secs. 16.2, 16.3) play a basic role, and other special functions (Sec. 16.4) can often be used. In the last section (Sec. 16.5) we discuss the idea of the Riemann surface.

Applications to potential theory follow in Chap. 17.

Prerequisites for this chapter: Chap. 12
Sections that may be omitted in a shorter course: Secs. 16.3, 16.4.
References: Appendix 1, Part D.
Answers to problems: Appendix 2.

16.1 Conformal Mapping

A complex-valued function

$$w = f(z) = u(x, y) + iv(x, y) \qquad (z = x + iy)$$

of a complex variable z gives a **mapping** of its domain of definition in the complex z-plane onto[1] its range of values in the complex w-plane. Examples are shown in Sec. 12.9, at which the reader may take a quick look, before going on. The most important mapping property of an *analytic* function is its conformality. This concept is defined as follows.

A mapping in the plane is said to be an *angle-preserving* or **conformal mapping** if it preserves angles between oriented curves in magnitude as well as in sense, that is, if the images of any two intersecting oriented curves, taken with their corresponding orientation, make the same angle of intersection as the curves, both in magnitude and direction. Here the **angle** between two oriented curves is defined to be the angle α ($0 \leqq \alpha \leqq \pi$) between their oriented tangents at the point of intersection (Fig. 369).

We want to show that a mapping $w = f(z)$ is conformal at every point at which $f(z)$ is analytic, except at points at which the derivative $f'(z)$ is zero. Such a point is called a **critical point**. For instance, in the case of $f(z) = z^2$ we have $f'(z) = 2z = 0$ at the point $z = 0$, at which the mapping is not conformal because angles are doubled there (cf. Example 2 in Sec. 12.9).

Theorem 1 (Conformal mapping)
The mapping defined by an analytic function $f(z)$ is conformal, except at critical points, at which the derivative $f'(z)$ is zero.

Proof. The idea is to consider a curve

$$(1) \qquad\qquad C: \quad z(t) = x(t) + iy(t)$$

in the domain of $f(z)$ and to show that the mapping $w = f(z)$ rotates the tangent to C at any point z_0 of C [with $f'(z_0) \neq 0$] through an angle that is independent of C, so that the tangents of *two* curves C_1 and C_2 passing through z_0 (Fig. 369) are rotated through the same angle, so that the images of these curves make the same angle in size and sense as the curves themselves, which means conformality, by definition. The details are as follows.

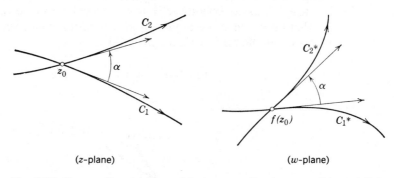

(z-plane)   (w-plane)

Fig. 369. Curves C_1 and C_2 and their respective images $C_1{}^*$ and $C_2{}^*$ under a conformal mapping

[1]The general terminology is as follows. A mapping of a set A into a set B is called **surjective** or a mapping of A **onto** B if every element of B is the image of at least one element of A. The mapping is said to be **injective** or **one-to-one** if different elements of A have different images in B. The mapping is said to be **bijective** if it is both surjective and injective.

We assume that C is a **smooth curve;** that is, $z(t)$ in (1) is differentiable and the derivative $\dot{z}(t) = dz/dt$ is continuous and nowhere zero. We claim that this means that C has a unique continuously turning tangent. In fact, by definition,

$$(2) \qquad \dot{z}(t_0) = \frac{dz}{dt}\bigg|_{t_0} = \lim_{\Delta t \to 0} \frac{z_1 - z_0}{\Delta t} = \lim_{\Delta t \to 0} \frac{z(t_0 + \Delta t) - z(t_0)}{\Delta t}$$

Now the numerator

$$z_1 - z_0 = z(t_0 + \Delta t) - z(t_0)$$

represents a chord of C (Fig. 370), and $(z_1 - z_0)/\Delta t$, where $\Delta t > 0$, has the same direction. As $\Delta t \to 0$, the point z_1 approaches z_0 along the curve, and $(z_1 - z_0)/\Delta t \to \dot{z}(t_0)$. This shows that $\dot{z}(t_0)$ is tangent to C at z_0. We say briefly that $\dot{z}(t_0)$ is a "tangent vector" to C at z_0. Our claim now follows.

Next we *orient* the tangent. We call the sense of increasing t in (1) the *positive sense* on C and on each of its tangents, so that each tangent is now oriented.

Consider now the image C^* of C under $w = f(z)$ (not a constant). C^* is a curve represented by

$$w(t) = f[z(t)].$$

The point $z_0 = z(t_0)$ corresponds to the point $w(t_0)$ of C^*, and $\dot{w}(t_0)$ represents a tangent vector to C^* at this point. Now by the chain rule,

$$(3) \qquad \frac{dw}{dt} = \frac{df}{dz} \frac{dz}{dt}.$$

Hence, if $f'(z_0) \neq 0$, we see that $\dot{w}(t_0) \neq 0$ and C^* has a unique tangent at $w(t_0)$. Now the angle between the tangent vector $\dot{w}(t_0)$ and the positive u-axis is $\arg \dot{w}(t_0)$, as follows directly from the polar form of complex numbers. Furthermore, since the argument of a product equals the sum of the arguments of the factors, we have from (3)

$$\arg \dot{w}(t_0) = \arg f'(z_0) + \arg \dot{z}(t_0).$$

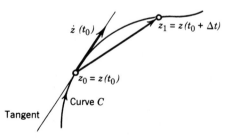

Fig. 370. Derivation of formula (2)

Thus under the mapping the tangent to C at z_0 is rotated through the angle

$$(4) \qquad\qquad \arg \dot{w}(t_0) - \arg \dot{z}(t_0) = \arg f'(z_0),$$

the angle between those two tangent vectors to C and C^*. Note that this angle is well-defined because $f'(z_0) \neq 0$ by assumption. Now comes the point: since the expression on the right is independent of the choice of C, we see that this angle is independent of C. We conclude that the mapping $w = f(z)$ rotates *all* the curves passing through z_0 through the *same* angle $\arg f'(z_0)$. Hence the images C_1^* and C_2^* of any two curves C_1 and C_2 through z_0 make the same angle as the curves themselves, in magnitude as well as in sense. This means conformality of $w = f(z)$ at z_0, and since z_0 was arbitrary in the domain of $f(z)$, the theorem is proved. ∎

EXAMPLE 1. Conformality of $w = z^n$ and $w = e^z$

The mapping $w = z^n$, $n = 2, 3, \cdots$ is conformal except at $z = 0$, where $w' = nz^{n-1} = 0$. For $w = z^2$ this is shown in Fig. 317 (Sec. 12.9), in which the image curves intersect at right angles, except at $z = 0$ where the angles are doubled under the mapping, since every ray $\arg z = c = const$ transforms into a ray $\arg w = 2c$.

The mapping $w = e^z$ is conformal for all z since $w' = e^z$ is not 0 for any z. ∎

Magnification Ratio. By the definition of the derivative we have

$$\lim_{z \to z_0} \left| \frac{f(z) - f(z_0)}{z - z_0} \right| = |f'(z_0)|.$$

Therefore, the mapping $w = f(z)$ magnifies (or shortens) the lengths of short lines by approximately the factor $|f'(z_0)|$. The image of a small figure *conforms* to the original figure in the sense that it has approximately the same shape. However, since $f'(z)$ varies from point to point, a *large* figure may have an image whose shape is quite different from that of the original figure.

More on the Condition $f'(z) \neq 0$. From (4) in Sec. 12.5 and the Cauchy–Riemann equations we obtain

$$(5') \quad |f'(z)|^2 = \left| \frac{\partial u}{\partial x} + i \frac{\partial v}{\partial x} \right|^2 = \left(\frac{\partial u}{\partial x} \right)^2 + \left(\frac{\partial v}{\partial x} \right)^2 = \frac{\partial u}{\partial x} \frac{\partial v}{\partial y} - \frac{\partial u}{\partial y} \frac{\partial v}{\partial x},$$

that is,

$$(5) \qquad\qquad |f'(z)|^2 = \begin{vmatrix} \dfrac{\partial u}{\partial x} & \dfrac{\partial u}{\partial y} \\[2mm] \dfrac{\partial v}{\partial x} & \dfrac{\partial v}{\partial y} \end{vmatrix} = \frac{\partial(u, v)}{\partial(x, y)}.$$

This determinant is the so-called **Jacobian** (cf. Sec. 9.2) of the transformation $w = f(z)$, written in real form

$$u = u(x, y), \qquad v = v(x, y).$$

Hence, the condition $f'(z_0) \neq 0$ implies that the Jacobian is not zero at z_0. This condition suffices that the restriction of the mapping $w = f(z)$ to a sufficiently small neighborhood N_0 of z_0 is **one-to-one** or injective, that is, different points in N_0 have different images; see Ref. [5] in Appendix 1.

EXAMPLE 2

The mapping $w = z^2$ is one-to-one in a sufficiently small neighborhood of any point $z \neq 0$. In a neighborhood of $z = 0$ it is not one-to-one. The full z-plane is mapped onto the w-plane so that each point $w \neq 0$ is the image of two points in the z-plane. For instance, the points $z = 1$ and $z = -1$ are both mapped onto $w = 1$, and, more generally, z_1 and $-z_1$ have the same image point $w = z_1^2$.   ∎

Harmonic functions remain harmonic under conformal mapping by analytic functions. This will be proved, discussed and applied to physical problems in Sec. 17.2, in connection with potential theory. (Harmonic functions were defined in Sec. 12.5.) More specifically, this will make conformal mapping a valuable tool in solving boundary value problems for the Laplace equation. For this, one needs to know what function to choose for mapping a specific given domain onto a simpler domain for which the problem can more easily be solved. This needs experience, and to gain it, we devote the remaining sections of this chapter to a discussion of the mappings by various functions. We begin with the so-called **linear fractional transformations,** which solve the problems of conformally mapping a half-plane onto a half-plane or a disk, and a disk onto a disk or a half-plane.

Problems for Sec. 16.1

1. Why do the images of the curves $|z| = const$ and $\arg z = const$ under a mapping by an analytic function intersect at right angles?

2. Why do the level curves $u = const$ and $v = const$ of an analytic function $w = u + iv = f(z)$ intersect at right angles at each point at which $f'(z) \neq 0$?

3. Does the mapping $w = \bar{z} = x - iy$ preserve angles in size as well as in sense?

Represent the following curves in the z-plane $(z = x + iy)$ in the form $z = z(t)$ and determine the corresponding tangent vector $\dot{z}(t)$.

4. $x^2 + 9y^2 = 9$ 5. $x^2 + y^2 = 25$ 6. $(x - 1)^2 + (y + 2)^2 = 16$

7. $y = 1/x$ 8. $x^2 - y^2 = 4$ 9. $y = ax + b$

Determine the points in the z-plane at which the mapping $w = f(z)$ fails to be conformal, where $f(z)$ equals

10. $\sin z$ 11. $z^2 + az + b$ 12. $\exp(z^5 - 5z)$

13. $z + z^{-1} \, (z \neq 0)$ 14. $\cosh z$ 15. $z^3 + 3z$

Verify (5) for the following functions $f(z) = u(x, y) + iv(x, y)$.

16. e^z 17. $\cos z$ 18. $z^2 - 3z$

19. **(Magnification of angles)** Let the function $f(z)$ be analytic at z_0. Suppose that $f'(z_0) = 0, \cdots, f^{(k-1)}(z_0) = 0$, whereas $f^{(k)}(z_0) \neq 0$. Then the mapping $w = f(z)$ magnifies angles with vertex at z_0 by a factor k. Illustrate this with examples for $k = 2, 3, 4$.

20. Prove the statement in Prob. 19 for arbitrary k. *Hint.* Use the Taylor series.

16.2 Linear Fractional Transformations

Linear fractional transformations (or **Möbius transformations**) are mappings

$$(1) \qquad \boxed{w = \frac{az + b}{cz + d}} \qquad (ad - bc \neq 0)$$

where a, b, c, d are complex or real numbers. These mappings are of practical importance in applications to boundary value problems since they are needed for mapping disks conformally onto half-planes or onto other disks, and conversely, as we shall see. They also provide further motivation of the extended complex plane (Sec. 14.10), which plays a role in fluid flow and other applications.

The condition $ad - bc \neq 0$ in (1) becomes clear if we differentiate:

$$w' = \frac{a(cz + d) - c(az + b)}{(cz + d)^2} = \frac{ad - bc}{(cz + d)^2}.$$

We see that $ad - bc \neq 0$ implies that w' is nowhere zero, hence the mapping (1) is everywhere conformal. From $ad - bc = 0$ we would obtain the uninteresting case that w' is identically zero, which we exclude once and for all. We begin our discussion with special cases of (1).

EXAMPLE 1. Translations, rotations, expansions, contractions

These are special cases of (1) of the form

$$(2) \qquad\qquad w = z + b \qquad\qquad (translation)$$

and

$$(3) \qquad\qquad w = az,$$

which is a *rotation* when $|a| = 1$, say, $a = e^{i\alpha}$ (α the angle of rotation), an *expansion* for real $a > 1$ and a *contraction* for $0 < a < 1$. These are also special cases of the **linear transformation**

$$w = az + b$$

obtained from (1) when $c = 0$ (except for the notation).

EXAMPLE 2. Mapping $w = 1/z$. Inversion

The mapping

$$(4) \qquad\qquad w = \frac{1}{z}$$

is an important special case of (1). It is best studied in terms of polar forms $z = re^{i\theta}$ and $w = Re^{i\phi}$. Then $w = 1/z$ becomes

$$Re^{i\phi} = \frac{1}{re^{i\theta}} \qquad \text{and gives} \qquad R = \frac{1}{r}, \qquad \phi = -\theta$$

by equating absolute values and arguments on both sides. We see from this that the image

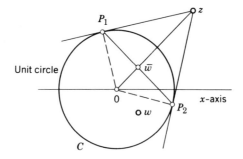

Fig. 371. Geometrical construction of $w = 1/z$. Here, $\overline{w}$ is the intersection of Oz and P_1P_2, where P_1 and P_2 are the points of contact of the tangents to the unit circle passing through z.

$w = 1/z$ of a $z \neq 0$ lies on the ray from the origin through $\overline{z}$, at the distance $1/|z|$. In particular, $z = e^{i\theta}$ on the unit circle $|z| = 1$ is mapped onto $w = e^{i\phi} = e^{-i\theta}$ on the unit circle $|w| = 1$.

We mention that $w = 1/z$ can be obtained geometrically from z by an *inversion in the unit circle* (Fig. 371) followed by a reflection in the x-axis. The reader may prove this, using similar triangles.

Figure 372 shows that $w = 1/z$ maps horizontal and vertical straight lines onto circles or straight lines. Even the following is true.

$w = 1/z$ maps every straight line or circle onto a circle or straight line.

Proof. Every straight line or circle in the z-plane can be written

$$A(x^2 + y^2) + Bx + Cy + D = 0 \qquad\qquad (A,\ B,\ C,\ D \text{ real}).$$

$A = 0$ gives a straight line and $A \neq 0$ a circle. In terms of z and $\overline{z}$, this equation becomes

$$A z\overline{z} + B\,\frac{z + \overline{z}}{2} + C\,\frac{z - \overline{z}}{2i} + D = 0.$$

Now $w = 1/z$. Substitution of $z = 1/w$ and multiplication by $w\overline{w}$ gives the equation

$$A + B\,\frac{\overline{w} + w}{2} + C\,\frac{\overline{w} - w}{2i} + Dw\overline{w} = 0$$

or, in terms of u and v,

$$A + Bu - Cv + D(u^2 + v^2) = 0.$$

This represents a circle (if $D \neq 0$) or a straight line (if $D = 0$) in the w-plane.   ∎

The proof in this example suggests the use of z and $\overline{z}$ instead of x and y, a general principle that is often quite useful in practice. Surprisingly, *every* linear fractional transformation has the property just proved:

Theorem 1 (Circles and straight lines)

Every linear fractional transformation (1) *maps the totality of circles and straight lines in the z-plane onto the totality of circles and straight lines in the w-plane.*

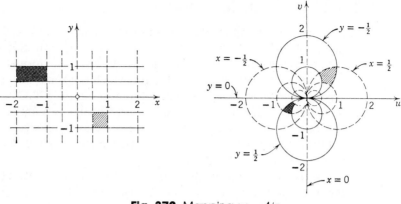

Fig. 372. Mapping $w = 1/z$

Proof. This is trivial for a translation or rotation, fairly obvious for a uniform expansion or contraction and true for $w = 1/z$, as just proved. Hence it also holds for composites of these special mappings. Now comes the key idea of the proof: represent (1) in terms of these special mappings. When $c = 0$, this is easy. When $c \neq 0$, the representation is

$$w = K \frac{1}{cz + d} + \frac{a}{c} \quad \text{where} \quad K = -\frac{ad - bc}{c}.$$

This can be verified by substituting K, taking the common denominator and simplifying; this yields (1). We can now set

$$w_1 = cz, \qquad w_2 = w_1 + d, \qquad w_3 = \frac{1}{w_2}, \qquad w_4 = Kw_3,$$

and see from the previous formula that then $w = w_4 + a/c$. This tells us that (1) is indeed a composite of those special mappings and completes the proof.   ∎

EXAMPLE 3. Image of a circle

Using Theorem 1, find the image of the unit circle $|z| = 1$ under the linear fractional transformation

$$w(z) = \frac{2iz - 2 - 2i}{(1 - i)z - 1}.$$

Solution. By Theorem 1, the image is a circle (or a straight line), hence determined by the images of three points 1, -1 and i on $|z| = 1$. We calculate these images, writing $w(z)$ more simply as

$$w(z) = 2i \frac{z - 1 + i}{(1 - i)z - 1}; \qquad \text{thus} \qquad w(1) = 2i \frac{1 - 1 + i}{1 - i - 1} = -2i;$$

similarly, $w(-1) = 2i$, so that the center of the circle must lie on the u-axis. Finally, $w(i) = -2 + 4i$. The perpendicular bisector of $2i$ and $-2 + 4i$ intersects the u-axis at -4 (make a sketch); this is the center of the circle. *Ans.* $|w + 4| = \sqrt{20}$.

We will develop a more powerful method for problems of this type in the next section, along with other applications of Theorem 1.   ∎

Extended Complex Plane

The extended complex plane was introduced in Sec. 14.10. It is the complex plane together with the point ∞ (*infinity*). We can now give it an even more natural motivation in terms of linear fractional transformations.

From (1) we see that to each z for which $cz + d \neq 0$ there corresponds precisely one complex number w. Suppose that $c \neq 0$. Then to $z = -d/c$, for which $cz + d = 0$, there does not correspond any number w. This suggests that to $z = -d/c$ we let $w = \infty$ correspond as the image.

Furthermore, when $c = 0$, we must have $a \neq 0$ and $d \neq 0$ (why?), and then we let $w = \infty$ be the image of $z = \infty$.

Finally, the inverse mapping of (1) is obtained by solving (1) for z; we find that it is again a linear fractional transformation:

$$(5) \qquad\qquad z = \frac{-dw + b}{cw - a} .$$

When $c \neq 0$, then $cw - a = 0$ for $w = a/c$, and we let a/c be the image of $z = \infty$. With these settings, the linear fractional transformation (1) is now a one-to-one conformal mapping of the extended z-plane onto the extended w-plane. We also say that every linear fractional transformation maps "the extended complex plane in a one-to-one and conformal manner onto itself."

Our present discussion suggests the following:

General Remark. If $z = \infty$, then the right-hand side of (1) becomes the meaningless expression $(a \cdot \infty + b)/(c \cdot \infty + d)$. We assign to it the value $w = a/c$ when $c \neq 0$ and $w = \infty$ when $c = 0$.

Fixed Points

Fixed points of a mapping $w = f(z)$ are points that are mapped onto themselves, are "kept fixed" under the mapping. Thus they are obtained from

$$w = f(z) = z.$$

The **identity mapping**

$$w = z$$

has every point as a fixed point. The mapping $w = \bar{z}$ has infinitely many fixed points, $w = 1/z$ has two, a rotation has one and a translation none in the finite plane. (Find them in each case.) For (1), the fixed-point condition $w = z$ is

$$z = \frac{az + b}{cz + d}$$

or

$$(6) \qquad\qquad cz^2 - (a - d)z - b = 0.$$

This is a quadratic equation in z whose coefficients all vanish if and only if the mapping is the identity mapping $w = z$ (in this case, $a = d \neq 0$, $b = c = 0$). Hence we have

Theorem 2 (Fixed points)

A linear fractional transformation, not the identity, has at most two fixed points. If a linear fractional transformation is known to have three or more fixed points, it must be the identity mapping $w = z$.

Special linear fractional transformations of practical importance and further general properties of linear fractional transformations will be discussed in the following section.

To make our present general discussion of linear fractional transformations really useful from a practical point of view, we extend it by further facts and typical examples, in the problem set as well as in the next section.

Problems for Sec. 16.2

Find the fixed points of the following mappings.

1. $w = z^2$

2. $w = z^5$

3. $w = z^2 + iz$

4. $w = (z - i)^2$

5. $w = \frac{1}{2}(z + 1)^2$

6. $w = z^5 + 5z^3 + 5z$

7. $w = \dfrac{5z + 4}{z + 5}$

8. $w = \dfrac{3z - 1}{z + 3}$

9. $w = \dfrac{2iz - 9}{z + 2i}$

Find a linear fractional transformation whose (only) fixed points are

10. 0, 1

11. $i, -i$

12. 0

Find all linear fractional transformations whose (only) fixed points are

13. $-i, i$

14. $-1, 1$

15. $0, \infty$

16. Find all linear fractional transformations without fixed points in the finite plane.

17. Show that a linear transformation with two or more fixed points in the finite plane must be the identity mapping.

18. If z_1 is a fixed point of a linear fractional transformation $w = f(z)$, it is clear that z_1 must also be a fixed point of the inverse $z = g(w)$. Give a formal proof of this.

Find the inverse by direct calculation, without using (5):

19. $w = \dfrac{z + 1}{z - 1}$

20. $w = \dfrac{z}{z + i}$

21. $w = \dfrac{3z + 4i}{iz + 5}$

22. Show the calculations of deriving (5) from (1).

23. Check (5) by substituting it into (1).

24. Prove Theorem 1 by substituting (1) in its given form into the equation of a circle or straight line. *Hint.* Show that the latter can be written ($w = u + iv$; $A = 0$ gives a straight line)

$$A w \overline{w} + B w + \overline{B} \overline{w} + C = 0.$$

25. If you are familiar with 2×2 matrices, calculate the product of the coefficient matrices of (1) and (5). Under what conditions will this be the 2×2 unit matrix?

16.3 Special Linear Fractional Transformations

We shall now learn how we can determine linear fractional transformations

$$(1) \qquad\qquad w = \frac{az + b}{cz + d} \qquad\qquad (ad - bc \neq 0)$$

for mapping certain simple domains onto others, and how we can discuss properties of (1). Four given numbers a, b, c, d determine a unique mapping (1), but we can drop or introduce a common factor without altering a given mapping (1). That is, (1) depends on *three* essential constants, namely, the ratios of any three of the a, b, c, d to the fourth. We conclude that we should get a unique mapping (1) if we impose three conditions, say, the conditions that three distinct points in the z-plane have specified distinct images in the w-plane. We show that this is the case and, more importantly, we give a formula that yields the mapping:

Theorem 1 (Three points and their images given)

Three given distinct points z_1, z_2, z_3 can always be mapped onto three prescribed distinct points w_1, w_2, w_3 by one, and only one, linear fractional transformation $w = f(z)$. This mapping is given implicitly by the equation

$$(2) \qquad \frac{w - w_1}{w - w_3} \cdot \frac{w_2 - w_3}{w_2 - w_1} = \frac{z - z_1}{z - z_3} \cdot \frac{z_2 - z_3}{z_2 - z_1}.$$

(If one of these points is the point ∞, the quotient of those two differences which contain this point must be replaced by 1.)

Proof. Equation (2) is of the form $F(w) = G(z)$, where F and G denote linear fractional functions of the respective variables. From this we readily obtain $w = f(z) = F^{-1}[G(z)]$, where F^{-1} denotes the inverse function of F. Since the inverse of a linear fractional transformation and the composite of linear fractional transformations are linear fractional transformations (cf. (5) in Sec. 16.2 and Prob. 1 at the end of this section), $w = f(z)$ is a linear fractional transformation. Furthermore, from (2) we see that

$$F(w_1) = 0, \qquad F(w_2) = 1, \qquad F(w_3) = \infty,$$

$$G(z_1) = 0, \qquad G(z_2) = 1, \qquad G(z_3) = \infty.$$

Hence, $w_1 = f(z_1)$, $w_2 = f(z_2)$, $w_3 = f(z_3)$. This proves the existence of a linear fractional transformation $w = f(z)$ that maps z_1, z_2, z_3 onto w_1, w_2, w_3, respectively.

We prove that $w = f(z)$ is uniquely determined. Suppose that $w = g(z)$ is another linear fractional transformation that maps z_1, z_2, z_3 onto w_1, w_2, w_3, respectively. Then its inverse $g^{-1}(w)$ maps w_1 onto z_1, w_2 onto z_2, and w_3 onto z_3. Consequently, the composite mapping $H = g^{-1}[f(z)]$ maps each of the points z_1, z_2, and z_3 onto itself; that is, it has three distinct fixed points z_1, z_2, z_3. From Theorem 2 in the preceding section it follows that H is the identity mapping, and, therefore, $g(z) \equiv f(z)$.

The last statement of the theorem follows from the general remark in the preceding section. This completes the proof.   ∎

Mapping of Half-Planes onto Disks

This is a task of practical interest, for instance in potential problems. Without loss of generality, let us map the upper half-plane $y \geqq 0$ onto the unit disk $|w| \leqq 1$. The boundary of this half-plane is the x-axis; clearly, it must be mapped onto the unit circle $|w| = 1$. This gives us an idea: to find a mapping, choose three points on the x-axis, prescribe their images on that circle and apply Theorem 1. Make sure that the half-plane $y \geqq 0$ is mapped onto the interior but not onto the exterior of that circle.

EXAMPLE 1. Mapping of a half-plane onto a disk
Find the linear fractional transformation (1) that maps $z_1 = -1$, $z_2 = 0$, $z_3 = 1$ onto $w_1 = -1$, $w_2 = -i$, $w_3 = 1$, respectively.

Solution. From (2) we obtain

$$\frac{w - (-1)}{w - 1} \cdot \frac{-i - 1}{-i - (-1)} = \frac{z - (-1)}{z - 1} \cdot \frac{0 - 1}{0 - (-1)}$$

or

(3)
$$w = \frac{z - i}{-iz + 1}.$$

Let us show that we can determine the specific properties of such a mapping without difficult calculations. The images of the lines $x = const$ and $y = const$ are obtained as follows. The point $z = i$ corresponds to $w = 0$, and $z = \infty$ corresponds to $w = i$. If $z = iy$, then we have $w = i(y - 1)/(y + 1)$; that is, the positive imaginary axis is mapped onto the segment $u = 0$, $-1 \leqq v \leqq 1$. Since the mapping is conformal and straight lines are mapped onto circles or straight lines, the lines $y = const$ are mapped onto circles through the image of $z = \infty$, that is, onto circles through $w = i$ and with center on the v-axis. It follows that, for the same reasons, the lines $x = const$ are mapped onto circles which are orthogonal to the image circles of the lines $y = const$ (Fig. 373). The lower half-plane corresponds to the exterior of the unit circle $|w| = 1$.

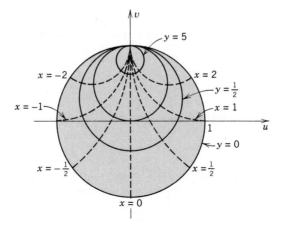

Fig. 373. Linear fractional transformation in Example 1

EXAMPLE 2. Occurrence of ∞

Determine the linear fractional transformation that maps $z_1 = 0$, $z_2 = 1$, $z_3 = \infty$ onto $w_1 = -1$, $w_2 = -i$, $w_3 = 1$, respectively.

Solution. From (2) we obtain the desired mapping

(4)
$$w = \frac{z - i}{z + i}.$$

This is sometimes called the *Cayley transformation.*[2] In this case, (2) gave at first the quotient $(1 - \infty)/(z - \infty)$, which we had to replace by 1. ∎

Mappings of Disks onto Half-Planes. This is quite similar to the case just discussed.

EXAMPLE 3. Mapping of the unit disk onto the right half-plane

Find the linear fractional transformation which maps $z_1 = -1$, $z_2 = i$, $z_3 = 1$ onto $w_1 = 0$, $w_2 = i$, $w_3 = \infty$, respectively. (Make a sketch of the disk and the half-plane.)

Solution. From (2) we obtain, after replacing $(i - \infty)/(w - \infty)$ by 1,

$$w = -\frac{z + 1}{z - 1}.$$
∎

Mappings of Half-Planes onto Half-Planes

This is another task of practical interest. We may map the upper half-plane $y \geq 0$ onto the upper half-plane $v \geq 0$, as a typical case. Then the x-axis must be mapped onto the u-axis.

EXAMPLE 4. Mapping of a half-plane onto a half-plane

Find the linear fractional transformation which maps the points $z_1 = -2$, $z_2 = 0$, $z_3 = 2$ onto the points $w_1 = \infty$, $w_2 = \frac{1}{4}$, $w_3 = \frac{3}{8}$, respectively.

Solution. From (2) we obtain

(5)
$$w = \frac{z + 1}{2z + 4},$$

as the reader may verify. What is the image of the x-axis? ∎

Mapping of Disks onto Disks

This is a third class of practical problems. We may map the unit disk in the z-plane onto the unit disk in the w-plane. It can be readily verified that the function

(6)
$$w = \frac{z - z_0}{cz - 1}, \qquad c = \bar{z}_0, \qquad |z_0| < 1$$

is of the desired type and maps the point z_0 onto the center $w = 0$ (cf. Prob. 4).

---

[2]ARTHUR CAYLEY (1821—1895), English mathematician and professor at Cambridge, is known for his important work in algebra, matrix theory and differential equations.

EXAMPLE 5. Mapping of the unit disk onto the unit disk

If we require that $z_0 = \frac{1}{2}$ be mapped onto $w = 0$, then (6) takes the form

$$w(z) = \frac{2z - 1}{z - 2}.$$

The real axes correspond to each other; in particular,

$$w(-1) = 1, \qquad w(0) = \tfrac{1}{2}, \qquad w(1) = -1.$$

Since the mapping is conformal and straight lines are mapped onto circles or straight lines and $w(\infty) = 2$, the images of the lines $x = const$ are circles through $w = 2$ with centers on the u-axis; the lines $y = const$ are mapped onto circles that are orthogonal to the aforementioned circles (cf. Fig. 374). ∎

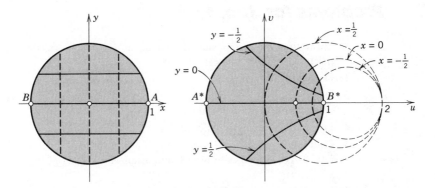

Fig. 374. Mapping in Example 5

Mappings of angular regions onto the unit disk may be obtained by combining linear fractional transformations and transformations of the form $w = z^n$, where n is an integer greater than 1.

EXAMPLE 6. Mapping of an angular region onto the unit disk

Map the angular region D: $-\pi/6 \leqq \arg z \leqq \pi/6$ onto the unit disk $|w| \leqq 1$.

Solution. We may proceed as follows. The mapping $Z = z^3$ maps D onto the right half of the Z-plane. Then we may apply a linear fractional transformation that maps this half-plane onto the unit disk, for example, the transformation

$$w = i \frac{Z - 1}{Z + 1}.$$

By inserting $Z = z^3$ into this mapping we find

$$w = i \frac{z^3 - 1}{z^3 + 1};$$

this mapping has the required properties (see Fig. 375 on the next page). ∎

This is the end of our discussion of linear fractional transformations. In the next section we turn to conformal mappings by other analytic functions (sine, cosine, etc.).

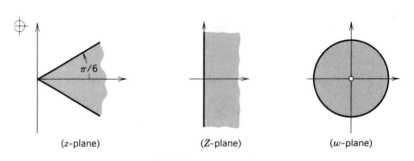

(z-plane)   (Z-plane)   (w-plane)

Fig. 375. Mapping in Example 6

Problems for Sec. 16.3

1. Show that the composite of two linear fractional transformations is a linear fractional transformation.
2. Derive (4) from (2).
3. Derive (5) from (2). Find the inverse of (5) and graph the curves corresponding to $u = const$ and $v = const$.
4. Prove the statement involving (6).
5. Find a linear fractional transformation that maps $|z| \leq 1$ onto $|w| \leq 1$ such that $z = i/2$ is mapped onto $w = 0$ and graph the images of the lines $x = const$ and $y = const$.
6. Find the inverse of (3). Show that (3) maps the lines $x = c = const$ onto circles with centers on the line $v = 1$.

Find the linear fractional transformation that maps:
7. $i, 0, 1$ onto $2 + i, 2, 3$, respectively
8. $\infty, 1, 0$ onto $0, 1, \infty$, respectively
9. $1, 0, -1$ onto $-1 - i/3, -3/4, -(3 - i)/5$, respectively
10. $0, 1, 2$ onto $1, 1/2, 1/3$, respectively
11. $1, 0, -1$ onto $\infty, -1, 0$, respectively
12. $i, -i, 0$ onto $0, \infty, -1$, respectively
13. $2i, i, 0$ onto $5i/2, 2i, \infty$, respectively
14. $1 + i, i, -1$ onto $2 + i, \infty, 0$, respectively
15. $1/2, 1, 3$ onto $\infty, 4, 6/5$, respectively

16. Find all linear fractional transformations $w(z)$ that map the x-axis onto the u-axis.
17. Determine the linear fractional transformation with fixed points -1 and 1 that maps 0 onto ip, where p is real. Discuss the cases $p = 0$ and $p = 1$.
18. Find an analytic function that maps the second quadrant of the z-plane onto the interior of the unit circle in the w-plane.
19. Find an analytic function $w = f(z)$ that maps the region $0 \leq \arg z \leq \pi/4$ onto the unit disk $|w| \leq 1$.
20. Find an analytic function $w = f(z)$ that maps the region $2 \leq y \leq x + 1$ onto the unit disk $|w| \leq 1$.

16.4 Mapping by Other Functions

Mappings by z^n, e^z and ln z were discussed in Sec. 12.9, which the student may want to study or review before going on with the present section.

Sine Function

We now consider the mapping (Sec. 12.7)

$$(1) \qquad w = u + iv = \sin z = \sin x \cosh y + i \cos x \sinh y$$

where

$$(2) \qquad u = \sin x \cosh y, \qquad v = \cos x \sinh y.$$

Since $\sin z$ is periodic with period 2π, the mapping (1) is certainly not one-to-one if we consider it in the full z-plane. We restrict z to the vertical infinite strip S defined by $-\frac{1}{2}\pi \leqq x \leqq \frac{1}{2}\pi$ (Fig. 376). Since $f'(z) = \cos z = 0$ at $z = \pm\frac{1}{2}\pi$, the mapping is not conformal at these two critical points. We can explore the properties of the mapping by determining the images of the vertical lines $x = const$ and the horizontal lines $y = const$.

When $x = 0$, then $u = 0$ and $v = \sinh y$ from (2). Hence the y-axis ($x = 0$) is mapped onto the v-axis.

When $x = \pm\frac{1}{2}\pi$, then $u = \pm\cosh y$ and $v = 0$ from (2). Since $\cosh y \geqq 1$, the vertical boundaries $x = \pm\frac{1}{2}\pi$ of the strip S are thus mapped onto the portions $u \leqq -1$ and $u \geqq 1$ of the u-axis, the images being "folded up" as indicated in Fig. 376.

When $x \neq 0$, $\pm\frac{1}{2}\pi$, using $\cosh^2 x - \sinh^2 x = 1$, we get from (2)

$$(3) \qquad \frac{u^2}{\sin^2 x} - \frac{v^2}{\cos^2 x} = 1.$$

For constant $x \neq 0$ these are hyperbolas. Hence these are the images of the vertical lines $x = const$.

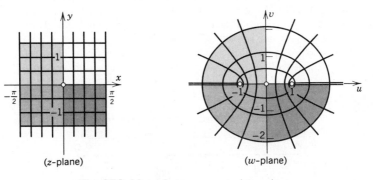

(z-plane)   (w-plane)

Fig. 376. Mapping $w = u + iv = \sin z$

When $y = 0$, then $\sinh y = 0$, hence $v = 0$ and $u = \sin x$ from (2). Hence the x-axis ($y = 0$) is mapped onto the segment $-1 \le u \le 1$ of the u-axis.

When $y \ne 0$, using $\cos^2 x + \sin^2 x = 1$, we get from (2)

(4)
$$\frac{u^2}{\cosh^2 y} + \frac{v^2}{\sinh^2 y} = 1.$$

For constant $y \ne 0$ these are confocal ellipses with foci ± 1. Hence these are the images of the horizontal lines $y = const$.

The image curves of $x = const$ and $y = const$ form an orthogonal net (intersect at right angles) because of conformality, except at the critical points $z = \pm\frac{1}{2}\pi$. Cf. Fig. 376.

From our discussion we see that $w = \sin z$ *maps the infinite vertical strip* $-\frac{1}{2}\pi < x < \frac{1}{2}\pi$ *onto the w-plane cut along the rays* $u \le -1$ *and* $u \ge 1$, $v = 0$, *and this mapping is one-to-one and conformal.*

EXAMPLE 1. Mapping of a rectangle onto an elliptical disk.
Find the image of the rectangle in Fig. 377 under $w = \sin z$.

Solution. The boundary of the rectangle corresponds to $x = \pm\pi/2$ and $y = \pm 1$. Hence the upper and lower edges are mapped onto the upper half and the lower half, respectively, of the ellipse (4) whose semiaxes are $\cosh 1 = 1.54$ and $\sinh 1 = 1.18$. The left edge is mapped onto the segment $-\cosh 1 \le u \le -1$ and the right edge onto the segment $1 \le u \le \cosh 1$ of the u-axis. Note that any two points $\pm y_0$ of those edges have the same image $-\cosh y_0$ (left edge) or $\cosh y_0$ (right edge) on the u-axis. Cf. Fig. 377.

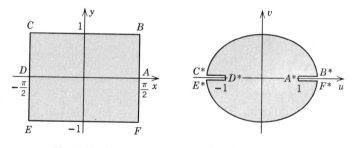

Fig. 377. Mapping by $w = \sin z$ in Example 1

EXAMPLE 2. Mapping of a rectangle onto an elliptical ring
Find the image of the rectangle R: $-\pi < x < \pi$, $1/2 < y < 1$ under $w = \sin z$. Cf. Fig. 378.

Solution. The upper edge of the rectangle is $y = 1$ and is mapped onto the ellipse (4) with semiaxes $\cosh 1$ and $\sinh 1$ (Fig. 378), the lower edge onto (4) with semiaxes $\cosh 1/2 = 1.13$ and $\sinh 1/2 = 0.52$. The lateral edges $x = \pm\pi$ are mapped onto the segment of the v-axis given by $-\sinh 1 \le v \le -\sinh 1/2$, as follows directly from (2) and $\sin(\pm\pi) = 0$, $\cos(\pm\pi) = -1$. Figure 378 shows the result.   ∎

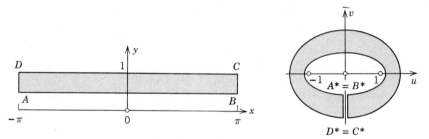

Fig. 378. Mapping by $w = \sin z$ in Example 2

Cosine and Tangent

Cosine Function. The mapping $w = \cos z$ could be discussed independently, but since

$$(5) \qquad w = \cos z = \sin (z + \tfrac{1}{2}\pi),$$

we see at once that this is the same mapping as $\sin z$ preceded by a translation to the right through $\tfrac{1}{2}\pi$ units.

Tangent Function. The idea of expressing mappings as composites of known mappings also helps in the case of $w = \tan z = \sin z / \cos z$. Expressing sine and cosine by exponential functions (Sec. 12.7) and multiplying both the numerator and the denominator by e^{iz}, we obtain

$$w = \tan z = \frac{\sin z}{\cos z} = \frac{(e^{iz} - e^{-iz})/i}{e^{iz} + e^{-iz}} = \frac{(e^{2iz} - 1)/i}{e^{2iz} + 1}.$$

Hence if we set $Z = e^{2iz}$ and use $1/i = -i$, we have

$$(6) \qquad w = \tan z = -iZ^*, \qquad Z^* = \frac{Z - 1}{Z + 1}, \qquad Z = e^{2iz}.$$

We now see that $w = \tan z$ is a linear fractional transformation preceded by an exponential mapping and followed by a clockwise rotation through an angle $\tfrac{1}{2}\pi$.

EXAMPLE 3. Mapping of an infinite strip onto a circular disk

Find the image of the infinite vertical strip S: $-\pi/4 < x < \pi/4$ (Fig. 379) under the mapping $w = \tan z$.

Solution. Since $Z = e^{2iz} = e^{-2y + 2ix}$, we obtain

$$|Z| = e^{-2y}, \qquad \text{Arg } Z = 2x$$

[cf. (8), Sec. 12.6]. Hence the vertical lines $x = -\pi/4$, 0, $\pi/4$ are mapped onto the rays Arg $Z = -\pi/2$, 0, $\pi/2$, respectively. Hence S is mapped onto the right Z-half-plane. Also $|Z| = e^{-2y} < 1$ when $y > 0$ and $|Z| > 1$ when $y < 0$. Hence the upper half of S is mapped inside the unit circle $|Z| = 1$ and the lower half of S outside $|Z| = 1$ (Fig. 379).

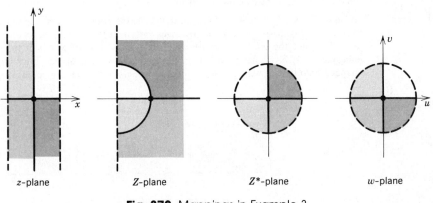

Fig. 379. Mappings in Example 3

Now comes

$$(7) \qquad\qquad Z^* = g(Z) = \frac{Z - 1}{Z + 1}.$$

For real Z this is real. Hence the real Z-axis is mapped onto the real Z^*-axis. Furthermore, the imaginary Z-axis is mapped onto the unit circle $|Z^*| = 1$ because for pure imaginary $Z = iY$ we get from (7)

$$|Z^*| = |g(iY)| = \left|\frac{iY - 1}{iY + 1}\right| = 1.$$

The right Z-half-plane is mapped inside this unit circle $|Z^*| = 1$, not outside, because $Z = 1$ has its image $g(1) = 0$ inside that circle. Finally, the unit circle $|Z| = 1$ is mapped onto the imaginary Z^*-axis, because this circle is $Z = e^{i\phi}$, so that (7) gives a pure imaginary expression, namely,

$$g(e^{i\phi}) = \frac{e^{i\phi} - 1}{e^{i\phi} + 1} = \frac{e^{i\phi/2} - e^{-i\phi/2}}{e^{i\phi/2} + e^{-i\phi/2}} = \frac{i \sin(\phi/2)}{\cos(\phi/2)}.$$

From the Z^*-plane we get to the w-plane simply by a clockwise rotation through $\pi/2$; cf. (6).

Ans. $w = \tan z$ maps S: $-\pi/4 < \mathrm{Re}\ z < \pi/4$ onto the unit disk $|w| = 1$, with the four quarters of S mapped as indicated in Fig. 379. This mapping is conformal and one-to-one.   ∎

Hyperbolic Functions

Hyperbolic functions can be treated independently in a similar fashion or reduced to the trigonometric functions just discussed. In particular, the **hyperbolic sine**

$$(8) \qquad\qquad w = \sinh z = -i \sin(iz)$$

defines a mapping that is a rotation $Z = iz$ followed by the mapping $Z^* = \sin Z$ and another rotation $w = -iZ^*$.

Similarly, the **hyperbolic cosine**

$$(9) \qquad\qquad w = \cosh z = \cos(iz)$$

defines a mapping that is a rotation $Z = iz$ followed by the mapping $w = \cos Z$.

EXAMPLE 4. Mapping of a semi-infinite strip onto a half-plane

Find the image of the semi-infinite strip $x \geq 0$, $0 \leq y \leq \pi$ (Fig. 380) under the mapping (9).

Solution. We set $w = u + iv$. Since $\cosh 0 = 1$, the point $z = 0$ is mapped onto $w = 1$. For real $z = x \geq 0$, $\cosh z$ is real and increases with increasing x in a monotone fashion, starting from 1. Hence the positive x-axis is mapped onto the portion $u \geq 1$ of the u-axis. For pure imaginary $z = iy$ we have $\cosh iy = \cos y$. It follows that the left boundary of the strip is mapped onto the segment $1 \geq u \geq -1$ of the u-axis, the point $z = \pi i$ corresponding to

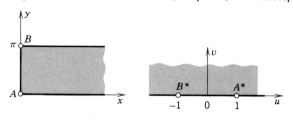

Fig. 380. Mapping in Example 4

$$w = \cosh i\pi = \cos \pi = -1.$$

On the upper boundary of the strip, $y = \pi$, and since $\sin \pi = 0$, $\cos \pi = -1$, it follows that this part of the boundary is mapped onto the portion $u \leqq -1$ of the u-axis. Hence the boundary of the strip is mapped onto the u-axis. It is not difficult to see that the interior of the strip is mapped onto the upper half of the w-plane, and the mapping is one-to-one.   ∎

In the next section we discuss **Riemann surfaces,** on which multivalued relations, such as $w = \sqrt{z}$ or $w = \ln z$, become single-valued, that is, functions in the usual sense. This idea is probably best understood if we explain it "geometrically," in the context of mappings, and for this reason we include it in the present chapter.

Problems for Sec. 16.4

1. Find an analytic function that maps the region bounded by the positive x and y axes and the hyperbola $xy = \pi/2$ in the first quadrant onto the upper half-plane. *Hint.* First map that region onto a horizontal strip.

Find and graph the images of the following regions under the mapping $w = \sin z$.

2. $0 < x < \pi/2, \quad 0 < y < 2$ 3. $-\pi/2 < x < \pi/2, \quad 1 < y < 2$
4. $-\pi/2 < x < \pi/2, \quad 0 < y < 1$ 5. $0 < x < 2\pi, \quad 1 < y < 2$

6. Find and plot the images of the lines $x = 0, \pm \pi/6, \pm \pi/3, \pm \pi/2$ under the mapping $w = \sin z$.
7. Determine all points at which the mapping $w = \sin z$ is not conformal.
8. Describe the transformation $w = \cosh z$ in terms of the transformation $w = \sin z$ and rotations and translations.
9. Find all points at which the mapping $w = \cosh z$ is not conformal.
10. Find all points at which the mapping $w = \cos z$ is not conformal. What happens to angles at these points?
11. Find the images of the lines $x = const$ under the mapping $w = \cos z$.
12. Find the images of the lines $y = const$ under the mapping $w = \cos z$.

Find and graph the images of the following regions under the mapping $w = \cos z$.

13. $0 < x < \pi, y < 0$ 14. $0 < x < \pi, y > 0$
15. $\pi < x < 2\pi, y < 0$ 16. $0 < x < \pi/2, 0 < y < 1$
17. $0 < x < \pi, 0 < y < 1$ 18. $0 < x < 2\pi, 1/2 < y < 1$

19. Find the image of the region $2 \leqq |z| \leqq 3$, $\pi/4 \leqq \theta \leqq \pi/2$ under the mapping $w = \text{Ln } z$.

20. Show that $w = \text{Ln } \dfrac{z-1}{z+1}$ maps the upper half-plane onto the horizontal strip $0 \leqq \text{Im } w \leqq \pi$ as shown in Fig. 381.

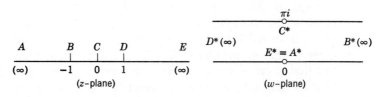

Fig. 381. Problem 20

16.5 Riemann Surfaces

We consider the mapping defined by

(1) $$w = u + iv = z^2$$

(cf. Sec. 12.9). This mapping is conformal, except at $z = 0$ where $w' = 2z$ is zero. The angles at $z = 0$ are doubled under the mapping. The right half of the z-plane (including the positive y-axis) is mapped onto the full w-plane cut along the negative half of the u-axis; the mapping is one-to-one. Similarly, the left half of the z-plane (including the negative y-axis) is mapped onto the cut w-plane in a one-to-one manner.

Obviously, the mapping of the full z-plane is not one-to-one, because every point $w \neq 0$ corresponds to precisely two points z. In fact, if z_1 is one of these points, then the other is $-z_1$. For example, $z = i$ and $z = -i$ have the same image, namely, $w = -1$, etc. Hence, the w-plane is "covered twice" by the image of the z-plane. We say that the full z-plane is mapped onto the *doubly covered* w-plane. We can still give our imagination the necessary support as follows.

We imagine one of the two previously obtained copies of the cut w-plane to be placed on the other so that the upper sheet is the image of the right half of the z-plane and the lower sheet is the image of the left half of the z-plane; we denote these half-planes by R and L, respectively. When we pass from R to L, the corresponding image point should pass from the upper to the lower sheet. For this reason we join the two sheets crosswise along the cut, that is, along the negative real axis. (This construction can be carried out only in imagination, since the penetration of the two sheets of a material model can be realized only imperfectly.) The two origins are fastened together. The configuration thus obtained is called a **Riemann surface.** On it every point $w \neq 0$ appears twice, at superposed positions, and the origin appears precisely once. The function $w = z^2$ now maps the full z-plane onto this Riemann surface in a one-to-one manner, and the mapping is conformal, except for the "winding point" or **branch point** at $w = 0$ (Fig. 382). Such a branch point connecting two sheets is said to be of the *first order*. (More generally, a branch point connecting n sheets is said to be of **order** $n - 1$.)

We now consider the double-valued relation

(2) $$\boxed{w = \sqrt{z}.}$$

To each $z \neq 0$ there correspond two values w, one of which is the principal value. If we replace the z-plane by the two-sheeted Riemann surface just considered, then each complex number $z \neq 0$ is represented by two points of the surface at superposed positions. We let one of these points correspond to the principal value—for example, the point in the upper sheet—and the other to the other value. Then (2) becomes single-valued, that is, (2) is a function of the points of the Riemann surface, and to any continuous motion of z on the surface there corresponds a continuous motion of the corresponding point in the w-plane. The function maps the sheet corresponding

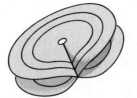

Fig. 382. Example of a Riemann surface

to the principal value onto the right half of the w-plane and the other sheet onto the left half of the w-plane.

Let us consider some further important examples.

EXAMPLE 1. Riemann surface of $\sqrt[n]{z}$
In the case of the relation

$$(3) \qquad\qquad w = \sqrt[n]{z} \qquad\qquad n = 3, 4, \cdots$$

we need a Riemann surface consisting of n sheets and having a branch point of order $n - 1$ at $z = 0$. One of the sheets corresponds to the principal value and the other $n - 1$ sheets to the other $n - 1$ values.

EXAMPLE 2. Riemann surface of the natural logarithm
For every $z \neq 0$ the relation

$$(4) \qquad\qquad w = \ln z = \text{Ln } z + 2n\pi i \qquad\qquad (n = 0, \pm 1, \pm 2, \cdots, \quad z \neq 0)$$

is infinitely many-valued. Hence (4) defines a function on a Riemann surface consisting of infinitely many sheets. The function $w = \text{Ln } z$ corresponds to one of these sheets. On this sheet the argument θ of z ranges in the interval $-\pi < \theta \leq \pi$ (cf. Sec. 12.8). The sheet is cut along the negative ray of the real axis, and the upper edge of the slit is joined to the lower edge of the next sheet, which corresponds to the interval $\pi < \theta \leq 3\pi$, that is, to the function $w = \text{Ln } z + 2\pi i$. In this way each value of n in (4) corresponds to precisely one of these infinitely many sheets. The function $w = \text{Ln } z$ maps the corresponding sheet onto the horizontal strip $-\pi < v \leq \pi$ in the w-plane. The next sheet is mapped onto the neighboring strip $\pi < v \leq 3\pi$, etc. The function $w = \ln z$ thus maps all the sheets of the corresponding Riemann surface onto the entire w-plane, the correspondence between the points $z \neq 0$ of the Riemann surface and those of the w-plane being one-to-one.

EXAMPLE 3. Mapping $w = z + z^{-1}$. Airfoils
Let us consider the mapping defined by

$$(5) \qquad\qquad w = z + \frac{1}{z} \qquad\qquad (z \neq 0)$$

which is important in aerodynamics (see below). Since this function has the derivative

$$w' = 1 - \frac{1}{z^2} = \frac{(z + 1)(z - 1)}{z^2},$$

the mapping is conformal except at the points $z = 1$ and $z = -1$; these points correspond to $w = 2$ and $w = -2$, respectively. From (5) we find

$$(6) \qquad\qquad z = \frac{w}{2} \pm \sqrt{\frac{w^2}{4} - 1} = \frac{w}{2} \pm \frac{1}{2}\sqrt{(w + 2)(w - 2)}.$$

Hence, the points $w = 2$ and $w = -2$ are branch points of the first order of $z = z(w)$. To any value w ($\neq 2$, $\neq -2$) there correspond two values of z. Consequently, (5) maps the z-plane onto a two-sheeted Riemann surface, the two sheets being connected crosswise from

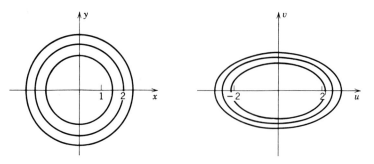

Fig. 383. Example 3

$w = -2$ to $w = 2$ (Fig. 383), and this mapping is one-to-one. We set $z = re^{i\theta}$ and determine the images of the curves $r = const$ and $\theta = const$. From (5) we obtain

$$w = u + iv = re^{i\theta} + \frac{1}{r}e^{-i\theta} = \left(r + \frac{1}{r}\right)\cos\theta + i\left(r - \frac{1}{r}\right)\sin\theta.$$

By equating the real and imaginary parts on both sides we have

(7) $$u = \left(r + \frac{1}{r}\right)\cos\theta, \qquad v = \left(r - \frac{1}{r}\right)\sin\theta.$$

From this we find

$$\frac{u^2}{a^2} + \frac{v^2}{b^2} = 1, \quad \text{where} \quad a = r + \frac{1}{r}, \qquad b = \left|r - \frac{1}{r}\right|.$$

The circles $r = const$ are thus mapped onto ellipses whose principal axes lie in the u and v axes and have the lengths $2a$ and $2b$, respectively. Since $a^2 - b^2 = 4$, independent of r, these ellipses are confocal, with foci at $w = -2$ and $w = 2$. The unit circle $r = 1$ is mapped onto the line segment from $w = -2$ to $w = 2$. For every $r \neq 1$ the two circles with radii r and $1/r$ are mapped onto the same ellipse in the w-plane, corresponding to the two sheets of the Riemann surface. Hence, the interior of the unit circle $|z| = 1$ corresponds to one sheet, and the exterior to the other.

Furthermore, from (7) we obtain

(8) $$\frac{u^2}{\cos^2\theta} - \frac{v^2}{\sin^2\theta} = 4.$$

The lines $\theta = const$ are thus mapped onto the hyperbolas that are the orthogonal trajectories of those ellipses. The real axis, that is, the rays $\theta = 0$ and $\theta = \pi$, are mapped onto the part of the real axis from $w = 2$ via ∞ to $w = -2$. The y-axis is mapped onto the v-axis. Any other pair of rays $\theta = \theta_0$ and $\theta = \theta_0 + \pi$ is mapped onto the two branches of the same hyperbola.

An exterior region of one of the above ellipses is free of branch points and corresponds to either the interior or the exterior of the corresponding circle in the z-plane, depending on the sheet of the Riemann surface to which the region belongs. In particular, the full w-plane corresponds to the interior or the exterior of the unit circle $|z| = 1$, as we mentioned before.

The mapping (5) transforms suitable circles into airfoils with a sharp trailing edge whose interior angle is zero; these airfoils are known as **Joukowski airfoils**.[3] Since the airfoil to be obtained has a sharp edge, it is clear that the circle to be mapped must pass through one of the points $z = \pm 1$ at which the mapping is not conformal. Let us choose a circle C through $z = -1$ and such that $z = 1$ lies inside C. The simplest way of determining the image is the graphical vector addition of the vectors corresponding to z and $1/z$ (where the latter can be obtained from the former as shown in Fig. 371 in Sec. 16.2). This leads to the result shown in Fig. 384. Further details are included in Ref. [D9] in Appendix 1.   ∎

[3]NIKOLAI JEGOROVICH JOUKOWSKI (1847—1921), Russian mathematician.

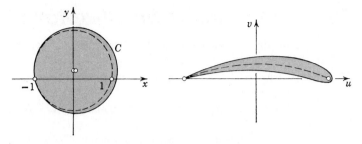

Fig. 384. Joukowski airfoil

A more elaborate treatment of Riemann surfaces can be found in Ref. [D2].

This is the end of Chap. 16 on conformal mapping. The next chapter, the last on complex analysis, is devoted to **potential problems** and includes applications of conformal mapping, along with various other ideas that are of practical importance to the engineer and physicist.

Problems for Sec. 16.5

1. Consider $w = \sqrt{z}$. Find the path of the image point w of a point z that moves twice around the unit circle, starting from the initial position $z = 1$.

2. Show that the Riemann surface of $w = \sqrt[3]{z}$ consists of three sheets and has a branch point of second order at $z = 0$. Find the path of the image point w of a point z that moves three times around the unit circle starting from the initial position $z = 1$.

3. Consider the Riemann surfaces of $w = \sqrt[4]{z}$ and $w = \sqrt[5]{z}$ in a fashion similar to that in Prob. 2.

4. Make a sketch, similar to Fig. 382, of the Riemann surfaces of $\sqrt[3]{z}$ and $\sqrt[4]{z}$.

5. Determine the path of the image of a point z under the mapping $w = \ln z$ as z moves several times around the unit circle.

6. Find the images of the annuli $\frac{1}{2} < |z| < 1$, $1 < |z| < 2$, and $2 < |z| < 3$ under the mapping $w = z + 1/z$.

7. Show that the Riemann surface of $w = \sqrt{(z - 1)(z - 4)}$ has branch points at $z = 1$ and $z = 4$ and consists of two sheets which may be cut along the line segment from 1 to 4 and joined crosswise. *Hint.* Introduce polar coordinates $z - 1 = r_1 e^{i\theta_1}$, $z - 4 = r_2 e^{i\theta_2}$.

8. Show that the Riemann surface of $w = \sqrt{(1 - z^2)(4 - z^2)}$ has four branch points and two sheets that may be joined crosswise along the segments $-2 \leqq x \leqq -1$ and $1 \leqq x \leqq 2$ of the x-axis.

Determine the location of the branch points and the number of sheets of the Riemann surfaces of the following functions.

9. $w = i\sqrt{z}$

10. $w = \sqrt[3]{z - i}$

11. $w = \sqrt{z - i}$

12. $\sqrt[3]{2z + 3i}$

13. $\sqrt{z(z - 1)(z + 1)}$

14. $\sqrt{z^2 + 1}$

15. $\ln(z - a)$

16. $\sqrt{(z - a)(z - b)}, a \neq b$

17. $1 + z + \sqrt{z}$

18. $\sqrt{\sqrt[3]{z} - 1}$

19. $e^{\sqrt{z}}$

20. $\sqrt{e^z}$

Review Problems for Chapter 16

1. How did we define the angle of intersection of two oriented curves, and what does it mean to say that a mapping is conformal?

2. At what points is a mapping $w = f(z)$ by an analytic function not conformal? Give examples.

3. What happens to angles at z_0 under a mapping $w = f(z)$ if $f'(z_0) = 0, f''(z_0) = 0,$ $f'''(z_0) \neq 0$?

4. What is a linear fractional transformation? Why are these mappings of great practical interest?

5. Why did we require that the coefficients of a linear fractional transformation satisfy $ad - bc \neq 0$?

6. What is the extended complex plane, and why did it occur in this chapter?

7. What is a fixed point of a mapping? Give examples of mappings by analytic functions that have no fixed points, precisely one fixed point, infinitely many fixed points.

8. What is a Riemann surface, and what purpose does it serve?

9. How many sheets does the Riemann surface of $w = \sqrt[4]{z}$ have? Where is its branch point?

10. Why is the mapping $w = z + 1/z$ of practical interest? How many sheets does the Riemann surface of the inverse of this mapping have?

Find and sketch the images of the following curves or regions under the mapping $w = u + iv = z^2$.

11. $y = 1, 2, 3$	12. $x = 1, 2, 3$	13. $|z| = 1.5, |\arg z| < \pi/4$
14. $xy = 1$	15. $\operatorname{Re} z > 0$	16. $y = x, y = -x$
17. $\pi/4 < \arg z < 3\pi/4$	18. $0 < y < 1$	19. $0 < x < 1$

Find and sketch the following curves or regions under the mapping $w = 1/z$.

20. $|z| \leq 1/3, y > 0$	21. $|\arg z| < \pi/4$	22. $x > 0, y > 0, |z| > 1$
23. $|z - 1/2| = 1/2$	24. $y = 1$	25. $y < 0, |z| > 4$

Find all points at which the following mappings $w = f(z)$ fail to be conformal, where $f(z)$ equals

26. $\sin 2\pi z^2$	27. $z^4 - z^2$	28. $\sinh z$
29. $\exp z^2$	30. $z^5 - 5z$	31. $z^2 + 1/z^2 \quad (z \neq 0)$

Find the linear fractional transformation that maps:

32. $-1, 0, 1$ onto $0, -1, \infty$, respectively.

33. $i, 0, -i$ onto $0, -1, \infty$, respectively.

34. $0, i, \infty$ onto $1, 1 + i, \infty$, respectively.

35. $0, \infty, -2$ onto $0, 1, \infty$, respectively.

36. $2i, 0, 4i$ onto $\infty, i/2, -i/2$, respectively.

37. $0, 1, 2$ onto $-i, (1 - i)/3, (2 - i)5$, respectively.

38. $0, 1, \infty$ onto $2 + i, (7 - i)/5, (1 - i)/2$, respectively.

39. $3, \infty, i$ onto $(1 - i)/4, -i/4, -i$, respectively.

40. $i/2, -i/2, i$ onto $-4/3, 0, -3$, respectively.

41. $i, 2i, 3i$ onto $-i/2, 0, i/8$, respectively.

Find an analytic function $w = u + iv = f(z)$ that maps:

42. The half-plane $x \geq 0$ onto the region $u \geq 2$ such that $z = 0$ has the image $w = 2 + i$

43. The region $0 < \arg z < \pi/3$ onto the region $u < 1$.

44. The interior of the unit circle $|z| = 1$ onto the exterior of the circle $|w + 1| = 5$.

45. The right half-plane onto the upper half-plane.

46. The square $1 \leq x \leq 2$, $1 \leq y \leq 2$ onto the region $e \leq |w| \leq e^2$, $1 \leq \arg w \leq 2$.

47. The infinite strip $0 < y < \pi/3$ onto the upper half-plane $v > 0$.

48. The semidisk $|z| < 1$, $x > 0$ onto the exterior of the unit circle $|w| = 1$.

49. The region $x > 0$, $y > 0$, $xy < k$ onto the strip $0 < v < 1$.

Find all fixed points of the mappings $w = f(z)$, where $f(z)$ equals

50. $z^2 + z - 1$

51. $z^4 + z - 81$

52. $z + \sin \pi z$

53. $\dfrac{z - 1}{z + 1}$

54. $\dfrac{3z + 2}{z - 1}$

55. $\dfrac{2iz - 1}{z + 2i}$

Summary of Chapter 16
Conformal Mapping

A complex function $w = f(z)$ gives a **mapping** of its domain of definition in the complex z-plane onto its range of values in the complex w-plane. If $f(z)$ is *analytic*, this mapping is **conformal**, that is, angle-preserving: the images of any two intersecting curves make the same angle of intersection, in both magnitude and sense, as the curves themselves (Sec. 16.1). **Harmonic functions** remain harmonic under conformal mapping (Sec. 16.1). For mapping properties of e^z, $\cos z$, $\sin z$, etc., see Secs. 12.9 and 16.4.

Linear fractional transformations, also called *Möbius transformations* (Secs. 16.2, 16.3)

$$(1) \qquad w = \frac{az + b}{cz + d} \qquad (ad - bc \neq 0)$$

map the extended complex plane (Sec. 16.2) conformally onto itself. They solve the problems of mapping half-planes onto half-planes or disks, and disks onto disks or half-planes. Prescribing the images of three points determines (1) uniquely (Sec. 16.3).

Riemann surfaces (Sec. 16.5) consist of several sheets connected at certain points called *branch points*. On them, multivalued relations become single-valued, that is, functions in the usual sense. *Examples.* For $w = \sqrt{z}$ we need two sheets since this relation is doubly valued. For $w = \ln z$ we need infinitely many sheets since this relation is infinitely many-valued (cf. Sec. 12.8).

Chapter 17

Complex Analysis Applied to Potential Theory

Laplace's equation $\nabla^2\Phi = 0$ is one of the most important partial differential equations in engineering mathematics, because it occurs in connection with gravitational fields (Sec. 8.8), electrostatic fields (Sec. 11.11), steady-state heat conduction (Secs. 9.7, 11.5), incompressible fluid flow, etc. The theory of the solutions of this equation is called **potential theory,** and solutions whose second partial derivatives are continuous are called **harmonic functions.**

In the "two-dimensional case" when Φ depends only on two Cartesian coordinates x and y, Laplace's equation becomes

$$\nabla^2\Phi = \Phi_{xx} + \Phi_{yy} = 0.$$

We know that then its solutions are closely related to complex analytic functions (cf. Sec. 12.5).[1] In this chapter we discuss this connection and its consequences in more detail and explain it in terms of practical problems taken from electrostatics (Secs. 17.1, 17.2), heat conduction (Sec. 17.3) and hydrodynamics (Sec. 17.4). In most cases this requires the solution of a **boundary value problem,** called the **Dirichlet problem** (or *first boundary value problem*) for the Laplace equation, that is, to find the solution of the Laplace equation in a given domain D assuming given values on the boundary of D. Such a problem can often be solved by the method of conformal mapping (Sec. 17.2), that is, by mapping D conformally onto a simpler domain (disk, half-plane etc.) for which the solution is known or can be obtained from a formula, such as Poisson's integral formula (Sec. 17.5). For *mixed boundary value problems* the situation is similar.

In the last section (Sec. 17.6) we show that various general properties of harmonic functions follow from results on analytic functions.

Prerequisites for this chapter: Chaps. 12, 13, 16.
References: Appendix 1, Part D.
Answers to problems: Appendix 2.

---

[1]No such close relation exists in the three-dimensional case.

On notation. We write Φ and later $\Phi + i\Psi$ since u and $u + iv$ will be needed in conformal mapping from Sec. 17.2 on.

17.1 Electrostatic Fields

The electrical force of attraction or repulsion between charged particles is governed by Coulomb's law. This force is the gradient of a function Φ, called the **electrostatic potential.** At any points free of charges, Φ is a solution of Laplace's equation

$$\nabla^2 \Phi = 0.$$

The surfaces $\Phi = const$ are called **equipotential surfaces.** At each point P the gradient of Φ is perpendicular to the surface $\Phi = const$ through P; that is, the electrical force has the direction perpendicular to the equipotential surface. (See also Secs. 8.8 and 11.11.)

The problems we shall discuss in this whole chapter are **two-dimensional** (for reasons just given); that is, they concern physical systems that lie in three-dimensional space, but are such that the potential Φ is independent of a third space coordinate but depends on two coordinates x, y only. Then Laplace's equation becomes

(1) $$\nabla^2 \Phi = \frac{\partial^2 \Phi}{\partial x^2} + \frac{\partial^2 \Phi}{\partial y^2} = 0.$$

Equipotential surfaces now appear as **equipotential lines** (curves) in the xy-plane.

EXAMPLE 1. Potential between parallel plates

Find the potential Φ of the field between two parallel conducting plates extending to infinity (Fig. 385), which are kept at potentials Φ_1 and Φ_2, respectively.

Solution. From the shape of the plates it follows that Φ depends only on x, and Laplace's equation becomes $\Phi'' = 0$. By integrating twice we obtain $\Phi = ax + b$, where the constants a and b are determined by the given boundary values of Φ on the plates. For example, if the plates correspond to $x = -1$ and $x = 1$, the solution is

$$\Phi(x) = \tfrac{1}{2}(\Phi_2 - \Phi_1)x + \tfrac{1}{2}(\Phi_2 + \Phi_1).$$

The equipotential surfaces are parallel planes.

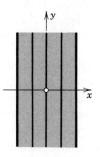

Fig. 385. Potential
in Example 1

EXAMPLE 2. Potential between coaxial cylinders

Find the potential Φ between two coaxial conducting cylinders which extend to infinity on both sides (Fig. 386) and are kept at potentials Φ_1 and Φ_2, respectively.

Solution. Here Φ depends only on $r = \sqrt{x^2 + y^2}$, for reasons of symmetry, and Laplace's equation becomes

$$r\Phi'' + \Phi' = 0 \qquad\qquad \text{[cf. (4) in Sec. 11.9].}$$

By separating variables and integrating we obtain

$$\frac{\Phi''}{\Phi'} = -\frac{1}{r}, \qquad \ln \Phi' = -\ln r + \tilde{a}, \qquad \Phi' = \frac{a}{r}, \qquad \Phi = a \ln r + b$$

and a and b are determined by the given values of Φ on the cylinders. Although no infinitely extended conductors exist, the field in our idealized conductor will approximate the field in a long finite conductor in that part which is far away from the ends of the two cylinders.

EXAMPLE 3. Potential in an angular region

Find the potential Φ between the conducting plates in Fig. 387, which are kept at potential Φ_1 (the lower plate) and Φ_2 and make an angle α, where $0 < \alpha \leq \pi$. (In the figure, $\alpha = 120° = 2\pi/3$.)

Solution. $\theta = \text{Arg } z$ ($z = x + iy \neq 0$) is constant on rays $\theta = const$. It is harmonic since it is the imaginary part of an analytic function, $\text{Ln } z$ (Sec. 12.8). Hence the solution is

$$\Phi(x, y) = a + b \text{ Arg } z$$

with a and b determined from the two boundary conditions (given values on the plates)

$$a + b(-\tfrac{1}{2}\alpha) = \Phi_1, \qquad a + b(\tfrac{1}{2}\alpha) = \Phi_2.$$

Thus $a = (\Phi_2 + \Phi_1)/2$, $b = (\Phi_2 - \Phi_1)/\alpha$. The answer is

$$\Phi(x, y) = \frac{1}{2}(\Phi_2 + \Phi_1) + \frac{1}{\alpha}(\Phi_2 - \Phi_1)\theta, \qquad \theta = \arctan \frac{y}{x}. \quad \blacksquare$$

Complex Potential

Let $\Phi(x, y)$ be harmonic in some domain D and $\Psi(x, y)$ a conjugate harmonic of Φ in D (Sec. 12.5). Then[2]

$$F(z) = \Phi(x, y) + i\Psi(x, y)$$

is an analytic function of $z = x + iy$. This function F is called the **complex potential** corresponding to the real potential Φ. Recall from Sec. 12.5 that for given Φ, a conjugate Ψ is uniquely determined except for an additive real constant. Hence we may say *the* complex potential, without causing misunderstandings.

The use of F has two advantages, a technical one and a physical one. Technically, F is easier to handle than real or imaginary parts, in connection with methods of complex analysis. Physically, Ψ has a meaning. By conformality, the curves $\Psi = const$ intersect the equipotential lines $\Phi = const$ at right angles [except where $F'(z) = 0$]. Hence they have the direction of the electrical force and, therefore, are called **lines of force.** They are the paths of moving charged particles (electrons in an electron microscope, etc.).

---

[2]We write $F = \Phi + i\Psi$, reserving $f = u + iv$ for conformal mapping, as needed from the next section on.

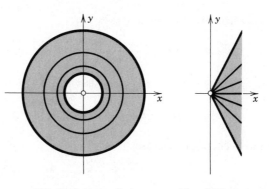

Fig. 386. Potential
in Example 2

Fig. 387. Potential
in Example 3

EXAMPLE 4. Complex potential

In Example 1, a conjugate is $\Psi = ay$. It follows that the complex potential is

$$F(z) = az + b = ax + b + iay,$$

and the lines of force are straight lines parallel to the x-axis.

EXAMPLE 5. Complex potential

In Example 2 we have

$$\Phi = a \ln r + b = a \ln |z| + b.$$

A conjugate is $\Psi = a \arg z$. The complex potential is

$$F(z) = a \ln z + b$$

and the lines of force are straight lines through the origin. $F(z)$ may also be interpreted as the complex potential of a source line whose trace in the xy-plane is the origin.

EXAMPLE 6. Complex potential

In Example 3, the complex potential is

$$F(z) = a - ib \ln z = a - ib(\ln |z| + i \arg z) = a + b \arg z - ib \ln |z|.$$

We see from this that the lines of force are concentric circles $|z| = const$. ∎

Superposition

More complicated potentials can often be obtained by superposition.

EXAMPLE 7. Potential of a pair of source lines

Determine the potential of a pair of oppositely charged source lines of the same strength at the points $z = c$ and $z = -c$ on the real axis.

Solution. From Examples 2 and 5 it follows that the potential of each of the source lines is

$$\Phi_1 = K \ln |z - c| \qquad \text{and} \qquad \Phi_2 = -K \ln |z + c|,$$

respectively. Here the constant K measures the strength (amount of charge). These are the real parts of the complex potentials

$$F_1(z) = K \ln (z - c) \qquad \text{and} \qquad F_2(z) = -K \ln (z + c).$$

Hence the complex potential of the combination of the two source lines is

(2) $$F(z) = F_1(z) + F_2(z) = K \ln \frac{z - c}{z + c}.$$

The equipotential lines are the curves

$$\Phi = \text{Re } F(z) \; K \ln \left| \frac{z-c}{z+c} \right| = const, \qquad \text{thus} \qquad \left| \frac{z-c}{z+c} \right| = const.$$

These are circles, as the reader may show by straightforward calculation. The lines of force are

$$\Psi = \text{Im } F(z) = K \arg \frac{z-c}{z+c} = K[\arg (z-c) - \arg (z+c)] = const.$$

We write this briefly (Fig. 388)

$$\Psi = K(\theta_1 - \theta_2) = const.$$

Now $\theta_1 - \theta_2$ is the angle between the line segments from z to c and $-c$ (Fig. 388). Hence the lines of force are the curves along each of which the line segment S: $-c \leqq x \leqq c$ appears under a constant angle. These curves are the totality of circular arcs over S, as is well known from elementary geometry. Hence the lines of force are circles. Figure 389 shows them together with the equipotential lines.

In addition to the interpretation as the potential of two source lines, this potential could also be thought of as the potential between two circular cylinders whose axes are parallel but do not coincide, or as the potential between two equal cylinders that lie outside each other, or as the potential between a cylinder and a plane wall. Explain this, using Fig. 389.　∎

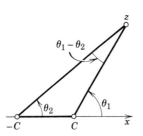

Fig. 388. Arguments in Example 7

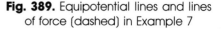

Fig. 389. Equipotential lines and lines of force (dashed) in Example 7

The idea of the complex potential as just explained is a key to a close relation of potential theory to complex analysis and will recur in heat flow and fluid flow. Another key is **conformal mapping** as applied to boundary value problems. We discuss this in the next section.

Problems for Sec. 17.1

1. Find and sketch the equipotential surfaces between two parallel plates at $x = -5$ and $x = 5$ having potentials 200 and 300 volts, respectively.

2. Find the complex potential in Prob. 1.

3. Find and sketch the equipotential surfaces between two parallel plates at $y = 0$ and $y = d$ having potentials $U_1 = 3 \text{ kV}$ (= 3000 volts) and $U_2 = 0$, respectively.

4. Find the potential Φ and the complex potential F between the plates $y = x$ and $y = x + 1$ having potentials 0 and 60 volts, respectively.

5. Find the potential in the first quadrant of the xy-plane between the axes (having potential 110 volts) and the hyperbola $xy = 4$ (having potential 220 volts).

Find the potential Φ between two infinite coaxial cylinders of radii r_1 and r_2 $(> r_1)$ having potentials U_1 and U_2, respectively, where

6. $r_1 = 1$, $r_2 = 5$, $U_1 = 0$, $U_2 = 100$ volts

7. $r_1 = 0.5$, $r_2 = 2$, $U_1 = -110$ volts, $U_2 = 110$ volts

8. $r_1 = 2$, $r_2 = 20$, $U_1 = 100$ volts, $U_2 = 200$ volts

9. $r_1 = 3$, $r_2 = 6$, $U_1 = 1000$ volts, $U_2 = 500$ volts

10. $r_1 = 10$, $r_2 = 100$, $U_1 = 10$ kV, $U_2 = 0$

Find the equipotential lines of the complex potential $F(z)$ and show these lines graphically, where

11. $F(z) = z^2$ **12.** $F(z) = -iz^2$ **13.** $F(z) = 1/z$ **14.** $F(z) = i/z$

15. $F(z) = i \operatorname{Ln} z$

16. Near each of the two source lines in Fig. 389 the circles $\Phi = const$ are almost concentric. Is this physically understandable?

17. Find the potential of two source lines at $z = a$ and $z = -a$ having the same charge.

18. Find the potential Φ between the infinite cylinders in Fig. 390, if on the left cylinder, $\Phi = -1$ and on the right, $\Phi = 1$. (Use the potential in Example 7.)

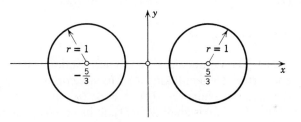

Fig. 390. Problem 18

19. Show that $F(z) = \cos^{-1} z$ may be interpreted as the complex potential of each of the three configurations in Fig. 391.

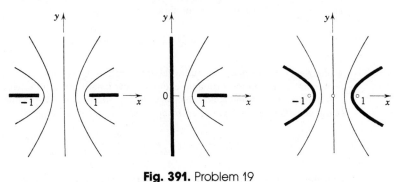

Fig. 391. Problem 19

20. Show that $F(z) = \cosh^{-1} z$ may be interpreted as the complex potential between two confocal elliptic cylinders.

17.2 Use of Conformal Mapping

In applications, we can often solve a boundary value problem for the Laplace equation in a complicated domain D in two steps. We first transform D by a conformal mapping $w = f(z)$ into a simpler domain D^* for which the solution $F^*(w)$ of the problem (corresponding to the transformed boundary conditions) is known or can be more readily obtained. Then we use $w = f(z)$ to transport the solution $F^*(w)$ to D, obtaining $F(z) = F^*(f(z))$. The key to this powerful method lies in the fact that under a conformal mapping, harmonic functions transform into harmonic functions. We shall utilize this fact in the following form.

Theorem 1 (Harmonic functions under conformal mapping)

If $\Phi^*(u, v)$ is harmonic in a domain D^* in the w-plane, and if an analytic function $w = u + iv = f(z)$ maps a domain D in the z-plane conformally onto D^*, then

$$(1) \qquad\qquad \Phi(x, y) = \Phi^*(u(x, y), v(x, y))$$

is harmonic in D.

Proof. The composite of analytic functions is analytic, as follows from the chain rule. Hence, taking a harmonic conjugate[3] $\Psi^*(u, v)$ of Φ^* and forming the analytic function $F^*(w) = \Phi^*(u, v) + i\Psi^*(u, v)$, we conclude that $F(z) = F^*(f(z))$ is analytic in D and its real part, $\Phi(x, y) = \operatorname{Re} F(z)$ is harmonic in D. ∎

EXAMPLE 1. Potential between noncoaxial cylinders
Find the potential between the cylinders C_1: $|z| = 1$ (having potential $U_1 = 0$) and C_2: $|z - 2/5| = 2/5$ (having potential $U_2 = 110$ volts).

Solution. We map the unit disk $|z| = 1$ onto the unit disk $|w| = 1$ in such a way that C_2 is mapped onto some cylinder C_2^*: $|w| = r_0$. By (6), Sec. 16.3, a linear fractional transformation mapping the unit disk onto the unit disk is

$$(2) \qquad\qquad w = \frac{z - b}{bz - 1}$$

where we have chosen $b = z_0$ real without restriction. z_0 is of no immediate help here because centers of circles do not map onto centers of the images, in general. However, we now have two free constants b and r_0 and shall succeed by imposing two reasonable conditions, namely, that 0 and 4/5 (Fig. 392) should be mapped onto r_0 and $-r_0$, respectively. This gives by (2)

$$r_0 = \frac{-b}{-1}, \qquad -r_0 = \frac{4/5 - b}{4b/5 - 1} = \frac{4/5 - r_0}{4r_0/5 - 1},$$

a quadratic equation in r_0 with solutions 2 (no good because $r_0 < 1$) and 1/2. Hence our mapping function (2) with $b = 1/2$ becomes

---

[3]Cf. Sec. 12.5. We mention without proof that if D^* is simply connected (Sec. 13.3), then a conjugate harmonic of Φ^* exists. Another proof without the use of a harmonic conjugate is given on p. 937 near the end of this chapter.

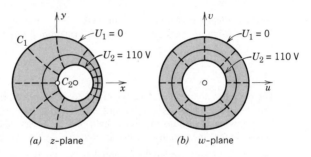

(a) z-plane   *(b) w-plane*

Fig. 392. Example 1

(3)
$$w = f(z) = \frac{2z - 1}{z - 2}.$$

From Example 5 in Sec. 17.1, writing w for z we have as the complex potential in the w-plane

$$F^*(w) = a \ln w + k$$

and from this the real potential

$$\Phi^*(u, v) = \text{Re } F^*(w) = a \ln |w| + k.$$

We determine a and k from the boundary conditions. When $|w| = 1$, then $\Phi^* = a \ln 1 + k = 0$, hence $k = 0$. When $|w| = 1/2$, then $\Phi^* = a \ln (1/2) = 110$, hence $a = 110/\ln (1/2) = -158.7$. Substitution of (3) now gives the desired solution in the given domain in the z-plane

$$F(z) = F^*(f(z)) = a \ln \frac{2z - 1}{z - 2}.$$

The real potential is

$$\Phi(x, y) = \text{Re } F(z) = a \ln \left| \frac{2z - 1}{z - 2} \right|, \qquad a = -158.7.$$

Can we "see" this result? Well, $\Phi(x, y) = const$ if and only if $|(2z - 1)/(z - 2)| = const$, that is, $|w| = const$ by (5). These circles correspond to circles in the z-plane, as follows from Theorem 1, Sec. 16.2, together with the fact that the inverse of a linear fractional transformation is linear fractional [see (5), Sec. 16.2]. Similarly for the rays arg $w = const$. Hence the equipotential lines $\Phi(x, y) = const$ are circles, and the lines of force are circular arcs (dashed in Fig. 392). These two families of curves intersect orthogonally, that is, at right angles, as shown in Fig. 392.

EXAMPLE 2. Potential between two semicircular plates

Find the potential between two semicircular plates P_1 and P_2 in Fig. 393a having potentials -3000 and 3000 volts, respectively. Use Example 3 in Sec. 17.1 and conformal mapping.

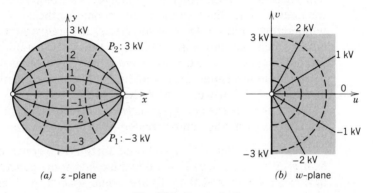

(a) z -plane   *(b) w-plane*

Fig. 393. Example 2

Solution. First step. We map the unit disk in Fig. 393a onto the right half of the w-plane (Fig. 393b) by using the linear fractional transformation in Example 3, Sec. 16.3:

$$w = f(z) = \frac{1 + z}{1 - z}.$$

The boundary $|z| = 1$ is mapped onto the boundary $u = 0$ (the v-axis), with $z = -1, i, 1$ going onto $w = 0, i, \infty$, respectively, and $z = -i$ onto $w = -i$. Hence the upper semicircle of $|z| = 1$ is mapped onto the upper half, and the lower semicircle onto the lower half of the v-axis, so that the boundary conditions in the w-plane are as indicated in Fig. 393b.

Second step. We determine the potential $\Phi^*(u, v)$ in the right half-plane of the w-plane. Example 3 in Sec. 17.1 with $\alpha = \pi$, $U_1 = -3000$ and $U_2 = 3000$ yields [with $\Phi^*(u, v)$ instead of $\Phi(x, y)$]

$$\Phi^*(u, v) = \frac{6000}{\pi} \varphi, \qquad\qquad \varphi = \text{arc tan } \frac{v}{u}.$$

On the positive half of the imaginary axis ($\varphi = \pi/2$), this equals 3000 and on the negative half -3000, as it should be. Φ^* is the real part of the complex potential

$$F^*(w) = -\frac{6000i}{\pi} \text{ Ln } w.$$

Third step. We substitute the mapping function into F^* to get the complex potential $F(z)$ in Fig. 393a in the form

$$F(z) = F^*(f(z)) = -\frac{6000i}{\pi} \text{ Ln } \frac{1 + z}{1 - z}.$$

The real part of this is the potential we wanted to determine:

$$\Phi(x, y) = \text{Re } F(z) = \frac{6000}{\pi} \text{ Im Ln } \frac{1 + z}{1 - z} = \frac{6000}{\pi} [\text{Arg } (1 + z) - \text{Arg } (1 - z)].$$

As in Example 1 we conclude that the equipotential lines $\Phi(x, y) = const$ are circular arcs because they correspond to Arg $[(1 + z)/(1 - z)] = const,$ hence to Arg $w = const.$ Also, Arg $w = const$ are rays from 0 to ∞, the images of $z = -1$ and $z = 1$, respectively. Hence the equipotential lines all have -1 and 1 (the points where the boundary potential jumps) as their endpoints (Fig. 393a). The lines of force are circular arcs, too, and since they must be orthogonal to the equipotential lines, their centers can be obtained as intersections of tangents to the unit circle with the x-axis. (Explain!)   ∎

Basic comment.

We formulated the examples in this section in terms of the electrostatic potential. It is quite important to realize that this is accidental. We could equally well have phrased everything in terms of (time-independent) heat flow; then instead of voltages, we would have had temperatures, the equipotential lines would have become isotherms (= lines of constant temperature) and the lines of the electrical force would have become lines along which heat flows from higher to lower temperatures. More on this in the next section. Or we could have talked about fluid flow; then the electrostatic equipotential lines would have become streamlines; more on this in Sec. 17.4. What we again see here is the **unifying power of mathematics:** different phenomena and systems from different areas in physics having the same types of model can be treated by the same mathematical methods.

This suggests our further program: Section 17.3 concerns two-dimensional heat problems. In Sec. 17.4 we first explain some general concepts needed in flow problems and then discuss some specific problems of practical interest.

Problems for Sec. 17.2

1. Verify Theorem 1 for $\Phi^*(u, v) = u^2 - v^2$ and $w = f(z) = e^z$.

2. Write down the formulas in the second proof of Theorem 1 and carry out all the steps in detail.

3. Find the potential Φ in the region R in the first quadrant of the z-plane bounded by the axes (having potential 0) and the hyperbola $y = 1/x$ (having potential U_2) in two ways: (i) directly, (ii) by mapping R onto a suitable infinite strip.

4. Verify the steps of the derivation of (3) from (2).

5. Applying a suitable conformal mapping, obtain from Fig. 393b the potential Φ in the angular region $-\pi/4 < \operatorname{Arg} z < \pi/4$, $\Phi = -3$ kV when $\operatorname{Arg} z = -\pi/4$ and $\Phi = 3$ kV when $\operatorname{Arg} z = \pi/4$.

6. Show that in Example 2, the y-axis is mapped onto the unit circle in the w-plane.

7. At $z = \pm 1$ in Fig. 393a the tangents to the equipotential lines shown make equal angles ($\pi/6$). Why?

8. In Example 2, set $z = Z^2$ ($Z = X + iY$) and show that the resulting potential $\Phi^*(X, Y) = \operatorname{Re} F(Z^2)$ is the potential in the portion of the unit disk $|Z| \leq 1$ in the first quadrant having the boundary values 0 on the axes and 3 on $|Z| = 1$.

9. Find the linear fractional transformation $z = g(Z)$ that maps $|Z| \leq 1$ onto $|z| \leq 1$ with $Z = i/2$ being mapped onto $z = 0$. Show that $Z_1 = (3 + 4i)/5$ is mapped onto $z = -1$, and $Z_2 = (-3 + 4i)/5$ onto $z = 1$, so that the equipotential lines of Example 2 look in $|Z| \leq 1$ as shown in Fig. 394.

10. Why are the equipotential lines in Prob. 9 circles?

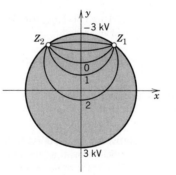

Fig. 394. Problem 9

17.3 Heat Problems

Heat conduction in a body of homogeneous material is governed by the **heat equation**

$$T_t = c^2 \nabla^2 T$$

where the function T is temperature, $T_t = \partial T/\partial t$, t is time and c^2 is a positive constant (depending on the material of the body). Hence if the problem is **"steady"** (that is, time-independent) and two-dimensional, the heat equation reduces to the two-dimensional Laplace equation

$$(1) \qquad\qquad \nabla^2 T = T_{xx} + T_{yy} = 0,$$

so that such a heat problem can be treated by our present methods. We call $T(x, y)$ the **heat potential.** The curves of constant temperature $T(x, y) = const$ are called **isotherms.** $T(x, y)$ is the real part of the complex potential

$$F(z) = T(x, y) + i\Psi(x, y).$$

The curves $\Psi(x, y) = const$ are the **heat flow lines,** the curves along which heat will flow from higher to lower temperatures.

It follows that all the examples considered so far (Secs. 17.1, 17.2) can now be reinterpreted as problems on heat flow. The electrostatic equipotential lines $\Phi(x, y) = const$ now become isotherms $T(x, y) = const,$ and the lines of electrical force become lines of heat flow. Mathematically, the calculations remain the same. New problems may arise, involving boundary conditions that would make no sense physically in electrostatics or would be of no practical interest there. Examples 3 and 4 (below) illustrate this.

Physically, to have a stationary problem, one must keep the boundaries of the domain in which the heat flow takes place at constant temperatures, by heating those which lose heat and cooling those toward which heat flows.

EXAMPLE 1. Temperature between parallel plates

Find the temperature between two parallel plates $x = 0$ and $x = d$ in Fig. 395 having temperatures 0 and 100 °C, respectively.

Solution. As in Sec. 17.1 we conclude that $T(x, y) = ax + b$. From the boundary conditions, $b = 0$ and $a = 100/d$. The answer is

$$T(x, y) = \frac{100}{d} x \ [\°C].$$

The corresponding complex potential is $F(z) = (100/d)z$. Heat flows horizontally, in the negative x-direction, along the lines $y = const$.

EXAMPLE 2. Temperature distribution between a wire and a cylinder

Find the temperature field around a long thin wire of radius $r_1 = 1$ mm that is electrically heated to $T_1 = 500$ °F and is surrounded by a circular cylinder of radius $r_2 = 100$ mm, which is kept at temperature $T_2 = 60$ °F by cooling it with air. Cf. Fig. 396.

Solution. T depends only on r, for reasons of symmetry. Hence, as in Sec. 17.1 (Example 2),

$$T(x, y) = a \ln r + b.$$

The boundary conditions are

$$T_1 = 500 = a \ln 1 + b, \qquad T_2 = 60 = a \ln 100 + b.$$

Hence $b = 500$ (since $\ln 1 = 0$) and $a = (60 - b)/\ln 100 = -95.54$. The answer is

$$T(x, y) = 500 - 95.54 \ln r \ [\°F].$$

The isotherms are concentric circles. Heat flows from the wire radially outward to the cylinder.

EXAMPLE 3. A mixed boundary value problem

Find the temperature distribution in the region in Fig. 397 (cross section of a solid cylinder), whose vertical portion of the boundary is at 20 °C, the horizontal portion at 50 °C, and the circular portion is insulated.

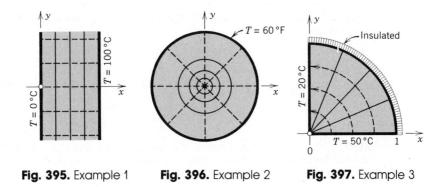

Fig. 395. Example 1   **Fig. 396.** Example 2   **Fig. 397.** Example 3

Solution. The insulated portion of the boundary must be a heat flow line, since the insulation prevents heat from crossing such a curve, hence it must flow along. Thus the isotherms must meet such a curve at right angles. Since T is constant along an isotherm, this means that

(2) $$\frac{\partial T}{\partial n} = 0 \qquad \text{along an isolated portion of the boundary,}$$

where n is a coordinate tangent to the isotherm and hence "normal" ($=$ perpendicular) to the insulated boundary curve. Such a problem in which T is prescribed on one portion of the boundary and $\partial T/\partial n$ on the other portion is called a **mixed boundary value problem.**

In our case, $n = r$ is normal to the circular insulated boundary, so that (2) is $\partial T/\partial r = 0$, meaning that along this curve the solution must not depend on r. Now Arg $z = \theta$ satisfies (1) as well as this condition, and is constant (0 and $\pi/2$) on the straight portions of the boundary. Hence the solution is of the form

$$T(x, y) = a\theta + b.$$

The boundary conditions yield $a(\pi/2) + b = 20$ and $a \cdot 0 + b = 50$. This gives

$$T(x, y) = 50 - \frac{60}{\pi}\,\theta, \qquad\qquad \theta = \arctan \frac{y}{x}.$$

The isotherms are portions of rays $\theta = const$. Heat flows from the x-axis along circles $r = const$ (dashed in Fig. 397) to the y-axis.

EXAMPLE 4. Another mixed boundary value problem in heat conduction

Find the temperature field in the upper half-plane when the x-axis is at $T = 0$ °C for $x < -1$, insulated for $-1 < x < 1$, and at $T = 20$ °C for $x > 1$ (Fig. 398a).

Solution. $w = \sin z$ maps the strip $-\pi/2 < x < \pi/2$, $y > 0$ onto the upper half-plane, with lines $x = const$ and $y = const$ being mapped onto hyperbolas and ellipses, respectively (Fig. 376 in Sec. 16.4). Hence $w = f(z) = \sin^{-1} z$ maps that half-plane onto that strip (Fig. 398b), the insulated segment being mapped onto the segment $-\pi/2 < u < \pi/2$ of the u-axis. There the potential $T^*(u, v)$ must satisfy $\partial T^*/\partial n = \partial T^*/\partial v = 0$ because v is a coordinate normal to that

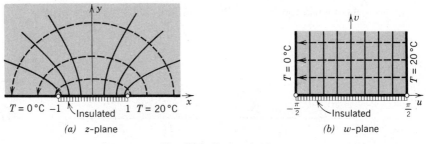

(a) z-plane   *(b) w-plane*

Fig. 398. Example 4

segment; that is, T^* must not depend on v. In that strip the solution satisfying the indicated boundary conditions is

$$T^*(u, v) = 10 + \frac{20}{\pi} u = \text{Re } F^*(w)$$

with the complex potential given by $F^*(w) = 10 + (20/\pi)w$. Hence the complex potential in the z-plane is

$$F(z) = F^*(f(z)) = 10 + \frac{20}{\pi} \sin^{-1} z$$

and $T(x, y) = \text{Re } F(z)$ is the solution. The isotherms are hyperbolas, and the heat flow lines are confocal ellipses. This follows from the property of $w = \sin z$ mentioned at the beginning. Cf. Fig. 398a.   ∎

This is the second time we have used complex potentials, and we shall meet them again in the next section on **fluid flow,** which is another area in which potential theory plays a role, provided we make suitable assumptions about the type of flow, which we state clearly and explicitly in the next section.

Problems for Sec. 17.3

1. Find the temperature between two parallel plates $y = 0$ and $y = d$ which are kept at temperatures 0 and 100 °C, respectively. (i) Proceed directly. (ii) Use Example 1 and a suitable mapping.
2. Find the temperature and the complex potential in an infinite plate with edges $y = x - 1$ and $y = x + 1$ kept at -20 and 40 °C, respectively.
3. Find the temperature in Fig. 397 if $T = 80$ °C on the y-axis, $T = -40$ °C on the x-axis and the circular portion of the boundary is insulated as before.
4. Find the temperature T in the sector $0 \leq \text{Arg } z \leq \pi/4$, $|z| \leq 1$, if $T = 100$ °C on the x-axis and $T = 20$ °C on $y = x$.
5. Find the temperature and the complex potential in the first quadrant of the z-plane when the y-axis is kept at 100 °C, the segment $0 < x < 1$ of the x-axis is insulated and the portion $x > 1$ of the x-axis is kept at 200 °C. *Hint.* Use Example 4.
6. Interpret Problem 9 in Sec. 17.2 as a heat flow problem. Along what curves does the heat flow?
7. Find the temperature T^* in the upper half-plane subject to the boundary conditions shown in Fig. 399.
8. Find the complex potential in Prob. 7.
9. Find the temperature in the upper half-plane in Fig. 400 satisfying the given boundary conditions.
10. Find the complex potential in Prob. 9.
11. Obtain the result of Prob. 7 from Prob. 9 by superposition.

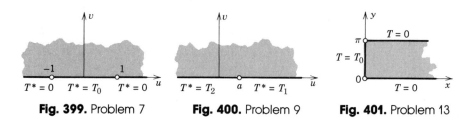

Fig. 399. Problem 7   **Fig. 400.** Problem 9   **Fig. 401.** Problem 13

12. What temperature in the first quadrant of the z-plane is obtained from Prob. 9 by the mapping $w = a + z^2$, and what are the transformed boundary conditions?

13. Using Prob. 7 and the mapping $w = \cosh z$ (cf. Example 4 in Sec. 16.4), show that the temperature in the infinite strip in Fig. 401 is

$$T(x, y) = \frac{T_0}{\pi} \operatorname{Arg} \frac{\cosh z - 1}{\cosh z + 1} = \frac{2T_0}{\pi} \operatorname{Arg} \left(\tanh \frac{z}{2} \right).$$

14. Show by direct calculation that $(\operatorname{Im} \tanh \frac{1}{2}z)/(\operatorname{Re} \tanh \frac{1}{2}z) = (\sin y)/(\sinh x)$, so that $T(x, y) = (2T_0/\pi)\arctan [(\sin y)/(\sinh x)]$ in Prob. 13.

Find the temperature $T(x, y)$ in the given thin metal plate whose faces are insulated and whose edges are kept at the temperatures shown in the figure.

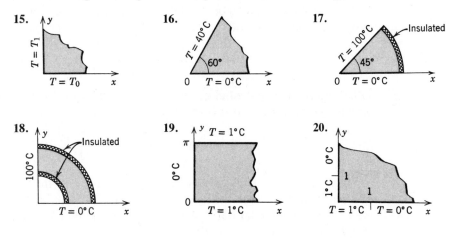

17.4 Fluid Flow

Harmonic functions also play an important role in hydrodynamics. To illustrate this, let us consider the two-dimensional steady motion of a nonviscous fluid. Here **"two-dimensional"** means that the motion of the fluid is the same in all planes parallel to the xy-plane, the velocity being parallel to this plane. It then suffices to consider the motion of the fluid in the xy-plane. **"Steady"** means that the velocity is independent of time.

At any point (x, y) the flow has a certain velocity which is determined by its magnitude and its direction and thus is a vector. Since in the complex plane any number a represents a vector (the vector from the origin to the point corresponding to a), we may represent the velocity of the flow by a complex variable, say,

(1) $$V = V_1 + iV_2.$$

Then V_1 and V_2 are the components of the velocity in the x and y directions, and V is tangential to the paths of the moving particles of the fluid. Such a path is called a **streamline** of the motion. Cf. Fig. 402 on the next page.

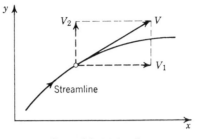

Fig. 402. Velocity

We shall show that under suitable assumptions, for a given flow there exists an analytic function

(2)
$$F(z) = \Phi(x, y) + i\Psi(x, y),$$

called the **complex potential** of the flow, such that the streamlines are given by $\Psi(x, y) = const,$ and the velocity is given by

(3)
$$V = V_1 + iV_2 = \overline{F'(z)}$$

where the bar denotes the complex conjugate. Ψ is called the **stream function.** The function Φ is called the **velocity potential.**[4] The curves $\Phi(x, y) = const$ are called **equipotential lines.** We shall also show that V is the **gradient** of Φ; by definition, this means that

(4)
$$V_1 = \frac{\partial \Phi}{\partial x}, \qquad V_2 = \frac{\partial \Phi}{\partial y}.$$

Furthermore, since $F(z)$ is analytic, Φ and Ψ satisfy Laplace's equation

(5)
$$\nabla^2 \Phi = \frac{\partial^2 \Phi}{\partial x^2} + \frac{\partial^2 \Phi}{\partial y^2} = 0, \qquad \nabla^2 \Psi = \frac{\partial^2 \Psi}{\partial x^2} + \frac{\partial^2 \Psi}{\partial y^2} = 0.$$

Whereas in electrostatics the boundaries (conducting plates) are equipotential lines, in fluid flow a boundary across which fluid cannot flow must be a streamline. Hence in fluid flow the stream function is of particular importance.

Before we discuss the conditions under which the statements involving (2)—(5) are valid, let us consider two flows of practical interest. Note that further flows are obtained by interpreting some of the examples in the previous sections as flow problems. Further flows are included in the problem set.

---

[4]Some authors use $-\Phi$ (instead of Φ) as the velocity potential.

EXAMPLE 1. Flow around a corner

The complex potential

$$F(z) = z^2 = x^2 - y^2 + 2ixy$$

describes a flow whose equipotential lines are the hyperbolas

$$\Phi = x^2 - y^2 = const$$

and whose streamlines are the hyperbolas

$$\Psi = 2xy = const.$$

From (3) we obtain the velocity vector

$$V = 2\bar{z} = 2(x - iy), \qquad \text{that is,} \qquad V_1 = 2x, \qquad V_2 = -2y.$$

The speed (magnitude of the velocity) is

$$|V| = \sqrt{V_1{}^2 + V_2{}^2} = 2\sqrt{x^2 + y^2}.$$

The flow may be interpreted as the flow in a channel bounded by the positive coordinates axes and a hyperbola, say, $xy = 1$ (Fig. 403). We note that the speed along a streamline S has a minimum at the point P where the cross section of the channel is large.

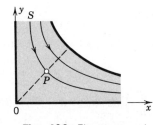

Fig. 403. Flow around a corner (Example 1)

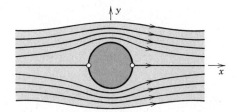

Fig. 404. Flow around a cylinder (Example 2)

EXAMPLE 2. Flow around a cylinder

Consider the complex potential

$$F(z) = \Phi(x, y) + i\Psi(x, y) = z + \frac{1}{z}.$$

Using the polar form $z = re^{i\theta}$, we obtain

$$F(z) = re^{i\theta} + \frac{1}{r}e^{-i\theta} = \left(r + \frac{1}{r}\right)\cos\theta + i\left(r - \frac{1}{r}\right)\sin\theta.$$

Hence the streamlines are

$$\Psi(x, y) = \left(r - \frac{1}{r}\right)\sin\theta = const.$$

In particular, $\Psi(x, y) = 0$ gives $r - 1/r = 0$ or $\sin\theta = 0$. Hence this streamline consists of the unit circle ($r = 1/r$ gives $r = 1$) and the x-axis ($\theta = 0$ and $\theta = \pi$). For large $|z|$ the term $1/z$ in $F(z)$ is small in absolute value, so that for these z the flow is nearly uniform and parallel to the x-axis. Hence we can interpret this as a flow around a long circular cylinder of unit radius. The flow has two **stagnation points** (i.e., points at which the velocity $V = 0$), at $z = \pm1$. This follows from $F'(z) = 1 - z^{-2}$ and (3). See Fig. 404. ∎

Assumptions and Theory Underlying (2)—(5)

If the domain of flow is simply connected and the flow is irrotational and incompressible, then the statements involving (2)—(5) hold. In particular, then the flow has a complex potential F(z), which is an analytic function. (Explanation of terms below.)

We prove this, along with a discussion of basic concepts related to fluid flow. Consider any smooth curve C in the z-plane, represented by $z(s) = x(s) + iy(s)$, where s is the arc length of C. Let the real variable V_t be the component of the velocity V tangent to C (Fig. 405). Then the value of the real line integral

$$(6) \qquad \int_C V_t \, ds$$

taken along C in the sense of increasing values of s is called the **circulation** of the fluid along C. Dividing the circulation by the length of C, we obtain the mean velocity[5] of the flow along the curve C. Now

$$V_t = |V| \cos \alpha \qquad \text{(cf. Fig. 405)}.$$

Hence V_t is the dot product (Sec. 6.5) of V and the tangent vector dz/ds of C (Sec. 16.1); thus in (6),

$$V_t \, ds = \left(V_1 \frac{dx}{ds} + V_2 \frac{dy}{ds} \right) ds = V_1 \, dx + V_2 \, dy.$$

The circulation (6) along C now becomes

$$(7) \qquad \int_C V_t \, ds = \int_C (V_1 \, dx + V_2 \, dy).$$

As the next idea, let C be a *closed* curve, namely, the boundary of a simply connected domain D, and suppose that V has continuous partial derivatives in a domain containing D and C. Then we can use Green's theorem (Sec. 9.3) to represent the circulation around C by a double integral,

$$(8) \qquad \oint_C (V_1 \, dx + V_2 \, dy) = \iint_D \left(\frac{\partial V_2}{\partial x} - \frac{\partial V_1}{\partial y} \right) dx \, dy.$$

---

[5]*Definitions:* $\dfrac{1}{b-a} \displaystyle\int_a^b f(x) \, dx = $ mean value of f on the interval $a \leq x \leq b$

$\dfrac{1}{L} \displaystyle\int_C f(s) \, ds = $ mean value of f on C $\quad$ ($L = $ length of C)

$\dfrac{1}{A} \displaystyle\iint_D f(x, y) \, dx \, dy = $ mean value of f on D $\quad$ ($A = $ area of D)

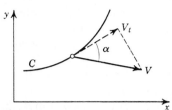

Fig. 405. Tangential component of the velocity
with respect to a curve C

The integrand of this double integral is called the **vorticity** of the flow, and the vorticity divided by 2,

$$(9) \qquad \omega(x, y) = \frac{1}{2} \left(\frac{\partial V_2}{\partial x} - \frac{\partial V_1}{\partial y} \right),$$

is called the **rotation.** If $\omega(x, y) = 0$ throughout the flow, we say that the flow is **irrotational.** We make this assumption; thus

$$(10) \qquad \frac{\partial V_2}{\partial x} - \frac{\partial V_1}{\partial y} = 0.$$

To understand the physical meaning of vorticity and rotation, take for C in (8) a circle of radius r. Then the circulation divided by the length $2\pi r$ of C is the mean velocity of the fluid along C. Hence by dividing this by r we obtain the mean *angular* velocity ω_0 of the fluid about the axis of the circle:

$$\omega_0 = \frac{1}{2\pi r^2} \iint_D \left(\frac{\partial V_2}{\partial x} - \frac{\partial V_1}{\partial y} \right) dx \, dy = \frac{1}{\pi r^2} \iint_D \omega(x, y) \, dx \, dy.$$

If we now let $r \to 0$, the limit of ω_0 is the value of ω at the center of C. Hence, $\omega(x, y)$ is the limiting angular velocity of a circular element of the fluid as the circle shrinks to the point (x, y). Roughly speaking, if a spherical element of the fluid were suddenly solidified and the surrounding fluid simultaneously annihilated, the element would rotate with the angular velocity ω.

Our second assumption is that the fluid is *incompressible.* (Examples are water and oil, whereas air is compressible.) Then

$$(11) \qquad \frac{\partial V_1}{\partial x} + \frac{\partial V_2}{\partial y} = 0 \qquad\qquad \text{[cf. (7), Sec. 8.9]}$$

in every region that is free of **sources** or **sinks,** that is, points at which fluid is produced or disappears. (The expression in (11) is called the *divergence* of V and is denoted by div V.)

If the domain D of the flow is simply connected and the flow is irrotational, then the line integral (7) is independent of path in D (by Theorem 3 in Sec. 9.9). Hence if we integrate from a fixed point (a, b) in D to a variable point (x, y) in D, the integral becomes a function of the point (x, y), say, $\Phi(x, y)$:

(12)
$$\Phi(x, y) = \int_{(a, b)}^{(x, y)} (V_1 \, dx + V_2 \, dy).$$

We claim that Φ is the desired velocity potential. To prove this, all we have to do is to show that (4) holds. Now since the integral (7) is independent of path, $V_1 \, dx + V_2 \, dy$ is an exact differential (Sec. 9.9), namely, the differential of Φ, that is,

$$V_1 \, dx + V_2 \, dy = \frac{\partial \Phi}{\partial x} \, dx + \frac{\partial \Phi}{\partial y} \, dy.$$

From this we see that $V_1 = \partial \Phi / \partial x$ and $V_2 = \partial \Phi / \partial y$, which gives (4).

We next show that Φ is harmonic. But this follows at once by substituting (4) into (11), giving the first Laplace's equation in (5).

We finally take a conjugate harmonic Ψ of Φ. Then the other equation in (5) holds. Also, assuming that the second partial derivatives of Φ and Ψ are continuous, we have that the complex function

$$F(z) = \Phi(x, y) + i\Psi(x, y)$$

is analytic in D. Since the curves $\Psi(x, y) = const$ are perpendicular to the equipotential curves $\Phi(x, y) = const$ (except where $F'(z) = 0$), we conclude that they are the streamlines. Hence Ψ is the stream function and $F(z)$ the complex potential of the flow.

Also, by the second Cauchy–Riemann equation and by (4),

$$F'(z) = \frac{\partial \Phi}{\partial x} + i \frac{\partial \Psi}{\partial x} = \frac{\partial \Phi}{\partial x} - i \frac{\partial \Phi}{\partial y} = V_1 - iV_2,$$

which yields (3). All our statements are now proved.   ∎

The first four sections of this chapter served to illustrate the use of complex analysis in various areas of physics, in connection with potential problems in electrostatics (Secs. 17.1, 17.2), heat flow (Sec. 17.3) and fluid flow (Sec. 17.4). Common to these sections was the use of analytic functions (the complex potential) and conformal mapping. In the next section we show that **complex integration** also plays a basic role in potential theory: from Cauchy's integral formula we derive a solution formula for the Dirichlet problem in a standard domain (a circular disk) in the form of an integral (**Poisson's integral formula**) and from it an interesting series development of the potential in the disk. This series will be related to *Fourier series*.

Problems for Sec. 17.4

1. **(Parallel flow)** Show that $F(z) = Kz$ (K positive real) describes a uniform flow to the right, which can be interpreted as a uniform flow between two parallel lines (between two parallel planes in three-dimensional space). Cf. Fig. 406. Find the velocity vector, the streamlines and the equipotential lines.

2. Obtain the flow in Example 1 from that in Prob. 1 by a conformal mapping of the first quadrant onto the upper half-plane.

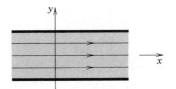

Fig. 406. Parallel flow in Problem 1

3. Show that $F(z) = iz^2$ describes a flow around a corner. Find and graph the streamlines and the equipotential lines. Find the velocity vector V.

4. Make a slight change of $F(z)$ in Example 2 so that you obtain a flow around a cylinder of radius r_0, that yields the flow in Prob. 1 as $r_0 \to 0$.

Consider the flow corresponding to the given complex potential $F(z)$. Show the streamlines and equipotential lines graphically. Find the velocity vector V. Determine all points at which V is parallel to the x-axis. (Assume k in Prob. 6 to be real.)

5. iz **6.** $-ikz$ **7.** $(1 + i)z$ **8.** $z^2 + z$ **9.** z^3 **10.** iz^3

11. (**Source and sink**) Consider the complex potential $F(z) = (c/2\pi) \ln z$, where c is positive real. Show that $V = (c/2\pi r^2)(x + iy)$, where $r = \sqrt{x^2 + y^2}$, and this implies that the flow is directed radially outward (cf. Fig. 407), so that this potential corresponds to a **point source** at $z = 0$ (that is, a **source line** $x = 0$, $y = 0$ in space). (The constant c is called the **strength** or **discharge** of the source. If c is negative real, the flow is said to have a **sink** at $z = 0$, it is directed radially inward and fluid disappears at the singular point $z = 0$ of the complex potential.)

12. (**Vortex line**) Show that $F(z) = -(iK/2\pi) \ln z$ with real positive K describes a flow circulating around the origin in the counterclockwise sense; cf. Fig. 408. (The point $z = 0$ is a **vortex**; the potential increases by K each time we travel around the vortex.)

13. Find the complex potential of a flow that has a point source of strength 1 at $z = -a$.

14. Find the complex potential of a flow that has a point sink of strength 1 at $z = a$.

15. Show that vector addition of the velocity vectors of two flows leads to a flow whose complex potential is obtained by adding those of the two flows.

16. Add the potentials in Probs. 13 and 14 and show the streamlines graphically.

17. Find the streamlines of the flow corresponding to $F(z) = 1/z$. Show that for small $|a|$ the streamlines in Prob. 16 look similar to those in the present problem.

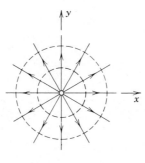

Fig. 407. Point source
in Problem 11

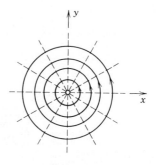

Fig. 408. Vortex flow
in Problem 12

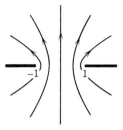

Fig. 409. Flow through an aperture in Problem 18

Fig. 410. Flow around a plate in Problem 19

18. Show that $F(z) = \cosh^{-1} z$ corresponds to a flow whose streamlines are confocal hyperbolas with foci at $z = \pm 1$, and the flow may be interpreted as a flow through an aperture. (Cf. Fig. 409.)

19. Show that $F(z) = \cos^{-1} z$ can be interpreted as the complex potential of a flow circulating around an elliptic cylinder or a plate (the straight segment from $z = -1$ to $z = 1$). Show that the streamlines are confocal ellipses with foci at $z = \pm 1$. (Cf. Fig. 410.)

20. **(Flow with circulation around a cylinder)** Show that addition of the potentials in Prob. 12 and Example 2 gives a flow such that the cylinder wall $|z| = 1$ is a streamline. Find the speed and show that the stagnation points are

$$ z = \frac{iK}{4\pi} \pm \sqrt{\frac{-K^2}{16\pi^2} + 1}; $$

when $K = 0$ they are at ± 1, as K increases they move up on the unit circle until they unite at $z = i$ ($K = 4\pi$, cf. Fig. 411), and when $K > 4\pi$ they lie on the imaginary axis (one lies in the field of flow and the other one lies inside the cylinder and has no physical meaning).

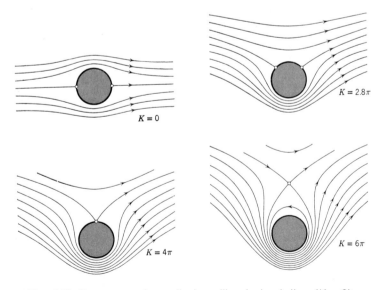

Fig. 411. Flow around a cylinder without circulation ($K = 0$) and with circulation (Problem 20)

17.5 Poisson's Integral Formula

If we want to solve the Dirichlet problem for the Laplace equation in a disk, or in a domain D for which we know the analytic function mapping D conformally onto a disk, we can use the Poisson formula.[6] This formula gives the potential in the disk in terms of given boundary values on the boundary circle of the disk. Let us derive this important formula.

We start from Cauchy's integral formula

$$(1) \qquad F(z) = \frac{1}{2\pi i} \oint_C \frac{F(z^*)}{z^* - z}\, dz^*.$$

Here we integrate over the circle

$$C: \qquad z^* = Re^{i\alpha} \qquad\qquad (0 \le \alpha \le 2\pi)$$

and we assume that the function $F(z)$ is analytic inside and on C.

Since $dz^* = iRe^{i\alpha}\, d\alpha = iz^*\, d\alpha$, we obtain from (1)

$$(2) \qquad F(z) = \frac{1}{2\pi} \int_0^{2\pi} F(z^*) \frac{z^*}{z^* - z}\, d\alpha \qquad\qquad (z^* = Re^{i\alpha}, \, z = re^{i\theta}).$$

On the other hand, if instead of z we take a point Z outside C, say, the point $Z = z^*\bar{z}^*/\bar{z}$ (whose absolute value is $R^2/r > R$), then the integrand in (1) is analytic in the disk $|z| \le R$, so that the integral is zero by Cauchy's theorem:

$$0 = \frac{1}{2\pi i} \oint_C \frac{F(z^*)}{z^* - Z}\, dz^* = \frac{1}{2\pi} \int_0^{2\pi} F(z^*) \frac{z^*}{z^* - Z}\, d\alpha.$$

If we insert $Z = z^*\bar{z}^*/\bar{z}$ and simplify the fraction, we find

$$0 = \frac{1}{2\pi} \int_0^{2\pi} F(z^*) \frac{\bar{z}}{\bar{z} - \bar{z}^*}\, d\alpha.$$

We subtract this from (2) and use

$$(3) \qquad \frac{z^*}{z^* - z} - \frac{\bar{z}}{\bar{z} - \bar{z}^*} = \frac{z^*\bar{z}^* - z\,\bar{z}}{(z^* - z)(\bar{z}^* - \bar{z})},$$

a formula that can be readily verified. We then obtain

$$(4) \qquad F(z) = \frac{1}{2\pi} \int_0^{2\pi} F(z^*) \frac{z^*\bar{z}^* - z\,\bar{z}}{(z^* - z)(\bar{z}^* - \bar{z})}\, d\alpha.$$

[6]SIMÉON DENIS POISSON (1781—1840), French mathematician and physicist, professor in Paris from 1809. His work includes potential theory, partial differential equations (Poisson equation; cf. Sec. 11.1) and probability.

From the polar representations of z and z^* we see that the quotient in the integrand is equal to

$$\frac{R^2 - r^2}{(Re^{i\alpha} - re^{i\theta})(Re^{-i\alpha} - re^{-i\theta})} = \frac{R^2 - r^2}{R^2 - 2Rr \cos (\theta - \alpha) + r^2}.$$

Consequently, writing

$$F(z) = \Phi(r, \theta) + i\Psi(r, \theta)$$

and taking the real part on both sides of formula (4), we obtain **Poisson's integral formula**

(5)
$$\Phi(r, \theta) = \frac{1}{2\pi} \int_0^{2\pi} \Phi(R, \alpha) \frac{R^2 - r^2}{R^2 - 2Rr \cos (\theta - \alpha) + r^2} \, d\alpha$$

which represents the harmonic function Φ in the disk $|z| \leq R$ in terms of its values $\Phi(R, \alpha)$ on the circle that bounds the disk.

It is of practical interest to note that in (5) we may use for $\Phi(R, \alpha)$ any function that is merely piecewise continuous on the interval of integration. Formula (5) then *defines* a function $\Phi(r, \theta)$ that is harmonic in the open disk $|z| < R$ and continuous on the circle $|z| = R$. On this circle this function is equal to $\Phi(R, \alpha)$, except at points where $\Phi(R, \alpha)$ is discontinuous. The proof can be found in Ref. [D1].

Series Representation of Potential in a Disk

From (5) we may obtain an important series development of Φ in terms of simple harmonic functions. We remember that the quotient in the integrand of (5) was derived from (3), and it is not difficult to see that the right-hand side of (3) is the real part of $(z^* + z)/(z^* - z)$. By the use of the geometric series we obtain

(6)
$$\frac{z^* + z}{z^* - z} = \frac{1 + (z/z^*)}{1 - (z/z^*)} = \left(1 + \frac{z}{z^*}\right) \sum_{n=0}^{\infty} \left(\frac{z}{z^*}\right)^n = 1 + 2 \sum_{n=1}^{\infty} \left(\frac{z}{z^*}\right)^n.$$

Since $z = re^{i\theta}$ and $z^* = Re^{i\alpha}$, we have

$$\text{Re} \left(\frac{z}{z^*}\right)^n = \text{Re} \left[\frac{r^n}{R^n} e^{in\theta} e^{-in\alpha}\right] = \left(\frac{r}{R}\right)^n \cos (n\theta - n\alpha).$$

On the right, $\cos (n\theta - n\alpha) = \cos n\theta \cos n\alpha + \sin n\theta \sin n\alpha$, so that from (6) we obtain

$$\text{Re} \frac{z^* + z}{z^* - z} = 1 + 2 \sum_{n=1}^{\infty} \left(\frac{r}{R}\right)^n (\cos n\theta \cos n\alpha + \sin n\theta \sin n\alpha).$$

This expression is equal to the quotient in (5), as we have mentioned before, and by inserting the series into (5) and integrating term by term we find

(7)
$$\Phi(r,\,\theta) = a_0 + \sum_{n=1}^{\infty} \left(\frac{r}{R}\right)^n (a_n \cos n\theta + b_n \sin n\theta)$$

where the coefficients are

$$a_0 = \frac{1}{2\pi} \int_0^{2\pi} \Phi(R,\,\alpha)\, d\alpha, \qquad a_n = \frac{1}{\pi} \int_0^{2\pi} \Phi(R,\,\alpha) \cos n\alpha\, d\alpha,$$

(8)

$$b_n = \frac{1}{\pi} \int_0^{2\pi} \Phi(R,\,\alpha) \sin n\alpha\, d\alpha, \qquad n = 1,\, 2,\, \cdots,$$

the Fourier coefficients of $\Phi(R,\,\alpha)$. Note that for $r = R$ the series (7) becomes the Fourier series of $\Phi(R,\,\alpha)$, and therefore the representation (7) will be valid whenever $\Phi(R,\,\alpha)$ can be represented by a Fourier series.

EXAMPLE 1. Dirichlet problem for the unit disk

Find the electrostatic potential $\Phi(r,\,\theta)$ in the unit disk $r < 1$ having the boundary values

$$\Phi(1,\,\alpha) = \begin{cases} -\alpha/\pi & \text{if} \quad -\pi < \alpha < 0 \\ \alpha/\pi & \text{if} \quad 0 < \alpha < \pi. \end{cases} \qquad \text{(Fig. 412)}$$

Solution. Since $\Phi(1,\,\alpha)$ is even, $b_n = 0$, and from (8) we obtain $a_0 = \frac{1}{2}$ and

$$a_n = \frac{1}{\pi} \left[-\int_{-\pi}^0 \frac{\alpha}{\pi} \cos n\alpha\, d\alpha + \int_0^{\pi} \frac{\alpha}{\pi} \cos n\alpha\, d\alpha \right] = \frac{2}{n^2 \pi^2} (\cos n\pi - 1).$$

Hence, $a_n = -4/n^2\pi^2$ if n is odd, $a_n = 0$ if $n = 2,\, 4,\, \cdots$, and the potential is

$$\Phi(r,\,\theta) = \frac{1}{2} - \frac{4}{\pi^2} \left[r \cos \theta + \frac{r^3}{3^2} \cos 3\theta + \frac{r^5}{5^2} \cos 5\theta + \cdots \right].$$

Figure 413 shows the unit disk and some of the equipotential lines (curves $\Phi = const$).   ∎

In the next section, the last in this chapter, we show how general properties of analytic functions lead to properties (notably mean value properties and the maximum modulus principle) that are common to all harmonic functions, regardless of their particular form.

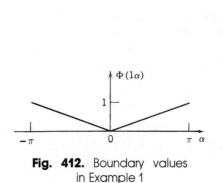

Fig. 412. Boundary values in Example 1

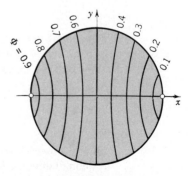

Fig. 413. Potential in Example 1

Problems for Sec. 17.5

1. Verify (3).
2. Show that each term in (7) is a harmonic function in the disk $r^2 < R^2$.

Using (7), find the potential $\Phi(r, \theta)$ in the unit disk $r < 1$ having the given boundary values $\Phi(1, \theta)$. Using the first few terms of the series, compute some values of Φ and sketch a figure of the equipotential lines.

3. $\Phi(1, \theta) = 2 - \cos \theta$
4. $\Phi(1, \theta) = 3 \sin \theta$
5. $\Phi(1, \theta) = \frac{1}{2} \sin 5\theta$
6. $\Phi(1, \theta) = \cos 4\theta - \cos 2\theta$
7. $\Phi(1, \theta) = \cos^2 \theta$
8. $\Phi(1, \theta) = \theta$ if $-\pi < \theta < \pi$
9. $\Phi(1, \theta) = 4 \sin^3 \theta$
10. $\Phi(1, \theta) = \theta$ if $0 < \theta < 2\pi$
11. $\Phi(1, \theta) = 1$ if $0 < \theta < \pi$ and 0 otherwise
12. $\Phi(1, \theta) = 1$ if $0 < \theta < \pi/2$, $\Phi(1, \theta) = -1$ if $\pi/2 < \theta < \pi$ and 0 otherwise
13. $\Phi(1, \theta) = \theta$ if $-\pi/2 < \theta < \pi/2$, $\Phi(1, \theta) = \pi - \theta$ if $\pi/2 < \theta < 3\pi/2$
14. $\Phi(1, \theta) = -\pi/2$ if $-\pi < \theta < -\pi/2$, $\Phi(1, \theta) = \theta$ if $-\pi/2 < \theta < \pi/2$, $\Phi(1, \theta) = \pi/2$ if $\pi/2 < \theta < \pi$

15. Using (8) in Sec. 14.6, show that the result of Prob. 12 may be written

$$\Phi(r, \theta) = \frac{1}{\pi} \text{ Im Ln } \frac{(1 + iz)(1 + z^2)}{(1 - iz)(1 - z^2)}.$$

16. Show that the potential in Prob. 8 may be written $\Phi(r, \theta) = 2 \text{ Im Ln } (1 + z)$.

17. Using (7), show that the potential $\Phi(r, \theta)$ in the unit disk $r < 1$ having the boundary values -1 if $-\pi < \theta < 0$ and 1 if $0 < \theta < \pi$ is given by the series

$$\Phi(r, \theta) = \frac{4}{\pi} \left(r \sin \theta + \frac{r^3}{3} \sin 3\theta + \frac{r^5}{5} \sin 5\theta + \cdots \right).$$

Compute some values of Φ by the use of the first few terms of this series and sketch some of the equipotential lines. Compare the result with Fig. 414. Graph the lines of force (orthogonal trajectories).

18. Using (8), Sec. 14.6, show that in Prob. 17,

$$\Phi(r, \theta) = \frac{2}{\pi} \text{ Im Ln } \frac{1 + z}{1 - z} = \frac{2}{\pi} [\text{Arg } (1 + z) - \text{Arg } (1 - z)].$$

19. Show that

$$F^*(w) = 1 + \frac{2}{\pi} \text{ Im Ln } \frac{w + 1}{w - 1} \qquad (w = u + iv)$$

is harmonic in the upper half-plane $v > 0$ and has the values -1 for $-1 < u < 1$ and $+1$ on the remaining part of the u-axis.

20. Show that the linear fractional transformation which maps $w_1 = -1$, $w_2 = 0$, $w_3 = 1$ onto $z_1 = -1$, $z_2 = -i$, $z_3 = 1$, respectively, is

$$z = \frac{w - i}{-iw + 1}.$$

Find the inverse $w = w(z)$, insert it into F^* in Prob. 19, and show that the resulting harmonic function is that in Prob. 18. Compare also with Example 2 in Sec. 17.2.

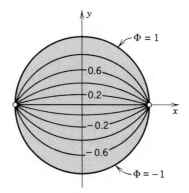

Fig. 414. Potential in Problem 17

17.6 General Properties of Harmonic Functions

In this section we show how results about complex analytic functions can be used for deriving general properties of harmonic functions.

If $\Phi(x, y)$ is harmonic in a domain D and D is simply connected, then Φ has a conjugate harmonic Ψ in D and

$$F(z) = \Phi(x, y) + i\Psi(x, y)$$

is analytic in D. (Cf. Sec. 12.5 and footnote 3 in Sec. 17.2.) Since an analytic function has derivatives of all orders, our first result is

Theorem 1 (Partial derivatives)

A function $\Phi(x, y)$ that is harmonic in a simply connected domain D has partial derivatives of all orders in D.

Furthermore, if $F(z)$ is analytic in a simply connected domain D, then, by Cauchy's integral formula (Sec. 13.5),

$$(1) \qquad F(z_0) = \frac{1}{2\pi i} \oint_C \frac{F(z)}{z - z_0} \, dz$$

where C is a simple closed path in D and z_0 lies inside C. Choosing for C a circle

$$z = z_0 + re^{i\alpha}$$

in D, we have

$$z - z_0 = re^{i\alpha}, \qquad dz = ire^{i\alpha} \, d\alpha$$

and the integral formula takes the following form:

(2)
$$F(z_0) = \frac{1}{2\pi} \int_0^{2\pi} F(z_0 + re^{i\alpha}) \, d\alpha.$$

The right-hand side is the mean value of F on the circle ($=$ value of the integral divided by the length of the interval of integration). This proves the following

Theorem 2 (Mean value property of analytic functions)

Let $F(z)$ be analytic in a simply connected domain D. Then the value of $F(z)$ at a point z_0 in D is equal to the mean value of $F(z)$ on any circle in D with center at z_0.

Another important property of analytic functions is the following one.

Theorem 3 (Maximum modulus theorem for analytic functions)

Let D be a bounded region and suppose that $F(z)$ is analytic and nonconstant in D and on the boundary of D. Then the absolute value $|F(z)|$ cannot have a maximum at an interior point of D. Consequently, the maximum of $|F(z)|$ is taken on the boundary of D. If $F(z) \neq 0$ in D, the same is true with respect to the minimum of $|F(z)|$.

Proof. We assume that $|F(z)|$ has a maximum at an interior point z_0 of D and show that this leads to a contradiction. Let $|F(z_0)| = M$ be this maximum. Since $F(z)$ is not constant, $|F(z)|$ is not constant, as follows from Example 4 in Sec. 12.5. Consequently, we can find a circle C of radius r with center at z_0 such that the interior of C is in D and $|F(z)|$ is smaller than M at some point P of C. Since $|F(z)|$ is continuous, it will be smaller than M on an arc C_1 of C which contains P, say, $|F(z)| \leq M - \epsilon$ ($\epsilon > 0$) for all z on C_1 (Fig. 415). If C_1 has the length L_1, the complementary arc C_2 of C has the length $2\pi r - L_1$. By applying the *ML*-inequality (Sec. 13.2) to (1) and noting that $|z - z_0| = r$, we thus obtain

$$M = |f(z_0)| \leq \frac{1}{2\pi} \left| \int_{C_1} \frac{f(z)}{z - z_0} \, dz \right| + \frac{1}{2\pi} \left| \int_{C_2} \frac{f(z)}{z - z_0} \, dz \right|$$

$$\leq \frac{1}{2\pi} \left(\frac{M - \epsilon}{r} \right) L_1 + \frac{1}{2\pi} \left(\frac{M}{r} \right) (2\pi r - L_1) = M - \frac{\epsilon L_1}{2\pi r} < M,$$

that is, $M < M$, which is impossible. Hence, our assumption is false and the first statement of the theorem is proved.

We prove the last statement. If $F(z) \neq 0$ in D, then $1/F(z)$ is analytic in D. From the statement already proved it follows that the maximum of $1/|F(z)|$ lies on the boundary of D. But this maximum corresponds to a minimum of $|F(z)|$. This completes the proof. ∎

From these theorems we shall now derive corresponding results about harmonic functions.

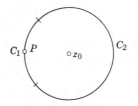

Fig. 415. Proof of Theorem 3

Theorem 4 (Harmonic functions)

Let D be a simply connected bounded domain and let C be its boundary curve. Then if $\Phi(x, y)$ is harmonic in a domain containing D and C, it has the following properties.

I. The value of $\Phi(x, y)$ at a point (x_0, y_0) in the domain D is equal to the mean value of $\Phi(x, y)$ on any circle in D with center at (x_0, y_0).

II. The value of $\Phi(x, y)$ at the point (x_0, y_0) is equal to the mean value of $\Phi(x, y)$ in any circular disk in D with center at (x_0, y_0). [Cf. footnote 5 in Sec. 17.4.]

III. **(Maximum principle)** *If $\Phi(x, y)$ is not constant, it has neither a maximum nor a minimum in D. Consequently, the maximum and the minimum are taken on the boundary of D.*

IV. If $\Phi(x, y)$ is constant on C, then $\Phi(x, y)$ is a constant.

V. If $h(x, y)$ is harmonic in D and on C and if $h(x, y) = \Phi(x, y)$ on C, then $h(x, y) = \Phi(x, y)$ everywhere in D.

Proof. Statement I follows from (2) by taking the real parts on both sides:

$$\Phi(x_0, y_0) = \operatorname{Re} F(x_0 + iy_0) = \frac{1}{2\pi} \int_0^{2\pi} \Phi(x_0 + r \cos \alpha, y_0 + r \sin \alpha) \, d\alpha.$$

If we multiply both sides by r and integrate over r from 0 to r_0 where r_0 is the radius of a circular disk in D with center at (x_0, y_0), then we obtain on the left-hand side $\frac{1}{2} r_0^2 \Phi(x_0, y_0)$ and therefore

$$\Phi(x_0, y_0) = \frac{1}{\pi r_0^2} \int_0^{r_0} \int_0^{2\pi} \Phi(x_0 + r \cos \alpha, y_0 + r \sin \alpha) r \, d\alpha \, dr.$$

This proves the second statement.

We prove statement III. Let $\Psi(x, y)$ be a conjugate harmonic function of $\Phi(x, y)$ in D. Then $F(z) = \Phi(x, y) + i\Psi(x, y)$ is analytic in D, and so is

$$G(z) = e^{F(z)}.$$

The absolute value is

$$|G(z)| = e^{\operatorname{Re} F(z)} = e^{\Phi(x, y)}.$$

From Theorem 3 it follows that $|G(z)|$ cannot have a maximum at an interior point of D. Since e^{Φ} is a monotone increasing function of the real variable Φ, statement III about the maximum of Φ follows. From this, the statement about the minimum follows by replacing Φ by $-\Phi$.

If Φ is constant on C, say, $\Phi = k$, then by III the maximum and the minimum of Φ are equal. From this, statement IV follows.

If h and Φ are harmonic in D and on C, then $h - \Phi$ is also harmonic in D and on C, and by assumption, $h - \Phi = 0$ everywhere on C. Hence, by IV, we have $h - \Phi = 0$ everywhere in D, and statement V is proved. This completes the proof of Theorem 4.   ∎

The last statement of Theorem 4 is very important. It means that a *harmonic function is uniquely determined in D by its values on the boundary of D*. Usually, $\Phi(x, y)$ is required to be harmonic in D and continuous on the boundary[7] of D. Under these circumstances the maximum principle (Theorem 4, III) is still applicable. The problem of determining $\Phi(x, y)$ when the boundary values are given is called the Dirichlet problem for the Laplace equation in two variables, as we know. From Theorem 4, V we thus have

Theorem 5 (Dirichlet problem)

If for a given region and given boundary values the Dirichlet problem for the Laplace equation in two variables has a solution, the solution is unique.

This is the end of Chap. 17 and Part D on complex analysis. We hope that the reader has gained an impression of the usefulness of complex analysis to the engineer and physicist, and of the mathematical beauty of this area, which probably will remain unsurpassed in applied mathematics. The physicist's demand for complex analysis seems to be increasing, due to recent "field theories" and related ideas involving Riemann surfaces.

Problems for Sec. 17.6

1. Verify Theorem 2 for $F(z) = (z + 2)^2$, $z_0 = 1$, and a circle of radius 1 with center at z_0.
2. Verify Theorem 2 for $F(z) = z^3$, $z_0 = 0$, and a circle of radius 1 about 0.
3. Integrate $F(z) = |z|$ around the unit circle. Does the result contradict Theorem 2?
4. Verify Theorem 3 for $F(z) = z^2$ and the rectangle $-2 < x < 2$, $-1 < y < 1$.
5. Verify Theorem 3 for $F(z) = e^z$ and any bounded domain.

Find the maximum of $|F(z)|$ in the disk $|z| \leq 1$ and the corresponding z, where

6. $F(z) = az + b$ 7. $F(z) = z^2 - 1$ 8. $F(z) = \sin z$

9. The function $F(z) = 1 + 3|z|^2$ is not zero in the disk $|z| \leq 2$ and has a minimum at an interior point of that disk. Does this contradict Theorem 3?
10. The *real* function $F(x) = \cos x$ has a maximum at $x = 0$. Using Theorem 3, conclude that this cannot be a maximum of the absolute value of the *complex* function $F(z) = \cos z$ in a domain containing $z = 0$.
11. Let $F(z)$ be analytic (not a constant) in the closed disk $|z| \leq 1$ and suppose that $|F(z)| = c = const$ on $|z| = 1$. Show that then $F(z)$ must have a zero in that disk.
12. If $F(z)$ is analytic (not a constant) in a simply connected domain D, and the curve

---

[7]That is, $\lim\limits_{\substack{x \to x_0 \\ y \to y_0}} \Phi(x, y) = \Phi(x_0, y_0)$, where (x_0, y_0) is on the boundary and (x, y) is in D.

given by $|F(z)| = c$ (c any fixed constant) lies in D and is closed, show that $F(z) = 0$ at a point in the interior of that curve. Give examples.

13. Verify statement I in Theorem 4 for $\Phi(x, y) = x^2 - y^2$ and a circle of radius 1 about $(x_0, y_0) = (1, 0)$.

14. Verify the maximum principle in Theorem 4 for $\Phi(x, y) = 5xy$ and the disk $x^2 + y^2 \leq 8$. Determine the maximum and minimum values of Φ in that disk and the points at which Φ takes those values.

15. Prove statement I of Theorem 4 by applying Poisson's integral formula (Sec. 17.5).

Further Proof in Chapter 17

Another proof of Theorem 1 in Sec. 17.2, without the use of a harmonic conjugate

We show that if $w = u + iv = f(z)$ is analytic and maps a domain D conformally onto a domain D^* and $\Phi^*(u, v)$ is harmonic in D^*, then

$$(1) \qquad\qquad \Phi(x, y) = \Phi^*(u(x, y), v(x, y))$$

is harmonic in D, that is, $\nabla^2\Phi = 0$ in D. We make no use of a harmonic conjugate of Φ^*, but use straightforward differentiation. By the chain rule,

$$\Phi_x = \Phi_u^* u_x + \Phi_v^* v_x.$$

We apply the chain rule again, underscoring the terms which will drop out when we form $\nabla^2\Phi$:

$$\Phi_{xx} = \underline{\Phi_u^* u_{xx}} + (\underline{\Phi_{uu}^* u_x} + \underline{\Phi_{uv}^* v_x})u_x$$

$$+ \underline{\Phi_v^* v_{xx}} + (\underline{\Phi_{vu}^* u_x} + \Phi_{vv}^* v_x)v_x.$$

Φ_{yy} is the same with each x replaced by y. We form the sum $\nabla^2\Phi$. In it, $\Phi_{vu}^* = \Phi_{uv}^*$ is multiplied by

$$u_x v_x + u_y v_y,$$

which is 0 by the Cauchy–Riemann equations. Also $\nabla^2 u = 0$ and $\nabla^2 v = 0$. There remains

$$\nabla^2\Phi = \Phi_{uu}^*(u_x^{\,2} + u_y^{\,2}) + \Phi_{vv}^*(v_x^{\,2} + v_y^{\,2}).$$

By the Cauchy–Riemann equations this becomes

$$\nabla^2\Phi = (\Phi_{uu}^* + \Phi_{vv}^*)(u_x^{\,2} + v_x^{\,2})$$

and is 0 since Φ^* is harmonic. ∎

Review Problems for Chapter 17

1. Why can complex analytic functions be used in solving two-dimensional potential problems? What do we mean by a "two-dimensional" problem?
2. What is a complex potential, and what advantage does it have?
3. What do we mean by the Dirichlet problem?
4. How is conformal mapping applied in solving the Dirichlet problem?
5. What is a mixed boundary value problem, and where in this chapter did it occur?
6. What is a streamline?
7. State the conditions a flow must satisfy if it is to be treated by methods of complex analysis.
8. Explain why steady-state heat flow is related to potential theory.
9. What do we mean by superposition of potentials?
10. List some of the remarkable properties common to all harmonic functions, regardless of their particular form.
11. Find the potential Φ in the first quadrant of the xy-plane if the y-axis has potential 110 volts and the x-axis is grounded (0 volt).
12. Find the potential between the plates Arg $z = \pi/4$, kept at 425 volts, and Arg $z = 3\pi/4$, kept at 275 volts.
13. Find the potential in the upper half-plane if the positive x-axis has potential 1000 volts and the negative x-axis has potential 1500 volts.
14. Find the potential on the ray $y = x$, $x > 0$ and on the positive x-axis if the positive y-axis is at 200 volts and the negative y-axis at 0.
15. Find the potential between the plates $y = x/2$, kept at 100 volts, and $y = x/2 + 4$, kept at 420 volts.
16. In Prob. 14, find a formula for the potential on the line $x = 1$. Find the potential at the point $x = 1$, $y = 2$.
17. Find the complex potential in Prob. 15.
18. Find the potential between the cylinders $|z| = 1$, having potential 10 volts, and $|z| = 10$, having potential 100 volts.
19. Find the complex potential in Prob. 11.
20. Find the complex potential in Prob. 18.
21. Find the equipotential lines of $F(z) = (1 - i)z$ and show that $F(z)$ can be interpreted as the complex potential between two parallel plates.
22. Find and sketch the equipotential lines of $F(z) = (1 + i)/z$.
23. Show that the equipotential lines of $F(z) = -iz^2 + 2z$ are hyperbolas.
24. Find the equipotential lines of $F(z) = 1/z^2$. Where is $\Phi(x, y) = 0$? What is Φ on the x-axis?
25. Find the equipotential lines of $F(z) = a \operatorname{Ln} (z - 1 - i)$, where a is real, and show that this is the complex potential between coaxial cylinders with axis at $1 + i$.
26. Find the potential in the upper half-plane H if the positive x-axis has potential 0 and the negative x-axis has potential Φ_1. Then show that $w = (z - i)/(-iz + 1)$ maps H onto the unit disk D: $|w| \leq 1$. Find the complex and real potentials in D. What is the potential on the v-axis?
27. Interpret Prob. 26 as a problem in heat flow.
28. Find and sketch the equipotential lines of $F(z) = 1/(1 - z)$.

29. Find the potential Φ in the unit disk $|w| \leqq 1$ in the form of an infinite series if the left semicircle of $|w| = 1$ has potential Φ_1 and the right semicircle has potential 0.

30. Find the complex potential in Prob. 29 in the form of a series. What is the sum of this series? *Hint.* Use Example 4 in Sec. 14.6.

Summary of Chapter 17
Complex Analysis Applied to
Potential Theory

Potential theory is the theory of solutions of **Laplace's equation**

$$(1) \qquad\qquad \nabla^2 \Phi = 0.$$

Solutions whose second partial derivatives are continuous are called **harmonic functions.** Equation (1) is the most important partial differential equation in physics, where it is of interest in two and three dimensions. It appears in electrostatics (Sec. 17.1), steady-state heat problems (Sec. 17.3), fluid flow (Sec. 17.4), gravity, etc. Whereas the three-dimensional case requires other methods (cf. Chap. 11), two-dimensional potential theory can be handled by complex analysis, since the real and imaginary parts of an analytic function are harmonic (Sec. 12.5). They remain harmonic under conformal mapping (Sec. 17.2), so that **conformal mapping** becomes a powerful tool in solving boundary value problems for (1), as is illustrated in this chapter. With a real potential Φ we can associate a **complex potential** (Sec. 17.1)

$$(2) \qquad\qquad F(z) = \Phi + i\Psi.$$

Then both families of curves $\Phi = const$ and $\Psi = const$ have a physical meaning. In electrostatics, they are equipotential lines and lines of electrical force of attraction or repulsion (Sec. 17.1). In heat problems, they are isotherms (curves of constant temperature) and lines of heat flow (Sec. 17.3). In fluid flow, they are equipotential lines of the velocity potential and streamlines (Sec. 17.4).

For the disk, the solution of the Dirichlet problem is given by the **Poisson formula** (Sec. 17.5) or by a series that on the boundary circle becomes the Fourier series of the given boundary values (Sec. 17.5).

Harmonic functions, like analytic functions, have a number of general properties; particularly important are the **mean value property** and the *maximum modulus property* (Sec. 17.6), which implies the uniqueness of the solution of the Dirichlet problem (Theorem 5 in Sec. 17.6).

PART E
NUMERICAL METHODS

No other field of mathematics has shown a recent increase in importance to the engineer comparable to that of numerical methods, nor has any other field developed as rapidly. Of course, the main reason for this evolution is the development of **computers,** from the microcomputer to the Cray, and we can see no end of it. Indeed, each new generation of computers presents new tasks, and even small improvements in algorithms may have great effects on time, storage demand, accuracy and stability. This opens up wide areas of research. A main goal is the development of good algorithms and corresponding software, along with well-structured software codes that take best possible advantage of all the power of computers of different sizes and structures.

Chapters 18–20 concern the development and investigation of **numerical methods,** which provide the transition from the mathematical model of a problem (the equations or functions obtained in calculus or algebra, etc.) to an **algorithm** which we can program (or use directly on a pocket calculator) to obtain the solution of the problem in the form of numbers. This includes the investigation of the range of applicability of numerical methods, and their error analysis, stability and properties in general.

We begin with numerical methods of a general nature in Chap. 18. In Chap. 19 we discuss numerical methods for problems in linear algebra, in particular the solution of systems of linear equations and algebraic eigenvalue problems. Chapter 20 is devoted to the numerical solution of ordinary and partial differential equations.

Actual programs, FORTRAN or other, are not included—they would allow the student to get results without understanding the underlying methods—but we give the algorithms in a form that seems best for showing how a method works and also is easy to program, even with little programming experience. The student is encouraged to program the given algorithms and try them out on the computer, and to become familiar with software packages as obtainable from IMSL (International Mathematical and Statistical Libraries: 2500 City West Boulevard, Houston, TX 77042-3020), which also serves as a distributor of good software from other sources. See also EISPACK and LINPACK (Refs. [E7], [E13] and [E30] in Appendix 1).

Chapter 18

Numerical Methods in General

Engineering mathematics ultimately comes down to numerical results, and engineering students should therefore include in their mathematical equipment a definite knowledge of some standard **numerical methods,** that is, methods of obtaining numerical results. The need for such processes for obtaining practical answers to given problems is obvious. Indeed, many problems of engineering mathematics can be solved only approximately. In other cases, an answer obtained from some theory may be almost useless for numerical purposes. Typical examples are the integral formula for the solution of a linear first-order differential equation (Sec. 1.7) if we cannot evaluate the integral by elementary methods, and Cramer's rule for solving systems of linear algebraic equations in terms of determinants (Secs. 7.9, 7.11). Sometimes it may also happen that the theory merely guarantees the existence (or merely characterizes a few general properties) of a solution, but gives no indication of how to actually obtain it.

The development of numerical methods is strongly influenced and determined by the **computer,** which has become indispensable in the mathematical work of every engineer, and by the tremendous progress in digital computation. A substantial part of engineering computing done today is "real-time" computing, in which the computation of results from given data is done virtually simultaneously with the event generating the data, as is needed in controlling ongoing chemical processes or nuclear reactions, in guiding airplanes or rockets and so on. Then questions of speed, storage demand and timing of portions of longer programs are of crucial importance.

It is clear that the high speed, reliability and flexibility of the computer provide access to practical problems that in precomputer times were not within the reach of numerical work. For instance, the solution of very large systems of linear equations is of that type. The present rapid development of numerical methods includes the creation of new methods, modification of existing methods to make them more efficient on the computer and a theoretical and practical analysis of algorithms for the standard computational processes, central tasks being the study of error and the elimination of arithmetic traps of all kinds that are ready for the unwary.

Prerequisite for this chapter: elementary calculus.
References: Appendix 1, Part E.
Answers to problems: Appendix 2.

18.1 Introduction

Errors in Computation

It is well known that the market offers an ever-increasing variety of **computers,** which range from (programmable or nonprogrammable) hand and desk calculators, microcomputers, minicomputers, etc., to powerful (and sometimes specialized) large-scale computers. Hand calculators may be useful in work involving not too much data and few operations, in quick spot checks of larger work or in gaining insight into a numerical method, by inspecting every detail of the computation. On the other hand, large-scale digital computers are very powerful in storing and handling enormous amounts of data, executing difficult programs and performing incredibly many arithmetic operations within a short time, and they rarely break down.

The solution of a problem involving numerical work consists of the following steps.

1. Modeling. Setting up a mathematical model, that is, formulating the problem in mathematical terms, taking into account the type of computer one wants to use.

2. Choice of numerical methods (algorithms), together with a preliminary error analysis (estimation of error, determination of steps size, etc., cf. below).

3. Programming. Usually this starts with a flowchart showing a block diagram of the procedures to be performed by the computer and then writing, say, a FORTRAN program.

4. Operation

5. Interpretation of results. This may also include decisions to rerun if further results are needed.

Steps 1 and 2 are related: a slight change in the model may often permit the use of a more efficient method.

In Step 3 the **program** consists of all relevant given data and a sequence of instructions that the computer will execute in a certain order, thereby eventually producing a numerical answer (or a graph, etc.) to the problem. The program is usually written in FORTRAN 77 (FORmula TRANslation), ALGOL (ALGOrithmic Language) or some other high-level language. A compiler then translates this program into a sequence of machine instructions that performs the desired task.

Our main concern in Chaps. 18—20 will be Step 2, that is, the explanation of basic general numerical methods.

Although **experience** cannot be *taught* but must be *gained,* a few general remarks may be in order.

All data to be entered into the computer should be carefully checked.

Brilliant programs do not compensate poor choices of methods; on the other hand, poor programs can spoil good methods, resulting in excessive time of computation and inaccuracy of results.

Some guidelines for detecting errors will be discussed near the end of this section.

The computer does not relieve the user of the responsibility of planning the work carefully; on the contrary, the computer demands much more careful planning.

Also, the computer does not reduce the need for a detailed understanding of the problem area or of the related mathematics.

The use of a package program from the library still requires knowledge of the purpose of the program and its limitations and whether it applies to the situation on hand. If a package program has been chosen for a given problem, we should first try it out on some simple problem whose solution we know. Furthermore, we should be aware that most problems cannot be solved simply by applying some standard program, but we must often devise new numerical methods by adapting standard methods to a new task.

Error

Since in computations we work with a finite number of digits and carry out finitely many steps, the methods of numerical analysis are **finite processes,** and a numerical result is an **approximate value** of the (unknown) exact result, except for the rare cases in which the exact answer is a sufficiently simple rational number and we can use a numerical method that gives the exact answer.

If $\tilde{a}$ is an approximate value of a quantity whose exact value is a, then the difference

$$\boxed{\epsilon = \tilde{a} - a}$$

is called the **error** of $\tilde{a}$. Hence[1]

$$\tilde{a} = a + \epsilon \qquad \text{Approximation = True value + error.}$$

For instance, if $\tilde{a} = 10.5$ is an approximation of $a = 10.2$, its error is $\epsilon = 0.3$, and the error of an approximation $\tilde{a} = 1.60$ of $a = 1.82$ is $\epsilon = -0.22$.

The **relative error** ϵ_r of $\tilde{a}$ is defined by

$$\epsilon_r = \frac{\epsilon}{a} = \frac{\tilde{a} - a}{a} = \frac{\text{Error}}{\text{True value}} \qquad (a \neq 0).$$

Clearly, $\epsilon_r \approx \epsilon/\tilde{a}$ if $|\epsilon|$ is much less than $|\tilde{a}|$. We may also introduce the quantity $\gamma = a - \tilde{a} = -\epsilon$ and call it the *correction*, thus[1]

$$a = \tilde{a} + \gamma \qquad \text{True value = Approximation + correction.}$$

Finally, an **error bound** for $\tilde{a}$ is a number β such that

$$|\tilde{a} - a| \leq \beta, \qquad \text{that is,} \qquad |\epsilon| \leq \beta.$$

Depending on the source of errors, we may distinguish between experi-

---

[1]In some books, $a - \tilde{a}$ is called the *error*. This differs from our definition by a factor of -1. It is rather irrelevant which of these two definitions is used, but one should be consistent.

mental errors, truncation errors, round–off errors and programming errors (blunders). (Moreover, hand computations may involve arithmetical blunders.) **Experimental errors** are errors of given data (probably arising from measurements). **Truncation errors** are errors caused by "truncating" (prematurely breaking off) a (finite or infinite) sequence of computational steps necessary for producing an exact result. These errors depend on the computational method used and must be dealt with individually for each method. **Round-off errors** are errors arising from the process of rounding off during computation. We discuss them, as well as programming errors, below.

Floating-Point Form of Numbers

In the decimal notation, every real number is represented by a finite or infinite sequence of decimal digits. For machine computation the number must be replaced by a number of finitely many digits. Most digital computers have two ways of representing numbers, called *fixed point* and *floating point*. In a **fixed-point** system all numbers are given with a *fixed number of decimal places,* e.g., 62.358, 0.013, 1.000. In a **floating-point** system numbers are given with a *fixed number of significant digits,* for instance,

$$0.6238 \times 10^3 \qquad 0.1714 \times 10^{-13} \qquad -0.2000 \times 10^1$$

also written[2]

$$0.6238E03 \qquad 0.1714E - 13 \qquad -0.2000E01.$$

Significant digit of a number c is any given digit of c, except possibly for zeros to the left of the first nonzero digit that serve only to fix the position of the decimal point. (Thus, any other zero is a significant digit of c.) For instance, each of the numbers 1360, 1.360, 0.001 360 has 4 significant digits.[3]

Most computers use (internally) the binary number system, whose base is 2 (cf. Probs. 3—5). A binary digit is briefly called a **bit.** By grouping bits, one can obtain octal (base 8) or hexadecimal (base 16) representations (the latter having been made popular by the IBM 360 and 370 series). In the computer, a number, represented in floating point, consists of the sign, a fractional part, called the *mantissa,*[4] and the exponential part, called the *characteristic*. For example, in the IBM 370 series, a (single-precision) floating-point number consists of 1 bit for the sign, 24 bits for the mantissa (corresponding to 6—7 decimal digits) and 7 bits for the (base 16) exponent, allowing values from -64 to 63. If in a computation a number greater than 16^{63} ($\approx 10^{76}$) occurs (which does happen!), this is called **overflow.** Then the computer halts. If a number less than 16^{-64} occurs, this is called **underflow.**

---

[2] One also uses a representation of the form 6.238×10^2, 1.1714×10^{-14}, etc.

[3] In tables of functions showing k significant digits, it is conventionally assumed that any given value $\tilde{a}$ deviates from the corresponding exact value a by at most ± 0.5 unit of the last given digit, unless otherwise stated; e.g., if $a = 1.1996$, then a table with 4 significant digits should show $\tilde{a} = 1.200$. Correspondingly, if 12 000 is correct to three digits only, we should write 120×10^2, etc. "Decimal" is abbreviated by D and "significant digit" by S. For example, 5D means 5 decimals, and 8S means 8 significant digits.

[4] This has nothing to do with "mantissa" as used in connection with logarithms. "**Single precision**" means the number of bits normally used in computations by the computer; "**double precision**" means twice as many bits, etc.

Similarly for a computer with another range of the size of numbers. In many computers, numbers causing underflow are set to 0.

The fixed-point form of numbers is impractical in most scientific work because of its limited range (explain!) and will not be of further concern to us.

Round-Off. Numerical Stability

An error is caused by **chopping** (= discarding all decimals from some decimal on) or **rounding.** The rule for rounding off a number to k decimals is as follows. (The rule for rounding off to k significant figures is the same, with "decimal" replaced by "significant figure.")

Round-off rule. Discard the $(k + 1)$th and all subsequent decimals. (a) If the number thus discarded is less than half a unit in the kth place, leave the kth decimal unchanged (*"rounding down"*). (b) If it is greater than half a unit in the kth place, add one to the kth decimal (*"rounding up"*). (c) If it is exactly half a unit, round off to the nearest *even* decimal. (Example: Rounding off 3.45 and 3.55 to 1 decimal gives 3.4 and 3.6, respectively.)

The last part of the rule is supposed to ensure that in discarding exactly half a decimal, rounding up and rounding down happens about equally often, on the average.

If we round off 1.253 5 to 3, 2, 1 decimals, we get 1.254, 1.25, 1.3, but if 1.25 is rounded off to one decimal, without further information, we get 1.2.

Chopping is not recommended since it introduces an error that is systematic and can be larger than an error in rounding off. Nevertheless, surprisingly many computers use chopping! A reason is that rounding is time-consuming, and manufacturers try all sorts of shortcuts to make the basic arithmetical operations as fast as possible. Most computers that use rounding off always round *up* in case (c) of the rule (or in the corresponding case for a base other than 10), since this is easier to realize technically.

Rounding errors may ruin a computation completely, even a small computation. In general, they are the more dangerous the more arithmetic operations (perhaps several millions!) we have to perform. It is therefore important to analyze computational programs for rounding errors to be expected and to find an arrangement of the computations such that the effect of rounding errors is as small as possible. We call a sequence of computations by well-defined rules an **algorithm,** and we call an algorithm **stable** if errors in intermediate results have little influence on the final result. If, however, intermediate errors have a very strong effect on the final result, we call the algorithm **unstable.** This *"numerical instability,"* which can be avoided by choosing a more suitable algorithm, must be carefully distinguished from *"mathematical instability"* of a given problem; the problem is then called *"ill-conditioned,"* a concept we discuss later (in Sec. 18.2).

We want to emphasize that a careful selection of most effective methods is at least as important in automatic computation as it is in using a desk calculator, in order to minimize time and cost as well as errors. Even for simple problems there may be several ways that differ by their efficiency. Let us illustrate this with an example.

EXAMPLE 1. Quadratic equation

Find the roots of each of the equations

(1) $\qquad\qquad$ (a) $x^2 - 4x + 2 = 0$ $\qquad$ and $\qquad$ (b) $x^2 - 40x + 2 = 0$.

using 4 significant digits in the computation.

Solution. A formula for the roots x_1, x_2 of a quadratic equation $ax^2 + bx + c = 0$ is

(2) $\qquad\qquad x_1 = \frac{1}{2a}(-b + \sqrt{b^2 - 4ac}), \qquad x_2 = \frac{1}{2a}(-b - \sqrt{b^2 - 4ac}).$

Furthermore, since $x_1 x_2 = c/a$, another formula for those roots is

(3) $\qquad\qquad\qquad\qquad x_1$ as before, $\qquad x_2 = \frac{c}{ax_1}.$

For (1a), formula (2) gives $x = 2 \pm \sqrt{2} = 2.000 \pm 1.414$, $x_1 = 3.414$, $x_2 = 0.586$, and (3) gives $x_1 = 3.414$, $x_2 = 2.000/3.414 = 0.5858$, in error by less than one unit in the last digit (cf. Prob. 24). For (1b), formula (2) gives $x = 20 \pm \sqrt{398} = 20.00 \pm 19.95$, $x_1 = 39.95$, $x_2 = 0.05$, which is poor, whereas (3) gives $x_1 = 39.95$, $x_2 = 2.000/39.95 = 0.05006$, in error by less than one unit in the last digit (cf. Prob. 24). $\qquad$ ∎

Programming Errors

In precomputer times, blunders occurred because of the fallibility of the human computer. In connection with a digital computer the situation is different. Since a computer rarely breaks down, **errors in programming** are now the main difficulty.

Errors of all types are collectively called **bugs.** The process of locating and removing bugs is called **debugging.** Various compilers provide *diagnostics* indicating all errors in a source program, except for errors in logic.

There are no general rules that guarantee the detection of all bugs in all programs. However, the following guidelines may be helpful. The most important way to determine whether a program has bugs is to run it with sets of data for which the answers are known (for instance, simplified data for which an answer can be computed by hand). Since user efficiency tends to fall off rapidly during an extended session, contemplation may often be more effective than attempting endless test runs. If test cases do not help us to detect an error whose existence seems obvious from nonsensical results, we may apply selective or full **tracing,** that is, printing out some or all intermediate computations and studying them, step by step, along with computations by hand or by a different computer program, looking for the first disagreement. In general, small errors in sophisticated programs (and their small but perhaps crucial effect) may be hard to detect. Consequently, programs should be prepared with greatest care and checked as far as possible *before* they reach the stage of machine operation. Detection of errors by means of such early checking is more efficient than reliance on programmed checks. In this connection it is important to realize that one must not assume a program to be correct merely because it can be translated and executed.

For general information on types of computers, computer components and arithmetic, programming, processing, etc., see Refs. [E10] or [E38] in Appendix 1.

Each of the next sections is devoted to some basic numerical task and methods for its solution. This begins in the next section with the **solution of equations,** a problem for which there are practically no formulas, except in a few simple cases, so that one depends almost entirely on numerical algorithms.

Problems for Sec. 18.1

1. **(Floating point)** Write 10 000, 736.128, 90.237, 0.0040506 in floating-point form, with 4 significant digits.
2. Write -0.53678, 1186.699, -0.00604, 23.9481, 1/3, 85/7 in floating-point form, with 5 significant digits.
3. **(Binary representation)** Most computers use the binary number system or a variant of it, such as base 8 or base 16. Writing the familiar base 10 representation in the form $(81.5)_{10}$, verify that

$$(81.5)_{10} = 8 \cdot 10^1 + 1 \cdot 10^0 + 5 \cdot 10^{-1} = 2^6 + 2^4 + 2^0 + 2^{-1} = (1010001.1)_2.$$

   Convert $(10)_{10}$, $(100)_{10}$, $(381.25)_{10}$ and $(0.044\ 921\ 875)_{10}$ to binary form.
4. Convert $(1987)_{10}$, $(5.25)_{10}$, $(15.664\ 062\ 5)_{10}$ and $(-24.875)_{10}$ to binary form.
5. Convert $(11100.1)_2$, $(0.00101)_2$, $(1.10111)_2$ and $(-11.01101)_2$ to base 10 form.
6. **(Small differences of large numbers)** Rounding errors can become particularly disadvantageous in expressions $a - b$ when a and b are close together. To illustrate this, calculate $0.36443/(17.862 - 17.798)$, first using the numbers as given, and then rounding stepwise to 4, 3, and 2 significant digits.
7. The quotient in Prob. 6 is of the form $a/(b - c)$. Write it in the form $a(b + c)/(b^2 - c^2)$ and compute it first with 5 significant digits, then rounding stepwise as in Prob. 6.
8. Solve $x^2 - 50x + 1 = 0$ by (2) and by (3), using 5 significant digits in the computation. Compare and comment.
9. Compute (a) $\sin^2 0.1$ and (b) $1 - \cos^2 0.1$ from Table A1 in Appendix 4, using 4 significant digits in the computation. Compare and comment.
10. What is a good way to compute $\sin a - \sin b$ when a and b are nearly equal?
11. Indicate how $\log a - \log b = \log (a/b)$ and $e^{a-b} = e^a/e^b$ may be used in computations to avoid loss of significant digits.
12. What is a good way to compute $f(x) = \sqrt{x^2 + 9} - 3$ for small $|x|$?
13. Answer the question in Prob. 12, when $f(x) = 1 - \cos x$.
14. **Approximations to π** = 3. 141 592 653 589 79 $\cdots$ are 22/7 and 355/113. Determine the corresponding errors and relative errors to 3 significant digits.
15. Compute π by Machin's approximation 16 arc tan (1/5) $-$ 4 arc tan (1/239) to 10 significant digits (Note that these digits are correct! For the first 100 000 digits of π, see D. Shanks and J. W. Wrench, *Mathematics of Computation,* vol. 16 (1962), pp. 76—99.)
16. Let 4.81 and 12.752 be correctly rounded to the number of digits shown. Determine the smallest interval in which the sum $s = 4.81 + 12.752$, using true instead of rounded values of the quantities, must lie.
17. Answer the question in Prob. 16, for the difference $d = 4.81 - 12.752$.
18. Let $a = 5$, $b = 4 \times 10^{-10}$, $c = 2 \times 10^{-10}$. Show that in floating-point arithmetic with 10 significant digits, $(a + b) + c = 5.000\ 000\ 000$, whereas $a + (b + c) = 5.000\ 000\ 001$. Comment. What would it indicate if your pocket calculator gave you 5.000 000 001 in both cases?

19. Given n numbers $a_1, \cdots, a_n$, where a_j is correctly rounded to D_j decimals. In computing the sum $a_1 + \cdots + a_n$, retaining $D = \min D_j$ decimals, is it essential whether we add first and then round the result or whether we first round each number to D decimals and then add?

20. If β_1 and β_2 are error bounds for $\tilde{a}_1$ and $\tilde{a}_2$, show that $\beta = \beta_1 + \beta_2$ is an error bound for the sum $\tilde{s} = \tilde{a}_1 + \tilde{a}_2$.

21. Show that β in Prob. 20 is an error bound for the difference $\tilde{d} = \tilde{a}_1 - \tilde{a}_2$, and illustrate with an example that in general β cannot be replaced by a smaller number.

22. (**Error propagation**) Show that for small relative errors the relative error of a product is approximately equal to the sum of the relative errors of the factors. Give a simple numerical example.

23. Show that for small relative errors the relative error of a quotient is approximately equal to the difference of the relative errors of the factors.

24. Using Prob. 23, show that in Example 1 the absolute value of the error of $x_2 = 2.000/3.414 = 0.5858$ is less than 0.0001, and the absolute value of the error of $x_2 = 2.000/39.95 = 0.050\ 06$ is less than 0.000 01.

25. Compute a two-decimal table[5] of $f(x) = x/16$, $x = 0(1)20$ and find out how the rounding error is distributed.

18.2 Solution of Equations by Iteration

In engineering mathematics we frequently have to find **solutions** of equations

(1)
$$\boxed{f(x) = 0,}$$

that is, numbers $x = s$ such that $f(s)$ equals zero; here f is a given function (and s should suggest "solution", but we shall also use other notations for solutions). Examples are $x^2 - 3x + 2 = 0$, $x^3 + x = 1$, $\sin x = 0.5x$, $\tan x = x$, $\cosh x = \sec x$, $\cosh x \cos x = -1$, which can all be written in the form (1). The first two are **algebraic equations** because the corresponding f is a polynomial, and in this case the solutions are also called **roots** of the equations. The other equations are **transcendental equations** because they involve transcendental functions. Engineering applications abound: some occur in Chaps. 2, 3, 7 (characteristic equations), 5 (partial fractions), 11 (eigenvalues, zeros of Bessel functions) and 16 (integration), but there are many, many others.

Formulas that give exact numerical values of the solution will exist only in very simple situations. In most cases we have to use an approximation method, in particular an **iteration method,** that is, a method in which we start from an initial guess x_0 (which may be poor) and compute step by step (in general better and better) approximations $x_1, x_2, \cdots$ of an unknown solution of (1), first

---

[5]The notation $x = a(h)b$ means $x = a, a + h, a + 2h, \cdots b$.

$$x_1 = g(x_0), \quad \text{then} \quad x_2 = g(x_1), \quad \text{then} \quad x_3 = g(x_2), \quad \text{etc.}$$

We discuss three such iteration methods for solving (1) that are of particular practical importance and mention two further such methods in the problem set.

In general, iteration methods are easy to program since they do the same sequence of steps over and over again, in the present case: the computation of $g(x)$ for $x_0, x_1, x_2, \cdots$. More importantly, if they converge, they generally are numerically stable (cf. Sec. 18.1).

Fixed-Point Iteration[6] for Solving Equations $f(x) = 0$

We may try to transform (1) *algebraically* into the form

$$(2) \qquad\qquad\qquad x = g(x).$$

Then we choose an x_0 and compute $x_1 = g(x_0)$, $x_2 = g(x_1)$, and in general

$$(3) \qquad\qquad\qquad \boxed{x_{n+1} = g(x_n)} \qquad\qquad (n = 0, 1, \cdots).$$

A solution of (2) is called a **fixed point** of g, motivating the name of the method, and this is a solution of (1), since from $x = g(x)$ we can return to the original form $f(x) = 0$. From (1) we may get several different forms of (2), and the behavior of corresponding iterative sequences $x_0, x_1, \cdots$ may differ accordingly, in particular with respect to their speed of convergence. Let us illustrate this with a simple example.

EXAMPLE 1. An iteration process (Fixed-point iteration)

Set up an iteration process for the equation $f(x) = x^2 - 3x + 1 = 0$. Since we know the solutions

$$x = 1.5 \pm \sqrt{1.25}, \quad \text{thus} \quad 2.618\ 034 \quad \text{and} \quad 0.381\ 966,$$

we can watch the behavior of the error as the iteration proceeds.

Solution. The equation may be written

$$(4a) \qquad\qquad x = g_1(x) = \tfrac{1}{3}(x^2 + 1), \quad \text{thus} \quad x_{n+1} = \tfrac{1}{3}(x_n^2 + 1),$$

and if we choose $x_0 = 1$, we obtain the sequence (cf. Fig. 416a on the next page)

$$x_0 = 1.000, \quad x_1 = 0.667, \quad x_2 = 0.481, \quad x_3 = 0.411, \quad x_4 = 0.390, \cdots$$

which seems to approach the smaller solution. If we choose $x_0 = 2$ the situation is similar. If we choose $x_0 = 3$, we obtain the sequence (cf. Fig. 416a, upper part)

$$x_0 = 3.000, \quad x_1 = 3.333, \quad x_2 = 4.037, \quad x_3 = 5.766, \quad x_4 = 11.415, \cdots$$

which seems to diverge.

Our equation may also be written

$$(4b) \qquad\qquad x = g_2(x) = 3 - \frac{1}{x}, \quad \text{thus} \quad x_{n+1} = 3 - \frac{1}{x_n},$$

[6]Our present use of the word "fixed point" has absolutely nothing to do with that in the previous section.

and if we choose $x_0 = 1$, we obtain the sequence (cf. Fig. 416b)

$$x_0 = 1.000, \quad x_1 = 2.000, \quad x_2 = 2.500, \quad x_3 = 2.600, \quad x_4 = 2.615, \cdots$$

which seems to approach the larger solution. Similarly, if we choose $x_0 = 3$, we obtain the sequence (cf. Fig. 416b)

$$x_0 = 3.000, \quad x_1 = 2.667, \quad x_2 = 2.625, \quad x_3 = 2.619, \quad x_4 = 2.618, \cdots.$$

Our figures show the following. In the lower part of Fig. 416a the slope of $g_1(x)$ is less than the slope of $y = x$, which is 1, thus $|g_1{}'(x)| < 1$, and we seem to have convergence. In the upper part, $g_1(x)$ is steeper ($g_1{}'(x) > 1$) and we have divergence. In Fig. 416b the slope of $g_2(x)$ is less near the intersection point ($x = 2.618$, fixed point of g_2, solution of $f(x) = 0$), and both sequences seem to converge. From all this we conclude that convergence seems to depend on the fact that in a neighborhood of a solution the curve of $g(x)$ is less steep than the straight line $y = x$, and we shall now see that this condition $|g'(x)| < 1$ (slope of $y = x$) is sufficient for convergence.   ∎

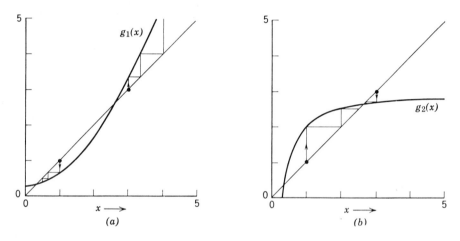

Fig. 416. Example 1, iterations (4a) and (4b)

An iteration process defined by (3) is said to be **convergent** for an x_0 if the corresponding sequence $x_0, x_1, \cdots$ is convergent.

A sufficient condition for convergence is given in the following theorem, which has various practical applications.

Theorem 1 (Convergence of fixed-point iteration)
Let $x = s$ be a solution of $x = g(x)$ and suppose that g has a continuous derivative in some interval J containing s. Then if $|g'(x)| \leq K < 1$ in J, the iteration process defined by (3) converges for any x_0 in J.

Proof. By the mean value theorem of differential calculus there is a t between x and s such that

$$g(x) - g(s) = g'(t)(x - s) \qquad\qquad (x \text{ in } J).$$

Since $g(s) = s$ and $x_1 = g(x_0)$, $x_2 = g(x_1)$, $\cdots$, we obtain

$$|x_n - s| = |g(x_{n-1}) - g(s)| = |g'(t)| \, |x_{n-1} - s|$$

$$\leqq K|x_{n-1} - s|$$

$$= K|g(x_{n-2}) - g(s)|$$

$$= K|g'(\tilde{t})| \, |x_{n-2} - s|$$

$$\leqq K^2|x_{n-2} - s|$$

$$\cdots \leqq K^n|x_0 - s|.$$

Since $K < 1$, we have $K^n \to 0$; hence $|x_n - s| \to 0$ as $n \to \infty$.   ∎

We mention that a function g satisfying the condition in Theorem 1 is called a **contraction** because $|g(x) - g(v)| \leqq K|x - v|$, where $K < 1$. Furthermore, K gives information on the speed of convergence. For instance, if $K = 0.5$, then the accuracy increases by at least 2 digits in only 7 steps because $0.5^7 < 0.01$.

EXAMPLE 2. An iteration process. Illustration of Theorem 1

Find a solution of $f(x) = x^3 + x - 1 = 0$ by iteration.

Solution. A rough sketch shows that a real solution lies near $x = 1$. We may write the equation in the form

$$x = g_1(x) = \frac{1}{1 + x^2}, \quad \text{so that} \quad x_{n+1} = \frac{1}{1 + x_n^2}.$$

Then $|g_1'(x)| = 2|x|/(1 + x^2)^2 < 1$ for any x, so that we have convergence for any x_0. Choosing $x_0 = 1$, we obtain (cf. Fig. 417)

$$x_1 = 0.500, \quad x_2 = 0.800, \quad x_3 = 0.610, \quad x_4 = 0.729, \quad x_5 = 0.653, \quad x_6 = 0.701, \cdots.$$

The solution exact to 6D is $s = 0.682\ 328$. The equation may also be written

$$x = g_2(x) = 1 - x^3. \quad \text{Then} \quad |g_2'(x)| = 3x^2$$

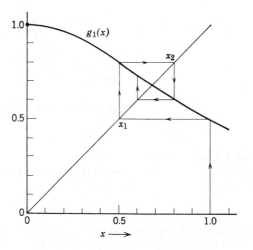

Fig. 417. Iteration in Example 2

and this is greater than 1 near the solution, so that we cannot expect convergence. The reader may try $x_0 = 1$, $x_0 = 0.5$, $x_0 = 2$ and see what happens.   ∎

Newton's Method for Solving Equations $f(x) = 0$

The **Newton method** or **Newton–Raphson**[7] **method** is another iteration method for solving equations $f(x) = 0$, where f is assumed to have a continuous derivative f'. The method is commonly used because of its simplicity and great speed. The underlying idea is that we approximate the graph of f by suitable tangents. Using an approximate value x_0 obtained from the graph of f, we let x_1 be the point of intersection of the x-axis and the tangent to the curve of f at x_0 (cf. Fig. 418). Then

$$\tan \beta = f'(x_0) = \frac{f(x_0)}{x_0 - x_1}, \qquad \text{hence} \qquad x_1 = x_0 - \frac{f(x_0)}{f'(x_0)}.$$

In the second step we compute $x_2 = x_1 - f(x_1)/f'(x_1)$, in the third step x_3 from x_2 again by the same formula, and so on. We thus have the algorithm shown in Table 18.1. Formula (5) in this algorithm can also be obtained from Taylor's formula (cf. Prob. 26).

If it happens that $f'(x_n) = 0$ for some n (see line 2 of the algorithm), then try another x_0. Line 3 is the heart of Newton's method. The inequality in line 4 is a **termination criterion**; if it holds, we have reached the desired accuracy of the approximation and stop. Line 5 gives another *termination criterion* and is needed since Newton's method may diverge or our initial guess x_0 may have been so poor that we will not reach the desired accuracy within a reasonable number of iterations. Then we try another x_0. If $f(x) = 0$ has more than one solution, different choices of x_0 may produce different solutions. Also, an iterative sequence may sometimes converge to a solution different from the expected one.

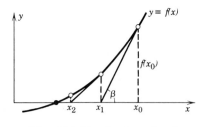

Fig. 418. Newton's method

EXAMPLE 3. Square root

Set up a Newton iteration for computing the square root x of a given positive number c and apply it to $c = 2$.

---

[7]JOSEPH RAPHSON (1648—1715), English mathematician, who published a method similar to Newton's method. For historical details, see Ref. [2], p. 203, listed in Appendix 1.

Table 18.1
Newton's Method for Solving Equations $f(x) = 0$

ALGORITHM NEWTON $(f, f', x_0, \epsilon, N)$

This algorithm computes a solution of $f(x) = 0$ given an initial approximation x_0 (starting value of the iteration). Here the function $f(x)$ is continuous and has a continuous derivative $f'(x)$.

INPUT: Initial approximation x_0, tolerance $\epsilon > 0$,
          maximum number of iterations N.

OUTPUT: Approximate solution x_n $(n \leq N)$
          or message of failure.

For $n = 0, 1, 2, \cdots, N - 1$ do:

1   Compute $f'(x_n)$.

2   If $f'(x_n) = 0$ then OUTPUT "Failure". Stop.
                [*Procedure completed unsuccessfully*]

3   Else compute

   (5) $$x_{n+1} = x_n - \frac{f(x_n)}{f'(x_n)}.$$

4   If $|x_{n+1} - x_n| \leq \epsilon$ then OUTPUT x_{n+1}. Stop.
                [*Procedure completed successfully*]
       End

5   OUTPUT "Failure". Stop.
          [*Procedure completed unsuccessfully*
          after N iterations]

End NEWTON

Solution. We have $x = \sqrt{c}$, hence $f(x) = x^2 - c = 0$, $f'(x) = 2x$, and (5) takes the form

$$x_{n+1} = x_n - \frac{x_n^2 - c}{2x_n} = \frac{1}{2}\left(x_n + \frac{c}{x_n}\right).$$

For $c = 2$, choosing $x_0 = 1$, we obtain

$$x_1 = 1.500\ 000, \qquad x_2 = 1.416\ 667, \qquad x_3 = 1.414\ 216, \qquad x_4 = 1.414\ 214, \cdots$$

x_4 is exact to 6D.

EXAMPLE 4. Iteration for a transcendental equation

Find the positive solution of $2 \sin x = x$.

Solution. Setting $f(x) = x - 2 \sin x$, we have $f'(x) = 1 - 2 \cos x$, and (5) gives

$$x_{n+1} = x_n - \frac{x_n - 2 \sin x_n}{1 - 2 \cos x_n} = \frac{2(\sin x_n - x_n \cos x_n)}{1 - 2 \cos x_n} = \frac{N_n}{D_n}.$$

From the graph of f we conclude that the solution is near $x_0 = 2$. We compute:

n	x_n	N_n	D_n	x_{n+1}
0	2.00000	3.48318	1.83229	1.90100
1	1.90100	3.12470	1.64847	1.89552
2	1.89552	3.10500	1.63809	1.89550
3	1.89550	3.10493	1.63806	1.89549

$x_4 = 1.89549$ is exact to 5D since the solution to 6D is 1.895 494.

EXAMPLE 5. Newton's method applied to an algebraic equation
Apply Newton's method to the equation $f(x) = x^3 + x - 1 = 0$.

Solution. From (5) we have

$$x_{n+1} = x_n - \frac{x_n^3 + x_n - 1}{3x_n^2 + 1} = \frac{2x_n^3 + 1}{3x_n^2 + 1}.$$

Starting from $x_0 = 1$, we obtain

$$x_1 = 0.750\,000, \qquad x_2 = 0.686\,047, \qquad x_3 = 0.682\,340, \qquad x_4 = 0.682\,328, \cdots$$

where x_4 is exact to 6D. A comparison with Example 2 show that the present convergence is much more rapid. This may motivate the concept of the *order of an iteration process*, which we shall now discuss. ∎

We shall see how we can characterize the quality of an iteration method by judging the speed of convergence, as follows.

Order of an Iteration Method

Let $x_{n+1} = g(x_n)$ define an iteration method, and let x_n approximate a solution s of $x = g(x)$. Then $x_n = s + \epsilon_n$, where ϵ_n is the error of x_n. Suppose that g is differentiable a number of times, so that the Taylor formula gives

$$\begin{aligned}
(6) \quad x_{n+1} = g(x_n) &= g(s) + g'(s)(x_n - s) + \tfrac{1}{2}g''(s)(x_n - s)^2 + \cdots \\
&= g(s) + g'(s)\epsilon_n + \tfrac{1}{2}g''(s)\epsilon_n^2 + \cdots .
\end{aligned}$$

The exponent of ϵ_n in the first nonvanishing term after $g(s)$ is called the **order** of the iteration process defined by g.

The order is a measure for the speed of convergence, because if we subtract $g(s) = s$ on both sides of (6), on the left we get the error $\epsilon_{n+1} = x_{n+1} - s$, and the remaining expression on the right equals about its first nonzero term (since $|\epsilon_n|$ is small in the case of convergence). Thus

$$(7) \quad \begin{aligned}
&\text{(a)} \quad \epsilon_{n+1} \approx g'(s)\epsilon_n && \text{in the case of first order,} \\
&\text{(b)} \quad \epsilon_{n+1} \approx \tfrac{1}{2}g''(s)\epsilon_n^2 && \text{in the case of second order,} && \text{etc.}
\end{aligned}$$

Hence in the case of second order the number of significant digits is about doubled in each iteration. (Explain!)

We shall prove that the situation just described applies to Newton's method. Indeed, the following theorem holds.

Theorem 2 (Second-order convergence of Newton's method)

If $f(x)$ is three times differentiable and f' and f'' are not zero at a solution
s of $f(x) = 0$, then for x_0 sufficiently close to s, Newton's method is of
second order.

Proof. By differentiating $g(x) = x - f(x)/f'(x)$ we obtain

$$g'(x) = 1 - \frac{f'(x)^2 - f(x)f''(x)}{f'(x)^2} = \frac{f(x)f''(x)}{f'(x)^2} .$$

Since $f(s) = 0$, this shows that also $g'(s) = 0$. Hence Newton's method is
at least of second order. Another differentiation shows that

$$(8) \qquad\qquad g''(s) = \frac{f''(s)}{f'(s)}$$

and gives the result.   ∎

Comment. Combining (7b) and (8) we have

$$(9) \qquad\qquad \epsilon_{n+1} \approx \frac{f''(s)}{2f'(s)} \epsilon_n^2.$$

EXAMPLE 6. Prior error estimate

Use $x_0 = 2$ and $x_1 = 1.901$ in Example 4 for estimating how many iterations we need to produce
the solution to 5D accuracy. This is an **a priori estimate** or **prior estimate** because we can
compute it after only one iteration, prior to further iterations.

Solution. $f''(s)/2f'(s) \approx f''(x_1)/2f'(x_1) = (2 \sin x_1)/2(1 - 2 \cos x_1) \approx 0.57$.
Hence (9) gives

$$\epsilon_{n+1} \approx 0.57\epsilon_n^2 \approx 0.57^2\epsilon_{n-1}^4 \approx \cdots \approx 0.57^{n+1}\epsilon_0^{2n+2} = (0.57\epsilon_0^2)^{n+1} \leq 5 \cdot 10^{-6}$$

We show below that $\epsilon_0 \approx 0.11$. Then by taking logarithms (and noting that these are negative,
so that the inequality sign is turned around) we get

$$n + 1 \geq \ln (5 \cdot 10^{-6})/\ln (0.57 \cdot 0.11^2) = 2.45.$$

Hence $n = 2$ is the smallest possible n, according to this crude estimate, in good agreement
with Example 4.

$\epsilon_0 \approx 0.11$ is obtained from $\epsilon_1 - \epsilon_0 = (\epsilon_1 + s) - (\epsilon_0 + s) = x_1 - x_0 \approx -0.10$, hence
$\epsilon_1 = \epsilon_0 - 0.10 \approx 0.57\epsilon_0^2$ or $0.57\epsilon_0^2 - \epsilon_0 + 0.10 \approx 0$.   ∎

Newton's method may cause difficulties when $|f'(x)|$ is very small near a
solution of $f(x) = 0$, and we may have to use double precision to get $f(x)$
and $f'(x)$ accurately enough. In this case we call the equation ill-conditioned.
That is, an equation $f(x) = 0$ is called **ill-conditioned** if a small function value

$$R(\tilde{s}) = f(\tilde{s})$$

can be produced by values $x = \tilde{s}$ considerably different from a solution s,
or if small changes of constants contained in f can produce large changes
in solutions. $R(\tilde{s}) = f(\tilde{s})$ is called the **residual** of the equation at $\tilde{s}$. If $R(\tilde{s})$
is small, the error of $\tilde{s}$ will be small only if the equation is not ill-conditioned.
If the equation is ill-conditioned, the error may then still be large.

Secant Method for Solving Equations $f(x) = 0$

The secant method is obtained from Newton's method by replacing the derivative $f'(x)$ by the difference quotient,

$$f'(x_n) \approx \frac{f(x_n) - f(x_{n-1})}{x_n - x_{n-1}}.$$

Then instead of (5) we have

(10)
$$x_{n+1} = x_n - f(x_n) \frac{x_n - x_{n-1}}{f(x_n) - f(x_{n-1})}.$$

Geometrically, in Newton's method x_{n+1} is the intersection of the x-axis and the tangent at x_n, and in the secant method x_{n+1} is the intersection of the x-axis and the secant of the curve of $f(x)$ corresponding to x_{n-1} and x_n (cf. Fig. 419). The secant method needs two starting values, x_0 and x_1, but avoids the evaluation of the derivatives. It can be shown (cf. Ref. [E26] in Appendix 1) that the secant method is somewhat inferior to Newton's method, but is still preferable when the work of computing a value of $f'(x)$ is more than 1/2 times the work of computing a value of $f(x)$. The algorithm is similar to that of Newton's method, as the reader may show. It is *not* advisable to write (10) in the form

$$x_{n+1} = \frac{x_{n-1} f(x_n) - x_n f(x_{n-1})}{f(x_n) - f(x_{n-1})},$$

since this may lead to trouble (loss of significant digits) when x_n and x_{n-1} are about equal. (Explain!)

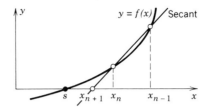

Fig. 419. Secant method

EXAMPLE 7. Secant method
Find the positive solution of $f(x) = x - 2 \sin x = 0$ by the secant method, starting from $x_0 = 2$, $x_1 = 1.9$.

Solution. Here, (10) is

$$x_{n+1} = x_n - \frac{(x_n - 2 \sin x_n)(x_n - x_{n-1})}{x_n - x_{n-1} + 2(\sin x_{n-1} - \sin x_n)} = x_n - \frac{N_n}{D_n}.$$

Numerical values are:

n	x_{n-1}	x_n	N_n	D_n	$x_{x+1} - x_n$
1	2.000 000	1.900 000	$-0.000\ 740$	$-0.174\ 005$	$-0.004\ 253$
2	1.900 000	1.895 747	$-0.000\ 002$	$-0.006\ 986$	$-0.000\ 252$
3	1.895 747	1.895 494	0		0

$x_3 = 1.895\ 494$ is exact to 6D. Cf. Example 4.   ∎

The **bisection method** and the **method of false position (regula falsi)** will be considered in the problem set.

Problems for Sec. 18.2

Fixed-Point Iteration

1. Why do we obtain a monotone sequence in Example 1, but not in Example 2?
2. Perform the iterations indicated at the end of Example 2. Sketch a figure similar to Fig. 417.
3. Find the solution of $x^5 = x + 0.2$ near $x = 0$ by transforming the equation algebraically into the form (2) and starting from $x_0 = 0$.
4. The equation in Prob. 3 has a solution near $x = 1$. Find this solution by writing the equation in the form $x = \sqrt[5]{x + 0.2}$ and iterating, starting with $x_0 = 1$.
5. Using iteration, show that the smallest positive solution of $x = \tan x$ is 4.49, approximately. *Hint.* Conclude from the graphs of x and $\tan x$ that a solution lies close to $x_0 = 3\pi/2$; write the equation in the form $x = \pi + \arctan x$. (Why?)
6. What happens in Prob. 4 if you write the equation in the form $x = x^5 - 0.2$ and start from $x_0 = 1$?
7. Show that if g is continuous in a closed interval I and its range lies in I, then the equation $x = g(x)$ has at least one solution in that interval. Illustrate that it may have more than one solution.
8. Show that in Example 2, $|g_1{}'(x)|$ is maximum at $\tilde{x} = \pm 1/\sqrt{3}$, the maximum value being $|g'(\tilde{x})| = 3\sqrt{3}/8 = 0.65$.
9. Of what order is the process given by $g_1(x)$ in Example 2?
10. Of what orders are the processes in Example 1?

Newton's Method

Find a solution (6 digits accuracy) by Newton's method, starting from the given x_0:

11. $x^4 - 4.00322x^3 + x^2 + 6.43105 = 0$, $x_0 = 3.5$
12. $x^3 - 1.2x^2 + 2x - 2.4 = 0$, $x_0 = 2$
13. $\sin x = \cot x$, $x_0 = 1$
14. $e^{-x} = \tan x$, $x_0 = 1$
15. $x^3 - 3.9x^2 + 4.79x - 1.881 = 0$, $x_0 = 1$

16. The solutions of the equation in Prob. 15 are 0.9, 1.1 and 1.9. Although $x_0 = 1$ lies close to 0.9 and 1.1, Newton's method does not yield one of these solutions. Why? Choose another x_0 such that the method yields approximations for the solution 1.1.

Find all real solutions of the following equations by Newton's method.

17. $\cos x = x$ 18. $x + \ln x = 2$ 19. $2x + \ln x = 1$
20. $x^4 - 0.1x^3 - 0.82x^2 - 0.1x - 1.82 = 0$

21. Calculate $\sqrt{5}$ by the iteration in Example 3, starting from $x_0 = 2$ and calculating $x_1, \cdots, x_3$. Calculate the error, using $\sqrt{5} = 2.236\ 068$.

22. In Example 3, show that $x_{n+1}^2 - c = (x_n - c/x_n)^2/4$, which measures accuracy. Show that $|x_n - \sqrt{c}| \approx |x_n - c/x_n|/2$ and apply this to Prob. 21.

23. Find an interval on the positive x-axis in which the iteration process in Example 3 with $c = 2$ satisfies the condition of Theorem 1.

24. Design a Newton iteration for the cube root. Calculate $\sqrt[3]{7}$ (6D), starting from $x_0 = 2$.

25. Design a Newton iteration for computing the kth root of a positive number c.

26. Obtain the formula for Newton's method by truncating the Taylor series.

Secant Method

Using the given x_0 and x_1, solve by the secant method:

27. Prob. 17; $x_0 = 0.5, x_1 = 1$ **28.** Prob. 14; $x_0 = 1, x_1 = 0.7$

29. Prob. 13; $x_0 = 1, x_1 = 0.5$ **30.** Prob. 18, $x_0 = 1, x_1 = 2$

Bisection Method and Method of False Position

31. (Bisection method) This simple but slowly convergent method for finding a solution of $f(x) = 0$ with continuous f is based on the **intermediate value theorem,** which states that if a continuous function f has opposite signs at some $x = a$ and $x = b$ ($> a$), that is, either $f(a) < 0$, $f(b) > 0$ or $f(a) > 0$, $f(b) < 0$, then f must be 0 somewhere on $[a, b]$. The solution is found by repeated bisection of the interval and in each iteration picking that half which also satisfies that sign condition. Write down an algorithm for the bisection method, solve $\cos x = x$, taking $a = 0$, $b = 1$, and compare the speed with that in Prob. 17.

Find a solution of the following equations by the bisection method (5 iterations), using the given a and b.

32. $e^{-x} = \ln x$, $a = 1, b = 2$ **33.** $x^2 = \ln x + 3$, $a = 1, b = 2$

34. $x^3 - 2x^2 + 6x = 10, a = 1.7, b = 1.8$ **35.** $e^x + x^4 + x = 2$, $a = 0, b = 1$

36. (Method of false position; regula falsi) Figure 420 shows the idea. We assume that f is continuous. Compute the x-intercept c_0 of the line through $(a_0, f(a_0))$, $(b_0, f(b_0))$. If $f(c_0) = 0$, we are done. If $f(a_0)f(c_0) < 0$ (as in Fig. 420), set $a_1 = a_0$, $b_1 = c_0$ and repeat to get c_1, etc. If $f(a_0)f(c_0) > 0$, then $f(c_0)f(b_0) < 0$ and we set $a_1 = c_0$, $b_1 = b_0$, etc. Show that

$$c_0 = \frac{a_0 f(b_0) - b_0 f(a_0)}{f(b_0) - f(a_0)}$$

and write an algorithm for the method. [The method is generally of first order and is good for starting, but it should not be used near a solution.]

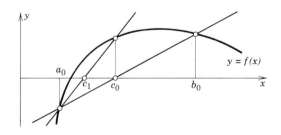

Fig. 420. Method of false position

Find all real solutions of the following equations by the method of false position.

37. $x^4 = 2$ **38.** $\cos x = \sqrt{x}$ **39.** $x^3 = 5x + 6$

40. In Prob. 37, the approximate values of the positive solution will always be somewhat smaller than the exact value of the solution. Why?

18.3 Interpolation

Given $n + 1$ data (pairs of numbers) (x_0, f_0), (x_1, f_1), $\cdots$, (x_n, f_n), with $x_0, x_1, \cdots, x_n$, all different, we want to find a polynomial $p_n(x)$ that at the x_j's takes the given values f_j, that is,

$$p_n(x_0) = f_0, \qquad p_n(x_1) = f_1, \qquad \cdots, \qquad p_n(x_n) = f_n,$$

and has degree n or less. Such a p_n is called an **interpolating polynomial.** The x_j's are often called **nodes.** The corresponding f_j's may be the values of some mathematical function $f(x)$ (ln x, sin x, a Bessel function, etc.), so that $f(x_j) = f_j$, and $p_n(x)$ is then an **approximation** of $f(x)$ with the property that p_n agrees with f at the nodes. Or the f_j's may be values empirically obtained in an experiment or observation. The polynomial $p_n(x)$ is used to get values for all x, be they approximate values of $f(x)$ or values at x's at which no measurement was taken. We call this **interpolation** if the x of interest lies between the nodes, and **extrapolation** if it does not lie between the nodes. In the latter case, the process is generally less accurate.

Such an interpolating polynomial $p_n(x)$ of degree n or less that fits given data exists—below we shall write it down explicitly—and is unique. Uniqueness follows by noting that the difference $d_n = p_n - q_n$ of two such polynomials is a polynomial of degree n or less and has at least $n + 1$ zeros [the nodes $x_0, \cdots, x_n$, at which $p_n(x) = q_n(x)$ by assumption], hence d_n must be identically zero, that is, $p_n(x) \equiv q_n(x)$, the asserted uniqueness.

Of course, the practical question is how to find $p_n(x)$ for given data. For this we consider some standard methods. For given data, these methods give the same polynomial, by the uniqueness just proved (which is thus of practical interest!), but in different forms, which differ in the amounts of computation involved.

Special Cases: Linear and Quadratic Interpolation

Linear interpolation is interpolation by means of the straight line through (x_0, f_0), (x_1, f_1), given by

(1) $$p_1(x) = f_0 + (x - x_0)f[x_0, x_1]$$

(cf. Fig. 421, p. 960) with the **first divided difference** $f[x_0, x_1]$ given by

(2) $$f[x_0, x_1] = \frac{f_1 - f_0}{x_1 - x_0}.$$

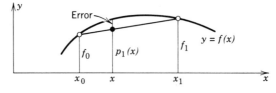

Fig. 421. Linear interpolation

Indeed, from (1) we see that $p_1(x_0) = f_0$, and from (1) and (2) we have $p_1(x_1) = f_1$, as claimed.

EXAMPLE 1. Linear interpolation

Estimate the population (in millions) of the United States in 1968 from

Year	1960	1970
Population	179.3	203.2

Solution. $p_1(1968) = 179.3 + \dfrac{203.2 - 179.3}{1970 - 1960}(1968 - 1960) = 198.4.$

EXAMPLE 2. Linear interpolation

Find ln 9.2 from ln 9.0 = 2.1972, ln 9.5 = 2.2513 by linear interpolation and determine the error from ln 9.2 = 2.2192 (4D).

Solution. $p_1(9.2) = 2.1972 + \dfrac{2.2513 - 2.1972}{9.5 - 9.0}(9.2 - 9.0) = 2.2188.$ Error $-0.0004.$ Hence linear interpolation is not sufficient to get 4D-accuracy; it would suffice for 3D-accuracy. ∎

Quadratic interpolation is interpolation by means of the polynomial $p_2(x)$ of at most second degree whose curve passes through three points (x_0, f_0), (x_1, f_1), (x_2, f_2),

(3) $\qquad p_2(x) = f_0 + (x - x_0)f[x_0, x_1] + (x - x_0)(x - x_1)f[x_0, x_1, x_2]$

with $f[x_0, x_1]$ as above and the **second divided difference** given by

(4) $\qquad\qquad f[x_0, x_1, x_2] = \dfrac{f[x_1, x_2] - f[x_0, x_1]}{x_2 - x_0}.$

Equation (3) shows that $p_2(x_0) = f_0$ because of the factors $x - x_0$, also

$$p_2(x_1) = f_0 + (x_1 - x_0)\frac{f_1 - f_0}{x_1 - x_0} = f_1$$

and $p_2(x_2) = f_2$ with a little more calculation. (Verify it.)

EXAMPLE 3. Quadratic interpolation

Find ln 9.2 from ln 8.0 = 2.0794, ln 9.0 = 2.1972, ln 9.5 = 2.2513.

Solution. We compute the divided differences by (2) and (4),

$x_0 = 8.0$	$f_0 = 2.0794$		
		$f[x_0, x_1] = 0.1178$	
$x_1 = 9.0$	$f_1 = 2.1972$		$f[x_0, x_1, x_2] = -0.0064.$
		$f[x_1, x_2] = 0.1082$	
$x_2 = 9.5$	$f_2 = 2.2513$		

and then from (3)

$$p_2(x) = 2.0794 + (x - 8.0) \cdot 0.1178 + (x - 8.0)(x - 9.0) \cdot (-0.0064)$$

$$= 0.6762 + 0.2266x - 0.0064x^2$$

so that we get the answer $p_2(9.2) = 2.2192$, which is exact to 4D.   ∎

Newton's Divided Difference Interpolation

We show next that (1) and (3) are particular cases of a general interpolation formula of great practical interest. Note first that $p_2 = p_1 +$ another term. This is good because it helps to save work if we suddenly realize that linear interpolation is not accurate enough and we want to go on to p_2. More generally, we want to have $p_n = p_{n-1} +$ another term, say,

$$(5) \qquad\qquad p_n(x) = p_{n-1}(x) + g_n(x)$$

where $p_{n-1}(x_0) = f_0, \cdots, p_{n-1}(x_{n-1}) = f_{n-1}$ and $p_n(x)$ has these same n values and, moreover, $p_n(x_n) = f_n$. We show that this idea leads to a reasonable general interpolation formula. We determine

$$(5') \qquad\qquad g_n(x) = p_n(x) - p_{n-1}(x).$$

Since p_n and p_{n-1} agree at $x_0, \cdots, x_{n-1}$, we see that g_n is zero there. Also, g_n will generally be a polynomial of nth degree because so is p_n, whereas p_{n-1} can be of degree $n - 1$ at most. Hence g_n must be of the form

$$(5'') \qquad\qquad g_n(x) = a_n(x - x_0)(x - x_1) \cdots (x - x_{n-1}).$$

We determine a_n. For this we set $x = x_n$ and solve (5'') algebraically for a_n. Replacing $g_n(x_n)$ according to (5') and using $p_n(x_n) = f_n$, we see that this gives

$$(6) \qquad\qquad a_n = \frac{f_n - p_{n-1}(x_n)}{(x_n - x_0)(x_n - x_1) \cdots (x_n - x_{n-1})}.$$

When $n = 1$, then $p_{n-1}(x_n) = p_0(x_1) = f_0$, so that (6) gives

$$a_1 = \frac{f_1 - p_0(x_1)}{x_1 - x_0} = \frac{f_1 - f_0}{x_1 - x_0} = f[x_0, x_1]$$

and (5) gives (1). When $n = 2$, then (6) with p_1 from (1) gives

$$a_2 = \frac{f_2 - p_1(x_2)}{(x_2 - x_0)(x_2 - x_1)} = \frac{f_2 - f_0 - (x_2 - x_0)f[x_0, x_1]}{(x_2 - x_0)(x_2 - x_1)} = f[x_0, x_1, x_2]$$

where the last equality follows by straightforward calculation and comparison with (4). (Verify it.) Hence (5) with $n = 2$ and $g_2(x)$ given by (5'') takes the form (3), as expected. By the same idea,

$$a_3 = f[x_0, x_1, x_2, x_3] = \frac{f[x_1, x_2, x_3] - f[x_0, x_1, x_2]}{x_3 - x_0}$$

and in general

(7) $$\boxed{a_k = f[x_0, \cdots, x_k] = \frac{f[x_1, \cdots, x_k] - f[x_0, \cdots, x_{k-1}]}{x_k - x_0}.}$$

$f[x_0, \cdots, x_k]$ is called the **kth divided difference.** Formula (5) with $n = k$ now becomes

(8) $$p_k(x) = p_{k-1}(x) + (x - x_0)(x - x_1) \cdots (x - x_{k-1}) f[x_0, \cdots, x_k].$$

With $p_0(x) = f_0$ by repeated application with $k = 1, \cdots, n$ this finally gives **Newton's divided difference interpolation formula**

(9) $$\boxed{\begin{aligned} f(x) &\approx f_0 + (x - x_0) f[x_0, x_1] + (x - x_0)(x - x_1) f[x_0, x_1, x_2] \\ &+ \cdots + (x - x_0) \cdots (x - x_{n-1}) f[x_0, \cdots, x_n]. \end{aligned}}$$

An algorithm is shown in Table 18.2. The first do-loop computes the divided differences and the second the desired value $p_n(\hat{x})$.

Example 4 shows how to arrange differences near the values from which they are obtained; the latter always stand a half-line above and a half-line below in the previous column. Such an arrangement is called a (divided) **difference table.**

EXAMPLE 4. Newton's Divided Difference Interpolation Formula

Find $f(9.2)$ from the given values.

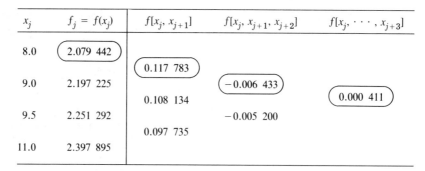

x_j	$f_j = f(x_j)$	$f[x_j, x_{j+1}]$	$f[x_j, x_{j+1}, x_{j+2}]$	$f[x_j, \cdots, x_{j+3}]$
8.0	2.079 442			
		0.117 783		
9.0	2.197 225		−0.006 433	
		0.108 134		0.000 411
9.5	2.251 292		−0.005 200	
		0.097 735		
11.0	2.397 895			

Solution. We compute the divided differences as shown. Sample computation: $(0.097\ 735 - 0.108\ 134)/(11 - 9) = -0.005\ 200$. The values which we need in (9) are circled. We have

$$f(x) \approx p_3(x) = 2.079\ 442 + 0.117\ 783(x - 8.0) - 0.006\ 433(x - 8.0)(x - 9.0)$$
$$+ 0.000\ 411(x - 8.0)(x - 9.0)(x - 9.5).$$

At $x = 9.2$,

$$f(9.2) \approx 2.079\ 442 + 0.141\ 340 - 0.001\ 544 - 0.000\ 030 = 2.219\ 208.$$

The value exact to 6D is $f(9.2) = \ln 9.2 = 2.219\ 203$. Note that we can nicely see how the accuracy increases from term to term:

$$p_1(9.2) = 2.220\ 782, \qquad p_2(9.2) = 2.219\ 238, \qquad p_3(9.2) = 2.219\ 203. \qquad ■$$

Table 18.2
Newton's Divided Difference Interpolation

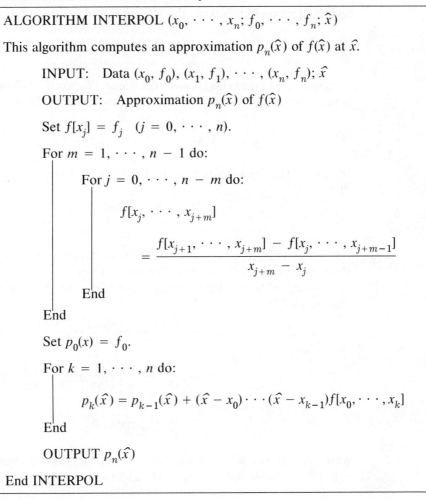

ALGORITHM INTERPOL $(x_0, \cdots, x_n; f_0, \cdots, f_n; \hat{x})$

This algorithm computes an approximation $p_n(\hat{x})$ of $f(\hat{x})$ at $\hat{x}$.

   INPUT: Data $(x_0, f_0), (x_1, f_1), \cdots, (x_n, f_n); \hat{x}$

   OUTPUT: Approximation $p_n(\hat{x})$ of $f(\hat{x})$

Set $f[x_j] = f_j$ $(j = 0, \cdots, n)$.

For $m = 1, \cdots, n - 1$ do:

    For $j = 0, \cdots, n - m$ do:

$$f[x_j, \cdots, x_{j+m}]$$
$$= \frac{f[x_{j+1}, \cdots, x_{j+m}] - f[x_j, \cdots, x_{j+m-1}]}{x_{j+m} - x_j}$$

    End

End

Set $p_0(x) = f_0$.

For $k = 1, \cdots, n$ do:

$$p_k(\hat{x}) = p_{k-1}(\hat{x}) + (\hat{x} - x_0) \cdots (\hat{x} - x_{k-1}) f[x_0, \cdots, x_k]$$

End

OUTPUT $p_n(\hat{x})$

End INTERPOL

Equal Spacing: Newton's Forward Difference Formula

Newton's formula (9) is valid for *arbitrarily spaced* nodes as they may occur in practice in experiments or observations, or as one wants them for functions that change rapidly near some x and slowly near others. However, in many applications the x_j's are *regularly spaced*—for instance, in function tables or in measurements at regular intervals of time. Then we can write

(10) $$\boxed{x_0, \quad x_1 = x_0 + h, \quad x_2 = x_0 + 2h, \quad \cdots, \quad x_n = x_0 + nh.}$$

We show how (7) and (9) simplify in this case.

To get started, let us define the *first forward difference* of f at x_j by

$$\Delta f_j = f_{j+1} - f_j,$$

the *second forward difference* of f at x_j by

$$\Delta^2 f_j = \Delta f_{j+1} - \Delta f_j$$

and, continuing in this way, the **kth forward difference** of f at x_j by

(11)
$$\boxed{\Delta^k f_j = \Delta^{k-1} f_{j+1} - \Delta^{k-1} f_j} \qquad (k = 1, 2, \cdots).$$

Examples and an explanation of the name "forward" follow below. What is the point of this? We show that when we have regular spacing (10), then

(12)
$$f[x_0, \cdots, x_k] = \frac{1}{k! h^k} \Delta^k f_0.$$

We prove (12) by induction. It is true for $k = 1$ because $x_1 = x_0 + h$, so that

$$f[x_0, x_1] = \frac{f_1 - f_0}{x_1 - x_0} = \frac{1}{h}(f_1 - f_0) = \frac{1}{1! h} \Delta f_0.$$

Assuming (12) to be true for all forward differences of order k, we show that it holds for $k + 1$. We use (7) with $k + 1$ instead of k, then $x_{k+1} = x_0 + (k + 1)h$, resulting from (10), and finally (11) with $j = 0$. This gives

$$f[x_0, \cdots, x_{k+1}] = \frac{f[x_1, \cdots, x_{k+1}] - f[x_0, \cdots, x_k]}{(k + 1)h}$$

$$= \frac{1}{(k + 1)h} \left[\frac{1}{k! h^k} \Delta^k f_1 - \frac{1}{k! h^k} \Delta^k f_0 \right]$$

$$= \frac{1}{(k + 1)! h^{k+1}} \Delta^{k+1} f_0$$

which is (12) with $k + 1$ instead of k. Formula (12) is proved.

In (9) we finally set $x = x_0 + rh$. Then $x - x_0 = rh$, $x - x_1 = (r - 1)h$ since $x_1 - x_0 = h$, and so on. With this and (12), formula (9) becomes **Newton's** (or *Gregory*[8]*–Newton's*) **forward difference interpolation formula**

(13)
$$\boxed{\begin{aligned} f(x) \approx p_n(x) &= \sum_{s=0}^{n} \binom{r}{s} \Delta^s f_0 \qquad (x = x_0 + rh, \quad r = (x - x_0)/h) \\ &= f_0 + r\Delta f_0 + \frac{r(r - 1)}{2!} \Delta^2 f_0 + \cdots + \frac{r(r - 1) \cdots (r - n + 1)}{n!} \Delta^n f_0 \end{aligned}}$$

where the **binomial coefficients** in the first line are defined by

(14) $\binom{r}{0} = 1, \quad \binom{r}{s} = \dfrac{r(r - 1)(r - 2) \cdots (r - s + 1)}{s!}$ ($s > 0$, integer)

and $s! = 1 \cdot 2 \cdots s$.

---

[8]JAMES GREGORY (1638—1675), Scots mathematician, professor at St. Andrews and Edinburgh.

The **error** involved in (13) in the case of a function f that is at least $n + 1$ times continuously differentiable is

(15)

$$\epsilon_n(x) = p_n(x) - f(x) = -\frac{1}{(n+1)!}(x - x_0) \cdots (x - x_n)f^{(n+1)}(t)$$

$$= -\frac{h^{n+1}}{(n+1)!}r(r-1) \cdots (r-n)f^{(n+1)}(t),$$

where $f^{(n+1)}$ is the $(n+1)$th derivative of f and t lies between x_0 and x_n (where also x is assumed to lie). We show in the next example how to use this formula in practice. For a proof, see Ref. [E22], p. 190, listed in Appendix 1.

Comments on accuracy.   **(A)** The error $\epsilon_n(x)$ is about of the order of magnitude of the next difference not used in $p_n(x)$.

   (B) One should choose $x_0, \cdots, x_n$ such that the x at which one interpolates is as well centered between $x_0, \cdots, x_n$ as possible.

   The reason for (A) is that in (15),

$$f^{n+1}(t) \approx \frac{\Delta^{n+1}f(t)}{h^{n+1}}, \qquad \frac{|r(r-1)\cdots(r-n)|}{1 \cdot 2 \cdots (n+1)} \leqq 1 \quad \text{when} \quad |r| \leqq 1$$

(and actually for any r as long as we do not *extrapolate*). The reason for (B) is that $|r(r-1) \cdots (r-n)|$ becomes smallest for that choice.

EXAMPLE 5. Newton's forward difference formula. Error estimation
Compute cosh 0.56 from (13) and the four values in the following table and estimate the error.

j	x_j	$f_j = \cosh x_j$	Δf_j	$\Delta^2 f_j$	$\Delta^3 f_j$
0	0.5	1.127 626			
			0.057 839		
1	0.6	1.185 465		0.011 865	
			0.069 704	0.012 562	0.000 697
2	0.7	1.255 169			
			0.082 266		
3	0.8	1.337 435			

Solution. We compute the forward differences as shown in the table. The values we need are circled. In (13) we have $r = (0.56 - 0.50)/0.1 = 0.6$, so that (13) gives

$$\cosh 0.56 \approx 1.127\ 626 + 0.6 \cdot 0.057\ 839 + \frac{0.6(-0.4)}{2} \cdot 0.011\ 865 + \frac{0.6(-0.4)(-1.4)}{6} \cdot 0.000\ 697$$

$$= 1.127\ 626 + 0.034\ 703 - 0.001\ 424 + 0.000\ 039 = 1.160\ 944.$$

Error estimate. From (15), since $\cosh^{(4)} t = \cosh t$,

$$\epsilon_3(0.56) = -\frac{0.1^4}{4!} \cdot 0.6(-0.4)(-1.4)(-2.4)\cosh t = A\cosh t,$$

where $A = 0.000\ 003\ 36$ and $0.5 \leqq t \leqq 0.8$. We don't know t, but we get an inequality by taking the smallest and largest cosh t in that interval:

$$A \cosh 0.5 \leqq \epsilon_3(0.62) \leqq A \cosh 0.8.$$

Since $f(x) = p_3(x) - \epsilon_3(x)$, this gives

$$p_3(0.56) - A \cosh 0.8 \leqq \cosh 0.56 \leqq p_3(0.56) - A \cosh 0.5.$$

Numerical values are

$$1.160\ 939 \leqq \cosh 0.56 \leqq 1.160\ 941.$$

The exact 6D-value is $\cosh 0.56 = 1.160\ 941$. It lies within these bounds. Such bounds are not always so tight. Also, we did not consider round-off errors, which will depend on the number of operations.   ■

This example also explains the name "*forward* difference formula": we see that the differences in the formula slope forward in the difference table.

Equal Spacing: Newton's Backward Difference Formula

Instead of forward-sloping differences we may also employ backward-sloping differences. The difference table remains the same as before (same numbers, in the same positions), except for a very harmless change of the running subscript j (which we explain in Example 6, below). Nevertheless, purely for reasons of convenience it is standard to introduce a second name and notation for differences as follows. We define the *first backward difference* of f at x_j by

$$\nabla f_j = f_j - f_{j-1},$$

the *second backward difference* of f at x_j by

$$\nabla^2 f_j = \nabla f_j - \nabla f_{j-1}$$

and, continuing in this way, the **kth backward difference** of f at x_j by

(16) $$\boxed{\nabla^k f_j = \nabla^{k-1} f_j - \nabla^{k-1} f_{j-1}} \qquad (k = 1, 2, \cdots).$$

A formula similar to (13) but involving backward differences is **Newton's** (or *Gregory–Newton's*) **backward difference interpolation formula**

(17) $$\boxed{\begin{aligned} f(x) \approx p_n(x) &= \sum_{s=0}^{n} \binom{r+s-1}{s} \nabla^s f_0 \qquad (x = x_0 + rh, r = (x - x_0)/h) \\ &= f_0 + r\nabla f_0 + \frac{r(r+1)}{2!}\nabla^2 f_0 + \cdots + \frac{r(r+1)\cdots(r+n-1)}{n!}\nabla^n f_0. \end{aligned}}$$

EXAMPLE 6. Newton's forward and backward interpolations

Compute a 7D-approximation of the Bessel function $J_0(x)$ for $x = 1.72$ from the four values in the following table, using (a) Newton's forward formula (13), (b) Newton's backward formula (17).

j_{for}	j_{back}	x_j	$J_0(x_j)$	1st Diff.	2nd Diff.	3rd Diff.
0	-3	1.7	0.397 9849			
				-0.057 9985		
1	-2	1.8	0.339 9864		-0.000 1693	
				-0.058 1678		0.000 4093
2	-1	1.9	0.281 8186		0.000 2400	
				-0.057 9278		
3	0	2.0	0.223 8908			

Solution. The computation of the differences is the same in both cases. Only their notation differs.

(a) **Forward.** In (13) we have $r = (1.72 - 1.70)/0.1 = 0.2$, and j goes from 0 to 3 (see first column). In each column we need the first given number, and (13) thus gives

$$J_0(1.72) \approx 0.397\ 9849 + 0.2(-0.057\ 9985) + \frac{0.2(-0.8)}{2}(-0.000\ 1693)$$

$$+ \frac{0.2(-0.8)(-1.8)}{6} \cdot 0.000\ 4093$$

$$= 0.397\ 9849 - 0.011\ 5997 + 0.000\ 0135 + 0.000\ 0196$$

$$= 0.386\ 4183,$$

which is exact to 6D, the exact 7D-value being 0.386 4185.

(b) **Backward.** For (17) we use j shown in the second column, and in each column the last number. Since $r = (1.72 - 2.00)/0.1 = -2.8$, we thus get from (17)

$$J_0(1.72) \approx 0.223\ 8908 - 2.8(-0.057\ 9278) + \frac{-2.8(-1.8)}{2} \cdot 0.000\ 2400$$

$$+ \frac{-2.8(-1.8)(-0.8)}{6} \cdot 0.000\ 4093$$

$$= 0.223\ 8908 + 0.162\ 1978 + 0.000\ 6048 - 0.000\ 2750$$

$$= 0.386\ 4184. \qquad \blacksquare$$

Central Difference Notation

There is a third notation for differences, which is useful in certain interpolation formulas (see, for instance, Prob. 19, below), in numerical differentiation to be discussed in Sec. 18.5 and in connection with differential equations (Chap. 20). This is the central difference notation. The *first central difference* of $f(x)$ at x_j is defined by

$$\delta f_j = f_{j+1/2} - f_{j-1/2}$$

and the **kth central difference** of $f(x)$ at x_j by

(18)
$$\boxed{\delta^k f_j = \delta^{k-1} f_{j+1/2} - \delta^{k-1} f_{j-1/2}} \qquad j = 2, 3, \cdots.$$

Thus in this notation a difference table, for example for f_{-1}, f_0, f_1, f_2, looks as shown on the next page.

$$
\begin{array}{lll}
x_{-1} & f_{-1} & \\
& & \delta f_{-1/2} & \\
x_0 & f_0 & & \delta^2 f_0 \\
& & \delta f_{1/2} & & \delta^3 f_{1/2} \\
x_1 & f_1 & & \delta^2 f_1 \\
& & \delta f_{3/2} & \\
x_2 & f_2 &
\end{array}
$$

Lagrangian Interpolation

Given $(x_0, f_0), \cdots, (x_n, f_n)$ with arbitrarily spaced x_j, Lagrange had the idea of multiplying each f_j by a polynomial that is 1 at x_j and 0 at the other n nodes, and then to take the sum of these $n + 1$ polynomials to get the unique interpolation polynomial of degree n or less. The resulting formula, called the **Lagrange interpolation formula,** is

(19a)
$$
f(x) \approx L_n(x) = \sum_{k=0}^{n} \frac{l_k(x)}{l_k(x_k)} f_k
$$

where

$$
l_0(x) = (x - x_1)(x - x_2) \cdots (x - x_n)
$$

(19b) $l_k(x) = (x - x_0) \cdots (x - x_{k-1})(x - x_{k+1}) \cdots (x - x_n), \qquad 0 < k < n$

$$
l_n(x) = (x - x_0)(x - x_1) \cdots (x - x_{n-1}).
$$

We can easily see that $L_n(x_k) = f_k$. Indeed, inspection of (19b) shows that $l_k(x_j) = 0$ when $j \neq k$, so that for $x = x_k$, the sum in (19a) reduces to the single term $(l_k(x_k)/l_k(x_k))f_k = f_k$.

EXAMPLE 7. An application of Lagrange's interpolation formula
Find ln 9.2, using (19) with $n = 3$ and the values

x	9.0	9.5	10.0	11.0
ln x	2.19722	2.25129	2.30259	2.39790

Solution. From (19) we have

$$
L_3(x) = \frac{l_0(x)}{l_0(x_0)} f_0 + \frac{l_1(x)}{l_1(x_1)} f_1 + \frac{l_2(x)}{l_2(x_2)} f_2 + \frac{l_3(x)}{l_3(x_3)} f_3
$$

where $l_0(x) = (x - 9.5)(x - 10)(x - 11)$, $l_1(x) = (x - 9)(x - 10)(x - 11)$, etc. From this we compute

$$
\ln 9.2 = \frac{-0.43200}{-1.00000} \cdot 2.19722 + \frac{0.28800}{0.37500} \cdot 2.25129 + \frac{0.10800}{-0.50000} \cdot 2.30259 + \frac{0.04800}{3.00000} \cdot 2.39790
$$

$$
= 2.21920 \text{ (exact to 5D).} \qquad \blacksquare
$$

The Lagrange polynomials $L_n(x)$ are not very practical in numerical work. Computations may become laborious and, more importantly, previous work is wasted in the transition to a polynomial of higher degree. However, they are of great interest in deriving other formulas. An important application of this type follows in Sec. 18.5 (the derivation of the famous Simpson rule of integration).

Inverse Interpolation

The problem of finding x for given $f(x)$ is known as **inverse interpolation.** If f is differentiable and df/dx is not zero near the point at which the inverse interpolation is to be effected, the inverse $x = F(y)$ of $y = f(x)$ exists locally near the given value of f and it may happen that F can be approximated in that neighborhood by a polynomial of moderately low degree. Then we may effect the inverse interpolation by tabulating F as a function of y and applying methods of direct interpolation to F. If $df/dx = 0$ near or at the desired point, it may be useful to solve $p(x) = \tilde{f}$ by iteration; here $p(x)$ is a polynomial that approximates $f(x)$ and $\tilde{f}$ is the given value.

In the next section we discuss another type of interpolation introduced in 1946 in order to avoid the disadvantage in working with high-degree interpolation polynomials, which often tend to oscillate between nodes. This is called **spline interpolation.**

Problems for Sec. 18.3

1. Verify the computations in Example 3.

Compute the quadratic interpolation:

2. $f(0.9)$ from $f(0.5) = 0.479$, $f(1.0) = 0.841$, $f(2.0) = 0.909$
3. $f(0.8)$ from the given values in Prob. 2
4. sinh 0.3 from sinh $(-0.5) = -0.521$, sinh $0 = 0$, sinh $1 = 1.175$
5. $\Gamma(1.01)$ and $\Gamma(1.03)$ from $\Gamma(1.00) = 1.000$, $\Gamma(1.02) = 0.989$, $\Gamma(1.04) = 0.978$
6. Find $f(6.5)$ from $f(6.0) = 0.1506$, $f(7.0) = 0.3001$, $f(7.5) = 0.2663$, $f(7.7) = 0.2346$ by cubic interpolation.
7. Using the values in the table below, find sin 0.26 by linear and by quadratic interpolation. How many decimals are exact? (sin 0.26 = 0.25708 exact to 5D.)

x	0.0	0.2	0.4	0.6	0.8	1.0
sin x	0.00000	0.19867	0.38942	0.56464	0.71736	0.84147

8. Compute sin 0.26 by (13) with $n = 3$ and $n = 4$. How many decimals are exact?
9. Apply the backward difference interpolation formula (17) (a) with $n = 1$, (b) with $n = 2$ to obtain sin 0.26. Observe that in both cases, the first two decimals will be exact. Hence the result in (b) is poorer than that in Prob. 7. Why?
10. Extend the given table so that you can apply (17) with (a) $n = 3$, (b) $n = 4$, (c) $n = 5$, for computing sin 0.26; show that you need sin x for the values $x = -0.6, -0.4, -0.2, 0, 0.2$, and the differences. What property of the sine function makes this extension very easy? Then apply (17), finding (a) 0.257 09, (b) 0.257 05, (c) 0.257 08 (the exact 5D-value). Why are the results poorer than in Prob. 8?

11. Using (13) and the values below, find $p_1(x)$, $p_2(x)$, $p_3(x)$ for $x = 2.05$ and observe how the accuracy increases. ($\sqrt{2.05} = 1.431\ 782$ to 6D.)

x	2.0	2.1	2.2	2.3	2.4
$\sqrt{x}$	1.414 214	1.449 138	1.483 240	1.516 575	1.549 193

12. Do the same task as in Prob. 11 when $x = 2.15$.

13. Using (15), show that for the error in linear interpolation we have $|\epsilon_1(x)| \leq \frac{1}{2}h^2|r(r-1)|M_2$, where M_2 is an upper bound for $|f''(x)|$, $x_0 \leq x \leq x_1$, that is, $|f''(x)| \leq M_2$ for these x. Apply this to Prob. 7 and compare it with the actual error.

14. Obtain from (15) an error estimate (similar to that in Prob. 13) for quadratic interpolation and apply it to Prob. 7.

15. Write the differences in the table of Example 5 in central difference notation.

16. Compute a difference table of $f(x) = x^3$ for[9] $x = 0(1)5$. Choose $x_0 = 2$ and write all occurring numbers in terms of the notations (a) for central differences, (b) for forward differences, (c) for backward differences.

17. Show that $\delta^n f_m = \Delta^n f_{m-n/2} = \nabla^n f_{m+n/2}$.

18. Show that $\delta^2 f_m = f_{m+1} - 2f_m + f_{m-1}$, $\delta^3 f_{m+1/2} = f_{m+2} - 3f_{m+1} + 3f_m - f_{m-1}$.

19. (**Everett interpolation formula**) There are formulas involving only even-order differences. A particularly useful such formula is the **Everett formula**

(20) $\quad f(x) \approx (1-r)f_0 + rf_1 + \dfrac{(2-r)(1-r)(-r)}{3!}\delta^2 f_0 + \dfrac{(r+1)r(r-1)}{3!}\delta^2 f_1$,

where $r = (x - x_0)/h$. Applying this formula, compute $e^{1.24}$ from $e^{1.1} = 3.004\ 166$, $e^{1.2} = 3.320\ 117$, $e^{1.3} = 3.669\ 297$, $e^{1.4} = 4.055\ 200$.

20. The error in the Everett formula is

(21) $\quad\quad\quad \epsilon(x) = p(x) - f(x) = -h^4 \begin{pmatrix} r+1 \\ 4 \end{pmatrix} f^{(4)}(t)$,

where $p(x)$ denotes the right-hand side of (20), and $x_0 - h < t < x_0 + 2h$. Use this formula to get error bounds in Prob. 19.

21. Using (20), compute $J_0(1.72)$ from $J_0(1.60) = 0.455\ 4022$ and $J_0(1.7)$, $J_0(1.8)$, $J_0(1.9)$ in Example 6. Note that the accuracy is higher than in Example 6, although the work is less. Can you explain why?

22. Compute $f(0.3)$ from the Everett formula (20) and the following table.

x	0.0	0.2	0.4	0.6	0.8
$f(x)$	0.50000	0.69867	0.88942	1.06464	1.21736

23. Compute $f(0.5)$ from (20) and the given values in Prob. 22.

24. Using $\ln 8.5 = 2.14007$ and $\ln 9.0$, $\ln 9.5$, $\ln 10$ as in Example 7, compute $\ln 9.2$ (a) by means of (13) with $n = 3$ (and $x_0 = 8.5$), (b) by means of (17) with $n = 3$ (and $x_0 = 10$).

25. Find $f(3)$, using Lagrange's formula (18) and $f(1) = 2$, $f(2) = 11$, $f(4) = 77$.

[9]The notation $x = a(h)b$ means $x = a, a + h, a + 2h, \cdots, b$.

18.4 Splines

One might expect the quality of interpolation to increase with increasing degree n of the polynomials used. Unfortunately, this is not generally true. Indeed, for various functions f the corresponding interpolation polynomials may tend to oscillate more and more between nodes as n increases. Hence we must be prepared for possible **numerical instability,** particularly if we cannot choose the x_j's freely (and compute or observe corresponding f_j's) but if they (and f_j's) are given, as will often be the case. A famous example is $f(x) = 1/(1 + x^2)$ for x on the interval $[-5, 5]$ and equidistant nodes, for which C. Runge[10] has shown that for $|x| > 3.64$ the maximum of the error approaches infinity as $n \to \infty$. (Proof in Ref. [E22], pp. 275—279. See also Ref. [E36], p. 148.) In the present section we discuss a popular and important method, the method of splines, which avoids that disadvantage. This approach was initiated in 1946 by I. J. Schoenberg (*Quarterly of Applied Mathematics* **4,** 1946, pp. 45—99, 112—141) and has found various applications.

   Spline interpolation is piecewise polynomial interpolation. This means that a function $f(x)$ is given on an interval $a \leqq x \leqq b$, and we want to approximate $f(x)$ on this interval by a function $g(x)$ that is obtained as follows. We *partition* $a \leqq x \leqq b$; that is, we subdivide it into subintervals with common endpoints (again called *nodes*)

(1) $$a = x_0 < x_1 < \cdots < x_n = b.$$

We now require that $g(x_0) = f(x_0), \cdots, g(x_n) = f(x_n)$, and that in these subintervals $g(x)$ be given by polynomials, one polynomial per subinterval, such that at those endpoints, $g(x)$ is several times differentiable. Hence, instead of approximating $f(x)$ by a single polynomial on the entire interval $a \leqq x \leqq b$, we now approximate $f(x)$ by n polynomials. In this way we may obtain approximating functions $g(x)$ that are more suitable in many problems of interpolation and approximation. In particular, this method will in general be numerically *stable;* these $g(x)$ will not be as oscillatory between nodes as a single polynomial on $a \leqq x \leqq b$ often is. Functions $g(x)$ thus obtained are called **splines.** This name is derived from thin rods, called *splines,* which engineers have used for a long time to fit curves through given points.

   The simplest continuous piecewise polynomial approximation would be by piecewise *linear* functions. But the graph of such a function has corners and would be of little practical interest—think of designing the body of a ship or a car. Hence it is preferable to use functions that have a certain number of derivatives *everywhere* on the interval $a \leqq x \leqq b$.

   We shall consider cubic splines, which are perhaps the most important ones from a practical point of view. By definition, a **cubic spline** $g(x)$ on $a \leqq x \leqq b$ corresponding to the partition (1) is a continuous function $g(x)$ that has continuous first and second derivatives everywhere in that interval and, in each subinterval of that partition, is represented by a polynomial of

---

[10]See the footnote in Sec. 20.1.

degree not exceeding three. Hence $g(x)$ consists of cubic polynomials, one in each subinterval.

If $f(x)$ is given on $a \leq x \leq b$ and a partition (1) has been chosen, we obtain a cubic spline $g(x)$ which approximates $f(x)$ by requiring that

(2) $g(x_0) = f(x_0) = f_0, \quad g(x_1) = f(x_1) = f_1, \quad \cdots, \quad g(x_n) = f(x_n) = f_n,$

as in the classical interpolation problem of the previous section. We claim that there is a cubic spline $g(x)$ satisfying these conditions (2). And if we also require that

(3) $$g'(x_0) = k_0, \qquad g'(x_n) = k_n$$

(k_0 and k_n given numbers), then we have a uniquely determined cubic spline. This is the content of the following existence and uniqueness theorem, whose proof will also pave the way for the practical determination of splines.

Theorem 1 (Cubic splines)

Let $f(x)$ be defined on the interval $a \leq x \leq b$, let a partition (1) *of the interval be given, and let k_0 and k_n be any two given numbers. Then there exists one and only one cubic spline $g(x)$ corresponding to* (1) *and satisfying* (2) *and* (3).

Proof. By definition, on every subinterval I_j given by $x_j \leq x \leq x_{j+1}$, the spline $g(x)$ must agree with a cubic polynomial $p_j(x)$ such that

(4) $p_j(x_j) = f(x_j), \qquad p_j(x_{j+1}) = f(x_{j+1}).$

We write $1/(x_{j+1} - x_j) = c_j$ and

(5) $p_j'(x_j) = k_j, \qquad p_j'(x_{j+1}) = k_{j+1},$

where k_0 and k_n are given, and $k_1, \cdots, k_{n-1}$ will be determined later. Equations (4) and (5) are four conditions for $p_j(x)$. By direct calculation we can verify that the unique cubic polynomial $p_j(x)$ satisfying (4) and (5) is

$$
\begin{aligned}
p_j(x) = \; & f(x_j)c_j^2(x - x_{j+1})^2[1 + 2c_j(x - x_j)] \\
& + f(x_{j+1})c_j^2(x - x_j)^2[1 - 2c_j(x - x_{j+1})] \\
& + k_j c_j^2(x - x_j)(x - x_{j+1})^2 \\
& + k_{j+1}c_j^2(x - x_j)^2(x - x_{j+1}).
\end{aligned}
$$

(6)

Differentiating twice, we obtain

(7) $p_j''(x_j) = -6c_j^2 f(x_j) + 6c_j^2 f(x_{j+1}) - 4c_j k_j - 2c_j k_{j+1}$

(8) $p_j''(x_{j+1}) = 6c_j^2 f(x_j) - 6c_j^2 f(x_{j+1}) + 2c_j k_j + 4c_j k_{j+1}.$

By definition, $g(x)$ has continuous second derivatives. This gives the conditions

$$p_{j-1}''(x_j) = p_j''(x_j) \qquad\qquad j = 1, \cdots, n - 1.$$

If we use (8) with j replaced by $j - 1$, and (7), these $n - 1$ equations become

$$(9) \quad c_{j-1}k_{j-1} + 2(c_{j-1} + c_j)k_j + c_jk_{j+1} = 3[c_{j-1}^2 \, \nabla f_j + c_j^2 \, \nabla f_{j+1}].$$

In these equations, $\nabla f_j = f(x_j) - f(x_{j-1})$ and $\nabla f_{j+1} = f(x_{j+1}) - f(x_j)$ and $j = 1, \cdots, n - 1$, as before. This system of $n - 1$ linear equations has a unique solution $k_1, \cdots, k_{n-1}$ because all the coefficients of the system are nonnegative, and each entry on the main diagonal is greater than the sum of the other entries in the corresponding row, so that the coefficient determinant cannot be zero.[11] Hence we are able to determine unique values $k_1, \cdots, k_{n-1}$ of the first derivative of $g(x)$ at the nodes. This completes the proof. ∎

We show how the formulas in this proof can be used for the actual determination of a spline. For simplicity we consider the case of equidistant nodes x_0, $x_1 = x_0 + h, \cdots, x_n = x_0 + nh$, the idea for nonequidistant nodes being the same (just the formulas would look somewhat more complicated). We then have $c_j = 1/(x_{j+1} - x_j) = 1/h$, so that (9), after multiplication by h and with our previous notation $f(x_j) = f_j$, becomes simply

$$(10) \qquad k_{j-1} + 4k_j + k_{j+1} = \frac{3}{h}(f_{j+1} - f_{j-1}), \qquad j = 1, \cdots, n - 1.$$

k_0 and k_n are given, for instance, $k_0 = f'(a)$, $k_n = f'(b)$ or otherwise, and $k_1, \cdots, k_{n-1}$ are obtained by solving the system (10) consisting of $n - 1$ linear equations. This was the first step.

The second step is the determination of the coefficients of the spline $g(x)$. On the interval $x_j \le x \le x_{j+1} = x_j + h$ the spline $g(x)$ is given by a cubic polynomial which we write in the form

$$(11) \qquad p_j(x) = a_{j0} + a_{j1}(x - x_j) + a_{j2}(x - x_j)^2 + a_{j3}(x - x_j)^3.$$

Using Taylor's formula, we obtain

$$(12) \quad
\begin{aligned}
a_{j0} &= p(x_j) = f_j & \text{by (2),} \\[4pt]
a_{j1} &= p_j'(x_j) = k_j & \text{by (5),} \\[4pt]
a_{j2} &= \frac{1}{2}p_j''(x_j) = \frac{3}{h^2}(f_{j+1} - f_j) - \frac{1}{h}(k_{j+1} + 2k_j) & \text{by (7),} \\[4pt]
a_{j3} &= \frac{1}{6}p_j'''(x_j) = \frac{2}{h^3}(f_j - f_{j+1}) + \frac{1}{h^2}(k_{j+1} + k_j)
\end{aligned}$$

[11]Cf. Sec. 19.7, Prob. 14.

with a_{j3} obtained by calculating $p_j''(x_{j+1})$ from (11) and equating the result to (8), that is,

$$p_j''(x_{j+1}) = 2a_{j2} + 6a_{j3}h = \frac{6}{h^2}(f_j - f_{j+1}) + \frac{2}{h}(k_j + 2k_{j+1}),$$

and now subtracting from this $2a_{j2}$ as given in (12) and simplifying.

EXAMPLE 1. Spline interpolation

Interpolate $f(x) = x^4$ on the interval $-1 \le x \le 1$ by the cubic spline $g(x)$ corresponding to the partition $x_0 = -1$, $x_1 = 0$, $x_2 = 1$ and satisfying $g'(-1) = f'(-1)$, $g'(1) = f'(1)$.

Solution. $n = 2$, $h = 1$, $f_0 = f(-1) = 1$, $f_1 = f(0) = 0$, $f_2 = f(1) = 1$, $k_0 = f'(-1) = -4$, $k_2 = f'(1) = 4$. Thus (10) is the single equation ($j = 1$)

$$-4 + 4k_1 + 4 = 3(1 - 1) = 0.$$

Hence $k_1 = 0$. From this and (12) with $j = 0$,

$$a_{00} = f_0 = 1$$
$$a_{01} = k_0 = -4$$
$$a_{02} = 3(0 - 1) - (0 - 8) = 5$$
$$a_{03} = 2(1 - 0) + 0 - 4 = 2.$$

This gives
$$p_0(x) = 1 - 4(x + 1) + 5(x + 1)^2 - 2(x + 1)^3 = -x^2 - 2x^3.$$

Similarly, from (12) with $j = 1$,

$$a_{10} = f_1 = 0$$
$$a_{11} = k_1 = 0$$
$$a_{12} = 3(1 - 0) - (4 + 2 \cdot 0) = -1$$
$$a_{13} = 2(0 - 1) + 4 + 0 = 2.$$

This gives
$$p_1(x) = -x^2 + 2x^3.$$

Figure 422 shows $f(x)$ and this spline

$$g(x) = \begin{cases} -x^2 - 2x^3 & \text{if} \quad -1 \le x \le 0 \\ -x^2 + 2x^3 & \text{if} \quad 0 \le x \le 1. \end{cases}$$

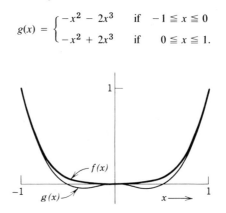

Fig. 422. Function $f(x) = x^4$ and cubic spline $g(x)$
in Example 1

EXAMPLE 2. Spline interpolation

Interpolate $f_0 = f(0) = 1$, $f_1 = f(2) = 9$, $f_2 = f(4) = 41$, $f_3 = f(6) = 41$ by the cubic spline satisfying $k_0 = 0$, $k_3 = -12$.

Solution. $n = 3$, $h = 2$, so that (10) is

$$k_0 + 4k_1 + k_2 \qquad = \tfrac{3}{2}(f_2 - f_0) = 60$$

$$k_1 + 4k_2 + k_3 = \tfrac{3}{2}(f_3 - f_1) = 48.$$

Since $k_0 = 0$ and $k_3 = -12$, the solution is $k_1 = 12$, $k_2 = 12$.

In (12) with $j = 0$ we have $a_{00} = f_0 = 1$, $a_{01} = k_0 = 0$,

$$a_{02} = \tfrac{3}{4}(9 - 1) - \tfrac{1}{2}(12 + 0) = 0$$

$$a_{03} = \tfrac{2}{8}(1 - 9) + \tfrac{1}{4}(12 + 0) = 1$$

From this and, similarly, from (12) with $j = 1$ and $j = 2$ we get the spline $g(x)$ consisting of the three polynomials (cf. Fig. 423)

$$p_0(x) = 1 + x^3 \qquad\qquad\qquad\qquad\qquad\qquad\qquad\qquad (0 \leq x \leq 2)$$

$$p_1(x) = 9 + 12(x - 2) + 6(x - 2)^2 - 2(x - 2)^3 = 25 - 36x + 18x^2 - 2x^3 \qquad (2 \leq x \leq 4)$$

$$p_2(x) = 41 + 12(x - 4) - 6(x - 4)^2 \qquad = -103 + 60x - 6x^2 \qquad (4 \leq x \leq 6). \quad \blacksquare$$

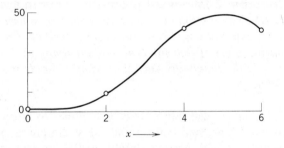

Fig. 423. Spline in Example 2

Splines have an interesting minimum property, which we shall now derive. Suppose that $f(x)$ in Theorem 1 is continuous and has continuous first and second derivatives on $a \leq x \leq b$. Assume that (3) is of the form

(13) $$g'(a) = f'(a), \qquad g'(b) = f'(b)$$

(as in Example 1). Then $f' - g'$ is zero at a and b. Thus, by integration by parts,

$$\int_a^b g''(x)[f''(x) - g''(x)] \, dx = -\int_a^b g'''(x)[f'(x) - g'(x)] \, dx.$$

Since $g'''(x)$ is constant on each subinterval of the partition, evaluating the integral on the right over a subinterval, we obtain $const \cdot [f(x) - g(x)]$, taken at the endpoints of the subinterval, which is zero by (2). Hence that integral on the right is zero. This proves the following formula on the next page:

$$\int_a^b f''(x)g''(x) \, dx = \int_a^b g''(x)^2 \, dx.$$

Consequently,

$$\int_a^b [f''(x) - g''(x)]^2 \, dx = \int_a^b f''(x)^2 \, dx - 2 \int_a^b f''(x)g''(x) \, dx + \int_a^b g''(x)^2 \, dx$$

$$= \int_a^b f''(x)^2 \, dx - \int_a^b g''(x)^2 \, dx.$$

Since the integrand on the left is nonnegative, so is the integral. This yields

(14)
$$\int_a^b f''(x)^2 \, dx \geqq \int_a^b g''(x)^2 \, dx.$$

Our result can be stated as follows.

Theorem 2 (Minimum property of cubic splines)

Let $f(x)$ be continuous and have continuous first and second derivatives on some interval $a \leqq x \leqq b$. Let $g(x)$ be the cubic spline corresponding to some partition (1) of that interval and satisfying (2) and (13). Then $f(x)$ and $g(x)$ satisfy the inequality (14), and equality holds if and only if $f(x)$ is the cubic spline $g(x)$.

Engineers use thin rods called *splines* to fit curves through given points, as was mentioned before, and the strain energy minimized by such splines is proportional, approximately, to the integral of the square of the second derivative of the spline. Hence inequality (14) explains the use of the term *splines* for the functions $g(x)$ considered in this section.

Cubic splines $g(x)$ that instead of (13) satisfy

(15)
$$g''(a) = 0, \qquad g''(b) = 0$$

are called **natural splines** because they have the following interesting minimum property. Among all twice differentiable functions $g(x)$ on $[a, b]$ that satisfy (2) and whose second derivative is continuous, natural splines are the functions that give the integral

$$\int_a^b [g''(x)]^2 \, dx$$

the smallest possible value. Condition (15) means that $g(x)$ has an almost linear graph near the ends a and b.

In the next section we discuss methods for another basic task of numerical analysis, namely, for numerical **integration** and **differentiation**.

Problems for Sec. 18.4

1. Verify that (6) satisfies (4) and (5).
2. Obtain (7) and (8) from (6) as indicated in the text.
3. Verify the derivation of (9) indicated in the text.
4. Derive (10) from (9).
5. Give the details of the derivation of a_{j2} and a_{j3} in (12).
6. Verify the computations in Example 1.
7. Derive $p_1(x)$ and $p_2(x)$ in Example 2 and verify that $g(x)$ in Example 2 has continuous first and second derivatives.
8. Confirm $g(x)$ in Example 1 by starting from cubic polynomials in the form

$$p_0(x) = b_0 + b_1 x + b_2 x^2 + b_3 x^3, \qquad p_1(x) = c_0 + c_1 x + c_2 x^2 + c_3 x^3$$

   and imposing the conditions a cubic spline must satisfy.
9. Compare the spline $g(x)$ in Example 1 with the quadratic interpolation polynomial $p(x)$ over the whole interval. What are the maximum deviations of $g(x)$ and $p(x)$ from $f(x)$? Comment.

Find the cubic spline to the given data, with k_0 and k_n as indicated.

10. $f_0 = f(0) = 4, f_1 = f(2) = 0, f_2 = f(4) = 4, f_3 = f(6) = 80, \quad k_0 = 0, k_3 = 0$
11. $f_0 = f(0) = 1, f_1 = f(1) = 0, f_2 = f(2) = -1, f_3 = f(3) = 0, \quad k_0 = 0, k_3 = -6$
12. $f_0 = f(0) = 0, f_1 = f(1) = 1, f_2 = f(2) = 6, f_3 = f(3) = 10, \quad k_0 = 0, k_3 = 0$
13. $f(-2) = f(-1) = f(1) = f(2) = 0, f(0) = 1, \quad k_0 = k_4 = 0$

14. Rewrite the polynomials in the answer to Prob. 13 in powers of x and show that $g(x)$ is an even function of x, that is, $g(-x) = g(x)$. Had this to be expected?
15. If we started from the piecewise linear function $f(x)$ in Fig. 424, we would obtain $g(x)$ in Prob. 13 as the spline satisfying $g'(-2) = f'(-2), g'(2) = f'(2)$. Find and sketch the corresponding interpolation polynomial of fourth degree and compare it with that spline (cf. Fig. 424). Comment.

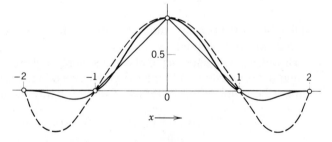

Fig. 424. Spline and interpolation polynomial in Problems 13—15

16. Show that for a given partition of the form (1) there exist $n + 1$ unique cubic splines $g_0(x), \cdots, g_n(x)$ such that $g_j{}'(a) = g_j{}'(b) = 0$ and

$$g_j(x_k) = \delta_{jk} = \begin{cases} 0 & \text{if } j \neq k \\ 1 & \text{if } j = k. \end{cases}$$

17. If a cubic spline is three times continuously differentiable (that is, it has continuous first, second and third derivatives), show that it must be a polynomial.

18. It may sometimes happen that a spline is represented by the same polynomial in adjacent subintervals of $a \leqq x \leqq b$. To illustrate this, find the cubic spline $g(x)$ for $f(x) = \sin x$ corresponding to the partition $x_0 = -\pi/2$, $x_1 = 0$, $x_2 = \pi/2$ of the interval $-\pi/2 \leqq x \leqq \pi/2$ and satisfying $g'(-\pi/2) = f'(-\pi/2)$ and $g'(\pi/2) = f'(\pi/2)$.

19. Show that the cubic splines corresponding to a given partition of a given interval form a vector space (cf. Sec. 6.4).

20. A possible geometric interpretation of (14) is that a cubic spline function minimizes the integral of the square of the curvature, at least approximately. Explain.

18.5 Numerical Integration and Differentiation

The problem of **numerical integration** is the numerical evaluation of a definite integral

$$J = \int_a^b f(x)\, dx$$

where a and b are given and f is a function given analytically by a formula or empirically by a table of values. Geometrically, J is the area under the curve of f between a and b (cf. Fig. 425).

We know that if f is such that we can find a differentiable function F whose derivative is f, then we can evaluate J by applying the familiar formula

$$J = \int_a^b f(x)\, dx = F(b) - F(a) \qquad [F'(x) = f(x)],$$

and tables of integrals, such as those in Ref. [3] in Appendix 1, may be helpful for that purpose.

However, in engineering applications there frequently occur integrals whose integrand is an empirical function given by a table, or is such that the integral cannot be represented in terms of finitely many elementary functions (ex-

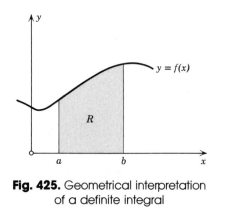

Fig. 425. Geometrical interpretation of a definite integral

amples in Appendix 3), or the explicit form of F is too complicated to be of practical value. Then we may use a numerical method of approximate integration.

Rectangular Rule. Trapezoidal Rule

Numerical integration methods may be obtained by approximating the integrand f by polynomials. To derive the simplest formula, we subdivide the interval of integration into n equal subintervals of length $h = (b - a)/n$ and approximate f in each such interval by a constant function $f(x_j^*)$, where x_j^* is the midpoint of the interval (cf. Fig. 426). Then f is approximated by a **step function** (piecewise constant function), the n rectangles in Fig. 426 have the areas $f(x_1^*)h, \cdots, f(x_n^*)h$, and we obtain the **rectangular rule**

$$(1) \qquad \boxed{J = \int_a^b f(x)\, dx \approx h[f(x_1^*) + f(x_2^*) + \cdots + f(x_n^*)]}$$

where $h = (b - a)/n$.

If we approximate f by a piecewise linear function (polygon of chords of the curve of f; cf. Fig. 427), we obtain the **trapezoidal rule**

$$(2) \qquad \boxed{J = \int_a^b f(x)\, dx \approx h[\tfrac{1}{2}f(a) + f(x_1) + f(x_2) + \cdots + f(x_{n-1}) + \tfrac{1}{2}f(b)],}$$

where $h = (b - a)/n$, and $x_0\ (= a),\ x_1,\ x_2,\ \cdots,\ x_{n-1},\ x_n\ (= b)$ are the endpoints of the intervals already used in (1); thus $x_j = x_0 + jh$. In fact, the n trapezoids in Fig. 427 have the areas

$$\tfrac{1}{2}[f(a) + f(x_1)]h, \qquad \tfrac{1}{2}[f(x_1) + f(x_2)]h, \qquad \cdots, \qquad \tfrac{1}{2}[f(x_{n-1}) + f(b)]h,$$

and their sum $\tilde{J}$ equals the right-hand side of (2).

EXAMPLE 1. Trapezoidal rule

Evaluate $J = \int_0^1 e^{-x^2}\, dx$ by means of (2) with $n = 10$.

Solution. $J \approx 0.1(0.5 \cdot 1.367\ 879 + 6.778\ 167) = 0.746\ 211$ from Table 18.3 on p. 980. ∎

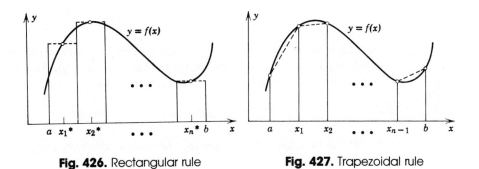

Fig. 426. Rectangular rule   **Fig. 427.** Trapezoidal rule

Table 18.3
Computation in Example 1

j	x_j	x_j^2	$e^{-x_j^2}$	
0	0	0	1.000 000	
1	0.1	0.01		0.990 050
2	0.2	0.04		0.960 789
3	0.3	0.09		0.913 931
4	0.4	0.16		0.852 144
5	0.5	0.25		0.778 801
6	0.6	0.36		0.697 676
7	0.7	0.49		0.612 626
8	0.8	0.64		0.527 292
9	0.9	0.81		0.444 858
10	1.0	1.00	0.367 879	
Sums			1.367 879	6.778 167

Error of the Trapezoidal Rule

By definition, the error ϵ of the approximation $\tilde{J}$ given by the trapezoidal rule (2) is (cf. Sec. 18.1)

$$\epsilon = \tilde{J} - J.$$

If $f(x)$ is a linear function, then $\epsilon = 0$. In this case f' is constant, f'' is zero for all x, and it seems plausible that for a general function f (having a continuous second derivative) we may obtain **error bounds** (bounds for ϵ) involving f''. For this purpose we replace b by a variable t and apply (2) with $n = 1$. Then the corresponding error is

$$\epsilon(t) = \frac{t - a}{2} [f(a) + f(t)] - \int_a^t f(x)\, dx.$$

We note that $\epsilon(a) = 0$, which is trivial. Differentiation gives

$$\epsilon'(t) = \tfrac{1}{2}[f(a) + f(t)] + \frac{t - a}{2} f'(t) - f(t).$$

We see that $\epsilon'(a) = 0$. Differentiating once more, we have

$$\epsilon''(t) = \tfrac{1}{2}(t - a)f''(t)$$

and can obtain bounds for ϵ'' by replacing f'' by its smallest and largest value in the interval $a \leqq t \leqq b$. Denoting these values by M_2^* and M_2, respectively, and noting that $t - a \geqq 0$, we obtain the inequality

$$\tfrac{1}{2}(t - a)M_2^* \leqq \epsilon''(t) \leqq \tfrac{1}{2}(t - a)M_2$$

for all t in that interval. Integrating this inequality from a to t, we have

$$\tfrac{1}{4}(t - a)^2 M_2{}^* \le \epsilon'(t) - \epsilon'(a) \le \tfrac{1}{4}(t - a)^2 M_2.$$

Using $\epsilon'(a) = 0$, $\epsilon(a) = 0$, integrating again, and setting $t = a + h$, we see that

$$\tfrac{1}{12}h^3 M_2{}^* \le \epsilon(a + h) \le \tfrac{1}{12}h^3 M_2.$$

For the errors corresponding to the other $n - 1$ subintervals we obtain $n - 1$ similar inequalities. Addition of all the n inequalities gives an inequality for the error ϵ in the trapezoidal rule (2) corresponding to the integration from a to b; since $h = (b - a)/n$, we obtain

(3)
$$\boxed{KM_2{}^* \le \epsilon \le KM_2 \qquad \text{where} \qquad K = \frac{(b - a)^3}{12n^2}}$$

and $M_2{}^*$ and M_2 are the smallest and largest values of the second derivative of f in the interval of integration.

EXAMPLE 2. Error estimate for the trapezoidal rule

Estimate the error of the approximate value in Example 1.

Solution. We use (3). By differentiation, $f''(x) = 2(2x^2 - 1)e^{-x^2}$. Also $f'''(x) > 0$ when $0 < x < 1$, so that the minimum and maximum occur at the ends of the interval. We calculate $M_2{}^* = f''(0) = -2$ and $M_2 = f''(1) = 0.735\ 759$. Furthermore, $K = 1/1200$, and (3) gives

$$-0.001\ 667 \le \epsilon \le 0.000\ 614.$$

Hence the exact value of J must lie between

$$0.746\ 211 - 0.000\ 614 = 0.745\ 597 \quad \text{and} \quad 0.746\ 211 + 0.001\ 667 = 0.747\ 878.$$

(Actually, $J = 0.746\ 824$, exact to 6D.) ∎

Simpson's Rule of Integration

Piecewise constant approximation of f led to the rectangular rule (1), piecewise linear approximation to the trapezoidal rule (2), and piecewise quadratic approximation will give Simpson's rule, which is of great practical importance because it is sufficiently accurate for most problems, but still sufficiently simple. To derive this formula, we subdivide the interval of integration $a \le x \le b$ into an *even* number of equal subintervals, say, into $2n$ subintervals of length $h = (b - a)/2n$, with endpoints $x_0\ (= a)$, x_1, $\cdots$, x_{2n-1}, $x_{2n}\ (= b)$; cf. Fig. 428. We consider the first two

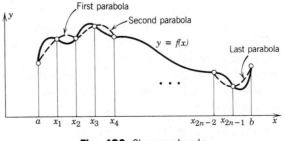

Fig. 428. Simpson's rule

subintervals and approximate $f(x)$ in the interval $x_0 \leqq x \leqq x_2 = x_0 + 2h$ by the Lagrange polynomial $L_2(x)$ through $(x_0, f_0), (x_1, f_1), (x_2, f_2)$, where $f_j = f(x_j)$. From (19a) in Sec. 18.3 we obtain

$$
(4^*) \quad L_2(x) = \frac{(x - x_1)(x - x_2)}{(x_0 - x_1)(x_0 - x_2)} f_0 + \frac{(x - x_0)(x - x_2)}{(x_1 - x_0)(x_1 - x_2)} f_1
$$
$$
+ \frac{(x - x_0)(x - x_1)}{(x_2 - x_0)(x_2 - x_1)} f_2.
$$

The denominators are $2h^2$, $-h^2$ and $2h^2$, respectively. Setting $s = (x - x_1)/h$, we have $x - x_0 = (s + 1)h, x - x_1 = sh, x - x_2 = (s - 1)h$, and obtain

$$
L_2(x) = \tfrac{1}{2} s(s - 1) f_0 - (s + 1)(s - 1) f_1 + \tfrac{1}{2}(s + 1) s f_2.
$$

We now integrate with respect to x from x_0 to x_2. This corresponds to integrating with respect to s from -1 to 1. Since $dx = h \, ds$, the result is

$$
\int_{x_0}^{x_2} f(x) \, dx \approx \int_{x_0}^{x_2} L_2(x) \, dx = h \left(\frac{1}{3} f_0 + \frac{4}{3} f_1 + \frac{1}{3} f_2 \right).
$$

A similar formula holds for the next two subintervals from x_2 to x_4, and so on. By summing all these n formulas we obtain **Simpson's rule**[12]

$$
(4) \quad \boxed{\int_a^b f(x) \, dx \approx \frac{h}{3}(f_0 + 4f_1 + 2f_2 + 4f_3 + \cdots + 2f_{2n-2} + 4f_{2n-1} + f_{2n}),}
$$

where $h = (b - a)/2n$ and $f_j = f(x_j)$. Table 18.4 shows an algorithm for Simpson's rule.

Bounds for the error ϵ_S in (4) can be obtained by a method similar to that in the case of the trapezoidal rule (2), assuming that the fourth derivative of f exists and is continuous in the interval of integration. The result is

$$
(5) \quad CM_4^* \leqq \epsilon_S \leqq CM_4, \quad \text{where} \quad C = \frac{(b - a)^5}{180(2n)^4}
$$

and M_4^* and M_4 are the smallest and the largest value of the fourth derivative of f in the interval of integration. It is trivial that $\epsilon_S = 0$ for a quadratic polynomial f, but (5) shows that $\epsilon_S = 0$ even for a cubic polynomial.

Simpson's rule is sufficiently accurate for most practical purposes and is preferred to more complicated formulas of higher precision, which can be obtained by the use of approximating polynomials of higher degree. A list of these so-called **Newton–Cotes formulas** is given on p. 316 of [E22] listed in Appendix 1.

---

[12]THOMAS SIMPSON (1710—1761), self-taught English mathematician, author of several popular textbooks. Simpson's rule was used much earlier by Torricelli, Gregory, (in 1668), and Newton (in 1676).

Table 18.4
Simpson's Rule of Integration

ALGORITHM SIMPSON $(x_j, f_j, j = 0, 1, \cdots, 2n)$

This algorithm computes the integral $J = \int_a^b f(x)\, dx$ from given values $f_j = f(x_j)$ at equidistant $x_0 = a, x_1 = x_0 + h, \cdots, x_{2n} = x_0 + 2nh = b$ by Simpson's rule (4), where $h = (b - a)/2n$.

INPUT: $a, b, n, f_0, \cdots, f_{2n}$

OUTPUT: Approximate value $\tilde{J}$ of J

Compute $s_0 = f_0 + f_{2n}$

$s_1 = f_1 + f_3 + \cdots + f_{2n-1}$

$s_2 = f_2 + f_4 + \cdots + f_{2n-2}$

$h = (b - a)/2n$

$\tilde{I} = \dfrac{h}{3}(s_0 + 4s_1 + 2s_2)$

OUTPUT $\tilde{J}$. Stop.

End SIMPSON

EXAMPLE 3. Simpson's rule. Error estimate

Evaluate $J = \displaystyle\int_0^1 e^{-x^2}\, dx$ by Simpson's rule with $2n = 10$ and estimate the error

Solution. Since $h = 0.1$, Table 18.5 gives

$$J \approx \frac{0.1}{3}(1.367\ 879 + 4 \cdot 3.740\ 266 + 2 \cdot 3.037\ 901) = 0.746\ 825.$$

Estimate of error. Differentiation gives $f^{\mathrm{IV}}(x) = 4(4x^4 - 12x^2 + 3)e^{-x^2}$. By considering the

Table 18.5
Computations in Example 3

j	x_j	x_j^2	$e^{-x_j^2}$		
0	0	0	1.000 000		
1	0.1	0.01		0.990 050	
2	0.2	0.04			0.960 789
3	0.3	0.09		0.913 931	
4	0.4	0.16			0.852 144
5	0.5	0.25		0.778 801	
6	0.6	0.36			0.697 676
7	0.7	0.49		0.612 626	
8	0.8	0.64			0.527 292
9	0.9	0.81		0.444 858	
10	1.0	1.00	0.367 879		
Sums			1.367 879	3.740 266	3.037 901

derivative of f^{IV} we find that the smallest value of f^{IV} in the interval of integration occurs at $x = x^* = 2.5 + 0.5\sqrt{10}$ and the largest value at $x = 0$. Computation gives the values $M_4^* = f^{IV}(x^*) = -7.359$ and $M_4 = f^{IV}(0) = 12$. Since $2n = 10$ and $b - a = 1$, we obtain the value $C = 1/1\,800\,000 = 0.000\,000\,56$. Therefore

$$-0.000\,004 \leqq \epsilon_S \leqq 0.000\,006, \qquad \text{and} \qquad 0.746\,818 \leqq J \leqq 0.746\,830,$$

which shows exactness of our approximation to at least 4D. Actually 0.746 825 is exact to 5D, because $J = 0.746\,824$ (exact to 6D).

Note that our present result is much better than that in Example 1 obtained by the trapezoidal rule, whereas the work is nearly the same in both cases.   ∎

Gaussian Integration Formulas

Our integration formulas discussed so far involve function values for equidistant x-values and give exact results for polynomials not exceeding a certain degree. More generally, we may set

(6) $$\int_{-1}^{1} f(x)\,dx \approx \sum_{j=1}^{n} A_j f_j \qquad\qquad [f_j = f(x_j)]$$

(choosing $a = -1$ and $b = 1$, which may be accomplished by a linear transformation of scale) and determine the $2n$ constants $A_1, \cdots, A_n$, $x_1, \cdots, x_n$ so that (6) gives exact results for polynomials of degree m as high as possible. Since $2n$ is the number of coefficients of a polynomial of degree $2n - 1$, it follows that $m \leqq 2n - 1$. Gauss has shown that exactness for polynomials of degree not exceeding $2n - 1$ (instead of $n - 1$ or n) can be attained if and only if the values $x_1, \cdots, x_n$ are the n zeros of the *Legendre polynomial* $P_n(x)$, where

$$P_0 = 1, \quad P_1(x) = x, \quad P_2(x) = \tfrac{1}{2}(3x^2 - 1), \quad P_3(x) = \tfrac{1}{2}(5x^3 - 3x), \cdots,$$

(see (11) in Sec. 4.3 for the general formula) and the coefficients A_j are suitably chosen. Then (6) is called a **Gaussian integration formula.** Some numerical values are (more values in Ref. [1], listed in Appendix 1)

n	Zeros		Coefficients
2	$\pm 1/\sqrt{3}$	$= \pm 0.57735\,02692$	1
3	0		8/9
	$\pm \sqrt{3/5}$	$= \pm 0.77459\,66692$	5/9
4	$\pm \sqrt{(15 - \sqrt{120})/35}$	$= \pm 0.33998\,10436$	0.65214 51549
	$\pm \sqrt{(15 + \sqrt{120})/35}$	$= \pm 0.86113\,63116$	0.34785 48451

EXAMPLE 4. Gaussian integration formula with $n = 3$

Evaluate the integral in Example 3 by the Gaussian formula (6) with $n = 3$.

Solution. We have to convert our integral from 0 to 1 into an integral from -1 to 1. We set $x = \tfrac{1}{2}(t + 1)$. Then $dx = \tfrac{1}{2}dt$, and (6) with $n = 3$ and the above values of the zeros and the coefficients yields

$$\int_0^1 \exp\left(-x^2\right) dx = \frac{1}{2} \int_{-1}^1 \exp\left(-\frac{1}{4}(t+1)^2\right) dt$$

$$\approx \frac{1}{2}\left[\frac{5}{9}\exp\left(-\frac{1}{4}\left(1-\sqrt{\frac{3}{5}}\right)^2\right) + \frac{8}{9}\exp\left(-\frac{1}{4}\right) + \frac{5}{9}\exp\left(-\frac{1}{4}\left(1+\sqrt{\frac{3}{5}}\right)^2\right)\right] = 0.746\,815$$

(exact to 6D: 0.746 825), which is almost as accurate as the Simpson result obtained in Example 3 with much more work.   ∎

The Gaussian integration formula (6) has the advantage of high accuracy. Its disadvantage is the irregular spacing of $x_1, \cdots, x_n$ and the somewhat inconvenient values of the coefficients, but this is not essential when the $f(x_j)$ are also computed (not taken from tables) or obtained from an experiment in which the x_j can be set once and for all. Formula (6) is hardly practical when $f(x_j)$ results from interpolation in a table giving f at equal intervals, because such an interpolation may outweigh any gain due to greater accuracy.

Since the endpoints -1 and 1 of the interval of integration in (6) are not zeros of P_n, they do not occur among $x_1, \cdots, x_n$, and the Gauss formula (6) is called, therefore, an **open formula,** in contrast to a **closed formula** in which the endpoints of the interval of integration are x_1 and x_n. [For example, (2) and (4) are closed formulas.]

We mention that, just as in the case of interpolation, there are numerical integration methods based on differences. A very efficient method uses the **central difference formula by Gauss**

(7) $$\int_{x_0}^{x_1} f(x) \, dx \approx \frac{h}{2}\left(f_0 + f_1 - \frac{\delta^2 f_0 + \delta^2 f_1}{12} + \frac{11(\delta^4 f_0 + \delta^4 f_1)}{720}\right).$$

For more details see Ref. [E36] in Appendix 1.

Numerical Differentiation

The problem of **numerical differentiation** is the determination of approximate values of the derivative of a function f that is given by a table. Numerical differentiation should be avoided whenever possible because approximate values of derivatives will in general be less accurate than the function values from which they are derived. In fact, the derivative is the limit of the difference quotient, and in the latter we normally subtract two large quantities and divide by a small one; furthermore, if a given function f is approximated by a polynomial p, the difference in function values may be small but the derivatives may differ considerably. Hence it is plausible that numerical differentiation is delicate, in contrast to numerical integration, which is not much affected by inaccuracies of function values, because integration is essentially a smoothing process.

It should also be noted that the formulas to be obtained will be important in the numerical solution of differential equations.

We use the notations $f_j' = f'(x_j)$, $f_j'' = f''(x_j)$, etc., and may obtain rough approximation formulas for derivatives by remembering that

$$f'(x) = \lim_{h \to 0} \frac{f(x+h) - f(x)}{h}.$$

This suggests

(8)
$$f'_{1/2} \approx \frac{\delta f_{1/2}}{h} = \frac{f_1 - f_0}{h} .$$

Similarly, for the second derivative we obtain

(9)
$$f_1'' \approx \frac{\delta^2 f_1}{h^2} = \frac{f_2 - 2f_1 + f_0}{h^2}$$

and so on.

More accurate approximations are obtained by differentiating suitable Lagrange polynomials. Differentiating (4*) and remembering that the denominators in (4*) are $2h^2$, $-h^2$, $2h^2$, we have

$$f'(x) \approx L_2'(x) = \frac{2x - x_1 - x_2}{2h^2} f_0 - \frac{2x - x_0 - x_2}{h^2} f_1 + \frac{2x - x_0 - x_1}{2h^2} f_2.$$

Evaluating this at x_0, x_1, x_2, we obtain the "three-point formulas"

(10)

(a) $\quad f_0' \approx \dfrac{1}{2h}(-3f_0 + 4f_1 - f_2)$

(b) $\quad f_1' \approx \dfrac{1}{2h}(-f_0 + f_2)$

(c) $\quad f_2' \approx \dfrac{1}{2h}(f_0 - 4f_1 + 3f_2).$

Applying the same idea to Lagrange's five-point formula, we obtain similar formulas, in particular

(11)
$$f_2' \approx \frac{1}{12h}(f_0 - 8f_1 + 8f_3 - f_4).$$

Further details and formulas are included in Ref. [E18] listed in Appendix 1.

Problems for Sec. 18.5

Review some integration formulas and methods by integrating

1. $\displaystyle\int \cos^2 \omega x \, dx$

2. $\displaystyle\int \sin^2 x \, dx$

3. $\displaystyle\int e^{ax} \sin bx \, dx$

4. $\displaystyle\int \tan kx \, dx$

5. $\displaystyle\int e^{ax} \cos bx \, dx$

6. $\displaystyle\int \ln x \, dx$

7. $\displaystyle\int \frac{dx}{\sqrt{a^2 - x^2}}$

8. $\displaystyle\int \frac{dx}{k^2 + x^2}$

9. $\displaystyle\int \frac{dx}{x^2(x^2 + 1)^2}$

10. Compute the integral in Example 1 by using (1) with $n = 5$.

11. To get a feeling for the increase in accuracy, compute $\int_0^1 x^2 \, dx$ by (1) with $h = 1, h = 0.5, h = 0.25$.

12. Perform the task in Prob. 11, using the trapezoidal rule (2) instead of (1). Also graph x^2 and the trapezoids.

13. Derive a formula that gives lower and upper bounds for J in connection with the rectangular rule (1).

14. Compute the integral in Prob. 11 from (1) with $h = 0.2$ and error bounds from the solution formula to Prob. 13. Compare with the actual error.

Using 5D-values of $\sin x$ (from your pocket calculator or from Table A1 in Appendix 4), evaluate $\int_0^1 \dfrac{\sin x}{x} \, dx$:

15. By the rectangular rule (1) with $n = 5$.

16. By the trapezoidal rule (2) with $n = 5$.

17. By (2) with $n = 10$.

18. By Simpson's rule with $2n = 2$ and with $2n = 10$.

19. Evaluate $\int_0^1 x^5 \, dx$ by Simpson's rule with $2n = 10$. What error bounds are obtained from (5)? What is the actual error of the result?

20. Find an approximate value of $\ln 2 = \int_1^2 \dfrac{dx}{x}$ by Simpson's rule with $2n = 4$. Estimate the error by (5).

21. Determine α and β so that $\int_{x_n}^{x_{n+1}} f(x) \, dx \approx h[\alpha f(x_n) + \beta f(x_{n+1})]$, where $h = x_{n+1} - x_n$, is exact for polynomials of first degree. What formula do you obtain?

22. Carry out the details of the derivation of (10).

23. Consider $f(x) = x^4$ for $x_0 = 0, x_1 = 0.2, x_2 = 0.4, x_3 = 0.6, x_4 = 0.8$. Calculate f_2' from (10a), (10b), (10c), (11). Determine the errors. Compare and comment.

24. A "four-point formula" for the derivative is

$$f_2' \approx \frac{1}{6h} (-2f_1 - 3f_2 + 6f_3 - f_4).$$

Apply it to $f(x) = x^4$ with $x_1, \cdots, x_4$ as in Prob. 23, determine the error, and compare it with that in the case of (11).

25. The derivative $f'(x)$ can also be approximated in terms of first and higher order differences:

$$f'(x_0) \approx \frac{1}{h} (\Delta f_0 - \frac{1}{2} \Delta^2 f_0 + \frac{1}{3} \Delta^3 f_0 - \frac{1}{4} \Delta^4 f_0 + - \cdots).$$

Obtain this formula by differentiating (13), Sec. 18.3, with respect to r, finding

$$hf'(x) \approx \Delta f_0 + \frac{2r - 1}{2!} \Delta^2 f_0 + \frac{3r^2 - 6r + 2}{3!} \Delta^3 f_0 + \cdots$$

where $x = x_0 + rh$, and setting $r = 0$. Apply the formula for obtaining approximations to $f'(0.4)$ in Prob. 23, using differences up to and including (a) first order, (b) second order, (c) third order, (d) fourth order.

18.6 Asymptotic Expansions

Asymptotic expansions are (in general divergent) series that are of great practical importance for computing values of a function $f(x)$ for large x. It is clear that the Maclaurin series of $f(x)$, if it exists and converges for large x, is not suitable for that purpose, since the number of terms needed to obtain a certain number of significant digits increases rapidly as x increases. For a Taylor series with center x_0 and large $|x - x_0|$, the situation is similar. We shall see that the larger x is, the fewer terms of an asymptotic expansion we need for obtaining a required accuracy. On the other hand, the accuracy is limited, and it decreases as x decreases so that asymptotic expansions can be used for large x only. These expansions were introduced by Poincaré.[13]

All variables and functions in this section are assumed to be real.

A series of the form

$$a_0 + \frac{a_1}{x} + \frac{a_2}{x^2} + \cdots \qquad (a_0, a_1, \cdots \text{constant})$$

(which need not converge for any value of x) *is called an* **asymptotic expansion,** *or* **asymptotic series,** *of a function* $f(x)$ *which is defined for every sufficiently large value of x if, for every fixed $n = 0, 1, 2, \cdots$,*

(1) $\quad \left[f(x) - \left(a_0 + \frac{a_1}{x} + \frac{a_2}{x^2} + \cdots + \frac{a_n}{x^n} \right) \right] x^n \rightarrow 0 \qquad \text{as } x \rightarrow \infty,$

and we shall then write

$$f(x) \sim a_0 + \frac{a_1}{x} + \frac{a_2}{x^2} + \cdots .$$

If a function $f(x)$ has an asymptotic expansion, then this expansion is unique, because its coefficients $a_0, a_1, \cdots$ are uniquely determined by (1). In fact, from (1) we obtain

(1*)
$$f(x) - a_0 \rightarrow 0 \qquad \text{or} \qquad a_0 = \lim_{x \rightarrow \infty} f(x),$$

$$\left[f(x) - a_0 - \frac{a_1}{x} \right] x \rightarrow 0 \qquad \text{or} \qquad a_1 = \lim_{x \rightarrow \infty} [f(x) - a_0]x, \quad \text{etc.}$$

On the other hand, different functions may have the same asymptotic expansion. In fact, let $f(x) = e^{-x}$. Then, since $e^{-x} \rightarrow 0$, $xe^{-x} \rightarrow 0$, etc., we see from (1*) that $a_0 = 0$, $a_1 = 0$, etc. Hence, $e^{-x} \sim 0 + 0/x + 0/x^2 \cdots$.

[13]JULES HENRI POINCARÉ (1854—1912), great French mathematician, professor at the Sorbonne in Paris from 1881, known by his numerous important publications in a wide range of subjects, notably in complex analysis, differential equations, topology, celestial mechanics, fluid flow, heat conduction, capillarity, and the foundations and philosophy of mathematics and mathematical physics.

Thus, if $g(x)$ has an asymptotic expansion, then $g(x) + e^{-x}$ has certainly the same asymptotic expansion.

For applications it is advantageous to *extend the definition* by writing

$$f(x) \sim g(x) + h(x) \left(a_0 + \frac{a_1}{x} + \frac{a_2}{x^2} + \cdots \right)$$

whenever

$$\cdot \frac{f(x) - g(x)}{h(x)} \sim a_0 + \frac{a_1}{x} + \frac{a_2}{x^2} + \cdots$$

in the sense of the preceding definition.

Only in rare cases may we determine the coefficients of an asymptotic expansion directly from (1*). In general, other methods will be more suitable, for example, successive integration by parts.

EXAMPLE 1. Asymptotic series of the error function

The error function erf x is defined by the integral (cf. Fig. 546 in Appendix 3)

(2a)
$$\operatorname{erf} x = \frac{2}{\sqrt{\pi}} \int_0^x e^{-t^2} \, dt$$

and its complementary function erfc x by

(2b)
$$\operatorname{erfc} x = 1 - \operatorname{erf} x = \frac{2}{\sqrt{\pi}} \int_x^\infty e^{-t^2} \, dt = \frac{1}{\sqrt{\pi}} \int_{x^2}^\infty e^{-\tau} \tau^{-1/2} \, d\tau,$$

where $t^2 = \tau$, and erf $\infty = 1$. Repeated integration by parts will lead to integrals of the form

(3)
$$F_n(x) = \int_{x^2}^\infty e^{-\tau} \tau^{-(2n+1)/2} \, d\tau, \qquad n = 0, 1, \cdots.$$

Note that erfc $x = F_0(x)/\sqrt{\pi}$. By integration by parts we obtain

$$F_n(x) = -e^{-\tau} \tau^{-(2n+1)/2} \bigg|_{x^2}^\infty - \frac{2n+1}{2} \int_{x^2}^\infty e^{-\tau} \tau^{-(2n+3)/2} \, d\tau.$$

The integral on the right is $F_{n+1}(x)$ and, therefore,

$$e^{x^2} F_n(x) = \frac{1}{x^{2n+1}} - \frac{2n+1}{2} e^{x^2} F_{n+1}(x), \qquad n = 0, 1, \cdots.$$

Repeated application of this formula yields

$$e^{x^2} F_0(x) = \frac{1}{x} - \frac{1}{2} e^{x^2} F_1(x)$$

$$= \frac{1}{x} - \frac{1}{2x^3} + \frac{1}{2} \cdot \frac{3}{2} e^{x^2} F_2(x)$$

(4)
$$\cdots \cdots \cdots \cdots \cdots \cdots$$

$$= \left[\frac{1}{x} - \frac{1}{2x^3} + \frac{1 \cdot 3}{2^2 x^5} - + \cdots + (-1)^{n-1} \frac{1 \cdot 3 \cdots (2n-3)}{2^{n-1} x^{2n-1}} \right]$$

$$+ (-1)^n \frac{1 \cdot 3 \cdots (2n-1)}{2^n} e^{x^2} F_n(x).$$

We show that the series obtained in this way is an asymptotic expansion,

(5) $$e^{x^2} F_0(x) \sim \frac{1}{x} - \frac{1}{2x^3} + \frac{1 \cdot 3}{2^2 x^5} - + \cdots .$$

Let S_{2n-1} denote the expression in the brackets in (4). Then from (4) we obtain

(6) $$[e^{x^2} F_0(x) - S_{2n-1}] x^{2n-1} = K_n e^{x^2} x^{2n-1} F_n(x),$$

where $K_n = (-2)^{-n} 1 \cdot 3 \cdots (2n - 1)$. We have to show that for each fixed $n = 1, 2, \cdots$ the expression on the right approaches zero as x approaches infinity. In (3) we have

$$\frac{1}{\tau^{(2n+1)/2}} \leq \frac{1}{x^{2n+1}} \qquad \text{for all } \tau \geq x^2.$$

Hence we obtain the inequality

(7) $$F_n(x) = \int_{x^2}^{\infty} \frac{e^{-\tau}}{\tau^{(2n+1)/2}} \, d\tau < \frac{1}{x^{2n+1}} \int_{x^2}^{\infty} e^{-\tau} \, d\tau = \frac{e^{-x^2}}{x^{2n+1}} .$$

From this we immediately see that

$$|K_n| e^{x^2} x^{2n-1} F_n(x) < \frac{|K_n|}{x^2} \;\to\; 0 \qquad\qquad (x \to \infty).$$

This proves that the series in (5) is an asymptotic expansion of the function on the left-hand side of (5). Since erf $x = 1 - $ erfc $x = 1 - F_0(x)/\sqrt{\pi}$, it follows that the desired asymptotic series of the error function is

(8) $$\text{erf } x \sim 1 - \frac{1}{\sqrt{\pi}} e^{-x^2} \left(\frac{1}{x} - \frac{1}{2x^3} + \frac{1 \cdot 3}{2^2 x^5} - \frac{1 \cdot 3 \cdot 5}{2^3 x^7} + - \cdots \right).$$

For large x we thus have the simple approximation

(8*) $$\text{erf } x \approx 1 - \frac{1}{\sqrt{\pi}\, x} e^{-x^2}.$$

From (6) and (7) we see that

$$|e^{x^2} F_0(x) - S_{2n-1}| = \frac{1 \cdot 3 \cdots (2n - 1)}{2^n} e^{x^2} F_n(x) < \frac{1 \cdot 3 \cdots (2n - 1)}{2^n} \frac{1}{x^{2n+1}} ,$$

and for sufficiently large x the expression on the right is very small. This shows that for such x, S_{2n-1} is a very good approximation to $e^{x^2} F_0(x)$, and so, for large x, the error function can be computed very accurately by means of (7). It turns out that even for relatively small $|x|$ the results are surprisingly accurate.

For instance, let us consider $x = 2$ and show that by taking suitably many terms in (8) we obtain three-digit accuracy of erf 2 (= 0.99532). Table 18.6 shows that the terms in (8) first decrease in absolute value and then (for $n > 5$) again increase (cf. Fig. 429). This is typical of an asymptotic expansion. In general, the greatest accuracy is obtained by taking the sum of the terms up to the smallest in absolute value. In our case, this is the sum of the first five terms (corresponding to $n = 1, \cdots, 5$); by taking more terms the error again increases; hence the accuracy is limited to three digits. This limited accuracy is in contrast to the situation in the case of a Maclaurin series where, for every fixed x for which the series converges, we can reach any degree of accuracy by taking the sum of sufficiently many terms. Hence, for fixed x, when using an asymptotic expansion we have a limited degree of accuracy. But the advantage of the asymptotic expansion is obvious. In our case we need only two terms (the constant term and the term involving $1/x$) to obtain the first three decimals of erf 2, and the error of this approximation is 1/2 unit of the third decimal (0.000 49). Table 18.7 shows the values of the terms of the Maclaurin series

(9) $$\frac{\sqrt{\pi}}{2} \text{ erf } x = x - \frac{x^3}{1! \, 3} + \frac{x^5}{2! \, 5} - \frac{x^7}{3! \, 7} + - \cdots$$

Table 18.6
Computation of erf x for $x = 2$ by Means of the Asymptotic Expansion

n	$A_1 = 1/x$ $A_n = (-1)^{n-1} \dfrac{1 \cdot 3 \cdots (2n-3)}{2^{n-1}x^{2n-1}}$	$1 - \dfrac{1}{\sqrt{\pi}} e^{-x^2}(A_1 + \cdots + A_n)$	Absolute Value of Error	Error Bound $k(2, n)$
1	0.500 000	0.994 83	0.000 49	0.000 65
2	−0.062 500	0.995 48	0.000 16	0.000 25
3	0.023 438	0.995 24	0.000 08	0.000 16
4	−0.014 648	0.995 39	0.000 07	0.000 14
5	0.012 817	0.995 26	0.000 06	0.000 15
6	−0.014 420	0.995 40	0.000 08	0.000 21
7	0.019 827	0.995 20	0.000 12	0.000 34
8	−0.032 219	0.995 53	0.000 21	0.000 63

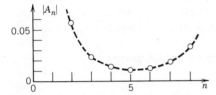

Fig. 429. Absolute values of the terms A_n of the asymptotic expansion (8) for $x = 2$

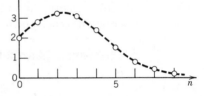

Fig. 430. Absolute values of the terms of the Maclaurin series (9) for $x = 2$

when $x = 2$ and Fig. 430 their absolute values. We see that to compute the first three decimals of erf 2 from (9) we must use the sum of the first 14 terms, the last one of which contains x^{27}. If four or more correct digits of erf 2 are needed, then we cannot use the asymptotic expansion, as we see from the error in Table 18.6.

We mention that we can obtain a bound for the absolute value of the error of our approximate values in Table 18.6 as follows. From (4) and

$$\text{erf } x = 1 - \text{erfc } x = 1 - \frac{1}{\sqrt{\pi}} F_0(x)$$

we see that the absolute value of the error is

$$\frac{1 \cdot 3 \cdots (2n-1)}{2^n \sqrt{\pi}} |F_n(x)|.$$

From (7) it follows that this expression is less than

Table 18.7
Terms of the Maclaurin Series (9) for $x = 2$

n	$\dfrac{(-1)^n 2^{2n+1}}{n!(2n+1)}$	n	$\dfrac{(-1)^n 2^{2n+1}}{n!(2n+1)}$	n	$\dfrac{(-1)^n 2^{2n+1}}{n!(2n+1)}$
0	2.000 00	5	−1.551 52	10	0.027 52
1	−2.666 67	6	0.875 21	11	−0.009 14
2	3.200 00	7	−0.433 44	12	0.002 80
3	−3.047 62	8	0.191 22	13	−0.000 80
4	2.370 37	9	−0.076 04	14	0.000 21

$$k(x, n) = \frac{1 \cdot 3 \cdots (2n - 1)}{2^n \sqrt{\pi}} \frac{e^{-x^2}}{x^{2n+1}} \qquad (x > 0, \, n = 1, 2, \cdots),$$

and this is the desired bound. Numerical values for $x = 2$ are shown in Table 18.6.

To understand our result concerning the asymptotic expansion, the reader should keep in mind that $x = 2$ is relatively small; for $x = 5$, for example, we obtain from (8*)

$$\text{erf } 5 \approx 1 - \frac{1}{\sqrt{\pi}} \frac{e^{-25}}{5} = 0.99999 \, 99999 \, 98433;$$

the first 13 decimals are correct, the error is three units of the fourteenth decimal. However, we wish to warn the reader that the present case may be misleading; for other asymptotic expansions a similarly favorable situation may arise not at $x = 2$ or 5, but only for much larger x, perhaps for $x = 20$ or $x = 100$.

Operations on Asymptotic Expansions

For practical applications it is useful to know that asymptotic series may be added, multiplied and under certain restrictions also integrated and differentiated term by term. Let us formulate these properties in a precise manner.

Theorem 1 (Addition and multiplication)
If

$$f(x) \sim \sum_{m=0}^{\infty} \frac{a_m}{x^m} \qquad and \qquad g(x) \sim \sum_{m=0}^{\infty} \frac{b_m}{x^m}$$

and A and B are any constants, then

$$(10) \qquad Af(x) + Bg(x) \sim \sum_{m=0}^{\infty} \frac{Aa_m + Bb_m}{x^m}$$

and

$$(11) \quad f(x)g(x) \sim \sum_{m=0}^{\infty} \frac{c_m}{x^m}, \qquad c_m = a_0 b_m + a_1 b_{m-1} + \cdots + a_m b_0.$$

Proof. The simple proof of (10) is left to the reader. We prove (11). We choose an arbitrary fixed n and write

$$(12) \quad f(x) = s_n(x) + \frac{h(x)}{x^n} \qquad where \qquad s_n(x) = a_0 + \frac{a_1}{x} + \cdots + \frac{a_n}{x^n},$$

$$g(x) = s_n^*(x) + \frac{l(x)}{x^n} \qquad where \qquad s_n^*(x) = b_0 + \frac{b_1}{x} + \cdots + \frac{b_n}{x^n}.$$

Then, by the definition of an asymptotic expansion,

$$(13) \qquad [f(x) - s_n(x)]x^n = h(x) \to 0,$$

$$[g(x) - s_n^*(x)]x^n = l(x) \to 0 \qquad \text{as } x \to \infty.$$

We set $s_n s_n^* = S_n + T_n$, where S_n is the sum of the terms involving the powers $1, 1/x, \cdots, 1/x^n$ and T_n is the sum of the other terms, which involve

$1/x^{n+1}, \cdots, 1/x^{2n}$; note that thus $T_n x^n \to 0$ as $x \to \infty$. Multiplying $s_n s_n^*$ term by term and collecting like powers, we can readily verify that

$$S_n(x) = c_0 + \frac{c_1}{x} + \cdots + \frac{c_n}{x^n}$$

with $c_0, \cdots, c_n$ as given in the theorem. By the definition of an asymptotic expansion we have to show that $(fg - S_n)x^n \to 0$ as $x \to \infty$. Now from (12),

$$fg = s_n s_n^* + \frac{h+l}{x^n} + \frac{hl}{x^{2n}} = S_n + T_n + \frac{h+l}{x^n} + \frac{hl}{x^{2n}}.$$

Remembering that $T_n x^n \to 0$ as $x \to \infty$ and using (13), we thus obtain the desired result which proves (11):

$$(fg - S_n)x^n = T_n x^n + h + l + \frac{hl}{x^n} \to 0. \qquad \blacksquare$$

We show next that termwise integration of an asymptotic expansion yields the asymptotic expansion of the integral of the given function. Of course, such an asymptotic expansion to be integrated cannot contain a constant term or a term involving $1/x$, terms whose integral would not remain finite as $x \to \infty$.

Theorem 2 (Integration)
Let $f(x)$ be continuous for all sufficiently large x, and let

$$f(x) \sim \sum_{m=2}^{\infty} \frac{a_m}{x^m} = \frac{a_2}{x^2} + \frac{a_3}{x^3} + \cdots.$$

Then for those x,

$$(14) \qquad \int_x^{\infty} f(t) \, dt \sim \sum_{m=2}^{\infty} \frac{a_m}{(m-1)x^{m-1}} = \frac{a_2}{x} + \frac{a_3}{2x^2} + \frac{a_4}{3x^3} + \cdots.$$

Proof. We denote the integral by $F(x)$ and use the notations

$$s_n(x) = \sum_{m=2}^{n} \frac{a_m}{x^m}, \qquad S_{n-1}(x) = \int_x^{\infty} s_n(t) \, dt = \sum_{m=2}^{n} \frac{a_m}{(m-1)x^{m-1}}.$$

Here we made use of the fact that we may integrate a finite sum term by term. From the definition of an asymptotic expansion and the continuity of f for large x it follows that for given $\epsilon > 0$ we can find an x_0 such that for all $x > x_0$ we have

$$|f(x) - s_n(x)|x^n < \epsilon, \qquad \text{hence} \qquad |f(x) - s_n(x)| < \epsilon/x^n.$$

This implies that for all $x > x_0$ we obtain

$$|F(x) - S_{n-1}(x)| = \left| \int_x^\infty f(t)\, dt - \int_x^\infty s_n(t)\, dt \right| = \left| \int_x^\infty [f(t) - s_n(t)]\, dt \right|$$

$$\leqq \int_x^\infty |f(t) - s_n(t)|\, dt < \epsilon \int_x^\infty \frac{dt}{t^n} = \frac{\epsilon}{(n-1)x^{n-1}}.$$

Multiplication by x^{n-1} gives $|F(x) - S_{n-1}(x)|x^{n-1} < \epsilon/(n-1)$ for all $x > x_0(\epsilon)$. Since we can choose ϵ as small as we please, we conclude that the left-hand side of the inequality approaches zero as $x \to \infty$. This completes the proof. ∎

If in the asymptotic expansion of a function $f(x)$ the terms a_0 and a_1/x are present, Theorem 2 can still be applied, namely, to $f(x) - a_0 - a_1/x$.

Differentiation is slightly more tricky, because if $f(x)$ has an asymptotic expansion, the same may not hold for the derivative. For example, from (1*) we obtain

$$f(x) = e^{-x} \sin (e^x) \sim 0 + \frac{0}{x} + \frac{0}{x^2} + \cdots.$$

But $f'(x) = -e^{-x} \sin (e^x) + \cos (e^x)$ does not have an asymptotic expansion. (Why?) So we have to be careful. However, if we assume that $f'(x)$ also has an asymptotic expansion, then we can in fact get it from that of $f(x)$ by termwise differentiation:

Theorem 3 (Differentiation)

Suppose that $f(x)$ has a continuous derivative $f'(x)$ and both f and f' have an asymptotic expansion. Then

$$(15) \quad \text{(a)} \quad f(x) \sim \sum_{m=0}^\infty \frac{a_m}{x^m} \quad \text{implies} \quad \text{(b)} \quad f'(x) \sim -\sum_{m=1}^\infty \frac{m a_m}{x^{m+1}}.$$

Proof. By assumption,

$$(16) \qquad\qquad f'(x) \sim b_0 + \frac{b_1}{x} + \frac{b_2}{x^2} + \cdots$$

and we have to show that this is identical to (15b). We first show that $b_0 = 0$. By calculus,

$$(17) \qquad\qquad f(x) = \int_{x_0}^x f'(t)\, dt + k \qquad\qquad [x_0\ (> 0)\text{ and } k \text{ constant}].$$

Now by the first line in (1*) applied to (15a) and (16) we have $a_0 = \lim f(x)$, $b_0 = \lim f'(x)$. Hence by (17),

$$a_0 = \lim_{x\to\infty} f(x) = \lim_{x\to\infty} \int_{x_0}^x f'(t)\, dt + k$$

but the limit of this integral would not exist if $b_0 = \lim f'(x)$ were not zero. Hence $b_0 = 0$. From the second line in (1*) we now have $b_1 = \lim xf'(x)$. By definition, this means that, given any $\epsilon > 0$, we have for all sufficiently large x,

(18) $b_1 - \epsilon < xf'(x) < b_1 + \epsilon$,   thus $\dfrac{b_1 - \epsilon}{x} < f'(x) < \dfrac{b_1 + \epsilon}{x}$.

Also, from the second line in (1*) applied to (15a), because of (17) we have

$$a_1 = \lim_{x\to\infty} (f(x) - a_0)x = \lim_{x\to\infty} \left(\int_{x_0}^{x} f'(t)\, dt + k - a_0 \right) x.$$

But (18) shows that for $b_1 \neq 0$ the limit on the right would not exist. Hence $b_1 = 0$, too, so that we can now apply Theorem 2 to $f'(x)$, obtaining

(19)
$$f(x) = \int_{x_0}^{\infty} f'(t)\, dt - \int_{x}^{\infty} f'(t)\, dt + k$$
$$\sim \int_{x_0}^{\infty} f'(t)\, dt + k - \frac{b_2}{x} - \frac{b_3}{2x^2} - \cdots,$$

the first integral on the right being a constant. If a given function has an asymptotic expansion, it is unique. We may thus compare corresponding terms in (15a) and (19), finding $b_2 = -a_1$, $b_3 = -2a_2$, etc. With these coefficients the series (15b) and (16) become identical, and the theorem is proved.   ∎

If we know that a function $f(x)$ satisfies the assumptions of Theorem 3 and is a solution of a first-order differential equation, then we may determine the coefficients of its asymptotic expansion by substituting (15) in the differential equation.

EXAMPLE 2. Exponential integral

The exponential integral Ei(x) is defined by the formula

$$\text{Ei}(x) = \int_{x}^{\infty} \frac{e^{-t}}{t}\, dt \qquad\qquad (x > 0).$$

To obtain the asymptotic expansion of Ei(x), we consider the function

$$y = f(x) = e^x\, \text{Ei}(x) = e^x \int_{x}^{\infty} \frac{e^{-t}}{t}\, dt \qquad\qquad (x > 0).$$

By differentiation we see that $f(x)$ satisfies the linear differential equation

(20) $y' - y + \dfrac{1}{x} = 0.$

It may be proved directly that this equation has only one solution y such that y and y' exist for positive x and have an asymptotic expansion. By substituting (15) into (20) and equating the coefficient of each power of x to zero we obtain

$$-c_0 = 0, \quad -c_1 + 1 = 0, \quad -c_1 - c_2 = 0, \quad \cdots, \quad -nc_n - c_{n+1} = 0, \quad \cdots,$$

that is,

$$c_0 = 0, \quad c_1 = 1, \quad c_2 = -1, \quad \cdots, \quad c_{n+1} = (-1)^n n!, \quad \cdots$$

and, therefore,

(21) $$\text{Ei}(x) = e^{-x} f(x) \sim e^{-x} \left(\frac{1}{x} - \frac{1}{x^2} + \frac{2!}{x^3} - \frac{3!}{x^4} + - \cdots \right).$$

Problems for Sec. 18.6

1. Show that the series $1 + \dfrac{1}{x} + \dfrac{1}{2! \, x^2} + \cdots$ [which converges for $|x| > 0$ and represents $e^{1/x}$] is an asymptotic expansion of $e^{1/x}$.

2. Show that $\sin \dfrac{1}{x} \sim \dfrac{1}{x} - \dfrac{1}{3! \, x^3} + \dfrac{1}{5! \, x^5} - + \cdots$.

Integrating by parts, find the asymptotic expansion of[14]

3. $\text{ci}(x) = \displaystyle\int_x^\infty \dfrac{\cos t}{t} \, dt$ (Cosine integral).

4. $\text{si}(x) = \displaystyle\int_x^\infty \dfrac{\sin t}{t} \, dt$ (Complementary sine integral).

5. $c(x) = \displaystyle\int_x^\infty \cos t^2 \, dt$ (Complementary Fresnel integral).

6. $s(x) = \displaystyle\int_x^\infty \sin t^2 \, dt$ (Complementary Fresnel integral).

7. $Q(\alpha, x) = \displaystyle\int_x^\infty e^{-t} t^{\alpha-1} \, dt$ (Incomplete gamma function).

Using the results of Probs. 4, 5, and 7, find the asymptotic expansion of

8. $\text{Si}(x) = \displaystyle\int_0^x \dfrac{\sin t}{t} \, dt$ (Sine integral).   Use $\text{si}(0) = \dfrac{\pi}{2}$.

9. $C(x) = \displaystyle\int_0^x \cos t^2 \, dt$ (Fresnel integral).   Use $c(0) = \dfrac{1}{2} \sqrt{\dfrac{\pi}{2}}$.

10. $P(\alpha, x) = \displaystyle\int_0^x e^{-t} t^{\alpha-1} \, dt$ (Incomplete gamma function).

11. Obtain the asymptotic expansion of erf x by showing that $y = \frac{1}{2} \sqrt{\pi} \, e^{x^2} \text{erfc } x$ satisfies $y' - 2xy + 1 = 0$.

12. Obtain the asymptotic expansion of $Q(\frac{1}{2}, x)$ by showing that $y = e^x \sqrt{x} \, Q(\frac{1}{2}, x)$ satisfies $y' = (1/2x + 1)y - 1$.

13. Obtain the asymptotic expansion of $Q(\alpha, x)$ from a differential equation for $y = e^x x^{1-\alpha} Q(\alpha, x)$.

---

[14]Some formulas for these and related functions are included in Appendix 3. Further formulas can be found in Ref. [7] (cf. Appendix 1) and references to tables of numerical values in Ref. [8]. In Prob. 3, *cosine integral* is the usual name since the integral from 0 to x does not exist. (Why?)

14. Show that

$$\text{Ei}(x) = e^{-x} \int_0^\infty \frac{e^{-u}}{u + x} \, du$$

and obtain from this the series in (21) by developing $1/(u + x)$ in powers of $1/x$ and termwise integration.

15. Show that $\text{Ei}(ix) = \text{ci}(x) - i \, \text{si}(x)$. Then replace x by ix in (21) and separate the real and imaginary parts on both sides; show that this leads to the asymptotic expansions of $\text{ci}(x)$ and $\text{si}(x)$ as obtained in Probs. 3 and 4.

Review Problems for Chapter 18

1. What do we mean by floating-point representation of numbers? What is its advantage over fixed-point representation?

2. To get the best possible result, in what order should we add given numbers in floating-point arithmetic?

3. What kinds of error will generally occur in computations?

4. Why is the choice of a "good" algorithm at least as important on a computer as it is when using a pocket calculator?

5. What do we mean by numerical instability of an algorithm? Give an example.

6. What is fixed-point iteration? State a condition sufficient for convergence.

7. What is the bisection method? What is its disadvantage? Can it diverge?

8. Can Newton's method for solving equations diverge?

9. What would you do if in Newton's method you reached an x at which $f'(x) = 0$?

10. What is the difference between the secant method and the method of false position? Between the method of false position and the bisection method?

11. What is an ill-conditioned equation? Why is this concept of practical importance?

12. Make a difference table of $f(x) = 1/x^2$, $x = 1.0(0.2)2.0$, 4D, and write the differences in all three notations.

13. What is the advantage of Newton's interpolation formulas over Lagrange's formula?

14. What is a cubic spline? Which of its derivatives may be discontinuous?

15. What is the advantage of spline interpolation over the usual polynomial interpolation?

16. Describe the general idea by which we obtained Simpson's rule. Why must we choose n even?

17. What is Gaussian integration? What are its advantages and disadvantages?

18. Why, in principle, is numerical differentiation more delicate than numerical integration?

19. What is an asymptotic expansion, and for what purpose is it useful?

20. Do there exist functions that have no asymptotic expansion?

21. Write 98.17, -100.988, 0.0047869, $-13\,800$ in floating-point form, with 4 significant digits.

22. Convert $(100)_{10}$, $(29.25)_{10}$, and $(3.75)_{10}$ to binary form.

23. Convert $(111.101)_2$, $(100010.1)_2$, and $(11.1011)_2$ to base 10 form.

24. Solve $x^2 + 30x - 1 = 0$ by (2) and by (3) in Sec. 18.1, using 6 significant figures in the computation. Compare and comment.

25. Let 12.03 and 9.2381 be correctly rounded to the number of digits shown. Determine the smallest interval in which the sum $s = 12.03 + 9.2381$, using true instead of rounded values of the quantities, must lie.

26. Answer the same question as in Prob. 25, for the difference $d = 12.03 - 9.2381$.

27. Write $f(x) = \sqrt{x^2 + 1} - 1$ in a form more suitable for computation for small $|x|$.

28. Do the same task as in Prob. 27, for $f(x) = \sqrt[4]{x^2 + 1} - 1$.

29. Compute a three-decimal table of $f(x) = x/3$, $x = 0(1)100$ and find out how the rounding error is distributed.

30. Explain the difficulties in obtaining a four-place table by rounding off the values (a) in a five-place table, (b) in a six-place table.

31. Let a and b be given and correctly rounded to $2D$ and D decimals, respectively, and let $|a| < |b| < 1$. If we want to compute a/b as a D-decimal quotient, show that dividing and then rounding is generally preferable to first rounding and then dividing. Give an example.

32. Find the solution of $x^4 = x + 0.15$ near $x = 0$ by fixed-point iteration, starting from $x_0 = 0$.

33. The equation in Prob. 32 has a solution near $x = 1$. Find this solution by writing the equation in the form $x = \sqrt[4]{x + 0.15}$ and iterating, starting with $x_0 = 1$.

34. What happens in Prob. 33 if you write the equation in the form $x = x^4 - 0.15$ and start from $x_0 = 1$?

35. Compute $\sqrt{7}$ by the iteration in Example 3, Sec. 18.2, starting from $x_0 = 2$ and computing x_1, x_2, x_3. Compute the error, using $\sqrt{7} = 2.645\,751$.

36. Find all real solutions of $\cos x = \sqrt{x}$ by Newton's method.

37. Do the same task as in Prob. 36, for $\sin x = x/2$.

38. Apply Newton's method to $x^4 - x^3 - 2x = 34$, starting from $x_0 = 3$.

39. Do the same task as in Prob. 38, for $x^3 - 5x + 3 = 0$, $x_0 = 2$.

40. Show that the equation in Prob. 39 can be written $x = \frac{1}{5}(x^3 + 3)$. Solve it by iteration (4 steps), starting from $x_0 = 1$. In what interval J is the condition of Theorem 1, Sec. 18.2, satisfied?

41. Find the real solutions of $x^4 = 3$ by the method of false position.

42. Do the same task as in Prob. 41, for $\cos x = x$.

43. Find $f(1.28)$ from $f(1.0) = 3.00000$, $f(1.2) = 2.98007$, $f(1.4) = 2.92106$, $f(1.6) = 2.82534$, $f(1.8) = 2.69671$, $f(2.0) = 2.54030$ by linear and by quadratic interpolation.

44. Do the same task as in Prob. 43, for $f(1.75)$.

45. In Prob. 43, compute $f(1.28)$ by cubic interpolation.

46. In Prob. 43, compute $f(1.75)$ by the Everett formula (cf. the problems for Sec. 18.3).

47. Compute $\sin 0.3$ and $\sin 0.5$ from 5D-values of $\sin 0.2$, $\sin 0.4$, $\sin 0.6$ and the Everett formula.

48. Show that $\Delta^k f_n = \binom{k}{0} f_{n+k} - \binom{k}{1} f_{n+k-1} + - \cdots + (-1)^k \binom{k}{k} f_n$.

49. Find the cubic spline for the data $f(0) = 1$, $f(1) = 0$, $f(2) = -3$, $k_0 = k_2 = 0$.

50. Find the cubic spline for the data $f(-1) = 3$, $f(1) = 1$, $f(3) = 23$, $f(5) = 45$, $k_0 = k_3 = 3$.

51. Using the rectangular rule (1) in Sec. 18.5 with $n = 5$, compute $\int_0^1 x^3\,dx$. What is the error?

52. Compute the integral in Prob. 51 by the trapezoidal rule (2) in Sec. 18.5 with $n = 5$. What error bounds are obtained from (3)? What is the actual error of the result? Why is this result larger than the exact value?

53. Compute the Fresnel integral $C(x) = \displaystyle\int_0^x \cos t^2 \, dt$ for $x = 1$ by Simpson's rule with $2n = 2$.

54. Compute the integral in Prob. 53 by Simpson's rule with $2n = 10$.

55. If $f(x)$ is a polynomial of second order, show that (8) in Sec. 18.5 is exact. What does this mean geometrically?

Summary of Chapter 18
Numerical Methods in General

In this chapter we discussed concepts that are relevant throughout numerical work as a whole and methods of a general nature, as opposed to methods for problems in linear algebra (Chap. 19) or in differential equations (Chap. 20).

In engineering computations we use the **floating-point** representation of numbers (Sec. 18.1); fixed-point representation is less suitable in most cases. Numerical methods give approximate values $\tilde{a}$ of quantities. The **error** ϵ of $\tilde{a}$ is

$$(1) \qquad\qquad \epsilon = \tilde{a} - a \qquad\qquad \text{(Sec. 18.1)}$$

where a is the exact value. The *relative error* of $\tilde{a}$ is ϵ/a. Errors arise from rounding off, inaccuracies of measured values, truncation (that is, replacement of integrals by sums, derivatives by difference quotients, series by partial sums) and so on. An algorithm is called **numerically stable** if errors in intermediate results have little influence on the final result (Sec. 18.1). Unstable algorithms are useless because errors may become so large that results will be very inaccurate. Numerical instability of algorithms must not be confused with mathematical instability of problems (*"ill-conditioned problems"*).

Fixed-point iteration is a method for solving equations $f(x) = 0$ in which the equation is first transformed algebraically to $x = g(x)$, an initial guess x_0 for the solution is made, and then approximations $x_1, x_2, \cdots$, are successively computed by iteration from (cf. Sec. 18.2)

$$(2) \qquad\qquad x_{n+1} = g(x_n) \qquad\qquad n = 0, 1, \cdots.$$

Newton's method for solving equations $f(x) = 0$ is an iteration

$$(3) \qquad\qquad x_{n+1} = x_n - \frac{f(x_n)}{f'(x_n)} \qquad\qquad \text{(Sec. 18.2)}.$$

x_{n+1} is the x-intercept of the tangent of the curve $y = f(x)$ at the point

x_n. This method is of second order (Theorem 2, Sec. 18.2). If we replace f' in (3) by a difference quotient (geometrically: we replace the tangent by a secant), we obtain the **secant method;** see (10) in Sec. 18.2. For the *bisection method* (which is slow) and the *method of false position,* see the problem set for Sec. 18.2.

Polynomial interpolation means the determination of an *interpolating polynomial,* that is, a polynomial $p_n(x)$ such that $p_n(x_j) = f_j$, where $j = 0, \cdots, n$ and $(x_0, f_0), \cdots, (x_n, f_n)$ are measured or observed values, values of a function, etc. For arbitrarily spaced $x_0, \cdots, x_n$, *Newton's divided difference* formula (9), Sec. 18.3, can be used. For regularly spaced $x_0, x_1 = x_0 + h, \cdots, x_n = x_0 + nh$, that formula becomes **Newton's forward difference formula** (Sec. 18.3)

$$
(4) \quad f(x) \approx p_n(x) = f_0 + r\Delta f_0 + \frac{r(r-1)}{2!} \Delta^2 f_0 + \cdots
$$

$$
\cdots + \frac{r(r-1) \cdots (r-n+1)}{n!} \Delta^n f_0
$$

where $r = (x - x_0)/h$ and the forward differences are $\Delta f_j = f_{j+1} - f_j$ and

$$
\Delta^k f_j = \Delta^{k-1} f_{j+1} - \Delta^{k-1} f_j.
$$

A similar formula is *Newton's backward difference interpolation formula* (Sec. 18.3). **Lagrange's interpolation formula** (19), Sec. 18.3, is helpful in deriving integration formulas etc., but it is impractical in numerical work.

Interpolation polynomials may become numerically unstable as n increases, and instead of interpolating and approximating by a single high-degree polynomial it is preferable to use a cubic **spline** $g(x)$, that is, a twice continuously differentiable interpolation function [thus, $g(x_j) = f_j$] which in each subinterval $x_j \leq x \leq x_{j+1}$ consists of a cubic polynomial $p_j(x)$; cf. Sec. 18.4.

Simpson's rule of numerical integration is (Sec. 18.5)

$$
(5) \quad \int_a^b f(x)\, dx \approx \frac{h}{3} (f_0 + 4f_1 + 2f_2 + 4f_3 + \cdots
$$

$$
+ 2f_{2n-2} + 4f_{2n-1} + f_{2n})
$$

where $h = (b - a)/2n$ (not n) and $f_j = f(x_j)$. For error estimates and other integration methods, see Sec. 18.5.

Section 18.6 deals with **asymptotic expansions;** these are in general divergent series, which are useful for computing functions $f(x)$ for large $|x|$.

Chapter 19

Numerical Methods in Linear Algebra

In this chapter we consider some of the most important numerical methods for solving systems of linear algebraic equations (Secs. 19.1—19.4), for fitting straight lines (Sec. 19.5) and for matrix eigenvalue problems (Secs. 19.6—19.10). These and similar methods are of great practical importance. Indeed, many engineering or other (for instance, statistical) problems lead to mathematical models whose solution requires methods of numerical linear algebra.

The present chapter is independent of Chap. 18 and can be studied immediately after Chap. 7.

Prerequisite for this chapter: Secs. 7.1—7.3, 7.12.
Section that may be omitted in a shorter course: Secs. 19.4—19.6, 19.9, 19.10.
References: Appendix 1, Part E.
Answers to Problems: Appendix 2.

19.1 Systems of Linear Equations: Gauss Elimination

A **system of n linear equations** (or *set of n simultaneous linear equations*) in *n unknowns* $x_1, \cdots, x_n$ is a set of equations $E_1, \cdots, E_n$ of the form

(1)

$$E_1: \quad a_{11}x_1 + \cdots + a_{1n}x_n = b_1$$

$$E_2: \quad a_{21}x_1 + \cdots + a_{2n}x_n = b_2$$

$$\cdots \cdots \cdots \cdots \cdots \cdots \cdots$$

$$E_n: \quad a_{n1}x_1 + \cdots + a_{nn}x_n = b_n$$

where the **coefficients** a_{jk} and the b_j are given numbers. The system is said to be **homogeneous** if all the b_j are zero; otherwise the system is said to be **nonhomogeneous.** Using matrix multiplication (Sec. 7.3), we can write (1) as a single vector equation

(2) $$\mathbf{Ax} = \mathbf{b}$$

where the **coefficient matrix** $\mathbf{A} = [a_{ik}]$ is the $n \times n$ matrix

$$\mathbf{A} = \begin{bmatrix} a_{11} & a_{12} & \cdots & a_{1n} \\ a_{21} & a_{22} & \cdots & a_{2n} \\ \cdot & \cdot & \cdots & \cdot \\ a_{n1} & a_{n2} & \cdots & a_{nn} \end{bmatrix}, \quad \text{and} \quad \mathbf{x} = \begin{bmatrix} x_1 \\ \cdot \\ \cdot \\ \cdot \\ x_n \end{bmatrix} \quad \text{and} \quad \mathbf{b} = \begin{bmatrix} b_1 \\ \cdot \\ \cdot \\ b_n \end{bmatrix}$$

are column vectors. The **augmented matrix** $\tilde{\mathbf{A}}$ of the system (1) is

$$\tilde{\mathbf{A}} = [\mathbf{A} \quad \mathbf{b}] = \begin{bmatrix} a_{11} & \cdots & a_{1n} & b_1 \\ a_{21} & \cdots & a_{2n} & b_2 \\ \cdot & \cdots & \cdot & \cdot \\ a_{n1} & \cdots & a_{nn} & b_n \end{bmatrix}$$

A **solution** of (1) is a set of numbers $x_1, \cdots, x_n$ that satisfy all the n equations, and a **solution vector** of (1) is a vector $\mathbf{x}$ whose components constitute a solution of (1).

The method of solving such a system by determinants (Cramer's rule in Sec. 7.11) is impracticable for large systems, even with efficient methods for evaluating the determinants.

A practical method for the solution of a system of linear equations is the so-called *Gauss elimination,* which we shall now discuss (proceeding independently of Sec. 7.5).

Gauss Elimination

This standard method for solving systems of linear equations (1) is a systematic process of elimination that reduces (1) to **"triangular form"** because then the system can be easily solved by **"back substitution."** For instance, a triangular system is

$$3x_1 + 5x_2 + 2x_3 = 8$$

$$8x_2 + 2x_3 = -7$$

$$6x_3 = 3$$

and back substitution gives $x_3 = 3/6 = 1/2$ from the third equation, then

$$x_2 = \tfrac{1}{8}(-7 - 2x_3) = -1$$

from the second equation, and finally from the first equation

$$x_1 = \tfrac{1}{3}(8 - 5x_2 - 2x_3) = 4.$$

How do we reduce a given system (1) to triangular form? In the first step we eliminate x_1 from equations E_2 to E_n in (1). This we do by subtracting suitable multiples of the first equation from the other equations. This first equation is called the **pivot equation** in this step, and a_{11} is called the **pivot entry.** This equation is left unaltered. In the second step we take the *new* second equation (which no longer contains x_1) as the pivot equation and use it to eliminate x_2 from equations E_3 to E_n, and so on. This gives a triangular system that can be solved by back substitution as just shown. In this way we obtain precisely all solutions of the *given* system (proof in Sec. 7.5). The pivot entries must be different from zero. To achieve this, we may have to change the order of equations. This is called **partial pivoting.**[1] Also, pivot entries should not be too small in absolute value, because of round-off errors, and this may be another reason for partial pivoting. We discuss this later in this section. Let us first consider a simple illustrative example.

EXAMPLE 1. Gauss elimination. Pivoting
Solve the system

$$8x_2 + 2x_3 = -7$$
$$3x_1 + 5x_2 + 2x_3 = 8$$
$$6x_1 + 2x_2 + 8x_3 = 26.$$

Solution. To get a pivot equation containing x_1, we have to reorder the equations, say, by interchanging the first equation and the first equation containing x_1 (that is, the second equation of our present system):

$$3x_1 + 5x_2 + 2x_3 = 8$$
$$8x_2 + 2x_3 = -7$$
$$6x_1 + 2x_2 + 8x_3 = 26.$$

First Step. Elimination of x_1
It would suffice to show the augmented matrix and operate on it. We show both the equations and the augmented matrix. In the first step, the first equation is the pivot equation. Thus

$$
\begin{array}{ll}
\text{Pivot} \rightarrow \boxed{3x_1} + 5x_2 + 2x_3 = 8 \\
\phantom{\text{Eliminate} \rightarrow} 8x_2 + 2x_3 = -7 \\
\text{Eliminate} \rightarrow \boxed{6x_1} + 2x_2 + 8x_3 = 26
\end{array}
\qquad
\begin{bmatrix}
3 & 5 & 2 & 8 \\
0 & 8 & 2 & -7 \\
6 & 2 & 8 & 26
\end{bmatrix}
$$

To eliminate x_1 from the other equations (here: from the third equation), do:

Subtract $6/3 = 2$ times the pivot equation from the third equation.

The result is shown below.

Second Step. Elimination of x_2
We leave the first equation untouched and take the *new* second equation as the pivot equation:

$$
\begin{array}{ll}
\phantom{\text{Pivot} \rightarrow} 3x_1 + 5x_2 + 2x_3 = 8 \\
\text{Pivot} \rightarrow \boxed{8x_2} + 2x_3 = -7 \\
\text{Eliminate} \rightarrow \boxed{-8x_2} + 4x_3 = 10
\end{array}
\qquad
\begin{bmatrix}
3 & 5 & 2 & 8 \\
0 & 8 & 2 & -7 \\
0 & -8 & 4 & 10
\end{bmatrix}
$$

---

[1] As opposed to **total pivoting,** which we discuss later in this section. For the Gauss—Jordan elimination, see Sec. 19.2.

To eliminate x_2 from the equations below the pivot equation (here: from the third equation), do:

Subtract $-8/8 = -1$ times the pivot equation from the third equation.

The resulting triangular system is shown below. This is the end of the forward elimination. Now comes the back substitution.

Back Substitution. Determination of x_3, x_2, x_1
The triangular system obtained in Step 2 is

$$
\begin{array}{rcr}
3x_1 + 5x_2 + 2x_3 &=& 8 \\
8x_2 + 2x_3 &=& -7 \\
6x_3 &=& 3
\end{array}
\qquad
\begin{bmatrix}
3 & 5 & 2 & 8 \\
0 & 8 & 2 & -7 \\
0 & 0 & 6 & 3
\end{bmatrix}
$$

From it, taking first the last equation, then the second equation and then the first equation, we compute the solution

$$
\begin{aligned}
x_3 &= \tfrac{3}{6} = \tfrac{1}{2} \\
x_2 &= \tfrac{1}{8}(-7 - 2x_3) = -1 \\
x_1 &= \tfrac{1}{3}(8 - 5x_2 - 2x_3) = 4.
\end{aligned}
$$

This agrees with the values given above. ∎

The general algorithm is shown in Table 19.1. To explain it, we have numbered some of its lines. b_j is denoted by $a_{j,n+1}$, for uniformity. m_{jk} in line 3 suggests *multiplier,* since these are the factors by which we have to multiply the pivot equation $E_k{}^*$ in Step k before subtracting it from an equation $E_j{}^*$ below $E_k{}^*$ from which we want to eliminate x_k. Here we have written $E_k{}^*$ and $E_j{}^*$ to indicate that after Step 1 these are no longer the given equations in (1), but these underwent a change in each step, as indicated in line 4. Accordingly, the possible pivoting in line 1 always refers to the most recent equations, and $j \geqq k$ indicates that we leave untouched all the equations that have served as pivot equations in previous steps. For $p = k$ in line 4 we get 0 on the right, as it should be in the elimination,

$$
a_{jk} - m_j a_{kk} = a_{jk} - \frac{a_{jk}}{a_{kk}} a_{kk} = 0.
$$

In line 5, if the last equation in the *triangular* system is $0 = b_n{}^* \neq 0$, we have no solution. If it is $0 = b_n{}^* = 0$, we have no unique solution because we then have fewer equations than unknowns.

EXAMPLE 2. Gauss elimination in Table 19.1, sample computation
In Example 1 we had $a_{11} = 0$ and $a_{21} \neq 0$, thus $j = 2$ and, by line 2 of the algorithm, interchanged the equations E_1 and E_2. Then we computed $m_{21} = 0/3 = 0$ and $m_{31} = 6/3 = 2$, so that line 4 with $p = 2, 3, 4$ gave no change in E_2, and in E_3 (labeling the new elements with an asterisk)

$$
\begin{aligned}
(p = 2) \quad & a_{32}^* = a_{32} - m_{31}a_{12} = 2 - 2 \cdot 5 = -8 \\
(p = 3) \quad & a_{33}^* = a_{33} - m_{31}a_{13} = 8 - 2 \cdot 2 = 4 \\
(p = 4) \quad & a_{34}^* = a_{34} - m_{31}a_{14} = b_3 - m_{31}b_1 = 26 - 2 \cdot 8 = 10
\end{aligned}
$$

as the coefficients of $E_3{}^*$ in Step 2. ∎

Table 19.1
Gauss Elimination

ALGORITHM GAUSS ($\widetilde{\mathbf{A}} = [a_{jk}] = [\mathbf{A} \quad \mathbf{b}]$)
This algorithm computes a unique solution $\mathbf{x} = [x_j]$ of the system (1) or indicates that (1) has no unique solution.

INPUT: Augmented $n \times (n + 1)$ matrix $\widetilde{\mathbf{A}} = [a_{jk}]$, where $a_{j,n+1} = b_j$

OUTPUT: Solution $\mathbf{x} = [x_j]$ of (1) or message that the system (1) has no unique solution

For $k = 1, \cdots, n - 1$, do:

1   Find the smallest $j \geq k$ such that $a_{jk} \neq 0$.
     If no such j exists then OUTPUT "No unique solution exists." Stop.
     [*Procedure completed unsuccessfully; A is singular*]

2   Else exchange the contents of rows j and k of $\widetilde{\mathbf{A}}$.

     For $j = k + 1, \cdots, n$, do:

3    $$m_{jk} := \frac{a_{jk}}{a_{kk}}$$

     For $p = k + 1, \cdots, n + 1$, do:

4    $$a_{jp} := a_{jp} - m_{jk} a_{kp}$$

     End
     End
     End

5   If $a_{nn} = 0$ then OUTPUT "No unique solution exists."
     Stop.
     Else,

6    $$x_n = \frac{a_{n,n+1}}{a_{nn}}$$ [*Start back substitution*]

     For $i = n - 1, \cdots, 1$, do:

7    $$x_i = \frac{1}{a_{ii}} \left(a_{i,n+1} - \sum_{j=i+1}^{n} a_{ij} x_j \right)$$

     End
     OUTPUT $x = [x_j]$. Stop.
End GAUSS

Operations Count

Quite generally, the quality of a numerical method is judged in terms of:

Amount of storage
Amount of time ($\equiv$ number of operations)
Effect of round-off error.

For the Gauss elimination, the operations count is as follows. In Step k we eliminate x_k from $n - k$ equations. This needs $n - k$ divisions in computing the m_{jk} (line 3) and $(n - k)(n - k + 1)$ multiplications and as many subtractions (both in line 4). Since we do $n - 1$ steps, k goes from 1 to $n - 1$ and thus the total number of operations in this forward elimination is

$$f(n) = \sum_{k=1}^{n-1} (n - k) + 2 \sum_{k=1}^{n-1} (n - k)(n - k + 1) \qquad \text{(write } n - k = s\text{)}$$

$$= \sum_{s=1}^{n-1} s + 2 \sum_{s=1}^{n-1} s(s + 1) = \tfrac{1}{2}(n - 1)n + \tfrac{2}{3}(n^2 - 1)n \approx \tfrac{2}{3}n^3$$

where $2n^3/3$ is obtained by dropping lower powers of n. We see that $f(n)$ grows about proportional to n^3. We say that $f(n)$ is of *order* n^3 and write

$$f(n) = O(n^3),$$

where O suggests **order**. The general definition of O is as follows. We write

$$f(n) = O(h(n))$$

if the quotient $|f(n)/h(n)|$ remains bounded (does not trail off to infinity) as $n \to \infty$. In our present case, $h(n) = n^3$ and, indeed, $f(n)/n^3 \to 2/3$ because the omitted terms divided by n^3 go to zero as $n \to \infty$. (See Sec. 22.2 for further examples, which are independent of the context in Sec. 22.2.)

In the back substitution of x_i we make $n - i$ multiplications and as many subtractions, as well as 1 division. Hence the number of operations in the back substitution is

$$b(n) = \sum_{i=1}^{n} (n - i) + n = \sum_{s=1}^{n} s + n = \tfrac{1}{2}n(n + 1) + n \approx \tfrac{1}{2}n^2 = O(n^2).$$

We see that it grows more slowly than the number of operations in the forward elimination of the Gauss algorithm, so that it is negligible for large systems because it is smaller by a factor n, approximately. For instance, if an operation takes 10^{-6} sec, then the times needed are:

n	Elimination	Back Substitution
100	0.7 sec	0.005 sec
1000	11 min	0.5 sec

More on Pivoting. Scaling

We know that a pivot entry a_{kk}^* must be different from zero. If it is zero, we *must* pivot (interchange equations). But this is not all: if a_{kk}^* is small in absolute value, we *should* pivot, because then we would have to subtract large multiples of the pivot equation from the other equations, thereby amplifying round-off errors. This may affect the accuracy of the result. Before discussing how to overcome this difficulty, let us illustrate it with a simple example.

EXAMPLE 3. Difficulty with small pivot entries

The solution of the system

$$0.0004x_1 + 1.402x_2 = 1.406$$

$$0.4003x_1 - 1.502x_2 = 2.501$$

is $x_1 = 10$, $x_2 = 1$. We solve this system by the Gauss elimination, using four-digit floating-point arithmetic.

(a) Picking the first equation as the pivot equation, we have to multiply this equation by $m = 0.4003/0.0004 = 1001$ and subtract the result from the second equation, obtaining

$$-1405x_2 = -1404.$$

Hence $x_2 = -1404/(-1405) = 0.9993$, and from the first equation, instead of $x_1 = 10$, we get

$$x_1 = \frac{1}{0.0004}(1.406 - 1.402 \cdot 0.9993) = \frac{0.005}{0.0004} = 12.5.$$

This failure occurs because $|a_{11}|$ is small compared to $|a_{12}|$, so that a small round-off error in x_2 led to a large error in x_1.

(b) Picking the second equation as the pivot equation, we have to multiply this equation by $0.0004/0.4003 = 0.000\,999\,3$ and subtract the result from the first equation, obtaining

$$1.404x_2 = 1.404.$$

Hence $x_2 = 1$, and from the second equation $x_1 = 10$. This success occurs because $|a_{21}|$ is not very small compared to $|a_{22}|$, so that a small round-off error in x_2 would not lead to a large error in x_1. Indeed, for instance, if we had $x_2 = 1.002$, we would still have from the second equation the good value $x_1 = (2.501 + 1.505)/0.4003 = 10.01$. ∎

In *partial pivoting,* in the first step we usually pick as the pivot equation an equation in which the coefficient of x_1 is largest in absolute value; similarly for x_2 in the second step, and so on. **Total pivoting** would mean that we look for a coefficient that is largest in absolute value *in the entire system* and start the elimination with the corresponding variable, using this coefficient as the pivot entry; similarly in the further steps. This is hardly done in practice, since it is much more expensive than partial pivoting.

There is a catch: we could magnify a coefficient by multiplying a whole equation, *but this would not change the computed solution.* Multiplication of an equation by a factor is called **row scaling;** here, one normally uses a power of 10 (or of the machine basis β) such that afterward the absolutely largest coefficients of the equation have absolute value between 0.1 and 1 (or β^{-1} and 1, respectively).

In practice, one uses scaled partial pivoting, that is, in the kth step $(k = 1, 2, \cdots)$ of the elimination one chooses as pivot equation an equation from the $n - k + 1$ available equations such that the quotient of the coefficient of x_k and the absolutely largest coefficient in the equation is maximum in absolute value. If there are several such equations, take the first of them.

EXAMPLE 4. Choice of pivot equations

Apply the method just mentioned to the selection of pivot equations in the following system and compute the solution.

$$3x_1 - 4x_2 + 5x_3 = -1$$
$$-3x_1 + 2x_2 + x_3 = 1$$
$$6x_1 + 8x_2 - x_3 = 35$$

Solution. We have

$$\frac{|a_{11}|}{\max |a_{1k}|} = \frac{3}{5}, \qquad \frac{|a_{21}|}{\max |a_{2k}|} = \frac{3}{3}, \qquad \frac{|a_{31}|}{\max |a_{3k}|} = \frac{6}{8}.$$

Accordingly, we select the second equation as the pivot equation and eliminate x_1, obtaining

$$-3x_1 + 2x_2 + x_3 = 1$$
$$-2x_2 + 6x_3 = 0$$
$$12x_2 + x_3 = 37$$

We now have

$$\frac{|a_{22}^*|}{\max |a_{2k}^*|} = \frac{2}{6}, \qquad \frac{|a_{32}^*|}{\max |a_{3k}^*|} = \frac{12}{12}.$$

Hence we select the last equation as the pivot equation and get the triangular system

$$-3x_1 + 2x_2 + x_3 = 1$$
$$12x_2 + x_3 = 37$$
$$\frac{37}{6} x_3 = \frac{37}{6}.$$

Back substitution now gives the solution $x_3 = 1$, $x_2 = 3$, $x_1 = 2$, in this order. ∎

No pivoting is needed for two important classes of matrices, namely, the **diagonally dominant matrices,** that is,

$$|a_{jj}| \geq \sum_{\substack{k=1 \\ k \neq j}}^{n} |a_{jk}| \qquad\qquad j = 1, \cdots, n$$

and the symmetric and positive definite matrices, that is,

$$\mathbf{A}^\mathsf{T} = \mathbf{A}, \qquad \mathbf{x}^\mathsf{T}\mathbf{A}\mathbf{x} > 0 \quad \text{for all } \mathbf{x} \neq \mathbf{0}.$$

Error estimates for the Gauss elimination are discussed in Ref. [E22], p. 37, listed in Appendix 1.

In the next section we discuss methods (by **Doolittle, Crout** and **Cholesky**) that are related to the Gauss elimination and result from the idea of factoring **A** as a product of two triangular matrices. This approach has the advantage that one avoids the solution of simultaneous equations but can deal with one equation at a time.

In the next section we shall also discuss the Gauss-Jordan elimination, which is of practical interest in connection with the determination of the inverse of a matrix. (For solving systems of equations, this method is inferior to the Gauss elimination, as we shall see.)

Problems for Sec. 19.1

Solve the following systems of linear equations by the Gauss elimination (with partial pivoting if necessary).

1. $2x_1 + 5x_2 = 2$
$6x_1 + 7x_2 = -10$

2. $-3x_1 + 8x_2 = -17$
$4x_1 + x_2 = 11$

3. $5x_1 + 2x_2 = -1.5$
$-12x_1 + 3x_2 = -12.0$

4. $4x_1 + 2x_2 - 2x_3 = 0$
$7x_2 + 4x_3 = 5$
$2x_3 = 6$

5. $x_2 - 4x_3 = 2$
$x_1 + 3x_2 - 2x_3 = 4$
$2x_1 + 5x_2 - 7x_3 = 6$

6. $2x_1 - 3x_2 + 4x_3 = 11.0$
$4x_1 - 6x_2 - x_3 = 17.5$
$3x_1 + 5x_2 + 9x_3 = 8.5$

7. $x_1 + x_2 + x_3 = 2$
$-x_1 + 2x_2 - 3x_3 = 32$
$3x_1 - 4x_3 = 17$

8. $3x_2 + 5x_3 = 1$
$2x_1 - 4x_2 - 8x_3 = 14$
$7x_1 + 2x_2 + x_3 = 52$

9. $2x_1 + 3x_3 = 15$
$4x_2 - x_3 = -13$
$3x_1 - x_2 + 5x_3 = 26$

10. $x_1 - 2x_2 + 4x_3 = 39$
$-2x_1 + x_2 - 5x_3 = -42$
$4x_1 - 5x_2 + x_3 = 36$

11. $9x_1 - 2x_2 = 1.5$
$2x_1 - 4x_2 + 3x_3 = 4.0$
$3x_2 + 7x_3 = 25.5$

12. $-3x_1 + x_2 + 5x_3 = -8$
$x_1 + 3x_2 - 4x_3 = 14$
$2x_1 + 2x_2 - 6x_3 = 16$

Solve the following systems of linear equations by the Gauss elimination with selection of pivot equations by the method in Example 4.

13. $5x_1 + 10x_2 - 2x_3 = -0.30$
$2x_1 - x_2 + x_3 = 1.91$
$3x_1 + 4x_2 = 1.16$

14. $x_1 - x_2 + 2x_3 = 3.8$
$4x_1 + 3x_2 - x_3 = -5.7$
$5x_1 + x_2 + 3x_3 = 2.8$

15. $x_2 - 3x_3 = -11.0$
$4x_1 + 5x_3 = 19.5$
$x_1 + x_2 = 0$

16. Discuss the following system in a fashion similar to that in Example 3.

$$0.0003x_1 + 3.0000x_2 = 2.0001$$
$$1.0000x_1 + 1.0000x_2 = 1.0000$$

17. Solve the following system by the Gauss elimination without pivoting.

$$\epsilon x_1 + x_2 = 1$$
$$x_1 + x_2 = 2$$

Show that, for any fixed machine word length and sufficiently small $\epsilon > 0$, the computer gives $x_2 = 1$ and then $x_1 = 0$. Solve the system exactly and show that for the exact solution, $x_1 \to 1$ and $x_2 \to 1$ as $\epsilon \to 0$. Comment.

18. Solve the system in Prob. 17 by the Gauss elimination with pivoting. Compare and comment.

19. Try to solve each of the following two systems by the Gauss elimination. Explain why the Gauss elimination fails when no solution exists.

$$
\begin{aligned}
x_1 + x_2 + x_3 &= 3 \\
4x_1 + 2x_2 - x_3 &= 5 \\
9x_1 + 5x_2 - x_3 &= 13
\end{aligned}
\qquad
\begin{aligned}
x_1 + x_2 + x_3 &= 3 \\
4x_1 + 2x_2 - x_3 &= 5 \\
9x_1 + 5x_2 - x_3 &= 12
\end{aligned}
$$

20. Verify the calculation of the operations count of the Gauss method in the text.

19.2 Systems of Linear Equations: LU-Factorization, Matrix Inversion

We continue our discussion of numerical methods for solving systems of n linear equations in n unknowns $x_1, \cdots, x_n$,

$$(1) \qquad\qquad \mathbf{Ax = b,}$$

where $\mathbf{A} = [a_{jk}]$ is the $n \times n$ coefficient matrix and $\mathbf{x}^\mathsf{T} = [x_1 \cdots x_n]$ and $\mathbf{b}^\mathsf{T} = [b_1 \cdots b_n]$. We present three related methods that are modifications of the Gauss elimination. They are named after Doolittle, Crout and Cholesky and use the idea of the LU-factorization of $\mathbf{A}$, which we explain first.

For any nonsingular (cf. Sec. 7.8) square matrix, the rows can be reordered so that the resulting matrix $\mathbf{A}$ has an **LU-factorization**

$$(2) \qquad\qquad \boxed{\mathbf{A = LU,}}$$

where $\mathbf{L}$ is lower triangular and $\mathbf{U}$ is upper triangular. This follows from the Gauss elimination in which $\mathbf{L}$ turns out to be the matrix of the multipliers m_{jk}, with main diagonal $1, \cdots, 1$, and $\mathbf{U}$ is the matrix of the triangular system at the end of the Gauss elimination. (Proof in Ref. [E4], pp. 155-156, listed in Appendix 1.) The crucial idea now is that $\mathbf{L}$ and $\mathbf{U}$ in (2) can be computed directly, without solving simultaneous equations (thus, without using the Gauss elimination). As a count shows, this needs about $n^3/3$ operations, about half as many as the Gauss elimination, which needs about $2n^3/3$ (see above). And once we have (2), we can use it for solving $\mathbf{Ax = b}$ in two steps, involving only about n^2 operations, simply by noting that $\mathbf{Ax = LUx = b}$ may be written

$$(3) \qquad \boxed{\text{(a)} \quad \mathbf{Ly = b} \qquad \text{where} \qquad \text{(b)} \quad \mathbf{Ux = y}}$$

and solving first (3a) for $\mathbf{y}$ and then (3b) for $\mathbf{x}$. This is called **Doolittle's method.** A similar method, **Crout's method** is obtained from (2) if $\mathbf{U}$ (instead of $\mathbf{L}$) is required to have main diagonal $1, \cdots, 1$. In either case, the factorization (2) is unique.

EXAMPLE 1. Doolittle's method
Solve the system in Example 1 of Sec. 19.1 by Doolittle's method.

Solution. The decomposition (2) is obtained from

$$
\mathbf{A} = [a_{jk}] = \begin{bmatrix} 3 & 5 & 2 \\ 0 & 8 & 2 \\ 6 & 2 & 8 \end{bmatrix} = \begin{bmatrix} 1 & 0 & 0 \\ m_{21} & 1 & 0 \\ m_{31} & m_{32} & 1 \end{bmatrix} \begin{bmatrix} u_{11} & u_{12} & u_{13} \\ 0 & u_{22} & u_{23} \\ 0 & 0 & u_{33} \end{bmatrix}
$$

by determining the m_{jk} and u_{jk}, using matrix multiplication. By going through $\mathbf{A}$ row by row we get successively

$a_{11} = 3 = u_{11}$	$a_{12} = 5 = u_{12}$	$a_{13} = 2 = u_{13}$

$a_{21} = 0 = m_{21}u_{11}$	$a_{22} = 8 = m_{21}u_{12} + u_{22}$	$a_{23} = 2 = m_{21}u_{13} + u_{23}$
$m_{21} = 0$	$u_{22} = 8$	$u_{23} = 2$

$a_{31} = 6 = m_{31}u_{11}$	$a_{32} = 2 = m_{31}u_{12} + m_{32}u_{22}$	$a_{33} = 8 = m_{31}u_{13} + m_{32}u_{23} + u_{33}$
$= m_{31} \cdot 3$	$= 2 \cdot 5 + m_{32} \cdot 8$	$= 2 \cdot 2 - 1 \cdot 2 + u_{33}$
$m_{31} = 2$	$m_{32} = -1$	$u_{33} = 6$

Thus the factorization (2) is

$$
\begin{bmatrix} 3 & 5 & 2 \\ 0 & 8 & 2 \\ 6 & 2 & 8 \end{bmatrix} = \begin{bmatrix} 1 & 0 & 0 \\ 0 & 1 & 0 \\ 2 & -1 & 1 \end{bmatrix} \begin{bmatrix} 3 & 5 & 2 \\ 0 & 8 & 2 \\ 0 & 0 & 6 \end{bmatrix}.
$$

We first solve $\mathbf{Ly} = \mathbf{b}$, that is,

$$
\begin{bmatrix} 1 & 0 & 0 \\ 0 & 1 & 0 \\ 2 & -1 & 1 \end{bmatrix} \begin{bmatrix} y_1 \\ y_2 \\ y_3 \end{bmatrix} = \begin{bmatrix} 8 \\ -7 \\ 26 \end{bmatrix}. \qquad \text{Solution} \qquad \mathbf{y} = \begin{bmatrix} 8 \\ -7 \\ 3 \end{bmatrix}.
$$

Then we solve $\mathbf{Ux} = \mathbf{y}$, determining x_3, then x_2, then x_1, that is,

$$
\begin{bmatrix} 3 & 5 & 2 \\ 0 & 8 & 2 \\ 0 & 0 & 6 \end{bmatrix} \begin{bmatrix} x_1 \\ x_2 \\ x_3 \end{bmatrix} = \begin{bmatrix} 8 \\ -7 \\ 3 \end{bmatrix}. \qquad \text{Solution} \qquad \mathbf{x} = \begin{bmatrix} 4 \\ -1 \\ 1/2 \end{bmatrix}.
$$

This agrees with the solution in Example 1 of Sec. 19.1. ∎

Our formulas in Example 1 suggest that for general n the elements of the matrices $\mathbf{L} = [m_{jk}]$ (with main diagonal $1, \cdots, 1$) and $\mathbf{U} = [u_{jk}]$ in the *Doolittle method* are computed from

$$u_{1k} = a_{1k} \qquad\qquad\qquad\qquad k = 1, \cdots, n$$

$$u_{jk} = a_{jk} - \sum_{s=1}^{j-1} m_{js}u_{sk} \qquad k = j, \cdots, n; \ \ j \geq 2$$

(4)

$$m_{j1} = \frac{a_{j1}}{u_{11}} \qquad\qquad\qquad\qquad j = 2, \cdots, n$$

$$m_{jk} = \frac{1}{u_{kk}}\left(a_{jk} - \sum_{s=1}^{k-1} m_{js}u_{sk} \right) \qquad j = k + 1, \cdots, n; \ \ k \geq 2.$$

The corresponding formulas for the LU-factorization in *Crout's method* are quite similar:

$$m_{j1} = a_{j1} \qquad\qquad j = 1, \cdots, n$$

$$m_{jk} = a_{jk} - \sum_{s=1}^{k-1} m_{js} u_{sk} \qquad\qquad j = k, \cdots, n; \quad k \geq 2$$

(5)

$$u_{1k} = \frac{a_{1k}}{m_{11}} \qquad\qquad k = 2, \cdots, n$$

$$u_{jk} = \frac{1}{m_{jj}} \left(a_{jk} - \sum_{s=1}^{j-1} m_{js} u_{sk} \right) \qquad k = j+1, \cdots, n; \quad j \geq 2.$$

Cholesky's Method

For a **symmetric, positive definite** matrix $\mathbf{A}$ (thus $\mathbf{A} = \mathbf{A}^\mathsf{T}$, $\mathbf{x}^\mathsf{T}\mathbf{A}\mathbf{x} > 0$ for all $\mathbf{x} \neq \mathbf{0}$) we can in (2) even choose $\mathbf{U} = \mathbf{L}^\mathsf{T}$, thus $u_{jk} = m_{kj}$ (but impose no conditions on the main diagonal entries). For example,

(6)
$$\begin{bmatrix} 4 & 2 & 14 \\ 2 & 17 & -5 \\ 14 & -5 & 83 \end{bmatrix} = \begin{bmatrix} 2 & 0 & 0 \\ 1 & 4 & 0 \\ 7 & -3 & 5 \end{bmatrix} \begin{bmatrix} 2 & 1 & 7 \\ 0 & 4 & -3 \\ 0 & 0 & 5 \end{bmatrix}.$$

The popular method of solving $\mathbf{A}\mathbf{x} = \mathbf{b}$ based on this factorization $\mathbf{A} = \mathbf{L}\mathbf{L}^\mathsf{T}$ is called **Cholesky's method.** The formulas for the factorization are

$$m_{11} = \sqrt{a_{11}}$$

$$m_{jj} = \sqrt{a_{jj} - \sum_{s=1}^{j-1} m_{js}^2} \qquad\qquad j = 2, \cdots, n$$

(7)

$$m_{j1} = \frac{a_{j1}}{m_{11}} \qquad\qquad j = 2, \cdots, n$$

$$m_{jk} = \frac{1}{m_{kk}} \left(a_{jk} - \sum_{s=1}^{k-1} m_{js} m_{ks} \right) \qquad j = k+1, \cdots, n; \quad k \geq 2.$$

If $\mathbf{A}$ is symmetric but not positive definite, this method can still be applied, but then leads to a *complex* matrix $\mathbf{L}$. Cf. Probs. 19, 20.

EXAMPLE 2. Cholesky's method
Solve by Cholesky's method:

$$4x_1 + 2x_2 + 14x_3 = 14$$
$$2x_1 + 17x_2 - 5x_3 = -101$$
$$14x_1 - 5x_2 + 83x_3 = 155.$$

Solution. From (7) or from the form of the factorization

$$
\begin{bmatrix} 4 & 2 & 14 \\ 2 & 17 & -5 \\ 14 & -5 & 83 \end{bmatrix} = \begin{bmatrix} m_{11} & 0 & 0 \\ m_{21} & m_{22} & 0 \\ m_{31} & m_{32} & m_{33} \end{bmatrix} \begin{bmatrix} m_{11} & m_{21} & m_{31} \\ 0 & m_{22} & m_{32} \\ 0 & 0 & m_{33} \end{bmatrix}
$$

we compute, in the given order,

$$
m_{11} = \sqrt{a_{11}} = 2 \qquad m_{21} = \frac{a_{21}}{m_{11}} = \frac{2}{2} = 1 \qquad m_{31} = \frac{a_{31}}{m_{11}} = \frac{14}{2} = 7
$$

$$
m_{22} = \sqrt{a_{22} - m_{21}{}^2} = \sqrt{17 - 1} = 4
$$

$$
m_{32} = \frac{1}{m_{22}}(a_{32} - m_{31}m_{21}) = \frac{1}{4}(-5 - 7 \cdot 1) = -3
$$

$$
m_{33} = \sqrt{a_{33} - m_{31}{}^2 - m_{32}{}^2} = \sqrt{83 - 7^2 - (-3)^2} = 5.
$$

This agrees with (6). We now have to solve $\mathbf{Ly} = \mathbf{b}$, that is,

$$
\begin{bmatrix} 2 & 0 & 0 \\ 1 & 4 & 0 \\ 7 & -3 & 5 \end{bmatrix} \begin{bmatrix} y_1 \\ y_2 \\ y_3 \end{bmatrix} = \begin{bmatrix} 14 \\ -101 \\ 155 \end{bmatrix}. \qquad \text{Solution} \qquad \mathbf{y} = \begin{bmatrix} 7 \\ -27 \\ 5 \end{bmatrix}.
$$

As the second step, we have to solve $\mathbf{Ux} = \mathbf{L}^T\mathbf{x} = \mathbf{y}$, that is,

$$
\begin{bmatrix} 2 & 1 & 7 \\ 0 & 4 & -3 \\ 0 & 0 & 5 \end{bmatrix} \begin{bmatrix} x_1 \\ x_2 \\ x_3 \end{bmatrix} = \begin{bmatrix} 7 \\ -27 \\ 5 \end{bmatrix}. \qquad \text{Solution} \qquad \mathbf{x} = \begin{bmatrix} 3 \\ -6 \\ 1 \end{bmatrix}. \qquad \blacksquare
$$

Gauss–Jordan Elimination. Matrix Inversion

Another variant of the Gauss elimination is the **Gauss-Jordan elimination,** introduced by W. Jordan in 1920, in which back substitution is avoided by additional computations that reduce the matrix to diagonal form, instead of the triangular form in the Gauss elimination. Because of those additional computations, the method seems to have no advantage in solving systems of equations on a computer. The situation is different for matrix inversion, where both the Gauss method and the Gauss-Jordan method need n^3 multiplications. The situation is as follows.

The **inverse** of a nonsingular square matrix $\mathbf{A}$ may be determined in principle by solving the n systems

$$
(8) \qquad\qquad \mathbf{Ax} = \mathbf{b}_j \qquad\qquad (j = 1, \cdots, n),
$$

where $\mathbf{b}_j$ is the jth column of the $n \times n$ unit matrix.

However, it is preferable to produce $\mathbf{A}^{-1}$ by operating on the unit matrix

I in the same way as the Gauss-Jordan algorithm, reducing **A** to **I**. A typical illustrative example of this method is given in Sec. 7.8.

In the next section we discuss the Gauss-Seidel method, which is an **iteration method** for solving systems of linear equations in certain special cases of practical interest.

Problems for Sec. 19.2

Solve the following systems of linear equations by Doolittle's method. Show also the LU-factorization.

1. $5x_1 + 3x_2 = 12$
$5x_1 + 9x_2 = 6$

2. $-4x_1 - 3x_2 = 8$
$28x_1 + 23x_2 = -64$

3. $2x_1 - 3x_2 = -6.5$
$8x_1 - 13x_2 = -28.5$

4. $2x_1 + x_2 + 2x_3 = 1$
$4x_1 + 6x_2 + 9x_3 = 1$
$12x_2 + 18x_3 = -6$

5. $x_1 + 2x_2 + 3x_3 = 17$
$2x_1 + 5x_2 + 8x_3 = 44$
$3x_1 + 8x_2 + 14x_3 = 76$

6. $2x_1 + 2x_2 + 4x_3 = -2$
$4x_1 + 5x_2 + 13x_3 = -7$
$10x_1 + 14x_2 + 43x_3 = -25$

7. $x_1 + 2x_2 + 5x_3 = -11$
$x_2 = 1$
$2x_1 + 9x_2 + 11x_3 = -20$

8. $3x_1 + 9x_2 + 6x_3 = 23$
$18x_1 + 48x_2 + 39x_3 = 136$
$9x_1 - 27x_2 + 42x_3 = 45$

9. $5x_1 + 4x_2 + x_3 = 3.4$
$10x_1 + 9x_2 + 4x_3 = 8.8$
$10x_1 + 13x_2 + 15x_3 = 19.2$

Solve the following systems of linear equations by Cholesky's method.

10. $4x_1 + 10x_2 + 8x_3 = 22$
$10x_1 + 26x_2 + 26x_3 = 64$
$8x_1 + 26x_2 + 61x_3 = 107$

11. $9x_1 + 6x_2 + 12x_3 = 17.4$
$6x_1 + 13x_2 + 11x_3 = 23.6$
$12x_1 + 11x_2 + 26x_3 = 30.8$

12. $4x_1 + 6x_2 + 8x_3 = 0$
$6x_1 + 34x_2 + 52x_3 = -160$
$8x_1 + 52x_2 + 129x_3 = -452$

Find the inverse of the given matrix by the Gauss-Jordan method.

13. $\begin{bmatrix} 6 & 2 \\ 4 & 8 \end{bmatrix}$

14. $\begin{bmatrix} -3 & 5 \\ 2 & 1 \end{bmatrix}$

15. $\begin{bmatrix} 1/\sqrt{2} & -1/\sqrt{2} \\ 1/\sqrt{2} & 1/\sqrt{2} \end{bmatrix}$

16. $\begin{bmatrix} 1.5 & 0 & -0.5 \\ 0 & 2.5 & 0.5 \\ 0.5 & 0 & -0.5 \end{bmatrix}$

17. $\begin{bmatrix} 6 & 12 & 3 \\ 3 & 6 & 3 \\ 9 & 12 & 6 \end{bmatrix}$

18. $\begin{bmatrix} 0.6 & 0.4 & 0.3 \\ 0.4 & 0.3 & 0.2 \\ 0.3 & 0.4 & 0.2 \end{bmatrix}$

If **A** in **Ax** = **b** is symmetric but not positive definite, Cholesky's method still applies but leads to a *complex* matrix **L**. Show this by finding **L** and solving:

19. $-x_1 + 3x_2 = 3$

$3x_1 - 13x_2 = -17$

20. $x_1 + 2x_2 + 3x_3 = 14$

$2x_1 + 3x_2 + 4x_3 = 20$

$3x_1 + 4x_2 + x_3 = 14$

19.3 Systems of Linear Equations: Solution by Iteration

The Gauss elimination and its variants discussed in the previous two sections belong to the **direct methods** for solving systems of linear equations; these are methods that yield solutions after an amount of computation that can be specified in advance. In contrast, in an **indirect** or **iterative method** we start from an approximation to the true solution and, if successful, obtain better and better approximations from a computational cycle repeated as often as may be necessary for achieving a required accuracy, so that the amount of arithmetic depends upon the accuracy required.

Iterative methods are used mainly in those problems for which convergence is known to be rapid, so that the solution is obtained with much less work than that of a direct method, and for systems of large order but with many zero coefficients, the so-called **"sparse" systems,** for which elimination methods would be relatively laborious and would need much storage. Such systems occur in connection with vibrational problems, partial differential equations and various other applications.

An iteration process of practical importance is the **Gauss-Seidel iteration,** which we shall explain in terms of an example.

EXAMPLE 1. Gauss-Seidel iteration

We consider the system of linear equations

(1)
$$
\begin{aligned}
x_1 - 0.25x_2 - 0.25x_3 &= 50 \\
-0.25x_1 + x_2 \qquad\qquad - 0.25x_4 &= 50 \\
-0.25x_1 \qquad\qquad + x_3 - 0.25x_4 &= 25 \\
- 0.25x_2 - 0.25x_3 + x_4 &= 25.
\end{aligned}
$$

(Equations of this form arise in the numerical solution of partial differential equations and in spline interpolation.) We write the system in the form

(2)
$$
\begin{aligned}
x_1 &= \qquad\quad 0.25x_2 + 0.25x_3 \qquad\quad + 50 \\
x_2 &= 0.25x_1 \qquad\qquad\qquad + 0.25x_4 + 50 \\
x_3 &= 0.25x_1 \qquad\qquad\qquad + 0.25x_4 + 25 \\
x_4 &= \qquad\quad 0.25x_2 + 0.25x_3 \qquad\quad + 25.
\end{aligned}
$$

We use these equations for iteration, that is, we start from a (possibly poor) approximation to the solution, say, $x_1^{(0)} = 100$, $x_2^{(0)} = 100$, $x_3^{(0)} = 100$, $x_4^{(0)} = 100$, and compute from (2) a presumably better approximation

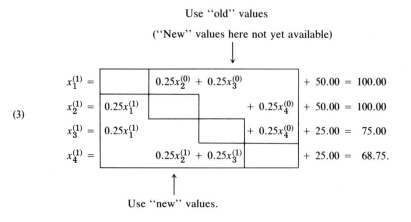

(3)

We see that these equations are obtained from (2) by substituting on the right the most recent approximations. In fact, corresponding elements replace previous ones as soon as they have been computed, so that in the second and third equations we use $x_1^{(1)}$ (not $x_1^{(0)}$), and in the last equation of (3) we use $x_2^{(1)}$ and $x_3^{(1)}$ (not $x_2^{(0)}$ and $x_3^{(0)}$). The next step yields

$$
\begin{aligned}
x_1^{(2)} &= \qquad\quad 0.25 x_2^{(1)} + 0.25 x_3^{(1)} &&\qquad + 50.00 = 93.75 \\
x_2^{(2)} &= 0.25 x_1^{(2)} &&+ 0.25 x_4^{(1)} + 50.00 = 90.62 \\
x_3^{(2)} &= 0.25 x_1^{(2)} &&+ 0.25 x_4^{(1)} + 25.00 = 65.62 \\
x_4^{(2)} &= \qquad\quad 0.25 x_2^{(2)} + 0.25 x_3^{(2)} &&\qquad + 25.00 = 64.06.
\end{aligned}
$$

In practice, one would do further steps and obtain a more accurate approximate solution.
The reader may show that the exact solution is $x_1 = x_2 = 87.5$, $x_3 = x_4 = 62.5$. ∎

To obtain an algorithm for the Gauss–Seidel iteration, let us derive the general formulas for this iteration.

We assume that $a_{jj} = 1$ for $j = 1, \cdots, n$. (Note that this can be achieved if we can rearrange the equations so that no diagonal coefficient is zero; then we may divide each equation by the corresponding diagonal coefficient.) We now write

(4) $$\mathbf{A} = \mathbf{I} + \mathbf{L} + \mathbf{U}$$

where $\mathbf{I}$ is the $n \times n$ unit matrix and $\mathbf{L}$ and $\mathbf{U}$ are respectively lower and upper triangular matrices with zero main diagonals. If we substitute (4) into $\mathbf{Ax} = \mathbf{b}$, we have

$$(\mathbf{I} + \mathbf{L} + \mathbf{U})\mathbf{x} = \mathbf{b}.$$

Taking $\mathbf{Lx}$ and $\mathbf{Ux}$ to the right, we obtain, since $\mathbf{Ix} = \mathbf{x}$,

(5) $$\mathbf{x} = \mathbf{b} - \mathbf{Lx} - \mathbf{Ux}.$$

Remembering from our computation in Example 1 that below the main diagonal we took "new" approximations and above the main diagonal "old" approximations, we obtain from (5) the desired iteration formulas

(6)
$$\mathbf{x}^{(m+1)} = \mathbf{b} - \mathbf{L}\mathbf{x}^{(m+1)} - \mathbf{U}\mathbf{x}^{(m)}$$

where $\mathbf{x}^{(m)} = [x_j^{(m)}]$ is the mth approximation and $\mathbf{x}^{(m+1)} = [x_j^{(m+1)}]$ is the $(m + 1)$st approximation. In components this gives the formula in line 1 of the algorithm in Table 19.2, where the factor $1/a_{jj}$ accomplishes our assumption in (4) and (5) that the main diagonal entries of $\mathbf{A}$ are all 1.

Table 19.2
Gauss–Seidel Iteration

ALGORITHM GAUSS–SEIDEL $(\mathbf{A}, \mathbf{b}, \mathbf{x}^{(0)}, \epsilon, N)$

This algorithm computes a solution $\mathbf{x}$ of the system $\mathbf{A}\mathbf{x} = \mathbf{b}$ given an initial approximation $\mathbf{x}^{(0)}$, where $\mathbf{A} = [a_{jk}]$ is an $n \times n$ matrix with $a_{jj} \neq 0$, $j = 1, \cdots, n$.

> INPUT: $\mathbf{A}, \mathbf{b}$, initial approximation $\mathbf{x}_0$, tolerance $\epsilon > 0$,
>   maximum number of iterations N
>
> OUTPUT: Approximate solution $\mathbf{x}^{(m)} = [x_j^{(m)}]$ or failure message
>   that $\mathbf{x}^{(N)}$ does not satisfy the tolerance condition

For $m = 0, \cdots, N - 1$, do:
> For $j = 1, \cdots, n$, do:

1
$$x_j^{(m+1)} = -\frac{1}{a_{jj}}\left(\sum_{k=1}^{j-1} a_{jk}x_k^{(m+1)} + \sum_{k=j+1}^{n} a_{jk}x_k^{(m)} - b_j\right)$$

> End

2   If $\max_j |x_j^{(m+1)} - x_j^{(m)}| < \epsilon$ then OUTPUT $\mathbf{x}^{(m+1)}$. Stop

> [*Procedure completed successfully*]

End

OUTPUT: "No solution satisfying the tolerance condition obtained after N iteration steps." Stop

> [*Procedure completed unsuccessfully*]

End GAUSS–SEIDEL

Convergence. Matrix Norms

An iteration method for solving $\mathbf{A}\mathbf{x} = \mathbf{b}$ is said to **converge** for an initial $\mathbf{x}^{(0)}$ if the corresponding iterative sequence $\mathbf{x}^{(0)}, \mathbf{x}^{(1)}, \mathbf{x}^{(2)}, \cdots$ converges to a solution of the given system. Convergence depends on the relation between $\mathbf{x}^{(m)}$ and $\mathbf{x}^{(m+1)}$. To get this relation for the Gauss–Seidel method, we use (6). We first have

$$(\mathbf{I} + \mathbf{L})\mathbf{x}^{(m+1)} = \mathbf{b} - \mathbf{U}\mathbf{x}^{(m)}$$

and by multiplying by $(\mathbf{I} + \mathbf{L})^{-1}$ from the left,

(7) $\mathbf{x}^{(m+1)} = \mathbf{C}\mathbf{x}^{(m)} + (\mathbf{I} + \mathbf{L})^{-1}\mathbf{b}$ where $\mathbf{C} = -(\mathbf{I} + \mathbf{L})^{-1}\mathbf{U}.$

The Gauss–Seidel iteration converges for every $\mathbf{x}^{(0)}$ if and only if all the eigenvalues (Sec. 7.12) of the "iteration matrix" $\mathbf{C}$ have absolute value less than 1. (Proof in Ref. [E4], p. 191, listed in Appendix 1.) *Caution!* If you want to get $\mathbf{C}$, first divide the rows of $\mathbf{A}$ by a_{jj} to have main diagonal $1, \cdots , 1$. If the **spectral radius** of $\mathbf{C}$ (= maximum of those absolute values) is small, then the convergence is rapid.

A sufficient condition for convergence is

(8)
$$\boxed{\|\mathbf{C}\| < 1.}$$

Here $\|\mathbf{C}\|$ is some **matrix norm,** such as

(9)
$$\|\mathbf{C}\| = \sqrt{\sum_{j=1}^{n} \sum_{k=1}^{n} c_{jk}^{2}}$$
(Frobenius norm)

or

(10)
$$\|\mathbf{C}\| = \max_{k} \sum_{j=1}^{n} |c_{jk}|$$
(Column "sum" norm)

or

(11)
$$\|\mathbf{C}\| = \max_{j} \sum_{k=1}^{n} |c_{jk}|$$
(Row "sum" norm).

These are the most frequently used matrix norms in numerical work. Let us explain our notations and reasoning in terms of a very simple example.

EXAMPLE 2. Test of convergence of the Gauss–Seidel iteration

Test whether the Gauss–Seidel iteration converges for the system

$$2x + y + z = 4 \qquad\qquad x = 2 - \tfrac{1}{2}y - \tfrac{1}{2}z$$
$$x + 2y + z = 4 \qquad \text{written} \qquad y = 2 - \tfrac{1}{2}x - \tfrac{1}{2}z$$
$$x + y + 2z = 4 \qquad\qquad z = 2 - \tfrac{1}{2}x - \tfrac{1}{2}y.$$

Solution. The decomposition

$$\begin{bmatrix} 1 & 1/2 & 1/2 \\ 1/2 & 1 & 1/2 \\ 1/2 & 1/2 & 1 \end{bmatrix} = \mathbf{I} + \mathbf{L} + \mathbf{U} = \mathbf{I} + \begin{bmatrix} 0 & 0 & 0 \\ 1/2 & 0 & 0 \\ 1/2 & 1/2 & 0 \end{bmatrix} + \begin{bmatrix} 0 & 1/2 & 1/2 \\ 0 & 0 & 1/2 \\ 0 & 0 & 0 \end{bmatrix}$$

shows that

$$\mathbf{C} = [c_{jk}] = -(\mathbf{I} + \mathbf{L})^{-1}\mathbf{U} = -\begin{bmatrix} 1 & 0 & 0 \\ -1/2 & 1 & 0 \\ -1/4 & -1/2 & 1 \end{bmatrix}\begin{bmatrix} 0 & 1/2 & 1/2 \\ 0 & 0 & 1/2 \\ 0 & 0 & 0 \end{bmatrix} = \begin{bmatrix} 0 & -1/2 & -1/2 \\ 0 & 1/4 & -1/4 \\ 0 & 1/8 & 3/8 \end{bmatrix}.$$

We compute the Frobenius norm of $\mathbf{C}$

$$\|\mathbf{C}\| = \left(\frac{1}{4} + \frac{1}{4} + \frac{1}{16} + \frac{1}{16} + \frac{1}{64} + \frac{9}{64}\right)^{1/2} = \left(\frac{50}{64}\right)^{1/2} = 0.884 < 1$$

and conclude from (8) that this Gauss–Seidel iteration would converge. It is interesting that the other two norms would permit no conclusion, as the student should verify. ∎

Jacobi Iteration

The Gauss–Seidel iteration is a method of **successive corrections** because we replace approximations by corresponding new ones as soon as the latter have been computed. A method is called a method of **simultaneous corrections** if no component of an approximation $\mathbf{x}^{(m+1)}$ is used until all the components of $\mathbf{x}^{(m+1)}$ have been computed. A method of this type is the **Jacobi iteration,** which is similar to the Gauss–Seidel iteration but consists in *not* using improved values until a step has been completed and then replacing $\mathbf{x}^{(m)}$ by $\mathbf{x}^{(m+1)}$ entirely for the next cycle. Hence, if we write $\mathbf{Ax} = \mathbf{b}$ in the form $\mathbf{x} = \mathbf{b} + (\mathbf{I} - \mathbf{A})\mathbf{x}$, the Jacobi iteration in matrix notation is

$$(12) \qquad \mathbf{x}^{(m+1)} = \mathbf{b} + (\mathbf{I} - \mathbf{A})\mathbf{x}^{(m)}.$$

This method is largely of theoretical interest. It converges for every choice of $\mathbf{x}^{(0)}$ if and only if the spectral radius of $\mathbf{I} - \mathbf{A}$ is less than 1; here we again assume $a_{jj} = 1$ for $j = 1, \cdots, n$.

Given a system $\mathbf{Ax} = \mathbf{b},$ we may introduce the **residual r** defined by

$$\mathbf{r} = \mathbf{b} - \mathbf{Ax}.$$

Clearly, $\mathbf{r} = \mathbf{0}$ if and only if $\mathbf{x}$ is a solution. Hence $\mathbf{r} \neq \mathbf{0}$ for an approximate solution. In the Gauss–Seidel iteration, at each stage we modify or *relax* a component of an approximate solution in order to reduce a component of $\mathbf{r}$ to zero. Hence the Gauss–Seidel iteration belongs to a class of methods often called **relaxation methods.** More about the residual follows in the next section.

Iteration for Computing the Inverse

The **inverse** of a nonsingular square matrix $\mathbf{A}$ may also be determined by an iteration method suggested by the following idea. The reciprocal x of a given number a satisfies $xa = 1$, and if we want to determine x without division, we may apply Newton's method (Sec. 18.2) to the function $f(x) = x^{-1} - a$. Since we have $f'(x) = -1/x^2$, the Newton iteration is

$$x_{m+1} = x_m - (x_m^{-1} - a)(-x_m^2) = x_m(2 - ax_m).$$

This suggests an analogous iteration formula for determining the inverse $\mathbf{X} = \mathbf{A}^{-1}$ of $\mathbf{A}$, namely,

$$(13) \qquad \mathbf{X}^{(m+1)} = \mathbf{X}^{(m)}(2\mathbf{I} - \mathbf{A}\mathbf{X}^{(m)}).$$

This process converges (produces $\mathbf{A}^{-1}$ as $m \to \infty$) if and only if an $\mathbf{X}^{(0)}$ is chosen so that every eigenvalue of $\mathbf{I} - \mathbf{A}\mathbf{X}^{(0)}$ is of absolute value less than 1 (cf. Ref. [E36], p. 223, listed in Appendix 1). The method is of practical interest if the occurring multiplications are simple, for instance, if $\mathbf{A}$ has many zeros. However, a suitable choice of $\mathbf{X}^{(0)}$ is generally difficult, and the method is mostly used for improving an inaccurate inverse obtained by another method.

In the next section, the last on linear equations, we give a systematic discussion of **ill-conditioned systems,** using vector and matrix **norms** and introducing the **condition number** of a matrix.

Problems for Sec. 19.3

Apply the Gauss–Seidel iteration to the following systems. Perform five steps, starting from 1, 1, 1. *Hint.* Make sure that at the beginning you solve each equation for the variable that has the largest coefficient. (Why?)

1. $6x_1 + x_2 + x_3 = 10.7$

$x_1 + 9x_2 - 2x_3 = 3.6$

$2x_1 - x_2 + 8x_3 = 12.1$

2. $10x_1 - x_2 + 2x_3 = 38$

$-x_1 + 15x_2 + 3x_3 = 26$

$2x_1 + 3x_2 + 12x_3 = 14$

3. $x_1 + 15x_2 - x_3 = 11$

$10x_1 + 3x_2 = -17$

$2x_1 - x_2 + 5x_3 = 5$

4. $4x_1 - x_2 = 5.5$

$4x_2 - x_3 = 0.4$

$-x_1 + 4x_3 = 11.2$

5. $6x_1 + x_2 - x_3 = 3$

$x_1 + 5x_2 + x_3 = 0$

$-x_1 + x_2 + 7x_3 = -17$

6. $2x_1 + 10x_2 - x_3 = -32$

$-x_1 + 2x_2 + 15x_3 = 17$

$10x_1 - x_2 + 2x_3 = 58$

7. $10x_1 + x_2 + x_3 = 6$

$x_1 + 10x_2 + x_3 = 6$

$x_1 + x_2 + 10x_3 = 6$

8. $9x_1 + 2x_3 = 25.7$

$12x_2 - x_3 = 16.7$

$2x_1 - x_2 + 9x_3 = 25.5$

9. $8x_1 + x_2 + x_3 - x_4 = 18$

$-2x_1 + 12x_2 - x_3 = -7$

$2x_1 + 16x_3 + 2x_4 = 54$

$x_2 + 2x_3 - 20x_4 = -14$

10. $7x_1 + x_3 - x_4 = 1.2$

$-x_1 + x_2 + 8x_3 = -8.3$

$10x_2 - x_3 + x_4 = 2.6$

$10x_1 + x_2 + 30x_4 = 22.1$

11. Apply the Gauss–Seidel iteration (3 steps) to the system in Prob. 7, starting from (a) 0, 0, 0, (b) 10, 10, 10. Compare and comment.

12. Apply the Gauss–Seidel and Jacobi iterations (3 steps each) to the system in Prob. 7, starting from 1, 1, 1. Compare and comment.

13. Verify the solution of (1) given in the text. Solve (1) by the Gauss elimination.

14. Apply the Jacobi iteration to the system (1) in the text; start from the vector $\mathbf{x}^{(0)} = [100 \ \ 100 \ \ 100 \ \ 100]^T$, perform 2 steps and compare the accuracy with that in the text.

15. Starting from 0, 0, 0, show that for the system in Example 2 the Jacobi iteration diverges.

16. It is plausible to think that the Gauss–Seidel iteration is better than the Jacobi iteration. Actually the methods are not comparable. Illustrate this surprising fact by showing that for the system

$$x_1 + x_3 = 2$$

$$-x_1 + x_2 = 0$$

$$x_1 + 2x_2 - 3x_3 = 0$$

the Jacobi iteration converges whereas the Gauss–Seidel iteration diverges. (*Hint.* Use eigenvalues.)

17. Find $\mathbf{C}$ [cf. (7)] for the system (1) in the text.

18. Compute $\mathbf{C}$ in Prob. 7 when x_1, x_2, x_3 are computed from the first, second and third equation, respectively, and conclude from (8) that this iteration converges.

19. Consider an approximation $\mathbf{X}^{(0)}$ to the inverse of a matrix $\mathbf{A}$, where

$$\mathbf{X}^{(0)} = \begin{bmatrix} 0.5 & -0.1 & 0.4 \\ 0 & 0.2 & 0 \\ -0.4 & 0.3 & -1.5 \end{bmatrix} \quad \text{and} \quad \mathbf{A} = \begin{bmatrix} 3 & 0 & 1 \\ 0 & 5 & 0 \\ -1 & 1 & -1 \end{bmatrix}$$

Calculate $\mathbf{X}^{(1)}$ by (13). Determine $\mathbf{A}^{-1}$ and show that each element of $\mathbf{X}^{(0)}$ deviates from the corresponding element of $\mathbf{A}^{-1}$ by at most 0.1, whereas for $\mathbf{X}^{(1)}$ that maximum deviation is 0.03.

20. Apply (13) with

$$\mathbf{X}^{(0)} = \begin{bmatrix} 2.9 & -0.9 \\ -4.9 & 1.9 \end{bmatrix} \quad \text{to the matrix} \quad \mathbf{A} = \begin{bmatrix} 2 & 1 \\ 5 & 3 \end{bmatrix},$$

verify that the condition for convergence is satisfied, perform two steps, and compare the results with the exact inverse.

Compute each of the norms (9), (10), (11) of the following matrices.

21. $\begin{bmatrix} 4 & 0 \\ 3 & 6 \end{bmatrix}$ **22.** $\begin{bmatrix} 0 & 0 \\ 0 & 0 \end{bmatrix}$ **23.** $\begin{bmatrix} 1 & 0 \\ 0 & 1 \end{bmatrix}$

24. $\begin{bmatrix} -1 & 3 & 2 \\ 0 & -4 & 4 \\ 6 & 0 & 0 \end{bmatrix}$ **25.** $\begin{bmatrix} 1 & 0 & 4 \\ 0 & 3 & -2 \\ 4 & -2 & 8 \end{bmatrix}$

19.4 Systems of Linear Equations: III-Conditioning, Norms

A system of linear equations is said to be **well-conditioned** if small errors in the coefficients or in the solving process have only a small effect on the solution. In this case the solution is relatively strongly indicated by the equations.

A system of linear equations is said to be **ill-conditioned** if small errors in the coefficients or in the solving process have a large effect on the solution. In this case the solution is relatively weakly indicated by the equations.

For instance, two linear equations in two unknowns represent two straight lines. Such a system is ill-conditioned if and only if the angle γ between the lines is small, that is, if and only if the lines are almost parallel. In fact, then a small change in a coefficient may cause a large displacement of the point of intersection of the lines (cf. Fig. 431, p. 1022). For larger systems of linear equations the situation is similar in principle, but no such simple geometrical

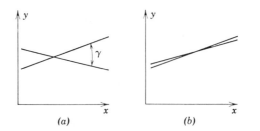

Fig. 431. (a) Well-conditioned and (b) ill-conditioned system
of two linear equations in two unknowns

interpretation is possible, and we would not be able to follow each detail of
the situation.

EXAMPLE 1. An ill-conditioned system

The reader may verify that the system

$$0.9999x - 1.0001y = 1$$

$$x - \quad y = 1$$

has the solution $x = 0.5$, $y = -0.5$, whereas the system

$$0.9999x - 1.0001y = 1$$

$$x - \quad y = 1 + \epsilon$$

has the solution $x = 0.5 + 5000.5\epsilon$, $y = -0.5 + 4999.5\epsilon$. This shows that the system is ill-
conditioned because a change on the right of magnitude ϵ produces a change in the solution of
magnitude 5000ϵ, approximately.   ∎

Ill-conditioning may be regarded as an approach toward singularity. It
manifests itself by a loss of significant figures during computation, so that
it is more difficult to get an accurate inverse or solution of the associated
equations. Several measures for ill-conditioning have been proposed, but
the common usage of the term is still qualitative.

In the case of ill-conditioning we must use a much larger number of decimal
places (double or triple precision) in intermediate computations necessary
to determine the solution to a specific number of decimal places (when round-
off error is present). If the coefficients of an ill-conditioned system as well
as the numbers on the right-hand side can be computed from a formula, we
can compute them, at some small effort, to an arbitrary degree of accuracy,
using single precision, double precision, triple precision, etc., so that the
situation is not too serious. A much greater handicap arises when those
coefficients and numbers are obtained experimentally with a necessarily
limited accuracy. Then we have to accept the fact that inaccuracies in the
given data permit the solution to be determined only within relatively wide
error limits, and we should perhaps try to reformulate the problem in terms
of another system of equations that is reasonably well-conditioned.

Some symptoms for ill-conditioning are as follows. $|\det \mathbf{A}|$ is small com-
pared with the size of the maximum of the $|a_{jk}|$'s and the terms on the right-
hand side of the system. Poor approximations of the solution may produce
merely very small residuals (see below). The entries in $\mathbf{A}^{-1}$ are large in

absolute value compared with the absolute values of the components of the solution.

If the main diagonal entries are large in absolute value compared with the absolute values of the other entries, the system is well-conditioned. If a square matrix with largest entries between 0.1 and 10 has an inverse with largest entries also of order unity, then the associated system of linear equations is well-conditioned.

In the case of ill-conditioning we may want to get an improved solution of

$$(1) \qquad\qquad\qquad \mathbf{Ax} = \mathbf{b}$$

from an approximate solution $\tilde{\mathbf{x}}$. To $\tilde{\mathbf{x}}$ there corresponds the **residual**

$$\mathbf{r} = \mathbf{b} - \mathbf{A}\tilde{\mathbf{x}}.$$

Hence

$$\mathbf{A}\tilde{\mathbf{x}} = \mathbf{b} - \mathbf{r}.$$

From this and (1),

$$(2) \qquad\qquad\qquad \mathbf{A}(\mathbf{x} - \tilde{\mathbf{x}}) = \mathbf{r}.$$

This shows that the correction to be applied to $\tilde{\mathbf{x}}$ to give the solution of (1) is the solution of (2).

One might think that a small residual $\mathbf{r}$ indicates that $\tilde{\mathbf{x}}$ is close to $\mathbf{x}$. However, this conclusion may be false, as Example 2 shows. What happens depends on $\mathbf{A}$ and will need further investigation.

EXAMPLE 2. Poor approximate solution with a small residual

The system

$$1.0001x_1 + \qquad x_2 = 2.0001$$

$$x_1 + 1.0001x_2 = 2.0001$$

has the exact solution $x_1 = 1$, $x_2 = 1$. Can you see this by inspection? The very poor approximation $\tilde{x}_1 = 2.0000$, $\tilde{x}_2 = 0.0001$ has the very small residual (to 4 decimals)

$$\mathbf{r} = \begin{bmatrix} 2.0001 \\ 2.0001 \end{bmatrix} - \begin{bmatrix} 1.0001 & 1.0000 \\ 1.0000 & 1.0001 \end{bmatrix} \begin{bmatrix} 2.0000 \\ 0.0001 \end{bmatrix} = \begin{bmatrix} 2.0001 \\ 2.0001 \end{bmatrix} - \begin{bmatrix} 2.0003 \\ 2.0001 \end{bmatrix} = \begin{bmatrix} -0.0002 \\ 0.0000 \end{bmatrix}.$$

From this, a naive person might draw the false conclusion that the approximation should be accurate to 3 or 4 decimals. ∎

To investigate this probably unexpected situation further, we need *norms* and some of their basic properties, which we discuss next. Norms play an increasing role in numerical work in general, in connection with error estimation, convergence of numerical methods, etc.

Vector Norms

A **vector norm** for all column vectors $\mathbf{x} = [x_j]$ with n components (n fixed) is a generalized length, is denoted by $\|\mathbf{x}\|$ and is defined by four properties of the usual length of vectors in three-dimensional space, namely,

(a) $\|\mathbf{x}\|$ is a nonnegative real number.

(b) $\|\mathbf{x}\| = 0$ if and only if $\mathbf{x} = \mathbf{0}$.

(3)

(c) $\|k\mathbf{x}\| = |k|\,\|\mathbf{x}\|$ for all k.

(d) $\|\mathbf{x} + \mathbf{y}\| \leq \|\mathbf{x}\| + \|\mathbf{y}\|$ (Triangle inequality).

If we use several norms, we label them by a subscript. Most important in numerical algebra is the *p-norm* defined by

$$(4) \qquad \|\mathbf{x}\|_p = (|x_1|^p + |x_2|^p + \cdots + |x_n|^p)^{1/p}$$

where p is a fixed number and $p \geq 1$. In practice, one usually takes $p = 1$ or 2 and, as a third norm, $\|\mathbf{x}\|_\infty$ (the latter as defined below), that is,

$$(5) \qquad \|\mathbf{x}\|_1 = |x_1| + \cdots + |x_n| \qquad (\text{``}l_1\text{-norm''})$$

$$(6) \qquad \|\mathbf{x}\|_2 = \sqrt{x_1^2 + \cdots + x_n^2} \qquad (\text{``Euclidean'' or ``}l_2\text{-norm''})$$

$$(7) \qquad \|\mathbf{x}\|_\infty = \max_j |x_j| \qquad (\text{``}l_\infty\text{-norm''}).$$

For $n = 3$ the l_2-norm is the usual length of a vector in three-dimensional space. The l_1-norm and l_∞-norm are usually more convenient in computation.

EXAMPLE 3. Vector norms
If $\mathbf{x}^T = [2 \ \ -3 \ \ 0 \ \ 1 \ \ -4]$, then

$$\|\mathbf{x}\|_1 = 10, \qquad \|\mathbf{x}\|_2 = \sqrt{30}, \qquad \|\mathbf{x}\|_\infty = 4. \qquad \blacksquare$$

In three-dimensional space, two points with position vectors $\mathbf{x}$ and $\tilde{\mathbf{x}}$ have distance $|\mathbf{x} - \tilde{\mathbf{x}}|$ from each other. This suggests, for a system of equations $\mathbf{Ax} = \mathbf{b}$, that we take $\|\mathbf{x} - \tilde{\mathbf{x}}\|$ as a measure of accuracy and call it the **distance** between an exact and an approximate solution, or the **error** of $\tilde{\mathbf{x}}$.

Matrix Norms

If $\mathbf{A}$ is an $n \times n$ matrix and $\mathbf{x}$ any vector with n components, then $\mathbf{Ax}$ is a vector with n components. We now take a vector norm and consider $\|\mathbf{x}\|$ and $\|\mathbf{Ax}\|$. One can prove (cf. Ref. [13], p. 95 listed in Appendix 1) that there is a number c (depending on $\mathbf{A}$) such that

$$(8) \qquad\qquad \|\mathbf{Ax}\| \leq c\|\mathbf{x}\| \qquad\qquad \text{for all } \mathbf{x}.$$

Let $\mathbf{x} \neq \mathbf{0}$. Then $\|\mathbf{x}\| > 0$ by (3b) and division gives

$$\frac{\|\mathbf{Ax}\|}{\|\mathbf{x}\|} \leq c.$$

We obtain the smallest possible c valid for *all* $\mathbf{x}$ ($\neq \mathbf{0}$) by taking the maximum[2] on the left. This smallest c is called the **matrix norm of A** *resulting from the vector norm we picked* and is denoted by $\|\mathbf{A}\|$. Thus

$$
(9) \qquad \boxed{\|\mathbf{A}\| = \max \frac{\|\mathbf{A}\mathbf{x}\|}{\|\mathbf{x}\|}} \qquad (\mathbf{x} \neq \mathbf{0})
$$

the maximum being taken over all $\mathbf{x} \neq \mathbf{0}$. Alternatively (Prob. 23),

$$
(10) \qquad \|\mathbf{A}\| = \max_{\|\mathbf{x}\| = 1} \|\mathbf{A}\mathbf{x}\|.
$$

Note well that $\|\mathbf{A}\|$ depends on the vector norm that we picked. In particular, one can show that for the l_1-norm (5) we get the matrix norm (10), Sec. 19.3, and for the l_∞-norm (7) the matrix norm (11), Sec. 19.3.

By taking our best possible (our smallest) $c = \|\mathbf{A}\|$ we have from (8)

$$
(11) \qquad \boxed{\|\mathbf{A}\mathbf{x}\| \leq \|\mathbf{A}\| \, \|\mathbf{x}\|.}
$$

This is the formula we need. Formula (9) also implies for two $n \times n$ matrices (cf. Ref. [13], p. 98)

$$
(12) \qquad \|\mathbf{A}\mathbf{B}\| \leq \|\mathbf{A}\| \, \|\mathbf{B}\|, \qquad \text{thus} \qquad \|\mathbf{A}^n\| \leq \|\mathbf{A}\|^n.
$$

See Refs. [13] and [E15] for other useful formulas on norms.

Before we go on, let us do a simple illustrative computation.

EXAMPLE 4. Matrix norms

Compute the matrix norms of the coefficient matrix $\mathbf{A}$ in Example 1 and of its inverse $\mathbf{A}^{-1}$ assuming that we use (a) the l_1-vector norm, (b) the l_∞-vector norm.

Solution. We use (4), Sec. 7.8, and then (10) and (11) in Sec. 19.3. Thus

$$
\mathbf{A} = \begin{bmatrix} 0.9999 & 1.0001 \\ 1.0000 & 1.0000 \end{bmatrix}, \qquad \mathbf{A}^{-1} = \begin{bmatrix} -5000.0 & 5000.5 \\ 5000.0 & -4999.5 \end{bmatrix}.
$$

Hence (a) $\|\mathbf{A}\| = 2.0001$, $\|\mathbf{A}^{-1}\| = 10\,000$; (b) $\|\mathbf{A}\| = 2$, $\|\mathbf{A}^{-1}\| = 10\,000.5$. We notice that $\|\mathbf{A}^{-1}\|$ is surprisingly large, which makes $\|\mathbf{A}\| \, \|\mathbf{A}^{-1}\|$ large (20 001). We shall see below that (and why) this is typical of an ill-conditioned system. ∎

Returning now to our actual problem of getting deeper insight into ill-conditioning, we introduce the following key concept.

Condition Number of a Matrix

The **condition number** $\kappa(\mathbf{A})$ of a (nonsingular) square matrix $\mathbf{A}$ is defined by

$$
(13) \qquad \boxed{\kappa(\mathbf{A}) = \|\mathbf{A}\| \, \|\mathbf{A}^{-1}\|.}
$$

---

[2]This maximum in (10) [hence in (9)] exists by Theorem 2.5-3 in Ref. [13], p. 77, since the norm is continuous (cf. Ref. [13], p. 60, listed in Appendix 1). The name "matrix *norm*" is justified since $\|\mathbf{A}\|$ satisfies (3) with $\mathbf{x}$ and $\mathbf{y}$ replaced by $\mathbf{A}$ and $\mathbf{B}$ (cf. Ref. [13], pp. 92-93).

Its role is seen from

Theorem 1 (Condition number)

A system of linear equations $\mathbf{Ax} = \mathbf{b}$ *whose condition number* (13) *is small is well-conditioned. A large condition number indicates ill-conditioning.*

Proof. $\mathbf{b} = \mathbf{Ax}$ and (11) give $\|\mathbf{b}\| \leq \|\mathbf{A}\| \, \|\mathbf{x}\|$. Let $\mathbf{b} \neq \mathbf{0}$ and $\mathbf{x} \neq \mathbf{0}$. Then by division,

$$(14) \qquad \frac{1}{\|\mathbf{x}\|} \leq \frac{\|\mathbf{A}\|}{\|\mathbf{b}\|} .$$

Multiplying (2) by $\mathbf{A}^{-1}$ from the left, we have $\mathbf{x} - \tilde{\mathbf{x}} = \mathbf{A}^{-1}\mathbf{r}$. From this, (11) and (14),

$$(15) \qquad \frac{\|\mathbf{x} - \tilde{\mathbf{x}}\|}{\|\mathbf{x}\|} = \frac{\|\mathbf{A}^{-1}\mathbf{r}\|}{\|\mathbf{x}\|} \leq \|\mathbf{A}^{-1}\| \, \|\mathbf{r}\| \frac{\|\mathbf{A}\|}{\|\mathbf{b}\|} = \kappa(\mathbf{A}) \frac{\|\mathbf{r}\|}{\|\mathbf{b}\|} .$$

Hence when $\kappa(\mathbf{A})$ is small, a small $\|\mathbf{r}\|/\|\mathbf{b}\|$ implies a small relative error $\|\mathbf{x} - \tilde{\mathbf{x}}\|/\|\mathbf{x}\|$, so that the system is well-conditioned. However, this does not hold when $\kappa(\mathbf{A})$ is large.   ∎

EXAMPLE 5. Condition number for an ill-conditioned system

From Example 4 we see that the condition number for the system in Example 1 is very large, $\kappa(\mathbf{A}) = \|\mathbf{A}\| \, \|\mathbf{A}^{-1}\| = 20\ 001$. This confirms that the system in Example 1 is very ill-conditioned.

EXAMPLE 6. Condition number for an ill-conditioned system

By (4) in Sec. 7.8 the inverse of the matrix $\mathbf{A}$ in Example 2 is

$$\mathbf{A}^{-1} = \frac{1}{0.0002} \begin{bmatrix} 1.0001 & -1.0000 \\ -1.0000 & 1.0001 \end{bmatrix} = \begin{bmatrix} 5000.5 & -5000.0 \\ -5000.0 & 5.000.5 \end{bmatrix}$$

From (10), Sec. 10.3, we thus get the very large value

$$\kappa(\mathbf{A}) = (1.0001 + 1.0000)(5000.5 + 5000.0) \approx 20\ 002;$$

similarly from (9) or (11) in Sec. 10.3. This shows that the system in Example 2 is very ill-conditioned and explains the surprising result in Example 2.   ∎

In practice, $\mathbf{A}^{-1}$ will not be known, so that in computing the condition number $\kappa(\mathbf{A})$, one must estimate $\|\mathbf{A}^{-1}\|$. A method for this proposed in 1979 is explained in Ref. [E15], p. 77, listed in Appendix 1.

$\kappa(\mathbf{A})$ can also be used for estimating the effect $\delta\mathbf{x}$ of an inaccuracy $\delta\mathbf{A}$ of $\mathbf{A}$ (errors of measurements of the a_{jk}, for instance), so that instead of $\mathbf{Ax} = \mathbf{b}$ we have

$$(\mathbf{A} + \delta\mathbf{A})(\mathbf{x} + \delta\mathbf{x}) = \mathbf{b}.$$

Multiplying out and subtracting $\mathbf{Ax} = \mathbf{b}$ on both sides, we have

$$\mathbf{A}\delta\mathbf{x} + \delta\mathbf{A}(\mathbf{x} + \delta\mathbf{x}) = \mathbf{0}.$$

Multiplication by A^{-1} from the left gives

$$\delta x = -A^{-1} \delta A (x + \delta x).$$

Two applications of (11) give

$$\|\delta x\| \leq \|A^{-1}\| \, \|\delta A(x + \delta x)\| \leq \|A^{-1}\| \, \|\delta A\| \, \|x + \delta x\|.$$

Now $\|A^{-1}\| = \kappa(A)/\|A\|$, so that division by $\|x + \delta x\|$ shows that the relative inaccuracy of x is related to that of A via the condition number by the inequality

$$(16) \qquad \frac{\|\delta x\|}{\|x\|} \approx \frac{\|\delta x\|}{\|x + \delta x\|} \leq \kappa(A) \, \frac{\|\delta A\|}{\|A\|}.$$

The reader may show that, similarly, when A is accurate, an inaccuracy δb of b causes an inaccuray δx satisfying

$$(17) \qquad \frac{\|\delta x\|}{\|x\|} \leq \kappa(A) \, \frac{\|\delta b\|}{\|b\|}.$$

As above we see that these $\|\delta x\|/\|x\|$ must remain relatively small whenever the condition number of A is small.   ∎

This is the end of our discussion of numerical methods for solving systems of linear equations. As an important application that leads to such systems we consider **curve fitting** in the next section.

Problems for Sec. 19.4

Compute each of the vector norms (5), (6), (7) of the following vectors.

1. $[3 \quad -3 \quad 0]^\mathsf{T}$ **2.** $[-10 \quad 4 \quad 10 \quad 3]^\mathsf{T}$ **3.** $[1 \quad -1 \quad 1 \quad -1]^\mathsf{T}$

4. $[0 \quad 1 \quad 0 \quad 0 \quad 0]^\mathsf{T}$ **5.** $[4 \quad 17 \quad -9]^\mathsf{T}$ **6.** $[3.2 \quad 0 \quad 4.6 \quad 0]^\mathsf{T}$

For the following matrices compute the matrix norm and the condition number corresponding to the l_∞-vector norm.

7. $\begin{bmatrix} 2 & 2 \\ 4 & 9 \end{bmatrix}$ **8.** $\begin{bmatrix} 1 & 0.1 \\ 0.2 & -1 \end{bmatrix}$ **9.** $\begin{bmatrix} 0 & 1 \\ 1 & 0 \end{bmatrix}$

10. $\begin{bmatrix} 100 & 0 & 0 \\ 0 & 0.1 & 0 \\ 0 & 0 & 9 \end{bmatrix}$ **11.** $\begin{bmatrix} 2 & 4 & 1 \\ 1 & 2 & 1 \\ 3 & 4 & 2 \end{bmatrix}$ **12.** $\begin{bmatrix} 30 & 20 & 15 \\ 20 & 15 & 10 \\ 15 & 20 & 10 \end{bmatrix}$

13. Verify (11) for $x^\mathsf{T} = [3 \quad 5 \quad 4]$ taken with the l_∞-norm and the matrix in Prob. 11.

14. Solve each of the following two systems, compare the solutions and comment.

$$2x_1 + 1.4x_2 = 1.4 \qquad\qquad 2x_1 + 1.4x_2 = 1.44$$

$$1.4x_1 + \quad x_2 = 1 \qquad\qquad 1.4x_1 + \quad x_2 = 1$$

15. Compute the condition number in Prob. 14 with respect to the l_1-vector norm and comment.

16. Solve each of the following two systems, compare the solutions and comment.

$$5x_1 - 7x_2 = -2 \qquad\qquad 5\,x_1 - 7x_2 = -2$$

$$-7x_1 + 10x_2 = \quad 3 \qquad\qquad -7x_1 + 10x_2 = \quad 3.1$$

17. Compute the condition numbers in Prob. 16 with respect to the l_1- and l_∞-vector norms.

18. Show that the solution of the system

$$6x_1 + 7x_2 + 8x_3 = 21$$

$$7x_1 + 8x_2 + 9x_3 = 24$$

$$8x_1 + 9x_2 + 9x_3 = 26$$

is $x_1 = 1$, $x_2 = 1$, $x_3 = 1$. Compute the determinant of the system and the residual corresponding to $x_1 = -0.8$, $x_2 = 2.9$, $x_3 = 0.7$. Comment.

19. Show that the angle γ between two lines represented by a system

$$a_{11}x_1 + a_{12}x_2 = b_1$$

$$a_{21}x_1 + a_{22}x_2 = b_2$$

is given by

$$\tan \gamma = (a_{11}a_{22} - a_{12}a_{21})/(a_{11}a_{21} + a_{12}a_{22})$$

and explain consequences of this formula in connection with ill-conditioning.

20. Using Prob. 19, calculate γ for the systems in Example 1 and Prob. 14 and comment.

21. **(Hilbert matrix)** Using the Gauss elimination, solve the system

$$x_1 + \tfrac{1}{2}x_2 + \tfrac{1}{3}x_3 = 1$$

$$\tfrac{1}{2}x_1 + \tfrac{1}{3}x_2 + \tfrac{1}{4}x_3 = 0$$

$$\tfrac{1}{3}x_1 + \tfrac{1}{4}x_2 + \tfrac{1}{5}x_3 = 0.$$

Computing the condition number (in the column sum norm), show that the system is ill-conditioned. (The coefficient matrix is called the 3×3 *Hilbert matrix*.)

22. By definition, the $n \times n$ Hilbert matrix $\mathbf{H}_n = [h_{jk}]$ has the elements $h_{jk} = 1/(j + k - 1)$. The elements of the inverse $\mathbf{H}_n{}^{-1}$ grow rapidly in absolute value as n increases. Illustrate this by computing $\mathbf{H}_2{}^{-1}$, $\mathbf{H}_3{}^{-1}$, $\mathbf{H}_4{}^{-1}$. ($\mathbf{H}_n$ is not an academic curiosity, but comparable matrices occur in connection with curve fitting by least squares, as we shall see in the next section.)

23. Derive (10) from (9).

24. Show that $\kappa(\mathbf{A}) \geq 1$ for the matrix norms (10), (11), Sec. 19.3, and $\kappa(\mathbf{A}) \geq \sqrt{n}$ for the Frobenius norm (9), Sec. 19.3.

25. Show that $\|\mathbf{x}\|_\infty \leq \|\mathbf{x}\|_1 \leq n\|\mathbf{x}\|_\infty$.

19.5 Method of Least Squares

In **curve fitting** we are given n points (pairs of numbers)

$$(x_1, y_1), \cdots, (x_n, y_n)$$

and we want to determine a function $f(x)$ such that $f(x_j) \approx y_j, j = 1, \cdots, n$. The type of function (for example, polynomials, exponential functions, sine and cosine functions) may be suggested by the nature of the problem (the underlying physical law, for instance), and in many cases a polynomial of a certain degree will be appropriate.

If we require strict equality $f(x_1) = y_1, \cdots, f(x_n) = y_n$ and use polynomials of sufficiently high degree, we may apply one of the methods discussed in Sec. 18.3 in connection with interpolation. However, in certain situations this would not be the appropriate solution of the actual problem. For instance, to the four points

$$(1) \quad (-1.0, 1.000), \quad (-0.1, 1.099), \quad (0.2, 0.808), \quad (1.0, 1.000)$$

there corresponds the Lagrange polynomial $f(x) = x^3 - x + 1$ (Fig. 432), but if we graph the points, we see that they lie nearly on a straight line. Hence if these values are obtained in an experiment and thus involve an experimental error, and if the nature of the experiment suggests a linear relation, we better fit a straight line through the points (Fig. 432). Such a line may be useful for predicting values to be expected for other values of x. In simple cases a straight line may be fitted by eye, but if the points are scattered, this becomes unreliable and we better use a mathematical principle. A widely used procedure of this type is the **method of least squares** by Gauss. In the present situation it may be formulated as follows.

Method of least squares. *The straight line*

$$y = a + bx$$

should be fitted through the given points $(x_1, y_1), \cdots, (x_n, y_n)$ so that the

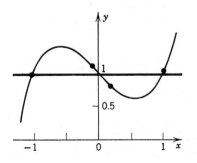

Fig. 432. The approximate fitting of a straight line

sum of the squares of the distances of those points from the straight line is minimum, where the distance is measured in the vertical direction (the y-direction).

The point on the line with abscissa x_j has the ordinate $a + bx_j$. Hence its distance from (x_j, y_j) is $|y_j - a - bx_j|$ (cf. Fig. 433) and that sum of squares is

$$q = \sum_{j=1}^{n} (y_j - a - bx_j)^2.$$

q depends on a and b. A necessary condition for q to be minimum is

(2)
$$\frac{\partial q}{\partial a} = -2 \sum (y_j - a - bx_j) = 0$$

$$\frac{\partial q}{\partial b} = -2 \sum x_j(y_j - a - bx_j) = 0$$

(where we sum over j from 1 to n). Writing each sum as three sums and taking one of them to the right, we obtain the result

(3)
$$\boxed{\begin{aligned} an \quad &+ b \sum x_j = \sum y_j \\ a \sum x_j &+ b \sum x_j^2 = \sum x_j y_j. \end{aligned}}$$

These equations are called the **normal equations** of our problem.

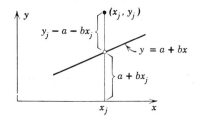

Fig. 433. Vertical distance of a point (x_j, y_j)
from a straight line $y = a + bx$

EXAMPLE 1. Straight line
Using the method of least squares, fit a straight line to the four points given in formula (1).

Solution. We obtain

$$n = 4, \quad \sum x_j = 0.1, \quad \sum x_j^2 = 2.05, \quad \sum y_j = 3.907, \quad \sum x_j y_j = 0.0517.$$

Hence the normal equations are
$$4a + 0.10b = 3.9070$$

$$0.1a + 2.05b = 0.0517$$

The solution is $a = 0.9773$, $b = -0.0224$, and we obtain the straight line (Fig. 432)

$$y = 0.9773 - 0.0224x.$$ ∎

Our approach of curve fitting can be generalized from a polynomial $y = a + bx$ to a polynomial degree m

$$p(x) = b_0 + b_1 x + \cdots + b_m x^m$$

where $m \leqq n - 1$. Then q takes the form

$$q = \sum_{j=1}^{n} (y_j - p(x_j))^2$$

and depends on $m + 1$ parameters $b_0, \cdots, b_m$. Instead of (2) we then have $m + 1$ conditions

$$\frac{\partial q}{\partial b_0} = 0, \quad \cdots, \quad \frac{\partial q}{\partial b_m} = 0$$

which give a system of $m + 1$ normal equations. The reader may show that in the case of a quadratic polynomial

$$(4) \qquad\qquad p(x) = b_0 + b_1 x + b_2 x^2$$

the normal equations are (summation from 1 to n)

$$(5) \qquad \begin{aligned} b_0 n \qquad + b_1 \sum x_j + b_2 \sum x_j^2 &= \sum y_j \\ b_0 \sum x_j + b_1 \sum x_j^2 + b_2 \sum x_j^3 &= \sum x_j y_j \\ b_0 \sum x_j^2 + b_1 \sum x_j^3 + b_2 \sum x_j^4 &= \sum x_j^2 y_j. \end{aligned}$$

Note that this system is symmetric. To solve it for the unknowns b_0, b_1, b_2, we may apply one of the methods discussed in Secs. 19.1–19.3.

Beginning in the next section, we shall turn to **matrix eigenvalue problems,** which are of great general interest in engineering and many other fields.

Problems for Sec. 19.5

In each case plot the given points in the xy-plane and fit a straight line (a) by eye, (b) by the method of least squares.

1. (2, 5), (3, 9), (4, 15), (5, 21)
2. (0, 2.3), (2, 4.1), (4, 5.7), (6, 6.9)
3. (4, 3), (15, 16), (30, 13), (100, 70), (200, 90)
4. (5, 5.0), (10, 3.9), (15, 3.2), (20, 2.0)
5. Density of ore x [grams/cm^3] 2.8 2.9 3.0 3.1 3.2 3.2 3.2 3.3 3.4
   Iron content y [percent]   32 28 35 33 35 37 39 38 35
6. Revolutions per minute x 400 500 600 700 750
   Power of a Diesel engine y [hp]   600 1050 1440 1900 2120

7. If a car is traveling along a straight road with constant speed $v = b_1$ [meters/sec], its position y[meters] at time t[sec] is $y = b_0 + b_1 t$. Suppose that measurements are

t	0	3	5	8	10
y	120	150	160	190	210

Graph these as points in the ty-plane. Fit a straight line (*a*) by eye, (*b*) by the least-squares method. What estimate of the speed can you obtain from it?

In each case determine the parabola (4) by the method of least squares.

8. $(-1, 2)$, $(0, 0)$, $(0, 1)$, $(1, 2)$

9. $(0, 3)$, $(1, 1)$, $(2, 0)$, $(4, 1)$, $(6, 4)$

10.
Worker's time on duty x [hours]	1	2	3	4	5	6
Worker's reaction time y [sec]	1.50	1.48	1.75	1.65	1.72	1.55

11.
Speed of a tractor x [km/hr]	1.4	1.8	2.3	3.0	4.0
Power of a tractor y [kg]	7400	7500	7600	7500	7200

12. Determine the normal equations in the case of a polynomial of the third degree.

13. In the least-squares principle involving a polynomial we seek to satisfy as well as possible

$$b_0 + b_1 x_j + b_2 x_j^2 + \cdots + b_m x_j^m = y_j \qquad (j = 1, \cdots, n).$$

Introduce a matrix $\mathbf{C}$ such that this can be written $\mathbf{Cb} = \mathbf{y}$ and show that then the normal equations can be written $\mathbf{C}^\mathsf{T}\mathbf{Cb} = \mathbf{C}^\mathsf{T}\mathbf{y}$.

14. The coefficient matrix in (5), call it $\mathbf{C}$, is symmetric and positive definite (definition in the problem set for Sec. 7.15), so that Cholesky's method (Sec. 19.2) is very satisfactory for solving (5) and, similarly, normal equations in the case of polynomials of higher degree. But $\mathbf{C}$ may be ill-conditioned. To understand this, reason as follows. Show that $\mathbf{C} = (c_{ik})$, where $c_{ik} = \Sigma x_j^{i+k-2}$ (sum over j from 1 to n). Assume that n is large and the x_j lie in the interval $0 \leq x \leq 1$ and are fairly uniformly distributed. Conclude from the definition of an integral that the integral of x^{i+k-2} over x from 0 to 1 equals about c_{ik}/n. Evaluating this integral, conclude that $\mathbf{C} = n\mathbf{H}_3$, where $\mathbf{H}_3$ is the 3×3 Hilbert matrix (Sec. 19.4, Prob. 21). Conclude that in the case of a polynomial of degree m one arrives at the Hilbert matrix $\mathbf{H}_{m+1}$, which is ill-conditioned for larger m.

15. In problems of growth it is often required to fit an exponential function $y = b_0 e^{bx}$ by the method of least squares. Show that by taking logarithms this task can be reduced to that of determining a straight line.

19.6 Matrix Eigenvalue Problems: Introduction

In the remaining sections of this chapter we discuss some of the most important ideas and numerical methods for matrix eigenvalue problems. This very extensive part of numerical linear algebra has great practical importance, with much research going on, particularly since 1945, and hundreds, if not thousands of papers published in various mathematical journals (see the references in [E13], [E15], [E21], [E30], [E33], [E39], [E40]). We begin with the concepts and general results we shall need in explaining and applying numerical methods for eigenvalue problems. (For typical applications, see Secs. 7.12–7.16.)

An **eigenvalue** or **characteristic value** (or *latent root*) of a given $n \times n$ matrix $\mathbf{A} = [a_{jk}]$ is a real or complex number λ such that the vector equation

$$\boxed{\mathbf{Ax} = \lambda\mathbf{x}} \tag{1}$$

has a nontrivial solution, that is, a solution $\mathbf{x} \neq \mathbf{0}$, which is then called an **eigenvector** or **characteristic vector** of $\mathbf{A}$ corresponding to that eigenvalue λ. The set of all eigenvalues of $\mathbf{A}$ is called the **spectrum** of $\mathbf{A}$. Equation (1) can be written

$$(\mathbf{A} - \lambda\mathbf{I})\mathbf{x} = \mathbf{0} \tag{2}$$

where $\mathbf{I}$ is the $n \times n$ unit matrix. This homogeneous system has a nontrivial solution if and only if the **characteristic determinant** det $(\mathbf{A} - \lambda\mathbf{I})$ is 0 (cf. Theorem 2 in Sec. 7.7). This gives (cf. Sec. 7.12)

Theorem 1 (Eigenvalues)

The eigenvalues of $\mathbf{A}$ *are the solutions of the* **characteristic equation**

$$(3) \quad \det(\mathbf{A} - \lambda\mathbf{I}) = \begin{vmatrix} a_{11} - \lambda & a_{12} & \cdots & a_{1n} \\ a_{21} & a_{22} - \lambda & \cdots & a_{2n} \\ \cdot & \cdot & \cdots & \cdot \\ a_{n1} & a_{n2} & \cdots & a_{nn} - \lambda \end{vmatrix} = 0.$$

Developing the characteristic determinant, we obtain the **characteristic polynomial** of $\mathbf{A}$, which is of degree n in λ. Hence $\mathbf{A}$ has at least one and at most n numerically different eigenvalues. If $\mathbf{A}$ is real, so are the coefficients of the characteristic polynomial. By familiar algebra it follows that then the roots (the eigenvalues of $\mathbf{A}$) are real or complex conjugates in pairs.

We shall usually denote the eigenvalues of $\mathbf{A}$ by

$$\boxed{\lambda_1, \lambda_2, \cdots, \lambda_n}$$

with the understanding that some (or all) of them may be numerically equal.

The sum of these n eigenvalues equals the sum of the entries on the main diagonal of $\mathbf{A}$, called the **trace** of $\mathbf{A}$; thus

$$\text{trace } \mathbf{A} = \sum_{j=1}^{n} a_{jj} = \sum_{k=1}^{n} \lambda_k. \tag{4}$$

Also

$$\det \mathbf{A} = \lambda_1\lambda_2\cdots\lambda_n. \tag{5}$$

Both formulas follow from the product representation of the characteristic polynomial $f(\lambda)$,

$$f(\lambda) = (-1)^n(\lambda - \lambda_1)(\lambda - \lambda_2) \cdots (\lambda - \lambda_n).$$

If we take equal factors together and denote the *numerically distinct* eigenvalues of **A** by $\lambda_1, \cdots, \lambda_r$ $(r \leqq n)$, then the product becomes

(6) $$f(\lambda) = (-1)^n(\lambda - \lambda_1)^{m_1}(\lambda - \lambda_2)^{m_2} \cdots (\lambda - \lambda_r)^{m_r}.$$

The exponent m_j is called the **algebraic multiplicity** of λ_j. The maximum number of linearly independent eigenvectors corresponding to λ_j is called the **geometric multiplicity** of λ_j. It is equal to or smaller than m_j.

Similarity. Spectral Shift. Special Matrices

An $n \times n$ matrix **B** is called **similar** to **A** if there is a matrix **T** such that

(7) $$\boxed{\mathbf{B} = \mathbf{T}^{-1}\mathbf{A}\mathbf{T}.}$$

Similarity is important for the following reason.

Theorem 2 (Similar matrices)
Similar matrices have the same eigenvalues. If **x** *is an eigenvector of* **A**, *then* $\mathbf{y} = \mathbf{T}^{-1}\mathbf{x}$ *is an eigenvector of* **B** *in* (7) *corresponding to the same eigenvalue.* (Proof in Sec. 7.15.)

Another theorem that has various applications in numerical work is

Theorem 3 (Spectral shift)
If **A** *has the eigenvalues* $\lambda_1, \cdots, \lambda_n$, *then* $\mathbf{A} - k\mathbf{I}$ *has the eigenvalues* $\lambda_1 - k, \cdots, \lambda_n - k$.

This theorem is a special case of the following **spectral mapping theorem**.

Theorem 4 (Polynomial matrices)
If λ *is an eigenvalue of* **A,** *then*

$$q(\lambda) = \alpha_s\lambda^s + \alpha_{s-1}\lambda^{s-1} + \cdots + \alpha_1\lambda + \alpha_0$$

is an eigenvalue of the polynomial matrix

$$q(\mathbf{A}) = \alpha_s\mathbf{A}^s + \alpha_{s-1}\mathbf{A}^{s-1} + \cdots + \alpha_1\mathbf{A} + \alpha_0\mathbf{I}.$$

Proof. $\mathbf{A}\mathbf{x} = \lambda\mathbf{x}$ implies $\mathbf{A}^2\mathbf{x} = \mathbf{A}\lambda\mathbf{x} = \lambda\mathbf{A}\mathbf{x} = \lambda^2\mathbf{x}$, $\mathbf{A}^3\mathbf{x} = \lambda^3\mathbf{x}$, etc. Thus

$$\mathbf{q}(\mathbf{A})\mathbf{x} = (\alpha_s\mathbf{A}^s + \alpha_{s-1}\mathbf{A}^{s-1} + \cdots)\mathbf{x}$$

$$= \alpha_s\mathbf{A}^s\mathbf{x} + \alpha_{s-1}\mathbf{A}^{s-1}\mathbf{x} + \cdots$$

$$= \alpha_s\lambda^s\mathbf{x} + \alpha_{s-1}\lambda^{s-1}\mathbf{x} + \cdots = q(\lambda)\mathbf{x}. \qquad \blacksquare$$

The eigenvalues of important special matrices can be characterized in a general way as follows.

Theorem 5 (Special matrices)

The eigenvalues of Hermitian matrices (i.e., $\overline{\mathbf{A}}^\mathsf{T} = \mathbf{A}$), hence of real symmetric matrices (i.e., $\mathbf{A}^\mathsf{T} = \mathbf{A}$), are real. The eigenvalues of skew-Hermitian matrices (i.e., $\overline{\mathbf{A}}^\mathsf{T} = -\mathbf{A}$), hence of real skew-symmetric matrices (i.e., $\mathbf{A}^\mathsf{T} = -\mathbf{A}$), are pure imaginary or 0. The eigenvalues of unitary matrices (i.e., $\overline{\mathbf{A}}^\mathsf{T} = \mathbf{A}^{-1}$), hence of orthogonal matrices (i.e., $\mathbf{A}^\mathsf{T} = \mathbf{A}^{-1}$), have absolute value 1. (Proof in Sec. 7.14.)

Choice of Numerical Methods

The choice of a numerical method for matrix eigenvalue problems depends essentially on two circumstances, on the kind of matrix (real symmetric, real general, complex, sparse or full) and on the kind of information to be obtained, that is, whether one wants to know all eigenvalues or merely specific ones, for instance, the largest eigenvalue, whether eigenvalues *and* eigenvectors are wanted, and so on. It is clear that we cannot enter into a systematic discussion of all these and further possibilities that arise in practice—look quickly into Ref. [E40] to get an idea—but we shall concentrate on some basic aspects and methods that will give us a good feeling and general understanding of this fascinating field.

19.7 Inclusion of Matrix Eigenvalues

The following interesting theorem by Gerschgorin yields a region consisting of closed circular disks and containing all the eigenvalues of a given matrix. Indeed, for each $j = 1, \cdots, n$ the inequality (1) in the theorem determines a closed circular disk in the complex λ-plane with center a_{jj} and radius given by the right-hand side of (1); and Theorem 1 states that each of the eigenvalues of $\mathbf{A}$ lies in one of these n disks.

Theorem 1 (Gerschgorin's theorem)

Let λ be an eigenvalue of an arbitrary $n \times n$ matrix $\mathbf{A} = [a_{jk}]$. Then for some integer j $(1 \leqq j \leqq n)$ we have

$$(1) \qquad |a_{jj} - \lambda| \leqq |a_{j1}| + |a_{j2}| + \cdots + |a_{j,j-1}| + |a_{j,j+1}| + \cdots + |a_{jn}|.$$

Proof. Let $\mathbf{x}$ be an eigenvector corresponding to that eigenvalue λ. Then

$$(2) \qquad\qquad \mathbf{Ax} = \lambda\mathbf{x} \qquad \text{or} \qquad (\mathbf{A} - \lambda\mathbf{I})\mathbf{x} = \mathbf{0}.$$

Let x_j be a component of $\mathbf{x}$ which is largest in absolute value. Then we have $|x_m/x_j| \leqq 1$ for $m = 1, \cdots, n$. The vector equation (2) is equivalent to a system of n equations for the n components of the vectors on both sides, and the jth of these n equations is

$$a_{j1}x_1 + \cdots + a_{j,j-1}x_{j-1} + (a_{jj} - \lambda)x_j + a_{j,j+1}x_{j+1} + \cdots + a_{jn}x_n = 0.$$

Division by x_j and reshuffling terms gives

$$a_{jj} - \lambda = -a_{j1}\frac{x_1}{x_j} - \cdots - a_{j,j-1}\frac{x_{j-1}}{x_j} - a_{j,j+1}\frac{x_{j+1}}{x_j} - \cdots - a_{jn}\frac{x_n}{x_j}.$$

By taking absolute values on both sides of this equation, applying the triangle inequality $|a + b| \leq |a| + |b|$ (where a and b are any complex numbers) and observing that

$$\left|\frac{x_1}{x_j}\right| \leq 1, \quad \cdots, \quad \left|\frac{x_n}{x_j}\right| \leq 1$$

we obtain (1), and the theorem is proved. ∎

EXAMPLE 1. Gerschgorin's theorem

For the eigenvalues of the matrix

$$\mathbf{A} = \begin{bmatrix} 0 & 1/2 & 1/2 \\ 1/2 & 5 & 1 \\ 1/2 & 1 & 1 \end{bmatrix}$$

we get the Gerschgorin disks (Fig. 434)

D_1: Center 0, radius 1

D_2: Center 5, radius 1.5

D_3: Center 1, radius 1.5.

Since $\mathbf{A}$ is symmetric, it follows from this and Theorem 5, Sec. 19.6, that the spectrum of $\mathbf{A}$ must actually lie in the intervals $[-1, 2.5]$ and $[3.5, 6.5]$ on the real axis. The student may confirm this by showing that, to 3 decimals, $\lambda_1 = -0.209$, $\lambda_2 = 5.305$, $\lambda_3 = 0.904$.

It is interesting that here the Gerschgorin disks form two disjoint sets, namely, $D_1 \cup D_3$, which contains two eigenvalues, and D_2, which contains one eigenvalue. This is typical, as the following theorem shows. ∎

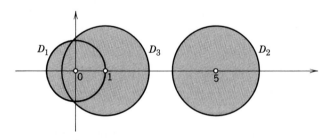

Fig. 434. Gerschgorin disks in Example 1

Theorem 2 (Extension of Gerschgorin's theorem)

*If p Gerschgorin disks form a set S that is disjoint from the n − p other disks of a given matrix **A**, then S contains precisely p eigenvalues of **A** (each counted with its algebraic multiplicity, as defined in Sec. 19.6).*

Proof. This is a "*continuity proof.*" Let $S = D_1 \cup D_2 \cup \cdots \cup D_p$ without restriction, where D_j is the Gerschgorin disk with center a_{jj}. We write $\mathbf{A} = \mathbf{B} + \mathbf{C}$, where $\mathbf{B} = \text{diag}(a_{jj})$ is the diagonal matrix with the main diagonal of $\mathbf{A}$ as its diagonal. We now consider

$$\mathbf{A}_t = \mathbf{B} + t\mathbf{C} \qquad\qquad \text{for } 0 \leqq t \leqq 1.$$

Then $\mathbf{A}_0 = \mathbf{B}$ and $\mathbf{A}_1 = \mathbf{A}$. Now by algebra, the roots of the characteristic polynomial $f_t(\lambda)$ of $\mathbf{A}_t$ (that is, the eigenvalues of $\mathbf{A}_t$) depend continuously on the coefficients of $f_t(\lambda)$, which in turn depend continuously on t. For $t = 0$, the eigenvalues are $a_{11}, \cdots, a_{nn}$. If we let t increase continuously from 0 to 1, the eigenvalues move continuously and, by Theorem 1, for each t lie in the Gerschgorin disks with centers a_{jj} and radii

$$tr_j \qquad \text{where} \qquad r_j = \sum_{k \neq j} |a_{jk}|.$$

Since at the end, S is disjoint from the other disks, the assertion follows. ∎

EXAMPLE 2. Another application of Gerschgorin's theorem. Similarity

Suppose that we have diagonalized a matrix by some numerical method that left us with some off-diagonal entries of size 10^{-5}, say,

$$\mathbf{A} = \begin{bmatrix} 2 & 10^{-5} & 10^{-5} \\ 10^{-5} & 2 & 10^{-5} \\ 10^{-5} & 10^{-5} & 4 \end{bmatrix}.$$

What can we conclude about deviations of the eigenvalues from the main diagonal entries?

Solution. By Theorem 2, one eigenvalue must lie in the disk of radius $2 \cdot 10^{-5}$ centered at 4 and two eigenvalues (or an eigenvalue of algebraic multiplicity 2) in the disk of radius $2 \cdot 10^{-5}$ centered at 2. Actually, since the matrix is symmetric, these eigenvalues must lie in the intersections of these disks and the real axis, by Theorem 5 in Sec. 19.6.

We show that and how an isolated disk can always be reduced in size by a similarity transformation. The matrix

$$\mathbf{B} = \mathbf{T}^{-1}\mathbf{A}\mathbf{T} = \begin{bmatrix} 1 & 0 & 0 \\ 0 & 1 & 0 \\ 0 & 0 & 10^{-5} \end{bmatrix} \begin{bmatrix} 2 & 10^{-5} & 10^{-5} \\ 10^{-5} & 2 & 10^{-5} \\ 10^{-5} & 10^{-5} & 4 \end{bmatrix} \begin{bmatrix} 1 & 0 & 0 \\ 0 & 1 & 0 \\ 0 & 0 & 10^{5} \end{bmatrix}$$

$$= \begin{bmatrix} 2 & 10^{-5} & 1 \\ 10^{-5} & 2 & 1 \\ 10^{-10} & 10^{-10} & 4 \end{bmatrix}$$

is similar to $\mathbf{A}$. Hence by Theorem 2, Sec. 19.6, it has the same eigenvalues as $\mathbf{A}$. From row 3 we get the smaller disk of radius $2 \cdot 10^{-10}$. Note that the other disks got bigger, approximately by a factor 10^5. And in choosing $\mathbf{T}$ we have to watch that the new disks do not overlap with the disk whose size we want to decrease. ∎

Further Inclusion Theorems

An **inclusion theorem** is a theorem that specifies a set which contains at least one eigenvalue of a given matrix. Thus, Theorems 1 and 2 are inclusion theorems; they even include the whole spectrum. We now discuss some famous theorems that yield inclusions of eigenvalues.

Theorem 3 (Schur's theorem[3])

Let $\mathbf{A} = [a_{jk}]$ *be an* $n \times n$ *matrix. Let* $\lambda_1, \cdots, \lambda_n$ *be its eigenvalues. Then*

$$(3) \qquad \sum_{i=1}^{n} |\lambda_i|^2 \leq \sum_{j=1}^{n} \sum_{k=1}^{n} |a_{jk}|^2 \qquad \textbf{(Schur's inequality).}$$

In (3) *the equality sign holds if and only if* $\mathbf{A}$ *is such that*

$$(4) \qquad \overline{\mathbf{A}}^{\mathsf{T}}\mathbf{A} = \mathbf{A}\overline{\mathbf{A}}^{\mathsf{T}}.$$

Matrices that satisfy (4) are called **normal matrices.** It is not difficult to see that Hermitian, skew-Hermitian and unitary matrices are normal, and so are real symmetric, skew-symmetric and orthogonal matrices.

Let λ_m be any eigenvalue of the matrix $\mathbf{A}$ in Theorem 3. Then $|\lambda_m|^2$ is less than or equal to the sum on the left-hand side of (3), and by taking square roots we obtain from (3)

$$(5) \qquad \boxed{|\lambda_m| \leq \sqrt{\sum_{j=1}^{n} \sum_{k=1}^{n} |a_{jk}|^2}.}$$

Note that the right-hand side is the Frobenius norm of $\mathbf{A}$ (cf. Sec. 19.3).

EXAMPLE 3. Bounds for eigenvalues obtained from Schur's inequality

For the matrix

$$\mathbf{A} = \begin{bmatrix} 26 & -2 & 2 \\ 2 & 21 & 4 \\ 4 & 2 & 28 \end{bmatrix}$$

we obtain from Schur's inequality

$$|\lambda| \leq \sqrt{1949} < 44.2.$$

(The eigenvalues of $\mathbf{A}$ are 30, 25 and 20; thus $30^2 + 25^2 + 20^2 = 1925 < 1949$; in fact, $\mathbf{A}$ is not normal.) ∎

The preceding theorems are valid for every real or complex square matrix. There are other theorems that hold for special classes of matrices only. The following theorem by Frobenius,[4] which we state without proof, is of this type.

---

[3]ISSAI SCHUR (1875—1941), German mathematician, professor in Berlin, also known by his important work in group theory.

[4]See the footnote in Sec. 4.4. For a proof of the theorem, see Ref. [B3], vol. II, pp. 53-62.

Theorem 4 (Perron–Frobenius's theorem)

Let **A** *be a real square matrix whose elements are all positive. Then* **A** *has at least one real positive eigenvalue* λ, *and the corresponding eigenvector can be chosen real and such that all its components are positive.*

From this theorem we may derive the following useful result by Collatz.[5]

Theorem 5 (Collatz's theorem)

Let **A** $= [a_{jk}]$ *be a real* $n \times n$ *matrix whose elements are all positive. Let* **x** *be any real vector whose components* $x_1, \cdots, x_n$ *are positive, and let* $y_1, \cdots, y_n$ *be the components of the vector* **y** $=$ **Ax**. *Then the closed interval on the real axis bounded by the smallest and the largest of the n quotients* $q_j = y_j/x_j$ *contains at least one eigenvalue of* **A**.

Proof. We have **Ax** $=$ **y** or

$$(6) \qquad\qquad \mathbf{y} - \mathbf{Ax} = \mathbf{0}.$$

The transpose $\mathbf{A}^{\mathsf{T}}$ satisfies the conditions of Theorem 4. Hence $\mathbf{A}^{\mathsf{T}}$ has a positive eigenvalue λ and, corresponding to this eigenvalue, an eigenvector **u** whose components u_j are all positive. Thus $\mathbf{A}^{\mathsf{T}}\mathbf{u} = \lambda\mathbf{u}$, and by taking the transpose we obtain $\mathbf{u}^{\mathsf{T}}\mathbf{A} = \lambda\mathbf{u}^{\mathsf{T}}$. From this and (6),

$$\mathbf{u}^{\mathsf{T}}(\mathbf{y} - \mathbf{Ax}) = \mathbf{u}^{\mathsf{T}}\mathbf{y} - \mathbf{u}^{\mathsf{T}}\mathbf{Ax} = \mathbf{u}^{\mathsf{T}}(\mathbf{y} - \lambda\mathbf{x}) = 0$$

or written out

$$\sum_{j=1}^{n} u_j(y_j - \lambda x_j) = 0.$$

Since all the components u_j are positive, it follows that

$$(7) \qquad \begin{array}{llll} y_j - \lambda x_j \geq 0, & \text{that is,} & q_j \geq \lambda & \text{for at least one } j, \\[2mm] y_j - \lambda x_j \leq 0, & \text{that is,} & q_j \leq \lambda & \text{for at least one } j. \end{array} \qquad \text{and}$$

Since **A** and $\mathbf{A}^{\mathsf{T}}$ have the same eigenvalues, λ is an eigenvalue of **A**, and from (7) the statement of the theorem follows.   ∎

EXAMPLE 4. Bounds for eigenvalues from Collatz's theorem

Let

$$\mathbf{A} = \begin{bmatrix} 8 & 1 & 1 \\ 1 & 5 & 2 \\ 1 & 2 & 5 \end{bmatrix}. \quad \text{Choose} \quad \mathbf{x} = \begin{bmatrix} 1 \\ 1 \\ 1 \end{bmatrix}. \quad \text{Then} \quad \mathbf{y} = \begin{bmatrix} 10 \\ 8 \\ 8 \end{bmatrix}.$$

Hence $q_1 = 10$, $q_2 = 8$, $q_3 = 8$, and Theorem 5 implies that one of the eigenvalues of **A** must lie in the interval $8 \leq \lambda \leq 10$. Of course, the length of such an interval depends on the choice of **x**. The student may show that $\lambda = 9$ is an eigenvalue of **A**.   ∎

[5]LOTHAR COLLATZ (born 1910), German mathematician, known for his work in numerical analysis.

In the next section we discuss the idea of computing eigenvalues by an iteration process, the **power method,** which is simple and natural (although sometimes somewhat slowly convergent).

Problems for Sec. 19.7

Using Theorem 1, determine and sketch disks that contain the eigenvalues of the following matrices.

1. $\begin{bmatrix} -1 & 2 & 3 \\ 2 & 2 & 6 \\ 3 & 6 & -1 \end{bmatrix}$
2. $\begin{bmatrix} 3.2 & 0.1 & 0 \\ 0.1 & 4.0 & -0.1 \\ 0 & -0.1 & 5.5 \end{bmatrix}$
3. $\begin{bmatrix} 26 & -2 & 2 \\ 2 & 21 & 4 \\ 4 & 2 & 28 \end{bmatrix}$

4. $\begin{bmatrix} 0 & 1 & 1 \\ 2 & 0 & 1 \\ 0 & 1 & 0 \end{bmatrix}$
5. $\begin{bmatrix} -2 & 10^{-3} & 10^{-3} \\ 10^{-4} & 1 & 10^{-3} \\ 10^{-4} & 10^{-4} & 4 \end{bmatrix}$
6. $\begin{bmatrix} 2 & 0.1i & -i \\ -0.1i & 0 & 0.1i \\ i & -0.1i & 4 \end{bmatrix}$

7. Show that the matrices in Probs. 1 and 4 have the eigenvalues 8, -2, -6 and 2, -1, -1, respectively.

8. Find a similarity transformation whose application to the following matrix **A** reduces the radius of the Gerschgorin circle with center 5 to 1/100 of its original value.

$$ A = \begin{bmatrix} 5 & 10^{-2} & 10^{-2} \\ 10^{-2} & 8 & 10^{-2} \\ 10^{-2} & 10^{-2} & 9 \end{bmatrix} $$

Using (5), obtain an upper bound for the absolute value of the eigenvalues of the matrix:

9. In Prob. 5
10. In Prob. 6
11. In Prob. 2

Apply Theorem 5 to the following matrices, choosing the given vectors as vectors **x**.

12. $\begin{bmatrix} 10 & 1 & 1 \\ 1 & 10 & 1 \\ 1 & 1 & 10 \end{bmatrix}$, $\begin{bmatrix} 1 \\ 2 \\ 1 \end{bmatrix}$, $\begin{bmatrix} 1 \\ 1 \\ 1 \end{bmatrix}$
13. $\begin{bmatrix} 40 & 8 & 1 \\ 8 & 41 & 8 \\ 1 & 8 & 40 \end{bmatrix}$, $\begin{bmatrix} 1 \\ 1 \\ 1 \end{bmatrix}$, $\begin{bmatrix} 1 \\ 2 \\ 1 \end{bmatrix}$, $\begin{bmatrix} 2 \\ 3 \\ 2 \end{bmatrix}$

14. **(Nonzero determinant)** If in each row of a determinant the entry on the main diagonal is larger in absolute value than the sum of the absolute values of the other entries in that row, show that the value of the determinant cannot be zero. What does this imply with respect to the solvability of systems of linear equations?

15. **(Inclusion set)** An *inclusion set* for a matrix **A** is a set in the complex plane containing at least one eigenvalue of **A**. What type of inclusion sets do we obtain for unitary matrices by combining Theorem 1 and Theorem 1c, Sec. 7.14?

16. Show that the matrix in Prob. 3 is not normal and has the eigenvalues 30, 25, 20.

17. Show that the matrix in Example 4 has the eigenvalues 9, 6, 3, and (3) holds with the equality sign.

18. Show that Hermitian, skew-Hermitian, and unitary matrices are normal.

19. Find a 2×2 matrix that is not normal.

20. Using Theorem 1 in Sec. 7.13 and Theorem 1 in Sec. 7.14, show that for a unitary matrix, Schur's inequality with the equality sign holds.

19.8 Eigenvalues by Iteration (Power Method)

A standard procedure for computing approximate values of the eigenvalues of an $n \times n$ matrix $\mathbf{A} = [a_{jk}]$ is the **power method.** In this method we start from any vector $\mathbf{x}_0$ ($\neq \mathbf{0}$) with n components and compute successively

$$\mathbf{x}_1 = \mathbf{A}\mathbf{x}_0, \qquad \mathbf{x}_2 = \mathbf{A}\mathbf{x}_1, \cdots, \qquad \mathbf{x}_s = \mathbf{A}\mathbf{x}_{s-1}.$$

For simplifying notation, we denote $\mathbf{x}_{s-1}$ by $\mathbf{x}$ and $\mathbf{x}_s$ by $\mathbf{y}$, so that $\mathbf{y} = \mathbf{A}\mathbf{x}$. If $\mathbf{A}$ is real symmetric, the following theorem gives an approximation and error bounds.

Theorem 1

Let $\mathbf{A}$ be an $n \times n$ real symmetric matrix. Let $\mathbf{x}$ ($\neq \mathbf{0}$) be any real vector with n components. Furthermore, let

$$\mathbf{y} = \mathbf{A}\mathbf{x}, \qquad m_0 = \mathbf{x}^\mathsf{T}\mathbf{x}, \qquad m_1 = \mathbf{x}^\mathsf{T}\mathbf{y}, \qquad m_2 = \mathbf{y}^\mathsf{T}\mathbf{y}.$$

Then the quotient

$$q = \frac{m_1}{m_0} \qquad \textbf{(Rayleigh quotient}[6]\textbf{)}$$

is an approximation for an eigenvalue[7] *λ of $\mathbf{A}$, and if we set $q = \lambda + \epsilon$, so that ϵ is the error of q, then*

(1)
$$|\epsilon| \leqq \sqrt{\frac{m_2}{m_0} - q^2}.$$

Proof. Let δ^2 denote the radicand in (1). Then, since $m_1 = qm_0$, we have

(2) $(\mathbf{y} - q\mathbf{x})^\mathsf{T}(\mathbf{y} - q\mathbf{x}) = m_2 - 2qm_1 + q^2m_0 = m_2 - q^2m_0 = \delta^2 m_0.$

---

[6]Cf. footnote 4 in the problem set for Sec. 3.2.

[7]Ordinarily that λ which is greatest in absolute value, but no general statements are possible.

Since $\mathbf{A}$ is real symmetric, it has an orthogonal set of n real unit eigenvectors $\mathbf{z}_1, \cdots, \mathbf{z}_n$ corresponding to the eigenvalues $\lambda_1, \cdots, \lambda_n$, respectively (some of which may be equal). (Proof in Ref. [B3], vol. 1, pp. 270-272, listed in Appendix 1.) Then $\mathbf{x}$ has a representation of the form

$$\mathbf{x} = a_1\mathbf{z}_1 + \cdots + a_n\mathbf{z}_n.$$

Now $\mathbf{A}\mathbf{z}_1 = \lambda_1\mathbf{z}_1$, etc., and we obtain

$$\mathbf{y} = \mathbf{A}\mathbf{x} = a_1\lambda_1\mathbf{z}_1 + \cdots + a_n\lambda_n\mathbf{z}_n$$

and, since the $\mathbf{z}_j$ are orthogonal unit vectors,

(3)
$$m_0 = \mathbf{x}^T\mathbf{x} = a_1{}^2 + \cdots + a_n{}^2.$$

It follows that in (2),

$$\mathbf{y} - q\mathbf{x} = a_1(\lambda_1 - q)\mathbf{z}_1 + \cdots + a_n(\lambda_n - q)\mathbf{z}_n.$$

Since the $\mathbf{z}_j$ are orthogonal unit vectors, we thus obtain from (2)

$$\delta^2 m_0 = a_1{}^2(\lambda_1 - q)^2 + \cdots + a_n{}^2(\lambda_n - q)^2.$$

Replacing each $(\lambda_j - q)^2$ by the smallest of these terms, we have from (3)

$$\delta^2 m_0 \geqq (\lambda_c - q)^2(a_1{}^2 + \cdots + a_n{}^2) = (\lambda_c - q)^2 m_0$$

where λ_c is an eigenvalue to which q is closest. Dividing this inequality by m_0 and taking square roots, we obtain (1), and the theorem is proved.   ∎

EXAMPLE 1. An application of Theorem 1
As in Example 4 of the preceding section, let us consider the real symmetric matrix

$$\mathbf{A} = \begin{bmatrix} 8 & 1 & 1 \\ 1 & 5 & 2 \\ 1 & 2 & 5 \end{bmatrix} \quad \text{and choose} \quad \mathbf{x}_0 = \begin{bmatrix} 1 \\ 1 \\ 1 \end{bmatrix}.$$

Then we obtain successively

$$\mathbf{x}_1 = \begin{bmatrix} 10 \\ 8 \\ 8 \end{bmatrix}, \quad \mathbf{x}_2 = \begin{bmatrix} 96 \\ 66 \\ 66 \end{bmatrix}, \quad \mathbf{x}_3 = \begin{bmatrix} 900 \\ 558 \\ 558 \end{bmatrix}, \quad \mathbf{x}_4 = \begin{bmatrix} 8316 \\ 4806 \\ 4806 \end{bmatrix}.$$

Taking $\mathbf{x} = \mathbf{x}_3$ and $\mathbf{y} = \mathbf{x}_4$, we have

$$m_0 = \mathbf{x}^T\mathbf{x} = 1\,432\,728, \quad m_1 = \mathbf{x}^T\mathbf{y} = 12\,847\,896, \quad m_2 = \mathbf{y}^T\mathbf{y} = 115\,351\,128.$$

From this we calculate

$$q = \frac{m_1}{m_0} = 8.967, \quad |\epsilon| \leqq \sqrt{\frac{m_2}{m_0} - q^2} = 0.311.$$

This shows that $q = 8.967$ is an approximation for an eigenvalue that must lie between 8.656 and 9.278. The student may show that an eigenvalue is $\lambda = 9$.   ∎

Suppose that we have determined an eigenvalue of a given matrix by the power method and want to obtain further eigenvalues of this matrix. What can we do? This problem will be discussed in the next section, devoted to **deflation.**

Problems for Sec. 19.8

Choosing $x_0 = [1\ \ 1]^T$ or $x_0 = [1\ \ 1\ \ 1]^T$, respectively, apply the power method (3 steps) to the following matrices, computing in each step the Rayleigh quotient and an error bound.

1.
$$\begin{bmatrix} 5 & 2 \\ 2 & 1 \end{bmatrix}$$

2.
$$\begin{bmatrix} 1 & 2 \\ 2 & 7 \end{bmatrix}$$

3.
$$\begin{bmatrix} 0.6 & 0.8 \\ 0.8 & -0.6 \end{bmatrix}$$

4.
$$\begin{bmatrix} 10 & 1 & 0 \\ 1 & 5 & 2 \\ 0 & 2 & 0 \end{bmatrix}$$

5.
$$\begin{bmatrix} -2 & 2 & 3 \\ 2 & 1 & 6 \\ 3 & 6 & -2 \end{bmatrix}$$

6.
$$\begin{bmatrix} 3 & 2 & 3 \\ 2 & 6 & 6 \\ 3 & 6 & 3 \end{bmatrix}$$

7. Compute the exact values of the eigenvalues in Prob. 1. Verify that the actual errors of the approximations are much smaller than the error bounds in Prob. 1. This is typical.

Apply the power method (3 steps) with different x_0 (below) to the matrix

$$A = \begin{bmatrix} 2 & -1 & 1 \\ -1 & 3 & 2 \\ 1 & 2 & 3 \end{bmatrix},$$

computing the Rayleigh quotient and error bound corresponding to the third step. Verify that A has the eigenvalues 5, 3, 0 and indicate which eigenvalue is approximated by the result.

8. $[0\ \ 1\ \ 0]^T$

9. $[1\ \ 1\ \ 1]^T$

10. $[0\ \ 1\ \ 1]^T$

Choosing $x_0 = [1\ \ 1\ \ 1\ \ 1]^T$, compute x_1, x_2 and approximations $q = x_1^T x_0 / x_0^T x_0$, $q = x_2^T x_1 / x_1^T x_1$ and corresponding error bounds for an eigenvalue of each of the following matrices.

11.
$$\begin{bmatrix} 1 & 0 & 0 & 1 \\ 0 & 2 & -1 & 0 \\ 0 & -1 & 3 & 0 \\ 1 & 0 & 0 & -1 \end{bmatrix}$$

12.
$$\begin{bmatrix} 3 & 2 & 0 & 1 \\ 2 & 0 & 5 & -1 \\ 0 & 5 & 2 & 1 \\ 1 & -1 & 1 & 4 \end{bmatrix}$$

13.
$$\begin{bmatrix} 2 & 0 & 1 & 0 \\ 0 & 0 & 3 & 1 \\ 1 & 3 & 4 & -2 \\ 0 & 1 & -2 & 0 \end{bmatrix}$$

14. To understand the importance of the error bound (1), consider the matrix

$$A = \begin{bmatrix} 3 & 4 \\ 4 & -3 \end{bmatrix}, \qquad \text{choose} \qquad x_0 = \begin{bmatrix} 3 \\ -1 \end{bmatrix}$$

and show that $q = 0$ for all s. Find the eigenvalues and explain what happened. Start again, choosing another x_0.

15. To understand why the Rayleigh quotient q is generally an approximation of that eigenvalue, λ_1, which is largest in absolute value, let

$$\mathbf{x}_0 = c_1 \mathbf{z}_1 + \cdots + c_n \mathbf{z}_n$$

(with $\mathbf{z}_1, \cdots, \mathbf{z}_n$ as in the proof of Theorem 1) and show that

$$\mathbf{x} = \mathbf{x}_{s-1} = c_1 \lambda_1^{s-1} \mathbf{z}_1 + \cdots + c_n \lambda_n^{s-1} \mathbf{z}_n,$$

$$\mathbf{y} = \mathbf{x}_s = c_1 \lambda_1^s \mathbf{z}_1 + \cdots + c_n \lambda_n^s \mathbf{z}_n,$$

$$q = \frac{m_1}{m_0} = \frac{c_1^2 \lambda_1^{2s-1} + \cdots}{c_1^2 \lambda_1^{2s-2} + \cdots} \approx \lambda_1.$$

Under what conditions will this be a good approximation?

16. **(Spectral shift)** Show that the matrix in Prob. 6 is $\mathbf{A} + 5\mathbf{I}$, where $\mathbf{A}$ is the matrix in Prob. 5. Verify that $\mathbf{A}$ has the eigenvalues $7, -3, -7$. What are the eigenvalues of $\mathbf{A} + 5\mathbf{I}$, and why can we expect an improvement of convergence by the "spectral shift" (cf. Sec. 19.6) from $\mathbf{A}$ to $\mathbf{A} + 5\mathbf{I}$?

17. Show that the matrix $\mathbf{A}$ in Example 1 has the eigenvalues $\lambda_1 = 9, \lambda_2 = 6, \lambda_3 = 3$. Apply the power method to $\mathbf{A} - 4.5\mathbf{I}$, starting from $\mathbf{x}_0 = [1 \quad 1 \quad 1]^\mathsf{T}$ and taking $\mathbf{x} = \mathbf{x}_3$ and $\mathbf{y} = \mathbf{x}_4$ in (1). Show that $q = 4.4995$, so that 8.9995 approximates λ_1, with an error $|\epsilon| \leq 0.0393$. Can you explain the reason for the great improvement over the result in Example 1? (For more details about improving the power method, see [E39] in Appendix 1.)

18. Let $\mathbf{A}$ be symmetric, with eigenvalues $\lambda_1 > \lambda_2 \geq \cdots \geq \lambda_{n-1} > \lambda_n$, assuming $|\lambda_1| > |\lambda_n|$, and let α be a good estimate of λ_1. Then the power method applied to $\mathbf{B} = \mathbf{A} - \alpha\mathbf{I}$ will in general yield approximations to λ_n. Why is this so, and what does "in general" mean? Apply the method to $\mathbf{A}$ in Probs. 8–10, using $\alpha = 4.9$ (as suggested by the answer to Prob. 9) and $\mathbf{x}_0^\mathsf{T} = [1 \quad 1 \quad 1]$ and taking $\mathbf{x}_1$ for $\mathbf{x}$ and $\mathbf{x}_2$ for $\mathbf{y}$ in (1).

19. Show that if $\mathbf{x}$ is an eigenvector, then $\epsilon = 0$ in (1).

20. Can we use (1) in connection with the matrices in Sec. 19.7, Probs. 1–3?

19.9 Deflation of a Matrix

In practice, it often happens that we have computed or guessed one of the eigenvalues of a given matrix $\mathbf{A}$, call it λ_1, and we want to find the further eigenvalues $\lambda_2, \cdots, \lambda_n$ of $\mathbf{A}$. We show that then we can obtain from $\mathbf{A}$ a matrix $\mathbf{A}_1$ whose spectrum consists of 0 and the (still unknown) eigenvalues $\lambda_2, \cdots, \lambda_n$ of $\mathbf{A}$. The process of replacing $\mathbf{A}$ by $\mathbf{A}_1$ in the further determination of those eigenvalues is called **deflation of A.** We discuss the popular Wielandt's method. A simpler but numerically poor method is mentioned in Prob. 15, below.

Wielandt's Deflation[8]

Let $\mathbf{A}$ be any real $n \times n$ matrix and λ_1 a known eigenvalue of $\mathbf{A}$, as before. Let $\mathbf{x}_1$ be an eigenvector of $\mathbf{A}$ corresponding to λ_1. Let $\mathbf{u}$ be any vector (to be determined later) such that

$$(1) \qquad \mathbf{u}^\mathsf{T}\mathbf{x}_1 = \lambda_1.$$

Consider

$$(2) \qquad \mathbf{A}_1 = \mathbf{A} - \mathbf{x}_1\mathbf{u}^\mathsf{T}.$$

We first show that 0 is an eigenvalue of $\mathbf{A}_1$ and $\mathbf{x}_1$ is a corresponding eigenvector. Indeed, by (1),

$$\mathbf{A}_1\mathbf{x}_1 = \mathbf{A}\mathbf{x}_1 - \mathbf{x}_1\mathbf{u}^\mathsf{T}\mathbf{x}_1 = \lambda_1\mathbf{x}_1 - \lambda_1\mathbf{x}_1 = 0.$$

Let λ_j, $j = 2, \cdots, n$, be the other eigenvalues of $\mathbf{A}$, with corresponding eigenvectors $\mathbf{x}_j$. We show that $\lambda_j \neq 0$ is an eigenvalue of $\mathbf{A}_1$, with corresponding eigenvector

$$(3) \qquad \mathbf{y}_j = \mathbf{x}_j - \frac{\mathbf{u}^\mathsf{T}\mathbf{x}_j}{\lambda_j}\mathbf{x}_1.$$

Using the definitions and multiplying out, we first get

$$\mathbf{A}_1\mathbf{y}_j = (\mathbf{A} - \mathbf{x}_1\mathbf{u}^\mathsf{T})\left(\mathbf{x}_j - \frac{\mathbf{u}^\mathsf{T}\mathbf{x}_j}{\lambda_j}\mathbf{x}_1\right)$$

$$= \mathbf{A}\mathbf{x}_j - \frac{\mathbf{u}^\mathsf{T}\mathbf{x}_j}{\lambda_j}\mathbf{A}\mathbf{x}_1 - \mathbf{x}_1(\mathbf{u}^\mathsf{T}\mathbf{x}_j) + \frac{\mathbf{u}^\mathsf{T}\mathbf{x}_j}{\lambda_j}\mathbf{x}_1\mathbf{u}^\mathsf{T}\mathbf{x}_1.$$

Now on the right, $\mathbf{A}\mathbf{x}_1 = \lambda_1\mathbf{x}_1$ in the second term, and $\mathbf{x}_1\mathbf{u}^\mathsf{T}\mathbf{x}_1 = \mathbf{x}_1\lambda_1$ in the fourth term [by (1)], so that these two terms cancel. Hence we are left with

$$\mathbf{A}_1\mathbf{y}_j = \lambda_j\mathbf{x}_j - \mathbf{x}_1(\mathbf{u}^\mathsf{T}\mathbf{x}_j) = \lambda_j\mathbf{y}_j.$$

This shows that λ_j is an eigenvalue of $\mathbf{A}_1$.

We show how to get $\mathbf{u}$ in practice, and we shall also see that our assumption $\lambda_j \neq 0$ is no handicap to the method. We have $\mathbf{A}\mathbf{x}_1 = \lambda_1\mathbf{x}_1$. We can easily find $\mathbf{v}$ such that $\mathbf{v}^\mathsf{T}\mathbf{x}_1 = 1$. Then

$$\mathbf{v}^\mathsf{T}\mathbf{A}\mathbf{x}_1 = \mathbf{v}^\mathsf{T}\lambda_1\mathbf{x}_1 = \lambda_1\mathbf{v}^\mathsf{T}\mathbf{x}_1 = \lambda_1.$$

Because of (1) this shows that we can take

$$\mathbf{u}^\mathsf{T} = \mathbf{v}^\mathsf{T}\mathbf{A}.$$

[8]HELMUT WIELANDT (born 1910), German mathematician.

Then (2) becomes

(4) $$\mathbf{A}_1 = \mathbf{A} - \mathbf{x}_1 \mathbf{v}^T \mathbf{A}$$

and in (3) we now get rid of λ_j, since $\mathbf{A}\mathbf{x}_j = \lambda_j \mathbf{x}_j$ and thus

$$\mathbf{y}_j = \mathbf{x}_j - \frac{\mathbf{v}^T \mathbf{A}\mathbf{x}_j}{\lambda_j} \mathbf{x}_j = \mathbf{x}_j - \mathbf{v}^T \mathbf{x}_j \mathbf{x}_1.$$

Furthermore, if $\mathbf{x}_1$ has a nonzero first component, we can normalize $\mathbf{x}_1$ to make its first component 1. Then we can simply take the vector $\mathbf{v}^T = \mathbf{e}_1{}^T = [1 \quad 0 \quad \cdots \quad 0]$. For this choice,

(5) $$\boxed{\mathbf{A}_1 = (\mathbf{I} - \mathbf{x}_1 \mathbf{e}_1{}^T)\mathbf{A}.}$$

EXAMPLE 1. Wielandt's deflation

Find the spectrum of the matrix

$$\mathbf{A} = \begin{bmatrix} 8 & -2 & 2 \\ -2 & 6 & -4 \\ 2 & -4 & 6 \end{bmatrix}$$

assuming that one of the eigenvalues, $\lambda_1 = 12$, is known.

Solution. We first compute an eigenvector $\mathbf{x}_1$ of $\mathbf{A}$ corresponding to $\lambda_1 = 12$ by the Gauss elimination, finding $\mathbf{x}_1 = [1 \quad -1 \quad 1]^T$. In Wielandt's method we can now take the vector $\mathbf{v}_1{}^T = \mathbf{e}_1{}^T = [1 \quad 0 \quad 0]$ since then $\mathbf{v}_1{}^T \mathbf{x}_1 = 1$. We compute

$$\mathbf{x}_1 \mathbf{e}_1{}^T = \begin{bmatrix} 1 \\ -1 \\ 1 \end{bmatrix} \begin{bmatrix} 1 & 0 & 0 \end{bmatrix} = \begin{bmatrix} 1 & 0 & 0 \\ -1 & 0 & 0 \\ 1 & 0 & 0 \end{bmatrix}$$

and from this, by (5),

$$\mathbf{A}_1 = (\mathbf{I} - \mathbf{x}_1 \mathbf{e}_1{}^T)\mathbf{A} = \begin{bmatrix} 0 & 0 & 0 \\ 1 & 1 & 0 \\ -1 & 0 & 1 \end{bmatrix} \begin{bmatrix} 8 & -2 & 2 \\ -2 & 6 & -4 \\ 2 & -4 & 6 \end{bmatrix} = \begin{bmatrix} 0 & 0 & 0 \\ 6 & 4 & -2 \\ -6 & -2 & 4 \end{bmatrix}$$

We can now determine the other two eigenvalues directly from $\det (\mathbf{A}_1 - \mathbf{I}) = 0$, where

$$\det (\mathbf{A}_1 - \lambda\mathbf{I}) = \begin{vmatrix} -\lambda & 0 & 0 \\ 6 & 4 - \lambda & -2 \\ -6 & -2 & 4 - \lambda \end{vmatrix} = -\lambda \begin{vmatrix} 4 - \lambda & -2 \\ -2 & 4 - \lambda \end{vmatrix} = -\lambda(\lambda^2 - 8\lambda + 12).$$

We obtain $\lambda_2 = 6$, $\lambda_3 = 2$.

In the case of a larger matrix, the next step would have been the computation of another eigenvalue (for instance, by the power method) followed by another application of Wielandt's method, and so on. ∎

In the next section we show that for determining the whole spectrum of a matrix there are better methods than the successive application of deflations.

Problems for Sec. 19.9

1. Verify that $\mathbf{A}_1\mathbf{x}_1 = \mathbf{0}$ in Example 1 in the text.

Apply Wielandt's method of deflation to the following matrices. Show the deflated matrix and determine the further eigenvalues.

2. $\begin{bmatrix} 2 & 1 & 1 \\ 1 & 2 & 1 \\ 1 & 1 & 2 \end{bmatrix}$, $\mathbf{x}_1 = \begin{bmatrix} 1 \\ 1 \\ 1 \end{bmatrix}$

3. $\begin{bmatrix} 1 & 2 & -1 \\ -2 & 3 & 1 \\ -3 & 8 & 1 \end{bmatrix}$, $\mathbf{x}_1 = \begin{bmatrix} 5 \\ 1 \\ 7 \end{bmatrix}$

4. $\begin{bmatrix} 9 & -1 & -5 \\ -4 & 5 & -2 \\ 18 & -5 & -6 \end{bmatrix}$, $\mathbf{x}_1 = \begin{bmatrix} 1 \\ 2 \\ 1 \end{bmatrix}$

5. $\begin{bmatrix} 31 & 16 & 72 \\ -24 & -12 & -57 \\ -8 & -4 & -19 \end{bmatrix}$, $\mathbf{x}_1 = \begin{bmatrix} 4 \\ -3 \\ -1 \end{bmatrix}$

6. $\begin{bmatrix} 8 & 1 & 1 \\ 2 & 8 & 1 \\ 0 & 1 & 8 \end{bmatrix}$, $\mathbf{x}_1 = \begin{bmatrix} 3 \\ 4 \\ 2 \end{bmatrix}$

7. $\begin{bmatrix} -1 & 2 & 3 \\ 2 & 2 & 6 \\ 3 & 6 & -1 \end{bmatrix}$, $\mathbf{x}_1 = \begin{bmatrix} 3 \\ 6 \\ 5 \end{bmatrix}$

8. $\begin{bmatrix} 6 & -2 & 2 \\ 2 & 1 & 4 \\ 4 & 2 & 8 \end{bmatrix}$, $\mathbf{x}_1 = \begin{bmatrix} 1 \\ 2 \\ 4 \end{bmatrix}$

9. $\begin{bmatrix} 11 & 7 & 7 \\ 7 & 11 & 7 \\ 7 & 7 & 11 \end{bmatrix}$, $\mathbf{x}_1 = \begin{bmatrix} 1 \\ 1 \\ 1 \end{bmatrix}$

10. $\begin{bmatrix} 1 & 4 & 4 & 1 \\ 4 & 1 & 1 & 4 \\ 4 & 1 & 1 & 4 \\ 1 & 4 & 4 & 1 \end{bmatrix}$, $\mathbf{x}_1 = \begin{bmatrix} 1 \\ 1 \\ 1 \\ 1 \end{bmatrix}$

11. $\begin{bmatrix} 4 & 4 & 1 & 1 \\ 4 & 4 & 1 & 1 \\ 1 & 1 & 3 & 2 \\ 1 & 1 & 2 & 3 \end{bmatrix}$, $\mathbf{x}_1 = \begin{bmatrix} 2 \\ 2 \\ 1 \\ 1 \end{bmatrix}$

12. $\begin{bmatrix} -6 & 4 & 4 & 4 \\ -6 & 4 & 4 & 4 \\ -4 & 0 & 6 & 4 \\ -2 & 0 & 0 & 8 \end{bmatrix}$, $\mathbf{x}_1 = \begin{bmatrix} 1 \\ 1 \\ 1 \\ 1 \end{bmatrix}$

13. $\begin{bmatrix} -12 & -8 & 16 & 40 \\ -12 & -8 & 16 & 40 \\ -8 & -16 & 20 & 40 \\ -4 & -8 & -8 & 56 \end{bmatrix}$, $\mathbf{x}_1 = \begin{bmatrix} 1 \\ 1 \\ 1 \\ 1 \end{bmatrix}$

14. Deflate the following matrix, using $\mathbf{x}_1 = [1 \quad -1 \quad 1 \quad 1]^T$. Then find by inspection that the resulting 3×3 matrix has an eigenvalue 0. Compute a corresponding eigenvector and deflate again.

$$\begin{bmatrix} 2 & -5 & 0 & 3 \\ 0 & 2 & -3 & -5 \\ 5 & -3 & 2 & 0 \\ 3 & 0 & 5 & 2 \end{bmatrix}$$

15. **(Hotelling's deflation)** Let $\mathbf{A}$ be symmetric, λ_1 a known eigenvalue of $\mathbf{A}$ and $\mathbf{x}_1$ a corresponding normalized eigenvector, $\mathbf{x}_1{}^T\mathbf{x}_1 = 1$. Show that

$$\mathbf{A}_1 = \mathbf{A} - \lambda_1\mathbf{x}_1\mathbf{x}_1{}^T \qquad \text{(Hotelling's deflation)}$$

has the eigenvalues $\lambda_1 = 0$ and the others equal to those of $\mathbf{A}$. (This method is numerically poor with respect to round-off.) *Hint.* Use Theorem 4, Sec. 7.15.

19.10 Householder Tridiagonalization and QR-Factorization

We consider the problem of computing *all* the eigenvalues of a *real symmetric* matrix $\mathbf{A} = [a_{jk}]$. Successive deflations would not be good, because of round-off error growth. We discuss a method widely used in practice. In the first stage we apply Householder's method,[9] which reduces the given matrix to a **tridiagonal matrix,** that is, a matrix having all its nonzero entries on the main diagonal and in the positions immediately adjacent to the main diagonal (such as $\mathbf{A}_3$ in Fig. 435). In the second stage, the tridiagonal matrix is factorized in the form **QR,** where $\mathbf{Q}$ is orthogonal and $\mathbf{R}$ upper triangular, and the eigenvalues are actually determined (approximately); we discuss this afterward. (For extensions to general matrices, see Ref. [E40] listed in Appendix 1.)

Householder's Method

This method reduces a given real *symmetric* $n \times n$ matrix $\mathbf{A} = [a_{jk}]$ by $n - 2$ successive similarity transformations (cf. Sec. 19.6) to tridiagonal form. The matrices $\mathbf{P}_1, \mathbf{P}_2, \cdots, \mathbf{P}_{n-2}$ are orthogonal and symmetric matrices. Hence $\mathbf{P}_1{}^{-1} = \mathbf{P}_1{}^T = \mathbf{P}_1$ and similarly for the others. The $n - 2$ similarity transformations that produce from the given $\mathbf{A}_0 = \mathbf{A} = [a_{jk}]$ successively the matrices $\mathbf{A}_1 = [a_{jk}^{(1)}]$, $\mathbf{A}_2 = [a_{jk}^{(2)}]$, etc. look as follows.

(1)
$$\mathbf{A}_1 = \mathbf{P}_1\mathbf{A}_0\mathbf{P}_1$$
$$\mathbf{A}_2 = \mathbf{P}_2\mathbf{A}_1\mathbf{P}_2$$
$$\cdots \cdots \cdots \cdots$$
$$\mathbf{B} = \mathbf{A}_{n-2} = \mathbf{P}_{n-2}\mathbf{A}_{n-3}\mathbf{P}_{n-2}.$$

These transformations create the necessary zeros, in the first step in row 1 and column 1, in the second step in row 2 and column 2, etc., as Fig. 435 illustrates for a 5×5 matrix. $\mathbf{B}$ is tridiagonal.

How to determine $\mathbf{P}_1, \mathbf{P}_2, \cdots$? All these $\mathbf{P}_r$ are of the form

(2)
$$\mathbf{P}_r = \mathbf{I} - 2\mathbf{v}_r\mathbf{v}_r{}^T \qquad r = 1, \cdots, n - 2$$

[9] *Journal of the Association for Computing Machinery* **5**, (1958), 335-342. See also Refs. [E37], [E39], etc. in Appendix 1.

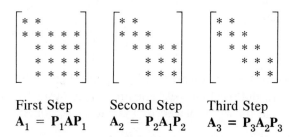

First Step   Second Step   Third Step
$$A_1 = P_1 A P_1 \qquad A_2 = P_2 A_1 P_2 \qquad A_3 = P_3 A_2 P_3$$

Fig. 435. Householder's method for a 5×5 matrix
Positions left blank are zeros created by the method.

where $\mathbf{v}_r = [v_{jr}]$ is a unit vector with its first r components 0; thus

$$(3) \qquad \mathbf{v}_1 = \begin{bmatrix} 0 \\ * \\ * \\ \vdots \\ * \end{bmatrix}, \qquad \mathbf{v}_2 = \begin{bmatrix} 0 \\ 0 \\ * \\ \vdots \\ * \end{bmatrix}, \qquad \cdots, \qquad \mathbf{v}_{n-2} = \begin{bmatrix} 0 \\ 0 \\ \vdots \\ * \\ * \end{bmatrix}$$

where the asterisks denote any other components. $\mathbf{v}_1$ has the components

$$(4) \qquad
\begin{cases}
\text{(a)} & v_{11} = 0 \\[2mm]
& v_{21} = \sqrt{\dfrac{1}{2}\left(1 + \dfrac{|a_{21}|}{S_1}\right)} \\[4mm]
\text{(b)} & v_{j1} = \dfrac{a_{j1}\ \text{sgn}\ a_{21}}{2 v_{21} S_1} \qquad j = 3, 4, \cdots, n \\[4mm]
& \text{where} \\[2mm]
\text{(c)} & S_1 = \sqrt{a_{21}{}^2 + a_{31}{}^2 + \cdots + a_{n1}{}^2}
\end{cases}$$

where $S_1 > 0$, and sgn $a_{21} = +1$ if $a_{21} \geqq 0$ and sgn $a_{21} = -1$ if $a_{21} < 0$. With this we compute $\mathbf{P}_1$ by (2) and then $\mathbf{A}_1$ by (1). This was the first step.

In the second step we compute $\mathbf{v}_2$ by (4) with all subscripts increased by 1 and the a_{jk} replaced by $a_{jk}^{(1)}$, the entries of $\mathbf{A}_1$ just computed. Thus [cf. also (3)]

$$v_{12} = v_{22} = 0$$

$$(4^*) \qquad v_{32} = \sqrt{\dfrac{1}{2}\left(1 + \dfrac{|a_{32}^{(1)}|}{S_2}\right)}$$

$$v_{j2} = \dfrac{a_{j2}^{(1)}\ \text{sgn}\ a_{32}^{(1)}}{2 v_{32} S_2} \qquad j = 4, 5, \cdots, n$$

where

$$S_2 = \sqrt{a_{32}^{(1)^2} + a_{42}^{(1)^2} + \cdots + a_{n2}^{(1)^2}}.$$

With this we compute $\mathbf{P}_2$ by (2) and then $\mathbf{A}_2$ by (1).

In the third step we compute $\mathbf{v}_3$ by (4*) with all subscripts increased by 1 and the $a_{jk}^{(1)}$ replaced by the entries $a_{jk}^{(2)}$ of $\mathbf{A}_2$, and so on.

EXAMPLE 1. Householder's method

Tridiagonalize the real symmetric matrix

$$\mathbf{A} = \mathbf{A}_0 = \begin{bmatrix} 6 & 4 & 1 & 1 \\ 4 & 6 & 1 & 1 \\ 1 & 1 & 5 & 2 \\ 1 & 1 & 2 & 5 \end{bmatrix}.$$

Solution. First Step. We compute $S_1^2 = 4^2 + 1^2 + 1^2 = 18$ from (4c). Since $a_{21} = 4 > 0$, we have sgn $a_{21} = +1$ in (4b) and get from (4) by straightforward computation

$$\mathbf{v}_1 = \begin{bmatrix} 0 \\ v_{21} \\ v_{31} \\ v_{41} \end{bmatrix} = \begin{bmatrix} 0 \\ 0.985\ 598\ 56 \\ 0.119\ 573\ 16 \\ 0.119\ 573\ 16 \end{bmatrix}.$$

From this and (2),

$$\mathbf{P}_1 = \begin{bmatrix} 1 & 0 & 0 & 0 \\ 0 & -0.942\ 809\ 04 & -0.235\ 702\ 26 & -0.235\ 702\ 26 \\ 0 & -0.235\ 702\ 26 & 0.971\ 404\ 52 & -0.028\ 595\ 48 \\ 0 & -0.235\ 702\ 26 & -0.028\ 595\ 48 & 0.971\ 404\ 52 \end{bmatrix}.$$

From the first line in (1) we now get

$$\mathbf{A}_1 = \mathbf{P}_1\mathbf{A}_0\mathbf{P}_1 = \begin{bmatrix} 6 & -\sqrt{18} & 0 & 0 \\ -\sqrt{18} & 7 & -1 & -1 \\ 0 & -1 & 9/2 & 3/2 \\ 0 & -1 & 3/2 & 9/2 \end{bmatrix}.$$

Second Step. From (4*) we compute $S_2^2 = 2$ and

$$\mathbf{v}_2 = \begin{bmatrix} 0 \\ 0 \\ v_{32} \\ v_{42} \end{bmatrix} = \begin{bmatrix} 0 \\ 0 \\ 0.923\ 879\ 53 \\ 0.382\ 683\ 43 \end{bmatrix}.$$

From this and (2),

$$\mathbf{P}_2 = \begin{bmatrix} 1 & 0 & 0 & 0 \\ 0 & 1 & 0 & 0 \\ 0 & 0 & -1/\sqrt{2} & -1/\sqrt{2} \\ 0 & 0 & -1/\sqrt{2} & 1/\sqrt{2} \end{bmatrix}.$$

The second line in (1) now gives

$$\mathbf{B} = \mathbf{A}_2 = \mathbf{P}_2\mathbf{A}_1\mathbf{P}_2 = \begin{bmatrix} 6 & -\sqrt{18} & 0 & 0 \\ -\sqrt{18} & 7 & \sqrt{2} & 0 \\ 0 & \sqrt{2} & 6 & 0 \\ 0 & 0 & 0 & 3 \end{bmatrix}.$$

This matrix **B** is tridiagonal. Since our given matrix has order $n = 4$, we needed $n - 2 = 2$ steps to accomplish this reduction, as claimed. (Do you see that we got more zeros than we can expect in general?) **B** is similar to **A**, as we now show in general. This is essential because **B** thus has the same spectrum as **A**, by Theorem 2 in Sec. 19.6.   ∎

We show that **B** in (1) is similar to $\mathbf{A} = \mathbf{A}_0$. The matrix $\mathbf{P}_r$ is symmetric,

$$\mathbf{P}_r^{\mathsf{T}} = (\mathbf{I} - 2\mathbf{v}_r\mathbf{v}_r^{\mathsf{T}})^{\mathsf{T}} = \mathbf{I}^{\mathsf{T}} - 2(\mathbf{v}_r\mathbf{v}_r^{\mathsf{T}})^{\mathsf{T}} = \mathbf{I} - 2\mathbf{v}_r\mathbf{v}_r^{\mathsf{T}} = \mathbf{P}_r.$$

Also, $\mathbf{P}_r$ is orthogonal because

$$\mathbf{P}_r\mathbf{P}_r^{\mathsf{T}} = \mathbf{P}_r^2 = (\mathbf{I} - 2\mathbf{v}_r\mathbf{v}_r^{\mathsf{T}})^2 = \mathbf{I} - 4\mathbf{v}_r\mathbf{v}_r^{\mathsf{T}} + 4\mathbf{v}_r\mathbf{v}_r^{\mathsf{T}}\mathbf{v}_r\mathbf{v}_r^{\mathsf{T}}$$

and $\mathbf{v}_r^{\mathsf{T}}\mathbf{v}_r = 1$ in the last term because $\mathbf{v}_r$ is a unit vector (see above), so that the right-hand side reduces to **I**. This gives $\mathbf{P}_r^{-1} = \mathbf{P}_r^{\mathsf{T}} = \mathbf{P}_r$. Hence from (1) we now obtain

$$\mathbf{B} = \mathbf{P}_{n-2}\mathbf{A}_{n-3}\mathbf{P}_{n-2} = \cdots$$

$$\cdots = \mathbf{P}_{n-2}\mathbf{P}_{n-3} \cdots \mathbf{P}_1\mathbf{A}\mathbf{P}_1 \cdots \mathbf{P}_{n-3}\mathbf{P}_{n-2}$$

$$= \mathbf{P}_{n-2}^{-1}\mathbf{P}_{n-3}^{-1} \cdots \mathbf{P}_1^{-1}\mathbf{A}\mathbf{P}_1 \cdots \mathbf{P}_{n-3}\mathbf{P}_{n-2}$$

$$= \mathbf{T}^{-1}\mathbf{A}\mathbf{T}$$

where $\mathbf{T} = \mathbf{P}_1\mathbf{P}_2 \cdots \mathbf{P}_{n-2}$. This proves our assertion.

QR-Factorization Method

In 1958 H. Rutishauser proposed the idea of using the LU-factorization (cf. Sec. 19.2; he called it LR-factorization) in solving eigenvalue problems. An improved version of Rutishauser's method (avoiding breakdown if certain submatrices become singular, etc., cf. Ref. [E39]) is the **QR-method,**[10] which is based on the factorization **QR**, where **R** is upper triangular as before but **Q** is orthogonal (instead of lower triangular). We discuss the QR-method for a real symmetric matrix. (For extensions to general real or complex matrices, see Refs. [E39] and [E40] in Appendix 1.) In this method we start from a real *symmetric* tridiagonal matrix $\mathbf{B}_0 = \mathbf{B}$ (as obtained from a real symmetric matrix **A** by Householder's method). We compute stepwise $\mathbf{B}_1, \mathbf{B}_2, \cdots$ according to this rule:

First Step. Factor

$$\mathbf{B}_0 = \mathbf{Q}_0\mathbf{R}_0$$

---

[10]Proposed independently by J.G.F. Francis, *Computer Journal* **4** (1961–62), 265–271, 332–345, and V.N. Kublanovskaya, *Zhurnal Vych. Mat. i Mat. Fiz.* **1** (1961), 555–570.

where $\mathbf{Q}_0$ is orthogonal and $\mathbf{R}_0$ is upper triangular. Then compute

$$\mathbf{B}_1 = \mathbf{R}_0\mathbf{Q}_0.$$

Second Step. Factor $\mathbf{B}_1 = \mathbf{Q}_1\mathbf{R}_1$. Then compute $\mathbf{B}_2 = \mathbf{R}_1\mathbf{Q}_1$.
General Step. Factor

(5)
$$\boxed{\mathbf{B}_s = \mathbf{Q}_s\mathbf{R}_s}$$

where $\mathbf{Q}_s$ is orthogonal and $\mathbf{R}_s$ is upper triangular. Then compute

(6)
$$\boxed{\mathbf{B}_{s+1} = \mathbf{R}_s\mathbf{Q}_s.}$$

The method of obtaining the factorization (5) will be explained below.
From (5) we have $\mathbf{R}_s = \mathbf{Q}_s^{-1}\mathbf{B}_s$. Substitution into (6) gives

(7)
$$\mathbf{B}_{s+1} = \mathbf{R}_s\mathbf{Q}_s = \mathbf{Q}_s^{-1}\mathbf{B}_s\mathbf{Q}_s.$$

Thus $\mathbf{B}_{s+1}$ is similar to $\mathbf{B}_s$. Hence $\mathbf{B}_{s+1}$ is similar to $\mathbf{B}_0 = \mathbf{B}$ for all s. By Theorem 2, Sec. 19.6, this implies that $\mathbf{B}_{s+1}$ has the same eigenvalues as $\mathbf{B}$.

Also, $\mathbf{B}_{s+1}$ is symmetric. This follows by induction. Indeed, $\mathbf{B}_0 = \mathbf{B}$ is symmetric. Assuming $\mathbf{B}_s$ to be symmetric and using $\mathbf{Q}_s^{-1} = \mathbf{Q}_s^{\mathsf{T}}$ (since $\mathbf{Q}_s$ is orthogonal), we get from (7)

$$\mathbf{B}_{s+1}^{\mathsf{T}} = (\mathbf{Q}_s^{\mathsf{T}}\mathbf{B}_s\mathbf{Q}_s)^{\mathsf{T}} = \mathbf{Q}_s^{\mathsf{T}}\mathbf{B}_s^{\mathsf{T}}\mathbf{Q}_s = \mathbf{Q}_s^{\mathsf{T}}\mathbf{B}_s\mathbf{Q}_s = \mathbf{B}_{s+1}$$

as claimed.

If the eigenvalues of $\mathbf{B}$ are different in absolute value, say, if we have $|\lambda_1| > |\lambda_2| > \cdots > |\lambda_n|$, then

$$\lim_{s\to\infty} \mathbf{B}_s = \mathbf{D}$$

where $\mathbf{D}$ is diagonal, with main diagonal entries $\lambda_1, \lambda_2, \cdots, \lambda_n$. (Proof in Ref. [E39] listed in Appendix 1.)

How to get the QR-factorization, say, $\mathbf{B} = \mathbf{B}_0 = [b_{jk}] = \mathbf{Q}_0\mathbf{R}_0$? The tridiagonal matrix $\mathbf{B}$ has $n - 1$ generally nonzero entries below the main diagonal. These are $b_{21}, b_{32}, \cdots, b_{n,n-1}$. We multiply $\mathbf{B}$ from the left by a matrix $\mathbf{C}_2$ such that $\mathbf{C}_2\mathbf{B} = [b_{jk}^{(2)}]$ has $b_{21}^{(2)} = 0$. This we multiply by a matrix $\mathbf{C}_3$ such that $\mathbf{C}_3\mathbf{C}_2\mathbf{B} = [b_{jk}^{(3)}]$ has $b_{32}^{(3)} = 0$, etc. After $n - 1$ such multiplications we are left with an upper triangular matrix $\mathbf{R}_0$, namely,

(8)
$$\mathbf{C}_n\mathbf{C}_{n-1} \cdots \mathbf{C}_3\mathbf{C}_2\mathbf{B}_0 = \mathbf{R}_0.$$

These $\mathbf{C}_j$ are very simple. $\mathbf{C}_j$ has the 2×2 submatrix

$$\begin{bmatrix} \cos\theta_j & \sin\theta_j \\ -\sin\theta_j & \cos\theta_j \end{bmatrix} \qquad (\theta_j \text{ suitable})$$

in rows $j - 1$ and j and columns $j - 1$ and j, entries 1 everywhere else on the main diagonal and all other entries 0. (This submatrix is the matrix of a plane rotation through the angle θ_j; cf. Prob. 19, Sec. 7.3.) For instance, when $n = 4$, writing $c_j = \cos \theta_j$, $s_j = \sin \theta_j$, we have

$$\mathbf{C}_2 = \begin{bmatrix} c_2 & s_2 & 0 & 0 \\ -s_2 & c_2 & 0 & 0 \\ 0 & 0 & 1 & 0 \\ 0 & 0 & 0 & 1 \end{bmatrix}, \qquad \mathbf{C}_3 = \begin{bmatrix} 1 & 0 & 0 & 0 \\ 0 & c_3 & s_3 & 0 \\ 0 & -s_3 & c_3 & 0 \\ 0 & 0 & 0 & 1 \end{bmatrix},$$

$$\mathbf{C}_4 = \begin{bmatrix} 1 & 0 & 0 & 0 \\ 0 & 1 & 0 & 0 \\ 0 & 0 & c_4 & s_4 \\ 0 & 0 & -s_4 & c_4 \end{bmatrix}.$$

These $\mathbf{C}_j$ are orthogonal. Hence their product in (8) is orthogonal, and so is the inverse of this product. We call this inverse $\mathbf{Q}_0$. Then from (8),

(9a) $$\mathbf{B}_0 = \mathbf{Q}_0 \mathbf{R}_0$$

where, with $\mathbf{C}_j^{-1} = \mathbf{C}_j^{\mathsf{T}}$,

(9b) $$\mathbf{Q}_0 = (\mathbf{C}_n \mathbf{C}_{n-1} \cdots \mathbf{C}_3 \mathbf{C}_2)^{-1} = \mathbf{C}_2^{\mathsf{T}} \mathbf{C}_3^{\mathsf{T}} \cdots \mathbf{C}_{n-1}^{\mathsf{T}} \mathbf{C}_n^{\mathsf{T}}.$$

This is our QR-factorization of $\mathbf{B}_0$. From it we have by (6) with $s = 0$

(10) $$\mathbf{B}_1 = \mathbf{R}_0 \mathbf{Q}_0 = \mathbf{R}_0 \mathbf{C}_2^{\mathsf{T}} \mathbf{C}_3^{\mathsf{T}} \cdots \mathbf{C}_{n-1}^{\mathsf{T}} \mathbf{C}_n^{\mathsf{T}}.$$

We do not need $\mathbf{Q}_0$ explicitly, but to get $\mathbf{B}_1$ from (10), we first compute $\mathbf{R}_0 \mathbf{C}_2^{\mathsf{T}}$, then $(\mathbf{R}_0 \mathbf{C}_2^{\mathsf{T}}) \mathbf{C}_3^{\mathsf{T}}$, etc. Similarly in the further steps that produce $\mathbf{B}_2, \mathbf{B}_3, \cdots$.

We show next how to determine $\cos \theta_2$ and $\sin \theta_2$ in $\mathbf{C}_2$ such that $b_{21}^{(2)} = 0$ in the product

$$\mathbf{C}_2 \mathbf{B} = \begin{bmatrix} c_2 & s_2 & 0 & \cdots \\ -s_2 & c_2 & 0 & \cdots \\ \cdot & \cdot & \cdot & \cdots \\ \cdot & \cdot & \cdot & \cdots \end{bmatrix} \begin{bmatrix} b_{11} & b_{12} & \cdots \\ b_{21} & b_{22} & \cdots \\ \cdot & \cdot & \cdots \\ \cdot & \cdot & \cdots \end{bmatrix}.$$

Now $b_{21}^{(2)}$ is obtained by multiplying the second row of $\mathbf{C}_2$ by the first column of $\mathbf{B}$, that is

$$b_{21}^{(2)} = -s_2 b_{11} + c_2 b_{21} = 0.$$

Hence $\tan \theta_2 = s_2/c_2 = b_{21}/b_{11}$, and

$$(11) \quad \cos \theta_2 = \frac{1}{\sqrt{1 + (b_{21}/b_{11})^2}}, \qquad \sin \theta_2 = \frac{b_{21}/b_{11}}{\sqrt{1 + (b_{21}/b_{11})^2}}.$$

Similarly for $\theta_3, \theta_4, \cdots$. The next example illustrates all this.

EXAMPLE 2. QR-factorization method

Compute all eigenvalues of the matrix

$$\mathbf{A} = \begin{bmatrix} 6 & 4 & 1 & 1 \\ 4 & 6 & 1 & 1 \\ 1 & 1 & 5 & 2 \\ 1 & 1 & 2 & 5 \end{bmatrix}.$$

Solution. We first reduce $\mathbf{A}$ to tridiagonal form. Applying Householder's method, we obtain (cf. Example 1)

$$\mathbf{A}_2 = \begin{bmatrix} 6 & -\sqrt{18} & 0 & 0 \\ -\sqrt{18} & 7 & \sqrt{2} & 0 \\ 0 & \sqrt{2} & 6 & 0 \\ 0 & 0 & 0 & 3 \end{bmatrix}.$$

From the characteristic determinant we see that $\mathbf{A}_2$, hence $\mathbf{A}$, has the eigenvalue 3. Hence it suffices to apply the QR-method to the 3×3 matrix

$$\mathbf{B}_0 = \mathbf{B} = \begin{bmatrix} 6 & -\sqrt{18} & 0 \\ -\sqrt{18} & 7 & \sqrt{2} \\ 0 & \sqrt{2} & 6 \end{bmatrix}.$$

First Step. We multiply $\mathbf{B}$ by

$$\mathbf{C}_2 = \begin{bmatrix} \cos \theta_2 & \sin \theta_2 & 0 \\ -\sin \theta_2 & \cos \theta_2 & 0 \\ 0 & 0 & 1 \end{bmatrix} \quad \text{and then } \mathbf{C}_2\mathbf{B} \text{ by} \quad \mathbf{C}_3 = \begin{bmatrix} 1 & 0 & 0 \\ 0 & \cos \theta_3 & \sin \theta_3 \\ 0 & -\sin \theta_3 & \cos \theta_3 \end{bmatrix}.$$

Here $(-\sin \theta_2) \cdot 6 + (\cos \theta_2)(-\sqrt{18}) = 0$ gives [cf. (11)]

$$\cos \theta_2 = 0.816\ 496\ 58, \qquad \sin \theta_2 = -0.577\ 350\ 27.$$

With these values, we compute

$$\mathbf{C}_2\mathbf{B} = \begin{bmatrix} 7.348\ 469\ 23 & -7.505\ 553\ 50 & -0.816\ 496\ 58 \\ 0 & 3.265\ 986\ 32 & 1.154\ 700\ 54 \\ 0 & 1.414\ 213\ 56 & 6.000\ 000\ 00 \end{bmatrix}.$$

In $\mathbf{C}_3$ we get from $(-\sin \theta_3) \cdot 3.265\ 986\ 32 + (\cos \theta_3) \cdot 1.414\ 213\ 56 = 0$ the values

$$\cos \theta_3 = 0.917\ 662\ 94, \qquad \sin \theta_3 = 0.397\ 359\ 71.$$

This gives

$$\mathbf{R}_0 = \mathbf{C}_3\mathbf{C}_2\mathbf{B} = \begin{bmatrix} 7.348\ 469\ 23 & -7.505\ 553\ 50 & -0.816\ 496\ 58 \\ 0 & 3.559\ 026\ 08 & 3.443\ 784\ 13 \\ 0 & 0 & 5.047\ 146\ 15 \end{bmatrix}.$$

From this we compute

$$\mathbf{B}_1 = \mathbf{R}_0\mathbf{C}_2{}^\mathsf{T}\mathbf{C}_3{}^\mathsf{T} = \begin{bmatrix} 10.333\ 333\ 33 & -2.054\ 804\ 67 & 0 \\ -2.054\ 804\ 67 & 4.035\ 087\ 72 & 2.005\ 532\ 51 \\ 0 & 2.005\ 532\ 51 & 4.631\ 578\ 95 \end{bmatrix}$$

which is symmetric and tridiagonal. The off-diagonal entries in $\mathbf{B}_1$ are large in absolute value. Hence we have to go on.

Second Step. We do the same computations as in the first step, with $\mathbf{B}_0 = \mathbf{B}$ replaced by $\mathbf{B}_1$. We obtain

$$\mathbf{R}_1 = \begin{bmatrix} 10.535\ 653\ 75 & -2.802\ 322\ 42 & -0.391\ 145\ 88 \\ 0 & 4.083\ 295\ 83 & 3.988\ 240\ 28 \\ 0 & 0 & 3.068\ 326\ 68 \end{bmatrix}$$

and from this

$$\mathbf{B}_2 = \begin{bmatrix} 10.879\ 879\ 88 & -0.796\ 379\ 19 & 0 \\ -0.796\ 379\ 18 & 5.447\ 386\ 64 & 1.507\ 025\ 00 \\ 0 & 1.507\ 025\ 00 & 2.672\ 733\ 48 \end{bmatrix}.$$

We see that the off-diagonal entries are somewhat smaller in absolute value than those of $\mathbf{B}_1$.

Third Step. We now obtain

$$\mathbf{R}_2 = \begin{bmatrix} 10.908\ 987\ 40 & -1.191\ 925\ 04 & -0.110\ 016\ 02 \\ 0 & 5.581\ 996\ 06 & 2.168\ 774\ 94 \\ 0 & 0 & 2.167\ 703\ 93 \end{bmatrix}$$

and from this

$$\mathbf{B}_3 = \begin{bmatrix} 10.966\ 892\ 94 & -0.407\ 497\ 542 & 0 \\ -0.407\ 497\ 54 & 5.945\ 898\ 56 & 0.585\ 235\ 81 \\ 0 & 0.585\ 235\ 82 & 2.987\ 208\ 51 \end{bmatrix}.$$

The reader may verify that the eigenvalues of $\mathbf{B}$ are 11, 6, 2. Hence the given matrix $\mathbf{A}$ has the spectrum 11, 6, 3, 2. ∎

Looking back at our discussion, we recognize that the purpose of applying Householder's method before the QR-factorization method is a substantial reduction of cost in each QR-factorization.

Convergence acceleration can be achieved by a **spectral shift** (cf. Sec. 19.6) in each step, that is, instead of $\mathbf{B}_s$ we take $\mathbf{B}_s - k_s\mathbf{I}$ with a suitable k_s. For instance, if we take $k_s = b_{nn}^{(s)}$ (the entry in the lower right-hand corner of $\mathbf{B}_s$), this will generally result in a more rapid decrease to 0 of the other entry in the nth row of $\mathbf{B}_s$, so that we can soon determine an eigenvalue with sufficient accuracy and reduce the size of the matrix in our further work.

(See Ref. [E39], p. 510, for a discussion of other choices of k_s.) It is fair to say that the QR-method, preceded by the Householder tridiagonalization, is a general-purpose method for symmetric matrices that is better than any other presently known method.

EXAMPLE 3. Spectral shift in the QR-method

In Example 2 we have $b_{nn} = b_{33} = 6$, so that the spectral shift with $k_0 = b_{nn} = 6$ gives the matrix

$$\hat{\mathbf{B}}_0 = \mathbf{B} - b_{33}\mathbf{I} = \mathbf{B} - 6\mathbf{I} = \begin{bmatrix} 0 & -\sqrt{18} & 0 \\ -\sqrt{18} & 1 & \sqrt{2} \\ 0 & \sqrt{2} & 0 \end{bmatrix}.$$

From this we get $(-\sin\theta_2) \cdot 0 + (\cos\theta_2)(-\sqrt{18}) = 0$, $\cos\theta_2 = 0$, $\sin\theta_2 = 1$ and

$$\mathbf{C}_2\hat{\mathbf{B}}_0 = \begin{bmatrix} -\sqrt{18} & 1 & \sqrt{2} \\ 0 & \sqrt{18} & 0 \\ 0 & \sqrt{2} & 0 \end{bmatrix}.$$

From this we get $(-\sin\theta_3)\sqrt{18} + (\cos\theta_3)\sqrt{2} = 0$, hence

$$\cos\theta_3 = 0.948\ 683\ 30, \qquad \sin\theta_3 = 0.316\ 227\ 77$$

and

$$\hat{\mathbf{R}}_0 = \mathbf{C}_3\mathbf{C}_2\hat{\mathbf{B}}_0 = \begin{bmatrix} -4.242\ 640\ 69 & 1.000\ 000\ 00 & 1.414\ 213\ 56 \\ 0 & 4.472\ 135\ 96 & 0 \\ 0 & 0 & 0 \end{bmatrix}.$$

Hence

$$\hat{\mathbf{B}}_1 = \hat{\mathbf{R}}_0\mathbf{C}_2{}^\mathsf{T}\mathbf{C}_3{}^\mathsf{T} = \begin{bmatrix} 1.000\ 000\ 00 & 4.472\ 135\ 96 & 0 \\ 4.472\ 135\ 96 & 0 & 0 \\ 0 & 0 & 0 \end{bmatrix}.$$

This shows that 0 is an eigenvalue of $\hat{\mathbf{B}}_0$. Hence 6 is an eigenvalue of **B**, by Theorem 3 in Sec. 19.6.

The result is **not** typical. Typically we could only expect that the off-diagonal entry in the nth row becomes rapidly smaller within the next few steps.

That we can now find the other eigenvalues of **B** from a quadratic equation is a consequence of the small size of our matrix, which we picked on purpose—we would not learn more from a monster. In practice, for a larger matrix we would now have to go on with QR (and shifts) as applied to a matrix of order $n - 1$. ∎

Nonsymmetric matrices **A** can also be treated by the QR-method. Instead of reducing **A** to tridiagonal form, we first reduce it to an **upper Hessenberg matrix B,** that is, **B** may have nonzero entries only where an upper triangular matrix or a tridiagonal matrix has them (cf. Fig. 436). For details, see Refs. [E39] or [E40] listed in Appendix 1.

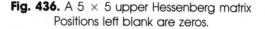

Fig. 436. A 5 × 5 upper Hessenberg matrix
Positions left blank are zeros.

Problems for Sec. 19.10

Tridiagonalize the following matrices by Householder's method.

1. $\begin{bmatrix} 0 & 1 & 1 \\ 1 & 0 & 1 \\ 1 & 1 & 0 \end{bmatrix}$

2. $\begin{bmatrix} 2 & -1 & 1 \\ -1 & 3 & 2 \\ 1 & 2 & 3 \end{bmatrix}$

3. $\begin{bmatrix} -1 & 2 & 3 \\ 2 & 2 & 6 \\ 3 & 5 & -1 \end{bmatrix}$

4. $\begin{bmatrix} 4 & 5 & 9 \\ 5 & 1 & 6 \\ 9 & 6 & 15 \end{bmatrix}$

5. $\begin{bmatrix} 1 & 2 & 3 \\ 2 & 4 & 6 \\ 3 & 6 & 9 \end{bmatrix}$

6. $\begin{bmatrix} 4 & 4 & 1 & 1 \\ 4 & 4 & 1 & 1 \\ 1 & 1 & 3 & 2 \\ 1 & 1 & 2 & 3 \end{bmatrix}$

7. Perform 1 step of the QR-method for the tridiagonal matrix in the answer to Prob. 3.

8. Determine the eigenvalues of the matrix in Prob. 4 by applying the QR-method to the tridiagonalized matrix in the answer to Prob. 4.

9. Perform 1 step of the QR-method (with spectral shift) for the matrix $\hat{\mathbf{B}}_0 = \mathbf{B}_0 - b_{33}\mathbf{I} = \mathbf{B}_0 + 4.461\ 538\ 46\mathbf{I}$, where $\mathbf{B}_0$ is the matrix in the answer to Prob. 3. Compare with the result in Prob. 7 and comment.

10. Apply another QR-step with spectral shift to the matrix in the answer to Prob. 9. Applying Gerschgorin's theorem to row 3 of the answer and using the symmetry of the matrix, find an inclusion interval and deduce from it that the given matrix $\mathbf{A}$ in Prob. 3 has an eigenvalue in the interval $-6.08 \leq \lambda = -5.92$. Verify directly that $\lambda = -6$ is an eigenvalue of $\mathbf{A}$.

Review Problems for Chapter 19

1. What is partial pivoting? When is it necessary? When advisable?

2. What do we mean by $O(n^3)$? By $O(2^n)$?

3. Which methods for solving systems of linear equations are based on an LU-factorization, and what is their advantage over the classical Gauss elimination?

4. What is the Gauss–Jordan method? Why is it not advisable for solving systems of linear equations? When is it useful?

5. What is a direct method for solving systems of linear equations, as opposed to an indirect method? Name an example of the latter. For what systems would one apply it?

6. What is an ill-conditioned system of linear equations? Why does one have to be particularly careful in the case of ill-conditioning? How can one recognize ill-conditioning?

7. What is a vector norm? A matrix norm? Give two examples (from memory if possible). Show that $\|\mathbf{x}\|_\infty \leq \|\mathbf{x}\|_1$.

8. What is the residual of an approximate solution of a system of linear equations? Does a small residual indicate that the solution is close to the actual solution?

9. What is the condition number of a matrix? What is its significance?

10. State from memory the definitions of an eigenvalue and an eigenvector of a matrix.

11. What are similarity transformations, and why are they of great interest in the numerical work on eigenvalue problems?

12. State Gerschgorin's theorem. Can you remember the proof?

13. What is the Rayleigh quotient? Can it be equal to an eigenvalue although the corresponding vector is not an eigenvector?

14. What are the advantages and disadvantages of the power method?

15. What is a spectral shift? What role does it play in the power method?

16. What do we mean by deflation of a matrix?

17. What is a tridiagonal matrix? What is the point of reducing a matrix to tridiagonal form before applying the QR-method?

18. What is the basic idea of the QR-method?

19. Can the inverse of a tridiagonal matrix be a full matrix?

20. What is the advantage of the QR-method over repeated deflation if one wants to obtain all eigenvalues of a matrix?

Solve the following systems of linear equations by the Gauss elimination.

21. $\begin{aligned} x_1 + x_2 + x_3 &= 5 \\ x_1 + 2x_2 + 2x_3 &= 6 \\ x_1 + 2x_2 + 3x_3 &= 8 \end{aligned}$

22. $\begin{aligned} x_1 + 2x_2 + 4x_3 &= 3.5 \\ 2x_1 + 13x_2 + 23x_3 &= 13.0 \\ 4x_1 + 23x_2 + 77x_3 &= 6.0 \end{aligned}$

23. $\begin{aligned} 2x_1 + x_2 + 2x_3 &= 5.6 \\ 8x_1 + 5x_2 + 13x_3 &= 20.9 \\ 6x_1 + 3x_2 + 12x_3 &= 11.4 \end{aligned}$

24. $\begin{aligned} 3x_2 + 4x_3 &= -31 \\ -x_1 \qquad - 2x_3 &= 13 \\ 2x_1 - 5x_2 - 6x_3 &= 59 \end{aligned}$

25. $\begin{aligned} 4x_2 - 3x_3 &= 11.8 \\ 5x_1 + 3x_2 + x_3 &= 34.2 \\ 6x_1 - 7x_2 + 2x_3 &= -3.1 \end{aligned}$

26. $\begin{aligned} 0.1x_1 + x_2 - 2x_3 &= -0.3 \\ 10x_1 - 3x_2 + 0.5x_3 &= 149.9 \\ -3x_1 + 4x_2 \qquad &= -44.2 \end{aligned}$

27. Solve Prob. 21 by Doolittle's method.
28. Solve Prob. 23 by Doolittle's method.
29. Solve Prob. 21 by Cholesky's method.
30. Solve Prob. 22 by Cholesky's method.

Compute the inverse:

31. $\begin{bmatrix} 0.5 & -0.5 & -1 \\ -1.5 & 0.5 & -0.5 \\ 0.5 & -1.5 & -2 \end{bmatrix}$ **32.** $\begin{bmatrix} 0 & -4 & 2 \\ 1 & 1 & -1 \\ -2 & 4 & 0 \end{bmatrix}$ **33.** $\begin{bmatrix} 6.8 & -4.1 & -1.7 & 1.0 \\ -4.1 & 2.5 & 1.0 & -0.6 \\ -1.7 & 1.0 & 0.5 & -0.3 \\ 1.0 & -0.6 & -0.3 & 0.2 \end{bmatrix}$

Apply the Gauss–Seidel iteration to the following systems. Perform three steps, starting from 1, 1, 1.

34. $10x - y - z = 13$
$x + 10y + z = 36$
$-x - y + 10z = 35$

35. $4x + y = -8$
$4y + z = 2$
$x + 2z = 2$

36. $4x + 2y + z = 14$
$x + 5y - z = 10$
$x + y + 8z = 20$

Compute the l_1-, l_2- and l_∞-norms of the vectors

37. $[-4 \ \ 1 \ \ 0 \ \ 2]^\mathsf{T}$

38. $[0 \ \ 1 \ \ 0]^\mathsf{T}$

39. $[3 \ \ 2 \ \ -4 \ \ -5]^\mathsf{T}$

40. $[6 \ \ 1 \ \ 0 \ \ 0]^\mathsf{T}$

41. $[3.2 \ \ 6.4 \ \ 0]^\mathsf{T}$

42. $[12 \ \ 4 \ \ 1 \ \ 8 \ \ 8]^\mathsf{T}$

Compute the matrix norm corresponding to the l_∞-vector norm, for the coefficient matrix of the system of equations:

43. In Prob. 25

44. In Prob. 26

45. In Prob. 24

Compute the condition number (corresponding to the l_1-vector norm) of the matrix

46. In Prob. 32

47. In Prob. 31

48. In Prob. 33

49. Fit a straight line through the data (0, 0), (2, 1.8), (4, 3.4), (6, 4.6) by the method of least squares.

50. Fit a quadratic parabola through the data

(1.09, 1.35), (1.28, 1.58), (1.36, 1.68), (1.44, 1.85), (1.60, 2.23), (1.65, 2.38)

by the method of least squares.

For each matrix find three circular disks that must contain all the eigenvalues.

51. $\begin{bmatrix} 10 & 0.1 & 0.1 \\ 0.1 & 4 & 0.2 \\ 0.1 & 0.2 & 3 \end{bmatrix}$

52. $\begin{bmatrix} 5 & 1 & 1 \\ 1 & 6 & 0 \\ 1 & 0 & 8 \end{bmatrix}$

53. $\begin{bmatrix} 11.4 & 2.0 & 0.6 \\ 2.0 & 14.4 & 1.2 \\ 0.6 & 1.2 & 14.6 \end{bmatrix}$

54. Apply 4 steps of the power method to the matrix in Prob. 52, starting from $[1 \ \ 1 \ \ 1]^\mathsf{T}$ and computing the Rayleigh quotients and error bounds.

55. Apply Wielandt's method of deflation to the matrix in Prob. 53, using that 16.4 is an eigenvalue.

Summary of Chapter 19
Numerical Methods
in Linear Algebra

This chapter deals with three large classes of numerical problems arising from linear algebra, namely, the numerical solution of systems of linear equations (Secs. 19.1–19.4), the fitting of straight lines or parabolas through given data (represented as points in the plane; Sec. 19.5) and the numerical solution of eigenvalue problems (Sec. 19.6–19.10).

Thus, in Secs. 19.1–19.4 we solve systems $\mathbf{Ax} = \mathbf{b}$, where $\mathbf{A} = [a_{jk}]$ is an $n \times n$ matrix, written out

$$E_1: \quad a_{11}x_1 + \cdots + a_{1n}x_n = b_1$$

$$E_2: \quad a_{21}x_1 + \cdots + a_{2n}x_n = b_2$$

(1)

$$\cdots\cdots\cdots\cdots\cdots\cdots\cdots\cdots\cdots$$

$$E_n: \quad a_{n1}x_1 + \cdots + a_{nn}x_n = b_n.$$

There are two types of numerical methods for this task, **direct methods,** such as the Gauss elimination, in which the amount of computation to get a solution can be specified in advance, and **indirect** or **iterative methods,** such as the Gauss–Seidel method, in which we start from a (possibly crude) approximation and improve it stepwise by repeatedly performing the same cycle of computation, with changing data (see below).

The **Gauss elimination** (Sec. 19.1) is a systematic elimination process that reduces (1) stepwise to triangular form. In step 1 we eliminate x_1 from equations E_2 to E_n by subtracting $(a_{21}/a_{11})\,E_1$ from E_2, then $(a_{31}/a_{11})\,E_1$ from E_3, etc. Equation E_1 is called the **pivot equation** in this step and a_{11} the **pivot entry.** In step 2 we take the new second equation as pivot equation and eliminate x_2, etc. When the triangular form is reached, we get x_n from the last equation, then x_{n-1} from the second last, etc. **Partial pivoting** (= interchange of equations) is *necessary* when pivot entries are zero, and *advisable* when they are small.

Variants of the Gauss elimination (**Doolittle's, Crout's, Cholesky's methods,** Sec. 19.2) use the idea of factoring

(2) $$\mathbf{A} = \mathbf{LU}$$

(**L** lower triangular, **U** upper triangular), setting $\mathbf{Ux} = \mathbf{y}$ and then solving $\mathbf{Ax} = \mathbf{LUx} = \mathbf{Ly} = \mathbf{b}$ by first solving $\mathbf{Ly} = \mathbf{b}$ for $\mathbf{y}$ and then $\mathbf{Ux} = \mathbf{y}$ for $\mathbf{x}$.

Iterative methods are obtained by writing $\mathbf{Ax} = \mathbf{b}$ as

(3) $$\mathbf{x} = \mathbf{b} - (\mathbf{A} - \mathbf{I})\mathbf{x}$$

and then substituting approximations on the right to get new approximations on the left. In particular, in the **Gauss–Seidel method** we first divide each equation to make $a_{11} = a_{22} = \cdots = a_{nn} = 1$, then write $\mathbf{A} = \mathbf{I} + \mathbf{L} + \mathbf{U}$, so that (3) becomes

$$\mathbf{x} = \mathbf{b} - \mathbf{Lx} - \mathbf{Ux}$$

and always take the most recent approximate x_j's on the right. This gives the iteration formula (Sec. 19.3)

(4) $$\mathbf{x}^{(m+1)} = \mathbf{b} - \mathbf{Lx}^{(m+1)} - \mathbf{Ux}^{(m)}.$$

If $\|\mathbf{C}\| < 1$, where $\mathbf{C} = -(\mathbf{I} + \mathbf{L})^{-1}\mathbf{U}$, then this process converges. Here, $\|\mathbf{C}\|$ denotes any matrix norm (cf. Sec. 19.3).

If the **condition number** $\kappa(\mathbf{A}) = \|\mathbf{A}\|\,\|\mathbf{A}^{-1}\|$ of $\mathbf{A}$ is large, then the system $\mathbf{Ax} = \mathbf{b}$ is **ill-conditioned** (Sec. 19.4), and a small **residual** $\mathbf{r} = \mathbf{b} - \mathbf{A\tilde{x}}$ does not imply that $\tilde{\mathbf{x}}$ is close to the exact solution.

The fitting of a polynomial $p(x) = b_0 + b_1 x + \cdots + b_m x^m$ through given data (points in the xy-plane) $(x_1, y_1), \cdots, (x_n, y_n)$ by the method of **least squares** is discussed in Sec. 19.5.

An **eigenvalue** of an $n \times n$ matrix $\mathbf{A} = [a_{jk}]$ is a number λ such that

$$(5) \qquad \mathbf{Ax} = \lambda\mathbf{x}, \qquad \text{thus} \qquad (\mathbf{A} - \lambda\mathbf{I})\mathbf{x} = \mathbf{0}$$

has a solution $\mathbf{x} \neq \mathbf{0}$, called an **eigenvector** of $\mathbf{A}$ corresponding to that λ. In Sec. 19.6 we list basic facts about the eigenvalue problem needed in numerical methods.

Section 19.7 concerns theorems that give **inclusion sets** (sets in the complex λ-plane that contain one or several eigenvalues of $\mathbf{A}$), notably the famous **Gerschgorin theorem,** which states that the **spectrum** ($=$ set of all eigenvalues) of $\mathbf{A}$ lies in the n disks

$$(6) \qquad |a_{jj} - \lambda| \leq \sum_{\substack{k=1 \\ k \neq j}}^{n} |a_{jk}| \qquad\qquad j = 1, \cdots, n$$

whose centers are $a_{11}, a_{22}, \cdots, a_{nn}$ and whose radii are given by the sums on the right.

The **power method** (Sec. 19.8) gives approximations

$$(7) \qquad q = \frac{(\mathbf{Ax})^{\mathsf{T}}\mathbf{x}}{\mathbf{x}^{\mathsf{T}}\mathbf{x}} \qquad\qquad (\textit{Rayleigh quotient}),$$

usually to the eigenvalue that is largest in absolute value, and, if $\mathbf{A}$ is symmetric, error bounds

$$(8) \qquad |\epsilon| \leq \sqrt{\frac{(\mathbf{Ax})^{\mathsf{T}}\mathbf{Ax}}{\mathbf{x}^{\mathsf{T}}\mathbf{x}} - q^2}.$$

Practically, one chooses any vector $\mathbf{x}_0 \neq \mathbf{0}$, computes the vectors $\mathbf{x}_1 = \mathbf{Ax}_0$, $\mathbf{x}_2 = \mathbf{Ax}_1$, $\cdots$, $\mathbf{x}_s = \mathbf{Ax}_{s-1}$ and takes $\mathbf{x} = \mathbf{x}_{s-1}$ and $\mathbf{Ax} = \mathbf{x}_s$ in (7) and (8).

If we know an eigenvalue λ_1 of $\mathbf{A}$, we can **deflate A,** that is, compute a matrix $\mathbf{A}_1$ with eigenvalues $0, \lambda_2, \cdots, \lambda_n$ (Sec. 19.9).

If we want to know all eigenvalues of a symmetric matrix, then better than repeated deflation is the **QR-method** preceded by **Householder's tridiagonalization** (Sec. 19.10). The QR-method is based on a factorization $\mathbf{A} = \mathbf{QR}$, where $\mathbf{Q}$ is orthogonal and $\mathbf{R}$ is upper triangular, and similarity transformations.

Chapter 20

Numerical Methods for Differential Equations

Numerical methods for differential equations are of great importance to the engineer and physicist because practical problems often lead to differential equations that cannot be solved by one of the methods in Chaps. 1–5 or similar methods, or to equations for which the solutions in terms of formulas are so complicated that one often prefers to compute a table of values by applying a numerical method to such an equation.

The present chapter includes basic methods for the numerical solution of ordinary differential equations (Secs. 20.1–20.3) and partial differential equations (Secs. 20.4–20.7).

Sections 20.1, 20.2 and 20.3 may also be studied immediately after Chaps. 1 and 2, respectively, since they are independent of Chaps. 18 and 19.

Sections 20.4–20.7 may also be studied immediately after Chap. 11, provided the reader has some knowledge in systems of linear algebraic equations.

Prerequisite for Secs. 20.1–20.3: Secs. 1.1–1.7, 2.1–2.3.
Prerequisite for Secs. 20.4–20.7: Secs. 11.1–11.3, 11.5, 11.11.
References: Appendix 1, Part E (see also Parts A and C).
Answers to problems: Appendix 2.

20.1 Methods for First-Order Differential Equations

From Chap. 1 we know that a *differential equation of the first order* is of the form $F(x, y, y') = 0$, and often it will be possible to write the equation in the *explicit form* $y' = f(x, y)$. An **initial value problem** consists of a differential equation and a condition the solution must satisfy (or several conditions referring to the same value of x if the equation is of higher order). In this section we shall consider initial value problems of the form

(1)
$$y' = f(x, y), \qquad y(x_0) = y_0$$

assuming f to be such that the problem has a unique solution on some interval containing x_0, and we shall discuss methods for computing numerical values of the solution.

If we can obtain a formula for the solution, we may evaluate it numerically, either directly or by the use of tables. In this approach the book by Kamke (Ref. [A11] in Appendix 1) and an index of tables (cf. Ref. [8] in Appendix 1) will be helpful. If that formula is too complicated or if no formula for the solution is available, we may apply one of the methods discussed in this section.

These methods are **step-by-step methods;** that is, we start from $y_0 = y(x_0)$ and proceed stepwise. In the first step we compute an approximate value y_1 of the solution y of (1) at $x = x_1 = x_0 + h$. In the second step we compute an approximate value y_2 of that solution at $x = x_2 = x_0 + 2h$, etc. Here h is a fixed number, for example, 0.2 or 0.1 or 0.01; principles for the choice of h will be discussed later in this section.

In each step the computations are done by the same formula. Such formulas are suggested by the Taylor series

$$y(x + h) = y(x) + hy'(x) + \frac{h^2}{2} y''(x) + \cdots .$$

From (1) we have $y' = f$. By differentiation, $y'' = f' = \partial f/\partial x + (\partial f/\partial y)y'$, etc., and the Taylor series becomes

$$(2) \qquad y(x + h) = y(x) + hf + \frac{h^2}{2} f' + \frac{h^3}{6} f'' + \cdots ,$$

where $f, f', f'', \cdots$ are evaluated at $(x, y(x))$.

For small values of h, the higher powers $h^2, h^3, \cdots$ in (2) will be very small. This suggests the crude approximation

$$y(x + h) \approx y(x) + hf$$

and the following iteration process. In the first step we compute

$$y_1 = y_0 + hf(x_0, y_0)$$

which approximates $y(x_1) = y(x_0 + h)$. In the second step we compute

$$y_2 = y_1 + hf(x_1, y_1)$$

which approximates $y(x_2) = y(x_0 + 2h)$, etc., and in general

$$(3) \qquad \boxed{y_{n+1} = y_n + hf(x_n, y_n)} \qquad (n = 0, 1, \cdots).$$

This is called the **Euler method** or **Euler–Cauchy method.** Geometrically it is an approximation of the curve of $y(x)$ by a polygon whose first side is tangent to the curve at x_0 (cf. Fig. 437).

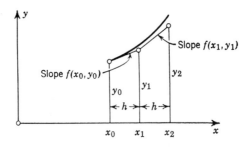

Fig. 437. Euler method

The method is called a **first-order method,** because in (2) we take only the constant term and the term containing the first power of h. The omission of the further terms in (2) causes an error, which is called the **truncation error** of the method. For small h, the third and higher powers of h will be small compared with h^2 in the first neglected term in (2), and we therefore say that the **truncation error per step** is of **order** h^2. In addition there are **round-off errors** in this and other methods, which may affect the accuracy of the values y_1, y_2, $\cdots$ more and more as n increases; we shall return to this point in the next section.

The practical value of the Euler method is limited, but since the method is simple, it may be helpful for understanding the basic idea of the methods in this section.

EXAMPLE 1. Euler method

Apply the Euler method to the following initial value problem, choosing $h = 0.2$ and computing $y_1, \cdots, y_5$:

$$(4) \qquad\qquad\qquad y' = x + y, \qquad y(0) = 0.$$

Solution. Here $f(x, y) = x + y$, and we see that (3) becomes

$$y_{n+1} = y_n + 0.2(x_n + y_n).$$

Table 20.1 shows the computations, the values of the exact solution

$$y(x) = e^x - x - 1$$

obtained from (4) in Sec. 1.7, and the error. In practice the exact solution is unknown, but an indication of the accuracy of the values can be obtained by applying the Euler method once more with step $2h = 0.4$ and comparing corresponding approximations. They differ by 0.040 ($x = 0.4$) and 0.110 ($x = 0.8$), as a simple calculation shows. Since the error is of order h^2, a switch from h to $2h$ corresponds to a multiplication by $2^2 = 4$, but since we need only half as many steps as before, the error will only be multiplied by $4/2 = 2$; hence those differences indicate the size of the error (cf. Table 20.1). ∎

Improved Euler Method (Heun's Method)

By taking more terms in (2) into account we obtain numerical methods of higher order and precision. The corresponding formulas can be represented in such a form that the complicated computation of derivatives of $f(x, y)$ is avoided and is replaced by computing f for one or several suitably chosen auxiliary values of (x, y). Let us discuss two such methods that are of practical importance.

Table 20.1
Euler Method Applied to (4) and Error

n	x_n	y_n	$0.2(x_n + y_n)$	Exact Values	Absolute Value of Error
0	0.0	0.000	0.000	0.000	0.000
1	0.2	0.000	0.040	0.021	0.021
2	0.4	0.040	0.088	0.092	0.052
3	0.6	0.128	0.146	0.222	0.094
4	0.8	0.274	0.215	0.426	0.152
5	1.0	0.489		0.718	0.229

The first method is the so-called **improved Euler method** or **improved Euler–Cauchy method** (sometimes also called **Heun's method**). In each step of this method we compute first the auxiliary value

(5a)
$$y_{n+1}^* = y_n + hf(x_n, y_n)$$

and then the new value

(5b)
$$y_{n+1} = y_n + \tfrac{1}{2}h[f(x_n, y_n) + f(x_{n+1}, y_{n+1}^*)].$$

This method has a simple geometric interpretation. In fact, we may say that in the interval from x_n to $x_n + \tfrac{1}{2}h$ we approximate the solution y by the straight line through (x_n, y_n) with slope $f(x_n, y_n)$, and then we continue along the straight line with slope $f(x_{n+1}, y_{n+1}^*)$ until x reaches x_{n+1} (cf. Fig. 438, where $n = 0$).

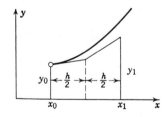

Fig. 438. Improved Euler method

The improved Euler–Cauchy method is a **predictor–corrector method,** because in each step we first predict a value by (5a) and then correct it by (5b).

In algorithmic form, using the notations $k_1 = hf(x_n, y_n)$ in (5a) and $k_2 = hf(x_{n+1}, y_{n+1}^*)$ in (5b) we can write this method as shown in Table 20.2 on the next page.

EXAMPLE 2. Improved Euler method
Apply the improved Euler method to the initial value problem (4), choosing $h = 0.2$, as before.

Table 20.2
Improved Euler Method (Heun's Method)

ALGORITHM EULER (f, x_0, y_0, h, N)

This algorithm computes the solution of the initial value problem $y' = f(x, y)$, $y(x_0) = y_0$ at equidistant points $x_1 = x_0 + h$, $x_2 = x_0 + 2h$, $\cdots, x_N = x_0 + Nh$; here f is such that this problem has a unique solution on the interval $[x_0, x_N]$ (cf. Sec. 1.11).

> INPUT: Initial values x_0, y_0, step size h, number of steps N
>
> OUTPUT: Approximation y_{n+1} to the solution $y(x_{n+1})$ at $x_{n+1} = x_0 + (n + 1)h$, where $n = 0, \cdots, N - 1$
>
> For $n = 0, 1, \cdots, N - 1$ do:
>
> > $x_{n+1} = x_n + h$
> >
> > $k_1 = hf(x_n, y_n)$
> >
> > $k_2 = hf(x_{n+1}, y_n + k_1)$
> >
> > $y_{n+1} = y_n + \frac{1}{2}(k_1 + k_2)$
> >
> > OUTPUT x_{n+1}, y_{n+1}
>
> End
>
> Stop

End EULER

Solution. For the present problem

$$k_1 = 0.2(x_n + y_n)$$

$$k_2 = 0.2(x_n + 0.2 + y_n + 0.2(x_n + y_n))$$

$$y_{n+1} = y_n + \frac{0.2}{2}(2.2x_n + 2.2y_n + 0.2)$$

$$= y_n + 0.22(x_n + y_n) + 0.02.$$

Table 20.3 shows that our present results are more accurate than those in Example 1; cf. also Table 20.6 on p. 1070. ∎

Table 20.3
Improved Euler Method Applied to (4) and Error

n	x_n	y_n	$0.22(x_n + y_n) + 0.02$	Exact Values	Error
0	0.0	0.0000	0.0200	0.0000	0.0000
1	0.2	0.0200	0.0684	0.0214	0.0014
2	0.4	0.0884	0.1274	0.0918	0.0034
3	0.6	0.2158	0.1995	0.2221	0.0063
4	0.8	0.4153	0.2874	0.4255	0.0102
5	1.0	0.7027		0.7183	0.0156

The improved Euler method is a **second-order method,** *because the truncation error per step is of order h^3.*

In fact, setting $\tilde{f}_n = f(x_n, y(x_n))$ and using (2), we have

(6a) $\qquad y(x_n + h) - y(x_n) = h\tilde{f}_n + \frac{1}{2}h^2\tilde{f}_n{}' + \frac{1}{6}h^3\tilde{f}_n{}'' + \cdots .$

Approximating the expression in the brackets in (5b) by $\tilde{f}_n + \tilde{f}_{n+1}$ and again using the Taylor expansion, we obtain from (5b)

(6b) $\qquad y_{n+1} - y_n \approx \frac{1}{2}h(\tilde{f}_n + \tilde{f}_n + h\tilde{f}_n{}' + \frac{1}{2}h^2\tilde{f}_n{}'' + \cdots).$

Subtraction of (6a) from (6b) gives the truncation error per step

$$\frac{h^3}{4}\tilde{f}_n{}'' - \frac{h^3}{6}\tilde{f}_n{}'' + \cdots = \frac{h^3}{12}\tilde{f}_n{}'' + \cdots .$$

This proves the assertion. ∎

We shall now discuss the **choice of the step** h, which is an important task in applying a step-by-step method. h should not be too small, because otherwise the number of steps and the round-off errors become large. On the other hand, h should not be too large, because a large h implies a large truncation error per step and, in addition, an error caused by the fact that f is evaluated at (x_n, y_n) instead of $(x_n, y(x_n))$. The latter would be zero if f were independent of y, and it will matter the more the faster f varies as y varies, that is, the larger the absolute value of the partial derivative $f_y = \partial f/\partial y$ is. More precisely, denoting that error by φ_n and applying the mean value theorem, we have

$$\varphi_n = f(x_n, y_n) - f(x_n, y(x_n)) = f_y(x_n, \bar{y})\eta_n$$

where $\eta_n = y_n - y(x_n)$ is the error of y_n and $\bar{y}$ lies between y_n and $y(x_n)$. Hence the contribution of φ_n to the error of y_{n+1} is approximately $h\varphi_n = hf_y(x_n, \bar{y})\eta_n$. This suggests to take a close upper bound K of $|f_y|$ in the region of interest and to choose h such that

$$\kappa = hK$$

is not too large. We see that if $|f_y|$ is large (strong dependence of f on y), then K is large and h must be small, which is understandable. (In Examples 1 and 2, $f_y = 1$, $K = 1$, $hK = 0.2$.) If f_y varies very much, we may choose a close upper bound K_n of $|f_y(x_n, \bar{y})|$ and choose two or even three different values of h in different regions, to keep

$$\kappa_n = hK_n$$

within a certain interval (e.g., $0.05 \leq \kappa_n \leq 0.1$), which depends on the desired accuracy; of course, because of the truncation error per step, we cannot let h increase beyond a certain value.

Runge–Kutta Method

A still more accurate method of great practical importance is the **Runge–Kutta method**,[1] shown in Table 20.4. We see that in each step we first compute four auxiliary quantities k_1, k_2, k_3, k_4 and then the new value y_{n+1}. These formulas look complicated at first sight, but they are in fact very easy to program.

It can be shown that the truncation error per step is of the order h^5 (cf. Ref. [E3] in Appendix 1) and the method is, therefore, a fourth-order method. Note that if f depends only on x, this method reduces to Simpson's rule of integration (Sec. 18.5).

In hand calculations, the frequent calculation of $f(x, y)$ is laborious. On a computer this does not matter too much, and the method is well suited because it needs no special starting procedure, makes light demands on the storage, requires no estimation, and uses the same straightforward computational procedure several times.

EXAMPLE 3. Runge–Kutta method

Apply the Runge–Kutta method to the initial value problem (4) in Example 1, choosing $h = 0.2$, as before and computing five steps.

Solution. For the present problem we have $f(x, y) = x + y$. Hence

$$k_1 = 0.2(x_n + y_n), \qquad\qquad k_2 = 0.2(x_n + 0.1 + y_n + 0.5k_1),$$

$$k_3 = 0.2(x_n + 0.1 + y_n + 0.5k_2), \qquad k_4 = 0.2(x_n + 0.2 + y_n + k_3).$$

Since these expressions are so simple, we find it convenient to insert k_1 into k_2, obtaining $k_2 = 0.22(x_n + y_n) + 0.02$, insert this into k_3, finding $k_3 = 0.222(x_n + y_n) + 0.022$, and finally insert this into k_4, finding $k_4 = 0.2444(x_n + y_n) + 0.0444$. If we use these expressions, the formula for y_{n+1} in Table 20.4 becomes

$$(7) \qquad\qquad y_{n+1} = y_n + 0.2214(x_n + y_n) + 0.0214.$$

Table 20.5 shows the corresponding computations, and from Table 20.6 on p. 1070 we see that the values are much more accurate than those in Examples 1 and 2. ∎

The **step length** h should not be greater than a certain value H which depends on the desired accuracy and should otherwise be such that

$$\kappa = hK \qquad\qquad (K \text{ a close upper bound for } |\partial f/\partial y|)$$

lies about between 0.01 and 0.05; this is similar to the case of the improved Euler method discussed before. It is an advantage of the Runge–Kutta method that we may control h by means of k_1, k_2, k_3, because from the definition of f_y we have

---

[1]Named after the German mathematicians CARL DAVID TOLMÉ RUNGE (1856—1927), professor of applied mathematics at Göttingen, and WILHELM KUTTA (1867—1944).

Without much historical justification, the improved Euler method is sometimes called a **second-order Runge–Kutta method.** The Runge–Kutta method discussed here is often called the **fourth-order Runge–Kutta method** because there are Runge–Kutta methods of still higher order based on the same principle of replacing the computation of derivatives by the computation of auxiliary values (values of f at certain points). For details, see Ref. [E32] in Appendix 1.

Table 20.4
Runge–Kutta Method (of Fourth Order)

ALGORITHM RUNGE–KUTTA (f, x_0, y_0, h, N).

This algorithm computes the solution of the initial value problem $y' = f(x, y)$, $y(x_0) = y_0$ at equidistant points

$$x_1 = x_0 + h, \; x_2 = x_0 + 2h, \; \cdots, \; x_N = x_0 + Nh;$$

here f is such that this problem has a unique solution on the interval $[x_0, x_N]$ (cf. Sec. 1.11).

INPUT: Initial values x_0, y_0, step size h, number of steps N

OUTPUT: Approximation y_{n+1} to the solution $y(x_{n+1})$ at $x_{n+1} = x_0 + (n + 1)h$, where $n = 0, 1, \cdots, N - 1$

For $n = 0, 1, \cdots, N - 1$ do:

$$k_1 = hf(x_n, y_n)$$

$$k_2 = hf(x_n + \tfrac{1}{2}h, y_n + \tfrac{1}{2}k_1)$$

$$k_3 = hf(x_n + \tfrac{1}{2}h, y_n + \tfrac{1}{2}k_2)$$

$$k_4 = hf(x_n + h, y_n + k_3)$$

$$x_{n+1} = x_n + h$$

$$y_{n+1} = y_n + \tfrac{1}{6}(k_1 + 2k_2 + 2k_3 + k_4)$$

OUTPUT x_{n+1}, y_{n+1}

End
Stop
End RUNGE–KUTTA

$$\kappa = hK \approx h|f_y| \approx h \left| \frac{f(x, y^*) - f(x, y^{**})}{y^* - y^{**}} \right|,$$

and if we choose $x = x_n + \tfrac{1}{2}h$, $y^* = y_n + \tfrac{1}{2}k_2$, $y^{**} = y_n + \tfrac{1}{2}k_1$, then we have $y^* - y^{**} = (k_2 - k_1)/2$, the numerator equals $|k_3 - k_2|/h$, and we obtain

Table 20.5
Runge–Kutta Method Applied to (4); Computations by the Use of (7)

n	x_n	y_n	$0.2214(x_n + y_n)$ $+ 0.0214$	Exact Values $y = e^x - x - 1$	$10^6 \times$ Error of y_n
0	0.0	0	0.021 400	0.000 000	0
1	0.2	0.021 400	0.070 418	0.021 403	3
2	0.4	0.091 818	0.130 289	0.091 825	7
3	0.6	0.222 107	0.203 414	0.222 119	11
4	0.8	0.425 521	0.292 730	0.425 541	20
5	1.0	0.718 251		0.718 282	31

Table 20.6
Comparison of the Accuracy of the Three Methods Under Consideration in the Case of the Initial Value Problem (4), with $h = 0.2$

x	$y = e^x - x - 1$	Absolute Value of Error		
		Euler Method (Table 20.1)	Improved Euler (Table 20.3)	Runge–Kutta (Table 20.5)
0.2	0.021 403	0.021	0.0014	0.000 003
0.4	0.091 825	0.052	0.0034	0.000 007
0.6	0.222 119	0.094	0.0063	0.000 011
0.8	0.425 541	0.152	0.0102	0.000 020
1.0	0.718 282	0.229	0.0156	0.000 031

$$(8) \qquad \kappa \approx 2 \left| \frac{k_3 - k_2}{k_2 - k_1} \right|.$$

We may now make provision to leave h unchanged if, say, $0.01 \leq \kappa_n \leq 0.05$, to decrease h by 50% if $\kappa_n > 0.05$, and to double h if $\kappa_n < 0.01$ (if doubling is possible without increasing h beyond a suitably chosen number H, which depends on the desired accuracy).

Another control of h results from performing the computation simultaneously with step $2h$, which corresponds to increasing the truncation error per step by a factor $2^5 = 32$, but since the number of steps decreases, the actual increase is by a factor $2^5/2 = 16$. Hence the error ϵ of an approximation $\tilde{y}$ obtained with step h equals about 1/15 times the difference $\delta = y^* - \tilde{y}$ of corresponding approximations obtained with steps $2h$ and h, respectively,

$$(9) \qquad \epsilon \approx \frac{1}{15} (y^* - \tilde{y}).$$

We may now choose a number ϵ (for example, 1 unit of the last digit that is supposed to be significant) and leave h unchanged if $0.2\epsilon \leq |\delta| \leq 10\epsilon$, decrease h by 50% if $|\delta| > 10\epsilon$, and double h if $|\delta| < 0.2\epsilon$; of course, in doubling we must take care that the step does not become larger than a suitable number H; this is as before.

Let us illustrate the error estimate (9) by a simple example.

EXAMPLE 4. Runge–Kutta method (of fourth order), error estimate

Solve the initial value problem $y' = (y - x - 1)^2 + 2$, $y(0) = 1$ by the Runge–Kutta method for $0 \leq x \leq 0.4$ with step $h = 0.1$ and estimate the error by (9).

Solution. The numerical results are shown in Table 20.7. They also illustrate how the accuracy increases with decreasing step (from $2h = 0.2$ to $h = 0.1$). The error estimates (9) are close to the actual error. Although we cannot always expect this, formula (9) will give information about the order of magnitude of the error.   ∎

It can be shown that the methods discussed in this section are *numerically stable* (definition in Sec. 18.1). They are **one-step methods** because in each step we use the data of just one preceding step. In the next section we discuss **multistep methods,** which in each step use data from several preceding steps.

Table 20.7
Runge–Kutta Method Applied to the Initial Value Problem in Example 4 and Error Estimate

x	$\tilde{y}$ (Step h)	y^* (Step $2h$)	Error Estimate (9)	Actual Error	Exact Solution (9D)
0.0	1.000 000 000	1.000 000 000	0.000 000 000	0.000 000 000	1.000 000 000
0.1	1.200 334 589			−0.000 000 083	1.200 334 672
0.2	1.402 709 878	1.402 707 341	−0.000 000 181	−0.000 000 157	1.402 710 036
0.3	1.609 336 039			−0.000 000 210	1.609 336 250
0.4	1.822 792 993	1.822 788 917	−0.000 000 291	−0.000 000 226	1.822 793 219

Problems for Sec. 20.1

Apply the Euler method to the following differential equations. (Carry out 10 steps.)

1. $y' = y$, $y(0) = 1$, $h = 0.1$ **2.** $y' = y$, $y(0) = 1$, $h = 0.01$

3. $y' = 2xy + 1$, $y(0) = 0$, $h = 0.1$ **4.** $y' = 2xy$, $y(0) = 1$, $h = 0.1$

5. Apply the improved Euler method to the initial value problem $y' = 2x$, $y(0) = 0$, choosing $h = 0.1$. Why are the errors zero?

6. Give some examples such that the improved Euler–Cauchy method yields the exact solution.

7. Repeat the computations in Example 1, choosing $h = 0.1$ (instead of $h = 0.2$), and note that the errors of the values thus obtained are about 50% of those in Example 1.

8. Do the same task as in Prob. 7, but with $h = 0.01$ (20 steps). Compare the error of the value corresponding to $x = 0.2$ with that in Example 1.

9. Repeat the computations in Example 2, choosing $h = 0.1$, and note that the errors of the values thus obtained are about 25% of those in Example 2.

10. Do the same task as in Prob. 9, but with $h = 0.05$ (8 steps). Compare the error of the value corresponding to $x = 0.4$ with those in Example 2 and Prob. 9.

11. Apply the Euler method and the improved Euler method with $h = 0.1$ and 10 steps to the initial value problem $y' = 2 - 2y$, $y(0) = 0$, and determine and compare the errors.

12. Apply the Runge–Kutta method with $h = 0.1$ and 10 steps to Prob. 11 and determine and compare the errors with those in Prob. 11.

13. Apply the Euler method to the initial value problem $y' = 1 + y^2$, $y(0) = 0$, choosing $h = 0.1$. Compute five steps and compare the results with the exact values.

14. Compute $y = e^x$ for $x = 0, 0.1, 0.2, \cdots, 1.0$ by applying the Runge–Kutta method to the initial value problem in Prob. 1. Show that the first five decimal places of the result are correct.

15. In Prob. 14, insert k_1 into k_2, then k_2 into k_3, etc., and show that the resulting formula becomes $y_{n+1} = 1.105\ 170\ 833 y_n$. Repeat the computations in Prob. 14, using this formula.

16. Apply the Euler method with $h = 0.1$ to $y' = -100y$, $y(0) = 1$. Show that it gives $y = (-9)^{10x}$, compare this and the exact solution, and comment.

17. Continue the computation of y^* in Example 4 with $h = 0.2$ to $x = 1.0$.

18. Solve the initial value problem in Example 4 analytically and verify the given exact values.

19. Another Euler–Cauchy type method is given by

$$y_{n+1} = y_n + hf(x_n + \tfrac{1}{2}h, y^*_{n+1}),$$

where $y^*_{n+1} = y_n + \tfrac{1}{2}hf(x_n, y_n)$. Give a geometric motivation of the method. Apply it to (4), choosing $h = 0.2$ and calculating 5 steps.

20. Kutta's third-order method is defined by

$$y_{n+1} = y_n + \frac{1}{6}(k_1 + 4k_2 + k_3^*)$$

where k_1 and k_2 are as in Table 20.4 and $k_3^* = hf(x_{n+1}, y_n - k_1 + 2k_2)$. Apply this method to (4) in Example 1. Choose $h = 0.2$ and do 5 steps. Compare with Table 20.6.

20.2 Multistep Methods

A **one-step method** is a method which in each step uses only values obtained in a single step, namely, in the preceding step. Examples are the Runge–Kutta method and all the other methods in the previous section. In contrast, a method that uses values from more than one preceding step is called a **multistep method.** We shall explain the idea of obtaining such methods in terms of the derivation of the **Adams–Moulton method,** which is of great practical importance. The initial value problem is as before:

(1)
$$\boxed{y' = f(x, y), \qquad y(x_0) = y_0}$$

where f is assumed to be such that the problem has a unique solution in some interval containing x_0.

Integrating this differential equation from x_n to $x_{n+1} = x_n + h$, we have

(2)
$$\int_{x_n}^{x_{n+1}} f(x, y(x))\, dx = \int_{x_n}^{x_{n+1}} y'(x)\, dx = y(x_{n+1}) - y(x_n).$$

Our notations are as before; in particular, $x_n = x_0 + nh$, and y_n denotes an approximate value of $y(x_n)$. In (2) we replace f by an interpolation polynomial $p_3(x)$ of third degree, so that we can later integrate. For $p_3(x)$ we take the polynomial that at $x_n, x_{n-1}, x_{n-2}, x_{n-3}$ has the values

(3)
$$f_n = f(x_n, y_n), \qquad f_{n-1} = f(x_{n-1}, y_{n-1}), \qquad f_{n-2} = f(x_{n-2}, y_{n-2}),$$
$$f_{n-3} = f(x_{n-3}, y_{n-3}),$$

respectively. (Actually, we could take a polynomial of higher degree, but a cubic polynomial is commonly used in practice.) We can obtain $p_3(x)$, for instance, from the Newton backward difference formula (17), Sec. 18.3:

$$p_3(x) = f_n + r\nabla f_n + \tfrac{1}{2}r(r + 1)\nabla^2 f_n + \tfrac{1}{6}r(r + 1)(r + 2)\nabla^3 f_n$$

where $r = (x - x_n)/h$. We integrate $p_3(x)$ over x from x_n to $x_{n+1} = x_n + h$. This corresponds to integrating over r from 0 to 1. (Note that $dx = h\,dr$, which introduces a common factor h in the result!) We obtain

$$(4) \qquad \int_{x_n}^{x_{n+1}} p_3(x)\,dx = h\left(f_n + \frac{1}{2}\nabla f_n + \frac{5}{12}\nabla^2 f_n + \frac{3}{8}\nabla^3 f_n\right).$$

It is practical to replace these differences by their expressions in terms of f:

$$\nabla f_n = f_n - f_{n-1}$$
$$\nabla^2 f_n = f_n - 2f_{n-1} + f_{n-2}$$
$$\nabla^3 f_n = f_n - 3f_{n-1} + 3f_{n-2} - f_{n-3}.$$

We substitute this into (4) and collect terms. By (2) we then obtain the multistep formula

$$(5) \qquad \boxed{y_{n+1}^* = y_n + \frac{h}{24}(55f_n - 59f_{n-1} + 37f_{n-2} - 9f_{n-3}).}$$

It expresses the new value y_{n+1}^* [approximation of the solution y of (1) at x_{n+1}] in terms of 4 values of f computed from the y-values obtained in the preceding 4 steps. The iteration based on (5) is called the **Adams–Bashford method.**

We wrote y_{n+1}^* in (5), instead of the usual y_{n+1}, because we want to extend the method, using (5) as a predictor and another formula [(6), below] as a corrector. A corrector is obtained by the same idea of integrating a cubic Newton–backward polynomial $\tilde{p}_3(x)$ which at x_{n+1}, x_n, x_{n-1}, x_{n-2} equals f_{n+1}, f_n, f_{n-1}, f_{n-2}, respectively; here,

$$f_{n+1} = f(x_{n+1}, y_{n+1}^*)$$

and the other f's are as in (3). The derivation is quite similar to that of (5). Indeed,

$$\tilde{p}_3(x) = f_{n+1} + r\nabla f_{n+1} + \tfrac{1}{2}r(r + 1)\nabla^2 f_{n+1} + \tfrac{1}{6}r(r + 1)(r + 2)\nabla^3 f_{n+1}$$

where $r = (x - x_{n+1})/h$. We integrate over x from x_n to x_{n+1} as before. This now corresponds to integrating over r from -1 to 0. We obtain

$$\int_{x_n}^{x_{n+1}} \tilde{p}_3(x)\,dx = h\left(f_{n+1} - \frac{1}{2}\nabla f_{n+1} - \frac{1}{12}\nabla^2 f_{n+1} - \frac{1}{24}\nabla^3 f_{n+1}\right).$$

Expressing differences in terms of values of f as before, we finally arrive at the corrector formula

(6)
$$y_{n+1} = y_n + \frac{h}{24}(9f^*_{n+1} + 19f_n - 5f_{n-1} + f_{n-2})$$

where $f^*_{n+1} = f(x_{n+1}, y^*_{n+1})$ and the other f's are as in (3). The predictor–corrector method (5), (6) is called the **Adams–Moulton method.** The program can provide for repeated application of the corrector for a fixed n, say, until the relative difference of successive values (for the same n) in absolute value becomes less than a small preassigned positive number.

Now comes an important feature of multistep methods. We see from (5) that *to get started,* we need f_0, f_1, f_2, f_3; that is, as (3) shows, we first have to compute y_1, y_2, y_3 by some other method, say, by the Runge–Kutta method, to get accurate values. Then we compute $y_4{}^*$ from (5) and y_4 from (6); in the second step we compute $y_5{}^*$ and y_5, and so on.

Indeed, a disadvantage of multistep formulas is that they are not self-starting. But on the other hand, (5), (6) is faster than the Runge–Kutta method since we now need only two new values of f per step, in contrast to the 4 values in each Runge–Kutta step. It can be shown that the present method is of fourth order, like the Runge–Kutta method, and that it is *numerically stable*.

Furthermore, predictor–corrector methods have the advantage that they provide an estimate of the error. Specifically, large $|y_n - y_n{}^*|$ indicates that the error of y_n is probably large in absolute value and calls for a reduction of h. On the other hand, if $|y_n - y_n{}^*|$ is very small, then h may be increased, say, by doubling it. (For more details, see Ref. [E22], pp. 388-393, listed in Appendix 1.)

EXAMPLE 1. Adams–Moulton method
Solve the initial value problem

(7)
$$y' = x + y, \qquad y(0) = 0$$

by the Adams–Moulton method on the interval $0 \leq x \leq 2$, choosing $h = 0.2$.

Solution. The problem is the same as in Examples 1–3, Sec. 20.1, so that we can compare the results. We compute starting values y_1, y_2, y_3 by the Runge–Kutta method. Then in each step we predict by (5) and make one correction by (6) before we execute the next step. The results are shown and compared with the exact values in Table 20.8. We see that the corrections improve the accuracy considerably. This is typical. ∎

This is the end of our discussion of methods for first-order differential equations. In the next section we turn to differential equations of **second order.**

Problems for Sec. 20.2

Solve the following initial value problems by the Adams–Moulton method with $h = 0.1$, starting values from Runge–Kutta, one correction per step, 10 steps. Determine the error of y_n by comparing with the exact solution, as well as the difference between predicted and corrected values.

   1. $y' = 2xy$, $y(0) = 1$
   2. $y' = x + y$, $y(0) = 0$
   3. $y' = 1 + y^2$, $y(0) = 0$

Table 20.8
Adams–Moulton Method Applied to the Initial Value Problem (7);
Predicted Values Computed by (5) and Corrected Values by (6)

n	x_n	Starting y_n	Predicted $y_n{}^*$	Corrected y_n	Exact Values	$10^6 \times$ Error of y_n
0	0.0	0.000 000			0.000 000	0
1	0.2	0.021 400			0.021 403	-3
2	0.4	0.091 818			0.091 825	-7
3	0.6	0.222 107			0.222 119	-12
4	0.8		0.425 361	0.425 529	0.425 541	-12
5	1.0		0.718 066	0.718 270	0.718 282	-12
6	1.2		1.119 855	1.120 106	1.120 117	-11
7	1.4		1.654 885	1.655 191	1.655 200	-9
8	1.6		2.352 653	2.353 026	2.353 032	-6
9	1.8		3.249 190	3.249 646	3.249 647	-1
10	2.0		4.388 505	4.389 062	4.389 056	6

4. Accurate starting values are important to the Adams–Moulton method. Illustrate this by replacing the starting values in Example 1 with the corresponding values obtained by the improved Euler–Cauchy method, calculating y for $x = 0.8, 1.0$ with 1 correction per step and comparing the results with those in Table 20.8.

5. Carry out the details of the calculations that lead to (5) and (6).

6. Verify numerically that in Table 20.8, the error of y_n does not exceed $\frac{1}{14}|y_n - y_n{}^*|$ in absolute value. Prove that for a function f that is at least 5 times continuously differentiable this is generally true, at least approximately; use the following facts. The Adams–Bashford and Adams–Moulton methods are of the fourth order and there are values s and t such that

$$y(x_{n+1}) - y_{n+1}^* = \frac{251}{720} h^5 y^{(5)}(s), \qquad x_{n-3} \leqq s \leqq x_{n+1},$$

$$y(x_{n+1}) - y_{n+1} = -\frac{19}{720} h^5 y^{(5)}(t), \qquad x_{n-3} \leqq t \leqq x_{n+1}.$$

Assume $y^{(5)}$ to be almost constant on $x_{n-3} \leqq x \leqq x_{n+1}$.

7. Show that by applying the method in the text to a polynomial of second degree we obtain the predictor and corrector formulas

$$y_{n+1}^* = y_n + \frac{h}{12}(23f_n - 16f_{n-1} + 5f_{n-2}),$$

$$y_{n+1} = y_n + \frac{h}{12}(5f_{n+1} + 8f_n - f_{n-1}).$$

8. Apply the method in Prob. 7 with $h = 0.2$ to the initial value problem $y' = x + y$, $y(0) = 0$; perform 5 steps and compare with the exact values.

9. Apply the method in Prob. 7 with $h = 0.1$, 10 steps, to the initial value problem $y' = 2xy$, $y(0) = 1$, using Runge–Kutta starting values. Compare with the exact solution.

10. Apply the method in Prob. 7 with $h = 0.2$, 5 steps, to the initial value problem $y' = (x + y - 4)^2$, $y(0) = 4$, using Runge–Kutta starting values. Compare with the exact solution.

20.3 Methods for Second-Order Differential Equations

This section may also be studied immediately after Chap. 2, since it is independent of the preceding sections on numerical methods in Chaps. 18–20.

An **initial value problem** for a second-order differential equation consists of that equation and two conditions (*initial conditions*) referring to the same point. In this section we shall consider two numerical methods for solving initial value problems of the form

(1)
$$y'' = f(x, y, y'), \qquad y(x_0) = y_0, \qquad y'(x_0) = y_0'$$

assuming f to be such that the problem has a unique solution on some interval containing x_0. The first method is simple (but inaccurate) and serves to illustrate the principle, whereas the second method is of great precision and practical importance.

In both methods we shall obtain approximate values of the solution $y(x)$ of (1) at equidistant points $x_1 = x_0 + h$, $x_2 = x_0 + 2h$, $\cdots$; these values will be denoted by $y_1, y_2, \cdots$, respectively. Similarly, approximate values of the derivative $y'(x)$ at those points will be denoted by $y_1', y_2', \cdots$, respectively.

The methods in the previous section were suggested by the Taylor expansion

(2)
$$y(x + h) = y(x) + hy'(x) + \frac{h^2}{2} y''(x) + \frac{h^3}{3!} y'''(x) + \cdots$$

which we shall now use for the same purpose, together with the expansion for the derivative

(3)
$$y'(x + h) = y'(x) + hy''(x) + \frac{h^2}{2} y'''(x) + \cdots .$$

The roughest numerical method is obtained by neglecting the terms containing y''' and the further terms in (2) and (3); this yields the approximations for $y(x + h)$ and the derivative $y'(x + h)$ given by the formulas

$$y(x + h) \approx y(x) + hy'(x) + \frac{h^2}{2} y''(x),$$

$$y'(x + h) \approx y'(x) + hy''(x).$$

In the first step of the method we compute

$$y_0'' = f(x_0, y_0, y_0')$$

from (1), then

$$y_1 = y_0 + hy_0' + \frac{h^2}{2} y_0''$$

which approximates $y(x_1) = y(x_0 + h)$, and furthermore

$$y_1' = y_0' + hy_0''$$

which will be needed in the next step. In the second step we compute

$$y_1'' = f(x_1, y_1, y_1')$$

from (1), then

$$y_2 = y_1 + hy_1' + \frac{h^2}{2} y_1''$$

which approximates $y(x_2) = y(x_0 + 2h)$, and furthermore

$$y_2' = y_1' + hy_1''.$$

In the $(n + 1)$th step we compute

$$y_n'' = f(x_n, y_n, y_n')$$

from (1), then the new value

(4a) $$y_{n+1} = y_n + hy_n' + \frac{h^2}{2} y_n''$$

which is an approximation for $y(x_{n+1})$, and furthermore

(4b) $$y_{n+1}' = y_n' + hy_n''$$

which is an approximation for $y'(x_{n+1})$ needed in the next step.

   Note that, geometrically speaking, this method is an approximation of the curve of $y(x)$ by portions of parabolas.

EXAMPLE 1. An application of the method defined by (4)
Apply (4) to the following initial value problem, choosing $h = 0.2$:

(5) $$y'' = \frac{1}{2}(x + y + y' + 2), \qquad y(0) = 0, \qquad y'(0) = 0.$$

Solution. For the present problem, formulas (4) become

$$y_{n+1} = y_n + 0.2y_n' + 0.02y_n''$$

$$\text{where} \qquad y_n'' = \frac{1}{2}(x_n + y_n + y_n' + 2).$$

$$y_{n+1}' = y_n' + 0.2y_n''$$

Table 20.9
Computations in Example 1

n	x_n	y_n	y_n'	y_n''	Exact (4D)	Error
0	0.0	0.0000	0.0000	1.0000	0.0000	0.0000
1	0.2	0.0200	0.2000	1.2100	0.0214	−0.0014
2	0.4	0.0842	0.4420	1.4631	0.0918	−0.0076
3	0.6	0.2019	0.7346	1.7682	0.2221	−0.0202
4	0.8	0.3842	1.0883	2.1362	0.4255	−0.0413
5	1.0	0.6446			0.7183	−0.0737

The computations are shown in Table 20.9. The student may verify that the exact solution is

$$y = e^x - x - 1.$$

We see that the errors of our approximate values are very large. This is typical, because in most practical cases our present method will be too inaccurate. ∎

Runge–Kutta–Nyström Method

A much more accurate method is the **Runge–Kutta–Nyström method,** which is a generalization of the Runge–Kutta method in Sec. 20.1. We mention without proof that this is a *fourth-order method,* which means that in the Taylor formulas for y and y' the first terms up to and including the term that contains h^4 are given exactly.

The computation in this method can be done as shown in Table 20.10. We see that in each step we compute four auxiliary quantities k_1, k_2, k_3, k_4 and then from them the new approximate value y_{n+1} of the solution y as well as an approximation of the derivative y' needed in the next step.

h can be controlled as described near the end of Sec. 20.1, where we now take for δ the larger of δ^* and δ^{**}; here δ^* is $\frac{1}{15}$ times the difference of corresponding values of y, and δ^{**} is $\frac{1}{15}$ times the difference of corresponding values of y'.

EXAMPLE 2. Runge–Kutta–Nyström method
Apply the Runge–Kutta–Nyström method to the initial value problem (5), choosing $h = 0.2$.

Solution. Here $f = 0.5(x + y + y' + 2)$. Hence

$$k_1 = 0.05(x_n + y_n + y_n' + 2)$$

$$k_2 = 0.05(x_n + 0.1 + y_n + K + y_n' + k_1 + 2) \qquad K = 0.1(y_n' + \tfrac{1}{2}k_1)$$

$$k_3 = 0.05(x_n + 0.1 + y_n + K + y_n' + k_2 + 2)$$

$$k_4 = 0.05(x_n + 0.2 + y_n + L + y_n' + 2k_3 + 2) \qquad L = 0.2(y_n' + k_3).$$

In the present case the differential equation is simple, and so are the expressions for k_1, k_2, k_3, k_4. Hence we may insert k_1 into k_2, then k_2 into k_3 and, finally, k_3 into k_4. The result of this simple calculation is

$$k_2 = 0.05[1.0525(x_n + y_n) + 1.152\,5y_n' + 2.205]$$

$$k_3 = 0.05[1.055\,125(x_n + y_n) + 1.160\,125y_n' + 2.215\,25]$$

$$k_4 = 0.05[1.116\,063\,75(x_n + y_n) + 1.327\,613\,75y_n' + 2.443\,677\,5].$$

Table 20.10
Runge–Kutta–Nyström Method

ALGORITHM R–K–N $(f, x_0, y_0, y_0', h, N)$.

This algorithm computes the solution of the initial value problem $y'' = f(x, y, y')$, $y(x_0) = y_0$, $y'(x_0) = y_0'$ at equidistant points $x_1 = x_0 + h$, $x_2 = x_0 + 2h, \cdots, x_N = x_0 + Nh$; here f is such that this problem has a unique solution on the interval $[x_0, x_N]$.

INPUT: Initial values x_0, y_0, y_0', step size h, number of steps N

OUTPUT: Approximation y_{n+1} to the solution $y(x_{n+1})$ at $x_{n+1} = x_0 + (n + 1)h$, where $n = 0, 1, \cdots, N - 1$

For $n = 0, 1, \cdots, N - 1$ do:

$$k_1 = \tfrac{1}{2}hf(x_n, y_n, y_n')$$

$$k_2 = \tfrac{1}{2}hf(x_n + \tfrac{1}{2}h, y_n + K, y_n' + k_1)$$

$$\text{where} \quad K = \tfrac{1}{2}h(y_n' + \tfrac{1}{2}k_1)$$

$$k_3 = \tfrac{1}{2}hf(x_n + \tfrac{1}{2}h, y_n + K, y_n' + k_2)$$

$$k_4 = \tfrac{1}{2}hf(x_n + h, y_n + L, y_n' + 2k_3)$$

$$\text{where} \quad L = h(y_n' + k_3)$$

$$x_{n+1} = x_n + h$$

$$y_{n+1} = y_n + h(y_n' + \tfrac{1}{3}(k_1 + k_2 + k_3))$$

OUTPUT x_{n+1}, y_{n+1}

        [Approximation to the solution at x_{n+1}]

$$y_{n+1}' = y_n' + \tfrac{1}{3}(k_1 + 2k_2 + 2k_3 + k_4)$$

        [Auxiliary value needed in the next step]

End
Stop
End R–K–N

From this we obtain

(6)
$$y_{n+1} = y_n + a(x_n + y_n) + by_n' + c$$

$$y_{n+1}' = y_n' + a^*(x_n + y_n) + b^*y_n' + c^*$$

where

$$a = 0.010\ 3588 \qquad b = 0.211\ 0421 \qquad c = 0.021\ 4008$$

$$a^* = 0.105\ 5219 \qquad b^* = 0.115\ 8811 \qquad c^* = 0.221\ 4030.$$

Table 20.11 shows the corresponding computations. The errors of the approximate values for $y(x)$ are much smaller than those in Example 1 (cf. Table 20.12).

Table 20.11
Runge–Kutta–Nyström Method (with $h = 0.2$) Applied to the Initial Value Problem (5); Five Steps Computed by the Use of (6)

n	x_n	y_n	$y_n{}'$	$a(x_n + y_n)$ $+ by_n{}' + c$	$a^*(x_n + y_n)$ $+ b^*y_n{}' + c^*$
0	0.0	0.000 0000	0.000 0000	0.021 4008	0.221 4030
1	0.2	0.021 4008	0.221 4030	0.070 4196	0.270 4220
2	0.4	0.091 8204	0.491 8250	0.130 2913	0.330 2940
3	0.6	0.222 1117	0.822 1190	0.203 4186	0.403 4219
4	0.8	0.425 5303	1.225 5409	0.292 7365	0.492 7403
5	1.0	0.718 2668	1.718 2812		

EXAMPLE 3. Runge–Kutta–Nyström method. Airy's equation. Airy function Ai(x)
Solve the initial value problem

$$y'' = xy, \qquad y(0) = 3^{-2/3}/\Gamma(2/3) = 0.355\ 028\ 05, \qquad y'(0) = -3^{-1/3}/\Gamma(1/3) = -0.258\ 819\ 40$$

by the Runge–Kutta–Nyström method with $h = 0.2$; do 5 steps. This is **Airy's equation,**[2] which arose in optics (see Ref. [A14], p. 188, listed in Appendix 1). Γ is the gamma function (see Appendix 3). The initial conditions are such that we obtain a standard solution, the **Airy function** Ai(x), which has been tabulated and investigated (see Ref. [A1], pp. 446, 475). Airy's equation is related to the Bessel equation (see Prob. 12, Sec. 4.6, with $k = i = \sqrt{-1}$).

Solution. For this equation and $h = 0.2$ we obtain from the general formulas in the algorithm

$$k_1 = 0.1x_ny_n,$$
$$k_2 = k_3 = 0.1(x_n + 0.1)(y_n + 0.1y_n{}' + 0.05k_1),$$
$$k_4 = 0.1(x_n + 0.2)(y_n + 0.2y_n{}' + 0.2k_2).$$

Table 20.13 shows the results. The error was determined by comparing with the values in Ref. [A1], p. 475, listed in Appendix 1. ∎

Our work in Example 3 also illustrates that methods for differential equations are often useful in the tabulation of "higher transcendental functions"

Table 20.12
Comparison of Accuracy of the Two Methods Under Consideration in the Case of the Initial Value Problem (5), with $h = 0.2$

x	$y =$ $e^x - x - 1$	Absolute Value of Error	
		Example 1	Table 20.11
0.2	0.021 4028	0.0014	0.000 0020
0.4	0.091 8247	0.0076	0.000 0043
0.6	0.222 1188	0.0202	0.000 0071
0.8	0.425 5409	0.0413	0.000 0106
1.0	0.718 2818	0.0737	0.000 0150

[2]Named after Sir GEORGE BIDELL AIRY (1801—1892), English mathematician, who is known for his work in elasticity and in partial differential equations.

Table 20.13
Runge–Kutta–Nyström Method Applied to Airy's Equation,
Computation of the Airy Function $y = \mathrm{Ai}(x)$

x_n	y_n	$y_n{}'$	$y(x)$ Exact (8D)	$10^8 \times$ Error of y_n
0.0	0.355 028 05	−0.258 819 40	0.355 028 05	0
0.2	0.303 703 04	−0.252 404 64	0.303 703 15	−11
0.4	0.254 742 11	−0.235 830 70	0.254 742 35	−24
0.6	0.209 799 74	−0.212 791 72	0.209 800 06	−32
0.8	0.169 845 99	−0.186 411 34	0.169 846 32	−33
1.0	0.135 292 18	−0.159 146 09	0.135 292 42	−24

given by some formula less practical in numerical work [in our example: a power series or integral representation of $\mathrm{Ai}(x)$].

The methods discussed in this section involve a truncation error. In addition, there are **round-off errors,** and we want to warn the reader that round-off errors may affect results to a large extent. For instance, the solution of the problem $y'' = y$, $y(0) = 1$, $y'(0) = -1$ is $y = e^{-x}$, but the round-off error will introduce a small multiple of the unwanted solution e^x, which may eventually (after sufficiently many steps) swamp the required solution. This is known as **building-up error.** In our simple example we may avoid it by starting from known values of e^{-x} and its derivative for a large x and compute in the reverse direction, but in more complicated cases considerable experience is needed to avoid that phenomenon.

In the remaining sections of this chapter we consider numerical methods for **partial differential equations.**

Problems for Sec. 20.3

1. Repeat the computation in Example 1, choosing $h = 0.1$, and compare the errors of the values thus obtained with those in Example 1.
2. Perform the same task as in Prob. 1, choosing $h = 0.05$.

Apply (4) to the following initial value problems (5 steps).
3. $y'' = -y$, $y(0) = 0$, $y'(0) = 1$, $h = 0.1$
4. $y'' = -y$, $y(0) = 1$, $y'(0) = 0$, $h = 0.1$
5. $y'' = -y$, $y(0) = 1$, $y'(0) = 0$, $h = 0.05$

6. Verify the formulas in Example 3 for the Airy equation.
7. Verify the computations in Example 3.
8. Verify, to 3D-accuracy, the values y_0 and $y_0{}'$ in Example 3, using (25) in Appendix A3.1 and Table A3 in Appendix 4.

Apply the Runge–Kutta–Nyström method to the following initial value problems.
9. The initial value problem in Prob. 3, $h = 0.2$, 5 steps. (Compare the results with the exact values.)
10. $y'' = xy' - 5y$, $y(0) = 0$, $y'(0) = 3$, $h = 0.2$,   5 steps
11. $y'' = xy' - 2y$, $y(0) = 1$, $y'(0) = 0$, $h = 0.2$,   5 steps

12. Show that the exact solution in Prob. 11 is $y = 1 - x^2$. Determine the errors. Comment.

13. Consider the initial value problem $(1 - x^2)y'' - 2xy' + 2y = 0$, $y(0) = 0$, $y'(0) = 1$. Show that (4) with $h = 0.1$ takes the form

$$y_{n+1} = y_n + 0.1y_n' + 0.01 \frac{x_n y_n' - y_n}{1 - x_n^2}, \qquad y_{n+1}' = y_n' + 0.2 \frac{x_n y_n' - y_n}{1 - x_n^2}.$$

Carry out five steps. Verify by substitution that the exact solution is $y = x$.

14. **(Boundary value problem)** Numerical methods for boundary value problems will not be considered in any generality, but the reader may show the following. To solve

$$y'' + f(x) y' + g(x) y = r(x), \qquad y(0) = 0, \quad y(b) = k$$

numerically, apply one of the previous methods to get a solution $Y(x)$ of the equation satisfying the initial conditions $Y(0) = 0$, $Y'(0) = 1$. Then determine by that method a solution $z(x)$ of the homogeneous equation satisfying the initial conditions $z(0) = 0$, $z'(0) = 1$. Show that

$$y(x) = Y(x) + cz(x)$$

with a suitable c satisfies the given problem. What is the condition for c?

15. To solve the boundary value problem

$$y'' + f(x) y' + g(x) y = r(x), \qquad y(a) = k_1, \quad y(b) = k_2$$

numerically, show that one may determine a solution $y_1(x)$ satisfying $y_1(a) = k_1$, $y_1'(a) = c_1$ and a solution $y_2(x)$ satisfying $y_2(a) = k_1$, $y_2'(a) = c_2$ and such that $y_1(b) \neq y_2(b)$ and then obtain the solution of the given problem in the form

$$y(x) = \frac{1}{y_1(b) - y_2(b)} [(k_2 - y_2(b)) y_1(x) + (y_1(b) - k_2) y_2(x)].$$

20.4 Numerical Methods for Elliptic Partial Differential Equations

The remaining sections of this chapter are devoted to numerical methods for partial differential equations, particularly for the Laplace, Poisson, heat and wave equations, which are basic in engineering and, at the same time, are model cases of elliptic, parabolic and hyperbolic equations. The definitions are as follows.

A partial differential equation is called **quasilinear** if it is linear in the highest derivatives. Hence a second-order quasilinear equation in two independent variables x, y can be written

(1) $$au_{xx} + 2bu_{xy} + cu_{yy} = F(x, y, u, u_x, u_y).$$

u is the unknown function. This equation is said to be of:

elliptic type   if $ac - b^2 > 0$ (example: *Laplace equation*)

parabolic type   if $ac - b^2 = 0$ (example: *heat equation*)

hyperbolic type if $ac - b^2 < 0$ (example: *wave equation*)

(where in the heat and wave equations, y is time t). Here, the *coefficients* a, b, c may be functions of x, y, so that the type of (1) may be different in different regions of the xy-plane. This classification is not merely a formal matter but is of great practical importance because the general behavior of solutions differs from type to type and so do the additional conditions (boundary and initial conditions) that must be taken into account.

Applications involving *elliptic* equations usually lead to boundary value problems in a region R, called a *first boundary value problem* or **Dirichlet problem** if u is prescribed on the boundary curve C of R, a *second boundary value problem* or **Neumann problem** if $u_n = \partial u/\partial n$ (normal derivative of u) is prescribed on C, and *third* or **mixed problem** if u is prescribed on a part of C and u_n on the remaining part. C usually is a closed curve (or sometimes consists of two or more such curves).

Difference Equations for the Laplace and Poisson Equations

In this section we consider the **Laplace equation**

$$(2) \qquad \nabla^2 u = u_{xx} + u_{yy} = 0$$

and the **Poisson equation**

$$(3) \qquad \nabla^2 u = u_{xx} + u_{yy} = f(x, y)$$

which are the most important elliptic equations in applications. To obtain methods of numerical solution, we replace the partial derivatives in a given equation by corresponding difference quotients, as follows. By the Taylor formula,

$$(a) \quad u(x + h, y) = u(x, y) + hu_x(x, y) + \tfrac{1}{2}h^2 u_{xx}(x, y) + \tfrac{1}{6}h^3 u_{xxx}(x, y) + \cdots$$

(4)

$$(b) \quad u(x - h, y) = u(x, y) - hu_x(x, y) + \tfrac{1}{2}h^2 u_{xx}(x, y) - \tfrac{1}{6}h^3 u_{xxx}(x, y) + \cdots.$$

Subtracting (4b) from (4a) and neglecting terms in h^3, h^4, $\cdots$, we obtain

$$(5a) \qquad u_x(x, y) \approx \frac{1}{2h} [u(x + h, y) - u(x - h, y)].$$

Similarly,

$$(5b) \qquad u_y(x, y) \approx \frac{1}{2k} [u(x, y + k) - u(x, y - k)].$$

We turn to second derivatives. Adding (4a) and (4b) and neglecting terms in h^4, h^5, $\cdots$, we obtain

$$u(x + h, y) + u(x - h, y) \approx 2u(x, y) + h^2 u_{xx}(x, y).$$

Solving for u_{xx}, we have

(6a) $$u_{xx}(x, y) \approx \frac{1}{h^2} [u(x + h, y) - 2u(x, y) + u(x - h, y)].$$

Similarly,

(6b) $$u_{yy}(x, y) \approx \frac{1}{k^2} [u(x, y + k) - 2u(x, y) + u(x, y - k)].$$

Also (cf. Prob. 3)

(6c) $$u_{xy}(x, y) \approx \frac{1}{4hk} [u(x + h, y + k) - u(x - h, y + k)$$
$$- u(x + h, y - k) + u(x - h, y - k)].$$

Figure 439 shows the points $(x + h, y)$, $(x - h, y)$, $\cdots$ occurring in (5) and (6). [Formula (6c) will not be needed in our further work.]

We now substitute (6a) and (6b) into the *Poisson equation* (3), choosing $k = h$ to obtain a simple formula:

(7)
$$u(x + h, y) + u(x, y + h) + u(x - h, y) + u(x, y - h) - 4u(x, y)$$
$$= h^2 f(x, y).$$

This is a difference equation corresponding to (3). Hence for the *Laplace equation* (2) the corresponding difference equation is

(8) $$u(x + h, y) + u(x, y + h) + u(x - h, y) + u(x, y - h) - 4u(x, y) = 0.$$

h is called the *mesh size*. Equation (8) relates u at (x, y) to u at the four neighboring points shown in Fig. 440. For convenience, these neighboring points are often called E (East), N (North), W (West), S (South). Then Fig. 440 takes the form of Fig. 441, and (7) becomes

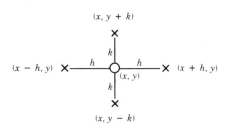

Fig. 439. Points in (5) and (6)

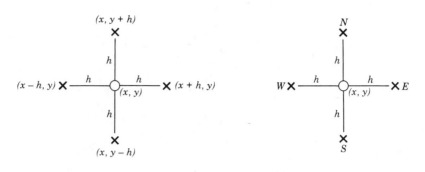

Fig. 440. Points in (7) and (8)   **Fig. 441.** Notation in (7*)

(7*) $$u(E) + u(N) + u(W) + u(S) - 4u(x, y) = h^2 f(x, y).$$

Our approximation of $h^2 \nabla^2 u$ in (7) and (8) is a 5-point approximation with the coefficient scheme or *pattern*

(9)
$$\begin{Bmatrix} & 1 & \\ 1 & -4 & 1 \\ & 1 & \end{Bmatrix}$$

Note that (8) has a remarkable interpretation: *u at (x, y) equals the mean of the values of u at the four neighboring mesh points.*

To grasp (7) at first sight, we may very conveniently represent it by that pattern:

$$\begin{Bmatrix} & 1 & \\ 1 & -4 & 1 \\ & 1 & \end{Bmatrix} u = h^2 f(x, y).$$

Dirichlet Problem

In the numerical solution of the Dirichlet problem (definition above) in a region R we first choose h and introduce in R a grid consisting of equidistant horizontal and vertical straight lines of distance h. Their intersections are called **mesh points** (or *nodes* or *lattice points*). Cf. Fig. 442. Then we use a difference equation approximating the given partial differential equation—formula (8) in the case of the Laplace equation—by which we relate the unknown values of u at the mesh points in R to each other and to the given boundary values, as will be discussed below. This yields a system of linear *algebraic* equations. By solving it we obtain approximations to the unknown values of u at the mesh points in R. We shall see that the number of equations equals the number of unknowns, that is, the number of mesh points in R. Since at each mesh point, u is only related to the values at the neighboring mesh points, the coefficients of the system form a **sparse matrix,** that is, a matrix with relatively few nonzero entries. In practice, this matrix will be

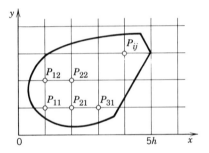

Fig. 442. Region in the *xy*-plane covered by a grid of mesh *h*, also showing mesh points $P_{11} = (h, h), \cdots, P_{ij} = (ih, jh), \cdots$

large, since for obtaining high accuracy one needs many mesh points, and a 500 × 500 or larger matrix may cause a storage problem.[3] Hence an indirect method (cf. Sec. 19.3) is preferable to a direct one. In particular, we may use the **Gauss–Seidel method**, which in the present context is also called **Liebmann's method.** We illustrate the whole approach by an example, in which we keep the number of equations small, for simplicity. As convenient *notations for mesh points and corresponding values of the solution* (and of approximate solutions) we use (cf. also Fig. 442)

(10) $$P_{ij} = (ih, jh), \qquad u_{ij} = u(ih, jh).$$

With this notation we can write (8) for any mesh point P_{ij} in the form

(11) $$\boxed{u_{i+1,j} + u_{i,j+1} + u_{i-1,j} + u_{i,j-1} - 4u_{ij} = 0.}$$

EXAMPLE 1. Laplace equation. Liebmann's method
The four sides of a square plate of side 12 cm made of homogeneous material are kept at constant temperature 0°C and 100°C as shown in Fig. 443*a*. Using a (very wide) grid of mesh 4 cm and applying Liebmann's method (that is, Gauss–Seidel iteration), find the (steady-state) temperature at the mesh points.

Solution. In the case of independence of time, the heat equation (cf. Sec. 9.7)

$$u_t = c^2(u_{xx} + u_{yy})$$

reduces to the Laplace equation. Hence our problem is a Dirichlet problem for this equation. We choose the grid shown in Fig. 443*b* and consider the mesh points in the order P_{11}, P_{21}, P_{12}, P_{22}. We use (11) and, in each equation, take to the right all the terms resulting from the given boundary values. Then we obtain the system

(12)
$$
\begin{aligned}
-4u_{11} + u_{21} + u_{12} \qquad\qquad &= -200 \\
u_{11} - 4u_{21} \qquad + u_{22} &= -200 \\
u_{11} \qquad - 4u_{12} + u_{22} &= -100 \\
u_{21} + u_{12} - 4u_{22} &= -100
\end{aligned}
$$

---

[3]The present matrix is *not* tridiagonal! (Definition below.) If it were, we could apply the Gauss elimination without causing that storage problem.

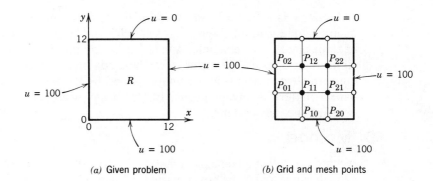

(a) Given problem   (b) Grid and mesh points

Fig. 443. Example 1

In practice, one would solve such a small system by the Gauss elimination, finding

$$u_{11} = u_{21} = 87.5, \qquad u_{12} = u_{22} = 62.5.$$

More exact values (exact to 1D) of the solution of our problem are 88.1 and 61.9, respectively. (These were obtained by using Fourier series.) Hence the error is about 1%, which is surprisingly accurate for a grid of such a large mesh size h. When the system of equations is large, one would solve it by an indirect method, such as Liebmann's method. For (12) this is as follows. We write (12) in the form

$$u_{11} = \qquad\qquad 0.25u_{21} + 0.25u_{12} \qquad\qquad\quad + 50$$
$$u_{21} = 0.25u_{11} \qquad\qquad\qquad\qquad\qquad + 0.25u_{22} + 50$$
$$u_{12} = 0.25u_{11} \qquad\qquad\qquad\qquad\qquad + 0.25u_{22} + 25$$
$$u_{22} = \qquad\qquad 0.25u_{21} + 0.25u_{12} \qquad\qquad\quad + 25$$

These equations are now used for the Gauss–Seidel iteration. They are identical with (2) in Sec. 19.3, where $u_{11} = x_1$, $u_{21} = x_2$, $u_{12} = x_3$, $u_{22} = x_4$, and the iteration is explained there, where 100, 100, 100, 100 are chosen as starting values. Using physical intuition on what the values at the mesh points might approximately be, one can save some work by choosing better starting values. The reader may verify that the exact solution of the system is $u_{11} = u_{21} = 87.5$, $u_{12} = u_{22} = 62.5$.

For a further idea that leads to substantial simplification, see Prob. 6.

Remark. It is interesting to note that if we choose mesh $h = L/n$ ($L =$ side of R) and consider the $(n - 1)^2$ mesh points row by row in the order

$$P_{11}, P_{21}, \cdots, P_{n1}, P_{12}, P_{22}, \cdots, P_{n2}, \cdots,$$

then the system of equations has the $(n - 1)^2 \times (n - 1)^2$ coefficient matrix

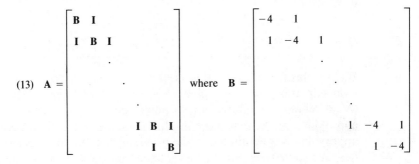

$$(13) \quad \mathbf{A} = \begin{bmatrix} \mathbf{B} & \mathbf{I} & & & \\ \mathbf{I} & \mathbf{B} & \mathbf{I} & & \\ & & \cdot & & \\ & & & \cdot & \\ & & \mathbf{I} & \mathbf{B} & \mathbf{I} \\ & & & \mathbf{I} & \mathbf{B} \end{bmatrix} \quad \text{where} \quad \mathbf{B} = \begin{bmatrix} -4 & 1 & & & \\ 1 & -4 & 1 & & \\ & & \cdot & & \\ & & & \cdot & \\ & & 1 & -4 & 1 \\ & & & 1 & -4 \end{bmatrix}$$

is an $(n - 1) \times (n - 1)$ matrix, and it can be shown that $\mathbf{A}$ is nonsingular. ∎

A matrix is called a **band matrix** if it has all its nonzero entries on the main diagonal and on sloping lines parallel to it (separated by sloping lines of zeros or not). For example, **A** in (13) is a band matrix. Although the Gauss elimination does not preserve zeros between bands, it does not introduce nonzero elements outside the limits defined by the original bands. Hence a band structure is advantageous. In (13) it has been achieved by carefully ordering the mesh points.

ADI Method

A matrix is called a **tridiagonal matrix** if it has all its nonzero entries on the main diagonal and in the positions immediately adjacent to the main diagonal. In this case the Gauss elimination is particularly simple.

This raises the question of whether in the solution of the Dirichlet problem for the Laplace or Poisson equations one could obtain a system of equations whose coefficient matrix is tridiagonal. The answer is yes, and a popular method of that kind, called the **ADI method** (*alternating direction implicit method*) was developed by Peaceman and Rachford. The idea is as follows. The pattern in (9) shows that we could obtain a tridiagonal matrix if there were only the three points in a row (or only the three points in a column). This suggests to write (11) in the form

(14a) $$u_{i-1,j} - 4u_{ij} + u_{i+1,j} = -u_{i,j-1} - u_{i,j+1}$$

so that the left-hand side belongs to y-row j and the right-hand side to x-column i. Of course, we can also write (11) in the form

(14b) $$u_{i,j-1} - 4u_{ij} + u_{i,j+1} = -u_{i-1,j} - u_{i+1,j}$$

so that the left-hand side belongs to column i and the right-hand side to row j. In the ADI method we proceed by iteration. At every mesh point we choose an arbitrary starting value $u_{ij}^{(0)}$. In each step we compute new values at all mesh points. In one step we use an iteration formula resulting from (14a) and in the next step an iteration formula resulting from (14b), and so on in alternating order.

In detail: suppose approximations $u_{ij}^{(m)}$ have been computed. Then, to obtain the next approximations $u_{ij}^{(m+1)}$, we substitute the $u_{ij}^{(m)}$ on the right-hand side of (14a) and solve for the $u_{ij}^{(m+1)}$ on the left-hand side; that is, we use

(15a) $$\boxed{u_{i-1,j}^{(m+1)} - 4u_{ij}^{(m+1)} + u_{i+1,j}^{(m+1)} = -u_{i,j-1}^{(m)} - u_{i,j+1}^{(m)}.}$$

We use this for a fixed j, that is, *for a fixed row j,* and for all internal mesh points in this row. This gives a system of N linear algebraic equations (N = number of internal mesh points per row) in N unknowns, the new approximations of u at these mesh points. Note that (15a) involves not only approximations computed in the previous step but also given boundary values. We solve the system (15a) (j fixed!) by the Gauss elimination. Then we go to the next row, obtain another system of N equations and solve it by

Gauss, and so on until all rows are done. In the next step we *alternate direction*, that is, we compute the next approximations $u_{ij}^{(m+2)}$ column by column from the $u_{ij}^{(m+1)}$ and the given boundary values, using a formula obtained from (14b) by substituting the $u_{ij}^{(m+1)}$ on the right:

$$\textbf{(15b)} \qquad \boxed{ u_{i,j-1}^{(m+2)} - 4u_{ij}^{(m+2)} + u_{i,j+1}^{(m+2)} = -u_{i-1,j}^{(m+1)} - u_{i+1,j}^{(m+1)}. }$$

For each fixed i, that is, *for each column*, this is a system of M equations (M = number of internal mesh points per column) in M unknowns, which we solve by the Gauss elimination. Then we go to the next column, and so on until all columns are done.

Let us consider an example that merely serves to explain the whole method. (In practice one would solve this problem directly by the Gauss elimination.) After this we shall discuss how to improve the convergence of the ADI method.

EXAMPLE 2. Dirichlet problem. ADI method

Explain the procedure and formulas of the ADI method in terms of the problem in Example 1, using the same grid and starting values 100, 100, 100, 100.

Solution. While working, we keep an eye on Fig. 443b, p. 1087, and the given boundary values. We obtain first approximations $u_{11}^{(1)}, u_{21}^{(1)}, u_{12}^{(1)}, u_{22}^{(1)}$ from (15a) with $m = 0$. We write boundary values contained in (15a) without an upper index, for better identification and to indicate that these given values remain the same during the iteration. From (15a) with $m = 0$ we have for $j = 1$ (first row) the system

$$(i = 1) \qquad u_{01} - 4u_{11}^{(1)} + u_{21}^{(1)} \qquad\quad = -u_{10} - u_{12}^{(0)}$$

$$(i = 2) \qquad\qquad u_{11}^{(1)} - 4u_{21}^{(1)} + u_{31} = -u_{20} - u_{22}^{(0)}.$$

The solution is $u_{11}^{(1)} = u_{21}^{(1)} = 100$. For $j = 2$ (second row) we obtain from (15a) the system

$$(i = 1) \qquad u_{02} - 4u_{12}^{(1)} + u_{22}^{(1)} \qquad\quad = -u_{11}^{(0)} - u_{13}$$

$$(i = 2) \qquad\qquad u_{12}^{(1)} - 4u_{22}^{(1)} + u_{32} = -u_{21}^{(0)} - u_{23}.$$

The solution is $u_{12}^{(1)} = u_{22}^{(1)} = 66.667$.

Second approximations $u_{11}^{(2)}, u_{21}^{(2)}, u_{12}^{(2)}, u_{22}^{(2)}$ are now obtained from (15b) with $m = 1$ by using the first approximations just computed and the boundary values. For $i = 1$ (first column) we obtain from (15b) the system

$$(j = 1) \qquad u_{10} - 4u_{11}^{(2)} + u_{12}^{(2)} \qquad\quad = -u_{01} - u_{21}^{(1)}$$

$$(j = 2) \qquad\qquad u_{11}^{(2)} - 4u_{12}^{(2)} + u_{13} = -u_{02} - u_{22}^{(1)}.$$

The solution is $u_{11}^{(2)} = 91.11$, $u_{12}^{(2)} = 64.44$. For $i = 2$ (second column) we obtain from (15b) the system

$$(j = 1) \qquad u_{20} - 4u_{21}^{(2)} + u_{22}^{(2)} \qquad\quad = -u_{11}^{(1)} - u_{31}$$

$$(j = 2) \qquad\qquad u_{21}^{(2)} - 4u_{22}^{(2)} + u_{23} = -u_{12}^{(1)} - u_{32}.$$

The solution is $u_{21}^{(2)} = 91.11$, $u_{22}^{(2)} = 64.44$.

In this example, which merely serves to explain the practical procedure in the ADI method, the accuracy of the second approximations is about the same as of those of two Gauss–Seidel steps in Sec. 19.3 (where $u_{11} = x_1$, $u_{21} = x_2$, $u_{12} = x_3$, $u_{22} = x_4$), as the table on the next page shows.

	u_{11}	u_{21}	u_{12}	u_{22}
ADI, 2nd approximations	91.11	91.11	64.44	64.44
Gauss–Seidel, 2nd approximations	93.75	90.62	65.62	64.06
Exact solution of (12)	87.50	87.50	62.50	62.50

Introducing a parameter p, we can also write (11) in the form

$$\text{(16a)} \quad u_{i-1,j} - (2 + p)u_{ij} + u_{i+1,j} = -u_{i,j-1} + (2 - p)u_{ij} - u_{i,j+1}$$

and

$$\text{(16b)} \quad u_{i,j-1} - (2 + p)u_{ij} + u_{i,j+1} = -u_{i-1,j} + (2 - p)u_{ij} - u_{i+1,j}.$$

This gives the more general ADI iteration formulas

$$\text{(17a)} \quad u_{i-1,j}^{(m+1)} - (2 + p)u_{ij}^{(m+1)} + u_{i+1,j}^{(m+1)} = -u_{i,j-1}^{(m)} + (2 - p)u_{ij}^{(m)} - u_{i,j+1}^{(m)}$$

and

$$\text{(17b)} \quad u_{i,j-1}^{(m+2)} - (2 + p)u_{ij}^{(m+2)} + u_{i,j+1}^{(m+2)} = -u_{i-1,j}^{(m+1)} + (2 - p)u_{ij}^{(m+1)} - u_{i+1,j}^{(m+1)}.$$

For $p = 2$, this is (15). The parameter p may be used for improving convergence. Indeed, one can show that the ADI method converges for positive p, and that the optimum value for maximum rate of convergence is

$$\text{(18)} \qquad\qquad p_0 = 2 \sin \frac{\pi}{K}$$

where K is the larger of $M + 1$ and $N + 1$ (see above). Even better results can be achieved by letting p vary from step to step. More details of the ADI method and variants are discussed in Ref. [E12] listed in Appendix 1.

Our discussion was concerned with the Dirichlet problem. In the next section we turn to the **Neumann problem** and the **mixed problem,** which require a new idea.

Problems for Sec. 20.4

1. Derive (5b).   **2.** Derive (6b).   **3.** Derive (6c).

4. Solve Example 1, choosing $h = 3$.

5. In Example 1, take $h = 6$, compute u_{11} and compare it with the exact value 75.

6. (**Use of symmetry**) Conclude from the boundary values in Example 1 that $u_{21} = u_{11}$ and $u_{22} = u_{12}$. Show that this leads to a system of two equations and solve it.

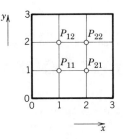

Fig. 444. Problems 7–13

Using the grid in Fig. 444, compute the potential at the four interior points for the following boundary values.

7. $u(1, 0) = -80$, $u(2, 0) = 400$, $u = 0$ on the three other edges of the boundary

8. $u = 110$ on the upper and lower edges, $u = 0$ on the left edge, $u = -120$ on the right edge

9. $u = x^4$ on the lower edge, $81 - 54y^2 + y^4$ on the right edge, $x^4 - 54x^2 + 81$ on the upper edge, y^4 on the left edge. (Verify that the exact solution is $x^4 - 6x^2y^2 + y^4$ and determine the error.)

10. $u = 0$ on the left edge, x^3 on the lower edge, $27 - 9y^2$ on the right edge, $x^3 - 27x$ on the upper edge

11. $u = \sin \frac{1}{3}\pi x$ on the upper edge, 0 on the other edges

12. Apply the ADI method (2 steps) to the Dirichlet problem in Prob. 11, using the grid in Fig. 444, as before and starting values zero.

13. What p_0 in (18) should we choose for Prob. 12? Apply the ADI formulas (17) with $p_0 = 1.7$ to Prob. 12, performing 1 step. Illustrate the improved convergence by comparing with the corresponding values 0.077, 0.308 after the first step in Prob. 12. (Use starting values zero.)

14. Find the potential in Fig. 445 using (a) the coarse grid, (b) the fine grid, and Gauss elimination. *Hint.* In (b), use symmetry; take $u = 0$ as the boundary value at the two points at which the potential has a jump.

15. How many Gauss–Seidel steps would be needed to obtain the answer to Prob. 14, coarse grid, to 5S (5 significant figures) if one started from 0, 0? In the case of the fine grid in Prob. 14, the Gauss–Seidel method converges more slowly. Can you see the reason by inspecting the system of equations?

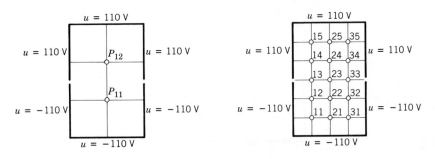

Fig. 445. Region and grids in Problem 14

20.5 Neumann and Mixed Problems. Irregular Boundary

We continue our consideration of the numerical solution of boundary value problems for elliptic equations in a region R in the xy-plane. The Dirichlet problem was studied in the previous section. In Neumann and mixed problems (defined in the previous section) we are confronted with a new situation, because there are boundary points at which the normal derivative $\partial u/\partial n$ of the solution is given, but u itself is unknown since it is not given. To handle such points we need a new idea. This idea is the same for Neumann and mixed problems. Hence we may explain it in connection with one of these two types of problem. We shall do so and consider a typical example as follows.

EXAMPLE 1. Mixed boundary value problem for a Poisson equation

Solve the mixed boundary value problem for the Poisson equation

$$\nabla^2 u = u_{xx} + u_{yy} = 12xy$$

shown in Fig. 446a.

Solution. We use the grid shown in Fig. 446b, where $h = 0.5$. From the formulas $u = 3y^3$ and $u_n = 6x$ given on the boundary we compute the boundary data

$$(1) \qquad u_{31} = 0.375, \qquad u_{32} = 3, \qquad \frac{\partial u_{12}}{\partial n} = \frac{\partial u_{12}}{\partial y} = 3, \qquad \frac{\partial u_{22}}{\partial n} = \frac{\partial u_{22}}{\partial y} = 6.$$

P_{11} and P_{21} are interior mesh points and can be handled as in the previous section. Indeed, from (7), Sec. 20.4, with $h^2 = 0.25$ and $f(x, y) = 12xy$, and from the given boundary values we obtain two equations corresponding to P_{11} and P_{21}:

$$(2a) \qquad \begin{aligned} -4u_{11} + u_{21} + u_{12} \phantom{+ u_{22}} &= 0.75 \\ u_{11} - 4u_{21} \phantom{+ u_{12}} + u_{22} &= 1.5 - 0.375 = 1.125. \end{aligned}$$

The only difficulty with these equations seems to be that they involve the unknown values u_{12} and u_{22} of u at P_{12} and P_{22} on the boundary, where the normal derivative $u_n = \partial u/\partial n = \partial u/\partial y$ is given, instead of u; but we shall overcome this difficulty as we go on.

We consider P_{12} and P_{22}. The idea that will help us here is as follows. We imagine the region R to be extended above to the first row of external mesh points (corresponding to $y = 1.5$),

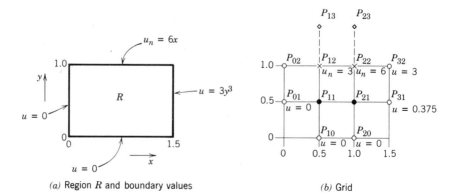

(a) Region R and boundary values   (b) Grid

Fig. 446. Mixed boundary value problem in Example 1

and we assume that the differential equation also holds in the extended region. Then we can write down two more equations as before (cf. Fig. 446b)

(2b)
$$u_{11} \qquad - 4u_{12} + \quad u_{22} + u_{13} \qquad = 1.5$$

$$u_{21} + \quad u_{12} - 4u_{22} \qquad + u_{23} = 3 - 3 = 0.$$

We remember that we have not yet used the boundary condition on the upper part of the boundary of R, and we also notice that in (2b) we have introduced two more unknowns u_{13}, u_{23}. But we can now use that condition and get rid of u_{13}, u_{23} by applying the central difference formula for u_y. From (1) we then obtain (cf. Fig. 446b)

$$3 = \frac{\partial u_{12}}{\partial y} = \frac{u_{13} - u_{11}}{2h} = u_{13} - u_{11}, \qquad \text{hence} \qquad u_{13} = u_{11} + 3$$

$$6 = \frac{\partial u_{22}}{\partial y} = \frac{u_{23} - u_{21}}{2h} = u_{23} - u_{21}, \qquad \text{hence} \qquad u_{23} = u_{21} + 6.$$

Substituting these results into (2b) and simplifying, we have

$$2u_{11} \qquad - 4u_{12} + \quad u_{22} = 1.5 - 3 = -1.5$$

$$2u_{21} + \quad u_{12} - 4u_{22} = 0 - 6 \quad = -6.$$

Together with (2a) this yields, written in matrix form,

(3)
$$\begin{bmatrix} -4 & 1 & 1 & 0 \\ 1 & -4 & 0 & 1 \\ 2 & 0 & -4 & 1 \\ 0 & 2 & 1 & -4 \end{bmatrix} \begin{bmatrix} u_{11} \\ u_{21} \\ u_{12} \\ u_{22} \end{bmatrix} = \begin{bmatrix} 0.750 \\ 1.125 \\ -1.500 \\ -6.000 \end{bmatrix}.$$

The solution is as follows; the exact values of the problem are given in parentheses.

$$u_{12} = 0.866 \quad \text{(exact 1)} \qquad u_{22} = 1.812 \quad \text{(exact 2)}$$

$$u_{11} = 0.077 \quad \text{(exact 0.125)} \qquad u_{21} = 0.191 \quad \text{(exact 0.25).} \qquad \blacksquare$$

Irregular Boundary

We continue our consideration of the numerical solution of boundary value problems for elliptic equations in a region R in the xy-plane. If R has a simple geometric shape, we can usually arrange for certain mesh points to lie on the boundary C of R, and then we can approximate partial derivatives as explained in the previous section. However, if C intersects the grid at points that are not mesh points, then at points close to the boundary we must proceed differently, as follows.

The mesh point O in Fig. 447 on the next page is of that kind. For O and its neighbors A and P we obtain from Taylor's theorem

(4)
$$\text{(a)} \quad u_A = u_O + ah \frac{\partial u_O}{\partial x} + \frac{1}{2}(ah)^2 \frac{\partial^2 u_O}{\partial x^2} + \cdots$$

$$\text{(b)} \quad u_P = u_O - h \frac{\partial u_O}{\partial x} + \frac{1}{2}h^2 \frac{\partial^2 u_O}{\partial x^2} + \cdots.$$

We disregard the terms marked by dots and eliminate $\partial u_O / \partial x$. Equation (4b) times a plus equation (4a) gives

$$u_A + au_P \approx (1 + a)u_O + \frac{1}{2} a(a + 1)h^2 \frac{\partial^2 u_O}{\partial x^2} .$$

We solve this algebraically for the derivative, obtaining

$$\frac{\partial^2 u_O}{\partial x^2} \approx \frac{2}{h^2} \left[\frac{1}{a(1 + a)} u_A + \frac{1}{1 + a} u_P - \frac{1}{a} u_O \right].$$

Similarly, by considering the points O, B and Q,

$$\frac{\partial^2 u_O}{\partial y^2} \approx \frac{2}{h^2} \left[\frac{1}{b(1 + b)} u_B + \frac{1}{1 + b} u_Q - \frac{1}{b} u_O \right].$$

By addition,

$$(5) \quad \nabla^2 u_O \approx \frac{2}{h^2} \left[\frac{u_A}{a(1 + a)} + \frac{u_B}{b(1 + b)} + \frac{u_P}{1 + a} + \frac{u_Q}{1 + b} - \frac{(a + b)u_O}{ab} \right].$$

For example, if $a = \frac{1}{2}$, $b = \frac{1}{2}$, instead of the pattern (cf. Sec. 20.4)

$$\left\{ \begin{matrix} & 1 & \\ 1 & -4 & 1 \\ & 1 & \end{matrix} \right\} \quad \text{we now have} \quad \left\{ \begin{matrix} & \frac{4}{3} & \\ \frac{2}{3} & -4 & \frac{4}{3} \\ & \frac{2}{3} & \end{matrix} \right\},$$

the sum of all five terms still being zero (which is useful for checking).

Using the same ideas, the reader may show that in the case of Fig. 448,

$$(6) \quad \nabla^2 u_O \approx \frac{2}{h^2} \left[\frac{u_A}{a(a + p)} + \frac{u_B}{b(b + q)} + \frac{u_P}{p(p + a)} + \frac{u_Q}{q(q + b)} - \frac{ap + bq}{abpq} u_O \right],$$

a formula that takes care of all conceivable cases.

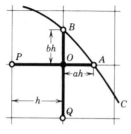

Fig. 447. Curved boundary C of a region R, a mesh point O near C and neighbors A, B, P, Q

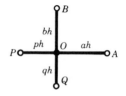

Fig. 448. Neighboring points A, B, P, Q of a mesh point O and notations in formula (6)

EXAMPLE 2. Dirichlet problem for the Laplace equation. Curved boundary

Find the potential u in the region in Fig. 449 that has the boundary values given in that figure; here the curved portion of the boundary is an arc of the circle of radius 10 about (0, 0). Use the grid in the figure.

Solution. u is a solution of the Laplace equation. From the given formulas for the boundary values $u = x^3$, $u = 512 - 24y^2$, $\cdots$ we compute the values at the points where we need them; the result is shown in the figure. For P_{11} and P_{12} we have the usual regular pattern, and for P_{21} and P_{22} we use (6), obtaining

$$(7) \quad P_{11}, P_{12}: \begin{Bmatrix} & 1 & \\ 1 & -4 & 1 \\ & 1 & \end{Bmatrix}, \quad P_{21}: \begin{Bmatrix} & 0.5 & \\ 0.6 & -2.5 & 0.9 \\ & 0.5 & \end{Bmatrix}, \quad P_{22}: \begin{Bmatrix} & 0.9 & \\ 0.6 & -3 & 0.9 \\ & 0.6 & \end{Bmatrix}.$$

We use this and the boundary values and take the mesh points in the order P_{11}, P_{21}, P_{12}, P_{22}. Then we obtain the system

$$-4u_{11} + u_{21} + u_{12} = 0 - 27 = -27$$

$$0.6u_{11} - 2.5u_{21} + 0.5u_{22} = -0.9 \cdot 296 - 0.5 \cdot 216 = -374.4$$

$$u_{11} - 4u_{12} + u_{22} = 702 + 0 = 702$$

$$0.6u_{21} + 0.6u_{12} - 3u_{22} = 0.9 \cdot 352 + 0.9 \cdot 936 = 1159.2.$$

In matrix form,

$$(8) \quad \begin{bmatrix} -4 & 1 & 1 & 0 \\ 0.6 & -2.5 & 0 & 0.5 \\ 1 & 0 & -4 & 1 \\ 0 & 0.6 & 0.6 & -3 \end{bmatrix} \begin{bmatrix} u_{11} \\ u_{21} \\ u_{12} \\ u_{22} \end{bmatrix} = \begin{bmatrix} -27 \\ -374.4 \\ 702 \\ 1159.2 \end{bmatrix}.$$

The Gauss elimination yields the (rounded) values

$$u_{11} = -55.6, \quad u_{21} = 49.2, \quad u_{12} = -298.5, \quad u_{22} = -436.3.$$

Clearly, from a grid with so few mesh points we cannot expect great accuracy. The exact values are

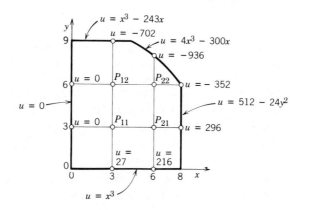

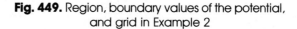

Fig. 449. Region, boundary values of the potential, and grid in Example 2

$$u_{11} = -54, \qquad u_{21} = 54, \qquad u_{12} = -297, \qquad u_{22} = -432.$$

In practice one would use a much finer grid and solve the resulting large system by an indirect method.   ∎

The concepts of elliptic, parabolic and hyperbolic equations were defined in Sec. 20.4. This section and the previous section were devoted to elliptic equations. In the next section we turn to **parabolic equations** and discuss an important numerical method for these equations.

Problems for Sec. 20.5

1. Check the values given at the end of Example 1 by solving the system (3) by the Gauss elimination.
2. Solve the mixed boundary value problem for the Laplace equation $\nabla^2 u = 0$ in the rectangle in Fig. 446 (using the grid in Fig. 446b) and the boundary conditions $u_x = 0$ on the left edge, $u_x = 6$ on the right edge, $u = 2x^2$ on the lower edge and $u = 2x^2 - 2$ on the upper edge.
3. Solve the mixed boundary value problem for the Poisson equation $\nabla^2 u = x^2 + y^2$ in the region and for the boundary conditions shown in Fig. 450, using the indicated grid.

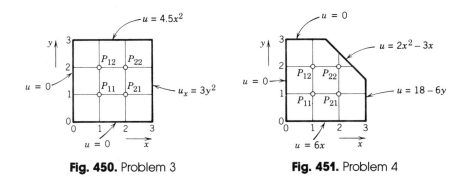

Fig. 450. Problem 3   **Fig. 451.** Problem 4

4. Solve the Laplace equation in the region and for the boundary values shown in Fig. 451, using the indicated grid. (The sloping portion of the boundary is $y = 4.5 - x$.)
5. Solve the Poisson equation $\nabla^2 u = 4$ in the region and for the boundary values shown in Fig. 452, using the grid also shown in the figure.
6. Verify the sample calculation right after (5).
7. Give a detailed derivation of (6).
8. Verify (7).
9. Solve (8) by the Gauss elimination.
10. Eliminating $\partial^2 u_0 / \partial x^2$ from (4), show that

$$\frac{\partial u_0}{\partial x} \approx \frac{1}{h}\left[\frac{1}{a(1+a)}u_A - \frac{1-a}{a}u_0 - \frac{a}{1+a}u_P\right].$$

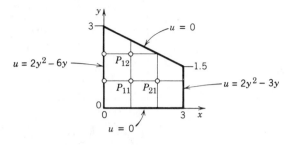

Fig. 452. Problem 5

20.6 Methods for Parabolic Equations

Solutions of equations of different type—elliptic, parabolic, hyperbolic—have a basically different general behavior, as was mentioned in Sec. 20.4, and this also reflects on numerical methods, as follows. For all three types, one replaces the equation by a corresponding difference equation, but for *parabolic* and *hyperbolic* equations this does not automatically guarantee the **convergence** of the approximate solution to the exact solution as the mesh $h \to 0$; in fact, it does not even guarantee convergence at all. For these two types of equation one needs additional conditions (inequalities) to assure convergence and **stability,** the latter roughly meaning that small perturbations in the initial data (or small errors at any time) remain small at later times.

In this section we consider the numerical solution of **parabolic equations,** concentrating on their prototype, the heat equation. We can study the situation just outlined as well as other facts in terms of the one-dimensional heat equation

$$u_t = c^2 u_{xx} \qquad (c \text{ constant}).$$

This equation is usually considered for x in some fixed interval, say, $0 \leqq x \leqq L$, and time $t \geqq 0$, and one prescribes the initial temperature $u(x, 0) = f(x)$ (f given) and boundary conditions at $x = 0$ and $x = L$ for all $t \geqq 0$, for instance $u(0, t) = 0$, $u(L, t) = 0$. We may assume $c = 1$ and $L = 1$; this can always be accomplished by a linear transformation of x and t (cf. Prob. 1). Then the **heat equation** and those conditions are

(1) $$u_t = u_{xx} \qquad\qquad 0 \leqq x \leqq 1, t \geqq 0$$

(2) $$u(x, 0) = f(x) \qquad\qquad \text{(initial condition)}$$

(3) $$u(0, t) = u(1, t) = 0 \qquad\qquad \text{(boundary conditions)}.$$

A simple finite difference approximation of (1) is [cf. (6a) in Sec. 20.4]

(4) $$\frac{1}{k} (u_{i,j+1} - u_{ij}) = \frac{1}{h^2} (u_{i+1,j} - 2u_{ij} + u_{i-1,j}).$$

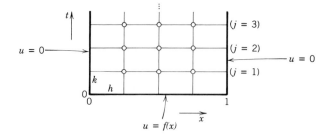

Fig. 453. Grid and mesh points corresponding to (4), (5)

Figure 453 shows a corresponding grid and mesh points. The mesh size is h in the x-direction and k in the t-direction. Formula (4) involves the four points shown in Fig. 454. On the left we used a *forward* difference quotient since we have no information for negative t at the start. From (4) we calculate $u_{i,j+1}$, which corresponds to time row $j + 1$, in terms of the three other u that correspond to time row j; solving (4) for $u_{i,j+1}$, we have

(5)
$$\boxed{u_{i,j+1} = (1 - 2r)u_{ij} + r(u_{i+1,j} + u_{i-1,j})} \qquad r = \frac{k}{h^2}.$$

Computations by means of this formula are simple and straightforward. However, it can be shown that crucial to the convergence of this method is the condition

(6)
$$r = \frac{k}{h^2} \leqq \frac{1}{2},$$

that is, that u_{ij} have a positive coefficient in (5) or (for $r = \frac{1}{2}$) is absent from (5). Intuitively, (6) means that we should not move too fast in the t-direction. An example is given below.

Crank–Nicolson Method

Condition (6) is a handicap in practice. Indeed, to attain sufficient accuracy, we have to choose h small, which makes k very small by (6). For example, if $h = 0.1$, then $k \leqq 0.005$. And a switch to $\frac{1}{2}h$ quadruples the number of time steps needed to reach a certain t-value. Accordingly, we should look for a method based on a more satisfactory discretization of the heat equation.

Such a method that imposes no restriction on $r = k/h^2$ is the **Crank–Nicolson method,** which uses values of u at the six points in Fig. 455. The idea of the method is the replacement of the difference quotient on the right-hand side of (4) by $\frac{1}{2}$ times the sum of two such difference quotients at two time rows

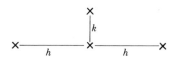

Fig. 454. The four points in (4) and (5)

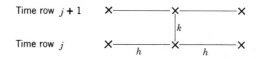

Fig. 455. The six points in the Crank–Nicolson formulas (7) and (8)

(cf. Fig. 455). Instead of (4) we then have

$$
(7) \qquad
\begin{aligned}
\frac{1}{k}(u_{i,j+1} - u_{ij}) &= \frac{1}{2h^2}(u_{i+1,j} - 2u_{ij} + u_{i-1,j}) \\[2mm]
&+ \frac{1}{2h^2}(u_{i+1,j+1} - 2u_{i,j+1} + u_{i-1,j+1}).
\end{aligned}
$$

Multiplying by $2k$ and writing $r = k/h^2$ as before, we can collect the three terms corresponding to time row $j + 1$ on the left and the three terms corresponding to time row j on the right; this gives

$$
(8) \quad (2 + 2r)u_{i,j+1} - r(u_{i+1,j+1} + u_{i-1,j+1}) = (2 - 2r)u_{ij} + r(u_{i+1,j} + u_{i-1,j}).
$$

How to use (8)? In general, the three values on the left are unknown, whereas the three values on the right are known. If we divide the x-interval $0 \leq x \leq 1$ in (1) into n equal intervals, we have $n - 1$ internal mesh points per time row (cf. Fig. 453, where $n = 4$). Then for $j = 0$ and $i = 1, \cdots, n - 1$, formula (8) gives a system of $n - 1$ linear equations for the $n - 1$ unknown values $u_{11}, u_{21}, \cdots, u_{n-1,1}$ in the first time row in terms of the initial values $u_{00}, u_{10}, \cdots, u_{n0}$ and the boundary values $u_{01}, u_{n1} (= 0)$. Similarly for $j = 1, j = 2$, and so on; that is, for each time row we have to solve such a system of $n - 1$ linear equations resulting from (8).

Although $r = k/h^2$ is no longer restricted, smaller r will still give better results. In practice, one chooses a k by which one can save a considerable amount of work, without making r too large. For instance, often a good choice is $r = 1$ (which would be impossible in the previous direct method). Then (8) becomes simply

$$
(9) \qquad 4u_{i,j+1} - u_{i+1,j+1} - u_{i-1,j+1} = u_{i+1,j} + u_{i-1,j}.
$$

EXAMPLE 1. Temperature in a rod. Crank–Nicolson method, direct method

Consider a laterally insulated metal rod of length 1 and such that $c^2 = 1$ in the heat equation. Suppose that the ends of the rod are kept at temperature $u = 0°C$ and the temperature in the rod at some instant—call it $t = 0$—is $f(x) = \sin \pi x$. Applying the Crank–Nicolson method with $h = 0.2$ and $r = 1$, find the temperature $u(x, t)$ in the rod for $0 \leq t \leq 0.2$. Compare the results with the exact solution. Also apply (5) with an r satisfying (6), say, $r = 0.25$, and with values not satisfying (6), say, $r = 1$ and $r = 2.5$.

Solution by Crank–Nicolson. Since $r = 1$, formula (8) takes the form (9). Since $h = 0.2$ and $r = k/h^2 = 1$, we have $k = h^2 = 0.04$. Hence we have to do 5 steps. Figure 456 on the next page shows the grid. We shall need the initial values

$$
u_{10} = \sin 0.2\pi = 0.587\ 785, \qquad u_{20} = \sin 0.4\pi = 0.951\ 057.
$$

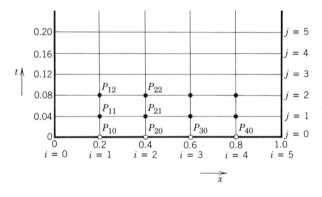

Fig. 456. Grid in Example 1

Also $u_{30} = u_{20}$ and $u_{40} = u_{10}$. (Recall that u_{10} means u at P_{10} in Fig. 456, etc.) In each time row in Fig. 456 there are 4 internal mesh points. Hence in each time step we would have to solve 4 equations in 4 unknowns. But since the initial temperature distribution is symmetric with respect to $x = 0.5$ and $u = 0$ at both ends for all t, we have $u_{31} = u_{21}$, $u_{41} = u_{11}$ in the first time row and similarly for the other rows. This reduces each system to 2 equations in 2 unknowns. By (9), since $u_{31} = u_{21}$, for $j = 0$ these equations are

$$4u_{11} - u_{21} \qquad = u_{00} + u_{20} = 0.951\ 057$$

$$-u_{11} + 4u_{21} - u_{21} = u_{10} + u_{20} = 1.538\ 842.$$

The solution is $u_{11} = 0.399\ 274$, $u_{21} = 0.646\ 039$. Similarly, for $j = 1$ we have the system

$$4u_{12} - u_{22} = u_{01} + u_{21} = 0.646\ 039$$

$$-u_{12} + 3u_{22} = u_{11} + u_{21} = 1.045\ 313.$$

The solution is $u_{12} = 0.271\ 221$, $u_{22} = 0.438\ 845$. And so on. This gives the temperature distribution (cf. Fig. 457):

t	$x = 0$	$x = 0.2$	$x = 0.4$	$x = 0.6$	$x = 0.8$	$x = 1$
0.00	0	0.588	0.951	0.951	0.588	0
0.04	0	0.399	0.646	0.646	0.399	0
0.08	0	0.271	0.439	0.439	0.271	0
0.12	0	0.184	0.298	0.298	0.184	0
0.16	0	0.125	0.202	0.202	0.125	0
0.20	0	0.085	0.138	0.138	0.085	0

Comparison with the exact solution. The present problem can be solved exactly by separating variables (cf. Sec. 11.5); the result is

(10) $$u(x, t) = \sin \pi x \, e^{-\pi^2 t}.$$

A comparison of numerical values is given below.

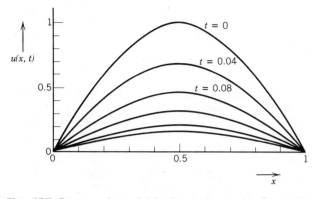

Fig. 457. Temperature distribution in the rod in Example 1

Solution by the direct method (5) with r = 0.25. For $h = 0.2$ and $r = k/h^2 = 0.25$ we have $k = rh^2 = 0.25 \cdot 0.04 = 0.01$. Hence we have to perform 4 times as many steps as with the Crank–Nicolson method! Formula (5) with $r = 0.25$ is

(11) $$u_{i,j+1} = 0.25(u_{i-1,j} + 2u_{ij} + u_{i+1,j}).$$

We can make use of the symmetry. For $j = 0$ we need $u_{00} = 0$, $u_{10} = 0.587\ 785$, $u_{20} = u_{30} = 0.951\ 057$ and compute

$$u_{11} = 0.25(u_{00} + 2u_{10} + u_{20}) = 0.531\ 657$$

$$u_{21} = 0.25(u_{10} + 2u_{20} + u_{30}) = 0.25(u_{10} + 3u_{20}) = 0.860\ 239.$$

Of course we can omit the boundary terms $u_{01} = 0$, $u_{02} = 0$, $\cdots$ from the formulas. For $j = 1$ we compute

$$u_{12} = 0.25(2u_{11} + u_{21}) = 0.480\ 888$$

$$u_{22} = 0.25(u_{11} + 3u_{21}) = 0.778\ 094$$

and so on. We have to perform 20 steps, but the numerical values show that the accuracy is only about the same as that of the Crank–Nicolson values CN (exact values given with 3 decimals):

t	$x = 0.2$			$x = 0.4$		
	CN	By (11)	Exact	CN	By (11)	Exact
0.04	0.399	0.393	0.396	0.646	0.637	0.641
0.08	0.271	0.263	0.267	0.439	0.426	0.432
0.12	0.184	0.176	0.180	0.298	0.285	0.291
0.16	0.125	0.118	0.121	0.202	0.191	0.196
0.20	0.085	0.079	0.082	0.138	0.128	0.132

Failure of (5) with r violating (6). Formula (5) with $h = 0.2$ and $r = 1$—which violates (6)—is

$$u_{i,j+1} = u_{i-1,j} - u_{ij} + u_{i+1,j}$$

and gives very poor values; some of them are as follows (see next page):

t	$x = 0.2$	Exact	$x = 0.4$	Exact
0.04	0.363	0.396	0.588	0.641
0.12	0.139	0.180	0.225	0.291
0.20	0.053	0.082	0.086	0.132

Formula (5) with an even larger $r = 2.5$ (and $h = 0.2$ as before) gives completely nonsensical results; some of them are:

t	$x = 0.2$	Exact	$x = 0.4$	Exact
0.1	0.0265	0.2191	0.0429	0.3545
0.3	0.0001	0.0304	0.0001	0.0492
0.5	0.0018	0.0042	-0.0011	0.0068

So far we have discussed numerical methods for elliptic and parabolic partial differential equations. In the next section, the last of this chapter, we discuss numerical methods for **hyperbolic equations.**

Problems for Sec. 20.6

1. **(Nondimensional form)** Show that the heat equation $\tilde{u}_{\tilde{t}} = c^2 \tilde{u}_{\tilde{x}\tilde{x}}$, $0 \leq \tilde{x} \leq L$, can be transformed to the "nondimensional" standard form $u_t = u_{xx}$, $0 \leq x \leq 1$, by setting $x = \tilde{x}/L$, $t = c^2\tilde{t}/L^2$, $u = \tilde{u}/u_0$, where u_0 is any constant temperature.

2. Derive the difference approximation (4) of the heat equation.

3. Using the direct method given by (5), with $h = 1$ and $k = 0.5$, find the temperature at $t = 2$ in a laterally insulated rod of length 10 whose ends at $x = 0$ and $x = 10$ are kept at zero temperature and whose initial temperature is $f(x) = 0.1x - 0.01x^2$.

4. Solve Prob. 3, choosing $k = 0.25$ (instead of 0.5), leaving everything else as before. To how many digits do the two solutions agree?

5. Extend the computation in Prob. 3 to $t = 4$. Then find the temperature at $t = 4$ and $x = 2, 4, 6, 8$ by the Crank–Nicolson formula (9) with $h = 2$ (thus $k = 4$) and compare the values.

6. Find the exact solution in Prob. 3 by the method in Sec. 11.5. Note that the series thus obtained converges rapidly, so that we can expect the sum of the first 2 terms to give 3D-values of the solution for $t = 2$ and $t = 4$. Compute $u(2, 2)$ and $u(4, 2)$ in this way and compare with the values in Prob. 3. Compute $u(2, 4)$, $u(4, 4)$ and compare with Prob. 5.

7. Solve the heat equation (1) for the initial condition $f(x) = x$ if $0 \leq x \leq \frac{1}{2}$, $f(x) = 1 - x$ if $\frac{1}{2} < x \leq 1$, and boundary condition (3) by the Crank–Nicolson formula (9) with $h = 0.2$ for $0 \leq t \leq 0.20$.

8. Solve Prob. 7 by the direct method with $h = 0.2$ and $r = 0.25$; do 8 steps and compare the last values with the Crank–Nicolson values 0.107, 0.175 (3D) and the values 0.108, 0.175 (exact to 3D).

9. Compute u in Prob. 7 for $x = 0.2, 0.4$ and $t = 0.04, 0.08, \cdots, 0.20$ from the series in Sec. 11.5, Example 3.

10. The accuracy of the results obtained by the direct method depends on r $(\leq \frac{1}{2})$. Illustrate this by reconsidering Prob. 8, choosing $r = \frac{1}{2}$ (and $h = 0.2$ as before), computing 4 steps and comparing the values for $t = 0.04$ and 0.08 with those in Prob. 8.

20.7 Methods for Hyperbolic Equations

In this section we consider the numerical solution of problems involving hyperbolic equations. We explain a standard method in terms of a typical setting for the prototype of a hyperbolic equation, the **wave equation:**

(1) $$u_{tt} = u_{xx} \qquad 0 \le x \le 1, t \ge 0$$

(2) $$u(x, 0) = f(x) \qquad \text{(Given initial displacement)}$$

(3) $$u_t(x, 0) = g(x) \qquad \text{(Given initial velocity)}$$

(4) $$u(0, t) = u(1, t) = 0 \qquad \text{(Boundary conditions)}.$$

Note that an equation $u_{tt} = c^2 u_{xx}$ and another x-interval can be reduced to the form (1) by a linear transformation of x and t. (This is similar to Sec. 20.6, Prob. 1.)

For instance, (1)–(4) is the model of a vibrating elastic string with fixed ends at $x = 0$ and $x = 1$ (cf. Sec. 11.2). Although an analytic solution of the problem is given in (7), Sec. 11.4, we use the problem for explaining basic ideas of the numerical approach that are also relevant for more complicated hyperbolic equations.

Replacing the derivatives by difference quotients as before, we obtain from (1) [cf. (6) in Sec. 20.4 with $y = t$]

(5) $$\frac{1}{k^2} (u_{i,j+1} - 2u_{ij} + u_{i,j-1}) = \frac{1}{h^2} (u_{i+1,j} - 2u_{ij} + u_{i-1,j})$$

where h is the mesh size in x, and k is the mesh size in t. This difference equation relates 5 points as shown in Fig. 458a. It suggests a rectangular grid similar to those for parabolic equations in the preceding section. We choose $r^* = k^2/h^2 = 1$. Then u_{ij} drops out and we have (cf. Fig. 458b)

(6) $$\boxed{u_{i,j+1} = u_{i-1,j} + u_{i+1,j} - u_{i,j-1}.}$$

It can be shown that for $0 < r^* \le 1$ the present method is stable, so that from (6) we may expect reasonable results for initial data that have no discontinuities. (For a hyperbolic equation, the latter would propagate into the solution domain—a phenomenon that would be difficult to deal with on our present grid; for details, see Ref. [E12] in Appendix 1.)

Equation (6) still involves 3 time steps $j - 1, j, j + 1$, whereas the formulas in the parabolic case involved only 2 time steps. Furthermore, we now have 2 initial conditions. So we ask how we get started and how we can use the initial condition (3). This can be done as follows. From $u_t(x, 0) = g(x)$ we

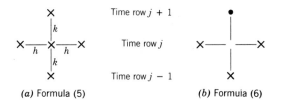

Fig. 458. Mesh points used in (5) and (6)

derive the difference formula

(7) $\qquad \dfrac{1}{2k}(u_{i1} - u_{i,-1}) = g_i,$ hence $\qquad u_{i,-1} = u_{i1} - 2kg_i$

where $g_i = g(ih)$. For $t = 0$, that is, $j = 0$, equation (6) is

$$u_{i1} = u_{i-1,0} + u_{i+1,0} - u_{i,-1}.$$

Into this we substitute (7), then take $-u_{i1}$ from the right to the left and divide by 2, finding

(8) $\qquad \boxed{u_{i1} = \tfrac{1}{2}(u_{i-1,0} + u_{i+1,0}) + kg_i}$

which expresses u_{i1} in terms of the initial data.

EXAMPLE 1. Vibrating string

Apply the present method of numerical solution with $h = k = 0.2$ to the problem (1)–(4), where

$$f(x) = \sin \pi x, \qquad g(x) = 0.$$

Solution. The grid is the same as in Fig. 456, Sec. 20.6, except for the values of t, which now are 0.2, 0.4, $\cdots$ (instead of 0.04, 0.08, $\cdots$). The initial values $u_{00}, u_{10}, \cdots$ are the same as in Example 1, Sec. 20.6. From (8) and $g(x) = 0$ we have

$$u_{i1} = \tfrac{1}{2}(u_{i-1,0} + u_{i+1,0}).$$

From this we compute

$$u_{11} = \tfrac{1}{2}(u_{00} + u_{20}) = \tfrac{1}{2} \cdot 0.951\ 057 = 0.475\ 528$$

$$u_{21} = \tfrac{1}{2}(u_{10} + u_{30}) = \tfrac{1}{2} \cdot 1.538\ 842 = 0.769\ 421,$$

and $u_{31} = u_{21}$, $u_{41} = u_{11}$ by symmetry as in Sec. 20.6, Example 1. From (6) with $j = 1$, using $u_{01} = u_{02} = \cdots = 0$, we now compute

$$u_{12} = u_{01} + u_{21} - u_{10} = 0.769\ 421 - 0.587\ 785 = 0.181\ 636$$

$$u_{22} = u_{11} + u_{31} - u_{20} = 0.475\ 528 + 0.769\ 421 - 0.951\ 057 = 0.293\ 892$$

and $u_{32} = u_{22}$, $u_{42} = u_{12}$ by symmetry; and so on. We thus obtain for the displacement $u(x, t)$ of the string over the first half-cycle the values:

t	$x = 0$	$x = 0.2$	$x = 0.4$	$x = 0.6$	$x = 0.8$	$x = 1$
0.0	0	0.588	0.951	0.951	0.588	0
0.2	0	0.476	0.769	0.769	0.476	0
0.4	0	0.182	0.294	0.294	0.182	0
0.6	0	-0.182	-0.294	-0.294	-0.182	0
0.8	0	-0.476	-0.769	-0.769	-0.476	0
1.0	0	-0.588	-0.951	-0.951	-0.588	0

These values are exact, the exact solution of the problem being (cf. Sec. 11.3)

$$u(x, t) = \sin \pi x \cos \pi t.$$

The reason for the exactness follows from d'Alembert's solution (6), Sec. 11.4. (Cf. Prob. 4, below.) ∎

This is the end of Chap. 20 on numerical methods for ordinary and partial differential equations. We have concentrated on some basic ideas and methods in this field, which is full of interesting open questions and is developing quite rapidly. This is also the end of Part E on numerical methods. The next part is devoted to optimization, an area that has many engineering, computer science and other applications, and is relatively young, particularly in its combinatorial methods involving graphs (Chap. 22).

Problems for Sec. 20.7

Solve the vibrating string problem (1)–(4) by the present numerical method with $h = k = 0.2$ on the given t-interval for initial velocity 0 and given initial deflection $f(x)$.

1. $0 \leq t \leq 2$, $f(x) = x(1 - x)$
2. $0 \leq t \leq 1$, $f(x) = x^2(1 - x^2)$
3. $0 \leq t \leq 1$, $f(x) = x$ if $0 \leq x \leq 0.2$, $f(x) = 0.25(1 - x)$ if $0.2 < x \leq 1$
4. Show that from d'Alembert's solution (6) in Sec. 11.4 with $c = 1$ it follows that (6) in the present section gives the exact value $u_{i,j+1} = u(ih, (j + 1)h)$.
5. If the string governed by (1) starts from its equilibrium position with initial velocity $g(x) = \sin \pi x$, what is its displacement at time $t = 0.4$ and $x = 0.2$, 0.4, 0.6, 0.8? (Use the present method with $h = 0.2$, $k = 0.2$. Compare with the exact values obtained from (7) in Sec. 11.4.)
6. Compute approximate values in Prob. 5, using a finer grid ($h = 0.1$, $k = 0.1$), and notice the increase in accuracy.
7. Illustrate the starting procedure for the present method in the case where both f and g are not identically zero, say,

$$f(x) = 1 - \cos 2\pi x, \qquad g(x) = x - x^2;$$

choose $h = k = 0.1$ and compute 2 time steps.
8. Show that (7) in Sec. 11.4 gives as another starting formula

$$u_{i1} = \frac{1}{2}(u_{i+1,0} + u_{i-1,0}) + \frac{1}{2}\int_{x_i-k}^{x_i+k} g(s)\, ds$$

(where one can evaluate the integral numerically if necessary). In what case is this identical with (8)?

9. Compute u in Prob. 7 for $t = 0.1$ and $x = 0.1, 0.2, \cdots, 0.9$, using the formula in Prob. 8, and compare the values.

10. Solve (1) numerically, subject to the conditions

$$u(x, 0) = x^2, \quad u_t(x, 0) = 2x, \quad u_x(0, t) = 2t, \quad u(1, t) = (1 + t)^2,$$

choosing $h = k = 0.2$ (5 time steps).

Review Problems for Chapter 20

1. What is a predictor–corrector method? Give an example.

2. Explain the similarities and differences of the Runge–Kutta and Runge–Kutta–Nyström methods.

3. Can you think of an extension of the method given by (4), Sec. 20.3, to third-order differential equations?

4. What is a one-step method for solving initial value problems? A multistep method? Give examples.

5. What are the advantages and disadvantages of multistep methods over one-step methods?

6. Can you recall the idea by which we got the Adams–Bashford method?

7. What is the difference between the Adams–Bashford method and the Adams–Moulton method?

8. What role did the Taylor series play in the derivation of some of the methods? Which methods?

9. What are the advantage and disadvantage in choosing a very small step size?

10. In this chapter we used finite differences a number of times. Explain why and how.

11. Can some of the methods discussed diverge in certain cases? Give examples.

12. What do we mean by an elliptic, parabolic or hyperbolic equation? Name and write down one equation of each type.

13. Why do we need different numerical methods for different types of partial differential equations?

14. How many initial conditions did we prescribe for the wave equation? For the heat equation?

15. In numerical methods for the Laplace equation $\nabla^2 u = 0$ in two variables, what formula relates u at a point P to u at the neighbors E, N, W, S?

16. Derive the formula in Prob. 15 from memory.

17. If we take a grid and a solution of Laplace's equation, say, $u = e^x \cos y$, can we expect the formula in Prob. 15 to hold exactly?

18. How did we handle boundary value problems on domains of irregular shape? Write down what you remember; only then look up more details in the text.

19. What was the idea in using given data for the normal derivative on the boundary of a region?

20. Describe the basic idea of the ADI method from memory as best as you can.

21. Solve $y' = 2x^{-1}\sqrt{y - \ln x} + x^{-1}$, $y(1) = 0$ for $1 \le x \le 1.8$ by Euler's method with $h = 0.1$. Verify that the exact solution is $y = (\ln x)^2 + \ln x$ and compute the error.

22. Solve Prob. 21 by the improved Euler method (Heun's method) with $h = 0.2$, determine the error, compare with Prob. 21 and comment. Note that this is a fair comparison because here we evaluate $f(x, y)$ eight times (4 steps with 2 evaluations each), just as in Prob. 21.

23. Solve Prob. 21 by the Runge–Kutta method with $h = 0.4$, determine the error and compare with Prob. 21. (Note that these 2 Runge–Kutta steps require 8 evaluations of $f(x, y)$, just as many as in Prob. 21.)

24. Solve Prob. 21 by the Runge–Kutta method with $h = 0.1$ and compare the error with that in Prob. 21.

25. Apply the Runge–Kutta method to the problem $y' = 1 + y^2$, $y(0) = 0$. Choose $h = 0.1$, compute 2 steps and compare with the exact 8D-values 0.100 334 67 and 0.202 710 04. Solve the problem analytically.

26. Show that the Euler method is also obtained by integrating $y' = f(x, y)$ from x_n to x_{n+1}, approximating the integrand $f[t, y(t)]$ by its value at x_n.

27. In Prob. 26, apply the trapezoidal rule to the integral, to get

$$y(x_n + h) \approx y(x_n) + \frac{h}{2} \{f[x_n, y(x_n)] + f[x_n + h, y(x_n + h)]\}.$$

Then approximate $y(x_n + h)$ on the right by the Euler–Cauchy method. Show that the result is the improved Euler–Cauchy method.

28. Solve $y' = (x + y - 4)^2$, $y(0) = 4$, by the Runge–Kutta method with step $h = 0.2$ for $x = 0, 0.2, \cdots, 1.4$.

29. Show that by applying the method in Sec. 20.2 to a polynomial of first degree we obtain the multistep predictor and corrector formulas

$$y_{n+1}^* = y_n + \frac{h}{2} (3f_n - f_{n-1}),$$

$$y_{n+1} = y_n + \frac{h}{2} (f_{n+1} + f_n).$$

30. Apply the multistep method in Prob. 29 to the initial value problem $y' = x + y$, $y(0) = 0$, choosing $h = 0.2$ and doing 5 steps. Compare with the exact values.

31. Solve $y' = (y - x - 1)^2 + 2$, $y(0) = 1$ for $0 \leq x \leq 1$ by the Adams–Moulton method (Sec. 20.2) with $h = 0.1$ and starting values from Example 4 in Sec. 20.1.

Apply the method given by (4), Sec. 20.3, with $h = 0.1$ to the given initial value problem (5 steps) and compare with the exact solution.

32. $y'' = y$, $y(0) = 1$, $y'(0) = 1$

33. $y'' = y$, $y(0) = 1$, $y'(0) = -1$

34. $y'' = xy' - 3y$, $y(0) = 0$, $y'(0) = -3$.   Exact solution $y = x^3 - 3x$

35. $y'' = xy' - 4y$, $y(0) = 3$, $y'(0) = 0$.   Exact solution $y = x^4 - 6x^2 + 3$

36. $(1 - x^2)y'' - 2xy' + 6y = 0$, $y(0) = -\frac{1}{2}$, $y'(0) = 0$.   Exact: $y = \frac{1}{2}(3x^2 - 1)$

37. Apply the Runge–Kutta–Nyström method to the initial value problem $y'' = -y$, $y(0) = 0$, $y'(0) = 1$; choose $h = 0.2$ and do 5 steps. Compare with the exact solution.

38. In Prob. 37, replace $h = 0.2$ by $h = 0.1$, do 4 steps and compare the results with those in Prob. 37 and with the exact 9D-values 0.099 833 417, 0.198 669 331, 0.295 520 207, 0.389 418 342.

39. Find rough approximate values of the electrostatic potential at P_{11}, P_{12}, P_{13} in Fig. 459 that lie in a field between conducting plates (in Fig. 459 appearing as sides of a rectangle) kept at potentials 0 and 220 volts as shown. (Use the indicated grid.)

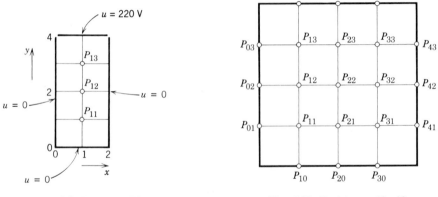

Fig. 459. Problem 39   **Fig. 460.** Problems 40–42

Find the potential in Fig. 460, using the given grid and the following boundary values:

40. $u = 100$ on the upper and left sides, $u = 0$ on the lower and right sides

41. $u(P_{10}) = u(P_{30}) = 320$, $u(P_{20}) = -160$, $u = 0$ elsewhere on the boundary

42. $u(P_{01}) = u(P_{03}) = u(P_{41}) = u(P_{43}) = 40$, $u(P_{10}) = u(P_{30}) = -80$, $u(P_{20}) = 320$, $u(P_{02}) = u(P_{42}) = u(P_{14}) = u(P_{24}) = u(P_{34}) = 0$

43. Verify (13) in Sec. 20.4 for the system (12) and show that $\mathbf{A}$ in (12) is nonsingular.

44. Formula (17) in Sec. 20.4 with $p = 0$ would not work. Show this by applying it to the initial value problem in Example 2 of Sec. 20.4, with grid and starting values as before; perform 2 steps. What happens?

45. Solve the heat equation (1), Sec. 20.6, for the initial condition $f(x) = x$ if $0 \leq x \leq 0.2$, $f(x) = 0.25(1 - x)$ if $0.2 < x \leq 1$ and boundary condition (3), Sec. 20.6, by the direct method [formula (5) in Sec. 20.6] with $h = 0.2$ and $k = 0.01$ so that you get values of the temperature at $t = 0.05$ as the answer.

46. A laterally insulated homogeneous rod with ends at $x = 0$ and $x = 1$ has initial temperature 0. Its left end is kept at 0, whereas the temperature at the right end varies sinusoidally according to

$$u(t, 1) = g(t) = \sin \tfrac{25}{3} \pi t.$$

Find the temperature $u(x, t)$ in the rod [solution of (1) in Sec. 20.6] by the direct method with $h = 0.2$ and $r = 0.5$ (one period, that is, $0 \leq t \leq 0.24$).

47. Find $u(x, 0.12)$ and $u(x, 0.24)$ in Prob. 46 if the left end of the rod is kept at $-g(t)$ (instead of 0), all the other data being as before.

48. Find out how the results of Prob. 46 can be used for obtaining the results in Prob. 47. Use the values 0.054, 0.172, 0.325, 0.406 ($t = 0.12$, $x = 0.2, 0.4, 0.6, 0.8$) and -0.009, -0.086, -0.252, -0.353 ($t = 0.24$) from the answer to Prob. 46 to check your answer to Prob. 47.

49. If the left end of a laterally insulated rod extending from $x = 0$ to $x = 1$ is insulated, the boundary condition at $x = 0$ is $u_n(0, t) = u_x(0, t) = 0$. Show that in the application of the direct method given by (5) in Sec. 20.6, we can compute $u_{0,j+1}$ by the formula

$$u_{0,j+1} = (1 - 2r)u_{0j} + 2ru_{1j}.$$

50. Applying the direct method with $h = 0.2$ and $r = 0.25$, determine the temperature $u(x, t)$, $0 \leq t \leq 0.12$ [solution of (1) in Sec. 20.6] in a laterally insulated rod extending from $x = 0$ to $x = 1$ if $u(x, 0) = 0$, the left end is insulated and the right end is kept at temperature $g(t) = \sin \tfrac{50}{3} \pi t$. *Hint.* Cf. Prob. 49.

Summary of Chapter 20
Numerical Methods
for Differential Equations

In this chapter we discussed numerical methods for ordinary differential equations (Secs. 20.1–20.3) and partial differential equations (Secs. 20.4–20.7). For first-order equations we considered initial value problems of the form (Sec. 20.1)

$$(1) \qquad\qquad y' = f(x, y), \qquad y(x_0) = y_0.$$

Numerical methods for solving such a problem can be obtained by truncating the Taylor series

$$y(x + h) = y(x) + hy'(x) + \frac{h^2}{2} y''(x) + \cdots$$

where, by (1), $y' = f$, $y'' = f' = \partial f/\partial x + (\partial f/\partial y)y'$, etc. Truncating after the term hy', we get the *Euler method*, in which we compute step by step

$$(2) \qquad\qquad y_{n+1} = y_n + hf(x_n, y_n) \qquad\qquad n = 0, 1, \cdots.$$

Taking one more term into account, we obtain the *improved Euler* or *Heun's method*. Both methods show the basic idea in a simple form, but are too inaccurate for most practical purposes. Truncating after the term in h^4, we get the important **Runge–Kutta method** (of fourth order). The crucial idea in this and similar methods is to replace the cumbersome evaluation of derivatives by the evaluation of $f(x, y)$ at suitable points (x, y); thus in each step we first compute four auxiliary quantities (Sec. 20.1)

$$(3\text{a}) \qquad
\begin{aligned}
k_1 &= hf(x_n, y_n) \\
k_2 &= hf(x_n + \tfrac{1}{2}h, y_n + \tfrac{1}{2}k_1) \\
k_3 &= hf(x_n + \tfrac{1}{2}h, y_n + \tfrac{1}{2}k_2) \\
k_4 &= hf(x_n + h, y_n + k_3)
\end{aligned}$$

and then from them the new value

$$(3\text{b}) \qquad\qquad y_{n+1} = y_n + \tfrac{1}{6}(k_1 + 2k_2 + 2k_3 + k_4).$$

This is a *one-step method*, since y_{n+1} is computed from data of a single step. A *multistep method* uses data from several preceding steps, thereby avoiding computations such as (3a). By integrating cubic interpolation polynomials we obtained as a multistep method the important Adams–Moulton method in which we first compute the *predictor* (Sec. 20.2)

(4a) $$y_{n+1}^* = y_n + \frac{1}{24} h(55f_n - 59f_{n-1} + 37f_{n-2} - 9f_{n-3})$$

where $f_j = f(x_j, y_j)$, and then from it the *corrector* (the actual new value)

(4b) $$y_{n+1} = y_n + \frac{1}{24} h(9f_{n+1}^* + 19f_n - 5f_{n-1} + f_{n-2})$$

where $f_{n+1}^* = f(x_{n+1}, y_{n+1}^*)$. Here, to get started, y_1, y_2, y_3 must be computed by the Runge–Kutta method or some other accurate method.

Section 20.3 concerned second-order ordinary differential equations, in particular, the **Runge–Kutta–Nyström method,** an extension of the Runge–Kutta method to these equations.

Numerical methods for **partial differential equations** are obtained by replacing partial derivatives by difference quotients. This leads to approximating difference equations, for the **Laplace equation** to (Sec. 20.4)

$$u_{i+1,j} + u_{i,j+1} + u_{i-1,j} + u_{i,j-1} - 4u_{ij} = 0,$$

for the **heat equation** to (Sec. 20.6)

$$\frac{1}{k}(u_{i,j+1} - u_{ij}) = \frac{1}{h^2}(u_{i+1,j} - 2u_{ij} + u_{i-1,j})$$

and for the **wave equation** to (Sec. 20.7)

$$\frac{1}{k^2}(u_{i,j+1} - 2u_{ij} + u_{i,j-1}) = \frac{1}{h^2}(u_{i+1,j} - 2u_{ij} + u_{i-1,j});$$

here h and k are the mesh sizes of a grid in the x- and y-directions, respectively. These equations are *elliptic, parabolic* and *hyperbolic,* respectively. Corresponding numerical methods differ, for the following reason. For elliptic equations we have boundary value problems, and we discussed for them the **Gauss–Seidel** or **Liebmann method** and the **ADI method** (Secs. 20.4, 20.5). For parabolic equations we are given one initial condition and boundary conditions, and we discussed a *direct method* and the **Crank–Nicolson method** (Sec. 20.6). For hyperbolic equations, the problems are similar but we are given a second initial condition; in Sec. 20.7 we explained how to handle such problems numerically.

PART F
OPTIMIZATION, GRAPHS

The ideas of optimization and the application of graphs and digraphs (= directed graphs) play an increasing role in engineering, computer science, systems theory, economics and other areas. In this part we explain some basic concepts, methods and results in unconstrained optimization (Sec. 21.1), linear programming (Sec. 21.1–21.4), graphs and digraphs and their application in combinatorial optimization (Chap. 22).

Chapter 21

Unconstrained Optimization, Linear Programming

> This chapter provides an introduction to the more important concepts, methods and results of optimization. Optimization principles are of increasing importance in modern design and system operation, engineering and other. The recent development has been affected by the modern computer capable of solving large-scale problems, and by the corresponding creation of new optimization techniques, so that the whole field is in the process of becoming a large area of its own.
>
> *Prerequisite:* A modest knowledge of systems of linear equations
> *References:* Appendix 1, Part F.
> *Answers to problems:* Appendix 2.

21.1 Basic Concepts. Unconstrained Optimization

In an **optimization problem,** the objective is to *optimize* (*maximize* or *minimize*) some function f. This function f is called the **objective function.**

For example, an objective function f to be *maximized* may be the revenue in a production of TV sets, the yield per minute in a chemical process, the mileage per gallon of a certain type of car, the hourly number of customers served in some office, the hardness of steel or the tensile strength of a rope.

Similarly, we may want to *minimize* f if f is the cost per unit of producing certain cameras, the operating cost of some power plant, the daily loss of heat in a heating system, the idling time of some lathe or the time needed to produce a fender.

In most optimization problems the objective function f depends on several variables

$$x_1, \cdots, x_n.$$

These are called **control variables** because we can "control" them, that is, choose their values.

For example, the yield of a chemical process may depend on pressure x_1

and temperature x_2. The efficiency of fluorescent lamps may depend on voltage x_1, room temperature x_2 and air motion x_3. The efficiency of a certain air conditioning system may depend on temperature x_1, air pressure x_2, moisture content x_3, cross-sectional area of outlet x_4 and so on.

Optimization theory develops methods for optimal choices of $x_1, \cdots, x_n$, which maximize (or minimize) the objective function f, that is, methods for finding optimal values of $x_1, \cdots, x_n$.

In many problems the choice of values of $x_1, \cdots, x_n$ is not entirely free but is subject to some **constraints,** that is, additional conditions arising from the nature of the problem and the variables.

For example, if x_1 is production cost, then $x_1 \geqq 0$, and there are many other variables (time, weight, distance traveled by a salesman, etc.) that can take nonnegative values only. Constraints can also have the form of equations (instead of inequalities).

Let us first consider **unconstrained optimization** in the case of a real-valued function $f(x_1, \cdots, x_n)$. We also write $\mathbf{x} = (x_1, \cdots, x_n)$ and $f(\mathbf{x})$, for convenience.

By definition, f has a **minimum** at a point $\mathbf{x} = \mathbf{X}_0$ in a region R where f is defined if $f(\mathbf{x}) \geqq f(\mathbf{X}_0)$ for all $\mathbf{x}$ in R. Similarly, f has a **maximum** at $\mathbf{X}_0$ if $f(\mathbf{x}) \leqq f(\mathbf{X}_0)$ for all $\mathbf{x}$ in R. Minima and maxima together are called **extrema.**

Furthermore, f is said to have a **local minimum** at $\mathbf{X}_0$ if $f(\mathbf{x}) \geqq f(\mathbf{X}_0)$ for all $\mathbf{x}$ in a neighborhood of $\mathbf{X}_0$, say, for all $\mathbf{x}$ satisfying

$$|\mathbf{x} - \mathbf{X}_0| = [(x_1 - X_1)^2 + \cdots + (x_n - X_n)^2]^{1/2} < r,$$

where $\mathbf{X}_0 = (X_1, \cdots, X_n)$ and $r > 0$ is sufficiently small. A *local maximum* is defined similarly.

If f is differentiable and has an extremum at a point $\mathbf{X}_0$ in the *interior of R* (that is, not on the boundary), then the partial derivatives $\partial f/\partial x_1, \cdots, \partial f/\partial x_n$ must be zero at $\mathbf{X}_0$. These are the components of a vector that is called the **gradient** of f and denoted by grad f or ∇f. (For $n = 3$ this agrees with Sec. 8.8.) Thus

$$(1) \qquad\qquad\qquad \nabla f(\mathbf{X}_0) = \mathbf{0}.$$

A point $\mathbf{X}_0$ at which (1) holds is called a **stationary point** of f.

Condition (1) is necessary for an extremum of f at $\mathbf{X}_0$ in the interior of R, but is not sufficient. Indeed, if $n = 1$, then for $y = f(x)$, condition (1) is $y' = f'(x) = 0$; and, for instance, $y = x^3$ satisfies $y' = 3x^2 = 0$ at $x = X_0 = 0$ where f has no extremum but a point of inflection. Similarly, for $f(\mathbf{x}) = x_1 x_2$ we have $\nabla f(0) = \mathbf{0}$, and f does not have an extremum but a saddle point at $\mathbf{0}$. Hence after solving (1), one must still find out whether one has obtained an extremum. In the case $n = 1$ the conditions $y'(X_0) = 0$, $y''(X_0) > 0$ guarantee a local minimum at X_0 and the conditions $y'(X_0) = 0$, $y''(X_0) < 0$ a local maximum, as is known from calculus. For $n > 1$ there exist similar criteria. However, in practice even solving (1) will often be difficult. For this reason, one generally prefers solution by iteration, by search processes that start at some point and move stepwise to points at which f is smaller (if a minimum of f is wanted) or bigger (in the case of a maximum).

Cauchy's **method of steepest descent** or **gradient method** is of this type. It was introduced in 1847 and has enjoyed popularity ever since.[1] The idea is to find a minimum of $f(\mathbf{x})$ by finding repeatedly minima of a function $g(t)$ of a single variable t, as follows. Suppose that f has a minimum at $\mathbf{X}_0$ and we start at a point $\mathbf{x}$. Then we look for a minimum of f closest to $\mathbf{x}$ along the straight line in the direction of

$$-\nabla f(\mathbf{x}),$$

the direction of steepest descent ($=$ direction of maximum decrease) of f at $\mathbf{x}$; that is, we determine the value of t and the corresponding point

$$(2) \qquad\qquad \mathbf{z}(t) = \mathbf{x} - t\nabla f(\mathbf{x})$$

at which the function

$$(3) \qquad\qquad g(t) = f(\mathbf{z}(t))$$

has a minimum. We take this $\mathbf{z}(t)$ as our next approximation to $\mathbf{X}_0$.

EXAMPLE 1. Method of steepest descent

Determine a minimum of

$$(4) \qquad\qquad f(\mathbf{x}) = x_1^2 + 3x_2^2,$$

starting from $\mathbf{x}_0 = 6\mathbf{i} + 3\mathbf{j}$ and applying the method of steepest descent.

Solution. Clearly, inspection shows that $f(\mathbf{x})$ has a minimum at $\mathbf{0}$. Knowing the solution gives us a better feeling of how the method works.

We obtain $\nabla f(\mathbf{x}) = 2x_1\mathbf{i} + 6x_2\mathbf{j}$ and from this

$$\mathbf{z}(t) = \mathbf{x} - t\nabla f(\mathbf{x}) = (1 - 2t)x_1\mathbf{i} + (1 - 6t)x_2\mathbf{j}$$

$$g(t) = f(\mathbf{z}(t)) = (1 - 2t)^2 x_1^2 + 3(1 - 6t)^2 x_2^2.$$

We now calculate $g'(t)$, which is a linear function of t, set $g'(t) = 0$ and solve for t, finding

Table 21.1
Method of Steepest Descent, Computations in Example 1

n	$\mathbf{x}$		t	$1 - 2t$	$1 - 6t$
0	6.000	3.000	0.210	0.581	-0.258
1	3.484	-0.774	0.310	0.381	-0.857
2	1.327	0.664	0.210	0.581	-0.258
3	0.771	-0.171	0.310	0.381	-0.857
4	0.294	0.147	0.210	0.581	-0.258
5	0.170	-0.038	0.310	0.381	-0.857
6	0.065	0.032			

[1]Convergence can sometimes be slow. For more refined methods and recent developments, see Ref. [F8] in Appendix 1.

$$t = \frac{x_1{}^2 + 9x_2{}^2}{2x_1{}^2 + 54x_2{}^2}.$$

Starting from $\mathbf{x_0} = 6\mathbf{i} + 3\mathbf{j}$, we compute the values in Table 21.1, which are shown in Fig. 461. ∎

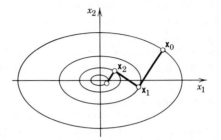

Fig. 461. Method of steepest descent in Example 1

Figure 461 suggests that in the case of slimmer ellipses ("a long narrow valley"), convergence would be poor. The reader may confirm this by replacing in (4) the coefficient 3 with a large coefficient. For more sophisticated descent and other methods, some of them also applicable to vector functions of vector variables, we refer to the references listed under F in Appendix 1.

In the present section we considered unconstrained optimization. In the next section we turn to **constrained optimization,** which plays a more important role in applications.

Problems for Sec. 21.1

1. What happens if you apply the method of steepest descent to $f(x) = x_1{}^2 + x_2{}^2$? (Explain!)

2. Apply the method of steepest descent to $f(x) = x_1{}^2 + 1.5x_2{}^2$, starting from $\mathbf{x_0} = 6\mathbf{i} + 3\mathbf{j}$ (3 steps). Why is the convergence faster than in Example 1?

3. In Prob. 2, start from $\mathbf{x_0} = 1.5\mathbf{i} + \mathbf{j}$. Show that the next approximations are $\mathbf{x_1} = k(1.5\mathbf{i} - \mathbf{j})$, $\mathbf{x_2} = k^2\mathbf{x_0}$, etc., where $k = 0.2$.

4. Verify that in Example 1, successive gradients are orthogonal (perpendicular). Is this just by chance?

5. Apply the method of steepest descent to $f(\mathbf{x}) = x_1{}^2 + cx_2{}^2$. Starting from $\mathbf{x_0} = c\mathbf{i} + \mathbf{j}$, show that subsequent approximations are $\mathbf{x_m} = a_m(c\mathbf{i} + (-1)^m\mathbf{j})$, where $a_m = (c - 1)^m/(c + 1)^m$. Conclude that for large c (slim ellipses $f(\mathbf{x}) = const$) the convergence becomes poor.

6. Apply the method of steepest descent to $f(\mathbf{x}) = x_1{}^4 + x_2{}^4 - 12$, starting from $2\mathbf{i} + 2\mathbf{j}$.

7. Sketch some curves $f(\mathbf{x}) = const$ in Prob. 6 for obtaining a qualitative picture of the behavior of the method of steepest descent for different choices of the starting point.

21.2 Linear Programming

Linear programming (or **linear optimization**) consists of methods for solving optimization problems in which the objective function f is a *linear* function of the control variables $x_1, \cdots, x_n$, and the domain of these variables is restricted by a system of linear inequalities. Problems of this type arise frequently, for instance in production, distribution of goods, economics and approximation theory. Let us illustrate this with a simple example.

EXAMPLE 1. Production plan
Suppose that in producing two types of containers K and L one uses two machines M_1, M_2. To produce a container K, one needs M_1 two minutes and M_2 four minutes. Similarly, L occupies M_1 eight minutes and M_2 four minutes. The net profit for a container K is $29 and for L it is $45. Determine the production plan that maximizes the net profit.

Solution. If we produce x_1 containers K and x_2 containers L per hour, the profit per hour is

$$f(x_1, x_2) = 29x_1 + 45x_2.$$

The constraints are

(1)
$$2x_1 + 8x_2 \leqq 60 \qquad \text{(resulting from machine } M_1\text{)}$$
$$4x_1 + 4x_2 \leqq 60 \qquad \text{(resulting from machine } M_2\text{)}$$
$$x_1 \qquad \geqq 0$$
$$x_2 \geqq 0.$$

Figure 462 shows these constraints. (x_1, x_2) must lie in the first quadrant and below or on the straight line $2x_1 + 8x_2 = 60$ as well as below or on the line $4x_1 + 4x_2 = 60$. Thus (x_1, x_2) is restricted to the quadrangle $OABC$. We have to find (x_1, x_2) in $OABC$ such that $f(x_1, x_2)$ is maximum. Now $f(x_1, x_2) = 0$ gives $x_2 = -(29/45)x_1$ (cf. Fig. 462). The lines $f(x_1, x_2) = const$ are parallel to that line. We see that B, that is, $x_1 = 10, x_2 = 5$, gives the optimum $f(10, 5) = 515$. Hence the answer is that the optimal production plan that maximizes the profit is achieved by producing containers K and L in the ratio $2:1$, the maximum profit being $515 per hour. ∎

Note that the problem in Example 1 or similar optimization problems *cannot* be solved by setting certain partial derivatives equal to zero, because crucial to such problems is the region in which the control variables can vary.

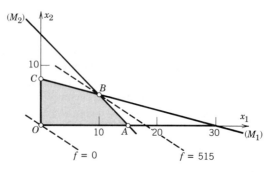

Fig. 462. Linear programming in Example 1

Furthermore, our "geometrical" or graphical method illustrated in Example 1 is confined to two variables x_1, x_2. However, most practical problems involve more variables, so that we need other methods of solution.

Normal Form of a Linear Programming Problem

To prepare for those methods, we show that constraints can be written more uniformly. Let us explain the idea in terms of the first inequality in (1),

$$2x_1 + 8x_2 \leqq 60.$$

This inequality implies $60 - 2x_1 - 8x_2 \geqq 0$ (and conversely), that is, the quantity

$$x_3 = 60 - 2x_1 - 8x_2$$

is nonnegative. Accordingly, our original inequality can now be written

$$2x_1 + 8x_2 + x_3 = 60,$$

$$x_3 \geqq 0.$$

x_3 is a nonnegative auxiliary variable introduced for the purpose of converting inequalities to equations. Such a variable is called a **slack variable.**

EXAMPLE 2. Conversion of inequalities to equations by using slack variables

With the help of two slack variables x_3, x_4 we can write the linear programming problem in Example 1 in the following form. *Maximize*

$$f = 29x_1 + 45x_2$$

under the constraints

(2)
$$\begin{aligned}
2x_1 + 8x_2 + x_3 \quad\;\; &= 60 \\
4x_1 + 4x_2 \quad\;\; + x_4 &= 60 \\
x_i \geqq 0 \quad (i = 1, &\cdots, 4).
\end{aligned}$$

We now have $n = 4$ variables and $m = 2$ (linearly independent) equations, so that two of the four variables, e.g., x_1, x_2, determine the others. Also note that each of the four sides of the quadrangle in Fig. 462 now has an equation of the form $x_i = 0$:

$$OA: x_2 = 0, \qquad AB: x_4 = 0, \qquad BC: x_3 = 0, \qquad CO: x_1 = 0.$$

A vertex of the quadrangle is the intersection of two sides. Hence at a vertex, $n - m = 4 - 2 = 2$ of the variables are zero and the others are nonnegative. Thus at A we have $x_2 = 0$, $x_4 = 0$, and so on. ∎

Our example suggests that a general linear optimization problem can be brought to the following **normal form.** *Maximize*

(3)
$$f = c_1x_1 + c_2x_2 + \cdots + c_nx_n$$

subject to the constraints

$$a_{11}x_1 + \cdots + a_{1n}\,x_n = b_1$$

$$a_{21}x_1 + \cdots + a_{2n}\,x_n = b_2$$

(4) $$\cdots \cdots \cdots \cdots \cdots \cdots \cdots$$

$$a_{m1}x_1 + \cdots + a_{mn}x_n = b_m$$

$$x_i \geqq 0 \qquad (i = 1, \cdots, n).$$

Here $x_1, \cdots, x_n$ include the slack variables (for which the c's in f are zero). We assume that the equations in (4) are linearly independent. Then, if we choose values for $n - m$ of the variables, the system uniquely determines the others.

Recall that our problem also includes the **minimization** of an objective function f since this corresponds to maximizing $-f$ and thus needs no separate consideration.

An n-tuple $(x_1, \cdots, x_n)$ that satisfies all the constraints in (4) is called a *feasible point* or **feasible solution**. A feasible solution is called an **optimal solution** if for it the objective function f becomes maximum, compared with the values of f at all feasible solutions.

Finally, by a **basic feasible solution** we mean a feasible solution for which at least $n - m$ of the variables $x_1, \cdots, x_n$ are zero. For instance, in Example 2 we have $n = 4$, $m = 2$, and the basic feasible solutions are the four vertices O, A, B, C in Fig. 462. Here B is an optimal solution (the only one in this example).

The following theorem is fundamental.

Theorem 1 (Optimal solution)

Some optimal solution of a linear programming problem (3), (4) is also a basic feasible solution of (3), (4).

For a proof, see Ref. [F12], Chap. 3 (listed in Appendix 1). A problem can have many optimal solutions and not all of them may be *basic* feasible solutions; but the theorem guarantees that we can find an optimal solution by searching through the basic feasible solutions. This is a great simplification, but since there are $\begin{pmatrix} n \\ n-m \end{pmatrix}$ different ways of equating $n - m$ of the n variables to zero, considering all these possibilities, dropping those which are not feasible and then searching through the rest would still involve very much work, even when n and m are relatively small. Hence a systematic search is needed. We shall explain an important method of this type in the next section.

Problems for Sec. 21.2

In each case, describe and graph the region in the first quadrant of the x_1x_2-plane determined by the given set of linear inequalities.

1. $x_1 + x_2 \leqq 5$
   $x_2 \leqq 3$
   $-x_1 + x_2 \leqq 2$

2. $x_1 - 2x_2 \leqq -2$
   $0.8x_1 + x_2 \leqq 5.5$

3. $x_1 - 2x_2 \geqq -4$
   $2x_1 + x_2 \leqq 12$
   $x_1 + x_2 \leqq 8$

4. $-x_1 + x_2 \geqq 0$

$x_1 + x_2 \leqq 4.7$

$-x_1 + x_2 \leqq 3$

5. $3x_1 + 4x_2 \leqq 5$

$x_1 - x_2 \leqq -8$

6. $x_1 + x_2 \leqq 1$

$x_1 - x_2 \leqq -1$

$3x_1 - 2x_2 \geqq 20$

7. $3x_1 - 7x_2 \geqq -28$

$x_1 + x_2 \geqq 6$

$9x_1 - 2x_2 \leqq 36$

$5x_1 + 6x_2 \leqq 48$

8. $3x_1 - 5x_2 \leqq 0$

$4x_1 + 3x_2 \leqq 28$

$5x_1 - 6x_2 \geqq -12$

$4x_1 + x_2 \geqq 4$

9. $-2x_1 + 3x_2 \leqq 9$

$5x_1 + x_2 \leqq 25$

$2x_1 + 3x_2 \geqq 3$

$x_1 - 4x_2 \leqq 4$

10. In Example 1, the solution is unique. Can we always expect uniqueness? (Give reason.)

11. Maximize the daily profit in manufacturing two alloys A_1, A_2 that are different mixtures of two metals M_1, M_2 as shown.

Metal	Proportion of Metal		Daily Supply (tons)
	In Alloy A_1	In Alloy A_2	
M_1	0.5	0.25	10
M_2	0.5	0.75	15
Net profit ($ per ton)	30	25	

12. Write Prob. 11 in normal form. What is the meaning of the two slack variables needed? Graph the quadrangle of feasible solutions.

13. Could one find a profit $f(x_1, x_2) = a_1 x_1 + a_2 x_2$ whose maximum is at an interior point of the quadrangle in Fig. 462? (Give reason for your answer.)

14. Maximize $f = x_1 + x_2$ subject to $x_1 \geqq 0$, $x_2 \geqq 0$ and

$$x_1 + 2x_2 \leqq 10, \qquad 2x_1 + x_2 \leqq 10, \qquad x_2 \leqq 4.$$

15. What is the meaning of the slack variables x_3, x_4 in Example 2 in terms of the problem in Example 1?

16. Maximize the total daily output $f = x_1 + x_2$ of a production in which one can choose from two production processes subject to the constraints

$$2x_1 + 4x_2 \leqq 200 \text{ (labor hours)}, \qquad 5x_1 + 2x_2 \leqq 150 \text{ (machine hours)}.$$

17. A company manufactures and sells two models of lamps L_1, L_2, the profit being $15 and $10, respectively. The process involves two workers W_1 and W_2 who are available for this kind of work 100 and 80 hours per month, respectively. W_1 assembles L_1 in 20 and L_2 in 30 minutes. W_2 paints L_1 in 20 and L_2 in 10 minutes. Assuming that all lamps made can be sold without difficulty, determine production figures that maximize the profit.

Minimize the given function f subject to $x_1 \geqq 0$, $x_2 \geqq 0$ and the given further constraints.

18. $f = 3x_1 + 2x_2$, $x_1 + x_2 \geqq 2$, $x_1 - x_2 \geqq -1$, $5x_1 + 3x_2 \leqq 15$

19. $f = 2x_1 + 8x_2$, $2x_1 + 4x_2 \geqq 8$, $2x_1 - 5x_2 \leqq 0$, $-x_1 + 5x_2 \leqq 5$

20. $f = -2x_1 + 10x_2$, $4 \leqq x_1 + x_2 \leqq 10$, $-6 \leqq x_1 - x_2 \leqq 0$

21.3 Simplex Method

From the previous section we recall the following. A linear optimization problem (linear programming problem) can be written in normal form; that is, *maximize*

$$(1) \qquad\qquad f = c_1 x_1 + \cdots + c_n x_n$$

subject to the constraints

$$
\begin{aligned}
a_{11} x_1 + \cdots + a_{1n}\, x_n &= b_1 \\
a_{21} x_1 + \cdots + a_{2n}\, x_n &= b_2
\end{aligned}
$$

$$(2) \qquad\qquad \cdots\cdots\cdots\cdots\cdots\cdots$$

$$a_{m1} x_1 + \cdots + a_{mn} x_n = b_m$$

$$x_i \geqq 0 \qquad (i = 1, \cdots, n).$$

For finding an optimal solution of this problem, we need to consider only the *basic feasible solutions* (defined in Sec. 21.2), but these are still so many that we have to follow a systematic search procedure. In 1948 G. B. Dantzig published an iterative method, called the **simplex method,** for that purpose. In this method, one proceeds stepwise from one basic feasible solution to another in such a way that the objective function f always increases its value. The method begins with an initial *Operation I_0* in which we find a basic feasible solution to start from. Each further step then consists of three operations

> *Operation O_1:* Test for optimality,
> *Operation O_2:* Location of a better basic feasible solution,
> *Operation O_3:* Transition to that better solution.

We describe the details and simultaneously illustrate them by a simple example (the example from Sec. 21.2, which can be solved by an elementary method, as we have seen).

Initial Operation I_0

Find any basic feasible solution (needed to begin the iteration). That is, divide the variables $x_1, \cdots, x_n$ into two classes by selecting m variables—call them **basic variables;** then the other $n - m$ variables are those which must be zero at a basic feasible solution; these are called **nonbasic variables** or **right-hand variables** because we shall write them on the right-hand side of our system (2), which we solve for the basic variables. For illustration:

  Maximize

$$(3a) \qquad\qquad f = 29 x_1 + 45 x_2$$

subject to the constraints

$$2x_1 + 8x_2 + x_3 \qquad = 60$$

(3b) $$\qquad 4x_1 + 4x_2 \qquad + x_4 = 60$$

$$x_i \geqq 0 \qquad (i = 1, \cdots, 4).$$

For instance, let us take x_3, x_4 as basic variables. We solve (3b) for these variables:

(4)
$$x_3 = 60 - 2x_1 - 8x_2$$
$$x_4 = 60 - 4x_1 - 4x_2.$$

In favorable cases, such as the present, we get the values of the variables of a basic feasible solution by setting the right-hand variables equal to zero. Indeed, $x_3 = 60$, $x_4 = 60$, and our basic feasible solution is the point O, the origin $x_1 = 0$, $x_2 = 0$ in Fig. 462 in the previous section.

If this method should give a negative value for some of the basic variables, we must try another set of basic variables.

1st Step. Operation O_1: Test for optimality
Find out whether all coefficients of f, expressed as a function of the present right-hand variables, are negative or zero.

If this optimality criterion is satisfied, then our basic feasible solution is optimal.

Indeed, then f cannot increase if we assign positive values (instead of zero) to the right-hand variables—remember that negative values are not allowed. Hence this condition is sufficient for optimality, and it can be shown that it is also necessary.

For the example, since by our choice the right-hand variables are x_1, x_2, we must consider f as *function of these*, that is, $f = 29x_1 + 45x_2$. The criterion shows that $x_1 = 0$, $x_2 = 0$ is not optimal, because 29 and 45 are positive.

1st Step. Operation O_2: Search for a better basic feasible solution
If the basic feasible solution just tested is not optimal, go to a neighboring basic feasible solution for which f is larger, as follows.

To go to a neighboring basic feasible solution means to go to a point at which another x_i is zero; that is, we have to make an **exchange:** the variable x_i which will now be zero leaves the set of basic variables, and one other variable becomes a basic variable instead. We explain this **method for exchange.**

Consider a right-hand variable x_R that has a positive coefficient in f. Keep the other right-hand variables at zero. Determine the largest increase Δx_R of x_R such that *all* the (old) basic variables are still nonnegative; also list the corresponding increase Δf of f.

In (4), for x_1 this looks as follows.

(5)
$$x_3 = 60 - 2x_1 - 8x_2, \qquad \Delta x_1 = \tfrac{60}{2} = 30, \qquad \Delta f = 29\,\Delta x_1 = 870$$
$$x_4 = 60 - \underline{4x_1} - 4x_2 \qquad \Delta x_1 = \tfrac{60}{4} = 15, \qquad \Delta f = 29\,\Delta x_1 = 435.$$

Underline $x_R = x_1$ in the equation that inhibited any further increase in x_R [as shown in (5)]. Do this for every right-hand variable x_R whose coefficient in f, *expressed as a function of the right-hand variables,* is positive.

Hence we must also consider $x_R = x_2$, underlining as before:

(6)
$$x_3 = 60 - 2x_1 - \underline{8x_2}, \qquad \Delta x_2 = \tfrac{60}{8} = 7.5, \qquad \Delta f = 45\,\Delta x_2 = 337.5$$
$$x_4 = 60 - 4x_1 - 4x_2, \qquad \Delta x_2 = \tfrac{60}{4} = 15, \qquad \Delta f = 45\,\Delta x_2 = 675.$$

1st Step. Operation O_3: Exchange variable
Consider the equations in which a variable is underlined. Thus, in (5), (6),

(7)
$$x_3 = 60 - 2x_1 - \underline{8x_2}, \qquad \Delta f = 337.5$$
$$x_4 = 60 - \underline{4x_1} - 4x_2, \qquad \Delta f = 435.$$

Exchange that right-hand variable x_R which gave the greatest Δf with the basic variable in the equation where x_R is underlined. Thus, by (7), we have to exchange x_1 and x_4, so that x_1 now becomes basic and x_4 right-hand. Solve the system for the new basic variables. Thus,

(8)
$$x_1 = 15 - x_2 - \tfrac{1}{4}x_4$$
$$x_3 = 30 - 6x_2 + \tfrac{1}{2}x_4.$$

2nd Step. Perform Operations O_1, O_2, O_3,
using (8). Thus, in Operation O_1 we must now express $f = 29x_1 + 45x_2$ in terms of the new right-hand variables x_2, x_4. Using the first equation in (8), we find

(9)
$$f = 435 + 16x_2 - 7.25x_4.$$

By the optimality criterion, the present basic feasible solution (point A in Fig. 462, Sec. 21.2) is not optimal because x_2 has a positive coefficient in (9).

In Operation O_2 we now consider $x_R = x_2$. From (8),

(10)
$$x_1 = 15 - x_2 - \tfrac{1}{4}x_4, \qquad \Delta x_2 = 15, \qquad \Delta f = 16\,\Delta x_2 = 240$$
$$x_3 = 30 - \underline{6x_2} + \tfrac{1}{2}x_4, \qquad \Delta x_2 = \tfrac{30}{6} = 5, \qquad \Delta f = 16\,\Delta x_2 = 80.$$

x_4 need not be considered because its coefficient in (9) is negative.

In Operation O_3 we exchange x_2 and x_3 and solve the system for the new basic variables x_1, x_2:

(11)
$$x_1 = 10 + \tfrac{1}{6}x_3 - \tfrac{1}{3}x_4$$
$$x_2 = 5 - \tfrac{1}{6}x_3 + \tfrac{1}{12}x_4.$$

3rd Step. Perform Operations O_1, O_2, O_3,
using (11). Thus, in Operation O_1 we must express $f = 29x_1 + 45x_2$ in terms
of x_3, x_4. By (11),

(12)
$$f = 515 - 2.667x_3 - 5.917x_4.$$

By the optimality criterion, this basic feasible solution is optimal. By (11),
the solution is $x_1 = 10$, $x_2 = 5$, and $f_{\text{opt}} = 515$ from (12), in agreement with
our result in Sec. 21.2.

Table 21.2 shows a tabular arrangement of calculations. In each step, the
right-hand variables are listed in the top line. Arrows point to variables that
enter or leave the basic variables. *Caution!* In the literature one sometimes
lists the *negative* of the right-hand variables ($-x_1$, $-x_2$, etc.); this introduces
a factor -1 in all coefficients. Since, however, this is a convention that adds
neither to the idea of the method as such nor to a better understanding of
it, we do not adopt it.

Table 21.2
Simplex Method, Calculations in Tabular Form

Basic Variables	Constants	↓ x_1	x_2
1st Step			
x_3	60	-2	-8
◀ x_4	60	$\underline{-4}$	-4
f	0	29	45

Basic Variables	Constants	↓ x_2	x_4
2nd Step			
x_1	15	-1	-0.25
◀ x_3	30	$\underline{-6}$	0.5
f	435	16	-7.25

Basic Variables	Constants	x_3	x_4
3rd Step			
x_1	10	0.167	-0.333
x_2	5	-0.167	0.083
f	515	-2.667	-5.917

Problems for Sec. 21.3

1. Maximize $f = 15x_1 + 10x_2$ subject to the constraints $x_1 \geq 0$, $x_2 \geq 0$,

$$3x_1 + 4x_2 \leq 60, \qquad 4x_1 + 3x_2 \leq 60, \qquad 5x_1 + x_2 \leq 60;$$

write the problem in normal form and use the simplex method, taking the slack variables as the initial basic variables.

2. Using the simplex method, maximize the daily output $f = x_1 + x_2$ in producing tables by two different processes subject to the constraints

$$5x_1 + 4x_2 \leq 1300 \text{ (machine hours)}, \qquad 3x_1 + 4x_2 \leq 1100 \text{ (labor).}$$

3. In the example discussed in the text, apply the simplex method, starting from x_2, x_4 as basic variables.

4. In the example in the text, replace f by $f = 15x_1 + 30x_2$ and apply the simplex method, starting from x_3, x_4 as basic variables. In Step 1, operation O_3, you will have a choice in exchanging variables. Explain why.

5. Using the simplex method, maximize the profit in the daily production of two kinds of metal frames F_1 (profit per frame \$135) and F_2 (profit per frame \$75) subject to the constraints (x_1, x_2 = numbers of F_1, F_2 produced per day)

$$x_1' + 3x_2 \leq 18 \qquad \text{(material)}$$

$$x_1 + x_2 \leq 10 \qquad \text{(machine hours)}$$

$$3x_1 + x_2 \leq 24 \qquad \text{(labor).}$$

6. Maximize $f = 4x_1 + x_2 + 2x_3$ subject to
   $x_1 \geq 0$, $x_2 \geq 0$, $x_3 \geq 0$, $x_1 + x_2 + x_3 \leq 1$, $x_1 + x_2 - x_3 \leq 0$.

7. Maximize $f = x_1 + 4x_2 - x_3$ subject to
   $x_1 \geq 0$, $x_2 \geq 0$, $x_3 \geq 0$, $x_1 + x_2 + x_3 \leq 4.8$, $10x_1 + x_3 \leq 9.9$, $x_2 - x_3 \leq 0.2$.

8. Maximize $f = 2x_1 + 3x_2 + x_3$ subject to the same constraints as in Prob. 7.

9. Maximize $f = 2x_1 + x_2 + 3x_3$ subject to
   $x_1 \geq 0$, $x_2 \geq 0$, $x_3 \geq 0$, $4x_1 + 3x_2 + 6x_3 \leq 12$.

10. Maximize $f = 4x_1 + x_2 + 3x_3 + 2x_4$ subject to $x_1 \geq 0$, $x_2 \geq 0$, $x_3 \geq 0$, $x_4 \geq 0$,
    $10x_1 + 15x_2 + 30x_3 + 6x_4 \geq 30$, $8x_1 + 10x_2 + 20x_3 + 5x_4 \leq 40$.

11. A company produces batteries I, II, choosing from four production processes P_1, P_2 (for I) and P_3, P_4 (for II). The profit per battery is \$10 for I and \$20 for II. Maximize the total profit $f = 10x_1 + 10x_2 + 20x_3 + 20x_4$ subject to the constraints

$$12x_1 + 8x_2 + 6x_3 + 4x_4 \leq 120 \qquad \text{(machine hours)}$$

$$3x_1 + 6x_2 + 12x_3 + 24x_4 \leq 180 \qquad \text{(labor hours).}$$

12. Suppose that for certain fuses the profit is \$3 per carton and one can choose from six processes of production $P_1, \cdots, P_6$. Maximize the total profit subject to the constraints (x_j = number of cartons of fuses produced by P_j)

$$10x_1 + 4x_2 + 16x_3 + 7.5x_4 + 5.6x_5 + 4.4x_6 \leq 400 \qquad \text{(machine } M_1)$$

$$2x_1 + 4x_2 + 1.6x_3 + 2.4x_4 + 2.8x_5 + 3.3x_6 \leq 120 \qquad \text{(machine } M_2).$$

21.4 Simplex Method: Degeneracy, Difficulties in Starting

We recall from the last section that in the simplex method we proceed stepwise from one basic feasible solution to another, thereby increasing the value of the objective function f until we reach an optimal solution. Each such transition is accomplished by an exchange in which a right-hand variable becomes a basic variable and vice versa.

It can happen, however, that a basic feasible solution is not optimal but one cannot increase f by exchanging a *single* variable without coming into conflict with the constraints. This situation can occur only for a basic feasible solution for which more than $n - m$ variables are zero (with m and n defined as in Sec. 21.3). Such a solution is called a **degenerate feasible solution.** In such a case one exchanges a right-hand variable with a basic variable which is zero for that solution, and then one proceeds as usual. In more complicated cases, one may even have to perform several such additional exchanges.

Theoretically, in the case of degeneracies there is a possibility that the simplex algorithm will get caught in an infinite loop, that is, will not terminate. Practically, the occurrence of such a loop has been extremely rare, so that it is sufficient to mention that techniques for handling degeneracies are further described in Refs. [F7] and [F12] listed in Appendix 1.

EXAMPLE 1. Simplex method, degenerate feasible solution

A foundry produces two kinds of iron I_1, I_2 by using three kinds of raw material R_1, R_2, R_3 (scrap iron and two kinds of ore) as shown. Maximize the daily net profit.

Raw Material	Raw Material Needed per Ton		Raw Material Available per Day (tons)
	Iron I_1	Iron I_2	
R_1	2	1	16
R_2	1	1	8
R_3	0	1	3.5
Net profit per ton	\$150	\$300	

Solution. Let x_1 and x_2 denote the amount (in tons) of iron I_1 and I_2, respectively, produced per day. Then our problem is as follows. Maximize

$$(1) \qquad\qquad f = 150x_1 + 300x_2$$

subject to the constraints $x_1 \geqq 0$, $x_2 \geqq 0$ and

$$2x_1 + x_2 \leqq 16$$

$$x_1 + x_2 \leqq 8$$

$$x_2 \leqq 3.5.$$

By introducing slack variables x_3, x_4, x_5 we obtain the normal form

(2)
$$
\begin{aligned}
2x_1 + x_2 + x_3 &= 16 \\
x_1 + x_2 \quad\ + x_4 &= 8 \\
x_2 \qquad\ + x_5 &= 3.5 \\
x_i \geq 0 \quad (i = 1, &\cdots, 5).
\end{aligned}
$$

Initial operation. To get started, we choose x_3, x_4, x_5 as basic variables and solve (2) for these:

(3)
$$
\begin{aligned}
x_3 &= 16 - 2x_1 - x_2 \\
x_4 &= 8 - x_1 - x_2 \\
x_5 &= 3.5 \qquad - x_2.
\end{aligned}
$$

Setting $x_1 = 0$, $x_2 = 0$, we get the positive values $x_3 = 16$, $x_4 = 8$, $x_5 = 3.5$, so that all the constraints are satisfied, that is, we have a basic feasible solution (point O in Fig. 463) and may begin with the iterative process.

1st Step. We test for optimality. In (3), the right-hand variables are x_1, x_2. In terms of these, f is given by (1). The optimality criterion (Sec. 21.3) shows that f is not optimal.

We prepare for an exchange of a variable. We have to consider both x_1 and x_2 since both x_1, x_2 have positive coefficients in f; cf. (1). We consider x_1. From (3) and (1) we obtain

$$
\begin{aligned}
x_3 &= 16 - \underline{2x_1} - x_2, & \Delta x_1 &= 8, & \Delta f &= 150\,\Delta x_1 = 1200 \\
x_4 &= 8 - \underline{x_1} - x_2, & \Delta x_1 &= 8, & \Delta f &= 150\,\Delta x_1 = 1200 \\
x_5 &= 3.5 \qquad - x_2,
\end{aligned}
$$

Similarly, for x_2 we obtain from (3) and (1)

$$
\begin{aligned}
x_3 &= 16 - 2x_1 - x_2, & \Delta x_2 &= 16 \\
x_4 &= 8 - x_1 - x_2, & \Delta x_2 &= 8 \\
x_5 &= 3.5 \qquad - \underline{x_2}, & \Delta x_2 &= 3.5, & \Delta f &= 300\Delta x_2 = 1050.
\end{aligned}
$$

We see that we have to exchange x_1 since then we get the maximum possible increase, $\Delta f = 1200$. From the underlining we see that we can exchange x_1 and x_3 or x_1 and x_4. We exchange x_1 and x_3. Solving for the new basic variables x_1, x_4, x_5, we obtain

(4)
$$
\begin{aligned}
x_1 &= 8 - 0.5x_2 - 0.5x_3 \\
x_4 &= \quad - 0.5x_2 + 0.5x_3 \\
x_5 &= 3.5 - x_2.
\end{aligned}
$$

Hence the present basic feasible solution is

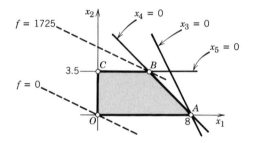

Fig. 463. Example 1, where A is degenerate

$$x_1 = 8, \quad x_2 = 0, \quad x_3 = 0, \quad x_4 = 0, \quad x_5 = 3.5$$

(point A in Fig. 463), and we note that it is **degenerate**, since it has more than $n - m = 5 - 3 = 2$ zeros, namely, 3 zeros.

2nd Step. We test for optimality. For this we have to express f in terms of the new right-hand variables x_2, x_3. From (1) and the first equation in (4) we obtain

$$(5) \qquad\qquad f = 1200 + 225x_2 - 75x_3.$$

The optimality criterion shows that the present basic feasible solution is not optimal.

We consider x_2 for an exchange. We do not consider x_3 since its coefficient in (5) is negative, for reasons explained in Sec. 21.3. From (4) we obtain

$$x_1 = 8 \quad - 0.5x_2 - 0.5x_3, \qquad \Delta x_2 = 16$$
$$x_4 = \qquad - \underline{0.5x_2} + 0.5x_3, \qquad \Delta x_2 = 0, \qquad \Delta f = 0,$$
$$x_5 = 3.5 - \quad x_2, \qquad\qquad \Delta x_2 = 3.5.$$

Because of the degeneracy we exchange x_2 and x_4 (which are both zero at A) without increasing f and without leaving A. We solve for the new basic variables x_1, x_2, x_5:

$$
\begin{aligned}
x_1 &= 8 \quad - x_3 + x_4 \\
x_2 &= \qquad x_3 - 2x_4 \\
x_5 &= 3.5 - x_3 + 2x_4.
\end{aligned}
\tag{6}
$$

3rd Step. From (1) and (6) we obtain f in terms of the new right-hand variables x_3, x_4:

$$(7) \qquad\qquad f = 1200 + 150x_3 - 450x_4.$$

By the optimality criterion, f is not optimal at A. To prepare for an exchange we consider x_3. We do not consider x_4 since its coefficient is negative. From (6) and (7),

$$x_1 = 8 \quad - x_3 + x_4, \qquad \Delta x_3 = 8$$
$$x_2 = \qquad x_3 - 2x_4$$
$$x_5 = 3.5 - \underline{x_3} + 2x_4, \qquad \Delta x_3 = 3.5, \qquad \Delta f = 150\,\Delta x_3 = 525.$$

We exchange x_3 and x_5 and solve for the new basic variables x_1, x_2, x_3:

$$
\begin{aligned}
x_1 &= 4.5 - \quad x_4 + x_5 \\
x_2 &= 3.5 \qquad\quad - x_5 \\
x_3 &= 3.5 + 2x_4 - x_5.
\end{aligned}
\tag{8}
$$

4th Step. From (1) and (8) we obtain f in terms of the new right-hand variables x_4, x_5:

$$(9) \qquad\qquad f = 1725 - 150x_4 - 150x_5.$$

The optimality criterion shows that our present basic feasible solution is optimal. From (8) we see that this solution is

$$x_1 = 4.5, \quad x_2 = 3.5, \quad x_3 = 3.5, \quad x_4 = 0, \quad x_5 = 0$$

(point B in Fig. 463), and $f_{opt} = 1725$ from (9).

Answer. The net profit is maximized if we produce $x_1 = 4.5$ tons of iron I_1 and $x_2 = 3.5$ tons of iron I_2 per day. This maximum amount is \$1725. Also $x_3 = 3.5$ tons of raw material R_1 remains unused in this plan.

Difficulties in Starting

It may sometimes be difficult to find a basic feasible solution to start from. In such a case the idea of an **artificial variable** (or several such variables) is helpful. We explain this method in terms of a typical example.

EXAMPLE 2. Simplex method: difficult start, artificial variable
Maximize

$$(10) \qquad\qquad f = 2x_1 + x_2$$

subject to the constraints $x_1 \geqq 0$, $x_2 \geqq 0$ and

$$x_1 - \tfrac{1}{2}x_2 \geqq 1$$
$$x_1 - x_2 \leqq 2$$
$$x_1 + x_2 \leqq 4.$$

Solution. By means of slack variables we achieve the normal form

$$(11) \qquad
\begin{aligned}
-x_1 + \tfrac{1}{2}x_2 + x_3 \qquad\qquad\qquad &= -1 \\
x_1 - x_2 \qquad + x_4 \qquad\quad &= 2 \\
x_1 + x_2 \qquad\qquad + x_5 &= 4 \\
\end{aligned}$$
$$x_i \geqq 0 \quad (i = 1, \cdots, 5).$$

We want to use x_3, x_4, x_5 as basic variables and start by solving accordingly:

$$(12) \qquad
\begin{aligned}
x_3 &= -1 + x_1 - \tfrac{1}{2}x_2 \\
x_4 &= 2 - x_1 + x_2 \\
x_5 &= 4 - x_1 - x_2.
\end{aligned}$$

But for $x_1 = 0$, $x_2 = 0$ we have the negative value $x_3 = -1$, so that we cannot proceed immediately. Now, instead of searching for other basic variables we can do the following. We introduce a new variable x_6, called an **artificial variable** and defined by

$$(13) \qquad\qquad x_3 = -1 + x_1 - \tfrac{1}{2}x_2 + x_6$$

with the constraint $x_6 \geqq 0$, and we take x_4, x_5, x_6 as basic variables:

$$(14) \qquad
\begin{aligned}
x_4 &= 2 - x_1 + x_2 \\
x_5 &= 4 - x_1 - x_2 \\
x_6 &= 1 - x_1 + \tfrac{1}{2}x_2 + x_3.
\end{aligned}$$

These are positive when the new right-hand variables x_1, x_2, x_3 are zero. Hence we have found a basic feasible solution $x_1 = x_2 = x_3 = 0$, $x_4 = 2$, $x_5 = 4$, $x_6 = 1$ for an **extended problem** involving 6 variables and 3 equations. For this new problem to yield a solution of the given problem we must take care that x_6 will eventually disappear. We accomplish this by the idea of modifying the objective function by adding a term $-Mx_6$, where M is very large; that is, we replace f in (10) by [cf. (10) and (14)]

$$(15) \qquad \hat{f} = f - Mx_6 = (M + 2)x_1 - (\tfrac{1}{2}M - 1)x_2 - Mx_3 - M.$$

Then a positive x_6 will make $\hat{f}$ very small, so that $x_6 = 0$ eventually.

1st Step. $\hat{f}$ is not maximum at our present basic feasible solution. For large M, the only positive coefficient in (15) is that of x_1; hence, by (14) and (15), our preparation for exchange is

$$x_4 = 2 - x_1 + x_2, \qquad \Delta x_1 = 2$$

$$x_5 = 4 - x_1 - x_2, \qquad \Delta x_1 = 4$$

$$x_6 = 1 - \underline{x_1} + \tfrac{1}{2}x_2 + x_3, \qquad \Delta x_1 = 1, \qquad \Delta \hat{f} = M + 2.$$

We exchange x_1 and x_6 and solve (14) and (15) accordingly:

$$x_1 = 1 + \tfrac{1}{2}x_2 + x_3 - x_6$$

(16) $$x_4 = 1 + \tfrac{1}{2}x_2 - x_3 + x_6$$

$$x_5 = 3 - \tfrac{3}{2}x_2 - x_3 + x_6$$

and

$$\hat{f} = 2 + 2x_2 + 2x_3 - (M + 2)x_6.$$

We see that the artificial variable x_6 has become a right-hand variable. We can now set $x_6 = 0$ since (16) then gives $x_1 = 1$, $x_2 = 0$, $x_3 = 0$, $x_4 = 1$, $x_5 = 3$ as a basic feasible solution of the *original* problem.

The further steps follow the usual pattern (cf. Sec. 21.3) and may be left to the reader.   ∎

Problems for Sec. 21.4

1. Maximize $f = 30x_1 + 50x_2$ subject to $x_i \geq 0$ $(i = 1, \cdots, 5)$,

$$2x_1 + 8x_2 + x_3 = 60, \qquad 4x_1 + 4x_2 + x_4 = 60, \qquad 2x_1 + x_2 + x_5 = 30,$$

choosing x_3, x_4, x_5 as basic variables and exchanging x_1 and x_5 in the first sep.

2. Maximize the daily output $f = x_1 + x_2$ in producing x_1 window frames by a process I and x_2 window frames by a process II subject to the constraints

$$2x_1 + 3x_2 \leq 260 \quad \text{(machine hours)}$$

$$3x_1 + 8x_2 \leq 600 \quad \text{(raw material supply)}$$

$$2x_1 + x_2 \leq 140 \quad \text{(labor hours)}.$$

3. Maximize the total output $f = x_1 + x_2 + x_3$ (production figures of three different production processes) subject to input constraints (limitation of machine time)

$$4x_1 + 5x_2 + 8x_3 \leq 6, \qquad 8x_1 + 5x_2 + 4x_3 \leq 6.$$

4. Maximize the weekly shipment $f = x_1 + x_2 + x_3$ of x_1 small containers of red paint, x_2 large containers of white paint and x_3 small containers of black paint, subject to the constraints (production limitation, space on trucks, space in separate storage rooms)

$$3x_1 + 2x_2 + x_3 \leq 250, \qquad x_1 + 3x_2 + x_3 \leq 180, \qquad 2x_1 + x_2 + 2x_3 \leq 210.$$

5. Maximize the total daily profit f of producing x_1, x_2, x_3 lamps of types I, II, III, respectively, if the profits per lamp are \$10, \$8, \$5, respectively, and the constraints are (daily metal supply, plastic and paper supply, glass supply)

$$3x_1 + x_2 \leq 450, \qquad 2x_2 + 3x_3 \leq 900, \qquad 2x_1 + x_2 \leq 350.$$

6. Maximize $f = 6x_1 + 6x_2 + 9x_3$ subject to $x_j \geq 0$ $(j = 1, \cdots, 5)$, and

$$x_1 + x_3 + x_4 = 1, \qquad x_2 + x_3 + x_5 = 1.$$

7. Continue and finish the solution process in Example 2.

8. Using an artificial variable, maximize $f = 2x_1 + x_2$ subject to

$$x_1 \geq 0, \qquad x_2 \geq 0, \qquad x_1 + 2x_2 \geq 4, \qquad x_1 + x_2 \leq 3.$$

9. Using an artificial variable, minimize $f = 2x_1 - x_2$ subject to

$$x_1 \geq 0, \qquad x_2 \geq 0, \qquad x_1 + x_2 \geq 5, \qquad -x_1 + x_2 \leq 1, \qquad 5x_1 + 4x_2 \leq 40.$$

10. If one uses the method of artificial variables in a problem without solution, this nonexistence will become apparent by the fact that one cannot get rid of the artificial variable. Illustrate this by trying to maximize $f = 2x_1 + x_2$ subject to $x_1 \geq 0, x_2 \geq 0$ and

$$2x_1 + x_2 \leq 2, \qquad x_1 + 2x_2 \geq 6, \qquad x_1 + x_2 \leq 4.$$

Review Problems for Chapter 21

1. What is the difference between constrained and unconstrained optimization?

2. Explain the idea of the method of steepest descent. On what will the speed of convergence depend?

3. Why does the function value decrease in each step of the method of steepest descent (unless we reach a point at which the gradient is the zero vector)?

4. What differentiability condition must we make in the method of steepest descent?

5. Write an algorithm for the method of steepest descent.

6. What assumptions about the object function do we make in linear programming?

7. Why can we not use methods of calculus for determining maxima or minima in a linear programming problem?

8. Explain the basic idea of the systematic search for a maximum or minimum in linear programming.

9. State from memory what we mean by the normal form of a linear programming problem and the idea by which we obtain this form.

10. Explain briefly some difficulties that may arise in a linear programming problem and discuss ways to overcome them.

11. Apply the method of steepest descent to $f(\mathbf{x}) = x_1^2 + x_2^2 - 2x_1 - 4x_2 + 7$, starting from $3\mathbf{i} + 3\mathbf{j}$.

12. Apply the method of steepest descent to $f(\mathbf{x}) = x_1^2 + 1.1x_2^2$, starting from $6\mathbf{i} + 3\mathbf{j}$ (3 steps).

13. Verify that in Prob. 12, successive gradients are orthogonal. What is the reason?

14. What result will the method of steepest descent give when you apply it to $f(\mathbf{x}) = x_1 x_2 + 1$, starting (a) from $4\mathbf{i} + 4\mathbf{j}$, (b) from $2\mathbf{i} + \mathbf{j}$ (1 step)? Compare and explain.

15. What does the method of steepest descent amount to in the case of a single variable?

16. Design a "method of steepest ascent" for determining maxima.

Describe and sketch the region in the first quadrant of the $x_1 x_2$-plane given by the following linear inequalities.

17.	18.	19.
$x_1 + x_2 \geq 2$	$5x_1 + 10x_2 \leq 50$	$-x_1 + x_2 \geq -1$
$2x_1 - 3x_2 \geq -12$	$x_1 + x_2 \leq 6$	$x_2 \leq 4$
$x_1 \leq 15$	$10x_1 + 5x_2 \leq 50$	$x_1 + 3x_2 \leq 15$

20. $x_1 - x_2 \geqq -3$ **21.** $x_1 + 2x_2 \geqq 2$ **22.** $x_1 - 3x_2 \geqq -3$

$\quad\quad x_1 + x_2 \geqq 3$ $\quad\quad x_1 \quad\quad \leqq 6$ $\quad\quad x_1 + x_2 \leqq 9$

$\quad\quad x_1 - 2x_2 \leqq 0$ $\quad\quad x_1 - 2x_2 \geqq 0$ $\quad\quad 2x_1 + 7x_2 \leqq 14$

$\quad\quad 5x_1 - x_2 \leqq 9$ $\quad\quad x_1 + 2x_2 \leqq 6$ $\quad\quad 3x_1 - x_2 \geqq 24$

Maximize the given objective function f subject to the given constraints.

23. $f = -5x_1 + x_2,$ $x_1 \geqq 0,$ $x_2 \geqq 0,$ $-x_1 + x_2 \geqq -1,$ $x_1 + x_2 \leqq 6,$ $x_2 \leqq 5$

24. $f = 2x_1 + 3x_2,$ $4x_1 + 3x_2 \geqq 12,$ $x_1 - x_2 \geqq -3,$ $x_2 \leqq 6,$ $2x_1 - 3x_2 \leqq 0$

25. $f = 3x_1 - 6x_2,$ $4x_1 + x_2 \geqq 4,$ $-x_1 + 2x_2 \geqq 6,$ $x_1 + 2x_2 \leqq 14$

26. Minimize f in Prob. 23.

27. Minimize f in Prob. 24.

28. Maximize $f = 12x_1 + 6x_2 + 4x_3$ subject to the constraints $x_1 \geqq 0, x_2 \geqq 0, x_3 \geqq 0,$

$$4x_1 + 2x_2 + x_3 \leqq 60, \quad\quad 2x_1 + 3x_2 + 3x_3 \leqq 50, \quad\quad x_1 + 3x_2 + x_3 \leqq 45.$$

29. A factory produces three kinds of gaskets G_1, G_2, G_3 with net profits of \$4, 6, 8, respectively. Maximize the total daily profit subject to the constraints (x_j = number of gaskets G_j produced per day)

$$x_1 + 2x_2 + 4x_3 \leqq 100 \quad \text{(machine hours)}, \quad\quad 2x_1 + x_2 \leqq 40 \quad \text{(labor)}.$$

30. Graph the region of feasible solutions in Prob. 29 and mark the optimal solution.

Summary of Chapter 21
Unconstrained Optimization,
Linear Programming

In optimization we maximize or minimize an **objective function** f depending on control variables x_j whose domain is either unrestricted ("**unconstrained optimization**," Sec. 21.1) or restricted by constraints in the form of inequalities or equations or both ("**constrained optimization**," Sec. 21.2).

If the objective function is *linear* and the constraints are *linear inequalities* in the control variables $x_1, \cdots, x_m$, by introducing **slack variables** $x_{m+1}, \cdots, x_n$ we can write the optimization problem in **normal form** with the objective function given by

$$(1) \qquad\qquad f_1 = c_1 x_1 + \cdots + c_n x_n$$

(where $c_{m+1} = \cdots = c_n = 0$) and the constraints given by

$$(2) \qquad
\begin{aligned}
a_{11}x_1 + a_{12}x_2 + \cdots + a_{1n}x_n &= b_1 \\
a_{21}x_1 + a_{22}x_2 + \cdots + a_{2n}x_n &= b_2 \\
\cdots\cdots\cdots\cdots\cdots\cdots\cdots\cdots\cdots& \\
a_{m1}x_1 + a_{m2}x_2 + \cdots + a_{mn}x_n &= b_m \\
x_1 \geqq 0, \cdots, x_n \geqq 0.&
\end{aligned}$$

In this case we can then apply the widely used **simplex method** (Sec. 21.3), a systematic stepwise search through a very much reduced subset of all feasible solutions. Section 21.4 shows how to overcome difficulties that may arise in connection with this method.

Chapter 22

Graphs and Combinatorial Optimization

Graphs and **digraphs** (= directed graphs) are presently developing into more and more powerful tools in electrical and civil engineering, communication networks, industrial management, operations research, computer science, marketing, economics, sociology and so on. An accelerating factor of this growth is the impact of fast large computers and their use in large-scale optimization problems that can be modeled in terms of graphs and solved by algorithms provided by graph theory. This approach yields models of general applicability and economic importance. It lies in the center of **"combinatorial optimization,"** a rather recently introduced term denoting optimization problems that have pronounced discrete or combinatorial structures.

The present chapter gives an introduction to this wide new area, which is full of new ideas as well as unsolved problems—in connection, for instance, with efficient computer algorithms and ideas of computational complexity. The classes of problems we shall consider often form the core of larger and more complex practical problems.

Prerequisites: None.
References: Appendix 1, Part F.
Answers to Problems: Appendix 2.

22.1 Graphs and Digraphs

Roughly, a *graph* consists of points, called *vertices,* and lines connecting them, called *edges.* For example, these may be four cities and five highways connecting them, as in Fig. 464. Or the points may represent some people, and we connect by an edge those who do business with each other. Or the vertices may represent workers and jobs, and we connect each worker with the jobs he can do. Let us now give a formal definition.

Definition of a graph

A **graph** G consists of two finite sets (sets having finitely many elements), a set V of points, called **vertices,** and a set E of connecting lines, called **edges,** such that each edge connects two vertices, called the *endpoints* of the edge. We write

$$G = (V, E).$$

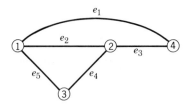
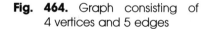

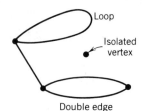
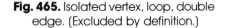

Fig. 464. Graph consisting of 4 vertices and 5 edges

Fig. 465. Isolated vertex, loop, double edge. (Excluded by definition.)

For simplicity we exclude[1] isolated vertices (vertices that are not end-points of any edge), *loops* (edges whose endpoints coincide) and *multiple edges* (edges that have both endpoints in common). Cf. Fig. 465. This is practical for our purpose. ∎

We denote vertices by letters, $u, v, \cdots$ or $v_1, v_2, \cdots$ or simply by numbers $1, 2, \cdots$ (as in Fig. 464). We denote edges by $e_1, e_2, \cdots$ or by their two endpoints; for instance, $e_1 = (1, 4)$, $e_2 = (1, 2)$ in Fig. 464.

We say that a vertex v_i is **incident** with an edge (v_i, v_j); similarly for v_j. The number of edges incident with a vertex v is called the **degree** of v. We call two vertices **adjacent** in G if they are connected by an edge in G (that is, if they are the endpoints of an edge in G).

We meet graphs in different fields under different names: as "networks" in electrical engineering, "structures" in civil engineering, "molecular structures" in chemistry, "organizational structures" in economics, "socio-grams," "road maps," "tele-communication networks" and so on.

Nets of one-way streets, pipeline networks, sequences of jobs in construction work, flows of computation in a computer, producer–consumer relations and many other applications suggest the idea of a "digraph" (= directed graph) in which each edge has as direction (indicated by an arrow, as in Fig. 466).

Definition of a digraph (directed graph)

A **digraph** $G = (V, E)$ is a graph in which each edge $e = (i, j)$ has a direction from its "*initial point*" i to its "*terminal point*" j.

Two edges connecting the same two points i, j are now permitted, provided they have opposite directions, that is, they are (i, j) and (j, i). *Example.* $(1, 4)$ and $(4, 1)$ in Fig. 466. ∎

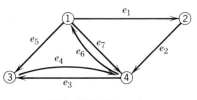

Fig. 466. Digraph

[1]As many authors do, but there is no uniformity, and one must be careful. Some authors permit multiple edges and call graphs without them **"simple graphs."** Others permit loops or isolated vertices, depending on the purpose.

A **subgraph** or subdigraph of a given graph or digraph $G = (V, E)$, respectively, is a graph or digraph obtained by deleting some of the edges and vertices of G, retaining the other edges of G (together with their pairs of endpoints). For instance, e_1, e_3 (together with the vertices 1, 2, 4) form a subgraph in Fig. 464, and e_3, e_4, e_5 (together with the vertices 1, 3, 4) form a subdigraph in Fig. 466.

Computer Representation of Graphs and Digraphs

Drawings of graphs are useful to people in explaining or illustrating specific situations. Here one should be aware that a graph may be sketched in various ways (cf. Fig. 467). For handling graphs and digraphs in computers, one uses matrices or lists, as follows.

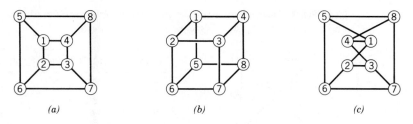

(a)   *(b)*   *(c)*

Fig. 467. Different sketches of the same graph

Adjacency matrix A $= [a_{ij}]$ of a graph G

$$a_{ij} = \begin{cases} 1 & \text{if } G \text{ has an edge } (i, j), \\ 0 & \text{else.} \end{cases}$$

Here, by definition, no vertex is considered to be adjacent to itself. **A** is symmetric. (Why?)

The adjacency matrix of a graph is generally much smaller than the so-called *incidence matrix* (cf. Probs. 21–24) and is preferred over the latter if one decides to store a graph in a computer in matrix form.

EXAMPLE 1. Adjacency matrix of a graph

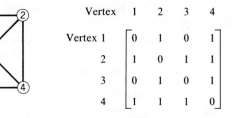

Vertex	1	2	3	4
Vertex 1	0	1	0	1
2	1	0	1	1
3	0	1	0	1
4	1	1	1	0

∎

Adjacency matrix A $= [a_{ij}]$ of a digraph G

$$a_{ij} = \begin{cases} 1 & \text{if } G \text{ has a directed edge } (i, j), \\ 0 & \text{else.} \end{cases}$$

EXAMPLE 2. Adjacency matrix of a digraph

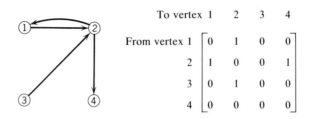

$$
\begin{array}{cccc}
\text{To vertex } 1 & 2 & 3 & 4 \\
\end{array}
$$

$$
\text{From vertex }
\begin{array}{c}
1 \\ 2 \\ 3 \\ 4
\end{array}
\begin{bmatrix}
0 & 1 & 0 & 0 \\
1 & 0 & 0 & 1 \\
0 & 1 & 0 & 0 \\
0 & 0 & 0 & 0
\end{bmatrix}
$$

Lists. The **vertex incidence list** of a graph shows for each vertex the incident edges. The **edge incidence list** shows for each edge its two endpoints. Similarly for a *digraph;* in the vertex list, outgoing edges then get a minus sign, and in the edge list we now have *ordered* pairs of vertices.

EXAMPLE 3. Vertex incidence list and edge incidence list of a graph
This graph is the same as in Example 1, except for notation.

Vertex	Incident edges	Edge	Endpoints
v_1	e_1, e_5	e_1	v_1, v_2
v_2	e_1, e_2, e_3	e_2	v_2, v_3
v_3	e_2, e_4	e_3	v_2, v_4
v_4	e_3, e_4, e_5	e_4	v_3, v_4
		e_5	v_1, v_4

"Sparse graphs" are graphs with few edges (far fewer than $n(n - 1)/2$, where n is the number of vertices). For these graphs, matrices are not efficient. *Lists* then have the advantage of needing much less storage and being easier to handle; they can be ordered, sorted or manipulated in various other ways directly within the computer. For instance, in tracing a "walk" (a connected sequence of edges with pairwise common endpoints), one can easily go back and forth between the two lists just discussed, instead of scanning a large column of a matrix for a single 1.

Computer science has developed more refined lists, which, in addition to the actual content, contain "pointers" indicating the preceding item or the next item to be scanned or both items (in the case of a "walk": the preceding edge or the subsequent one). For details, see Ref. [F16], vol. 1, or Ref. [F14] in Appendix 1.

The present section was devoted to basic concepts and notations needed throughout this chapter. We shall now consider some of the most important classes of combinatorial optimization problems, beginning with **shortest path problems** in the next section. All these problems will concern graphs or digraphs as they arise in applications.

Comment. For grasping the basic ideas, it will be practical to explain these problems and algorithms in terms of *small* graphs, so that one can often see solutions by inspection, but the student should keep in mind that real-life problems are often much larger and may sometimes involve thousands of vertices and edges (think of continent-wide electrical networks, worldwide air travel, companies that have stores or representatives in all large cities in the country), and then a reliable and efficient algorithm is an absolute necessity—solution by inspection would no longer work, even if "nearly optimal" solutions were acceptable.

Problems for Sec. 22.1

1. Sketch the graph consisting of the vertices and edges of a tetrahedron.
2. Worker W_1 can do jobs J_1 and J_3, worker W_2 job J_4, worker W_3 jobs J_2 and J_3. Represent this by a graph.
3. Explain how the following may be regarded as graphs or digraphs: air connections between given cities; memberships of some persons in some committees; relations between chapters of a book; a tennis tournament; countries on a map.
4. How would you represent a net consisting of one-way and two-way streets by a digraph?
5. Give further examples that could be represented by a graph or a digraph.
6. Find the adjacency matrix of the graph in Fig. 464.

In each case find the adjacency matrix of the given graph or digraph.

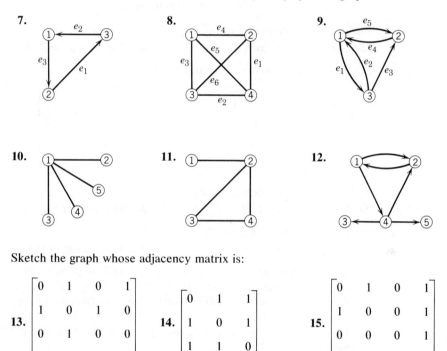

7.   8.   9.

10.   11.   12.

Sketch the graph whose adjacency matrix is:

13. $\begin{bmatrix} 0 & 1 & 0 & 1 \\ 1 & 0 & 1 & 0 \\ 0 & 1 & 0 & 0 \\ 1 & 0 & 0 & 0 \end{bmatrix}$ 14. $\begin{bmatrix} 0 & 1 & 1 \\ 1 & 0 & 1 \\ 1 & 1 & 0 \end{bmatrix}$ 15. $\begin{bmatrix} 0 & 1 & 0 & 1 \\ 1 & 0 & 0 & 1 \\ 0 & 0 & 0 & 1 \\ 1 & 1 & 1 & 0 \end{bmatrix}$

16. Show that the adjacency matrix of a graph is symmetric.
17. Under what condition will the adjacency matrix of a digraph G be symmetric?
18. What are the column sums of the adjacency matrix of a graph?

Sketch the digraph whose adjacency matrix is:
19. The matrix in Prob. 14.   20. The matrix in Prob. 15.

Incidence matrix B of a graph. The definition is $\mathbf{B} = [b_{jk}]$, where

$$b_{jk} = \begin{cases} 1 & \text{if vertex } j \text{ is an endpoint of edge } e_k, \\ 0 & \text{otherwise.} \end{cases}$$

Find the incidence matrix of:
21. The graph in Fig. 464.   22. The graph in Prob. 8.

Sketch the graph whose incidence matrix is:

23.
$$\begin{bmatrix} 1 & 0 & 1 \\ 1 & 1 & 0 \\ 0 & 1 & 0 \\ 0 & 0 & 1 \end{bmatrix}$$

24.
$$\begin{bmatrix} 0 & 1 & 1 & 0 & 1 \\ 1 & 0 & 0 & 1 & 1 \\ 0 & 1 & 0 & 1 & 0 \\ 1 & 0 & 1 & 0 & 0 \end{bmatrix}$$

Incidence matrix $\tilde{\mathbf{B}}$ of a digraph. The definition is $\tilde{\mathbf{B}} = (\tilde{b}_{jk})$, where

$$\tilde{b}_{jk} = \begin{cases} -1 & \text{if edge } e_k \text{ leaves vertex } j, \\ 1 & \text{if edge } e_k \text{ enters vertex } j, \\ 0 & \text{otherwise.} \end{cases}$$

Find the incidence matrix of:

25. The digraph in Prob. 7.   **26.** The digraph in Prob. 9.

Sketch the digraph whose incidence matrix is:

27.
$$\begin{bmatrix} 1 & -1 & 0 & 0 \\ -1 & 1 & -1 & 1 \\ 0 & 0 & 1 & -1 \end{bmatrix}$$

28.
$$\begin{bmatrix} 1 & 1 & -1 \\ 0 & -1 & 0 \\ -1 & 0 & 0 \\ 0 & 0 & 1 \end{bmatrix}$$

Make the vertex incidence list of:

29. The graph in Prob. 8.   **30.** The digraph in Prob. 7.

Make the edge incidence list of:

31. The graph in Fig. 464.   **32.** The digraph in Prob. 9.

33. (Complete graph) Show that a graph G with n vertices can have at most $n(n-1)/2$ edges, and G has exactly $n(n-1)/2$ edges if G is *complete*, that is, if every pair of vertices of G is joined by an edge. (Recall that loops and multiple edges are excluded.)

34. How many diagonals (chords) does an octogon have? An n-gon? What does this have to do with Prob. 33?

35. Show that in any graph, the number of vertices of odd degree must be even.

22.2 Shortest Path Problems. Complexity

In a graph $G = (V, E)$ we can walk from a vertex v_1 along some edges to some other vertex v_k. Here we can make no restrictions, or require that each edge of G be traversed at most once, or that each vertex be visited at most once. These three possibilities suggest three related concepts. If there are no restrictions, we call this a **walk**. Thus a walk from v_1 to v_k is of the form

(1) $$(v_1, v_2), (v_2, v_3), \cdots, (v_{k-1}, v_k),$$

where some of these edges or vertices may be the same. If we are permitted to go along an *edge* at most once, we call the walk a **trail.** If we are permitted to visit each *vertex* at most once, except that we may end our trail at the vertex at which we started, we call the trail a **path.** A walk, trail or path is called **closed** if its last vertex coincides with the first; thus $v_k = v_1$ in (1). A closed path is called a **cycle.** Note that since we do not permit double edges (cf. Sec. 22.1), a cycle has *at least three edges*. Figure 468 illustrates these concepts.

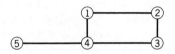

Fig. 468. Walk, trail, path, cycle

$$1 - 2 - 3 - 2 \text{ is a walk (not a trail).}$$
$$4 - 1 - 2 - 3 - 4 - 5 \text{ is a trail (not a path).}$$
$$1 - 2 - 3 - 4 - 5 \text{ is a path (not a cycle).}$$
$$1 - 2 - 3 - 4 - 1 \text{ is a cycle.}$$

In this section we consider a graph $G = (V, E)$ in which each edge (v_i, v_j) has a given "length" $l_{ij} > 0$. We ask for a **shortest path** $v_1 \to v_k$ (v_1 and v_k fixed), that is, a path (1) such that the sum of the lengths of its edges

$$l_{12} + l_{23} + l_{34} + \cdots + l_{k-1,k}$$

is minimum (as small as possible among all paths from v_1 to v_k). Similarly, a **longest path** $v_1 \to v_k$ is one for which that sum is maximum.

Shortest (and longest) path problems are among the most important optimization problems. Here, "length" l_{ij} (often also called "cost" or "weight") can be an actual length measured in miles or travel time or gasoline expenses, but it may also be something entirely different.

For instance, the "*traveling salesman problem*" requires the determination of a shortest **Hamiltonian**[2] **cycle** in a given graph, that is, a cycle that contains all the vertices of the graph.

As another example, by choosing the "most profitable" route $v_1 \to v_k$, a salesman may want to maximize Σl_{ij}, where l_{ij} is his expected commission minus his travel expenses for going from town i to town j.

In an investment problem, i may be the day an investment is made, j the day it matures and l_{ij} the resulting profit, and one gets a graph by considering the various possibilities of investing and reinvesting over a given period of time.

---

[2] WILLIAM ROWAN HAMILTON (1805—1865), Irish mathematician known for his work in dynamics.

Shortest Path If All Edges Have Length 1

Obviously, if all edges have length 1, then a shortest path $v_1 \to v_k$ is one that has the smallest number of edges among all paths $v_1 \to v_k$ in a given graph G. For this problem we discuss a BFS algorithm. BFS stands for **Breadth First Search.** This means that in each step the algorithm visits *all neighboring* (all adjacent) vertices of a vertex reached, as opposed to a DFS algorithm (**Depth First Search** algorithm), which makes a long trail (as in a maze).

We want to find a shortest path in G from a vertex s (*start*) to a vertex t (*terminal*). To guarantee that there is a path from s to t, we make sure that G does not consist of separate portions. Thus we assume that G is **connected,** that is, for any two vertices v and w there is a path $v \to w$ in G. For our problem, the following algorithm is widely used. (Recall that a vertex v is called *adjacent* to a vertex u if there is an edge (u, v) in G.)

Table 22.1
Moore's BFS for Shortest Path (All Lengths One)[3]

ALGORITHM MOORE $[G = (V, E), s, t]$

This algorithm determines a shortest path in a connected graph $G = (V, E)$ from a vertex s to a vertex t.

INPUT:　Connected graph $G = (V, E)$, in which one vertex is denoted by s and one by t, and each edge (i, j) has length $l_{ij} = 1$. Initially all vertices are unlabeled.

OUTPUT:　A shortest path $s \to t$ in $G = (V, E)$

1. Label s with 0.

2. Set $i = 0$.

3. Find all *unlabeled* vertices adjacent to a vertex labeled i.

4. Label the vertices just found with $i + 1$.

5. If vertex t is labeled, then "backtracking" gives the shortest path

$$k \ (= \text{label of } t), \ k - 1, \ k - 2, \ \cdots, \ 0$$

OUTPUT $k, k - 1, k - 2, \cdots, 0$. Stop

Else increase i by 1. Go to Step 3.

End MOORE

EXAMPLE 1. Application of Moore's BFS algorithm
Find a shortest path $s \to t$ in the graph G shown in Fig. 469.

Solution. Figure 469 shows the labels. The heavy edges form a shortest path (length 4). There is another shortest path $s \to t$. (Can you find it?) Hence in the program we may wish to introduce a rule that makes backtracking unique. For instance, assuming that the vertices are numbered $1, \cdots, n$, we may require that in each step of backtracking, the vertex with the smallest possible number (not label!) be taken.　　　∎

---

[3]*Proceedings of the International Symposium for Switching Theory.* Cambridge: Harvard University Press, 1959. Part II, pp. 285-292.

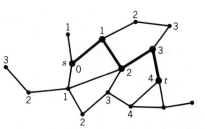

Fig. 469. Example 1, given graph and result of labeling

Complexity of an Algorithm

Complexity *of Moore's algorithm.* To find the vertices to be labeled 1, we have to scan all edges incident with s. Next, when $i = 1$, we have to scan all edges incident with vertices labeled 1, etc. Hence each edge is scanned twice. These are $2m$ operations (m = number of edges of G). This is a function $c(m)$. Whether it is $2m$ or $5m + 3$ or $12m$ is not so essential; it *is* essential that $c(m)$ is proportional to m (not m^2, for example); it is of the "order" m. We write for any function $am + b$ simply $O(m)$, for any function $am^2 + bm + d$ simply $O(m^2)$, and so on; here, O suggests "order." The underlying idea and practical aspect are as follows.

In judging an algorithm, we are mostly interested in its behavior for very large problems (large m in the present case), since these are going to determine the limits of the applicability of the algorithm. Thus, the essential item is the fastest growing term (am^2 in $am^2 + bm + d$, etc.) since it will overwhelm the others when m is large enough. Also, a constant factor in this term is not very essential; for instance, the difference between two algorithms of orders, say, $5m^2$ and $8m^2$ is generally not very essential and can be made irrelevant by a modest increase in the speed of computers. However, it does make a great practical difference whether an algorithm is of order m or m^2 or of a still higher power m^p. And the biggest difference occurs between these "polynomial orders" and "exponential orders," such as 2^m.

For instance, on a computer that does 10^6 operations per second, a problem of size $m = 50$ will take 5 minutes with an algorithm that requires m^5 operations, but 36 years with an algorithm that requires 2^m operations. But this is not our only reason for regarding polynomial orders as good and exponential orders as bad. Another reason is the gain in using a faster computer. For example, since $1000 = 31.6^2 = 2^{9.97}$, an increase in speed by a factor 1000 will permit us to do per hour problems 1000 and 31.6 times as big if the algorithms are $O(m)$ and $O(m^2)$, respectively, but when an algorithm is $O(2^m)$, all we gain is a relatively modest increase of 10 in problem size since $2^{9.97} \cdot 2^m = 2^{m+9.97}$.

The **symbol** O is quite practical and commonly used whenever the order of growth is essential, but not the specific form of a function. Thus if a function $g(m)$ is of the form

$$g(m) = kh(m) + \text{more slowly growing terms} \qquad (k \text{ constant}),$$

we say that $g(m)$ is of the *order* $h(m)$ and write

$$g(m) = O(h(m)).$$

For instance, $am + b = O(m)$,

$$am^2 + bm + d = O(m^2), \qquad 5 \cdot 2^m + 3m^2 = O(2^m),$$

and so on.

We want an algorithm $\mathcal{A}$ to be "efficient," that is, "good" with respect to

> (i) *Time* (number $c_{\mathcal{A}}(m)$ of computer operations)

or

> (ii) *Space* (storage needed in the internal memory)

or both. Here $c_{\mathcal{A}}$ suggests "**complexity**" of $\mathcal{A}$. Two popular choices for $c_{\mathcal{A}}$ are

> (*Worst case*) $c_{\mathcal{A}}(m) = $ longest time $\mathcal{A}$ takes for a problem of size m.

> (*Average case*) $c_{\mathcal{A}}(m) = $ average time $\mathcal{A}$ takes for a problem of size m.

In problems on graphs, the "size" will often be m (number of edges) or n (number of vertices). For our present simple algorithm, $c_{\mathcal{A}}(m) = 2m$ in both cases.

For a "good" algorithm $\mathcal{A}$, we want that $c_{\mathcal{A}}(m)$ does not grow too fast. Accordingly, we call $\mathcal{A}$ **efficient** if $c_{\mathcal{A}}(m) = O(m^k)$ for some integer $k \geq 0$; that is, $c_{\mathcal{A}}$ may contain only powers of m (or functions that grow even more slowly, such as $\ln m$), but no exponential functions. Furthermore, we call $\mathcal{A}$ **polynomially bounded** if $\mathcal{A}$ is efficient when we choose the "worst case" $c_{\mathcal{A}}(m)$.

These conventional concepts have intuitive appeal, as our discussion shows. But for moderate m they do not always tell the full story because then $O(2^m)$ or $O(e^m)$ may be smaller than $O(m^2)$, say; cf. Prob. 11.

Complexity should be investigated for every algorithm, so that one can also compare different algorithms for the same task. This may often exceed the level in this chapter; accordingly, we shall confine ourselves to a few occasional comments in this direction.

In the next section we turn to more general shortest path problems in which edges can have arbitrary positive lengths and discuss **Bellman's optimality principle** as well as a resulting algorithm for solving these general problems.

Problems for Sec. 22.2

1. A shortest path $s \rightarrow t$ for given s and t need not be unique. Illustrate this by finding another shortest path $s \rightarrow t$ in Example 1.

2. Show the nonuniqueness of a shortest path $s \rightarrow t$ by the simplest possible graph G. How many edges must G have?

3. How many edges can a shortest path between any two vertices in a graph with n vertices at most have? Give reason. In a complete graph with all edges of length 1?

In each graph find a shortest path P: $s \to t$ by Moore's BFS algorithm; sketch the graph with the labels and indicate P by heavier lines (as in Fig. 469).

4.   **5.**   **6.**

7. **(Moore's algorithm)** Show that if vertex v has label $\lambda(v) = k$, then there is a path $s \to v$ of length k.

8. **(Moore's algorithm)** Call the length of a shortest path $s \to v$ briefly the *distance* of v from s. Show that if v has distance l, it has label $\lambda(v) = l$.

9. **(Order)** Show that $O(m^2) + O(m^2) = O(m^2)$ and $kO(m^p) = O(m^p)$.

10. Show that $\sqrt{1 + m^2} = O(m)$, $0.02e^m + 100m^2 = O(e^m)$.

11. Show that $0.001e^m = O(e^m)$, $1000m^2 = O(m^2)$, and determine the positive integers m for which the first of these functions is smaller than the second. What does this illustrate?

12. If we switch from one computer to another that is 100 times as fast, what is our gain in problem size per hour in the use of an algorithm that is $O(m)$, $O(m^2)$, $O(m^5)$, $O(e^m)$?

13. **(Hamiltonian cycle)** Find and sketch a Hamiltonian cycle in the graph in Prob. 6.

14. Does the graph in Prob. 4 have a Hamiltonian cycle? (Give reason.)

15. Find and sketch a Hamiltonian cycle in Fig. 467, Sec. 22.1.

16. Find and sketch a Hamiltonian cycle in the graph of a dodecahedron, which has 12 pentagonal faces and 20 vertices (Fig. 470). This is a problem Hamilton considered.

17. Divide a square into 5×5 congruent squares and find in the resulting graph a Hamiltonian cycle with 36 vertices.

18. **(Euler graph)** An *Euler graph* G is a graph that has a closed Euler trail. An **Euler trail** is a trail that contains every edge of G exactly once. Which subgraph with four edges of the graph in Example 1, Sec. 22.1, is an Euler graph?

19. Find four different closed Euler trails in Fig. 471.

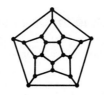

Fig. 470. Problem 16   **Fig. 471.** Problem 19

20. **(Postman problem)** The *postman problem* (or *Chinese postman problem*[4]) is the problem of finding a closed walk W: $s \to s$ (s is the post office) in a graph G with edges (i, j) of lengths $l_{ij} > 0$ such that every edge of G is traversed at least once and the length of W is minimum. Explain some other applications in which this problem is essential.

---

[4]Since it was first considered in the journal *Chinese Mathematics* **1** (1962), 273-77.

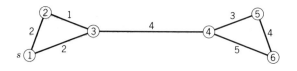

Fig. 472. Problem 21

21. Find a solution of the postman problem in Fig. 472 by inspection.
22. Is the graph in Prob. 21 an Euler graph? (Give reason.)
23. What is the difference between the postman problem and the traveling salesman problem mentioned in the text?
24. Show that the length of a shortest postman trail is the same for every starting vertex.
25. Give an example of a digraph (as simple as possible) for which the postman problem has no solution.

 Bellman's Optimality Principle. Dijkstra's Algorithm

We continue our discussion of the shortest path problem in a graph G in which every edge (i, j) has assigned a length l_{ij}. Section 22.2 was devoted to the special case $l_{ij} = 1$ for all edges in G. We now turn to the general case of any $l_{ij} > 0$, writing $l_{ij} = \infty$ for any edge (i, j) that does not exist in G (and setting, as usual, $\infty + a = \infty$ for any number a).

We consider the problem of finding shortest paths from a given vertex, denoted by 1 and called the *origin,* to *all* other vertices $2, 3, \cdots, n$ of G. We let L_j denote the length of a shortest path $1 \rightarrow j$ in G.

Bellman's minimality principle (or optimality principle)
If P: $1 \rightarrow j$ is a shortest path from 1 to j in G and (i, j) is the last edge of P (Fig. 473), then P_i: $1 \rightarrow i$ [obtained by dropping (i, j) from P] is a shortest path $1 \rightarrow i$.

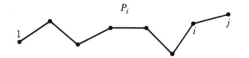

Fig. 473. Paths P and P_i in Bellman's minimality principle

Proof. Suppose that the conclusion is false. Then there is a path P_i^*: $1 \rightarrow i$ which is shorter than P_i. Hence if we now add (i, j) to P_i^*, we get a path $1 \rightarrow j$ that is shorter than P. This contradicts our assumption that P is shortest. ∎

From Bellman's principle we can derive basic equations as follows. For fixed j we may obtain various paths $1 \rightarrow j$ by taking shortest paths P_i for various i and adding to them an edge (i, j). These paths $1 \rightarrow j$ obviously have lengths $L_i + l_{ij}$ (L_i = length of P_i). We can now take the minimum over i, that is, pick an i for which $L_i + l_{ij}$ is smallest. By the Bellman principle, this gives a shortest path $1 \rightarrow j$. It has the length

$$
\begin{array}{|l|}
\hline
L_1 = 0 \\[2mm]
L_j = \min_{i \neq j} (L_i + l_{ij}), \qquad\qquad j = 2, \cdots, n. \\
\hline
\end{array}
$$

These are the **Bellman equations.** Since $l_{ii} = 0$ by definition, instead of $\min_{i \neq j}$ we can simply write $\min_i$. These equations suggest the idea of one of the best known algorithms for the shortest path problem, as follows.

Dijkstra's Algorithm for Shortest Paths

Table 22.2
Dijkstra's Algorithm for Shortest Path[5] (Explanation see next page)

ALGORITHM DIJKSTRA [$G = (V, E)$, $V = \{1, \cdots, n\}$, l_{ij} for all (i, j) in E]

Given a connected graph $G = (V, E)$ with vertices $1, \cdots, n$ and edges (i, j) having lengths $l_{ij} > 0$, this algorithm determines the lengths of shortest paths from vertex 1 to the vertices $2, \cdots, n$.

INPUT: Number of vertices n, edges (i, j) and lengths l_{ij}
OUTPUT: Lengths L_j of shortest paths $1 \rightarrow j$, $j = 2, \cdots, n$

1. *Initial step*
   Vertex 1 gets PL: $L_1 = 0$.
   Vertex j ($= 2, \cdots, n$) gets TL: $\tilde{L}_j = l_{1j}$ ($= \infty$ if there is no edge $(1, j)$ in G).
   Set $\mathscr{PL} = \{1\}$, $\mathscr{TL} = \{2, 3, \cdots, n\}$.

2. *Fixing a permanent label*
   Find a k in $\mathscr{TL}$ for which $\tilde{L}_k$ is minimum, set $L_k = \tilde{L}_k$. Take the smallest k if there are several. Delete k from $\mathscr{TL}$ and include it in $\mathscr{PL}$.
   If $\mathscr{TL} = \varnothing$ (that is, $\mathscr{TL}$ is empty) then
         OUTPUT $L_2, \cdots, L_n$. Stop
   Else continue (i.e., go to Step 3).

3. *Updating tentative labels*
   For all j in $\mathscr{TL}$, set[6] $\tilde{L}_j = \min_k \{\tilde{L}_j, L_k + l_{kj}\}$.
   Go to Step 2.

End DIJKSTRA

---

[5]*Numerische Mathematik* **1** (1959), 269-271.
[6]That is, take the number $\min_k \{\tilde{L}_j, L_k + l_{kj}\}$ as your new $\tilde{L}_j$.

This algorithm is shown in Table 22.2. It is a labeling procedure. At each stage of the computation, each vertex v gets a label, either

(PL) a *permanent label* = length L_v of a shortest path $1 \to v$

or

(TL) a *temporary label* = upper bound $\tilde{L}_v$ for the length of a shortest path $1 \to v$.

We denote by $\mathcal{PL}$ and $\mathcal{TL}$ the sets of vertices with a permanent label and with a temporary label, respectively. The algorithm has an initial step in which vertex 1 gets the permanent label $L_1 = 0$ and the other vertices get temporary labels, and then the algorithm alternates between Steps 2 and 3. In Step 2 the idea is to pick k "minimally." In Step 3 the idea is that the upper bounds will in general improve (decrease) and must be updated accordingly.

EXAMPLE 1. Application of Dijkstra's algorithm

Applying Dijkstra's algorithm to the graph in Fig. 474a, find shortest paths from vertex 1 to vertices 2, 3, 4.

Solution. We list the steps and computations.

1. $L_1 = 0, \tilde{L}_2 = 8, \tilde{L}_3 = 5, \tilde{L}_4 = 7,$ $\mathcal{PL} = \{1\},$ $\mathcal{TL} = \{2, 3, 4\}$

2. $L_3 = \min\{\tilde{L}_2, \tilde{L}_3, \tilde{L}_4\} = 5, k = 3,$ $\mathcal{PL} = \{1, 3\},$ $\mathcal{TL} = \{2, 4\}$

3. $\tilde{L}_2 = \min\{8, L_3 + l_{32}\} = \min\{8, 5 + 1\} = 6$

   $\tilde{L}_4 = \min\{7, L_3 + l_{34}\} = \min\{7, \infty\} = 7$

2. $L_2 = \min\{\tilde{L}_2, \tilde{L}_4\} = 6, k = 2,$ $\mathcal{PL} = \{1, 2, 3\},$ $\mathcal{TL} = \{4\}$

3. $\tilde{L}_4 = \min\{7, L_2 + l_{24}\} = \min\{7, 6 + 2\} = 7$

2. $L_4 = 7, k = 4$ $\mathcal{PL} = \{1, 2, 3, 4\},$ $\mathcal{TL} = \varnothing.$ ∎

Figure 474b shows the resulting shortest paths, of lengths $L_2 = 6, L_3 = 5, L_4 = 7$.

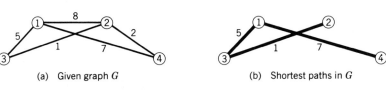

(a)   Given graph G (b)   Shortest paths in G

Fig. 474. Example 1

Complexity. *Dijkstra's algorithm is $O(n^2)$.*

Proof. Step 2 requires comparison of elements, first $n - 2$, the next time $n - 3$, etc., a total of $(n - 2)(n - 1)/2$. Step 3 requires the same number of comparisons, a total of $(n - 2)(n - 1)/2$, as well as additions, first $n - 2$, the next time $n - 3$, etc., again a total of $(n - 2)(n - 1)/2$. Hence the total number of operations is $3(n - 2)(n - 1)/2 = O(n^2)$.   ∎

This section and the preceding one were devoted to shortest path problems. In the next section we begin our discussion of **trees,** which are particularly important graphs, and corresponding optimization problems that arise quite often in practice.

Problems for Sec. 22.3

1. The net of roads in Fig. 475 connecting four villages is to be reduced to minimum length, but so that one can still reach every village from every other village. Which of the roads should be retained? Find the solution (a) by inspection, (b) by Dijkstra's algorithm.

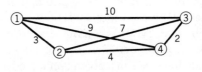

Fig. 475. Problem 1

Applying Dijkstra's algorithm, find shortest paths for the following graphs.

2.   **3.**   **4.**

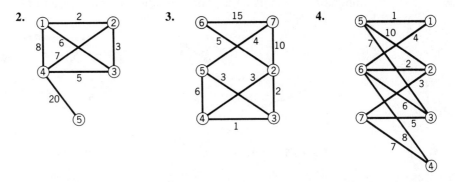

5. Show that in Dijkstra's algorithm, at each instant the demand on storage is light (data for less than n edges).

6. Show that in Dijkstra's algorithm, for L_k there is a path $P: 1 \to k$ of length L_k.

7. What is the complexity of the algorithm for finding all shortest paths (beginning at any vertex)?

Negative lengths. We do not generally consider the case of negative lengths, but we want to indicate that negative lengths may cause difficulties, particularly when they lead to cycles of negative length.

8. Find shortest paths in Fig. 476 by inspection and confirm them by Bellman's equations.

9. Find shortest paths in Fig. 477 by inspection and show that Bellman's equations do not have a unique solution.

10. Show by inspection that there are no unique shortest paths in Fig. 478.

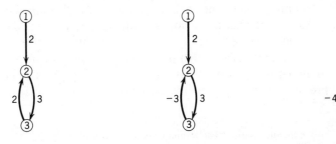

Fig. 476. Problem 8   **Fig. 477.** Problem 9   **Fig. 478.** Problem 10

22.4 Shortest Spanning Trees. Kruskal's Greedy Algorithm

By definition, a **tree** T is a graph that is connected and has no cycles. "**Connected**" means that there is a path from any vertex in T to any other vertex in T. A **cycle**[7] is a path $s \to t$ of at least three edges which is closed ($t = s$); cf. also Sec. 22.2. Figure 479a shows an example.

A **spanning tree** T in a given connected graph $G = (V, E)$ is a tree containing *all* the n vertices of G. Cf. Fig. 479b. Such a tree has $n - 1$ edges. (Proof?)

A **shortest spanning tree** T in a connected graph G whose edges (i, j) have lengths $l_{ij} > 0$ is a spanning tree for which Σl_{ij} (sum over all edges of T) is minimum compared to Σl_{ij} for any other spanning tree in G.

(a) A cycle   *(b)* A spanning tree

Fig. 479. Example of (a) a cycle, (b) a spanning tree in a graph

Trees are among the most important types of graphs, and they occur in various applications. Familiar examples are family trees and organization charts. Trees can be used to exhibit, organize or analyze electrical networks, producer–consumer and other business relations, information in database systems, syntactic structure of computer programs, etc. We mention a few specific applications that need no lengthy additional explanations.

The set of shortest paths from vertex 1 to the vertices 2, $\cdots$, n in the last section forms a spanning tree.

Railway lines connecting a number of cities (the vertices) can be set up in the form of a spanning tree, the "length" of a line (edge) being the construction cost, and one wants to minimize the total construction cost. Similarly for bus lines, where "length" may be the average annual operating cost. Or for steamship lines (freight lines), where "length" may be profit and the goal is the maximization of total profit. Or in a network of telephone lines between some cities, a shortest spanning tree may simply represent a selection of lines that connect all the cities at minimal cost.

As a somewhat more sophisticated example, consider a private communication network G, let p_{ij} be the probability of intercepting line (i, j) by an outsider, and suppose that one wants to communicate a confidential message to all participants (vertices) along a spanning tree T in G that minimizes the product of the p_{ij} of all the edges (lines) of T, that is, assuming stochastic

---

[7]Or **circuit**. Caution! The terminology varies considerably.

independence, the total probability of interception; equivalently, a tree T that minimizes the sum of the logarithms of the p_{ij} or, better, the sum of $l_{ij} = \ln \tilde{p}_{ij}$, where $\tilde{p}_{ij} = Kp_{ij}$ with K so large that $\tilde{p}_{ij} > 1$ for each edge of the network G, thus $l_{ij} > 0$.

In addition to these examples from transportation and communication networks, one could mention others from distribution networks, and so on.

We shall now discuss a simple algorithm for the shortest spanning tree problem, which is particularly suitable for sparse graphs (graphs with very few edges; cf. Sec. 22.1). This algorithm is shown in Table 22.3.

Table 22.3
Kruskal's Greedy Algorithm for Shortest Spanning Trees [8]

> ALGORITHM KRUSKAL [$G = (V, E)$, l_{ij} for all (i, j) in E]
> Given a connected graph $G = (V, E)$ with edges (i, j) having length $l_{ij} > 0$, the algorithm determines a shortest spanning tree T in G.
>
> INPUT: Edges (i, j) of G and their lengths l_{ij}
> OUTPUT: Shortest spanning tree T in G
>
> **1.** Order the edges of G in ascending order of length.
> **2.** Choose them in this order as edges of T, rejecting an edge only
>   if it forms a cycle with edges already chosen.
>   If $n - 1$ edges have been chosen, then
>   OUTPUT T ($=$ the set of edges chosen). Stop
> End KRUSKAL

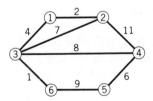

Fig. 480. Graph in Example 1

EXAMPLE 1. Application of Kruskal's algorithm
Find a shortest spanning tree in the graph in Fig. 480.

Solution. See Table 22.4. In some of the intermediate stages the edges chosen form a *discon-nected* graph (cf. Fig. 481); this is typical. We stop after $n - 1 = 5$ choices since a spanning tree has $n - 1$ edges. In our problem the edges chosen are in the upper part of the list. This is typical of problems of any size; in general, edges farther down in the list have a smaller chance of being chosen.   ■

[8]*Proceedings of the American Mathematical Society* **7** (1956), 48-50.

Table 22.4
Solution in Example 1

Edge	Length	Choice
(3, 6)	1	1st
(1, 2)	2	2nd
(1, 3)	4	3rd
(4, 5)	6	4th
(2, 3)	7	Reject
(3, 4)	8	5th
(5, 6)	9	
(2, 4)	11	

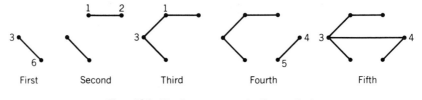

Fig. 481. Choice process in Example 1

The efficiency of Kruskal's method is greatly increased by

Double labeling of vertices

Each vertex i carries a double label (r_i, p_i), where

   r_i = *Root of the subtree to which i belongs,*
   p_i = *Predecessor of i in its subtree*
   p_i = *0 for roots.*

This simplifies

Rejecting. *If (i, j) is next in the list to be considered, reject (i, j) if $r_i = r_j$* (that is, i and j are in the same subtree, so that they are already joined by edges and (i, j) would thus create a cycle). *If $r_i \neq r_j$, include (i, j) in T.*

If there are several choices for r_i, choose the smallest. If subtrees merge (become a single tree), retain the smallest root as the root of the new subtree.
For Example 1, the double-label list is shown in Table 22.5. In storing it, at each instant one may retain only the latest double label. We show all double labels in order to exhibit the process in all its stages. Labels that remain unchanged are not listed again. Underscored are the two 1s that are the common root of vertices 2 and 3, the reason for rejecting the edge (2, 3). By reading for each vertex the latest label we can read from this list that 1 is the vertex we have chosen as a root and the tree is as shown in the last part of Fig. 481. This is made possible by the predecessor label that each vertex carries. Also, for accepting or rejecting an edge we have to make only one comparison (the roots of the two endpoints of the edge).

Table 22.5
List of Double Labels in Example 1

Vertex	Choice 1 (3, 6)	Choice 2 (1, 2)	Choice 3 (1, 3)	Choice 4 (4, 5)	Choice 5 (3, 4)
1		(1, 0)			
2		($\underline{1}$, 1)			
3	(3, 0)		($\underline{1}$, 1)		
4				(4, 0)	(1, 3)
5				(4, 4)	(1, 4)
6	(3, 3)		(1, 3)		

Ordering is the more expensive part of the algorithm. It is a standard process in data processing for which various methods have been suggested (cf. Ref. [F16] in Appendix 1). For a complete list of m edges, an algorithm would be $O(m \log_2 m)$, but since the $n - 1$ edges of the tree are most likely be found earlier, by inspecting the $q \, (< m)$ topmost edges, for such a list of q edges one would have $O(q \log_2 m)$.

Figure 481 shows that Kruskal's algorithm will only eventually give a tree. In the next section we discuss **Prim's algorithm** for shortest spanning trees which gives a tree in each step. This tree has at first one edge and is grown stepwise until it becomes spanning (and is shortest).

Problems for Sec. 22.4

Applying Kruskal's algorithm, find a shortest spanning tree for the following graphs.

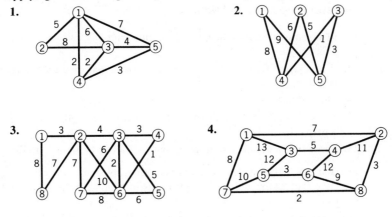

5. The graph in Prob. 4, Sec. 22.3

6. Find a shortest spanning tree in the complete graph of all possible 30 air connections between the six cities given (distances in miles, rounded). Can you think of a practical application of the result? (See the table on the next page.)

Table for Problem 6

	Dallas	Denver	Los Angeles	New York	Washington, DC
Chicago	800	900	1800	700	650
Dallas		650	1300	1350	1200
Denver			850	1650	1500
Los Angeles				2500	2350
New York					200

7. In the graph in Fig. 482 find (a) shortest paths from 1 to 2 and from 1 to 3, (b) a shortest spanning tree. Compare the sum of the lengths of the edges used in (a) with that in (b) and comment.

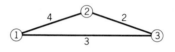

Fig. 482. Problem 7

8. Design an algorithm for obtaining longest spanning trees.

9. Apply the algorithm in Prob. 8 to the graph in Example 1. Compare with the result in Example 1.

10. To get a minimum spanning tree, instead of adding shortest edges, one could think of deleting longest edges. For what graphs would this be feasible? Describe an algorithm for this.

11. Apply the method suggested in Prob. 10 to the graph in Example 1. Do you get the same tree?

General Properties of Trees

12. **(Uniqueness)** Show that the path connecting any two vertices u and v in a tree T is unique.

13. If in a graph G, any two vertices are connected by a unique path, show that G is a tree.

14. Show that a tree with n vertices has $n - 1$ edges.

15. Show that a tree T with exactly two vertices of degree 1 (cf. Sec. 22.1) must be a path.

16. Show that a connected graph G with n vertices and $n - 1$ edges is a tree.

17. **(Forest)** A (not necessarily connected) graph without cycles is called a *forest*. Give typical examples of applications in which graphs occur that are forests or trees.

18. Show that if a graph G has no cycles, then G must have at least 2 vertices of degree 1.

19. Show that a graph G with n vertices is a tree if and only if G has $n - 1$ edges and has no cycles.

20. Show that if one joins two vertices in a tree T by a new edge, then a cycle is formed.

22.5 Prim's Algorithm for Shortest Spanning Trees

Table 22.6
Prim's Algorithm for Shortest Spanning Tree[9]

ALGORITHM PRIM [$G = (V, E)$, $V = \{1, \cdots, n\}$, l_{ij} for all (i, j) in E]
Given a connected graph $G = (V, E)$ with vertices $1, 2, \cdots, n$ and edges (i, j) having length $l_{ij} > 0$, this algorithm determines a shortest spanning tree T in G and its length $L(T)$.

INPUT: n, edges (i, j) of G and their lengths l_{ij}
OUTPUT: Edge set S of a shortest spanning tree T in G; $L(T)$
[Initially, all vertices are unlabeled.]

1. *Initial step*
   Set $i(k) = 1$, $U = \{1\}$, $S = \varnothing$.
   Label vertex k ($= 2, \cdots, n$) with $\lambda_k = l_{ik}$ [$= \infty$ if G has no edge $(1, k)$].

2. *Addition of an edge to the tree T*
   Let λ_j be the smallest λ_k for k not in U. Include vertex j in U and edge $(i(j), j)$ in S.
   If $U = V$ then compute
   $$L(T) = \Sigma l_{ij} \quad \text{(sum over all edges in } S)$$
   OUTPUT S, $L(T)$. Stop
      [S is the edge set of a shortest spanning tree T in G.]
   Else continue (i.e., go to Step 3).

3. *Label updating*
   For every k not in U, if $l_{jk} < \lambda_k$ then set $\lambda_k = l_{jk}$ and $i(k) = j$.
   Go to Step 2.

End PRIM

Prim's algorithm shown in Table 22.6 is another popular algorithm for the shortest spanning tree problem (cf. Sec. 22.4), which at each stage gives a tree T. Starting from any single vertex, which we call 1, we "grow" the tree

---

[9]*Bell System Technical Journal* **36** (1957), 1389-1401. For an improved version of the algorithm, see Cheriton and Tarjan (*SIAM Journal on Computation* **5** (1976), 724-742). For his work, in 1983 Robert Tarjan was awarded the first Nevanlinna prize, named in honor of ROLF NEVANLINNA (1895—1980), an outstanding Finnish mathematician (and close personal friend of this author).

T by adding edges to it, one at a time, according to some rule (below) until *T* finally becomes a *spanning* tree, which is shortest.

We denote by *U* the set of vertices of the growing tree *T* and by *S* the set of its edges. Thus, initially $U = \{1\}$ and $S = \emptyset$; at the end, $U = V$, the vertex set of the given graph $G = (V, E)$, whose edges (i, j) have length $l_{ij} > 0$, as before.

Thus at the beginning (Step 1) the labels $\lambda_2, \cdots, \lambda_n$ of the vertices $2, \cdots, n$ are the lengths of the edges connecting them to vertex 1 (or ∞ if there is no such edge in *G*). And we pick (Step 2) the shortest of these as the first edge of the growing tree *T* and include its other end *j* in *U* (choosing the smallest *j* if there are several, to make the process unique). Updating labels in Step 3 (at this stage and at any later stage) concerns each vertex *k* not yet in *U*. Vertex *k* has label $\lambda_k = l_{i(k),k}$ from before. If $l_{jk} < \lambda_k$, this means that *k* is closer to the new member *j* just included in *U* than *k* is to its old "closest neighbor" *i(k)* in *U*. Then we update the label of *k*, replacing $\lambda_k = l_{i(k),k}$ by $\lambda_k = l_{jk}$ and setting $i(k) = j$. If, however, $l_{jk} \geq \lambda_k$ (the *old* label of *k*), we don't touch the old label. Thus the label λ_k always identifies the closest neighbor of *k* in *U*, and this is updated in Step 3 as *U* and the tree *T* grow. From the final labels we can backtrack the final tree, and from their numerical values we compute the total length (sum of the lengths of the edges) of this tree.

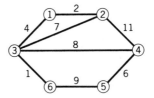

Fig. 483. Graph in Example 1

EXAMPLE 1. Application of Prim's algorithm
Find a shortest spanning tree in the graph in Fig. 483 (which is the same as in Example 1, Sec. 22.4, so that we can compare).

Solution. The steps are as follows.

1. $i(k) = 1$, $U = \{1\}$, $S = \emptyset$, initial labels see Table 22.7

2. $\lambda_2 = l_{12} = 2$ is smallest. $U = \{1, 2\}$, $S = \{(1, 2)\}$

3. Update labels as shown in Table 22.7, column (I).

2. $\lambda_3 = l_{13} = 4$ is smallest. $U = \{1, 2, 3\}$, $S = \{(1, 2), (1, 3)\}$

3. Update labels as shown in Table 22.7, column (II).

2. $\lambda_6 = l_{36} = 1$ is smallest. $U = \{1, 2, 3, 6\}$, $S = \{(1, 2), (1, 3), (3, 6)\}$

3. Update labels as shown in Table 22.7, column (III).

2. $\lambda_4 = l_{34} = 8$ is smallest. $U = \{1, 2, 3, 4, 6\}$, $S = \{(1, 2), (1, 3), (3, 4), (3, 6)\}$

3. Update labels as shown in Table 22.7, column (IV).

2. $\lambda_5 = l_{45} = 6$ is smallest. $U = V$, $S = (1, 2), (1, 3), (3, 4), (3, 6), (4, 5)$. Stop.

The tree is the same as in Example 1, Sec. 22.4. Its length is 21.

Table 22.7
Labeling of Vertices in Example 1

Vertex	Initial Label	Relabeling			
		(I)	(II)	(III)	(IV)
2	$l_{12} = 2$	—	—	—	—
3	$l_{13} = 4$	$l_{13} = 4$	—	—	—
4	∞	$l_{24} = 11$	$l_{34} = 8$	$l_{34} = 8$	—
5	∞	∞	∞	$l_{65} = 9$	$l_{45} = 6$
6	∞	∞	$l_{36} = 1$	—	—

So far we have considered shortest path problems (Secs. 22.2, 22.3) and shortest spanning trees (Secs. 22.4, 22.5). In the next two sections we discuss flow problems as they arise in **networks** (electrical, water, communication, traffic, business connection, etc.).

Problems for Sec. 22.5

Applying Prim's algorithm, find a shortest spanning tree S in the following graphs and draw a sketch of S.

1. The graph in Prob. 1, Sec. 22.4

2. The graph in Prob. 2, Sec. 22.4

3.

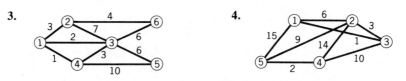

4.

5. In what case will at the end $S = E$ in Prim's algorithm?

6. (**Complexity**) Show that Prim's algorithm has complexity $O(n^2)$.

7. What would the result be if one applied Prim's algorithm to a graph that is not connected?

8. For a complete graph (or one that is almost complete), if our data is an $n \times n$ distance table (as in Prob. 6, Sec. 22.4), show that the present algorithm [which is $O(n^2)$] cannot easily be replaced by an algorithm of order less than $O(n^2)$.

9. How does Prim's algorithm prevent the formation of cycles as one grows T?

10. (**Distance, eccentricity**) Call the length of a shortest path $u \to v$ in a graph $G = (V, E)$ the *distance* $d(u, v)$ from u to v. For fixed u, call the greatest $d(u, v)$ as v ranges over V the *eccentricity* $\epsilon(u)$ of u. Find the eccentricity of vertices 1, 2, 3 in the graph in Prob. 3.

11. (**Diameter, radius, center**) The *diameter* $d(G)$ of a graph $G = (V, E)$ is the maximum of $d(u, v)$ (cf. Prob. 10) as u and v vary over V, and the *radius* $r(G)$ is the smallest eccentricity $\epsilon(v)$ of the vertices v. A vertex v with $\epsilon(v) = r(G)$ is called a *central vertex*. The set of all central vertices is called the *center* of G. Find $d(G)$, $r(G)$ and the center of the graph in Prob. 3.

12. What are diameter, radius and center of the spanning tree in Example 1?

13. Explain how the idea of a center can be used in setting up an emergency service facility on a transportation network. In setting up a fire station, a shopping center. How would you generalize the concepts in the case of two or more such facilities?

14. Show that a tree T whose edges all have length 1 has a center consisting of either one vertex or two adjacent vertices.

15. Set up an algorithm of complexity $O(n)$ for finding the center of a tree T.

22.6 Networks. Flow Augmenting Paths

In this section we consider flows in networks. A **network** is a digraph $G = (V, E)$ (cf. Sec. 22.1) in which each edge (i, j) has assigned to it a "capacity" $c_{ij} > 0$ [= maximum possible flow along (i, j)], and at one vertex, s, called the *source*, a flow is produced that flows along the edges to another vertex, t, called the *target* or *sink*, where the flow disappears.

In applications, this may be the flow of electricity in wires, of water in pipes, of cars on roads, of people in a public transportation system, of goods from a producer to consumers, of letters from senders to recipients and so on. (Note that in many cases there will be more than one source or sink.)

We denote the flow along a (directed!) edge (i, j) by f_{ij} and impose two conditions:

1. For each edge (i, j) in G,

(1) $\qquad 0 \leqq f_{ij} \leqq c_{ij}$ $\qquad$ ("*Edge condition*")

2. For each vertex, not s or t,

$\qquad$ Inflow = Outflow $\qquad$ ("*Vertex condition*," "*Kirchhoff's law*");

in a formula,

(2) $\qquad \underbrace{\sum_k f_{ki}}_{\text{Inflow}} - \underbrace{\sum_j f_{ij}}_{\text{Outflow}} = \begin{cases} 0 \text{ if vertex } i \neq s, i \neq t, \\ -f \text{ at the source } s, \\ f \text{ at the target (sink) } t. \end{cases}$

Here f is the total flow. Figure 484 illustrates the notation.

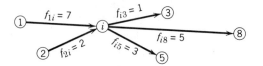

Fig. 484. Notation in (2): inflow and outflow for a vertex i (not s or t)

By a **path** $v_1 \to v_k$ from a vertex v_1 to a vertex v_k in a digraph G we mean a sequence of *undirected edges*

$$(v_1, v_2), (v_2, v_3), \cdots, (v_{k-1}, v_k)$$

that forms a path as defined in Sec. 22.2. Thus, when we travel from v_1 to v_k we may travel through some edge *in* its given direction—then we call it a **forward edge**—or *opposite to* its given direction—then we call it a **backward edge.** Figure 485 shows a forward edge (u, v) and a backward edge (w, v) of a path $v_1 \to v_k$

Fig. 485. Forward edge (u, v) and backward edge (w, v) of a path $v_1 \to v_k$

Caution! Each edge in a network has a given direction, *which we cannot change.* Accordingly, if (u, v) is a forward edge in a path $v_1 \to v_k$, then (u, v) can become a backward edge only in another path $x_1 \to x_j$ in which it is an edge and is traversed in the opposite direction as one goes from x_1 to x_j; cf. Fig. 486. This should be kept in mind, to avoid misunderstandings.

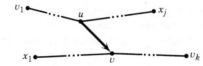

Fig. 486. Edge (u, v) as forward edge in the path $v_1 \to v_k$ and as backward edge in the path $x_1 \to x_j$

Flow Augmenting Paths

To pursue the goal of maximizing the flow f in a network, we should look for methods of increasing an existing flow. The idea then is to find a path $P: s \to t$ all of whose edges are not yet fully used, so that we can push additional flow through P by

    (i) increasing the flow on forward edges, and

    (ii) decreasing the flow on backward edges.

This suggests the following

Definition

A **flow augmenting path** in a network with a given flow f_{ij} on each edge (i, j) is a path $P: s \to t$ such that

    (i) no forward edge is used to capacity; thus $f_{ij} < c_{ij}$ for these;

    (ii) no backward edge has flow 0; thus $f_{ij} > 0$ for these.

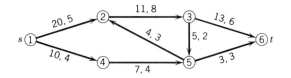

Fig. 487. Network in Example 1
First number = Capacity, Second number = Given flow

EXAMPLE 1. Flow augmenting paths

Find flow augmenting paths in the network in Fig. 487, where the first number is the capacity and the second number a given flow.

Solution. In practical problems, networks are large and one needs a systematic method for augmenting flows. In our small network, which should help to illustrate and clarify the concepts and ideas, we can find flow augmenting paths by inspection and augment the existing flow $f = 9$ in Fig. 487. We use the notation

$$\Delta_{ij} = c_{ij} - f_{ij} \quad \text{for forward edges}$$

$$\Delta_{ij} = f_{ij} \quad \text{for backward edges}$$

$$\Delta = \min \Delta_{ij} \quad \text{taken over all edges of a path.}$$

From Fig. 487 we see that a flow augmenting path P_1: $s \rightarrow t$ is P_1: $1 - 2 - 3 - 6$ (Fig. 488), with $\Delta_{12} = 20 - 5 = 15$, etc., and $\Delta = 3$. Hence we can use P_1 to increase the given flow 9 to $f = 9 + 3 = 12$. All three edges of P_1 are forward edges. We augment the flow by 3. Then the flow in each of the edges of P_1 is increased by 3, so that we now have $f_{12} = 8$ (instead of 5), $f_{23} = 11$ (instead of 8) and $f_{36} = 9$ (instead of 6). Edge (2, 3) is now used to capacity. The flow in the other edges remains as before.

We shall now try to increase the flow in this network (Fig. 487) beyond $f = 12$.

There is another flow augmenting path P_2: $s \rightarrow t$, namely, P_2: $1 - 4 - 5 - 3 - 6$ (Fig. 488). It shows how a backward edge comes in and how it is handled. Edge (3, 5) is a backward edge. It has flow 2, so that $\Delta_{35} = 2$. We compute $\Delta_{14} = 10 - 4 = 6$, etc. (Fig. 488) and $\Delta = 2$. Hence we can use P_2 for another augmentation to get $f = 12 + 2 = 14$. The new flow is shown in Fig. 489. No further augmentation is possible. We shall confirm later that $f = 14$ is maximum. (The "cut" in Fig. 489 will be explained below.) ▮

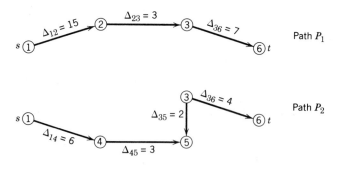

Fig. 488. Flow-augmenting paths in Example 1

Cut Sets

A "cut set" is a set of edges in a network. The underlying idea is simple and natural. If we want to find out what is flowing from s to t in a network, we may cut the network somewhere between s and t (Fig. 489 shows an example) and see what is flowing in the edges hit by the cut, because any flow from s to t must sometimes pass through some of these edges. These form what is called a **cut set.** [In Fig. 489, the cut set consists of the edges

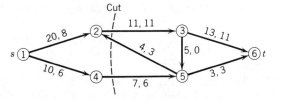

Fig. 489. Maximum flow in Example 1
(The "cut" is explained after Example 1.)

(2, 3), (5, 2), (4, 5).] We denote this cut set by (S, T). Here S is the set of vertices on that side of the cut on which s lies ($S = \{s, 2, 4\}$ for the cut in Fig. 489) and T is the set of the other vertices ($T = \{3, 5, t\}$ in Fig. 489). We say that a cut "*partitions*" the vertex set V into two parts S and T. Obviously, the corresponding cut set (S, T) consists of all the edges in the network with one end in S and the other end in T.

By definition, the **capacity** cap (S, T) of a cut set (S, T) is the sum of the capacities of all **forward edges** in (S, T) (forward edges only!), that is, the edges that are directed *from S to T*,

$$(3) \qquad \text{cap } (S, T) = \Sigma c_{ij} \qquad \text{[sum over the forward edges of } (S, T)]$$

Thus, cap $(S, T) = 11 + 7 = 18$ in Fig. 489.

The other edges (directed *from T to S*) are called **backward edges** of the cut set (S, T), and by the **net flow** through a cut set we mean the sum of the flows in the forward edges minus the sum of the flows in the backward edges of the cut set.

Caution! Distinguish well between forward and backward edges in a cut set and in a path: (5, 2) in Fig. 489 is a backward edge for the cut shown but a forward edge in the path $1 - 4 - 5 - 2 - 3 - 6$.

For the cut in Fig. 489 the net flow is $11 + 6 - 3 = 14$. For the same cut in Fig. 487 (not indicated there), the net flow is $8 + 4 - 3 = 9$. In both cases it equals the flow f. We claim that this is not just by chance, but cuts do serve the purpose for which we have introduced them:

Theorem 1 (Net flow in cut sets)
Any given flow in a network G is the net flow through any cut set (S, T) of G.

Proof. By Kirchhoff's law (2), at a vertex i,

$$(4) \qquad \underbrace{\sum_j f_{ij}}_{\text{Outflow}} - \underbrace{\sum_l f_{li}}_{\text{Inflow}} = \begin{cases} 0 \text{ if } i \neq s, t, \\ f \text{ if } i = s. \end{cases}$$

Here we can sum over j and l from 1 to n ($=$ number of vertices) by putting $f_{ij} = 0$ for $j = i$ and also for edges without flow or nonexisting edges; hence we can write the two sums as one,

$$\sum_j (f_{ij} - f_{ji}) = \begin{cases} 0 \quad \text{if } i \neq s, t, \\ f \quad \text{if } i = s. \end{cases}$$

We now sum over all i in S. Since $s \in S$, this sum equals f:

$$\text{(5)} \qquad \sum_{i \in S} \sum_{j \in V} (f_{ij} - f_{ji}) = f.$$

We claim that in this sum, only the edges belonging to the cut set contribute. Indeed, edges with both ends in T cannot contribute, since we sum only over i in S; but edges (i, j) with both ends in S contribute $+f_{ij}$ at one end and $-f_{ij}$ at the other, a total contribution of 0. Hence the left side of (5) equals the net flow through the cut set. By (5), this is equal to the flow f and proves the theorem. ∎

This theorem has the following consequence, which we shall need below.

Theorem 2 (Upper bound for flows)
A flow f in a network G cannot exceed the capacity of any cut set (S, T) in G.

Proof. By Theorem 1, the flow f equals the net flow through the cut set, $f = f_1 - f_2$, where f_1 is the sum of the flows through the forward edges and $f_2 \; (\geqq 0)$ is the sum of the flows through the backward edges of the cut set. Thus $f \leqq f_1$. Now f_1 cannot exceed the sum of the capacities of the forward edges; but this sum equals the capacity of the cut set, by definition. Together, $f \leqq \text{cap} \, (S, T)$, as asserted. ∎

Cut sets will now bring out the full importance of augmenting paths:

Main Theorem (Augmenting path theorem for flows)
A flow from s to t in a network G is maximum if and only if there does not exist a flow augmenting path $s \to t$ in G.

Proof. (a) If there is a flow augmenting path $P: s \to t$, we can use it to push through it an additional flow. Hence the given flow cannot be maximum.

(b) On the other hand, suppose that there is no flow augmenting path $s \to t$ in G. Let S_0 be the set of all vertices i (including s) such that there is a flow augmenting path $s \to i$, and let T_0 be the set of the other vertices in G. Consider any edge (i, j) with i in S_0 and j in T_0. Then we have a flow augmenting path $s \to i$ since i is in S_0, but $s \to i \to j$ is not flow augmenting because j is not in S_0. Hence we must have

$$\text{(6)} \qquad f_{ij} = \begin{cases} c_{ij} \\ 0 \end{cases} \text{ if } (i, j) \text{ is a } \begin{cases} \text{forward} \\ \text{backward} \end{cases} \text{edge of the path } s \to i \to j.$$

Otherwise we could use (i, j) to get a flow augmenting path $s \to i \to j$. Now (S_0, T_0) defines a cut set (since t is in T_0; why?). Since by (6), forward edges are used to capacity and backward edges carry no flow, the net flow through the cut set (S_0, T_0) equals the sum of the capacities of the forward edges, which is cap (S_0, T_0) by definition. This net flow equals the given flow f by Theorem 1. Thus $f = \text{cap} \, (S_0, T_0)$. Also $f \leqq \text{cap} \, (S_0, T_0)$ by Theorem 2. Hence f must be maximum since we have reached equality. ∎

The end of this proof yields another basic result (by Ford and Fulkerson, *Canadian Journal of Mathematics* **8** (1956), 399-404), namely, the so-called

Max-flow min-cut theorem
The maximum flow[10] in any network G equals the capacity of a "minimum cut set" (= a cut set of minimum capacity) in G.

Proof. We have just seen that $f = \text{cap}(S_0, T_0)$ for a maximum flow f and a suitable cut set (S_0, T_0). Now by Theorem 2 we also have $f \le \text{cap}(S, T)$ for this f and any cut set (S, T) in G. Together, $\text{cap}(S_0, T_0) \le \text{cap}(S, T)$. Hence (S_0, T_0) is a minimum cut set, and the theorem is proved.   ∎

The two basic tools in connection with networks are flow augmenting paths and cut sets. In the next section we show how flow augmenting paths can be used as the basic tool in an algorithm for maximum flows, the **Ford–Fulkerson algorithm,** in which a given flow (for instance, zero flow in all edges) is stepwise increased until it is maximum.

Problems for Sec. 22.6

In Fig. 487, find T and cap (S, T) if S equals
  1. {1, 2, 3}   **2.** {1, 2, 4, 5}   **3.** {1, 3, 5}

  4. Find a minimum cut set in Fig. 487 and verify that its capacity equals the maximum flow $f = 14$.
  5. In Fig. 487, is $1 - 4 - 5 - 2 - 3 - 6$ a flow augmenting path? (Give reason for your answer.)

Find the maximum flow by inspection. (The given numbers are capacities.)

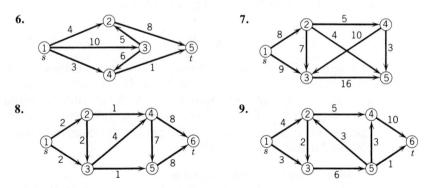

  10. Find a minimum cut set in Prob. 7 and its capacity.
  11. Find examples of flow augmenting paths and the maximum flow in the network in Fig. 490. (First number = capacity; second number = given flow).
  12. Find a minimum cut set in Prob. 11 and its capacity.
  13. Give a simple example of a network that has a cut set of capacity zero.

---

[10]The existence of a maximum flow follows in the case of rational capacities from the algorithm in the next section and in the case of arbitrary capacities from a modification of this algorithm mentioned in footnote 11 in the next section.

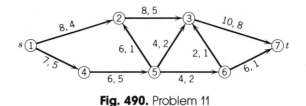

Fig. 490. Problem 11

14. Why are backward edges not considered in the definition of the capacity of a cut set?

15. In which case can an edge (i, j) be used as a forward as well as a backward edge of a path in a network with a given flow?

22.7 Ford–Fulkerson Algorithm for Maximum Flow

The underlying idea of this algorithm is a stepwise construction of flow augmenting paths, one at a time, until no further such paths can be constructed, which happens precisely when the flow is maximum. The algorithm is shown in Table 22.8.

In Step 1, an initial flow may be given. In Step 3, a vertex j can be labeled if there is an edge (i, j) with i labeled and

$$c_{ij} > f_{ij} \qquad\qquad (\text{``}forward\ edge\text{''})$$

or if there is an edge (j, i) with i labeled and

$$f_{ji} > 0 \qquad\qquad (\text{``}backward\ edge\text{''}).$$

To **scan** a labeled vertex i means to label every unlabeled vertex j adjacent to i that can be labeled. Before scanning a labeled vertex i, scan all the vertices that got labeled before i. This BFS (Breadth First Search) strategy was suggested by Edmonds and Karp[11] in 1972. It has the effect that one gets shortest possible augmenting paths. The computational advantage of this is illustrated in Prob. 13.

EXAMPLE 1. Ford–Fulkerson algorithm
Applying the Ford–Fulkerson algorithm, determine the maximum flow for the network in Fig. 491 (which is the same as that in Example 1, Sec. 22.6, so that we can compare).

[11]*Journal of the Association of Computing Machinery* **19** (1972), 248-64. This also removes a convergence difficulty that Ford and Fulkerson (Ref. [F11], p. 21, listed in Appendix 1) had described and avoided by requiring integer (or rational) capacities. The algorithm is of complexity $O(n^3 m)$, n and m the numbers of vertices and edges, respectively. A more advanced algorithm of complexity $O(n^2 m)$ was designed by Dinic (*Soviet Mathematics Doklady* **11** (1970), 1277-80) and another algorithm of complexity $O(n^3)$ by Malhotra et al. (Computer Science Program, Indian Institute of Technology, Kanpur, India, 1978).

The Ford-Fulkerson algorithm was first published in *Canadian Journal of Mathematics* **9** (1957), 210-218.

Table 22.8
Ford–Fulkerson Algorithm for Maximum Flow

ALGORITHM FORD–FULKERSON

$[G = (V, E),$ vertices $1 (= s), \cdots, n (= t),$ edges $(i, j), c_{ij}]$

This algorithm computes the maximum flow in a network G with source s, sink t and capacities $c_{ij} > 0$ of the edges (i, j).

INPUT: $n, s = 1, t = n,$ edges (i, j) of G, c_{ij}
OUTPUT: Maximum flow f in G

1. Assign an initial flow f_{ij} (for instance, $f_{ij} = 0$ for all edges), compute f.

2. Label s by $\varnothing$. Mark the other vertices "*unlabeled.*"

3. Find a labeled vertex i that has not yet been scanned. Scan i as follows.

   For every unlabeled adjacent vertex j, if $c_{ij} > f_{ij}$, compute

$$\Delta_{ij} = c_{ij} - f_{ij} \quad \text{and} \quad \Delta_j = \begin{cases} \Delta_{1i} & \text{if } i = 1 \\ \min (\Delta_i, \Delta_{ij}) & \text{if } i > 1 \end{cases}$$

   and label j with a "*forward label*" (i^+, Δ_j); or if $f_{ji} > 0$, compute

$$\Delta_j = \min (\Delta_i, f_{ji})$$

   and label j by a "backward label" (i^-, Δ_j).
   If no such j exists then OUTPUT f. Stop
       [*f is the maximum flow.*]
   Else continue (i.e., go to Step 4).

4. Repeat Step 3 until t is reached.
       [*This gives a flow augmenting path P: $s \to t$.*]
   If it is impossible to reach t then OUTPUT f. Stop
       [*f is the maximum flow.*]
   Else continue (i.e., go to Step 5).

5. Backtrack the path P, using the labels.

6. Using P, augment the existing flow by Δ_t. Set $f = f + \Delta_t$.

7. Remove all labels from vertices $2, \cdots, n$. Go to Step 3.

End FORD–FULKERSON

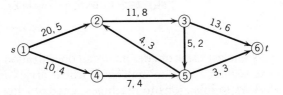
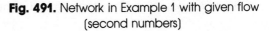

Fig. 491. Network in Example 1 with given flow
(second numbers)

Solution. The algorithm proceeds as follows.

1. An initial flow $f = 9$ is given.

2. Label s (= 1) by $\varnothing$. Mark 2, 3, 4, 5, 6 "unlabeled."

3. Scan 1.

    Compute $\Delta_{12} = 20 - 5 = 15 = \Delta_2$. Label 2 by $(1^+, 15)$.

    Compute $\Delta_{14} = 10 - 4 = 6 = \Delta_4$. Label 4 by $(1^+, 6)$.

4. Scan 2.

    Compute $\Delta_{23} = 11 - 8 = 3$, $\Delta_3 = \min(\Delta_2, 3) = 3$. Label 3 by $(2^+, 3)$.

    Compute $\Delta_5 = \min(\Delta_2, 3) = 3$. Label 5 by $(2^-, 3)$.

    Scan 3.

    Compute $\Delta_{36} = 13 - 6 = 7$, $\Delta_6 = \Delta_t = \min(\Delta_3, 7) = 3$. Label 6 by $(3^+, 3)$.

5. P: $1 - 2 - 3 - 6 (= t)$ is a flow augmenting path.

6. $\Delta_t = 3$. Augmentation gives $f_{12} = 8$, $f_{23} = 11$, $f_{36} = 9$, other f_{ij} unchanged. Augmented flow $f = 9 + 3 = 12$.

7. Remove labels on vertices 2, $\cdots$, 6. Go to Step 3.

3. Scan 1.

    Compute $\Delta_{12} = 20 - 8 = 12 = \Delta_2$. Label 2 by $(1^+, 12)$.

    Compute $\Delta_{14} = 10 - 4 = 6 = \Delta_4$. Label 4 by $(1^+, 6)$.

4. Scan 2.

    Compute $\Delta_5 = \min(\Delta_2, 3) = 3$. Label 5 by $(2^-, 3)$.

    Scan 4. [*No vertex left for labeling.*]

    Scan 5.

    Compute $\Delta_3 = \min(\Delta_5, 2) = 2$. Label 3 by $(5^-, 2)$.

    Scan 3.

    Compute $\Delta_{36} = 13 - 9 = 4$, $\Delta_6 = \min(\Delta_3, 4) = 2$. Label 6 by $(3^+, 2)$.

5. P: $1 - 2 - 5 - 3 - 6 (= t)$ is a flow augmenting path.

6. $\Delta_t = 2$. Augmentation gives $f_{12} = 10$, $f_{52} = 1$, $f_{35} = 0$, $f_{36} = 11$, other f_{ij} unchanged. Augmented flow $f = 12 + 2 = 14$.

7. Remove labels on vertices 2, $\cdots$, 6. Go to Step 3.

One can now scan 1 and then scan 2, as before, but in scanning 4 and then 5 one finds that no vertex is left for labeling. Thus one can no longer reach t. Hence the flow obtained (Fig. 492) is maximum, in agreement with our earlier result in the previous section.   ∎

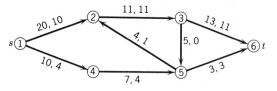

Fig. 492. Maximum flow in Example 1

As the last class of combinatorial optimization problems we discuss in the next section *bipartite* **matching**, which arises in assignment problems (of workers to jobs, jobs to machines, goods to storage, ships to piers, classes to classrooms, exams to time periods) and in other applications.

Problems for Sec. 22.7

1. Carry out the details of the further computations indicated in Example 1.
2. Apply the Ford–Fulkerson algorithm to Example 1 with initial flow 0. Comment on the amount of work compared to that in Example 1.
3. Which are the "bottleneck" edges by which the flow in Example 1 is actually limited? Hence which capacities could be decreased without affecting the maximum flow?

Applying the Ford–Fulkerson algorithm and starting from zero flow, find the maximum flow:

4. In Prob. 6, Sec. 22.6.   5. In Prob. 7, Sec. 22.6.

6. How can one see from the algorithm that Ford and Fulkerson follow a BFS technique?
7. Are the consecutive flow augmenting paths produced by the Ford–Fulkerson algorithm unique?
8. What is the (simple) reason that in the augmentation of a flow by the use of a flow augmenting path, Kirchhoff's law is preserved?
9. How does the Ford–Fulkerson algorithm prevent the formation of cycles?
10. **(Integer flow theorem)** Show that if the capacities in a network G are integers, there is a maximum flow that is an integer.
11. **(Several sources and sinks)** If a network has several sources $s_1, \cdots, s_k$, show that it can be reduced to the case of a single-source network by introducing a new vertex s and connecting s to $s_1, \cdots, s_k$ by k edges of capacity ∞. Similarly if there are several sinks. Illustrate this idea by a network with two sources and two sinks.
12. Find the maximum flow in the network in Fig. 493 with two sources (factories) and two sinks (consumers).

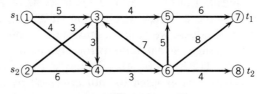

Fig. 493. Problem 12

13. How many augmentations would you need to obtain the maximum flow in the network in Fig. 494 if you started from zero flow and alternatingly used the paths $P_1:\ s\ -\ 2\ -\ 3\ -\ t$ and $P_2:\ s\ -\ 3\ -\ 2\ -\ t$? How does the Ford–Fulkerson algorithm prevent this poor choice?

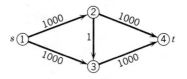

Fig. 494. Problem 13

14. If the Ford–Fulkerson algorithm stops without reaching t, show that the edges with one end labeled and the other end unlabeled form a cut set (S, T) whose capacity equals the maximum flow.

15. Find a minimum cut set in Fig. 491 and its capacity.

16. Show that in a network G with all $c_{ij} = 1$, the maximum flow equals the number of edge-disjoint paths $s \rightarrow t$.

17. In Prob. 15, the cut set contains precisely all forward edges used to capacity by the maximum flow (Fig. 492). Is this just by chance?

18. Show that in a network G with capacities all equal to 1, the capacity of a minimum cut set (S, T) equals the minimum number q of edges whose deletion destroys all directed paths $s \rightarrow t$. (A **directed path** $v \rightarrow w$ is a path in which each edge has the direction in which it is traversed in going from v to w.)

22.8 Bipartite Matching

This is another practically important task in connection with graphs. To explain the problem, we need the following concepts.

A **bipartite graph** $G = (V, E)$ is a graph in which the vertex set V is partitioned into two sets S and T (without common elements, by the definition of a partition) such that every edge of G has one end in S and the other in T, so that there are no edges in G that have both ends in S or both ends in T. Such a graph $G = (V, E)$ is also written $G = (S, T; E)$.

Figure 495 shows an illustration, in which V consists of seven elements, three workers a, b, c, making up the set S, and four jobs 1, 2, 3, 4, making up the set T. The edges indicate that worker a can do the jobs 1 and 2, worker b the jobs 1, 2, 3, etc. and the problem is to assign one job to each worker so that every worker gets one job to do. This suggests the next concept:

A **matching** in $G = (S, T; E)$ is a set M of edges of G such that no two of them have a vertex in common. If M consists of the greatest possible number of edges, we call it a **maximum cardinality matching**[12] in G.

For instance, a matching in Fig. 495 is $M_1 = \{(a, 2), (b, 1)\}$. Another is $M_2 = \{(a, 1), (b, 3), (c, 4)\}$; obviously, this is of maximum cardinality.

A vertex v is **exposed** (or *not covered*) by a matching M if v is not an endpoint of an edge of M. This concept, which always refers to some matching, will be of interest when we begin to augment given matchings (below).

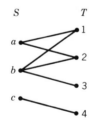

Fig. 495. Bipartite graph in the assignment
of a set $S = \{a, b, c\}$ of workers
to a set $T = \{1, 2, 3, 4\}$ of jobs

---

[12]Or simply a **maximum matching,** but this term is sometimes also used in a different sense, which does not interest us here.

If a matching leaves no vertex exposed, we call it a *complete matching*. Obviously, a complete matching can exist only if S and T have the same number of vertices.

We now want to show how one can stepwise increase the cardinality of a matching M until it becomes maximum. Central in this task is the concept of an augmenting path:

An **alternating path** is a path that consists alternately of edges in M and not in M (Fig. 496A). An **augmenting path** is an alternating path both of whose endpoints (a and b in Fig. 496B) are exposed. By dropping from the matching M the edges which are on an augmenting path P (two edges in Fig. 495B) and adding to M the other edges of P (three in the figure), we get a new matching, with one more edge than M. This is how we use an augmenting path in augmenting a given matching by one edge. We assert that this will always lead, after a number of steps, to a maximum cardinality matching; indeed the basic role of augmenting paths is expressed in the following theorem.

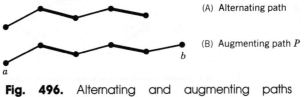

(A) Alternating path

(B) Augmenting path P

Fig.   496.   Alternating   and   augmenting   paths
Heavy edges are those belonging to a matching M.

Augmenting path theorem for bipartite matching

A matching M in a bipartite graph G = (S, T; E) is of maximum cardinality if and only if there does not exist an augmenting path P with respect to M.

Proof. (a) We show that if such a path P exists, then M is not of maximum cardinality. Let P have q edges belonging to M. Then P has $q + 1$ edges not belonging to M. (In Fig. 496B we have $q = 2$.) The endpoints a and b of P are exposed, and all the other vertices on P are endpoints of edges in M, by the definition of an alternating path. Hence if an edge of M is not an edge of P, it cannot have an endpoint on P since then M would not be a matching. Consequently, the edges of M not on P, together with the $q + 1$ edges of P not belonging to M form a matching of cardinality one more than the cardinality of M because we omitted q edges from M and added $q + 1$ instead. Hence M cannot be of maximum cardinality.

(b) We now show that if there is no augmenting path for M, then M is of maximum cardinality. Let M^* be a maximum cardinality matching and consider the graph H consisting of all edges that belong to M or to M^*, but not to both. Then it is possible that two edges of H have a vertex in common, but three edges cannot have a vertex in common since then two of the three would have to belong to M (or to M^*), violating that M and M^* are matchings. So every v in V can be in common with two edges of H or with one or none. Hence we can characterize each "component" (= maximal *connected* subset) of H; a component can be:

(A) A closed path with an *even* number of edges (in the case of an *odd* number, two edges from M or two from M^* would meet, violating the matching property). See (A) in Fig. 497.

(B) An open path P with the same number of edges from M and edges from M^*, for the following reason. P must be alternating, that is, an edge of M is followed by an edge of M^*, etc. (since M and M^* are matchings). Now if P had an edge more from M^*, then P would be augmenting for M [see (B2) in Fig. 497], contradicting our assumption that there is no augmenting path. If P had an edge more from M, it would be augmenting for M^* [see (B3) in Fig. 497], violating the maximum cardinality of M^*, by part (a) of this proof. Hence in each component of H, the two matchings have the same number of edges. Adding to this the number of edges that belong to both M and M^* (which we left aside when we made up H), we conclude that M and M^* must have the same number of edges. M^* being of maximum cardinality, this shows that the same holds for M, as we wanted to prove.   ∎

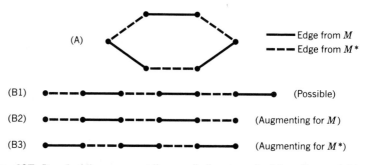

Fig. 497. Proof of the augmenting path theorem for bipartite matching

This theorem suggests an algorithm for obtaining augmenting paths in which vertices are labeled for the purpose of backtracking paths. Such a label is *in addition* to the number of the vertex, which is also retained. Clearly, to get an augmenting path, one must start from an *exposed* vertex, and then trace an alternating path until one arrives at another *exposed* vertex. Table 22.9 shows such an algorithm. After Step 3 all vertices in S are labeled. In Step 4, the set T contains at least one exposed vertex, since otherwise we would have stopped at Step 1.

EXAMPLE 1. Maximum cardinality matching

Is the matching M_1 in Fig. 498a of maximum cardinality? If not, augment it until maximum cardinality will be reached.

Solution. We apply the algorithm.

1. Label 1 and 4 with $\varnothing$.

2. Label 7 with 1. Label 5, 6, 8 with 3.

3. Label 2 with 6, and 3 with 7.

  [*All vertices are now labeled as shown in Fig. 498a.*]

4. P_1: 1 — 7 ■ 3 — 5. [*By backtracking. P_1 is augmenting.*]

  P_2: 1 — 7 ■ 3 — 8. [*P_2 is augmenting.*]

Table 22.9
Bipartite Maximum Cardinality Matching

ALGORITHM MATCHING [$G = (S, T; E)$, M, n]

This algorithm determines a maximum cardinality matching M in a bipartite graph G by augmenting a given matching in G.

INPUT: Bipartite graph $G = (S, T; E)$ with vertices $1, \cdots, n$, matching M in G (for instance, $M = \varnothing$)

OUTPUT: Maximum cardinality matching M in G

1. If there is no exposed vertex in S then
   OUTPUT M. Stop
   [*M is of maximum cardinality in G.*]
   Else label all *exposed* vertices *in S* with $\varnothing$.

2. For each i in S and edge (i, j) *not* in M, label j with i, unless already labeled.

3. For each *nonexposed j* in T, label i with j, where i is the other end of the unique edge (i, j) in M.

4. Backtrack the alternating paths P ending on an exposed vertex in T by using the labels on the vertices.

5. If no P in Step 4 is augmenting then
   OUTPUT M. [*M is of maximum cardinality in G.*]

   Else augment M by using an augmenting path P. Remove all labels. Go to Step 1

End MATCHING

---

5. Augment M_1 by using P_1, dropping $(3, 7)$ from M and including $(1, 7)$ and $(3, 5)$. Remove all labels. Go to Step 1.
   Figure 498b shows the resulting matching $M_2 = \{(1, 7), (2, 6), (3, 5)\}$.

1. Label 4 with $\varnothing$.

2. Label 7 with 2.   Label 6 and 8 with 3.

3. Label 1 with 7, and 2 with 6, and 3 with 5.

4. P_3: 5 — 3 — 8.   [P_3 *is alternating but not augmenting.*]

5. Stop. M_2 *is of maximum cardinality (namely, 3).*   ∎

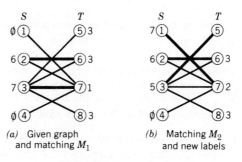

(a) Given graph   *(b)* Matching M_2
   and matching M_1 and new labels

Fig. 498. Example 1

This is the end of Chap. 22 and of Part F on optimization, an area that is new, is full of unsolved problems, has various applications (many of which have not yet been explored!) and is of rapidly increasing interest to the engineer and computer scientist, as well as in management, micro- and macroeconomics and other fields.

The next and last part of the book, Part G, is devoted to **mathematical statistics** and its foundation, which is **probability theory,** areas that have become indispensable to the engineer and more recently in computer science (pattern recognition, robotic vision, etc.), not to mention medicine, biology, agriculture, metereology, geography, sociology, economics, demography and other applications.

Problems for Sec. 22.8

Are the following graphs bipartite? If your answer is yes, find S and T.

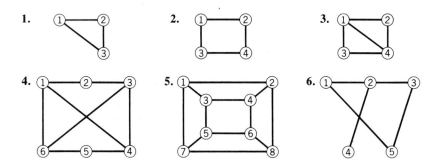

7. Can you obtain the answer to Prob. 3 from the answer to Prob. 1?
8. If a graph $G = (V, E) = (S, T; E)$ is bipartite, what special form does its adjacency matrix have? *Hint.* Think of listing the vertices in special ways.

Find an augmenting path in the following graphs.

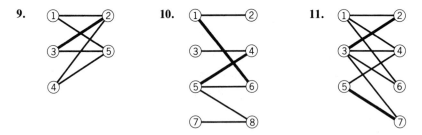

By augmenting the given matching, find a maximum cardinality matching of the graph:
12. In Prob. 10.   13. In Prob. 9.   14. In Prob. 11.

15. **(Complete bipartite graphs)** A bipartite graph $G = (S, T; E)$ is called *complete* if every vertex in S is joined to every vertex in T by an edge, and is denoted by K_{n_1,n_2}, where n_1 and n_2 are the numbers of vertices in S and T, respectively. How many edges does this graph have?

16. What is the limit of the ratio of the edges of $K_{n,n}$ and of the edges of K_{2n} (the complete graph with $2n$ vertices)? Can you find the result by a simple argument, without calculations?

17. Show that in a bipartite graph $G = (S, T; E)$ with n vertices, the number m of edges cannot exceed $n^2/4$. In what case is $m = n^2/4$?

18. **(Timetabling and matching)** Three teachers x_1, x_2, x_3 teach four classes y_1, y_2, y_3, y_4 for these numbers of periods:

	y_1	y_2	y_3	y_4
x_1	1	0	1	1
x_2	1	1	1	1
x_3	0	1	1	1

Show that this arrangement can be represented by a bipartite graph G and that a teaching schedule for one period corresponds to a matching in G. Set up a teaching schedule with the smallest possible number of periods.

19. **(Planar graph)** A *planar graph* is a graph that can be drawn on a sheet of paper so that no two edges cross. Show that the complete graph K_4 with four vertices is planar. The complete graph K_5 with five vertices is not planar. Make this plausible by attempting to draw K_5 so that no edges cross. Interpret the result in terms of a net of roads between five cities.

20. **(Bipartite graph $K_{3,3}$ not planar)** Three factories 1, 2, 3 are each supplied underground by water, gas and electricity, from points A, B, C, respectively. Show that this can be represented by $K_{3,3}$ (the complete bipartite graph $G = (S, T; E)$ with S and T consisting of three vertices each) and that eight of the nine supply lines (edges) can be laid out without crossing. Make it plausible that $K_{3,3}$ is not planar by attempting to draw the ninth line without crossing the others.

21. **(Vertex coloring and exam scheduling)** What is the smallest number of exam periods for six subjects a, b, c, d, e, f if some of the students take a, b, f, some c, d, e, some a, c, e and some c, e? Solve this as follows. Sketch a graph with six vertices $a, \cdots, f$ and join vertices if they represent subjects simultaneously taken by some students. Color the vertices so that adjacent vertices receive different colors. (Use numbers 1, 2, $\cdots$ instead of actual colors if you want.) What is the minimum number of colors you need? For any graph G, this minimum number is called the (vertex-) **chromatic number** $\chi_v(G)$. Why is this the answer to the problem? Write down a possible schedule.

22. How many colors do you need in vertex coloring the graph in Prob. 5?

23. Show that all trees can be vertex colored with two colors.

24. In some computation, temporary storage of frequently used variables $v_1, \cdots, v_6$ during overlapping time intervals $(0, 3)$, $(2, 4)$, $(3, 6)$, $(1, 4)$, $(5, 7)$, $(3, 6)$, respectively, is required. How many index registers (storage locations) are needed? *Hint.* Join v_i and v_j by an edge if their intervals overlap. Then color vertices.

25. What would be the answer to Prob. 24 if not only overlapping but also common endpoints were excluded?

26. **(Four (vertex-) color theorem)** The famous *four-color theorem* states that one can color the vertices of any *planar* graph (so that adjacent vertices get different colors) with at most four colors. It had been conjectured for a long time and was eventually proved in 1976 by Appel and Haken [*Bulletin of the American Mathematical Society* **82** (1976), 711-712]. Can you color the complete graph K_5 with four colors? Does the result contradict the four color theorem? (For more details, see Chap. 12 of Ref. [F15] in Appendix 1.)

27. **(Edge coloring)** The *edge chromatic number* $\chi_e(G)$ of a graph G is the minimum number of colors needed for coloring the edges of G so that incident edges get different colors. Clearly, $\chi_e(G) \geqq \max d(u)$, where $d(u)$ is the degree of vertex u. If $G = (S, T; E)$ is bipartite, the equality sign holds. Prove this for $K_{n,n}$.

28. Prove the last statement in Prob. 27 for arbitrary bipartite graphs $G = (S, T; E)$.

29. Vizing's theorem states that for any graph G (without multiple edges!), max $d(u) \leq \chi_e(G) \leq$ max $d(u) + 1$. Give an example of a graph for which $\chi_e(G)$ does exceed max $d(u)$.

30. (Vertex cover of edges) A *vertex cover* K_v *of a set* L *of edges* in a graph G is a set of vertices of G such that at least one endpoint of each edge of L is in K_v. How many police officers (at least) are needed to cover all blocks (edges) in the graph in Fig. 499? Where should they be placed?

31. How many police officers (at least) are needed to keep every corner in Fig. 499 under surveillance (i.e., an officer being at most one block away from every corner)? Where should they be placed?

32. (Edge cover of vertices) An *edge cover* K_e of a set M of vertices in a graph G is a set of edges such that each vertex in M is endpoint of at least one edge in K_e. Find a minimum edge cover of the graph in Fig. 499 (consisting of the smallest possible number of edges).

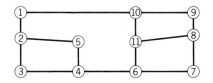

Fig. 499. Problems 30–32

33. (König's theorem) A famous theorem of D. König states that in a bipartite graph G, the number of edges in a maximum cardinality matching equals the number of vertices in a minimum vertex cover of G. Illustrate this by an example. Show that it is not true for a nonbipartite graph (with "*matching*" defined as in the bipartite case).

34. Euler's polyhedron formula is $n - m + f = 2$, where n, m, f are the numbers of vertices, edges and faces, respectively, of a polyhedron. Verify this for the cube. This holds for a connected planar graph (drawn so that no edges cross), where "face" now means a plane region bounded by edges and such that any two points in the region can be joined by a continuous curve that meets no edges or vertices; here the exterior region that extends to infinity must be counted. Verify this for Fig. 467a in Sec. 22.1.

35. Use the Euler formula in Prob. 34 to prove that the complete graph K_5 with five vertices is not planar. *Hint.* Use that each region is bounded by at least three edges, but each edge is a part of the boundary of at most two faces.

Review Problems for Chapter 22

1. What is a graph? A digraph? A tree?

2. What is a path? What do we mean by a shortest path problem?

3. What do we mean by saying that an algorithm is $O(n^2)$?

4. What is BFS? DFS? In what connection did these concepts occur in this chapter?

5. What is the "traveling salesman problem"?

6. What is the intuitive idea of Bellman's optimality principle, and how is the principle used in Dijkstra's algorithm?

7. Name some applications in which spanning trees play a role.

8. What is the basic idea of Kruskal's greedy algorithm?

9. What is a network? What kinds of optimization problem are connected with it?

10. There is a famous theorem on cut sets. Can you remember it?

11. What is a flow augmenting path and why is this concept important?

12. Can a forward edge in one path be a backward edge in another path? In a cut set? Explain.

13. What is a bipartite graph? Give some typical applications that motivate this concept.

14. What kind of optimization problems did we consider in connection with bipartite graphs?

15. What is an augmenting path in bipartite matching? How is it used in obtaining a maximum cardinality matching?

Find the adjacency matrices of the following graphs.

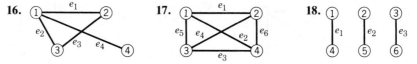

16.   17.   18.

19. **(Complete graph)** A *complete graph G* is a graph that has a maximum number of edges (i.e., any two vertices are connected by an edge). Show that if G has n vertices and is complete, it has $n(n - 1)/2$ edges. (Remember that we excluded multiple edges!)

20. In what case are all the off-diagonal elements of the adjacency matrix of a graph G equal to 1?

Find the adjacency matrices of the following digraphs.

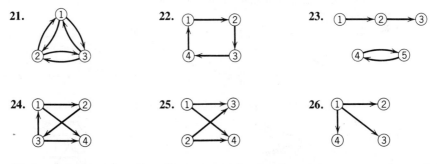

21.   22.   23.

24.   25.   26.

27. In what cases does the adjacency matrix of a digraph have a row of zeros? A column of zeros?

Make a vertex incidence list of the graph:

28. In Prob. 16.   29. In Prob. 17.   30. In Prob. 18.

31. The graph in Prob. 18 is sparse. Compare its adjacency matrix and its vertex incidence list and comment.

32. Company A has offices in Chicago, Los Angeles and New York, Company B in Boston and New York, Company C in Chicago, Dallas and Los Angeles. Represent this by a bipartite graph.

Sketch the graph whose adjacency matrix is:

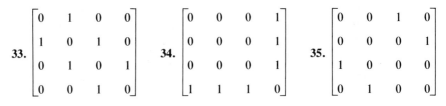

33. $\begin{bmatrix} 0 & 1 & 0 & 0 \\ 1 & 0 & 1 & 0 \\ 0 & 1 & 0 & 1 \\ 0 & 0 & 1 & 0 \end{bmatrix}$ **34.** $\begin{bmatrix} 0 & 0 & 0 & 1 \\ 0 & 0 & 0 & 1 \\ 0 & 0 & 0 & 1 \\ 1 & 1 & 1 & 0 \end{bmatrix}$ **35.** $\begin{bmatrix} 0 & 0 & 1 & 0 \\ 0 & 0 & 0 & 1 \\ 1 & 0 & 0 & 0 \\ 0 & 1 & 0 & 0 \end{bmatrix}$

Find a shortest path and its length by Moore's BFS algorithm, assuming that all the edges have length 1:

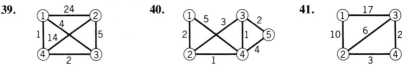

36.   **37.**   **38.**

Find shortest paths by Dijkstra's algorithm:

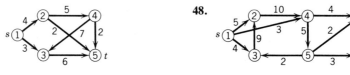

39.   **40.**   **41.**

Find a shortest spanning tree for the graph:

42. In Prob. 39.   **43.** In Prob. 40.   **44.** In Prob. 41.

45. Cayley's theorem states that the number of spanning trees in a complete graph (cf. Prob. 19) with n vertices is n^{n-2}. Verify this for $n = 2, 3, 4$.

46. Show that $O(m^3) + O(m^2) = O(m^3)$.

Find the maximum flow in the following networks, where the given numbers are capacities.

47.   **48.**

By augmenting the given matching, find a maximum cardinality matching:

49.   **50.**

Summary of Chapter 22
Graphs and Combinatorial
Optimization

Combinatorial optimization concerns optimization problems of a discrete or combinatorial structure. It uses graphs and digraphs (Sec. 22.1) as basic tools.

A **graph** $G = (V, E)$ consists of a set V of **vertices** $v_1, v_2, \cdots$ (often simply denoted by 1, 2, $\cdots$, n) and a set E of **edges** $e_1, e_2, \cdots$ each of which connects two vertices. We also write (i, j) for an edge with vertices i and j as endpoints. A **digraph** (= directed graph) is a graph in which each edge has a direction (indicated by an arrow). For handling graphs and digraphs in computers, one can use *matrices* or *lists* (Sec. 22.1).

This chapter concerns important classes of optimization problems for graphs that all arise from practical applications, and corresponding algorithms, as follows.

In a **shortest path problem** (Sec. 22.2) we determine a path of minimum length (consisting of edges) from a vertex s to a vertex t in a graph whose edges (i, j) have a "length" $l_{ij} > 0$, which may be an actual length or a travel time or cost or an electrical resistance [if (i, j) is a wire in a net], and so on. **Dijkstra's algorithm** (Sec. 22.3) or, when all $l_{ij} = 1$, **Moore's algorithm** (Sec. 22.2) are suitable for these problems.

A **tree** is a graph that is connected and has no **cycles** (no closed paths). Trees are very important in practice. A *spanning tree* in a graph G is a tree containing *all* the vertices of G. If the edges of G have lengths, we can determine a **shortest spanning tree,** for which the sum of the lengths of all its edges is minimum. Corresponding algorithms are those by **Kruskal** (Sec. 22.4) and by **Prim** (Sec. 22.5).

A **network** (Sec. 22.6) is a digraph in which each edge (i, j) has a *capacity* $c_{ij} > 0$ [= maximum possible flow along (i, j)] and at one vertex, the *source* s, a flow is produced that flows along the edges to a vertex t, the *sink* or *target,* where the flow disappears. The problem is to maximize the flow, for instance, by applying the **Ford–Fulkerson algorithm** (Sec. 22.7), which uses flow **augmenting paths** (Sec. 22.6). Another related concept is that of a **cut set,** as defined in Sec. 22.6.

A **bipartite graph** $G = (V, E)$ (Sec. 22.8) is a graph whose vertex set V consists of two parts S and T such that every edge of G has one end in S and the other in T, so that there are no edges connecting vertices in S or vertices in T. A **matching** in G is a set of edges, no two of which have an endpoint in common. The problem then is to find a **maximum cardinality matching** in G, that is, a matching M that has a maximum number of edges. For an algorithm, see Sec. 22.8.

PART G

PROBABILITY AND STATISTICS

Chapter 23   Probability Theory
Chapter 24   Mathematical Statistics

Probability theory and mathematical statistics are important to the engineer in testing of materials, acceptance sampling, quality control, general performance tests of systems, automatization and robotics, planning of production processes, optimization problems, marketing and so on. To this we could add a long list of other applications of probability and statistics in agriculture, biology, economics, geography, management of natural resources, medicine, meteorology, politics, psychology, sociology, traffic control, etc.

Probability theory (Chap. 23) provides mathematical models of processes in which "chance effects" play an essential role; these are effects of factors that we cannot control or predict with certainty. Probability theory also gives the mathematical foundation of methods of mathematical statistics, which we consider in Chap. 24. These will be methods of *statistical inference*, that is, methods of conclusions from samples about unknown properties of the population from which the samples are taken.

Chapter 23

Probability Theory

The concept of probability (Sec. 23.2) had its origin in connection with games of chance, such as flipping coins, rolling dice or playing cards. Nowadays it yields mathematical models of chance processes ("random experiments," "random observations": Sec. 23.1). In any such experiment we observe a "random variable" X (a function whose values in the experiment occur "by chance"; Sec. 23.4), which is characterized by a probability distribution (Secs. 23.5–23.7). Or we observe more than one random variable (e.g., height and weight of persons, hardness and tensile strength of steel; Sec. 23.8). The last section (Sec. 23.8) will also be basic in justifying the statistical methods in Chap. 24.

Prerequisite for this chapter: elementary differential and integral calculus.
References: Appendix 1, Part G.
Answers to problems: Appendix 2.

23.1 Random Experiments, Outcomes, Events

In this section we introduce basic concepts we shall need throughout this chapter and the next.

Games of chance such as rolling a die or flipping a coin are examples of "random experiments." If we pick eight screws at random from a box of 100 screws and count the number of defective screws among the eight, this is another "random experiment." Other examples are the random selection of five students in a class and the measurement of their height and weight, the measurement of the yield of a chemical process at a certain time and so on. This suggests the following concept.

A *random experiment* or *random observation*—briefly, an **experiment** or **observation**—is a process that has the following properties.

1. It is performed according to a set of rules that determines the performance completely.

2. It can be repeated arbitrarily often.

3. The result of each performance depends on "chance" (that is, on influences that we cannot control) and therefore cannot be uniquely predicted.

We call a single performance a **trial** and the result the **outcome** of the trial. We call the set of all possible outcomes of an experiment the **sample space** *S* of the experiment, and each outcome an *element* or a *point* of *S*. All these are standard terms.

EXAMPLE 1. Sample space

In flipping a coin, the possible outcomes are *H* (head) and *T* (tail), so that

$$S = \{H, T\}.$$

In rolling a die,

$$S = \{1, 2, 3, 4, 5, 6\};$$

thus *S* has six elements corresponding to the six faces of the die. In measuring the humidity (in percent) in an air-conditioned room, any outcome between 0 and 100 is theoretically possible, so that *S* is the interval from 0 to 100 and thus is an infinite set (that is, consists of infinitely many points). ∎

Events

In most practical problems we are not so much interested in the individual outcomes but in whether an outcome belongs (or does not belong) to a certain set of outcomes. Clearly, each such set *A* is a subset of the sample space *S*. It is called an **event.**

Since an outcome is a subset of *S*, it is an event, but a rather special one, sometimes called an *elementary event*. Similarly, the entire space *S* is another special event.

EXAMPLE 2. Events

If we draw two gaskets from a set of five gaskets (numbered from 1 to 5), the sample space consists of the 10 possible outcomes

$$1, 2 \quad 1, 3 \quad 1, 4 \quad 1, 5 \quad 2, 3 \quad 2, 4 \quad 2, 5 \quad 3, 4 \quad 3, 5 \quad 4, 5,$$

but we may be interested in the number of defective gaskets we get in the draws and thus distinguish among the three events

$$A: \ No\ defective\ gasket, \quad B: \ 1\ defective\ gasket, \quad C: \ 2\ defective\ gaskets.$$

Assuming that three gaskets, say 1, 2, 3, are defective, we see that

A occurs, if we draw   4, 5

B occurs, if we draw   1, 4;   1, 5;   2, 4;   2, 5;   3, 4;   or 3, 5

C occurs, if we draw   1, 2;   1, 3;   or 2, 3. ∎

A sample space *S* and the events of an experiment can be represented graphically by a **Venn diagram,**[1] as follows. Suppose that the set of points inside the rectangle in Fig. 500 on p. 1180 represents *S*. Then the interior of a closed curve inside the rectangle represents an event which we denote by *E*. The set of all the elements (outcomes) not in *E* is called the **complement** *of E in S* and is denoted[2] by E^C.

---

[1]JOHN VENN (1834—1923), English mathematician.

[2]Or by $\overline{E}$, but we shall not use this notation because it is used in set theory for another purpose (to denote the closure of a set).

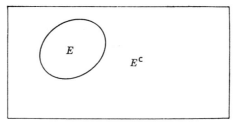

Fig. 500. Venn diagram representing a sample space *S* and events *E* and *E*ᶜ

For example, in the experiment of rolling a die the complement of

E: *The die turns up an even number*

is the event

*E*ᶜ: *The die turns up an odd number.*

An event containing no element is called the **impossible event** or *empty event* and is denoted by ∅.

Let *A* and *B* be any two events in an experiment. Then the event consisting of all the elements of the sample space *S* that belong to *A* or *B*, or both, is called the **union** of *A* and *B* and is denoted by

$$A \cup B.$$

The event consisting of all the elements in *S* that belong to both *A* and *B* is called the **intersection** of *A* and *B* and is denoted by

$$A \cap B.$$

Figure 501 illustrates how to represent unions and intersections by Venn diagrams.

If *A* and *B* have no elements in common, then $A \cap B = \varnothing$, and *A* and *B* are called **disjoint events;** they are also called **mutually exclusive events** because the occurrence of one of them in a trial excludes the possible occurrence of the other in the same trial.

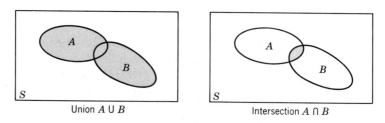

Union *A* ∪ *B*   Intersection *A* ∩ *B*

Fig. 501. Venn diagrams representing two events *A* and *B* in a sample space *S* and their union *A* ∪ *B* (shaded) and intersection *A* ∩ *B* (shaded)

For instance, in Example 2, $B \cap C = \varnothing$ and $B \cup C$: 1 *or* 2 *defective gaskets.*

EXAMPLE 3. Union and intersection of events
In the random experiment of rolling a die, the events

A: *The die turns up a number not smaller than* 4

B: *The die turns up a number divisible by* 3

have the union $A \cup B = \{3, 4, 5, 6\}$ and the intersection $A \cap B = \{6\}$ (cf. Fig. 502).   ∎

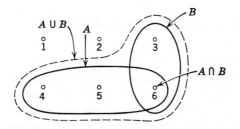

Fig. 502. Venn diagram in Example 3

If all the elements of an event A also belong to an event B, then A is called a **subevent** of B, and we write[3]

$$A \subset B \quad \text{or} \quad B \supset A.$$

Clearly, if $A \subset B$, then if A occurs, B necessarily occurs. For instance, the event $D = \{4, 6\}$ is a subevent of the event $E = \{2, 4, 6\}$ that the die turns up an even number.

There is a practical need for extending the concepts of union and intersection to more than two elements. This is what we do next.

Let $A_1, \cdots, A_m$ be several events in a sample space S. Then the event consisting of all elements each of which belongs to one or more of these m events is called the **union** of $A_1, \cdots, A_m$ and is denoted by

$$A_1 \cup A_2 \cup \cdots \cup A_m, \quad \text{or more briefly} \quad \bigcup_{j=1}^{m} A_j.$$

The event consisting of all elements each of which belongs to all those m events is called the **intersection** of $A_1, \cdots, A_m$ and is denoted by

$$A_1 \cap A_2 \cap \cdots \cap A_m, \quad \text{or more briefly} \quad \bigcap_{j=1}^{m} A_j.$$

---

[3]$A \subset B$ includes $A = B$ as a special case, by the definition of $\subset$, which is sometimes written $\subseteq$. The notation is not uniform.

More generally, let $A_1, A_2, \cdots, A_m, \cdots$ be infinitely many elements in S. Then the **union**

$$A_1 \cup A_2 \cup \cdots, \qquad \text{or more briefly} \qquad \bigcup_{j=1}^{\infty} A_j$$

is defined to be the event consisting of all elements each of which belongs to at least one of those events, and the **intersection**

$$A_1 \cap A_2 \cap \cdots, \qquad \text{or more briefly} \qquad \bigcap_{j=1}^{\infty} A_j$$

is the event consisting of all elements each of which belongs to all those events.

If events $A_1, A_2, \cdots, A_m \cdots$ are such that the occurrence of any of them makes the simultaneous occurrence of any other of them impossible, then we have $A_j \cap A_k = \varnothing$ for any j and $k \neq j$, and the events are called **mutually exclusive events** or **disjoint events.**

For instance, the events A, B and C in Example 2 are mutually exclusive.

EXAMPLE 4. Unions and Intersections of 3 events
In rolling a die, consider the events

> A: *Number greater than* 3
>
> B: *Number less than* 6
>
> C: *Even number*

Then $A \cap B = \{4, 5\}$, $B \cap C = \{2, 4\}$, $C \cap A = \{4, 6\}$, $A \cap B \cap C = \{4\}$. Can you sketch a Venn diagram of this? Furthermore, $A \cup B = S$, hence $A \cup B \cup C = S$ (why?), etc.   ∎

In the next section we introduce the concept of **probability,** which will be central in our further work, and discuss its most basic properties.

Problems for Sec. 23.1

Graph a sample space for the following random experiments.

1. Tossing two coins
2. Rolling two dice
3. Drawing two screws from a box containing right-handed and left-handed screws
4. Tossing three coins
5. Recording the lifetime of each of three electronic components
6. Drawing four razor blades from a production, distinguishing between sharp and blunt blades

7. In testing a car, find the complement of the event E: *the car uses at least* 5.0 *and at most* 5.5 *gallons per* 100 *miles.*
8. In Prob. 2, circle and mark the events:

> A: *Faces are equal.* B: *Sum of faces exceeds* 9.
>
> C: *Sum of faces equals* 7. D: *Sum of faces is even.*

Which of these events are mutually exclusive? What is A^C, B^C, $A \cup B$, $A \cap B$, $A \cup C$, $A \cap C$?

9. Two persons are randomly selected from a group. Describe the sample space, assuming that we distinguish between F: *female* and M: *male*.

10. Two balls are drawn from a box containing five blue, four green balls and one yellow ball; assume that like colored balls cannot be distinguished. Describe the sample space.

11. Using Venn diagrams, graph and check the rules

$$A \cup (B \cap C) = (A \cup B) \cap (A \cup C)$$

$$A \cap (B \cup C) = (A \cap B) \cup (A \cap C)$$

12. **(De Morgan's laws)** Using Venn diagrams, graph and check *De Morgan's laws*

$$(A \cup B)^C = A^C \cap B^C$$

$$(A \cap B)^C = A^C \cup B^C$$

13. Show that, by the definition of a complement,

$$(A^C)^C = A, \quad S^C = \varnothing, \quad \varnothing^C = S, \quad A \cup A^C = S, \quad A \cap A^C = \varnothing.$$

Here, A is any subset of a sample space S.

14. Using a Venn diagram, show that $A \subset B$ if and only if $A \cup B = B$.

15. Find a condition for $A \subset B$ in terms of the intersection $A \cap B$.

23.2 Concept of Probability

Experience shows that most random experiments exhibit *statistical regularity;* that is, the **relative frequency** of an event E (= proportion of times of occurrence of E) in a long sequence of trials is about the same if we perform several such long sequences. Table 23.1 shows classical results confirming this for the experiment of flipping a coin. Figure 503 concerns

Table 23.1
Coin Tossing

Experiments by	Number of Throws	Number of Heads	Relative Frequency of Heads
BUFFON	4,040	2,048	0.5069
K. PEARSON	12,000	6,019	0.5016
K. PEARSON	24,000	12,012	0.5005

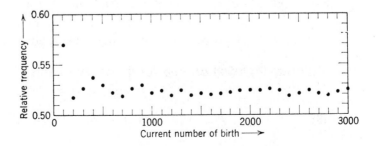

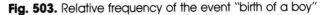

Fig. 503. Relative frequency of the event "birth of a boy"

the random experiment "*birth of a single child*" (data from a city of 250 000 inhabitants) and shows that in the long run the event "*birth of a boy*" seems to have a relative frequency of about 0.52. Another example is the relative frequency of defective items in a production under reasonably constant conditions.

Since most random experiments show statistical regularity, we may assert that for every event E in a random experiment there is a number $P(E)$ such that the relative frequency of E in a large number of performances of the experiment is approximately equal to $P(E)$.

For this reason we now postulate the existence of a number $P(E)$, which is called the **probability** *of the event E in that random experiment*. Note that this number is not an absolute property of E but refers to a certain random experiment with corresponding sample space S.

The statement "E has the probability $P(E)$" then means that if we perform the experiment very often, it is practically certain that the relative frequency of E is approximately equal to $P(E)$. (Here the term "approximately equal" can be characterized quite well, as we shall see in Sec. 23.7.)

Since probability is the counterpart of relative frequency, we define it by the following properties.

Axioms of mathematical probability

1. *If E is any event in a sample space S, then*

(1) $$0 \leq P(E) \leq 1.$$

2. *The entire sample space S has the probability*

(2) $$P(S) = 1.$$

3. *If A and B are mutually exclusive events* (cf. Sec. 23.1), *then*

(3) $$P(A \cup B) = P(A) + P(B).$$

If the sample space is infinite, we must replace Axiom 3 by

3*. *If $E_1, E_2, \cdots$ are mutually exclusive events, then*

(3*) $$P(E_1 \cup E_2 \cup \cdots) = P(E_1) + P(E_2) + \cdots.$$

From Axiom 3 we obtain by induction the following

Theorem 1 (Addition rule for mutually exclusive events)
If $E_1, \cdots, E_m$ are mutually exclusive events, then

(4) $$P(E_1 \cup E_2 \cup \cdots \cup E_m) = P(E_1) + P(E_2) + \cdots + P(E_m).$$

Another basic property of probability is the following.

Theorem 2 (Addition rule for arbitrary events)
If A and B are any events in a sample space S, then

(5) $$P(A \cup B) = P(A) + P(B) - P(A \cap B).$$

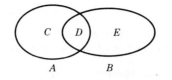

Fig. 504. Proof of Theorem 2

Proof. C, D, E in Fig. 504 are disjoint, and $A = C \cup D$, $B = D \cup E$. Hence by Axiom 3,

$$P(A) = P(C) + P(D), \qquad P(B) = P(D) + P(E).$$

By addition,

$$P(A) + P(B) = [P(C) + P(D) + P(E)] + P(D).$$

Now $A \cup B = C \cup D \cup E$, so that $[\cdot \cdot \cdot]$ is $P(A \cup B)$, by Theorem 1. Hence subtraction of $P(D)$ on both sides gives (5) because $D = A \cap B$.   ∎

Furthermore, an event E and its complement E^C (Sec. 23.1) are mutually exclusive, and $E \cup E^C = S$. Using Axioms 3 and 2, we thus have

$$P(E \cup E^C) = P(E) + P(E^C) = 1.$$

This yields

Theorem 3 (Complementation rule)
The probabilities of an event E and its complement E^C in a sample space S are related by the formula

(6) $$P(E) = 1 - P(E^C).$$

We shall use (6) whenever $P(E^C)$ is easier to compute than $P(E)$, in Example 2 (below) and later on.

Assignment of Probabilities. Equally Likely Outcomes

How can we assign probabilities to the various events in a sample space S?
   *If S is finite, consisting of k elements, and the nature of the experiment shows that these k outcomes are **equally likely** to occur, then we may assign the same probability to each outcome, and because of Axiom 2 this probability must equal $1/k$.* In this case the computation of probabilities of events reduces to that of counting the elements that make up the events.

EXAMPLE 1. Fair die
We consider a **fair die,** that is, a die of homogeneous material and strictly cubical form, and roll it once. In this experiment, $S = \{1, 2, 3, 4, 5, 6\}$. We thus have the probabilities $P(1) = 1/6$, $P(2) = 1/6$, $\cdots$, $P(6) = 1/6$. From this and Theorem 1 we see that, for instance, the event

A: *An even number turns up*

has the probability $P(A) = P(2) + P(4) + P(6) = 1/2$, the event

B: A number greater than 4 turns up

has the probability $P(B) = P(5) + P(6) = 1/3$, etc. Formulas for more complicated cases and further examples will be included in the next section.

EXAMPLE 2. Coin tossing

Five coins are tossed simultaneously. Find the probability of the event *A*: *At least one head turns up*. Assume that the coins are fair.

Solution. Since each coin can turn up heads or tails, the sample space consists of $2^5 = 32$ elements. Since the coins are fair, we may assign the same probability (1/32) to each outcome. Then the event A^C (*no heads turn up*) consists of only 1 outcome. Hence $P(A^C) = 1/32$, and the answer is $P(A) = 1 - P(A^C) = 31/32$.   ∎

If the nature of the experiment does not show that the finitely many outcomes are equally likely to occur, or if the sample space is not finite, we must assign probabilities by using relative frequencies observed in long sequences of trials in such a fashion that the axioms of probability are satisfied.

In this way we obtain only approximate values, but this does not matter. It resembles the situation in classical physics, where we postulate that each body has a certain mass but cannot determine that mass exactly. This is no handicap for developing the theory.

If we doubt whether we assigned the probabilities in the right fashion, we may apply a statistical test which we shall explain later (in Sec. 24.10).

Conditional Probability. Independent Events

Often it is required to find the probability of an event *B* under the condition that an event *A* occurs. This probability is called the **conditional probability** *of B given A* and is denoted by $P(B|A)$. In this case *A* serves as a new (reduced) sample space, and that probability is the fraction of $P(A)$ which corresponds to $A \cap B$. Thus

(7)
$$P(B|A) = \frac{P(A \cap B)}{P(A)}$$
$[P(A) \neq 0]$.

Similarly, the *conditional probability of A given B* is

(8)
$$P(A|B) = \frac{P(A \cap B)}{P(B)}$$
$[P(B) \neq 0]$.

Solving (7) and (8) for $P(A \cap B)$, we obtain

Theorem 4 (Multiplication rule)

If A and B are events in a sample space S and $P(A) \neq 0$, $P(B) \neq 0$, then

(9)
$$P(A \cap B) = P(A)P(B|A) = P(B)P(A|B).$$

EXAMPLE 3. Multiplication rule

In producing screws, let A mean "screw too slim" and B "screw too short." Let $P(A) = 0.1$ and let the conditional probability that a slim screw is also too short be $P(B|A) = 0.2$. What is the probability that a screw that we pick randomly from the lot produced will be both too slim and too short?

Solution. $P(A \cap B) = P(A)P(B|A) = 0.1 \cdot 0.2 = 0.02 = 2\%$, by Theorem 4.   ∎

If the events A and B are such that

(10)
$$\boxed{P(A \cap B) = P(A)P(B),}$$

they are called **independent events.** Assuming $P(A) \neq 0$, $P(B) \neq 0$, we see from (7)–(9) that in this case

$$P(A|B) = P(A), \qquad P(B|A) = P(B)$$

which means that the probability of A does not depend on the occurrence or nonoccurrence of B, and conversely. This justifies the terminology.

Similarly, m events $A_1, \cdots, A_m$ are said to be **independent** if for any k events $A_{j_1}, A_{j_2}, \cdots, A_{j_k}$ (where $1 \leqq j_1 < j_2 < \cdots < j_k \leqq m$ and $k = 2, 3, \cdots, m$)

(11)
$$P(A_{j_1} \cap A_{j_2} \cap \cdots \cap A_{j_k}) = P(A_{j_1})P(A_{j_2}) \cdots P(A_{j_k}).$$

Accordingly, three events A, B, C are independent if

$$P(A \cap B) = P(A)P(B),$$

$$P(B \cap C) = P(B)P(C),$$

(12)
$$P(C \cap A) = P(C)P(A),$$

$$P(A \cap B \cap C) = P(A)P(B)P(C).$$

To prepare for another example, we first note that there are two ways of drawing objects (one at a time!) for obtaining a sample from a given set of objects, briefly referred to as **sampling from a population,** as follows.

1. In **sampling with replacement,** the object that was drawn at random is placed back to the given set and the set is mixed thoroughly. Then we draw the next object at random.

2. In **sampling without replacement** the object that was drawn is put aside.

EXAMPLE 4. Sampling with and without replacement

A box contains 10 screws, three of which are defective. Two screws are drawn at random. Find the probability that none of the two screws is defective.

Solution. We consider the events

A: First drawn screw nondefective.

B: Second drawn screw nondefective.

Clearly, $P(A) = \frac{7}{10}$ because 7 of the 10 screws are nondefective and we sample at random, so that each screw has the same probability ($\frac{1}{10}$) of being picked. If we sample with replacement, the situation before the second drawing is the same as at the beginning, and $P(B) = \frac{7}{10}$. The events are independent, and the answer is

$$P(A \cap B) = P(A)P(B) = 0.7 \cdot 0.7 = 0.49 = 49\%.$$

If we sample without replacement, then $P(A) = \frac{7}{10}$, as before. If A has occurred, then there are 9 screws left in the box, 3 of which are defective. Thus $P(B \mid A) = \frac{6}{9} = \frac{2}{3}$, and Theorem 4 yields the answer

$$P(A \cap B) = \frac{7}{10} \cdot \frac{2}{3} \approx 47\%. \qquad\blacksquare$$

We have seen that in random experiments in which there are finitely many equally likely outcomes, the assignment of probabilities to events amounts to counting the outcomes (= points in the sample space S) that make up the various events. Since the number of possible outcomes is often very large in practice, **systematic counting** is advisable, to prevent us from forgetting a point or counting it twice. We discuss methods for this purpose in the next section.

Problems for Sec. 23.2

1. What is the probability of obtaining at least one head in tossing six fair coins?
2. What is the probability of a sum 17 or 18 in rolling three fair dice?
3. What is the probability of obtaining a sum divisible by 4 if we roll two fair dice?
4. In throwing three fair dice, find the probability of the event E that at least two of the three numbers obtained will be different.
5. Three screws are drawn at random from a lot of 50 screws, 10 of which are defective. Find the probability of the event that all three screws drawn are nondefective, assuming that we draw (a) with replacement, (b) without replacement.
6. If a certain kind of tire has a life exceeding 30 000 miles with probability 0.9, what is the probability that a set of these tires on a car will last longer than 30 000 miles?
7. In Prob. 6, what is the probability that at least one of the tires will not last for 30 000 miles?
8. Three boxes contain five chips each, numbered from 1 to 5, and one chip is drawn from each box. Find the probability of the event E that the sum of the numbers on the drawn chips is greater than 3.
9. A batch of 200 iron rods consists of 50 oversized rods, 50 undersized rods and 100 rods of the desired length. If two rods are drawn at random without replacement, what is the probability of obtaining (a) two rods of the desired length, (b) one of the desired length, (c) none of the desired length, (d) two undersized rods?
10. A process of producing spark plugs has been operated for a long time and the fraction of defective items has remained constant at 2%. Every half-hour, two plugs just produced are inspected. What is the probability of obtaining (a) no defectives, (b) 1 defective, (c) 2 defectives? What is the sum of the probabilities in (a), (b), (c)?
11. Suppose that certain sheets of paper are inspected by drawing three sheets without replacement from every lot of 100 sheets. What is the probability of obtaining three clean sheets although 10% of the sheets contain impurities?

12. An automatic pressure control apparatus contains six electronic tubes. The apparatus will not work unless all tubes are operative. If the probability of failure of each tube during some interval of time is 0.05, what is the corresponding probability of failure of the apparatus?

13. Suppose that we draw cards repeatedly and with replacement from a file of 200 cards, 100 of which refer to male and 100 to female persons. What is the probability of obtaining the third "male" card before the second "female" card?

14. Given $P(A^C) = 0.5$, $P(B) = 0.4$ and $P(A \cap B^C) = 0.3$, find $P(B \mid A \cup B^C)$. (*Hint.* Use a Venn diagram.)

15. A box of 100 gaskets contains 10 gaskets with type A defects, 5 gaskets with type B defects, and 2 gaskets with both types of defects. Find the probability that a gasket to be drawn has a type B defect under the condition that it has a type A defect.

16. Find the probability that in throwing two fair dice the sum of the faces exceeds seven, given that one of the faces is a five.

17. Show that if B is a subset of A, then $P(B) \leqq P(A)$.

18. In the case of mutually exclusive events, Theorem 2 reduces to Axiom 3. Does this mean that we have proved Axiom 3 (so that we could delete it)?

19. Prove Theorem 1.

20. Extending Theorem 4, show that $P(A \cap B \cap C) = P(A)P(B \mid A)P(C \mid A \cap B)$.

23.3 Permutations and Combinations

From the last section we know that in the case of a finite sample space S consisting of k equally likely outcomes, each outcome has probability $1/k$, and the probability of an event A is obtained by counting the number of outcomes of which A consists. Clearly, if that number is m, then

$$P(A) = \frac{m}{k}.$$

For the counting of outcomes the formulas that follow will often be useful.

Suppose that there are given n things (*elements* or *objects*). We may lay them out in a row *in any order*. Each such arrangement is called a **permutation** of those things.

Theorem 1 (Permutations)

The number of permutations of n different things taken all at a time is

(1) $$n! = 1 \cdot 2 \cdot 3 \cdots n$$ (read "*n factorial*").

In fact, there are n possibilities for filling the first place in the row; then $n - 1$ objects are still available for filling the second place, etc. Similarly, if not all given things are different, we obtain for their number of permutations the following theorem.

Theorem 2 (Permutations)

If n given things can be divided into c classes such that things belonging to the same class are alike while things belonging to different classes are different, then the number of permutations of these things taken all at a time is

(2)
$$\frac{n!}{n_1!n_2! \cdots n_c!} \qquad\qquad (n_1 + n_2 + \cdots + n_c = n)$$

where n_j is the number of things in the jth class.

EXAMPLE 1. Illustration of Theorems 1 and 2

If there are 10 different screws in a box that are needed in a certain order for assembling a certain product, and these screws are drawn at random from the box, the probability P of picking them in the desired order is very small, namely (cf. Theorem 1),

$$P = 1/10! = 1/3628\ 800 \approx 0.000\ 03\%.$$

If the box contains six alike right-handed screws and four alike left-handed screws and the six right-handed ones are needed first and the four left-handed ones are needed afterward, the probability P that random drawing will yield the screws in the desired order is (cf. Theorem 2)

$$P = 6!4!/10! = 1/210 \approx 0.5\%. \qquad\qquad ■$$

A **permutation of *n* things taken *k* at a time** is a permutation containing only *k* of the *n* given things. Two such permutations consisting of the same *k* elements, in a different order, are different, by definition. For example, there are 6 different permutations of the three letters *a*, *b*, *c*, taken two letters at a time, namely *ab*, *ac*, *bc*, *ba*, *ca*, *cb*.

A **permutation of *n* things taken *k* at a time with repetitions** is an arrangement obtained by putting any given thing in the first position, any given thing, including a repetition of the one just used, in the second, and continuing until *k* positions are filled. For example, there are $3^2 = 9$ different such permutations of *a*, *b*, *c*, taken 2 letters at a time, namely, the preceding 6 permutations and *aa*, *bb*, *cc*. The reader may prove the following theorem (cf. Prob. 15).

Theorem 3 (Permutations)

The number of different permutations of n different things taken k at a time without repetitions is

(3a)
$$n(n - 1)(n - 2) \cdots (n - k + 1) = \frac{n!}{(n - k)!}$$

and with repetitions is

(3b)
$$n^k.$$

EXAMPLE 2. Illustration of Theorem 3

In a coded telegram the letters are arranged in groups of five letters, called *words*. From (3b) we see that there are

$$26^5 = 11\ 881\ 376$$

different such words. From (3a) it follows that there are

$$26!/(26 - 5)! = 26 \cdot 25 \cdot 24 \cdot 23 \cdot 22 = 7\ 893\ 600$$

different such words that contain each letter no more than once.   ∎

Combinations

In the case of a permutation not only are the things themselves essential, but so also is their order. A **combination** of given things is any selection of one or more of the things *without regard to order*. There are two types of combinations, as follows.

The number of **combinations of** *n* **different things, taken** *k* **at a time, without repetitions** is the number of sets that can be made up from the *n* given things, each set containing *k* different things and no two sets containing exactly the same *k* things.

The number of **combinations of** *n* **different things, taken** *k* **at a time, with repetitions** is the number of sets that can be made up of *k* things chosen from the given *n* things, each being used as often as desired.

For example, there are three combinations of the three letters *a*, *b*, *c*, taken two letters at a time, without repetitions, namely, *ab*, *ac*, *bc*, and six such combinations with repetitions, namely, *ab*, *ac*, *bc*, *aa*, *bb*, *cc*.

Theorem 4 (Combinations)

The number of different combinations of n different things, k at a time, without repetitions, is

(4a)
$$\binom{n}{k} = \frac{n!}{k!(n-k)!} = \frac{n(n-1)\cdots(n-k+1)}{1 \cdot 2 \cdots k},$$

and the number of those combinations with repetitions is

(4b)
$$\binom{n+k-1}{k}.$$

The statement involving (4a) follows from the first part of Theorem 3 by noting that there are *k*! permutations of *k* things from the given *n* things that differ by the order of the elements (cf. Theorem 1), but there is only a single combination of those *k* things of the type characterized in the first statement of Theorem 4. The last statement of Theorem 4 can be proved by induction (cf. Prob. 14).

EXAMPLE 3. Illustration of Theorem 4

The number of samples of five light bulbs that can be selected from a lot of 500 bulbs is [cf. (4a)]

$$\binom{500}{5} = \frac{500!}{5!495!} = \frac{500 \cdot 499 \cdot 498 \cdot 497 \cdot 496}{1 \cdot 2 \cdot 3 \cdot 4 \cdot 5} = 255\ 244\ 687\ 600.$$ ∎

Factorial Function

Let us include some remarks about the **factorial function.** We define

(5)
$$0! = 1$$

and may compute further values by the relation

(6)
$$(n + 1)! = (n + 1)n!.$$

For large n the function is very large (cf. Table A4 in Appendix 4). A convenient approximation for large n is the **Stirling formula**[4]

(7)
$$n! \sim \sqrt{2\pi n} \left(\frac{n}{e}\right)^n$$
$\qquad (e = 2.718 \cdots)$

where $\sim$ is read "*asymptotically equal*" and means that the ratio of the two sides of (7) approaches 1 as n approaches infinity.

EXAMPLE 4. Stirling formula

To get a feeling for the Stirling formula, verify the following approximations:

$n!$	By (7)	Exact Value	Relative Error
4!	23.5	24	1.4%
10!	3 598 696	3 628 800	0.8%
20!	$2.422\ 79 \cdot 10^{18}$	2 432 902 008 176 640 000	0.4%

Binomial Coefficients

The binomial coefficients are defined by the formula

(8)
$$\binom{a}{k} = \frac{a(a - 1)(a - 2) \cdots (a - k + 1)}{k!}$$
$\qquad (k \geqq 0, \text{ integer}).$

The numerator has k factors. Furthermore, we define

(9)
$$\binom{a}{0} = 1, \quad \text{in particular} \quad \binom{0}{0} = 1.$$

For integer $a = n$ we obtain from (8)

(10)
$$\binom{n}{k} = \binom{n}{n - k}$$
$\qquad (n \geqq 0, 0 \leqq k \leqq n).$

Binomial coefficients may be computed recursively, because

---

[4]JAMES STIRLING (1692—1770), Scots mathematician.

(11) $$\binom{a}{k} + \binom{a}{k+1} = \binom{a+1}{k+1}$$ $\quad(k \geq 0,\ \text{integer}).$

Formula (8) also yields

(12) $$\binom{-m}{k} = (-1)^k \binom{m+k-1}{k}$$ $\quad\begin{array}{l}(k \geq 0,\ \text{integer})\\ (m > 0).\end{array}$

There are numerous further relations; we mention

(13) $$\sum_{s=0}^{n-1} \binom{k+s}{k} = \binom{n+k}{k+1}$$ $\quad\begin{array}{l}(k \geq 0,\ n \geq 1,\\ \text{both integer})\end{array}$

and

(14) $$\sum_{k=0}^{r} \binom{p}{k} \binom{q}{r-k} = \binom{p+q}{r}.$$

In the next section we introduce the basic concept of a **random variable** (and its *"probability distribution"*). Random variables are functions whose values are observed in a random experiment and depend on "chance" in the sense that we cannot predict with certainty which of the possible values will occur in a future trial. The number on a die is an example; the number of defective items in a production is another.

Problems for Sec. 23.3

1. List all permutations of four digits 1, 2, 3, 4, taken all at a time.
2. List all permutations of five letters a, e, i, o, u, taken two at a time.
3. In how many ways can a committee of four be chosen from 16 persons?
4. Find the number of samples of four screws that can be drawn from a lot of 50 screws.
5. How many different license plates showing five symbols, namely, two letters followed by three digits, could be made?
6. Of a lot of 12 items, three are defective. (a) Find the number of different samples of four. Find the number of samples of four containing (b) no defectives, (c) one defective, (d) two defectives, (e) three defectives.
7. A box contains four blue, two green and three red balls. We draw a ball at random and put it aside. Then we draw the next ball, and so on. Find the probability of drawing at first all the blue balls, then all the green ones, and finally the red ones.
8. If five different inks are available, in how many ways can we select two colors for a printing job? Three colors?
9. In how many ways can we assign 10 workmen to 10 different jobs?
10. In how many ways can four executives each choose a car from a pool of nine cars?
11. If a cage contains 100 mice, three of which are male, what is the probability that the three male mice will be included if 10 mice are randomly selected?

12. What is the probability that in a group of 20 people (that include no twins) at least two have the same birthday, if we assume that the probability of having birthday at a given day is 1/365 for every day. First guess, then calculate.

13. Determine the number of different bridge hands. (A bridge hand consists of 13 cards selected from a full deck of 52 cards.)

14. Prove the last statement of Theorem 4. *Hint.* Use (13).

15. Prove Theorem 3.

16. For the same number of different items taken six at a time without repetition, how will the number of combinations be related to the number of permutations?

17. Using (7), compute approximate values of 6! and 8! and determine the absolute and relative errors.

18. **(Binomial theorem)** By the binomial theorem,

$$(a + b)^n = \sum_{k=0}^{n} \binom{n}{k} a^k b^{n-k},$$

so that $a^k b^{n-k}$ has the coefficient $\binom{n}{k}$. Can you conclude this from Theorem 4 or is this merely a coincidence?

19. Derive (11) from (8).

20. Prove (14) by applying the binomial theorem (Prob. 18) to

$$(1 + b)^p (1 + b)^q = (1 + b)^{p+q}.$$

23.4 Random Variables. Discrete and Continuous Distributions

If we roll two dice, we know that the sum X of the two numbers that turn up must be an integer between 2 and 12, but we cannot predict which value of X will occur in the next trial, and we may say that X depends on "chance." Similarly, if we want to draw 5 bolts from a lot of bolts and measure their diameter, we cannot predict how many will be defective, that is, will not meet given requirements; hence $X = $ *number of defectives* is again a function that depends on "chance." The lifetime X of a light bulb to be drawn at random from a lot of bulbs also depends on "chance," and so does the content X of a bottle of lemonade filled by a filling machine and selected at random from a given lot.

Roughly speaking, a **random variable** X (also called **stochastic variable** or **variate**) is a function associated with a random experiment whose values are real numbers and their occurrence in the trials (the performances of the experiment) depends on "chance"; more precisely, it is a function X that has the following properties:

1. *X is defined on the sample space S of the experiment, and its values are real numbers.*

2. *Let a be any real number, and let I be any interval. Then the set of all outcomes in S for which X = a has a well-defined probability, and the same is true for the set consisting of all outcomes in S for which the values of X are in I. These probabilities are in agreement with the axioms in Sec. 23.2.*

Although this definition is quite general and includes many functions, we shall see that in practice the number of important types of random variables and corresponding "probability distributions" is rather small.

If we perform a random experiment and the event corresponding to a number a occurs, then we say that in this trial the random variable X corresponding to that experiment has **assumed** the value a. We also say that we have **observed** the value $X = a$. Instead of "the event corresponding to a number a," we say, more briefly, "the event $X = a$." The corresponding probability is denoted by $P(X = a)$. We list this notation along with standard notations for other events in the following table:

Event E	Probability of E
X assumes a given value a	$P(X = a)$
X assumes any value in an interval $a < X < b$	$P(a < X < b)$
X assumes any value not exceeding a given c	$P(X \leq c)$
X assumes any value greater than a given c	$P(X > c)$

The last two events in this table are mutually exclusive. From Axiom 3 in Sec. 23.2 we thus obtain

$$P(X \leq c) + P(X > c) = P(-\infty < X < \infty).$$

Axiom 2 implies that the right-hand side equals 1, because $-\infty < X < \infty$ corresponds to the whole sample space. This yields the important formula

(1) $$\boxed{P(X > c) = 1 - P(X \leq c)}$$ (c arbitrary).

EXAMPLE 1. Illustration of the notations in the table

Let X be the number that turns up in rolling a fair die. Then $P(X = 1) = \frac{1}{6}$, $P(X = 2) = \frac{1}{6}$, etc., $P(1 < X < 2) = 0$, $P(1 \leq X \leq 2) = \frac{1}{3}$, $P(0 \leq X \leq 3.2) = \frac{1}{2}$, $P(X > 4) = \frac{1}{3}$, $P(X \leq 0.5) = 0$, etc. ∎

In most random experiments of practical interest in which we observe a single quantity, either we *count* (cars on a road, defective items in a production, deaths caused by cancer, days of sunshine in a city, etc.) or we *measure* (temperature, voltage, carbon content of steel, water level in a canal, etc.). If we *count*, we call the corresponding random variable X *discrete*. If we measure, we call X *continuous* (or of continuous type). Let us discuss these two important concepts one after the other.

Discrete Random Variables and Distributions

A random variable X and the corresponding distribution are said to be **discrete,** if X has the following properties.

1. The number of values for which X has a probability different from 0 is finite or at most countably infinite.

2. If an interval $a < X \leqq b$ does not contain such a value, then $P(a < X \leqq b) = 0$.

Let

$$x_1, \quad x_2, \quad x_3, \quad \cdots$$

be the values for which X has a positive probability, and let

$$p_1, \quad p_2, \quad p_3, \quad \cdots$$

be the corresponding probabilities. Then $P(X = x_1) = p_1$, etc. We now introduce the so-called **probability function** of X, defined by

(2) $$f(x) = \begin{cases} p_j & \text{when } x = x_j \quad (j = 1, 2, \cdots) \\ 0 & \text{otherwise.} \end{cases}$$

Since $P(S) = 1$ (cf. Axiom 2 in Sec. 23.2), we must have

(3) $$\sum_{j=1}^{\infty} f(x_j) = 1.$$

If we know the probability function of a discrete random variable X, then we may readily compute the probability $P(a < X \leqq b)$ corresponding to any interval $a < X \leqq b$. In fact,

(4) $$\boxed{P(a < X \leqq b) = \sum_{a < x_j \leqq b} f(x_j) = \sum_{a < x_j \leqq b} p_j.}$$

This is the sum of all the probabilities $f(x_j) = p_j$ for which the x_j lie in that interval. For a closed, open, or infinite interval the situation is quite similar. We express this by saying that the probability function $f(x)$ determines the **probability distribution** or, briefly, the **distribution** of the random variable X in a unique fashion.

If X is any random variable, **_not necessarily discrete,_** then for any real number x there exists the probability $P(X \leqq x)$ corresponding to

$$X \leqq x \quad (X \text{ assumes any value smaller than } x \text{ or equal to } x).$$

Clearly, $P(X \leqq x)$ depends on the choice of x; it is a function of x, which is called the **distribution function**[5] of X and is denoted by $F(x)$. Thus

---

[5]We note that some authors call $F(x)$ the **cumulative distribution function,** in particular those who use the term *distribution function* for the probability function $f(x)$.

(5)
$$F(x) = P(X \leq x).$$

Since for any a and $b\ (> a)$ we have

$$P(a < X \leq b) = P(X \leq b) - P(X \leq a),$$

it follows that

(6)
$$P(a < X \leq b) = F(b) - F(a).$$

This shows that the distribution function determines the distribution of X uniquely and that it can be used for computing probabilities.

*Suppose that X is a **discrete** variable.* Then we may represent the distribution function $F(x)$ in terms of the probability function $f(x)$. In fact, by inserting (4) (with $a = -\infty$ and $b = x$) we have

(7)
$$F(x) = \sum_{x_j \leq x} f(x_j)$$

where the right-hand side is the sum of all those $f(x_j)$ for which $x_j \leq x$.

The values $x_1, x_2, x_3, \cdots$ for which a discrete random variable X has positive probability (and only these values) are called the **possible values** of X. In each interval that contains no possible value the distribution function $F(x)$ is constant. Hence $F(x)$ is a **step function** (piecewise constant function) that has an upward jump of magnitude $p_j = P(X = x_j)$ at $x = x_j$ and is constant between two subsequent possible values. Let us illustrate these concepts and facts by typical examples.

EXAMPLE 2. Probability function and distribution function
Figure 505 on the next page shows the probability function $f(x)$ and the distribution function $F(x)$ of the discrete random variable

$$X = \textit{Number that a fair die turns up.}$$

X has the possible values $x = 1, 2, 3, 4, 5, 6$ with probability 1/6 each, and precisely at these x the distribution function has upward jumps of magnitude 1/6. Hence from the graph of $f(x)$ we can construct the graph of $F(x)$, and conversely.

EXAMPLE 3. Probability function and distribution function
The random variable

$$X = \textit{Sum of the two numbers two fair dice turn up}$$

is discrete and has the possible values $2\ (= 1 + 1), 3, 4, \cdots, 12\ (= 6 + 6)$. There are $6 \cdot 6 = 36$ equally likely outcomes $(1, 1), (1, 2), \cdots, (6, 6)$, where the first number is that shown on the first die and the second number that on the other die. Each such outcome has probability 1/36. Now $X = 2$ occurs in the case of the outcome $(1, 1)$; $X = 3$ in the case of the two outcomes $(1, 2)$ and $(2, 1)$; $X = 4$ in the case of the three outcomes $(3, 1), (2, 2), (3, 1)$, and so on. Hence $f(x) = P(X = x)$ has the values (see Fig. 506)

x	2	3	4	5	6	7	8	9	10	11	12
$f(x)$	1/36	2/36	3/36	4/36	5/36	6/36	5/36	4/36	3/36	2/36	1/36

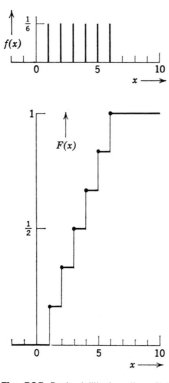

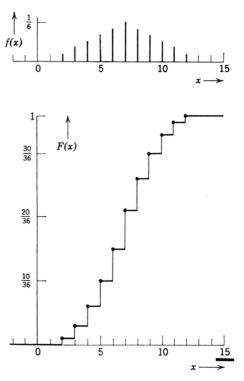

Fig. 505. Probability function $f(x)$ and distribution function $F(x)$ of the random variable X = *Number obtained in rolling a fair die once*

Fig. 506. Probability function $f(x)$ and distribution function $F(x)$ of the random variable X = *Sum of the two numbers obtained in rolling two fair dice once*

Figure 506 shows a bar chart of this function and the graph of the distribution function, which is again a step function, with jumps at the possible values of X.   ∎

Continuous Random Variables and Distributions

We shall now define and consider continuous random variables. A random variable X and the corresponding distribution are said to be *of continuous type* or, briefly, **continuous** if the distribution function $F(x) = P(X \leq x)$ of X can be represented by an integral in the form[6]

$$(8) \qquad F(x) = \int_{-\infty}^{x} f(v) \, dv$$

---

[6]$F(x)$ is continuous, but continuity of $F(x)$ does not imply the existence of a representation of the form (8). Since continuous distribution functions that cannot be represented in the form (8) are rare in practice, the widely accepted terms "continuous random variable" and "continuous distribution" should not be confusing.

where the integrand is continuous, possibly except for at most finitely many values of v, and is nonnegative. The integrand f is called the *probability density* or, briefly, the **density** of the distribution. Differentiating (8) we see that

$$F'(x) = f(x)$$

for every x at which $f(x)$ is continuous. In this sense *the density is the derivative of the distribution function.*

From (8) and Axiom 2, Sec. 23.2, we also have

$$(9) \qquad \int_{-\infty}^{\infty} f(v) \, dv = 1.$$

Furthermore, from (6) and (8) we obtain the important formula

$$(10) \qquad \boxed{P(a < X \leq b) = F(b) - F(a) = \int_{a}^{b} f(v) \, dv.}$$

Hence this probability equals the area under the curve of the density $f(x)$ between $x = a$ and $x = b$, as shown in Fig. 507.

Obviously, for any fixed a and $b\ (> a)$, in the case of a *continuous* random variable X the probabilities corresponding to the intervals $a < X \leq b$, $a < X < b$, $a \leq X < b$, and $a \leq X \leq b$ are all the same. This is different from the case of a discrete distribution. (Explain why!)

Examples of continuous distributions are included in the next problem set and in later sections.

The distribution function (and equally well the probability function or density, respectively) determines all the properties of a random variable completely. In addition, we may express some of the most important properties, for instance, the average value (**"mean"**) and the spread (**"variance"**) of the values, by certain numbers that can be computed from the distribution function and often give sufficient information for practical purposes. We discuss this idea in the next section and define such *"parameters"* of a distribution.

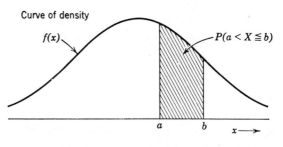

Fig. 507. Example illustrating formula (10)

Problems for Sec. 23.4

1. Graph the probability function $f(x) = x^2/14$ ($x = 1, 2, 3$) and the distribution function.

2. Graph f and F when $f(0) = f(3) = 1/6$, $f(1) = f(2) = 1/3$. Can f have further positive values?

3. Let X be the number of years before a particular type of machine will need replacement. Assume that X has the probability function $f(1) = 0.1$, $f(2) = 0.3$, $f(3) = 0.6$. Graph f and F.

4. If X has the probability function $f(2) = 1/2$, $f(3) = 1/4$, $f(4) = f(5) = 1/8$, what is the probability that X will assume any value less than 4? Greater than 4?

5. Graph the density $f(x) = \frac{1}{4}$ ($1 < x < 5$) and the distribution function.

6. The daily demand for gasoline [gallons] at a certain gas station is a random variable X. Assume that X has the density $f(x) = k$ if $2000 < x < 6000$ and 0 otherwise. Find k. Find and graph the distribution function $F(x)$.

7. Find and graph the probability function $f(x)$ of the random variable $X = sum$ *of the three numbers obtained in rolling three fair dice.*

8. Let $f(x) = ke^{-2x}$ when $x > 0$, $f(x) = 0$ when $x < 0$. Find k. Graph f and F.

9. Show that the random variable $X = $ *number of times a fair coin is tossed until a head appears* has the probability function $f(x) = 2^{-x}$ ($x = 1, 2, \cdots$). Show that $f(x)$ satisfies (3).

10. Let X [millimeters] be the thickness of washers a machine turns out. Assume that X has the density $f(x) = kx$ if $1.8 < x < 2.2$ and 0 otherwise. Find k. What is the probability that a washer will have thickness between 1.95 and 2.05 mm?

11. Consider the random variable $X = $ *Number of times a fair die is rolled until the first 6 appears*. Find the probability function of X and show that it satisfies (3).

12. Let X be the ratio of sales to profits of some company. Assume that X has the distribution function $F(x) = 0$ if $x < 5$, $F(x) = (x^2 - 25)/75$ if $5 \leqq x < 10$, $F(x) = 1$ if $x \geqq 10$. Find and graph the density. Find the probability that X is between 8 (12.5% profit) and 10 (10% profit).

13. Find the probability function of the random variable $X = $ *Number of times a fair die is rolled until the first 5 or 6 appears.*

14. Suppose that in an automatic process of filling oil into cans, the content of a can [gallons] is $Y = 100 + X$, where X is a random variable with density $f(x) = 1 - |x|$ if $|x| \leqq 1$ and 0 if $|x| > 1$. Graph $f(x)$ and $F(x)$. In a lot of 1000 cans, about how many will contain 100 gallons or more? What is the probability that a can will contain less than 99.5 gallons? Less than 99 gallons?

15. Find the probability that none of three bulbs in a traffic signal will have to be replaced during the first 1200 hours of operation if the lifetime X of a bulb is a random variable with the density $f(x) = 6[0.25 - (x - 1.5)^2]$ when $1 \leqq x \leqq 2$ and $f(x) = 0$ otherwise, where x is measured in multiples of 1000 hours.

16. Suppose that certain bolts have length $L = 200 + X$ mm, where X is a random variable with density $f(x) = \frac{3}{4}(1 - x^2)$ when $-1 \leqq x \leqq 1$ and 0 otherwise. Determine c so that with a probability of 95% a bolt will have any length between $200 - c$ and $200 + c$.

17. Let X have the density $f(x) = kx$ if $0 \leqq x \leqq 3$. Find k. Find x such that $P(X \leqq x) = 10\%$. Find x such that $P(X \leqq x) = 95\%$.

18. Let $f(x) = kx^2$ when $0 \leqq x \leqq 1$ and 0 otherwise. Find k. Find the number c such that $P(X \leqq c) = 72.9\%$.

19. Let X have the density $f(x) = 0$ if $x < 0$ and $f(x) = e^{-x}$ if $x > 0$. Find c_1 such that $P(X < c_1) = 50\%$ and c_2 such that $P(X \leqq c_2) = 95\%$.

20. Show that $b < c$ implies $P(X \leqq b) \leqq P(X \leqq c)$.

23.5 Mean and Variance of a Distribution

The *mean value* or **mean** of a distribution is denoted by μ and is defined by

(1)

$$\textbf{(a)} \quad \mu = \sum_j x_j f(x_j) \qquad \text{(Discrete distribution)}$$

$$\textbf{(b)} \quad \mu = \int_{-\infty}^{\infty} x f(x)\, dx \qquad \text{(Continuous distribution).}$$

In (1a) the function $f(x)$ is the probability function of the random variable X considered, and we sum over all possible values (cf. Sec. 23.4). In (1b) the function $f(x)$ is the density of X. The mean is also known as the *mathematical expectation of X* and is sometimes denoted by $E(X)$. By definition it is assumed that the series in (1a) converges absolutely and the integral of $|x| f(x)$ from $-\infty$ to ∞ exists. If this does not hold, we say that the distribution does not have a mean; this case will rarely occur in engineering applications.

A distribution is said to be **symmetric** with respect to a number $x = c$ if for every real x,

(2)
$$f(c + x) = f(c - x).$$

The reader may prove

Theorem 1 (Mean of a symmetric distribution)
If a distribution is symmetric with respect to $x = c$ and has a mean μ, then $\mu = c$.

The **variance** of a distribution is denoted by σ^2 and is defined by the formula

(3)

$$\textbf{(a)} \quad \sigma^2 = \sum_j (x_j - \mu)^2 f(x_j) \qquad \text{(Discrete distribution)}$$

$$\textbf{(b)} \quad \sigma^2 = \int_{-\infty}^{\infty} (x - \mu)^2 f(x)\, dx \qquad \text{(Continuous distribution)}$$

where, by definition, it is assumed that the series in (3a) converges absolutely and the integral in (3b) exists.

In the case of a discrete distribution with $f(x) = 1$ at a point and $f = 0$ otherwise, we have $\sigma^2 = 0$. This case is of no practical interest. In any other case

(4)
$$\sigma^2 > 0.$$

The positive square root of the variance is called the **standard deviation** and is denoted by σ.

Roughly speaking, the variance measures the spread or dispersion of the values that the corresponding random variable X can assume.

EXAMPLE 1. Mean and variance
The random variable

$$X = \textit{Number of heads in a single toss of a fair coin}$$

has the possible values $X = 0$ and $X = 1$ with probabilities $P(X = 0) = \frac{1}{2}$ and $P(X = 1) = \frac{1}{2}$. From (1a) we thus obtain the mean value $\mu = 0 \cdot \frac{1}{2} + 1 \cdot \frac{1}{2} = \frac{1}{2}$, and (3a) yields

$$\sigma^2 = (0 - \tfrac{1}{2})^2 \cdot \tfrac{1}{2} + (1 - \tfrac{1}{2})^2 \cdot \tfrac{1}{2} = \tfrac{1}{4}.$$

EXAMPLE 2. Uniform distribution
The distribution with the density

$$f(x) = \frac{1}{b - a} \qquad \text{when} \qquad a < x < b$$

and $f = 0$ otherwise is called the *uniform distribution* on the interval $a < x < b$. From Theorem 1 or from (1b) we find that $\mu = (a + b)/2$, and (3b) yields the variance

$$\sigma^2 = \int_a^b \left(x - \frac{a + b}{2} \right)^2 \frac{1}{b - a}\, dx = \frac{(b - a)^2}{12}.$$

Figure 508 shows special cases illustrating that σ^2 measures the spread.   ∎

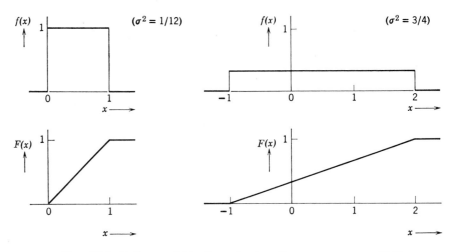

Fig. 508. Uniform distributions having the same mean (0.5)
but different variances σ^2

Theorem 2 (Linear transformation)
If a random variable X has mean μ and variance σ^2, the random variable $X^* = c_1 X + c_2 \ (c_1 \neq 0)$ has the mean

$$(5) \qquad\qquad \mu^* = c_1 \mu + c_2$$

and the variance

$$(6) \qquad\qquad \sigma^{*2} = c_1^2 \sigma^2.$$

Proof. We prove (5) in the continuous case, first assuming that $c_1 > 0$. For corresponding x and $x^* = c_1 x + c_2$ the densities $f(x)$ of X and $f^*(x^*)$ of X^* satisfy the relation $f^*(x^*) = f(x)/c_1$, because to a small interval of length Δx on the X-axis there corresponds the probability $f(x)\,\Delta x$ (approximately) and this must equal $f^*(x^*)\,\Delta x^*$, where $\Delta x^* = c_1\Delta x$ is the length of the corresponding interval on the X^*-axis. Since $dx^*/dx = c_1$, $dx^* = c_1\,dx$, we thus have $f^*(x^*)\,dx^* = f(x)\,dx$. Hence

$$\mu^* = \int_{-\infty}^{\infty} x^* f^*(x^*)\,dx^* = \int_{-\infty}^{\infty} (c_1 x + c_2) f(x)\,dx$$

$$= c_1 \int_{-\infty}^{\infty} x f(x)\,dx + c_2 \int_{-\infty}^{\infty} f(x)\,dx.$$

The last integral equals 1, cf. (9), Sec. 23.4, and formula (5) is proved. Since

$$x^* - \mu^* = (c_1 x + c_2) - (c_1 \mu + c_2) = c_1 x - c_1 \mu,$$

the definition of the variance yields

$$\sigma^{*2} = \int_{-\infty}^{\infty} (x^* - \mu^*)^2 f^*(x^*)\,dx^* = \int_{-\infty}^{\infty} (c_1 x - c_1 \mu)^2 f(x)\,dx = c_1^2 \sigma^2.$$

If $c_1 < 0$, the results remain the same, because we get two additional minus signs, one from changing the direction of integration in x (note that $x^* = -\infty$ corresponds to $x = \infty$) and the other one from $f^*(x^*) = f(x)/(-c_1)$; here $-c_1 > 0$ is needed since densities are nonnegative.

For a discrete distribution, the proof of Theorem 2 is similar.   ∎

From (5) and (6) we readily obtain

Theorem 3 (Standardized variable)
If a random variable X has mean μ and variance σ^2, then the corresponding random variable

$$\boxed{Z = \frac{X - \mu}{\sigma}}$$

has the mean 0 *and the variance* 1.

Z is called the **standardized variable** corresponding to X. Applications of this concept follow in Sec. 23.7.

If X is any random variable and $g(X)$ is any continuous function defined for all real X, then the number

(7)

$$\text{(a)}\quad E(g(X)) = \sum_{j} g(x_j) f(x_j) \qquad\qquad (X \text{ discrete})$$

$$\text{(b)}\quad E(g(X)) = \int_{-\infty}^{\infty} g(x) f(x)\,dx \qquad\qquad (X \text{ continuous})$$

is called the **mathematical expectation** of $g(X)$. Here f is the probability function or the density, respectively.

Taking $g(X) = X^k$ ($k = 1, 2, \cdots$) in (7), we obtain the expressions

$$(8) \qquad E(X^k) = \sum_j x_j{}^k f(x_j) \quad \text{and} \quad E(X^k) = \int_{-\infty}^{\infty} x^k f(x)\, dx$$

respectively. $E(X^k)$ is called the **kth moment** of X. Taking $g(X) = (X - \mu)^k$ in (7), we have

$$(9) \qquad E([X - \mu]^k) = \sum_j (x_j - \mu)^k f(x_j) \quad \text{and} \quad \int_{-\infty}^{\infty} (x - \mu)^k f(x)\, dx$$

respectively. This expression is called the **kth central moment** of X. The reader may show that

$$(10) \qquad\qquad\qquad E(1) = 1$$

$$(11) \qquad\qquad\qquad \mu = E(X)$$

$$(12) \qquad\qquad\qquad \sigma^2 = E([X - \mu]^2).$$

It has been said in Sec. 23.4 that the definition of a random variable is quite general, but in practice we need only a handful of special random variables and their distributions. We discuss three important discrete distributions in the next section and the most important continuous distribution, the normal distribution, in Sec. 23.7.

Problems for Sec. 23.5

1. Find the mean and variance of the discrete random variable X having the probability function $f(0) = 0.1$, $f(1) = 0.8$, $f(2) = 0.1$.
2. Do the same task as in Prob. 1, when $f(0) = 0.001$, $f(1) = 0.027$, $f(2) = 0.243$, $f(3) = 0.729$.
3. Let X have the density $f(x) = 2x$ if $0 \leq x \leq 1$ and 0 otherwise. Show that X has the mean 2/3 and the variance 1/18.
4. Find the mean and the variance of $Y = -2X + 3$, where X is the random variable in Prob. 3.
5. Find the mean and variance of $5X - 5$ with X as in Prob. 1.
6. Find the mean and the variance of the distribution having the density $f(x) = \frac{1}{2} e^{-|x|}$.
7. Let X [cm] be the diameter of bolts in a production. Assume that X has the density $f(x) = k(x - 0.9)(1.1 - x)$ if $0.9 < x < 1.1$ and 0 otherwise. Determine k, graph $f(x)$ and find μ and σ^2.
8. Suppose that in Prob. 7, a bolt is regarded as being defective if its diameter deviates from 1.00 cm by more than 0.06 cm. What percentage of defective bolts should we then expect?
9. A small filling station is supplied with gasoline every Saturday afternoon. Assume that its volume X of sales in thousands of gallons has the probability density $f(x) = 6x(1 - x)$ if $0 \leq x \leq 1$ and 0 otherwise. Determine the mean and the variance.

10. What capacity must the tank in Prob. 9 have for the probability that the tank will be emptied in a given week to be 5%?

11. Assume that the mileage (in thousands of miles) car owners get with a certain kind of tires is a random variable X having the density

$$f(x) = \theta e^{-\theta x} \quad \text{if } x > 0$$

and 0 otherwise. Here, $\theta \, (> 0)$ is a parameter. (*a*) What mileage can a car owner expect to get with one of these tires? (*b*) Let $\theta = 0.05$. Find the probability that the tire will last at least 30 000 miles.

12. Prove Theorem 1.

13. Derive Theorem 3 from (5) and (6).

14. Show that $E(X - \mu) = 0$ and $\sigma^2 = E(X^2) - \mu^2$.

15. Let X have the density $f(x) = 2x$ if $0 < x < 1$ and 0 otherwise. Find all the moments. Calculate σ^2 by the formula in Prob. 14.

16. $E([X - c]^2)$ is called the *second moment of X with respect to c*. If it exists, show that it is minimum when $c = \mu$.

17. Find the moments of the uniform distribution on the interval $0 \leqq x \leqq 1$.

18. (**Skewness**) The number

$$\gamma = \frac{1}{\sigma^3} E([X - \mu]^3)$$

is called the *skewness* of X. Justify this term by showing that if X is symmetric with respect to μ and the third central moment exists, this moment is zero.

19. Find the skewness of $X = $ *number of 6's in rolling two fair dice*.

20. (**Moment generating function**) The so-called *moment generating function* of a discrete or continuous random variable X is given by the formulas

$$G(t) = E(e^{tX}) = \sum_j e^{tx_j} f(x_j) \quad \text{and} \quad G(t) = E(e^{tX}) = \int_{-\infty}^{\infty} e^{tx} f(x) \, dx,$$

respectively. Assuming that differentiation under the summation sign and the integral sign is permissible, show that $E(X^k) = G^{(k)}(0)$, in particular $\mu = G'(0)$, where $G^{(k)}(t)$ is the kth derivative of G with respect to t.

Binomial, Poisson and Hypergeometric Distributions

We shall now consider special discrete distributions that are particularly important in probability theory and statistics.

Binomial Distribution

We start with the binomial distribution, which is obtained if we are interested in the number of times an event A occurs in n independent performances of an experiment, assuming that A has probability $P(A) = p$ in a single trial. Then $q = 1 - p$ is the probability that in a single trial the event A does not occur. We assume that the experiment is performed n times and consider the random variable

$$X = \textit{Number of times } A \textit{ occurs.}$$

Then X can assume the values $0, 1, \cdots, n$, and we want to determine the corresponding probabilities. For this purpose we consider any of these values, say, $X = x$, which means that in x of the n trials A occurs and in $n - x$ trials A does not occur. This may look as follows:

(1)
$$\underbrace{AA \cdots A}_{x \text{ times}} \; \underbrace{BB \cdots B.}_{n-x \text{ times}}$$

Here $B = A^C$, that is, A does not occur. We assume that the trials are *independent,* that is, do not influence each other. Then, since $P(A) = p$ and $P(B) = q$, we see that to (1) there corresponds the probability

$$\underbrace{pp \cdots p}_{x \text{ times}} \; \underbrace{qq \cdots q}_{n-x \text{ times}} = p^x q^{n-x}.$$

Clearly, (1) is merely *one* order of arranging x A's and $n - x$ B's, and the probability $P(X = x)$ thus equals $p^x q^{n-x}$ times the number of different arrangements of x A's and $n - x$ B's, as follows from Theorem 1 in Sec. 23.2. We may number the n trials from 1 to n and pick x of these numbers corresponding to those trials in which A happens. Since the order in which we pick the x numbers does not matter, we see from (4a) in Sec. 23.3 that we can pick those x numbers from the n numbers in $\binom{n}{x}$ different ways. Hence the probability $P(X = x)$ corresponding to $X = x$ equals

(2)
$$\boxed{f(x) = \binom{n}{x} p^x q^{n-x}} \qquad (x = 0, 1, \cdots, n)$$

and $f(x) = 0$ for any other value of x. This is the probability that in n independent trials an event A occurs precisely x times, where p is the probability of A in a single trial and $q = 1 - p$. The distribution determined by the probability function (2) is called the **binomial distribution** or *Bernoulli distribution.* The occurrence of A is called *success,* and the nonoccurrence is called *failure.* p is called the *probability of success in a single trial.* Figure 509 shows illustrative examples of $f(x)$.

The binomial distribution has the mean (cf. Prob. 21)

(3)
$$\mu = np$$

and the variance (cf. Prob. 22)

(4)
$$\sigma^2 = npq.$$

Note that when $p = 0.5$, the distribution is symmetric with respect to μ.

A table of the binomial distribution is included in Appendix 4. For more extended tables see [G17] and [G14], vol. 1, p. 131, listed in Appendix 1.

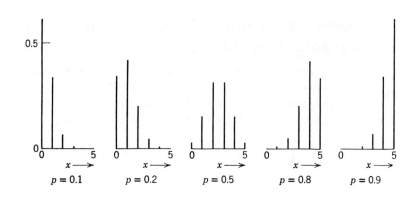

Fig. 509. Probability function (2) of the binomial distribution for $n = 5$ and various values of p

Poisson Distribution

The discrete distribution with the probability function

$$(5) \qquad f(x) = \frac{\mu^x}{x!} e^{-\mu} \qquad (x = 0, 1, \cdots)$$

is called the **Poisson distribution,** named after S. D. Poisson (cf. Sec. 17.5). Figure 510 shows (5) for some values of μ. It can be proved that this distribution may be obtained as a limiting case of the binomial distribution, if we let $p \to 0$ and $n \to \infty$ so that the mean $\mu = np$ approaches a finite value. The Poisson distribution has the mean μ and the variance (cf. Prob. 23)

$$(6) \qquad \sigma^2 = \mu.$$

This distribution provides probabilities, for instance, of given numbers of cars passing an intersection per unit interval of time, of given numbers of defects per unit length of wire or unit area of paper or textile, and so on.

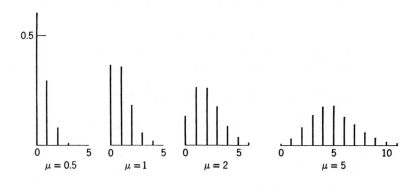

Fig. 510. Probability function (5) of the Poisson distribution for various values of μ

Sampling With and Without Replacement. Hypergeometric Distribution

The binomial distribution is important in **sampling with replacement** (cf. Example 4, Sec. 23.2), as follows. Suppose that a box contains N things, for example, screws, M of which are defective. If we want to draw a screw at random, the probability of obtaining a defective screw is

$$p = \frac{M}{N}.$$

Hence in drawing a sample of n screws *with replacement,* the probability of obtaining precisely x defective screws is [cf. (2)]

(7) $$f(x) = \binom{n}{x} \left(\frac{M}{N}\right)^x \left(1 - \frac{M}{N}\right)^{n-x} \qquad (x = 0, 1, \cdots, n).$$

In **sampling without replacement** that probability is

(8) $$f(x) = \frac{\binom{M}{x}\binom{N - M}{n - x}}{\binom{N}{n}} \qquad (x = 0, 1, \cdots, n).$$

This distribution with the probability function (8) is called the **hypergeometric distribution.**[7]

To verify our statement we first note that, by (4a) in Sec. 23.3, there are

(a) $\binom{N}{n}$ different ways of picking n things from N,

(b) $\binom{M}{x}$ different ways of picking x defectives from M,

(c) $\binom{N - M}{n - x}$ different ways of picking $n - x$ nondefectives from $N - M$,

and each way in (b) combined with each way in (c) gives the total number of mutually exclusive ways of obtaining x defectives in n drawings without replacement. Since (a) is the total number of outcomes and we draw at random, each such way has the probability $1 \Big/ \binom{N}{n}$. From this, (8) follows.

The hypergeometric distribution has the mean (cf. Prob. 25)

(9) $$\mu = n\frac{M}{N}$$

---

[7]Because the moment generating function (cf. Prob. 20, Sec. 23.5) of this distribution can be represented in terms of the hypergeometric function.

and the variance

(10)
$$\sigma^2 = \frac{nM(N - M)(N - n)}{N^2(N - 1)} .$$

EXAMPLE 1. Sampling with and without replacement

We want to draw random samples of two gaskets from a box containing 10 gaskets, three of which are defective. Find the probability function of the random variable

$$X = Number\ of\ defectives\ in\ the\ sample.$$

Solution. We have $N = 10$, $M = 3$, $N - M = 7$, $n = 2$. For sampling with replacement, (7) yields

$$f(x) = \binom{2}{x}\left(\frac{3}{10}\right)^x \left(\frac{7}{10}\right)^{2-x}, \qquad f(0) = 0.49, \quad f(1) = 0.42, \quad f(2) = 0.09.$$

For sampling without replacement we have to use (8), finding

$$f(x) = \binom{3}{x}\binom{7}{2 - x} \Big/ \binom{10}{2}, \qquad f(0) = f(1) = \frac{21}{45} \approx 0.47, \quad f(2) = \frac{3}{45} \approx 0.07. \quad \blacksquare$$

If N, M, and $N - M$ are large compared with n, then it does not matter too much whether we sample with or without replacement, and in this case the hypergeometric distribution may be approximated by the binomial distribution (with $p = M/N$), which is somewhat simpler.

*Hence in sampling from an indefinitely large population (**"infinite population"**) we may use the binomial distribution, regardless of whether we sample with or without replacement.*

The binomial, Poisson and hypergeometric distributions discussed in this section are the most important *discrete* distributions. In the next section we discuss the **normal distribution,** which is continuous. This distribution is probably the most important of all distributions from the viewpoint of applications, and the student should study the next section with particular care.

Problems for Sec. 23.6

1. Five fair coins are tossed simultaneously. Find the probability function of the random variable X = *Number of heads* and compute the probabilities of obtaining no heads, precisely one head, at least one head, and not more than four heads.

2. If, in an experiment, positive and negative values are equally likely, what is the probability of obtaining at most one negative value in nine trials?

3. If the probability of hitting a target is 25% and four shots are fired independently, what is the probability that the target will be hit at least once?

4. Would the probability be less than, equal to or greater than it is in Prob. 3 if the probability of hitting were 1/16 and we fired 16 shots? First guess, then compute.

5. Let $p = 2\%$ be the probability that a certain type of light bulb will fail in a 24-hour test. Find the probability that a sign consisting of 15 such bulbs will burn for 24 hours with no bulb failures.

6. Answer the same question as in Prob. 5, for a sign consisting of 100 bulbs. First guess how much less that probability is. Then compute.

7. Suppose that 4% of bolts made by a machine are defective, the defectives occurring at random during production. If the bolts are packaged 100 per box, what is the Poisson approximation of the probability that a given box will contain x defectives?

8. Suppose that in the production of 60-ohm radio resistors, nondefective items are those that have a resistance between 55 and 65 ohms and the probability of a resistor's being defective is 0.5%. The resistors are sold in lots of 200, with the guarantee that all resistors are nondefective. What is the probability that a given lot will violate this guarantee? (Use the Poisson distribution.)

9. Classical experiments by E. Rutherford and H. Geiger in 1910 showed that the number of alpha particles emitted per second in a radioactive process is a random variable X having a Poisson distribution. If X has mean 0.5, what is the probability of observing two or more particles during any given second?

10. Suppose that a telephone switchboard handles 240 calls on the average during a rush hour, and that the board can make at most eight connections per minute. Using the Poisson distribution and Table A7 in Appendix 4, estimate the probability that the board will be overtaxed during a given minute.

11. Let X be the number of cars per minute passing a certain point of some road between 8 A.M. and 10 A.M. on a Sunday. Assume that X has a Poisson distribution with mean 4. Find the probability of observing two or fewer cars during any given minute.

12. A distributor sells rubber bands in packages of 100 and guarantees that at most 10% will be defective. A consumer controls each package by drawing 10 bands without replacement. If the sample contains no defective rubber bands, he accepts the package. Otherwise he rejects it. Find the probability that in this process the consumer will reject a package that contains 10 defective bands (thus still satisfying the guarantee).

13. In Problem 12 find the probability that any given package will be accepted although it contains 20 defective rubber bands.

14. A process of manufacturing screws is checked every hour by inspecting n screws selected at random from that hour's production. If one or more screws are defective, the process is halted and carefully examined. How large should n be if the manufacturer wants the probability to be about 95% that the process will be halted when 8% of the screws being produced are defective? (Assume independence of the quality of any item of that of the other items.)

15. **(Multinomial distribution)** Suppose a trial can result in precisely one of k mutually exclusive events $A_1, \cdots, A_k$ with probabilities $p_1, \cdots, p_k$, respectively, where $p_1 + \cdots + p_k = 1$. Suppose that n independent trials are performed. Show that the probability of getting x_1 A_1's, $\cdots$, x_k A_k's is

$$f(x_1, \cdots, x_k) = \frac{n!}{x_1! \cdots x_k!} p_1^{x_1} \cdots p_k^{x_k}$$

where $0 \leqq x_j \leqq n$, $j = 1, \cdots, k$, and $x_1 + \cdots + x_k = n$. The distribution having this probability function is called the *multinomial distribution*.

16. What distribution do we obtain in Prob. 15 when $k = 2$?

17. Suppose that in a production of electrical resistors the probabilities of producing a resistor of resistance $R < 299$ ohms or $R > 303$ ohms are 3 and 2%, respectively. Find the probability that a random sample of 10 resistors contains precisely x_1 resistors with $R < 299$ ohms and precisely x_2 resistors with $R > 303$ ohms.

18. Let X have a binomial distribution with $p = 0.5$. Find the probability that X will assume a value in the interval $\mu - \sigma < X < \mu + \sigma$, assuming that (a) $n = 1$, (b) $n = 2$, (c) $n = 3$, (d) $n = 4$, and (e) $n = 5$.

19. Show that the distribution function of the Poisson distribution satisfies $F(\infty) = 1$.

20. Using the binomial theorem, show that the binomial distribution has the moment generating function (cf. Prob. 20, Sec. 23.5)

$$G(t) = \sum_{x=0}^{n} e^{tx} \binom{n}{x} p^x q^{n-x} = \sum_{x=0}^{n} \binom{n}{x} (pe^t)^x q^{n-x} = (pe^t + q)^n.$$

21. Using Prob. 20 and $p + q = 1$, prove (3).

22. Prove (4).

23. Show that the Poisson distribution has the moment-generating function

$$G(t) = e^{-\mu} e^{\mu e^t}$$

and prove (6).

24. Show that $E([X - \mu]^3) = E(X^3) - 3\mu E(X^2) + 2\mu^3$. Using this and Prob. 23, show that the Poisson distribution has the skewness $\gamma = 1/\sqrt{\mu}$ and conclude that for large μ the distribution is almost symmetric (cf. Fig. 510).

25. Prove (9).

23.7 Normal Distribution

The continuous distribution having the density

(1)
$$f(x) = \frac{1}{\sigma\sqrt{2\pi}}\, e^{-\frac{1}{2}\left(\frac{x-\mu}{\sigma}\right)^2} \qquad\qquad (\sigma > 0)$$

is called the **normal distribution** or *Gauss distribution*. A random variable having this distribution is said to be **normal** or *normally distributed*. This distribution is very important, because many random variables of practical interest are normal or approximately normal or can be transformed into normal random variables in a relatively simple fashion. Furthermore, the normal distribution is a useful approximation of more complicated distributions. It also appears in the mathematical proofs of various statistical tests.

In (1), μ is the mean and σ is the standard deviation of the distribution. The curve of $f(x)$ is called the *bell-shaped curve*. It is symmetric with respect to μ. Figure 511 on p. 1212 shows $f(x)$ for $\mu = 0$. For $\mu > 0$ ($\mu < 0$) the curves have the same shape, but are shifted $|\mu|$ units to the right (to the left). The smaller σ^2 is, the higher is the peak at $x = 0$ in Fig. 511 and the steeper are the descents on both sides. This agrees with the meaning of the variance.

From (1) we see that the normal distribution has the distribution function

(2)
$$F(x) = \frac{1}{\sigma\sqrt{2\pi}} \int_{-\infty}^{x} e^{-\frac{1}{2}\left(\frac{v-\mu}{\sigma}\right)^2} dv.$$

From this and (10) in Sec. 23.4 we obtain

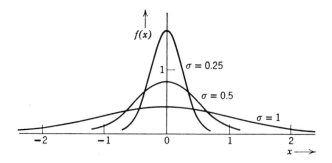

Fig. 511. Density (1) of the normal distribution with $\mu = 0$
for various values of σ

$$(3) \qquad P(a < X \leqq b) = F(b) - F(a) = \frac{1}{\sigma\sqrt{2\pi}} \int_a^b e^{-\frac{1}{2}\left(\frac{v-\mu}{\sigma}\right)^2} dv.$$

The integral in (2) cannot be evaluated by elementary methods, but can be represented in terms of the integral (Fig. 512)

$$(4) \qquad \Phi(z) = \frac{1}{\sqrt{2\pi}} \int_{-\infty}^z e^{-u^2/2} \, du.$$

This is the distribution function of the **standardized normal distribution,** that is, the normal distribution with mean 0 and variance 1. It has been tabulated (Table A8 in Appendix 4).

To express (2) in terms of (4), we set

$$\frac{v - \mu}{\sigma} = u. \qquad \text{Then} \qquad \frac{du}{dv} = \frac{1}{\sigma}, \qquad dv = \sigma \, du$$

and we now have to integrate with respect to u from $-\infty$ to $z = (x - \mu)/\sigma$. From (2) we thus obtain

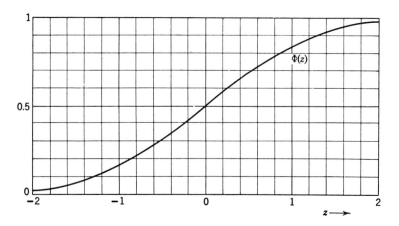

Fig. 512. Distribution function $\Phi(z)$ of the normal distribution
with mean 0 and variance 1

$$F(x) = \frac{1}{\sigma\sqrt{2\pi}} \int_{-\infty}^{(x-\mu)/\sigma} e^{-u^2/2} \, \sigma \, du.$$

σ drops out, and the expression on the right equals (4) where $z = (x - \mu)/\sigma$:

(5)
$$F(x) = \Phi\left(\frac{x - \mu}{\sigma}\right).$$

From this important formula and (3) we obtain another important formula:

(6)
$$P(a < X \le b) = F(b) - F(a) = \Phi\left(\frac{b - \mu}{\sigma}\right) - \Phi\left(\frac{a - \mu}{\sigma}\right).$$

In particular, when $a = \mu - \sigma$ and $b = \mu + \sigma$, the right-hand side equals $\Phi(1) - \Phi(-1)$; to $a = \mu - 2\sigma$ and $b = \mu + 2\sigma$ there corresponds the value $\Phi(2) - \Phi(-2)$, etc. Using Table A8 in Appendix 4 we thus find (Fig. 513)

(7)

(a) $P(\mu - \sigma < X \le \mu + \sigma) \approx 68\%,$

(b) $P(\mu - 2\sigma < X \le \mu + 2\sigma) \approx 95.5\%,$

(c) $P(\mu - 3\sigma < X \le \mu + 3\sigma) \approx 99.7\%.$

Hence we may expect that a large number of observed values of a normal random variable X will be distributed as follows:

(a) *About $\frac{2}{3}$ of the values will lie between $\mu - \sigma$ and $\mu + \sigma$.*

(b) *About 95% of the values will lie between $\mu - 2\sigma$ and $\mu + 2\sigma$.*

(c) *About $99\frac{3}{4}\%$ of the values will lie between $\mu - 3\sigma$ and $\mu + 3\sigma$.*

This may also be expressed as follows.

A value that deviates more than σ from μ will occur about once in 3 trials. A value that deviates more than 2σ or 3σ from μ will occur about once in 20 or 400 trials, respectively. Practically speaking, this means that all the values will lie between $\mu - 3\sigma$ and $\mu + 3\sigma$; these two numbers are called **three-sigma limits.** In a similar fashion we obtain

(8)

(a) $P(\mu - 1.96\sigma < X \le \mu + 1.96\sigma) = 95\%,$

(b) $P(\mu - 2.58\sigma < X \le \mu + 2.58\sigma) = 99\%,$

(c) $P(\mu - 3.29\sigma < X \le \mu + 3.29\sigma) = 99.9\%.$

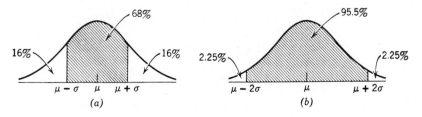

Fig. 513. Illustration of formula (7)

The following typical examples should help the reader to understand the practical use of Tables A8 and A9 in Appendix 4.

EXAMPLE 1

Determine the probabilities

$$\text{(a) } P(X \leq 2.44), \quad \text{(b) } P(X \leq -1.16), \quad \text{(c) } P(X \geq 1), \quad \text{(d) } P(2 \leq X \leq 10)$$

where X is assumed to be normal with mean 0 and variance 1.

Solution. Since $\mu = 0$ and $\sigma^2 = 1$, we may obtain the desired values directly from Table A8:

(a) 0.9927, (b) 0.1230, (c) $1 - P(X \leq 1) = 1 - 0.8413 = 0.1587$ [cf. (6), Sec. 23.2],

(d) $\Phi(10) = 1.0000$ (why?), $\Phi(2) = 0.9772$, $\Phi(10) - \Phi(2) = 0.0228$.

EXAMPLE 2

Determine the probabilities in the previous example, assuming that X is normal with mean 0.8 and variance 4.

Solution. From (6) and Table A8 we obtain

$$\text{(a) } \quad F(2.44) = \Phi\left(\frac{2.44 - 0.80}{2}\right) = \Phi(0.82) = 0.7939$$

$$\text{(b) } \quad F(-1.16) = \Phi(-0.98) = 0.1635$$

$$\text{(c) } \quad 1 - P(X \leq 1) = 1 - F(1) = 1 - \Phi(0.1) = 0.4602$$

$$\text{(d) } \quad F(10) - F(2) = \Phi(4.6) - \Phi(0.6) = 1 - 0.7257 = 0.2743.$$

EXAMPLE 3

Let X be normal with mean 0 and variance 1. Determine the constant c such that

$$\text{(a) } \quad P(X \geq c) = 10\%, \qquad \text{(b) } \quad P(X \leq c) = 5\%,$$

$$\text{(c) } \quad P(0 \leq X \leq c) = 45\%, \qquad \text{(d) } \quad P(-c \leq X \leq c) = 99\%.$$

Solution. From Table A9 in Appendix 4 we obtain

$$\text{(a) } \quad 1 - P(X \leq c) = 1 - \Phi(c) = 0.1, \; \Phi(c) = 0.9, \; c = 1.282$$

$$\text{(b) } \quad c = -1.645$$

$$\text{(c) } \quad \Phi(c) - \Phi(0) = \Phi(c) - 0.5 = 0.45, \; \Phi(c) = 0.95, \; c = 1.645$$

$$\text{(d) } \quad c = 2.576.$$

EXAMPLE 4

Let X be normal with mean -2 and variance 0.25. Determine c such that

(a) $P(X \geq c) = 0.2$ (b) $P(-c \leq X \leq -1) = 0.5$

(c) $P(-2 - c \leq X \leq -2 + c) = 0.9$ (d) $P(-2 - c \leq X \leq -2 + c) = 99.6\%$.

Solution. Using Table A9 in Appendix 4, we obtain the following results.

$$\text{(a) } \quad 1 - P(X \leq c) = 1 - \Phi\left(\frac{c + 2}{0.5}\right) = 0.2,$$

$$\Phi(2c + 4) = 0.8, \; 2c + 4 = 0.842, \; c = -1.579$$

$$\text{(b) } \quad \Phi\left(\frac{-1 + 2}{0.5}\right) - \Phi\left(\frac{-c + 2}{0.5}\right) = 0.9772 - \Phi(4 - 2c) = 0.5,$$

$$\Phi(4 - 2c) = 0.4772, \; 4 - 2c = -0.057, \; c = 2.03$$

(c) $\Phi\left(\dfrac{-2 + c + 2}{0.5}\right) - \Phi\left(\dfrac{-2 - c + 2}{0.5}\right)$

$= \Phi(2c) - \Phi(-2c) = 0.9,\ 2c = 1.645,\ c = 0.823$

(d) $\Phi(2c) - \Phi(-2c) = 99.6\%,\ 2c = 2.878,\ c = 1.439.$

EXAMPLE 5

Suppose that iron plates in production are required to have a certain thickness, and the machine work is done by a shaper. In any case industrial products will differ slightly from each other because the properties of the material and the behavior of the machines and tools used show slight random variations caused by small disturbances we cannot predict. We may therefore regard the thickness X [mm] of the plates as a random variable. We assume that for a certain setting the variable X is normal with mean $\mu = 10$ mm and standard deviation $\sigma = 0.02$ mm. We want to determine the percentage of defective plates to be expected, assuming that defective plates are (a) plates thinner than 9.97 mm, (b) plates thicker than 10.05 mm, (c) plates that deviate more than 0.03 mm from 10 mm. (d) How should we choose numbers $10 - c$ and $10 + c$ to ensure that the expected percentage of defectives will not be greater than 5%? (e) How does the percentage of defectives in part (d) change if μ is shifted from 10 mm to 10.01 mm?

Solution. Using Table A8 in Appendix 4 and (6), we obtain the following solutions.

(a) $P(X \leqq 9.97) = \Phi\left(\dfrac{9.97 - 10.00}{0.02}\right) = \Phi(-1.5) = 0.0668 \approx 6.7\%$

(b) $P(X \geqq 10.05) = 1 - P(X \leqq 10.05) = 1 - \Phi\left(\dfrac{10.05 - 10.00}{0.02}\right)$

$= 1 - \Phi(2.5) = 1 - 0.9938 \approx 0.6\%$

(c) $P(9.97 \leqq X \leqq 10.03) = \Phi\left(\dfrac{10.03 - 10.00}{0.02}\right) - \Phi\left(\dfrac{9.97 - 10.00}{0.02}\right)$

$= \Phi(1.5) - \Phi(-1.5) = 0.8664.$ *Answer* $1 - 0.8664 \approx 13\%.$

(d) From (8a) we obtain $c = 1.96\sigma = 0.039.$ *Answer* 9.961 and 10.039 mm.

(e) $P(9.961 \leqq X \leqq 10.039) = \Phi\left(\dfrac{10.039 - 10.010}{0.02}\right) - \Phi\left(\dfrac{9.961 - 10.010}{0.02}\right)$

$= \Phi(1.45) - \Phi(-2.45) = 0.9265 - 0.0071 \approx 92\%;$

hence the answer is 8%, and we see that this slight change in the adjustment of the tool bit causes a considerable increase of the percentage of defectives.   ∎

Under a linear transformation, a normal random variable transforms into a normal random variable. In fact, using (5), the reader may prove the following theorem.

Theorem 1 (Linear transformation)

If X is normal with mean μ and variance σ^2, then $X^* = c_1 X + c_2\ (c_1 \neq 0)$ is normal with mean $\mu^* = c_1\mu + c_2$ and variance $\sigma^{*2} = c_1^2\sigma^2$.

The normal distribution may also be used for approximating the binomial distribution when n is large, so that the binomial coefficients and powers in the probability function

(9) $$f(x) = \binom{n}{x} p^x q^{n-x} \qquad (x = 0, 1, \cdots, n)$$

(cf. Sec. 23.6) of the binomial distribution become very inconvenient. In fact, the following important theorem holds.

Theorem 2 (Limit theorem of De Moivre and Laplace)

For large n,

$$f(x) \sim f^*(x) \qquad (x = 0, 1, \cdots, n)$$

where f is given by (9),

(10)
$$f^*(x) = \frac{1}{\sqrt{2\pi}\sqrt{npq}} e^{-z^2/2}, \qquad z = \frac{x - np}{\sqrt{npq}}$$

*is the density of the normal distribution with mean $\mu = np$ and variance $\sigma^2 = npq$ (the mean and variance of the binomial distribution), and the symbol $\sim$ (read **asymptotically equal**) means that the ratio of both sides approaches 1 as n approaches ∞. Furthermore, for any nonnegative integers a and b ($> a$),*

(11)
$$P(a \leq X \leq b) = \sum_{x=a}^{b} \binom{n}{x} p^x q^{n-x} \sim \Phi(\beta) - \Phi(\alpha),$$

$$\alpha = \frac{a - np - 0.5}{\sqrt{npq}}, \qquad \beta = \frac{b - np + 0.5}{\sqrt{npq}}.$$

A proof of this theorem can be found in [G4] or [G15] listed in Appendix 1. The proof shows that the term 0.5 in α and β is a correction caused by the change from a discrete to a continuous distribution.

In the next section, the last of this chapter, we discuss **distributions of several random variables,** which occur in random experiments in which we observe several quantities (for instance, the yield strength, tensile strength, elongation and Brinell hardness of brass, height and weight of persons, etc.); these distributions will also be needed in the mathematical justification of the statistical methods in Chap. 24.

Problems for Sec. 23.7

1. Let X be normal with mean 10 and variance 4. Find $P(X > 12)$, $P(X < 10)$, $P(X < 11)$ and $P(9 < X < 13)$.

2. Let X be normal with mean 3.5 and variance 0.16. Find $P(X > 3.9)$, $P(X \geq 3.9)$, $P(X \leq 3.9)$, $P(2.5 < X < 4.5)$.

3. Let X be normal with mean 50 and variance 9. Determine c such that $P(X < c) = 5\%$, $P(X > c) = 1\%$ and $P(50 - c < X < 50 + c) = 50\%$.

4. Let X be normal with mean 28.3 and variance 1.44. Find $P(X < c) = 95\%$, $P(X \leq c) = 99\%$, $P(27.1 < X \leq c) = 68\%$.

5. What percentage of defective iron plates can we expect in part (c) of Example 5 if we use a better shaper so that $\sigma = 0.01$ mm?

6. In Example 5, part (c), what value should σ have in order to reduce the percentage of defective plates to 1%?

7. If the lifetime X of a certain kind of automobile battery is normally distributed with a mean of 3 years and a standard deviation of 1 year, and the manufacturer wishes to guarantee the battery for 2 years, about what percentage of the batteries will he have to replace under the guarantee?

8. What is the probability of obtaining at least 2048 heads if a coin is tossed 4040 times and heads and tails are equally likely? (Cf. Table 23.1 in Sec. 23.2.)

9. A manufacturer produces airmail envelopes whose weight is normal with mean $\mu = 1.95$ grams and standard deviation $\sigma = 0.05$ gram. The envelopes are sold in lots of 1000. How many envelopes in a lot will be heavier than 2 grams?

10. In Prob. 9, how many envelopes weighing 2.05 grams or more can be expected in a given package of 100 envelopes?

11. A manufacturer knows from experience that the resistance of resistors he produces is normal with mean $\mu = 100$ ohms and standard deviation $\sigma = 2$ ohms. What percentage of the resistors will have resistance between 98 and 102 ohms? Between 95 and 105 ohms?

12. Specifications for a certain job call for bolts with a diameter of 0.260 ± 0.005 cm. If the diameters of the bolts made by some manufacturer are normally distributed with $\mu = 0.259$ and $\sigma = 0.003$, what percentage of these bolts will meet specifications?

13. How does the answer to Prob. 12 change if $\sigma = 0.003$ is changed to $\sigma = 0.030$ (lower quality of production)?

14. A producer sells electric bulbs in cartons of 1000 bulbs. Using (11), find the probability that any given carton contains not more than 1% defective bulbs, assuming the production process to be a Bernoulli experiment with $p = 1\%$ ($=$ probability that any given bulb will be defective).

15. The breaking strength X [kg] of a certain type of plastic block is normally distributed with a mean of 1000 kg and a standard deviation of 75 kg. What is the maximum load such that we can expect no more than 5% of the blocks to break?

16. **(Bernoulli's law of large numbers)** In a random experiment let an event A have probability p $(0 < p < 1)$, and let X be the number of times A happens in n independent trials. Show that for any given $\epsilon > 0$,

$$P \left(\left| \frac{X}{n} - p \right| < \epsilon \right) \to 1 \qquad\qquad \text{as } n \to \infty.$$

17. Considering $\Phi^2(\infty)$ and introducing polar coordinates in the double integral, prove

(12) $$\Phi(\infty) = \frac{1}{\sqrt{2\pi}} \int_{-\infty}^{\infty} e^{-u^2/2} \, du = 1.$$

18. Using (12) and integration by parts, show that σ in (1) is the standard deviation of the normal distribution.

19. Show that $\Phi(-z) = 1 - \Phi(z)$.

20. Prove Theorem 1.

23.8 Distributions of Several Random Variables

If in a random experiment we observe a single quantity, we have to associate with that experiment a single random variable, call it X. From Sec. 23.4 we remember that the corresponding distribution function $F(x) = P(X \leq x)$ determines the distribution completely, because for each interval $a < X \leq b$,

$$P(a < X \leq b) = F(b) - F(a).$$

If in a random experiment we observe two quantities, we have to associate with the experiment two random variables, say, X and Y. For example, X may correspond to the Rockwell hardness and Y to the carbon content of steel, if we measure these two quantities. Each performance of the experiment yields a pair of numbers $X = x$, $Y = y$, briefly (x, y), which may be plotted as a point in the XY-plane. We may now consider a rectangle $a_1 < X \leq b_1$, $a_2 < Y \leq b_2$ (Fig. 514). If for each such rectangle we know the corresponding probability

$$P(a_1 < X \leq b_1, a_2 < Y \leq b_2),$$

then we say that the **two-dimensional probability distribution** of the random variables X and Y or of the **two-dimensional random variable** (X, Y) is known. The function

(1)
$$\boxed{F(x, y) = P(X \leq x, Y \leq y)}$$

is called the **distribution function** of this distribution or of (X, Y). It determines the distribution uniquely, because (cf. Prob. 4)

(2)
$$P(a_1 < X \leq b_1, a_2 < Y \leq b_2)$$
$$= F(b_1, b_2) - F(a_1, b_2) - F(b_1, a_2) + F(a_1, a_2).$$

Discrete Two-Dimensional Distributions

The distribution and the variable (X, Y) are said to be **discrete** if (X, Y) has the following properties.

(X, Y) can assume only finitely many or at most countably infinitely many pairs of values (x, y), the corresponding probabilities being positive. To every domain containing no such pairs there corresponds the probability[8] 0.

Let (x_i, y_j) be any such pair and let $P(X = x_i, Y = y_j) = p_{ij}$ (where we admit that p_{ij} may be 0 for certain pairs of subscripts i, j). The function

---

[8]Note that this does not follow from the first of these two conditions.

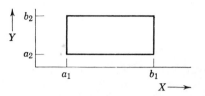

Fig. 514. Notion of a two-dimensional distribution

(3) $f(x, y) = p_{ij}$ when $x = x_i$, $y = y_j$ and $f(x, y) = 0$ otherwise

is called the **probability function** of (X, Y); here $i = 1, 2, \cdots$ and $j = 1, 2, \cdots$ independently. In analogy to (7), Sec. 23.4 we now have for the distribution function the formula

(4)
$$F(x, y) = \sum_{x_i \leq x} \sum_{y_j \leq y} f(x_i, y_j),$$

and instead of formula (3) in Sec. 23.4 we now have the condition

(5)
$$\sum_i \sum_j f(x_i, y_j) = 1.$$

EXAMPLE 1. Two-dimensional discrete distribution
If we simultaneously toss a dime and a nickel and consider

$$X = \textit{Number of heads the dime turns up,}$$

$$Y = \textit{Number of heads the nickel turns up,}$$

then X and Y can have the values 0 or 1, and the probability function is

$$f(0, 0) = f(1, 0) = f(0, 1) = f(1, 1) = \tfrac{1}{4}, \quad f(x, y) = 0 \text{ otherwise.} \qquad \blacksquare$$

Continuous Two-Dimensional Distributions

(X, Y) and its distribution are said to be **continuous** if the corresponding distribution function $F(x, y)$ can be represented by a double integral

(6)
$$F(x, y) = \int_{-\infty}^{y} \int_{-\infty}^{x} f(x^*, y^*) \, dx^* \, dy^*$$

where $f(x, y)$ is defined, nonnegative and bounded in the entire plane, possibly except for finitely many continuously differentiable curves. $f(x, y)$ is called the **probability density** of the distribution. It follows that

(7)
$$P(a_1 < X \leq b_1, a_2 < Y \leq b_2) = \int_{a_2}^{b_2} \int_{a_1}^{b_1} f(x, y) \, dx \, dy.$$

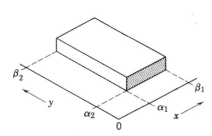
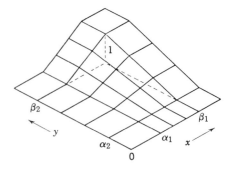

Fig. 515. Probability density function (8) of the uniform distribution

Fig. 516. Distribution function of the uniform distribution defined by (8)

EXAMPLE 2. Two-dimensional uniform distribution in a rectangle
The density (cf. Fig. 515)

(8) $$f(x, y) = 1/k \text{ if } (x, y) \text{ is in } R, \quad f(x, y) = 0 \text{ otherwise}$$

defines the so-called *uniform distribution in the rectangle R*; here k is the area of R, that is, $k = (\beta_1 - \alpha_1)(\beta_2 - \alpha_2)$. The distribution function is shown in Fig. 516.   ∎

Marginal Distributions of a Discrete Distribution

In the case of a discrete random variable (X, Y) with probability function $f(x, y)$ we may ask for the probability $P(X = x, Y \text{ arbitrary})$ that X assume a value x, while Y may assume any value, which we ignore. This probability is a function of x, say $f_1(x)$, and we have the formula

(9) $$f_1(x) = P(X = x, Y \text{ arbitrary}) = \sum_y f(x, y)$$

where we sum all the values of $f(x, y)$ that are not 0 for that x. Clearly, $f_1(x)$ is the probability function of the probability distribution of a single random variable. This distribution is called the **marginal distribution** *of X with respect to the given two-dimensional distribution.* It has the distribution function

(10) $$F_1(x) = P(X \leq x, Y \text{ arbitrary}) = \sum_{x^* \leq x} f_1(x^*).$$

Similarly, the probability function

(11) $$f_2(y) = P(X \text{ arbitrary}, Y = y) = \sum_x f(x, y)$$

determines the so-called **marginal distribution** *of Y with respect to the given two-dimensional distribution.* In (11) we sum all the values of $f(x, y)$ that are not 0 for the corresponding y. The distribution function of this distribution is

(12) $$F_2(y) = P(X \text{ arbitrary}, Y \leq y) = \sum_{y^* \leq y} f_2(y^*).$$

Obviously, both marginal distributions of a discrete random variable (X, Y) are discrete.

EXAMPLE 3. Marginal distributions of a discrete two-dimensional random variable
Suppose that we want to draw three cards with replacement from a bridge deck and observe the two-dimensional discrete random variable (X, Y), where

$$X = Number\ of\ queens,$$

$$Y = Number\ of\ kings\ or\ aces.$$

Since the 52 cards of the deck contain four queens, four kings and four aces, the probability of obtaining a queen in drawing a single card is $4/52 = 1/13$, and a king or ace is obtained with probability $8/52 = 2/13$. Hence (X, Y) has the probability function

$$f(x, y) = \frac{3!}{x!\,y!\,(3 - x - y)!} \left(\frac{1}{13}\right)^x \left(\frac{2}{13}\right)^y \left(\frac{10}{13}\right)^{3-x-y} \qquad (x + y \leq 3)$$

and $f(x, y) = 0$ otherwise. Table 23.2 shows the values of $f(x, y)$ in the center and, on the right and lower margins, the values of the probability functions $f_1(x)$ and $f_2(y)$ of the marginal distributions of X and Y, respectively. ∎

Table 23.2
Values of the Probability Functions $f(x, y)$, $f_1(x)$, $f_2(y)$ in Drawing Three Cards with Replacement from a Bridge Deck, where X is the Number of Queens Drawn and Y is the Number of Kings or Aces Drawn

x \ y	0	1	2	3	$f_1(x)$
0	$\frac{1000}{2197}$	$\frac{600}{2197}$	$\frac{120}{2197}$	$\frac{8}{2197}$	$\frac{1728}{2197}$
1	$\frac{300}{2197}$	$\frac{120}{2197}$	$\frac{12}{2197}$	0	$\frac{432}{2197}$
2	$\frac{30}{2197}$	$\frac{6}{2197}$	0	0	$\frac{36}{2197}$
3	$\frac{1}{2197}$	0	0	0	$\frac{1}{2197}$
$f_2(y)$	$\frac{1331}{2197}$	$\frac{726}{2197}$	$\frac{132}{2197}$	$\frac{8}{2197}$	

Marginal Distributions of a Continuous Distribution

Similarly, in the case of a continuous random variable (X, Y) with density $f(x, y)$ we may consider

$$(X \leq x,\ Y\ arbitrary) \quad \text{or} \quad (X \leq x,\ -\infty < Y < \infty).$$

Obviously, the corresponding probability is

$$F_1(x) = P(X \leq x,\ -\infty < Y < \infty) = \int_{-\infty}^{x} \left(\int_{-\infty}^{\infty} f(x^*, y)\, dy \right) dx^*.$$

Setting

(13)
$$f_1(x) = \int_{-\infty}^{\infty} f(x, y)\, dy,$$

we may write

(14)
$$F_1(x) = \int_{-\infty}^{x} f_1(x^*)\, dx^*.$$

$f_1(x)$ is called the *density* and $F_1(x)$ the *distribution function* of the **marginal distribution** *of X with respect to the given continuous distribution*. The function

(15)
$$f_2(y) = \int_{-\infty}^{\infty} f(x, y)\, dx$$

is called the *density* and

(16)
$$F_2(y) = \int_{-\infty}^{y} f_2(y^*)\, dy^* = \int_{-\infty}^{y} \int_{-\infty}^{\infty} f(x, y^*)\, dx\, dy^*$$

is called the *distribution function* of the **marginal distribution** *of Y with respect to the given two-dimensional distribution.* We see that both marginal distributions of a continuous distribution are continuous.

Independence of Random Variables

The two random variables X and Y of a two-dimensional (X, Y)-distribution with distribution function $F(x, y)$ are said to be **independent** if

(17)
$$F(x, y) = F_1(x)F_2(y)$$

holds for all (x, y). Otherwise these variables are said to be **dependent.**

Suppose that X and Y are either both discrete or both continuous. Then X and Y are independent if and only if the corresponding probability functions or densities $f_1(x)$ and $f_2(y)$ satisfy

(18)
$$f(x, y) = f_1(x)f_2(y)$$

for all (x, y); cf. Prob. 16. For example, the variables in Table 23.2 are dependent. The variables X = *Number of heads on a dime*, Y = *Number of heads on a nickel* in tossing a dime and a nickel once may assume the values 0 or 1 and are independent.

The notions of independence and dependence may be extended to the n

random variables of an n-dimensional $(X_1, \cdots, X_n)$-distribution with distribution function

$$F(x_1, \cdots, x_n) = P(X_1 \leqq x_1, \cdots, X_n \leqq x_n).$$

These random variables are said to be **independent** if for all $(x_1, \cdots, x_n)$,

(19) $$F(x_1, \cdots, x_n) = F_1(x_1)F_2(x_2) \cdots F_n(x_n),$$

where $F_j(x_j)$ is the distribution function of the marginal distribution of X_j, that is,

$$F_j(x_j) = P(X_j \leqq x_j, X_k \text{ arbitrary}, k \neq j).$$

Otherwise these random variables are said to be **dependent.**

Functions of Random Variables

Let (X, Y) be a random variable with probability function or density $f(x, y)$ and distribution function $F(x, y)$, and let $g(x, y)$ be any continuous function that is defined for all (x, y) and is not constant. Then $Z = g(X, Y)$ is a random variable, too. For example, if we roll two dice and X is the number that the first die turns up whereas Y is the number that the second die turns up, then $Z = X + Y$ is the sum of those two numbers (cf. Fig. 506 in Sec. 23.4).

If $(X_1, \cdots, X_n)$ is an n-dimensional random variable and $g(x_1, \cdots, x_n)$ is a continuous function that is defined for all $(x_1, \cdots, x_n)$ and is not constant, then $Z = g(X_1, \cdots, X_n)$ is a random variable, too.

In the case of a *discrete* random variable (X, Y) we may obtain the probability function $f(z)$ of $Z = g(X, Y)$ by summing all $f(x, y)$ for which $g(x, y)$ equals the value of z considered; thus

(20) $$f(z) = P(Z = z) = \sum_{g(x,y)=z} \sum f(x, y).$$

The distribution function of Z is

(21) $$F(z) = P(Z \leqq z) = \sum_{g(x,y)\leqq z} \sum f(x, y)$$

where we sum all values of $f(x, y)$ for which $g(x, y) \leqq z$.

In the case of a *continuous* random variable (X, Y) we similarly have

(22) $$F(z) = P(Z \leqq z) = \iint_{g(x,y)\leqq z} f(x, y) \, dx \, dy$$

where for each z we integrate over the region $g(x, y) \leqq z$ in the xy-plane.

Addition of Means and Variances

The number

$$
(23) \quad E(g(X, Y)) = \begin{cases} \displaystyle\sum_x \sum_y g(x, y)f(x, y) & [(X, Y) \text{ discrete}] \\[2em] \displaystyle\int_{-\infty}^{\infty} \int_{-\infty}^{\infty} g(x, y)f(x, y)\, dx\, dy & [(X, Y) \text{ continuous}] \end{cases}
$$

is called the **mathematical expectation** or, briefly, the **expectation of** $g(X, Y)$. Here it is assumed that the double series converges absolutely and the integral of $|g(x, y)|f(x, y)$ over the xy-plane exists. Since summation and integration are linear processes, we have from (23)

$$
(24) \qquad E(ag(X, Y) + bh(X, Y)) = aE(g(X, Y)) + bE(h(X, Y)).
$$

An important special case is $E(X + Y) = E(X) + E(Y)$, and by induction we have the following result.

Theorem 1 (Addition of means)
The mean (expectation) of a sum of random variables equals the sum of the means (expectations), that is,

$$
(25) \quad \boxed{E(X_1 + X_2 + \cdots + X_n) = E(X_1) + E(X_2) + \cdots + E(X_n).}
$$

Furthermore we readily obtain

Theorem 2 (Multiplication of means)
*The mean (expectation) of the product of **independent** random variables equals the product of the means (expectations), that is,*

$$
(26) \qquad E(X_1 X_2 \cdots X_n) = E(X_1)E(X_2) \cdots E(X_n).
$$

Proof. If X and Y are independent random variables (both discrete or both continuous), then $E(XY) = E(X)E(Y)$. In fact, in the discrete case we have

$$
E(XY) = \sum_x \sum_y xy f(x, y) = \sum_x x f_1(x) \sum_y y f_2(y) = E(X)E(Y),
$$

and in the continuous case the proof of the relation is similar. Extension to n independent random variables gives (26), and Theorem 2 is proved. ∎

We shall now discuss the **addition of variances.** Let $Z = X + Y$ and let μ and σ^2 denote the mean and variance of Z. We first have (cf. Prob. 14 in Sec. 23.5)

$$
\sigma^2 = E([Z - \mu]^2) = E(Z^2) - [E(Z)]^2.
$$

From (24) we see that the first term on the right equals

$$E(Z^2) = E(X^2 + 2XY + Y^2) = E(X^2) + 2E(XY) + E(Y^2),$$

and for the second term on the right, we obtain from Theorem 1

$$[E(Z)]^2 = [E(X) + E(Y)]^2 = [E(X)]^2 + 2E(X)E(Y) + [E(Y)]^2.$$

By substituting these expressions in the formula for σ^2 we have

$$\sigma^2 = E(X^2) - [E(X)]^2 + E(Y^2) - [E(Y)]^2$$
$$+ 2[E(XY) - E(X)E(Y)].$$

From Prob. 14, Sec. 23.5, we see that the expression in the first line on the right is the sum of the variances of X and Y, which we denote by $\sigma_1{}^2$ and $\sigma_2{}^2$, respectively. The quantity

(27) $$\sigma_{XY} = E(XY) - E(X)E(Y)$$

is called the **covariance** of X and Y. Consequently, our result is

(28) $$\sigma^2 = \sigma_1{}^2 + \sigma_2{}^2 + 2\sigma_{XY}.$$

If X and Y are independent, then $E(XY) = E(X)E(Y)$; hence $\sigma_{XY} = 0$, and

(29) $$\sigma^2 = \sigma_1{}^2 + \sigma_2{}^2.$$

Extension to more than two variables yields

Theorem 3 (Addition of variances)
*The variance of the sum of **independent** random variables equals the sum of the variances of these variables.*

Caution! In the numerous applications of Theorems 1 and 3 we must always remember that Theorem 3 holds only for *independent* variables.

This is the end of Chap. 23 on probability theory. Most of the concepts, methods and special distributions discussed in this chapter will play a fundamental role in the next chapter, which deals with methods of **statistical inference,** that is, conclusions from samples to populations, whose unknown properties we want to know, and try to discover by looking at suitable properties of samples which we have obtained.

Problems for Sec. 23.8

1. Let $f(x, y) = k$ when $8 \leq x \leq 12$ and $0 \leq y \leq 2$ and zero elsewhere. Find k. Find $P(X \leq 11, 1 \leq Y \leq 1.5)$ and $P(9 \leq X \leq 13, Y \leq 1)$.

2. Let $f(x, y) = 2$ if $x > 0, y > 0, x + y < 1$ and 0 otherwise. Graph f and the corresponding distribution function.

3. Let $f(x, y) = k$ if $x > 0, y > 0, x + y < 3$ and 0 otherwise. Find k. Graph $f(x, y)$. Find $P(X + Y \leq 1)$, $P(Y > X)$.

4. Prove (2).

5. Find $P(X > Y)$ when (X, Y) has the density $f(x, y) = e^{-(x+y)}$ if $x \geq 0$, $y \geq 0$ and 0 otherwise.

6. Find the densities of the marginal distributions in Prob. 5.

7. Find the marginal distributions of the distribution in Figs. 515 and 516 on p. 1220.

8. Let X [cm] and Y [cm] be the diameters of a pin and a hole, respectively. Suppose that (X, Y) has the density

$$f(x, y) = 625 \quad \text{if} \quad 0.98 < x < 1.02, \quad 1.00 < y < 1.04$$

and 0 otherwise. (a) Find the marginal distributions. (b) What is the probability that a pin chosen at random will fit a hole whose diameter is 1.00?

9. If certain sheets of paper have a mean weight of 2 grams each, with a standard deviation of 0.03 gram, what are the mean weight and the standard deviation of a pack of 10 000 sheets?

10. An electronic device consists of two components. Let X and Y [years] be the times to failure of the first and second components, respectively. Assume that (X, Y) has the probability density

$$f(x, y) = 4e^{-2(x+y)} \quad \text{if } x > 0 \text{ and } y > 0$$

and 0 otherwise. (a) Are X and Y dependent or independent? (b) Find the densities of the marginal distributions. (c) What is the probability that the first component will have a lifetime of 2 years or longer?

11. If the weight of certain (empty) containers has mean 5 lb and standard deviation 0.2 lb, and if the filling of the containers has mean weight 100 lb and standard deviation 0.5 lb, what are the mean weight and the standard deviation of filled containers?

12. A five-gear assembly is put together with spacers between the gears. The mean thickness of the gears is 5.030 cm with a standard deviation of 0.008 cm. The mean thickness of the spacers is 0.140 cm with a standard deviation of 0.005 cm. Find the mean and the standard deviation of the assembled units consisting of five randomly selected gears and four randomly selected spacers.

13. What are the mean thickness and the standard deviation of transformer cores each consisting of 50 layers of sheet metal and 49 insulating paper layers if the metal sheets have mean thickness 0.5 mm each with a standard deviation of 0.05 mm and the paper layers have mean 0.05 mm each with a standard deviation of 0.02 mm?

14. Give an example of two different discrete distributions that have the same marginal distributions.

15. Show that the random variables with the densities $f(x, y) = x + y$ and $g(x, y) = (x + \frac{1}{2})(y + \frac{1}{2})$ when $0 \leq x \leq 1$, $0 \leq y \leq 1$, have the same marginal distributions.

16. Prove the statement involving (18).

17. Let (X, Y) have the probability function given by $f(0, 0) = f(1, 1) = 0.3$, $f(0, 1) = f(1, 0) = 0.2$. Are X and Y independent?

18. Using Theorem 1, obtain the formula for the mean μ of the binomial distribution.

19. Using Theorem 3, obtain the formula for the variance σ^2 of the binomial distribution.

20. Using Theorem 1, obtain the formula for the mean of the hypergeometric distribution. Can you use Theorem 3 to obtain the variance of that distribution?

Review Problems for Chapter 23

1. If $P(A) = P(B)$ and $A \subset B$, can $A \neq B$?

2. Give some simple examples of random experiments in which it is not possible to use probabilities based on equally likely cases.

3. If $E \neq S$ (= the sample space), can $P(E) = 1$?

4. What is the difference between mutually exclusive events and independent events?

5. What is the difference between the concepts of a permutation and a combination?

6. List all permutations and all combinations of the letters a, b, c, d taken two at a time.

7. What do we mean by conditional probability? Give a typical example.

8. What distributions correspond to sampling with replacement and without replacement?

9. Under what conditions will it practically make no difference whether we sample with or without replacement?

10. What are the complements of the events $X \leq b$, $X < b$, $X \geq c$, $X > c$, $b \leq X \leq c$, $b < X \leq c$?

11. In what kind of experiment do we have a discrete random variable and in what kind a continuous random variable?

12. What is the sum of the probabilities of all the possible values of a discrete random variable? What is the analog of this for a continuous random variable?

13. Can the probability function of a discrete random variable have infinitely many possible values? (Give reason for your answer.)

14. If X is a continuous random variable, what is $P(X = a)$, where a is any given number?

15. How is the density of a continuous random variable related to the distribution function?

16. What property of a distribution does the variance characterize? The mean?

17. Write down the transformation formulas for mean and variance under a linear transformation (from memory). Why are these formulas important?

18. Under what conditions does the addition formula for variances hold? The addition formula for means?

19. Show that the inflection points of the density curve of the normal distribution correspond to $x = \mu \pm \sigma$.

20. Write down the formula for the distribution function $F(x)$ of an arbitrary normal random variable in terms of the distribution function of the standardized normal distribution. Why is this practically important?

21. A box contains 50 screws, five of which are defective. Find the probability function of the random variable $X = $ *Number of defective screws in drawing two screws without replacement* and compute its values.

22. Find the values of the distribution function in Prob. 21.

23. Find the probability function of $X = $ *Number of times of rolling a fair die until the first even number appears*.

24. In Prob. 23, find the mean.

25. If X has the probability function $f(2) = 0.3$, $f(3) = 0.5$, $f(4) = f(5) = 0.1$, what is the probability that X will assume any value less than 4? Greater than 4?

26. Of a lot of 10 items, two are defective. (a) Find the number of different samples of four. Find the number of samples of four containing (b) no defectives, (c) one defective, (d) two defectives.

27. Find the number of samples of three light bulbs that can be drawn from a box containing 24 bulbs.

28. In how many ways can a committee of four persons be chosen from 20 persons?

29. Compute 5! by the Stirling formula and find the absolute and relative errors.

30. If we know that a process of producing bolts is running smoothly, with a fraction defective of 4%, and we inspect the process by taking samples of two bolts, what is the probability of no defectives? One defective? Two defectives?

31. A motor drives an electric generator. During a 60-day period, the motor needs repair with probability 5% and the generator needs repair with probability 6%. What is the probability that during a given period, the entire apparatus will need repair?

32. In flipping three fair coins, what is the probability of three heads under the condition that one obtain at least one head?

33. Let $f(x) = kx^2$ when $0 \leq x \leq 1$ and 0 otherwise. Find k. Find the number c such that $P(X \leq c) = 95\%$.

34. Find the mean and variance of a discrete random variable X having the probability function $f(0) = \frac{1}{4}$, $f(1) = \frac{1}{2}$, $f(2) = \frac{1}{4}$.

35. Do the same task as in Prob. 34, if $f(0) = 0.512$, $f(1) = 0.384$, $f(2) = 0.096$, $f(3) = 0.008$.

36. Let X have the density $f(x) = 0.5x$ if $0 \leq x \leq 2$ and 0 otherwise. Determine the mean and variance and the standardized random variable corresponding to X.

37. Let X be as in Prob. 36. Find the mean and variance of $Y = -2X + 5$.

38. Find the skewness of the distribution with density $f(x) = 2(1 - x)$ if $0 < x < 1$, $f(x) = 0$ otherwise.

39. Show that $E(ag(X) + bh(X)) = aE(g(X)) + bE(h(X))$; [$a$ and b are constants].

40. Find the skewness of the distribution with density $f(x) = xe^{-x}$ if $x > 0$ and $f(x) = 0$ otherwise. Graph $f(x)$.

41. If the number of cars per second passing over a bridge at 5 P.M. on a weekday has a Poisson distribution with mean $\mu = 5$, what is the probability of observing four or more cars during any given second?

42. If in a production of rubber bands, 3% are defective and the bands are sold in packs of 50, what is the Poisson approximation that a given pack will contain x defectives?

43. In Prob. 42, let 0.2% be defective. If the bands are sold in packs of 100 and with the guarantee that all the bands in a pack are nondefective, what is the probability that a given pack will violate this guarantee?

44. Suppose that a test for extrasensory perception consists of naming (in any order) three cards randomly drawn from a deck of 13 cards. Find the probability that by chance alone, the person will correctly name (a) no cards, (b) one card, (c) two cards, (d) three cards.

45. If in a production of washers, 3% are too thin and 5% are too thick, what is the probability that a sample of 20 washers will contain precisely x_1 washers that are too thin and x_2 that are too thick?

46. If string contains, on the average, 2 defects per 100 ft, what is the probability that a roll of 300 ft will contain (a) x defects, (b) no defect?

47. Find the probability of obtaining x dull drill bits in a sample of three drawn without replacement from a lot of 20 bits, five of which are dull.

48. If the breaking strength of rope is normal with mean 1250 kg and standard deviation 55 kg, what is the maximum load such that we can expect at most 5% of the ropes to break?

49. Let X be normal with $\mu = 0.279$ and $\sigma = 0.001$. Find $P(0.278 < X < 0.282)$.

50. If the weight of bags of cement is normal with mean 50 kg and standard deviation 1 kg, what is the probability that 100 bags will be heavier than 5030 kg?

Summary of Chapter 23
Probability Theory

A **random experiment** is an experiment or observation in which the result ("**outcome**") depends on "chance" (effects of factors unknown to us). Examples are games of chance with dice or cards, measuring the hardness of steel, observing weather conditions, counting the number of telephone calls in an office or recording the number of accidents in a city. (Thus the word "experiment" is used here in a wider sense than in common language.) The outcomes are regarded as points (elements) of a set S, called the **sample space**, whose subsets are called **events**. For events E we define a **probability** $P(E)$ by the axioms (Sec. 23.2)

$$0 \leqq P(E) \leqq 1$$

$$(1) \qquad\qquad P(S) = 1$$

$$P(E_1 \cup E_2 \cup \cdots) = P(E_1) + P(E_2) + \cdots \qquad (E_j \cap E_k = \varnothing).$$

The complement E^C of E then has the probability

$$(2) \qquad\qquad P(E^C) = 1 - P(E).$$

The **conditional probability** of an event B under the condition that an event A happens is

$$(3) \qquad\qquad P(B \,|\, A) = \frac{P(A \cap B)}{P(A)}.$$

A and B are called **independent** if

$$(4) \qquad\qquad P(A \cap B) = P(A)P(B).$$

With a random experiment we associate a **random variable** X; this is a function that is defined on S, whose values are real numbers, and X is such that the probability $P(X \leqq a)$ that X assume any value a, and the probability $P(a \leqq X < b)$ that X assume any value in any interval are defined (Sec. 23.4). The probability distribution of X is determined by the distribution function (Sec. 23.4)

$$(5) \qquad\qquad F(x) = P(X \leqq x).$$

There are two practically important kinds of random variables: those of **discrete** type, which appear if we count (defective items, customers in a bank, etc.) and those of **continuous** type, which appear if we measure (length, speed, temperature, weight, etc.).

A discrete random variable has a **probability function**

$$(6) \qquad\qquad f(x) = P(X = x).$$

Its **mean** μ and **variance** σ^2 are (Sec. 23.5)

$$(7) \qquad = \sum_j x_j f(x_j), \qquad \text{and} \qquad \sigma^2 = \sum_j (x_j - \mu)^2 f(x_j)$$

where the x_j are the values for which X has a positive probability. Important discrete random variables and distributions are the *binomial, Poisson* and *hypergeometric distributions* (Secs. 23.6).
    A continuous random variable has a **density**

$$(8) \qquad\qquad f(x) = F'(x) \qquad\qquad\qquad \text{[cf. (5)].}$$

Its mean and variance are (Sec. 23.5)

$$(9) \qquad \mu = \int_{-\infty}^{\infty} x f(x)\, dx \qquad \text{and} \qquad \sigma^2 = \int_{-\infty}^{\infty} (x - \mu)^2 f(x)\, dx.$$

Very important is the **normal distribution** whose density is (Sec. 23.7)

$$(10) \qquad\qquad f(x) = \frac{1}{\sqrt{2\pi}} \exp\left[-\frac{1}{2}\left(\frac{x - \mu}{\sigma}\right)\right]^2$$

and whose distribution function is (Sec. 23.7; Tables A8, A9 in Appendix 4)

$$(11) \qquad\qquad F(x) = \Phi\left(\frac{x - \mu}{\sigma}\right).$$

A **two-dimensional random variable** (X, Y) occurs if we simultaneously observe two quantities (e.g., height X and weight Y of persons). Its distribution function is (Sec. 23.8)

$$(12) \qquad\qquad F(x, y) = P(X \leq x, Y \leq y).$$

X and Y have the distribution functions (Sec. 23.8)

$$(13) \quad F_1(x) = P(X \leq x, Y \text{ arbitrary}), \quad F_2(y) = P(x \text{ arbitrary}, Y \leq y);$$

their distributions are called *marginal distributions.* If both X, and Y are discrete, then (X, Y) has a probability function

$$f(x, y) = P(X = x, Y = y).$$

If both X and Y are continuous, then (X, Y) has a density $f(x, y)$.

Chapter 24

Mathematical Statistics

In **probability theory** we set up mathematical models of processes and systems that are affected by "chance." In mathematical statistics or, briefly, **statistics,** we check these models against the reality, to determine whether they are faithful and accurate enough for practical purposes. This is done by drawing **samples,** inspecting their properties and then making conclusions about properties of the population from which the samples were drawn. These conclusions are true, not absolutely, but with some (usually high) probability which we can choose (95%, for instance) or at least compute. Such conclusions result from methods of statistical inference that we shall develop in this chapter, using probability theory as the foundation of our approach.

We talk about this approach in general terms in Sec. 24.1, and then we discuss samples and their means (Secs. 24.2–24.4). **Statistical methods** begin in Sec. 24.5 with point and interval **estimation of parameters** (Secs. 24.5, 24.6) and continue with **hypothesis testing** and applications (Secs. 24.7–24.9), *goodness of fit* (test for distribution functions, Sec. 24.10) and some nonparametric tests (Sec. 24.11). The last section deals with *pairs of measurements* and **regression analysis.**

Prerequisites for this chapter: Chap. 23.
Sections that may be omitted in a shorter course: 24.3, 24.8, 24.9, 24.11.
References: Appendix 1, Part G.
Answers to problems: Appendix 2.

24.1 Nature and Purpose of Statistics

In engineering statistics we are concerned with methods for designing and evaluating experiments to obtain information about practical problems, for example, the inspection of quality of raw material or manufactured products, the comparison of machines and tools or methods used in production, the output of workers, the reaction of consumers to new products, the yield of a chemical process under various conditions, the relation between the iron content of iron ore and the density of the ore, the efficiency of air-conditioning systems under various temperatures, the relation between the Rockwell hardness and the carbon content of steel, and so on.

For instance, in a process of mass production (of screws, bolts, light bulbs, typewriter keys, etc.) we ordinarily have *nondefective items,* that is, articles that satisfy the quality requirements, and *defective items,* which do not satisfy the requirements. Such requirements include maximum and minimum diameters of axle shafts, minimum lifetimes of light bulbs, limiting values of the resistance of resistors in radios and TV sets, maximum weights of airmail envelopes, minimum contents of automatically filled bottles, maximum reaction times of switches and minimum values of strength of yarn.

The reason for the differences in the quality of products is *variation* due to numerous factors (in the raw material, function of automatic machinery, workmanship, etc.) whose influence cannot be predicted, so that the variation must be regarded as a *random variation.* In the case of evaluating the efficiency of production methods and in the other preceding examples the situation is similar.

In most cases the inspection of *each* item of a production would be too expensive and time-consuming. It may even be impossible if it leads to the destruction of the item. Hence instead of inspecting all the manufactured items, just a few of them (a "**sample**") are inspected and from this inspection conclusions can be drawn about the totality (the "**population**"). If we draw 100 screws from a lot of 10,000 screws and find that 4 of the 100 screws are defective, we are inclined to conclude that *about* 4% of the lot are defective, provided the screws were drawn "**at random,**" that is, so that each screw of the lot had the same "chance" of being drawn. It is clear that such a conclusion is not absolutely certain; that is, we cannot say that the lot contains *precisely* 4% defectives, but in most cases such a precise statement would not be of particular practical interest anyway. We also feel that the conclusion is the more dependable the more (randomly selected!) screws we inspect. Using the theory of mathematical probability we shall see that these somewhat vague intuitive notions and ideas can be made precise. Furthermore, this theory will also yield measures for the reliability of conclusions about populations obtained from samples by statistical methods. Hence probability theory forms the basis of statistical methods.

Similarly, to obtain information about the iron content μ of iron ore, we may pick a certain number n of specimens at random and measure their iron content. This yields a sample of n numbers $x_1, \cdots, x_n$ (the results of the n measurements) whose average $\bar{x} = (x_1 + x_2 + \cdots + x_n)/n$ is an approximate value for μ.

Problems of differing natures may require different methods and techniques, but the steps leading from the formulation of a problem to its solution are the same in almost all cases. They are as follows.

1. *Formulation of the problem.* It is important to formulate the problem in a precise fashion and to limit the investigation, so that we can expect a useful answer within a prescribed interval of time, taking into account the cost of the statistical investigation, the skill of the investigators and the facilities available. This step must also include the creation of a mathematical model based on clear concepts. (For example, we must define what we mean by a defective item in a specific situation.)

2. *Design of experiment.* This step includes the choice of the statistical method to be used in the last step, the sample size n (number of items to be drawn and inspected or number of experiments to be made, etc.), and the physical methods and techniques to be used in the experiment. The goal is to obtain a maximum of information using a minimum of cost and time.

3. *Experimentation or collection of data.* This step should adhere to strict rules.

4. *Tabulation.* In this step we arrange the experimental data in a clear and simple tabular form, and we may represent them graphically by diagrams, bar charts, etc. We also compute numbers that characterize the average size and the spread of the sample values.

5. *Statistical inference.* In this step we use the sample and apply a suitable statistical method for drawing conclusions about the unknown properties of the population so that we obtain the answer to our problem.

It seems clear that **samples** will play a crucial role in the whole approach. Accordingly, we should make this intuitive notion rigorous. We do this as our next task in the coming section.

24.2 Random Sampling. Random Numbers

Probability theory helps to create mathematical models of populations, and the statistical methods to be discussed later in this chapter yield relations between the theory (the models) and the observable reality; they permit conclusions about populations by means of samples (statistical inference; cf. Sec. 24.1). Hence we should first focus our attention on the concept of a sample.

So far it has been sufficient to know that a sample from a population is a selection taken from a population (for examples, see Sec. 24.1), but from now on we must define this notion in a precise fashion. In fact, to obtain meaningful information about populations from samples, a sample must be a **random selection;** that is, each element of the population must have a known probability of being taken into the sample. This condition must be satisfied (at least approximately); otherwise the methods to be discussed may yield completely meaningless and misleading results.

If the sample space is infinite, the sample values will be **independent;** that is, the results of the n performances of a random experiment made for obtaining n sample values will not influence each other. This certainly applies to samples from a normal population. If the sample space is finite, the sample values will still be independent if we sample with replacement; they will be *practically* independent if we sample without replacement but keep the size of the sample small compared to the size of the population (for instance, samples of 5 or 10 values from a population of 1000 values). However, if we sample without replacement and take large samples from a finite population, the dependence will matter considerably.

Random Numbers

It is not so easy to satisfy the requirement that a sample be a random selection, because there are many and subtle factors of various types that can bias results of sampling. For example, if an interested purchaser wants to draw and inspect a sample of 10 items from a lot of 80 items before he decides whether to purchase the lot, how should he select physically those 10 items so that he can be reasonably sure that all possible

$$\binom{80}{10} \approx 1.6 \cdot 10^{12}$$

samples of size 10 would be equally probable?

Methods to solve this problem have been developed, and we shall now describe such a procedure, which is frequently used.

We number the items of that lot from 1 to 80. Then we select 10 items using Table A10 in Appendix 4, which contains *random numbers* made up of sets of **random digits,** as follows. We first select a row number from 0 to 99 at random. This can be done by tossing a fair coin seven times denoting heads by 1 and tails by 0, thus generating a seven-place binary number that will represent $0, 1, \cdots, 127$ with equal probabilities. We use this number if it is $0, 1, \cdots,$ or 99. Otherwise we ignore it and repeat this process. Then we select a column number from 0 to 9 by generating a four-place binary number in a similar fashion. Suppose that we obtained 0011010 ($= 26$) and 0111 ($= 7$), respectively. In row 26 and column 7 of Table A10 we find 44973. We keep the first two digits, that is, 44. We move down the column, starting with 44973 and keeping only the first two digits. In this way we find

$$44 \quad 44 \quad 83 \quad 91 \quad 55 \quad \text{etc.}$$

We omit numbers which are greater than 80 or occur for the second time and continue until 10 numbers are obtained. This yields

$$44 \quad 55 \quad 53 \quad 03 \quad 52 \quad 61 \quad 67 \quad 78 \quad 39 \quad 54.$$

The 10 items having these numbers represent the desired selection.

For a larger table of random digits see Ref. [G18], Appendix 1.

However, for large samples such tables may become cumbersome. For this reason, procedures for generating numbers with similar properties have been developed and are available in a so-called **random number generator** in many computer languages and subroutine libraries.

The next section deals with samples from the data processing point of view (their tabular and graphical representation). Conceptually, this leads to the notion of a **sample frequency function** $\tilde{f}(x)$, which in a sense is an empirical counterpart of the probability function or density $f(x)$ of the population from which the sample is taken.

Problems for Sec. 24.2

1. Suppose that in the example explained in the text we would start from row 43 and column 5 of Table A10 in Appendix 4, and move upward. What items would then be included in a sample of size 10?

2. Using Table A10 in Appendix 4, select a sample of size 15 from a lot of 250 given items.

3. How can fair dice be used in connection with random selection?

4. Consider a random variable Y having the uniform density $f(y) = 1$ if $0 < y < 1$, and 0 otherwise. We can easily simulate Y (that is, sample the values of Y) with the use of random digits. For instance, to obtain 20 values rounded to two decimals, consider any of the 10 columns in Table A10, start with some randomly chosen row and move downward, taking the first two digits of the five digits given in the column and put a decimal point to the left of the first digit. Suppose your choice was column 3 and row 36. Show that you get the following sample and graph a dot frequency diagram of it.

| 0.89 | 0.40 | 0.67 | 0.86 | 0.87 | 0.86 | 0.06 | 0.20 | 0.38 | 0.12 |
| 0.68 | 0.50 | 0.53 | 0.10 | 0.08 | 0.90 | 0.19 | 0.85 | 0.53 | 0.98 |

5. How would you modify the method in Prob. 4 if you want a sample of size 20 from the uniform distribution on the interval $3 < y < 5$?

6. Random digits can also be used to simulate *any* continuous random variable X. For this end we graph the distribution function of X, use random digits to obtain values of the variable Y described in Prob. 4, plot these values on the vertical axis and read off the corresponding values of X. Illustrate this procedure for a normal random variable X with mean 0 and variance 1, using the sample in Prob. 4.

7. Write the method in Prob. 6 in terms of $\Phi(z)$, so that you could program it on a computer.

8. Extend the method in Prob. 6 to the normal distribution with mean 6 and variance 4. What do you then get from 0.89, 0.40, 0.67 in Prob. 4?

9. What sample from a random variable with density $f(x) = e^{-x}$ if $x > 0$, $f(x) = 0$ if $x < 0$, would you get from the sample in Prob. 4? Compute the first three values.

10. The simulation technique described in Prob. 6 applies also to *discrete* random variables. Explain how you would proceed if X is the sum of the two numbers obtained in rolling two fair dice (cf. Fig. 506 in Sec. 23.4).

24.3   Processing of Samples

In this section we discuss methods for handling samples and representing them by diagrams (graphical representations); more fundamentally, we also introduce concepts related to samples that we shall need throughout this chapter.

In the course of a statistical experiment we usually obtain a sequence of observations (numbers in most cases). These should always be recorded in the order in which they occur. They are called **sample values;** their number is called the *sample* **size** and is denoted by n. As a typical example, Table 24.1 shows a sample of size $n = 100$ values. These data were obtained by making standard test cylinders (diameter 6 inches, length 12 inches) of concrete and splitting them 28 days later.

Table 24.1
Sample of 100 Values of the Splitting Tensile Strength (lb/in.²)
of Concrete Cylinders

320	380	340	410	380	340	360	350	320	370
350	340	350	360	370	350	380	370	300	420
370	390	390	440	330	390	330	360	400	370
320	350	360	340	340	350	350	390	380	340
400	360	350	390	400	350	360	340	370	420
420	400	350	370	330	320	390	380	400	370
390	330	360	380	350	330	360	300	360	360
360	390	350	370	370	350	390	370	370	340
370	400	360	350	380	380	360	340	330	370
340	360	390	400	370	410	360	400	340	360

   D. L. IVEY, Splitting tensile tests on structural lightweight aggregate concrete. Texas Transportation Institute, College Station, Texas.

Representation of a Sample in a Frequency Table

We show next how to arrange a sample in tabular form from which we can get an impression of the information contained in a sample. It is sufficient to explain this in terms of our sample in Table 24.1. The point is that we arrange the occurring sample values in increasing order, taking together equal values. For larger samples this is best done on a computer. (Sorting subroutines are available.) Columns 1 and 3 of Table 24.2 show the result. For smaller samples or in the field we may do sorting by hand, running through Table 24.1 row by row and **tallying** as column 2 in Table 24.2 shows.

   We call **absolute frequency** or simply **frequency** of a number x in a sample the number of times this x occurs in the sample. (Thus $x = 330$ has the absolute frequency 6 and $x = 346.2$ the absolute frequency 0.) Dividing the frequency by the sample size n, we get the **relative frequency** of x in the sample, which we denote by $\tilde{f}(x)$. Thus $x = 330$ has the relative frequency $\tilde{f}(330) = 6/n = 6/100 = 0.06$; see column 4 in Table 24.2. Also, $\tilde{f}(346.2) = 0$.

   When $\tilde{f}(x)$ is considered as a function of x for all x, we call it the **frequency function** of the sample. Thus this function will be zero most of the time and positive only for numbers x that actually occur in a sample (in Table 24.2 only for the 15 numbers 300, 310, $\cdots$, 440). This function $\tilde{f}(x)$ shows how the values of the sample are distributed. For this reason we say that $\tilde{f}(x)$ determines the **frequency distribution** of the sample.

   Clearly, $\tilde{f}(x) = 0$ for any number x not in the sample, and $\tilde{f}(x) = n/n = 1$ if all the sample values happen to be equal. Since these are the two extreme possibilities, we have

Theorem 1 (Relative frequency)
The relative frequency satisfies

(1)
$$\boxed{0 \leq \tilde{f}(x) \leq 1.}$$

Table 24.2
Frequency Table of the Sample in Table 24.1

1 Tensile Strength x [lb/in.^{2}]	2 Absolute Frequency	3	4 Relative Frequency	5 Cumulative Absolute Frequency	6 Cumulative Relative Frequency
	Tallies				
300	||	2	0.02	2	0.02
310		0	0.00	2	0.02
320	||||	4	0.04	6	0.06
330	⅂H |	6	0.06	12	0.12
340	⅂H ⅂H |	11	0.11	23	0.23
350	⅂H ⅂H ||||	14	0.14	37	0.37
360	⅂H ⅂H ⅂H |	16	0.16	53	0.53
370	⅂H ⅂H ⅂H	15	0.15	68	0.68
380	⅂H |||	8	0.08	76	0.76
390	⅂H ⅂H	10	0.10	86	0.86
400	⅂H |||	8	0.08	94	0.94
410	||	2	0.02	96	0.96
420	|||	3	0.03	99	0.99
430		0	0.00	99	0.99
440	|	1	0.01	100	1.00

This is the empirical counterpart of Axiom 1 for probability in Sec. 23.2. Furthermore, the sum of all absolute frequencies in a sample of size n must be n (why?), so that division by n gives

Theorem 2 (Sum of all relative frequencies)
The sum of all relative frequencies in a sample equals 1,

$$(2) \qquad\qquad \sum_x f(x) = 1.$$

This is the empirical counterpart of Axiom 2 in Sec. 23.2. (The analog of Axiom 3 would be of no great interest.) In (2) we have a finite sum, which we can write in more orderly form if we first denote the *numerically different* values in a sample of size n by

$$x_1, \qquad x_2, \qquad \cdots, \qquad x_m \qquad\qquad (m \leqq n).$$

(Thus $n = 100$ but $m = 15$ in Table 24.2, and $x_1 = 330$, $x_2 = 340$, $\cdots$, $x_{15} = 440$.) Then (2) becomes

$$(2^*) \qquad\qquad \boxed{\sum_{j=1}^{m} \widetilde{f}(x_j) = 1}$$

(summation from 1 to m, not n). Verify (2^*) for Table 24.2.

Graphical Representations of Samples

Figures 517–520 show possibilities of representing samples by graphs. These are self-explanatory, but the following comment is of practical interest. In Fig. 519 the area of each rectangle is equal to the corresponding relative frequency. Hence the ordinate should be labeled "*relative frequency per unit interval.*" Since in the present case the rectangles are equally wide, those values on the ordinate are proportional to $\tilde{f}(x)$ and we may label the ordinate in terms of $\tilde{f}(x)$. However, this would no longer be true if the rectangles were of different width. In Fig. 520 the situation is similar.

Sample Distribution Function

The frequency function $\tilde{f}(x)$ of a sample is an empirical counterpart or analog of the probability function or density $f(x)$ of the corresponding population—

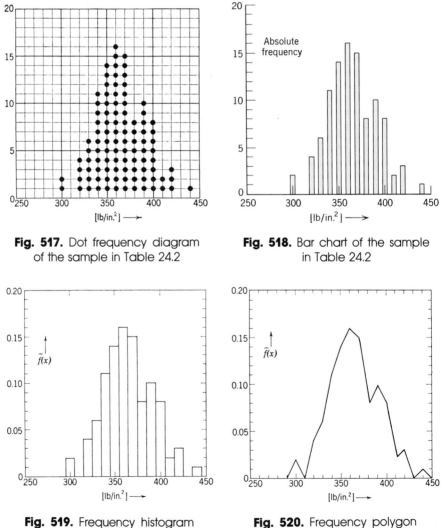

Fig. 517. Dot frequency diagram of the sample in Table 24.2

Fig. 518. Bar chart of the sample in Table 24.2

Fig. 519. Frequency histogram of the sample in Table 24.2

Fig. 520. Frequency polygon of the sample in Table 24.2

although these functions are conceptually quite different; most obviously: a population has *one* $f(x)$, but if we take 10 samples from the same population, we shall generally get 10 different sample frequency functions $\tilde{f}(x)$. (Why?) In our further work we shall also need an analog of the distribution function $F(x)$ of a population. We define

(3) $\tilde{F}(x)$ = *Sum of the relative frequencies of all the values that are smaller than x or equal to x.*

This function is called the **cumulative frequency function of the sample** or the **sample distribution function.** An example is shown in Fig. 521.

$\tilde{F}(x)$ is a *step function* (piecewise constant function) having jumps of magnitude $\tilde{f}(x)$ precisely at those x at which $\tilde{f}(x) \neq 0$. The first jump is at the smallest sample value and the last at the largest. Afterwards $\tilde{F}(x) = 1$.

The relation between $\tilde{f}(x)$ and $\tilde{F}(x)$ is

(4)
$$\tilde{F}(x) = \sum_{t \leq x} \tilde{f}(t)$$

where $t \leq x$ means that for an x we have to sum all those $\tilde{f}(t)$ for which t is less than x or equal to x.

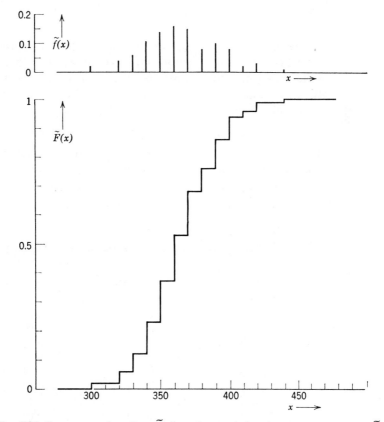

Fig. 521. Frequency function $\tilde{f}(x)$ and cumulative frequency function $\tilde{F}(x)$ of the sample in Table 24.2

Grouping of Samples

If a sample consists of too many numerically different sample values, then its tabular and graphical representations are unnecessarily complicated but may be simplified by the process of **grouping,** as follows.

A sample being given, we choose an interval I that contains all the sample values. We subdivide I into subintervals, called **class intervals.** The midpoints of these intervals are called **class midpoints** or **class marks.** The sample values in each such interval are said to form a **class.** Their number is called the corresponding **absolute class frequency.** Division by the sample size n gives the **relative class frequency.** This frequency, considered as a function $\tilde{f}(x)$ of the class marks, is called the **frequency function of the grouped sample.** The corresponding cumulative relative class frequency, considered as a function $\tilde{F}(x)$ of the class marks, is called the **distribution function of the grouped sample.** An example is shown in Tables 24.3 and 24.4.

Table 24.3
Strength of 50 Lots of Cotton (lb required to break a skein)

114	118	86	107	87	94	82	81	98	84
120	126	98	89	114	83	94	106	96	111
123	110	83	118	83	96	96	74	91	81
102	107	103	80	109	71	96	91	86	129
130	104	86	121	96	96	127	94	102	87

The fewer classes we choose, the simpler the distribution of the grouped sample becomes but the more information we lose, because the original sample values no longer appear explicitly. Grouping should be done so that only unessential details are eliminated. Unnecessary complications in the later use of a grouped sample are avoided by obeying the following rules.

1. All the class intervals should have the same length.

2. The class intervals should be chosen so that the class marks correspond to simple numbers (numbers with few nonzero digits).

3. If a sample value x_j coincides with the common endpoint of two class intervals, take it into the class interval that extends from x_j to the right.

Table 24.4
Frequency Table of the Sample in Table 24.3 (Grouped)

Class Interval	Class Mark x	Absolute Frequency	$\tilde{f}(x)$	$\tilde{F}(x)$
65– 75	70	2	0.04	0.04
75– 85	80	8	0.16	0.20
85– 95	90	11	0.22	0.42
95–105	100	12	0.24	0.66
105–115	110	8	0.16	0.82
115–125	120	5	0.10	0.92
125–135	130	4	0.08	1.00
	Sum	50	1.00	

In the case of populations we have defined numbers, called parameters (the mean μ and variance σ^2 above all) that characterize important properties of the distribution. For samples we can do the same, as we show in the next section, where we introduce the **sample mean** $\bar{x}$ and **sample variance** s^2.

Problems for Sec. 24.3

In each case make a frequency table of the given sample and represent the sample by a dot frequency diagram, a bar chart and a histogram.

1. Resistance [ohms] of resistors

99   100   102   101   98   103   100   102   99   101
100   100   99   101   100   102   99   101   98   100

2. Numbers that turned up on a die

6   2   4   1   2   4   3   3   2   1   6   5   6   3   4

3. Release time [sec] of a relay

1.3   1.4   1.1   1.5   1.4   1.3   1.2   1.4   1.5   1.3
1.2   1.3   1.5   1.4   1.4   1.6   1.3   1.5   1.1   1.4

4. Carbon content [%] of coal

87   86   85   87   86   87   86   81   77   85
86   84   83   83   82   84   83   79   82   73

5. Tensile strength [kg/mm^2] of sheet steel

44   43   41   41   44   44   43   44   42   45   43   43   44   45   46
42   45   41   44   44   43   44   46   41   43   45   45   42   44   44

6. Number of sheets of paper over and under the desired number of 100 sheets per package in a packaging process

0   -1 0   0   1   1   2   0   1   0

7. Miles per gallon of gasoline required by six cars of the same make

15.0   15.5   14.5   15.0   15.5   15.0

8. Weight of filled bags [grams] in an automatic filling process

200   203   199   198   201   200   201   201

9. Waiting time [min, rounded] of a commuter for a train in a certain subway

3   4   1   0   2   2   3   1   5   3

10. Graph the cumulative frequency function of the sample in Prob. 3.

11. Graph the bar chart, the histogram and the frequency polygon of the grouped sample in Table 24.4.

12. Graph the histogram of the following sample of lifetimes [hours] of light bulbs.

Lifetime	Absolute Frequency	Lifetime	Absolute Frequency	Lifetime	Absolute Frequency
950-1050	4	1350-1450	51	1750-1850	20
1050-1150	9	1450-1550	58	1850-1950	9
1150-1250	19	1550-1650	53	1950-2050	3
1250-1350	36	1650-1750	37	2050-2150	1

13. Group the sample in Table 24.1, using class intervals with midpoints 300, 320, 340, $\cdots$. Make up the corresponding frequency table. Graph the histogram and compare it with Fig. 519. Graph the cumulative frequency function.

14. Group the sample shown in Table 24.3, using class intervals with midpoints 75, 85, 95, $\cdots$. Make up the corresponding frequency table. Graph the histogram and compare it with that in Prob. 11.

15. The smallest of 1500 measurements was 10.8 cm and the largest was 11.9 cm. Suggest class intervals for grouping these data.

24.4 Sample Mean, Sample Variance

The frequency function (or equally well the distribution function) characterizes a given sample in detail. From this function we may compute numbers that characterize *certain properties* of the sample, such as the average size of the sample values, the "spread," the "asymmetry," etc. In the present section we consider the two most important such quantities, which are called the sample mean and the sample variance.

The *mean value of a sample* $x_1, x_2, \cdots, x_n$ or, briefly, **sample mean**, is denoted by $\bar{x}$ and is defined by the formula

(1)
$$ \bar{x} = \frac{1}{n} \sum_{j=1}^{n} x_j = \frac{1}{n}(x_1 + x_2 + \cdots + x_n). $$

It is the sum of all the sample values, divided by the size n of the sample. Obviously, it measures the average size of the sample values.

The *variance of a sample* $x_1, x_2, \cdots, x_n$ or, briefly, **sample variance**, is denoted by s^2 and is defined by the formula

(2)
$$ s^2 = \frac{1}{n-1} \sum_{j=1}^{n} (x_j - \bar{x})^2 = \frac{1}{n-1}[(x_1 - \bar{x})^2 + \cdots + (x_n - \bar{x})^2]. $$

It is the sum of the squares of the deviations of the sample values from the mean $\bar{x}$, divided by $n - 1$. It measures the spread or dispersion of the sample values and is positive, except for the rare case when all the sample values are equal (and are then equal to $\bar{x}$). The positive square root of the sample variance s^2 is called the **standard deviation** of the sample and is denoted by s.

EXAMPLE 1. Sample mean and sample variance
Ten randomly selected nails had the lengths [in.]

$$ 0.80 \quad 0.81 \quad 0.81 \quad 0.82 \quad 0.81 \quad 0.82 \quad 0.80 \quad 0.82 \quad 0.81 \quad 0.81. $$

Find the mean and the variance of this sample.

Solution. From (1) we see that the sample mean is

$$ \bar{x} = \tfrac{1}{10}(0.80 + 0.81 + 0.81 + 0.82 + \cdots + 0.81) = 0.811 \text{ [in.]}. $$

Applying (2), we thus obtain the sample variance

$$ s^2 = \tfrac{1}{9}[(0.800 - 0.811)^2 + \cdots + (0.810 - 0.811)^2] = 0.000\ 054 \text{ [in.}^2]. $$

This calculation becomes simpler if we take equal sample values together. Then

$$\bar{x} = \tfrac{1}{10}(2 \cdot 0.80 + 5 \cdot 0.81 + 3 \cdot 0.82) = 0.811.$$

In the parentheses we have the sum of the three *numerically different* sample values $x_1 = 0.80$, $x_2 = 0.81$, $x_3 = 0.82$, each multiplied by its absolute frequency. Similarly,

$$s^2 = \tfrac{1}{9}[2(0.800 - 0.811)^2 + 5(0.810 - 0.811)^2 + 3(0.820 - 0.811)^2] = 0.000\ 054. \quad ∎$$

The example illustrates how we may compute $\bar{x}$ and s^2 by the use of the frequency function $\tilde{f}(x)$ of the sample. If a sample of n values contains precisely m **numerically different** sample values

$$x_1, \qquad\qquad x_2, \qquad\qquad \cdots, \qquad\qquad x_m$$

(where $m \leq n$), the corresponding relative frequencies are

$$\tilde{f}(x_1), \qquad\qquad \tilde{f}(x_2), \qquad\qquad \cdots, \qquad\qquad \tilde{f}(x_m)$$

and the corresponding absolute frequencies needed in the computation are

$$a(x_1) = n\tilde{f}(x_1), \qquad a(x_2) = n\tilde{f}(x_2), \qquad \cdots, \qquad a(x_m) = n\tilde{f}(x_m).$$

We see that (1) now takes the form

(3)
$$\bar{x} = \frac{1}{n} \sum_{j=1}^{m} x_j a(x_j) \qquad\qquad a(x_j) = n\tilde{f}(x_j)$$

and (2) takes the form

(4)
$$s^2 = \frac{1}{n-1} \sum_{j=1}^{m} (x_j - \bar{x})^2 a(x_j).$$

Note that in (1) and (2) we sum over *all* the sample values, whereas now we sum over the *numerically different* sample values. The absolute frequencies $a(x_j)$ are integers, while the relative frequencies $\tilde{f}(x_j)$ may be messy, for example if $n = 23$ or $n = 84$, etc.

This has been the last of four sections on the general idea of mathematical statistics and on handling given samples (without drawing any conclusions about populations). In the remaining sections of this chapter we discuss important basic tasks and statistical methods, beginning with **point estimation of parameters** in the next section.

Problems for Sec. 24.4

1. Compute the mean and variance of the sample in Prob. 2, Sec. 24.3, using (1), (2) or (3), (4).
2. Compute the mean and variance of the sample 6, 8, 20, 8, 8. Notice the large contribution of 20 to s^2 and comment.
3. Compute the mean and variance of the sample in Prob. 4, Sec. 24.3.

4. Graph a histogram of the sample 2, 1, 4, 5 and guess $\bar{x}$ and s by inspecting the histogram. Then compute $\bar{x}$, s^2 and s.

5. To illustrate that s^2 measures the spread, compute s^2 for the two samples 108, 110, 112 and 100, 110, 120, and compare.

6. **(Working origin)** If $x_j = x_j^* + c$, where $j = 1, \cdots, n$ and c is any constant, show that

$$\bar{x} = c + \bar{x}^* \qquad \left(\bar{x}^* = \frac{1}{n} \sum_{j=1}^{n} x_j^* \right), \qquad s^2 = s^{*2}$$

where s^{*2} is the variance of the x_j^*'s. (In practice, c is chosen so that the x_j^*'s are small in absolute value. Geometrically this is a shift of the origin and is called the *method of working origin*.)

7. Apply the method of working origin to the sample in Example 1.

8. **(Full coding)** If $x_j = c_1 x_j^* + c_2$, where $j = 1, \cdots, n$ and c_1 and c_2 are constants, show that

$$\bar{x} = c_1 \bar{x}^* + c_2, \qquad s^2 = c_1^2 s^{*2}$$

where $\bar{x}^*$ and s^{*2} have the same meaning as in Prob. 6. (This is called the *method of full coding*. Obviously it is valuable in computations with a pocket calculator, for instance for quick checks.)

9. Using the method of working origin, compute the mean of the sample in Prob. 1 of Sec. 24.3.

10. Apply full coding to the sample in Example 1.

11. Prove that $s^2 = 0$ if and only if all sample values are equal.

12. Prove that $\bar{x}$ lies between the smallest and the largest sample values.

13. **(Range of a sample)** The difference between the largest and the smallest values in a sample is called the *range of the sample*. Find the range of the sample in Example 1.

14. An advantage of the range is that it can be computed more easily than s^2. Can you think of a disadvantage?

15. **(Percentile, median)** The pth **percentile** of a sample is a number Q_p such that at least $p\%$ of the sample values are smaller than or equal to Q_p and also at least $(100 - p)\%$ of those values are larger than or equal to Q_p. If there is more than one such number (in which case there will be an interval of them), the pth percentile is defined as the average of the numbers (midpoint of that interval). In particular, Q_{50} is called the **middle quartile** or **median** and is denoted by $\tilde{x}$. Find $\tilde{x}$ for the sample in Table 24.2 (Sec. 24.3).

16. The percentiles Q_{25} and Q_{75} of a sample are called the **lower** and **upper quartiles** of the sample, and $Q_{75} - Q_{25}$, which is a measure for the spread, is called the **interquartile range**. Find Q_{25}, Q_{75}, and $Q_{75} - Q_{25}$ for the sample in Table 24.2.

17. Do the same tasks as in Probs. 15 and 16 for the sample in Table 24.3.

18. **(Mode)** A *mode* of a sample is a sample value that occurs most frequently in the sample. Find the mean, median and mode of the following sample. Comment.

Total market value of stocks owned ($):	100	1000	100,000
Absolute frequency (= number of owners):	100	90	20

19. Compute the mean and the variance of the ungrouped sample in Table 24.3 (Sec. 24.3) and the grouped sample in Table 24.4 and compare the results.

20. If a sample is being grouped, its mean will change, in general. Show that the change cannot exceed $L/2$ where L is the length of each class interval.

24.5 Estimation of Parameters

Quantities appearing in distributions, such as p in the binomial distribution and μ and σ in the normal distribution, are called **parameters.**

A **point estimate** of a parameter is a number (point on the real line), which is computed from a given sample and serves as an approximation of the unknown exact value of the parameter. An **interval estimate** is an interval ("*confidence interval*") obtained from a sample; such estimates will be considered in the next section. Estimation of parameters is an important practical problem.

As an approximation of the mean μ of a population we may take the mean $\bar{x}$ of a corresponding sample. This gives the estimate $\hat{\mu} = \bar{x}$ for μ, that is,

$$\text{(1)} \qquad \hat{\mu} = \bar{x} = \frac{1}{n}(x_1 + \cdots + x_n)$$

where n is the size of the sample. Similarly, an estimate $\hat{\sigma}^2$ for the variance of a population is the variance s^2 of a corresponding sample, that is,

$$\text{(2)} \qquad \hat{\sigma}^2 = s^2 = \frac{1}{n-1}\sum_{j=1}^{n}(x_j - \bar{x})^2.$$

Clearly, (1) and (2) are estimates of parameters for distributions in which μ or σ^2 appear explicitly as parameters, such as the normal and Poisson distributions. For the binomial distribution, $p = \mu/n$ [cf. (3) in Sec. 23.6]. In this case, $x_j = 1$ in (1) if the event A whose probability is p occurred in the jth trial, and $x_j = 0$ if A did not occur in that trial. From (1) we thus obtain for p the estimate

$$\text{(3)} \qquad \hat{p} = \frac{\bar{x}}{n}.$$

We mention that (1) is a special case of the so-called **method of moments.** In this method the parameters to be estimated are expressed in terms of the moments of the distribution (cf. Sec. 23.5). In the resulting formulas those moments are replaced by the corresponding moments of the sample. This gives the estimates. Here the **kth moment of a sample** $x_1, \cdots, x_n$ is

$$\boxed{m_k = \frac{1}{n}\sum_{j=1}^{n} x_j^k.}$$

Maximum Likelihood Method

Another method for obtaining estimates is the so-called **maximum likelihood method** of R. A. Fisher (*Messenger Math.* **41,** 1912, 155–160). To explain it, we consider a discrete (or continuous) random variable X whose probability function (or density) $f(x)$ depends on a single parameter θ and take a corresponding sample of n independent values $x_1, \cdots, x_n$. Then in the discrete case the probability that a sample of size n consists precisely of those n values is

(4) $$l = f(x_1)f(x_2) \cdots f(x_n).$$

In the continuous case the probability that the sample consists of values in the small intervals $x_i \leqq x \leqq x_i + \Delta x$ $(i = 1, 2, \cdots, n)$ is

(5) $$f(x_1) \, \Delta x f(x_2) \, \Delta x \cdots f(x_n) \, \Delta x = l(\Delta x)^n.$$

Since $f(x_i)$ depends on θ, the function l depends on $x_1, \cdots, x_n$ and θ. We imagine $x_1, \cdots, x_n$ to be given and fixed. Then l is a function of θ, which is called the **likelihood function.** The basic idea of the maximum likelihood method is very simple, as follows. We choose that approximation for the unknown value of θ for which l is as large as possible. If l is a differentiable function of θ, a necessary condition for l to have a maximum (not at the boundary) is

(6) $$\frac{\partial l}{\partial \theta} = 0.$$

(We write a *partial* derivative, because l depends also on $x_1, \cdots, x_n$.) A solution of (6) depending on $x_1, \cdots, x_n$ is called a **maximum likelihood estimate** for θ. We may replace (6) by

(7) $$\frac{\partial \ln l}{\partial \theta} = 0,$$

because $f(x) \geqq 0$, a maximum of f is in general positive, and $\ln l$ is a monotone increasing function of l. This often simplifies calculations.

If the distribution of X involves r parameters $\theta_1, \cdots, \theta_r$, then instead of (6) we have the r conditions $\partial l/\partial \theta_1 = 0, \cdots, \partial l/\partial \theta_r = 0$, and instead of (7) we have

(8) $$\frac{\partial \ln l}{\partial \theta_1} = 0, \qquad \cdots, \qquad \frac{\partial \ln l}{\partial \theta_r} = 0.$$

EXAMPLE 1. Normal distribution
Find maximum likelihood estimates for μ and σ in the case of the normal distribution.

Solution. From (1), Sec. 23.7, and (4) we obtain

$$l = \left(\frac{1}{\sqrt{2\pi}}\right)^n \left(\frac{1}{\sigma}\right)^n e^{-h} \quad \text{where} \quad h = \frac{1}{2\sigma^2} \sum_{i=1}^{n} (x_i - \mu)^2.$$

Taking logarithms, we have

$$\ln l = -n \ln \sqrt{2\pi} - n \ln \sigma - h.$$

The first equation in (8) is $\partial \ln l/\partial \mu = 0$, written out

$$\frac{\partial \ln l}{\partial \mu} = -\frac{\partial h}{\partial \mu} = \frac{1}{\sigma^2} \sum_{i=1}^{n} (x_i - \mu) = 0, \qquad \text{hence} \qquad \sum_{i=1}^{n} x_i - n\mu = 0.$$

The solution is the desired estimate $\hat{\mu}$ for μ; we find

$$\hat{\mu} = \frac{1}{n} \sum_{i=1}^{n} x_i = \bar{x}.$$

The second equation in (8) is $\partial \ln l/\partial \sigma = 0$, written out

$$\frac{\partial \ln l}{\partial \sigma} = -\frac{n}{\sigma} - \frac{\partial h}{\partial \sigma} = -\frac{n}{\sigma} + \frac{1}{\sigma^3} \sum_{i=1}^{n} (x_i - \mu)^2 = 0.$$

Replacing μ by $\hat{\mu}$ and solving for σ^2, we obtain the estimate

$$\bar{\sigma}^2 = \frac{1}{n} \sum_{i=1}^{n} (x_i - \bar{x})^2$$

which we shall use in Sec. 24.10. Note that this differs from (2). We cannot discuss criteria for the goodness of estimates, but we want to mention that for small n, formula (2) is preferable.

In the next section we discuss interval estimates of parameters ("**confidence intervals**"). This is important not only for practical reasons but also as one of the basic ideas in modern statistical thinking, and the student should follow the next section with particular care.

Problems for Sec. 24.5

1. Find the maximum likelihood estimate for the parameter μ of a normal distribution with known variance $\sigma^2 = \sigma_0^2$.
2. Apply the maximum likelihood method to the normal distribution with $\mu = 0$.
3. (**Binomial distribution**) Derive a maximum likelihood estimate for p.
4. Extend Prob. 3 as follows. Suppose that m times n trials were made and in the first n trials A happened k_1 times, in the second n trials A happened k_2 times, $\cdots$, in the mth n trials A happened k_m times. Find a maximum likelihood estimate of p based on this information.
5. (**Poisson distribution**) Apply the maximum likelihood method to the Poisson distribution.
6. (**Uniform distribution**) Show that in the case of the parameters a and b of the uniform distribution (cf. Sec. 23.5), the maximum likelihood estimate cannot be obtained by equating the first derivative to zero. How can we obtain maximum likelihood estimates in this case?
7. Consider $X = $ *Number of independent trials until an event A occurs.* Show that X has the probability function $f(x) = pq^{x-1}$, $x = 1, 2, \cdots$, where p is the probability of A in a single trial and $q = 1 - p$. Find the maximum likelihood estimate of p corresponding to a single observed value x of X.
8. In Prob. 7, find the maximum likelihood estimate of p resulting from a sample $x_1, \cdots, x_n$.
9. Find the maximum likelihood estimate of θ in the density $f(x) = \theta e^{-\theta x}$ if $x \geq 0$ and $f(x) = 0$ if $x < 0$.
10. In Prob. 9, find the mean μ, substitute it in $f(x)$, find the maximum likelihood estimate of μ, and show that it is identical with the estimate for μ which can be obtained from that for θ in Prob. 9.

11. Compute θ in Prob. 9 from the sample 1.9, 0.4, 0.7, 0.6, 1.4. Graph the sample distribution function $\tilde{F}(x)$ and the distribution function $F(x)$ of the random variable, with $\theta = \hat{\theta}$, on the same axes. Do they agree reasonably well? (We consider goodness of fit systematically in Sec. 24.10.)

12. Using Table A10 in Appendix 4 and the method explained in Prob. 6 of Sec. 24.2, obtain a sample of size 20 from the distribution with density $f(x) = 0.5e^{-0.5x}$ if $x > 0$, $f(x) = 0$ if $x < 0$, and compare the distribution functions of the sample and of the population (as in Prob. 11).

24.6 Confidence Intervals

The last section was devoted to point estimates of parameters, and we shall now discuss **interval estimates,** starting with a general motivation.

Whenever we use mathematical approximation formulas, we should try to find out how much the approximate value can at most deviate from the unknown true value. For example, in the case of numerical integration methods there exist "error formulas" from which we can compute the maximum possible error (that is, the difference between the approximate and the true value). Suppose that in a certain case we obtain 2.47 as an approximate value of a given integral and ± 0.02 as the maximum possible deviation from the unknown exact value. Then we are sure that the values $2.47 - 0.02 = 2.45$ and $2.47 + 0.02 = 2.49$ "*include*" the unknown exact value; that is, 2.45 is smaller than or equal to that value and 2.49 is larger than or equal to that value.

In estimating a parameter θ, the corresponding problem would be the determination of two numerical quantities that depend on the sample values and include the unknown value of the parameter with certainty. However, we already know that from a sample we cannot draw conclusions about the corresponding population that are 100% certain. So we have to be more modest and modify our problem, as follows.

Choose a probability γ close to 1 (for example, $\gamma = 95\%$, 99%, or the like). Then determine two quantities Θ_1 and Θ_2 such that the probability that Θ_1 and Θ_2 include the exact unknown value of the parameter θ is equal to γ.

Here the idea is that we replace the impossible requirement "with certainty" by the attainable requirement "with a preassigned probability close to 1."

Numerical values of those two quantities should be computed from a given sample $x_1, \cdots, x_n$. The n sample values may be regarded as observed values of n random variables $X_1, \cdots, X_n$. Then Θ_1 and Θ_2 are functions of these random variables and therefore random variables, too. Our requirement above may thus be written

$$P(\Theta_1 \leqq \theta \leqq \Theta_2) = \gamma.$$

If we know such functions Θ_1 and Θ_2 and a sample is given, we may compute a numerical value θ_1 of Θ_1 and a numerical value θ_2 of Θ_2. The interval with endpoints θ_1 and θ_2 is called a **confidence interval**[1] or *interval estimate* for the unknown parameter θ, and we shall denote it by

$$\boxed{\text{CONF}\,\{\theta_1 \leqq \theta \leqq \theta_2\}.}$$

The values θ_1 and θ_2 are called **lower** and **upper confidence limits** for θ. The number γ is called the **confidence level.** One chooses $\gamma = 95\%$, 99%, sometimes 99.9%.

Clearly, if we intend to obtain a sample and determine a corresponding confidence interval, then γ is the probability of getting an interval that will include the unknown exact value of the parameter.

For example, if we choose $\gamma = 95\%$, then we can expect that *about* 95% of the samples that we may obtain will yield confidence intervals that do include the value θ, whereas the remaining 5% do not. Hence the statement "the confidence interval includes θ" will be correct in *about* 19 out of 20 cases, while in the remaining case it will be false.

Choosing $\gamma = 99\%$ instead of 95%, we may expect that statement to be correct even in *about* 99 out of 100 cases. But we shall see later that the intervals corresponding to $\gamma = 99\%$ are longer than those corresponding to $\gamma = 95\%$. This is the disadvantage of increasing γ.

What value of γ should we choose in a concrete case? This is not a mathematical question, but one that must be answered from the viewpoint of the application, taking into account what risk of making a false statement we can afford.

It is clear that the uncertainty involved in the present method as well as in the methods yet to be discussed comes from the sampling process, so that the statistician should be prepared for his share of mistakes. However, he is no worse off than a judge or a banker, who is subject to the laws of chance. Quite the contrary, he has the advantage that he can *measure* his chances of making a mistake.

Normal Distribution: Confidence Intervals for μ and σ^2

We shall now consider methods for obtaining confidence intervals for the mean (Tables 24.5, 24.6) and the variance (Table 24.7) of the **normal distribution.** The corresponding theory will be explained in the last part of this section.

EXAMPLE 1. Confidence interval for μ of the normal distribution with known σ^2

Determine a 95% confidence interval for the mean of a normal distribution with variance $\sigma^2 = 9$, using a sample of $n = 100$ values with mean $\bar{x} = 5$.

Solution.   1st Step. $\gamma = 0.95$ is required.

2nd Step. The corresponding c equals 1.960; see Table 24.5, p. 1250.

3rd Step. $\bar{x} = 5$ is given.

[1]The modern theory and terminology of confidence intervals were developed by J. Neyman (*Annals of Mathematical Statistics* **6,** 1935, 111-116). See also footnote 3 in the next section.

In mathematics, $\theta_1 \leqq \theta \leqq \theta_2$ means that θ lies between θ_1 and θ_2, and, to avoid misunderstandings, it seems worthwhile to characterize a confidence interval by a special symbol, such as CONF.

Table 24.5
Determination of a Confidence Interval for the Mean μ
of a Normal Distribution with Known Variance σ^2

1st Step. Choose a confidence level γ (95%, 99% or the like).
2nd Step. Determine the corresponding c:

γ	0.90	0.95	0.99	0.999
c	1.645	1.960	2.576	3.291

3rd Step. Compute the mean $\bar{x}$ of the sample $x_1, \cdots, x_n$.
4th Step. Compute $k = c\sigma/\sqrt{n}$. The confidence interval for μ is

(1) $$\mathrm{CONF}\ \{\bar{x} - k \leqq \mu \leqq \bar{x} + k\}.$$

4th Step. We need $k = 1.960 \cdot 3/\sqrt{100} = 0.588$. Hence $\bar{x} - k = 4.412$, $\bar{x} + k = 5.588$ and the confidence interval is

$$\mathrm{CONF}\ \{4.412 \leqq \mu \leqq 5.588\}.$$

Sometimes this is written $\mu = 5 \pm 0.588$, but we shall not use this notation, which can be misleading.

EXAMPLE 2. Sample size needed for a confidence interval of prescribed length
How large must n in the last example be if we want to obtain a 95% confidence interval of length $L = 0.4$?

Solution. The interval (1) has the length $L = 2k = 2c\sigma/\sqrt{n}$. Solving for n, we obtain

$$n = (2c\sigma/L)^2.$$

In the present case the answer is $n = (2 \cdot 1.960 \cdot 3/0.4)^2 \approx 870$.

Figures 522 and 523 show that L decreases as n increases. The shorter the confidence interval should be, the larger the sample size n must be chosen. ∎

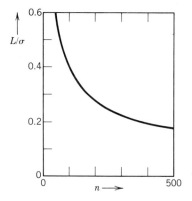

Fig. 522. Length of the confidence interval (1) with $\gamma = 95\%$ (measured in multiples of σ) as a function of the sample size n

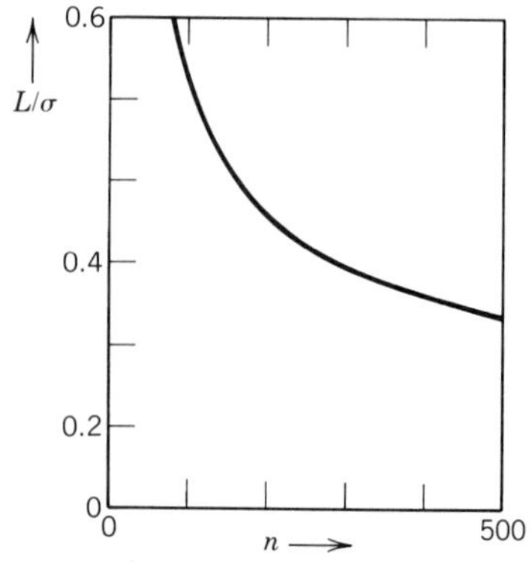

Fig. 523. Length of the confidence interval (1) with $\gamma = 99\%$ (measured in multiples of σ) as a function of the sample size n

Table 24.6 shows how to determine a confidence interval for the mean μ of a normal distribution with unknown variance σ^2. The steps are similar to those in Table 24.5, but k in Table 24.6 is different from that in Table 24.5. Moreover, c depends on n and must be determined from Table A11 in Appendix 4, which contains values z corresponding to given values of the distribution function

$$F(z) = K_m \int_{-\infty}^{z} \left(1 + \frac{u^2}{m}\right)^{-(m+1)/2} du$$

of the so-called **t-distribution** of Student.[2] Here K_m is a constant, namely, $K_m = \Gamma(\tfrac{1}{2}m + \tfrac{1}{2})/[\sqrt{m\pi}\,\Gamma(\tfrac{1}{2}m)]$, and $\Gamma(\alpha)$ is the gamma function [cf. (24) in Appendix 3]. $m\ (= 1, 2, \cdots)$ is a parameter called the **number of degrees of freedom** of the distribution.

EXAMPLE 3. Confidence interval for μ of the normal distribution with unknown σ^2
Using the sample in Table 24.2, Sec. 24.3, determine a 99% confidence interval for the mean μ of the corresponding population, assuming that the population is normal. (This assumption will be justified in Sec. 24.10.)

Solution.   1st Step. $\gamma = 0.99$ is required.

2nd Step. Since $n = 100$, we obtain $c = 2.63$.

3rd Step. Computation gives $\bar{x} = 364.70$ and $s = \sqrt{720.1} = 26.83$.

4th Step. We find $k = 26.83 \cdot 2.63/10 = 7.06$. Hence the confidence interval is

$$\text{CONF } \{357.64 \leq \mu \leq 371.76\}.$$

For comparison, if σ were known and equal to 26.83, Table 24.5 would give the value $k = 2.576 \cdot 26.83/\sqrt{100} = 6.91$, and CONF $\{357.79 \leq \mu \leq 371.61\}$. This differs but little from the preceding result because n is large. For smaller n the difference would be considerable, as Figs. 524 and 525 illustrate.   ∎

Table 24.6
Determination of a Confidence Interval for the Mean μ
of a Normal Distribution with Unknown Variance σ^2

1st Step. Choose a confidence level γ (95%, 99% or the like).

2nd Step. Determine the solution c of the equation

$$(2) \qquad F(c) = \tfrac{1}{2}(1 + \gamma)$$

from the table of the t-distribution with $n - 1$ degrees of freedom (Table A11 in Appendix 4; n = sample size).

3rd Step. Compute the mean $\bar{x}$ and the variance s^2 of the sample $x_1, \cdots, x_n$.

4th Step. Compute $k = sc/\sqrt{n}$. The confidence interval is

$$(3) \qquad \text{CONF } \{\bar{x} - k \leq \mu \leq \bar{x} + k\}.$$

---

[2]Pseudonym for WILLIAM SEALY GOSSET (1876—1937), English statistician, who discovered the t-distribution in 1907—1908.

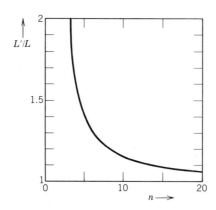

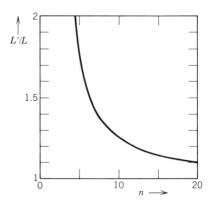

Fig. 524. Ratio of the lengths L' and L of the confidence intervals (3) and (1) with $\gamma = 95\%$ as a function of the sample size n for equal s and σ

Fig. 525. Ratio of the lengths L' and L of the confidence intervals (3) and (1) with $\gamma = 99\%$ as a function of the sample size n for equal s and σ

The steps in Table 24.7 for obtaining a confidence interval for the variance of a normal distribution are similar to those in Tables 24.5 and 24.6, but we now have to determine two numbers c_1 and c_2. Both numbers are obtained from Table A12 in Appendix 4, which contains values z corresponding to given values of the distribution function $F(z) = 0$ if $z < 0$ and

$$F(z) = C_m \int_0^z e^{-u/2} u^{(m-2)/2} \, du \quad \text{if} \quad z \geqq 0.$$

This is the distribution function of the so-called χ^2-**distribution** (*chi-square distribution*); here $C_m = 1/[2^{m/2}\Gamma(\tfrac{1}{2}m)]$ and $m \ (= 1, 2, \cdots)$ is a parameter that is called the *number of degrees of freedom* of the distribution.

Table 24.7
Determination of a Confidence Interval for the Variance σ^2 of a Normal Distribution, Whose Mean Need Not Be Known

---

1st Step. Choose a confidence level γ (95%, 99% or the like).

2nd Step. Determine solutions c_1 and c_2 of the equations

(4) $\qquad F(c_1) = \tfrac{1}{2}(1 - \gamma), \qquad F(c_2) = \tfrac{1}{2}(1 + \gamma)$

from the table of the chi-square distribution with $n - 1$ degrees of freedom (Table A12 in Appendix 4; $n = $ sample size).

3rd Step. Compute $(n - 1)s^2$, where s^2 is the variance of the sample $x_1, \cdots, x_n$.

4th Step. Compute $k_1 = (n - 1)s^2/c_1$ and $k_2 = (n - 1)s^2/c_2$. The confidence interval is

(5) $\qquad\qquad\qquad \text{CONF } \{k_2 \leqq \sigma^2 \leqq k_1\}.$

EXAMPLE 4. Confidence interval for the variance of the normal distribution
Using the sample in Table 24.2, Sec. 24.3, determine a 95% confidence interval for the variance of the corresponding population.

Solution.   1st Step. $\gamma = 0.95$ is required.

2nd Step. Since $n = 100$, we find $c_1 = 73.4$ and $c_2 = 128$.

3rd Step. From Table 24.2 we compute $99s^2 = 71\ 291$.

4th Step. The confidence interval is

$$\text{CONF }\{556 \leqq \sigma^2 \leqq 972\}. \qquad\blacksquare$$

Other distributions. Confidence intervals for the mean and variance of other distributions may be obtained by using the previous methods and *sufficiently large samples. Practically* speaking, if the sample indicates that the skewness of the unknown distribution is small, one should take samples of size $n = 20$ at least for obtaining confidence intervals for μ and samples of size $n = 50$ at least for obtaining confidence intervals for σ^2. The reason for this method will be explained at the end of this section.

Theoretical Basis of the Methods in Tables 24.5–24.7

We shall now discuss the theory which justifies our methods for obtaining confidence intervals, using the following simple but very important idea.

So far we have regarded the values $x_1, \cdots, x_n$ of a sample as n observed values of a single random variable X. We may equally well regard these n values as single observations of n random variables $X_1, \cdots, X_n$ that have the same distribution (the distribution of X) and are independent because the sample values are assumed to be independent.

For deriving (1) in Table 24.5 we need

Theorem 1 (Sum of independent normal random variables)
Suppose that $X_1, X_2, \cdots, X_n$ are independent normal random variables with means $\mu_1, \mu_2, \cdots, \mu_n$ and variances $\sigma_1^2, \sigma_2^2, \cdots, \sigma_n^2$, respectively. Then the random variable

$$X = X_1 + X_2 + \cdots + X_n$$

is normal with the mean

$$\mu = \mu_1 + \mu_2 + \cdots + \mu_n$$

and the variance

$$\sigma^2 = \sigma_1^2 + \sigma_2^2 + \cdots + \sigma_n^2.$$

The statements about μ and σ follow directly from Theorems 1 and 3 in Sec. 23.8. The proof that X is normal can be found in Refs. [G4] or [G15] listed in Appendix 1.

From this theorem, Theorem 1 in Sec. 23.7 and Theorem 3 in Sec. 23.5 we obtain the following theorem about the distribution of the sum of independent identically distributed normal random variables.

Theorem 2

If $X_1, \cdots, X_n$ are independent normal random variables each of which has mean μ and variance σ^2, then the random variable

$$(6) \qquad \boxed{\overline{X} = \frac{1}{n}(X_1 + \cdots + X_n)}$$

is normal with the mean μ and the variance σ^2/n, and the random variable

$$(7) \qquad \boxed{Z = \sqrt{n}\,\frac{\overline{X} - \mu}{\sigma}}$$

is normal with the mean 0 and the variance 1.

Let us derive (1). From the general motivation at the beginning of this section we recall that our goal is to find two random variables Θ_1 and Θ_2 such that

$$(8) \qquad P(\Theta_1 \le \mu \le \Theta_2) = \gamma,$$

where γ is chosen, and the sample gives observed values θ_1 of Θ_1 and θ_2 of Θ_2, which then yield a confidence interval CONF $\{\theta_1 \le \mu \le \theta_2\}$. In the present case this can be done as follows. We choose a number γ between 0 and 1 and determine c from Table A9, Appendix 4, such that $P(-c \le Z \le c) = \gamma$. (When $\gamma = 0.90$, etc., we obtain the values c in Table 24.5.) The inequality $-c \le Z \le c$ with Z given by (7) is

$$-c \le \sqrt{n}(\overline{X} - \mu)/\sigma \le c$$

and can be transformed into an inequality for μ. In fact, multiplication by $\sigma/\sqrt{n}$ yields $-k \le \overline{X} - \mu \le k$, where $k = c\sigma/\sqrt{n}$. Multiplying by -1 and adding $\overline{X}$, we get

$$(9) \qquad \overline{X} + k \ge \mu \ge \overline{X} - k.$$

Thus $P(-c \le Z \le c) = \gamma$ is equivalent to $P(\overline{X} - k \le \mu \le \overline{X} + k) = \gamma$. This is of the form (8) with $\Theta_1 = \overline{X} - k$ and $\Theta_2 = \overline{X} + k$. Under our assumptions it means that with probability γ the random variables $\overline{X} - k$ and $\overline{X} + k$ will assume values that include the unknown mean μ. Regarding the sample values $x_1, \cdots, x_n$ in Table 24.5 as observed values of n independent normal random variables $X_1, \cdots, X_n$, we see that the sample mean $\overline{x}$ is an observed value of (6), and by inserting this value into (9) we obtain (1). ∎

For deriving (3) in Table 24.6 we need

Theorem 3

Let $X_1, \cdots, X_n$ be independent normal random variables with the same mean μ and the same variance σ^2. Then the random variable

(10)
$$T = \sqrt{n}\,\frac{\bar{X} - \mu}{S}$$

has a t-distribution (cf. p. 1251) with $n - 1$ *degrees of freedom; here* $\bar{X}$ *is given by* (6) *and*

(11)
$$S^2 = \frac{1}{n - 1} \sum_{j=1}^{n} (X_j - \bar{X})^2.$$

Proof in Refs. [G4] or [G15] in Appendix 1.

The derivation of (3) is similar to that of (1). We choose a number γ between 0 and 1 and determine a number c from Table A11, Appendix 4, with $n - 1$ degrees of freedom such that

(12)
$$P(-c \leqq T \leqq c) = F(c) - F(-c) = \gamma.$$

Since the *t*-distribution is symmetric, we have $F(-c) = 1 - F(c)$, and (12) assumes the form (2). Transforming $-c \leqq T \leqq c$ in (12) as before, we get

(13)
$$\bar{X} - K \leqq \mu \leqq \bar{X} + K \qquad \text{where} \qquad K = cS/\sqrt{n},$$

and (12) becomes $P(\bar{X} - K \leqq \mu \leqq \bar{X} + K) = \gamma$. By inserting the observed values $\bar{x}$ of $\bar{X}$ and s^2 of S^2 into (13) we obtain (3).   ∎

For deriving (5) in Table 24.7 we need

Theorem 4
Under the assumptions in Theorem 3 the random variable

(14)
$$Y = (n - 1)\,\frac{S^2}{\sigma^2}$$

where S^2 *is given by* (11), *has a chi-square distribution (cf. p. 1252) with* $n - 1$ *degrees of freedom.*

Proof in Refs. [G4] or [G15] in Appendix 1.

The derivation of (5) is similar to that of (1) and (3). We choose a number γ between 0 and 1 and determine c_1 and c_2 from Table A12, Appendix 4, such that [cf. (4)]

$$P(Y \leqq c_1) = F(c_1) = \tfrac{1}{2}(1 - \gamma), \qquad P(Y \leqq c_2) = F(c_2) = \tfrac{1}{2}(1 + \gamma).$$

Subtraction yields

$$P(c_1 \leqq Y \leqq c_2) = P(Y \leqq c_2) - P(Y \leqq c_1) = \gamma.$$

Transforming $c_1 \leqq Y \leqq c_2$ with Y given by (14) into an inequality for σ^2, we obtain

$$\frac{n-1}{c_2} S^2 \leqq \sigma^2 \leqq \frac{n-1}{c_1} S^2.$$

By inserting the observed value s^2 of S^2 we obtain (5).   ∎

Confidence Intervals for Other Distributions

In the case of other distributions we may also obtain confidence intervals by the methods in Tables 24.5 and 24.7, but in this case we must use large samples. This follows from the basic

Theorem 5 (Central limit theorem)

Let $X_1, \cdots, X_n, \cdots$ be independent random variables that have the same distribution function and therefore the same mean μ and the same variance σ^2. Let $Y_n = X_1 + \cdots + X_n$. Then the random variable

(15)
$$\boxed{Z_n = \frac{Y_n - n\mu}{\sigma\sqrt{n}}}$$

*is **asymptotically normal** with mean 0 and variance 1; that is, the distribution function $F_n(x)$ of Z_n satisfies*

$$\lim_{n\to\infty} F_n(x) = \Phi(x) = \frac{1}{\sqrt{2\pi}} \int_{-\infty}^{x} e^{-u^2/2} \, du.$$

A proof can be found in Ref. [G4] listed in Appendix 1.

We know that if $X_1, \cdots, X_n$ are independent random variables with the same mean μ and the same variance σ^2, then their sum $X = X_1 + \cdots + X_n$ has the following properties.

(A) X has the mean $n\mu$ and the variance $n\sigma^2$ (cf. Theorems 1 and 3 in Sec. 23.8).

(B) If those variables are normal, then X is normal (cf. Theorem 1).

If those variables are not normal, then (B) fails to hold, but if n is large, then X is approximately normal (cf. Theorem 5) and this justifies the application of methods for the normal distribution to other distributions, but in such a case we have to use large samples.

The idea of confidence intervals that we have just discussed and applied is characteristic of modern statistics, and the other equally (or even more) important idea is that of **statistical tests,** to which we turn next.

Problems for Sec. 24.6

1. Find a 95% confidence interval for the mean μ of a normal population with standard deviation 2.5, using the sample 16, 12, 19, 10, 15.

2. Obtain a 95% confidence interval for the mean μ of a normal population with variance $\sigma^2 = 16$, using a sample of size 300 with mean 87.

3. Find a 99% confidence interval for the mean μ of a normal population with variance $\sigma^2 = 1.69$, using a sample of size 36 with mean 18.4.

4. What sample size would be needed to produce a 95% confidence interval (1) of length (a) 2σ, (b) σ?

Assuming that the populations from which the following samples are taken are normal, determine a 95% confidence interval for the mean μ of the population.

5. Nitrogen content [%] of steel 0.74, 0.75, 0.73, 0.75, 0.74, 0.72.

6. A sample of diameters of 10 gaskets with mean 4.37 cm and standard deviation 0.157 cm.

7. Density [g/cm^3] of coke 1.40, 1.45, 1.39, 1.44, 1.38.

8. What sample size should we use in Prob. 6 if we want to obtain a confidence interval of length 0.1?

9. The specific heat of iron was measured 41 times at a temperature of 25 °C. The sample had the mean 0.106 [cal/g °C] and the standard deviation 0.002 [cal/g °C]. What can we assert with a probability of 99% about the possible size of the error if we use that sample mean to estimate the true specific heat of iron?

10. Find a 95% confidence interval for the percentage of cars on a certain highway that have poorly adjusted brakes, using a random sample of 1000 cars stopped at a roadblock on that highway, 188 of which had poorly adjusted brakes.

Assuming that the populations from which the following samples are taken are normal, determine a 99% confidence interval for the variance σ^2 of the population.

11. A sample of size $n = 128$ with variance $s^2 = 1.921$.

12. Rockwell hardness of tool bits 64.9, 64.1, 63.8, 64.0.

13. The sample in Prob. 5.

Assuming that the populations from which the following samples are taken are normal, determine a 95% confidence interval for the variance σ^2 of the population.

14. Carbon monoxide emission [grams per mile] of a certain type of passenger car [cruising at 55 mph]: 17.3, 17.8, 18.0, 17.7, 18.2, 17.4, 17.6, 18.1

15. Mean energy [keV] of delayed neutron group [Group 3, half-life 6.2 sec.] for uranium U^{235} fission: 435, 451, 430, 444, 438

16. Ultimate tensile strength [kpsi] of alloy steel (Maraging H) at room temperature: 251, 255, 258, 253, 253, 252, 250, 252, 255, 256

17. If X is normal with mean 12 and variance 25, what distributions do $-X$, $2X$ and $4X - 1$ have?

18. If the weight X of bags of sand is normally distributed with a mean of 40 kg and a standard deviation of 2 kg, how many bags can a delivery truck carry so that the probability of the total load exceeding 2000 kg will be 5%?

19. If X_1 and X_2 are independent normal random variables with mean 6 and -2 and variance 4 and 6, respectively, what distribution does $4X_1 - X_2$ have?

20. Find out in what fraction of random matches of holes and bolts the bolts would fit, assuming that the diameter D_1 of the holes is normal with mean $\mu_1 = 1$ in. and standard deviation $\sigma_1 = 0.01$ in. and the diameter D_2 of the bolts is normal with mean $\mu_2 = 0.99$ in. and standard deviation $\sigma_2 = 0.01$ in. (Use Theorem 1.)

24.7 Testing of Hypotheses, Decisions

A statistical **hypothesis** is an assumption about the distribution of a random variable, for example, that a certain distribution has mean 20.3, etc. A statistical **test**[3] of a hypothesis is a procedure in which a sample is used to find out whether we may **"not reject"** (**"accept"**) the hypothesis, that is, act as though it is true, or whether we should **"reject"** it, that is, act as though it is false.

These tests are applied quite frequently, and we may ask why they are important. Often we have to make decisions in situations where chance variation plays a role. If we have a choice between, say, two possibilities, the decision between them might be based on the result of some test.

For example, if we want to use a certain lathe for producing bolts whose diameter should lie between given limits and we allow at most 2% defective bolts, we may take a sample of 100 bolts produced on that lathe and use it for testing the hypothesis $\sigma^2 = \sigma_0^2$ that the variance σ^2 of the corresponding population has a certain value σ_0^2, which we choose such that we can expect to obtain not more than 2% defectives. A meaningful *alternative* in this case is $\sigma^2 > \sigma_0^2$. Depending on the result of the test, we either do not reject the hypothesis $\sigma^2 = \sigma_0^2$ (and then use that lathe) or we reject it, assert that $\sigma^2 > \sigma_0^2$, and use a better lathe. In the latter case we say that the test indicates a **significant deviation** of σ^2 from σ_0^2, that is, a deviation that is not caused merely by the unavoidable influence of chance factors but by the lack of precision of the lathe.

In other cases we may want to compare two things, for example, two different medications, two methods of performing a certain work, the accuracy of two methods of measurement, the quality of products produced by two different tools, etc. Depending on the result of a suitable test, we decide to use one of the two medications, to introduce the better method of working, etc.

Typical sources for hypotheses are as follows.

1. The hypothesis comes from a quality requirement. (Experience about attainable quality may be gained by producing a large number of items with special care.)

2. The hypothesis is based on values known from previous experience.

3. The hypothesis results from a theory one wants to verify.

4. The hypothesis is a pure guess caused by occasional observations.

Let us start with a simple introductory example.

EXAMPLE 1. Test of a hypothesis

The birth of a single child may be regarded as a random experiment with two possible outcomes, namely, B: *Birth of a boy* and G: *Birth of a girl*. Intuitively we should feel that both outcomes are about equally likely. However, in the literature it is often claimed that births of boys are somewhat more frequent than births of girls. On the basis of this situation we want to test the

---

[3]A systematic theory of tests was developed by the American statistician JERZY NEYMAN (1894—1981) and the English statistician EGON SHARPE PEARSON (1895—1980), the son of Karl Pearson (cf. footnote 6), beginning around 1930.

hypothesis that the two outcomes B and G have the same probability. If we let p denote the probability of the outcome B, our hypothesis to be tested is $p = 50\% = 0.5$. Because of those claims we choose the alternative $p > 0.5$.

For the test we use a sample of $n = 3000$ babies from a city of about 250 000 inhabitants; 1578 of these babies were boys.

If the hypothesis is true, we expect that in a sample of $n = 3000$ births there will be *about* 1500 boys. If the alternative holds, then we expect more than 1500 boys, on the average. Hence if the number of boys actually observed is much larger than 1500, we can use this as an indication that the hypothesis may be false, and we reject it.

To perform our test, we proceed as follows. We first determine a critical value c. Because of the alternative, c will be greater than 1500. (A method for determining c will be given below.) Then, if the observed number of boys is greater than c, we reject the hypothesis. If that number is not greater than c, we do not reject the hypothesis.

The basic question now is how we should choose c, that is, where we should draw the line between small random deviations and large significant deviations. Different people may have different opinions, and to answer the question we must use mathematical arguments. In the present case these are very simple, as we now show.

We determine c such that if the hypothesis is true, the probability of observing more than c boys in a sample of 3000 single births is a very small number, call it α. It is customary to choose $\alpha = 1\%$ or 5%. Choosing $\alpha = 1\%$ (or 5%) we risk about once in 100 cases (in 20 cases, respectively) rejecting a hypothesis even though it is true. We shall return to this point later. Let us choose $\alpha = 1\%$ and consider the random variable

$$X = Number\ of\ boys\ in\ 3000\ births.$$

Assuming that the hypothesis is true, we obtain the critical value c from the equation

(1) $$P(X > c)_{p=0.5} = \alpha = 0.01.$$

(That assumption is indicated by the subscript $p = 0.5$.) If the observed value 1578 is greater than c, we reject the hypothesis. If $1578 \leqq c$, we do not reject the hypothesis.

To determine c from (1) we must know the distribution of X. For our purpose the binomial distribution is a sufficiently accurate model. Hence if the hypothesis is true, X has a binomial distribution with $p = 0.5$ and $n = 3000$. This distribution can be approximated by the normal distribution with mean $\mu = np = 1500$ and variance $\sigma^2 = npq = 750$; cf. Sec. 23.7. (For the sake of simplicity we shall disregard the term 0.5 in (11), Sec. 23.7.) The curve of the density is shown in Fig. 526. Using (1), we thus obtain

$$P(X > c) = 1 - P(X \leqq c) \approx 1 - \Phi\left(\frac{c - 1500}{\sqrt{750}}\right) = 0.01.$$

Table A9 in Appendix 4 yields $(c - 1500)/\sqrt{750} = 2.326$. Hence $c = 1564$. Since $1578 > c$, we reject the hypothesis and assert that $p > 0.5$. This completes the test.   ∎

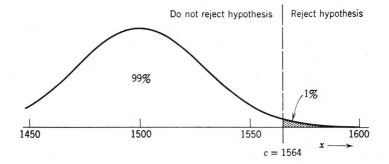

Fig. 526. (Approximate) density of X in Example 1 if the hypothesis is true.
Critical value $c = 1564$

Notion of an Alternative. Kinds of Alternatives

The hypothesis to be tested is sometimes called the *null hypothesis,* and a counterassumption (such as $p > 0.5$ in Example 1) is called an *alternative hypothesis* or, briefly, an **alternative.** The number α (or $100\alpha\%$) is called the **significance level** of the test. c is called the *critical value.* The region containing the values for which we reject the hypothesis is called the **rejection region** or **critical region.** The region of values for which we do not reject the hypothesis is called the **acceptance region.** A frequent choice for α is 5%.

Let θ be an unknown parameter in a distribution, and suppose that we want to test the hypothesis $\theta = \theta_0$. There are three main kinds of alternatives, namely

(2) $\theta > \theta_0$

(3) $\theta < \theta_0$

(4) $\theta \neq \theta_0$.

(2) and (3) are called **one-sided alternatives,** and (4) is called a **two-sided alternative.** (2) is of the type considered in Example 1 (where $\theta_0 = p = 0.5$ and $\theta = p > 0.5$); c lies to the right of θ_0 and the rejection region extends from c to ∞ (Fig. 527, upper part). The test is called a **right-sided test.** In the case of (3) the number c lies to the left of θ_0, the rejection region extends from c to $-\infty$ (Fig. 527, middle part) and the test is called a **left-sided test.** Tests of both kinds are called **one-sided tests.** In the case of (4) we have two critical values c_1 and c_2 ($> c_1$), the rejection region extends from c_1 to $-\infty$ and from c_2 to ∞, and the test is called a **two-sided test.**

All three kinds of alternatives are of practical importance. For example, (3) may appear in connection with testing strength of material. θ_0 may then be the required strength, and the alternative characterizes an undesirable weakness. The case that the material may be stronger than required is acceptable, of course, and does therefore not need special attention. (4) may be important, for example, in connection with the diameter of an axle shaft. Then θ_0 is the required diameter, and slimmer axle-shafts are as bad as thicker ones, so that one has to watch for undesirable deviations from θ_0 in both directions.

Types of Errors in Tests

We shall now consider the **risks of making false decisions** in a test of a hypothesis $\theta = \theta_0$ against an alternative which, for the sake of simplicity, is a single number[4] θ_1. Let $\theta_1 > \theta_0$ so that we have a right-sided test, the consideration for a left-sided or a two-sided test being similar. From a given sample $x_1, \cdots, x_n$ we compute a value $\hat{\theta} = g(x_1, \cdots, x_n)$. If $\hat{\theta} > c$, the hypothesis is rejected (as in Example 1). If $\hat{\theta} \leqq c$, the hypothesis is not rejected. The value $\hat{\theta}$ can be regarded as an observed value of the random variable

---

[4]This standard notation has absolutely nothing to do with the use of the notation θ_1 in connection with confidence intervals in Sec. 24.6.

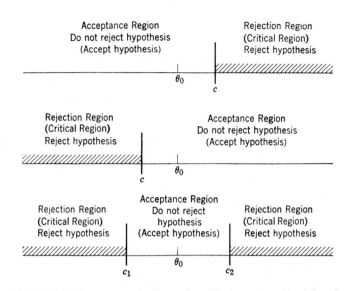

Fig. 527. Test in the case of alternative (2) (upper part of the figure), alternative (3) (middle part), and alternative (4)

$$\hat{\Theta} = g(X_1, \cdots, X_n)$$

because x_j may be regarded as an observed value of X_j, $j = 1, \cdots, n$. In this test there are two possibilities of making an error, as follows.

Type I error (cf. Table 24.8 on p. 1262). The hypothesis is true but is rejected (hence the alternative is accepted) because Θ assumes a value $\hat{\theta} > c$. Obviously, the probability of making such an error equals

(5)
$$P(\hat{\Theta} > c)_{\theta = \theta_0} = \alpha,$$

the significance level of the test.

Type II error (cf. Table 24.8). The hypothesis is false but is accepted (is not rejected) because $\hat{\Theta}$ assumes a value $\hat{\theta} \le c$. The probability of making such an error is denoted by β; thus

(6)
$$P(\hat{\Theta} \le c)_{\theta = \theta_1} = \beta.$$

$\eta = 1 - \beta$ is called the **power** of the test. Obviously, this is the probability of avoiding a Type II error.

Formulas (5) and (6) show that both α and β depend on c, and we would like to choose c so that these probabilities of making errors are as small as possible. But Fig. 528 (p. 1262) shows that these are conflicting requirements because, to let α decrease we must shift c to the right, but then β increases. In practice we first choose α (5%, sometimes 1%), then determine c and

Table 24.8
Type I and Type II Errors in Testing a Hypothesis $\theta = \theta_0$
Against an Alternative $\theta = \theta_1$

		Unknown Truth	
		$\theta = \theta_0$	$\theta = \theta_1$
Accepted	$\theta = \theta_0$	True decision $P = 1 - \alpha$	Type II error $P = \beta$
	$\theta = \theta_1$	Type I error $P = \alpha$	True decision $P = 1 - \beta$

finally compute β. If β is large so that the power $\eta = 1 - \beta$ is small, we should repeat the test, choosing a larger sample, for reasons that will appear shortly.

If the alternative is not a single number but is of the form (2)–(4), then β becomes a function of θ. This function $\beta(\theta)$ is called the **operating characteristic** (OC) of the test and its curve the **OC curve.** Clearly, in this case $\eta = 1 - \beta$ also depends on θ, and this function $\eta(\theta)$ is called the **power function** of the test.

Of course, from a test that leads to the acceptance of a certain hypothesis θ_0 it does *not* follow that this is the only possible hypothesis or the best possible hypothesis. Hence the terms **"not reject"** or **"fail to reject"** are perhaps better than the term **"accept."**

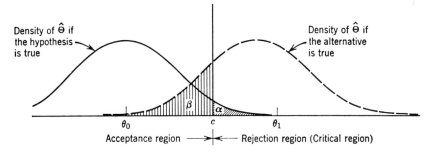

Fig. 528. Illustration of Type I and II errors in testing a hypothesis $\theta = \theta_0$ against an alternative $\theta = \theta_1 \, (> \theta_0)$

Tests in the Case of the Normal Distribution

The following examples will explain tests of practically important hypotheses.

EXAMPLE 2. Test for the mean of the normal distribution with known variance
Let X be a normal random variable with variance $\sigma^2 = 9$. Using a sample of size $n = 10$ with mean $\bar{x}$, test the hypothesis $\mu = \mu_0 = 24$ against the three kinds of alternatives, namely,

$$(a) \quad \mu > \mu_0 \quad (b) \quad \mu < \mu_0 \quad (c) \quad \mu \neq \mu_0.$$

Solution. We choose the significance level $\alpha = 0.05$. An estimate of the mean will be obtained from

$$\bar{X} = \frac{1}{n}(X_1 + \cdots + X_n).$$

If the hypothesis is true, $\bar{X}$ is normal with mean $\mu = 24$ and variance $\sigma^2/n = 0.9$, cf. Theorem 2, Sec. 24.6. Hence we may obtain the critical value c from Table A9 in Appendix 4.

Case (a). We determine c from $P(\bar{X} > c)_{\mu=24} = \alpha = 0.05$, that is,

$$P(\bar{X} \leq c)_{\mu=24} = \Phi\left(\frac{c - 24}{\sqrt{0.9}}\right) = 1 - \alpha = 0.95.$$

Table A9 in Appendix 4 yields $(c - 24)/\sqrt{0.9} = 1.645$, and $c = 25.56$, which is greater than μ_0, as in the upper part of Fig. 527. If $\bar{x} \leq 25.56$, the hypothesis is not rejected. If $\bar{x} > 25.56$, it is rejected. The power of the test is (cf. Fig. 529)

(7)
$$\eta(\mu) = P(\bar{X} > 25.56)_\mu = 1 - P(\bar{X} \leq 25.56)_\mu$$

$$= 1 - \Phi\left(\frac{25.56 - \mu}{\sqrt{0.9}}\right) = 1 - \Phi(26.94 - 1.05\mu).$$

Case (b). The critical value c is obtained from the equation

$$P(\bar{X} \leq c)_{\mu=24} = \Phi\left(\frac{c - 24}{\sqrt{0.9}}\right) = \alpha = 0.05.$$

Table A9 in Appendix 4 yields $c = 24 - 1.56 = 22.44$. If $\bar{x} \geq 22.44$, we do not reject the hypothesis. If $\bar{x} < 22.44$, we reject it. The power of the test is

(8)
$$\eta(\mu) = P(\bar{X} \leq 22.44)_\mu = \Phi\left(\frac{22.44 - \mu}{\sqrt{0.9}}\right) = \Phi(23.65 - 1.05\mu).$$

Case (c). Since the normal distribution is symmetric, we choose c_1 and c_2 equidistant from $\mu = 24$, say, $c_1 = 24 - k$ and $c_2 = 24 + k$, and determine k from

$$P(24 - k \leq \bar{X} \leq 24 + k)_{\mu=24} = \Phi\left(\frac{k}{\sqrt{0.9}}\right) - \Phi\left(-\frac{k}{\sqrt{0.9}}\right) = 1 - \alpha = 0.95.$$

Table A9 in Appendix 4 gives $k/\sqrt{0.9} = 1.960$, $k = 1.86$. Hence $c_1 = 24 - 1.86 = 22.14$ and $c_2 = 24 + 1.86 = 25.86$. If $\bar{x}$ is not smaller than c_1 and not greater than c_2, we do not reject the hypothesis. Otherwise we reject it. The power of the test is (cf. Fig. 529)

$$\eta(\mu) = P(\bar{X} < 22.14)_\mu + P(\bar{X} > 25.86)_\mu = P(\bar{X} < 22.14)_\mu + 1 - P(\bar{X} \leq 25.86)_\mu$$

(9)
$$= 1 + \Phi\left(\frac{22.14 - \mu}{\sqrt{0.9}}\right) - \Phi\left(\frac{25.86 - \mu}{\sqrt{0.9}}\right)$$

$$= 1 + \Phi(23.34 - 1.05\mu) - \Phi(27.26 - 1.05\mu).$$

Consequently, the operating characteristic $\beta(\mu) = 1 - \eta(\mu)$ (see before) is (cf. Fig. 530)

$$\beta(\mu) = \Phi(27.26 - 1.05\mu) - \Phi(23.34 - 1.05\mu).$$

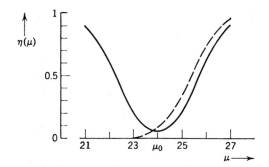

Fig. 529. Power $\eta(\mu)$ in Example 2, case (a) (dashed) and case (c)

If we take a larger sample, say, of size $n = 100$ (instead of 10), then $\sigma^2/n = 0.09$ (instead of 0.9) and the critical values are $c_1 = 23.41$ and $c_2 = 24.59$, as can be readily verified. Then the operating characteristic of the test is

$$\beta(\mu) = \Phi\left(\frac{24.59 - \mu}{\sqrt{0.09}}\right) - \Phi\left(\frac{23.41 - \mu}{\sqrt{0.09}}\right)$$

$$= \Phi(81.97 - 3.33\mu) - \Phi(78.03 - 3.33\mu).$$

Figure 530 shows that the corresponding OC curve is steeper than that for $n = 10$. This means that the increase of n has led to an improvement of the test. In any practical case, n is chosen as small as possible but so large that the test brings out deviations between μ and μ_0 that are of practical interest. For instance, if deviations of ± 2 units are of interest, we see from Fig. 530 that $n = 10$ is much too small because when $\mu = 24 - 2 = 22$ or $\mu = 24 + 2 = 26$, then β is almost 50%. On the other hand, we see that $n = 100$ will be sufficient for that purpose.

EXAMPLE 3. Test for the mean of the normal distribution with unknown variance

The tensile strength of a sample of $n = 16$ manila ropes (diameter 3 in.) was measured. The sample mean was $\bar{x} = 4482$ kg, and the sample standard deviation was $s = 115$ kg (N. C. Wiley, 41st Annual Meeting of the American Society for Testing Materials). Assuming that the tensile strength is a normal random variable, test the hypothesis $\mu_0 = 4500$ kg against the alternative $\mu_1 = 4400$ kg. Here μ_0 may be a value given by the manufacturer, while μ_1 may result from previous experience.

Solution. We choose the significance level $\alpha = 5\%$. If the hypothesis is true, it follows from Theorem 3 in Sec. 24.6, that the random variable

$$T = \sqrt{n}\,\frac{\bar{X} - \mu_0}{S} = 4\,\frac{\bar{X} - 4500}{S}$$

has a t-distribution with $n - 1 = 15$ degrees of freedom. The critical value c is obtained from

$$P(T < c)_{\mu_0} = \alpha = 0.05.$$

Table A11 in Appendix 4 yields $c = -1.75$. As an observed value of T we obtain from the sample $t = 4(4482 - 4500)/115 = -0.626$. We see that $t > c$ and do not reject the hypothesis. For obtaining numerical values of the power of the test, we would need tables called noncentral Student t-tables; we shall not discuss this question here.

EXAMPLE 4. Test for the variance of the normal distribution

Using a sample of size $n = 15$ and sample variance $s^2 = 13$ from a normal population, test the hypothesis $\sigma^2 = \sigma_0^2 = 10$ against the alternative $\sigma^2 = \sigma_1^2 = 20$.

Solution. We choose the significance level $\alpha = 5\%$. If the hypothesis is true, then

$$Y = (n - 1)\,\frac{S^2}{\sigma_0^2} = 14\,\frac{S^2}{10} = 1.4S^2$$

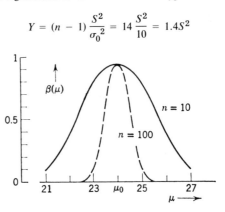

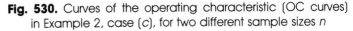

Fig. 530. Curves of the operating characteristic (OC curves) in Example 2, case (c), for two different sample sizes n

has a chi-square distribution with $n - 1 = 14$ degrees of freedom, cf. Theorem 4, Sec. 24.6. From

$$P(Y > c) = \alpha = 0.05, \qquad \text{that is,} \qquad P(Y \leq c) = 0.95,$$

and Table A12 in Appendix 4 with 14 degrees of freedom we obtain $c = 23.68$. This is the critical value of Y. Hence to $S^2 = \sigma_0^2 Y/(n - 1) = 0.714Y$ there corresponds the critical value $c^* = 0.714 \cdot 23.68 = 16.91$. Since $s^2 < c^*$, we do not reject the hypothesis.

If the alternative is true, the variable

$$Y_1 = 14 \frac{S^2}{\sigma_1^2} = 0.7S^2$$

has a chi-square distribution with 14 degrees of freedom. Hence our test has the power

$$\eta = P(S^2 > c^*)_{\sigma^2 = 20} = P(Y_1 > 0.7c^*)_{\sigma^2 = 20} = 1 - P(Y_1 \leq 11.84)_{\sigma^2 = 20} \approx 62\%$$

and we see that the Type II risk is very large, namely, 38%. To make this risk smaller, we would have to increase the sample size.

EXAMPLE 5. Comparison of the means of two normal distributions

Using a sample $x_1, \cdots, x_{n_1}$ from a normal distribution with unknown mean μ_1 and a sample $y_1, \cdots, y_{n_2}$ from another normal distribution with unknown mean μ_2, we want to test the hypothesis that the means are equal, $\mu_1 = \mu_2$, against an alternative, say, $\mu_1 > \mu_2$. The variances need not be known but are assumed to be equal.[5] Two cases are of practical importance:

Case A. The samples have the same size. Furthermore, each value of the first sample corresponds to precisely one value of the other, because corresponding values result from the same person or thing (**paired comparison**); for example, two measurements of the same thing by two different methods or two measurements from the two eyes of the same animal; more generally, they may result from pairs of *similar* individuals or things, for example, identical twins, pairs of used front tires from the same car, etc. Then we should form the differences of corresponding values and test the hypothesis that the population corresponding to the differences has mean 0, using the method in Example 3. If we have a choice, this method is better than the following.

Case B. The two samples are independent and not necessarily of the same size. Then we may proceed as follows. Suppose that the alternative is $\mu_1 > \mu_2$. We choose a significance level α. We compute the sample means $\bar{x}$ and $\bar{y}$ and $(n_1 - 1)s_1^2$, $(n_2 - 1)s_2^2$, where s_1^2 and s_2^2 are the sample variances. Using Table A11 in Appendix 4 with $n_1 + n_2 - 2$ degrees of freedom, we determine c from

$$(10) \qquad\qquad\qquad P(T \leq c) = 1 - \alpha.$$

We finally compute

$$(11) \qquad\qquad t_0 = \sqrt{\frac{n_1 n_2 (n_1 + n_2 - 2)}{n_1 + n_2}} \frac{\bar{x} - \bar{y}}{\sqrt{(n_1 - 1)s_1^2 + (n_2 - 1)s_2^2}}.$$

It can be shown that this is an observed value of a random variable which has a t-distribution with $n_1 + n_2 - 2$ degrees of freedom, provided the hypothesis is true. If $t_0 \leq c$, the hypothesis is not rejected. If $t_0 > c$, it is rejected.

If the alternative is $\mu_1 \neq \mu_2$, then (10) must be replaced by

$$(10^*) \qquad\qquad P(T \leq c_1) = 0.5\alpha, \qquad P(T \leq c_2) = 1 - 0.5\alpha.$$

Note that for samples of equal size $n_1 = n_2 = n$, formula (11) reduces to

---

[5]If the test in the next example shows that the variances differ significantly, then choose two not too small samples of the same size $n_1 = n_2 = n$ (> 30, say), use the fact that (12) is an observed value of an approximately normal random variable with mean 0 and variance 1, and proceed as in Example 2.

(12)
$$t_0 = \sqrt{n}\,\frac{\bar{x} - \bar{y}}{\sqrt{s_1^2 + s_2^2}}.$$

To illustrate the numerical work, let us consider the two samples

and

| 105 | 108 | 86 | 103 | 103 | 107 | 124 | 105 |
| 89 | 92 | 84 | 97 | 103 | 107 | 111 | 97 |

showing the relative output of tin plate workers under two different working conditions (J.J.B. Worth, *Journal of Industrial Engineering* **9** (1958), 249-253). Assuming that the corresponding populations are normal and have the same variance, let us test the hypothesis $\mu_1 = \mu_2$ against the alternative $\mu_1 \neq \mu_2$. (Equality of variances will be tested in the next example.)

Solution. We find

$$\bar{x} = 105.125, \quad \bar{y} = 97.500, \quad s_1^2 = 106.125, \quad s_2^2 = 84.000.$$

We choose the significance level $\alpha = 5\%$. From (10*) with $0.5\alpha = 2.5\%$, $1 - 0.5\alpha = 97.5\%$ and Table A11 in Appendix 4 with 14 degrees of freedom we obtain $c_1 = -2.15$ and $c_2 = 2.15$. Formula (12) with $n = 8$ gives the value

$$t_0 = \sqrt{8} \cdot 7.625/\sqrt{190.125} = 1.56.$$

Since $c_1 \leq t_0 \leq c_2$, we do not reject the hypothesis $\mu_1 = \mu_2$ that under both conditions the mean output is the same.

Case A applies to the example, because the two first sample values correspond to a certain type of work, the next two were obtained in another kind of work, etc. So we may use the differences

| 16 | 16 | 2 | 6 | 0 | 0 | 13 | 8 |

of corresponding sample values and the method in Example 3 to test the hypothesis $\mu = 0$, where μ is the mean of the population corresponding to the differences. As a logical alternative we take $\mu \neq 0$. The sample mean is $\bar{d} = 7.625$, and the sample variance is $s^2 = 45.696$. Hence

$$t = \sqrt{8}\,(7.625 - 0)/\sqrt{45.696} = 3.19.$$

From $P(T \leq c_1) = 2.5\%$, $P(T \leq c_2) = 97.5\%$ and Table A11 in Appendix 4 with $n - 1 = 7$ degrees of freedom we obtain $c_1 = -2.37$, $c_2 = 2.37$ and reject the hypothesis because $t = 3.19$ does not lie between c_1 and c_2. Hence our present test, in which we used more information (but the same samples), shows that the difference in output is significant.

EXAMPLE 6. Comparison of the variances of two normal distributions

Using the two samples in the last example, test the hypothesis $\sigma_1^2 = \sigma_2^2$; assume that the corresponding populations are normal and the nature of the experiment suggests the alternative $\sigma_1^2 > \sigma_2^2$.

Solution. We find $s_1^2 = 106.125$, $s_2^2 = 84.000$. We choose the significance level $\alpha = 5\%$. Using $P(V \leq c) = 1 - \alpha = 95\%$ and Table A13 in Appendix 4, with $(n_1 - 1, n_2 - 1) = (7, 7)$ degrees of freedom, we determine $c = 3.79$. We finally compute $v_0 = s_1^2/s_2^2 = 1.26$. Since $v_0 \leq c$, we do not reject the hypothesis. If $v_0 > c$, we would reject it.

This test is justified by the fact that v_0 is an observed value of a random variable that has a so-called **F-distribution** with $(n_1 - 1, n_2 - 1)$ degrees of freedom, provided the hypothesis is true. (Proof in Ref. [G4] listed in Appendix 1.) The F-distribution with (m, n) degrees of freedom was introduced by R.A. Fisher[6] and has the distribution function $F(z) = 0$ when $z < 0$ and

---

[6]After the pioneering work of the English statistician and biologist, KARL PEARSON (1857—1936), the founder of the English school of statistics, and W. S. GOSSET (cf. footnote in Sec. 24.6), the English statistician Sir RONALD AYLMER FISHER (1890—1962), professor of eugenics in London (1933—1943) and professor of genetics in Cambridge, England (1943—1957), had great influence on the further development of modern statistics.

(13) $$F(z) = K_{mn} \int_0^z t^{(m-2)/2}(mt + n)^{-(m+n)/2} \, dt \qquad (z \geq 0),$$

where $K_{mn} = m^{m/2}n^{n/2}\Gamma(\frac{1}{2}m + \frac{1}{2}n)/\Gamma(\frac{1}{2}m)\Gamma(\frac{1}{2}n)$. (For Γ see Appendix 3.)   ∎

This section contained the basic ideas and concepts on testing, along with typical applications, and the student may perhaps want to review it quickly before going on. In the next section we apply testing to **quality control,** an important task in modern production.

Problems for Sec. 24.7

1. Test $\mu = 0$ against $\mu > 0$, assuming normality and using the sample 1, -1, 1, 3, -8, 6, 0 (deviations of the azimuth [multiples of 0.01 radian] in some revolution of a satellite).

2. Using the data of Buffon in Table 23.1 (Sec. 23.2), test the hypothesis that the coin is fair, that is, that heads and tails have the same probability of occurrence, against the alternative that heads are more likely than tails. (Choose $\alpha = 5\%$.)

3. Do the same test as in Prob. 2, using the data of Pearson in Table 23.1.

4. A firm sells paint in cans containing 1 kg of paint per can and is interested to know whether the mean weight differs significantly from 1 kg, in which case the filling machine must be adjusted. Set up a hypothesis and an alternative and perform the test, assuming normality and using a sample of 20 fillings having a mean of 991 g and a standard deviation of 8 g. (Choose $\alpha = 5\%$.)

5. Choosing $\alpha = 5\%$, test the hypothesis that there is no significant difference between two methods of measuring the starch content of potatoes, assuming normality and using the sample [differences of corresponding values from 16 potatoes, measured in multiples of 0.1%]

$$2 \quad 0 \quad 0 \quad 1 \quad 2 \quad 2 \quad 3 \quad -3 \quad 1 \quad 2 \quad 3 \quad 0 \quad -1 \quad 1 \quad -2 \quad 1.$$

6. Assuming normality and known variance $\sigma^2 = 4$, test the hypothesis $\mu = 15.0$ against the alternative (a) $\mu = 12.0$, (b) $\mu = 15.8$, using a sample of size 10 with mean $\bar{x} = 14$ and choosing $\alpha = 5\%$.

7. How does the result in Prob. 6 change if we use a larger sample, say, of size 100, the other data ($\bar{x} = 14$, $\alpha = 5\%$, etc.) remaining as before?

8. Determine the power of the test in Prob. 6(a).

9. What is the rejection region in Prob. 6 in the case of a two-sided test with $\alpha = 5\%$?

10. If a sample of 100 tires of a certain kind has a mean life of 26 000 km and a standard deviation of 2000 km, can the manufacturer claim that the true mean life of such tires is greater than 25 000 km? Set up and test a corresponding hypothesis, assuming normality and choosing $\alpha = 1\%$.

11. Assuming normality and equal variance and using independent samples with $n_1 = 9$, $\bar{x} = 12$, $s_1 = 2$, $n_2 = 9$, $\bar{y} = 15$, $s_2 = 2$, test H_0: $\mu_1 = \mu_2$ against $\mu_1 \neq \mu_2$; choose $\alpha = 5\%$.

12. Three specimens of high-quality concrete had compressive strength 357, 359, 413 [kg/cm^2], and for three specimens of ordinary concrete the values were 346, 358, 302. Test for equality of the population means, $\mu_1 = \mu_2$, against the alternative $\mu_1 > \mu_2$. (Assume normality and equality of variances. Choose $\alpha = 5\%$.)

13. For checking the reliability of a certain electrical method for measuring temperatures, two samples were obtained. Assuming normality and equality of variances of the corresponding populations, test the hypothesis that the population means are equal. Choose $\alpha = 5\%$. The samples are [°C]

$$106.9 \quad 106.3 \quad 107.0 \quad 106.0 \quad 104.9$$

$$106.5 \quad 106.7 \quad 106.8 \quad 106.1 \quad 105.6$$

14. If a standard medication cures about 75% of patients with a certain disease and a new medication cured 157 of the first 200 patients on whom it was tried, can we conclude that the new medication is better? (Choose $\alpha = 5\%$.)

15. Using samples of sizes 10 and 5 with variances $s_1^2 = 50$ and $s_2^2 = 20$ and assuming normality of the corresponding populations, test the hypothesis H_0: $\sigma_1^1 = \sigma_2^2$ against the alternative $\sigma_1^2 > \sigma_2^2$. Choose $\alpha = 5\%$.

16. Using two samples (weight of dust in certain tubes, measured in mg) obtained in an experiment by the British Coal Utilisation Research Association and assuming normality of the corresponding populations, find whether the variances of the populations differ significantly from each other. Choose $\alpha = 5\%$. The samples are

$$75 \quad 20 \quad 70 \quad 70 \quad 85 \quad 90 \quad 100 \quad 40 \quad 35 \quad 65 \quad 90 \quad 35$$

$$20 \quad 35 \quad 55 \quad 50 \quad 65 \quad 40$$

17. Suppose that brand I and brand II light bulbs have the same price and are of the same quality, perhaps except for their lifetime. A buyer tested 100 bulbs of each brand and found that I had a mean lifetime of 1120 hr with a standard deviation of 75 hr, and for II the corresponding values were 1064 hr and 82 hr. Is the difference in lifetime significant? (Assume normality; choose $\alpha = 5\%$.)

18. Construct two simple samples such that the hypothesis of equal population means is not rejected although the difference between the sample means is large.

19. Show that for a normal distribution the two types of errors in a test of a hypothesis H_0: $\mu = \mu_0$ against an alternative H_1: $\mu = \mu_1$ can be made as small as one pleases (not zero) by taking the sample sufficiently large.

20. Graph the OC curves in Example 2, cases (a) and (b).

24.8 Quality Control

No production process is so perfect that all the products are completely alike. There is always a small variation that is caused by a great number of small, uncontrollable factors and must therefore be regarded as a chance variation. It is important to make sure that the products have required values (for example, length, strength, or whatever property may be of importance in a particular case). For this purpose one makes a test of the hypothesis that the products have the required property, say, $\mu = \mu_0$, where μ_0 is a required value. If this is done after an entire lot has been produced (for example, a lot of 100,000 screws), the test will tell us how good or how bad the products are, but it is obviously too late to alter undesirable results. It is much better to test during the production run. This is done at regular intervals of time (for example, every hour or half-hour) and is called **quality control**. Each time a sample of the same size is taken, in practice 3 to 10 items. If the hypothesis is rejected, we stop the production process and look for the trouble that causes the deviation.

If we stop the production process even though it is progressing properly, we make a Type I error. If we do not stop the process even though something is not in order, we make a Type II error (cf. Sec. 24.7).

The result of each test is marked in graphical form on what is called a **control chart.** This was proposed by W. A. Shewhart in 1924 and makes quality control particularly effective.

Control Chart for the Mean

An illustration and example of a control chart is given in the upper part of Fig. 531. This control chart for the mean shows the **lower control limit** LCL, the **center control line** CL and the **upper control limit** UCL. The two **control**

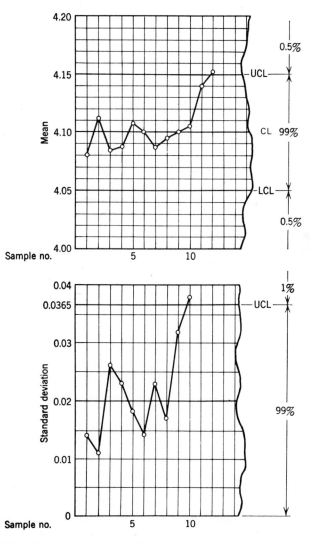

Fig. 531. Control charts for the mean (upper part of figure) and the standard deviation in the case of the sample in Table 24.9, p. 1270

Table 24.9
Twelve Samples of Five Values Each
(Diameter of Small Cylinders, Measured in Millimeters)

Sample Number	Sample Values					$\bar{x}$	s	R
1	4.06	4.08	4.08	4.08	4.10	4.080	0.014	0.04
2	4.10	4.10	4.12	4.12	4.12	4.112	0.011	0.02
3	4.06	4.06	4.08	4.10	4.12	4.084	0.026	0.06
4	4.06	4.08	4.08	4.10	4.12	4.088	0.023	0.06
5	4.08	4.10	4.12	4.12	4.12	4.108	0.018	0.04
6	4.08	4.10	4.10	4.10	4.12	4.100	0.014	0.04
7	4.06	4.08	4.08	4.10	4.12	4.088	0.023	0.06
8	4.08	4.08	4.10	4.10	4.12	4.096	0.017	0.04
9	4.06	4.08	4.10	4.12	4.14	4.100	0.032	0.08
10	4.06	4.08	4.10	4.12	4.16	4.104	0.038	0.10
11	4.12	4.14	4.14	4.14	4.16	4.140	0.014	0.04
12	4.14	4.14	4.16	4.16	4.16	4.152	0.011	0.02

limits correspond to the critical values c_1 and c_2 in case (c) of Example 2 in Sec. 24.7. As soon as a sample mean falls outside the range between the control limits, we reject the hypothesis and assert that the production process is "out of control"; that is, we assert that there has been a shift in process level. Action is called for whenever a point exceeds the limits.

If we choose control limits that are too loose, we shall not detect process shifts. On the other hand, if we choose control limits that are too tight, we shall be unable to run the process because of frequent searches for non-existent trouble. The usual significance level is $\alpha = 1\%$. From Theorem 2 in Sec. 24.6 and Table A9 in Appendix 4 we see that in the case of the normal distribution the corresponding control limits for the mean are

$$(1) \qquad \text{LCL} = \mu_0 - 2.58 \frac{\sigma}{\sqrt{n}} \qquad \text{and} \qquad \text{UCL} = \mu_0 + 2.58 \frac{\sigma}{\sqrt{n}}.$$

Here σ is assumed to be known. If σ is unknown, we may calculate the standard deviations of the first 20 or 30 samples and take their arithmetic mean as an approximation of σ. The broken line connecting the means in Fig. 531 is merely to display the results effectively.

Control Chart for the Variance

In addition to the mean, one often controls the variance, the standard deviation or the range. To set up a control chart for the variance in the case of a normal distribution, we may employ the method in Example 4 of Sec. 24.7 for determining control limits. It is customary to use only one control limit, namely, an upper control limit. From Example 4 of Sec. 24.7 we see that this limit is

$$(2) \qquad \text{UCL} = \frac{\sigma^2 c}{n - 1}$$

where c is obtained from the equation

$$P(Y > c) = \alpha, \quad \text{that is,} \quad P(Y \leq c) = 1 - \alpha$$

and the table of the chi-square distribution (Table A12 in Appendix 4) with $n - 1$ degrees of freedom; here α (5% or 1%, say) is the probability that an observed value s^2 of S^2 in a sample is greater than the upper control limit.

If we wanted a control chart for the variance with both an upper control limit UCL and a lower control limit LCL, these limits would be

$$(3) \qquad\qquad \text{LCL} = \frac{\sigma^2 c_1}{n-1} \quad \text{and} \quad \text{UCL} = \frac{\sigma^2 c_2}{n-1}$$

where c_1 and c_2 are obtained from the equations

$$(4) \qquad\qquad P(Y \leq c_1) = \frac{\alpha}{2} \quad \text{and} \quad P(Y \leq c_2) = 1 - \frac{\alpha}{2}$$

and Table A12 in Appendix 4 with $n - 1$ degrees of freedom.

Control Chart for the Standard Deviation

Similarly, to set up a control chart for the standard deviation, we need an upper control limit

$$(5) \qquad\qquad\qquad \text{UCL} = \frac{\sigma \sqrt{c}}{\sqrt{n-1}}$$

obtained from (2). For example, in Table 24.9 we have $n = 5$. Assuming that the corresponding population is normal with standard deviation $\sigma = 0.02$ and choosing $\alpha = 1\%$, we obtain from the equation

$$P(Y \leq c) = 1 - \alpha = 99\%$$

and Table A12 in Appendix 4 with 4 degrees of freedom the critical value $c = 13.28$ and from (5) the corresponding value

$$\text{UCL} = \frac{0.02\sqrt{13.28}}{\sqrt{4}} = 0.0365,$$

which is shown in the lower part of Fig. 531.

A control chart for the standard deviation with both an upper and a lower control limit is obtained from (3).

Control Chart for the Range

Instead of the variance or standard deviation, one often controls the **range** R (= largest sample value minus smallest sample value). It can be shown that in the case of the normal distribution, the standard deviation σ is proportional to the expectation of the random variable R^* for which R is an observed value, say, $\sigma = \lambda_n E(R^*)$, where the factor of proportionality λ_n depends on the sample size n and has the values

n	2	3	4	5	6	7	8	9	10
$\lambda_n = \sigma/E(R^*)$	0.89	0.59	0.49	0.43	0.40	0.37	0.35	0.34	0.32

n	12	14	16	18	20	30	40	50
$\lambda_n = \sigma/E(R^*)$	0.31	0.29	0.28	0.28	0.27	0.25	0.23	0.22

Since R depends on two sample values only, it gives less information about a sample than s does. Clearly, the larger the sample size n is, the more information we shall lose in using R instead of s. A practical rule is to use s when n is larger than 10.

In the next section we discuss **acceptance sampling,** another important application of the ideas of hypothesis testing.

Problems for Sec. 24.8

1. Suppose a machine for filling cans with lubricating oil is set so that it will generate fillings which form a normal population with mean 1 gallon and standard deviation 0.02 gallon. Set up a control chart of the type shown in Fig. 531 for controlling the mean (that is, find LCL and UCL), assuming the sample size to be 4.

2. **(Three-sigma control chart)** Show that in Prob. 1, the requirement of the significance level $\alpha = 0.3\%$ leads to LCL $= \mu - 3\sigma/\sqrt{n}$ and UCL $= \mu + 3\sigma/\sqrt{n}$, and find the corresponding numerical values.

3. What sample size should we choose in Prob. 1 if we want LCL and UCL somewhat closer together, say, UCL $-$ LCL $= 0.02$, without changing the significance level?

4. What effect on UCL $-$ LCL does it have if we double the sample size? If we switch from $\alpha = 1\%$ to $\alpha = 5\%$?

5. Eight samples of size 2 were taken from a production lot of screws. The values (length in inches) are

Sample No.	1	2	3	4	5	6	7	8
Length	3.49	3.48	3.52	3.50	3.51	3.49	3.52	3.53
	3.50	3.47	3.49	3.51	3.48	3.50	3.50	3.49

Assuming that the population is normal with mean 3.500 and variance 0.0004 and using (1), set up a control chart for the mean and graph the sample means on the chart.

6. Graph the means of the following 10 samples (thickness of gaskets, coded values) on a control chart for means, assuming that the population is normal with mean 5 and standard deviation 1.16.

Time	10:00	11:00	12:00	13:00	14:00	15:00	16:00	17:00	18:00	19:00
	4	5	6	5	5	3	3	5	7	7
Sample	4	6	4	5	2	4	6	2	5	3
values	3	4	6	6	5	8	6	5	4	6
	6	6	4	4	3	4	8	6	4	5

7. Graph the ranges of the samples in Prob. 6 on a control chart for ranges.

8. Graph $\lambda_n = \sigma/E(R^*)$ as a function of n. Why is λ_n a monotone decreasing function of n?

9. **(Number of defectives)** Find formulas for the UCL, CL and LCL (corresponding to 3σ-limits) in the case of a control chart for the number of defectives, assuming that in a state of statistical control the fraction of defectives is p.

10. **(Attribute control charts)** Twenty samples of size 100 were taken from a production of containers. The numbers of defectives (leaking containers) in those samples (in the order observed) were

$$3 \quad 7 \quad 6 \quad 1 \quad 4 \quad 5 \quad 4 \quad 9 \quad 7 \quad 0 \quad 5 \quad 6 \quad 13 \quad 4 \quad 9 \quad 0 \quad 2 \quad 1 \quad 12 \quad 8$$

From previous experience it was known that the average fraction defective is $p = 5\%$, provided the production process is running properly. Using the binomial distribution, set up a *fraction defective chart* (also called a **p-chart**); that is, choose the LCL $= 0$ and determine the UCL for the fraction defective [%] by the use of 3σ-limits, where σ^2 is the variance of the random variable $\overline{X}$ = *fraction defective in a sample of size* 100. Is the process in control?

11. **(Number of defects per unit)** A so-called *c-chart* or *defects-per-unit chart* is used for the control of the number X of defects per unit (for instance, the number of defects per 100 meters of paper, the number of rivets missing from an airplane wing, etc.). (*a*) Set up formulas for CL and LCL, UCL corresponding to $\mu \pm 3\sigma$, assuming that X has a Poisson distribution. (*b*) Compute CL, LCL and UCL in a control process of the number of imperfections in sheet glass; assume that this number is 3.6 per sheet on the average when the process is under statistical control.

12. How does the meaning of the control limits (1) change if we apply a control chart with these limits in the case of a population that is not normal?

13. Since the presence of a point outside control limits for the mean indicates trouble, how often would we be making the mistake of looking for nonexistent trouble if we used (*a*) 1-sigma limits, (*b*) 2-sigma limits? (Assume normality.)

14. What LCL and UCL should we use instead of (1) if we use the sum $x_1 + \cdots + x_n$ of the sample values instead of $\bar{x}$? Determine these limits in the case of Fig. 531.

15. How would progressive tool wear in an automatic lathe operation be indicated by a control chart for the mean? Answer the same question for a sudden change in the position of the tool in that operation.

24.9 Acceptance Sampling

Acceptance sampling is applied in mass production when a *producer* is supplying to a *consumer* lots of N items. In such a situation the decision to accept or reject an individual lot must be made. This decision is often based on the result of inspecting a sample of size n from the lot and determining the number of **defective items,** briefly called **defectives,** that is, items that do not meet the specifications (size, color, strength, or whatever may be important). If the number of defective items x in the sample is not greater than a specified number c ($< n$), the lot is accepted. If $x > c$, the lot is rejected. c is called the *allowable number of defectives* or the **acceptance number.** It is clear that the producer and the consumer must agree on a certain **sampling plan,** that is, on a certain sample size n and an allowable number c. Such a plan is called a **single sampling plan** because it is based on a single sample. *Double sampling plans,* in which two samples are used, will be mentioned later.

Let A be the event that a lot is accepted. It is clear that the corresponding probability $P(A)$ depends not only on n and c but also on the number of defectives in the lot. Let M denote this number, let the random variable X be the number of defectives in a sample and suppose that we sample without replacement. Then (cf. Sec. 23.6)

$$(1) \qquad P(A) = P(X \leqq c) = \sum_{x=0}^{c} \binom{M}{x}\binom{N-M}{n-x} \Big/ \binom{N}{n}.$$

If $M = 0$ (no defectives in the lot), then X must assume the value 0, and

$$P(A) = \binom{0}{0}\binom{N}{n} \Big/ \binom{N}{n} = 1.$$

For fixed n and c and increasing M the probability $P(A)$ decreases. If $M = N$ (all items of the lot defective), then X must assume the value n, and we have $P(A) = P(X \leqq c) = 0$ because $c < n$.

The ratio $\theta = M/N$ is called the **fraction defective** in the lot. Note that $M = N\theta$, and (1) may be written

$$(2) \qquad \boxed{P(A;\, \theta) = \sum_{x=0}^{c} \binom{N\theta}{x}\binom{N-N\theta}{n-x} \Big/ \binom{N}{n}.}$$

Since θ can have one of the $N + 1$ values 0, $1/N$, $2/N$, $\cdots$, N/N, the probability $P(A)$ is defined for these values only. For fixed n and c we may plot $P(A)$ as a function of θ. These are $N + 1$ points. Through these points we may then draw a smooth curve, which is called the **operating characteristic curve** (OC curve) of the sampling plan considered.

EXAMPLE 1. Single sampling plan

Suppose that certain tool bits are packaged 20 to a box, and the following single sampling plan is used. A sample of two tool bits is drawn, and the corresponding box is accepted if and only if both bits in the sample are good. In this case, $N = 20$, $n = 2$, $c = 0$, and (2) takes the form

$$P(A;\, \theta) = \binom{20\theta}{0}\binom{20-20\theta}{2} \Big/ \binom{20}{2} = \frac{(20-20\theta)(19-20\theta)}{380}.$$

Numerical values are

θ	0.00	0.05	0.10	0.15	0.20	$\cdots$
$P(A;\, \theta)$	1.00	0.90	0.81	0.72	0.63	$\cdots$

The OC curve is shown in Fig. 532.   ∎

In most practical cases θ will be small (less than 10%). In many cases the lot size N will be very large (1000, 10,000 etc.), so that we may approximate the hypergeometric distribution in (1) and (2) by the binomial distribution with $p = \theta$. Then if n is such that $n\theta$ is moderate (say, less than 20), we may approximate that distribution by the Poisson distribution with mean $\mu = n\theta$. From (2) we then have

$$(3) \qquad\qquad P(A; \theta) \sim e^{-\mu} \sum_{x=0}^{c} \frac{\mu^x}{x!} \qquad\qquad (\mu = n\theta).$$

EXAMPLE 2. Single sampling plan. Poisson distribution

Suppose that for large lots the following single sampling plan is used. A sample of size $n = 20$ is taken. If it contains not more than one defective, the lot is accepted. If the sample contains two or more defectives, the lot is rejected. In this plan, we obtain from (3)

$$P(A; \theta) \sim e^{-20\theta}(1 + 20\theta).$$

The corresponding OC curve is shown in Fig. 533.   ∎

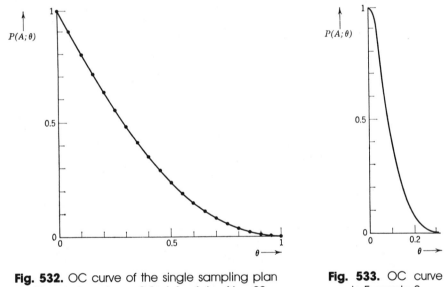

Fig. 532. OC curve of the single sampling plan with $n = 2$ and $c = 0$ for lots of size $N = 20$

Fig. 533. OC curve in Example 2

Errors in Acceptance Sampling

We shall now discuss the two types of possible errors in acceptance sampling and the related problem of choosing n and c. In acceptance sampling the producer and the consumer have different interests. The producer may require the probability of rejecting a "good" or "acceptable" lot to be a small number, call it α. The consumer (buyer) may want the probability of accepting a "bad" or "unacceptable" lot to be a small number β. More precisely, suppose that the two parties agree that a lot for which θ does not exceed a certain number θ_0 is an *acceptable lot* whereas a lot for which θ is greater than or equal to a certain number θ_1 is an *unacceptable lot*. Then

α is the probability of rejecting a lot with $\theta \leqq \theta_0$ and is called **producer's risk.** This corresponds to a Type I error in testing a hypothesis (Sec. 24.7). β is the probability of accepting a lot with $\theta \geqq \theta_1$ and is called **consumer's risk.** This corresponds to a Type II error in Sec. 24.7. Figure 534 shows an illustrative example. θ_0 is called the **acceptable quality level** (AQL), and θ_1 is called the **lot tolerance percent defective** (LTPD) or the **rejectable quality level** (RQL). A lot with $\theta_0 < \theta < \theta_1$ may be called an *indifferent lot*.

From Fig. 534 we see that the points $(\theta_0, 1 - \alpha)$ and (θ_1, β) lie on the OC curve. It can be shown that for large lots we can choose θ_0, θ_1 $(> \theta_0)$, α, β and then determine n and c such that the OC curve runs very close to those prescribed points. Sampling plans for specified α, β, θ_0, and θ_1 have been published; cf. Ref. [G6] in Appendix 1.

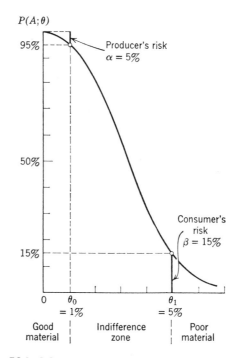

Fig. 534. OC curve, producer's and consumer's risks

There is a close relationship between sampling inspection and testing a hypothesis, as follows:

Sampling Inspection	Hypothesis Testing
Acceptable quality level (AQL) $\theta = \theta_0$	Hypothesis $\theta = \theta_0$
Lot tolerance percent defectives (LTPD) $\theta = \theta_1$	Alternative $\theta = \theta_1$
Allowable number of defectives c	Critical value c
Producer's risk α of rejecting a lot with $\theta \leqq \theta_0$	Probability α of making a Type I error (significance level)
Consumer's risk β of accepting a lot with $\theta \geqq \theta_1$	Probability β of making a Type II error

The sampling procedure by itself does not protect the consumer sufficiently well. In fact, if the producer is permitted to resubmit a rejected lot without telling that the lot has already been rejected, then even bad lots will eventually be accepted. To protect the consumer against this and other possibilities, the producer may agree with the consumer that a rejected lot is **rectified,** that is, is inspected 100%, item by item, and all defective items in the lot are removed and replaced by nondefective items.[7] Suppose that a plant produces $100\theta\%$ defective items and rejected lots are rectified. Then K lots of size N contain KN items, $KN\theta$ of which are defective. $KP(A; \theta)$ of the lots are accepted; these contain a total of $KPN\theta$ defective items. The rejected and rectified lots contain no defective items. Hence after the rectification the fraction defective in the K lots equals $KPN\theta/KN = \theta P(A; \theta)$. This function of θ is called the **average outgoing quality** (AOQ) and is denoted by AOQ(θ). Thus

(4)
$$\boxed{\text{AOQ}(\theta) = \theta P(A; \theta).}$$

A sampling plan being given, this function and its graph, the *average outgoing quality curve* (AOQ curve), can readily be obtained from $P(A; \theta)$ and the OC curve. An example is shown in Fig. 535.

Clearly AOQ(0) = 0. Also, AOQ(1) = 0 because $P(A; 1) = 0$. From this and AOQ $\geqq 0$ we conclude that this function must have a maximum at some $\theta = \theta^*$. The corresponding value AOQ(θ^*) is called the **average outgoing quality limit** (AOQL). This is the worst average quality that may be expected to be accepted under the rectifying procedure.

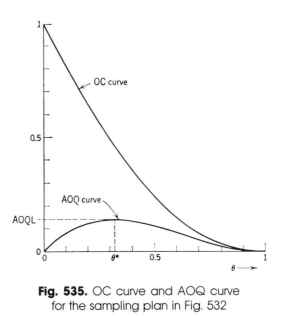

Fig. 535. OC curve and AOQ curve
for the sampling plan in Fig. 532

[7]Of course, rectification is impossible if the inspection is destructive, or is not worthwhile if it is too expensive, compared with the value of the lot. The rejected lot may then be sold at a cut-rate price or scrapped.

It turns out that several single sampling plans may correspond to the same AOQL; see Ref. [G6] in Appendix 1. Hence if the AOQL is all the consumer cares about, the producer has some freedom in choosing a sampling plan and may select a plan that minimizes the amount of sampling, that is, the number of inspected items per lot. This number is

$$nP(A; \theta) + N(1 - P(A; \theta))$$

where the first term corresponds to the accepted lots and the last term to the rejected and rectified lots; in fact, rectification requires the inspection of all N items of the lot, and $1 - P(A; \theta)$ is the probability of rejecting a lot.

We mention that inspection work may be saved by using a **double sampling plan,** in which the sample of size n is broken up into two samples of sizes n_1 and n_2 (where $n_1 + n_2 = n$). If the lot is very good or very bad, it may then be possible to decide about acceptance or rejectance, using one sample only, so that the other sample will be necessary in the case of a lot of intermediate quality only. Reference [G6] contains double sampling plans using rectifying inspection of the following type (where x_1 and x_2 are the numbers of defectives in those two samples).

1. If $x_1 \leqq c_1$, accept the lot. If $x_1 > c_2$, reject the lot.

2. If $c_1 < x_1 \leqq c_2$, use the second sample, too. If $x_1 + x_2 \leqq c_2$, accept the lot. If $x_1 + x_2 > c_2$, reject the lot.

So far we have considered tests of hypotheses for parameters (μ, σ, θ etc.). These are of practical interest if we already know the kind of distribution or do not care about it but have a large sample, so that by the central limit theorem (Sec. 24.6) we can apply methods designed for the normal distribution to other distributions (with proper caution!). In the next section we turn to a different task, namely, the testing of the hypothesis that a given function is the distribution function of some population that interests us. This is called **goodness of fit:** How well does the sample distribution function fit the hypothetical distribution function?

Problems for Sec. 24.9

1. Large lots of razor blades are inspected by a single sampling plan that uses a sample of size 20 and the acceptance number $c = 1$. What are the probabilities of accepting a lot with 1%, 2%, 10% defectives (dull blades)? Use Table A7 in Appendix 4. Graph the OC curve.

2. What are the probabilities in Prob. 1 if the sample size is increased to 50? Graph the OC curve and compare with Prob. 1.

3. Will the probabilities in Prob. 1 increase or decrease if we change the acceptance number to $c = 0$, retaining sample size 20? First think about, then determine them.

4. What are the producer's and consumer's risks in Prob. 1 if the AQL is 1.5% and the RQL is 7.5%?

5. Calculate $P(A; \theta)$ in Example 1 when the sample size is increased from $n = 2$ to $n = 3$, the other data remaining as before. Compute $P(A; 0.10)$ and $P(A; 0.20)$ and compare with Example 1.

6. Find the binomial approximation of the hypergeometric distribution in Example 1 and compare the approximate and the accurate values.

7. In Example 1, what are the producer's and consumer's risks if the AQL is 0.05 and the RQL is 0.45?

8. Suppose in a single sampling plan for inspecting screws we have sample size 40 and $c = 3$, and we want the producer's risk to be 5%. What is the AQL? (*Hint.* Use the De Moivre-Laplace theorem.)

9. Samples of five screws are drawn from a lot with fraction defective θ. The lot is accepted if the sample contains (a) no defective screws, (b) at most one defective screw. Using the binomial distribution, find, graph and compare the OC curves.

10. Lots of copper pipes are inspected according to a single sample plan that uses sample size 30 and acceptance number 1. Graph the OC curve of the plan, using the Poisson approximation.

11. Graph the AOQ curve in Prob. 10. Determine the AOQL, assuming that rectification is applied.

12. Do the work required in Prob. 10 when $n = 35$ and $c = 0$.

13. Graph and compare sampling plans with $c = 0$ and increasing values of n, say, $n = 2, 3, 4$. (Use the binomial distribution.)

14. Samples of three flashlight bulbs are drawn from lots and a lot is accepted if in the corresponding sample we find no more than one defective bulb. Criticize this sampling plan. In particular, find the probability of accepting a lot that is 50% defective. (Use the binomial distribution.)

15. Why is it impossible for an OC curve to have a vertical portion separating good from poor quality?

24.10 Goodness of Fit. χ^2-Test

Using a sample $x_1, \cdots, x_n$, we want to test the hypothesis that a certain function $F(x)$ is the distribution function of the population from which the sample was taken. Clearly, the sample distribution function $\tilde{F}(x)$ is an approximation of $F(x)$, and if it approximates $F(x)$ "sufficiently well," we shall not reject the hypothesis that $F(x)$ is the distribution function of the population. If $\tilde{F}(x)$ deviates "too much" from $F(x)$, we shall reject the hypothesis.

To decide in this fashion, we have to know how much $\tilde{F}(x)$ can differ from $F(x)$ if the hypothesis is true. Hence we must first introduce a quantity that measures the deviation of $\tilde{F}(x)$ from $F(x)$, and we must know the probability distribution of this quantity under the assumption that the hypothesis is true. Then we proceed as follows. We determine a number c such that if the hypothesis is true, a deviation greater than c has a small preassigned probability. If, nevertheless, a deviation larger than c occurs, we have reason to doubt that the hypothesis is true and we reject it. On the other hand, if the deviation does not exceed c, so that $\tilde{F}(x)$ approximates $F(x)$ sufficiently well, we do not reject the hypothesis. Of course, if we do not reject the hypothesis, this means that we have insufficient evidence to reject it and does not exclude the possibility that there are other functions that would not be rejected in the test. In this respect the situation is quite similar to that in Sec. 24.7.

Table 24.10 shows a test of this type, which was introduced by R. A. Fisher. This test is justified by the fact that if the hypothesis is true, then χ_0^2 is an observed value of a random variable whose distribution function approaches that of the chi-square distribution with $K - 1$ degrees of freedom (or $K - r - 1$ degrees of freedom if r parameters are estimated) as n approaches infinity. The requirement that at least five sample values lie in each interval in Table 24.10 results from the fact that for finite n that random variable has only approximately a chi-square distribution. A proof can be found in Ref. [G4] listed in Appendix 1. If the sample is so small that the requirement cannot be satisfied, one may continue with the test, but the result should be used with great caution.

Table 24.10
Chi-square Test for the Hypothesis That $F(x)$ is the Distribution Function of a Population from Which a Sample $x_1, \cdots, x_n$ Is Taken

---

1st Step. Subdivide the x-axis into K intervals $I_1, I_2, \cdots, I_K$ such that each interval contains at least 5 values of the given sample $x_1, \cdots, x_n$. Determine the number b_j of sample values in the interval I_j, where $j = 1, \cdots, K$. If a sample value lies at a common boundary point of two intervals, add 0.5 to each of the two corresponding b_j.

2nd Step. Using $F(x)$, compute the probability p_j that the random variable X under consideration assumes any value in the interval I_j, where $j = 1, \cdots, K$. Compute

$$e_j = np_j.$$

(This is the number of sample values theoretically expected in I_j if the hypothesis is true.)

3rd Step. Compute the deviation

$$\text{(1)} \qquad \chi_0^2 = \sum_{j=1}^{K} \frac{(b_j - e_j)^2}{e_j}.$$

4th Step. Choose a significance level (5%, 1% or the like).

5th Step. Determine the solution c of the equation

$$P(\chi^2 \leqq c) = 1 - \alpha$$

from the table of the chi-square distribution with $K - 1$ degrees of freedom (Table A12, Appendix 4). If r parameters of $F(x)$ are unknown and their maximum likelihood estimates (Sec. 24.5) are used, then use $K - r - 1$ degrees of freedom (instead of $K - 1$). If $\chi_0^2 \leqq c$, do not reject the hypothesis. If $\chi_0^2 > c$, reject the hypothesis.

EXAMPLE 1. Test of normality
Test whether the population from which the sample in Table 24.2, Sec. 24.3, was taken is normal.

Solution. The maximum likelihood estimates for μ and σ^2 are $\hat{\mu} = \bar{x} = 364.7$ and $\tilde{\sigma}^2 = 712.9$. The computation in Table 24.11 yields $\chi_0^2 = 2.790$. We choose $\alpha = 5\%$. Since $K = 10$ and we estimated $r = 2$ parameters, we have to use Table A12, Appendix 4, with $K - r - 1 = 7$ degrees of freedom, finding $c = 14.07$ as the solution of $P(\chi^2 \leqq c) = 95\%$. Since $\chi_0^2 < c$, we do not reject the hypothesis that the population is normal.   ∎

Table 24.11
Computations in Example 1

x_j	$\dfrac{x_j - 364.7}{26.7}$	$\Phi\left(\dfrac{x_j - 364.7}{26.7}\right)$	e_j	b_j	Terms in (1)
$-\infty \cdots 325$	$-\infty \cdots -1.49$	$0.0000 \cdots 0.0681$	6.81	6	0.096
$325 \cdots 335$	$-1.49 \cdots -1.11$	$0.0681 \cdots 0.1335$	6.54	6	0.045
$335 \cdots 345$	$-1.11 \cdots -0.74$	$0.1335 \cdots 0.2296$	9.61	11	0.201
$345 \cdots 355$	$-0.74 \cdots -0.36$	$0.2296 \cdots 0.3594$	12.98	14	0.080
$355 \cdots 365$	$-0.36 \cdots 0.00$	$0.3594 \cdots 0.5000$	14.06	16	0.268
$365 \cdots 375$	$0.00 \cdots 0.39$	$0.5000 \cdots 0.6517$	15.17	15	0.002
$375 \cdots 385$	$0.39 \cdots 0.76$	$0.6517 \cdots 0.7764$	12.47	8	1.602
$385 \cdots 395$	$0.76 \cdots 1.13$	$0.7764 \cdots 0.8708$	9.44	10	0.033
$395 \cdots 405$	$1.13 \cdots 1.51$	$0.8708 \cdots 0.9345$	6.37	8	0.417
$405 \cdots \infty$	$1.51 \cdots \infty$	$0.9345 \cdots 1.0000$	6.55	6	0.046

$$\chi_0^2 = 2.790$$

Problems for Sec. 24.10

1. If 100 flips of a coin result in 40 heads and 60 tails, can we assert on the 5% level that the coin is fair?

2. If in 10 flips of a coin we get the same ratio as in Prob. 1 (4 heads and 6 tails), is the conclusion the same as in Prob. 1? First conjecture, then compute.

3. If in rolling a die 180 times we get $1, \cdots, 6$ with the absolute frequencies 25, 31, 33, 27, 29, 35, can we claim on the 5% level that the die is fair?

4. Solve Prob. 3 if the sample is 39, 22, 41, 26, 20, 32.

5. In a classical experiment, R. Wolf rolled a die 20 000 times and obtained $1, \cdots, 6$ with the absolute frequencies 3407, 3631, 3176, 2916, 3448, 3422. Test whether the die was fair, using $\alpha = 5\%$.

6. Can we assert on the 5% level that a random variable X assumes each of three values with probability 1/3 if we observe in 198 trials the absolute frequencies 87, 49, 62? On the 1% level?

7. In one of G. Mendel's classical experiments the result was 355 yellow peas and 123 green peas. Is this in agreement with Mendel's theory according to which in the present case the ratio of yellow peas to green peas should be 3:1?

8. A manufacturer claims that in a process of producing razor blades, only 2.5% of the blades are dull. Test the claim against the alternative that more than 2.5% of the blades are dull, using a sample of 400 blades containing 17 dull blades. (Use $\alpha = 5\%$.)

9. If in some week the numbers of customers in a bank from Monday to Friday were 2680, 1600, 2020, 2250, 3650, can we assert on the 1% level that the random variable $X = $ *Number of customers per day* has the same probability (0.2) for each weekday?

10. Repeat the test in Prob. 9 for the more reasonable model that the five probabilities are 0.22, 0.14, 0.16, 0.18, 0.30, respectively, choosing $\alpha = 1\%$.

11. Using the given sample, test that the corresponding population has a Poisson distribution. x is the number of alpha particles per 7.5-second intervals observed by E. Rutherford and H. Geiger in one of their classical experiments in 1910, and $a(x)$ is the absolute frequency ($=$ number of time periods during which exactly x particles were observed).

x	$a(x)$	x	$a(x)$	x	$a(x)$
0	57	5	408	10	10
1	203	6	273	11	4
2	383	7	139	12	2
3	525	8	45	$\geqq 13$	0
4	532	9	27		

12. Test the hypothesis that the population from which the given sample of 1000 sheets of paper was taken has a Poisson distribution. x is the number of dark spots on a sheet and $a(x)$ is the observed absolute frequency (number of sheets with x spots). Use (a) $\alpha = 5\%$, (b) $\alpha = 1\%$.

x	0	1	2	3	4	$\geqq 5$
$a(x)$	419	352	154	56	19	0

13. Was the sample in Table 24.4, Sec. 24.3, taken from a normal population?

14. Test for normality of the population from which the following sample was taken, where x [kg/mm^2] is the tensile strength of steel sheets 0.3 mm thick.

x	Absolute Frequency	x	Absolute Frequency
<42.0	15	43.5–44.0	22.5
42.0–42.5	11	44.0–44.5	19.5
42.5–43.0	15	44.5–45.0	12
43.0–43.5	14	>45.0	19

15. The following sample consists of values of the mass of cotton material [g/m^2]. Test the hypothesis that the corresponding population is normal.

Class mark	96	98	100	102	104	106	108	110	112
Absolute class frequency	1	0	1	2	8	19	28	30	41

Class mark	114	116	118	120	122	124	126	128	130
Absolute class frequency	66	50	27	8	5	3	0	1	1

24.11 Nonparametric Tests

The tests in Sec. 24.7 refer to normal populations. In many applications the distributions of populations are known to be not normal or are even completely unknown. Then we may apply a so-called **nonparametric test** or **distribution-free test,** which is based on order statistics and therefore is valid for any *continuous* distribution. These tests are simple. However, for a normal distribution the tests in Sec. 24.7 yield better results. To illustrate the basic idea of a nonparametric test, let us consider two typical examples.

EXAMPLE 1. Sign test for the median

A **median** is a solution $x = \tilde{\mu}$ of the equation $F(x) = 0.5$, where F is the distribution function. Using the sample of differences in Example 5, Sec. 24.7, that is,

$$16 \quad 16 \quad 2 \quad 6 \quad 0 \quad 0 \quad 13 \quad 8,$$

we shall test the hypothesis $\tilde{\mu} = 0$, that is, the outputs corresponding to the two different working conditions do not differ significantly.

Solution. We choose the alternative $\tilde{\mu} > 0$ and the significance level $\alpha = 5\%$. If the hypothesis is true, the probability p of a positive difference is the same as that of a negative difference. Hence in this case, $p = 0.5$, and the random variable

$$X = Number\ of\ positive\ values\ among\ n\ values$$

has a binomial distribution with $p = 0.5$. Our sample has eight values. We omit the values 0, which do not contribute to the decision. Then six values are left, all of which are positive. Since

$$P(X = 6) = \binom{6}{6} (0.5)^6 (0.5)^0 = 0.0156 = 1.56\% < \alpha,$$

we reject the hypothesis.

EXAMPLE 2. Test for arbitrary trend

A certain machine is used for cutting lengths of wire. Five successive pieces had the lengths

$$29 \quad 31 \quad 28 \quad 30 \quad 32.$$

Using this sample, test the hypothesis that there is *no trend,* that is, the machine does not have the tendency to produce longer and longer pieces or shorter and shorter pieces. Assume that the type of machine suggests the alternative that there is *positive trend,* that is, there is the tendency of successive pieces to get longer.

Solution. We count the number of **transpositions** in the sample, that is, the number of times a larger value precedes a smaller value:

$$29 \text{ precedes } 28 \qquad (1 \text{ transposition}),$$

$$31 \text{ precedes } 28 \text{ and } 30 \qquad (2 \text{ transpositions}).$$

The remaining three sample values follow in ascending order. Hence in the sample there are $1 + 2 = 3$ transpositions. We now consider the random variable

$$T = Number\ of\ transpositions.$$

If the hypothesis is true (no trend), then each of the $5! = 120$ permutations of five elements 1 2 3 4 5 has the same probability ($1/120$). We arrange these permutations according to their number of transpositions:

$T = 0$	$T = 1$	$T = 2$	$T = 3$	
1 2 3 4 5	1 2 3 5 4	1 2 4 5 3	1 2 5 4 3	
	1 2 4 3 5	1 2 5 3 4	1 3 4 5 2	
	1 3 2 4 5	1 3 2 5 4	1 3 5 2 4	
	2 1 3 4 5	1 3 4 2 5	1 4 2 5 3	
		1 4 2 3 5	1 4 3 2 5	
		2 1 3 5 4	1 5 2 3 4	
		2 1 4 3 5	2 1 4 5 3	
		2 3 1 4 5	2 1 5 3 4	etc.
		3 1 2 4 5	2 3 1 5 4	
			2 3 4 1 5	
			2 4 1 3 5	
			3 1 2 5 4	
			3 1 4 2 5	
			3 2 1 4 5	
			4 1 2 3 5	

From this we obtain

$$P(T \leq 3) = \tfrac{1}{120} + \tfrac{4}{120} + \tfrac{9}{120} + \tfrac{15}{120} = \tfrac{29}{120} = 24\%.$$

Hence we do not reject the hypothesis.

Values of the distribution function of T in the case of no trend are shown in Table A14, Appendix 4. Our method and those values refer to *continuous* distributions. Theoretically we may then expect that all the values of a sample are different. Practically, some sample values may still be equal, because of rounding off. If m values are equal, add $m(m - 1)/4$ (= mean value of the transpositions in the case of the permutations of m elements), that is, $\tfrac{1}{2}$ for each pair of equal values, $\tfrac{3}{2}$ for each triple, etc.   ∎

So far we have discussed random experiments in which we observe a single quantity and get samples whose values are single numbers. In the next section we consider **regression analysis,** in which we get samples of *pairs* of numbers. We shall begin with the famous **least-squares principle** for fitting straight lines through given data (pairs of numbers) and then turn to random variables Y whose mean μ depends on an ordinary variable x (examples: the yield of a chemical process on temperature, the wear of tires on mileage).

Problems for Sec. 24.11

1. Apply the sign test to the sample in Prob. 1, Sec. 24.7.
2. Does a process of producing plastic pipes of length $\tilde{\mu} = 2$ meters need adjustment if in a sample, 4 pipes have the exact length and 15 are shorter and 3 longer than 2 meters? (Use the normal approximation of the binomial distribution; cf. Sec. 23.7.)
3. Do the calculations in Prob. 2 without the use of the DeMoivre–Laplace limit theorem.
4. Each of 10 patients were given two different sedatives A and B. The following table shows the effect (increase of sleeping time, measured in hours). Using the sign test, find out whether the difference is significant.

A	1.9	0.8	1.1	0.1	−0.1	4.4	5.5	1.6	4.6	3.4
B	0.7	−1.6	−0.2	−1.2	−0.1	3.4	3.7	0.8	0.0	2.0
Difference	1.2	2.4	1.3	1.3	0.0	1.0	1.8	0.8	4.6	1.4

5. Assuming that the populations corresponding to the samples in Prob. 4 are normal, apply the test explained in Example 3, Sec. 24.7.
6. Apply the sign test to a sample of size 19 consisting of 4 positive and 10 negative values and 5 zeros. (The values of the binomial distribution needed here are not contained in Table A6, Appendix 4, but must be computed.)
7. Test whether a thermostatic switch is properly set to 25°C against the alternative that its setting is too low. Use a sample of eight values, seven of which are less than 25°C and one is greater than 25°C.
8. Set up a sign test for the lower quartile q_{25} (defined by the condition $F(q_{25}) = 0.25$).
9. How would you proceed in the sign test if the hypothesis is $\tilde{\mu} = \tilde{\mu}_0$ (any number) instead of $\tilde{\mu} = 0$?
10. Test the hypothesis that for a certain type of voltmeter, readings are independent of temperature T against the alternative that they tend to increase with T. Use a sample of values obtained by applying a constant voltage:

Temperature T [°C]	20	25	30	35	40
Reading V [volts]	110.2	111.4	110.6	111.0	112.1

11. In a swine feeding experiment, the following gains in weight [kg] of 10 animals (ordered according to increasing amounts of food given per day) were recorded:

20   17   19   18   23   16   25   28   24   22

Test for no trend against positive trend.

12. Does the amount of fertilizer increase the yield of wheat X [kg/plot]? Use a sample of values ordered according to increasing amounts of fertilizer:

41.4   43.3   39.6   43.0   44.1   45.6   44.5   46.7

13. Does the yield of some chemical process depend on temperature? (Test against positive trend.)

Temperature [°C]	0	15	30	45	60	75
Yield [kg/min]	0.8	1.1	0.9	1.6	1.2	1.8

14. Do increases in the price of milk lead to a decrease in sales if daily sales figures [gallons] (ordered according to increasing daily prices) are as follows?

82   76   40   60   65   50   35   55   25

15. Apply the test in Example 2 to the following data (x = disulfide content of a certain type of wool, measured in percent of the content in unreduced fibers; y = saturation water content of the wool, measured in percent).

x	10	15	30	40	50	55	80	100
y	50	46	43	42	36	39	37	33

24.12 Pairs of Measurements. Fitting Straight Lines

We shall now discuss experiments in which we observe or measure two quantities simultaneously. In practice we may distinguish between two kinds of experiment, as follows.

1. In **correlation analysis** both quantities are random variables and we are interested in relations between them. (We shall not discuss this branch of statistics.)

2. In **regression analysis** one of the two variables, call it x, can be regarded as an ordinary variable, that is, can be measured without appreciable error. The other variable, Y, is a random variable. x is called the *independent* (sometimes the *controlled*) *variable,* and one is interested in the dependence of Y on x. Typical examples are the dependence of the blood pressure Y on the age x of a person or, as we shall now say, the regression of Y on x, the regression of the gain of weight Y of certain animals on the daily ration of food x, the regression of the heat conductivity Y of cork on the specific weight x of the cork, etc.

In the experiment the experimenter first selects n values $x_1, \cdots, x_n$ of x and then observes Y at those values of x, so that he obtains a sample of the form $(x_1, y_1), (x_2, y_2), \cdots, (x_n, y_n)$. In regression analysis the mean μ of Y is assumed to depend on x, that is, it is a function $\mu = \mu(x)$ in the ordinary sense. The curve of $\mu(x)$ is called the *regression curve of Y on x*. In the present section we shall discuss the simplest case, when $\mu(x)$ is a linear function, $\mu(x) = \alpha + \beta x$. Then we may want to plot the sample values as n points in the xY-plane, fit a straight line through them and use this line for estimating $\mu(x)$ for given values of x, so that we know what values of Y we can expect if we choose certain values of x. If the points are scattered, fitting "by eye" becomes unreliable and we need a mathematical method for fitting lines that yields a unique result depending only on the points. A widely used procedure is the **method of least squares** by Gauss. In our present situation it may be formulated as follows.

The straight line should be fitted through the given points so that the sum of the squares of the distances of those points from the straight line is minimum, where the distance is measured in the vertical direction (the y-direction).

General assumption (A1)
The x-values $x_1, \cdots, x_n$ of our sample $(x_1, y_1), \cdots, (x_n, y_n)$ are not all equal.

Consider a sample $(x_1, y_1), \cdots, (x_n, y_n)$ of size n. The vertical distance (distance measured in the y-direction) of a sample value (x_j, y_j) from a straight line $y = a + bx$ is $|y_j - a - bx_j|$; cf. Fig. 536. Hence the sum of the squares of these distances is

(1)
$$q = \sum_{j=1}^{n} (y_j - a - bx_j)^2.$$

In the method of least squares we choose a and b such that q is minimum. q depends on a and b, and a necessary condition for q to be minimum is

(2)
$$\frac{\partial q}{\partial a} = 0 \quad \text{and} \quad \frac{\partial q}{\partial b} = 0.$$

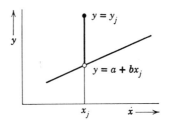

Fig. 536. Vertical distance of a point (x_j, y_j)
from a straight line $y = a + bx$

We shall see that from this condition we obtain the formula

(3)
$$y - \bar{y} = b(x - \bar{x})$$

where

(4) $\bar{x} = \dfrac{1}{n}(x_1 + \cdots + x_n)$ and $\bar{y} = \dfrac{1}{n}(y_1 + \cdots + y_n).$

The straight line (3) is called the **regression line** *of the y-values of the sample on the x-values of the sample*. Its slope b is called the **regression coefficient** *of y on x,* and we shall see that

(5)
$$b = \frac{s_{xy}}{s_1{}^2}.$$

Here,

(6) $s_1{}^2 = \dfrac{1}{n-1} \displaystyle\sum_{j=1}^{n} (x_j - \bar{x})^2 = \dfrac{1}{n-1}\left[\sum_{j=1}^{n} x_j^2 - \dfrac{1}{n}\left(\sum_{j=1}^{n} x_j\right)^2 \right]$

and

(7) $s_{xy} = \dfrac{1}{n-1} \displaystyle\sum_{j=1}^{n} (x_j - \bar{x})(y_j - \bar{y}) = \dfrac{1}{n-1}\left[\sum_{j=1}^{n} x_j y_j - \dfrac{1}{n}\left(\sum_{i=1}^{n} x_i\right)\left(\sum_{j=1}^{n} y_j\right) \right].$

In (5) and (7), s_{xy} is called the **covariance** of the sample. Obviously, the regression line (3) passes through the point $(\bar{x}, \bar{y})$.

To derive (3), we use (1) and (2), finding

$$\frac{\partial q}{\partial a} = -2 \sum (y_j - a - bx_j) = 0$$

$$\frac{\partial q}{\partial b} = -2 \sum x_j(y_j - a - bx_j) = 0$$

(where we sum over j from 1 to n). Thus

$$na + b \sum x_j = \sum y_j$$

$$a \sum x_j + b \sum x_j^2 = \sum x_j y_j.$$

Because of Assumption (A1), the determinant

$$n \sum x_j^2 - \left(\sum x_j\right)^2 = n(n-1)s_1{}^2$$

[cf. (6)] of this system of linear equations is not zero, and the system has a unique solution [cf. (4), (6), (7)]

$$
(8) \qquad a = \bar{y} - b\bar{x}, \qquad b = \frac{n \sum x_j y_j - \sum x_i \sum y_j}{n(n-1)s_1^2}.
$$

This yields (3) with b given by (5)-(7). (The equality of the two expressions for s_1^2 in (6) may be shown by the reader (cf. Prob. 15); similarly for (7).)

Hand calculations can be simplified by *coding*, that is, by setting

$$
(9) \qquad x_j = c_1 x_j^* + l_1, \qquad y_j = c_2 y_j^* + l_2
$$

and choosing the constants c_1, c_2, l_1, l_2 such that the transformed values x_j^* and y_j^* are as simple as possible. We first compute the values $\bar{x}^*$, $\bar{y}^*$, x_1^{*2}, s_{xy}^* corresponding to the transformed values and then

$$
(10) \qquad
\begin{aligned}
\bar{x} &= c_1 \bar{x}^* + l_1, & \bar{y} &= c_2 \bar{y}^* + l_2 \\
s_1^2 &= c_1^2 s_1^{*2}, & s_{xy} &= c_1 c_2 s_{xy}^*.
\end{aligned}
$$

Table 24.12
Regression of the Decrease of Volume y [%]
of Leather on the Pressure x [Atmospheres]

Given values		Auxiliary values	
x_j	y_j	x_j^2	$x_j y_j$
4,000	2.3	16,000,000	9,200
6,000	4.1	36,000,000	24,600
8,000	5.7	64,000,000	45,600
10,000	6.9	100,000,000	69,000
28,000	19.0	216,000,000	148,400

(C. E. WEIR, *Journ. Res. Nat. Bureau of Standards* **45** (1950), 468.)

EXAMPLE 1. Regression line
The decrease of volume y [%] of leather for certain fixed values of high pressure x [atmospheres] was measured. The results are shown in the first two columns of Table 24.12. Find the regression line of y on x.

Solution. We see that $n = 4$ and obtain the values $\bar{x} = 28\,000/4 = 7000$, $\bar{y} = 19.0/4 = 4.75$,

$$
s_1^2 = \frac{1}{3}\left(216\,000\,000 - \frac{28\,000^2}{4}\right) = \frac{20\,000\,000}{3}
$$

$$
s_{xy} = \frac{1}{3}\left(148\,400 - \frac{28\,000 \cdot 19}{4}\right) = \frac{15\,400}{3}.
$$

Hence $b = 15\,400/20\,000\,000 = 0.000\,77$, and the regression line is

$$
y - 4.75 = 0.000\,77(x - 7000) \qquad \text{or} \qquad y = 0.000\,77x - 0.64. \qquad \blacksquare
$$

We now make the following two assumptions.

Assumption (A2)

For each fixed x the random variable Y is normal with mean

$$(11) \qquad \qquad \mu(x) = \alpha + \beta x$$

and variance σ^2 where the latter is independent of x.

Assumption (A3)

The n performances of the experiment by which we obtain a sample $(x_1, y_1), \cdots , (x_n, y_n)$ are independent (cf. p. 1233).

β in (11) is called the **regression coefficient** *of the population,* because it can be shown that under Assumptions (A1)–(A3) the maximum likelihood estimate of β is the sample regression coefficient b in (5).

Under Assumptions (A1)–(A3) we may obtain a confidence interval for β, as shown in Table 24.13.

Table 24.13
Determination of a Confidence Interval for β in (11) under Assumptions (A1)-(A3)

1st Step. Choose a confidence level γ (95%, 99% or the like).

2nd Step. Determine the solution c of the equation

$$(12) \qquad \qquad F(c) = \tfrac{1}{2}(1 + \gamma)$$

from the table of the t-distribution with $n - 2$ degrees of freedom (Table A11 in Appendix 4; n = sample size).

3rd Step. Using a sample $(x_1, y_1), \cdots , (x_n, y_n)$, compute $(n - 1)s_1^2$ from (6), $(n - 1)s_{xy}$ from (7), b from (5),

$$(13) \qquad (n - 1)s_2^2 = \sum_{j=1}^{n} y_j^2 - \frac{1}{n} \left(\sum_{j=1}^{n} y_j \right)^2$$

and

$$(14) \qquad \qquad q_0 = (n - 1)(s_2^2 - b^2 s_1^2).$$

4th Step. Compute $k = c\sqrt{q_0/(n - 2)(n - 1)s_1^2}$. The confidence interval is

$$(15) \qquad \qquad \text{CONF } \{b - k \leqq \beta \leqq b + k\}.$$

EXAMPLE 2. Confidence interval for the regression coefficient

Using the sample in Table 24.12, determine a confidence interval for β by the method in Table 24.13.

Solution. *1st Step.* We choose $\gamma = 0.95$.

2nd Step. Equation (12) takes the form $F(c) = 0.975$, and Table A11 in Appendix 4 with $n - 2 = 2$ degrees of freedom yields $c = 4.30$.

3rd Step. From Example 1 we have $3s_1^2 = 20\,000\,000$ and $b = 0.000\,77$. From Table 24.12 we compute

$$3s_2^2 = 102.2 - \frac{19^2}{4} = 11.95, \qquad q_0 = 11.95 - 20\,000\,000 \cdot 0.000\,77^2 = 0.092.$$

4th Step. We thus obtain $k = 4.30\sqrt{0.092/2 \cdot 20\,000\,000} = 0.000\,206$ and

$$\text{CONF } \{0.000\,56 \leqq \beta \leqq 0.000\,98\}.$$

Problems for Sec. 24.12

1. Fit a straight line by eye. Estimate the stopping distance of a car traveling at 35 miles per hour.

$x = $ Speed (mph)	20	30	40	50
$y = $ Stopping distance (ft)	50	95	150	210

2. Obtain the results in Example 1 by coding; set $x_j = 2000x_j^* + 4000$, $y_j = 0.1y_j^* + 5$.

In each case find and graph the sample regression line of y on x.

3. $(1, 1), (2, 1.7), (3, 3)$ 4. $(2, 3.5), (4, 2), (6, 4.5), (8, 3)$

5. $(1, 1), (2, 2), (3, 4), (4, 2), (5, 4), (6, 5)$

6. $(21, 12), (25, 8), (27, 6), (30, -1), (32, 0)$

7. The sample in Prob. 15, Sec. 24.11

8. Number of revolutions (per minute) x and power y [hp] of a Diesel engine

x	400	500	600	700	750
y	580	1030	1420	1880	2100

9. Deformation x [mm] and Brinell hardness y [kg/mm^2] of a certain type of steel

x	6	9	11	13	22	26	28	33	35
y	68	67	65	53	44	40	37	34	32

10. Mortality y [%] of termites (*Reticulitermes lucifugus* Rossi) as a function of the concentration X [%] of chloronaphthalene

x	0.04	0.15	0.30	1.00	2.00
y	3	16	13	70	90

In each case find a 95% confidence interval for the regression coefficient β, using the given sample and supposing that Assumptions (A2) and (A3) are satisfied.

11. $(1, 1), (2, 2 + a), (3, 3)$, where a is a constant

12. The sample in Prob. 8.

13. The sample in Prob. 9.

14. $x = $ humidity of air [%], $y = $ expansion of gelatin [%]

x	10	20	30	40
y	0.8	1.6	2.3	2.8

15. Derive the second expression for s_1^2 in (6) from the first one. *Hint.* Perform the square, substitute the defining expression for $\bar{x}$ and simplify.

Review Problems for Chapter 24

1. Explain the role of probability theory in statistics.
2. Why do we take samples?
3. Can we draw absolutely certain conclusions about populations from samples?
4. Why is the sample distribution function in the continuous case a step function, whereas the random variable has a continuous distribution function?
5. What do we mean by grouping of samples? Why and when do we group?
6. What is a histogram of a sample? Give a simple example.
7. What is a point estimate? An interval estimate? Why is the latter more useful in most cases?
8. Explain the basic idea of a maximum likelihood estimate for a single parameter from memory.
9. What errors are involved in testing?
10. Does a test *prove* that a hypothesis is true or false? Explain.
11. If we do not reject a hypothesis, are there other hypotheses that also would not be rejected in the same test?
12. On what facts does the choice of α depend in a specific application?
13. Why can we generally obtain better results in testing by taking larger samples?
14. Under what conditions can we (cautiously) apply tests for the normal distribution in the case of other distributions?
15. What is quality control? Give a simple example of your own.
16. What is a single sampling plan for controlling the mean? Give an example.
17. What is a nonparametric test? What are its advantages and disadvantages?
18. What is the intuitive idea of the chi-square test?
19. What is Gauss's least-squares principle (which Gauss found when he was 18 years old)?
20. Why do we not simply fit a straight line through given points by eye?
21. Graph a histogram of the sample 8, 2, 4, 10 and guess $\bar{x}$ and s by inspecting the histogram. Then calculate $\bar{x}$, s^2, and s.

Find the mean, variance and range of the samples:

22. 4.3   5.1   5.0   4.7   4.6
23. 210   220   200   190   200   210   230   210   190   200
24. 16   18   17   15   14   12   17   16

25. Find the maximum likelihood estimates of mean and variance of a normal distribution, using the sample

    16   14   17   10   18   15   16   14   20   18   12   16   15   17   16   15

26. Determine a 95% confidence interval for the mean μ of a normal population with variance $\sigma^2 \doteq 9$, using a sample of size 100 with mean 38.25.
27. What will happen to the length of the interval in Prob. 26 if we reduce the sample size to 25?
28. Determine a 99% confidence interval for the mean of a normal population with standard deviation 2.2, using the sample 28, 24, 31, 27, 22.
29. Obtain a 99% confidence interval for the mean of a normal population with variance $\sigma^2 = 0.36$ from Fig. 524 in Sec. 24.6, using a sample of size 290 with mean 16.30.

30. Using the sample 145.3, 145.1, 145.4, 146.2 and assuming normality, find a 95% confidence interval for σ^2.

31. Do the same task as in Prob. 30, using a sample of size 30 with variance 0.00070.

32. A machine fills boxes weighing Y lb with X lb of salt, where X and Y are normal with means 100 and 5 lb and standard deviations 1.0 and 0.5 lb, respectively. What percent of filled boxes weighing between 104 and 106 lb are to be expected?

Assuming that the populations from which the following samples are taken are normal, determine a 99% confidence interval for the mean μ of the population.

33. 325, 320, 325, 335

34. A sample of lengths of 20 screws with mean 10.20 cm and variance 0.04 cm^2.

35. 124, 127, 126, 122, 124

36. If X is normal with mean 27 and variance 16, what distributions do $-X$, $3X$, and $5X - 2$ have?

37. If X_1 and X_2 are independent normal random variables with means 23 and 4 and variances 3 and 1, respectively, what distribution does $4X_1 - X_2$ have?

38. Find a 99% confidence interval for the parameter p of the binomial distribution, using Pearson's result in the last row of Table 23.1 in Sec. 23.2.

39. Using the sample in Example 1, Sec. 24.4, test the hypothesis $\mu = 0.80$ in. (the length indicated on the box) against the alternative $\mu \neq 0.80$ in. (Assume normality.)

40. Suppose that in operating battery-powered electrical equipment, it is less expensive to replace all batteries at fixed intervals than to replace each battery individually when it breaks down, provided the standard deviation of the lifetime is less than a certain limit, say, less than 5 hr. Set up and apply a suitable test, using a sample of 28 values of lifetimes with standard deviation $s = 3.5$ hr and assuming normality; choose $\alpha = 5\%$.

41. Suppose that in the past the standard deviation of weights of certain 25.0-oz packages filled by a machine was 0.4 oz. Test the hypothesis H_0: $\sigma = 0.4$ against the alternative H_1: $\sigma > 0.4$ (an undesirable increase), using a sample of 10 packages with standard deviation 0.5 oz and assuming normality. (Choose $\alpha = 5\%$.)

42. If simultaneous measurements of electric voltage by two different types of voltmeter yield the differences 0.8, 0.2, -0.3, 0.1, 0.0, 0.5, 0.7, 0.2 [volts], can we assert at the 5% level that there is no significant difference in the calibration of the two types of instrument? Assume normality.

43. Brand A gasoline was used in nine similar automobiles under identical conditions. The corresponding sample of nine values [miles per gallon] had mean 20.2 and standard deviation 0.5. Under the same conditions, high-power brand B gasoline gave a sample of 10 values with mean 21.8 and standard deviation 0.6. Is the mileage of B significantly better than that of A? Assume normality. Choose $\alpha = 5\%$.

44. The two samples 50, 90, 100, 90, 110, 80 and 110, 110, 120, 110, 130, 110, 120 are values of the differences of temperatures [°C] of iron at two stages of casting, taken from two different crucibles. Is the variance of the first population larger than that of the second? Assume normality. Choose $\alpha = 5\%$.

45. Using samples of sizes 10 and 16 with variances $s_1^2 = 50$ and $s_2^2 = 30$ and assuming normality of the corresponding populations, test the hypothesis H_0: $\sigma_1^1 = \sigma_2^2$ against the alternative $\sigma_1^2 > \sigma_2^2$. Choose $\alpha = 5\%$.

46. Set up a control chart (find LCL and UCL) for the mean of a normal distribution (length of screws) with mean 5.00 cm and standard deviation 0.03 cm, for the sample size $n = 8$.

47. (**Three-sigma control chart**) Show that for a control chart for the mean, the requirement of the significance level $\alpha = 0.3\%$ leads to LCL $= \mu - 3\sigma/\sqrt{n}$ and UCL $= \mu + 3\sigma/\sqrt{n}$.

48. How should we change the sample size in controlling the mean of a normal population if we want UCL − LCL to decrease to half its original value?

49. Assuming that the following 10 samples of size two come from a normal distribution (filling of oil) with mean 27.5 lb and variance 0.024 lb^2, set up a control chart for the mean and graph the sample means on the chart.

27.4	27.4	27.5	27.3	27.9	27.6	27.6	27.8	27.5	27.3
27.6	27.4	27.7	27.4	27.5	27.5	27.4	27.3	27.4	27.7

50. Graph the OC curve and the AOQ curve for the single sampling plan for large lots with $n = 5$ and $c = 0$, and find the AOQL.

51. Find θ_0 in Prob. 50 from the curve if the producer's risk is 5%. Find θ_1 in Prob. 50 if the consumer's risk is 10%.

52. Do the same task as in Prob. 50 when $n = 4$ and $c = 1$.

53. If in a single sampling plan for large lots of spark plugs, the sample size is 100 and we want the AQL to be 5% and the producer's risk 2%, what acceptance number c should we choose? (Use the normal approximation.)

54. What is the consumer's risk in Prob. 53 if we want the RQL to be 12%?

55. Find the risks in the single sampling plan with $n = 5$ and $c = 0$, assuming that the AQL is $\theta_0 = 1\%$ and the RQL is $\theta_1 = 15\%$.

56. In a table of random digits, even and odd digits should be *about* equally frequent. Test this, using as a sample the 50 digits in row 0 of Table A10 in Appendix 4. (Choose $\alpha = 5\%$.)

57. If it is known that 25% of certain steel rods produced by a standard process will break when subjected to a load of 5000 lb, can we claim that a new process yields the same breakage rate if we find that in a sample of 80 rods produced by the new process, 27 rods broke when subjected to that load?

58. Can we assert that the traffic on the three lanes of an expressway (in one direction) is about the same on each lane if a count gives 910, 850, 720 cars on the right, middle and left lanes, respectively, during the same interval of time?

59. Are air filters of type A better than type B filters if in 10 trials, A gave cleaner air than B in seven cases, B gave cleaner air than A in one case, whereas in two of the trials the results for A and B were practically the same?

60. Under what condition can we use the sign test as a test for the mean of a continuous distribution?

Summary of Chapter 24
Mathematical Statistics

With a random experiment in which we observe some quantity (number of defectives, height of persons, etc.) there is associated a random variable X whose probability distribution is determined by a distribution function (Sec. 23.4)

(1) $$F(x) = P(X \leqq x)$$

which for each x gives the probability that X assumes any value not exceeding x. In statistics we take random samples $x_1, \cdots, x_n$ of size n by performing that experiment n times (Secs. 24.2, 24.3). The purpose is to draw conclusions from properties of samples about properties of the distribution of the corresponding X (Sec. 24.1). We do this by calculating *point estimates* or *confidence intervals* or by performing a *test* for **parameters** (μ and σ^2 in the normal distribution, p in the binomial distribution, etc.) or by a test for distribution functions.

A **point estimate** (Sec. 24.5) is an approximate value for a parameter in the distribution of X obtained from a sample. For instance, the **sample mean** (Sec. 24.4)

$$(2) \qquad \bar{x} = \frac{1}{n}(x_1 + \cdots + x_n)$$

is an estimate of the mean μ of X, and the **sample variance** (Sec. 24.4)

$$(3) \qquad s^2 = \frac{1}{n-1}[(x_1 - \bar{x})^2 + \cdots + (x_n - \bar{x})^2]$$

is an estimate of the variance σ^2 of X. Point estimation can be done by the basic **maximum likelihood method** (Sec. 24.5).

Confidence intervals (Sec. 24.6) are intervals $\theta_1 \leq \theta \leq \theta_2$ with endpoints calculated from a sample such that with a high probability γ, which we can choose (95%, for instance), we obtain an interval that contains the unknown true value of the parameter θ in the distribution of X. We denote such an interval by CONF $\{\theta_1 \leq \theta \leq \theta_2\}$.

In a **test** for a parameter we test a *hypothesis* $\theta = \theta_0$ against an *alternative* $\theta = \theta_1$ [this notation has nothing to do with the θ_1 above] and then, on the basis of a sample, reject (or do not reject) the hypothesis in favor of the alternative (Sec. 24.7). Like any conclusion about X from samples, this may involve errors leading to a false decision, as follows. There is a small probability α (which we can choose, 5% or 1%, for instance) that we reject a true hypothesis, and there is a probability β (which we can compute and decrease by taking larger samples) that we do not reject a false hypothesis. α is called the **significance level** and $1 - \beta$ the **power** of the test. Among many applications, testing is used in **quality control** (Sec. 24.8) and **acceptance sampling** (Sec. 24.9).

If not merely a parameter but the kind of distribution of X is unknown, we can use the **chi-square test** (Sec. 24.10) for testing the hypothesis that some function $F(x)$ is the unknown distribution function of X. This is done by determining the discrepancy between $F(x)$ and the distribution function $\tilde{F}(x)$ of a given sample.

"Distribution-free" or **nonparametric tests** are tests that apply to any distribution, since they are based on combinatorial ideas. Two such tests are discussed in Sec. 24.11.

Finally, Sec. 24.12 deals with samples consisting of pairs of values, as they arise if we simultaneously consider two quantities, and gives an introduction to **regression analysis.**

Appendix 1

References

General References

[1] Abramowitz, M. and I. A. Stegun (eds.), *Handbook of Mathematical Functions.* 10th printing, with corrections. Washington, DC: National Bureau of Standards, 1972. (Also New York: Dover, 1965.)

[2] Cajori, F., *A History of Mathematics.* 3rd ed. New York: Chelsea, 1980.

[3] CRC Handbook of Mathematical Sciences. 5th ed. West Palm Beach, FL: CRC Press, 1978.

[4] Courant, R. and D. Hilbert, *Methods of Mathematical Physics.* 2 vols. New York: Wiley-Interscience, 1953, 1962.

[5] Courant, R. and F. John, *Introduction to Calculus and Analysis.* 2 vols. New York: Wiley-Interscience, 1965, 1974.

[6] *Encyclopedia of Mathematics.* 10 vols. Hingham, MA: Reidel, to appear 1987–1990.

[7] Erdélyi, A., W. Magnus, F. Oberhettinger and F. G. Tricomi, *Higher Transcendental Functions.* 3 vols. New York: McGraw-Hill, 1953, 1955.

[8] Fletcher, A., J. C. P. Miller, L. Rosenhead and L. J. Comrie, *An Index of Mathematical Tables.* Oxford: Blackwell, 1962.

[9] Fulks, W., *Advanced Calculus.* 3rd ed. New York: Wiley, 1978.

[10] Hildebrand, F. B., *Advanced Calculus for Applications.* 2nd ed. Englewood Cliffs, NJ: Prentice-Hall, 1976.

[11] Itô, K. (ed.), *Encyclopedic Dictionary of Mathematics.* 4 vols. 2nd ed. Cambridge, MA: MIT Press, 1987.

[12] Jahnke, E., F. Emde and F. Lösch, *Tables of Higher Functions.* 7th ed. Stuttgart: Teubner, 1966.

[13] Kreyszig, E., *Introductory Functional Analysis with Applications.* New York: Wiley, 1978.

[14] Luke, Y. L., *The Special Functions and their Approximations.* 2 vols. New York: Academic Press, 1969.

[15] Magnus, W., F. Oberhettinger and R. P. Soni, *Formulas and Theorems for the Special Functions of Mathematical Physics.* 3rd ed. New York: Springer, 1966.

[16] Pearson, C. E. (ed.), *Handbook of Applied Mathematics.* 2nd ed. New York: Van Nostrand Reinhold, 1983.

[17] Rudin, W., *Principles of Mathematical Analysis.* 3rd ed. New York: McGraw-Hill, 1976.

[18] Thomas, G. B. and R. L. Finney, *Calculus and Analytic Geometry.* 6th ed. Reading, MA: Addison-Wesley, 1984.

[19] Whittaker, E. T. and G. N. Watson, *A Course of Modern Analysis.* 4th ed. Cambridge: Harvard University Press, 1927 (reprinted 1965).

Part A.   Ordinary Differential Equations (Chaps. 1-5)

See also Part E: Numerical Analysis.

[A1] Bhatia, N. P. and G. P. Szegö, *Stability Theory of Dynamical Systems.* New York: Springer, 1970.

[A2] Birkhoff, G. and G.-C. Rota, *Ordinary Differential Equations*. 3rd ed. New York: Wiley, 1978.

[A3] Cesari, L., *Asymptotic Behavior and Stability Problems in Ordinary Differential Equations*. 3rd ed. New York: Springer, 1971.

[A4] Churchill, R. V., *Operational Mathematics*. 3rd ed. New York: McGraw-Hill, 1972.

[A5] Coddington, E. A. and N. Levinson, *Theory of Ordinary Differential Equations*. New York: McGraw-Hill, 1955.

[A6] Director, S. W., *Circuit Theory*. New York: Wiley, 1975.

[A7] Erdélyi, A., W. Magnus, F. Oberhettinger and F. Tricomi, *Tables of Integral Transforms*. 2 vols. New York: McGraw-Hill, 1954.

[A8] Hartman, P., *Ordinary Differential Equations*. New York: Wiley, 1964.

[A9] Hille, E., *Ordinary Differential Equations in the Complex Domain*. New York: Wiley-Interscience, 1976.

[A10] Ince, E. L., *Ordinary Differential Equations*. New York: Dover, 1956.

[A11] Kamke, E., *Differentialgleichungen, Lösungsmethoden und Lösungen. I. Gewöhnliche Differentialgleichungen*. 3rd ed. New York: Chelsea, 1948. (This extremely useful book contains a systematic list of more than 1500 differential equations and their solutions.)

[A12] McLachlan, N. W., *Ordinary Non-Linear Differential Equations in Engineering and Physical Sciences*. 2nd ed. Oxford: Clarendon, 1956.

[A13] Oberhettinger, F. and L. Badii, *Tables of Laplace Transforms*. New York: Springer, 1973.

[A14] Watson, G. N., *A Treatise on the Theory of Bessel Functions*. 2nd ed. Cambridge: University Press, 1944. (Reprinted 1966.)

[A15] Widder, D. V., *The Laplace Transform*. Princeton, NJ: Princeton University Press, 1941.

Part B. Linear Algebra, Vector Calculus (Chaps. 6-9)

Books on *numerical* linear algebra see also under Part E: Numerical Analysis.

[B1] Bellman, R., *Introduction to Matrix Analysis*. 2nd ed. New York: McGraw-Hill, 1970.

[B2] Faddeev, D. K. and V. N. Faddeeva, *Computational Methods of Linear Algebra*. San Francisco: Freeman, 1963.

[B3] Gantmacher, F. R., *The Theory of Matrices*. 2 vols. New York: Chelsea, 1960.

[B4] Herstein, I. N., *Topics in Algebra*. 2nd ed. Lexington, MA: Xerox College Publishing, 1975.

[B5] Hoffman, K. and R. Kunze, *Linear Algebra*. 2nd ed. Englewood Cliffs, NJ: Prentice-Hall, 1971.

[B6] Kreyszig, E., *Introduction to Differential Geometry and Riemannian Geometry*. Toronto: University of Toronto Press, 1975.

[B7] McDuffee, C. C., *The Theory of Matrices*. New York: Chelsea, 1946.

[B8] Nering, E. D., *Linear Algebra and Matrix Theory*. 2nd ed. New York: Wiley, 1970.

[B9] Wilkinson, J. H., *The Algebraic Eigenvalue Problem*. Oxford: Clarendon, 1965.

[B10] Wilkinson, J. H. and C. Reinsch, *Linear Algebra*. New York: Springer, 1971.

Part C. Fourier Analysis and Partial Differential Equations (Chaps. 10, 11)

Books on *numerical* methods for partial differential equations see also under Part E: Numerical Analysis.

[C1] Bateman, H., *Partial Differential Equations of Mathematical Physics*. Cambridge: University Press, 1944 (reprinted 1964).

[C2] Churchill, R. V. and J. W. Brown, *Fourier Series and Boundary Value Problems*. 3rd ed. New York: McGraw-Hill, 1978.

[C3] Duff, G. F. D. and D. Naylor, *Differential Equations of Applied Mathematics*. New York: Wiley, 1966.

[C4] Erdélyi, A., W. Magnus, F. Oberhettinger and F. Tricomi, *Tables of Integral Transforms*. 2 vols. New York: McGraw-Hill, 1954.

[C5] Friedman, A., *Partial Differential Equations*. Reprint with corrections. Huntington, NY: Krieger, 1976.

[C6] Garabedian, P. R., *Partial Differential Equations*. New York: Wiley, 1964.

[C7] Gilbarg, D. and N. S. Trudinger, *Elliptic Partial Differential Equations of Second Order*. New York: Springer, 1977.

[C8] Hellwig, G., *Partial Differential Equations*. 2nd ed. Stuttgart: Teubner, 1977.

[C9] John, F., *Partial Differential Equations*. New York: Springer, 1971.

[C10] Kellogg, O. D., *Foundations of Potential Theory*. New York: Dover, 1953.

[C11] Rogosinski, W., *Fourier Series*. 2nd ed. New York: Chelsea, 1959.

[C12] Sneddon, I. N., *Elements of Partial Differential Equations*. New York: McGraw-Hill, 1957.

[C13] Sneddon, I. N., *The Use of Integral Transforms*. New York: McGraw-Hill, 1972.

[C14] Sommerfeld, A., *Partial Differential Equations in Physics*. New York: Academic Press, 1949.

[C15] Szegö, G., *Orthogonal Polynomials*. 3rd ed. New York: American Mathematical Society, 1967.

[C16] Titchmarsh, E. C., *Introduction to the Theory of Fourier Integrals*. 2nd ed. Oxford: Clarendon, 1948.

[C17] Trèves, F., *Basic Linear Partial Differential Equations*. New York: Academic Press, 1975.

[C18] Widder, D. V., *The Heat Equation*. New York: Academic Press, 1975.

[C19] Zauderer, E., *Partial Differential Equations of Applied Mathematics*. New York: Wiley, 1983.

[C20] Zygmund, A., *Trigonometric Series*. 2nd ed., reprinted with corrections. Cambridge: University Press, 1977.

Part D.   Complex Analysis (Chaps. 12-17)

[D1] Ahlfors, L. V., *Complex Analysis*. 3rd ed. New York: McGraw-Hill, 1979.

[D2] Ahlfors, L. V. and L. Sario, *Riemann Surfaces*. Princeton, NJ: Princeton University Press, 1960.

[D3] Bieberbach, L., *Conformal Mapping*. New York: Chelsea, 1964.

[D4] Hayman, W. K., *Meromorphic Functions*. Oxford: Clarendon Press, 1975.

[D5] Hille, E., *Analytic Function Theory*. 2 vols. Boston: Ginn, 1959, 1962.

[D6] Knopp, K., *Theory of Functions*. 2 vols. New York: Dover, 1945.

[D7] Nehari, Z., *Conformal Mapping*. New York: McGraw-Hill, 1952.

[D8] Nevanlinna, R., *Analytic Functions*. New York: Springer, 1970.

[D9] Rothe, R., F. Ollendorf and K. Pohlhausen, *Theory of Functions as Applied to Engineering Problems*. New York: Dover, 1961.

[D10] Rudin, W., *Real and Complex Analysis*. 2nd ed. New York: McGraw-Hill, 1974.

[D11] Saks, S. and A. Zygmund, *Analytic Functions*. New York: Elsevier, 1971.

[D12] Titchmarsh, E. C., *The Theory of Functions*. 2nd ed. London: Oxford University Press, 1939 (reprinted 1975).

[D13] Weyl, H., *The Concept of a Riemann Surface*. Reading, MA: Addison-Wesley, 1964.

Part E. Numerical Analysis (Chaps. 18-20)

[E1] Aho, A. V., J. E. Hopcroft and J. D. Ullman, *Data Structures and Algorithms*. Reading, MA: Addison-Wesley, 1983.

[E2] Birkhoff, G. and R. E. Lynch, *Numerical Solution of Elliptic Problems*. Philadelphia: SIAM Publications, 1984.

[E3] Collatz, L., *The Numerical Treatment of Differential Equations*. 3rd ed. New York: Springer, 1966.

[E4] Dahlquist, G. and Å. Björck, *Numerical Methods*. Englewood Cliffs, NJ: Prentice-Hall, 1974.

[E5] Davis, P. and P. Rabinowitz, *Methods of Numerical Integration*. 2nd ed. New York: Academic Press, 1984.

[E6] DeBoor, C., *A Practical Guide to Splines*. New York: Springer, 1978.

[E7] Dongerra, J., J. R. Bunch, C. B. Moler and G. W. Stewart, *LINPACK Users Guide*. Philadelphia: SIAM Publications, 1978.
EISPACK see [E30].
EISPACK 2 see [E13].

[E8] Evans, D. J. (ed.), *Sparsity and Its Applications*. Cambridge: Cambridge University Press, 1985.

[E9] Faddeev, D. K. and V. N. Faddeeva, *Computational Methods of Linear Algebra*. San Francisco: Freeman, 1963.

[E10] Feingold, C., *Introduction to Data Processing*. 3rd ed. Dubuque, IA: Brown, 1980.

[E11] Forsythe, G. E., M. A. Malcolm and C. B. Moler, *Computer Methods for Mathematical Computations*. Englewood Cliffs, NJ: Prentice-Hall, 1977.

[E12] Forsythe, G. E. and W. R. Wasow, *Finite-Difference Methods for Partial Differential Equations*. New York: Wiley, 1960.

[E13] Garbow, B. S., J. M. Boyle, J. J. Dongerra and C. B. Moler, *Matrix Eigensystem Routines: EISPACK Guide Extension*. New York: Springer, 1972.

[E14] Gear, C. W., *Numerical Initial Value Problems in Ordinary Differential Equations*. Englewood Cliffs, NJ: Prentice-Hall, 1971.

[E15] Golub, G. H. and C. F. Van Loan, *Matrix Computations*. Baltimore: Johns Hopkins University Press, 1984.

[E16] Gotlieb, C. C. and L. R. Gotlieb, *Data Types and Structures*. Englewood Cliffs, NJ: Prentice-Hall, 1978.

[E17] Hennie, F., *Introduction to Computability*. Reading, MA: Addison-Wesley, 1977.

[E18] Hildebrand, F. B., *Introduction to Numerical Analysis*. 2nd ed. New York: McGraw-Hill, 1974.

[E19] Horowitz, E. and S. Sahni, *Fundamentals of Computer Algorithms*. Rockville, MD: Computer Science Press, 1985.

[E20] Hull, T. E., W. H. Enright, B. M. Fellen and A. E. Sedgewick, Comparing numerical methods for ordinary differential equations. *SIAM Journal on Numerical Analysis* **9** (1972), 603-637.

[E21] IMSL (International Mathematical and Statistical Libraries), *FORTRAN Subroutines for Mathematics and Statistics. User's Manual*. 4 vols. Houston, TX: IMSL, 1985.

[E22] Isaacson, E. and H. B. Keller, *Analysis of Numerical Methods*. New York: Wiley, 1966.

[E23] Knuth, D. E., *The Art of Computer Programming*. 3 vols. 2nd ed. Reading, MA: Addison-Wesley, 1973.

[E24] Lambert, J. D., *Computational Methods in Ordinary Differential Equations*. New York: Wiley, 1983.

[E25] Lapidus, L. and W. Schiesser (eds.), *Numerical Methods for Differential*

Equations: Recent Developments in Algorithms, Software, New Applications. New York: Academic Press, 1976.

LINPACK see [E7].

[E26] Ostrowski, A., *Solution of Equations in Euclidean and Banach Spaces.* 3rd ed. New York: Academic Press, 1973.

[E27] Parlett, B. N., *The Symmetric Eigenvalue Problem.* Englewood-Cliffs, NJ: Prentice-Hall, 1980.

[E28] Schumaker, L. L., *Spline Functions: Basic Theory.* New York: Wiley, 1981.

[E29] Shampine, L. and M. Gordon, *Computer Solution of Ordinary Differential Equations.* San Francisco: Freeman, 1975.

[E30] Smith, B. T., J. M. Boyle, J. J. Dongarra, B. S. Garbow, Y. Ikebe, V. C. Klema and C. B. Moler, *Matrix Eigensystem Routines–EISPACK Guide.* 2nd ed. New York: Springer, 1976.

[E31] Smith, G. D., *Numerical Solution of Partial Differential Equations.* London: Oxford University Press, 1974.

[E32] Stetter, H. J., *Analysis of Discretization Methods for Ordinary Differential Equations.* New York: Springer, 1973.

[E33] Stewart, G. W., *Introduction to Matrix Computations.* New York: Academic Press, 1973.

[E34] Strang, G. and G. J. Fix, *An Analysis of the Finite Element Method.* Englewood Cliffs, NJ: Prentice-Hall, 1973.

[E35] Tikhonov, A. and V. Arsenin. *Solutions of Ill-Posed Problems.* New York: Wiley, 1977.

[E36] Todd, J. (ed.), *Survey of Numerical Analysis.* New York: McGraw-Hill, 1962.

[E37] Todd, J., *Basic Numerical Mathematics.* 2 vols. New York: Academic Press, 1978–1980.

[E38] Wakerly, J. F., *Microcomputer Architecture and Programming.* New York: Wiley, 1981.

[E39] Wilkinson, J. H., *The Algebraic Eigenvalue Problem.* Oxford: Clarendon, 1965.

[E40] Wilkinson, J. H. and C. Reinsch, *Linear Algebra.* New York: Springer, 1971.

[E41] Young, D. M., Iterative Solution of Large Linear Systems. New York: Academic Press, 1971.

Part F.   Optimization (Chaps. 21, 22)

[F1] Bellman, R. E. and R. Kalaba, *Dynamic Programming and Modern Control Theory.* New York: Academic Press, 1971.

[F2] Berge, C., *Graphs and Hypergraphs.* 2nd ed. New York: North-Holland, 1976.

[F3] Bondy, J. A. and U. S. R. Murty, *Graph Theory with Applications.* New York: North-Holland, 1976.

[F4] Christofides, N., *Graph Theory.* New York: Academic Press, 1975.

[F5] Christofides, N., A. Mingozzi, P. Toth and C. Sandi (eds.), *Combinatorial Optimization.* New York: Wiley, 1979.

[F6] Chvátal, V., *Linear Programming.* San Francisco: Freeman, 1983.

[F7] Dantzig, G. B., *Linear Programming and Extensions.* 2nd ed. Princeton, NJ: Princeton University Press, 1963.

[F8] Dixon, L. C. W., *Nonlinear Optimization.* London: English Universities Press, 1972.

[F9] Even, S., *Graph Algorithms.* Rockville, MD: Computer Science Press, 1979.

[F10] Fletcher, R., *Practical Methods of Optimization.* New York: Wiley, 1980.

[F11] Ford, L. R., Jr. and D. R. Fulkerson, *Flows in Networks.* Princeton, NJ: Princeton University Press, 1962.

[F12] Gass, S. I., *Linear Programming, Methods and Applications*. 3rd ed. New York: McGraw-Hill, 1969.

[F13] Gondran, M. and M. Minoux, *Graphs and Algorithms*. New York: Wiley, 1984.

[F14] Gotlieb, C. C. and L. R. Gotlieb, *Data Types and Structures*. Englewood Cliffs, NJ: Prentice-Hall, 1978.

[F15] Harari, F., *Graph Theory*. Reading, MA: Addison-Wesley, 1969.

[F16] Knuth, D. E., *The Art of Computer Programming*. 3 vols. 2nd ed. Reading, MA: Addison-Wesley, 1973.

[F17] Künzi, H. P. and W. Krelle, *Nonlinear Programming*. Waltham, MA: Blaisdell, 1966.

[F18] Lawler, E. L., *Combinatorial Optimization: Networks and Matroids*. New York: Holt, Rinehart & Winston, 1977.

[F19] Murty, K. G., *Linear and Combinatorial Programming*. New York: Wiley, 1976.

[F20] Papadimitriou, C. H. and K. Steiglitz, *Combinatorial Optimization: Algorithms and Complexity*. Englewood Cliffs, NJ: Prentice-Hall, 1982.

Part G. Probability and Statistics (Chaps. 23, 24)

[G1] Anderson, T. W., *An Introduction to Multivariate Statistical Analysis*. 2nd ed. New York: Wiley, 1984.

[G2] Cochran, W. G., *Sampling Techniques*. 3rd ed. New York: Wiley, 1977.

[G3] Cox, D. R., *Planning of Experiments*. New York: Wiley, 1958.

[G4] Cramér, H., *Mathematical Methods of Statistics*. Princeton: Princeton University Press, 1946. (Reprinted 1961.)

[G5] David, H. A., *Order Statistics*. 2nd ed. New York: Wiley, 1981.

[G6] Dodge, H. F. and H. G. Romig, *Sampling Inspection Tables*. 2nd ed. New York: Wiley, 1959.

[G7] Feller, W., *An Introduction to Probability Theory and Its Applications*. Vol. I. 3rd ed. New York: Wiley, 1968.

[G8] Gibbons, J. D., *Nonparametric Statistical Inference*. 2nd ed. New York: Dekker, 1985.

[G9] Harter, H. L. and D. B. Owen (eds.), *Selected Tables in Mathematical Statistics*. Vol. 1. Chicago: Markham, 1970.

[G10] Huber, P. J., *Robust Statistics*. New York: Wiley, 1981.

[G11] Huff, D., *How to Lie with Statistics*. New York: Norton, 1954.

[G12] IMSL, *FORTRAN Subroutines for Mathematics and Statistics. User's Manual*. 4 vols. Houston, TX: IMSL, 1985.

[G13] Johnson, N. L. and F. C. Leone, *Statistics and Experimental Design in Engineering and the Physical Sciences*. 2 vols. 2nd ed. New York: Wiley, 1977.

[G14] Kendall, M. and A. Stuart, *The Advanced Theory of Statistics*. 3 vols. 4th ed. New York: Macmillan, 1977, 1979, 1983. (Vol. 3 coauthored by K. Ord.)

[G15] Kreyszig, E., *Introductory Mathematical Statistics. Principles and Methods*. New York: Wiley, 1970.

[G16] Maindonald, J. H., *Statistical Computation*. New York: Wiley, 1984.

[G17] Pearson, E. S. and H. O. Hartley, *Biometrika Tables for Statisticians*. Vol. I. 3rd ed. Cambridge: University Press, 1966.

[G18] Rand Corporation, *A Million Random Digits with 100,000 Normal Deviates*. Glencoe, IL: Free Press, 1955.

[G19] Randles, R. H. and D. A. Wolfe, *Introduction to the Theory of Nonparametric Statistics*. New York: Wiley, 1979.

[G20] Scheffé, H., *The Analysis of Variance*. New York: Wiley, 1959.

Appendix 2

Answers to Odd-Numbered Problems

SECTION 1.1, page 10

1. First order
3. Second order
5. Second order
11. $y = -2e^{-x/2} + c$
13. $y = -\frac{1}{4} \sin 2x + ax + b$
15. $y' + y \tan x = 0$
17. $xy' = 4y$
19. $y = \frac{1}{2}e^{x^2}$
21. $y = -\frac{1}{2}x^3$
23. About 1600 years
27. 7.42 km = 4.61 mi
29. $y = y_0 \exp(-0.000\ 116\ x)$

SECTION 1.2, page 14

1. Suppose you forgot $\tilde{c}$ in (4), wrote $\ln|y| = -x^2$, transformed this to $y = e^{-x^2}$ and afterward added a constant c. Then $y = e^{-x^2} + c$, which is not a solution of $y' = -2xy$.
3. $y = ce^{kx}$
5. $y = c \sin x$
7. $y = c \exp(x^3)$
9. $y = ce^{-ax} - b/a$
11. $y = (c - 2e^x)^{-1/2}$
13. $y = \sin(x + c)$
15. $y = c \ln|x|$
17. $y = c\sqrt{\sin 2x}$
19. $xy = 1$
21. $x^2 - y^2 = 4$
23. $r = -4 \sin^2 \theta$
25. $y = 1.4 \cosh x$
27. $I = I_0 e^{-(R/L)t}$

SECTION 1.3, page 22

1. $v^2 = 2gR^2/r$ at $r_0 = R + 1000 = 7372$ gives $v(r_0) = 10.39$ km/sec.
3. $pV = c = const$. This is the law empirically found by Boyle (1662) and Mariotte (1676).
5. $-v_0$
7. 70.7%, 25%
9. $2^7 p_0 = 128 p_0$
11. 10 hr
13. $y(t) = 0.01e^{-0.275t}$
15. $T(t) = 23 - 8e^{-kt}$, $T(1) = 19$ gives $k = \ln 2$, $t = 6.32$ min.
19. $5\frac{1}{2}$ times, which is surprisingly little.
21. 1.21 m
23. $y = cx$
25. $y^2 - x^2 = c$

SECTION 1.4, page 26

1. $y = x(\ln|x| + c)$
3. $y = x + x(c - x^2)^{-1/2}$
5. $y = x + x/(c - \ln|x|)$
7. $y = x \arcsin(x + c)$
9. $y = x(e^x + c)^{1/4}$
11. $y = x(\ln x - 7.4)$

13. $y = x(x^2 + 5)^{1/2}$

15. $y^2 - 2xy - x^2 = 4$

17. $y = x^{-1} \ln (x + c)$

19. $y = x + \dfrac{1 + ce^{2x}}{1 - ce^{2x}}$

21. $(y - x)^2 + 10y - 2x = c$

23. $u' = a + by' = a + bf(u), \ du/(a + bf(u)) = dx$

25. $y = [(c - 15x)^{-1/3} - 2]/5$

27. $y = x\sqrt{\ln x^2}$

29. $y = ax, \ y/x = a = const, \ y' = g(a) = const$

SECTION 1.5, page 30

1. $2x \, dx + 18y \, dy = 0$, ellipses

3. $2x \, dx - 2y \, dy = 0$, hyperbolas

5. $(y \, dx + x \, dy)e^{xy} = 0$, hyperbolas

7. $-2x^{-3}y \, dx + x^{-2} \, dy = 0$, parabolas

9. $(2xy^{-1} \, dx - x^2y^{-2} \, dy)e^{x^2/y} = 0$, parabolas

11. $xy = c$

13. $xy^2 = c$

15. $(e^x - e^y)x = c$

17. $\cosh x \cos y = c$

19. $re^{-\theta} = c$

21. No, $y = 1/(3 \ln |x| + c)$

23. Yes, $\sinh y \cos 2x = c$

25. No, $y = c - x^2$

27. Yes, $y = c \sin 2x$

29. No, $y = c/\cosh x$

31. $y = cx$

33. $b = k, \ ax^2 + 2kxy + ly^2 = c$

35. $\partial f/\partial y = g'(x)h(y)$

37. $y = cx^3$

39. $x^2 + y/x = c$

41. Hyperbola $(x - 2)(y + 3) = 4$

43. $x^3(y - 1)^2 = -18$

45. $\sin \pi x \cos 2\pi y = 1$

47. Hyperbola $(x - 2)(y - 2) = 4$

49. Ellipse $x^2 + 4(y - 1)^2 = 4$

SECTION 1.6, page 33

1. $x^2y = c$

3. $y = cx$

5. $e^x \sin y = c$

7. $ye^{xy} = c$

9. $F = e^x, \ 2e^x - e^y = c$

11. $F = 1/x^2, \ \sinh y = cx$

13. $F = 1/(x + 1)^2, \ y = c(x + 1) - 1$

17. $F = e^y, \ e^y \sin x = c$

19. $F = \sinh x, \ \sinh^2 x \cos y = c$

21. $F = x, \ x^2 \cos y + x^4 = c$

23. $F = x, \ x^3 e^y + x^2 y = c$

25. $F = e^{x^2}, \ e^{x^2} \tan y = c$

27. $\partial P/\partial y = \partial Q/\partial x + (a/x)Q$

29. $\partial Q/\partial x = \partial P/\partial y + P$

31. $H(u)F, \ u = x^2 y^3, \ H$ arbitrary

33. $H(u)F, \ u = x^3 \csc^2 y, \ H$ arbitrary

SECTION 1.7, page 37

3. $y = ce^x - 2$

5. $y = ce^{2x} + x$

7. $y = cx^{-1} + x$

9. $y = (c + x)e^{-kx}$

11. $y = cx^2 + x^2 e^x$

13. $y = ce^{4x} + \frac{1}{2}x^2$

15. $y = (c + e^{x^2})x^{-2}$

17. $y = ce^{-x^2} - 3$

19. $y = -e^{-x} + x^2 + 1$

21. $y = (x + 1)e^{2x} - 2 - 0.25e^{-2x}$

23. $y = e^x x^3 - x$

25. $y = 3 \cosh 2x + 1$

27. The continuity of p and r implies the existence of the integrals in (4), so that (4) determines y explicitly, with c uniquely determined by $y(x_0) = y_0$.

37. $T(t) = T_1 + (T_0 - T_1)e^{-kt}$

39. $dv/dt = 0$ gives $v = (W - B)/k$. The speed would increase without bound.

41. $x = (c + y)e^{-3y}$ **43.** $x = (c + y)e^y$

45. $x = cy^2 + y^2e^y$

47. $u = y^{1-a}, u' = (1 - a)y^{-a}y' = (1 - a)y^{-a}(gy^a - py) = (1 - a)(g - pu)$

49. $y^2 = u, u = ce^{-2x} + x - 0.5$ **51.** $y = (ce^x - 2x - 1)^{-1/3}$

53. $y^2 = 1 + ce^{-x^2}$ **55.** $\sin y = z, z = 2 + ce^{-x^2/2}$

57. $y = (\frac{1}{2}xe^x + cxe^{-x})^{1/2}$

SECTION 1.8, page 45

7. $L = 0.348$ henry   **9.** It will decrease.

11. $\tau_L = 0.025$ sec, $t = 0.02$ sec   **17.** $I(t) = ce^{-(R/L)t} + t/R - L/R^2$

19. $I = I_1 = 1 - \frac{1}{2}e^{-t}$ if $0 \leq t \leq 3, I = I_2 = c_2e^{-(t-3)}$,
  $c_2 = I_2(3) = I_1(3) = 1 - \frac{1}{2}e^{-3} = 0.975$

21. $I_1 = 1 + c_1e^{-0.05t}, I_2 = (1 + c_1e^{-0.5})e^{-0.05(t-10)}$,
  $I_2(20) = (1 + c_1e^{-0.5})e^{-0.5} = 1 + c_1 = I(0)$

29. $RQ' + Q/C = E_0, Q(t) = E_0C(1 - e^{-t/RC})$,
  $V(t) = Q(t)/C = 12(1 - e^{-0.05t})$

31. $Q = ce^{-2t} + 0.001(1 - 2e^{-t})$

33. $Q = 0.000011 \cos 314t + 0.007006 \sin 314t$

35. $Q = \sin \pi t - 2\pi \cos \pi t$

SECTION 1.9, page 51

1. Parallel straight lines (slope $\frac{1}{4}$)   **3.** Congruent parabolas

5. Semicubical parabolas   **7.** Hyperbolas

9. Ellipses   **11.** $x^2/c^2 + y^2/(c^2 + 4) - 1 = 0$

13. $x^2 + (y - c)^2 - 1 - c^2 = 0$ **15.** $y' = y$

17. $y' = 3y/x$ **19.** $y' = 1 + y^2$

21. $y' = y \cot x$ **23.** $y' = \tanh x \cot y$

25. $y = \ln |x| + c^*$ **27.** $y = c^*x^2$

29. $y = \sqrt{c^* - 2x}$ **31.** $xy = c^*$

33. $y = c^*x$ **35.** $2x^2 + y^2 = c^*$

37. $y' = y/x, y = c^*x, x = 0$ **39.** $x^2 - y^2 = c^*$

41. $y = c^*e^{-2x}$

43. $dv = 0$ gives $dy/dx = -v_x/v_y$ and this must equal u_y/u_x (cf. Prob. 42).

45. $u_x = e^x \cos y = v_y, v = e^x \sin y + k(x)$,
  $v_x = e^x \sin y + k'(x) = -u_y = e^x \sin y, k'(x) = 0, e^x \sin y = c^*$

SECTION 1.10, page 57

1. $xy = c$ **3.** $y = ce^{-x^2/2}$ **5.** $x^2 + 4y^2 = c$

7. $v(t) = 5(1 - e^{-4t})/(1 + e^{-4t})$, $5(1 + \frac{1}{6}e^{-4t})/(1 - \frac{1}{6}e^{-4t})$

13. $y_1(1) = 2$, $y_2(1) = 2.5$, $y_3(1) = 2.667$

15. $y_n = 1 + 2\int_0^x y_{n-1}(t)\, dt$, $y_0 = 1$, $y_1 = 1 + 2x$, $y_2 = 1 + 2x + (2x)^2/2!$, etc., $y = e^{2x}$

17. $y_0 = 1$, $y_1 = 1 + x$, $y_2 = 1 + x + x^2 + \frac{1}{3}x^3$, etc.; exact $y = 1/(1 - x)$

19. $y_0 = 0$, $y_1 = x^2 - \frac{1}{4}x^4$, $y_2 = x^2 - \dfrac{1}{4 \cdot 6}x^6$, etc., $y = x^2$

SECTION 1.11, page 64

1. General solution $y = cx^3$. No; $f(x, y) = 3y/x$ is not defined when $x = 0$.

3. $y = cx^2$, c arbitrary. No; $f(x, y) = 2y/x$ is not defined when $x = 0$.

5. $y = (x - 1)^2$, $y \equiv 0$

7. This would violate the uniqueness.

9. $\frac{1}{2}$, the largest value of $b/(1 + b^2)$, where $1 + b^2$ is the smallest K.

11. $\alpha = \frac{1}{4}$, the largest value of $b/(b^2 + 4)$, where $b^2 + 4$ is the smallest K. The solution $y = 2 \tan 2x$ exists for $|x| < \pi/4$.

13. $|f(x, y_2) - f(x, y_1)| = |y_2 + y_1||y_2 - y_1| \leq 2b|y_2 - y_1|$

15. $y = (x - 1)^2$, $y \equiv 0$

17. $|\sin y_2 - \sin y_1| = 2\left|\sin\dfrac{y_2 - y_1}{2}\right|\left|\cos\dfrac{y_2 + y_1}{2}\right| \leq |y_2 - y_1|$; cf. (12), Appendix 3.

19. $y \equiv 0$, $y = ce^{x^2/2}$ $(c > 0)$, $y = ce^{-x^2/2}$ $(c < 0)$

CHAPTER 1 (REVIEW PROBLEMS), page 65

11. $y = cx^4 + 3$

13. $e^{-x}\cos 2y = c$

15. $y = \tan(\frac{1}{2}x^2 + x + c)$

17. $y^2 = 1 + ce^{-x^2}$

19. $y = x^{-2}(c + \frac{1}{3}\cosh 3x)$

21. $2\sin x + \sin y = c$

23. $e^y = ce^x + x^2$

25. $y = \tan(x + c) - x$

27. $y = (e^x + \sin x)x^3 + 2x^2$

29. $4\sin x + \sin y = 0$

31. $y = 2\sin x$

33. $y = \arccos(x - 3.1001)$

35. $y = (4.5x - x^2)^{-1}$

37. $2(y - 1)^2 + x^2 = c^*$

39. $(y - 1)^2 - (x - 1)^2 = c^*$

41. 1.7 days, 2.7 days

43. 6.58 days, 43.7 days

45. $I = I_0 e^{-(t - t_2)/RC}$ with $I_0 = -0.99E_0/R$ from (7), Sec. 1.8

47. $v(t) = 37.19(1 - e^{-0.1089t})$, $s(t) = 37.19[t - 9.183(1 - e^{-0.1089t})]$, $s(2.22) = 9.2$ meters

49. $a = 9 \cdot 10^6$ meters/sec^2, 5.5 meters

51. $y^2 - x^2 = c^*$

53. $y = c\sqrt{x}$

55. $r^2 = a^2\cos 2\theta$

SECTION 2.1, page 73

3. Factor c out.

5. Since $y'' + py' + qy = cr$.

7. Since $y'' + py' + qy = r - r$.

11. $y = c_1 e^x + c_2$

13. $y = c_1 \sinh x + c_2$

15. $y = c_1^{-1} \ln |c_1 x + 1| + c_2$

17. $y = (c_1 x + c_2)^{-1}$

19. $x = e^y + c_1 y + c_2$

21. $x = -\cos y + c_1 y + c_2$

23. $y = \frac{1}{2} e^{3t} + \frac{1}{2}, t \approx 1$

25. $y = [(2t + 4)^{3/2} - 2]/3, y(6) = 62/3, y'(6) = 4$

27. $y = e^x - 1$

29. $y = x^4$

SECTION 2.2, page 76

1. e^x, e^{-x}

3. e^{3x}, e^{-3x}

5. $e^{-x/2}, e^{x/4}$

7. $e^{-\pi x}, e^{\pi x}$

9. $e^{0.4x}, e^{-0.2x}$

11. e^{2ix}, e^{-2ix}

13. $e^{-(2+i)x}, e^{-(2-i)x}$

15. $1, e^x$

17. (a) $e^{-4x}, 1$;   (b) $c_1 e^{-4x} + c_2$

19. $(\lambda - \lambda_1)(\lambda - \lambda_2) = \lambda^2 - (\lambda_1 + \lambda_2)\lambda + \lambda_1 \lambda_2 = \lambda^2 + a\lambda + b$.
Now compare.

21. $y'' - 2y' = 0$

23. $y'' + 2y' + 5y = 0$

25. $y'' + 2\alpha y' + (\alpha^2 + \omega^2)y = 0$

SECTION 2.3, page 81

1. Linearly independent

3. Linearly independent

5. Linearly dependent

7. Linearly independent

9. Linearly dependent

11. $c_1 e^{5x} + c_2 e^{-5x}$

13. $c_1 e^{2x} + c_2 e^{-x}$

15. $(c_1 + c_2 x)e^{4x}$

17. $c_1 e^{3ix} + c_2 e^{-3ix}$

19. $(c_1 + c_2 x)e^{x/3}$

21. $y = 3e^{2x} - 2e^{-2x}$

23. $y = 3e^{-x/2}$

25. $y = (2 + 3x)e^{-1.5x}$

27. $y = 0.8 e^{x/2}$

29. $y = e^{0.5x} - e^{-x}$

31. $y'' + y' - 6y = 0$

33. $y'' + \omega^2 y = 0$

35. $y'' + 6y' + 10y = 0$

37. $y_2 = x \ln x$

39. $y_2 = x^2 + x$

41. $y_2 = x^{-1} \cos x$

45. Impossible, since the functions of a basis are not proportional.

SECTION 2.4, page 88

1. $A \cos 5x + B \sin 5x$

3. $e^x(A \cos x + B \sin x)$

5. $e^{kx}(A \cos 2\pi x + B \sin 2\pi x)$

7. I, $c_1 e^{2x} + c_2 e^{-2x}$

9. II, $e^{-0.3x}(A \cos x + B \sin x)$

11. II, $e^{-2x}(A \cos \frac{1}{2}x + B \sin \frac{1}{2}x)$

13. III, $(c_1 + c_2 x)e^{-4.5x}$

15. I, $c_1 e^{1.6x} + c_2 e^{-1.6x}$

17. $y = (1 - x)e^{x/4}$

19. $y = e^{0.1x}(\cos 10x - \sin 10x)$

21. $y = 4$

23. $y = (3 + 2x)e^{-10x}$

25. $y = e^x(\cos \pi x - \sin \pi x)$

27. $y = 2e^{-4x} + 4e^{x/4}$

29. $y = e^{-0.5x}$

31. $y = (c_1 + c_2 x)e^x$

37. $y = 3e^{4x}$

39. $y = e^{2x} - 1$

41. $y = e^x - e^{-2x}$

43. $y = -3e^x \cos x$

45. If $y \neq 0$, $y(P_1) = y(P_2) = 0$ and y_1 satisfies (1), (9), so does $y_2 = y_1 + y$, and $y_2 \neq y_1$. Conversely, if y_1 and y_2 are different solutions of (1), (9), then $y = y_1 - y_2 \neq 0$ is a solution of (1) and $y(P_1) = y(P_2) = 0$.

SECTION 2.5, page 91

1. $9e^{3x}$, $10e^{2x} - 2e^{-x}$, 0

3. 0, $3e^{2x}$, 0, $-3e^{-x}$

5. $y = c_1 e^{2x} + c_2 e^{-x}$

7. $y = c_1 e^{x/2} + c_2 e^{-x/3}$

9. $y = c_1 e^{0.2x} + c_2 e^{-2.2x}$

11. $y = e^{0.9x}(A \cos 2.1x + B \sin 2.1x)$

13. $y = (c_1 + c_2 x)e^{2x/\pi}$

15. $y = e^{-x} \cos x$

17. $y = (1 - 3x)e^{-x/3}$

19. $y = 3e^{-x/2} - \frac{1}{2}e^{x/4}$

21. $y = 25xe^{-0.5x}$

SECTION 2.6, page 99

1. $y = (y_0^2 + v_0^2 \omega_0^{-2})^{1/2} \cos (\omega_0 t - \arctan (v_0/y_0\omega_0))$

3. 3.52 cycles/sec, 0.284 sec

5. Lower by $\sqrt{2}$, higher, unchanged

7. $1/2\pi$, $1/\pi$, $3/2\pi$, $2/\pi$

9. $k = k_1 + k_2$, 0.225, 0.318, 0.390

11. $mL\theta'' = -mg \sin \theta \approx -mg\theta$ (the tangential component of $w = mg$), $\theta'' + \omega_0^2 \theta = 0$, $\omega_0^2 = g/L$. Ans. $\sqrt{g/L}/2\pi$.

13. $T = 2\pi \sqrt{\dfrac{mL}{mg + (k_1 + k_2)L}}$

15. $\theta(t) = 0.2618 \cos 3.7t + 0.0472 \sin 3.7t$ [rad]

17. $my'' = -a\gamma y$, where $m = 1$ kg, $ay = \pi \cdot 0.01^2 \cdot 2y$ meter^3 is the volume of the water which causes the restoring force $a\gamma y$ with $\gamma = 9800$ nt ($=$ weight/meter^3). $y'' + \omega_0^2 y = 0$, $\omega_0^2 = a\gamma/m = a\gamma = 0.000628\gamma$. Frequency $\omega_0/2\pi = 0.4$ [sec^{-1}].

21. If c_1 and c_2 have the same sign.

27. (a) $c = 0.282\sqrt{mk}$, $0.6245\sqrt{mk}$; (b) $c = 0.283\sqrt{mk}$, $0.632\sqrt{mk}$

29. The positive solutions of $\tan t = 1$, that is, $\pi/4$ (max), $5\pi/4$ (min), etc.

31. Cycle: $t = 1$, $e^{-10\alpha} = \frac{1}{2}$, $\alpha = (\ln 2)/10$, $c = 2m\alpha = \alpha = 0.0693$ [kg/sec] by (8).

33. $y = [(v_0 + \alpha y_0)t + y_0]e^{-\alpha t}$

35. $y = A \cos \gamma x + B \sin \gamma x$, $\gamma^2 = F/EI$, $A = 0$ since $y(0) = 0$, and $\cos \gamma l = 0$, $\gamma l = \pi/2$ since $y'(l) = 0$; this yields F_{crit} and $y = B \sin (\pi x/2l)$. Note that further forces and solutions satisfying the boundary conditions are obtained from $\gamma l = 3\pi/2$, $5\pi/2$, etc.

SECTION 2.7, page 106

5. $y = c_1 x^{-4} + c_2 x^5$

7. $y = (c_1 + c_2 \ln x)x^4$

9. $y = c_1 x^{-1/2} + c_2 x^{-3/2}$

11. $y = x^{-1}[A \cos (2 \ln x) + B \sin (2 \ln x)]$

13. $y = (c_1 + c_2 \ln x)x^{-1.8}$

15. $y = c_1 x^{-0.2} + c_2 x^{0.5}$

17. $y = (c_1 + c_2 \ln x)x^{-3}$

19. $y = c_1 x^{-3} + c_2 x^{1.5}$

21. $y = (c_1 + c_2 \ln x)x^{0.6}$

23. $y = -2/x$

25. $y = 2 \cos (3 \ln x)$

27. $y = x^{0.1}$

29. $y = -x \cos (\ln x)$

31. $x = e^t$, $t = \ln x$; by the chain rule, $y' = \dot{y}t' = \dot{y}/x$, $y'' = \ddot{y}/x^2 - \dot{y}/x^2$, hence $\ddot{y} - \dot{y} + a\dot{y} + by = 0$.

35. $2z - 3 = x$, $y = c_1(2z - 3)^{-2} + c_2(2z - 3)^{-1/2}$

SECTION 2.8, page 111

1. $W = (\lambda_2 - \lambda_1) \exp (\lambda_1 x + \lambda_2 x)$

3. $W = e^{-ax}\omega$

5. $W = \nu x^{2\mu - 1}$

7. $y'' - 2y' + y = 0$

9. $y'' + 16y = 0$

11. $x^2 y'' - 3xy' + 4y = 0$

13. $y'' - 4y = 0$

15. $y'' + 4y = 0$

17. $W = 0$, since $y_1' = 0$, $y_2' = 0$ at such an x, by calculus.

19. $z_1 = (y_1 + y_2)/2$, $z_2 = (y_1 - y_2)/2$, det $[a_{jk}] = -1/2$

SECTION 2.9, page 116

15. $y = 4x^2 - 4$

17. $y = x^{1/2} - 2x^{-1/2} + 1$

19. $y = -5 \sin 2x$

21. Linearly independent

23. Linearly independent

25. Linearly dependent

27. Linearly independent

29. Linearly dependent

33. Suppose $y_1 = ky_2$. Then (4) holds if we choose any $k_1 \neq 0$, $k_2 = -kk_1$, and $k_3 = 0$, $k_4 = 0$, $\cdots$.

SECTION 2.10, page 120

1. $y^{IV} - 6y''' + 9y'' = 0$

3. $y^{IV} - 2y'' + y = 0$

5. $y''' + 6y'' + 12y' + 8y = 0$

7. $y^{IV} - 2y'' + y = 0$

9. $c_1 + c_2 e^{-x} + c_3 e^x$

11. $c_1 e^x + c_2 e^{-x} + c_3 e^{2x} + c_4 e^{-2x}$

13. $c_1 e^x + c_2 e^{-x} + c_3 \cos x + c_4 \sin x$

15. $(c_1 + c_2 x + c_3 x^2)e^{-x}$

17. $y = 5 - \cos 3x$

19. $y = 2e^x + 3 \cos x$

21. $y = -e^x + \cos 2x$

23. $c_1 e^x + A \cos 2x + B \sin 2x$

25. $c_1 e^{-x} + (c_2 + c_3 x)e^{-x/2}$

27. $c_1 e^{-3x} + e^x(A \cos 3x + B \sin 3x)$

33. $y = c_1 x^{-1} + c_2 + c_3 x$

35. $y = c_1 x^{-2} + c_2 + c_3 x^2$

SECTION 2.11, page 124

1. $c_1 e^{-x} + c_2 e^{-2x} + x^2 - 3x + 8$

3. $A \cos x + B \sin x + 2.5 \sin 3x$

5. $c_1 e^{x/2} + c_2 e^{x/4} + 2e^x + 3x + 2$

7. $c_1 e^x + c_2 e^{-4x} + 2xe^x$

9. $(c_1 + c_2 x)e^{3x} + x^2 e^{3x}$

11. $y = -\cos x + 6 \sin x + 2x$

13. $y = 5e^x - e^{2x} + x^2$

15. $y = -e^x + xe^x$

17. $y = \sin x - x \cos x$ **19.** $y = \ln x$

23. $A \cos x + B \sin x + e^x + x \sin x$

SECTION 2.12, page 128

1. $x^2 + x - 2$ **3.** $-x \cos x$

5. $-5 \cos 2x + \sin 2x$ **7.** $1.5x^4 - 5x^3 + 9.5x^2 - 11x + 6$

9. $2x^3 - 3x^2 + 15x - 8$ **11.** $A \cos x + B \sin x - x^2 - x + 2$

13. $c_1 e^x + c_2 e^{-x} + 0.5xe^x$ **15.** $c_1 e^{3x} + c_2 e^x + 0.5xe^{3x}$

17. $c_1 e^{-2x} + c_2 e^x + xe^x/3$ **19.** $c_1 e^x + c_2 e^{2x} + c_3 e^{-2x} + e^{-x}$

21. $5 \cos 5x - \sin 5x + 0.2x$ **23.** $0.3e^x + 2x^2$

25. $e^{-x} - e^{2x} + xe^{2x}$ **27.** $\cos 2x + 2 \sin 2x + e^{-2x} + x^2$

29. $e^{-x} \cos 3x + x^2$

31. $D(D - s)y = kx^n$, $D^{n+2}(D - s)y = kD^{n+1}[x^n] = 0$,
$y = K_0 + K_1 x + \cdots + K_{n+1} x^{n+1} + ce^{sx}$

33. $ce^{-x} + x^4 - 4x^3 + 12x^2 - 24x + 24$

35. $ce^{-x} + \cos x + \sin x$

SECTION 2.13, page 135

1. $y = A \cos 3t + B \sin 3t + \frac{1}{8} \sin t$

3. $y = \sin t - 3 \cos t$

5. $y = 8 \sin 2t - \cos 2t$

7. $y = -\cos t + \sin t + \cos 2t + 5 \sin 2t$

9. $y = e^{-t}(A \cos t + B \sin t) + 6 \sin 3t - 7 \cos 3t$

11. $y = e^{-t/2}(A \cos \frac{1}{2}\sqrt{5}t + B \sin \frac{1}{2}\sqrt{5}t) - 5 \cos 3t + 2 \sin 3t$.

13. $y = e^{-2t}(A \cos 4t + B \sin 4t) + \frac{1}{19} \sin t$

15. $y = A \cos t + B \sin t + (1 - \omega^2)^{-1} \cos \omega t$

17. $y = \cos 5t + \sin t$

19. $y = 6e^{-t/4} \sin \frac{1}{4}\sqrt{31}t + 4 \cos t + 2 \sin t$

21. $y = 3te^{-t/2} + 2 \sin t - \sin 2t$

25. $y = (1 - \omega^2)^{-1}(\cos \omega t - \cos t)$

SECTION 2.14, page 140

1. $I = \cos t + 2 \sin t$ **3.** $I = 10(\cos 5t + \sin 5t)$

5. $I = e^{-2t}(A \cos 4t + B \sin 4t) + 10 \cos 4t + 40 \sin 4t$

7. $I = 30e^{-2t} \sin t$

9. $I = e^{-3t}(3 \sin 4t - 4 \cos 4t) + 4 \cos 5t$

13. $S = 0$, that is, $C = 1/\omega^2 L$ **15.** $I = A \cos 10t + B \sin 10t$

17. $I = 5 \sin 5t$ **19.** $I = 5(\cos 4t - \cos 10t)$

21. The discontinuity of $I'(t)$ follows from $LI' + Q/C = E$ and the fact that the charge $Q(t)$ in the capacitor is a continuous function of t. Continuity of $I(t)$ follows by integrating that equation, because $\int E \, dt$ is continuous.

23. $I = \begin{cases} \frac{1}{2}(e^{-t} - \cos t + \sin t) & \text{if } 0 < t < \pi \\ -\frac{1}{2}(1 + e^{-\pi}) \cos t + \frac{1}{2}(3 - e^{-\pi}) \sin t & \text{if } t > \pi \end{cases}$

25. $(1/2\pi)\sqrt{1/LC}$

29. $I = 1 - e^{-t}$ if $t < a$, $I = [(1 - a)e^a - 1]e^{-t}$ if $t > a$

SECTION 2.15, page 144

1. $I_p = \text{Re}[(-0.1 - 0.2i)(\cos 2t + i \sin 2t)] = -0.1 \cos 2t + 0.2 \sin 2t$

3. $I_p = (50 \sin 4t - 110 \cos 4t)/73$ **5.** $I_p = (360 \sin 3t - 300 \cos 3t)/61$

7. $y_p = -4 \cos 2t$ **9.** $y_p = \cos 3t$

SECTION 2.16, page 147

1. $y = (c_1 + c_2 x + x \ln x - x)e^{2x}$

3. $y = (c_1 + c_2 x - \sin x)e^x$ **5.** $y = (c_1 + c_2 x - \cos x)e^{-x}$

7. $y = (c_1 + c_2 x + \frac{4}{35}x^{7/2})e^x$ **9.** $y = (c_1 + c_2 x - \ln x)e^{-2x}$

11. $y = (c_1 + c_2 x + x^{-2})e^{2x}$ **13.** $y = (c_1 + c_2 x + (2x)^{-1})e^x$

15. $y = c_1 x^{-1} + c_2 x - 4$ **17.** $y = c_1 x + c_2 x^2 + \frac{1}{12}x^{-2}$

19. $y = c_1 x + c_2 x^2 + x^4/6$ **21.** $y = c_1 + c_2 x^2 + (2x - 2)e^x$

23. $y = c_1 x^2 + c_2 x^3 + \frac{1}{42}x^{-4}$ **25.** $y = c_1 x + c_2 x^{-1} + 2x^3$

CHAPTER 2 (REVIEW PROBLEMS), page 150

17. $(c_1 + c_2 x)e^x + \frac{1}{2}e^x/x$ **19.** $(c_1 + c_2 x)e^{-2x} + \frac{1}{2}e^x + \frac{9}{2}e^{-x}$

21. $c_1 x + c_2 x^2 - 10x \cos x$ **23.** $c_1\sqrt{x} + c_2 x^{5/2}$

25. $(c_1 + c_2 \ln x)x^3$ **27.** $c_1 e^{-x} + c_2 e^{-3x} + \frac{1}{2}\sin x$

29. $c_1 x + c_2 x^2 + 4x^{-2}$ **31.** $4 \cos 1.7x - 3.2 \sin 1.7x + 12$

33. $5 \cos 2x + (1/\sqrt{15})e^{-x/2} \sin \frac{1}{2}\sqrt{15}x$

35. $\cosh 2x - 2 \cos x$

37. $3 + (2 - x)e^{-x} + x^2$ **39.** $12.8x^{-1} + 12.5x - 26$

41. $-4e^x - 2e^{-2x} - x^2 + 6$ **43.** $e^{-x}(\cos x - \cos 2x \sec x)$

45. $3e^{2x} - 4e^{-2x} + 2 \cos x$

47. $c_1 e^{-x} + c_2 e^{-3x} + 8 \sin 2x - \cos 2x$

49. $c_1 e^{-1999.87t} + c_2 e^{-0.125\,008t}$ **51.** $\frac{1}{73}(50 \sin 4t - 110 \cos 4t)$

53. $\frac{1}{2}\sin 2t$ **55.** $e^{-t} \cos \sqrt{5}t + \frac{1}{2}\sin 2t$

SECTION 3.1, page 160

1. $x = c_1 e^t + c_2 e^{-t}$, $y = c_1 e^t - c_2 e^{-t}$

3. $x = c_1 e^t + c_2 e^{3t}$, $y = \frac{1}{3}(-c_1 e^t + c_2 e^{3t})$

5. $x = c_1 e^{3t} + c_2 e^{-t}$, $y = 2(c_1 e^{3t} - c_2 e^{-t})$

7. $x = \frac{1}{2}(\sin t - \cos t) + c_1$, $y = \frac{1}{2}(\sin t + \cos t) + c_2$

9. $x = -2c_1 e^{-t} - 3c_2 e^{-2t}$, $y = c_1 e^{-t} + c_2 e^{-2t}$

11. $x = \frac{1}{8}e^{4t} + c_1$, $y = -\frac{1}{8}e^{-4t} + c_2$

13. $x = c_1 e^t + c_2 e^{-t} + \cos t + \sin t$, $y = -c_1 e^t + c_2 e^{-t}$

15. $x = \frac{1}{2}t^2 + c_1 t + c_2$, $y = \frac{1}{6}t^3 + \frac{1}{2}c_1 t^2 + (c_2 - c_1)t + c_3$

17. $x = \frac{1}{2}(e^t - e^{-t}) = \sinh t$, $y = \frac{1}{2}(e^t + e^{-t}) = \cosh t$

19. $x = \sin 2t - \cos 2t + 2.2$, $y = \sin 2t + \cos 2t - 1$

21. $x = e^{3t}$, $y = e^{3t}$
23. $x = 2e^{5t} - e^{-5t}$, $y = e^{5t} + 2e^{-5t}$
25. $x = e^{5t} - \frac{5}{3}e^{2t}$, $y = e^{5t} - \frac{2}{3}e^{2t}$
27. $I_1 = 2 - 2e^{-300t}$, $I_2 = 2 + e^{-300t}$, steady-state currents 2 amp, 2 amp.
29. $I_1 = -\frac{200}{3}e^{-2t} + \frac{125}{3}e^{-0.8t} + 25$, $I_2 = -\frac{100}{3}e^{-2t} + \frac{100}{3}e^{-0.8t}$

SECTION 3.2, page 166

1. $\sqrt{A^2 + B^2}$
5. Ellipses
7. Ellipses $y^2 + 4v^2 = const$
9. Hyperbolas $9y^2 - v^2 = const$
11. $v = 1 + y^2$
13. $v = 4y + const$
15. Hyperbolas $yv = const$
17. $v^2 + y^2 = const$
19. $vy^{1/2} = const$
21. Consider $dv/dy = 0$.

SECTION 3.3, page 176

7. $x = Ae^{\tau}$, $y = Be^{2\tau}$, same parabolas, traced in the opposite sense.
9. $x = Ae^t$, $y = Be^{2t}$, $y = (B/A^2)x^2$ if $A \neq 0$, arcs of parabolas and of the y-axis. Unstable improper node at $(0, 0)$.
11. $x'' = -x$, $x^2 + y^2 = const$, orientation clockwise; center at $(0, 0)$.
13. $x = Ae^t + Be^{3t}$, $y = \frac{1}{3}(-Ae^t + Be^{3t})$, unstable node at $(0, 0)$
15. $x = Ae^{3t} + Be^{-t}$, $y = 2Ae^{3t} - 2Be^{-t}$, saddle
19. $(0, 0)$, $(-3, 0)$, $(3, 0)$
21. $(0, 0)$ saddle point, $(-2, 0)$ and $(2, 0)$ centers. *Hint.* To linearize at $(-2, 0)$, set $x = -2 + u$ to get $u' = y$, $y' = -8u + 6u^2 - u^3$.
23. Saddle point as before, stable and attractive spiral points instead of centers. Note the similarity to the situation for the undamped and damped pendulum equations.
25. Unstable node or spiral point when $\Delta \geqq 0$ or $\Delta < 0$, respectively.

CHAPTER 3 (REVIEW PROBLEMS), page 176

11. $x = c_1 e^t + c_2 e^{3t}$, $y = -3c_1 e^t + 3c_2 e^{3t}$
13. $x = c_1 e^t + c_2 e^{2t}$, $y = 4c_1 e^t + 5c_2 e^{2t}$
15. $x = c_1 \cos t + c_2 \sin t + t + 1$,
   $y = (c_2 - c_1)\cos t - (c_2 + c_1)\sin t - t$
17. $x = 2e^{-\sqrt{2}t}$, $y = -2(\sqrt{2} + 1)e^{-\sqrt{2}t}$
19. $x = (1 + t)e^t + \sin t$, $y = te^t + 3\sin t - \cos t$
21. $m_1 y_1'' + k_1 y_1 - k_2(y_2 - y_1) = 0$, $m_2 y_2'' + k_2(y_2 - y_1) = 0$,
   $y_1 = c_1 \cos t + c_2 \sin t + c_3 \cos 3t + c_4 \sin 3t$,
   $y_2 = \frac{3}{2}c_1 \cos t + \frac{3}{2}c_2 \sin t - \frac{1}{2}c_3 \cos 3t - \frac{1}{2}c_4 \sin 3t$
23. $y_1 = 2\sin t$, $y_2 = 3\sin t$
25. $I_1 = (-19 - 62.5t)e^{-5t} - 19\cos t + 62.5\sin t$,
   $I_2 = (-6 + 62.5t)e^{-5t} + 6\cos t + 2.5\sin t$
27. Unstable spiral point
29. Unstable node

SECTION 4.2, page 189

1. $y = -2 + a_3 x^3$

3. $y = a_0(1 + x^2 + x^4 + \cdots) = a_0/(1 - x^2)$

5. $y = a_0(1 + 3x + 4x^2 + \frac{10}{3}x^3 + 2x^4 + \frac{14}{15}x^5 + \cdots) = a_0(x + 1)e^{2x}$

7. $y = a_0(1 - \frac{1}{6}x^2 - \frac{1}{27}x^3 + \frac{1}{216}x^4 + \frac{1}{270}x^5 + \cdots)$

9. $y = a_0 + a_1 x + (\frac{3}{2}a_1 - a_0)x^2 + (\frac{7}{6}a_1 - a_0)x^3 + \cdots$. Setting $a_0 = A + B$ and $a_1 = A + 2B$, we obtain $y = Ae^x + Be^{2x}$. This illustrates the fact that even if the solution of an equation is a known function, the power series method may not yield it immediately in the usual form. Of course, cases really of practical interest in applications are mostly those in which one gets *new* functions.

11. $y = a_1 x + a_0(1 - x^2 - \frac{1}{3}x^4 - \frac{1}{5}x^6 - \frac{1}{7}x^8 - \cdots)$. (This is a particular case of Legendre's equation ($n = 1$), which we consider in Sec. 4.3.)

13. $(t + 1)\dot{y} - y = t + 1$, $y = a_0(1 + t) + (1 + t)(t - \frac{1}{2}t^2 + - \cdots)$
$$= a_0 x + x \ln x \qquad (\dot{y} = dy/dt)$$

15. $y = a_0\left(1 - \dfrac{t^2}{2!} + \dfrac{t^4}{4!} - + \cdots\right) + a_1\left(t - \dfrac{t^3}{3!} + - \cdots\right)$
$$= a_0 \cos(x - 1) + a_1 \sin(x - 1)$$

17. 3

19. 1

21. 3

23. $\sqrt{|k|}$

25. ∞

27. $\sqrt{5/7}$

29. $m = p + 1 = s + 2$

31. $\displaystyle\sum_{m=1}^{\infty} \dfrac{(m + 1)(m + 2)}{(m + 1)^2 + 1} x^m$; 1

33. $\displaystyle\sum_{m=1}^{\infty} \dfrac{m^3}{(m + 1)!} x^m$; ∞

35. $\displaystyle\sum_{m=2}^{\infty} \dfrac{5^m}{m + 1} x^m$; $\frac{1}{5}$

SECTION 4.3, page 194

9. $y''/y' = 2x/(1 - x^2)$, $\ln y' = -\ln(1 - x^2) + c_1$, etc.

17. Set $x = -1$ and use the formula for the sum of a geometric series.

19. Set $x = 0$ and use $(1 + u^2)^{-1/2} = \displaystyle\sum \binom{-1/2}{m} u^{2m}$.

SECTION 4.4, page 203

1. $y_1 = x^{-1} \cos 2x$, $y_2 = x^{-1} \sin 2x$

3. $y_1 = e^x$, $y_2 = e^x \ln x$

5. $y_1 = (1 - x)^{-1}$, $y_2 = x^{-1}$

7. $y_1 = x^3$, $y_2 = x^3 \ln x$

9. $y_1 = \sqrt{x}$, $y_2 = 1 + x$

11. $y_1 = x + 2$, $y_2 = (x + 2)^{-1}$

13. $y_1 = (x - 1)^2$, $y_2 = (x - 1)^{-2}$

15. $y_1 = (1 - x)^{-1}$, $y_2 = x^{-1/2}(1 - x)^{-1}$

17. $y_1 = x$, $y_2 = x \ln x + x^2$

19. $y_1 = \sqrt{x}\, e^x$, $y_2 = e^{2x}$

35. $y = AF(1, \frac{1}{2}, -\frac{1}{2}; x) + Bx^{3/2}F(\frac{5}{2}, 2, \frac{5}{2}; x)$

37. $y = AF(\frac{1}{2}, \frac{1}{2}, \frac{1}{2}; x) + B\sqrt{x}F(1, 1, \frac{3}{2}; x)$

39. $y = AF(1, \frac{3}{2}, \frac{3}{2}; x) + Bx^{-1/2}F(\frac{1}{2}, 1, \frac{1}{2}; x)$
$\quad = A(1 - x)^{-1} + Bx^{-1/2}(1 - x)^{-1}$

41. $y = AF(1, -\frac{1}{3}, \frac{1}{3}; x) + Bx^{2/3}F(\frac{5}{3}, \frac{1}{3}, \frac{5}{3}; x)$

43. $y = c_1F(-1, \frac{1}{3}, \frac{1}{3}; t + 1) + c_2(t + 1)^{2/3}F(-\frac{1}{3}, 1, \frac{5}{3}; t + 1)$

45. $y = C_1F(-\frac{1}{2}, -\frac{1}{2}, \frac{1}{2}; t - 1) + C_2(t - 1)^{1/2}$

SECTION 4.5, page 210

7. $\frac{1}{4}\pi + k\pi$, $k = 1, 2, 3, 4$; 2.5%, 0.8%, 0.4%, 0.2%

9. Perform the indicated product differentiations in (20) and (21), multiply the last formula by $x^{2\nu}$, subtract it from the first and divide the result by x^ν.

11. Start from (21), with ν replaced by $\nu - 1$, and replace $J_{\nu-1}$ by using (20).

17. $J_3 = (8x^{-2} - 1)J_1 - 4x^{-1}J_0$; 0.1289, 0.1623, 0.1981, 0.2353, 0.2726, exact to 4D, except $J_3(2.8) = 0.2727$.

19. Let $J_n(x_1) = J_n(x_2) = 0$. Then $x_1^{-n}J_n(x_1) = x_2^{-n}J_n(x_2) = 0$, and $[x^{-n}J_n(x)]' = 0$ somewhere between x_1 and x_2 by Rolle's theorem. Now use (21). Then use (20) with $\nu = n + 1$.

23. $-2J_2(x) - J_0(x) + c$ **25.** $-2J_4 - 2J_2 - J_0 + c$

SECTION 4.6, page 217

1. $AJ_2(x) + BY_2(x)$ **3.** $AJ_0(\sqrt{x}) + BY_0(\sqrt{x})$

5. $AJ_{1/4}(x^2) + BY_{1/4}(x^2)$ **7.** $x^2[AJ_2(x^2) + BY_2(x^2)]$

9. $\sqrt{x}[AJ_{1/2}(\sqrt{x}) + BJ_{-1/2}(\sqrt{x})] = x^{1/4}[\tilde{A}\sin\sqrt{x} + \tilde{B}\cos\sqrt{x}]$

11. $\sqrt{x}[AJ_{1/4}(\frac{1}{2}x^2) + BY_{1/4}(\frac{1}{2}x^2)]$ **13.** $\sqrt{x}[AJ_{1/4}(\frac{1}{2}kx^2) + BY_{1/4}(\frac{1}{2}kx^2)]$

15. 1.1

17. Set $H_\nu^{(1)} = kH_\nu^{(2)}$, use (10), obtain a contradiction.

19. For $x \neq 0$ all the terms of the series (13) are real and positive.

SECTION 4.7, page 224

1. 1/4, $1/\sqrt{3}$, $1/\sqrt{5}$, 2/15   **3.** 0, $1/\sqrt{2}$, $1/\sqrt{2}$, 0

5. $1/\sqrt{2\pi}$, $(\cos x)/\sqrt{\pi}$, $(\cos 2x)/\sqrt{\pi}$, $\cdots$

7. $\sin\pi x$, $\sin 2\pi x$, $\cdots$ **9.** $1/\sqrt{2}$, $\cos\pi x$, $\cos 2\pi x$, $\ldots$

11. $(1/\sqrt{L})\sin(n\pi x/L)$ **13.** $P_0/\sqrt{2}$, $\sqrt{\frac{3}{2}}P_1(x)$, $\sqrt{\frac{5}{2}}P_2(x)$

15. Set $x = ct + k$.   **17.** $c = k = \frac{1}{2}\pi$

SECTION 4.8, page 230

5. $\lambda = [(2n + 1)\pi/2]^2$; $n = 0, 1, \cdots$, $y_n(x) = \sin(\frac{1}{2}(2n + 1)\pi x)$

7. $\lambda = [(2n + 1)\pi/2L]^2$, $n = 0, 1\cdots$; $y_n(x) = \sin((2n + 1)\pi x/2L)$

9. $\lambda = n^2$, $n = 0, 1, \cdots$; $y_n(x) = \cos nx$

11. $\lambda = (n\pi/L)^2$, $n = 0, 1, \cdots$; $y_n(x) = \cos(n\pi x/L)$

13. $\lambda = 4n^2$, $n = 0, 1, \cdots$, $y_0(x) = 1$, $y_n(x) = \cos nx$, $\sin nx$ $(n \geq 1)$

15. $\lambda = ((2n + 1)\pi/2)^2$, $n = 0, 1, \cdots$; $y_n(x) = \sin((2n + 1)\frac{\pi}{2}\ln|x|)$

17. $\lambda = n^2\pi^2$, $n = 1, 2, \cdots$; $\quad y_n(x) = x \sin(n\pi \ln|x|)$

19. $y'' + \lambda y = 0$, $\quad y(0) = y(2L)$, $\quad y'(0) = y'(2L)$

SECTION 4.9, page 233

1. $x^2 = \frac{1}{3}P_0 + \frac{2}{3}P_2$, $\quad x^4 = \frac{1}{5}P_0 + \frac{4}{7}P_2 + \frac{8}{35}P_4$

3. $2(-P_0 + P_1 - P_2 + P_3)$ **5.** $\frac{1}{4}P_0 + \frac{1}{2}P_1 + \frac{5}{16}P_2 + \cdots$

7. $\frac{1}{2}P_0 + \frac{5}{8}P_2 + \cdots$

13. $G_x = \sum a_n'(x)t^n = \sum He_n'(x)t^n/n! = tG = \sum He_{n-1}(x)t^n/(n-1)!$,

etc.

15. Write $e^{-x^2/2} = v$, $v^{(n)} = d^nv/dx^n$ etc., integrate by parts, use the formula $He_n' = nHe_{n-1}$ (Prob. 13): then, for $n > m$,

$$\int_{-\infty}^{\infty} vHe_mHe_n \, dx = (-1)^n\int_{-\infty}^{\infty} He_m v^{(n)} \, dx = (-1)^{n-1}\int_{-\infty}^{\infty} He_m' v^{(n-1)} \, dx$$

$$= (-1)^{n-1}m \int_{-\infty}^{\infty} He_{m-1}v^{(n-1)} \, dx = \cdots$$

$$= (-1)^{n-m}m! \int_{-\infty}^{\infty} He_0 v^{(n-m)} \, dx = 0.$$

21. Since the highest power in L_m is x^m, it suffices to show that $\int e^{-x}x^kL_n \, dx = 0$ for $k < n$:

$$\int_0^{\infty} e^{-x}x^kL_n(x) \, dx = \frac{1}{n!} \int_0^{\infty} x^k \frac{d^n}{dx^n}(x^ne^{-x}) \, dx = \cdots$$

$$= (-1)^k\frac{k!}{n!} \int_0^{\infty} \frac{d^{n-k}}{dx^{n-k}}(x^ne^{-x}) \, dx = -\frac{k}{n!} \int_0^{\infty} x^{k-1}\frac{d^{n-1}}{dx^{n-1}}(x^ne^{-x}) \, dx = 0.$$

25. By (20) in the problem set to Sec. 4.5, with $\nu = 1$,

$$a_m = \frac{2}{R^2J_1^2(\alpha_{m0})} \int_0^R xJ_0\left(\frac{\alpha_{m0}}{R}x\right) dx = \frac{2}{\alpha_{m0}^2J_1^2(\alpha_{m0})} \int_0^{\alpha_{m0}} wJ_0(w) \, dw$$

$$= \frac{2}{\alpha_{m0}J_1(\alpha_{m0})}, f = 2\left(\frac{J_0(\lambda_{10}x)}{\alpha_{10}J_1(\alpha_{10})} + \frac{J_0(\lambda_{20}x)}{\alpha_{20}J_1(\alpha_{20})} + \cdots\right)$$

27. $a_m = \dfrac{2akJ_1(\alpha_{m0}a/R)}{\alpha_{m0}RJ_1^2(\alpha_{m0})}$ **29.** $a_m = \dfrac{4J_2(\alpha_{m0})}{\alpha_{m0}^2J_1^2(\alpha_{m0})}$

31. $a_m = \dfrac{2R^2}{\alpha_{m0}J_1(\alpha_{m0})}\left[1 - \dfrac{2J_2(\alpha_{m0})}{\alpha_{m0}J_1(\alpha_{m0})}\right]$

35. $x^3 = 16\left[\dfrac{J_3(\alpha_{13}x/2)}{\alpha_{13}J_4(\alpha_{13})} + \dfrac{J_3(\alpha_{23}x/2)}{\alpha_{23}J_4(\alpha_{23})} + \cdots\right]$

CHAPTER 4 (REVIEW PROBLEMS), page 239

17. $a_0 \cos 2x + a_1 \sin 2x$ **19.** $a_0e^{-x} + (a_2 - \frac{1}{2}a_0)x^2$

21. $a_0 \cosh x^2 + a_2 \sinh x^2$ **23.** $a_1x \cosh x + a_2x \sinh x$

25. $a_0(x^{-1} + 1)$ (from $r_1 = -1$); $A_0x^{-2} + A_1(x^{-1} + 1)$ (from $r_2 = -2$)

27. $x[AJ_1(x) + BY_1(x)]$　　　　　**29.** $[AJ_1(x) + BY_1(x)]/x$
31. $(\sin nx)/\sqrt{\pi}$
33. $\sqrt{2/\pi}$, $(2/\sqrt{\pi})\cos 4nx$, $(2/\sqrt{\pi})\sin 4nx$
35. $\frac{1}{2}P_0(\frac{1}{2}x), \frac{1}{2}\sqrt{3}\,P_1(\frac{1}{2}x), \frac{1}{2}\sqrt{5}\,P_2(\frac{1}{2}x)$　　**37.** $\lambda = n^2$, $y_n(x) = \sin nx$
39. $\lambda = n^2\pi^2$, $y_n(x) = \cos(n\pi \ln x)$

SECTION 5.1, page 248

1. $1/s^2 + 4/s$　　　　　　　　　　　**3.** $a/s + b/s^2 + 2c/s^3$
5. $k(1 - e^{-cs})/s$　　　　　　　　　**7.** $-e^{-s}/s + (1 - e^{-s})/s^2$
9. $24s^{-5} + 2s^{-3} + \frac{1}{4}s^{-1}$　　　**11.** $\pi/(s^2 + \pi^2)$
13. $(s \cos\theta - \omega \sin\theta)/(s^2 + \omega^2)$　　**15.** $1/2s + s/(2s^2 + 8)$
17. $\cos^2 \omega t = \frac{1}{2} + \frac{1}{2}\cos 2\omega t$.　$Ans.$ $1/2s + s/(2s^2 + 8\omega^2)$
19. $\frac{1}{2}[s/(s^2 - 16) - 1/s]$　　　　　**21.** $\frac{1}{3}\sin 3t$
23. $\cosh 2t - 2\sinh 2t$　　　　　　　**25.** $0.8e^{-1.3t}$
27. $2 + e^{-2t}$　　　　　　　　　　　**29.** $\frac{1}{24}t^4 + t$
31. $1 - e^{-t}$　　　　　　　　　　　　**33.** $\frac{1}{3}(1 - e^{-3t})$
35. $(e^{at} - e^{bt})/(a - b)$
45. Let $F = \mathscr{L}(f)$, $G = \mathscr{L}(g)$. Then $aF + bG = a\mathscr{L}(f) + b\mathscr{L}(g) = \mathscr{L}(af + bg)$. Apply $\mathscr{L}^{-1}$: $\mathscr{L}^{-1}(aF + bG) = \mathscr{L}^{-1}\mathscr{L}(af + bg) = af + bg = a\mathscr{L}^{-1}(F) + b\mathscr{L}^{-1}(G)$, since $f = \mathscr{L}^{-1}(F), g = \mathscr{L}^{-1}(G)$. Note that we have proved much more: if a *linear* transformation has an inverse, this inverse is itself linear.

SECTION 5.2, page 254

1. $\dfrac{s^2 - 2}{s(s^2 - 4)}$　　**3.** $\dfrac{1}{(s - 1)^2}$　　**5.** $\dfrac{2\omega^2}{s(s^2 + 4\omega^2)}$　　**7.** $\dfrac{s^2 - 1}{(s^2 + 1)^2}$
19. $\mathscr{L}(f'') = \mathscr{L}(-\omega^2 \cos \omega t) = s^2\mathscr{L}(\cos \omega t) - s$, etc.
21. $1 - e^{-t}$　　　　　　　　　　　**23.** $10(1 - \cos 3t)/9$
25. $t + e^{-t} - 1$　　　　　　　　　**27.** $2(1 - e^{-t}) - t$
29. $2e^{3t} - 9t^2 - 6t - 2$　　　　　**31.** $1 + t - \cos t - \sin t$
33. $\frac{2}{3}\sin 3t$　　　　　　　　　　**35.** $A \cos \omega t + (B/\omega)\sin \omega t$
37. $-e^{-t} + 2e^{3t}$　　　　　　　　**39.** $\cos\frac{1}{2}t - 4\sin\frac{1}{2}t$
43. $f' \equiv 0$, $\mathscr{L}(f') = 0 = s\mathscr{L}(f) - 0 - (k - 0)e^{-s} - (0 - k)e^{-2s}$.
　　　$Ans.$ $k(e^{-s} - e^{-2s})/s$
45. $(3 - 2e^{-5s})/s$

SECTION 5.3, page 260

1. $0.4/(s - 2.5)^2$　　　　　　　　　**3.** $(s - 1)/(s^2 - 2s + 2)$
5. $n\pi/[(s + 2)^2 + n^2\pi^2]$　　　　**7.** $(s + 1)/(s^2 + 2s)$
9. $s/(s^2 + 6s + 13)$　　　　　　　　**11.** $3te^{-t}$
13. $0.3e^{-t}\sin t$　　　　　　　　　**15.** $2e^{2t}\sinh 3t$
17. $\frac{5}{3}e^{-\pi t}t^3$　　　　　　　　　**19.** $e^{-t}(\cos \omega t - 2\sin \omega t)$
25. $y = 2te^{2t}$　　　　　　　　　　**27.** $y = (t + 2)e^{-2t}$

29. $y = e^{-t/2} \cos t$ **31.** $k(e^{-as} - e^{-bs})/s$

33. $1/s(1 + e^{-s})$ **35.** e^{-s}/s^2

37. $2e^{-s}/s^3$ **39.** $-se^{-\pi s}/(s^2 + 1)$

41. $e^{-s-2}/(s + 2)$ **43.** $1/s^2 - e^{-s}/s^2 - e^{-s}/s$

45. $2s^{-3} - e^{-s}(2s^{-3} + 2s^{-2} + s^{-1})$ **47.** $K(e^{-2\pi s} - e^{-4\pi s})/(s^2 + 1)$

49. $Ks(1 + e^{-\pi s/\omega})/(s^2 + \omega^2)$ **51.** -1 if $1 < t < 3$

53. $u(t - 2)$ **55.** $e^{2(t-2)}u(t - 2)$

57. $-u(t - 1)\pi^{-1} \sin \pi t$ **59.** $e^{3(t-1)}u(t - 1)$

61. $q(t) = CV_0 e^{-t/RC} \ (t \geqq 0)$

63. $v = 1 - u(t - a)$, $sI + I/s = 1/s - e^{-as}/s$. *Ans.* $\sin t$ if $0 < t < a$, and $\sin t - \sin (t - a)$ if $t > a$

65. $(e^{-t} - \cos t + \sin t)/2$ if $0 < t < \pi$, $[-(1 + e^{-\pi}) \cos t + (3 - e^{-\pi}) \sin t]/2$ if $t > \pi$

SECTION 5.4, page 267

1. $i = 0 \ (t < 1)$, $\quad i = e^{-0.1(t-1)} \ (1 < t < 2)$, $i = e^{-0.1(t-1)} - e^{-0.1(t-2)} \ (t > 2)$

3. $i = 0 \ (t < 2)$, $\quad i = (10e^{-(t-2)} - e^{-0.1(t-2)})/900e^2 \ (t > 2)$

5. $i = 0 \ (t < 1)$, $\quad i = 1 - e^{-0.1(t-1)} \ (t > 1)$

7. $y = k(1 - \cos t) \ (t < a)$, $\quad y = k[\cos (t - a) - \cos t] \ (t > a)$

9. $y = \frac{2}{3} \sin t - \frac{1}{3} \sin 2t \ (0 < t < \pi)$, $\quad y = \frac{4}{3} \sin t \ (t > \pi)$

11. $y = 0 \ (t < 1)$, $\quad y = \frac{1}{3} \sin (3t - 3) \ (t > 1)$

13. $y = \sin t \ (0 < t < \pi)$, $\quad y = 0 \ (\pi < t < 2\pi)$, $\quad y = -\sin t \ (t > 2\pi)$

15. $y = e^{-t} \cos t \ (0 < t < 2\pi)$, $\quad y = e^{-t}(\cos t + e^2 \sin t) \ (t > 2\pi)$

17. $y = 2e^{-3t} + e^{-t} \ (0 < t < \frac{1}{2})$, $y = 2e^{-3t} + e^{-t} + \frac{1}{4}[e^{-3(t-1/2)} - e^{t-1/2}] \ (t > \frac{1}{2})$

19. $y = 5t - 2 \ (0 < t < \pi)$, $\quad y = 5t - 2 - \frac{1}{2}e^{-(t-\pi)} \sin 2t \ (t > \pi)$

SECTION 5.5, page 270

1. $\dfrac{1}{(s - 1)^2}$ **3.** $\dfrac{2}{(s + 1)^3}$ **5.** $\dfrac{4s}{(s^2 - 4)^2}$

7. $\dfrac{2s(s^2 - 3)}{(s^2 + 1)^3}$ **9.** $\dfrac{s^2 + 2s + 5}{(s^2 + 2s - 3)^2}$ **11.** $\dfrac{2s(s^2 - 3\omega^2)}{(s^2 + \omega^2)^3}$

13. $t \sin t$ **15.** $t \sinh 3t$ **17.** $(2 \sinh 5t)/t$

19. $(e^t - 1)/t$ **21.** $(\sin \omega t)/t$

SECTION 5.6, page 276

1. t **3.** $\sin t$

5. te^{kt} **7.** $-\dfrac{t}{2} \cos \omega t + \dfrac{1}{2\omega} \sin \omega t$

9. $\dfrac{t}{2} \cos \omega t + \dfrac{1}{2\omega} \sin \omega t$ **11.** $e^t - 1$

13. $\frac{2}{3} \sin t - \frac{1}{3} \sin 2t$ **15.** $0 \ (t < \pi)$, $-\sin t \ (t > \pi)$

17. $\frac{1}{2}e^{2t} - \frac{1}{2}$ **19.** $6t^2e^{0.4t}$

21. $\frac{1}{2}(\sin t - t \cos t)$ **23.** $(\omega t - \sin \omega t)/\omega^3$

25. $(t \sin \omega t)/2\omega$ **27.** $t \cos \omega t$

29. Set $t - \tau = \sigma$. Then $\displaystyle\int_0^t f(\tau)g(t - \tau)\, d\tau = \int_t^0 g(\sigma)f(t - \sigma)(-d\sigma).$

33. $\mathcal{L}(\delta * f) = \mathcal{L}(\delta)\mathcal{L}(f) = 1 \cdot \mathcal{L}(f) = \mathcal{L}(f)$, since $\mathcal{L}\{\delta(t)\} = 1$ by (4), Sec. 5.4.

39. $-\frac{1}{2} \sin 2t + \sin t$ **41.** $0.2e^{-t} + \cos 5t - 2 \sin 5t$

43. $\frac{1}{2} - \frac{1}{2}e^{-2t}$ if $0 < t < 1$, $e^{-t+1} - \frac{1}{2}(1 + e^2)e^{-2t}$ if $t > 1$

45. $y = e^t$ **47.** $y = \sqrt{2} \sin \sqrt{2}t$

49. $y = \cosh t$

SECTION 5.7, page 286

1. $(e^{at} - e^{bt})/(a - b)$ **3.** $1 + 3 \sinh 3t$ **5.** $(1 - t)e^{-t}$

7. $(t + 1)e^{-t}$ **9.** $te^{-t} \cos t$ **11.** $e^{-t}(1 - t - t^2)$

13. $2e^{2t} + te^t$ **15.** $te^{2t}(\cos t - \sin t)$ **17.** $te^t - 0.5t^2e^{2t}$

25. $2 \cos t + t \sin t$ **27.** $2e^{3t} + e^t$ **29.** $5e^{2t} - 2t^2$

31. $2e^{2t} + e^{-t}$ **33.** $e^t \cos 2t + 2 \sin t$ **35.** $y_1 = \cos t, y_2 = \sin t$

37. $y_1 = e^t + e^{2t}, y_2 = e^{2t}$

39. $y_1 = e^t, y_2 = e^{-t}, y_3 = e^t - e^{-t}$

SECTION 5.8, page 294

1. $\dfrac{1}{s^2} - \dfrac{2\pi e^{-2\pi s}}{s(1 - e^{-2\pi s})}$ **3.** $\dfrac{e^{2(1-s)\pi} - 1}{(1 - s)(1 - e^{-2\pi s})}$

5. $\dfrac{2(e^{2\pi s} - 1 - 2\pi s - 2\pi^2 s^2)}{s^3(e^{2\pi s} - 1)}$ **7.** $\dfrac{-(1 - e^{-\pi s})^2}{s(1 - e^{-2\pi s})}$

9. $[s^{-2}(e^{-\pi s} - e^{-2\pi s}) - \pi s^{-1}e^{-2\pi s}]/(1 - e^{-2\pi s})$

15. $\dfrac{\omega}{s^2 + \omega^2} \coth \dfrac{\pi s}{2\omega}$ **21.** $y = \dfrac{P(1 - (1 + \alpha t)e^{-\alpha t})}{K}$

23. For $0 < t < 1$ the solution is as in Prob. 22. When $t > 1$,

$$v = \omega - \frac{P}{M_1 + M_2} + \frac{P}{k(M_1 + M_2)} [\sin kt - \sin (k(t - 1))],$$

$$k^2 = c \left(\frac{1}{M_1} + \frac{1}{M_2} \right), \quad c = \text{stiffness of shaft.}$$

27. $i(t) = \begin{cases} a \cos \omega t + b \sin \omega t - ae^{-t/RC} & \text{if } 0 < t < \pi/\omega \\ -a(1 + e^{\pi/\omega RC})e^{-t/RC} & \text{if } t > \pi/\omega \end{cases}$

    where $a = \omega CK, b = \omega^2 RC^2 K, K = 1/[1 + (\omega RC)^2]$

31. $i(t) = (V_0/\omega^* L)e^{-\alpha t} \sin \omega^* t$ if $0 < t < a$, $\alpha = R/2L, \omega^{*2} = 1/LC - \alpha^2$,

    $i(t) = (V_0/\omega^* L)[e^{-\alpha t} \sin \omega^* t - e^{-\alpha(t-a)} \sin \{\omega^*(t - a)\}]$ if $t > a$

33. $i(t) = R^{-1}t + R^{-2}L(e^{-tR/L} - 1)$ $(0 < t < 1)$,

    $i(t) = (R^{-1}e^{R/L}(1 - R^{-1}L) + R^{-2}L)e^{-tR/L}$ $(t > 1)$

35. Initial current 2 amperes, initial charge 0, $i(t) = e^{-t/5}(2 \cos \frac{2}{5}t + \sin \frac{2}{5}t)$

39. $i_1 = V_0 \left[e^{-\alpha t} \left(-\dfrac{1}{R_1} \cos \omega^* t + \dfrac{1}{\omega^*} \left(\dfrac{L_2}{A} - \dfrac{\alpha}{R_1} \right) \sin \omega^* t \right) + \dfrac{1}{R_1} \right]$

CHAPTER 5 (REVIEW PROBLEMS), page 301

11. $(s + 1)/(s^2 + 2s + 17)$
13. $(s^{-2} + 3s^{-1})e^{-3s}$
15. $(s^2 - 2)/(s^3 - 4s)$
17. arc tan $(1/s)$
19. $2s^2/(s^2 + 1)^2$
21. $(e^{3t} + e^{-4t})/7$
23. $\frac{1}{2} \sinh^2 t$
25. $e^{2t} - 1 - 2t - 2t^2$
27. $2 + e^t - 2e^{2t}$
29. $\frac{1}{3}[1 - \cos \sqrt{3}(t - 4)]u(t - 4)$
31. 0 if $0 < t < 3$, $0.2[1 - \cos (3t - 9)]$ if $t > 3$
33. $\frac{1}{2}e^{-3t} + \frac{1}{2}e^{-t} - e^{-2t}$
35. $2e^t + e^{-2t} + \frac{1}{2}t^2 - 3$
37. $1.5e^t \cos t + te^t \sin t$
39. $2te^{2t} + (20 + 12t + 3t^2)e^t$
41. t
43. e^{-3t}
45. $i(t) = e^{-4t}(\frac{3}{26} \cos 3t - \frac{10}{39} \sin 3t) - \frac{3}{26} \cos 10t + \frac{8}{65} \sin 10t$
49. $y_1 = \cos \sqrt{2}\, t$, $y_2 = \cos \sqrt{2}\, t$
55. $i_1 = -8e^{-2t} + 5e^{-0.8t} + 3$, $i_2 = -4e^{-2t} + 4e^{-0.8t}$

SECTION 6.2, page 311

1. $2, 2, -1, |\mathbf{v}| = 3$
3. $-14, 5, 9, |\mathbf{v}| = \sqrt{302}$
5. $3, -2, -7, |\mathbf{v}| = \sqrt{62}$
7. $-4, 1, -2, |\mathbf{v}| = \sqrt{21}$
9. $-8, 2, -1, |\mathbf{v}| = \sqrt{69}$
11. Q: $(6, 12, 0)$, $|\mathbf{v}| = \sqrt{13}$
13. Q: $(1.6, 2.7, 2.1)$, $|\mathbf{v}| = \sqrt{4.2}$
15. Q: $(12, -15, -1)$, $|\mathbf{v}| = \sqrt{385}$
17. Q: $(8, 8, 8)$, $|\mathbf{v}| = \sqrt{48}$
19. Q: $(3, 2, 1)$, $|\mathbf{v}| = \frac{1}{2}\sqrt{105}$
21. P: $(-\frac{3}{2}, -\frac{1}{3}, 0)$, $|\mathbf{v}| = \frac{1}{12}\sqrt{61}$
23. P: $(0, 0, 0)$, $|\mathbf{v}| = \sqrt{170}$
25. P: $(-4, 0, 2)$, $|\mathbf{v}| = \sqrt{20}$
27. P: $(-6, -12, -14)$, $|\mathbf{v}| = \sqrt{94}$
29. P: $(-1, 9, -7)$, $|\mathbf{v}| = \sqrt{35}$
31. $[0.80, 1.39]$
33. $[1.73, -1.73]$
35. $[-8.02, 2.15]$

SECTION 6.3, page 315

1. $[3, 3, 1]$
3. $[3, -2, 1]$
5. $[-4, -48, 12]$
7. $\sqrt{59}, \sqrt{14} - 5$
9. $\sqrt{131}$
11. $[2\sqrt{14}, 6\sqrt{14}, 4\sqrt{14}]$
15. $[1, -1, 0], \sqrt{2}$
17. $[3, -2, 18], \sqrt{337}$
19. $\mathbf{p} = [0, -3, 2]$
21. $\mathbf{p} = [-2, 0], \mathbf{q} = [0, 2]$
23. $1 \leq |\mathbf{p} + \mathbf{q}| \leq 5$; nothing about the direction
25. $\pi - (2\pi/n)$ is the interior angle at a vertex of the given n-gon, and so is the angle between the sides of the broken line L obtained by geometrically adding those vectors, since at the center, adjacent ones make an angle $2\pi/n$. Hence L closes, that is, the vector sum is $\mathbf{0}$.

SECTION 6.4, page 321

1. $[2, 4, -6, -2]$
3. $[7, -10, 13, 0]$
5. $[11, -13, 16, -1]$

7. a, b in the set S implies **a** + **b** in S and c**a** in S. Dimension 2. Basis [1, 0, 0], [0, 1, 0].

9. 1; [3, −1] **11.** 2; [1, 0, −1], [0, 1, −1]

13. 2; [1, 0, 0], [0, 1, 1] **15.** 2; [1, 2, 3, 0], [0, 0, 0, 1]

17. 2; cos 2x, sin 2x **19.** 3; e^x, e^{2x}, e^{-x}

23. No **25.** No **27.** No **29.** No

31. $\overrightarrow{OP}$ = k(**a** + **b**); also $\overrightarrow{OP}$ = **a** + l(**b** − **a**); see figure. Hence k(**a** + **b**) − (**a** + l(**b** − **a**)) = 0, k − 1 + l = 0, k − l = 0, since {**a, b**} is linearly independent; $k = l = \frac{1}{2}$.

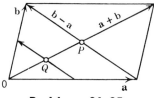

Problems 31, 35

35. $\overrightarrow{OQ}$ = k(**a** + **b**) = $\frac{1}{2}$**a** + $l\frac{1}{2}$(**b** − **a**), hence $k = \frac{1}{2} - \frac{1}{2}l$, $k = \frac{1}{2}l = \frac{1}{4}$.

37. If $S_0 = \{\mathbf{a}_{(1)}, \cdots, \mathbf{a}_{(k)}\}$, then $c_1\mathbf{a}_{(1)} + \cdots + c_k\mathbf{a}_{(k)} = \mathbf{0}$ for $c_1, \cdots, c_k$ not all 0, and $c_1\mathbf{a}_{(1)} + \cdots + c_k\mathbf{a}_{(k)} + \cdots + c_m\mathbf{a}_{(m)} = \mathbf{0}$ for those $c_1, \cdots, c_k$ and $c_{k+1} = \cdots = c_m = 0$.

39. i = [1, 0, 0], **j** = [0, 1, 0], **k** = [0, 0, 1], as mentioned in the text.

SECTION 6.5, page 328

1. 5, 5 **3.** $\sqrt{31.5}$, $\sqrt{31.5}$ **5.** 30, 30 **7.** 16, −16

9. $\sqrt{62}$, $\sqrt{62}$ **11.** 27 **15.** 2, 0, −2 **17.** −4

19. 6

23. Let $\overrightarrow{AB}$ = **a**, $\overrightarrow{BC}$ = **b**. Then W = **a**•**p** + **b**•**p** = (**a** + **b**)•**p**.

25. 2 **27.** −5 **29.** 0 **31.** $-\frac{8}{3}$

35. $[v_1, -\frac{1}{2}v_1, v_3]$; v_1, v_3 arbitrary. Yes. **39.** c = −9

41. If and only if all four sides are equal.

43. $|\mathbf{a} + \mathbf{b}|^2 = |\mathbf{a} - \mathbf{b}|^2$ implies **a**•**b** = 0.

45. −23/26 **47.** $\sqrt{3}/2$ **49.** 90° **51.** 122.1°

53. 55°, 79°, 46°, approximately **55.** 55.5°

57. $|\mathbf{a} - \mathbf{b}|^2 = (\mathbf{a} - \mathbf{b})\cdot(\mathbf{a} - \mathbf{b}) = |\mathbf{a}|^2 + |\mathbf{b}|^2 - 2|\mathbf{a}||\mathbf{b}| \cos \gamma$, etc.

59. $|\mathbf{a} + \mathbf{b}|^2 = (\mathbf{a} + \mathbf{b})\cdot(\mathbf{a} + \mathbf{b}) \leqq |\mathbf{a}|^2 + 2|\mathbf{a}||\mathbf{b}| + |\mathbf{b}|^2 = (|\mathbf{a}| + |\mathbf{b}|)^2$

SECTION 6.8, page 338

1. ±[6, −3, 0], 5 **3.** [0, 0, 3], 3

5. [−12, 6, −6], [12, −6, 6] **7.** [3, 3, 6]

9. [3, −6, −3], [−3, 6, 3] **11.** [−5, 5, 0], −9

15. [0, 12, 6] **17. 0**

19. [−12, −12, −6] **21.** [24, −18, 0]

23. $[-20, -12, -12]$ **25.** Orthogonal
27. Parallel   **29.** No
31. $x + y + z = 5$ **33.** $x - 2y + z = 0$
35. $11x - 7y - 4z = 13$ **37.** $2x - y - z = 0$
39. $3y + 2z = 5$ **41.** $\sqrt{5}$ **43.** $\sqrt{2016}$ **45.** $\sqrt{21}$
47. 4   **49.** 11   **51.** $\frac{1}{2}\sqrt{3}$ **53.** $\frac{1}{2}$
55. $\frac{1}{2}\sqrt{27}$ **57.** $\pm[\frac{2}{3}, \frac{2}{3}, -\frac{1}{3}]$
59. $[2, 3, 4] \times [1, -1, 1] = [7, 2, -5]$

SECTION 6.9, page 343

1. 1   **3.** 1   **5.** 39   **7.** -17 **9.** -4
11. Dependent   **13.** Independent   **15.** Dependent
17. Dependent   **19.** Independent   **21.** No
23. 10/7   **25.** 96   **27.** 9   **29.** 3   **31.** 7/3   **33.** 1/6
35. 1/3   **37.** $[6, 6, -7], 9$ **39.** $[6, 3, -3], [5, 5, 0]$
41. $[17, -26, -27]$
43. $\mathbf{a}\cdot[\mathbf{b} \times (\mathbf{c} \times \mathbf{d})]$ equals $(\mathbf{a}\ \ \mathbf{b}\ \ [\mathbf{c} \times \mathbf{d}]) = (\mathbf{a} \times \mathbf{b})\cdot(\mathbf{c} \times \mathbf{d})$ as well as $(\mathbf{a}\cdot\mathbf{c})(\mathbf{b}\cdot\mathbf{d}) - (\mathbf{a}\cdot\mathbf{d})(\mathbf{b}\cdot\mathbf{c})$.
45. Formula (8) follows from (6) with $\mathbf{b}$ replaced by $\mathbf{a} \times \mathbf{b}$.

CHAPTER 6 (REVIEW PROBLEMS), page 345

17. $\mathbf{0}$ **19.** $[28, 16, 0]$ **21.** $\sqrt{33}, 3 - \sqrt{14}$
23. $(1/\sqrt{14})[-1, 2, -3]$ **25.** $-5, 5$
27. $[-8, 14, 12]$ **29.** $[7, -1, -3], [-3, 8, 3]$
31. 0   **33.** $-15, 30$ **35.** $[12, -21, 35]$ **37.** $[-3, 5, -6]$
39. $55.5°$ **41.** 10   **43.** 0   **45.** 24
47. $[0, 0, 10]$ **49.** $[-9, 0, 0]$ **51.** $[30, -7, -1]$ **53.** 26
55. 7/3   **57.** $\sqrt{20}$ **59.** $\frac{1}{2}\sqrt{3}$
61. $7x - 8y + 6z = 5$ **63.** 1
65. $-6/\sqrt{35}$ **67.** $[-5, -5], [-2, 2]$

SECTIONS 7.1 and 7.2, page 355

1. $\begin{bmatrix} 0 & 1 \\ 4 & 7 \end{bmatrix}$ **3.** $\begin{bmatrix} 12 & 4 \\ -2 & 4 \end{bmatrix}$ **5.** $\begin{bmatrix} 30 & 48 & 0 \\ 24 & 30 & 24 \end{bmatrix}$

7. Undefined   **9.** $\begin{bmatrix} \frac{5}{2} & \frac{1}{2} \\ -\frac{7}{4} & -\frac{3}{2} \end{bmatrix}$ **11.** $-2\mathbf{E}_{11} + 3\mathbf{E}_{21} + 4\mathbf{E}_{22}$

15. $\begin{bmatrix} 0 & -1 & -4 \\ -9 & 1 & 0 \\ 0 & -3 & -1 \end{bmatrix}$ **17.** $\begin{bmatrix} 0 & 3 & 12 \\ 27 & -3 & 0 \\ 0 & 9 & 3 \end{bmatrix}$ **19.** $\begin{bmatrix} -4 & -4 & -16 \\ -34 & 2 & 0 \\ -6 & -4 & -4 \end{bmatrix}$

21. 9   **23.** No   **25.** 1

27.
$$\begin{bmatrix} 1 & -1 & 0 & 0 \\ -1 & 1 & 1 & 1 \\ 0 & 0 & -1 & 0 \end{bmatrix}$$

29.
$$\begin{bmatrix} -1 & 1 & 1 & 0 & 0 & -1 \\ 1 & -1 & 0 & 1 & 0 & 0 \\ 0 & 0 & -1 & 0 & 1 & 0 \\ 0 & 0 & 0 & -1 & -1 & 1 \end{bmatrix}$$

31.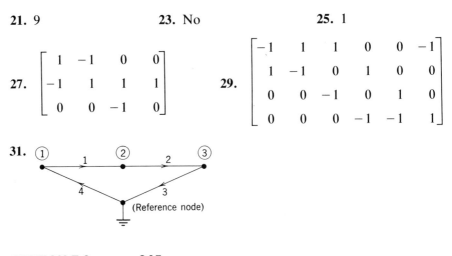

(Reference node)

SECTION 7.3, page 365

1.
$$\begin{bmatrix} 4 & 3 & 0 \\ 0 & 0 & 0 \\ 8 & 6 & 0 \end{bmatrix}, \begin{bmatrix} 4 \end{bmatrix}$$

3. $\mathbf{CB} = \begin{bmatrix} 2 & -3 \\ 3 & 3 \\ 7 & 1 \end{bmatrix}$

5. $\mathbf{dC} = \begin{bmatrix} 4 & 2 & 9 \end{bmatrix}$

7. $\begin{bmatrix} 17 & -3 \end{bmatrix}$

9. $\begin{bmatrix} 22 \\ 0 \\ 44 \end{bmatrix}$

11. $\begin{bmatrix} 1 & 0 \\ 0 & 0 \end{bmatrix}, \begin{bmatrix} 1 & 1 \\ 0 & 0 \end{bmatrix}$, etc.

13. $\begin{bmatrix} 0 & 1 \\ 0 & 0 \end{bmatrix}$, etc.

17. $\begin{bmatrix} y_1 \\ y_2 \end{bmatrix} = \begin{bmatrix} 1 & 3 \\ 1 & -3 \end{bmatrix} \begin{bmatrix} w_1 \\ w_2 \end{bmatrix}$

21. $\begin{bmatrix} 1/\sqrt{2} & 1/\sqrt{2} \\ -1/\sqrt{2} & 1/\sqrt{2} \end{bmatrix}$

23. Uniform stretch or contraction, same in all directions.
25. (a) C_1 to J_2, C_3 to J_1, J_3. (b) C_1 to J_2, C_2 to J_1, C_3 to J_3
27. His bill would be 7.70 at I, 7.30 at II, 5.60 at III.

29.
$$\begin{bmatrix} 17500 & 12500 & 17000 & 16500 \\ 95000 & 65000 & 90000 & 85000 \end{bmatrix}$$

SECTION 7.4, page 370

1. $\mathbf{a}^{\mathsf{T}}\mathbf{a} = \begin{bmatrix} 26 \end{bmatrix}$, $\mathbf{a}\mathbf{a}^{\mathsf{T}} = \begin{bmatrix} 1 & 3 & 4 \\ 3 & 9 & 12 \\ 4 & 12 & 16 \end{bmatrix}$

3. $\begin{bmatrix} 0 & 3 & 7 \\ 3 & 0 & 2 \\ 7 & 2 & 8 \end{bmatrix}$

5. $\mathbf{Ca} = \begin{bmatrix} 11 \\ 14 \\ 18 \end{bmatrix}$, $\mathbf{C}^\mathsf{T}\mathbf{a} = \begin{bmatrix} 26 \\ -3 \\ 27 \end{bmatrix}$ **7.** $\begin{bmatrix} 14 & 32 \end{bmatrix}$, $\begin{bmatrix} 14 \\ 32 \end{bmatrix}$

13. 10   **15.** $\frac{1}{2}n(n-1)$ **19. 0**

27. Write $\mathbf{A}^2 = [c_{jk}]$. Then

$$\sum_{k=1}^{n} c_{jk} = \sum_{k=1}^{n}\sum_{l=1}^{n} a_{jl}a_{lk} = \sum_{l=1}^{n} a_{jl}\left(\sum_{k=1}^{n} a_{lk}\right) = \sum_{l=1}^{n} a_{jl} \cdot 1 = 1.$$

29. Married (single) 4700 (1300), 4490 (1510), 4343 (1657)

SECTION 7.5, page 379

1. $x = 1, y = 2$ **3.** $x = \frac{1}{7}, y = -\frac{1}{7}$
5. $x = z, y = -z$ **7.** $x = -2, y = 0, z = 1$
9. $x = \frac{1}{2}, y = 2, z = \frac{3}{2}$ **11.** $x = 0, z = 3 - 2y$
13. $x = y + 2, z = 2y - 1$ **15.** $x = -2z, y = \frac{1}{4}$
17. $x = -1, y = 2, z = -3$ **19.** $w = 2, x = 0, y = 1, z = 3$
21. $I_1 = 1, I_2 = 3, I_3 = 4$ (amperes)
23. $I_1 = (R_1 + R_2)E_0/R_1R_2, I_2 = E_0/R_1, I_3 = E_0/R_2$
27. $P = 4, D = S = 5$

SECTION 7.6, page 384

1. 1   **3.** 1   **5.** 2   **7.** 1   **9.** 1   **11.** 2   **13.** 1   **15.** 3
17. Linearly independent   **19.** Linearly dependent
21. Linearly dependent   **23.** Linearly independent
25. Use Theorem 1.   **27.** Use Theorem 1.
29. Use the definition of rank and Theorem 1.

SECTION 7.8, page 394

3. $\begin{bmatrix} 3 & -1 \\ -5 & 2 \end{bmatrix}$ **5.** $\begin{bmatrix} 0.2 & 0 \\ 0 & 2.5 \end{bmatrix}$ **7.** $\begin{bmatrix} 0 & 1/b \\ 1/a & 0 \end{bmatrix}$

9. $\begin{bmatrix} 0.25 & 0 & 0 \\ 0 & 2 & 0 \\ 0 & 0 & -0.2 \end{bmatrix}$ **11.** $\begin{bmatrix} 1 & 0 & 0 \\ -\frac{1}{2} & 1 & 0 \\ \frac{3}{4} & -\frac{5}{2} & \frac{1}{2} \end{bmatrix}$

13. $\begin{bmatrix} 3 & -1 & 1 \\ -15 & 6 & -5 \\ 5 & -2 & 2 \end{bmatrix}$ **15.** $\begin{bmatrix} 0 & 1 & 0 \\ 1 & 0 & 0 \\ 0 & 0 & 1 \end{bmatrix}$

17. $\begin{bmatrix} 0 & 0 & 1/a \\ 0 & 1/b & 0 \\ 1/c & 0 & 0 \end{bmatrix}$

19. $x = 2x^* + y^*$
$\quad\; y = 5x^* + 3y^*$

21. $x = \;\;19x^* + 2y^* - 9z^*$
$\quad\; y = -4x^* - \;\;y^* + 2z^*$
$\quad\; z = -2x^* \qquad\;\; + \;\;z^*$

23. $x = -2y^* + z^*$
$\quad\; y = \frac{1}{2}(x^* + y^* - z^*)$
$\quad\; z = -x^* + 2y^*$

25. $x = 2x^* \qquad\;\; - \;\;z^*$
$\quad\; y = 5x^* + y^*$
$\quad\; z = \qquad\quad y^* + 3z^*$

27. Use (7).
29. Use Prob. 26.

SECTION 7.9, page 401

1. 14   **3.** 1   **5.** -474 **7.** -45 **9.** 0   **11.** -38
13. $a^3 + b^3 + c^3 - 3abc$ **15.** $4a^2b^2c^2$

SECTION 7.10, page 408

1. 90   **3.** -17 **5.** 0   **7.** 72   **9.** 120   **11.** -116

SECTION 7.11, page 413

1. 2   **3.** 2   **5.** 2   **7.** 2   **9.** 3

11. $\begin{bmatrix} 1 & 2 \\ 3 & 4 \end{bmatrix}$

13. $\begin{bmatrix} 3 & -1 & 1 \\ -15 & 6 & -5 \\ 5 & -2 & 2 \end{bmatrix}$

15. $\begin{bmatrix} 0.2 & -0.08 & -0.144 \\ 0 & 0.2 & -0.04 \\ 0 & 0 & 0.2 \end{bmatrix}$

17. $\begin{bmatrix} 0 & 0 & 1/c \\ 0 & 1/b & 0 \\ 1/a & 0 & 0 \end{bmatrix}$

19. $\begin{bmatrix} 6 & 4 & 3 \\ 4 & 3 & 4 \\ 3 & 2 & 2 \end{bmatrix}$

21. $\begin{bmatrix} 2 & 4 & 1 \\ 1 & 2 & 1 \\ 3 & 4 & 2 \end{bmatrix}$

23. $x = 2, y = 4$
27. $x = -1, y = 0, z = 1$
37. $18x - 25y - 22z + 19 = 0$

25. $x = 2, y = 1, z = -3$
33. Use Theorem 4.

SECTION 7.12, page 422

1. 6, $\begin{bmatrix} 2 \\ 1 \end{bmatrix}$, 4, $\begin{bmatrix} 1 \\ 1 \end{bmatrix}$

3. 1, $\begin{bmatrix} 1 \\ 0 \end{bmatrix}$, 2, $\begin{bmatrix} 0 \\ 1 \end{bmatrix}$

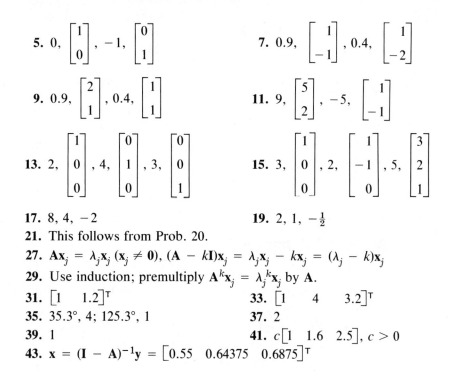

5. $0,\ \begin{bmatrix} 1 \\ 0 \end{bmatrix},\ -1,\ \begin{bmatrix} 0 \\ 1 \end{bmatrix}$ **7.** $0.9,\ \begin{bmatrix} 1 \\ -1 \end{bmatrix},\ 0.4,\ \begin{bmatrix} 1 \\ -2 \end{bmatrix}$

9. $0.9,\ \begin{bmatrix} 2 \\ 1 \end{bmatrix},\ 0.4,\ \begin{bmatrix} 1 \\ 1 \end{bmatrix}$ **11.** $9,\ \begin{bmatrix} 5 \\ 2 \end{bmatrix},\ -5,\ \begin{bmatrix} 1 \\ -1 \end{bmatrix}$

13. $2,\ \begin{bmatrix} 1 \\ 0 \\ 0 \end{bmatrix},\ 4,\ \begin{bmatrix} 0 \\ 1 \\ 0 \end{bmatrix},\ 3,\ \begin{bmatrix} 0 \\ 0 \\ 1 \end{bmatrix}$ **15.** $3,\ \begin{bmatrix} 1 \\ 0 \\ 0 \end{bmatrix},\ 2,\ \begin{bmatrix} 1 \\ -1 \\ 0 \end{bmatrix},\ 5,\ \begin{bmatrix} 3 \\ 2 \\ 1 \end{bmatrix}$

17. $8, 4, -2$ **19.** $2, 1, -\frac{1}{2}$

21. This follows from Prob. 20.

27. $\mathbf{A}\mathbf{x}_j = \lambda_j\mathbf{x}_j\ (\mathbf{x}_j \neq \mathbf{0}),\ (\mathbf{A} - k\mathbf{I})\mathbf{x}_j = \lambda_j\mathbf{x}_j - k\mathbf{x}_j = (\lambda_j - k)\mathbf{x}_j$

29. Use induction; premultiply $\mathbf{A}^k\mathbf{x}_j = \lambda_j^k\mathbf{x}_j$ by $\mathbf{A}$.

31. $\begin{bmatrix} 1 & 1.2 \end{bmatrix}^\mathsf{T}$ **33.** $\begin{bmatrix} 1 & 4 & 3.2 \end{bmatrix}^\mathsf{T}$

35. $35.3°, 4; 125.3°, 1$ **37.** 2

39. 1 **41.** $c\begin{bmatrix} 1 & 1.6 & 2.5 \end{bmatrix}, c > 0$

43. $\mathbf{x} = (\mathbf{I} - \mathbf{A})^{-1}\mathbf{y} = \begin{bmatrix} 0.55 & 0.64375 & 0.6875 \end{bmatrix}^\mathsf{T}$

SECTION 7.13, page 427

1. Skew-symmetric, orthogonal   **3.** Orthogonal

5. Orthogonal   **9.** No. Yes, e.g., $\mathbf{I}$.

13. $|\mathbf{v}| = |\mathbf{w}| = \sqrt{20},\ |\mathbf{x}| = |\mathbf{y}| = \sqrt{5}.\ 126.87°.$ Theorem 4.

15. $(\mathbf{AB})^{-1} = \mathbf{B}^{-1}\mathbf{A}^{-1} = \mathbf{B}^\mathsf{T}\mathbf{A}^\mathsf{T} = (\mathbf{AB})^\mathsf{T}$

17. $\mathbf{A} = \mathbf{H} + \mathbf{S}$, where $\mathbf{H} = (\mathbf{A} + \overline{\mathbf{A}}^\mathsf{T})/2$ is Hermitian and $\mathbf{S} = (\mathbf{A} - \overline{\mathbf{A}}^\mathsf{T})/2$ is skew-Hermitian.

19. $\mathbf{C}^\mathsf{T}$ must equal $-\overline{\mathbf{C}}$, which implies a, b real.

21. $\mathbf{AB} = \overline{(\mathbf{AB})}^\mathsf{T} = \overline{\mathbf{B}}^\mathsf{T}\overline{\mathbf{A}}^\mathsf{T} = \mathbf{BA}$ **23.** $\begin{bmatrix} 0 & -i \\ -i & 0 \end{bmatrix}$

25. $\begin{bmatrix} -0.8i & -0.6i \\ 0.6i & -0.8i \end{bmatrix}$ **27.** $\begin{bmatrix} 1/\sqrt{2} & \frac{1}{2} - \frac{1}{2}i \\ \frac{1}{2} + \frac{1}{2}i & -1/\sqrt{2} \end{bmatrix}$

29. Let $\mathbf{A}$ be unitary. Set $\mathbf{A}^{-1} = \mathbf{B}$. Then
$\mathbf{B}^\mathsf{T} = (\mathbf{A}^{-1})^\mathsf{T} = (\mathbf{A}^\mathsf{T})^{-1} = (\overline{\mathbf{A}}^{-1})^{-1} = \overline{\mathbf{B}}^{-1}.$

31. $1, [1\ \ 1]^\mathsf{T}, -1, [1\ \ -1]^\mathsf{T}; 1, [1\ \ i]^\mathsf{T}, -1, [1\ \ -i]^\mathsf{T};$
$1, [1\ \ 0]^\mathsf{T}, -1, [0\ \ 1]^\mathsf{T}$

SECTION 7.14, page 432

3. $\pm i$, Theorems 2(b), 2(c)   **5.** $\pm|a|i$, Theorem 1(b)

7. $\cos\theta \pm i\sin\theta$, Theorem 2(c)   **9.** ± 1, Theorem 1(c)

11. $1 \pm \sqrt{30}$, Theorem 1(a)   **13.** $-1, (1 \pm i)/\sqrt{2}$, Theorem 2(c)

15. 4   **17.** -2 **19.** 2

21. 0   **23.** $2i\,|x_1|^2 + 8i\,\text{Im}\,\bar{x}_1 x_2$

25. $\begin{bmatrix} 6 & -2 \\ -2 & 2 \end{bmatrix}$ **27.** $\begin{bmatrix} 5 & -1 \\ -1 & 1 \end{bmatrix}$

29. $\begin{bmatrix} 1 & 1 & 0 \\ 1 & 1 & 0 \\ 0 & 0 & -1 \end{bmatrix}$ **31.** $\begin{bmatrix} 1 & -1 & 1 \\ -1 & 1 & -1 \\ 1 & -1 & 1 \end{bmatrix}$

SECTION 7.15, page 439

1. $\begin{bmatrix} 0.92 & 1.44 \\ 1.44 & 0.08 \end{bmatrix}$, 2, $\begin{bmatrix} 4 \\ 3 \end{bmatrix}$, $\mathbf{x} = \begin{bmatrix} 5 \\ 0 \end{bmatrix}$; -1, $\begin{bmatrix} 3 \\ -4 \end{bmatrix}$, $\mathbf{x} = \begin{bmatrix} 0 \\ -5 \end{bmatrix}$

3. $\begin{bmatrix} 4 & 0 \\ 6 & 6 \end{bmatrix}$, 6, $\begin{bmatrix} 0 \\ 1 \end{bmatrix}$, $\mathbf{x} = \begin{bmatrix} 2 \\ 1 \end{bmatrix}$; 4, $\begin{bmatrix} 1 \\ -3 \end{bmatrix}$, $\mathbf{x} = \begin{bmatrix} -1 \\ -1 \end{bmatrix}$

5. $\begin{bmatrix} 0.9 & 0 \\ 0 & 0.4 \end{bmatrix}$, 0.9, $\begin{bmatrix} 1 \\ 0 \end{bmatrix}$, $\mathbf{x} = \begin{bmatrix} 1 \\ -1 \end{bmatrix}$; 0.4, $\begin{bmatrix} 0 \\ 1 \end{bmatrix}$, $\mathbf{x} = \begin{bmatrix} 1 \\ -2 \end{bmatrix}$

7. $\begin{bmatrix} -10 & 16 \\ -6 & 10 \end{bmatrix}$, 2, $\begin{bmatrix} 4 \\ 3 \end{bmatrix}$, $\mathbf{x} = \begin{bmatrix} 1 \\ 0 \end{bmatrix}$; -2, $\begin{bmatrix} 2 \\ 1 \end{bmatrix}$, $\mathbf{x} = \begin{bmatrix} 0.4 \\ -0.2 \end{bmatrix}$

9. $\begin{bmatrix} 2 \\ 3 \end{bmatrix}$, $\begin{bmatrix} 2 \\ -3 \end{bmatrix}$, $\begin{bmatrix} 6 & 0 \\ 0 & -6 \end{bmatrix}$ **11.** $\begin{bmatrix} 1 \\ 0 \end{bmatrix}$, $\begin{bmatrix} 1 \\ -2 \end{bmatrix}$, $\begin{bmatrix} 1 & 0 \\ 0 & -1 \end{bmatrix}$

13. $\begin{bmatrix} 1 \\ 1 \end{bmatrix}$, $\begin{bmatrix} 1 \\ -1 \end{bmatrix}$, $\begin{bmatrix} 2 & 0 \\ 0 & 0 \end{bmatrix}$

15. $\begin{bmatrix} 1 \\ 0 \\ \sqrt{3} \end{bmatrix}$, $\begin{bmatrix} 0 \\ 1 \\ 0 \end{bmatrix}$, $\begin{bmatrix} 1 \\ 0 \\ -\sqrt{3} \end{bmatrix}$, $\begin{bmatrix} \sqrt{3} & 0 & 0 \\ 0 & 2 & 0 \\ 0 & 0 & -\sqrt{3} \end{bmatrix}$

17. $\begin{bmatrix} 1 & -1 \\ -1 & 1 \end{bmatrix}$ **19.** $\begin{bmatrix} 0 & -\frac{9}{2} \\ -\frac{9}{2} & 3 \end{bmatrix}$

21. $\begin{bmatrix} 1 & 1 & 1 \\ 1 & 1 & 1 \\ 1 & 1 & 1 \end{bmatrix}$ **23.** $\begin{bmatrix} 36 & 12 & 4 \\ 12 & 4 & 0 \\ 4 & 0 & 0 \end{bmatrix}$

25. Hyperbola $y_1^2 - y_2^2 = 1$; $x_1 = 0.8y_1 - 0.6y_2$, $x_2 = 0.6y_1 + 0.8y_2$

27. Ellipse $4y_1^2 + 9y_2^2 = 1$; $x_1 = (12y_1 - 5y_2)/13$, $x_2 = (5y_1 + 12y_2)/13$

29. Parallel straight lines $y_1 = \pm 1$; $x_1 = (y_1 - 3y_2)/\sqrt{10}$, $x_2 = (3y_1 + y_2)/\sqrt{10}$

31. Hyperbola $2y_1^2 - 3y_2^2 = 1$; $x_1 = 0.8y_1 - 0.6y_2$, $x_2 = 0.6y_1 + 0.8y_2$

37. Compare the coefficient of λ^{n-1} in

$$(-1)^n(\lambda - \lambda_1) \cdots (\lambda - \lambda_n)$$
$$= (-1)^n\lambda^n + (-1)^{n-1}(\lambda_1 + \cdots + \lambda_n)\lambda^{n-1} + \cdots$$

with that of λ^{n-1} in the development of the characteristic determinant.

39. $\det (\mathbf{T}^{-1}\mathbf{A}\mathbf{T} - \lambda\mathbf{I}) = \det (\mathbf{T}^{-1}(\mathbf{A} - \lambda\mathbf{I})\mathbf{T}) = \det \mathbf{T}^{-1} \det (\mathbf{A} - \lambda\mathbf{I}) \det \mathbf{T}$
$$= \det(\mathbf{A} - \lambda\mathbf{I})$$

SECTION 7.16, page 446

1. $\mathbf{y} = c_1\mathbf{x}_1 e^{2t} + c_2\mathbf{x}_2 e^{-2t}$, $\mathbf{x}_1^{\mathsf{T}} = \begin{bmatrix} 1 & 1 \end{bmatrix}$, $\mathbf{x}_2^{\mathsf{T}} = \begin{bmatrix} 1 & -3 \end{bmatrix}$

3. $\mathbf{y} = c_1\mathbf{x}_1 e^{9t} + c_2\mathbf{x}_2 e^{4t}$, $\mathbf{x}_1^{\mathsf{T}} = \begin{bmatrix} 2 & 1 \end{bmatrix}$, $\mathbf{x}_2^{\mathsf{T}} = \begin{bmatrix} 1 & 1 \end{bmatrix}$

5. $\mathbf{y} = c_1\mathbf{x}_1 e^{3t} + c_2\mathbf{x}_2 e^{-2t}$, $\mathbf{x}_1^{\mathsf{T}} = \begin{bmatrix} 6 & -11 \end{bmatrix}$, $\mathbf{x}_2^{\mathsf{T}} = \begin{bmatrix} 1 & -1 \end{bmatrix}$

7. $\mathbf{y} = c_1\mathbf{x}_1 e^{4t} + c_2\mathbf{x}_2 e^{-3t} + c_3\mathbf{x}_3 e^{2t}$,

$\mathbf{x}_1^{\mathsf{T}} = \begin{bmatrix} 1 & 1 & 0 \end{bmatrix}$, $\mathbf{x}_2^{\mathsf{T}} = \begin{bmatrix} 1 & -6 & 0 \end{bmatrix}$, $\mathbf{x}_3^{\mathsf{T}} = \begin{bmatrix} 0 & 0 & 1 \end{bmatrix}$

9. $\mathbf{y} = \mathbf{x}_1 e^{3t} + \mathbf{x}_2 e^{2t}$, $\mathbf{x}_1^{\mathsf{T}} = \begin{bmatrix} 2 & 1 \end{bmatrix}$, $\mathbf{x}_2^{\mathsf{T}} = \begin{bmatrix} 1 & 1 \end{bmatrix}$; thus

$y_1 = 2e^{3t} + e^{2t}$, $y_2 = e^{3t} + e^{2t}$

11. $y_1 = e^{3t}$, $y_2 = e^{3t}$

13. $y_1 = 2e^{3t}$, $y_2 = -10e^{3t}$, $y_3 = -27e^{3t}$

17. $y_1 = \sin \sqrt{2}t$, $y_2 = 2 \sin \sqrt{2}t$

21. $\mathbf{y} = \mathbf{x}_1 \cos \sqrt{k}t + \mathbf{x}_2 \sin \sqrt{3k}t$

23. $\mathbf{y}'' = \mathbf{A}\mathbf{y}$, $\mathbf{A} = \begin{bmatrix} -(k_1 + k_2) & k_2 & 0 \\ k_2 & -(k_2 + k_3) & k_3 \\ 0 & k_3 & -(k_3 + k_4) \end{bmatrix}$, $\mathbf{y} = \begin{bmatrix} y_1 \\ y_2 \\ y_3 \end{bmatrix}$

CHAPTER 7 (REVIEW PROBLEMS), page 450

17. $\begin{bmatrix} -10 & 20 & -10 \\ -30 & -15 & -30 \\ -5 & 5 & 65 \end{bmatrix}$ **19.** $\begin{bmatrix} 6 \\ -13 \\ 6 \end{bmatrix}$

21. $\begin{bmatrix} 57 \\ 46 \\ -98 \end{bmatrix}$ **23.** $\begin{bmatrix} 5 & 2 & 3 \\ 2 & 5 & -4 \\ 3 & -4 & 13 \end{bmatrix}$

25. $\begin{bmatrix} 0.6 & -0.2 & -0.2 \\ -0.2 & 0.4 & 0.4 \\ 0.3 & -0.6 & -0.1 \end{bmatrix}$ **27.** 1

29. $\begin{bmatrix} 1 & 0 & -2 \\ 0 & 0 & 0 \\ -2 & 0 & 4 \end{bmatrix}$, $\begin{bmatrix} 5 \end{bmatrix}$ **31.** $\begin{bmatrix} 23 & -12 & -14 \\ -12 & 5 & 24 \\ 21 & -36 & -26 \end{bmatrix}$

33. $x = \begin{bmatrix} 2 & 1 & 3 \end{bmatrix}^{\mathsf{T}}$ **35.** $x = 3, y = -2$

37. $x = 1, y = 0, z = 2$ **39.** $x = y + 3, z = y + 3$

41. $x = 3, y = -2, z = 1$ **43.** $y = 2x + 4, z = x + 3$

45. $\begin{bmatrix} -1 & 4 \\ -4 & 6 \end{bmatrix}$ **47.** $\begin{bmatrix} 0.8 & -0.6 \\ 0.6 & 0.8 \end{bmatrix}$

49. No inverse

51. $\begin{bmatrix} 1 & -3 & 1 \\ 0 & 1 & -2 \\ 0 & 0 & 1 \end{bmatrix}$ **53.** $\begin{bmatrix} -1 & 1 & 2 \\ 3 & -1 & 1 \\ -1 & 3 & 4 \end{bmatrix}$

55. $\begin{bmatrix} 0.20 & 0.04 & -0.12 \\ 0.55 & 0.46 & -0.38 \\ 0.15 & -0.02 & 0.06 \end{bmatrix}$ **57.** $2, \begin{bmatrix} 3 \\ -1 \end{bmatrix}; -2, \begin{bmatrix} 1 \\ -1 \end{bmatrix}$

59. $\sqrt{8}, \begin{bmatrix} \sqrt{2} \\ 1 \end{bmatrix}; -\sqrt{8}, \begin{bmatrix} \sqrt{2} \\ -1 \end{bmatrix}$ **61.** $1, \begin{bmatrix} 1 \\ -i \end{bmatrix}, -1, \begin{bmatrix} 1 \\ i \end{bmatrix}$

63. $1 + i, \begin{bmatrix} 1 \\ 1 \end{bmatrix}; 1 - i, \begin{bmatrix} 1 \\ -1 \end{bmatrix}$ **65.** $3, \begin{bmatrix} 2 \\ 3 \\ 3 \end{bmatrix}; 0, \begin{bmatrix} 1 \\ 0 \\ -1 \end{bmatrix}; -5, \begin{bmatrix} 2 \\ -5 \\ 3 \end{bmatrix}$

67. $2i, \begin{bmatrix} 2 \\ 0 \\ 3 \end{bmatrix}; i, \begin{bmatrix} 0 \\ 1 \\ 0 \end{bmatrix}; 0, \begin{bmatrix} 3 \\ 0 \\ 4 \end{bmatrix}$

69. $y_1 = c_1 e^{7t} + c_2 e^{-7t}, \quad y_2 = c_1 e^{7t} - c_2 e^{-7t}$

71. $y_1 = c_1 e^{3t} + 2c_2, \quad y_2 = 2c_1 e^{3t} + c_2$

73. $y_1 = 5c_1 e^{-2t} + c_2 e^{-5t}, \quad y_2 = 12c_1 e^{-2t} + 3c_2 e^{-5t}$

75. $y_1 = 2c_1 e^{6t} + c_2 + 2c_3 e^{-10t}, \quad y_2 = 3c_1 e^{6t} - 5c_3 e^{-10t},$
$y_3 = 3c_1 e^{6t} - c_2 + 3c_3 e^{-10t}$

77. $I_1 = 7.5, I_2 = 5, I_3 = 2.5$ [amps]   **79.** $I_1 = 4, I_2 = 5, I_3 = 1$ [amps]

81. $\mathbf{A} = \dfrac{1}{t_{12}} \begin{bmatrix} t_{22} & -\det \mathbf{T} \\ 1 & -t_{11} \end{bmatrix}$

SECTION 8.1, page 460

11. Ellipses

13. The region between the ellipses $9x^2 + y^2 = 1$ and $x^2 + \frac{1}{9}y^2 = 1$

15. Spheres with center 0
17. Circular cylinders with the z-axis as axis
19. Spheres with center 0
21. Cones of revolution $z = \sqrt{x^2 + y^2} + c$
23. Ellipsoids   25. $2x^2 + 2xy + 5y^2$
35. $x^2 + y^2 = const$, $y/x = const$ 37. $x^4 + y^4 = const$, $x/y = const$
39. $x^2 + y^2 = const$

SECTION 8.2, page 463

1. **b, 0, |b|, 0**   3. $\mathbf{i} + 4\mathbf{j}$, **0**, $\sqrt{17}$, **0**
5. $2t\mathbf{i} + 6t\mathbf{j}$, $2\mathbf{i} + 6\mathbf{j}$, $|t|\sqrt{40}$, $\sqrt{40}$
7. $-\sin t\,\mathbf{i} + \cos t\,\mathbf{j}$, $-\cos t\,\mathbf{i} - \sin t\,\mathbf{j}$, 1, 1
9. $-\sin t\,\mathbf{i} + \cos t\,\mathbf{j} + \mathbf{k}$, $-\cos t\,\mathbf{i} - \sin t\,\mathbf{j}$, $\sqrt{2}$, 1
11. $\mathbf{i} + 2t\mathbf{j} + 3t^2\mathbf{k}$, $2\mathbf{j} + 6t\mathbf{k}$, $\sqrt{1 + 4t^2 + 9t^4}$, $2\sqrt{1 + 9t^2}$
13. $6t^2$ 15. $6t^5 - 10t^4 - 3t^2$
17. $1 + 3t^2 + 12t^3$ 19. $6 + 48t^3$
21. $-(4t^3 + 5t^4)\mathbf{i} + (4t^3 + 14t^6)\mathbf{j} - (12t^5 - 2t)\mathbf{k}$
23. **i, 2j, 0**   25. $y \cos x\,\mathbf{j}$, $\sin x\,\mathbf{j} + e^y\mathbf{k}$, **i**
27. **j + k, j − k, 0**   29. **0**, $-z \sin yz\,\mathbf{i}$, $-y \sin yz\,\mathbf{i}$
35. $(\mathbf{u}\cdot\mathbf{v})'' = \mathbf{u}''\cdot\mathbf{v} + 2\mathbf{u}'\cdot\mathbf{v}' + \mathbf{u}\cdot\mathbf{v}''$,
    $(\mathbf{u}\times\mathbf{v})'' = \mathbf{u}''\times\mathbf{v} + 2\mathbf{u}'\times\mathbf{v}' + \mathbf{u}\times\mathbf{v}''$

SECTION 8.3, page 467

1. $\mathbf{r}(t) = t\mathbf{i} + t\mathbf{j}$
3. $\mathbf{r}(t) = 2t\mathbf{i} + (-5 + t)\mathbf{j} + (3 - t)\mathbf{k}$
5. $\mathbf{r}(t) = (-3 + \frac{1}{2}t)\mathbf{i} + 8\mathbf{j} - 7t\mathbf{k}$ 7. $\mathbf{r}(t) = t(\mathbf{i} + \mathbf{j} + \mathbf{k})$
9. $\mathbf{r}(t) = 2t\mathbf{i} + \mathbf{j} + 3\mathbf{k}$ 11. $\mathbf{r}(t) = (4 + 3t)\mathbf{i} + (2 - 5t)\mathbf{j}$
13. $\mathbf{r}(t) = t(\mathbf{i} + \mathbf{j} + \mathbf{k})$
15. $\mathbf{r}(t) = t\mathbf{i} + (1 + t)\mathbf{j} + (-3 + 2t)\mathbf{k}$
17. Hyperbola $xy = 2$, $z = 0$ 19. Ellipse $y^2 + \frac{1}{4}z^2 = 1$, $x = 0$
21. Hyperbola $x^2 - y^2 = 1$, $z = 0$ 23. Hyperbola $y = 1/x$ $(x > 0)$, $z = 0$
25. $2 \cos t\,\mathbf{i} + 2 \sin t\,\mathbf{j} + 5\mathbf{k}$ 27. $\cosh t\,\mathbf{i} + \frac{1}{4}\sinh t\,\mathbf{j} + \mathbf{k}$
29. $t\mathbf{i} + (1 - t^2)\mathbf{j} - 2\mathbf{k}$ 31. $6 \cos t\,\mathbf{i} + 6 \sin t\,\mathbf{j} + 6 \cos t\,\mathbf{k}$
33. $y = x^4$, $z = 0$, $t^{*3}\mathbf{i} + t^{*12}\mathbf{j}$

SECTION 8.4, page 471

1. (a) $\mathbf{i} + 2t\mathbf{j}$, $(1/\sqrt{1 + 4t^2})(\mathbf{i} + 2t\mathbf{j})$, (b) $\mathbf{i} + 2\mathbf{j}$, $(1/\sqrt{5})(\mathbf{i} + 2\mathbf{j})$,
   (c) $(1 + w)\mathbf{i} + (1 + 2w)\mathbf{j}$
3. (a), (b) $3\mathbf{i} - \mathbf{k}$, $(1/\sqrt{10})(3\mathbf{i} - \mathbf{k})$, (c) $(13 + 3w)\mathbf{i} - w\mathbf{k}$
5. (a) $\sinh t\,\mathbf{i} + \cosh t\,\mathbf{j}$, $(1/\sqrt{\sinh^2 t + \cosh^2 t})(\sinh t\,\mathbf{i} + \cosh t\,\mathbf{j})$,
   (b) $t = \cosh^{-1}\frac{5}{3}$, $\frac{4}{3}\mathbf{i} + \frac{5}{3}\mathbf{j}$, $(1/\sqrt{41})(4\mathbf{i} + 5\mathbf{j})$, (c) $(\frac{5}{3} + \frac{4}{3}w)\mathbf{i} + (\frac{4}{3} + \frac{5}{3}w)\mathbf{j}$
7. (a) $-\sin t\,\mathbf{i} + \cos t\,\mathbf{j} + \mathbf{k}$, $(1/\sqrt{2})(-\sin t\,\mathbf{i} + \cos t\,\mathbf{j} + \mathbf{k})$, (b) $\mathbf{j} + \mathbf{k}$,
   $(1/\sqrt{2})(\mathbf{j} + \mathbf{k})$, (c) $\mathbf{i} + w\mathbf{j} + w\mathbf{k}$

9. (a) $-3 \sin t\,\mathbf{i} - 3 \cos t\,\mathbf{j}$, $-\sin t\,\mathbf{i} - \cos t\,\mathbf{j}$, (b) $t = \frac{3}{2}\pi$, $3\mathbf{i}$, $\mathbf{i}$,
  (c) $3w\mathbf{i} + 3\mathbf{j}$

11. $\sinh 1$

13. $8(\sqrt{1000} - 1)/27$

15. $\frac{1}{2}\pi^2$

17. Start from $\mathbf{r}(t) = t\mathbf{i} + f(t)\mathbf{j}$.

21. $\sqrt{2}(e^\pi - 1)$

23. $2\sqrt{2}$

SECTION 8.5, page 476

1. $\mathbf{i}$, 1, $\mathbf{0}$

3. $(2 - 2t)\mathbf{i}$, $|2 - 2t|$, $-2\mathbf{i}$

5. $2\mathbf{i} - 2\mathbf{j} + \mathbf{k}$, 3, $\mathbf{0}$

7. $xy = 1\ (x > 0)$, $e^t\mathbf{i} - e^{-t}\mathbf{j}$, $\sqrt{2}\cosh 2t$, $e^t\mathbf{i} + e^{-t}\mathbf{j}$

9. $-3 \sin t\,\mathbf{i} + 3 \cos t\,\mathbf{j} + 2\mathbf{k}$, $\sqrt{13}$, $-3 \cos t\,\mathbf{i} - 3 \sin t\,\mathbf{j}$

15. $[3 \sin t \cos t/(\sin^2 t + 4 \cos^2 t)](\sin t\,\mathbf{i} - 2 \cos t\,\mathbf{j})$

17. $|\mathbf{a}| = \omega^2 R = |\mathbf{v}|^2/R = 30^2/(30 \cdot 86400 \cdot 365/2\pi) = 5.98 \cdot 10^{-6}$ [km/sec^2]

19. $\mathbf{r}(t) = \frac{1}{2}\mathbf{a}_0 t^2 + \mathbf{v}_0 t + \mathbf{r}_0$ ($\mathbf{a}_0$, $\mathbf{v}_0$, $\mathbf{r}_0$ constant vectors)

SECTION 8.6, page 479

3. $\mathbf{r}(t) = [t,\ \ y(t),\ \ 0]$, $\mathbf{r}' = [1,\ \ y',\ \ 0]$, $\mathbf{r}' \cdot \mathbf{r}' = 1 + y'^2$, $\mathbf{r}'' = [0,\ \ y'',\ \ 0]$
  etc.

5. $2/(1 + 4x^2)^{3/2}$

7. $a/(a^2 + c^2)$

9. $6/\sqrt{t}(4 + 9t)^{3/2}$

15. $3/(1 + 9t^2 + 9t^4)$

17. $2/(2e^t + e^{-t})$

19. $-6 \cos 2t/(5 + 12 \sin^2 2t)$

SECTION 8.7, page 484

1. $4t^3 + 2/t$

3. $(g'h - gh')/h^2$

7. $-t(1 + t^2)^{-3/2}$

9. $\cosh 2t$

11. $4u$, $4v$

13. $e^{2u} \sin 2v$, $e^{2u} \cos 2v$

15. $4u^3 - 12uv^2$, $-12u^2 v + 4v^3$

SECTION 8.8, page 491

1. $y\mathbf{i} + x\mathbf{j}$

3. $2x\mathbf{i} + 2y\mathbf{j}$

5. $-8x\mathbf{i} + \mathbf{j}$

7. $2x\mathbf{i} + 8y\mathbf{j}$

9. $2(x\mathbf{i} + y\mathbf{j})(x^2 + y^2)^{-1}$

11. $e^x \cos y\,\mathbf{i} - e^x \sin y\,\mathbf{j}$

13. $yz\mathbf{i} + xz\mathbf{j} + xy\mathbf{k}$

15. $\frac{1}{2}x\mathbf{i} - 8y\mathbf{j} + z\mathbf{k}$

17. $(yz\mathbf{i} + xz\mathbf{j} + xy\mathbf{k})e^{xyz}$

19. $x + y + z$

21. $x^2 - 3y^2 + z$

23. xyz

25. $xy + yz$

27. e^{xy}

33. $-2/5$

35. $-7/125\sqrt{3}$

37. $-18/\sqrt{13}$

39. 0

41. $0.8\mathbf{i} - 0.6\mathbf{j}$

43. $0.6\mathbf{i} + 0.8\mathbf{j}$

45. $(-3\mathbf{i} + 2\mathbf{j})/\sqrt{13}$

47. $(3\mathbf{i} + 4\mathbf{j} - 5\mathbf{k})/\sqrt{50}$

49. $(\mathbf{i} + \mathbf{j})/\sqrt{2}$

51. The direction of $-3\mathbf{i} + 4\mathbf{j}$

53. The direction of $6\mathbf{i} - 5\mathbf{j}$

55. The direction of $-5\mathbf{i} + 2\mathbf{j}$

SECTION 8.9, page 495

1. 2   **3.** 0   **5.** 0   **7.** $2e^x \cos y$

9. $y + x + xy$ **11.** 0   **13.** 6   **15.** -6

17. $2xz/y^3$ **19.** $2(\mathbf{i} + \mathbf{j} + \mathbf{k})$ **21.** $2\mathbf{i}$ **23.** 0

25. 272

35. $v_1 = w_2 z - w_3 y$ is independent of x, thus $\partial v_1/\partial x = 0$, etc.

SECTION 8.10, page 498

1. $-2\mathbf{k}$ **3.** $-(xe^{xy} + ye^{-xy})\mathbf{k}$

5. $-\mathbf{i} - \mathbf{j} - \mathbf{k}$ **7.** 0   **9.** $-(z \sin y + y \cos z)\mathbf{i}$

11. curl $\mathbf{v} = -3y^2\mathbf{k}$, div $\mathbf{v} = 0$, incompressible, $\mathbf{r} = (c_2{}^3 t + c_1)\mathbf{i} + c_2\mathbf{j} + c_3\mathbf{k}$

13. curl $\mathbf{v} = -2\mathbf{k}$, incompressible, $x' = y$, $y' = -x$, $x^2 + y^2 = c_1$, $z = c_2$

21. 0, $z(x - y)\mathbf{i} + x(y - z)\mathbf{j} + y(z - x)\mathbf{k}$

23. $(y^2 + 2xz)\mathbf{i} + (z^2 + 2xy)\mathbf{j} + (x^2 + 2yz)\mathbf{k}$, $2x + 2y + 2z$

CHAPTER 8 (REVIEW PROBLEMS), page 508

11. $6\mathbf{i} - 18\mathbf{j}$ **13.** 28

15. $72xy^3z^2$ **17.** $3y^2z^2$, $6xy^2z$, $2y^3z$

19. $2x^2z + 18y(y - z)^2 + 16xyz^2$ **21.** $2x^2z$

23. -11 **25.** 0

27. $-x\mathbf{i} - 2(y - z)\mathbf{j} - (2y - z)\mathbf{k}$ **29.** $18/\sqrt{5}$

31. 0   **33.** $2y - z + 2xy + 2x^2z$

35. $4xz\mathbf{i} + 2x^2\mathbf{k}$ **37.** $y^3z^2 + 3xy^2z^2 + 2xy^3z$

39. $-2xy^3z\mathbf{i} + y^3z^2\mathbf{k}$ **41.** Parabola, $\mathbf{i} + 12\mathbf{j}$, $\sqrt{145}$, $6\mathbf{j}$

43. Circle, $-2\mathbf{j}$, 2, $4\mathbf{i}$ **45.** Straight line, $\mathbf{i} + 3\mathbf{j} + \mathbf{k}$, $\sqrt{11}$, 0

47. Hyperbola, $\mathbf{i} - \frac{1}{4}\mathbf{j}$, $\frac{1}{4}\sqrt{17}$, $\frac{1}{4}\mathbf{j}$ **49.** Helix, $2\mathbf{j} + \mathbf{k}$, $\sqrt{5}$, $-4\mathbf{i}$

51. $2/(1 + 4t^2)^{3/2}$, 0   **53.** $2/|t|^3(1 + t^{-4})^{3/2}$, 0

55. $2(1 + 9t^2 + 9t^4)^{1/2}/(1 + 4t^2 + 9t^4)^{3/2}$, $3/(1 + 9t^2 + 9t^4)$

57. $2(1 + 9t^{-4} + t^{-6})^{1/2}/(1 + t^{-4} + 4t^2)^{3/2}$, $3/(t^4 + 9 + t^{-2})$

59. $(1 + 2e^{-2t})^{1/2}/(1 + e^{-2t})^{3/2}$, $-2e^{-t}/(1 + 2e^{-2t})$

SECTION 9.1, page 518

1. 1   **3.** $\frac{28}{3}$ **5.** $-\frac{4}{3}$ **7.** -12.8 **9.** 0

11. 0   **13.** $\frac{7}{5}$ **15.** $\frac{3}{2}$ **17.** $-\frac{25}{2}$

19. 0   **21.** $\frac{5}{3}\sqrt{5}$ **23.** 4π **25.** $\frac{5}{3}$

SECTION 9.2, page 526

1. 2/3   **3.** 0   **5.** $\frac{1}{2} - \frac{1}{2}e^{-4}$ **7.** $(8 + e^3)/9$

9. 67/120   **11.** 156   **13.** 2/3   **15.** 1/6

17. 1/12   **19.** 1/3   **21.** 72   **23.** $32\pi/3$

25. $e^{-1} - e^{-4}$ **27.** 8/3, 4/3   **29.** $8/5\pi$, $8/5\pi$

31. $I_x = bh^3/12$, $I_y = 7b^3h/48$
33. $I_x = h^3(3b + a)/12$, $I_y = h(a^4 - b^4)/48(a - b)$
35. 1   **37.** 1   **39.** ab

SECTION 9.3, page 533

1. -8 **3.** $17/60$ **5.** $-\pi/2$ **7.** -4 **9.** 4
11. $-1/3$ **13.** 0   **15.** $16/5$ **17.** πab **19.** $3\pi a^2$
21. 32π **23.** 60π **25.** 10π **27.** 0   **29.** 16
31. $2e^2 - 2$ **33.** 16π **35.** 3

SECTION 9.4, page 538

1. Straight lines, $\mathbf{k}$ **3.** Straight lines, $-2\mathbf{k}$
5. Straight lines, ellipses, $2 \cos u\mathbf{i} + 4 \sin u\mathbf{j}$
7. $z = c\sqrt{x^2 + y^2}$, circles, straight lines, $-cu \cos v\mathbf{i} - cu \sin v\mathbf{j} + u\mathbf{k}$
9. Helices, horizontal straight lines, $\sin v\mathbf{i} - \cos v\mathbf{j} + u\mathbf{k}$
11. $x^2 + y^2 + \frac{1}{4}z^2 = 1$, ellipses, circles,
 $2 \cos^2 v \cos u\mathbf{i} + 2 \cos^2 v \sin u\mathbf{j} + \cos v \sin v\mathbf{k}$
13. $z = x^2/a^2 - y^2/b^2$, hyperbolas, parabolas,
 $-2bu^2 \cosh v\mathbf{i} + 2au^2 \sinh v\mathbf{j} + abu\mathbf{k}$
15. Catenaries, circles, $(v^2 - 1)^{-1/2}v(\cos u\mathbf{i} + \sin u\mathbf{j}) - v\mathbf{k}$
17. $u\mathbf{i} + v\mathbf{j} + v\mathbf{k}$, $-\mathbf{j} + \mathbf{k}$ **19.** $u\mathbf{i} + u^2\mathbf{j} + v\mathbf{k}$, $2u\mathbf{i} - \mathbf{j}$
21. $2 \cos u\mathbf{i} + v\mathbf{j} + 3 \sin u\mathbf{k}$, $-3 \cos u\mathbf{i} - 2 \sin u\mathbf{k}$
23. $u \cos v\mathbf{i} + u \sin v\mathbf{j} + 4u^2\mathbf{k}$, $-8u^2 \cos v\mathbf{i} - 8u^2 \sin v\mathbf{j} + u\mathbf{k}$
25. $u \cos v\mathbf{i} + u \sin v\mathbf{j} + u^4\mathbf{k}$, $-4u^4 \cos v\mathbf{i} - 4u^4 \sin v\mathbf{j} + u\mathbf{k}$
27. $\mathbf{r}(u, v) = h(u, v)\mathbf{i} + u\mathbf{j} + v\mathbf{k}$, $\mathbf{i} - h_u\mathbf{j} - h_v\mathbf{k}$
29. $u\mathbf{i} + v\mathbf{j} + 4(u^2 + v^2)\mathbf{k}$, $-8u\mathbf{i} - 8v\mathbf{j} + \mathbf{k}$
31. $(1/\sqrt{50})(4\mathbf{i} - 5\mathbf{j} + 3\mathbf{k})$ **33.** $(1/5)(y\mathbf{j} + z\mathbf{k})$
35. $(1/a)(x\mathbf{i} + y\mathbf{j} + z\mathbf{k})$ **37.** $(x^2 + y^2 + 1)^{-1/2}(y\mathbf{i} + x\mathbf{j} - \mathbf{k})$
39. $(2 + 4x^2)^{-1/2}(2x\mathbf{i} - \mathbf{j} - \mathbf{k})$ **45.** $x^* + y^* - z^* = 1$
47. $4x^* - z^* = 4$ **49.** $-3x^* + y^* = 20$

SECTION 9.5, page 547

1. -8 **3.** $-4/3$ **5.** 4   **7.** $-59/180$
9. 32   **11.** $\cosh 1 - 2$
13. $30 + 2e^3 - 2e \approx 64.73$ **15.** $\pi/4 + \pi^3/48 \approx 1.431$
17. $2 \cosh^3 2 - 2 \approx 104.5$ **19.** $16/3 + 16\pi/5 \approx 15.39$
21. $\sqrt{3}$ **23.** $\sqrt{3}$ **25.** 6π **27.** 4π
29. 0   **31.** $\frac{1}{9}(5^{3/2} - 1)$ **33.** $8\pi\sqrt{2}$ **35.** 8π
39. $I_{x=y} = \displaystyle\iint_S [\frac{1}{2}(x - y)^2 + z^2]\sigma \, dA$

41. $h\pi(1 + h^2/6)$ **43.** $2\pi^2ab(2a^2 + 3b^2)$
45. $h\pi(1 + 2h^2/3)$ **47.** Use $d\mathbf{r} = \mathbf{r}_u \, du + \mathbf{r}_v \, dv$.
53. $E = (a + b \cos v)^2$, $F = 0$, $G = b^2$, etc.

SECTION 9.6, page 555

1. 1/8   **3.** 24π **5.** 1/4   **7.** 4   **9.** 16/3
11. $\frac{1}{12}abc(b^2 + c^2)$ **13.** $\pi ha^4/2$ **15.** $8\pi a^5/15$
17. $2a^2(e^a - 1)$ **19.** 160π **21.** $8\pi^2$ **23.** 24
25. 0   **27.** $384\pi/5$ **29.** 1/6   **31.** $3\pi c^2 h$
33. 0   **35.** 24

SECTION 9.7, page 561

3. Put $f = g$ in (10).
5. $h = f - g$ is harmonic, and $\partial h/\partial n = 0$ on S. Thus $h = const$ in T (cf. Prob. 4).
9. We have div $\mathbf{u} = 3$ and $\mathbf{u} \cdot \mathbf{n} = |\mathbf{u}|\,|\mathbf{n}| \cos \theta$, $|\mathbf{u}| = \sqrt{x^2 + y^2 + z^2} = r$, $|\mathbf{n}| = 1$. Thus $\mathbf{u} \cdot \mathbf{n} = r \cos \theta$, and the formula follows.

SECTION 9.8, page 567

1. ± 4 **3.** $\pm 3\pi/2$ **5.** $\pm 16/3$ **7.** $\pm 2(1 - e^2)$
11. $\cosh 1 - 1$ **13.** $-2/3$ **15.** -10 **17.** 6π
19. -4π **21.** 0   **23.** 2

SECTION 9.9, page 576

1. $x^2 y + z^3$ **3.** $\frac{1}{4}x^4 + \frac{1}{2}y^2 + \sin z$ **5.** Not exact
7. $xz - yz$ **9.** Not exact   **11.** $\frac{1}{2}[x^2 + (y - z)^2]$
13. $xe^y + ze^x$ **15.** $-(y + z)/x$ **17.** Not exact   **19.** $x \sin^2 y$
21. 4   **23.** 0   **25.** 12   **27.** 2
29. $64e^2 \approx 472.9$

CHAPTER 9 (REVIEW PROBLEMS), page 577

17. 2/3   **19.** 0   **21.** -9π **23.** 45/4
25. 64   **27.** 0   **29.** 0   **31.** 0
33. $1 - \cosh 1$ **35.** 1   **37.** 0   **39.** 5π
41. $\frac{8}{3}\pi(1 + \pi^2)$ **43.** 1/8   **45.** 159/20   **47.** $\pi/2, \pi/8$
49. 4/5, 8/15
51. $\mathbf{r} = u\mathbf{i} + v\mathbf{j} + (1 - 3u - 2v)\mathbf{k}, \quad 3\mathbf{i} + 2\mathbf{j} + \mathbf{k}$
53. $\mathbf{r} = u \cos v\,\mathbf{i} + u\mathbf{j} + u \sin v\,\mathbf{k}, \quad u \cos v\,\mathbf{i} - u\mathbf{j} + u \sin v\,\mathbf{k}$
55. $\mathbf{r} = u\mathbf{i} + v\mathbf{j} + u^2\mathbf{k}, \quad -2u\mathbf{i} + \mathbf{k}$
57. $\mathbf{r} = 2 \cosh u\,\mathbf{i} + \sinh u\,\mathbf{j} + v\mathbf{k}, \quad \cosh u\,\mathbf{i} - 2 \sinh u\,\mathbf{j}$
59. $\mathbf{r} = u \cos v\,\mathbf{i} + 2u \sin v\,\mathbf{j} + 2u\mathbf{k}, \quad -4u \cos v\,\mathbf{i} - 2u \sin v\,\mathbf{j} + 2u\mathbf{k}$
61. 5/12   **63.** 640/3   **65.** -98.4 **67.** 40
69. $-4/3 + 8\pi/3$

SECTION 10.1, page 585

1. $2\pi, 2\pi, \pi, \pi, 2, 2, 1, 1$ **23.** 0

25. 0 (n even), $\quad 2/n$ (n odd)

27. $[(-1)^n e^{\pi} - 1]/(1 + n^2)$

29. $n[(-1)^n e^{-\pi} - 1]/(1 + n^2)$

SECTION 10.2, page 592

1. $\dfrac{1}{2} + \dfrac{2}{\pi} \left(\cos x - \dfrac{1}{3} \cos 3x + \dfrac{1}{5} \cos 5x - + \cdots \right)$

3. $\dfrac{1}{2} + \dfrac{2}{\pi} \left(\sin x + \dfrac{1}{3} \sin 3x + \dfrac{1}{5} \sin 5x + \cdots \right)$

5. $-\dfrac{4k}{\pi} \left(\cos x - \dfrac{1}{3} \cos 3x + \dfrac{1}{5} \cos 5x - + \cdots \right)$

7. $2 \left(\sin x - \dfrac{1}{2} \sin 2x + \dfrac{1}{3} \sin 3x - \dfrac{1}{4} \sin 4x + - \cdots \right)$

9. $\pi - 2 \left(\sin x + \dfrac{1}{2} \sin 2x + \dfrac{1}{3} \sin 3x + \cdots \right)$

11. $\dfrac{\pi^2}{3} - 4 \left(\cos x - \dfrac{1}{4} \cos 2x + \dfrac{1}{9} \cos 3x - \dfrac{1}{16} \cos 4x + - \cdots \right)$

13. $\dfrac{2}{\pi} \sin x + \dfrac{1}{2} \sin 2x - \dfrac{2}{9\pi} \sin 3x - \dfrac{1}{4} \sin 4x + \dfrac{2}{25\pi} \sin 5x + \cdots$

15. $2 \left[\left(\pi - \dfrac{4}{\pi} \right) \sin x - \dfrac{\pi}{2} \sin 2x + \left(\dfrac{\pi}{3} - \dfrac{4}{3^2 \pi} \right) \sin 3x - \dfrac{\pi}{4} \sin 4x + \cdots \right]$

19. Set $x = x^* - \pi/2$.

SECTION 10.3, page 596

1. $\dfrac{1}{2} + \dfrac{2}{\pi} \left(\cos \dfrac{\pi x}{2} - \dfrac{1}{3} \cos \dfrac{3\pi x}{2} + \dfrac{1}{5} \cos \dfrac{5\pi x}{2} - + \cdots \right)$

3. $\dfrac{1}{2} + \dfrac{2}{\pi} \left(\sin \dfrac{\pi x}{2} + \dfrac{1}{3} \sin \dfrac{3\pi x}{2} + \dfrac{1}{5} \sin \dfrac{5\pi x}{2} + \cdots \right)$

5. $\dfrac{8}{\pi} \left(\sin \dfrac{\pi x}{4} - \dfrac{1}{2} \sin \dfrac{2\pi x}{4} + \dfrac{1}{3} \sin \dfrac{3\pi x}{4} - \dfrac{1}{4} \sin \dfrac{4\pi x}{4} + - \cdots \right)$

7. $\dfrac{2}{\pi^2} (\cos \pi x + \tfrac{1}{9} \cos 3\pi x + \cdots) - \dfrac{1}{\pi} (2 \sin \pi x - \tfrac{1}{2} \sin 2\pi x + - \cdots)$

9. $\dfrac{1}{\pi} (\sin 4x - \tfrac{1}{9} \sin 12x + \tfrac{1}{25} \sin 20x - \tfrac{1}{49} \sin 28x + - \cdots)$

11. $\dfrac{1}{3} - \dfrac{4}{\pi^2} (\cos \pi x - \tfrac{1}{4} \cos 2\pi x + \tfrac{1}{9} \cos 3\pi x - \tfrac{1}{16} \cos 4\pi x + - \cdots)$

13. $\dfrac{4}{\pi} \left(\dfrac{1}{2} - \dfrac{1}{1 \cdot 3} \cos 2\pi x - \dfrac{1}{3 \cdot 5} \cos 4\pi x - \dfrac{1}{5 \cdot 7} \cos 6\pi x - \cdots \right)$

15. $\dfrac{2}{\pi} + \cos 100\pi t$

$\qquad + \dfrac{4}{\pi} \left(\dfrac{1}{1 \cdot 3} \cos 200\pi t - \dfrac{1}{3 \cdot 5} \cos 400\pi t + \dfrac{1}{5 \cdot 7} \cos 600\pi t - + \cdots \right)$

SECTION 10.4, page 600

1. Even: $|x^3|$, $x^2 \cos nx$, $\cosh x$.   Odd: $x \cos nx$, $\sinh x$, $x|x|$

3. Odd   **5.** Neither one   **7.** Even   **9.** Odd

11. Neither one   **17.** $|f(-x)| = |-f(x)| = |f(x)|$, etc.

19. $\cosh x + \sinh x$ **21.** $x^2(1 - x^2)^{-1} + x(1 - x^2)^{-1}$

23. $f \equiv 0$

27. $\dfrac{\pi}{2} - \dfrac{4}{\pi}\left(\cos x + \dfrac{1}{9}\cos 3x + \dfrac{1}{25}\cos 5x + \cdots\right)$

29. $\dfrac{\pi^2}{12} - \cos x + \dfrac{1}{4}\cos 2x - \dfrac{1}{9}\cos 3x + \dfrac{1}{16}\cos 4x - + \cdots$

31. $\dfrac{\pi^3}{4} + 6\left[\left(-\dfrac{\pi}{1^2} + \dfrac{4}{1^4\pi}\right)\cos x + \dfrac{\pi}{2^2}\cos 2x\right.$

$$\left. + \left(-\dfrac{\pi}{3^2} + \dfrac{4}{3^4\pi}\right)\cos 3x + \dfrac{\pi}{4^2}\cos 4x + \cdots\right]$$

33. $\dfrac{1}{2} + \dfrac{2}{\pi}\left(\cos x - \dfrac{1}{3}\cos 3x + \dfrac{1}{5}\cos 5x - + \cdots\right)$

35. $\dfrac{\pi^2}{6} - \dfrac{4}{\pi}\cos x - \dfrac{2}{2^2}\cos 2x + \dfrac{4}{3^3\pi}\cos 3x + \dfrac{2}{4^2}\cos 4x - \dfrac{4}{5^3\pi}\cos 5x + \cdots$

SECTION 10.5, page 604

1. $f(x) = k$

3. $\dfrac{1}{2} - \dfrac{2}{\pi}\left(\cos\dfrac{\pi x}{L} - \dfrac{1}{3}\cos\dfrac{3\pi x}{L} + \dfrac{1}{5}\cos\dfrac{5\pi x}{L} - + \cdots\right)$

5. $\dfrac{L^2}{3} - \dfrac{4L^2}{\pi^2}\left(\cos\dfrac{\pi x}{L} - \dfrac{1}{4}\cos\dfrac{2\pi x}{L} + \dfrac{1}{9}\cos\dfrac{3\pi x}{L} - \dfrac{1}{16}\cos\dfrac{4\pi x}{L} + - \cdots\right)$

7. $\dfrac{L^3}{4} + \dfrac{6L^3}{\pi^2}\left[\left(\dfrac{4}{\pi^2} - 1\right)\cos\dfrac{\pi x}{L} + \dfrac{1}{2^2}\cos\dfrac{2\pi x}{L} + \left(\dfrac{4}{3^4\pi^2} - \dfrac{1}{3^2}\right)\cos\dfrac{3\pi x}{L} + \cdots\right]$

9. $\dfrac{2}{\pi} - \dfrac{4}{\pi}\left(\dfrac{1}{1\cdot 3}\cos\dfrac{2\pi x}{L} + \dfrac{1}{3\cdot 5}\cos\dfrac{4\pi x}{L} + \dfrac{1}{5\cdot 7}\cos\dfrac{6\pi x}{L} + \cdots\right)$

11. $\dfrac{4k}{\pi}\left(\sin\dfrac{\pi x}{L} + \dfrac{1}{3}\sin\dfrac{3\pi x}{L} + \dfrac{1}{5}\sin\dfrac{5\pi x}{L} + \cdots\right)$

13. $\dfrac{2L}{\pi}\left(\sin\dfrac{\pi x}{L} + \dfrac{1}{2}\sin\dfrac{2\pi x}{L} + \dfrac{1}{3}\sin\dfrac{3\pi x}{L} + \cdots\right)$

15. $\dfrac{4}{\pi}\left(\sin\dfrac{\pi x}{L} - \dfrac{1}{2}\sin\dfrac{2\pi x}{L} + \dfrac{1}{3}\sin\dfrac{3\pi x}{L} + \dfrac{1}{5}\sin\dfrac{5\pi x}{L} - \dfrac{1}{6}\sin\dfrac{6\pi x}{L} + \dfrac{1}{7}\sin\dfrac{7\pi x}{L} + \cdots\right)$

17. $\dfrac{4L}{\pi^2}\left(\dfrac{1}{1^2}\sin\dfrac{\pi x}{L} - \dfrac{1}{3^2}\sin\dfrac{3\pi x}{L} + \dfrac{1}{5^2}\sin\dfrac{5\pi x}{L} - + \cdots\right)$

19. $\dfrac{2L^2}{\pi}\left[\left(1 - \dfrac{4}{\pi^2}\right)\sin\dfrac{\pi x}{L} - \dfrac{1}{2}\sin\dfrac{2\pi x}{L} + \left(\dfrac{1}{3} - \dfrac{4}{3^3\pi^2}\right)\sin\dfrac{3\pi x}{L} - \dfrac{1}{4}\sin\dfrac{4\pi x}{L} + \cdots\right]$

23. $i\displaystyle\sum_{\substack{n=-\infty \\ n\neq 0}}^{\infty} \dfrac{(-1)^n}{n} e^{inx}$

25. $\dfrac{\sinh \pi}{\pi}\displaystyle\sum_{n=-\infty}^{\infty} (-1)^n \dfrac{1 + in}{1 + n^2} e^{inx}$

SECTION 10.6, page 610

11. $12 \left(\sin x - \dfrac{1}{2^3} \sin 2x + \dfrac{1}{3^3} \sin 3x - + \cdots \right)$

15. No

SECTION 10.7, page 613

3. $y = C_1 \cos \omega t + C_2 \sin \omega t + A(\omega) \sin t, \quad A(\omega) = 1/(\omega^2 - 1),$
$A(0.5) = -1.33, \quad A(0.7) = -0.20, \quad A(0.9) = -5.3, \quad A(1.1) = 4.8,$
$A(1.5) = 0.8, \quad A(2) = 0.33, \quad A(10) = 0.01$

5. $y = C_1 \cos \omega t + C_2 \sin \omega t + \displaystyle\sum_{n=1}^{N} \dfrac{b_n}{\omega^2 - n^2} \sin nt$

7. $y = C_1 \cos \omega t + C_2 \sin \omega t$
$\qquad + \dfrac{1}{2\omega^2} - \dfrac{1}{1 \cdot 3(\omega^2 - 4)} \cos 2t - \dfrac{1}{3 \cdot 5(\omega^2 - 16)} \cos 4t - \cdots$

9. $y = C_1 \cos \omega t + C_2 \sin \omega t$
$\qquad + \dfrac{4}{\pi} \left(\dfrac{1}{\omega^2 - 1} \sin t - \dfrac{1}{9(\omega^2 - 9)} \sin 3t + \dfrac{1}{25(\omega^2 - 25)} \sin 5t - + \cdots \right)$

11. $y = -\dfrac{K}{c} \cos t$

13. $y = \dfrac{1 - n^2}{D} a_n \cos nt + \dfrac{nc}{D} a_n \sin nt, \quad D = (1 - n^2)^2 + n^2 c^2$

15. $y = A_1 \cos t + B_1 \sin t + A_3 \cos 3t + B_3 \sin 3t + \cdots$
$\qquad$ where $A_n = -ncb_n/D, \quad B_n = (1 - n^2)b_n/D, \quad D = (1 - n^2)^2 + n^2 c^2,$
$\qquad b_1 = 1, b_2 = 0, b_3 = -\frac{1}{9}, b_4 = 0, b_5 = \frac{1}{25}, \cdots$

17. $I = \displaystyle\sum_{n=1}^{\infty} (A_n \cos nt + B_n \sin nt),$

$\qquad A_n = \dfrac{80(10 - n^2)}{\pi n^2 D_n}, \quad B_n = \dfrac{800}{n \pi D_n} \quad (n \text{ odd}),$

$\qquad A_n = 0, B_n = 0 \quad (n \text{ even}), \qquad D_n = (10 - n^2)^2 + 100\, n^2,$
$\qquad I = 1.266 \cos t + 1.406 \sin t + 0.003 \cos 3t + 0.094 \sin 3t + \cdots$

SECTION 10.8, page 617

1. $F = \dfrac{4}{\pi} \left[\sin x + \dfrac{1}{3} \sin 3x + \cdots + \dfrac{1}{N} \sin Nx \right] \qquad (N \text{ odd})$

5. $2 \left(\sin x - \dfrac{1}{2} \sin 2x + \cdots + \dfrac{(-1)^{N+1}}{N} \sin Nx \right),$
$\qquad E^* \approx 8, 5, 3.6, 2.8, 2.3$

7. $\dfrac{\pi^2}{3} - 4 \left(\cos x - \dfrac{1}{4} \cos 2x + \dfrac{1}{9} \cos 3x - \cdots + \dfrac{(-1)^{N+1}}{N^2} \cos Nx \right),$
$\qquad E^* \approx 4.14, 1.00, 0.38, 0.18, 0.10$

SECTION 10.9, page 625

9. $\dfrac{2}{\pi} \displaystyle\int_0^\infty \left[\dfrac{a \sin aw}{w} + \dfrac{\cos aw - 1}{w^2} \right] \cos xw \; dw$

11. $\dfrac{6}{\pi} \displaystyle\int_0^\infty \dfrac{2 + w^2}{4 + 5w^2 + w^4} \cos xw \; dw$

13. $\dfrac{2}{\pi} \displaystyle\int_0^\infty \left[\left(a^2 - \dfrac{2}{w^2} \right) \sin aw + \dfrac{2a}{w} \cos aw \right] \dfrac{\cos wx}{w} \; dw$

15. $f(ax) = \displaystyle\int_0^\infty A(w) \cos axw \; dw = \int_0^\infty A\left(\dfrac{p}{a} \right) \cos xp \dfrac{dp}{a}$ where $wa = p$

SECTION 10.10, page 630

1. No   **5.** $e^{-w} \sqrt{\pi/2}$

7. $\sqrt{\pi/2}$ if $0 < w < \pi$, 0 if $w > \pi$

9. $\sqrt{\pi/2} \cos w$ if $|w| < \pi/2$, 0 if $|w| > \pi/2$

SECTION 10.11, page 635

1. $1/(1 + iw)\sqrt{2\pi}$ **3.** $\sqrt{2/\pi}(2 - w)^{-1} \sin (2 - w)$

5. $(e^{1-iw} - e^{-1+iw})/(1 - iw)\sqrt{2\pi}$

CHAPTER 10 (REVIEW PROBLEMS), page 640

21. $\dfrac{1}{2} - \dfrac{2}{\pi} \left(\sin x + \dfrac{1}{3} \sin 3x + \dfrac{1}{5} \sin 5x + \cdots \right)$

23. $\dfrac{2}{\pi} \left(\sin x - \dfrac{2}{2} \sin 2x + \dfrac{1}{3} \sin 3x + \dfrac{1}{5} \sin 5x - \dfrac{2}{6} \sin 6x + \cdots \right)$

25. $\dfrac{\pi}{2} - \dfrac{4}{\pi} \left(\cos x + \dfrac{1}{9} \cos 3x + \dfrac{1}{25} \cos 5x + \cdots \right)$

27. $-\sin x + \dfrac{1}{2} \sin 2x - \dfrac{1}{3} \sin 3x + \dfrac{1}{4} \sin 4x - + \cdots$

29. $1 + \dfrac{1}{3} \pi^2 - 4 \left(\cos x - \dfrac{1}{4} \cos 2x + \dfrac{1}{9} \cos 3x - + \cdots \right)$

31. $\dfrac{\pi^2}{12} - \dfrac{2}{\pi} \cos x - \dfrac{1}{2^2} \cos 2x + \dfrac{2}{3^3 \pi} \cos 3x + \dfrac{1}{4^2} \cos 4x - \cdots$

33. $2 \sin x + \dfrac{\pi}{2} \sin 2x - \dfrac{2}{9} \sin 3x - \dfrac{\pi}{4} \sin 4x + \dfrac{2}{25} \sin 5x + \cdots$

35. $\dfrac{2}{a(\pi - a)} \left[\dfrac{\sin a}{1^2} \sin x + \dfrac{\sin 2a}{2^2} \sin 2x + \dfrac{\sin 3a}{3^2} \sin 3x + \cdots \right]$

37. $k + \dfrac{2k}{\pi} \left(\sin x - \dfrac{2}{2} \sin 2x + \dfrac{1}{3} \sin 3x + \dfrac{1}{5} \sin 5x - \dfrac{2}{6} \sin 6x + \cdots \right)$

39. $\dfrac{8}{\pi}\left(\sin x + \dfrac{1}{3^3}\sin 3x + \dfrac{1}{5^3}\sin 5x + \cdots\right)$

41. $\dfrac{4}{\pi}\left(\sin \pi x + \dfrac{1}{3}\sin 3\pi x + \dfrac{1}{5}5\pi x + \cdots\right)$

43. $\dfrac{2}{\pi}\left(\sin \pi x - \dfrac{1}{2}\sin 2\pi x + \dfrac{1}{3}3\pi x - + \cdots\right)$

45. $-\dfrac{1}{4} + \dfrac{2}{\pi^2}\left(\cos \pi x + \dfrac{1}{9}\cos 3\pi x + \cdots\right) + \dfrac{1}{\pi}\left(\sin \pi x - \dfrac{1}{2}\sin 2\pi x + - \cdots\right)$

47. $-\dfrac{4}{\pi^2}\left(\cos \pi x + \dfrac{1}{9}\cos 3\pi x + \cdots\right) + \dfrac{2}{\pi}\left(2\sin \pi x - \dfrac{1}{2}\sin 2\pi x + - \cdots\right)$

49. $\dfrac{4}{3} - \dfrac{4}{\pi^2}\left(\cos \pi x - \dfrac{1}{4}\cos 2\pi x + \dfrac{1}{9}\cos 3\pi x - \dfrac{1}{16}\cos 4\pi x + - \cdots\right)$

51. $\cosh 3x + \sinh 3x$ **57.** $\pi/4$ **59.** $\pi^3/32$

65. $y = C_1 \cos \omega t + C_2 \sin \omega t$

$+ \dfrac{1}{\omega^2 - 1}\sin t - \dfrac{1}{2^3(\omega^2 - 4)}\sin 2t + \dfrac{1}{3^3(\omega^2 - 9)}\sin 3t - + \cdots$

SECTION 11.1, page 646

27. By integration, $u = f(y)$ **29.** $u_x = f(y)$, $u = xf(y) + g(y)$

31. $u = A(y)e^x + B(y)e^{-2x}$ **33.** $u = v(x) + w(y)$

35. $u = c = const$ **37.** $u = cx + g(y)$

SECTION 11.3, page 655

3. $u = k \cos 2t \sin 2x$

5. $u = \dfrac{0.08}{\pi}\left(\cos t \sin x + \dfrac{1}{3^3}\cos 3t \sin 3x + \dfrac{1}{5^3}\cos 5t \sin 5x + \cdots\right)$

7. $u = 12k\left[\left(\dfrac{\pi}{1^3} - \dfrac{8}{1^5\pi}\right)\cos t \sin x + \left(\dfrac{\pi}{3^3} - \dfrac{8}{3^5\pi}\right)\cos 3t \sin 3x + \cdots\right]$

9. $u = \dfrac{4}{5\pi}\left(\dfrac{1}{4}\cos 2t \sin 2x - \dfrac{1}{36}\cos 6t \sin 6x + \dfrac{1}{100}\cos 10t \sin 10x - + \cdots\right)$

11. $27{,}960/\pi^6 \approx 0.9986$

13. $u = 0.05 \sin 2x \sin 2t$ **15.** $u = 0.1 \sin x (\cos t - 2 \sin t)$

17. $u = ke^{c(x-y)}$ **19.** $u = ke^{x^2 + y^2 + c(x-y)}$

21. $u = x^k y^k$ **23.** $u = ke^{cx + y/c}$

25. $u(0, t) = 0$, $w(0, t) = 0$, $u(L \cdot t) = h(t)$, $w(L, t) = h(t)$. The simplest w satisfying these conditions is $w(x, t) = xh(t)/L$. Then

$$v(x, 0) = f(x) - xh(0)/L = f_1(x),$$
$$v_t(x, 0) = g(x) - xh'(0)/L = g_1(x),$$
$$v_{tt} - c^2 v_{xx} = -xh''/L.$$

SECTION 11.4, page 659

7. $u = f_1(y) + f_2(x + y)$ **9.** $u = yf_1(x + y) + f_2(x + y)$

11. $u = f_1(x + y) + f_2(2y - x)$

15. $y'^2 - 2y' + 1 = (y' - 1)^2 = 0$, $y = x + c$, $\Psi(x, y) = x - y$, $v = x$,
$z = x - y$

19. $u = \dfrac{8L^2}{\pi^3} \left(\cos c\left(\dfrac{\pi}{L}\right)^2 t \sin \dfrac{\pi x}{L} + \dfrac{1}{3^3} \cos c\left(\dfrac{3\pi}{L}\right)^2 t \sin \dfrac{3\pi x}{L} + \cdots \right)$

21. $u(0, t) = 0$, $u(L, t) = 0$, $u_x(0, t) = 0$, $u_x(L, t) = 0$

23. $\beta L \approx \frac{3}{2}\pi, \frac{5}{2}\pi, \frac{7}{2}\pi, \cdots$ (more exactly 4.730, 7.853, 10.996, . . .)

25. $\beta L \approx \frac{1}{2}\pi, \frac{3}{2}\pi, \frac{5}{2}\pi, \cdots$ (more exactly 1.875, 4.694, 7.855, . . .)

SECTION 11.5, page 669

1. The solutions of the wave equation are periodic in t.

5. $u = \sin 0.1\pi x\, e^{-1.752\pi^2 t/100}$

7. $u = \dfrac{40}{\pi^2} \left(\sin 0.1\pi x\, e^{-0.01752\pi^2 t} - \dfrac{1}{9} \sin 0.3\pi x\, e^{-0.01752(3\pi)^2 t} + - \cdots \right)$

9. $u = \dfrac{800}{\pi^3} \left(\sin 0.1\pi x\, e^{-0.01752\pi^2 t} + \dfrac{1}{3^3} \sin 0.3\pi x\, e^{-0.01752(3\pi)^2 t} + \cdots \right)$

11. Since the temperatures at the ends are kept constant, the temperature will approach a steady-state (time-independent) distribution $u_I(x)$ as $t \to \infty$, and $u_I = U_1 + (U_2 - U_1)x/L$, the solution of (1) with $\partial u/\partial t = 0$ satisfying the boundary conditions.

15. $u = k$

17. $u = \cos 2x\, e^{-4t}$

19. $u = \frac{1}{2} + \frac{1}{2} \cos 2x\, e^{-4t}$

21. $u = \dfrac{\pi}{8} + \left(1 - \dfrac{2}{\pi}\right) \cos x\, e^{-t} - \dfrac{1}{\pi} \cos 2x\, e^{-4t} - \left(\dfrac{1}{3} + \dfrac{2}{9\pi}\right) \cos 3x\, e^{-9t} + \cdots$

25. $w = e^{-\beta t}$

27. $u = \dfrac{440}{\pi} \displaystyle\sum_{n=0}^{\infty} \dfrac{\sin \frac{1}{20}(2n + 1)\pi x \, \sinh \frac{1}{40}(2n + 1)\pi y}{(2n + 1) \sinh (2n + 1)\pi}$

29. $u_y(x, 0, t) = u_y(x, \pi, t) = 0$, $u = \sin mx \cos ny \exp(-c^2(m^2 + n^2)t)$

SECTION 11.6, page 674

7. $\dfrac{1}{\sqrt{\pi}} \displaystyle\int_{(a-x)/\tau}^{(b-x)/\tau} e^{-w^2}\, dw - \dfrac{1}{\sqrt{\pi}} \displaystyle\int_{(a+x)/\tau}^{(b+x)/\tau} e^{-w^2}\, dw$

SECTION 11.8, page 685

1. c increases and so does the frequency.

5. $c\pi \sqrt{260}$ (corresponding eigenfunctions $F_{4,16}$, $F_{16,14}$), etc.

13. $B_{mn} = (-1)^{m+1}8/mn\pi$ (n odd), 0 (n even)

15. $B_{mn} = (-1)^{m+n}4/mn$

17. $4(\cos m\pi/2 - (-1)^m)(\cos n\pi/2 - (-1)^n)/mn\pi^2$

19. $B_{mn} = (-1)^{m+n} \dfrac{4ab}{mn\pi^2}$

21. $B_{mn} = 0$ (m or n even), $B_{mn} = \dfrac{64a^2b^2}{\pi^6 m^3 n^3}$ (m, n both odd)

23. $B_{mn} = (-1)^{m+n} \dfrac{144a^3b^3}{m^3 n^3 \pi^6}$ **25.** $u = k \cos \pi \sqrt{5}\, t \sin \pi x \sin 2\pi y$

27. $u = \dfrac{0.64}{\pi^6} \displaystyle\sum_{\substack{m=1 \\ m,n \ \ odd}}^{\infty} \sum_{n=1}^{\infty} \dfrac{1}{m^3 n^3} \cos \left(\pi t \sqrt{m^2 + n^2}\right) \sin m\pi x \sin n\pi y$

29. $u = \dfrac{144k}{\pi^6} \displaystyle\sum_{m=1}^{\infty} \sum_{n=1}^{\infty} \dfrac{(-1)^{m+n}}{m^3 n^3} \cos \left(\pi t \sqrt{m^2 + n^2}\right) \sin m\pi x \sin n\pi y$

SECTION 11.9, page 687

7. $a^2 u_{x^*x^*} + c^2 u_{y^*y^*}$ **9.** $5 + 5r^2 \cos 2\theta$

11. $u = \dfrac{4}{\pi} \left(r \sin \theta + \dfrac{1}{3} r^3 \sin 3\theta + \dfrac{1}{5} r^5 \sin 5\theta + \cdots \right)$

13. $u = \dfrac{2}{\pi} r \sin \theta + \dfrac{1}{2} r^2 \sin 2\theta - \dfrac{2}{9\pi} r^3 \sin 3\theta - \dfrac{1}{4} r^4 \sin 4\theta + - \cdots$

15. $u = \dfrac{\pi}{2} - \dfrac{4}{\pi} \left(r \cos \theta + \dfrac{1}{9} r^3 \cos 3\theta + \dfrac{1}{25} r^5 \cos 5\theta + \cdots \right)$

17. $u = \dfrac{880}{\pi} \left(r \sin \theta + \dfrac{1}{3^3} r^3 \sin 3\theta + \dfrac{1}{5^3} r^5 \sin 5\theta + \cdots \right)$

19. $u \approx \dfrac{\pi}{2} \pm \dfrac{4}{\pi} \left(r + \dfrac{1}{9} r^3 + \dfrac{1}{25} r^5 \right)$; $\quad 2.60, 2.23, 1.89, 1.57, 1.25, 0.91, 0.54$

SECTION 11.10, page 694

11. $u = 4k \displaystyle\sum_{m=1}^{\infty} \dfrac{J_2(\alpha_m)}{\alpha_m^2 J_1^2(\alpha_m)} \cos \alpha_m t\, J_0(\alpha_m r)$ **13.** No

25. $\alpha_{11}/2\pi \approx 0.6099$ (cf. Table A2 in Appendix 4)

SECTION 11.11, page 698

1. $a^2 u_{x^*x^*} + b^2 u_{y^*y^*} + c^2 u_{z^*z^*}$ **11.** $u = 275/r - 27.5$

13. $u = 110(\ln 10 - \ln r)/\ln 5 \approx 157.4 - 68.3 \ln r$

17. $u = (u_1 - u_0)(\ln r) / \ln (r_1/r_0) + (u_0 \ln r_1 - u_1 \ln r_0) / \ln (r_1/r_0)$

SECTION 11.12, page 704

7. $u = 1$

9. $\cos 2\phi = 2 \cos^2 \phi - 1$, $2x^2 - 1 = \frac{4}{3} P_2(x) - \frac{1}{3}$, $u = \frac{4}{3} r^2 P_2(\cos \phi) - \frac{1}{3}$

11. $x^3 = \frac{2}{5} P_3(x) + \frac{3}{5} P_1(x)$, $u = \frac{2}{5} r^3 P_3(\cos \phi) + \frac{3}{5} r P_1(\cos \phi)$

13. $u = 4r^3 P_3(\cos \phi) - 2r^2 P_2(\cos \phi) + r P_1(\cos \phi) - 2$

25. $u = U_0 \cos (\pi t/l \sqrt{LC}) \sin (\pi x/l)$

SECTION 11.13, page 708

5. $U(x, s) = \dfrac{c(s)}{x^s} + \dfrac{x}{s^2(s + 1)}$, $U(0, s) = 0$, $c(s) = 0$,

$u(x, t) = x(t - 1 + e^{-t})$

9. Set $x^2/4c^2\tau = z^2$. Use z as a new variable of integration. Use $\operatorname{erf}(\infty) = 1$.

CHAPTER 11 (REVIEW PROBLEMS), page 713

21. $u = A(y) \cos 3x + B(y) \sin 3x$ **23.** $u = A(x)e^{-4y} + B(x)e^y - 2$

25. $u - \exp [k(x^2 + y^2)]$ **27.** $u = -0.1 \cos 3t \sin 3x$

29. $\dfrac{3}{4} \cos t \sin x - \dfrac{1}{4} \cos 3t \sin 3x$ **37.** $u = xf_1(x - y) + f_2(x - y)$

39. $u = f_1(x) + f_2(xy)$ **41.** $u = \sin \dfrac{\pi x}{50} e^{-0.004572t}$

43. $u = \dfrac{3}{4} \sin \dfrac{\pi x}{10} e^{-0.1143t} - \dfrac{1}{4} \sin \dfrac{3\pi x}{10} e^{-1.029t}$

45. $u = \dfrac{200}{\pi^2} \left(\sin \dfrac{\pi x}{50} e^{-0.004572t} - \dfrac{1}{9} \sin \dfrac{3\pi x}{50} e^{-0.04115t} + \cdots \right)$

47. $u = 95 \cos 2x \, e^{-4t}$

49. $u = \pi - \dfrac{32}{\pi} \left(\dfrac{1}{4} \cos 2x \, e^{-4t} + \dfrac{1}{36} \cos 6x \, e^{-36t} + \cdots \right)$

SECTION 12.1, page 725

3. $7/29 + (26/29)i$ **5.** $7/50 - (13/25)i$

7. $-12 + 16i$ **9.** $-2 + 5i$ **11.** $2 - i$ **13.** $10 + i$

15. $-1/5$ **17.** $2/25$ **19.** $4x^3y - 4xy^3, \; 4x^2y^2$

21. $(x^2 - y^2)/(x^2 + y^2)$

SECTION 12.2, page 732

3. $3/2$ **5.** 10 **7.** 1 **9.** $1/\sqrt{49 + \pi^2}$

11. π **13.** $-\pi/3$ **15.** $\sqrt{32} \left(\cos \dfrac{3\pi}{4} + i \sin \dfrac{3\pi}{4} \right)$

17. $7(\cos \pi + i \sin \pi)$ **19.** $\frac{1}{3}(\cos \frac{1}{4}\pi + i \sin \frac{1}{4}\pi)$

21. $(3/\sqrt{2})(\cos \pi + i \sin \pi)$ **23.** $2 + 2i$ **25.** $-5 + 5i$

27. $\pm(1 - i)/\sqrt{2}$ **29.** $\pm(\sqrt{3}/2 - i/\sqrt{2})$

31. $\pm(2 + i)$ **33.** $\pm(1 - 3i)$

35. $-1, \cos \dfrac{\pi}{5} \pm i \sin \dfrac{\pi}{5}, \cos \dfrac{3\pi}{5} \pm i \sin \dfrac{3\pi}{5}$

37. $\pm 1, \pm i, \pm \dfrac{1}{\sqrt{2}} (1 \pm i)$ **39.** $2 + i, \; 1 - i$

41. $\pm(1 + i), \pm(2 + i)$

43. $|z| = \sqrt{x^2 + y^2} \geq |x|, \; |z| = \sqrt{x^2 + y^2} \geq |y|$

45. Equation (5) holds when $z_1 + z_2 = 0$. Let $z_1 + z_2 \neq 0$ and
$c = a + ib = z_1/(z_1 + z_2)$. By (19) in Prob. 43, $|a| \leq |c|$,
$|a - 1| \leq |c - 1|$. Thus $|a| + |a - 1| \leq |c| + |c - 1|$. Clearly
$|a| + |a - 1| \geq 1$. Together we have the inequality below; multiply
by $|z_1 + z_2|$ to get (5).

$$1 \leq |c| + |c - 1| = \left| \frac{z_1}{z_1 + z_2} \right| + \left| \frac{z_2}{z_1 + z_2} \right|$$

SECTION 12.3, page 735

1. Circle of radius 2 and center at $2i$.

3. The region between the two branches of the hyperbola $x^2 - y^2 = 1$.

5. Horizontal strip bounded by $y = -\pi$ and $y = \pi$, with $y = \pi$ included.

7. $(x + 1)^2 + y^2 = (x - 1)^2 + y^2$; the y-axis.

9. Circle of radius 3/8 and center at $(1, -3/8)$ and its exterior.

SECTION 12.4, page 739

1. $11 + 13i$, $-27 + 3i$, $41 - 23i$ **3.** $2 - i$, $(4 - 3i)/5$, $(8 - i)/13$

5. $(1 - x)/[(1 - x)^2 + y^2]$, $y/[(1 - x)^2 + y^2]$

7. Re $w \geq 0$ **9.** $|\arg w| \leq \frac{2}{3}\pi$ **11.** Yes   **13.** Yes

15. No   **17.** $-5i/z^6$ **19.** $6z(z^2 + i)^2$ **21.** 0

23. $\frac{1}{2}(1 - i)$ **25.** $-\frac{1}{8}(1 + i)$ **27.** $-i/8$

29. The quotient in (4) is $\Delta x/\Delta z$, which is 0 if $\Delta x = 0$ but 1 if $\Delta y = 0$, so that it has no limit as $\Delta z \to 0$.

31. Use Re $f(z) = [f(z) + \overline{f(z)}]/2$, Im $f(z) = [f(z) - \overline{f(z)}]/2i$.

33. By continuity, for any $\epsilon > 0$ there is a $\delta > 0$ so that $|f(z) - f(a)| < \epsilon$ when $|z - a| < \delta$. Now $|z_n - a| < \delta$ for all sufficiently large n since $\lim z_n = a$. Thus $|f(z_n) - f(a)| < \epsilon$ for these n.

SECTION 12.5, page 745

1. Yes   **3.** Yes   **5.** No   **7.** Yes   **9.** No   **11.** No

13. $f(z) = -z^2$ **15.** $f(z) = -1/z$ **17.** No

19. $f(z) = iz^3$

21. $f(z) = \sin x \cosh y + i \cos x \sinh y$

23. $a = 2$, $e^{2x} \sin 2y$ **25.** $a = 1$, $-\sin x \sinh y$

27. $a = 0$, $\frac{1}{2}b(y^2 - x^2)$

35. $f'(z) = u_x + iv_x = 0$, $u_x = v_x = 0$, hence $v_y = u_y = 0$ by (1), $u = const$, $v = const$, $f = u + iv = const$.

SECTION 12.6, page 749

3. $0.540 - 0.841i$, 1   **5.** -0.135, 0.135

7. -15.154, 15.154   **9.** $0.260 + 0.260i$, 0.368

11. $e^{2\pi x} \cos 2\pi y$, $e^{2\pi x} \sin 2\pi y$ **13.** $e^{x-y} \cos(x + y)$, $e^{x-y} \sin(x + y)$

15. $e^{\pi i}$, $\sqrt{2}\,e^{3\pi i/4}$ **17.** $10e^{-0.927i}$ **19.** $(\frac{1}{2} + 2n)\pi i$
21. $\pm\sqrt{n\pi}\,(1 \pm i)$ **23.** $y = \frac{1}{2}n\pi$, n integer
25. All z **27.** $\pi i e^{\pi i z}$ **29.** $-e^{1/z}/z^2$
31. $(g/f)' = (g'f - gf')/g^2 = 0$, since $g' = g$, $f' = f$; $g/f = k = const$
    by Prob. 35, Sec. 12.5. $g(0) = f(0) = e^0 = 1$ gives $k = 1$, $g = f$.
33. $re^{-i\theta}$, $re^{i(\theta+\pi)}$, $re^{-i(\theta+\pi)}$

SECTION 12.7, page 752

3. $-z^{-2}\cos(1/z)$ **5.** $2z\cosh z^2$
7. $0.531 - 3.591i$ **9.** $-\cosh\pi$
11. $26.974 - 4.256i$
13. $\frac{1}{2}(\pm 4n + 1)\pi \pm 2i$ $(n = 0, 1, 2, \ldots)$
15. $\sin x \cos x/(\cos^2 x + \sinh^2 y)$ **17.** $\cos x \cosh y/(\cos^2 x + \sinh^2 y)$
21. $i\sinh 1 = 1.175i$
33. $\sin x \sinh y = 0$, $\sin x = 0$, since $\cosh 0 = 1$ and $\cos x = 5$ is impossible,
    $x = \pm 2n\pi$ to make $\cos x > 0$, $\cosh y = 5$ when $y = \pm 2.292$.
35. Use (15) and Prob. 34, or Prob. 18 only.

SECTION 12.8, page 757

In these answers, $n = 0, 1, 2, \ldots$.

5. $1.946 + 3.142i$ **7.** $0.693 - 1.571i$
9. $0.693 \pm 2n\pi i$ **11.** $-1.099 + (-\frac{1}{2} \pm 2n)\pi i$
13. $(2 \pm 2n\pi)i$ **15.** $1.193 + (-0.308 \pm 2n\pi)i$
17. $-i$ **19.** $10.85 - 16.90i$
21. $(1 + i)/\sqrt{2}$ **23.** $0.117 + 0.262i$
25. $36.84 - 72.14i$ **27.** $7\,083\,319 + 1\,103\,646i$
31. $\cosh w = \frac{1}{2}(e^w + e^{-w}) = z$, $(e^w)^2 - 2ze^w + 1 = 0$,
    $e^w = z + \sqrt{z^2 - 1}$

SECTION 12.9, page 761

1. $v = -u + 2x - 2$, $v = -u - 2$, $-u$, $-u + 2$, $-u + 4$
3. $|w - 2i| \leqq 2\sqrt{2}$ **5.** $|w| \leqq 16$ **7.** $\frac{1}{4} < |w| < 2$
9. $-\pi/2 < \arg w < \pi$ **11.** $2xy = const$ **13.** $v^2 = 4k^2(k^2 + u)$
15. $|\arg w| < \pi/4$ **17.** $|w - \frac{1}{2}| = \frac{1}{2}$ **19.** $|w| < e$, $w \neq 0$

CHAPTER 12 (REVIEW PROBLEMS), page 763

21. $-4 - 3i$ **23.** $-46 + 9i$ **25.** $2 + 3i$ **27.** 4
29. $-46/13$ **31.** 2   **33.** 19   **35.** 6
37. $3\sqrt{2}e^{3\pi i/4}$ **39.** $64e^{\pi i}$ **45.** $6\pi i e^{-9\pi^2}$ **47.** $1 + i/8$
49. $(4 + 5i)/41$ **51.** $-3\cosh 3$ **53.** $z^2 + 2z$ **55.** e^{z^2}
57. $e^{\pi x}\sin\pi y$ **59.** $-y/(x^2 + y^2)$

SECTION 13.1 and 13.2, page 776

1. $z = (1 + 2i)t,\ 0 \le t \le 1$ **3.** $z = 1 + i + (2 - 5i)t,\ 0 \le t \le 1$
5. $z = -2 + i + it,\ 0 \le t \le 3$ **7.** $z = 1 - 2i + 3e^{it},\ 0 \le t \le 2\pi$
9. $z = t + i/t,\ 1 \le t \le 4$ **11.** $z = 3 \cos t + i \sin t,\ 0 \le t \le 2\pi$
13. Upper half of the circle $x^2 + (y - 3)^2 = 9$, $\dot{z}(t) = 3ie^{it}$
15. Parabola $y = 3x^2$ from $(-1, 3)$ to $(2, 12)$, $\dot{z}(t) = 1 + 6ti$
17. Ellipse $4x^2 + y^2 = 4$, $\dot{z}(t) = -\sin t + 2i \cos t$
19. $-2\pi i$ **21.** $-3\pi i/2$ **23.** $-2\pi i$ **25.** $r^2 \pi i$
27. $2b(1 + i)$ **29.** 0 **35.** $32/3 + 64i/3,\ 32i,\ 8 + 128i/5$
37. $i,\ 2i,\ 2i$ **39.** $\cos 2 - 1$ **41.** $-\frac{1}{2} \sin \pi^2$
43. $(\pi - \frac{1}{2} \sinh 2\pi)i$ **45.** $2\pi i$
47. $5e^3$ **49.** 5

SECTION 13.3, page 783

5. No, yes   **7.** 0, yes   **9.** $-\pi$, no   **11.** 0, yes   **13.** 0, no
15. 0, no   **19.** $2\pi i,\ \pi i$ **21.** $-\pi,\ \pi$ **23.** 0 **25.** $-2\pi i,\ 0$
27. 0 **29.** $\pi i/2$

SECTION 13.4, page 787

1. $2 + 2i$ **3.** $-1566i/35$ **5.** $-2e$ **7.** 0
9. $(\pi - \frac{1}{2} \sinh 2\pi)i = -130.7i$ **11.** 0
13. $F(z) = \text{Ln } z$ is suitable.   **15.** Yes, 0

SECTION 13.5, page 791

1. π **3.** 0 **5.** $\pi i/2$ **7.** $\pi/2$
9. $\pi i/2$ **11.** $i\pi e^{-1/4}$ **13.** 0 **15.** $2\pi i$
17. $\pi/2$ **19.** $-2\pi \tanh 1$
21. $\pi e^{-2+2i} = -0.1769 + 0.3866i$ **23.** $2\pi i \text{ Ln } 4 = 8.710i$
25. Reduction by partial fractions is simpler than the method in Prob. 24.

SECTION 13.6, page 795

1. $\pi i/2$ **3.** $3\pi/8$ **5.** $2\pi i$
7. $2\pi i$ **9.** $\pi(\pi + 4i)(1 - i)/32\sqrt{2}$
11. $2\pi i$ **13.** $(-1)^n 2\pi i/(2n)!$ **15.** $2\pi i$
17. $2\pi i$ **19.** πi **21.** $\frac{16}{25}\pi e^2 i$
23. $(-5 + 8i)\pi$ **25.** $\frac{9}{2}\pi e^{-4}i$

CHAPTER 13 (REVIEW PROBLEMS), page 798

11. $\frac{1}{2}(1 - \cosh \pi^2)$ **13.** 0 **15.** 0

17. $i/2$ **19.** $-2 + 8i$ **21.** $2\pi i$ **23.** 0

25. $1 - \cosh 1$ **27.** 0 **29.** 0 **31.** 0

33. $\pi i/3$ **35.** $-2\pi i$ **37.** $2 \sin 1$ **39.** 0

SECTION 14.1, page 805

1. $i, -\frac{1}{8}, -i/27, \frac{1}{64}, \cdots$

3. $\frac{1}{2} + \frac{1}{2}i, -\frac{8}{5} + \frac{4}{5}i, -\frac{9}{10} - \frac{27}{10}i, \cdots$

5. $(1 + i)/\sqrt{2}, i, (-1 + i)/\sqrt{2}, \cdots$ **7.** Yes, 0

9. No   **11.** No   **13.** Yes, 0   **15.** Yes, $-\pi i/2$

SECTION 14.2, page 811

1. Yes, no, ± 1 **3.** Yes, no, $\pm 1, \pm i$ **5.** Yes, yes, 0

7. No, no, none   **9.** Yes, no, $\pm 1 + 2i$ **17.** Convergent

19. Convergent   **21.** Divergent   **23.** Divergent

25. Convergent   **27.** Divergent   **29.** 4, 8, 66

SECTION 14.3, page 816

1. $-2i, 1$ **3.** $0, 3$ **5.** $2 + i, 2$ **7.** $-i, 1/3$

9. $-5i, 1/5$ **11.** $0, 1/4$ **13.** $0, \infty$ **15.** $3 - \pi i, 0$

17. $0, 0$ **19.** $0, 1/4$ **21.** $\sqrt{2}$ **23.** ∞

25. 3   **27.** Start from Σz^{n+m}.

29. $\Sigma a_n z^{2n} = \Sigma a_n (z^2)^n, \quad |z^2| < R$, etc.

SECTION 14.4, page 822

3. $1/4$ **5.** 1 **7.** 3 **9.** 1 **11.** ∞ **13.** $1/4$

SECTION 14.6, page 828

9. $1/(1 + z)$

11. $-1 - (z - \pi i) - (z - \pi i)^2/2! - \cdots, \quad R = \infty$

13. $1 - 2z + (2z)^2/2! - (2z)^3/3! + - \cdots, \quad R = \infty$

15. $-1 - (z + 1) - (z + 1)^2 - (z + 1)^3 - \cdots, \quad R = 1$

17. $1 + (z + i)^2/2! + (z + i)^4/4! + \cdots, \quad R = \infty$

19. $2^{-1/2}[1 + (z + \frac{1}{4}\pi) - (z + \frac{1}{4}\pi)^2/2! - (z + \frac{1}{4}\pi)^3/3! + + - - \cdots],$
$R = \infty$

21. $-27i - 27(z - 2i) + 9i(z - 2i)^2 + (z - 2i)^3$

23. $z + \frac{1}{2}\pi - (z + \frac{1}{2}\pi)^3/3! + (z + \frac{1}{2}\pi)^5/5! - + \cdots, \quad R = \infty$

25. $\cos^2 z = \frac{1}{2}(1 + \cos 2z) = \frac{1}{2}[2 - (2^2/2!)z^2 + (2^4/4!)z^4 - + \cdots],$
$R = \infty$

27. $(z - 1) - \frac{1}{2}(z - 1)^2 + \frac{1}{3}(z - 1)^3 - + \cdots, \quad R = 1$

29. $z + \frac{1}{3}z^3 + \frac{2}{15}z^5 + \cdots, \quad R = \frac{1}{2}\pi$

31. $1 - \frac{1}{3}z^2 - \frac{1}{45}z^4 - \frac{2}{945}z^6 - \cdots, \quad R = \pi$

33. $\cosh^2 z = \cos^2 iz = 1 + z^2 + \frac{1}{3}z^4 + \cdots$, $R = \infty$

35. $z + \dfrac{z^2}{2!2} + \dfrac{z^3}{3!3} + \dfrac{z^4}{4!4} + \cdots$, $R = \infty$

37. $\dfrac{2}{\sqrt{\pi}}\left(z - \dfrac{z^3}{3} + \dfrac{z^5}{2!5} - \dfrac{z^7}{3!7} + - \cdots\right)$, $R = \infty$

39. $\dfrac{z^3}{1!3} - \dfrac{z^7}{3!7} + \dfrac{z^{11}}{5!11} - + \cdots$, $R = \infty$

SECTION 14.7, page 832

1. $z - z^4 + z^7 - + \cdots$, $R = 1$

3. $1 - 2z^2 + 3z^4 - 4z^6 + - \cdots$, $R = 1$

5. $z + z^5/2! + z^9/3! + \cdots$, $R = \infty$

7. $1 + 3z + 6z^2 + 10z^3 + \cdots$, $R = 1$

9. $k^2 - 2k^3 z + 3k^4 z^2 - 4k^5 z^3 + - \cdots$, $k = (3 + 4i)/25$, $R = 5$

11. $z^2 + z^4 + \frac{1}{3}z^6 + \cdots$, $R = \infty$

13. $e(1 + 2z + 6z^2 + \frac{52}{3}z^3 + \cdots)$, $R = 1/2$

15. $e[1 + z + z^2 + \frac{5}{6}z^3 + \cdots]$, $R = \infty$

19. $k + 3k^2(z - 1 - i) + 3^2 k^3 (z - 1 - i)^2 + \cdots$, $k = (1 + 3i)/\sqrt{10}$, $R = \frac{1}{3}\sqrt{10}$

21. $k - k^2(z - 2 - 3i) + k^3(z - 2 - 3i)^2 - + \cdots$, $k = (2 - 3i)/13$, $R = \sqrt{13}$

23. $1/3! - (z + 1)^2/5! + (z + 1)^4/7! - + \cdots$, $R = \infty$

25. $1 + 2(z - \frac{1}{4}\pi) + 2(z - \frac{1}{4}\pi)^2 + \frac{8}{3}(z - \frac{1}{4}\pi)^3 + \cdots$, $R = \frac{1}{4}\pi$

SECTION 14.8, page 840

1. Use Theorem 1.

3. $R = 4$; use Theorem 1.

5. $R = \sqrt{2/\pi} > 0.79$; use Theorem 1.

7. By Theorem 1, since $R = 3$.

9. $|\cos^n |z|| \leqq 1$; Σn^{-2} converges.

11. $|z - i| \leqq \sqrt{2} - \delta$, $\delta > 0$

13. Everywhere

15. No. Why?

17. Convergence follows from the comparison test (Sec. 14.2). Let $R_n(z)$ and R_n^* be the remainders of (1) and (5), respectively. Since (5) converges, for given $\epsilon > 0$ we can find an $N(\epsilon)$ such that $R_n^* < \epsilon$ for all $n > N(\epsilon)$. Since $|f_n(z)| \leqq M_n$ for all z in the region G, we also have $|R_n(z)| \leqq R_n^*$ and therefore $|R_n(z)| < \epsilon$ for all $n > N(\epsilon)$ and all z in the region G. This proves that the convergence of (1) in G is uniform.

SECTION 14.9, page 847

1. $\dfrac{1}{z^3} + \dfrac{2}{z^2} + \dfrac{2}{z} + \dfrac{4}{3} + \dfrac{2}{3}z + \cdots$, $R = \infty$

3. $z - \dfrac{1}{2z} + \dfrac{1}{24z^3} - + \cdots$, $R = \infty$

5. $-\dfrac{1}{3z^2} - \dfrac{1}{9z} - \dfrac{1}{27} - \dfrac{z}{81} - \cdots$, $R = 3$

7. $\dfrac{1}{z^2} - \dfrac{2}{z} + 3 - 4z + - \cdots, \quad R = 1$

9. $\dfrac{2}{z} - \dfrac{1}{2} + \dfrac{1}{2}z - \dfrac{1}{8}z^2 + - \cdots, \quad R = 2$

11. Let $\displaystyle\sum_{-\infty}^{\infty} a_n(z - z_0)^n$ and $\displaystyle\sum_{-\infty}^{\infty} c_n(z - z_0)^n$ be two Laurent series of the same function $f(z)$ in the same annulus. We multiply both series by $(z - z_0)^{-k-1}$ and integrate along a circle with center at z_0 in the interior of the annulus. Since the series converge uniformly, we may integrate term by term. This yields $2\pi i a_k = 2\pi i c_k$. Thus, $a_k = c_k$ for all $k = 0, \pm 1, \cdots$.

13. $z^3 + \dfrac{z}{2!} + \dfrac{1}{4!z} + \dfrac{1}{6!z^3} + \cdots, \quad R = \infty$

15. $-\displaystyle\sum_{n=0}^{\infty} \left(\dfrac{i}{2}\right)^{n+1} (z - i)^{n-1}, \quad R = 2$

17. $\dfrac{1}{z + i} + \dfrac{1}{(z + i)^2}$

19. $-\displaystyle\sum_{n=0}^{\infty} \dfrac{(z + i)^{n-1}}{(2i)^{n+1}}, \quad R = 2$

21. $\displaystyle\sum_{n=0}^{\infty} z^{2n}, \quad |z| < 1; \qquad -\displaystyle\sum_{n=0}^{\infty} \dfrac{1}{z^{2n+2}}, \quad |z| > 1$

23. $\displaystyle\sum_{n=0}^{\infty} (-1)^n(z - 1)^n, \quad 0 < |z - 1| < 1; \qquad \displaystyle\sum_{n=0}^{\infty} \dfrac{(-1)^n}{(z - 1)^{n+1}}, \quad |z - 1| > 1$

25. $4 \displaystyle\sum_{n=0}^{\infty} z^{4n}, \quad |z| < 1, \qquad -\displaystyle\sum_{n=0}^{\infty} \dfrac{4}{z^{4n+4}}, \quad |z| > 1$

27. $(1 - 4z) \displaystyle\sum_{n=0}^{\infty} z^{4n}, \quad |z| < 1, \qquad \left(\dfrac{4}{z^3} - \dfrac{1}{z^4}\right) \displaystyle\sum_{n=0}^{\infty} \dfrac{1}{z^{4n}}, \quad |z| > 1$

29. $\displaystyle\sum_{n=0}^{\infty} \dfrac{(-1)^{n+1}(z + \frac{1}{2}\pi)^{2n-1}}{(2n)!}, \quad |z + \frac{1}{2}\pi| > 0$

SECTION 14.10, page 853

1. $\pm 1, \pm i$ (simple)   **3.** $-i$ (third order)

5. $\frac{1}{2}(\pm 2n + 1)\pi i, n = 0, 1, \ldots$ (second order)

7. $\pm 2n\pi i, n = 1, 2, \ldots$ (simple)   **9.** $\pm 1, \pm 2, \ldots$ (third order)

11. $f(z) = (z - z_0)^n g(z)$ by (3), and $g(z_0) \neq 0$. Hence $1/g(z_0)$ is analytic at $z = z_0$. Let its Taylor series be

$$\dfrac{1}{g(z)} = c_0 + c_1(z - z_0) + \cdots. \text{ Then } \dfrac{1}{f(z)} = \dfrac{c_0 + c_1(z - z_0) + \cdots}{(z - z_0)^n},$$

which proves the first statement. Multiplication by $h(z)$ does not change this.

13. $0, \infty$ (simple poles)

15. $0, \pm\pi, \pm2\pi, \ldots$ (simple poles), ∞ (essential singularity)
17. $0, \infty$ (essential singularities) **19.** $\pm i, \infty$ (simple poles)
21. π (essential singularity)
23. $\frac{1}{4}\pi \pm 2n\pi, \frac{5}{4}\pi \pm 2n\pi, n = 0, 1, \ldots$ (simple poles), ∞ (essential singularity)
27. No, yes, no **29.** Yes **31.** Yes
33. No, no **35.** No

CHAPTER 14 (REVIEW PROBLEMS), page 856

21. 1/2 **23.** 3 **25.** $1/\pi^2$ **27.** ∞ **29.** $1/\sqrt{3}$

31. $\sum\limits_{n=0}^{\infty} \frac{(-1)^n}{n!} z^n, \quad \infty$ **33.** $\ln 2 + \sum\limits_{n=1}^{\infty} \frac{(-1)^{n+1}}{2^n n}(z-2)^n, \quad 2$

35. $\sum\limits_{n=0}^{\infty} (-1)^n z^{4n}, \quad 1$ **37.** $\sum\limits_{n=0}^{\infty} \frac{(z-2i)^n}{(1-2i)^{n+1}}, \quad \sqrt{5}$

39. $\sum\limits_{n=0}^{\infty} \frac{(-1)^n}{n!(2n+1)} z^{2n+1}, \quad \infty$

41. $\sum\limits_{n=0}^{\infty} z^{n-4}, \quad 0 < |z| < 1$, pole of fourth order

43. $\sum\limits_{n=0}^{\infty} \frac{1}{(2n)! z^{4n+8}}, \quad |z| > 0$, essential singularity

45. $\sum\limits_{n=0}^{\infty} \frac{(-1)^{n+1}}{(2n+1)!}\left(z - \frac{\pi}{2}\right)^{2n-2}, \quad \left|z - \frac{\pi}{2}\right| > 0$, pole of third order

47. $\sum\limits_{n=1}^{\infty} \frac{(-1)^{n+1}}{n}(z-1)^{n-5}, \quad 0 < |z-1| < 1$, pole of fifth order

49. $\sum\limits_{n=1}^{\infty} \frac{1}{n!n} z^{n-2}, \quad |z| > 0$, pole of second order

SECTION 15.1, page 864

1. -6 (at $z = 1$) **3.** 2 (at $z = -1$)
5. $-\frac{1}{4}, \frac{1}{4}$ (at $z = 1, -1$) **7.** -1 (at $z = \pm2n\pi i$)
9. -1 (at $z = (2n+1)\pi/2$) **11.** $-5/4$ (at $z = -1$)
13. -2 (at $z = 0$) **15.** 2, 5 (at $z = 0, 1$)
19. $2\pi i$ **21.** 0
23. $5\pi/2$ **25.** $-\pi^4 i/3$
27. $-2\pi i$ **29.** $-\pi i$

SECTION 15.2, page 867

1. $-4\pi i/3$ **3.** $2\pi i/3$ **5.** $-6\pi i$ **7.** $\pi i/2$ **9.** 0
11. $10\pi i$ **13.** $-\pi \sin\frac{1}{2}$ **15.** $4\pi i/5$ **17.** $-4i$ **19.** 0
21. πi **23.** 0 **25.** $2\pi i/9$ **27.** 4 **29.** $-2\pi/3$

SECTION 15.3, page 872

1. $2\pi/\sqrt{3}$ **3.** $\pi/\sqrt{k^2 - 1}$ **5.** $\pi/\sqrt{2}$ **7.** $4\pi/15$

9. 0

11. Use $\cos 2\theta = \frac{1}{2}(z^2 + z^{-2})$. *Ans.* $\pi/20$

13. $\displaystyle\int_C \frac{1}{iz} \cdot \frac{1}{2^{2n}} \sum_{j=0}^{2n} \binom{2n}{j} z^j \left(\frac{1}{z}\right)^{2n-j} dz$, and $j = 2n - j$, that is, $j = n$, gives the constant term $K = \binom{2n}{n} = (2n)!/(n!)^2$ and the residue is $K/(i2^{2n})$. Multiply by $2\pi i$.

17. π **19.** $\pi/2$ **21.** $2\pi/3$ **23.** $3\pi/8$ **25.** $\pi/12$ **27.** 0

SECTION 15.4, page 877

3. $5\pi/16e^4$ **5.** $\pi e^{-s}/2$ **7.** $\pi/10e^2 - \pi/15e^3$

9. $-(2\pi e^{-\sqrt{3}} \sin 1)/\sqrt{3}$ **11.** $x = t/k$, $a = ks$

15. π **17.** $-\sqrt{3}\pi/6$ **19.** $-\pi/2$ **21.** $\pi/2$

23. $-\frac{1}{4}\pi[(1 + \pi)e^{-\pi} + 1]$

CHAPTER 15 (REVIEW PROBLEMS), page 879

11. $8\pi i$, yes   **13.** 0, yes   **15.** $4\pi i$, yes

17. 0, no   **19.** $2 \sin 1 + \cos 1 - 4 \sin 3 - \cos 3$, no

21. $-64\pi i$, yes   **23.** $-\pi^2(\pi^2 + 7)i$, yes   **25.** $8\pi i$, yes

27. $4\pi/\sqrt{3}$ **29.** $2\pi/\sqrt{4\pi^2 - 1}$ **31.** $2\pi/35$

33. $2\pi/15$ **35.** $\pi/2$ **37.** $\pi/6$

39. π/k

SECTION 16.1, page 886

3. No, the size is preserved, but the sense is reversed.

5. $5e^{it}$, $5ie^{it}$ **7.** $t + it^{-1}$, $1 - it^2$

9. $t + i(at + b)$, $1 + ia$ **11.** $-a/2$

13. ± 1 **15.** $\pm i$

SECTION 16.2, page 891

1. 0, 1   **3.** 0, $1-i$ **5.** $\pm i$ **7.** ± 2 **9.** $\pm 3i$

11. $w = -1/z$ **13.** $w = (az + b)/(a - bz)$

15. $w = az/d$ **19.** $z = (w + 1)/(w - 1)$

21. $z = (-5w + 4i)/(iw - 3)$ **25.** $bc - ad = 1$

SECTION 16.3, page 896

3. $z = (-4w + 1)/(2w - 1)$ **5.** $w = (2z - i)/(-iz - 2)$

7. $w = z + 2$ **9.** $w = (iz + 3)/(z - 4)$

11. $w = (z + 1)/(z - 1)$ **13.** $w = (3iz + 1)/z$

15. $w = (z + 3)/(2z - 1)$ **17.** $w = (z + ip)/(ipz + 1)$

19. $w = (z^4 - i)/(-iz^4 + 1)$

SECTION 16.4, page 901

1. $t = z^2$ maps the given region onto the strip $0 < \text{Im } t < \pi$, and $w = e^t$ maps this strip onto the upper half-plane. *Ans.* $w = e^{z^2}$.

3. The region in the upper half-plane bounded by the ellipses
$$\frac{u^2}{\cosh^2 1} + \frac{v^2}{\sinh^2 1} = 1 \quad \text{and} \quad \frac{u^2}{\cosh^2 2} + \frac{v^2}{\sinh^2 2} = 1$$

5. Elliptical annulus bounded by the ellipses in Prob. 3 and cut along the positive imaginary axis

7. $\pm(2n + 1)\pi/2, \; n = 0, 1, \cdots$ **9.** $\pm n\pi i, \; n = 0, 1, \cdots$

11. Hyperbolas $u^2/(\cos^2 x) - v^2/(\sin^2 x) = 1$

13. Upper half-plane $v > 0$ **15.** Lower half-plane $v < 0$

17. Lower half of the interior of the ellipse $u^2/(\cosh^2 1) + v^2/(\sinh^2 1) = 1$

19. Rectangle $\ln 2 \leqq u \leqq \ln 3, \; \pi/4 \leqq v \leqq \pi/2$

SECTION 16.5, page 905

1. w moves once around the unit circle $|w| = 1$.

3. 4 and 5 sheets, respectively, branch point at $z = 0$

9. 2 sheets, branch point at $z = 0$ **11.** 2 sheets, branch point at $z = i$

13. $0, 1, -1, 2$ sheets   **15.** a, infinitely many sheets

17. $0, 2$ sheets   **19.** $0, 2$ sheets

CHAPTER 16 (REVIEW PROBLEMS), page 906

11. $u = \frac{1}{4}v^2 - 1, \frac{1}{16}v^2 - 4, \frac{1}{36}v^2 - 9$ **13.** $|w| = 2.25, \; u > 0$

15. The w-plane except for the negative real axis

17. $\pi/2 < \arg w < 3\pi/2$

19. The domain to the left of the parabola $v^2 = 4(1 - u)$, except for the negative u-axis

21. $|\arg w| < \pi/4$ **23.** $u = 1$ **25.** $v > 0, \; |w| < 1/4$

27. $0, \; \pm 1/\sqrt{2}$ **29.** 0 **31.** $\pm 1, \; \pm i$

33. $w = \dfrac{z - i}{z + i}$ **35.** $w = \dfrac{z}{z + 2}$ **37.** $w = \dfrac{z - i}{2z + 1}$

39. $w = \dfrac{z + 3i}{4iz}$ **41.** $w = \dfrac{iz + 2}{3z - i}$ **43.** $w = iz^3 + 1$

45. $w = iz$ **47.** $w = e^{3z}$ **49.** $w = z^2/2k$

51. $\pm 3, \; \pm 3i$ **53.** $\pm i$ **55.** $\pm i$

SECTION 17.1, page 912

1. $10x + 250 = const$ **3.** $3(1 - y/d) = const$

5. $\Phi = 110 + 27.5xy$ **7.** $\Phi = 110 \ln r/\ln 2$

9. $\Phi = 500(\ln 12 - \ln r)/\ln 2$ 11. $x^2 - y^2 = const$
13. $(x - 1/2c)^2 + y^2 = 1/4c^2$ 15. Arg $z = const$
17. $u = c$ Re ln $(z^2 - a^2) = c$ ln $|z^2 - a^2|$

SECTION 17.2, page 917

3. $\Phi(x, y) = U_2 xy$. $w = u + iv = iz^2$ maps R onto $-2 \leq u \leq 0$.
5. Apply $w = z^2$.
7. Corresponding rays in the w-plane make equal angles, and the mapping is conformal.
9. $z = (2Z - i)/(-iZ - 2)$

SECTION 17.3, page 920

1. $(100/d)y$. Apply a rotation through $\pi/2$.   3. $-40 + 240\theta/\pi$
5. Re $F(z) = 100 + (200/\pi)$ Re $(\sin^{-1}z)$ 7. $T_0[$Arg $(w - 1) -$ Arg $(w + 1)]/\pi$
9. $T_1 + \pi^{-1}(T_2 - T_1)$ Arg $(w - a)$ 15. $T_0 + 2\pi^{-1}(T_1 - T_0)$ Arg z
17. $400($Arg $z)/\pi$ 19. $1 - (2/\pi)$ arc tan $[(\sin y)/(\sinh x)]$

SECTION 17.4, page 926

1. $V = V_1 = K$, $Ky = const$, $Kx = const$
3. $x^2 - y^2 = const$, $xy = const$, $V = -2i\bar{z} = -2y - 2ix$
5. Parallel flow in the negative y-direction, $V = -i$
7. Parallel flow in the direction of $y = -x$, $V = 1 - i$
9. $V = 3(x^2 - y^2) - 6xyi$, $V_2 = 0$ on the coordinate axes
13. $F(z) = [\ln (z + a)]/2\pi$
17. The streamlines are circles $y/(x^2 + y^2) = c$ or $x^2 + (y - k)^2 = k^2$.

SECTION 17.5, page 932

3. $\Phi = 2 - r \cos \theta$ 5. $\Phi = \frac{1}{2}r^5 \sin 5\theta$
7. $\Phi = \frac{1}{2} + \frac{1}{2}r^2 \cos 2\theta$ 9. $\Phi = 3r \sin \theta - r^3 \sin 3\theta$
11. $\Phi = \frac{1}{2} + (2/\pi)(r \sin \theta + \frac{1}{3}r^3 \sin 3\theta + \frac{1}{5}r^5 \sin 5\theta + \cdots)$
13. $\Phi = (4/\pi)(r \sin \theta - \frac{1}{9}r^3 \sin 3\theta + \frac{1}{25}r^5 \sin 5\theta - + \cdots)$

SECTION 17.6, page 936

5. Use that $|e^z| = e^x$ is monotone in x.
7. 2, at $z = \pm i$ 11. Use Theorem 3.
15. Set $r = 0$ in that formula.

CHAPTER 17 (REVIEW PROBLEMS), page 938

11. $(220/\pi)$ Arg z 13. $1000 + (500/\pi)$ Arg z
15. $100 + 80(y - x/2)$ 17. $100 - 80(\frac{1}{2} + i)z$ 19. $-(220i/\pi)$ Ln z

21. $x + y = const$ **23.** $2x(y + 1) = const$ **25.** $|z - 1 - i| = const$

29. $\dfrac{1}{2}\,\Phi_1 - \dfrac{2}{\pi}\,\Phi_1\left(r\cos\theta - \dfrac{1}{3}r^3\cos 3\theta + \dfrac{1}{5}r^5\cos 5\theta - + \cdots\right)$

SECTION 18.1, page 947

1. 0.1000E05, 0.7361E03, 0.9024E02, 0.4051E-02

3. $(1010)_2$, $(1100100)_2$, $(101111101.01)_2$, $(0.000010111)_2$

5. 28.5, 0.15625, 1.71875, -3.40625 **7.** 5.6945, 5.697, 5.70, 5.7

9. 0.009966, 0.010 **13.** $f(x) = 2\sin^2\frac{1}{2}x$

17. $-7.9475 \leqq d \leqq -7.9365$ **19.** Add, then round.

21. $|\tilde{d} - d| = |\epsilon_1 - \epsilon_2| \leqq |\epsilon_1| + |\epsilon_2| \leqq \beta_1 + \beta_2 = \beta$, $a_1 = 1$, $\tilde{a}_1 = 1.1$,

$a_2 = 1$, $\tilde{a}_2 = 0.9$ gives $|\epsilon| = |\epsilon_1| + |\epsilon_2|$.

23. $\dfrac{\tilde{a}_1}{\tilde{a}_2} = \dfrac{a_1 + \epsilon_1}{a_2 + \epsilon_2} = \dfrac{a_1 + \epsilon_1}{a_2}\left(1 - \dfrac{\epsilon_2}{a_2} + \dfrac{\epsilon_2^{\,2}}{a_2^{\,2}} - + \cdots\right) \approx \dfrac{a_1}{a_2} + \dfrac{\epsilon_1}{a_2} - \dfrac{\epsilon_2}{a_2}\cdot\dfrac{a_1}{a_2}$,

$\left(\dfrac{\tilde{a}_1}{\tilde{a}_2} - \dfrac{a_1}{a_2}\right)\Big/\dfrac{a_1}{a_2} \approx \dfrac{\epsilon_1}{a_1} - \dfrac{\epsilon_2}{a_2}$

SECTION 18.2, page 957

3. $x_1 = -0.2$, $x_2 = -0.200\,32$, $x_3 = -0.200\,323$, ...

7. This follows from the intermediate value theorem of calculus.

9. First order **11.** 3.584 631 **13.** 0.904 557

15. $x_1 = 1.9$ (exact 1.9) **17.** 0.739 085 **19.** 0.687 411

21. Errors 0.236 068, 0.013 932, 0.000 043, 0.000 000

23. $x \geqq \sqrt{2/(1 + 2\alpha)}$, $\alpha < 1$.

25. $f(x) = x^k - c$, $x_{n+1} = (1 - k^{-1})x_n + k^{-1}cx_n^{\,1-k}$

27. 0.739 085 **29.** 0.904 557

31. ALGORITHM BISECT (f, a_0, b_0, N) **Bisection Method**

This algorithm computes an interval $[a_n, b_n]$ containing a solution of $f(x) = 0$ (f continuous), or it computes a solution c_n, given an initial interval $[a_0, b_0]$ such that $f(a_0)f(b_0) < 0$.

INPUT: Initial interval $[a_0, b_0]$, maximum number of iterations N.

OUTPUT: Interval $[a_N, b_N]$ containing a solution, or a solution c_n.

For $n = 0, 1, \cdots, N - 1$ do:

> Compute $c_n = \frac{1}{2}(a_n + b_n)$.
> If $f(c_n) = 0$ then OUTPUT c_n. Stop. [*Procedure completed*]
> Else continue.
> If $f(a_n)f(c_n) < 0$ then $a_{n+1} = a_n$ and $b_{n+1} = c_n$.
> Else set $a_{n+1} = c_n$ and $b_{n+1} = b_n$.

End

OUTPUT $[a_N, b_N]$. Stop.

[*Procedure completed*]

End BISECT

33. [1.90625, 1.93750] **35.** [0.40625, 0.43750] **37.** ± 1.189 **39.** 2.689

SECTION 18.3, page 969

3. 0.722

5. 0.994, 0.984

7. 0.25590, 0.25753

9. (a) 0.258 27, (b) 0.258 27

11. 1.431 676, 1.431 779, 1.431 782

13. $|\epsilon_1(0.26)| \leq \frac{1}{2} \cdot 0.04 \cdot 0.3 \cdot 0.7 \cdot 0.38942 = 0.00164$. Actual error 0.00118

15. $\delta f_{1/2} = 0.057\ 839$, $\delta f_{3/2} = 0.069\ 704$, etc.

19. 3.455606, exact to 5D

21. 0.386 4185, exact to 7D

23. 0.97941

25. $8x^2 - 15x + 9$; 36

SECTION 18.4, page 977

9. 1/16, 1/4

11. $1 - x^2$, $-2(x - 1) - (x - 1)^2 + 2(x - 1)^3$,
$-1 + 2(x - 2) + 5(x - 2)^2 - 6(x - 2)^3$

13. $-\frac{3}{4}(x + 2)^2 + \frac{3}{4}(x + 2)^3$, $\frac{3}{4}(x + 1) + \frac{3}{2}(x + 1)^2 - \frac{5}{4}(x + 1)^3$,
$1 - \frac{9}{4}x^2 + \frac{5}{4}x^3$, $-\frac{3}{4}(x - 1) + \frac{3}{2}(x - 1)^2 - \frac{3}{4}(x - 1)^3$

15. $1 - 5x^2/4 + x^4/4$

SECTION 18.5, page 986

11. 0.25, 0.3125, 0.328 125

13. $A \leq J \leq B$, $A = h \Sigma A_j$, $B = h \Sigma B_j$, A_j and B_j being lower and upper bounds for f in the jth subinterval.

15. 0.9466 (0.9461 exact to 4D)

17. 0.9458

19. 0.16670, $0 \leq \epsilon_S \leq 0.000\ 067$

21. $\alpha = \beta = \frac{1}{2}$, trapezoidal rule

23. 0.08, 0.32, 0.176, 0.256 (exact)

25. 0.52, 0.080, 0.304, 0.256

SECTION 18.6, page 996

1. Use (1).

3. $\mathrm{ci}(x) \sim \sin x \left(-\dfrac{1}{x} + \dfrac{2!}{x^3} - \dfrac{4!}{x^5} + - \cdots \right)$

$$+ \cos x \left(\dfrac{1}{x^2} - \dfrac{3!}{x^4} + \dfrac{5!}{x^6} - + \cdots \right)$$

5. $c(x) \sim -A(x) \sin x^2 + B(x) \cos x^2$, where

$$A(x) = \frac{1}{2} \left(\frac{1}{x} - \frac{1 \cdot 3}{4x^5} + \frac{1 \cdot 3 \cdot 5 \cdot 7}{16x^9} - + \cdots \right),$$

$$B(x) = \frac{1}{2} \left(\frac{1}{2x^3} - \frac{1 \cdot 3 \cdot 5}{8x^7} + - \cdots \right).$$

7. $Q(\alpha, x) \sim x^\alpha e^{-x} \left[\dfrac{1}{x} - \dfrac{1 - \alpha}{x^2} + - \cdots \right.$

$$\left. + (-1)^{n-1} \frac{(1 - \alpha)(2 - \alpha) \cdots (n - 1 - \alpha)}{x^n} + \cdots \right]$$

9. $C(x) \sim \pi^{1/2}2^{-3/2} + A(x) \sin x^2 - B(x) \cos x^2$ with A and B as in Prob. 5.

13. $y' = x^{-1}(1 - \alpha + x)y - 1$, etc.

CHAPTER 18 (REVIEW PROBLEMS), page 997

21. 0.9817 E 02, -0.1010 E 03, 0.4787 E -02, -0.1380 E 05

23. $(7.625)_{10}$, $(34.5)_{10}$, $(3.6875)_{10}$ **25.** $21.26305 \leq s \leq 21.27315$

27. $x^2/(\sqrt{x^2 + 1} + 1)$

33. $1.035\ 558$, $1.043\ 472$, $1.045\ 209$, $1.045\ 589$, $x_5 = 1.045\ 672$

37. ± 1.8955 **39.** $1.857\ 143$, $1.834\ 787$, $1.834\ 244$

41. 1.316 **43.** 2.95647, 2.96087

45. 2.96111 **47.** 0.29551, 0.47941

49. $1 - x^3$ if $0 \leq x \leq 1$, $-3(x - 1) - 3(x - 1)^2 + 3(x - 1)^3$ if $1 \leq x \leq 2$

51. 0.245, $\epsilon = -0.005$ **53.** 0.9027 (0.9045 exact to 4D)

53. The secant corresponding to $x - h/2$, $x + h/2$ is parallel to the tangent at x.

SECTION 19.1, page 1009

1. $x_1 = -4$, $x_2 = 2$ **3.** $x_1 = \frac{1}{2}$, $x_2 = -2$

5. $x_1 = -2$, $x_2 = 2$, $x_3 = 0$ **7.** $x_1 = -1$, $x_2 = 8$, $x_3 = -5$

9. $x_1 = 6$, $x_2 = -3$, $x_3 = 1$ **11.** $x_1 = \frac{1}{2}$, $x_2 = \frac{3}{2}$, $x_3 = 3$

13. $x_1 = 0.4$, $x_2 = -0.01$, $x_3 = 1.1$ **15.** $x_1 = 0.5$, $x_2 = -0.5$, $x_3 = 3.5$

17. $(1 - 1/\epsilon)x_2 = 2 - 1/\epsilon$ eventually becomes $x_2/\epsilon \approx 1/\epsilon$, $x_2 \approx 1$,

  $x_1 = (1 - x_2)/\epsilon \approx 0$.

19. $x_1 = x_2 = x_3 = 1$; no solution

SECTION 19.2, page 1014

1. $\begin{aligned} x_1 &= 3, \\ x_2 &= -1, \end{aligned}$ $\begin{bmatrix} 1 & 0 \\ 1 & 1 \end{bmatrix} \begin{bmatrix} 5 & 3 \\ 0 & 6 \end{bmatrix}$

3. $\begin{aligned} x_1 &= \frac{1}{2}, \\ x_2 &= \frac{5}{2}, \end{aligned}$ $\begin{bmatrix} 1 & 0 \\ 4 & 1 \end{bmatrix} \begin{bmatrix} 2 & -3 \\ 0 & -1 \end{bmatrix}$

5. $\begin{aligned} x_1 &= 2, \\ x_2 &= 0, \\ x_3 &= 5, \end{aligned}$ $\begin{bmatrix} 1 & 0 & 0 \\ 2 & 1 & 0 \\ 3 & 2 & 1 \end{bmatrix} \begin{bmatrix} 1 & 2 & 3 \\ 0 & 1 & 2 \\ 0 & 0 & 1 \end{bmatrix}$

7. $\begin{aligned} x_1 &= 2, \\ x_2 &= 1, \\ x_3 &= -3, \end{aligned}$ $\begin{bmatrix} 1 & 0 & 0 \\ 0 & 1 & 0 \\ 2 & 5 & 1 \end{bmatrix} \begin{bmatrix} 1 & 2 & 5 \\ 0 & 1 & 0 \\ 0 & 0 & 1 \end{bmatrix}$

9. $\begin{aligned} x_1 &= 0.2, \\ x_2 &= 0.4, \\ x_3 &= 0.8, \end{aligned}$ $\begin{bmatrix} 1 & 0 & 0 \\ 2 & 1 & 0 \\ 2 & 5 & 1 \end{bmatrix} \begin{bmatrix} 5 & 4 & 1 \\ 0 & 1 & 2 \\ 0 & 0 & 3 \end{bmatrix}$

11. $x_1 = 0.6,$ $x_2 = 1.2,$ $x_3 = 0.4,$ $\begin{bmatrix} 3 & 0 & 0 \\ 2 & 3 & 0 \\ 4 & 1 & 3 \end{bmatrix} \begin{bmatrix} 3 & 2 & 4 \\ 0 & 3 & 1 \\ 0 & 0 & 3 \end{bmatrix}$

13. $\begin{bmatrix} 0.2 & -0.05 \\ -0.1 & 0.15 \end{bmatrix}$

15. $\begin{bmatrix} 1/\sqrt{2} & 1/\sqrt{2} \\ -1/\sqrt{2} & 1/\sqrt{2} \end{bmatrix}$

17. $\begin{bmatrix} 0 & -\frac{2}{3} & \frac{1}{3} \\ \frac{1}{6} & \frac{1}{6} & -\frac{1}{6} \\ -\frac{1}{3} & \frac{2}{3} & 0 \end{bmatrix}$

19. $x_1 = 3,$ $x_2 = 2,$ $\begin{bmatrix} i & 0 \\ -3i & -2i \end{bmatrix} \begin{bmatrix} i & -3i \\ 0 & -2i \end{bmatrix}$

SECTION 19.3, page 1020

1. Exact 1.5, 0.5, 1.2   **3.** Exact $-$ 2, 1, 2   **5.** Exact 0, 0.5, -2.5

7. Exact 0.5, 0.5, 0.5   **9.** Exact 2, 0, 3, 1

15. The Jacobi iteration yields 0, 0, 0, then 2, 2, 2, then 0, 0, 0, etc.

17. $\begin{bmatrix} 0 & 0.25 & 0.25 & 0 \\ 0 & 0.0625 & 0.0625 & 0.25 \\ 0 & 0.0625 & 0.0625 & 0.25 \\ 0 & 0.03125 & 0.03125 & 0.125 \end{bmatrix}$

19. $\mathbf{X}^{(1)} = \begin{bmatrix} 0.49 & -0.1 & 0.51 \\ 0 & 0.2 & 0 \\ -0.51 & 0.3 & -1.47 \end{bmatrix}$, $\mathbf{A}^{-1} = \begin{bmatrix} 0.5 & -0.1 & 0.5 \\ 0 & 0.2 & 0 \\ -0.5 & 0.3 & -1.5 \end{bmatrix}$

21. $\sqrt{61}$, 7, 9   **23.** $\sqrt{2}$, 1, 1   **25.** $\sqrt{114}$, 14, 14

SECTION 19.4, page 1027

1. 6, $\sqrt{18}$, 3   **3.** 4, 2, 1   **5.** 30, $\sqrt{386}$, 17

7. 13, 14.3   **9.** 1, 1   **11.** 9, 27

13. $\|\mathbf{Ax}\| = 37 < 9 \cdot 5$ **15.** 289   **17.** 289, 289

21. $x_1 = 9$, $x_2 = -36$, $x_3 = 30$; $\kappa(\mathbf{A}) = 1.83 \cdot 408 = 746.64$

23. Let $\mathbf{x} \neq \mathbf{0}$. Set $\mathbf{x} = \|\mathbf{x}\|\mathbf{y}$. Then $\|\mathbf{y}\| = \|\mathbf{x}\|/\|\mathbf{x}\| = 1$. Also $\mathbf{Ax} = \mathbf{A}(\|\mathbf{x}\|\mathbf{y}) = \|\mathbf{x}\|\mathbf{Ay}$ since $\|\mathbf{x}\|$ is a number. Hence $\|\mathbf{Ax}\|/\|\mathbf{x}\| = \|\mathbf{Ay}\|$, and in (9), instead of taking the maximum over *all* $\mathbf{x} \neq \mathbf{0}$, since $\|\mathbf{y}\| = 1$ we only take the maximum over all $\mathbf{y}$ of norm 1. Write $\mathbf{x}$ for $\mathbf{y}$ to get (10) from this.

25. Let $\mathbf{x}_1$ be largest in absolute value. Then

$$\|\mathbf{x}\|_\infty = |x_1| \leq \Sigma |x_j| = \|\mathbf{x}\|_1 \leq n|x_1| = n\|\mathbf{x}\|_\infty.$$

SECTION 19.5, page 1031

1. $y = 5.4x - 6.4$ **3.** $y = 6.497 + 0.457x$ **5.** $y = -3.05 + 12.1x$

7. $y = 120.127 + 8.822\,t$. Speed 8.822 [meters/sec]

9. $n = 5$, $\Sigma x_i = 13$, $\Sigma x_i^2 = 57$, $\Sigma x_i^3 = 289$, $\Sigma x_i^4 = 1569$,
$\Sigma y_i = 9$, $\Sigma x_i y_i = 29$, $\Sigma x_i^2 y_i = 161$,
$$y = 2.825\ 159 - 2.049\ 040x + 0.377\ 398x^2$$

11. $y = 6642 + 762.3x - 156.1x^2$

13. $\mathbf{C} = [c_{jk}]$, $c_{jk} = x_j^{k-1}$, $\mathbf{b}^\mathsf{T} = [b_0 \cdots b_m]$

15. $y^* = a^* + bx$, where $y^* = \ln y$ and $a^* = \ln b_0$

SECTION 19.7, page 1040

1. $-1, 2, -1$ radii 5, 8, 9   **3.** 26, 21, 28, radii 4, 6, 6

5. $-2, 1, 4$, radii 0.002, 0.0011, 0.0002

9. $\sqrt{21.\ 000\ 0003\ 03}$ **11.** $\sqrt{56.53}$

13. $49 \leqq \lambda \leqq 57$, $49 \leqq \lambda \leqq 57$, $51.66 \leqq \lambda \leqq 53$

15. Circular arcs

SECTION 19.8, page 1043

1. $q = 5, 5.8276, 5.828\ 426$; $|\epsilon| \leqq 2, 0.069, 0.0021$

3. $q = 0.8, 0.8, \ldots$; $|\epsilon| \leqq 0.6, 0.6, \ldots$

5. $q = 6.333, 6.871, 6.976$; $|\epsilon| \leqq 2.5, 1.2, 0.49$

7. $3 \pm \sqrt{8} = 5.828\ 427\ 124, 0.171\ 572\ 875$

9. $q = 4.91718$, $|\epsilon| \leqq 0.39848$

11. $q = 5/4, 14/9$; $|\epsilon| \leqq \sqrt{11}/4, \sqrt{101}/9$

13. $q = 3, 5.774$; $|\epsilon| \leqq 2.55, 1.337$

17. $\mathbf{A} - 4.5\mathbf{I}$ has the eigenvalues 4.5, 1.5, -1.5, and $4.5/1.5 = 3$, whereas for $\mathbf{A}$ we have $9/6 = 1.5$ and slower convergence.

SECTION 19.9, page 1047

3. $\begin{bmatrix} 0 & 0 & 0 \\ -2.2 & 2.6 & 1.2 \\ -4.4 & 5.2 & 2.4 \end{bmatrix}$, 0, 0, 5   **5.** $\begin{bmatrix} 0 & 0 & 0 \\ -0.75 & 0 & -3 \\ -0.25 & 0 & -1 \end{bmatrix}$, 1, 0, -1

7. $\begin{bmatrix} 0 & 0 & 0 \\ 4 & -2 & 0 \\ 14/3 & 8/3 & -6 \end{bmatrix}$, 8, -2, -6 **9.** $\begin{bmatrix} 0 & 0 & 0 \\ -4 & 4 & 0 \\ -4 & 0 & 4 \end{bmatrix}$, 25, 4, 4

11. $\begin{bmatrix} 0 & 0 & 0 & 0 \\ 0 & 0 & 0 & 0 \\ -1 & -1 & 2.5 & 1.5 \\ -1 & -1 & 1.5 & 2.5 \end{bmatrix}$, 9, 4, 1, 0

13. $\begin{bmatrix} 0 & 0 & 0 & 0 \\ 0 & 0 & 0 & 0 \\ 4 & -8 & 4 & 0 \\ 8 & 0 & -24 & 16 \end{bmatrix}$, 36, 16, 4, 0

15. $A_1x_1 = Ax_1 - \lambda_1x_1x_1^Tx_1 = \lambda_1x_1 - \lambda_1x_1 = 0$. For an eigenvector x_j of A corresponding to λ_j we get $A_1x_j = Ax_j = \lambda_1x_1x_1^Tx_j = \lambda_jx_j$ since $x_1^Tx_j = 0$.

SECTION 19.10, page 1057

1. $\begin{bmatrix} 0 & -\sqrt{2} & 0 \\ -\sqrt{2} & 1 & 0 \\ 0 & 0 & -1 \end{bmatrix}$

3. $\begin{bmatrix} -1 & -3.605\ 551\ 28 & 0 \\ -3.605\ 551\ 28 & 5.461\ 538\ 46 & 3.692\ 307\ 69 \\ 0 & 3.692\ 307\ 69 & -4.461\ 538\ 46 \end{bmatrix}$

5. $\begin{bmatrix} 1 & -\sqrt{13} & 0 \\ -\sqrt{13} & 13 & 0 \\ 0 & 0 & 0 \end{bmatrix}$

7. $\begin{bmatrix} 3.142\ 857\ 14 & 5.938\ 459\ 91 & 0 \\ 5.938\ 459\ 91 & 0.190\ 476\ 19 & -2.494\ 438\ 26 \\ 0 & -2.494\ 438\ 26 & -3.333\ 333\ 33 \end{bmatrix}$

9. $\begin{bmatrix} 10.426\ 483\ 98 & -4.072\ 844\ 87 & 0 \\ -4.072\ 844\ 87 & 4.223\ 244\ 41 & -1.093\ 611\ 69 \\ 0 & -1.093\ 611\ 69 & -1.265\ 113\ 01 \end{bmatrix}$

CHAPTER 19 (REVIEW PROBLEMS), page 1057

21. $\begin{bmatrix} 4 & -1 & 2 \end{bmatrix}^T$ 23. $\begin{bmatrix} 2.2 & 3.0 & -0.9 \end{bmatrix}^T$

25. $\begin{bmatrix} 3.9 & 4.3 & 1.8 \end{bmatrix}^T$

27. All entries of the triangular matrices are 1.

31. $\begin{bmatrix} 1.4 & -0.4 & -0.6 \\ 2.6 & 0.4 & -1.4 \\ -1.6 & -0.4 & 0.4 \end{bmatrix}$ 33. $\begin{bmatrix} 50 & 70 & 60 & 50 \\ 70 & 100 & 80 & 70 \\ 60 & 80 & 100 & 90 \\ 50 & 70 & 90 & 100 \end{bmatrix}$

35. $\begin{bmatrix} -2.25 & 0.25 & 2.125 \end{bmatrix}$, $\begin{bmatrix} -2.06250 & -0.03125 & 2.03125 \end{bmatrix}$, $\cdots$; exact $\begin{bmatrix} -2 & 0 & 2 \end{bmatrix}$

37. 7, $\sqrt{21}$, 4   **39.** 14, $\sqrt{54}$, 5   **41.** 9.6, $\sqrt{51.2}$, 6.4

43. 15 **45.** 13   **47.** $3.5 \cdot 5.6 = 19.6$ **49.** $y = 0.14 + 0.77x$

51. Centers 10, 4, 3; radii 0.2, 0.3, 0.3

53. Centers 11.4, 14.4, 14.6; radii 2.6, 3.2, 1.8

55. $\begin{bmatrix} 0 & 0 & 0 \\ -20.8 & 10.4 & 0 \\ -18.4 & -2.133 & 13,6 \end{bmatrix}$, eigenvalues 16.4, 10.4, 13.6

SECTION 20.1, page 1071

3. 0.1, 0.202, 0.31008, 0.42869, etc. Exact solution: $y = e^{x^2} \displaystyle\int_0^x e^{-t^2}\, dt$

11. For instance, $x = 0.5$, $y = 0.672$ (error 0.040), 0.6293 (error -0.0029); $x = 1$, $y = 0.893$ (error 0.028), 0.8626 (error -0.0020)

13. $y = \tan x$

17. 2.284 133 308, 2.829 636 493, 3.557 351 471

SECTION 20.2, page 1074

1. $y_4 = 1.173\ 518$, $y_5 = 1.284\ 044$, $y_6 = 1.433\ 364$, $y_7 = 1.632\ 374$, $y_8 = 1.896\ 572$, $y_9 = 2.248\ 046$, $y_{10} = 2.718\ 486$

3. $y_4 = 0.422\ 798$, $y_5 = 0.546\ 315$, $y_6 = 0.684\ 161$, $y_7 = 0.842\ 332$, $y_8 = 1.029\ 714$, $y_9 = 1.260\ 288$, $y_{10} = 1.555\ 763$

9. $y_1 = 1.010\ 050$, $y_2 = 1.040\ 811$, $y_3 = 1.094\ 224$, $y_4 = 1.173\ 622$, $y_5 = 1.284\ 219$, $y_6 = 1.433\ 636$, $y_7 = 1.632\ 782$, $y_8 = 1.897\ 074$, $y_9 = 2.248\ 543$, $y_{10} = 2.718\ 780$

SECTION 20.3, page 1081

11. $y_1 = 0.96$, $y_2 = 0.84$, $y_3 = 0.64$, $y_4 = 0.36$, $y_5 = 0$

SECTION 20.4, page 1090

3. $u_{xy}(x, y) \approx \dfrac{1}{2k} [u_x(x, y + k) - u_x(x, y - k)]$. Now substitute

$$u_x(x, y \pm k) \approx \frac{1}{2h} [u(x + h, y \pm k) - u(x - h, y \pm k)].$$

5. 75

7. $u(P_{11}) = u(P_{12}) = 10$, $u(P_{21}) = 110$, $u(P_{22}) = 30$

9. $-2, -5, -5, -62$

11. $u_{11} = u_{21} = 0.1083$, $u_{12} = u_{22} = 0.3248$; exact to 4D: 0.0937, 0.2999.

13. $\sqrt{3}$. $u_{11} = u_{21} = 0.0859$, $u_{12} = u_{22} = 0.3180$. (0.1083, 0.3248 are 4D-values of the solution of the linear equations of the problem.)

15. Only 5 steps

SECTION 20.5, page 1096

3. $u_{11} = 1/2$, $u_{21} = u_{12} = 2$, $u_{22} = 8$

5. $-4u_{11} + u_{21} + u_{12} = 8$, $u_{11} - 4u_{21} = 5$, $\frac{4}{3}u_{11} - 6u_{12} = 8$, $u_{11} = -3$,
$u_{21} = -2$, $u_{12} = -2$

SECTION 20.6, page 1102

3. 0, 0.06625, 0.12500, 0.17125, 0.20000, 0.21000, 0.20000, etc.

5. 0. 0.05352, 0.10156 [0.10182 by (9)], 0.13984, 0.16406 [0.16727 by (9)],
0.17266, etc.

7. Calculations to 6D, values rounded off to 3D

t	$x = 0.2$	Exact	$x = 0.4$	Exact
0.04	0.164	0.159	0.255	0.260
0.08	0.107	0.108	0.175	0.175
0.12	0.073	0.073	0.119	0.118
0.16	0.050	0.049	0.081	0.079
0.20	0.034	0.033	0.055	0.054

9. The values are given as exact values in the answer to Prob. 7. The series
converges rapidly.

SECTION 20.7, page 1105

1. For $x = 0.2$, 0.4 we obtain 0.12, 0.2 ($t = 0.2$), 0.04, 0.08 ($t = 0.4$),
-0.04, -0.08 ($t = 0.6$), etc.

3. $u(x, 1) = 0$, -0.05, -0.10, -0.15, -0.20, 0

5. 0.190, 0.308, 0.308, 0.190 (0.178, 0.288, 0.288, 0.178 exact to 3D)

7. 0, 0.354, 0.766, 1.271, 1.679, 1.834, $\cdots$ ($t = 0.1$); 0, 0.575, 0.935, 1.135,
1.296, 1.357, $\cdots$ ($t = 0.2$)

CHAPTER 20 (REVIEW PROBLEMS), page 1106

21. $y = 0$, 0.1000, 0.2034, 0.3109, 0.4217, 0.5348, 0.6494, 0.7649, 0.8806;
error 0, -0.0044, -0.0122, $\cdots$, -0.0527

23. $y = 0$, 04352, 0.9074; error -0.0145, -0.0258 (about 50% of that in
Prob. 21)

25. $y = \tan x$

31. $y(0.4) = 1.822\ 798$, $y(0.5) = 2.046\ 315$, $y(0.6) = 2.284\ 155$, $y(0.7) = 2.542\ 325$, $y(0.8) = 2.829\ 706$, $y(0.9) = 3.160\ 277$, $y(1.0) = 3.557\ 611$

39. 3.93, 15.71, 58.93

41. $u(P_{11}) = u(P_{31}) = 90$, $u(P_{21}) = u(P_{13}) = u(P_{23}) = u(P_{33}) = 10$, $u(P_{12}) = u(P_{32}) = 30$, $u(P_{22}) = 20$

45. 0.06279, 0.09336, 0.08364, 0.04707

47. 0, -0.352, -0.153, 0.153, 0.352, 0 if $t = 0.12$ and 0, 0.344, 0.166,
-0.166, -0.344, 0 if $t = 0.24$

SECTION 21.2, page 1118

5. Contradictory, no solution
11. $x_1 = 15$ tons of A_1, $x_2 = 10$ tons of A_2, daily profit \$700
13. No
15. x_3, $x_4 =$ unused time on M_1, M_2, respectively
17. $f_{max} = f(210, 60) = 3750$
19. $f(20/9, 8/9) = 104/9 \approx 11.56$

SECTION 21.3, page 1124

1. $f_{max} = f(\frac{120}{11}, \frac{60}{11}) = \frac{2400}{11}$ 5. $f_{max} = f(7, 3) = 1170$
7. $f_{max} = f(0, 2.5, 2.3) = 7.7$
9. $f_{max} = 6$ on the segment with endpoints $(3, 0, 0)$ and $(0, 0, 2)$
11. $f_{max} = f(2.857, 0, 14.286, 0) = 314$

SECTION 21.4, page 1129

1. $f_{max} = f(10, 5) = 550$ 3. $f_{max} = f(0, 1.2, 0) = 1.2$
5. $f_{max} = f(100, 150, 200) = 3200$
7. Exchange x_2, x_5, then x_3, x_4, $f_{max} = f(3, 1) = 7$.
9. $f_{min} = -f_{max} = -f(2, 3) = 1$

CHAPTER 21 (REVIEW PROBLEMS), page 1130

11. The first step gives $z = i + 2j$, where $f(x)$ has a minimum.
23. $f_{max} = f(0, 5) = 5$
25. $f_{max} = -18$ at every point of the segment with endpoints $(2/9, 28/9)$ and $(4, 5)$
27. $f_{min} = f(2, 4/3) = 8$ 29. $f_{max} = f(0, 40, 5) = 280$

SECTION 22.1, page 1137

7. $\begin{bmatrix} 0 & 1 & 0 \\ 0 & 0 & 1 \\ 1 & 0 & 0 \end{bmatrix}$ 9. $\begin{bmatrix} 0 & 1 & 1 \\ 1 & 0 & 0 \\ 1 & 1 & 0 \end{bmatrix}$ 11. $\begin{bmatrix} 0 & 1 & 0 & 0 \\ 1 & 0 & 1 & 1 \\ 0 & 1 & 0 & 1 \\ 0 & 1 & 1 & 0 \end{bmatrix}$

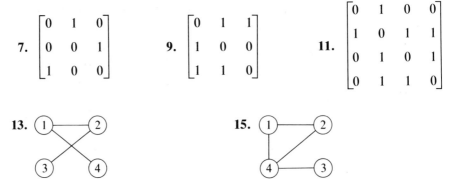

17. If G, with any (i, j) also contains (j, i).

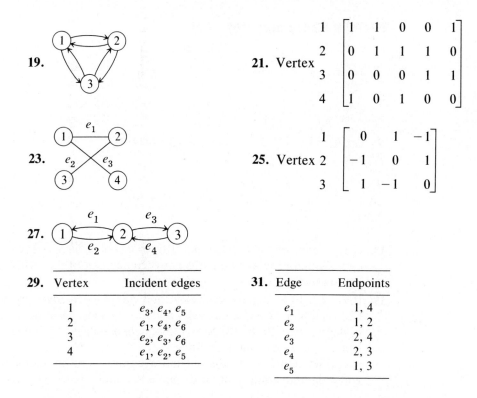

19.

21. Vertex
$$\begin{array}{c} 1 \\ 2 \\ 3 \\ 4 \end{array} \begin{bmatrix} 1 & 1 & 0 & 0 & 1 \\ 0 & 1 & 1 & 1 & 0 \\ 0 & 0 & 0 & 1 & 1 \\ 1 & 0 & 1 & 0 & 0 \end{bmatrix}$$

23.

25. Vertex
$$\begin{array}{c} 1 \\ 2 \\ 3 \end{array} \begin{bmatrix} 0 & 1 & -1 \\ -1 & 0 & 1 \\ 1 & -1 & 0 \end{bmatrix}$$

27.

29.

Vertex	Incident edges
1	e_3, e_4, e_5
2	e_1, e_4, e_6
3	e_2, e_3, e_6
4	e_1, e_2, e_5

31.

Edge	Endpoints
e_1	1, 4
e_2	1, 2
e_3	2, 4
e_4	2, 3
e_5	1, 3

35. Since each edge has 2 endpoints, the sum of the degrees of *all* vertices must be even. Now use that the sum of the even degrees *is* even.

SECTION 22.2, page 1142

3. $n - 1$. If it had more, a vertex would appear more than once and the corresponding cycle could be omitted. One edge.

5. 4

7. The idea is to go backward. There is a v_{k-1} adjacent to v_k and labeled $k - 1$, etc. Now the only vertex labeled 0 is s. Hence $\lambda(v_0) = 0$ implies $v_0 = s$, so that $v_0 - v_1 - \cdots - v_{k-1} - v_k$ is a path $s \to v_k$ that has length k.

11. $m \leqq 19$

19. $1 - 2 - 3 - 4 - 5 - 3 - 1$, $1 - 3 - 4 - 5 - 3 - 2 - 1$ and these two trails traversed in the opposite sense.

21. $1 - 2 - 3 - 4 - 5 - 6 - 4 - 3 - 1, L = 25$

25. The digraph consisting of just one edge

SECTION 22.3, page 1147

1. $(1, 2), (2, 4), (4, 3)$; $L_2 = 3, L_3 = 9, L_4 = 7$

3. $L_2 = 9, L_3 = 7, L_4 = 8, L_5 = 4, L_6 = 14$

7. $O(n^3)$ **9.** $L_2 = 2, L_3 = 5$

SECTION 22.4, page 1151

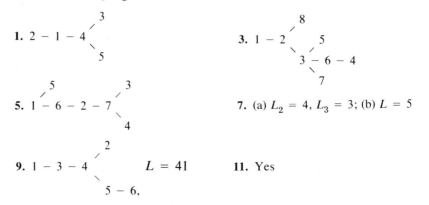

1. $2 - 1 - 4 \nearrow\!\!\!\!\! \begin{array}{c} 3 \\ \searrow\ 5 \end{array}$

3. $1 - 2 \nearrow\!\!\!\!\! \begin{array}{c} 8 \\ 5 \\ 3 - 6 - 4 \\ 7 \end{array}$

5. $1 - 6 - 2 - 7 \nearrow\!\!\!\!\! \begin{array}{c} 5 \qquad 3 \\ \searrow\ 4 \end{array}$

7. (a) $L_2 = 4$, $L_3 = 3$; (b) $L = 5$

9. $1 - 3 - 4 \nearrow\!\!\!\!\! \begin{array}{c} 2 \\ \searrow\ 5 - 6, \end{array}$ $L = 41$ **11.** Yes

13. G is connected. If G were not a tree, it would have a cycle, but this cycle would provide two paths between any pair of its vertices, contradicting the uniqueness.

19. If G is a tree, it has no cycles, and has $n - 1$ edges by Prob. 14. Conversely, let G have no cycles and $n - 1$ edges. Then G has 2 vertices of degree 1 by Prob. 18. Now prove connectedness by induction. True when $n = 2$. Assume true for $n = k - 1$. Let G with k vertices have no cycles and $k - 1$ edges. Omit a vertex v and its incident edge e, apply the induction hypothesis and add e and v back on.

SECTION 22.5, page 1155

1. $(1, 2)$, $(1, 4)$, $(3, 4)$, $(4, 5)$, $L = 12$

3. $(1, 2)$, $(1, 3)$, $(1, 4)$, $(2, 6)$, $(3, 5)$, $L = 16$

5. If G is a tree.

7. A shortest spanning tree of the largest connected graph that contains vertex 1.

11. $d(G) = 12$, $r(G) = 6 = \epsilon(3)$, center $\{3\}$

15. Choose a vertex u and find a farthest v_1. From v_1 find a farthest v_2. Find w such that $d(w, v_1)$ is as close as possible to being equal to $\frac{1}{2}d(v_1, v_2)$.

SECTION 22.6, page 1161

1. $\{4, 5, 6\}$, 28   **3.** $\{2, 4, 6\}$, 50   **5.** Yes   **7.** $f = 17$ **9.** $f = 7$

11. $1 - 2 - 3 - 7$, $\Delta f = 2$; $1 - 4 - 5 - 6 - 7$, $\Delta f = 1$;
$1 - 2 - 3 - 6 - 7$, $\Delta f = 1$; $f_{max} = 14$

15. If $f_{ij} < c_{ij}$ as well as $f_{ij} > 0$.

SECTION 22.7, page 1165

5. $f = 17$ **13.** 2000

15. $S = \{1, 2, 4, 5\}$, $T = \{3, 6\}$, cap $(S, T) = 14$

SECTION 22.8, page 1170

1. No   **3.** No   **5.** Yes, $S = \{1, 4, 5, 8\}$

7. Yes; a graph is not bipartite if it has a nonbipartite subgraph.

9. $1 - 2 - 3 - 5$ 　　　　　　　　　　**11.** $1 - 2 - 3 - 7 - 5 - 4$

13. $(1, 5), (2, 3)$ 　　　　　　　　　　**15.** $n_1 n_2$

17. $m \leq n_1(n - n_1) =$ number of edges in a complete bipartite graph, and this is maximum when $n_1 = n/2$, and then equal to $n^2/4$.

21. 3 　　　　　　　　　　　　　　　　**25.** 5

27. max $d(u) = n$. Let $u_1, \cdots, u_n$ and $v_1, \cdots, v_n$ denote the vertices of S and T, respectively. Color edges $(u_1, v_1), \cdots, (u_1, v_n)$ by colors $1, \cdots, n$, respectively, then edges $(u_2, v_1), \cdots, (u_2, v_n)$ by colors $2, \cdots, n, 1$, respectively, etc., cyclicly permuted.

29. K_3 　　　　　　　　　　**31.** 3; at 1, 4, 8 　　**33.** K_3

35. $n = 5, m = 10, 2m = 20 \geq 3f, f \leq 6$ by the hint, but $f = 2 - 5 + 10 = 7$ by Euler, a contradiction.

CHAPTER 22 (REVIEW PROBLEMS), page 1172

17. $\begin{bmatrix} 0 & 1 & 1 & 1 \\ 1 & 0 & 1 & 1 \\ 1 & 1 & 0 & 1 \\ 1 & 1 & 1 & 0 \end{bmatrix}$

21. $\begin{bmatrix} 0 & 1 & 1 \\ 1 & 0 & 1 \\ 1 & 1 & 0 \end{bmatrix}$

23. $\begin{bmatrix} 0 & 1 & 0 & 0 & 0 \\ 0 & 0 & 1 & 0 & 0 \\ 0 & 0 & 0 & 0 & 0 \\ 0 & 0 & 0 & 0 & 1 \\ 0 & 0 & 0 & 1 & 0 \end{bmatrix}$

25. $\begin{bmatrix} 0 & 0 & 1 & 1 \\ 0 & 0 & 1 & 1 \\ 0 & 0 & 0 & 0 \\ 0 & 0 & 0 & 0 \end{bmatrix}$

29.

1	e_1, e_2, e_5
2	e_1, e_4, e_6
3	e_3, e_4, e_5
4	e_2, e_3, e_6

33.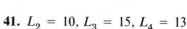

35. ①——③　②——④ 　　　　　　**37.** $L = 4$

39. $L_2 = 8, L_3 = 3, L_4 = 1$ 　　　　**41.** $L_2 = 10, L_3 = 15, L_4 = 13$

43. $1 - 2 - 4 - 3 - 5$ 　　　　　　　**47.** $f = 7$

49. $(1, 3), (2, 4)$

SECTION 23.1, page 1182

5. The space of all ordered triples of nonnegative numbers

7. Less than 5.0 or more than 5.5 gallons

15. $A \subset B$ if and only if $A \cap B = A$.

SECTION 23.2, page 1188

1. $31/32 \approx 0.97$ **3.** $1/4$

5. (a) 0.512, (b) About 0.504   **7.** 0.3439

9. (a), (c) 0.2487, (b) 0.5025, (d) 0.0606

11. 0.72653   **13.** 5/16

15. $P(E_B|E_A) = P(E_A \cap E_B)/P(E_A) = 0.02/0.10 = 20\%$

17. $A = (A \cap B) \cup (A \cap B^C) = B \cup (A \cap B^C)$ and Axiom 3 imply
$P(A) = P(B) + P(A \cap B^C) \geqq P(B)$ because $P(A \cap B^C) \geqq 0$.

SECTION 23.3, page 1193

3. 1820   **5.** $26^2 \cdot 10^3 = 676\ 000$

7. 1/1260   **9.** $10! = 3\ 628\ 800$

11. $\dbinom{97}{7} \Big/ \dbinom{100}{10} = 0.074\%$ **13.** $\dbinom{52}{13} = 635\ 013\ 559\ 600$

15. The idea of proof is the same as that of Theorem 1, but instead of filling n places we now have to fill only k places. If repetitions are permitted, we have n elements for filling each of the k places.

17. 710.08, 9.92, 1.4%; 39 902, 418, 1%

SECTION 23.4, page 1200

5. $F(x) = 0, (x - 1)/4, 1$ if $x < 1, 1 \leqq x < 5, x \geqq 5$, respectively

7. $f(3) = 1/216, f(4) = 3/216$, etc.

11. $f(x) = (5/6)^{x-1}/6, x = 1, 2, \cdots$;
$\Sigma f(x) = (1/6)/(1 - 5/6) = 1$ (geometric series)

13. $f(x) = (1/3)(2/3)^{x-1}, x = 1, 2, \cdots$

15. $P(X > 1200) = \displaystyle\int_{1.2}^{2} 6[0.25 - (x - 1.5)^2]\, dx = 0.896.$

     Ans. $0.896^3 = 72\%$

17. $k = 2/9, x = 0.9487, 2.9240$ **19.** $x = \ln 2 = 0.693, \ln 20 = 2.996$

SECTION 23.5, page 1204

1. $1, 0.2$ **5.** $0, 5$ **7.** $k = 750; 1, 0.002$

9. $\frac{1}{2}, \frac{1}{20}$ **11.** $\mu = 1/\theta; 22\%$

13. Set $c_1 = 1/\sigma$ and $c_2 = -\mu/\sigma$.   **15.** $E(X^k) = 2/(k + 2), \sigma^2 = \frac{1}{18}$

17. $E(X^k) = 1/(k + 1)$ **19.** $2\sqrt{2/5} \approx 1.2649$

SECTION 23.6, page 1209

1. $f(x) = \dbinom{5}{x} 0.5^5, f(0) = 0.03125, f(1) = 0.15625, 1 - f(0) = 0.96875,$

     $1 - f(5) = 0.96875$

3. 68.4%   **5.** $(0.98)^{15} \approx 74\%$ **7.** $f(x) = 4^x e^{-4}/x!$

9. $f(x) = 0.5^x e^{-0.5}/x!, f(0) + f(1) = e^{-0.5}(1.0 + 0.5) = 0.91.$ *Ans.* 9%

11. 23.81%   **13.** $f(0) = \dbinom{20}{0}\dbinom{80}{10}\bigg/\dbinom{100}{10} = \dfrac{80 \cdot 79 \cdots 71}{100 \cdot 99 \cdots 91} = 9.5\%$

17. $f(x_1, x_2) = 10! \cdot 0.03^{x_1} 0.02^{x_2} 0.95^{10 - x_1 - x_2}/x_1! x_2! (10 - x_1 - x_2)!$

23. $G(0) = 1$, $G'(t) = e^{-\mu} \exp[\mu e^t] \mu e^t = \mu e^t G(t)$, $G''(t) = \mu e^t [G(t) + G'(t)]$,
$E(X^2) = G''(0) = \mu + \mu^2$, $\sigma^2 = E(X^2) - \mu^2 = \mu$

25. Use $x \dbinom{M}{x} = M \dbinom{M-1}{x-1}$ and (14) in Sec. 23.3 with $p = M - 1$,

$k = x - 1$, $q = N - M$, $r - k = n - x$.

SECTION 23.7, page 1216

1. 0.1587, 0.5, 0.6915, 0.6247   **3.** 45.065, 56.978, 2.022
5. 0.27%   **7.** 16%
9. About 160   **11.** 68%, 98.8%
13. 13.1%   **15.** 876 kg

17. $\Phi^2(\infty) = \dfrac{1}{2\pi} \displaystyle\int_{-\infty}^{\infty} \int_{-\infty}^{\infty} e^{-u^2/2} e^{-v^2/2} \, du \, dv = \dfrac{1}{2\pi} \int_0^{2\pi} \int_0^{\infty} e^{-r^2/2} r \, dr \, d\theta = 1$

$(u = r \cos \theta, v = r \sin \theta, du \, dv = r \, dr \, d\theta)$

SECTION 23.8, page 1225

1. 1/8, 3/16, 3/8   **3.** 2/9, 1/9, 1/2   **5.** 50%
7. $f_1(x) = 1/(\beta_1 - \alpha_1)$ if $\alpha_1 < x < \beta_1$, $f_2(y) = 1/(\beta_2 - \alpha_2)$ if $\alpha_2 < y < \beta_2$
9. 20 kg, 3 grams   **11.** 105 lb, 0.539 lb   **13.** 27.45 mm, 0.38 mm
17. No

CHAPTER 23 (REVIEW PROBLEMS), page 1227

21. $f(0) = 0.80816$, $f(1) = 0.18367$, $f(2) = 0.00816$
23. $f(x) = 2^{-x}$, $x = 1, 2, \cdots$ **25.** 0.8, 0.1
27. 2024   **29.** 118.019, 1.98, 1.65%   **31.** 10.7%
33. $k = 3$, $c = 0.983$ **35.** 0.6, 0.48   **37.** 7/3, 8/9
41. 73.5%   **43.** $1 - e^{-0.2}$

45. $f(x_1, x_2) = \dfrac{20!}{x_1! \, x_2! \, (20 - x_1 - x_2)!} \, 0.03^{x_1} 0.05^{x_2} 0.92^{20 - x_1 - x_2}$

47. $f(x) = \dbinom{5}{x}\dbinom{15}{3-x}\bigg/\dbinom{20}{3}$; $f(0) = 91/228$, $f(1) = 35/76$, $f(2) = 5/38$,

$f(3) = 1/114$
49. 84%

SECTION 24.2, page 1234

1. 41, 13, 79, 57, 21, 44, 3, 25, 65, 63
7. $x = 0.89 = \Phi(z)$ gives $z = 1.227$ (Table A9), etc.
9. 2.207, 0.511, 1.109

SECTION 24.4, page 1243

1. $\bar{x} = 3.47$, $s^2 = 2.98$ **3.** $\bar{x} = 83.3$, s^2 13.17
5. 4, 100   **9.** 100.25
15. 360   **17.** 96, 86, 110, 24
19. $\bar{x} = 99.2$, $s^2 = 234.7$; grouped; $\bar{x} = 99.4$, $s^2 = 254.7$

SECTION 24.5, page 1247

3. $l = p^k(1 - p)^{n-k}$, $\hat{p} = k/n$, k = number of successes in n trials.
5. $\hat{\mu} = \bar{x}$
7. $\dfrac{\partial \ln l}{\partial p} = \dfrac{\partial \ln f}{\partial p} = \dfrac{1}{p} - \dfrac{x - 1}{1 - p} = 0$ gives $\hat{p} = \dfrac{1}{x}$.
9. $\hat{\theta} = n/\Sigma\, x_j = 1/\bar{x}$ **11.** $\hat{\theta} = 1$

SECTION 24.6, page 1256

1. CONF $\{12.2 \leqq \mu \leqq 16.6\}$ **3.** CONF $\{17.84 \leqq \mu \leqq 18.96\}$
5. CONF $\{0.726 \leqq \mu \leqq 0.751\}$ **7.** CONF $\{1.373 \leqq \mu \leqq 1.451\}$
9. CONF $\{0.105 \leqq \mu \leqq 0.107\}$; that error is 0.001, approximately (0.000 8 more exactly).
11. $c_1 = (\sqrt{253} - 2.58)^2/2 = 88.8$, $c_2 = 170.9$, CONF $\{1.42 \leqq \sigma^2 \leqq 2.75\}$
13. $c_1 = 0.41$, $c_2 = 16.7$, CONF $\{0.000\ 04 \leqq \sigma^2 \leqq 0.001\ 67\}$
15. CONF $\{23 \leqq \sigma^2 \leqq 553\}$
17. Normal distributions, means -12, 24, 47, variances 25, 100, 400
19. Normal distribution with mean 26 and variance 70

SECTION 24.7, page 1267

1. $t = \sqrt{7}(0.286 - 0)/4.31 = 0.18 < c = 1.94$ ($\alpha = 5\%$); do not reject the hypothesis.
3. $c = 6090 > 6019$, $c = 12\ 127 > 12\ 012$; do not reject the hypothesis.
5. $n = 16$, $\bar{x} = 0.75$, $s^2 = 2.87$, $t = 1.77 < c = 2.13$ [$\alpha = 5\%$, alternative $\mu \neq 0$, c obtained from $P(-c \leqq T \leqq c) = 1 - \alpha$]. The hypothesis is not rejected.
7. (a) $c = 14.7 > 12.0$; reject as before.
   (b) $c = 15.3 < 15.8$; reject the hypothesis.
9. $\mu < 13.76$ or $\mu > 16.24$.
11. $t_0 = \sqrt{9}(12 - 15)/\sqrt{4 + 4} = -3.18 < -2.12$ (Table A11 with 16 degrees of freedom); reject the hypothesis.
13. Hypothesis $\mu_1 = \mu_2$. Alternative $\mu_1 \neq \mu_2$, $\bar{x} = 106.22$; $\bar{y} = 106.34$, $s_1{}^2 = 0.717$, $s_2{}^2 = 0.243$, $t_0 = -0.27$ lies between $c_1 = -2.31$ and $c_2 = 2.31$ ($\alpha = 5\%$, 8 degrees of freedom). The hypothesis is not rejected.
15. $v_0 = 2.5 < 6.0$ [(9, 4) degrees of freedom]; do not reject the hypothesis.

17. Hypothesis $\mu_I = \mu_{II}$, alternative $\mu_I \neq \mu_{II}$, and

$$t_0 = \sqrt{100}\,(1120 - 1064)/\sqrt{75^2 + 82^2} = 5.04 > c_2 = 1.97$$

($\alpha = 5\%$, $c_1 = -1.97$, 198 degrees of freedom). Reject the hypothesis.

SECTION 24.8, page 1272

1. LCL $= 1 - 2.58 \cdot 0.02/2 = 0.974$, UCL $= 1.026$ **3.** 27
5. $2.58\sqrt{0.0004}/\sqrt{2} = 0.036$, UCL $= 3.536$, LCL $= 3.464$
9. UCL $= np + 3\sqrt{np(1 - p)}$, CL $= np$, LCL $= np - 3\sqrt{np(1 - p)}$
11. CL $= \mu = 3.6$, UCL $= \mu + 3\sqrt{\mu} \approx 9.3$, LCL $= \mu - 3\sqrt{\mu}$ is negative in (b) and we set LCL $= 0$.
13. In 30% (5%) of the cases, approximately.
15. Trend of sample means to increase. Abrupt change of sample means.

SECTION 24.9, page 1278

1. 0.9825, 0.9384, 0.4060   **3.** 0.8187, 0.6703, 0.1353

5. $P(A; \theta) = \dbinom{20\theta}{0}\dbinom{20 - 20\theta}{3}\Big/\dbinom{20}{3} = \dfrac{(20 - 20\theta)(19 - 20\theta)(18 - 20\theta)}{6 \cdot 1140}$

gives 0.72, 0.49.
7. About 10%, 30%   **9.** $(1 - \theta)^5$, $(1 - \theta)^5 + 5\theta(1 - \theta)^4$
11. 0.028 (at $\theta = 0.054$)   **13.** $P(A; \theta) = (1 - \theta)^n$
15. Because n is finite.

SECTION 24.10, page 1281

1. $\chi_0^2 = (40 - 50)^2/50 + (60 - 50)^2/50 = 4 > c = 3.84$. No.
3. $\chi_0^2 = 2.33 < c = 11.07$. Yes.   **5.** $\chi_0^2 = 94.19 > c = 11.07$. Reject.
7. $\chi_0^2 = 3.5^2/358.5 + 3.5^2/119.5 = 0.137 < c = 3.84$. Yes.
9. $\chi_0^2 = 999.9 > c = 13.28$. No.
11. Taking the last three rows together, we have $K - r - 1 = 9$, where $r = 1$, since the mean was estimated. $\chi_0^2 = 12.8 < c = 16.92$ ($\alpha = 5\%$). Do not reject the hypothesis.
13. $\bar{x} = 99.4$, $\tilde{\sigma} = 15.8$, $K = 5$ (endpoints $-\infty$, 85, 95, 105, 115, ∞), $\chi_0^2 = 0.7 < c = 5.99$ ($\alpha = 5\%$). Hypothesis not rejected.
15. $\bar{x} = 113$, $\tilde{\sigma} = 4.6$. We choose $K = 12$, the endpoints of the intervals being $-\infty$, 103, 105, . . . , 123, ∞. Then $\chi_0^2 = 18.14 > c = 16.92$ ($\alpha = 5\%$, 9 degrees of freedom). Hypothesis rejected.

SECTION 24.11, page 1284

1. $P(X \leq 2) = 0.5^6(1 + 6 + 15) = 34\%$. Do not reject the hypothesis $\tilde{\mu} = 0$.

3. 0.38% is the exact probability of obtaining 3 or fewer rods that are longer than 2 meters; this is somewhat smaller than the approximate value in Prob. 2.

5. Hypothesis $\mu = 0$. Alternative $\mu > 0$, $\bar{x} = 1.58$,
   $t = \sqrt{10} \cdot 1.58/1.23 = 4.06 > c = 1.83$ ($\alpha = 5\%$). Hypothesis rejected.

7. $P(X \geq 1) = 0.5^8(1 + 8) = 3.5\% < \alpha = 5\%$. Reject the hypothesis.

9. Consider $y_j = x_j - \tilde{\mu}_0$.   11. $P(T \leq 14) = 7.8\%$. Do not reject.

13. $P(T \leq 2) = 2.8\%$. Reject.   15. $P(T \leq 2) = 0.1\%$. Reject.

SECTION 24.12, page 1290

1. About 120 feet   3. $y = x - 0.1$

5. $y - 3 = 0.685(x - 3.5)$ 7. $y - 40.75 = -0.17(x - 47.5)$

9. $y - 48.89 = -1.32(x - 20.33)$

11. $2s_1^2 = 2$, $2s_{xy} = 2$, $b = 1$, $2s_2^2 = 2 + 2p^2/3$, $q_0 = 2p^2/3$,
   $k = 12.7a/\sqrt{3} = 7.3a$ ($\gamma = 95\%$), CONF $\{1 - 7.3a \leq \beta \leq 1 + 7.3a\}$

13. $q_0 = 76$, $k = 2.37\sqrt{76/(7 \cdot 944)} = 0.254$, CONF $\{-1.58 \leq \beta \leq -1.06\}$

CHAPTER 24 (REVIEW PROBLEMS), page 1291

21. $\bar{x} = 6$, $s^2 = 13.3$ 23. $\bar{x} = 206$, $s^2 = 160$, $r = 40$

25. $\hat{\mu} = 15.5625$, $\hat{\sigma}^2 = 5.3711$ 27. It will double.

29. CONF $\{16.21 \leq \mu \leq 16.39\}$ 31. CONF $\{0.00044 \leq \sigma^2 \leq 0.00127\}$

33. CONF $\{307 \leq \mu \leq 345\}$ 35. CONF $\{120.6 \leq \mu \leq 128.6\}$

37. Normal distribution with mean 88 and variance 49

39. $\alpha = 5\%$, $t = \sqrt{10}(0.811 - 0.800)/\sqrt{0.000054} = 4.73$, $P(-c \leq T \leq c)$
   $= 0.95$, $c = 2.26$ ($n - 1 = 9$ degrees of freedom). Reject the hypothesis.

41. $\alpha = 5\%$, $c = 16.92 > 9 \cdot 0.5^2/0.4^2 = 14.06$; do not reject the hypothesis.

43. $t_0 = \sqrt{10 \cdot 9 \cdot 17/19}\ (21.8 - 20.2)/\sqrt{9 \cdot 0.6^2 + 8 \cdot 0.5^2} = 6.3 > c = 1.74$
   (17 degrees of freedom). Reject the hypothesis and assert that B is better.

45. $v_0 = 50/30 = 1.67 < c = 2.59$ [(9, 15) degrees of freedom]; do not reject the hypothesis.

49. LCL $= 27.217$, UCL $= 27.783$ 51. $\theta_0 \approx 0.01$, $\theta_1 \approx 0.37$

53. 9   55. $\alpha = 5\%$, $\beta = 44\%$

57. $\chi_0^2 = 49/20 + 49/60 = 3.27 < c = 3.84$ (1 degree of freedom, $\alpha = 5\%$), which supports the claim.

59. Hypothesis: A and B are of equal quality. Then the event of 7 or 8 trials favorable to A in 8 trials has probability 3.5%. Reject the hypothesis.

Appendix 3

Auxiliary Material

A3.1 Formulas for Special Functions

Tables of numerical values see Appendix 4.

Exponential function e^x (Fig. 537)

$$e = 2.71828\ 18284\ 59045\ 23536\ 02874\ 71353$$

(1) $\qquad e^x e^y = e^{x+y}, \qquad e^x/e^y = e^{x-y}, \qquad (e^x)^y = e^{xy}$

Natural logarithm (Fig. 538)

(2) $\quad \ln (xy) = \ln x + \ln y, \qquad \ln (x/y) = \ln x - \ln y, \qquad \ln (x^a) = a \ln x$

$\ln x$ is the inverse of e^x, and $e^{\ln x} = x$, $e^{-\ln x} = e^{\ln (1/x)} = 1/x$.

Logarithm of base ten $\log_{10} x$ or simply $\log x$

(3) $\quad \log x = M \ln x, \quad M = \log e = 0.43429\ 44819\ 03251\ 82765\ 11289\ 18917$

(4) $\quad \ln x = \dfrac{1}{M} \log x, \quad \dfrac{1}{M} = 2.30258\ 50929\ 94045\ 68401\ 79914\ 54684$

$\log x$ is the inverse of 10^x, and $10^{\log x} = x$, $10^{-\log x} = 1/x$.

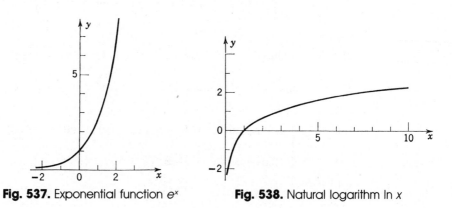

Fig. 537. Exponential function e^x **Fig. 538.** Natural logarithm $\ln x$

Sine and cosine functions (Figs. 539, 540). In calculus, angles are measured in radians, so that sin x and cos x have period 2π.

sin x is odd, sin $(-x) = -\sin x$, and cos x is even, cos $(-x) = \cos x$.

$$1° = 0.01745\ 32925\ 19943 \text{ radian}$$

$$1 \text{ radian} = 57°\ 17'\ 44.80625''$$

$$= 57.29577\ 95131°$$

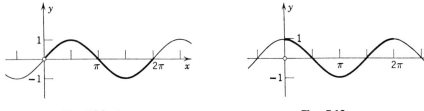

Fig. 539. sin x **Fig. 540.** cos x

(5) $$\sin^2 x + \cos^2 x = 1$$

(6)
$$
\begin{cases}
\sin (x + y) = \sin x \cos y + \cos x \sin y \\
\sin (x - y) = \sin x \cos y - \cos x \sin y \\
\cos (x + y) = \cos x \cos y - \sin x \sin y \\
\cos (x - y) = \cos x \cos y + \sin x \sin y
\end{cases}
$$

(7) $$\sin 2x = 2 \sin x \cos x, \qquad \cos 2x = \cos^2 x - \sin^2 x$$

(8)
$$
\begin{cases}
\sin x = \cos \left(x - \dfrac{\pi}{2} \right) = \cos \left(\dfrac{\pi}{2} - x \right) \\[2mm]
\cos x = \sin \left(x + \dfrac{\pi}{2} \right) = \sin \left(\dfrac{\pi}{2} - x \right)
\end{cases}
$$

(9) $$\sin (\pi - x) = \sin x, \qquad \cos (\pi - x) = -\cos x$$

(10) $$\cos^2 x = \tfrac{1}{2}(1 + \cos 2x), \qquad \sin^2 x = \tfrac{1}{2}(1 - \cos 2x)$$

(11)
$$
\begin{cases}
\sin x \sin y = \tfrac{1}{2}[-\cos (x + y) + \cos (x - y)] \\
\cos x \cos y = \tfrac{1}{2}[\cos (x + y) + \cos (x - y)] \\
\sin x \cos y = \tfrac{1}{2}[\sin (x + y) + \sin (x - y)]
\end{cases}
$$

$$(12) \begin{cases} \sin u + \sin v = 2 \sin \dfrac{u+v}{2} \cos \dfrac{u-v}{2} \\[2mm] \cos u + \cos v = 2 \cos \dfrac{u+v}{2} \cos \dfrac{u-v}{2} \\[2mm] \cos v - \cos u = 2 \sin \dfrac{u+v}{2} \sin \dfrac{u-v}{2} \end{cases}$$

$$(13) \quad A \cos x + B \sin x = \sqrt{A^2 + B^2} \cos (x \pm \delta), \quad \tan \delta = \frac{\sin \delta}{\cos \delta} = \mp \frac{B}{A}$$

$$(14) \quad A \cos x + B \sin x = \sqrt{A^2 + B^2} \sin (x \pm \delta), \quad \tan \delta = \frac{\sin \delta}{\cos \delta} = \pm \frac{A}{B}$$

Tangent, cotangent, secant, cosecant (Figs. 541, 542)

$$(15) \quad \tan x = \frac{\sin x}{\cos x}, \quad \cot x = \frac{\cos x}{\sin x}, \quad \sec x = \frac{1}{\cos x}, \quad \csc x = \frac{1}{\sin x}$$

$$(16) \quad \tan (x+y) = \frac{\tan x + \tan y}{1 - \tan x \tan y}, \quad \tan (x-y) = \frac{\tan x - \tan y}{1 + \tan x \tan y}$$

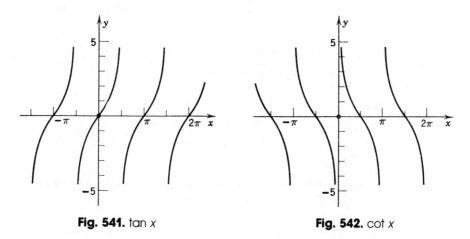

Fig. 541. tan x **Fig. 542.** cot x

Hyperbolic functions (hyperbolic sine sinh x, etc.; Figs. 543, 544, A76)

$$(17) \qquad \sinh x = \tfrac{1}{2}(e^x - e^{-x}), \qquad \cosh x = \tfrac{1}{2}(e^x + e^{-x})$$

$$(18) \qquad \tanh x = \frac{\sinh x}{\cosh x}, \qquad \coth x = \frac{\cosh x}{\sinh x}$$

$$(19) \qquad \cosh x + \sinh x = e^x, \qquad \cosh x - \sinh x = e^{-x}$$

$$(20) \qquad \cosh^2 x - \sinh^2 x = 1$$

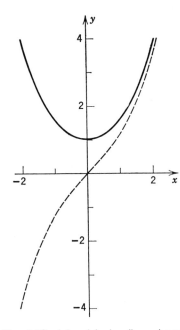

Fig. 543. sinh x (dashed) and cosh x

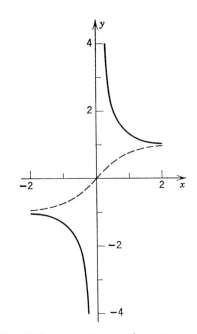

Fig. 544. tanh x (dashed) and coth x

(21) $\sinh^2 x = \frac{1}{2}(\cosh 2x - 1),$ $\cosh^2 x = \frac{1}{2}(\cosh 2x + 1)$

(22) $\begin{cases} \sinh (x \pm y) = \sinh x \cosh y \pm \cosh x \sinh y \\ \cosh (x \pm y) = \cosh x \cosh y \pm \sinh x \sinh y \end{cases}$

(23) $\tanh (x \pm y) = \dfrac{\tanh x \pm \tanh y}{1 \pm \tanh x \tanh y}$

Gamma function (cf. Fig. 545 and Table A3 in Appendix 4). The gamma function $\Gamma(\alpha)$ is defined by the integral

(24) $\Gamma(\alpha) = \displaystyle\int_0^\infty e^{-t} t^{\alpha-1}\, dt$ $(\alpha > 0)$

which is meaningful only if $\alpha > 0$ (or, if we consider complex α, for those α whose real part is positive). Integration by parts gives the important *functional relation of the gamma function,*

(25) $\Gamma(\alpha + 1) = \alpha\Gamma(\alpha).$

From (24) we readily have $\Gamma(1) = 1$; hence if α is a positive integer, say k, then by repeated application of (25) we obtain

(26) $\Gamma(k + 1) = k!$ $(k = 0, 1, \cdots).$

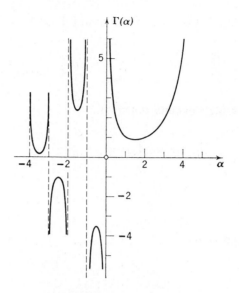

Fig. 545. Gamma function

This shows that *the gamma function can be regarded as a generalization of the elementary factorial function.* [Sometimes the notation $(\alpha - 1)!$ is used for $\Gamma(\alpha)$, even for noninteger values of α, and the gamma function is also known as the **factorial function.**]

By repeated application of (25) we obtain

$$\Gamma(\alpha) = \frac{\Gamma(\alpha + 1)}{\alpha} = \frac{\Gamma(\alpha + 2)}{\alpha(\alpha + 1)} = \cdots = \frac{\Gamma(\alpha + k + 1)}{\alpha(\alpha + 1)(\alpha + 2) \cdots (\alpha + k)}$$

and we may use this relation

$$(27) \qquad \Gamma(\alpha) = \frac{\Gamma(\alpha + k + 1)}{\alpha(\alpha + 1) \cdots (\alpha + k)} \qquad (\alpha \neq 0, -1, -2, \cdots)$$

for defining the gamma function for negative α ($\neq -1, -2, \cdots$), choosing for k the smallest integer such that $\alpha + k + 1 > 0$. *Together with* (24), *this then gives a definition of* $\Gamma(\alpha)$ *for all α not equal to zero or a negative integer* (Fig. 545).

It can be shown that the gamma function may also be represented as the limit of a product, namely, by the formula

$$(28) \qquad \Gamma(\alpha) = \lim_{n \to \infty} \frac{n! \, n^{\alpha}}{\alpha(\alpha + 1)(\alpha + 2) \cdots (\alpha + n)} \qquad (\alpha \neq 0, -1, \cdots).$$

From (27) or (28) we see that, for complex α, the gamma function $\Gamma(\alpha)$ is a meromorphic function that has simple poles at $\alpha = 0, -1, -2, \cdots$.

An approximation of the gamma function for large positive α is given by the **Stirling formula**

$$(29) \qquad \Gamma(\alpha + 1) \approx \sqrt{2\pi\alpha} \left(\frac{\alpha}{e}\right)^{\alpha}$$

where e is the base of the natural logarithm. We finally mention the special value

(30)
$$\Gamma(\tfrac{1}{2}) = \sqrt{\pi}.$$

Incomplete gamma functions

(31) $P(\alpha, x) = \displaystyle\int_0^x e^{-t} t^{\alpha-1}\, dt, \qquad Q(\alpha, x) = \displaystyle\int_x^\infty e^{-t} t^{\alpha-1}\, dt \qquad (\alpha > 0)$

(32)
$$\Gamma(\alpha) = P(\alpha, x) + Q(\alpha, x)$$

Beta function

(33) $B(x, y) = \displaystyle\int_0^1 t^{x-1}(1 - t)^{y-1}\, dt \qquad\qquad (x > 0,\, y > 0)$

Representation in terms of gamma functions:

(34)
$$B(x, y) = \frac{\Gamma(x)\Gamma(y)}{\Gamma(x + y)}$$

Error function (cf. Fig. 546 and Table A5 in Appendix 4; cf. also Sec. 18.6)

(35)
$$\operatorname{erf} x = \frac{2}{\sqrt{\pi}} \int_0^x e^{-t^2}\, dt$$

(36)
$$\operatorname{erf} x = \frac{2}{\sqrt{\pi}} \left(x - \frac{x^3}{1!3} + \frac{x^5}{2!5} - \frac{x^7}{3!7} + - \cdots \right)$$

$\operatorname{erf}(\infty) = 1$, *complementary error function*

(37)
$$\operatorname{erfc} x = 1 - \operatorname{erf} x = \frac{2}{\sqrt{\pi}} \int_x^\infty e^{-t^2}\, dt$$

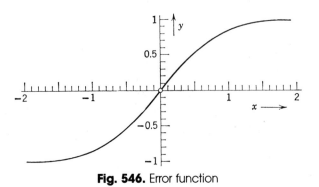

Fig. 546. Error function

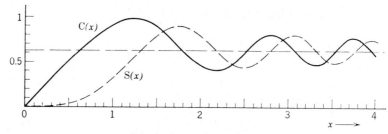

Fig. 547. Fresnel integrals

Fresnel integrals[1] (Fig. 547)

$$(38) \qquad C(x) = \int_0^x \cos (t^2)\, dt, \qquad S(x) = \int_0^x \sin (t^2)\, dt$$

$C(\infty) = \sqrt{\pi/8}$, $S(\infty) = \sqrt{\pi/8}$, *complementary functions*

$$(39) \qquad \begin{aligned} c(x) &= \sqrt{\frac{\pi}{8}} - C(x) = \int_x^\infty \cos (t^2)\, dt \\[2mm] s(x) &= \sqrt{\frac{\pi}{8}} - S(x) = \int_x^\infty \sin (t^2)\, dt \end{aligned}$$

Sine integral (cf. Fig. 548 and Table A5 in Appendix 4)

$$(40) \qquad Si(x) = \int_0^x \frac{\sin t}{t}\, dt$$

$Si(\infty) = \pi/2$, *complementary function*

$$(41) \qquad si(x) = \frac{\pi}{2} - Si(x) = \int_x^\infty \frac{\sin t}{t}\, dt$$

Cosine integral (cf. Table A5 in Appendix 4)

$$(42) \qquad ci(x) = \int_x^\infty \frac{\cos t}{t}\, dt \qquad\qquad (x > 0)$$

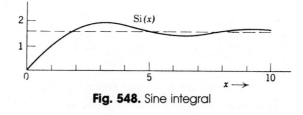

Fig. 548. Sine integral

[1]AUGUSTIN FRESNEL (1788—1827), French physicist and mathematician. For tables see Refs. [8], [12]; cf. also *Acta mathematica,* vol. **85** (1951), p. 180, and vol. **89** (1953), p. 130.

Exponential integral (cf. Sec. 18.6)

$$(43) \qquad \text{Ei}(x) = \int_x^\infty \frac{e^{-t}}{t}\, dt \qquad (x > 0)$$

Logarithmic integral

$$(44) \qquad \text{li}(x) = \int_0^x \frac{dt}{\ln t}$$

A3.2 Partial Derivatives

Differentiation formulas see inside of front cover.

Let $z = f(x, y)$ be a real function of two independent real variables, x and y. If we keep y constant, say, $y = y_1$, and think of x as a variable, then $f(x, y_1)$ depends on x alone. If the derivative of $f(x, y_1)$ with respect to x for a value $x = x_1$ exists, then the value of this derivative is called the **partial derivative** *of* $f(x, y)$ *with respect to* x *at the point* (x_1, y_1) and is denoted by

$$\frac{\partial f}{\partial x}\bigg|_{(x_1, y_1)} \qquad \text{or by} \qquad \frac{\partial z}{\partial x}\bigg|_{(x_1, y_1)}$$

Other notations are

$$f_x(x_1, y_1) \qquad \text{and} \qquad z_x(x_1, y_1);$$

these may be used when subscripts are not used for another purpose and there is no danger of confusion.

We thus have, by the definition of the derivative,

$$(1) \qquad \frac{\partial f}{\partial x}\bigg|_{(x_1, y_1)} = \lim_{\Delta x \to 0} \frac{f(x_1 + \Delta x, y_1) - f(x_1, y_1)}{\Delta x}.$$

The partial derivative of $z = f(x, y)$ with respect to y is defined similarly; we now keep x constant, say, equal to x_1, and differentiate $f(x_1, y)$ with respect to y. Thus

$$(2) \qquad \frac{\partial f}{\partial y}\bigg|_{(x_1, y_1)} = \frac{\partial z}{\partial y}\bigg|_{(x_1, y_1)} = \lim_{\Delta y \to 0} \frac{f(x_1, y_1 + \Delta y) - f(x_1, y_1)}{\Delta y}.$$

Other notations are $f_y(x_1, y_1)$ and $z_y(x_1, y_1)$.

It is clear that the values of those two partial derivatives will in general depend on the point (x_1, y_1). Hence the partial derivatives $\partial z/\partial x$ and $\partial z/\partial y$ at a variable point (x, y) are functions of x and y. The function $\partial z/\partial x$ is obtained as in ordinary calculus by differentiating $z = f(x, y)$ with respect to x, *treating y as a constant*, and $\partial z/\partial y$ is obtained by differentiating z with respect to y, *treating x as a constant*.

EXAMPLE 1
Let $z = f(x, y) = x^2 y + x \sin y$. Then

$$\frac{\partial f}{\partial x} = 2xy + \sin y, \qquad \frac{\partial f}{\partial y} = x^2 + x \cos y. \qquad \blacksquare$$

The partial derivatives $\partial z/\partial x$ and $\partial z/\partial y$ of a function $z = f(x, y)$ have a very simple **geometric interpretation.** The function $z = f(x, y)$ can be represented by a surface in space. The equation $y = y_1$ then represents a vertical plane intersecting the surface in a curve, and the partial derivative $\partial z/\partial x$ at a point (x_1, y_1) is the slope of the tangent (i.e., $\tan \alpha$ where α is the angle shown in Fig. 549) to the curve. Similarly, the partial derivative $\partial z/\partial y$ at (x_1, y_1) is the slope of the tangent to the curve $x = x_1$ on the surface $z = f(x, y)$ at (x_1, y_1).

The partial derivatives $\partial z/\partial x$ and $\partial z/\partial y$ are called *first partial derivatives* or *partial derivatives of first order*. By differentiating these derivatives once more, we obtain the four *second partial derivatives* (or *partial derivatives of second order*)[1]

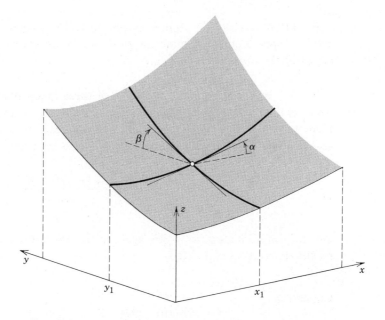

Fig. 549. Geometrical interpretation of first partial derivatives

[1]Caution! In the subscript notation the subscripts are written in the order in which we differentiate, whereas in the "∂" notation the order is opposite.

$$\frac{\partial^2 f}{\partial x^2} = \frac{\partial}{\partial x}\left(\frac{\partial f}{\partial x}\right) = f_{xx}$$

$$\frac{\partial^2 f}{\partial x\,\partial y} = \frac{\partial}{\partial x}\left(\frac{\partial f}{\partial y}\right) = f_{yx}$$

(3)

$$\frac{\partial^2 f}{\partial y\,\partial x} = \frac{\partial}{\partial y}\left(\frac{\partial f}{\partial x}\right) = f_{xy}$$

$$\frac{\partial^2 f}{\partial y^2} = \frac{\partial}{\partial y}\left(\frac{\partial f}{\partial y}\right) = f_{yy}.$$

It can be shown that if all the derivatives concerned are continuous, then the two mixed partial derivatives are equal, so that then the order of differentiation does not matter (see Ref. [5] in Appendix 1),

(4)
$$\frac{\partial^2 z}{\partial x\,\partial y} = \frac{\partial^2 z}{\partial y\,\partial x}.$$

EXAMPLE 2
For the function in Example 1,

$$f_{xx} = 2y, \qquad f_{xy} = 2x + \cos y = f_{yx}, \qquad f_{yy} = -x\sin y. \qquad \blacksquare$$

By differentiating the second partial derivatives again with respect to x and y, respectively, we obtain the *third partial derivatives* or *partial derivatives of the third order* of f, etc.

If we consider a function $f(x, y, z)$ of **three independent variables,** then we have the three first partial derivatives $f_x(x, y, z)$, $f_y(x, y, z)$ and $f_z(x, y, z)$. Here f_x is obtained by differentiating f with respect to x, *treating both y and z as constants.* Thus, in analogy to (1), we now have

$$\left.\frac{\partial f}{\partial x}\right|_{(x_1,y_1,z_1)} = \lim_{\Delta x \to 0} \frac{f(x_1 + \Delta x, y_1, z_1) - f(x_1, y_1, z_1)}{\Delta x},$$

etc. By differentiating f_x, f_y, f_z again in this fashion we obtain the second partial derivatives of f, etc.

EXAMPLE 3
Let $f(x, y, z) = x^2 + y^2 + z^2 + xy\,e^z$. Then

$$f_x = 2x + y\,e^z, \qquad f_y = 2y + x\,e^z, \qquad f_z = 2z + xy\,e^z,$$

$$f_{xx} = 2, \qquad f_{xy} = f_{yx} = e^z, \qquad f_{xz} = f_{zx} = y\,e^z,$$

$$f_{yy} = 2, \qquad f_{yz} = f_{zy} = x\,e^z, \qquad f_{zz} = 2 + xy\,e^z.$$

A3.3 Sequences and Series

See also Chap. 14.

Monotone Real Sequences

We call a real sequence $x_1, x_2, \cdots, x_n, \cdots$ a **monotone sequence** if it is either **monotone increasing,** that is,

$$x_1 \leqq x_2 \leqq x_3 \leqq \cdots$$

or **monotone decreasing,** that is,

$$x_1 \geqq x_2 \geqq x_3 \geqq \cdots,$$

We call $x_1, x_2, \cdots$ a **bounded sequence** if there is a positive constant K such that $|x_n| < K$ for all n.

Theorem 1
If a real sequence is bounded and monotone, it converges.

Proof. Let $x_1, x_2, \cdots$ be a bounded monotone increasing sequence. Then its terms are smaller than some number B and, since $x_1 \leqq x_n$ for all n, they lie in the interval $x_1 \leqq x_n \leqq B$, which will be denoted by I_0. We bisect I_0; that is, we subdivide it into two parts of equal length. If the right half (together with its endpoints) contains terms of the sequence, we denote it by I_1. If it does not contain terms of the sequence, then the left half of I_0 (together with its endpoints) is called I_1. This is the first step.

In the second step we bisect I_1, select one half by the same rule, and call it I_2, and so on (see Fig. 550).

In this way we obtain shorter and shorter intervals $I_0, I_1, I_2, \cdots$ with the following properties. Each I_m contains all I_n for $n > m$. No term of the sequence lies to the right of I_m, and, since the sequence is monotone increasing, all x_n with n greater than some number N lie in I_m; of course, N will depend on m, in general. The lengths of the I_m approach zero as m approaches infinity. Hence there is precisely one number, call it L, which

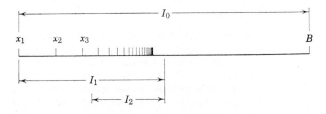

Fig. 550. Proof of Theorem 1

lies in all those intervals,[2] and we may now easily prove that the sequence is convergent with the limit L.

In fact, given an $\epsilon > 0$, we choose an m such that the length of I_m is less than ϵ. Then L and all the x_n with $n > N(m)$ lie in I_m, and, therefore, $|x_n - L| < \epsilon$ for all those n. This completes the proof for an increasing sequence. For a decreasing sequence the proof is the same, except for a suitable interchange of "left" and "right" in the construction of those intervals. ∎

Real Series

Leibniz test for real series

Let x_1, x_2, $\cdots$ be real and monotone decreasing to zero, that is,

$$(1) \qquad \text{(a) } x_1 \geqq x_2 \geqq x_3 \geqq \cdots, \qquad \text{(b) } \lim_{m \to \infty} x_m = 0.$$

Then the series with terms of alternating signs

$$x_1 - x_2 + x_3 - x_4 + - \cdots$$

converges, and for the remainder R_n after the nth term we have the estimate

$$(2) \qquad |R_n| \leqq x_{n+1}.$$

Proof. Let s_n be the nth partial sum of the series. Then, because of (1a),

$$s_1 = x_1, \qquad\qquad s_2 = x_1 - x_2 \leqq s_1,$$

$$s_3 = s_2 + x_3 \geqq s_2, \qquad s_3 = s_1 - (x_2 - x_3) \leqq s_1,$$

so that $s_2 \leqq s_3 \leqq s_1$. Proceeding in this fashion, we conclude that (Fig. 551)

$$(3) \qquad s_1 \geqq s_3 \geqq s_5 \geqq \cdots \geqq s_6 \geqq s_4 \geqq s_2$$

which shows that the odd partial sums form a bounded monotone sequence, and so do the even partial sums. Hence, by Theorem 1, both sequences converge, say,

$$\lim_{n \to \infty} s_{2n+1} = s, \qquad \lim_{n \to \infty} s_{2n} = s^*.$$

---

[2]This statement seems to be obvious, but actually it is not; it may be regarded as an axiom of the real number system in the following form. Let $J_1, J_2, \cdots$ be closed intervals such that each J_m contains all J_n with $n > m$, and the lengths of the J_m approach zero as m approaches infinity. Then there is precisely one real number that is contained in all those intervals. This is the so-called **Cantor–Dedekind axiom,** named after the German mathematicians GEORG CANTOR (1845—1918), the creator of set theory, and RICHARD DEDEKIND (1831—1916), known for his fundamental work in number theory. For further details see Ref. [2] in Appendix 1. (An interval I is said to be **closed** if its two endpoints are regarded as points belonging to I. It is said to be **open** if the endpoints are not regarded as points of I.)

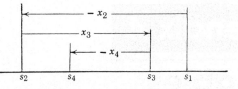

Fig. 551. Proof of the Leibniz test

Now, since $s_{2n+1} - s_{2n} = x_{2n+1}$, we readily see that (1b) implies

$$s - s^* = \lim_{n \to \infty} s_{2n+1} - \lim_{n \to \infty} s_{2n} = \lim_{n \to \infty} (s_{2n+1} - s_{2n}) = \lim_{n \to \infty} x_{2n+1} = 0.$$

Hence $s^* = s$, and the series converges with the sum s.

We prove the estimate (2) for the remainder. Since $s_n \to s$, it follows from (3) that

$$s_{2n+1} \geqq s \geqq s_{2n} \qquad \text{and also} \qquad s_{2n-1} \geqq s \geqq s_{2n}.$$

By subtracting s_{2n} and s_{2n-1}, respectively, we obtain

$$s_{2n+1} - s_{2n} \geqq s - s_{2n} \geqq 0, \qquad 0 \geqq s - s_{2n-1} \geqq s_{2n} - s_{2n-1}.$$

In these inequalities, the first expression is equal to x_{2n+1}, the last is equal to $-x_{2n}$, and the expressions between the inequality signs are the remainders R_{2n} and R_{2n-1}. Thus the inequalities may be written

$$x_{2n+1} \geqq R_{2n} \geqq 0, \qquad 0 \geqq R_{2n-1} \geqq -x_{2n}$$

and we see that they imply (2). This completes the proof.   ∎

Appendix 4

Tables

Tables of Laplace transforms in Secs. 5.9 and 5.10
Tables of Fourier transforms in Sec. 10.12

Table A1
Some Elementary Functions

We leave this table in as a "first-aid kit" if no pocket calculator is available.

x	$\sin x$	$\cos x$	$\tan x$	e^x	$\sinh x$	$\cosh x$
0.0	0.00000	1.00000	0.00000	1.00000	0.00000	1.00000
0.1	0.09983	0.99500	0.10033	1.10517	0.10017	1.00500
0.2	0.19867	0.98007	0.20271	1.22140	0.20134	1.02007
0.3	0.29552	0.95534	0.30934	1.34986	0.30452	1.04534
0.4	0.38942	0.92106	0.42279	1.49182	0.41075	1.08107
0.5	0.47943	0.87758	0.54630	1.64872	0.52110	1.12763
0.6	0.56464	0.82534	0.68414	1.82212	0.63665	1.18547
0.7	0.64422	0.76484	0.84229	2.01375	0.75858	1.25517
0.8	0.71736	0.69671	1.02964	2.22554	0.88811	1.33743
0.9	0.78333	0.62161	1.26016	2.45960	1.02652	1.43309
1.0	0.84147	0.54030	1.55741	2.71828	1.17520	1.54308
1.1	0.89121	0.45360	1.96476	3.00417	1.33565	1.66852
1.2	0.93204	0.36236	2.57215	3.32012	1.50946	1.81066
1.3	0.96356	0.26750	3.60210	3.66930	1.69838	1.97091
1.4	0.98545	0.16997	5.79788	4.05520	1.90430	2.15090
1.5	0.99749	0.07074	14.10142	4.48169	2.12928	2.35241
1.6	0.99957	−0.02920	−34.23253	4.95303	2.37557	2.57746
1.7	0.99166	−0.12884	−7.69660	5.47395	2.64563	2.82832
1.8	0.97385	−0.22720	−4.28626	6.04965	2.94217	3.10747
1.9	0.94630	−0.32329	−2.92710	6.68589	3.26816	3.41773
2.0	0.90930	−0.41615	−2.18504	7.38906	3.62686	3.76220

x	$\ln x$	x	$\ln x$	x	$\ln x$	x	$\ln x$
1.0	0.00000	2.0	0.69315	3.0	1.09861	5	1.60944
1.1	0.09531	2.1	0.74194	3.1	1.13140	7	1.94591
1.2	0.18232	2.2	0.78846	3.2	1.16315	11	2.39790
1.3	0.26236	2.3	0.83291	3.3	1.19392	13	2.56495
1.4	0.33647	2.4	0.87547	3.4	1.22378	17	2.83321
1.5	0.40547	2.5	0.91629	3.5	1.25276	19	2.94444
1.6	0.47000	2.6	0.95551	3.6	1.28093	23	3.13549
1.7	0.53063	2.7	0.99325	3.7	1.30833	29	3.36730
1.8	0.58779	2.8	1.02962	3.8	1.33500	31	3.43399
1.9	0.64185	2.9	1.06471	3.9	1.36098	37	3.61092

Table A1 (*continued*)

$\dfrac{y}{x}$	arc tan $\dfrac{y}{x}$	$\dfrac{y}{x}$	arc tan $\dfrac{y}{x}$	$\dfrac{y}{x}$	arc tan $\dfrac{y}{x}$	$\dfrac{y}{x}$	arc tan $\dfrac{y}{x}$
0.0	0.00000	1.0	0.78540	2.0	1.10715	4.0	1.32582
0.1	0.09967	1.1	0.83298	2.2	1.14417	4.5	1.35213
0.2	0.19740	1.2	0.87606	2.4	1.17601	5.0	1.37340
0.3	0.29146	1.3	0.91510	2.6	1.20362	5.5	1.39094
0.4	0.38051	1.4	0.95055	2.8	1.22777	6.0	1.40565
0.5	0.46365	1.5	0.98279	3.0	1.24905	7.0	1.42890
0.6	0.54042	1.6	1.01220	3.2	1.26791	8.0	1.44644
0.7	0.61073	1.7	1.03907	3.4	1.28474	9.0	1.46014
0.8	0.67474	1.8	1.06370	3.6	1.29985	10.0	1.47113
0.9	0.73282	1.9	1.08632	3.8	1.31347	11.0	1.48014

Table A2
Bessel Functions

For more extensive tables see Ref. [12] in Appendix 1.

x	$J_0(x)$	$J_1(x)$	x	$J_0(x)$	$J_1(x)$	x	$J_0(x)$	$J_1(x)$
0.0	1.0000	0.0000	3.0	-0.2601	0.3991	6.0	0.1506	-0.2767
0.1	0.9975	0.0499	3.1	-0.2921	0.3009	6.1	0.1773	-0.2559
0.2	0.9900	0.0995	3.2	-0.3202	0.2613	6.2	0.2017	-0.2329
0.3	0.9776	0.1483	3.3	-0.3443	0.2207	6.3	0.2238	-0.2081
0.4	0.9604	0.1960	3.4	-0.3643	0.1792	6.4	0.2433	-0.1816
0.5	0.9385	0.2423	3.5	-0.3801	0.1374	6.5	0.2601	-0.1538
0.6	0.9120	0.2867	3.6	-0.3918	0.0955	6.6	0.2740	-0.1250
0.7	0.8812	0.3290	3.7	-0.3992	0.0538	6.7	0.2851	-0.0953
0.8	0.8463	0.3688	3.8	-0.4026	0.0128	6.8	0.2931	-0.0652
0.9	0.8075	0.4059	3.9	-0.4018	-0.0272	6.9	0.2981	-0.0349
1.0	0.7652	0.4401	4.0	-0.3971	-0.0660	7.0	0.3001	-0.0047
1.1	0.7196	0.4709	4.1	-0.3887	-0.1033	7.1	0.2991	0.0252
1.2	0.6711	0.4983	4.2	-0.3766	-0.1386	7.2	0.2951	0.0543
1.3	0.6201	0.5220	4.3	-0.3610	-0.1719	7.3	0.2882	0.0826
1.4	0.5669	0.5419	4.4	-0.3423	-0.2028	7.4	0.2786	0.1096
1.5	0.5118	0.5579	4.5	-0.3205	-0.2311	7.5	0.2663	0.1352
1.6	0.4554	0.5699	4.6	-0.2961	-0.2566	7.6	0.2516	0.1592
1.7	0.3980	0.5778	4.7	-0.2693	-0.2791	7.7	0.2346	0.1813
1.8	0.3400	0.5815	4.8	-0.2404	-0.2985	7.8	0.2154	0.2014
1.9	0.2818	0.5812	4.9	-0.2097	-0.3147	7.9	0.1944	0.2192
2.0	0.2239	0.5767	5.0	-0.1776	-0.3276	8.0	0.1717	0.2346
2.1	0.1666	0.5683	5.1	-0.1443	-0.3371	8.1	0.1475	0.2476
2.2	0.1104	0.5560	5.2	-0.1103	-0.3432	8.2	0.1222	0.2580
2.3	0.0555	0.5399	5.3	-0.0758	-0.3460	8.3	0.0960	0.2657
2.4	0.0025	0.5202	5.4	-0.0412	-0.3453	8.4	0.0692	0.2708
2.5	-0.0484	0.4971	5.5	-0.0068	-0.3414	8.5	0.0419	0.2731
2.6	-0.0968	0.4708	5.6	0.0270	-0.3343	8.6	0.0146	0.2728
2.7	-0.1424	0.4416	5.7	0.0599	-0.3241	8.7	-0.0125	0.2697
2.8	-0.1850	0.4097	5.8	0.0917	-0.3110	8.8	-0.0392	0.2641
2.9	-0.2243	0.3754	5.9	0.1220	-0.2951	8.9	-0.0653	0.2559

$J_0(x) = 0$ for $x = 2.405,\ 5.520,\ 8.654,\ 11.792,\ 14.931,\ \cdots$

$J_1(x) = 0$ for $x = 0,\ 3.832,\ 7.016,\ 10.173,\ 13.324,\ \cdots$

Table A2 (*continued*)

x	$Y_0(x)$	$Y_1(x)$	x	$Y_0(x)$	$Y_1(x)$	x	$Y_0(x)$	$Y_1(x)$
0.0	$(-\infty)$	$(-\infty)$	2.5	0.498	0.146	5.0	−0.309	0.148
0.5	−0.445	−1.471	3.0	0.377	0.325	5.5	−0.339	−0.024
1.0	0.088	−0.781	3.5	0.189	0.410	6.0	−0.288	−0.175
1.5	0.382	−0.412	4.0	−0.017	0.398	6.5	−0.173	−0.274
2.0	0.510	−0.107	4.5	−0.195	0.301	7.0	−0.026	−0.303

Table A3
Gamma Function (cf. (24) in Appendix 3)

α	$\Gamma(\alpha)$	α	$\Gamma(\alpha)$	α	$\Gamma(\alpha)$	α	$\Gamma(\alpha)$	α	$\Gamma(\alpha)$
1.00	1.000 000	1.20	0.918 169	1.40	0.887 264	1.60	0.893 515	1.80	0.931 384
1.02	0.988 844	1.22	0.913 106	1.42	0.886 356	1.62	0.895 924	1.82	0.936 845
1.04	0.978 438	1.24	0.908 521	1.44	0.885 805	1.64	0.898 642	1.84	0.942 612
1.06	0.968 744	1.26	0.904 397	1.46	0.885 604	1.66	0.901 668	1.86	0.948 687
1.08	0.959 725	1.28	0.900 718	1.48	0.885 747	1.68	0.905 001	1.88	0.955 071
1.10	0.951 351	1.30	0.897 471	1.50	0.886 227	1.70	0.908 639	1.90	0.961 766
1.12	0.943 590	1.32	0.894 640	1.52	0.887 039	1.72	0.912 581	1.92	0.968 774
1.14	0.936 416	1.34	0.892 216	1.54	0.888 178	1.74	0.916 826	1.94	0.976 099
1.16	0.929 803	1.36	0.890 185	1.56	0.889 639	1.76	0.921 375	1.96	0.983 743
1.18	0.923 728	1.38	0.888 537	1.58	0.891 420	1.78	0.926 227	1.98	0.991 708
1.20	0.918 169	1.40	0.887 264	1.60	0.893 515	1.80	0.931 384	2.00	1.000 000

Table A4
Factorial Function

n	$n!$	$\log(n!)$	n	$n!$	$\log(n!)$	n	$n!$	$\log(n!)$
1	1	0.000 000	6	720	2.857 332	11	39 916 800	7.601 156
2	2	0.301 030	7	5 040	3.702 431	12	479 001 600	8.680 337
3	6	0.778 151	8	40 320	4.605 521	13	6 227 020 800	9.794 280
4	24	1.380 211	9	362 880	5.559 763	14	87 178 291 200	10.940 408
5	120	2.079 181	10	3 628 800	6.559 763	15	1 307 674 368 000	12.116 500

Table A5
Error Function, Sine and Cosine Integrals (cf. (35), (40), (42) in Appendix 3)

x	erf x	Si(x)	ci(x)	x	erf x	Si(x)	ci(x)
0.0	0.0000	0.0000	∞	2.0	0.9953	1.6054	−0.4230
0.2	0.2227	0.1996	1.0422	2.2	0.9981	1.6876	−0.3751
0.4	0.4284	0.3965	0.3788	2.4	0.9993	1.7525	−0.3173
0.6	0.6039	0.5881	0.0223	2.6	0.9998	1.8004	−0.2533
0.8	0.7421	0.7721	−0.1983	2.8	0.9999	1.8321	−0.1865
1.0	0.8427	0.9461	−0.3374	3.0	1.0000	1.8487	−0.1196
1.2	0.9103	1.1080	−0.4205	3.2	1.0000	1.8514	−0.0553
1.4	0.9523	1.2562	−0.4620	3.4	1.0000	1.8419	0.0045
1.6	0.9763	1.3892	−0.4717	3.6	1.0000	1.8219	0.0580
1.8	0.9891	1.5058	−0.4568	3.8	1.0000	1.7934	0.1038
2.0	0.9953	1.6054	−0.4230	4.0	1.0000	1.7582	0.1410

Table A6
Binomial Distribution

Probability function $f(x)$ (cf. (2), Sec. 23.6) and distribution function $F(x)$

n	x	$p = 0.1$ $f(x)$	$F(x)$	$p = 0.2$ $f(x)$	$F(x)$	$p = 0.3$ $f(x)$	$F(x)$	$p = 0.4$ $f(x)$	$F(x)$	$p = 0.5$ $f(x)$	$F(x)$
		0.		**0.**		**0.**		**0.**		**0.**	
1	0	9000	0.9000	8000	0.8000	7000	0.7000	6000	0.6000	5000	0.5000
	1	1000	1.0000	2000	1.0000	3000	1.0000	4000	1.0000	5000	1.0000
	0	8100	0.8100	6400	0.6400	4900	0.4900	3600	0.3600	2500	0.2500
2	1	1800	0.9900	3200	0.9600	4200	0.9100	4800	0.8400	5000	0.7500
	2	0100	1.0000	0400	1.0000	0900	1.0000	1600	1.0000	2500	1.0000
	0	7290	0.7290	5120	0.5120	3430	0.3430	2160	0.2160	1250	0.1250
3	1	2430	0.9720	3840	0.8960	4410	0.7840	4320	0.6480	3750	0.5000
	2	0270	0.9990	0960	0.9920	1890	0.9730	2880	0.9360	3750	0.8750
	3	0010	1.0000	0080	1.0000	0270	1.0000	0640	1.0000	1250	1.0000
	0	6561	0.6561	4096	0.4096	2401	0.2401	1296	0.1296	0625	0.0625
	1	2916	0.9477	4096	0.8192	4116	0.6517	3456	0.4752	2500	0.3125
4	2	0486	0.9963	1536	0.9728	2646	0.9163	3456	0.8208	3750	0.6875
	3	0036	0.9999	0256	0.9984	0756	0.9919	1536	0.9744	2500	0.9375
	4	0001	1.0000	0016	1.0000	0081	1.0000	0256	1.0000	0625	1.0000
	0	5905	0.5905	3277	0.3277	1681	0.1681	0778	0.0778	0313	0.0313
	1	3281	0.9185	4096	0.7373	3602	0.5282	2592	0.3370	1563	0.1875
5	2	0729	0.9914	2048	0.9421	3087	0.8369	3456	0.6826	3125	0.5000
	3	0081	0.9995	0512	0.9933	1323	0.9692	2304	0.9130	3125	0.8125
	4	0005	1.0000	0064	0.9997	0284	0.9976	0768	0.9898	1563	0.9688
	5	0000	1.0000	0003	1.0000	0024	1.0000	0102	1.0000	0313	1.0000
	0	5314	0.5314	2621	0.2621	1176	0.1176	0467	0.0467	0156	0.0156
	1	3543	0.8857	3932	0.6554	3025	0.4202	1866	0.2333	0938	0.1094
	2	0984	0.9841	2458	0.9011	3241	0.7443	3110	0.5443	2344	0.3438
6	3	0146	0.9987	0819	0.9830	1852	0.9295	2765	0.8208	3125	0.6563
	4	0012	0.9999	0154	0.9984	0595	0.9891	1382	0.9590	2344	0.8906
	5	0001	1.0000	0015	0.9999	0102	0.9993	0369	0.9959	0938	0.9844
	6	0000	1.0000	0001	1.0000	0007	1.0000	0041	1.0000	0156	1.0000
	0	4783	0.4783	2097	0.2097	0824	0.0824	0280	0.0280	0078	0.0078
	1	3720	0.8503	3670	0.5767	2471	0.3294	1306	0.1586	0547	0.0625
	2	1240	0.9743	2753	0.8520	3177	0.6471	2613	0.4199	1641	0.2266
7	3	0230	0.9973	1147	0.9667	2269	0.8740	2903	0.7102	2734	0.5000
	4	0026	0.9998	0287	0.9953	0972	0.9712	1935	0.9037	2734	0.7734
	5	0002	1.0000	0043	0.9996	0250	0.9962	0774	0.9812	1641	0.9375
	6	0000	1.0000	0004	1.0000	0036	0.9998	0172	0.9984	0547	0.9922
	7	0000	1.0000	0000	1.0000	0002	1.0000	0016	1.0000	0078	1.0000
	0	4305	0.4305	1678	0.1678	0576	0.0576	0168	0.0168	0039	0.0039
	1	3826	0.8131	3355	0.5033	1977	0.2553	0896	0.1064	0313	0.0352
	2	1488	0.9619	2936	0.7969	2965	0.5518	2090	0.3154	1094	0.1445
	3	0331	0.9950	1468	0.9437	2541	0.8059	2787	0.5941	2188	0.3633
8	4	0046	0.9996	0459	0.9896	1361	0.9420	2322	0.8263	2734	0.6367
	5	0004	1.0000	0092	0.9988	0467	0.9887	1239	0.9502	2188	0.8555
	6	0000	1.0000	0011	0.9999	0100	0.9987	0413	0.9915	1094	0.9648
	7	0000	1.0000	0001	1.0000	0012	0.9999	0079	0.9993	0313	0.9961
	8	0000	1.0000	0000	1.0000	0001	1.0000	0007	1.0000	0039	1.0000

Table A7
Poisson Distribution

Probability function $f(x)$ (cf. (5), Sec. 23.6) and distribution function $F(x)$

x	$\mu = 0.1$		$\mu = 0.2$		$\mu = 0.3$		$\mu = 0.4$		$\mu = 0.5$	
	$f(x)$	$F(x)$	$f(x)$	$F(x)$	$f(x)$	$F(x)$	$f(x)$	$F(x)$	$f(x)$	$F(x)$
	0.		**0.**		**0.**		**0.**		**0.**	
0	9048	0.9048	8187	0.8187	7408	0.7408	6703	0.6703	6065	0.6065
1	0905	0.9953	1637	0.9825	2222	0.9631	2681	0.9384	3033	0.9098
2	0045	0.9998	0164	0.9989	0333	0.9964	0536	0.9921	0758	0.9856
3	0002	1.0000	0011	0.9999	0033	0.9997	0072	0.9992	0126	0.9982
4	0000	1.0000	0001	1.0000	0003	1.0000	0007	0.9999	0016	0.9998
5							0001	1.0000	0002	1.0000

x	$\mu = 0.6$		$\mu = 0.7$		$\mu = 0.8$		$\mu = 0.9$		$\mu = 1$	
	$f(x)$	$F(x)$	$f(x)$	$F(x)$	$f(x)$	$F(x)$	$f(x)$	$F(x)$	$f(x)$	$F(x)$
	0.		**0.**		**0.**		**0.**		**0.**	
0	5488	0.5488	4966	0.4966	4493	0.4493	4066	0.4066	3679	0.3679
1	3293	0.8781	3476	0.8442	3595	0.8088	3659	0.7725	3679	0.7358
2	0988	0.9769	1217	0.9659	1438	0.9526	1647	0.9371	1839	0.9197
3	0198	0.9966	0284	0.9942	0383	0.9909	0494	0.9865	0613	0.9810
4	0030	0.9996	0050	0.9992	0077	0.9986	0111	0.9977	0153	0.9963
5	0004	1.0000	0007	0.9999	0012	0.9998	0020	0.9997	0031	0.9994
6			0001	1.0000	0002	1.0000	0003	1.0000	0005	0.9999
7									0001	1.0000

x	$\mu = 1.5$		$\mu = 2$		$\mu = 3$		$\mu = 4$		$\mu = 5$	
	$f(x)$	$F(x)$	$f(x)$	$F(x)$	$f(x)$	$F(x)$	$f(x)$	$F(x)$	$f(x)$	$F(x)$
	0.		**0.**		**0.**		**0.**		**0.**	
0	2231	0.2231	1353	0.1353	0498	0.0498	0183	0.0183	0067	0.0067
1	3347	0.5578	2707	0.4060	1494	0.1991	0733	0.0916	0337	0.0404
2	2510	0.8088	2707	0.6767	2240	0.4232	1465	0.2381	0842	0.1247
3	1255	0.9344	1804	0.8571	2240	0.6472	1954	0.4335	1404	0.2650
4	0471	0.9814	0902	0.9473	1680	0.8153	1954	0.6288	1755	0.4405
5	0141	0.9955	0361	0.9834	1008	0.9161	1563	0.7851	1755	0.6160
6	0035	0.9991	0120	0.9955	0504	0.9665	1042	0.8893	1462	0.7622
7	0008	0.9998	0034	0.9989	0216	0.9881	0595	0.9489	1044	0.8666
8	0001	1.0000	0009	0.9998	0081	0.9962	0298	0.9786	0653	0.9319
9			0002	1.0000	0027	0.9989	0132	0.9919	0363	0.9682
10					0008	0.9997	0053	0.9972	0181	0.9863
11					0002	0.9999	0019	0.9991	0082	0.9945
12					0001	1.0000	0006	0.9997	0034	0.9980
13							0002	0.9999	0013	0.9993
14							0001	1.0000	0005	0.9998
15									0002	0.9999
16									0000	1.0000

Table A8
Normal Distribution

Values of the distribution function $\Phi(z)$ (cf. (4), Sec. 23.7)
$\Phi(-z) = 1 - \Phi(z)$, $\Phi(0) = 0.5000$

z	$\Phi(z)$	z	$\Phi(z)$	z	$\Phi(z)$	z	$\Phi(z)$	z	$\Phi(z)$	z	$\Phi(z)$
	0.		**0.**		**0.**		**0.**		**0.**		**0.**
0.01	5040	0.51	6950	1.01	8438	1.51	9345	2.01	9778	2.51	9940
0.02	5080	0.52	6985	1.02	8461	1.52	9357	2.02	9783	2.52	9941
0.03	5120	0.53	7019	1.03	8485	1.53	9370	2.03	9788	2.53	9943
0.04	5160	0.54	7054	1.04	8508	1.54	9382	2.04	9793	2.54	9945
0.05	5199	0.55	7088	1.05	8531	1.55	9394	2.05	9798	2.55	9946
0.06	5239	0.56	7123	1.06	8554	1.56	9406	2.06	9803	2.56	9948
0.07	5279	0.57	7157	1.07	8577	1.57	9418	2.07	9808	2.57	9949
0.08	5319	0.58	7190	1.08	8599	1.58	9429	2.08	9812	2.58	9951
0.09	5359	0.59	7224	1.09	8621	1.59	9441	2.09	9817	2.59	9952
0.10	5398	0.60	7257	1.10	8643	1.60	9452	2.10	9821	2.60	9953
0.11	5438	0.61	7291	1.11	8665	1.61	9463	2.11	9826	2.61	9955
0.12	5478	0.62	7324	1.12	8686	1.62	9474	2.12	9830	2.62	9956
0.13	5517	0.63	7357	1.13	8708	1.63	9484	2.13	9834	2.63	9957
0.14	5557	0.64	7389	1.14	8729	1.64	9495	2.14	9838	2.64	9959
0.15	5596	0.65	7422	1.15	8749	1.65	9505	2.15	9842	2.65	9960
0.16	5636	0.66	7454	1.16	8770	1.66	9515	2.16	9846	2.66	9961
0.17	5675	0.67	7486	1.17	8790	1.67	9525	2.17	9850	2.67	9962
0.18	5714	0.68	7517	1.18	8810	1.68	9535	2.18	9854	2.68	9963
0.19	5753	0.69	7549	1.19	8830	1.69	9545	2.19	9857	2.69	9964
0.20	5793	0.70	7580	1.20	8849	1.70	9554	2.20	9861	2.70	9965
0.21	5832	0.71	7611	1.21	8869	1.71	9564	2.21	9864	2.71	9966
0.22	5871	0.72	7642	1.22	8888	1.72	9573	2.22	9868	2.72	9967
0.23	5910	0.73	7673	1.23	8907	1.73	9582	2.23	9871	2.73	9968
0.24	5948	0.74	7704	1.24	8925	1.74	9591	2.24	9875	2.74	9969
0.25	5987	0.75	7734	1.25	8944	1.75	9599	2.25	9878	2.75	9970
0.26	6026	0.76	7764	1.26	8962	1.76	9608	2.26	9881	2.76	9971
0.27	6064	0.77	7794	1.27	8980	1.77	9616	2.27	9884	2.77	9972
0.28	6103	0.78	7823	1.28	8997	1.78	9625	2.28	9887	2.78	9973
0.29	6141	0.79	7852	1.29	9015	1.79	9633	2.29	9890	2.79	9974
0.30	6179	0.80	7881	1.30	9032	1.80	9641	2.30	9893	2.80	9974
0.31	6217	0.81	7910	1.31	9049	1.81	9649	2.31	9896	2.81	9975
0.32	6255	0.82	7939	1.32	9066	1.82	9656	2.32	9898	2.82	9976
0.33	6293	0.83	7967	1.33	9082	1.83	9664	2.33	9901	2.83	9977
0.34	6331	0.84	7995	1.34	9099	1.84	9671	2.34	9904	2.84	9977
0.35	6368	0.85	8023	1.35	9115	1.85	9678	2.35	9906	2.85	9978
0.36	6406	0.86	8051	1.36	9131	1.86	9686	2.36	9909	2.86	9979
0.37	6443	0.87	8078	1.37	9147	1.87	9693	2.37	9911	2.87	9979
0.38	6480	0.88	8106	1.38	9162	1.88	9699	2.38	9913	2.88	9980
0.39	6517	0.89	8133	1.39	9177	1.89	9706	2.39	9916	2.89	9981
0.40	6554	0.90	8159	1.40	9192	1.90	9713	2.40	9918	2.90	9981
0.41	6591	0.91	8186	1.41	9207	1.91	9719	2.41	9920	2.91	9982
0.42	6628	0.92	8212	1.42	9222	1.92	9726	2.42	9922	2.92	9982
0.43	6664	0.93	8238	1.43	9236	1.93	9732	2.43	9925	2.93	9983
0.44	6700	0.94	8264	1.44	9251	1.94	9738	2.44	9927	2.94	9984
0.45	6736	0.95	8289	1.45	9265	1.95	9744	2.45	9929	2.95	9984
0.46	6772	0.96	8315	1.46	9279	1.96	9750	2.46	9931	2.96	9985
0.47	6808	0.97	8340	1.47	9292	1.97	9756	2.47	9932	2.97	9985
0.48	6844	0.98	8365	1.48	9306	1.98	9761	2.48	9934	2.98	9986
0.49	6879	0.99	8389	1.49	9319	1.99	9767	2.49	9936	2.99	9986
0.50	6915	1.00	8413	1.50	9332	2.00	9772	2.50	9938	3.00	9987

Table A9
Normal Distribution

Values of z for given values of $\Phi(z)$ (cf. (4), Sec. 23.7) and $D(z) = \Phi(z) - \Phi(-z)$
Example: $z = 0.279$ if $\Phi(z) = 61\%$; $z = 0.860$ if $D(z) = 61\%$.

%	$z(\Phi)$	$z(D)$	%	$z(\Phi)$	$z(D)$	%	$z(\Phi)$	$z(D)$
1	−2.326	0.013	41	−0.228	0.539	81	0.878	1.311
2	−2.054	0.025	42	−0.202	0.553	82	0.915	1.341
3	−1.881	0.038	43	−0.176	0.568	83	0.954	1.372
4	−1.751	0.050	44	−0.151	0.583	84	0.994	1.405
5	−1.645	0.063	45	−0.126	0.598	85	1.036	1.440
6	−1.555	0.075	46	−0.100	0.613	86	1.080	1.476
7	−1.476	0.088	47	−0.075	0.628	87	1.126	1.514
8	−1.405	0.100	48	−0.050	0.643	88	1.175	1.555
9	−1.341	0.113	49	−0.025	0.659	89	1.227	1.598
10	−1.282	0.126	50	0.000	0.674	90	1.282	1.645
11	−1.227	0.138	51	0.025	0.690	91	1.341	1.695
12	−1.175	0.151	52	0.050	0.706	92	1.405	1.751
13	−1.126	0.164	53	0.075	0.722	93	1.476	1.812
14	−1.080	0.176	54	0.100	0.739	94	1.555	1.881
15	−1.036	0.189	55	0.126	0.755	95	1.645	1.960
16	−0.994	0.202	56	0.151	0.772	96	1.751	2.054
17	−0.954	0.215	57	0.176	0.789	97	1.881	2.170
18	−0.915	0.228	58	0.202	0.806	97.5	1.960	2.241
19	−0.878	0.240	59	0.228	0.824	98	2.054	2.326
20	−0.842	0.253	60	0.253	0.842	99	2.326	2.576
21	−0.806	0.266	61	0.279	0.860	99.1	2.366	2.612
22	−0.772	0.279	62	0.305	0.878	99.2	2.409	2.652
23	−0.739	0.292	63	0.332	0.896	99.3	2.457	2.697
24	−0.706	0.305	64	0.358	0.915	99.4	2.512	2.748
25	−0.674	0.319	65	0.385	0.935	99.5	2.576	2.807
26	−0.643	0.332	66	0.412	0.954	99.6	2.652	2.878
27	−0.613	0.345	67	0.440	0.974 ˙	99.7	2.748	2.968
28	−0.583	0.358	68	0.468	0.994	99.8	2.878	3.090
29	−0.553	0.372	69	0.496	1.015	99.9	3.090	3.291
30	−0.524	0.385	70	0.524	1.036			
31	−0.496	0.399	71	0.553	1.058	99.91	3.121	3.320
32	−0.468	0.412	72	0.583	1.080	99.92	3.156	3.353
33	−0.440	0.426	73	0.613	1.103	99.93	3.195	3.390
34	−0.412	0.440	74	0.643	1.126	99.94	3.239	3.432
35	−0.385	0.454	75	0.674	1.150	99.95	3.291	3.481
36	−0.358	0.468	76	0.706	1.175	99.96	3.353	3.540
37	−0.332	0.482	77	0.739	1.200	99.97	3.432	3.615
38	−0.305	0.496	78	0.772	1.227	99.98	3.540	3.719
39	−0.279	0.510	79	0.806	1.254	99.99	3.719	3.891
40	−0.253	0.524	80	0.842	1.282			

Table A10
Random Digits

Cf. Sec. 24.2.

Row No.	Column Number									
	0	1	2	3	4	5	6	7	8	9
0	87331	82442	28104	26432	83640	17323	68764	84728	37995	96106
1	33628	17364	01409	87803	65641	33433	48944	64299	79066	31777
2	54680	13427	72496	16967	16195	96593	55040	53729	62035	66717
3	51199	49794	49407	10774	98140	83891	37195	24066	61140	65144
4	78702	98067	61313	91661	59861	54437	77739	19892	54817	88645
5	55672	16014	24892	13089	00410	81458	76156	28189	40595	21500
6	18880	58497	03862	32368	59320	24807	63392	79793	63043	09425
7	10242	62548	62330	05703	33535	49128	66298	16193	55301	01306
8	54993	17182	94618	23228	83895	73251	68199	64639	83178	70521
9	22686	50885	16006	04041	08077	33065	35237	02502	94755	72062
10	42349	03145	15770	70665	53291	32288	41568	66079	98705	31029
11	18093	09553	39428	75464	71329	86344	80729	40916	18860	51780
12	11535	03924	84252	74795	40193	84597	42497	21918	91384	84721
13	35066	73848	65351	53270	67341	70177	92373	17604	42204	60476
14	57477	22809	73558	96182	96779	01604	25748	59553	64876	94611
15	48647	33850	52956	45410	88212	05120	99391	32276	55961	41775
16	86857	81154	22223	74950	53296	67767	55866	49061	66937	81818
17	20182	36907	94644	99122	09774	29189	27212	79000	50217	71077
18	83687	31231	01133	41432	54542	60204	81618	09586	34481	87683
19	81315	12390	46074	47810	90171	36313	95440	77583	28506	38808
20	87026	52826	58341	76549	04105	66191	12914	55348	07907	06978
21	34301	76733	07251	90524	21931	83695	41340	53581	64582	60210
22	70734	24337	32674	49508	49751	90489	63202	24380	77943	09942
23	94710	31527	73445	32839	68176	53580	85250	53243	03350	00128
24	76462	16987	07775	43162	11777	16810	75158	13894	88945	15539
25	14348	28403	79245	69023	34196	46398	05964	64715	11330	17515
26	74618	89317	30146	25606	94507	98104	04239	44973	37636	88866
27	99442	19200	85406	45358	86253	60638	38858	44964	54103	57287
28	26869	44399	89452	06652	31271	00647	46551	83050	92058	83814
29	80988	08149	50499	98584	28385	63680	44638	91864	96002	87802
30	07511	79047	89289	17774	67194	37362	85684	55505	97809	67056
31	49779	12138	05048	03535	27502	63308	10218	53296	48687	61340
32	47938	55945	24003	19635	17471	65997	85906	98694	56420	78357
33	15604	06626	14360	79542	13512	87595	08542	03800	35443	52823
34	12307	27726	21864	00045	16075	03770	86978	52718	02693	09096
35	02450	28053	66134	99445	91316	25727	89399	85272	67148	78358
36	57623	54382	35236	89244	27245	90500	75430	96762	71968	65838
37	91762	78849	93105	40481	99431	03304	21079	86459	21287	76566
38	87373	31137	31128	67050	34309	44914	80711	61738	61498	24288
39	67094	41485	54149	86088	10192	21174	39948	67268	29938	32476
40	94456	66747	76922	87627	71834	57688	04878	78348	68970	60048
41	68359	75292	27710	86889	81678	79798	58360	39175	75667	65782
42	52393	31404	32584	06837	79762	13168	76055	54833	22841	98889
43	59565	91254	11847	20672	37625	41454	86861	55824	79793	74575
44	48185	11066	20162	38230	16043	48409	47421	21195	98008	57305
45	19230	12187	86659	12971	52204	76546	63272	19312	81662	96557
46	84327	21942	81727	68735	89190	58491	55329	96875	19465	89687
47	77430	71210	00591	50124	12030	50280	12358	76174	48353	09682
48	12462	19108	70512	53926	25595	97085	03833	59806	12351	64253
49	11684	06644	57816	10078	45021	47751	38285	73520	08434	65627

Table A10
Random Digits (*continued*)

Row No.	Column Number									
	0	1	2	3	4	5	6	7	8	9
50	12896	36576	68686	08462	65652	76571	70891	09007	04581	01684
51	59090	05111	27587	90349	30789	50304	70650	06646	70126	15284
52	42486	67483	65282	19037	80588	73076	41820	46651	40442	40718
53	88662	03928	03249	85910	97533	88643	29829	21557	47328	36724
54	69403	03626	92678	53460	15465	83516	54012	80509	55976	46115
55	56434	70543	38696	98502	32092	95505	62091	39549	30117	98209
56	58227	62694	42837	29183	11393	68463	25150	86338	95620	39836
57	41272	94927	15413	40505	33123	63218	72940	98349	57249	40170
58	36819	01162	30425	15546	16065	68459	35776	64276	92868	07372
59	31700	66711	26115	55755	33584	18091	38709	57276	74660	90392
60	69855	63699	36839	90531	97125	87875	62824	03889	12538	24740
61	44322	17569	45439	41455	34324	90902	07978	26268	04279	76816
62	62226	36661	87011	66267	78777	78044	40819	49496	39814	73867
63	27284	19737	98741	72531	52741	26699	98755	19657	08665	16818
64	88341	21652	94743	77268	79525	44769	66583	30621	90534	62050
65	53266	18783	51903	56711	38060	69513	61963	80470	88018	86510
66	50527	49330	24839	42529	03944	95219	88724	37247	84166	23023
67	15655	07852	77206	35944	71446	30573	19405	57824	23576	23301
68	62057	22206	03314	83465	57466	10465	19891	32308	01900	67484
69	41769	56091	19892	96253	92808	45785	52774	49674	68103	65032
70	25993	72416	44473	41299	93095	17338	69802	98548	02429	85238
71	22842	57871	04470	37373	34516	04042	04078	35336	34393	97573
72	55704	31982	05234	22664	22181	40358	28089	15790	33340	18852
73	94258	18706	09437	96041	90052	80862	20420	24323	11635	91677
74	74145	20453	29657	98868	56695	53483	87449	35060	98942	62697
75	88881	12673	73961	89884	73247	97670	69570	88888	58560	72580
76	01508	56780	52223	35632	73347	71317	46541	88023	36656	76332
77	92069	43000	23233	06058	82527	25250	27555	20426	60361	63525
78	53366	35249	02117	68620	39388	69795	73215	01846	16983	78560
79	88057	54097	49511	74867	32192	90071	04147	46094	63519	07199
80	85492	82238	02668	91854	86149	28590	77853	81035	45561	16032
81	39453	62123	69611	53017	34964	09786	24614	49514	01056	18700
82	82627	98111	93870	56969	69566	62662	07353	84838	14570	14508
83	61142	51743	38209	31474	96095	15163	54380	77849	20465	03142
84	12031	32528	61311	53730	89032	16124	58844	35386	45521	59368
85	31313	59838	29147	76882	74328	09955	63673	96651	53264	29871
86	50767	41056	97409	44376	62219	35439	70102	99248	71179	26052
87	30522	95699	84966	26554	24768	72247	84993	85375	92518	16334
88	74176	19870	89874	64799	03792	57006	57225	36677	46825	14087
89	17114	93248	37065	91346	04657	93763	92210	43676	44944	75798
90	53005	11825	64608	87587	05742	31914	55044	41818	29667	77424
91	31985	81539	79942	49471	46200	27639	94099	42085	79231	03932
92	63499	60508	77522	15624	15088	78519	52279	79214	43623	69166
93	30506	42444	99047	66010	91657	37160	37408	85714	21420	80996
94	78248	16841	92357	10130	68990	38307	61022	56806	81016	38511
95	64996	84789	50185	32200	64382	29752	11876	00664	54547	62597
96	11963	13157	09136	01769	30117	71486	80111	09161	08371	71749
97	44335	91450	43456	90449	18338	19787	31339	60473	06606	89788
98	42277	11868	44520	01113	11341	11743	97949	49718	99176	42006
99	77562	18863	58515	90166	78508	14864	19111	57183	85808	59385

Table A11
t-Distribution

Values of z for given values of the distribution function $F(z)$ (cf. p. 1251)
Example: For 9 degrees of freedom, $z = 1.83$ when $F(z) = 0.95$.

$F(z)$	Number of Degrees of Freedom									
	1	2	3	4	5	6	7	8	9	10
0.5	0.00	0.00	0.00	0.00	0.00	0.00	0.00	0.00	0.00	0.00
0.6	0.33	0.29	0.28	0.27	0.27	0.27	0.26	0.26	0.26	0.26
0.7	0.73	0.62	0.58	0.57	0.56	0.55	0.55	0.55	0.54	0.54
0.8	1.38	1.06	0.98	0.94	0.92	0.91	0.90	0.89	0.88	0.88
0.9	3.08	1.89	1.64	1.53	1.48	1.44	1.42	1.40	1.38	1.37
0.95	6.31	2.92	2.35	2.13	2.02	1.94	1.90	1.86	1.83	1.81
0.975	12.7	4.30	3.18	2.78	2.57	2.45	2.37	2.31	2.26	2.23
0.99	31.8	6.97	4.54	3.75	3.37	3.14	3.00	2.90	2.82	2.76
0.995	63.7	9.93	5.84	4.60	4.03	3.71	3.50	3.36	3.25	3.17
0.999	318.3	22.3	10.2	7.17	5.89	5.21	4.79	4.50	4.30	4.14

$F(z)$	Number of Degrees of Freedom									
	11	12	13	14	15	16	17	18	19	20
0.5	0.00	0.00	0.00	0.00	0.00	0.00	0.00	0.00	0.00	0.00
0.6	0.26	0.26	0.26	0.26	0.26	0.26	0.26	0.26	0.26	0.26
0.7	0.54	0.54	0.54	0.54	0.54	0.54	0.53	0.53	0.53	0.53
0.8	0.88	0.87	0.87	0.87	0.87	0.87	0.86	0.86	0.86	0.86
0.9	1.36	1.36	1.35	1.35	1.34	1.34	1.33	1.33	1.33	1.33
0.95	1.80	1.78	1.77	1.76	1.75	1.75	1.74	1.73	1.73	1.73
0.975	2.20	2.18	2.16	2.15	2.13	2.12	2.11	2.10	2.09	2.09
0.99	2.72	2.68	2.65	2.62	2.60	2.58	2.57	2.55	2.54	2.53
0.995	3.11	3.06	3.01	2.98	2.95	2.92	2.90	2.88	2.86	2.85
0.999	4.03	3.93	3.85	3.79	3.73	3.69	3.65	3.61	3.58	3.55

$F(z)$	Number of Degrees of Freedom									
	22	24	26	28	30	40	50	100	200	∞
0.5	0.00	0.00	0.00	0.00	0.00	0.00	0.00	0.00	0.00	0.00
0.6	0.26	0.26	0.26	0.26	0.26	0.26	0.26	0.25	0.25	0.25
0.7	0.53	0.53	0.53	0.53	0.53	0.53	0.53	0.53	0.53	0.52
0.8	0.86	0.86	0.86	0.86	0.85	0.85	0.85	0.85	0.84	0.84
0.9	1.32	1.32	1.32	1.31	1.31	1.30	1.30	1.29	1.29	1.28
0.95	1.72	1.71	1.71	1.70	1.70	1.68	1.68	1.66	1.65	1.65
0.975	2.07	2.06	2.06	2.05	2.04	2.02	2.01	1.98	1.97	1.96
0.99	2.51	2.49	2.48	2.47	2.46	2.42	2.40	2.37	2.35	2.33
0.995	2.82	2.80	2.78	2.76	2.75	2.70	2.68	2.63	2.60	2.58
0.999	3.51	3.47	3.44	3.41	3.39	3.31	3.26	3.17	3.13	3.09

Table A12
Chi-square Distribution

Values of x for given values of the distribution function $F(z)$ (cf. p. 1252)
Example: For 3 degrees of freedom, $z = 11.34$ when $F(z) = 0.99$.

$F(z)$	Number of Degrees of Freedom									
	1	2	3	4	5	6	7	8	9	10
0.005	0.00	0.01	0.07	0.21	0.41	0.68	0.99	1.34	1.73	2.16
0.01	0.00	0.02	0.11	0.30	0.55	0.87	1.24	1.65	2.09	2.56
0.025	0.00	0.05	0.22	0.48	0.83	1.24	1.69	2.18	2.70	3.25
0.05	0.00	0.10	0.35	0.71	1.15	1.64	2.17	2.73	3.33	3.94
0.95	3.84	5.99	7.81	9.49	11.07	12.59	14.07	15.51	16.92	18.31
0.975	5.02	7.38	9.35	11.14	12.83	14.45	16.01	17.53	19.02	20.48
0.99	6.63	9.21	11.34	13.28	15.09	16.81	18.48	20.09	21.67	23.21
0.995	7.88	10.60	12.84	14.86	16.75	18.55	20.28	21.96	23.59	25.19

$F(z)$	Number of Degrees of Freedom									
	11	12	13	14	15	16	17	18	19	20
0.005	2.60	3.07	3.57	4.07	4.60	5.14	5.70	6.26	6.84	7.43
0.01	3.05	3.57	4.11	4.66	5.23	5.81	6.41	7.01	7.63	8.26
0.025	3.82	4.40	5.01	5.63	6.26	6.91	7.56	8.23	8.91	9.59
0.05	4.57	5.23	5.89	6.57	7.26	7.96	8.67	9.39	10.12	10.85
0.95	19.68	21.03	22.36	23.68	25.00	26.30	27.59	28.87	30.14	31.41
0.975	21.92	23.34	24.74	26.12	27.49	28.85	30.19	31.53	32.85	34.17
0.99	24.73	26.22	27.69	29.14	30.58	32.00	33.41	34.81	36.19	37.57
0.995	26.76	28.30	29.82	31.32	32.80	34.27	35.72	37.16	38.58	40.00

$F(z)$	Number of Degrees of Freedom									
	21	22	23	24	25	26	27	28	29	30
0.005	8.0	8.6	9.3	9.9	10.5	11.2	11.8	12.5	13.1	13.8
0.01	8.9	9.5	10.2	10.9	11.5	12.2	12.9	13.6	14.3	15.0
0.025	10.3	11.0	11.7	12.4	13.1	13.8	14.6	15.3	16.0	16.8
0.05	11.6	12.3	13.1	13.8	14.6	15.4	16.2	16.9	17.7	18.5
0.95	32.7	33.9	35.2	36.4	37.7	38.9	40.1	41.3	42.6	43.8
0.975	35.5	36.8	38.1	39.4	40.6	41.9	43.2	44.5	45.7	47.0
0.99	38.9	40.3	41.6	43.0	44.3	45.6	47.0	48.3	49.6	50.9
0.995	41.4	42.8	44.2	45.6	46.9	48.3	49.6	51.0	52.3	53.7

$F(z)$	Number of Degrees of Freedom							
	40	50	60	70	80	90	100	>100 (Approximation)
0.005	20.7	28.0	35.5	43.3	51.2	59.2	67.3	$\frac{1}{2}(h - 2.58)^2$
0.01	22.2	29.7	37.5	45.4	53.5	61.8	70.1	$\frac{1}{2}(h - 2.33)^2$
0.025	24.4	32.4	40.5	48.8	57.2	65.6	74.2	$\frac{1}{2}(h - 1.96)^2$
0.05	26.5	34.8	43.2	51.7	60.4	69.1	77.9	$\frac{1}{2}(h - 1.64)^2$
0.95	55.8	67.5	79.1	90.5	101.9	113.1	124.3	$\frac{1}{2}(h + 1.64)^2$
0.975	59.3	71.4	83.3	95.0	106.6	118.1	129.6	$\frac{1}{2}(h + 1.96)^2$
0.99	63.7	76.2	88.4	100.4	112.3	124.1	135.8	$\frac{1}{2}(h + 2.33)^2$
0.995	66.8	79.5	92.0	104.2	116.3	128.3	140.2	$\frac{1}{2}(h + 2.58)^2$

In the last column, $h = \sqrt{2m - 1}$ where m is the number of degrees of freedom.

Table A13
F-Distribution with (*m*, *n*) Degrees of Freedom

Values of z for which the distribution function $F(z)$ (cf. (13), Sec. 24.7) has the value **0.95**
Example: For (7, 4) degrees of freedom, $z = 6.09$ if $F(z) = 0.95$.

n	*m* = 1	*m* = 2	*m* = 3	*m* = 4	*m* = 5	*m* = 6	*m* = 7	*m* = 8	*m* = 9
1	161	200	216	225	230	234	237	239	241
2	18.5	19.0	19.2	19.2	19.3	19.3	19.4	19.4	19.4
3	10.1	9.55	9.28	9.12	9.01	8.94	8.89	8.85	8.81
4	7.71	6.94	6.59	6.39	6.26	6.16	6.09	6.04	6.00
5	6.61	5.79	5.41	5.19	5.05	4.95	4.88	4.82	4.77
6	5.99	5.14	4.76	4.53	4.39	4.28	4.21	4.15	4.10
7	5.59	4.74	4.35	4.12	3.97	3.87	3.79	3.73	3.68
8	5.32	4.46	4.07	3.84	3.69	3.58	3.50	3.44	3.39
9	5.12	4.26	3.86	3.63	3.48	3.37	3.29	3.23	3.18
10	4.96	4.10	3.71	3.48	3.33	3.22	3.14	3.07	3.02
11	4.84	3.98	3.59	3.36	3.20	3.09	3.01	2.95	2.90
12	4.75	3.89	3.49	3.26	3.11	3.00	2.91	2.85	2.80
13	4.67	3.81	3.41	3.18	3.03	2.92	2.83	2.77	2.71
14	4.60	3.74	3.34	3.11	2.96	2.85	2.76	2.70	2.65
15	4.54	3.68	3.29	3.06	2.90	2.79	2.71	2.64	2.59
16	4.49	3.63	3.24	3.01	2.85	2.74	2.66	2.59	2.54
17	4.45	3.59	3.20	2.96	2.81	2.70	2.61	2.55	2.49
18	4.41	3.55	3.16	2.93	2.77	2.66	2.58	2.51	2.46
19	4.38	3.52	3.13	2.90	2.74	2.63	2.54	2.48	2.42
20	4.35	3.49	3.10	2.87	2.71	2.60	2.51	2.45	2.39
22	4.30	3.44	3.05	2.82	2.66	2.55	2.46	2.40	2.34
24	4.26	3.40	3.01	2.78	2.62	2.51	2.42	2.36	2.30
26	4.23	3.37	2.98	2.74	2.59	2.47	2.39	2.32	2.27
28	4.20	3.34	2.95	2.71	2.56	2.45	2.36	2.29	2.24
30	4.17	3.32	2.92	2.69	2.53	2.42	2.33	2.27	2.21
32	4.15	3.30	2.90	2.67	2.51	2.40	2.31	2.24	2.19
34	4.13	3.28	2.88	2.65	2.49	2.38	2.29	2.23	2.17
36	4.11	3.26	2.87	2.63	2.48	2.36	2.28	2.21	2.15
38	4.10	3.24	2.85	2.62	2.46	2.35	2.26	2.19	2.14
40	4.08	3.23	2.84	2.61	2.45	2.34	2.25	2.18	2.12
50	4.03	3.18	2.79	2.56	2.40	2.29	2.20	2.13	2.07
60	4.00	3.15	2.76	2.53	2.37	2.25	2.17	2.10	2.04
70	3.98	3.13	2.74	2.50	2.35	2.23	2.14	2.07	2.02
80	3.96	3.11	2.72	2.49	2.33	2.21	2.13	2.06	2.00
90	3.95	3.10	2.71	2.47	2.32	2.20	2.11	2.04	1.99
100	3.94	3.09	2.70	2.46	2.31	2.19	2.10	2.03	1.97
150	3.90	3.06	2.66	2.43	2.27	2.16	2.07	2.00	1.94
200	3.89	3.04	2.65	2.42	2.26	2.14	2.06	1.98	1.93
1000	3.85	3.00	2.61	2.38	2.22	2.11	2.02	1.95	1.89
∞	3.84	3.00	2.60	2.37	2.21	2.10	2.01	1.94	1.88

Table A13
F-Distribution with (*m, n*) Degrees of Freedom (*continued*)

Values of z for which the distribution function $F(z)$ (cf. (13), Sec. 24.7) has the value **0.95**

n	m = 10	m = 15	m = 20	m = 30	m = 40	m = 50	m = 100	∞
1	242	246	248	250	251	252	253	254
2	19.4	19.4	19.4	19.5	19.5	19.5	19.5	19.5
3	8.79	8.70	8.66	8.62	8.59	8.58	8.55	8.53
4	5.96	5.86	5.80	5.75	5.72	5.70	5.66	5.63
5	4.74	4.62	4.56	4.50	4.46	4.44	4.41	4.37
6	4.06	3.94	3.87	3.81	3.77	3.75	3.71	3.67
7	3.64	3.51	3.44	3.38	3.34	3.32	3.27	3.23
8	3.35	3.22	3.15	3.08	3.04	3.02	2.97	2.93
9	3.14	3.01	2.94	2.86	2.83	2.80	2.76	2.71
10	2.98	2.85	2.77	2.70	2.66	2.64	2.59	2.54
11	2.85	2.72	2.65	2.57	2.53	2.51	2.46	2.40
12	2.75	2.62	2.54	2.47	2.43	2.40	2.35	2.30
13	2.67	2.53	2.46	2.38	2.34	2.31	2.26	2.21
14	2.60	2.46	2.39	2.31	2.27	2.24	2.19	2.13
15	2.54	2.40	2.33	2.25	2.20	2.18	2.12	2.07
16	2.49	2.35	2.28	2.19	2.15	2.12	2.07	2.01
17	2.45	2.31	2.23	2.15	2.10	2.08	2.02	1.96
18	2.41	2.27	2.19	2.11	2.06	2.04	1.98	1.92
19	2.38	2.23	2.16	2.07	2.03	2.00	1.94	1.88
20	2.35	2.20	2.12	2.04	1.99	1.97	1.91	1.84
22	2.30	2.15	2.07	1.98	1.94	1.91	1.85	1.78
24	2.25	2.11	2.03	1.94	1.89	1.86	1.80	1.73
26	2.22	2.07	1.99	1.90	1.85	1.82	1.76	1.69
28	2.19	2.04	1.96	1.87	1.82	1.79	1.73	1.65
30	2.16	2.01	1.93	1.84	1.79	1.76	1.70	1.62
32	2.14	1.99	1.91	1.82	1.77	1.74	1.67	1.59
34	2.12	1.97	1.89	1.80	1.75	1.71	1.65	1.57
36	2.11	1.95	1.87	1.78	1.73	1.69	1.62	1.55
38	2.09	1.94	1.85	1.76	1.71	1.68	1.61	1.53
40	2.08	1.92	1.84	1.74	1.69	1.66	1.59	1.51
50	2.03	1.87	1.78	1.69	1.63	1.60	1.52	1.44
60	1.99	1.84	1.75	1.65	1.59	1.56	1.48	1.39
70	1.97	1.81	1.72	1.62	1.57	1.53	1.45	1.35
80	1.95	1.79	1.70	1.60	1.54	1.51	1.43	1.32
90	1.94	1.78	1.69	1.59	1.53	1.49	1.41	1.30
100	1.93	1.77	1.68	1.57	1.52	1.48	1.39	1.28
150	1.89	1.73	1.64	1.53	1.48	1.44	1.34	1.22
200	1.88	1.72	1.62	1.52	1.46	1.41	1.32	1.19
1000	1.84	1.68	1.58	1.47	1.41	1.36	1.26	1.08
∞	1.83	1.67	1.57	1.46	1.39	1.35	1.24	1.00

Table A13
F-Distribution with (*m*, *n*) Degrees of Freedom (*continued*)

Values of z for which the distribution function $F(z)$ (cf. (13), Sec. 24.7) has the value **0.99**

n	*m* = 1	*m* = 2	*m* = 3	*m* = 4	*m* = 5	*m* = 6	*m* = 7	*m* = 8	*m* = 9
1	4052	4999	5403	5625	5764	5859	5928	5982	6022
2	98.5	99.0	99.2	99.3	99.3	99.3	99.4	99.4	99.4
3	34.1	30.8	29.5	28.7	28.2	27.9	27.7	27.5	27.3
4	21.2	18.0	16.7	16.0	15.5	15.2	15.0	14.8	14.7
5	16.3	13.3	12.1	11.4	11.0	10.7	10.5	10.3	10.2
6	13.7	10.9	9.78	9.15	8.75	8.47	8.26	8.10	7.98
7	12.2	9.55	8.45	7.85	7.46	7.19	6.99	6.84	6.72
8	11.3	8.65	7.59	7.01	6.63	6.37	6.18	6.03	5.91
9	10.6	8.02	6.99	6.42	6.06	5.80	5.61	5.47	5.35
10	10.0	7.56	6.55	5.99	5.64	5.39	5.20	5.06	4.94
11	9.65	7.21	6.22	5.67	5.32	5.07	4.89	4.74	4.63
12	9.33	6.93	5.95	5.41	5.06	4.82	4.64	4.50	4.39
13	9.07	6.70	5.74	5.21	4.86	4.62	4.44	4.30	4.19
14	8.86	6.51	5.56	5.04	4.70	4.46	4.28	4.14	4.03
15	8.68	6.36	5.42	4.89	4.56	4.32	4.14	4.00	3.89
16	8.53	6.23	5.29	4.77	4.44	4.20	4.03	3.89	3.78
17	8.40	6.11	5.18	4.67	4.34	4.10	3.93	3.79	3.68
18	8.29	6.01	5.09	4.58	4.25	4.01	3.84	3.71	3.60
19	8.18	5.93	5.01	4.50	4.17	3.94	3.77	3.63	3.52
20	8.10	5.85	4.94	4.43	4.10	3.87	3.70	3.56	3.46
22	7.95	5.72	4.82	4.31	3.99	3.76	3.59	3.45	3.35
24	7.82	5.61	4.72	4.22	3.90	3.67	3.50	3.36	3.26
26	7.72	5.53	4.64	4.14	3.82	3.59	3.42	3.29	3.18
28	7.64	5.45	4.57	4.07	3.75	3.53	3.36	3.23	3.12
30	7.56	5.39	4.51	4.02	3.70	3.47	3.30	3.17	3.07
32	7.50	5.34	4.46	3.97	3.65	3.43	3.26	3.13	3.02
34	7.44	5.29	4.42	3.93	3.61	3.39	3.22	3.09	2.98
36	7.40	5.25	4.38	3.89	3.57	3.35	3.18	3.05	2.95
38	7.35	5.21	4.34	3.86	3.54	3.32	3.15	3.02	2.92
40	7.31	5.18	4.31	3.83	3.51	3.29	3.12	2.99	2.89
50	7.17	5.06	4.20	3.72	3.41	3.19	3.02	2.89	2.79
60	7.08	4.98	4.13	3.65	3.34	3.12	2.95	2.82	2.72
70	7.01	4.92	4.08	3.60	3.29	3.07	2.91	2.78	2.67
80	6.96	4.88	4.04	3.56	3.26	3.04	2.87	2.74	2.64
90	6.93	4.85	4.01	3.54	3.23	3.01	2.84	2.72	2.61
100	6.90	4.82	3.98	3.51	3.21	2.99	2.82	2.69	2.59
150	6.81	4.75	3.92	3.45	3.14	2.92	2.76	2.63	2.53
200	6.76	4.71	3.88	3.41	3.11	2.89	2.73	2.60	2.50
1000	6.66	4.63	3.80	3.34	3.04	2.82	2.66	2.53	2.43
∞	6.63	4.61	3.78	3.32	3.02	2.80	2.64	2.51	2.41

Table A13
F-Distribution with (m, n) Degrees of Freedom (continued)

Values of z for which the distribution function $F(z)$ (cf. (13), Sec. 24.7) has the value **0.99**

n	m = 10	m = 15	m = 20	m = 30	m = 40	m = 50	m = 100	∞
1	6056	6157	6209	6261	6287	6300	6330	6366
2	99.4	99.4	99.4	99.5	99.5	99.5	99.5	99.5
3	27.2	26.9	26.7	26.5	26.4	26.4	26.2	26.1
4	14.5	14.2	14.0	13.8	13.7	13.7	13.6	13.5
5	10.1	9.72	9.55	9.38	9.29	9.24	9.13	9.02
6	7.87	7.56	7.40	7.23	7.14	7.09	6.99	6.88
7	6.62	6.31	6.16	5.99	5.91	5.86	5.75	5.65
8	5.81	5.52	5.36	5.20	5.12	5.07	4.96	4.86
9	5.26	4.96	4.81	4.65	4.57	4.52	4.42	4.31
10	4.85	4.56	4.41	4.25	4.17	4.12	4.01	3.91
11	4.54	4.25	4.10	3.94	3.86	3.81	3.71	3.60
12	4.30	4.01	3.86	3.70	3.62	3.57	3.47	3.36
13	4.10	3.82	3.66	3.51	3.43	3.38	3.27	3.17
14	3.94	3.66	3.51	3.35	3.27	3.22	3.11	3.00
15	3.80	3.52	3.37	3.21	3.13	3.08	2.98	2.87
16	3.69	3.41	3.26	3.10	3.02	2.97	2.86	2.75
17	3.59	3.31	3.16	3.00	2.92	2.87	2.76	2.65
18	3.51	3.23	3.08	2.92	2.84	2.78	2.68	2.57
19	3.43	3.15	3.00	2.84	2.76	2.71	2.60	2.49
20	3.37	3.09	2.94	2.78	2.69	2.64	2.54	2.42
22	3.26	2.98	2.83	2.67	2.58	2.53	2.42	2.31
24	3.17	2.89	2.74	2.58	2.49	2.44	2.33	2.21
26	3.09	2.82	2.66	2.50	2.42	2.36	2.25	2.13
28	3.03	2.75	2.60	2.44	2.35	2.30	2.19	2.06
30	2.98	2.70	2.55	2.39	2.30	2.25	2.13	2.01
32	2.93	2.66	2.50	2.34	2.25	2.20	2.08	1.96
34	2.89	2.62	2.46	2.30	2.21	2.16	2.04	1.91
36	2.86	2.58	2.43	2.26	2.17	2.12	2.00	1.87
38	2.83	2.55	2.40	2.23	2.14	2.09	1.97	1.84
40	2.80	2.52	2.37	2.20	2.11	2.06	1.94	1.80
50	2.70	2.42	2.27	2.10	2.01	1.95	1.82	1.68
60	2.63	2.35	2.20	2.03	1.94	1.88	1.75	1.60
70	2.59	2.31	2.15	1.98	1.89	1.83	1.70	1.54
80	2.55	2.27	2.12	1.94	1.85	1.79	1.66	1.49
90	2.52	2.24	2.09	1.92	1.82	1.76	1.62	1.46
100	2.50	2.22	2.07	1.89	1.80	1.73	1.60	1.43
150	2.44	2.16	2.00	1.83	1.73	1.66	1.52	1.33
200	2.41	2.13	1.97	1.79	1.69	1.63	1.48	1.28
1000	2.34	2.06	1.90	1.72	1.61	1.54	1.38	1.11
∞	2.32	2.04	1.88	1.70	1.59	1.52	1.36	1.00

Table A14
Distribution Function $F(x) = P(T \leq x)$ of the Random Variable T in Section 24.11

If $n = 3$, then $F(2) = 1 - 0.167 = 0.833$.
If $n = 4$, then $F(3) = 1 - 0.375 = 0.625$, $F(4) = 1 - 0.167 = 0.833$, etc.

$n = 3$ (values 0.)

x	
0	167
1	500

$n = 4$ (values 0.)

x	
0	042
1	167
2	375

$n = 5$ (values 0.)

x	
0	008
1	042
2	117
3	242
4	408

$n = 6$ (values 0.)

x	
0	001
1	008
2	028
3	068
4	136
5	235
6	360
7	500

$n = 7$ (values 0.)

x	
1	001
2	005
3	015
4	035
5	068
6	119
7	191
8	281
9	386
10	500

$n = 8$ (values 0.)

x	
2	001
3	003
4	007
5	012
6	031
7	054
8	089
9	138
10	199
11	274
12	360
13	452

$n = 9$ (values 0.)

x	
4	001
5	003
6	006
7	012
8	022
9	038
10	060
11	090
12	130
13	179
14	238
15	306
16	381
17	460

$n = 10$ (values 0.)

x	
6	001
7	002
8	005
9	008
10	014
11	023
12	036
13	054
14	078
15	108
16	146
17	190
18	242
19	300
20	364
21	431
22	500

$n = 11$ (values 0.)

x	
8	001
9	002
10	003
11	005
12	008
13	013
14	020
15	030
16	043
17	060
18	082
19	109
20	141
21	179
22	223
23	271
24	324
25	381
26	440
27	500

$n = 20$ (values 0.)

x	
50	001
51	002
52	002
53	003
54	004
55	005
56	006
57	007
58	008
59	010
60	012
61	014
62	017
63	020
64	023
65	027
66	032
67	037
68	043
69	049
70	056
71	064
72	073
73	082
74	093
75	104
76	117
77	130
78	144
79	159
80	176
81	193
82	211
83	230
84	250
85	271
86	293
87	315
88	339
89	362
90	387
91	411
92	436
93	462
94	487

$n = 19$ (values 0.)

x	
43	001
44	002
45	002
46	003
47	003
48	004
49	005
50	006
51	008
52	010
53	012
54	014
55	017
56	021
57	025
58	029
59	034
60	040
61	047
62	054
63	062
64	072
65	082
66	093
67	105
68	119
69	133
70	149
71	166
72	184
73	203
74	223
75	245
76	267
77	290
78	314
79	339
80	365
81	391
82	418
83	445
84	473
85	500

$n = 18$ (values 0.)

x	
38	001
39	002
40	003
41	003
42	004
43	005
44	007
45	009
46	011
47	013
48	016
49	020
50	024
51	029
52	034
53	041
54	048
55	056
56	066
57	076
58	088
59	100
60	115
61	130
62	147
63	165
64	184
65	205
66	227
67	250
68	275
69	300
70	327
71	354
72	383
73	411
74	441
75	470
76	500

$n = 17$ (values 0.)

x	
32	001
33	002
34	002
35	003
36	004
37	005
38	007
39	009
40	011
41	014
42	017
43	021
44	026
45	032
46	038
47	046
48	054
49	064
50	076
51	088
52	102
53	118
54	135
55	154
56	174
57	196
58	220
59	245
60	271
61	299
62	328
63	358
64	388
65	420
66	452
67	484

$n = 16$ (values 0.)

x	
27	001
28	002
29	002
30	003
31	004
32	006
33	008
34	010
35	013
36	016
37	021
38	026
39	032
40	039
41	048
42	058
43	070
44	083
45	097
46	114
47	133
48	153
49	175
50	199
51	225
52	253
53	282
54	313
55	345
56	378
57	412
58	447
59	482

$n = 15$ (values 0.)

x	
23	001
24	002
25	003
26	004
27	006
28	008
29	010
30	014
31	018
32	023
33	029
34	037
35	046
36	057
37	070
38	084
39	101
40	120
41	141
42	164
43	190
44	218
45	248
46	279
47	313
48	349
49	385
50	423
51	461
52	500

$n = 14$ (values 0.)

x	
18	001
19	002
20	002
21	003
22	005
23	007
24	010
25	013
26	018
27	024
28	031
29	040
30	051
31	063
32	079
33	096
34	117
35	140
36	165
37	194
38	225
39	259
40	295
41	334
42	374
43	415
44	457
45	500

$n = 13$ (values 0.)

x	
14	001
15	001
16	002
17	003
18	005
19	007
20	011
21	015
22	021
23	029
24	038
25	050
26	064
27	082
28	102
29	126
30	153
31	184
32	218
33	255
34	295
35	338
36	383
37	429
38	476

$n = 12$ (values 0.)

x	
11	001
12	002
13	003
14	004
15	007
16	010
17	016
18	022
19	031
20	043
21	058
22	076
23	098
24	125
25	155
26	190
27	230
28	273
29	319
30	369
31	420
32	473

INDEX

Page numbers A1, A2, A3, . . . refer to Appendix 1 to Appendix 4 at the end of the book.

H

I